Conversion Factors

Mass

$1 \text{ g} = 10^{-3} \text{ kg}$
$1 \text{ kg} = 10^3 \text{ g}$
$1 \text{ u} = 1.66 \times 10^{-24} \text{ g} = 1.66 \times 10^{-27} \text{ kg}$
$1 \text{ metric ton} = 1000 \text{ kg}$

Length

$1 \text{ nm} = 10^{-9} \text{ m}$
$1 \text{ cm} = 10^{-2} \text{ m} = 0.394 \text{ in.}$
$1 \text{ m} = 10^{-3} \text{ km} = 3.28 \text{ ft} = 39.4 \text{ in.}$
$1 \text{ km} = 10^3 \text{ m} = 0.621 \text{ mi}$
$1 \text{ in.} = 2.54 \text{ cm} = 2.54 \times 10^{-2} \text{ m}$
$1 \text{ ft} = 0.305 \text{ m} = 30.5 \text{ cm}$
$1 \text{ mi} = 5280 \text{ ft} = 1609 \text{ m} = 1.609 \text{ km}$

Area

$1 \text{ cm}^2 = 10^{-4} \text{ m}^2 = 0.1550 \text{ in}^2$
$\quad = 1.08 \times 10^{-3} \text{ ft}^2$
$1 \text{ m}^2 = 10^4 \text{ cm}^2 = 10.76 \text{ ft}^2 = 1550 \text{ in}^2$
$1 \text{ in}^2 = 6.94 \times 10^{-3} \text{ ft}^2 = 6.45 \text{ cm}^2$
$\quad = 6.45 \times 10^{-4} \text{ m}^2$
$1 \text{ ft}^2 = 144 \text{ in}^2 = 9.29 \times 10^{-2} \text{ m}^2 = 929 \text{ cm}^2$

Volume

$1 \text{ cm}^3 = 10^{-6} \text{ m}^3 = 3.35 \times 10^{-5} \text{ ft}^3$
$\quad = 6.10 \times 10^{-2} \text{ in}^3$
$1 \text{ m}^3 = 10^6 \text{ cm}^3 = 10^3 \text{ L} = 35.3 \text{ ft}^3$
$\quad = 6.10 \times 10^4 \text{ in}^3 = 264 \text{ gal}$
$1 \text{ liter} = 10^3 \text{ cm}^3 = 10^{-3} \text{ m}^3 = 1.056 \text{ qt}$
$\quad = 0.264 \text{ gal} = 0.0353 \text{ ft}^3$
$1 \text{ in}^3 = 5.79 \times 10^{-4} \text{ ft}^3 = 16.4 \text{ cm}^3$
$\quad = 1.64 \times 10^{-5} \text{ m}^3$
$1 \text{ ft}^3 = 1728 \text{ in}^3 = 7.48 \text{ gal} = 0.0283 \text{ m}^3$
$\quad = 28.3 \text{ L}$
$1 \text{ qt} = 2 \text{ pt} = 946 \text{ cm}^3 = 0.946 \text{ L}$
$1 \text{ gal} = 4 \text{ qt} = 231 \text{ in}^3 = 0.134 \text{ ft}^3 = 3.785 \text{ L}$

Time

$1 \text{ h} = 60 \text{ min} = 3600 \text{ s}$
$1 \text{ day} = 24 \text{ h} = 1440 \text{ min} = 8.64 \times 10^4 \text{ s}$
$1 \text{ y} = 365 \text{ days} = 8.76 \times 10^3 \text{ h}$
$\quad = 5.26 \times 10^5 \text{ min} = 3.16 \times 10^7 \text{ s}$

Angle

$1 \text{ rad} = 57.3°$

$1° = 0.0175 \text{ rad}$	$60° = \pi/3 \text{ rad}$
$15° = \pi/12 \text{ rad}$	$90° = \pi/2 \text{ rad}$
$30° = \pi/6 \text{ rad}$	$180° = \pi \text{ rad}$
$45° = \pi/4 \text{ rad}$	$360° = 2\pi \text{ rad}$

$1 \text{ rev/min} = (\pi/30) \text{ rad/s} = 0.1047 \text{ rad/s}$

Speed

$1 \text{ m/s} = 3.60 \text{ km/h} = 3.28 \text{ ft/s}$
$\quad = 2.24 \text{ mi/h}$
$1 \text{ km/h} = 0.278 \text{ m/s} = 0.621 \text{ mi/h}$
$\quad = 0.911 \text{ ft/s}$
$1 \text{ ft/s} = 0.682 \text{ mi/h} = 0.305 \text{ m/s}$
$\quad = 1.10 \text{ km/h}$
$1 \text{ mi/h} = 1.467 \text{ ft/s} = 1.609 \text{ km/h}$
$\quad = 0.447 \text{ m/s}$
$60 \text{ mi/h} = 88 \text{ ft/s}$

Force

$1 \text{ N} = 0.225 \text{ lb}$
$1 \text{ lb} = 4.45 \text{ N}$
Equivalent weight of a mass of 1 kg
$\quad$ on Earth's surface $= 2.2 \text{ lb} = 9.8 \text{ N}$

Pressure

$1 \text{ Pa (N/m}^2) = 1.45 \times 10^{-4} \text{ lb/in}^2$
$\quad = 7.5 \times 10^{-3} \text{ torr (mm Hg)}$
$1 \text{ torr (mm Hg)} = 133 \text{ Pa (N/m}^2)$
$\quad = 0.02 \text{ lb/in}^2$
$1 \text{ atm} = 14.7 \text{ lb/in}^2 = 1.013 \times 10^5 \text{ N/m}^2$
$\quad = 30 \text{ in. Hg} = 76 \text{ cm Hg}$
$1 \text{ lb/in}^2 = 6.90 \times 10^5 \text{ Pa (N/m}^2)$
$1 \text{ bar} = 10^5 \text{ Pa}$
$1 \text{ millibar} = 10^2 \text{ Pa}$

Energy

$1 \text{ J} = 0.738 \text{ ft·lb} = 0.239 \text{ cal}$
$\quad = 9.48 \times 10^{-4} \text{ Btu} = 6.24 \times 10^{18} \text{ eV}$
$1 \text{ kcal} = 4186 \text{ J} = 3.968 \text{ Btu}$
$1 \text{ Btu} = 1055 \text{ J} = 778 \text{ ft·lb} = 0.252 \text{ kcal}$
$1 \text{ cal} = 4.186 \text{ J} = 3.97 \times 10^{-3} \text{ Btu}$
$\quad = 3.09 \text{ ft·lb}$
$1 \text{ ft·lb} = 1.36 \text{ J} = 1.29 \times 10^{-3} \text{ Btu}$
$1 \text{ eV} = 1.60 \times 10^{-19} \text{ J}$
$1 \text{ kWh} = 3.6 \times 10^6 \text{ J}$

Power

$1 \text{ W} = 0.738 \text{ ft·lb/s} = 1.34 \times 10^{-3} \text{ hp}$
$\quad = 3.41 \text{ Btu/h}$
$1 \text{ ft·lb/s} = 1.36 \text{ W} = 1.82 \times 10^{-3} \text{ hp}$
$1 \text{ hp} = 550 \text{ ft·lb/s} = 745.7 \text{ W}$
$\quad = 2545 \text{ Btu/h}$

Mass–Energy Equivalents

$1 \text{ u} = 1.66 \times 10^{-27} \text{ kg} \leftrightarrow 931.5 \text{ MeV}$
$1 \text{ electron mass} = 9.11 \times 10^{-31} \text{ kg}$
$\quad = 5.49 \times 10^{-4} \text{ u} \leftrightarrow 0.511 \text{ MeV}$
$1 \text{ proton mass} = 1.67262 \times 10^{-27} \text{ kg}$
$\quad = 1.007\,276 \text{ u} \leftrightarrow 938.27 \text{ MeV}$
$1 \text{ neutron mass} = 1.67493 \times 10^{-27} \text{ kg}$
$\quad = 1.008\,665 \text{ u} \leftrightarrow 939.57 \text{ MeV}$

Temperature

$T_F = \frac{9}{5} T_C + 32$
$T_C = \frac{5}{9}(T_F - 32)$
$T_K = T_C + 273$

cgs Force

$1 \text{ dyne} = 10^{-5} \text{ N} = 2.25 \times 10^{-6} \text{ lb}$

cgs Energy

$1 \text{ erg} = 10^{-7} \text{ J} = 7.38 \times 10^{-6} \text{ ft·lb}$

HW Slot #17

College Physics

College Physics

Fifth Edition

Jerry D. Wilson

Lander University
Greenwood, SC

Anthony J. Buffa

California Polytechnic State University
San Luis Obispo, CA

with Bo Lou

Ferris State University
Big Rapids, MI

Pearson Education, Inc.
Upper Saddle River, NJ 07458

Library of Congress Cataloging-in-Publication Data

Wilson, Jerry D.
  College physics.—5th ed. / Jerry D. Wilson, Anthony J. Buffa.
    p. cm.
  Includes bibliographical references and index.
  ISBN 0-13-067644-6
    1. Physics. I. Buffa, Anthony J.

QC21.3 .W35 2003
530—dc21   2002016998

Senior Editor: Erik Fahlgren
Editor in Chief, Physical Sciences: John Challice
Vice President of Production and Manufacturing: David W. Riccardi
Executive Managing Editor: Kathleen Schiaparelli
Assistant Managing Editor: Beth Sweeten
Development Editors: Mary Catherine Hager and Karen Karlin
Editor in Chief, Development: Carol Trueheart
Executive Marketing Manager: Mark Pfaltzgraff
Manufacturing Manager: Trudy Pisciotti
Assistant Manufacturing Manager: Michael Bell
Director of Creative Services: Paul Belfanti
Director of Design: Carole Anson
Art Director: Jonathan Boylan
Managing Editor, Visual Assets and Production: Patricia Burns
Art Editor: Adam Velthaus
Art Studio: Imagineering
Interior and Cover Designer: John Christiana
Cover Photograph: Sean Justice/Getty Images–Image Bank "Swimmer doing butterfly stroke"
Photo Researcher: Yvonne Gerin
Editorial Assistants: Eileen Nee, Nancy Bauer
Assistant Editor: Christian Botting
Production Supervision/Composition: WestWords, Inc.

© 2003, 2000, 1997, 1994, 1990 by Pearson Education, Inc.
Pearson Education, Inc.
Upper Saddle River, New Jersey 07458

Printed in the United States of America
10 9 8 7 6 5 4 3 2 1

ISBN 0-13-067644-6 (college edition)
ISBN 0-13-048459-8 (school edition)

Pearson Education LTD., *London*
Pearson Education Australia PTY, Limited, *Sydney*
Pearson Education Singapore, Pte. Ltd
Pearson Education North Asia Ltd, *Hong Kong*
Pearson Education Canada, Ltd., *Toronto*
Pearson Educacíon de Mexico, S.A. de C.V.
Pearson Education — Japan, *Tokyo*
Pearson Education Malaysia, Pte. Ltd

About the Authors

Jerry D. Wilson, a native of Ohio, is Emeritus Professor of Physics and former Chair of the Division of Biological and Physical Sciences at Lander University in Greenwood, South Carolina. He received his B.S. degree from Ohio University, M.S. degree from Union College, and, in 1970, a Ph.D. from Ohio University. He earned his M.S. degree while employed as a Materials Behavior Physicist by the General Electric Co.

As a doctoral graduate student, Professor Wilson held the faculty rank of Instructor and began teaching physical-science courses. During this time, he coauthored a physical-science text that is now in its 10th edition. In conjunction with his teaching career, Professor Wilson continued his writing and has authored or coauthored six titles. Having retired from full-time teaching, he continues to write, producing, among other works, *The Curiosity Corner*, a weekly column for local newspapers that can also be found on the Internet.

With several competitive books available, one may wonder why I chose to cowrite another algebra-based physics text. Having taught introductory physics many times, I was well aware of the needs of students and the difficulties they have in mastering the subject. I decided to write a text that presents the basic physics principles in a clear and concise manner, with illustrative examples that help resolve the major difficulty in learning physics: problem solving. Also, I wanted to write a text that is relevant to real-life situations, so as to show students how physics applies in their everyday world, how things work, and why things happen. Once the basics are learned, an understanding of such applications follows naturally.

—Jerry Wilson

Anthony J. Buffa received his B.S. degree in physics from Rensselaer Polytechnic Institute and both his M.S. and Ph.D. degrees in physics from the University of Illinois, Urbana–Champaign. In 1970, Professor Buffa joined the faculty at California Polytechnic State University, San Luis Obispo, where he is currently Professor of Physics, and has been a research associate with the Radioanalytical Facility of the department of physics since 1980.

Professor Buffa's main interest continues to be teaching. He has taught courses at Cal Poly ranging from introductory physical science to quantum mechanics, has developed and revised many laboratory experiments, and has taught elementary physics to local teachers in an NSF-sponsored workshop. Combining physics with his interests in art and architecture, Dr. Buffa develops his own artwork and sketches, which he uses to increase his effectiveness in teaching physics.

I try to teach my students the crucial role physics plays in understanding all aspects of the world around them—whether it be technology, biology, astronomy, or any other field. In that regard, I emphasize conceptual understanding before number crunching. To this end, I rely heavily on visual methods. I hope the artwork and other pedagogical features in this book assist you in achieving your own teaching goals for your students.

—Tony Buffa

Brief Contents

Contents

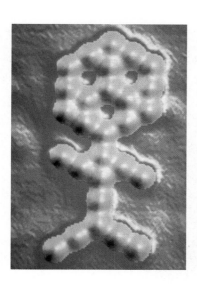

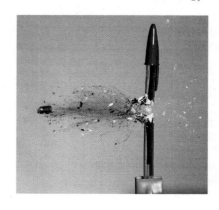

6 Linear Momentum and Collisions   179

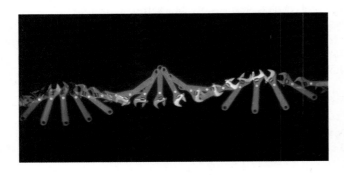

7 Circular Motion and Gravitation   219

8 Rotational Motion and Equilibrium   260

9 Solids and Fluids   305

10 Temperature and Kinetic Theory   345

11 Heat 373

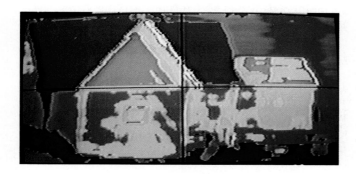

12 Thermodynamics 404

13 Vibrations and Waves 444

14 Sound 478

15 Electric Charge, Forces, and Fields 513

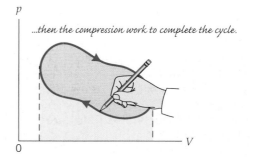

...then the compression work to complete the cycle.

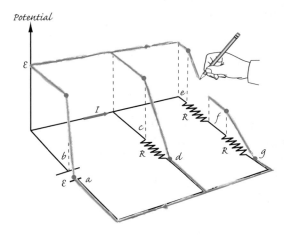

Potential

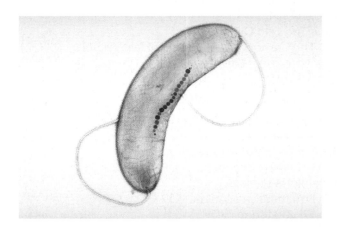

Learn by Drawing

Applications (Insights appear in **boldface**, and "(bio)" indicates a biomedical application)

Physlet® Illustrations

Preface

We believe that there are two basic goals in any introductory physics course: (1) to impart an understanding of the basic concepts of physics and (2) to enable students to use these concepts to solve a variety of problems.

These goals are linked. We want students to apply concepts to the problems that they are trying to solve. However, they often begin the problem-solving process by searching for an equation. There is the temptation to try and plug numbers into equations before visualizing the situation or considering the physical concepts that could be used to solve the problem.

Research in physics education has shown that a surprising number of students who learn to solve typical problems well enough to pass examinations do so without ever arriving at a real understanding of the most elementary physical concepts. Simply put, they can solve quantitative problems and get the right answer, but they do not know why it is right. In addition, students often do not check their numerical answer to see if it matches their understanding of the relevant physical concept.

Our Goals—Features of the Fifth Edition

Our goals for the Fifth Edition of this text are simple, yet challenging. With the goals of the course in mind, we identified areas in need of improvement and made efforts to further enhance the strengths of the book.

First, we asked a trusted colleague to contribute to our efforts. Bo Lou, of Ferris State University, has been an important part of *College Physics* since the Third Edition. He has authored the *Instructor's Solutions Manual* and the *Student Study Guide* and has played an important role as a member of AZTEC (Absolutely Zero Tolerance for Errors Club). In this edition, his expertise in optics has been used to update the chapters dealing with that topic (Chapters 22–25). Also, Professor Lou was responsible for updating the end-of-chapter exercises. His Ph.D., in condensed-matter physics, is from Emory University.

We feel, and many users have agreed, that the strengths of this textbook are as follows:

Conceptual Basis. We believe that giving students a secure grasp of physical principles will almost invariably enhance their problem-solving abilities. Central to this belief is an approach to the development of problem-solving skills that stresses an understanding of basic concepts, rather than the mechanical and rote use of equations, as the essential foundation. Throughout the writing of *College Physics*, we have organized discussions and incorporated pedagogical tools to ensure that conceptual insight drives the development of practical skills.

Concise Coverage. To maintain a sharp focus on essential concepts, a textbook should emphasize the basics and minimize superfluous material. In this text, topics of marginal interest have been avoided, as have those that present formal or mathematical difficulties for students. Similarly, we have not wasted space on deriving relationships when they shed no additional light on the principle involved. It is usually more important for students in a course such as this book is geared toward to understand what a relationship means and how it can be used, rather than the mathematical or analytical techniques employed to derive it.

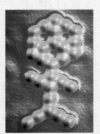

INSIGHT

Nanotechnology

A prefix for metric units refers to the order of the dimension of measurement used. The prefix *micro-* (10^{-6}) has been in common use for some time—e.g., in microscope, microchip, microbiology, microchemistry, and so on. Now the prefix *nano-* is coming into use, and you will probably be hearing a great deal about it with respect to nanotechnology (*nanotech* for short).

In general, nanotechnology is any technology done on the nanometer scale. A nanometer is one billionth (10^{-9}) of a meter, about the width of three to four atoms. Basically, nanotechnology involves the manufacture or building of things one atom or molecule at a time, so the nanometer is the appropriate scale. One atom or molecule at a time? That may sound a bit farfetched, but it's not. This possibility was advanced by physicist Richard Feynman in 1956: "The principles of physics, as far as I can see, do not speak against the possibility of maneuvering things atom by atom." This has been accomplished by using a special microscope at very low temperatures and "pushing around" single atoms (Fig. 1).

In a sense, nanotechnology occurs naturally in the cells of our bodies. Ribosomes (tiny particles in all living cells) are the sites at which information carried by the genetic code is converted into protein molecules. Ribosomes are like tiny "machines," no larger then a few nanometers long, that read DNA instructions on how to build enzymes and other proteins molecule by molecule.

Perhaps you can now see the potential of nanotechnology. The chemical properties of atoms and molecules are well understood. For example, rearranging the atoms in coal can produce a diamond. (We can already do this task without nanotechnology, using heat and pressure.) Nanotechnology presents the possibility of constructing novel molecular devices or "machines" with extraordinary properties and

FIGURE 1 Molecular Man This figure was crafted by moving 28 molecules, one at a time. Each of the gold-colored peaks is the image of a carbon monoxide molecule. The molecules rest on a single crystal platinum surface. "Molecular Man" measures 5 nm tall and 2.5 nm wide (hand-to-hand). More than 20000 figures, linked hand-to-hand, would be needed to span a single human hair. The molecules in the figure were positioned using a special microscope at very low temperatures.

PHYSLET®
ILLUSTRATION

Scanning tunneling microscope

Applications. *College Physics* is known for the strong mix of applications related to medicine, science, technology, architecture, and everyday life in its text narrative and Insight boxes. While the Fifth Edition continues to have a wider range of applications than do most texts, we have also increased the number of biological applications, in recognition of the high percentage of premed and allied health majors who take the course for which it is used. Some examples of topics discussed in biology-oriented Insights are nanotechnology, weight-lessness and its effects on the human body, the physics of ear popping, desirable and undesirable resonance, body-fat analysis, cornea surgery, and bioengineering. A complete list of applications discussed, with page references, is found on page xiii.

Learn by Drawing

A Lens Ray Diagram (See Example 23.5.)

1 **Parallel ray**

2 **Chief (central) ray**

3 **Locating image**

4 **Can also use focal ray**

The following pedagogical features have been enhanced in the Fifth Edition:

Learn by Drawing Boxes. Visualization is one of the most important problem-solving tools in physics. In many cases, if students can make a sketch of a problem, they can solve it. "Learn by Drawing" features offer students specific help on making certain types of sketches and graphs that will provide key insights into a variety of physical situations.

Integrated Learning Objectives. Specific learning objectives, located at the beginning of each chapter section, help students structure their reading and facilitate review of the material.

Suggested Problem-Solving Procedure. An extensive section (Section 1.7) provides a framework for thinking about problem solving. This section includes

- An overview of problem-solving strategies;
- A seven-step procedure that is general enough to apply to most problems in physics, but is easily used in specific situations;
- Three Examples that illustrate the detailed problem-solving process, showing how the general procedure is applied in practice.

Problem-Solving Strategies and Hints.

The initial treatment of problem solving is followed up throughout *College Physics* with an abundance of suggestions, tips, cautions, shortcuts, and useful techniques for solving specific kinds of problems. These strategies and hints help students apply general principles to specific contexts, as well as avoid common pitfalls and misunderstandings.

Conceptual Examples.

College Physics was among the first physics texts to include examples that are conceptual in nature, in addition to quantitative ones. Our Conceptual Examples ask students to think about a physical situation and choose the correct prediction out of a set of possible outcomes, on the basis of an understanding of relevant principles. The discussion that follows ("Reasoning and Answer") explains clearly how the correct answer can be identified, as well as why the other answers are wrong.

Worked Examples.

We have tried to make the solutions to in-text Examples as clear and detailed as possible. The aim is not merely to show students which equations to use, but to explain the strategy being employed and the role of each step in the overall plan. Students are encouraged to learn the "why" of each step along with the "how." This technique will make it easier for students to apply the demonstrated techniques to other problems that are not identical in structure. Each worked Example also includes the following:

- *Thinking It Through Step.* This section, which follows the statement of the problem and precedes the solution, focuses students on the critical thinking and analysis they should undertake before beginning to use equations.
- *Follow-up Exercise.* The Follow-up Exercise at the end of each Conceptual Example and each regular worked Example further reinforces the importance of conceptual understanding and offers additional practice. (Answers to Follow-up Exercises are given at the back of the text.)

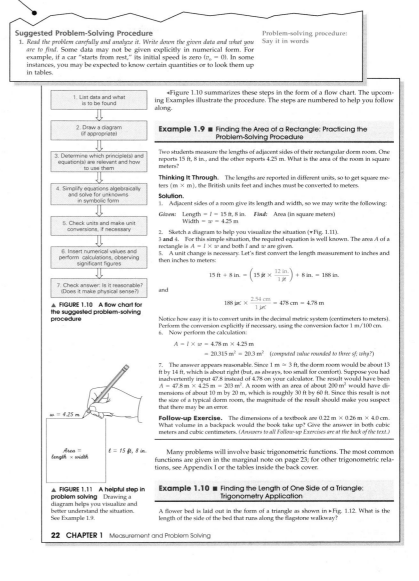

Suggested Problem-Solving Procedure

1. *Read the problem carefully and analyze it. Write down the given data and what you are to find.* Some data may not be given explicitly in numerical form. For example, if a car "starts from rest," its initial speed is zero ($v_o = 0$). In some instances, you may be expected to know certain quantities or to look them up in tables.

Problem-solving procedure:
Say it in words

1. List data and what is to be found
2. Draw a diagram (if appropriate)
3. Determine which principle(s) and equation(s) are relevant and how to use them
4. Simplify equations algebraically and solve for unknowns in symbolic form
5. Check units and make unit conversions, if necessary
6. Insert numerical values and perform calculations, observing significant figures
7. Check answer: Is it reasonable? (Does it make physical sense?)

▲ **FIGURE 1.10** A flow chart for the suggested problem-solving procedure

◄ Figure 1.10 summarizes these steps in the form of a flow chart. The upcoming Examples illustrate the procedure. The steps are numbered to help you follow along.

Example 1.9 ■ Finding the Area of a Rectangle: Practicing the Problem-Solving Procedure

Two students measure the lengths of adjacent sides of their rectangular dorm room. One reports 15 ft, 8 in., and the other reports 4.25 m. What is the area of the room in square meters?

Thinking It Through. The lengths are reported in different units, so to get square meters (m × m), the British units feet and inches must be converted to meters.

Solution.
1. Adjacent sides of a room give its length and width, so we may write the following:

Given: Length = l = 15 ft, 8 in.   *Find:* Area (in square meters)
       Width = w = 4.25 m

2. Sketch a diagram to help you visualize the situation (▼ Fig. 1.11).
3 **and** 4. For this simple situation, the required equation is well known. The area A of a rectangle is $A = l \times w$ and both l and w are given.
5. A unit change is necessary. Let's first convert the length measurement to inches and then inches to meters:

$$15\ \text{ft} + 8\ \text{in.} = \left(15\ \text{ft} \times \frac{12\ \text{in.}}{1\ \text{ft}} \right) + 8\ \text{in.} = 188\ \text{in.}$$

and

$$188\ \text{in.} \times \frac{2.54\ \text{cm}}{1\ \text{in.}} = 478\ \text{cm} = 4.78\ \text{m}$$

Notice how easy it is to convert units in the decimal metric system (centimeters to meters). Perform the conversion explicitly if necessary, using the conversion factor 1 m/100 cm.
6. Now perform the calculation:

$$A = l \times w = 4.78\ \text{m} \times 4.25\ \text{m}$$
$$= 20.315\ \text{m}^2 = 20.3\ \text{m}^2 \quad (computed\ value\ rounded\ to\ three\ sf;\ why?)$$

7. The answer appears reasonable. Since 1 m ≈ 3 ft, the dorm room would be about 13 ft by 14 ft, which is about right (but, as always, too small for comfort). Suppose you had inadvertently input 47.8 instead of 4.78 on your calculator. The result would have been $A = 47.8\ \text{m} \times 4.25\ \text{m} = 203\ \text{m}^2$. A room with an area of about 200 m² would have dimensions of about 10 m by 20 m, which is roughly 30 ft by 60 ft. Since this result is not the size of a typical dorm room, the magnitude of the result should make you suspect that there may be an error.

Follow-up Exercise. The dimensions of a textbook are 0.22 m × 0.26 m × 4.0 cm. What volume in a backpack would the book take up? Give the answer in both cubic meters and cubic centimeters. (*Answers to all Follow-up Exercises are at the back of the text.*)

Many problems will involve basic trigonometric functions. The most common functions are given in the marginal note on page 23; for other trigonometric relations, see Appendix I or the tables inside the back cover.

Example 1.10 ■ Finding the Length of One Side of a Triangle: Trigonometry Application

A flower bed is laid out in the form of a triangle as shown in ▶ Fig. 1.12. What is the length of the side of the bed that runs along the flagstone walkway?

$w = 4.25\ m$

$Area = length \times width$ $l = 15\ ft,\ 8\ in.$

▲ **FIGURE 1.11** A helpful step in problem solving Drawing a diagram helps you visualize and better understand the situation. See Example 1.9.

22 CHAPTER 1 Measurement and Problem Solving

Integration of Conceptual and Quantitative Exercises. To help break down the artificial barrier between conceptual questions and quantitative problems, we do not separate these categories in the end-of-chapter exercises. Instead, each section begins with a series of multiple-choice and short-answer questions that provide review of the chapter's content, test students' conceptual understanding, and ask students to reason from principles. The aim is to show students that the same kind of conceptual insight is required regardless of whether the desired answer involves words, equations, or numbers. The conceptual questions are marked by a bold CQ in the text for easy reference when assigning questions. *College Physics* offers short answers to all odd-numbered conceptual questions (as well as to all odd-numbered quantitative problems) at the back of the text, so that students can check their understanding of those problems.

Paired Exercises. Most numbered sections include at least one set of paired Exercises that deal with similar situations. The first problem in a pair is solved in the *Student Study Guide and Solutions Manual;* the second problem, which explores a similar situation to that presented in the first prob-lem, has only an answer at the back of the book, thereby encouraging students to work out the problem on their own.

Additional Exercises. Each chapter includes a supplemental section of Additional Exercises drawn from all sections of the chapter, to ensure that students can synthesize concepts.

New features to the Fifth Edition include the following:

Physlet® Illustrations. Physlet® Illustrations are short Java applets that clearly illustrate, through animation, a concept from the text. Available on the Wilson/Buffa Companion Web site, Physlet® Illustrations are followed by a series of questions that ask students to think critically about the concept at hand. Physlet® Illustrations are denoted by an icon in the margin of the text.

Integrated Examples. In order to further emphasize the connection between conceptual understanding and quantitative problem-solving, we have developed Integrated Examples for each chapter. These Examples work through a physical situation both qualitatively and quantitatively. Integrated Examples demonstrate how conceptual understanding and numerical calculations go hand in hand in understanding and solving problems.

Integrated Exercises. Like the Integrated Examples in the chapter, Integrated Exercises ask students to solve a problem quantitatively as well as answer a conceptual question dealing with the

---

$(\mathbf{F}_{net} = m\mathbf{a})$ and inserting the expression for centripetal acceleration from Eq. 7.9, we can write

$$F_c = ma_c = \frac{mv^2}{r} \qquad \begin{array}{l}\textit{magnitude of}\\\textit{centripetal force}\end{array} \qquad (7.11)$$

The centripetal force, like the centripetal acceleration, is directed radially toward the center of the circular path.

Keep in mind that, in general, a net force applied at an angle to the direction of motion of an object produces changes in the magnitude *and* direction of the veloc-ity. However, when a net force of constant magnitude is continuously applied at an angle of 90° to the direction of motion (as is centripetal force), only the direc-tion of the velocity changes. Also notice that because the centripetal force is al-ways perpendicular to the direction of motion, this force does no work. (Why?) Therefore, by the work–energy theorem, a centripetal force does not change the kinetic energy or speed of the object.

Note that the centripetal force in the form $F = mv^2/r$ is not really a new indi-vidual force, but rather the cause of the centripetal acceleration supplied by a real force or forces. In Example 7.5, the force supplying the centripetal acceleration was gravity. In Conceptual Example 7.7, it was the tension in the string. Another force that often supplies centripetal acceleration is friction. Suppose that an auto-mobile moves into a level, circular curve. To negotiate the curve, the car must have a centripetal acceleration, which is supplied by the force of friction between the tires and the road.

However, this (static; why?) friction has a maximum limiting value. If the speed of the car is high enough, the friction will not be sufficient to supply the necessary centripetal acceleration, and the car will skid outward from the cen-ter of the curve. If the car moves onto a wet or icy spot, the friction between the tires and the road may be reduced, allowing the car to skid at an even lower speed. (Banking a curve also helps vehicles negotiate the curve. See Exercises 45 and 59.)

PHYSLET® ILLUSTRATION

Centripetal Force

230 CHAPTER 7 Circular Motion and Gravitation

Integrated Example 11.3 ■ Cooking Class 101: Studying Specific Heats While Boiling Water

To prepare pasta, you bring a pot of water from room temperature (20°C) to its boil-ing point (100°C). The pot has a mass of 0.900 kg, is made of steel, and holds 3.00 L of water. (a) Which of the following is true? (1) The pot requires more heat, (2) the water requires more heat, or (3) they require the same amount of heat. (b) Determine the re-quired heat for both the water and the pot, and the ratio Q_w/Q_{pot}.

(a) Conceptual Reasoning. The temperature increase is the same for the water and the pot. Thus, the only factors that determine the difference in required heat are mass and specific heat. Since a liter of water has a mass of 1.0 kg, we have 3.0 kg of water to heat. This mass is more than three times the mass of the pot. From Table 11.1, the specif-ic heat of water is about nine times larger than that of steel. Thus, both factors indicate that the water will require significantly more heat than the pot, so the answer is (2).

(b) Thinking It Through. The heats can be found using Eq. 11.1, after looking up the specific heats and converting the volume of water to mass. The temperature change is easily determined from the initial and final values.

We list the data given and find the water's mass from its volume and density:

Given: $m_{pot} = 0.900$ kg   **Find:** The heat for the water and the
        $m_w = 3.00$ kg   pot and the heat ratio Q_w/Q_{pot}
        $c_{pot} = 460$ J/kg·C° (From Table 11.1)
        $c_w = 4186$ J/kg·C° (From Table 11.1)

In general, the amount of heat required is given by $Q = mc\Delta T$. The temperature in-crease for both objects is 80 C°. Thus, the heat required for the water is

$$Q_w = m_w c_w \Delta T_w$$
$$= (3.00 \text{ kg})(4186 \text{ J/kg·C°})(80 \text{ C°}) = 1.00 \times 10^6 \text{ J}$$

and the heat required for the pot is

$$Q_{pot} = m_{pot} c_{pot} \Delta T_{pot}$$
$$= (0.900 \text{ kg})(460 \text{ J/kg·C°})(80 \text{ C°}) = 3.31 \times 10^4 \text{ J}$$

The water requires over 30 times the heat, since

$$\frac{Q_w}{Q_{pot}} = \frac{1.00 \times 10^6 \text{ J}}{3.31 \times 10^4 \text{ J}} = 30.2$$

Follow-up Exercise. (a) In this Example, if the pot were made of the same mass of aluminum, would you expect the ratio of heat (water to pot) to be smaller or larger than the answer for the steel pot? Explain. (b) Verify your choice by calculating this ratio for the case of the aluminum pot.

Exercise. By answering both parts, students can see if their numerical answer matches their conceptual understanding.

Figure Reference Icon.   In the Fifth Edition, we have placed an arrow next to each in-text figure reference as well as next to each figure caption. These "placeholders" point the student in the direction of the appropriate figure and are easily located when the student returns to the sentence.

Chapter Review.   The Important Concepts and Equations section is integrated into the new Chapter Review section of each chapter. Key concepts are in bold and defined in words as well as symbolically. This new format provides a quick study reference for students.

We have continued to ensure accuracy through the Absolutely Zero Tolerance for Errors Club (The AZTECs). This team approach to accuracy checking worked quite well in the third and fourth editions, so we did it again. Bo Lou of Ferris State University, the author of our *Instructor's Solutions Manual*, headed the AZTEC team and was supported by the text's authors and two other accuracy checkers, Bill Mc-Corkle of West Liberty State University and Dave Curott of the University of North Alabama. Each member of the team individually and independently worked all end-of-chapter Exercises. The results were then collected, and any discrepancies were re-solved by a team discussion. All data in the chapters, as well as the answers at the back of the book, were checked and rechecked in first- and second-page proofs. In addition, five other physics teachers—Xiaochun He of Georgia State University, Jerry Shi of Pasadena City College, John Walkup of California Polytechnic State University at San Luis Obispo, William Dabby of Edison Community College, and Donald El-liott of Carroll College—read pages in detail, checking for errors in the chapter nar-rative, worked Examples, and text art. Although it is almost certainly not humanly possible to produce a physics text with absolutely no errors, that was our goal; we worked very hard to make the book as error free as possible.

The Fifth Edition is supplemented by a state-of-the-art Media and Print Ancil-lary package developed to address the needs of both students and instructors.

Companion Web Site.   Our Web site (http://www.prenhall.com/wilson), which hosts contributions from leaders in physics education research, provides students with a variety of interactive explorations of each chapter's topics, easily accommodating differences in learning styles. Student tools provided on the Web site include Physlet® Illustrations by Steve Mellema and Chuck Niederriter (Gustavus Adolphus College); Warm-Ups, Puzzles, and "What Is Physics Good For?" applications by Gregor Novak and Andy Gavrin (Indiana University–Purdue University, Indianapolis); award-winning Java-based Physlet® problems by Wolfgang Christian (Davidson College); algorithmically generated numerical Practice Problems, multiple-choice Practice Questions, and on-line destinations by Carl Adler (East Carolina University); Ranking Task Exercises edited by Tom O'Kuma (Lee College), David Maloney (Indiana University–Purdue University, Fort Wayne), and Curtis Hieggelke (Joliet Junior College); Chapter Objectives and Solutions to Select Exercises by Bo Lou (Ferris State University); and MCAT Questions by Glen Terrell (University of Texas at Arlington) and from ARCO's MCAT Supercourse. Using the Preferences module on the opening page of the site or the tool in the "Results reporter" part of each module, students can, at a professor's request, have the results of their work on the Companion Web site e-mailed to the professor or teaching assistant. Instructor tools include on-line grading capabilities and a Syllabus Manager. See pp. **xxx–xxxi** for further infor-mation about the modules in this site.

For the Instructor

Annotated Instructor's Edition (0-13-047193-3). The margins of the *Annotated Instructor's Edition* (AIE) contain an abundance of suggestions for classroom demonstrations and activities, along with teaching tips (points to emphasize, discussion suggestions, and common misunderstandings to avoid). In addition, the *AIE* contains

- Icons that identify each illustration reproduced as a transparency in the *Transparency Pack* and
- Answers to end-of-chapter Exercises (following each Exercise).

Instructor's Resource Manual (0-13-047180-1). Written by Kathy Whatley and Judy Beck (both of University of North Carolina–Asheville), the IRM, new to this edition, provides teaching suggestions, lecture outlines, notes, demonstrations, sample syllabi, and additional references and resources.

Instructor's Solutions Manual (0-13-047194-1). Prepared by Bo Lou of Ferris State University, the *Instructor's Solutions Manual* supplies answers with complete, worked-out solutions to all end-of-chapter exercises. Each solution has been checked for accuracy by a minimum of five instructors. This manual is also available electronically on both Windows (*0-13-047203-4*) and Macintosh (*0-13-047202-6*) platforms.

Test Item File (0-13-047196-8). Fully revised by Dave Curott of the University of North Alabama, the *Test Item File* now offers more than 2600 Multiple-Choice, Essay, True/False, and Fill-in-the-Blank questions. The questions are organized and referenced by chapter section and by question type.

Test Generator EQ (0-13-047778-8). New to the Fifth Edition, TestGenEQ is an easy-to-use, fully networkable software program for creating tests ranging from short quizzes to long exams. Questions from the *Test Item File,* including algorithmic versions, are supplied, and professors can use the Question Editor to modify existing questions or create new questions.

Transparency Pack (0-13-047199-2). The *Transparency Pack* contains more than 300 full-color acetates of text illustrations useful for class lectures. It is available upon adoption of the text.

Media Portfolio CD-ROM (0-13-047190-9). Prepared by Sue Willis (Northern Illinois University), this CD-ROM, new to the Fifth Edition, contains all the art from the text in JPEG format, for easy incorporation into presentation software. It is also available as a password-protected module on the Companion Web site.

"Physics You Can See" Video Demonstrations (0-205-12393-7). Each segment, 2–5 minutes long, demonstrates a classical physics experiment. Eleven segments are included, such as "Coin & Feather" (acceleration due to gravity), "Monkey & Gun" (rate of vertical free fall), "Swivel Hips" (force pairs), and "Collapse a Can" (atmospheric pressure).

Peer Instruction (0-13-656441-6). Authored by Eric Mazur (Harvard University), this manual explains peer instruction, an interactive teaching style that actively involves students in the learning process by focusing their attention on underlying concepts through interactive "ConcepTests," reading quizzes, and conceptual exam questions. Results are assessed though scores on the Force Concept Inventory and final exams, showing that students better understand concepts and perform more highly on conventional problems in this environment. Peer instruction can be easily adapted to fit individual lecture styles and used in a variety of settings.

Just-in-Time Teaching: Blending Active Learning with Web Technology (0-13-085034-9). Just-in-Time Teaching (JiTT) is an exciting teaching and learning methodology designed to engage students. Using feedback from preclass Web assignments, instructors can adjust classroom lessons so that students receive rapid response to the specific questions and problems they are having—instead of more generic lectures that may or may not address topics with which students actually need help. Many teachers have found that this process makes students become active and interested learners. In this resource book for educators, authors Gregor Novak (Indiana University–Purdue University, Indianapolis), Evelyn Patterson (United States Air Force Academy), Andrew Gavrin (Indiana University–Purdue University, Indianapolis), and Wolfgang Christian (Davidson College) more fully explain what Just-in-Time Teaching is, its underlying goals and philosophies, and how to implement it. They also provide an extensive section of tested resource materials that can be used in introductory physics courses with the JiTT approach.

Ranking Task Exercises in Physics (0-13-022355-7). This book, by Thomas L. O'Kuma (Lee College), David P. Maloney (Indiana University–Purdue University, Fort Wayne), and Curtis J. Hieggelke (Joliet Junior College), describes ranking tasks, which are an innovative type of conceptual exercise that asks students to make comparative judgments about a set of variations on a particular physical situation. This text is a unique resource for physics instructors who are looking for tools to incorporate more conceptual analysis in their courses. This supplement contains approximately 200 Ranking Task Exercises that cover all classical physics topics (with the exception of optics).

Physlets®: Teaching Physics with Interactive Curricular Material (0-13-029341-5). Authored by Wolfgang Christian and Mario Belloni (both of Davidson College), this text is a teacher's resource book with an accompanying CD for instructors who are interested in incorporating Physlets® into their physics courses. The book and CD discuss the pedagogy behind the use of Physlets® and provide instructors with information on how to author their own interactive curricular material, using Physlets®.

For the Student

Student Study Guide and Solutions Manual (0-13-047195-X). Updated by Bo Lou of Ferris State University, the *Student Study Guide and Solutions Manual* presents chapter-by chapter reviews, chapter summaries, additional worked examples, and solutions to paired and selected exercises.

Student Pocket Guide (0-13-047192-5). Written by Biman Das (State University of New York–Potsdam), this easy-to-carry 5" $\times$ 7" paperback contains a summary of the entire text, including all key concepts and equations, as well as tips and hints. Perfect for carrying to lectures and taking notes in.

MCAT Physics Study Guide (0-13-627951-1). This study resource, by Joseph Boone of California Polytechnic State University–San Luis Obispo, references all of the physics topics on the MCAT to the appropriate sections in the text.

Since most MCAT questions require more thought and reasoning than simply plugging numbers into an equation, this study guide is designed to refresh students' memory about the topics they've covered in class. Additional review, practice problems, and review questions are included.

Tutorials in Introductory Physics (0-13-097069-7). Authored by Lillian C. McDermott, Peter S. Schaffer, and the Physics Education Group at the University of Washington, this landmark book presents a series of physics tutorials designed by a leading physics education research group. Emphasizing the development of concepts and scientific reasoning skills, the tutorials focus on the specific conceptual and reasoning difficulties that students tend to encounter. The tutorials cover a range of topics in Mechanics, E & M, and Waves and Optics.

Interactive Physics Player Workbook (0-13-067108-8). Written by Cindy Schwarz of Vassar College, this highly interactive workbook and software package contains simulation projects of varying difficulty. Each includes a physics review, simulation details, hints, an explanation of results, math help, and a self-test.

Acknowledgments

We would like to acknowledge the generous assistance we received from many people during the preparation of the Fifth Edition. First, our sincere thanks go to Bo Lou of Ferris State University for his vital contributions to the chapters on optics. His meticulous, conscientious help with checking solutions and answers to problems, as well as preparing the *Instructor's Solutions Manual*, the answer keys for the back of the book, and the *Student Study Guide and Solutions Manual* are greatly appreciated. We are similarly grateful to Dave Curott of the University of North Alabama for preparing the *Test Item File* as well as for his participation as an accuracy checker for all solutions to end-of-chapter exercises.

Indeed, all the members of AZTEC—Bo Lou, Dave Curott, and Bill McCorkle (West Liberty State University)—as well as the reviewers of first-and second-page proofs—William Dabby (Edison Community College), Donald Elliott (Carroll College), Xiaochun He (Georgia State University), Jerry Shi (Pasadena City College), John Walkup (California Polytechnic State University at San Luis Obispo)—deserve more than a special thanks for their tireless, timely, and extremely thorough review of all materials in the book for scientific accuracy.

Dozens of other colleagues, listed in the upcoming section, helped us with reviews of the Fourth Edition to help us plan the Fifth Edition, as well as with reviews of manuscript as it was developed. We are indebted to them, as their thoughtful and constructive suggestions benefited the book greatly.

The editorial staff of Prentice Hall continued to be particularly helpful. First, we would like to extend a heartfelt thanks to our former editor, Alison Reeves, for initiating this revision and providing insightful direction. We are grateful to Mary Catherine Hager, Development Editor; Patrick Burt, Project Manager for the book; and Beth Sturla Sweeten, Assistant Managing Editor, who kept the whole complex endeavor moving forward, while designer Jonathan Boylan made sure that the ultimate physical presentation would be both visually engaging and clean and easy to use. We also thank Mark Pfaltzgraff, Executive Marketing Manager; Christian Botting, Assistant Editor, for his extensive work on the supplements, media program, and review program; Erik Fahlgren, Acquisitions Editor, and Eileen Nee, Editorial Assistant, for their help in coordinating all of these facets; and John Challice, Editor in Chief, for his support and encouragement.

In addition, I (Tony Buffa) once again extend many thanks to my coauthor, Professor Jerry Wilson, for his cheerful helpfulness and professional approach to the work on this edition. I am also indebted to Professor Bo Lou, who contributed many good comments and ideas that helped enormously. As always, several colleagues of mine at Cal Poly gave of their time for fruitful discussions. Among them are Professors Joseph Boone, Ronald Brown, Theodore Foster, Richard Frankel, and John Walkup. My family—my wife, Connie, and daughters, Jeanne and Julie—was, as always, a continuous and welcomed source of support. I also acknowledge the support from my father, Anthony Buffa, Sr., and my aunt, Dorothy Abbott. Last, I thank the students in my classes who contributed excellent ideas over the past few years.

Finally, both of us would like to urge anyone using the book—student or instructor—to pass on to us any suggestions that you have for its improvement. We look forward to hearing from you.

—*Jerry D. Wilson*
jwilson@ais-gwd.com
—*Anthony J. Buffa*
abuffa@calpoly.edu

Reviewers of the Fifth Edition

Alice Hawthorne Allen
Virginia Tech

Anand Batra
Howard University

Michael Berger
Indiana University

James Borgardt
Juniata College

Jeffrey Braun
University of Evansville

Michael Browne
University of Idaho

Mike Broyles
Collin County Community College

James Carroll
Eastern Michigan State University

Robert Coakley
University of Southern Maine

Sergio Conetti
University of Virginia, Charlottesville

William Dabby
Edison Community College

Purna Das
Purdue University

Donald Elliott
Carroll College

Lewis Ford
Texas A&M University

Gary Hastings
Georgia State University

Xiaochun He
Georgia State University

Andy Hollerman
University of Louisiana, Layfayette

Randall Jones
Loyola University

Kevin Lee
University of Nebraska

Paul Lee
California State University, Northridge

William McCorkle
West Liberty State University

Paul Morris
Abilene Christian University

K. W. Nicholson
Central Alabama Community College

Erin O'Connor
Allan Hancock College

Anthony Pitucco
Glendale Community College

William Pollard
Valdosta State University

David Rafaelle
Glendale Community College

Robert Ross
University of Detroit–Mercy

Craig Rottman
North Dakota State University

Om Rustgi
Buffalo State College

Cindy Schwarz
Vassar College

Bartlett Sheinberg
Houston Community College

Jerry Shi
Pasadena City College

Christopher Sirola
Tri-County Technical College

Gene Skluzacek
St. Petersburg College

Ross Spencer
Brigham Young University

Mark Sprague
East Carolina University

Gabriel Umerah
Florida Community College–Jacksonville

Lorin Vant-Hull
University of Houston

Karl Vogler
Northern Kentucky University

John Walkup
California Polytechnic State University

Larry Weinstein
Old Dominion University

Jeffery Wragg
College of Charleston

Rob Wylie
Carl Albert State University

Reviewers of Previous Editions

William Achor
Western Maryland College

Arthur Alt
College of Great Falls

Zaven Altounian
McGill University

Frederick Anderson
University of Vermont

Charles Bacon
Ferris State College

Ali Badakhshan
University of Northern Iowa

William Berres
Wayne State University

Hugo Borja
Macomb Community College

Bennet Brabson
Indiana University

Michael Browne
University of Idaho

David Bushnell
Northern Illinois University

Lyle Campbell
Oklahoma Christian University

Aaron Chesir
Lucent Technologies

Lowell Christensen
American River College

Philip A. Chute
University of Wisconsin–Eau Claire

Lawrence Coleman
University of California–Davis

Lattie F. Collins
East Tennessee State University

James Cook
Middle Tennessee State University

David M. Cordes
Belleville Area Community College

James R. Crawford
Southwest Texas State University

J. P. Davidson
University of Kansas

Donald Day
Montgomery College

Richard Delaney
College of Aeronautics

James Ellingson
College of DuPage

Arnold Feldman
University of Hawaii

John Flaherty
Yuba College

Rober J. Foley
University of Wisconsin–Stout

Donald Foster
Wichita State University

Donald R. Franceschetti
Memphis State University

Frank Gaev
ITT Technical Institute–Ft. Lauderdale

Rex Gandy
Auburn University

Simon George
California State–Long Beach

Barry Gilbert
Rhode Island College

Richard Grahm
Ricks College

Tom J. Gray
University of Nebraska

Douglas Al Harrington
Northeastern State University

J. Erik Hendrickson
University of Wisconsin–Eau Claire

Al Hilgendorf
University of Wisconsin–Stout

Joseph M. Hoffman
Frostburg State University

Jacob W. Huang
Towson University

Omar Ahmad Karim
University of North Carolina–Wilmington

S. D. Kaviani
El Camino College

Victor Keh
ITT Technical Institute–Norwalk, California

John Kenny
Bradley University

James Kettler
Ohio University, Eastern Campus

Dana Klinck
Hillsborough Community College

Chantana Lane
University of Tennessee–Chattanooga

Phillip Laroe
Carroll College

Rubin Laudan
Oregon State University

Bruce A. Layton
Mississippi Gulf Coast Community College

R. Gary Layton
Northern Arizona University

Federic Liebrand
Walla Walla College

Mark Lindsay
University of Louisville

Bryan Long
Columbia State Community College

Multimedia Explorations of Physics

Companion Website www.prenhall.com/wilson

This text-specific Website for introductory physics provides students and instructors with a wealth of innovative on-line materials for use with *College Physics, Fifth Edition.*

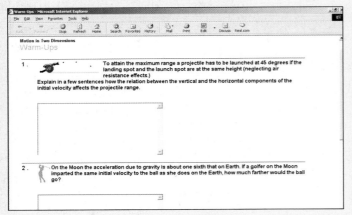

Warm-Ups & Puzzles
by Gregor Novak and Andrew Gavrin (Indiana University–Purdue University, Indianapolis)
Warm-Up and Puzzle questions are real-world short-answer questions based on important concepts in the text chapters. Both types of questions get students' attention, often refer to current events, and are good discussion starters. The **Warm-Ups** are designed to help introduce a topic, whereas the **Puzzles** are more complex and often require the integration of more than one concept. Thus, professors can assign Warm-Up questions after students have read a chapter but before the class lecture on that topic, and Puzzle questions as follow-up assignments submitted after class.

Physlet® Illustrations
by Steve Mellema and Chuck Niederritter (Gustavus Adolphus College)
New to the fifth edition, **Physlet® Illustrations** are short interactive Java applets that clearly illustrate, through animation, a concept from the text. Available on the Wilson/Buffa Companion Website, Physlet® Illustrations are often followed by a series of questions that ask students to think critically about the concept at hand. Physlet® Illustrations are denoted by an icon in the margin of the text:

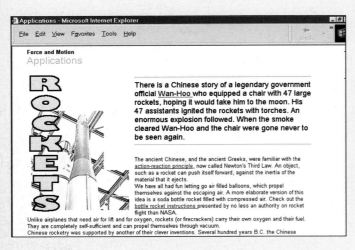

Applications
by Gregor Novak and Andrew Gavrin (Indiana University–Purdue University, Indianapolis)
The **Applications** modules answer the question "What is physics good for?" by connecting physics concepts to real-world phenomena and new developments in science and technology. These illustrated essays include embedded Web links to related sites, one for each chapter. Each essay is followed by short-answer/essay questions, which professors can assign for extra credit.

Physlet® Problems

by Wolfgang Christian (Davidson College)

Physlet® Problems are multimedia-focused problems based on Wolfgang Christian's award-winning Java applets for physics, called Physlets®. With these problems, students use multimedia elements to help solve a problem by observing, applying appropriate physics concepts, and making measurements of parameters they deem important. No numbers are given, so students are required to consider a problem qualitatively instead of plugging numbers into formulas.

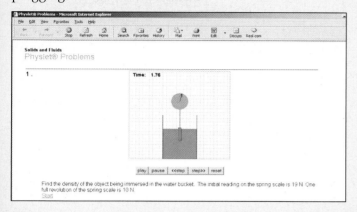

Practice Questions

by Carl Adler (East Carolina University)

Two modules of 10 to15 multiple-choice **Practice Questions** are available for review with each chapter.

MCAT Study Guide

by Glen Terrell (University of Texas at Arlington) and ARCO's MCAT Supercourse

For all relevant chapters, the MCAT Study Guide module provides students with an average of 25 multiple-choice questions on topics and concepts covered on the MCAT exam. As with all multiple-choice modules, the computer automatically grades and scores student responses and provides cross-references to corresponding text sections.

Practice Problems

by Carl Adler (East Carolina University)

Ten algorithmically generated numerical **Practice Problems** per chapter allow students to get multiple iterations of each problem set for practice.

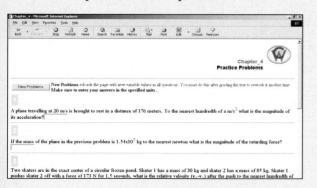

Destinations

Destinations are links to relevant Websites for each chapter, either about the physics topic in the chapter or about related applications.

Solutions to Selected Exercises

by Bo Lou (Ferris State University)

Solutions to six selected exercises per chapter from Bo Lou's *Student Study Guide and Solutions Manual*.

Syllabus Manager

Wilson/Buffa's **Syllabus Manager** provides instructors with an easy, step-by-step process for creating and revising a class syllabus with direct links to the text's Companion Website and other on-line content. Through this on-line syllabus, instructors can add assignments and send announcements to the class with the click of a button. The completed syllabus is hosted on Prentice Hall's servers, allowing the syllabus to be updated from any computer with Internet access.

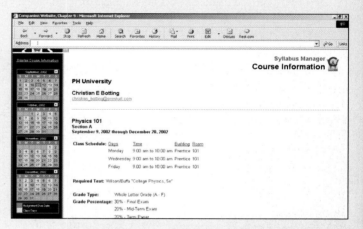

Ranking Task Exercises

edited by Thomas O'Kuma (Lee College), David Maloney (Indiana University–Purdue University, Fort Wayne), and Curtis Hieggelke (Joliet Junior College)

Available for most text chapters as PDF files downloadable from the Wilson/Buffa Website, **Ranking Tasks** are conceptual exercises that require students to rank a number of situations or variations of a situation. Engaging in this process of making comparative judgments helps students reason about physical situations and often gives them new insights into relationships among various concepts and principles.

On-line Grading

Scoring for all objective questions and problems, as well as responses to essay questions, can be e-mailed back to the instructor.

Course and Homework Management Tools
for *College Physics*, Fifth Edition

Blackboard

Blackboard ™ is a comprehensive and flexible e-Learning software platform that delivers a course management system, customizable institution-wide portals, online communities, and an advanced architecture that allows for Web-based integration of multiple administrative systems.

Features:

- Course Management System including progress tracking, class and student management, grade book, communication, assignments, and reporting tools.
- Testing programs with the capability for instructors to create online quizzes and tests, and automatically grade and track results. All tests can be entered into the grade book for easy course management.
- Communications Tools such as the Virtual Classroom™ (chat rooms, Whiteboard, slides), document sharing, and bulletin boards.

CourseCompass
Powered by **Blackboard**

With the highest level of service, support, and training available today, CourseCompass combines tested, quality online course resources with easy-to-use online course management tools. CourseCompass is designed to address the individual needs of instructors, who will be able to create an online course without any special technical skills or training.

Features:

- Greater ease of use—you will be able to create an online course in 15 minutes or less.
- Higher flexibility—Professors can adapt Prentice Hall content to match their own teaching goals, with little or no outside assistance needed.
- Assessment, customization, class administration, and communication tools.
- Point-and-click access—Prentice Hall's vast educational resources are available to instructors at a click of the mouse.
- A nationally hosted and fully supported system that relieves individuals and institutions of the burdens of trouble-shooting and maintenance.

WebCT—WebCT offers a powerful set of tools that enables you to create practical Web-based educational programs- the ideal resources to enhance a campus course or to construct one entirely online. The WebCT shell and tools, integrated with the *College Physics, Fifth Edition* content, results in a versatile, course-enhancing teaching and learning system.

Features:

- Page tracking, progress tracking, class and student management, grade book, communication, calendar, reporting tools, and more.
- Communication tools including chat rooms, bulletin boards, private e-mail, and WhiteBoard.
- Testing tools that help create and administer timed online quizzes and tests, and automatically grade and track all results.

WebAssign

WebAssign unlocks the door to a new way of teaching—to a classroom without walls, where time isn't a boundary and record keeping is no longer a constraint. WebAssign's unparalleled homework delivery service harnesses the power of the Internet and puts it to work for you. Collecting and grading homework becomes the province of the WebAssign service, which gives you the freedom to get back to teaching. Create assignments from a database of exercises from *College Physics, Fifth Edition*, or write and customize your own exercises. You have complete control over the homework your students receive, including due date, content, feedback, and question formats.

Features:

- Create, post, and review assignments 24 hours a day, 7 days a week.
- Deliver, collect, grade, and record assignments instantly.
- Offer more practice exercises, quizzes, homework, labs, and tests.
- Randomize numerical values or phrases to create unique questions.
- Assess student performance to keep abreast of individual progress.
- Grade algebraic formulas with enhanced math-type display.
- Capture the attention of your online, distance learning students.

Measurement and Problem Solving

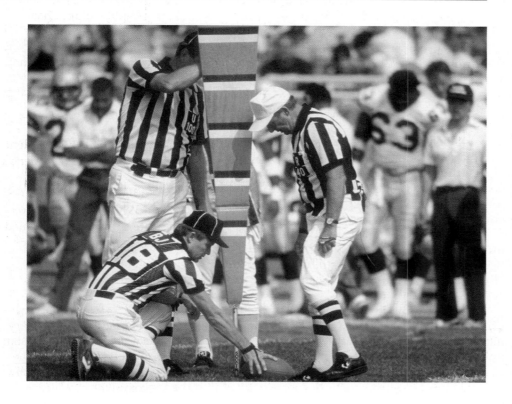

Is it first and 10? A measurement is needed, as with many other things in our lives. Length measurements tell us how far it is between cities, how tall you are, and as in the photo, if it's first and 10. Time measurements tell you how long it is until the class ends, when the semester or quarter begins, and how old you are. Drugs taken because of illnesses are given in measured doses. Lives depend on various measurements made by doctors, medical technologists, and pharmacists in the diagnosis and treatment of disease.

Measurements allow us to compute quantities and solve problems. For example, suppose you were asked to find the volume of a cylindrical container. How would you do it? By measuring the diameter (d) and height (h) of the container, and computing the volume from the standard equation $V = \pi r^2 h$, where r, the radius, is one-half the measured diameter. Your problem is solved. Units also come into such problem solving. If you had measured the cylinder's dimensions in inches, the volume would be in in.^{3} (cubic inches); if in centimeters, then cm^3 (cubic centimeters).

Measurement and problem solving are part of our lives. They play a particularly central role in our attempts to describe and understand the physical world, as we shall see in this chapter.

1.1 Why and How We Measure

OBJECTIVE: **To distinguish standard units and systems of units.**

Imagine that someone is giving you directions to her house. Would you find it helpful to be told, "Drive along Elm Street for a little while, and turn right at one of the lights. Then keep going for quite a long way."? Or would you want to deal with a bank that sent you a statement at the end of the month saying, "You still have some money left in your account. Not a great deal, though."?

Measurement is important to all of us. It is one of the concrete ways in which we deal with our world. This concept is particularly true in physics. *Physics is concerned with the description and understanding of nature,* and measurement is one of its most important tools.

There certainly are ways of describing the physical world that do not involve measurement. For instance, we might talk about the color of a flower or a dress. But perception of color is subjective: It may vary from one person to another. Indeed, many people are color-blind and cannot tell certain colors apart. Light can also be described in terms of wavelengths and frequencies. Different wavelengths are associated with different colors because of the physiological response of our eyes to light. But unlike the sensations or perceptions of color, the wavelengths can be measured. They are the same for everyone. In other words, they are *objective. Physics attempts to describe nature in an objective way through measurements.*

Standard Units

Measurements are expressed in unit values, or units. As you are probably aware, a large variety of units is used to express measured values. Some of the earliest units of measurement, such as the foot, were originally referenced to parts of the human body. (Even today, the hand is still used as a unit to measure the height of horses.) If a unit becomes officially accepted, it is called a **standard unit**. Traditionally, a government or international body establishes standard units.

A group of standard units and their combinations is called a **system of units**. Two major systems of units are in use today—the metric system and the British system. The latter is still widely used in the United States, but has virtually disappeared in the rest of the world, having been replaced by the metric system.

Different units in the same system or units from different systems can be used to describe the same thing. For example, your height can be expressed in inches, feet, centimeters, meters—or even miles, for that matter (although this unit would not be very convenient). It is always possible to convert from one unit to another, and such conversion is sometimes necessary. It is best, however, and certainly most practical, to work consistently within the same system of units, as you will see.

1.2 SI Units of Length, Mass, and Time

OBJECTIVES: To (a) describe the SI and (b) specify the references for the three main base quantities of this system.

Length, mass, and time are fundamental physical quantities that describe a great many objects and phenomena. In fact, the topics of mechanics (the study of motion and force) covered in the first part of this book require *only* these physical quantities. The system of units used by scientists to represent these and other quantities is based on the metric system

Historically, the metric system was the outgrowth of proposals for a more uniform system of weights and measures in France during the 17th and 18th centuries. The modern version of the metric system is called the **International System of Units**, officially abbreviated as **SI** (from the French *Système International des Unités*).

The SI includes *base quantities* and *derived quantities*, which are described by base units and derived units, respectively. **Base units**, such as the meter and the kilogram, are represented by standards. Other quantities that may be expressed in terms of combinations of base units are called **derived units**. (Think of how we commonly measure the length of a trip in miles and the amount of time the trip takes in hours. To express how fast we travel, we use the derived unit of miles per hour, which represents distance per unit of time, or length per time.)

One of the refinements of the SI was the adoption of new standard references for some base units, including those of length and of time.

Length

Length is the base quantity used to measure distances or dimensions in space. We commonly say that length is the distance between two points. But the distance between any two points depends on how the space between them is traversed, which may be in a straight or a curved path.

The SI unit of length is the **meter (m)**. The meter was originally defined as 1/10 000 000 of the distance from the North Pole to the equator along a meridian running through Paris (▼Fig. 1.1a).* A portion of this meridian between Dunkirk, France, and Barcelona, Spain, was surveyed to establish the standard length, which was assigned the name *metre*, from the Greek word *metron*, meaning "a measure." (The American spelling is *meter*.) A meter is 39.37 inches—slightly longer than a yard.

The length of the meter was initially preserved in the form of a material standard: the distance between two marks on a metal bar (made of a platinum–iridium alloy) that was stored under controlled conditions and called the Meter of the Archives. However, it is not desirable to have a reference standard that changes with external conditions, such as temperature. In 1983, the meter was redefined in terms of a more accurate standard, an unvarying property of light: the length of the path traveled by light in a vacuum during an interval of 1/299 792 458 of a second (Fig. 1.1b). In other words, light travels 299 792 458 meters in a second, and the speed of light in a vacuum is defined to be 299 792 458 meters per second. Note that the length standard is referenced to time, which can be measured with great accuracy.

▼ **FIGURE 1.1   The SI length standard: the meter**   (a) The meter was originally defined as 1/10 000 000 of the distance from the North Pole to the Equator along a meridian running through Paris, of which a portion was measured between Dunkirk and Barcelona. A metal bar (called the Meter of the Archives) was constructed as a standard. **(b)** The meter is currently defined in terms of the speed of light.

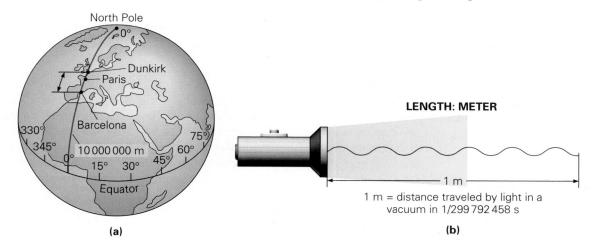

(a)   (b)

*Note that this book and most physicists have adopted the practice of writing large numbers with a thin space for three-digit groups—for example, 10 000 000 (not 10,000,000). This is done to avoid confusion with the European practice of using a comma as a decimal point. For instance, 3.141 in the United States would be written as 3,141 in Europe. Large decimal numbers, such as 0.537 84, may also be separated, for consistency. Spaces are generally used for numbers with more than four digits on either side of the decimal point.

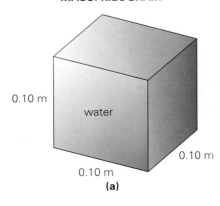

0.10 m

water

0.10 m

0.10 m

(a)

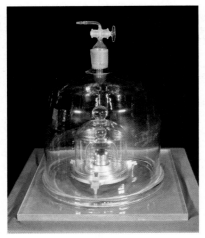

(b)

▲ **FIGURE 1.2 The SI mass standard: the kilogram (a)** The kilogram was originally defined in terms of a specific volume of water, that of a cube 0.10 m on a side, thereby associating the mass standard with the length standard. The standard kilogram is now defined by a metal cylinder. **(b)** The international prototype of the kilogram is kept at the French Bureau of Weights and Measures. It was manufactured in the 1880s of an alloy of 90% platinum and 10% iridium. Copies have been made for use as 1 kg national prototypes, one of which is the mass standard for the United States. It is kept at the National Institute of Standards and Technology (NIST) in Washington, D.C.

Mass

Mass is the base quantity used to describe amounts of matter. The more massive an object, the more matter it contains. (We will encounter more precise definitions of mass in Chapters 4 and 7.)

The SI unit of mass is the **kilogram** (**kg**). The kilogram was originally defined in terms of a specific volume of water, but is now referenced to a specific material standard: the mass of a prototype platinum–iridium cylinder kept at the International Bureau of Weights and Measures in Sèvres, France (◀Fig. 1.2). The United States has a duplicate of the prototype cylinder. The duplicate serves as a reference for secondary standards that are used in everyday life and commerce. (The kilogram may eventually be referenced to something less subject to time and chance than a piece of metal.)

You may have noticed that the phrase "weights and measures" is generally used instead of "masses and measures." In the SI, mass is a base quantity, but in the more familiar British system, weight is used instead to describe amounts of mass—for example, weight in pounds instead of mass in kilograms. The weight of an object is the gravitational attraction that the Earth exerts on the object. For example, when you weigh yourself on a scale, your weight is a measure of the downward gravitational force exerted on you by the Earth. We can use weight in this way because near the Earth's surface, mass and weight are directly proportional to each other: They differ only by a particular constant.

But treating weight as a base quantity creates some problems. A base quantity is naturally most useful if its value is the same everywhere. This is the case with mass—an object has the same mass, or amount of matter, regardless of its location. *But it is not true of weight.* For example, the weight of an object on the Moon is less than its weight on the Earth. This is because the Moon is less massive than the Earth, and the gravitational attraction exerted on an object by the Moon (i.e., the object's weight) is less than that exerted by the Earth. That is, an object with a given amount of mass has a particular weight on the Earth, but on the Moon, the same amount of mass will weigh only about one sixth as much. Similarly, the weight of an object would vary on the surfaces of different planets.

For now, keep in mind that *weight is related to mass, but they are not the same.* Since the weight of an object of a certain mass can vary with location, it is much more useful to take mass as the base quantity, as the SI does. Base quantities should remain the same regardless of where they are measured, under normal or standard conditions. The distinction between mass and weight will be more fully explained in a later chapter. Our discussion until then will be chiefly concerned with mass.

Time

Time is a difficult concept to define. (Try it.) A common definition is that time is the forward flow of events. This statement is not so much a definition as an observation that time has never been known to run backward, as it might appear to do when you view a film run backward in a projector. Time is sometimes said to be a fourth dimension, accompanying the three dimensions of space (x, y, z, t). Thus, if something exists in space, it also exists in time. In any case, events can be used to mark time measurements. The events are analogous to the marks on a meterstick used for measurements of length.

The SI unit of time is the **second** (**s**). The solar "clock" was originally used to define the second. A solar day is the interval of time that elapses between two successive crossings of the same longitude line (meridian) by the Sun at its highest point in the sky at that longitude. A second was fixed as $1/86\,400$ of this apparent solar day (1 day = 24 h = 1440 min = 86 400 s). However, the elliptical path of the Earth's motion around the Sun causes apparent solar days to vary in length.

As a more precise standard, an average, or mean, solar day was computed from the lengths of the apparent solar days during a solar year. In 1956, the second

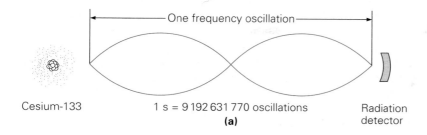

One frequency oscillation

Cesium-133   1 s = 9 192 631 770 oscillations   Radiation
                              (a)   detector

(b)

▲ **FIGURE 1.3   The SI time standard: the second**   The second was once defined in terms of the average solar day. **(a)** It is now defined by the frequency of the radiation associated with an atomic transition. **(b)** The atomic fountain "clock" shown here, at NIST, is the time standard for the United States. The variation of this "timepiece" is less than one second per 20 million years.

was referenced to this mean solar day. But the mean solar day is not exactly the same for each yearly period, because of minor variations in the Earth's motions and a steady slowing of its rate of rotation, due to tidal friction. So scientists kept looking for something better.

In 1967, an atomic standard was adopted as a better reference. The second was defined by the radiation frequency of the cesium-133 atom. This "atomic clock" used a beam of cesium atoms to maintain our time standard, with a variation of about one second in 300 years. In 1999, another cesium-133 atomic clock was adopted, the atomic fountain clock, which, as the name implies, is based on the radiation frequency of a fountain of cesium atoms rather than a beam (▲Fig. 1.3). The variation of this timepiece is less than one second per 20 million years!*

SI Base Units

The complete SI has seven base quantities and base units. In addition to the meter, kilogram, and second for (1) length, (2) mass, and (3) time, respectively, we measure (4) electric current (charge/second) in amperes (A), (5) temperature in kelvins (K), (6) amount of substance in moles (mol), and (7) luminous intensity in candelas (cd). See Table 1.1.

The foregoing quantities are thought to compose the smallest number of base quantities needed for a full description of everything observed or measured in nature. There are also two supplemental units for angular measure: the two-dimensional plane angle and the three-dimensional solid angle. Scientists have not reached general agreement about whether these geometric units are base or derived units.

---

*An even more precise clock, the all-optical atomic clock, is under development. It is so named because it uses laser technology and measures the shortest time interval ever recorded—100 000 times shorter than that of the best current clocks. The new clock does not use cesium atoms, but rather a single cooled ion of liquid mercury linked to a laser oscillator. The frequency of the mercury ion is 100 000 times higher than the frequency of cesium atoms, hence the shorter, more precise time interval.

TABLE 1.1   The Seven Base Units of the SI

Name of Unit (abbreviation)	Property Measured
meter (m)	length
kilogram (kg)	mass
second (s)	time
ampere (A)	electric current
kelvin (K)	temperature
mole (mol)	amount of substance
candela (cd)	luminous intensity

PHYSLET®
ILLUSTRATION

Introduction to Physlet Illustrations

1.3 More about the Metric System

OBJECTIVES: To use common (a) metric prefixes and (b) nonstandard metric units.

The metric system involving the standard units of length, mass, and time, now incorporated in the SI was once called the **mks system** (for *meter–kilogram–second*). Another metric system that has been used in dealing with relatively small quantities is the **cgs system** (for *centimeter–gram–second*). In the United States, the system still generally in use is the British (or English) engineering system, in which the standard units of length, mass, and time are foot, slug, and second, respectively. You may not have heard of the slug, because, as we mentioned earlier, gravitational force (weight) is commonly used instead of mass—pounds instead of slugs—to describe quantities of matter. As a result, the British system is sometimes called the **fps system** (for *foot–pound–second*).

The metric system is predominant throughout the world and is coming into increasing use in the United States. Primarily because it is simple mathematically, it is the preferred system of units for science and technology. SI units are used throughout most of this book. All quantities can be expressed in SI units. However, some units from other systems are accepted for limited use as a matter of practicality—for example, the time unit of hour and the temperature unit of degree Celsius. British units will sometimes be used in the early chapters for comparison purposes, since these units are still employed in everyday activities and for many practical applications.

The increasing worldwide use of the metric system means that you should be familiar with it. One of the greatest advantages of the metric system is that it is a decimal, or base-10, system. This means that larger or smaller units are obtained by multiplying or dividing, respectively, a base unit by powers of 10. A list of some multiples and corresponding prefixes for metric units is given in Table 1.2. In decimal measurements, the prefixes milli-, centi-, and kilo- are the ones most commonly used. The decimal characteristic makes it convenient to change measurements from one size of metric unit to another. With the familiar British system, different conversion factors must be used, such as 16 for converting pounds to ounces and 12 for converting feet to inches. The British system developed historically and not very scientifically.

TABLE 1.2 Some Multiples and Prefixes for Metric Units*

Multiple†	Prefix (and Abbreviation)	Pronunciation	Multiple†	Prefix (and Abbreviation)	Pronunciation
10^{12}	tera- (T)	ter'a (as in *terra*ce)	10^{-2}	centi- (c)	sen'ti (as in *senti*mental)
10^{9}	giga- (G)	jig'a (*jig* as in *jig*gle, *a* as in *a*bout)	10^{-3}	milli- (m)	mil'li (as in *mili*tary)
10^{6}	mega- (M)	meg'a (as in *mega*phone)	10^{-6}	micro- (μ)	mi'kro (as in *micro*phone)
10^{3}	kilo- (k)	kil'o (as in *kilo*watt)	10^{-9}	nano- (n)	nan'oh (*an* as in *ann*ual)
10^{2}	hecto- (h)	hek'to (*heck-toe*)	10^{-12}	pico- (p)	pe'ko (*peek-oh*)
10	deka- (da)	dek'a (*deck* plus *a* as in *a*bout)	10^{-15}	femto- (f)	fem'toe (*fem* as in *femi*nine)
10^{-1}	deci- (d)	des'i (as in *deci*mal)			

*For example, 1 gram (g) multiplied by 1000, or 10^3, is 1 kilogram (kg); 1 gram multiplied by 1/1000, or 10^{-3}, is 1 milligram (mg).

†The most commonly used prefixes are printed in blue. Note that the abbreviations for the multiples 10^6 and greater are capitalized, whereas the abbreviations for the smaller multiples are lowercased.

You are already familiar with one base-10 system—U.S. currency. Just as a meter can be divided into 10 decimeters, 100 centimeters, or 1000 millimeters, the "base unit" of the dollar can be broken down into 10 "decidollars" (dimes), 100 "centidollars" (cents), or 1000 "millidollars" (tenths of a cent, or mills, used in figuring property taxes and bond levies). Since all the metric prefixes are powers of 10, there are no metric analogues for quarters or nickels.

The official metric prefixes can help eliminate confusion. In the United States, a billion is a thousand million (10^9); in Great Britain, a billion is a million million (10^{12}). The use of metric prefixes eliminates any confusion, since giga- indicates 10^9 and tera- stands for 10^{12}. You may be hearing more about nano, a prefix that indicates 10^{-9}, and the nanometer. See the Insight on *nano*technology.

Volume

In the SI, the standard unit of volume is the cubic meter (m^3)—the three-dimensional derived unit of the meter base unit. Because this unit is rather large, it is often more convenient to use the nonstandard unit of volume (or capacity) of a cube 10 cm (centimeters) on a side. This volume was given the name *litre*, which is spelled **liter**, and abbreviated as L in the United States. The volume of a liter is $1000 \ cm^3 (10 \ cm \times 10 \ cm \times 10 \ cm)$. Since 1 L = 1000 mL (milliliters), it follows that $1 \ mL = 1 \ cm^3$. (The cubic centimeter is sometimes abbreviated as cc, particularly in chemistry and biology.) See ▶ Fig. 1.4a.

Note: Liter is sometimes abbreviated as a lowercase "ell" (l), but a capital "ell" (L) is preferred in the United States so that the abbreviation is less likely to be confused with the numeral one. (Isn't 1 L clearer than 1 l ?)

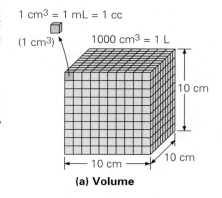

(a) Volume

Example 1.1 ■ The Metric Ton (or Tonne): Another Unit of Mass

As we have seen, the metric unit of mass was originally related to the length standard, with a liter ($1000 \ cm^3$) of water having a mass of 1 kg. The standard metric unit of volume is the cubic meter (m^3) and this volume of water was used to define a larger unit of mass called the *metric ton* (or *tonne*, as it is sometimes spelled). A metric ton is equivalent to how many kilograms?

Thinking It Through. A cubic meter is a relatively large volume and holds a large amount of water (more than a cubic yard; why?). The key is to find how many cubic volumes measuring 10 cm on a side (liters) are in a cubic meter. We expect, therefore, a large number.

Solution. Each liter of water has a mass of 1 kg, so we must find out how many liters there are in $1 \ m^3$. Since there are 100 cm in a meter, we can visualize a cubic meter as simply a cube with sides 100 cm in length. Therefore, a cubic meter ($1 \ m^3$) has a volume of $10^2 \ cm \times 10^2 \ cm \times 10^2 \ cm = 10^6 \ cm^3$. Since $1 \ L = 10^3 \ cm^3$, there must be $(10^6 \ cm^3)/(10^3 \ cm^3/L) = 1000 \ L$ in $1 \ m^3$. Thus, 1 metric ton is equivalent to 1000 kg.

Note that this entire line of reasoning can be expressed very concisely in a single calculation:

$$\frac{1 \ m^3}{1 \ L} = \frac{100 \ cm \times 100 \ cm \times 100 \ cm}{10 \ cm \times 10 \ cm \times 10 \ cm} = 1000 \quad \text{or} \quad 1 \ m^3 = 1000 \ L$$

Follow-up Exercise. One kilogram has an equivalent weight of 2.2 lb. Using this relationship, how does the metric ton compare with the British ton? (*Answers to all Follow-up Exercises are at the back of the text.*)*

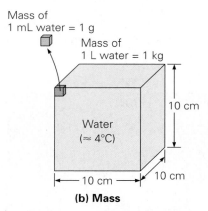

(b) Mass

▲ **FIGURE 1.4 The liter and the kilogram** Other metric units are derived from the meter. **(a)** A unit of volume (capacity) was taken to be the volume of a cube 10 cm, or 0.01 m, on a side and was given the name *liter* (L). **(b)** The mass of a liter of water (at its maximum density) was defined to be 1 kg. Note that the decimeter cube contains $1000 \ cm^3$, or 1000 mL. Thus, $1 \ cm^3$, or 1 mL, of water has a mass of 1 g.

Recall that the standard unit of mass, the kilogram, was originally defined to be the mass of a cubic volume of water 10 cm, or 0.10 m, on a side, or the volume of one liter of water (at its maximum density, at about 4°C). The density of water

*The Answers to Follow-up Exercises section after the appendices contains the answers—and, for Conceptual Exercises, the reasoning—for all Follow-up Exercises in this book.

Nanotechnology

A prefix for metric units refers to the order of the dimension of measurement used. The prefix *micro-* (10^{-6}) has been in common use for some time—e.g., in microscope, microchip, microbiology, microchemistry, and so on. Now the prefix *nano-* is coming into use, and you will probably be hearing a great deal about it with respect to nanotechnology (*nanotech* for short).

In general, nanotechnology is any technology done on the nanometer scale. A nanometer is one billionth (10^{-9}) of a meter, about the width of three to four atoms. Basically, nanotechnology involves the manufacture or building of things one atom or molecule at a time, so the nanometer is the appropriate scale. One atom or molecule at a time? That may sound a bit farfetched, but it's not. This possibility was advanced by physicist Richard Feynman in 1956: "The principles of physics, as far as I can see, do not speak against the possibility of maneuvering things atom by atom." This has been accomplished by using a special microscope at very low temperatures and "pushing around" single atoms (Fig. 1).

In a sense, nanotechnology occurs naturally in the cells of our bodies. Ribosomes (tiny particles in all living cells) are the sites at which information carried by the genetic code is converted into protein molecules. Ribosomes are like tiny "machines," no larger then a few nanometers long, that read DNA instructions on how to build enzymes and other proteins molecule by molecule.

Perhaps you can now see the potential of nanotechnology. The chemical properties of atoms and molecules are well understood. For example, rearranging the atoms in coal can produce a diamond. (We can already do this task without nanotechnology, using heat and pressure.) Nanotechnology presents the possibility of constructing novel molecular devices or "machines" with extraordinary properties and

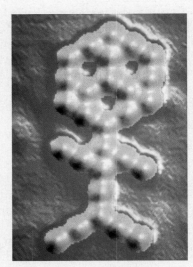

FIGURE 1 Molecular Man This figure was crafted by moving 28 molecules, one at a time. Each of the gold-colored peaks is the image of a carbon monoxide molecule. The molecules rest on a single crystal platinum surface. "Molecular Man" measures 5 nm tall and 2.5 nm wide (hand-to-hand). More than 20 000 figures, linked hand-to-hand, would be needed to span a single human hair. The molecules in the figure were positioned using a special microscope at very low temperatures.

Scanning tunneling microscope

does not vary greatly with temperature, so we can say with fairly good accuracy that *1 L of water has a mass of 1 kg*, an approximation that we will use throughout this book (Fig. 1.4b). Also, since 1 kg = 1000 g and 1 L = 1000 cm^3, then *1 cm^3 (or 1 mL) of water has a mass of 1 g*.

You are probably more familiar with the liter than you think. The use of the liter is becoming quite common in the United States, as ◀Fig. 1.5 indicates.

Because the metric system is coming into increasing use in the United States, you may find it helpful to have an idea of how metric and British units compare. The relative sizes of some units are illustrated in ▶ Fig. 1.6. The mathematical conversion from one unit to another will be discussed shortly.

1.4 Dimensional Analysis and Unit Analysis

OBJECTIVES: To explain the advantages of and apply (a) dimensional analysis and (b) unit analysis.

The fundamental, or base, quantities used in physical descriptions are called *dimensions*. For example, length, mass, and time are dimensions. You could measure the distance between two points and express it in units of meters, centimeters, or feet, but the quantity would still have the dimension of length.

▲ **FIGURE 1.5 Two, three, one, and one-half liters** The liter is now a common volume unit for soft drinks.

abilities. Here are a few potential applications that scientists hope to develop:

- *Medical delivery*. Nanostructures might be injected into the body to go to a particular site, such as a cancerous growth, and deliver a drug directly. Other organs of the body would then be spared any effects of the drug. (This process might be considered nanochemotherapy.)

- *Nanocomputing*. A new technology will soon be needed to make smaller or faster computer chips. As more and more transistors are packed onto a chip, the microscopic "wires" that connect the transistors must be made smaller and thinner. (They are currently down to a thickness of a few hundred atoms.) When down to a thickness of several atoms, the wires explode when electrical signals are passed through them. To save the day, enter the "nanotube," a nanostructure composed of carbon atoms and shaped like a hollow tube, which can carry the electrical signals without being destroyed. Nanotubes have already been developed, and research on their application is underway.

- *Fabric improvement*. A research company is now developing nanostructures that can be attached to fibers. One type, when attached to cotton fibers, causes spilled liquids to bead up on the surface so they can be quickly wiped off, making the fabric both water-repellent and resistant to stains. Another type applies to synthetic fibers, such as polyester. The nanostructure allows the fabric to act as natural fibers in clothing do, moving moisture away from a person's body to the surface for quick evaporation and more comfort. (See the discussion of latent heat in Chapter 11.)

To illustrate the new technology, scientists have made a "nanoguitar" (Fig. 2). The world's smallest guitar has been carved out of crystalline silicon. How is this procedure done? The key is electron-beam technology. Electron-beam lithography is a technique for creating extremely fine patterns, much smaller than can be seen with the unaided eye. It was developed from early electron-beam microscope technology. (See the Insight in Chapter 28 on the electron microscope.) The nanoguitar was made for fun and illustration—it doesn't play. (Perhaps the scientists haven't been able to fashion a "nanopick" to use on the strings.) Actually, the strings, which are only 50 nanometers (nm) wide, or the width of about 100 atoms, could be plucked by special microscope techniques and would vibrate, but at inaudible frequencies. (See Chapter 14.)

It is difficult for us to grasp or visualize the new concept of nanotechnolgy. Even so, keep in mind that a nanometer is one billionth of a meter. The diameter of a human hair is about 200000 nanometers—huge compared with the new nanoapplications. The future should be an exciting nanotime.

FIGURE 2   Nanoguitar   The world's smallest guitar is 10 μm long—about the size of a single cell—with six strings, each about 50 nm (or 100 atoms) wide. The guitar was formed from crystalline silicone.

▼ **FIGURE 1.6   Comparison of some SI and British units**   The bars illustrate the relative magnitudes of each pair of units. (*Note*: The comparison scales are different in each case.)

Volume

1 L = 1.06 qt

1 qt = 0.947 L

Length

1 cm = 0.394 in.

1 in. = 2.54 cm

1 m = 1.09 yd

1 yd = 0.914 m

1 km = 0.621 mi

1 mi = 1.61 km

Mass

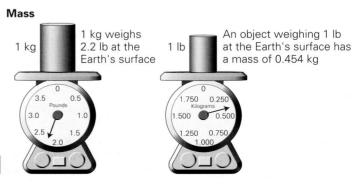

1 kg weighs 2.2 lb at the Earth's surface

An object weighing 1 lb at the Earth's surface has a mass of 0.454 kg

It is common to express dimensional quantities by bracketed symbols, such as [L], [M], and [T] for length, mass, and time, respectively. (The dimensions and units of some common quantities are given in Table 1.3.) Derived quantities are combinations of dimensions; for example, velocity (v) has the dimensions [L]/[T] (think of miles/hour or meters/second), and volume (V) has the dimensions [L] $\times$ [L] $\times$ [L], or [L^3]. Addition and subtraction can be done only with quantities that have the same dimensions; for example, 10 s + 20 s = 30 s, or [T] + [T] = [T].

Dimensional analysis is a procedure by which the dimensional consistency of any equation may be checked. You have used equations and know that an equation is a mathematical equality. Since physical quantities used in equations have dimensions, *the two sides of an equation must be equal not only in numerical value, but also in dimensions.* For example, suppose you had the length quantities $a = 3.0$ m and $b = 4.0$ m. Inserting these values into the equation $a \times b = c$, gives 3.0 m $\times$ 4.0 m = 12 m^2. Both sides of the equation are numerically equal (3 $\times$ 4 = 12), and both sides are dimensionally equal ([L] $\times$ [L] = [L^2]). Notice that we are not algebraically manipulating the dimension symbol [L] in the same way you normally would a variable, such as L. (Note also that variables are italicized.)

The expression [L] $\times$ [L] = [L^2] is read, "when you take a length dimension and multiply it by a length dimension, the result has the dimension of length squared." A second example will help clarify the point: If a and b are inserted into the equation $a + b = c$, we get 3.0 m + 4.0 m = 7.0 m. Comparing both sides of the equation dimensionally, we would say [L] + [L] = [L], or a length plus a length gives another length, but *not* [L] + [L] = 2[L].

One of the major advantages of dimensional analysis is its usefulness in checking an equation to see if it has the correct form. For example, suppose that you think you can solve a problem about the distance an object travels by using the equation $x = at$, where x represents distance, or length, and has the dimension [L]; a represents acceleration, which, as we will see in Chapter 2, has the dimensions [L]/[T^2]; and t represents time and has the dimension [T]. First, before trying to use numbers with the equation, you can check to see if it is dimensionally correct:

$$x = at$$

is expressed dimensionally as

$$[L] = \frac{[L]}{[T^2]} \times [T] \qquad \text{or} \qquad [L] = \frac{[L]}{[T]}$$

which is not true. Thus, $x = at$ cannot be a correct equation.

Dimensional analysis will tell you if an equation is dimensionally incorrect, but a dimensionally consistent equation does not necessarily express correctly the real relationship of quantities. For example, in terms of dimensions,

$$x = at^2$$

is

$$[L] = \frac{[L]}{[T^2]} \times [T^2] \qquad \text{or} \qquad [L] = [L]$$

This equation is dimensionally correct. But, as you will see in Chapter 2, it is not physically correct. The correct form of the equation—both dimensionally and physically—is $x = \frac{1}{2}at^2$. (The fraction $\frac{1}{2}$ has no dimensions; it is a dimensionless number.) Thus, you must be sure that you have the correct form of an equation before you use it to find solutions to problems.

TABLE 1.3 Some Dimensions and Units of Common Quantities

Quantity	Dimension	Unit
mass	[M]	kg
time	[T]	s
length	[L]	m
area	[L^2]	m^2
volume	[L^3]	m^3
velocity (v)	$\dfrac{[L]}{[T]}$	$\dfrac{m}{s}$
acceleration (a or g)	$\dfrac{[L]}{[T^2]}$	$\dfrac{m}{s^2}$

Doing dimensional analysis with the symbols [L], [T], and [M] is fine, but in practice, it is often more convenient to use specific units, such as m, s, and kg. Units can also be treated as algebraic quantities and, like symbols, can be canceled. Using units instead of symbols in dimensional analysis is called **unit analysis**. If an equation is correct by unit analysis, it must be dimensionally correct. Example 1.2 demonstrates the use of unit analysis.

Example 1.2 ■ Checking Dimensions: Unit Analysis

A professor puts two equations on the board: (a) $v = v_o + at$ and (b) $x = v/2a$, where x is a distance in meters (m); v and v_o are velocities in meters/second (m/s); a is acceleration in (meters/second)/second, or meters/second² (m/s^2); and t is time in seconds (s). Are the equations dimensionally correct? Use unit analysis to find out.

Thinking It Through. Simply insert the units for the quantities in each equation, cancel, and check the units on both sides.

Solution.
(a) The equation is

$$v = v_o + at$$

Inserting units for the physical quantities gives

$$\frac{m}{s} = \frac{m}{s} + \left(\frac{m}{s^2} \times s\right) \qquad \text{or} \qquad \frac{m}{s} = \frac{m}{s} + \left(\frac{m}{s \times s} \times s\right)$$

Notice that units cancel like numbers in a fraction. Then, we have

$$\frac{m}{s} = \frac{m}{s} + \frac{m}{s} \qquad \begin{bmatrix} \textit{dimensionally} \\ \textit{correct} \end{bmatrix}$$

The equation is dimensionally correct, since the units on each side are meters per second. (The equation is also a correct relationship, as we shall see in Chapter 2.)
(b) By unit analysis, the equation

$$x = \frac{v}{2a}$$

is

$$m = \frac{\left(\dfrac{m}{s}\right)}{\left(\dfrac{m}{s^2}\right)} = \frac{m}{s} \times \frac{s^2}{m} \qquad \text{or} \qquad m = s \qquad \begin{bmatrix} \textit{not dimensionally} \\ \textit{correct} \end{bmatrix}$$

Meters (m) cannot equal seconds (s), so in this case, the equation is dimensionally incorrect and therefore not physically correct.

Follow-up Exercise. Is the equation $x = v^2/a$ dimensionally correct? (*Answers to all Follow-up Exercises are at the back of the text.*)

Another useful application of unit analysis is to determine the units of a quantity from a correct equation. For example, $x = \frac{1}{2}at^2$ is a correct expression relating distance (x), acceleration (a), and time (t), and say you want to know the units of a. Then,

$$a = \frac{2x}{t^2} \qquad \text{or} \qquad a = 2\left(\frac{m}{s^2}\right)$$

Thus, we find that the units of a are m/s^2. (The 2 is a numerical factor and is not associated with units.)

Mixed Units

Unit analysis also allows you to check for mixed units. In general, when working problems, you should always use the same unit for a given dimension throughout a problem. For example, suppose that you wanted to buy a new carpet to fit a rectangular room. To compute the area of the floor, you would measure the lengths of two adjacent sides of the room and multiply them. Would you measure the sides in different units, such as 10 ft × 3.0 m, and express the area as 30 ft · m? Probably not. Such mixed units are not usually very useful.

Let's look at mixed units in an equation. Suppose that you used centimeters as the unit for x in the equation

$$v^2 = v_0^2 + 2ax$$

and the units for the other quantities as in Example 1.2. In terms of units, this equation would give

$$\left(\frac{m}{s}\right)^2 = \left(\frac{m}{s}\right)^2 + \left(\frac{m \times cm}{s^2}\right)$$

or

$$\frac{m^2}{s^2} = \frac{m^2}{s^2} + \frac{m \times cm}{s^2}$$

which is dimensionally correct. That is, it is dimensionally equivalent to

$$\frac{[L^2]}{[T^2]} = \frac{[L^2]}{[T^2]} + \frac{[L^2]}{[T^2]}$$

But the units are mixed (m and cm). The terms on the right-hand side should not be added together without centimeters first being converted to meters.

Determining the Units of Quantities

Another aspect of unit analysis that is very important in physics is the determination of the units of quantities from defining equations. For example, **density** (represented by the Greek letter rho, ρ) is defined by the equation

$$\rho = \frac{m}{V} \tag{1.1}$$

where m is mass and V is volume. (Density is the mass of an object or substance per unit volume and is a measure of the compactness of the mass of that object or substance.) What are the units of density? In SI units, mass is measured in kilograms and volume in cubic meters. Hence, the defining equation

$$\rho = \frac{m}{V} \left(\frac{kg}{m^3}\right)$$

gives the derived SI unit for density as kilograms per cubic meter (kg/m^3).

What are the units of π? The relationship between the circumference (c) and the diameter (d) of a circle is given by the equation $c = \pi d$, so $\pi = c/d$. If length is measured in meters, then

$$\pi = \frac{c}{d} \left(\frac{\not{m}}{\not{m}}\right)$$

Thus, the constant π has no units, because they cancel out. It is unitless, or a dimensionless, constant.

OBJECTIVES: **To (a) explain conversion-factor relationships and (b) apply them in converting units within a system or from one system of units to another.**

Because units in different systems, or even different units in the same system, can express the same quantity, it is sometimes necessary to convert the units of a quantity from one unit to another. For example, we may need to convert feet to yards or inches to centimeters. You already know how to do many unit conversions. If a room is 12 ft long, what is its length in yards? Your immediate answer is 4 yd.

How did you do this conversion? Well, you must have known a relationship between the units of foot and yard. That is, you know that 1 yd = 3 ft. This is what we call an *equivalence statement*. As we saw in Section 1.4, the numerical values and units on both sides of an equation must be the same. In equivalence statements, we commonly use an equal sign to indicate that 1 yd and 3 ft stand for the *same*, or *equivalent*, *length*. The numbers are different because they stand for different *units* of length.

Mathematically, to change units, we use **conversion factors**, which are simply equivalence statements expressed in the form of ratios—for example, 1 yd/3 ft or 3 ft/1 yd. (The "1" is often omitted in the denominators of such ratios for convenience—for example, 3 ft/yd.) To understand why such ratios are useful, note that dividing the expression 1 yd = 3 ft by 3 ft (or 3 ft = 1 yd by 1 yd) on both sides gives, respectively,

$$\frac{(1 \text{ yd} = 3 \text{ ft})}{3 \text{ ft}} = \frac{1 \text{ yd}}{3 \text{ ft}} = \frac{3 \text{ ft}}{3 \text{ ft}} = 1 \quad \text{or} \quad \frac{(3 \text{ ft} = 1 \text{ yd})}{1 \text{ yd}} = \frac{3 \text{ ft}}{1 \text{ yd}} = \frac{1 \text{ yd}}{1 \text{ yd}} = 1$$

As you can see from these examples, a conversion factor always has an actual value of unity—and you can multiply any quantity by one without changing its value or size. Thus, *a conversion factor simply lets you express a quantity in terms of other units without changing its physical value or size.*

The manner in which 12 feet is converted to yards may be expressed mathematically as follows:

$$12 \text{ ft} \times \frac{1 \text{ yd}}{3 \text{ ft}} = 4 \text{ yd} \quad (\textit{units cancel})$$

Using the appropriate conversion-factor form, the units cancel, as shown by the slash marks, giving the correct unit analysis, yd = yd.

Suppose you are asked to convert 12.0 inches to centimeters. You may not know the conversion factor in this case, but you can get it from a table (such as the one that appears inside the front cover of this book) that gives the needed relationships: 1 in. = 2.54 cm or 1 cm = 0.394 in. It makes no difference which of these equivalence statements is used. The question, once you have expressed the equivalence statement as a conversion factor, is whether to divide or multiply by that factor to make the conversion. *In doing unit conversions, take advantage of unit analysis*—that is, let the units determine the appropriate form of conversion factor, so to speak.

Note that the equivalence statement 1 in. = 2.54 cm can give rise to two forms of the conversion factor: 1 in./2.54 cm or 2.54 cm/1 in. When changing in. to cm, the appropriate form for multiplying is 2.54 cm/in. When changing cm to in. use the form 1 in./2.54 cm. (The inverse forms could be used in each case, but the quantities would have to be *divided* by the conversion factors for proper unit cancellation.) In general, the multiplication form of conversion factors will be used throughout this book.

A few commonly used equivalence statements are not dimensionally correct; for example, consider 1 kg = 2.2 lb, which is used for conversions. The kilogram is a unit of mass, and the pound is a unit of weight. This means that 1 kilogram is *equivalent* to 2.2 pounds; that is, a 1-kg *mass* has a *weight* of 2.2 lb, but only on the Earth's surface. Since mass and weight are directly proportional, differing only by

Note: 1 kg of mass has an equivalent weight of 2.2 lb near the surface of the Earth.

▲ FIGURE 1.7   Unit conversion
Signs sometimes list both the British and metric units, as shown here for elevation and distance.

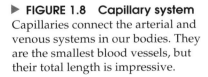

a particular constant, we can use the dimensionally incorrect conversion factor 1 kg/2.2 lb (but *only* near the Earth's surface).

Example 1.3 ■ Converting Units: Use of Conversion Factors

(a) A basketball player is 6.5 ft tall. What is the player's height in meters? (b) How many seconds are there in a 30-day month? (c) What is 50 mi/h in meters per second? (See table of conversion factors inside front cover of this book.)

Thinking It Through.   If we use the correct conversion factors, the rest is arithmetic.

Solution.
(a)   From the conversion table, we have 1 ft = 0.305 m, so

$$6.5 \text{ ft} \times \frac{0.305 \text{ m}}{1 \text{ ft}} = 2.0 \text{ m}$$

Another foot–meter conversion is shown in ◄Fig. 1.7. Is it correct? You should be able to do this one in your head.
(b)   The conversion factor for days and seconds is available from the table (1 day = 86 400 s), but you may not always have a table handy. You can always use several better known conversion factors to get the result:

$$30 \frac{\text{days}}{\text{month}} \times \frac{24 \text{ h}}{\text{day}} \times \frac{60 \text{ min}}{\text{h}} \times \frac{60 \text{ s}}{\text{min}} = \frac{2.6 \times 10^6 \text{ s}}{\text{month}}$$

Note how unit analysis checks the conversion factors for you. The rest is simple arithmetic.
(c)   In this case, from the conversion table, we have 1 mi = 1609 m and 1 h = 3600 s. (The latter is easily computed.) These ratios are used to cancel the units that are to be changed, leaving behind the ones that are wanted:

$$\frac{50 \text{ mi}}{1 \text{ h}} \times \frac{1609 \text{ m}}{1 \text{ mi}} \times \frac{1 \text{ h}}{3600 \text{ s}} = 22 \text{ m/s}$$

Follow-up Exercise.   (a) Convert 50 mi/h directly to meters per second by using a single conversion factor, and (b) show that this single conversion factor can be derived from those in part (c) of this Example. (*Answers to all Follow-up Exercises are at the back of the text.*)

Example 1.4 ■ More Conversions: A Really Long Capillary System

Capillaries, the smallest blood vessels of the body, connect the arterial system with the venous system and supply our tissues with oxygen and nutrients (▼Fig. 1.8). It is estimated that if all of the capillaries of an average adult were unwound and spread out end

► FIGURE 1.8   Capillary system
Capillaries connect the arterial and venous systems in our bodies. They are the smallest blood vessels, but their total length is impressive.

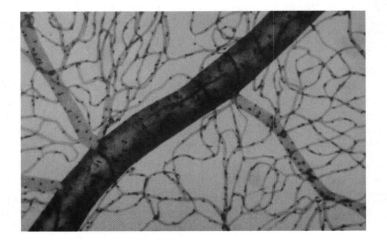

to end, they would extend to a length of about 64 000 km. (a) How many miles is this length? (b) Compare this length with the circumference of the Earth.

Thinking It Through. (a) This conversion is straightforward—just use the appropriate conversion factor. (b) How do we calculate the circumference of a circle or sphere? There is an equation to do so, and you must know the radius or diameter of the Earth. (If you do not remember one of these values, see the solar system data table inside the back cover of this book.)

Solution.

(a) We see in the conversion table that 1 km = 0.621 mi, so

$$64\,000 \ \cancel{\text{km}} \times \frac{0.621 \ \text{mi}}{1 \ \cancel{\text{km}}} = 40\,000 \ \text{mi} \quad (\textit{rounded off})$$

(b) A length of 40 000 mi is substantial. To see how this length compares with the circumference (c) of the Earth, recall that the radius of the Earth is about 4000 mi, so the diameter (d) is 8000 mi. The circumference of a circle is given by $c = \pi d$, as we saw in Section 1.4. Therefore, we have

$$c = \pi d \approx 3 \times 8000 \ \text{mi} = 24\,000 \ \text{mi}$$

[To make a general comparison, we round $\pi \, (=3.14 \ldots)$ off to 3.] So,

$$\frac{\text{capillary length}}{\text{Earth's circumference}} = \frac{40\,000 \ \text{mi}}{24\,000 \ \text{mi}} = 1.7$$

The capillaries of your body have a total length that would extend 1.7 times around the world. Wow!

Follow-up Exercise. Taking the average distance between the east and west coasts of the continental United States to be 4800 km, how many times would the total length of your body's capillaries cross the country? (*Answers to all Follow-up Exercises are at the back of the text.*)

Example 1.5 ■ Converting Units of Area: Choosing the Correct Conversion Factor

A hall bulletin board has an area of 2.5 m². What is this area in square centimeters (cm²)?

Thinking It Through. This problem is a conversion of area units, and we know that 1 m = 100 cm. So, some squaring must be done to get square meters and square centimeters.

Solution. A common error in such conversions is the use of incorrect conversion factors. Because 1 m = 100 cm, it is sometimes assumed that 1 m² = 100 cm², which is wrong. The correct area conversion factor may be obtained directly from the correct linear conversion factor, 100 cm/1 m, or 10^2 cm/1 m, by squaring the linear conversion factor:

$$\left(\frac{10^2 \ \text{cm}}{1 \ \text{m}} \right)^2 = \frac{10^4 \ \text{cm}^2}{1 \ \text{m}^2}$$

Hence, 1 m² = 10^4 cm²(= 10 000 cm²). We can therefore write the following:

$$2.5 \ \text{m}^2 \times \left(\frac{10^2 \ \text{cm}}{1 \ \text{m}} \right)^2 = 2.5 \ \cancel{\text{m}^2} \times \frac{10^4 \ \text{cm}^2}{1 \ \cancel{\text{m}^2}} = 2.5 \times 10^4 \ \text{cm}^2$$

Follow-up Exercise. How many cubic centimeters are there in one cubic meter? (*Answers to all Follow-up Exercises are at the back of the text.*)

Throughout this textbook, you will be presented with various Conceptual Examples. These examples show the reasoning used in applying particular concepts, often with little or no mathematics.

Conceptual Example 1.6 ■ Which Is Faster? Comparison Using Unit Conversions

Two students disagree on which speed is faster, (a) 1 km/h or (b) 1 m/s. Which would you choose? *Clearly establish the reasoning used in determining your answer before you check it below. That is, **why** did you select your answer?*

Reasoning and Answer. To answer this, the quantities should be compared in the same units, so unit conversion is involved, and one looks for the easiest conversions. Observing the prefix *kilo-*, we know that 1 km is 1000 m, so that conversion is easy. Also, one hour is quickly expressed as 3600 s. (Why?) Then we have 1 km/h = 1000 m/3600 s, which is less than 1 m/s, so the answer is (b).

Follow-up Exercise. An American and a European are comparing the gas mileage they get with their RVs. The American calculates that he gets 10 mi/gal, and the European gives his as 10 km/L. Who is getting the better gas mileage? *(Answers to all Follow-up Exercises are at the back of the text.)*

Some examples of the importance of unit conversion are given in the accompanying Insight.

1.6 Significant Figures

OBJECTIVES: To (a) determine the number of significant figures in a numerical value and (b) report the proper number of significant figures after performing simple calculations.

Most of the time, you will be given numerical data when asked to solve a problem. In general, such data are either exact numbers or measured numbers (quantities). **Exact numbers** are numbers without any uncertainty or error. This category includes numbers such as the "100" used to calculate a percentage and the "2" in the equation $r = d/2$ relating the radius and diameter of a circle. **Measured numbers** are numbers obtained from measurement processes and thus generally have some degree of uncertainty or error.

When calculations are done with measured numbers, the error of measurement is propagated, or carried along, by the mathematical operations. A question of how to report a result arises. For example, suppose that you are asked to find time (t) from the formula $x = vt$ and are given that $x = 5.3$ m and $v = 1.67$ m/s. Then

$$t = \frac{x}{v} = \frac{5.3 \text{ m}}{1.67 \text{ m/s}} = ?$$

▲ **FIGURE 1.9 Significant figures and insignificant figures** For the division operation 5.3/1.67, a calculator with a floating decimal point gives many digits. A calculated quantity can be no more accurate than the least accurate quantity involved in the calculation, so this result should be rounded off to two significant figures—that is, 3.2.

Doing the division operation on a calculator yields a result such as 3.173 652 695 (◄Fig. 1.9). How many figures, or digits, should you report in the answer?

The error or uncertainty of the result of a mathematical operation may be computed by statistical methods. A simpler, widely used procedure for estimating this uncertainty involves the use of **significant figures** (**sf**), sometimes called *significant digits*. The degree of accuracy of a measured quantity depends on how finely divided the measuring scale of the instrument is. For example, you might measure the length of an object as 2.5 cm with one instrument and 2.54 cm with another; the second instrument provides more significant figures and a greater degree of accuracy.

Is Unit Conversion Important? You Be The Judge

The answer to this question is, you bet! Here are a couple of cases in point. In 1999, the $125 million Mars Climate Orbiter was making a trip to the Red Planet to investigate its atmosphere (Fig. 1). The spacecraft approached the planet in September, but suddenly contact between the Orbiter and personnel on Earth was lost, and the Orbiter was never heard from again.

Investigations showed that the Orbiter had approached Mars at a far lower altitude than planned. Instead of passing 147 km (87 mi) above the Martian surface, tracking data showed that the Orbiter was on a trajectory that would have taken it as close as 57 km (35 mi) to the surface—some 80 km (50 mi) closer to the planet than intended. As a result, the spacecraft either burned up in the Martian atmosphere or crashed into the surface.

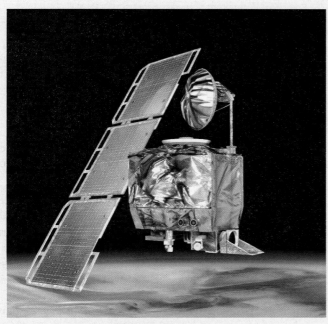

FIGURE 1 Mars Climate Orbiter An artist's conception of the Orbiter near the surface of Mars. The actual Orbiter either burned up in the Martian atmosphere or crashed into the surface. The cause was attributed to a mix-up in units, and a $125 million spacecraft was lost.

How could this have happened? Investigations showed that the failure of the Orbiter was primarily a problem of unit conversions, or a lack thereof. At Lockheed Martin Astronautics, which built the spacecraft, the engineers calculated the navigational information in British units. When scientists at NASA's Jet Propulsion Laboratory received the data they assumed that the information was in metric units, as called for in the mission specifications. These unit conversions weren't made, and a $125 million spacecraft was lost on the Red Planet—causing a few red faces.

A less well-remembered flight closer to Earth is that of the "Gimli Glider" in 1983. Air Canada Flight 143 was on course from Montréal to Edmonton, Canada, with 61 passengers in a new Boeing 767, at the time the most advanced jetliner in the world. But, almost halfway into the flight, a warning light came on for a fuel pump, then for another, and finally for all four pumps. The engines quit, and this advanced plane was now a glider, about 100 miles from the nearest major airport, at Winnipeg.

Without engines, Flight 143's descent would bring it down 10 miles short of the airport, so it was diverted to an old Royal Canadian Air Force landing field at Gimli, which had been converted into an auto racetrack with end barriers. The pilot maneuvered the powerless plane to a landing, stopping just short of the barrier, hence the name Gimli Glider. Did the plane have bad fuel pumps? No, the plane had run out of fuel!

This near-disaster was caused by another conversion problem. It seems that the fuel computers weren't working properly in the first place, so the mechanics had used the old procedure of measuring the fuel in the tanks with a dipstick. The length of the stick that is wet gives the mechanics information with which to determine the volume of fuel by using conversion values in tables. Air Canada had for years computed the amount of fuel in pounds, whereas the new 767's fuel consumption was expressed in kilograms. Even worse, the dipstick procedure gave the amount of fuel onboard in liters instead of pounds or kilograms.

These incidents underscore the importance of using appropriate units, making correct unit conversions, and working consistently in the same system of units. Several problems at the end of the chapter will challenge you to develop your skills in accurate unit conversions (See Exercises, Section 1.5 Unit Conversions).

Basically, *the significant figures in any measurement are the digits that are known with certainty, plus one digit that is uncertain.* This set of digits is usually defined as all of the digits that can be read directly from the instrument used in making the measurement, plus one uncertain digit that is obtained by estimating the fraction of the smallest division of the instrument's scale.

The quantities 2.5 cm and 2.54 cm have two and three significant figures, respectively. This information is rather evident. However, some confusion may arise when a quantity contains one or more zeros. For example, how many significant figures does the quantity 0.0254 m have? What about 104.6 m? 2705.0 m? In such cases, we will use the following rules:

1. Zeros at the beginning of a number are not significant. They merely locate the decimal point. For example,

0.0254 m has three significant figures (2, 5, 4)

2. Zeros within a number are significant. For example,

104.6 m has four significant figures (1, 0, 4, 6)

3. Zeros at the end of a number after the decimal point are significant. For example,

2705.0 m has five significant figures (2, 7, 0, 5, 0).

4. In whole numbers without a decimal point that end in one or more zeros (trailing zeros)—for example, 500 kg—the zeros may or may not be significant. In such cases, it is not clear which zeros serve only to locate the decimal point and which are actually part of the measurement. That is, if the first zero from the left (5<u>0</u>0 kg) is the estimated digit in the measurement, then only two digits are reliably known, and there are only two significant figures. Similarly, if the last zero is the estimated digit (50<u>0</u> kg), then there are three significant figures. This ambiguity may be removed by using scientific (powers-of-ten) notation:

5.0×10^2 kg has two significant figures
5.00×10^2 kg has three significant figures

This notation is helpful in expressing the results of calculations with the proper numbers of significant figures, as we shall see shortly. (Appendix I includes a review of scientific notation.)

Note: To avoid confusion regarding numbers having trailing zeroes used as given quantities in text examples and exercises, we will consider the trailing zeroes to be significant. For example, assume that a time of 20 s has two significant figures, even if it is not written out as 2.0×10^1 s.

It is important to report the results of mathematical operations with the proper number of significant figures. This is accomplished by using rules for (1) multiplication and division, and (2) addition and subtraction. To obtain the proper number of significant figures, the results are rounded off. Here are some general rules that will be used for mathematical operations and rounding.

Significant Figures in Calculations

1. When multiplying and dividing quantities, leave as many significant figures in the answer as there are in the quantity with the least number of significant figures.
2. When adding or subtracting quantities, leave the same number of decimal places (rounded) in the answer as there are in the quantity with the least number of decimal places.

Rules for Rounding*

1. If the first digit to be dropped is less than 5, leave the preceding digit as is.
2. If the first digit to be dropped is 5 or greater, increase the preceding digit by one.

What the rules for significant figures mean is that the result of a calculation can be no more accurate than the least accurate quantity used. That is, you cannot gain

---

*It should be noted that these rounding rules give an approximation of accuracy, as opposed to the results provided by more advanced statistical methods.

accuracy in performing mathematical operations. Thus, the result that should be reported for the division operation discussed at the beginning of this section is

$$\frac{\overset{(2\ sf)}{5.3\ \text{m}}}{\underset{(3\ sf)}{1.67\ \text{m/s}}} = 3.2\ \text{s}\ \ (2\ sf)$$

The result is rounded off to two significant figures. (See Fig. 1.9.)
Applications of these rules are shown in the following Examples.

Example 1.7 ■ Using Significant Figures in Multiplication and Division: Rounding Applications

The following operations are performed and the results rounded off to the proper number of significant figures:

Multiplication

$$\underset{(2\ sf)}{2.4\ \text{m}} \times \underset{(3\ sf)}{3.65\ \text{m}} = 8.76\ \text{m}^2 = 8.8\ \text{m}^2 \quad (\textit{rounded to two sf}\,)$$

Division

$$\frac{\overset{(4\ sf)}{725.0\ \text{m}}}{\underset{(3\ sf)}{0.125\ \text{s}}} = 5800\ \text{m/s} = 5.80 \times 10^3\ \text{m/s} \quad (\textit{represented with three sf; why?})$$

Follow-up Exercise. Perform the following operations, and express the answers in the standard powers-of-ten notation (one digit to the left of the decimal point) with the proper number of significant figures: (a) $(2.0 \times 10^5\ \text{kg})(0.035 \times 10^2\ \text{kg})$ and (b) $(148 \times 10^{-6}\ \text{m})/(0.4906 \times 10^{-6}\ \text{m})$. *(Answers to all Follow-up Exercises are at the back of the text.)*

Example 1.8 ■ Using Significant Figures in Addition and Subtraction: Application of Rules

The following operations are performed by finding the number that has the least number of decimal places. (Units have been omitted for convenience.)

Addition
In the numbers to be added, note that 23.1 has the least number of decimal places:

$$
\begin{array}{r}
23.1 \\
0.546 \\
\underline{1.45} \\
25.096
\end{array}
\xrightarrow{\ (\textit{rounding off}\,)\ } 25.1
$$

Subtraction
The same rounding procedure is used. Here, the "157" has the least number of decimal places (none).

$$
\begin{array}{r}
157 \\
\underline{-\ 5.5} \\
151.5
\end{array}
\xrightarrow{\ (\textit{rounding off}\,)\ } 152
$$

Follow-up Exercise. Given the numbers 23.15, 0.546, and 1.058, (a) add the first two numbers and (b) subtract the last number from the first. *(Answers to all Follow-up Exercises are at the back of the text.)*

Suppose that you must deal with mixed operations—multiplication and/or division *and* addition and/or subtraction. What do you do in this case? Just follow your regular rules for order of algebraic operations, and observe significant figures as you go.

The number of digits reported in a result depends on the number of digits in the given data. The rules for rounding will generally be observed for examples in this book. However, there will be exceptions that may make a difference, as explained in the following hint.

Problem-Solving Hint: The "Correct" Answer

When working problems, you naturally strive to get the correct answer and will probably want to check your answers against those listed in the Answers to Odd-Numbered Exercises section in the back of the book. However, on occasion, you may find that your answer differs slightly from that given, even though you have solved the problem correctly. There are several reasons why this could happen.

As stated previously, it is best to round off only the final result of a multipart calculation, but this practice is not always convenient in elaborate calculations. Sometimes, the results of intermediate steps are important in themselves and need to be rounded off to the appropriate number of digits as if each were a final answer. Similarly, Examples in this book are often worked in steps to show the stages in the *reasoning* of the solution. The results obtained when the results of intermediate steps are rounded off may differ slightly from those obtained when only the final answer is rounded.

Rounding differences may also occur when using conversion factors. For example, in changing 5.0 mi to kilometers using the conversion factor listed in the front of the book in different forms,

$$5.0 \text{ mi} \left(\frac{1.609 \text{ km}}{1 \text{ mi}} \right) = (8.045 \text{ km}) = 8.0 \text{ km} \quad (\textit{two significant figures})$$

and

$$5.0 \text{ mi} \left(\frac{1 \text{ km}}{0.621 \text{ mi}} \right) = (8.051 \text{ km}) = 8.1 \text{ km} \quad (\textit{two significant figures})$$

The difference arises because of rounding of the conversion factors. Actually, 1 km = 0.6214 mi, so 1 mi = (1/0.6214) km = 1.609 269 km ≈ 1.609 km. (Try repeating these conversions with the unrounded factors, and see what you get.) To avoid rounding differences in conversions, we will generally use the multiplication form of a conversion factor, as in the first of the foregoing equations, unless there is a convenient exact factor, such as 1 min/60 s.

Slight differences in answers may occur when different methods are used to solve a problem, because of rounding differences. Keep in mind that when solving a problem (a general procedure for which is given in Section 1.7), *if your answer differs from that in the text in only the last digit, the disparity is most likely the result of a rounding difference for an alternative method of solution being used.*

1.7 Problem Solving

OBJECTIVES: **To (a) establish a problem-solving procedure and (b) apply it to typical problems.**

An important aspect of physics is problem solving. In general, problem solving involves the application of physical principles and equations to data from a particular situation in order to find some unknown or wanted quantity. There is no universal method for approaching a problem that will automatically produce a solution.

A few general points, though, are worth keeping in mind:

- *Make sure you understand the problem.* The foundation of successful problem solving is having a thorough grasp of the problem. Before starting to work a problem, make it a habit to list everything that is given or known (as we will do for most of the Examples in this text). It's all too easy to overlook a critical piece of information.

- To proceed from the given data to the solution, you may have to *devise a strategy or plan.* Many problems cannot be solved merely by finding one equation and "plugging in" the given quantities to get the solution. Often, you will need to perform intermediate steps, each of which will bring you closer to the final answer.

- *Remember that equations are expressions of physical principles.* Solving problems involves translating principles into the "language" of equations, but you should also be able to look at an equation and determine what physical relationship it embodies.

- *A common cause of failure in problem solving is the misapplication of physical principles or equations.* Principles and equations are generally subject to certain conditions or are limited in their applicability to physical situations.

- *Many problems can be solved by more than one method.* If you understand the problem and the relevant principles well, you can probably figure out how to solve the problem in the fastest and easiest way.

Although there is no magic formula for problem solving, there are some sound practices that can be very useful. The steps in the following procedure are intended to provide you with a framework that can be applied to solving most of the problems that you will encounter in this text. We generally will use these steps in dealing with the Example problems throughout this text. Additional helpful problem-solving hints will be given where appropriate in subsequent chapters.

Suggested Problem-Solving Procedure

1. *Read the problem carefully, and analyze it. Write down the given data and what you are to find.* Some data may not be given explicitly in numerical form. For example, if a car "starts from rest," its initial speed is zero ($v_o = 0$). In some instances, you may be expected to know certain quantities or to look them up in tables.

2. *Draw a diagram as an aid in visualizing and analyzing the physical situation of the problem where appropriate.* This step may not be necessary in every case, but it is usually helpful.

3. *Determine which principle(s) and equation(s) are applicable to this situation and how they can be used to get from the information given to what is to be found.* You may have to devise a strategy that involves several steps.

4. *Simplify mathematical expressions as much as possible through algebraic manipulation before inserting actual values.* Trigonometric relationships (summarized in Appendix I) can sometimes be used to simplify equations. The less calculation you do, the less likely you are to make a mistake—so *don't put the numbers in until you have to.*

5. *Check units before doing calculations.* Make unit conversions if necessary so that all units are in the same system (preferably standard units). This practice avoids mixed units and is helpful in unit analysis. (Unit checking and conversions are often done when writing the data in Step 1.)

6. *Substitute given quantities into equation(s), and perform calculations. Report the result with the proper units and the proper number of significant figures.*

7. *Consider whether the result is reasonable.* Does the answer have an appropriate magnitude? (This means, is it in the right ballpark?) For example, if a person's calculated mass turns out to be 4.60×10^2 kg, the result should be questioned, since 460 kg corresponds to a weight of 1010 lb.

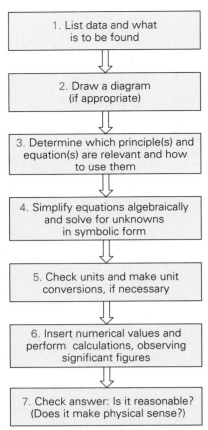

1. List data and what is to be found

2. Draw a diagram (if appropriate)

3. Determine which principle(s) and equation(s) are relevant and how to use them

4. Simplify equations algebraically and solve for unknowns in symbolic form

5. Check units and make unit conversions, if necessary

6. Insert numerical values and perform calculations, observing significant figures

7. Check answer: Is it reasonable? (Does it make physical sense?)

▲ FIGURE 1.10 A flow chart for the suggested problem-solving procedure

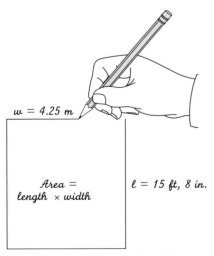

$w = 4.25$ m

Area = length × width

$l = 15$ ft, 8 in.

▲ FIGURE 1.11 A helpful step in problem solving Drawing a diagram helps you visualize and better understand the situation. See Example 1.9.

◄Figure 1.10 summarizes these steps in the form of a flow chart. The upcoming Examples illustrate the procedure. The steps are numbered to help you follow along.

Example 1.9 ■ Finding the Area of a Rectangle: Practicing the Problem-Solving Procedure

Two students measure the lengths of adjacent sides of their rectangular dorm room. One reports 15 ft, 8 in., and the other reports 4.25 m. What is the area of the room in square meters?

Thinking It Through. The lengths are reported in different units, so to get square meters (m × m), the British units feet and inches must be converted to meters.

Solution.
1. Adjacent sides of a room give its length and width, so we may write the following:

Given: Length = l = 15 ft, 8 in.   *Find:* Area (in square meters)
Width = w = 4.25 m

2. Sketch a diagram to help you visualize the situation (▼Fig. 1.11).
3 and 4. For this simple situation, the required equation is well known. The area A of a rectangle is $A = l \times w$ and both l and w are given.
5. A unit change is necessary. Let's first convert the length measurement to inches and then inches to meters:

$$15 \text{ ft} + 8 \text{ in.} = \left(15 \text{ ft} \times \frac{12 \text{ in.}}{1 \text{ ft}} \right) + 8 \text{ in.} = 188 \text{ in.}$$

and

$$188 \text{ in.} \times \frac{2.54 \text{ cm}}{1 \text{ in.}} = 478 \text{ cm} = 4.78 \text{ m}$$

Notice how easy it is to convert units in the decimal metric system (centimeters to meters). Perform the conversion explicitly if necessary, using the conversion factor 1 m/100 cm.
6. Now perform the calculation:

$$A = l \times w = 4.78 \text{ m} \times 4.25 \text{ m}$$
$$= 20.315 \text{ m}^2 = 20.3 \text{ m}^2 \quad (\textit{computed value rounded to three sf; why?})$$

7. The answer appears reasonable. Since 1 m ≈ 3 ft, the dorm room would be about 13 ft by 14 ft, which is about right (but, as always, too small for comfort). Suppose you had inadvertently input 47.8 instead of 4.78 on your calculator. The result would have been $A = 47.8 \text{ m} \times 4.25 \text{ m} = 203 \text{ m}^2$. A room with an area of about 200 m² would have dimensions of about 10 m by 20 m, which is roughly 30 ft by 60 ft. Since this result is not the size of a typical dorm room, the magnitude of the result should make you suspect that there may be an error.

Follow-up Exercise. The dimensions of a textbook are 0.22 m × 0.26 m × 4.0 cm. What volume in a backpack would the book take up? Give the answer in both cubic meters and cubic centimeters. (*Answers to all Follow-up Exercises are at the back of the text.*)

Many problems will involve basic trigonometric functions. The most common functions are given in the marginal note on page 23; for other trigonometric relations, see Appendix I or the tables inside the back cover.

Example 1.10 ■ Finding the Length of One Side of a Triangle: Trigonometry Application

A flower bed is laid out in the form of a triangle as shown in ▶Fig. 1.12. What is the length of the side of the bed that runs along the flagstone walkway?

Thinking It Through. The length of one side of a right triangle and an angle are given. Basic trigonometry applies.

Solution.

1 and 2. In some problems, we are given diagrams that contain data. Here, we have the following information:

Given: $x = 5.4$ m   *Find:* r (long side of triangle)
$\theta = 40°$

3 and 4. Noting in the figure that the flower bed is a right triangle (as indicated by the right-angle symbol at the large angle), we can use a trigonometric function to find the hypotenuse r. Recalling that x and r are associated with the cosine for a standard right triangle, we can write

$$\cos \theta = \frac{x}{r} \quad \text{or} \quad r = \frac{x}{\cos \theta}$$

5 and 6. The units are fine. (Since x is in meters and $\cos \theta$ is dimensionless, r will be in meters.) So putting in the data:

$$r = \frac{x}{\cos \theta} = \frac{5.4 \text{ m}}{\cos 40°} = \frac{5.4 \text{ m}}{0.766} = 7.0 \text{ m}$$

7. The magnitude of r seems reasonable compared with the value of x. Note that r must be greater than x (why?), and it is.

Follow-up Exercise. To keep animals out of the flower bed in Fig. 1.12, the owner decides to fence in the perimeter of the bed. What will be the total length of the fence? *(Answers to all Follow-up Exercises are at the back of the text.)*

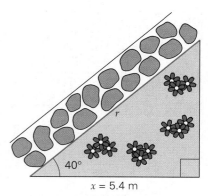

▲ **FIGURE 1.12 The length of a triangle's side** See Example 1.10.

Trigonometric functions:

$$\cos \theta = \frac{x}{r} \left(\frac{\text{side adjacent}}{\text{hypotenuse}} \right)$$

$$\sin \theta = \frac{y}{r} \left(\frac{\text{side opposite}}{\text{hypotenuse}} \right)$$

$$\tan \theta = \frac{\sin \theta}{\cos \theta} = \frac{y}{x} \left(\frac{\text{side opposite}}{\text{side adjacent}} \right)$$

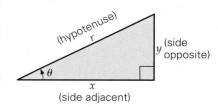

It is understood that you do not need to write out the steps of your problem-solving procedure each time. However, it is good practice to run through them mentally when working a problem. In the examples in upcoming chapters, the problem-solving steps will not always be listed as was done here for illustration. Even so, you should be able to see the general pattern outlined in the foregoing examples.

The main point of this section is that in solving problems, you should have some systematic procedure for analyzing the situation and extracting the information wanted. The suggested problem-solving procedure given here is one option.

In short, this procedure is (1) a pictorial representation, (2) a physical representation, (3) a mathematical representation, and (4) an evaluation of the results. To emphasize the aspects of problem solving, an occasional Integrated Example will be presented, in which a sketch or reasoning of the situation is specifically called for as part of the exercise. Integrated Exercises are also given in the end-of-chapter exercises.) Here is an Integrated Example.

Integrated Example 1.11 ■ Around the World—at Different Latitudes

If we consider the Earth to be a sphere with a radius of $R_E = 4000$ mi, it is easy to find the distance around the Earth at the equator (0° latitude), or the equatorial circumference ($c = 2\pi R_E$). However, a student wants to know the distance around the Earth at her latitude of 40° N (about the midlatitude of the conterminous 48 states). (a) Is the circumference radius at this latitude given by (1) $R_{40} = R_E \cos 40°$, (2) $R_{40} = R_E \sin 40°$, (3) $R_{40} = R_E \cos 50°$, or (4) $R_{40} = R_E \sin 50°$? (b) What is the distance around the Earth at 40° N latitude in km?

(a) Conceptual Reasoning. Recalling (or looking up) that the latitude angle is measured from the center of the Earth relative to the equator, we observe that the distance

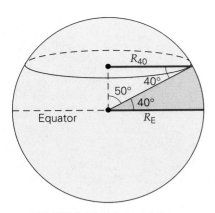

▲ FIGURE 1.13 Around the World Not so far around at 40° latitude as at the equator (0°). See Integrated Example 1.11.

around the Earth at 40° N latitude is that along a circular path (called a *parallel*) at this latitude. The radius of this circle (R_{40}) involves a cosine or sine function, but it is difficult to visualize which and whether the angle would be the latitude angle (40°) or its complementary angle (50° = 90° − 40°). A quick sketch as in ◄Fig. 1.13 helps, and it is seen that the answer is (1).

(b) Thinking It Through. The distance around the parallel at 40° N (the circumference) is given by $c = 2\pi R_{40}$, which can be readily computed.

Given: $R_E = 4000$ mi $(1.61 \text{ km/mi}) = 6440$ km   *Find:* Distance around the Earth at
$\quad\quad\quad \theta = 40°$ latitude 40° N

[Note: After reading in part (b) of the question that the distance is to be in km, we do the conversion prior to working the rest of the problem.] Then,

$$c = 2\pi R_{40} = 2\pi R_E \cos 40°$$
$$= 2\pi(6440 \text{ km})(0.766) = 3.10 \times 10^4 \text{ km}$$

Follow-up Exercise. At what latitude would the distance around a parallel be 3.71×10^4 km? *Hint*: Should you have two answers? (*Answers to all Follow-up Exercises are at the back of the text.*)

Approximation and Order-of-Magnitude Calculations

At times when solving a problem, you may not be interested in an exact answer, but want only an estimate, or a "ballpark" figure. Approximations can be made by rounding off quantities so as to make the calculations easier and, perhaps, obtainable without the use of a calculator. For example, suppose you want to get an idea of the area of a circle with a radius $r = 9.5$ cm. Then rounding 9.5 cm $\approx$ 10 cm, and $\pi \approx 3$ instead of 3.14. And,

$$A = \pi r^2 \approx 3(10 \text{ cm})^2 = 300 \text{ cm}^2$$

(Note that significant figures are not a concern in calculations involving approximations.) The answer is not exact, but it is a good approximation. Compute the exact answer and see.

Powers-of-ten, or scientific, notation is particularly convenient in making estimates or approximations in what are called **order-of-magnitude calculations**. *Order of magnitude* means that we express a quantity to the power of 10 closest to the actual value. For example, in the foregoing calculation, approximating 9.5 cm $\approx$ 10 cm is expressing 9.5 as 10^1, and we say that the radius is *on the order of* 10 cm. Expressing a distance of 75 km $\approx 10^2$ km indicates that the distance is on the order of 10^2 km. The radius of the Earth is 6.4×10^3 km $\approx 10^4$ km, or on the order of 10^4 km. A nanostructure with a width of 8.2×10^{-9} m is on the order of 10^{-8} m, or 10 nm. (Why a −8 exponent?)

An order-of-magnitude calculation gives only an estimate, of course. But this estimate may be enough to provide you with a better grasp or understanding of a physical situation. Usually, the result of an order-of-magnitude calculation is precise within a power of 10, or *within an order of magnitude*. That is, the prefix to the power of 10 is somewhere between 1 and 10. For example, if we got a time result of 10^5 s, we would expect the exact answer to be somewhere between 1×10^5 s and 10×10^5 s.

Example 1.12 ■ Order-of-Magnitude Calculation: Drawing Blood

A medical technologist draws 15 cc of blood from a patient's vein. Back in the lab, it is determined that this volume of blood has a mass of 16 g. Estimate the density of the blood, in standard units.

Thinking It Through. The data are given in cgs (centimeter–gram–second) units, which are often used for practicality when dealing with small, whole-number quantities in some situations. The cc abbreviation is commonly used in the medical and chemistry fields for cm^3. Density (ρ) is mass per unit volume, where $\rho = m/V$ (Section 1.4).

Solution.

Given: $\quad m = 16 \, g \left(\dfrac{1 \, kg}{1000 \, g}\right) = 1.6 \times 10^{-2} \, kg \approx 10^{-2} \, kg \qquad$ *Find:* $\quad \rho$ (density)

$$V = 15 \, cm^3 \left(\dfrac{1 \, m}{10^2 \, cm}\right)^3 = 1.5 \times 10^{-5} \, m^3 \approx 10^{-5} \, m^3$$

So, we have

$$\rho = \frac{m}{V} \approx \frac{10^{-2} \, kg}{10^{-5} \, m^3} = 10^3 \, kg/m^3$$

This result is quite close to the average density of whole blood, $1.05 \times 10^3 \, kg/m^3$.

Follow-up Exercise. A patient receives 750 cc of whole blood. Estimate the mass of the blood, in standard units. *(Answers to all Follow-up Exercises are at the back of the text.)*

Example 1.13 ■ How Many Cells in Your Blood?

The blood volume in the human body varies with a person's age, body size, and sex. On average, this volume is about 5 L. A typical value of red blood cells (erythrocytes) per volume is 5 000 000 per mm^3. Estimate how many "red cells" you have in your body.

Thinking It Through. The red blood cell count in cells/mm^3 is sort of a red blood cell "density." Multiplying this figure by the total volume of blood [(cells/volume) $\times$ total volume] will give the total number of cells. But note that we must have the volumes in the same units.

Solution.

Given: $\qquad V = 5 \, L \qquad\qquad$ *Find:* the approximate number of red cells in the body

$$= 5 \, L \left(10^{-3} \frac{m^3}{L}\right)$$

$$= 5 \times 10^{-3} \, m^3 \approx 10^{-2} \, m^3$$

$$\text{cells/volume} = 5 \times 10^6 \, \frac{cells}{mm^3} \approx 10^7 \, \frac{cells}{mm^3}$$

Then, changing to m^3,

$$\frac{cells}{volume} \approx 10^7 \, \frac{cells}{mm^3} \left(\frac{10^3 \, mm}{1 \, m}\right)^3 = 10^{16} \, \frac{cells}{m^3}$$

Note: The conversion factor for L to m^3 was obtained directly from the conversion tables, but there is no conversion factor given for converting mm^3 to m^3, so we just use a conversion we know and cube it. So, we have

$$\left(\frac{cells}{volume}\right)(\text{total volume}) \approx \left(10^{16} \, \frac{cells}{m^3}\right)(10^{-2} \, m^3) = 10^{14} \, \text{red cells}$$

Follow-up Exercise. The average number of white cells (leukocytes) in human blood is normally 5000 to 10 000 cells per mm^3. Estimate the number of white blood cells you have in your body. *(Answers to all Follow-up Exercises are at the back of the text.)*

Chapter Review

Important Concepts and Equations

- **SI units of length, mass, and time.** The meter (m), the kilogram (kg), and the second (s), respectively.
- **Liter (L).** A volume of 1000 mL, or 1000 cm^3. To a good approximation, a liter of water has a mass of 1 kg.
- **Dimensional analysis and/or unit analysis.** Either can be used to determine if an equation has the correct form. Unit analysis can be used to find the unit of a quantity.
- **Significant figures (digits).** The digits that are known with certainty, plus one digit that is uncertain, in a measured value.

- **Problem solving.** Problems should be worked using a consistent procedure. Order-of-magnitude calculations may be done when only an estimated value is desired.
- **Density (ρ).** The mass per unit volume of an object or substance, which is a measure of the compactness of the material it contains:

$$\rho = \frac{m}{V} \quad \left(\frac{\text{mass}}{\text{volume}}\right) \tag{1.1}$$

Exercises*

*Exercises designated **CQ** are Conceptual Questions; those designated **IE** are Integrated Exercises, which involve conceptual reasoning and calculations. Throughout the text, many exercise sections will include "paired" exercises. These exercise pairs, identified with red numbers, are intended to assist you in problem solving and learning. In a pair, the first exercise (even numbered) is worked out in the Study Guide so that you can consult it should you need assistance in solving it. The second exercise (odd numbered) is similar in nature, and its answer is given at the back of the book.*

1.2 SI Units of Length, Mass, and Time
and
1.3 More about the Metric System

1. The only SI standard represented by an artifact is the (a) meter, (b) kilogram, (c) second, or (d) electric charge.

2. Which of the following is *not* an SI base unit? (a) length, (b) mass, (c) weight, or (d) time.

3. Which one of the following is the SI unit of mass? (a) pound, (b) gram, (c) kilogram, or (d) ton.

4. **CQ** Is each of the following statements reasonable? (Justify your answers.) (a) It took 300 L of gasoline to fill up the car's tank. (b) The center on the basketball team is 225 cm tall. (c) The area of a dorm room is 120 m^2.

5. The prefix micro- (μ) means (a) 10^6, (b) 10^{-6}, (c) 10^3, or (d) 10^{-3}.

6. **CQ** If a fellow student tells you that he saw a 3-cm-long ladybug in his vegetable garden, would you believe him? How about if another student says she caught a 10-kg salmon?

*Keep in mind here and throughout the text that your answer to an odd-number exercise may differ slightly from that given at the back of the book because of rounding. See Problem-Solving Hint: The "Correct Answer" in this chapter.

7. ■ The metric system is a decimal (base-10) system, and the British system is, in part, a duodecimal (base-12) system. Discuss the ramifications if our monetary system had a duodecimal base. What would be the possible values of our coins if this were the case?

8. ■ (a) In the British system, 16 oz = 1 pt and 16 oz = 1 lb. Is there something wrong here? Explain. (b) Here's an old one: A pound of feathers weighs more than a pound of gold. How can that be? (*Hint*: Look up *ounce* in the dictionary.)

9. ■■ A sailor tells you that if his ship is traveling at 25 knots (nautical miles per hour), it is moving faster than the 25 miles per hour your car travels. How can that be?

1.4 Dimensional Analysis and Unit Analysis**

10. Both sides of an equation are equal in (a) numerical value, (b) units, (c) dimensions, or (d) all of the preceding.

11. Unit analysis of an equation cannot tell you if (a) the equation is dimensionally correct, (b) the equation is physically correct, (c) the numerical value is correct, or (d) both (b) and (c).

**Dimensions and/or units of velocity and acceleration are given in the chapter.

12. **CQ** Can dimensional analysis tell you whether you have used the correct equation in solving a problem? Explain.

13. If an equation has the same numerical value on both sides, the equation (a) is dimensionally correct, (b) is correct, (c) may be correct, or (d) both (a) and (c).

14. If meters/second (m/s) is divided by second (s), what is the resulting unit?

15. **CQ** Discuss the differences between dimensional analysis and unit analysis.

16. ■ Show that the equation $x = x_o + vt$, where v is velocity and x and x_o are lengths, is dimensionally correct.

17. ■ If x refers to distance, v_o and v to speeds, a to acceleration, and t to time, which of the following equations is dimensionally correct? (a) $x = v_o t + at^3$, (b) $v^2 = v_o^2 + 2at$, (c) $x = at + vt^2$, or (d) $v^2 = v_o^2 + 2ax$.

18. ■ Show that $t = \sqrt{2x/a}$ is dimensionally correct. (a is acceleration, t is time, and x is length.)

19. ■■ Use SI unit analysis to show that the equation $A = 4\pi r^2$, where A is the area and r is the radius of a sphere, is dimensionally correct.

20. ■■ You are told that the volume of a sphere is given by $V = \pi d^3/4$, where V is the volume and d is the diameter of the sphere. Is this equation dimensionally correct? (Use SI unit analysis to find out.)

21. ■■ The correct equation for the volume of a sphere is $V = 4\pi r^3/3$, where r is the radius of the sphere. Is the equation in Exercise 20 correct? If not, what should it be when expressed in terms of d?

22. ■■ If $x = gt^2/2$, where x is length and t is time, is dimensionally correct, what are the SI units of the constant g?

23. ■■ Is the equation $v = v_o \sin\theta - gt^2$ dimensionally correct? Use SI unit analysis to find out. (v and v_o are velocities, θ is an angle, t is time, and g is acceleration.)

24. ■■ Density is defined as the mass of an object divided by the volume of the object. Using SI unit analysis, determine the SI unit for density. (See Section 1.4 for units of mass and volume.)

25. ■■ Is the equation for the area of a trapezoid, $A = \frac{1}{2}a(b_1 + b_2)$, where a is the height and b_1 and b_2 are the bases, dimensionally correct? (▶Fig. 1.14) If not, what should be changed to correct it?

26. ■■ One student, using unit analysis, says that the equation $v = \sqrt{2ax}$ is dimensionally correct. Another says it isn't. With whom do you agree, and why?

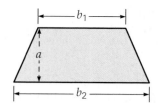

◀ **FIGURE 1.14 The area of a trapezoid** See Exercise 25.

27. ■■ The equation for the frequency f of oscillation of a simple pendulum is

$$f = \frac{1}{2\pi}\sqrt{\frac{g}{L}}$$

where L is the length of the pendulum string and g is the acceleration due to gravity. Frequency is commonly expressed in units of hertz (Hz). What is a hertz unit in terms of SI base units?

28. ■■■ Newton's second law of motion (Chapter 4) is expressed by the equation $F = ma$, where F represents force, m is mass, and a is acceleration. (a) The SI unit of force is, appropriately, called the newton (N). What are the units of the newton in terms of base quantities? (b) An equation for force associated with uniform circular motion (Chapter 7) is $F = mv^2/r$, where v is speed and r is the radius of the circular path. Does this equation give the same units for the newton?

29. ■■■ Einstein's famous mass–energy equivalence is expressed by the equation $E = mc^2$, where E is energy, m is mass, and c is the speed of light. (a) What are the SI base units of energy? (b) Another equation for energy is $E = mgh$, where m is mass, g is the acceleration due to gravity, and h is height. Does this equation give the same units as in (a)?

1.5 Unit Conversions*

30. A good way to ensure proper unit conversion is to (a) use another measurement instrument, (b) always work in one system of units, (c) use unit analysis, or (d) use dimensional analysis.

31. You often see 1 kg = 2.2 lb. This expression means that (a) 1 kg is equivalent to 2.2 lb, (b) this is a true equation, (c) 1 lb = 2.2 kg, or (d) none of the preceding

32. ■ Figure 1.7 (top) shows the elevation of a location in both feet and meters. If a town is 130 ft above sea level, what is the elevation in meters?

33. **IE** ■ (a) If you wanted to express your height with the largest number, you would use (1) meters, (2) feet, (3) inches, or (4) centimeters? Why? (b) If you are 6.00 ft tall, what is your height in centimeters?

34. ■ What is the length in feet of (a) a 100-m dash and (b) a 2.4-m high jump?

*Conversion factors are listed inside the front cover of the text.

35. ■ If the capillaries of an average adult were unwound and spread out end to end, they would extend to a length over 40 000 mi (Fig. 1.8). If you are 1.75 m tall, how many times your height would the capillary length equal?

36. ■ Standing at 452 m, the Petronas Twin Towers in Malaysia is one of the tallest buildings in the world. What is its height in feet?

37. ■ The largest airplane, the Airbus A380, has a length of 239 ft, 6 in; a wingspan of 261 ft, 10 in; and a height of 79 ft, 1 in. What are these dimensions in meters?

38. IE ■ (a) Compared with a two-liter soda bottle, a half-gallon soda bottle holds (1) more, (2) the same amount of, or (3) less soda? Why? (b) Verify your answer for (a).

39. ■ A commuting student wants to buy 18 gal of gas, but the gas station has installed new pumps that are measured in liters. How many liters of gas (rounded off to a whole number) should he ask for?

40. ■ (a) A football field is 300 ft long and 160 ft wide. What are the field's dimensions in meters? (b) A football is 11.0 to $11\frac{1}{4}$ in. long. What is its length in centimeters?

41. ■ Suppose that when the United States goes completely metric, the dimensions of a football field are established as 100 m by 54 m. Which would be larger, the metric football field or a current football field (see Exercise 40a), and what would be the difference between the areas?

42. ■■ If blood flows with an average speed of 0.35 m/s in the human circulatory system, how many miles does a blood cell travel in 1 h?

43. ■■ Driving a jet-powered car, Royal Air Force pilot Andy Green broke the sound barrier on land for the first time and achieved a record land speed of more than 763 mi/h in Black Rock Desert, NV, on Oct. 15, 1997 (▼Fig. 1.15). (a) What is this speed expressed in m/s? (b) How long would it take the jet-powered car to travel the length of a 300-ft football field at this speed?

▲ FIGURE 1.15   Record run   See Exercise 43.

44. IE ■■ (a) Which one of the following represents the greatest speed? (1) 1 m/s, (2) 1 km/h, (3) 1 ft/s, or (4) 1 mi/h. (b) Express the speed 15.0 m/s in mi/h.

45. ■■ An automobile speedometer is shown in ▼Fig. 1.16. (a) What would be the equivalent scale readings (for each empty box) in kilometers per hour? (b) What would be the 70-mi/h speed limit in kilometers per hour?

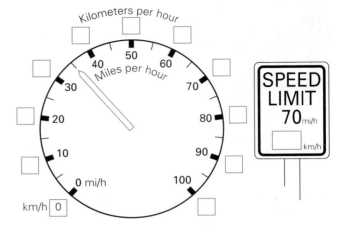

▲ FIGURE 1.16   Speedometer readings   See Exercise 45.

46. ■■ A student has a car that gets, on average, 25.0 mi/gal of gasoline. She plans to spend a year in Europe and take the car with her. (a) What should she expect the car's average gas mileage to be in kilometers per liter? (b) During the year there, she drove 6000 km. Assuming that gas costs $5.00/gal in Europe, how much did she spend on fuel? (Compute to the nearest dollar.)

47. ■■ Some common product labels are shown in ▼Fig. 1.17. From the units on the labels, find (a) the number of milliliters in 2 fl. oz and (b) the number of ounces in 100 g.

▲ FIGURE 1.17   Conversion factors   See Exercise 47.

48. ■■ ▶Fig. 1.18 is a picture of red blood cells seen under a scanning electron microscope. Normally, women possess about 4.5 million of these cells in each cubic millimeter of blood. If the blood flow through the heart is 250 milliliters per minute, how many red blood cells flow through a woman's heart each second?

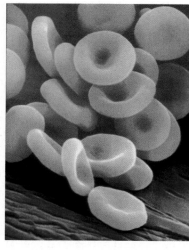

◀ **FIGURE 1.18 Red blood cells** See Exercise 48.

49. ■■ A student was 18 in. long when she was born. She is now 5 ft 6 in. tall and 20 years old. How many centimeters a year did she grow on average?

50. ■■ A 19-in. TV has a diagonal length of the TV tube of 19 inches. Assuming that the face of the tube is flat and rectangular and that the diagonal makes a 37° angle with the base of the tube, what is the area of the tube face in (a) square inches and (b) square centimeters? (*Hint*: A diagram, as suggested in the problem-solving procedure, is helpful here.)

51. ■■ The width and length of a room are 3.2 yd and 4.0 yd, respectively. If the height of the room is 8.0 ft, what is the volume of the room in (a) cubic meters and (b) cubic feet?

52. ■■■ The density of metal mercury is 13.6 g/cm³. (a) What is this density as expressed in kg/m³? (b) How many kilograms of mercury would be required to fill a 0.250-L container? (a)

53. ■■■ Engineers often express the density of a substance as a weight density, for example, in pounds per cubic foot. (a) What is the weight density of water? (b) What is the weight of one gallon of water?

54. ■■■ In the Bible, Noah is instructed to build an ark 300 cubits long, 50.0 cubits wide, and 30.0 cubits high (▶Fig. 1.19). Historical records indicate a cubit is equal to half of a yard. (a) What would the dimensions of the ark be in meters? (b) What would the ark's volume be in cubic meters? To approximate, assume that the ark is to be rectangular.

1.6 Significant Figures

55. Which of the following has the greatest number of significant figures? (a) 103.07, (b) 124.5, (c) 0.099 16, or (d) 5.408 × 10⁵.

56. In a multiplication and/or division operation involving the numbers 15 437, 201.08, and 408.0 × 10⁵, the result should be rounded to how many significant figures? (a) 3, (b) 4, (c) 5, or (d) any number.

▲ **FIGURE 1.19 Noah and his ark** See Exercise 54.

57. ■ Express the length 50 500 μm (micrometers) in centimeters, decimeters, and meters, to three significant figures.

58. ■ Using a meterstick, a student measures a length and reports it to be 0.8755 m. What is the smallest division on the meterstick scale?

59. ■ If a measured length is reported as 25.483 cm, could this length have been measured with an ordinary meterstick whose smallest division is millimeters? Discuss in terms of significant figures.

60. ■ Determine the number of significant figures in the following measured numbers: (a) 1.007 m; (b) 8.03 cm; (c) 16.272 kg; (d) 0.015 μs (microseconds).

61. ■ Express each of the numbers in Exercise 60 with two significant figures.

62. ■ Which of the following quantities has three significant figures? (a) 305.0 cm, (b) 0.0500 mm, (c) 1.000 81 kg, or (d) 8.06 × 10⁴ m².

63. ■ Express each of the following numbers to only three significant figures: (a) 10.072 m; (b) 775.4 km; (c) 0.002 549 kg; (d) 93 000 000 mi.

64. ■■ The cover of your physics book measures 0.274 m long and 0.222 m wide. What is its area in m²?

65. ■■ A compact disc (CD) has a diameter of approximately 12 cm. What is its area in m²?

66. ■■ The side of a cube is measured, and its volume is reported to be 2.5 × 10² cm³. What was the measured length of the side of the cube?

67. IE ■■ The outside dimensions of a cylindrical soda can are reported as 12.559 cm for the diameter and 5.62 cm for

the height. (a) How many significant figures will the total outside area have, (1) two, (2) three, (3) four, or (4) five? Why? (b) What is the total outside surface area of the can in cm²?

68. **IE ■■■** In doing a problem, a student adds 46.9 m and 5.72 m and then subtracts 38 m from the result. (a) How many decimal places will the final answer have, (1) zero, (2) one, or (3) two? Why? (b) What is the final answer?

69. **■■■** Work this exercise by the two given procedures as directed, commenting on and explaining any difference in the answers. Use your calculator for the calculations. Compute $p = mv$, where $v = x/t$. Given: $x = 8.5$ m, $t = 2.7$ s, and $m = 0.66$ kg. (a) First compute v and then p. (b) Compute $p = mx/t$ without an intermediate step. (c) Are the results the same? If not, why?

1.7 Problem Solving

70. An important step in problem solving before mathematically solving an equation is (a) checking units, (b) checking significant figures, (c) consulting with a friend, or (d) checking to see if the result is reasonable.

71. An important final step in problem solving before reporting an answer is (a) reading the problem again, (b) saving your calculations, (c) seeing if the answer is reasonable, or (d) checking your results with another student.

72. **CQ** When you do order-of-magnitude calculations, should you be concerned about significant figures? Explain.

73. **■** A corner construction lot has the shape of a right triangle. If the two sides perpendicular to each other are 37 m long and 42.3 m long, respectively, what is the length of the hypotenuse?

74. **■** The Earth has a mass of 6.0×10^{24} kg and a volume of 1.1×10^{21} m³. What is the Earth's average density?

75. **■** The lightest solid material is silica aerogel, which has a typical density of only about 0.10 g/cm³. The molecular structure of silica aerogel is typically 95% empty space. What is the mass of 1 m³ of silica aerogel?

76. **■** Estimate the area of a circle if its radius is 3.1×10^{-4} m.

77. **■■** Nutrition Facts labels now appear on most foods. An abbreviated label concerned with fat is shown in ▶Fig. 1.20. When burned in the body, each gram of fat supplies 9 Calories. (A food Calorie is really a kilocalorie, as we shall see in Chapter 11.) (a) What percentage of the Calories in one serving is supplied by fat? (b) You may notice that our answer doesn't agree with the listed Total Fat percentage in Fig. 1.20. This is because the given Percent Daily Values are the percentages of the maximum recommended amounts of nutrients (in grams) contained in a 2000-Calorie diet. What are the maximum recommended amounts of total fat and saturated fat for a 2000-Calorie diet?

Nutrition Facts
Serving Size: 1 can
Calories: 310

Amount Per Serving	**% Daily Value***
Total Fat 18 g	28%
Saturated Fat 7g	35%

* Percent Daily Values are based on a 2,000 Calorie diet.

◄ **FIGURE 1.20 Nutrition Facts** See Exercise 77.

78. **■■** The thickness of the total of numbered pages of a textbook is measured to be 3.75 cm. (a) If the last page of the book is numbered 860, what is the average thickness of a page? (b) Repeat the calculation by using order-of-magnitude calculations.

79. **■■** A light-year is a unit of distance corresponding to the distance light travels in a vacuum in 1 year. If the speed of light is 3.00×10^8 m/s, what is the length of a light-year in meters?

80. **IE ■■** To go to a football stadium from your house, you first drive 1000 m north, then 500 m west, and finally 1500 m south. (a) Relative to your home, the football stadium is (1) north of west, (2) south of east, (3) north of east, or (4) south of west. (b) What is the straight-line distance from your house to the stadium?

81. **■■** Two chains of length 1.0 m are used to support a lamp, as shown in ▼Fig. 1.21. The distance between the two chains is 1.0 m along the ceiling. What is the vertical distance from the lamp to the ceiling?

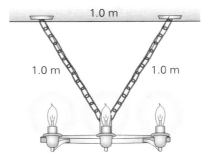

1.0 m

1.0 m   1.0 m

◄ **FIGURE 1.21 Support the lamp** See Exercise 81.

82. **■■** Tony's Pizza Palace sells a medium 9.0-in. (diameter) pizza for $7.95, and a large 12-in. pizza for $13.50. Which pizza is the better buy?

83. **■■** In ▶Fig. 1.22 which black region has the greater area, the center circle or the outer ring?

84. **■■** The shortest distance between the bases in a baseball field is 90 ft. What is the straight-line distance in meters between first base and third base?

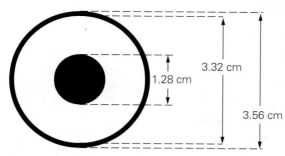

▲ **FIGURE 1.22 Which black area is greater?** See Exercise 83.

85. ■■ The Channel Tunnel, or "Chunnel" which runs under the English Channel between Great Britain and France, is 31 mi long. (There are actually three separate tunnels.) A shuttle train that carries passengers through the tunnel travels with an average speed of 75 mi/h. On average, how long, in minutes, does it take the shuttle to make a one-way trip through the Chunnel?

86. ■■■ Approximately 118 mi wide and 307 mi long and averaging 279 ft in depth, Lake Michigan is the second-largest Great Lake by volume. Estimate its volume of water in m³.

87. ■■■ A student wants to determine the distance of a small island from the lakeshore (▼Fig. 1.23). He first draws a 50-m line parallel to the shore. Then, he goes to the ends of the line and measures the angles of the lines of sight from the island relative to the line he has drawn. The angles are 30° and 40°. How far is the island from the shore?

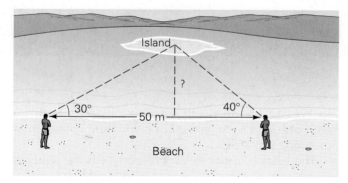

▲ **FIGURE 1.23 Measuring with lines of sight** See Exercise 87.

Additional Exercises

88. Express the following calculations to the proper number of significant figures: (a) $12.634 + 2.1$; (b) $13.5 - 2.134$; (c) $\pi(0.25 \text{ m})^2$; (d) $\sqrt{2.37/3.5}$.

89. Suppose you are paying $1.20 for 1 gal of gas. Then the United States switches to SI units, and you find that gas costs $0.32/L. Which is the greater cost for gas?

90. On average, the human heart beats 70 times a minute (called the *pulse rate*). On average, how many times does the heart beat in a 70-year lifetime?

91. The base and the height of a right triangle are 11.2 cm long and 7.5 cm high, respectively. What is the area of the triangle?

92. The general equation for a parabola is $y = ax^2 + bx + c$, where a, b, and c are constants. What are the units of each constant if y and x are in meters?

93. A rectangular block has the dimensions 2.8 cm, 9.5 cm, and 8.7 cm. Estimate the volume of the block in cubic centimeters.

94. The radius of a solid sphere is 12 cm. What is the surface area of the sphere in (a) square centimeters and (b) square meters? (c) If the mass of the sphere is 9.0 kg, what is the sphere's density in kilograms per cubic meter?

95. A cylindrical drinking glass has an inside diameter of 8.0 cm and a depth of 12 cm. If a person drinks a completely full glass of water, how much water (in liters) will be consumed?

96. IE The top of a rectangular table measures 1.245 m by 0.760 m. (a) The smallest division on the scale of the measurement instrument is (1) m, (2) cm, or (3) mm. Why? (b) What is the area of the tabletop?

97. When computing the average speed of a cross-country runner, a student gets 25 m/s. Is this result reasonable? Justify your answer.

98. The average density of the Moon is 3.36 g/cm³, and the Moon's diameter is 2160 mi. What is the total mass of the Moon in kilograms? (Compare your answer with the value given inside the back cover of this text.)

99. IE A car is driven 13 miles east and then a certain distance due north and ends up at a position of 25 degrees north of east. (a) The distance traveled by the car due north is (1) less then, (2) equal to, or (3) greater than 13 miles. Why? (b) What distance does the car go due north?

100. An airplane flies 100 mi south from city A to city B, 200 mi east from city B to city C, and then 300 mi north from city C to city D. (a) What is the straight-line distance from city A to city D? (b) What is the direction of city D relative to city A?

101. A hollow spherical ball of radius 12 cm is filled with water. What is the mass of water inside the sphere, in kilograms?

102. Tampa, Florida, is at a latitude of about 28° North. What is the distance traveled by a building in Tampa in one day (in space) as a result of the Earth's rotation? The equatorial radius of the Earth is 6.38×10^6 m.

2

Kinematics: Description of Motion

The cheetah is running at full stride. This fastest of all land animals is capable of attaining speeds up to 113 km/h, or 70 mi/h. The sense of motion in this photograph is so strong that you can almost feel the air rushing by you. And yet, this sense of motion is an illusion. Motion takes place in time, but the photo can only "freeze" a single instant. You'll find that, without the dimension of time, you can hardly describe motion at all.

The description of motion involves the representation of a restless world. Nothing is ever perfectly still. You may sit, apparently at rest, but your blood flows, and air moves into and out of your lungs. The air is composed of gas molecules moving at different speeds and in different directions. And, while you experience stillness, you, your chair, the building you are in, and the air

you breathe are all revolving through space with the Earth, part of a solar system in a spiraling galaxy in an expanding universe.

The branch of physics concerned with the study of motion and what produces and affects motion is called **mechanics**. The roots of mechanics and of human interest in motion go back to early civilizations. The study of the motions of heavenly bodies, or *celestial mechanics*, grew out of the need to measure time and location. Several early Greek scientists, notably Aristotle, put forth theories of motion that were useful descriptions, but were later proved to be incomplete or incorrect. Our currently accepted concepts of motion were formulated in large part by Galileo (1564–1642) and Isaac Newton (1642–1727).

Mechanics is usually divided into two parts: (1) kinematics and (2) dynamics. **Kinematics** deals with the *description* of the motion of objects, without consideration of what causes the motion. **Dynamics** analyzes the *causes* of motion. This chapter covers kinematics and reduces the description of motion to its simplest terms by considering motion in a straight line. We'll learn to analyze changes in motion—speeding up, slowing down, stopping. Along the way, we'll deal with a particularly interesting case of accelerated motion: free fall under the influence only of gravity. Chapter 3 focuses on motion in two dimensions (which can easily be extended to three dimensions). Chapter 4 investigates dynamics to show what causes changes in motion.

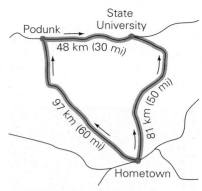

2.1 Distance and Speed: Scalar Quantities

OBJECTIVES: To (a) define distance and calculate speed and (b) explain what is meant by a scalar quantity.

Distance

We can observe motion in many instances around us. What is motion? This question seems simple but you might have some difficulty giving an immediate answer (and it's not fair to use forms of the verb "to move" to describe motion). After a little thought, you should be able to conclude that **motion** (or moving) involves the changing of position. Motion can be described in part by specifying *how far* something travels in changing position—that is, the distance it travels. **Distance** is simply the *total path length* traversed in moving from one location to another. For example, you may drive to school from your hometown and express the distance traveled in miles or kilometers. In general, the distance between two points depends on the path traveled (▶ Fig. 2.1).

Along with many other quantities in physics, distance is a scalar quantity. A **scalar quantity** is a quantity with only magnitude, or size. That is, a *scalar* has only a numerical value, such as 160 km or 100 mi. (Note that the magnitude includes units.) Distance is a scalar quantity; it tells you the magnitude only—how far, but not how far in any direction. Other examples of scalars are quantities such as 10 s (time), 3.0 kg (mass), and 20°C (temperature).

▲ **FIGURE 2.1   Distance—total path length**   In driving to State University from Hometown, one student may take the shortest route and travel a distance of 81 km (50 mi). Another student takes a longer route in order to visit a friend in Podunk before returning to school. The longer trip is in two segments, but the distance traveled is the total length, 97 km + 48 km = 145 km (90 mi).

Note: A scalar quantity has magnitude, but no direction.

Speed

When something is in motion, its position changes with time. That is, it moves a certain distance in a given amount of time. Both length and time are therefore important quantities in describing motion. For example, imagine a car and a pedestrian moving down a street and traveling a distance (length) of one block. You would expect the car to travel faster, and thus to cover the distance in a shorter time, than the person does. This relation can be expressed by using length and time to give the rate at which distance is traveled, or the **speed**, for each.

Average speed ($\bar{s}$) is the distance *d* traveled, i.e., the actual length of the path, divided by the total time Δt elapsed in traveling that distance:

Definition of: Average Speed

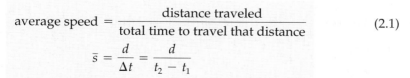

$$\text{average speed} = \frac{\text{distance traveled}}{\text{total time to travel that distance}} \qquad (2.1)$$

$$\bar{s} = \frac{d}{\Delta t} = \frac{d}{t_2 - t_1}$$

SI unit of speed: meters per second (m/s)

Note: A symbol with a bar over it is often used to denote an average. The Greek letter Δ is used to represent a change or difference in a quantity, in this case the change in time between the beginning (t_1) and end (t_2) of a trip, or the elapsed time.

The SI standard unit of speed is meters per second (length/time), although kilometers per hour is used in many everyday applications. The British standard unit is feet per second, but we often use miles per hour.

Since distance is a scalar (as is time), speed is also a scalar. The distance does not have to be in a straight line. (See Fig. 2.1.) For example, you probably have computed the average speed of an automobile trip by using the distance obtained from the starting and ending odometer readings. Suppose these readings were 17 455 km and 17 775 km, respectively, for a four-hour trip. (We'll assume that you have a car with odometer readings in kilometers.) Subtracting the readings gives a traveled distance d of 320 km, so the average speed of the trip is $d/t = 320$ km/4.0 h = 80 km/h (or about 50 mi/h).

Average speed gives a general description of motion over a time interval Δt. In the case of the auto trip with an average speed of 80 km/h, the car's speed wasn't *always* 80 km/h. With various stops and starts on the trip, the car must have been moving more slowly than the average speed part of the time. It therefore had to be moving more rapidly than the average speed another part of the time. With an average speed, you really don't know how fast the car was moving at any particular time during the trip. Similarly, the average test score of a class doesn't tell you the score of any particular student.

If the time interval Δt considered becomes smaller and smaller and approaches zero, the speed calculation gives an **instantaneous speed**. This quantity is how fast something is moving *at a particular instant of time*. The speedometer of a car gives an approximate instantaneous speed. For example, the speedometer shown in ◄Fig. 2.2 indicates a speed of about 44 mi/h, or 70 km/h. If the car travels with constant speed (so the speedometer reading does not change), then the average and instantaneous speeds will be equal. (Do you agree? Think of the average test score analogy.)

▲ **FIGURE 2.2   Instantaneous speed**   The speedometer of a car gives the speed over a very short interval of time, so its reading approaches the instantaneous speed.

Example 2.1 ■ Slow Motion: *Sojourner* Moves Along

On July 4, 1997, the *Pathfinder Lander* touched down on the surface of Mars. Out rolled the rover named *Sojourner* (◄Fig. 2.3).

Sojourner could move at a maximum speed of 0.60 m/min. At this speed, what is the shortest time it takes the rover to travel 3.0 m to get to another rock to analyze?

Thinking It Through.   Knowing the average speed and distance, we can compute the time from the equation for average speed.

Solution.   Listing the data and what is to be found in symbol form:

Given: $\bar{s} = 0.60 \dfrac{m}{min} \left(\dfrac{1 \ min}{60 \ s} \right)$ *Find:* Δt (time to travel distance d)
      $= 0.010$ m/s
   $d = 3.0$ m

In the given data, the unit meters per minute was converted to the standard meters per second.
   From Eq. 2.1, we have

$$\bar{s} = \frac{d}{\Delta t}$$

Rearranging,

$$\Delta t = \frac{d}{\bar{s}} = \frac{3.0 \ m}{0.010 \ m/s} = 3.0 \times 10^2 \ s \ (= 5.0 \ min)$$

Follow-up Exercise.   (a) Was it necessary to convert meters per minute to meters per second? Explain. (b) Suppose *Sojourner* took 15.0 min to travel the 3.00 m. What would be the rover's average speed in this case? (*Answers to all Follow-up Exercises are at the back of the text.*)

▲ **FIGURE 2.3   Away we go!** Sojourner speeding at 0.60 m/min (0.010 m/s) along the Martian surface. See Example 2.1.

PHYSLET®
ILLUSTRATION

Determination of Average Speed

2.2 One-Dimensional Displacement and Velocity: Vector Quantities

OBJECTIVES: To (a) define displacement and calculate velocity, and (b) explain the difference between scalar and vector quantities.

Displacement

For straight-line, or linear, motion, it is convenient to specify position by using the familiar two-dimensional Cartesian coordinate system, with x- and y-axes at right angles. A straight-line path can be in any direction, but for convenience, we usually orient the coordinate axes so that the motion is along one of them. (See Learn by Drawing (LDB).)

As we have seen, distance is a scalar quantity with only magnitude (and units). However, often when we describe motion, more information can be given by adding a *direction*. This information is particularly convenient for a change of position in a straight line. We define **displacement** as the straight-line distance between two points, along with the *direction* from the starting point to the final position. Unlike distance (a scalar), displacement can have either positive or negative values, with the signs indicating the directions along a coordinate axis.

As such, displacement is a **vector quantity**. A *vector* has both magnitude and direction. For example, when we describe the displacement of an airplane as 25 km north, we are giving a *vector* description (magnitude and direction). Other vector quantities include velocity and acceleration.

Algebra applies to vectors, but we have to know how to specify and deal with the direction part of the vector. This process is relatively simple in one dimension. To illustrate this with respect to finding displacements, consider the situation shown in ▼Fig. 2.4, where x_1 and x_2 indicate positions on the x-axis. A student moves in a straight line from the lockers to the physics lab. As can be seen in Fig. 2.4a, the scalar distance between the two points is 8.0 m. To specify displacement (a vector) between x_1 and x_2, we use the expression

$$\Delta x = x_2 - x_1 \tag{2.2}$$

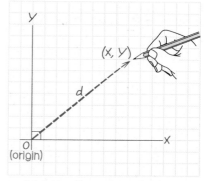
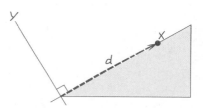
◀ **FIGURE 2.4 Distance (scalar) and displacement (vector)** **(a)** The distance (straight-line path) between the student and the physics lab is 8.0 m and is a scalar quantity. **(b)** To indicate displacement, x_1 and x_2 specify the initial and final positions, respectively. The displacement is then $\Delta x = x_2 - x_1 = 9.0 \text{ m} - 1.0 \text{ m} = +8.0$ m—that is, 8.0 m in the $+x$-direction.

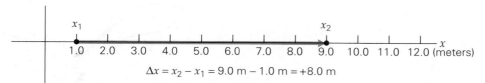

(a) Distance (magnitude or numerical value)

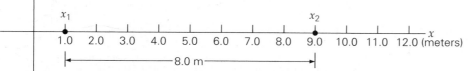

$$\Delta x = x_2 - x_1 = 9.0 \text{ m} - 1.0 \text{ m} = +8.0 \text{ m}$$

(b) Displacement (magnitude and direction)

where Δ is again used to represent a change or difference in a quantity. Then, as in Fig. 2.4b, we have

$$\Delta x = x_2 - x_1 = 9.0 \text{ m} - 1.0 \text{ m} = +8.0 \text{ m}$$

Hence, the student's displacement (magnitude and direction) is 8.0 m in the positive x-direction, as indicated by the positive (+) result in Fig. 2.4b. (As in "regular" mathematics, the plus sign is often omitted, as it is understood, so this displacement can be written as $\Delta x = 8.0$ m instead of $\Delta x = +8.0$ m.)

Vector quantities in this book are usually indicated by boldface type; for example, a velocity vector is indicated by **v**. However, when working in one dimension, this notation is not needed and can be simplified by using plus and minus signs to indicate the only two possible directions. The x-axis is commonly used for horizontal motions, and a plus (+) sign is taken to indicate the direction to the right, or in the "positive x-direction," and a minus (−) sign indicates the direction to the left, or in the "negative x-direction." Keep in mind that these signs only "point" in *particular directions*. An object moving along the negative x-axis toward the origin would be moving in the + direction, even though its x-position is negative. How about an object moving along the +x-axis toward the origin? If you said the minus (−) direction, you are correct.

Suppose the other student in Fig. 2.4 walks from the physics lab (the initial position is different and $x_1 = 9.0$ m) to the end of the lockers (the final position is now $x_2 = 1.0$ m). Her displacement would be

$$\Delta x = x_2 - x_1 = 1.0 \text{ m} - 9.0 \text{ m} = -8.0 \text{ m}$$

The minus sign indicates that the direction of the displacement was in the negative x-direction. In this case, we say that the two students' displacements are equal (in magnitude) and opposite (in direction).

Velocity

PHYSLET®
ILLUSTRATION

Velocity and Speed

As we have seen, speed, like the distance it incorporates, is a scalar quantity—it has magnitude only. Another quantity used to describe motion is *velocity*. Speed and velocity are often used synonymously in everyday conversation, but the terms have different meanings in physics. Speed is a scalar, and velocity is a vector—it has both magnitude and direction. Unlike speed (but like displacement) velocity can have both positive and negative values, indicating directions.

Velocity tells you how fast something is moving *and* in which direction it is moving. And just as we can speak of average and instantaneous speeds, we have average and instantaneous velocities involving vector displacements. The **average velocity** is the displacement divided by the total travel time.

Definition of: Average velocity

$$\text{average velocity} = \frac{\text{displacement}}{\text{total travel time}} \qquad (2.3)*$$

$$\bar{v} = \frac{\Delta x}{\Delta t} = \frac{x_2 - x_1}{t_2 - t_1}$$

SI unit of velocity: meters per second (m/s),

*Another common form of this equation is

$$\bar{v} = \frac{\Delta x}{\Delta t} = \frac{(x_2 - x_1)}{(t_2 - t_1)} = \frac{(x - x_o)}{(t - t_o)} = \frac{(x - x_o)}{t},$$

or, after rearranging

$$x = x_o + \bar{v}t, \qquad (2.3)$$

where x_o is the initial position, x is the final position, and $\Delta t = t$ with $t_o = 0$. See Section 2.3 for more on this notation.

In this equation, Δx is the displacement and Δt is the time interval, where t_1 is the initial time and t_2 is the final time.

In the case of more than one displacement (successive displacements), the average velocity for all the displacements is equal to the total or net displacement divided by the total time. The total displacement is found by adding the displacements algebraically according to the directional signs.

You might be wondering whether there is a relationship between average speed and average velocity. A quick look at Fig. 2.4 will show you that if all the motion is in one direction, that is, no reversal of direction, the distance is equal to the magnitude of the displacement, and the average speed is equal to the magnitude of the average velocity. However, be careful. This set of relationships is not true if there is a reversal of direction, as Example 2.2 shows.

Example 2.2 ■ There and Back: Average Velocities

A jogger jogs from one end to the other of a straight 300-m track in 2.50 min and then jogs back to the starting point in 3.30 min. What was the jogger's average velocity (a) in jogging to the far end of the track, (b) coming back to the starting point, and (c) for the total jog?

Thinking It Through.   The average velocities are computed from the defining equation. Note that the times given are the Δt's associated with the particular displacements.

Solution.   From the problem, we have:

Given: $\Delta x_1 = 300$ m (taking the initial direction as positive)   *Find:*   Average velocities
$\Delta x_2 = -300$ m (taking the direction of the return   for (a) the first leg
                 trip as negative)   of the jog,
$\Delta t_1 = 2.50$ min $(60\ \text{s/min}) = 150$ s $\Big\}$ (conversion to   (b) the return jog,
$\Delta t_2 = 3.30$ min $(60\ \text{s/min}) = 198$ s $\Big\}$ standard units)   and
                                                                                       (c) the total jog

(a)   The jogger's average velocity for the trip down the track is found from Eq. 2.3:

$$\bar{v}_1 = \frac{\Delta x_1}{\Delta t_1} = \frac{300\ \text{m}}{150\ \text{s}} = +2.00\ \text{m/s}$$

(b)   Similarly, for the return trip, we have

$$\bar{v}_2 = \frac{\Delta x_2}{\Delta t_2} = \frac{-300\ \text{m}}{198\ \text{s}} = -1.52\ \text{m/s}$$

(c)   For the total trip, there are two displacements to consider, down and back, so these are added together to get the total displacement, and then divided by the total time;

$$\bar{v}_3 = \frac{\Delta x_1 + \Delta x_2}{\Delta t_1 + \Delta t_2} = \frac{300\ \text{m} + (-300\ \text{m})}{150\ \text{s} + 198\ \text{s}} = 0\ \text{m/s}$$

The average velocity for the total trip is zero! Do you see why? Recall from the definition of displacement that the magnitude of displacement is the straight-line distance between two points. The displacement from one point back to the same point is zero, hence the average velocity is zero. (See ▶Fig. 2.5.)

The total displacement could have been found by simply taking $\Delta x = x_{\text{final}} - x_{\text{initial}} = 0 - 0 = 0$, but it was done in parts here for illustration purposes.

Follow-up Exercise.   Find the jogger's average speed for each of the cases in this Example, and compare it with the respective average velocities. [Will the average speed for (c) be zero?] *(Answers to all Follow-up Exercises are at the back of the text.)*

▲ **FIGURE 2.5   Back home again!**   Despite having covered nearly 110 m on the base paths, at the moment the runner slides through the batter's box (his original position) into home plate, his displacement is zero—at least, if he is a right-handed batter. No matter how fast he ran the bases, his average velocity for the round trip is also zero.

As Example 2.2 shows, average velocity provides only a very general description of motion. One way to take a closer look at motion is to take smaller time intervals, that is, to let the observation time (Δt) become smaller and smaller. As with speed, when Δt approaches zero, we obtain the **instantaneous velocity**, which describes how fast something is moving and in which direction at a particular instant of time.

Note: for displacements in both the + and − directions (reversal of direction), the distance is *not* the magnitude of the total displacement.

Instantaneous velocity is defined mathematically as

$$v = \lim_{\Delta t \to 0} \frac{\Delta x}{\Delta t} \qquad (2.4)$$

This expression is read as "the instantaneous velocity is equal to the limit of $\Delta x/\Delta t$ as Δt goes to zero." The time interval does not ever equal zero (why?), but *approaches* zero. Instantaneous velocity is an average velocity, but over such a small Δt that it is essentially an average "at an instant in time," which is why we call it the instantaneous velocity.

In one dimension, **uniform motion** means motion with a constant velocity (constant magnitude *and* constant direction). For example, the car in ▼Fig. 2.6 has a uniform velocity (as well as a uniform speed). It travels the same distance in equal time intervals (50 km each hour), and the direction of its motion does not change.

Note: The word *uniform* means constant.

Graphical Analysis

Graphical analysis is often helpful in understanding motion and its related quantities. For example, the motion of the car in Fig. 2.6a may be represented on a plot of position versus time, or x versus t. As can be seen from Fig. 2.6b, a straight line is obtained for a uniform, or constant, velocity on such a graph.

Recall from Cartesian graphs of y versus x that the slope of a straight line is given by $\Delta y/\Delta x$. Here, with a plot of x versus t, the slope of the line, $\Delta x/\Delta t$, is

▶ **FIGURE 2.6 Uniform linear motion—constant velocity** In uniform linear motion, an object travels at a constant velocity, covering the same distance in equal time intervals. **(a)** Here, a car travels 50 km each hour. **(b)** An x-versus-t plot is a straight line, since equal displacements are covered in equal times. The numerical value of the slope of the line is equal to the magnitude of the velocity, and the sign of the slope gives its direction. (The average velocity equals the instantaneous velocity in this case. Why?)

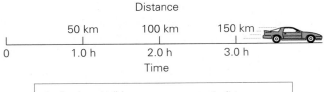

Δx (km)	Δt (h)	$\Delta x/\Delta t$
50	1.0	50 km/1.0 h = 50 km/h
100	2.0	100 km/2.0 h = 50 km/h
150	3.0	150 km/3.0 h = 50 km/h

(a)

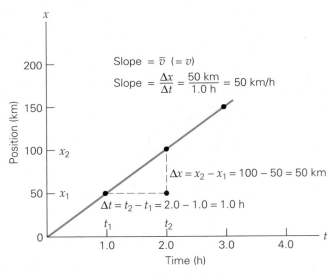

Uniform velocity

(b)

equal to the average velocity $\bar{v} = \Delta x/\Delta t$. For uniform motion, this value is equal to the instantaneous velocity. That is, $\bar{v} = v$. (Why?) The numerical value of the slope is the magnitude of the velocity, and the sign of the slope gives the direction. A positive slope indicates that x increases with time, so the motion is in the positive x-direction. (The plus sign is often omitted, as being understood.)

Suppose that a plot of position versus time for a car's motion was a straight line with a negative slope, as in ▼Fig. 2.7. What does this slope indicate? As the figure shows, the position (x) values get smaller with time at a constant rate, indicating that the car was traveling in uniform motion in the negative x-direction.

In most instances, the motion of an object is nonuniform, meaning that different distances are covered in equal intervals of time. An x-versus-t plot for such motion in one dimension is a curved line, as illustrated in ▼Fig. 2.8. The average velocity of the object at a particular interval of time is the slope of a straight line between the two points on the curve that correspond to the starting and ending times of the interval. In the figure, the average velocity of the total trip is the slope of the straight line joining the beginning and ending points of the curve (t_1 and t_2).

The instantaneous velocity is equal to the slope of a straight line tangent to the curve at a specific point. Five tangent lines are shown in Fig. 2.8. At (1), the slope is positive, and the motion is in the positive x-direction. At (2), the slope of a horizontal tangent line is zero, so there is no motion. That is, the object has instantaneously stopped ($v = 0$). At (3), the slope is negative, so the object is moving in the negative x-direction. Thus, the object stopped and changed direction at point (2). What is happening at points (4) and (5)?

Drawing various tangent lines along the curve, we see that their slopes vary, indicating that the instantaneous velocity is changing with time. An object in

PHYSLET®
ILLUSTRATION

Graphical Determination of Velocity

▼ **FIGURE 2.8 Position-versus-time graph for an object in nonuniform linear motion** For a nonuniform velocity, an x-versus-t plot is a curved line. The slope of the line between two points is the average velocity between those positions, and the instantaneous velocity is the slope of a line tangent to the curve at any point. Five tangent lines are shown, with the value of $\Delta x/\Delta t$ given for the fifth. Can you describe the object's motion in words?

▼ **FIGURE 2.7 Position-versus-time graph for an object in uniform motion in the $-x$-direction** A straight line on an x-versus-t plot with a negative slope indicates uniform motion in the $-x$-direction. Note that the object's location changes at a constant rate. At $t = 4.0$ h, the object is at $x = 0$. How would the graph look if the motion continues for $t > 4.0$ h?

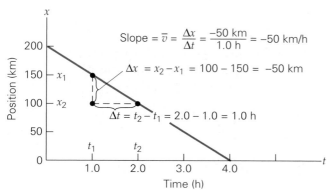

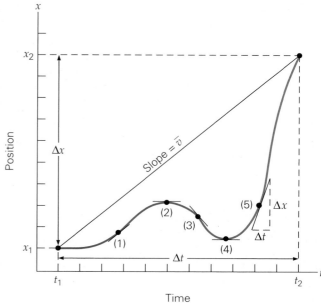

nonuniform motion can speed up, slow down, or change direction. How we describe motion with changing velocity is the topic of Section 2.3.

2.3 Acceleration

OBJECTIVES: To (a) explain the relationship between velocity and acceleration and (b) perform graphical analyses of acceleration.

The basic description of motion involves the time rate of change of position, which we call *velocity*. Going one step further, we can consider how this *rate of change* changes. Suppose that something is moving at a constant velocity and then the velocity changes. Such a change in velocity is called an *acceleration*. The gas pedal on an automobile is commonly called the *accelerator*. When you press down on the accelerator, the car speeds up; when you let up on the accelerator, the car slows down. In either case, there is a change in velocity with time. We define **acceleration** as the time rate of change of velocity.

Analogous to average velocity is the **average acceleration**, or the change in velocity divided by the time taken to make the change:

$$\text{average acceleration} = \frac{\text{change in velocity}}{\text{time to make the change}} \tag{2.5}$$

$$\bar{a} = \frac{\Delta v}{\Delta t}$$

$$= \frac{v_2 - v_1}{t_2 - t_1} = \frac{v - v_\mathrm{o}}{t - t_\mathrm{o}}$$

SI unit of acceleration: meters per second squared $(\mathrm{m/s^2})$.

Note that we have changed the initial and final variables to a more commonly used notation. v_o and t_o are the initial or original velocity and time, respectively, and v and t are the general velocity and time at some time in the future, such as when you want to know the velocity v at a particular time t.

The dimensions of acceleration are $([L]/[T])/[T]$, from $\Delta v/\Delta t$. The SI units of acceleration are therefore meters per second per second, that is, $(\mathrm{m/s})/\mathrm{s}$, or $\mathrm{m/s \cdot s}$, commonly expressed as meters per second squared $(\mathrm{m/s^2})$. In the British system, the units are feet per second squared $(\mathrm{ft/s^2})$.

Because velocity is a vector quantity, so is acceleration, since acceleration represents a change in velocity. Being a vector quantity, velocity has both magnitude and direction, and a change in velocity may thus involve either or both of these factors. An acceleration, therefore, may result from a change in *speed* (magnitude), a change in *direction*, or a change in *both*, as illustrated in ▶ Fig. 2.9.

For straight-line, linear motion, plus and minus signs will be used to indicate the directions of velocity and acceleration, as was done for linear displacements. Eq. 2.5 is commonly simplified and written as

$$\bar{a} = \frac{v - v_\mathrm{o}}{t} \tag{2.6}$$

where t_o is taken to be zero. (v_o may not be zero, so it cannot generally be omitted.)

Analogous to instantaneous velocity, **instantaneous acceleration** is the acceleration at a particular instant of time. This quantity is expressed mathematically as

$$a = \lim_{\Delta t \to 0} \frac{\Delta v}{\Delta t} \tag{2.7}$$

The conditions of the time interval approaching zero are the same here as described for instantaneous velocity.

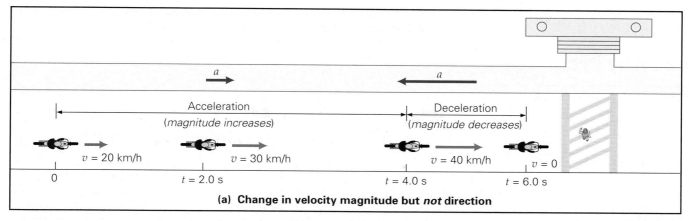

(a) **Change in velocity magnitude but *not* direction**

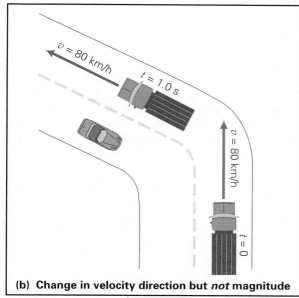

(b) **Change in velocity direction but *not* magnitude**

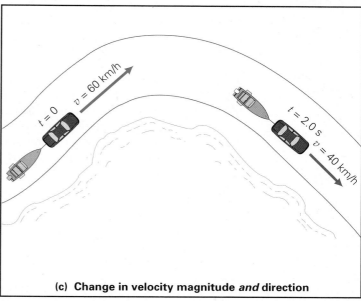

(c) **Change in velocity magnitude *and* direction**

▲ **FIGURE 2.9   Acceleration—the time rate of change of velocity**   Since velocity is a vector quantity, with magnitude and direction, an acceleration can occur when there is **(a)** a change in magnitude, but not direction, **(b)** a change in direction, but not magnitude, or **(c)** a change in both magnitude and direction.

Example 2.3 ■ Slowing It Down: Average Acceleration

A couple in their sport utility vehicle (SUV) is traveling 90 km/h down a straight high-way. They see an accident in the distance, so the driver slows down to 40 km/h in 5.0 s. What is the average acceleration of the SUV?

Thinking It Through.   To find the average acceleration, the variables as defined in Eq. 2.6 must be given, and they are.

Solution.   From the statement of the problem, we have the following data:

Given: $v_\text{o} = (90 \text{ km/h}) \left(\dfrac{0.278 \text{ m/s}}{1 \text{ km/h}} \right)$ *Find:* $\bar{a}$ (average acceleration)

$\quad\quad = 25 \text{ m/s}$

$\quad\quad v = (40 \text{ km/h}) \left(\dfrac{0.278 \text{ m/s}}{1 \text{ km/h}} \right)$

$\quad\quad = 11 \text{ m/s}$

$\quad\quad t = 5.0 \text{ s}$

[With the motion in a straight line, the instantaneous velocities are assumed to be in the positive direction, and conversions to standard units (kilometers per second to meters

a positive
v positive
Result: Faster in +x direction

−x   +x

a negative
v positive
Result: Slower in +x direction

−x   +x

a positive
v negative
Result: Slower in −x direction

−x   +x

a negative
v negative
Result: Faster in −x direction

−x   +x

per second) are made right away, since it is noted that the time is given in seconds. In general, we always work with acceleration in standard units.]

Given the initial and final velocities and the time interval, the average acceleration can be found by using Eq. 2.6:

$$\bar{a} = \frac{v - v_o}{t} = \frac{11 \text{ m/s} - 25 \text{ m/s}}{5.0 \text{ s}} = -2.8 \text{ m/s}^2$$

The minus sign indicates the direction of the (vector) acceleration. In this case, the acceleration is opposite to the direction of the motion ($+v$), and the car slows. Such an acceleration is sometimes called a *deceleration*, since the car is slowing.

Follow-up Exercise. Does a negative acceleration necessarily mean that a moving object is slowing down (decelerating) or that its speed is decreasing? *Hint*: see LBD. (*Answers to all Follow-up Exercises are at the back of the text.*)

Constant Acceleration

Although acceleration can vary with time, our study of motion will generally be restricted to constant accelerations for simplicity. (An important constant acceleration is the acceleration due to gravity near the Earth's surface, which will be considered in the next section.) Since for a constant acceleration, the average is equal to the constant value ($\bar{a} = a$), the bar over the acceleration in Eq. 2.6 may be omitted. Thus, for a constant acceleration, the equation relating velocity, acceleration, and time is commonly written (rearranging Eq. 2.6) as follows:

$$v = v_o + at \quad \text{(constant acceleration only)} \quad (2.8)$$

(Note that the term at represents the *change* in velocity that occurs, since $at = v - v_o = \Delta v$.)

Example 2.4 ■ Fast Start, Slow Stop: Motion with Constant Acceleration

A drag racer starting from rest accelerates in a straight line at a constant rate of 5.5 m/s² for 6.0 s. (a) What is the racer's velocity at the end of this period of time? (b) If a parachute deployed at this time causes the racer to slow down uniformly at a rate of 2.4 m/s², how long will it take the racer to come to a stop?

Thinking It Through. The racer first speeds up and then slows down, so close attention must be given to the directional signs of the vector quantities. Choose a coordinate system with the positive direction in the direction of the initial velocity. (Draw a sketch of the situation for yourself.) The answers can then be found by using the appropriate equations.

Solution. Taking the initial motion to be in the positive direction, we have the following data:

Given: (a) $v_o = 0$ (at rest)   *Find:*   (a) v (final velocity)
            $a = 5.5 \text{ m/s}^2$ (b) t (time)
            $t = 6.0 \text{ s}$
     (b) $v_o = v$ [from part (a)]
          $v = 0$ (comes to stop)
          $a = -2.4 \text{ m/s}^2$ (opposite direction of v_o).

The data have been listed in two parts. This practice helps avoid confusion with symbols. Note that the final velocity v that is to be found in part (a) becomes the initial velocity v_o for part (b).

(a) To find the final velocity v, we use Eq. 2.8 directly:

$$v = v_o + at = 0 + (5.5 \text{ m/s}^2)(6.0 \text{ s}) = 33 \text{ m/s}$$

(b) Here, we want to find time, so solving Eq. 2.6 for t and using $v_o = 33$ m/s from part (a), we have

$$t = \frac{v - v_o}{a} = \frac{0 - 33 \text{ m/s}}{-2.4 \text{ m/s}^2} = 14 \text{ s}$$

Note that the time comes out positive, as it should. We start implicitly at zero time (taking $t_o = 0$ when the parachute is deployed), and time goes forward, or in a "positive direction."

Follow-up Exercise. What is the racer's instantaneous velocity 10 seconds after the parachute is deployed? *(Answers to all Follow-up Exercises are at the back of the text.)*

Motions with constant accelerations are easy to represent graphically by plotting instantaneous velocity versus time. A v-versus-t plot is a straight line whose slope is equal to the acceleration, as illustrated in ▼Fig. 2.10. Note that Eq. 2.8 can be written as $v = at + v_o$, which, as you may recognize, has the form of an equation of a straight line, $y = mx + b$ (slope m and intercept b). In Fig. 2.10a, the motion is in the positive direction, and the acceleration adds to the velocity

PHYSLET® ILLUSTRATION

Graphical Determination of Acceleration

▼ **FIGURE 2.10   Velocity-versus-time graphs for motions with constant accelerations**   The slope of a v-versus-t plot is the acceleration. **(a)** A positive slope indicates an increase in the velocity in the positive direction. The vertical arrows to the right indicate how the acceleration adds velocity to the initial velocity v_o. **(b)** A negative slope indicates a decrease in the initial velocity v_o, or a deceleration. **(c)** Here a negative slope indicates a negative acceleration, but the initial velocity is in the negative direction, $-v_o$, so the speed of the object increases in that direction. **(d)** The situation here is initially similar to that of (b) but ends up resembling that in (c). Can you explain what happened at time t_1?

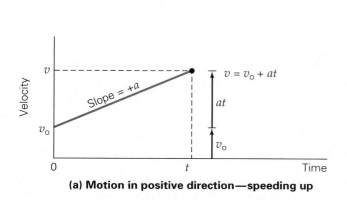

(a) Motion in positive direction—speeding up

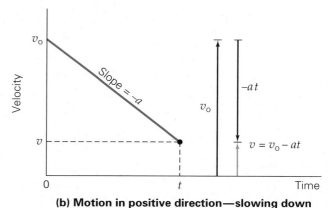

(b) Motion in positive direction—slowing down

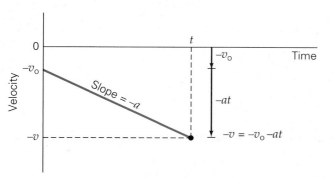

(c) Motion in negative direction—speeding up

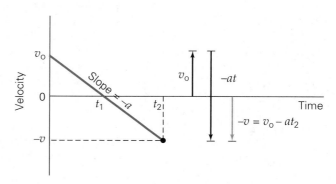

(d) Changing direction

for a time t, as illustrated by the vertical arrows at the right of the graph. Here, the slope is positive, $+a$. In Fig. 2.10b, the negative slope $(-a)$ indicates a negative acceleration that produces a slowing down, or deceleration. However, Fig. 2.10c illustrates how a negative acceleration can speed things up (for motion in the negative direction). The situation in Fig. 2.10d is slightly more complex. Can you explain what is happening there?

When an object moves at a constant acceleration, its velocity changes by the same amount in each time unit. For example, if the acceleration is 10 m/s^2 in the same direction as that of the initial velocity, the object's velocity increases by 10 m/s in each second. Suppose that the object has an initial velocity v_0 of 20 m/s at $t_0 = 0$. Then, for $t = 0, 1.0, 2.0, 3.0,$ and 4.0 s, the velocities are 20, 30, 40, 50, and 60 m/s, respectively. The average velocity over the four-second interval is $\bar{v} = 40$ m/s.

This average velocity may be computed in the regular manner (Eq. 2.3), or you may immediately recognize that the uniformly increasing series of numbers 20, 30, 40, 50, and 60 has an average value of 40 (the midway value of the series). Note that the average of the extreme (initial and final) values also gives the average of the series—that is, $(20 + 60)/2 = 40$. When the velocity changes at a uniform rate because of a constant acceleration, $\bar{v}$ is the average of the initial and final velocities:

$$\bar{v} = \frac{v + v_0}{2} \qquad \text{(constant acceleration only)} \qquad (2.9)$$

Example 2.5 ■ On the Water: Using Multiple Equations

A motorboat starting from rest on a lake accelerates in a straight line at a constant rate of 3.0 m/s^2 for 8.0 s. How far does the boat travel during this time?

Thinking It Through. We have only one equation for distance (Eq. 2.3, $x = x_0 + \bar{v}t$), but this equation cannot be used directly. The average velocity must first be found, so multiple equations are involved.

Solution. Reading the problem and summarizing the given data and what is to be found, we have the following:

Given: $x_0 = 0$ *Find:* x (distance)
$v_0 = 0$
$a = 3.0$ m/s^2
$t = 8.0$ s

(Note that all of the units are standard.)

In analyzing the problem, we might reason as follows: To find x, we will need to use Eq. 2.3, $x_0 + \bar{v}t$. (The average velocity $\bar{v}$ must be used because the velocity is changing and thus not constant.) With time t given, the solution to the problem then involves finding $\bar{v}$. By Eq. 2.9, $\bar{v} = (v + v_0)/2$, and with $v_0 = 0$, we need only find the final velocity v to solve the problem. Equation 2.8, $v = v_0 + at$, enables us to calculate v from the given data. So, we have the following:

The velocity of the boat at the end of 8.0 s is

$$v = v_0 + at = 0 + (3.0 \text{ m/s}^2)(8.0 \text{ s}) = 24 \text{ m/s}$$

The average velocity over that time interval is

$$\bar{v} = \frac{v + v_0}{2} = \frac{24 \text{ m/s} + 0}{2} = 12 \text{ m/s}$$

Finally, the magnitude of the displacement, which in this case is the same as the distance traveled, is given by Eq. 2.3 (with $x_0 = 0$):

$$x = \bar{v}t = (12 \text{ m/s})(8.0 \text{ s}) = 96 \text{ m}$$

Follow-up Exercise. (Sneak preview.) In Section 2.4, the following equation will be derived, $x = v_{\mathrm{o}}t + \frac{1}{2}at^2$. Use the data in this Example to see if this equation gives the distance traveled. (*Answers to all Follow-up Exercises are at the back of the text.*)

2.4 Kinematic Equations (Constant Acceleration)

OBJECTIVES: To (a) explain the kinematic equations of constant acceleration and (b) apply them to physical situations.

The description of motion in one dimension with constant acceleration requires only three basic equations. From previous sections, these equations are

$$x = x_{\mathrm{o}} + \bar{v}t \tag{2.3}$$

$$\bar{v} = \frac{v + v_{\mathrm{o}}}{2} \qquad \textit{(constant acceleration only)} \tag{2.9}$$

$$v = v_{\mathrm{o}} + at \qquad \textit{(constant acceleration only)} \tag{2.8}$$

(Keep in mind that the first equation, Eq. 2.3, is general and is not limited to situations in which there is constant acceleration, as the latter two equations are.)

However, as Example 2.5 showed, the description of motion in some instances requires multiple applications of these equations, which may not be obvious at first. It would be helpful if there were a way to reduce the number of operations in solving kinematic problems, and there is—by combining equations algebraically.

For instance, suppose we want an expression that gives location x in terms of time and acceleration rather than in terms of time and average velocity (as in Eq. 2.3). We can eliminate v from Eq. 2.3 by substituting for v from Eq. 2.9 into Eq. 2.3:

$$x = x_{\mathrm{o}} + \bar{v}t = x_{\mathrm{o}} + \left(\frac{v + v_{\mathrm{o}}}{2}\right)t$$

and

$$x = x_{\mathrm{o}} + \tfrac{1}{2}(v + v_{\mathrm{o}})t \qquad \textit{(constant acceleration only)} \tag{2.10}$$

Then, substituting for v from Eq. 2.8 gives

$$x = x_{\mathrm{o}} + \tfrac{1}{2}(v_{\mathrm{o}} + at + v_{\mathrm{o}})t$$

Simplifying,

$$x = x_{\mathrm{o}} + v_{\mathrm{o}}t + \tfrac{1}{2}at^2 \qquad \textit{(constant acceleration only)} \tag{2.11}$$

Essentially, this series of steps was done in Example 2.5. The combined equation allows the distance traveled by the motorboat in that Example to be computed directly:

$$x - x_{\mathrm{o}} = \Delta x = v_{\mathrm{o}}t + \tfrac{1}{2}at^2 = 0 + \tfrac{1}{2}(3.0 \text{ m/s}^2)(8.0 \text{ s})^2 = 96 \text{ m}$$

Much easier, isn't it?

Perhaps we want an expression that gives velocity as a function of position x rather than time (as in Eq. 2.8). We can eliminate t from Eq. 2.8 by using Eq. 2.3 in the form $t = (x - x_{\mathrm{o}})/\bar{v}$:

$$v = v_{\mathrm{o}} + a\left(\frac{x - x_{\mathrm{o}}}{\bar{v}}\right)$$

Note: $\Delta x = x - x_{\mathrm{o}}$ is displacement, but with $x_{\mathrm{o}} = 0$, as it often is, then $\Delta x = x$, and the value of the x position is the same as that of the displacement, which saves writing $\Delta x = x - x_{\mathrm{o}}$ every time.

Then we replace $\bar{v}$, using Eq. 2.9:

$$v = v_0 + \frac{a(x - x_0)}{\left(\dfrac{v + v_0}{2}\right)} \quad \text{or} \quad v - v_0 = \frac{2a(x - x_0)}{v + v} \quad \text{or} \quad (v + v_0)(v - v_0) = 2a(x - x_0)$$

Simplifying by using the algebraic relationship $(v + v_0)(v - v_0) = v^2 - v_0^2$, we have

$$v^2 = v_0^2 + 2a(x - x_0) \quad \text{(constant acceleration only)} \quad (2.12)$$

Problem-Solving Hint

Students in introductory physics courses are sometimes overwhelmed by the various kinematic equations. Keep in mind that equations and mathematics are the tools of physics. As any mechanic or carpenter will tell you, tools make your work easier so long as you are familiar with them and know how to use them. The same is true for physics tools.

In Sections 2.3 and 2.4, we here presented the kinematic equations. The following set of equations for linear motion with *constant* acceleration is used to solve the majority of kinematic problems:

$$v = v_0 + at \tag{2.8}$$

$$x = x_0 + \tfrac{1}{2}(v + v_0)t \tag{2.10}$$

$$x = x_0 + v_0 t + \tfrac{1}{2}at^2 \tag{2.11}$$

$$v^2 = v_0^2 + 2a(x - x_0) \tag{2.12}$$

(Occasionally, we are interested in average speed or velocity, but, as noted earlier, averages generally don't tell you a great deal.) Note that each of the equations in the list has four or five variables. All but one of the variables in an equation must be known in order to be able to solve for what you are trying to find. That is, for Eq. 2.8, three variables must be known to be able to solve for the fourth, unknown variable, and, for the rest of the equations, four variables must be known to be able to solve for a fifth, unknown variable.

Always try to understand and visualize a problem. Listing the data as described in the Suggested Problem-Solving Procedure in Chapter 1 may help you decide which equation to use, by determining the known and unknown variables. Remember this approach as you work through the remaining Examples in the chapter. Also, don't overlook any *implied data*, an error illustrated by Example 2.6.

Integrated Example 2.6 ■ Moving Apart: Where Are They Now?

Two riders on dune buggies sit 10 m apart on a long, straight track, facing in opposite directions. Starting at the same time, both riders accelerate at a constant rate of 2.0 m/s². (a) Make a problem-solving sketch of the situation. (b) How far apart will the dune buggies be at the end of 3.0 s?

(a) Conceptual Reasoning. We know only that the dune buggies are initially 10 m apart, so they can be positioned anywhere on the x-axis. It is convenient to place one at the origin so that one initial position (x_0) is zero. A sketch of the situation is shown in ▶ Fig. 2.11.

(b) Thinking It Through. Referring to the sketch, we have the following data:

Given: $x_{0_A} = 0$ *Find:* separation distance at $t = 3.0$ s
$a_A = -2.0$ m/s²
$t = 3.0$ s
$x_{0_B} = 10$ m
$a_B = 2.0$ m/s²

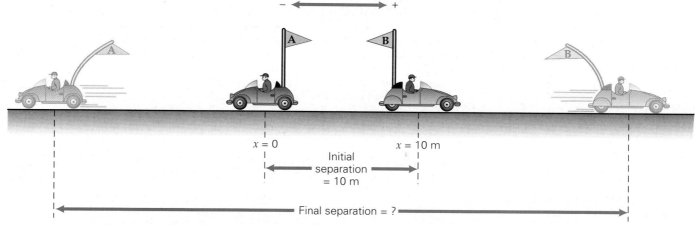

$x = 0$

$x = 10$ m

Initial
separation
= 10 m

Final separation = ?

▲ **FIGURE 2.11   Away they go!**   Two dune buggies accelerate away from each other. How far are they apart at a later time? See Integrated Example 2.6.

The distance each vehicle travels is given by Eq. 2.11 [the only displacement (x) equation with acceleration (a)]: $x = x_o + v_o t + \frac{1}{2}at^2$. But wait, we don't have v_o in the Given list. Some implied data may have been missed. We quickly note that $v_o = 0$ for both vehicles, so

$$x_A = x_{o_A} + v_{o_A}t + \tfrac{1}{2}a_A t^2 = 0 + 0 + \tfrac{1}{2}(-2.0 \text{ m/s}^2)(3.0 \text{ s})^2 = -9 \text{ m}$$

and

$$x_B = x_{o_B} + v_{o_A}t + \tfrac{1}{2}a_B t^2 = 10 \text{ m} + 0 + \tfrac{1}{2}(2.0 \text{ m/s}^2)(3.0 \text{ s})^2 = 19 \text{ m}$$

What does $x_A = -9$ m tell us? That vehicle A is at the -9 m position from the origin on the $-x$-axis, whereas vehicle B is at a position of 19 m on the $+x$-axis. Hence, the separation distance between the two dune buggies is 28 m.

Follow-up Exercise.   Would it make any difference in the separation distance if vehicle B had been initially put at the origin instead of vehicle A? (*Answers to all Follow-up Execises are at the back of the text.*)

Conceptual Example 2.7 ■ Two Race Cars: The Effect of Squared Quantities

During some time trials, race car A, starting from rest, accelerates uniformly along a straight, level track for a particular interval of time. Race car B, also starting from rest, accelerates at the same rate, but for twice the time. At the ends of their respective acceleration periods, which of these statements is true: (a) Car A has traveled a greater distance; (b) car B has traveled twice as far as car A; (c) car B has traveled four times as far as car A; (d) both cars have traveled the same distance? *Clearly establish the reasoning and physical principle(s) used in determining your answer before checking it below. That is, **why** did you select your answer?*

Reasoning and Answer.   It is given that $v_o = 0$, and the acceleration is the same for both cars. To find the distances traveled in a time t, you would use Eq. 2.11, which, with $v_o = 0$ and taking $x_o = 0$, becomes $x = \frac{1}{2}at^2$. The important thing to note here is that the distance increases as t^2. That is, if you double the amount of time, the distance quadruples (i.e., increases by a factor of four).

In this example, car B accelerates for twice as long as car A, or $t_B = 2t_A$, so car B travels four times as far as car A, and the answer is (c). Expressed mathematically, the solution is

$$x_A = \tfrac{1}{2}at_A^2 \quad \text{and} \quad x_B = \tfrac{1}{2}at_B^2 = \tfrac{1}{2}a(2t_A)^2 = \tfrac{1}{2}a(4t_A^2) = 4(\tfrac{1}{2}at_A^2) = 4x_A$$

How do the cars' distances compare if $t_B = 3t_A$?

Follow-up Exercise.   In this Example, how do the speeds of the cars compare at the ends of the acceleration periods? (*Answers to all Follow-up Exercises are at the back of the text.*)

Example 2.8 ■ Putting on the Brakes: Vehicle Stopping Distance

The stopping distance of a vehicle is an important factor in road safety. This distance depends on the initial speed $(+v_0)$ and the braking capacity, or deceleration, $-a$, which is assumed to be constant. (Recall that the minus sign indicates that the acceleration is in the negative direction. In this case, the sign of acceleration is opposite that of the velocity, which is taken to be positive. Thus, the car slows to a stop.) Express the stopping distance x in terms of these quantities.

Thinking It Through. Again, a kinematic equation is required, and the appropriate one is determined by listing what is given and what is to be found. Notice that the distance x is wanted, and time is not involved.

▶ **FIGURE 2.12 Vehicle stopping distance** A sketch to help visualize the situation in Example 2.8.

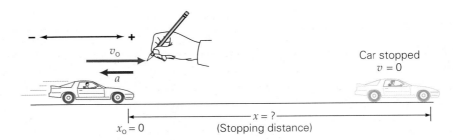

Solution. Here, we are working with variables, so we can represent quantities only in symbolic form.

Given: v_0 (positive direction)   *Find:* x (in terms of the given
  $-a$ (opposite direction of v_0)   variables)
  $v = 0$ (car comes to stop)
  $x_0 = 0$ (car taken to be initially at the origin)

Again, it is helpful to make a sketch of the situation, particularly when directional vector quantities are involved (▲Fig. 2.12). (Since Eq. 2.12 has the variables we want, it should allow us to find the stopping distance x. Expressing the negative acceleration explicitly and assuming $x_0 = 0$ gives

$$v^2 = v_0^2 + 2(-a)x = v_0^2 - 2ax$$

Since the vehicle comes to a stop $(v = 0)$, we can solve for x:

$$x = \frac{v_0^2}{2a}$$

This equation gives us x expressed in terms of the vehicle's initial speed and stopping acceleration.

Notice that the stopping distance x is proportional to the *square* of the initial speed. Doubling the initial speed therefore increases the stopping distance by a factor of 4 (for the same deceleration). That is, if the stopping distance is x_1 for an initial speed of v_1, then for a twofold increase in the initial speed $(v_2 = 2v_1)$, the stopping distance would increase fourfold:

$$x_1 = \frac{v_1^2}{2a}$$

$$x_2 = \frac{v_2^2}{2a} = \frac{(2v_1)^2}{2a} = 4\left(\frac{v_1^2}{2a}\right) = 4x_1$$

We can get the same result by directly using ratios:

$$\frac{x_2}{x_1} = \frac{v_2^2}{v_1^2} = \left(\frac{v_2}{v_1}\right)^2 = 2^2 = 4$$

Do you think this consideration is important in setting speed limits, for example, in school zones? (The driver's reaction time should also be considered. A method for approximating a person's reaction time is given in Section 2.5.)

Follow-up Exercise. Tests have shown that the Chevy Blazer has an average braking deceleration of 7.5 m/s^2, while that of a Toyota Celica is 9.2 m/s^2. Suppose two of these vehicles are being driven down a straight, level road at 97 km/h (60 mi/h), with the Celica in front of the Blazer. A cat runs across the road ahead of them, and both drivers apply their brakes at the same time and come to safe stops (not hitting the cat). Assuming the same reaction times for both drivers, what is the minimum safe tailgating distance for the Blazer so that there won't be a rear-end collision with the Celica when the two vehicles came to a stop? (*Answers to all Follow-up Exercises are at the back of the text.*)

Graphical Analysis of Kinematic Equations

As was shown in Fig. 2.10, plots of v versus t gave straight-line graphs where the slopes were constant accelerations. There is another interesting aspect of v-versus-t graphs. Consider the one shown in ▸Fig. 2.13a, particularly the shaded area under the curve. Suppose we calculate the area of the shaded triangle, where, in general, $A = \frac{1}{2}ab$ [Area $= \frac{1}{2}$(altitude)(base)].

For the graph in Fig. 2.13a, the altitude is v and the base is t, so $A = \frac{1}{2}vt$. But, from the equation $v = v_0 + at$, we have $v = at$, where $v_0 = 0$ (zero intercept on graph). Therefore,

$$A = \tfrac{1}{2}vt = \tfrac{1}{2}(at)t = \tfrac{1}{2}at^2 = \Delta x = x$$

(The equation $x = \frac{1}{2}at^2$ arises from Eq. 2.11 with $v_0 = 0$ and $x_0 = 0$, with the object initially at the origin.) Hence, x, the distance covered, is equal to the area under a v-versus-t curve.

Now take a look at Fig. 2.13b. Here, there is a nonzero value of v_0 at $t = 0$, so the object is initially moving. Consider the shaded areas. We know that the area of the triangle is $A_2 = \frac{1}{2}at^2$, and the area of the rectangle can be seen (with $x_0 = 0$) to be $A_1 = v_0t$. Adding these areas to get the total area yields

$$A_1 + A_2 = v_0t + \tfrac{1}{2}at^2 = x$$

Eq. 2.11 is on the right, and again x, the distance covered, is equal to the area under the v-versus-t curve.

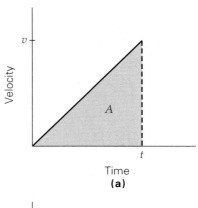

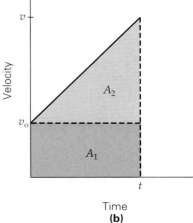

▲ **FIGURE 2.13** *v-versus-t graphs, one more time* **(a)** In the straight-line plot for a constant acceleration, the area under the curve is equal to x, the distance covered. **(b)** If v_0 is not zero, the distance is still given by the area under the curve, here divided into two parts, areas A_1 and A_2.

2.5 Free Fall

OBJECTIVE: **To use the kinematic equations to analyze free fall.**

One of the more common cases of constant acceleration is the acceleration due to gravity near the Earth's surface. When an object is dropped, its initial velocity (at the instant it is released) is zero. At a later time while falling, it has a nonzero velocity. There has been a change in velocity and thus, by definition, an acceleration. This **acceleration due to gravity** (g) has an approximate magnitude of

$$g = 9.80 \text{ m/s}^2, \quad acceleration\ due\ to\ gravity$$

(or 980 cm/s^2) and is directed downward (toward the center of the Earth). In British units, the value of g is about 32.2 ft/s^2.

The values given here for g are only approximate because the acceleration due to gravity varies slightly at different locations as a result of differences in elevation and regional average mass density of the Earth. These small variations will be ignored in this book unless otherwise noted. (Gravitation is studied in more detail in Chapter 7.) Air resistance is another factor that affects the acceleration of a falling object, but it, too, will be ignored here for simplicity. (The frictional effect of air resistance will be considered in Chapter 4.)

*Objects in motion solely under the influence of gravity are said to be in **free fall**.* The words "free fall" bring to mind dropped objects that are moving downward under the influence of gravity ($g = 9.80 \text{ m/s}^2$ in the absence of air resistance). However, the term can be applied in general to any vertical motion under the sole influence of gravity. Objects released from rest or thrown upward or downward are in free fall once they are released. That is, after $t = 0$ (the time of release), only gravity is acting and influencing the motion. (Even when an object projected upward is traveling upward, *it is still accelerating downward*.) Thus, the set of equations for motion in one dimension can be used to describe generalized free fall.

The acceleration due to gravity, g, is the *constant* acceleration for all free-falling objects, regardless of their mass or weight. It was once thought that heavier bodies fall faster than lighter bodies. This concept was part of Aristotle's theory of motion. You can easily observe that a coin falls faster than a sheet of paper when dropped simultaneously from the same height. But in this case, air resistance plays a noticeable role. If the paper is crumpled into a compact ball, it gives the coin a better race. Similarly, a feather "floats" down much more slowly than a coin falls. However, in a near-vacuum, where there is negligible air resistance, the feather and the coin fall with the same acceleration—the acceleration due to gravity (▼ Fig. 2.14).

Astronaut David Scott performed a similar experiment on the Moon in 1971 by simultaneously dropping a feather and a hammer from the same height. He did not need a vacuum pump: The Moon has no atmosphere and therefore no air resistance. The hammer and the feather reached the lunar surface together, but both fell at a slower rate than on Earth. The acceleration due to gravity near the Moon's surface is about one sixth of that near the Earth's surface ($g_M \approx g/6$).

Currently accepted ideas about the motion of falling bodies are due in large part to Galileo. He challenged Aristotle's theory and experimentally investigated the motion of objects. Legend has it that Galileo studied the accelerations of falling bodies by dropping objects of different weights from the top of the Leaning Tower of Pisa. (See the accompanying Insight on Galileo.)

▶ **FIGURE 2.14 Free fall and air resistance** **(a)** When dropped simultaneously from the same height, a feather falls more slowly than a coin, because of air resistance. But when both objects are dropped in an evacuated container with a good partial vacuum, where air resistance is negligible, the feather and the coin fall together with a constant acceleration. **(b)** An actual demonstration with multiflash photography: An apple and a feather are released simultaneously through a trap door into a large vacuum chamber, and they fall together—almost. Because the chamber has a partial vacuum, there is still some air resistance. (How can you tell?)

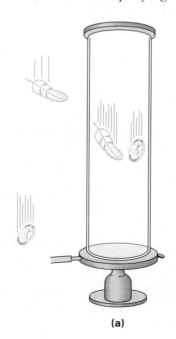

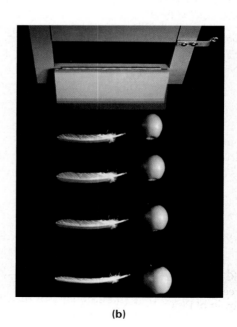

(a)   (b)

Galileo Galilei and the Leaning Tower of Pisa

Galileo Galilei (Fig. 1) was born in Pisa, Italy, in 1564 during the Renaissance. Today, he is known throughout the world by his first name and is often referred to as the father of modern science or the father of modern mechanics and experimental physics, which attests to the magnitude of his scientific contributions.

One of Galileo's greatest contributions to science was the establishment of the scientific method—that is, investigation through experiment. In contrast, Aristotle's approach was based on logical deduction. By the scientific method, for a theory to be valid, it must predict or agree with experimental results. If it doesn't, it is invalid or requires modification. Galileo said, "I think that in the discussion of natural problems we ought not to begin at the authority of places of Scripture, but at sensible experiments and necessary demonstrations."*

Probably the most popular and well-known legend about Galileo is that he performed experiments with falling bodies by dropping objects from the Leaning Tower of Pisa (Fig. 2). There is some debate as to whether Galileo actually did this, but there is little doubt that he questioned Aristotle's view on the motion of falling objects. In 1638, Galileo wrote,

> Aristotle says that an iron ball of one hundred pounds falling from a height of one hundred cubits reaches the ground before a one-pound ball has fallen a single cubit. I say that they arrive at the same time. You find, on making the experiment, that the larger outstrips the smaller by two finger-breadths, that is, when the larger has reached the ground, the other is short of it by two finger-breadths; now you would not hide behind these two fingers the ninety-nine cubits of Aristotle.[†]

*From *Growth of Biological Thought: Diversity, Evolution & Inheritance*, by F. Meyr (Cambridge, MA: Harvard University Press, 1982).

[†]From *Aristotle, Galileo, and the Tower of Pisa*, by L. Cooper (Ithaca, NY: Cornell University Press, 1935).

This and other writings show that Galileo was aware of the effect of air resistance.

The experiments at the Tower of Pisa were supposed to have taken place around 1590. In this writings of about that time, Galileo mentions dropping objects from a high tower, but never specifically names the Tower of Pisa. A letter written to Galileo from another scientist in 1641 describes the dropping of a cannon ball and a musket ball from the Tower of Pisa. The first account of Galileo doing a similar experiment was written a dozen years after his death by Vincenzo Viviani, his last pupil and first biographer. It is not known whether Galileo told this story to Viviani in his declining years or Viviani created this picture of his former teacher.

The important point is that Galileo recognized (and probably experimentally showed) that free-falling objects fall with the same acceleration regardless of their mass or weight. (See Fig. 2.14.) Galileo gave no reason as to why all objects in free fall have the same acceleration, but Newton did, as you will learn in a later chapter. (For more on the current status of the Leaning Tower, see the Insight in Chapter 8.)

FIGURE 2 The Leaning Tower of Pisa The tower, constructed as a belfry for a nearby cathedral, was built on shifting subsoil. Construction began in 1173, and the tower started to shift one way and then the other, before inclining to its present direction. Today, the tower leans about 5 m (16 ft) from the vertical at the top. It was closed in 1990, and efforts were made to stabilize the leaning. (See the related Insight in Chapter 8.)

FIGURE 1   Galileo Galileo is alleged to have performed free-fall experiments by dropping objects off the Leaning Tower of Pisa.

It is customary to use y to represent the vertical direction and to take upward as positive (as with the vertical y-axis of Cartesian coordinates). Because the acceleration due to gravity is always downward, then it is in the negative y-direction. This negative acceleration, $a = -g = -9.80 \text{ m/s}^2$, should be substituted into the equations of motion. However, the relationship $a = -g$ may be expressed explicitly in the equations for linear motion for convenience:

$$v = v_o - gt \tag{2.8'}$$

$$y = y_o + v_o t - \tfrac{1}{2}gt^2 \qquad \begin{array}{l}\textit{Free-fall equations with}\\ -g \textit{ expressed explicitly}\end{array} \tag{2.11'}$$

$$v^2 = v_o^2 - 2g(y - y_o) \tag{2.12'}$$

Equation 2.9 applies as well, but it does not contain g:

$$y = y_o + \tfrac{1}{2}(v + v_o)t \tag{2.10'}$$

The origin ($y = 0$) of the frame of reference is usually taken to be at the initial position of the object. Since upward is generally taken to be in the positive direction ($+y$-axis on a graph), writing $-g$ explicitly in the equations reminds you of directional differences, and the value of g is inserted as 9.80 m/s^2. However, the choice is arbitrary. The equations can be written with $a = g$, for example, $v = v_o + gt$, with the directional minus sign associated directly with g. In this case, a value of -9.80 m/s^2 must be substituted for g each time.

Note that you have to be explicit about the directions of vector quantities. The location y and the velocities v and v_o may be positive or negative, depending on the direction of motion. The use of these equations and the sign convention are illustrated in the following Examples.

PHYSLET®
ILLUSTRATION

Free Fall

Example 2.9 ■ A Stone Thrown Downward: The Kinematic Equations Revisited

A boy on a bridge throws a stone vertically downward with an initial velocity of 14.7 m/s toward the river below. If the stone hits the water 2.00 s later, what is the height of the bridge above the water?

Thinking It Through. This is a free-fall problem, but note that the initial velocity is downward, or negative. It is important to express this factor explicitly.

Solution. As usual, we first write down what is given and what is to be found:

Given: $v_o = -14.7 \text{ m/s}$ (downward taken as   *Find:* y (height)
     $t = 2.00 \text{ s}$ the negative direction)
     $g \, (= 9.80 \text{ m/s}^2)$

Notice that g is taken as a positive number, since the directional minus sign has already been put into the previous equations of motion. After a while, you will probably just write down the symbol g, since you will be familiar with its numerical value. This time, draw a sketch of your own to help you analyze the situation.

Which equation(s) will provide the solution using the given data? It should be evident that the distance the stone travels in an amount of time t is given directly by Eq. 2.11 ' (taking $y_o = 0$):

$$y = v_o t - \tfrac{1}{2}gt^2 = (-14.7 \text{ m/s})(2.00 \text{ s}) - \tfrac{1}{2}(9.80 \text{ m/s}^2)(2.00 \text{ s})^2$$
$$= -29.4 \text{ m} - 19.6 \text{ m} = -49.0 \text{ m}$$

The minus sign indicates that the displacement is downward, which agrees with what you know from the statement of the problem. Thus the height is 49.0 m.

Follow-up Exercise. How much longer would it take for the stone to reach the river if the boy in this Example had dropped the ball rather than thrown it? (*Answers to all Follow-up Exercises are at the back of the text.*)

Example 2.10 ■ Measuring Reaction Time: Free Fall

Reaction time is the time it takes a person to notice, think, and act in response to a situation—for example, the time between first observing and then responding to an obstruction on the road ahead while you are driving an automobile. Reaction time varies with the complexity of the situation (and with the individual). In general, the largest part of a person's reaction time is spent thinking, but practice in dealing with a given situation can reduce this time.

A person's reaction time can be measured by having another person drop a ruler (without warning) even with and through the first person's thumb and forefinger, as shown in ▶Fig. 2.15. The first person grasps the falling ruler as quickly as possible, and the length of the ruler below the top of the finger is noted. If, on average, the ruler descends 18.0 cm before it is caught, what is the person's average reaction time?

Thinking It Through. Both distance and time are involved. This observation indicates which kinematic equation should be used.

Solution. Notice that only the distance of fall is given. However, we know a couple of other things, such as v_o and g, so, taking $y_o = 0$:

Given: $y = -18.0 \text{ cm} = -0.180 \text{ m}$ *Find:* t (reaction time)
$v_o = 0$
$g (= 9.80 \text{ m/s}^2)$

(Note that the distance y has been converted directly to meters.) We can see that Eq. 2.11′ applies here, giving us

$$y = v_o t - \tfrac{1}{2}gt^2$$

or

$$y = -\tfrac{1}{2}gt^2 \quad (\text{with } v_o = 0)$$

Solving for t gives

$$t = \sqrt{\frac{2y}{-g}} = \sqrt{\frac{2(-0.180 \text{ m})}{-9.80 \text{ m/s}^2}} = 0.192 \text{ s}$$

Try this experiment with a fellow student, and measure your reaction time. Why do you think another person besides you should drop the ruler?

Follow-up Exercise. A popular trick is to substitute a crisp dollar bill lengthwise for the ruler in Fig. 2.15, telling the person that he or she can have the dollar if able to catch it. Is this proposal a good deal? (The length of a dollar is 15.7 cm.) (*Answers to all Follow-up Exercises are at the back of the text.*)

▲ **FIGURE 2.15 Reaction time**
A person's reaction time can be measured by having the person grasp a dropped ruler. See Example 2.10.

Example 2.11 ■ Free Fall Up and Down: Using Implicit Data

A worker on a scaffold on a billboard throws a ball straight up. The ball has an initial velocity of 11.2 m/s when it leaves the worker's hand at the top of the billboard (▶Fig. 2.16). (a) What is the maximum height the ball reaches relative to the top of the billboard? (b) How long does it take the ball to reach this height? (c) What is the position of the ball at $t = 2.00 \text{ s}$?

Thinking It Through. In (a), only the upward part of the motion has to be considered. Note that the ball stops (zero velocity) at the maximum height, which allows us to determine this height. (b) Knowing the maximum height, we can determine the upward time of flight. In (c), the distance–time equation (Eq. 2.11′) applies for any time and gives the position (y) of the ball relative to the launch point.

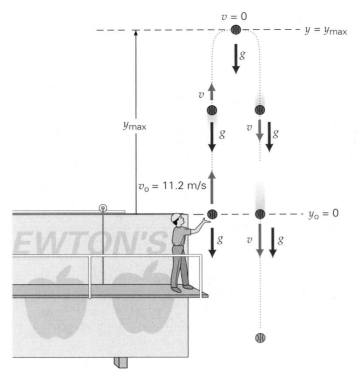

FIGURE 2.16 Free fall up and down Note the lengths of the velocity and acceleration vectors at different times. (The upward and downward paths of the ball are horizontally displaced for illustration purposes.) See Example 2.11.

Solution. It might appear that all that is given in the general problem is the initial velocity v_o. However, a couple of other pieces of information are implied, because they should be understood. One is the acceleration g, and the other is the velocity at the maximum height where the ball stops. Here, in changing direction, the velocity of the ball is momentarily zero, so we have (again taking $y_o = 0$):

Given: $v_o = 11.2$ m/s   **Find:** (a) y_{max} (maximum height)
$\quad\quad\quad$ $g\ (= 9.80$ m/s$^2)$ $\quad\quad\quad\quad$ (b) t_u $\quad$ (time upward)
$\quad\quad\quad$ $v = 0$ (at y_{max}) $\quad\quad\quad\quad$ (c) y $\quad$ (at $t = 2.00$ s)
$\quad\quad\quad$ $t = 2.00$ s [for part(c)]

(a) Notice that we reference the height ($y = 0$) to the top of the billboard. For this part of the problem, we need be concerned with only the upward motion—a ball is thrown upward and stops at its maximum height y_{max}. With $v = 0$ at this height, y_{max} may be found directly from Eq. 2.12′ as

$$v^2 = 0 = v_o^2 - 2gy_{max}$$

So,

$$y_{max} = \frac{v_o^2}{2g} = \frac{(11.2 \text{ m/s})^2}{2(9.80 \text{ m/s}^2)} = 6.40 \text{ m}$$

relative to the top of billboard ($y_o = 0$; see Fig. 2.16).

(b) The time the ball travels upward is designated t_u. This is the time it takes for the ball to reach y_{max}, where $v = 0$. Since we know v_o and v, and can find the time t_u directly from Eq. 2.8′ as

$$v = 0 = v_o - gt_u$$

So,

$$t_u = \frac{v_o}{g} = \frac{11.2 \text{ m/s}}{9.80 \text{ m/s}^2} = 1.14 \text{ s}$$

(c) The height of the ball at $t = 2.00$ s is given directly by Eq. 2.11′:

$$y = v_o t - \tfrac{1}{2}gt^2$$
$$= (11.2 \text{ m/s})(2.00 \text{ s}) - \tfrac{1}{2}(9.80 \text{ m/s}^2)(2.00 \text{ s})^2 = 22.4 \text{ m} - 19.6 \text{ m} = 2.8 \text{ m}$$

Note that this height is 2.8 m above, or measured upward from, the reference point ($y_o = 0$). The ball has reached its maximum height and is on its way back down.

Considered from another reference point, this situation is like dropping a ball from a height of y_{max} above the top of the billboard with $v_o = 0$ and asking how far it falls in a time $t = 2.00$ s $- t_u = 2.00$ s $- 1.14$ s $= 0.86$ s. The answer is (with $y_o = 0$ at the maximum height)

$$y = v_o t - \tfrac{1}{2}gt^2 = 0 - \tfrac{1}{2}(9.80 \text{ m/s}^2)(0.86 \text{ s})^2 = -3.6 \text{ m}$$

This height is the same as the position found previously, but is measured with respect to the maximum height as the reference point; that is,

$$y_{max} - 3.6 \text{ m} = 6.4 \text{ m} - 3.6 \text{ m} = 2.8 \text{ m}$$

Follow-up Exercise. At what height does the ball in this Example have a speed of 5.00 m/s? (*Hint*: The ball attains this height twice—once on the way up, and once on the way down.) (*Answers to all Follow-up Exercises are at the back of the text.*)

Here are a couple of interesting facts about the vertical projectile motion of an object thrown upward in the absence of significant air resistance. First, the times of flight upward and downward are the same. That is, the time it takes the object to reach its maximum height is the same as the time it takes the object to fall from the maximum height back to the initial starting point. Note that at the very top of the trajectory, the object's velocity is zero for an instant, but the acceleration (even at the top) remains a constant 9.8 m/s^2 downward. If the acceleration went to zero, the object would remain there, and gravity would be turned off!

Second, the object returns to the starting point with the same speed as that at which it was launched. (The velocities have the same magnitude, but are opposite in direction.)

These facts can be shown mathematically. See Exercise 89.

Problem-Solving Hint

When working vertical projection problems involving motions up and down, it is often convenient to divide the problem into two parts and consider each part separately. As seen in Example 2.11, for the upward part of the motion, the velocity is zero at the maximum height. A quantity of zero simplifies the calculations. Similarly, the downward part of the motion is analogous to that of an object dropped from a height where the initial velocity is zero.

However, as Example 2.11 shows, the appropriate equations may be used directly for any position or time of the motion. For instance, note in part (c) that the height was found directly for a time *after* the ball had reached the maximum height. The velocity of the ball at that time could also have been found directly from Eq. 2.8', $v = v_o - gt$.

Also, note that the initial position was consistently taken as $y_o = 0$. This assumption is generally the case and is accepted for convenience when the situation involves only one object (then $y_o = 0$ at $t_o = 0$). Using this convention can save a lot of time in writing and solving equations.

The same is true with only one object in horizontal motion: You can usually take $x_o = 0$ at $t_o = 0$. There are a couple of exceptions to this case, however: first, if the problem specifies the object to be initially located at a position other than $x_o = 0$; and second, if the problem involves two objects, as in Integrated Example 2.7. In the latter case, if one object is taken to be initially at the origin, the other's initial position is not zero.

In our study, these situations do not arise very often, but should be kept in mind. At times, we write the kinematic equations without x_o and y_o, assuming they are taken to be at the origin and equal to zero at $t_o = 0$.

As a final example in this chapter, consider the following Example.

Example 2.12 ■ Free Fall on Mars

The Mars Polar Lander was launched in January 1999 and was lost near the Martian surface in December, 1999. What became of the spacecraft is unknown. Let's assume the retrorockets were fired and shut off, and then the Lander came to a stop, and fell to the surface from a height of 40 m. (Highly unlikely, but let's assume this is correct.) Considering the spacecraft to be in free fall, with what speed did it hit the surface?

Thinking It Through. This appears to be analogous to a simple problem of dropping an object from a height. And it is, but this situation takes place on Mars. It was noted in this section that the acceleration due to gravity on the surface of the Moon is one sixth of that on the Earth. Acceleration due to gravity also varies for other planets. So, we need to know g_{Mars}. Try looking in Appendix III. (The Appendices contain a lot of useful information, so don't forget to check them out.)

Solution.

Given: $y = -40$ m ($y_o = 0$ again)   *Find:* v (magnitude, speed)
$v_o = 0$
$g_{Mars} = (0.379)g = (0.379)(9.8 \text{ m/s}^2)$
$= 3.7 \text{ m/s}^2$ (from Appendix III)

Then Eq. 2.12' can be used:

$$v^2 = v_o^2 - 2g_{Mars}y = 0 - 2(3.7 \text{ m/s}^2)(-40 \text{ m}).$$

So,

$$v^2 = 296 \text{ m}^2/\text{s}^2 \quad \text{and} \quad v = \sqrt{296 \text{ m}^2/\text{s}^2} = \pm 17 \text{ m/s}$$

This is the velocity, which we know is downward, so the negative root is selected, and $v = -17$ m/s. Since the speed is the magnitude of the velocity, the speed is 17 m/s.

Follow-up Exercise. From the 40-m height, how long did the Lander's descent take? Compute this using two different kinematic equations, and compare the answers. *(Answers to all Follow-up Exercises are at the back of the text.)*

Chapter Review

Important Concepts and Equations

- **Motion** involves a change of position; it can be described in terms of the distance moved (a scalar) or the displacement (a vector).

- A **scalar** quantity has magnitude (value and units) only; a **vector** quantity has magnitude *and* direction.

- **Speed** (a scalar) is the time rate of change of position:

$$\text{average speed} = \frac{\text{distance traveled}}{\text{total time to travel that distance}}$$

or

$$\bar{s} = \frac{d}{\Delta t} = \frac{d}{t_2 - t_1} \qquad (2.1)$$

- **Average velocity** (a vector) is the displacement divided by the total travel time:

$$\text{average velocity} = \frac{\text{displacement}}{\text{total travel time}}$$

$$\bar{v} = \frac{\Delta x}{\Delta t} = \frac{x_2 - x_1}{t_2 - t_1} \quad \text{or} \quad x = x_o + \bar{v}t \qquad (2.3)$$

- **Instantaneous velocity** (a vector) describes how fast something is moving and in what direction at a particular instant of time.

- **Acceleration** is the time rate of change of velocity and hence is a vector quantity:

$$\text{average acceleration} = \frac{\text{change in velocity}}{\text{time to make the change}}$$

$$\bar{a} = \frac{\Delta v}{\Delta t} = \frac{v_2 - v_1}{t_2 - t_1} \qquad (2.5)$$

- **The kinematic equations for constant acceleration:**

$$\bar{v} = \frac{v + v_o}{2} \qquad (2.9)$$

$$v = v_o + at \qquad (2.8)$$

$$x = x_o + \tfrac{1}{2}(v + v_o)t \qquad (2.10)$$

$$x = x_o + v_o t + \tfrac{1}{2}at^2 \qquad (2.11)$$

$$v^2 = v_o^2 + 2a(x - x_o) \qquad (2.12)$$

- An object in **free fall** has a constant acceleration of magnitude $g = 9.80 \text{ m/s}^2$ (acceleration due to gravity) near the surface of the Earth.

- Expressing $a = -g$ in the kinematic equations for constant acceleration in the y-direction yields the following:

$$v = v_\text{o} - gt \qquad (2.8')$$

$$y = y_\text{o} + \tfrac{1}{2}(v + v_\text{o})t \qquad (2.10')$$

$$y = y_\text{o} + v_\text{o}t - \tfrac{1}{2}gt^2 \qquad (2.11')$$

$$v^2 = v_\text{o}^2 - 2g(y - y_\text{o}) \qquad (2.12')$$

Exercises

2.1 Distance and Speed: Scalar Quantities
and
2.2 One-Dimensional Displacement and Velocity: Vector Quantities

1. Identify each of the following as a scalar or vector quantity: (a) distance, (b) velocity, (c) speed, and (d) displacement.

2. CQ Two people choose different reference points to specify an object's position. Does this difference affect their descriptions of the object's coordinates? How about their assessment of the object's displacement? Explain.

3. A scalar quantity has (a) only magnitude, (b) only direction, or (c) both direction and magnitude.

4. A vector quantity has (a) only magnitude, (b) only direction, or (c) both direction and magnitude.

5. CQ Can the displacement of a person's trip be zero, yet the distance involved in the trip be nonzero? How about the reverse situation? Explain.

6. CQ You are told that a person has walked 500 m. What can you safely say about the person's final position relative to the starting point?

7. CQ An object travels at a constant velocity. What is the relationship of the object's speed to its velocity?

8. CQ Speed is the magnitude of velocity. Is average speed the magnitude of average velocity? Explain.

9. CQ If the displacement of an object is 300 m north, what can you say about the distance traveled by the object?

10. ■ What is the magnitude of the displacement of a car that travels half a lap along a circle that has a radius of 150 m? How about when the car travels a full lap?

11. ■ A student throws a rock straight upward at shoulder level, which is 1.65 m above the ground. What is the displacement of the rock when it hits the ground?

12. ■ In 1999, the Moroccan runner Hicham El Guerrouj ran the 1-mile race in 3 min, 43.13 s. What was his average speed during the race?

13. ■ A bus travels at an average speed of 90 km/h. On average, how far does the bus travel in 20 min? Is this distance the magnitude of the actual displacement of the bus? Explain.

14. ■ A motorist drives 150 km from one city to another in 2.5 h, but makes the return trip in only 2.0 h. What are the average speeds for (a) each half of the round-trip and (b) the total trip?

15. ■ A senior citizen walks 0.30 km in 10 min, going around a shopping mall. (a) What is her average speed in meters per second? (b) If she wants to increase her average speed by 20% in walking a second lap, what would her travel time in minutes have to be?

16. IE ■ A race car travels a complete lap on a circular track of radius 500 m in 50 s. (a) The average velocity of the race car is (1) zero, (2) 100 m/s, (3) 200 m/s, or (4) none of the preceding. Why? (b) What is the average speed of the race car?

17. IE ■■ A student runs 30 m east, 40 m north, and 50 m west. (a) The magnitude of the student's net displacement is (1) between 0 and 20 m, (2) between 20 m and 40 m, or (3) between 40 m and 60 m. (b) What is his net displacement?

18. ■■ A student throws a ball vertically upward such that it travels 7.1 m to its maximum height. If the ball is caught at the initial height 2.4 s after being thrown, (a) what is the ball's average speed, and (b) what is its average velocity?

19. ■■ An insect crawls along the edge of a rectangular swimming pool of length 27 m and width 21 m (▼Fig. 2.17). If it crawls from corner A to corner B in 30 min, (a) what is its average speed, and (b) what is the magnitude of its average velocity?

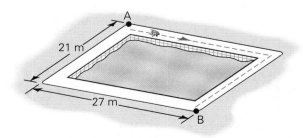

▲ **FIGURE 2.17 Speed versus velocity**   See Exercise 19. (Not drawn to scale, insect is displaced for clarity.)

20. ■■ The distance of one lap around an oval dirt-bike track is 1.50 km. If a rider going at a constant speed makes one lap in 1.10 min, what is the speed of the bike and rider in meters per second? Is the velocity of the bike also constant? Explain.

21. ■■ (a) Given that the speed of sound is 340 m/s and the speed of light is 3.00×10^8 m/s (186 000 mi/s), how much time will elapse between a lightning flash and the resulting thunder if the lightning strikes 2.50 km away from the observer? (b) Does your answer change if you assume that light travels with an infinite speed (i.e., the lightning flash is seen instantaneously)?

22. ■■ A plot of position versus time is shown in ▼Fig. 2.18 for an object in linear motion. (a) What are the average velocities for the segments AB, BC, CD, DE, EF, FG, and BG? (b) State whether the motion is uniform or nonuniform in each case. (c) What is the instantaneous velocity at point D?

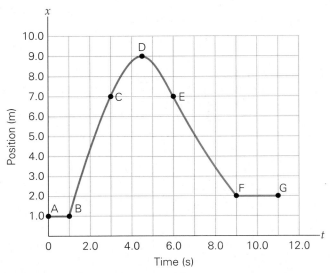

▲ **FIGURE 2.18 Position versus time** See Exercise 22.

23. ■■ In demonstrating a dance step, a person moves in one dimension, as shown in ▼Fig. 2.19. What are (a) the average speed and (b) the average velocity for each

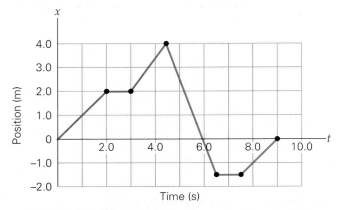

▲ **FIGURE 2.19 Position versus time** See Exercise 23.

phase of the motion? (c) What are the instantaneous velocities at $t = 1.0$ s, 2.5 s, 4.5 s, and 6.0 s? (d) What is the average velocity for the interval between $t = 4.5$ s and $t = 9.0$ s? [*Hint*: Recall that the overall displacement is the displacement between the starting point and the ending point.]

24. ■■ You can determine the speed of a car by measuring the time it takes to travel between mile markers on a highway. (a) How many seconds should elapse between two consecutive mile markers if the car's average speed is 65 mi/h? (b) What is the car's average speed if it takes 65 s to travel between the mile markers?

25. ■■ An earthquake releases two types of traveling waves, called "transverse" and "longitudinal." (See Chapter 13.) The average speeds of transverse and longitudinal seismic waves in rock are 8.9 km/s and 5.1 km/s, respectively. A seismograph records the arrival of the transverse waves 73 s before that of the longitudinal waves. Assuming that the waves travel in straight lines, how far away is the center of the earthquake from the seismograph?

26. ■■ An airline company operates two types of aircraft. The faster type can cruise at 565 mi/h; the slower type can travel only at 505 mi/h. If it takes the faster aircraft 4.50 hours to cover an established route, how many more minutes will it take the slower one to complete the same trip?

27. ■■■ A student driving home for the holidays starts at 8:00 AM to make the 675-km trip, practically all of which is on nonurban interstate highway. If she wants to arrive home no later than 3:00 PM, what must be her minimum average speed? Will she have to exceed the 65-mi/h speed limit?

28. ■■■ Two runners approaching each other on a straight track have constant speeds of 4.50 m/s and 3.50 m/s, respectively, when they are 100 m apart (▼Fig. 2.20). How long will it take for the runners to meet, and at what position will they meet if they maintain these speeds?

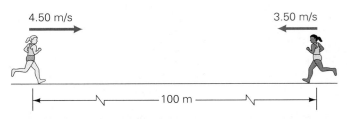

▲ **FIGURE 2.20 When and where do they meet?** See Exercise 28.

29. ■■■ In driving the usual route to school, a student computes his average speed to be 30 km/h. In a hurry to get home that afternoon, he wants to average 60 km/h for the round trip. What would his average speed for the return trip over the same route have to be in order to do this?

2.3 Acceleration

30. **CQ** The gas pedal of an automobile is commonly referred to as the *accelerator*. Which of the following might also be called an acceleration? (a) the brakes; (b) the steering wheel; (c) the gear shift; (d) all of the preceding. Explain.

31. **CQ** A car is traveling at a constant speed of 55 mi/h on a circular track. Is the car accelerating? Explain.

32. On a position-versus-time plot for an object that has a constant acceleration, the graph is (a) a horizontal line, (b) a nonhorizontal and nonvertical straight line, (c) a vertical line, or (d) a curve.

33. **CQ** An object traveling at a constant velocity v_o experiences a constant acceleration in the same direction for a period of time t. Then, an acceleration of equal magnitude is experienced in the opposite direction of v_o for the same period of time t. What is the object's final velocity?

34. **CQ** Does a fast-moving object always have higher acceleration than a slower object? Give a few examples, and explain.

35. **CQ** Can an object have a positive velocity, yet a negative acceleration? Support your answer with an example.

36. **CQ** Describe the motions of the two objects that have the velocity-versus-time plots shown in ▼Fig. 2.21.

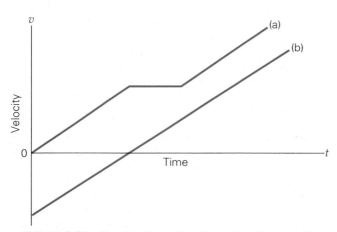

▲ **FIGURE 2.21   Description of motion**   See Exercise 36.

37. ■ An automobile traveling at 25.0 km/h along a straight, level road accelerates to 65.0 km/h in 6.00 s. What is the magnitude of the auto's average acceleration?

38. ■ A sports car can accelerate from 0 to 60 mi/h in 3.9 s. What is the magnitude of the average acceleration of the car in meters per second squared?

39. ■ If the sports car in Exercise 38 can accelerate at a rate of $7.2 \, \text{m/s}^2$, how long does it take for the car to accelerate from 0 to 60 mi/h?

40. **IE** ■ A couple is traveling by car 40 km/h down a straight highway. They see an accident in the distance, so the driver applies the brakes, and in 5.0 s the car slows down uniformly to rest. (a) The direction of the acceleration vector is (1) in the same direction as, (2) opposite to, or (3) at 90° relative to the velocity vector. Why? (b) By how much must the velocity change each second from the start of braking to the car's complete stop?

41. ■■ A boat starting from rest on a lake accelerates in a straight line at a constant rate of $2.0 \, \text{m/s}^2$ for 6.0 s. How far does the boat travel during this time?

42. ■■ A car with an initial velocity of 3.5 m/s coasts with the engine off and the transmission in neutral. If the combined effect of air resistance and rolling friction causes a deceleration of $0.50 \, \text{m/s}^2$, after how long a time will the car come to a stop?

43. ■■ What is the acceleration for each graph segment in ▼Fig. 2.22? Describe the motion of the object over the total time interval.

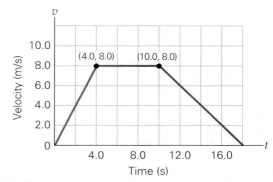

▲ **FIGURE 2.22   Velocity versus time**   See Exercises 43 and 66.

44. ■■ ▼Figure 2.23 shows a plot of velocity versus time for an object in linear motion. (a) Compute the acceleration

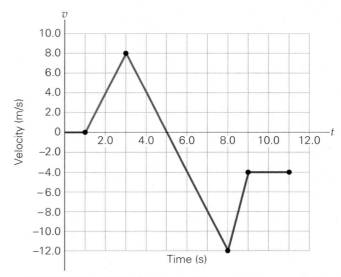

▲ **FIGURE 2.23   Velocity versus time**   See Exercises 44 and 67.

for each phase of motion. (b) Describe how the object moves during the last time segment.

45. ■■■ A train normally travels at a uniform speed of 72 km/h on a long stretch of straight, level track. On a particular day, the train must make a 2.0-min stop at a station along this track. If the train decelerates at a uniform rate of $1.0\,\text{m/s}^2$ and, after the stop, accelerates at a rate of $0.50\,\text{m/s}^2$, how much time is lost because of stopping at the station?

2.4 Kinematic Equations (Constant Acceleration)

46. For a constant linear acceleration, the velocity-versus-time graph is (a) a horizontal line, (b) a vertical line, (c) a nonhorizontal and nonvertical straight line, or (d) a curved line.

47. For a constant linear acceleration, the position-versus-time graph would be (a) a horizontal line, (b) a vertical line, (c) a nonhorizontal and nonvertical straight line, or (d) a curve.

48. CQ If an object's velocity-versus-time graph is a horizontal line, what can you say about the object's acceleration?

49. An object accelerates uniformly from rest for t seconds. The object's average speed for this time interval is (a) $\frac{1}{2}at$, (b) $\frac{1}{2}at^2$, (c) $2at$, or (d) $2at^2$.

50. CQ A classmate states that a negative acceleration always means that a moving object is decelerating. Is this statement true? Explain.

51. ■ At a sports-car rally, a car starting from rest accelerates uniformly at a rate of $9.0\,\text{m/s}^2$ over a straight-line distance of 100 m. The time to beat in this event is 4.5 s. Does the driver do it? What must the minimum acceleration be to do so?

52. ■ A car accelerates from rest at a constant rate of $2.0\,\text{m/s}^2$ for 5.0 s. (a) What is the speed of the car at the end of that time? (b) How far does the car travel in this time?

53. ■ A car traveling at 35 mi/h is to stop on a 35-m long shoulder of the road. (a) What is the required magnitude of the minimum acceleration? (b) How much time will elapse during this minimum deceleration until the car stops?

54. ■ A motorboat traveling on a straight course slows down uniformly from 75 km/h to 40 km/h in a distance of 50 m. What is the boat's acceleration?

55. ■■ The driver of a pickup truck going 100 km/h applies the brakes, giving the truck a uniform deceleration of $6.50\,\text{m/s}^2$ while it travels 20.0 m. (a) What is the speed of the truck in kilometers per hour at the end of this distance? (b) How much time has elapsed?

56. ■■ An experimental rocket car starting from rest reaches a speed of 560 km/h after a straight 400-m run on a level

salt flat. Assuming that acceleration is constant, (a) what was the time of the run, and (b) what is the magnitude of the acceleration?

57. ■■ A rocket car is traveling at a constant speed of 250 km/h on a salt flat. The driver gives the car a reverse thrust, and the car experiences a continuous and constant deceleration of $8.25\,\text{m/s}^2$. How much time elapses until the car is 175 m from the point where the reverse thrust is applied? Describe the situation(s) for your answer(s).

58. ■■ Two identical cars capable of accelerating at $3.00\,\text{m/s}^2$ are racing on a straight track with running starts. Car A has an initial speed of 2.50 m/s; car B starts with an initial speed of 5.00 m/s. (a) What is the separation of the two cars after 10 s? (b) Which car is moving faster after 10 s?

59. IE ■■ An object moves in the $+x$-direction at a speed of 40 m/s. As it passes through the origin, it starts to experience a constant acceleration of $3.5\,\text{m/s}^2$ in the $-x$-direction. (a) What will happen next? (1) The object will reverse its direction of travel at the origin. (2) The object will keep traveling in the $+x$-direction. (3) The object will travel in the $+x$-direction and then reverse its direction. Why? (b) How much time elapses before the object returns to the origin? (c) What is the velocity of the object when it returns to the origin?

60. ■■ A rifle bullet with a muzzle speed of 330 m/s is fired directly into a special, dense material that stops the bullet in 30 cm. Assuming the bullet's deceleration to be constant, what is its magnitude?

61. ■■ A bullet traveling horizontally at a speed of 350 m/s hits a board perpendicular to the surface, passes through it, and emerges on the other side at a speed of 210 m/s. If the board is 4.00 cm thick, how long does the bullet take to pass through it?

62. ■■ A rock hits the ground at a speed of 10 m/s and leaves a hole 25 cm deep. What is the magnitude of the deceleration of the rock?

63. ■■ A jet aircraft being launched from an aircraft carrier is accelerated from rest along a 94-m track for 2.5 s. (a) What is the acceleration of the aircraft, assuming it is constant? (b) What is the launch speed of the jet?

64. ■■ The speed limit in a school zone is 40 km/h (about 25 mi/h). A driver traveling at this speed sees a child run onto the road 13 m ahead of his car. He applies the brakes, and the car decelerates at a uniform rate of $8.0\,\text{m/s}^2$. If the driver's reaction time is 0.25 s, will the car stop before hitting the child?

65. ■■ Assuming a reaction time of 0.50 s for the driver in Exercise 64, will the car stop before hitting the child?

66. ■■ (a) Show that the area under the curve of a velocity-versus-time plot for a constant acceleration is equal to the displacement. [*Hint*: The area of a triangle is $ab/2$, or one

half the altitude times the base.] (b) Compute the distance traveled for the motion represented by Fig. 2.22.

67. ■■■ Figure 2.23 shows a plot of velocity versus time for an object in linear motion. (a) What are the instantaneous velocities at $t = 8.0$ s and $t = 11.0$ s? (b) Compute the final displacement of the object. (c) Compute the total distance the object travels. (See Exercise 66(a).)

68. IE ■■■ (a) A car traveling at a speed of v can brake to an emergency stop in a distance x. Assuming all other driving conditions are all similar, if the traveling speed of the car doubles, the stopping distance will be (1) $\sqrt{2}x$, (2) $2x$, or (3) $4x$: (b) A driver traveling at 40.0 km/h in a school zone can brake to an emergency stop in 3.00 m. What would be the braking distance if the car were traveling at 60.0 km/h?

69. ■■■ List the four kinematics equations in the text. For each equation, discuss the following questions: (a) What does each symbol represent? (b) Under what conditions does the equation hold?

70. ■■■ An object moves in the positive x-direction with a constant acceleration. At $x = 5.0$ m, its speed is 10 m/s. At 2.5 s later, the object is at $x = 65$ m. What is its acceleration?

2.5 Free Fall (Neglect air resistance)

71. An object is thrown vertically upward. Which one of the following statements is true? (a) Its velocity changes nonuniformly; (b) its maximum height is independent of the initial velocity; (c) its travel time upward is slightly greater than its travel time downward; (d) the speed on returning to its starting point is the same as its initial speed.

72. The "free fall" motion described in this section applies to (a) an object dropped from rest, (b) an object thrown vertically downward, (c) an object thrown vertically upward, or (d) all of the preceding.

73. CQ When a ball is thrown upward, what are its velocity and acceleration at its highest point?

74. CQ A dropped object in free fall (a) falls 9.8 m each second, (b) falls 9.8 m during the first second, (c) has an increase in speed of 9.8 m/s each second, or (d) has an increase in acceleration of 9.8 m/s each second.

75. CQ Imagine you are in space far away from any planet, and you throw a ball as you would on Earth. Describe the ball's motion.

76. CQ From the window of a building, you drop a stone. After a second, you drop another stone. Does the distance separating the two stones (a) increase, (b) decrease, or (c) stay the same, or (d) there is not enough information given to determine the answer. Explain.

77. ■ If a dropped object falls 19.6 m in 2.00 s, how far will it fall in 4.00 s?

78. ■ A student drops a ball from the top of a tall building; it takes 2.8 s for the ball to reach the ground. (a) What was the ball's speed just before hitting the ground? (b) What is the height of the building?

79. IE ■ The time it takes for an object dropped from the top of cliff A to hit the water in the lake below is twice the time it takes for another object dropped from the top of cliff B to reach the lake. (a) The height of cliff A is (1) one half, (2) two, or (3) four times that of cliff B. (b) If it takes 1.80 s for the object to fall from cliff A to the water, what are the heights of cliffs A and B?

80. ■ For the motion of a dropped object in free fall, sketch the general forms of the graphs of (a) v versus t and (b) y versus t.

81. ■ You can perform a popular trick by dropping a dollar bill (lengthwise) through the thumb and forefinger of a fellow student. Tell your fellow student to grab the dollar bill as fast as possible, and he or she can have the dollar if able to catch it. (The length of a dollar is 15.7 cm, and the average human reaction time is about 0.2 s. See Fig. 2.15.). Is this proposal a good deal? Justify your answer.

82. ■ A boy throws a stone straight upward with an initial speed of 15 m/s. What maximum height will the stone reach before falling back down?

83. ■ In Exercise 82, what would be the maximum height of the stone if the boy and the stone were on the surface of the Moon, where the acceleration due to gravity is only $1.67 \, \text{m/s}^2$?

84. ■■ A spring-loaded gun shoots a 0.0050-kg bullet vertically upward with an initial velocity of 21 m/s. (a) What is the height of the bullet 3.0 s after the gun fires? (b) At what times is the bullet 12 m above the muzzle of the gun?

85. ■■ The ceiling of a classroom is 3.75 m above the floor. A student tosses an apple vertically upward, releasing it 0.50 m above the floor. What is the maximum initial speed that can be given to the apple if it is not to touch the ceiling?

86. ■■ The Petronas Twin Towers in Malaysia and the Chicago Sears Tower have heights of about 452 m and 443 m, respectively. If objects were dropped from the top of each, what would be the difference in the time it takes the objects to reach the ground?

87. ■■ A stone is thrown vertically downward at an initial speed of 14 m/s from a height of 65 m above the ground. (a) How far does the stone travel in 2.0 s? (b) What is its velocity just before it hits the ground?

88. ■■ An arrow is shot vertically upward. Three seconds later, it is at a height of 35 m. (a) What was the initial speed of the arrow? (b) How long is the arrow in flight from launch to returning to the initial height? (Assume the arrow falls vertically downward.)

89. ■■ Referring to the thrown ball in Example 2.11 and Fig. 2.16, (a) compare the travel time upward (t_u) with the travel time downward (t_d) required for the ball to return to its starting point, and (b) compare the initial velocity of the ball with its velocity upon its return to the starting point.

90. ■■ You throw a stone vertically upward, with an initial speed of 6.0 m/s, from a third-story office window. If the window is 12 m above the ground, find (a) the time the stone is in the air, and (b) the speed of the stone just before it hits the ground.

91. IE ■■ A Superball™ is dropped from a height of 4.00 m. Assuming the ball rebounds with 95% of its impact speed, (a) would the ball bounce to (1) less than 95%, (2) equal to 95.0%, or (3) more than 95% of the initial height? (b) How high will the ball go?

92. ■■ Two balls are thrown vertically, both at the initial speed of 10.0 m/s from a height of 60.0 m above the ground. Ball A is thrown upward, while ball B is projected downward. (a) What is the time lapse between the times two balls hit the ground? (b) Do the masses of the balls affect the result?

93. ■■■ In ▼Fig. 2.24, a student at a window on the second floor of a dorm sees his math professor walking on the sidewalk beside the building. He drops a water balloon from 18.0 m above the ground when the prof is 1.00 m from the point directly beneath the window. If the prof. is 170 cm tall and walks at a rate of 0.450 m/s, does the balloon hit her? If not, how close does it come?

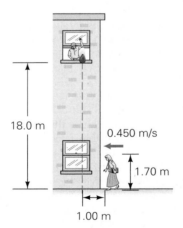

▲ FIGURE 2.24 **Hit the prof** See Exercise 93. (This figure is not drawn to scale.)

94. ■■■ A photographer in a helicopter ascending vertically at a constant rate of 12.5 m/s accidentally drops a camera out the window when the helicopter is 60.0 m above the ground. (a) How long will it take the camera to reach the ground? (b) What will its speed be when it hits?

95. IE ■■■ The acceleration due to gravity on the Moon is about one sixth of that on Earth. (a) If an object were dropped from the same height on the Moon and on the Earth, the time it would take to reach the surface on the Moon is (1) $\sqrt{6}$, (2) 6, or (3) 36 times that it would take on the Earth. (b) For a projectile with an initial velocity of 18.0 m/s upward, what would be the maximum height and the total time of flight on the Moon and on the Earth?

96. ■■■ It takes 0.210 s for a dropped object to pass a window that is 1.35 m tall. From what height above the top of the window was the object released? (See ▼Fig. 2.25.)

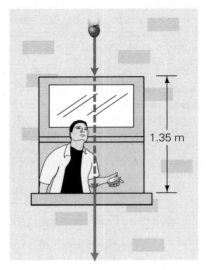

1.35 m

▲ FIGURE 2.25 **From where did it come?** See Exercise 96.

Additional Exercises

97. In throwing an object vertically upward at a speed of 7.25 m/s from the top of a tall building, a student leans over the edge of the building so that the object will not strike the building on the return trip. (a) What is the velocity of the object when it has traveled a total distance of 25.0 m? (b) How long does it take to travel this distance?

98. After landing, a jet liner on a straight runway taxis to a stop at an average velocity of −35.0 km/h. If the plane takes 7.00 s to come to rest, what are the plane's initial velocity and acceleration?

99. A vertically moving projectile reaches a maximum height of 23 m above its starting position. (a) What was the projectile's initial speed? (b) What is its height above the starting point at $t = 1.3$ s?

100. A drag racer traveling at a speed of 200 km/h on a straight track deploys a parachute and slows uniformly to a speed of 20 km/h in 12 s. (a) What is the racer's acceleration? (b) How far does the racer travel in the 12-s interval?

101. On a cross-country trip, a couple drives 500 mi in 10 h on the first day, 380 mi in 8.0 h on the second day, and 600 mi in 15 h on the third day. What was the average speed for the whole trip?

102. A car traveling at 25 mi/h has 1.5 s to come to a stop when approaching a red light. The magnitude of the maximum deceleration of the car is 7.0 m/s^2. Will the car stop safely before it reaches the red light?

103. IE A car travels three quarters of a lap on a circular track of radius R. (a) The magnitude of the displacement is (1) less than R; (2) greater than R, but less than $2R$; or (3) greater than $2R$. (b) If $R = 50$ m, what is the magnitude of the displacement?

104. From behind a billboard a police officer is watching for speeders at an intersection ($x = 0$). At some instant ($t = 0$), an obvious speeder comes by at 30 m/s (67 mph). The police officer starts a chase at the same instant, maintaining a constant acceleration of 3.5 m/s^2 as he tries to catch the speeder, in the $+x$-direction. The speeder maintains his speed throughout the pursuit. (a) On an x-versus-t graph, plot the position for each car. (b) How long does it take for the officer to catch the speeder?

105. At what speed must an object be projected vertically upward for it to reach a maximum height of 14.0 m above its starting point?

106. A train on a straight, level track has an initial speed of 45.0 km/h. A uniform acceleration of 1.50 m/s^2 is applied while the train travels 200 m. (a) What is the speed of the train at the end of this distance? (b) How long did it take for the train to travel the 200 m?

107. A car going 85 km/h on a straight road is brought uniformly to a stop in 10 s. How far does the car travel during that time?

108. A car and a motorcycle start from rest at the same time on a straight track, but the motorcycle is 25.0 m behind the car (▼Fig. 2.26). The car accelerates at a uniform rate of 3.70 m/s^2 and the motorcycle at a uniform rate of 4.40 m/s^2. (a) How much time elapses before the motorcycle overtakes the car? (b) How far will each have traveled during that time? (c) How far ahead of the car will the motorcycle be 2.00 s later? (Both vehicles are still accelerating.)

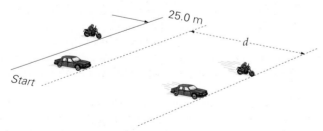

▲ **FIGURE 2.26 A tie race** See Exercise 108. (This figure is not drawn to scale.)

109. A person throws a stone straight upward at an initial speed of 15 m/s on a bridge that is 25 m above the surface of the water. If the stone just misses the bridge on the way down, (a) what is the speed of the stone just before it hits the water? (b) What is the total time the stone is in the air?

110. An object initially at rest experiences an acceleration of 1.5 m/s^2 for 6.0 s and then travels at that constant velocity for another 8.0 s. What is the object's average velocity over the 14-s interval?

111. IE To find the depth of the surface of water in a well, a person drops a stone from the top of the well and simultaneously starts a stopwatch. The watch is stopped when a splash is heard, giving a reading of 3.65 s. The speed of sound is 340 m/s. (a) Why is the speed of sound considered here? (b) Find the depth of the water surface below the top of the well. Take the person's reaction time for stopping the watch to be 0.250 s.

112. A test rocket containing a probe to determine the composition of the upper atmosphere is fired vertically upward from an initial position at ground level. During the time t while its fuel lasts, the rocket ascends with a constant upward acceleration of magnitude $2g$. Assume that the rocket travels a small enough height that the Earth's gravitational force can be considered constant. (a) What are the speed and height of the rocket when its fuel runs out? (b) What is the maximum height the rocket reaches? (c) If $t = 30.0$ s, calculate the rocket's maximum height.

113. A car starting from rest and traveling in only one direction has velocities of 5.0 m/s, 10 m/s, 15 m/s, 20 m/s, and 25 m/s at 1.0 s, 2.0 s, 3.0 s, 4.0 s, and 5.0 s, respectively. (a) What is the magnitude of the car's acceleration? (b) What is the car's average velocity for the 5-s interval? (c) Sketch a graph of v versus t. (d) Sketch a graph of x versus t.

114. On a walk in the country, a couple goes 1.80 km east along a straight road in 20.0 min and then 2.40 km directly north in 35.0 min. (a) What is the average velocity for each segment of the couple's walk and for the total walk? (b) What is the average speed for the total walk? (c) If the couple takes a straight-line path that gets it back to its original starting place in 25.0 min, what are the couple's average speed and average velocity for the total trip?

115. A student calculates that if he skis at 10 km/h, he will arrive at his cabin in the woods at 1:00 PM. If he skis at 15 km/h, he will arrive at the cabin at 11:00 AM. How fast must he ski to arrive at the cabin at noon?

CHAPTER 3

Motion in Two Dimensions

You *can* get there from here! It's just a matter of knowing which way to head at the crossroads. But did you ever wonder why so many roads meet at right angles? There's a good reason. Living on the Earth's surface, we are used to describing locations in two dimensions, and one of the easiest ways to do this is by referring to a pair of mutually perpendicular axes. When you want to tell someone how to get to a particular place in the city, you might say, "Go uptown for four blocks and then crosstown for three more blocks." In the country, it's "Go south for five miles and then another half mile east." Either way, though, you need to know how far to go in each of *two* directions that are 90° apart.

You can use the same approach to describe motion—and the motion doesn't have to be in a straight line. As you will shortly see, we can use vectors, introduced in Chapter 2, to describe motion in curved paths as well. Such analysis of *curvilinear* motion will eventually allow you to analyze the behavior of batted balls, planets circling the Sun, and even electrons in atoms.

Curvilinear motion can be analyzed by using rectangular components of motion. Essentially, you break down, or *resolve*, the curved motion into rectangular (*x* and *y*) components and look at the motion in both dimensions simultaneously. You can apply to those components the kinematic equations introduced in Chapter 2. For an object moving in a curved path, for example, the *x*- and *y*-coordinates of the motion at any time give the object's position as the point (*x*, *y*).

3.1 Components of Motion

OBJECTIVES: To (a) analyze motion in terms of its components and (b) apply the kinematic equations to components of motion.

An object moving in a straight line was considered in Chapter 2 to be moving along one of the Cartesian axes (x or y). But what if the motion is not along an axis? For example, consider the situation illustrated in ▼Fig. 3.1. Here, three balls are moving uniformly across a tabletop. The ball rolling in a straight line along the side of the table, designated as the x-direction, is moving in one dimension. That is, its motion can be described with a single coordinate, x, as was done for motions in Chapter 2. Similarly, the motion of the ball rolling in the y-direction can be described by a single y-coordinate. However, both x- and y-coordinates are needed to describe the motion of the ball rolling diagonally across the table. We say that this motion is *in two dimensions*.

You might observe that if the diagonally moving ball were the only object you had to consider, the x-axis could be chosen to be in the direction of that ball's motion, and the motion would thereby be reduced to one dimension. This observation is true, but once the coordinate axes are fixed, motions not along the axes must be described with two coordinates (x, y), or in two dimensions. Also, keep in mind that not all motions in a plane (two dimensions) are in straight lines. Think about the path of a ball you toss to another person. The path is curved for such projectile motion; and this concept will be considered in Section 3.4.

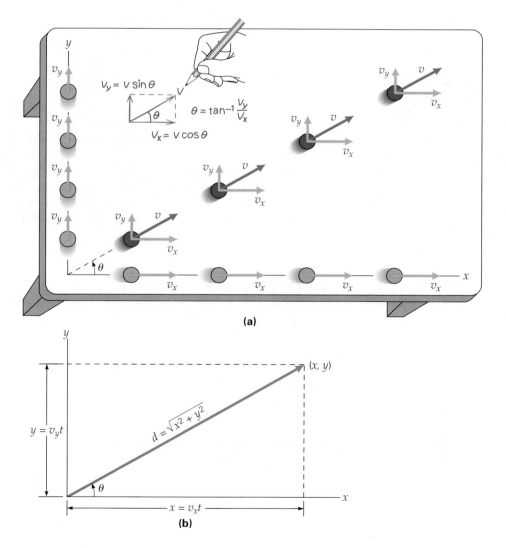

(a)

(b)

◄ **FIGURE 3.1 Components of motion** **(a)** The velocity (and displacement) for uniform, straight-line motion—that of the dark purple ball—may have x- and y-components (v_x and v_y as shown in the pencil drawing), because of the chosen orientation of the coordinate axes. Note that the velocity and displacement of the ball in the x-direction are exactly the same as those that a ball rolling along the x-axis with a uniform velocity of v_x would have. A comparable relationship holds true for the ball's motion in the y-direction. Since the motion is uniform, the ratio v_y/v_x (and therefore θ) is constant. **(b)** The coordinates (x, y) of the ball's position and the distance d the ball has traveled from the origin can be found at any time t.

In considering the motion of the ball moving diagonally across the table in Fig. 3.1a, we can think of the ball as moving in the x- and y-directions simultaneously. That is, it has a velocity in the x-direction (v_x) *and* a velocity in the y-direction (v_y) at the same time. The combined velocity components describe the actual motion of the ball. If the ball has a constant velocity v in a direction at an angle θ relative to the x-axis, then the velocities in the x- and y-directions are obtained by resolving, or breaking down, the velocity vector into **components of motion** in these directions, as shown in the pencil drawing in Fig. 3.1a. As this drawing shows, the v_x and v_y components have magnitudes of

$$v_x = v \cos \theta \tag{3.1a}$$

and

$$v_y = v \sin \theta \tag{3.1b}$$

respectively. (Notice that $v = \sqrt{v_x^2 + v_y^2}$, so v is a combination of the velocities in the x- and y-directions.)

You are familiar with the use of two-dimensional length components in finding the x- and y-coordinates in a Cartesian system. For the ball rolling on the table, its position (x, y), or the distance traveled from the origin in each of the component directions at time t, is given by

$$x = x_o + v_x t \quad \text{\textit{Magnitudes of displacement}} \tag{3.2a}$$
$$\text{\textit{components (under conditions of}}$$
$$y = y_o + v_y t \quad \text{\textit{constant velocity and zero acceleration})} \tag{3.2b}$$

respectively, just as it is for motion in the x- and y-directions separately. (Here, the x_o and y_o are the ball's coordinates at $t = 0$, which may be other than zero.) The ball's straight-line distance from the origin is then $d = \sqrt{x^2 + y^2}$ (Fig. 3.1b).

Note that $\tan \theta = v_y/v_x$, so the direction of the motion relative to the x-axis is given by $\theta = \tan^{-1}(v_y/v_x)$. (See the hand-drawn sketch in Fig. 3.1a.) Also, $\theta = \tan^{-1}(y/x)$. Why?

In this introduction to components of motion, the velocity vector has been taken to be in the first quadrant $(0 < \theta < 90°)$, where both the x- and y-components are positive. But, as will be shown in more detail in the next section, vectors may be in any quadrant, and their components can be negative. Can you tell in which quadrants the v_x or v_y components would be negative?

PHYSLET®
ILLUSTRATION

Motion in Two Dimensions

Example 3.1 ■ On a Roll: Using Components of Motion

If the diagonally moving ball in Fig. 3.1a has a constant velocity of 0.50 m/s at an angle of 37° relative to the x-axis, find how far it travels in 3.0 s by using x- and y-components of its motion.

Thinking It Through. Given the magnitude and direction (angle) of the velocity of the ball, we can find the x- and y-components of the velocity. Then we can compute the distance in each direction. Since the x- and y-axes are at right angles to each other, the Pythagorean theorem gives the distance of the straight-line path of the ball as shown in Fig. 3.1b. (Note the procedure: Separate the motion into components, calculate what is needed in each direction, and recombine if necessary.)

Solution. Organizing the data, we have

Given: $v = 0.50$ m/s   *Find:* d (distance traveled)
        $\theta = 37°$
        $t = 3.0$ s

The distance traveled by the ball in terms of its x- and y-components is given by $d = \sqrt{x^2 + y^2}$. To find x and y as given by Eq. 3.2, we must first compute the velocity components v_x and v_y (Eq. 3.1):

$$v_x = v\cos 37° = (0.50 \text{ m/s})(0.80) = 0.40 \text{ m/s}$$
$$v_y = v\sin 37° = (0.50 \text{ m/s})(0.60) = 0.30 \text{ m/s}$$

Then, the component distances are

$$x = v_x t = (0.40 \text{ m/s})(3.0 \text{ s}) = 1.2 \text{ m}$$

and

$$y = v_y t = (0.30 \text{ m/s})(3.0 \text{ s}) = 0.90 \text{ m}$$

and the actual distance of the path is

$$d = \sqrt{x^2 + y^2} = \sqrt{(1.2 \text{ m})^2 + (0.90 \text{ m})^2} = 1.5 \text{ m}$$

Follow-up Exercise. Suppose that a ball were rolling diagonally across a table with the same speed as in Example 3.1, but from the lower right corner, which is taken as the origin of the coordinate system, toward the upper left corner at an angle of 37° relative to the $-x$-axis. What would be the velocity components in this case? (Would the distance change?) *(Answers to all Follow-up Exercises are at the back of the text.)*

Problem-Solving Hint

Note that for this simple case, the distance can also be obtained directly from $d = vt = (0.50 \text{ m/s})(3.0 \text{ s}) = 1.5 \text{ m}$. However, we have solved this Example in a more general way to illustrate the use of components of motion. The direct solution would have been evident if the equations had been combined algebraically before calculation, that is, as

$$x = v_x t = (v\cos\theta)t$$

and

$$y = v_y t = (v\sin\theta)t$$

from which it follows that

$$d = \sqrt{x^2 + y^2} = \sqrt{(v\cos\theta)^2 t^2 + (v\sin\theta)^2 t^2} = \sqrt{v^2 t^2(\cos^2\theta + \sin^2\theta)} = vt$$

Before embarking on the first solution strategy that occurs to you, pause for a moment to see whether there might be an easier or more direct way of approaching the problem.

Kinematic Equations for Components of Motion

Example 3.1 involved two-dimensional motion in a plane. With a constant velocity (constant components v_x and v_y), the motion is in a straight line. The motion may also be accelerated. For motion in a plane *with a constant acceleration* that has components a_x and a_y, the displacement and velocity components are given by the kinematic equations of Chapter 2 for the x- and y-directions:

$$x = x_o + v_{x_o}t + \tfrac{1}{2}a_x t^2 \qquad (3.3a)$$

$$y = y_o + v_{y_o}t + \tfrac{1}{2}a_y t^2 \qquad (3.3b)$$

$$v_x = v_{x_o} + a_x t \qquad (3.3c)$$

$$v_y = v_{y_o} + a_y t \qquad (3.3d)$$

(constant acceleration only)

Kinematic equations for displacement and velocity components

If an object is initially moving with a constant velocity and suddenly experiences an acceleration in the direction of the velocity ($0°$) or opposite to it ($180°$), it will continue in a straight-line path, either speeding up or slowing down, respectively.

If, however, the acceleration vector is at some angle other than $0°$ or $180°$ to the velocity vector, the motion is along a curved path. For the motion of an object to be *curvilinear*—that is, to vary from a straight-line path—an acceleration is required. For a curved path, the ratio of the velocity components varies with time. That is, the direction of the motion, $\theta = \tan^{-1}(v_y/v_x)$, varies with time, because one or both of the velocity components do, so the motion is not in a straight line.

Consider a ball initially moving along the *x*-axis, as illustrated in ▾Fig. 3.2. Assume that, starting at a time $t_o = 0$, the ball receives a constant acceleration a_y in the *y*-direction. The magnitude of the *x*-component of the ball's displacement is given by $x = v_x t$; the $\frac{1}{2}a_x t^2$ term of Eq. 3.3a drops out, since there is no acceleration in the *x*-direction. Prior to t_o, the motion is in a straight line along the *x*-axis. But at

▶ **FIGURE 3.2 Curvilinear motion** An acceleration not parallel to the instantaneous velocity produces a curved path. Here, an acceleration a_y is applied at $t_o = 0$ to a ball initially moving with a constant velocity v_x. The result is a curved path with the velocity components as shown. Notice how v_y increases with time, while v_x remains constant.

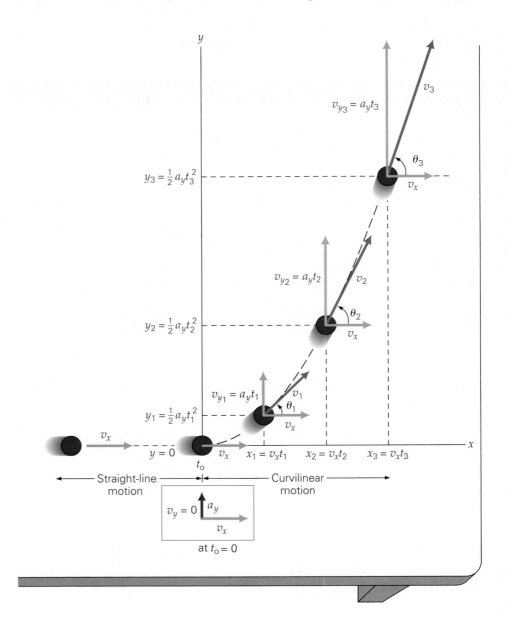

any time after t_o, there is a y-displacement with a magnitude of $y = \frac{1}{2}a_yt^2$ (as given by Eq. 3.3b with $y_o = 0$ and $v_{y_o} = 0$). The result is a curved path for the ball.

Note that the length (magnitude) of the velocity component v_y changes with time, while that of the v_x component remains constant. The total velocity vector *at any time* is tangent to the curved path of the ball. It is at an angle θ relative to the positive x-axis, given by $\theta = \tan^{-1}(v_y/v_x)$, which now changes with time, as we see in Fig. 3.2 and in Example 3.2.

Example 3.2 ■ A Curving Path: Vector Components

Suppose that the ball in Fig. 3.2 has an initial velocity of 1.50 m/s along the x-axis and, starting at $t_o = 0$, receives an acceleration of 2.80 m/s^2 in the y-direction. (a) What is the position of the ball 3.00 s after t_o? (b) What is the velocity of the ball at that time?

Thinking It Through.　Keep in mind that the motions in the x- and y-directions can be analyzed independently. That is, both component motions occur at the same time. For (a), simply compute the x- and y-positions at the given time, taking into account the acceleration in the y-direction. For (b), find the component velocities, and vectorially combine them to get the total velocity.

Solution.　Referring to Fig. 3.2, we have the following:

Given:　$v_{x_o} = v_x = 1.50$ m/s　**Find:**　(a) (x, y) (position coordinates)
$\qquad v_{y_o} = 0$　　　　　　　　　　(b) v (velocity)
$\qquad a_x = 0$
$\qquad a_y = 2.80$ m/s^2
$\qquad t = 3.00$ s

(a)　At 3.00 s after $t_o = 0$, Eqs. 3.3a and 3.3b tell us that the ball has traveled the following distances from the origin $(x_o = y_o = 0)$ in the x- and y-directions, respectively:

$$x = v_{x_o}t + \tfrac{1}{2}a_xt^2 = (1.50 \text{ m/s})(3.00 \text{ s}) + 0 = 4.50 \text{ m}$$
$$y = v_{y_o}t + \tfrac{1}{2}a_yt^2 = 0 + \tfrac{1}{2}(2.80 \text{ m/s}^2)(3.00 \text{ s})^2 = 12.6 \text{ m}$$

Thus, the position of the ball is $(x, y) = (4.50 \text{ m}, 12.6 \text{ m})$. If you had computed the distance $d = \sqrt{x^2 + y^2}$, what would you have gotten? (Note that this quantity is not the actual distance the ball has traveled in 3.00 s, but rather the magnitude of the *displacement*, or straight-line distance, from the origin at $t = 3.00$ s.)

Note: Don't confuse the direction of the velocity with the direction of the displacement from the origin. The direction of the velocity is always tangent to the path.

(b)　The x-component of the velocity is given by Eq. 3.3c:

$$v_x = v_{x_o} + a_xt = 1.50 \text{ m/s} + 0 = +1.50 \text{ m/s}$$

(This component is constant, since there is no acceleration in the $+x$-direction.) Similarly, the y-component of the velocity is given by Eq. 3.3d:

$$v_y = v_{y_o} + a_yt = 0 + (2.80 \text{ m/s}^2)(3.00 \text{ s}) = 8.40 \text{ m/s}$$

The velocity therefore has a magnitude of

$$v = \sqrt{v_x^2 + v_y^2} = \sqrt{(1.50 \text{ m/s})^2 + (8.40 \text{ m/s})^2} = 8.53 \text{ m/s}$$

and its direction relative to the $+x$-axis is

$$\theta = \tan^{-1}\left(\frac{v_y}{v_x}\right) = \tan^{-1}\left(\frac{8.40 \text{ m/s}}{1.50 \text{ m/s}}\right) = 79.9°$$

Follow-up Exercise.　Suppose that the ball in this Example also received an acceleration of 1.00 m/s^2 in the $+x$-direction starting at t_o. What would be the position of the ball 3.00 s after t_o in this case? (*Answers to all Follow-up Exercises are at the back of the text.*)

Problem-Solving Hint

When using the kinematic equations, it is important to note that motion in the x- and y-directions can be analyzed independently—the factor connecting them is time t. That is, both components of motion occur at the same time, so that you find (x, y) and/or (v_x, v_y) at a given time t. Also, keep in mind that we often set $x_o = 0$ and $y_o = 0$, which means that we locate the object at the origin at $t_o = 0$. If the object is actually elsewhere at $t_o = 0$, then an x_o and/or a y_o would have to be used in the appropriate equations. (See Eqs. 3.3a and b.)

3.2 Vector Addition and Subtraction

OBJECTIVES: To (a) learn vector notation, (b) be able to add and subtract vectors graphically and analytically, and (c) use vectors to describe motion in two dimensions.

Many physical quantities, including those describing motion, have a direction associated with them—that is, they are vectors. You have already worked with a few such quantities related to motion (displacement, velocity, and acceleration) and will encounter more during this course of study. A very important technique in the analysis of many physical situations is the addition (and subtraction) of vectors. By adding or combining such quantities (**vector addition**), you can obtain the overall, or net, effect that occurs—the *resultant*, as we call the vector sum.

You have already been adding vectors. In Chapter 2, we added displacements to get the net displacement. In this chapter, we will add vector components of motion to get net effects. Notice that, in Example 3.2, we combined the velocity components v_x and v_y to get the resultant velocity.

In this section, we will look at vector addition and subtraction in general, along with common vector notation. As you will learn, these operations are not the same as scalar, or numerical, addition and subtraction, with which you are already familiar. Vectors have magnitudes *and* directions, so different rules apply.

In general, there are geometric (graphical) methods and analytical (computational) methods of vector addition. The geometric methods are useful in helping you visualize the concepts of vector addition, particularly with a quick sketch. Analytical methods are more commonly used, however, because they are faster and more precise.

Vector Addition: Geometric Methods

Note: In vector notation, vectors are represented by boldface symbols, e.g., **A** and **B**, and their magnitudes by italic symbols, *A* and *B*. When handwritten, an arrow over the symbol is commonly used to denote a vector, e.g., $\vec{A}$ and $\vec{B}$, and the magnitudes by the symbols without arrows, A and B.

Triangle Method To add two vectors—say, to add **B** to **A** (that is, to find **A** + **B**) by the **triangle method**—you first draw **A** on a sheet of graph paper to some scale (▶ Fig. 3.3a). For example, if **A** is a displacement in meters, a convenient scale is 1 cm : 1 m, or 1 cm of vector length on the graph corresponding to 1 m of displacement. As shown in Fig. 3.3b, the direction of the **A** vector is specified as being at an angle θ_A relative to a coordinate axis, usually the x-axis.

Next, draw **B** with its tail starting at the tip of **A**. (Thus, this method is also called the *tip-to-tail method*.) The vector from the tail of **A** to the tip of **B** is then the vector sum **R**, or the resultant of the two vectors: **R** = **A** + **B**.

If the vectors had been drawn to scale, the magnitude of **R** can be found by measuring its length and using the scale conversion. In such a graphical approach, the direction angle θ_R is measured with a protractor. If we know the magnitudes and directions (angles θ) of **A** and **B**, we can also find the magnitude and direction of **R** analytically by using trigonometric methods. For the nonright triangle in Fig. 3.3b, the laws of sines and cosines could be used. (See Appendix I.)

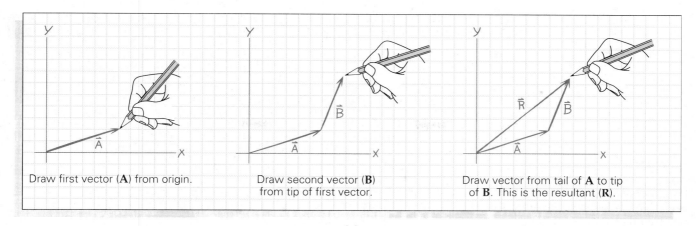

Draw first vector (**A**) from origin.

Draw second vector (**B**) from tip of first vector.

Draw vector from tail of **A** to tip of **B**. This is the resultant (**R**).

(a)

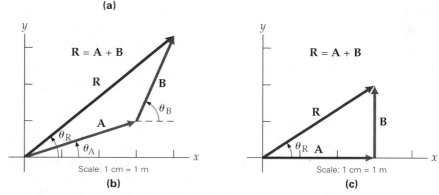

(b)

(c)

▲ **FIGURE 3.3   Triangle method of vector addition**   (a) The vectors **A** and **B** are placed tip to tail. The vector that extends from the tail of **A** to the tip of **B**, forming the third side of the triangle, is the resultant or sum **R** = **A** + **B**. **(b)** When the vectors are drawn to scale, the magnitude of **R** can be found by measuring the length of **R** and using the scale conversion, and the direction angle θ_R can be measured with a protractor. Analytical methods can also be used. For a nonright triangle, as in (b), the laws of sines and cosines can be used to determine the magnitude of **R** and θ (Appendix I). **(c)** If the vector triangle is a right triangle, **R** is easily obtained via the Pythagorean theorem, and the direction angle is given by an inverse trigonometric function.

The resultant of the vector right triangle in Fig. 3.3c would be much easier to find, by using the Pythagorean theorem for the magnitude and an inverse trigonometric function to find the direction angle. (Notice that **R** is made up of x- and y-components **A** and **B**.)

Parallelogram Method   Another graphical method of vector addition, similar to the triangle method, is the **parallelogram method**. In ▶Fig. 3.4, **A** and **B** are drawn tail to tail, and a parallelogram is formed as shown. The resultant vector **R** lies along the diagonal of the parallelogram. By drawing the diagram to scale with proper orientations, the magnitude and direction of **R** can be measured directly from the diagram as in the triangle method.

Notice that **B** could be moved to the other side of the parallelogram, forming the **A** + **B** triangle (and demonstrating why the triangle and parallelogram methods are equivalent). In general, a vector (arrow) can be moved around in vector addition methods. As long as you don't change its length (magnitude) or direction, you don't change the vector. In Fig. 3.4, this shifting of vector arrows shows that **A** + **B** = **B** + **A**, that is, the vectors can be added in either order.

Polygon Method   The triangle method can be extended to include the addition of any number of vectors. The method is then called the **polygon method**, because the resulting graphical figure is a polygon. This method is illustrated for four vectors in ▶Fig. 3.5, where **R** = **A** + **B** + **C** + **D**. Note that

Note: In general, a vector (arrow) can be moved around in vector addition methods. As long as you don't change its length (magnitude) or direction, you don't change the vector.

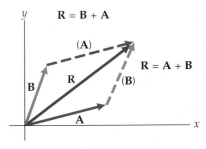

▲ FIGURE 3.4 Parallelogram method of vector addition
The diagonal of the parallelogram formed after **A** and **B** are drawn with their tails at the origin is the resultant $R = A + B$. If the vectors are drawn to scale, the magnitude and direction of **R** can be measured directly from the diagram as in a triangle method. The parallelogram method is equivalent to the triangle method. Shifting **B** to the right forms the same $A + B$ vector triangle as shown in Fig. 3.3b. Shifting **A** up has the same effect, with $R = B + A$.

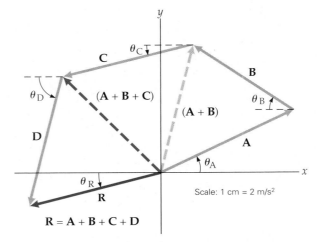

▲ FIGURE 3.5 Polygon method of vector addition
The vectors to be added are placed tip to tail. The resultant **R** is the vector from the tail of the first vector **A** to the tip of the last vector **D** and **R** completes the polygon. This method is essentially composed of multiple applications of the triangle method: $(A + B), (A + B) + C$, and so on.

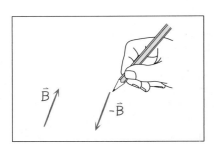

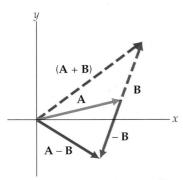

▲ FIGURE 3.6 Vector subtraction Vector subtraction is a special case of vector addition; that is, $A - B = A + (-B)$, where $-B$ has the same magnitude as **B**, but is in the opposite direction. (See the sketch.) Thus, $A + B$ is not the same as $B - A$, either in length or direction. Can you show that $B - A = -(A - B)$ geometrically?

this addition is essentially three applications of the triangle method. The length and direction of the resultant could be found analytically by successive applications of the laws of sines and cosines (see Appendix I), but an easier analytical method, the component method, will be described shortly. As in the parallelogram method, the four vectors (or any number of vectors) can be added in any order.

Vector Subtraction Vector subtraction is a special case of vector addition:

$$A - B = A + (-B)$$

That is, to subtract **B** from **A**, a *negative* **B** is added to **A**. In Chapter 2, you learned that a minus sign simply means that the direction of a vector is opposite that of one with a plus sign (for example, $+x$ and $-x$). The same is true with vectors represented by bold face notation. The vector $-B$ has the same magnitude as the vector **B**, but is in the opposite direction (◄Fig. 3.6). The vector diagram in Fig 3.6 provides a graphical representation of $A - B$.

Vector Components and the Analytical Component Method

Probably the most widely used analytical method for adding multiple vectors is the **component method**. It will be used again and again throughout the course of our study, so a basic understanding of the method is *essential*. Learn this section well.

Adding Rectangular Vector Components By *rectangular components*, we mean vector components at right (90°) angles to each other, usually taken in the rectangular-coordinate x- and y-directions. You have already had an introduction to the addition of such components in the discussion of the velocity components of motion in Section 3.1. For the general case, suppose that **A** and **B**, two vectors at right angles, are added, as illustrated in ►Fig. 3.7a. The right angle makes the math easy. The magnitude of **C** is given by the Pythagorean theorem:

$$C = \sqrt{A^2 + B^2} \tag{3.4a}$$

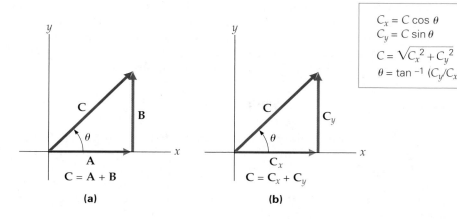

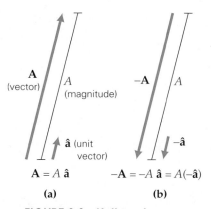

◀ FIGURE 3.7 Vector components (a) The vectors **A** and **B** along the x- and y-axes, respectively, add to give **C**. (b) A vector **C** may be resolved into rectangular components $\mathbf{C}_x$ and $\mathbf{C}_y$.

The orientation of **C** relative to the x-axis is given by the angle

$$\theta = \tan^{-1}\left(\frac{B}{A}\right) \qquad (3.4b)$$

This notation is how a resultant is expressed in **magnitude–angle form**.

Resolving a Vector into Rectangular Components; Unit Vectors Resolving a vector into rectangular components is essentially the reverse of adding the rectangular components of the vector. Given a vector **C**, Fig. 3.7b illustrates how it may be resolved into x and y vector components $\mathbf{C}_x$ and $\mathbf{C}_y$. Simply complete the vector triangle with x- and y-components. As the diagram shows, the magnitudes, or vector lengths, of these components are given by

$$C_x = C\cos\theta \qquad \textit{magnitudes of} \qquad (3.5a)$$
$$C_y = C\sin\theta \qquad \textit{components} \qquad (3.5b)$$

respectively (similar to $v_x = v\cos\theta$ and $v_y = v\sin\theta$ in Example 3.1).* The angle of direction of **C** can also be expressed in terms of the components, since $\tan\theta = C_y/C_x$, or

$$\theta = \tan^{-1}\left(\frac{C_y}{C_x}\right) \qquad \textit{direction of vector from} \atop \textit{magnitudes of components} \qquad (3.6)$$

A general notation for expressing the magnitude and direction of a vector involves the use of unit vectors. For example, as illustrated in ▶ Fig. 3.8, a vector **A** can be written as $\mathbf{A} = A\hat{\mathbf{a}}$. The numerical magnitude is represented by A, and $\hat{\mathbf{a}}$ is called a **unit vector**. That is, it has a magnitude of unity, or one, but no units and thus simply indicates the vector's direction. For example, a velocity along the x-axis might be written $\mathbf{v} = (4.0\ \text{m/s})\,\hat{\mathbf{x}}$ (that is, 4.0 m/s magnitude in the $+x$-direction).

Note in Fig. 3.8 how $-\mathbf{A}$ would be represented in this notation. Although the minus sign is sometimes put in front of the numerical magnitude, this quantity is an absolute number; the minus actually goes with the unit vector: $-\mathbf{A} = -A\hat{\mathbf{a}} = A(-\hat{\mathbf{a}})$.† That is, the unit vector is in the $-\hat{\mathbf{a}}$ direction (opposite $\hat{\mathbf{a}}$). A velocity of $\mathbf{v} = (-4.0\ \text{m/s})\,\hat{\mathbf{x}}$ has a magnitude of 4.0 m/s in the $-x$ direction; that is, $\mathbf{v} = (4.0\ \text{m/s})\,(-\hat{\mathbf{x}})$.

Note: The notation $\tan^{-1} x$ stands for arctangent x, or the angle whose tangent is x.

Magnitude–angle form of a vector

▲ FIGURE 3.8 Unit vectors
(a) A unit vector $\hat{\mathbf{a}}$ has a magnitude of unity, or one, and thereby simply indicates a vector's direction. Written with the magnitude A, it represents the vector **A**, and $\mathbf{A} = A\hat{\mathbf{a}}$. (b) For the vector $-\mathbf{A}$, the unit vector is $-\hat{\mathbf{a}}$, and $-\mathbf{A} = -A\hat{\mathbf{a}} = A(-\hat{\mathbf{a}})$. The magnitude A is a positive number—that is, it is $|A|$.

---

*Figure 3.7b illustrates only a vector in the first quadrant, but the equations hold for all quadrants when vectors are referenced to either the positive or negative x-axis. The directions of the components are indicated by $+$ and $-$ signs, as will be shown shortly.

†The notation is sometimes written with an absolute value, $\mathbf{A} = |A|\hat{\mathbf{a}}$, or $-\mathbf{A} = -|A|\hat{\mathbf{a}}$, so as to clearly show that the magnitude of **A** is a positive quantity.

This notation can be used to express explicitly the rectangular components of a vector. For example, the ball's displacement from the origin in Example 3.2 could be written $\mathbf{d} = (4.50 \text{ m}) \, \hat{\mathbf{x}} + (12.6 \text{ m}) \, \hat{\mathbf{y}}$, where $\hat{\mathbf{x}}$ and $\hat{\mathbf{y}}$ are unit vectors in the x- and y-directions, respectively. In some instances, it may be more convenient to express a general vector in this unit-vector **component form**:

$$\mathbf{C} = C_x \, \hat{\mathbf{x}} + C_y \, \hat{\mathbf{y}} \tag{3.7}$$

Vector Addition Using Components The **analytical component method** of vector addition involves resolving the vectors into rectangular components and adding the components for each axis independently. This method is illustrated graphically in ▼Fig. 3.9 for two vectors $\mathbf{F}_1$ and $\mathbf{F}_2$.* The sums of the x- and y-components of the vectors being added are then equal to the corresponding components of the resultant vector.

The same principle applies if you are given three (or more) vectors to add. You could find the resultant by applying the graphical tip-to-tail method, as illustrated in ▶Fig. 3.10a. However, this technique involves drawing the vectors to scale and using a protractor to measure angles, which is time consuming. Note that the magnitude of the directional angle θ is not obvious in the figure and would have to be measured (or computed trigonometrically).

However, if you use the component method, you do not have to draw the vectors tip to tail. In fact, it is usually more convenient to put all of the tails together at the origin, as shown in Fig. 3.10b. Also, the vectors do not have to be drawn to scale, since the approximate sketch is just a visual aid in applying the analytical method.

In the component method, you resolve the vectors to be added into their x- and y-components, add the respective components, and recombine to find the resultant. The resultant is shown in Fig. 3.10c. By looking at the x-components, it can be seen that the vector sum of these components is in the $-x$-direction. Similarly, the sum of the y-components is in the $+y$-direction. (Note that $\mathbf{v}_2$ is in the y-direction and has a zero x-component, just as a vector in the x-direction would have a zero y-component.)

▼ **FIGURE 3.9 Component addition** **(a)** In adding vectors by the component method, each vector is first resolved into its x- and y-components. **(b)** The sums of the x- and y-components of vectors $\mathbf{F}_1$ and $\mathbf{F}_2$ are $\mathbf{F}_x = \mathbf{F}_{x_1} + \mathbf{F}_{x_2}$ and $\mathbf{F}_y = \mathbf{F}_{y_1} + \mathbf{F}_{y_2}$, respectively.

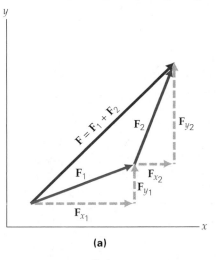

(a)

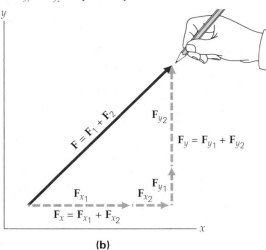

(b)

*The symbol $\mathbf{F}$ is commonly used to denote force, a very important vector quantity that you will study in Chapter 4. Here, $\mathbf{F}$ is employed as a general vector, but its use provides familiarity with the notation used in the next chapter, where a knowledge of the addition of forces is essential.

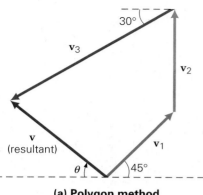

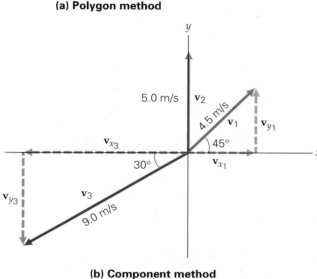

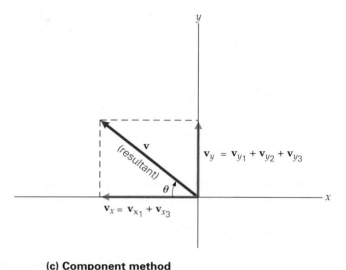

◀ FIGURE 3.10 Component method of vector addition **(a)** Several vectors may be added graphically to find the resultant **v**, but this technique is time consuming and less accurate than the component method. **(b)** In the analytical component method, all the vectors to be added (**v**$_1$, **v**$_2$, and **v**$_3$) are first placed with their tails at the origin so that they may be easily resolved into rectangular components. **(c)** The respective summations of all the x-components and all the y-components are then added to give the components of the resultant **v**.

(a) Polygon method

(b) Component method
(resolving into components)

(c) Component method
(adding x- and y-components, shown as offset dashed arrows, and finding resultant)

In using the plus and minus notation to indicate directions, we write the x- and y-components of the resultant: $v_x = v_{x_1} - v_{x_3}$ and $v_y = v_{y_1} + v_{y_2} - v_{y_3}$, as shown in Fig. 3.10c. When the numerical values of the vector components are computed and put into these equations, you will have values for $-v_x$ and $+v_y$.

Notice also in Fig. 3.10c that the directional angle θ of the resultant is referenced to the x-axis, as are the individual vectors in Fig. 3.10b. *In adding vectors by the component method, we will reference all vectors to the nearest x-axis—that is, the +x-axis or −x-axis.* This policy eliminates angles greater than 90° (as occurs when we customarily measure angles counterclockwise from the +x-axis) and the use of double-angle formulas, such as $\cos(\theta + 90°)$. This restriction greatly simplifies calculations. The recommended procedures for adding vectors analytically by the component method can be summarized as follows:

Procedures for Adding Vectors by the Component Method

1. Resolve the vectors to be added into their x- and y-components. Use the acute angles (angles less than 90°) between the vectors and the x-axis, and indicate the directions of the components by plus and minus signs (▶ Fig. 3.11).

2. Add all of the x-components together, and all of the y-components together vectorially to obtain the x- and y-components of the resultant, or vector sum.

3. Express the resultant vector, using:
   (a) the component form—for example, $\mathbf{C} = C_x\,\hat{\mathbf{x}} + C_y\,\hat{\mathbf{y}}$—or
   (b) the magnitude–angle form.

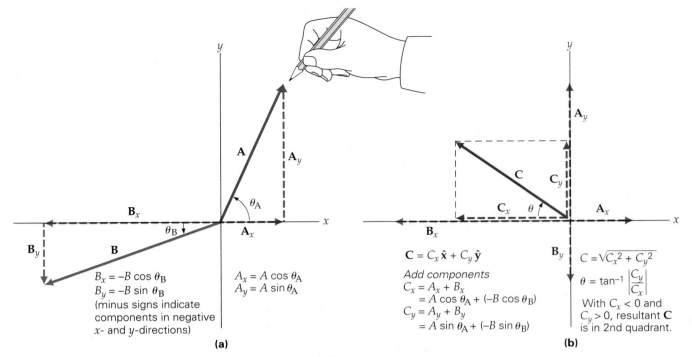

▲ FIGURE 3.11 Vector addition by the analytical component method (a) Resolve the vectors into their x- and y-components. (b) Add all of the x-components and all of the y-components together vectorially to obtain the x- and y-components $\mathbf{C}_x$ and $\mathbf{C}_y$, respectively, of the resultant. Express the resultant in either component form or magnitude–angle form. All angles are referenced to the $+x$- or $-x$-axis to keep them less than $90°$.

For the latter notation, find the magnitude of the resultant by using the summed x- and y-components and the Pythagorean theorem:

$$C = \sqrt{C_x^2 + C_y^2}$$

Find the angle of direction (relative to the x-axis) by taking the inverse tangent $(\tan^{-1})$ of the *absolute value* (that is, the positive value, ignoring any minus signs) of the ratio of the y- and x-components :

$$\theta = \tan^{-1}\left|\frac{C_y}{C_x}\right|$$

Note: The absolute value indicates that minus signs are ignored (for example, $|-3| = 3$). This operation is done to avoid negative values and angles greater than $90°$.

Designate the quadrant in which the resultant lies. This information is obtained from the signs of the summed components or from a sketch of their addition via the triangle (or parallelogram) method. (See Fig. 3.11.) The angle θ is the angle between the resultant and the x-axis in that quadrant.

Example 3.3 ■ Applying the Analytical Component Method:
Separating and Combining x- and y-components

Let's apply the procedural steps of the component method to the addition of the vectors in Fig. 3.10b. The vectors with units of meters per second represent velocities.

Thinking It Through. Follow and learn the steps of the procedure. Basically, you resolve the vectors into components and add the respective components to get the components of the resultant, which then may be expressed in component form or magnitude–angle form.

Solution
1. The rectangular components of the vectors are shown in Fig. 3.10b.

2. Summing these components gives

$$\mathbf{v} = v_x\,\hat{\mathbf{x}} + v_y\,\hat{\mathbf{y}} = (v_{x_1} + v_{x_2} + v_{x_3})\,\hat{\mathbf{x}} + (v_{y_1} + v_{y_2} + v_{y_3})\,\hat{\mathbf{y}}$$

where

$$v_x = v_{x_1} + v_{x_2} + v_{x_3} = v_1\cos 45° + 0 - v_3\cos 30°$$
$$= (4.5\text{ m/s})(0.707) - (9.0\text{ m/s})(0.866) = -4.6\text{ m/s}$$

and

$$v_y = v_{y_1} + v_{y_2} + v_{y_3} = v_1\sin 45° + v_2 - v_3\sin 30°$$
$$= (4.5\text{ m/s})(0.707) + (5.0\text{ m/s}) - (9.0\text{ m/s})(0.50) = 3.7\text{ m/s}$$

In tabular form, the components are provided as follows:

	x-components		y-components	
v_{x_1}	$+v_1\cos 45° = +3.2$ m/s	v_{y_1}	$+v_1\sin 45° = +3.2$ m/s	
v_{x_2}	$= 0$ m/s	v_{y_2}	$= +5.0$ m/s	
v_{x_3}	$-v_3\cos 30° = -7.8$ m/s	v_{y_3}	$-v_3\sin 30° = -4.5$ m/s	
Sums:	$v_x = -4.6$ m/s		$v_y = +3.7$ m/s	

The directions of the components are indicated by signs. (The $+$ sign is sometimes omitted as being understood.) In this case, v_2 has no x-component. Note that in general, for the analytical component method, the x-components are cosine functions and the y-components are sine functions, as long as we reference to the nearest x-axis.

3. In component form, the resultant vector is

$$\mathbf{v} = (-4.6\text{ m/s})\,\hat{\mathbf{x}} + (3.7\text{ m/s})\,\hat{\mathbf{y}}$$

In magnitude–angle form, the resultant velocity has a magnitude of

$$v = \sqrt{v_x^2 + v_y^2} = \sqrt{(-4.6\text{ m/s})^2 + (3.7\text{ m/s})^2} = 5.9\text{ m/s}$$

Since the x-component is negative and the y-component is positive, the resultant lies in the second quadrant at an angle of

$$\theta = \tan^{-1}\left|\frac{v_y}{v_x}\right| = \tan^{-1}\left(\frac{3.7\text{ m/s}}{4.6\text{ m/s}}\right) = 39°$$

above the negative x-axis (See Fig. 3.10c.)

Follow-up Exercise. Suppose in this Example that there were an additional velocity vector $\mathbf{v}_4 = (+4.6\text{ m/s})\,\hat{\mathbf{x}}$. What would be the resultant of all four vectors in this case? *(Answers to all Follow-up Exercises are at the back of the text.)*

Although our discussion is limited to motion in two dimensions (in a plane), the component method is easily extended to three dimensions. For a velocity in three dimensions, the vector has x-, y-, and z-components: $\mathbf{v} = v_x\,\hat{\mathbf{x}} + v_y\,\hat{\mathbf{y}} + v_z\,\hat{\mathbf{z}}$ and magnitude $v = \sqrt{v_x^2 + v_y^2 + v_z^2}$

Conceptual Example 3.4 ■ Vector Components in Action: Sailing "into" the Wind

A sailboat (or wind surfer) on a lake travels in the direction the wind is blowing and then returns. Traveling "into" the wind, how does the boat get back home? *Clearly establish the reasoning and physical principle(s) used in determining your answer before checking it here. That is, **why** did you select your answer?*

Reasoning and Answer. Sailing into the wind is called *tacking*. This method is not a direct mode of sailing, but is a wise use of vector components. Here, we will consider components of force, which is a vector quantity. As you know from experience, a single force (or force component), such as a push, gives rise to motion in the direction in which

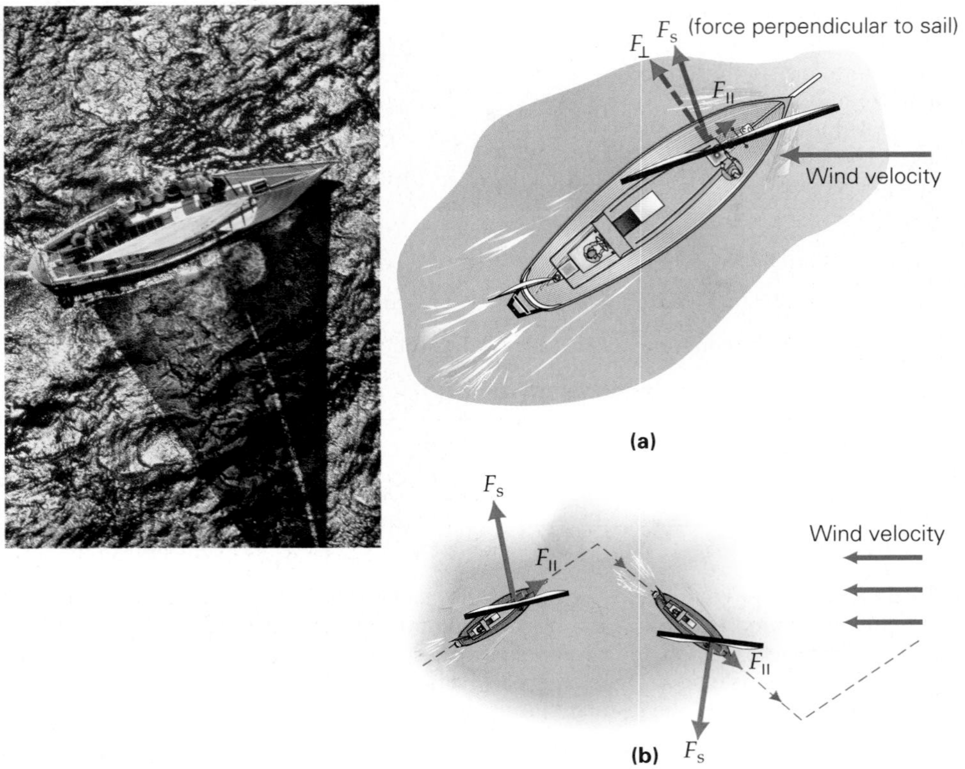

▶ **FIGURE 3.12 Let's go tacking**
(a) The wind filling the sail exerts a force perpendicular to the sail ($\mathbf{F_s}$). We can resolve this force vector into components. The one parallel to the motion of the boat ($\mathbf{F_{\parallel}}$) has an upwind component. **(b)** By changing the direction of the sail, the sailor can "tack" upwind. See Conceptual Example 3.4.

the force is applied. (We shall learn much more about forces in Chapter 4.) For simplicity, let's assume that the wind fills the sail and exerts a force perpendicular to the sail, $\mathbf{F}_s$, as illustrated in ▲ Fig. 3.12a. (Ignore friction and water current effects.)

Note that this force is resolved into components. The force ($\mathbf{F}_{\parallel}$) in the direction of the boat's velocity is at an acute angle ($<90°$) to the direction of the wind and is an "upwind" force on the boat. But, before the boat heads too far to the northeast (as in Fig. 3.12a), the sail is turned so that the direction of $\mathbf{F}_{\parallel}$ is changed by about 90° (Fig. 3.12b). The boat then comes back more in line with the desired upwind course. Using this zigzag process, the boat sails "into" the wind and eventually gets home.

Follow-up Exercise.   The zigzag, straight-line path in Fig. 3.12b was drawn for simplicity. Would this path be the actual path of the boat? Explain. [*Hint*: Consider $\mathbf{F}_{\perp}$.] (*Answers to all Follow-up Exercises are at the back of the text.*)

Integrated Example 3.5 ■ Find the Vector: Add Them Up

You are given two displacement vectors: **A**, with a magnitude of 8.0 m in a direction 45° below the $+x$-axis, and **B**, which has an x-component of $+2.0$ m and a y-component of $+4.0$ m. (a) Sketch the vectors as accurately as you can, using x–y coordinates. (b) Find a vector **C** so that $\mathbf{A} + \mathbf{B} + \mathbf{C}$ equals a vector **D** that has a magnitude of 6.0 m in the $+y$-direction.

(a) Conceptual Reasoning.   In drawing a vector, the length of the vector arrow is proportional to the magnitude of the vector. This length is usually set to some scale—for example, 1 cm per meter (as will be used here). The scale lengths could be measured with a ruler, but in practice, one usually makes a sketch, estimating the vector lengths as best one can. [*See the Learn by Drawing (LBD) figure on the next page.*]

(b) Thinking It Through.   Here again, a sketch helps to understand the situation and gives a general idea of the attributes of **C**. Note in the second LBD figure that both **A** and **B** have $+x$-components, so **C** would have to have a $-x$ component to cancel these components out. (Resultant **D** points only in the $+y$-direction.) $\mathbf{B}_y$ and **D** are in the $+y$-direction, but the $\mathbf{A}_y$-component is larger in the $-y$-direction, so **C** would have to have a $+y$-component. With this information, we see that **C** would lie in the second quadrant if its tail were placed at the origin. A polygon sketch (shown in the LBD figure) confirms this observation.

So, we know that **C** has second quadrant components and that it has a relatively large magnitude (from the lengths of the vectors in the polygon drawing). This information gives us an idea of what we are looking for, making it easer to see if the results from the analytic solution are reasonable.

Given: **A**: 8.0 m, 45° below the $-x$-axis   **Find:** (b) **C** such that
         (fourth quadrant) $\mathbf{A} + \mathbf{B} + \mathbf{C} = \mathbf{D} = (+6.0 \text{ m})\,\hat{\mathbf{y}}$
         $\mathbf{B}_x = (+2.0 \text{ m})\,\hat{\mathbf{x}}$
         $\mathbf{B}_y = (+4.0 \text{ m})\,\hat{\mathbf{y}}$

Let's set up the components in tabular form again so they can be easily seen:

x-components	y-components
$A_x = A\cos 45° = (8.0 \text{ m})(0.707) = +5.7 \text{ m}$	$A_y = A\sin 45° = (8.0 \text{ m})(0.707)$
$B_x = +2.0 \text{ m}$	$= -5.7 \text{ m}$
$C_x = ?$	$B_y = +4.0 \text{ m}$ $C_y = ?$
$D_x = 0$	$D_y = +6.0 \text{ m}$

To find the components of **C**, where $\mathbf{A} + \mathbf{B} + \mathbf{C} = \mathbf{D}$, we sum the x- and y-components of the vectors:

$$x: \quad \mathbf{A}_x + \mathbf{B}_x + \mathbf{C}_x = \mathbf{D}_x$$

or

$$+5.7 \text{ m} + 2.0 \text{ m} + C_x = 0 \quad \text{and} \quad C_x = -7.7 \text{ m}$$

$$y: \quad \mathbf{A}_y + \mathbf{B}_y + \mathbf{C}_y = \mathbf{D}_y$$

or

$$-5.7 \text{ m} + 4.0 \text{ m} + C_y = 6.0 \text{ m} \quad \text{and} \quad C_y = +7.7 \text{ m}$$

So,

$$\mathbf{C} = (-7.7 \text{ m})\,\hat{\mathbf{x}} + (7.7 \text{ m})\,\hat{\mathbf{y}}$$

We can also express the result in magnitude–angle form:

$$C = \sqrt{C_x^2 + C_y^2} = \sqrt{(-7.7 \text{ m})^2 + (7.7 \text{ m})^2} = 11 \text{ m}$$

and

$$\theta = \tan^{-1}\left|\frac{C_y}{C_x}\right| = \tan^{-1}\left|\frac{7.7 \text{ m}}{-7.7 \text{ m}}\right| = 45° \text{ (above the } -x \text{ axis)}$$

Follow-Up Example. Suppose **D** pointed in the opposite direction $[\mathbf{D} = (-6.0 \text{ m})\,\mathbf{y}]$. What would **C** be in this case?

3.3 Relative Velocity: Applying Vector Addition

OBJECTIVE: **To determine relative velocities through vector addition and subtraction.**

Measurements must be made with respect to some reference. This reference is usually taken to be the origin of a coordinate system. The point you designate as the origin of a set of coordinate axes is arbitrary and entirely a matter of choice. For example, you may "attach" the coordinate system to the road or the ground and then measure the displacement or velocity of a car relative to these axes. You may then change the origin of coordinate axes to reflect a different perspective of the car's position.

Learn by Drawing

Make a Sketch and Add Them Up

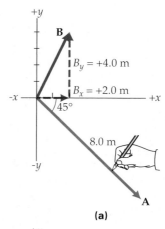

(a)

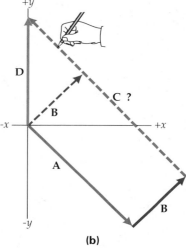

(b)

(a) A sketch is made for the vectors **A** and **B**. In a vector drawing, the vector lengths are usually set to some scale—for example, 1 cm per meter—but in a quick sketch, the vector lengths are estimated. **(b)** By shifting **B** to the tip of **A** and putting in **D**, the vector **C** can be found from $\mathbf{A} + \mathbf{B} + \mathbf{C} = \mathbf{D}$.

We can analyze a situation from any frame of reference. For example, the origin of the coordinate axes may be attached to a car moving along a highway. In analyzing motion from another reference frame, you do not change the physical situation or what is taking place, only the point of view from which you describe it. Hence, we say that motion is *relative* (to some reference frame), and we refer to **relative velocity**. Since velocity is a vector, vector addition and subtraction are helpful in determining relative velocities.

Relative Velocities in One Dimension

When the velocities are linear (along a straight line) in the same or opposite directions and all have the same reference (such as the ground), we can find relative velocities by using vector subtraction. As an illustration, consider cars moving with constant velocities along a straight, level highway, as in ▼Fig. 3.13. The velocities shown in the figure are *relative to the Earth, or the ground*, as indicated by the reference set of coordinate axes in Fig. 3.13a, with motions along the *x*-axis. They are also relative to the stationary observers standing by the highway and sitting in the parked car A. That is, these observers see the cars as moving with velocities $\mathbf{v}_B = +90$ km/h and $\mathbf{v}_C = -60$ km/h. The relative velocity of two objects is given by the velocity (vector) difference between them. For example, the velocity of car B *relative to car A* is given by

$$\mathbf{v}_{BA} = \mathbf{v}_B - \mathbf{v}_A = (+90 \text{ km/h}) \, \hat{\mathbf{x}} - 0 = (+90 \text{ km/h}) \, \hat{\mathbf{x}}$$

Thus, a person sitting in car A would see car B move away (in the positive *x*-direction) with a speed of 90 km/h. For this linear case, the directions of the velocities are indicated by plus and minus signs (in addition to the minus sign in the equation).

Similarly, the velocity of car C relative to an observer in car A is

$$\mathbf{v}_{CA} = \mathbf{v}_C - \mathbf{v}_A = (-60 \text{ km/h}) \, \hat{\mathbf{x}} - 0 = (-60 \text{ km/h}) \, \hat{\mathbf{x}}$$

Note: Use subscripts carefully!
$\mathbf{v}_{AB}$ = velocity of *A* relative to *B*.

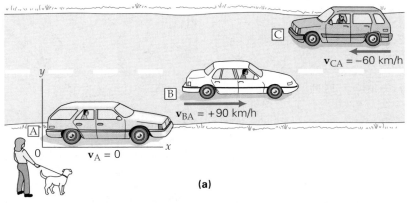

(a)

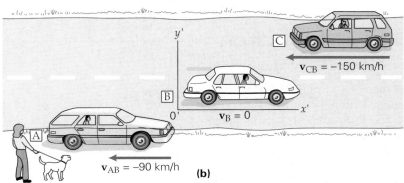

$\mathbf{v}_{AB} = -90$ km/h **(b)**

◀ **FIGURE 3.13 Relative velocity** The observed velocity of a car depends on, or is relative to, the frame of reference. The velocities shown in **(a)** are relative to the ground or to the parked car. In **(b)**, the frame of reference is with respect to car *B*, and the velocities are those that a driver of car *B* would observe. **(c)** These aircraft, performing air-to-air refueling, are normally described as traveling at hundreds of kilometers per hour. To what frame of reference do these velocities refer? What is their velocity relative to each other?

(c)

The person in car A would see car C approaching (in the negative x-direction) with a speed of 60 km/h.

But suppose that you want to know the velocities of the other cars *relative to car B* (that is, from the point of view of an observer in car B) or relative to a set of coordinate axes with the origin fixed to car B (Fig. 3.13b). Relative to those axes, car B is not moving; it acts as the fixed reference point. The other cars are moving relative to car B. The velocity of car C relative to car B is

$$\mathbf{v}_{CB} = \mathbf{v}_C - \mathbf{v}_B = (-60 \text{ km/h}) \, \hat{\mathbf{x}} - (+90 \text{ km/h}) \, \hat{\mathbf{x}} = (-150 \text{ km/h}) \, \hat{\mathbf{x}}$$

Similarly, car A has a velocity relative to car B of

$$\mathbf{v}_{AB} = \mathbf{v}_A - \mathbf{v}_B = 0 - (+90 \text{ km/h}) \, \hat{\mathbf{x}} = (-90 \text{ km/h}) \, \hat{\mathbf{x}}$$

Notice that relative to B, the other cars are both moving in the negative x-direction. That is, C is approaching B with a velocity of 150 km/h in the $-x$-direction, and A appears to be receding from B with a velocity of 90 km/h in the $-x$-direction. (Imagine yourself in car B, and take that position as stationary. Car C would appear to be coming toward you at a high rate of speed, and car A would be getting farther and farther away, as though it were moving backward relative to you.) Note that, in general,

$$\mathbf{v}_{AB} = -\mathbf{v}_{BA}$$

What about the velocities of cars A and B relative to car C? From the point of view (or reference point) of car C, cars A and B would both appear to be approaching or moving in the positive x-direction. For the velocity of B relative to C, we have

$$\mathbf{v}_{BC} = \mathbf{v}_B - \mathbf{v}_C = (90 \text{ km/h}) \, \hat{\mathbf{x}} - (-60 \text{ km/h}) \, \hat{\mathbf{x}} = (+150 \text{ km/h}) \, \hat{\mathbf{x}}$$

Can you show that $\mathbf{v}_{AC} = +60$ km/h? Also note the situation in Fig. 3.13c.

In some instances, we may need to work with velocities that do not all have the same reference point. In such cases, relative velocities can be found by means of vector addition. To solve problems of this kind, *it is essential to identify the velocity references with care*.

Let's look first at a one-dimensional (linear) example. Suppose that a straight moving walkway in a major airport moves with a velocity of $\mathbf{v}_{wg} = (+1.0 \text{ m/s}) \, \hat{\mathbf{x}}$, where the subscripts indicate the velocity of the walkway (w) relative to the ground (g). A passenger (p) on the walkway (w) trying to make a flight connection walks with a velocity of $\mathbf{v}_{pw} = (+2.0 \text{ m/s}) \, \hat{\mathbf{x}}$ relative to the walkway. What is the passenger's velocity relative to an observer standing next to the walkway (that is, relative to the ground)?

The velocity we are seeking, $\mathbf{v}_{pg}$, is given by

$$\mathbf{v}_{pg} = \mathbf{v}_{pw} + \mathbf{v}_{wg} = (2.0 \text{ m/s}) \, \hat{\mathbf{x}} + (1.0 \text{ m/s}) \, \hat{\mathbf{x}} = (3.0 \text{ m/s}) \, \hat{\mathbf{x}}$$

Thus, the stationary observer sees the passenger as traveling with speed of 3.0 m/s down the walkway. (Make a sketch, and show how the vectors add.)

Problem-Solving Hint

Notice the pattern of the subscripts in this example. On the right side of the equation, the two inner subscripts out of the four total subscripts are the same (w). The outer subscripts (p and g) are sequentially the same as those for the relative velocity on the left side of the equation. When adding relative velocities, always check to make sure that the subscripts have this relationship—it indicates that you have set up the equation correctly.

What if a passenger got on the walkway going in the opposite direction and walked with the same speed as that of the walkway? Now it is essential to indicate the direction in which the passenger is walking by means of a minus sign: $\mathbf{v}_{pw} = (-1.0 \text{ m/s}) \, \hat{\mathbf{x}}$. In this case, relative to the stationary observer,

$$\mathbf{v}_{pg} = \mathbf{v}_{pw} + \mathbf{v}_{wg} = (-1.0 \text{ m/s}) \, \hat{\mathbf{x}} + (1.0 \text{ m/s}) \, \hat{\mathbf{x}} = 0$$

so the passenger is stationary with respect to the ground, and the walkway acts as a treadmill. (Excellent physical exercise!)

Relative Velocities in Two Dimensions

Of course, velocities are not always in the same or opposite directions. However, if we know how to use rectangular components to add or subtract vectors, we can solve problems involving relative velocities in two dimensions, as Examples 3.6 and 3.7 show.

Example 3.6 ■ Across and Down the River: Relative Velocity and Components of Motion

The current of a 500-m-wide straight river has a flow rate of 2.55 km/h. A motorboat that travels with a constant speed of 8.00 km/h in still water crosses the river (▼Fig. 3.14). (a) If the boat's bow points directly across the river toward the opposite shore, what is the velocity of the boat relative to the stationary observer sitting at the corner of the bridge? (b) How far downstream will the boat's landing point be from the point directly opposite its starting point? (c) What is the distance traveled by the boat in crossing the river? (Assume that the boat comes instantaneously to rest on a grassy shore.)

Thinking It Through. Careful designation of the given quantities is very important— the velocity of what, relative to what? Once this is done, part (a) should be straightforward. (See the previous Problem-Solving Hint.) For parts (b) and (c), we use kinematics, where the time it takes the boat to cross the river is the key.

Solution. As indicated in Fig. 3.14, we take the river's flow velocity ($\mathbf{v}_{rs}$, river to shore) to be in the x-direction and the boat's velocity ($\mathbf{v}_{br}$, boat to river) to be in the y-direction. Note that the river's flow velocity is *relative to the shore* and that the boat's velocity is *relative to the river*, as indicated by the subscripts. Listing the data, we have:

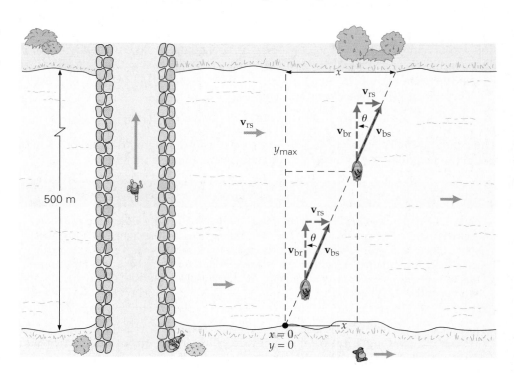

▶ **FIGURE 3.14 Relative velocity and components of motion** As the boat moves across the river, it is carried downstream by the current. See Example 3.6.

Given: y_{max} = 500 m (river width)
$\mathbf{v}_{rs}$ = (2.55 km/h) $\hat{\mathbf{x}}$
   = (0.709 m/s) $\hat{\mathbf{x}}$
   (velocity of river
   relative to shore)
$\mathbf{v}_{br}$ = (8.00 km/h) $\hat{\mathbf{y}}$
   = (2.22 m/s) $\hat{\mathbf{y}}$
   (velocity of boat
   relative to river)

Find: (a) $\mathbf{v}_{bs}$ (velocity of boat
      relative to shore)
  (b) x (distance downstream)
  (c) d (distance traveled by
      boat)

Notice that as the boat moves toward the opposite shore, it is also carried downstream by the current. These velocity components would be clearly apparent relative to the jogger crossing the bridge and to the person sauntering downstream in Fig. 3.14. If both observers stay even with the boat, the velocity of each will match one of the components of the boat's velocity. Since the velocity components are constant, the boat travels in a straight line diagonally across the river (much like the ball rolling across the table in Example 3.1).

(a) The velocity of the boat relative to the shore ($\mathbf{v}_{bs}$) is given by vector addition. In this case, we have

$$\mathbf{v}_{bs} = \mathbf{v}_{br} + \mathbf{v}_{rs}$$

Since the velocities are not along one axis, their magnitudes cannot be added directly. Notice in Fig. 3.14 that the vectors form a right triangle, so we can apply the Pythagorean theorem to find the magnitude of $\mathbf{v}_{bs}$:

$$v_{bs} = \sqrt{v_{br}^2 + v_{rs}^2} = \sqrt{(2.22 \text{ m/s})^2 + (0.709 \text{ m/s})^2}$$
$$= 2.33 \text{ m/s}$$

The direction of this velocity is defined by

$$\theta = \tan^{-1}\left(\frac{v_{rs}}{v_{br}}\right) = \tan^{-1}\left(\frac{0.709 \text{ m/s}}{2.22 \text{ m/s}}\right) = 17.7°$$

(b) To find the distance x that the current carries the boat downstream, we use components. Note that in the y-direction, $y_{max} = v_{br}t$, and

$$t = \frac{y_{max}}{v_{br}} = \frac{500 \text{ m}}{2.22 \text{ m/s}} = 225 \text{ s}$$

which is the time it takes the boat to cross the river.
During this time, the boat is carried downstream by the current a distance of

$$x = v_{rs}t = (0.709 \text{ m/s})(225 \text{ s}) = 160 \text{ m}$$

(c) We could find the distance d the boat travels by using x and y_{max} with the Pythagorean theorem again, but let's use the magnitude of the relative velocity and time instead:

$$d = v_{bs}t = (2.33 \text{ m/s})(225 \text{ s}) = 524 \text{ m}$$

Follow-up Exercise. In part (c), find the distance d by using the Pythagorean theorem. (The answer may be slightly different. Why?) (*Answers to all Follow-up Exercises are at the back of the text.*)

**PHYSLET®
ILLUSTRATION**

Relative Motion

Example 3.7 ■ Flying Into the Wind: Relative Velocity

An airplane with an air speed of 200 km/h (its speed in still air) flies in a direction such that with a west wind of 50.0 km/h blowing, it travels in a straight line northward. (Wind direction is specified by the direction *from* which the wind blows, so a west wind blows from west to east.) To maintain its course due north, the plane must fly at an angle, as illustrated in ▶ Fig. 3.15. What is the speed of the plane along its northward path?

Thinking It Through. Here again, the velocity designations are important, but Fig. 3.15 shows that the velocity vectors form a right triangle, and the magnitude of the unknown velocity can be found by using the Pythagorean theorem.

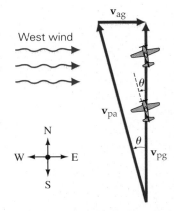

▲ **FIGURE 3.15 Flying into the wind** To fly directly north, the plane's heading (θ-direction) must be west of north. See Example 3.7.

Solution. As always, it is important to identify the reference frame to which the given velocities are relative.

Given: $\mathbf{v}_{pa}$ = 200 km/h at angle θ *Find:* v_{pg} (ground speed of plane)
  (velocity of plane with
  respect to still air = air speed)
$\mathbf{v}_{ag}$ = 50.0 km/h east
  (velocity of air with respect to the
  Earth, or ground, = wind speed)
Plane flies due north with
velocity $\mathbf{v}_{pg}$

The speed of the plane with respect to the Earth, or the ground, v_{pg}, is called the plane's *ground speed*; v_{pa} is its air speed. Vectorially, the respective velocities are related by

$$\mathbf{v}_{pg} = \mathbf{v}_{pa} + \mathbf{v}_{ag}$$

If there were no wind blowing $(v_{ag} = 0)$, the ground speed and air speed would be equal. However, a head wind (a wind blowing directly toward the plane) would cause a slower ground speed, and a tail wind would cause a faster ground speed. The situation is analogous to that of a boat going upstream versus downstream.

Here, $\mathbf{v}_{pg}$ is the resultant of the other two vectors, which can be added by the triangle method. We use the Pythagorean theorem to find v_{pg}, noting that v_{pa} is the hypotenuse of the triangle:

$$v_{pg} = \sqrt{v_{pa}^2 - v_{ag}^2} = \sqrt{(200 \text{ km/h})^2 - (50.0 \text{ km/h})^2} = 194 \text{ km/h}$$

(Note that it was convenient to use the units of kilometers per hour, since the calculation did not involve any other units.)

Follow-up Exercise. What must be the plane's heading (θ-direction) in this Example for the plane to fly directly north? (*Answers to all Follow-up Exercises are at the back of the text.*)

3.4 Projectile Motion

OBJECTIVES: **To analyze projectile motion to find (a) position, (b) time of flight, and (c) range.**

A familiar example of two-dimensional, curvilinear motion is the motion of objects that are thrown or projected by some means. The motion of a stone thrown across a stream or a golf ball driven off a tee is **projectile motion**. A special case of projectile motion in one dimension occurs when an object is projected vertically upward. This case was treated in Chapter 2 in terms of free fall (air resistance neglected). We will treat projectile motion as free fall, too, so that the only acceleration of a projectile is due to gravity.

Note: Review Section 2.5 (free fall).

We can use vector components to analyze projectile motion. We simply break up the motion into its x- and y-components and treat them separately.

Horizontal Projections

It is worthwhile to analyze first the special case of the motion of an object projected horizontally, or parallel to a level surface. Suppose that you throw an object horizontally with an initial velocity v_{x_o} (▶Fig. 3.16a). Projectile motion is analyzed beginning at the instant of release ($t = 0$). Once the object is released, there is no longer a horizontal acceleration ($a_x = 0$), so throughout the object's path, the horizontal velocity remains constant: $v_x = v_{x_o}$.

According to the equation $x = x_o + v_x t$ (Eq. 3.2a), the projected object would continue to travel in the horizontal direction indefinitely. However, you know that this is not what happens. As soon as the object is projected, it is in free fall in the vertical direction, with $v_{y_o} = 0$ (as though it were dropped) and $a_y = -g$. In other words, the projected object travels at a uniform velocity in the horizontal direction while *at the same time* undergoing acceleration in the downward direction under the

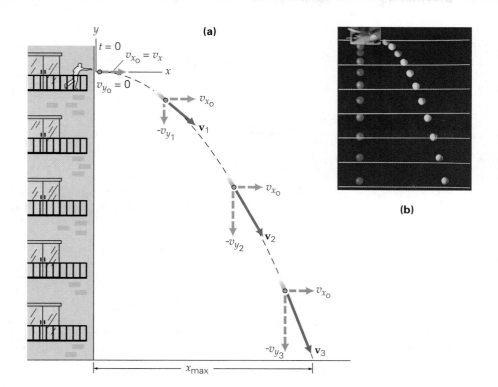

(a)

(b)

◀ **FIGURE 3.16 Horizontal projection** **(a)** The velocity components of a projectile launched horizontally show that the projectile travels to the right as it falls downward. **(b)** A multiflash photograph shows the paths of two golf balls. One was projected horizontally at the same time that the other was dropped straight down. The horizontal lines are 15 cm apart, and the interval between flashes was $\frac{1}{30}$ s. The vertical motions of the balls are the same. Why? Can you describe the horizontal motion of the yellow ball?

influence of gravity. The result is a curved path, as illustrated in Fig. 3.16. (Compare the motions in Fig. 3.16 and Fig. 3.2. Do you see any similarities?) If there were no horizontal motion, the object would simply drop to the ground in a straight line. In fact, the time of flight of the projected object is *exactly the same as if it were falling vertically.*

Note the components of the velocity vector in Fig. 3.16a. The length of the horizontal component of the velocity vector remains the same, but the length of the vertical component increases with time. What is the instantaneous velocity at any point along the path? (Think in terms of vector addition, covered in Section 3.3.) The photo in Fig. 3.16b shows the actual motions of a horizontally projected golf ball and one that is simultaneously dropped from rest. The horizontal reference lines show that the balls fall vertically at the same rate. The only difference is that the horizontally projected ball also travels to the right as it falls.

Example 3.8 ■ Starting at the Top: Horizontal Projection

Suppose that the ball in Fig. 3.16a is projected from a height of 25.0 m above the ground and is thrown with an initial horizontal velocity of 8.25 m/s. (a) How long is the ball in flight before striking the ground? (b) How far from the building does the ball strike the ground?

Thinking It Through. In looking at the components of motion, we find that part (a) involves the time it takes the ball to fall vertically, analogous to a ball dropped from that height. This time is also the time the ball travels in the horizontal direction. The horizontal speed is constant, so we can find the horizontal distance, requested in part (b).

Solution. Writing the data with the origin chosen as the point from which the ball is thrown and downward taken as the negative direction, we have

Given: $y = -25.0 \text{ m}$ 　　　　*Find:*　(a) t (time of flight)
　　　　$v_{x_0} = 8.25 \text{ m/s}$ 　　　　　　　(b) x (horizontal distance)
　　　　$a_x = 0$
　　　　$v_{y_0} = 0$
　　　　$a_y = -g$
　　　　$(x_0 = 0$ and $y_0 = 0$ because of
　　　　our choice of axes location.)

(a) As noted previously, the time of flight is the same as the time it takes for the ball to fall vertically to the ground. To find this time, we can use the equation $y = y_o + v_{y_o}t - \frac{1}{2}gt^2$, in which the negative direction of g is expressed explicitly, as was done in Chapter 2. With $v_{y_o} = 0$, we have

$$y = -\tfrac{1}{2}gt^2$$

So,

$$t = \sqrt{\frac{2y}{-g}} = \sqrt{\frac{2(-25.0 \text{ m})}{-9.80 \text{ m/s}^2}} = 2.26 \text{ s}$$

(b) The ball travels in the x-direction for the same amount of time it travels in the y-direction (that is, 2.26 s). Since there is no acceleration in the horizontal direction, the ball travels in this direction with a uniform velocity. Thus, with $a_x = 0$, we have

$$x = v_{x_o}t = (8.25 \text{ m/s})(2.26 \text{ s}) = 18.6 \text{ m}$$

Follow-up Exercise. (a) Choose the axes to be at the base of the building, and show that the resulting equation is the same as in the Example. (b) What is the velocity (in component form) of the ball just before it strikes the ground? *(Answers to all Follow-up Exercises are at the back of the text.)*

Projections at Arbitrary Angles

The general case of projectile motion involves an object projected at an arbitrary angle θ relative to the horizontal—for example, a golf ball struck by a club (▼ Fig. 3.17). During projectile motion, the object travels up and down while traveling horizontally with a constant velocity. (Does the ball have acceleration? Yes. At each point of the motion, gravity acts, and $\mathbf{a} = -g\,\hat{\mathbf{y}}$.)

This motion is also analyzed by using its components. As before, upward is taken as the positive direction and downward as the negative direction. The initial velocity v_o is first resolved into rectangular components:

$$v_{x_o} = v_o \cos \theta \qquad \text{(3.8a)}$$
$$v_{y_o} = v_o \sin \theta \qquad \text{(3.8b)}$$

initial velocity components $(t_o = 0)$

▼ **FIGURE 3.17 Projection at an angle** The velocity components of the ball are shown for various times. Note that $v_y = 0$ at the top of the arc, or at y_{max}. The range R is the maximum horizontal distance, or x_{max}.

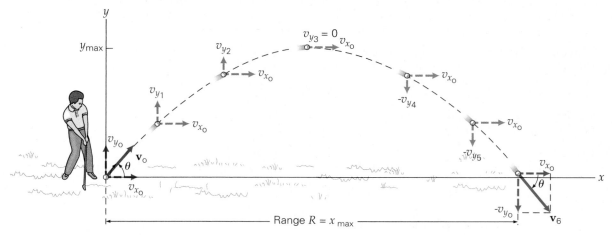

Since there is no horizontal acceleration and gravity acts in the negative y-direction, the x-component of the velocity is constant, and the y-component varies with time (see Eq. 3.3):

$$v_x = v_{x_o} = v_o \cos \theta \qquad \text{(3.9a)}$$

$$v_y = v_{y_o} - gt = v_o \sin \theta - gt \qquad \text{(3.9b)}$$

projectile motion velocity components

The components of the instantaneous velocity at various times are illustrated in Fig. 3.17. The instantaneous velocity is the sum of these components and is tangent to the curved path of the ball at any point. Notice that the ball strikes the ground at the same speed as it was launched (but with $-v_{y_o}$) and at the same angle below the horizontal.

Similarly, the displacement components are given by

$$x = v_{x_o}t = (v_o \cos \theta)t \qquad \text{(3.10a)}$$

$$y = v_{y_o}t - \tfrac{1}{2}gt^2 = (v_o \sin \theta)t - \tfrac{1}{2}gt^2 \qquad \text{(3.10b)}$$

projectile motion displacement components

The curve produced by these equations, or the path of motion of the projectile, is called a **parabola**. The path of projectile motion is often referred to as a *parabolic arc*. Such arcs are commonly observed (▶Fig. 3.18).

Note that, as in the case of horizontal projection, *time is the common feature shared by the components of motion*. Aspects of projectile motion that may be of interest in various situations include the time of flight, the maximum height reached, and the **range** (R), which is the maximum horizontal distance traveled.

▲ **FIGURE 3.18 Parabolic arcs** Sparks of hot metal from welding describe parabolic arcs.

Example 3.9 ■ Teeing Off: Projection at an Angle

Suppose a golf ball is hit off the tee with an initial velocity of 30.0 m/s at an angle of 35° to the horizontal, as in Fig. 3.17. (a) What is the maximum height reached by the ball? (b) What is its range?

Thinking It Through. The maximum height involves the y-component; the procedure for finding it is like that for finding the maximum height of a ball projected vertically upward. The ball travels in the x-direction for the same amount of time it would take for the ball to go up and down.

Solution.

Given: $v_o = 30.0$ m/s *Find:* (a) y_{max}
$\quad\quad\quad \theta = 35°$ (b) $R = x_{max}$
$\quad\quad\quad a_y = -g$
$\quad\quad\quad (x_o$ and $y_o = 0$
$\quad\quad\quad$ *and final* $y_f = 0)$

Let us compute v_{x_o} and v_{y_o} explicitly so that we can use simplified kinematic equations:

$$v_{x_o} = v_o \cos 35° = (30.0 \text{ m/s})(0.819) = 24.6 \text{ m/s}$$

$$v_{y_o} = v_o \sin 35° = (30.0 \text{ m/s})(0.574) = 17.2 \text{ m/s}$$

(a) Just as for an object thrown vertically upward, $v_y = 0$ at the maximum height (y_{max}). Thus, we can find the time to reach the maximum height (t_u) by using Eq. 3.9b with v_y set equal to zero:

$$v_y = 0 = v_{y_o} - gt_u$$

Solving for t_u, we have

$$t_u = \frac{v_{y_o}}{g} = \frac{17.2 \text{ m/s}}{9.80 \text{ m/s}^2} = 1.76 \text{ s}$$

Projectile Motion

(Note that t_u represents the amount of time the ball moves upward.)

The maximum height y_{max} is then obtained by substituting t_u into Eq. 3.10b:

$$y_{max} = v_{y_o}t_u - \tfrac{1}{2}gt_u^2 = (17.2 \text{ m/s})(1.76 \text{ s}) - \tfrac{1}{2}(9.80 \text{ m/s}^2)(1.76 \text{ s})^2 = 15.1 \text{ m}$$

The maximum height could also be obtained directly from Eq. 2.11′, $v_y^2 = v_{y_o}^2 - 2gy$, with $y = y_{max}$ and $v_y = 0$. However, the method of solution used here illustrates how the time of flight is obtained.

(b) As in the case of vertical projection, the time in going up is equal to the time in coming down, so the total time of flight is $t = 2t_u$ (to return to the elevation from which the object was projected, $y - y_o = v_{y_o}t - \tfrac{1}{2}gt^2 = 0$, and $t = 2v_{y_o}/g = 2t_u$.)

The range R is equal to the horizontal distance traveled (x_{max}), which is easily found by substituting the total time of flight $t = 2t_u = 2(1.76 \text{ s}) = 3.52 \text{ s}$ into Eq. 3.10a:

$$R = x_{max} = v_xt = v_{x_o}(2t_u) = (24.6 \text{ m/s})(3.52 \text{ s}) = 86.6 \text{ m}$$

Follow-up Exercise. How would the values of maximum height (y_{max}) and the range (x_{max}) compare with those found in this Example if the golf ball had been similarly teed off on the surface of the Moon? (*Hint:* $g_M = g/6$; that is, acceleration due to gravity on the Moon is one sixth of that on Earth.) Do not do any numerical calculations. Find the answers by "sight reading" the equations. (*Answers to all Follow-up Exercises are at the back of the text.*)

The range of a projectile is an important consideration in various applications. This factor is particularly important in sports in which a maximum range is desired, such as golf and javelin throwing.

In general, what is the range of a projectile launched with velocity v_o at an angle θ? In order to answer this question, we must to consider the equation used in Example 3.9 to calculate the range, $R = v_xt$. First let's look at the expressions for v_x and t. Since there is no acceleration in the horizontal direction, we know that

$$v_x = v_{x_o} = v_o \cos \theta$$

and the total time t (as shown in Example 3.9) is

$$t = \frac{2v_{y_o}}{g} = \frac{2v_o \sin \theta}{g}$$

Then,

$$R = v_xt = (v_o \cos \theta)\left(\frac{2v_o \sin \theta}{g}\right) = \frac{2v_o^2 \sin \theta \cos \theta}{g}$$

Using the trigonometric identity $\sin 2\theta = 2 \cos \theta \sin \theta$ (see Appendix I), we have

$$R = \frac{v_o^2 \sin 2\theta}{g} \quad \begin{array}{l} \textit{projectile range } x_{max} \\ (\textit{only for } y_{initial} = y_{final}) \end{array} \quad (3.11)$$

Note that the range depends on the magnitude of the initial velocity (or speed), v_o, and the angle of projection, θ, and g is assumed to be constant. Keep in mind that this equation applies only to the *special*, but common, case of $y_{initial} = y_{final}$—that is, when the landing point is at the same height as the launch point.

Example 3.10 ■ A Throw from the Bridge

A young girl standing on a bridge throws a stone with an initial velocity of 12 m/s at a downward angle of 45° to the horizontal, in an attempt to hit a block of wood floating in the river below (▶ Fig. 3.19). If the stone is thrown from a height of 20 m and it

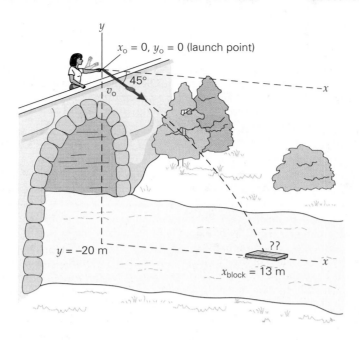

$x_0 = 0, y_0 = 0$ (launch point)

45°

v_0

$y = -20$ m

??

$x_{block} = 13$ m

◀ **FIGURE 3.19 A throw from the bridge—hit or miss?** See Example 3.10.

reaches the river when the block is 13 m from the bridge, does the stone hit the block? (Assume that the block does not move appreciably and that it is in the plane of the throw.)

Thinking It Through.   The question is, what is the range of the stone? If this range is the same as the distance between the block and the bridge, then the stone hits the block. To find the range of the stone, we need to find the time of descent (from the y-component of motion) and then use this time to find the distance x_{max}. (Time is the connecting factor.)

Solution.

Given: $v_0 = 12$ m/s
　　　 $\theta = 45°$ 　　$v_{x_0} = v_0 \cos 45° = 8.5$ m/s
　　　 $y = -20$ m 　$v_{y_0} = -v_0 \sin 45° = -8.5$ m/s
　　 $x_{block} = 13$ m
　　　 $(x_0 = y_0 = 0)$

Find: Range or x_{max} of stone from bridge. (Is it the same as the block's distance from the bridge?)

To find the time for upward projections, we have previously used $v_y = v_{y_0} - gt$, where $v_y = 0$ at the top of the arc. However, in this case, v_y is not zero when the stone reaches the river, so to use this equation, we need to find v_y. This value may be found from the kinematic equation Eq. 2.11′,

$$v_y^2 = v_{y_0}^2 - 2gy$$

as

$$v_y = \sqrt{(-8.5 \text{ m/s})^2 - 2(9.8 \text{ m/s}^2)(-20 \text{ m})} = -22 \text{ m/s}$$

(minus root because v_y is downward).
　　　Then solving $v_y = v_{y_0} - gt$ for t,

$$t = \frac{v_{y_0} - v_y}{g} = \frac{-8.5 \text{ m/s} - (-22 \text{ m/s})}{9.8 \text{ m/s}^2} = 1.4 \text{ s}$$

The stone's horizontal distance from the bridge at this time is

$$x_{max} = v_{x_0} t = (8.5 \text{ m/s})(1.4 \text{ s}) = 12 \text{ m}$$

So the girl's throw falls short by a meter (block at 13 m).

Note that Eq. 3.10b, $y = y_o + v_{y_o}t - \frac{1}{2}gt^2$, could have been used to find the time, but this calculation would have involved solving a quadratic equation.

Follow-up Exercise. (a) Why was it assumed that the block was in the plane of the throw? (b) Why wasn't Eq. 3.11 used in this Example to find the range? Show that Eq. 3.11 works in Example 3.9, but not in Example 3.10, by computing the range in each case and comparing your results with the answers found in the Examples. (*Answers to all Follow-up Exercises are at the back of the text.*)

Conceptual Example 3.11 ■ Which Has the Greater Velocity?

Consider two balls, both thrown with the same initial speed v_o, but one at an angle of 45° above the horizontal and the other at an angle of 45° below the horizontal (▼Fig. 3.20). Determine whether, on reaching the ground, (a) the ball projected upward will have the greater speed, (b) the ball projected downward will have the greater speed, or (c) both balls will have the same speed. *Clearly establish the reasoning and physical principle(s) used in determining your answer before checking it below. That is, **why** did you select your answer?*

Reasoning and Answer. At first, you might think the answer is (b), because this ball is projected downward. But the ball projected upward falls from a greater maximum height, perhaps the answer is (a). To solve this dilemma, look at the horizontal line in Fig. 3.20 between the two velocity vectors that extends beyond the upper trajectory. From this diagram, you should be able to see that the trajectories for both balls are the same below this line. Moreover, the downward velocity of the upper ball on reaching this line is v_o at an angle of 45° below the horizontal. (See Fig. 3.17.) Therefore, relative to the horizontal line and below, the conditions are identical, with the same y-component and same, constant x-component. So, the answer is (c).

Follow-Up Exercise. Suppose the ball thrown downward was thrown at an angle of −40°. Which ball would hit the ground with the greater speed in this case? (*Answers to all Follow-up Exercises are at the back of the text.*)

▶ **FIGURE 3.20 Which has the greater velocity?** See Example 3.11.

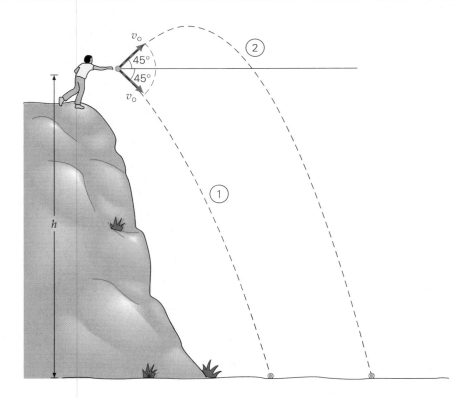

The range of a downward projectile, as in Fig. 3.20, is found as illustrated in Example 3.10. But what about the range of the upward projectile? This task might be thought of as an "extended-range" problem. One way to solve it is to divide the trajectory into two parts—(1) the arc above the horizontal line and (2) the downward part below the horizontal line—such that $x_{max} = x_1 + x_2$. You know how to find x_1 (Example 3.9) and x_2 (Example 3.10). Another way to solve the problem is to use $y = y_o + v_{y_o}t - \frac{1}{2}gt^2$, where y is the final position of the projectile, and solve for t, the total time of flight. You would then use that value in the equation $x = v_{x_o}t$.

Equation 3.11 allows us to compute the range for a particular projection angle and initial velocity. However, we are sometimes interested in the maximum range for a given initial velocity—for example, the maximum range of an artillery piece that fires a projectile with a particular muzzle velocity. Is there an optimum angle that gives the maximum range? Under ideal conditions, the answer is yes.

For a particular v_o, the range is a maximum (R_{max}) when $\sin 2\theta = 1$, since this value of θ yields the maximum value of the sine function (which varies from 0 to 1). Thus,

$$R_{max} = \frac{v_o^2}{g} \quad (y_{initial} = y_{final}) \qquad (3.12)$$

Because this maximum range is obtained when $\sin 2\theta = 1$ and because $\sin 90° = 1$, we have

$$2\theta = 90° \quad \text{or} \quad \theta = 45°$$

for the maximum range for a given initial speed when the projectile returns to the elevation from which it was projected. At a greater or smaller angle, for a projectile with the same initial speed, the range will be less as illustrated in ▼Fig. 3.21. Also, the range is the same for angles equally above and below 45°, such as 30° and 60°.

Thus, to get the maximum range, a projectile *ideally* should be projected at an angle of 45°. However, up to now, we have neglected air resistance. In actual situations, such as when a ball or object is thrown or hit hard, this factor may have a significant effect. Air resistance reduces the speed of the projectile, thereby

PHYSLET®
ILLUSTRATION

Range of a Projectile

▼ **FIGURE 3.21 Range** For a projectile with a given initial speed, the maximum range is ideally attained with a projection of 45° (no air resistance). For projection angles above and below 45°, the range is shorter, and it is equal for angles equally different from 45° (for example, 30° and 60°).

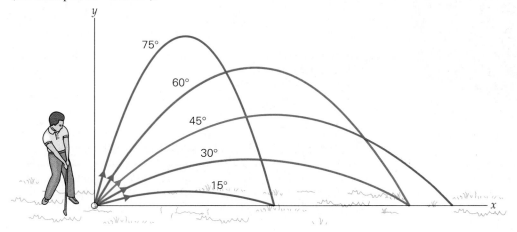

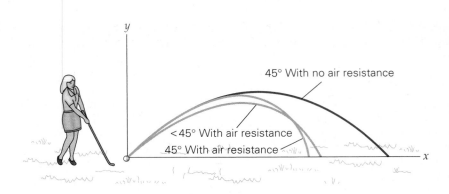

FIGURE 3.22 Air resistance and range When air resistance is a factor, the angle of projection for maximum range is less than 45°.

45° With no air resistance

<45° With air resistance

45° With air resistance

reducing the range. As a result, when air resistance is a factor, the angle of projection for maximum range is less than 45°, which gives a greater initial horizontal velocity (▲ Fig. 3.22). Other factors, such as spin and wind, may also affect the range of a projectile. For example, backspin on a driven golf ball provides lift, and the projection angle for the maximum range may be considerably less than 45°.

Keep in mind that for the maximum range to occur at a projection angle of 45°, the components of initial velocity must be equal—i.e., $\tan^{-1}(v_{y_o}/v_{x_o}) = 45°$ and $\tan 45° = 1$, so that $v_{y_o} = v_{x_o}$. However, this condition may not always be physically possible, as Conceptual Example 3.12 shows.

Conceptual Example 3.12 ■ The Longest Jump: Theory and Practice

In a long-jump event, the jumper normally has a launch angle of (a) less than 45°, (b) exactly 45°, or (c) greater than 45°? *Clearly establish the reasoning and physical principle(s) used in determining your answer before checking it below. That is, **why** did you select your answer?*

Reasoning and Answer. Air resistance is not a major factor here (although wind speed is taken into account for record setting in track-and-field events). Therefore, it would seem that, to achieve maximum range, the jumper would take off at an angle of 45°. But there is another physical consideration. Let's look more closely at the jumper's initial velocity components.

To maximize a long jump, the jumper runs as fast as possible and then pushes upward as strongly as possible to maximize the velocity components. The initial vertical velocity component v_{y_o} depends on the upward push of the jumper's legs, whereas the initial horizontal velocity component v_{x_o} depends mostly on the running speed toward the jump point. In general, a greater velocity can be achieved by running than by jumping, so $v_{x_o} > v_{y_o}$. Then, since $\theta = \tan^{-1}(v_{y_o}/v_{x_o})$, we have $\theta < 45°$, where $v_{y_o}/v_{x_o} < 1$ in this case. Hence, the answer is (a)—it certainly could not be (c). A typical launch angle for a long jump is 20° to 25°. (If a jumper increased her launch angle to be closer to the ideal 45°, then her running speed would have to decrease, resulting in a decrease in range.)

Follow-up Exercise. When jumping to score, basketball players seem to be suspended momentarily, or to "hang" in the air (◄ Fig. 3.23). Explain the physics of this effect. (*Answers to all Follow-up Exercises are at the back of the text.*)

▲ FIGURE 3.23 Hanging in there Basketball players seem to "hang" in the air at the peak of their jump. Why is this? See the Follow-up Exercise for Conceptual Example 3.12.

Integrated Example 3.13 ■ A "Slap Shot": Is It Good?

A hockey player hits a "slap shot" in practice (with no goalie present) when he is 15.0 m directly in front of the net. The net is 1.20 m high, and the puck is initially hit at an angle of 5.00° above the ice with a speed of 35.0 m/s. (a) Make a sketch of the situation using

x–y coordinates, assuming that the puck is at the origin at the time it is hit. Be sure to locate the net in the sketch and show its height. (b) Determine if the puck makes it into the net. If it does, determine whether the puck is rising or falling vertically as it crosses the front plane of the net.

(a) Conceptual Reasoning. A sketch of the situation is shown in ▼Fig. 3.24. Note that the launch angle is exaggerated. An angle of 5.00° is quite small, but then again, the top of the net is not overly high (1.2 m).

(b) Thinking It Through. To determine if the shot is of goal quality, we need to know if the puck's trajectory takes it above the net or into the net. That is, what is the puck's height (y) when its horizontal distance is $x = 15$ m? Whether or not the puck is rising or falling at this horizontal distance depends on when the puck reaches its maximum height. The appropriate equation(s) should tell us this information; we must keep mind that time is the connecting factor between the x- and y-components. Listing the data as usual, we have

Given: $x = 15.0$ m, $x_o = 0$
$\qquad y_{net} = 1.20$ m, $y_o = 0$
$\qquad \theta = 5.00°$
$\qquad v_o = 35.0$ m/s
$\qquad v_{x_o} = v_o \cos 5.00° = 34.9$ m/s
$\qquad v_{y_o} = v_o \sin 5.00° = 3.05$ m/s

Find: (b) If the puck goes into the
$\qquad$ net and if so, if it is
$\qquad$ rising or falling

The vertical location of the puck at any time t is given by $y = v_{y_o}t - \frac{1}{2}gt^2$, so we need to know how long it takes for the puck to travel the 15.0 m to the net. The connecting factor of the components is time, so this time can be found from the x motion:

$$x = v_{x_o}t \quad \text{or} \quad t = \frac{x}{v_{x_o}} = \frac{15.0 \text{ m}}{34.9 \text{ m/s}} = 0.430 \text{ s}$$

So, on reaching the front of the net, the puck is at a height of

$$y = v_{y_o}t - \frac{1}{2}gt^2 = (3.05 \text{ m/s})(0.430 \text{ s}) - \frac{1}{2}(9.80 \text{ m/s}^2)(0.430 \text{ s})^2$$
$$= 1.31 \text{ m} - 0.906 \text{ m} = 0.40 \text{ m}$$

Goal!

The time (t_u) for the puck to reach its maximum height is given by $v_y = v_{y_o} - gt_u$, where $v_y = 0$ and

$$t_u = \frac{v_{y_o}}{g} = \frac{3.05 \text{ m/s}}{9.80 \text{ m/s}^2} = 0.311 \text{ s}$$

and with the puck reaching the net in 0.430 s, it is descending.

Follow-up Exercise. At what distance from the net did the puck start to descend? *(Answers to all Follow-up Exercises are at the back of the text.)*

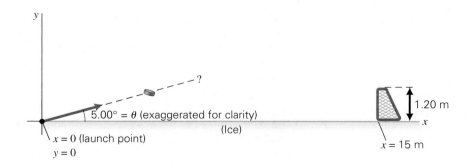

◀ **FIGURE 3.24 Slap shot** Is it a goal? See Integrated Example 3.13.

Chapter Review

Important Concepts and Equations

- Motion in two dimensions is analyzed by considering the motion of linear components. The connecting factor between components is time.

  Components of Initial Velocity:

  $$v_{x_o} = v_o \cos \theta \tag{3.1a}$$

  $$v_{y_o} = v_o \sin \theta \tag{3.1b}$$

  Components of Displacement (constant acceleration only):

  $$x = x_o + v_{x_o}t + \tfrac{1}{2}a_x t^2 \tag{3.3a}$$

  $$y = y_o + v_{y_o}t + \tfrac{1}{2}a_y t^2 \tag{3.3b}$$

  Component of Velocity (constant acceleration only):

  $$v_x = v_{x_o} + a_x t \tag{3.3c}$$

  $$v_y = v_{y_o} + a_y t \tag{3.3d}$$

- Of the various methods of vector addition, the component method is most useful. A resultant vector can be expressed in **magnitude–angle form** or in **unit-vector component form**.

Vector Representation:

$$\left. \begin{aligned} C &= \sqrt{C_x^2 + C_y^2} \\[6pt] \theta &= \tan^{-1}\left|\frac{C_y}{C_x}\right| \end{aligned} \right\} \quad \textit{magnitude–angle form} \tag{3.4a}$$

$$\mathbf{C} = C_x\,\hat{\mathbf{x}} + C_y\,\hat{\mathbf{y}} \quad \textit{component form} \tag{3.7}$$

- **Relative velocity** is expressed *relative* to a particular reference frame.

- **Projectile motion** is analyzed by considering horizontal and vertical components separately—constant velocity in the horizontal direction and an acceleration due to gravity, g, in the downward vertical direction. (The foregoing equations for constant acceleration then have an acceleration of $a = -g$ instead of a.)

Exercises

3.1 Components of Motion

1. On Cartesian axes, the x-component of a vector is generally associated with a (a) cosine, (b) sine, (c) tangent, or (d) none of the foregoing.

2. **CQ** Can the x-component of a vector be greater than the magnitude of the vector? How about the y-component? Explain.

3. **CQ** Is it possible for an object's velocity to be perpendicular to the object's acceleration? If so, describe the motion.

4. For an object in curvilinear motion, (a) the object's velocity components are constant, (b) the y-velocity component is necessarily greater than the x-velocity component, (c) there is an acceleration nonparallel to the object's path, or (d) the velocity and acceleration vectors must be at right angles (90°).

5. **CQ** Describe the motion of an object that is initially traveling with a constant velocity and then receives an acceleration of constant magnitude (a) in a direction parallel to the initial velocity, (b) in a direction perpendicular to the initial velocity, and (c) that is always perpendicular to the instantaneous velocity or direction of motion.

6. **IE** ■ A golf ball is hit with an initial speed of 35 m/s at an angle less than 45° above the horizontal. (a) The horizontal velocity component is (1) greater than, (2) equal to, or (3) less than the vertical velocity component. Why? (b) If the ball is hit at an angle of 37°, what are the initial horizontal- and vertical-velocity components?

7. **IE** ■ The x- and y-components of an acceleration vector are 3.0 m/s² and 4.0 m/s², respectively. (a) The magnitude of the acceleration vector is (1) less than 3.0 m/s², (2) between 3.0 m/s² and 4.0 m/s², (3) between 4.0 m/s² and 7.0 m/s², or (4) equal to 7.0 m/s². (b) What are the magnitude and direction of the acceleration vector?

8. ■ If the magnitude of a velocity vector is 7.0 m/s and the x-component is 3.0 m/s, what is the y-component?

9. ■■ The x-component of a velocity vector that has an angle of 37° to the $+x$-axis has a magnitude of 4.8 m/s. (a) What is the magnitude of the velocity? (b) What is the magnitude of the y-component of the velocity?

10. **IE** ■■ A student walks 100 m west and 50 m south. (a) To get back to the starting point, the student must walk in a general direction of (1) south of west, (2) north of east,

(3) south of east, or (4) north of west. (b) What displacement will bring the student back to the starting point?

11. ■■ A student strolls diagonally across a level rectangular campus plaza, covering the 50-m distance in 1.0 min (▼Fig. 3.25). (a) If the diagonal route makes a 37° angle with the long side of the plaza, what would be the distance if the student had walked halfway around the outside of the plaza instead of along the diagonal route? (b) If the student had walked the outside route in 1.0 min at a constant speed, how much time would she have spent on each side?

▲ FIGURE 3.25   Which way?   See Exercise 11.

12. ■■ The displacement vector of a moving object initially at the origin has a magnitude of 12.5 cm and is at an angle of 30° below the −x-axis at a particular instant. What are the coordinates of the object at that instant?

13. ■■ A ball rolls at a constant velocity of 1.50 m/s at an angle of 45° below the +x-axis in the fourth quadrant. If we take the ball to be at the origin at $t = 0$, what are its coordinates (x, y) 1.65 s later?

14. ■■ A ball rolling on a table has a velocity with rectangular components $v_x = 0.60$ m/s and $v_y = 0.80$ m/s. What is the displacement of the ball in an interval of 2.5 s?

15. ■■ A ball has an initial velocity of 1.30 m/s along the +y-axis and, starting at t_o, receives an acceleration of 2.10 m/s^2 in the +x-direction. (a) What is the position of the ball 2.50 s after t_o? (b) What is the velocity of the ball at that time?

16. ■■ A small plane takes off at a constant velocity of 150 km/h at an angle of 37°. At 3.00 s, (a) how high is the plane above the ground, and (b) what horizontal distance has the plane traveled from the liftoff point?

17. ■■ A ball rolls diagonally across a tabletop from corner to corner with a constant speed of 0.75 m/s. The tabletop is 3.0 m wide and 4.0 m long. Another ball, starting at the same time as the first one and from the same corner, rolls along the longer edge of the table. What constant speed must the second ball have in order to reach the same corner at the same time as the ball traveling diagonally?

18. ■■ A particle moves at a speed of 2.5 m/s in the +x-direction. Upon reaching the origin, the particle receives a continuous constant acceleration of 0.75 m/s^2 in the −y-direction. What is the position of the particle 4.0 s later?

19. ■■■ An automobile travels, at a constant speed of 60 km/h, 800 m along a straight highway that is inclined 5.0° to the horizontal. An observer notes only the vertical motion of the car. What is the car's (a) vertical velocity magnitude and (b) vertical travel distance?

3.2 Vector Addition and Subtraction*

20. CQ Two vectors of magnitudes 3 and 4, respectively, are added. The magnitude of the resultant vector is (a) 1, (b) 7, or (c) between 1 and 7.

21. CQ In Exercise 20, under what condition would the magnitude of the resultant equal 1? How about 7? How about 5?

22. CQ The resultant of $\mathbf{A} - \mathbf{B}$ is the same as (a) $\mathbf{B} - \mathbf{A}$, (b) $-\mathbf{A} + \mathbf{B}$, (c) $-(\mathbf{A} + \mathbf{B})$, or (d) $-(\mathbf{B} - \mathbf{A})$.

23. CQ Can a nonzero vector have a zero x-component? Explain.

24. CQ Is it possible to add a vector quantity to a scalar quantity?

25. ■ Find the rectangular components of a velocity vector that has a magnitude of 10.0 m/s and is oriented at an angle of 30° above the +x-axis.

26. ■ Using the triangle method, show graphically that (a) $\mathbf{A} + \mathbf{B} = \mathbf{B} + \mathbf{A}$ and (b) if $\mathbf{A} - \mathbf{B} = \mathbf{C}$, then $\mathbf{A} = \mathbf{B} + \mathbf{C}$.

27. IE ■ (a) Is vector addition associative? That is, does $(\mathbf{A} + \mathbf{B}) + \mathbf{C} = \mathbf{A} + (\mathbf{B} + \mathbf{C})$? (b) Justify your answer graphically.

28. ■ A vector has an x-component of -2.5 m and a y-component of 4.2 m. Express the vector in magnitude–angle form.

29. ■ Consider yourself to be in boat A traveling along a straight path on a lake with a speed of $v_A = 30$ km/h. Another boat, boat B, travels at a speed of $v_B = 45$ km/h. Find the difference in the velocities, $\mathbf{v}_{BA} = \mathbf{v}_B - \mathbf{v}_A$, when (a) the other boat travels in the same direction in front of you and (b) the other boat is approaching you from the opposite direction.

30. ■ For the two vectors $\mathbf{x}_1 = (20 \text{ m}) \, \hat{\mathbf{x}}$ and $\mathbf{x}_2 = (15 \text{ m}) \, \hat{\mathbf{x}}$, compute and show graphically (a) $\mathbf{x}_1 + \mathbf{x}_2$, (b) $\mathbf{x}_1 - \mathbf{x}_2$, and (c) $\mathbf{x}_2 - \mathbf{x}_1$.

*There are a few exercises in this section that use force vectors (**F**). These vectors should be treated as vectors to be added, just as you would treat velocity vectors. The SI unit of force is the newton (N). A force vector might be written as $\mathbf{F} = 50$ N at an angle of 20°, similar to how we would specify a velocity vector $\mathbf{v} = 30$ m/s at an angle of 40°. Some familiarity with **F** vectors will be helpful in Chapter 4.

31. ■ During a takeoff (in still air), an airplane moves at a speed of 120 mi/h at an angle of 25° above the ground. What is the ground speed of the plane?

32. ■■ A dog walks northeast for 15 m and then east for 25 m. Find the resultant (or sum) displacement vector by (a) the graphical method and (b) the component method.

33. ■■ An airplane flies northwest for 250 mi and then west for 150 mi. Find the resultant (or sum) displacement vector by (a) the graphical method and (b) the component method.

34. ■■ Two boys are pulling a box across a horizontal floor as shown in ▶Fig 3.26. If F_1 = 50.0 N and F_2 = 100 N, find the resultant (or sum) force by (a) the graphical method and (b) the component method.

35. ■■ For each of the given vectors, give a vector that, when added to it, yields a *null vector* (a vector with a magnitude of zero). Express the vector in the form other than that in which it is given (component or magnitude–angle). (a) **A** = 4.5 cm, 40° above the +x-axis; (b) **B** = (2.0 cm) $\hat{x}$ − (4.0 cm) $\hat{y}$; (c) **C** = 8.0 cm at an angle of 60° above the −x-axis.

36. IE ■■ (a) If each of the two components (x and y) of a vector are doubled, (1) the vector's magnitude doubles, but the direction remains unchanged; (2) the vector's magnitude remains unchanged, but the direction angle doubles; or (3) both the vector's magnitude and direction angle double. (b)

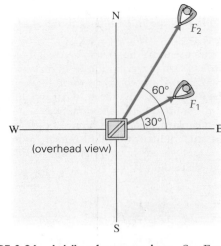

▲ **FIGURE 3.26 Adding force vectors** See Exercises 34 and 51.

If the x- and y-components of a vector of 10 m at 45° are tripled, what is the new vector?

37. ■■ (a) Find the resultant (or sum) of the vectors $\mathbf{F}_1$ and $\mathbf{F}_2$ in ▼Fig. 3.27. (b) If $\mathbf{F}_1$ in the figure were at an angle of 27° instead of 37° with the +x-axis, what would be the resultant (or sum) of $\mathbf{F}_1$ and $\mathbf{F}_2$?

38. ■■ Given two vectors **A**, which has a length of 10.0 and makes an angle of 45° below the −x-axis, and **B**, which has an x-component of +2.0 and a y-component of +4.0, (a) sketch the vectors on x–y axes, with all their "tails" starting at the origin, and (b) calculate **A** + **B**.

▶ **FIGURE 3.27 Vector addition** See Exercise 37.

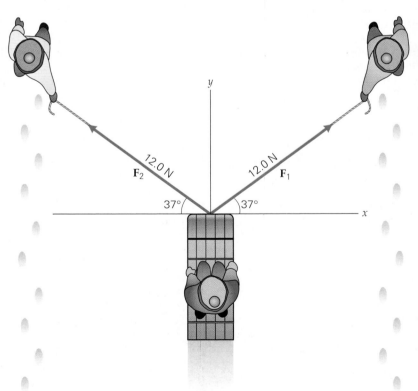

39. ■■ For the velocity vectors shown in ▼Fig. 3.28, determine **A** + **B** + **C**.

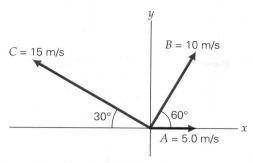

▲ **FIGURE 3.28   Adding velocity vectors**   See Exercises 39 and 40.

40. ■■ For the velocity vectors shown in Fig. 3.28, determine **A** − **B** − **C**.

41. ■■ Given two vectors **A** and **B** with magnitude A and B, respectively, you can subtract **B** from **A** to get a third vector **C** = **A** − **B**. If the magnitude of **C** is equal to $C = A + B$, what is the relative orientation of vectors **A** and **B**?

42. ■■ Two force vectors $\mathbf{F}_1 = (3.0\,\mathrm{N})\,\hat{\mathbf{x}} − (3.0\,\mathrm{N})\,\hat{\mathbf{y}}$ and $\mathbf{F}_2 = (−6.0\,\mathrm{N})\,\hat{\mathbf{x}} + (4.5\,\mathrm{N})\,\hat{\mathbf{y}}$ are applied to a particle. What third force $\mathbf{F}_3$ would make the net, or resultant, force on the particle zero?

43. ■■ Two force vectors $\mathbf{F}_1 = 8.0\,\mathrm{N}$ at an angle of 60° above the +x-axis and $\mathbf{F}_2 = 5.5\,\mathrm{N}$ at an angle of 45° below the +x-axis are applied to a particle at the origin. What third force $\mathbf{F}_3$ would make the net, or resultant, force on the particle zero?

44. ■■ A student works three problems involving the addition of two different vectors $\mathbf{F}_1$ and $\mathbf{F}_2$. He states that the magnitudes of the three resultants are given by (a) $F_1 + F_2$, (b) $F_1 − F_2$, and (c) $\sqrt{F_1^2 + F_2^2}$. Are these results possible? If so, describe the vectors in each case.

45. ■■ A block weighing 50 N rests on an inclined plane. Its weight is a force directed vertically downward, as illustrated in ▼Fig. 3.29. Find the components of the force

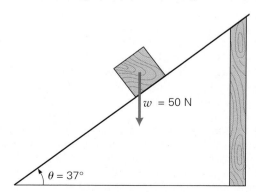

▲ **FIGURE 3.29   Block on an inclined plane**   See Exercise 45.

parallel to the surface of the plane and perpendicular to it.

46. ■■■ A roller coaster at an amusement park starts out on a level track 50.0 m long and then goes up a 25.0-m incline at an angle of 30° to the horizontal. It then goes down a 15.0-m ramp with an incline of 40° to the horizontal. When the roller coaster has reached the bottom of the ramp, what is its displacement from its starting point?

47. ■■■ A person walks from point A to point B as shown in ▼Fig. 3.30. What is the person's displacement relative to A?

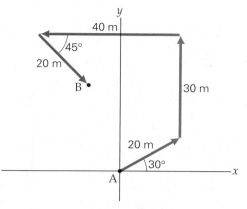

▲ **FIGURE 3.30   Adding displacement vectors**   See Exercise 47.

48. IE ■■■ A meteorologist tracks the movement of a thunderstorm with Doppler radar. At 8:00 PM, the storm was 60 mi northeast of her station. At 10:00 PM, the storm is at 75 mi north. (a) The general direction of the thunderstorm's velocity is (1) south of east, (2) north of west, (3) north of east, or (4) south of west. (b) What is the average velocity of the storm?

49. IE ■■■ A flight controller determines that an airplane is 20.0 mi south of him. Half an hour later, the same plane is 35.0 mi northwest of him. (a) The general direction of the airplane's velocity is (1) east of south, (2) north of west, (3) north of east, or (4) west of south. (b) If the plane is flying with constant velocity, what is its velocity during this time?

50. ■■■ A ship is seen on a radar screen to be 10 km east of the radar site. Some time later, the ship is at 15 km northwest. What is the displacement of the ship?

51. ■■■ Two students are pulling a box as shown in Fig. 3.26. If $F_1 = 100\,\mathrm{N}$ and $F_2 = 150\,\mathrm{N}$ and a third student wants to stop the box, what force should he apply?

3.3 Relative Velocity

52. A student walks on a treadmill moving at 4.0 m/s and remains at the same place in the gym. (a) What is the student's velocity relative to the gym floor? (b) What is the student's speed relative to the treadmill?

53. **CQ** You are running in the rain along a straight sidewalk to your dorm. If the rain is falling vertically downward relative to the ground, how should you hold your umbrella so as to minimize the rain landing on you? Explain.

54. **CQ** When driving to the basket for a layup, a basketball player usually tosses the ball gently upward relative to herself. Explain why?

55. **CQ** When you are riding in a fast-moving car, in what direction would you throw an object up so it will return to your hand? Explain.

56. ■ While you are traveling in a car on a straight, level interstate highway at 90 km/h, another car passes you in the same direction; its speedometer reads 120 km/h. (a) What is your velocity relative to the other driver? (b) What is the other car's velocity relative to you?

57. ■ A shopper is in a hurry to catch a bargain in a department store. She walks up the escalator, rather than letting it carry her, at a speed of 1.0 m/s relative to the escalator. If the escalator is 20 m long and moves at a speed of 0.50 m/s, how long does it take for the shopper to get to the next floor?

58. ■ A person riding in the back of a pickup truck traveling at 70 km/h on a straight, level road throws a ball with a speed of 15 km/h relative to the truck in the direction opposite to the truck's motion. What is the velocity of the ball (a) relative to a stationary observer by the side of the road, and (b) relative to the driver of a car moving in the same direction as the truck at a speed of 90 km/h?

59. ■ In Exercise 58, what are the relative velocities if the ball is thrown in the direction of the truck?

60. ■ A boat heads upstream for 30 s at 5.0 m/s relative to still water. If the speed of the current is 3.0 m/s, what distance does the boat travel? How about if the boat were heading downstream?

61. ■■ In a 500-m stretch of a river, the speed of the current is a steady 5.0 m/s. How long does it take for a boat to finish a round-trip (upstream and downstream) if the speed of the boat is 7.5 m/s relative to still water?

62. ■■ Suppose you are climbing up the mast of a sailboat at a speed of 0.20 m/s. If the sailboat is moving with a speed of 0.60 m/s in a lake, what is your relative velocity to the lake?

63. ■■ A moving walkway in an airport is 75 m long and moves at a speed of 0.30 m/s. A passenger, after traveling 25 m while standing on the walkway, starts to walk at a speed of 0.50 m/s relative to the surface of the walkway. How long does it take her to travel the total distance of the walkway?

64. **IE** ■■ A swimmer swims north at 0.15 m/s relative to still water across a river that flows at a rate of 0.20 m/s

from west to east. (a) The general direction of the swimmer's velocity, relative to the river bank, is (1) north of east, (2) south of west, (3) north of west, or (4) south of east. (b) Calculate the swimmer's velocity relative to the riverbank.

65. ■■ A swimmer maintains a speed of 0.15 m/s relative to the water when he swims directly toward the opposite shore of a river. The river has a current that flows at 0.75 m/s. (a) How far downstream is he carried in 1.5 min? (b) What is his velocity relative to an observer on shore?

66. ■■ A boat that travels at a speed of 6.75 m/s in still water is to go directly across a river and back (▼Fig. 3.31). The current flows at 0.50 m/s. (a) At what angle(s) must the boat be steered? (b) How long does it take to make the round-trip? (Assume that the boat's speed is constant at all times, and neglect turnaround time.)

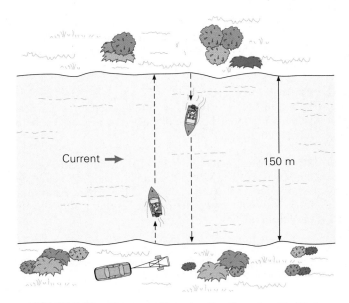

Current →   150 m

▲ **FIGURE 3.31** **Over and back** See Exercise 66. (Not drawn to scale.)

67. **IE** ■■ It is raining, and there is no wind. When you are sitting in a stationary car, the rain falls straight down relative to the car and the ground. But when you're driving, the rain appears to hit the windshield at an angle. (a) As the velocity of the car increases, this angle (1) also increases, (2) remains the same, or (3) decreases. Why? (b) If the raindrops fall straight down at a speed of 10 m/s, but appear to make an angle of 25° to the vertical, what is the speed of the car?

68. ■■ Swimming at 0.15 m/s relative to still water, a swimmer heads directly across a 100-m-wide river. He arrives 50 m downstream from a point directly across the river from his starting point. (a) What is the speed of the current in the river? (b) In what direction should the swimmer head so as to arrive at a point directly opposite his starting point?

69. ■■■ An airplane is flying at 150 mi/h (its speed in still air) in a direction such that with a wind of 60.0 mi/h blowing from east to west, the airplane travels in a straight line southward. (a) What must be the plane's heading direction for it to fly directly south? (b) If the plane has to go 200 mi in the southward direction, how long does it take?

3.4 Projectile Motion*

70. If air resistance is neglected, the motion of an object projected at an angle consists of a uniform downward acceleration combined with (a) an equal horizontal acceleration, (b) a uniform horizontal velocity, (c) a constant upward velocity, or (d) an acceleration that is always perpendicular to the path of motion.

71. CQ A golf ball is hit on a level fairway. When it lands, its velocity vector has rotated through an angle of 90°. What was the launch angle of the golf ball? [*Hint*: See Fig. 3.17.]

72. CQ Figure 3.16b shows a multiflash photograph of one ball dropping from rest and, at the same time, another ball projected horizontally at the same height. The two balls hit the ground at the same time. Why? Explain.

73. CQ In ▼Fig. 3.32, a spring-loaded "cannon" on a wheeled car fires a metal ball vertically. The car is given a push and set in motion horizontally with constant velocity. A pin is pulled with a string to launch the ball, which travels upward and then falls back into the moving cannon every time. Why does the ball always fall back into the cannon? Explain.

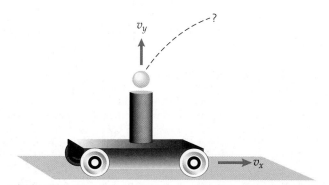

▲ **FIGURE 3.32   A ballistics car**   See Exercises 73 and 83.

74. ■ A ball with a horizontal speed of 1.5 m/s rolls off a bench 2.0 m high. (a) How long will it take the ball to reach the floor? (b) How far from a point on the floor directly below the edge of the bench will the ball land?

75. ■ An electron is ejected horizontally at a speed of 1.5×10^6 m/s from the electron gun of a computer monitor. If the viewing screen is 35 cm away from the end of

*Assume angles to be exact for significant figure purposes.

the gun, how far will the electron travel in the vertical direction before hitting the screen? Based on your answer, do you think designers need to worry about this gravitational effect?

76. ■ A ball is thrown horizontally, with a speed of 15 m/s, from the top of a 6.0-m tall hill. How far from the point on the ground directly below the launch point does the ball strike the ground?

77. ■ If Exercise 76 were to take place on the surface of the Moon, where the acceleration due to gravity is only 1.67 m/s², what would be the answer?

78. ■ A ball rolls horizontally with a speed of 7.6 m/s off the edge of a tall platform. If the ball lands 8.7 m from the point on the ground directly below the edge of the platform, what is the height of the platform?

79. ■ A golf ball is hit at a speed of 30 m/s at an angle of 30° above the horizontal. What are the horizontal and vertical components of the ball's velocity?

80. ■■ A pitcher throws a fastball horizontally at a speed of 140 km/h toward home plate, 18.4 m away. (a) If the batter's combined reaction and swing times total 0.350 s, how long can the batter watch the ball after it has left the pitcher's hand before swinging? (b) In traveling to the plate, how far does the ball drop from its original horizontal line?

81. IE ■■ Ball A rolls at a constant speed of 0.25 m/s on a table 0.95 m above the floor, and ball B rolls on the floor directly under the first ball and with the same speed and direction. (a) When ball A rolls off the table and hits the floor, (1) ball B is ahead of ball A, (2) ball B collides with ball A, or (3) ball A is ahead of ball B. Why? (b) When ball A hits the floor, how far from the point directly below the edge of the table will both balls be?

82. ■■ A package of supplies is to be dropped from an airplane so that it hits the ground at a designated spot near some campers. The airplane, moving horizontally at a constant velocity of 140 km/h, approaches the spot at an altitude of 0.500 km above level ground. Having the designated point in sight, the pilot prepares to drop the package. (a) What should the angle be between the horizontal and the pilot's line of sight when the package is released? (b) What is the location of the plane when the package hits the ground?

83. ■■ A wheeled car with a spring-loaded cannon fires a metal ball vertically (◄Fig. 3.32). If the vertical initial speed of the ball is 5.0 m/s as the cannon moves horizontally at a speed of 0.75 m/s, (a) how far from the launch point does the ball fall back into the cannon, and (b) what would happen if the cannon were accelerating?

84. ■■ A soccer player kicks a stationary ball, giving it a speed of 20.0 m/s at an angle of 15.0° to the horizontal. (a) What is the maximum height reached by the ball?

(b) What is the ball's range? (c) How could the range be increased?

85. ■■ A rifle fires a bullet at a speed of 250 m/s at an angle of 37° above the horizontal. (a) What height does the bullet reach? (b) How long is the ball in the air? (c) What is the ball's horizontal range?

86. ■■ A golf ball, hit off a tee on level ground, lands 62 m away 3.0 s later. What was the initial velocity of the golf ball?

87. ■■ An arrow has an initial launch speed of 18 m/s. If it must strike a target 31 m away at the same elevation, what should be the projection angle?

88. ■■ An astronaut on the Moon fires a projectile from a launcher on a level surface so as to get the maximum range. If the launcher gives the projectile a muzzle velocity of 25 m/s, what is the range of the projectile? [*Hint*: The acceleration due to gravity on the Moon is only one sixth of that on the Earth.]

89. ■■ A stone thrown off a bridge 20 m above a river has an initial velocity of 12 m/s at an angle of 45° above the horizontal (▼ Fig. 3.33). (a) What is the range of the stone? (b) At what velocity does the stone strike the water?

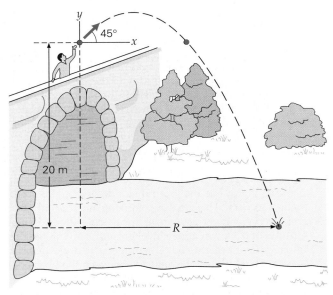

▲ FIGURE 3.33 **A view from the bridge** See Exercise 89.

90. ■■ William Tell is said to have shot an apple off his son's head with an arrow. If the arrow was shot with an initial speed of 55 m/s and the boy was 15 m away, at what launch angle did Bill aim the arrow? (Assume that the arrow and apple are initially at the same height above the ground.)

91. ■■■ This time, William Tell is shooting at an apple that hangs on a tree (▶ Fig. 3.34). The apple is a horizontal dis-

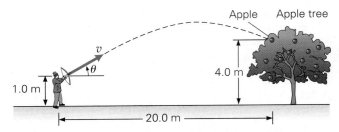

▲ FIGURE 3.34 **Hit the apple** See Exercise 91. (Not drawn to scale.)

tance of 20.0 m away and at a height of 4.00 m above the ground. If the arrow is released from a height of 1.00 m above the ground and hits the apple 0.500 s later, what is the arrow's initial velocity?

92. ■■■ A ditch 2.5 m wide crosses a trail bike path (▼ Fig 3.35). An upward incline of 15° has been built up on the approach so that the top of the incline is level with the top of the ditch. What is the minimum speed a trail bike must be moving to clear the ditch? (Add 1.4 m

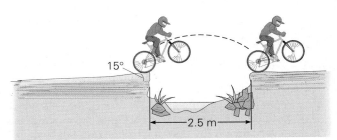

▲ FIGURE 3.35 **Clear the ditch** See Exercise 92. (Not drawn to scale.)

to the range for the back of the bike to clear the ditch safely.)

93. ■■■ A quarterback passes a football—at a velocity of 50 ft/s at an angle of 40° to the horizontal—toward an intended receiver 30 yd downfield. The pass is released 5.0 ft above the ground. Assume that the receiver is stationary and that he will catch the ball if it comes to him. Will the pass be completed? If not, will the throw be long or short?

94. ■■■ A 2.05-m-tall basketball player takes a shot when he is 6.02 m from the basket (at the three-point line). If the launch angle is 25° and the ball was launched at the level of the player's head, what must be the release speed of the ball for the player to make the shot? The basket is 3.05 m above the floor.

95. ■■■ The hole on a level, elevated golf green is a horizontal distance of 150 m from the tee and at an elevation of 12.0 m above the tee. A golfer hits a ball at an angle 10.0° greater than that of the hole above the tee and makes a hole in one! (a) Sketch the situation. (b) What

was the initial speed of the ball? (c) Suppose the next golfer hits her ball toward the hole at the same speed, but at an angle of 10.5° greater than that of the hole above the tee. Does the ball go in the hole, or is the shot too long or too short?

Additional Exercises

96. The apparatus for a popular lecture demonstration is shown in ▼Fig. 3.36. A gun is aimed directly at a can, which is simultaneously released when the gun is fired. This gun won't miss as long as the initial speed of the bullet is sufficient to reach the falling target before the target hits the floor. Verify this statement, using the figure. [*Hint*: Note that $y_o = x \tan \theta$.]

97. What is the resultant velocity vector of the following three velocity vectors? $\mathbf{v}_1 = (3.0 \text{ m/s}) \hat{\mathbf{x}} + (4.0 \text{ m/s}) \hat{\mathbf{y}}$; $\mathbf{v}_2 = (-4.0 \text{ m/s}) \hat{\mathbf{x}} + (5.0 \text{ m/s}) \hat{\mathbf{y}}$; $\mathbf{v}_3 = (5.0 \text{ m/s}) \hat{\mathbf{x}} - (7.0 \text{ m/s}) \hat{\mathbf{y}}$. Express your result in both component and magnitude–angle forms.

98. A motorboat travels at a speed of 40.0 km/h in a straight path on a still lake. Suddenly, a strong, steady wind pushes the boat at a speed of 15.0 km/h perpendicularly to its straight-line path for 5.00 s. Relative to its position just when the wind started to blow, where is the boat located at the end of this time?

99. You are traveling and start out 200 mi directly south of your destination. If you first go 150 mi at 25° west of north to visit your friends, what must be your second displacement in order for you to get to your destination?

100. If the flow rate of the current in a straight river is greater than the speed of a boat in the water, the boat cannot make a trip *directly across* the river. Prove this statement.

101. A student throws a softball horizontally from a dorm window 15.0 m above the ground. Another student standing 10.0 m away from the dorm catches the ball at a height of 1.50 m above the ground. What is the initial velocity of the ball?

102. In two attempts, a javelin is thrown at angles of 35° and 60°, respectively, to the horizontal from the same height and at the same speed in each case. For which throw does the javelin go farther, and how many times farther? (Assume that the landing place is at the same height as the launching place.)

103. If the maximum height reached by a projectile launched on level ground is equal to half the projectile's range, what is the launch angle?

104. In a movie, a monster climbs to the top of a building 30 m above the ground and hurls a boulder downward with a speed of 25 m/s at an angle of 45° below the horizontal. How far from the building does the boulder land?

105. The shells fired from an artillery piece have a muzzle speed of 150 m/s, and the target is at a horizontal distance of 2.00 km. (a) At what angle relative to the horizontal should the gun be aimed? (b) Could the gun hit a target 3.00 km away?

106. A hockey player hits a "slap shot" in practice at a horizontal distance of 15 m from the net (with no goalie present). The net is 1.2 m high, and the puck is initially hit at an angle of 5.0° above the horizontal with a speed of 50 m/s. Does the puck make it into the net?

107. A ball is thrown horizontally from the top of a building at a height of 32.5 m above the ground and hits the level ground 56.0 m from the base of the building. (a) What is the initial speed of the ball? (b) What is the velocity of the ball just before it hits the ground?

108. A field goal is attempted when the football is at the center of the field, 40 yd from the goalposts. If the kicker gives the ball a velocity of 70 ft/s toward the goalposts at an angle of 45° to the horizontal, will the kick be good? (The crossbar of the goalposts is 10 ft above the ground, and the ball must be higher than the crossbar when it reaches the goalposts for the field goal to be good.)

▶ **FIGURE 3.36 A sure shot** See Exercise 96. (Not drawn to scale.)

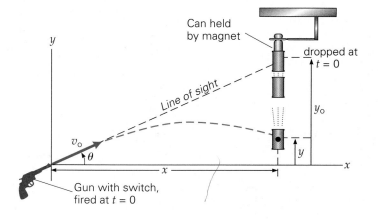

109. (a) An object moving at a speed of 10 m/s in a straight line has a velocity with an x-component of $+6.0$ m/s. In what directions could the the object be moving? (b) What is the value(s) of the y-component of the velocity?

110. At a track-and-field meet, the best long jump is measured as 8.20 m. The jumper took off at an angle of 37° to the horizontal. (a) What was the jumper's initial speed? (b) If there were another meet on the Moon and the same jumper could attain only half of the initial speed he had on the Earth, what would be the maximum jump there? (Air resistance does not have to be neglected in part (b). Why?)

111. An airplane with a speed of 150 km/h heads directly north while a west wind blows at a constant speed of 30 km/h. How far does the plane travel during the 4-hour flight?

Force and Motion

Y ou don't have to understand any physics to know what's needed to get the car in the picture (or anything else) moving: a push or a pull. If the frustrated motorist (or the tow truck that he will soon call) can apply enough *force*, the car will move.

But what's keeping the car stuck in the snow? A car's engine can generate plenty of force—so why doesn't the motorist just put the car into reverse and back out? For a car to move, another force is needed besides that exerted by the engine: *friction*. Here, the problem is most likely that there is not enough friction between the tires and the snow.

In Chapters 2 and 3, we learned how to analyze motion in terms of kinematics. Now we need to know more about the *dynamics* of motion—that is, what *causes* motion and changes in motion? This inquiry leads us to the concept of force and inertia.

The study of force and motion occupied many early scientists. It was the English scientist Isaac Newton (1642–1727) (▶ Fig. 4.1) who summarized the various relationships and principles of those early scientists into three statements, or laws, which, not surprisingly, are known as *Newton's laws of motion*. These laws sum up the concepts of dynamics. In this chapter, you'll learn what Newton had to say about force and motion.

4.1 The Concepts of Force and Net Force

OBJECTIVES: **To (a) relate force and motion and (b) explain what is meant by a net or unbalanced force.**

Let's first take a closer look at the meaning of force. It is easy to give examples of forces, but how would you generally define this concept? An operational definition of force is based on observed effects. That is, a force is described in terms of what it does. From your own experience, you

▲ FIGURE 4.1 Isaac Newton
Newton (1642–1727), one of the
greatest scientific minds of all time,
made fundamental contributions
to mathematics, astronomy, and
several branches of physics,
including optics and mechanics.
He formulated the laws of motion
and universal gravitation (Chapter 7)
and was one of the inventors of
calculus. He did some of his most
profound work when he was in his
midtwenties.

Note: In the notation ΣF_i, the Greek
letter sigma means the "sum of" the
individual forces, as indicated by the i
subscript: $\Sigma F_i = F_1 + F_2 + F_3 + F_4$, that
is, a vector sum. The i subscripts are
sometimes omitted as being
understood, and we write ΣF.

know that *forces can produce changes in motion.* A force can set a stationary object into
motion. It can also speed up or slow down a moving object or change the direction of
its motion. In other words, a force can produce a change in velocity (speed and/or
direction)—that is, an acceleration. Therefore, an observed change in motion, including motion starting from rest, is evidence of a force. This concept leads to a common
definition of **force**:

> A force is something that is capable of changing an object's state of motion (its
> velocity).

The word "capable" is very significant here. It takes into account the fact that a
force may be acting on an object, but its capability to produce a change in motion
may be balanced, or canceled, by one or more other forces. The net effect is then
zero. Thus, a force may *not necessarily* produce a change in motion. However, it follows that if a force acts *alone*, the object on which it acts *will* accelerate.

Since a force can produce an acceleration—a vector quantity—force must itself be a vector quantity, with both magnitude and direction. When several forces
act on an object, you will often be interested in their combined effect—the net
force. The **net force**, $\mathbf{F}_{net}$, is the vector sum $\Sigma \mathbf{F}_i$, or resultant, of all the forces acting on an object or system. (See the note in the margin.) Consider the opposite
forces illustrated in ▼ Fig. 4.2a. The net force is zero when forces of equal magnitude act in opposite directions (Fig. 4.2b). Such forces are said to be *balanced forces.*
A nonzero net force is referred to as an unbalanced force (Fig. 4.2c). In this case,
the situation can be analyzed as though only one force equal to the net force were acting.
An unbalanced, or nonzero, net force produces an acceleration. In some instances, an applied unbalanced force may also deform an object, that is, change
its size and/or shape (as we shall see in Chapter 9). A deformation involves a
change in motion for some part of an object; hence, there is an acceleration.

Forces are sometimes divided into two types or classes. The more familiar of
these classes is *contact forces.* Such forces arise because of physical contact between
objects. For example, when you push on a door to open it or throw or kick a ball,
you exert a contact force on the door or ball.

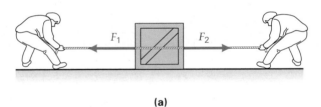

(a)

► FIGURE 4.2 Net force
(a) Opposite forces are applied to a
crate. **(b)** If the forces are of equal
magnitude, the vector resultant, or
the net force acting on the crate in
the x-direction, is zero. The forces
acting on the crate are said to be
balanced. **(c)** If the forces are
unequal in magnitude, the resultant
is not zero. A nonzero net force, or
an unbalanced force, then acts on
the crate, producing an acceleration
(for example, setting the crate in
motion if it was initially at rest).

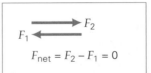

(b) Zero net force (balanced forces)

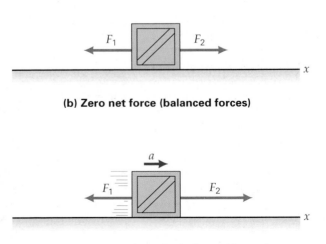

(c) Nonzero net force (unbalanced forces)

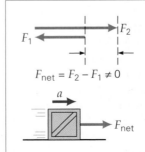

The other class of forces is called *action-at-a-distance forces*. Examples of these forces include gravity, the electrical force between two charges, and the magnetic force between two magnets. The Moon is attracted to the Earth and maintained in orbit by a gravitational force, but there seems to be nothing physically transmitting that force. In Chapter 30, you will learn the modern view of how such action-at-a-distance forces are thought to be transmitted.

Now, with a better understanding of the concept of force, let's see how force and motion are related through Newton's laws.

4.2 Inertia and Newton's First Law of Motion

OBJECTIVES: To (a) state and explain Newton's first law of motion, and
(b) describe inertia and its relationship to mass.

The groundwork for Newton's first law of motion was laid by Galileo. In his experimental investigations, Galileo dropped objects to observe motion under the influence of gravity. (See the related Insight in Chapter 2.) However, the relatively large acceleration due to gravity causes dropped objects to move quite fast and quite far in a short time. From the kinematic equations in Chapter 2, you can see that 3.0 s after being dropped (if we neglect air resistance), an object in free fall has a speed of about 29 m/s (64 mi/h) and has fallen a distance of 44 m (about 48 yd, or almost half the length of a football field). Thus, experimental measurements of free-fall distance versus time were particularly difficult to make with the instrumentation available in Galileo's time.

To slow things down so that he could study motion, Galileo used balls rolling on inclined planes. He allowed a ball to roll down one inclined plane and then up another with a different degree of incline (▼Fig. 4.3). Galileo noted that the ball rolled to approximately the same height in every case, but it rolled farther in the horizontal direction when the angle of incline was smaller. When allowed to roll onto a horizontal surface, the ball traveled a considerable distance and went even farther when the surface was made smoother. Galileo wondered how far the ball would travel if the horizontal surface could be made perfectly smooth (frictionless). Although this situation was impossible to attain experimentally, Galileo reasoned that in this ideal case with an infinitely long surface, the ball would continue to travel (actually slide) indefinitely with straight-line, uniform motion, since there would be nothing (no net force) to cause its motion to change.

According to Aristotle's theory of motion, which had been accepted for about 1500 years prior to Galileo's time, the normal state of a body was to be at rest (with the exception of celestial bodies, which were thought to be naturally in motion). Aristotle probably observed that objects moving on a surface tend to slow down and come to rest, so this conclusion would have seemed logical to him. However, from his experiments, Galileo concluded that bodies in motion exhibit the behavior of maintaining that motion and that if an object is initially at rest, it will remain so, unless something causes it to move.

Galileo called this tendency of an object to maintain its initial state of motion **inertia.** That is,

Definition of: Inertia

Inertia is the natural tendency of an object to maintain a state of rest or to remain in uniform motion in a straight line (constant velocity).

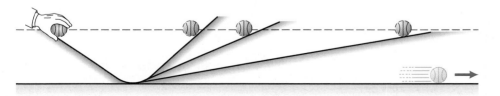

◀ **FIGURE 4.3 Galileo's experiment** A ball rolls farther along the upward incline as the angle of incline is decreased. On a smooth, horizontal surface, the ball rolls a greater distance before coming to rest. How far would the ball travel on an ideal, perfectly smooth surface?

For example, if you've ever tried to stop a slowly rolling automobile by pushing on it, you felt its resistance to a change in motion, to slowing down. Physicists describe the property of inertia in terms of observed behavior, as they do for all physical phenomena. A comparative example of inertia is illustrated in ◄Fig. 4.4. If the two punching bags have the same density (mass per unit volume; see Chapter 1), the larger one has more mass and therefore more inertia, as you would quickly notice when you try to punch both bags.

Newton related the concept of inertia to mass. Originally, he called mass a quantity of matter, but he later redefined it as follows:

Relationship of mass and inertia

| Mass is a quantitative measure of inertia. |

That is, a massive object has more inertia, or more resistance to a change in motion, than does a less massive object. For example, a car has more inertia than a bicycle.

Newton's first law—the law of inertia

Newton's first law of motion, sometimes called the *law of inertia*, summarizes these observations:

Note: Inertia is *not* a force.

| In the absence of an unbalanced applied force ($F_{net} = 0$), a body at rest remains at rest, and a body already in motion remains in motion with a constant velocity (constant speed and direction). |

That is, if the net force acting on an object is zero, then its acceleration is zero.

4.3 Newton's Second Law of Motion

OBJECTIVES: To (a) state and explain Newton's second law of motion, (b) apply it to physical situations, and (c) distinguish between weight and mass.

A change in motion, or an acceleration (i.e., a change in speed and/or direction), is evidence of a net force. All experiments indicate that an acceleration is directly proportional to, and in the direction of, the applied net force; that is,

$$\mathbf{a} \propto \mathbf{F}_{net}$$

where the boldface symbols indicate vector quantities. For example, if you hit an identical second ball twice as hard as the first one (i.e., you applied twice as much force), you would expect the acceleration of the second ball to be twice as great as that of the first ball (but still in the direction of the force).

However, as Newton recognized, the inertia or mass of the object also plays a role. For a given net force, the more massive the object, the *less* its acceleration will be. That is, the magnitude of the acceleration and mass (m) of an object are inversely proportional:

$$a \propto \frac{1}{m}$$

For example, if you hit two balls of different masses with the same force, the less massive ball would experience a greater acceleration.

Then combining these relationships, we have

$$\mathbf{a} \propto \frac{\mathbf{F}_{net}}{m}$$

or, in words,

| The acceleration of an object is directly proportional to the net force acting on it and inversely proportional to its mass. The direction of the acceleration is in the direction of the applied net force. |

▲ **FIGURE 4.4 A difference in inertia** The larger punching bag has more mass and hence more inertia, or resistance to a change in motion.

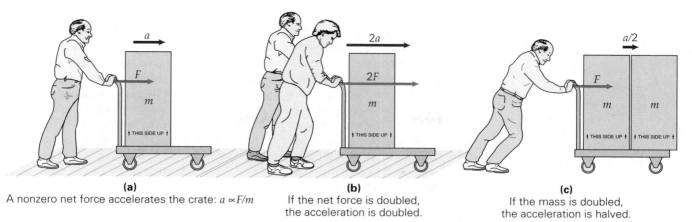

(a)
A nonzero net force accelerates the crate: $a \propto F/m$

(b)
If the net force is doubled, the acceleration is doubled.

(c)
If the mass is doubled, the acceleration is halved.

▲ **FIGURE 4.5 Newton's second law** The relationships among force, acceleration, and mass shown here are expressed by Newton's second law of motion (assuming no friction).

▲Figure 4.5 presents some illustrations of this principle.

With $\mathbf{F}_{net} \propto m\mathbf{a}$, **Newton's second law of motion** is commonly expressed in equation form as

Newton's second law—force and acceleration

$$\mathbf{F}_{net} = m\mathbf{a} \qquad \textit{Newton's second law} \qquad (4.1)$$

SI unit of force: newton (N) or kilogram-meter per second squared $(kg \cdot m/s^2)$

where $\mathbf{F}_{net} = \Sigma\mathbf{F}_i$. Equation 4.1 defines the SI unit of force, which is appropriately called the newton (N).

Equation 4.1 also shows that (by unit analysis) a newton in base units is defined as $1\,N = 1\,kg \cdot m/s^2$. That is, a net force of 1 N gives a mass of 1 kg an acceleration of $1\,m/s^2$ (▶Fig. 4.6). The British-system unit of force is the pound (lb). One pound is equivalent to about 4.5 N (actually, 4.448 N). An average apple weighs about 1 N.

Thus, if the net force acting on an object is zero, the object's acceleration is zero, and it remains at rest or in uniform motion, which is consistent with the first law. For a nonzero net force (an unbalanced force), the resulting acceleration is in the same direction as the force.*

Weight

Equation 4.1 can be used to relate mass and weight. Recall from Chapter 1 that weight is the gravitational force of attraction that a celestial body exerts on an object. For us, this force is the gravitational attraction of the Earth. Its effects are easily demonstrated: When you drop an object, it falls (accelerates) toward the Earth. Since there is only one force acting on the object, its **weight (w)** is the net force $\mathbf{F}_{net}$, and the acceleration due to gravity (**g**) can be substituted for **a** in Eq. 4.1. We can therefore write, in terms of magnitude,

$$w = mg \qquad (4.2)$$

$$(F_{net} = ma)$$

The magnitude of the weight of 1.0 kg of mass is $w = mg = (1.0\,kg)(9.8\,m/s^2) = 9.8\,N$.

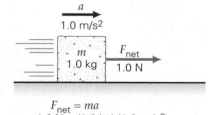

$F_{net} = ma$
$1.0\,N = (1.0\,kg)\,(1.0\,m/s^2)$

▲ **FIGURE 4.6 The newton (N)**
A net force of 1.0 N acting on a mass of 1.0 kg produces an acceleration of $1.0\,m/s^2$ (on a frictionless surface).

---

*It may appear that Newton's first law is a special case of Newton's second law, but this is not so. The first law *defines* what is called an *inertial reference system* (as we shall see in Chapter 26): a system in which there is no net force, that is not accelerating, or in which an isolated object is stationary or moves with a constant velocity. If Newton's first law holds, then the second law in the form $\mathbf{F}_{net} = m\mathbf{a}$ applies to the system.

Thus, 1.0 kg of mass has a weight of approximately 9.8 N, or 2.2 lb near the Earth's surface. But, although weight and mass are simply related through Eq. 4.2, keep in mind that *mass is the fundamental property*. Mass doesn't depend on the value of g, but weight does. As pointed out previously, the acceleration due to gravity on the Moon is about one sixth that on the Earth. The weight of an object on the Moon would thus be one sixth of its weight on the Earth, but its mass, which reflects the quantity of matter it contains and its inertia, would be the same in both places.

Newton's second law also explains why all objects in free fall have the same acceleration. Consider, for example, two falling objects, one with twice the mass of the other. The object with twice as much mass would have twice as much weight, or two times as much gravitational force acting on it. But the more massive object also has twice the inertia, so twice as much force is needed to give it the same acceleration. Expressing this relationship mathematically, for the smaller mass (m), we can write $F_{net}/m = mg/m = g$, and for the larger mass ($2m$), we have the same acceleration: $F_{net}/m = 2mg/2m = g$ (▶Fig. 4.7). Some other effects of g, which you may have experienced, are discussed in the following Insight.

INSIGHT

g's of Force and Effects on the Human Body

The value of g at the Earth's surface is referred to as the standard acceleration and is sometimes used as a nonstandard unit. For example, when a spacecraft lifts off, astronauts are said to experience an acceleration of "several g's." This expression means that the astronauts' acceleration is several times the standard acceleration g. Since $g = w/m$, we can also think of g as the (weight) *force per unit mass*. Thus, the term **g's of force** is sometimes used for the force corresponding to multiples of the standard acceleration.

To help you better understand this nonstandard unit of force, let's look at some examples. During the takeoff of a jet airliner, you experience an average horizontal force of about 0.20 g. This means that as the plane accelerates down the runway, the seat back exerts on you a horizontal force of about one fifth of your weight (to accelerate you along with the plane), but you experience a feeling of being pushed back into the seat. On takeoff at an angle of 30°, the force increases to about 0.70 g, with a component of gravity helping to "push" you back into the seat.

In situations where a person is subjected to several g's vertically, blood can begin to pool in the lower extremities, which may cause blood vessels to distend or capillaries to rupture. Under such conditions, the heart has a difficult time pumping blood throughout the body. At a force of about four g's, the pooling of blood in the lower body deprives the head of sufficient oxygen. Lack of blood circulation to the eyes can cause temporary blindness, and if the brain is deprived of oxygen, a person becomes disoriented and eventually "blacks out" or loses consciousness. The average person can withstand several g's of force only for a short period of time. You may have experienced the feeling of these effects on a roller coaster when it is climbing out of a big dip.

The maximum force on astronauts in a space shuttle upon blastoff is about three g's. But jet-fighter pilots are subjected to as much as nine g's when pulling out of a downward dive. However, these folks wear "g-suits," which are specially designed to prevent blood pooling. The common g-suit is inflated by compressed air and applies pressure to the pilot's lower body to prevent the blood from accumulating there. (Work is being done on the development of a hydrostatic g-suit that contains liquid, which is less restrictive than air. When the number of g's increases, the liquid, like the blood in the body, flows into the lower part of the suit and applies pressure to the legs.) Also, although astronauts are in a reclined position on blastoff, the seating in a jet fighter is slightly horizontal. When a pilot is positioned horizontally, his or her blood flows from front to back and does not pool in the lower extremities. This configuration helps the heart keep pumping oxygenated blood throughout the body. (See the Insight on blood pressure in Chapter 9.)

Meanwhile, back on Earth, where there is only 1 g, a partial "g-suit" of sorts is being used to prevent blood clots in patients—most of whom are over the age of 65—who have undergone hip replacement surgery. It is estimated that each year 400 to 800 people die in the first three months after such surgery. The patients die primarily because of blood clots forming in a leg, breaking off into the blood stream, and finally lodging in the lungs—giving rise to a condition called *pulmonary embolism*. In other cases, a blood clot in the leg may slow the flow of blood to the heart. These complications arise more often after hip replacement surgery than after almost any other surgery and occur after the patient has left the hospital.

Studies have shown that pneumatic (operated by air) compression of the legs during the hospital stay reduces these risks. A plastic, thigh-high leg cuff inflates every few minutes, forcing the blood from the ankle to the thigh. This mechanical massaging is thought to prevent blood from pooling in the veins and clotting. By using both this technique and anticlotting drug therapy, it is hoped that many of the postoperative deaths can be prevented.

Newton's second law allows us to analyze dynamic situations. In using this equation, you should keep in mind that F_{net} is the *magnitude of the net force* and *m* is the *total mass of the system*. The boundaries defining a system may be real or imaginary. For example, a system might consist of all the gas molecules in a particular sealed vessel. But you might also define a system to be all the gas molecules in an arbitrary cubic meter of air. In studying dynamics, we often have occasion to work with systems made up of one or more discrete masses—the Earth and Moon, for instance, or a series of blocks on a tabletop, or a tractor and wagon, as in Example 4.1.

Note: In $F_{net} = ma$, *m* is the total mass of the system.

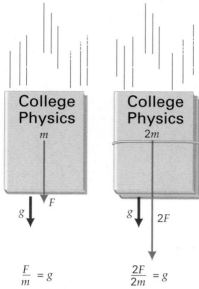

$$\frac{F}{m} = g \qquad \frac{2F}{2m} = g$$

▲ **FIGURE 4.7   Newton's second law and free fall**   In free fall, all objects fall with the same constant acceleration *g*. An object with twice the mass of another has twice as much gravitational force acting on it. But with twice the mass, the object also has twice as much inertia, so twice as much force is needed to give it the same acceleration.

Example 4.1 ■ Newton's Second Law: Finding Acceleration

A tractor pulls a loaded wagon on a level road with a constant force of 440 N (▼Fig. 4.8). If the total mass of the wagon and its contents is 275 kg, what is the wagon's acceleration? (Ignore any frictional forces.)

Thinking It Through.   This problem is a direct application of Newton's second law. Note that the total mass is given; we treat the two separate masses (wagon and contents) as one and look at the whole system.

Solution.   Listing the data, we have

Given: $F = 440$ N   *Find:*   *a* (acceleration)
          $m = 275$ kg

In this case, *F* is the net force, and the acceleration is given by Eq. 4.1, $F_{net} = ma$. Solving for the magnitude of *a*,

$$a = \frac{F_{net}}{m} = \frac{440 \text{ N}}{275 \text{ kg}} = 1.60 \text{ m/s}^2$$

and the direction of *a* is that in which the tractor is pulling.

Note that *m* is the *total* mass of the wagon and its contents. If the masses of the wagon and its contents had been given separately—say, $m_1 = 75$ kg and $m_2 = 200$ kg, respectively—they would have been added together in Newton's law: $F = (m_1 + m_2)a$. Also, in reality, there would be an opposing force of friction. Suppose there were an effective frictional force of $f = 140$ N. In this case, the net force would be the vector sum of the force exerted by the tractor and the frictional force, and the acceleration would be (using directional signs)

$$a = \frac{F_{net}}{m} = \frac{F - f}{m_1 + m_2} = \frac{440 \text{ N} - 140 \text{ N}}{275 \text{ kg}} = 1.09 \text{ m/s}^2$$

Again, the direction of *a* would be the direction in which the tractor is pulling.

With a constant net force, the acceleration is also constant, so the kinematic equations of Chapter 2 can be applied. Suppose the wagon started from rest ($v_0 = 0$). Could you find how far it traveled in 4.00 s? Using the appropriate kinematic equation (Eq. 2.11, with $x_0 = 0$) for the case with friction, we have

$$x = v_0 t + \tfrac{1}{2} a t^2 = 0 + \tfrac{1}{2} (1.09 \text{ m/s}^2)(4.00 \text{ s})^2 = 8.72 \text{ m}$$

◄ **FIGURE 4.8   Force and acceleration**   See Example 4.1.

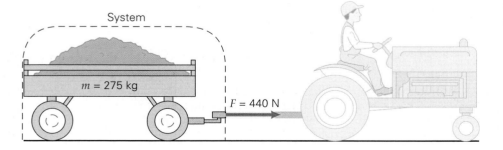

Follow-up Exercise. Suppose the applied force on the wagon is 550 N. With the same frictional force, what would be the wagon's velocity 4.0 s after starting from rest? *(Answers to all Follow-up Exercises are at the back of the text.)*

Example 4.2 ■ Newton's Second Law: Finding Mass

A student weighs 588 N. What is her mass?

Thinking It Through. Newton's second law allows us to determine an object's mass if we know the object's weight (force), since g is known.

Solution.

Given: $w = 588$ N   *Find:* m (mass)

Recall that weight is a (gravitational) force and that Newton's second law, $F_{net} = ma$, can be written in the form $w = mg$ (Eq. 4.2), where g is the acceleration due to gravity (9.80 m/s²). Rearranging the equation, we have

$$m = \frac{w}{g} = \frac{588 \text{ N}}{9.80 \text{ m/s}^2} = 60.0 \text{ kg}$$

On the surface of the Earth, this is equivalent to 60.0 kg (2.2 lb/kg) = 132 lb. In countries that use the metric system, the kilogram unit of mass, rather than a force unit, is used to express "weight." It would be said that this student weighs 60.0 "kilos."

Follow-up Exercise. (a) A person in Europe is a bit overweight and would like to lose 5.0 "kilos." What would be the equivalent loss in pounds? (b) What is your "weight" in kilos? *(Answers to all Follow-up Exercises are at the back of the text.)*

A dynamic system may consist of more than one object. In applications of Newton's second law, it is often advantageous, and sometimes necessary, to isolate a given object within a system. This isolation is possible because the motion of any part of a system is also described by Newton's second law, as Example 4.3 shows.

Example 4.3 ■ Newton's Second Law: All or Part of the System?

Two blocks with masses $m_1 = 2.5$ kg and $m_2 = 3.5$ kg rest on a frictionless surface and are connected by a light string (▶Fig. 4.9).* A horizontal force (F) of 12.0 N is applied to m_1, as shown in the figure. (a) What is the magnitude of the acceleration of the masses (that is, of the total system)? (b) What is the magnitude of the force (T) in the string? [When a rope or string is stretched taut, it is said to be under tension, which is represented by the magnitude of the force acting at any point. For a very light string, the force at the right end of the string has the same magnitude (T) as the force at the left end.]

Thinking It Through. It is important to remember that Newton's second law may be applied to a total system or any part of it (a subsystem, so to speak). This capability allows for the analysis of a particular component of a system, if desired. Identification of the acting forces is critical, as this Example shows. We then apply $F_{net} = ma$ to each subsystem or component.

Solution. Carefully listing the data and what we want to find, we have the following information:

Given: $m_1 = 2.5$ kg   *Find:* (a) a (acceleration)
         $m_2 = 3.5$ kg   (b) T (tension, a force)
         $F = 12.0$ N

Given an applied force, such as a light string, the acceleration of the masses can be found from Newton's second law. In using Newton's second law, it is important to keep in mind that it applies to the total system *or to any part of it*—that is, to the total mass

*When an object is described as being "light," its mass can be ignored in analyzing the situation given in the problem. That is, its mass is negligible relative to the other masses.

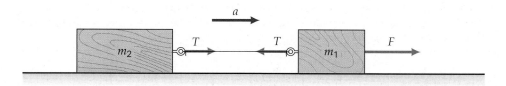

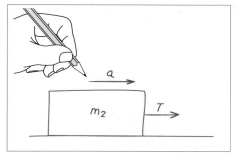

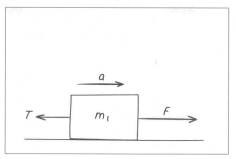

Isolating the masses

◀ **FIGURE 4.9 An accelerated system** See Example 4.3.

$(m_1 + m_2)$, to m_1 individually, or to m_2 individually. However, *we must be sure to identify correctly the appropriate force or forces in each case.* The net force acting on the combined masses, for example, is not the same as the net force acting on m_2 considered separately, as will be seen.

(a) First, taking the system as a whole (i.e., considering both m_1 and m_2), we see that the net force acting on this system is F. Note that in considering the total system, we are concerned only about the net external force acting on it. The *internal* equal and opposite T forces are not a consideration in this case. Representing the total mass as M, we can thus write

$$a = \frac{F_{net}}{M} = \frac{F}{m_1 + m_2} = \frac{12.0 \text{ N}}{2.5 \text{ kg} + 3.5 \text{ kg}} = 2.0 \text{ m/s}^2$$

The acceleration is in the direction of the applied force, as the figure indicates. Note that M is the *total* mass of the system, or all the mass that is accelerated. (The mass of the light string is small enough to be ignored.)

(b) Under tension, a force is exerted on an object by flexible strings (or ropes or wires) and is directed along the string. Note in the figure that we are assuming the tension to be transmitted *undiminished* through the string. That is, the tension is the same everywhere in the string. Thus, the magnitude of **T** acting on m_2 is the same as that acting on m_1. This is actually true only if the string has zero mass. Only such idealized *light* (i.e., of negligible mass) strings or ropes will be considered in this book.

So, there is a force of magnitude T on each of the masses, because of tension in the connecting string. To find the value of T, we must consider a *part* of the system that is affected by this force.

Each block may be considered as a separate system to which Newton's second law applies. In these subsystems, the tension comes into play explicitly. Looking at the sketch of the isolated m_2 in Fig. 4.9, we see that the only force acting to accelerate this mass is T. From the values of m_2 and a, the magnitude of this force is given directly by

$$F_{net} = T = m_2 a = (3.5 \text{ kg})(2.0 \text{ m/s}^2) = 7.0 \text{ N}$$

An isolated sketch of m_1 is also shown in Fig. 4.9, and Newton's second law can equally well be applied to this block to find T. We must add the forces vectorially to get the net force on m_1 that produces its acceleration. Recalling that vectors in one dimension can be written with directional signs and magnitudes, we have

$$F_{net} = F - T = m_1 a \quad \text{(direction of } \mathbf{F} \text{ taken to be positive)}$$

Then, solving for the magnitude of **T**,

$$T = F - m_1 a$$
$$= 12.0 \text{ N} - (2.5 \text{ kg})(2.0 \text{ m/s}^2) = 12.0 \text{ N} - 5.0 \text{ N} = 7.0 \text{ N}$$

Follow-up Exercise. Suppose there were an additional horizontal force to the left of 3.0 N applied to m_2 in Fig. 4.9. What would be the tension in the connecting string in this case? *(Answers to all Follow-up Exercises are at the back of the text.)*

The Second Law in Component Form

Not only does Newton's second law hold for any part of a system, but it also applies to the components of motion. For example, a force may be expressed in component notation in two dimensions as follows:

$$\Sigma \mathbf{F}_i = m\mathbf{a}$$

and

$$\Sigma (F_x \hat{\mathbf{x}} + F_y \hat{\mathbf{y}}) = m(a_x \hat{\mathbf{x}} + a_y \hat{\mathbf{y}}) = ma_x \hat{\mathbf{x}} + ma_y \hat{\mathbf{y}} \qquad (4.3a)$$

Hence, to satisfy both x and y together, we have the components

$$\Sigma F_x = ma_x \quad \text{and} \quad \Sigma F_y = ma_y \qquad (4.3b)$$

and Newton's second law applies separately to each component of motion. Note that *both* equations must be true. Example 4.4 demonstrates how the second law is applied using components.

Example 4.4 ■ Newton's Second Law: Components of Force

A block of mass 0.50 kg travels with a speed of 2.0 m/s in the x-direction on a flat, frictionless surface. On passing through the origin, the block experiences a constant force of 3.0 N at an angle of 60° relative to the x-axis for 1.5 s (▼Fig. 4.10). What is the velocity of the block at the end of this time?

Thinking It Through. With the force at an angle to the initial motion, it would appear that the solution is complicated. But note in the insert in Fig. 4.10 that the force can be resolved into components. The motion can then be analyzed in each component direction.

Solution. First, we write the given data and what is to be found:

Given: $m = 0.50$ kg   *Find:* $\mathbf{v}$ (velocity at the end of 1.5 s)
$v_{x_o} = 2.0$ m/s
$v_{y_o} = 0$
$F = 3.0$ N, $\theta = 60°$
$t = 1.5$ s

► **FIGURE 4.10 Off the straight and narrow** A force is applied to a moving block when it reaches the origin, and the block deviates from its straight-line path. See Example 4.4.

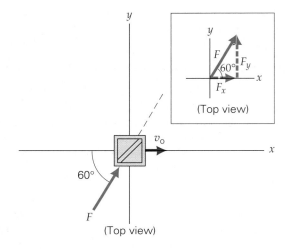

Let's find the magnitudes of the forces in the component (x and y) directions:

$$F_x = F \cos 60° = (3.0 \text{ N})(0.500) = 1.5 \text{ N}$$
$$F_y = F \sin 60° = (3.0 \text{ N})(0.866) = 2.6 \text{ N}$$

Then, applying Newton's second law to each direction to find the components of acceleration, we get

$$a_x = \frac{F_x}{m} = \frac{1.5 \text{ N}}{0.50 \text{ kg}} = 3.0 \text{ m/s}^2$$
$$a_y = \frac{F_y}{m} = \frac{2.6 \text{ N}}{0.50 \text{ kg}} = 5.2 \text{ m/s}^2$$

Next, from the kinematic equation relating velocity and acceleration (Eq. 2.8), the velocity components of the block are given by

$$v_x = v_{x_o} + a_x t = 2.0 \text{ m/s} + (3.0 \text{ m/s}^2)(1.5 \text{ s}) = 6.5 \text{ m/s}$$
$$v_y = v_{y_o} + a_y t = 0 + (5.2 \text{ m/s}^2)(1.5 \text{ s}) = 7.8 \text{ m/s}$$

At the end of the 1.5 s, the velocity of the block is

$$\mathbf{v} = v_x \hat{\mathbf{x}} + v_y \hat{\mathbf{y}} = (6.5 \text{ m/s}) \hat{\mathbf{x}} + (7.8 \text{ m/s}) \hat{\mathbf{y}}$$

Follow-up Exercise.   (a) What is the direction of motion at the end of the 1.5 s? (b) If the force were applied at an angle of 30° (rather than 60°) relative to the x-axis, how would the results of this Example be different? *(Answers to all Follow-up Exercises are at the back of the text.)*

4.4 Newton's Third Law of Motion

OBJECTIVES: **To (a) state and explain Newton's third law of motion and (b) identify action–reaction force pairs.**

Newton formulated a third law that is as far reaching in its physical significance as the first two laws. For a simple introduction to the third law, consider the forces involved in seatbelt safety. When the brakes are suddenly applied when you are riding in a moving car, you continue to move forward. (The frictional force on the seat of your pants is not enough to stop you.) In doing so, you exert forces on the seatbelt and shoulder strap. The belt and strap exert corresponding reaction forces on you, causing you to slow down with the car. If you haven't buckled up, you may keep on going (Newton's first law) until another force, such as that applied by the dashboard or windshield, slows you down.

We commonly think of forces as occurring singly. However, Newton recognized that it is impossible to have a single force. He observed that in any application of force, there is always a mutual interaction, and forces always occur in pairs. An example given by Newton was the following: If you press on a stone with a finger, then the finger is also pressed by, or receives a force from, the stone.

Newton termed the paired forces *action* and *reaction*, and **Newton's third law of motion** is as follows:

Newton's third law—action and reaction

| For every force (action), there is an equal and opposite force (reaction). |

In symbol notation, Newton's third law is

$$\mathbf{F}_{12} = -\mathbf{F}_{21}$$

That is, $\mathbf{F}_{12}$ is the force exerted *on* object 1 *by* object 2, and $-\mathbf{F}_{21}$ is the equal and opposite force exerted *on* object 2 *by* object 1. (The minus sign indicates the opposite direction.) *Which force is considered the action or the reaction is arbitrary;* $\mathbf{F}_{21}$ may be the reaction to $\mathbf{F}_{12}$ or vice versa.

Newton's third law at first glance may seem to contradict Newton's second law: If there are always equal and opposite forces, how can there be a nonzero net force? An important thing to remember about the force pair of the third law is that

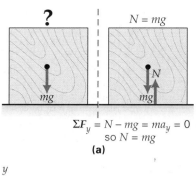

$$\Sigma F_y = N - mg = ma_y = 0$$
$$\text{so } N = mg$$

(a)

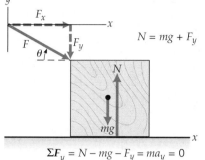

$$\Sigma F_y = N - mg - F_y = ma_y = 0$$

(b)

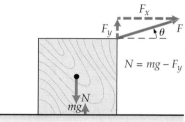

$$\Sigma F_y = N - mg + F_y = ma_y = 0$$

(c)

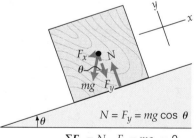

$$\Sigma F_y = N - F_y = ma_y = 0$$

(d)

▲ **FIGURE 4.11 Distinctions between Newton's second and third laws** Newton's second law deals with the forces acting on a particular object (or system). Newton's third law deals with the force pair that acts on different objects. This distinction is illustrated for various situations. (See the text for descriptions.)

the action–reaction forces do not act on the same object. The second law is concerned with force(s) acting on a particular object (or system). The opposing forces of the third law act on *different* objects. Hence, the forces cannot cancel each other out or have a vector sum of zero when we apply the second law to the individual objects.

To illustrate this distinction, consider the situations shown in ◄Fig. 4.11. We often tend to forget the reaction force. For example, in the left portion of Fig. 4.11a, the obvious force that acts on a block sitting on a table is the Earth's gravitational attraction, which is expressed by the weight mg. But, there *has to be another force* acting on the block. For the block not to accelerate, the table must exert an upward force $\mathbf{N}$ whose magnitude is equal to the block's weight. Thus, $\Sigma\mathbf{F}_y = +N - mg = ma_y = 0$, where the direction of the vectors on the right side of the equation are indicated by plus and minus signs.

In reaction to $\mathbf{N}$, the block exerts a downward force on the table, $-\mathbf{N}$, whose magnitude is the same as the block's weight, mg. However, $-\mathbf{N}$ is *not* the object's weight. Weight and $-\mathbf{N}$ have two different origins: Weight is the action-at-a-distance gravitational force, and $-\mathbf{N}$ is a contact force between the two surfaces.

You could easily demonstrate that this upward force on the block is there by placing the block on your hand and holding it stationary—you would exert an upward force on the block. (And you would feel a reaction force of $-\mathbf{N}$ on your hand.) If you applied a greater force, that is, $N > mg$, then the block would accelerate upward.

We call the force that a surface exerts on an object a *normal* force and use the symbol N to denote the force. *Normal* means *perpendicular*. The **normal force** that a surface exerts on an object is always perpendicular to the surface. In Fig. 4.11a, the normal force is equal and opposite to the weight of the block.

The normal force is not always equal and opposite to an object's weight, however. The normal force is a "reaction" force; it reacts to the situation. If a force is applied to a block downward at an angle, as shown in Fig. 4.11b, then a downward component (F_y) acts on the block in addition to the block's weight; thus, the normal force must balance both forces ($\Sigma\mathbf{F}_y = +N - mg - F_y = ma_y = 0$), and the magnitude of $N = mg + F_y > mg$.

In Fig. 4.11b, the block could accelerate to the right. (Why?) However, this horizontal component (F_x) might be balanced by a frictional force, as it often is. We'll bring friction into the picture in a later section; the normal force will be quite important there.

For the situation in Fig. 4.11c, the normal force is reduced to less than w. Can you see why? Note that the F_y component of the applied force is opposite in direction to the weight. In this situation, $\Sigma\mathbf{F}_y = +N - mg + F_y = ma_y = 0$, and in magnitude, $N = mg - F_y < mg$.

Inclined planes are important in both physics and the real world. For a block on an inclined plane (Fig. 4.11d), the normal force on the block is *perpendicular to the surface of the plane*. This force is the reaction to the block's force on the surface of the plane, and since $a_y = 0$ (i.e., there is no acceleration in the y-direction), we have $\Sigma\mathbf{F}_y = N - F_y = ma_y = 0$, and in magnitude, $N = F_y = mg \cos\theta$. The weight component (F_x) accelerates the block down the plane in the absence of an equal opposing friction force between the block and the surface of the plane.

As another example, consider the situation in ▶Fig. 4.12a. Two third-law force pairs are acting when the person is holding the briefcase. First, there is a pair of contact forces: The person's hand exerts an upward force on the handle, and the handle exerts an equal downward force on the hand. That is, $+F_1 = -F_1$. This force pair is an action–reaction pair, with the forces acting on different objects. The other third-law force pair consists of action-at-a-distance forces associated with gravitational attraction: The Earth attracts the briefcase (its weight), and the briefcase attracts the Earth, or $+F_2 = -F_2$.

Concentrating on the briefcase in isolation, we can see that only two of the four forces in Fig. 4.12a act on it—the upward force on the handle ($+F_1$) and the

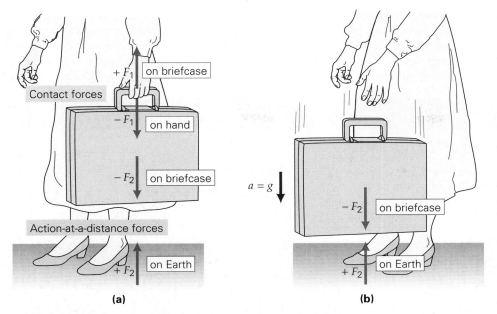

◀ FIGURE 4.12 Force pairs of
Newton's third law (a) When a
person holds a briefcase, there are
two force pairs: a contact pair ($+F_1$
and $-F_1$) and an action-at-a-distance
(gravity) pair ($+F_2$ and $-F_2$). The
net force acting on the briefcase is
zero: The upward contact force
($+F_1$) balances the downward
weight force. Note, however, that
the upward contact force and
downward weight force are *not* a
third-law pair. (b) When the
briefcase is released, there is an
unbalanced, or nonzero net, force
acting on the briefcase (its weight
force), and it accelerates downward
(at g for free fall).

downward force of the briefcase's weight ($-F_2$). These two forces *cannot* constitute a
third-law force pair, however, because they act on the same object. Since the briefcase
does not accelerate, these forces must be equal and opposite. Thus, the net force on the
isolated stationary briefcase is zero, as required by Newton's first and second laws.

When the person drops the briefcase (Fig. 4.12b), there is then a nonzero net
force (the unbalanced force of gravity) on the briefcase, and it accelerates toward the
Earth. But there is still a pair of third-law forces *acting on different objects*—the briefcase and the Earth. (The Earth actually accelerates upward toward the briefcase as
well, but the amount of the acceleration is infinitesimal. Can you explain why?)

Jet propulsion is yet another example of Newton's third law in action. In the case
of a rocket, the rocket and exhaust gases exert equal and opposite forces on each other.
As a result, the exhaust gases are accelerated away from the rocket, and the rocket is
accelerated in the opposite direction. When a big rocket "blasts off," as in a space-
shuttle launch, there is a fiery release of exhaust from the rocket. A common misconception is that the exhaust gases "push" against the launch pad to accelerate the
rocket. If this interprets were true, there would be no space travel, since there is nothing to "push" against in space. The correct explanation is one of action (gases exerting
a force on the rocket) and reaction (rocket exerting an opposite force on the gases).

Another action–reaction pair is given in the Insight on p. 116.

4.5 More on Newton's Laws: Free-Body Diagrams and Translational Equilibrium

OBJECTIVES: To (a) apply Newton's laws in analyzing various situations, using free-
body diagrams, and (b) understand the concept of translational
equilibrium.

Now that you have been introduced to Newton's laws and some applications in
analyzing motion, the importance of these laws should be evident. They are so
simply stated, yet so far reaching. The second law is probably the most often applied, because of its mathematical relationship. However, the first and third laws
are often used in qualitative analysis, as our continuing study of the different
areas of physics will reveal.

The simple relationship expressed by Newton's second law, $\mathbf{F}_{net} = m\mathbf{a}$, allows
the quantitative analysis of force and motion. We can think of it as a cause-and-effect
relationship, with force being the cause and acceleration being the motional effect.

In general, we will be concerned with applications that involve constant forces. Constant forces result in constant accelerations and allow us to use the kinematic equations from Chapter 2 in analyzing the motion. When there is a variable force, Newton's second law holds for the *instantaneous* force and acceleration, but the acceleration will vary with time, requiring calculus to analyze. We will generally limit our-

INSIGHT

Action and Reaction: The Harrier "Jump" Jet

The action–reaction principle of Newton's third law is used to power jet aircraft (and propeller-driven planes), usually for propulsion rather than directly for lift. However, there is one type of aircraft that uses jet propulsion to take off vertically—no runway needed, like a helicopter: the Harrier "Jump" jet, originally developed by the British (Fig. 1). This plane has four engine nozzles that can be swiveled to point at any angle from vertical to horizontal. These engines provide the power for both vertical and horizontal flight.

A Harrier takeoff is illustrated in Fig. 2. (Other small jets on the plane that help control and stabilize the aircraft are not shown.) As shown in Fig. 2a, the nozzles are directed downward for takeoff, and the reaction forces raise the plane. After the plane has achieved a proper altitude, the nozzles are swiveled, and a forward component of the force gives the plane a forward acceleration (Fig. 2b). Finally, with the jets horizontal, the planes flies conventionally (Fig. 2c).

But wait! Did you notice something missing in Fig. 2c? The plane has weight, or a downward gravitational force acting on it, that is not shown. What is the counterbalancing force to this weight that keeps the plane from falling? As you will learn in Chapter 9, this *lift* force results from the curved wing surface and fluid (air) dynamics.

FIGURE 1 **"Jump" jet** A Harrier "Jump" jet lifting off vertically.

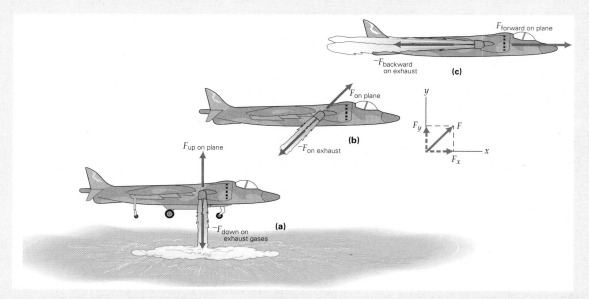

FIGURE 2 **Up, up, and away** **(a)** The reaction force to the downward exhaust force from the engines raises the plane. Only one nozzle is shown, for simplicity. **(b)** At the appropriate altitude, the engine nozzles are swiveled, and a forward force component gives the plane a forward acceleration (while the plane still gains altitude—why?). **(c)** With the nozzles swiveled 90° from their position in (a), the plane flies conventionally.

selves to constant accelerations and forces. This section presents several examples of applications of Newton's second law so that you can become familiar with its use. This small, but powerful, equation will be used again and again throughout this text.

Newton's second law gives the acceleration resulting from an applied force, but it can also be used to find the force from motional effects, as Example 4.5 shows.

Example 4.5 ■ A Braking Car: Finding a Force from Motional Effects

A car traveling at 72.0 km/h along a straight, level road is brought uniformly to a stop in a distance of 40.0 m. If the car weighs 8.80×10^3 N, what is the braking force?

Thinking It Through. We know that the car's velocity changes, so there is an acceleration. Given a distance, we might surmise that first the acceleration is found by using a kinematic equation, and then the acceleration is used to compute the force.

Solution. In bringing the car to a stop, the braking force caused an acceleration (actually a deceleration), as illustrated in ▼Fig. 4.13. Listing what is given and what must be found, we have

Given: $v_0 = 72.0 \text{ km/h} = 20.0 \text{ m/s}$ *Find:* F_b (braking force)
$v = 0$
$x = 40.0 \text{ m}, x_0 = 0$
$w = 8.80 \times 10^3 \text{ N}$

We know that $F_{net} = ma$, so we can easily calculate F_b if we can find m and a. The car's mass m can be obtained from the car's weight, which is given. The other given quantities should remind you of a kinematic equation from Chapter 2 from which the acceleration a can be found. Since the car is brought uniformly to a stop, the acceleration is constant, and we may use Eq. 2.11, $v^2 = v_0^2 + 2ax$, to find a:

$$a = \frac{v^2 - v_0^2}{2x} = \frac{0 - (20.0 \text{ m/s})^2}{2(40.0 \text{ m})} = -5.00 \text{ m/s}^2$$

The minus sign indicates that the acceleration is opposite to $+v_0$, as expected for a braking force, which slows the car.

The mass of the car is obtained from the car's weight: $w = mg$, or $m = w/g$. Using this expression along with the acceleration obtained previously gives a braking force of

$$F_{net_x} = F_b = ma = \left(\frac{w}{g}\right)a = \left[\frac{8.80 \times 10^3 \text{ N}}{9.80 \text{ m/s}^2}\right](-5.00 \text{ m/s}^2) = -4.49 \times 10^3 \text{ N}$$

Follow-up Exercise. A 1000-kg car starting from rest travels in a straight line and uniformly reaches a speed of 10 m/s in 5.0 s. What is the magnitude of the net force that accelerates the car? (*Answers to all Follow-up Exercises are at the back of the text.*)

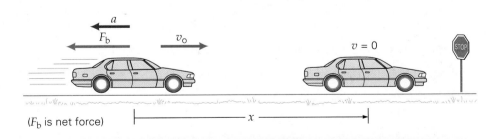

(*F_b is net force*)

◀ **FIGURE 4.13 Finding force from motional effects** See Example 4.5.

Problem-Solving Strategy: Free-Body Diagrams

In illustrations of physical situations, sometimes called *space diagrams*, force vectors may be drawn at different locations to indicate their points of application. However, because we are presently concerned only with linear motions, vectors in free-body

Learn by Drawing

Drawing a Free-Body Diagram

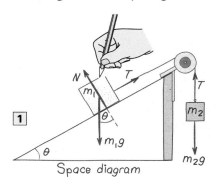

1

Space diagram

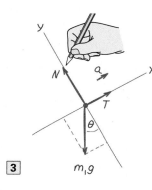

2

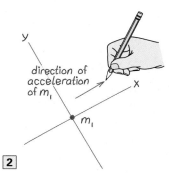

3

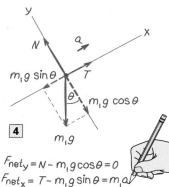

4

$F_{net_y} = N - m_1 g \cos\theta = 0$
$F_{net_x} = T - m_1 g \sin\theta = m_1 a$

---

diagrams may be shown as emanating from a common point, which is chosen as the origin of the x–y axes. One of the axes is generally chosen along the direction of the net force acting on a body, since that is the direction in which the body will accelerate. Also, it is often important to resolve force vectors into components, and properly chosen x–y axes simplify this task.

In a free-body diagram, the vector arrows do not have to be drawn exactly to scale. However, it should be made apparent if there is a net force, and whether forces balance each other in a particular direction. When the forces aren't balanced, we know from Newton's second law that there must be an acceleration.

In summary, the general steps in constructing and using free-body diagrams are as follows (refer to the accompanying Learn by Drawing as you read):

1. Make a sketch, or space diagram, of the situation (if one is not already available), and identify the forces acting on each body of the system. (A space diagram is an illustration of the physical situation that identifies the force vectors.)

2. Isolate the body for which the free-body diagram is to be constructed. Draw a set of Cartesian axes, with the origin at a point through which the forces act and with one of the axes along the direction of the body's acceleration. (The acceleration will be in the direction of the net force, if there is one.)

3. Draw properly oriented force vectors (including angles) on the diagram, emanating from the origin of the axes. If there is an unbalanced force, assume a direction of acceleration, and indicate it with an acceleration vector. Be sure to include only those forces that act on the isolated body.

4. Resolve any forces that are not directed along the x- or y-axes into x- or y-components (use plus and minus signs to indicate direction). Use the free-body diagram to analyze the forces in terms of Newton's second law of motion. (*Note*: If you assume that the acceleration is in one direction, and in the solution it comes out with the opposite sign, then the acceleration is actually in the opposite direction from that assumed. For example, if you assume that **a** is in the +x-direction, but you get a negative answer, then **a** is in the −x-direction.)

Free-body diagrams are a particularly useful way of following one of the Suggested Problem-Solving Procedures in Chapter 1: Draw a diagram as an aid in visualizing and analyzing the physical situation of the problem. *Make it a practice to draw free-body diagrams for force problems, as is done in the following Examples.*

Example 4.6 ■ Up or Down?: Motion on a Frictionless Inclined Plane

Two masses are connected by a light string running over a light pulley of negligible friction as illustrated in the Learn By Drawing diagrams. One mass ($m_1 = 5.0$ kg) is on a frictionless 20° inclined plane, and the other ($m_2 = 1.5$ kg) is freely suspended. What is the acceleration of the masses? (Only the free-body diagram for m_1 is shown in the Learn By Drawing diagram. You need to draw the free-body diagram for m_2.)

Thinking It Through. Apply the preceding Problem-Solving Strategy.

Solution. Following our usual procedure, we write

Given: $m_1 = 5.0$ kg   *Find:* **a** (acceleration)
$m_2 = 1.5$ kg
$\theta = 20°$

To help us visualize the forces involved, we isolate m_1 and m_2 and draw free-body diagrams for each mass. For mass m_1, there are three concurrent forces (forces acting through a common point). These forces are T, its weight $m_1 g$, and N, where T is the tension force of the string on m_1 and N is the normal force of the plane on the block (LBD-3). The forces are shown as emanating from their common point of action. (Recall that a vector can be moved as long as its direction and magnitude are not changed.)

We will start by assuming that m_1 accelerates up the plane, which is taken to be in the $+x$-direction. (It makes no difference whether it is assumed that m_1 accelerates up or down the plane, as we shall see shortly.) Notice that m_1g (the weight) is broken down into components. The x-component is opposite to the assumed direction of acceleration, and the y-component acts perpendicularly to the plane and is balanced by the normal force N. (There is no acceleration in the y-direction, so there is no net force in this direction.)

Then, applying Newton's second law in component form (Eq. 4.3b) to m_1, we have

$$\Sigma F_{x_1} = T - m_1g \sin\theta = m_1a$$
$$\Sigma F_{y_1} = N - m_1g \cos\theta = m_1a_y = 0 \quad (a_y = 0, \text{ no net force, so the forces cancel})$$

And for m_2,

$$\Sigma F_{y_2} = m_2g - T = m_2a_y = m_2a$$

where the masses of the string and pulley have been neglected. Since the accelerations of m_1 and m_2 have the same magnitudes, we can use $a_x = a_y = a$.

Adding the first and last equations to eliminate T, we have

$$m_2g - m_1g \sin\theta = (m_1 + m_2)g$$

(net force = *total* mass × acceleration)

(Note that this is the equation that would be obtained by applying Newton's second law to the system as a whole, because in the system of both blocks, the $\pm T$ forces are internal forces and cancel.)

Then, solving for a:

$$a = \frac{m_2g - m_1g \sin 20°}{m_1 + m_2}$$

$$= \frac{(1.5 \text{ kg})(9.8 \text{ m/s}^2) - (5.0 \text{ kg})(9.8 \text{ m/s}^2)(0.342)}{5.0 \text{ kg} + 1.5 \text{ kg}}$$

$$= -0.32 \text{ m/s}^2$$

The minus sign indicates that the acceleration is opposite to the assumed direction. That is, m_1 actually accelerates down the plane, and m_2 accelerates upward. As this example shows, if you assume the acceleration to be in the wrong direction, the sign on the result will give you the correct direction anyway.

Could you find the tension force T in the string if you were asked to do so? How this task could be done should be quite evident from the free-body diagram.

Follow-up Exercise. (a) In this Example, what is the minimum amount of mass for m_2 that would cause m_1 barely to accelerate up the plane? (b) Keeping the masses the same as in the Example, how should the angle of incline be adjusted so that m_1 would barely accelerate up the plane? (*Answers to all Follow-up Exercises are at the back of the text.*)

PHYSLET®
ILLUSTRATION

Block Sliding on a Incline

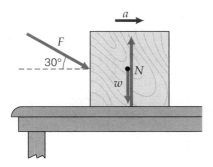

Example 4.7 ■ Components of Force and Free-Body Diagrams

A force of 10.0 N is applied at an angle of 30° to the horizontal on a 1.25-kg block at rest on a frictionless surface, as illustrated in ▶Fig. 4.14. (a) What is the magnitude of the resulting acceleration of the block? (b) What is the magnitude of the normal force?

Thinking It Through. The applied force may be resolved into components. The horizontal component accelerates the block. The vertical component affects the normal force. (Review Fig. 4.11.)

Solution. First we write down the given data and what is to be found:

Given: $F = 10.0$ N   *Find:* (a) a (acceleration)
        $m = 1.25$ kg   (b) N (normal force)
        $\theta = 30°$
        $v_0 = 0$

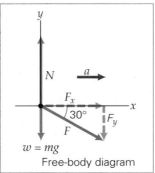

▲ **FIGURE 4.14 Newton's second law and components of force** See Example 4.7

Then we draw a free-body diagram for the block, as in Fig. 4.14.

(a) The acceleration of the block can be calculated using Newton's second law. We choose our axes so that a is in the $+x$-direction. As the free-body diagram shows, only a component (F_x) of the applied force F acts in this direction. The component of F in the direction of motion is $F_x = F \cos \theta$. We apply Newton's second law in the $+x$-direction to calculate the acceleration:

$$F_x = F \cos 30° = ma_x$$
$$a_x = \frac{F \cos 30°}{m} = \frac{(10.0 \text{ N})(0.866)}{1.25 \text{ kg}} = 6.93 \text{ m/s}^2$$

(b) The acceleration found in part (a) is the total acceleration of the block, since the block moves only in the x-direction (i.e., it does not accelerate in the y-direction). With $a_y = 0$, the sum of the forces in the y-direction must then be zero. That is, the downward component of F acting on the block, F_y, and its downward weight force w must be balanced by the upward normal force N that the surface exerts on the block. If this were not the case, then there would be a net force and an acceleration in the y-direction.

We sum the forces in the y-direction with upward taken as positive and

$$N - F_y - w = 0$$

or

$$N - F \sin 30° - mg = 0$$

and

$$N = F \sin 30° + mg = (10.0 \text{ N})(0.500) + (1.25 \text{ kg})(9.80 \text{ m/s}^2) = 17.3 \text{ N}$$

The surface then exerts a force of 14.9 N upward on the block, which balances the sum of the downward forces acting on it, $w = mg$ and F_y.

Follow-up Exercise. (a) Suppose the applied force on the block is applied for only a short time. What is the magnitude of the normal force after the applied force is removed? (b) If the block slides off the edge of the table, what would be the net force on the block just after it leaves the table (with the applied force removed)? *(Answers to all Follow-up Exercises are at the back of the text.)*

Problem-Solving Hint

There is no single fixed way to go about solving a problem. However, there are general strategies or procedures that are helpful in solving problems involving Newton's second law. When using our Suggested Problem-Solving Procedures introduced in Chapter 1, you might include the following steps when solving problems involving force applications:

- Draw a free-body diagram for each individual body, showing all of the forces acting on that body.
- Depending on what is to be found, apply Newton's second law either to the system as a whole (in which case internal forces cancel) or to a part of the system. Basically, *you want to obtain an equation (or set of equations) containing the quantity for which you want to solve.* Review Example 4.3. (If there are two unknown quantities, application of Newton's second law to two parts of the system may give you two equations and two unknowns. See Example 4.6.)
- Keep in mind that Newton's second law may be applied to components of motion and that forces may be resolved into components to do this. Review Example 4.7.

Translational Equilibrium

Forces may act on an object without producing an acceleration. In such a case, with $\mathbf{a} = 0$, we know from Newton's second law that

$$\Sigma \mathbf{F}_i = 0 \tag{4.4}$$

That is, the vector sum of the forces, or the net force, is zero, so the object either remains at rest (as in ▶Fig. 4.15) or moves with a constant velocity. In such cases, objects are said to be in **translational equilibrium**. When remaining at rest, an object is said to be in *static translational equilibrium*.

It follows that the sums of the rectangular components of the forces for an object in translational equilibrium are also zero (why?):

$$\Sigma F_x = 0 \quad \text{(translational}$$
$$\Sigma F_y = 0 \quad \text{equilibrium only)} \tag{4.5}$$

For three-dimensional problems, we note that $\Sigma F_z = 0$. However, we will restrict our discussion of forces to two dimensions.

Equations 4.5 give what is often referred to as the **condition for translational equilibrium**. (A condition for rotational considerations will be given in Chapter 8.) Let's apply this translational-equilibrium condition to a case involving static equilibrium.

Example 4.8 ■ A Hanging Sign: Static Translational Equilibrium

A 3.0-kg sign hangs in a hall in the physics department as shown in ▶Fig. 4.16a. (a) What is the minimum tensile strength necessary for the cord used to hang the sign? (The minimum tensile strength of the cord is the amount of tension the cord must be able to support without breaking.) (b) Suppose the sign were hung off-center and one segment of the support cords was longer than the other. Would this have any effect?

Thinking It Through. The sign hangs motionless on the wall, so it is in static translational equilibrium. Hence, the sum of the forces is zero, and here we see that there will be components in two dimensions.

Solution.

Given: $m = 3.0 \text{ kg}$ *Find:* T_1 and T_2 (tensions in cord segments)
$\quad\quad\quad \theta_1 = \theta_2 = 45°$

(a) With a cord of sufficient strength, the sign will hang in static equilibrium. We want to find the values of T_1 and T_2 that will just support the weight (mg) of the sign.

The rectangular components of the tensions are shown in the free-body diagram (Fig. 4.16b). Applying the component conditions for static translational equilibrium (Eq. 4.5), we have

$$\Sigma F_x = T_{1_x} - T_{2_x} = 0$$

or

$$T_{1_x} = T_{2_x}$$

Hence, the magnitudes of the x-components of the tensions are equal, as you might have expected from symmetry. (They are the *only* forces in the x-direction, and they must balance each other, since $a_x = 0$.) However, this relationship gives no direct information about the magnitude of the tensions.

Since angles θ_1 and θ_2 are equal, we also know that $T_{1_y} = T_{2_y}$. (Why?) We now apply the other component condition:

$$\Sigma F_y = T_{1_y} + T_{2_y} - mg = 2T_{1_y} - mg = ma_y = 0$$

Therefore,

$$2(T_1 \sin 45°) - mg = 0$$

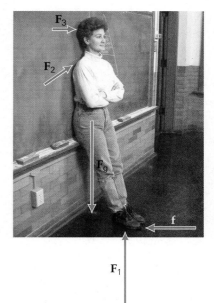

(a)

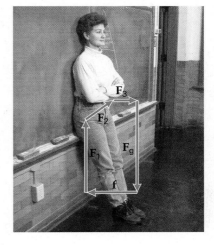

(b)

▲ **FIGURE 4.15 Many forces, no acceleration** (a) At least five different external forces act on this physics professor. (Here, **f** is the force of friction.) Nevertheless, she experiences no acceleration. Why? (b) Adding the force vectors by the polygon method reveals that the vector sum of the forces is zero. The professor is in static translational equilibrium. (She is also in static rotational equilibrium; we'll see why in Chapter 8.)

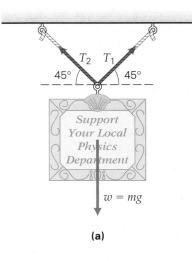
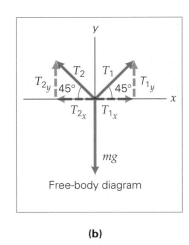

Free-body diagram

(a)   **(b)**

▶ **FIGURE 4.16 Static translational equilibrium** See Example 4.8.

$w = mg$

Learn by Drawing

Vector Diagram

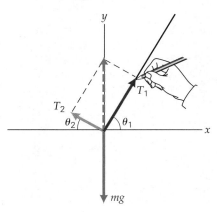

Vector diagram for Example 4.8

Solving for T_1, we obtain

$$T_1 = \frac{mg}{2 \sin 45°} = \frac{(3.0 \text{ kg})(9.8 \text{ m/s}^2)}{2(0.707)} = 21 \text{ N}$$

Thus, $T_1 = T_2 = 21$ N, so the sign should be hung from a cord with a tensile strength of at least 21 N.

(b) If the sign were hung off-center, as illustrated in the Learn By Drawing diagram, there would be a greater tension in the shorter cord segment, and if a 21-N test cord were used, this segment would break.

Follow-up Exercise. Suppose the angles in the LBD figure were 60° and 30°. Show that the tension in the shorter segment (T_1) is greater than 21 N.

Example 4.9 ■ On Your Toes: In Static Equilibrium

An 80-kg person stands on one foot with the heel elevated (▶ Fig. 4.17a). This gives rise to a tibia force F_1 and an Achilles-tendon force F_2, as illustrated in Fig. 4.17b. Typical angles are $\theta_1 = 15°$ and $\theta_2 = 21°$, respectively. (a) Find general equations for F_1 and F_2, and show that θ_2 must be greater than θ_1 to prevent damage to the Achilles tendon. (b) Compare the force of the Achilles tendon with the weight of the person.

Thinking It Through. This is a case of static translational equilibrium, so we can sum the x- and y-components to get equations for F_1 and F_2.

Solution. Listing what is given and what is to be found, we have

Given: $m = 80$ kg   *Find:*   (a) general equations for F_1 and F_2
         $F_1 =$ tibia force   (b) comparison of F_2 and the
         $F_2 =$ tendon "pull"   person's weight
         $\theta_1 = 15°, \theta_2 = 21°$
         $N =$ normal force of floor on foot

(a) It is assumed that the person is at rest, standing on one foot. Then, summing the force components, we have

$$\Sigma F_x = +F_1 \sin \theta_1 - F_2 \sin \theta_2 = 0$$
$$\Sigma F_y = +N - F_1 \cos \theta_1 + F_2 \cos \theta_2 - m_f g = 0$$

Where m_f is the mass of the foot. From the F_x equation, we have,

$$F_1 = F_2 \left(\frac{\sin \theta_2}{\sin \theta_1} \right) \tag{1}$$

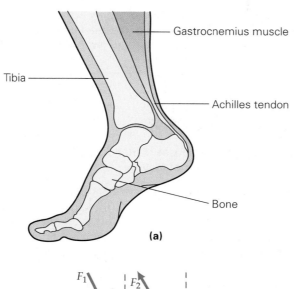

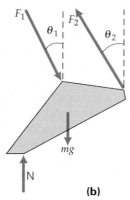

(a)

(b)

◀ **FIGURE 4.17** **On your toes**
(a) A person stands on one foot
with the heel elevated. (b) The
forces involved for this position
(not to scale). See Example 4.9.

And substituting into the F_y equation, we obtain

$$N - F_2\left(\frac{\sin\theta_2}{\sin\theta_1}\right)\cos\theta_1 + F_2\cos\theta_2 - m_f g = 0$$

solving for F_2 yields

$$F_2 = \frac{N - m_f g}{\left(\dfrac{\sin\theta_2}{\tan\theta_2}\right) - \cos\theta_2} = \frac{N}{\cos\theta_2\left(\dfrac{\tan\theta_2}{\tan\theta_1} - 1\right)} \qquad (2)$$

Then, examining the F_2 in equation (Eq. 2), we see that if $\theta_2 = \theta_1$, or $\tan\theta_2 = \tan\theta_1$, then F_2 is very large—infinite! (Why?) So, to have a finite force, we must have $\tan\theta_2 > \tan\theta_1$ or $\theta_2 > \theta_1$, and $21° > 15°$, so Nature obviously knows her physics.
(b) The weight of the person is $w = mg = (80\text{ kg})(9.8\text{ m/s}^2) = 7.8 \times 10^2\text{ N}$. To compare F_2 (Eq. 2) with this quantity, we need to know the normal force N. Thinking big, we find that the y-component equilibrium condition for the *total* body would be

$$\Sigma F_y = 0 = N - w \qquad \text{or} \qquad N = w = mg$$

where m is the mass of the person's body and $m \gg m_f$. Then

$$F_2 = \frac{w - m_f g}{\cos\theta_2\left(\dfrac{\tan\theta_2}{\tan\theta_1} - 1\right)} \approx \frac{w}{\cos 21°\left(\dfrac{\tan 21°}{\tan 15°} - 1\right)} = 2.5\,w$$

The Achilles tendon force is thus approximately 2.5 times the person's weight. No wonder folks stretch or tear this tendon, even without jumping!

Follow-up Exercise. (a) Compare the tibia force with the weight of the person. (b) Suppose the person jumped upward from the one-foot toe position (as in taking a running jump shot in basketball). How would this jump affect F_1 and F_2? (*Answer to all Follow-up Exercises are at the back of the text.*)

4.6 Friction

OBJECTIVES: To explain (a) the causes of friction and (b) how friction is described by using coefficients of friction.

Friction refers to the ever-present resistance to motion that occurs whenever two materials, or media, are in contact with each other. This resistance occurs for all types of media—solids, liquids, and gases—and is characterized as the **force of friction** (f). For simplicity, we have up to now generally ignored all kinds of friction (including air resistance) in examples and problems. Now that you know how to describe motion, we are ready to consider situations that are more realistic, in that the effects of friction are included.

In some real situations, we want to increase friction—for example, by putting sand on an icy road or sidewalk to improve traction. This might seem contradictory, since an increase in friction presumably would increase the resistance to motion. However, consider the forces involved in walking, as illustrated in ◄Fig. 4.18. Without friction, the foot would slip backward. (Think about walking on a slippery surface.) The force of friction prevents this slipping and sometimes needs to be increased on slippery surfaces (►Fig. 4.19a). In other situations, we try to reduce friction (Fig. 4.19b). For instance, we lubricate moving machine parts to allow them to move more freely, thereby lessening wear and reducing the expenditure of energy. Automobiles would not run without friction-reducing oils and greases.

This section is concerned chiefly with friction between solid surfaces. All surfaces are microscopically rough, no matter how smooth they appear or feel. It was originally thought that friction was due primarily to the mechanical interlocking of surface irregularities, or *asperities* (high spots). However, research has shown that friction between the contacting surfaces of ordinary solids (metals in particular) is due mostly to local adhesion. When surfaces are pressed together, local welding or bonding occurs in a few small patches where the largest asperities make contact. To overcome this local adhesion, a force great enough to pull apart the bonded regions must be applied. Once contacting surfaces are in relative motion, another form of friction may result when the asperities of a harder material dig into a softer material, with a "plowing" effect.

Friction between solids is generally classified into three types: static, sliding (kinetic), and rolling. **Static friction** includes all cases in which the frictional force is sufficient to prevent relative motion between surfaces. Suppose you want to move a large desk. You push it, but the desk doesn't move. The force of static friction between the desk's legs and the floor opposes and equals the horizontal force you are applying, so there is no motion—a static condition.

Sliding friction, or **kinetic friction**, occurs when there is relative (sliding) motion at the interface of the surfaces in contact. In pushing on the desk, you eventually get it sliding, but there is still a great deal of resistance between the desk's legs and the floor—kinetic friction.

Rolling friction occurs when one surface rotates as it moves over another surface, but does not slip or slide at the point or area of contact. Rolling friction, such as occurs between a train wheel and a rail, is attributed to small local deformations in the contact region. This type of friction is somewhat difficult to analyze.

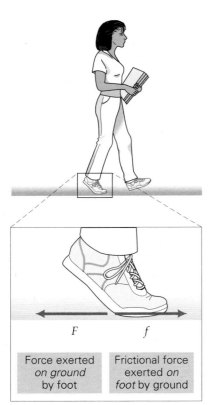

F	f
Force exerted *on ground* by foot	Frictional force exerted *on foot* by ground

▲ **FIGURE 4.18 Friction and walking** The force of friction, *f*, is shown in the direction of the walking motion. This direction may seem wrong at first glance, but it's not. The force of friction prevents the foot from slipping backward while the other foot is brought forward. If you walk on a deep-pile rug, *F* is evident in that the pile will be bent backward.

Frictional Forces and Coefficients of Friction

In this subsection, we will consider the forces of friction on stationary and sliding objects. These forces are called the *force of static friction* and the *force of kinetic* (or *sliding*) *friction*, respectively. Experimentally, it has been found that the force of friction depends on both the nature of the two surfaces and the *load*, or the contact force that presses the surfaces together. For an object on a horizontal surface, this force is equal in magnitude to the object's weight. (Why?) However, as was shown in Example 4.6, on an inclined plane, only a component of the weight force contributes to the load.

Thus, to avoid confusion, remember that the force of friction is proportional to the normal force N in magnitude; that is, $f \propto N$. As we learned earlier, the normal force always acts perpendicular to, and away from, the surface. (It is the force exerted *by* the surface *on* the object.) In the absence of other perpendicular forces, the normal force is equal in magnitude to the component of the weight force acting perpendicular to the surface.

The force of static friction (f_s) between parallel surfaces in contact acts in the direction that opposes the initiation of relative motion between the surfaces. The magnitude has different values such that

$$f_s \leq \mu_s N \quad \text{(static conditions)} \quad (4.6)$$

where μ_s is the **coefficient of static friction**. ("μ" is the Greek letter mu. Note that it is a dimensionless constant. How do you know this from the equation?)

The less-than-or-equal-to sign ($\leq$) indicates that the force of static friction may have different values or magnitudes up to some maximum. To understand this concept, look at ▶Fig. 4.20. In Fig. 4.20a, one person pushes on a file cabinet, but it doesn't move. With no acceleration, the net force on the cabinet is zero, and $F - f_s = 0$, or $F = f_s$. Suppose that a second person also pushes, and the file cabinet still doesn't budge (Fig. 4.20b). Then f_s must now be larger, since the applied force has been increased. Finally, if the applied force is made large enough to overcome the static friction, motion occurs (Fig. 4.20c). The greatest, or maximum, force of static friction is exerted just before the cabinet starts to slide (Fig. 4.20b), and for this case, Eq. 4.6 can be written with an equal sign:

$$f_{s_{max}} = \mu_s N \quad (4.7)$$

Once an object is sliding, the force of friction changes to kinetic friction (f_k). This force acts in the direction opposite to the direction of motion and has a magnitude of

$$f_k = \mu_k N \quad \text{(sliding conditions)} \quad (4.8)$$

where μ_k is the **coefficient of kinetic friction** (sometimes called the *coefficient of sliding friction*). Note that Eqs. 4.6 and 4.8 are *not* vector equations, since f and N are in different directions. Generally, the coefficient of kinetic friction is less than the coefficient of static friction ($\mu_k < \mu_s$), which means that the force of kinetic friction is less than $f_{s_{max}}$, as illustrated in Fig. 4.20. The coefficients of friction between some common materials are listed in Table 4.1.

Note that the force of static friction (f_s) exists in response to an applied force. The magnitude of f_s and its direction depend on the magnitude and direction of the applied force. Up to its maximum value, the force of static friction is equal in magnitude and opposite in direction to the applied force (F), since there is no acceleration ($F - f_s = ma = 0$). Thus, if the person in Fig. 4.20a were to push on the cabinet in the opposite direction, f_s would change to oppose the new push. If there were no applied force F, then f_s would be zero. When F exceeds $f_{s_{max}}$, the

(a)

(b)

▲ **FIGURE 4.19 Increasing and decreasing friction** **(a)** To get a fast start, drag racers need to make sure that their wheels don't slip when the starting light goes on and they floor the accelerator. They therefore try to maximize the friction between their tires and the track by "burning in" the tires just before the start of the race. This "burn in" is done by spinning the wheels with the brakes on until the tires are extremely hot. The rubber becomes so sticky that it almost welds itself to the surface of the road. **(b)** Water serves as a good lubricant to reduce friction in rides such as this one.

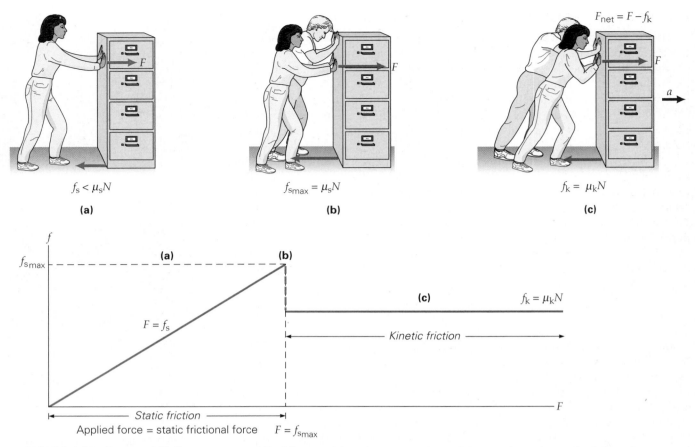

▲ FIGURE 4.20 Force of friction versus applied force (a) In the static region of the graph, as the applied force F increases, so does f_s; that is, $f_s = F$ and $f_s < \mu_s N$. **(b)** When the applied force F exceeds $f_{s_{max}} = \mu_s N$, the heavy file cabinet is set into motion. **(c)** Once the cabinet is moving, the frictional force is decreased, since kinetic friction is less than static friction ($f_k < f_{s_{max}}$). Thus, if the applied force is maintained, there is a net force, and the cabinet is accelerated. For the cabinet to move with constant velocity, the applied force must be reduced to equal the kinetic friction force: $f_k = \mu_k N$.

TABLE 4.1 Approximate Values for Coefficients of Static and Kinetic Friction between Certain Surfaces

Friction between Materials	μ_s	μ_k
aluminum on aluminum	1.90	1.40
glass on glass	0.94	0.35
rubber on concrete		
dry	1.20	0.85
wet	0.80	0.60
steel on aluminum	0.61	0.47
steel on steel		
dry	0.75	0.48
lubricated	0.12	0.07
Teflon on steel	0.04	0.04
Teflon on Teflon	0.04	0.04
waxed wood on snow	0.05	0.03
wood on wood	0.58	0.40
lubricated ball bearings	<0.01	<0.01
synovial joints (at the ends of most long		
bones—for example, elbows and hips)	0.01	0.01

cabinet begins moving (accelerates), and kinetic friction comes into effect, with $f_k = \mu_k N$. If F is reduced to f_k, the cabinet will slide with a constant velocity; if F is maintained as greater than f_k, the cabinet will continue to accelerate.

It has been experimentally determined that the coefficients of friction (and therefore the forces of friction) are nearly independent of the size of the contact area between metal surfaces. This means that the force of friction between a brick-shaped metal block and a metal surface is the same regardless of whether the block is lying on a larger side or a smaller side. The observation is generally not valid for other surfaces, such as wood, and does not apply to plastic or polymer surfaces.

Finally, keep in mind that although the equation $f = \mu N$ holds in general for frictional forces, friction may not be linear over a wide range. That is, μ is not always constant. For example, the coefficient of kinetic friction varies somewhat with the relative speed of the surfaces. However, for speeds up to several meters per second, the coefficients are relatively constant. Thus, this discussion will neglect any variations due to speed (or area), and the forces of static and kinetic friction will be assumed to depend only on the load (N) and the nature of the materials as expressed by the given coefficients of friction.

PHYSLET® ILLUSTRATION

Static and Kinetic Friction

Example 4.10 ■ Pulling a Crate: Static and Kinetic Forces of Friction

(a) In ▼Fig. 4.21, if the coefficient of static friction between the 40.0-kg crate and the floor is 0.650, with what minimum horizontal force must the worker pull to get the crate moving? (b) If the worker maintains that force once the crate starts to move and the coefficient of kinetic friction between the surfaces is 0.500, what is the magnitude of the acceleration of the crate?

Thinking It Through. This scenario involves applications of the forces of friction. In (a), the maximum force of static friction must be calculated. In (b), if the worker maintains an applied force of this magnitude after the crate is in motion, there will be an acceleration, since $f_k < f_{s_{max}}$.

Solution. Listing the given data and what we want to find, we have

Given: $m = 40.0$ kg   *Find:* (a) F (minimum force necessary to move crate)
       $\mu_s = 0.650$ (b) a (acceleration)
       $\mu_k = 0.500$

(a) The crate will not move until the applied force F slightly exceeds the maximum static frictional force $f_{s_{max}}$. So we must find $f_{s_{max}}$ to see what force the worker must apply.

▼ **FIGURE 4.21 Forces of static and kinetic friction** See Example 4.10.

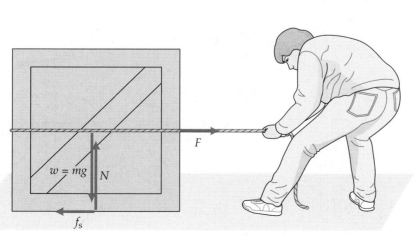

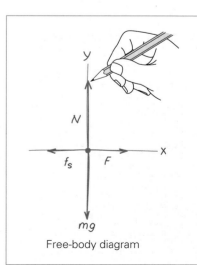

Free-body diagram

The weight of the crate and the normal force are equal in magnitude in this case (see the free-body diagram in Fig. 4.21), so the maximum force of static friction is

$$f_{s_{max}} = \mu_s N = \mu_s (mg)$$
$$= (0.650)(40.0 \text{ kg})(9.80 \text{ m/s}^2) = 255 \text{ N}$$

The crate moves if the applied force F exceeds 255 N.

(b) Now the crate is in motion, and the worker maintains a constant applied force $F = f_{s_{max}} = 255$ N. The force of kinetic friction f_k acts on the crate, but this force is smaller than F, because $\mu_k < \mu_s$. Hence, there is a net force, and the acceleration of the crate may be found by using Newton's second law in the x-direction:

$$\Sigma F_x = +F - f_k = F - \mu_k N = ma_x$$

Solving for a_x, we obtain

$$a_x = \frac{F - \mu_k N}{m} = \frac{F - \mu_k(mg)}{m}$$
$$= \frac{255 \text{ N} - (0.500)(40.0 \text{ kg})(9.80 \text{ m/s}^2)}{40.0 \text{ kg}} = 1.5 \text{ m/s}^2$$

Follow-up Exercise. On the average, by what factor does μ_s exceed μ_k for nonlubricated, metal-on-metal surfaces? (See Table 4.1.) (*Answers to all Follow-up Exercises are at the back of the text.*)

Let's look at another worker with the same crate, but this time assume that the worker applies the force at an angle (▼ Fig. 4.22).

Example 4.11 ■ Pulling at an Angle: A Closer Look at the Normal Force

A worker pulling a crate applies a force at an angle of 30° to the horizontal, as shown in Fig. 4.22. How large a force must he apply to move the crate? (Before looking at the solution, would you expect that the force needed in this case would be greater or lesser than that in Example 4.10?)

Thinking It Through. We see from the figure that the applied force is at an angle to the horizontal surface, so the vertical component will affect the normal force. (See Fig. 4.11). This change in the normal force will, in turn, affect the maximum force of static friction.

▼ **FIGURE 4.22 Normal force** See Example 4.11.

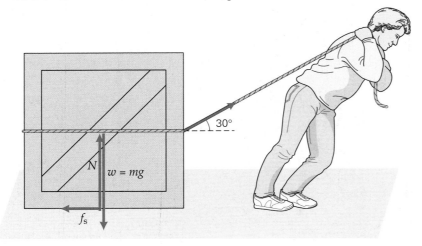

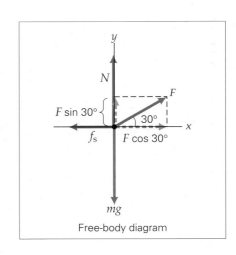

Free-body diagram

Solution. The data are the same as in Example 4.10, except that the force is applied at an angle.

Given: $\theta = 30°$ **Find:** F (minimum force necessary to move crate)

In this case, the crate will move when the *horizontal component* of the applied force, $F \cos 30°$, slightly exceeds the maximum static friction force. So, we may write the following for the maximum friction:

$$F \cos 30° = f_{s_{max}} = \mu_s N$$

However, the magnitude of the normal force is not equal to the weight of the crate here, because of the upward component of the applied force. (See the free-body diagram in Fig. 4.22.) By Newton's second law, since $a_y = 0$, we have

$$\Sigma F_y = +N + F \sin 30° - mg = 0$$

or

$$N = mg - F \sin 30°$$

In effect, the applied force partially supports the weight of the crate. Substituting this expression for N into the first equation gives

$$F \cos 30° = \mu_s(mg - F \sin 30°)$$

Solving for F gives

$$F = \frac{mg}{(\cos 30°/\mu_s) + \sin 30°}$$

$$= \frac{(40.0 \text{ kg})(9.80 \text{ m/s}^2)}{(0.866/0.650) + 0.500} = 214 \text{ N}$$

Thus, less applied force is needed in this case, reflecting the fact that the frictional force is less, because of the reduced normal force.

Follow-up Exercise. Note that in this Example, applying the force at an angle produces two effects. As the angle between the applied force and the horizontal increases, the horizontal component of the applied force is reduced. However, the normal force also gets smaller, resulting in a lower $f_{s_{max}}$. Does one effect always outweigh the other? That is, does the applied force F necessary to move the crate always decrease with increasing angle? (*Hint:* Investigate F for different angles. For example, compute F for 20° and 50°. You already have a value for 30°. What do the results tell you?) (*Answers to all Follow-up Exercises are at the back of the text.*)

Pulling a Block

Integrated Example 4.12 ■ No Slip, No Slide: Static Friction

A crate sits in the middle of the bed on a flatbed truck that is traveling at a speed of 80 km/h on a straight, level road. The coefficient of static friction between the crate and the truck bed is 0.40. When the truck comes uniformly to a stop, the crate does not slide, but remains stationary on the truck. (a) Draw a free-body diagram for the crate while the truck is stopping. (b) What is the minimum stopping distance for the truck so the crate does not slide on the truck bed?

Solution.

(a) Conceptual Reasoning. There are three forces on the crate, as shown in the free-body diagram in ▶Fig. 4.23 (assuming that the truck is initially traveling in the +x-direction). But wait. There is a net force in the −x-direction, and hence there should be an acceleration in that direction ($-a_x$). What does this mean? It means that relative to the ground, the crate is decelerating at the same rate as the truck, which is necessary for the crate not to slide—the crate and the truck slow down uniformly together.

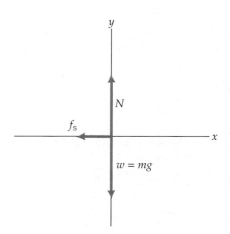

(Free-body diagram)

▲ **FIGURE 4.23 Free-body diagram** See Example 4.12.

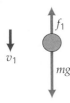

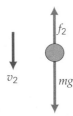

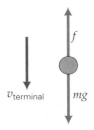

(a) As v increases, so does f.

(b) When $f = mg$, the object falls with a constant (terminal) velocity.

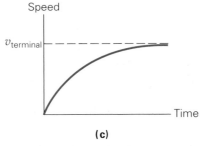

(c)

▲ **FIGURE 4.25   Air resistance and terminal velocity**   (a) As the speed of a falling object increases, so does the frictional force of air resistance. (b) When this force of friction equals the weight of the object, the net force is zero, and the object falls with a constant (terminal) velocity. (c) A plot of speed versus time, showing these relationships.

▶ **FIGURE 4.24   Air foil**   The air foil at the top of the truck's cab makes the truck streamlined and therefore reduces air resistance.

(b) Thinking It Through.   The force creating this acceleration for the crate is the static force of friction. The acceleration is found using Newton's second law, and then used in one of the kinematic equations to find the distance.

Given: $v_{x_o} = 80 \text{ km/h} = 22 \text{ m/s}$ *Find:*   (b) minimum stopping distance
$\mu_s = 0.40$

Applying Newton's second law:

$$\Sigma F_x = -f_s = -\mu_s N = -\mu_s mg = ma_x$$

Solving for a_x, we obtain

$$a_x = -\mu_s g = -(0.40)(9.8 \text{ m/s}^2) = -3.9 \text{ m/s}^2$$

which is the maximum deceleration of the truck so the crate does not slide. Otherwise, $ma_x > f_{s_{max}}$, and kinetic friction would take over (sliding). Hence, the minimum stopping distance (x) for the truck is given by Eq. 2.11, where $v_x = 0$ and x_o is taken to be zero. So,

$$v_x^2 = 0 = v_{x_o}^2 + 2(a_x)x$$

Solving for x, we obtain

$$x = \frac{v_{x_o}^2}{-2a} = \frac{(22 \text{ m/s})^2}{-2(-3.9 \text{ m/s}^2)} = +62 \text{ m}$$

Is the answer reasonable? This length is about two thirds of a football field.

Follow-up Exercise.   Draw a free-body diagram, and describe what happens in terms of accelerations and coefficients of friction if the crate starts to slide forward when the truck is braking to a stop (in other words, if a_x exceeds -3.9 m/s^2). (*Answers to all Follow-up Exercises are at the back of the text.*)

Air Resistance

Air resistance refers to the resistance force acting on an object as it moves through air. In other words, air resistance is a type of frictional force. In analyses of falling objects, you can generally ignore the effect of air resistance and still get valid approximations for falling relatively short distances. However, for longer distances, air resistance cannot be ignored.

Air resistance occurs when a moving object collides with air molecules. Therefore, air resistance depends on the object's shape and size (which determine the area of the object that is exposed to collisions) as well as its speed. The larger the object and the faster it moves, the more collisions there will be with air molecules. (Air density is also a factor, but this quantity can be assumed to be constant near the Earth's surface.) To reduce air resistance (and fuel consumption), automobiles are made more "streamlined," and air foils are used on trucks and campers (▲ Fig. 4.24).

Consider now a falling object. Since air resistance depends on speed, as a falling object accelerates under the influence of gravity, the retarding force of air resistance increases (◀Fig. 4.25a). Eventually, the magnitude of the retarding force equals that of the object's weight force (Fig. 4.25b), so the net force on the object is zero. The object then falls with a maximum constant velocity, which is called the **terminal velocity**.

This can be easily seen from Newton's second law. For the falling object, we have

$$F_{net} = ma$$

or

$$mg - f = ma$$

where downward has been taken as positive for convenience. Solving for a, we obtain

$$a = g - \frac{f}{m}$$

where a is the magnitude of the instantaneous acceleration.

Notice that the acceleration for a falling object when air resistance is included is less than g; that is, $a < g$, or $a < 9.8$ m/s². As the object continues to fall, its speed increases, and thus the force of air resistance, f, increases (since it is speed dependent) until $a = 0$, when $f = mg$ and $f - mg = 0$. The object then falls at its constant terminal velocity.

For a sky diver with an unopened parachute, the terminal velocity is about 200 km/h (about 125 mi/h). To reduce the terminal velocity so that it can be reached sooner and the time of fall extended, a sky diver will try to increase exposed body area to a maximum by assuming a spread-eagle position (▸Fig. 4.26). This position takes advantage of the dependence of air resistance on the size and shape of the falling object. Once the parachute is open (giving a larger exposed area and a shape that catches the air), the additional air resistance slows the diver down to about 40 km/h (or 25 mi/h), which is preferable for landing.

PHYSLET®
ILLUSTRATION

Air Resistance

▲ **FIGURE 4.26   Terminal velocity**   Sky divers assume a spread-eagle position to maximize air resistance. This causes them to reach terminal velocity more quickly and prolongs the time of fall.

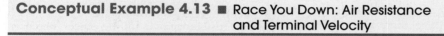
Conceptual Example 4.13 ■ Race You Down: Air Resistance and Terminal Velocity

From a high altitude, a balloonist simultaneously drops two balls of identical size, but appreciably different in weight. Assuming that both balls reach terminal velocity during the fall, which of the following is true? (a) The heavier ball reaches terminal velocity first; (b) the balls reach terminal velocity at the same time; (c) the heavier ball hits the ground first; (d) the balls hit the ground at the same time? *Clearly establish the reasoning and physical principle(s) used in determining your answer before checking it below. That is, **why** did you select your answer?*

Reasoning and Answer.   Terminal velocity is reached when the weight of a ball is balanced by the frictional air resistance. Both balls initially experience the same acceleration, g, and their speeds and the retarding forces of air resistance increase at the same rate. The weight of the lighter ball will be balanced first, so (a) and (b) are incorrect. The lighter ball reaches terminal velocity ($a = 0$) first, but the heavier ball continues to accelerate, speeding up, and pulls ahead of the lighter ball. Hence, the heavier ball hits the ground first, and the answer is (c), and (d) does not apply.

Follow-up Exercise.   Suppose the heavier ball were much larger in size than the lighter ball. How might this difference affect the outcome? *(Answers to all Follow-up Exercises are at the back of the text.)*

You see an example of terminal velocity quite often. Why do clouds stay seemingly suspended in the sky? Certainly the water droplets or ice crystals (high clouds) should fall—and they do. However, they are so small that their terminal velocity is reached quickly, and the very slow rate of their descent goes unnoticed. Also, there may be some helpful updrafts that keep the water and ice from reaching the ground.

Chapter Review

Important Concepts and Equations

- A **force** is something that is capable of changing an object's state of motion. To produce a change in motion, there must be a nonzero net, or unbalanced, force:

$$\mathbf{F}_{net} = \Sigma \mathbf{F}_i$$

- **Newton's first law of motion** is also called the *law of inertia*, where inertia is the natural tendency of an object to maintain its state of motion. It states that in the absence of a net applied force, a body at rest remains at rest, and a body in motion remains in motion with constant velocity.

- **Newton's second law** relates the net force acting on an object or system to the (total) mass and the resulting acceleration. It defines the cause-and-effect relationship between force and acceleration:

$$\Sigma \mathbf{F}_i = \mathbf{F}_{net} = m\mathbf{a} \qquad (4.1)$$

The equation for **weight** in terms of mass is a form of Newton's second law:

$$w = mg \qquad (4.2)$$

The component form of Newton's second law:

$$\Sigma(F_x\hat{\mathbf{x}} + F_y\hat{\mathbf{y}}) = m(a_x\hat{\mathbf{x}} + a_y\hat{\mathbf{y}}) = ma_x\hat{\mathbf{x}} + ma_x\hat{\mathbf{y}} \quad (4.3a)$$

and

$$\Sigma F_x = ma_x \qquad \text{and} \qquad \Sigma F_y = ma_y \qquad (4.3b)$$

- **Newton's third law** states that for every force, there is an equal and opposite reaction force. The opposing forces of a third-law pair always act on different objects.

- An object is said to be in **translational equilibrium** when it either is at rest or moves with a constant velocity. When remaining at rest, an object is said to be in *static translational equilibrium*. The condition for translational equilibrium is represented as

$$\Sigma \mathbf{F}_i = 0 \qquad (4.4)$$

or

$$\Sigma F_x = 0 \quad \text{and} \quad \Sigma F_y = 0 \qquad (4.5)$$

- **Friction** is the resistance to motion that occurs between contacting surfaces. (In general, friction occurs for all types of media—solids, liquids, and gases.)

- The frictional force between surfaces is characterized by coefficients of friction (μ), one for the static case and one for the kinetic (moving) case. In many cases, $f = \mu N$, where N is the normal force—the force perpendicular to the surface (i.e., the force exerted *by* the surface *on* the object). As a ratio of forces (f/N), μ is unitless.

Force of Static Friction:

$$f_s \leq \mu_s N \qquad (4.6)$$

$$f_{s_{max}} = \mu_s N \qquad (4.7)$$

Force of Kinetic (Sliding) Friction:

$$f_k = \mu_k N \qquad (4.8)$$

- The force of air resistance on a falling object increases with increasing speed. It eventually attains a constant velocity, called the *terminal velocity*.

Exercises

Note: Unless otherwise stated, all objects are located near the Earth's surface, where $g = 9.80$ m/s^2.

4.1 The Concept of Force and Net Force
and
4.2 Inertia and Newton's First Law of Motion

1. **CQ** If an object is at rest, there must be no force acting on it. Is this statement correct? Explain.

2. The tendency of an object to maintain its state of motion is called (a) Newton's second law, (b) Galileo's principle, or (c) inertia.

3. **CQ** When on a jet airliner that is taking off, you feel that you are being "pushed" back into the seat. Use Newton's first law to explain why.

4. **CQ** The net force on an object is zero. Can you conclude that the object is at rest? Explain.

5. If an object is moving at constant velocity, (a) there must be a force in the direction of the velocity, (b) there must be no force in the direction of the velocity, (c) there must be no net force, or (d) there must be a net force in the direction of the velocity.

6. If the net force on an object is zero, the object could (a) be at rest, (b) be in motion at a constant velocity, (c) have zero acceleration, or (d) all of the preceding.

7. **CQ** An object weighs 300 N on Earth and 50 N on the Moon. Does the object also have less inertia on the Moon?

8. **CQ** Consider an air-bubble level that is sitting on a horizontal surface (▶ Fig. 4.27). Initially, the air bubble is in the middle of the horizontal glass tube. (a) If the level is pushed and a force is applied to accelerate it, which way would the

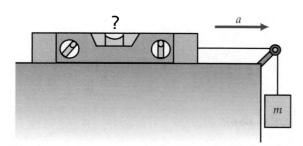

▲ FIGURE 4.27 An air-bubble level/accelerometer
See Exercise 8.

bubble move? Which way would the bubble move if the force is then removed and the level slows down, due to friction? (b) Such a level is sometimes used as an "accelerometer" to indicate the direction of the acceleration. Explain the principle involved. [*Hint*: Think about pushing a pan of water.]

9. **CQ** As a follow-up to Exercise 8, consider the situation of a child holding a helium balloon in a closed car at rest. What would the child observe when the car (a) accelerates from rest and (b) brakes to a stop? (The balloon does not touch the roof of the car.)

10. **CQ** Objects have no appreciable weight in deep space. How can you then distinguish their masses?

11. **CQ** The following is an old trick: If a tablecloth is yanked out very quickly, the dishes on top of it will barely move (▼Fig. 4.28). Why?

▲ FIGURE 4.28 No more dinner? See Exercise 11.

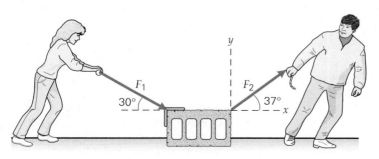

12. **CQ** If your car is moving around a circular track at a constant speed of 55 mi/h, is there a net force acting on your car? Explain.

13. ■ Which has more inertia, 20 cm^3 of water or 10 cm^3 of aluminum, and how many times more? (See Table 9.2.)

14. **IE** ■■ (a) You are told that an object has zero acceleration. Which of the following is true? (1) The object is at rest; (2) the object is moving with constant velocity; (3) either (1) or (2) is possible; or (4) neither (1) nor (2) is possible. (b) Two forces on the object are F_1 = 3.6 N at 74° below the +x-axis and F_2 = 3.6 N at 34° above the −x-axis. Is there a third force on the object, and why? If yes, what is it?

15. ■■ A 5.0-kg block at rest on a frictionless surface is acted on by forces F_1 = 5.5 N and F_2 = 3.5 N, as illustrated in ▼Fig. 4.29. What additional horizontal force will keep the block at rest?

16. ■■■ A 1.5-kg object moves up the y-axis at a constant speed. When it reaches the origin, the forces F_1 = 5.0 N at 37° above the +x-axis, F_2 = 2.5 N in the +x-direction, F_3 = 3.5 N at 45° below the −x-axis, and F_4 = 1.5 N in the −y-direction are applied to it. (a) Will the object continue to move along the y-axis? (b) If not, what simultaneously applied force will keep it moving along the y-axis at a constant speed?

4.3 Newton's Second Law of Motion

17. The newton unit of force is equivalent to (a) kg·m/s, (b) kg·m/s^2, (c) kg·m^2/s, or (d) none of the preceding.

18. **CQ** In general, this chapter has considered forces that are applied to objects of constant mass. What would be the situation if mass were added to or lost from a system while a force was being applied to the system? Give examples of situations in which this set of events might happen.

19. **CQ** The engines of most rockets produce a constant thrust (forward force). However, when a rocket is fired into space, its acceleration increase with time as the engine continues to operate. Is this situation a violation of Newton's second law? Explain.

20. **CQ** An astronaut has a mass of 70 kg when measured on Earth. What is his weight in deep space, far from any celestial body? What is his mass there?

◀ FIGURE 4.29 Two applied forces See Exercises 15 and 94.

21. **CQ** In football, good wide receivers usually have "soft" hands for catching balls (▼Fig. 4.30). How would you interpret this description on the basis of Newton's second law?

▲ **FIGURE 4.30** **Soft hands** See Exercise 21.

22. ■ A 3.0-N net force is applied to a 1.5-kg mass. What is the object's acceleration?

23. ■ What is the mass of an object that accelerates at 3.0 m/s^2 under the influence of a 5.0-N net force?

24. ■ A worker pushes on a crate that experiences a net force of 75 N. If the crate also experiences an acceleration of 0.50 m/s^2, what is its weight?

25. ■ A loaded Boeing 747 jumbo jet has a mass of 2.0×10^5 kg. What net force is required to give the plane an acceleration of 3.5 m/s^2 down the runway for takeoffs?

26. **IE** ■ A 6.0-kg object is brought to the Moon, where the acceleration due to gravity is only one sixth of that on the Earth. (a) The mass of the object on the Moon is (1) zero, (2) 1.0 kg, (3) 6.0 kg, or (4) 36 kg. Why? (b) What is the weight of the object on the Moon?

27. ■ What is the mass of a person weighing 740 N on the Earth?

28. ■ What is the net force acting on a 1.0-kg object in free fall?

29. ■ What is the weight of an 8.0-kg mass in newtons? How about in pounds?

30. ■■ What is the weight of a 150-lb person in newtons? What is his mass in kilograms?

31. **IE** ■■ ▶Fig. 4.31 shows a product label. (a) This label is correct (1) on the Earth; (2) on the Moon, where the acceleration due to gravity is only one sixth of that on the

▲ **FIGURE 4.31** **Correct label?** See Exercise 31.

Earth; (3) in deep space, where there is little gravity; or (4) all of the preceding. (b) What mass of lasagne would a label show for the amount that weighs 2 lb on the Moon?

32. ■■ In a college homecoming competition, 16 students lift a sports car. While holding the car off the ground, each student exerts an upward force of 400 N. (a) What is the mass of the car in kilograms? (b) What is its weight in pounds?

33. ■■ A horizontal force of 12 N acts on an object that rests on a level, frictionless surface on the Earth, where the object has a weight of 98 N. (a) What is the magnitude of the acceleration of the object? (b) What would be the acceleration of the same object in a similar situation on the Moon?

34. ■■ The engine of a 1.0-kg toy plane exerts a 15-N forward force. If the air exerts a 8.0-N resistive force on the plane, what is the magnitude of the acceleration of the plane?

35. ■■ When a horizontal force of 300 N is applied to a 75.0-kg box, the box slides on a level floor, opposed by a force of kinetic friction of 120 N. What is the magnitude of the acceleration of the box?

36. **IE** ■■ (a) A horizontal force acts on an object on a frictionless horizontal surface. If the force is halved and the mass of the object is doubled, the acceleration will be (1) four times, (2) two times, (3) one half, or (4) one fourth as great. (b) If the acceleration of the object is 1.0 m/s^2, and the force on it is doubled and its mass is halved, what is the new acceleration?

37. ■■ A stalled 1500-kg automobile is pushed toward a gas station by a man and a woman on a level road. The applied horizontal forces are 200 N for the woman and 300 N for the man. (a) If there is an effective force of friction of 300 N on the car as it moves, what is its acceleration? (b) Once the car is moving appreciably, what would be an appropriate combined applied force, and why?

38. ■■ In an emergency stop to avoid an accident, a shoulder-strap seat belt holds a 60-kg passenger firmly in place. If the car were initially traveling at 90 km/h and came to a stop in 5.5 s along a straight, level road, what was the average force applied to the passenger by the seatbelt?

39. ■■ A jet catapult on an aircraft carrier accelerates a 2000-kg plane uniformly from rest to a launch speed of 320 km/h in 2.0 s. What is the magnitude of the net force on the plane?

40. ■■ In serving, a tennis player accelerates a 56-g tennis ball horizontally from rest to a speed of 35 m/s. Assuming that the acceleration is uniform when the racquet is applied over a distance of 0.50 m, what is the magnitude of the force exerted on the ball by the racquet?

4.4 Newton's Third Law of Motion

41. A brick hits a glass window. The brick breaks the glass, so (a) the magnitude of the force of the brick on the glass is greater than the magnitude of the force of the glass on the brick, (b) the magnitude of the force of the brick on the glass is smaller than the magnitude of the force of the glass on the brick, (c) the magnitude of the force of the brick on the glass is equal to the magnitude of the force of the glass on the brick, or (d) none of the previous.

42. A freight truck collides head on with a passenger car, causing a lot more damage to the car than to the truck. From this condition, we can say that (a) the magnitude of the force of the truck on the car is greater than the magnitude of the force of the car on the truck, (b) the magnitude of the force of the truck on the car is smaller than the magnitude of the force of the car on the truck, (c) the magnitude of the force of the truck on the car is equal to the magnitude of the force of the car on the truck, or (d) none of the preceding.

43. CQ Here is a story of a horse and a farmer: One day, the farmer attaches a heavy cart to the horse and demands that the horse pull the cart. "Well," says the horse, "I cannot pull the cart, because, according to Newton's third law, if I apply a force to the cart, the cart will apply an equal and opposite force on me. The net result will be that I cannot pull the cart, since all the forces will cancel. Therefore, it is impossible for me to pull this cart." The farmer was very upset! What could he say to convince the horse to move?

44. The force pair of Newton's third law (a) consists of forces that are always opposite, but not always equal; (b) always cancels when the second law is applied to a body; (c) always acts on the same object; or (d) consists of forces that are equal and opposite, but act on different objects.

45. CQ Is there something wrong with the following statements? When a baseball is hit with a bat, there are equal and opposite forces on the bat and baseball. The forces then cancel, and there is no motion.

46. IE A book is sitting on a horizontal surface. (a) There is (are) (1) one, (2) two, or (3) three force(s) acting on the book. (b) Identify the reaction force to each force on the book.

47. CQ By using the right technique, a karate master can exert huge forces on objects. If a brick is hit by a fist with a force of 800 N, what else do we know, according to Newton's third law? Is this something you should try at home?

48. A person pushes on a block of wood that has been placed against a wall. Draw a free-body diagram and identify the reaction forces to all the forces on the block.

49. ■■ In an Olympic figure-skating event, a 60-kg male skater pushes a 45-kg female skater, causing her to accelerate at a rate of 2.0 m/s². At what rate will the male skater accelerate? What is the direction of his acceleration?

50. ■■ Jane and John, with masses of 50 kg and 60 kg, respectively, stand on a frictionless surface 10 m apart. John pulls on a rope that connects him to Jane, giving Jane an acceleration of 0.92 m/s² toward him. (a) What is John's acceleration? (b) If the pulling force is applied constantly, where will Jane and John meet?

4.5 More on Newton's Laws: Free-Body Diagrams and Translational Equilibrium

Note: In Exercises with strings and pulleys, "ideal conditions" means that the masses of the string(s) and pulley(s), as well as the friction of the pulley(s), should be neglected.

51. Draw a free-body diagram of a car coasting (with its engine off) up a long, straight ramp. Clearly mark the forces, and identify their sources.

52. IE ■ (a) When an object is on a inclined plane, the normal force exerted by the inclined plane on the object is (1) less than, (2) equal to, or (3) more than the weight of the object. Why? (b) For a 10-kg object on a 30° inclined plane, what are the object's weight and the normal force exerted on the object by the inclined place?

53. IE ■■ The weight of a 500-kg object is 4900 N. (a) When the object is on a moving elevator, its measured weight could be (1) zero, (2) between zero and 4900 N, (3) more than 4900 N, or (4) all of the preceding. Why? (b) Describe the motion if the object's measured weight is only 4000 N in a moving elevator.

54. ■■ A 75.0-kg person is standing on a scale in an elevator. What is the reading of the scale in newtons if the elevator is (a) at rest, (b) moving up at a constant velocity of 2.00 m/s, and (c) accelerating up at 2.00 m/s²?

55. ■■ In Exercise 54, what if the elevator is accelerating down?

56. ■■ Two boats pull a 75.0-kg water skier, as illustrated in ▼Fig. 4.32. (a) If each boat pulls with a force of 600 N and the skier travels at a constant velocity, what is the magnitude of the retarding force between the water and the skis? (b) Assuming that the retarding force remains constant, if each boat pulls with a force of 700 N, what is the magnitude of the acceleration of the skier?

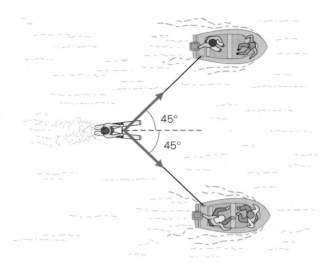

▲ **FIGURE 4.32   Double tow**   See Exercise 56.

57. ■■ (a) A 65-kg water skier is pulled by a boat with a horizontal force of 400 N due east with a water drag on the skis of 300 N. A sudden gust of wind supplies another horizontal force of 50 N on the skier at an angle of 60° north of east. At that instant, what is the skier's acceleration? (b) What would be the skier's acceleration if the wind force were in the opposite direction to that in part (a)?

58. ■■ A boy pulls a box of mass 30 kg with a force of 25 N in the direction shown in ▼Fig. 4.33. (a) Ignoring friction, what is the acceleration of the box? (b) What is the normal force exerted on the box by the ground?

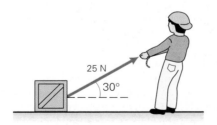

▲ **FIGURE 4.33   Pulling a box**   See Exercises 58 and 89.

59. ■■ A girl pushes a 25-kg lawn mower as shown in ▶Fig. 4.34. If $F = 30$ N and $\theta = 37°$, (a) what is the acceleration of the mower, and (b) what is the normal force exerted on the mower by the lawn? Ignore friction.

▲ **FIGURE 4.34   Mowing the lawn**   See Exercise 59.

60. IE ■■ (a) An Olympic skier coasts down a slope with an angle of inclination of 37°. Neglecting friction, there is (are) (1), one, (2) two, or (3) three force(s) acting on the skier. (b) What is the acceleration of the skier? (c) If the skier has a speed of 5.0 m/s at the top of the slope, what is his speed when he reaches the bottom of the 35-m-long slope?

61. ■■ A car coasts (engine off) up a 30° grade. If the speed of the car is 25 m/s at the bottom of the grade, what is the distance traveled by the car before it comes to rest?

62. ■■ A horizontal force of 40 N acting on a block on a frictionless level surface produces an acceleration of 2.5 m/s². A second block, with a mass of 4.0 kg, is dropped onto the first. What is the magnitude of the acceleration of the combination of blocks if the same force continues to act? (Assume that the second block does not slide on the first block.)

63. IE ■■ A rope is fixed at both ends on two trees, and a bag is hung in the middle of the rope, causing the rope to sag vertically. (a) The tension in the rope depends on (1) only the tree separation, (2) only the sag, (3) both the tree separation and sag, or (4) neither the tree separation nor the sag. (b) If the tree separation is 10 m, the mass of the bag is 5.0 kg, and the sag is 0.20 m, what is the tension in the line?

64. ■■ A 50-kg gymnast hangs vertically from a pair of parallel rings. (a) If the ropes supporting the rings are attached to the ceiling directly above, what is the tension in each rope? (b) If the ropes are supported so that they make an angle of 45° with the ceiling, what is the tension in each rope?

65. ■■ A 3000-kg truck tows a 1500-kg car by a chain. If the net forward force on the truck by the ground is 3200 N, (a) what is the acceleration of the car, and (b) what is the tension in the connecting chain?

66. ■■ Three blocks are pulled along a frictionless surface by a horizontal force as shown in ▶Fig. 4.35. (a) What is the acceleration of the system? (b) What are the tension forces in the light strings? (*Hint*: Can T_1 equal T_2? Investigate by drawing free-body diagrams of each block separately.)

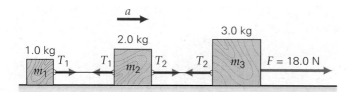

▲ FIGURE 4.35 Three-block system See Exercises 66 and 95.

67. ■■ Assume ideal conditions for the apparatus illustrated in ▼Fig. 4.36. What is the acceleration of the system if (a) $m_1 = 0.25$ kg, $m_2 = 0.50$ kg, and $m_3 = 0.25$ kg, and (b) $m_1 = 0.35$ kg, $m_2 = 0.15$ kg, and $m_3 = 0.50$ kg?

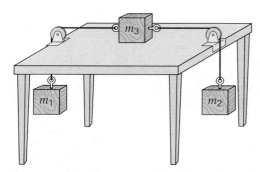

▲ FIGURE 4.36 Which way will they accelerate? See Exercises 67, 97, and 98.

68. ■■ The *Atwood machine* consists of two masses suspended from a fixed pulley, as shown in ▼Fig. 4.37. It is named after British scientist George Atwood (1746–1807), who

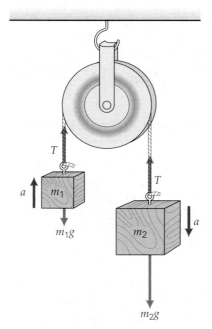

▲ FIGURE 4.37 Atwood machine See Exercises 68, 69, 70, and 107.

used it to study motion and to measure the value of g. If $m_1 = 0.55$ kg and $m_2 = 0.80$ kg, (a) what is the acceleration of the system, and (b) what is the magnitude of the tension in the string?

69. ■■ An Atwood machine (see Fig. 4.37) has suspended masses of 0.25 kg and 0.20 kg. Under ideal conditions, what will be the acceleration of the smaller mass?

70. ■■■ One mass, $m_1 = 0.215$ kg , of an ideal Atwood machine (see Fig. 4.37) rests on the floor 1.10 m below the other mass, $m_2 = 0.255$ kg. (a) If the masses are released from rest, how long does it take m_2 to reach the floor? (b) How high will mass m_1 ascend from the floor? [*Hint*: When m_2 hits the floor, m_1 continues to move upward.]

71. ■■■ A 0.20-kg ball is released from a height of 10 m above the beach; the impression the ball makes in the sand is 5.0 cm deep. What is the average force acting on the ball by the sand?

72. ■■■ In the ideal apparatus shown in ▼Fig. 4.38, $m_1 = 2.0$ kg. What is m_2 if both masses are at rest? How about if both masses are moving at constant velocity?

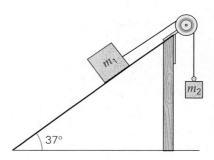

▲ FIGURE 4.38 Inclined Atwood machine See Exercises 72, 73, and 99.

73. ■■■ In the ideal setup shown in Fig. 4.38, $m_1 = 3.0$ kg and $m_2 = 2.5$ kg. (a) What is the acceleration of the masses? (b) What is the tension in the string?

4.6 Friction

Note: Neglect air resistance, unless otherwise stated.

74. In general, the frictional force (a) is greater for smooth than rough surfaces, (b) depends on sliding speeds, (c) is proportional to the normal force, or (d) depends significantly on the surface area of contact.

75. **CQ** Identify the direction of the friction force in the following cases: (a) a book sitting on a table; (b) a box sliding on a horizontal surface; (c) a car making a turn on a flat road; (d) the initial motion of a machine part delivered on a conveyor belt in an assembly line.

76. The coefficient of kinetic friction, μ_k (a) is usually greater than the coefficient of static friction, μ_s; (b) usually equals μ_s; (c) is usually smaller than μ_s; or (d) equals the applied force that exceeds the maximum static force.

77. **CQ** The purpose of a car's antilock brakes is to prevent the wheels from locking up so as to keep the car rolling rather than sliding. Why would rolling decrease the stopping distance as compared with sliding?

78. **CQ** Is it easier to push or pull a lawn mower at an angle? Draw a free-body diagram, and explain.

79. **CQ** ▼Fig. 4.39 shows the front and rear wings of an Indy racing car. These wings generate *down force*, the vertical downward force produced by the air moving over the car. Why is such a down force desired? An Indy car can create a down force equal to twice its weight. Why not simply make the cars heavier?

▲ **FIGURE 4.39 Down force** See Exercise 79.

80. **CQ** (a) We commonly say that friction opposes motion. Yet, when we walk, the frictional force is in the direction of our motion (Fig. 4.18). Is there an inconsistency in terms of Newton's second law? Explain. (b) What effects would wind have on air resistance? [*Hint*: The wind can blow in different directions.]

81. **CQ** Why are drag-racing tires wide and smooth, whereas passenger-car tires are narrower and have tread (▶Fig. 4.40)? Are there frictional and/or safety considerations? Does this difference between the tires contradict the fact that friction is independent of surface area?

82. ■ In moving a 35.0-kg desk from one side of a classroom to the other, a professor finds that a horizontal force of 275 N is necessary to set the desk in motion, and a force of 195 N is necessary to keep it in motion at a constant speed. What are the coefficients of (a) static and (b) kinetic friction between the desk and the floor?

83. ■ A 40-kg crate is at rest on a level surface. If the coefficient of static friction between the crate and the surface is 0.69, what horizontal force is required to get the crate moving?

▲ **FIGURE 4.40 Racing tires versus passenger-car tires: safety** See Exercise 81.

84. **IE** ■ A 20-kg box sits on a rough horizontal surface. When a horizontal force of 120 N is applied, the object accelerates at 1.0 m/s^2. (a) If the applied force is doubled, the acceleration will (1) increase, but less than double; (2) also double; or (3) increase, but more than double. Why? (b) Calculate the acceleration to prove your answer to (a).

85. ■ The coefficients of static and kinetic friction between a 50-kg box and a horizontal surface are 0.60 and 0.40, respectively. (a) What is the acceleration of the object if a 250-N horizontal force is applied to the box? (b) What is the acceleration if the applied force is 350 N?

86. ■■ A 1500-kg automobile travels at a speed of 90 km/h along a straight concrete highway. Faced with an emergency situation, the driver jams on the brakes, and the car skids to a stop. What will be the car's stopping distance for (a) dry pavement and (b) wet pavement, respectively?

87. ■■ A hockey player hits a puck with his stick, giving the puck an initial speed of 5.0 m/s. If the puck slows uniformly and comes to rest in a distance of 20 m, what is the coefficient of kinetic friction between the ice and the puck?

88. ■■ A packing crate is placed on a 20° inclined plane. If the coefficient of static friction between the crate and the plane is 0.65, will the crate slide down the plane? Justify your answer.

89. ■■ In Exercise 58 and Fig. 4.33, if the coefficient of kinetic friction between the box and the surface is 0.03 (waxed wood box on snow), what is the acceleration of the box?

90. ■■ Suppose the slope conditions for the skier shown in ▶Fig. 4.41 are such that the skier travels at a constant velocity. From the photo, could you find the coefficient of kinetic friction between the snowy surface and the skis? If so, describe how this would be done.

91. ■■ A 5.0-kg wooden block is placed on an adjustable wooden inclined plane. (a) What is the angle of incline above which the block will *start* to slide down the plane?

▲ **FIGURE 4.41 A downslope run** See Exercise 90.

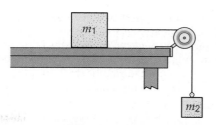

▲ **FIGURE 4.43 Friction and motion** See Exercise 96.

(b) At what angle of incline will the block then slide down the plane at a constant speed?

92. ■■ A block that has a mass of 2.0 kg and is 10 cm wide on each side just begins to slide down an inclined plane with a 30° angle of incline (▼Fig. 4.42). Another block of the same height and same material has base dimensions of 20 cm × 10 cm and thus a mass of 4.0 kg. (a) At what critical angle will the more massive block start to slide down the plane? Why? (b) Estimate the coefficient of static friction between the block and the plane.

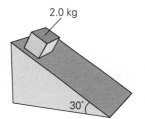
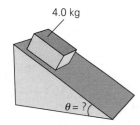

2.0 kg   4.0 kg

30° $\theta = ?$

▲ **FIGURE 4.42 At what angle will it begin to slide?** See Exercise 92.

93. ■■ While being unloaded from a truck, a 10-kg suitcase is placed on a flat ramp inclined at 37°. When released from rest, the suitcase accelerates down the ramp at 0.15 m/s². What is the coefficient of kinetic friction between the suitcase and the ramp?

94. ■■ For the situation shown in Fig. 4.29, what is the minimum coefficient of static friction between the block and the surface that will keep the block from moving? ($F_1 = 5.0$ N, $F_2 = 4.0$ N , and $m = 5.0$ kg.)

95. ■■ For the system illustrated in Fig. 4.35, if $\mu_s = 0.45$ and $\mu_k = 0.35$ between the blocks and the surface, what applied forces will (a) set the blocks in motion and (b) move the blocks at a constant velocity?

96. ■■ In the apparatus shown in ▶Fig. 4.43, $m_1 = 10$ kg and the coefficients of static and kinetic friction between m_1 and the table are 0.60 and 0.40, respectively. (a) What mass of m_2 will set the system in motion? (b) After the system moves, what is the acceleration?

97. ■■■ For the apparatus in Fig. 4.36, what is the minimum value of the coefficient of static friction between the block (m_3) and the table that would keep the system at rest if

$m_1 = 0.25$ kg, $m_2 = 0.50$ kg , and $m_3 = 0.75$ kg? (Assume ideal conditions for the string and pulleys.)

98. ■■■ If the coefficient of kinetic friction between the block and the table in Fig. 4.36 is 0.560, and $m_1 = 0.150$ kg and $m_2 = 0.250$ kg, (a) what should m_3 be if the system is to move with a constant speed? (b) If $m_3 = 0.100$ kg, what is the magnitude of the acceleration of the system? (Assume ideal conditions for the string and pulleys.)

99. ■■■ In the apparatus shown in Fig. 4.38, $m_1 = 2.0$ kg and the coefficients of static and kinetic friction between m_1 and the inclined plane are 0.30 and 0.20, respectively. (a) What is m_2 if both masses are at rest? (b) What is m_2 if both masses are moving at constant velocity? Neglect all friction.

Additional Exercises

100. ■■ At the end of most landing runways in airports, an extension of the runway is constructed using a special substance called *formcrete*. Formcrete can support the weight of cars, but crumbles under the weight of airplanes, so as to slow them down if they run off the end of a runway. If a plane of mass 2.00×10^5 kg is to stop from a speed of 25.0 m/s on a 100-m-long stretch of formcrete, what is the average force exerted on the plane by the formcrete?

101. A rifle weighs 50.0 N, and its barrel is 0.750 m long. It shoots a 25.0-g bullet, which leaves the barrel at a speed (muzzle velocity) of 300 m/s after being uniformly accelerated. What is the magnitude of the force exerted on the rifle by the bullet?

102. The maximum load that can safely be supported by a rope in an overhead hoist is 400 N. What is the maximum acceleration that can safely be given to a 25-kg object being hoisted vertically upward?

103. The coefficient of static friction between a 9.0-kg object and a horizontal surface is 0.45. Would a force of 35 N applied 20° above the horizontal cause the object to move from rest? If so, what would be the object's acceleration?

104. A 2.0-kg object travels at a constant velocity of 4.8 m/s northward. It is then acted on by forces of 6.5 N to the north and 8.5 N to the south. (a) How far will the object travel before coming to rest? (b) What will be the object's position 1.5 s after it comes momentarily to rest, with the forces still acting?

105. In the operation of a machine, a 5.0-kg steel part is to move on a horizontal steel surface. Force is applied to the part downward at an angle of 30° from the horizontal. (a) What is the magnitude of the applied force required to set the part in motion if the surface is dry? (b) If the surface is lubricated, by what factor is the force reduced?

106. A crate weighing 9.80×10^3 N is pulled up a 37° incline by a force parallel to the plane. If the coefficient of kinetic friction between the crate and the surface of the plane is 0.750, what is the magnitude of the applied force required to move the crate at a constant velocity?

107. For an Atwood machine (see Fig. 4.37) with suspended masses of 0.30 kg and 0.40 kg, the acceleration of the masses is measured as 0.95 m/s^2. What is the effective force of friction for the system?

108. A loaded jet plane with a weight of 2.75×10^6 N is ready for takeoff. If its engines supply 6.35×10^6 N of net thrust, how long a runway will the plane need to reach its minimum takeoff speed of 285 km/h?

109. A 0.45-kg shuffleboard puck is given an initial speed of 4.5 m/s down the flat playing surface. If the coefficient of sliding friction between the puck and the surface is 0.20, how far will the puck slide before coming to rest?

110. A 135-m-long ramp is to be built for a ski jump. If a skier starting from rest at the top is to have a speed no faster than 24 m/s at the bottom, what should be the maximum angle of inclination? Ignore friction.

111. A school bus pulls into an intersection as a car approaches at a speed of 25 km/h on an icy street. Seeing the bus from 26 m away, the driver of the car steps on and inadvertently locks the brakes, causing the car to slide toward the intersection. (The car does not have antilock brakes.) If the coefficient of kinetic friction between the car's tires and the icy road is 0.10, does the car hit the bus? Justify your answer.

112. While catching a baseball that is traveling horizontally at a speed of 15 m/s, a player's glove and arm move straight backward 25 cm from the time of contact to the time the ball comes to rest. If the ball has a mass of 0.14 kg, what is the average force on the ball during that interval?

Work and Energy

A description of pole vaulting might be as follows: An athlete runs with a pole, plants it into the ground, and tries to vault his body over a bar set at a certain height. However, a physicist might give a different description: The athlete has chemical potential energy stored in his body. He uses this potential energy to do work in running down the path to gain speed, or kinetic energy. When he plants the pole, most of his kinetic energy goes into elastic potential energy of the bent pole as shown in the photo. This potential energy is used to lift the vaulter or to do work against gravity, and is partially converted into gravitational potential energy. At the top, there is just enough kinetic energy left to carry the vaulter over the bar. On the way down, the gravitational potential energy is converted back to kinetic energy, which is absorbed by the mat in doing work to stop the fall. The pole vaulter participates in a game of work–energy give and take (see Fig. 5.28).

This chapter centers on these two concepts that are important in both science and everyday life—*work* and *energy*. We commonly think of work as being associated with doing or accomplishing something. Because work makes us physically (and sometimes mentally) tired, we have invented machines that we use to decrease the amount of effort we expend personally. Thinking about energy tends to bring to mind the cost of fuel for transportation and heating, or perhaps the food that supplies the energy our bodies need to carry out life processes and to do work.

Although these notions do not really define work and energy, they point us in the right direction. As you might have guessed, work and energy are closely related. In physics, as in everyday life, when something possesses energy, it has the ability to do work. For example, water rushing through the sluices of a dam has energy of motion, and this energy allows the water to do the work

141

of driving a turbine or dynamo. Conversely, no work can be performed without energy.

Energy exists in various forms: There is mechanical energy, chemical energy, electrical energy, heat energy, nuclear energy, and so on. A transformation from one form to another may take place, but the total amount of energy is *conserved*, or always remains the same. This is the point that makes the concept of energy so useful. When a physically measurable quantity is conserved, it not only gives us an insight that leads to a better understanding of nature, but also usually provides another approach to practical problems. (You will be introduced to other conserved quantities during the course of our study of physics.)

5.1 Work Done by a Constant Force

OBJECTIVES: To (a) define mechanical work and (b) compute the work done in various situations.

The word *work* is commonly used in a variety of ways: We go to work; we work on projects; we work at our desks or on computers; we work problems. In physics, however, *work* has a very specific meaning. Mechanically, work involves force and displacement, and we use the word *work* to describe quantitatively what is accomplished when a force moves an object through a distance. In the simplest case of a *constant* force acting on an object, work is defined as follows:

> The **work** done by a constant force acting on an object is equal to the product of the magnitudes of the displacement and the component of the force parallel to that displacement.

Work—involves force and displacement

Note: The product of two vectors (force and displacement) in this case is a special type of vector multiplication and yields a scalar quantity equal to $(F\cos\theta)d$. Thus, work is a scalar—it does not have direction. It can, however, be positive, zero, or negative, depending on the angle.

Work then involves moving an object through a distance. A force may be applied, as in ▾Fig. 5.1a, but *if there is no motion (no displacement), then no work is done*. For a constant force F acting *in the same direction* as the displacement d (Fig. 5.1b), the work W is defined as the product of their magnitudes:

$$W = Fd \qquad (5.1)$$

(As you might expect, when work is done in Fig. 5.1b, energy is expended. We shall discuss the relationship between work and energy in Section 5.3.)

In general, work is done on an object only by a force, or force *component*, parallel to the line of motion or displacement of the object (Fig. 5.1c). That is, if the force acts at an angle θ to the object's displacement, then $F_\parallel = F\cos\theta$ is the component of the force parallel to the displacement. Thus, a more general equation for work done by a constant force is

PHYSLET® ILLUSTRATION

Work and Angles

$$W = F_\parallel d = (F\cos\theta)d \qquad \textit{work done by a constant force} \quad (5.2)$$

▼ **FIGURE 5.1 Work done by a constant force—the product of the magnitudes of the parallel component of force and the displacement** **(a)** If there is no displacement, no work is done: $W = 0$. **(b)** For a constant force in the same direction as the displacement, $W = Fd$. **(c)** For a constant force at an angle to the displacement, $W = (F\cos\theta)d$.

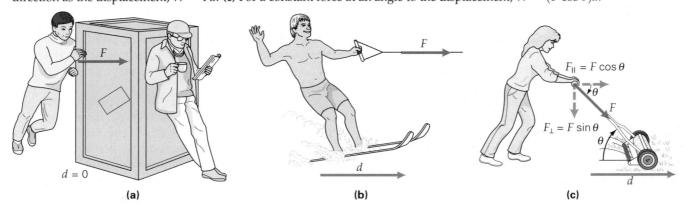

Notice that θ is the angle *between* the force and the displacement vectors. To remind yourself of this factor, you can write $\cos \theta$ between the magnitudes of the force and displacement, as in Eq. 5.2. If $\theta = 0°$ (i.e., force and displacement are in the same direction as in Fig. 5.1b), then $W = F(\cos 0°)d = Fd$, so Eq. 5.2 reduces to Eq. 5.1. The perpendicular component of the force, $F_\perp = F \sin \theta$, does no work, since there is no displacement in this direction.

The units of work can be determined from the equation $W = Fd$. With force in newtons and displacement in meters, work has the SI unit of newton-meter ($N \cdot m$). This unit is given the special name *joule* (J):

$$Fd = W$$
$$1\,N \cdot m = 1\,J$$

For example, the work done by a force of 25 N on an object as the object moves through a parallel displacement of 2.0 m is $W = Fd = (25\,N)(2.0\,m) = 50\,N \cdot m$, or 50 J.

From the previous displayed equation, we also see that in the British system, work would have the unit pound-foot. However, this name is commonly written in reverse: The British standard unit of work is the *foot-pound* ($ft \cdot lb$). One $ft \cdot lb$ is equal to 1.36 J.

We can analyze work graphically. Suppose a constant force F acts on an object as it moves a distance x. Then $W = Fx$, and if F versus x is plotted, a straight-line graph is obtained such as shown in the Learn by Drawing figure. The area under the line is Fx, so this area is equal to the work done by the force over the given distance. We will consider a nonconstant, or variable, force later.*

Remember that *work is a scalar quantity* and, as such, may have a positive or negative value. In Fig. 5.1b, the work is positive, because the force acts in the same direction as the displacement (and $\cos 0°$ is positive). The work is also positive in Fig. 5.1c, because a force component acts in the direction of the displacement (and $\cos \theta$ is positive).

However, if the force, or a force component, acts in the opposite direction of the displacement, the work is negative, since the cosine term is negative. For example, for $\theta = 180°$ (force opposite to the displacement), $\cos 180° = -1$, so the work is negative: $W = Fd = (F \cos 180°)d = -Fd$. An example is a braking force that slows down or decelerates an object. See Learn by Drawing on p. 144.

Learn by Drawing

Work: Area Under the F versus x Curve

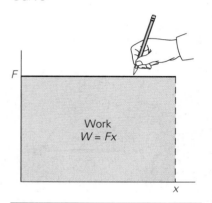

Example 5.1 ■ Applied Psychology: Mechanical Work

A student holds her psychology textbook, which has a mass of 1.5 kg, out of a second-story dormitory window until her arm is tired; then she releases it (▶Fig. 5.2). (a) How much work is done on the book by the student in simply holding it out the window? (b) How much work is done by the force of gravity during the time in which the book falls 3.0 m?

Thinking It Through. Analyze the situations in terms of the definition of work, keeping in mind that force and displacement are the key factors.

Solution. Listing the data, we have

Given: $v_o = 0$ (initially at rest) *Find:* (a) W (work done by student in holding)
$m = 1.5\,kg$ (b) W (work done by gravity in falling)
$d = 3.0\,m$

(a) Even though the student gets tired (because work is performed within the body to maintain muscles in a state of tension), she does *no work on the book* in merely holding it stationary. She exerts an upward force on the book (equal in magnitude to its weight), but the displacement is zero in this case ($d = 0$). Thus, $W = Fd = F \times 0 = 0\,J$.
(b) While the book is falling, the only force acting on it is the force of gravity, which is equal in magnitude to the weight of the book: $F = w = mg$ (neglecting air resistance).

*Work is the area under the F versus x curve even if the curve is not a straight line. Finding the work in such cases generally requires advanced mathematics.

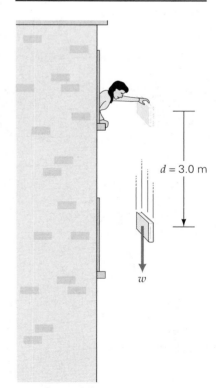

▲ **FIGURE 5.2 Mechanical work requires motion** See Example 5.1.

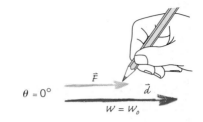

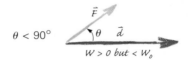

$\theta = 0°$

$W = W_o$

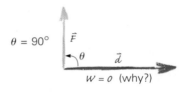

$\theta < 90°$

$W > 0$ but $< W_o$

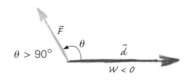

$\theta = 90°$

$W = 0$ (why?)

$\theta > 90°$

$W < 0$

$\theta = 180°$

$W = -W_o$

Definition of net work

The displacement is in the same direction as the force ($\theta = 0°$) and has a magnitude of $d = 3.0$ m, so the work done by gravity is

$$W = F(\cos 0°)d = (mg)d = (1.5 \text{ kg})(9.8 \text{ m/s}^2)(3.0 \text{ m}) = +44 \text{ J}$$

Follow-up Exercise. A 0.20-kg ball is thrown upward. How much work is done on the ball by gravity as the ball rises between heights of 2.0 m and 3.0 m? *(Answers to all Follow-up Exercises are at the back of the text.)*

Example 5.2 ■ Yard Work: Parallel Component of a Force

If the person in Fig. 5.1c pushes on the lawn mower with a constant force of 90.0 N at an angle of 40° to the horizontal, how much work does she do in pushing it a horizontal distance of 7.50 m?

Thinking It Through. Only the component of the force parallel to the displacement does work on the mower. Clearly identify the angle between the force and the displacement.

Solution.

Given: $F = 90.0$ N   *Find:* W (work done by person pushing the mower)
$\theta = 40°$
$d = 7.50$ m

Here, the horizontal component of the applied force, $F \cos \theta$, is parallel to the displacement, so Eq. 5.2 applies:

$$W = (F \cos 40°)d = (90.0 \text{ N})(0.766)(7.50 \text{ m}) = +517 \text{ J}$$

Work is done by the horizontal component of the force, but the vertical component does no work, because there is no vertical displacement.

Follow-up Exercise. When you push a wheelbarrow on a level surface, the force you apply has an upward component. Is work done on the wheelbarrow by this component?

We commonly specify which force is doing work *on* which object. For example, the force of gravity does work on a falling object, such as the book in Example 5.1. Also, when you lift an object, *you* do work *on* the object. We sometimes describe this as doing work *against* gravity, because the force of gravity acts in the direction opposite that of the applied lift force and opposes it. For example, an average-sized apple has a weight of about 1 N. So, if you lifted such an apple a distance of 1 m with a force equal to its weight, you would have done 1 J of work against gravity $[W = Fd = (1 \text{ N})(1 \text{ m}) = 1 \text{ J}]$. This example gives you an idea of how much work 1 J represents.

In both Examples 5.1 and 5.2, work was done by a single constant force. If more than one force acts on an object, the work done by each can be calculated separately:

The *total,* or *net, work* is the work done by all the forces, or the scalar sum of those quantities of work.

This concept is illustrated in Example 5.3.

Example 5.3 ■ Total or Net Work

A 0.75-kg block slides with a uniform velocity down a 20° inclined plane (►Fig. 5.3). (a) How much work is done by the force of friction on the block as it slides the total length of the plane? (b) What is the net work done on the block? (c) Discuss the net work done if the angle of incline is adjusted so that the block accelerates down the plane.

Thinking It Through. (a) The length of the plane can be found using trigonometry, so this part boils down to finding the force of friction. (b) The net work is the sum of all the

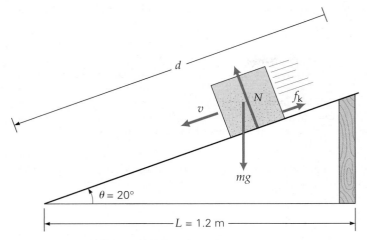

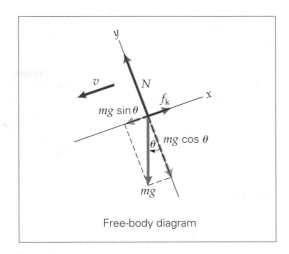

Free-body diagram

▲ **FIGURE 5.3 Total or net work**   See Example 5.3.

work done by the individual forces. (*Note:* Since the block has a uniform, or constant, velocity, the net force on it is zero. This observation should tell you the answer, but it will be shown explicitly in the solution.) (c) If there is acceleration, Newton's second law applies, which involves a net force, so there will be net work.

Solution.   We list the information that is given. In addition, it is equally important that we list specifically what is to be found.

Given: $m = 0.75\text{kg}$ *Find:*   (a) W_f (work done on the block by friction)
$\quad\quad\theta = 20°$ (b) W_{net} (net work on the block)
$\quad\quad L = 1.2$ m (from Fig. 5.3)   (c) W (discuss net work with block
$\quad\quad\quad\quad\quad\quad\quad\quad\quad\quad\quad\quad\quad\quad\quad\quad$ accelerating)

(a)   Note from Fig. 5.3 that only two forces do work, because there are only two forces parallel to the motion: f_k, the force of kinetic friction, and $mg \sin\theta$, the component of the block's weight acting down the plane. The normal force N and $mg \cos\theta$, the components of the block's weight acting perpendicular to the plane, do no work on the block. (Why?)
We first find the work done by the frictional force:

$$W_f = f_k(\cos 180°)d = -f_k d = -\mu_k N d$$

Note: Recall the discussion of friction in Section 4.6.

The angle 180° indicates that the force and displacement are in opposite directions. (It is common in such cases to write $W_f = -f_k d$ directly, since kinetic friction typically opposes motion.) The distance d the block slides down the plane can be found by using trigonometry. Note that $\cos\theta = L/d$, so

$$d = \frac{L}{\cos\theta}$$

We know that $N = mg \cos\theta$, but what is μ_k? It would appear that we are lacking some information. When this situation occurs, you should look for another approach to solve the problem. As noted earlier, there are only two forces parallel to the motion, and they are opposite, so with a constant velocity their magnitudes are equal, $f_k = mg \sin\theta$. Thus,

$$W_f = -f_k d = -(mg \sin\theta)\left(\frac{L}{\cos\theta}\right) = -mgL \tan 20°$$
$$= -(0.75 \text{ kg})(9.8 \text{ m/s}^2)(1.2 \text{ m})(0.364) = -3.2 \text{ J}$$

(b)   To find the net work, we need to calculate the work done by gravity and then add it to our result in part (a). Since $F_\parallel$ for gravity is just $mg \sin\theta$, we have

$$W_g = F_\parallel d = (mg \sin\theta)\left(\frac{L}{\cos\theta}\right) = mgL \tan 20° = +3.2 \text{ J}$$

where the calculation is the same as in (a) except for the sign. Then,

$$W_{net} = W_g + W_f = +3.2 \text{ J} + (-3.2 \text{ J}) = 0$$

Remember that work is a scalar quantity, so scalar addition is used to find net work.

(c) If the block accelerates down the plane, then from Newton's second law, we have $F_{net} = mg \sin \theta - f_k = ma$. The component of the gravitational force ($mg \sin \theta$) is greater than the opposing frictional force (f_k), so there is net work done on the block, because now $|W_g| > |W_f|$. You may be wondering what the effect of nonzero net work is. As you will learn shortly, nonzero net work causes a change in the amount of energy an object has.

Follow-up Exercise. In part (c) of this Example, is it possible for the frictional work to be greater in magnitude than the gravitational work? What would this condition mean in terms of the block's speed?

Work on an Incline

Problem-Solving Hint

Note how in part (a) of Example 5.3, the equation for W_f was simplified by using the algebraic expressions for N and d instead of by computing these quantities initially. It is a good rule of thumb not to plug numbers into an equation until you have to. Simplifying an equation through cancellation is easier with symbols and saves computation time.

5.2 Work Done by a Variable Force

OBJECTIVES: **To (a) differentiate between work done by constant and variable forces and (b) compute the work done by a spring force.**

The discussion in the preceding section was limited to work done by constant forces. In general, however, forces are variable; that is, they change in magnitude and/or angle with time and/or position. For example, someone might push harder and harder on an object to overcome the force of static friction, until the applied force exceeds $f_{s_{max}}$. However, the force of static friction does no work, because there is no motion or displacement.

An example of a variable force that does work is illustrated in ▶Fig. 5.4, which depicts a spring being stretched. As the spring is stretched (or compressed) farther and farther, its restoring force (the force that opposes the stretching or compression) becomes greater, and an increased applied force is required. For most springs, the applied force F is directly proportional to the change in length of the spring from its unstretched length. In equation form, this relationship is expressed as

$$F = k \Delta x = k(x - x_o)$$

or, if we choose $x_o = 0$, then

$$F = kx$$

where x now represents the distance the spring is stretched (or compressed) from its unstretched length. As can be seen, the force varies with x. We describe this relationship by saying that the *force is a function of position*.

The k in this equation is a constant of proportionality and is commonly called the **spring constant**, or **force constant**. The greater the value of k, the stiffer or stronger is the spring. As you should be able to prove to yourself, the SI unit of k is newton per meter (N/m).

Note: In Fig. 5.4, the hand applies a variable force F in stretching the spring. At the same time, the spring exerts an equal and opposite force F_s on the hand.

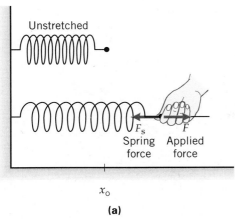

Unstretched

F_s F
Spring Applied
force force

x_o

(a)

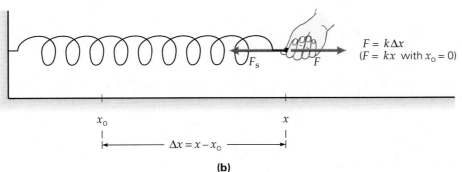

$F = k\Delta x$
$(F = kx$ with $x_o = 0)$

F_s F

x_o x

$\Delta x = x - x_o$

(b)

◀ **FIGURE 5.4 Spring force**
(a) An applied force F stretches the spring, and the spring exerts an equal and opposite force F_s on the hand. **(b)** The magnitude of the force depends on the change Δx spring's length. This change is often referenced to the end of the unstretched spring, x_o.

The relationship expressed by the equation $F = kx$ holds only for ideal springs. Real springs approximate this linear relationship between force and displacement within certain limits. If a spring is stretched beyond a certain point, called its *elastic limit*, the spring will be permanently deformed, and $F = kx$ will no longer apply.

Note that a spring exerts a force that is equal and opposite to the external applied force F. Thus,

$$F_s = -k\,\Delta x = -k(x - x_o)$$

or, if $x_o = 0$,

$$F_s = -kx \qquad \text{ideal spring force} \qquad (5.3)$$

The minus sign indicates that the spring force acts in the direction opposite to the displacement when the spring is either stretched or compressed. Equation 5.3 is a form of what is known as *Hooke's law*, named after Robert Hooke, a contemporary of Newton.

To compute the work done by variable forces generally requires calculus. But we are fortunate in that the spring force is a special case that can be computed by using the average force. We examined an analogous case involving average velocity, $\bar{v} = (v + v_o)/2$, in our study of kinematics in Chapter 2, and it is instructive to explore this analogy graphically.

Plots of F versus x and of v versus t are shown in ▶Fig. 5.5. The graphs have straight-line slopes of k and a, respectively, with $F = kx$ and $v = at$. [The applied force F, rather than the spring force F_s, is plotted to avoid a negative slope $(F_s = -kx)$ and thereby allow analogous graphs.]

By the same reasoning given for v in Section 2.3, the average force $\bar{F}$ can be expressed as

$$\bar{F} = \frac{F + F_o}{2}$$

► **FIGURE 5.5 Work done by a uniformly variable force** The work done by a uniformly varying force of the form $F = kx$ is $W = \frac{1}{2}kx^2$. A plot of this special case of a variable force is graphically analogous to a plot of v versus t for a uniformly varying velocity starting at rest ($v_o = 0$): $v = at$. The work W and the distance y are equal to the areas under the respective straight lines.

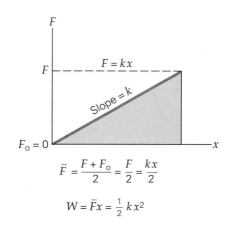

$$\bar{F} = \frac{F + F_o}{2} = \frac{F}{2} = \frac{kx}{2}$$

$$W = \bar{F}x = \frac{1}{2}kx^2$$

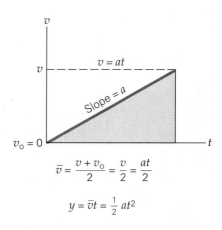

$$\bar{v} = \frac{v + v_o}{2} = \frac{v}{2} = \frac{at}{2}$$

$$y = \bar{v}t = \frac{1}{2}at^2$$

or, since $F_o = 0$ (why?), as

$$\bar{F} = \frac{F}{2}$$

Thus, the work done in stretching or compressing the spring an amount x from its unstretched length is

$$W = \bar{F}x = \frac{Fx}{2}$$

Since $F = kx$, the work done is

$$W = \tfrac{1}{2}kx^2 \qquad \begin{array}{l} \textit{work done in stretching} \\ \textit{(or compressing)} \\ \textit{a spring from } x_o = 0 \end{array} \qquad (5.4)$$

Note that the work done is just the area under the curve in Fig. 5.5.

Example 5.4 ■ Determining the Spring Constant

A 0.15-kg mass is attached to a vertical spring and descends a distance of 4.6 cm below its original position. It then hangs at rest (►Fig. 5.6). An additional 0.50-kg mass is then suspended from the first mass. What is the total extension of the spring? (Neglect the mass of the spring.)

Thinking It Through. The spring constant k appears in Eq. 5.3. Therefore, to find the value of k for a particular instance, the spring force and distance the spring is stretched (or compressed) must be known.

Solution. The data given are as follows:

Given: $m_1 = 0.15$ kg   *Find:* x (total stretch distance)
$\qquad x_1 = 4.6$ cm $= 0.046$ m
$\qquad m_2 = 0.50$ kg

The total stretch distance is given by $x = F/k$, where F is the applied force, which in this case is the weight of the mass suspended on the spring. However, the spring constant k is not given. This quantity may be found from the data pertaining to the suspension of m_1 and resulting displacement x_1. (This method is commonly used to determine spring constants.) As seen in Fig. 5.6a, the magnitudes of the weight force and the restoring spring force are equal, since $a = 0$, so we may equate them (with the directional minus signs omitted):

$$F_s = kx_1 = m_1g$$

Solving for k, we obtain

$$k = \frac{m_1 g}{x_1} = \frac{(0.15 \text{ kg})(9.8 \text{ m/s}^2)}{0.046} = 32 \text{ N/m}$$

Then, knowing k, we find the total extension of the spring from the balanced-force situation shown in Fig. 5.6b:

$$F_s = (m_1 + m_2)g = kx$$

Thus,

$$x = \frac{(m_1 + m_2)g}{k} = \frac{(0.15 \text{ kg} + 0.50 \text{ kg})(9.8 \text{ m/s}^2)}{32 \text{ N/m}} = 0.20 \text{ m (or 20 cm)}$$

Follow-up Exercise. How much work is done by gravity in stretching the spring through both displacements in Example 5.4?

Problem-Solving Hint

The reference position x_o for the change in length of a spring is arbitrary and is usually chosen for convenience. *The important quantity in computing work is the difference in position, Δx, or the net change in the length of the spring from its unstretched length.* As shown in ▼Fig. 5.7 for a mass suspended on a spring, x_o can be referenced to the unloaded length of the spring or to the loaded position, which may be taken as the zero position for convenience. In Example 5.4, x_o was referenced to the end of the unloaded spring. Also, with the displacement only in one direction, downward can be designated as the $+x$-direction in order to avoid minus signs.

When the net force on the suspended mass is zero, the mass is said to be at its *equilibrium position* (as in Fig. 5.7a with m_1 suspended). This position, rather than the unloaded length, may be taken as a zero reference ($x = x_o = 0$; see Fig. 5.7b). The equilibrium position is a convenient reference point for cases in which the mass oscillates up and down on the spring. (We will describe this motion in Chapter 13.) Notice that, in general, there are both positive and negative directions. The chosen signs are arbitrary, but must be adhered to for the entire solution to the problem. Also, since the displacement is in the vertical direction, the x's are often replaced by y's.

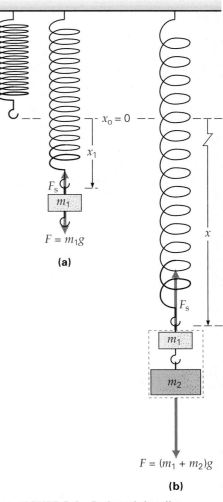

▲ **FIGURE 5.6 Determining the spring constant and the work done in stretching a spring** See Example 5.4.

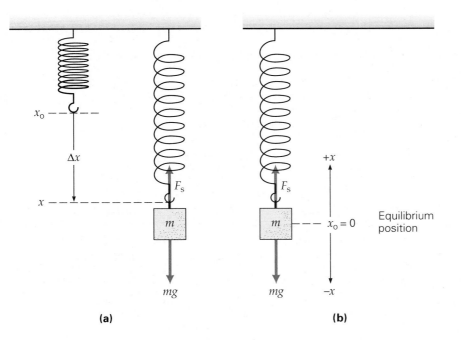

◄ **FIGURE 5.7 Displacement reference** The reference position x_o is arbitrary and is usually chosen for convenience. It may be **(a)** at the end of the spring at its unloaded position or **(b)** at the equilibrium position when a mass is suspended on the spring. The latter is particularly convenient in cases in which the mass oscillates up and down on the spring.

5.3 The Work–Energy Theorem: Kinetic Energy

OBJECTIVES: To (a) study the work–energy theorem and (b) apply it in solving problems.

Now that we have an operational definition of work, we are ready to look at how work is related to energy. Energy is one of the most important concepts in science. We describe it as a quantity that objects or systems possess. Basically, work is something that is *done on* objects, whereas energy is something that objects *have*—the ability to do work.

One form of energy that is closely associated with work is *kinetic energy*. (Another form of energy, *potential energy*, will be described in Section 5.4.) Consider an object at rest on a frictionless surface. Let a horizontal force act on the object and set it in motion. Work is done *on* the object, but where does the work "go," so to speak? It goes into setting the object into motion, or changing its *kinetic* conditions. Because of its motion, we say the object has gained energy—kinetic energy, which gives it the capability to do work.

For a constant net force doing work on a moving object, as illustrated in ▼Fig. 5.8, the force does an amount of work $W = Fx$. But what are the kinematic effects? The force causes the object to accelerate, and from Eq. 2.11, $v^2 = v_o^2 + 2ax$ (with $x_o = 0$),

$$a = \frac{v^2 - v_o^2}{2x}$$

where v_o may or may not be zero. Writing the magnitude of the force in Newton's second law form ($F = ma$, where $F = F_{net}$) and then substituting in the expression for a from the previous equation gives

$$F = ma = m\left(\frac{v^2 - v_o^2}{2x}\right)$$

Using this expression in the equation for work, we have

$$W = Fx = m\left(\frac{v^2 - v_o^2}{2x}\right)x$$
$$= \tfrac{1}{2}mv^2 - \tfrac{1}{2}mv_o^2$$

Definition of kinetic energy—the energy of motion

It is convenient to define $\tfrac{1}{2}mv^2$ as the **kinetic energy** K of the moving object:

$$K = \tfrac{1}{2}mv^2 \quad \text{kinetic energy} \tag{5.5}$$

SI unit of energy: joule (J)

▶ **FIGURE 5.8 The relationship of work and kinetic energy** The work done on a block in moving it along a horizontal frictionless surface is equal to the change in the block's kinetic energy: $W = \Delta K$.

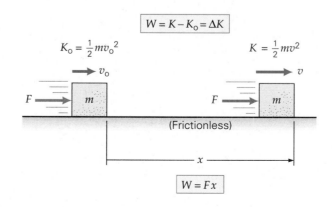

$W = K - K_o = \Delta K$

$K_o = \tfrac{1}{2}mv_o^2$ $K = \tfrac{1}{2}mv^2$

(Frictionless)

$W = Fx$

Kinetic energy is often called the *energy of motion*. Note that it is directly proportional to the square of the (instantaneous) speed of a moving object, and therefore cannot be negative.

Then, in terms of kinetic energy, the previous expression for work can be written as

$$W = \tfrac{1}{2}mv^2 - \tfrac{1}{2}mv_o^2 = K - K_o = \Delta K$$

Work–energy theorem

or

$$W = \Delta K \qquad (5.6)$$

where it is understood that W is the net work if more than one force acts on the object, as shown in Example 5.3. This equation is called the **work–energy theorem**; it relates the work done on an object to the change in the object's kinetic energy. That is, *the net work done on a body by all the forces acting on it is equal to the change in kinetic energy of the body*. Both work and energy have the units of joules, and both are *scalar* quantities. Keep in mind that the work–energy theorem is true in general for varying forces and not just for the special case considered in deriving Eq. 5.6.

To illustrate that net work is equal to the change in kinetic energy, recall that in Example 5.1 the force of gravity did $+44$ J of work on a book that fell from rest through a distance of $y = 3.0$ m. At that position and instant, the falling book had 44 J of kinetic energy. Since $v_o = 0$ in this case, $\tfrac{1}{2}mv^2 = mgy$, and $gy = v^2/2$. Substituting this expression into the equation for the work done on the falling book by gravity, we get

$$W = Fd = mgy = \frac{mv^2}{2} = K = \Delta K$$

where $K_o = 0$. Thus the kinetic energy gained by the book is equal to the net work done on it: 44 J in this case. (As an exercise, confirm this fact by calculating the speed of the book and evaluating its kinetic energy.)

What the work–energy theorem tells us is that when work is done, there is a change in or a transfer of energy. In general, then, we might say that *work is a measure of the transfer of kinetic energy*. For example, a force doing work on an object that causes the object to speed up gives rise to an increase in the object's kinetic energy. Conversely, (negative) work done by the force of kinetic friction may cause a moving object to slow down and decrease its kinetic energy. So, for an object to have a change in its kinetic energy, there must be net work done on the object, as Eq. 5.6 tells us.

When an object is in motion, it possesses kinetic energy and has the capability to do work. For example, a moving automobile has kinetic energy and can do work in crumpling a fender in a fenderbender—not *useful* work in that case, but still work. Another example of work done by kinetic energy is shown in ▶Fig. 5.9.

▲ **FIGURE 5.9 Kinetic energy and work** A moving object, such as a wrecking ball, possesses kinetic energy and thus can do work.

Example 5.5 ■ A Game of Shuffleboard: The Work–Energy Theorem

A shuffleboard player (▶Fig. 5.10) pushes a 0.25-kg puck that is initially at rest such that a constant horizontal force of 6.0 N acts on it through a distance of 0.50 m. (Neglect friction.) (a) What are the kinetic energy and the speed of the puck when the force is removed? (b) How much work would be required to bring the puck to rest?

Thinking It Through. Apply the work–energy theorem. If you can find the amount of work done, you know the change in kinetic energy, and vice versa.

Solution. Listing the given data as usual, we have

Given: $m = 0.25$ kg   *Find:* (a) K (kinetic energy)
       $F = 6.0$ N v (speed)
       $d = 0.50$ m   (b) W (work done in stopping puck)
       $v_o = 0$

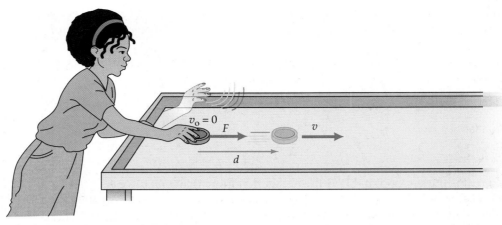

▲ **FIGURE 5.10   Work and kinetic energy**   See Example 5.5.

(a)   Since the speed is not known, we cannot compute the kinetic energy ($K = \frac{1}{2}mv^2$) directly. However, kinetic energy is related to work by the work–energy theorem. The work done on the puck is

$$W = Fd = (6.0\,\text{N})(0.50\,\text{m}) = +3.0\,\text{J}$$

Then, by the work–energy theorem, we obtain

$$W = \Delta K = K - K_\text{o} = +3.0\,\text{J}$$

But $K_\text{o} = \frac{1}{2}mv_\text{o}^2 = 0$, because $v_\text{o} = 0$, so

$$K = 3.0\,\text{J}$$

The speed can be found from the kinetic energy. Since $K = \frac{1}{2}mv^2$, we have

$$v = \sqrt{\frac{2K}{m}} = \sqrt{\frac{2(3.0\,\text{J})}{0.25\,\text{kg}}} = 4.9\,\text{m/s}$$

(b)   As you might guess, the work required to bring the puck to rest is equal to the puck's kinetic energy (i.e., the amount of energy that must be "removed" from the puck to stop its motion). To confirm this equality, we essentially perform the reverse of the previous calculation, with $v_\text{o} = 4.9\,\text{m/s}$ and $v = 0$:

$$W = K - K_\text{o} = 0 - K_\text{o} = -\tfrac{1}{2}mv_\text{o}^2 = -\tfrac{1}{2}(0.25\,\text{kg})(4.9\,\text{m/s})^2 = -3.0\,\text{J}$$

The minus sign indicates that the puck loses energy as it slows down. The work is done *against* the motion of the puck; that is, the opposing force is in a direction opposite that of the motion. (In a real-life situation, the opposing force could be friction.)

Follow-up Exercise.   Suppose the puck in this Example had twice the final speed when released. Would it then take twice as much work to stop the puck? Justify your answer numerically.

Problem-Solving Hint

Notice how work–energy considerations were used to find speed in Example 5.5. This operation can be done in another way as well. First, the acceleration could be found from $a = F/m$, and then the kinematic equation $v^2 = v_\text{o}^2 + 2ax$ could be used to find v (where $x = d = 0.50\,\text{m}$).

The point is that many problems can be solved in different ways, and finding the fastest and most efficient way is often the key to success. As our discussion of energy progresses, you will see how useful and powerful the notions of work and energy are, both as theoretical concepts and as practical tools for solving many kinds of problems.

Conceptual Example 5.6 ■ Kinetic Energy: Mass versus Speed

In a football game, a 140-kg guard runs at a speed of 4.0 m/s, and a 70-kg free safety moves at 8.0 m/s. In this situation, it is correct to say that (a) both players have the same kinetic energy, (b) the safety has twice as much kinetic energy as the guard, (c) the guard has twice as much kinetic energy as the safety, or (d) the safety has four times as much kinetic energy as the guard.

Reasoning and Answer.   The kinetic energy of a body depends on both its mass and its speed. You might think that, with half the mass but twice the speed, the safety would have the same kinetic energy as the guard, but this is not the case. As we can observe from the relationship $K = \frac{1}{2}mv^2$, kinetic energy is directly proportional to the mass, but it is also proportional to the *square* of the speed. Thus, halving the mass decreases the kinetic energy by a factor of two; so if the two athletes had equal speeds, the safety would have had half as much kinetic energy as the guard.

However, doubling the speed increases the kinetic energy, not by a factor of 2 but by a factor of 2^2, or 4. Thus, the safety, with half the mass but twice the speed, would have $\frac{1}{2} \times 4 = 2$ times as much kinetic energy as the guard, and so the answer is (b).

Note that to answer this question, it was not necessary to calculate the kinetic energy of each player. We can do so, however, to verify our conclusions:

$$K_{\text{safety}} = \tfrac{1}{2}m_s v_s^2 = \tfrac{1}{2}(70 \text{ kg})(8.0 \text{ m/s})^2 = 2.2 \times 10^3 \text{ J}$$
$$K_{\text{guard}} = \tfrac{1}{2}m_g v_g^2 = \tfrac{1}{2}(140 \text{ kg})(4.0 \text{ m/s})^2 = 1.1 \times 10^3 \text{ J}$$

Thus, we see explicitly that our answer was correct.

Follow-up Exercise.   Suppose that the safety's speed were only 50 percent greater than the guard's, or 6.0 m/s. Which athlete would then have the greater kinetic energy, and how many times greater?

Problem-Solving Hint

Note that the work–energy theorem relates the work done to the *change* in the kinetic energy. Often, we have $v_o = 0$ and $K_o = 0$, so $W = \Delta K = K$. But take care! You *cannot* simply use the square of the change in speed, $(\Delta v)^2$, to calculate ΔK, as you might at first think. In terms of speed, we have

$$W = \Delta K = K - K_o = \tfrac{1}{2}mv^2 - \tfrac{1}{2}mv_o^2 = \tfrac{1}{2}m(v^2 - v_o^2)$$

But $v^2 - v_o^2$ is not the same as $(v - v_o)^2 = (\Delta v)^2$, since $(v - v_o)^2 = v^2 - 2vv_o + v_o^2$. Hence, the work, or change in kinetic energy, is *not* equal to $\tfrac{1}{2}m(v - v_o)^2 = \tfrac{1}{2}m(\Delta v)^2$.

What this observation means is that to calculate work, or the charge in kinetic energy, you must compute the kinetic energy of an object at one point or time (using the instantaneous speed to get the instantaneous kinetic energy) and also at another point or time. Then the quantities are subtracted to find the change in kinetic energy, or the work. Alternatively, you can find the difference of the *squares* of the speeds $(v^2 - v_o^2)$ first in computing the change, but do not use the square of the difference of the speeds. To see this hint in action, take a look at Conceptual Example 5.7.

Conceptual Example 5.7 ■ An Accelerating Car: Speed and Kinetic Energy

A car traveling at 5.0 m/s speeds up to 10 m/s, with an increase in kinetic energy that requires work W_1. Then the car's speed increases from 10 m/s to 15 m/s, requiring additional work W_2. Which of the following relationships accurately compares the two works (a) $W_1 > W_2$; (b) $W_1 = W_2$; (c) $W_2 > W_1$.

(a)

(b)

▲ **FIGURE 5.11 Potential energy**
Potential energy has many forms.
(a) Work must be done to bend the bow, giving it potential energy. That energy is converted into kinetic energy when the arrow is released.
(b) Gravitational potential energy is converted into kinetic energy when an object falls. (Where did the gravitational potential energy of the water and the diver come from?)

Reasoning and Answer. As noted previously, the work–energy theorem relates the work done to the *change* in the kinetic energy. Since the speeds have the same increment in each case ($\Delta v = 5.0$ m/s), it might appear that (b) would be the answer. However, keep in mind that the work is equal to the *change* in kinetic energy and involves $v_2^2 - v_1^2$, *not* $(\Delta v)^2 = (v_2 - v_1)^2$.

So, the greater the speed of an object, the greater is its kinetic energy, and we would expect the *difference* in kinetic energy in changing speeds (or the work required to change speed) to be greater for higher speeds for the same Δv. Thus, (c) is the answer.

The main point is that the Δv values are the same, but more work is required to increase the kinetic energy of an object at higher speeds.

Follow-up Exercise. Suppose the car speeds up a third time, from 15 m/s to 20 m/s, a change requiring work W_3. How does the work done in this increment compare with W_2? Justify your answer numerically. (*Hint*: Use a ratio.)

5.4 Potential Energy

OBJECTIVES: **To (a) define and understand potential energy and (b) learn about gravitational potential energy.**

An object in motion has kinetic energy. However, whether an object is in motion or not, it may have another form of energy—potential energy. As the name implies, an object having potential energy has the *potential* to do work. You can probably think of many examples: a compressed spring, a drawn bow, water held back by a dam, a wrecking ball poised to drop. In all such cases, the potential to do work derives from the *position* or *configuration* of bodies. The spring has energy because it is compressed, the bow because it is drawn, the water and the ball because they have been lifted above the surface of the Earth (◀Fig. 5.11). Consequently, **potential energy** U, is often called the energy of position (and/or configuration).

In a sense, potential energy can be thought of as stored work, just as kinetic energy can. You have already seen an example of potential energy in Section 5.2 when work was done in compressing a spring from its equilibrium position. Recall that the work done in such a case is $W = \frac{1}{2}kx^2$ (with $x_\text{o} = 0$). Note that the amount of work done depends on the amount of compression (x). Because work is done, there is a *change* in potential energy (ΔU), which is equal to the work done *by the applied force* in compressing (or stretching) the spring:

$$W = \Delta U = U - U_\text{o} = \tfrac{1}{2}kx^2 - \tfrac{1}{2}kx_\text{o}^2$$

Thus, with $x_\text{o} = 0$ and $U_\text{o} = 0$, as they are commonly taken for convenience, the *potential energy of a spring* is

$$U = \tfrac{1}{2}kx^2 \quad \textit{potential energy of a spring} \quad (5.7)$$

SI unit of energy: joule (J)

[*Note*: Since the potential energy varies as x^2, the previous problem-solving hint also applies. That is, when $x_\text{o} \neq 0$, then $x^2 - x_\text{o}^2 \neq (x - x_\text{o})^2$.]

Perhaps the most common type of potential energy is **gravitational potential energy**. In this case, position refers to the height of an object above some reference point, such as the floor or the ground. Suppose that an object of mass m is lifted a distance Δy (▶Fig. 5.12). Work is done against the force of gravity, and an applied force at least equal to the object's weight is necessary to lift the object: $F = w = mg$. The work done in lifting is then equal to the change in potential

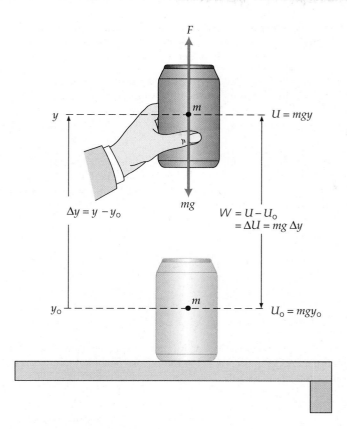

◀ **FIGURE 5.12 Gravitational potential energy** The work done in lifting an object is equal to the change in gravitational potential energy: $W = F\Delta y = mg(y - y_o)$.

energy. Expressing this relationship in equation form, since there is no overall change in kinetic energy, we have

work done by external force = change in gravitational potential energy

or

$$W = F\Delta y = mg(y - y_o) = mgy - mgy_o = \Delta U = U - U_o$$

where y is used as the vertical coordinate and, with the common choices of $y_o = 0$ and $U_o = 0$, the **gravitational potential energy** is

$$U = mgy \qquad (5.8)$$

SI unit of energy: joule (J)

Definition of: Gravitational potential energy

(Eq. 5.8 represents the gravitational potential energy on or near the Earth's surface where g is considered to be constant. A more general form of gravitational potential energy shall be given in Section 7.5.)

Example 5.8 ■ A Thrown Ball: Kinetic Energy and Gravitational Potential Energy

A 0.50-kg ball is thrown vertically upward with an initial velocity of 10 m/s (▶Fig. 5.13). (a) What is the change in the ball's kinetic energy between the starting point and the ball's maximum height? (b) What is the change in the ball's potential energy between the starting point and the ball's maximum height? (Neglect air resistance.)

Thinking It Through. Kinetic energy is lost and gravitational potential energy is gained as the ball travels upward.

Solution. Studying Fig. 5.13 and listing the data, we have

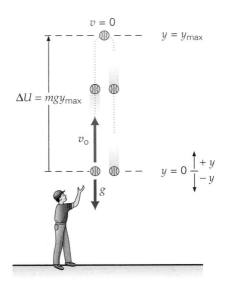

▲ FIGURE 5.13 Kinetic and potential energies See Example 5.8. (The ball is displaced sideways for clarity.)

Given: $m = 0.50$ kg   **Find:**   (a) ΔK (the change in kinetic energy)
$v_0 = 10$ m/s   (b) ΔU (the change in potential energy
$a = g$ between y_0 and y_{max})

(a) To find the *change* in kinetic energy, we first compute the kinetic energy at each point. We know the initial velocity v_0, and at the maximum height, $v = 0$, so $K = 0$. Thus,

$$\Delta K = K - K_0 = 0 - K_0 = -\tfrac{1}{2}mv_0^2 = -\tfrac{1}{2}(0.50 \text{ kg})(10 \text{ m/s})^2 = -25 \text{ J}$$

That is, the ball loses 25 J of kinetic energy as negative work is done on it by the force of gravity. (The gravitational force and the ball's displacement are in opposite directions.)
(b) To find the change in potential energy, we need to know the ball's height above its starting point when $v = 0$. Using Eq. 2.11', $v^2 = v_0^2 - 2gy$ (with $y_0 = 0$) to find y_{max},

$$y_{max} = \frac{v_0^2}{2g} = \frac{(10 \text{ m/s})^2}{2(9.8 \text{ m/s}^2)} = 5.1 \text{ m}$$

Then, with $y_0 = 0$ and $U_0 = 0$, $\Delta U = U - U_0 = U - 0$, and

$$\Delta U = U = mgy_{max} = (0.50 \text{ kg})(9.8 \text{ m/s}^2)(5.1 \text{ m}) = +25 \text{ J}$$

The potential energy increases by 25 J, as might be expected. Notice that this value is the change in potential energy with respect to the release point, which was taken as the zero reference point ($y_0 = 0$).

Follow-up Exercise. In this Example, what are the overall changes in the ball's kinetic and potential energies when the ball returns to the starting point?

Zero Reference Point

An important point is illustrated in Example 5.8, namely, the choice of a zero reference point. Potential energy is the energy of *position*, and the potential energy at a particular position (U) is referenced to the potential energy at some other position (U_0). The reference position or point is arbitrary, as is the origin of a set of coordinate axes for analyzing a system. Reference points are usually chosen with convenience in mind—for example, $y_0 = 0$. The value of the potential energy at a particular position depends on the reference point used. However, the *difference, or change, in potential energy associated with two positions is the same regardless of the reference position.*

If, in Example 5.8, ground level had been taken as the zero reference point, then U_0 at the release point would not have been zero. However, U at the maximum height would have been greater, and $\Delta U = U - U_0$ would have been the same. This concept is illustrated in ▸Fig. 5.14. Note that the potential energy can be negative. When an object has a negative potential energy, it is said to be in a potential-energy *well*, which is analogous to being in an actual well: Work is needed to raise the object to a higher position in the well or to get it out of the well.

Also, for gravitational potential energy, the path by which an object is raised (or lowered) makes no difference (▸Fig. 5.15). That is, *the change in gravitational potential energy is independent of path.* As illustrated in Fig. 5.15, an object raised to a height y has a change in potential energy of $\Delta U = mgy$ no matter whether it is lifted vertically or moved along an inclined plane. The change in potential energy is the same for both cases because the force of gravity always acts downward, and only vertical displacement (or the vertical component) is involved in doing work against gravity and changing the potential energy.

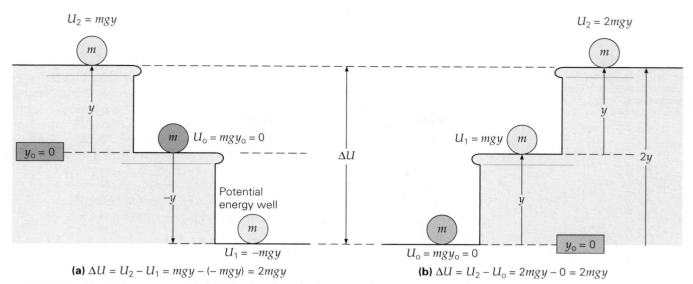

(a) $\Delta U = U_2 - U_1 = mgy - (-mgy) = 2mgy$ **(b)** $\Delta U = U_2 - U_o = 2mgy - 0 = 2mgy$

▲ **FIGURE 5.14 Reference point and change in potential energy** **(a)** The choice of a reference point (zero height) is arbitrary and may give rise to a negative potential energy. An object is said to be in a potential-energy well in this case. **(b)** The well may be avoided by selecting a new zero reference. Note that the difference, or *change*, in potential energy (ΔU) associated with the two positions is the same, regardless of the reference point. There is no physical difference, even though there are two coordinate systems and two different zero reference points.

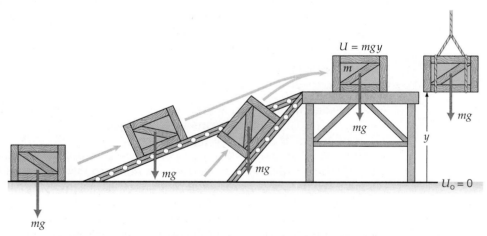

▲ **FIGURE 5.15 Path independence of the change in gravitational potential energy** When it is resting on the table, the crate has the same potential energy relative to the floor, regardless of how it got on the table. Only the vertical component of the force that moves the crate does work against gravity in lifting the crate vertically or moving it up an inclined plane or ramp.

5.5 The Conservation of Energy

OBJECTIVES: **To (a) distinguish between conservative and nonconservative forces and (b) explain their effects on the conservation of energy.**

Conservation laws are the cornerstones of physics, both theoretically and practically. Most scientists would probably name the conservation of energy as the most profound and far reaching of these important laws. When we say that a physical quantity is *conserved*, we mean that it is constant, or has a constant value. Because so many things continually change in physical processes, conserved

Note: A system is a physical situation with real or imaginary boundaries. A classroom might be considered a system, and so might an arbitrary cubic meter of air.

quantities are extremely helpful in our attempts to understand and describe the universe. Keep in mind, though, that quantities are generally conserved only under special conditions.

One of the most important conservation laws is that concerning conservation of energy. (You may have seen this topic coming in Example 5.8.) A familiar statement is that the total energy of the universe is conserved. This statement is true, because the whole universe is taken to be a system. A *system* is defined as a definite quantity of matter enclosed by boundaries, either real or imaginary. In effect, the universe is the largest possible closed, or isolated, system we can imagine. Within a *closed system*, particles can interact with each other, but have absolutely no interaction with anything outside. In general, then, the amount of energy in a system remains constant when no mechanical work is done on or by the system, and no energy is transmitted to or from the system (including thermal energy and radiation).

Thus, the **law of conservation of total energy** may be stated as follows:

Conservation of total energy

| The total energy of an isolated system is always conserved. |

Within such a system, energy may be converted from one form to another, but the total amount of all forms of energy is constant, or unchanged. Total energy can never be created or destroyed.

Conservative and Nonconservative Forces

Note: Friction is discussed in Section 4.6.

We can make a general distinction among systems by considering two categories of forces that may act within them: conservative and nonconservative forces. You have already been introduced to a couple of conservative forces: the force due to gravity and the spring force. We considered a classic nonconservative force, friction, in Chapter 4. A conservative force is defined as follows:

Conservative force—work independent of path

| A force is said to be conservative if the work done by or against it in moving an object is independent of the object's path. |

What this definition means is that the work done by a **conservative force** depends only on the initial and final positions of an object.

The concept of conservative and nonconservative forces is sometimes difficult to comprehend at first. Because this concept is so important in the conservation of energy, let's consider some illustrative examples to increase our understanding.

First, what does *independent of path* mean? An example of path independence was given in Fig. 5.15, where work was done against the *conservative force of gravity*. The figure illustrates that the work done in moving a crate onto a table does not depend on the *path* of the crate, but only on the initial and final positions of the crate. The magnitude of the work done is equal to the change in potential energy (under frictionless conditions only), and in fact, *the concept of potential energy is associated only with conservative forces*. A change in potential energy can be defined in terms of the work done by a conservative force.

Conversely, a **nonconservative force** *does* depend on path.

Nonconservative force—work dependent on path

| A force is said to be nonconservative if the work done by or against it in moving an object does depend on the object's path. |

Friction is a nonconservative force. For example, given the same frictional conditions for each case in Fig. 5.15, it would take more work (against friction) to move the crate up the longer ramp, so the work would depend on path—the longer the path, the more frictional work is done. It should therefore be evident that the *total work* done is not equal to a change in potential energy. Some work was done against a nonconservative force, in addition to the conservative gravitational force. In this case, the energy associated with the work done against friction would be converted to heat energy. Hence, in a sense, a conservative force allows you to conserve or store all of the energy as potential energy, whereas a nonconservative force does not.

To further highlight this idea, consider moving the crate around the tabletop. Here, work is done only against the nonconservative frictional force. (Why?) Certainly the work done in moving the crate between two points depends on the path taken. The work would be greater, for example, if we were to move the crate around the table's edges before coming to the destination point than if we were to take a straight-line path.

Another approach to explain the distinction between conservative and nonconservative forces is through an equivalent statement of the previous definition of conservative force:

| A force is conservative if the work done by or against it in moving an object | Another way of describing a
| through a round-trip is zero. | conservative force

Consider a book resting on a table. It has gravitational potential energy $U = mgy$, relative to some reference point $y_o = 0$. You could drop the book on the floor, pick it up, and place it back at its original position; or, you could pick it up from the table, carry it around with you all day, and then place it back at its original position. Both sets of circumstances are round-trips, and the potential energy of the book is the same when it is returned to its original position as it was before the book was moved. Thus, the change in the book's potential energy, or the work done by the conservative force of gravity, is $\Delta U = W = 0$. However, if you were to push the book around on the tabletop and eventually back to its original position, the work done against the nonconservative force of friction would depend on the path (i.e., the longer the path, the more work is done). The work done would not be stored, but would instead be lost as heat and sound.

Notice that for the *conservative* gravitational force, the force and displacement are sometimes in the same direction (in which case positive work is done by the force) and sometimes in opposite directions (in which case negative work is done by the force) during a round-trip. Think of the simple case of the book falling to the floor and being placed back on the table. With positive and negative work, the total work done by gravity can be zero.

However, for only a *nonconservative* force like that of kinetic friction, which always opposes the motion or is in the opposite direction to the displacement, the total work done in a round-trip can *never* be zero and is always negative (i.e., energy is lost). But don't get the idea that nonconservative forces only take energy away from a system. On the contrary, we often supply nonconservative pushes and pulls (forces) that add to the energy of a system, such as when you push a stalled car.

Conservation of Total Mechanical Energy

The idea of a conservative force allows us to extend the conservation of energy to the special case of mechanical energy, which greatly helps us better analyze many physical situations. The sum of the kinetic and potential energies is called the **total mechanical energy**:

$$
\underset{\substack{total \\ mechanical \\ energy}}{E} = \underset{\substack{kinetic \\ energy}}{K} + \underset{\substack{potential \\ energy}}{U}
$$

Total mechanical energy—kinetic plus potential

(5.9)

For a **conservative system** (i.e., a system in which only conservative forces do work) the total mechanical energy is constant, or conserved; that is,

$$E = E_o$$

Substituting for E and E_o from Eq. 5.9,

$$K + U = K_o + U_o \qquad (5.10a)$$

or

$$\tfrac{1}{2}mv^2 + U = \tfrac{1}{2}mv_o^2 + U_o \qquad (5.10b)$$

Conservation of mechanical energy

Equation 5.10b is a mathematical statement of the **law of the conservation of mechanical energy**:

| In a conservative system, the sum of all types of kinetic energy and potential energy is constant and equals the total mechanical energy of the system. |

The kinetic and potential energies in a conservative system may change, but their sum is always constant. This concept is illustrated in ▼Fig. 5.16a.

Notice for a conservative system that when work is done and energy is transferred within a system, we can write Eq. 5.10a as

$$(K - K_o) + (U - U_o) = 0 \qquad (5.11a)$$

or as

$$\Delta K + \Delta U = 0 \qquad (5.11b)$$
(for a conservative system)

This expression tells us that these quantities are related in a seesaw fashion: If there is a decrease in potential energy, then the kinetic energy must increase by an equal amount to keep the sum of the changes equal to zero. However, in a nonconservative system, mechanical energy is usually lost (for example, to the heat of friction), and thus $\Delta K + \Delta U < 0$. Such a situation is illustrated in Fig. 5.16b. In terms of the total mechanical energy, $\Delta E = E - E_o < 0$, where ΔE is the amount of energy lost from the system. But, keep in mind, as pointed out previously, a nonconservative force may instead add energy to a system (or have no effect at all).

Examples 5.9 through 5.11 illustrate the conservation of mechanical energy for some conservative systems.

▼ **FIGURE 5.16   Conservative and nonconservative systems**   (a) The work done by a conservative force exchanges energy between kinetic and potential forms; that is, $\Delta K + \Delta U = 0$. This relationship means that $K_o + U_o = K + U$, or $E_o = E$, and the total mechanical energy is conserved (i.e., no mechanical energy is lost or gained). (b) For a nonconservative force, not all the work goes into the exchange of mechanical energy; some energy (ΔE) is lost, for example, if friction acts. The mechanical energy is not conserved in this case. (c) A graphical energy summary for the nonconservative case. (Keep in mind that a nonconservative force doing work may instead add energy to a system, rather than subtracting it.) How would the energy summary for a conservative system look?

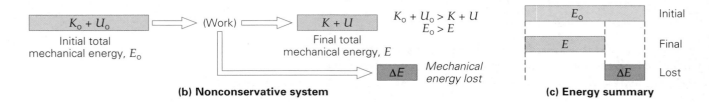

Example 5.9 ■ Look Out Below! Conservation of Mechanical Energy

A painter on a scaffold drops a 1.50-kg can of paint from a height of 6.00 m. (a) What is the kinetic energy of the can when the can is at a height of 4.00 m? (b) With what speed will the can hit the ground? (Neglect air resistance.)

Thinking It Through. Total mechanical energy is conserved, since only the conservative force of gravity acts on the system (the can). The initial total mechanical energy can be found, and potential energy decreases as kinetic energy (as well as speed) increases.

Solution. Listing what is given and what we are to find, we have:

Given: $m = 1.50$ kg   *Find:* (a) K (kinetic energy at $y = 4.00$ m)
$y_o = 6.00$ m   (b) v (speed hitting the ground)
$y = 4.00$ m
$v_o = 0$

(a) First, it is convenient to find the can's total mechanical energy, since this quantity is conserved while the can is falling (why?). Initially, with $v_o = 0$, the can's total mechanical energy is all potential energy. Taking the ground as the zero reference point, we have

$$E = K_o + U_o = 0 + mgy_o = (1.50 \text{ kg})(9.80 \text{ m/s}^2)(6.00 \text{ m}) = 88.2 \text{ J}$$

The relation $E = K + U$ continues to hold while the can is falling, but now we know what E is. Rearranging the equation, we have $K = E - U$ and can find U at $y = 4.00$ m:

$$K = E - U = E - mgy = 88.2 \text{ J} - (1.50 \text{ kg})(9.80 \text{ m/s}^2)(4.00 \text{ m}) = 29.4 \text{ J}$$

Alternatively, we could have computed the change in (in this case, the loss of) potential energy, ΔU. Whatever potential energy was lost must have been gained as kinetic energy (Eq. 5.11). Then,

$$\Delta K + \Delta U = 0$$
$$(K - K_o) + (U - U_o) = (K - K_o) + (mgy - mgy_o) = 0$$

With $K_o = 0$ (since $v_o = 0$), we obtain

$$K = mg(y_o - y) = (1.50 \text{ kg})(9.8 \text{ m/s}^2)(6.00 \text{ m} - 4.00 \text{ m}) = 29.4 \text{ J}$$

(b) Just before the can strikes the ground ($y = 0, U = 0$), the total mechanical energy is all kinetic energy, or

$$E = K = \tfrac{1}{2}mv^2$$

Thus,

$$v = \sqrt{\frac{2E}{m}} = \sqrt{\frac{2(88.2 \text{ J})}{1.50 \text{ kg}}} = 10.8 \text{ m/s}$$

Basically, all of the potential energy of a free-falling object released from some height y is converted into kinetic energy just before the object hits the ground, so

$$|\Delta K| = |\Delta U|$$

Thus,

$$\tfrac{1}{2}mv^2 = mgy$$

or

$$v = \sqrt{2gy}$$

Note that the mass cancels and is not a consideration. This result is also obtained from the kinematic equation $v^2 = 2gy$ (Eq. 2.11′), with $v_o = 0$ and $y_o = 0$.

Follow-up Exercise. A fellow painter on the ground wishes to toss a paintbrush vertically upward a distance of 5.0 m to his partner on the scaffold. Use methods of conservation of mechanical energy to determine the minimum speed that he must give to the brush.

Roller Coaster

Conceptual Example 5.10 ■ A Matter of Direction? Speed and Conservation of Energy

Three balls of equal mass m are projected with the same speed in different directions, as shown in ▼Fig. 5.17. If air resistance is neglected, which ball would you expect to strike the ground with the greatest speed? (a) ball 1; (b) ball 2; (c) ball 3; (d) all balls strike with the same speed.

Reasoning and Answer. All of the balls have the same initial kinetic energy $K_o = \frac{1}{2}mv_o^2$. (Recall that energy is a scalar quantity, and the different directions of projection do not produce any difference in the kinetic energies.) Regardless of their trajectories, all of the balls ultimately descend a distance y relative to their common starting point, so they all lose the same amount of potential energy. (Recall that U is energy of *position* and thus is *independent* of path—see Fig. 5.15.)

By the law of conservation of mechanical energy, the amount of potential energy each ball loses is equal to the amount of kinetic energy it gains. Since all of the balls start with the same amount of kinetic energy and gain the same amount of kinetic energy, all three will have equal kinetic energies just before striking the ground. This means that their speeds must be equal, so the answer is (d).

Note that although balls 1 and 2 are projected at 45° angles, this factor is not relevant. Since the change in potential energy is independent of path, it is independent of the projection angle. The vertical distance between the starting point and the ground is the same (y) for projectiles at any angle. (*Note*: Although the strike speeds are equal, the *times* it takes the balls to reach the ground are different. Refer to Conceptual Example 3.11 for another approach.)

Follow-up Exercise. Would the balls strike the ground with different speeds if their masses were different? (Neglect air resistance.)

▶ **FIGURE 5.17 Speed and energy** See Conceptual Example 5.10.

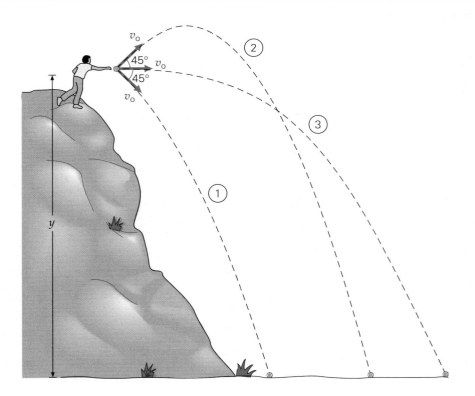

Example 5.11 ■ Conservative Forces: Mechanical Energy of a Spring

A 0.30-kg block sliding on a horizontal frictionless surface with a speed of 2.5 m/s, as depicted in ▶Fig. 5.18, strikes a light spring that has a spring constant of 3.0×10^3 N/m. (a) What is the total mechanical energy of the system? (b) What is the kinetic energy K_1 of the block when the spring is compressed a distance $x_1 = 1.0$ cm? (Assume that no energy is lost in the collision.)

Thinking It Through. (a) Initially, the total mechanical energy is all kinetic energy. (b) The total energy is the same as in (a), but it is now divided between kinetic energy and spring potential energy.

Solution.

Given: $m = 0.30$ kg   *Find:* (a) E (total mechanical energy)
$\quad\quad\ v_o = 2.5$ m/s $\quad\quad\quad$ (b) K_1 (kinetic energy)
$\quad\quad\ k = 3.0 \times 10^3$ N/m
$\quad\quad\ x_1 = 1.0$ cm $= 0.010$ m

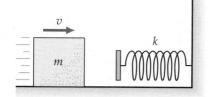

▲ **FIGURE 5.18 Conservative force and the mechanical energy of a spring** See Example 5.11.

(a) Before the block makes contact with the spring, the total mechanical energy of the system is all in the form of kinetic energy; therefore,

$$E = K_o = \tfrac{1}{2}mv_o^2 = \tfrac{1}{2}(0.30 \text{ kg})(2.5 \text{ m/s})^2 = 0.94 \text{ J}$$

Since the system is conservative (i.e., no mechanical energy is lost), this quantity is the total mechanical energy at any time.

(b) When the spring is compressed a distance x_1, it has potential energy $U_1 = \tfrac{1}{2}kx_1^2$, and

$$E = K_1 + U_1 = K_1 + \tfrac{1}{2}kx_1^2$$

Solving for K_1, we have

$$K_1 = E - \tfrac{1}{2}kx_1^2$$

$$= 0.94 \text{ J} - \tfrac{1}{2}(3.0 \times 10^3 \text{ N/m})(0.010 \text{ m})^2 = 0.94 \text{ J} - 0.15 \text{ J} = 0.79 \text{ J}$$

Follow-up Exercise. How far will the spring in Example 5.11 be compressed when the block comes to a stop? (Solve using energy principles.)

See the Learn by Drawing on the next page for another example of energy exchange.

Total Energy and Nonconservative Forces

In the preceding examples, we ignored the force of friction, which is probably the most common nonconservative force. In general, both conservative and nonconservative forces can do work on objects. However, as you know, when some nonconservative forces do work, the total mechanical energy is not conserved. Mechanical energy is "lost" through the work done by nonconservative forces, such as friction.

You might think that we can no longer use an energy approach to analyze problems involving such nonconservative forces, since mechanical energy can be lost or dissipated (▶Fig. 5.19). However, in some instances, we can use the total energy to find out how much energy was lost to the work done by a nonconservative force. Suppose an object initially has mechanical energy and that nonconservative

▲ **FIGURE 5.19 Nonconservative force and energy loss** Friction is a nonconservative force—when friction is present and does work, mechanical energy is not conserved. Can you tell from the photo what is happening to the work being done by the motor on the grinding wheel after the work is converted into rotational kinetic energy? (Note that the worker is wisely wearing a face shield rather than just goggles as the sign in the background suggests.)

Learn by Drawing

Energy Exchanges: A Falling Ball

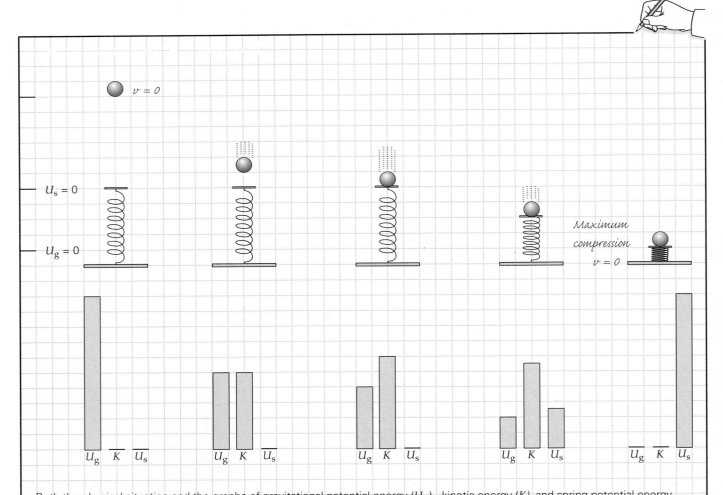

Both the physical situation and the graphs of gravitational potential energy (U_g), kinetic energy (K), and spring potential energy (U_s) are drawn to scale. (Air resistance, the mass of the spring, and any energy loss in the collision are assumed to be negligible.) Why is the spring energy only one-quarter of the total when the spring is halfway compressed?

forces do an amount of work W_{nc} on it. Starting with the work–energy theorem, we have

$$W = \Delta K = K - K_o$$

In general, the net work (W) may be done by both conservative forces (W_c) and nonconservative forces (W_{nc}), so we may write

$$W_c + W_{nc} = K - K_o \qquad (5.12)$$

But recall that the work done by conservative forces is equal to $-\Delta U$, or $W_{nc} = U_o - U$, and Eq. 5.12 then becomes

$$W_{nc} = K - K_o - (U_o - U)$$
$$= (K + U) - (K_o + U_o)$$

Therefore,

$$W_{nc} = E - E_o = \Delta E \qquad (5.13)$$

Hence, the work done by the nonconservative forces acting on a system is equal to the change in mechanical energy. Notice that for dissipative forces, $E_o > E$. Thus, the change is negative, indicating a decrease in mechanical energy. This condition agrees in sign with W_{nc}, which, for friction, would also be negative. Example 5.12 illustrates this concept.

Example 5.12 ■ Downhill Racer: Nonconservative Force

A skier with a mass of 80 kg starts from rest at the top of a slope and skis down from an elevation of 110 m (▼Fig. 5.20). The speed of the skier at the bottom of the slope is 20 m/s. (a) Show that the system is nonconservative. (b) How much work is done by the nonconservative force of friction?

Thinking It Through. (a) If the system is nonconservative, then $E_o \neq E$, and these quantities can be computed. (b) We cannot determine the work from force–distance considerations, but W_{nc} is equal to the difference in total energies (Eq. 5.13).

Solution.

Given: $m = 80 \text{ kg}$ *Find:*   (a) Show that E is not constant.
$\quad\quad\quad v_o = 0$ (b) W_{nc} (work done by friction)
$\quad\quad\quad v = 20 \text{ m/s}$
$\quad\quad\quad y_o = 110 \text{ m}$

(a) If the system is conservative, the total mechanical energy is constant. Taking $U_o = 0$ at the bottom of the hill, we find the initial energy at the top of the hill to be

$$E_o = U = mgy_o = (80 \text{ kg})(9.8 \text{ m/s}^2)(110 \text{ m}) = 8.6 \times 10^4 \text{ J}$$

Then we find the energy at the bottom of the slope to be

$$E = K = \tfrac{1}{2}mv^2 = \tfrac{1}{2}(80 \text{ kg})(20 \text{ m/s})^2 = 1.6 \times 10^4 \text{ J}$$

Therefore, $E_o \neq E$, so this system is not conservative.

(b) The amount of work done by the nonconservative force of friction is equal to the change in the mechanical energy, or to the amount of mechanical energy lost (Eq. 5.13):

$$W_{nc} = E - E_o = (1.6 \times 10^4 \text{ J}) - (8.6 \times 10^4 \text{ J}) = -7.0 \times 10^4 \text{ J}$$

This quantity is over 80% of the initial energy. (Where did this energy actually go?)

Follow-up Exercise. In free fall, air resistance is neglected, but for skydivers, air resistance has a very practical effect. Typically, a skydiver descends about 450 m before reaching a terminal velocity (Section 4.6) of 60 m/s. (a) What is the percentage of energy

$v = 20 \text{ m/s}$

110 m

◀ **FIGURE 5.20 Work done by a nonconservative force** See Example 5.12.

loss to nonconservative forces during this descent? (b) Show that after terminal velocity is reached, the rate of energy loss is given by $60mg$ J/s, where m is the mass of the skydiver.

Integrated Example 5.13 ■ Nonconservative Force: One More Time

A 0.75-kg block slides on a frictionless surface with a speed of 2.0 m/s. It then slides over a rough area 1.0 m in length and onto another frictionless surface. The coefficient of kinetic friction between the block and the rough surface is 0.17. (a) Make a sketch of the situation, with the regions labelled with regard to energy, work, and speed. (b) What is the speed of the block after it passes across the rough surface?

(a) Conceptual Reasoning. The block originally has energy E_o and speed v_o, loses energy (W_{nc}), and slows down in the rough area, and emerges with energy E and speed v. Assuming that the motion is in the $+x$-direction, the sketch would look like that shown in ▼Fig. 5.21.

(b) Thinking It Through. The task of finding the final speed implies that we use equations involving kinetic energy, where the final kinetic energy can be found by using the conservation of *total* energy. Note that the initial and final energies are kinetic energies, since there is no change in gravitational potential energy. Listing the data as usual, we have the following:

Given: $m = 0.75$ kg **Find:** v (final speed of block)
$x = 1.0$ m
$\mu_k = 0.17$
$v_o = 2.0$ m/s

For this nonconservative system we have from Eq. 5.13,

$$W_{nc} = E - E_o = K - K_o$$

In the rough area, the block loses energy, because of the work done against friction (W_{nc}), and thus

$$W_{nc} = -f_k x = -\mu_k N x = -\mu_k mgx$$

[negative because f_k and the displacement x are in opposite directions, i.e., $(f_k \cos 180°)x = -f_k x$].

Then, rearranging the energy equation and writing the terms out in detail, we have

$$K = K_o + W_{nc}$$

or

$$\tfrac{1}{2}mv^2 = \tfrac{1}{2}mv_o^2 - \mu_k mgx$$

Simplifying yields

$$v = \sqrt{v_o^2 - 2\mu_k gx} = \sqrt{(2.0 \text{ m/s})^2 - 2(0.17)(9.8 \text{ m/s}^2)(1.0 \text{ m})} = 0.82 \text{ m/s}$$

Note that the mass of the block was not needed. And, it can be easily shown that the block lost over 80% of its energy to friction.

Follow-up Exercise. Suppose the coefficient of kinetic friction between the block and the rough surface were 0.25. What would happen to the block in this case?

▶ **FIGURE 5.21 A nonconservative rough spot**
See Integrated Example 5.13.

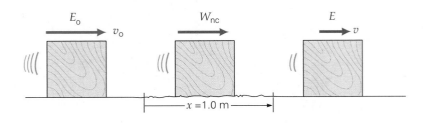

Note that in a nonconservative system, the *total energy* (*not* the total mechanical energy) is conserved (including nonmechanical forms of energy, such as heat), but not all of it is available for mechanical work. For a conservative system, you get back what you put in, so to speak. That is, if you do work on the system, the transferred energy is available to do work. But keep in mind that conservative systems are idealizations, because most all real systems are nonconservative to some degree. However, working with ideal conservative systems gives us an understanding of the conservation of energy.

Total energy is always conserved. During this course of study, you will learn about other forms of energy, such as thermal, electrical, nuclear, and chemical energies. In general, on the microscopic and submicroscopic levels, these forms of energy can be described in terms of kinetic energy and potential energy. Also, you will learn that mass is a form of energy and that the law of the conservation of energy must take this form into account in order to be applied to the analysis of nuclear reactions.

PHYSLET® ILLUSTRATION

Work Done by Friction

5.6 Power

OBJECTIVES: **To (a) define power and (b) describe mechanical efficiency.**

A particular task may require a certain amount of work, but that work might be done over different lengths of time or at different rates. For example, suppose that you have to mow a lawn. This task takes a certain amount of work, but you might do the job in a half hour, or you might take an hour or two. There's a practical distinction to be made here. There is usually not only an interest in the amount of work done, but also an interest in how fast it is done—that is, the rate at which it is done. *The time rate of doing work* is called **power**.

The average power is the work done divided by the time it takes to do the work, or work per unit of time:

Definition of power: the time rate of doing work)

$$\overline{P} = \frac{W}{t} \tag{5.14}$$

If we are interested in the work (and power) done by a constant force of magnitude F acting while an object moves through a parallel displacement of magnitude d, then

$$\overline{P} = \frac{W}{t} = \frac{Fd}{t} = F\left(\frac{d}{t}\right) = F\overline{v} \tag{5.15}$$

SI unit of power: J/s or watt (W)

where it is assumed that the force is in the direction of the displacement. Here, $\overline{v}$ is the magnitude of the average velocity. If the velocity is constant, then $\overline{P} = P = Fv$. If the force and displacement are not in the same direction, then we can write

$$\overline{P} = \frac{F(\cos\theta)d}{t} = F\,\overline{v}\cos\theta \tag{5.16}$$

where θ is the angle between the force and the displacement.

As you can see from Eq. 5.15, the SI unit of power is joules per second (J/s), but this unit is given another name, the **watt (W)**:

$$1\ \text{J/s} = 1\ \text{watt (W)}$$

The SI unit of power is named in honor of James Watt (1736–1819), a Scottish engineer who developed one of the first practical steam engines. A common unit of electrical power is the *kilowatt* (kW).

The British unit of power is foot-pound per second (ft·lb/s). However, a larger unit, the **horsepower** (**hp**), is more commonly used:

$$1 \text{ hp} = 550 \text{ ft} \cdot \text{lb/s} = 746 \text{ W}$$

Power tells you how fast work is being done *or* how fast energy is transferred. For example, motors have power ratings commonly given in horsepower. A 2-hp motor can do a given amount of work in half the time that a 1-hp motor would take, or twice the work in the same amount of time. That is, a 2-hp motor is twice as "powerful" as a 1-hp motor.

▲ **FIGURE 5.22 Power delivery**
See Example 5.14.

Example 5.14 ■ A Crane Hoist: Work and Power

A crane hoist like the one shown in ◄Fig. 5.22 lifts a load of 1.0 metric ton a vertical distance of 25 m in 9.0 s at a constant velocity. How much useful work is done by the hoist each second?

Thinking It Through. The useful work done each (i.e., per) second is the power output, so this quantity is what is to be found.

Solution.

Given: m = 1.0 metric ton   *Find:* W per second (= power, P)
         = 1.0×10^3 kg
     y = 25 m
     t = 9.0 s

Keep in mind that the work per unit time (work per second) is power, so this quantity is what we need to compute. Since the load moves with a constant velocity, $\overline{P} = P$. (Why?) The work is done against gravity, so $F = mg$, and

$$P = \frac{W}{t} = \frac{Fd}{t} = \frac{mgy}{t}$$

$$= \frac{(1.0 \times 10^3 \text{ kg})(9.8 \text{ m/s}^2)(25 \text{ m})}{9.0 \text{ s}} = 2.7 \times 10^4 \text{ W (or 27 kW)}$$

Thus, since a watt (W) is a joule per second (J/s), the hoist did 2.7×10^4 J of work each second. Note that the velocity has a magnitude of $v = d/t = 25 \text{ m}/9.0 \text{ s} = 2.8 \text{ m/s}$, and the power could be found using $P = Fv$.

Follow-up Exercise. If the hoist motor of the crane in this Example is rated at 70 hp, what percentage of this power output goes into useful work?

Example 5.15 ■ Cleaning Up: Work and Time

The motors of two vacuum cleaners have net power outputs of 1.00 hp and 0.500 hp, respectively. (a) How much work in joules can each motor do in 3.00 min? (b) How long does it take for each motor to do 97.0 kJ of work?

Thinking It Through. (a) Since power is work/time ($P = W/t$), the work can be computed. Note that power is given in horsepower units. (b) This part of the problem is another application of Eq. 5.15.

Solution.

Given: P_1 = 1.00 hp = 746 W   *Find:* (a) W (work for each)
      P_2 = 0.500 hp = 373 W   (b) t (time for each)
        t = 3.00 min = 180 s
     W = 97.0 kJ = 97.0×10^3 J

(a) Since $P = W/t$,

$$W_1 = P_1t = (746 \text{ W})(180 \text{ s}) = 1.34 \times 10^5 \text{ J}$$

and

$$W_2 = P_2t = (373 \text{ W})(180 \text{ s}) = 0.67 \times 10^5 \text{ J}$$

Note that in the same amount of time, the smaller motor does half the work as the larger one, as you would expect.

(b) The times are given by $t = W/P$, and for the same amount of work,

$$t_1 = \frac{W}{P_1} = \frac{97.0 \times 10^3 \text{ J}}{746 \text{ W}} = 130 \text{ s}$$

and

$$t_2 = \frac{W}{P_2} = \frac{97.0 \times 10^3 \text{ J}}{373 \text{ W}} = 260 \text{ s}$$

Note that the smaller motor takes twice as long as the larger one to do the same amount of work.

Follow-up Exercise. (a) A 10-hp motor breaks down and is temporarily replaced with a 5-hp motor. What can you say about the rate of work output? (b) Suppose the situation were reversed—a 5-hp motor is replaced with a 10-hp motor. What can you say about the rate of work output for this case?

A real-life example of power, energy, and physics is given in the Insight at the end of this section.

Efficiency

Machines and motors are commonly used items in our daily lives, and we often talk about their efficiency. Efficiency involves work, energy, and/or power. Both simple and complex machines that do work have mechanical parts that move, so some input energy is always lost because of friction or some other cause (perhaps in the form of sound). Thus, not all of the input energy goes into doing useful work.

Mechanical efficiency is essentially a measure of what you get out for what you put in—that is, the *useful* work output compared with the energy input. **Efficiency** ε is given as a fraction (or percentage):

$$\varepsilon = \frac{\text{work output}}{\text{energy input}} (\times 100\%) = \frac{W_{out}}{E_{in}} (\times 100\%) \qquad (5.17)$$

Efficiency is a unitless quantity

For example, if a machine has a 100-joule (energy) input and a 40-joule (work) output, then its efficiency is

$$\varepsilon = \frac{W_{out}}{E_{in}} = \frac{40 \text{ J}}{100 \text{ J}} = 0.40 \, (\times 100\%) = 40\%$$

An efficiency of 0.40, or 40%, means that 60% of the energy input is lost because of friction or some other cause and doesn't serve its intended purpose. Note that if both terms of the ratio in Eq. 5.17 are divided by time t, we obtain $W_{out}/t = P_{out}$ and $E_{in}/t = P_{in}$. So, we can also write efficiency in terms of power P:

$$\varepsilon = \frac{P_{out}}{P_{in}} (\times 100\%) \qquad (5.18)$$

Example 5.16 ■ Home Improvement: Mechanical Efficiency and Work Output

The motor of an electric drill with an efficiency of 80% has a power input of 600 W. How much useful work is done by the drill in a time of 30 s?

Thinking It Through. This example is an application of Eq. 5.18 and the definition of power.

Solution.

Given: $\varepsilon = 80\% = 0.80$ *Find:* W_{out} (work output)
$P_{in} = 600$ W
$t = 30$ s

Given the efficiency and power input, we can readily find the power output P_{out} from Eq. 5.18, and this quantity is related to the work output ($P_{out} = W_{out}/t$). First, we re-arrange Eq. 5.18:

$$P_{out} = \varepsilon P_{in} = (0.80)(600 \text{ W}) = 4.8 \times 10^2 \text{ W}$$

Then, substituting this value into the equation relating power output and work output, we obtain

$$W_{out} = P_{out}t = (4.8 \times 10^2 \text{ W})(30 \text{ s}) = 1.4 \times 10^4 \text{ J}$$

Follow-up Exercise. (a) Is it possible to have a mechanical efficiency of 100%? (b) What would an efficiency of greater than 100% imply?

TABLE 5.1 Typical Efficiencies of Some Machines

Machine	Efficiency (approximate %)
Compressor	85
Electric motor	70–95
Automobile	20
Human muscle*	20–25
Steam locomotive	5–10

*Technically not a machine, but used to perform work.

Table 5.1 lists the typical efficiencies of some machines. You may be surprised by the relatively low efficiency of the automobile. Much of the energy input (from gasoline combustion) is lost as exhaust heat and through the cooling system (more than 60%), and friction accounts for a great deal more. About 20% of the input energy is converted to useful work that goes into propelling the vehicle. Air conditioning, power steering, radio, and tape and CD players are nice, but they also use energy and also contribute to the car's decrease in efficiency.

INSIGHT

More Broken Records: The Clap Skate

At the 1998 Winter Olympics in Nagano, Japan, speed skating records were broken and new ones set—largely because of an innovation in ice skates (and applied physics). This new skate technology, invented by Dutch researchers in biomechanics, is commonly called the *clap skate*. The skates are designed to increase the amount of time they are in contact with the ice and therefore also the length of the skater's stride.

The new skates have a spring-loaded hinge on the toe that allows the blade to pull away from the heel (Fig. 1). With a longer stride, the amount of work done by the skater's leg muscle is increased, providing more kinetic energy and speed. Toward the end of the stride, the blade returns to the boot with a "clap" sound when the foot is lifted off the ice—hence, the name *clap skate*. (The skates' inventors called the skates "slap skates," because the skates enable a skater to "slap on" an extra amount of work with each stride.)

Traditional skates require skaters to push from side to side. However, with claps, a new skating technique must be learned, with an emphasis on pushing straight back during the stride to keep the blade on the ice longer. Will the clap skate replace the traditional skate? Probably not completely. On a long track against the clock, the clapping is no more

than an annoyance. But on a relatively short curved track, where skaters compete against each other, a surprise move to pass another skater would be well announced—clap, clap, clap—no surprise.

Related Exercise: 35

FIGURE 1 Clap, clap, clap The new clap skate is helping skaters set records. A hinge allows the heel to lift off the blade so that the blade stays on the ice longer, thus providing the skater with more kinetic energy and speed.

Chapter Review

Important Concepts and Equations

- **Work done by a constant force** is the product of the magnitude of the displacement and the component of the force parallel to the displacement:

$$W = (F \cos \theta)d \qquad (5.2)$$

- Calculating work done by a variable force requires advanced mathematics. An example of a variable force is the **spring force**, given by *Hooke's law*:

$$F_s = -kx \qquad (5.3)$$

The **work done by a spring force** is given by

$$W = \tfrac{1}{2}kx^2 \qquad (5.4)$$

- **Kinetic energy** is the energy of motion and is given by

$$K = \tfrac{1}{2}mv^2 \qquad (5.5)$$

- By the **work–energy theorem**, the net work done on an object is equal to the change in the kinetic energy of the object:

$$W = K - K_o = \Delta K \qquad (5.6)$$

- **Potential energy** is the energy of position and/or configuration. The elastic **potential energy of a spring** is given by

$$U = \tfrac{1}{2}kx^2 \qquad \text{(with } x_o = 0\text{)} \qquad (5.7)$$

The most common type of potential energy is **gravitational potential energy**, associated with the gravitational attraction near the Earth's surface.

$$U = mgy \qquad \text{(with } y_o = 0\text{)} \qquad (5.8)$$

- **Conservation of energy**: The total energy of the universe or of an isolated system is always conserved.

  Conservation of mechanical energy: The total mechanical energy (kinetic plus potential) is constant in a conservative system:

$$\tfrac{1}{2}mv^2 + U = \tfrac{1}{2}mv_o^2 + U_o \qquad (5.10b)$$

- In systems with **nonconservative forces**, where mechanical energy is lost, the work done by a nonconservative force is given by

$$W_{nc} = E - E_o = \Delta E \qquad (5.13)$$

- **Power** is the time rate of doing work (or expending energy). **Average power** is given by

$$\overline{P} = \frac{W}{t} = \frac{Fd}{t} = F\overline{v} \qquad (5.15)$$

(constant force in direction of *d* and *v*)

$$\overline{P} = \frac{F(\cos \theta)d}{t} = F\overline{v} \cos \theta \qquad (5.16)$$

(constant force acts at an angle θ between *d* and *v*)

- **Efficiency** relates work output to energy (work) input as a percent:

$$\varepsilon = \frac{W_{out}}{E_{in}} \ (\times 100\%) \qquad (5.17)$$

$$\varepsilon = \frac{P_{out}}{P_{in}} \ (\times 100\%) \qquad (5.18)$$

Exercises

5.1 Work Done by a Constant Force

1. The units of work are (a) $N \cdot m$, (b) $kg \cdot m^2/s^2$, (c) J, or (d) all of the preceding.

2. **CQ** If you push against the wall of a building, are you doing any work? Explain.

3. **CQ** When you catch a basketball, are you doing positive, negative, or zero work on the ball? Explain.

4. **CQ** Can the work done by a frictional force on an object be positive? If yes, give an example.

5. **CQ** (a) As a weightlifter strains to lift a barbell from the floor (▶Fig. 5.23a), is he doing work? Why or why not? (b) In raising the barbell above his head, is he doing work? Explain. (c) In holding the barbell above his head (Fig. 5.23b), is he doing more work, less work, or the same amount of work as in lifting the barbell? Explain. (d) If the weightlifter drops the barbell, is work done on the barbell? Explain what happens in this situation.

(a)   (b)

▲ **FIGURE 5.23 Man at work?** See Exercise 5.

6. **CQ** You are carrying a backpack across campus. What is the work done by your vertical carrying force on the backpack? Explain.

7. **CQ** A jet plane flies in a vertical circular loop. In what regions of the loop is the work done by the plane's weight positive and/or negative? Is the work constant? If not, are there maximum and minimum instantaneous values? Explain.

8. ■ If a person does 50 J of work in moving a 30-kg box over a 10-m distance on a horizontal surface, what is the minimum force required?

9. ■ A 5.0-kg box slides a 10-m distance on ice. If the coefficient of kinetic friction is 0.20, what is the work done by the friction force?

10. ■ A passenger at an airport pulls a rolling suitcase by its handle. If the force used is 10 N and the handle makes an angle of 25° to the horizontal, what is the work done by the pulling force after the passenger walks 200 m?

11. ■ A college student earning some summer money pushes a lawn mower on a level lawn with a constant force of 250 N at an angle of 30° downward from the horizontal. How far does the student push the mower in doing 1.44×10^3 J of work?

12. ■■ A 3.00-kg block slides down a frictionless plane inclined 20° to the horizontal. If the length of the plane's surface is 1.50 m, how much work is done, and by what force?

13. ■■ Suppose the coefficient of kinetic friction between the block and the plane in Exercise 12 is 0.275. What would be the net work done in this case?

14. ■■ The formula for work is sometimes written $W = F_\parallel d$, where $F_\parallel$ is the component of the force parallel to the displacement. Show that work is also given by $F d_\parallel$, where $d_\parallel$ is the component of the displacement parallel to the force.

15. **IE** ■■ A hot-air balloon ascends at a constant rate. (a) The weight of the balloon does (1) positive work, (2) negative work, or (3) no work. Why? (b) A hot-air balloon with a mass of 500 kg ascends at a constant rate of 1.50 m/s for 20.0 s. How much work is done by the upward buoyant force? (Neglect air resistance.)

16. ■■ A father pulls his young daughter on a sled with a constant velocity on a level surface through a distance of 10 m, as illustrated in ▶Fig. 5.24a. If the total mass of the sled and the girl is 35 kg and the coefficient of kinetic friction between the sled runners and the snow is 0.20, how much work does the father do?

17. ■■ A father pushes horizontally on his daughter's sled to move it up a snowy incline, as illustrated in Fig. 5.24b. If the sled moves up the hill with a constant velocity, how

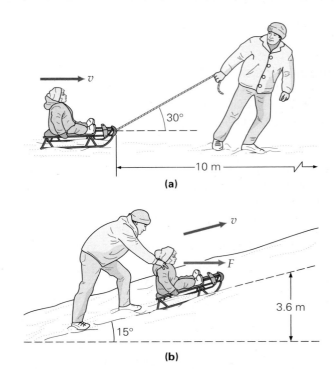

▲ **FIGURE 5.24 Fun and work** See Exercises 16 and 17.

much work is done by the father in moving it from the bottom to the top of the hill? (Some necessary data are given in Exercise 16.)

18. ■■ A boy pulls a 20-kg box with a 50-N force at 37° above a horizontal surface. The coefficient of kinetic friction between the box and the horizontal surface is 0.15, and the box is pulled over a distance of 25 m. (a) What is the work done by the boy? (b) What is the work done by the frictional force? (c) What is the net work done on the box?

19. ■■■ A 500-kg helicopter ascends from the ground with an acceleration of 2.00 m/s². Over a 5.00-s interval, what is (a) the work done by the lifting force, (b) the work done by the gravitational force, and (c) the net work done on the helicopter?

20. **IE** ■■■ A student could either pull or push, at an angle of 30° from the horizontal, a 50-kg crate on a horizontal surface, where the coefficient of kinetic friction between the crate and surface is 0.20. The crate is to be moved a horizontal distance of 15 m. (a) Compared with pushing, pulling requires the student to do (1) less, (2) the same, or (3) more work. (b) Calculate the minimum work required for both pulling and pushing.

5.2 Work Done by a Variable Force

21. The work done by a variable force of the form $F = kx$ is equal to (a) kx^2, (b) kx, (c) $\frac{1}{2}kx^2$, or (d) none of the preceding.

22. **CQ** Does it take the same amount of work to stretch a spring two centimeters from its equilibrium position as it does to stretch it one centimeter? Explain.

23. CQ If a spring is compressed 2.0 cm from its equilibrium position and then compressed an additional 2.0 cm, how much more work is done in the second compression than in the first? Explain.

24. ■ To measure the spring constant of a certain spring, a student applies a 4.0-N force, and the spring stretches by 5.0 cm. What is the spring constant?

25. ■ If a 10-N force is used to compress a spring with a spring constant of 4.0×10^2 N/m, what is the resulting spring compression?

26. ■ A spring has a spring constant of 40 N/m. How much work is required to stretch the spring 2.0 cm from its equilibrium position?

27. ■ If it takes 400 J of work to stretch a spring 8.00 cm, what is the spring constant?

28. IE ■ A certain amount of work is required to stretch a spring from its equilibrium position. (a) If twice the work is performed on the spring, the spring will stretch more by a factor of (1) $\sqrt{2}$, (2) 2, (3) $1/\sqrt{2}$, or (4) $\frac{1}{2}$? Why? (b) If 100 J of work is done to pull a spring 1.0 cm, what work is required to stretch it 3.0 cm?

29. ■■ When a 75-g mass is suspended from a vertical spring, the spring is stretched from a length of 4.0 cm to a length of 7.0 cm. If the mass is then pulled downward an additional 10 cm, what is the total work done against the spring force in joules?

30. ■■ A particular spring has a force constant of 2.5×10^3 N/m. (a) How much work is done in stretching the relaxed spring by 6.0 cm? (b) How much more work is done in stretching the spring an additional 2.0 cm?

31. ■■ For the spring in Exercise 30, how much mass would have to be suspended from the vertical spring to stretch it (a) the first 6.0 cm and (b) the additional 2.0 cm?

32. ■■ A particular force is described by the equation $\mathbf{F} = (60 \text{ N/m})x\,\hat{\mathbf{x}}$. How much work is done when this force pushes a box horizontally between (a) $x_\circ = 0$ and $x = 0.15$ m, and (b) $x_\circ = 0.15$ m and $x = 0.25$ m?

33. ■■ Compute the work done by the variable force in the graph of F versus x in ▶Fig. 5.25. [*Hint*: The area of a triangle is $A = \frac{1}{2}$ altitude $\times$ base.]

5.3 The Work–Energy Theorem: Kinetic Energy

34. If the angle between the net force and the displacement of an object is greater than 90°, (a) kinetic energy increases, (b) kinetic energy decreases, (c) kinetic energy remains the same, or (d) the object stops.

35. CQ ▶Fig. 5.26 shows a close-up of the clap skate discussed in the Insight in this chapter. The skates have a spring-loaded hinged toe at the front of the boot and a

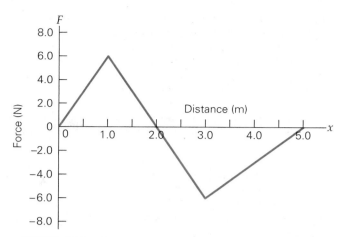

▲ **FIGURE 5.25 How much work is done?** See Exercise 33.

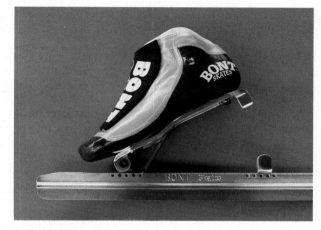

▲ **FIGURE 5.26 Clap to records** See Exercise 35.

plunger system at the heel. As the skater strides, the heel lifts from the blade—unlike with traditional skates—lengthening the skater's stride. Explain how these skates can improve performance.

36. CQ Which of the following objects has the smallest kinetic energy? (a) an object of mass $4m$ and speed v; (b) an object of mass $3m$ and speed $2v$; (c) an object of mass $2m$ and speed $3v$; (d) an object of mass m and speed $4v$.

37. CQ You want to decrease the kinetic energy of an object as much as you can, and you can do so by either reducing the mass by half or reducing the speed by half. Which option should you pick, and why?

38. CQ A certain amount of work W is required to accelerate a car from rest to a speed v. How much work is required to accelerate the car from rest to a speed of $2v$?

39. CQ A certain amount of work W is required to accelerate a car from rest to a speed v. If an amount of work equal to $2W$ is done on the car, what is the car's speed?

40. CQ Two identical cars traveling at 55 mph collide head on. A third identical car crashes into a wall at 55 mph. Which car has more damage? Explain.

41. **IE ■** A 0.20-kg object with a horizontal speed of 10 m/s 10 m/s hits a wall and bounces directly back with only half the original speed. (a) The percentage of lost kinetic energy compared with the object's original kinetic energy is (1) 25%, (2) 50%, or (3) 75%. (b) How much kinetic energy is lost in the ball's collision with the wall?

42. **■** A 1200-kg automobile travels at a speed of 90 km/h. (a) What is its kinetic energy? (b) What is the net work that would be required to bring it to a stop?

43. **■** A constant net force of 75 N acts on an object initially at rest through a parallel distance of 0.60 m. (a) What is the final kinetic energy of the object? (b) If the object has a mass of 0.20 kg, what is its final speed?

44. **■** A 3.0-g bullet traveling at 350 m/s hits a tree and slows down uniformly to a stop while penetrating a distance of 12 cm into the tree's trunk. What was the force exerted on the bullet in bringing it to rest?

45. **■■** The stopping distance of a vehicle is an important safety factor. Assuming a constant braking force, use the work–energy theorem to show that a vehicle's stopping distance is proportional to the square of its initial speed. If an automobile traveling at 45 km/h is brought to a stop in 50 m, what would be the stopping distance for an initial speed of 90 km/h?

46. **IE ■■** A large car of mass $2m$ travels at speed v, and a small car of mass m travels with a speed $2v$. Both skid to a stop with the same coefficient of friction. (a) The small car will have (1) a longer, (2) the same, or (3) a shorter stopping distance. (b) Calculate the ratio of the stopping distance of the small car to that of the large car. (Use the work–energy theorem, not Newton's laws.)

47. **■■■** If the work required to speed a car up from 10 km/h to 20 km/h is 5.0×10^3 J, what would be the work required to increase the car's speed from 20 km/h to 30 km/h?

5.4 Potential Energy

48. A change in gravitational potential energy (a) is always positive, (b) depends on the reference point, (c) depends on the path, or (d) depends only on the initial and final positions.

49. **CQ** If a spring changes its position from x_o to x, the change in potential energy is then proportional to what? (Express the quantity in terms of x_o and x.)

50. Sketch a plot of U versus x for a mass oscillating on a spring between the limits of $-A$ and $+A$.

51. **■** What is the gravitational potential energy, relative to the ground, of a 1.0-kg box at the top of a 50-m building?

52. **■** To store exactly 1.0 J of potential energy in a spring for which $k = 2.0 \times 10^4$ N/m, how much would the spring

have to be stretched beyond its equilibrium length? How about to store 4.0 J?

53. **■** How much more gravitational potential energy does a 1.0-kg hammer have when it is on a shelf 1.5 m high than when it is on a shelf 0.90 m high?

54. **IE ■** You are told that the gravitational potential energy of a 2.0-kg object has decreased by 10 J. (a) With this information, you can determine (1) the object's initial height, (2) the object's final height, (3) both the initial and the final height, or (4) only the difference between the two heights. Why? (b) What can you say has physically happened to the object?

55. **■■** A 0.20-kg stone is thrown vertically upward with an initial velocity of 7.5 m/s from a starting point 1.2 m above the ground. (a) What is the potential energy of the stone at its maximum height relative to the ground? (b) What is the change in the potential energy of the stone between its launch point and its maximum height?

56. **■■** A 60-kg diver dives off a board that is 5.0 m above the surface of the water in a swimming pool; he touches the bottom of the pool 3.0 m below the water's surface. (a) What are the respective potential energies of the diver relative to the surface of the water when he is on the board and at the bottom of the swimming pool? (b) What is the change in the diver's potential energy relative to the board, to the surface of the water, and to the bottom of the swimming pool?

57. **IE ■■** The floor of the basement of a house is 3.0 m below ground level, and the floor of the attic is 4.5 m above ground level. (a) If an object in the attic were brought to the basement, the change in potential energy will be greatest relative to which floor, (1) attic, (2) ground, (3) basement, or (4) all the same? Why? (b) What are the respective potential energies of 1.5-kg objects in the basement and attic, relative to ground level? (c) What is the change in potential energy if the object in the attic is brought to the basement?

58. **■■■** A student has six textbooks, each with a thickness of 4.0 cm and a weight of 30 N. What is the minimum work the student would have to do to place all the books in a single vertical stack, starting with all the books on the surface of the table?

5.5 The Conservation of Energy

59. If a nonconservative force acts on an object, (a) the object's kinetic energy is conserved, (b) the object's potential is conserved, (c) the mechanical energy is conserved, or (d) the mechanical energy is not conserved.

60. The speed of a pendulum is greatest (a) when the pendulum's kinetic energy is a minimum, (b) when the pendulum's acceleration is a maximum, (c) when the pendulum's potential energy is a minimum, or (d) none of the preceding.

61. **CQ** For a classroom demonstration, a bowling ball suspended from a ceiling is displaced from the vertical position to one side and released from rest just in front of the nose of a student (▼Fig. 5.27). If the student doesn't move, why won't the bowling ball hit his nose?

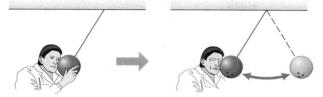

▲ **FIGURE 5.27 In the face?** See Exercise 61.

62. **CQ** Discuss all the different energy conversions that are involved in a pole vault, as shown in ▼Fig. 5.28. (Include the vaulter's running start. Where does the energy for this action come from?)

▲ **FIGURE 5.28 Energy conversion(s)** See Exercise 62.

63. **CQ** Here's an energy-transfer question: When jumping straight upward from the ground, you can achieve only the same maximum height with each jump. However, on a trampoline, you can jump higher and higher with each bounce. Why? However, there's a limit on the trampoline, too. What determines this limit?

64. **CQ** When you throw an object into the air, is its initial velocity the same as its velocity just before it returns to your hand? Explain by applying the concept of the conservation of mechanical energy.

65. **CQ** A rubber ball dropped on a floor will bounce back to a height lower than its original height. Is this phenomenon a violation of the conservation of energy? Discuss some of the energy conversions that take place in the process.

66. ■ A person standing on a bridge at a height of 115 m above a river drops a 0.250-kg rock. (a) What is the rock's mechanical energy at the time of release relative to the surface of the river? (b) What are the rock's kinetic, potential, and mechanical energies after it has fallen 75.0 m? (c) Just before the rock hits the water, what are its speed and total mechanical energy? (d) Answer parts (a)–(c) for a reference point ($y = 0$) at the elevation where the rock is released. (Neglect air resistance.)

67. ■ A 0.300-kg ball is thrown vertically upward with an initial speed of 10.0 m/s. If the initial potential energy is taken as zero, find the ball's kinetic, potential, and mechanical energies (a) at its initial position, (b) at 2.50 m above the initial position, and (c) at its maximum height.

68. ■ What is the maximum height reached by the ball in Exercise 67?

69. ■■ A 0.50-kg ball thrown vertically upward has an initial kinetic energy of 80 J. (a) What are its kinetic and potential energies when it has traveled three fourths of the distance to its maximum height? (b) What is the ball's speed at this point? (c) What is its potential energy at the maximum height? (Assume a reference is chosen to be zero at the launch point.)

70. **IE** ■■ A girl swings back and forth on a swing with ropes that are 4.00 m long. The maximum height she reaches is 2.00 m above the ground. At the lowest point of the swing, she is 0.500 m above the ground. (a) The girl attains the maximum speed (1) at the top, (2) in the middle, or (3) at the bottom of the swing. Why? (b) What is the girl's maximum speed?

71. ■■ When a certain rubber ball is dropped from a height of 1.25 m onto a hard surface, it loses 18.0% of its mechanical energy on each bounce. (a) How high will the ball bounce on the first bounce? (b) How high will it bounce on the second bounce? (c) With what speed would the ball have to be thrown downward to make it reach its original height on the first bounce?

72. ■■ A skier coasts down a very smooth, 10-m-high slope similar to the one shown in Fig. 5.20. If the speed of the skier on the top of the slope is 5.0 m/s, what is his speed at the bottom of the slope?

73. ■■ A roller coaster travels on a frictionless track as shown in ▶Fig. 5.29. (a) If the speed of the roller coaster at point A is 5.0 m/s, what is its speed at point B? (b) Will it reach point C? (c) What speed at point A is required for the roller coaster to reach point C?

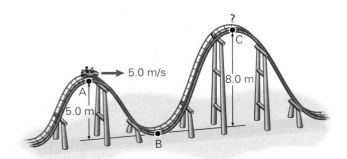

▲ FIGURE 5.29 **Enough energy?** See Exercise 73.

74. ■■ A simple pendulum has a length of 0.75 m and a bob whose mass is 0.15 kg. The bob is released from an angle of 25° relative to a vertical reference line (▼Fig. 5.30). (a) Show that the vertical height of the bob when it is released is $h = L(1 - \cos 25°)$. (b) What is the kinetic energy of the bob when the string is at an angle of 9.0°? (c) What is the speed of the bob at the bottom of the swing? (Neglect friction and the mass of the string.)

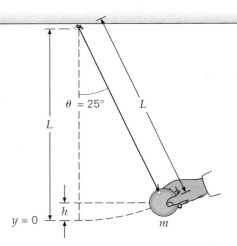

▲ FIGURE 5.30 **A pendulum swings** See Exercise 74.

75. ■■ Suppose the simple pendulum in Exercise 74 were released from an angle of 60°. (a) What would be the speed of the bob at the bottom of the swing? (b) To what height would the bob swing on the other side? (c) What angle of release would give half the speed of that for the 60° release angle at the bottom of the swing?

76. ■■ A 1.5-kg box that is sliding on a frictionless surface with a speed of 12 m/s approaches a horizontal spring. (See Fig. 5.18.) The spring has a spring constant of 2000 N/m. (a) How far will the spring be compressed in stopping the box? (b) How far will the spring be compressed when the box's speed is reduced to half of its initial speed?

77. ■■ A 28-kg child slides down a playground slide from a height of 3.0 m above the bottom of the slide. If her speed at the bottom is 2.5 m/s, what is the work done by nonconservative forces?

78. IE ■■ A 50-kg student on a sled starts from rest at a vertical height of 20 m above the horizontal base of a hill and slides down. (a) If the sled and the student have a speed of 10 m/s at the bottom of the hill, this system is (1) conservative, (2) nonconservative, or (3) none of the preceding. Why? (b) What is the work done by the nonconservative force?

79. ■■■ In Exercise 72, if the skier has a mass of 60 kg and the force of friction retards his motion by doing 2500 J of work, what is his speed at the bottom of the slope?

5.6 Power

80. Which of the following is not a unit of power? (a) J/s; (b) W·s; (c) W; (d) hp.

81. CQ If you check your electricity bill, you will note that you are paying the power company for so many kilowatt-hours (kWh). Are you really paying for power? Explain. Also, convert 2.5 kWh to J.

82. CQ (a) Does efficiency describe how fast work is done? Explain. (b) Does a more powerful machine always perform more work than a less powerful one? Explain.

83. CQ Two students who weigh the same start at the same ground-floor location at the same time to go to the same classroom on the third floor by different routes. If they arrive at different times, which student will have expended more power? Explain.

84. ■ What is the power in watts of a motor rated at $\frac{1}{4}$ hp?

85. ■ A girl consumes 8.4×10^6 J (2000 food calories) of energy per day while maintaining a constant weight. What is the average power she produces in a day?

86. ■ A 1500-kg race car can go from 0 to 90 km/h in 5.0 s. What average power is required to do this?

87. ■ The two 0.50-kg weights of a cuckoo clock descend 1.5 m in a three-day period. At what rate is gravitational potential energy decreased?

88. ■ A 60-kg woman runs up a staircase 15 m high (vertically) in 20 s. (a) How much power does she expend? (b) What is her horsepower rating?

89. ■■ An electric motor with a 2.0-hp output drives a machine with an efficiency of 45%. What is the energy output of the machine per second?

90. ■■ Water is lifted out of a well 30.0 m deep by a motor rated at 1.00 hp. Assuming 90% efficiency, how many kilograms of water can be lifted in 1 min?

91. ■■ The Gossamer Albatross, a human-powered aircraft that requires the pilot to pedal, flew across the English Channel on June 12, 1979, in 2 h and 49 min . If the aver-

age power needed to keep the aircraft flying is 0.30 hp, how much energy did the pilot use during the flight?

92. ■■■ A 3250-kg aircraft takes 12.5 min to achieve its cruising altitude of 10.0 km and cruising speed of 850 km/h. If the plane's engines deliver, on average, 1500 hp of power during this time, what is the efficiency of the engines?

93. ■■■ A sleigh and driver with a total mass of 120 kg is pulled up a hill with a 15° incline by a horse, as illustrated in ▼Fig. 5.31. (a) If the overall retarding frictional force is 950 N and the sled moves up the hill with a constant velocity of 5.0 km/h, what is the power output of the horse? (Express in horsepower, of course. Note the magnitude of your answer, and explain.) (b) Suppose that in a spurt of energy, the horse accelerates the sled uniformly from 5.0 km/h to 20 km/h in 5.0 s. What is the horse's maximum instantaneous power output? Assume the same force of friction.

▲ FIGURE 5.31   A one-horse open sleigh   See Exercise 93.

Additional Exercises

94. It is estimated that a 60-kg Olympic sprinter in the 100-m dash can achieve a kinetic energy of 3.0×10^3 J during the race. What is the sprinter's speed at this time?

95. A large electric motor with an efficiency of 75% has a power output of 1.5 hp. If the motor is run steadily and the cost of electricity is $0.12/kWh, how much does it cost, to the nearest penny, to run the motor for 2.0 h?

96. In planing a piece of wood 35 cm long, a carpenter applies a force of 40 N to a plane at a downward angle of 25° to the horizontal. How much work is done by the carpenter?

97. IE A spring with a force constant of 50 N/m is to be stretched from 0 to 20 cm. (a) The work required to stretch the spring from 10 cm to 20 cm is (1) more than, (2) the same as, or (3) less than that required to stretch it from 0 to 10 cm. (b) Compare the two work values to prove your answer to (a).

98. A 120-kg sleigh is pulled by one horse at a constant velocity for a distance of 0.75 km on a level snowy sur-

face. The coefficient of kinetic friction between the sleigh runners and the snow is 0.25. (a) Calculate the work done by the horse. (b) Calculate the work done by friction.

99. A water slide has a height of 4.0 m. The people coming down the slide shoot out horizontally at the bottom, which is a distance of 1.5 m above the surface of the water in the swimming pool. (a) If a person starts down the slide from rest, neglecting frictional losses, how far from a point directly below the bottom of the slide does the person land? (b) Does it make any difference whether the person is a small child or an adult?

100. A hiker plans to swing on a rope across a ravine in the mountains, as illustrated in ▼Fig. 5.32, and to drop when she is just above the far edge. (a) At what horizontal speed should she be moving when she starts to swing? (b) At what point would she be in danger of falling into the ravine? Explain.

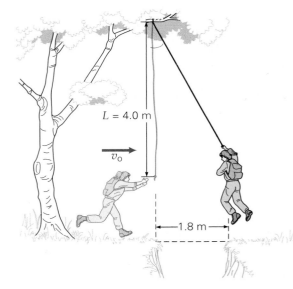

▲ FIGURE 5.32   Can she make it?   See Exercise 100.

101. A tractor pulls a wagon from rest with a constant force of 700 N, eventually giving the wagon a constant speed of 20.0 km/h. (a) How much work is done by the tractor in 3.50 min ? (b) What is the tractor's power output?

102. A sports car weighs one third as much as a large luxury car. (a) If the sports car is traveling at a speed of 90 km/h, at what speed would the larger car have to travel to have the same kinetic energy? (b) Suppose that the large car is traveling at a speed that gives it half the kinetic energy of the sports car. What is the speed of the more massive car?

103. A ball with a mass of 0.360 kg is dropped from a height of 1.20 m above the top of a fixed vertical spring whose force constant is 350 N/m. (a) What is the maximum distance the spring is compressed by the ball? (Neglect energy loss due to the collision.) (b) What is the speed of the ball when the spring has been compressed 5.00 cm?

104. A constant horizontal force of 30 N moves a box with a constant speed along a rough surface. If the force does work at a rate of 50 W, (a) what is the box's speed, and (b) how much work is done by the force in 2.5 s?

105. In a time of 10 s, a 70-kg student runs up two flights of stairs whose combined vertical height is 8.0 m. Compute the student's power output in doing work against gravity in (a) watts and (b) horsepower.

106. ■■■ A 50-kg crate slides down a 5.0-m loading ramp that is inclined at an angle of 25° to the horizontal. A worker pushes on the crate parallel to the surface of the ramp so that the crate slides down with a constant velocity. If the coefficient of kinetic friction between the crate and the ramp is 0.33, how much work is done by (a) the worker, (b) the force of friction, and (c) the force of gravity? (d) What is the net work done on the crate?

Linear Momentum and Collisions

Tomorrow, the sportscasters may say that the momentum of the entire game changed as a result of this clutch hit. One team gained momentum and went on to win the game, while its opponents lost it. But regardless of the effect on the team, it's clear that the momentum of the *ball* must have changed dramatically in the instant before this photograph was taken. The ball was traveling from right to left, probably at a pretty good rate of speed—and thus with lots of momentum. But a collision with several pounds of hardwood—with plenty of momentum of its own—changed the ball's trajectory in a fraction of a second. A fan might say that the batter turned the ball around. After studying Chapter 4, you might say that the force he applied gave it a large negative acceleration, reversing its velocity (momentum) vector. Yet, if you summed up the momentum of the ball and bat just before the collision and just afterward, you'd discover that although both the ball and the bat had momentum changes, the total momentum never changed!

If you were bowling and the ball bounced off the pins and rolled back toward you, you would probably be very surprised. But why? What leads us to expect that the ball will send the pins flying and continue on its way, rather than rebounding? You might say that the momentum of the ball carries it onward even after the collision (and you would be right)—but what does that really mean? In this chapter, you will study the concept of *momentum* and learn how it is particularly useful in analyzing motion and collisions.

6.1 Linear Momentum

OBJECTIVE: To compute linear momentum and the components of momentum.

The term *momentum* may bring to mind a football player running down the field, knocking down players who are trying to stop him. Or you might have heard someone say that a team lost its momentum (and so lost the game). Such everyday usages give some insight into the meaning of momentum. They suggest the idea of mass in motion and therefore of inertia. We tend to think of heavy or massive objects in motion as having a great deal of momentum, even if they move very slowly. However, according to the technical definition of momentum, a light object can have just as much momentum as a heavier one, and sometimes more.

Newton referred to what modern physicists term **linear momentum** as "the quantity of motion . . . arising from velocity and the quantity of matter conjointly." In other words, the momentum of a body is proportional to both its mass and its velocity. By definition,

Definitiion of: linear momentum

Note: The momentum vector of a single object is in the direction of the object's velocity.

the linear momentum of an object is the product of its mass and velocity:

$$\mathbf{p} = m\mathbf{v} \qquad (6.1)$$

SI unit of momentum: kilogram-meter per second $(\text{kg} \cdot \text{m/s})$

It is common to refer to linear momentum as simply *momentum*. From Eq. 6.1, you can see that the SI units for momentum are kilogram-meters per second. Momentum is a vector quantity that has the same direction as the velocity, and x–y components with magnitudes of $p_x = mv_x$ and $p_y = mv_y$, respectively.

Equation 6.1 expresses the momentum of a single object or particle. For a system of more than one particle, the **total linear momentum** of the system is the vector sum of the momenta (plural of *momentum*) of the individual particles:

Note: Total linear momentum—a vector sum

$$\mathbf{P} = \mathbf{p}_1 + \mathbf{p}_2 + \mathbf{p}_3 + \mathbf{p}_4 = \Sigma\mathbf{p}_i \qquad (6.2)$$

(*Note:* **P** signifies the *total* momentum, while **p** signifies an *individual* momentum.)

Example 6.1 ■ Momentum: Mass *and* Velocity

A 100-kg football player runs with a velocity of 4.0 m/s straight down the field. A 1.0-kg artillery shell leaves the barrel of a gun with a muzzle velocity of 500 m/s. Which has the greater momentum (magnitude), the football player or the shell?

Thinking It Through. Given the mass and velocity of an object, the momentum can be calculated from Eq. 6.1.

Solution. As usual, we first list the given data and what we are to find, using the subscripts "p" and "s" to refer to the player and shell, respectively.

Given: $m_p = 100 \text{ kg}$ *Find:* p_p and p_s (magnitudes of the momenta)
$\qquad\quad v_p = 4.0 \text{ m/s}$
$\qquad\quad m_s = 1.0 \text{ kg}$
$\qquad\quad v_s = 500 \text{ m/s}$

The magnitude of the momentum of the football player is

$$p_p = m_p v_p = (100 \text{ kg})(4.0 \text{ m/s}) = 4.0 \times 10^2 \text{ kg} \cdot \text{m/s}$$

and that of the shell is

$$p_s = m_s v_s = (1.0 \text{ kg})(500 \text{ m/s}) = 5.0 \times 10^2 \text{ kg} \cdot \text{m/s}$$

Thus, the less massive shell has the greater momentum. Remember, the magnitude of momentum depends on *both* the mass *and* the magnitude of the velocity.

Follow-up Exercise. What would the football player's speed have to be for his momentum to have the same magnitude as the artillery shell's momentum? Would this speed be realistic? *(Answers to all Follow-up Exercises are at the back of the text.)*

Integrated Example 6.2 ■ Linear Momentum: Some Ballpark Comparisons

Consider the three objects shown in ▶Fig. 6.1—a .22-caliber bullet, a cruise ship, and a glacier. Assuming each to be moving at its normal speed, (a) which would you expect to have the (1) greatest linear momentum and (2) the least linear momentum? (b) Estimate the masses and velocities and compute order-of-magnitude values of the linear momentum of the objects.

(a) Conceptual Reasoning. Certainly the bullet travels the fastest and the glacier the slowest, with the cruise ship in between. But, momentum, $p = mv$, is equally dependent on mass and velocity. The fast bullet has a tiny mass compared with that of the ship and the glacier. The slow glacier has a huge mass that greatly overshadows that of the bullet, but not so much that of the ship. The cruise ship weighs a great deal and thus has considerable mass. Which object has the greater momentum also depends on the relative speeds. The glacier "creeps" along compared with the ship, so the very slow speed of the glacier counterbalances its huge mass to make its momentum less than might be expected. Assuming the speed difference to be greater than the mass difference for the ship and glacier, the ship would have the larger momentum. Similarly, because of the fast bullet's relatively tiny mass, it would be expected to have the least momentum. So, with this reasoning, the largest momentum goes to the ship and the smallest momentum to the bullet.

(b) Thinking It Through. With no physical data given, you are asked to estimate the masses and velocities (speeds) of the objects so as to be able to compute their momenta [which will verify the reasoning in part (a)]. As is often the case in real-life problems, you may have difficulty estimating the values, so you would try to look up approximate values for the various quantities. For this example, we will provide these estimates. (Note that the units given in references vary, and it is important to convert units correctly.)

Given: Estimates (given below) of weight (mass) and speed for the bullet, cruise ship, and glacier.

Find: The approximate magnitudes of the momenta for the bullet (p_b), cruise ship (p_s), and glacier (p_g).

Bullet: A typical .22-caliber bullet has a weight of about 30 grains and a muzzle velocity of about 1300 ft/s. (A grain, abbreviated gr, is an old British unit. It was once commonly used for pharmaceuticals, such as 5-grain aspirin tablets; 1 lb = 7 000 gr.)

Ship: A ship like the one shown in Fig. 6.1 would have a weight of about 70 000 tons and a speed of about 20 knots. (A knot is another old unit, still commonly used in nautical contexts; 1 knot = 1.15 mi/h.)

Glacier: The glacier might be 1 km wide, 10 km long, and 250 m deep and move at a rate of 1 m per day. (There is much variation among glaciers. Therefore, these figures must involve more assumptions and rougher estimates than those for the bullet or ship. For example, we are assuming a uniform, rectangular cross-sectional area for the glacier. The depth is particularly difficult to estimate from a photograph; a minimum value is given by the fact that glaciers must be at least 50–60 m thick before they can "flow." Observed speeds range from a few centimeters to as much as 40 m a day for valley glaciers such as the one shown in Fig. 6.1c. The value chosen here is considered a typical one.)

(a)

(b)

(c)

▲ **FIGURE 6.1 Three moving objects: a comparison of momenta and kinetic energies** (a) A .22-caliber bullet shattering a ballpoint pen; (b) a cruise ship; (c) a glacier, Glacier Bay, Alaska. See Example 6.2.

Then, converting the data to metric units and giving orders of magnitude yields the following:

Bullet: $\quad m_b = 30 \text{ gr}\left(\dfrac{1 \text{ lb}}{7000 \text{ gr}}\right)\left(\dfrac{1 \text{ kg}}{2.2 \text{ lb}}\right) = 0.0019 \text{ kg} \approx 10^{-3} \text{ kg}$

$\quad v_b = 1.3 \times 10^3 \text{ ft/s}\left(\dfrac{0.305 \text{ m/s}}{\text{ft/s}}\right) = 4.0 \times 10^2 \text{ m/s} \approx 10^2 \text{ m/s}$

Ship: $\quad m_s = 7.0 \times 10^4 \text{ ton}\left(\dfrac{2.0 \times 10^3 \text{ lb}}{\text{ton}}\right)\left(\dfrac{1 \text{ kg}}{2.2 \text{ lb}}\right) = 6.4 \times 10^7 \text{ kg} \approx 10^8 \text{ kg}$

$\quad v_s = 20 \text{ knots}\left(\dfrac{1.15 \text{ mi/h}}{\text{knot}}\right)\left(\dfrac{0.447 \text{ m/s}}{\text{mi/h}}\right) = 10 \text{ m/s} = 10^1 \text{ m/s}$

Glacier: width $\approx 10^3$ m, length $\approx 10^4$ m, depth $\approx 10^2$ m

$\quad v_g = 1 \text{ m/day}\left(\dfrac{1 \text{ day}}{86\,400 \text{ s}}\right) = 1.2 \times 10^{-5} \text{ m/s} \approx 10^{-5} \text{ m/s}$

We have all the speeds and masses except for m_g, the mass of the glacier. To compute this value, we need to know the density of ice, since $m = \rho V$ (Eq. 1.1). The density of ice is less than that of water (ice floats in water), but the two are not very different, so we will use the density of water, 1.0×10^3 kg/m^3, to simplify the calculations. (The actual density of ice is 0.92×10^3 kg/m^3, but most of our other data in this Example are also approximations or estimates, so this shortcut should produce results that are good enough for our present purposes.)

Thus, the mass of the glacier is approximated as

$$m_g = \rho V = \rho(l \times w \times d)$$
$$\approx (10^3 \text{ kg/m}^3)[(10^4 \text{m})(10^3 \text{m})(10^2 \text{m})] = 10^{12} \text{ kg}$$

Then, calculating the magnitudes of the momenta of the objects, we have

Bullet: $\quad p_b = m_b v_b \approx (10^{-3} \text{ kg})(10^2 \text{ m/s}) = 10^{-1} \text{ kg} \cdot \text{m/s}$

Ship: $\quad p_s = m_s v_s \approx (10^8 \text{ kg})(10^1 \text{ m/s}) = 10^9 \text{ kg} \cdot \text{m/s}$

Glacier: $\quad p_g = m_g v_g \approx (10^{12} \text{ kg})(10^{-5} \text{ m/s}) = 10^7 \text{ kg} \cdot \text{m/s}$

So, the ship does have the largest momentum, and the bullet has the smallest.

Follow-up Exercise. Which of the objects in Example 6.2 has (1) the greatest kinetic energy and (2) the least kinetic energy? Justify your choices using order-of-magnitude calculations. (Notice here that the dependence is on the *square* of the velocity, $K = \frac{1}{2}mv^2$.)

Example 6.3 ■ Total Momentum: A Vector Sum

What is the total momentum for each of the systems of particles illustrated in ▶Fig. 6.2?

Thinking It Through. The total momentum is the vector sum of the individual momenta (Eq. 6.2). This quantity can be computed using the components of each vector.

Solution.

Given: Magnitudes and directions of momenta from Fig. 6.2   *Find:* (a) Total momentum (**P**) for Fig. 6.2a
                                                (b) Total momentum (**P**) for Fig. 6.2b

(a) The total momentum of a system is the vector sum of the momenta of the individual particles, so

$$\mathbf{P} = \mathbf{p}_1 + \mathbf{p}_2 = (2.0 \text{ kg} \cdot \text{m/s})\hat{\mathbf{x}} + (3.0 \text{ kg} \cdot \text{m/s})\hat{\mathbf{x}} = (5.0 \text{ kg} \cdot \text{m/s})\hat{\mathbf{x}} \quad (+x\text{-direction})$$

(b) Computing the total momenta in the x- and y-directions gives

$$\mathbf{P}_x = \mathbf{p}_1 + \mathbf{p}_2 = (5.0 \text{ kg} \cdot \text{m/s})\hat{\mathbf{x}} + (-8.0 \text{ kg} \cdot \text{m/s})\hat{\mathbf{x}}$$
$$= -(3.0 \text{ kg} \cdot \text{m/s})\hat{\mathbf{x}} \quad (-x\text{-direction})$$

$$\mathbf{P}_y = \mathbf{p}_3 = (4.0 \text{ kg} \cdot \text{m/s})\hat{\mathbf{y}} \quad (+y\text{-direction})$$

Note: Review Example 4.4 and Fig. 4.10.

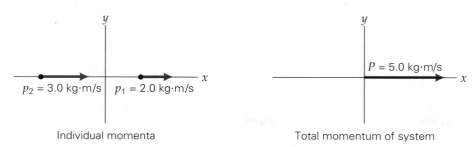

(a) $\mathbf{P} = \mathbf{p}_1 + \mathbf{p}_2$

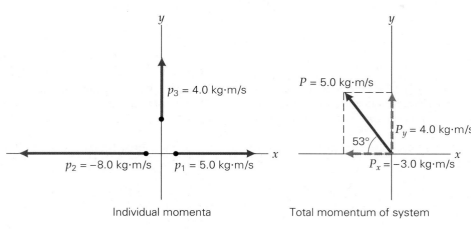

(b) $\mathbf{P} = \mathbf{p}_1 + \mathbf{p}_2 + \mathbf{p}_3$

◄ **FIGURE 6.2 Total momentum**
The total momentum of a system of particles is the vector sum of the particles' individual momenta. See Example 6.3.

Then

$$\mathbf{P} = \mathbf{P}_x + \mathbf{P}_y = (-3.0\,\text{kg} \cdot \text{m/s})\hat{\mathbf{x}} + (4.0\,\text{kg} \cdot \text{m/s})\hat{\mathbf{y}}$$

Follow-up Exercise. In this Example, if $\mathbf{p}_1$ and $\mathbf{p}_2$ in part (a) were added to $\mathbf{p}_2$ and $\mathbf{p}_3$ in part (b), what would be the total momentum?

In Example 6.3a, the momenta were along the coordinate axes and thus were added straightforwardly. If the motion of one (or more) of the particles is not along an axis, its momentum vector may be broken up, or resolved, into rectangular components, and individual components can then be added to find the components of the total momentum, just as you learned to do with force components in Chapter 4.

Since momentum is a vector, a change in momentum can result from a change in magnitude and/or direction. Examples of changes in the momenta of particles because of changes of direction on collision are illustrated in ▶Fig. 6.3. In the figure, the magnitude of a particle's momentum is taken to be the same both before and after collision (as indicated by the arrows of equal length). Figure 6.3a illustrates a direct rebound—a 180° change in direction. Note that the change in momentum ($\Delta\mathbf{p}$) is the vector difference and that directional signs for the vectors are important. Figure 6.3b shows a glancing collision, for which the change in momentum is given by analyzing the x- and y-components.

Force and Momentum

As you know from Chapter 4, if an object has a change in velocity (an acceleration), a net force must be acting on it. Similarly, since momentum is directly related to velocity (as well as mass), a change in momentum also requires a force. In fact, Newton originally expressed his second law of motion in terms of momentum rather than acceleration. The force–momentum relationship may be seen by

PHYSLET®
ILLUSTRATION

Two-Dimensional Collisions

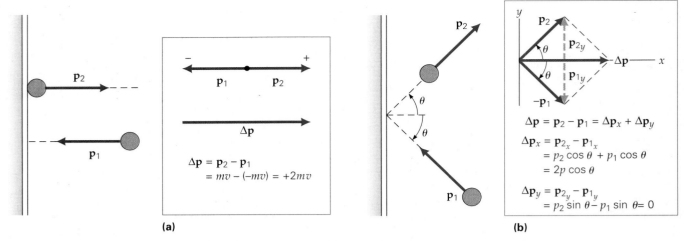

(a)

$$\Delta p = p_2 - p_1$$
$$= mv - (-mv) = +2mv$$

(b)

$$\Delta \mathbf{p} = \mathbf{p}_2 - \mathbf{p}_1 = \Delta \mathbf{p}_x + \Delta \mathbf{p}_y$$

$$\Delta \mathbf{p}_x = \mathbf{p}_{2_x} - \mathbf{p}_{1_x}$$
$$= p_2 \cos \theta + p_1 \cos \theta$$
$$= 2p \cos \theta$$

$$\Delta \mathbf{p}_y = \mathbf{p}_{2_y} - \mathbf{p}_{1_y}$$
$$= p_2 \sin \theta - p_1 \sin \theta = 0$$

▲ **FIGURE 6.3 Change in momentum** The change in momentum is given by the *difference* in the momentum vectors. **(a)** Here, the vector sum is zero, but the vector *difference*, or change in momentum, is not. (The particles are displaced for convenience.) **(b)** The change in momentum is found by computing the change in the components.

starting with $\mathbf{F}_{net} = m\mathbf{a}$ and using $\mathbf{a} = (\mathbf{v} - \mathbf{v}_o)/\Delta t$, where the mass is assumed to be constant. Thus,

$$\mathbf{F}_{net} = m\mathbf{a} = \frac{m(\mathbf{v} - \mathbf{v}_o)}{\Delta t} = \frac{m\mathbf{v} - m\mathbf{v}_o}{\Delta t} = \frac{\mathbf{p} - \mathbf{p}_o}{\Delta t} = \frac{\Delta \mathbf{p}}{\Delta t}$$

or

Newton's second law of motion in
terms of momentum

$$\mathbf{F}_{net} = \frac{\Delta \mathbf{p}}{\Delta t} \tag{6.3}$$

where $\mathbf{F}_{net}$ is the *average* net force on the object if the acceleration is not constant (or the *instantaneous* net force if Δt goes to zero).

Expressed in this form, Newton's second law states that *the net external force acting on an object is equal to the time rate of change of the object's momentum.* It is easily seen from the development of Eq. 6.3 that the equations $\mathbf{F}_{net} = m\mathbf{a}$ and $\mathbf{F}_{net} = \Delta \mathbf{p}/\Delta t$ are equivalent if the mass is constant. In some situations, however, the mass may vary. This factor will not be a consideration here in our discussion of particle collisions, but a special case will be given later in the chapter. The more general form of Newton's second law, Eq. 6.3, is true even if the mass varies.

Just as the equation $\mathbf{F}_{net} = m\mathbf{a}$ indicates that an acceleration is evidence of a net force, the equation $\mathbf{F}_{net} = \Delta \mathbf{p}/\Delta t$ indicates that *a change in momentum is evidence of a net force.* For example, as illustrated in ▼Fig. 6.4, the momentum of a projectile is tangential to the projectile's parabolic path and changes in both magnitude and direction. The change in momentum indicates that there is a net force acting on the projectile, which you know is the force of gravity. Changes in momentum were il-

▶ **FIGURE 6.4 Change in the momentum of a projectile**
The total momentum vector of a projectile is tangential to the projectile's path (as is its velocity); this vector changes in magnitude and direction, because of the action of an external force (gravity). The x-component of the momentum is constant. (Why?)

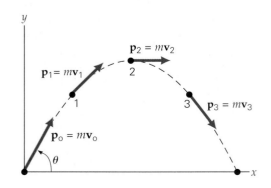

lustrated in Fig. 6.3. Can you identify the forces in these cases? Think in terms of Newton's third law.

6.2 Impulse

OBJECTIVES: **To relate (a) impulse and momentum, and (b) kinetic energy and momentum.**

When two objects—such as a hammer and a nail, a golf club and a golf ball, or even two cars—collide, they can exert large forces on one another for a short period of time (▶Fig. 6.5a). The force is not constant in this case. However, Newton's second law in momentum form is still useful for analyzing such situations by using average values. Written in this form, the law states that the net *average* force is equal to the time rate of change of momentum: $\overline{\mathbf{F}} = \Delta\mathbf{p}/\Delta t$ (Eq. 6.3). Rewriting the equation to express the change in momentum, we have (with only one force acting on the object)

$$\overline{\mathbf{F}}\Delta t = \Delta\mathbf{p} = \mathbf{p} - \mathbf{p}_o \qquad (6.4)$$

The term $\overline{\mathbf{F}}\Delta t$ is known as the **impulse** of the force:

$$\text{Impulse} = \overline{\mathbf{F}}\Delta t = \Delta\mathbf{p} = m\mathbf{v} - m\mathbf{v}_o \qquad (6.5)$$

SI unit of impulse and momentum: newton-second (N·s)

Thus, *the impulse exerted on a body is equal to the change in the body's momentum.* This statement is referred to as the **impulse–momentum theorem**. Impulse has units of newtons-second (N·s), which are also units of momentum by Eq. 6.4 and Eq. 6.1 (1 kg·m/s = 1 N·s).

In Chapter 5, you learned that by the work–energy theorem ($W_{net} = F_{net}\Delta x = \Delta K$), the area under an F_{net} versus x curve is equal to the net work, or change in kinetic energy. Similarly, the area under an F_{net} versus t curve is equal to the impulse, or the change in momentum (Fig. 6.5b). An impulse force usually varies with time and is therefore not a constant force. However, in general, it is convenient to talk about the equivalent *constant* average force $\overline{\mathbf{F}}$ acting over a time interval Δt to give the same impulse (same area under the force-versus-time curve), as shown in ▶Fig. 6.6.

(a)

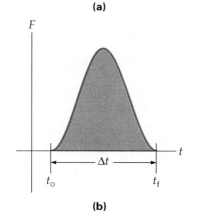

(b)

▲ **FIGURE 6.5 Collision impulse** **(a)** Collision impulse causes the football to be deformed. **(b)** The impulse is the area under the curve of an *F*-versus-*t* graph. Note that the impulse force on the ball is not constant, but rises to a maximum.

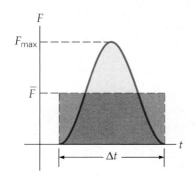

▲ **FIGURE 6.6 Average impulse force** The area under the average-force curve ($\overline{F}\Delta t$, within the dashed red lines) is the same as the area under the *F* versus *t* curve, which is usually difficult to evaluate.

Example 6.4 ■ Teeing Off: The Impulse–Momentum Theorem

A golfer drives a 0.10-kg ball from an elevated tee, giving the ball an initial horizontal speed of 40 m/s (about 90 mi/h). The club and the ball are in contact for 1.0 ms (millisecond). What is the average force exerted by the club on the ball during this time?

Thinking It Through. The average force is equal to the time rate of change of momentum (Eq. 6.5 rearranged).

Solution.

Given: $m = 0.10$ kg   *Find:* $\overline{F}$ (average force)
$v = 40$ m/s
$v_o = 0$
$\Delta t = 1.0$ ms $= 1.0 \times 10^{-3}$ s

Notice that the mass and the initial and final velocities are given, so the change in momentum can be found. Then, the average force can be computed from the impulse–momentum theorem:

$$\overline{F}\Delta t = p - p_o = mv - mv_o$$

Thus,

$$\overline{F} = \frac{mv - mv_o}{\Delta t} = \frac{(0.10\text{ kg})(40\text{ m/s}) - 0}{1.0 \times 10^{-3}\text{ s}} = 4000\text{ N (or 900 lb)}$$

$\overline{F}\,\Delta t = mv_0$

(a)

$\overline{F}\,\Delta t = mv_0$

(b)

▲ FIGURE 6.7 Adjust the impulse (a) The change in momentum in catching the ball is a constant mv_0. If the ball is stopped quickly (small Δt), the impulse force is large (big $\overline{F}$) and stings the catcher's bare hands. (b) Increasing the contact time (large Δt) by moving the hands with the ball reduces the impulse force and makes catching more enjoyable.

[This a very large force compared with the weight of the ball, $w = mg = (0.10\,\text{kg})$ $(9.8\,\text{m/s}^2) = 0.98\,\text{N}$.] The force is in the direction of the acceleration and is the *average* force. The instantaneous force is even greater near the midpoint of the time interval of the collision (Δt in Fig. 6.6).

Follow-up Exercise.   Suppose the golfer in this Example drives the ball with the same average force, but "follows through" on the swing so as to increase the contact time to 1.5 ms. What effect would this change have on the initial horizontal speed of the drive?

Example 6.4 illustrates the large forces that colliding objects can exert on one another during short contact times. However, in some instances, the impulse may be manipulated to reduce the force. Suppose there is a fixed change in momentum in a given situation. Then, since $\Delta p = \overline{F}\,\Delta t$, if Δt could be made longer, the average impulse force $\overline{F}$ would be reduced.

You have probably tried to minimize the impulse force on occasion. For example, when jumping from a height onto a hard surface, you try not to land stiff legged. The abrupt stop (small Δt) would apply a large impulse force to your leg bones and joints and could cause injury. If you bend your knees as you land, the impulse is vertically upward, opposite your velocity ($\overline{F}\,\Delta t = \Delta p = -mv_0$, with the final velocity being zero). Thus, increasing the time interval Δt makes the impulse force smaller.

Similarly, in catching a hard, fast-moving ball, you quickly learn not to catch it with your arms rigid, but rather to move your hands with the ball. This movement increases the contact time and reduces the impulse force and the "sting" (◀Fig. 6.7).

Another example in which the contact time is increased to decrease the impulse force is given in the Insight in this chapter.

In other instances, the *applied* impulse force may be relatively constant and the contact time (Δt) deliberately increased to produce a greater impulse, and thus a greater change in momentum ($\overline{F}\,\Delta t = \Delta p$). This is the principle of "following through" in sports, for example, when hitting a ball with a bat or racquet, or driving a golf ball. In the latter case (▶ Fig. 6.8a), assuming that the golfer supplies the same average force with each swing, the longer the contact time, the greater will be the impulse or the momentum the ball receives. That is, with $\overline{F}\,\Delta t = mv$ (since $v_0 = 0$), the greater the value of Δt, the greater will be the final velocity of the ball. (This principle is illustrated in the Follow-up Exercise in Example 6.4.) As you learned in Section 3.4, a greater projection velocity gives a projectile a greater range. (Ideally, at what angle of projection should a golfer try to achieve when driving the ball on a level fairway? Remember from Chapter 3?)

In some instances, a long follow-through that increases the contact time may be used to improve control of the ball's direction (Fig. 6.8b).

The word *impulse* implies that the impulse force acts only briefly or quickly (like an "impulsive" person), and this is true in many instances. However, the definition of *impulse* places no limit on the time interval over which the force may act. Technically, a comet at its closest approach to the Sun is involved in a collision, because in physics, collision forces do not have to be contact forces. Basically, a **collision** is an interaction of objects in which there is an exchange of momentum and energy. As you might expect from the work–energy theorem and the impulse–momentum theorem, momentum and kinetic energy are directly related. A little algebraic manipulation of the equation for kinetic energy (Eq. 5.5) allows us to express kinetic energy in terms of the *magnitude* of momentum:

$$K = \tfrac{1}{2}mv^2 = \frac{(mv)^2}{2m} = \frac{p^2}{2m} \qquad (6.6)$$

Thus, kinetic energy and momentum are intimately related. The conservation of these quantities is important in collisions, as we shall see later in the chapter.

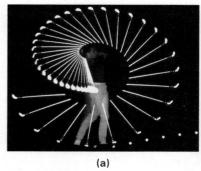

(a)

(b)

▲ **FIGURE 6.8 Increasing the contact time**
(a) A golfer follows through on a drive swing. One
reason he does so is to increase the contact time so
that the ball receives greater impulse and momentum.
(b) The follow-through on a long putt increases the
contact time for greater impulse and momentum,
but the main reason is for directional control.
Notice that the putter is in contact with the ball for
a time equivalent to about four flash intervals.

6.3 The Conservation of Linear Momentum

OBJECTIVES: To **(a)** explain the condition for the conservation of linear momentum and **(b)** apply them to physical situations.

Like total mechanical energy, the momentum of a body or system is a conserved
quantity under certain conditions. This fact allows us to analyze a wide range of situations and solve many problems readily. The conservation of momentum is one of
the most important principles in physics. In particular, it is used to analyze the collision of objects ranging from subatomic particles to automobiles in traffic accidents.

For the linear momentum of an object to be conserved (i.e., to remain constant with time), one condition must hold that is apparent from the momentum
form of Newton's second law (Eq. 6.3). If the net force acting on a particle is zero,
that is,

$$\mathbf{F}_{net} = \frac{\Delta \mathbf{p}}{\Delta t} = 0$$

then

$$\Delta \mathbf{p} = 0 = \mathbf{p} - \mathbf{p}_o$$

where $\mathbf{p}_o$ is the initial momentum and $\mathbf{p}$ is the momentum at some later time.
Since these two values are equal, the momentum is conserved:

$$\mathbf{p} = \mathbf{p}_o \qquad (6.7)$$

or

$$m\mathbf{v} = m\mathbf{v}_o$$

Note that this conservation is consistent with Newton's first law: An object remains at rest ($\mathbf{p} = 0$), or in motion with a *uniform* velocity (constant $\mathbf{p}$), unless
acted on by a net external force.

The conservation of momentum can be extended to a system of particles if we
write Newton's second law in terms of the net force acting on the system and of
the momenta of the particles: $\mathbf{F}_{net} = \Sigma \mathbf{F}_i$ and $\mathbf{P} = \Sigma \mathbf{p}_i = \Sigma m \mathbf{v}_i$.

Since $\mathbf{F}_{net} = \Delta \mathbf{P}/\Delta t$, and if there is no net external force acting *on the system*,
then $\mathbf{F}_{net} = 0$, and $\Delta \mathbf{P} = 0$; so $\mathbf{P} = \mathbf{P}_o$, and the total momentum is conserved.
This generalized condition is referred to as the law of **conservation of linear
momentum**:

$$\mathbf{P} = \mathbf{P}_o$$

Note: A lowercase "**p**" means
individual momentum. A capital "**P**"
means total system momentum. Both
are vectors.

Conservation of momentum—no net
external force

The Automobile Air Bag

A dark, rainy night—a car goes out of control and hits a big tree head on! But the driver walks away with only minor injuries, because he had his seatbelt buckled and was in a car equipped with air bags. Air bags, along with seatbelts, are safety devices designed to prevent (or lessen) injuries to passengers in the front seat in automobile collisions.

When a car collides with something basically immovable, such as a tree or a bridge abutment, or has a head-on collision with another vehicle, it stops almost instantaneously. If the front-seat passengers have not buckled up (and there are no air bags), they keep moving until acted on by an external force (Newton's first law). For the driver, this force is supplied by the steering wheel and column, and for the passenger by the dashboard and/or windshield.

Even when everyone has buckled up, there can be injuries. Seatbelts absorb energy by stretching, and they spread the force over a wider area. However, if a car is going fast enough and hits something truly immovable, there may be too much energy for the belts to absorb. This is where the air bag comes in. The bag inflates automatically on hard impact (Fig. 1), cushioning the driver (and front-seat passenger if both sides are equipped with air bags). In terms of impulse, the air bag increases the stopping contact time—the fraction of a second it takes your head to sink into the inflated bag is many times longer than the instant in which your head would be stopped by the dashboard. A longer contact time means a reduced average impact force and thus much less likelihood of an injury. (Because the bag is large, the total impact force is also spread over a greater area of the body, so the force on any one part of the body is also less.)

An interesting point is the inflating mechanism of an air bag. Think of how little time elapses between the front-end impact and the driver hitting the steering column in the collision of a fast-moving automobile. We say such a collision

FIGURE 1 Impulse and safety   An automobile air bag increases the contact time over what a person in a crash would experience with the dashboard or windshield, thereby decreasing the impulse force that could cause injury.

takes place instantaneously, yet during this time the air bag must be inflated! How is this done?

First, the air bag is equipped with sensors that detect the sharp deceleration associated with a head-on collision the instant it begins. If the deceleration exceeds the sensors' threshold settings, an electric current in an igniter in the air bag sets off a chemical explosion that generates gas to inflate the bag. The complete process from sensing to full inflation takes only on the order of 25 *thousandths* of a second (0.025 s).

The sensors' signals go first to a control unit, which determines whether a frontal collision rather than a system malfunction is occurring. (Accidental deployment could be dangerous as well as costly.) Typically, the control unit com-

---

Thus, the total linear momentum of a system, $\mathbf{P} = \Sigma \mathbf{p}_i$, is conserved if the net external force acting on the system is zero.

There are other ways to express this condition. For example, recall from Chapter 5 that a *closed*, or *isolated*, system is one on which no net external force acts, so the total linear momentum of an isolated system is conserved.

Note: Third-law force pairs were discussed in Section 4.4.

Within a system, internal forces are acting, for example, when particles collide. These are force pairs of Newton's third law, and there is a good reason that such forces are not explicitly referred to in the condition for the conservation of momentum: By Newton's third law, these internal forces are equal and opposite and vectorially cancel each other. Thus, the net internal force of a closed system is always zero.

An important point to understand, however, is that the momenta of *individual* particles or objects within a system may change. But in the absence of a net external force, the *vector sum* of all the momenta (the total system momentum **P**) remains the same. If the objects are initially at rest (i.e., the total momentum is zero) and then are set in motion as the result of internal forces, the total momentum must still add to zero. This principle is illustrated in ▶Fig. 6.9 and analyzed in Example 6.5. Objects in an isolated system may transfer momentum among them-

pares signals from two different sensors for collision verification. The unit is equipped with its own power source, since the car's battery and alternator are usually destroyed in a hard front-end collision. Sensing a collision, the control unit completes the circuit to the air-bag igniter. It initiates a chemical reaction in a sodium-azide (NaN_3) propellant. Gas (mostly nitrogen) is generated at an explosive rate, which inflates the air bag. The bag itself is made of thin nylon that is covered with cornstarch. The cornstarch acts as a lubricant to help the bag unfold smoothly on inflation.

Not only is sodium azide explosive, but it is also poisonous. Thus, you should not open a collapsed air bag, and if a car is junked, undeployed air bags must be disposed of properly. Other, nonpoisonous compounds are being developed for use as air-bag inflators.

In general, air bags offer protection only if the occupants are thrown forward, because the bags are designed to deploy only in front-end collisions. They are of little use in side-impact crashes. Side air bags, however, are now available in many new automobile models.

Recent Concerns

In some cases, however, the deployment of air bags has been shown to cause injuries and deaths. An air bag is not a soft, fluffy pillow. When activated, it is ejected out of its compartment at speeds up to 320 km/h (200 mi/h) and could hit a person close by with enough force to cause severe injury and even death. Adults are advised to sit at least 13 cm (10 in.) from the air-bag compartment to allow a margin of safety from the 5–8 cm (2–3-in.) injury "risk zone." Seats should be adjusted to allow the proper safety distance.*

*Guidelines from the National Highway Traffic Safety Administration (www.nhtsa.dot.gov).

FIGURE 2   The wrong way

The most serious concern involves children. Children may get too close to the dashboard if they are not buckled in or not buckled in securely with an adjustable shoulder harness. Using a rear-facing child seat in the front passenger seat is also dangerous if the air bag inflates (Fig. 2). According to government recommendations, (1) *children 12 years old and younger should ride buckled up in the back seat*, and (2) *a rear-facing child seat should never be put in front of an air bag*. The next generation of "smart" air bags will be designed to sense the severity of an accident and inflate accordingly, and to base inflation on the weight of the occupant. For example, if sensors detect the weight of a child, or a child in a safety seat, in the passenger seat, the air bag won't deploy.

There may be specific problems in some instances, but air bags do save many lives. All new passenger cars are now required to have dual air bags. But even if your car is equipped with air bags, *always* remember to buckle up. (Maybe we should make that Newton's fourth law of motion!)

Related Exercise: 41

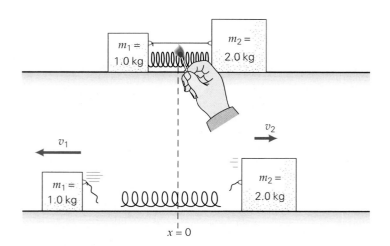

◄ **FIGURE 6.9   An internal force and the conservation of momentum**   The spring force is an internal force, so the momentum of the system is conserved. See Example 6.5.

selves, but the total momentum after the changes must add up to the initial value, assuming that friction is zero.

The conservation of momentum is often a powerful and convenient tool for analyzing situations involving motion. Its application is illustrated in the following

Examples. (Notice that conservation of momentum, in many cases, bypasses the need to know the forces involved.)

Example 6.5 ■ Before and After: Conservation of Momentum

Two masses, $m_1 = 1.0$ kg and $m_2 = 2.0$ kg, are held on either side of a light compressed spring by a light string joining them, as shown in Fig. 6.9. The string is burned (negligible external force), and the masses move apart on the frictionless surface, with m_1 having a velocity of 1.8 m/s to the left. What is the velocity of m_2?

Thinking It Through. With no net external force (the weights are canceled by normal forces), the total momentum of the system is conserved. It is initially zero, so after the string is burned, the momentum of m_2 must be *equal to* and opposite that of m_1. (Vector addition gives zero total momentum. Also, note that the term *light* indicates that the masses of the spring and string can be ignored.)

Solution. Listing the masses and speed given, we have

Given: $m_1 = 1.0$ kg   *Find:* v_2 (velocity—speed and direction)
$\qquad m_2 = 2.0$ kg
$\qquad v_1 = 1.8$ m/s (left)

Here, the system consists of the two masses and the spring. Since the spring force is internal to the system, the momentum of the system is conserved. It should be apparent that the initial total momentum of the system ($\mathbf{P}_o$) is zero, and therefore the final momentum must also be zero. Thus, we may write

$$\mathbf{P}_o = \mathbf{P} = 0 \qquad \text{and} \qquad \mathbf{P} = \mathbf{p}_1 + \mathbf{p}_2 = 0$$

(The momentum of the spring does not come into the equations, because its velocity is zero.) Thus,

$$\mathbf{p}_2 = -\mathbf{p}_1$$

which means that the momenta of m_1 and m_2 are equal and opposite. Using directional signs (with $+x$ indicating the direction to the right in the figure), we get

$$p_2 - p_1 = m_2 v_2 - m_1 v_1 = 0$$

or, in terms of magnitudes,

$$m_2 v_2 = m_1 v_1$$

and

$$v_2 = \left(\frac{m_1}{m_2}\right) v_1 = \left(\frac{1.0 \text{ kg}}{2.0 \text{ kg}}\right)(1.8 \text{ m/s}) = 0.90 \text{ m/s}$$

Thus, the velocity of m_2 is 0.90 m/s in the positive x-direction, or to the right in the figure. This value is half of v_1, as you might have expected, since m_2 has twice the mass of m_1.

Follow-up Exercise. (a) Suppose that the large block in Fig. 6.9 were attached to the Earth's surface so that the block could not move when the string was burned. Would momentum be conserved in this case? Explain. (b) Two girls, each having a mass of 50 kg, stand at rest on skateboards with negligible friction. The first girl tosses a 2.5-kg ball to the second. If the speed of the ball is 10 m/s, what is the speed of each girl after the ball is caught, and what is the momentum of the ball before it is tossed, while it is in the air, and after it is caught?

Integrated Example 6.6 ■ Conservation of Linear Momentum: Fragments and Components

A 30-g bullet with a speed of 400 m/s strikes a glancing blow to a target brick of mass 1.0 kg. The brick breaks into two fragments. The bullet deflects at an angle of 30° and has a reduced speed of 100 m/s. One piece of the brick (with mass 0.75 kg) goes off to the right, or in the initial direction of the bullet, with a speed of 5.0 m/s. (a) Make "before" and "after" sketches of the situation, along with a momentum vector diagram for the after-collision condition. (b) Determine the speed and direction of the other piece of the brick immediately after collision (where gravity can be neglected).

(a) Conceptual Reasoning. The idea here is to apply the conservation of linear momentum because there is no net external force (gravity neglected) on the system, bullet + brick. Adding the linear momentum vectors should be facilitated by a sketch. Could you have come up with something like ▼Fig. 6.10?

(b) Thinking It Through. There is one momentum before collision (that of the bullet), and three momenta afterwards (for the bullet and two fragments). By the conservation of linear momentum, the total (vector) momentum after collision equals that before collision. This quantity may be expressed in component form (Fig. 6.10), which should allow the velocity (speed and direction) of the fragment to be determined.

Given: $m_b = 30\text{ g} = 0.030\text{ kg}$ Find: v_2 (speed of the smaller brick
$\quad\quad v_{b_o} = 400\text{ m/s}$ (initial bullet speed) $\quad\quad\quad\quad$ fragment) and θ_2 (direction
$\quad\quad v_b = 100\text{ m/s}$ (final bullet speed) $\quad\quad\quad\quad$ of the fragment relative to
$\quad\quad \theta_b = 30°$ (final bullet angle) $\quad\quad\quad\quad$ the original direction of the
$\quad\quad M = 1.0\text{ kg}$ (brick mass) $\quad\quad\quad\quad$ bullet)
$\quad\quad m_1 = 0.75\text{ kg}$ and $\theta_1 = 0°$ (mass and
$\quad\quad\quad\quad$ angle of the large fragment)
$\quad\quad v_1 = 5.0\text{ m/s}$
$\quad\quad m_2 = 0.25\text{ kg}$ (mass of small fragment)

With no external forces (gravity neglected), the linear total momentum is conserved. Therefore, we can write both the x- and y-components of the total momentum, before and after, as follows (see Fig. 6.10):

$$\text{before} \quad\quad\quad\quad \text{after}$$
$$x: \quad m_b v_{b_o} = m_b v_b \cos\theta_b + m_1 v_1 + m_2 v_2 \cos\theta_2$$
$$y: \quad 0 = m_b v_b \sin\theta_b - m_2 v_2 \sin\theta_2$$

The x equation can be rearranged to solve for the magnitude of the x velocity of the smaller fragment:

$$x: \quad v_2 \cos\theta_2 = \frac{m_b v_{b_o} - m_b v_b \cos\theta_b - m_1 v_1}{m_2}$$

$$= \frac{(3.0 \times 10^{-2}\text{ kg})(4.0 \times 10^2\text{ m/s}) - (3.0 \times 10^{-2}\text{ kg})(10^2\text{ m/s})(0.866) - (0.75\text{ kg})(5.0\text{ m/s})}{0.25\text{ kg}} = 23\text{ m/s}$$

Similarly, the y equation can be solved for the magnitude of the y velocity component of the smaller fragment:

$$y: \quad v_2 \sin\theta_2 = \frac{m_b v_b \sin\theta_b}{m_2} = \frac{(3.0 \times 10^{-2}\text{ kg})(10^2\text{ m/s})(0.50)}{0.25\text{ kg}} = 6.0\text{ m/s}$$

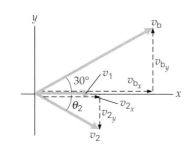

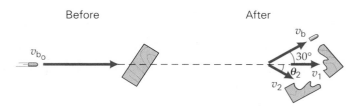

Before   After

◀ **FIGURE 6.10   A glancing collision**   Momentum is conserved in an isolated system. The motion in two dimensions may be analyzed in terms of the components of momentum, which are also conserved. See Integrated Example 6.6.

Forming a ratio of the equations ($y:/x:$),

$$\frac{v_2 \sin \theta_2}{v_2 \cos \theta_2} = \frac{6.0 \text{ m/s}}{23 \text{ m/s}} = 0.26 = \tan \theta_2 \quad \left(\text{where the } v_2 \text{ terms cancel, and } \frac{\sin \theta_2}{\cos \theta_2} = \tan \theta_2 \right)$$

and

$$\theta_2 = \tan^{-1}(0.26) = 15°$$

Then from the x equation,

$$v_2 = \frac{23 \text{ m/s}}{\cos 15°} = \frac{23 \text{ m/s}}{0.97} = 24 \text{ m/s}$$

Follow-up Exercise.   Is the kinetic energy conserved for the collision in Exercise 6.6? If not, where did the energy go?

Example 6.7 ■ Physics on Ice

A physicist is lowered from a helicopter to the middle of a smooth, level, frozen lake, the surface of which has negligible friction, and challenged to make her way off the ice. Walking is out of the question. (Why?) As she stands there pondering her predicament, she decides to use the conservation of momentum by throwing her heavy, identical mittens, which will provide her with the momentum to get herself to shore. To get to the shore more quickly, which should this sly physicist do, throw both mittens at once or throw them separately, one after the other?

Thinking It Through.   The initial momentum of the system (physicist and mittens) is zero. With no net external force, by the conservation of momentum, this remains zero, so if the mittens are thrown in one direction, she will go in the opposite direction (because momenta vectors in opposite directions can add to zero). So, which way of throwing gives greater speed? If both the mittens were thrown together, the magnitude of their momentum would be $2mv$, where v is relative to the ice and m is the mass of one mitten.

When thrown separately, the first mitten would have a momentum of mv. The physicist and the second mitten would then be in motion, and throwing the second mitten would add some more momentum to the physicist and increase her speed, but would the speed now be greater than that if both mittens were thrown simultaneously? Let's analyze the conditions of the second throw. After throwing the first mitten, the physicist "system" would have less mass. With less mass, the second throw would produce a greater acceleration and speed things up. But, on the other hand, after the first throw, the second mitten is moving with the person, and when thrown in the opposite direction, the mitten would have a velocity less than v relative to the ice (or to a stationary observer).

So, which effect would be greater? What do you think? Sometimes situations are difficult to analyze intuitively, and you must apply scientific principles to figure them out.

Solution.

Given: m = mass of single mitten   *Find:*   Which method of mitten throwing
          M = mass of physicist   gives the physicist the greater
          v = velocity of thrown mitten(s)   speed
          V_p = velocity of physicist

When the mittens are thrown together, by the conservation of momentum (V_p positive and v negative),

$$0 = 2m(-v) + MV_p \quad \text{and} \quad V_p = \frac{2mv}{M} \quad \text{(Thrown together)} \quad (1)$$

When they are thrown separately,

First throw: $0 = m(-v) + (M + m)V_{p_1}$ and $V_{p_1} = \frac{mv}{M + m}$ (Thrown separately) (2)

Second throw: $(M + m)V_{p_1} = m(V_{p_1} - v) + MV_{p_2}$

Note that in the m term, the velocity of the mitten is that relative to the ice. With an initial velocity of $+V_{p_1}$ when the first mitten is thrown in the negative direction, we have $V_{p_1} - v$. (Recall relative velocities from Chapter 3.)

Solving for V_{p_2}

$$V_{p_2} = V_{p_1} + \left(\frac{m}{M}\right)v = \frac{mv}{M+m} + \left(\frac{m}{M}\right)v = \left(\frac{m}{M+m} + \frac{m}{M}\right)v \qquad (3)$$

where Eq. 2 was substituted for V_{p_1} of the first throw.

Now, when the mittens are thrown together,

$$V_p = \left(\frac{2m}{M}\right)v$$

(Eq. 1), so the question is whether the result of Eq. 3 is greater or less than that of Eq. 1. Notice that with a greater denominator for the $m/(M+m)$ term in Eq. 3, is less than the m/M term. So,

$$\left(\frac{m}{M+m} + \frac{m}{M}\right) < \frac{2m}{M}$$

and $V_p > V_{p_2}$, or (thrown together) > (thrown separately).

Follow-up Exercise. Suppose the second throw were in the direction of the physicist's velocity from the first throw. Would this throw bring her to a stop?

The conservation of momentum is one of the most important principles in physics. As mentioned previously, it is used to analyze the collisions of objects ranging from subatomic particles to automobiles in traffic accidents. In many instances, however, external forces may be acting on the objects, which means that the momentum is not conserved.

But, as you will learn in the next section, the conservation of momentum often allows a good approximation *over the short time of a collision*, because the internal forces, for which momentum is conserved, are much greater than the external forces. For example, external forces such as gravity and friction also act on colliding objects, but are often relatively small compared with the internal forces. (This concept was implied in Example 6.6.) Therefore, if the objects interact for only a brief time, the effects of the external forces may be negligible compared with those of the large internal forces during that time.

6.4 Elastic and Inelastic Collisions

OBJECTIVE: **To describe the conditions on kinetic energy and momentum in elastic and inelastic collisions.**

Taking a closer look at collisions in terms of the conservation of momentum is simpler if we consider an isolated system, such as a system of particles (or balls) involved in head-on collisions. For simplicity, we will consider only collisions in one dimension. Such collisions can also be analyzed in terms of the conservation of energy. On the basis of what happens to the total kinetic energy, we can define two types of collisions: *elastic* and *inelastic*.

In an **elastic collision**, the total kinetic energy is conserved. That is, the *total* kinetic energy of all the objects of the system after the collision is the same as their *total* kinetic energy before the collision (►Fig. 6.11a). Kinetic energy may be traded between objects of a system, but the total kinetic energy in the system remains constant. Therefore,

Elastic collision—total kinetic energy conserved, as is momentum

total K after = total K before
$$K_f = K_i \qquad (\textit{elastic collision}) \qquad (6.8)$$

During such a collision, some or all of the initial kinetic energy is temporarily converted to potential energy as the objects are deformed. But, after the maximum

| (a) | (b) |

▶ **FIGURE 6.11 Collisions**
(a) Approximate elastic collisions.
(b) An inelastic collision.

Inelastic collision—total kinetic energy
not conserved, but momentum is

deformations occur, the objects *elastically* spring back to their original shapes, and the system regains all of its original kinetic energy. For example, two steel balls or two billiard balls may have a nearly elastic collision, with each ball having the same shape afterward as before; that is, there is no permanent deformation.

In an **inelastic collision** (Fig. 6.11b), total kinetic energy is *not* conserved. For example, one or more of the colliding objects may not spring back to its original shape, or heat or sound may be generated. In such interactions, work is done by nonconservative forces (Section 5.5), such as friction, and some kinetic energy is lost:

$$\text{total } K \text{ after} = \text{total } K \text{ before}$$
$$K_f < K_i \qquad \text{(condition for an inelastic collision)} \qquad (6.9)$$

Note: In reality, only atoms and subatomic particles have truly elastic collisions, but some larger hard objects have nearly elastic collisions in which the kinetic energy is approximately conserved.

For example, a hollow aluminum ball that collides with a solid steel ball may be dented. Permanently deformation of the ball takes work, and that work is done at the expense of the original kinetic energy of the system. Everyday collisions are inelastic.

For isolated systems, momentum is conserved in both elastic and inelastic collisions. *For an inelastic collision, only an amount of kinetic energy consistent with the conservation of momentum may be lost.* It may seem strange that kinetic energy can be lost and momentum still be conserved, but this fact provides insight into the difference between scalar and vector quantities.

Energy and Momentum in Inelastic Collisions

To see how momentum can remain constant while the kinetic energy changes (decreases) in inelastic collisions, consider the examples illustrated in ▶Fig. 6.12. In Fig. 6.12a, two balls of equal mass ($m_1 = m_2$) approach each other with equal and opposite velocities ($v_{1_o} = -v_{2_o}$). Hence, the total momentum before the collision is (vectorially) zero, but the (scalar) total kinetic energy is *not* zero. After the collision, the balls are stuck together and stationary, so the total momentum is unchanged—still zero. Momentum is conserved because the forces of collision are internal to the system of the two balls; thus, there is no net external force on the system. The total kinetic energy, however, has decreased to zero. In this case, some of the kinetic energy went into the work done in permanently deforming the balls. Some energy may also have gone into doing work against friction (producing heat) or may have been lost in some other way (for example, in producing sound).

It should be noted that the balls need not come to rest after collision. In a less inelastic collision, the balls may recoil in opposite directions at reduced, but equal, speeds. The momentum would be conserved (still equal to zero—why?), but the kinetic energy would again not be conserved. Under all conditions, the amount of kinetic energy lost must be consistent with the conservation of momentum.

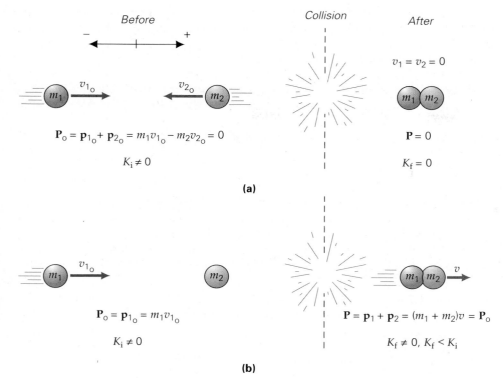

◄ FIGURE 6.12 Inelastic collisions In inelastic collisions, momentum is conserved, but kinetic energy is not. Collisions like the ones shown here, in which the objects stick together, are called *completely* or *totally inelastic collisions*. The maximum amount of kinetic energy lost is consistent with the law of conservation of momentum. (The equations are first provided in boldface (vector) notation and then in sign-magnitude notation.)

In Fig. 6.12b, one ball is initially at rest as the other approaches. The balls stick together after collision, but are still in motion. Both of these cases are examples of a **completely inelastic collision**, in which the objects stick together, and hence both objects have the same velocity after colliding. The coupling of colliding railroad cars is a practical example of a completely inelastic collision.

Assume that the balls in Fig. 6.12b have different masses. Since the momentum is conserved even in inelastic collisions,

$$\overset{\textit{before}}{m_1 v_{1_o}} = \overset{\textit{after}}{(m_1 + m_2)v}$$

and

$$v = \left(\frac{m_1}{m_1 + m_2} \right)v_{1_o} \qquad \begin{array}{l}\textit{(m}_2 \textit{ initially at rest,}\\ \textit{completely inelastic}\\ \textit{collision only)}\end{array} \qquad (6.10)$$

Thus, v is less than v_{1_o}, since $m_1/(m_1 + m_2)$ must be less than one. Now let us consider how much kinetic energy has been lost. Initially, $K_i = \frac{1}{2} m_1 v_o^2$; finally, after the collision,

$$K_f = \tfrac{1}{2}(m_1 + m_2)v^2$$

Substituting for v from Eq. 6.10 and simplifying the result, we have

$$K_f = \tfrac{1}{2}(m_1 + m_2)\left(\frac{m_1 v_{1_o}}{m_1 + m_2} \right)^2 = \frac{\frac{1}{2}m_1^2 v_{1_o}^2}{m_1 + m_2} = \left(\frac{m_1}{m_1 + m_2} \right)\tfrac{1}{2}m_1 v_{1_o}^2 = \left(\frac{m_1}{m_1 + m_2} \right)K_i$$

and

$$\frac{K_f}{K_i} = \frac{m_1}{m_1 + m_2} \qquad \begin{array}{l}\textit{(m}_2 \textit{ initially at rest,}\\ \textit{completely inelastic}\\ \textit{collision only)}\end{array} \qquad (6.11)$$

Equation 6.11 gives the fractional amount of the initial kinetic energy that the system has after a completely inelastic collision. For example, if the masses of the balls are equal $(m_1 = m_2)$, then $m_1/(m_1 + m_2) = \frac{1}{2}$ and $K_f/K_i = \frac{1}{2}$, or $K_f = K_i/2$. That is, only half of the initial kinetic energy is lost.

Note that not all of the kinetic energy can be lost in this case, no matter what the masses of the balls are. The total momentum after collision cannot be zero, since it was not zero initially. Thus, after the collision, the balls must be moving and must have some kinetic energy $(K_f \neq 0)$. *In a completely inelastic collision, the maximum amount of kinetic energy is lost, consistent with the conservation of momentum.*

Example 6.8 ■ Stuck Together: Completely Inelastic Collision

A 1.0-kg ball with a speed of 4.5 m/s strikes a 2.0-kg stationary ball. If the collision is completely inelastic, (a) what are the speeds of the balls after the collision? (b) What percentage of the initial kinetic energy do the balls have after the collision? (c) What is the total momentum after the collision?

Thinking It Through. Recall the definition of a *completely inelastic collision*. The balls stick together after collision; kinetic energy is *not* conserved, but total momentum is.

Solution. Using the labeling as in the preceding discussion, we have

Given: $m_1 = 1.0$ kg   *Find:* (a) v (speed after collision)
$m_2 = 2.0$ kg
$v_0 = 4.5$ m/s   (b) $\dfrac{K_f}{K_i}$ $(\times\ 100\%)$

(c) P_f (total momentum after collision)

(a) The momentum is conserved and

$$\mathbf{P}_f = \mathbf{P}_o$$

or

$$(m_1 + m_2)v = m_1 v_o$$

The balls stick together and have the same speed after collision. This speed is then

$$v = \left(\frac{m_1}{m_1 + m_2}\right)v_o = \left(\frac{1.0\ \text{kg}}{1.0\ \text{kg} + 2.0\ \text{kg}}\right)(4.5\ \text{m/s}) = 1.5\ \text{m/s}$$

(b) The fractional part of the initial kinetic energy that the balls have after the completely inelastic collision is given by Eq. 6.11. Notice that this fraction, as given by the masses, is the same as that for the speeds (Eq. 6.10) in this special case. By inspection, we can write

$$\frac{K_f}{K_i} = \frac{m_1}{m_1 + m_2} = \frac{1.0\ \text{kg}}{1.0\ \text{kg} + 2.0\ \text{kg}} = \frac{1}{3} = 0.33\ (\times\ 100\%) = 33\%$$

Let's show this relationship explicitly:

$$\frac{K_f}{K_i} = \frac{\frac{1}{2}(m_1 + m_2)v^2}{\frac{1}{2}m_1 v_o^2} = \frac{\frac{1}{2}(1.0\ \text{kg} + 2.0\ \text{kg})(1.5\ \text{m/s})^2}{\frac{1}{2}(1.0\ \text{kg})(4.5\ \text{m/s})^2} = 0.33\ (=33\%)$$

Keep in mind that Eq. 6.11 applies *only* to *completely* inelastic collisions. For other types of collisions, the initial and final values of the kinetic energy must be computed explicitly.

(c) The total momentum is conserved in all collisions (in the absence of external forces), so the total momentum after collision is the same as before collision. That value is the momentum of the incident ball, with a magnitude of

$$P_f = p_{1_o} = m_1 v_o = (1.0\ \text{kg})(4.5\ \text{m/s}) = 4.5\ \text{kg} \cdot \text{m/s}$$

and the same direction as that of the incoming ball. Also, $P_f = (m_1 + m_2)v = 4.5\ \text{kg} \cdot \text{m/s}$.

Follow-up Exercise. A small hard-metal ball of mass m collides with a larger, stationary, soft-metal ball of mass M. A *minimum* amount of work W is required to make a dent in the larger ball. If the smaller ball initially has kinetic energy $K = W$, will the larger ball be dented in a completely inelastic collision between the two balls?

Energy and Momentum in Elastic Collisions

For a general *elastic* collision of two objects,

$$\text{Conservation of } K: \quad \overset{before}{\tfrac{1}{2}m_1 v_{1_o}^2 + \tfrac{1}{2}m_2 v_{2_o}^2} = \overset{after}{\tfrac{1}{2}m_1 v_1^2 + \tfrac{1}{2}m_2 v_2^2} \qquad (6.12a)$$

Linear momentum is conserved even if the collision is inelastic, so we can write

$$\text{Conservation of } \mathbf{P}: \quad m_1 \mathbf{v}_{1_o} + m_2 \mathbf{v}_{2_o} = m_1 \mathbf{v}_1 + m_2 \mathbf{v}_2 \qquad (6.12b)$$

Thus, for head-on elastic collisions, knowing the masses of the objects and the initial velocities allows you to find the final velocities.

A common collision situation is that in which one of the objects is initially stationary and then is hit head on by the other (▼Fig. 6.13). Assume that the motion of both of the balls after the collision is to the right, or in the positive direction, as shown in the figure. If this is not the case, the math will tell you. (How?)

This situation is one dimensional. The condition for an elastic, head-on collision in this situation is

$$\tfrac{1}{2}m_1 v_{1_o}^2 = \tfrac{1}{2}m_1 v_1^2 + \tfrac{1}{2}m_2 v_2^2 \qquad (6.13a)$$

and conservation of linear momentum gives

$$m_1 v_{1_o} = m_1 v_1 + m_2 v_2 \qquad (6.13b)$$

Rearranging these equations gives

$$m_1(v_{1_o}^2 - v_1^2) = m_2 v_2^2 \qquad (6.13a')$$

and

$$m_1(v_{1_o} - v_1) = m_2 v_2 \qquad (6.13b')$$

Then, using the algebraic relationship $x^2 - y^2 = (x + y)(x - y)$ and dividing Eq. 6.13a' by Eq. 6.13b', we get

$$\frac{m_1(v_{1_o} + v_1)(v_{1_o} - v_1)}{m_1(v_{1_o} - v_1)} = \frac{m_2 v_2^2}{m_2 v_2} \qquad \frac{(\text{Eq. } 6.13a')}{(\text{Eq. } 6.13b')} \quad (divide)$$

or

$$v_{1_o} + v_1 = v_2 \qquad (6.14)$$

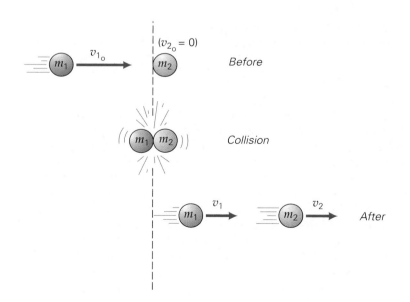

◄ FIGURE 6.13 Elastic collision
For an elastic collision between two bodies, one of which is initially at rest, the velocities after the collision depend on the relative masses of the bodies.

Equation 6.14 can now be used to eliminate v_1 or v_2 from Eq. 6.13b. Thus, each of these final velocities can be expressed in terms of the initial velocity (v_{1_o}):

$$v_1 = \left(\frac{m_1 - m_2}{m_1 + m_2}\right)v_{1_o}$$

(*final velocities for*   (6.15)
elastic, head-on collision

$$v_2 = \left(\frac{2m_1}{m_1 + m_2}\right)v_{1_o}$$

with m_2 initially   (6.16)
stationary)

Note that the final velocities depend on mass *ratios* of the objects. Looking at Eq. 6.15, you can see that if m_1 is greater than m_2, then v_1 is positive, or in the same direction as the initial velocity of the incoming ball, as was assumed in Fig. 6.13. However, if m_2 is greater than m_1, then v_1 is negative, and the incoming ball recoils in the opposite direction after collision. Equation 6.16 shows that v_2 is always in the same direction as the velocity of the incoming ball.

You can also get some general ideas about what happens after such a collision by considering three possible situations for the relative masses of the objects, illustrated in ▶ Fig. 6.14:

Case 1. $m_1 = m_2$ (Fig. 6.14a). From Eqs. 6.15 and 6.16,

$$v_1 = 0 \qquad \text{and} \qquad v_2 = v_{1_o}$$

That is, if the masses of the colliding objects are equal, the objects simply exchange momentum and energy. The incoming ball is stopped on collision, and the originally stationary ball moves off with the same velocity as that of the incoming ball, thus obviously conserving the system's kinetic energy and momentum. (A real-world example would be two billiard balls colliding along a straight line.)

Case 2. $m_1 \gg m_2$ (m_1 very much greater than m_2; Fig. 6.14b). In this case, m_2 can be ignored in the addition and subtraction with m_1 in Eqs. 6.15 and 6.16, and

$$v_1 \approx v_{1_o} \qquad \text{and} \qquad v_2 \approx 2v_{1_o}$$

This pair of relationships tells you that if a very massive object collides with a stationary light object, the massive object is slowed down only slightly by the collision, and the light object is knocked away with a velocity almost twice that of the initial velocity of the massive object. (Think of a bowling ball hitting a pin.)

Case 3. $m_1 \ll m_2$ (m_1 very much less than m_2; Fig. 6.14c). Here, m_1 can be ignored in the addition and subtraction with m_2 in Eqs. 6.15 and 6.16, and

$$v_1 \approx -v_{1_o} \qquad \text{and} \qquad v_2 \approx 0$$

(In the second equation, the approximation $m_1/m_2 \approx 0$ is made.) Thus, if a light object collides with a massive stationary one, the massive object remains *almost* stationary, and the light object recoils backward with approximately the same speed it had before collision. An extreme case of this type is similar to a particle striking a solid, immovable wall. (See Fig. 6.3.) If the wall is anchored to the ground, then the Earth and wall recoil together, but this effect is totally unnoticeable. (Why?) Total momentum, however, is still conserved. For the case in Fig. 6.14c, the massive ball must move a bit after the collision to conserve momentum. (Think of throwing a small rubber ball at a stationary bowling ball.)

Remember, these special cases give you a general idea about what might happen for similar, but not exact, situations. Here are some examples that demonstrate such cases.

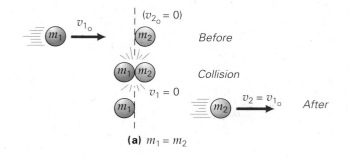

(a) $m_1 = m_2$

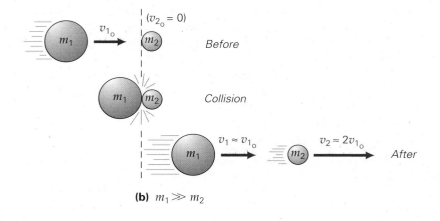

(b) $m_1 \gg m_2$

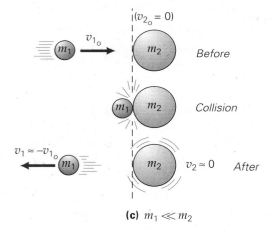

(c) $m_1 \ll m_2$

◀ **FIGURE 6.14 Special cases of head-on elastic collisions** **(a)** When a moving object collides elastically with a stationary object of equal mass, there is a complete exchange of momentum and energy. **(b)** When a very massive moving object collides elastically with a much less massive stationary object, the very massive object continues to move essentially as before, and the less massive object is given a velocity of almost twice the initial velocity of the large mass. **(c)** When a moving object of small mass collides elastically with a very massive stationary object, the incoming object recoils in the opposite direction with approximately the same speed as its initial speed, and the very massive object remains essentially stationary.

Example 6.9 ■ Elastic Collision: Conservation of Momentum and Kinetic Energy

A 0.30-kg object with a speed of 2.0 m/s in the positive x-direction has a head-on elastic collision with a stationary 0.70-kg object located at $x = 0$. What is the distance separating the objects 2.5 s after the collision?

Thinking It Through.   The incoming object is less massive than the stationary one. However, it is not extremely less massive, so we might expect the objects to separate in opposite directions after collision, as in Case 1. Equations 6.15 and 6.16 will give the velocities; knowing the time, we can find the distances traveled or the separation distance after collision.

Solution.   Using the notation used previously, we have

PHYSLET®
ILLUSTRATION

Elastic and Inelastic Collisions

Given: $m_1 = 0.30$ kg **Find:** $\Delta x = x_2 - x_1$ (separation distance
$\quad\quad\quad v_{1_o} = 2.0$ m/s $\quad\quad\quad\quad\quad\quad\quad\quad$ 2.5 s after collision)
$\quad\quad\quad m_2 = 0.70$ kg
$\quad\quad\quad v_{2_o} = 0$
$\quad\quad\quad\quad t = 2.5$ s

From Eqs. 6.15 and 6.16, the velocities after collision are

$$v_1 = \left(\frac{m_1 - m_2}{m_1 + m_2}\right)v_{1_o} = \left(\frac{0.30 \text{ kg} - 0.70 \text{ kg}}{0.30 \text{ kg} + 0.70 \text{ kg}}\right)(2.0 \text{ m/s}) = -0.80 \text{ m/s}$$

$$v_2 = \left(\frac{2m_1}{m_1 + m_2}\right)v_{1_o} = \left[\frac{2(0.30 \text{ kg})}{0.30 \text{ kg} + 0.70 \text{ kg}}\right](2.0 \text{ m/s}) = 1.2 \text{ m/s}$$

Here, m_1 is less than m_2, but not *so much* less that it could be ignored, as was done in Fig. 6.14c.

The objects are separating after collision, and their positions relative to the collision point ($x_o = 0$) are

$$x_1 = v_1 t = (-0.80 \text{ m/s})(2.5 \text{ s}) = -2.0 \text{ m}$$
$$x_2 = v_2 t = (1.2 \text{ m/s})(2.5 \text{ s}) = 3.0 \text{ m}$$

So,

$$\Delta x = x_2 - x_1 = 3.0 \text{ m} - (-2.0 \text{ m}) = 5.0 \text{ m}$$

The objects are 5.0 m apart at that time.

Follow-up Exercise. Suppose the objects in this Example had equal masses. What would be the separation distance 2.5 s after collision in this case?

Example 6.10 ■ Spare! Conservation of Momentum and Kinetic Energy

A 7.1-kg bowling ball with a speed of 6.0 m/s has a head-on elastic collision with a stationary 1.6-kg pin. (a) What is the velocity of each object after the collision? (b) What is the total momentum after the collision?

Thinking It Through. If the ball and the pin have an elastic collision, then Eqs. 6.15 and 6.16 apply and will give the velocities. (The ball is quite a bit more massive than the pin, so one might expect that the ball and the pin will go in same direction after collision, as in Fig. 6.14b.) The total momentum is conserved, so we can calculate it from the initial total momentum. (Why?)

Solution. Listing the data, we have

Given: $m_1 = 7.1$ kg **Find:** (a) $\mathbf{v}_1$ and $\mathbf{v}_2$ (velocities after collision)
$\quad\quad\quad v_{1_o} = 6.0$ m/s $\quad\quad\quad\quad\quad$ (b) $\mathbf{P}$ (total momentum after collision)
$\quad\quad\quad m_2 = 1.6$ kg
$\quad\quad\quad v_{2_o} = 0$

(a) The velocities are given directly by Eqs. 6.15 and 6.16, and the direction of the incoming ball is taken as positive:

$$v_1 = \left(\frac{m_1 - m_2}{m_1 + m_2}\right)v_{1_o} = \left(\frac{7.1 \text{ kg} - 1.6 \text{ kg}}{7.1 \text{ kg} + 1.6 \text{ kg}}\right)(6.0 \text{ m/s}) = 3.8 \text{ m/s}$$

$$v_2 = \left(\frac{2m_1}{m_1 + m_2}\right)v_{1_o} = \left[\frac{2(7.1 \text{ kg})}{7.1 \text{ kg} + 1.6 \text{ kg}}\right](6.0 \text{ m/s}) = 9.8 \text{ m/s}$$

So both objects move in the same direction. Here, m_1 is greater than m_2, but not *so much* greater that m_2 could be ignored, as was done in Fig. 6.14b.
(b) Momentum is conserved in elastic (and inelastic) collisions, so the total momentum afterward is the same as that before the collision. Thus, the total momentum is that of the incident ball, in the same direction and with a magnitude of

$$P_f = p_{1_o} = m_1 v_{1_o} = (7.1 \text{ kg})(6.0 \text{ m/s}) = 43 \text{ kg} \cdot \text{m/s}$$

Follow-up Exercise. In general, for head-on, elastic collisions with one of the objects initially stationary, is it possible for the velocities of the objects to be the same after collision as they were before it? Explain.

Example 6.11 ■ An Elastic Collision on a Small Scale

An electron has an elastic, head-on collision with a stationary hydrogen atom. What fraction of the electron's initial kinetic energy is transferred to the hydrogen atom? (The mass of the hydrogen atom is 1840 times the mass of the electron.)

Thinking It Through. This is certainly a case of an object hitting a much more massive, stationary object (similar to Fig. 6.14c). You may notice immediately that we are not given any initial velocity, only a ratio of masses. Right away, this should make you think of a solution involving a ratio—that of the kinetic energies, which would be a fractional representation.

Solution.

Given: $M_H = 1840 \, m_e$ or $M_H/m_e = 1840$ *Find:* Fraction of kinetic energy transfer

Since no initial velocity for the electron is given, we take it to be v_{1_o}. After the collision, the hydrogen atom would have a velocity given by Eq. 6.16. These values may be used to compute kinetic energies. Then, the ratio K_H/K_{e_o} (K of the H atom after collision/K of the electron before collision) is the fraction of the kinetic energy transferred to the hydrogen atom, since initially, the total kinetic energy is that of the electron. So, using Eq. 6.16, we have

$$\frac{K_H}{K_{e_o}} = \frac{\frac{1}{2} M_H v^2}{\frac{1}{2} m_e v_{1_o}^2} = \frac{\frac{1}{2} M_H \left(\dfrac{2m_e}{m_e + M_H}\right)^2 v_{1_o}^2}{\frac{1}{2} m_e v_{1_o}^2} = \left(\frac{M_H}{m_e}\right)\left[\frac{2}{1 + \left(\dfrac{M_H}{m_e}\right)}\right]^2$$

$$= (1840)\left[\frac{4}{(1 + 1840)^2}\right] \approx \left(\frac{4}{1840}\right) = 0.0022$$

where an approximation was made in the last step. (Is it reasonable?) So, a little over $\frac{2}{1000}$ of the electron's kinetic energy is transferred to the hydrogen atom, which means that the more massive atom doesn't move very fast after collision.

Follow-up Exercise. What would be the kinetic-energy transfer if a moving hydrogen atom collided with a stationary electron?

Conceptual Example 6.12 ■ Two In, One Out?

A novelty collision device, as shown in ▶ Fig. 6.15, consists of five identical metal balls. When one ball swings in, after multiple collisions, one ball swings out at the other end of the row of balls. When two balls swing in, two swing out; when three swing in, three swing out, and so on—always the same number out as in.

Suppose that two balls, each of mass m, swing in at velocity v and collide with the next ball. Why doesn't one ball swing out at the other end with a velocity $2v$?

Reasoning and Answer. The collisions along the horizontal row of balls are approximately elastic. The case of two balls swinging in and one ball swinging out with twice the velocity wouldn't violate the conservation of momentum: $(2m)v = m(2v)$. However, there's another condition that applies if we assume elastic collisions—the conservation of kinetic energy. Let's check to see if this condition is upheld for this case:

▲ **FIGURE 6.15 One in, one out**
See Conceptual Example 6.12.

$$K_i = K_f$$

$$\tfrac{1}{2}(2m)v^2 \stackrel{?}{=} \tfrac{1}{2}m\,(2v)^2$$

$$mv^2 \neq 2mv^2$$

Hence, the kinetic energy would *not* be conserved if this happened, and the equation is telling us that this situation violates established physical principles and does not occur. Note that there's a big violation—more energy out than in.

Follow-up Exercise. Suppose the first ball of mass m were replaced with a ball of mass $2m$. When this ball is pulled back and allowed to swing in, how many balls will swing out? [*Hint*: Think about the analogous situation in Fig. 6.14, and remember that the balls in the row are actually colliding. It may help to think of them as being separated.]

6.5   Center of Mass

OBJECTIVES: To (a) explain the concept of the center of mass and compute its location for simple systems, and (b) describe how the center of mass and center of gravity are related.

The conservation of total momentum gives us a method of analyzing a "system of particles." Such a system may be virtually anything—for example, a volume of gas, water in a container, or a baseball. Another important concept, the center of mass, allows us to analyze the overall motion of a system of particles. It involves representing the whole system as a single particle or point mass. This concept will be introduced here and applied in more detail in the upcoming chapters.

We have seen that if no net external force acts on a particle, the particle's linear momentum is constant. Similarly, if no net external force acts on a *system* of particles, the linear momentum of the system is constant. This similarity implies that a system of particles might be represented by an *equivalent* single particle. Moving rigid objects, such as balls, automobiles, and so forth, are essentially systems of particles and can be effectively represented by equivalent single particles when we analyze motion. Such representation is done through the concept of the **center of mass (CM)**:

The center of mass is the point at which all of the mass of an object or system may be considered to be concentrated, for the purposes of linear or translational motion only.

Even if a rigid object is rotating, an important result (beyond the scope of this text to derive) is that the center of mass moves as though it were a particle (►Fig. 6.16). The center of mass is sometimes described as the *balance point* of a solid object. For example, if you balance a meterstick on your finger, the center of mass of the stick is located directly above your finger, and all of the mass (or weight) seems to be concentrated there.

An expression similar to Newton's second law for a single particle applies to a *system* when the center of mass is used:

$$\mathbf{F}_{net} = M\mathbf{A}_{CM} \tag{6.17}$$

Here, $\mathbf{F}_{net}$ is the net external force on the system, M is the total mass of the system or the sum of the masses of the particles of the system ($M = m_1 + m_2 + m_3 + \cdots + m_n$, where the system has n particles), and $\mathbf{A}_{CM}$ is the acceleration of the center of mass of the system. In words, Eq. 6.17 says that the *center of mass* of a system of particles moves as though all the mass of the system were concentrated there and acted on by the resultant of the external forces. Note that the movement of the individual parts of the system is *not* predicted by Eq. 6.17.

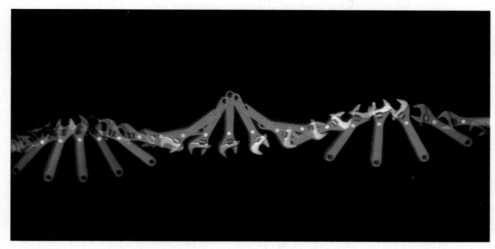

▲ **FIGURE 6.16 Center of mass** The center of mass of this sliding wrench moves in a straight line as though it were a particle. Note the white dot on the wrench that marks the center of mass.

It follows that *if the net external force on a system is zero*, the total linear momentum of the center of mass is conserved (i.e., stays constant), because

$$\mathbf{F}_{net} = M\mathbf{A}_{CM} = M(\Delta\mathbf{V}_{CM}/\Delta t) = \Delta(M\mathbf{V}_{CM})/\Delta t = \Delta\mathbf{P}/\Delta t = 0 \qquad (6.18)$$

Then, $\Delta\mathbf{P}/\Delta t = 0$, which means that there is no change in $\mathbf{P}$ during a time Δt, or the total momentum of the system, $\mathbf{P} = M\mathbf{V}_{CM}$, is constant (but not necessarily zero). Since M is constant (why?), $\mathbf{V}_{CM}$ is a constant in this case. Thus, the center of mass either moves with a constant velocity or remains at rest.

Although you may more readily visualize the center of mass of a solid object, the concept of the center of mass applies to any system of particles or objects, even a quantity of gas. For a system of n particles arranged in one dimension, along the x-axis (▶ Fig. 6.17), the location of the center of mass is given by

$$X_{CM} = \frac{m_1\mathbf{x}_1 + m_2\mathbf{x}_2 + m_3\mathbf{x}_3 + \cdots + m_n\mathbf{x}_n}{m_1 + m_2 + m_3 + \cdots + m_n} \qquad (6.19)$$

That is, X_{CM} is the x-coordinate of the center of mass of a system of particles. In shorthand notation (using signs to indicate vector directions in one dimension), this relationship is expressed as

$$X_{CM} = \frac{\Sigma m_i x_i}{M} \qquad (6.20)$$

where Σ is the summation of the products $m_i x_i$ for n particles ($i = 1, 2, 3, \ldots, n$). If $\Sigma m_i x_i = 0$, then $X_{CM} = 0$, and the center of mass of the one-dimensional system is located at the origin.

Other coordinates of the center of mass for systems of particles are similarly defined. For a two-dimensional distribution of masses, the coordinates of the center of mass are (X_{CM}, Y_{CM}).

Example 6.13 ■ Finding the Center of Mass: A Summation Process

Three masses, 2.0 kg, 3.0 kg, and 6.0 kg, are located at positions (3.0, 0), (6.0, 0), and (−4.0, 0), respectively, in meters from the origin (Fig. 6.17). Where is the center of mass of this system?

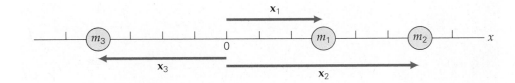

► **FIGURE 6.17 System of particles in one dimension** Where is the system's center of mass? See Example 6.13.

Thinking It Through. The masses and the positions are given, so the definition of X_{CM} (Eq. 6.20) applies. However, keep in mind that the positions are located by vector displacements from the origin and are indicated in one dimension by the appropriate signs (+ or −).

Solution.

Given: $m_1 = 2.0$ kg *Find:* X_{CM} (CM coordinate)
$m_2 = 3.0$ kg
$m_3 = 6.0$ kg
$x_1 = 3.0$ m
$x_2 = 6.0$ m
$x_3 = -4.0$ m

Then, simply performing the summation as indicated in Eq. 6.20 yields

$$X_{CM} = \frac{\Sigma_i m_i x_i}{M}$$

$$= \frac{(2.0 \text{ kg})(3.0 \text{ m}) + (3.0 \text{ kg})(6.0 \text{ m}) + (6.0 \text{ kg})(-4.0 \text{ m})}{2.0 \text{ kg} + 3.0 \text{ kg} + 6.0 \text{ kg}} = 0$$

The center of mass is at the origin.

Follow-up Exercise. Describe a single particle that could replace the system in this Example in order for us to be able to analyze the motion of the CM.

Example 6.14 ■ A Dumbbell: Center of Mass Revisited

A dumbbell (▼Fig. 6.18) has a connecting bar of negligible mass. Find the location of the center of mass (a) if m_1 and m_2 are each 5.0 kg and (b) if m_1 is 5.0 kg and m_2 is 10.0 kg.

Thinking It Through. This Example shows how the location of the center of mass depends on the distribution of mass. In (b), you might expect the center of mass to be located closer to the more massive end of the dumbbell.

Solution. Listing the data, with the coordinates from Eq. 6.18

Given: $x_1 = 0.20$ m *Find:* (a) (X_{CM}, Y_{CM}) (CM coordinates), with $m_1 = m_2$
$x_2 = 0.90$ m   (b) (X_{CM}, Y_{CM}), with $m_1 \neq m_2$
$y_1 = y_2 = 0.10$ m
(a) $m_1 = m_2 = 5.0$ kg
(b) $m_1 = 5.0$ kg
$m_2 = 10.0$ kg

Note that each mass is considered to be a particle located at the center of the sphere (its center of mass).

► **FIGURE 6.18 Location of the center of mass** See Example 6.14.

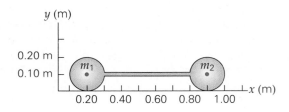

(a) X_{CM} is given by

$$X_{CM} = \frac{m_1 x_1 + m_2 x_2}{m_1 + m_2}$$

$$= \frac{(5.0\ kg)(0.20\ m) + (5.0\ kg)(0.90\ m)}{5.0\ kg + 5.0\ kg} = 0.55\ m$$

Similarly, we find that $Y_{CM} = 0.10\ m$. (You might have seen this right away, since each center of mass is at this height.) The center of mass of the dumbbell is then located at $(X_{CM}, Y_{CM}) = (0.55\ m, 0.10\ m)$, or midway between the end masses.

(b) With $m_2 = 10.0\ kg$,

$$X_{CM} = \frac{m_1 x_1 + m_2 x_2}{m_1 + m_2}$$

$$= \frac{(5.0\ kg)(0.20\ m) + (10.0\ kg)(0.90\ m)}{5.0\ kg + 10.0\ kg} = 0.67\ m$$

which is two thirds of the way between the masses. (Note that the distance of the CM from the center of m_1 is $\Delta x = 0.67\ m - 0.20\ m = 0.47\ m$. With the distance $L = 0.70\ m$ between the centers of the masses, $\Delta x/L = 0.47\ m/0.70\ m = 0.67$, or $\frac{2}{3}$.) You might expect the balance point of the dumbbell in this case to be closer to m_2. The y-coordinate of the center of mass is again $Y_{CM} = 0.10\ m$, as you can prove for yourself.

Follow-up Exercise. In part (b) of this Example, take the origin of the coordinate axes to be at the point where m_1 touches the x-axis. What are the coordinates of the CM in this case, and how does its location compare with that found in the Example?

In Example 6.14, when the value of one of the masses changed, the x-coordinate of the center of mass changed. You might have expected the y-coordinate to change also. However, the centers of the end masses were still at the same height, and Y_{CM} remained the same. To increase Y_{CM}, one or both of the end masses would have to be in a higher position.

Now let's see how the concept of the center of mass can be applied to a realistic situation.

Example 6.15 ■ Internal Motion: Where's the Center of Mass?

A 75.0-kg man stands in the far end of a 50.0-kg boat 100 m from the shore, as illustrated in ▼Fig. 6.19. If he walks to the other end of the 6.00-m-long boat, how far is he from the shore? Neglect friction, and assume that the center of mass of the boat is at its midpoint.

Thinking It Through. The answer is *not* $100\ m - 6.00\ m = 94.0\ m$, because the boat moves as the man walks. Why? With no net external force, the acceleration of the center of mass of the man–boat system is zero (Eq. 6.17), and so is the total momentum by Eq. 6.18

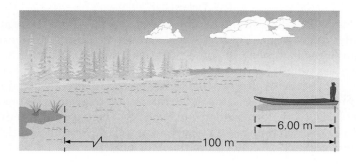

◀ **FIGURE 6.19 Walking toward shore** See Example 6.15.

$(\mathbf{P} = M\mathbf{V}_{CM} = 0)$. Hence, the velocity of the center of mass of the system is zero, or the center of mass is stationary and remains so to conserve system momentum; that is, X_{CM} (initial) $= X_{CM}$ (final).

Solution. Taking the shore as the origin ($x = 0$), we have

Given: $m_m = 75.0$ kg   *Find:* x_{m_f} (distance of man
 $x_{m_i} = 100$ m from shore)
 $m_b = 50.0$ kg
 $x_{b_i} = 94.0$ m $+ 3.00$ m $= 97.0$ m (CM position)
 (where the location of the boat is taken
 at its center of mass)

Note that if we take the man's final position to be a distance x_{m_f} from the shore, then the final position of the boat's center of mass will be $x_{b_f} = x_{m_f} + 3.00$ m, since the man will be at the front of the boat.
Then initially,

$$X_{CM_i} = \frac{m_m x_{m_i} + m_b x_{b_i}}{m_m + m_b}$$

$$= \frac{(75.0 \text{ kg})(100 \text{ m}) + (50.0 \text{ kg})(97.0 \text{ m})}{75.0 \text{ kg} + 50.0 \text{ kg}} = 98.8 \text{ m}$$

And finally,

$$X_{CM_f} = \frac{m_m x_{m_f} + m_b x_{b_f}}{m_m + m_b}$$

$$= \frac{(75.0 \text{ kg})x_{m_f} + (50.0 \text{ kg})(x_{m_f} + 3.00 \text{ m})}{75.0 \text{ kg} + 50.0 \text{ kg}} = 98.8 \text{ m}$$

Here, $X_{CM_i} = 98.8$ m $= X_{CM_f}$, since the CM does not move. Then, solving for x_{m_f}, we get

$$(125 \text{ kg})(98.8 \text{ m}) = (125 \text{ kg})x_{m_f} + (50.0 \text{ kg})(3.00 \text{ m})$$

and

$$x_{m_f} = 97.6 \text{ m}$$

from the shore.

PHYSLET®
ILLUSTRATION

Where Is the Center of Mass

Follow-up Exercise. Suppose the man then walks back to his original position at the opposite end of the boat. Would he then be 100 m from shore again?

Center of Gravity

As you know, mass and weight are related. Closely associated with the concept of the center of mass is the concept of the **center of gravity (CG)**, the point where all of the weight of an object may be considered to be concentrated when the object is represented as a particle. If the acceleration due to gravity is constant both in magnitude and direction over the extent of the object, Eq. 6.20 can be rewritten as (with all $g_i = g$)

$$MgX_{CM} = \Sigma_i m_i g x_i \qquad (6.21)$$

Then, the weight Mg acts as if it were concentrated at X_{CM}, and the center of mass and the center of gravity coincide. As you may have noticed, the location of the center of gravity was implied in some previous figures in Chapter 4, where the vector arrows for weight ($\mathbf{w} = m\mathbf{g}$) were drawn from a point at or near the center of an object. For practical purposes, the center of gravity is considered to coincide with the center of mass. That is, the acceleration due to gravity is constant for all parts of the object. (Note the constant g in Eq. 6.21.) There would be a difference in the locations of the two points if an object were so large that the acceleration due to gravity were different at different parts of the object.

In some cases, the center of mass or the center of gravity of an object may be located by symmetry. For example, for a spherical object that is homogeneous (i.e., the mass is distributed evenly throughout), the center of mass is at the geometrical center (or center of symmetry). In Example 6.14(a), where the end masses of the dumbbell were equal, it was probably apparent that the center of mass was midway between them.

The location of the center of mass or center of gravity of an irregularly shaped object is not so evident and is usually difficult to calculate (even with advanced mathematical methods that are beyond the scope of this book). In some instances, the center of mass may be located experimentally. For example, the center of mass of a flat, irregularly shaped object can be determined experimentally by suspending it freely from different points (▼Fig. 6.20). A moment's thought should convince you that the center of mass (or center of gravity) always lies vertically below the point of suspension. Since the center of mass is defined as the point at which all the mass of a body can be considered to be concentrated, this is analogous to a particle of mass suspended from a string. Suspending the object from two or more points and marking the vertical lines on which the center of mass must lie locates the center of mass as the intersection of the lines.

The center of mass (or center of gravity) of an object may lie outside the body of the object (▶Fig. 6.21). For example, the center of mass of a homogeneous ring is at the ring's center. The mass in any section of the ring is compensated for by the mass in an equivalent section directly across the ring, and by symmetry, the center of mass is at the center of the ring. For an L-shaped object with equal legs, the center of mass lies on a line that makes a 45° angle with both legs. Its location can easily be determined by suspending the L from a point on one of the legs and noting where a vertical line from that point intersects the diagonal line.

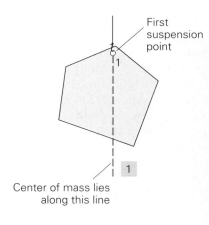

First suspension point

Center of mass lies along this line

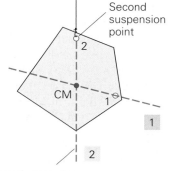

Second suspension point

CM

Center of mass also lies along this line

(a)

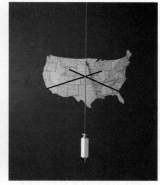

(b)

▲ **FIGURE 6.20 Location of the center of mass by suspension**   **(a)** The center of mass of a flat, irregularly shaped object can be found by suspending the object from two or more points. The CM (and CG) lies on a vertical line under any point of suspension, so the intersection of two such lines marks its location midway through the thickness of the body. The sheet could be balanced horizontally at this point. Why? **(b)** The process is illustrated with a cutout map of the United States. Note that a plumb line dropped from any other point (third photo) does in fact pass through the CM as located in the first two photos.

(a)

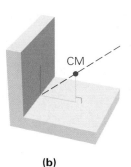

(b)

▲ **FIGURE 6.21 The center of mass may be located outside of a body** The center of mass (and center of gravity) may lie either inside or outside of a body, depending on the distribution of that object's mass. **(a)** For a uniform ring, the center of mass is at the center of the ring. **(b)** For an L-shaped object, if the mass distribution is uniform and the legs are of equal length, the center of mass lies on the diagonal between the legs.

▲ **FIGURE 6.22 Center of gravity** By arching his body, this high jumper can get over the bar even though his center of gravity passes beneath it.

Keep in mind that the location of the center of mass or center of gravity of an object depends on the distribution of mass. Therefore, for a flexible object such as the human body, the position of the center of gravity changes as the object changes configuration (distribution of mass). For example, when a person raises both arms overhead, his or her center of gravity is raised several centimeters. For a high jumper going over a bar, the center of gravity lies outside the arched body (▼Fig. 6.22). In fact, the center of gravity passes *beneath* the bar. This configuration is made purposefully, because work must be done to raise the center of gravity, and only the jumper's body has to clear the bar, not the CG.

*6.6 Jet Propulsion and Rockets

OBJECTIVE: **To apply the conservation of momentum in the explanation of jet propulsion and the operation of rockets.**

The word *jet* is sometimes used to refer to a stream of liquid or gas emitted at a high speed—for example, a jet of water from a fountain or a jet of air from an automobile tire. **Jet propulsion** is the application of such jets to the production of motion. This concept usually brings to mind jet planes and rockets, but squid and octopi propel themselves by squirting jets of water.

You have probably tried the simple application of blowing up a balloon and releasing it. Lacking any guidance or rigid exhaust system, the balloon zigzags around, driven by the escaping air. In terms of Newton's third law, the air is forced out by the contraction of the stretched balloon—that is, the balloon exerts a force on the air. Thus, there must be an equal and opposite reaction force exerted by the air on the balloon. It is this force that propels the balloon on its erratic path.

Jet propulsion is explained by Newton's third law, and in the absence of external forces, the conservation of momentum also applies. You may understand this concept better by considering the recoil of a rifle, taking the rifle and the bullet as an isolated system (▶Fig. 6.23). Initially, the total momentum of this system is zero. When the rifle is fired (by remote control to avoid external forces), the expansion of the gases from the exploding charge accelerates the bullet down the barrel. These gases push backward on the rifle as well, producing a recoil force (the "kick" experienced by a person firing a weapon). Since the initial momentum of the system is zero and the force of the expanding gas is an internal force, the momenta of the bullet and of the rifle must be exactly equal and opposite at any instant. After the bullet leaves the barrel, there is no propelling force, so the bullet and the rifle move with constant velocities (unless acted on by a net external force such as gravity or air resistance).

Similarly, the thrust of a rocket is created by exhausting the gas from burning fuel out the rear of the rocket. The expanding gas exerts a net force on the rocket that propels the rocket in the forward direction (▶Fig. 6.24a and b). The rocket exerts a reaction force on the gas, so all of the gas is directed out the exhaust nozzle. If the rocket is at rest when the engines are turned on and there are no external forces (as in deep space, where friction is zero and gravitational forces are negligible), then the instantaneous momentum of the exhaust gas is equal and opposite to that of the rocket. The numerous exhaust-gas molecules have small masses and high velocities, and the rocket has a much larger mass and a smaller velocity.

Unlike a rifle firing a single shot, which has negligible mass, a rocket continuously loses mass when burning fuel. (The rocket is more like a machine gun). Thus, the rocket is a system for which the mass is not constant. As the mass of the rocket decreases, it accelerates more easily. Multistage rockets take advantage of this fact. The hull of a burnt-out stage is jettisoned to give a further in-flight reduction in mass (Fig. 6.24c). The payload (cargo) is typically a very small part of the initial mass of rockets for space flights.

Suppose that the purpose of a space flight is to land a payload on the Moon. At some point on the journey, the gravitational attraction of the Moon will become

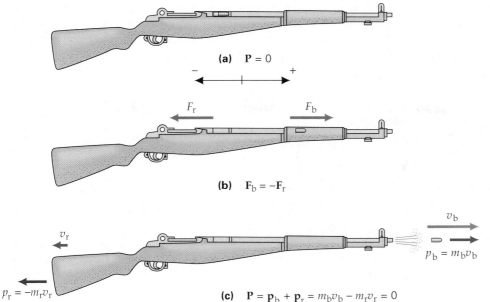

(a)   P = 0

(b) $\mathbf{F}_b = -\mathbf{F}_r$

v_b

$p_b = m_b v_b$

v_r

$p_r = -m_r v_r$

(c) $\mathbf{P} = \mathbf{p}_b + \mathbf{p}_r = m_b v_b - m_r v_r = 0$

◀ **FIGURE 6.23 Conservation of momentum** **(a)** Before the rifle is fired, the total momentum of the rifle and bullet (as an isolated system) is zero. **(b)** During firing, there are equal and opposite internal forces, and the instantaneous total momentum of the rifle–bullet system remains zero (neglecting external forces, such as arise when a rifle is being held). **(c)** When the bullet leaves the barrel, the total momentum of the system is still zero. (The vector equation is written in boldface (vector) notation and then in sign–magnitude notation so as to indicate directions.)

(b)

(c)

greater than that of the Earth, and the spacecraft will accelerate toward the Moon. A soft landing is desirable, so the spacecraft must be slowed down enough to go into orbit around the Moon. This slowing down is accomplished by using the rocket engines to apply a *reverse thrust*, or braking thrust. The spacecraft is maneuvered through a 180° angle, or turned around, which is quite easy to do in space. The rocket engines are then fired, expelling the exhaust gas toward the Moon and supplying a braking action.

You have experienced a reverse-thrust effect if you have flown in a commercial jet. In this instance, however, the craft is not turned around. Instead, after touchdown, the jet engines are revved up, and a braking action can be felt. Ordinarily, revving up the engines accelerates the plane forward. The reverse thrust is accomplished by activating thrust reversers in the engines that deflect the exhaust gases forward (▶Fig. 6.25). The gas experiences an impulse force and a change in momentum in the forward direction (see Fig. 6.3b), and the engine and the aircraft have an equal and opposite momentum change and braking impulse force.

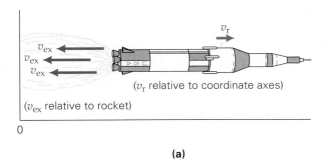

v_{ex}

v_{ex}

v_{ex}

v_r

(v_r relative to coordinate axes)

(v_{ex} relative to rocket)

0

(a)

▲ **FIGURE 6.24 Jet propulsion and mass reduction** **(a)** A rocket burning fuel is continuously losing mass and thus becomes easier to accelerate. The resulting force on the rocket (the thrust) depends on the product of the rate of change of its mass with time and the velocity of the exhaust gases: $(\Delta m/\Delta t)\mathbf{v}_{ex}$. Since the mass is decreasing, $\Delta m/\Delta t$ is negative, and the thrust is opposite $\mathbf{v}_{ex}$. **(b)** The space shuttle uses a multistage rocket. Both the two booster rockets and the huge external fuel tank are jettisoned in flight. **(c)** The first and second stages of a *Saturn V* rocket separating after 148 s of burn time.

Question: There are no end-of-chapter exercises on the material covered in this section, so test your knowledge with this one: Astronauts use handheld maneuvering devices (small rockets) to move around on space walks. Describe how these rockets are used. Is there any danger on an untethered space walk?

Rocket

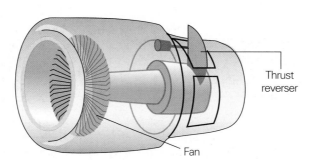

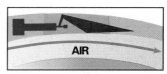

Normal operation

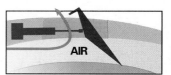

Thrust reverser activated

▶ **FIGURE 6.25 Reverse thrust**
Thrust reversers are activated on jet engines during landing to help slow the plane. The gas experiences an impulse force and a change in momentum in the forward direction, and the plane experiences an equal and opposite momentum change and a braking impulse force.

Thrust reverser

Fan

Chapter Review

Important Concepts and Equations

- The **linear momentum (p)** of a particle is a vector and is defined as the product of mass and velocity.

$$\mathbf{p} = m\mathbf{v} \tag{6.1}$$

- The **total linear momentum (P)** of a system is the vector sum of the momenta of the individual particles:

$$\mathbf{P} = \mathbf{p}_1 + \mathbf{p}_2 + \mathbf{p}_3 + \cdots = \Sigma\mathbf{p}_i \tag{6.2}$$

- **Newton's second law in terms of momentum (for a particle):**

$$\mathbf{F}_{net} = \frac{\Delta\mathbf{p}}{\Delta t} \tag{6.3}$$

- **Conservation of linear momentum:** In the absence of a net external force, the total linear momentum of a system is conserved.

- The **impulse–momentum theorem** relates the impulse acting on an object to its change in momentum:

$$\text{Impulse} = \overline{\mathbf{F}}\,\Delta t = \Delta\mathbf{p}_o = m\mathbf{v} - m\mathbf{v}_o \tag{6.5}$$

- **In an elastic collision, the total kinetic energy of the system is conserved.**

- **Momentum is conserved in both elastic and inelastic collisions.** In a completely inelastic collision, objects stick together after impact.

- **Conditions for an elastic collision:**

$$\begin{aligned} \mathbf{P}_f &= \mathbf{P}_i \\ K_f &= K_i \end{aligned} \tag{6.8}$$

- **Conditions for an inelastic collision:**

$$\begin{aligned} \mathbf{P}_f &= \mathbf{P}_i \\ K_f &< K_i \end{aligned} \tag{6.9}$$

- **Final velocities in head-on, two-body elastic collisions** ($v_{2_o} = 0$):

$$v_1 = \left(\frac{m_1 - m_2}{m_1 + m_2}\right)v_{1_o} \tag{6.15}$$

$$v_2 = \left(\frac{2m_1}{m_1 + m_2}\right)v_{1_o} \tag{6.16}$$

- The **center of mass** is the point at which all of the mass of an object or system may be considered to be concentrated. (The **center of gravity** is the point at which all the weight may be considered to be concentrated.)

Coordinates of the center of mass (using signs for directions):

$$X_{CM} = \frac{\Sigma_i m_i x_i}{M} \tag{6.20}$$

Exercises

6.1 Linear Momentum

1. Linear momentum has units of (a) N/m, (b) kg·m/s, (c) N/s, or (d) all of the preceding.

2. Linear momentum is (a) always conserved, (b) a scalar quantity, (c) a vector quantity, or (d) unrelated to force.

3. **CQ** Does a fast-running running back always have more linear momentum than a slow-moving, more massive lineman? Explain.

4. **CQ** Both the kinetic energy and the linear momentum of an object depend on the mass and velocity of the object. Explain the differences between kinetic energy and linear momentum.

5. **CQ** If two objects have the same linear momentum, will they have the same kinetic energy? Explain.

6. ■ If a 60-kg woman is riding in a car traveling at 90 km/h, what is her linear momentum relative to (a) the ground and (b) the car?

7. ■ The linear momentum of a runner in a 100-m dash is 7.5×10^2 kg·m/s. If the runner's speed is 10 m/s, what is his mass?

8. ■ Find the magnitude of the linear momentum of (a) a 7.1-kg bowling ball traveling at 12 m/s and (b) a 1200-kg automobile traveling at 90 km/h.

9. **IE** ■ In a football game, a lineman usually has more mass than a running back. (a) Will a lineman always have greater linear momentum than a running back? Why? (b) Who has greater linear momentum, a 75-kg running back running at 8.5 m/s or a 120-kg lineman moving at 5.0 m/s?

10. ■ How fast would a 1200-kg car travel if it had the same linear momentum as a 1500-kg truck traveling at 90 km/h?

11. ■ A ball of mass 3.0 kg has a linear momentum of 12 kg·m/s. What is the kinetic energy of the ball?

12. ■■ A 0.150-kg baseball traveling with a horizontal speed of 4.50 m/s is hit by a bat and then moves with a speed of 34.7 m/s in the opposite direction. What is the change in the ball's momentum?

13. ■■ A 15.0-g rubber bullet hits a wall with a speed of 150 m/s. If the bullet bounces straight back with a speed of 120 m/s, what is the change in momentum of the bullet?

14. **IE** ■■ Two protons approach each other with different speeds. (a) Will the magnitude of the total momentum of the two-proton system be (1) greater than the magnitude of the momentum of either proton, (2) equal to the difference between the magnitudes of momenta of the two protons, or (3) equal to the sum of the magnitudes of momenta of the two protons? Why? (b) If the speeds of the two protons are 340 m/s and 450 m/s, respectively, what is the total momentum of the two-proton system? [*Hint:* Find the mass of a proton in one of the tables inside the backcover.]

15. ■■ If a 0.50-kg ball is dropped from a height of 10 m, what is the momentum of the ball (a) 0.75 s after being released and (b) just before it hits the ground?

16. ■■ Taking the density of air to be 1.29 kg/m³, what is the magnitude of the linear momentum of a cubic meter of air moving with a wind speed of (a) 36 km/h and (b) 74 mi/h (the wind speed at which a tropical storm becomes a hurricane)?

17. ■■ Two runners of mass 70 kg and 60 kg, respectively, have a total linear momentum of 350 kg·m/s. The heavier runner is running at 2.0 m/s. Determine the possible speeds of the lighter runner.

18. ■■ A 0.20-kg billiard ball traveling at a speed of 15 m/s strikes the side rail of a pool table at an angle of 60° (▼Fig. 6.26). If the ball rebounds at the same speed and angle, what is the change in its momentum?

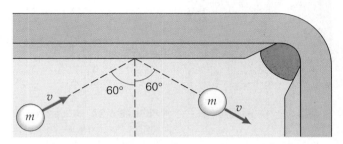

▲ **FIGURE 6.26 Glancing collision** See Exercises 18, 19, and 40.

19. ■■ Suppose that the billiard ball in Fig. 6.26 approaches the rail at a speed of 15 m/s and an angle of 60°, as shown, but rebounds at a speed of 10 m/s and an angle of 50°. What is the change in momentum in this case? [*Hint:* Use components.]

20. ■■ A person pushes a 10-kg box from rest and accelerates it to a speed of 4.0 m/s with a constant force. If the box is pushed for a time of 2.5 s, what is the force exerted by the person?

21. ■■ A loaded tractor-trailer with a total mass of 5000 kg traveling at 3.0 km/h hits a loading dock and comes to a stop in 0.64 s. What is the magnitude of the average force exerted on the truck by the dock?

22. ■■ A 2.0-kg mud ball drops from rest at a height of 15 m. If the impact between the ball and the ground lasts 0.50 s, what is the average net force exerted by the ball on the ground?

23. ■■■ At a basketball game, a 120-lb cheerleader is tossed vertically upward with a speed of 4.50 m/s by a male cheerleader. (a) What is the cheerleader's change in momentum from the time she is released to just before being caught if she is caught at the height at which she was released? (b) Would there be any difference if she were caught 0.25 m below the point of release? If so, what is the change then?

6.2 Impulse

24. Impulse is equal to (a) $F\Delta x$, (b) the change in kinetic energy, (c) the change in momentum, or (d) $\Delta p/\Delta t$.

25. CQ "Follow-through" is very important in many sports, such as in serving a tennis ball. Explain how follow-through can increase the speed of the tennis ball when it is served.

26. CQ A karate student tries *not* to follow through in order to break a board, as shown in ▼Fig. 6.27. How can the abrupt stop of the hand (with no follow-through) generate so much force?

◄ **FIGURE 6.27 A karate punch** See Exercise 26 and 33.

27. CQ Explain the difference for each of the following pairs of actions in terms of impulse: (a) a golfer's long drive and a short chip shot; (b) a boxer's jab and knock-out punches; (c) a baseball player's bunting action and home-run swing.

28. CQ Explain the principle behind (a) the use of Styrofoam as packing material to prevent objects from breaking, (b) the use of shoulder pads of football players to prevent injuries, and (c) the thicker glove used by a baseball catcher than by his teammates in the field.

29. CQ Will a greater force always generate a greater impulse? Explain.

30. ■ When tossed upward and hit horizontally by a batter, a 0.20-kg softball receives an impulse of 3.0 N·s. With what horizontal speed does the ball move away from the bat?

31. ■ An automobile with a linear momentum of 3.0×10^4 kg · m/s is brought to a stop in 5.0 s. What is the magnitude of the average braking force?

32. ■ A pool player imparts an impulse of 3.2 N·s to a stationary 0.25-kg cue ball with a cue stick. What is the speed of the ball just after impact?

33. ■■ For the karate punch in Exercise 26, assume that the hand has a mass of 0.35 kg and that the speeds of the hand just before and just after hitting the board are 10 m/s and 0, respectively. What is the average force exerted by the fist on the board if (a) the fist follows

through, so the contact time is 3.0 ms, and (b) the fist stops abruptly, so the contact time is only 0.30 ms?

34. IE ■■ When bunting, a baseball player uses the bat to change both the speed and direction of the baseball. (a) Will the magnitude of the change in momentum of the baseball before and after the bunt be (1) greater than the magnitude of the momentum of the baseball either before or after the bunt, (2) equal to the difference between the magnitudes of momenta of the baseball before and after the bunt, or (3) equal to the sum of the magnitudes of momenta of the baseball before and after the bunt? Why? (b) The baseball has a mass of 0.16 kg; its speeds before and after the bunt are 15 m/s and 10 m/s, respectively; and the bunt lasts 0.025 s. What is the change in momentum of the baseball? (c) What is the average force on the ball by the bat?

35. IE ■■ A volleyball is traveling toward you. (a) Which action will require a greater force on the volleyball, your catching the ball or your hitting the ball back? Why? (b) A 0.45-kg volleyball travels with a horizontal velocity of 4.0 m/s over the net. You jump up and hit the ball back with a horizontal velocity of 7.0 m/s. If the contact time is 0.040 s, what was the average force on the ball?

36. ■■ A 1.0-kg ball is thrown horizontally with a velocity of 15 m/s against a wall. If the ball rebounds horizontally with a velocity of 13 m/s and the contact time is 0.020 s, what is the force exerted on the ball by the wall?

37. ■■ A boy catches—with bare hands and his arms rigidly extended—a 0.16-kg baseball coming directly toward him at a speed of 25 m/s. He emits an audible "ouch!", because the ball stings his hands. He learns quickly to move his hands with the ball as he catches it. If the contact time of the collision is increased from 3.5 ms to 8.5 ms in this way, how do the magnitudes of the average impulse forces compare?

38. ■■ A one-dimensional impulse force acts on a 3.0-kg object as diagrammed in ▼Fig. 6.28. Find (a) the magnitude of the impulse given to the object, (b) the magnitude of the average force, and (c) the final speed if the object had an initial speed of 6.0 m/s.

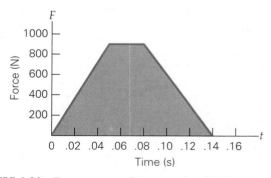

▲ **FIGURE 6.28 Force versus time graph** See Exercise 38.

39. ■■ A 0.35-kg piece of putty is dropped from a height of 2.5 m above a flat surface. When it hits the surface, the putty comes to rest in 0.30 s. What is the average force exerted on the putty by the surface?

40. ■■ If the billiard ball in Fig. 6.26 is in contact with the rail for 0.010 s, what is the magnitude of the average force exerted on the ball? (See Exercise 18.)

41. ■■■ In a simulated head-on crash test, a car impacts a wall at 25 mi/h (40 km/h) and comes abruptly to rest. A 120-lb passenger dummy (with a mass of 55 kg), without a seat belt, is stopped by an air bag, which exerts a force on the dummy of 2400 lb. How long was the dummy in contact with the air bag while coming to a stop?

6.3 The Conservation of Linear Momentum

42. The linear momentum of an object is conserved if (a) the force acting on the object is conservative; (b) there is a single, unbalanced internal force acting on the object; (c) the mechanical energy is conserved; or (d) none of the preceding.

43. Internal forces do not affect the conservation of momentum because (a) they cancel each other, (b) their effects are canceled by external forces, (c) they can never produce a change in velocity, or (d) Newton's second law is not applicable to them.

44. CQ An airboat of the type used in swampy and marshy areas is shown in ▼Fig. 6.29. Explain the principle of its propulsion. Using the concept of conservation of linear momentum, determine what would happen to the boat if a sail were installed behind the fan?

▲ **FIGURE 6.29 Fan propulsion** See Exercise 44.

45. CQ Imagine yourself standing in the middle of a frozen lake. The ice is so smooth that it is frictionless. How could you get to shore? (You couldn't walk. Why?)

46. CQ A stationary object receives a direct hit by another object moving toward it. Is it possible for both objects to be at rest after the collision? Explain.

47. CQ When a golf ball is driven off the tee, its speed is often much greater than the speed of the golf club. Explain how this situation can happen.

48. ■ A 60-kg astronaut floating at rest in space outside a space capsule throws his 0.50-kg hammer such that it moves with a speed of 10 m/s relative to the capsule. What happens to the astronaut?

49. ■ In a pairs figure-skating competition, a 65-kg man and his 45-kg female partner stand facing each other on skates on the ice. If they push apart and the woman has a velocity of 1.5 m/s eastward, what is the velocity of her partner? (Neglect friction.)

50. ■ To get off a frozen, frictionless lake, a 70.0-kg person takes off a 0.150-kg shoe and throws it horizontally, directly away from the shore with a speed of 2.00 m/s. If the person is 5.00 m from the shore, how long does it take for him to reach it?

51. ■■ A 100-g bullet is fired horizontally into a 14.9-kg block of wood resting on a horizontal surface, and the bullet becomes embedded in the block. If the muzzle velocity of the bullet is 250 m/s, what is the speed of the block immediately after the impact? (Neglect surface friction.)

52. IE ■■ An object initially at rest explodes and splits into three fragments. The first fragment flies off to the west, and the second fragment flies off to the south. The third fragment will fly off toward a general direction of (1) southwest, (2) north of east, or (3) either due north or due east. Why? (b) If the object has a mass of 3.0 kg, the first fragment has a mass of 0.50 kg and a speed of 2.8 m/s, and the second fragment has a mass of 1.3 kg and a speed of 1.5 m/s, what are the speed and direction of the third fragment?

53. ■■ Suppose that the 3.0-kg object in Exercise 52 is initially traveling at a speed of 2.5 m/s in the positive x-direction. What will be the speed and direction of the third fragment in this case?

54. ■■ Two identical cars hit each other and lock bumpers. In each of the following cases, what are the speeds of the cars immediately after coupling bumpers? (a) a car moving with a speed of 90 km/h approaches a stationary car; (b) two cars approach each other with speeds of 90 km/h and 120 km/h, respectively; (c) two cars travel in the same direction with speeds of 90 km/h and 120 km/h, respectively.

55. ■■ A 1200-kg car moving to the right with a speed of 25 m/s collides with a 1500-kg truck and locks bumpers with the truck. Calculate the velocity of the combination after the collision if the truck is initially (a) at rest, (b) moving to the right with a speed of 20 m/s, and (c) moving to the left with a speed of 20 m/s.

56. ■■ A 10-g bullet moving horizontally at 400 m/s penetrates a 3.0-kg wood block resting on a horizontal surface. If the bullet slows down to 300 m/s after emerging from the block, what is the speed of the block immediately after the bullet emerges (▼ Fig. 6.30)?

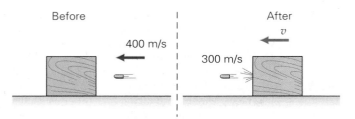

▲ **FIGURE 6.30** **Momentum transfer?** See Exercise 56.

57. ■■ A 1600-kg (empty) truck rolls with a speed of 2.5 m/s under a loading bin, and a mass of 3500 kg is deposited in the truck. What is the truck's speed immediately after loading?

58. ■■■ A projectile that is fired from a gun has an initial velocity of 90.0 km/h at an angle of 60.0° above the horizontal. When the projectile is at the top of its trajectory, an internal explosion causes it to separate into two fragments of equal mass. One of the fragments falls straight downward as though it had been released from rest. How far from the gun does the other fragment land?

59. ■■■ A moving shuffleboard puck has a glancing collision with a stationary puck of the same mass, as shown in ▼ Fig. 6.31. If friction is negligible, what are the speeds of the pucks after the collision?

60. ■■■ A *ballistic pendulum* is a device used to measure the velocity of a projectile—for example, the muzzle velocity of a rifle bullet. The projectile is shot horizontally into, and becomes embedded in, the bob of a pendulum, as illustrated in ▶ Fig. 6.32. The pendulum swings upward to some height h, which is measured. The masses of the block and the bullet are known. Using the laws of momentum and energy, show that

the initial velocity of the projectile is given by $v_o = [(m + M)/m]\sqrt{2gh}$.

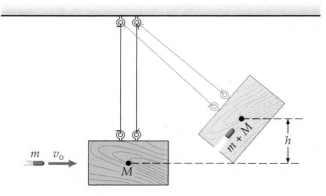

▲ **FIGURE 6.32** **A ballistic pendulum** See Exercises 60, 78, and 79.

6.4 Elastic and Inelastic Collisions

61. Which of the following is *not* conserved in an inelastic collision? (a) momentum, (b) mass, (c) kinetic energy, or (d) total energy.

62. In a head-on elastic collision, mass m_1 strikes a stationary mass m_2. There is a complete transfer of energy if (a) $m_1 = m_2$, (b) $m_1 \gg m_2$, (c) $m_1 \ll m_2$, or (d) the masses stick together.

63. **CQ** Since $K = p^2/2m$, how can kinetic energy be lost in an inelastic collision while the total momentum is still conserved? Explain.

64. **CQ** Discuss the common and different characteristics of an elastic collision and an inelastic collision.

65. **CQ** If a rubber ball hits a wall and bounces back with the same speed, is the collision elastic or inelastic? Explain.

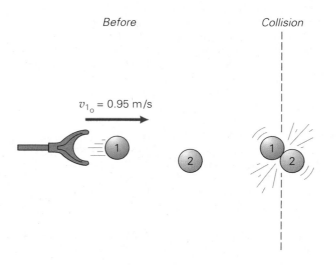

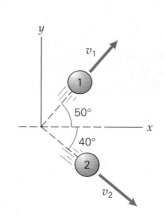

◀ **FIGURE 6.31** **Another glancing collision** See Exercise 59.

66. ■■ A 4.0-kg ball with a velocity of 4.0 m/s in the $+x$-direction collides head on elastically with a stationary 2.0-kg ball. What are the velocities of the balls after the collision?

67. ■■ A ball with a mass of 0.10 kg is traveling with a velocity of 0.50 m/s in the $+x$-direction and collides head on with a 5.0-kg ball that is at rest. Find the velocities of the balls after the collision. Assume that the collision is elastic.

68. IE ■■ For the apparatus in Fig. 6.15, one ball swinging in at a speed of $2v_0$ will not cause two balls to swing out with speeds v_0. (a) Which law of physics precludes this situation from happening, the law of conservation of momentum or the law of conservation of mechanical energy? (b) Prove this law mathematically.

69. ■■ A proton of mass m moving with a speed of 3.0×10^6 m/s undergoes a head-on elastic collision with an alpha particle of mass $4m$, which is initially at rest. What are the velocities of the two particles after the collision?

70. ■■ Two balls with masses of 2.0 kg and 6.0 kg travel toward each other at speeds of 12 m/s and 4.0 m/s, respectively. If the balls have a head-on, inelastic collision and the 2.0-kg ball recoils with a speed of 8.0 m/s, how much kinetic energy is lost in the collision?

71. IE ■■ ▼Fig. 6.33 shows a bird catching a fish. Assume that initially the fish jumps up and that the bird coasts horizontally and does not touch water with its feet or flap its wings. (a) Is this kind of collision (1) elastic, (2) inelastic, or (3) completely inelastic? Why? (b) If the mass of the bird is 5.0 kg, the mass of the fish is 0.80 kg, and the bird coasts with a speed of 6.5 m/s before grabbing, what is the speed of the bird after grabbing the fish?

▲ **FIGURE 6.33 Elastic or inelastic?** See Exercise 71.

72. ■■ Two balls approach each other as shown in ▶Fig. 6.34, where $m = 2.0$ kg, $v = 3.0$ m/s, $M = 4.0$ kg, and $V =$

5.0 m/s. If the balls collide and stick together at the origin, (a) what are the components of the velocity v of the balls after collision, and (b) what is the angle θ?

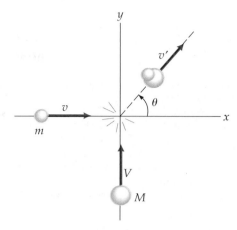

▲ **FIGURE 6.34 A completely inelastic collision** See Exercise 72.

73. IE ■■ A car traveling east and a minivan traveling south collide in a completely inelastic collision at a perpendicular intersection. (a) Right after the collision, will the car and minivan move toward a general direction (1) southeast, (2) north of west, or (3) either due south or due east? Why? (b) If the initial speed of the 1500-kg car was 90.0 km/h and the initial speed of the 3000-kg minivan was 60.0 km/h, what is the velocity of the vehicles immediately after collision?

74. ■■ In a pool game, a cue ball traveling at 0.75 m/s hits the stationary eight ball. The eight ball moves off with a velocity of 0.25 m/s at an angle of 37° relative to the cue ball's initial direction. Assuming that the collision is inelastic, at what angle will the cue ball be deflected, and what will be its speed?

75. ■■ A fellow student states that the total momentum of a three-particle system ($m_1 = 0.25$ kg, $m_2 = 0.20$ kg, and $m_3 = 0.33$ kg) is initially zero, and he calculates that after an inelastic triple collision, the particles have velocities of 4.0 m/s at 0°, 6.0 m/s at 120°, and 2.5 m/s at 230°, respectively, measured from the $+x$-direction. Do you agree with his calculations? If not, assuming the first two answers to be correct, what should be the momentum of the third particle so the total momentum is zero?

76. ■■ A freight car with a mass of 25 000 kg rolls down an inclined track through a vertical distance of 2.5 m. At the bottom of the incline, on a level track, the car collides and couples with an identical freight car that was at rest. What percentage of the initial kinetic energy is lost in the collision?

77. ■■■ In an elastic head-on collision with a stationary target particle, a moving particle recoils at one third of its incident speed. (a) What is the ratio of the particles' masses (m_1/m_2)? (b) What is the speed of the target particle after the collision in terms of the initial speed of the incoming particle?

78. ■■■ Show that the fraction of kinetic energy lost in a ballistic-pendulum collision (as in Fig. 6.32) is equal to $M/(m + M)$.

79. ■■■ A 10-g bullet is fired horizontally into, and becomes embedded in, a suspended block of wood whose mass is 0.890 kg. (See Fig. 6.32.) (a) How does the speed of the block with the embedded bullet immediately after the collision compare with the initial speed of the bullet (v_o)? (b) If the block with the embedded bullet swings upward and its center of mass is raised 0.40 m, what was the initial speed of the bullet? (c) Was the collision elastic? If not, what percentage of the initial kinetic energy was lost?

80. ■■■ A moving billiard ball collides with an identical stationary one, and the incoming ball is deflected at an angle of 45° from its original direction. Show that if the collision is elastic, both balls will have the same speed afterward and will move at a right angle (90°) relative to each other.

81. ■■■ (a) For an elastic, two-body head-on collision, show that, in general, $v_2 - v_1 = -(v_{2_o} - v_{1_o})$. That is, the relative speed of recession after the collision is the same as the relative speed of approach before it. (b) In general, a collision is either completely inelastic, completely elastic, or somewhere in between. The degree of elasticity is sometimes expressed as the *coefficient of restitution* (e), which is defined as the ratio of the relative velocities of recession and approach: $v_2 - v_1 = -e(v_{2_o} - v_{1_o})$. What are the values of e for an elastic collision and a completely inelastic collision?

82. ■■■ The coefficient of restitution (see Exercise 81) for steel colliding with steel is 0.95. If a steel ball is dropped from a height h_o above a steel plate, to what height will the ball rebound?

6.5 Center of Mass

83. The center of mass of an object (a) always lies at the center of the object, (b) is at the location of the most massive particle in the object, (c) always lies within the object, or (d) none of the preceding.

84. CQ ▶ Figure 6.35 shows a performer walking on a tightrope. The long pole he holds curves down at the ends. What effect does this pole have on the center of mass of the performer–pole system? [In Chapter 8, you will learn why this effect and another effect caused by the long pole make walking on a tightrope easier.]

85. CQ ▶ Figure 6.36 shows a flamingo standing on one of its two legs, with its other leg lifted. What can you say about

▲ FIGURE 6.35 Tightrope walking See Exercise 84.

▲ FIGURE 6.36 Delicate balance See Exercise 85.

the location of the flamingo's center of mass?

86. CQ A spacecraft is initially at rest in free space, and then its rocket engines are fired. Describe the motion of the center of mass of the system after the firing.

87. ■ (a) The center of mass of a system consisting of two 0.10-kg particles is located at the origin. If one of the particles is at (0, 0.45 m), where is the other? (b) If the masses are moved so their center of mass is located at (0.25 m, 0.15 m), can you tell where the particles are located?

88. ■ The centers of a 4.0-kg sphere and a 7.5-kg sphere are separated by a distance of 1.5 m. Where is the center of mass of the two-sphere system?

89. ■ (a) Find the center of mass of the Earth–Moon system. [*Hint*: Use data from tables on the inside cover of the book, and consider the distance between the Earth and Moon to be measured from their centers.] (b) Where is that center of mass relative to the surface of the Earth?

90. ■■ Find the center of mass of a system composed of three spherical objects with masses of 3.0 kg, 2.0 kg, and 4.0 kg and centers located at (−6.0 m, 0), (1.0 m, 0), and (3.0 m, 0), respectively.

91. ■■ Rework Exercise 58, using the concept of the center of mass, and compute the distance the other fragment landed from the gun.

92. IE ■■ A 3.0-kg rod of length 5.0 m has at opposite ends point masses of 4.0 kg and 6.0 kg. (a) Will the center of mass of this system be (1) nearer to the 4.0-kg mass, (2) nearer to the 6.0-kg mass, or (3) at the center of the rod? Why? (b) Where is the center of mass of the system?

93. ■■ A piece of uniform sheet metal measures 25 cm by 25 cm. If a circular piece with a radius of 5.0 cm is cut from the center of the sheet, where is the sheet's center of mass now?

94. ■■ Locate the center of mass of the system shown in ▼Fig. 6.37 (a) if all of the masses are equal; (b) if $m_2 = m_4 = 2m_1 = 2m_3$; (c) if $m_1 = 1.0$ kg, $m_2 = 2.0$ kg, $m_3 = 3.0$ kg, and $m_4 = 4.0$ kg.

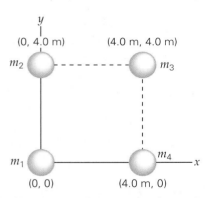

▲ FIGURE 6.37 **Where's the center of mass?** See Exercise 94.

95. ■■ A system of two masses has a center of mass given by X_{CM1}. A different system of three masses has a center of mass given by X_{CM2}. Show that if all five masses are considered to be one system, the center of mass of the combined system is not $X_{CM} = X_{CM1} + X_{CM2}$.

96. ■■ A 100-kg astronaut (mass includes space gear) on a space walk is 5.0 m from a 3000-kg space capsule and at the full length of her safety cord. To return to the capsule, she pulls herself along the cord. Where do the astronaut and capsule meet?

97. ■■ Two skaters with masses of 65 kg and 45 kg, respectively, stand 8.0 m apart, each holding one end of a piece of rope. (a) If they pull themselves along the rope until they meet, how far does each skater travel? (Neglect fric-

tion.) (b) If only the 45-kg skater pulls along the rope until she meets her friend (who just holds onto the rope), how far does each skater travel?

98. ■■■ Three particles, each with a mass of 0.25 kg, are located at (−4.0 m, 0), (2.0 m, 0), and (0, 3.0 m) and are acted on by forces $\mathbf{F}_1 = (−3.0\text{ N})\,\hat{\mathbf{y}}$, $\mathbf{F}_2 = (5.0\text{ N})\,\hat{\mathbf{y}}$, and $\mathbf{F}_3 = (4.0\text{ N})\,\hat{\mathbf{x}}$, respectively. Find the acceleration (magnitude and direction) of the center of mass of the system. [*Hint*: Consider the components of the acceleration.]

Additional Exercises

99. **CQ** Two objects have the same momentum. (a) Will they always have the same kinetic energy? (b) What can you say definitely about their motion?

100. A 1.0-kg object moving at 10 m/s collides with a stationary 2.0-kg object as shown in ▼Fig. 6.38. If the collision is perfectly inelastic, how far along the inclined plane will the combined system travel? Neglect friction.

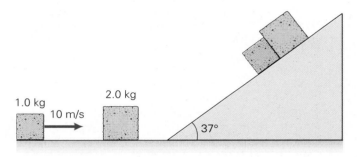

▲ FIGURE 6.38 **How far is up?** See Exercise 100.

101. A 1500-kg truck moving with a speed of 25 m/s runs into the rear of a 1200-kg stopped car. If the collision is perfectly inelastic, what is the kinetic energy lost in the collision?

102. Two balls of equal mass (0.50 kg) approach the origin along the positive x- and y-axes at the same speed (3.3 m/s). (a) What is the total momentum of the system? (b) Will the balls necessarily collide at the origin? What is the total momentum of the system after both balls have passed through the origin?

103. Two identical billiard balls approach each other at the same speed (2.0 m/s). At what speeds do they rebound after a head-on elastic collision?

104. A truck with a mass of 2400 kg travels at a constant speed of 90 km/h. (a) What is the magnitude of the truck's linear momentum? (b) What average force would be required to stop the truck in 8.0 s?

105. A 15 000-N automobile travels at a speed of 45 km/h northward along a street, and a 7500-N sports car travels at a speed of 60 km/h eastward along an intersecting street.

(a) If neither driver brakes and the cars collide at the intersection and lock bumpers, what will the velocity of the cars be immediately after the collision? (b) What percentage of the initial kinetic energy will be lost in the collision?

106. For a movie scene, a 75-kg stunt man drops from a tree onto a 50-kg sled that is moving on a frozen lake with a velocity of 10 m/s toward the shore. (a) What is the speed of the sled after the stunt man is on board? (b) If the sled hits the bank and stops, but the stunt man keeps on going, with what speed does he leave the sled? (Neglect friction.)

107. IE During a snowball fight, a snowball traveling horizontally hits a student in the back of the head and sticks there. (a) Is this collision elastic or inelastic? Why? (b) If the mass of the snowball is 0.15 kg and its initial speed is 14 m/s, what is the impulse? (c) If the contact time is 0.10 s, what is the average force on the student's head?

108. A 2.5-kg block sliding with a constant velocity of 6.0 m/s on a frictionless horizontal surface approaches a stationary 6.5-kg block. (a) If the blocks suffer a completely inelastic collision, what is their velocity after the collision? (b) How much mechanical energy is lost in the completely inelastic collision?

109. A 90-kg astronaut is stranded in space at a point 6.0 m from his spaceship, and he needs to get back in 4.0 minutes to control the spaceship. To get back, he throws a 0.50-kg piece of equipment so that it moves at a speed of 4.0 m/s directly away from the spaceship. (a) Does he get back in time? (b) How fast must he throw the piece of equipment so he gets back in time?

110. In nuclear reactors, subatomic particles called *neutrons* are slowed down by allowing them to collide with the atoms of a moderator material, such as carbon, which is 12 times more massive than neutrons. (a) In a head-on elastic collision with a carbon atom, what percentage of a neutron's energy is lost? (b) If the neutron has an initial speed of 1.5×10^7 m/s, what will be its speed after collision?

111. A 70-kg athlete achieves a height of 2.25 m in a high jump. Considering the jumper and the Earth as an isolated system, with what speed does the Earth initially move as the jumper launches himself upward?

112. A uniform, flat piece of metal is shaped like an equilateral triangle with sides that are 30 cm long. What are the coordinates of the center of mass in the xy-plane if one apex is at the origin and one side is along the y-axis?

Circular Motion and Gravitation

INSIGHTS

- The Centrifuge: Separating Blood Components
- Space Exploration: Gravity Assists
- "Weightlessness": Effects on the Human Body

Learn by Drawing

- The Small-Angle Approximation

People often say that rides like this one "defy gravity." Of course, you know that in reality, gravity cannot be defied; it commands respect. There is nothing that will shield you from it and no place in the universe where you can go to be entirely free of it. What, then, keeps these people in their seats, despite the tug of Earth's ever-present gravity? You may be surprised to find that if you can answer this question, you'll also be able to understand what prevents satellites from falling to the Earth and keeps the Earth in orbit around the Sun.

Circular motion is everywhere, from atoms to galaxies, from flagella of bacteria to Ferris wheels. Two terms are frequently used to describe such motion. In general, we say that an object *rotates* when the axis of rotation lies within the body and that it *revolves* when the axis is out-side the body. Thus, the Earth rotates on its axis and re-volves about the Sun.

When a solid body rotates on its axis, all the particles of the body move in circular paths about the body's axis of rotation. For example, all the particles that make up a compact disc travel in circles about the hub of the CD player. In fact, as a "particle" on the Earth, you are contin-uously in circular motion about the Earth's rotational axis.

Circular motion is motion in two dimensions, and so it can be described by rectangular components as used in Chapter 3. However, it is usually more convenient to de-scribe circular motion in terms of angular quantities that will be introduced in this chapter. Being familiar with the description of circular motion will make the study of rotating rigid bodies much easier, as you will find in Chapter 8.

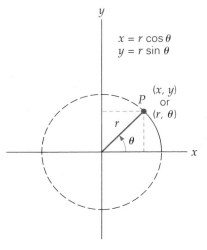

▲ FIGURE 7.1 Polar coordinates
A point may be described by polar coordinates instead of Cartesian coordinates—that is, by (r, θ) instead of (x, y). For a circle, θ is the angular distance and r is the radial distance. The two types of coordinates are related by the transformation equations $x = r \cos \theta$ and $y = r \sin \theta$.

Gravity plays a large role in determining the motions of the planets, since it supplies the force necessary to maintain their nearly circular orbits. This chapter will consider Newton's law of gravitation, which describes this fundamental force, and will analyze planetary motion in terms of this and related basic laws. The same considerations will help you understand the motions of the Earth's satellites, which include one natural one (the Moon) and many artificial ones.

7.1 Angular Measure

OBJECTIVES: **To (a) define units of angular measure and (b) show how angular measure is related to circular arc length.**

Motion is described as a time rate of change of position (Section 2.1). As you might guess, *angular speed* and *angular velocity* also involve a time rate of change of position, which is expressed by an *angular change*. Consider a particle traveling in a circular path, as shown in ◀Fig. 7.1. At a particular instant, the particle's position (P) may be designated by the Cartesian coordinates x and y. However, the position may also be designated by the polar coordinates r and θ. The distance r extends from the origin, and the angle θ is commonly measured counterclockwise from the x-axis. The transformation equations that relate one set of coordinates to the other are

$$x = r \cos \theta \qquad (7.1a)$$
$$y = r \sin \theta \qquad (7.1b)$$

as can be seen from the x- and y-components of point P in Fig. 7.1.

Note that r is the same for any point on a given circle. As a particle travels in a circle, the value of r is constant, and only θ changes with time. Thus, circular motion can be described by using one polar coordinate (θ) that changes with time, instead of two Cartesian coordinates (x and y), both of which change with time.

Analogous to linear displacement is **angular displacement**, the magnitude of which is given by

$$\Delta\theta = \theta - \theta_{o} \qquad (7.2)$$

or simply $\Delta\theta = \theta$ when we choose $\theta_{o} = 0°$. (The direction of the angular displacement will be explained in the next section, on angular velocity.) A unit commonly used to express angular displacement (or angles) is the degree (°); there are 360° in one complete circle, or revolution.*

It is important to be able to relate the angular description of circular motion to the orbital or tangential description, that is, to relate the angular displacement to the arc length s. The *arc length* is the distance traveled along the circular path, and the angle θ is said to *subtend* (define) the arc length. A unit that is very convenient for relating angle to arc length is the **radian (rad)**, which is defined as the angle subtending an arc length (s) that is equal to the radius r (◀Fig. 7.2).

The number of radians subtended by an arbitrary arc length s is equal to the number of radii that fits into s, or the number of radians $\theta = s/r$. Thus, we can write

$$s = r\theta \qquad (7.3)$$

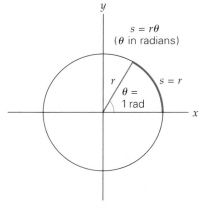

▲ FIGURE 7.2 Radian measure
Angular displacement may be measured either in degrees or in radians (rad). An angle θ is subtended by an arc length s. When $s = r$, the angle subtending s is defined to be 1 rad. More generally, $\theta = s/r$, where θ is in radians. One radian is equal to 57.3°.

which is an important relationship between the circular arc length s and the radius of the circle, r. Notice that since $\theta = s/r$, the angle in radians is the ratio of two lengths. This means that a radian measure is a pure number, that is, it is dimensionless and has no units.

---

*A degree may be divided into the smaller units of minutes (1 degree = 60 minutes) and seconds (1 minute = 60 seconds). These divisions have nothing to do with time units.

To get a general relationship between radians and degrees, let's consider the distance around a complete circle (360°). For one full circle, with $s = 2\pi r$ (the circumference), there is a total of $\theta = s/r = 2\pi r/r = 2\pi$ rad, or

$$2\pi \text{ rad} = 360°$$

This relationship can be used to obtain convenient conversions of common angles (Table 7.1). Thus,

$$1 \text{ rad} = 360°/2\pi = 57.3°$$

(to three significant figures). Notice in Table 7.1 that the angles in radians are expressed in terms of π explicitly, for convenience.

TABLE 7.1 Equivalent Degree and Radian Measures	
Degrees	*Radians*
360°	2π
180°	π
90°	$\pi/2$
60°	$\pi/3$
57.3°	1
45°	$\pi/4$
30°	$\pi/6$

Example 7.1 ■ Finding Arc Length: Using Radian Measure

A spectator standing at the center of a circular running track observes a runner start a practice race 256 m due east of her own position (▶Fig. 7.3). The runner runs on the same track to the finish line, which is located due north of the observer's position. What is the distance of the run?

Thinking It Through. Note that the subtending angle of the section of circular track is $\theta = 90°$. The arc length (s) can be found, since the radius r of the circle is known.

Solution. Listing what is given and what is to be found, we have

Given: $r = 256$ m $\qquad$ *Find:* s (arc length)
$\qquad \theta = 90° = \pi/2$ rad

Simply using Eq. 7.3 to find the arc length, we get

$$s = r\theta = (256 \text{ m})\left(\frac{\pi}{2}\right) = 402 \text{ m}$$

Note that the rad unit is omitted, and the equation is dimensionally correct. Why?

Follow-up Exercise. What would be the distance of one complete lap around the track in this Example? *(Answers to all Follow-up Exercises are at the back of the text.)*

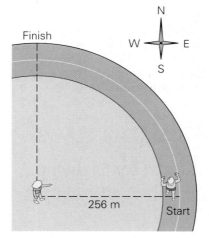

▲ **FIGURE 7.3 Arc length—found by means of radians**
See Example 7.1.

Example 7.2 ■ How Far Away? A Useful Approximation

A sailor measures the length of a distant tanker as an angular measure of 1.15° with a divided circle, as illustrated in ▶Fig. 7.4a. He knows from the shipping charts that the tanker is 150 m in length. Approximately how far away is the tanker?

Thinking It Through. Note in the Learn by Drawing feature on p. 222 that for small angles, the arc length approximates the y length of the triangle, or $s \approx y$. Hence, if we know the length and the angle, we can find the radial distance, which is approximately equal to the tanker's distance from the sailor.

Solution. To approximate the distance, we take the ship's length to be nearly equal to the arc length subtended by the measured angle. This approximation is good for small angles. The data are then as follows:

Given: $\theta = 1.15°$ $(1 \text{ rad}/57.3°) = 0.0201$ rad $\qquad$ *Find:* r (radial distance)
$\qquad s = 150$ m

Knowing the arc length and angle, we can use Eq. 7.3 to find r (note the unitless rad is omitted):

$$r = \frac{s}{\theta} = \frac{150 \text{ m}}{0.0201} = 7.46 \times 10^3 \text{ m} = 7.46 \text{ km}$$

As noted, the distance r is an approximation, obtained by assuming that for small angles, the arc length s and the straight-line chord length L are very nearly the same length

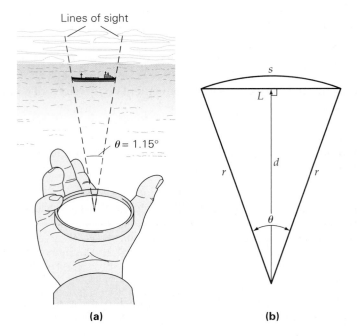

Lines of sight

$\theta = 1.15°$

(a)

(b)

▶ **FIGURE 7.4 Angular distance**
For small angles, the arc length is approximately a straight line, or the chord length. Knowing the length of the tanker, we can find how far away it is by measuring its angular size. See Example 7.2. (Drawing not to scale for clarity.)

Learn by Drawing

The Small-Angle Approximation

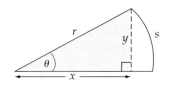

θ *not* small:

$$\theta \ (\text{in rad}) = \frac{s}{r}$$

$$\sin \theta = \frac{y}{r} \qquad \tan \theta = \frac{y}{x}$$

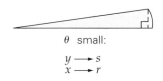

θ small:

$$\begin{aligned} y &\longrightarrow s \\ x &\longrightarrow r \end{aligned}$$

$$\theta \ (\text{in rad}) = \frac{s}{r} \approx \frac{y}{r} \approx \frac{y}{x}$$

$$\theta \ (\text{in rad}) \approx \sin \theta \approx \tan \theta$$

(Fig. 7.4b). How good is this approximation? To see, let's compute the perpendicular distance d to the ship. From the geometry, we have $\tan (\theta/2) = (L/2)/d$, so

$$d = \frac{L}{2 \tan (\theta/2)} = \frac{150 \ \text{m}}{2 \tan (1.15°/2)} = 7.47 \times 10^3 \ \text{m} = 7.47 \ \text{km}$$

The first calculation is a pretty good approximation—the values derived by the two methods are nearly equal.

Follow-up Exercise.   As pointed out, the approximation used in this Example is for *small* angles. You might wonder what is small. To investigate this question, what would be the percentage error of the approximated distance to the tanker for angles of 10° and 20°?

Problem-Solving Hint

In computing trigonometric functions such as $\tan \theta$ or $\sin \theta$, the angle may be expressed in degrees or radians; for example, $\sin 30° = \sin [(\pi/6) \ \text{rad}] = \sin (0.524 \ \text{rad}) = 0.500$. When finding trig functions with a calculator, note that there is usually a way to change the angle entry between "deg" and "rad" modes. Calculators commonly are set in the degree mode, so if you want to find the value of, say, $\sin (1.22 \ \text{rad})$, first change to the "rad" mode and enter sin 1.22, and $\sin (1.22 \ \text{rad}) = 0.939$.

Your calculator may have a third mode, "grad." The grad is a little-used angular unit. A grad is $1/100$ of a right (90°) angle; that is, there are 100 grads in a right angle.

7.2 Angular Speed and Velocity

OBJECTIVES: **To (a)** describe and compute angular speed and velocity and **(b)** explain their relationship to tangential speed.

The description of circular motion in angular form is analogous to the description of linear motion. In fact, you'll notice that the equations are almost mathematically identical, with different symbols being used to indicate that the quantities have different meanings. The lowercase Greek letter omega with a bar over it ($\bar{\omega}$) is used to represent **average angular speed**, the magnitude of the angular displacement divided by the total time to travel the distance:

$$\bar{\omega} = \frac{\Delta\theta}{\Delta t} = \frac{\theta - \theta_o}{t - t_o} \quad \textit{average angular speed} \qquad (7.4)$$

(As with other instantaneous quantities, we say that the units of angular speed are radians per second. Technically, they are s^{-1}, since the radian is unitless, but it is useful to keep the rad in there to indicate the quantity is angular speed. The *instantaneous angular speed* is given by considering a very small time interval, that is, as Δt approaches zero.)

Taking θ_o and t_o to be zero in Eq. 7.4, we write

$$\bar{\omega} = \frac{\theta}{t} \quad \text{or} \quad \theta = \bar{\omega}t \qquad (7.5)$$

SI unit of angular speed: radians per second (rad/s or s^{-1})

As in the linear case, if the angular speed is *constant*, then $\bar{\omega} = \omega$.

Another common descriptive unit for angular speed is revolutions per minute (rpm); for example, a CD (compact disc) rotates at a speed of 200–500 rpm (depending on the location of the track). This nonstandard unit of revolutions per minute is readily converted to radians per second, since 1 revolution = 2π rad.

The **average and instantaneous angular velocities** are analogous to their linear counterparts. Angular velocity is associated with angular displacement. Both are vectors and thus have direction; however, this directionality is, by convention, specified in a special way. In one-dimensional, or linear, motion, a particle can go only in one direction or the other (+ or −), so the displacement and velocity vectors can have only these two directions. In the angular case, a particle moves one way or the other, but the motion is along its *circular path*. Thus, the angular-displacement and angular velocity vectors of a particle in circular motion can have only two directions, which correspond to going around the circular path with either increasing or decreasing angular displacement from θ_o—that is, clockwise or counterclockwise. Let's focus on the angular velocity vector $\boldsymbol{\omega}$. (The direction of the angular displacement will be the same as that of the angular velocity. Why?)

The *direction* of the angular velocity vector is given by the *right-hand rule*, illustrated in ▶ Fig. 7.5a. When the fingers of your right hand are curled in the direction of circular motion, your extended thumb points in the direction of $\boldsymbol{\omega}$. Note that circular motion can be in only one of two circular *senses*, clockwise or counterclockwise. Plus and minus signs can be used to distinguish circular rotation directions relative to the angular velocity vector. It is customary to take a counterclockwise rotation as positive (+), since positive angular distance (and displacement) is conventionally measured counterclockwise from the positive x-axis.

Why not just designate the direction of the angular velocity vector to be either clockwise or counterclockwise? This designation is not made because clockwise (cw) and counterclockwise (ccw) are directional senses or indications rather than actual directions. These rotational senses are like right and left. If you faced another person and each of you were asked if some object is on the right or left, your answers would disagree. Similarly, if you held this book up toward a person facing you and rotated it, would it be rotating cw or ccw?

We can use cw and ccw to indicate rotational "directions" when they are specified relative to a reference, for example, the positive x-axis as in the preceding discussion. Referring to Fig. 7.5, imagine yourself being first on one side of one of the rotating disks and then on the other. Which way is the disk rotating from each vantage point, cw or ccw? Then, apply the right-hand rule on both sides. You should find that the direction of the angular velocity vector is the same for both locations (because it is referenced to the right hand). Relative to this vector—for example,

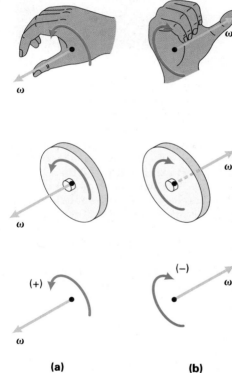

▲ **FIGURE 7.5 Angular velocity** The direction of the angular-velocity vector for an object in rotational motion is given by the right-hand rule: When the fingers of the right hand are curled in the direction of the rotation, the extended thumb points in the direction of the angular-velocity vector. Circular senses or directions are commonly indicated by **(a)** plus and **(b)** minus signs.

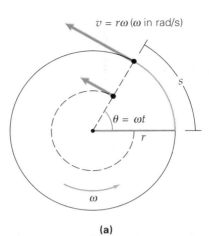

$v = r\omega$ (ω in rad/s)

$\theta = \omega t$

r

s

ω

(a)

(b)

▲ **FIGURE 7.6** **Tangential and angular speeds** **(a)** Tangential and angular speeds are related by $v = r\omega$, where ω is in radians per second. Note that all of the particles of an object rotating about a fixed axis travel in circles. All the particles have the same angular speed ω, but particles at different distances from the axis of rotation have different tangential speeds. **(b)** Sparks from a grinding wheel provide a graphic illustration of instantaneous tangential velocity. (Why do the paths curve slightly?)

Note: Whenever a tangential or linear quantity is calculated from an angular quantity, the angular unit radian is dropped in the final answer. When an angular quantity is asked for, the unit radian is usually included in the final answer, for clarity.

looking at the tip—there is no ambiguity in using $+$ and $-$ to indicate rotational senses or directions.

Relationship between Tangential and Angular Speeds

A particle moving in a circle has an instantaneous velocity tangential to its circular path. For a constant angular velocity and speed, the particle's orbital speed, or **tangential speed**, v (the magnitude of the tangential velocity) is also constant. How the angular and tangential speeds are related is revealed by starting with Eq. 7.3 ($s = r\theta$) and Eq. 7.5 ($\theta = \omega t$):

$$s = r\theta = r(\omega t)$$

The arc length, or distance, is also given by

$$s = vt$$

Combining the equations for s gives

$$v = r\omega \qquad \textit{tangential speed relation to angular speed for circular motion} \qquad (7.6)$$

where ω is in radians per second. Equation 7.6 holds in general for instantaneous tangential and angular speeds for solid- or rigid-body rotation about a fixed axis, even when ω might vary with time.

Note that all the particles of an object rotating with constant angular velocity have the same angular speed, but the tangential speeds are different at different distances from the axis of rotation (◄Fig. 7.6).

Example 7.3 ■ Merry-Go-Rounds: Do Some Go Faster Than Others?

An amusement-park merry-go-round at its constant operational speed makes one complete rotation in 45 s. Two children are on horses, one at 3.0 m from the center of the ride and the other farther out, 6.0 m from the center. What are (a) the angular speed and (b) the tangential speed of each child?

Thinking It Through. The angular speed of each child is the same, since both children make a complete rotation in the same time. However, the tangential speeds will be different, because the radii are different. That is, the child at the greater radius travels in a larger circle during the rotation time and thus must travel faster.

Solution.

Given: $\theta = 2\pi$ rad (one rotation)   *Find:* (a) ω_1 and ω_2 (angular speeds)
       $t = 45$ s   (b) v_1 and v_2 (tangential speeds)
       $r_1 = 3.0$ m
       $r_2 = 6.0$ m

(a) As noted, $\omega_1 = \omega_2$—that is, both riders rotate at the same angular speed. All points on the merry-go-round travel through 2π rad in the time it takes to make one rotation. The angular speed can be found from Eq. 7.5 (constant ω) as

$$\omega = \frac{\theta}{t} = \frac{2\pi \text{ rad}}{45 \text{ s}} = 0.14 \text{ rad/s}$$

Hence, $\omega = \omega_1 = \omega_2 = 0.14$ rad/s.

(b) The tangential speed is different at different locations on the merry-go-round. All of the "particles" making up the merry-go-round go through one rotation in the same amount of time. Therefore, the farther a particle is from the center, the longer its circular path will be, and the greater is its tangential speed, as Eq. 7.6 indicates. (See also Fig. 7.6a.) Thus,

$$v_1 = r_1\omega = (3.0 \text{ m})(0.14 \text{ rad/s}) = 0.42 \text{ m/s}$$

and

$$v_2 = r_2\omega = (6.0 \text{ m})(0.14 \text{ rad/s}) = 0.84 \text{ m/s}$$

(Note that the unitless radian has been dropped from the answer.)

Thus, a rider on the outer part of the ride has a greater tangential speed than a rider closer to the center.

Follow-up Exercise. (a) On an old 45-rpm record, the beginning track is 8.0 cm from the center, and the end track is 5.0 cm from the center. What are the angular speeds and the tangential speeds at these distances when the record is spinning at 45 rpm? (b) Why, on oval racetracks, do inside and outside runners have different starting points (called a "staggered" start), such that some runners start "ahead" of others?

Period and Frequency

Some other quantities commonly used to describe circular motion are period and frequency. The time it takes for an object in circular motion to make one complete revolution, or *cycle*, is called the **period** (T). For example, the period of revolution of the Earth about the Sun is one year, and the period of the Earth's axial rotation is 24 hours. The standard unit of period is the second (s). Descriptively, the period is sometimes given in seconds per revolution (s/rev) or seconds per cycle (s/cycle).

Closely related to the period is the **frequency** (f), which is the number of revolutions, or cycles, made in a given time, generally a second. For example, if a particle traveling uniformly in a circular orbit makes 5.0 revolutions in 2.0 s, the frequency (of revolution) is $f = 5.0 \text{ rev}/2.0 \text{ s} = 2.5 \text{ rev/s}$, or 2.5 cycles/s (cps, or cycles per second). *Revolution* and *cycle* are merely descriptive terms used for convenience and are *not* units. Without these descriptive terms, we see that the unit of frequency is inverse seconds (1/s, or s^{-1}), which is called the **hertz** (Hz) in the SI.

Since the units for frequency and period are inverses of one another (1/s and s), it follows that the two quantities are related by

The hertz (Hz), a unit of frequency, is named for Heinrich Hertz (1857–1894), a German physicist and pioneering investigator of electromagnetic waves, which are also characterized by frequency.

$$f = \frac{1}{T} \quad \text{frequency and period} \quad (7.7)$$

SI unit of frequency: hertz (Hz, 1/s or s^{-1})

Note: Frequency (f) and period (T) are inversely related.

where the period is in seconds and the frequency is in hertz, or inverse seconds.

The frequency can also be related to the angular speed. For uniform circular motion, the orbital speed can be written as $v = 2\pi r/T$—that is, the distance traveled in one revolution divided by the time for one revolution (one period). Similarly, for the angular case, since a distance of 2π rad is traveled in one period (by definition of the period), we have

$$\omega = \frac{2\pi}{T} = 2\pi f \quad \begin{array}{l}\textit{angular speed in terms} \\ \textit{of period and frequency}\end{array} \quad (7.8)$$

Notice that ω and f have the same units, that is, $\omega = \text{rad/s} = 1/\text{s}$ and $2\pi f = \text{rad/s} = 1/\text{s}$. This notation can easily cause confusion, which is why the unitless radian term is often added.

Example 7.4 ■ Frequency and Period: An Inverse Relationship

A compact disc (CD) rotates in a player at a constant speed of 200 rpm. What are the CD's (a) frequency and (b) period of revolution?

Thinking It Through. We can use the relationships among the frequency (f), the period (T), and the angular frequency ω, expressed in Eqs. 7.7 and 7.8.

Solution. The angular speed is not in standard units and thus must be converted. Revolutions per minute (rpm) can be converted to radians per second (rad/s).

Given: $\omega = 200 \text{ rev/min} \left(\dfrac{2\pi \text{ rad}}{1 \text{ rev}}\right)\left(\dfrac{1 \text{ min}}{60 \text{ s}}\right)$ *Find:*   (a) f (frequency)
(b) T (period)

$= 20.9 \text{ rad/s}$

(A convenient conversion factor is $1 \text{ rev/min} = \pi/30 \text{ rad/s}$. Can you see how this relationship is obtained?)

(a)   Rearranging Eq. 7.8 and solving for f, we get

$$f = \frac{\omega}{2\pi} = \frac{20.9 \text{ rad/s}}{2\pi} = 3.33 \text{ Hz}$$

The units of 2π are rad/cycle or revolution, so the result is in cycles/second or inverse seconds, which is the hertz.

(b)   Equation 7.8 could be used to find T, but Eq. 7.7 is a bit simpler:

$$T = \frac{1}{f} = \frac{1}{3.33 \text{ Hz}} = 0.300 \text{ s}$$

Thus, it takes 0.300 s for the CD to make one revolution. (Notice that with $\text{Hz} = 1/\text{s}$, the equation is dimensionally correct.)

Follow-up Exercise.   If the period of a particular CD is 0.500 s, what is the CD's angular speed in revolutions per minute?

7.3 Uniform Circular Motion and Centripetal Acceleration

OBJECTIVES: To (a) explain why there is a centripetal acceleration in constant or uniform circular motion, and (b) compute centripetal acceleration.

Note: Review the discussion of curvilinear motion in Section 3.1.

A simple, but important, type of circular motion is **uniform circular motion**, which occurs when an object moves at a constant speed in a circular path. An example of this movement is a car going around a circular track (▼Fig. 7.7). The motion of the Moon around the Earth is approximated by uniform circular motion. Such motion is curvilinear, so you know from the discussion in Chapter 3 that there must be an acceleration. But what are its magnitude and direction?

Centripetal Acceleration

The acceleration of uniform circular motion is not in the same direction as the instantaneous velocity (which is tangent to the circular path at any point). If it were, the object would speed up, and the motion wouldn't be uniform. Recall

▶ **FIGURE 7.7   Uniform circular motion**   The speed of an object in uniform circular motion is constant, but the object's velocity changes in the direction of motion. Thus, there is an acceleration.

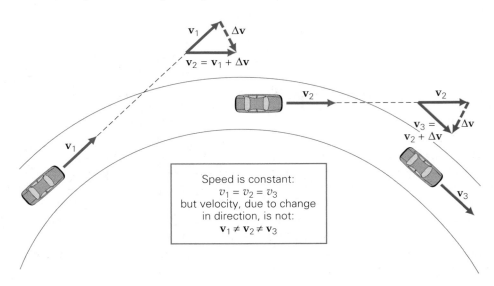

Speed is constant:
$v_1 = v_2 = v_3$
but velocity, due to change in direction, is not:
$\mathbf{v}_1 \neq \mathbf{v}_2 \neq \mathbf{v}_3$

that acceleration is the time rate of change of velocity and that velocity has both *magnitude* and *direction*. In uniform circular motion, the direction of the velocity is continuously changing, which is a clue to the direction of the acceleration. (See Fig. 7.7.)

The velocity vectors at the beginning and end of a time interval give the change in velocity, or $\Delta\mathbf{v}$, via vector subtraction. All of the instantaneous-velocity vectors have the same magnitude or length (constant speed), but they differ in direction. Note that since $\Delta\mathbf{v}$ is not zero, there must be an acceleration ($\mathbf{a} = \Delta\mathbf{v}/\Delta t$).

As illustrated in ▶Fig. 7.8, as Δt (or $\Delta\theta$) becomes smaller, $\Delta\mathbf{v}$ points more and more toward the center of the circular path. As Δt approaches zero, the instantaneous change in the velocity, and therefore the acceleration, points exactly toward the center of the circle. As a result, the acceleration in uniform circular motion is called **centripetal acceleration**, which means center-seeking acceleration (from Latin *centri*, "center," and *petere*, "to fall toward" or "to seek").

Without the centripetal acceleration, the motion would not be in a curved path, but rather in a straight line. The centripetal acceleration must be directed radially inward, that is, with no component in the direction of the perpendicular (tangential) velocity, or else the magnitude of that velocity would change (▶Fig. 7.9). Note that for an object in uniform circular motion, the direction of the centripetal acceleration is continuously changing. In terms of x- and y-components, a_x and a_y are not constant. Can you describe how this set of conditions differs from that for projectile motion?

The magnitude of the centripetal acceleration can be deduced from the small shaded triangles in Fig. 7.8. (For very short time intervals, the arc length of Δs is almost a straight line—the chord.) These two triangles are similar, because each has a pair of equal sides surrounding the same angle $\Delta\theta$. (Note that the velocity vectors have the same magnitude.) Thus, Δv is to v as Δs is to r, which we can write as

$$\frac{\Delta v}{v} \approx \frac{\Delta s}{r}$$

The arc length Δs is the distance traveled in time Δt; thus, $\Delta s = v\Delta t$, so

$$\frac{\Delta v}{v} \approx \frac{\Delta s}{r} = \frac{v\Delta t}{r}$$

and

$$\frac{\Delta v}{\Delta t} \approx \frac{v^2}{r}$$

Then, as Δt approaches zero, this approximation becomes exact. The instantaneous centripetal acceleration, $\Delta v/\Delta t = a_c$, thus has a magnitude of

$$a_c = \frac{v^2}{r} \quad \begin{array}{l}\textit{magnitude of centripetal acceleration}\\ \textit{in terms of tangential speed}\end{array} \quad (7.9)$$

Using Eq. 7.6 ($v = r\omega$), we can also write the equation for centripetal acceleration in terms of the angular speed:

$$a_c = \frac{v^2}{r} = \frac{(r\omega)^2}{r} = r\omega^2 \quad \begin{array}{l}\textit{magnitude of centripetal acceleration}\\ \textit{in terms of angular speed}\end{array} \quad (7.10)$$

Orbiting satellites have centripetal accelerations (Example 7.5), and a down-to-Earth practical application of centripetal acceleration is discussed in the Insight on p. 229 and illustrated in Example 7.6.

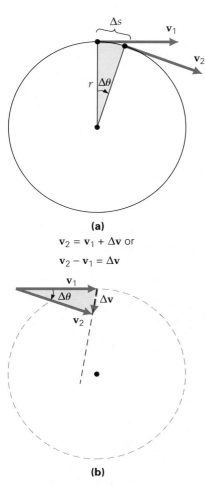

(a)

$$\mathbf{v}_2 = \mathbf{v}_1 + \Delta\mathbf{v} \text{ or}$$

$$\mathbf{v}_2 - \mathbf{v}_1 = \Delta\mathbf{v}$$

(b)

▲ **FIGURE 7.8 Analysis of centripetal acceleration** **(a)** The velocity vector of an object in uniform circular motion is constantly changing direction. **(b)** As Δt, the time interval for $\Delta\theta$, is taken to be smaller and smaller and approaches zero, Δv (the change in the velocity, and therefore an acceleration) is directed toward the center of the circle. The centripetal, or center-seeking, acceleration has a magnitude of $a_c = v^2/r$.

Note: Centripetal acceleration depends on tangential speed (v) and radius (r).

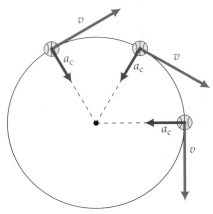

▲ FIGURE 7.9 Centripetal acceleration For an object in uniform circular motion, the centripetal acceleration is directed radially inward. There is no acceleration component in the tangential direction; if there were, the magnitude of the velocity (tangential *speed*) would change.

Example 7.5 ■ A Satellite in Orbit: Centripetal Acceleration

A space station is in a circular orbit about the Earth at an altitude h of 5.0×10^2 km. If the station makes one revolution every 95 min, what are its (a) orbital speed and (b) centripetal acceleration?

Thinking It Through. From the given data, we can find the length of the circular path (by calculating the radius); and, using the period, we can find the orbital speed. The centripetal acceleration can then be computed by using Eq. 7.9.

Solution.

Given: $h = 5.0 \times 10^2$ km   *Find:* (a) v (tangential, or orbital, speed)
        $= 5.0 \times 10^5$ m   (b) $\mathbf{a}_c$ (centripetal acceleration)
        $t = T = 95$ min $= 5.7 \times 10^3$ s

(a) The radius of the circular orbit is not h, but $R_E + h$, where R_E is the (average) radius of the Earth, 6.4×10^6 m (see Appendix III):

$$r = R_E + h = (6.4 \times 10^6 \text{ m}) + (5.0 \times 10^5 \text{ m}) = 6.9 \times 10^6 \text{ m}$$

The station travels the circumference of its circular orbit $(2\pi r)$ in the given time, so the tangential speed is

$$v = \frac{2\pi r}{T} = \frac{2\pi(6.9 \times 10^6 \text{ m})}{5.7 \times 10^3 \text{ s}} = 7.6 \times 10^3 \text{ m/s}$$

(b) Then the centripetal acceleration is (from Eq. 7.9)

$$a_c = \frac{v^2}{r} = \frac{(7.6 \times 10^3 \text{ m/s})^2}{6.9 \times 10^6 \text{ m}} = 8.4 \text{ m/s}^2$$

directed toward the center of the orbit. Alternatively, we could have computed the angular speed $\omega = 2\pi/T$ and used Eq. 7.10, $a_c = r\omega^2$.

As you might suspect, this centripetal acceleration is supplied by the Earth's gravitational force—it is the acceleration due to gravity at an altitude of 500 km.

Follow-up Exercise. (a) How does the acceleration found in this Example compare percentagewise with the acceleration due to gravity on the Earth's surface? (b) Would an astronaut in the space station be "weightless"?

▲ FIGURE 7.10 Centrifuge Centrifuges are used to separate particles of different sizes and densities suspended in liquids. For example, red and white blood cells can be separated from each other and from the plasma that makes up the liquid portion of the blood.

Example 7.6 ■ A Centrifuge: Centripetal Acceleration

A laboratory centrifuge like that shown in ◄Fig. 7.10 operates at a rotational speed of 12 000 rpm. (a) What is the magnitude of the centripetal acceleration of a red blood cell at a radial distance of 8.00 cm from the centrifuge's axis of rotation? (b) How does this acceleration compare with g?

Thinking It Through. Here, the angular speed and the radius are given, so the magnitude of the centripetal acceleration can be computed directly from Eq. 7.10. The result can be compared with g by using $g = 9.80$ m/s^2.

Solution. The data are as follows:

Given: $\omega = (1.20 \times 10^4 \text{ rpm})\left[\dfrac{(\pi/30) \text{ rad/s}}{\text{rpm}}\right]$ *Find:* (a) a_c
                                                                          (b) how a_c compares with g
        $= 1.26 \times 10^3$ rad/s
        $r = 8.00$ cm $= 0.0800$ m

Note that we converted the angular speed to radians per second.

(a) The centripetal acceleration is found from Eq. 7.10:

$$a_c = r\omega^2 = (0.0800 \text{ m})(1.26 \times 10^3 \text{ rad/s})^2 = 1.27 \times 10^5 \text{ m/s}^2$$

The Centrifuge: Separating Blood Components

The centrifuge is a rotating machine used to separate particles of different sizes and densities suspended in a liquid (or a gas). For example, cream is separated from milk by centrifuging, and blood components are separated in centrifuges in medical laboratories (see Fig. 7.10).

Blood components will eventually settle toward the bottom of a tube in layers—a process called *sedimentation*—under the influence of normal gravity alone. The viscous drag of the plasma on the particles is analogous to (but much greater than) the air resistance that determines the terminal velocity of falling objects (Section 4.6). Red blood cells settle in the bottom layer of the plasma, because they reach a greater terminal velocity than do the white blood cells and platelets and thus reach the bottom sooner. However, gravitational sedimentation is generally a very slow process.

Since clinicians cannot afford to wait a long time to see the fractional volume of red cells in the blood, centrifugation is used to speed up the sedimentation process. Tubes are spun horizontally. The resistance of the fluid medium on the particles supplies the centripetal acceleration that keeps them moving in slowly widening circles as they settle toward the bottom of the tube. The bottom of the tube itself must exert a strong force on the contents as a whole and must be strong enough so as not to break.

Laboratory centrifuges commonly operate at speeds sufficient to produce centripetal accelerations thousands of times larger than g. (See Example 7.6.) Since the principle of the centrifuge involves centripetal acceleration, perhaps "centripuge" would be a more descriptive name.

Related Exercises: 30, 36, and 103

(b) Using the relationship $1g = 9.80 \text{ m/s}^2$ to express a_c in terms of g, we get

$$a_c = (1.27 \times 10^5 \text{ m/s}^2)\left(\frac{1\,g}{9.80 \text{ m/s}^2}\right) = 1.30 \times 10^4\, g\,(= 13\,000\,g!)$$

Follow-up Exercise. (a) What angular speed in revolutions per minute would give a centripetal acceleration of $1\,g$ at the radial distance in this Example, and, taking gravity into account, what would be the resultant acceleration? (b) Compare the effect of gravity on the spinning tubes at the rotational speeds in the Example and in part (a) of this Follow-up Exercise.

Conceptual Example 7.7 ■ Breaking Away

A ball attached to a string is swung with uniform motion in a horizontal circle above a person's head (▶Fig. 7.11a). If the string breaks, which of the trajectories shown in Fig. 7.11b (viewed from above) would the ball follow?

Reasoning and Answer. When the string breaks, the centripetal force goes to zero. There is no force in the outward direction, so the ball could not follow trajectory (a). Newton's first law states that if no force acts on an object in motion, the object will continue to move in a straight line. This factor rules out trajectories b, d, and e.

It should be evident from the previous discussion that at any instant (including the instant when the string breaks), the isolated ball has a horizontal, tangential velocity. The downward force of gravity acts on it, but this force affects only its vertical motion, which is not visible in Fig. 7.11b. The ball thus flies off tangentially and is essentially a horizontal projectile (with $v_{x_o} = v$, $v_{y_o} = 0$, and $a_y = -g$). Viewed from above, the ball would follow the path labeled c.

Follow-up Exercise. If you swing a ball in a horizontal circle about your head, can the string be exactly horizontal? (See Fig. 7.11a.) Explain your answer. *Hint*: Analyze the forces acting on the ball.

Centripetal Force

For there to be an acceleration, there must be a net force. Thus, for there to be a centripetal (inward) acceleration, we must have a **centripetal force** (net inward force). Expressing the magnitude of this force in terms of Newton's second law

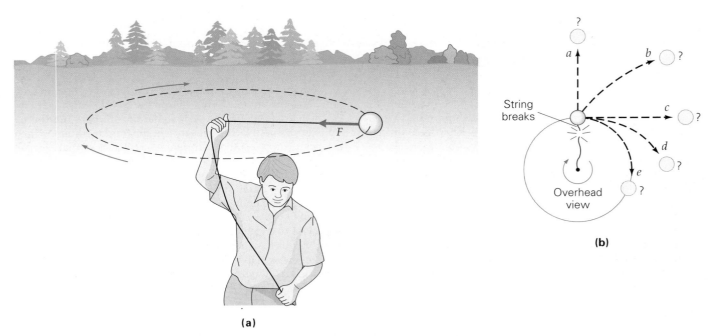

▲ FIGURE 7.11 Centripetal force (a) A ball is swung in a horizontal circle. (b) If the string breaks and the centripetal force goes to zero, what happens to the ball? See Conceptual Example 7.7.

($\mathbf{F}_{net} = m\mathbf{a}$) and inserting the expression for centripetal acceleration from Eq. 7.9, we can write

$$F_c = ma_c = \frac{mv^2}{r} \qquad \begin{array}{l} \textit{magnitude of} \\ \textit{centripetal force} \end{array} \qquad (7.11)$$

Centripetal Force

The centripetal force, like the centripetal acceleration, is directed radially toward the center of the circular path.

Keep in mind that, in general, a net force applied at an angle to the direction of motion of an object produces changes in the magnitude *and* direction of the velocity. However, when a net force of constant magnitude is continuously applied at an angle of 90° to the direction of motion (as is centripetal force), only the direction of the velocity changes. Also notice that because the centripetal force is always perpendicular to the direction of motion, this force does no work. (Why?) Therefore, by the work–energy theorem, a centripetal force does not change the kinetic energy or speed of the object.

Note that the centripetal force in the form $F = mv^2/r$ is not really a new individual force, but rather the cause of the centripetal acceleration supplied by a real force or forces. In Example 7.5, the force supplying the centripetal acceleration was gravity. In Conceptual Example 7.7, it was the tension in the string. Another force that often supplies centripetal acceleration is friction. Suppose that an automobile moves into a level, circular curve. To negotiate the curve, the car must have a centripetal acceleration, which is supplied by the force of friction between the tires and the road.

However, this (static; why?) friction has a maximum limiting value. If the speed of the car is high enough, the friction will not be sufficient to supply the necessary centripetal acceleration, and the car will skid outward from the center of the curve. If the car moves onto a wet or icy spot, the friction between the tires and the road may be reduced, allowing the car to skid at an even lower speed. (Banking a curve also helps vehicles negotiate the curve. See Exercises 45 and 59.)

Example 7.8 ■ Where the Rubber Meets the Road: Friction and Centripetal Force

A car approaches a level, circular curve with a radius of 45.0 m. If the concrete pavement is dry, what is the maximum speed at which the car can negotiate the curve at a constant speed?

Thinking It Through.   The car is in uniform circular motion on the curve, so there must be a centripetal force. This force is supplied by friction, so the maximum frictional force provides the centripetal force when the car is at its maximum tangential speed.

Solution.   Writing down what is given and what is to be found, we have

Given: $r = 45.0$ m   *Find:* v (maximum speed)

To go around the curve at a particular speed, the car must have a centripetal acceleration, and therefore a centripetal force must act on it. This inward force is supplied by static friction between the tires and the road. (The tires are not slipping or skidding relative to the road.)
  Recall from Chapter 4 that the maximum frictional force is given by $f_{s_{max}} = \mu_s N$ (Eq. 4.7), where N is the magnitude of the normal force on the car and is equal to the weight of the car, mg, on the level road (why?). We may set this expression to be equal in magnitude to the expression for centripetal force ($F_c = mv^2/r$) in order to find the maximum speed. To find $f_{s_{max}}$, we will need the coefficient of friction between rubber and concrete; from Table 4.1, it is $\mu_s = 1.20$. Then, we can write

$$f_{s_{max}} = F_c$$

$$\mu_s N = \mu_s mg = \frac{mv^2}{r}$$

So

$$v = \sqrt{\mu_s rg} = \sqrt{(1.20)(45.0\text{ m})(9.80\text{ m/s}^2)} = 23.0\text{ m/s}$$

(about 83 km/h, or 52 mi/h).

Follow-up Exercise.   Would the centripetal force be the same for all types of vehicles in this Example?

The proper safe speed for driving on a highway curve is an important consideration. The coefficient of friction between tires and the road may vary, depending on weather, road conditions, the design of the tires, the amount of tread wear, and so on. When a curved road is designed, safety may be promoted by banking, or inclining, the roadway. This design reduces the chances of skidding because the normal force exerted on the car by the road then has a component toward the center of the curve that reduces the need for friction. In fact, for a circular curve with a given banking angle and radius, there is one speed for which no friction is required at all. This condition is used in banking design. (See Exercise 59.)
  Let's look at one more example of centripetal force, this time with two objects in uniform circular motion. This Example will help you understand the motions of satellites in circular orbits, discussed in a later section.

Example 7.9 ■ Strung Out: Centripetal Force and Newton's Second Law

Suppose that two masses, $m_1 = 2.5$ kg and $m_2 = 3.5$ kg, are connected by light strings and are in uniform circular motion on a horizontal frictionless surface as illustrated in ▶Fig. 7.12, where $r_1 = 1.0$ m and $r_2 = 1.3$ m, respectively. The forces acting on the masses are $T_1 = 4.5$ N and $T_2 = 2.9$ N, which are the tensions in the strings, respectively. Find (a) the

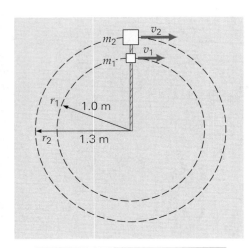

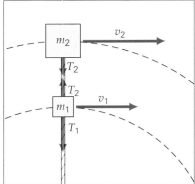

▲ FIGURE 7.12 Centripetal force and Newton's second law See Example 7.9.

centripetal accelerations and (b) the magnitudes of the tangential velocities (tangential speeds) of the masses.

Thinking It Through. The centripetal forces on the masses are supplied by the tensions (T_1 and T_2) in the strings. By isolating the masses, we can find a_c for each mass, because the net force on a mass is equal to the mass' centripetal force ($F_c = ma_c$).

The tangential speeds can then be found, since the radii are known ($a_c = v^2/r$).

Solution.

Given: $r_1 = 1.0$ m and $r_2 = 1.3$ m $\quad$ *Find:* (a) $\mathbf{a}_{c_1}$ and $\mathbf{a}_{c_2}$ (centripetal accelerations)
$m_1 = 2.5$ kg and $m_2 = 3.5$ kg $\qquad\qquad$ (b) v_1 and v_2 (tangential speeds)
$T_1 = 4.5$ N
$T_2 = 2.9$ N

By isolating m_2 in the figure, we can see that the centripetal force is provided by the tension in the string. (T_2 is the only force acting on m_2 toward the center of its circular path.) Thus,

$$T_2 = m_2 a_{c_2}$$

and

$$a_{c_2} = \frac{T_2}{m_2} = \frac{2.9 \text{ N}}{3.5 \text{ kg}} = 0.83 \text{ m/s}^2$$

The acceleration is toward the center of the circle. You can find the tangential speed of m_2 from $a_c = v^2/r$:

$$v_2 = \sqrt{a_{c_2} r_2} = \sqrt{(0.83 \text{ m/s}^2)(1.3 \text{ m})} = 1.0 \text{ m/s}$$

The situation is a bit different for m_1. In this case, there are two radial forces acting on m_1: the string tensions $+T_1$ and $-T_2$. Also, by Newton's second law, in order to have a centripetal acceleration, there must be a net force, which is given by the difference in the two tensions, so we expect $T_1 > T_2$. Thus, we have

$$F_{\text{net}_1} = +T_1 + (-T_2) = m_1 a_{c_1} = \frac{m_1 v_1^2}{r_1}$$

where the radial direction (toward the center of the circular path) is taken to be positive. Then

$$a_{c_1} = \frac{T_1 - T_2}{m_1} = \frac{4.5 \text{ N} + (-2.9 \text{ N})}{2.5 \text{ kg}} = 0.64 \text{ m/s}^2$$

and

$$v_1 = \sqrt{a_{c_1} r_1} = \sqrt{(0.64 \text{ m/s}^2)(1.0 \text{ m})} = 0.80 \text{ m/s}$$

Follow-up Exercise. Notice in this Example that the centripetal acceleration of m_2 is greater than that of m_1, yet $r_2 > r_1$, and $a_c \propto 1/r$. Is there something wrong here? Explain.

Integrated Example 7.10 ■ Center-Seeking Force: One More Time

A 1.0-m cord is used to suspend a 0.50-kg tetherball from the top of the pole. After being hit several times, the ball goes around the pole in uniform circular motion with a tangential speed of 1.1 m/s at an angle of 20° relative to the pole. (a) The force that supplies the centripetal acceleration is (1) the gravitational weight of the ball, (2) a component of the tension force in the string, or (3) the total tension in the string. (b) What is the magnitude of the centripetal force?

(a) Conceptual Reasoning. The centripetal force, being a "center-seeking" force, is directed perpendicularly toward the pole, about which the ball is in circular motion. As suggested in the problem-solving procedures provided in Section 1.7, it is most always helpful to sketch a diagram, such as that in ▶Fig. 7.13. Immediately, it can be seen that choices (1) and (3) are not correct, as these forces are not directly toward the pole. (mg and T_y are equal and opposite, and there is no acceleration in the y-direction.) The answer is obviously (2), with a component of the tension force, T_x, supplying the centripetal force.

(b) Thinking It Through. T_x supplies the centripetal force, and the given data are for the dynamical form of the centripetal force, that is, $T_x = F_c = mv^2/r$ (Eq. 7.11).

Given: $L = 1.0$ m   **Find:** The magnitude of the centripetal force
$v = 1.1$ m/s
$m = 0.50$ kg
$\theta = 20°$

As pointed out previously, the magnitude of the centripetal force may be found, using Eq. 7.11:

$$F_c = T_x = \frac{mv^2}{r}$$

But, we need the radial distance r. From the figure, this quantity can be seen to be $r = L \sin 20°$, so

$$F_c = \frac{mv^2}{L \sin 20°} = \frac{(0.50 \text{ kg})(1.1 \text{ m/s})^2}{(1.0 \text{ m})(0.342)} = 1.8 \text{ N}$$

Follow-up Exercise.   What is the magnitude of the tension T in the string?

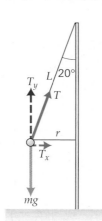

▲ **FIGURE 7.13   Ball on a string** See Integrated Example 7.10.

7.4   Angular Acceleration

OBJECTIVES: **To (a) define angular acceleration and (b) analyze rotational kinematics.**

As you might have guessed, another type of acceleration in angular motion is angular acceleration. This quantity is the time rate of change of angular velocity. In the case of circular motion, if there were an angular acceleration, the motion would not be uniform, because the speed would be changing. Analogous to the linear case, the magnitude of the **average angular acceleration** ($\bar{\alpha}$) is given by

$$\bar{\alpha} = \frac{\Delta\omega}{\Delta t}$$

Definition of: Average Angular Acceleration

where the bar over the alpha indicates that it is an average value, as usual. Taking $t_o = 0$, and if the angular acceleration is constant, so that $\bar{\alpha} = \alpha$, we have

$$\alpha = \frac{\omega - \omega_o}{t} \quad \text{(constant angular acceleration)}$$

SI unit of angular acceleration: radians per second squared (rad/s^2)

or

$$\omega = \omega_o + \alpha t \quad \text{(constant acceleration only)} \quad (7.12)$$

Angular Acceleration

No boldface symbols are used in Eq. 7.12, because, plus and minus signs are used to indicate angular directions, as described earlier. As in the case of linear motion, if the angular acceleration increases the angular velocity, both quantities have the same sign, meaning that their vector directions are the same (i.e., α is in the same direction as ω as given by the right-hand rule). If the angular acceleration decreases the angular velocity, then the two quantities have opposite signs, meaning that their vectors are opposed (i.e., α is in the direction opposite to ω as given by the right-hand rule, or is an angular deceleration, so to speak).

Example 7.11 ■ A Rotating CD: Angular Acceleration

A CD accelerates uniformly from rest to its operational speed of 500 rpm in 3.50 s. What is the angular acceleration of the CD (a) during this time? (b) What is the angular acceleration of the CD after this time? (c) If the CD comes uniformly to a stop in 4.50 s, what is its angular acceleration?

Thinking It Through. (a) We are given the initial and final angular velocities; the constant (uniform) angular acceleration can thus be calculated (Eq. 7.12), since we know the amount of time during which the CD accelerates. (b) Keep in mind that the operational speed is constant. (c) Everything is given for Eq. 7.12, but a negative result should be expected. (Why?)

Solution.

Given: $\omega_{o} = 0$

$$\omega = (500\ \text{rpm}) \left[\frac{(\pi/30)\ \text{rad/s}}{1\ \text{rev/min}} \right] = 52.4\ \text{rad/s}$$

$t = 3.50\ \text{s}$ (starting up)
$t = 4.50\ \text{s}$ (coming to a stop)

Find: (a) α (during start-up)
(b) α (in operation)
(c) α (in coming to a stop)

(a) Using Eq. 7.12, we get

$$\alpha = \frac{\omega - \omega_{o}}{t} = \frac{52.4\ \text{rad/s} - 0}{3.50\ \text{s}} = 15.0\ \text{rad/s}^2$$

in the direction of the angular velocity.

(b) After the CD reaches its operational speed, the angular velocity remains constant, so $\alpha = 0$.

(c) Again using Eq. 7.12, but this time with $\omega_{o} = 500\ \text{rpm}$ and $\omega = 0$, we get

$$\alpha = \frac{\omega - \omega_{o}}{t} = \frac{0 - 52.4\ \text{rad/s}}{4.50\ \text{s}} = -11.6\ \text{rad/s}^2$$

where the minus sign indicates that the angular acceleration is in the direction opposite that of the angular velocity (which is taken as $+$).

Follow-up Exercise. (a) What are the directions of the $\boldsymbol{\omega}$ and $\boldsymbol{\alpha}$ vectors in part (a) of this Example if the CD rotates clockwise when viewed from above? (b) Do the directions of these vectors change in part (c) of the Example?

As with arc length and angle ($s = r\theta$) and tangential and angular speeds ($v = r\omega$), there is a relationship between the tangential acceleration and the angular acceleration. The **tangential acceleration** is associated with changes in tangential speed and hence continuously changes direction. The magnitudes of the tangential and angular accelerations are related by a factor of r. For circular motion with a constant radius r,

$$a_{t} = \frac{\Delta v}{\Delta t} = \frac{\Delta (r\omega)}{\Delta t} = \frac{r\Delta \omega}{\Delta t} = r\alpha$$

$$a_t = r\alpha \qquad \begin{array}{l}\textit{magnitude of} \\ \textit{tangential acceleration}\end{array} \qquad (7.13)$$

The tangential acceleration (a_t) is written with a subscript t to distinguish it from the radial, or *centripetal, acceleration* (a_c). Centripetal acceleration is necessary for circular motion, but tangential acceleration is not. For uniform circular motion, there is no angular acceleration ($\alpha = 0$) or tangential acceleration, as can be seen from Eq. 7.13; there is only centripetal acceleration (▶Fig. 7.14a).

However, when there is an angular acceleration α (and therefore a tangential acceleration $a_t = r\alpha$), there is a change in *both* the angular *and* the tangential velocities ($v = r\omega$). As a result, the centripetal acceleration $a_c = v^2/r = r\omega^2$ must increase or decrease if the object is to maintain the same circular orbit (that is, if r is to stay the same). When there are both tangential and centripetal accelerations, the total instantaneous acceleration is their vector sum (Fig. 7.14b). The tangential-acceleration vector and the centripetal-acceleration vector are perpendicular to each other at any instant, and the total acceleration is $\mathbf{a} = a_t\hat{\mathbf{t}} + a_c\hat{\mathbf{r}}$, where $\hat{\mathbf{t}}$ and $\hat{\mathbf{r}}$ are unit vectors directed tangentially and radially inward, respectively. You should be able to find the magnitude of $\mathbf{a}$ and the angle it makes relative to $\mathbf{a}_t$ by using trigonometry (Fig. 7.14b).

The other angular equations can be derived, as was done for the linear equations in Chapter 2. That development will not be shown here; the set of angular equations with their linear counterparts for constant accelerations are listed in Table 7.2. A quick review of Chapter 2 (with a change of symbols) will show you how the angular equations are derived.

Example 7.12 ■ Even Cooking: Rotational Kinematics

A microwave oven has a rotating plate 30 cm in diameter for even cooking. The plate accelerates from rest at a uniform rate of 0.87 rad/s² for 0.50 s before reaching its constant operational speed. (a) How many revolutions does the plate make before reaching its operational speed? (b) What are the operational angular speed of the plate and the operational tangential speed at its rim?

Thinking It Through. This Example involves the use of the angular kinematic equations (Table 7.2). In (a), the angular distance θ will give the number of revolutions. For (b), we want to find ω and then $v = r\omega$.

Solution. Listing what is given and what we are to find, we have

Given: $d = 30$ cm, $r = 15$ cm $= 0.15$ m (radius) *Find:* (a) θ (in revolutions)
$\omega_o = 0$ (at rest)   (b) ω and v (angular and
$\alpha = 0.87$ rad/s²   tangential speeds,
$t = 0.50$ s   respectively)

(a) To find the angular distance θ *in radians*, we can use Eq. 4 from Table 7.2:

$$\theta = \omega_o t + \tfrac{1}{2}\alpha t^2 = 0 + \tfrac{1}{2}(0.87 \text{ rad/s}^2)(0.50 \text{ s})^2 = 0.11 \text{ rad}$$

Since 2π rad $= 1$ rev (or 360°), (1 rev/2π rad) $= 0.018$ rev/rad, so the plate reaches its operational speed in only a small fraction of a revolution.

(b) From Table 7.2, we see that Eq. 3 gives the angular speed, and

$$\omega = \omega_o + \alpha t = 0 + (0.87 \text{ rad/s}^2)(0.50 \text{ s}) = 0.44 \text{ rad/s}$$

Then, Eq. 7.6 gives the tangential speed at the rim radius:

$$v = r\omega = (0.15 \text{ m})(0.44 \text{ rad/s}) = 0.066 \text{ m/s}$$

Follow-up Exercise. When the oven is turned off, the plate makes half of a revolution before stopping. What is the plate's angular acceleration during this period?

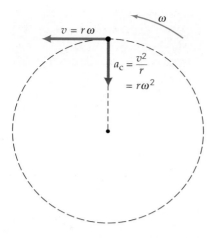

(a) Uniform circular motion
$(a_t = r\alpha = 0)$

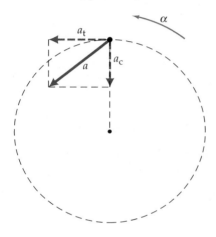

(b) Nonuniform circular motion
$(\mathbf{a} = \mathbf{a}_t + \mathbf{a}_c)$

▲ **FIGURE 7.14 Acceleration and circular motion (a)** In uniform circular motion, there is centripetal acceleration, but no angular acceleration ($\alpha = 0$) or tangential acceleration ($a_t = r\alpha = 0$). **(b)** In nonuniform circular motion, there are angular and tangential accelerations, and the total acceleration is the vector sum of the tangential and centripetal components.

Linear	Angular	
$x = \bar{v}t$	$\theta = \bar{\omega}t$	(1)
$\bar{v} = \dfrac{v + v_o}{2}$	$\bar{\omega} = \dfrac{\omega + \omega_o}{2}$	(2)
$v = v_o + at$	$\omega = \omega_o + \alpha t$	(3)
$x = v_o t + \frac{1}{2}at^2$	$\theta = \omega_o t + \frac{1}{2}\alpha t^2$	(4)
$v^2 = v_o^2 + 2ax$	$\omega^2 = \omega_o^2 + 2\alpha\theta$	(5)

*For these equations, $x_o = 0$, $\theta_o = 0$, and $t_o = 0$. The first equation in each column is general, that is, is not limited to situations where the acceleration is constant.

7.5 Newton's Law of Gravitation

OBJECTIVES: To (a) describe Newton's law of gravitation and how it relates to the acceleration due to gravity and (b) investigate how the law applies to gravitational potential energy.

Another of Isaac Newton's many accomplishments was the formulation of what is called the **universal law of gravitation**. This law is very powerful and fundamental. Without it, for example, we would not understand the cause of tides or know how to put satellites into particular orbits around the Earth. This law allows us to analyze the motions of planets and comets, stars and galaxies. The word *universal* in the name indicates that we believe it to apply everywhere in the universe. (This term highlights the importance of the law, but for brevity, it is common to refer simply to *Newton's law of gravitation* or the *law of gravitation*.)

Newton's law of gravitation in mathematical form gives a simple relationship for the gravitational interaction between two particles, or point masses, m_1 and m_2 separated by a distance r (▶Fig. 7.15a). Basically, every particle in the universe has an attractive interaction with every other particle. The forces of mutual interaction are equal and opposite, forming a force pair as described by Newton's third law (Chapter 4) that is, $F_{21} = -F_{12}$ in Fig. 7.15a.

The gravitational attraction, or force (F), decreases as the the square of the distance (r^2) between two point masses increases; that is, the magnitude of the gravitational force and the distance separating the two particles are related as follows:

$$F \propto \frac{1}{r^2}$$

(This type of relationship is called an *inverse-square law*—F is inversely proportional to r^2.)

Newton's law also correctly postulates that the gravitational force, or attraction of a body, depends on the body's mass—the greater the mass, the greater is the attraction. However, because gravity is a mutual interaction between masses, it should be directly proportional to both masses—that is, to their product ($F \propto m_1 m_2$).

Hence, Newton's law of gravitation has the form $F \propto m_1 m_2/r^2$. Expressed as an equation with a constant of proportionality, the magnitude of the mutually attractive gravitational force between two masses is given by

$$F_g = \frac{Gm_1 m_2}{r^2}$$

(7.14)

Newton's universal law of gravitation

where G is a constant called the **universal gravitational constant** and has a value of

$$G = 6.67 \times 10^{-11} \, \text{N} \cdot \text{m}^2/\text{kg}^2$$

This constant is often referred to as "big G" to distinguish it from "little g," the acceleration due to gravity. Note from Eq. 7.14 that F approaches zero only when r is infinitely large. Thus, the gravitational force has, or acts over, an *infinite range*.

How did Newton come to his conclusions about the force of gravity? Legend has it that his insight came after he observed an apple fall from a tree to the ground. Newton had been wondering what supplied the centripetal force to keep the Moon in orbit and might have had this thought: "If gravity attracts an apple toward the Earth, perhaps it also attracts the Moon, and the Moon is falling, or accelerating toward the Earth, under the influence of gravity" (▶Fig. 7.16).

Whether or not the legendary apple did the trick, Newton assumed that the Moon and the Earth were attracted to each other and could be treated as point masses, with their total masses concentrated at their centers (Fig 7.15b). The inverse-square relationship had been speculated on by some of his contemporaries. Newton's achievement was demonstrating that the relationship could be deduced from one of Johannes Kepler's laws of planetary motion (Section 7.6).

Newton expressed Eq. 7.14 as a proportion ($F \propto m_1 m_2 / r^2$) because he did not know the value of G. It was not until 1798 (71 years after Newton's death) that the value of the universal gravitational constant was experimentally determined by an English physicist, Henry Cavendish. Cavendish used a sensitive balance to measure the gravitational force between separated spherical masses (as illustrated in Fig. 7.15b). If F, r, and the m's are known, big G can be computed from Eq. 7.14.

As mentioned earlier, Newton considered the nearly spherical Earth and Moon to be point masses located at their respective centers. It took him some years, using mathematical methods he developed, to prove that this condition is the case only for only spherical, *homogeneous* objects.* The general concept is illustrated in ▶ Fig. 7.17.

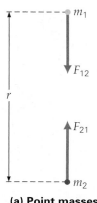

(a) Point masses

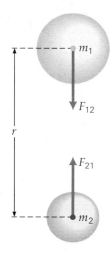

(b) Homogeneous spheres

$$F = \frac{Gm_1 m_2}{r^2}$$

▲ **FIGURE 7.15 Universal law of gravitation** **(a)** Any two particles, or point masses, are gravitationally attracted to each other with a force that has a magnitude given by Newton's universal law of gravitation. **(b)** For homogeneous spheres, the masses may be considered to be concentrated at their centers.

Note: The symbol g is reserved for the acceleration due to gravity at the Earth's surface; a_g is more generally the acceleration due to gravity at some greater radial distance.

Example 7.13 ■ Gravitational Attraction between the Earth and the Moon: A Centripetal Force

Compute the magnitude of the mutual gravitational force between the Earth and the Moon. (You can assume that the Earth and the Moon are homogeneous spheres.)

Thinking It Through. This Example is an application of Eq. 7.14 for which the masses and the distance must be looked up. (See the Appendices.)

Solution. No data are given, so they must be available from references.

Given: (from tables inside the back cover of this text)   *Find:* F_g (gravitational force)
$M_E = 6.0 \times 10^{24}$ kg
$m_M = 7.4 \times 10^{22}$ kg
$r_{EM} = 3.8 \times 10^8$ m

The average distance from the Earth to the Moon (r_{EM}) is taken to be the distance from the center of one to the center of the other. Using Eq. 7.14, we get

$$F_g = \frac{Gm_1 m_2}{r^2} = \frac{GM_E m_M}{r_{EM}^2}$$
$$= \frac{(6.67 \times 10^{-11}\ \text{N} \cdot \text{m}^2/\text{kg}^2)(6.0 \times 10^{24}\ \text{kg})(7.4 \times 10^{22}\ \text{kg})}{(3.8 \times 10^8\ \text{m})^2}$$
$$= 2.1 \times 10^{20}\ \text{N}$$

This quantity is the magnitude of the centripetal force that keeps the Moon revolving in its orbit around the Earth. It is a very large force, but the Moon is a very massive object, with a correspondingly large inertia to overcome. Because of the inward or radial component, it is sometimes said that the Moon is "falling" toward the Earth. This motion, combined with the tangential motion, results in the Moon's nearly circular orbit about the Earth.

Follow-up Exercise. With what acceleration is the Moon "falling" toward the Earth?

The acceleration due to gravity at a particular distance from a planet can also be investigated by using Newton's second law of motion and his law of gravitation. The magnitude of the acceleration due to gravity, which we will generally write as a_g, at a distance r from the center of a spherical mass M is found by setting the force of gravitational attraction due to that spherical mass equal to ma_g, the net force on an object of mass m at a distance r:

$$ma_g = \frac{GmM}{r^2}$$

*For a homogeneous sphere, the equivalent point mass is located at the center of mass. However, this is a special case. The center of gravitational force and the center of mass of a configuration of particles or an object do not generally coincide.

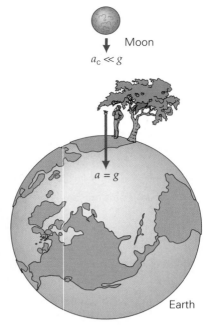

Moon

$a_c \ll g$

$a = g$

Earth

▲ **FIGURE 7.16 Gravitational insight?** Newton developed his law of gravitation while studying the orbital motion of the Moon. According to legend, his thinking was spurred when he observed an apple falling from a tree. He supposedly wondered whether the force causing the apple to accelerate toward the ground could extend to the Moon and cause it to fall toward Earth, that is, supply its orbital centripetal acceleration.

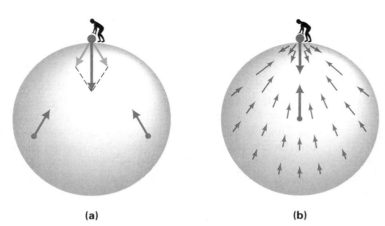

(a)   **(b)**

▲ **FIGURE 7.17 Uniform spherical masses** **(a)** Gravity acts between any two particles. The resultant gravitational force exerted on an object outside a homogeneous sphere by two particles at symmetric locations within the sphere is directed toward the center of the sphere. **(b)** Because of the sphere's symmetry and uniform distribution of mass, the net effect is as though all the mass of the sphere were concentrated as a particle at its center. For this special case, the gravitational center of force and center of mass coincide, but this condition is generally not true for other objects. (Only a few of the red force arrows are shown because of space considerations.)

Then, the acceleration due to gravity at any distance r from the planet's center is

$$a_g = \frac{GM}{r^2} \tag{7.15}$$

Notice that a_g is proportional to $1/r^2$, so the farther away an object is from the planet, the smaller is its acceleration due to gravity and the smaller is the attractive force (ma_g) on the object. The force or weight vector is directed toward the center of the planet.

Equation 7.15 can be applied to the Moon or any planet. For example, taking the Earth to be a point mass M_E located at its center and R_E as its radius, we obtain the acceleration due to gravity ($a_{g_E} = g$) at the Earth's surface by setting the distance r to be equal to R_E.

$$a_{g_E} = g = \frac{GM_E}{R_E^2} \tag{7.16}$$

This equation has several interesting implications. First, it reveals that taking g to be constant everywhere on the surface of the Earth involves the assumption that the Earth has a homogeneous distribution of mass and that the distance from the center of the Earth to any location on its surface is the same. These two assumptions are not exactly true. Therefore, taking g to be a constant is only an approximation, but one that works pretty well for most situations.

Also, you can see why the acceleration due to gravity is the same for all free-falling objects, that is, independent of the mass of the object. The mass of the object doesn't appear in Eq. 7.16, so all objects in free fall accelerate at the same rate.

Finally, if you're observant, you'll notice that Eq. 7.16 can be used to compute the mass of the Earth. All of the other quantities in the equation are measurable, and their values are known, so M_E can readily be calculated. This is what Cavendish did after he determined the value of G experimentally. Then he also found the average density of the Earth. (What do you think this value showed? Think of the difference between the Earth's crust and its metallic core. How does the density of a crustal rock compare with Cavendish's average density of the Earth?)

The acceleration due to gravity does vary with altitude. At a distance h above the Earth's surface, we have $r = R_E + h$. The acceleration is then given by

$$a_g = \frac{GM_E}{(R_E + h)^2} \qquad (7.17)$$

Problem-Solving Hint

When comparing accelerations due to gravity or gravitational forces, you will often find it convenient to work with ratios. For example, comparing a_g with g (Eqs. 7.15 and 7.16) for the Earth gives

$$\frac{a_g}{g} = \frac{GM_E/r^2}{GM_E/R_E^2} = \frac{R_E^2}{r^2} = \left(\frac{R_E}{r}\right)^2 \quad \text{or} \quad \frac{a_g}{g} = \left(\frac{R_E}{r}\right)^2$$

Note how the constants cancel out. Taking $r = R_E + h$, you can easily compute a_g/g, or the acceleration due to gravity at some altitude h above the Earth compared with g on the Earth's surface (9.80 m/s^2).

Because R_E is very large compared with readily attainable everyday altitudes above the Earth's surface, the acceleration due to gravity does not decrease very rapidly as we ascend. At an altitude of 16 km (10 mi), $a_g/g = 0.99$, and thus a_g is still 99% of the value of g at the Earth's surface. At an altitude of 160 km (100 mi), a_g is 95% of g.

Another aspect of the decrease of g with altitude concerns potential energy. In Chapter 5, you learned that $U = mgh$ for an object at a height h above some zero reference point, since g is essentially constant near the Earth's surface. This potential energy is equal to the work done in raising the object a distance h above the Earth's surface in a *uniform* gravitational field. But what if the change in altitude is so large that g cannot be considered constant while work is done in moving an object, such as a satellite? In this case, the equation $U = mgh$ doesn't apply. In general, it can be shown (using mathematical methods that are beyond the scope of this book) that the **gravitational potential energy** of two point masses separated by a distance r is given by

Definition of: Gravitational potential energy

$$U = -\frac{Gm_1m_2}{r} \qquad (7.18)$$

The minus sign in Eq. 7.18 arises from the choice of the zero reference point (the point where $U = 0$), which is $r = \infty$.

In terms of the Earth and a mass m at an altitude h above the Earth's surface,

$$U = -\frac{Gm_1m_2}{r} = -\frac{GmM_E}{R_E + h} \qquad (7.19)$$

Note: Review Fig. 5.14, which illustrates negative potential energy.

where r is the distance separating the Earth's center and the mass. What this equation means is that on the Earth we are in a negative gravitational potential energy well (►Fig. 7.18) that extends to infinity, because the force of gravity has an infinite range. As h increases, so does U. That is, U becomes *less negative*, or gets closer to zero, corresponding to a higher position in the potential energy well. The same is true for the finite-well approximation given by $U = mgh$, but over long distances, for which Eq. 7.19 applies, the change in U is not linear; it varies as $1/r$.

Thus, when gravity does negative work (an object moves higher in the well) or gravity does positive work (an object falls lower in the well), there is a *change* in potential energy. As with finite potential energy wells, this *change* in energy is always what's important in analyzing situations.

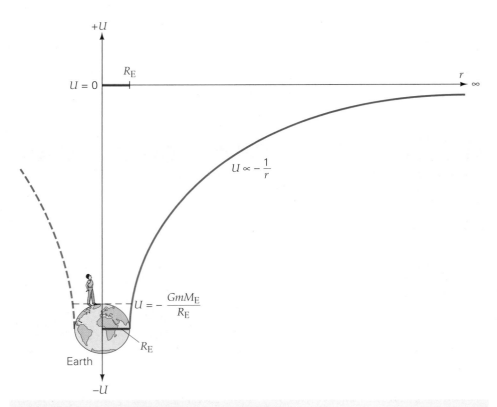

FIGURE 7.18 Gravitational potential energy well On the Earth, we are in a negative gravitational potential energy well. As with an actual well or hole in the ground, work must be done against gravity to get higher in the well. The potential energy of an object increases as the object moves higher in the well. This means that the value of U becomes less negative. The top of the Earth's gravitational well is at infinity, where the gravitational potential energy is, by choice, zero.

Example 7.14 ■ Different Orbits: Change in Gravitational Potential Energy

Two 50-kg satellites move in circular orbits about the Earth at altitudes of 1000 km (about 620 mi) and 36 000 km (about 22 400 mi), respectively. The lower one monitors particles about to enter the atmosphere, and the higher, geosynchronous one (synchronous with the Earth's rotation) takes weather pictures from its stationary position with respect to the Earth's surface. What is the difference in the gravitational potential energies of the two satellites in their respective orbits?

Thinking It Through.   The potential energies of the satellites are given by Eq. 7.19, where the greater the altitude (h), the smaller or less negative is U. Thus, the satellite with the greater h is higher in the gravitational-potential-energy well and has more potential energy.

Solution.   Listing the data so that we can better see what's given, we have (with two significant figures)

Given: $m = 50$ kg
$\quad\quad h_1 = 1000$ km $= 1.0 \times 10^6$ m
$\quad\quad h_2 = 36\,000$ km $= 36 \times 10^6$ m
$\quad\quad M_E = 6.0 \times 10^{24}$ kg (from a table inside
$\quad\quad\quad\quad\quad\quad\quad\quad$ the back cover of the book)
$\quad\quad R_E = 6.4 \times 10^6$ m

Find: ΔU (difference in
$\quad\quad\quad$ potential energy)

The difference in the gravitational potential energy can be computed directly from Eq. 7.19. Keep in mind that the potential energy is the energy of position, so we compute the potential energies for each position or altitude and subtract one from the other. Thus, we have

$$\Delta U = U_2 - U_1 = -\frac{GmM_E}{R_E + h_2} - \left(-\frac{GmM_E}{R_E + h_1}\right) = GmM_E\left(\frac{1}{R_E + h_1} - \frac{1}{R_E + h_2}\right)$$

$$= (6.67 \times 10^{-11}\,\text{N}\cdot\text{m}^2/\text{kg}^2)(50\,\text{kg})(6.0 \times 10^{24}\,\text{kg})$$

$$\times \left[\frac{1}{6.4 \times 10^6\,\text{m} + 1.0 \times 10^6\,\text{m}} - \frac{1}{6.4 \times 10^6\,\text{m} + 36 \times 10^6\,\text{m}}\right]$$

$$= +2.2 \times 10^9\,\text{J}$$

Note: For communications purposes, many satellites are launched into a circular orbit above the equator at an altitude of about 36 000 km. Satellites there are synchronous to the Earth's rotation. That is, they remain "fixed" over one point on the equator; to an observer on the Earth, they are always seen in the same position in the sky.

Because ΔU is positive, m_2 is higher in the gravitational potential energy well than m_1. Note that even though both U_1 and U_2 are negative, U_2 is "more positive," or "less negative," and closer to zero. Thus, you have to put in more energy to get a satellite farther from the Earth.

Follow-up Exercise. (a) Suppose that the altitude of the higher satellite in this Example were doubled, to 72 000 km. Would the difference in the gravitational potential energies of the two satellites then be twice as great? Justify your answer. (b) Note that U_2 has a *smaller* negative value than U_1 (since $h_2 > h_1$). What does this mean?

Substituting the gravitational potential energy (Eq. 7.18) into the equation for the total mechanical energy gives the equations a different form than it had in Chapter 5. For example, the total mechanical energy of a mass m_1 moving near a stationary mass m_2 is

$$E = K + U = \tfrac{1}{2} m_1 v^2 - \frac{G m_1 m_2}{r} \tag{7.20}$$

(Note that the potential energy $U = -G m_1 m_2/r$ is *not* written as mgh.) This equation and the principle of the conservation of energy can be applied to the Earth's motion about the Sun, by neglecting other gravitational forces. The Earth's orbit is not quite circular, but slightly elliptical. At *perihelion* (the point of the Earth's closest approach to the Sun), the mutual gravitational potential energy is less (a larger negative number) than it is at *aphelion* (the point farthest from the Sun). Therefore, as can be seen from Eq. 7.20 in the form $\tfrac{1}{2} m_1 v^2 = E + G m_1 m_2/r$, where E is constant, the Earth's kinetic energy and orbital speed are greatest at perihelion (the smallest value of r) and least at aphelion (the greatest value of r). Or, in general, the Earth's orbital speed is greater when it is nearer the Sun than when it is farther away.

Mutual gravitational potential energy also applies to a group, or *configuration*, of more than two masses. That is, there is gravitational potential energy due to the masses being in a configuration, which is the case because work was done in bringing the masses together. Suppose that there is a single fixed mass m_1, and another mass m_2 is brought close to m_1 from an infinite distance (where $U = 0$). The work done against the attractive force of gravity is negative (why?) and equal to the change in the mutual potential energy of the masses, which are now separated by a distance r_{12}; that is, $U_{12} = -G m_1 m_2/r_{12}$.

If a third mass m_3 is brought close to the other two fixed masses, there are then two forces of gravity acting on m_3, so $U_{13} = -G m_1 m_3/r_{13}$ and $U_{23} = -G m_2 m_3/r_{23}$. The total gravitational potential energy of the configuration is therefore

$$U = U_{12} + U_{13} + U_{23}$$
$$= -\frac{G m_1 m_2}{r_{12}} - \frac{G m_1 m_3}{r_{13}} - \frac{G m_2 m_3}{r_{23}} \tag{7.21}$$

A fourth mass could be brought in to further prove the point, but this development should be sufficient to suggest that the total gravitational potential energy of a configuration of particles is equal to the sum of the individual potential energies for all pairs of particles.

Example 7.15 ■ Total Gravitational Potential Energy: Energy of Configuration

Three masses are in a configuration as shown in ▶Fig. 7.19. What is their total gravitational potential energy?

Thinking It Through. Equation 7.21 applies, but be sure to keep your masses and their distances distinct.

Solution. From the figure, we have

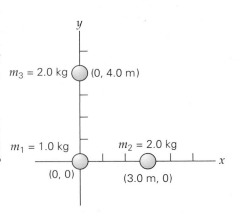

▲ **FIGURE 7.19** Total gravitational potential energy
See Example 7.15.

Space Exploration: Gravity Assists

As you read this text, the *Cassini* spacecraft is on its way to Saturn, after having made two Venus flybys, a Jupiter flyby, and one Earth flyby (Fig. 1).* What is going on here? Why was the spacecraft launched toward Venus, an inner planet, in order to go to Saturn, an outer planet?

Although space probes can be launched from the Earth with the current rocket technology, there are limitations— namely, fuel versus payload: The more fuel, the smaller is the payload. Using rockets alone, planetary spacecraft are realistically limited to visiting Venus, Mars, and Jupiter. The other planets could not be reached by a spacecraft of reasonable size without taking decades to get there.

So, how will *Cassini* get to Saturn in 2004, almost 7 years after its 1997 launch? By using gravity in a clever scheme called *gravity assist*. By using gravity assists, missions to all of the planets in our solar system are possible. Rocket energy is needed to get a spacecraft to the first planet, and after that,

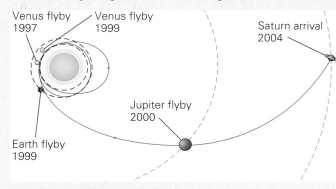

FIGURE 1 *Cassini* trajectory

*Cassini was the French–Italian astronomer who studied Saturn, discovering four of its moons and that Saturn's rings are separated into two parts by a narrow gap, now called the *Cassini Division*. The *Cassini* spacecraft will release a "Huygens" probe to Saturn's moon Titan. This moon was discovered by the Dutch scientist Christian Huygens.

the energy is more or less "free." Basically, during a planetary swing-by, there is an exchange of energy between the planet and the spacecraft, which enables the spacecraft to increase its speed to the Sun. (This phenomenon is sometimes called a *slingshot effect*.)

Gravity assists, or swing-bys, are a key factor that has made exploration of the solar system possible. The first swing-by was in 1973 for the *Mariner 10* mission, which flew by Venus on its way to Mercury. Since then, there have been many planetary missions with gravity assists—for example, the *Voyagers* to the outer planets and *Galileo* to Jupiter. Perhaps most spectacular of all was the *Voyager 2* spacecraft, which made three planetary swingbys (of Jupiter, Saturn, and Uranus) on the way to its encounter with Neptune in 1989. After the spacecraft reached Neptune, there was another swing-by in order for it to gain momentum and trajectory that sent it out of the solar system—and it is still going (Newton's first law).

Let's take a brief look at the physics of this ingenious use of gravity. Imagine the *Cassini* spacecraft making a swing-by of Jupiter, as illustrated in Fig. 2, where the momenta vectors of the motion are shown. Recall from Chapter 6 that a collision is an interaction of objects in which there is an exchange of momentum and energy. Technically, in a swing-by, a spacecraft is having a "collision" with a planet.

As shown in Fig. 2, when the spacecraft approaches from "behind" the planet and leaves in "front" (relative to the planet's motional direction), the gravitational interaction gives rise to a change in momentum—that is, $|\mathbf{p}_2| > |\mathbf{p}_1|$, and the direction is different. Looking at the vector triangle, we see that there is a $\Delta\mathbf{p}$ in the general "forward" direction of the spacecraft. Since $\Delta\mathbf{p} \propto \mathbf{F}$, there is a force acting on the craft that gives it a "kick" of energy in that direction. So, there is positive net work done ($W > 0$) and a change in kinetic energy ($\Delta K > 0$, from the work–energy theorem), and

Given: $m_1 = 1.0$ kg *Find:* U (total gravitational potential energy)
$\quad\quad\quad m_2 = 2.0$ kg
$\quad\quad\quad m_3 = 2.0$ kg
$\quad\quad\quad r_{12} = 3.0$ m; $r_{13} = 4.0$ m; $r_{23} = 5.0$ m (3–4–5 right triangle)

We can use Eq. 7.21 directly, since only three masses are used in this example. (Note that Eq. 7.21 can be extended to any number of masses.)

$$U = U_{12} + U_{13} + U_{23}$$

$$= -\frac{Gm_1m_2}{r_{12}} - \frac{Gm_1m_3}{r_{13}} - \frac{Gm_2m_3}{r_{23}}$$

$$= (6.67 \times 10^{-11}\ \text{N}\cdot\text{m}^2/\text{kg}^2)$$

$$\times \left[-\frac{(1.0\ \text{kg})(2.0\ \text{kg})}{3.0\ \text{m}} - \frac{(1.0\ \text{kg})(2.0\ \text{kg})}{4.0\ \text{m}} - \frac{(2.0\ \text{kg})(2.0\ \text{kg})}{5.0\ \text{m}} \right]$$

$$= -1.3 \times 10^{-10}\ \text{J}$$

Follow-up Exercise. Explain what the *negative* potential energy in this Example means in physical terms.

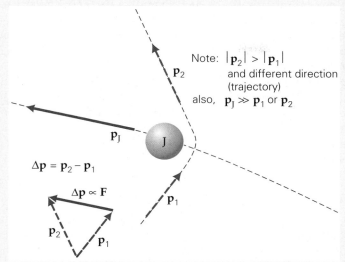

FIGURE 2 Spacecraft on a planetary swing-by The gravitational interaction of Jupiter (J) gives the spacecraft a change in momentum (greater magnitude and different direction), as the vector diagram shows. With $\Delta \mathbf{p} \propto \mathbf{F}$, positive work is done on the craft: It has more energy and a greater speed on leaving the region than on entering. By the conservation of momentum, the planet also has a change in momentum, but the motional effects on it are negligible, because of its large mass.

the spacecraft leaves with more energy, a greater speed, and a new direction. (If the flyby occurred in the opposite direction, the spacecraft would slow down.)

Momentum and energy are conserved in this elastic collision, and the planet gets an equal and opposite change in momentum, giving it a retarding effect. But because the planet's mass is so much larger than that of the spacecraft, the effect on the planet is negligible.

To help you grasp the idea of a gravity assist, consider the analogous roller derby "slingshot maneuver" illustrated in Fig. 3. The skaters interact, and skater S comes out of the "flyby" with increased speed. Here, the change in momentum on the "slinger," skater J, would probably be noticeable, but that would not be so on Jupiter or any other planet.

Related Exercise: 93

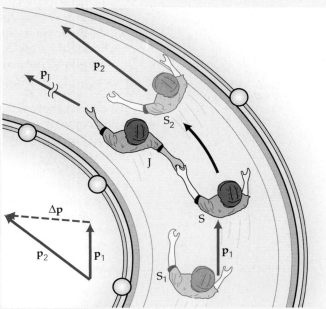

FIGURE 3 Skating swing-by As an analogy to a planetary swing-by, consider a roller derby "slingshot maneuver." Skater J slings skater S, who comes out of the "flyby" with greater speed than she had before. In this case, the change in momentum on skater J, the slinger, would probably be noticeable. (Why?)

The effects of gravity are familiar to us, perhaps indirectly. When we lift an object, we may think of it as being heavy, but we are working against gravity. Gravity causes rocks to tumble down and causes mud slides. But we often put gravity to use. For example, fluids from bottles for intravenous infusions flow because of gravity. An extraterrestrial application of how we put gravity to work is given in the Insight on Space Exploration.

7.6 Kepler's Laws and Earth Satellites

OBJECTIVES: To (a) state and explain Kepler's laws of planetary motion and (b) describe the orbits and motions of satellites.

The force of gravity determines the motions of the planets and satellites and holds the solar system (and galaxy) together. A general description of planetary motion had been set forth shortly before Newton's time by the German astronomer and mathematician Johannes Kepler (1571–1630). Kepler was able to formulate three empirical laws from observational data gathered during a 20-year period by the Danish astronomer Tycho Brahe (1546–1601).

Kepler went to Prague to assist Brahe, who was the official mathematician at the court of the Holy Roman Emperor. Brahe died the next year, and Kepler succeeded him, inheriting his records of the positions of the planets. Analyzing these data, Kepler announced the first two of his three laws in 1609 (the year Galileo built his first telescope). These laws were applied initially only to Mars. Kepler's third law came 10 years later.

Interestingly enough, Kepler's laws of planetary motion, which took him about 15 years to deduce from observed data, can now be derived theoretically with a page or two of calculations. These three laws apply not only to planets, but also to any system composed of a body revolving about a more massive body, to which the inverse-square law of gravitation applies (such as the Moon, artificial Earth satellites, and solar-bound comets).

Kepler's first law (the law of orbits) says the following:

Kepler's First Law

> Planets move in elliptical orbits, with the Sun at one of the focal points.

An ellipse, shown in ▼Fig. 7.20a, has, in general, an oval shape, resembling a flattened circle. In fact, a circle is a special case of an ellipse in which the focal points, or *foci* (plural of *focus*), are at the same point (the center of the circle). Although the orbits of the planets are elliptical, most do not deviate very much from circles (Mercury and Pluto are notable exceptions; see Appendix III, "Eccentricity"). For example, the difference between the perihelion and aphelion of the Earth (its closest and farthest distances from the Sun, respectively) is about 5 million km. This distance may sound like a lot, but it is only a little over three percent of 150 million km, which is the average distance between the Earth and the Sun.

Kepler's second law (the law of areas) says the following:

Kepler's Second Law

> A line from the Sun to a planet sweeps out equal areas in equal lengths of time.

This law is illustrated in Fig. 7.20b. Since the time to travel the different orbital distances (s_1 and s_2) is the same such that the areas swept out (A_1 and A_2) are equal,

▶ **FIGURE 7.20 Kepler's first and second laws of planetary motion** (a) In general, an ellipse has an oval shape. The sum of the distances from the focal points F to any point on the ellipse is constant: $r_1 + r_2 = 2a$. Here, $2a$ is the length of the line joining the two points on the ellipse at the greatest distance from its center, called the *major axis*. (The line joining the two points closest to the center is b, the *minor axis*.) Planets revolve about the Sun in elliptical orbits for which the Sun is at one of the focal points and nothing is at the other. **(b)** A line joining the Sun and a planet sweeps out equal areas in equal times. Since $A_1 = A_2$, a planet travels faster along s_1 than along s_2.

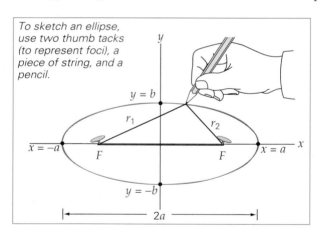

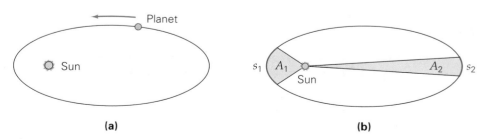

(a)   (b)

this law tells you that the orbital speed of a planet varies in different parts of its orbit. Because a planet's orbit is elliptical, its orbital speed is greater when it is closer to the Sun than when it is farther away. We used the conservation of energy in Section 7.5 (Eq. 7.20) to deduce this relationship for the Earth.

Kepler's third law (the law of periods) says the following:

Kepler's third law

> The square of the orbital period of a planet is directly proportional to the cube of the average distance of the planet from the Sun; that is, $T^2 \propto r^3$.

Kepler's third law is easily derived for the special case of a planet with a circular orbit, using Newton's law of gravitation. Since the centripetal force is supplied by the force of gravity, the expressions for these forces can be set equal

$$\underset{\substack{\text{centripetal} \\ \text{force}}}{\frac{m_p v^2}{r}} = \underset{\substack{\text{gravitational} \\ \text{force}}}{\frac{G m_p M_S}{r^2}}$$

or

$$v = \sqrt{\frac{GM_S}{r}}$$

In these equations, m_p and M_S are the masses of the planet and the Sun, respectively, and v is the planet's orbital speed. But $v = 2\pi r/T$ (circumference/period), so

$$\frac{2\pi r}{T} = \sqrt{\frac{GM_S}{r}}$$

PHYSLET®
ILLUSTRATION

Kepler's Laws

Squaring both sides and solving for T^2 gives

$$T^2 = \left(\frac{4\pi^2}{GM_S}\right) r^3$$

or

$$T^2 = Kr^3 \tag{7.22}$$

The constant K for solar-system planetary orbits is easily evaluated from orbital data (for T and r) for the Earth: $K = 2.97 \times 10^{-19} \text{ s}^2/\text{m}^3$. (As an exercise, you might wish to convert K to the more useful units of y^2/km^3.) Notice that Eq. 7.22 can be used to find the value of K, and if G is known, then the mass of the Sun can be determined. (*Note*: This value of K does *not* apply to Earth satellites. Why?)

Earth's Satellites

We are only about a half a century into the space age. Since the 1950s, numerous unmanned satellites have been put into orbit about the Earth, and now astronauts regularly spend weeks or months in orbiting space laboratories.

Putting a spacecraft into orbit about the Earth (or any planet) is an extremely complex task. However, you can get a basic understanding of the method from fundamental principles of physics. First, suppose that a projectile could be given the initial speed required to take it just to the top of the Earth's potential-energy well. At the exact top of the well, which is an infinite distance away ($r = \infty$), the potential energy is zero. By the conservation of energy and Eq. 7.18,

$$\underset{\text{initial}}{K_o + U_o} = \underset{\text{final}}{K + U}$$

or

$$\tfrac{1}{2}mv_{esc}^2 - \frac{GmM_E}{R_E} = \underset{\text{initial}}{0} + \underset{\text{final}}{0}$$

where v_{esc} is the **escape speed**—that is, the initial speed needed to escape (that is, so as not to come back) from the surface of the Earth. The final energy is zero, since the projectile stops at the top of the well (at very large distances, and it is barely moving), and $U = 0$ there. Solving for v_{esc} gives

$$\tfrac{1}{2}mv_{esc}^2 = \frac{GmM_E}{R_E}$$

and

$$v_{esc} = \sqrt{\frac{2GM_E}{R_E}} \qquad (7.23)$$

Since $g = GM_E/R_E^2$ (Eq. 7.17), it is convenient to write

$$v_{esc} = \sqrt{2gR_E} \qquad (7.24)$$

Although derived here for the Earth, this equation may be used generally to find the escape speeds for other planets and our Moon.

Example 7.16 ■ No Return: Escape Speed

What is the escape speed at the Earth's surface?

Thinking It Through. This Example is a direct application of Eq. 7.24, using the Earth's radius, found inside the back cover of this book (or in Appendix III).

Solution. Equation 7.24 can be used with known data:

$$v_{esc} = \sqrt{2gR_E} = \sqrt{2(9.80 \text{ m/s}^2)(6.4 \times 10^6 \text{ m})} = 11 \times 10^3 \text{ m/s} = 11 \text{ km/s (about 7 mi/s)}$$

Thus, if a projectile or spacecraft could be given an initial upward speed of 11 km/s (about 40 000 km/h, or 25 000 mi/h), it would leave the Earth and not return. Note from the equation that the escape speed is independent of the mass of the spacecraft; this quantity is the initial speed required to send *any* object to the top of the Earth's potential-energy well.

Such an initial speed is not attainable instantly at launch, but can be acheived in multistage steps. For example, in the 1980s, probes (e.g., the *Voyagers*) to the outer planets left the Earth, never to return, and are currently beyond the limits of the solar system.

Follow-up Exercise. How many times greater is the escape speed of the Earth than that of the Moon?

A tangential speed less than the escape speed is required for a satellite to orbit. Consider the centripetal force of a satellite in circular orbit about the Earth. Since the centripetal force on the satellite is supplied by the gravitational attraction between the satellite and the Earth, we may again write

$$F_c = \frac{mv^2}{r} = \frac{GmM_E}{r^2}$$

Then

$$v = \sqrt{\frac{GM_E}{r}} \qquad (7.25)$$

where $r = R_E + h$. For example, suppose that a satellite is in a circular orbit at an altitude of 500 km (about 300 mi); its tangential speed is

$$v = \sqrt{\frac{GM_E}{r}} = \sqrt{\frac{GM_E}{R_E + h}} = \sqrt{\frac{(6.67 \times 10^{-11} \text{ N} \cdot \text{m}^2/\text{kg}^2)(6.0 \times 10^{24} \text{ kg})}{6.4 \times 10^6 \text{ m} + 5.0 \times 10^5 \text{ m}}}$$

$$= 7.6 \times 10^3 \text{ m/s} = 7.6 \text{ km/s} \quad (\text{about } 4.7 \text{ mi/s})$$

This speed is about 27 000 km/h, or 17 000 mi/h. As can be seen from Eq. 7.25, the required circular orbital speed *decreases* with altitude.

In practice, a satellite is given a tangential speed by a component of the thrust from a rocket stage (▶Fig. 7.21a). The inverse-square relationship of Newton's law of gravitation means that the satellite orbits that are possible about a large mass are ellipses, of which a circular orbit is a special case. This condition is illustrated in Fig. 7.21b for the Earth, using the previously calculated values. If a satellite is not given a sufficient tangential speed, it will fall back to the Earth (and possibly be burned up while falling through the atmosphere). If the tangential speed reaches the escape speed, the satellite will leave its orbit and go off into space.

Finally, the total energy of an orbiting satellite in circular orbit is

$$E = K + U = \tfrac{1}{2}mv^2 - \frac{GmM_E}{r} \tag{7.26}$$

Substituting the expression for v from Eq. 7.25 into the kinetic energy term in Eq. (7.26) gives

$$E = \frac{GmM_E}{2r} - \frac{GmM_E}{r}$$

Thus,

$$E = -\frac{GmM_E}{2r} \qquad \begin{array}{l} \textit{total energy of} \\ \textit{an orbiting satellite} \end{array} \tag{7.27}$$

Note that the total energy of the satellite is negative: More work is required to put a satellite into a higher orbit, where it has more potential and total energy. The total energy E increases as its *numerical value* becomes smaller—i.e., less negative—as the satellite goes to a higher orbit toward the zero potential at the top of the well. That is, the farther a satellite is from Earth, the greater is its total energy.

To help understand why the total energy increases when its value becomes less negative, think of a change in energy from, say, 5.0 J to 10 J. This change would be considered an increase in energy. Similarly, a change from −10 J to −5.0 J would be an increase in energy, even though the *absolute* value has decreased:

$$\Delta U = U - U_o = -5.0 \text{ J} - (-10 \text{ J}) = +5.0 \text{ J}$$

The relationship of speed and energy to orbital radius is summarized in Table 7.3.

TABLE 7.3 Relationship of Radius, Speed, and Energy for Circular Orbital Motion

	Increasing r (larger orbit)	Decreasing r (smaller orbit)
ω	decreases	increases
v	decreases	increases
K	decreases	increases
U	increases (smaller negative value)	decreases (larger negative value)
$E \; (= K + U)$	increases (smaller negative value)	decreases (larger negative value)

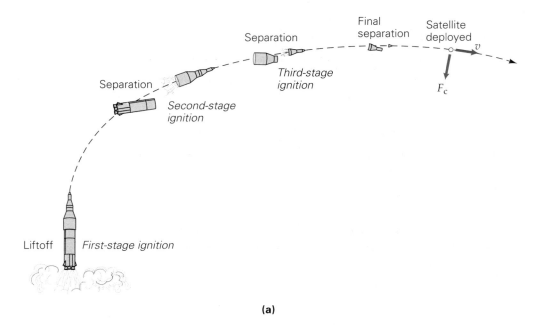

(a)

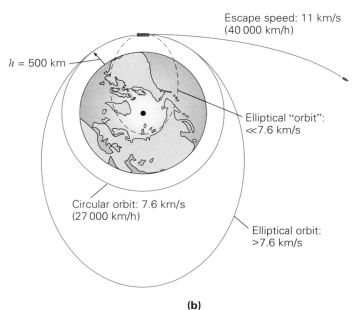

(b)

▲ **FIGURE 7.21 Satellite orbits** **(a)** A satellite is put into orbit by giving it a tangential speed sufficient for maintaining an orbit at a particular altitude. The higher the orbit, the smaller is the tangential speed. **(b)** At an altitude of 500 km, a tangential speed of 7.6 km/s is required for a circular orbit. With a tangential speed between 7.6 km/s and 11 km/s (the escape speed), the satellite would move out of the circular orbit. Since it would not have the escape speed, it would "fall" around the Earth in an elliptical orbit, with the Earth's center at one focal point. A tangential speed less than 7.6 km/s would also give an elliptical path about the center of the Earth, but because the Earth is not a point mass, a certain minimum speed would be needed to keep the satellite from striking the Earth's surface.

Also note from the development of Eq. 7.27 that the kinetic energy of an orbiting satellite is equal to the absolute value of the satellite's total energy:

$$K = \frac{GmM_{\mathrm{E}}}{2r} = |E| \qquad (7.28)$$

Conceptual Example 7.17 ■ Applying the Brakes to Speed Up: Conservation of Energy

A spacecraft is in a circular orbit about the Earth, well outside the atmosphere. Describe what happens to the spacecraft if its retrorockets are fired. (Retrorockets point in the direction of the spacecraft's motion and produce a reverse thrust.)

Reasoning and Answer.   The retrorockets apply a braking force, or reverse thrust, to the spacecraft. *Negative* work is done (because the force and displacement are in opposite directions), and the spacecraft loses some energy (from the work–energy theorem). That is, the total energy of the spacecraft,

Note: The work–energy theorem is discussed in Section 5.3.

$$E = -\frac{GmM_E}{2r}$$

is reduced. If E decreases (that is, it takes on a greater negative value), we can see from the foregoing equation that r must also decrease. Thus, the spacecraft will spiral inward toward the Earth. From Eq. 7.28,

$$K = \tfrac{1}{2}mv^2 = \frac{GmM_E}{2r} \qquad \text{or} \qquad v^2 = \frac{GM_E}{r}$$

So, if r decreases, v increases, and the spacecraft actually speeds up; even though both U and E decrease, K increases!

Follow-up Exercise.   Suppose the retrorockets of an orbiting spacecraft were fired until the craft was within the top layers of the Earth's atmosphere, where air resistance is appreciable. If the rockets were then shut off, what would happen to the spacecraft?

Reverse thrust, provided by the engines of docked cargo ships, was used to put the Russian space station *Mir* into lower orbits and ultimately led to its final destruction in March 2001. A final thrust sent the station into a decaying orbit and into our atmosphere. Most of the 120-ton *Mir* burned up in the atmosphere; however, some pieces did fall into the Pacific Ocean.

Conceptual Example 7.18 ■ Forward Thrust: A Change in Energy

The engines of a spacecraft in circular orbit about the Earth are fired so as to give the craft a *forward* thrust. How will the spacecraft's energy be changed after the firing, assuming that another circular orbit is achieved? (a) K increased and U decreased, (b) K decreased and U increased, (c) both K and U increased, or (d) the total energy decreased.

Reasoning and Answer.   You might think that a forward thrust would cause the spacecraft to speed up and increase its kinetic energy. However, an increase in v would require an increased centripetal force to maintain the spacecraft in orbit. This force is supplied by gravitational attraction and thus could increase only if r were reduced (that is, if the spacecraft moved to a smaller orbit)—something that would not result from the application of a forward thrust. (Draw a diagram to confirm this condition; in what direction would the thrust act with respect to the circular orbit?)

A forward thrust (in contrast to reverse thrust, described in Conceptual Example 7.17) will send the craft into a larger, rather than a smaller, orbit, because the forward thrust does positive work on the spacecraft, causing it to rise higher in its potential-energy well and thus increasing its potential energy. With an increase in r, the kinetic energy *decreases* (Eq. 7.27) while the total energy increases by becoming less negative (Eq. 7.27). Hence, the answer is (b). (In actuality, if the engine "burn" is brief, the spacecraft would go into an elliptical orbit, with its lowest altitude equal to the radius of the initial circular orbit.)

Follow-up Exercise.   Imagine yourself in a spacecraft in a circular orbit, well behind a space station, but in the same radial orbit. Wishing to dock with the station, what maneuvers would you execute to catch up with it?

The advent of the space age and the use of orbiting satellites have brought us the terms *weightlessness* and *zero gravity*, because astronauts appear to "float" about in orbiting spacecraft (▼Fig. 7.22a). However, the terms are misnomers. As mentioned earlier in the chapter, gravity is an infinite-range force, and Earth's gravity acts on a spacecraft and astronauts, supplying the centripetal force necessary to keep them in orbit. Gravity there is not zero, so there must be weight.

A better term to describe the floating effect of astronauts in orbiting spacecraft would be *apparent weightlessness*. The astronauts "float" because both the astronauts and the spacecraft are centripetally accelerating (or falling) toward the Earth at the same rate. To help you understand this effect, consider the analogous situation of a person standing on a scale in an elevator (Fig. 7.22b). The "weight" measurement that the scale registers is actually the normal force N of the scale on the person. In a nonaccelerating elevator ($a = 0$), we have $N = mg = w$, and N is

(a)

▶ **FIGURE 7.22 Apparent weightlessness** (a) An astronaut "floats" in a spacecraft, seemingly in a weightless condition. (b) In a stationary elevator (top), a scale reads the passenger's true weight. The weight reading is the reaction force N of the scale on the person. If the elevator is descending with an acceleration $a < g$ (middle), the reaction force and apparent weight are less than the true weight. If the elevator were in free fall ($a = g$; bottom), the reaction force and indicated weight would be zero, since the scale would be falling as fast as the person.

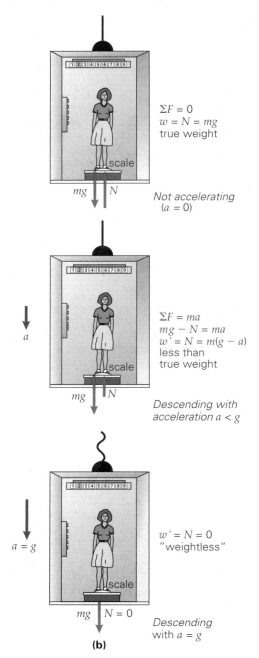

$\Sigma F = 0$
$w = N = mg$
true weight

Not accelerating
($a = 0$)

$\Sigma F = ma$
$mg - N = ma$
$w' = N = m(g - a)$
less than
true weight

Descending with
acceleration $a < g$

$w' = N = 0$
"weightless"

Descending
with $a = g$

(b)

equal to the true weight of the individual. However, suppose the elevator is descending with an acceleration a, where $a < g$. As the vector diagram in the figure shows,

$$mg - N = ma$$

and the *apparent* weight w' is

$$w' = N = m(g - a) < mg$$

where the downward direction is taken as positive in this instance. With a downward acceleration a, we see that N is less than mg, hence the scale indicates that the person weighs less than his or her true weight. Note that the *apparent acceleration* due to gravity is $g' = g - a$.

Now suppose the elevator were in free fall, with $a = g$. As you can see, N (and thus the apparent weight w') would be zero. Essentially, the scale is accelerating, or falling, at the same rate as the person. The scale may indicate a "weightless" condition ($N = 0$), but gravity still acts, as would be noted by the sudden stop at the bottom of the shaft. (See the Insight on the effects of "weightlessness" on the human body in space.)

INSIGHT

"Weightlessness": Effects on the Human Body

Astronauts spend weeks and months in orbiting spacecraft and space stations. Although gravity acts on them, the astronauts experience long durations of "zero gravity" (zero-g)*, due to the centripetal motion. On Earth, gravity provides the force that causes our muscles and bones to develop to the proper strength so we may function in our environment. That is, our muscles and bones must be strong enough for us to be able to walk, lift objects, and so on. And we exercise and eat properly to maintain our ability to function against the pull of gravity.

However, physical conditions are different in a zero-g environment. Muscle atrophy occurs quickly, because the body perceives no need for muscles. That is, muscles lose mass if there is no need for them to respond to gravity. In zero-g, muscle mass may deplete as much as 5 percent per week. Bone loss also occurs at a rate of about 1 percent per month. Models show that the total bone loss could reach 40 to 60 percent.

The circulatory system is affected too. On Earth, gravity causes blood to pool in our feet. When we stand, the blood pressure in our feet (about 200 mm Hg) is much greater than that in our heads (60—80 mm Hg), because of the head-to-toe gravity gradient. (See Section 9.2 for a discussion of the measurement of blood pressure.) In the zero-g experienced by astronauts, this gradient is not there, and the blood pressure equalizes throughout the body at about 100 mm Hg. This condition causes fluid to flow from the legs to the head, and the face gets puffy and the legs thin.

Even more serious, the condition of greater-than-normal blood pressure in the head is interpreted by the brain as indi-

cating that there is too much blood in the body, and blood production is signaled to slow down. Astronauts can lose up to 22 percent of their blood as a result of the equalized body pressure in zero-g. Also, with equalized blood pressure, the heart doesn't work as hard, and the heart muscles may atrophy.

All of these phenomena explain why astronauts undergo rigorous physical-fitness programs before going into space. On returning to Earth, their bodies have to readjust to a "9.8 m/s^2 g" environment. Each of the bodily losses requires different recovery times. Blood volume is typically restored in a few days, with astronauts drinking lots of liquids. Most muscles are regenerated in a month or so, depending on the length of stay in zero-g. Bone recovery takes much longer. Astronauts spending three to six months in space may require two or three years to regain the lost bone, if it is regained at all. Exercise and nutrition are very important in all the recovery processes.

Several devices have been developed to mimic gravity, which would allow astronauts to exercise in space. These devices range from a Russian method of providing "artificial gravity," using bungee cords on a treadmill, to the application of negative pressure to the lower part of the body to restore a pressure gradient in space. The effectiveness of such devices is still being studied.

There is much to learn about the effects of zero-g—or even reduced-g. Unmanned spacecraft have visited Mars, with the aim of one day sending astronauts to the Red Planet. This task would involve perhaps a six-month trip in zero-g and, on arrival, a Martian surface gravity that is only 38 percent of the Earth's gravity. No one yet understands completely the effects that such a space journey would have on an astronaut's body.

Related Exercises: 75 and 85

*This term will be used here for description, with the understanding it is "apparent" zero-g.

Space has been called the *final frontier*. Someday, instead of brief stays in Earth-orbiting spacecraft, we may have permanent space colonies with "artificial" gravity. One proposal is to have a huge, rotating space colony in the form of a wheel—somewhat like an automobile tire, with the inhabitants living inside the tire. As you know, centripetal force is necessary to keep an object in rotational circular motion. On the rotating Earth, that force is supplied by gravity, and we refer to it as *weight*. We exert a force on the ground, and the normal force (by Newton's third law) upward on our feet gives us the feeling of "having our feet on solid ground."

In a rotating space colony, the situation is somewhat reversed. The rotating colony would supply the centripetal force on the inhabitants, and the centripetal force would be perceived as a normal force sensation on the soles of our feet, or artificial gravity. Rotation at the proper speed would produce a simulation of "normal" gravity ($g = 9.80$ m/s^2) within the colony wheel. Note that in the colonists' world, "down" would be outward, toward the periphery of the space station, and "up" would always be inward, toward the axis of rotation.

Chapter Review

Important Concepts and Equations

- The **radian** (rad) is a measure of angle; 1 rad is the angle of a circle subtended by an arc length (s) equal to the radius (r):

  Arc Length (*angle in radians*):

  $$s = r\theta \qquad (7.3)$$

Angular Kinematic Equations (*for $\theta_o = 0$ and $t_o = 0$; see Table 7.2 for linear analogues*):

$$\theta = \overline{\omega}t \quad \begin{matrix}(\textit{general, not limited to}\\ \textit{constant acceleration})\end{matrix} \qquad (7.5)$$

$$\left.\begin{matrix}\overline{\omega} = \dfrac{w + w_o}{2} \\[2mm] w = w_o + \alpha t \\[2mm] \theta = w_o t + \frac{1}{2}\alpha t^2 \\[2mm] w^2 = w_o{}^2 + 2\alpha\theta \end{matrix}\right\} \begin{matrix}\textit{constant} \\ \textit{acceleration} \\ \textit{only}\end{matrix} \quad \begin{matrix}(2,\,\text{Table 7.2})\\[2mm](7.12)\\[2mm](4,\,\text{Table 7.2})\\[2mm](5,\,\text{Table 7.2})\end{matrix}$$

- **Tangential speed** (v) and **angular speed** (ω) for circular motion are directly proportional, with the radius r being the constant of proportionality:

  $$v = r\omega \qquad (7.6)$$

- The **frequency** (f) and **period** (T) are inversely related:

  $$f = \frac{1}{T} \qquad (7.7)$$

- **Angular speed** (*with uniform circular motion*) in terms of period (T) and frequency (f):

  $$\omega = \frac{2\pi}{T} = 2\pi f \qquad (7.8)$$

- In uniform circular motion, a **centripetal acceleration** (a_c) is required and is always directed toward the center of the circular path, and its magnitude is given by:

$$a_c = \frac{v^2}{r} = r\omega^2 \qquad (7.10)$$

- A **centripetal force**, F_c, (the net force directed toward the center of a circle) is a requirement for circular motion:

$$F_c = ma_c = \frac{mv^2}{r} \qquad (7.11)$$

- **Angular acceleration** (α) is the time rate of change of angular velocity and is related to the tangential acceleration (a_t) by

$$a_t = r\alpha \qquad (7.13)$$

- According to **Newton's law of gravitation**, every particle attracts every other particle in the universe with a force that is proportional to the masses of both particles and inversely proportional to the square of the distance between them:

$$F_g = \frac{Gm_1 m_2}{r^2} \qquad (7.14)$$

$$(G = 6.67 \times 10^{-11}\ \text{N} \cdot \text{m}^2/\text{kg}^2)$$

- *Acceleration due to gravity at an altitude h:*

$$a_g = \frac{GM_E}{(R_E + h)^2} \qquad (7.17)$$

- *Gravitational potential energy of two particles:*

$$U = -\frac{Gm_1 m_2}{r} \qquad (7.18)$$

- *Kepler's law of periods:*

$$T^2 = Kr^3 \qquad (7.22)$$

(*K depends on the mass of the object being orbited; for objects orbiting the Sun, $K = 2.97 \times 10^{-19}$ s^2/m^3.*)

- **Escape speed** (*from the Earth*) is

$$v_{esc} = \sqrt{\frac{2GM_E}{R_E}} = \sqrt{2gR_E} \qquad (7.23, 7.24)$$

- Earth's satellites are in a negative potential-energy well; the higher the object is in the well, the greater is the object's potential energy and the less is its kinetic energy.

- **Energy of a satellite orbiting Earth**

$$E = -\frac{GmM_E}{2r} \qquad (7.27)$$

$$K = |E| \qquad (7.28)$$

Exercises

7.1 Angular Measure

1. The radian unit is equivalent to a ratio of (a) degree/time, (b) length, (c) length/length, or (d) length/time.

2. A proper conversion factor for degrees to radians would be (a) $(\pi/7 \text{ rad})/270°$, (b) $(\pi/2 \text{ rad})/60°$, (c) $(3\pi \text{ rad})/540°$, or (d) $(0.50 \text{ rad})/180°$.

3. **CQ** A wheel rotates about a rigid axis through its center. Do all points on the wheel travel the same distance? How about the same *angular* distance?

4. ■ The Cartesian coordinates of a point on a circle are (1.5 m, 2.0 m). What are the polar coordinates (r, θ) of this point?

5. ■ The polar coordinates of a point are (5.3 m, 32°). What are the point's Cartesian coordinates?

6. ■ In Cartesian coordinates, the equation of a circle is $x^2 + y^2 = a^2$, where a is the radius of the circle. What is that equation in polar coordinates?

7. ■ Convert the following angles from degrees to radians, to three significant figures: (a) 15°, (b) 45°, (c) 90°, and (d) 120°.

8. ■ Convert the following angles from radians to degrees: (a) $\pi/6$ rad, (b) $5\pi/12$ rad, (c) $3\pi/4$ rad, and (d) π rad.

9. ■ You can determine the diameter of the Sun by measuring the angle subtended by it. If the angle is 0.535° and the average distance from the Earth to the Sun is 1.5×10^{11} m, what is the diameter of the Sun?

10. ■ You measure the length of a distant car to be subtended by an angular distance of 2.0°. If the car is actually 5.0 m long, approximately how far away is the car?

11. ■ A jogger on a circular track that has a radius of 0.250 km runs a distance of 1.00 km. What angular distance does the jogger cover in (a) radians and (b) degrees?

12. ■ Assume that the Earth's orbit around the Sun is circular. In kilometers, what is the approximate orbital distance the Earth travels in three months?

13. ■ Using the ratio equation $s/\theta = c/360°$, where c is the circumference of a circle, (a) show that, by definition, 1 rad is equivalent to 57.3°, and (b) show that 2π rad is equivalent to 360°.

14. ■■ The hour, minute, and second hands on a clock are 0.25 m, 0.30 m, and 0.35 m long, respectively. What are the distances traveled by the tips of the hands in a 30-min interval?

15. ■■ In Europe, a large circular walking track with a diameter of 0.900 km is marked in angular distances in radians. An American tourist who walks 3.00 mi daily goes to the track. How many radians should he walk per day to maintain his daily routine?

16. ■■ You ordered a 12-in. pizza for a party of five. For the pizza to be distributed evenly, how should it be cut in triangular pieces? (▼Fig. 7.23)?

▲ **FIGURE 7.23 Tough pizza to cut** See Exercise 16.

17. ■■ In degrees and radians, what is the angular width (a) of a full Moon as viewed from the Earth? (b) of a full Earth as viewed from the Moon? [*Hint*: Use data from Appendix III.]

18. IE ■■ To attend the 2000 Summer Olympics, a fan flew from Mosselbaai, South Africa (34°S, 22°E) to Sydney, Australia (34°S, 151°E). (a) What is the smallest angular distance the fan has to travel, (1) 34°, (2) 12°, (3) 117°, or (4) 129°? Why? (b) Determine the approximate shortest flight distance, in kilometers.

19. IE ■■■ (a) Could a circular pie be cut such that all of the wedge-shaped pieces have an arc length along the outer crust equal to the pie's radius? (b) If not, how many such pieces could you cut, and what would be the angular dimension of the final piece?

20. ■■■ Electrical wire with a diameter of 0.75 cm is wound on a spool with a radius of 30 cm and a height of 24 cm. (a) Through how many radians must the spool be turned to wrap one even layer of wire? (b) What is the length of this wound wire?

7.2 Angular Speed and Velocity

21. Viewed from above, a turntable is observed to rotate counterclockwise. The angular velocity vector is then (a) tangential to the turntable's rim, (b) out of the plane of the turntable, (c) counterclockwise, or (d) none of the preceding.

22. The frequency unit of hertz is equivalent to (a) that of the period, (b) that of the cycle, (c) radian/s, or (d) s^{-1}.

23. CQ Do all points on a wheel rotating about a fixed axis through its center have the same angular velocity? The same tangential speed? Explain.

24. CQ When "clockwise" or "counterclockwise" is used to describe rotational motion, why is a phrase such as "viewed from above" added?

25. CQ A wheel is spinning at a constant angular velocity about an axis through its center. Which point(s) on the wheel have the greatest and smallest tangential speeds? Which point(s) on the wheel have the greatest and smallest angular speeds?

26. ■ A computer DVD-ROM has a variable angular speed from 200 rpm to 500 rpm. Express this range of angular speed in radians per second.

27. ■ A satellite in a circular orbit has a period of 10 h. What is the satellite's frequency in revolutions per day?

28. ■ A race car makes two and a half laps around a circular track in 3.0 min. What is the car's average angular speed?

29. ■ If a particle is rotating with an angular speed of 3.5 rad/s, how long does it take for the particle to go through one revolution?

30. ■ What is the period of revolution for (a) a 12 000-rpm centrifuge and (b) a 10 000-rpm computer hard-disk drive?

31. ■■ Determine which has the greater angular speed: particle A, which travels 160° in 2.00 s, or particle B, which travels 4π rad in 8.00 s.

32. ■■ The tangential speed of a particle on a rotating wheel is 3.0 m/s. If the particle is 0.20 m from the axis of rotation, how long will it take for the particle to go through one revolution?

33. ■■ A merry-go-round makes 24 revolutions in a 3.0-min ride. (a) What is its average angular speed in rad/s? (b) What are the tangential speeds of two people 4.0 m and 5.0 m from the center, or axis of rotation?

34. ■■ In 1.5 min, a car moving at constant speed travels halfway around a circular track that has a diameter of 1.0 km. What are the car's (a) angular speed and (b) tangential speed?

35. IE ■■ The Earth rotates on its own axis once a day and revolves around the Sun once a year. (a) Which is greater, the rotating angular speed or the revolving angular speed? Why? (b) Calculate both angular speeds in rad/s.

36. ■■ A medical centrifuge is rated at 15 000 rpm. How long will it take the centrifuge to make one revolution?

37. ■■■ The driver of a car sets the cruise control and ties the steering wheel so that the car travels at a uniform speed of 15 m/s in a circle with a diameter of 120 m. (a) Through what angular distance does the car move in 4.00 min? (b) What linear distance does it travel in this time?

7.3 Uniform Circular Motion and Centripetal Acceleration

38. In uniform circular motion, there is a (a) constant velocity, (b) constant angular velocity, (c) zero acceleration, or (d) nonzero tangential acceleration.

39. If the centripetal force on a particle in uniform circular motion is increased, (a) the tangential speed will remain constant, (b) the tangential speed will decrease, (c) the radius of the circular path will increase, or (d) the tangential speed will increase and/or the radius will decrease.

40. CQ Explain why mud flies off a fast-spinning wheel.

41. CQ Is it possible for an object in uniform circular motion to have nonzero tangential acceleration, yet zero centripetal acceleration? Explain.

42. CQ The spin cycle of a washing machine is used to extract water from recently washed clothes. Explain the physical principle(s) involved here.

43. CQ The apparatus illustrated in ▼Fig. 7.24 is used to demonstrate forces in a rotating system. The floats are in jars of water. When the arm is rotated, which way will the floats move? (Compare this apparatus with the accelerometer shown in Fig. 4.27.) Does it make a difference which way the arm is rotated?

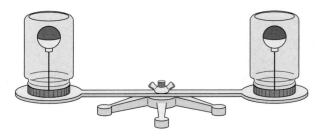

▲ **FIGURE 7.24   When set in motion, a rotating system** See Exercise 43.

44. CQ When rounding a curve in a fast-moving car, we experience a feeling of being thrown outward (▶Fig. 7.25). It

▲ **FIGURE 7.25   A center-fleeing force?**   See Exercise 44.

is sometimes said that this effect occurs because of an outward centrifugal (center-fleeing) force. However, in terms of Newton's laws, this pseudo, or false, force doesn't really exist. Analyze the situation in the figure to show that this is the case (i.e., that the force does not exist). [*Hint*: Start with Newton's first law.]

45. **CQ** Many racetracks have banked turns, which allow the cars to travel faster around the curves than if the curves were flat. Actually, cars could also make turns on these banked curves if there were no friction at all. Explain this statement using the free-body diagram shown in ▼Fig. 7.26.

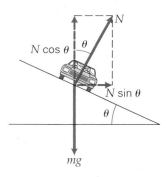

▲ **FIGURE 7.26   Banking safety**   See Exercises 45 and 59.

46. **CQ** Two students are about to ride a roller-coaster ride, but are arguing about one thing: One student says that it is not necessary to wear a safety harness when the roller coaster goes upside down, while the other insists that wearing a safety harness is required throughout the ride. Who is correct? Explain.

47. ■ A rotating cylinder about 16 km in length and 8.0 km in diameter is designed to be used as a space colony. With what angular speed must it rotate so that the residents on it will experience the same acceleration due to gravity on Earth?

48. ■ An Indy car with a speed of 120 km/h goes around a level, circular track with a radius of 1.00 km. What is the centripetal acceleration of the car?

49. ■ A wheel of radius 1.5 m rotates at a uniform speed. If a point on the rim of the wheel has a centripetal acceleration of 1.2 m/s², what is the point's tangential speed?

50. ■■ The Moon revolves around the Earth in 27.3 days in a nearly circular orbit with a radius of 3.8×10^5 km. Assuming that the Moon's orbital motion is a uniform circular motion, what is the Moon acceleration as it "falls" toward the Earth?

51. ■■ A car with a constant speed of 83.0 km/h enters a circular flat curve with a radius of curvature of 0.400 km. If the friction between the road and the car's tires can supply a centripetal acceleration of 1.25 m/s², does the car negotiate the curve safely? Justify your answer.

52. **IE** ■■ Imagine that you swing about your head a ball attached to the end of a string. The ball moves at a constant speed in a horizontal circle. (a) Can the string be exactly horizontal? Why? (b) If the mass of the ball is 0.250 kg, the radius of the circle is 1.50 m, and it takes 1.20 s for the ball to make one revolution, what is the ball's tangential speed? (c) What centripetal force are you imparting to the ball via the string?

53. ■■ In Exercise 52, if you supplied a tension force of 12.5 N to the string, what angle would the string make relative to the horizontal?

54. **IE** ■■ A student is to swing a bucket of water in a vertical circle without spilling any (▼Fig. 7.27). (a) Explain how this task is possible. (b) If the distance from his shoulder to the center of mass of the bucket of water is 1.0 m, what is the minimum speed required to keep the water from coming out of the bucket at the top of the swing?

▲ **FIGURE 7.27   Weightless water?**   See Exercise 54.

55. ■■ A car of mass 1.5×10^3 kg rounds a circular turn of radius 20 m. If the road is flat and the coefficient of static friction between the tires and the road is 0.50, how fast can the car travel without skidding?

56. **IE** ■■ A jet pilot puts an aircraft with a constant speed into a vertical circular loop. (a) Which is greater, the normal force exerted on the seat by the pilot at the bottom of

the loop or that at the top of the loop? Why? (b) If the speed of the aircraft is 700 km/h and the radius of the circle is 2.0 km, calculate the normal forces exerted on the seat by the pilot at the bottom and top of the loop. Express your answer in terms of the pilot's weight *mg*.

57. ■■■ A block of mass *m* slides down an inclined plane into a loop-the-loop of radius *r* (▼Fig. 7.28). (a) Neglecting friction, what is the minimum speed the block must have at the highest point of the loop in order to stay in the loop? [*Hint*: What force must act on the block at the top of the loop to keep the block on a circular path?] (b) At what vertical height on the inclined plane (in terms of the radius of the loop) must the block be released if it is to have the required minimum speed at the top of the loop?

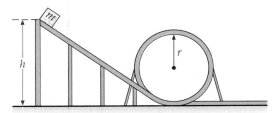

▲ **FIGURE 7.28 Loop-the-loop** See Exercise 57.

58. ■■■ For a scene in a movie, a stunt driver drives a 1.50×10^3-kg pickup truck with a length of 4.25 m around a circular curve with a radius of curvature of 0.333 km (▼Fig. 7.29). The truck is to curve off the road, jump across a gully 10.0 m wide, and land on the other side 2.96 m below the initial side. What is the minimum centripetal acceleration the truck must have in going around the circular curve to clear the gully and land on the other side?

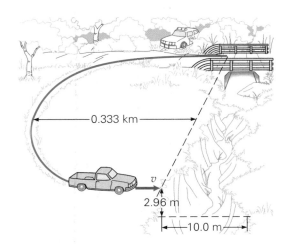

▲ **FIGURE 7.29 Over the gully** See Exercise 58.

59. ■■■ For the banked curve in Fig. 7.26, the horizontal component of the normal force toward the center of the curve reduces the need for friction to prevent skidding. In fact, for a circular curve banked at a given angle, there is one speed for which no frictional force is required—

the centripetal force is supplied completely by the inward component of the normal force. This relationship is used in designing banked curves. (a) Show that the banking angle θ for which no friction is required for a car to negotiate a circular curve safely is given by $\tan \theta = v^2/(gr)$, where *v* is the car's speed and *r* is the radius of curvature. (b) How would the banking angle differ for a more massive truck? (c) Show how friction affects the situation.

60. IE ■■■ In a theme park, you walk into a giant can-shaped room that can rotate. When the angular speed of the "room" reaches a certain value, the bottom "floor" drops out (▼Fig. 7.30). However, you don't fall down! Instead, you "stick" to the wall. (a) Explain why in terms of the forces involved. (b) If the coefficient of static friction between you and the side wall of the curved room is 0.30 and the radius of the room is 2.5 m, what is the minimum rotational speed of the "can" so that you can "stick" to the wall of the bottomless room?

▲ **FIGURE 7.30 Bottomless room** See Exercise 60.

7.4 Angular Acceleration

61. The angular acceleration in circular motion (a) is equal in magnitude to the tangential acceleration divided by the radius, (b) increases the angular velocity if both angular velocity and angular acceleration are in the same direction, (c) has units of s^{-2}, or (d) all of the preceding.

62. CQ A car increases its speed when it is on a circular track. Does the car have centripetal acceleration? How about angular acceleration? Explain.

63. CQ Is it possible for a car traveling on a circular track to have angular acceleration, but not centripetal acceleration?

64. ■ During an acceleration, the angular speed of an engine increases from 700 rpm to 3000 rpm in 3.0 s. What is the average angular acceleration of the engine?

65. ■ A merry-go-round accelerating uniformly from rest achieves its operating speed of 2.5 rpm in five revolutions. What is the magnitude of its angular acceleration?

66. ■■ A $33\frac{1}{3}$-rpm record on a turntable uniformly reaches its operating speed in 2.45 s once the record player is turned on. (a) What is the angular distance traveled during this time? (b) What is the corresponding arc length, in feet, on the circumference of a 12-in.-diameter record?

67. ■■ A bicycle being repaired is turned upside down, and one wheel is rotated at a rate of 60 rpm. If the wheel slows uniformly to a stop in 15 s, how many revolutions does it make during this time?

68. ■■ A Ferris wheel with a diameter of 35.0 m starts from and achieves its maximum operational tangential speed of 2.20 m/s in a time of 15.0 s. (a) What is the magnitude of the wheel's angular acceleration? (b) What is the magnitude of the tangential acceleration after the maximum operational speed is reached?

69. ■■ The blades of a fan running at low speed turn at 250 rpm. When the fan is switched to high speed, the rotation rate increases uniformly to 350 rpm in 5.75 s. (a) What is the magnitude of the angular acceleration of the blades? (b) How many revolutions do the blades go through while the fan is accelerating?

70. ■■ In landing, the rotor of a helicopter slows from an angular speed of 4500 rpm to rest in 5.0 s. How many revolutions does the rotor travel during this time?

71. IE ■■ A car on a circular track accelerates from rest. (a) The car experiences (1) only angular acceleration, (2) only centripetal acceleration, or (3) both angular and centripetal accelerations. Why? (b) If the radius of the track is 0.30 km and the magnitude of the constant angular acceleration is 4.5×10^{-3} rad/s², how long does it take the car to make one lap around the track? (c) What is the total (vector) acceleration of the car when it has completed half of a lap?

72. ■■■ A student investigating circular motion places a dime 10 cm from the center of a $33\frac{1}{3}$-rpm record on a turntable. The record player can accelerate at 1.42 rad/s². The student notes that the dime slides outward 2.25 s after she has switched on the turntable. (a) Why does the dime slide outward? (b) What is the coefficient of static friction between the dime and the record?

7.5 Newton's Law of Gravitation

73. The gravitational force is (a) a linear function of distance, (b) an inverse function of distance, (c) an inverse function of distance squared, or (d) sometimes repulsive.

74. The acceleration due to gravity of an object on the Earth's surface (a) is a universal constant, like G; (b) does not depend on the Earth's mass; (c) is directly proportional to

the Earth's radius; or (d) does not depend on the object's mass.

75. CQ Astronauts in a spacecraft orbiting the Earth or out for a "space walk" (▼Fig. 7.31) are seen to "float" in midair. This phenomenon is sometimes referred to as *weightlessness* or *zero gravity* (zero-*g*). Are these terms correct? Explain why an astronaut appears to float in or near an orbiting spacecraft.

▲ **FIGURE 7.31** **Out for a walk** Why does this astronaut seem to "float"? See Exercise 75.

76. CQ If the cup in ▼Fig. 7.32 were dropped, no water would run out. Explain.

▲ **FIGURE 7.32** **Let it go** See Exercise 76.

77. CQ Can you determine the mass of the Earth simply by measuring the gravitational acceleration near the Earth's surface? If yes, give the details.

78. CQ (a) A spring scale calibrated in kilograms on the Earth is taken to the Moon to make measurements. Will the scale read correctly? Explain. (b) For a given mass, will the scale's reading be greater or less at the Earth's equator than at the North Pole? Explain.

79. ■ From the known mass and radius of the Moon (see the tables inside the back cover of the book), compute the value of the acceleration due to gravity, g_M, at the surface of the Moon.

80. ■ Calculate the gravitational force between the Earth and the Moon.

81. **IE** ■■ A solar eclipse occurs when the Moon is between the Sun and the Earth, and a lunar eclipse occurs when the Earth is between the Sun and the Moon. (a) Is the gravitational force exerted on the Moon by the Earth and the Sun greater during a solar eclipse or during a lunar eclipse? Why? (b) Calculate the two forces.

82. ■■ For a spacecraft going directly from the Earth to the Moon, beyond what point will lunar gravity begin to dominate? That is, where will the lunar gravitational force be equal in magnitude to the Earth's gravitational force? Are the astronauts onboard a spacecraft at this point truly weightless?

83. ■■ Four identical masses of 2.5 kg each are located at the corners of a square with 1.0-m sides. What is the net force on any one of the masses?

84. ■■ What is the acceleration due to gravity on the top of Mt. Everest? The summit is about 8.80 km above sea level. (Report to three significant figures.)

85. ■■ A 75-kg man weighs 735 N on the Earth's surface. How far above the surface of the Earth would he have to go to "lose" 10% of his body weight?

86. **IE** ■■ Two objects are attracting each other with a certain gravitational force. (a) If the distance between the objects is reduced to half, the new gravitational force will (1) increase by a factor of 2, (2) increase by a factor of 4, (3) decrease by a factor of 2, or (4) decrease by a factor of 4. Why? (b) If the original force between the two objects is 0.90 N, and the distance is tripled, what is the new gravitational force between the objects?

87. ■■ Which is greater, the gravitational force exerted on the Earth by the Sun or that exerted on the Earth by the Moon? (Compare these attractions by forming a ratio and giving a factor of how many times greater or smaller one is than the other.)

88. ■■ Assuming that we could obtain enough soil to add a uniform outer layer onto the Earth, approximately how thick would this layer have to be to have $g = 10.0 \text{ m/s}^2$ exactly? (Take the average density of the Earth to be 5.52 g/cm^3 and $R_E = 6.40 \times 10^3 \text{ km}$.)

89. ■■■ (a) What is the mutual gravitational potential energy of the configuration shown in ▶ Fig. 7.33 if all the masses are 1.0 kg? (b) What is the gravitational force per unit mass at the center of the configuration?

7.6 Kepler's Laws and Earth's Satellites

90. A new planet is discovered and its period determined. The new planet's distance from the Sun could then be found by using Kepler's (a) first law, (b) second law, or (c) third law.

91. As a planet moves in its elliptical orbit, (a) its speed is constant. (b) its distance from the Sun is constant, (c) it moves faster when it is closer to the Sun, or (d) it moves slower when it is closer to the Sun.

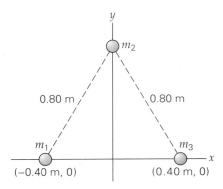

▲ **FIGURE 7.33 Gravitational potential, gravitational force, and center of mass** See Exercise 89.

92. **CQ** (a) In one revolution, how much work does the centripetal force do on a satellite in circular orbit about the Earth? (b) A person in a freely falling elevator thinks he can avoid injury by jumping upward just before the elevator strikes the floor. Would this strategy work?

93. **CQ** (a) In putting a satellite into orbit, over the equator, should the rocket be launched eastward or westward? Why? (b) In the United States, such satellites are launched from Florida. Why not from California, which generally has better weather conditions?

94. **CQ** As you pilot your spaceship in orbit, you see a piece of equipment ahead of you in the same orbit. (a) Can you speed up your spaceship with one rocket burst in order to catch the equipment? Explain. (b) What do you have to do to catch the equipment? [*Hint*: See Conceptual Example 7.17.]

95. ■ An instrument package is projected vertically upward to collect data at the top of the Earth's atmosphere (at an altitude of about 800 km). (a) What initial speed is required at the Earth's surface for the package to reach this height? (b) What percentage of the escape speed is this initial speed?

96. ■■ Using a development similar to Kepler's law of periods for planets orbiting the Sun, find the required altitude of geosynchronous satellites above the Earth. [*Hint*: The period of such satellites is the same as that of the Earth.]

97. ■■ Venus has a rotational period of 243 days. What would be the altitude of a synchronous satellite for this planet (similar to geosynchronous satellite on the Earth)?

98. ■■ The asteroid belt that lies between Mars and Jupiter may be the debris of a planet that broke apart or that was not able to form as a result of Jupiter's strong gravitation. The asteroid belt has a period of about 5.0 years. Approximately how far from the Sun would this "fifth" planet have been?

Additional Exercises

99. At sunset, the Sun has an angular width of about 0.50°. From the time the lower edge of the Sun just touches the

horizon, about how long does it take the Sun, in minutes, to disappear?

100. CQ Ocean tides are produced primarily by the gravitational attraction of the Moon. Explain how this attraction gives rise to two tidal "bulges" on opposite sides of the Earth, resulting in two daily high tides and two daily low tides (▼Fig. 7.34). [*Hint*: Consider the inverse-square relationship of the distance in the gravitational force acting on the water on opposite sides of the Earth and on the Earth itself.] (You may want to go to the library for a little help on this one.)

(a)   **(b)**

▲ **FIGURE 7.34 Tides** **(a)** Low tide and **(b)** high tide at the same location off the California coast. See Exercise 100.

101. A 60-kg woman prepares to swing out over a pond on a rope (▼Fig. 7.35). What tension in the rope is required if she is to maintain a circular arc (a) when she starts her swing by stepping off the platform, (b) when she is 5.0 m above the surface of the pond, and (c) at the lowest point

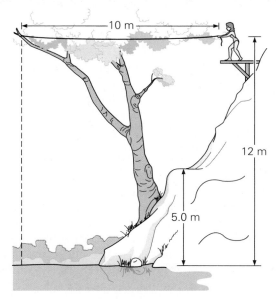

▲ **FIGURE 7.35 A swinging time** See Exercise 101.

of her swing? [*Hint*: Treat the woman and the rope as a simple pendulum.]

102. CQ When a candle is lit in an orbiting spaceship, it quickly goes out. Why?

103. An ultracentrifuge (a very fast centrifuge) operates at a speed of 5.00×10^6 rpm. It contains test tubes filled with viruses. (a) What is the centripetal acceleration on a virus at a radial distance of 4.00 cm from the centrifuge's axis of rotation? (b) How does this acceleration compare with g, the acceleration due to gravity?

104. A garden hose wound on a rotating storage cylinder has an outer layer at a distance of 0.45 m from the center of the cylinder. As this layer of wound hose is pulled out, the cylinder makes three revolutions. What is the length of the unwound hose?

105. What is the escape speed of a satellite at an altitude of 750 km in a circular orbit about the Earth?

106. An astronaut has a weight of 735 N on Earth. What would be the force of gravity on the astronaut in a spacecraft in a circular orbit at an altitude of 450 km?

107. The value of the Kepler constant is $K = 2.97 \times 10^{-19}$ s²/m³, and $T^2 = (2.97 \times 10^{-19}$ s²/m³$)r^3$. With nonstandard units, K can be made equal to unity, and the equation can be written as $T^2 = r^3$. Using data on the Earth (from the tables on the inside back cover of the book), find the units of K that make it equal to unity. [*Hint*: The average, or mean, distance of the Earth from the Sun is used in astronomy as a unit called an *astronomical unit* (AU).]

108. A pendulum swinging in a circular arc under the influence of gravity, as shown in ▼Fig. 7.36, has both centripetal and tangential components of acceleration. (a) If the pendulum bob has a speed of 2.7 m/s when the cord makes an angle of $\theta = 15°$ with the vertical, what are the magnitudes of the components at this time? (b) Where is the centripetal acceleration a maximum? What is the value of the tangential acceleration at that location?

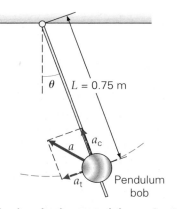

▲ **FIGURE 7.36 A swinging pendulum** See Exercise 108.

Rotational Motion and Equilibrium

INSIGHTS

- The Leaning Tower of Pisa
- Stability in Action
- Slide or Roll to a Stop?—Antilock Brakes

It's always a good idea to keep your equilibrium—but it's more important in some situations than in others! When you look at a photo like this, your first reaction is probably to wonder how these acrobats keep from falling. Presumably, the poles must help—but in what way? If you think about the picture and about what you have already learned, a less obvious, but in some ways more interesting, question may come to mind: Why do we think they're in danger in the first place? After all, the wire is strong enough to support their weight.

We might say that the acrobats are in equilibrium. Translational equilibrium ($\Sigma \mathbf{F}_i = 0$) was discussed in Chapter 4, but here there is another consideration: rotation. Should the performers start to fall (and we hope they won't), there would initially be a sideways rotation about the wire. To avoid this calamity, another condition must be met: rotational equilibrium, which we will discuss in this chapter.

These tightrope riders are striving to avoid rotational motion. But rotational motion is very important in physics, because rotating objects are all around us: wheels on vehicles (like the ones on the bicycles in the photo), gears and pulleys in machinery, planets in our solar system, and even many bones in the human body. (Can you think of a few bones that rotate in sockets?)

Fortunately, the equations describing rotational motion can be written as almost direct analogues of those for translational (linear) motion. In Chapter 7, this similarity was pointed out with respect to the linear and angular kinematic equations. With the addition of equations describing rotational dynamics, you will be able to analyze the general motions of real objects.

8.1 Rigid Bodies, Translations, and Rotations

OBJECTIVES: To **(a)** distinguish between pure translational and pure rotational motions of a rigid body and **(b)** state the condition(s) for rolling without slipping.

In previous chapters, it was convenient to consider motions of objects with the understanding that the motion of an object can be represented by a particle located at the center of mass of the object. Rotation, or spinning, was not a consideration then, because a particle, a point mass, has no physical dimensions. Rotational motion becomes relevant when we analyze the motion of solid extended objects or *rigid bodies*, the focus of this chapter.

| A **rigid body** is an object or a system of particles in which the distances between particles are fixed and remain constant.

A quantity of liquid water is not a rigid body, but the ice that would form if the water were frozen would be. The discussion of rigid body rotation is thus restricted to solids. Actually, the concept of a rigid body is an idealization. In reality, the particles (atoms and molecules) of a solid vibrate constantly. Also, solids can undergo elastic (and inelastic) deformations (Chapter 6). Even so, most solids can be considered to be rigid bodies for purposes of analyzing rotational motion.

A rigid body may be subject to either or both of two types of motions: translational and rotational. Translational motion is basically the linear motion we studied in previous chapters. If an object has only **translational motion** (▼Fig. 8.1a), every particle in it has the same instantaneous velocity, which means that the object is not rotating. (Why?)

An object may have only **rotational motion**—motion about a fixed axis (Fig. 8.1b). In this case, all of the particles of the object have the same instantaneous angular velocity and travel in circles about the axis of rotation. Although the axis of

Note: The words "rotation" and "revolution," are commonly used synonymously. In general, this book uses "rotation" when the axis of rotation goes through the body (e.g., the Earth's rotation on its axis, in a period of 24 h) and "revolution" when the axis is outside the body (e.g., the revolution of the Earth about the Sun, in a period of 365 days).

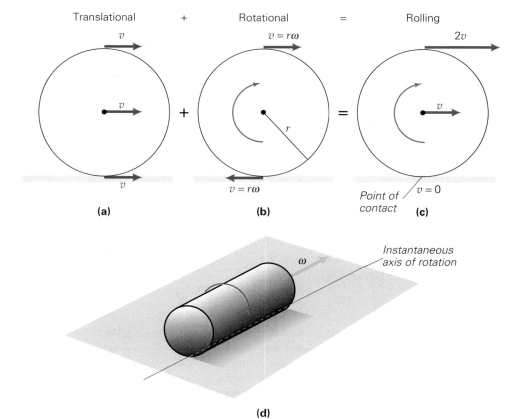

◀ **FIGURE 8.1   Rolling—a combination of translational and rotational motions** **(a)** In pure translational motion, all the particles of an object have the same instantaneous velocity. **(b)** In pure rotational motion, all the particles of an object have the same instantaneous angular velocity. **(c)** Rolling is a combination of translational and rotational motions. Summing the velocity vectors for these two motions shows that the point of contact (for a sphere) or the line of contact (for a cylinder) is instantaneously at rest. **(d)** The line of contact (or, for a sphere, a line through the point of contact) is called the *instantaneous axis of rotation*. Note that the center of mass of a rolling object on a level surface moves linearly and remains over the point or line of contact.

rotation of an object is commonly taken through the object's center of mass, this is not always the case. For example, you might pivot a meterstick through one end and have rotational motion about an axis through that end.

General rigid-body motion is a combination of translational and rotational motions. When you throw a ball, the translational motion is described by the motion of its center of mass (as in projectile motion). But the ball may also spin, or rotate, and usually does. A common example of rigid-body motion involving both translation and rotation is rolling, as illustrated in Fig. 8.1c. The combined motion of any point or particle is given by the vector sum of the particle's instantaneous velocity vectors. (Three points or particles are shown in the figure— one at the top, one in the middle, and one at the bottom, of the object.) At each instant, a rolling object rotates about an **instantaneous axis of rotation** through the point of contact of the object with the surface it is rolling on (for a sphere) or along the line of contact of the object with the surface (for a cylinder, Fig. 8.1d). The location of this axis changes with time. However, note in Fig.8.1c that the point or line of contact of the body with the surface is instantaneously at rest (and thus has zero velocity), as can be seen from the vector addition of the combined motions at that point. Also, the point on the top has twice the tangential speed ($2v$) of the middle (center-of-mass) point (v), because the top point is twice as far away from the instantaneous axis of rotation as the middle point is. (With a radius r, for the middle point, $r\omega = v$, and for the top point, $2r\omega = 2v$.)

When an object rolls without slipping (which was the case for the spheres and cylinder in Fig. 3.1), its translational and rotational motions are related simply as discussed in Chapter 7. For example, when a uniform ball (or cylinder) rolls in a straight line on a flat surface, it turns through an angle θ, and a point on the object that was initially in contact with the surface moves through an arc distance s (◄Fig. 8.2). From Chapter 7, $s = r\theta$ (Eq. 7.3). The center of mass of the ball is directly over the point of contact and moves a linear distance s. Then

$$v_{CM} = \frac{s}{t} = \frac{r\theta}{t} = r\omega$$

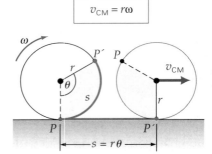

$$v_{CM} = r\omega$$

▲ **FIGURE 8.2 Rolling without slipping** As an object rolls without slipping, the length of the arc between two points of contact on the circumference is equal to the linear distance traveled. (Think of paint coming off a roller.) This distance is $s = r\theta$. The speed of the center of mass is $v_{CM} = r\omega$.

where $\omega = \theta/t$. In terms of the speed of the center of mass and the angular speed ω, the *condition for rolling without slipping* is

$$v_{CM} = r\omega \quad \text{(rolling, no slipping)} \qquad (8.1)$$

The condition for rolling without slipping is also expressed by

$$s = r\theta \qquad (8.1a)$$

where s is the distance the object rolls (the distance the center of mass moves).

Carrying Eq. 8.1 one step further, we can write an expression for the time rate of change of the velocity. Assuming it started from rest, $\Delta v_{CM} = v_{CM}/t = (r\omega)/t$. This yields an equation for *accelerated rolling without slipping*:

$$a_{CM} = \frac{v_{CM}}{t} = \frac{r\omega}{t} = r\alpha \qquad (8.1b)$$

where $\alpha = \omega/t$ (for a constant α).

PHYSLET®
ILLUSTRATION

Rolling Motion

Example 8.1 ■ Slipping or Not Slipping? That Is the Question

A cylinder with a radius of 12 cm rolls with an instantaneous angular speed of 0.75 rad/s down an inclined plane. (a) If the center of mass of the cylinder travels at a speed of 0.10 m/s at this time, is the cylinder rolling without slipping? (b) The cylinder then goes onto a level surface and rolls without slipping with the same speed of the center of mass as in part (a). Assuming that this speed remains constant for 2.0 s, through what angle does the cylinder turn during that time?

Thinking It Through. Part (a) is a simple test of Eq. 8.1, which must be satisfied for rolling without slipping. In part (b), Eq. 8.1 is valid and can be used to find the angular speed. With the speed and time, we can find the angular distance—that is, the angle through which the cylinder rotated.

Solution. We list the data as usual:.

Given: $r = 12$ cm $= 0.12$ m   Find: (a) if cylinder is slipping
$\omega = 0.75$ rad/s   (b) θ (angle of rotation)
$v_{CM} = 0.10$ m/s
$t = 2.0$ s

(a) If the data satisfy Eq. 8.1, the cylinder will be rolling without slipping; thus,

$$v_{CM} = r\omega = (0.12 \text{ m})(0.75 \text{ rad/s}) = 0.090 \text{ m/s} \neq 0.10 \text{ m/s}$$

Hence, the cylinder is slipping as it goes down the plane. The center of mass is traveling too fast for the nonslipping, rolling condition.
(b) For rolling without slipping on the level surface, the angular speed is given by Eq. 8.1:

$$\omega = \frac{v_{CM}}{r} = \frac{0.10 \text{ m/s}}{0.12 \text{ m}} = 0.83 \text{ rad/s}$$

Since this speed is constant the entire interval, we have

$$\theta = \omega t = (0.83 \text{ rad/s})(2.0 \text{ s}) = 1.7 \text{ rad}$$

The cylinder makes a little over one-quarter of a rotation. (Right?) Check it yourself.

Follow-up Exercise. How far does the cylinder travel linearly in part (b) of this Example? Find the distance by using two different methods, translational and rotational. (*Answers to all Follow-up Exercises are at the back of the text.*)

Conceptual Example 8.2 ■ On a Roll, but How Far?

One end of a ruler is placed on the side of a soda can, and the other end is held horizontally so that the ruler is level (▶Fig. 8.3). With a slight pressure on the ruler so that there is no slipping, the ruler is moved over the can, thereby causing the can to roll uniformly on the horizontal surface until the other end of the ruler is on the top of the can. Will the can roll (a) a distance twice the length of the ruler, (b) the same distance as the length of the ruler, or (c) half the distance of the length of the ruler?

Reasoning and Answer. Since there is no slipping between the ruler and can, the answer can't be (a). Instead, the can would roll such that the length of its circumference in contact with the can would be the same as the ruler's length. This might make you think that (b) is the answer, but let's look at the situation more closely.
 The ruler moves with the same speed as that of a point on top of the can. The contact line between the can and the surface is instantaneously at rest and is the instantaneous axis of rotation. Then, a point on top of the can would have a tangential velocity with a magnitude of $v_{top} = 2R\omega$. This is twice the magnitude of the velocity of the center point (the center of mass), $v_{CM} = R\omega$. Therefore, during the duration of the roll, the center of mass of the can moves half the distance that the top of the can moves (the length of the ruler), and the answer is (c). (Get a ruler and a can and try it yourself!)

Follow-up Exercise. Suppose a can with a larger radius were used. Would this have any effect on the result?

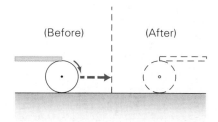

(Before)   (After)

▲ **FIGURE 8.3 Rolling on** See Conceptual Example 8.2.

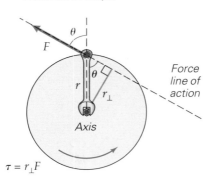

$$\tau = r_\perp F$$

(a) Counterclockwise torque

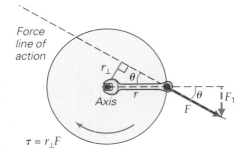

$$\tau = r_\perp F$$

(b) Smaller clockwise torque

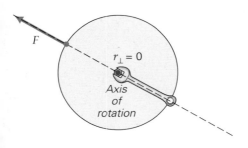

(c) Zero torque

▲ **FIGURE 8.4 Torque and moment arm (a)** The perpendicular distance $r_\perp$ from the axis of rotation to the line of action of a force is called the *moment arm* (or *lever arm*) and is equal to $r \sin \theta$. The torque, or twisting force, that produces rotational motion is $\tau = r_\perp F$. **(b)** The same force in the opposite direction with a smaller moment arm produces a smaller torque in the opposite direction. Note that $r_\perp F = rF_\perp$, or $(r \sin \theta)F = r(F \sin \theta)$. **(c)** When a force acts through the axis of rotation, $r_\perp = 0$ and $\tau = 0$.

8.2 Torque, Equilibrium, and Stability

OBJECTIVES: To **(a)** define torque, **(b)** apply the conditions for mechanical equilibrium, and **(c)** describe the relationship between the location of the center of gravity and stability.

Torque

As with translational motion, a force is necessary to produce a change in rotational motion. The rate of change of motion depends not only on the magnitude of the force, but also on the perpendicular distance of its line of action from the axis of rotation, $r_\perp$ (◄Fig. 8.4a, b). The line of action of a force is an imaginary line extending through the force vector arrow—that is, the line along which the force acts.

Figure 8.4 shows that $r_\perp = r \sin \theta$, where r is the straight-line distance between the axis of rotation and the point at which the force acts and θ is the angle between the line of r and the force vector **F**. This perpendicular distance $r_\perp$ is called the **moment arm** or **lever arm**.

The product of the force and the lever arm is called **torque** τ (from the Latin *torquere*, meaning "to twist"). The magnitude of the torque provided by the force is

$$\tau = r_\perp F = rF \sin \theta \tag{8.2}$$

SI unit of torque: meter-newton (m · N)

(The symbolism $r_\perp F$ is commonly used to denote torque, but note from Fig. 8.4 that $r_\perp F = rF_\perp$.) The SI units of torque are meters × newtons (m · N), the same as the units of work, $W = Fd$ (N · m, or J). However, we will write the units of torque in reverse order as m · N to avoid confusion. Keep in mind, though, that torque is *not* work.

Rotational acceleration is not *always* produced when a force acts on a stationary rigid body. From Eq. 8.2, we see that $\tau = 0$ when $\theta = 0°$—that is, when the force acts through the axis of rotation (Fig. 8.4c). Also, when $\theta = 90°$, the torque is maximum and the force acts perpendicularly to r, just as when you push a door open. Thus, the angular acceleration depends on *where* a perpendicular force is applied (and therefore on the length of the lever arm). As a practical example, think of applying a force to a heavy glass door that swings in and out. Where you apply the force makes a great difference in how easily the door opens or rotates (through the hinges) on its axis. Have you ever tried to open such a door and inadvertently pushed on the side near the hinges? This force produces a small torque and little or no rotational acceleration.

We can think of torque in rotational motion as the analogue of force in translational motion. An unbalanced or net force changes translational motion, and an unbalanced or net torque changes rotational motion. Torque, the product of a force and a moment arm—both vectors—is itself a vector. Its direction is always perpendicular to the plane of the force and moment arm vectors and is given by a *right-hand rule* similar to that for angular velocity given in Section 7.2. If the fingers of the right hand are curled around the axis of rotation in the direction that the torque would produce a rotational (angular) acceleration, the extended thumb points in the direction of the torque. A sign convention, as in the case of linear motion, can be used to represent torque directions, as will be shown shortly.

Example 8.3 ■ Lifting and Holding: Muscle Torque at Work

In our bodies, torques produced by the contraction of our muscles cause some bones to rotate at joints. For example, when you lift something with your forearm, a torque is applied on the lower arm by the biceps muscle (▶Fig. 8.5). With the axis of rotation through the elbow joint and the muscle attached 4.0 cm from the joint, what are the magnitudes of the muscle torques for cases (a) and (b) in Fig. 8.5 if the muscle exerts a force of 600 N?

Thinking It Through. As in many rotational situations, it is important to know the orientations of the *r* and *F* vectors so that we can find the angle *between* them to determine the lever arm. Note in the inset in Fig. 8.5a that if the tails of the **r** and **F** vectors were put together, the angle between them would be greater than 90°, that is, $30° + 90° = 120°$. In Fig. 8.5b, the angle is 90°.

Solution. First we list the data given here and in the figure. This Example demonstrates an important point. Recall that θ is the angle *between* the radial vector **r** and the force **F**.

Given: $r = 4.0 \text{ cm} = 0.040 \text{ m}$ **Find:** (a) τ (muscle torque magnitude) for Fig. 8.5a
$F = 600 \text{ N}$ (b) τ (muscle torque magnitude) for Fig. 8.5b
(a) $\theta = 30° + 90° = 120°$
(b) $\theta = 90°$

(a) In this case, **r** is directed along the forearm, so the angle between the **r** and **F** vectors is $\theta = 120°$. Using Eq. 8.2, we have

$$\tau = rF \sin(120°) = (0.040 \text{ m})(600 \text{ N})(0.8660) = 21 \text{ m} \cdot \text{N}$$

at the instant in question.

(b) Here, the distance vector **r** and the line of action of the force are perpendicular ($\theta = 90°$), and $r_\perp = r \sin 90° = r$. Then,

$$\tau = r_\perp F = rF = (0.040 \text{ m})(600 \text{ N}) = 24 \text{ m} \cdot \text{N}$$

The torque is greater in (b). This is to be expected because the maximum value of the torque (τ_{max}) occurs when $\theta = 90°$.

Follow-up Exercise. In (a) of this Example, there must have been a net torque, since the ball was accelerated upward by a rotation of the forearm. In (b), the ball is just being held and there is no rotational acceleration, so there is no net torque on the system. Identify the other torque(s) in each case.

▼ **FIGURE 8.5   Human torque**   See Example 8.2.

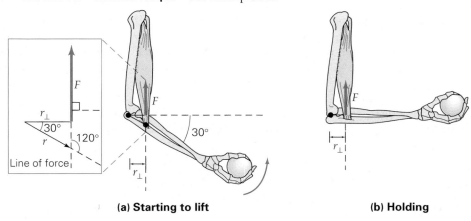

(a) **Starting to lift**   (b) **Holding**

Before considering rotational dynamics with net torques and rotational motions, let's look at a situation in which the forces and torques acting on a body are balanced, or in equilibrium.

Equilibrium

In general, equilibrium means that things are in balance or are stable. This definition applies in the mechanical sense to forces and torques. Unbalanced forces produce translational accelerations, but *balanced* forces produce the condition we call

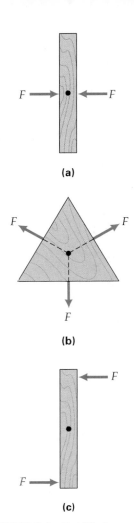

(a)

(b)

(c)

▲ **FIGURE 8.6 Equilibrium and forces** Forces with lines of action through the same point are said to be *concurrent*. The resultants of the concurrent forces acting on the objects in **(a)** and **(b)** are zero, and the objects are in equilibrium, because the net torque *and* net force are zero. In **(c)**, the object is in *translational* equilibrium, but it will undergo angular acceleration; thus, the object is *not* in rotational equilibrium.

translational equilibrium. Similarly, unbalanced torques produce rotational accelerations, and *balanced* torques produce *rotational equilibrium.*

According to Newton's first law of motion, when the sum of the forces acting on a body is zero, the body remains either at rest (static) or in motion with a constant velocity. In either case, the body is said to be in **translational equilibrium**. Stated another way, the *condition for translational equilibrium* is that the net force on a body is zero; that is, $\mathbf{F}_{net} = \Sigma\mathbf{F}_i = 0$. It should be apparent that this condition is satisfied for the situations illustrated in ◄Fig. 8.6a, b. Forces with lines of action through the same point are called **concurrent forces**. When these forces vectorially add to zero, as in Figs. 8.6a and b, the body is in translational equilibrium.

But what about the situation pictured in Fig. 8.6c? Here, $\Sigma\mathbf{F}_i = 0$, but the opposing forces will cause the object to rotate, and it will clearly not be in a state of static equilibrium. (Such a pair of equal and opposite forces that do not have the same line of action is called a *couple.*) Thus, the condition $\Sigma\mathbf{F}_i = 0$ is a necessary, but *not sufficient*, condition for static equilibrium.

Since $\mathbf{F}_{net} = \Sigma\mathbf{F}_i = 0$ is the condition for translational equilibrium, you might predict (and correctly so) that $\tau_{net} = \Sigma\tau_i = 0$ is the *condition for rotational equilibrium.* That is, if the sum of the *torques* acting on an object is zero, then the object is in **rotational equilibrium**—it remains rotationally at rest or rotates with a constant angular velocity.

Thus, we see that there are actually *two* equilibrium conditions. Taken together, they define what is called **mechanical equilibrium**:

A body is said to be in mechanical equilibrium when the conditions for both translational and rotational equilibrium are satisfied:

$$\mathbf{F}_{net} = \Sigma\mathbf{F}_i = 0 \quad (\textit{for translational equilibrium}) \quad (8.3)$$

$$\tau_{net} = \Sigma\tau_i = 0 \quad (\textit{for rotational equilibrium})$$

A rigid body in mechanical equilibrium may be either at rest or moving with a constant linear or angular velocity. An example of the latter is an object rolling without slipping on a level surface, with the center of mass of the object having a constant velocity. However, this is an ideal condition. Of greater practical interest is **static equilibrium**, the condition that exists when a rigid body remains at rest—that is, a body for which $v = 0$ and $\omega = 0$. There are many instances in which we do not want things to move, and this absence of motion can occur only if the equilibrium conditions are satisfied. It is particularly comforting to know, for example, that a bridge you are crossing is in static equilibrium and not subject to translational or rotational motion.

Let's consider examples of static translational equilibrium and static rotational equilibrium separately and then an example in which both apply.

Example 8.4 ■ Translational Static Equilibrium: No Translational Acceleration or Motion

A picture hangs motionless on a wall as shown in ►Fig. 8.7a. If the picture has a mass of 5.0 kg, what are the tension forces in the wires?

Thinking It Through. Since the picture remains motionless, it must be in static equilibrium, so applying the conditions for mechanical equilibrium should give equations that yield the tensions. Note that all the forces (tension and weight force) are concurrent; that is, their lines of action pass through a common point, the nail. Because of this, the condition for rotational equilibrium ($\Sigma\tau_i = 0$) is automatically satisfied: With respect to an axis of rotation at the nail, the moment arms ($r_\perp$) of the forces are zero, and therefore the torques are zero. Thus, we need consider only translational equilibrium.

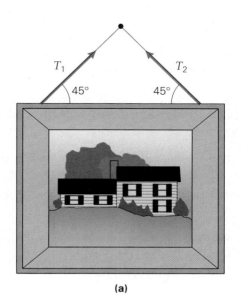

(a)

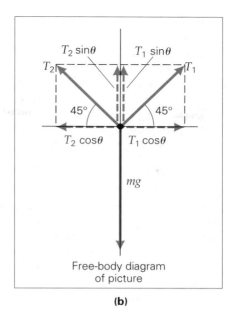

Free-body diagram
of picture

(b)

◀ **FIGURE 8.7 Translational static equilibrium (a)** Since the picture hangs motionless on the wall, the sum of the forces acting on it must be zero. The forces are concurrent, with their lines of action passing through a common point at the nail. **(b)** In the free-body diagram, all the forces are represented as acting at the common point at the nail. T_1 and T_2 have been moved to this point for convenience. Note, however, that the forces shown are acting on the *picture*, not on the nail. See Example 8.4.

Solution.

Given: $\theta = 45°$ **Find:** T_1 and T_2
   $m = 5.0 \text{ kg}$

You will find it helpful to isolate the forces acting on the picture in a free-body diagram, as was done in Chapter 4 for force problems (Fig. 8.7b). The diagram shows the concurrent forces acting through their common point. Note that we have moved all the force vectors to that point, which is taken as the origin of the coordinate axes. The weight force mg acts downward.

  With the system in static equilibrium, the net force on the picture is zero; that is, $\Sigma \mathbf{F}_i = 0$. Thus, the sums of the rectangular components are also zero: $\Sigma \mathbf{F}_{x_i} = 0$ and $\Sigma \mathbf{F}_{y_i} = 0$. Then (using $\pm$ for direction), we get

$$\Sigma \mathbf{F}_{x_i}: \quad +T_1 \cos 45° - T_2 \cos 45° = 0$$

This equation tells you that the x components of the forces are equal and opposite and that $T_1 = T_2 = T$. For the y components, with two upward components of $T \sin 45°$, we have

$$\Sigma \mathbf{F}_{y_i}: \quad +T \sin 45° + T \sin 45° - mg = 0$$

or

$$2T \sin 45° - mg = 0$$

Thus,

$$T = \frac{mg}{2 \sin 45°} = \frac{(5.0 \text{ kg})(9.8 \text{ m/s}^2)}{2(0.707)} = 35 \text{ N}$$

Follow-up Exercise. Analyze the situation in Fig. 8.7 that would result if the wires were shortened such that the angles were decreased but kept equal. Carry your analysis to the limit where the angles approach zero. Is the answer realistic?

  As pointed out earlier, torque is a vector and therefore has direction. Similar to the way we treated linear motion (Chapter 2), in which we used plus and minus signs to express opposite directions (for example, $+x$ and $-x$), we designate torque directions as being plus and minus, depending on the rotational acceleration they tend to produce. The rotational "directions" are taken as clockwise or counterclockwise around the axis of rotation. A torque that tends to produce a counterclockwise rotation will be taken as positive (+), and a torque that tends to produce a clockwise rotation will be taken as negative (−). (See the right-hand rule in Section 7.2.) To illustrate, let's apply our convention to the situation shown in ▼Fig. 8.8.

Example 8.5 ■ Rotational Static Equilibrium: No Rotational Motion

Three masses are suspended from a meterstick as shown in Fig. 8.8a. How much mass must be suspended on the right side for the system to be in static equilibrium? (Neglect the mass of the meterstick.)

Thinking It Through. As the free-body diagram (Fig. 8.8b) shows, the translational equilibrium condition will be satisfied with the upward reaction force N balancing the downward weight forces, so long as the stick is horizontal. But N is not known if m_3 is unknown, so applying the condition for rotational equilibrium should give the required value of m_3. (Note that the lever arms are measured from the pivot point, the center of the meterstick.)

Solution. From the figure, we have

Given: $m_1 = 25$ g   *Find:* m_3 (unknown mass)
           $r_1 = 50$ cm
           $m_2 = 75$ g
           $r_2 = 30$ cm
           $r_3 = 35$ cm

Because the condition for translational equilibrium ($\Sigma F_i = 0$) is satisfied (there is no $\mathbf{F}_{net}$ in the y direction), $N - Mg = 0$, or $N = Mg$, where M is the total mass. This is true no matter what the total mass may be—that is, regardless of how much mass we add for m_3. However, unless the proper mass for m_3 is placed on the right side, the stick will experience a net torque and begin to rotate.

▶ **FIGURE 8.8 Rotational static equilibrium** For the meterstick to be in rotational equilibrium, the sum of the torques acting about any selected axis must be zero. (The mass of the meterstick is considered negligible.) See Example 8.5.

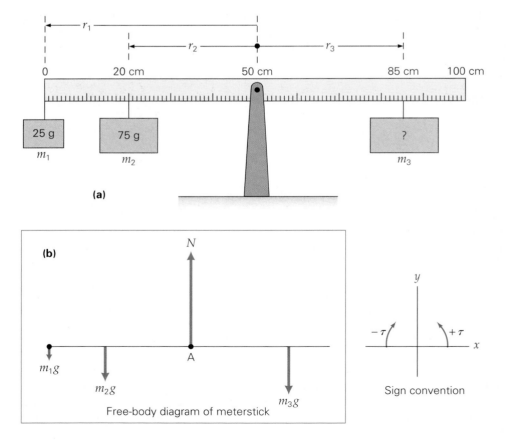

Notice that the masses on the left side produce torques that would tend to rotate the stick counterclockwise, and the mass on the right side produces a torque that would tend to rotate the stick clockwise. We apply the condition for rotational equilibrium by summing the torques about an axis. Let's take this to be through the center of the stick at the 50-cm position, or point A in Fig. 8.8b. Then, noting that N passes through the axis of rotation ($r_\perp = 0$) and produces no torque, we have

$$\Sigma\tau_i : \quad \tau_1 + \tau_2 + \tau_3 = +r_1 F_1 + r_2 F_2 - r_3 F_3 \qquad \textit{(using our sign}$$
$$= r_1(m_1 g) + r_2(m_2 g) - r_3(m_3 g) = 0 \quad \begin{array}{l}\textit{convention for}\\ \textit{torque vectors)}\end{array}$$

Noting that the g's cancel out and solving for m_3, we have

$$m_3 = \frac{m_1 r_1 + m_2 r_2}{r_3} = \frac{(25\ \text{g})(50\ \text{cm}) + (75\ \text{g})(30\ \text{cm})}{35\ \text{cm}} = 100\ \text{g}$$

where it was convenient not to convert to standard units. (The mass of the stick was neglected. If the stick is uniform, however, its mass will not affect the equilibrium, as long as the pivot point is at the 50-cm mark. Why?)

Follow-up Exercise. The axis of rotation could have been taken through any point along the stick. That is, if a system is in static rotational equilibrium, the condition $\Sigma\tau_i = 0$ holds for *any* axis of rotation. Show that the preceding statement is true for the system in this Example by taking the axis of rotation through the left end of the stick ($x = 0$).

In general, the conditions for both translational and rotational equilibrium need to be written explicitly to solve a statics problem. Example 8.6 is one such case.

Example 8.6 ■ Static Equilibrium: No Translation, No Rotation

A ladder with a mass of 15 kg rests against a smooth wall (▼Fig. 8.9a). A painter who has a mass of 78 kg stands on the ladder as shown in the figure. What frictional force must act on the bottom of the ladder to keep it from slipping?

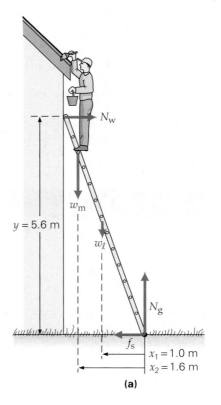

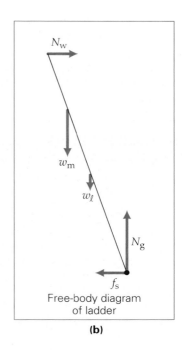

◀ FIGURE 8.9   Static equilibrium For the painter's sake, the ladder has to be in static equilibrium; that is, both the sum of the forces and the sum of the torques must be zero. See Example 8.6.

Thinking It Through. Here we have a variety of forces and torques. However, the ladder will not slip as long as the conditions for static equilibrium are satisfied. Summing both the forces and torques to zero should enable us to solve for the necessary frictional force. Also, as we will see, choosing a convenient axis of rotation, such that one or more τ's are zero in the summation of the torques, can simplify the torque equation.

Solution.

Given: $m_\ell = 15$ kg   *Find:* f_s (force of static friction)
$m_m = 78$ kg
Distances given in figure

Because the wall is smooth, there is negligible friction between it and the ladder, and only the normal reaction force of the wall (N_w) acts on the ladder at this point (Fig. 8.9b).

In applying the conditions for static equilibrium, you are free to choose any axis of rotation for the rotational condition. (The conditions must hold for all parts of a system that is in static equilibrium; that is, there can't be motion in any part of the system.) Note that choosing an axis at the end of the ladder where it touches the ground eliminates the torques due to f_s and N_g, since the moment arms are zero. Then you can write three equations (using mg for w):

$$\Sigma F_{x_i}: \quad N_w - f_s = 0$$
$$\Sigma F_{y_i}: \quad N_g - m_m g - m_\ell g = 0$$

and

$$\Sigma \tau_i: \quad (m_\ell g)x_1 + (m_m g)x_2 + (-N_w y) = 0$$

The weight of the ladder is considered to be concentrated at its center of gravity. Solving the third equation for N_w and substituting the given values for the masses and distances gives

$$N_w = \frac{(m_\ell g)x_1 + (m_m g)x_2}{y}$$

$$= \frac{(15\text{ kg})(9.8\text{ m/s}^2)(1.0\text{ m}) + (78\text{ kg})(9.8\text{ m/s}^2)(1.6\text{ m})}{5.6\text{ m}} = 2.4 \times 10^2\text{ N}$$

From the first equation, then,

$$f_s = N_w = 2.4 \times 10^2\text{ N}$$

Follow-up Exercise. In this Example, would the frictional force between the ladder and the ground (call it f_{s_1}) remain the same if there were friction between the wall and the ladder (call it f_{s_2})? Justify your answer.

Problem-Solving Hint

As the preceding Examples have shown, a good procedure to follow in working problems involving static equilibrium is as follows:

1. Sketch a space diagram of the problem.
2. Draw a free-body diagram, showing and labeling all external forces and, if necessary, resolving the forces into x and y components.
3. Apply the equilibrium conditions. Sum the forces: $\Sigma F_i = 0$, usually in component form, $\Sigma F_{x_i} = 0$ and $\Sigma F_{y_i} = 0$. Sum the torques: $\Sigma \tau_i = 0$. Remember to select an appropriate axis of rotation to reduce the number of terms as much as possible. Use $\pm$ sign conventions for both $\mathbf{F}$ and τ.
4. Solve for the unknown quantities.

Stability and Center of Gravity

The equilibrium of a particle or a rigid body can be stable or unstable in a gravitational field. For rigid bodies, these categories of equilibria are conveniently analyzed in terms of the center of gravity of the body. Recall from Chapter 6 that the **center of gravity** is the point at which all the weight of an object may be considered to be acting as if the object were a particle. When the acceleration due to gravity is constant, the center of gravity and the center of mass coincide.

If an object is in **stable equilibrium**, any small displacement results in a restoring force or torque, which tends to return the object to its original equilibrium position. As illustrated in ▼Fig. 8.10a, a ball in a bowl is in stable equilibrium. Analogously, the center of gravity of an extended body in stable equilibrium is essentially in a potential-energy bowl. Any slight displacement raises its center of gravity, and a restoring gravitational force tends to return it to the position of minimum potential energy. This force actually produces a restoring torque that is due to a component of the weight force and that tends to rotate the object about a pivot point back to its original position.

For an object in **unstable equilibrium**, any small displacement from equilibrium results in a torque that tends to rotate the object farther away from its equilibrium position. This situation is illustrated in Fig. 8.10b. Note that the center of gravity of the object is at the top of an overturned, or inverted, potential-energy bowl; that is, the potential energy is at a maximum in this case.

Small displacements or slight disturbances have profound effects on objects that are in unstable equilibrium: It doesn't take much to cause such an object to change its position. Yet, even if the angular displacement of an object in stable equilibrium is quite substantial, the object will still be restored to its equilibrium

Note: Stable equilibrium—a restoring torque

Note: Unstable equilibrium—a torque that topples

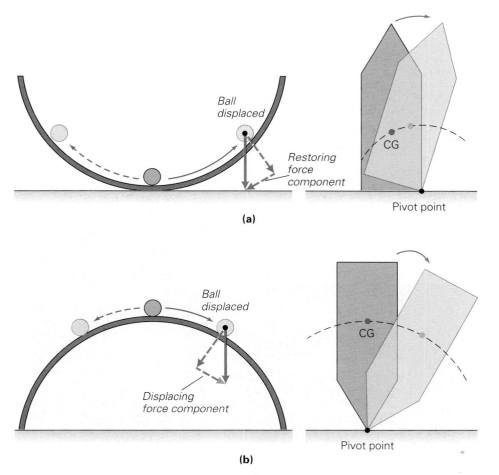

(a)

(b)

◀ **FIGURE 8.10 Stable and unstable equilibria** **(a)** When an object is in stable equilibrium, any small displacement from an equilibrium position results in a force or torque that tends to return the object to that position. A ball in a bowl (left) returns to the bottom after being displaced. Analogously, the center of gravity (CG) of an extended object (right) can be thought of as being in a potential-energy bowl: A small displacement raises the CG, increasing the object's potential energy. **(b)** For an object in unstable equilibrium, any small displacement from its equilibrium position results in a force or torque that tends to take the object farther away from that position. The ball on top of an overturned bowl (left) is in unstable equilibrium. For an extended object (right), the CG can be thought of as being on an inverted potential-energy bowl: A small displacement lowers the CG, decreasing the object's potential energy.

position. An object lying on its long side can be rotated through quite a distance and will still fall back to that original position. This is another way of expressing the *condition of stable equilibrium*:

> An object is in stable equilibrium as long as its center of gravity after a small displacement still lies above and inside the object's original base of support. That is, the line of action of the weight force of the center of gravity intersects the original base of support.

When this is the case, there will always be a restoring gravitational torque (▼Fig. 8.11a). However, when the center of gravity or center of mass falls outside the base of support, over goes the object—because of a gravitational torque that rotates it away from its equilibrium position (Fig. 8.11b).

Rigid bodies with wide bases and low centers of gravity are therefore most stable and least likely to tip over. This relationship is evident in the design of high-speed race cars, which have wide wheel bases and centers of gravity close to the ground (▼Fig. 8.12). SUVs, on the other hand, can roll over more easily. Why?

The location of the center of gravity of the human body has an effect on certain physical abilities. For example, women can generally bend over and touch their toes or touch their palms to the floor more easily than can men, who often fall over trying. On the average, men have higher centers of gravity (larger shoulders) than do women (larger pelvises), so it is more likely that a man's center of gravity will be outside his base of support when he bends over. See the Insight on p. 273 for another real-life example of equilibrium and stability.

▼ **FIGURE 8.11   Examples of stable and unstable equilibria**   **(a)** When the center of gravity is above and inside an object's base of support, the object is in stable equilibrium. (There is a restoring torque.) Note how the line of action of the weight of the center of gravity (CG) intersects the original base of support after the displacement. **(b)** When the center of gravity lies outside the base of support or the line of action of the weight does not intersect the original base of support, the object is unstable. (There is a displacing torque.)

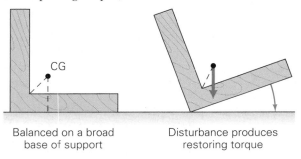

| Balanced on a broad base of support | Disturbance produces restoring torque | | Balanced carefully on narrow base of support (point) | Disturbance produces displacing torque |

(a) Stable Equilibrium   **(b) Unstable Equilibrium**

▶ **FIGURE 8.12   Stable and unstable**   **(a)** Race cars are very stable because of their wide wheel bases and low center of gravity. **(b)** The acrobat's base of support is very narrow: the small area of head-to-head contact. As long as his center of gravity remains above this area, he is in equilibrium, but a displacement of only a few inches would probably be enough to topple him. (Why he is in a spread-eagle position will become clearer in Section 8.3.)

(a)   **(b)**

The Leaning Tower of Pisa: An Exercise in Stability

The Leaning Tower of Pisa is a famous Italian landmark from which Galileo allegedly performed experiments. (See the Chapter 2 Insight on p. 51.) Recently, the Tower of Pisa has been in the spotlight—for attempts made to keep it from falling. The Tower started leaning before its completion in A.D. 1350 because of soft subsoil underneath it. It was closed to the public in 1990, with a lean of about 5.5° from the verticle (about 5 m, or 17 ft, at the top) and an average increase in lean of about 1.2 mm a year.

Cement was injected into the base in the 1930s, but the lean continued to increase. In the past decade, there has been an international effort to save the tower. From 1993 to 1995, a concrete ring was built at the base, and 620 tons of lead counterweight were placed on it (Fig. 1). Within months, the incline was corrected by 2 to 3 cm. But the construction of a second ring jolted the tower, and overnight it tilted 2.5 mm. With that big scare, an additional 230 tons of lead weight were added.

A significant improvement was made by removing soil from beneath the high side of the tower. Augers drilled diagonally below the foundation, creating small cavities from which soil could be removed. This allowed the tower to settle back into a straighter position (and keep the center of gravity more inside the base of support). Even so, steel cables were attached to the tower to support it should there be a shift in the wrong direction (Fig. 2). However, with the drilling technique, the tower settled back to about a 5° lean, or a shift of 40 cm at the top.

The Leaning Tower of Pisa opened its doors for tourists in December, 2001 for the first time after 12 years of work to reduce the lean. Before the tower was closed in 1990, about a million visitors a year climbed the 293-step, spiral staircase inside to the top. However, now, this will be limited to parties of 30 people for a 40-minute guided tour. This will slow things down a bit and cause long lines.

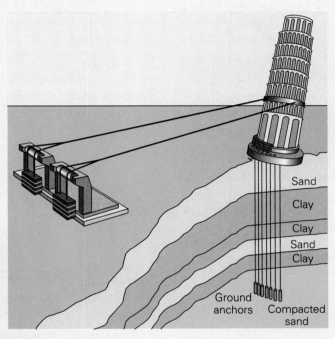

FIGURE 2 Keep the center of gravity over the base of support. Efforts were made to stabilize the tower by removing soil from beneath the high side. For safety in case of sudden shifts, steel support cables were attached.

FIGURE 1 Hold it stable! Tons of lead counterweight being used to help correct the tower's lean.

Example 8.7 ■ Stack Them Up: Center of Gravity

Uniform, identical bricks 20 cm long are stacked so that 4.0 cm of each brick extends beyond the brick beneath, as shown in ▶ Fig. 8.13a. How many bricks can be stacked in this way before the stack falls over?

Thinking It Through. As each brick is added, the center of mass (or center of gravity) of the stack moves to the right. The stack will be stable as long as the combined center of mass (CM) is over the base of support—the bottom brick. All of the bricks have the same mass, and the center of mass of each is located at its midpoint. So the horizontal location of the stack's CM must be computed as bricks are added, until the CM extends beyond the base. The location of the CM was discussed in Chapter 6 (see Eq. 6.19).

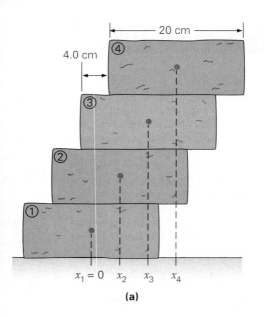

20 cm

4.0 cm

④
③
②
①

$x_1 = 0$ x_2 x_3 x_4

(a)

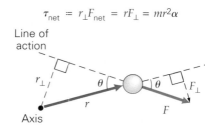

(b)

▲ **FIGURE 8.13   Stack them up!**
(a) How many bricks can be stacked like this before the stack falls? See Example 8.7. **(b)** Try a similar experiment with books.

Solution.

Given: length = 20 cm
   displacement of each brick = 4.0 cm

Find: maximum number of bricks that yields stability

Taking the origin to be at the center of the bottom brick, we find that the horizontal coordinate of the center of mass (or center of gravity) for the first two bricks in the stack is given by Eq. 6.19, where $m_1 = m_2 = m$ and x_2 is the displacement of the second brick:

$$X_{CM_2} = \frac{mx_1 + mx_2}{m + m} = \frac{m(x_1 + x_2)}{2m} = \frac{x_1 + x_2}{2} = \frac{0 + 4.0 \text{ cm}}{2} = 2.0 \text{ cm}$$

The masses of the bricks cancel out (since they are all the same). For three bricks,

$$X_{CM_3} = \frac{m(x_1 + x_2 + x_3)}{3m} = \frac{0 + 4.0 \text{ cm} + 8.0 \text{ cm}}{3} = 4.0 \text{ cm}$$

For four bricks,

$$X_{CM_4} = \frac{m(x_1 + x_2 + x_3 + x_4)}{4m} = \frac{0 + 4.0 \text{ cm} + 8.0 \text{ cm} + 12.0 \text{ cm}}{4} = 6.0 \text{ cm}$$

and so on.

This series of results shows that the center of mass of the stack moves horizontally 2.0 cm for each brick added to the bottom one. For a stack of six bricks, the center of mass is 10 cm from the origin and directly over the edge of the bottom brick (2.0 cm × 5 *added* bricks = 10 cm, which is half the length of the bottom brick), so the stack is just at unstable equilibrium. The stack may not topple if the sixth brick is positioned very carefully, but it is doubtful that this could be done in practice. A seventh brick would definitely cause the stack to fall.

Follow-up Exercise.   If the bricks in this Example were stacked so that, alternately, 4.0 cm and 6.0 cm extended beyond the brick beneath, how many bricks could be stacked before the stack toppled?

For another case of stability, see the Insight "Stability in Action," on p. 275.

8.3   Rotational Dynamics

OBJECTIVES: To **(a)** describe the moment of inertia of a rigid body and **(b)** apply the rotational form of Newton's second law to physical situations.

Moment of Inertia

Torque is the rotational analogue of force in linear motion, and a net torque produces rotational motion. To analyze this relationship, consider a constant net force acting on a particle of mass m (◄Fig. 8.14). The magnitude of the torque on the particle is

$$\tau_{net} = r_\perp F_{net} = rF_\perp = rma_\perp = mr^2\alpha \quad \text{torque on a particle} \quad (8.4)$$

where $a_\perp = a_t = r\alpha$ is the tangential acceleration (a_t, Eq. 7.13). For the rotation of a rigid body about a fixed axis, this equation can be applied to each particle and the results summed over the entire body (n particles) to find the total torque. Since all the particles of a rotating rigid body have the same angular acceleration, we can simply add the individual torque magnitudes:

$$\begin{aligned}
\tau_{net} = \Sigma\tau_i &= \tau_1 + \tau_2 + \tau_3 + \cdots + \tau_n \\
&= m_1 r_1^2\alpha + m_2 r_2^2\alpha + m_3 r_3^2\alpha + \cdots + m_n r_n^2\alpha \\
&= (m_1 r_1^2 + m_2 r_2^2 + m_3 r_3^2 + \cdots + m_n r_n^2)\alpha
\end{aligned}$$

$$\Sigma\tau_{net} = \left(\Sigma m_i r_i^2\right)\alpha \quad (8.5)$$

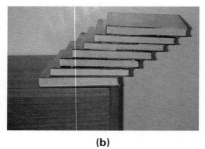

$\tau_{net} = r_\perp F_{net} = rF_\perp = mr^2\alpha$

Line of action

$r_\perp$

θ θ $F_\perp$

r F

Axis

▲ **FIGURE 8.14   Torque on a particle**   The magnitude of the torque on a particle of mass m is $\tau = mr^2\alpha$.

Stability in Action

When riding a bicycle and going around a curve or making a turn on a level surface, a rider instinctively leans into the curve. Why? We might think that leaning over, rather than remaining upright, is more likely to cause a spill. However, leaning really does increase stability—it's all a matter of torques.

When a vehicle goes around a level circular curve, a centripetal force is needed to keep the vehicle on the road, as we learned in Chapter 7. This force is generally supplied by the force of static friction between the tires and the road. As illustrated in Fig. 1a, the reaction force $\mathbf{R}$ of the ground on the bicycle provides the required centripetal force ($\mathbf{R}_x = \mathbf{F}_c = \mathbf{f}_s$) to round the curve, and the normal force $\mathbf{R}_y = \mathbf{N}$.

Suppose the rider tried to go around the curve with these forces operative while remaining upright, as shown in Fig. 1a. Note that the line of action of $\mathbf{R}$ does not go through the system's center of gravity (indicated by a dot). With an axis of rotation through the center of gravity, there would be a counterclockwise torque that would tend to rotate the bicycle

in such a way that the wheels would slide inward underneath the rider. However, if the rider leans inward at the proper angle (Fig. 1b), both the line of action of $\mathbf{R}$ and the weight force go through the center of gravity, and there is no rotational instability (as the gentleman on the bicycle well knew).

There is still a torque on the rider, however. Indeed, when the rider leans into the curve, the weight force gives rise to a torque about an axis through the point of contact with the ground. This torque, along with the turning of the handlebars, causes the bicycle to turn. If the bicycle were not moving, there would be a rotation about this axis, and the bicycle and rider would fall over.

The need to lean into a curve is readily apparent in bicycle and motorcycle races on level tracks. Things can be made easier for the riders if tracks or roadways are banked to provide a natural lean. (Recall Section 7.3 and Exercises 7.45 and 7.59.)

Related Exercise: 41

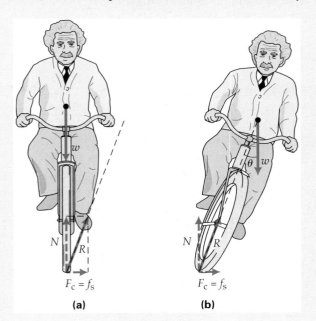

(a)   (b)

FIGURE 1 Leaning into a curve When rounding a curve or making a turn, a bicycle rider must lean into the curve. (This rider could have told you why.)

But for a rigid body, the masses (m_i's) and the distances from the axis of rotation (r_i's) do not change. Therefore, the quantity in the parentheses in Eq. 8.5 is constant, and it is called the **moment of inertia**, I (for a given axis):

Definition of: moment of inertia

$$I = \Sigma m_i r_i^2 \quad \textit{moment of inertia} \quad (8.6)$$

SI unit of moment of inertia: kilogram-meters squared ($\text{kg} \cdot \text{m}^2$)

Thus, the magnitude of the net torque can be conveniently written as

$$\tau_{\text{net}} = I\alpha \quad \textit{net torque on a rigid body} \quad (8.7)$$

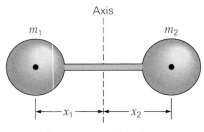

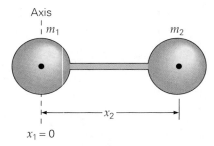

(a) $m_1 = m_2 = 30$ kg
$x_1 = x_2 = 0.50$ m

(b) $m_1 = 40$ kg, $m_2 = 10$ kg
$x_1 = x_2 = 0.50$ m

(c) $m_1 = m_2 = 30$ kg
$x_1 = x_2 = 1.5$ m

(d) $m_1 = m_2 = 30$ kg
$x_1 = 0$, $x_2 = 3.0$ m

(e) $m_1 = 40$ kg, $m_2 = 10$ kg
$x_1 = 0$, $x_2 = 3.0$ m

▲ **FIGURE 8.15 Moment of inertia** The moment of inertia depends on the distribution of mass relative to a particular axis of rotation and, in general, has a different value for each axis. This difference reflects the fact that objects are easier or more difficult to rotate about certain axes. See Example 8.8.

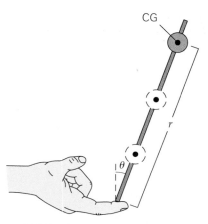

▲ **FIGURE 8.16 Greater stability with a higher center of gravity?** See Integrated Example 8.9.

This is the *rotational form of Newton's second law* ($\tau_{net} = I\alpha$, in vector form). Keep in mind that, like *net* forces, *net* torques (τ_{net}) are necessary to produce accelerations.

As you might infer by comparing the rotational form of Newton's second law with the translational form ($\mathbf{F}_{net} = m\mathbf{a}$), the moment of inertia I is a measure of *rotational inertia*, or a body's tendency to resist change in its rotational motion. Although I is constant for a rigid body and is the rotational analogue of mass, you must keep in mind that, unlike the mass of a particle, the moment of inertia of a body is referenced to a particular axis and can have different values for different axes.

The moment of inertia also depends on the mass distribution of the body *relative* to its axis of rotation. It is easier (i.e., it takes less torque) to give an object an angular acceleration about some axes than about others. The following Example illustrates this point.

Example 8.8 ■ Rotational Inertia: Mass Distribution and Axis of Rotation

Find the moment of inertia about the axis indicated for each of the one-dimensional dumbbell configurations in ◄Fig. 8.15. (Consider the mass of the connecting bar to be negligible, and state your answer to three figures for comparison.)

Thinking It Through. This is a direct application of Eq. 8.6 to cases with different masses and distances. It will show that the moment of inertia of an object depends on the axis of rotation and on the mass distribution relative to the axis of rotation. The sum for I will include only two terms (two masses).

Solution.

Given: Values of m and r from the figure.   *Find:* $I = \Sigma m_i r_i^2$

With $I = m_1 r_1^2 + m_2 r_2^2$:
(a) $I = (30 \text{ kg})(0.50 \text{ m})^2 + (30 \text{ kg})(0.50 \text{ m})^2 = 15.0 \text{ kg} \cdot \text{m}^2$
(b) $I = (40 \text{ kg})(0.50 \text{ m})^2 + (10 \text{ kg})(0.50 \text{ m})^2 = 12.5 \text{ kg} \cdot \text{m}^2$
(c) $I = (30 \text{ kg})(1.5 \text{ m})^2 + (30 \text{ kg})(1.5 \text{ m})^2 = 135 \text{ kg} \cdot \text{m}^2$
(d) $I = (30 \text{ kg})(0 \text{ m})^2 + (30 \text{ kg})(3.0 \text{ m})^2 = 270 \text{ kg} \cdot \text{m}^2$
(e) $I = (40 \text{ kg})(0 \text{ m})^2 + (10 \text{ kg})(3.0 \text{ m})^2 = 90.0 \text{ kg} \cdot \text{m}^2$

This Example clearly shows how the moment of inertia depends on the mass *and* its distribution relative to a particular axis of rotation. In general, the moment of inertia is larger the farther the mass is from the axis of rotation. This principle is important in the design of flywheels, which are used in automobiles to keep the engine running smoothly between cylinder firings. The mass of a flywheel is concentrated near the rim, giving a large moment of inertia, which resists changes in motion.

Follow-up Exercise. In parts (d) and (e) of this Example, would the moments of inertia be different if the axis of rotation went through m_2? Explain.

Integrated Example 8.9 ■ Balancing Act: Locating the Center of Gravity

(a) A rod with a movable ball, like that shown in ◄Fig. 8.16, is more easily balanced if the ball is in a higher position. Is this because, when the ball in a higher position, (1) the system has a higher center of gravity and more stability; (2) the center of gravity is off the vertical, and there is less torque and a smaller angular acceleration; (3) the center of gravity is closer to the axis of rotation; or (4) the moment of inertia about the axis of rotation is larger? (b) Suppose the distance of the ball from the finger for the farthest position in Fig. 8.16 is 60 cm and the distance to the closest position is 20 cm. When the rod rotates, how many times greater is the angular acceleration of the rod with the ball at the closest position than that with the ball at the farthest position? (Neglect the mass of the rod.)

(a) Conceptual Reasoning. With the ball at any position and the rod vertical, the system is in unstable equilibrium. We saw in Section 8.2 that rigid bodies with wide bases and *low* centers of gravity are more stable, so answer (1) isn't correct. Any slight movement would cause the rod to rotate about an axis through its point of contact. With the center of gravity (CG) at a higher position and off the vertical, there would be a greater lever arm (and thus a *greater* torque), so (2) is also incorrect. With the ball in a higher position, the center of gravity is *farther* from the axis of rotation, which makes (3) incorrect. This leaves (4) by a process of elimination, but let's justify it as the correct answer.

Moving the CG farther from the axis of rotation has an interesting consequence: a greater moment of inertia, or resistance to change in rotational motion, and hence a smaller angular acceleration. With the ball in a higher position, as the rod starts to rotate there is a greater torque, but the increased moment of inertia produces an even greater resistance to rotational motion and hence a smaller angular acceleration. [Note that the torque ($\tau = rF \sin \theta$) varies as r, whereas the moment of inertia ($I = mr^2$) varies as r^2 and so has a larger increase with increasing r. What effect does $\sin \theta$ have?] Then, the smaller the angular acceleration, the more time there is available to adjust your hand under the rod to balance it by bringing the finger and the axis of rotation under the center of gravity. The torque is then zero and the rod is again in equilibrium, albeit unstable. Thus, the answer is (4).

(b) Thinking It Through. Being asked how many times greater or less something is compared to something else usually implies the use of a ratio in which some quantity (or quantities) that is not given cancels. Note that the mass of the ball is not given, which would be needed to compute the torque (τ). Also, the angle θ is not given. So it is best to start with basic equations and see what happens.

Given: $r_1 = 20$ cm   **Find:** How many times greater the rod's angular acceleration
$r_2 = 60$ cm   is with the ball at r_1 compared to when it is at r_2

Angular acceleration is given by Eq. 8.7, $\alpha = \tau_{net}/I$. So attention turns to the torque τ_{net} and the moment of inertia I. From the basic equations of the chapter, $\tau_{net} = r_\perp F = rF \sin \theta$ (Eq. 8.2), or $\tau_{net} = rmg \sin \theta$, where $F = mg$ in this case, with m being the mass of the ball. Similarly, $I = mr^2$ (Eq. 8.6). Thus,

$$\alpha = \frac{\tau_{net}}{I} = \frac{rmg \sin \theta}{mr^2} = \frac{g \sin \theta}{r}$$

(Note that the angular acceleration α is inversely proportional to the lever arm r; that is, the greater the lever arm, the smaller is the angular acceleration.) Sin θ is still there, but note what happens when the ratio of the angular accelerations is formed:

$$\frac{\alpha_1}{\alpha_2} = \frac{g \sin \theta / r_1}{g \sin \theta / r_2} = \frac{r_2}{r_1} = \frac{60 \text{ cm}}{20 \text{ cm}} = 3 \quad \text{or} \quad \alpha_2 = \frac{\alpha_1}{3}$$

Hence, the angular acceleration of the rod with the ball at the upper position is one-third that with the ball at the lower position.

Follow-up Exercise. When walking on a thin bar or rail, such as a railroad rail, you have probably found that it helps to hold your arms outstretched. Similarly, tightrope walkers often carry long poles, as in the chapter-opening photo. How does this posture help a performer to maintain balance? *(Answers to all Follow-up Exercises are at the back of the text.)*

As Integrated Example 8.9 shows, the moment of inertia is an important consideration in rotational motion. By changing the axis of rotation and the relative mass distribution, the value of I can be changed and the motion affected. You were probably told to do this when playing softball or baseball as a child. When at bat, children are often instructed to "choke up" on the bat—to move their hands farther up on the handle.

Now you know why. In doing so, the child moves the axis of rotation of the bat closer to the more massive end of the bat (or its center of mass). Hence, the moment of inertia of the bat is decreased (smaller r in the mr^2 term). Then, when a swing is taken, the angular acceleration is greater. The bat gets around quicker, and the chance of hitting the ball before it goes past is greater. A batter has only a fraction of a second to swing, and with $\theta = \frac{1}{2}\alpha t^2$, a larger α allows the bat to rotate more quickly (swing faster).

Parallel-Axis Theorem

Calculations of the moments of inertia of most extended rigid bodies require math that is beyond the scope of this book. The results for some common shapes are given in ▼Fig. 8.17. The rotational axes are generally taken along axes of symmetry—that is, axes running through the center of mass so as to give a symmetrical mass distribution. An exception is the rod with an axis of rotation through one end (Fig. 8.17c). This axis is parallel to an axis of rotation through the center of mass of the rod (Fig. 8.17b). The moment of inertia about such a parallel axis is given by a useful theorem called the **parallel axis theorem**, namely,

$$I = I_{CM} + Md^2 \qquad (8.8)$$

Note: $I = I_{CM}$, the minimum I value, when $d = 0$.

▼ **FIGURE 8.17   Moments of inertia of some uniform-density objects with common shapes**

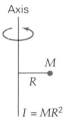

$I = MR^2$

(a) Particle

$I = \frac{1}{12}ML^2$

(b) Thin rod

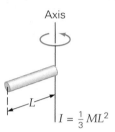

$I = \frac{1}{3}ML^2$

(c) Thin rod

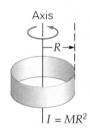

$I = MR^2$

(d) Thin cylindrical shell, hoop, or ring

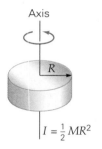

$I = \frac{1}{2}MR^2$

(e) Solid cylinder or disk

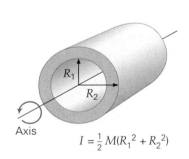

$I = \frac{1}{2}M(R_1{}^2 + R_2{}^2)$

(f) Annular cylinder

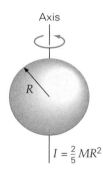

$I = \frac{2}{5}MR^2$

(g) Solid sphere about any diameter

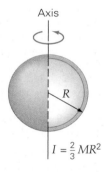

$I = \frac{2}{3}MR^2$

(h) Thin spherical shell

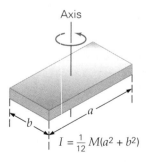

$I = \frac{1}{12}M(a^2 + b^2)$

(i) Rectangular plate

$I = \frac{1}{12}ML^2$

(j) Thin rectangular sheet

$I = \frac{1}{3}ML^2$

(k) Thin rectangular sheet

where I is the moment of inertia about an axis that is parallel to one through the center of mass and at a distance d from it, I_{CM} is the moment of inertia about an axis through the center of mass, and M is the total mass of the body (▶Fig. 8.18). For the axis through the end of the rod (Fig. 8.17c), the moment of inertia is obtained by applying the parallel-axis theorem to the thin rod in Fig. 8.17b:

$$I = I_{CM} + Md^2 = \tfrac{1}{12}ML^2 + M\left(\frac{L}{2}\right)^2 = \tfrac{1}{12}ML^2 + \tfrac{1}{4}ML^2 = \tfrac{1}{3}ML^2$$

Applications of Rotational Dynamics

The rotational form of Newton's second law allows us to analyze dynamic rotational situations. Examples 8.10 and 8.11 illustrate how this is done. In such situations, it is very important to make certain that all the data are properly listed to help with the increasing number of variables.

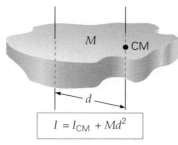

▲ **FIGURE 8.18 Parallel-axis theorem** The moment of inertia about an axis parallel to another through the center of mass of a body is $I = I_{CM} + Md^2$, where M is the total mass of the body and d is the distance between the two axes.

Example 8.10 ■ Opening the Door: Torque in Action

A student opens a 12-kg door by applying a constant force of 40 N at a perpendicular distance of 0.90 m from the hinges (▶Fig. 8.19). If the door is 2.0 m in height and 1.0 m wide, what is the magnitude of its angular acceleration? (Assume that the door rotates freely on its hinges.)

Thinking It Through. From the given information, we can calculate the applied net torque. To find the angular acceleration of the door, we need to know its moment of inertia. This can be calculated, since we know the door's mass and dimensions.

Solution. From the information set forth in the problem, we can list the following:

Given: $M = 12\,\text{kg}$
$\quad\quad\;\; F = 40\,\text{N}$
$\quad\quad\;\; r_\perp = r = 0.90\,\text{m}$
$\quad\quad\;\; h = 2.0\,\text{m}$ (door height)
$\quad\quad\;\; w = 1.0\,\text{m}$ (door width)

Find: α (magnitude of angular acceleration)

We need to apply the rotational form of Newton's second law (Eq. 8.7), $\tau_{net} = I\alpha$, where I is about the hinge axis. The problem boils down to finding the moment of inertia of the door.

Looking at Fig. 8.17, we see that (case k) applies to a door (treated as a uniform rectangle) rotating on hinges, so $I = \tfrac{1}{3}ML^2$, where $L = w$, the width of the door. We must simply do the calculations:

$$\tau_{net} = I\alpha$$

or

$$\alpha = \frac{\tau_{net}}{I} = \frac{r_\perp F}{\tfrac{1}{3}ML^2} = \frac{3rF}{Mw^2} = \frac{3(0.90\,\text{m})(40\,\text{N})}{(12\,\text{kg})(1.0\,\text{m})^2} = 9.0\,\text{rad/s}^2$$

Follow-up Exercise. In this Example, if the constant torque were applied through an angular distance of 45° and then removed, how long would it take the door to swing completely open (90°)?

▲ **FIGURE 8.19 Torque in action** See Example 8.10.

In problems involving pulleys in Chapter 4, the mass (as well as the inertia) of the pulley was always neglected in order to simplify things. Now we know how to include those quantities and can treat pulleys more realistically.

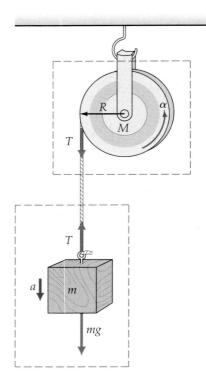

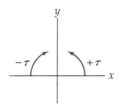

▲ FIGURE 8.20 **Pulley with inertia** Taking the mass, or rotational inertia, of a pulley into account allows a more realistic description of the motion. See Example 8.11.

PHYSLET®
ILLUSTRATION

Torque and Moment of Inertia

Example 8.11 ■ Pulleys Have Mass, Too: Taking Account of Pulley Inertia

A block of mass m hangs from a string wrapped around a frictionless, disk-shaped pulley of mass M and radius R, as shown in ◄Fig. 8.20. If the block descends from rest under the influence of gravity, what is the magnitude of its linear acceleration? (Neglect the mass of the string.)

Thinking It Through. Real pulleys have mass and rotational inertia, which affect their motion. The suspended mass (via the string) applies a torque to the pulley. Here we use the rotational form of Newton's second law to find the angular acceleration of the pulley and then its tangential acceleration, which is the same in magnitude as the linear acceleration of the block. (Why?)

Solution. The linear acceleration of the block depends on the angular acceleration of the pulley, so we look at the pulley system first. The pulley is treated as a disk and thus has a moment of inertia $I = \frac{1}{2}MR^2$ (Fig. 8.17e). A torque due to the tension force in the string (T) acts on the pulley. With $\tau = I\alpha$ (considering only the upper dashed box in Fig. 8.20), we obtain

$$\tau_{\text{net}} = r_\perp F = RT = I\alpha = (\tfrac{1}{2}MR^2)\alpha$$

so that

$$\alpha = \frac{2T}{MR}$$

The linear acceleration of the block and the angular acceleration of the pulley are related by $a = R\alpha$, where a is the tangential acceleration, and

$$a = R\alpha = \frac{2T}{M} \qquad (1)$$

But T is unknown. Looking at the descending mass (the lower dashed box) and summing the forces in the vertical direction (positive in the direction of motion) gives

$$mg - T = ma$$

or

$$T = mg - ma \qquad (2)$$

Using Eq. 2 to eliminate T from Eq. 1 yields

$$a = \frac{2T}{M} = \frac{2(mg - ma)}{M}$$

Solving for a, we get

$$a = \frac{2mg}{(2m + M)} \qquad (3)$$

Note that if $M \to 0$ (as in the case of ideal, massless pulleys in previous chapters), then $I \to 0$ and $a = g$ (from Eq. 3). Here, however, $M \neq 0$, so we have $a < g$. (Why?)

Follow-up Exercise. Pulleys can be analyzed even more realistically. In this Example, friction was neglected, but practically, a frictional torque (τ_f) exists and should be included. What would be the form, as in Eq. 3, of the angular acceleration in this case? Show that your result is dimensionally correct.

In pulley problems, we also neglect the mass of the string, an approach that still gives a good approximation if the string is relatively light. Taking the mass of the string into account would give a continuously varying mass hanging on the pulley, thus producing a variable torque. Such a problem is beyond the scope of this book.

Suppose you had masses suspended from each side of a pulley. Here, you'd have to compute the net torque. If you didn't know the values of the masses or which way the pulley would rotate, then you could simply assume a direction. As in the linear case, if the result came out with the opposite sign, it would indicate that you had assumed the wrong direction.

Problem-Solving Hint

For problems such as those of Examples 8.10 and 8.11, dealing with coupled rotational and translational motions, keep in mind that, with no rope slippage, the magnitudes of the accelerations are usually related by $a = r\alpha$, while $v = r\omega$ relates the magnitudes of the velocities at any instant of time. Applying Newton's second law (in rotational or linear form) to different parts of the system gives equations that can be combined by using such relationships. Also, for rolling without slipping, $a = r\alpha$ and $v = r\omega$ relate the angular quantities to the linear motion of the center of mass.

Another application of rotational dynamics is the analysis of motion on an inclined plane. Previously, we worked with block-shaped objects sliding down planes. Now we can generalize to include objects that can roll.

Integrated Example 8.12 ■ Down It Goes: The Yo-Yo

Most everyone has played with a yo-yo, a very popular toy. If you stop and think about it, the yo-yo's operation involves forces, torques, and rotational dynamics. There is a lot of physics in a yo-yo that can be quite complicated. Here, we will consider the simple case of a descending yo-yo.

A yo-yo descends from rest, increasing its rotational speed as it goes. The forces involved are the tension in the string (**T**) and the yo-yo's weight. (a) Make a sketch of the descending yo-yo, showing the forces. (b) Determine the magnitudes of the tension in the string and the yo-yo's downward translational (linear) acceleration, using general principles. (Neglect the mass of the string.)

(a) Conceptual Reasoning. Let's make a sketch to help better understand the situation and "learn by drawing" (Fig. 8.21). Notice that the tension **T** produces a torque that causes the yo-yo to rotate.

(b) Thinking It Through. The answers can be expressed in symbol equation form. (Note that no numerical values are given.) Looking at the sketch (▸ Fig. 8.21), we can see that there are vertical forces (in opposite directions) and a torque due to the tension force in the string, with a lever arm of r. In analyzing motion and forces, Newton's second law is always a good place to start—but remember, we now have two forms of the second law, one for translational (linear) motion, $\mathbf{F}_{net} = m\mathbf{a}$, and one for rotational motion, $\tau_{net} = I\alpha$. To apply these two forms of Newton's second law, we sum the forces (T and mg) and the torques.

Given: Force–torque situation (Fig. 8.21).   *Find:* T (tension in the string)
a (linear acceleration of yo-yo)

Both forces are vertical and there is only one torque, so (choosing downward as positive for convenience)

$$\Sigma \mathbf{F}_y: \quad mg - T = ma \tag{1}$$

$$\Sigma \tau: \quad rT = I\alpha \tag{2}$$

(where I is about an axis through the yo-yo's center of mass)

Since we want to find the linear acceleration, α in (2) can be converted to linear tangential acceleration using $\alpha = a/r$:

$$rT = I(a/r) \quad \text{or} \quad r^2 T = Ia$$

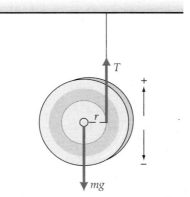

▲ **FIGURE 8.21   Down it goes!**
See Integrated Example 8.12

Substituting T from (1) yields

$$r^2(mg - ma) = Ia$$

and

$$a = \frac{mr^2g}{I + mr^2} = \left(\frac{mr^2}{I + mr^2}\right)g \qquad (3)$$

where the equation has been written so you can compare a with g. Then, solving for T from (1), we obtain

$$T = mg - ma = mg - m\left(\frac{mr^2g}{I + mr^2}\right) = mg\left(1 - \frac{mr^2}{I + mr^2}\right) = mg\left(\frac{I}{I + mr^2}\right)$$

and

$$T = mg\left(\frac{I}{I + mr^2}\right) \qquad (4)$$

which shows that T is less than mg.

Writing Eqs. (3) and (4) in these forms allows the results to be checked to see if they are reasonable. From Eq. (3), it can be seen that $a < g$. That is, the translational acceleration of the yo-yo descending on the string is smaller than it would be in free fall, which indicates that the velocity would be smaller also, at any time. This would be expected, but the changes in potential energy would be the same in each case. The yo-yo descending on the string would have less kinetic energy than if it were in free fall. Where did this energy go? Don't worry, energy is still conserved. Some of the potential energy goes into the *rotational* kinetic energy of the yo-yo, which has both translational and rotational kinetic energies, as will be discussed in detail in the next section. (This was a sneak preview.)

From Eq. (4), it can be seen that $T < mg$, which also would be as expected, since the CM of the yo-yo accelerates downward. This would be the case until the yo-yo gets to the end of the string and is in a "sleeper" mode (spinning in place in the loop of the string, with $T = mg$). With a lot of rotational energy, various "sleeper" tricks can be performed. (The motions of a yo-yo are much more complicated than this simplified case. For example, it is very difficult to explain getting the yo-yo to return to the hand by giving it a quick pull so that the string rewinds.)

Follow-up Exercise. Initially, a yo-yo is usually "thrown" downward. What effect does this have?

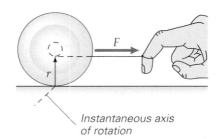

Instantaneous axis of rotation

▲ **FIGURE 8.22 Pulling the yo-yo's string** See Conceptual Example 8.13.

Conceptual Example 8.13 ■ Applying a Torque One More Time: Which Way Does the Yo-yo Roll?

The string of a yo-yo sitting on a level surface is pulled as shown in ◄Fig. 8.22. Will the yo-yo roll (a) toward the person or (b) away from the person?

Reasoning and Answer. Apply the physics we have just studied to the situation. Note that the instantaneous axis of rotation is along the line of contact of the yo-yo with the surface. If you had a stick standing vertically in place of the **r** vector and pulled on a string attached to the top of the stick in the direction of **F**, which way would the stick rotate? Of course, it would rotate clockwise (about its instantaneous axis of rotation). The yo-yo reacts similarly; that is, it rolls in the direction of the pull, so the answer is (a). (Get a yo-yo and try it if you're a nonbeliever.)

There is more interesting physics in our yo-yo situation. The pull force is not the only force acting on the yo-yo; there are three others. Do they contribute torques? Let's identify these forces. There's the weight of the yo-yo and the normal force from the surface. Also, there is a horizontal force of static friction between the yo-yo and the surface. (Otherwise the yo-yo would slide rather than roll.) But these three forces act through the line of contact or through the instantaneous axis of rotation, so there are no torques here. (Why?)

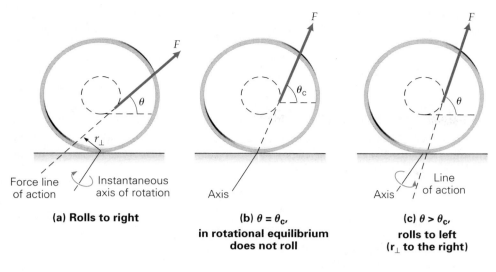

(a) Rolls to right

Force line of action

Instantaneous axis of rotation

$r_\perp$

θ

F

(b) $\theta = \theta_c$,
in rotational equilibrium
does not roll

Axis

θ_c

F

(c) $\theta > \theta_c$,
rolls to left
($r_\perp$ to the right)

Axis

Line of action

θ

F

◀ **FIGURE 8.23 The angle makes a difference** **(a)** With the line of action to the left of the instantaneous axis, the yo-yo rolls to the right. **(b)** At a critical angle θ_c, the line of action passes through the axis, and the yo-yo is in equilibrium. **(c)** When the line of action is to the right of the axis, the yo-yo rolls to the left. See Conceptual Example 8.13.

What would happen if we increased the angle of the string or pull force (relative to the horizontal) as illustrated in ▲Fig. 8.23a? The yo-yo would still roll to the right. As can be seen in Fig. 8.23b, at some critical angle θ_c the line of force goes through the axis of rotation, and the net torque on the yo-yo is zero, so the yo-yo does not roll.

If we exceed this critical angle (Fig. 8.23c), the yo-yo will begin to roll counterclockwise, or to the left. Note that the line of action of the force is on the other side of the axis of rotation than that in Fig. 8.23a and that the lever arm ($r_\perp$) has changed directions, resulting in a reversed net torque direction.

Follow-up Exercise.   Suppose you set the yo-yo string at the critical angle, with the string over a round, horizontal bar at the appropriate height, and you suspend a weight on the end of the string to supply the force for the equilibrium condition. What will happen if you then pull the yo-yo toward you, away from its equilibrium position, and release it?

8.4 Rotational Work and Kinetic Energy

OBJECTIVES: **To discuss, explain, and use the rotational forms of (a) work, (b) kinetic energy, and (c) power.**

This section gives the rotational analogues of various equations of linear motion associated with work and kinetic energy for constant torques. Because their development is similar to that given for their linear counterparts, detailed discussion is not needed. As in Chapter 5, it is understood that W is the net work if more than one force or torque acts on an object.

Rotational Work We can go directly from work done by a force to work done by a torque, since the two are related ($\tau = r_\perp F$). For rotational motion, the **rotational work** $W = Fs$ done by a single force F acting tangentially along an arc length s is

Definition of: Rotational work

$$W = Fs = F(r_\perp \theta) = \tau \theta$$

where θ is in radians. Thus, for a single torque acting through an angle of rotation, θ,

$$W = \tau \theta \quad \text{(single force)} \qquad (8.9)$$

In this book, both the torque (τ) and angular displacement (θ) vectors are almost always along the fixed axis of rotation, so you will not need to be concerned about parallel components, as you were for translational work. The torque and angular displacement may be in opposite directions, in which case the torque does negative work and slows down the rotation of the body. Negative rotational work is analogous to F and d being in opposite directions for translational motion.

Rotational Power An expression for **rotational power**, the rotational analogue of power (the time rate of doing work), is easily obtained from Eq. 8.9:

$$P = \frac{W}{t} = \tau\left(\frac{\theta}{t}\right) = \tau\omega \tag{8.10}$$

The Work–Energy Theorem and Kinetic Energy

The relationship between the net rotational work done on a rigid body and the change in rotational kinetic energy of the body can be derived as follows, starting with the equation for rotational work:

$$W_{net} = \tau\theta = I\alpha\theta$$

Since, by hypothesis, our torques are due only to constant forces, we get a constant α. But from rotational kinematics in Chapter 7, we know that for a constant angular acceleration, $\omega^2 = \omega_0^2 + 2\alpha\theta$, and

$$W_{net} = I\left(\frac{\omega^2 - \omega_0^2}{2}\right) = \tfrac{1}{2}I\omega^2 - \tfrac{1}{2}I\omega_0^2$$

Using Eq. 5.6 (work–energy), we know that $W_{net} = \Delta K$. Therefore,

$$W_{net} = \tfrac{1}{2}I\omega^2 - \tfrac{1}{2}I\omega_0^2 = K - K_0 = \Delta K \tag{8.11}$$

Then the expression for **rotational kinetic energy**, K, is

$$K = \tfrac{1}{2}I\omega^2 \tag{8.12}$$

Thus, *the net rotational work done on an object is equal to the change in the rotational kinetic energy of the object* (with zero linear kinetic energy). Consequently, to change the rotational kinetic energy of an object, a net torque must be applied.

It is possible to derive the expression for the kinetic energy of a rotating rigid body (on a fixed axis) directly. Summing the instantaneous kinetic energies of the body's individual particles relative to the fixed axis gives

$$K = \tfrac{1}{2}\Sigma m_i v_i^2 = \tfrac{1}{2}(\Sigma m_i r_i^2)\omega^2 = \tfrac{1}{2}I\omega^2$$

where, for each particle of the body, $v_i = r_i\omega$. Thus, Eq. 8.12 doesn't represent a new form of energy; rather, it is simply another expression for kinetic energy, in a form that is more convenient for rigid-body rotation.

A summary of translational and rotational analogues is given in Table 8.1. (The table also contains momentum, which we shall discuss in Section 8.5.)

When an object has both translational and rotational motion, its total kinetic energy may be divided into parts to reflect the two kinds of motion. For example,

TABLE 8.1 Translational and Rotational Quantities and Equations

Translational		Rotational	
Force:	$\mathbf{F}$	Torque (magnitude):	$\tau = rF\sin\theta$
Mass (inertia):	m	Moment of inertia:	$I = \Sigma m_i r_i^2$
Newton's second law:	$\mathbf{F}_{net} = m\mathbf{a}$	Newton's second law:	$\tau_{net} = I\boldsymbol{\alpha}$
Work:	$W = Fd$	Work:	$W = \tau\theta$
Power:	$P = Fv$	Power:	$P = \tau\omega$
Kinetic energy:	$K = \tfrac{1}{2}mv^2$	Kinetic energy:	$K = \tfrac{1}{2}I\omega^2$
Work–energy theorem:	$W_{net} = \tfrac{1}{2}mv^2 - \tfrac{1}{2}mv_0^2 = \Delta K$	Work–energy theorem:	$W_{net} = \tfrac{1}{2}I\omega^2 - \tfrac{1}{2}I\omega_0^2 = \Delta K$
Linear momentum:	$\mathbf{p} = m\mathbf{v}$	Angular momentum:	$\mathbf{L} = I\boldsymbol{\omega}$

for a cylinder rolling without slipping on a level surface, the motion is purely rotational relative to the instantaneous axis of rotation (the point or line of contact), which is instantaneously at rest. The total kinetic energy of the rolling cylinder is

$$K = \tfrac{1}{2}I_i\,\omega^2$$

where I_i is the moment of inertia about the instantaneous axis. This moment of inertia about the point of contact (our axis) is given by the parallel-axis theorem (Eq. 8.8), $I_i = I_{CM} + MR^2$, where R is the radius of the cylinder. Then

$$K = \tfrac{1}{2}I_i\,\omega^2 = \tfrac{1}{2}(I_{CM} + MR^2)\omega^2 = \tfrac{1}{2}I_{CM}\,\omega^2 + \tfrac{1}{2}MR^2\,\omega^2$$

But since there is no slipping, $v_{CM} = R\omega$, and

$$\underset{\substack{total \\ KE}}{K} = \underset{\substack{rotational \\ KE}}{\tfrac{1}{2}I_{CM}\,\omega^2} + \underset{\substack{translational \\ KE}}{\tfrac{1}{2}Mv_{CM}^2} \qquad (rolling,\ no\ slipping) \quad (8.13)$$

Total kinetic energy

Note that although a cylinder was used as an example here, this is a general result and applies to any object that is rolling without slipping.

Thus, *the total kinetic energy of such an object is the sum of two contributions: the translational kinetic energy of the object's center of mass and the rotational kinetic energy of the object relative to a horizontal axis through its center of mass.*

Note: A rolling body has both translational kinetic energy and rotational kinetic energy.

Example 8.14 ■ Division of Energy: Rotational and Translational

A uniform, solid 1.0-kg cylinder rolls without slipping at a speed of 1.8 m/s on a flat surface. (a) What is the total kinetic energy of the cylinder? (b) What percentage of this total is rotational kinetic energy?

Thinking It Through. The cylinder has both rotational and translational kinetic energies, so Eq. 8.13 applies, and its terms are related by the condition of rolling without slipping.

Solution.

Given: $M = 1.0\,\text{kg}$

$v_{CM} = 1.8\,\text{m/s}$

$I_{CM} = \tfrac{1}{2}MR^2$ (from Fig. 8.17e)

Find: (a) K (total kinetic energy)

(b) $\dfrac{K_r}{K}$ ($\times 100\%$) (percentage of rotational energy)

(a) The cylinder rolls without slipping, so the condition $v_{CM} = R\omega$ applies. Then the total kinetic energy is the sum of the rotational kinetic energy K_r and the translational kinetic energy of the center of mass, K_{CM} (Eq. 8.13):

$$K = \tfrac{1}{2}(\tfrac{1}{2}MR^2)\left(\dfrac{v_{CM}}{R}\right)^2 + \tfrac{1}{2}Mv_{CM}^2 = \tfrac{1}{4}Mv_{CM}^2 + \tfrac{1}{2}Mv_{CM}^2$$

$$= \tfrac{3}{4}Mv_{CM}^2 = \tfrac{3}{4}(1.0\,\text{kg})(1.8\,\text{m/s})^2 = 2.4\,\text{J}$$

(b) The rotational kinetic energy K_r of the cylinder is the first term of the preceding equation, so, forming a ratio in symbol form, we get

$$\dfrac{K_r}{K} = \dfrac{\tfrac{1}{4}Mv_{CM}^2}{\tfrac{3}{4}Mv_{CM}^2} = \tfrac{1}{3}(\times 100\%) = 33\%$$

Thus, the total kinetic energy of the cylinder is made up of rotational and translational parts, with one-third being rotational.

Note that in part (b) the radius of the cylinder was not needed and neither were the mass or any numerical values. Because we used a ratio, these quantities canceled out. *Don't* think that this exact division of energy is a general result, however: It is easy to show that the percentage is different for objects with different moments of inertia.

For example, you should expect a rolling sphere to have a smaller percentage of rotational kinetic energy than a cylinder has, because the sphere has a smaller moment of inertia ($I = \frac{2}{5}MR^2$).

Follow-up Exercise. Potential energy can be brought into the act by applying the conservation of energy to an object rolling up or down an inclined plane. In this Example, suppose that the cylinder rolled up a 20° inclined plane without slipping. (a) At what vertical height (measured by the vertical distance of its CM) on the plane does the cylinder stop? (b) To find the height in part (a), you probably equated the initial total kinetic energy to the final gravitational potential energy. That is, the total kinetic energy was reduced by the work done by gravity. However, a frictional force also acts (to prevent slipping). Is there not work done here, too?

Example 8.15 ■ Rolling Down versus Sliding Down: Which Is Faster?

A uniform cylindrical hoop is released from rest at a height of 0.25 m near the top of an inclined plane (◀Fig. 8.24). If the cylinder rolls down the plane without slipping and there is no energy loss due to friction, what is the linear speed of the cylinder's center of mass at the bottom of the incline?

Thinking It Through. Here, gravitational potential energy is converted into kinetic energy—both rotational and translational. The conservation of (mechanical) energy applies, since W_f is zero.

Solution.

Given: $h = 0.25$ m   *Find:* v_{CM} (speed of CM)
$I_{CM} = MR^2$ (from Fig. 8.17d)

Because the total mechanical energy of the cylinder is conserved, you can write

$$E_o = E$$

or, since $v_o = 0$ at the top of the incline and assuming that $U = 0$ at the bottom,

$$U_o = K$$
$$\underbrace{Mgh}_{\text{at rest}} = \underbrace{\tfrac{1}{2}I_{CM}\omega^2 + \tfrac{1}{2}Mv_{CM}^2}_{\text{at bottom of incline}}$$

Using the rolling condition $v_{CM} = R\omega$ gives

$$Mgh = \tfrac{1}{2}(MR^2)\left(\frac{v_{CM}}{R}\right)^2 + \tfrac{1}{2}Mv_{CM}^2 = Mv_{CM}^2$$

Solving for v_{CM}, we get

$$v_{CM} = \sqrt{gh} = \sqrt{(9.8 \text{ m/s}^2)(0.25 \text{ m})} = 1.6 \text{ m/s}$$

Again, not much numerical information was needed here. Note that the hoop rolls down from the same height that the solid cylinder rolled up in the Example 8.14 Follow-up Exercise, yet the speed of the hoop is less than that of the cylinder at the bottom of the incline. Why? Because of differences in the moment of inertia.

Follow-up Exercise. Suppose the inclined plane in this Example were frictionless and the hoop slid down the plane instead of rolling. How would the speed at the bottom compare in this case? Why are the speeds different?

A Fixed Race As Example 8.15 shows, v_{CM} is independent of M and R. The masses and radii cancel out, so all objects of a particular shape (with the same equation for the moment of inertia) roll with the same speed, regardless of their size or density. But the rolling speed does vary with the moment of inertia, which varies with the shape. Therefore, rigid bodies with different shapes roll with different speeds. For

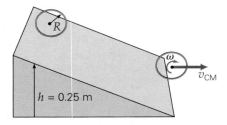

▲ **FIGURE 8.24 Rolling motion and energy** When an object rolls down an inclined plane, potential energy is converted to translational *and* rotational kinetic energy. This makes the rolling slower than frictionless sliding. See Example 8.15.

PHYSLET®
ILLUSTRATION

Rolling on an Incline

example, if you released a cylindrical hoop, a solid cylinder, and a uniform sphere at the same time from the top of an inclined plane, the sphere would win the race to the bottom, followed by the cylinder, with the hoop coming in last—every time!

You can try this experiment with a couple of cans of food or other cylindrical containers—one full of some solid material (in effect, a rigid body) and one empty and with the ends cut out—and a smooth, solid ball. Remember that the masses and the radii make no difference. You might think that an annular cylinder (a hollow cylinder with inner and outer radii that vary appreciably—Fig. 8.17f) would be a possible front-runner, or "front-roller," in such a race, but it wouldn't win. The rolling race down an incline is fixed even when you vary the masses and the radii. (See Exercise 84.)

Another aspect of rolling is discussed in the Insight "Slide or Roll to a Stop."

8.5 Angular Momentum

OBJECTIVES: To (a) define angular momentum and (b) apply the conservation of angular momentum to physical situations.

Another important quantity in rotational motion is angular momentum. Recall from Section 6.1 how the linear momentum of an object is changed by a force. Analogously, changes in angular momentum are associated with torques. As we

INSIGHT

Slide or Roll to a Stop?—Antilock Brakes

While driving, in an emergency you may instinctively jam on the brakes, trying to come to a quick stop—that is, to stop in the shortest distance. But with the wheels locked, the car skids, or slides, to a stop, often out of control. In this case, the force of sliding friction is acting on the wheels.

To prevent skidding, you may have learned to pump the brakes in order to roll rather than slide to a stop, particularly on wet or icy roads. Many newer automobiles have a computerized antilock braking system (ABS) that does this automatically. When the brakes are applied firmly and the car begins to slide, sensors in the wheels note the sliding motion, and a computer takes over control of the braking system. It momentarily releases the brakes and then varies the brake-fluid pressure with a pumping action (up to thirteen times per second!) so that the wheels will continue to roll without slipping.

In the absence of sliding, both rolling friction and static friction act. In many cases, however, the force of rolling friction is small, and only static friction need be taken into account, as will be done here. The ABS works to keep static friction near the maximum, $f_s \approx f_{s_{max}}$, which you can't do easily by foot.

Does sliding instead of rolling make a big difference in an automobile's stopping distance? We can calculate the difference by assuming that rolling friction is negligible. Although the external force of static friction does no work to dissipate energy in slowing a car (this is done internally by friction on the brake pads), it does determine whether the wheels roll or slide.

In Example 2.8, a vehicle's stopping distance was given by

$$x = \frac{v_o^2}{2a}$$

By Newton's second law, the net force in the horizontal direction is

$$F = f = \mu N = \mu mg = ma$$

and the stopping acceleration is then

$$a = \mu g$$

Thus,

$$x = \frac{v_o^2}{2\mu g} \qquad (1)$$

But, as was noted in Chapter 4, the coefficient of sliding (kinetic) friction is generally less than that of static friction; that is, $\mu_k < \mu_s$. The general difference between rolling and sliding stops can be seen by using the same initial velocity v_o for both cases. Then, using Eq. 1 to form a ratio, we obtain

$$\frac{x_{roll}}{x_{slide}} = \frac{\mu_k}{\mu_s} \qquad or \qquad x_{roll} = \left(\frac{\mu_k}{\mu_s}\right) x_{slide}$$

From Table 4.1, the value of μ_k for rubber on wet concrete is 0.60, and the value of μ_s for these surfaces is 0.80. Using these values for a comparison of the stopping distances gives

$$x_{roll} = \left(\frac{0.60}{0.80}\right) x_{slide} = (0.75) x_{slide}$$

Thus, the car comes to a rolling stop in 75 percent of the distance required for a sliding stop—for example, 15 m instead of 20 m. Although this may vary for different conditions, it could be an important, perhaps lifesaving, difference.

have seen, torque is the product of a moment arm and a force. In a similar manner, **angular momentum (L)** is the product of a moment arm and a linear momentum. For a particle of mass m, the magnitude of the linear momentum is $p = mv$, where $v = r\omega$. The magnitude of the angular momentum is

$$L = r_\perp p = mr_\perp v = mr_\perp^2 \omega \qquad \begin{array}{l} \textit{single-particle} \\ \textit{angular momentum} \end{array} \qquad (8.14)$$

SI unit of angular momentum: kilogram-meters squared per second $(\text{kg} \cdot \text{m}^2/\text{s})$

where v is the speed of the particle, $r_\perp$ is the moment arm, and ω is the angular speed.

For circular motion, $r_\perp = r$, since **v** is perpendicular to **r**. For a system of particles making up a rigid body, all the particles travel in circles, and the total magnitude of the angular momentum is

$$L = (\Sigma m_i r_i^2)\omega = I\omega \qquad \begin{array}{l} \textit{rigid-body} \\ \textit{angular momentum} \end{array} \qquad (8.15)$$

which, for rotation about a fixed axis, is (in vector notation)

$$\mathbf{L} = I\boldsymbol{\omega} \qquad (8.16)$$

Definition of: Angular momentum

Thus, **L** is in the direction of the angular velocity vector $(\boldsymbol{\omega})$. This direction is given by the right-hand rule.

Note: Review Eq. 6.3, Section 6.1.

For linear motion, the change in the total linear momentum of a system is related to the net external force by $\mathbf{F}_{\text{net}} = \Delta\mathbf{P}/\Delta t$. Angular momentum is analogously related to net torque:

$$\tau_{\text{net}} = I\alpha = \frac{I\Delta\omega}{\Delta t} = \frac{\Delta(I\omega)}{\Delta t} = \frac{\Delta\mathbf{L}}{\Delta t}$$

That is,

Net torque: The time rate of change of angular momentum

$$\tau_{\text{net}} = \frac{\Delta\mathbf{L}}{\Delta t} \qquad (8.17)$$

Thus, the net torque is equal to *the time rate of change of angular momentum*. In other words, a net torque causes a *change* in angular momentum.

Conservation of Angular Momentum

Equation 8.17 was derived by using $\tau_{\text{net}} = I\alpha$, which applies to a rigid system of particles or a rigid body having a constant moment of inertia. However, Eq. 8.17 is a general equation that also applies to a nonrigid system of particles. In such a system, there may be a change in the mass distribution and a change in the moment of inertia. As a result, there may be an angular acceleration even in the absence of a net torque. How can this be?

If the net torque on a system is zero, then, by Eq. 8.17, $\tau_{\text{net}} = \Delta\mathbf{L}/\Delta t = 0$, and

$$\Delta\mathbf{L} = \mathbf{L} - \mathbf{L}_\text{o} = I\omega - I_\text{o}\omega_\text{o} = 0$$

or

conservation of angular momentum

$$I\omega = I_\text{o}\omega_\text{o} \qquad (8.18)$$

Note: Angular momentum is conserved when the net torque is zero. (L stays fixed.) This is the third conservation law in mechanics.

Thus, the condition for the **conservation of angular momentum** is as follows:

In the absence of an external, unbalanced torque, the total (vector) angular momentum of a system is conserved (remains constant).

As with total linear momentum, the internal torques arising from internal forces cancel out.

For a rigid body with a constant moment of inertia (i.e., $I = I_o$), the angular speed remains constant ($\omega = \omega_o$) in the absence of a net torque. But it is possible for the moment of inertia to change in some systems, giving rise to a change in the angular speed, as the following Example illustrates.

Example 8.16 ■ Pull It Down: Conservation of Angular Momentum

A small ball at the end of a string that passes through a tube is swung in a circle, as illustrated in ▼Fig. 8.25. When the string is pulled downward through the tube, the angular speed of the ball increases. (a) Is the increase in angular speed caused by a torque due to the pulling force? (b) If the ball is initially swung at a speed of 2.8 m/s in a circle with a radius of 0.30 m, what will be its tangential speed if the string is pulled down far enough to reduce the radius of the circle to 0.15 m? (Neglect the mass of the string.)

Thinking It Through. (a) A force is applied to the ball via the string, but consider the axis of rotation. (b) In the absence of a net torque, the angular momentum is conserved (Eq. 8.18), and the tangential speed is related to the angular speed by $v = r\omega$.

Solution.

Given: $r_1 = 0.30$ m   *Find:* (a) Cause of the increase in angular speed
$r_2 = 0.15$ m   (b) v_2 (final tangential speed)
$v_1 = 2.8$ m/s

(a) The change in the angular velocity, or an angular acceleration, is not caused by a torque due to the pulling force. The force on the ball, as transmitted by the string (tension), acts through the axis of rotation, and therefore the torque is zero. Because the rotating portion of the string is shortened, the moment of inertia of the ball ($I = mr^2$, from Fig. 8.17a) decreases. This causes the ball to speed up, since, in the absence of an external torque, the angular momentum of the ball is conserved.

(b) Because the angular momentum is conserved, we can equate the magnitudes of the angular momenta:

$$I_o\omega_o = I\omega$$

Then, using $I = mr^2$ and $\omega = v/r$ gives

$$mr_1v_1 = mr_2v_2$$

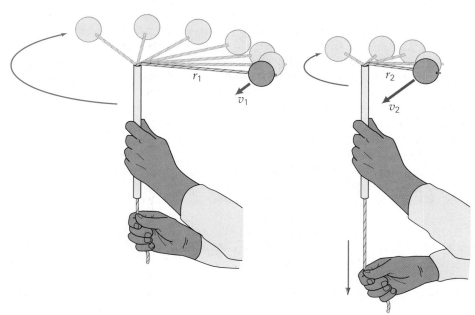

◀ **FIGURE 8.25 Conservation of angular momentum** When the string is pulled downward through the tube, the revolving ball speeds up. See Example 8.16.

and

$$v_2 = \left(\frac{r_1}{r_2}\right)v_1 = \left(\frac{0.30\ \text{m}}{0.15\ \text{m}}\right)2.8\ \text{m/s} = 5.6\ \text{m/s}$$

When the radial distance is shortened, the ball speeds up.

Follow-up Exercise. Let's look at the situation in this Example in terms of work and energy. If the initial speed is the same and the vertical pulling force is 7.8 N, what is the final speed of the 0.10-kg ball?

Example 8.16 should help you understand Kepler's law of equal areas (Chapter 7) from another viewpoint. A planet's angular momentum is conserved to a good approximation, since the Sun exerts no torque on it (why?) and we can neglect the torques from other planets (why?). Thus, when a planet is closer to the Sun in its elliptical orbit and so has a shorter moment arm, its speed is greater, by the conservation of angular momentum. Similarly, when an orbiting satellite's altitude varies during the course of an elliptical orbit about a planet, the satellite speeds up or slows down in accordance with the same principle.

Real-Life Angular Momentum

A popular demonstration of the conservation of angular momentum is shown in ▸Fig. 8.26a. A person sitting on a stool that rotates holds weights with his arms outstretched and is started slowly rotating. An external torque to start this rotation must be supplied by someone else, because the person on the stool cannot initiate the motion by himself. (Why not?) Once rotating, if the person brings his arms inward, the angular speed increases and he spins much faster. Extending his arms again slows him down. Can you explain this phenomenon?

If L is constant, what happens to ω when I is made smaller by reducing r? The angular speed must increase to compensate and keep L constant. Ice skaters do dizzying spins by pulling in their arms to reduce their moment of inertia (Fig. 8.26b). Similarly, a diver spins during a high dive by tucking in the body and limbs, greatly decreasing his or her moment of inertia. The enormous wind speeds of tornadoes and hurricanes represent another example of the same effect (Fig. 8.26c).

Angular momentum also plays a role in ice-skating jumps in which the skater spins in the air, such as a triple axel or triple lutz. A torque applied on the jump gives the skater angular momentum, and the arms and legs are drawn into the body, which, as in spinning on one's toes, decreases the moment of inertia and increases the angular speed so that multiple spins can be made during the jump. To land with a smaller rate of spin, the skater opens the arms and the nonlanding leg. You may have noticed that most jump landings proceed in a curved arc, which allows the skater to gain control.

Angular momentum, **L**, is a vector, and when it is conserved or constant, its magnitude *and* direction must remain unchanged. Thus, when no external torques act, the direction of **L** is fixed in space. This is the principle behind passing a football accurately, as well as that behind the movement of a gyrocompass (▸Fig. 8.27). A football is normally passed with a spiraling rotation. This spin, or gyroscopic action, stabilizes the ball's spin axis in the direction of motion. Similarly, rifle bullets are set spinning by the rifling in the barrel for directional stability.

The **L** vector of a spinning gyroscope in the compass is set in a particular direction (usually north). In the absence of external torques, the compass direction remains fixed, even though its carrier (an airplane or ship, for example) changes directions. You may have played with a toy gyroscope that is set spinning and placed on a pedestal. In a "sleeping" condition, the gyro stands straight up with its angular-momentum vector fixed in space for some time. The gyro's center of gravity is on the axis of rotation, so there is no net torque due to its weight.

(a)

(b)

(c)

◀ **FIGURE 8.26 Change in moment of inertia**
(a) Spinning slowly with masses in outstretched arms, the man's moment of inertia is relatively large. (The masses are farther from the axis of rotation.) Note that he is isolated, with no external torques (neglecting friction) acting on him, so his angular momentum, $L = I\omega$, is conserved. Pulling his arms inward decreases his moment of inertia. (Why?) Consequently, ω must increase, and he goes into a dizzying spin. **(b)** Ice skaters change their moment of inertia to increase ω in doing spins. **(c)** The same principle helps explain the violence of the winds that spiral around the center of a hurricane. As air rushes in toward the center of the storm, where the pressure is low, its rotational velocity must increase for angular momentum to be conserved.

However, the gyroscope eventually slows down because of friction, causing **L** to tilt. In watching this motion, you may have noticed how the spin axis revolves, or *precesses*, about the vertical axis. It revolves tilted over, so to speak (▶Fig. 8.28a). Since the gyroscope precesses, the angular-momentum vector **L** is no longer constant in direction, indicating that a torque must be acting to produce a change (Δ**L**) with time. As can be seen from the figure, the torque arises from the vertical component of the weight force, since the center of gravity no longer lies directly above the point of support or on the vertical axis of rotation. The instantaneous torque is such that the gyroscope's axis moves or precesses about the vertical axis.

In a similar manner, the Earth's rotational axis precesses. The Earth's spin axis is tilted $23\frac{1}{2}°$ with respect to a line perpendicular to the plane of its revolution about the Sun; the axis precesses about this line (Fig. 8.28b). The precession is due to slight gravitational torques exerted on the Earth by the Sun and the Moon. The torques arise because the Earth is not perfectly spherical—it has an equatorial bulge.

The period of the precession of the Earth's axis is about 26 000 years, so the precession has little day-to-day effect. However, it does have an interesting long-term effect. Polaris will not always be (nor has it always been) the North Star—that is, the star toward which the Earth's axis of rotation points. About 5000 years ago, Alpha Draconis was the North Star, and 5000 years from now it will be Alpha Cephei, which is at an angular distance of about 68° away from Polaris on the circle described by the precession of the Earth's axis.

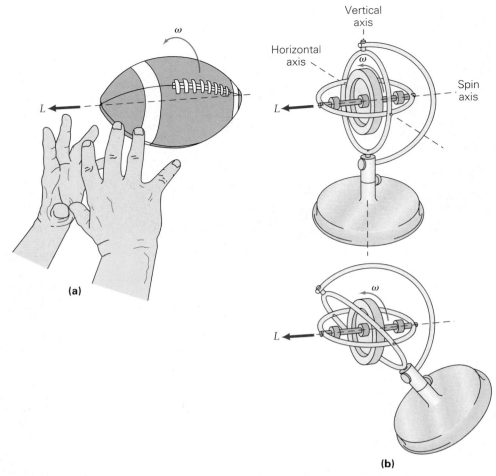

(a)

(b)

▲ **FIGURE 8.27   Constant direction of angular momentum**   When angular momentum is conserved, its direction is constant in space. **(a)** This principle can be demonstrated by a passed football. **(b)** Gyroscopic action also occurs in a gyroscope, a rotating wheel that is universally mounted on gimbals (rings) so that it is free to turn about any axis. When the frame moves, the wheel maintains its direction. This is the principle of the gyrocompass.

▶ **FIGURE 8.28   Precession**   An external torque causes a change in angular momentum. **(a)** For a spinning gyroscope, this change is directional, and the axis of rotation precesses at angular acceleration ω_p about a vertical line. (The torque due to the weight force would point out of the page as drawn here, as would $\Delta\mathbf{L}$.) Note that although there is a torque that would topple a nonspinning gyroscope, a spinning gyroscope doesn't fall. **(b)** Similarly, the Earth's axis precesses because of gravitational torques caused by the Sun and the Moon. We don't notice this motion because the period of precession is about 26 000 years.

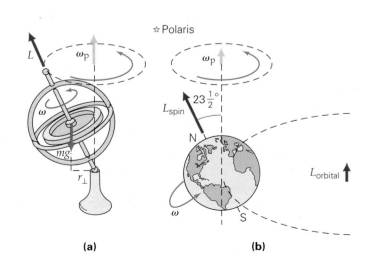

(a)   **(b)**

There are some other long-term torque effects on the Earth and the Moon. Did you know that the Earth's daily spin rate is slowing down and hence the days are getting longer? Also, that the Moon is receding, or getting farther away, from the Earth? This is due primarily to ocean tidal friction, which gives rise to a torque. As a result, the Earth's spin angular momentum, and therefore its rate of rotation, is changing. The slowing rate of rotation causes the average day to be longer; this century will be about 25 seconds longer than the previous.

But the rate we are talking about is an average rate. At times, the Earth's rotation speeds up for relatively short periods. This increase is thought to be associated with the rotational inertia of the liquid layer of the Earth's core. (See the Insight in Chapter 13.)

The tidal torque on the Earth results chiefly from the Moon's gravitational attraction, which is the main cause of ocean tides. This torque is *internal* to the Earth–Moon system, so the total angular momentum of that system is conserved. Since the Earth is losing angular momentum, the Moon must be gaining angular momentum to keep the total angular momentum of the system constant. The Earth loses rotational (spin) angular momentum, and the Moon gains orbital angular momentum. As a result, the Moon drifts slightly farther from Earth and its orbital speed decreases. The Moon moves away from the Earth at about 1 cm per lunar month (calculated from the rate of decrease of the Earth's rotation). Thus, the Moon moves in a slowly widening spiral.

Finally, a common example in which angular momentum is an important consideration is the helicopter. What would happen if a helicopter had a single rotor? Since the motor supplying the torque is internal, the angular momentum would be conserved. Initially, $\mathbf{L} = 0$; hence, to conserve the total angular momentum of the system (rotor plus body), the separate angular momenta of the rotor and body would have to be in opposite directions to cancel. Thus, on takeoff, the rotor would rotate one way and the helicopter body the other, which is not a desirable situation.

To prevent this situation, helicopters have two rotors. Large helicopters have two overlapping rotors (▾Fig. 8.29a). The oppositely rotating rotors cancel each other's angular momenta, so the helicopter body does not have to rotate to provide

▼ **FIGURE 8.29 Different rotors** See text for description.

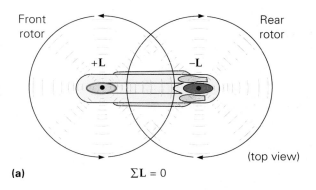

Front rotor

Rear rotor

$+\mathbf{L}$ $-\mathbf{L}$

(top view)

(a) $\Sigma \mathbf{L} = 0$

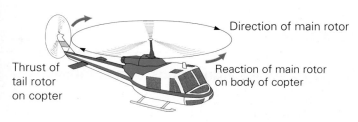

Direction of main rotor

Thrust of tail rotor on copter

Reaction of main rotor on body of copter

(b)

canceling angular momentum. The rotors are offset at different heights so that they do not collide.

Small helicopters with a single overhead rotor have small "anti-torque" tail rotors (Fig. 8.29b). The tail rotor produces a thrust like a propeller and supplies the torque to counterbalance the torque produced by the overhead rotor. The tail rotor also helps in steering the craft. By increasing or decreasing the tail rotor's thrust, the helicopter turns (rotates) one way or the other.

Chapter Review

Important Concepts and Equations

- **In pure translational motion,** all of the particles that make up a rigid object have the same instantaneous velocity.

- **In pure rotational motion (about a fixed axis),** all of the particles that make up a rigid object have the same instantaneous angular velocity.

  Condition for rolling without slipping:
  $$v_{CM} = r\omega$$
  $$\text{(or} \quad s = r\theta \quad \text{or} \quad a_{CM} = r\alpha) \tag{8.1}$$

- **Torque,** the rotational analogue of force, is the product of a force and a moment arm, or lever arm.

  Torque (magnitude):
  $$\tau = r_\perp F = rF \sin\theta \tag{8.2}$$
  (Direction is given by right-hand rule.)

- **Mechanical equilibrium** requires that the net force, or summation of the forces, be zero (translational equilibrium) and that the net torque, or summation of the torques, be zero (rotational equilibrium).

  Conditions for translational and rotational mechanical equilibrium, respectively:
  $$\mathbf{F}_{net} = \Sigma \mathbf{F}_i = 0 \quad \text{and} \quad \tau_{net} = \Sigma \tau_i = 0 \tag{8.3}$$

- An object is in **stable equilibrium** as long as its center of gravity, upon small displacement, lies above and inside the object's original base of support.

- **Moment of inertia** (*I*) is the rotational analogue of mass and is given by

  Rotational form of Newton's second law:
  $$I = \Sigma m_i r_i^2 \tag{8.6}$$
  $$\tau_{net} = I\alpha \tag{8.7}$$

Parallel-Axis Theorem:
$$I = I_{CM} + Md^2 \tag{8.8}$$

Rotational work:
$$W = \tau\theta \tag{8.9}$$

Rotational power:
$$P = \tau\omega \tag{8.10}$$

Work–Energy Theorem (rotational):
$$W_{net} = \Delta K = \tfrac{1}{2}I\omega^2 - \tfrac{1}{2}I\omega_o^2 \tag{8.11}$$

Rotational kinetic energy:
$$K = \tfrac{1}{2}I\omega^2 \tag{8.12}$$

Kinetic energy of a rolling object:
$$K = \tfrac{1}{2}I_{CM}\omega^2 + \tfrac{1}{2}Mv_{CM}^2 \tag{8.13}$$

- **Angular momentum,** the product of a moment arm and linear momentum, as well as the product of a moment of inertia and angular velocity, is conserved in the absence of an external, unbalanced torque.

Angular momentum of a particle in circular motion (magnitude):
$$L = r_\perp p = mr_\perp v = mr_\perp^2 \omega \tag{8.14}$$

Angular momentum of a rigid body:
$$\mathbf{L} = I\boldsymbol{\omega} \tag{8.16}$$

Torque as change in angular momentum:
$$\tau_{net} = \frac{\Delta \mathbf{L}}{\Delta t} \tag{8.17}$$

Conservation of angular momentum (with $\tau_{net} = 0$):
$$I\omega = I_o\omega_o \tag{8.18}$$

Exercises

8.1 Rigid Bodies, Translations, and Rotations

1. In pure rotational motion of a rigid body, (a) all the particles of the body have the same angular velocity, (b) all the particles of the body have the same tangential velocity, (c) acceleration is always zero, or (d) there are always two simultaneous axes of rotation.

2. The condition for rolling without slipping is (a) $a_c = r\omega^2$, (b) $v_{CM} = r\omega$, (c) $F = ma$, or (d) $a_c = v^2/r$.

3. **CQ** Suppose someone in your physics class says that it is possible for a rigid body to have translational motion and rotational motion at the same time. Would you agree? If so, give an example.

4. **CQ** For a rolling cylinder, what would happen if the tangential speed v were less than $r\omega$? Is it possible for v to be greater than $r\omega$? Explain.

5. For the tires on your skidding car, (a) $v_{CM} = r\omega$, (b) $v_{CM} > r\omega$, (c) $v_{CM} < r\omega$, or (d) none of the above.

6. **CQ** If the top of your automobile tire is moving with a speed of v, what is the reading of your speedometer?

7. **CQ** What is the instantaneous speed of the point on a car's tire that makes contact with the ground when the car is moving with a speed of v?

8. ■ A rope goes over a circular pulley with a radius of 6.5 cm. If the pulley makes four revolutions without the rope slipping, what length of rope passes over the pulley?

9. ■ A wheel rolls five revolutions on a horizontal surface without slipping. If the center of the wheel moves 3.2 m, what is the radius of the wheel?

10. ■ A wheel rolls uniformly on level ground without slipping. A piece of mud on the wheel flies off when it is at the 9:00 o'clock position (rear of wheel). Describe the subsequent motion of the mud.

11. ■ A circular disk with a radius of 0.25 m rolls without slipping on a level surface with an angular speed of 2.0 rad/s. (a) What is the linear speed of the center of mass of the disk? (b) What is the instantaneous tangential speed of the top of the disk?

12. ■■ A ball with a radius of 15 cm rolls on a level surface, and the translational speed of the center of mass is 0.25 m/s. What is the angular speed about the center of mass if the ball rolls without slipping?

13. **IE** ■■ (a) When a disk rolls without slipping, should the product $r\omega$ be (1) greater than, (2) equal to, or (3) less than v_{CM}? (b) A disk with a radius of 0.15 m rotates through 270° as it travels 0.71 m. Does the disk roll without slipping? Prove your answer.

14. ■■ Show that $a_{CM} = r\alpha$ is also a condition for rolling without slipping.

15. ■■■ A cylinder with a diameter of 20 cm rolls with an angular speed of 0.50 rad/s on a level surface. If the cylinder experiences a uniform tangential acceleration of 0.018 m/s² without slipping until its angular speed is 1.25 rad/s, through how many complete revolutions does the cylinder rotate during the time it accelerates?

8.2 Torque, Equilibrium, and Stability

16. It is possible to have a net torque when (a) all forces act through the axis of rotation, (b) $\Sigma \mathbf{F}_i = 0$, (c) an object is in rotational equilibrium, or (d) an object remains in unstable equilibrium.

17. If an object in unstable equilibrium is displaced slightly, (a) its potential energy will decrease, (b) the center of gravity is directly above the axis of rotation, (c) no gravitational work is done, or (d) stable equilibrium follows.

18. **CQ** Why is a doorstop most effective when placed at the side of the door away from the hinge?

19. **CQ** When lifting an object with the use of the back rather than the legs, we often experience back pain. (See ▼Fig. 8.30.) Why?

Ouch!

▲ **FIGURE 8.30 Back pain** See Exercise 19.

20. **CQ** Explain the balancing acts in ▼Fig. 8.31. Where are the centers of gravity?

▲ **FIGURE 8.31 Balancing acts** See Exercise 20. *Left*: A toothpick on the rim of the glass supports a fork and spoon. *Right*: A toy bird balances on its beak.

21. **CQ** ►Figure 8.32 shows a toy clown walking on a tightrope. Explain how this brainless toy can perform such a difficult task. Is the clown in stable equilibrium? What happens if the weights are removed? [*Hint*: Think of center of gravity.]

22. ■ The drain plug on a car's engine has been tightened to a torque of 25 m · N. If a 0.15-m-long wrench is used

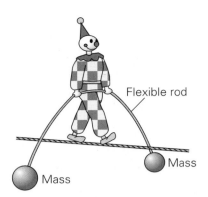

▲ **FIGURE 8.32 Tightrope walking** See Exercise 21.

to change the oil, what is the minimum force needed to loosen the plug?

23. ■ In Exercise 22, due to limited work space, you must crawl under the car. The force thus cannot be applied perpendicularly to the length of the wrench. If the applied force makes a 30° angle with the length of the wrench, what is the force required to loosen the drain plug?

24. ■ In Fig. 8.5a, if the arm makes a 37° angle with the horizontal and a torque of $18 \, \text{m} \cdot \text{N}$ is to be produced, what force must the biceps muscle supply?

25. ■ How many different positions of stable equilibrium and unstable equilibrium are there for a cube? Consider each surface, edge, and corner to be a different position.

26. ■ The pedals of a bicycle rotate in a circle with a diameter of 40 cm. What is the maximum torque a 55-kg rider can apply by putting all of her weight on one pedal?

27. IE ■ Two children are sitting on opposite ends of a uniform seesaw of negligible mass. (a) Can the seesaw be balanced if the masses of the children are different? How? (b) If a 35-kg child is 2.0 m from the pivot point (or fulcrum), how far from the pivot point will her 30-kg playmate have to sit on the other side for the seesaw to be in equilibrium?

28. ■ A uniform meterstick pivoted at its center, as in Example 8.5, has a 100-g mass suspended at the 25.0-cm position. (a) At what position should a 75.0-g mass be suspended to put the system in equilibrium? (b) What mass would have to be suspended at the 90.0-cm position for the system to be in equilibrium?

29. ■■ Show that the balanced meterstick in Example 8.5 is in static rotational equilibrium about a horizontal axis through the 100-cm end of the stick.

30. IE ■■ Telephone and electrical lines are allowed to sag between poles so that the tension will not be too great when something hits or sits on the line. (a) Is it possible to have the lines perfectly horizontal? Why or why not? (b) Suppose that a line were stretched almost perfectly horizontally between two poles that are 30 m apart. If a

0.25-kg bird perches on the wire midway between the poles and the wire sags 1.0 cm, what would be the tension in the wire?

31. ■■ In ▼Fig. 8.33, what is the force F_m supplied by the deltoid muscle so as to hold up the outstretched arm if the mass of the arm is 3.0 kg? (F_j is the joint force on the bone of the upper arm—the humerus.)

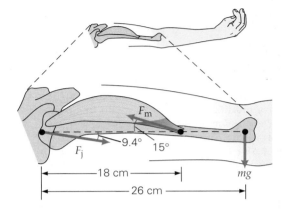

▲ **FIGURE 8.33 Arm in static equilibrium** See Exercise 31.

32. ■■ A variation of Russell traction (▼Fig. 8.34) supports the lower leg in a cast. Suppose that the patient's leg and cast have a combined mass of 15.0 kg and m_1 is 4.50 kg. (a) What is the reaction force of the leg muscles to the traction? (b) What must m_2 be to keep the leg horizontal?

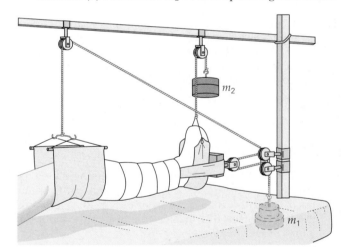

▲ **FIGURE 8.34 Static traction** See Exercise 32.

33. ■■ A 5.0-N uniform meterstick is pivoted so that it can rotate about a horizontal axis through one end (▶Fig. 8.35). If a 0.15-kg mass is suspended 75 cm from the pivoted end, what is the tension in the string?

34. ■■ In Example 8.6, if the mass of the painter is 65 kg, what frictional force must act on the bottom of the ladder to keep it from slipping?

35. ■■ An artist wishes to construct a birds-and-bees mobile, as shown in ▶Fig. 8.36. If the mass of the bee on the lower left is 0.10 kg and each vertical support string has a

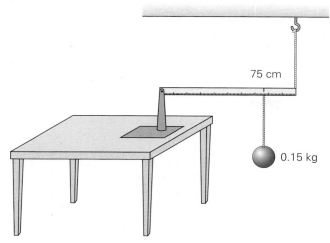

▲ FIGURE 8.35 **What's the tension?** See Exercise 33.

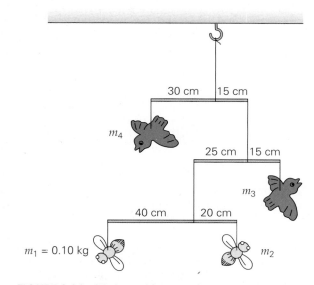

▲ FIGURE 8.36 **Birds and bees** See Exercise 35.

length of 30 cm, what are the masses of the other birds and bees? (Neglect the masses of the bars and strings.)

36. ■■ (a) How many uniform, identical textbooks of width 25.0 cm can be stacked on top of each other on a level surface without the stack falling over if each successive book is displaced 3.00 cm in width relative to the book below it? (b) If the books are 5.00 cm thick, what will be the height of the center of mass of the stack above the level surface?

37. ■■ If four metersticks were stacked on a table with 10 cm, 15 cm, 30 cm, and 50 cm, respectively, hanging over the edge, as shown in ▶Fig. 8.37, would the top meterstick remain on the table?

38. ■■ A 10.0-kg solid uniform cube with 0.500-m sides rests on a level surface. What is the minimum amount of work necessary to put the cube into an unstable equilibrium position?

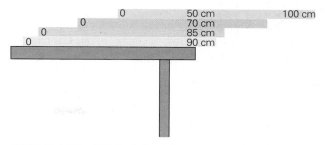

▲ FIGURE 8.37 **Will they fall off?** See Exercise 37.

39. ■■■ While standing on a long board resting on a scaffold, a 70-kg painter paints the side of a house, as shown in ▼Fig. 8.38. If the mass of the board is 15 kg, how close to the end can the painter stand without tipping the board over?

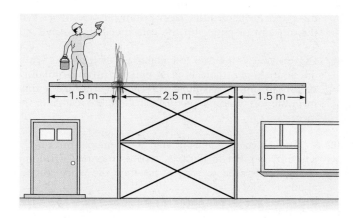

▲ FIGURE 8.38 **Not too far!** See Exercise 39 and 40.

40. ■■■ Suppose that the board in Fig. 8.38 were suspended from vertical ropes attached to each end instead of resting on scaffolding. If the painter stood 1.5 m from one end of the board, what would the tensions in the ropes be? (See Exercise 39 for additional data.)

41. IE ■■■ The forces acting on Einstein and the bicycle (Fig. 1 of Insight on p. 275) are the total weight of Einstein and the bicycle (mg) at the center of gravity of the system, the normal force (N) exerted by the road, and the force of static friction (f_s) acting on the tires due to the road. (a) If Einstein is to maintain balance, should the tangent of the lean angle θ (tan θ) be (1) greater than, (2) equal to, or (3) less than f_s/N? (b) The angle θ in the picture is about 11°. What is the minimum coefficient of static friction μ_s between the road and the tires? (c) If the radius of the circle is 6.5 m, what is the maximum speed of Einstein's bicycle? [*Hint*: The net torque about the center of gravity must be zero for rotational equilibrium.]

8.3 Rotational Dynamics

42. The moment of inertia of a rigid body (a) depends on the axis of rotation, (b) cannot be zero, (c) depends on mass distribution, or (d) all of the preceding.

43. Which of the following best describes the physical quantity called torque? (a) rotational analogue of force, (b) energy due to rotation, (c) rate of change of linear momentum, or (d) force that is tangent to a circle

44. CQ (a) Does the moment of inertia of a rigid body depend in any way on the center of mass of the body? Explain. (b) Can a moment of inertia have a negative value? If so, explain what this would mean.

45. CQ Why does the moment of inertia of a rigid body have different values for different axes of rotation? What does this mean physically?

46. CQ When a hard-boiled egg lying on a table is given a quick torque (spin), it will rise up and spin on one end like a top. A raw egg will not. Why the difference?

47. CQ Why does jerking a paper towel from a roll cause the paper to tear better than pulling it smoothly? Will the amount of paper on the roll affect the results?

48. CQ (a) Why is it easier to balance a meterstick vertically on your finger than it is a pencil? (b) A softball, a volleyball, and a basketball are released at the same time from the top of an inclined plane. Give the results of the race.

49. CQ If a bicycle wheel is rotated about an axis through its rim and parallel to its axle, will its moment of inertia change compared to what it is when the wheel rotates about its axle?

50. CQ What happens to a rigid body if the net torque acting on it is zero?

51. ■ For the system of masses shown in ▼Fig. 8.39, find the moment of inertia about (a) the x axis, (b) the y axis, and (c) an axis through the origin and perpendicular to the page (z axis). Neglect the masses of the connecting rods.

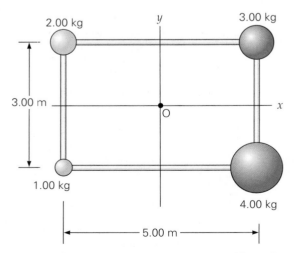

▲ FIGURE 8.39 Moments of inertia about different axes
See Exercise 51.

52. ■ A fixed 0.15-kg solid-disk pulley with a radius of 0.075 m is acted on by a net torque of $6.4 \text{ m} \cdot \text{N}$. What is the angular acceleration of the pulley?

53. ■ What net torque is required to give a uniform 20-kg solid ball with a radius of 0.20 m an angular acceleration of 2.0 rad/s^2?

54. ■ A light meterstick is loaded with masses of 2.0 kg and 4.0 kg at the 30-cm and 75-cm positions, respectively. (a) What is the moment of inertia about an axis through the 0-cm end of the meterstick? (b) What is the moment of inertia about an axis through the center of mass of the system? (c) Use the parallel-axis theorem to find the moment of inertia about the axis through the 0-cm end of the stick, and compare the result with the result of part (a).

55. IE ■■ Two objects of different masses are joined by a light rod. (a) Is the moment of inertia about the center of mass the minimum or the maximum? Why? (b) If the two masses are 3.0 kg and 5.0 kg and the length of the rod is 2.0 m, find the moments of inertia of the system about an axis perpendicular to the rod, through the center of the rod and the center of mass.

56. ■■ A 2000-kg Ferris wheel accelerates from rest to an angular speed of 2.0 rad/s in 12 s. Approximate the Ferris wheel as a circular disk with a radius of 30 m. What is the net torque on the wheel?

57. ■■ A 15-kg uniform sphere with a radius of 15 cm rotates about an axis tangent to its surface at 3.0 rad/s. A constant torque of $10 \text{ m} \cdot \text{N}$ then increases the rotational speed to 7.5 rad/s. Through what angle does the sphere rotate while accelerating?

58. ■■ A 10-kg solid disk of radius 0.50 m is rotated about an axis through its center. If the disk accelerates from rest to an angular speed of 3.0 rad/s while rotating 2.0 revolutions, what net torque is required?

59. ■■ If the Earth rotated about a north–south axis tangent to the equator, how much more moment of inertia would it have than if it rotated about a similarly oriented axis through its center?

60. ■■ A 0.25-kg disk-shaped machine part with a radius of 6.0 cm rotates eccentrically (off center) about an axis that is normal to its flat surface and that is located two-thirds of the distance from the center to the circumference. (a) What torque is required to rotate the part about this off-center axis with an angular acceleration of 2.0 rad/s^2? (b) How much greater is this torque than the torque required to rotate the part with the same angular acceleration about a parallel axis through the center?

61. ■■ Two masses are suspended from a pulley as shown in ▶Fig. 8.40 (the Atwood machine revisited, see Chapter 4, Exercise 68). The pulley itself has a mass of 0.20 kg, a radius of 0.15 m, and a constant torque of $0.35 \text{ m} \cdot \text{N}$ due to the friction between the rotating pulley and its axle. What

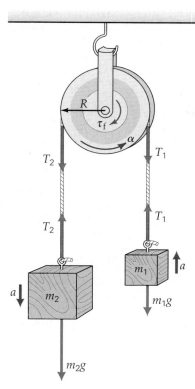

▲ FIGURE 8.40 The Atwood machine revisited See Exercise 61.

is the magnitude of the acceleration of the suspended masses if $m_1 = 0.40$ kg and $m_2 = 0.80$ kg? (Neglect the mass of the string.)

62. ■■ For the system shown in ▼Fig. 8.41, $m_1 = 8.0$ kg, $m_2 = 3.0$ kg, $\theta = 30°$, and the radius and mass of the pulley are 0.10 m and 0.10 kg, respectively. (a) What is the acceleration of the masses? (Neglect friction and the string's mass.) (b) If the pulley has a constant frictional torque of 0.050 m · N when the system is in motion, what is the acceleration of the masses? [*Hint:* Isolate the forces. The tensions in the strings are different. Why?]

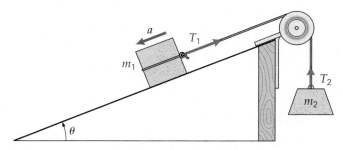

▲ FIGURE 8.41 Inclined plane and pulley See Exercise 62.

63. ■■ What is the magnitude of the tangential force that must be applied to the rim of a 2.0-kg wheel that is disk shaped and that has a radius of 0.50 m in order to give the wheel an angular acceleration of 4.8 rad/s²? (Neglect friction.)

64. ■■ A man starts his lawn mower by using the starter rope to apply a constant tangential force of 150 N to the 0.30-kg disk-shaped flywheel. The diameter of the flywheel is 18 cm. What is the angular speed of the wheel after it has turned through one revolution? (Neglect friction and motor compression.)

65. ■■ A meterstick pivoted about a horizontal axis through the 0-cm end is held in a horizontal position and let go. (a) What is the tangential acceleration of the 100-cm position? Are you surprised by this result? (b) Which position has a tangential acceleration equal to the acceleration due to gravity?

66. ■■ Pennies are placed every 10 cm on a meterstick. One end of the stick is put on a table and the other end is held horizontally with a finger, as shown in ▼Fig. 8.42. If the finger is pulled away, what happens to the pennies?

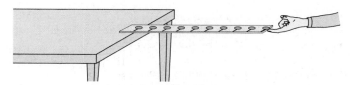

▲ FIGURE 8.42 Money left behind? See Exercise 66.

67. ■■■ A uniform 2.0-kg cylinder of radius 0.15 m is suspended by two strings wrapped around it (▼Fig. 8.43). As the cylinder descends, the strings unwind from it. What is the acceleration of the center of mass of the cylinder? (Neglect the mass of the string.)

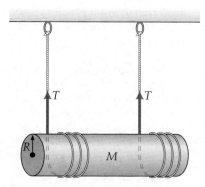

▲ FIGURE 8.43 Unwinding with gravity See Exercise 67.

68. ■■■ A uniform hoop rolls without slipping down a 15° inclined plane. What is the acceleration of the hoop's center of mass?

69. IE ■■■ A ball of radius R and mass M rolls down an incline of angle θ. (a) For the ball to roll without slipping, should the tangent of the maximum angle of incline $(\tan \theta)$ be equal to (1) $3\mu_s/2$, (2) $5\mu_s/2$, (3) $7\mu_s/2$, or (4) $9\mu_s/2$? Here, μ_s is the coefficient of static friction. (b) If the ball is made of wood and the surface is also wood, what is the maximum angle of incline? [*Hint:* See Table 4.1.]

8.4 Rotational Work and Kinetic Energy

70. From $W = \tau\theta$, the unit of rotational work is the (a) watt, (b) N · m, (c) kg · rad/s², (d) N · rad.

71. A bowling ball rolls without slipping on a flat surface. The ball has (a) rotational kinetic energy, (b) translational kinetic energy, (c) both translational and rotational kinetic energy, or (d) neither translational nor rotational kinetic energy.

72. CQ Can you increase the rotational kinetic energy of a wheel without changing its translational kinetic energy? Explain.

73. CQ In order to produce fuel-efficient vehicles, automobile manufacturers want to minimize rotational kinetic energy and maximize translational kinetic energy when a car is traveling. If you were the designer of wheels of a certain diameter, how would you design them?

74. ■ A constant retarding torque of 12 m · N stops a rolling wheel of diameter 0.80 m in a distance of 15 m. How much work is done by the torque?

75. ■ A person opens a door by applying a 15-N force perpendicular to it at a distance 0.90 m from the hinges. The door is pushed wide open (to 120°) in 2.0 s. (a) How much work was done? (b) What was the average power delivered?

76. ■ A constant torque of 10 m · N is applied to a 10-kg uniform disk of radius 0.20 m. What is the angular speed of the disk about an axis through its center after it rotates 2.0 revolutions from rest?

77. ■ A 2.5-kg pulley of radius 0.15 m is pivoted about an axis through its center. What constant torque is required for the pulley to reach an angular speed of 25 rad/s after rotating 3.0 revolutions, starting from rest?

78. IE ■ In Fig. 8.20, a mass m descends a vertical distance from rest. (Neglect friction and the mass of the string.) (a) From the conservation of mechanical energy, will the linear speed of the descending mass be (1) greater than, (2) equal to, or (3) less than $\sqrt{2gh}$? Why? (b) If $m = 1.0$ kg, $M = 0.30$ kg, and $R = 0.15$ m, what is the linear speed of the mass after it has descended a vertical distance of 2.0 m from rest?

79. ■■ A sphere with a radius of 15 cm rolls on a level surface with a constant angular speed of 10 rad/s. To what height on a 30° inclined plane will the sphere roll before coming to rest? (Neglect frictional losses.)

80. ■■ A thin 1.0-m-long rod pivoted at one end falls (rotates) from a horizontal position, starting from rest and with no friction. What is the angular speed of the rod when it is vertical? [*Hint*: Consider the center of mass and use the conservation of mechanical energy.]

81. ■■ A uniform sphere and a uniform cylinder with the same mass and radius roll at the same velocity side by side on a level surface without slipping. If the sphere and the cylinder approach an inclined plane and roll up it without slipping, will they be at the same height on the plane when they come to a stop? If not, what will be the percentage difference of the heights?

82. ■■ A hoop starts from rest at a height 1.2 m above the base of an inclined plane and rolls down under the influence of gravity. What is the linear speed of the hoop's center of mass just as the hoop leaves the incline and rolls onto a horizontal surface? (Neglect friction.)

83. ■■ An industrial flywheel with a moment of inertia of 4.25×10^2 kg · m² rotates with a speed of 7500 rpm. (a) How much work is required to bring the flywheel to rest? (b) If this work is done uniformly in 1.5 min, how much power is expended?

84. ■■ A cylindrical hoop, a cylinder, and a sphere of equal radius and mass are released at the same time from the top of an inclined plane. Using the conservation of mechanical energy, show that the sphere always gets to the bottom of the incline first with the fastest speed and that the hoop always arrives last with the slowest speed.

85. ■■ For the following objects, which all roll without slipping, determine the rotational kinetic energy about the center of mass as a percentage of the total kinetic energy: (a) a solid sphere, (b) a thin spherical shell, and (c) a thin cylindrical shell.

86. ■■ A 0.050-kg phonograph record with a radius of 0.15 m drops onto a turntable and is soon rotating at $33\frac{1}{3}$ rpm. How much work must be supplied to get the record to rotate at this speed, and what supplies the work?

87. ■■■ A steel ball rolls down an incline into a loop-the-loop of radius R (▶Fig. 8.44a). (a) What minimum speed must the ball have at the top of the loop in order to stay on the track? (b) At what vertical height (h) on the incline, in terms of the radius of the loop, must the ball be released in order for it to have the required minimum speed at the top of the loop? (Neglect frictional losses.) (c) Figure 8.44b shows the loop-the-loop of a roller coaster. What are the sensations of the riders if the roller coaster has the minimum or a greater speed at the top of the loop? [*Hint*: In case the speed is below the minimum, seat and shoulder straps hold the riders in.

8.5 Angular Momentum

88. The units of angular momentum are (a) N · m, (b) kg · m/s², (c) kg · m²/s, or (d) J · m.

89. CQ A child stands on the edge of a rotating playground merry-go-round (the hand-driven type). He then starts to walk toward the center of the merry-go-round. This results in a dangerous situation. Why?

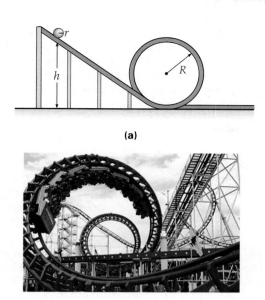

(a)

(b)

▲ FIGURE 8.44   Loop-the-loop and rotational speed
See Exercise 87.

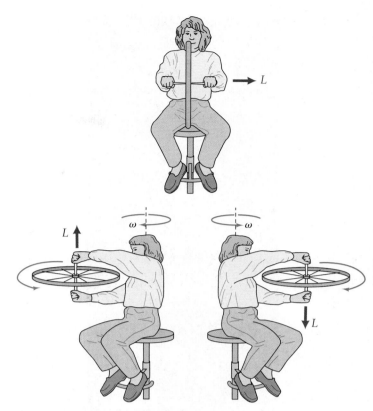

▲ FIGURE 8.46   A double rotation   See Exercise 92.

90. **CQ** The release of vast amounts of carbon dioxide may result in an increase in the Earth's average temperature through the so-called greenhouse effect and cause melting of the polar ice caps. If this occurred and the ocean level rose substantially, what effect would it have on the Earth's rotation and on the length of the day?

91. **CQ** An ice skater goes into a fast spin by tucking her arms in (▼Fig. 8.45a), and a high platform diver often draws her legs up against her chest (Fig. 8.45b). Why are the arms and legs put into these positions in each case?

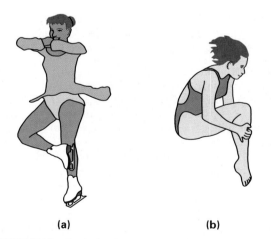

(a)   (b)

▲ FIGURE 8.45   Faster rotation   See Exercise 91.

92. **CQ** In the classroom demonstration illustrated in ▶Fig. 8.46, a person on a rotating stool holds a rotating bicycle wheel by handles attached to the wheel. When the wheel is held horizontally, she rotates one way (clock-

wise as viewed from above). When the wheel is turned over, she rotates in the opposite direction. Explain why this occurs. [*Hint*: Consider angular-momentum vectors.]

93. **CQ** Cats usually land on their feet when they fall, even if held upside down when dropped (▶Fig. 8.47). While a cat is falling, there is no external torque and its center of mass falls as a particle. How can cats turn themselves over while falling?

94. **CQ** Why are you able to ride a bicycle without having your hands on the handlebar?

95. **CQ** Explain why the larger helicopters, such as the one shown in Fig. 8.29(a), use two overhead rotors instead of one.

96. ■ What is the angular momentum of a 2.0-g particle moving counterclockwise (as viewed from above) with an angular speed of 5π rad/s in a horizontal circle of radius 15 cm? (Give the magnitude and direction.)

97. ■ A 10-kg rotating disk of radius 0.25 m has an angular momentum of 0.45 kg·m²/s. What is the angular speed of the disk?

98. ■■ Compute the ratio of the magnitudes of the Earth's orbital angular momentum and its rotational angular momentum. Are these momenta in the same direction?

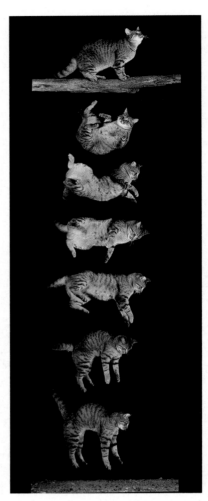

▲ **FIGURE 8.47   One down, eight to go**   See Exercise 93.

99. ■■ The period of the Moon's rotation is the same as the period of its revolution: 27.3 days (sidereal). What is the angular momentum for each rotation and revolution? (Because the periods are equal, we see only one side of the Moon from Earth.)

100. IE ■■ Circular disks are used in automobile clutches and transmissions. When a rotating disk couples to a stationary one through frictional force, the energy from the rotating disk can transfer to the stationary one. (a) Is the angular speed of the coupled disks (1) greater than, (2) less than, or (3) the same as the angular speed of the original rotating disk? Why? (b) If a disk rotating at 800 rpm couples to a stationary disk with three times the moment of inertia, what is the angular speed of the combination?

101. ■■ During part of its life cycle, a rotating spherical star expands to six times its normal volume. Assuming that the mass of the star remains constant and is uniformly distributed, how is the period of rotation affected?

102. ■■ A skater has a moment of inertia of $100 \, \text{kg} \cdot \text{m}^2$ when his arms are outstretched and a moment of inertia of $75 \, \text{kg} \cdot \text{m}^2$ when his arms are tucked in close to his chest. If he starts to spin at an angular speed of 2.0 rps (revolutions per second) with his arms outstretched, what will his angular speed be when they are tucked in?

103. ■■ An ice skater spinning with outstretched arms has an angular speed of 4.0 rad/s. She tucks in her arms, decreasing her moment of inertia by 7.5%. (a) What is the resulting angular speed? (b) By what factor does the skater's kinetic energy change? (Neglect any frictional effects.) (c) Where does the extra kinetic energy come from?

104. ■■■ A comet approaches the Sun as illustrated in ▼Fig. 8.48 and is deflected by the Sun's gravitational attraction. This event is considered a collision, and b is called the impact parameter. Find the distance of closest approach (d) in terms of the impact parameter and the velocities (v_o at large distances and v at closest approach). Assume that the radius of the Sun is negligible compared to d. (As the figure shows, the tail of a comet always "points" away from the Sun.)

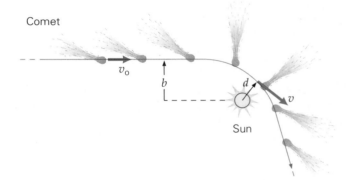

▲ **FIGURE 8.48   A comet "collision"**   See Exercise 104.

105. IE ■■■ A kitten stands on the edge of a lazy Susan (a turntable). Assume that the lazy Susan has frictionless bearings and is initially at rest. (a) If the kitten starts to walk around the edge of the lazy Susan, the Susan will (1) remain lazy and stationary, (2) rotate in the direction opposite that in which the cat is walking, or (3) rotate in the direction the cat is walking. Explain. (b) The mass of the kitten is 0.50 kg, and the lazy Susan has a mass of 1.5 kg and a radius of 0.30 m. If the kitten walks at a speed of 0.25 m/s relative to the ground, what will be the angular speed of the lazy Susan? (c) When the kitten has walked completely around the edge and is back at its starting point, will that point be above the same point on the ground as it was at the start? If not, where is the kitten relative to the starting point? (Speculate on what might happen if everyone on Earth suddenly started to run eastward. What effect might this have on the length of a day?)

Additional Exercises

106. A 10000-kg bridge of length 10 m is supported at both ends. If a 2000-kg car is parked on the bridge 3.0 m from the left support, what are the supporting forces at the left and right ends?

107. A spring is twisted 60° about its linear axis (axis of symmetry) by a torque of 100 m · N. How much torsional, or rotational, energy is stored in the spring? [*Hint*: Let the restoring torque of the spring be $\tau = k\theta$, the rotational analogue of $F = kx$ (Hooke's law—see Chapter 5) with the minus sign omitted.]

108. Prove that a thin hoop or ring starting from rest will roll more slowly down an inclined plane than an annular cylinder will. (Remember that masses and radii do not have to be taken into account.)

109. A 10-kg pulley of radius 0.50 m is connected to a 5.0-kg mass through a string as shown in ▼Fig. 8.49. (a) If the mass is let go from rest, what is its linear acceleration? (b) What is the angular acceleration of the pulley?

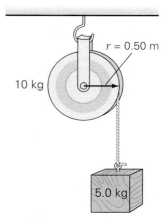

▲ FIGURE 8.49 Linear and angular accelerations See Exercise 109.

110. A bicycle wheel has a diameter of 0.80 m and a mass of 2.0 kg. (a) If the bicycle travels with a linear speed of 1.5 m/s, what is the angular speed of the wheel? (b) What is the angular momentum of the wheel? (Consider the wheel to be a thin ring.)

111. A piece of machinery with a moment of inertia of 2.6 kg · m² rotates with a constant angular speed of 4.0 rad/s when a frictional torque of 0.56 m · N acts on it. What is the net torque acting on the piece?

112. IE The location of a person's center of gravity relative to his or her height can be found by using the arrangement shown in ▶Fig. 8.50. The scales are initially adjusted to zero with the board alone. (a) Would you expect the location of the center of gravity to be (1) midway between the scales, (2) toward the scale at the person's head, or (3) to-

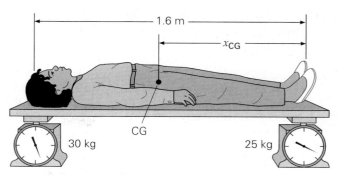

▲ FIGURE 8.50 Locating the center of gravity See Exercise 112.

ward the scale at the person's feet? Why? (b) Locate the center of gravity of the person relative to the horizontal dimension.

113. Using bricks identical to those in Example 8.7, you are asked to create a stack of nine bricks (maximum) without its toppling, with each added brick displaced an equal distance horizontally in the same direction. (a) What is the maximum displacement that will allow this arrangement? (b) What is the height of the center of mass of the stack if the uniform bricks are 8.0 cm thick?

114. The *radius of gyration* (k) is defined as the distance from an axis of rotation at which all the mass of an object would have to be concentrated to give the same moment of inertia as the actual mass distribution would have; that is, $I = Mk^2$. Take the object to be a rotating particle of mass M. From the axes shown in Fig. 8.17, determine the radius of gyration for (a) a uniform sphere, (b) a uniform cylinder, and (c) a particle.

115. A 25-kg child jumps onto the rim of a 100-kg rotating disk of radius 2.0 m. If the angular speed of the disk before the child's jump was 2.0 rad/s, what is the angular speed of the disk–child system?

116. A mass is suspended by two cords as shown in ▼Fig. 8.51. What are the tensions in the cords?

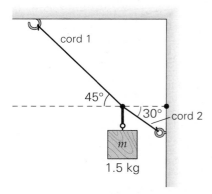

▲ FIGURE 8.51 Strung-out equilibrium See Exercises 116 and 117.

117. If the cord attached to the vertical wall in Fig. 8.51 were horizontal (instead of at a 30° angle), what would the tensions in the cords be?

118. The outer back wheels of an oversize tractor–trailer rig are 3.66 m apart. When the trailer is loaded, its center of gravity is equidistant from its sides and 3.58 m above the road surface. At what angle with the road will the parked trailer be in unstable equilibrium?

119. A 0.50-kg thin rectangular plate has dimensions of 30 cm by 40 cm. For which axis of rotation does the plate have the greater rotational inertia, an axis tangent to a smaller side at its midpoint or an axis tangent to a larger side at its midpoint? How many times more? (Both axes are perpendicular to the surface of the plate.)

Solids and Fluids

INSIGHTS

- Feat of Strength or Knowledge of Materials?
- An Atmospheric Effect: Possible Earaches
- Blood Pressure and Its Measurement
- Throwing a Curveball

Like the person who is hang gliding, we exist at the intersection of three realms. We walk on the solid surface of the Earth and in our daily lives use solid objects of all sorts, from scissors to computers. But we are surrounded by fluids—liquids and gases—on which we are dependent. Without the water that we drink, we could survive only for a few days at most; without the oxygen in the air we breathe, we could not live for more than a few minutes. Indeed, we ourselves are not nearly as solid as we think. By far the most abundant substance in our bodies is water, and it is in the watery environment of our cells that all chemical processes on which life depends take place.

On the basis of general physical distinctions, matter is commonly divided into three phases: solid, liquid, and gas. A solid has a definite shape and volume. A liquid has a fairly definite volume, but assumes the shape of its container. A gas takes on the shape and volume of its container. Solids and liquids are sometimes called *condensed*

matter. We will use a different classification scheme and consider matter in terms of solids and fluids. Liquids and gases are referred to collectively as fluids. A **fluid** is a substance that can flow; liquids and gases qualify, but solids do not.

A simplistic description of solids is that they are made up of particles called atoms that are held rigidly together by interatomic forces. In Chapter 8, the concept of an ideal rigid body was used to describe rotational motion. Real solid bodies are not absolutely rigid and can be elastically deformed by external forces. Elasticity usually brings to mind a rubber band or spring that will resume its original dimensions even after being greatly deformed. In fact, all materials—even very hard steel— are elastic to some degree. But, as you will learn, such deformation has an *elastic limit*.

Fluids, however, have little or no elastic response to a force. Instead, the force merely causes an unconfined fluid to flow. This chapter pays particular attention to the behavior of fluids, shedding light on such questions as how hydraulic lifts work, why icebergs and ocean liners float, and what the 10W–30 on a can of motor oil means. You'll also discover why the person in the photo can neither float like a helium balloon nor fly like a hummingbird, yet, with the aid of a suitably shaped piece of plastic, can soar like an eagle.

Because of their fluidity, liquids and gases have many properties in common, and it is convenient to study them together. There are important differences as well. For example, liquids are not very compressible, whereas gases are easily compressed, as we will see shortly.

9.1 Solids and Elastic Moduli

OBJECTIVES: **To (a) distinguish between stress and strain and (b) use elastic moduli to compute dimensional changes.**

As stated previously, all solid materials are elastic to some degree. That is, a body which is slightly deformed by an applied force will return to its original dimensions or shape when the force is removed. The deformation may not be noticeable for many materials, but it's there.

You may be able to visualize why materials are elastic if you think in terms of the simplistic model of a solid in ◄Fig. 9.1. The atoms of the solid substance are imagined to be held together by springs. The elasticity of the springs represents the resilient nature of the interatomic forces. The springs resist permanent deformation, as do the forces between atoms. The elastic properties of solids are commonly discussed in terms of stress and strain. **Stress** is a measure of the force causing a deformation. **Strain** is a relative measure of the deformation a stress causes. Quantitatively, *stress is the applied force per unit cross-sectional area:*

$$\text{stress} = \frac{F}{A} \tag{9.1}$$

SI unit of stress: newton per square meter (N/m^2)

Here, F is the magnitude of the applied force normal (perpendicular) to the cross-sectional area. Equation 9.1 shows that the SI units for stress are newtons per square meter (N/m^2).

As illustrated in ▼Fig. 9.2, a force applied to the ends of a rod gives rise to either a *tensile stress* (an elongating tension, $+\Delta L$) or a *compressional stress* (a shortening tension, $-\Delta L$), depending on the direction of the force. In both these cases,

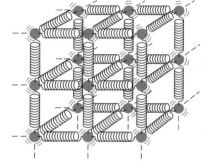

▲ FIGURE 9.1 A springy solid
The elastic nature of interatomic forces is indicated by simplistically representing them as springs, which, like the forces, resist deformation.

► FIGURE 9.2 Tensile and compressional stress and strain
Tensile and compressional stresses are due to forces applied normally to the surface area of the ends of bodies. **(a)** A tension, or tensile stress, tends to increase the length of an object. **(b)** A compressional stress tends to shorten the length. $\Delta L = L - L_o$ can be positive, as in (a), or negative, as in (b). The sign is not needed in Eq. 9.2, so the absolute value, $|\Delta L|$, is used.

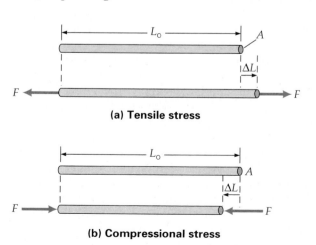

(a) Tensile stress

(b) Compressional stress

the *tensile strain* is the ratio of the change in length ($\Delta L = L - L_o$) to the original length (L_o), without regard to the sign, so we use the absolute value, $|\Delta L|$:

$$\text{strain} = \frac{\text{change in length}}{\text{original length}} = \frac{|\Delta L|}{L_o} = \frac{|L - L_o|}{L_o} \qquad (9.2)$$

Strain is a positive unitless quantity

Thus the strain is the *fractional change* in length. For example, if the strain is 0.05, the length of the material has changed by 5% of the original length.

As might be expected, the resulting strain is proportional to the applied stress; that is, strain $\propto$ stress. For relatively small stresses, this is a direct proportion. The constant of proportionality, which depends on the nature of the material, is called the **elastic modulus**. Thus,

Definition of elastic modulus

$$\text{stress} = \text{elastic modulus} \times \text{strain}$$

or

$$\text{elastic modulus} = \frac{\text{stress}}{\text{strain}} \qquad (9.3)$$

SI unit of elastic modulus: newton per square meter (N/m^2)

That is, the elastic modulus is the stress divided by the strain, and the elastic modulus has the same units as stress. (Why?)

Three general types of elastic moduli (plural of modulus) are associated with stresses that produce changes in length, shape, or volume. These are called *Young's modulus*, the *shear modulus*, and the *bulk modulus*, respectively.

Change in Length: Young's Modulus

▼Figure 9.3 is a graph of the tensile stress versus the strain for a typical metal rod. The curve is a straight line up to a point called the *proportional limit*. Beyond this point, the strain begins to increase more rapidly to another critical point called the **elastic limit**. If the tension is removed at this point, the material will return to its original length. If the tension is applied beyond the elastic limit and then removed, the material will recover somewhat, but will retain some permanent deformation.

The straight-line part of the graph shows a direct proportionality between stress and strain. This relationship, first formalized by Robert Hooke in 1678, is

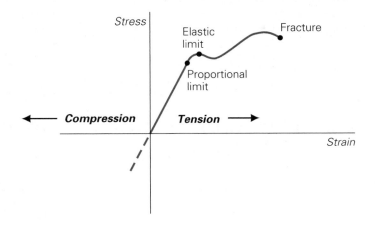

◀ **FIGURE 9.3   Stress versus strain**
A plot of stress versus strain for a typical metal rod is a straight line up to the proportional limit. Then elastic deformation continues until the elastic limit is reached. Beyond that, the rod will be permanently deformed and will eventually fracture or break.

now known as *Hooke's law*. (It is the same general relationship as that given for a spring in Section 5.2—see Fig. 5.5.) The elastic modulus for a tension or a compression is called **Young's modulus** (Y):

$$\frac{F}{A} = Y\left(\frac{\Delta L}{L_\text{o}}\right) \quad \text{or} \quad Y = \frac{F/A}{\Delta L/L_\text{o}} \qquad (9.4)$$

$$\underset{\text{stress}}{\underbrace{\phantom{\frac{F}{A}}}} \quad \underset{\text{strain}}{\underbrace{\phantom{\frac{\Delta L}{L}}}}$$

SI unit of Young's modulus: newton per square meter (N/m^2)

The units of Young's modulus are the same as those of stress, newtons per square meter (N/m^2), since the strain is unitless. Some typical values of Young's modulus are given in Table 9.1.

To obtain a conceptual or physical understanding of Young's modulus, let's solve Eq. 9.4 for ΔL:

$$\Delta L = \left(\frac{FL_\text{o}}{A}\right)\frac{1}{Y} \quad \text{or} \quad \Delta L \propto \frac{1}{Y}$$

Hence, the larger the modulus of a material, the smaller its change in length (other parameters being equal).

Young's Modulus and Tensile Strength

Example 9.1 ■ Pulling My Leg: Under a Lot of Stress

The femur (upper leg bone) is the longest and strongest bone in the body. Taking a typical femur to be approximately circular with a radius of 2.0 cm, how much force would be required to extend the bone by 0.010%?

Thinking It Through. We can see that Eq. 9.4 should apply, but where does the percentage increase fit in? We can answer this question as soon as we recognize that the $\Delta L/L_\text{o}$ term is the *fractional* increase in length. For example, if you had a spring with a length of 10 cm (L_o), and you stretched it 1.0 cm (ΔL), then $\Delta L/L_\text{o} =$ 1.0 cm/10 cm = 0.10. This ratio can readily be changed to a percentage, and we would say the spring's length was increased by 10%. So the percentage increase is really just the value of the $\Delta L/L_\text{o}$ term (multiplied by 100%).

Solution. Listing the data, we have the following:

Given: $r = 2.0 \text{ cm} = 0.020 \text{ m}$ **Find:** F (tensile force)
$\Delta L/L_\text{o} = 0.010\% = 1.0 \times 10^{-4}$
$Y = 1.5 \times 10^{10} \text{ N/m}^2$ (for bone from Table 9.1)

TABLE 9.1 Elastic Moduli for Various Materials (in N/m^2)			
Substance	*Young's Modulus (Y)*	*Shear Modulus (S)*	*Bulk Modulus (B)*
Solids			
Aluminum	7.0×10^{10}	2.5×10^{10}	7.0×10^{10}
Bone (limb)	1.5×10^{10}	8.0×10^{10}	
Brass	9.0×10^{10}	3.5×10^{10}	7.5×10^{10}
Copper	11×10^{10}	3.8×10^{10}	12×10^{10}
Glass	5.7×10^{10}	2.4×10^{10}	4.0×10^{10}
Iron	15×10^{10}	6.0×10^{10}	12×10^{10}
Steel	20×10^{10}	8.2×10^{10}	15×10^{10}
Liquids			
Alcohol, ethyl			1.0×10^{9}
Glycerin			4.5×10^{9}
Mercury			26×10^{9}
Water			2.2×10^{9}

Using Eq. 9.4, we have

$$F = Y(\Delta L/L_o)A = Y(\Delta L/L_o)\pi r^2$$
$$= (1.5 \times 10^{10} \text{ N/m}^2)(1.0 \times 10^{-4})\pi(0.020 \text{ m})^2 = 1.9 \times 10^3 \text{ N}$$

How much force is this? Quite a bit—in fact, more than 400 lb. The femur is a pretty strong bone.

Follow-up Exercise. A total mass of 16 kg is suspended from a 0.10-cm-diameter steel wire. (a) By what percentage does the length of the wire increase? (b)The tensile or ultimate strength of a material is the maximum stress the material can support before breaking or fracturing. If the tensile strength of the steel wire in (a) is $4.9 \times 10^8 \text{ N/m}^2$, how much mass could be suspended before the wire would break? (*Answers to all Follow-up Exercises are at the back of the text.*)

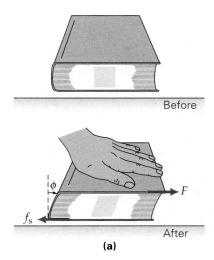

Before

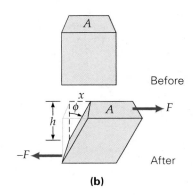

After

(a)

Change in Shape: Shear Modulus

Another way an elastic body can be deformed is by a *shear stress*. In this case, the deformation is due to an applied force that is tangential to the surface area (▶Fig. 9.4). A change in shape results without a change in volume. The *shear strain* is given by x/h, where x is the relative displacement of the faces and h is the distance between them.

The shear strain is sometimes defined in terms of the *shear angle* ϕ. As Fig. 9.4 shows, $\tan \phi = x/h$. But the shear angle is usually quite small, so a good approximation is $\tan \phi \approx \phi \approx x/h$, where ϕ is in radians. (If $\phi = 10°$, for example, there is only 1.0% difference between ϕ and $\tan \phi$.) The **shear modulus** (sometimes called the *modulus of rigidity*) is then

Before

After

(b)

$$S = \frac{F/A}{x/h} \approx \frac{F/A}{\phi} \tag{9.5}$$

SI unit of shear modulus: newton per square meter (N/m^2)

▲ **FIGURE 9.4 Shear stress and strain** A shear stress is produced when a force is applied tangentially to a surface area. The strain is measured in terms of the relative displacement of the object's faces, or the shear angle ϕ.

Note in Table 9.1 that the shear modulus is generally less than Young's modulus. In fact, S is approximately $Y/3$ for many materials, which indicates that there is a greater response to a shear stress than to a tensile stress. Note also the inverse relationship $\phi \approx 1/S$, similar to that pointed out previously for Young's modulus.

A shear stress may be of the torsional type, resulting from the twisting action of a torque. For example, a torsional shear stress may shear off the head of a bolt that is being tightened.

Liquids do not have shear moduli (or Young's moduli)—hence the gaps in Table 9.1. A shear stress cannot be effectively applied to a liquid or a gas. It is often said that *fluids cannot support a shear.* (Why?)

Change in Volume: Bulk Modulus

Suppose that a force directed inward acts over the entire surface of a body (▶Fig. 9.5). Such a *volume stress* is often applied by pressure transmitted by a fluid. An elastic material will be compressed by a volume stress; that is, the material will show a change in volume, but not in general shape, in response to a pressure change Δp. (Pressure is force per unit area, as we shall see in Section 9.2.) The change in pressure is equal to the volume stress, or $\Delta p = F/A$. The *volume strain* is the ratio of the volume change (ΔV) to the original volume (V_o). The **bulk modulus** is then

$$B = \frac{F/A}{-\Delta V/V_o} = -\frac{\Delta p}{\Delta V/V_o} \tag{9.6}$$

SI unit of bulk modulus: newton per square meter (N/m^2)

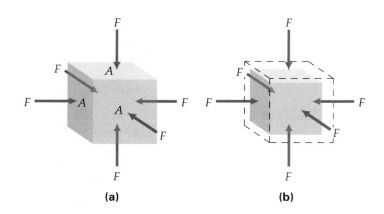

▶ **FIGURE 9.5 Volume stress and strain** (a) A volume stress is applied when a normal force acts over an entire surface area, as shown here for a cube. This type of stress most commonly occurs in gases. (b) The resulting strain is a change in volume.

(a)   (b)

The minus sign is introduced to make B a positive quantity, since $\Delta V = V - V_o$ is negative for an increase in external pressure (when Δp is positive). Similarly to the previous moduli relationships, $\Delta V \propto 1/B$.

Bulk moduli of solids and liquids are listed in Table 9.1. Gases also have bulk moduli, since they can be compressed. For a gas, it is common to talk about the reciprocal of the bulk modulus, which is called the **compressibility** (k):

$$k = \frac{1}{B} \quad \text{(compresibility for gases)} \tag{9.7}$$

The change in volume ΔV is thus directly proportional to the compressibility k.

Solids and liquids are relatively incompressible and thus have small values of compressibility. Conversely, gases are easily compressed and have large compressibilities, which vary with pressure and temperature.

Example 9.2 ■ Compressing a Liquid: Volume Stress and Bulk Modulus

By how much should the pressure on a liter of water be changed to compress it by 0.10%?

Thinking It Through. Similarly to the fractional change in length, $\Delta L/L_o$, the fractional change in volume is given by $-\Delta V/V_o$, which may be expressed as a percentage. The pressure change can then be found from Eq. 9.6. Compression implies a negative ΔV.

Solution. Listing the data, we have the following:

Given: $-\Delta V/V_o = 0.0010$ (or 0.10%)   *Find:* Δp
$V_o = 1.0 \text{ L} = 1000 \text{ cm}^3$
$B_{H_2O} = 2.2 \times 10^9 \text{ N/m}^2$ (from Table 9.1)

Note that $-\Delta V/V_o$ is the *fractional* change in the volume. With $V_o = 1000 \text{ cm}^3$, the reduction in volume is

$$-\Delta V = 0.0010 V_o = 0.0010(1000 \text{ cm}^3) = 1.0 \text{ cm}^3$$

However, the change in volume is not needed. The fractional change, as listed in the given data, can be used directly in Eq. 9.6 to find the increase in pressure:

$$\Delta p = B\left(\frac{-\Delta V}{V_o}\right) = (2.2 \times 10^9 \text{ N/m}^2)(0.0010) = 2.2 \times 10^6 \text{ N/m}^2$$

(This increase is about 22 times normal atmospheric pressure!)

Follow-up Exercise. If an extra $1.0 \times 10^6 \text{ N/m}^2$ of pressure above normal atmospheric pressure is applied to a half liter of water, what is the change in the water's volume?

INSIGHT

Feat of Strength or Knowledge of Materials?

The breaking of wooden boards, concrete blocks, or similar materials with a bare hand or foot is an impressive demonstration often performed by martial arts experts (Fig. 1). The physics of this feat can be analyzed in terms of the properties of the materials involved.

In delivering the blow, the expert imparts a large impulse force (Chapter 6) to the top board or block, which bends under the pressure. Note that the objects are struck midway between the end supports. At the same time, the bones of the hand are being compressed as well. Fortunately, human bone can withstand more compressive force than can wood, tile, or concrete, which is why the expert's bones aren't damaged. The ultimate compressive strength of bone is at least four times greater than that of concrete.*

When each board or block is hit, the upper surface is compressed and the lower surface is elongated, or subjected to a tension force. Wood, tile, and concrete are weaker under tension than under compression. (The ultimate tensile strength of concrete is only about a twentieth of its ultimate compressive strength.) The board or block therefore begins to crack at the bottom surface first. The crack propagates from the underside *toward* the hand, widens, and becomes a complete break. Thus, the hand never actually "cuts through" the board or block.

The amount of force required to break a board or block in this way depends on several factors. For a wooden board, these include the type of wood, the width and thickness of the board or block, and the distance between the end supports. Also, the edge of the hand must strike the board parallel to the grain of the wood. Similar considerations apply for a tile or a concrete block. Because tile and concrete are more rigid than wood, more force is generally required to break these substances.

*Ultimate strength is the maximum strength a material can stand before it breaks or fractures.

FIGURE 1 Feat of strength or knowledge of materials? Breaking wooden boards, concrete blocks, or roofing tiles (as shown here) with a karate blow depends on both the physical strength of the expert and the strength of the material. Wood, concrete, and tile have different maximum tensile and compressional stresses, but the maximum compressional strength of bone is greater than any of these. (This photo, from a high-speed photography sequence, was made toward the end of the blow.)

Some karate experts are able to break through a stack of boards or tiles and more than one concrete block. The force required does not increase by a factor equal to the number of objects. High-speed photography shows that the hand makes contact with only one or two boards or tiles at the top of the stack. Then, as each object breaks, it collides with and breaks through the one below it.

Even though the properties of the materials seem to guarantee the success of this demonstration, you should not attempt the feat unless you are an expert and know what you're doing. The board or block might win.

9.2 Fluids: Pressure and Pascal's Principle

OBJECTIVES: To (a) explain the pressure–depth relationship and (b) state Pascal's principle and describe how it is used in practical applications.

A force can be applied to a solid at a point of contact, but this won't work with a fluid, since a fluid cannot support a shear. With fluids, a force must be applied over an area. Such an application of force is expressed in terms of **pressure**, or the *force per unit area*:

Definition of pressure—force per unit area

$$p = \frac{F}{A} \qquad (9.8a)$$

SI unit of pressure: newton per square meter (N/m^2), or pascal (Pa)

The force in this equation is understood to be acting normally (perpendicularly) to the surface area. F may be the perpendicular component of a force that acts at an angle to the surface (▶Fig. 9.6).

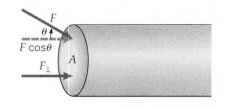

$$p = \frac{F_\perp}{A} = \frac{F \cos\theta}{A}$$

▲ **FIGURE 9.6 Pressure**
Pressure is usually written
$p = F/A$, where it is understood
that F is the force or component
of force normal to the surface. In
general, then, $p = (F \cos\theta)/A$.

As Fig. 9.6 shows, in the more general case we should write

$$p = \frac{F_\perp}{A} = \frac{F \cos\theta}{A} \tag{9.8b}$$

Pressure is a scalar quantity (with magnitude only), even though the force producing it or the force produced within a fluid by a pressure does have direction and so is a vector.

Pressure has SI units of newton per square meter (N/m^2), or **pascal (Pa)**, in honor of the French scientist and philosopher Blaise Pascal (1623–1662), who studied fluids and pressure. By definition,

$$1\ \text{Pa} = 1\ N/m^2$$

In the British system, a common unit of pressure is pound per square inch (lb/in^2, or psi). Other units, some of which will be introduced later, are used in special applications. Before going on, here's a "solid" example of the relationship between force and pressure.

Conceptual Example 9.3 ■ Force and Pressure: Taking a Nap on a Bed of Nails

Suppose you are getting ready to take a nap, and you have a choice of lying stretched out on your back on (a) a bed of nails, (b) a hardwood floor, or (c) a couch. Which one would you choose, and *why*? (See Fig. 9.27.)

Reasoning and Answer. The comfortable choice is quite apparent—the couch. But here, the conceptual question is *why*.

First let's look at the prospect of lying on a bed of nails, an old trick that originated in India and used to be demonstrated in carnival sideshows. There is really no trick here, just physics—namely, force and pressure. It is the force per unit area, or pressure $(p = F/A)$, that determines whether a nail will pierce the skin. The force is determined by the weight of the person lying on the nails. The area is determined by the *effective* area of the nails in contact with the skin (neglecting one's clothes). If there were only one nail, the person's weight on the area of its tip (pressure) would be very great—a situation in which the lone nail would pierce the skin. However, when a bed of nails is used, the same force (weight) is distributed over hundreds of nails, which gives a relatively large effective area of contact. The pressure is then reduced to a level at which the nails do not pierce the skin.

When you are lying on a hardwood floor, the area in contact with your body is appreciable and the pressure is reduced, but it still may be uncomfortable. Parts of your body, such as your neck and the small of your back, are *not* in contact with a surface, but they would be on a soft couch, making for a comfortable pressure—the lower the pressure, the more comfort (same force over a larger area). So (c) is the answer.

Follow-up Exercise. What are a couple of important considerations in constructing a bed of nails to lie on?

Now, let's take a quick review of density, which is an important consideration in the study of fluids. Recall from Chapter 1 that the density (ρ) of a substance is defined as (Eq. 1.1)

$$\text{density} = \frac{\text{mass}}{\text{volume}}$$

$$\rho = \frac{m}{V}$$

SI unit of density: kilogram per cubic meter (kg/m^3)

(common cgs unit: gram per cubic centimeter, or g/cm^3)

TABLE 9.2 Densities of Some Common Substances (in kg/m³)

Solids	Density (ρ)	Liquids	Density (ρ)	Gases*	Density (ρ)
Aluminum	2.7×10^3	Alcohol, ethyl	0.79×10^3	Air	1.29
Brass	8.7×10^3	Alcohol, methyl	0.82×10^3	Helium	0.18
Copper	8.9×10^3	Blood, whole	1.05×10^3	Hydrogen	0.090
Glass	2.6×10^3	Blood plasma	1.03×10^3	Oxygen	1.43
Gold	19.3×10^3	Gasoline	0.68×10^3	Water vapor (100°C)	0.63
Ice	0.92×10^3	Kerosene	0.82×10^3		
Iron (and steel)	7.8×10^3 (general value)	Mercury	13.6×10^3		
Lead	11.4×10^3	Seawater (4°C)	1.03×10^3		
Silver	10.5×10^3	Water, fresh (4°C)	1.00×10^3		
Wood, oak	0.81×10^3				

*At 0°C and 1 atm, unless otherwise specified.

That is, density is mass per *unit* volume. The densities of some common substances are given in Table 9.2.

Water has a density of 1.00×10^3 kg/m³ (or 1.00 g/cm³), from the original definition of the kilogram (Chapter 1). Mercury has a density of 13.6×10^3 kg/m³ (or 13.6 g/cm³). Hence, mercury is 13.6 times denser than water. Gasoline, however, is less dense than water. (See Table 9.2.) (*Note*: Be careful not to confuse the symbol for density, ρ, with that for pressure, p.)

We say that density is a measure of the compactness of the matter of a substance: The greater the density, the more matter or mass there is in a given volume. Notice how density quantifies the amount or mass per unit volume. It would be incorrect to say that mercury is "heavier" than water, because you could have a large volume of water that would be heavier than some much smaller volume of mercury.

Pressure and Depth

If you have gone scuba diving, you well know that pressure increases with depth, having felt the increased pressure on your eardrums. An opposite effect is commonly felt when you fly in a plane or ride in a car going up a mountain: With increasing altitude, your ears may "pop" because of *reduced* external air pressure.

How the pressure in a fluid varies with depth can be demonstrated by considering a container of liquid at rest. Imagine that you can isolate a rectangular column of water, as shown in ▼Fig. 9.7. Then the force on the bottom of the container below the column (or the hand) is equal to the weight of the liquid making up the

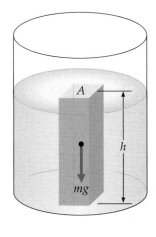
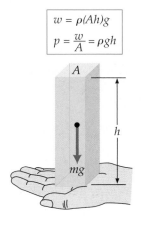

$$w = \rho(Ah)g$$
$$p = \frac{w}{A} = \rho gh$$

◀ **FIGURE 9.7 Pressure and depth** The extra pressure at a depth h in a liquid is due to the weight of the liquid above: $p = \rho gh$, where ρ is the density of the liquid (assumed to be constant). This is shown for an imaginary rectangular column of liquid.

column: $F = w = mg$. Since density is $\rho = m/V$, the mass in the column is equal to the density times the volume; that is, $m = \rho V$. (The liquid is assumed incompressible, so ρ is constant.)

The volume of the isolated liquid column is equal to the height of the column times the area of its base, or $V = hA$. Thus, we can write

$$F = w = mg = \rho V g = \rho g h A$$

With $p = F/A$, the pressure at a depth h due to the weight of the column is

$$p = \rho g h \qquad (9.9)$$

This is a general result for incompressible liquids. The pressure is the same everywhere on a horizontal plane at a depth h (with ρ and g constant). Note that Eq. 9.9 is independent of the base area of the rectangular column: We could have taken the whole cylindrical column of the liquid in the container in Fig. 9.7 and gotten the same result.

The derivation of Eq. 9.9 did not take into account pressure being applied to the open surface of the liquid. This factor adds to the pressure at a depth h to give a *total* pressure of

Pressure–depth relationship
$$p = p_o + \rho g h \qquad \begin{array}{l}\textit{(incompressible fluid}\\\textit{at constant density)}\end{array} \qquad (9.10)$$

where p_o is the pressure applied to the liquid surface (that is, the pressure at $h = 0$). For an open container, p_o is atmospheric pressure, or the weight (force) per unit area due to the gases in the atmosphere above the liquid's surface. The average atmospheric pressure at sea level is sometimes used as a unit, called an **atmosphere (atm)**:

$$1 \text{ atm} \equiv 101.325 \text{ kPa} = 1.013\,25 \times 10^5 \text{ N/m}^2 \approx 14.7 \text{ lb/in}^2.$$

The measurement of atmospheric pressure is described shortly.

Example 9.4 ■ A Scuba Diver: Pressure and Force

(a) What is the total pressure on the back of a scuba diver in a lake at a depth of 8.00 m?
(b) What is the force on the diver's back due to the water alone, taking the surface of the back to be a rectangle 60.0 cm by 50.0 cm?

Thinking It Through.   (a) This is a direct application of Eq. 9.10 in which p_o is taken as the atmospheric pressure p_a. (b) Knowing the area and the pressure due to the water, the force can be found from the definition of pressure, $p = F/A$.

Solution.

Given: $h = 8.00$ m *Find:* (a) p (total pressure)
  $A = 60.0 \text{ cm} \times 50.0 \text{ cm}$ (b) F (force due to water)
    $= 0.600 \text{ m} \times 0.500 \text{ m} = 0.300 \text{ m}^2$
  $\rho_{H_2O} = 1.00 \times 10^3 \text{ kg/m}^3$ (from Table 9.2)
  $p_a = 1.01 \times 10^5 \text{ N/m}^2$

(a)   The total pressure is the sum of the pressure due to the water and the atmospheric pressure (p_a). By Eq. 9.10, this is

$$p = p_a + \rho g h$$

$$= (1.01 \times 10^5 \text{ N/m}^2) + (1.00 \times 10^3 \text{ kg/m}^3)(9.80 \text{ m/s}^2)(8.00 \text{ m})$$

$$= (1.01 \times 10^5 \text{ N/m}^2) + (0.784 \times 10^5 \text{ N/m}^2) = 1.79 \times 10^5 \text{ N/m}^2 \text{ (or Pa)}$$

$$\text{(expressed in atmospheres)} \approx 1.8 \text{ atm}$$

(b)   The pressure p_{H_2O} due to the water alone is the $\rho g h$ portion of the preceding equation, so $p_{H_2O} = 0.784 \times 10^5 \text{ N/m}^2$.

Then, $p_{H_2O} = F/A$, and

$$F = p_{H_2O} A = (0.784 \times 10^5 \, N/m^2)(0.300 \, m^2)$$

$$= 2.35 \times 10^4 \, N \, (\text{or } 5.29 \times 10^3 \, lb\text{—about 2.6 tons!})$$

Follow-up Exercise.   You might question the answer to part (b) of this Example—how could the diver support such a force? To get a better idea of the forces our bodies can support, what would be the force on the diver's back at the water surface from atmospheric pressure alone? How do you suppose our bodies can support such forces or pressures?

Pascal's Principle

When the pressure (for example, air pressure) is increased on the entire open surface of an incompressible liquid at rest, the pressure at any point in the liquid or on the boundary surfaces increases by the same amount. The effect is the same if pressure is applied to any surface of an enclosed fluid by means of a piston (►Fig. 9.8). The transmission of pressure in fluids was studied by Pascal, and the observed effect is called **Pascal's principle**:

Pascal's principle

| Pressure applied to an enclosed fluid is transmitted undiminished to every point in the fluid and to the walls of the container.

For an incompressible liquid, the change in pressure is transmitted essentially instantaneously. For a gas, a change in pressure will generally be accompanied by a change in volume or temperature (or both), but after equilibrium has been reestablished, Pascal's principle remains valid.

Common practical applications of Pascal's principle include the hydraulic braking systems used on automobiles. Through tubes filled with brake fluid, a force on the brake pedal transmits a force to the wheel brake cylinder. Similarly, hydraulic lifts and jacks are used to raise automobiles and other heavy objects (▼Fig. 9.9).

Using Pascal's principle, we can show how such systems not only allow us to transmit force from one place to another, but also to multiply that force. The input pressure p_i supplied by compressed air for a garage lift, for example, gives an input force F_i on a small piston area A_i (Fig. 9.9). The full magnitude of the pressure is transmitted to the output piston, which has an area A_o. Since $p_i = p_o$, it follows that

$$\frac{F_i}{A_i} = \frac{F_o}{A_o}$$

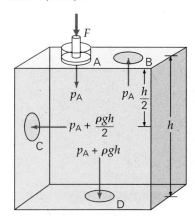

▲ **FIGURE 9.8   Pascal's principle**
The pressure applied at point A is fully transmitted to all parts of the fluid and to the walls of the container. There is also pressure due to the weight of the fluid above at different depths (for instance, $\rho gh/2$ at C and ρgh at D).

▼ **FIGURE 9.9   The hydraulic lift and shock absorbers**   **(a)** Because the input and output pressures are equal (Pascal's principle), a small input force gives a large output force proportional to the ratio of the piston areas. **(b)** A simplified exposed view of one type of shock absorber. (See Follow-up Exercise 9.5 for description.)

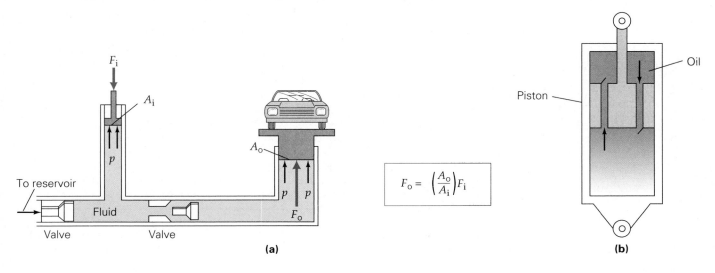

$$F_o = \left(\frac{A_o}{A_i}\right)F_i$$

(a)

(b)

and

$$F_o = \left(\frac{A_o}{A_i}\right)F_i \qquad \begin{array}{l}\textit{hydraulic force}\\ \textit{multiplication}\end{array} \qquad (9.11)$$

With A_o larger than A_i, F_o will be larger than F_i. The input force is greatly multiplied if the input piston has a relatively small area.

Example 9.5 ■ The Hydraulic Lift: Pascal's Principle

A garage lift has input and lift (output) pistons with diameters of 10 cm and 30 cm, respectively. The lift is used to hold up a car with a weight of 1.4×10^4 N. (a) What is the force on the input piston? (b) What pressure is applied to the input piston?

Thinking It Through. (a) Pascal's principle, as expressed in the hydraulic Eq. 9.11, has four variables, and three are given (areas via diameters). (b) The pressure is simply $p = F/A$.

Solution.

Given: $\quad d_i = 10$ cm $= 0.10$ m $\quad$ *Find:* (a) F_i (input force)
$\qquad d_o = 30$ cm $= 0.30$ m $\qquad\qquad$ (b) p_i (input pressure)
$\qquad F_o = 1.4 \times 10^4$ N

(a) Rearranging Eq. 9.11 and using $A = \pi r^2 = \pi d^2/4$ for the circular piston $(r = d/2)$ gives

$$F_i = \left(\frac{A_i}{A_o}\right)F_o = \left(\frac{\pi d_i^2/4}{\pi d_o^2/4}\right)F_o = \left(\frac{d_i}{d_o}\right)^2 F_o$$

or

$$F_i = \left(\frac{0.10 \text{ m}}{0.30 \text{ m}}\right)^2 F_o = \frac{F_o}{9} = \frac{1.4 \times 10^4 \text{ N}}{9} = 1.6 \times 10^3 \text{ N}$$

The input force is one-ninth of the output force; in other words, the force was multiplied by 9 (i.e., $F_o = 9F_i$).

(Note that we didn't really need to write the complete expressions for the areas. We know that the area of a circle is proportional to the square of the diameter of the circle. If the ratio of the piston diameters is 3 to 1, the ratio of their areas must therefore be 9 to 1, and we could have used this ratio directly in Eq. 9.11.)

(b) Then we apply Eq. 9.8a:

$$p_i = \frac{F_i}{A_i} = \frac{F_i}{\pi r_i^2} = \frac{F_i}{\pi(d_i/2)^2} = \frac{1.6 \times 10^3 \text{ N}}{\pi(0.10 \text{ m})^2/4}$$

$$= 2.0 \times 10^5 \text{ N/m}^2 \, (= 200 \text{ kPa})$$

This pressure is about 30 lb/in², a common pressure used in automobile tires and about twice atmospheric pressure (which is approximately 100 kPa, or 15 lb/in².)

Follow-up Exercise. Pascal's principle is used in shock absorbers on automobiles and on the landing gear of airplanes. (The polished steel piston rods can be seen above the wheels on aircraft.) In these devices, a large force (the shock produced on hitting a bump in the road or an airport runway at high speed) must be reduced to a safe level by removing energy. Basically, fluid is forced by the motion of a large-diameter piston through small channels in the piston on each stroke cycle (Fig. 9.9b).

Note that the valves allow for fluid through the channel, which creates resistance to the motion of the piston (effectively the reverse of the situation in Fig. 9.9a). The piston goes up and down, dissipating the energy of the shock. This is called *damping* (Section 13.2). Suppose that the input piston of a shock absorber on a jet plane has a diameter of 8.0 cm. What would be the diameter of an output channel that would reduce the force by a factor of 10?

As Example 9.5 shows, we can relate forces produced by pistons directly to their diameters: $F_i = (d_i/d_o)^2 F_o$ or $F_o = (d_o/d_i)^2 F_i$. By making $d_o \gg d_i$, we can get huge factors of force multiplication, as is typical for hydraulic presses, jacks, and earth-moving equipment. (The shiny input piston rods are often visible on front loaders and backhoes.) Inversely, we can get a force reduction by making $d_i > d_o$, as in Follow-up Exercise 9.5.

However, don't think that you are getting something for nothing with large force multiplications: Energy is still a factor, and it can never be multiplied by a machine. (Why not?) Looking at the work involved and assuming that the work output is equal to the work input, $W_o = W_i$ (an ideal condition—why?), we have, from Eq. 5.1,

$$F_o x_o = F_i x_i$$

or

$$F_o = \left(\frac{x_i}{x_o}\right) F_i$$

where x_o and x_i are the output and input distances moved by the respective pistons.

Thus, the output force can be much greater than the input force only if the input distance is much greater than the output distance. For example, if $F_o = 10F_i$, then $x_i = 10x_o$, and the input piston must travel 10 times the distance of the output piston. We say that *force is multiplied at the expense of distance.*

Pressure Measurement

Pressure can be measured by mechanical devices that are often spring loaded (e.g., a tire gauge). Another type of instrument, called a manometer, uses a liquid—usually mercury—to measure pressure. An *open-tube manometer* is illustrated in ▼Fig. 9.10a. One end of the U-shaped tube is open to the atmosphere, and the other is connected to the container of gas whose pressure is to be measured. The

▼ **FIGURE 9.10 Pressure measurement**   (a) For an open-tube manometer, the pressure of the gas in the container is balanced by the pressure of the liquid column and atmospheric pressure acting on the open surface of the liquid. The absolute pressure of the gas equals the sum of the atmospheric pressure (p_a) and ρgh, the gauge pressure. (b) A tire gauge measures gauge pressure, the difference between the pressure in the tire and atmospheric pressure: $p_{gauge} = p - p_a$. Thus, if a tire gauge reads 200 kPa (30 lb/in²), the actual pressure within the tire is 1 atm higher, or 300 kPa. (c) A barometer is a closed-tube manometer that is exposed to the atmosphere and thus reads only atmospheric pressure.

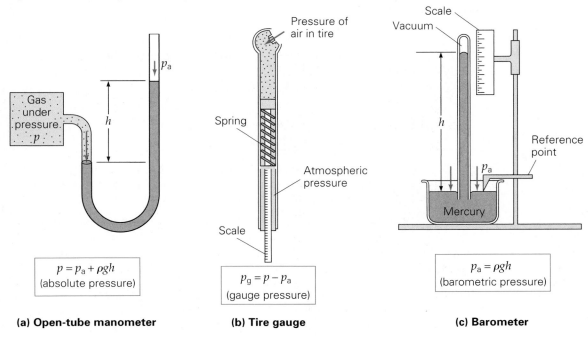

(a) Open-tube manometer

$$p = p_a + \rho gh$$
(absolute pressure)

(b) Tire gauge

$$p_g = p - p_a$$
(gauge pressure)

(c) Barometer

$$p_a = \rho gh$$
(barometric pressure)

liquid in the U-tube acts as a reservoir through which pressure is transmitted according to Pascal's principle.

The pressure of the gas (p) is balanced by the weight of the column of liquid (of height h, the difference in the heights of the columns) and the atmospheric pressure (p_a) on the open liquid surface:

$$p = p_a + \rho g h \qquad (9.12)$$

The pressure p is called the **absolute pressure**.

You may have measured pressure using pressure gauges; a tire gauge used to measure air pressure in automobile tires is a common example (Fig. 9.10b). Such gauges, quite appropriately, measure **gauge pressure**: A pressure gauge registers only the pressure *above* atmospheric pressure. Hence, to get the absolute pressure (p), you have to add the atmospheric pressure (p_a) to the gauge pressure (p_g):

$$p = p_a + p_g$$

For example, suppose your tire gauge reads a pressure of 200 kPa ($\approx$ 30 lb/in^2). The absolute pressure within the tire is then $p = p_a + p_g = 101$ kPa $+ 200$ kPa $= 301$ kPa, where normal atmospheric pressure is about 101 kPa (14.7 lb/in^2), as will be shown shortly.

The gauge pressure of a tire keeps the tire rigid or operational. In terms of the more familiar pounds per square inch (psi, or lb/in^2), a tire with a gauge pressure of 30 psi has an absolute pressure of about 45 psi ($30 + 15$, with atmospheric pressure $\approx$15 psi). Hence, the pressure on the inside of the tire is 45 psi, and that on the outside is 15 psi. The Δp of 30 psi keeps the tire inflated. If you open the valve or get a puncture, the internal and external pressures equalize and you have a flat!

Atmospheric pressure itself can be measured with a *barometer*. The principle of a mercury barometer is illustrated in Fig. 9.10c. The device was invented by Evangelista Torricelli (1608–1647), Galileo's successor as professor of mathematics at the academy in Florence. A simple barometer consists of a tube filled with mercury that is inverted into a reservoir. Some mercury runs from the tube into the reservoir, but a column supported by the air pressure on the surface of the reservoir remains in the tube. This device can be considered to be a *closed-tube manometer*, and the pressure it measures is just the atmospheric pressure, since the gauge pressure (the pressure *above* atmospheric pressure) is zero.

The atmospheric pressure is then equal to the pressure due to the weight of the column of mercury, or

$$p_a = \rho g h \qquad (9.13)$$

A *standard atmosphere* is defined as the pressure supporting a column of mercury exactly 76 cm in height at sea level and at 0°C. (For a common biological atmospheric effect, see the Insight on page 319.)

Example 9.6 ■ Standard Atmospheric Pressure: Converting to Pascals

If a standard atmosphere supports a column height of exactly 76 cm of mercury (chemical symbol Hg), what is the standard atmospheric pressure in pascals? (The density of mercury is 13.5951 $\times$ 10^3 kg/m^3 at 0°C, and $g = 9.80665$ m/s^2.)

Thinking It Through. This is a direct application of Eq. 9.13 to find standard atmospheric pressure in metric units (pascals, or newtons per square meter). Note that the height of the column is given in centimeters.

Solution.

Given: $h = 76$ cm $= 0.76$ m (exact)   *Find:* p_a (atmospheric pressure)
        $\rho_{Hg} = 13.5951 \times 10^3$ kg/m^3
         $g = 9.80665$ m/s^2

An Atmospheric Effect: Possible Earaches

Variations in atmospheric pressure can have a common physiological effect: changes in pressure in the ears with a change in altitude. This "plugging up and popping" of the ears is frequently experienced in ascents and descents on mountain roads or on airplanes. The eardrum, so important to your hearing, is a membrane that separates the middle ear from the outer ear. [See Fig. 2 in the Chapter 14 (Sound) Insight on Speech and Hearing, on p. 494, to view the anatomy of the ear.] The middle ear is connected to the throat by the Eustachian tube, the end of which is normally closed. The tube opens during swallowing or yawning to permit air to escape, so the internal and external pressures are equalized.

However, when you climb relatively quickly in a car in a hilly or mountainous region, the air pressure on the outside of the ear may be less than that in the middle ear. This difference in pressure would force the eardrum outward. If the outward pressure were not relieved, you might soon have an earache. But the pressure is relieved by a "pushing" of air through the Eustachian tube into the throat, which produces a "popping" sound. We often swallow or yawn to assist this process. Similarly, when we descend a hill or mountain, the higher outside pressure at lower altitudes needs to be equalized with the lower pressure in the middle ear.

Nature takes care of us, but it is important to understand what is going on. Suppose you have a throat infection. Then there may be a swelling of the opening of the Eustachian tube to the throat, partially blocking the tube. You may be tempted to hold your nose and blow with your mouth closed in order to clear your ears. Don't do it! You may blow infectious mucus into the inner ear and cause a painful inner-ear infection. Instead, swallow hard several times and give some big yawns to help open the Eustachian tube and equalize the pressure.

Using Eq. 9.13, we have

$$p_a = \rho_{Hg}gh = (13.5951 \times 10^3 \text{ kg/m}^3)(9.80665 \text{ m/s}^2)(0.760000 \text{ m})$$
$$= 101325 \text{ N/m}^2 = 1.01325 \times 10^5 \text{ Pa} \quad \text{(or 101.325 kPa)}$$

Follow-up Exercise. What would be the height of a barometer column for 1 standard atmosphere if water were used instead of mercury?

Changes in atmospheric pressure can be observed as changes in the height of a column of mercury. These changes are due primarily to high- and low-pressure) air masses that travel across the country. Atmospheric pressure is commonly reported in terms of the height of the barometer column, and weather forecasters say that the barometer is rising or falling. That is,

$$1 \text{ atm} = 76 \text{ cm Hg} = 760 \text{ mm Hg}$$
$$= 29.92 \text{ in. Hg} \quad \text{(about 30 in. Hg)}$$

In honor of Torricelli, a pressure supporting 1 mm of mercury is given the name *torr*:

$$1 \text{ mm Hg} \equiv 1 \text{ torr}$$

and

$$1 \text{ atm} = 760 \text{ torr}$$

Because mercury is highly toxic, it is sealed inside a barometer. A safer and less expensive device that is widely used to measure atmospheric pressure is the *aneroid* ("without fluid") *barometer*. In an aneroid barometer, a sensitive metal diaphragm on an evacuated container (something like a drumhead) responds to pressure changes, which are indicated on a dial. This is the kind of barometer you frequently find in homes in decorative wall mountings.

Since air is compressible, the atmospheric density and pressure are greatest at the Earth's surface and decrease with altitude. We live at the bottom of the atmosphere, but don't notice its pressure very much in our daily activities. Remember that our bodies are composed largely of fluids, which exert a matching

Note: Another unit sometimes used in weather reports is the millibar (mb). By definition, 1 atm = 1.01325 × 10^5 N/m^2 = 1.01325 bar = 1013.25 mb. Normal atmospheric pressures are around 1000 mb.

INSIGHT

Blood Pressure and Its Measurement

Basically, a pump is a machine that transfers mechanical energy to a fluid, thereby increasing the pressure and causing the fluid to flow. There is a wide variety of pumps, but one which is of interest to everyone is the heart, a muscular pump that drives blood through the body's circulatory network of arteries, capillaries, and veins. With each pumping cycle, the human heart's interior chambers enlarge and fill with freshly oxygenated blood from the lungs (Fig. 1).

When the chambers called ventricles contract, blood is forced out through the arteries. Smaller and smaller arteries branch off from the main ones, until the very small capillaries are reached. There, food and oxygen being carried by the blood are exchanged with the surrounding tissues, and wastes are picked up. The blood then flows into the veins to complete the circuit back to the heart.

The arterial blood pressure rises and falls in response to the cardiac cycle. That is, when the ventricles contract, forcing blood into the arterial system, the pressure in the arteries increases sharply. The maximum pressure achieved during the ventricular contraction is called the *systolic pressure*. When the ventricles relax, the arterial pressure drops, and the lowest pressure before the next contraction, called *diastolic pressure*, is reached. (These pressures are named after two parts of the pumping cycle, the *systole* and the *diastole*.)

The walls of the arteries have considerable elasticity and expand and contract with each pumping cycle. This alternate expansion and contraction can be felt as a *pulse* in an artery near the surface of the body. For example, the radial artery near the surface of the wrist is commonly used to measure a person's pulse. The pulse rate is equal to the ventricle contraction rate, and hence the pulse rate indicates the heart rate.

Taking a person's blood pressure involves measuring the pressure of the blood on the arterial walls. This is done with a *sphygmomanometer*. (The Greek word *sphygmo* means "pulse.") An inflatable cuff is used to shut off the blood

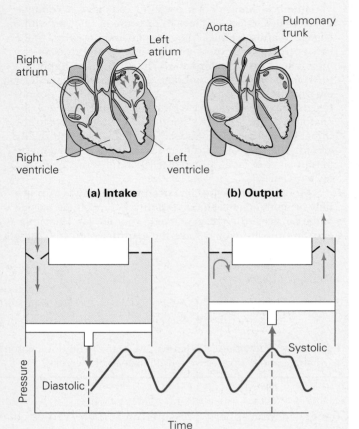

(a) Intake   **(b) Output**

FIGURE 1   The heart as a pump   The human heart is analogous to a mechanical force pump. Its pumping action, consisting of **(a)** intake and **(b)** output, gives rise to variations in blood pressure.

outward pressure. Indeed, the external pressure of the atmosphere is so important to our normal functioning that we take it with us wherever we can. The pressurized suits worn by astronauts in space or on the Moon are needed not only to supply oxygen, but also to provide an external pressure similar to that on the Earth's surface.

A very important gauge pressure reading is discussed in the Insight on blood pressure. Read it before going on to Example 9.7.

Example 9.7 ■ An IV: A Gravity Assist

An IV (*intra*venous injection) is quite a different type of gravity assist from that discussed for space probes in Chapter 7. Consider a hospital patient who receives an IV under gravity flow, as seen in ◄Fig. 9.11. If the blood gauge pressure in the vein is 20.0 mm Hg, above what height should the bottle be placed for the IV to function properly?

Thinking It Through.   The fluid gauge pressure at the bottom of the IV tube must be greater than the pressure in the vein and can be computed from Eq. 9.9. (The liquid is assumed to be incompressible.)

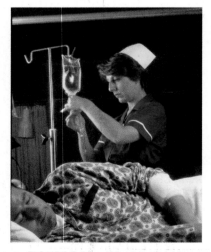

▲ **FIGURE 9.11   What height is needed?**   See Example 9.7.

flow temporarily. The cuff pressure is slowly released, and the artery is monitored with a stethoscope (Fig. 2). A point is reached where blood is just forced through the constricted artery. This flow is turbulent and gives rise to a specific sound with each heartbeat. When the sound is first heard, the (systolic) pressure is noted on the gauge, which is nor-

mally about 120 mm Hg. (The gauge in Fig. 2 is an aneroid type; older types of sphygmomanometers used a mercury column to measure blood pressure.) When the turbulent beats disappear because of smooth blood flow, the diastolic reading is taken. The pressure at this point is normally about 80 mm Hg. Blood pressure is commonly reported by giving the systolic and diastolic pressures, separated by a slash—for example, 120/80 (read as "120 over 80"). Normal systolic blood pressure ranges between 100 and 140, normal diastolic between 70 and 90. (Blood pressure is a gauge pressure. Why?)

Away from the heart, blood vessels branch to smaller and smaller diameters. The pressure in the blood vessels decreases as their diameter decreases. In small arteries, such as those in the arm, the blood pressure is on the order of 10 to 20 mm Hg, and there is no systolic–diastolic variation.

High blood pressure is a common health problem. The elastic walls of the arteries expand under the hydraulic force of the blood pumped from the heart. Their elasticity may diminish with age, however. Fatty deposits (of cholesterols) can narrow and roughen the arterial passageways, impeding the blood flow and giving rise to a form of arteriosclerosis, or hardening of the arteries. Because of these defects, the driving pressure must increase to maintain a normal blood flow. The heart must work harder, which places a greater demand on its muscles. A relatively slight decrease in the effective cross-sectional area of a blood vessel has a rather large effect (an increase) on the flow rate, as will be shown in Section 9.4.

Related Exercises: 28, 79, and 89.

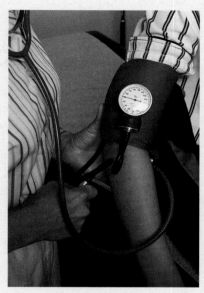

FIGURE 2 Measuring blood pressure The pressure is indicated on the gauge in millimeters Hg, or torr.

Solution.

Given: p_v = 20.0 mm Hg (vein gauge pressure)   *Find:* h (height for $p_v > 20$ mm Hg)
$\rho = 1.05 \times 10^3$ kg/m³ (whole blood from Table 9.2)

First, we need to change the common medical unit of mm Hg (or torr) to the SI unit of pascal (Pa, or N/m²):

$$p_v = (20.0 \text{ mm Hg})[133 \text{ Pa/(mm Hg)}] = 2.66 \times 10^3 \text{ Pa}$$

Then, for $p > p_v$,

$$p = \rho g h > p_v$$

or

$$h > \frac{p_v}{\rho g} = \frac{2.66 \times 10^3 \text{ Pa}}{(1.05 \times 10^3 \text{ kg/m}^3)(9.80 \text{ m/s}^2)} = 0.259 \text{ m } (\approx 26 \text{ cm})$$

The IV bottle needs to be at least 26 cm above the injection site.

Follow-up Exercise. The normal (gauge) blood pressure range is commonly reported as 120/80 (in millimeters Hg). Why is the blood pressure of 20 mm Hg in this Example so low?

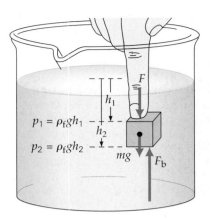

$p_1 = \rho_f g h_1$

$p_2 = \rho_f g h_2$

$\Delta p = \rho_f g(h_2 - h_1)$

(a)

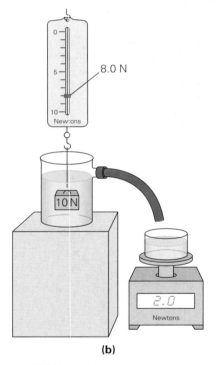

8.0 N

10 N

2.0

Newtons

(b)

▲ **FIGURE 9.13 Buoyancy and Archimedes' principle** (a) A buoyant force arises from the difference in pressure at different depths. The pressure on the bottom of the submerged block (p_2) is greater than that on the top (p_1), so there is a (buoyant) force directed upward. (It is shifted for clarity.) (b) Archimedes' principle: The buoyant force on the object is equal to the weight of the volume of fluid displaced. (The scale is set to read zero when the container is empty.)

9.3 Buoyancy and Archimedes' Principle

OBJECTIVES: To (a) relate the buoyant force to Archimedes' principle and (b) tell whether an object will float in a fluid, on the basis of relative densities.

When placed in a fluid, an object will either sink or float. This is most commonly observed with liquids; for example, objects float or sink in water. But the same effect occurs in gases: A falling object sinks in the atmosphere, and other bodies float (▶ Fig. 9.12).

Things float because they are buoyant, or are buoyed up. For example, if you immerse a cork in water and release it, the cork will be buoyed up to the surface and float there. From your knowledge of forces, you know that such motion requires an upward net force on the object. That is, there must be an upward force acting on the object that is greater than the downward force of its weight. The forces are equal when the object floats in equilibrium. The upward force resulting from an object being wholly or partially immersed in a fluid is called the **buoyant force**.

How the buoyant force comes about can be seen by considering a buoyant object being held under the surface of a fluid (◀Fig. 9.13a). The pressures on the upper and lower surfaces of the block are $p_1 = \rho_f g h_1$ and $p_2 = \rho_f g h_2$, respectively, where ρ_f is the density of the fluid. Thus, there is a pressure difference $\Delta p = p_2 - p_1 = \rho_f g (h_2 - h_1)$ between the top and bottom of the block, which gives an upward force (the buoyant force) F_b. This force is balanced by the applied force and the weight of the block.

It is not difficult to derive an expression for the magnitude of the buoyant force. We know that pressure is force per unit area. Thus, if both the top and bottom areas of the block are A, the magnitude of the net buoyant force in terms of the pressure difference is

$$F_b = p_2 A - p_1 A = (\Delta p)A = \rho_f g(h_2 - h_1)A$$

Since $(h_2 - h_1)A$ is the volume of the block, and hence the volume of fluid displaced by the block, V_f, we can write the expression for F_b as

$$F_b = \rho_f g V_f$$

But $\rho_f V_f$ is simply the mass of the fluid displaced by the block, m_f. Thus, we can write the expression for the buoyant force as $F_b = m_f g$: The magnitude of the buoyant force is equal to the weight of the fluid displaced by the block (Fig. 9.13b). This general result is known as **Archimedes' principle**:

A body immersed wholly or partially in a fluid experiences a buoyant force equal in magnitude to the weight of the *volume of fluid* that is displaced:

$$F_b = m_f g = \rho_f g V_f \qquad (9.14)$$

Archimedes (287 B.C.–212 B.C.) was given the task of determining whether a crown made for a certain king was pure gold or contained some other, cheaper metal. Legend has it that the solution to the problem came to him when he was bathing, perhaps from seeing the water level rise when he got into the tub and experiencing the buoyant force on his limbs. In any case, it is said that he was so excited that he ran through the streets of the city shouting "Eureka!" (Greek for "I have found it".) Although Archimedes' solution to the problem involved density and volume (see Exercises 57 and 61), it presumably got him thinking about buoyancy.

Example 9.8 ■ Lighter Than Air: Buoyant Force

What is the buoyant force in air on a spherical helium balloon with a radius of 30 cm if $\rho_{air} = 1.29 \ kg/m^3$? (Neglect the weight of the balloon.)

Thinking It Through. Helium is termed "lighter" (less dense) than air. Since the balloon displaces air, there is a buoyant force F_b on the balloon. This force is equal to the weight of the volume of air that the balloon displaces. To find that weight, we first find the balloon's volume and then use the density of air to find the air's mass and weight.

Solution.

Given: $r = 30\text{ cm} = 0.30\text{ m}$ *Find:* F_b (buoyant force)
$\rho_{\text{air}} = 1.29\text{ kg/m}^3$

The volume of the air in the balloon, assuming that the balloon is a sphere, is

$$V = \tfrac{4}{3}\pi r^3 = \tfrac{4}{3}\pi(0.30\text{ m})^3 = 0.11\text{ m}^3$$

Then, by Eq. 9.14, the weight of the air displaced by the balloon's volume, or the magnitude of the upward buoyant force, is

$$F_b = m_{\text{air}}g = (\rho_{\text{air}}V)g = (1.29\text{ kg/m}^3)(0.11\text{ m}^3)(9.8\text{ m/s}^2) = 1.4\text{ N}$$

Note that the buoyant force depends on the *density of the fluid* and *the volume of the body.* Shape makes no difference.

Follow-up Exercise. Would the buoyant force on the balloon be greater or lesser if the balloon were submerged in water, and by what factor? (Assume the same volume.)

▲ **FIGURE 9.12 Fluid buoyancy** The air is a fluid in which objects such as these dirigibles float. The helium inside the blimps is lighter or less dense than the surrounding air. The blimps are supported by the resulting buoyant forces.

Integrated Example 9.9 ■ Weight and Buoyant Force: Archimedes' Principle

A container of water with an overflow tube, similar to that shown in Fig. 9.13b, sits on a scale that reads 40 N. The water level is just below the exit tube in the side of the container. (a) An 8.0-N cube of wood is placed in the container. The water displaced by the floating cube runs out the exit tube into another container that is not on the scale. Will the scale reading then be (1) exactly 48 N, (2) between 40 N and 48 N, (3) exactly 40 N, or (4) less than 40 N? (b) Suppose you pushed down on the cube with your finger such that the top surface of the cube was even with the water level. How much force would have to be applied if the wooden cube measured 10 cm on a side?

(a) Conceptual Reasoning. By Archimedes' principle, the block is buoyed upward with a force equal in magnitude to the weight of the water displaced. Since the block floats, the upward buoyant force must balance the weight of the cube and so has a magnitude of 8.0 N. Thus, a volume of water weighing 8.0 N is displaced from the container as 8.0 N of weight is added to the container. The scale still reads 40 N, so the answer is (3).

Note that the upward buoyant force and the block's weight act *on the block.* The reaction force (pressure) of the block *on the water* is transmitted to the bottom of the container (Pascal's principle) and is registered on the scale.

(b) Thinking It Through. Here there are three forces acting on the stationary cube: the buoyant force upward and the weight and the force applied by the finger downward. The weight of the cube is known, so to find the applied finger force, we need to determine the buoyant force on the cube.

Given: $\ell = 10\text{ cm} = 0.10\text{ m}$ (side length of cube)   *Find:*   downward applied force
$w = 8.0\text{ N}$ (weight of cube)   necessary to put cube even
with water level

The summation of the forces acting on the cube is $\Sigma F_i = +F_b - w - F_f = 0$, where F_b is the upward buoyant force and F_f is the downward force applied by the finger. Hence, $F_f = F_b - w$. As we know, the magnitude of the buoyant force is equal to the weight of the water the cube displaces, which is given by $F_b = \rho_f g V_f$ (Eq. 9.14). The density of the fluid is that of water, which is known ($1.0 \times 10^3\text{ kg/m}^3$, Table 9.2), so

$$F_b = \rho_f g V_f = (1.0 \times 10^3\text{ kg/m}^3)(9.8\text{ m/s}^2)(0.10\text{ m})^3 = 9.8\text{ N}$$

Thus,

$$F_f = F_b - w = 9.8\text{ N} - 8.0\text{ N} = 1.8\text{ N}$$

Follow-up Exercise. In part (a), would the scale still read 40 N if the object had a density greater than that of water? In part (b), what would the scale read?

Buoyancy and Density

We commonly say that helium and hot-air balloons float because they are lighter than air. To be technically correct, we should say they are *less dense than air*. An object's density will tell you whether it will sink or float in a fluid, as long as you also know the density of the fluid. Consider a solid uniform object that is totally immersed in a fluid. The weight of the object is

$$w_o = m_o g = \rho_o V_o g$$

The weight of the volume of fluid displaced, or the magnitude of the buoyant force, is

$$F_b = w_f = m_f g = \rho_f V_f g$$

If the object is *completely submerged*, $V_f = V_o$. Dividing the second equation by the first gives

$$\frac{F_b}{w_o} = \frac{\rho_f}{\rho_o} \quad \text{or} \quad F_b = \left(\frac{\rho_f}{\rho_o}\right)w_o \quad \begin{matrix}\text{(completely}\\\text{submerged)}\end{matrix} \quad (9.15)$$

Thus, if ρ_o is less than ρ_f, then F_b will be greater than w_o, and the object will be buoyed to the surface and float. If ρ_o is greater than ρ_f, then F_b will be less than w_o, and the object will sink. If ρ_o equals ρ_f, then F_b will be equal to w_o, and the object will remain in equilibrium at any submerged depth (as long as the density of the fluid is constant). If the object is not uniform, so that its density varies over its volume, then the density of the object in Eq. 9.15 means average density.

Expressed in words, these three conditions are as follows:

An object will float in a fluid if the average density of the object is less than the density of the fluid ($\rho_o < \rho_f$).

An object will sink in a fluid if the average density of the object is greater than the density of the fluid ($\rho_o > \rho_f$).

An object will be in equilibrium at any submerged depth in a fluid if the average density of the object and the density of the fluid are equal ($\rho_o = \rho_f$).

See ◄Fig. 9.14 for an example of the last condition.

A quick look at Table 9.2 will tell you whether an object will float in a fluid, regardless of the shape or volume of the object. The three conditions just stated also apply to a fluid in a fluid, provided that the two are immiscible (do not mix). For example, you might think that cream is "heavier" than skim milk, but that's not so: Since cream floats on milk, it is less dense than milk.

In general, the densities of objects or fluids will be assumed to be uniform and constant in this book. (The density of the atmosphere does vary with altitude, but is relatively constant near the surface of the Earth.) In any event, in practical applications it is the *average* density of an object that often matters with regard to floating and sinking. For example, an ocean liner is, on average, less dense than water, even though it is made of steel. Most of its volume is occupied by air, so the liner's average density is less than that of water. Similarly, the human body has air-filled spaces, so most of us float in water. The surface depth at which a person floats depends on his or her density. (Why?)

In some instances, the overall density of an object is purposefully varied. For example, a submarine submerges by flooding its tanks with seawater (called "taking on ballast"), which increases its average density. When the sub is to surface, the water is pumped out of the tanks, so the average density of the sub becomes less than that of the surrounding seawater.

PHYSLET® ILLUSTRATION

Archimedes' Principle

Float or sink? Depends on densities of object and fluid

▲ **FIGURE 9.14   Equal densities and buoyancy**   This soft drink contains colored gelatin beads that remain suspended for months with no change. What is the density of the beads compared to the density of the drink?

Example 9.10 ■ Float or Sink? Comparison of Densities

A uniform solid cube of material 10 cm on each side has a mass of 700 g. (a) Will the cube float in water? (b) If so, how much of its volume will be submerged?

Thinking It Through. (a) The question is whether the density of the material the cube is made of greater or less than that of water, so we compute the cube's density. (b) If the cube floats, then the buoyant force and the cube's weight are equal. Both of these forces are related to the cube's volume, so we can write them in terms of that volume and equate them.

Solution.

Given: $m = 700\text{ g} = 0.700\text{ kg}$
$L = 10\text{ cm}$
$\rho_{H_2O} = 1.00 \times 10^3\text{ kg/m}^3 = 1.00\text{ g/cm}^3$
(Table 9.2)

Find: (a) Whether the cube will float in water
(b) The percentage of the volume submerged if the cube does float

It is sometimes convenient to work in cgs units in comparing quantities, particularly when one is working with ratios. For densities in g/cm³, drop the "$\times 10^3$" from the values given in Table 9.2 for solids and liquids, and add "$\times 10^{-3}$" for gases.

(a) The density of the cube is

$$\rho_c = \frac{m}{V_c} = \frac{m}{L^3} = \frac{700\text{ g}}{(10\text{ cm})^3} = 0.70\text{ g/cm}^3 < \rho_{H_2O} = 1.00\text{ g/cm}^3$$

Since ρ_c is less than ρ_{H_2O} the cube will float.

(b) The weight of the cube is $w_c = \rho_c g V_c$. When the cube is floating, it is in equilibrium, which means that its weight is balanced by the buoyant force. That is, $F_b = \rho_{H_2O} g V_{H_2O}$, where V_{H_2O} is the volume of water the submerged part of the cube displaces. Equating the expressions for weight and buoyant force gives

$$\rho_{H_2O} g V_{H_2O} = \rho_c g V_c$$

or

$$\frac{V_{H_2O}}{V_c} = \frac{\rho_c}{\rho_{H_2O}} = \frac{0.70\text{ g/cm}^3}{1.00\text{ g/cm}^3} = 0.70$$

Thus, $V_{H_2O} = 0.70 V_c$, and 70% of the cube is submerged.

Follow-up Exercise. Most of an iceberg floating in the ocean (▶Fig. 9.15) is submerged. What is seen is the proverbial "tip of the iceberg." What percentage of an iceberg's volume is seen above the surface? [Note: Icebergs are frozen *fresh* water floating in cold (salty) water.]

▲ **FIGURE 9.15 The tip of the iceberg** The vast majority of an iceberg's bulk is underneath the water. (Does the submerged ice have to be directly below the exposed tip of the iceberg?)

A quantity called specific gravity is related to density. It is commonly used for liquids, but also applies to solids. The **specific gravity** (*sp. gr.*) of a substance is equal to the ratio of the density of the substance (ρ_s) to the density of water (ρ_{H_2O}) at 4°C, the temperature for maximum density:

$$sp.\,gr. = \frac{\rho_s}{\rho_{H_2O}}$$

Because it is a ratio of densities, specific gravity has no units. In cgs units, $\rho_{H_2O} = 1.00\text{ g/cm}^3$, so

$$sp.\,gr. = \frac{\rho_s}{1.00} = \rho_s$$

That is, the specific gravity of a substance is equal to the numerical value of its density *in cgs units*. For example, if a liquid has a density of 1.5 g/cm³, its specific gravity is 1.5, which tells you that it is 1.5 times denser than water. To get density values in grams per cubic centimeter, divide the value in Table 9.2 by 10^3. The

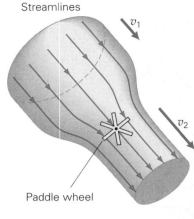

Streamlines

v_1

v_2

Paddle wheel

(a)

(b)

▲ **FIGURE 9.16 Streamline flow**
(a) Streamlines never cross and are closer together in regions of greater fluid velocity. The stationary paddle wheel indicates that the flow is irrotational, or without whirlpools and eddy currents. **(b)** The smoke from an extinguished candle begins to rise in nearly streamline flow, but quickly becomes rotational and turbulent.

specific gravities of automobile coolant and battery electrolyte (both water solutions) are measured so as to determine their relative concentrations of antifreeze and sulfuric acid, respectively.

9.4 Fluid Dynamics and Bernoulli's Equation

OBJECTIVES: To **(a)** identify the simplifications used in describing ideal fluid flow and **(b)** use the continuity equation and Bernoulli's equation to explain common effects of ideal fluid flow.

In general, fluid motion is difficult to analyze. For example, think of trying to describe the motion of a particle (a molecule, as an approximation) of water in a rushing stream. The overall motion of the stream may be apparent, but a mathematical description of the motion of any one particle of it may be virtually impossible because of eddy currents (small whirlpool motions), the gushing of water over rocks, frictional drag on the stream bottom, and so on. A basic description of fluid flow is conveniently obtained by ignoring such complications and considering an ideal fluid. Actual fluid flow can then be approximated with reference to this simpler theoretical model.

In this simplified approach to fluid dynamics, it is customary to consider four characteristics of an **ideal fluid**. In such a fluid, flow is (1) *steady*, (2) *irrotational*, (3) *nonviscous*, and (4) *incompressible*.

Condition 1: *Steady flow* means that all the particles of a fluid have the same velocity as they pass a given point.

Steady flow might be called smooth or regular flow. The path of steady flow can be depicted in the form of **streamlines** (◄Fig. 9.16). Every particle that passes a particular point moves along a streamline. That is, every particle moves along the same path (streamline) as particles that passed by earlier. Streamlines never cross; if they did, a particle would have alternative paths and abrupt changes in its velocity, in which case the flow would not be steady.

Steady flow requires low velocities. For example, steady flow is approximated by the flow relative to a canoe that is gliding slowly through still water. When the flow velocity is high, eddies tend to appear, especially near boundaries, and the flow becomes turbulent.

Streamlines also indicate the relative magnitude of the velocity of a fluid. The velocity is greater where the streamlines are closer together. Notice this effect in Fig. 9.16a. The reason for it will be explained shortly.

Condition 2: *Irrotational flow* means that a fluid element (a small volume of the fluid) has no net angular velocity, which eliminates the possibility of whirlpools and eddy currents. (The flow is nonturbulent.)

Consider the small paddle wheel in Fig. 9.16a. With a zero net torque, the wheel does not rotate. Thus, the flow is irrotational.

Condition 3: *Nonviscous flow* means that viscosity is negligible.

Viscosity refers to a fluid's internal friction, or resistance to flow. (For example, honey has a much greater viscosity than water.) A truly nonviscous fluid would flow freely with no energy lost within it. Also, there would be no frictional drag between the fluid and the walls containing it. In reality, when a liquid flows through a pipe, the speed is lower near the walls because of frictional drag and is higher toward the center of the pipe. (Viscosity is discussed in more detail in Section 9.5.)

Condition 4: *Incompressible flow* means that the fluid's density is constant.

Liquids can usually be considered incompressible. Gases, by contrast, are quite compressible. Sometimes, however, gases approximate incompressible flow—for example, air flowing relative to the wings of an airplane traveling at low speeds.

Theoretical or ideal fluid flow is not characteristic of most real situations, but the analysis of ideal flow provides results that approximate, or generally describe, a variety of applications. Usually, this analysis is derived, not from Newton's laws, but instead from two basic principles: conservation of mass and conservation of energy.

Equation of Continuity

If there are no losses of fluid within a uniform tube, the mass of fluid flowing into the tube in a given time must be equal to the mass flowing out of the tube in the same time (by the conservation of mass). For example, in ▼Fig. 9.17a, the mass (Δm_1) entering the tube during a short time (Δt) is

$$\Delta m_1 = \rho_1 \Delta V_1 = \rho_1(A_1 \Delta x_1) = \rho_1(A_1 v_1 \Delta t)$$

where A_1 is the cross-sectional area of the tube at the entrance and, in a time Δt, a fluid particle moves a distance equal to $v_1 \Delta t$. Similarly, the mass leaving the tube in the same interval is (Fig. 9.17b)

$$\Delta m_2 = \rho_2 \Delta V_2 = \rho_2(A_2 \Delta x_2) = \rho_2(A_2 v_2 \Delta t)$$

Since the mass is conserved, $\Delta m_1 = \Delta m_2$, and it follows that

$$\rho_1 A_1 v_1 = \rho_2 A_2 v_2 \quad \text{or} \quad \rho A v = \text{constant} \quad (9.16)$$

This general result is called the **equation of continuity**.
    For an incompressible fluid, the density ρ is constant, so

Equation of continuity or flow rate equation

$$A_1 v_1 = A_2 v_2 \quad \text{or} \quad A v = \text{constant} \quad (\textit{for an incompressible fluid}) \quad (9.17)$$

This is sometimes called the **flow rate equation**, since Av has the SI units of cubic meters per second (m³/s, or volume/time). (In the British system, the

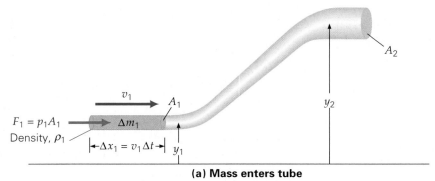

(a) Mass enters tube

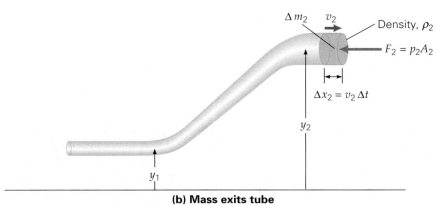

(b) Mass exits tube

◀ **FIGURE 9.17 Flow continuity**
Ideal fluid flow can be described in terms of the conservation of mass by the equation of continuity.

▲ FIGURE 9.18 Flow rate By the flow rate equation, the speed of a fluid is greater when the cross-sectional area of the tube through which the fluid is flowing is smaller. Think of a hose that is equipped with a nozzle such that the cross-sectional area of the hose is made smaller.

unit gallons per minute is often used.) Note that the flow rate equation shows that the fluid velocity is greater where the cross-sectional area of the tube is smaller. That is,

$$v_2 = \left(\frac{A_1}{A_2}\right)v_1$$

and v_2 is greater than v_1 if A_2 is less than A_1. This effect is evident in the common experience that the speed of water is greater from a hose fitted with a nozzle than from the same hose without a nozzle (◄Fig. 9.18).

The flow rate equation can be applied to the flow of blood in your body. Blood flows from the heart into the aorta. It then makes a circuit through the circulatory system, passing through arteries, arterioles (small arteries), capillaries, and venules (small veins) and back to the heart through veins. The speed is lowest in the capillaries. Is this a contradiction? No: The *total* area of the capillaries is much larger than that of the arteries or veins, so the flow rate equation holds.

Example 9.11 ■ Blood Flow: Cholesterol and Plaque

High cholesterol in the blood can cause fatty deposits called plaques to form on the walls of blood vessels. Suppose a plaque reduces the effective radius of an artery by 25%. How does this partial blockage affect the speed of blood through the artery?

Thinking It Through.   The flow rate equation (Eq. 9.17) applies, but note that no values of area or speed are given. This indicates that we should use ratios.

Solution.   Taking the unclogged artery to have a radius r_1, we can say that the plaque then reduces the effective radius to r_2.

Given: $r_2 = 0.75r_1$ (for a 25% reduction)   *Find:* v_2

Writing the flow rate equation in terms of the radii, we have

$$A_1 v_1 = A_2 v_2$$
$$(\pi r_1^2)v_1 = (\pi r_2^2)v_2$$

Rearranging and canceling, we get

$$v_2 = \left(\frac{r_1}{r_2}\right)^2 v_1$$

From the given information, $r_1/r_2 = 1/0.75$, so

$$v_2 = (1/0.75)^2\, v_1 = 1.8 v_1$$

Hence, the speed through the clogged artery increases by 80%.

Follow-up Exercise.   By how much would the effective radius of an artery have to be reduced to have a 50% increase in the speed of the blood flowing through it?

Bernoulli's Equation

The conservation of energy or the general work–energy theorem leads to another relationship that has great generality for fluid flow. This relationship was first derived in 1738 by the Swiss mathematician Daniel Bernoulli (1700–1782) and is named for him.

Let's look again at the ideal fluid flowing in the tube in Fig. 9.17. Work is done by the external forces at the ends of the tube; F_1 does positive work (in the same direction as the fluid's motion) and F_2 does negative work (opposite to the fluid motion). The net work done on the system by these forces is then

$$W_{\text{net}} = F_1 \Delta x_1 - F_2 \Delta x_2 = (p_1 A_1)(v_1 \Delta t) - (p_2 A_2)(v_2 \Delta t)$$

The flow rate equation (Eq. 9.17) requires that $A_1v_1 = A_2v_2$, so we may write the work equation as

$$W_{\text{net}} = A_1v_1\Delta t(p_1 - p_2)$$

Recall from the preceding derivation of the equation of continuity that $\Delta m_1 = \rho_1\Delta V_1 = \rho_1(A_1v_1\Delta t)$ and $\Delta m_1 = \Delta m_2$, so we may write, in general,

$$W_{\text{net}} = \frac{\Delta m}{\rho}(p_1 - p_2)$$

The net work done on the system by the external forces (nonconservative work) must be equal to the change in total mechanical energy. That is, $W_{\text{net}} = \Delta E = \Delta K + \Delta U$. Looking at the change in kinetic energy of an element of mass Δm, we have

$$\Delta K = \tfrac{1}{2}\Delta m(v_2^2 - v_1^2)$$

The corresponding change in gravitational potential energy is

$$\Delta U = \Delta mg(y_2 - y_1)$$

Thus,

$$W_{\text{net}} = \Delta K + \Delta U$$

$$\frac{\Delta m}{\rho}(p_1 - p_2) = \tfrac{1}{2}\Delta m(v_2^2 - v_1^2) + \Delta mg(y_2 - y_1)$$

Canceling each Δm and rearranging gives the common form of **Bernoulli's equation**:

Bernoulli's equation

$$p_1 + \tfrac{1}{2}\rho v_1^2 + \rho gy_1 = p_2 + \tfrac{1}{2}\rho v_2^2 + \rho gy_2 \qquad (9.18)$$

or

$$p + \tfrac{1}{2}\rho v^2 + \rho gy = \text{constant}$$

Note that in working with a fluid, the terms in Bernoulli's equation are work or energy per unit volume (J/m^3). That is, $W = F\,\Delta x = p(A\,\Delta x) = p\,\Delta V$ and therefore $p = W/\Delta V$ (work/volume). Similarly, with $\rho = m/V$, we have $\tfrac{1}{2}\rho v^2 = \tfrac{1}{2}mv^2/V$ (energy/volume) and $\rho gy = mgy/V$ (energy/volume).

Note: Compare the derivation of Eq. 5.10 in Section 5.5.

    Bernoulli's equation, or principle, can be applied to many situations. If there is horizontal flow ($y_1 = y_2$), then $p + \tfrac{1}{2}\rho v^2 = \text{constant}$, which indicates that the pressure decreases if the speed of the fluid increases (and vice versa). This effect is illustrated in ▼Fig. 9.19, where the difference in flow heights through the pipe is considered negligible (so the ρgy term drops out).

PHYSLET®
ILLUSTRATION

Blood Flow in Arteries

▼ **FIGURE 9.19 Flow rate and pressure** Taking the horizontal difference in flow heights to be negligible in a constricted pipe, we obtain, for Bernoulli's equation, $p + \tfrac{1}{2}\rho v^2 = \text{constant}$. In a region of smaller cross-sectional area, the flow speed is greater (see flow rate equation); from Bernoulli's equation, the pressure in that region is lower than in other regions.

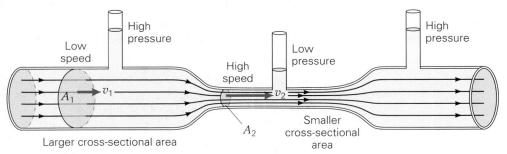

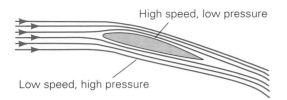

High speed, low pressure

Low speed, high pressure

▶ **FIGURE 9.20 Airplane lift—Bernoulli's principle in action** Because of the shape and orientation of an airfoil or airplane wing, the air streamlines are closer together, and the air speed is greater above the wing than below it. By Bernoulli's principle, the resulting pressure difference supplies an upward force, or lift.

Chimneys and smokestacks are tall in order to take advantage of the more consistent and higher wind speeds at greater heights. The faster the wind blows over the top of a chimney, the lower is the pressure, and the greater is the pressure difference between the bottom and top of the chimney. Thus, the chimney draws exhaust out better. Bernoulli's equation and the continuity equation (Av = constant) also tell you that if the cross-sectional area of a pipe is reduced, so that the velocity of the fluid passing through it is increased, then the pressure is reduced.

The Bernoulli effect (as it is sometimes called) is also partially responsible for the lift of an airplane. Ideal airflow over an airfoil or wing is shown in ▲Fig. 9.20. (Turbulence is neglected.) The wing is curved on the top side and is angled relative to the incident streamlines. As a result, the streamlines above the wing are closer together than those below, which causes a higher air speed and lower pressure above the wing. With a higher pressure on the bottom of the wing, there is a net upward force, or *lift*. Note in the figure that the streamlines leaving the wing curve downward. This curving reflects the fact that, since the wing is acquiring upward momentum, the air molecules must acquire an equal downward component of momentum.

Suppose a fluid is at rest ($v_2 = v_1 = 0$). Bernoulli's equation then becomes

$$p_2 - p_1 = \rho g(y_1 - y_2)$$

This is the pressure–depth relationship derived earlier (Eq. 9.10).

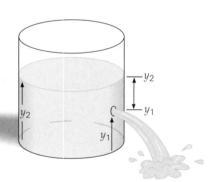

y_2

y_2

y_1

y_1

▲ **FIGURE 9.21 Fluid flow from a tank** The flow rate is given by Bernoulli's equation. See Example 9.12.

Example 9.12 ■ Flow Rate from a Tank: Bernoulli's Equation

A cylindrical tank containing water has a small hole punched in its side below the water level, and water runs out (◀Fig. 9.21). What is the approximate initial flow rate of water out of the tank?

Thinking It Through. Equation 9.17 ($A_1v_1 = A_2v_2$) is the flow rate equation, where Av has units of m³/s, or volume/time. The v terms can be related by Bernoulli's equation, which also contains y, which can be used to find differences in height. The areas are not given, so relating the v terms might require some sort of approximation, as will be seen. (Note that the *approximate* initial flow rate is wanted.)

Solution.

Given: No specific values are given,   *Find:* An expression for the approximate
so symbol variables will be used.   initial water flow rate from the hole.

Bernoulli's equation,

$$p_1 + \tfrac{1}{2}\rho v_1^2 + \rho g y_1 = p_2 + \tfrac{1}{2}\rho v_2^2 + \rho g y_2$$

can be used to write $y_2 - y_1$, which is the height of the surface of the liquid above the hole. The atmospheric pressures acting on the open surface and at the hole, p_1 and p_2, respectively, are essentially equal and cancel from the equation, as does the density, so

$$v_1^2 - v_2^2 = 2g(y_2 - y_1)$$

By the equation of continuity (the flow rate equation, Eq. 9.17), $A_1v_1 = A_2v_2$, where A_2 is the cross-sectional area of the tank and A_1 is that of the hole. Since A_2 is much greater than A_1, v_1 is much greater than v_2 (initially, $v_2 \approx 0$). So, to a good approximation,

$$v_1^2 = 2g(y_2 - y_1) \qquad \text{or} \qquad v_1 = \sqrt{2g(y_2 - y_1)}$$

The flow rate (volume/time) is then

$$\text{flow rate} = A_1 v_1 = A_1 \sqrt{2g(y_2 - y_1)}$$

Given the area of the hole and the height of the liquid above it, you can find the initial speed of the water coming from the hole and the flow rate. (What happens as the water level falls?)

Follow-up Exercise. What would be the percentage change in the initial flow rate from the tank in this Example if the diameter of the small circular hole were increased by 30.0%?

Another example of the Bernoulli effect is given in the Insight on p. 332.

Conceptual Example 9.13 ■ A Stream of Water: Smaller and Smaller

You have probably observed that a steady stream of water flowing out of a faucet gets smaller the farther the water gets from the faucet. Why does that happen?

Reasoning and Answer. This effect can be explained by Bernoulli's principle. As the water falls, it accelerates and its speed increases. Then, by Bernoulli's principle, the internal liquid pressure inside the stream decreases. (See Fig. 9.19.) A pressure difference between that inside stream and the atmospheric pressure on the outside is thus created. As a result, there is an increasing inward force as the stream falls, so it becomes smaller. Eventually, the stream may get so thin that it breaks up into individual droplets.

Follow-up Exercise. The equation of continuity can also be used to explain this stream effect. Give this explanation.

*9.5 Surface Tension, Viscosity, and Poiseuille's Law

OBJECTIVES: To (a) describe the source of surface tension and its effects and (b) discuss fluid viscosity.

Surface Tension

The molecules of a liquid exert small attractive forces on each other. Even though molecules are electrically neutral overall, there is often some slight asymmetry of charge that gives rise to attractive forces between them (called *van der Waals forces*). Within a liquid, any molecule is completely surrounded by other molecules, and the net force is zero (▶Fig. 9.22a). However, for molecules at the surface of the liquid, there is no attractive force acting from above the surface. (The effect of air molecules is small and considered negligible.) As a result, net forces act upon the molecules of the surface layer, due to the attraction of neighboring molecules just below the surface. This inward pull on the surface molecules causes the surface of the liquid to contract and to resist being stretched or broken, a property called **surface tension**.

If a sewing needle is carefully placed on the surface of a bowl of water, the surface acts like an elastic membrane under tension. There is a slight depression in the surface, and molecular forces along the depression act at an angle to the surface (Fig. 9.22b). The vertical components of these forces balance the weight (mg) of the needle, and the needle "floats" on the surface. Similarly, surface tension supports the weight of a water strider (Fig. 9.22c).

Throwing a Curveball

In baseball, a pitcher can cause the ball to curve as it moves toward home plate by giving it an appropriate spin. Why a curveball curves can be understood in terms of Bernoulli's equation and the viscous properties of air.

If air were an ideal fluid, the spin of a pitched ball would have no effect in changing the direction of the ball's motion. (In Fig. 1a, the streamlines are those which would be seen by an observer moving with the ball.) However, because air is viscous, friction between it and the ball causes a thin (boundary) layer of air to be dragged around by the spinning ball. This effect is illustrated in Fig. 1b for a relatively slow airstream in which the flow is generally smooth. The speed of the ball relative to the air is thus greater on one side than on the other, because the velocities of the ball and of the air are in the same direction on one side (top in Fig. 1b) and in opposite directions on the other. By Bernoulli's equation, the low-velocity side (v') has a greater pressure than the high-velocity side (v''). Thus, there is a net force on the ball toward the low-pressure side, and the ball is deflected. (Its path curves.)

However, baseballs are usually pitched very fast, and another effect applies. With a high airflow speed, the rotation of the ball causes the boundary layer to separate at different points on either side of the ball. This separation gives rise to a turbulent wake that is deflected in the direction of the ball's spin (Fig. 2). A pressure difference results, and the ball is deflected or curves the opposite way. This is called the Magnus effect (after Gustav Magnus, who observed the effect in 1852). In terms of Newton's third law, the turbulent wake is deflected in one direction and the reaction force deflects the ball in the opposite direction.

Note that the ball in Fig. 2 is rotating counterclockwise and is deflected to the left (toward the top, in this overhead view). To have the ball curve to the right requires a clockwise rotation. In general, the curve is in the direction that the front part of the ball is rotating. With the spin axis in the horizontal plane, the deflecting force may be up or down, which can cause the ball not to drop as much under the influence of gravity or to "sink" even faster. The spin axis can be oriented in any direction to produce a variety of effects with the deflecting force and gravity.

Incidentally, golf balls are dimpled to create turbulence, which is usually a bad thing in trying to get distance in a resistive medium (air). A swimmer, for example, wants as little turbulence (and resistance) as possible. However, with golf balls, localized turbulence can reduce drag. This involves the Magnus effect and is a bit complicated, but it gives golf balls a greater range.

Related Exercise: 75.

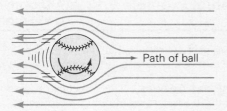

(a) No viscosity: same airflow on both sides

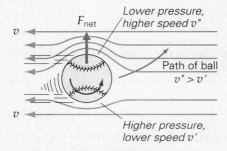

(b) With viscosity: faster airflow on one side of the ball creates a pressure difference

FIGURE 1   Bernoulli effect   (a) If air were an ideal fluid (without viscosity) and spin were applied, the airflow would be the same on both sides of the ball. **(b)** Because air has viscosity, some air is dragged around with the ball, so the speed of the ball with respect to the air is greater on one side (top view shown). Pressure on that side is lower, resulting in a net force in that direction.

FIGURE 2   Magnus effect
The turbulent wake of a fast-moving ball causes the ball to curve. In terms of Newton's third law, the wake is deflected in one direction, and the ball is deflected in the opposite direction by the reaction force.

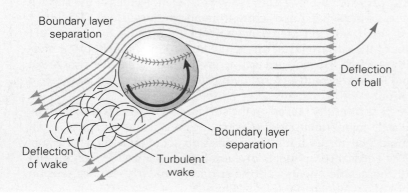

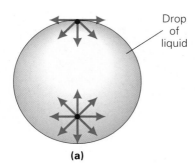

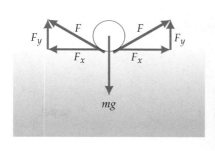

(a)

(b)

▲ **FIGURE 9.22 Surface tension** **(a)** The net force on a molecule in the interior of a liquid is zero, because the molecule is surrounded by other molecules. However, a nonzero net force acts on a molecule at the surface, due to the attractive forces of the neighboring molecules just below the surface. **(b)** For an object such as a needle to form a depression on the surface, work must be done, since more interior molecules must be brought to the surface to increase its area. As a result, the surface area acts like a stretched elastic membrane, and the weight of the object is supported by the upward components of the surface tension. Insects such as this water strider can walk on water because of the upward components of the surface tension, much as you might walk on a large trampoline. Note the depressions in the surface of the liquid where the legs touch it.

The net effect of surface tension is to make the surface area of a liquid as small as possible. That is, a given volume of liquid tends to assume the shape that has the least surface area. As a result, drops of water and soap bubbles have spherical shapes, because a sphere has the smallest surface area for a given volume (▼Fig. 9.23). In forming a drop or bubble, surface tension pulls the molecules together to minimize the surface area.

Viscosity

All real fluids have an internal resistance to flow, or **viscosity**, which can be considered to be friction between the molecules of a fluid. In liquids, viscosity is caused by short-range cohesive forces, and in gases, it is caused by collisions between molecules. (See the discussion of air resistance in Section 4.6.) The viscous drag for both liquids and gases depends on their velocity and may be directly proportional to it in some cases. However, the relationship varies with the conditions; for example, the drag is approximately proportional to v^2 or v^3 in turbulent flow.

Internal friction causes the layers of a fluid to move relative to each other in response to a shear stress. This layered motion, called *laminar flow*, is characteristic

▼ **FIGURE 9.23 Surface tension at work** Because of surface tension, **(a)** water droplets and **(b)** soap bubbles tend to assume the shape that minimizes their surface area—that of a sphere.

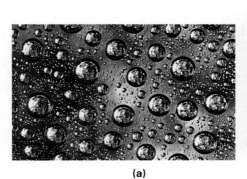

(a)

(b)

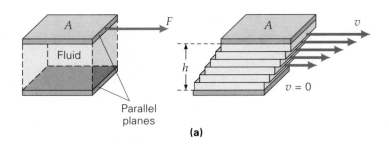

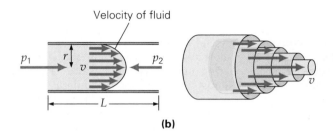

(b)

FIGURE 9.24 Laminar flow
(a) A shear stress causes layers of a fluid to move over each other in laminar flow. The shear force and the flow rate depend on the viscosity of the fluid. (b) For laminar flow through a pipe, the speed of the fluid is less near the walls of the pipe than near the center because of frictional drag between the walls and the fluid.

Viscosity: resistance to flow

Note: *SAE* stands for *Society of Automotive Engineers*, an organization that designates the grades of motor oils based on their viscosity.

of steady flow for viscous liquids at low velocities (▲ Fig. 9.24a). At higher velocities, the flow becomes rotational, or *turbulent*, and difficult to analyze.

Since there are shear stresses and shear strains (deformation) in laminar flow, the viscous property of a fluid can be described by a coefficient, like the elastic moduli discussed in Section 9.1. Viscosity is characterized by a *coefficient of viscosity*, η (the Greek letter eta), commonly referred to as simply the viscosity.

The coefficient of viscosity is, in effect, the ratio of the shear stress to the rate of change of the shear strain (since motion is involved). Unit analysis shows that the SI unit of viscosity is the pascal-second (Pa·s). This combined unit is called the *poiseuille* (Pl), in honor of the French scientist Jean Poiseuille (1799–1869), who studied the flow of liquids, particularly blood. (Poiseuille's law on flow rate will be presented shortly.) The cgs unit of viscosity is the *poise* (P). A smaller multiple, the *centipoise* (cP), is widely used because of its convenient size; 1 Pl = 10^2 cP.

The viscosities of some fluids are listed in Table 9.3. The greater the viscosity of a liquid, which is easier to visualize than that of a gas, the greater is the shear stress required to get the layers of the liquid to slide along each other. Note, for example, the large viscosity of glycerin compared to that of water.*

As you might expect, viscosity, and thus fluid flow, varies with temperature, which is evident from the old saying, "slow as molasses in January." A familiar application is the viscosity grading of motor oil used in automobiles. In winter, a low-viscosity, or relatively thin, oil should be used (such as SAE grade 10W or 20W), because it will flow more readily, particularly when the engine is cold at start-up. In summer, a higher viscosity, or thicker, oil is used (SAE 30, 40, or even 50). Seasonal changes in the grade of motor oil are not necessary if you use the multigrade year-round oils. These oils contain additives called viscosity improvers, which are polymers whose molecules are long, coiled chains. An increase in temperature causes the molecules to uncoil and intertwine. Thus, the normal decrease in viscosity is counteracted. The action is reversed on cooling, and the oil maintains a relatively small viscosity range over a large temperature range. Such motor oils are graded, for example, as SAE 10W–30 (or 10W–30, for short).

---

*If you want to think about a very large viscosity, consider that of glass. It has been said that the glass in the stained-glass windows of medieval churches has "flowed" over time, such that the panes are now thicker at the bottom than at the top. A recent analysis indicates that window glass may even flow over incredibly long periods that exceed the limits of human history. On human time scales, such a flow would not be evident. [See E. D. Zanotto, *Am. J. Phys.* **66** (May, 1998), 392–395.]

TABLE 9.3 Viscosities of Various Fluids*

Fluid	Viscosity (η) Poiseuille (Pl)
Liquids	
Alcohol, ethyl	1.2×10^{-3}
Blood, whole (37°C)	1.7×10^{-3}
Blood plasma (37°C)	2.5×10^{-3}
Glycerin 1.5	1.5×10^{3}
Mercury	1.55×10^{-3}
Oil, light machine	1.1
Water	1.00×10^{-3}
Gases	
Air	1.9×10^{-5}
Oxygen	2.2×10^{-5}

*At 20°C unless otherwise indicated.

Poiseuille's Law Viscosity makes analyzing fluid flow difficult. For example, when a fluid flows through a pipe, there is frictional drag between the liquid and the walls, and the fluid velocity is greater toward the center of the pipe (Fig. 9.24b). In practice, this effect makes a difference in a fluid's *average flow rate* $Q = A\bar{v} = \Delta V / \Delta t$ (see Eq. 9.17), which describes the volume (ΔV) of fluid flowing past a given point during a time Δt. The SI unit of flow rate is cubic meters per second (m^3/s). The flow rate depends on the properties of the fluid and the dimensions of the pipe, as well as on the pressure difference (Δp) between the ends of the pipe.

Jean Poiseuille studied flow in pipes and tubes, assuming a constant viscosity and steady or laminar flow. He derived the following relationship, known as **Poiseuille's law**, for the flow rate:

Poiseuille's law

$$Q = \frac{\Delta V}{\Delta t} = \frac{\pi r^4 \Delta p}{8 \eta L} \qquad (9.19)$$

Here, r is the radius of the pipe and L is its length.

As expected, the flow rate is inversely proportional to the viscosity (η) and the length of the pipe. Also as expected, the flow rate is directly proportional to the pressure difference Δp between the ends of the pipe. Somewhat surprisingly, however, the flow rate is proportional to r^4, which makes it more highly dependent on the radius of the tube than we might have thought.

An application of fluid flow in a medical IV was examined in Example 9.7. However, Poiseuille's law, which incorporates the flow rate, affords more reality to this application, as the next Example shows.

Example 9.14 ■ Poiseuille's Law: A Blood Transfusion

A hospital patient needs a blood transfusion, which will be administered through a vein in the arm via a gravity IV. The physician wishes to have 500 cc of whole blood delivered over a period of 10 min by an 18-gauge needle with a length of 50 mm and an inner diameter of 1.0 mm. At what height above the arm should the bag of blood be hung? (Assume a venous blood pressure of 15 mm Hg.)

Thinking It Through.   This is an application of Poiseuille's law (Eq. 9.19) to find the pressure needed at the inlet of the needle that will provide the required flow rate (Q). Note that $\Delta p = p_{in} - p_{out}$ (inlet pressure minus outlet pressure). Knowing the inlet pressure, we can find the required height of the bag as in Example 9.7. (*Caution*: There are a lot of nonstandard units here, and some quantities are assumed to be known from tables.)

Solution. First we write the given (and known) quantities, converting to standard SI units as we go:

Given: $\Delta V = 500 \text{ cc} = 500 \text{ cm}^3 \, (1 \text{ m}^3/10^6 \text{ cm}^3) = 5.00 \times 10^{-4} \text{ m}^3$ *Find:* h (height
$\Delta t = 10 \text{ min} = 600 \text{ s} = 6.00 \times 10^2 \text{ s}$ of bag)
$L = 50 \text{ mm} = 5.0 \times 10^{-2} \text{ m}$
$d = 1.0 \text{ mm, or } r = 0.50 \text{ mm} = 5.0 \times 10^{-4} \text{ m}$
$p_{\text{out}} = 15 \text{ mm Hg} = 15 \text{ torr } (133 \text{ Pa/torr}) = 2.0 \times 10^3 \text{ Pa}$
$\eta = 1.7 \times 10^{-3} \text{ Pl (whole blood, from Table 9.3)}$

The flow rate is

$$Q = \frac{\Delta V}{\Delta t} = \frac{5.00 \times 10^{-4} \text{ m}^3}{6.00 \times 10^2 \text{ s}} = 8.33 \times 10^{-7} \text{ m}^3/\text{s}$$

We insert this number into Eq. 9.19 and solve for Δp:

$$\Delta p = \frac{8\eta L Q}{\pi r^4} = \frac{8(1.7 \times 10^{-3} \text{ Pl})(5.0 \times 10^{-2} \text{ m})(8.33 \times 10^{-7} \text{ m}^3/\text{s})}{\pi (5.0 \times 10^{-4} \text{ m})^4} = 2.9 \times 10^3 \text{ Pa}$$

With $\Delta p = p_{\text{in}} - p_{\text{out}}$, we have

$$p_{\text{in}} = \Delta p + p_{\text{out}} = (2.9 \times 10^3 \text{ Pa}) + (2.0 \times 10^3 \text{ Pa}) = 4.9 \times 10^3 \text{ Pa}$$

Then, to find the height of the bag that will deliver this amount of pressure, we use $p_{\text{in}} = \rho g h$ (where $\rho_{\text{whole blood}} = 1.05 \times 10^3 \text{ kg/m}^3$, from Table 9.2). Thus,

$$h = \frac{p_{\text{in}}}{\rho g} = \frac{4.9 \times 10^3 \text{ Pa}}{(1.05 \times 10^3 \text{ kg/m}^3)(9.80 \text{ m/s}^2)} = 0.48 \text{ m}$$

Hence, for the prescribed flow rate, the bag of blood should be hung about 48 cm above the needle in the arm.

Follow-up Exercise. Suppose the physician wants to follow up the blood transfusion with 500 cc of saline solution at the same rate of flow. At what height should the saline bag be placed? (The *isotonic* saline solution administered by IV is a 0.85% aqueous salt solution, which has the same salt concentration as do body cells. To a good approximation, saline has the same density as water.)

Gravity-flow IVs are still used, but with modern technology, the flow rates of IVs are now often controlled and monitored by machines (◄Fig. 9.25).

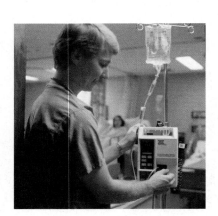

▲ **FIGURE 9.25 IV technology**
The mechanism of intravenous injection is still a gravity assist, but IV flow rates are now commonly controlled and monitored by machines.

Chapter Review

Important Concepts and Equations

- In the deformation of elastic solids, **stress** is a measure of the force causing the deformation:

$$\text{stress} = \frac{F}{A} \qquad (9.1)$$

strain is a relative measure of the deformation a stress causes:

$$\text{strain} = \frac{\text{change in length}}{\text{original length}} = \frac{|\Delta L|}{L_o} = \frac{|L - L_o|}{L_o} \qquad (9.2)$$

- An **elastic modulus** is the ratio of stress to strain.

Young's Modulus:

$$Y = \frac{F/A}{\Delta L/L_o} \qquad (9.4)$$

Shear Modulus:

$$S = \frac{F/A}{x/h} \approx \frac{F/A}{\phi} \qquad (9.5)$$

Bulk Modulus:

$$B = \frac{F/A}{-\Delta V/V_o} = -\frac{\Delta p}{\Delta V/V_o} \qquad (9.6)$$

- Pressure is the force per unit area.

$$p = \frac{F}{A} \qquad (9.8a)$$

- **Pascal's principle**. Pressure applied to an enclosed fluid is transmitted undiminished to every point in the fluid and to the walls of the container.

  Pressure–Depth Equation (for an incompressible fluid at constant density):

$$p = p_o + \rho g h \qquad (9.10)$$

- **Archimedes' principle**. A body immersed wholly or partially in a fluid is buoyed up by a force equal in magnitude to the weight of the volume of fluid displaced.

  Buoyant force:

$$F_b = m_f g = \rho_f g V_f \qquad (9.14)$$

- An object will float in a fluid if the average density of the object is less than the density of the fluid. If the average density of the object is greater than the density of the fluid, the object will sink.

- For an ideal fluid, the flow is (1) steady, (2) irrotational, (3) nonviscous, and (4) incompressible. The following equations describe such a flow:

Equation of Continuity:

$$\rho_1 A_1 v_1 = \rho_2 A_2 v_2 \qquad \text{or} \qquad \rho A v = \text{constant} \qquad (9.16)$$

Flow Rate Equation (for an incompressible fluid):

$$A_1 v_1 = A_2 v_2 \qquad \text{or} \qquad A v = \text{constant} \qquad (9.17)$$

Bernoulli's Equation (for an incompressible fluid):

$$p_1 + \tfrac{1}{2}\rho v_1^2 + \rho g y_1 = p_2 + \tfrac{1}{2}\rho v_2^2 + \rho g y_2$$

or

$$p + \tfrac{1}{2}\rho v^2 + \rho g y = \text{constant} \qquad (9.18)$$

- Bernoulli's equation is a statement of the conservation of energy for a fluid.

- **Viscosity** is a fluid's internal resistance to flow. All real fluids have a nonzero viscosity.

  *Poiseuille's Law (flow rate in pipes and tubes for fluids with constant viscosity and steady or laminar flow):

$$Q = \frac{\pi r^4 \Delta p}{8 \eta L} \qquad (9.19)$$

Exercises

9.1 Solids and Elastic Moduli

(Use as many significant figures as you need to show small changes.)

1. The pressure on an elastic body is described by (a) a modulus, (b) work, (c) stress, or (d) strain.

2. Shear moduli are not zero for (a) solids, (b) liquids, (c) gases, or (d) all of these.

3. CQ Which has a greater Young's modulus, a steel wire or a rubber band? Explain.

4. CQ Why are scissors sometimes called shears? Is this a descriptive name in the physical sense?

5. ■ Write the general form of Hooke's law, and find the units of the "spring constant" for elastic deformation.

6. ■ Suppose you use the tip of one finger to support a 5.0-kg object. If your finger has a diameter of 2.0 cm, what is the stress on your finger?

7. ■ A 5.0-m-long rod is stretched 0.10 m by a force. What is the strain in the rod?

8. ■ A 250-N force is applied at a 37° angle to the surface of the end of a square bar. The surface is 4.0 cm on a side.

What are (a) the compressional stress and (b) the shear stress on the bar?

9. ■■ A metal wire 1.0 mm in diameter and 2.0 m long hangs vertically with a 6.0-kg object suspended from it. If the wire stretches 1.4 mm under the tension, what is the value of Young's modulus for the metal?

10. ■■ A 5.0-kg object is supported by an aluminum wire of length 2.0 m and diameter 2.0 mm. How much will the wire stretch?

11. ■■ A copper wire has a length of 5.0 m and a diameter of 3.0 mm. Under what load will its length increase by 0.3 mm?

12. IE ■■ When railroad tracks are installed, gaps are left between the rails. (a) Should a greater gap be used if the rails are installed on (1) a cold day or (2) a hot day? Or (3) Does it make any difference? Why? (b) Each steel rail is 8.0 m long and has a cross-sectional area of 0.0025 m². On a hot day, each rail thermally expands as much as 3.0×10^{-3} m. If there were no gaps between the rails, what would be the force on the ends of each rail?

13. ■■ A rectangular steel column (20.0 cm × 15.0 cm) supports a load of 12.0 metric tons. If the column is 2.00 m in length before being stressed, what is the decrease in length?

14. IE ■■ A bimetallic rod as illustrated in ▼Fig 9.26 is composed of brass and copper. (a) If the rod is subjected to a compressive force, will the rod bend toward the brass or the copper? Why? (b) Justify your answer mathematically if the compressive force is 5.00×10^4 N.

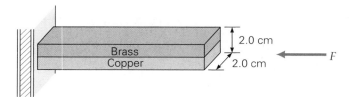

▲ **FIGURE 9.26** **Bimetallic rod and mechanical stress** See Exercise 14.

15. ■■ A 500-N shear force is applied to one face of a cube of aluminum measuring 10 cm on each side. What is the displacement of that face relative to the opposite face?

16. IE ■■ Two same-size metal posts, one aluminum and one copper, are subjected to equal shear stresses. (a) Which post will show the larger deformation angle, (1) the copper post or (2) the aluminum post? Or (3) Is the angle the same for both? Why? (b) By what factor is the deformation angle of one post greater than the other?

17. ■■ A rectangular block of gelatin of length, width, and height 10 cm, 8.0 cm, and 4.0 cm, respectively, is subjected to a 0.40-N shear force on its upper surface. If the top surface is displaced 0.30 mm relative to the bottom surface, what is the shear modulus of the gelatin?

18. ■■ Two metal plates are held together by two steel rivets, each of diameter 0.20 cm and length 1.0 cm. How much force must be applied parallel to the plates to shear off both rivets?

19. IE ■■ (a) Which of the liquids in Table 9.1 has the greatest compressibility? Why? (b) For equal volumes of ethyl alcohol and water, which would require more pressure to be compressed by 0.10%, and how many times more?

20. ■■■ A brass cube 6.0 cm on each side is placed in a pressure chamber and subjected to a pressure of 1.2×10^7 N/m^2 on all of its surfaces. By how much will each side be compressed under this pressure?

21. ■■■ A 45-kg traffic light is suspended from two steel cables of equal length and radii 0.50 cm. If each cable makes a 15° angle with the horizontal, what is the fractional increase in their length due to the weight of the light?

9.2 Fluids: Pressure and Pascal's Principle

22. ▶Figure 9.27 shows a famous magician's "trick." The magician lies on a bed of thousands of nails, and a sledgehammer is used to crack a concrete block on his chest. The nails will not pierce the magician's skin. Explain why.

▲ **FIGURE 9.27** **The nail mattress** See Exercise 22.

23. For the pressure–depth relationship for a fluid ($p = \rho g h$), it is assumed that (a) the pressure decreases with depth, (b) a pressure difference depends on the reference point, (c) the fluid density is constant, or (d) the relationship applies only to liquids.

24. CQ Two dams form artificial lakes of equal depth. However, one lake backs up 15 km behind the dam, and the other backs up 50 km behind. What effect does the difference in length have on the pressures on the dams?

25. CQ A water dispenser for pets has an inverted plastic bottle, as shown in ▼Fig. 9.28. (The water is dyed blue for contrast.) When a certain amount of water is drunk from the bowl, more water flows automatically from the bottle into the bowl. The bowl never overflows. Explain the operation of the dispenser. Does the height of the water in the bottle depend on the surface area of the water in the bowl?

◀ **FIGURE 9.28** **Pet barometer** See Exercise 25.

26. CQ (a) Liquid storage cans, such as gasoline cans, generally have capped vents. What is the purpose of the vents, and what happens if you forget to remove the cap before you pour the liquid? (b) Explain how a medicine

dropper works. (c) Explain how we breathe (inhalation and exhalation).

27. **CQ** Automobile tires are inflated to about 30 lb/in², whereas thin bicycle tires are inflated to 90 to 115 lb/in²— at least three times higher pressure! Why?

28. **CQ** Blood pressure is usually measured at the arm. However, suppose the pressure reading were taken on the calf of the leg of a standing person. Would there be a difference, in principle? Explain.

29. **CQ** Atmospheric pressure can be tremendous. Why is it that our bodies are not crushed under this enormous pressure?

30. **CQ** Explain in terms of fluid pressure how you are able to use a straw to drink soda from a bottle.

31. **IE** ■ In his original barometer, Pascal used water instead of mercury. (a) Water is less dense than mercury, so the water barometer would have (1) a higher height than, (2) a lower height than, or (3) the same height as the mercury barometer. Why? (b) How high would the water column have been?

32. ■ If you dive to 15 m below the surface of a lake, (a) what is the pressure due to the water alone? (b) What is the total or absolute pressure at that depth?

33. **IE** ■ In an open U-tube, the pressure of a water column on one side is balanced by the pressure of a column of gasoline on the other side. (a) Compared to the height of the water column, the gasoline column will have (1) a higher, (2) a lower, or (3) the same height. Why? (b) If the height of the water column is 15 cm, what is the height of the gasoline column?

34. ■ A 75-kg athlete does a single-hand handstand. If the area of the hand in contact with the floor is 125 cm², what pressure is exerted on the floor?

35. ■ The gauge pressure in both tires of a bicycle is 690 kPa. If the bicycle and the rider have a combined mass of 90.0 kg, what is the area of contact of *each* tire with the ground? (Assume that each tire supports half the total weight of the bicycle.)

36. ■■ In a sample of seawater taken from an oil spill, an oil layer 4.0 cm thick floats on 55 cm of water. If the density of the oil is 0.75×10^3 kg/m³, what is the absolute pressure on the bottom of the container?

37. **IE** ■■ In a lecture demonstration, an empty can is used to demonstrate the force exerted by air pressure (▶ Fig. 9.29). A small quantity of water is poured into the can, and the water is brought to a boil. Then, the can is sealed with a rubber stopper. As you watch, the can is slowly crushed with sounds of metal bending. (Why is a rubber stopper used as a safety precaution?) (a) This is because of (1) thermal expansion and contraction, (2) a higher

▲ **FIGURE 9.29 Air pressure** See Exercise 37.

steam pressure inside the can, or (3) a lower pressure inside the can as steam condenses. Why? (b) Assuming the dimensions of the can are 0.24 m × 0.16 m × 0.10 m and the inside of the can is in a perfect vacuum, what is the total force exerted on the can by the air pressure?

38. ■■ What is the fractional decrease in pressure when a barometer is raised 35 m to the top of a building? (Assume that the density of air is constant over that distance.)

39. ■■ A student decides to compute the standard barometric reading on top of Mt. Everest (29 028 ft) by assuming that the density of air has the same constant density as at sea level. Try this yourself. What does the result tell you?

40. ■■ Here is a demonstration Pascal used to show the importance of a fluid's pressure on the fluid's depth (▼ Fig. 9.30): An oak barrel with a lid of area 0.20 m² is filled with water. A long, thin tube of cross-sectional area 5.0×10^{-5} m² is inserted into a hole at the center of the lid, and water is poured into the tube. When the water reaches 12 m high, the barrel bursts. (a) What was the weight of the water in the tube? (b) What was the pressure

▲ **FIGURE 9.30 Pascal and the bursting barrel** See Exercise 40.

of the water on the lid of the barrel? (c) What was the net force on the lid due to the water pressure?

41. ■■ The door and the seals on an aircraft are subject to a tremendous amount of force during flight. At an altitude of 10 000 m (about 33 000 ft), the air pressure outside the airplane is only $2.7 \times 10^4 \, N/m^2$, while the inside is still at normal atmospheric pressure, due to pressurization of the cabin. Calculate the net force due to the air pressures on a door of area 3.0 m².

42. ■■ The pressure exerted by a person's lungs can be measured by having the person blow as hard as possible into one side of a manometer. If a person blowing into one side of an open-tube manometer produces an 80-cm difference between the heights of the columns of water in the manometer arms, what is the gauge pressure of the lungs?

43. ■■ In 1960, the U.S. Navy's bathyscaphe *Trieste* (a submersible) descended to a depth of 10 912 m (about 35 000 ft) into the Marianas Trench in the Pacific Ocean. (a) What was the pressure at that depth? (Assume that seawater is incompressible.) (b) What was the force on a circular observation window with a diameter of 15 cm?

44. ■■ The output piston of a hydraulic press has a cross-sectional area of 0.20 m². (a) How much pressure on the input piston is required for the press to generate a force of $1.5 \times 10^6 \, N$? (b) What force is applied to the input piston if it has a diameter of 5.0 cm?

45. ■■ A hydraulic lift in a garage has two pistons: a small one of cross-sectional area 4.00 cm² and a large one of cross-sectional area 250 cm². If this lift is designed to raise a 3000-kg car, what minimum force must be applied to the small piston? If the force is applied through compressed air, what must be the minimum air pressure applied to the small piston?

46. ■■■ A hypodermic syringe has a plunger of area 2.5 cm² and a 5.0×10^{-3}-cm² needle. (a) If a 1.0-N force is applied to the plunger, what is the gauge pressure in the syringe's chamber? (b) If a small obstruction is at the end of the needle, what force does the fluid exert on it? (c) If the blood pressure in a vein is 50 mm Hg, what force must be applied on the plunger so that fluid can be injected into the vein?

47. ■■■ A hydraulic balance used to detect small changes in mass is shown in ▸Fig. 9.31. If a mass *m* of 0.25 g is placed on the balance platform, by how much will the height of the water in the smaller, 1.0-cm-diameter cylinder have changed when the balance comes to equilibrium?

9.3 Buoyancy and Archimedes' Principle

48. A wood block floats in a swimming pool. The buoyant force exerted on the block by water depends on (a) the volume of water in the pool, (b) the volume of the wood

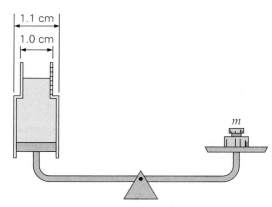

▲ **FIGURE 9.31   A hydraulic balance**   See Exercise 47.

block, (c) the volume of the wood block under water, or (d) all of the above.

49. If a submerged object displaces an amount of liquid of greater weight than its own and is then released, the object will (a) rise to the surface and float, (b) sink, or (c) remain in equilibrium at its submerged position.

50. CQ (a) What is the most important factor in constructing a life jacket that will keep a person afloat? (b) Why is it so easy to float in Utah's Great Salt Lake?

51. CQ An ice cube floats in a glass of water. As the ice melts, how does the level of the water in the glass change? Would it make any difference if the ice cube were hollow? Explain.

52. CQ Oceangoing ships in port are loaded to the so-called *Plimsoll mark*, which is a line indicating the maximum safe loading depth. However, in New Orleans, located at the mouth of the Mississippi River, where the water is brackish (partly salty and partly fresh), ships are loaded until the Plimsoll mark is somewhat below the water line. Why?

53. CQ Two blocks of equal volume, one iron and one aluminum, are dropped into a body of water. Which block will experience the greater buoyant force? Why?

54. CQ How do you think balloonists control the altitude of a hot-air balloon?

55. CQ A bucket of water is sitting on a flat scale. Will the reading of the scale change if you dip a finger in the water without touching the bucket? Explain.

56. IE ■ (a) If the density of an object is exactly equal to the density of a fluid, the object will (1) float, (2) sink, or (3) stay at any height in the fluid, as long as it is totally immersed. Why? (b) A cube 8.5 cm on each side has a mass of 0.65 kg. Will the cube float or sink in water? Prove your answer.

57. ■ Suppose that Archimedes found that the king's crown had a mass of 0.750 kg and a volume of 3.980×10^{-5} m³. (a) What simple approach did Archimedes use to determine the crown's volume? (b) Was the crown pure gold?

58. ■ A rectangular boat, as illustrated in ▼Fig. 9.32, is overloaded such that the water level is just 1.0 cm below the top of the boat. What is the combined mass of the people and the boat?

▲ **FIGURE 9.32   An overloaded boat**   (Not drawn to scale.) See Exercise 58.

59. ■ An aluminum cube 0.15 m on each side is completely submerged in water. What is the buoyant force on it? How does the answer change if the cube is made of steel?

60. ■■ An object has a weight of 8.0 N in air. However, it apparently weighs only 4.0 N when it is completely submerged in water. What is the density of the object?

61. ■■ When a 0.80-kg crown is submerged in water, its apparent weight is measured to be 7.3 N. Is the crown pure gold?

62. IE ■■ (a) Given a piece of metal with a light string attached, a scale, and a container of water in which the piece of metal can be submersed, how could you find the volume of the piece without using the variation in the water level? (b) An object has a weight of 0.882 N. It is suspended from a scale, which reads 0.735 N when the piece is submerged in water. What are the volume and density of the piece of metal?

63. ■■ A flat-bottomed rectangular boat is 4.0 m long and 1.5 m wide. If the load is 2000 kg (including the mass of the boat), how much of the boat will be submerged when it floats in a lake?

64. ■■ A block of iron quickly sinks in water, but ships constructed of iron float. A solid cube of iron 1.0 m on each side is made into sheets. To make these sheets into a hollow cube that will not sink, what should be the minimum length of the sides of the sheets?

65. ■■ Plans are being made to bring back the zeppelin, a lighter-than-air airship like the Goodyear blimp that carries passengers and cargo, but is filled with helium, not flammable hydrogen (as was used in the ill-fated *Hindenburg*). One design calls for the ship to be 110 m long and to have a total mass (without helium) of 30.0 metric tons. Assuming the ship's "envelope" to be cylindrical, what would its diameter have to be so as to lift the total weight of the ship and the helium?

66. ■■■ A girl floats in a lake with 97% of her body beneath the water. What are (a) her mass density and (b) her weight density?

67. ■■■ A block of wood made of oak is held under the water's surface in a swimming pool. At the instant the block is released, what is its acceleration?

9.4   Fluid Dynamics and Bernoulli's Equation

68. If the speed at some point in a fluid changes with time, the fluid flow is *not* (a) steady, (b) irrotational, (c) incompressible, (d) nonviscous.

69. CQ The speed of blood flow is greater in arteries than in capillaries. However, the flow rate equation ($Av = $ constant) seems to predict that the speed should be greater in the smaller capillaries. Can you resolve this apparent inconsistency?

70. CQ Explain why water shoots out farther from a hose if you put your finger over the tip of the hose.

71. According to Bernoulli's equation, if the pressure on the liquid in Fig. 9.19 is increased, (a) the flow speed always increases, (b) the height of the liquid always increases, (c) both the flow speed and the height of the liquid may increase, or (d) none of these.

72. CQ When you suddenly turn on a water faucet in a shower, the shower curtain moves inward. Why?

73. CQ If an Indy car had a flat bottom, it would be highly unstable (like an airplane wing) due to the lift it gets when it moves at a high speed. To increase friction and stability of the car, the bottom has a concave section called the *Venturi tunnel* (▼Fig. 9.33). In terms of Bernoulli's

▲ **FIGURE 9.33   Venturi tunnel**   See Exercise 73.

equation, explain how this concavity supplies extra downward force to the car in addition to that supplied by the front and rear wings (Exercise 4.79).

74. **CQ** Here are two common demonstrations of Bernoulli effects: (a) If you hold a narrow strip of paper in front of your mouth and blow over the top surface, the strip will rise (▼Fig. 9.34a). (Try it.) Why? (b) A plastic egg is supported vertically by a stream of air from a tube (Fig. 9.34b). The egg will not move away from the midstream position. Why not?

(a)   (b)

▲ **FIGURE 9.34 Bernoulli effects** See Exercise 74.

75. **CQ** A "split-finger" fastball sinks faster than a regular fastball. How should the ball be spinning for this to happen? Explain.

76. ■ An ideal fluid is moving at 3.0 m/s in a section of a pipe of radius 0.20 m. If the radius in another section is 0.35 m, what is the flow speed there?

77. **IE** ■ (a) If the radius of a pipe narrows to half of its original size, will the flow speed in the narrow section (1) increase by a factor of 2, (2) increase by a factor of 4, (3) decrease by a factor of 2, or (4) decrease by a factor of 4? Why? (b) If the radius widens to three times its original size, what is the ratio of the flow speed in the wider section to that in the narrow section?

78. ■■ Show that the static pressure–depth equation can be derived from Bernoulli's equation.

79. ■■ The speed of blood in a major artery of diameter 1.0 cm is 4.5 cm/s. (a) What is the flow rate in the artery? (b) If the capillary system has a total cross-sectional area of 2500 cm², the average speed of blood through the capillaries is what percentage of that through the major artery? (c) Why must blood flow at low speed through the capillaries?

80. ■■ A room measures 3.0 m by 4.5 m by 6.0 m. If the heating and air-conditioning ducts to and from the room are circular with diameter 0.30 m and all the air in the room is to be exchanged every 12 min, (a) what is the average flow rate? (b) What is the necessary flow

speed in the duct? (Assume that the density of the air is constant.)

81. ■■ The spout heights in the container in ▼Fig. 9.35 are 10 cm, 20 cm, 30 cm, and 40 cm. The water level is maintained at a 45-cm height by an outside supply. (a) What is the speed of the water out of each hole? (b) Which water stream has the greatest range relative to the base of the container? Justify your answer.

▲ **FIGURE 9.35 Streams as projectiles** See Exercise 81.

82. ■■■ In an industrial cooling process, water is circulated through a system. If the water is pumped with a speed of 0.45 m/s under a pressure of 400 torr from the first floor through a 6.0-cm diameter pipe, what will be the pressure on the next floor 4.0 m above in a pipe with a diameter of 2.0 cm?

83. ■■■ Water flows at a rate of 25 L/min through a horizontal 7.0-cm-diameter pipe under a pressure of 6.0 Pa. At one point, calcium deposits reduce the cross-sectional area of the pipe to 30 cm². What is the pressure at this point? (Consider the water to be an ideal fluid.)

84. ■■■ A Venturi meter can be used to measure the flow speed of a liquid. A simple such device is shown in ▼Fig. 9.36. Show that the flow speed of an ideal fluid is given by
$$v_1 = \sqrt{\frac{2g\,\Delta h}{(A_1^2/A_2^2) - 1}}.$$

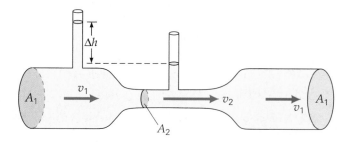

▲ **FIGURE 9.36 A flow speed meter** See Exercise 84.

*9.5 Surface Tension, Viscosity, and Poiseuille's Law

85. Water droplets and soap bubbles tend to assume the shape of a sphere. This effect is due to (a) viscosity,

(b) surface tension, (c) laminar flow, or (d) none of the above.

86. Some insects can walk on water because (a) the density of water is greater than that of the insect, (b) water is viscous, (c) water has surface tension, or (d) none of the above.

87. The viscosity of a fluid is due to (a) forces causing friction between the molecules, (b) surface tension, (c) density, or (d) none of the above.

88. **CQ** A motor oil is labeled 10W–40. What do the numbers 10 and 40 measure? How about the "W"?

89. ■■ The pulmonary artery, which connects the heart to the lungs, is about 8.0 cm long and has an inside diameter of 5.0 mm. If the flow rate in it is to be 25 mL/s, what is the required pressure difference over its length?

90. ■■ A hospital patient receives a quick 500-cc blood transfusion through a needle with a length of 5.0 cm and an inner diameter of 1.0 mm. If the blood bag is suspended 0.85 m above the needle, how long does the transfusion take? (Neglect the viscosity of the blood flowing in the plastic tube between the bag and the needle.)

Additional Exercises

91. A copper wire 100.00 cm long is stretched to 100.02 cm when it supports a certain load. If an aluminum wire of the same diameter is used to support the same load, what should its initial length be if its stretched length is to be 100.02 cm also?

92. A steel cube 0.25 m on each side is suspended from a scale and immersed in water. What will the scale read?

93. A wood cube 0.30 m on each side has a density of 700 kg/m^3 and floats levelly in water. (a) What is the distance from the top of the wood to the water surface? (b) What mass has to be placed on top of the wood so that its top is just at the water level?

94. A 60-kg woman balances herself on the heel of one of her high-heeled shoes. If the heel is a square of sides 1.5 cm, what is the pressure exerted on the floor? (Express your answer in Pa, lb/in^2, and atm.)

95. What is the total inward force the atmosphere exerts on a surface area of 1.30 m^2 of a person's body? (Express your answer in both newtons and pounds.)

96. A scuba diver dives to a depth of 12 m in a lake. If the circular glass plate on the diver's face mask has a diameter of 18 cm, what is the force on it due to the water only?

97. Oil is poured into the open side of an open-tube manometer containing mercury. What is the density of the oil if a column of mercury 5.0 cm high supports a column of oil 80 cm high?

98. If the atmosphere had a constant density equal to its density at the Earth's surface, how high would it extend? What does this tell you?

99. A cylinder has a diameter of 15 cm (▼Fig. 9.37). The water level in the cylinder is maintained at a constant height of 0.45 m. If the diameter of the spout pipe is 0.50 cm, how high is h, the vertical stream of water? (Assume the water to be an ideal fluid.)

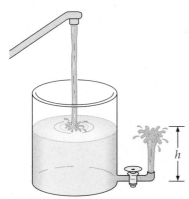

▲ **FIGURE 9.37 How high a fountain?** See Exercise 99.

100. A vertical steel beam with a rectangular cross-sectional area of 24 cm^2 is used to support a sagging floor in a building. If the beam supports a load of 10 000 N, by what percentage is the beam compressed?

101. A container 100 cm deep is filled with water. If a small hole is punched in its side 25 cm from the top, at what initial speed will the water flow from the hole?

102. Show that specific gravity is equivalent to a ratio of densities, given that its strict definition is the ratio of the weight of a given volume of a substance to the weight of an equal volume of water.

103. The flow rate of water through a garden hose is 66 cm^3/s, and the hose and nozzle have cross-sectional areas of 6.0 cm^2 and 1.0 cm^2, respectively. (a) If the nozzle is held 10 cm above the spigot, what are the flow speeds through the spigot and the nozzle? (b) What is the pressure difference between these points? (Consider the water to be an ideal fluid.)

104. **CQ** Ancient stonemasons sometimes split huge blocks of rock by inserting wooden pegs into holes drilled in the rock and then pouring water on the pegs. Can you explain the physics that underlies this technique? [*Hint*: Think about sponges and paper towels.]

105. A crane lifts a rectangular iron bar from the bottom of a lake. The dimensions of the bar are 0.25 m × 0.20 m × 10 m. What is the minimum upward force the crane must

supply when the bar is (a) in the water and (b) out of the water?

106. A submarine has a mass of 10 000 metric tons. What weight of water must be displaced for the sub to be in equilibrium just below the ocean surface?

107. ▶ Figure 9.38 shows a simple laboratory experiment. Calculate (a) the volume and (b) the density of the suspended sphere. (Assume that the density of the sphere is uniform and that the liquid in the beaker is water.) (c) Would you be able to make the same determinations if the liquid in the beaker were mercury? (See Table 9.2.) Explain.

▲ **FIGURE 9.38   Dunking a sphere**   See Exercise 107.

Temperature and Kinetic Theory

INSIGHTS

- Human Body Temperature
- Physiological Diffusion in Life Processes

Learn by Drawing

✎ Thermal Area Expansion

L ike sailboats, hot-air balloons are low-tech devices in a high-tech world. You can equip a balloon with the latest satellite-linked, computerized navigational system and attempt to fly across the Pacific, but the basic principles that keep you aloft were known and understood centuries ago.

But why must the air in a hot-air balloon be hot? More fundamentally, what do we mean by "hot"? How does hot air differ from cool air? Temperature and heat are frequent subjects of conversation, but if you had to state what the words really mean, you might find yourself at a loss. We use thermometers of all sorts to record temperatures, which provide an objective equivalent for our sensory experience of hot and cold. Also, we know that a temperature change generally results from the application or removal of heat. Temperature, therefore, is related to heat. But what is heat? In this chapter, you'll find that the an-

swers to such questions lead to an understanding of some far-reaching physical principles.

An early theory of heat considered it to be a fluidlike substance called caloric (from the Latin word *calor*, meaning "heat") that could be made to flow into and out of a body. Even though this theory has been abandoned, we still speak of heat as flowing from one object to another. Heat is now known to be energy in transit, and temperature and thermal properties are explained by considering the atomic and molecular behavior of substances. This and the next two chapters examine the nature of temperature and heat in terms of microscopic (molecular) theory and macroscopic observations. Here, you'll explore the nature of heat and the ways in which we measure temperature. You'll also encounter the gas laws, which explain not only the behavior of hot-air balloons, but also more important phenomena, such as how our lungs supply us with the oxygen we need to live.

10.1 Temperature and Heat

OBJECTIVE: To distinguish between temperature and heat.

A good way to begin studying thermal physics is with definitions of temperature and heat. **Temperature** is a relative measure, or indication, of hotness or coldness. A hot stove is said to have a high temperature and an ice cube to have a low temperature. An object that has a higher temperature than another object is said to be hotter. Note that *hot* and *cold* are relative terms, like *tall* and *short*. We can perceive temperature by touch. However, this temperature sense is somewhat unreliable, and its range is too limited to be useful for scientific purposes.

Heat is related to temperature and describes the process of energy transfer from one object to another. That is, **heat** is *the net energy transferred from one object to another because of a temperature difference.* Thus, heat is energy in transit, so to speak. Once transferred, the energy becomes part of the total energy of the molecules of the object or system, its **internal energy**. So heat (energy) transfers between objects can result in internal energy changes.

On a microscopic level, temperature is associated with molecular motion. We will show that in kinetic theory (Section 10.5), which treats gas molecules as point particles, temperature is a measure of the average random *translational* kinetic energy of the molecules. However, diatomic molecules and other real substances, besides having such translational "temperature" energy, also may have kinetic energy due to linear vibration and rotation, as well as potential energy due to the attractive forces between molecules. These energies do not contribute to the temperature of the gas, but are definitely part of its internal energy, which is the sum of all such energies (▶Fig. 10.1).

Note that a higher temperature does not necessarily mean that one system has a greater internal energy than another. For example, in a classroom on a cold day, the air temperature is relatively high compared to that of the outdoor air. But all that cold air outside the classroom has far more internal energy than does the warm air inside, simply because there is so much *more* of it. If this were not the case, heat pumps would not be practical (Chapter 12). In other words, the internal energy of a system also depends on its mass, or the number of molecules in the system.

When heat is transferred between two objects, regardless of whether they are touching, the objects are said to be in *thermal contact.* When there is no longer a net heat transfer between objects in thermal contact, they have come to the same temperature and are said to be in *thermal equilibrium.*

10.2 The Celsius and Fahrenheit Temperature Scales

OBJECTIVES: To (a) explain how a temperature scale is constructed and (b) convert temperatures from one scale to another.

A measure of temperature is obtained by using a **thermometer**, a device constructed to make use of some property of a substance that changes with temperature. Fortunately, many physical properties of materials change sufficiently with temperature to be used as the bases for thermometers. By far the most obvious and commonly used property is **thermal expansion** (Section 10.4), a change in the dimensions or volume of a substance that occurs when the temperature changes.

Almost all substances expand with increasing temperature, but they do so to different extents. Most substances also contract with decreasing temperature. (Thermal expansion refers to both expansion and contraction; contraction is considered to be a negative expansion.) Because some metals expand more than oth-

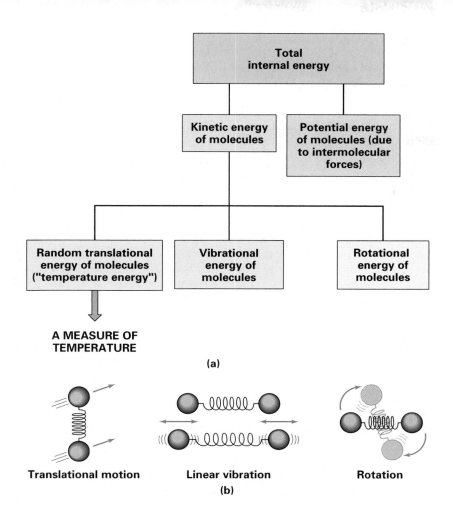

Total internal energy

Kinetic energy of molecules

Potential energy of molecules (due to intermolecular forces)

Random translational energy of molecules ("temperature energy")

Vibrational energy of molecules

Rotational energy of molecules

A MEASURE OF TEMPERATURE

(a)

Translational motion

Linear vibration

Rotation

(b)

◀ **FIGURE 10.1 Molecular motions** **(a)** Temperature is associated with random translational motion; the internal energy of a system is its total energy. **(b)** A molecule may move as a whole in translational motion or may have vibrational or rotational motions (or any combination of the three kinds of motion).

ers, a bimetallic strip (a strip made of two different metals bonded together) can be used to measure temperature changes. As heat is added, the composite strip will bend away from the side made of the metal that expands more (▲Fig. 10.2). Coils formed from such strips are used in dial thermometers and in common household thermostats (▶Fig. 10.3).

A common thermometer is the liquid-in-glass type, which is based on the thermal expansion of a liquid. A liquid in a glass bulb expands into a glass

▼ **FIGURE 10.2 Thermal expansion** **(a)** A bimetallic strip is made of two strips of different metals bonded together. **(b)** When such a strip is heated, it bends because of unequal expansions of the two metals. Here, brass expands more than iron, so the deflection is toward the iron. The deflection of the end of a strip could be used to measure temperature.

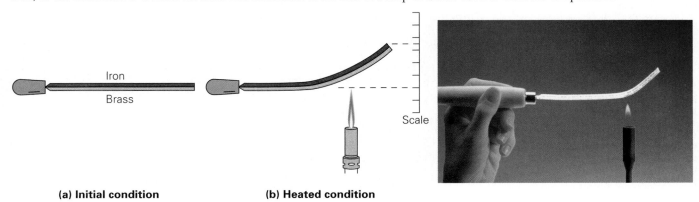

Iron

Brass

Scale

(a) Initial condition

(b) Heated condition

(a)

(b)

▲ **FIGURE 10.3 Bimetallic coil**
Bimetallic coils are used in **(a)** dial thermometers (the coil is in the center) and **(b)** household thermostats (the coil is to the right). Thermostats are used to regulate a heating or cooling system, turning off and on as the temperature of the room changes. The expansion and contraction of the coil causes the tilting of a glass vial containing mercury, which makes and breaks electrical contact.

Note: For distinction, a particular temperature measurement, such as $T = 20°C$, is written with °C (pronounced 20 degrees Celsius), whereas a temperature interval, such as $\Delta T = 80°C - 60°C = 20 \; C°$, is written with C° (pronounced 20 Celsius degrees).

stem, rising in a capillary bore (a thin tube). Mercury and alcohol (usually dyed red to make it more visible) are the liquids used in most liquid-in-glass thermometers. These substances are chosen because of their relatively large thermal expansion and because they remain liquids over normal temperature ranges.

Thermometers are calibrated so that a numerical value can be assigned to a given temperature. For the definition of any standard scale or unit, two fixed reference points are needed. The ice point and the steam point of water at standard atmospheric pressure are two convenient fixed points. More commonly known as the freezing and boiling points, these are the temperatures at which pure water freezes and boils, respectively, under a pressure of 1 atm (standard pressure).

The two most familiar temperature scales are the **Fahrenheit temperature scale** (used in the United States) and the **Celsius temperature scale** (used in the rest of the world). As shown in ▶ Fig. 10.4, the ice and steam points have values of 32°F and 212°F, respectively, on the Fahrenheit scale and 0°C and 100°C, respectively, on the Celsius scale. On the Fahrenheit scale, there are 180 equal intervals, or degrees (F°), between the two reference points; on the Celsius scale, there are 100 degrees (C°). Therefore, since $180/100 = 9/5 = 1.8$, a Celsius degree is almost twice as large as a Fahrenheit degree. (See margin note for difference between °C and C°.)

A relationship for converting between the two scales can be obtained from a graph of Fahrenheit temperature (T_F) versus Celsius temperature (T_C), such as the one in ▶ Fig. 10.5. The equation of the straight line (in slope–intercept form, $y = mx + b$) is $T_F = (180/100)T_C + 32$, and

$$T_F = \tfrac{9}{5}T_C + 32 \quad \text{or} \quad T_F = 1.8T_C + 32 \qquad \text{\textit{Celsius-to-Fahrenheit}} \atop \text{\textit{conversion}} \qquad (10.1)$$

where $\tfrac{9}{5}$ or 1.8 is the slope of the line and 32 is the intercept on the vertical axis. Thus, to change from a Celsius temperature (T_C) to its equivalent Fahrenheit temperature (T_F), you simply multiply the Celsius reading by $\tfrac{9}{5}$ and add 32.

The equation can be solved for T_C to convert from Fahrenheit to Celsius:

$$T_C = \tfrac{5}{9}(T_F - 32) \qquad \text{\textit{Fahrenheit-to-Celsius}} \atop \text{\textit{conversion}} \qquad (10.1)$$

Example 10.1 ■ Converting Temperature Scale Readings: Fahrenheit and Celsius

What are (a) the typical room temperature of 20°C and a cold temperature of −18°C on the Fahrenheit scale; and (b) another cold temperature of −10°F and normal body temperature, 98.6°F, on the Celsius scale?

Thinking It Through. This is a direct application of Eqs. 10.1 and 10.2.

Solution. We wish to make the following conversions:

Given: (a) $T_C = 20°C$ and $T_C = -18°C$ *Find:* for each temperature,
       (b) $T_F = -10°F$ and $T_F = 98.6°F$ (a) T_F
                                         (b) T_C

(a) Equation 10.1 is for changing Celsius readings to Fahrenheit:

20°C: $T_F = \tfrac{9}{5}T_C + 32 = \tfrac{9}{5}(20) + 32 = 68°F$

−18°C: $T_F = \tfrac{9}{5}T_C + 32 = \tfrac{9}{5}(-18) + 32 = 0°F$

(This typical room temperature of 20°C is a good one to remember.)

(b) Equation 10.2 changes Fahrenheit to Celsius:

$-10°F:$ $\quad T_C = \frac{5}{9}(T_F - 32) = \frac{5}{9}(-10 - 32) = 23°C$

$98.6°F:$ $\quad T_C = \frac{5}{9}(T_F - 32) = \frac{5}{9}(98.6 - 32) = 37.0°C$

From the last calculation, we see that normal body temperature has a whole-number value on the Celsius scale. Keep in mind that a Celsius degree is 1.8 times (almost twice) as large as a Fahrenheit degree, so a temperature elevation of several degrees on the Celsius scale makes a big difference. For example, a temperature of 40.0°C represents an elevation of 3.0 C° over normal body temperature. However, on the Fahrenheit scale, this is an increase of $3.0 \times 1.8 = 5.4$ F°, or a temperature of $98.6 + 5.4 = 104.0°F$. (For more on "normal" body temperature, see the Insight on p. 350.)

Follow-up Exercise. Convert the following temperatures: (a) −40°F to Celsius and (b) −40°C to Fahrenheit. *(Answers to all Follow-up Exercises are at the back of the text.)*

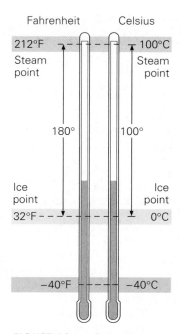

▲ **FIGURE 10.4 Celsius and Fahrenheit temperature scales** Between the ice and steam fixed points, there are 100 degrees on the Celsius scale and 180 degrees on the Fahrenheit scale. Thus, a Celsius degree is 1.8 times larger than a Fahrenheit degree.

Problem-Solving Hint

Because Eqs. 10.1 and 10.2 are so similar, it is easy to miswrite them. Since they are equivalent, you need to know only one of them—say, Celsius to Fahrenheit (Eq. 10.1, $T_F = \frac{9}{5}T_C + 32$). Solving this equation for T_C algebraically gives Eq. 10.2. A good way to make sure that you have written the conversion equation correctly is to test it with a known temperature, such as the boiling point of water. For example, $T_C = 100°$, so

$$T_F = \frac{9}{5}T_C + 32 = \frac{9}{5}(100) + 32 = 212°F$$

Thus, we know the equation is correct.

Liquid-in-glass thermometers are adequate for many temperature measurements, but problems arise when highly accurate determinations are needed. A material may not expand uniformly over a wide temperature range. When calibrated to the ice and steam points, an alcohol thermometer and a mercury thermometer have the same readings at those points, but because alcohol and mercury have different expansion properties, the thermometers will not have exactly the same reading at an intermediate temperature, such as room temperature. For very sensitive temperature measurements and to define intermediate temperatures precisely, some other type of thermometer must be used. One such thermometer, a *gas thermometer*, is discussed next.

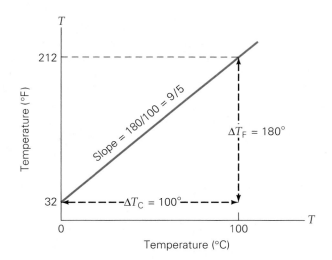

◀ **FIGURE 10.5 Fahrenheit versus Celsius** A plot of Fahrenheit temperature versus Celsius temperature gives a straight line of the general form $y = mx + b$, where $T_F = \frac{9}{5}T_C + 32$.

Human Body Temperature

We commonly take "normal" human body temperature to be 98.6°F (or 37.0°C). The source of this value is a study of human temperature readings done in 1868—more than 130 years ago! A more recent study, conducted in 1992, notes that the 1868 study used thermometers that were not as accurate as modern electronic thermometers. The new study has some interesting results.

The "normal" human body temperature from oral measurements varies among individuals over a range of about 96°F to 101°F, with an average temperature of 98.2°F. After strenuous exercise, the oral temperature can rise as high as 103°F. When the body is exposed to cold, oral temperatures can fall below 96°F. A rapid drop in temperature of 2 to 3 F° produces uncontrollable shivering. There is a contraction not only of the skeletal muscles, but also of the tiny muscles attached to the hair follicles. The result is "goose bumps."

Your body temperature is typically lower in the morning, after you have slept and your digestive processes are at a low point. "Normal" body temperature generally rises during the day to a peak and then recedes. The 1992 study also indicated that women have a slightly higher average body temperature than do men (98.4°F versus 98.1°F).

What about the extremes? A fever temperature is typically between 102°F and 104°F. A body temperature above 106°F is extremely dangerous. At such temperatures, the enzymes that take part in certain chemical reactions in the body begin to be inactivated, and a total breakdown of body chemistry can result.

On the cold side, decreasing the body temperature results in memory lapse and slurred speech, muscular rigidity, erratic heartbeats, and loss of consciousness. Below 78°F,

death occurs due to heart failure. However, mild hypothermia (lower-than-normal body temperature) can be beneficial. A decrease in body temperature slows down the body's chemical reactions, and cells use less oxygen than they normally do. This effect is applied in some surgeries (Fig. 1). A patient's body temperature may be lowered significantly to avoid damage to the heart, which must be stopped during such procedures, and to the brain.

Related Exercise: 10

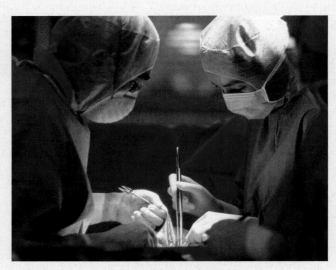

FIGURE 1 Lower than normal During some surgeries, the patient's body temperature is lowered to slow down the body's chemical reactions and to reduce the need for blood to supply oxygen to the tissues.

10.3 Gas Laws and Absolute Temperature

OBJECTIVES: To (a) describe the ideal gas law, (b) explain how it is used to determine absolute zero, and (c) understand the Kelvin temperature scale.

Whereas different liquid-in-glass thermometers show slightly different readings for temperatures other than fixed points because of the liquids' different expansion properties, a thermometer that uses a gas gives the same readings regardless of the gas used. The reason is that at very low densities all gases exhibit the same expansion behavior.

The variables that describe the behavior of a given quantity (mass) of gas are pressure, volume, and temperature (p, V, and T). When temperature is held constant, the pressure and volume of a quantity of gas are related as follows:

$$pV = \text{constant} \qquad \text{or} \qquad p_1V_1 = p_2V_2 \quad (\textit{at constant temperature}) \quad (10.3)$$

That is, the product of pressure and volume is a constant. This relationship is known as *Boyle's law*, after Robert Boyle (1627–1691), the English chemist who discovered it.

When the pressure is held constant, the volume of a quantity of gas is related to the *absolute* temperature (to be defined shortly):

$$\frac{V}{T} = \text{constant} \qquad \text{or} \qquad \frac{V_1}{T_1} = \frac{V_2}{T_2} \qquad \textit{(at constant pressure)} \quad (10.4)$$

That is, the ratio of the volume to the temperature is a constant. This relationship is known as *Charles's law*, named for the French scientist Jacques Charles (1747–1823), who made early hot-air balloon flights and was therefore quite interested in the relationship between the volume and temperature of a gas. A popular demonstration of Charles's law is shown in ▶ Fig. 10.6.

Low-density gases obey these laws, which may be combined into a single relationship. Since pV = constant and V/T = constant for a given quantity of gas, pV/T must also equal a constant. This relationship is the **ideal gas law**:

$$\frac{pV}{T} = \text{constant} \qquad \text{or} \qquad \frac{P_1V_1}{T_1} = \frac{P_2V_2}{T_2} \qquad \begin{array}{l}\textit{ideal gas law}\\ \textit{(ratio form)}\end{array} \quad (10.5)$$

That is, the ratio pV/T at one time (t_1) is the same as at another time (t_2), or at any other time, as long as the quantity (or mass) of gas does not change.

This relationship can be written in a more general form that applies not just to a given quantity of a single gas, but to any quantity of any low-pressure, dilute gas. With a quantity of gas determined by the number of molecules (N) in the gas (i.e., $pV/T \propto N$), it follows that

$$\frac{pV}{T} = Nk_B \qquad \text{or} \qquad pV = Nk_BT \qquad \textit{ideal gas law} \quad (10.6)$$

where k_B is a constant of proportionality known as *Boltzmann's constant*: $k_B = 1.38 \times 10^{-23}$ J/K. The K stands for temperature on the Kelvin scale, discussed shortly. (Can you show that the units are correct?) Note that the mass of the sample does not appear explicitly in Eq. 10.6. However, the number of molecules N in a sample of a gas is proportional to the total mass of the gas. The ideal gas law, sometimes called the *perfect gas law*, applies to gases with low pressures and densities and describes the behavior of most gases fairly accurately at normal densities.

Macroscopic Form of the Ideal Gas Law

Equation 10.6 is a "microscopic" (*micro* means small) form of the ideal gas law in that it refers specifically to the number of molecules, N. However, the law can be rewritten in a "macroscopic" (*macro* means large) form, which involves quantities that can be measured with everyday laboratory equipment. In this form, we have

$$pV = nRT \qquad \textit{ideal gas law} \quad (10.7)$$

using nR rather than Nk_B for convenience. Here, n is the number of moles (mol) of the gas, a quantity defined next, and R is called the *universal gas constant*:

$$R = 8.31 \text{ J/(mol·K)}$$

In chemistry, a **mole** (abbreviated mol) of a substance is defined as the quantity that contains **Avogadro's number** (N_A) of molecules:

$$N_A = 6.02 \times 10^{23} \text{ molecules/mol}$$

▲ **FIGURE 10.6 Charles's law in action** Demonstrations of the relationship between the volume and the temperature of a quantity of gas. A weighted balloon, initially at room temperature, is placed in a beaker of water. **(a)** When ice is placed in the beaker and the temperature falls, the balloon's volume is reduced. **(b)** When the water is heated and the temperature rises, the balloon's volume increases.

Ludwig Boltzmann was a 19th-century Austrian physicist who used stastical mechanics to investigate atomic phenomena.

Note: The temperature *T* in the ideal gas law is absolute (Kelvin) temperature.

Note: *N* is the total number of molecules; N_A is Avogadro's number; $n = N/N_A$ is the number of moles.

Thus, n and N in the two forms of the ideal gas law are related by $N = nN_A$. From Eq. 10.7, it can be shown that 1 mol of *any* gas occupies 22.4 L at 0°C and 1 atm. These conditions are known as *standard temperature and pressure (STP)*.

It is important to note what these equations for the macroscopic (Eq. 10.7) and microscopic (Eq. 10.6) forms of the ideal gas law represent. For the macroscopic form of the ideal gas law, the constant $R = pV/(nT)$ has units of J/(mol·K). For the microscopic form of the law, $k_B = pV/(NT)$, with units of J/(molecule·K). Note the difference between the macroscopic and microscopic forms of the ideal gas law is moles versus molecules, and we can measure moles.

Equation 10.7 is a practical form of the ideal gas law, because we generally work with measured (macroscopic or laboratory) quantities or moles (n) of gases rather than the number of molecules (N). To use Eq. 10.7, we need to know the number of moles of a gas. The mass of 1 mol of any substance is its *formula mass*, expressed in grams. The formula mass is determined from the chemical formula and the atomic masses of the atoms. (The latter are listed in Appendix IV and are commonly rounded to the nearest one-half.) For example, water, H_2O, with two hydrogen atoms and one oxygen atom, has a formula mass of $2m_H + 1m_O = 2(1.0) + 1(16.0) = 18.0$, because the atomic mass of each hydrogen atom is 1.0 g and that of an oxygen atom is 16.0 g. Thus, 1 mol of water has a formula mass of 18.0 g. Similarly, the oxygen we breathe, O_2, has a formula mass of $2 \times 16.0 = 32.0$. Hence, a mole of oxygen, with a mass of 32 g, would occupy 22.4 L at STP.

It is interesting to note that Avogadro's number allows you to compute the mass of a particular type of molecule. For example, suppose you want to know the mass of a water molecule (H_2O). As we have just seen, the formula mass of 1 mol of water is 18.0 g, or 18.0 g/mol. The *molecular mass* (m) is then given by

$$m = \frac{\text{formula mass (in kilograms)}}{N_A}$$

and, converting grams to kilograms, we have

$$m_{H_2O} = \frac{(18.0 \text{ g/mol})(10^{-3} \text{ kg/g})}{6.02 \times 10^{23} \text{ molecules/mol}} = 2.99 \times 10^{-26} \text{ kg/molecule}$$

Thus, the *formula mass* (in kilograms) is also equivalent to the mass of 1 mol (in kilograms).

Absolute Zero and the Kelvin Temperature Scale

The product of the pressure and the volume of a sample of ideal gas is directly proportional to the temperature of the gas: $pV \propto T$. This relationship allows a gas to be used to measure temperature in a *constant-volume gas thermometer*. Holding the volume of the gas constant, which can be done easily in a rigid container (► Fig. 10.7), means that $p \propto T$. Thus, using a constant-volume gas thermometer, one reads the temperature in terms of pressure. A plot of pressure versus temperature gives a straight line in this case (► Fig. 10.8a).

As can be seen in Fig. 10.8b, measurements of real gases (plotted data points) deviate from the values predicted by the ideal gas law at very low temperatures. This is because the gases liquefy at such temperatures. However, the relationship is linear over a large temperature range, and it looks as though the pressure might reach zero with decreasing temperature if the gas were to continue to be gaseous (ideal or perfect).

The absolute minimum temperature for an ideal gas is therefore inferred by extrapolating, or extending the straight line to the axis, as in Fig. 10.8b. This temperature is found to be −273.15°C and is designated as **absolute zero**. Absolute zero is believed to be the lower limit of temperature, but it has never been attained. In fact, there is a law of thermodynamics that says it never can be (Section 12.5). There is no

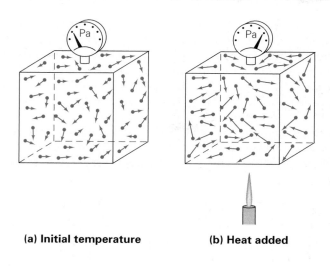

(a) Initial temperature　　**(b) Heat added**

◀ **FIGURE 10.7 Constant-volume gas thermometer**
Such a thermometer indicates temperature as a function of pressure, since, for a low-density gas, $p \propto T$. **(a)** At some initial temperature, the pressure reading has a certain value. **(b)** When the gas thermometer is heated, the pressure (and temperature) reading is higher, because, on average, the gas molecules are moving faster.

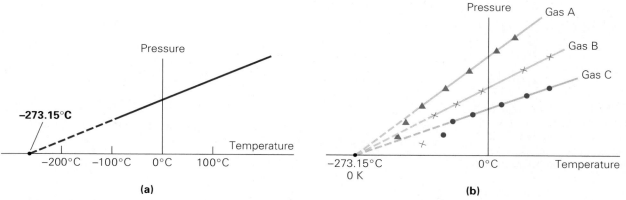

(a)　　　　**(b)**

▲ **FIGURE 10.8 Pressure versus temperature (a)** A low-density gas kept at a constant volume gives a straight line on a graph of p versus T, that is, $p = (Nk_B/V)T$. When the line is extended to the zero pressure value, a temperature of $-273.15°C$ is obtained, which is taken to be absolute zero. **(b)** Extrapolation of lines for all low-density gases indicates the same absolute zero temperature. The actual behavior of gases deviates from this straight-line relationship at low temperatures because the gases start to liquefy.

known upper limit to temperature. For example, the temperatures at the centers of some stars are estimated to be greater than 100 million degrees.

Absolute zero is the foundation of the **Kelvin temperature scale**, named after the British scientist Lord Kelvin.* On this scale, $-273.15°C$ is taken as the zero point—that is, as 0 K (▶Fig. 10.9). The size of a single unit of Kelvin temperature is the same as that of the Celsius degree, so temperatures on these scales are related by

$$T_K = T_C + 273.15 \quad \textit{Celsius-to-Kelvin conversion} \quad (10.8)$$

where T_K is the temperature in **kelvins** (*not* degrees Kelvin; for example, 300 kelvins). The kelvin is abbreviated as K (*not* °K). For general calculations, it is common to round the 273.15 in the Eq. 10.8 to 273, that is,

$$T_K = T_C + 273 \quad (\textit{for general calculations}) \quad (10.8a)$$

*Lord Kelvin, born William Thomson (1824–1907), developed devices to improve telegraphy and the compass and was involved in the laying of the first transatlantic cable. When he received his title, it is said that he considered choosing Lord Cable or Lord Compass as the title, but decided on Lord Kelvin, after a river that runs near the University of Glasgow in Scotland, where he was a professor of physics for 50 years.

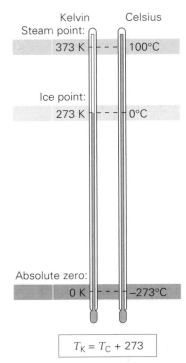

Kelvin   Celsius
Steam point:
  373 K ┤---├ 100°C

Ice point:
  273 K ┤---├ 0°C

Absolute zero:
  0 K ┤---├ −273°C

$$T_K = T_C + 273$$

▲ **FIGURE 10.9 The Kelvin temperature scale** The lowest temperature on the Kelvin scale (corresponding to −273.15°C) is absolute zero. A unit interval on the Kelvin scale, called a kelvin and abbreviated K, is equivalent to a temperature change of 1 C°; thus, $T_K = T_C + 273.15$. (The constant is usually rounded to 273 for convenience.) For example, a temperature of 0°C is equal to 273 kelvins.

The absolute Kelvin scale is the official SI temperature scale; however, the Celsius scale is used in most parts of the world for everyday temperature readings. The absolute temperature in kelvins is used primarily in scientific applications.

Hint: Keep in mind that Kelvin temperatures *must* be used with the ideal gas law. It is a common mistake to use Celsius or Fahrenheit temperatures in that equation. Suppose you used a Celsius temperature of $T = 0°C$ in the gas law. You would have $pV = 0$, which makes no sense.

Note that there can be no negative temperatures on the Kelvin scale if absolute zero is the lowest possible temperature. That is, the Kelvin scale doesn't have an arbitrary zero temperature somewhere within the scale—zero K is absolute zero, period.

Example 10.2 ■ Deepest Freeze: Absolute Zero on the Fahrenheit Scale

What is absolute zero on the Fahrenheit scale?

Thinking It Through. We need to convert 0 K to the Fahrenheit scale. Let's make the first conversion to the Celsius scale. (Why?)

Solution.

Given: $T_K = 0 \text{ K}$ *Find:* T_F

Temperatures on the Kelvin scale are related directly to Celsius temperatures by $T_K = T_C + 273.15$ (Eq. 10.8), so first we convert 0 K to a Celsius value:

$$T_C = T_K - 273.15 = 0 - 273.15 = -273.15°C$$

(We use −273.15°C for absolute zero to give a more accurate value of absolute zero on the Fahrenheit scale.) Then, converting to Fahrenheit (Eq. 10.1) gives

$$T_F = \tfrac{9}{5}T_C + 32 = \tfrac{9}{5}(-273.15) + 32 = -459.67°F$$

Thus, absolute zero is about −460°F.

Follow-up Exercise. There is an absolute temperature scale associated with the Fahrenheit temperature scale called the Rankine scale. A Rankine degree is the same size as a Fahrenheit degree, and absolute zero is taken as 0°R (zero degrees Rankine). Write the conversion equations between (a) the Rankine and the Fahrenheit scales, (b) the Rankine and the Celsius scales, and (c) the Rankine and the Kelvin scales.

Initially, gas thermometers were calibrated by using the ice and steam points. The Kelvin scale uses absolute zero, and a second fixed point adopted in 1954 by the International Committee on Weights and Measures. This second fixed point is the **triple point of water**, at which water coexists simultaneously in equilibrium as a solid (ice), liquid (water), and gas (water vapor). The triple point occurs at a unique set of values for temperature and pressure—a temperature of 0.01°C and a pressure of 4.58 mm of Hg—and provides a reproducible reference temperature for the Kelvin scale. The temperature of the triple point on the Kelvin scale was assigned a value of 273.16 K. The SI kelvin unit is then defined as 1/273.16 of the temperature at the triple point of water.*

Now let's use the ideal gas law, which requires absolute temperatures.

Example 10.3 ■ The Ideal Gas Law: Using Absolute Temperatures

A quantity of low-density gas in a rigid container is initially at room temperature (20°C) and a particular pressure (p_1). If the gas is heated to a temperature of 60°C, by what factor does the pressure change?

---

*The 273.16 value given here for the triple point temperature and the −273.15 value determined in Fig. 10.8 indicate different things. The −273.15°C is taken as 0 K. The 273.16 K (or 0.01°C) is a different reading on a different temperature scale.

Thinking It Through. A "factor" of change implies a ratio (p_2/p_1), so Eq. 10.5 should apply. Note that the container is rigid, which means that $V_1 = V_2$.

Solution.

Given: $T_1 = 20°C$ **Find:** p_2/p_1 (pressure ratio or factor)
            $T_2 = 60°C$
            $V_1 = V_2$

Since we want the factor by which the pressure changes, we write p_2/p_1 as a ratio. For example, if $p_2/p_1 = 2$, then $p_2 = 2p_1$, or the pressure would change (increase) by a factor of 2. The ratio also indicates that we should use the ideal gas law in ratio form. The law requires *absolute* temperatures, so we first change the Celsius temperatures to kelvins:

$$T_1 = 20°C + 273 = 293 \text{ K}$$
$$T_2 = 60°C + 273 = 333 \text{ K}$$

Observe that a rounded value of 273 was used in Eq. 10.8 for convenience. Then, using the ideal gas law (Eq. 10.5) in the form $p_2V_2/T_2 = p_1V_1/T_1$, we have, since $V_1 = V_2$,

$$p_2 = \left(\frac{T_2}{T_1}\right)p_1 = \left(\frac{333 \text{ K}}{293 \text{ K}}\right)p_1 = 1.14\, p_1$$

Thus, p_2 is 1.14 times p_1; that is, the pressure increases by a factor of 1.14, or 14 percent. (What would the factor be if the Celsius temperatures were *incorrectly* used? It would be much larger: $60°C/20°C = 3$, or $p_2 = 3p_1$.)

Follow-up Exercise. If the gas in this Example is heated at room temperature, so that the pressure increases by a factor of 1.26, what is the final Celsius temperature?

Note: *Always* use Kelvin (absolute) temperatures with the ideal gas law.

Because of its absolute nature, the Kelvin temperature scale has special significance. As we shall see in Section 10.5, the absolute temperature is directly proportional to the internal energy of an ideal gas and so can be used as an indication of that energy. There are no negative values on the absolute scale. Negative absolute temperatures would imply negative internal energy for the gas, a meaningless concept.

Suppose you were asked to double the temperatures of, say, $-10°C$ and $0°C$. What would you do? The following Integrated Example should help.

Integrated Example 10.4 ■ Some Like It Hot: Doubling the Temperature

The evening weather report gives the day's high temperature as $10°C$ and predicts the next day's high to be $20°C$. (a) A father tells his son that this means it will be twice as warm tomorrow, but the son says it does not. With whom do you agree? (b) What is the factor increase for the temperatures on the Kelvin scale, and what does it indicate?

(a) Conceptual Reasoning. Keep in mind that temperature gives a relative *indication* of hotness or coldness. Certainly, $20°C$ would be warmer than $10°C$. But just because the numerical value of the higher temperature is twice as great (or greater by a factor of 2, because $20°C/10°C = 2$) does not necessarily mean it is twice as warm, but only that the air temperature is 10 degrees higher and therefore relatively warmer. So the son wins.

(b) Thinking It Through. The Kelvin temperatures can be computed directly from Eq. 10.8, and a ratio of these temperatures will give the factor of increase. As noted, the absolute temperature is directly proportional to the internal energy of an ideal gas. If we consider a volume of air to be an ideal gas at these absolute temperatures, a ratio of the temperatures would give the factor of increase in internal energy.

Given: $T_{C_1} = 10°C$ **Find:** T_{K_2}/T_{K_1}
         $T_{C_2} = 20°C$

The equivalent absolute temperatures are

$$T_{K_1} = T_{C_1} + 273 = 10°C + 273 = 283 \text{ K}$$
$$T_{K_2} = T_{C_2} + 273 = 20°C + 273 = 293 \text{ K}$$

and

$$\frac{T_{K_2}}{T_{K_1}} = \frac{293 \text{ K}}{283 \text{ K}} = 1.04$$

So there is an increase of 0.04, or 4%, in the internal energy of an ideal gas in going from 10°C to 20°C, and it would not be twice as warm.

Follow-up Exercise. The weather report gives the day's high temperature as 0°C. If the next day's temperature were double that, what would the temperature be in degrees Celsius? Would this be environmentally possible?

10.4 Thermal Expansion

OBJECTIVE: To understand and be able to calculate the thermal expansions of solids and liquids.

Changes in the dimensions and volumes of materials are common thermal effects. As you learned earlier, thermal expansion provides a means of measuring temperature. The thermal expansion of gases is generally described by the ideal gas law and is very obvious. Less dramatic, but by no means less important, is the thermal expansion of solids and liquids.

Note: Solids are discussed in Section 9.1.

Thermal expansion results from a change in the average distance separating the atoms of a substance. The atoms are held together by bonding forces, which can be simplistically represented as springs in a simple model of a solid. (See Fig. 9.1.) The atoms vibrate back and forth; with increased temperature (i.e., more internal energy), they become increasingly active and vibrate over greater distances. With wider vibrations in all dimensions, the solid expands as a whole.

The change in one dimension of a solid (length, width, or thickness) is called *linear* expansion. For small temperature changes, linear expansion is approximately proportional to ΔT, or $T - T_o$ (▼Fig. 10.10a). The *fractional* change* in length is

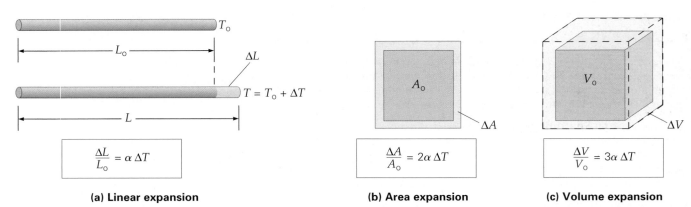

(a) Linear expansion **(b) Area expansion** **(c) Volume expansion**

▲ **FIGURE 10.10   Thermal expansion**   (a) Linear expansion is proportional to the temperature change; that is, the change in length ΔL is proportional to ΔT, and $\Delta L/L_o = \alpha \Delta T$, where α is the thermal coefficient of linear expansion. **(b)** For isotropic expansion, the thermal coefficient of area expansion is approximately 2α. **(c)** The thermal coefficient of volume expansion for solids is about 3α.

*A fractional change may also be expressed as a percent. For example, by analogy, if you invested $100 ($_o$) and made $10 ($\Delta$ $$), then the fractional change would be $\Delta \$/\$_o = 10/100 = 0.10$, or an increase of 10%.

$(L - L_o)/L_o$, or $\Delta L/L_o$, where L_o is the original length of the solid at the initial temperature. This ratio is related to the change in temperature by

$$\frac{\Delta L}{L_o} = \alpha \Delta T \qquad \text{or} \qquad \Delta L = \alpha L_o \Delta T \tag{10.9}$$

where α is the **thermal coefficient of linear expansion**. Note that the unit of α is inverse temperature: inverse Celsius degrees ($1/C^\circ$, or $C^{\circ -1}$). Values of α for some materials are given in Table 10.1.

A solid may have different coefficients of linear expansion for different directions, but for simplicity, this book will assume that the same coefficient applies to all directions (in other words, that solids show *isotropic* expansion). Also, the coefficient of expansion may vary slightly for different temperature ranges. Since this variation is negligible for most common applications, α will be considered to be constant and independent of temperature.

Equation 10.9 can be rewritten to give the final length (L) after a change in temperature:

$$\Delta L = \alpha L_o \Delta T$$
$$L - L_o = \alpha L_o \Delta T$$
$$L = L_o + \alpha L_o \Delta T$$

or

$$L = L_o(1 + \alpha \Delta T) \tag{10.10}$$

We can then use Eq. 10.10 to compute the thermal expansion of *areas* of flat objects. Since area (A) is length squared (L^2) for a square, we get

$$A = L^2 = L_o^2(1 + \alpha \Delta T)^2 = A_o(1 + 2\alpha \Delta T + \alpha^2 \Delta T^2)$$

where A_o is the original area. Because the values of α for solids are much less than 1 ($\sim 10^{-5}$, as shown in Table 10.1), the second-order term (containing $\alpha^2 \approx (10^{-5})^2 = 10^{-10} \ll 10^{-5}$) can be dropped with negligible error. As a first-order approximation, then, and with the understanding that the change in area, $\Delta A = A - A_o$, we have

$$A = A_o(1 + 2\alpha \Delta T) \qquad \text{or} \qquad \frac{\Delta A}{A_o} = 2\alpha \Delta T \tag{10.11}$$

Thus, the **thermal coefficient of area expansion** (Fig. 10.10b) is twice as large as the coefficient of linear expansion. (That is, it is equal to 2α.) This relationship is valid for all flat shapes. (See the Learn by Drawing feature.)

Learn by Drawing

Thermal Area Expansion

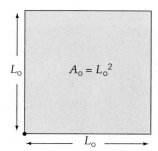

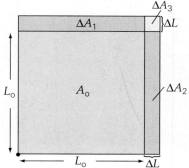

$\Delta A = \Delta A_1 + \Delta A_2 + \Delta A_3$
$\Delta A_1 = \Delta A_2 = L_o \Delta L$
$\qquad = L_o (\alpha L_o \Delta T) = \alpha A_o \Delta T$
Since ΔA_3 is very small
compared to ΔA_1 and ΔA_2,
$\qquad \Delta A \approx 2\alpha A_o \Delta T$

Thermal Expansion

Material	Coefficient of linear expansion (α)	Material	Coefficient of volume expansion (β)
Aluminum	24×10^{-6}	Alcohol, ethyl	1.1×10^{-4}
Brass	19×10^{-6}	Gasoline	9.5×10^{-4}
Brick or concrete	12×10^{-6}	Glycerin	4.9×10^{-4}
Copper	17×10^{-6}	Mercury	1.8×10^{-4}
Glass, window	9.0×10^{-6}	Water	2.1×10^{-4}
Glass, Pyrex	3.3×10^{-6}		
Gold	14×10^{-6}	Air (and most other gases	3.5×10^{-3}
Ice	52×10^{-6}	at 1 atm)	
Iron and steel	12×10^{-6}		

TABLE 10.1 Values of Thermal Expansion Coefficients (in $C^{\circ -1}$) for Some Materials at 20°C

Similarly, a first-order expression for thermal *volume* expansion is

$$V = V_0(1 + 3\alpha \, \Delta T) \quad \text{or} \quad \frac{\Delta V}{V_0} = 3\alpha \, \Delta T \tag{10.12}$$

The **thermal coefficient of volume expansion** (Fig. 10.10c) is equal to 3α (for isotropic solids and liquids).

The equations for thermal expansions are approximations. (Why?) Even though an equation is a description of a physical relationship, always keep in mind that it may be only an approximation of physical reality or may apply only in certain situations.

The thermal expansion of materials is an important consideration in construction. Seams are put in concrete highways and sidewalks to allow room for expansion and to prevent cracking. Expansion gaps in large bridges and between railroad rails are necessary to prevent damage. Similarly, expansion loops are found in oil pipelines (◀Fig. 10.11). The thermal expansion of steel beams and girders can produce tremendous pressures, as the following Example shows.

Example 10.5 ■ Temperature Rising: Thermal Expansion and Stress

A steel beam is 5.0 m long at a temperature of 20°C (68°F). On a hot day, the temperature rises to 40°C (104°F). (a) What is the change in the beam's length due to thermal expansion? (b) Suppose that the ends of the beam are initially in contact with rigid vertical supports. How much force will the expanded beam exert on the supports if the beam has a cross-sectional area of 60 cm²?

Thinking It Through. (a) This is a direct application of Eq. 10.9. (b) As the constricted beam expands, it applies a stress, and hence a force, to the supports. For linear expansion, Young's modulus (Section 9.1) should come into play.

Solution.

Given: $L_0 = 5.0$ m
$T_0 = 20°C$
$T = 40°C$
$\alpha = 12 \times 10^{-6} \, C^{o-1}$ (from Table 10.1)

Find: (a) ΔL (change in length)
(b) F (force)

$$A = 60 \text{ cm}^2 \left(\frac{1 \text{ m}}{100 \text{ cm}} \right)^2 = 6.0 \times 10^{-3} \text{ m}^2$$

(a) Using Eq. 10.9 to find the change in length with $\Delta T = T - T_0 = 40°C - 20°C = 20 \, C°$, we get

$$\Delta L = \alpha L_0 \Delta T = (12 \times 10^{-6} \, C^{o-1})(5.0 \text{ m})(20 \, C°) = 1.2 \times 10^{-3} \text{ m} = 1.2 \text{ mm}$$

This may not seem like much of an expansion, but it can give rise to a great deal of force if the beam is constrained and kept from expanding, as part (b) will show.

(b) By Newton's third law, if the beam is kept from expanding, the force the beam exerts on its constraint supports is equal to the force exerted by the supports to prevent the beam from expanding by a length ΔL. This is the same as the force that would be required to compress the beam by that length. Using the Young's modulus form of Hooke's law (Chapter 9) with $Y = 20 \times 10^{10} \text{ N/m}^2$ (Table 9.1), we calculate the stress on the beam as

$$\frac{F}{A} = \frac{Y \Delta L}{L_0} = \frac{(20 \times 10^{10} \text{ N/m}^2)(1.2 \times 10^{-3} \text{ m})}{5.0 \text{ m}} = 4.8 \times 10^7 \text{ N/m}^2$$

The force is then

$$F = (4.8 \times 10^7 \text{ N/m}^2)A = (4.8 \times 10^7 \text{ N/m}^2)(6.0 \times 10^{-3} \text{ m}^2)$$
$$= 2.9 \times 10^5 \text{ N (about 65 000 lb, or 32.5 tons!)}$$

(a)

(b)

▲ **FIGURE 10.11 Expansion gaps** (a) Expansion gaps are built into bridge roadways to prevent contact stresses produced by thermal expansion. (b) These loops in oil pipelines serve a similar purpose. As hot oil passes through them, the pipes expand, and the loops take up the extra length. The loops also accommodate expansions resulting from day–night temperature variations.

Follow-up Exercise. Expansion gaps between identical steel beams laid end to end are specified to be 0.060% of the length of a beam at the installation temperature. With this specification, what is the temperature range for noncontact expansion?

Conceptual Example 10.6 ■ Larger or Smaller? Area Expansion

A circular piece is cut from a flat metal sheet (▶ Fig. 10.12a). If the sheet is then heated in an oven, the size of the hole will (a) become larger, (b) become smaller, (c) remain unchanged.

Reasoning and Answer. It is a common misconception to think that the area of the hole will shrink because the metal expands inwardly around it. To counter this misconception, think of the piece of metal removed from the hole rather than of the hole itself. This piece would expand with increasing temperature. The metal in the heated sheet reacts as if the piece that is removed were still part of it. (Think of putting the piece of metal back into the hole after heating, as in Fig. 10.12b, or consider drawing a circle on an uncut metal sheet and heating it.) So the answer is (a).

Follow-up Exercise. A circular ring of iron has a tight-fitting metal bar inside it, across its diameter. If the ring is heated in an oven to a high temperature, would it be distorted or bent out of shape, or would it remain circular?

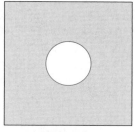

(a) Metal plate with hole

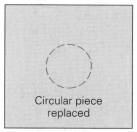

(b) Metal plate without hole

▲ **FIGURE 10.12   A larger or smaller hole?**   See Conceptual Example 10.6.

Fluids (liquids and gases), like solids, normally expand with increasing temperature. Because fluids have no definite shape, only volume expansion (and not linear or area expansion) is meaningful. The expression is

$$\frac{\Delta V}{V_\text{o}} = \beta \, \Delta T \quad \textit{fluid volume expansion} \quad (10.13)$$

where β is the coefficient of volume expansion for fluids. Note in Table 10.1 that the values of β for fluids are typically larger than the values of 3α for solids.

Unlike most liquids, water exhibits an anomalous expansion in volume near its freezing point. The volume of a given amount of water decreases as it is cooled from room temperature, until its temperature reaches 4°C (▼Fig. 10.13a). Below 4°C, the volume increases, and therefore the density decreases (Fig. 10.13b). This means that water has its maximum density ($\rho = m/V$) at 4°C (actually, 3.98°C).

◀ **FIGURE 10.13   Thermal expansion of water**   Water exhibits nonlinear expansion behavior near its freezing point. **(a)** Above 4°C (actually, 3.98°C), water expands with increasing temperature, but from 4°C down to 0°C, it expands with decreasing temperature. **(b)** As a result, water has its maximum density near 4°C.

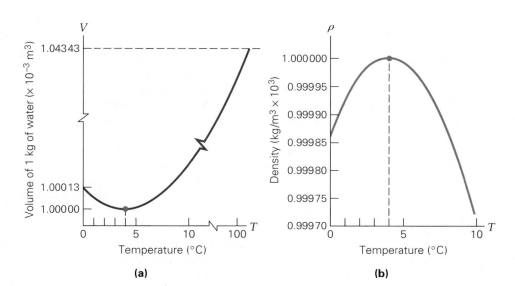

(a)

(b)

When water freezes, its molecules form a hexagonal (six-sided) lattice pattern. (This is why snowflakes have hexagonal shapes.) It is the open structure of this lattice that gives water its almost unique property of expanding on freezing and being less dense as a solid than as a liquid. (This is why ice floats in water and frozen water pipes burst.) The variation in the density of water over the temperature range from 4°C to 0°C indicates that the open lattice structure is beginning to form at about 4°C rather than exactly at the freezing point.

This property has an important environmental effect: Bodies of water such as lakes and ponds freeze at the top first, and the ice that forms floats. As a lake cools toward 4°C, water near the surface loses energy to the atmosphere, becomes denser, and sinks. The warmer, less dense water near the bottom rises. However, once the colder water on top reaches temperatures below 4°C, it becomes less dense and remains at the surface, where it freezes. If water did not have this property, lakes and ponds would freeze from the bottom up, which would destroy much of their animal and plant life (and would make ice skating a lot less popular). There would also be no oceanic ice caps at the polar regions. Instead, there would be a thick layer of ice at the bottom of the ocean, covered by a layer of water.

Conceptual Example 10.7 ■ Quick Chill: Temperature and Density

Ice is put into a container of water at room temperature. For faster cooling, the ice should (a) be allowed to float naturally in the water or (b) be pushed to the bottom of the container with a stick and held there.

Reasoning and Answer. When the ice melts, the water in its vicinity is cooled and hence becomes denser (Fig. 10.13b). If the ice is permitted to float at the top, the denser water sinks and the warmer, less dense water rises. This mixing causes the water to cool quickly. If the ice were at the bottom of the container, however, the colder, denser water would remain there, and cooling of the top layer of warm water would be slowed, so the answer is (a).

Follow-up Exercise. Suppose that the density-versus-temperature curve for water (Fig. 10.13b) were inverted, so that it dipped downward. What would this imply for the situation in this Example and for the freezing of lakes? Explain.

10.5 The Kinetic Theory of Gases

OBJECTIVES: To (a) relate kinetic theory and temperature and (b) explain the process of diffusion.

If the molecules of a sample of gas are viewed as colliding particles, the laws of mechanics can be applied to each molecule of that gas. We should then be able to describe the gas's microscopic characteristics, such as pressure, internal energy, and so on, in terms of molecular motion. Because of the large number of particles involved, however, a statistical approach is employed for such a microscopic description.

One of the major accomplishments of theoretical physics was to do exactly that: derive the ideal gas law from mechanical principles. This derivation led to a new interpretation of temperature in terms of the translational kinetic energy of the gas molecules. As a theoretical starting point, the molecules of an ideal gas are viewed as point masses in random motion with relatively large distances separating them.

In this section, we will consider primarily the kinetic theory of *mon*atomic (single-atom) gases, such as He, and learn about the internal energy of such a gas. In the next section, the internal energy of *di*atomic (two-atom molecules) gases, such as O_2, will be considered. In either case, the vibrational and rotational motions can be ignored with regard to temperature and pressure, since these quantities depend only on *linear* motion.

According to the **kinetic theory of gases**, the molecules of a gas undergo perfectly elastic collisions with the walls of its container. (With the molecules taken to be point particles molecular collisions can be neglected.) From Newton's laws of motion, the force on the walls of the container can be calculated from the change in momentum of the gas molecules when they collide with the walls (▶Fig. 10.14). If this force is expressed in terms of pressure (force/area), the following equation is obtained (see Appendix II for derivation):

Note: Elastic collisions are discussed in Section 6.4.

$$pV = \tfrac{1}{3} Nmv_{rms}^2 \qquad (10.14)$$

Here, V is the volume of the container or gas, N is the number of gas molecules in the closed container, m is the mass of a gas molecule, and the speed v_{rms} is the average speed of the molecules, but a special kind of average. It is obtained by averaging the squares of the speeds and then taking the square root of the average—that is, $\sqrt{\overline{v^2}} = v_{rms}$. As a result, v_{rms} is called the *root-mean-square (rms)* speed.

Solving Eq. 10.6 for pV and equating the resulting expression with Eq. 10.14 shows how temperature came to be interpreted as a measure of translational kinetic energy:

$$pV = Nk_BT = \tfrac{1}{3} Nmv_{rms}^2 \quad \text{or} \quad \tfrac{1}{2} mv_{rms}^2 = \tfrac{3}{2} k_BT \qquad \begin{array}{l}\textit{(for all ideal} \\ \textit{gases)}\end{array} \quad (10.15)$$

Thus, the temperature of a gas (and that of the walls of the container or a thermometer bulb in thermal equilibrium with the gas) is directly proportional to its average random kinetic energy (per molecule), since $\overline{K} = \tfrac{1}{2} mv_{rms}^2 = \tfrac{3}{2} k_BT$. (Don't forget that T is the absolute temperature in kelvins.)

Average molecular kinetic energy and temperature

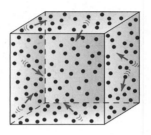

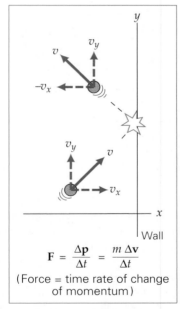

$$\mathbf{F} = \frac{\Delta \mathbf{p}}{\Delta t} = \frac{m \, \Delta \mathbf{v}}{\Delta t}$$

(Force = time rate of change of momentum)

▲ **FIGURE 10.14 Kinetic theory of gases** The pressure a gas exerts on the walls of a container is due to the force resulting from the change in momentum of the gas molecules that collide with the wall. The force exerted by an individual molecule is equal to the time rate of change of momentum; that is, $\mathbf{F} = \Delta\mathbf{p}/\Delta t = m\,\Delta\mathbf{v}/\Delta t$, where $\mathbf{p} = m\mathbf{v}$. The sum of the instantaneous normal components of the collision forces gives rise to the average pressure on the wall.

Example 10.8 ■ Molecular Speed: Relation to Absolute Temperature

What is the average (rms) speed of a helium atom (He) in a helium balloon at room temperature? (Take the mass of the helium atom to be 6.65×10^{-27} kg.)

Thinking It Through. All the data we need to solve for the average speed in Eq. 10.15 are known.

Solution.

Given: $m = 6.65 \times 10^{-27}$ kg *Find:* v_{rms} (rms speed)
$\quad\quad\quad$ $T = 20°C$ (room temperature)
$\quad\quad\quad$ $k_B = 1.38 \times 10^{-23}$ J/K (known)

Eq. 10.15 will be used, so we list k_B among the given quantities.

The Celsius temperature must be changed to kelvins, and note that k_B has units of J/K, so

$$T_K = T_C + 273 = 20°C + 273 = 293 \text{ K}$$

Rearranging Eq. 10.15, we have

$$v_{rms} = \sqrt{\frac{3k_BT}{m}} = \sqrt{\frac{3(1.38 \times 10^{-23} \text{ J/K})(293 \text{ K})}{6.65 \times 10^{-27} \text{ kg}}} = 1.35 \times 10^3 \text{ m/s} = 1.35 \text{ km/s}$$

This is over 3000 mi/h—pretty fast!

Follow-up Exercise. In this Example, if the temperature of the gas were increased by $10 C°$, what would be the corresponding percentage increases in the average (rms) speed and in the average kinetic energy?

Interestingly, Eq. 10.15 predicts that at absolute zero ($T = 0$ K) all translational molecular motion of a gas would cease. According to classical theory, this would correspond to absolute zero energy. However, modern quantum theory says that there would still be some zero-point motion, and a corresponding minimum *zero-point energy*. Basically, absolute zero is the temperature at which all the energy that *can* be removed from an object has been removed.

Internal Energy of Monatomic Gases

Because the "particles" in an ideal monatomic gas do not vibrate or rotate, as explained previously, the total kinetic energy of all the molecules is equal to the total internal energy of the gas. That is, the gas's internal energy is all "temperature" energy (Section 10.1). With N molecules in a system, we can use Eq. 10.15, expressing the energy per molecule, to write an equation for the total internal energy U:

$$U = N(\tfrac{1}{2}mv_{rms}^2) = \tfrac{3}{2}Nk_BT = \tfrac{3}{2}nRT \qquad \text{(for ideal monatomic \quad (10.16) \\ gases only)}$$

Thus, we see that the internal energy of an ideal monatomic gas is directly proportional to its absolute temperature. (In Section 10.6, we will see that this is true regardless of the molecular structure of the gas. However, the expression for U will be a bit different for gases that are not monatomic.) This means that if the absolute temperature of a gas is doubled (by heat transfer), for example, from 200 K to 400 K, then the internal energy of the gas is also doubled.

Diffusion

We depend on our sense of smell to detect odors, such as the smell of smoke from something burning. That you can smell something from a distance implies that molecules get from one place to another in the air—from the source to your nose. This process of random molecular mixing in which particular molecules move from a region where they are present in higher concentration to one where they are in lower concentration is called **diffusion**. Diffusion also occurs readily in liquids; think about what happens to a drop of ink in a glass of water (▼ Fig. 10.15). It even occurs to some degree in solids.

The rate of diffusion for a particular gas depends on the rms speed of its molecules. Even though gas molecules have large average speeds (Example 10.8), their

▼ **FIGURE 10.15 Diffusion in liquids** Random molecular motion would eventually distribute the dye throughout the water. Here there is some distribution due to mixing, and the ink colors the water after a few minutes. The distribution would take more time by diffusion only.

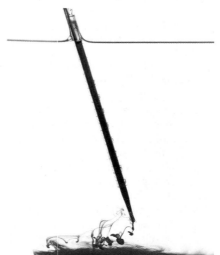
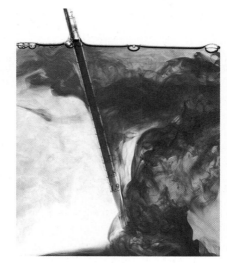
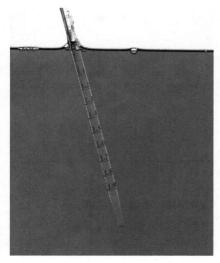

average positions change slowly, and the molecules do not fly from one side of a room to the other. Instead, there are frequent collisions, and as a result, the molecules "drift" rather slowly. For example, suppose someone opened a bottle of ammonia on the other side of a closed room. It would take some time for the ammonia to diffuse across the room until you could smell it. (Much of the movement that people commonly attribute to diffusion is actually due to air currents.)

Gases can also diffuse through porous materials or permeable membranes. (This process is sometimes referred to as *effusion*.) Energetic molecules enter the material through the pores (openings), and, colliding with the pore walls, they slowly meander through the material. Such gaseous diffusion can be used to physically separate the different gases making up a mixture.

The kinetic theory of gases says that the average translational kinetic energy (per molecule) of a gas is proportional to the absolute temperature of the gas: $\frac{1}{2}mv_{rms}^2 = \frac{3}{2}k_BT$. So, on the average, the molecules of different gases (having different masses) move at different speeds at a given temperature. As you might expect, because they move faster, lighter gas molecules diffuse through the tiny openings of a porous material faster than do heavier gas molecules.

For instance, at a particular temperature, molecules of oxygen (O_2) move faster on the average than do the more massive molecules of carbon dioxide (CO_2). Because of this difference in molecular speed, oxygen can diffuse through a barrier faster than carbon dioxide can. Suppose that a mixture of equal volumes of oxygen and carbon dioxide is contained on one side of a porous barrier (▼Fig. 10.16). After a while, some O_2 molecules and some CO_2 molecules will have diffused through the barrier, but more oxygen than carbon dioxide. Repeating the process on this diffused gas mixture would cause the oxygen concentration to become even greater on the far side of the barrier. Almost pure oxygen can be obtained by repeating the separation process many times. Separation by gaseous diffusion is a key process in obtaining enriched uranium, which was used in the first atomic bomb and in nuclear reactors that generate electricity (Chapter 30).

Fluid diffusion is very important to organisms. In plant photosynthesis, carbon dioxide from the air diffuses into leaves, and oxygen and water vapor diffuse out. The diffusion of liquid water across a permeable membrane down a concentration gradient (a concentration difference) is called **osmosis**, a process that is vital in living cells. Osmotic diffusion is also important to kidney functioning: Tubules in the kidneys concentrate waste matter from the blood in much the same way that oxygen is removed from mixtures. (See the Insight on page 364 for other examples of diffusion.)

Osmosis is the tendency for the solvent of a solution, such as water, to diffuse across a semipermeable membrane from the side where the solvent is at higher concentration to the side where it is at lower concentration. When pressure is applied to the side with the lower concentration, the diffusion is reversed—a process called *reverse osmosis*. Reverse osmosis is used in desalination plants to provide freshwater from seawater in dry coastal regions.

Reverse osmosis is also used in water purification. You may have drunk such purified water. One of the popular bottled waters is purified "using state of the art treatment by reverse osmosis," according to the label.

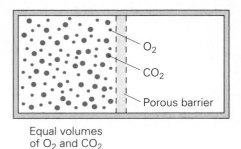

Equal volumes
of O_2 and CO_2

O_2
CO_2
Porous barrier

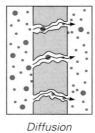

*Diffusion
through barrier*

◀ **FIGURE 10.16 Separation by gaseous diffusion** The molecules of both gases diffuse (or effuse) through the porous barrier, but because oxygen molecules have the greater average speed, more of them pass through. Thus, over time, there is a greater concentration of oxygen molecules on the other side of the barrier.

Physiological Diffusion in Life Processes

Diffusion plays a central role in many life processes. For example, consider a cell membrane in the lung. Such a membrane is permeable to a number of substances, any of which will diffuse through the membrane from a region where its concentration is high to a region where its concentration is low. Most importantly, the lung membrane is permeable to oxygen (O_2), and the transfer of O_2 across the membrane occurs because of a concentration gradient.

The blood carried to the lungs is low in O_2, having given up the oxygen during its circulation through the body to tissues requiring O_2 for metabolism. Conversely, the air in the lungs is high in O_2, because there is a continuous exchange of fresh air in the breathing process. As a result of this concentration difference, or gradient, O_2 diffuses from the lung volume into the blood that flows through the lung tissue, and the blood leaving the lungs is high in O_2.

Exchanges between the blood and the tissues occur across capillary walls, and diffusion again is a major factor. The chemical composition of arterial blood is regulated to maintain the proper concentrations of particular solutes (substances dissolved in the blood solution), so diffusion takes place in the appropriate directions across capillary walls. For example, as cells take up O_2 and glucose (blood sugar), the blood continuously brings in fresh supplies of the substances to maintain the concentration gradient needed for diffusion to the cells. The continuous production of carbon dioxide

(CO_2) and metabolic wastes in the cells produces concentration gradients in the opposite direction for these substances. They therefore diffuse out of the cells into the blood, to be carried away from the tissues by the circulatory system.

During periods of physical exertion, cellular activity increases. More O_2 is used up and more CO_2 is produced, thereby increasing the concentration gradients and the diffusion rates. How do the lungs respond to satisfy an increased demand for O_2 to the blood? As you might expect, the rate of diffusion depends on the surface area and thickness of the lung membrane. Deeper breathing during exercise causes the alveoli (small air sacs in the lungs) to increase in volume. Such stretching increases the alveolar surface area and decreases the thickness of the membrane wall, allowing more rapid diffusion.

Also, the heart works harder during exercise, and the blood pressure is raised. The increased pressure forces open capillaries that are normally closed during rest or normal activity. As a result, the total exchange area between the blood and cells is increased. Each of these changes helps expedite the exchange of gases during exercise.

Now you can understand why people with emphysema (a disease involving the breakdown of alveoli walls) or pneumonia (a condition characterized by fluid accumulations within or around the lungs) have difficulty providing enough oxygen to their tissues.

*10.6 Kinetic Theory, Diatomic Gases, and the Equipartition Theorem

OBJECTIVES: To understand (a) the difference between monatomic and diatomic gases, (b) the meaning of the equipartition theorem, and (c) the expression for the internal energy of a diatomic gas.

In the real world, most of the gases we deal with are *not* monatomic gases. Recall from chemistry that monatomic gases are elements known as *noble* or *inert* gases, because they do not readily combine with other atoms. These elements are found on the far right side of the periodic chart: helium, neon, argon, krypton, xenon, and radon.

However, the mixture of gases we breathe (collectively known as "air") consists mainly of diatomic molecules of nitrogen (N_2, 78% by volume) and oxygen (O_2, 21% by volume). Each of these gases has two identical atoms chemically bonded together to form a single molecule. How do we deal with these realistic, more complicated molecules in terms of the kinetic theory of gases? [There are even more complicated gas molecules consisting of more than two atoms, such as carbon dioxide (CO_2). However, because of the complexity of such gas molecules, we will limit our discussion to diatomic molecules.]

The Equipartition Theorem

As we saw in Section 10.5, only the translational kinetic energy of a gas is determined by the gas's temperature. Thus, for any type of gas, regardless of how many atoms make up its molecules, it is *always true* that the average *translational* kinetic energy per molecule is still proportional to the temperature of the gas (Eq. 10.15): $\frac{1}{2}mv_{rms}^2 = \frac{3}{2}k_BT$ (for all gases).

Recall that for monatomic gases, the total internal energy U consists solely of translational kinetic energy. For diatomic molecules, this is not true, because a diatomic molecule is free to rotate and vibrate in addition to moving linearly. Therefore, these extra forms of energy must be taken into account. The expression given in Eq. 10.16 ($U = \frac{3}{2}Nk_BT$) for monatomic gases, which assumes that the total energy is due only to translational kinetic energy, therefore does *not* hold for diatomic gases.

Scientists wondered exactly how the expression for the internal energy of a diatomic gas might differ from that for a monatomic gas. In looking at the derivation of Eq. 10.16 from the kinetic theory, they realized that the factor of 3 in that equation was due to the fact that the gas molecules had three independent linear ways (dimensions) of moving. Thus, for each molecule, there were three independent ways of possessing kinetic energy: with x, y, and z linear motion. Each independent way a molecule has for possessing energy is called a **degree of freedom**.

According to this scheme, a monatomic gas has only three degrees of freedom, since its molecules can move only linearly and can possess kinetic energy in three dimensions. Scientists reasoned that, quite possibly, a diatomic gas could vibrate (see Fig. 10.1), thus having vibrational kinetic and potential energies (two additional degrees of freedom). A diatomic molecule might also rotate.

Consider a symmetric diatomic molecule—for example, O_2. A classical model describes such a diatomic molecule as though the molecules were particles connected by a rigid rod (▶Fig. 10.17). The rotational moment of inertia, I, has the same value about each of the axes (x and y) that pass perpendicularly through the center of the rod. The moment of inertia about the z-axis is essentially zero. (Why?) Thus, there are only two degrees of freedom associated with the kinetic energies of diatomic molecule rotation.

On the basis of the understanding of monatomic gases and their three degrees of freedom, the **equipartition theorem** was proposed. (As the name implies, the total energy of a gas or molecule is "partitioned," or divided, equally for each degree of freedom.) That is,

> On average, the total internal energy U of an ideal gas is divided equally among each degree of freedom its molecules possess. Furthermore, each degree of freedom contributes $\frac{1}{2}Nk_BT$ (or $\frac{1}{2}nRT$) to the total internal energy of the gas.

The equipartition theorem fits the special case of a monatomic gas, since the theorem predicts that $U = \frac{3}{2}Nk_BT$, which we know to be true. With three degrees of freedom, we have $U = 3(\frac{1}{2}Nk_BT)$, in agreement with the monatomic result given earlier (Eq. 10.16).

The Internal Energy of a Diatomic Gas

Exactly how does the equipartition theorem enable us to calculate the internal energy of a diatomic gas such as oxygen? To perform such a calculation, we must realize that U now includes all the available degrees of freedom. In addition to the translational degrees of freedom, what other motions are the molecules undergoing? The analysis is complicated and beyond the scope of this text, so we give only the general results. *Typically, for normal (room) temperatures, quantum theory predicts (and experiment verifies) that only the rotational motions are important in the degrees of freedom.*

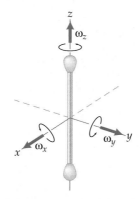

▲ **FIGURE 10.17 Model of a diatomic gas molecule** A dumbbell-like molecule can rotate about three axes. The moment of inertia, I, about the x- and y-axes is the same. The masses (molecules) on the ends of the rod are pointlike particles, so the moment of inertia about the z-axis is zero.

Definition of: Equipartition theorem

So the total internal energy of a diatomic gas is composed of the internal energies due to the three linear degrees of freedom and the two rotational degrees of freedom, for a total of five degrees of freedom. Hence, we can write

$$U = K_{trans} + K_{rot} = 3(\tfrac{1}{2}nRT) + 2(\tfrac{1}{2}nRT)$$
$$= \tfrac{5}{2}nRT = \tfrac{5}{2}Nk_BT$$

(*for diatomic gases near room temperature*) (10.17)

Thus, a monatomic sample of gas at normal room temperature has 40% less internal energy than a diatomic sample at the same temperature. Or, equivalently, the monatomic sample possesses only 60% of the internal energy of the diatomic sample.

Example 10.9 ■ Monatomic versus Diatomic: Are Two Atoms Better Than One?

More than 99% of the air we breathe consists of diatomic gases, mainly nitrogen (N_2, 78%) and oxygen (O_2, 21%). There are traces of other gases, one of which is radon (Ra), a monatomic gas arising from radioactive decay of uranium in the ground. [Radon itself is radioactive, which is irrelevant here, but this fact does make radon a possible health danger when concentrated inside a house (as we shall see in Chapter 29).] (a) Calculate the total internal energy of 1.00-mol samples each of oxygen and radon at room temperature (20°C). (b) For each sample, calculate the amount of internal energy associated with molecular *translational* kinetic energy.

Thinking It Through. (a) We have to consider the number of degrees of freedom in a monatomic gas and a diatomic gas in computing the internal energy U. (b) Only three linear degrees of freedom contribute to the translational kinetic energy portion (U_{trans}) of the internal energy.

Solution. We list the data and convert to kelvins right away because we know that the internal energy is expressed in terms of absolute temperature:

Given: $n = 1.00$ mol
$T = 20°C + 273 = 293$ K

Find: (a) U for O_2 and Ra samples at room temperature
(b) U_{trans} for O_2 and Ra samples at room temperature

(a) Let's compute the total internal energy of the (monatomic) radon sample first, using Eq. 10.16:

$$U_{Ra} = \tfrac{3}{2}nRT = \tfrac{3}{2}(1.00 \text{ mol})[8.31 \text{ J/(mol} \cdot \text{K)}](293 \text{ K}) = 3.65 \times 10^3 \text{ J}$$

Since this sample is at room temperature, the (diatomic) oxygen will also include internal energy stored as two extra degrees of freedom, due to rotation. Thus, we have

$$U_{O_2} = \tfrac{5}{2}nRT = \tfrac{5}{2}(1.00 \text{ mol})[8.31 \text{ J/(mol} \cdot \text{K)}](293 \text{ K}) = 6.09 \times 10^3 \text{ J}$$

As we have seen, even though there is the same number of molecules in each sample and the temperature is the same, the oxygen sample has almost 70% more total internal energy.
(b) For (monatomic) radon, all the internal energy is in the form of translational kinetic energy; hence, the answer is the same as in (a):

$$U_{trans} = U_{Ra} = 3.65 \times 10^3 \text{ J}$$

For (diatomic) oxygen, only $\tfrac{3}{2}nRT$ of the total internal energy ($\tfrac{5}{2}nRT$) is in the form of translational kinetic energy, so the answer is the same as for radon; that is, $U_{trans} = 3.65 \times 10^3$ J for both gas samples.

Follow-up Exercise. (a) In this Example, how much energy is associated with the rotational motion of the oxygen molecules? (b) Which sample has the highest rms speed? (*Note:* The mass of one radon atom is about seven times the mass of an oxygen molecule.) Explain your reasoning.

Chapter Review

Important Concepts and Equations

Celsius-Fahrenheit conversion:

$$T_F = \tfrac{9}{5}T_C + 32 \quad \text{or} \quad T_F = 1.8T_C + 32 \quad (10.1)$$

$$T_C = \tfrac{5}{9}(T_F - 32) \quad (10.2)$$

- **Heat** is the net energy transferred from one object to another because of temperature differences. Once transferred, the energy becomes part of the internal energy of the object (or system).

- The **ideal (or perfect) gas law** relates the pressure, volume, and absolute temperature of an ideal or dilute gas.

Ideal (or perfect) gas law (always use absolute temperatures):

$$\frac{p_1 V_1}{T_1} = \frac{p_2 V_2}{T_2} \quad \text{or} \quad pV = Nk_B T \quad (10.5, 10.6)$$

or

$$pV = nRT \quad (10.7)$$

$$k_B = 1.38 \times 10^{-23}\,\text{J/K}$$

$$R = 8.31\,\text{J/(mol·K)}$$

- **Absolute zero (0 K)** corresponds to $-273.15°C$.

Kelvin-Celsius conversion:

$$T_K = T_C + 273.15 \quad (10.8)$$

$$T_K = T_C + 273 \quad (\text{for general calculations}) \quad (10.8a)$$

- **Thermal coefficients of expansion** relate the fractional change in dimension(s) to a change in temperature.

Thermal expansion of solids:

$$\text{linear:} \quad \frac{\Delta L}{L_o} = \alpha\,\Delta T \quad \text{or} \quad L = L_o(1 + \alpha\,\Delta T) \quad (10.9, 10.10)$$

$$\text{area:} \quad \frac{\Delta A}{A_o} = 2\alpha\,\Delta T \quad \text{or} \quad A = A_o(1 + 2\alpha\,\Delta T) \quad (10.11)$$

$$\text{volume:} \quad \frac{\Delta V}{V_o} = 3\alpha\,\Delta T \quad \text{or} \quad V = V_o(1 + 3\alpha\,\Delta T) \quad (10.12)$$

Thermal volume expansion of fluids:

$$\frac{\Delta V}{V_o} = \beta\,\Delta T \quad (10.13)$$

- According to the **kinetic theory of gases**, the absolute temperature of a gas is directly proportional to the average random kinetic energy per molecule.

Results of kinetic theory of gases:

$$pV = \tfrac{1}{3}Nmv_{rms}^2 \quad (10.14)$$

$$\tfrac{1}{2}mv_{rms}^2 = \tfrac{3}{2}k_B T \quad (\text{all gases}) \quad (10.15)$$

$$U = \tfrac{3}{2}Nk_B T = \tfrac{3}{2}nRT \quad \begin{matrix}(\text{ideal monatomic} \\ \text{gases only})\end{matrix} \quad (10.16)$$

$$U = \tfrac{5}{2}Nk_B T = \tfrac{5}{2}nRT \quad \begin{matrix}(\text{for diatomic gases near} \\ \text{room temperature})\end{matrix} \quad (10.17)$$

Exercises*

10.1 Temperature and Heat
and
10.2 The Celsius and Fahrenheit Temperature Scales

1. Which one of the following is the closest to 15°C? (a) 8.3°F, (b) 27°F, (c) 40°F, or (d) 50°F

2. **CQ** Which temperature scale has the smaller degree interval, Celsius or Fahrenheit?

3. **CQ** Heat flows spontaneously from a body at a higher temperature to one at a lower temperature that is in thermal contact with it. Does heat always flow from a body with more internal energy to one with less internal energy? Explain.

4. **CQ** What is the hottest (highest temperature) item in a home? (*Hint:* Think about this one, and maybe a light will come on.)

*Assume all temperatures to be exact, and neglect significant figures for small changes in dimension.

5. **CQ** The tires of commercial jumbo jets are inflated with nitrogen, not air. Why?

6. **CQ** When temperature changes during the day, which scale, Celsius or Fahrenheit, will read a smaller change? Explain.

7. ■ Convert the following to Celsius readings: (a) 1000°F, (b) 0°F, (c) −20°F, and (d) −40°F.

8. ■ Convert the following to Fahrenheit readings: (a) 150°C, (b) 32°C, (c) −25°C, and (d) −273°C.

9. ■ The coldest inhabited village in the world is Oymyakon, a town located in eastern Siberia, where it gets as cold as −94°F. What is this temperature on the Celsius scale?

10. ■ A person running a fever has a body temperature of 39.4°C. What is this temperature on the Fahrenheit scale?

11. ■ The highest and lowest recorded air temperatures in the United States are, respectively, 134°F (Death Valley, California, 1913) and −80°F (Prospect Creek, Alaska, 1971). What are these temperatures on the Celsius scale?

12. ■ Which is the lower temperature? (a) 245°C or 245°F. (b) 200°C or 375°F.

13. ■ The highest and lowest recorded air temperatures in the world are, respectively, 58°C (Libya, 1922) and −89°C (Antarctica, 1983). What are these temperatures on the Fahrenheit scale?

14. IE ■■ There is one temperature at which the Celsius and Fahrenheit scales have the same reading. (a) To find that temperature, would you set (1) $5T_F = 9T_C$, (2) $9T_F = 5T_C$, or (3) $T_F = T_C$? Why? (b) Find the temperature.

15. ■■■ (a) The largest temperature drop recorded in the United States in one day occurred in Browning, Montana, in 1916, when the temperature went from 7°C to −49°C. What is the corresponding change on the Fahrenheit scale? (b) On the Moon, the average surface temperature is 127°C during the day and −183°C during the night. What is the corresponding change on the Fahrenheit scale?

16. IE ■■■ Fig. 10.5a is a plot of Fahrenheit temperature versus Celsius temperature. (a) Is the value of the y-intercept found by setting (1) $T_F = T_C$, (2) $T_C = 0$, or (3) $T_F = 0$? Why? (b) Compute the value of the y-intercept. (c) What would be the slope and y-intercept if the graph were plotted the opposite way (Celsius versus Fahrenheit)?

10.3 Gas Laws and Absolute Temperature

17. The temperature used in the ideal gas law must be expressed on which scale? (a) Celsius, (b) Fahrenheit, (c) Kelvin, or (d) any of the preceding.

18. When the temperature of a quantity of gas is increased, (a) the pressure must increase, (b) the volume must increase, (c) both the pressure and volume must increase, or (d) none of the preceding.

19. The temperature of a quantity of gas is decreased. How is the density affected (a) if the pressure is held constant? (b) if the volume is held constant?

20. CQ A type of constant-volume gas thermometer is shown in ▶ Fig. 10.18. Describe how it operates.

21. CQ Describe how a constant-pressure gas thermometer might be constructed.

22. CQ In terms of the idea gas law, what would a temperature of absolute zero imply? How about a negative absolute temperature?

23. CQ Excited about a New Year's Eve party in Times Square, you pump up 10 balloons in your warm apart-

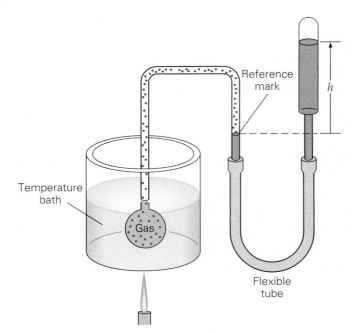

▲ FIGURE 10.18 A type of constant-volume gas thermometer See Exercise 20.

ment and take them to the cold square. However, you are very disappointed with your decorations. Why?

24. CQ Which has more molecules, 1 mole of oxygen or 1 mole of nitrogen? Explain.

25. CQ "Oxygen will start to flow even though the bag may not inflate." Heard that before? This is part of the safety instructions given by a flight crew before the plane takes off. Actually, the bag inflates fully at high altitude, but only slightly at lower altitude, when the cabin pressure is lost. Explain.

26. ■ Convert the following temperatures to absolute temperatures in kelvins: (a) 0°C, (b) 100°C, (c) 20°C, and (d) −35°C.

27. ■ Convert the following temperatures to degrees Celsius: (a) 0 K, (b) 250 K, (c) 273 K, and (d) 325 K.

28. ■ What are the absolute temperatures in kelvins for (a) −40°F and (b) −40°C?

29. ■ Which is the lower temperature, 300°F or 300 K?

30. ■ When lightning strikes, it can heat the air around it to more than 30 000 K, five times the surface temperature of the Sun. (a) What is this temperature on the Fahrenheit and Celsius scales? (b) The temperature is sometimes reported to be 30 000°C. Assuming that 30 000 K is correct, what is the percentage error of this Celsius value?

31. ■ How many moles are there in (a) 40 g of water, (b) 245 g of H_2SO_4 (sulfuric acid), (c) 138 g of NO_2 (nitrogen dioxide), and (d) 56 L of SO_2 (sulfur dioxide) at STP (stan-

dard temperature of exactly 0°C and pressure of exactly 1 atm)?

32. **IE ■** (a) In a constant-volume gas thermometer, if the pressure of the gas decreases, will the temperature of the gas (a) increase, (2) decrease, or (3) remain the same? Why? (b) The initial absolute pressure of a gas is 1000 Pa at room temperature (20°C). If the pressure increases to 1500 Pa, what is the new Celsius temperature?

33. **■** The air in a balloon of volume 0.10 m³ exerts a pressure of 1.4×10^5 Pa. If the volume of the balloon increases to 0.12 m³ at constant temperature, what is the pressure?

34. **■** Show that 1.00 mol of ideal gas under STP occupies a volume of 0.0224 m³ = 22.4 L.

35. **■** What is the volume occupied by 4.0 g of hydrogen under a pressure of 2.0 atm and a temperature of 300 K?

36. **■■** An athlete has a large lung capacity, 7.5 L. Assuming air to be an ideal gas, how many molecules of air are in the athlete's lung when the air temperature in the lungs is 37°C under normal atmospheric pressure?

37. **IE ■■** (a) If the temperature of an ideal gas increases and its volume decreases, will the pressure of the gas (1) increase, (2) remain the same, or (3) decrease? Why? (b) The Kelvin temperature of an ideal gas is doubled and its volume is halved. How is the pressure affected?

38. **■■** On a warm day (92°F), an air-filled balloon occupies a volume of 0.20 m³ and is under a pressure of 20.0 lb/in.² If the balloon is cooled to 32°F in a refrigerator while its pressure is reduced to 14.7 lb/in.², what is the volume of the air in the container? (Assume that the air behaves as an ideal gas.)

39. **■■** A steel-belted radial automobile tire is inflated to a gauge pressure of 30.0 lb/in.² when the temperature is 61°F. Later in the day, the temperature rises to 100°F. Assuming that the volume of the tire remains constant, what is the tire's pressure at the elevated temperature? (*Hint:* Remember that the ideal gas law uses absolute pressure.)

40. **■■** If 2.4 m³ of a gas initially at STP is compressed to 1.6 m³ and its temperature raised to 30°C, what is the final pressure?

41. **IE ■■** The pressure on a low-density gas in a cylinder is kept constant as its temperature is increased. (a) Does the volume of the gas (1) increase, (2) decrease, or (3) remain the same? Why? (b) If the temperature is increased from 10°C to 40°C, what is the percentage change in the volume of the gas?

42. **■■■** A diver releases an air bubble of volume 2.0 cm³ from a depth of 15 m below the surface of a lake, where the temperature is 7.0°C. What is the volume of the bubble when it reaches just below the surface of the lake, where the temperature is 20°C?

10.4 Thermal Expansion

43. Are the units of the thermal coefficient of linear expansion (a) m/C°, (b) m²/C°, (c) m·C°, or (d) 1/C°?

44. Is the thermal coefficient of volume expansion for a solid (a) α, (b) 2α, (c) 3α, or (d) α^3?

45. **CQ** A cube of ice sits on a bimetallic strip at room temperature (▼Fig. 10.19). What will happen if (a) the upper strip is aluminum and the lower strip brass, or (b) the upper strip is iron and the lower strip copper? (c) If the cube is made of a hot metal rather than ice and the two strips are brass and copper, should the brass or copper be on top to keep the cube from falling off?

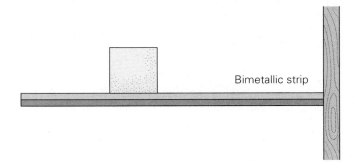

▲ FIGURE 10.19   Which way will the cube go?
See Exercise 45.

46. **CQ** A demonstration of thermal expansion is shown in ▶Fig. 10.20. (a) Initially, the ball goes through the ring made of the same metal. When the ball is heated (b), it does not go through the ring (c). If both the ball and the ring are heated, the ball again goes through the ring. Explain what is being demonstrated.

47. **CQ** What happens to a volume of water if it is cooled from 4°C to 2°C?

48. **CQ** A solid metal disk rotates freely, so the conservation of angular momentum applies (Chapter 8). If the disk is heated while it is rotating, will there be any effect on the rate of rotation (the angular speed)?

49. **CQ** A circular ring of iron has a tight-fitting iron bar across its diameter, as illustrated in ▶Fig. 10.21. If the arrangement is heated in an oven to a high temperature, will the circular ring be distorted? What if the bar is made of aluminum?

50. **CQ** In terms of thermal expansion, which would be more accurate, a steel tape measure or an aluminum one? Explain.

51. **CQ** We often use hot water to loosen tightly sealed metal lids on glass jars. Explain why this works.

52. **■** A steel beam 10 m long is installed in a structure at 20°C. What are the beam's changes in length at the temperature extremes of −30°C to 45°C?

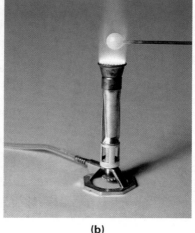
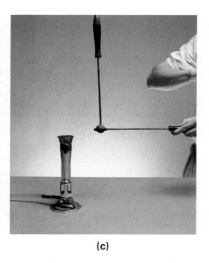

(a)   (b)   (c)

▲ **FIGURE 10.20 Ball-and-ring expansion** See Exercises 46 and 59.

▲ **FIGURE 10.21 Stress out of shape?** See Exercise 49.

53. IE ■ An aluminum tape measure is accurate at 20°C. (a) If the tape measure is placed in a freezer, would it read (a) high, (2) low, or (3) the same? Why? (b) If the temperature of the freezer is −5.0°C, what would be the stick's percentage error because of thermal contraction?

54. ■ Concrete highway slabs are poured in lengths of 10.0 m. How wide should the expansion gaps between the slabs be at a temperature of 20°C to ensure that there will be no contact between adjacent slabs over a temperature range of −25°C to 45°C?

55. ■ A man's gold wedding ring has an inner diameter of 2.4 cm at 20°C. If the ring is dropped into boiling water, what will be the change in the inner diameter of the ring?

56. ■■ What temperature change would cause a 0.10% increase in the volume of a quantity of water that was initially at 20°C?

57. ■■ A piece of copper tubing used in plumbing has a length of 60.0 cm and an inner diameter of 1.50 cm at 20°C. When hot water at 85°C flows through the tube, what are (a) the tube's new length and (b) the change in its cross-sectional area? Does the latter affect the flow speed?

58. IE ■■ A circular piece is cut from an aluminum sheet at room temperature. (a) When the sheet is then placed in an oven, will the hole (1) get larger, (2) get smaller, or (3) remain the same? Why? (b) If the diameter of the hole is 8.00 cm at 20°C and the temperature of the oven is 150°C, what will be the new area of the hole?

59. IE ■■ In Fig 10.20, the steel ring of diameter 2.5 cm is 0.10 mm smaller in diameter than the steel ball at 20°C. (a) For the ball to go through the ring, should you heat (1) the ring, (2) the ball, or (3) both so that the ball will go through the ring? Why? (b) What is the minimum required temperature?

60. ■■ A circular steel plate of radius 0.10 m is cooled from 350°C to 20°C. By what percentage does the plate's area decrease?

61. IE ■■ One morning, an employee at a rental car company fills a car's steel gas tank to the top and then parks the car a short distance away. (a) That afternoon, when the temperature increases, will there be any gas spill or not? Why? (b) If the temperatures in the morning and afternoon are, respectively, 10°C and 30°C and the gas tank can hold 25 gal in the morning, how much gas will be lost? (Neglect the expansion of the tank.)

62. ■■ Show that the density of a solid substance varies with temperature as $\rho = \rho_o(1 - 3\alpha \, \Delta T)$.

63. ■■ The density of mercury is $13.6 \times 10^3 \text{ kg/m}^3$ at 0°C. Calculate its density at 100°C.

64. ■■ A copper block has an internal spherical cavity with a 10-cm diameter (▶Fig. 10.22). The block is heated in an oven from 20°C to 500 K. (a) Does the cavity get larger or smaller? (b) What is the change in the cavity's volume?

65. ■■■ A brass rod has a circular cross section of radius 0.500 cm. The rod fits into a circular hole in a copper sheet with a clearance of 0.010 mm completely around it when

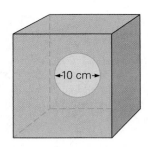

▲ FIGURE 10.22 A hole in a block See Exercise 64.

both it and the sheet are at 20°C. (a) At what temperature will the clearance be zero? (b) Would such a tight fit be possible if the sheet were brass and the rod were copper?

66. ■■■ A Pyrex beaker that has a capacity of 1000 cm³ at 20°C contains 990 cm³ of mercury at that temperature. Is there some temperature at which the mercury will completely fill the beaker? Justify your answer. (Assume that no mass is loss by vaporization.)

10.5 The Kinetic Theory of Gases

67. If the average kinetic energy of the molecules in an ideal gas initially at 20°C doubles, what is the final temperature of the gas? (a) 10°C, (b) 40°C, (c) 313°C, or (d) 586°C.

68. If the temperature of a quantity of ideal gas is raised from 100 K to 200 K, is the internal energy of the gas (a) doubled, (b) halved, (c) unchanged, or (d) none of the preceding?

69. CQ Equal volumes of helium gas (He) and neon gas (Ne) at the same temperature (and pressure) are on opposite sides of a porous membrane (▼Fig. 10.23). Describe what happens after a period of time, and why.

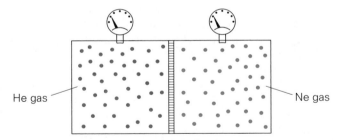

▲ FIGURE 10.23 What happens as time passes?
See Exercise 69.

70. CQ Natural gas is odorless; to alert people to gas leaks, the gas company inserts an additive that has a distinctive scent. When there is a gas leak, the additive reaches your nose before the gas does. What can you conclude about the masses of the additive molecules and gas molecules?

71. ■ What is the average kinetic energy per molecule in an ideal gas at (a) 20°C and (b) 100°C?

72. IE ■ If the Celsius temperature of an ideal gas is doubled, (a) will the internal energy of the gas (1) double, (2) increase by less than a factor of 2, (3) be half as much, or (4) decrease by less than a factor of 2? Why? (b) If the temperature is raised from 20°C to 40°C, by how much will the internal energy of 2.0 mol of an ideal gas change?

73. ■ (a) What is the average kinetic energy per molecule of an ideal gas at a temperature of 27°C? (b) What is the average (rms) speed of the molecules if the gas is helium? (A helium molecule consists of a single atom of mass 6.65×10^{-27} kg.)

74. ■ What is the average speed of the molecules in low-density oxygen gas at 0°C? (The mass of an oxygen molecule, O_2, is 5.31×10^{-26} kg.)

75. ■■ If the temperature of an ideal gas increases from 300 K to 600 K, what happens to the rms speed of the gas molecules?

76. ■■ At a given temperature, which would be greater, the rms speed of oxygen (O_2) or of ozone (O_3), and how many times greater?

77. ■■ If the temperature of an ideal gas is raised from 25°C to 100°C, how many times faster is the new average (rms) speed of the gas molecules?

78. ■■ A quantity of an ideal gas is at 0°C. An equal quantity of another ideal gas is twice as hot. What is its temperature?

10.6 Kinetic Theory, Diatomic Gases, and the Equipartition Theorem

79. Is the temperature of a diatomic molecule like O_2 a measure of its (a) translational kinetic energy, (b) rotational kinetic energy, (c) vibrational kinetic energy, or (d) all of the above?

80. On the average, is the total internal energy of a gas divided equally among (a) each atom, (b) each degree of freedom, (c) linear motion, rotational motion, and vibrational motion, or (d) none of the above?

81. CQ Why does a sample of gas with diatomic molecules have more internal energy than a similar sample with monatomic molecules at the same temperature?

82. ■ If 1.0 mol of ideal monoatomic gas has a total internal energy of 5.0×10^3 J at a certain temperature, what is the total internal energy of 1.0 mol of a diatomic gas at the same temperature?

83. ■ What is the total internal energy of 1.0 mol of O_2 gas at 20°C?

84. ■■ For an average molecule of N_2 gas at 0°C, what are its (a) translational kinetic energy, (b) rotational kinetic energy, and (c) total energy?

Additional Exercises

85. In Exercise 61, if the expansion of the steel gas tank is *not* ignored, what is the volume of gas spilled?

86. To conserve energy, thermostats in an office building are set at 78°F in the summer and 65°F in the winter. What would the settings be if the thermostats had a Celsius scale?

87. A copper wire 0.500 m long is at 20°C. If the temperature is increased to 100°C, what is the change in the wire's length?

88. If 2.0 mol of oxygen gas is confined in a 10-L bottle under a pressure of 6.0 atm, what is the average kinetic energy of an oxygen molecule?

89. A mercury thermometer has a uniform capillary bore whose cross-sectional area is 0.012 mm². The volume of the mercury in the thermometer bulb at 10°C is 0.130 cm³. If the temperature is increased to 50°C, how much will the height of the mercury column in the capillary bore change? (Neglect the expansion of the mercury in the bore and of the glass of the thermometer.)

90. If the rms speed of the molecules in an ideal gas at 20°C increases by a factor of two, what is the new temperature?

91. What temperature increase would produce a stress of 8.0×10^7 N/m² on a rigidly held steel beam?

92. Using the ideal-gas law, express the coefficient of volume expansion in terms of temperature and pressure. (*Hint:* Use different temperatures and pressures.)

93. An ideal gas occupies a container of volume 0.75 L at STP. Find (a) the number of moles and (b) the number of molecules of the gas. (c) If the gas is carbon monoxide (CO), what is its mass?

94. An aluminum tape measure is used to measure a 1.0-m-long steel beam. At 0°C, the measurement is accurate. What will the instrument measure if the temperature increases to 40°C? (Ignore significant figures.)

95. (a) If the temperature drops by 10 C°, what is the corresponding temperature change on the Fahrenheit scale?

(b) If the temperature rises by 10 F°, what is the corresponding change on the Celsius scale?

96. A new metal alloy is made into a 50-cm rod. (a) Determine the alloy's coefficient of linear expansion if after being heated from 20°C to 250°C, the rod has elongated by 0.44 mm. (b) Is the new metal a potentially valuable alloy in terms of linear expansion? Explain.

97. Calculate the number of gas molecules in a container of volume 0.20 m³ filled with gas under a partial vacuum of pressure 20 Pa at 20°C.

98. Is there a temperature that has the same numerical value on the Kelvin and Fahrenheit scales? Justify your answer.

99. Show that the coefficient of area expansion for flat, square solids is approximately equal to 2α.

100. In the troposphere (the lowest part of the atmosphere), the temperature decreases rather uniformly with altitude at a so-called "lapse" rate of about 6.5 C°/km. What are the temperatures (a) near the top of the troposphere (which has an average thickness of 11 km) and (b) outside a commercial aircraft flying at a cruising altitude of 34 000 ft? (Assume that the ground temperature is normal room temperature.)

101. A mercury thermometer has a bulb volume of 0.200 cm³ and a capillary bore diameter of 0.065 mm. How far up the bore will the column of mercury move if the overall temperature of the thermometer is increased by 30 C°? (Neglect the expansion of the glass of the thermometer.)

102. On a hot, 40°C day, copper electric wires are used to connect a power box to a home 150 m away. Assuming the wire would break when the temperature dropped to 10°C, what should the minimum extra length of the wire be to avoid breaking?

103. A quantity of an ideal gas initially at atmospheric pressure is maintained at a constant temperature while it is compressed to half of its volume. What is the final pressure of the gas?

Heat

INSIGHTS

- Phase Changes and Ice Skating
- Physics, the Construction Industry, and Energy Conservation
- The Greenhouse Effect

Most of us know to be very careful about handling anything that has recently been in contact with a flame or other source of heat. Yet, while the metal handle of a pot on the stove may be almost as hot as the bottom of the pot, a wooden pot handle is probably safe to touch. In other words, not only does heat flow through objects, but the rate of flow can vary widely (as the photograph shows in dramatic fashion). Direct contact isn't always necessary for heat to be transmitted: Think of the temperature of the dashboard or the metal seat-belt buckle of a car left in direct sunlight for a few hours. Clearly, they were warmed without direct physical contact with the Sun, but how was heat transferred?

Heat is also crucial to our existence. Our bodies must precisely balance heat loss and gain to stay within the narrow temperature range necessary for life. On a larger scale, the average temperature of the Earth, so critical to our environment and the survival of the organisms that inhabit it, is maintained through a similar balance. We are warmed by the Sun's energy that comes to us across a 150-million-km void. Each day, vast quantities of solar energy reach our planet's atmosphere and surface only to be radiated into space.

These thermal balances are delicate, and any disturbance can have serious consequences. In an individual, sickness can disrupt the thermal balance, producing a chill or fever. Similarly, we worry that an atmospheric buildup of "greenhouse" gases, a product of our industrial society, could significantly raise our planet's average temperature. This change could have a negative effect on all life.

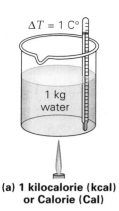

(a) 1 kilocalorie (kcal) or Calorie (Cal)

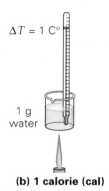

(b) 1 calorie (cal)

1 British thermal unit (Btu)

▲ **FIGURE 11.1 Units of heat**
(a) A kilocalorie raises the temperature of 1 kg of water by 1 C°. **(b)** A calorie raises the temperature of 1 g of water by 1 C°. **(c)** A Btu raises the temperature of 1 lb of water by 1 F°. (Not drawn to scale.)

In this chapter, you'll learn what heat is and how it is measured and study the various mechanisms by which heat passes from one body to another. This knowledge will allow you to explain many everyday phenomena as well as provide a basis for understanding the conversion of thermal energy into useful mechanical work.

11.1 Units of Heat

OBJECTIVES: **To (a) distinguish the various units of heat and (b) define the mechanical equivalent of heat.**

Like work, heat involves a transfer of energy. In the 1800s, it was thought that heat describes the amount of energy an object has, but this is not true. Rather, **heat** is the name we use to describe a type of energy *transfer*. When we refer to "heat" or "heat energy," we mean the amount of energy added to, or removed from, the internal energy of an object due to a temperature difference.

Because heat is energy *in transit*, it is described by standard energy units. As usual, we will use SI units, but for completeness, we define other commonly used units of heat. A chief one is the **kilocalorie (kcal)** (◄Fig. 11.1a):

> One kilocalorie (kcal) is defined as the amount of heat needed to raise the temperature of 1 kg of water by 1 C° (from 14.5°C to 15.5°C).

The kilocalorie is technically known as the 15° kilocalorie. The temperature range is specified because the energy needed varies slightly with temperature—a variation so small that it can be ignored.

For smaller quantities, the **calorie (cal)** is sometimes used (1 kcal = 1000 cal). *One calorie is the amount of heat needed to raise the temperature of 1 g of water by 1 C° (from 14.5°C to 15.5°C)* (Fig. 11.1b).

A familiar use of the larger unit, the kilocalorie, is for specifying the energy values of foods. In this context, the word is usually shortened to *Calorie* (Cal). That is, people on diets really count kilocalories. This quantity refers to the energy that is available for conversion to heat, for mechanical movement, or to increase body mass. (See Example 11.1.) The capital C distinguishes the larger kilogram-Calorie, or kilocalorie, from the smaller gram-calorie, or calorie. They are sometimes referred to as "big Calorie" and "little calorie." (In some countries, the joule is used for food values—see ▶Fig. 11.2.)

A unit of heat sometimes used in American industry is the **British thermal unit (Btu)**. *One Btu is the amount of heat needed to raise the temperature of 1 lb of water by 1 F° (from 63°F to 64°F)* (Fig. 11.1c), and 1 Btu = 252 cal = 0.252 kcal. If you buy an air conditioner or an electric heater, you will find that it is rated in Btu per hour. For example, window air conditioners range from 4000 to 25 000 Btu/hr. This specifies the rate at which the appliance can remove heat.

The Mechanical Equivalent of Heat

The idea that heat is actually a transfer of energy is the result of work by many scientists. Some early observations were made by Benjamin Thompson (Count Rumford) while he was supervising the boring of cannon barrels in Germany. Rumford noticed that water put into the bore of the cannon (to prevent overheating during drilling) boiled away and had to be replenished. The theory of heat at that time pictured it as a "caloric fluid," which flowed from hot objects to colder ones. Rumford did several experiments to detect "caloric fluid" by measuring changes in the weights of heated substances. Since no weight change was ever detected, he concluded that mechanical work was responsible for the heating of the water.

This conclusion was later proven quantitatively by the English scientist James Joule. Using the apparatus shown in ▶Fig. 11.3, Joule demonstrated that when a given amount of mechanical work was done, the water was heated, as indicated by an increase in its temperature. He found that for every 4186 J of work done, the temperature of the water rose 1 C° per kg, or that 4186 J was equivalent to 1 kcal:

$$1 \text{ kcal} = 4186 \text{ J} = 4.186 \text{ kJ} \quad \text{or} \quad 1 \text{ cal} = 4.186 \text{ J}$$

This relationship is called the **mechanical equivalent of heat**. Example 11.1 illustrates an everyday use of these conversion factors.

Example 11.1 ■ Working Off That Birthday Cake: Mechanical Equivalent of Heat to the Rescue

At a birthday party, a student ate a piece of cake (400 Cal). To prevent this energy from being stored as fat, she took a stationary-bicycle workout class the next day. This exercise requires the body to do work at an average rate of 200 watts. How long must the student bicycle to achieve her goal of "working off" the cake's energy?

Thinking It Through. Power is the rate at which the student does work, and the watt (W) is its SI unit (1 W = 1 J/s). To find the time to do this work, we express the Caloric content in joules and use the definition of average power, $\overline{P} = W/t$ (work/time).

Solution. The work required to "burn up" the energy content of the cake is 400 Cal. We list the data given while converting to SI units. (Remember that Cal means kcal.) Thus,

Given: $W = (400 \text{ kcal})\left(\dfrac{4186 \text{ J}}{\text{kcal}}\right) = 1.67 \times 10^6 \text{ J}$ *Find:* Time t to "burn up" 400 Cal

$\overline{P} = 200 \text{ W} = 200 \text{ J/s}$

Rearranging the equation for power, we have

$$t = \frac{W}{\overline{P}} = \frac{1.67 \times 10^6 \text{ J}}{200 \text{ J/s}} = 8.35 \times 10^3 \text{ s} = 139 \text{ min}$$

Follow-up Exercise. If the 400 Cal in this Example were used to increase the students gravitational potential energy, how high would she rise? (Assume her mass is 70 kg.) *(Answers to all Follow-up Exercises are at the back of the text.)*

11.2 Specific Heat

OBJECTIVES: To (a) define specific heat and (b) explain how the specific heats of materials are measured using the technique of calorimetry.

Specific Heats of Solids and Liquids

Recall from Chapter 10 that when heat is added to a solid or liquid, the energy may go toward increasing molecular kinetic energy (temperature change) and also toward increasing the potential energy associated with the molecular bonds. Different substances have different molecular configurations and bonding patterns. Thus, if equal amounts of heat are added to equal masses of different substances, the resulting temperature changes will *not* generally be the same.

The amount of heat (Q) required to change the temperature of a substance is proportional to the mass (m) of the substance and to the change in its temperature (ΔT). That is, $Q \propto m\,\Delta T$, or, in equation form,

$$Q = mc\,\Delta T \quad \text{or} \quad c = \frac{Q}{m\,\Delta T} \quad \textit{specific heat} \quad (11.1)$$

Here, $\Delta T = T_f - T_i$ is the temperature change and c is the *specific heat capacity*, or the **specific heat**. The SI units of specific heat are J/(kg · K) or J/(kg · C°) It is characteristic of the substance and independent of its mass. Thus, the specific heat gives us an indication of a material's internal molecular configuration and bonding.

▲ **FIGURE 11.2   It's a joule!** In Australia, diet drinks are labeled as being "low joule." In Germany, the labeling is a bit more specific: "Less than 4 kilojoules (1 kcal) in 0.3 Liter." How does this labeling compare to that for diet drinks in the United States?

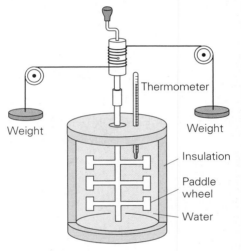

▲ **FIGURE 11.3   Joule's apparatus for determining the mechanical equivalent of heat** As the weights descend, the paddle wheels churn the water, and the mechanical energy, or work, is converted into heat energy, raising the internal energy of the water. For every 4186 J of work done, the temperature of the water rises 1 C° per kilogram. Thus, 4186 J is equivalent to 1 kcal.

TABLE 11.1 Specific Heats of Various Substances (Solids and Liquids) at 20°C and 1 atm

Substance	Specific heat (c)	
	$J/(kg \cdot C°)$	$kcal/(kg \cdot C°)[or\ cal/(g \cdot C°)]$
Solids		
Aluminum	920	0.220
Copper	390	0.0932
Glass	840	0.201
Ice ($-10°C$)	2100	0.500
Iron or steel	460	0.110
Lead	130	0.0311
Soil (average)	1050	0.251
Wood (average)	1680	0.401
Liquids		
Ethyl alcohol	2450	0.585
Glycerin	2410	0.576
Mercury	139	0.0332
Water ($15°C$)	4186	1.000

Note that the *specific heat (in SI) is the heat required to raise the temperature of 1 kg of a substance by 1 C°*. The specific heats of some common substances are given in Table 11.1. Specific heats vary slightly with temperature, but at everyday temperatures can be considered constant.

The greater the specific heat of a substance, the more energy must be transferred to or taken from it to change the temperature of a given mass of the substance. That is, a substance with a higher specific heat has a greater capacity for storage of transferred heat, for a given temperature change (and mass). In Table 11.1, notice that metals have specific heats considerably lower than that of water. Thus it takes only a small amount of heat to produce a large temperature increase in a metal object.

Water has a large specific heat of 4186 $J/(kg \cdot C°)$, or 1.00 kcal/(kg·C°). (The kilocalorie value is exactly 1.00, because the definition of the kilocalorie states that 1 kcal raises the temperature of 1 kg of water by 1 C°.) You have been the victim of the high specific heat of water if you have ever burned your mouth on a baked potato or the cheese of a pizza. These foods have a high water content and a large heat capacity, so they don't cool off as quickly as some other foods do.

Note from Eq. 11.1 that when there is a temperature increase, ΔT is positive ($T_f > T_i$) and Q is positive. This condition corresponds to energy being *added* to a system or object. Conversely, ΔT and Q are negative when energy is *removed from* a system or object. This sign convention will be used throughout this book.

Sign Convention:
$Q > 0$ or $+Q$ means heat added;
$Q < 0$ or $-Q$ means heat removed.

Example 11.2 ■ Cooling Down with Water: Specific Heat

A half-liter of water at 30°C is cooled, with the removal of 63 kJ of heat. What is the final temperature of the water?

Thinking It Through.   Recall that 1.0 L of water has a mass of 1.0 kg; hence the mass of the water is half that, or 0.50 kg. We want T_f—note that it is part of ΔT in Eq. 11.1.

Solution.

Given: $m = 0.50$ kg   *Find:* T_f (final temperature)
$\qquad\ \ T_i = 30$ C°
$\qquad\ \ Q = -63$ kJ (heat is removed)
$\qquad\ \ \ c = 4.186$ kJ/(kg·C°) (from Table 11.1)

With relatively large numbers, it can be convenient to work in kilojoules rather than the standard joules. Writing out the ΔT term of Eq. 11.1 gives

$$Q = mc\,\Delta T = mc\,(T_f - T_i)$$

Solving for T_f gives

$$T_f = \frac{Q}{mc} + T_i = \frac{-63\text{ kJ}}{(0.50\text{ kg})[4.186\text{ kJ}/(\text{kg}\cdot\text{C}^\circ)]} + 30^\circ\text{C} = -30^\circ\text{C} + 30^\circ\text{C} = 0^\circ\text{C}$$

The (liquid) water is at its freezing point; the removal of more heat would cause the water to freeze (Section 11.3).

Follow-up Exercise.   (a) Suppose in this Example that 30 kJ of heat were initially removed and then 60 kJ were added. What would be the final temperature of the water? (b) Determine the heat required to raise the temperature of 2.0 L of water initially at 20°C to the boiling point (at 1 atm).

Integrated Example 11.3 ■ Cooking Class 101: Studying Specific Heats While Boiling Water

To prepare pasta, you bring a pot of water from room temperature (20°C) to its boiling point (100°C). The pot has a mass of 0.900 kg, is made of steel, and holds 3.00 L of water. (a) Which of the following is true? (1) The pot requires more heat, (2) the water requires more heat, or (3) they require the same amount of heat. (b) Determine the required heat for both the water and the pot, and the ratio Q_w/Q_{pot}.

(a) Conceptual Reasoning.   The temperature increase is the same for the water and the pot. Thus, the only factors that determine the difference in required heat are mass and specific heat. Since a liter of water has a mass of 1.0 kg, we have 3.0 kg of water to heat. This mass is more than three times the mass of the pot. From Table 11.1, the specific heat of water is about nine times larger than that of steel. Thus, both factors indicate that the water will require significantly more heat than the pot, so the answer is (2).

(b) Thinking It Through.   The heats can be found using Eq. 11.1, after looking up the specific heats and converting the volume of water to mass. The temperature change is easily determined from the initial and final values.

We list the data given and find the water's mass from its volume and density:

Given: $m_{pot} = 0.900$ kg   *Find:*   The heat for the water and the
  $m_w = 3.00$ kg   pot and the heat ratio Q_w/Q_{pot}
  $c_{pot} = 460$ J/kg·C° (From Table 11.1)
  $c_w = 4186$ J/kg·C° (From Table 11.1)

In general, the amount of heat required is given by $Q = mc\,\Delta T$. The temperature increase for both objects is 80 C°. Thus, the heat required for the water is

$$Q_w = m_w c_w \Delta T_w$$
$$= (3.00\text{ kg})(4186\text{ J}/\text{kg}\cdot\text{C}^\circ)(80\text{ C}^\circ) = 1.00 \times 10^6\text{ J}$$

and the heat required for the pot is

$$Q_{pot} = m_{pot} c_{pot} \Delta T_{pot}$$
$$= (0.900\text{ kg})(460\text{ J}/\text{kg}\cdot\text{C}^\circ)(80\text{ C}^\circ) = 3.31 \times 10^4\text{ J}$$

The water requires over 30 times the heat, since

$$\frac{Q_w}{Q_{pot}} = \frac{1.00 \times 10^6\text{ J}}{3.31 \times 10^4\text{ J}} = 30.2$$

Follow-up Exercise.   (a) In this Example, if the pot were made of the same mass of aluminum, would you expect the ratio of heat (water to pot) to be smaller or larger than the answer for the steel pot? Explain. (b) Verify your choice by calculating this ratio for the case of the aluminum pot.

▲ FIGURE 11.4 Calorimetry apparatus The calorimetry cup (center, with black insulating ring), goes into the larger container. The cover with the thermometer and stirrer is seen at the right. Metal shot or pieces of metal are heated in the small cup (with the handle) that is inserted into the hole at the top of the steam generator on the tripod.

Calorimetry

Calorimetry is the quantitative measure of heat exchange. Such measurements are made by using a *calorimeter* (cal-oh-RIM-i-ter), an insulated container that allows little heat loss to the environment (ideally none). A simple laboratory calorimeter is shown in ◄Fig. 11.4.

The specific heat of a substance can be determined by measuring the quantities in Eq. 11.1 ($Q = mc\Delta T$)* other than c. A substance of known mass and temperature is put into a quantity of water in a calorimeter. The water is at a different temperature than that of the mass, usually a lower one. The principle of the conservation of energy is then applied to determine c. This procedure is called the *method of mixtures*. Example 11.4 illustrates the use of this procedure. Such heat-exchange problems are just a matter of "thermal accounting." Because of the conservation of energy, if something loses heat $(-Q)$, something else must gain an equal, amount of heat $(+Q)$ —that is, $\Sigma Q_i = 0$.

Example 11.4 ■ Calorimetry: The Method of Mixtures

Students in a physics lab are to determine the specific heat of copper experimentally. They heat 0.150 kg of copper shot to 100°C and then carefully pour the hot shot into a calorimeter cup (Fig. 11.4) containing 0.200 kg of water at 20°C. The final temperature of the mixture in the cup is measured to be 25°C. If the aluminum cup has a mass of 0.045 kg, what is the specific heat of copper? (Assume that there is no heat loss to the surroundings.)

Thinking It Through.   The conservation of heat energy is involved: $\Sigma Q_i = 0$, taking into account the correct ± signs. In calorimetry problems, it is important to identify and label all of the quantities with proper signs. Identification of the heat gains and losses is crucial. You will probably use this method in the laboratory.

Solution.   We will use the subscripts Cu, w, and Al to refer to the copper, water, and aluminum calorimeter cup, respectively, and the subscripts *h, i,* and *f* to refer to the temperature of the *h*ot metal shot, the water (and cup) *i*nitially at room temperature, and the *f*inal temperature of the system, respectively. With this notation, we have

Given: $m_{Cu} = 0.150$ kg   *Find:* c_{Cu} (specific heat)
$m_w = 0.200$ kg
$c_w = 4186$ J/(kg·C°) (from Table 11.1)
$m_{Al} = 0.045$ kg
$c_{Al} = 920$ J/(kg·C°) (from Table 11.1)
$T_h = 100°C$, $T_i = 20°C$, and $T_f = 25°C$

If no heat is lost from the system, the system's total energy is conserved: $\Sigma Q_i = 0$, and

$$\Sigma Q_i = Q_w + Q_{Al} + Q_{Cu} = 0$$

or

$$Q_w + Q_{Al} = -Q_{Cu}$$

Substituting the relationship in Eq. 11.1 for these heats, we have

$$m_w c_w \Delta T_w + m_{Al} c_{Al} \Delta T_{Al} = -m_{Cu} c_{Cu} \Delta T_{Cu}$$

or

$$m_w c_w (T_f - T_i) + m_{Al} c_{Al} (T_f - T_i) = -m_{Cu} c_{Cu} (T_f - T_h)$$

Here, the water and aluminum cup, initially at T_i, are heated to T_f, so $\Delta T_w = \Delta T_{Al} = (T_f - T_i)$; and the copper initially at T_h is cooled to T_f, so $\Delta T_{Cu} = (T_f - T_h)$. Solving for c_{Cu} gives

PHYSLET®
ILLUSTRATION

Specific Heat

---

*In this section, calorimetry will *not* involve "phase changes" such as ice melting or water boiling. These effects are discussed in Section 11.3.

$$c_{\text{Cu}} = \frac{(m_w c_w + m_{A1} c_{A1})(T_f - T_i)}{-m_{\text{Cu}}(T_f - T_h)}$$

$$= \frac{\{(0.200 \text{ kg})[4186 \text{ J/(kg} \cdot \text{C°)}] + (0.045 \text{ kg})[920 \text{ J/(kg} \cdot \text{C°)}]\}(25°C - 20°C)}{(-0.150 \text{ kg})(25°C - 100°C)}$$

$$= 390 \text{ J/(kg} \cdot \text{C°)}$$

Notice that the proper use of signs resulted in a positive answer for c_{Cu}, as required. If, for example, the Q_{Cu} term in the second equation had not had the correct sign, the answer would have been negative—a big clue that you had an initial sign error.

Follow-up Exercise. In this example, what would the final equilibrium temperature be if the calorimeter (water and cup) initially had been at a warmer 30°C?

Specific Heat of Gases

When heat is added to or taken from most materials, the materials expand or contract. During expansion, for example, the materials do work on the atmosphere. Fortunately, for most solids and liquids, this work is negligible, which is why we didn't include it in our discussion of specific heat of solids and liquids.

However, for gases, expansion and contraction can be significant. It is therefore important to specify the *conditions* under which heat is transferred to a gas. If heat is added to a gas at constant volume (a *rigid* container), the gas does no work. (Why?) In this case, all of the heat goes into increasing the gas's internal energy and, therefore, to increasing its temperature. However, if the same amount of heat is added at constant pressure (a *nonrigid* container), a portion of the heat is converted to work as the gas expands. Thus, not all of the heat will go into the gas's internal energy. This process results in a smaller temperature change than occurred during the constant-volume process.

To designate these physical quantities that are held constant while heat is added or removed from a gas, we use a subscript notation: c_p means "specific heat under conditions of constant pressure (p)," and c_v means "specific heat under conditions of constant volume (v)."

For solids and liquids, we use mass to designate different amounts. However, for gases, we commonly use the number of moles (n) to describe the amount of gas. Specific heats for gases are on a *per mole* basis. (See Section 10.3.) Mathematically, the specific heats for a gas are defined as

$$c_p = \left(\frac{Q}{n \Delta T}\right)_p \quad \textit{(molar specific heat at constant pressure)} \quad \text{(11.2a)}$$

and

$$c_v = \left(\frac{Q}{n \Delta T}\right)_v \quad \textit{(molar specific heat at constant volume)} \quad \text{(11.2b)}$$

From these definitions, the SI units of molar specific heat are joules per mole per kelvin [J/(mol·K)].

Table 11.2 gives values for molar specific heats for common gases. Even a quick look shows that c_p is greater than c_v.* This is true because, at *constant volume*, all of the heat goes into internal energy. Thus, ΔT_v is as large as it can be. At *constant pressure*, however, a gas's temperature rise is less, or $\Delta T_p < \Delta T_v$. Thus, c_p is greater than c_v. As a practical application of the specific heats of gases, consider the following Example.

*The observant student will notice in Table 11.2 that the specific heats depend on the number of atoms per molecule. This is not a coincidence, but its explanation is beyond the scope of this book.

TABLE 11.2 Molar Specific Heats of Various Gases [J/(mol·K)] at 20°C and 1 atm

Gas	c_p	c_v
Monatomic Gases		
Helium	20.8	12.5
Argon	20.8	12.5
Neon	20.8	12.6
Krypton	20.8	12.4
Diatomic Gases		
Hydrogen (H_2)	28.8	20.4
Nitrogen (N_2)	29.1	20.8
Oxygen (O_2)	29.4	21.1
Carbon monoxide (CO)	34.7	25.7
Triatomic Gases		
Carbon dioxide (CO_2)	37.1	28.5
Sulfur dioxide (SO_2)	40.4	31.4
Water vapor (H_2O)	35.4	27.0

Example 11.5 ■ Warming a Room for Free?: Solving the Energy Crisis with Human Heat

The average human body gives off heat at a rate of about 100 W. A law office with three clerks measures 4.0 m wide, 6.0 m long, and 3.0 m high. Assume that only one third of the radiated body heat in this office goes into warming the air and that the air was a chilly 15°C (59°F) when work began at 8 AM. Assume further that the office remains at normal atmospheric pressure the entire day and that no heating systems are turned on before 10 AM.

Thinking It Through. Air is composed chiefly of nitrogen and oxygen, both diatomic molecules, so we can estimate the molar specific heat (at constant pressure) of air from Table 11.2. The room's volume enables us to estimate the number of moles of gas in the room at 8 AM, and we will neglect leakage as the air warms and expands. From the specific heat and number of moles, we can determine the temperature rise

Solution. We list the data given, converting the dimensions into room volume:

Given: $V = 4.0\text{ m} \times 6.0\text{ m} \times 3.0\text{ m} = 72\text{ m}^3$
$P = \frac{1}{3} \times 300\text{ W} = 100\text{ J/s}$
$t = 2.0\text{ h} = 7.2 \times 10^3\text{ s}$
$c_p = 29.2\text{ J/mol·K}$ (from Table 11.2, weighing toward the nitrogen)
$T_i = 15°\text{C} = 288\text{ K}$

Find: T_f, the final air temperature after two hours

The amount of heat energy absorbed in the 2-hour interval will be

$$Q = P \times t = (100\text{ J/s})(7.2 \times 10^3\text{ s}) = 7.2 \times 10^5\text{ J}$$

We can use $pV = nRT$ to solve for the number of moles, or

$$n = \frac{pV}{RT} = \frac{(1.01 \times 10^5\text{ N/m}^2)(72\text{ m}^3)}{(8.31\text{ J/mol·K})(288\text{ K})} = 3.0 \times 10^3\text{ mol}$$

Finally, from $c_p = (Q/n\Delta T)_p$, we solve for ΔT (recall 1 K = 1 C°):

$$\Delta T = \frac{Q}{nc_p} = \frac{7.2 \times 10^5\text{ J}}{(3.0 \times 10^3\text{ mol})(29.2\text{ J/mol·K})} = 8.2\text{ K} = 8.2\text{ C}°$$

Hence, $T_f = 15°\text{C} + 8\text{ C}° = 23°\text{C}$, or about 73°F. So by 10 AM, the room may even be a bit warm for most peoples' preferences—all from body heat!

Follow-up Exercise. In the office in this Example, suppose one of the workers had used a copier for 15 minutes. If the copier was rated at 200 W, and half of its energy was absorbed by the air, how much more would the room's temperature have risen?

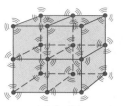

(a) Solid

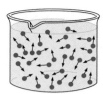

(b) Liquid

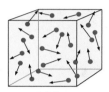

(c) Gas

11.3 Phase Changes and Latent Heat

OBJECTIVES: **To (a) compare and contrast the three common phases of matter and (b) relate latent heat to phase changes.**

Matter normally exists in one of three *phases*: solid, liquid, or gas (►Fig. 11.5). The phase that a substance is in depends on the substance's internal energy (as indicated by its temperature) and the pressure on it. However, it is likely that you more readily think of adding or removing heat to change the phase of a substance.

In the **solid phase**, molecules are held together by attractive forces, or bonds. Simplistically, these bonds can be represented as springs (Fig. 11.5a). Adding heat causes increased motion about the molecular-equilibrium positions. If enough heat is added to provide sufficient energy to break the intermolecular bonds, the solid undergoes a phase change and becomes a liquid. The temperature at which this phase change occurs is called the **melting point**. The temperature at which a liquid becomes a solid is called the **freezing point**. In general, these temperatures are the same for a given substance, but they can differ slightly.

In the **liquid phase**, molecules of a substance are relatively free to move and a liquid assumes the shape of its container (Fig. 11.5b). In certain liquids, there may be some locally ordered structure, giving rise to liquid crystals, such as those used in LCDs (liquid crystal displays) of calculators and clocks (Chapter 24).

Adding even more heat increases the motion of the molecules of a liquid. When they have enough energy to become separated, the liquid changes to the **gaseous** (or **vapor***) **phase**. This change may occur slowly, by *evaporation* (p. 387), or rapidly at a particular temperature called the **boiling point**. The temperature at which a gas condenses into a liquid is the **condensation point**.

Some solids, such as dry ice (solid carbon dioxide), mothballs, and certain air fresheners, change directly from the solid to the gaseous phase. This process is called **sublimation**. Like the rate of evaporation, the rate of sublimation increases with temperature. A phase change from a gas to a solid is called *deposition*. Frost, for example, is solidified water vapor deposited on grass, car windows, and other objects. Frost is not frozen dew, as is sometimes mistakenly assumed.

Latent Heat

In general, when heat is transferred to a substance, the substance's temperature increases as the average kinetic energy per molecule increases. However, when heat added (or removed) causes only a phase change, the temperature of the substance does *not* change. For example, if heat is added to a quantity of ice at −10°C, the temperature of the ice increases until it reaches its melting point of 0°C. At this point, the addition of more heat does not increase the temperature, but causes the ice to melt, or change phase. (The heat must be added slowly so that the ice and melted water remain in thermal equilibrium.) Only after the ice is completely melted does adding more heat cause the temperature of the water to rise. A similar situation occurs during the liquid–gas phase change at the boiling point. Adding more heat to boiling water only causes more vaporization. A temperature increase occurs only *after* the water is completely boiled, resulting in *superheated steam*.

▲ **FIGURE 11.5 Three phases of matter** **(a)** The molecules of a solid are held together by bonds; consequently, a solid has a definite shape and volume. **(b)** The molecules of a liquid can move more freely, so a liquid has a definite volume and assumes the shape of its container. **(c)** The molecules of a gas interact weakly and are separated by relatively large distances; thus, a gas has no definite shape or volume, unless it is confined in a container.

Note: Solid, liquid, and gas are sometimes referred to as states of matter rather than phases of matter, but the *state* of a system has a different meaning in physics, as you will learn in Chapter 12.

Note: Keep in mind that ice can be colder than 0°C.

*The terms *vapor* and *vapor phase* are sometimes used interchangeably with the term *gaseous phase*. Strictly speaking, *vapor* refers to the gaseous phase of a substance in contact with its liquid phase.

During a phase change, the heat goes into breaking bonds and separating molecules (increasing their potential, rather than kinetic, energies) rather than into increasing the temperature. The heat in a phase change is called the **latent heat** (L)*, which is defined as the magnitude of the heat needed per unit mass:

$$L = \frac{|Q|}{m} \quad latent\ heat \tag{11.3}$$

where m is the mass of the substance. Latent heat has units of joules per kilogram (J/kg) in the SI, or kilocalories per kilogram (kcal/kg).

The latent heat for a solid–liquid phase change is called the **latent heat of fusion** (L_f), and that for a liquid–gas phase change is called the **latent heat of vaporization** (L_v). These quantities are often referred to as simply the *heat of fusion* and the *heat of vaporization*. The latent heats of some substances, along with their melting and boiling points, are given in Table 11.3. (The latent heat for the less common solid–gas phase change is called the *latent heat of sublimation* and is symbolized by L_s). As you might expect from the conservation of energy, the latent heat (in joules per kilogram) is the amount of energy per kilogram given up when the phase change is in the opposite direction, from liquid to solid or gas to liquid.

A more useful form of Eq. 11.3 is given by solving for Q and including a plus/minus sign for the two possible directions of heat flow:

$$Q = \pm mL \quad (signs\ with\ latent\ heat) \tag{11.3}$$

This equation is more practical for problem solving, because in calorimetry problems, you are typically interested in applying conservation of energy in the form $\Sigma Q_i = 0$. The plus/minus sign $(\pm)$ is explicitly expressed because heat can flow either into $(+)$ or out of $(-)$ the object or system of interest.

When solving calorimetry problems involving phase changes, you must be careful to use the correct sign, in agreement with our sign conventions (▶Fig. 11.6). For example, if water is condensing from steam into liquid droplets, removal of heat is involved, necessitating the choice of the minus sign.

Problem-Solving Hint

Recall in Section 11.2 that where there were no phase changes, the expression for heat $Q = mc\,\Delta T$ *automatically* gave the correct sign for Q from the sign of ΔT. But there is no ΔT during a phase change. *Choosing the correct sign is up to you!*

TABLE 11.3 Temperatures of Phase Changes and Latent Heats for Various Substances (at 1 atm)

Substance	Melting point	L_f J/kg	kcal/kg	Boiling point	L_v J/kg	kcal/kg
Alcohol, ethyl	−114°C	1.0×10^5	25	78°C	8.5×10^5	204
Gold	1063°C	0.645×10^5	15.4	2660°C	15.8×10^5	377
Helium†	—	—	—	−269°C	0.21×10^5	5
Lead	328°C	0.25×10^5	5.9	1744°C	8.67×10^5	207
Mercury	−39°C	0.12×10^5	2.8	357°C	2.7×10^5	65
Nitrogen	−210°C	0.26×10^5	6.1	−196°C	2.0×10^5	48
Oxygen	−219°C	0.14×10^5	3.3	−183°C	2.1×10^5	51
Tungsten	3410°C	1.8×10^5	44	5900°C	48.2×10^5	1150
Water	0°C	3.33×10^5	80	100°C	22.6×10^5	540

†Not a solid at a pressure of 1 atm; melting point is −272°C at 26 atm.

Latent is Latin for "hidden"; Q seems to be disappearing with no change in temperature.

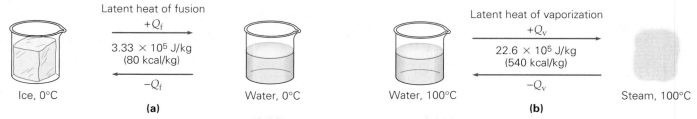

(a)

(b)

▲ **FIGURE 11.6   Phase changes and latent heats   (a)** At 0°C, 3.33 × 10⁵ J must be added to 1 kg of ice or removed from 1 kg of liquid water to change its phase. **(b)** At 100°C, 22.6 × 10⁵ J must be added to 1 kg of liquid water or removed from 1 kg of steam to change its phase.

It is helpful to focus on the fusion and vaporization of water. A plot of temperature versus heat for 1 kg of water is shown in ▼Fig. 11.7. When heat is added (or removed) at the temperature of a phase change, 0°C or 100°C, the temperature remains constant. Once the phase change is complete, adding more heat causes the temperature to increase. The slopes of the lines in Fig. 11.7 are not all the same, which indicates that the specific heats of the various phases are not the same. (Why?)

For water, the latent heats of fusion and vaporization are

$$L_f = 3.33 \times 10^5 \text{ J/kg} \quad \text{(or about 80 kcal/kg)}$$
$$L_v = 22.6 \times 10^5 \text{ J/kg} \quad \text{(or about 540 kcal/kg)}$$

Example 11.6 shows explicitly the two types of heat terms that must be calculated in the general situation when an object undergoes a temperature change either preceding or following a phase change.

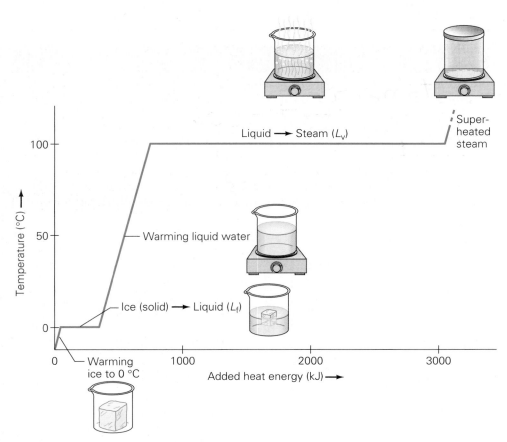

◀ **FIGURE 11.7   Temperature versus heat added for water**   As heat is added to the various phases of water, the temperature of the water increases. During a phase change, however, the heat energy does work to separate the molecules (break bonds), and the temperature remains constant. Note the different slopes of the phase lines, which indicate different values of specific heat (ice versus liquid water versus superheated steam). (Drawn to scale for 1 kg of water.)

Example 11.6 ■ Temperature Changes and Phase Changes: Specific Heat and Latent Heat

Heat is added to 0.500 kg of water at room temperature (20°C). How much heat is required to change the water to superheated steam at 110°C?

Thinking It Through. Three steps are involved: heating water (specific heat), a phase change (latent heat), and heating steam (specific heat).

Solution.

Given: $m = 0.500$ kg   *Find:* Q (total heat in joules)
         $T_i = 20°C$
         $T_f = 110°C$
         $L_v = 22.6 \times 10^5$ J/kg   (from Table 11.3)
         (specific heats from Table 11.1)

The temperature intervals and the heats required in the process are shown in ▼Fig. 11.8. First, the water is heated to the boiling point (100°C); next, a phase change occurs; and finally, the steam is heated to 110°C. The Q's for the various steps are

(*heating water*) $Q_1 = mc_w\Delta T_1 = (0.500 \text{ kg})[4186 \text{ J}/(\text{kg}\cdot\text{C°})](100°C - 20°C)$

                     $= +1.67 \times 10^5$ J

(*phase change*) $Q_v = +mL_v = (0.500 \text{ kg}) (22.6 \times 10^5 \text{ J/kg}) = +11.3 \times 10^5$ J

(*heating steam*) $Q_2 = mc_s\Delta T_2 = (0.500 \text{ kg})[2010 \text{ J}/(\text{kg}\cdot\text{C°})](110°C - 100°C)$

                     $= +0.101 \times 10^5$ J

The total heat required is

$$Q = \Sigma Q_i = (1.7 + 11.3 + 0.1) \times 10^5 \text{ J} = +13.1 \times 10^5 \text{ J}$$

Follow-up Exercise. How much heat must a refrigerator freezor remove from liquid water (initally at 20°C) to create 0.250 kg of ice at −10°C?

Problem-Solving Hint

Note that you must compute the latent heat at each phase change. It is a common error to use the specific-heat equation with a temperature interval that includes a phase change. Also, you cannot assume a complete phase change until you have checked for it numerically. (See Example 11.7.)

Note that the freezing and boiling points of 0°C and 100°C, respectively, for water apply at 1 atm of pressure. The phase-change temperatures vary with pressure. For example, the boiling point of water decreases with decreasing pressure.

▶ **FIGURE 11.8 Steps in changing phases** In going from liquid water at 20°C to steam at 110°C, three heat-adding steps are involved. See Example 11.6.

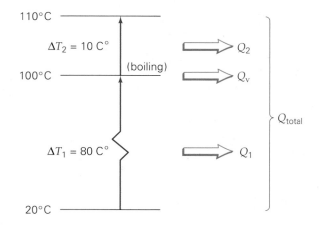

At high altitudes, where there is lower atmospheric pressure, the boiling point of water is lowered. For example, at Pike's Peak, Colorado, at an elevation of about 4300 m, the atmospheric pressure is about 0.79 atm and water boils at about 94°C rather than at 100°C. The lower temperature lengthens the cooking time of food. A pressure cooker can be used to reduce the cooking time (at Pike's Peak or at sea level)—by increasing the pressure, a pressure cooker raises the boiling point.

Conversely, the freezing point of water decreases with increasing pressure. This inverse relationship is characteristic of only a very few substances, including water (Section 10.4), that expand when they freeze. Pressure and ice skating are discussed in the Insight on page 386.

Example 11.7 ■ Practical Calorimetry: Using Phase Changes to Save a Life

Organ transplants are becoming commonplace. Many times, the procedure involves removing a healthy organ from a deceased person and flying it to the recipient. During that time, to prevent its deterioration, the organ is packed in ice in an insulated container. Assume that the liver has a mass of 0.500 kg and is initially at 29°C. The specific heat of the human liver is 3500 J/(kg·C°). The liver is surrounded by 2.00 kg of ice initially at −10°C. Calculate the final equilibrium temperature.

Thinking It Through. Clearly, the liver will cool, and the ice will warm. However, it is not clear what temperature the ice will reach. If it gets to the freezing point, it will begin to melt, and we must then consider a phase change. If all of it melts, then we must in addition consider the heat required to warm that water to a temperature above 0°C. Thus, care must be taken here, since it *cannot* be assumed that all the ice melts, or even that the ice reaches its melting point. Hence, we cannot write down the calorimetry equation (conservation of energy) until we determine which terms are in it. First wse need to review the *possible* heat transfers. Only then can the final temperature be determined.

Solution. Listing the data given and the information obtained from tables, we have

Given: $m_\ell = 0.500$ kg   *Find:* The final temperature
$m_{ice} = 2.00$ kg   of the system
$c_\ell = 3500$ J/(kg·C°)
$c_{ice} = 2100$ J/(kg·C°) (from Table 11.1)
$L_f = 3.33 \times 10^5$ J/(kg·C°) (from Table 11.3)

The amount of heat required to bring the ice to 0°C from −10°C is

$$Q_{ice} = m_{ice}c_{ice}\Delta T_{ice} = (2.00 \text{ kg})(2100 \text{ J/kg·C°})(+10 \text{ C°}) = +4.20 \times 10^4 \text{ J}$$

Since this heat must come from the liver, we need to calculate the maximum heat available from the liver, that is, if its temperature drops all the way from 29°C to 0°C:

$$Q_{\ell,max} = m_\ell c_\ell \Delta T_{\ell,max} = (0.500 \text{ kg})(3500 \text{ J/kg·C°})(-29 \text{ C°}) = -5.08 \times 10^4 \text{ J}$$

This is enough to bring the ice to 0°C. If 4.20×10^4 J flows into the ice (bringing it to 0°C), the liver is still not at 0°C. How much ice melts? This depends on how much more heat can be transferred from the liver.

How much more heat Q' would be extracted from the liver if its temperature were to drop to 0°C? This value is just the maximum amount minus the heat that went into warming the ice, or

$$Q' = |Q_{\ell,max}| - 4.20 \times 10^4 \text{ J}$$
$$= 5.08 \times 10^4 \text{ J} - 4.20 \times 10^4 \text{ J} = 8.8 \times 10^3 \text{ J}$$

Compare this with the magnitude of the heat needed to melt the ice completely ($|Q_{melt}|$) to decide if this can, in fact, happen. The heat required to melt all the ice is

$$|Q_{melt}| = +m_{ice}L_{ice} = +(2.00 \text{ kg})(3.33 \times 10^5 \text{ J/kg}) = +6.66 \times 10^5 \text{ J}$$

Phase Changes and Ice Skating

It was once thought that the lowering of the freezing point of water by pressure provided the mechanism that makes ice skating possible. Supposedly, the pressure of the narrow skate blade on the ice would lower the melting point below the ambient temperature. The skater would thus glide on a thin film of water, which quickly refroze when the pressure was removed. For water, a pressure of about 120 atm is needed to lower the freezing point by 1 C°.

A typical skate blade is about 25 cm long and is "hollow ground." This means that each blade actually consists of two very narrow (0.1 mm) wide "runners." (See Fig. 1.) Thus, the total area of one skate in contact with the ice is $2 \times L \times w = 2 \times (0.25 \, \text{m}) \times (10^{-4} \, \text{m}) = 5 \times 10^{-5} \, \text{m}^2$. For a 70-kg skater with all the weight on one blade, the pressure would be about

$$p = \frac{F}{A} = \frac{(70 \, \text{kg})(9.8 \, \text{m/s}^2)}{(5 \times 10^{-5} \, \text{m}^2)} = 1.4 \times 10^7 \, \text{Pa}$$

or about 140 atm. This pressure would lower the melting point of the ice by a bit more than 1C°. However, since outdoor rinks (and even most indoor ones) are kept at well below −1°C, this temperature drop alone cannot explain the phenomenon of low-friction skating.

Frictional heating between the skate blade and the ice does contribute to melting. For high-friction surfaces, such as skis on snow, this condition is the main mechanism allowing rapid movement. However, for ice skating, another factor contributes to the low coefficient of friction of ice. This factor is called *surface melting*, proposed by the English scientist Michael Faraday (1791–1867; (see Chapter 20). He

suggested that a thin layer of liquid normally exists on the surface of a solid even at temperatures well below the solid's melting point. Modern techniques have shown this phenomenon to be the case for most solids. For ice, the thickness of the water film is about 40 nm near 0°C and about 0.50 nm near −35°C. Recall that $1 \, \text{nm} = 10^{-9} \text{m}$. Atomic diameters are on the order of 0.1 nm; thus, this film is only hundreds of atoms thick. (See Exercise 87, at the end of the chapter.)

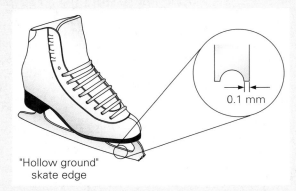

"Hollow ground"
skate edge

FIGURE 1   Where the blade meets the ice   Actual skates are "hollow ground." (See expanded inset.) This term means that only a very small area (with a width of perhaps only 0.10 mm) actually touches the ice, much smaller than you would think from looking at the thickness of the blade.

Since this amount of heat is much larger than the amount available from the liver, only part of the ice melts. In the process, the liver has dropped to 0°C, and the remainder of the ice is at 0°C. Since everything in the "calorimeter" is at the same temperature, heat flow stops, and our final temperature is 0°C. Thus the final result is that the liver is in a bag containing ice and some liquid water, all at 0°C. Since the container is a very good insulator, it will prevent any inward heat flow, which could raise the liver's temperature. We therefore expect that the liver will arrive at its destination in good shape.

Follow-up Exercise.   (a) In this Example, how much of the ice melts? (b) If the ice originally had been at its melting point (0°C), what would the equilibrium temperature have been?

Problem-Solving Hint

Notice in Example 11.7 that numbers were *not* plugged directly into $\Sigma Q_i = 0$, assuming that all the ice melts. In fact, if this step had been done, we would have been on the wrong track. For calorimetry problems *involving phase changes*, a careful step-by-step numerical "accounting" procedure must be followed until all of the pieces of the system are at the same temperature. At that point, the problem is over, because no more heat exchanges can happen.

Evaporation

The **evaporation** of water from an open container becomes evident only after a relatively long period of time. This phenomenon can be explained in terms of the kinetic theory (Section 10.5). The molecules in a liquid are in motion at different speeds. A faster moving molecule near the surface may momentarily leave the liquid. If its speed is not too large, the molecule will return to the liquid, because of the attractive forces exerted by the other molecules. Occasionally, however, a molecule has a large enough speed to leave the liquid entirely. The higher the temperature of the liquid, the more likely this phenomenon is to occur.

The escaping molecules take their energy with them. Since those molecules with greater than average energy are the ones most likely to escape, the average molecular energy, and thus the temperature of the remaining liquid, will be reduced. Thus, *evaporation is a cooling process* for the object from which the molecules escape. You have probably noticed this phenomenon when drying off after a bath or shower.

Although evaporation is a relatively slow process, it is often important in preventing our bodies from overheating. Usually, radiation, natural convection, conduction (discussed in Section 11.4), and possibly slight perspiration are sufficient to maintain a rate of heat loss that keeps us comfortable, given the temperature difference between our bodies and our surroundings. However, when the air gets hot, the temperature difference that enables these mechanisms to work narrows (or disappears), and we start to perspire.

The evaporation of perspiration lowers the temperature of perspiration on our bodies, which can then draw heat from our skin and thus cool our bodies. On a hot summer day, a person may stand in front of a fan and remark how cool the blowing air feels. But the fan is merely blowing hot air from one place to another. The air feels cool because it is relatively dry and its flow promotes evaporation, which removes latent-heat energy. Of course, evaporation depends on the humidity (the amount of moisture already in the air). Without a fan, we would be less comfortable on hot, humid days, because evaporation is reduced. (Why?)

11.4 Heat Transfer

OBJECTIVES: To (a) describe the three methods of heat transfer and (b) give practical and/or environmental examples of each.

The transfer of heat is an important topic and has many practical applications. Heat can move from place to place by three different mechanisms: conduction, convection, or radiation.

Conduction

You can keep a pot of coffee hot on a hot stove because heat is conducted through the bottom of the coffee pot from the burner. The process of **conduction** is visualized as resulting from molecular interactions. Molecules in one part of a body at a higher temperature vibrate faster. They collide with, and transfer some of their energy to, the less energetic molecules located toward the cooler part of the body. In this way, energy is conductively transferred from a higher temperature region to a lower temperature region—transfer as a result of a temperature difference.

Solids can be divided into two general categories: metals and nonmetals. Metals are good conductors of heat, or **thermal conductors**. Metals have a large number of electrons that are free to move around (not permanently bound to a particular molecule or atom). These free electrons (rather than the interaction between adjacent atoms) are primarily responsible for the good heat conduction in metals. Nonmetals, such as wood and cloth, have relatively few free electrons. The absence of this transfer mechanism makes them poor heat conductors relative to metals. A poor heat conductor is called a **thermal insulator**.

In general, the ability of a substance to conduct heat depends on the substance's phase. Gases are poor thermal conductors; their molecules are relatively far apart, and collisions are therefore infrequent. Liquids and solids are better thermal conductors than gases, because their molecules are closer together and can interact more readily.

Heat conduction can be described quantitatively as the time rate of heat flow ($\Delta Q/\Delta t$) in a material for a given temperature difference (ΔT), as illustrated in ▼Fig. 11.9. Experiment has established that the rate of heat flow through a substance depends on the temperature difference between its boundaries. Heat conduction also depends on the size and shape of the object. In our analysis of heat flow, we will use a uniform slab of the substance.

Experimentally, it is found that the heat flow rate ($\Delta Q/\Delta t$ in J/s) through a slab of material is directly proportional to the material's surface area (A) and inversely proportional to its thickness (d). That is,

$$\frac{\Delta Q}{\Delta t} \propto \frac{A\,\Delta T}{d}$$

The ratio $\Delta T/d$ is called the *thermal gradient* (the change in temperature per unit of length). Using a constant of proportionality allows us to write the relation as an equation:

$$\frac{\Delta Q}{\Delta t} = \frac{kA\,\Delta T}{d} \qquad (conduction\ only) \qquad (11.4)$$

The constant k, called the **thermal conductivity**, characterizes the heat-conducting ability of a material. The greater the value of k for a material, the more rapidly it will conduct heat. The units of k are $J/(m \cdot s \cdot C°) = W/(m \cdot C°)$ or $kcal/(m \cdot s \cdot C°)$. The thermal conductivities of various substances are listed in Table 11.4. These values vary slightly over different temperature ranges, but can be considered to be constant over normal temperature ranges.

Compare the relatively large thermal conductivities of the good thermal conductors, the metals, with the relatively small thermal conductivities of some good thermal insulators, such as Styrofoam™ and wood. Some cooking pots have copper coatings on the bottoms (▶Fig. 11.10). Being a good conductor of heat, the copper

▶ **FIGURE 11.9 Thermal conduction** Heat conduction is characterized by the time rate of heat flow ($\Delta Q/\Delta t$) in a material with a temperature difference across it of ΔT. For a slab of material, $\Delta Q/\Delta t$ is directly proportional to the cross-sectional area (A) and the thermal conductivity (k) of the material; it is inversely proportional to the thickness of the slab (d).

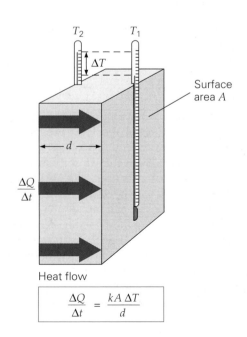

TABLE 11.4 Thermal Conductivities of Some Substances

Substance	Thermal conductivity k	
	$J/(m \cdot s \cdot C°)$ or $W/(m \cdot C°)$	$kcal/(m \cdot s \cdot C°)$
Metals		
Aluminum	240	5.73×10^{-2}
Copper	390	9.32×10^{-2}
Iron and steel	46	1.1×10^{-2}
Silver	420	10×10^{-2}
Liquids		
Transformer oil	0.18	4.3×10^{-5}
Water	0.57	14×10^{-5}
Gases		
Air	0.024	0.57×10^{-5}
Hydrogen	0.17	4.1×10^{-5}
Oxygen	0.024	0.57×10^{-5}
Other materials		
Brick	0.71	17×10^{-5}
Concrete	1.3	31×10^{-5}
Cotton	0.075	1.8×10^{-5}
Fiberboard	0.059	1.4×10^{-5}
Floor tile	0.67	16×10^{-5}
Glass (typical)	0.84	20×10^{-5}
Glass wool	0.042	1.0×10^{-5}
Human tissue (average)	0.20	4.8×10^{-5}
Ice	2.2	53×10^{-5}
Styrofoam™	0.042	1.0×10^{-5}
Wood, oak	0.15	3.6×10^{-5}
Wood, pine	0.12	2.9×10^{-5}
Vacuum	0	0

▲ **FIGURE 11.10 Copper-bottomed pots** A layer of copper is coated on the bottoms of some pots and saucepans. The high heat conductivity of this metal ensures the rapid and even spread of heat from the burner, reducing the likelihood of local "hot spots." (Contrast this photo to the chapter-opening photo; what can you say about that tile's conductivity compared with that of copper?)

promotes the distribution of heat over the bottom of a pot for even cooking. Conversely, plastic foams are good insulators mainly because they contain small trapped pockets of air, thus reducing conduction and convection (p. 392) losses.

Thermal Conductivity

Example 11.8 ■ Thermal Insulation: Helping to Conserve Energy

A room with a pine ceiling that measures 3.0 m × 5.0 m × 2.0 cm thick has a layer of glass-wool insulation above it that is 6.00 cm thick (▶ Fig. 11.11a). On a cold day, the temperature inside the room at ceiling height is 20°C, and the temperature in the attic above the insulation layer is 8°C. Assuming that the temperatures remain constant with a steady heat flow, how much energy does the layer of insulation save in one hour? Assume that losses are due to conduction only.

Thinking It Through.   Here, we have two materials, so we must consider Eq. 11.4 for two different thermal conductivities (k). We want to find $\Delta Q/\Delta t$ for the combination so that we can get ΔQ for $\Delta t = 1.0$ h. The situation is a bit complicated, because the heat flows through two materials. But we know that at a steady rate, *the heat flows must be the same through both* (why?). To find the energy saved in one hour, we need to compute how much heat is conducted in this time with and without the layer of insulation.

Solution.   Computing some of the quantities in Eq. 11.4 and making conversions as we list the data, we have

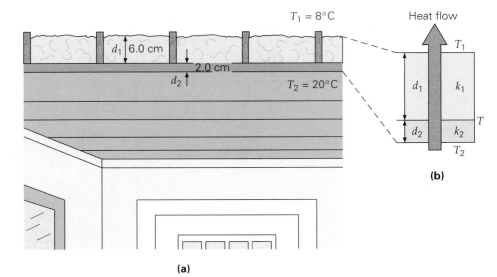

$T_1 = 8°C$

Heat flow

d_1 6.0 cm

2.0 cm

d_2

$T_2 = 20°C$

T_1

d_1 k_1

T

d_2 k_2

T_2

(b)

(a)

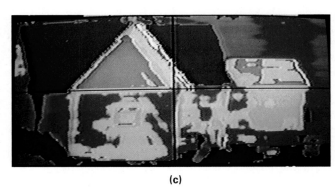

(c)

► **FIGURE 11.11 Insulation and thermal conductivity (a), (b)** Attics should be insulated to prevent loss of heat by the mechanism of conduction. See Example 11.8 and the Insight on physics, construction, and energy conservation on p. 392. **(c)** This thermogram of a house allows us to visualize the house's heat loss. Blue represents the areas that have the lowest rate of heat leaking; white, pink, and red indicate areas with increasingly larger heat losses. (Red areas have the most loss). What recommendations would you make to the owner of this house to save both money and energy? (Compare this figure with Fig. 11.15.)

Given: $A = 3.0$ m $\times 5.0$ m $= 15$ m²   **Find:** Energy saved in one hour
$d_1 = 6.0$ cm $= 0.060$ m
$d_2 = 2.0$ cm $= 0.020$ m
$\Delta T = T_2 - T_1 = 20°C - 8.0°C = 12$ C°
$\Delta t = 1.0$ h $= 3.6 \times 10^3$ s

$k_1 = 0.042$ J/(m·s·C°) (glass wool)
$k_2 = 0.12$ J/(m·s·C°) (wood) $\Big\}$ (from Table 11.4)

(In working such problems with lots of given quantities, it is especially important to label all the data correctly.)

First, let's consider how much heat would be conducted in one hour through the wooden ceiling with no insulation present. Since we know Δt, we can rearrange Eq. 11.4 * to find ΔQ_c (heat conducted through the wooden ceiling alone, assuming the same ΔT):

$$\Delta Q_c = \left(\frac{k_2 A \Delta T}{d_2}\right)\Delta t = \left\{\frac{[0.12 \text{ J/(m·s·C°)}](15 \text{ m}^2)(12 \text{ C°})}{0.020 \text{ m}}\right\}(3.6 \times 10^3 \text{ s}) = 3.9 \times 10^6 \text{ J}$$

Now we need to find the heat conducted through the ceiling *and* the insulation layer together. Let T be the temperature at the interface of the materials and T_2 and T_1 be the warmer and cooler temperatures, respectively (Fig. 11.11b). Then

$$\frac{\Delta Q_1}{\Delta t} = \frac{k_1 A(T - T_1)}{d_1} \quad \text{and} \quad \frac{\Delta Q_2}{\Delta t} = \frac{k_2 A(T_2 - T)}{d_2}$$

*Equation 11.4 can be extended to any number of layers or slabs of materials: $\Delta Q/\Delta t = A(T_2 - T_1)/\Sigma(d_i/k_i)$. (See the Insight involving insulation in building construction, p. 392.)

We don't know T, but when the conduction is steady, the flow rates are the same for both materials; that is, $\Delta Q_1/\Delta t = \Delta Q_2/\Delta t$, or

$$\frac{k_1 A(T - T_1)}{d_1} = \frac{k_2 A(T_2 - T)}{d_2}$$

The A's cancel, and solving for T gives

$$T = \frac{k_1 d_2 T_1 + k_2 d_1 T_2}{k_1 d_2 + k_2 d_1}$$

$$= \frac{[0.042\,\text{J}/(\text{m}\cdot\text{s}\cdot\text{C}°)](0.020\,\text{m})(8.0°\text{C}) + [0.12\,\text{J}/(\text{m}\cdot\text{s}\cdot\text{C}°)](0.060\,\text{m})(20.0°\text{C})}{[0.042\,\text{J}/(\text{m}\cdot\text{s}\cdot\text{C}°)](0.020\,\text{m}) + [0.12\,\text{J}/(\text{m}\cdot\text{s}\cdot\text{C}°)](0.060\,\text{m})}$$

$$= 18.7°\text{C}$$

Since the flow rate is the same through the wood and the insulation, we can use the expression for either material to calculate it. Let's use the expression for the wood ceiling. Here, care must be taken to use the correct ΔT. The temperature at the wood–insulation interface is 18.7°C; thus,

$$\Delta T_{\text{wood}} = |T_2 - T| = |20°\text{C} - 18.7°\text{C}| = 1.3°\text{C}$$

Therefore, the heat flow rate is

$$\frac{\Delta Q_2}{\Delta t} = \frac{k_2 A |\Delta T_{\text{wood}}|}{d_2} = \frac{[(0.12\,\text{J}/(\text{m}\cdot\text{s}\cdot\text{C}°)](15\,\text{m}^2)(1.3\,\text{C}°)}{0.020\,\text{m}} = 1.2 \times 10^2\,\text{J/s (or W)}$$

In one hour, the heat loss with insulation in place is

$$\Delta Q_2 = \frac{\Delta Q_2}{\Delta t} \times \Delta t = (1.2 \times 10^2\,\text{J/s})(3600\,\text{s}) = 4.3 \times 10^5\,\text{J}$$

This value represents a decreased heat loss of

$$\Delta Q_c - \Delta Q_2 = 3.9 \times 10^6\,\text{J} - 4.3 \times 10^5\,\text{J} = 3.5 \times 10^6\,\text{J}$$

Ideally, this amount represents a savings of $\dfrac{3.5 \times 10^6\,\text{J}}{3.9 \times 10^6\,\text{J}} \times (100\%) = 90\%$!

Follow-up Exercise. Verify that the heat flow rate is the same through the insulation as through the wood ($1.2 \times 10^2\,\text{J/s}$) in this Example.

Convection

In general, compared with solids, liquids, and gases are not good thermal conductors. However, the mobility of molecules in fluids permits heat transfer by another process—convection. **Convection** is heat transfer as a result of mass transfer, which can be natural or forced.

Natural convection cycles occur in liquids and gases. For example, when cold water is in contact with a hot object, such as the bottom of a pot on a stove, the object transfers heat to the water adjacent to the pot by conduction. But the water carries the heat away with it by natural convection, and a cycle is set up in which upper, colder (more dense) water replaces the rising warm (less dense) water. Such cycles are important in atmospheric processes, as illustrated in ▶ Fig. 11.12. During the day, the ground heats up more quickly than do large bodies of water, as you may have noticed if you have been to the beach. This phenomenon occurs both because the water has a greater specific heat than that of the land and because convection currents disperse the absorbed heat throughout the great volume of water. The air in contact with the warm ground is heated and expands, becoming less dense. As a result, the warm air rises (air currents) and, to fill the space, other air moves horizontally (winds)—creating a sea breeze near a large body of water. Cooler air descends, and a thermal convection cycle is set up, which transfers heat away from the land. At night, the ground loses its heat more quickly than the water, and the surface of the water is warmer than the land. As a result, the cycle is reversed.

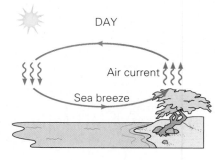

DAY

Land warmer than water

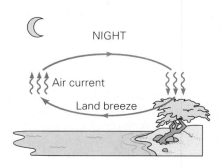

NIGHT

Water warmer than land

▲ **FIGURE 11.12 Convection cycles** During the day, natural-convection cycles give rise to sea breezes near large bodies of water. At night, the pattern of circulation is reversed, and the land breezes blow. The temperature differences between land and water are the result of their specific-heat differences. Water has a much larger specific heat, so the land warms up more quickly during the day. At night, the land cools more quickly, while the water remains warmer, because of its larger specific heat.

INSIGHT

Physics, the Construction Industry, and Energy Conservation

In the last few decades, many home owners have found it cost effective to retrofit their homes with better insulation (or, in some cases, to add insulation not previously there). This task typically includes cutting small holes in walls and blowing foam insulation into the air spaces, or laying thick slabs of insulation above the ceiling in the attic. To quantify the insulation properties of various materials, the insulation and construction industries do not use thermal conductivity k. Rather, they use a quantity called the *thermal resistance*, which is related to the inverse of k.

To see how these two quantities are related, consider Eq. 11.4 rewritten as

$$\frac{\Delta Q}{\Delta t} = \left(\frac{k}{d}\right)A\Delta T = \left(\frac{1}{R_t}\right)A\Delta T$$

where the *thermal resistance* is $R_t = d/k$. Note that R_t depends not only on the material's properties (expressed in the thermal conductivity k), but also on its thickness. R_t is a measure of how "resistant" to heat flow the slab of material is.

$\Delta Q/\Delta t$ is the *thermal conduction current* (I_t), defined as the rate of heat conduction, or $I_t = \Delta Q/\Delta t$. Expressed in these terms, the preceding equation becomes

$$I_t = \frac{A\Delta T}{R_t}$$

Note that the thermal conduction current is proportional to the area of the material and the temperature difference. More area means more heat conduction, and, of course, temperature differences are the fundamental cause of the heat flow in the first place. But also note that the thermal current is inversely related to the thermal resistance: More resistance results in less heat flow.

For the home owners, the lesson is clear. To reduce the heat flow (and thus minimize energy loss in the winter and gain in the summer), they should reduce areas of low thermal resistance, like windows or, if that is not possible, at least increase their resistance by switching to double or triple panes. Similarly for walls, increasing the thermal resistance by adding or upgrading insulation is the way to go. Lastly, changing interior temperature requirements (changing $\Delta T = |T_{\text{exterior}} - T_{\text{interior}}|$) can make a big difference. In summer, home owners should raise the thermostat setting on the air conditioning (lowering ΔT by increasing T_{interior}), and in winter, they should lower the thermostat setting on the heating system (lowering ΔT by decreasing T_{interior}).

Insulation and building materials are classified according to their "R-values," that is, their thermal resistance values. In the United States, the units of R_t are $\text{ft}^2 \cdot \text{h} \cdot \text{F}°/\text{Btu}$. While these units may seem awkward, the important point is that they are proportional to the thermal resistance of the material. Thus, wall insulation with a value of R-31 is about 1.6 times (or $\frac{31}{19}$) less conductive than insulation with a value of R-19. A comparison photo of various types of insulation is shown in Fig. 1.

FIGURE 1   Differences in R-values   For insulation blankets made of identical materials, the R-values are proportional to the materials' thickness.

In *forced convection*, the fluid is moved mechanically. This condition results in transfer without a temperature difference. In fact, we can transfer heat energy from a low-temperature region to a high-temperature region this way, as in the case of the forced convection of a refrigerator coolant removing energy from the inside of the refrigerator. (The circulating coolant carries heat energy from the inside of the refrigerator, and this heat is given up to the environment, as we will see in Section 12.4.)

Other common examples of forced convection systems are forced-air heating systems in homes (▶ Fig. 11.13), the human circulatory system, and the cooling system of an automobile engine. The human body does not use all of the energy obtained from food; a great deal is lost. (There's usually a temperature difference between your body and your surroundings.) So that body temperature will stay normal, the internally generated heat energy is transferred close to the surface of the skin by blood circulation. From the skin, the energy is conducted to the air or lost by radiation (the other heat-transfer mechanism, to be discussed shortly).

Water or some other coolant is circulated (pumped) through most automobile cooling systems. (Some engines are air-cooled.) The coolant carries engine heat to the radiator (a form of *heat exchanger*), where forced air flow produced by the fan and car movement carries it away. The *radiator* of an automobile is actually misnamed—most of the heat is transferred from it by forced convection rather than by radiation.

Conceptual Example 11.9 ■ Foam Insulation: Better Than Air?

Polymer foam insulation is sometimes blown into the space between the inner and outer walls of a house. Since air is a good thermal insulator, why is the foam insulation needed: (a) To prevent loss of heat by conduction, (b) to prevent loss of heat by convection, or (c) for fireproofing?

Reasoning and Answer. Polymer foams will generally burn, so (c) isn't likely to be the answer. Air is a poor thermal conductor, even poorer than plastic foam (Styrofoam™; see Table 11.4), so the answer can't be (a). However, as a gas, air is subject to convection within the wall space. In the winter, the air near the warm inner wall is heated and rises, thus setting up a convection cycle in the space and transferring heat to the cold outer wall. In the summer, with air conditioning, the heat-loss cycle is reversed. Foam blocks the movement of air and thus stops such convection cycles. Hence, the answer is (b).

Follow-up Exercise. Thermal underwear and thermal blankets are loosely knitted with lots of small holes. Wouldn't they be more effective if the material were denser?

Radiation

Conduction and convection require some material as a transport medium. The third mechanism of heat transfer needs no medium; it is called **radiation**, which refers to energy transfer by electromagnetic waves (Chapter 20). Heat is transferred to the Earth from the Sun through empty space by radiation. Visible light and other forms of electromagnetic radiation are commonly referred to as *radiant energy*.

You have experienced heat transfer by radiation if you've ever stood near an open fire (▼Fig. 11.14a). You can feel the heat on your exposed hands and face. This heat transfer is not due to convection or conduction, since heated air rises and air is a poor conductor. Visible radiation is emitted from the burning material, but most of the heating effect comes from the invisible **infrared radiation** emitted by the glowing embers or coals. You feel this radiation because it is absorbed by water molecules in your skin. (Body tissue is about 85% water.) The water molecule has an internal vibration whose frequency coincides with that of infrared radiation,

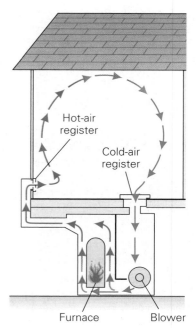

▲ **FIGURE 11.13 Forced convection** Houses are commonly heated by forced convection. Registers or gratings in the floors or walls allow heated air to enter and cooler air to return to the heat source. (Can you explain why the registers are located near the floor?) In older homes, hot water runs through pipes that are along the wall baseboards, and natural convection distributes the heat vertically upward.

Note: Resonance is discussed in Section 13.5.

▼ **FIGURE 11.14 Heating by conduction, convection, and radiation** (a) The hands at left are warmed by the convection of rising hot air (and some radiation). The gloved hand at the upper right is warmed by conduction. The hands at the lower right are warmed by radiation. (b) A practical application of heat transfer by radiation. A Tibetan teakettle is heated by focusing sunlight, using a metal reflector.

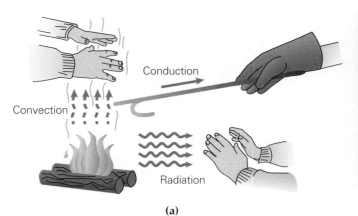

(a)

(b)

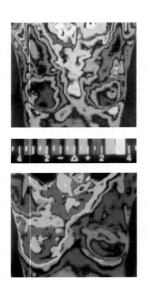

▲ **FIGURE 11.15 Applied Thermography** Thermograms can be used to detect breast cancer by detecting tumor regions that are higher in temperature than normal. The upper photo shows a thermogram scan of a woman without breast cancer. The lower photo shows the result for a woman with breast cancer. The "hot spots" in this scan tells the physician where the cancer resides.

which is therefore readily absorbed. (This effect is called *resonance absorption*. The electromagnetic wave drives the molecular vibration, and energy is transferred to the molecule, somewhat like pushing a swing.) Heat transfer by radiation can play a practical role in daily living (Fig. 11.14b).

Infrared radiation is sometimes referred to as "heat rays." You may have noticed the red infrared lamps used to keep food warm in cafeterias. Heat transfer by infrared radiation is also important in maintaining our planet's warmth by a mechanism known as the *greenhouse effect*. This important environmental topic is discussed in the Insight on p. 395.

Although infrared radiation is invisible to the human eye, it can be detected by other means. The frequency of the infrared radiation is proportional to the temperature of its source. This relationship is the basis for infrared thermometers, which, using infrared detectors, can measure temperature remotely. Also, some cameras can use special infrared film. A picture taken with this film is an image consisting of contrasting light and dark areas, corresponding to regions of higher and lower temperatures, respectively. Special instruments that apply such *thermography* are used in industry and medicine; the images they produce are called *thermograms* (◄Fig. 11.15).

A new application of thermograms is for security. The system consists of an infrared camera and a computer that identifies an individual by means of the unique heat pattern emitted by the facial blood vessels. The camera takes a picture of the radiation from a person's face, which is compared with an earlier image stored in the computer memory. Such a system reportedly can even distinguish between identical twins, whose facial features are slightly different. Also, changes in body temperature from weather conditions or a fever do not affect the identification, as the relative patterns of radiation remain the same.

The rate at which an object radiates energy has been found to be proportional to the fourth power of the objective's absolute temperature (T^4). This relationship is expressed in an equation known as **Stefan's law**, expressed as

$$P = \frac{\Delta Q}{\Delta t} = \sigma A e T^4 \quad \textit{(radiation only)} \quad (11.5)$$

where P is the power radiated in watts (W), or joules per second (J/s). The symbol σ (the Greek letter sigma) is the *Stefan–Boltzmann constant*: $\sigma = 5.67 \times 10^{-8}$ W/(m$^2 \cdot$K^4). The radiated power is also proportional to the surface area (A) of the object. The **emissivity** (e) is a unitless number between 0 and 1 that is characteristic of the material. Dark surfaces have emissivities close to 1, and shiny surfaces have emissivities close to 0. The emissivity of human skin is about 0.70.

Dark surfaces are not only better emitters of radiation, but they are also good absorbers. This must be the case, because to maintain a constant temperature, the incident energy absorbed must equal the emitted energy. *Thus, a good absorber is also a good emitter.* An ideal, or perfect, absorber (and emitter) is referred to as a **black-body** ($e = 1.0$). Shiny surfaces are poor absorbers, since most of the incident radiation is reflected. This fact can be demonstrated easily, as shown in ◄Fig. 11.16. (Can you see why it is better to wear light-colored clothes in the summer and dark-colored clothes in the winter?)

When an object is in thermal equilibrium with its surroundings, its temperature is constant; thus, it must be emitting and absorbing radiation at the same rate. However, if the temperatures of the object and its surroundings are different, there will be a net flow of radiant energy. If an object is at a temperature T and its surroundings are at a temperature T_s, the net rate of energy loss or gain per unit time (power) is given by

$$P_{\text{net}} = \sigma A e (T_s^4 - T^4) \quad (11.6)$$

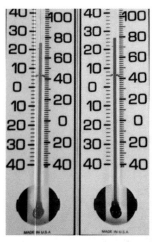

▲ **FIGURE 11.16 Good absorber** Black objects are generally good absorbers of radiation. The bulb of the thermometer on the right has been painted black. Note the difference in temperature readings.

The Greenhouse Effect

You may have heard or read about the greenhouse effect in connection with global warming. The *greenhouse effect* helps regulate the Earth's long-term average temperature, which has been fairly constant. A portion of the solar radiation we receive reaches and warms the Earth's surface. The Earth, in turn, reradiates energy in the form of infrared (IR) radiation. The balance between absorption and radiation is a major factor in stabilizing the Earth's temperature.

This balance is affected by the concentration of *greenhouse gases*—primarily water vapor and carbon dioxide (CO_2)—in the atmosphere. As the reradiated infrared radiation passes back through the atmosphere, some of the radiation is absorbed by the greenhouse gases. These gases are selective absorbers: They absorb radiation at certain wavelengths (IR), but not at others (Fig. 1a).

If IR radiation is absorbed, the atmosphere warms, warming the Earth (heat transfer by radiation). This rise in surface temperature causes a shift in the wavelength of the emitted radiation. The wavelength is eventually shifted to a "window" in the absorption spectrum where little or no absorption takes place, and the terrestrial radiation passes through the atmosphere into space. Thus, the Earth loses energy, and its surface temperature decreases.

But with a temperature decrease, the terrestrial radiation shifts to a longer wavelength and is again absorbed by the greenhouse gases. So, under normal conditions, we have a "feedback" situation that keeps the temperature near an equilibrium value. There is a "turning on and off," so to speak, similar to the action of a thermostat. Hence, the selective absorption of atmospheric gases plays an important role in maintaining the Earth's average temperature.

Why is this phenomenon called the *greenhouse effect*? The reason is that the atmosphere functions like the glass in a greenhouse. That is, the absorption and transmission properties of glass are similar to those of the atmospheric greenhouse gases—in general, visible radiation is transmitted, but infrared radiation is selectively absorbed (Fig. 1b). We have all observed the warming effect of sunlight passing through glass, for example, in a closed car on a sunny, cold day. Similarly, a greenhouse heats up by absorbing sunlight and trapping the reradiated infrared radiation. Thus, it is quite warm inside on a sunny day, even in winter. (The glass enclosure also keeps warm air from escaping upward. In practice, this elimination of heat loss by convection is the chief factor in maintaining an elevated temperature.)

The problem is that on Earth, human activities may accentuate greenhouse warming. With the combustion of fuels for heating and industrial processes, vast amounts of CO_2 and other greenhouse gases are vented into the atmosphere, where they might trap increasingly more IR. There is concern that the result of this trend will be—or already is—*global warming*: an increase in the Earth's average temperature that could dramatically affect the environment. For example, the climate in many parts of the globe would be altered, with effects on agricultural production and world food supplies that are very difficult to predict. A general rise in temperature could cause partial melting of the polar ice caps. Sea levels would rise, flooding low-lying regions and endangering coastal ports and population centers.

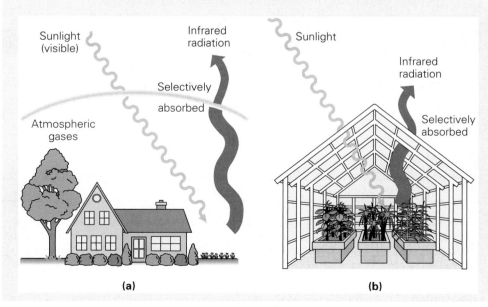

(a)   (b)

FIGURE 1   The greenhouse effect   (a) The greenhouse gases of the atmosphere, particularly water vapor and carbon dioxide, are selective absorbers with absorption properties similar to those of glass used in greenhouses. Visible light is transmitted and heats the Earth's surface, while some of the infrared radiation that is reemitted is absorbed and trapped in the Earth's atmosphere. **(b)** A greenhouse operates in a similar way.

Note that if T_s is less than T, then P (or $\Delta Q/\Delta t$) will be negative, indicating a net energy loss, in keeping with our heat-flow sign convention. *Keep in mind that the temperatures used in calculating radiated power are the absolute temperatures in kelvins.*

You may have noticed in Chapter 10 that we defined *heat* as the net energy transfer because of a temperature difference. The word *net* here is important. It is possible to have energy transfer between an object and its surroundings, or between objects, at the same temperature. Note that if $T_s = T$ (i.e., there is no temperature difference), there is a continuous exchange of radiant energy (Eq. 11.5 still holds), but there is no *net* change of internal energy of the object.

Example 11.10 ■ Body Heat: Radiant-Heat Transfer

Suppose that your skin has an emissivity of 0.70 and that its exposed area is 0.27 m². How much net energy will be radiated per second from this area if the ambient room temperature is 20°C? Assume your skin temperature to be the same as the normal body temperature, 37°C. (The average skin temperature is actually less.)

Thinking it Through. Everything is given for us to find P_{net} from Eq. 11.6. The net radiant-energy transfer is between the skin's surface and the walls, because air is transparent to visible radiation. We must remember to work in kelvins.

Solution.

Given: $T_s = 20°C + 273 = 293$ K   *Find:* P_{net} (net power)
$T = 37°C + 273 = 310$ K
$e = 0.70$
$A = 0.27$ m²
$\sigma = 5.67 \times 10^{-8}$ W/(m²·K⁴) (known)

Using Eq. 11.6 directly, we get

$$P_{net} = \sigma A e(T_s^4 - T^4)$$
$$= [5.67 \times 10^{-8} \text{W}/(\text{m}^2 \cdot \text{K}^4)](0.27 \text{ m}^2)(0.70)[(293 \text{ K})^4 - (310 \text{ K})^4]$$
$$= -20 \text{ W} \quad (\text{or } -20 \text{ J/s})$$

Thus, 20 J of energy is radiated, or *lost* (as indicated by the minus sign), each second.

Follow-up Exercise. (a) In this Example, suppose the skin had been exposed to an ambient room temperature of only 10°C. What would the rate of heat loss be? (b) Elephants have huge body masses and daily caloric food intakes. Can you explain how their huge ear flaps (large surface area) might help stabilize their body temperature?

Problem-Solving Hint

Note that in Example 11.10, the fourth powers of the temperatures were found first, and then their difference was found. It is *not* correct to find the temperature difference and then raise it to the fourth power: $T_s^4 - T^4 \neq (T_s - T)^4$.

Let's consider a practical example of heat transfer.

Conceptual Example 11.11 ■ Solar Panels: Reducing the Heat Transfer

Solar panels are used to collect solar energy to heat water, which may then be used directly to heat a home at night. The panel boxes have black interiors (why?) through which the piping runs to carry the water, and the top is covered with glass. However, ordinary glass absorbs most of the Sun's ultraviolet radiation. This absorption reduces the heating effect. Wouldn't it be better to leave the glass off the panel boxes?

Reasoning and Answer. Some energy is absorbed by the glass, but the glass serves a useful purpose and saves a lot more energy than it absorbs. As the black interior and piping of the panel box heat up, there could be heat loss by radiation (infrared) and convection. The glass prevents this loss from occurring via the greenhouse effect. It absorbs much of the infrared radiation and keeps the convection inside the solar-panel box. (See the Insight on the greenhouse effect, p. 395.)

Follow-up Exercise. Is there any practical reason for window drapes (other than privacy)?

Let's look at a few more real-life examples of heat transfer. In the spring, a late frost could kill the buds on fruit trees in an orchard. To reduce heat transfer, some growers spray water on the trees to form ice before a hard frost occurs. Using ice to save buds? Ice is a relatively poor (and inexpensive) conductor of heat, so it has an insulating effect. It will maintain the buds' temperature at 0°C, not going below that value, and therefore protects the buds.

Another method to protect orchards from freezing is the use of smudge pots, containers in which material is burned to create a dense cloud of smoke. At night, when the Sun-warmed ground cools off by radiation, the cloud absorbs this heat and reradiates it back to the ground. Thus, the ground takes longer to cool, hopefully without reaching freezing temperatures before the Sun comes up. (Recall that frost is the direct condensation of water vapor in the air to ice—not frozen dew.)

Take a look at ▸Fig. 11.17. Why would anyone wear a dark robe in the desert? We have learned that dark objects absorb radiation (Fig. 11.16). Wouldn't a white robe be better? A dark robe definitely absorbs more radiant energy and warms the air inside near the body. But note that the robe is open at the bottom. The warm air rises (since it is less dense) and exits at the neck area, and outside cooler air enters the robe at the bottom—a natural-convection air circulation!

Finally, consider some of the thermal factors involved in "passive" solar house design (▾Fig. 11.18). The term "passive" means that the design elements require no "active" use of energy to conserve energy. Their initial cost can usually be quickly recovered by the savings in energy bills.

▲ **FIGURE 11.17 A dark robe in the desert?** Dark objects absorb more radiation than do lighter ones, and they become hotter. What's going on here? See the text for an explanation.

South

Shade

Summer

(a)

South

Sunlight

Winter

(b)

◀ **FIGURE 11.18 Aspects of passive solar design** **(a)** In summer, with the sun angle high, large overhangs on the south-facing side of the house provide shade. The leaves of deciduous trees provide additional shade. The walls of the house are thick to reduce conductive heat flow to the interior, and the windows and doors are equipped with drapes, venetian blinds, and shades (not shown) to keep the reflected sunlight out. **(b)** In winter, the sun angle is low. The deciduous trees have dropped their leaves, allowing sunlight to come in, even at low angles. The sunlight streams through the large south-facing windows and doors, heating tile floors, which will reradiate the energy to the interior during the cold evening hours. The thick walls prevent heat flow to the exterior at night. Drawing the drapes helps reduce heat leaks to the outside during the night. Small north-facing windows help reduce heat leaks during cold winter days.

Chapter Review

Important Concepts and Equations

- **Heat** (Q) is the (atomic-level) energy exchanged between objects, most commonly because they are at different temperatures.

- **The specific heat** (c) for solids and liquids tells how much heat is needed to raise the temperature of 1 kilogram of a particular material by 1 C°. It is characteristic of the type of material and is defined by

$$c = \frac{Q}{m\,\Delta T} \quad \text{or} \quad Q = mc\,\Delta T \quad (11.1)$$

- **The (molar) specific heats** (c_p and c_v) for gases tell how much heat is needed to raise the temperature of 1 mole of a gas by 1 C°. It is a characteristic of the type of gas and depends on whether the heat is added at constant pressure (c_p) or volume (c_v).

- **Calorimetry** is a technique that uses heat transfer between objects, most commonly to measure specific heats of materials. It is based on conservation of energy written as $\Sigma Q_i = 0$, assuming no heat losses or gains to the outside environment.

- **Latent heat** (L) is that heat required to change the phase of an object *per kilogram* of mass. During the phase change, the temperature of the system does not change. Its general definition is

$$L = \frac{|Q|}{m} \quad \text{or} \quad Q = \pm mL \quad (11.3)$$

- Heat transfer due to direct contact of objects that have different temperatures is called **heat conduction**. The rate of heat flow by conduction is given by

$$\frac{\Delta Q}{\Delta t} = \frac{kA\,\Delta T}{d} \quad (11.4)$$

- **Heat convection** refers to heat transfer due to mass movement of gas or liquid molecules. *Natural convection* is driven by density differences caused by temperature differences. In *forced convection*, the movement is driven by mechanical means.

- **Heat radiation** refers to heat transferred by electromagnetic radiation (light) between objects that have different temperatures. The rate of transfer is given by

$$P_{net} = \sigma Ae(T_s^4 - T^4) \quad (11.6)$$

where σ is the Stefan–Boltzmann constant, whose value is $5.67 \times 10^{-8} \text{ W/(m}^2 \cdot \text{K}^4)$.

Exercises*

11.1 Units of Heat

1. The SI unit of heat energy is the (a) calorie, (b) kilocalorie, (c) Btu, or (d) joule.

2. Which of the following is the largest unit of heat energy? (a) calorie, (b) Btu, (c) joule, or (d) kilojoule.

3. ■ A heater supplies 240 Btu of energy. What is this quantity in joules?

4. ■ A person goes on a 1500 Cal/day diet to lose weight. What is the equivalent daily allowance expressed in joules?

5. ■ A window air conditioner has a rating of 20000 Btu (per hour). What is this rating in watts?

6. ■ What is the mechanical equivalent of heat expressed in British thermal units?

7. ■■ A student ate a Thanksgiving dinner that totaled 2800 Cal. He wants to use up all that energy by lifting a 20-kg

mass up a distance of 1.0 m. (a) How many times must he lift the mass? (b) If he can lift the mass once every 5.0 s, how long does this exercise take (neglecting time taken to lower the mass)?

11.2 Specific Heat

8. The amount of heat necessary to change the temperature of 1 kg of a substance by 1 C° is called the substance's (a) specific heat, (b) latent heat, (c) heat of combustion, or (d) mechanical equivalent of heat.

9. CQ When you swim in the ocean or a lake at night, the water may feel pleasantly warm even when the air is quite cool. Why?

10. CQ Is it possible to have a negative specific heat? Explain.

11. CQ Is it possible for an object to be described with a negative value of heat? If yes, what is the significance of the negative sign?

12. CQ Near a large body of water, what would you expect the direction of a breeze to be during the night?

*Neglect heat losses to the external environment in the exercises unless instructed otherwise, and consider all temperatures to be exact.

13. **CQ** Does heat flow between two objects depend on their temperatures or just the difference between their temperatures?

14. **CQ** The British thermal unit is defined as the heat required to increase the temperature of 1 lb of water by 1 F°. An astronaut in space asks the following question: "How many Btu's does it take to increase the temperature of 3 lb of water from 60°F to 212°F?" How would you respond?

15. **CQ** Equal amounts of heat are added to two different objects at the same initial temperature. What factors can cause the final temperature of the two objects to be different?

16. **CQ** Two identical objects of the same mass and at the same initial temperature cool off. If object A cools off faster than object B, what can you say about the specific heats of the two objects? Explain.

17. **CQ** For gases, is c_p always greater than c_v? Explain.

18. ■ A 5.0-g pellet of aluminum at 20°C gains 200 J of heat. What is its final temperature?

19. ■ How many joules of heat must be added to 5.0 kg of water at 20°C to bring it to the boiling point?

20. **IE** ■ The temperature of a lead block and a copper block, both 1.0 kg and at 110 K, is to be raised to 190 K. (a) Copper requires (1) more heat, (2) the same heat or (3) less heat than lead? Why? (b) Calculate the difference between the heat required for the two blocks to prove your answer to (a).

21. ■■ How many kilograms of aluminum will experience the same temperature rise as 3.00 kg of copper when the same amount of heat is added to each?

22. ■■ A 0.250-kg coffee cup at 20° C is filled with 0.250 kg of boiling coffee. The cup and the coffee come to thermal equilibrium at 80° C. If no heat is lost to the environment, what is the specific heat of the cup material? [*Hint*: Consider the coffee essentially to be boiling water.]

23. ■■ An aluminum spoon at 100°C is placed in a Styrofoam™ cup containing 0.200 kg of water at 20°C. If the final equilibrium temperature is 30°C and no heat is lost to the cup itself, what is the mass of the aluminum spoon?

24. **IE** ■■ Equal amounts of heat are added to different quantities of copper and lead. The temperature of the copper increases by 5.0 C° and the temperature of the lead by 10 C°. (a) Lead has (1) a greater mass than copper, (2) the same amount of mass as copper, or (3) less mass than copper. (b) Calculate the mass ratio of lead to copper to prove your answer to (a).

25. ■■ A camper heats 30 L of water to boiling to purify it for bathing. What volume of water from a 15°C stream must he then add to get the bath water to 45°C? (Neglect any losses.)

26. ■■ To determine the specific heat of a new metal alloy, 0.150 kg of the substance is heated to 400°C and then placed in a 0.200-kg aluminum calorimeter cup containing 0.400 kg of water at 10.0°C. If the final temperature of the mixture is 30.5°C, what is the specific heat of the alloy? (Ignore the calorimeter stirrer and thermometer.)

27. **IE** ■■ In a calorimetry experiment, 0.50 kg of a metal at 100°C is added to 0.50 kg of water at 20°C in an aluminum calorimeter cup. The cup has a mass of 0.250 kg. (a) If some water splashed out of the cup when the metal was added, the measured specific heat will appear to be (1) higher, (2) the same, or (3) lower than the value calculated for the case in which the water does not splash out. Why? (b) If the final temperature of the mixture is 25°C, and no water splashed out, what is the specific heat of the metal?

28. ■■■ An electric immersion heater has a power rating of 1500 W. If the heater is placed in a liter of water at 20°C, how many minutes will it take to bring the water to a boil? (Assume that there is no heat loss except to the water itself.)

29. ■■■ A 0.100-kg piece of aluminum at 90°C is immersed in 1.00 kg of water at 20°C. Assuming that no heat is lost to the surroundings or the container, what is the temperature of the metal and water when they reach thermal equilibrium?

30. ■■■ At what average rate would heat have to be removed from 1.5 L of (a) water and (b) mercury to reduce the liquid's temperature from 20°C to its freezing point in 3.0 min?

11.3 Phase Changes and Latent Heat

31. The SI units of latent heat are (a) $1/C°$, (b) $J/(kg \cdot C°)$, (c) $J/C°$, or (d) J/kg.

32. Latent heat is always (a) part of the specific heat, (b) related to the specific heat, (c) the same as the mechanical equivalent of heat, or (d) involved in a phase change.

33. **CQ** Why do different substances have different freezing and boiling points? Would you expect the latent heats to be different for different substances? Explain.

34. **CQ** You are monitoring the temperature of some cold ice cubes ($-5.0°C$) in a cup as the ice and cup are heated. Initially, the temperature rises, but it stops at 0°C. After a while, it begins rising again. Is there anything wrong with the thermometer? Explain.

35. **CQ** In general, you can get a more severe burn from steam at 100°C than from the same mass of hot water at 100°C. Why?

36. CQ When you breathe out in the winter, your breath looks like water vapor. Explain.

37. CQ Explain how a fan can make you feel cool in the summer.

38. ■ How much heat is required to boil away 0.500 kg of water that is initially at 100°C?

39. IE ■ (a) Converting 1.0 kg of water at 100°C to steam at 100°C requires (1) more heat, (2) the same amount of heat, or (3) less heat than converting 1.0 kg of ice at 0°C to water at 0°C. Why? (b) Calculate the difference in heat required to prove your answer to (a).

40. ■ How much heat must be added to 0.75 kg of lead at 20°C to cause it to melt completely?

41. ■ How much heat is required to boil away 0.50 L of liquid nitrogen at −196°C. (Take the density of liquid nitrogen to be $0.80 \times 10^3 \ kg/m^3$.)

42. ■ How much heat is required to boil away 0.50 kg of water that is initially at 50°C?

43. ■ A mercury diffusion vacuum pump contains 0.015 kg of mercury vapor at a temperature of 630 K. Suppose that you wanted to condense the vapor. How much heat would have to be removed?

44. ■■ (a) A 0.500-kg piece of ice at −20°C is converted to steam at 115°C. How much heat must be supplied to do this operation? (b) To convert the steam back to ice at −5°C, how much heat would have to be removed?

45. ■■ How much ice (at 0°C) must be added to 1.0 kg of water at 100°C so as to end up with all liquid at 20°C?

46. ■■ 0.60 kg of ice at −10°C is placed in 0.30 kg of water at 50°C. How much liquid is left when the system reaches thermal equilibrium?

47. ■■ If 0.050 kg of ice at 0°C is added to 0.300 kg of water at 25°C in a 0.100-kg aluminum calorimeter cup, what is the final temperature of the water?

48. ■■ Steam at 100°C is bubbled into 0.250 kg of water at 20°C in a calorimeter cup. How much steam will have been added when the water in the cup reaches 60°C? (Ignore the effect of the cup.)

49. ■■ Ice (initially at 0°C) is added to 0.75 L of tea at 20°C to make the coldest possible iced tea. If enough ice is added so the mixture is all liquid, how much liquid is in the pitcher when this condition occurs?

50. ■■ A volume of 0.50 L of water at 16°C is put into an aluminum ice-cube tray of mass 0.250 kg at the same temperature. How much energy must be removed from this system by the refrigerator to turn the water into ice at −8.0°C?

51. IE ■■ Evaporation of water from our skin is a very important mechanism for controlling body temperature. (a) This is because (1) water has a high specific heat, (2) water has a high heat of vaporization, (3) water contains more heat when hot, or (4) water is a good heat conductor. (b) If an athlete loses 0.50 kg of water through evaporation, estimate the amount of heat lost in the process to cool off his body in the phase change part of the vaporization.

52. ■■■ A kilogram of a substance gives a T-versus-Q graph as shown in ▼Fig. 11.19. (a) What are the melting and boiling points? In SI units, what are (b) the specific heats of the substance during its various phases and (c) the latent heats of the substance at the various phase changes?

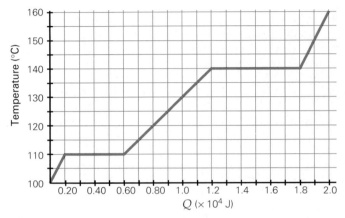

▲ FIGURE 11.19 Temperature versus heat See Exercise 52.

53. ■■■ Some ceramic materials will become superconducting if immersed in liquid nitrogen. In an experiment, a 0.150-kg piece of such material at 20°C is placed in liquid nitrogen at its boiling point to cool in a perfectly insulated flask, which allows the gaseous N_2 immediately to escape. How many liters of liquid nitrogen will be boiled away in doing this operation? (Take the specific heat of the ceramic material to be the same as that of glass, and take the density of liquid nitrogen to be $0.80 \times 10^3 \ kg/m^3$.)

11.4 Heat Transfer

54. The warming of the atmosphere involves (a) conduction, (b) convection, (c) radiation, or (d) all of the preceding.

55. CQ Water is a very poor heat conductor. Why can a pot of water be heated relatively quickly?

56. CQ A plastic ice-cube tray and a metal ice-cube tray are removed from the same freezer, at the same initial temperature. However, once your hands touch both, the metal one feels cooler. Why?

57. CQ (a) You blow on a spoonful of hot soup to cool it. Yet, your breath is warm. How does blowing cause cooling of the soup? (b) On hot, humid days, ice can build up on the cooling coils of a window air conditioner.

Does this phenomenon help with the task of cooling the room? Explain.

58. **CQ** Why is the warning shown on the highway road sign in ▼Fig. 11.20 necessary?

▲ **FIGURE 11.20 A cold warning** See Exercise 58.

59. **CQ** What is the purpose of the fins on a motorcycle radiator or an automobile radiator?

60. **CQ** Why do many aluminum pots and saucepans have a layer of copper on the bottom?

61. **CQ** A Thermos bottle (▼Fig. 11.21) keeps cold beverages cold and hot ones hot. It consists of a double-walled, partially evacuated container with silvered walls (mirrored interior). The bottle is constructed to minimize all three mechanisms of heat transfer. Explain how.

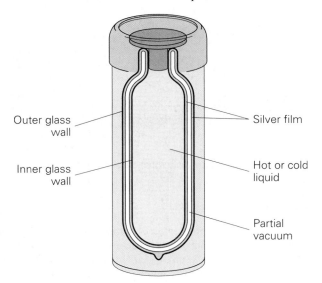

Outer glass wall

Inner glass wall

Silver film

Hot or cold liquid

Partial vacuum

▲ **FIGURE 11.21 Thermal insulation** The Thermos bottle uses all three methods of heat transfer. See Exercise 61.

62. **CQ** If you put your hand in a hot oven, the heat is uncomfortable, but tolerable. Yet, if you touch a pan in the oven, you burn your hand. Why the difference?

63. **CQ** Polar bears have an excellent heat insulation system. (Sometimes infrared cameras cannot even detect them.) Polar-bear hairs are actually hollow inside. Explain how

this helps the bears maintain their body temperature in the cold winter.

64. **CQ** The SR-71 Blackbird (whose color is black) is one of the world's fastest planes, flying at over 2350 mph. During flight, air resistance makes the plane's surface very hot. If the airplane were painted white rather than black, what would happen to the temperature of the plane?

65. **IE** ■ Assume that a tile floor and an oak floor each have the same temperature and thickness. (a) Compared with the oak floor, the tile floor will conduct heat from your bare feet (1) faster, (2) at the same rate, or (3) slower. Why? (b) Calculate the ratio of the rate of heat flow of the tile floor to that of the oak floor.

66. ■■ Assume that your skin has an emissivity of 0.70, a normal temperature of 34°C, and a total exposed area of 0.25 m². How much heat energy per second do you lose due to radiation per second if the outside temperature is 22°C?

67. ■■ The glass pane in a window has dimensions of 2.00 m × 1.50 m and is 4.00 mm thick. How much heat will flow through the glass in 1.00 h if there is a temperature difference of 2 C° between the inner and outer surfaces? (Consider conduction only.)

68. ■■ Assuming that the human body has a 1.0-cm-thick layer of skin tissue and a surface area of 0.30 m², estimate the rate at which heat is conducted from inside the body to the surface if the skin temperature is 33°C. (Assume a normal body temperature of 37°C for the temperature of the interior.)

69. ■■ A copper teakettle with a circular bottom 30 cm in diameter has a uniform thickness of 2.5 mm. It sits on a burner whose temperature is 150°C. (a) If the teakettle is full of boiling water, what is the rate of heat conduction through its bottom? (b) Assuming that the heat from the burner is the only heat input, how much water is boiled away in 5.0 min? Is your answer reasonable? If not, explain why.

70. **IE** ■■ An aluminum bar and a copper bar of identical cross-sectional area have the same temperature difference between their ends and conduct heat at the same rate. (a) The copper bar is (1) longer, (2) of the same length, or (3) shorter than the aluminum bar. Why? (b) Calculate the ratio of the length of the copper bar to that of the aluminum bar.

71. ■■ A large Styrofoam™ cooler has a surface area of 1.0 m² and a thickness of 2.5 cm. If 5.0 kg of ice at 0°C is stored inside and the outside temperature is a constant 35°C, how long does it take for all the ice to melt? (Consider conduction only.)

72. ■■ The thermal insulation used in building is commonly rated in terms of its *R-value*, defined as d/k, where d is the thickness of the insulation in inches and k is the thermal

conductivity. (See the Insight on the construction industry and energy conservation on p. 394.) For example, 3.0 in. of foam plastic would have an R-value of 3.0/0.30 = 10, where, in British units, $k = 0.30$ Btu·in./(ft²·h·F°). This value is expressed as R-10. (a) What does the R-value tell you? That is, how does thermal insulation vary with R-value? (b) What thicknesses of (1) fiberboard and (2) brick would give an R-value of R-10? (*Hint*: Use known ratios for unit conversion.)

73. ■■ Pine wood that is 14 in. thick has an R-value of 19. What thickness of (a) glass wool and (b) fiberboard would have the same R-value? (See Exercise 72 and the Insight on the construction industry and energy conservation on p. 392.)

74. ■■ A large picture window measures 2.0 m by 3.0 m. At what rate will heat be conducted through the window when the room temperature is 20°C and the outside temperature is 0°C (a) if the window has a single pane of glass 4.0 mm thick and (b) if the window has a double pane of glass (a thermopane), where each pane is 2.0 mm thick, with an intervening air space of 1.0 mm? (Assume that there is a constant temperature difference and consider conduction only.)

75. IE ■■ The emissivity of an object is 0.60. (a) Compared with a perfect blackbody at the same temperature, this object would radiate (1) more, (2) the same amount of, or (3) less power. Why? (b) Calculate the ratio of the power radiated by the blackbody to that radiated by the object.

76. ■■ The wall of a house is composed of a solid concrete block with outside brick veneer and is faced on the inside with fiberboard, as illustrated in ▼Fig. 11.22. If the outside temperature on a cold day is −10°C and the inside temperature is 20°C, how much energy is conducted through a wall with dimensions of 3.5 m × 5.0 m in 1.0 h?

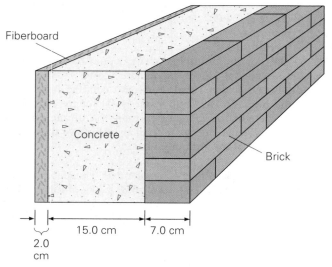

▲ **FIGURE 11.22 Thermal conductivity and heat loss** See Exercise 76.

Fiberboard

Concrete

Brick

15.0 cm   7.0 cm
2.0 cm

77. ■■ Suppose you wished to cut the heat loss through the wall in Exercise 76 by 50% by installing insulation. What thickness of Styrofoam™ would have to be placed between the fiberboard and concrete block to accomplish this goal?

78. IE ■■ (a) If the Kelvin temperature of an object is doubled, the power radiated increases (1) 2, (2) 4, (3) 8, or (4) 16 times. Why? (b) If the temperature is increased from 20°C to 40°C, how is the power radiated changed?

79. ■■ A steel cylinder of radius 5.0 cm and length 4.0 cm is placed in end-to-end thermal contact with a copper cylinder of the same dimensions. If the free ends of the two cylinders are maintained at constant temperatures of 95°C (steel) and 15°C (copper), how much heat will flow through the cylinders in 20 min?

80. ■■■ For the metal cylinders in Exercise 79, what is the temperature at the interface of the cylinders?

81. ■■■ Solar heating takes advantage of solar collectors such as the type shown in ▼Fig. 11.23. About 50% of the solar radiation received at the top of the atmosphere reaches the Earth during daylight hours. (The rest is reflected, scattered, absorbed, and so on.) How much heat energy would be collected, on average, by the cylindrical collector shown in the figure during 10 h of collection? [*Hint*: The average intensity of solar radiation is about 1400 J/(m²·s).]

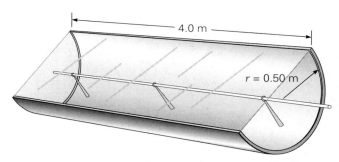

4.0 m

$r = 0.50$ m

▲ **FIGURE 11.23 Solar collector and solar heating** See Exercise 81.

Additional Exercises

82. A modern engine of alloy construction consists of 25 kg of aluminum and 80 kg of iron. How much heat does the engine absorb as its temperature increases from 20°C to 120°C as it warms up to operating temperature?

83. A 0.200-kg glass cup at 20°C is filled with 0.40 kg of hot water at 90°C. Neglecting any loss, what is the final equilibrium temperature of the water?

84. A lamp filament radiates 100 W of power when the temperature of the surroundings is 20°C, and it radiates 99.5 W of power when the temperature of the surroundings is 30°C. If the temperature of the filament is the same in each case, what is that temperature on the Celsius scale?

85. A student doing an experiment pours 0.150 kg of heated copper shot into a 0.375-kg aluminum calorimeter cup containing 0.200 kg of water at 25°C. The mixture (and the cup) comes to thermal equilibrium at 28°C. What was the initial temperature of the shot?

86. A student mixes 1.0 L of water at 40°C with 1.0 L of ethyl alcohol at 20°C. Assuming that there is no heat lost to the container or the surroundings, what is the final temperature of the mixture? [*Hint*: See Table 11.1.]

87. After performing a barrel jump, a 65-kg ice skater traveling at 25 km/h glides to a stop. If 40% of the frictional heat generated by the skate blades goes into melting ice at 0°C, how much ice is momentarily melted? Where does the other 60% of the energy go? (See the related Insight, "Phase Change and Ice Skating," on p. 386.)

88. A 0.030-kg lead bullet hits a steel plate, both initially at 20°C. The bullet melts and splatters on impact. (This action has been photographed.) Assuming that the bullet receives 80% of its kinetic energy as heat energy, at what minimum speed must it be traveling to melt on impact?

89. IE Initially at 20°C, 0.50 kg of aluminum and 0.50 kg of iron are heated to 100°C. (a) Aluminum gains (1) more heat than iron, (2) the same amount of heat as iron, or (3) less heat than iron. Why? (b) Calculate the difference in heat required to prove your answer to (a).

90. A waterfall is 75 m high. If all of the gravitational potential energy of the water were converted into heat energy, by how much would the temperature of the water increase in going from the top to the bottom of the falls? [*Hint*: Consider a kilogram of water going over the falls.]

91. How much heat is released when ice at −10°C is formed from 0.75 kg of steam at 120°C?

92. CQ Water is most dense at 4°C. Fish can survive in severe winter temperatures in modestly deep lakes and ponds because the water temperature near the bottom is either at or slightly above 4°C. Can you explain why?

93. Newton's law of cooling (Sir Isaac was a busy man) states that, in general, the heat-loss rate from an object is proportional to the temperature difference between it and the surroundings: $\Delta Q/\Delta t = K\Delta T$. K is a constant that includes losses by conduction, convection, and radiation. Apply this law to the following question: Would a cup of hot coffee stay hotter longer if you put cream in it right away or waited until a later time, when you were ready to drink it?

CHAPTER 12

Thermodynamics

At certain locations on the Earth, water from hot springs deep in the interior rises to the surface. In Yellowstone National Park, this produces boiling pools and geysers such as Old Faithful, shown here. On Iceland, the hot water warms the ocean and can create warm lagoons surrounded by glaciers. Such intriguing settings are popular vacation spots.

But the uses of such hot springs extend beyond recreation. Iceland's capital city of Reykjavik is heated by piping the hot-spring water to homes and businesses. Furthermore, whenever there is a temperature difference, the potential exists for obtaining useful work. For instance, geothermal power plants draw on the energy of the geysers as a renewable resource to generate electric energy while producing virtually no pollution. In this chapter, you'll learn under what conditions, and with what efficiency, heat can be exploited to perform work, in machines as different as automobile engines and home freezers. You'll find that the laws governing such energy conversions include some of the most general and far-reaching laws in all of physics.

As the word implies, **thermodynamics** deals with the transfer (dynamics) of heat (the Greek word for "heat" is *therme*). The development of thermodynamics started about 200 years ago and grew out of efforts to develop heat engines. The steam engine was one of the first such devices, designed to convert heat to mechanical work. Steam engines in factories and locomotives powered the Industrial Revolution that changed the world. Although our study is primarily concerned with heat and work, thermodynamics is a broad and comprehensive science that includes a great deal more than heat-engine theory.

In this chapter, you will learn about the laws on which thermodynamics is based, as well as the concept of entropy.

12.1 Thermodynamic Systems, States, and Processes

OBJECTIVES: To (a) define thermodynamic systems and states of systems, and (b) explain how thermal processes affect such systems.

Thermodynamics is a field that describes systems with so many particles—think of the number of molecules in a gas sample—that using ordinary dynamics (Newton's laws) to keep track of them is impossible. Therefore, even though the underlying physics is the same as for other systems, we generally use alternative (macroscopic) variables, such as pressure and temperature, to describe thermodynamic systems as a whole. Because of this difference in "language," it is important to become familiar with the terms and definitions at the outset.

The term **system**, as used in thermodynamics, refers to a definite quantity of matter enclosed by boundaries or surfaces, either real or imaginary. For example, a quantity of gas in the piston cylinder of an engine has real boundaries, and imaginary boundaries enclose a cubic meter of air in a room. The boundaries of a system need not have a definite shape nor enclose a fixed volume. For example, a cylinder of gas experiences a volume change when the piston is moved.

Occasionally, it is necessary to consider systems between which matter is transferred. However, for the most part, we will consider systems of constant mass. More important will be the interchange of energy between a system and its surroundings. This exchange may occur through a transfer of heat and/or the performance of mechanical work. For example, if a balloon is warmed (meaning that heat is transferred into it), it can expand and do work on its containing surface (its outside latex "skin") by exerting a force through a distance, as discussed in Chapter 5.

If heat transfer into or out of the system is impossible, the system is said to be a **thermally isolated system**. However, work may be done on a thermally isolated system, thus transferring energy to it. For example, a thermally isolated cylinder (perhaps surrounded by heavy insulation) filled with gas can be compressed by an external force exerted on a piston cover. As such, work is done on the system. As we know, work is a way of transferring energy.

When heat does enter or leave a system, it is absorbed from, or given up to, the surroundings or to heat reservoirs. A **heat reservoir** is a large separate system assumed to have unlimited heat capacity. Any amount of heat can be withdrawn from or added to a heat reservoir without appreciably changing its temperature. For example, pouring a bottle of warm water into a cold lake does not noticeably raise the lake's temperature. The lake is a low-temperature heat reservoir.

State of a System

Just as there are kinematic equations to describe the motion of an object, there are **equations of state** to describe the conditions of thermodynamic systems. Such an equation expresses a mathematical relationship between the thermodynamic variables of a system. The ideal gas law, $pV = nRT$ (Section 10.3), is an example of an equation of state. This expression establishes a relationship among the pressure (p), volume (V), absolute temperature (T), and number of moles (n, or equivalently, N, the number of molecules, since from Section 10.3, we know that $N = nN_A$) of a gas. These ideal gas quantities are examples of *state variables*. Clearly, then, different states have different sets of values for these variables.

For an ideal gas, a set of these three variables (p, V, and T) that satisfies the ideal gas law specifies its state completely as long as the system is in thermal equilibrium and has a uniform temperature. It is convenient to plot the states accord-

Note: Because p, V, and T are connected by the ideal gas law, specifying the values of any two of these variables automatically tells you the value of the third variable.

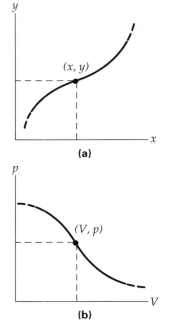

(a)

(b)

▲ **FIGURE 12.1 Graphing states**
(a) On a Cartesian graph, the coordinates (x, y) represent an individual point. **(b)** Similarly, on a $p–V$ graph or diagram, the coordinates (V, p) represent a particular state of a system. [It is common to say $p–V$, rather than $V–p$, diagram to match the order of the coordinates (p, V, T) in the ideal gas law.]

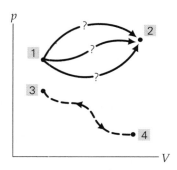

▲ **FIGURE 12.2 Paths of reversible and irreversible processes** If a gas quickly goes from state 1 to state 2, the process is irreversible, since we do not know the "path." If, however, the gas is taken through many closely spaced equilibrium states (as in going from state 3 to state 4), the process is reversible (from state 4 back to state 3), in principle. Reversible means "exactly retraceable."

ing to the thermodynamic coordinates (p, V, T), much as we plot graphs using Cartesian coordinates (x, y, z). A general two-dimensional illustration of such a plot is shown in ◄Fig. 12.1.

Just as the coordinates (x, y) specify individual points on a Cartesian graph, the coordinates (V, p) specify individual *states* on the $p–V$ graph or diagram. This is because the ideal gas law, $pV = nRT$, can be solved for the unique temperature of a gas if we know the gas's pressure and volume and the number of molecules or moles in the sample. In other words, on a $p–V$ diagram, each "coordinate" gives the pressure and volume of a gas directly and the temperature of the gas indirectly. Thus, to describe a gas completely, only a $p–V$ plot is necessary. In some cases, however, it can be instructive to refer to other plots, such as $p–T$ or $T–V$ plots. (Notice that Fig. 12.1b could illustrate a phenomenon that you might be familiar with—reduction of the pressure of a gas resulting in its expansion.)

Processes

A **process** is any *change* in the state, or the thermodynamic coordinates, of a system. For instance, when an ideal gas undergoes a process, its state variables p, V, and T will, in general, all change. Suppose a gas initially in state 1, described by state variables (p_1, V_1, T_1), changes to a second state, state 2. State 2 will, in general, be described by a different set of state variables (p_2, V_2, T_2). A system that has undergone a change of state has been subjected to a *thermodynamic process*.

Processes are classified as either reversible or irreversible. Suppose that a system of gas in equilibrium (with known p, V, and T values) is allowed to expand quickly when the pressure on it is reduced. The state of the system will change rapidly and unpredictably, but eventually the system will return to a different state of equilibrium, with another set of thermodynamic coordinates. On a $p–V$ diagram (◄Fig. 12.2), we would be able to show the initial and final states (labeled 1 and 2, respectively), but *not* what happened in between them. This type of process is called an **irreversible process**—a process for which the intermediate steps are nonequilibrium states. "Irreversible" does not mean that the system can't be taken back to the initial state; it means only that the process path can't be retraced, because of the nonequilibrium conditions that existed.

If, however, the gas changes state very, very slowly, passing from one equilibrium state to a neighboring one and eventually arriving at the final state (see Fig. 12.2, initial and final states 3 and 4, respectively), then the process path is known. In such a situation, the system could be brought back to its initial conditions by "traveling" the path in the opposite direction, re-creating every intermediate state (again, in many small steps) along the way. Such a process is called a **reversible process**. In practice, a perfectly reversible process cannot be achieved. All real thermodynamic processes are irreversible to some degree, because they follow complicated paths with many intermediate nonequilibrium states. However, the concept of an ideal reversible process is useful and will be our primary tool in discussing the thermodynamics of an ideal gas.

12.2 The First Law of Thermodynamics

OBJECTIVES: To **(a)** explain the relationship among internal energy, heat, and work as expressed by the first law of thermodynamics, and **(b)** learn the technique for calculating work done by gases.

Recall from mechanics (Chapter 5) that work describes the transfer of energy from one object to another by application of a force. For example, when you push on a chair initially at rest, some of the work you do (exerting a force through a distance) on the chair goes into increasing its kinetic energy. At the same time, you lose an amount of stored (chemical) energy in your body equal to the amount of work you do. This type of work is done in an *orderly* way, that is, by applying

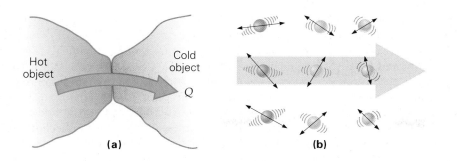

FIGURE 12.3 Heat flow (via conduction) on the atomic scale (a) Macroscopically, heat is transferred by conduction from the hot object to the cold one. (b) On the atomic scale, heat conduction is explained as the energy transfer from the more energetic atoms (in the hotter object) to the less energetic atoms (in the colder object). This transfer of energy from an atom to its neighbor results in the heat transfer we observe in part (a).

various forces, in well-defined directions, on an object of interest. For example, when a gas (enclosed in a cylinder and fitted with a piston) is allowed to expand, it does work on the piston at the expense of some of its internal energy. From Chapter 10, we now know there is a second way to change the energy of a system—by adding or removing heat energy. Thus, internal energy is lost by a hot object when the heat is transferred to a cold object, which then gains internal energy. This process changes both objects' internal energies, but in opposite ways.

Although we can't see the actual process, heat transfer is really the same concept as the work we know from mechanics, but on a microscopic (atomic) level. During a conduction process, for example, energy is transferred from a hot solid object to a colder solid object, because the faster vibrating atoms of the warmer object do work on the slower molecules of the colder object (▲Fig. 12.3). This energy is then transferred further into the volume of the cold object as more work is done on the neighboring (slower vibrating) atoms. This ongoing process is the "flow" or "transfer" of energy we observe macroscopically as heat transfer.

The first law of thermodynamics describes how work and heat are related to a system's internal energy. This law is a restatement of energy conservation in terms of thermodynamic variables. It relates the change in internal energy (ΔU) *of a system* to the work (W) done *by that system* and the heat energy transferred (Q) *to or from that system*. Depending on the conditions, heat transfer Q can result in a change in that system's internal energy, ΔU. However, because of the heat transfer, the system might do work on the environment. Thus, heat transferred to a system can end up in two places: a change in the internal energy of the system and/or work done by the system. Therefore, the first law of thermodynamics is usually written as

$$Q = \Delta U + W \qquad (12.1)$$

First Law of Thermodynamics

As always, it is important to remember what the symbols mean and what their sign conventions denote (shown in ▼Fig. 12.4). Q is the heat added to or removed

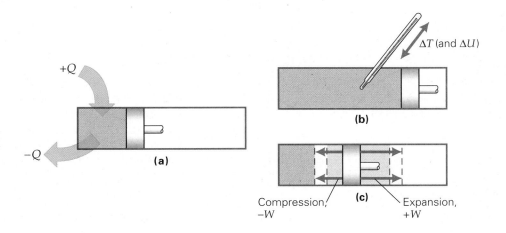

◀ FIGURE 12.4 Sign conventions for Q, W, and ΔU (a) If heat flows into a system, it is positive. Heat flowing out is designated as negative. (b) The experimental way to tell if a gas's internal energy changes is to take its temperature. Since internal energy is determined by temperature, a rise or fall in one of these quantities implies a similar rise or fall in the other. (c) If a gas expands (pushes in the direction its piston moves), the work it does is positive. If the gas is compressed, the work done by the gas is designated as negative.

from the system, ΔU is the change in internal energy *of the system* and W is the work done *by the system* (on the environment).*

The first law of thermodynamics can be compared with a financial transaction you might make at an automatic teller machine (ATM). Suppose you have a savings account (stored money), which is analogous to internal energy (stored energy). In addition, at the same bank, you have a checking account (which, analogous to work, enables you to do things). Q would be analogous to deposits of money to or removals of money from either account. For example, if you deposit a check for $1000, you can choose to put $600 in savings and $400 in checking. Then, with the money in checking, you would be able to pay some of your bills (i.e., do some useful work).

Similarly, in thermodynamics, a gas might absorb 1000 J of heat and do 400 J of work on the environment, thus leaving 600 J as the increase in the gas's internal energy. If the gas were to do more than 400 J of work, less energy would be stored in the internal energy "account." Thus, the first law does *not* tell you how much energy goes into ΔU or W. These amounts depend, as we shall see, on the system's conditions (constant pressure, constant volume, etc.) when the heat energy is transferred (Section 12.3).

Consider the application of the first law of thermodynamics to weight loss.

Example 12.1 ■ Energy Balancing: Dieting Using Physics

A man usually eats an average of 3000 kcal (of food energy) per day. With this intake, he has maintained his weight. However, he is overweight, and his physician suggests that he lose 50 lbs. Assume that the weight to be lost is entirely in the form of fat and that the person must reduce his caloric intake to 2000 kcal per day, while maintaining his current exercise and work routines. "Burning up" a gram of fat requires about 9.3 kcal. How long will it take the patient to lose the 50 lbs of fat?

Thinking It Through. Since the patient's weight has been steady, it must be that the 3000 kcal have *not* been going into increasing internal energy (fat-cell storage)—(i.e., $\Delta U = 0$). All of the food-intake energy (3000 kcal) is converted into heat to keep his body temperature constant and/or into mechanical work (W), such as walking, running, typing, and talking. On the diet, his body will tap into the internal energy storage in fat cells to make up the input difference. This will result in a loss of fat (mass).

Solution. The values given are listed below. Note the use of subscripts: A subscript 1 refers to the current intake, and a subscript 2 signifies intake under the restricted diet.

Given: $Q_{in,1} = 3000$ kcal/d

$Q_{in,2} = 2000$ kcal/d

energy value of fat $= 9.3$ kcal/g

$\qquad = 9.3 \times 10^3$ kcal/kg

$m_{fat} = (50.0 \text{ lb})\left(\dfrac{1 \text{ kg}}{2.20 \text{ lb}}\right) = 22.7$ kg

Find: Time t to burn up the energy content in 50 lbs of fat

Energy conservation can be used to describe the patient's situation. Let's denote Q_{out} as the normal heat loss, W as all types of mechanical work done, and $Q_{in,1}$ as the dietary-intake energy, all on a per-day basis, when the patient is on his normal diet. Since $\Delta U_1 = 0$ we have

$$Q_{in,1} = Q_{out} + W$$

Under the restricted diet, ΔU_2 must be negative. Energy conservation for the restricted diet can be written as

$$Q_{in,2} = Q_{out} + \Delta U_2 + W$$

*In some chemistry and engineering books, the first law of thermodynamics is written as $Q = \Delta U - W'$. The two equations are the same, but have a different emphasis. In this expression, W' means the work done *by the environment on the system* and is thus the negative of our work W (why?), or $W = -W'$. The first law was discovered by researchers interested in building heat engines (Sections 12.5 and 12.6). Their emphasis was on finding the work done *by* the system, W, not W'. Since we want to understand heat engines, we adopt the historical definition: W *means the work done by the system.*

or, after we substitute for $Q_{out} + W$ from the first equation, the result is

$$\Delta U_2 = Q_{in, 2} - Q_{in, 1}$$
$$= 2000 \text{ kcal/d} - 3000 \text{ kcal/d}$$
$$= -1000 \text{ kcal/d}$$

As expected, the patient will, with less intake, eliminate the fat-energy equivalent of 1000 kcal per day. The total amount of energy to be removed is

$$(9.3 \times 10^3 \text{ kcal/kg}) \times (22.7 \text{ kg}) = 2.1 \times 10^5 \text{ kcal}$$

The amount of time is therefore

$$t = \frac{2.1 \times 10^5 \text{ kcal}}{1000 \text{ kcal/d}} = 2.1 \times 10^2 \text{ d} \approx 7 \text{ months}$$

Follow-up Exercise.   In this Example, by how much would the time t have been reduced if, in addition to trimming his dietary intake, the patient had burned an additional 100 kcal per day? (*Answers to all Follow-up Exercises are at the back of the text.*)

When applying the first law of thermodynamics, the proper use of signs (shown in Fig. 12.4) cannot be overemphasized. The signs for work are easy to remember if you keep in mind that positive work is done by a force that acts generally in the direction of the displacement, such as when a gas expands. Similarly, negative work means that the force acts generally opposite to the direction of the displacement as when a gas contracts.

But how do you compute the work done by the gas? To answer this question, consider a cylindrical piston with end area A, containing a known sample of gas (▶Fig. 12.5). Let us imagine that the gas is allowed to expand over a very small distance Δx. If the volume of the gas does not change appreciably, then the pressure remains constant. In moving the piston slowly and steadily outward, the gas does positive work on the piston. Thus from the definition of work,

$$W = F \Delta x \cos 0° = F \Delta x$$

In terms of pressure, $p = F/A$, or $F = pA$. Substituting for F, we have

$$W = pA \Delta x$$

But $A \Delta x$ is the volume of a right cylinder with end area A and height Δx. Here, that volume represents the change in volume of the gas, or $\Delta V = A \Delta x$. Thus,

$$W = p \Delta V$$

Note that the work done in Fig. 12.5 is positive because ΔV is positive. If the gas contracts, the work is negative because the volume change is negative ($\Delta V = V_2 - V_1 < 0$).

Of course, gases don't always change their volumes by small amounts and aren't usually subject to constant pressure. In fact, the changes in volume and pressure can be significant. How do we handle the calculation of work under these circumstances? The answer is seen in ▶Fig. 12.6. Here, we have a reversible path on a p–V diagram. Notice that during each small step, the pressure remains approximately constant. Therefore, for each step, we approximate the work done by $p \Delta V$. Graphically, this quantity is just the area of a small narrow rectangle, extending from the process curve to the V-axis. However, as the volume changes, so does the pressure. Therefore, to approximate the *total* work, we add up these small amounts of work or $W \approx \Sigma p \Delta V$. To get an exact value, think of the area as made up of a large number of very thin rectangles. As the number of rectangles becomes infinitely large, each rectangle's thickness approaches zero. This process involves calculus and is beyond the scope of this book. However, it should be clear that the following is true:

The work done by a system is equal to the area under the process curve on a p–V diagram.

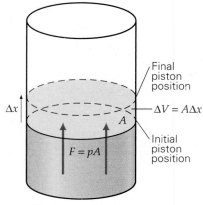

$$W = F \Delta x = pA \Delta x = p \Delta V$$

▲ **FIGURE 12.5   Work in thermodynamic terms**   If a gas expands a very small amount, its pressure remains constant. The small amount of work done by the gas is $p \Delta V$.

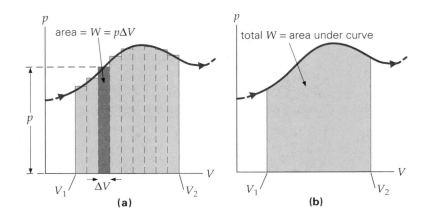

▶ FIGURE 12.6 Thermodynamic work as the area under the process curve (a) If a gas expands by a significant amount, the work done can be computed by treating the expansion in little steps, each one yielding a small amount of work. The total work is determined (approximately) by adding up the many rectangular strips. **(b)** If the number of rectangular strips becomes large, and each one becomes very thin, the calculation of the area becomes exact. The work done is equal to the area between the process curve and the V-axis.

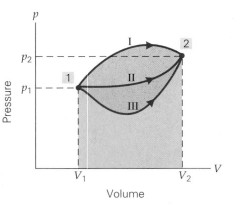

▲ FIGURE 12.7 Thermodynamic work depends on the process curve This graph shows the work done by a gas as it expands the same amount, but by three different processes. The work done during process I is larger than process II's work, which in turn is larger than process III's work. Fundamentally, applying a larger force (pressure) through the same distance (volume change) requires more work. Process I includes the blue, green and pink areas; process II includes just the green and pink areas; and process III includes just the pink area.

Note: *Isothermal means constant temperature.*

Before we discuss specific types of processes, note that there is a fundamental difference between U and both Q and W. Any system "contains" a certain amount of internal energy U. However, it is wrong to say that a system "possesses" certain "amounts" of heat or work, as these quantities represent energy *transfers*, not total energies. A further distinction is that both Q and W depend on how the gas gets from its initial to its final state, whereas ΔU does not. The heat added to, or removed from, a system *depends on the conditions* under which this transfer is done (Section 11.2). Similarly, from the area-under-the-curve representation, the work *depends on the path* (◀Fig. 12.7). For example, more work is done if the process takes place at higher pressures. This situation is represented as a larger area, with more work done by a larger force with the same volume change.

Contrast these properties with those of ΔU for an ideal gas. To find ΔU, we need know only the internal energies at the ends of the path. This is because for an ideal gas (with a fixed number of moles), the internal energy depends only upon the absolute temperature of the gas. For that case (see Chapter 10), $U \propto nRT$; thus, $\Delta U = U_f - U_i$ depends only on ΔT. To summarize, the change in the internal energy, ΔU, is *independent* of the process path, whereas Q and W both *depend* on the path.

12.3 Thermodynamic Processes for an Ideal Gas

OBJECTIVES: To (a) describe and understand the four fundamental thermodynamic processes using an ideal gas and (b) analyze the work done, heat flow, and change in internal energy that occurs during each of these processes.

The first law of thermodynamics can be applied to several processes for a system consisting of an ideal gas. Note that in three of the processes, one thermodynamic variable is kept constant. Such processes have names that begin with *iso-* (from the Greek *isos*, meaning "equal").

Isothermal Process. An **isothermal process** is a constant-temperature process (*iso* for equal, *thermal* for temperature). In this case, the process path is called an *isotherm*, or a line of constant temperature. (See ▶ Fig. 12.8.) The ideal gas law may be rewritten as $p = nRT/V$. Since the gas remains at constant temperature, nRT is a constant. Therefore, p is inversely proportional to V, that is, $p \propto 1/V$, which is a hyperbola. (Recall that a hyperbola is written as $y = a/x$ or $y \propto 1/x$, and it plots as a downward-curving line on a set of x–y axes.)

In the expansion from state 1 (initial) to state 2 (final) in Fig. 12.8, heat is added to the system, while both the pressure and volume vary in such a way as to keep the temperature constant. Positive work is done by the expanding gas. On an isotherm, $\Delta T = 0$; therefore, $\Delta U = 0$. The heat added to the gas is exactly equal to the amount of work done by the gas, and none of the heat goes into increasing the gas's internal

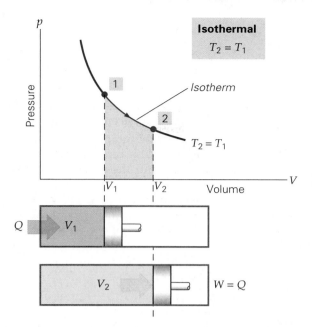

energy. See the Learn by Drawing feature "Leaning on Isotherms," on p. 412. (What would be the ATM analogy here? See p. 408.)

In terms of the first law of thermodynamics, we can write

$$Q = \Delta U + W = 0 + W$$

or

$$Q = W \quad \text{(ideal gas isothermal process)} \quad (12.2)$$

The magnitude of the work done on the gas is equal to the area under the curve (requiring calculus to compute). We simply state it as follows:

$$W_{\text{isothermal}} = nRT \ln \left(\frac{V_2}{V_1} \right) \quad \text{(ideal gas isothermal process)} \quad (12.3)$$

Work Done at Constant Temperature

Since the product nRT is a constant along a given isotherm, the work done depends on the ratio of the end-point volumes.

Problem-Solving Hint

In Eq. 12.3, the function "ln" stands for *natural logarithm*. Recall that "regular," or "common," logarithms are referenced to the base 10. For this type, the exponent of the base 10 is the logarithm of the number in question. For example, since $100 = 10^2$, the logarithm of 100 is 2, or, in equation terms, $\log 100 = 2$. In general, if $y = 10^x$, then x is the logarithm of y, or $x = \log y$. The *natural logarithm* is similar, except it uses a different base, e, which is an irrational number ($e \approx 2.7183$ K). On most calculators, there is a button labeled "ln" for the natural logarithm, and it is suggested that you familiarize yourself with it before proceeding. As a check, find the natural logarithm of 100. (The answer is $\ln 100 = 4.605$ to four significant figures.)

Isobaric Process A constant-pressure process is called an **isobaric process** (*iso* for equal, and *bar* for pressure). An isobaric process for an ideal gas is illustrated in ► Fig. 12.9. On a $p–V$ diagram, an isobaric process is represented by a horizontal line called an *isobar*. When heat is added to or removed from an ideal gas at constant pressure, the ratio V/T remains constant (since $V/T = nR/p$ = constant). As the heated gas expands, its temperature must increase, and the gas crosses to higher temperature isotherms. This temperature increase means that the internal energy of the gas increases, since $\Delta U \propto \Delta T$.

Note: *Isobaric* means *constant pressure.*

Learn by Drawing

When you are analyzing thermodynamic processes, it is sometimes hard to keep track of the signs of heat flow (Q), work (W), and internal-energy change (ΔU). One method that can help with this bookkeeping is to superimpose a series of isotherms on the p–V plot you are working with (as in Figs. 12.8 through 12.12). This method is useful even if the situation you are studying does not involve isothermal processes.

Before starting, recall that an isothermal process is one in which the temperature remains constant:

1. In an isothermal process for an ideal gas, ΔU is zero. (Why?)
2. Since T is constant, pV must also be constant, since, from the ideal gas law (Eq. 10.3), $pV = nRT =$ constant. You may recall from algebra that $p = k/V$ is the equation of a hyperbola. Thus, on a p–V diagram, an isothermal process is described by a hyperbola. The farther from the axes the hyperbola is, the higher is the temperature it represents (Fig. 1).

To take advantage of these properties, follow these steps:

- Sketch a set of isotherms for a series of increasing temperatures on the p–V plot (Fig. 1).
- Then sketch the process you are analyzing—for example, the isobar shown in Fig. 2. [Take a moment to recall what the names mean: isomet = constant volume (vertical line), isobar = constant pressure (hori-

zontal line), and adiabat = no heat flow (downward sloping curve, steeper than an isotherm)].

- Next, use the graphs to determine the signs of W and ΔU. W is represented by the area under the p–V curve for the process represented, and its sign is determined by whether the gas expanded ($+$) or was compressed ($-$). The sign of ΔT will be clear from the isotherms, since they serve as a temperature scale, with progressively higher temperatures represented by isotherms further away from the axes. For example, a rise in T implies an increase in U.

- Last, determine the sign of Q from the first law of thermodynamics, $Q = \Delta U + W$. From the sign of Q, it should be clear if heat was transferred into or out of the system.

The example in Fig. 2 shows the power of this visual approach. Here, we are to decide whether heat flows into or out of the gas during an isobaric expansion. Expansion implies positive work done by the gas. But what is the direction of the heat flow (or is it zero)? Sketching the isobar, we see that it crosses from lower temperature isotherms to higher temperature ones. Hence, there is a temperature increase, and ΔU is positive. From $Q = \Delta U + W$, we see that Q is the sum of two positive quantities, ΔU and W. Therefore, Q must be positive, which means that heat enters the gas.

As an exercise, try analyzing an isometric process, using this graphical approach. See also Integrated Examples 12.2 and 12.5.

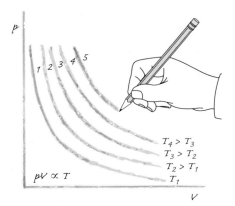

FIGURE 1   Isotherms on a p–V plot

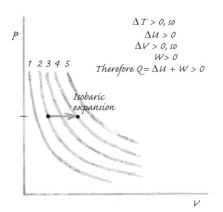

FIGURE 2   An isobaric expansion

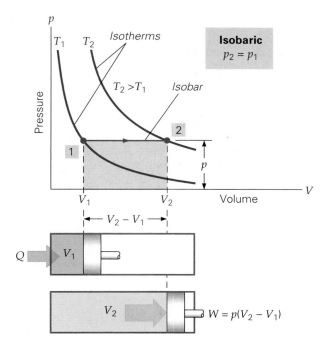

▶ **FIGURE 12.9 Isobaric (constant-pressure) process** The heat added to the gas in the frictionless piston goes into work done by the gas and into changing the internal energy of the gas: $Q = \Delta U + W$. The work is equal to the area under the isobar (from state 1 to state 2 here, shown as shaded) on the p–V diagram. Note the two isotherms. They are not part of the isobaric process, but they help show us that the temperature rises during the isobaric expansion.

As you can see from the isobar in Fig. 12.9, the area representing the work is rectangular. Thus, the work is relatively easy to compute (just length times width):

$$W_{\text{isobar}} = p(V_2 - V_1) = p\Delta V \qquad \text{(ideal gas isobaric process)} \qquad (12.4)$$

For example, when heat is added to or removed from a gas under isobaric conditions, the gas's internal energy changes *and* the gas expands or contracts, doing positive or negative works respectively. (See Integrated Example 12.2 for the signs.) We can write this relationship, using the first law of thermodynamics, with the work expression appropriate for isobaric conditions (Eq. 12.4):

$$Q = \Delta U + W = \Delta U + p\Delta V \qquad \text{(ideal gas isobaric process)} \qquad (12.5)$$

For example, if heat is added, the internal energy increases, and the gas does positive (expansional) work. To see a detailed comparison of an isobaric process and an isothermal process, consider the following Integrated Example.

Integrated Example 12.2 ■ Isotherms versus Isobars: What Is Your Sign?

Two moles of a monatomic ideal gas, initially at 0°C and 1.00 atm, is expanded to twice its original volume, using two different processes. It is expanded first isothermally and then, starting in the same initial state, isobarically. (a) During which process does the gas do more work, (1) the isothermal process, (2) the isobaric process, or (3) it does the same amount of work during both processes? Explain. (b) To prove your answer, determine the work done by the gas in each case.

(a) Conceptual Reasoning. As shown in ▶Fig. 12.10, both processes involve expansion. The isobar is horizontal, and the isotherm is a decreasing hyperbola. Thus, the gas does more work during the isobaric expansion (more area under the curve). Fundamentally, this is because the isobaric process is done at higher (constant) pressure than is the isothermal process (where the pressure drops as the gas expands). In both cases, the work is positive. (How do we know this?) Thus, the correct answer to part (a) is (2): The isobaric process does more work.

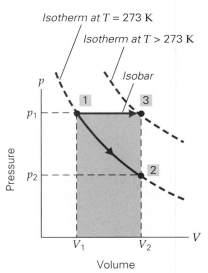

▲ **FIGURE 12.10 Comparing work** In Integrated Example 12.2, the gas does positive work while expanding. It does more work under isobaric conditions (from state 1 to state 3) than under isothermal conditions (from state 1 to state 2), because the pressure remains constant on the isobar, but decreases along the isotherm. (Compare areas under the curves.)

(b) Thinking It Through.   We can use Eqs. 12.3 and 12.4, if we know the volumes. These quantities can be calculated from the ideal gas law.

Solution.   Listing the data, we have the following:

Given: $p_1 = 1.00\,\text{atm} = 1.01 \times 10^5\,\text{N/m}^2$ *Find:* The work done during the
  $T_1 = 0°C = 273\,\text{K}$ isothermal and isobaric processes
  $n = 2.00\,\text{mol}$
  $V_2 = 2V_1$

For the isotherm, use Eq. 12.3 (the natural logarithm of the volume ratio is $\ln 2 = 0.693$):

$$W_{\text{isothermal}} = nRT \ln\left(\frac{V_2}{V_1}\right) = (2.00\,\text{mol})[8.31\,\text{J/(mol}\cdot\text{K)}](273\,\text{K})(\ln 2)$$

$$= +3.14 \times 10^3\,\text{J}$$

For the isobar, we need to know the two volumes. Using the ideal gas law, we obtain

$$V_1 = \frac{nRT_1}{p_1} = \frac{(2.00\,\text{mol})[8.31\,\text{J/(mol}\cdot\text{K)}](273\,\text{K})}{1.01 \times 10^5\,\text{N/m}^2}$$

$$= 4.49 \times 10^{-2}\,\text{m}^3$$

and therefore

$$V_2 = 2V_1 = 8.98 \times 10^{-2}\,\text{m}^3$$

The work is given by Eq. 12.4 as

$$W_{\text{isobar}} = p(V_2 - V_1)$$
$$= (1.01 \times 10^5\,\text{N/m}^2)(8.98 \times 10^{-2}\,\text{m}^3 - 4.49 \times 10^{-2}\,\text{m}^3) = +4.53 \times 10^3\,\text{J}$$

This amount is larger than the isothermal work, as expected from part (a).

Follow-up Exercise.   In this Example, what is the heat flow in each processes?

Note: *Isometric* means *constant volume.*

Isometric Process An **isometric process** (short for *isovolumetric*, or constant-volume, process), sometimes called an *isochoric process*, is a constant-volume process. As illustrated in ▼Fig. 12.11, the process path on a *p–V* diagram is a vertical line, commonly called an *isomet*. No work is done, since the area under such a curve is zero. (There is no displacement, so there is no change in volume.) Since the gas can do no work, if heat is added it must go completely into increasing the gas's internal energy and, therefore, its temperature. In terms of the first law of thermodynamics,

$$Q = \Delta U + W = \Delta U + 0 = \Delta U$$

▶ **FIGURE 12.11 Isometric (constant-volume) process** All of the heat added to the gas goes into increasing the gas's internal energy, since there is no work done ($W = 0$); thus, $Q = \Delta U$. (Notice the locking nut on the piston, which prevents any movement from occurring.) Again, the isotherms, although not part of the isometric process, tell us visually that the temperature of the gas rises.

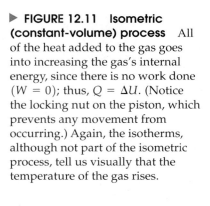

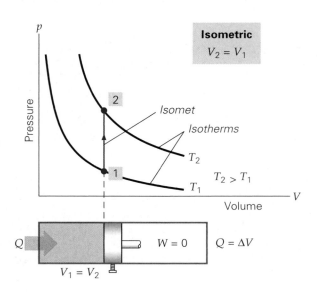

and thus

$$Q = \Delta U \quad \text{(ideal gas isometric process)} \quad (12.6)$$

Consider the following example of an isovolumetric process in action.

Example 12.3 ■ A Practical Isometric Exercise:
How NOT to Recycle a Spray Can

Many "empty" aerosol cans contain remnant propellant gases under approximately 1 atm of pressure (assume 1.00 atm) at 20°C. They display the warning "Do not dispose of this can in an incinerator or open fire." (a) Explain why it is dangerous to throw such a can into a fire. (b) Calculate the heat added to 0.50 L of such a gas when the can is thrown in a fire, assuming that it is an ideal diatomic gas initially at 20°C and comes to equilibrium at the fire's temperature of 2000°F. (c) What is the final pressure of the gas?

Thinking It Through. This process is isovolumetric; hence, all the heat goes into increasing the gas's internal energy. A pressure rise is expected, which is where the danger lies. The number of moles can be determined from the ideal gas law. Combining this with the temperature change, we can calculate the heat transfer, since we know the specific heat (how?). Lastly, the final pressure can be obtained using the ideal gas law.

Solution. The specific heats for gases at constant volume are listed in Table 11.2. We convert the given temperatures into kelvins. (Again, for qualitative reasoning, you may want to refer to the "Leaning on Isotherms" drawing feature on p. 417.)

Given: $p_1 = 1 \text{ atm} = 1.01 \times 10^5 \text{ N/m}^2$ *Find:* (a) Explain the danger in
$V_1 = 0.500 \text{ L} = 5.00 \times 10^{-4} \text{ m}^3$ heating the can.
$T_1 = 20°C = 293 \text{ K}$ (b) Q (heat added to gas)
$T_2 = 2000°F = 1.09 \times 10^3 °C = 1.37 \times 10^3 \text{ K}$ (c) p_2 (final pressure of
$c_v = 20.8 \text{ J/(mol} \cdot \text{K)}$ (from Table 11.1)   gas)

(a) When heat is added, it all goes into increasing the gas's internal energy. Because at consistant volume, pressure is proportional to temperature, the final pressure will be greater than 1 atm. The danger is that the container could explode into metallic fragments like a grenade if its maximum design pressure is exceeded.

(b) To calculate the heat added, we need the number of moles, n. We use the ideal gas law (Eq. 10.7), and solve for n:

$$n = \frac{pV}{RT} = \frac{(1.01 \times 10^5 \text{ N/m}^2)(5.00 \times 10^{-4} \text{ m}^3)}{[8.31 \text{ J/(mol} \cdot \text{K)}](293\text{K})} = 2.07 \times 10^{-2} \text{ mol}$$

From the molar specific heat at constant volume, we find Q:

$$Q = nc_v \Delta T = (2.07 \times 10^{-2} \text{ mol})[20.8 \text{ J/(mol} \cdot \text{K)}](1.37 \times 10^3 \text{ K} \cdot 293\text{K}) = +462 \text{ J}$$

(c) The final pressure of the gas is determined directly from the ideal gas law:

$$\frac{p_2 V_2}{T_2} = \frac{p_1 V_1}{T_1} \quad \text{or} \quad p_2 = p_1 \left(\frac{V_1}{V_2}\right)\left(\frac{T_2}{T_1}\right) = (1 \text{ atm})\left(\frac{V_1}{V_1}\right)\left(\frac{1.37 \times 10^3 \text{ K}}{293 \text{ K}}\right) = 4.68 \text{ atm}$$

Follow-up Exercise. (a) In this Example, what is the change in internal energy of the gas? (b) Suppose the can were designed to withstand pressures up to 3.5 atm. What would be the highest temperature it could sustain without exploding? (c) In part (b) how much heat energy is needed to bring the gas to that point?

Adiabatic Process In an **adiabatic process**, no heat is transferred into or out of the system. That is, $Q = 0$ (► Fig. 12.12). (The Greek word *adiabatos* means "impass-able.") This condition is satisfied for a thermally isolated system, one completely surrounded by "perfect" insulation. This is ideal situation, since there is some heat transfer even with the best of materials, if you wait long enough. Thus, for real-life conditions, we can only approximate adiabatic processes. For example, nearly

Note: In practice, adiabatic means *quickly*, before significant amounts of heat can flow in or out ($Q = 0$).

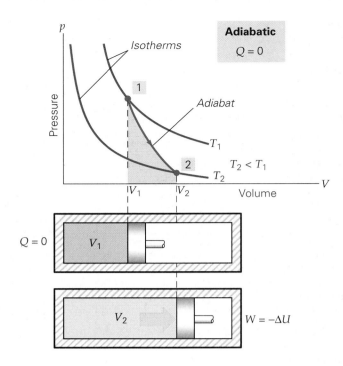

► **FIGURE 12.12 Adiabatic (no heat transfer) process** In an adiabatic process (shown here for a cylinder with heavy insulation), no heat is added to or removed from the system; thus, $Q = 0$. During expansion (shown here), positive work is done by the gas at the expense of its internal energy: $W = -\Delta U$. The pressure, volume, and temperature all change in the process. The work done by the gas is the shaded area between the adiabat and the V-axis.

adiabatic processes can take place if the changes occur rapidly enough so that there isn't time for significant heat to flow into or out of the system.

The curve for this process is called an *adiabat*. During an adiabatic process, all three thermodynamic coordinates (p, V, T) change. For example, if the pressure on a gas is reduced, the gas expands. However, no heat flows into the gas. Without a compensating input of heat, work is done at the expense of the gas's internal energy. Therefore, ΔU must be negative. Since the internal energy, and thus the temperature, both decrease, such an expansion is a cooling process. Similarly, an adiabatic compression is a warming process (temperature increase).

From the first law of thermodynamics, an adiabatic process can be described by,

$$Q = 0 = \Delta U + W$$

or

$$\Delta U = -W \quad \text{(adiabatic process)} \qquad (12.7)$$

For completeness, we state some other relationships in an adiabatic process. In such a process, an important factor is the ratio of the gas's molar specific heats, defined by a dimensionless quantity $\gamma = c_p/c_v$. We can determine this ratio from the values in Table 11.2, depending on the type of gas. For the two common types of gas molecules, monatomic and diatomic, the values of γ are about 1.66 and 1.40, respectively. (Check these values yourself.) The volume and pressure at any two points on an adiabat are related by

$$p_1 V_1^{\gamma} = p_2 V_2^{\gamma} \quad \text{(ideal gas adiabatic process)} \quad (12.8)$$

The work done by an ideal gas during an adiabatic process is

$$W_{\text{adiabatic}} = \frac{p_1 V_1 - p_2 V_2}{\gamma - 1} \quad \begin{matrix}\text{(ideal gas} \\ \text{adiabatic process)}\end{matrix} \qquad (12.9)$$

Conceptual Example 12.4 ■ Exhaling: Blowing Hot and Cold?

The air in your lungs is warm. This can be shown by putting your bare upper arm near your mouth and blowing air on your arm with your mouth opened widely. If you repeat this with your lips puckered, the air would feel (a) warmer, (b) cooler, or (c) the same.

Reasoning and Answer. We'll go from answer to reasoning on this one, since it is easy to determine the correct answer experimentally. Try it, and you might be surprised to find that the answer is (b).

The interesting thing is why this situation occurs. When you blow on your arm with mouth open, you get a gush of warm air (roughly at body temperature). However, when you blow with your lips puckered, the stream of air is compressed. Then, as it emerges, the air expands, doing positive work against the atmosphere. The process is approximately adiabatic, since it takes place quickly. From the first law, since $Q = 0$, $\Delta U = -W$; therefore, ΔU is negative, and the temperature drops. Eventually, the cooled air absorbs heat from the surrounding air, and its temperature rises.

Follow-up Exercise. Even during winter days with snow on the slopes, it is not uncommon in the Rocky Mountains to experience blasts of warm air coming down the slopes. (These gusts are called *Chinook winds*.) Explain how these winds could experience a significant rise in temperature while there is still snow and ice on the ground.

To clear up confusion that often occurs between isotherms and abiabats, see the following Integrated Example.

Integrated Example 12.5 ■ Adiabats versus Isotherms: Two Different Processes That Are Often Confused

A sample of helium expands to triple its initial volume adiabatically in one case and isothermally in another. In both cases, it starts from the same initial state. The sample contains 2.00 moles of helium, initially at 20°C and 1.00 atm. (a) During which process does the gas do more work, (1) the adiabatic process, (2) the isothermal process, or (3) the work done is the same during both processes? (b) Calculate the work done during each process to verify your reasoning in part (a).

(a) Conceptual Reasoning. To determine graphically which process involves more work, we look at the areas under the process curves (see Fig. 12.12). The area under the isothermal process curve is larger; thus, the gas does more work during its isothermal expansion, and the correct answer is (2). Physically, the isothermal expansion involves more work, because the pressures are always higher during the isothermal expansion than during the adiabatic one.

(b) Thinking It Through. To determine the isothermal work, we need the initial and final volumes, which can be determined using the ideal gas law. For the adiabatic work, the ratio of specific heats, γ, is important, as are the final pressure and volume. γ can be determined using the specific heats in Table 11.2. The final pressure can be found using Eq. 12.8, and the ideal gas law enables us to determine the final volume.

Solution. We list the values given and convert the given temperature into kelvins as follows:

Given: $p_1 = 1.00 \text{ atm} = 1.01 \times 10^5 \text{ N/m}^2$ *Find:* Work done during each
       $n = 2.00 \text{ mol}$ process
       $T_1 = 20°C + 273 = 293 \text{ K}$
       $V_2 = 3V_1$
       $c_v = 12.5 \text{ J/(mol·K)}$ (from Table 11.2)
       $c_p = 20.8 \text{ J/(mol·K)}$ (from Table 11.2)

The data necessary to calculate the isothermal work from Eq. 12.3 are given. Since the volume ratio is 3 and ln 3 = 1.10, we have

$$W_{\text{isothermal}} = nRT \ln\left(\frac{V_2}{V_1}\right)$$

$$= (2.00 \text{ mol})[8.31 \text{ J/(mol} \cdot \text{K)}](293 \text{ K})(\ln 3) = +5.35 \times 10^3 \text{ J}$$

For the adiabatic process, the specific heat ratio is

$$\gamma = \frac{c_p}{c_v} = \frac{20.8 \text{ J/(mol} \cdot \text{K)}}{12.5 \text{ J/(mol} \cdot \text{K)}} = 1.66$$

We can determine the work from Eq. 12.9, but first we need the final pressure and volume. The final pressure can be determined from a ratio form of Eq. 12.8:

$$p_2 = p_1\left(\frac{V_1}{V_2}\right)^{\gamma} = p_1\left(\frac{V_1}{3V_1}\right)^{\gamma} = p_1\left(\frac{1}{3}\right)^{1.66} = 0.161\, p_1$$

$$= (0.161)(1.01 \times 10^5 \text{ N/m}^2) = 1.63 \times 10^4 \text{ N/m}^2$$

The initial volume is determined from the ideal gas law:

$$V_1 = \frac{nRT_1}{p_1} = \frac{(2.00 \text{ mol})[8.31 \text{ J/(mol} \cdot \text{K)}](293 \text{ K})}{1.01 \times 10^5 \text{ N/m}^2}$$

$$= 4.82 \times 10^{-2} \text{ m}^3$$

Therefore, $V_2 = 3V_1 = 1.45 \times 10^{-1} \text{ m}^3$. Now, applying Eq. 12.9, we have

$$W_{\text{adiabatic}} = \frac{p_1 V_1 - p_2 V_2}{\gamma - 1}$$

$$= \frac{(1.01 \times 10^5 \text{ N/m}^2)(4.82 \times 10^{-2} \text{ m}^3) - (1.63 \times 10^4 \text{ N/m}^2)(1.45 \times 10^{-1} \text{ m}^3)}{1.66 - 1}$$

$$= +3.80 \times 10^3 \text{ J}$$

As expected, this result is less than the isothermal work.

Follow-up Exercise. In this Example, (a) calculate the final temperature of the gas in the adiabatic expansion. (b) During the adiabatic expansion, determine the gas's change in internal energy by using the expression for the internal energy of a monatomic gas. Does it equal the negative of the work done (as calculated in the Example)? Explain.

12.4 The Second Law of Thermodynamics and Entropy

OBJECTIVES: **To (a) state and explain the second law of thermodynamics in several forms and (b) explain the concept of entropy.**

Suppose that a piece of hot metal is placed in an insulated container of cool water. Heat will be transferred from the metal to the water, and the two will come to thermal equilibrium at some intermediate temperature. For a thermally isolated system, the system's total energy remains constant. Could heat have been transferred from the cooler water to the hot metal instead? This process would not happen naturally. But if it did, the total energy of the system would remain constant, and this "impossible" inverse process would *not* violate energy conservation.

Clearly, there must be another principle that specifies the *direction* in which a process can take place. This principle is embodied in the **second law of thermodynamics**, which says that certain processes do not take place, or have never been observed to take place, even though they may be consistent with the first law.

There are many equivalent statements of the second law, which are worded according to their application. One applicable to the aforementioned situation is

Heat will not flow spontaneously from a colder body to a warmer body.

An equivalent alternative statement of the second law involves thermal cycles. A *thermal cycle* typically consists of several separate thermal processes after which the system ends up back at its starting conditions. If the system is a gas, this means the same *p-V-T* state from which it started. The second law, stated in terms of a thermal cycle (operating as a heat engine; see Section 12.5):

| In a thermal cycle, heat energy cannot be completely transformed into mechanical work.

In general, the second law of thermodynamics applies to all forms of energy. It is considered to be true because no one has ever found an exception to it. If it were not valid, a perpetual-motion machine could be built. Such a machine could first transform heat completely into work and motion (mechanical energy), with no energy loss. The mechanical energy could then be transformed back into heat and be used to reheat the reservoir from where the heat came originally (again with no loss). Since the processes could be repeated indefinitely, the machine would run perpetually, just shifting energy back and forth. All of the energy is accounted for, so this situation does *not* violate the first law. However, it is obvious that real machines are always less than 100% efficient, that is, the work output is always less than the energy input. Another statement of the second law is therefore as follows:

Note: Real machines can be made to be virtually frictionless, but their efficiency is still less than 100%. The second-law limit does not refer to frictional losses.

| It is impossible to construct an operational perpetual motion machine.

Attempts have been made to construct such machines, with no success.*

It would be convenient to have some way of expressing the *direction* of a process in terms of the thermodynamic properties of a system. One such property is temperature. In analyzing a conductive heat-transfer process, you need to know the temperatures of the system and its surroundings. Knowing the temperature difference between the two allows you to state the direction in which the heat transfer will spontaneously take place. Another useful quantity, particularly during the discussion of heat engines, is entropy.

Entropy

A property that indicates the *natural direction* of a process was first described by Rudolf Clausius (1822–1888), a German physicist. This property is **entropy**. Entropy is a multifaceted concept, with many different physical interpretations:

- Entropy is a measure of a system's ability to do useful work. As a system loses the ability to do work, its entropy increases.
- Entropy determines the direction of time. It is "time's arrow" that points out the forward flow of events, distinguishing past events from future ones.
- Entropy is a measure of disorder. A system naturally moves toward greater disorder, or disarray. The more order there is, the lower is the system's entropy.

All of these statements (and others) turn out to be equally valid interpretations of entropy and are physically equivalent, as you will see in the upcoming discussions. First, however, we introduce the definition of entropy. The change in a system's entropy (ΔS) when an amount of heat (Q) is added or removed by a reversible process at a constant temperature is

Note: ΔS is positive if a system absorbs heat ($+Q$) and negative if heat leaves ($-Q$). The sign of ΔS is determined by the sign of Q.

$$\Delta S = \frac{Q}{T} \quad \text{(change in entropy at constant temperature)} \quad (12.10)$$

SI unit of entropy: joule per kelvin (J/K)

*Although perpetual motion *machines* cannot exist, (very nearly) perpetual motion is known to exist—for example, the planets have been in motion around the Sun for about 5 billion years.

Note: **Always use kelvins when computing entropy!** What would happen if you used the Celsius temperature to calculate ΔS for a phase change from ice to water at $0°C$?

If the temperature changes during the process, calculating the change in entropy requires advanced mathematics. Our discussions will be limited to isothermal processes or those involving small temperature changes. For the latter, we can approximate entropy changes by using average temperatures, as Example 12.7. But first, let's look at a change in entropy and how it is interpreted.

Example 12.6 ■ Change in Entropy: An Isothermal Process

What is the change in entropy of 0.25 kg of ethyl alcohol when it vaporizes at its boiling point of 78°C (latent heat of vaporization $L_v = 1.0 \times 10^5$ J/kg)?

Thinking It Through. A phase change occurs at constant temperature; hence, we can apply Eq. 12.10 ($\Delta S = Q/T$) after converting to kelvins. From Eq. 11.2 ($Q = mL_v$) we can compute the amount of heat added.

Solution. From the statement of the problem, we have

Given: $m = 0.25$ kg   *Find:* ΔS (change in entropy)
$T = 78°C + 273 = 351$ K
$L_v = 1.0 \times 10^5$ J/kg

Since a phase change occurs, latent heat is absorbed by the alcohol:

$$Q = mL_v = (0.25 \text{ kg})(1.0 \times 10^5 \text{ J/kg}) = 2.5 \times 10^4 \text{ J}$$

Then

$$\Delta S = \frac{Q}{T} = \frac{+2.5 \times 10^4 \text{ J}}{351 \text{ K}} = +71 \text{ J/K}$$

Q is positive, because heat is added to the system. The change in entropy, then, is also positive, and the entropy of the alcohol increases. This outcome is reasonable, because a gaseous state is more random (disordered) than a liquid state.

Follow-up Exercise. What is the change in entropy of a 1.00-kg water sample when it freezes to form ice at 0°C?

Example 12.7 ■ Warm Spoon into Cool Water: System Entropy Increase or Decrease?

A metal spoon at 24°C is immersed in 1.00 kg of water at 18°C. The system (spoon and water) is thermally isolated and comes to equilibrium at a temperature of 20°C. (a) Find the approximate change in the entropy of the system. (b) Repeat the calculation, assuming, although this can't happen, that the water temperature dropped to 16°C and the spoon's rose to 28°C. Comment on how entropy tells us that the situation in part (b) cannot happen.

Thinking It Through. The system is thermally isolated, so there is only heat exchange between the spoon and the water, that is, $Q_m + Q_w = 0$, where the subscripts "w" and "m" stand for water and metal, respectively. We can determine Q_w from the known water mass, specific heat, and temperature change. Therefore, we can determine both Q values (equal, but opposite, signs). Strictly speaking, we cannot use Eq. 12.10, because it is applicable only for constant-temperature processes. However, here the temperature changes are small, so we can get a good approximation for ΔS by using each object's *average* temperature $\overline{T}$.

Solution. Using "i" and "f" to stand for "initial" and "final," respectively, we have

Given: $T_{m,i} = 24°C$ *Find:* (a) ΔS (change in entropy
$T_{w,i} = 18°C$ of the system in a
$m_w = 1.00$ kg   realistic situation)
$c_w = 4186$ J/(kg·C°) (from Table 11.1)   (b) ΔS (change in entropy
(a) $T_f = 20°C$ of the system in an
(b) $T_f = 16°C$ unrealistic situation)

(a) We need to find the amount of heat transferred (Q) in order to solve for ΔS. With $\Delta T_w = T_f - T_{w,i} = 20°C - 18°C = +2.0\,C°$, the heat gained by the water is

$$Q_w = m_w c_w \Delta T = (1.00\text{ kg})[4186\text{ J}/(\text{kg}\cdot C°)](2.0\,C°) = +8.37 \times 10^3\text{ J}$$

from Eq. 11.1. This quantity is also the magnitude of the heat *lost* by the metal. Therefore,

$$Q_m = -8.37 \times 10^3\text{ J}$$

The average temperatures are:

$$\overline{T}_w = \frac{T_{w,i} + T_f}{2} = \frac{18°C + 20°C}{2} = 19°C = 292\text{ K}$$

$$\overline{T}_m = \frac{T_{m,i} + T_f}{2} = \frac{24°C + 20°C}{2} = 22°C = 295\text{ K}$$

We then use these average temperatures and Eq. 12.10 to compute the approximate entropy changes for the water and the metal:

$$\Delta S_w \approx \frac{Q_w}{\overline{T}_w} = \frac{+8.37 \times 10^3\text{ J}}{292\text{ K}} = +28.7\text{ J/K}$$

$$\Delta S_m \approx \frac{Q_m}{\overline{T}_m} = \frac{-8.37 \times 10^3\text{ J}}{295\text{ K}} = -28.4\text{ J/K}$$

The change in the entropy of the *system* is the sum of these, or

$$\Delta S = \Delta S_w + \Delta S_m \approx +28.7\text{ J/K} - 28.4\text{ J/K} = +0.3\text{ J/K}$$

The entropy of the metal decreased, because heat was lost. The entropy of the water increased by a greater amount, so overall, the system's entropy increased.

(b) Although this situation does conserve energy, it violates the second law of thermodynamics. To see this violation in terms of entropy, let's repeat the foregoing calculation, using the second set of numbers. With $\Delta T_w = T_f - T_{w,i} = 16°C - 18°C = -2.0\,C°$, the heat lost by the water is

$$Q_w = m_w c_w \Delta T = (1.00\text{ kg})[4186\text{ J}/(\text{kg}\cdot C°)](-2.0\,C°) = -8.37 \times 10^3\text{ J}$$

We again use the average temperatures, $\overline{T}_w = 17°C = 290\text{ K}$ and $\overline{T}_m = 26°C = 299\text{ K}$, to compute the approximate entropy changes for the water and the metal spoon:

$$\Delta S_w \approx \frac{Q_w}{\overline{T}_w} = \frac{-8.37 \times 10^3\text{ J}}{290\text{ K}} = -28.9\text{ J/K}$$

$$\Delta S_m \approx \frac{Q_m}{\overline{T}_m} = \frac{+8.37 \times 10^3\text{ J}}{299\text{ K}} = +28.0\text{ J/K}$$

The change in the entropy of the *system* is:

$$\Delta S = \Delta S_w + \Delta S_m \approx -28.9\text{ J/K} + 28.0\text{ J/K} = -0.9\text{ J/K}$$

In this unrealistic scenario, the entropy of the metal increased, but the entropy of the water decreased by a greater amount, and the total system entropy decreased.

Follow-up Exercise. What should the initial temperatures in this Example be to make the overall system entropy change zero? Explain in terms of heat transfers.

Note how the entropy change of the system in Example 12.7a is positive, because the process is a *natural* one. That is, it is a process that is always observed to occur. In general, the direction of any process is toward an increase in total system entropy. That is, *the entropy of an isolated system never decreases.* Another way to state this observation is to say that *the entropy of an isolated system increases for every natural process* ($\Delta S > 0$). In coming to an intermediate temperature, the water and spoon in Example 12.7a are undergoing a natural process. In part (b) of the same Example, the process would never be observed, and the decrease in system entropy indicates this. Similarly, water at room temperature in an isolated ice-cube tray will not naturally (spontaneously) turn into ice.

Note: For every natural process, the entropy of a closed system increases.

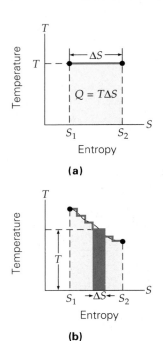

(a)

(b)

▲ **FIGURE 12.13** *T–S* **(temperature–entropy) diagrams** When temperature (T) is plotted versus entropy (S), the area under the process path is equal to the heat (Q) transferred in the process. **(a)** This *T–S* diagram is for an isothermal process. (How can you tell?) Note that the area under the path is a rectangle and thus is easy to compute. **(b)** This *T–S* diagram is for a more general heat-transfer process in which the temperature varies. The area under the curve still represents the heat transferred, but in this case, the area must be broken up into rectangular strips and added using calculus, much like the computation of the area under the *p–V* curve.

Note: Total energy can be neither created nor destroyed; total entropy can be created, but not destroyed.

However, if a system is *not* isolated, it may undergo a decrease in entropy. For example, if the ice-cube tray filled with water is instead put into a freezer compartment, the water will freeze, undergoing a decrease in entropy. But there will be a *larger increase in entropy somewhere else* in the universe. In this case, the refrigerator freezer warms the kitchen as it freezes the ice, and the total entropy of the system (ice + kitchen) actually increases.

Thus, a statement of the second law of thermodynamics in terms of entropy (for naturally occurring processes) is as follows:

The total entropy of the universe increases in every *natural* process.

The entropy of a system is a function of its state. Each state of a system has a particular value of entropy, and a change in entropy depends only on the initial and final states for a process, or $\Delta S = S_2 - S_1$. (This relationship is analogous to the change in internal energy for an ideal gas, $\Delta U = U_2 - U_1$, which depends only on the initial and final states and *not* on the details of the process.) An entropy change can be represented on a *T–S* diagram, as shown in ◄Fig. 12.13. The area under a *T–S* curve is equal to the heat flow, which is easily seen for the isothermal process in Fig. 12.13 (since $Q = T \Delta S$). Heat transferred in a process with variable temperature is also equal to the area under the *T–S* curve. The expression $T \Delta S$ is interpreted as the area of the small rectangles (Fig. 12.13b). The total heat flow during the process is determined by summing up the areas of all the rectangles, taking them to have very narrow widths, as $Q \approx \Sigma T \Delta S$. Since this calculation requires advanced mathematical techniques, we will restrict ourselves to straight-line processes on the *T–S* plots.

Processes exist for which the entropy is constant. One obvious such process is any adiabatic process, since $Q = 0$. In this case, $\Delta S = Q/T = 0$. Similarly, any reversible isothermal expansion that is followed immediately by an isothermal compression along the same path has zero net entropy change. This last example is true because the two heat flows are the same, but opposite in sign, and the temperatures are also the same; thus $\Delta S = Q/T + (-Q/T) = 0$. With the realization that, under some circumstances, it *is* possible to have $\Delta S = 0$, we generalize the previous statement of the second law of thermodynamics to include all possible processes. It now reads as follows:

During *any* process, the entropy of the universe can only increase or remain constant ($\Delta S \geq 0$).

To appreciate one of the many alternative (and equivalent) interpretations of entropy, consider the foregoing statements rewritten in terms of order and disorder. Here, entropy is interpreted as a measure of the disorder of a system. Thus, a larger value for entropy means more disorder (or, equivalently, less order):

All naturally occurring processes move toward a state of greater disorder or disarray.

A working definition of order and disorder may be extracted from everyday observations. Suppose you are making a pasta salad and have chopped tomatoes ready to toss into the cooked pasta. Before you mix the pasta and tomatoes, there is a relative amount of order, that is, the ingredients are separate and unmixed. Upon mixing, the separate ingredients become one dish, and there is less order (or more disorder, if you prefer). The pasta salad, once mixed, will never separate into the individual ingredients on its own (that is, by a natural process—one that happens on its own). Of course, you could go in and pick out the individual tomato pieces, but that would not be a natural process. Similarly, broken eyeglasses do not, on their own, go back together. Nor do mixed gases suddenly separate back into their separate species. Disorder is sometimes described as a measure of the randomness in a system. In the pasta dish after the ingredients are thoroughly mixed, the pieces are distributed randomly throughout the salad.

To better understand this interpretation of entropy, consider the Insight entitled "Life, Order, and the Second Law" on p. 423 and Conceptual Example 12.8.

INSIGHT

Life, Order, and the Second Law

The second law of thermodynamics is one of the foundations of physics and operates universally—the total entropy of the universe increases in every natural process. Certainly, life is a natural process. Yet clearly there is an increase in order during the development of the human embryo into a mature adult. Moreover, in virtually all life processes, larger, more complex molecules and cellular structures are continuously being synthesized from simpler components. Does this mean that living things are exempt from the second law? (We encountered a similar situation in Conceptual Example 12.8.)

To answer this question, we must realize that cells are *not* closed systems. They are involved in continuous exchanges of matter and energy with their surroundings. Living cells are constantly occupied in the exchange and conversion of energy from one form to another, which is governed by the first law of thermodynamics. For example, cells convert chemical potential energy into kinetic energy (movement) and electrical energy (the basis of nerve impulses). Many plant cells convert sunlight energy into chemical energy through photosynthesis (Fig. 1a). We have taken a hint from nature in attempting to harness the energy in sunlight (Fig. 1b).

In all biological processes, living systems maintain or even increase their organization while at the same time causing a larger decrease in the order of their surroundings. The second law is not violated by a *local* decrease in entropy as long as the *total* entropy of the universe increases in the process.

As an example of how cells create order while remaining within the confines of the second law, consider how they use chemical raw materials such as the sugar glucose ($C_6H_{12}O_6$). Cells use carbon atoms from glucose (and other organic compounds) to build up more complex molecules—a decrease in entropy. But they get the energy for these reactions by "burning" a great many glucose molecules as well—oxidizing them to produce many smaller, simpler molecules of water (H_2O) and carbon dioxide (CO_2). This process involves an increase in entropy. The molecules produced, being less highly structured than glucose, have greater entropy. Moreover, heat is released by the oxidation process, raising the entropy of the cell's surroundings. Thus, a cell "pays" for its increase in order by increasing the disorder of the rest of the universe.

But, you may ask, Where did the complex glucose molecule, with its chemical potential energy, come from in

the first place? As we mentioned previously, photosynthetic cells use the energy of sunlight to make glucose and related compounds. Thus, we can look at the flow of energy through organisms in a more general way. As we saw in Chapter 11 Insight on the greenhouse effect, life on the Earth is dependent on receiving radiant energy in the form of sunlight and reradiating infrared energy back into space. In effect, the Sun is a high-temperature source, and space is a low-temperature sink. The flow of energy from one to another allows living things to decrease entropy locally, creating this island of order that we call the biosphere. Thus, biological organisms do not violate the second law, as is sometimes thought; they merely take advantage of a loophole in it.

(a) **(b)**

FIGURE 1 Solar collectors (a) Green plants capture huge amounts of solar energy every day and have been doing so for billions of years. The chemical energy they store in the form of sugars and other complex molecules is used by nearly all organisms on the Earth to maintain and increase the highly organized state that we call life. **(b)** Humans have developed devices (such as these reflective solar-energy concentrators) to capture some of the energy of sunlight, although the technology is not yet in widespread use.

Conceptual Example 12.8 ■ More Order in the System: Entropy Decreasing?

Which of the following involves system undergoing a *decrease* in entropy? (a) A hot fudge sundae is left uneaten, so the ice cream melts and the fudge solidifies; (b) a green plant combines water and carbon-dioxide molecules in photosynthesis to make a larger, more complex molecule of sugar; (c) you drop your term paper on the way up the library stairs and find that, when you have gathered up all the pages, they are no longer in sequence; (d) perfume in an opened bottle evaporates and fills the room with scent; (e) a shortage of library personnel makes it increasingly difficult to find the books you want in their proper locations on the shelves.

Reasoning and Answer. It is not difficult to see that both cases (c) and (e) involve a decrease in order and thus an increase in entropy. The same is true, in a slightly less obvious way, of case (d). When the perfume molecules are no longer confined within the restricted space of the bottle, but instead are distributed at random throughout the larger volume of air in the room, there is a loss of order—the positions of individual perfume molecules cannot be specified as accurately.

Similarly, in case (a), the system (fudge plus ice cream) is less highly ordered when the speeds of all the molecules are randomized than when they are divided into two distinct populations, one hot and one cold. (We showed in another way in Example 12.7 how the flow of heat from a warmer body to a colder one entails an increase in entropy.) Hence, the answer is (b). In this instance, a more complex, highly ordered structure was created from simpler components—a process that will not occur naturally without an input of energy and a larger increase in entropy somewhere else in the universe.

Follow-up Exercise. In a garden compost pile, newly clipped grass blades are gradually broken down into a natural fertilizer. Does this process represent an increase or decrease in the total entropy of the grass blades? of the universe?

Notice that care must be taken to define the "system." For part (b) of Conceptual Example 12.8 in particular, the system is stated to be the plant, water, and carbon dioxide. However, this system is *not* isolated from the universe. (How can you tell?) Thus, the overall entropy of the universe suffers an increase in entropy, even though a nonisolated part of it managed an entropy decrease. This point is an often misunderstood one and is brought home in the "Life, Order, and the Second Law" Insight with a discussion of the building blocks of life, which, on the surface, may seem to decrease the entropy of the universe.

12.5 Heat Engines and Thermal Pumps

OBJECTIVES: To (a) explain the concept of a heat engine and compute thermal efficiency and (b) explain the concept of a thermal pump and compute the coefficient of performance.

A **heat engine** is any device that converts heat energy into work. Since the second law of thermodynamics prohibits perpetual-motion machines, some of the heat supplied to a heat engine will necessarily be lost. (In our study, we will not be concerned with the details of the mechanical components of an engine, such as pistons and cylinders.)

For our purposes, a heat engine is a device that takes heat from a high-temperature source (a hot reservoir), converts some of it to useful work, and transfers the rest to its surroundings (a cold, or low-temperature, reservoir). For example, most turbines that generate electricity (Chapter 20) are heat engines, using input heat from various sources. These sources include the burning of chemical ("fossil") fuels (oil, gas, or coal); nuclear reactions (Chapter 30); and the thermal energy already present beneath the Earth's surface (see the chapter-opening photo). They might be cooled by river water, for example, thus losing heat to this low-temperature reservoir. A generalized heat engine is represented in ▶ Fig. 12.14a.

A few reminders about our sign convention are in order before we start analyzing heat engines. For engines, we are interested primarily in the work W done *by the gas on the environment*. During expansion, the gas does positive work. Similarly, during a compression, the work done *by the gas* is negative. Also, we will assume that the "working substance" (the material that absorbs the heat and does the work) behaves like an ideal gas. The fundamental physics upon which heat engines are based is the same regardless of the working substance. However, using ideal gases makes the mathematics easier.

Adding heat to a gas can produce work. But since a *continuous* output is usually wanted, practical heat engines operate in a **thermal cycle**, or a series of processes

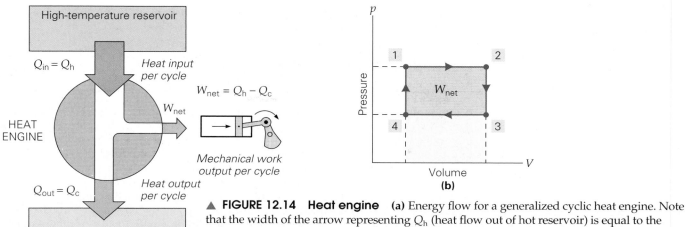

(a)

(b)

▲ **FIGURE 12.14 Heat engine** **(a)** Energy flow for a generalized cyclic heat engine. Note that the width of the arrow representing Q_h (heat flow out of hot reservoir) is equal to the combined widths of the arrows representing W_{net} and Q_c (heat flow into cold reservoir), reflecting the conservation of energy: $Q_h = Q_c + W_{net}$. **(b)** This specific cyclic process consists of two isobars and two isomets. The net work output per cycle is the area of the rectangle formed by the process paths. (See Example 12.11 for the analysis of this particular cycle.)

that brings the system back to its original condition. Cyclic heat engines include steam engines and internal combustion engines, such as automobile engines.

An idealized, rectangular thermodynamic cycle is shown in Fig. 12.14b. It consists of two isobars and two isomets. When these processes occur in the sequence indicated, the system goes through a cycle (1–2–3–4–1), returning to its original condition. When the gas expands, it does (positive) work equal to the area under the isobar. Doing positive work is exactly the desired output of an engine. (Think of a leaf blower pushing air out, or a car engine piston moving the crankshaft.) *However*, there must be a compression (during 3 to 4) of the gas to bring it back to its initial conditions. During this phase, the work done by the gas is negative, which is *not* the object of an engine. In a sense, some of the positive work done by the gas is "cancelled" by the negative work done during the compression.

From this discussion, you can see that the important quantity in engine design is not the amount of expansional work, but rather the *net work* W_{net}. This quantity is represented graphically as the area enclosed by the process curves that make up the cycle. In Fig 12.14, the area is rectangular. This "subtraction" of the two areas to form the net work is shown more generally in the Learn By Drawing feature "Representing Work in Thermal Cycles" on this page. When the paths are not straight lines, numerical calculations of the areas may be difficult, but the concept is the same.

Learn by Drawing

Representing Work in Thermal Cycles

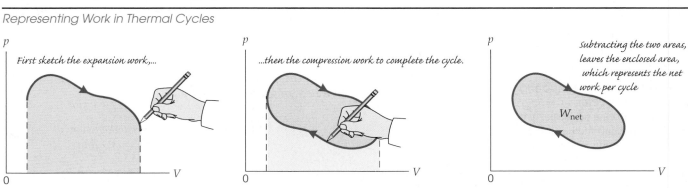

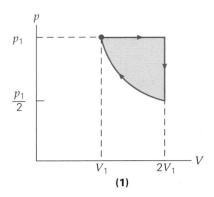

(1)

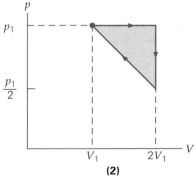

(2)

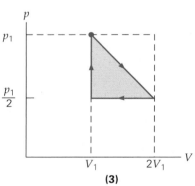

(3)

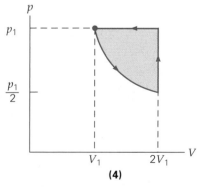

(4)

▲ **FIGURE 12.15 Comparing heat-engine cycles** Comparing the net work done per cycle, using graphical (visual) methods. (See the Learn By Drawing feature on p. 425 and Conceptual Example 12.9 for details.)

Before discussing engine efficiency, let's take a look at a conceptual example that uses these graphical techniques.

Conceptual Example 12.9 ■ Net Work: A Matter of Comparing Areas

A group of engineers is charged with designing a heat engine. They have narrowed their choices to four cycles (◄Fig. 12.15). They must decide which will generate the most net work. Can you help them decide? Order the given cycles from *least* net work to *most* net work per cycle. Assume in all cases that the gas starts in the same initial condition and that all expansions involve a doubling of the initial volume.

The cycles under consideration are as follows: (1) The gas expands isobarically and then has its pressure halved isometrically. Finally, it is isothermally compressed to its original state. (2) The gas expands (again) isobarically and then has its pressure halved isometrically. Lastly, it is compressed along a straight-line path (on a *p–V* diagram) to its original state. (3) The gas has its pressure halved (while doubling its volume) along a straight-line path (on a *p–V* diagram). Then it is isobarically compressed to its original volume. Finally, its pressure is raised isometrically to bring it back to its original state. (4) A gas is isothermally expanded, then its pressure is doubled isometrically, and last it is brought back to its original state isobarically.

Reasoning and Answer. The key here is to translate the words into graphical areas, as illustrated in Fig. 12.15. There each cycle is sketched (to scale) and then the enclosed areas are compared.

1. An isobaric expansion is represented by a horizontal line to the right and represents positive work done by the gas. Isometric compression is represented by a vertical line and does no work. At this point, the temperature is the same as the initial value, since the pressure is halved and the volume is doubled, since $p_f V_f = (p_i/2)(2V_i) = p_i V_i = nRT_i$. Thus, an isothermal process takes the gas back to its initial state.
2. This is the same as (1) for the first two processes, but the return is *not* isothermal.
3. During the first process, the gas does positive work. It expands along a straight line that ends up at the initial temperature. (How can you tell?) The isobaric compression is represented by a horizontal line to the left. The final process is a vertical isometric pressure increase back to initial conditions, during which zero work is done.
4. First there is an isothermal expansion to halve the initial pressure and double the initial volume. Here, the gas does positive work. Then a pressure increase brings the gas back to its initial pressure, but requires no work. Finally, the isobaric compression restores the gas to its initial conditions.

First, it should be clear that cycle (4) is not a heat engine, since it takes more work to compress it than the work it does on the environment, and the person who suggested this cycle should go back to school!

Cycle (1) does more net work than (2), so $W_{net, 1} > W_{net, 2}$. It is also clear that cycle 2 and cycle 3 do the same amount of net work, although they do it in a different way. Thus, $W_{net, 2} = W_{net, 3}$ and the winner is cycle (a), with (b) and (c) tied for second (or last?) place.

Follow-up Exercise. (a) In this Example, describe what changes you would make to cycle (2) to make it the winner. (*Hint:* There are many ways to achieve this goal.)

Thermal Efficiency

Thermal efficiency is an important parameter used to rate heat engines. The **thermal efficiency** (ε) of a heat engine is defined as

$$\varepsilon = \frac{\text{net work out}}{\text{heat in}} = \frac{W_{net}}{Q_{in}} \qquad \begin{array}{l}\textit{thermal efficiency}\\\textit{of heat engine}\end{array} \qquad (12.11)$$

The efficiency tells us how much useful work (W_{net}) the engine does in comparison with the input heat it receives (Q_{in}). For example, modern automobile engines have an efficiency of about 20 to 25%.* This means that only about one-fourth of the heat generated by igniting the air–gas mixture actually turns into mechanical work, which turns the car wheels, etc. Alternatively, you could say that the engine wastes about three fourths of the heat, eventually transferring it to the atmosphere through the hot exhaust system, radiator system, and metal engine.

For one cycle of an ideal gas heat engine, W_{net} is determined by applying the first law of thermodynamics to the complete cycle. Recall that our heat sign convention designates Q_{out} as negative. For our discussion of heat engines and pumps, *all heat symbols Q will represent magnitude only*. Therefore, Q_{out} is written as $-Q_c$ (the negative of a positive quantity to indicate flow *out* of the engine into a cold reservoir). Q_{in} is positive by our sign convention and is shown as $+Q_h$.

Note: Here, all Q's are positive; heat flow in or out will be shown by a $+$ or $-$ sign, respectively.

Applying the first law of thermodynamics to the expansional part of the cycle and showing the work done by the gas as $W = +W_{exp}$, we have $\Delta U_h = +Q_h - W_{exp}$. For the compression part of the cycle, the work done by the gas is shown explicitly as being negative ($W = -W_{comp}$ and $\Delta U_c = -Q_c + W_{comp}$). Adding these equations and realizing that for an ideal gas, $\Delta U_{cycle} = \Delta U_h + \Delta U_c = 0$ (why?), we obtain

$$0 = (Q_h - Q_c) + (W_{comp} - W_{exp})$$

or

$$W_{exp} - W_{comp} = Q_h - Q_c$$

However, $W_{net} = W_{exp} - W_{comp}$, so the final result is

$$W_{net} = Q_h - Q_c \quad \text{(Q represents magnitude here)}$$

So the thermal efficiency of a heat engine can be rewritten in terms the heat flows as

$$\varepsilon = \frac{W_{net}}{Q_h} = \frac{Q_h - Q_c}{Q_h} = 1 - \frac{Q_c}{Q_h} \quad \text{\textit{efficiency of an ideal gas heat engine}} \quad (12.12)$$

Like mechanical efficiency, thermal efficiency is a dimensionless fraction and is commonly expressed as a percentage. Equation 12.12 indicates that a heat engine could have 100% efficiency if Q_c were zero. This condition would mean that no heat energy would be lost and all the input (hot reservoir) heat would be converted to useful work. However, this situation is impossible according to the second law of thermodynamics. In 1851, this observation led Lord Kelvin to state the second law in yet another, but physically equivalent, manner:

No cyclic heat engine can convert its heat input completely to work.

Second Law of Thermodynamics

From Eq. 12.12 we see that to maximize the work output per cycle of a heat engine, we must minimize Q_c/Q_h, which increases the efficiency. The higher compression and pressure present in an automotive diesel engine, for example, give it a higher thermal efficiency than that of a gasoline engine. For more about engines, see the Insight "The Otto Cycle" on p. 428.

Example 12.10 ■ Thermal Efficiency: What You Get Out of What You Put In

The small, gasoline-powered engine of a leaf blower absorbs 800 J of heat energy from a high-temperature reservoir (the ignited gas–air mixture) and exhausts 700 J to a low-temperature reservoir (the outside air, through its cooling fins). What is the engine's thermal efficiency?

*The thermal efficiency of a heat engine in an automobile depends on many parameters, including the operation of features such as air conditioning. Thus, these values are only approximate.

Thinking It Through. We can use the definition of thermal efficiency of a heat engine (Eq. 12.11) if W_{net} can be determined. (Keep in mind that the Q's mean heat magnitudes.)

Solution.

Given: $Q_h = 800 \text{ J}$ *Find:* ε (thermal efficiency)
$\qquad\quad\ Q_c = 700 \text{ J}$

The net work done by the engine per cycle is

$$W_{net} = Q_h - Q_c = 800 \text{ J} - 700 \text{ J} = 100 \text{ J}$$

Therefore, the thermal efficiency is

$$\varepsilon = \frac{W_{net}}{Q_h} = \frac{100 \text{ J}}{800 \text{ J}} = 0.125$$

or, in terms of percentages,

$$\varepsilon = 0.125 \times 100\% = 12.5\%$$

Follow-up Exercise. (a) What would be the net work per cycle of the engine in this Example if the efficiency were raised to 15% and the input heat per cycle were raised to 1000? (b) How much heat would be exhausted in this case?

INSIGHT

The Otto Cycle

The internal combustion engine is a cyclic heat engine employed in automobiles and other devices. Most automobile engines use a four-stroke cycle. An approximation to this important cycle involves the steps shown in Fig. 1, along with a *p–V* diagram of the thermodynamic processes that make up the cycle. This theoretical cycle is called the *Otto cycle*, for the German engineer Nickolas Otto (1832–1891), who built one of the first successful gasoline engines.

During the intake stroke (1–2), an isobaric expansion, the air–fuel mixture is admitted at atmospheric pressure through the open valve as the piston drops. This mixture is compressed adiabatically (quickly) on the compression stroke (2–3). This step is followed by fuel ignition (3–4, when the spark plug fires, giving an isometric pressure rise). Next, an adiabatic expansion occurs during the power stroke (4–5). Following this step is an isometric cooling of the system when the piston is at its lowest position (5–2). The final, exhaust stroke is along the isobaric leg of the Otto cycle (2–1). Notice that it takes two cycles of the piston to produce one power stroke.

Related Exercise: 73

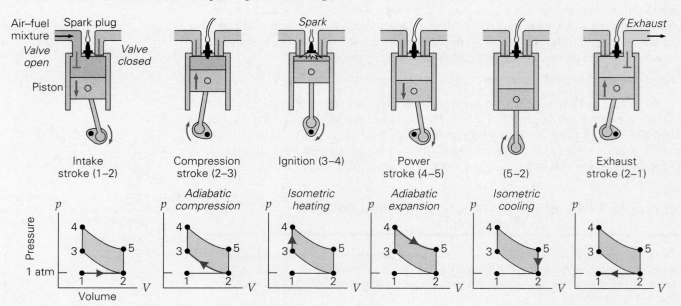

FIGURE 1 The four-stroke cycle of a heat engine The process steps for the four-stroke Otto cycle. The piston moves up and down twice each cycle, making a total of four strokes per cycle.

To see how the efficiency of an ideal gas cycle is calculated using the law of thermodynamics, consider the following Example.

Example 12.11 ■ Thermal Efficiency: Applying the Basic Definition

Assume you have 0.100 mole of an ideal monatomic gas that follows the cycle given in Fig. 12.14b and that the pressure and volume at the lower left-hand corner of that figure are 1.00 atm and 20°C, respectively. Further assume that the pressure doubles during the isometric pressure increase and the volume doubles during the isobaric expansion. What would be the thermal efficiency of this cycle?

Thinking It Through.   We can apply the definition of thermal efficiency of a heat engine (Eq. 12.11). However, we have to be careful, because heat exchanges can occur during more than one of the processes in the cycle. To determine heat input during the isobaric expansion, we need to know the change in internal energy and thus the change in temperature. So it seems likely that the temperatures at the four corners of the cycle will also be needed.

Solution.   Let us label the four corners with numbers, as shown in the figure. Listing the data given and converting to SI units, we have

Given: $p_4 = p_3 = 1.00 \text{ atm} = 1.01 \times 10^5 \text{ N/m}^2$ **Find:** ε (thermal efficiency)
$n = 0.100 \text{ mol}$
$T_4 = 20°C = 293 \text{ K}$
$p_1 = p_2 = 2.00 \text{ atm} = 2.02 \times 10^5 \text{ N/m}^2$
$V_2 = V_3 = 2V_4 = 2V_2$

First, let us compute the volumes and temperatures at the corners, using the ideal gas law:

$$V_4 = V_1 = \frac{nRT_1}{p_1} = \frac{(0.100 \text{ mol})[8.31 \text{ J/(mol} \cdot \text{K)}](293 \text{ K})}{1.01 \times 10^5 \text{ N/m}^2} = 2.41 \times 10^{-3} \text{ m}^3$$

Therefore,

$$V_2 = V_3 = 2V_1 = 4.82 \times 10^{-3} \text{ m}^3$$

During the isometric processes, the temperature is directly proportional to the pressure, and during the isobaric processes, the temperature is directly proportional to the volume. Therefore,

$$T_1 = 2T_4 = 586 \text{ K}$$
$$T_2 = 2T_1 = 1172 \text{ K}$$
$$T_3 = \tfrac{1}{2}T_2 = 586 \text{ K}$$

Now we are in position to calculate the heat transfers. $W = 0$ during the 4–1 process, and for a monatomic gas, $\Delta U = 3nR\,\Delta T/2$; Therefore

$$Q_{41} = \Delta U_{41} = \tfrac{3}{2}nR\,\Delta T_{41} = \tfrac{3}{2}(0.100 \text{ mol})[8.31 \text{ J/(mol} \cdot \text{K)}](586 \text{ K} - 293 \text{ K}) = +365 \text{ J}$$

During the 1–2 process, the gas expands and its internal energy increases. The work done by the gas is

$$W_{12} = p_1\Delta V_{12} = (2.02 \times 10^5 \text{ N/m}^2)(4.82 \times 10^{-3} \text{ m}^3 - 2.41 \times 10^{-3} \text{ m}^3) = +487 \text{ J}$$

Since work was done *and* the internal energy went up we have, a

$$Q_{12} = \Delta U_{12} + W_{12} = \tfrac{3}{2}nR\,\Delta T_{12} + 487 \text{ J}$$
$$= \tfrac{3}{2}(0.100 \text{ mol})[8.31 \text{ J/(mol} \cdot \text{K)}](1172 \text{ K} - 586 \text{ K}) + 487 \text{ J}$$
$$= +730 \text{ J} + 487 \text{ J} = +1.22 \times 10^3 \text{ J}$$

Thus the input heat Q_h is just

$$Q_h = Q_{41} + Q_{12} = 1.59 \times 10^3 \text{ J}$$

To find the net work, we need the area enclosed by the cycle. Therefore,

$$W_{net} = (\Delta p_{23})(\Delta V_{12}) = (1.01 \times 10^5 \text{ N/m}^2)(2.41 \times 10^{-3} \text{ m}^3) = +243 \text{ J}$$

and the efficiency is

$$\varepsilon = \frac{W_{net}}{Q_h} = \frac{243\ J}{1.59 \times 10^3\ J} = 0.153 \text{ or } 15.3\%$$

Follow-up Exercise. Calculate the total output heat (Q_c) in this Example by calculating the Q's involved in processes 2–3 and 3–4 and adding them. Your answer should be in agreement with the value of Q_c found from simple subtraction, using the results in the Example. Do you get such agreement?

Thermal Pumps: Refrigerators, Air Conditioners, and Heat Pumps

Note: As used here, the term *thermal pump* is general and does not refer specifically to the devices that are commonly used for heating buildings. These devices are called "heat pumps."

The function performed by a thermal pump is basically the reverse of that of a heat engine. The name **thermal pump** is a generic term for *any* device, including refrigerators, air conditioners, and heat pumps, that transfers heat energy from a low-temperature reservoir to a high-temperature reservoir (▼ Fig. 12.16). For such a transfer to occur, there must be work input. Since the second law of thermodynamics says that heat will not *spontaneously* flow from a cold body to a hot body, we must provide the means for this process to happen, that is, using work from the environment.

A familiar example of a thermal pump is a refrigerator. By using input work from electrical energy, heat is transferred from inside the refrigerator (low-temperature reservoir) to the surroundings (high-temperature reservoir). The knowledge you have gained about thermodynamic processes will allow you to understand the refrigerator's basic operation (▶ Fig. 12.17).

▼ **FIGURE 12.16 Thermal pumps** (a) An energy-flow diagram for a generalized cyclic thermal pump. The width of the arrow representing Q_h, the heat transferred into the high-temperature reservoir, is equal to the combined widths of the arrows representing W_{in} and Q_c, reflecting the conservation of energy: $Q_h = Q_c + W_{in}$. **(b)** An air conditioner is an example of a thermal pump. Using the input work, it transfers heat (Q_c) from a low-temperature reservoir (inside the house) to a high-temperature reservoir (outside). Thermodynamically, it operates like a refrigerator, cooling the interior while warming the exterior, that is, pumping heat "uphill."

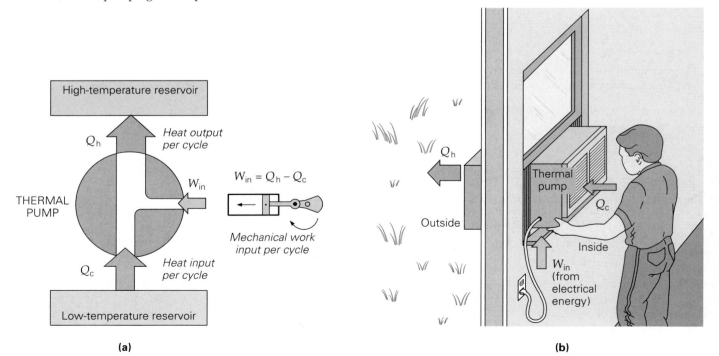

(a)

(b)

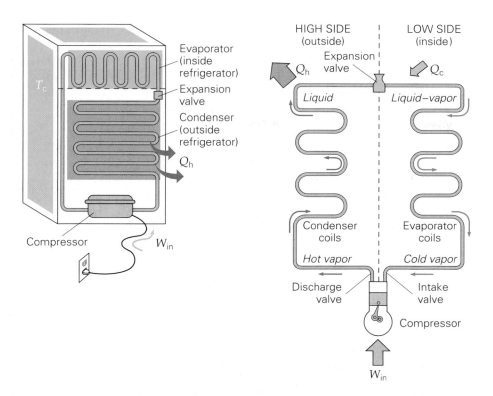

◀ FIGURE 12.17 Refrigerator operation Heat (Q_c) is carried away from the interior by the refrigerant as latent heat. This heat energy and that of the work input (W_{in}) are discharged from the condenser to the surroundings (Q_h). A refrigerator can be thought of as a remover of heat (Q_c) from an already cold region (its interior) *or* as a heat pump that adds heat (Q_h) to an already warm area (the kitchen).

The refrigerant, or heat-transferring medium, is a substance with a relatively low boiling point. Until recently, Freon (boiling point of $-29.8°C$ at 1 atm) was used in most domestic refrigerators. Freon has now been replaced by still other refrigerants, because of its destruction of the Earth's ozone layer. (See the related Insight in Chapter 20.)

It is convenient to think of the system diagrammed in Fig. 12.17 as having high and low (temperature and pressure) sides. On the low side, heat is transferred to the evaporator coils from inside the refrigerator. The heat causes the refrigerant to boil and is carried away as latent heat of vaporization. The gaseous vapor is drawn into the compressor chamber on the downstroke of the piston and is compressed on the upstroke (where the work input is done by the compressor motor). The compression increases the temperature of the gas, which is discharged from the compressor as a superheated vapor (on the high side). This vapor condenses in the cooler condenser unit, where circulating air (or water in larger units) carries away the latent heat of condensation and the internal energy caused by the work done during compression. The condensed liquid then collects in a receiver.

An expansion valve maintains a balance between the high and low sides, thereby controlling the rate at which the refrigerant passes back to the low side. On being admitted to the low side, the liquid immediately boils and vaporizes, because of the low pressure. This stage is a cooling process, because the heat of vaporization is supplied by the internal energy of the liquid. The cooled refrigerant is drawn into the evaporator, and the cycle begins again.

In essence, a refrigerator (or air conditioner, as in Fig. 12.16b) pumps heat up a temperature gradient, or "hill." (Think of pumping water up an actual hill against the force of gravity.) The cooling efficiency of this operation is based upon the amount of heat *extracted* from the cold-temperature reservoir (the refrigerator, the freezer, or the inside of a house), Q_c, compared with the work W_{in} needed to do so. Since a practical refrigerator operates in a cycle to provide a continuous removal of heat, $\Delta U = 0$ for the cycle. Then, by the conservation of energy (the first law of thermodynamics), $Q_c + W_{in} = Q_h$, where Q_h is the heat ejected to the high-temperature reservoir or the outside.

The measure of a refrigerator's or air conditioner's performance is defined differently than that for a heat engine, because of the difference in their functions. For the cooling appliances, the efficiency is expressed as a **coefficient of performance (COP)**. Since the purpose is to extract the most heat (Q_c to make or keep things cold) per unit of work input (W_{in}), the coefficient of performance for a refrigerator or air conditioner (COP_{ref}), is the ratio of these two quantities:

$$COP_{ref} = \frac{Q_c}{W_{in}} = \frac{Q_c}{Q_h - Q_c} \quad \begin{matrix} \textit{(refrigerator or} \\ \textit{air conditioner)} \end{matrix} \quad (12.13)$$

Thus, the greater the COP, the better is the performance—that is, more heat is extracted for each unit of work done. For normal refrigerator operation, the work input is less than the heat removed, so the COP is greater than 1. The COPs of typical refrigerators range from 3 to 5, depending on operating conditions. This range means that the amount of heat removed from the cold reservoir (the refrigerator, freezer, or house interior) is three to five times the amount of work needed to remove it.

Any machine that transfers heat in the opposite direction to which it would naturally flow is called a *thermal pump*. The term **heat pump** is specifically applied to commercial devices used to cool homes and offices in the summer and to heat them in the winter. The summer operation is that of an air conditioner. In this mode, it cools the interior of the house and heats the outdoors. Operating in its winter heating mode, a heat pump heats the interior and cools the outdoors, usually by taking heat energy from the cold air or ground.

For a heat pump in its heating mode, the heat *output* (to warm something up or keep something warm) is the item of interest, so the COP for this case is defined differently than that of a refrigerator or air conditioner. As you might guess, it is the ratio of Q_h to W_{in} (the heating you get for the work put in), or

$$COP_{hp} = \frac{Q_h}{W_{in}} = \frac{Q_h}{Q_h - Q_c} \quad \begin{matrix} \textit{(heat pump} \\ \textit{in heating mode)} \end{matrix} \quad (12.14)$$

where, again, we have used $Q_c + W_{in} = Q_h$. Typical COPs for heat pumps range between 2 and 4, again depending on the operating conditions.

Compared with electrical heating, heat pumps are efficient. For each unit of electric energy consumed, a heat pump typically pumps in from 1.5 to 3 times as much heat as direct electric heating systems provide. However, if the outside temperature is very low, a heat pump may not be efficient enough to heat a building adequately. Then it must be supplemented by a conventional heating system. Some heat pumps use water from underground reservoirs, wells, or buried loops of pipe as a low-temperature heat reservoir. These heat pumps are more efficient than the ones that use the outside air, because water has a larger specific heat than air, and the average temperature difference between the water and the inside air is usually smaller.

Example 12.12 ■ Air Conditioner/Heat Pump: Thermal Switch Hitting

An air conditioner operating in summer extracts 100 J of heat from the interior of the house for every 40 J of electric energy required to operate it. Determine (a) the air conditioner's COP and (b) its COP if it runs as a heat pump in the winter. Assume it is capable of moving the same amount of heat for the same amount of electric energy, regardless of the direction in which it runs.

Thinking It Through. We know the input work and input heat in part (a), so we apply the definition of COP for a refrigerator (Eq. 12.13). For the reverse operation, it is the output heat that is important, so we must find this quantity from the conservation of energy.

Solution.

Given: $Q_c = 100\text{ J}$ **Find:** (a) COP_{ref}
$W_{\text{in}} = 40\text{ J}$ (b) COP_{hp}

(a) From Eq. 12.13, the COP for this engine operating as an air conditioner is

$$\text{COP}_{\text{ref}} = \frac{Q_c}{W_{\text{in}}} = \frac{100\text{ J}}{40\text{ J}} = 2.5$$

(b) When the engine operates as a heat pump, the relevant heat is the output heat, which can be calculated from the conservation of energy:

$$Q_h = Q_c + W_{\text{in}} = 100\text{ J} + 40\text{ J} = 140\text{ J}$$

Thus, the COP for this engine operating as a heat pump in winter is, from Eq. 12.14,

$$\text{COP}_{\text{hp}} = \frac{Q_h}{W_{\text{in}}} = \frac{140\text{ J}}{40\text{ J}} = 3.5$$

Follow-up Exercise. (a) Suppose you redesigned the engine in this Example to perform the same operation, but with 25% less work input. What would be the new values of the two COPs? (b) Which COP would have the larger percentage increase?

12.6 The Carnot Cycle and Ideal Heat Engines

OBJECTIVES: To (a) explain how the Carnot cycle applies to heat engines, (b) compute the ideal Carnot efficiency, and (c) state the third law of thermodynamics.

Lord Kelvin's statement of the second law of thermodynamics says that any *cyclic* heat engine, regardless of its design, must always exhaust some heat energy (Section 12.5). But how much heat must be lost in the process? In other words, what is the *maximum* possible efficiency of a heat engine? In designing heat engines, engineers strive to make them as efficient as possible, but there must be some theoretical limit, and, according to the second law, it must be less than 100%.

Sadi Carnot (1796–1832), a French engineer, studied this limit. The first thing he sought was the thermodynamic cycle an *ideal* heat engine would use, that is, the most efficient cycle. Carnot found that the ideal heat engine absorbs heat from a *constant* high-temperature reservoir (T_h) and exhausts it to a *constant* low-temperature reservoir (T_c). These processes are ideally reversible isothermal processes and may be represented as two isotherms on a *p–V* diagram. But what are the processes that complete the cycle? Carnot showed that these processes are reversible adiabatic processes. As we saw in Section 12.3, the curves on a *p–V* diagram are called adiabats and are steeper than isotherms (►Fig. 12.18a).

Thus, the ideal **Carnot cycle** consists of two isotherms and two adiabats and is conveniently represented on a *T–S* diagram, where it forms a rectangle (Fig. 12.18b). The area under the upper isotherm (1–2) is the heat added to the system from the high-temperature reservoir: $Q_h = T_h \Delta S$. Similarly, the area under the lower isotherm (3–4) is the heat exhausted: $Q_c = T_c \Delta S$. Here, Q_h and Q_c are the heat transfers at *constant* temperatures (T_h and T_c, respectively). There is no heat transfer ($Q = 0$) during the adiabatic legs of the cycle. (Why?)

The difference between these heat transfers is the work output, which is equal to the area enclosed by the process paths (the shaded areas on the diagrams):

$$W_{\text{net}} = Q_h - Q_c = (T_h - T_c)\Delta S$$

Since ΔS is the same for both isotherms (see Fig. 12.18b, processes 1–2 and 3–4), we can use the expressions to relate the temperatures and heats. That is, since

$$\Delta S = \frac{Q_h}{T_h} \quad \text{and} \quad \Delta S = \frac{Q_c}{T_c}$$

PHYSLET®
ILLUSTRATION

Carnot Cycle

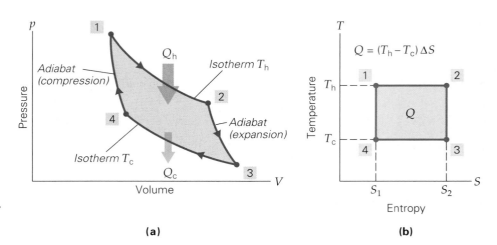

► FIGURE 12.18 The Carnot cycle (a) The Carnot cycle consists of two isotherms and two adiabats. Heat is absorbed during the isothermal expansion and exhausted during the isothermal compression. (b) On a *T–S* diagram, the Carnot cycle forms a rectangle, the area of which is equal to *Q*.

(a)

(b)

we have

$$\frac{Q_h}{T_h} = \frac{Q_c}{T_c} \quad \text{or} \quad \frac{Q_c}{Q_h} = \frac{T_c}{T_h}$$

This equation can be used to express the efficiency of an ideal heat engine in terms of temperature. From Eq. 12.12, this ideal **Carnot efficiency** (ε_C) is

$$\varepsilon_C = 1 - \frac{Q_c}{Q_h} = 1 - \frac{T_c}{T_h}$$

or

$$\varepsilon_C = 1 - \frac{T_c}{T_h} \qquad \begin{array}{l} \textit{Carnot efficiency} \\ \textit{(ideal heat engine)} \end{array} \qquad (12.15)$$

where the fractional efficiency is often expressed as a percentage. Note that T_c and T_h must be expressed in kelvins.

Note: The temperatures in the Carnot efficiency are the absolute, or Kelvin, temperatures.

The Carnot efficiency expresses the theoretical upper limit on the thermodynamic efficiency of a cyclic heat engine operating between two known temperature extremes. In practice, this limit can never be achieved. A true Carnot engine cannot be built, because the necessary reversible processes can only be approximated.

However, the Carnot efficiency does illustrate a general idea: The greater the difference in the temperatures of the heat reservoirs, the greater is the efficiency. For example, if T_h is twice T_c, or $T_c/T_h = 0.50$, the Carnot efficiency is

$$\varepsilon_C = 1 - \frac{T_c}{T_h} = 1 - 0.50 = 0.50\,(\times 100\%) = 50\%$$

However, if T_h is four times T_c, or $T_c/T_h = 0.25$, then

$$\varepsilon_C = 1 - \frac{T_c}{T_h} = 1 - 0.25 = 0.75\,(\times 100\%) = 75\%$$

Note: The Carnot efficiency, which can never be attained, is an ideal upper limit.

Since a heat engine can never attain 100% thermal efficiency, it is useful to compare its actual efficiency ε with its theoretical maximum efficiency, that of a Carnot cycle, ε_C. To see the importance of this concept in more detail, study the next Example carefully.

Example 12.13 ■ Carnot Efficiency: The True Measure of Efficiency for Any Real Engine

---

An engineer is designing a cyclic heat engine to operate between temperatures of 150°C and 27°C. (a) What is the maximum theoretical efficiency that can be achieved? (b) Suppose the engine, when built, does 100 J of work per cycle for every 500 J of input heat per cycle. What is its efficiency, and how close is it to the Carnot efficiency?

Thinking It Through. The maximum efficiency for specific high and low temperatures is given by Eq. 12.15. Remember, we must convert absolute temperatures. In part (b), we calculate the actual efficiency and compare it with our answer in part (a).

Solution.

Given: $T_h = 150°C + 273 = 423$ K   *Find:* (a) ε_C (Carnot efficiency)
$T_c = 27°C + 273 = 300$ K   (b) ε (actual efficiency) and compare
$W_{net} = 100$ J   it with ε_C
$Q_h = 500$ J

(a) Using Eq. 12.15 to find the maximum theoretical efficiency, we get

$$\varepsilon_C = 1 - \frac{T_c}{T_h} = 1 - \frac{300 \text{ K}}{423 \text{ K}} = 0.291 \, (\times 100\%) = 29.1\%$$

(b) The actual efficiency is, from Eq. 12.12,

$$\varepsilon = \frac{W_{net}}{Q_h} = \frac{100 \text{ J}}{500 \text{ J}} = 0.200 \quad \text{or} \quad 20.0\%$$

Thus,

$$\frac{\varepsilon}{\varepsilon_C} = \frac{0.200}{0.291} = 0.687 \quad \text{or} \quad 68.7\%$$

In other words, the heat engine is operating at 68.7% of its theoretical maximum. That's not bad!

Follow-up Exercise. If the operating high temperature of the engine in this Example were increased to 200°C, what would be the change in the theoretical efficiency?

There are also Carnot COPs for refrigerators and heat pumps. (See Exercise 98.)

The Third Law of Thermodynamics

Another inference might be drawn from the expression for the Carnot efficiency (Eq. 12.15). It would seem possible to have ε_C equal to 100%, if only T_c could be absolute zero. (See Section 10.3.) However, absolute zero has never been achieved, although ultralow-temperature (cryogenic) experiments have gotten within 20 nK (2×10^{-8} K) of it. Apparently, reducing the temperature of a system already close to absolute zero, in a finite number of steps is impossible. The **third law of thermodynamics**, simply stated, is as follows:

| It is impossible to reach absolute zero in a finite number of thermal processes. | Third Law of Thermodynamics

Chapter Review

Important Concepts and Equations

- **The first law of thermodynamics** is a statement of the conservation of energy for a thermodynamic system. Expressed in equation form, it relates the change in a system's internal energy to the heat flow and the work done on it and is given by

$$Q = \Delta U + W \qquad (12.1)$$

- Some **thermodynamic processes** for gases are

  isothermal: a process that occurs at constant temperature
  isobaric: a process that occurs at constant pressure
  isometric: a process that occurs at constant volume
  adiabatic: a process involving no heat flow

- The expressions for **thermodynamic work**, are done by an ideal gas

$$W_{isothermal} = nRT \ln\left(\frac{V_2}{V_1}\right) \quad \begin{matrix}\textit{(ideal gas isothermal} \\ \textit{process)}\end{matrix} \quad (12.3)$$

$$W_{isobar} = p(V_2 - V_1) = p\Delta V \quad \begin{matrix}\textit{(ideal gas isobaric} \\ \textit{process)}\end{matrix} \quad (12.4)$$

$$W_{adiabatic} = \frac{p_1V_1 - p_2V_2}{\gamma - 1} \quad \begin{matrix}\textit{(ideal gas adiabatic} \\ \textit{process)}\end{matrix} \quad (12.9)$$

(In the adiabatic process , $\gamma = c_p/c_v$ is the ratio of specific heats.)

- The **second law of thermodynamics** states whether a process can take place naturally or, alternatively, specifies the direction a process can take.

- **Entropy** (S) is a measure of the disorder of a system.

- The **change in entropy** of an object at constant temperature is given by

$$\Delta S = \frac{Q}{T} \qquad (12.10)$$

The total entropy of the universe increases in every natural process.

- A **heat engine** is a device that converts heat into work. Its **thermal efficiency** ε is the ratio of work output to heat input, or

$$\varepsilon = \frac{W_{net}}{Q_h} = \frac{Q_h - Q_c}{Q_h} = 1 - \frac{Q_c}{Q_h} \qquad (12.12)$$

- A **thermal pump** is a device that transfers heat energy from a low-temperature reservoir to a high-temperature reservoir. The coefficient of performance (COP) is the ratio of heat transferred to the input work. The COP differs depending on whether the thermal pump is used as a heat pump or an air conditioner/refrigerator.

- A **Carnot cycle** is a theoretical heat-engine cycle consisting of two isotherms and two adiabats. Its efficiency is the highest possible efficiency that any heat engine could have, operating between two temperature extremes. The efficiency of a Carnot cycle is

$$\varepsilon_C = 1 - \frac{T_c}{T_h} \qquad (12.15)$$

Exercises*

12.1 Thermodynamic Systems, States, and Processes

1. There may be an exchange of heat with the surroundings for (a) a thermally isolated system, (b) a completely isolated system, (c) a heat reservoir, or (d) none of the preceding.

2. Only initial and final states are known for irreversible processes on (a) p–V diagrams, (b) p–T diagrams, (c) V–T diagrams, or (d) all of the preceding.

3. On a p–V diagram, a reversible process is a process (a) whose path is known, (b) whose path is unknown, (c) for which the intermediate steps are nonequilibrium states, or (d) none of the preceding.

4. **CQ** Explain why the process shown in Fig. 12.1b is not that of an ideal gas at constant temperature.

12.2 The First Law of Thermodynamics
and
12.3 Thermodynamic Processes for an Ideal Gas

5. **CQ** On a p–V diagram, sketch a cyclic process that consists of an isothermal expansion, an isobaric compression, and an isometric process—in that order.

6. **CQ** On a p–T diagram, sketch the general paths for the following reversible processes for an ideal gas: (a) isothermal, (b) isobaric, (c) isometric, and (d) adiabatic.

*Take temperatures to be exact—that is, with no uncertainty.

7. There is no heat flow into or out of the system in an (a) isothermal process, (b) adiabatic process, (c) isobaric process, or (d) isometric process.

8. According to the first law of thermodynamics, if work is done on a system, then (a) the internal energy of the system must change, (b) heat must be transferred from the system, (c) the internal energy of the system must change and/or heat must be transferred from the system, or (d) heat must be transferred to the system.

9. When heat is added to a system of ideal gas during an isothermal expansion process, (a) work is done on the system, (b) the internal energy decreases, (c) the effect is the same as for an isometric process, or (d) none of the preceding.

10. **CQ** (a) Why does the body of a hand pump become hot when it is used to pump up a tire? (Neglect friction. The plunger is usually lubricated to prevent air leakage.) (b) Why does the valve of a tire become cold when air is released from the tire?

11. **CQ** In ▶Fig. 12.19, the plunger of a syringe is pushed in quickly, and the small pieces of paper in the syringe catch fire. Explain this phenomenon, using the first law of thermodynamics. (Similarly, in a diesel engine, there are no spark plugs. How can the air–fuel mixture ignite?)

12. **CQ** Discuss heat, work, and the change in internal energy of your body when you play a game of basketball.

13. **CQ** In an adiabatic process, there is no heat exchange, but the temperature of an ideal gas changes. How can that be? Explain.

▲ **FIGURE 12.19   Fire syringe**   See Exercise 11.

14. **CQ** An ideal gas initially at temperature T_o, pressure P_o, and volume V_o is compressed to one-half its initial volume. As shown in ▼ Fig. 12.20, process 1 is adiabatic, 2 is isothermal, and 3 is isobaric. Rank the work done on the gas, from high to low, of the three processes, and explain your ranking.

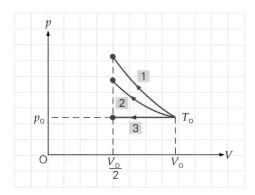

▲ **FIGURE 12.20   Thermodynamic processes**   See Exercises 14 and 15.

15. **CQ** In Exercise 14, rank the final temperatures, from low to high, of the three processes, and explain your ranking.

16. **IE** ■ A rigid container contains 1.0 mole of an ideal gas that slowly receives 2.0×10^4 J of heat. (a) The work done by the gas is (1) positive, (2) zero, or (3) negative. Why? (b) What is the change in the internal energy of the gas?

17. **IE** ■ A quantity of ideal gas goes through a cyclic process and does 400 J of net work. (a) The temperature of the gas at the end of the cycle is (1) higher than, (2) the same as, or (3) less than when it started? Why? (b) Is heat added to or removed from the system, and how much heat is involved?

18. ■ While playing in a basketball game, you lost 6.5×10^5 J of heat, and your internal energy also decreased by 1.2×10^6 J. How much work did you do in the game?

19. **IE** ■ While doing 500 J of work, a system of ideal gas expands adiabatically to 1.5 times its volume. (a) The temperature of the gas (1) increases, (2) remains the same, or (3) decreases. Why? (b) What is the change in the internal energy of the gas?

20. **IE** ■ An ideal gas system expands from 1.0 m³ to 3.0 m³ at atmospheric pressure while absorbing 5.0×10^4 J of heat in the process. (a) Since the system absorbs heat, (1) the temperature of the system must increase, (2) the internal energy of the system must increase, or (3) none of the preceding. Why? (b) What is the change in internal energy of the system?

21. ■■ A gas at low density has an initial pressure of 1.65×10^4 Pa and occupies a volume of 0.20 m³. The slow addition of 8.4×10^3 J of heat to the gas causes it to expand isobarically to a volume of 0.40 m³. (a) How much work is done by the gas in the process? (b) Does the internal energy of the gas change? If so, by how much?

22. ■■ An Olympic weight lifter lifts 145 kg a vertical distance of 2.1 m. In doing so, his internal energy decreases by 6.0×10^4 J. How much heat, in kilocalories, flows, and in what direction?

23. **IE** ■■ An ideal gas is taken through the reversible processes shown in ▼ Fig. 12.21. (a) Is the overall change in the internal energy of the gas (1) positive, (2) zero, or (3) negative? Why? (b) In terms of state variables p and V, how much work is done by or on the gas, and (c) what is the net heat transfer in the overall process?

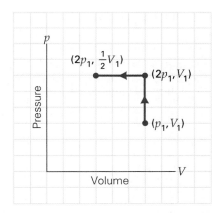

▲ **FIGURE 12.21   A p–V diagram for an ideal gas**   See Exercise 23.

24. ■■ The temperature of 2.0 moles of ideal gas is increased from 150°C to 250°C by two different processes. In one, 2500 J of heat is added to the gas; in the other, 3000 J of heat is added. (a) What is the change in the internal energy of the gas in each case? (*Hint*: See Eq. 10.16.) (b) In which case is more work done, and how much more?

25. ■■ 2.0 moles of an ideal gas expands isothermally from a volume of 20 L to 40 L while its temperature is 300 K. (a) Is

work done by the gas or on the gas? (b) What is the magnitude of the work?

26. ■■ A fixed quantity of gas undergoes the reversible changes illustrated in the p–V diagram in ▼Fig. 12.22. How much work is done in each process?

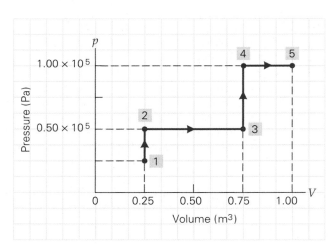

▲ **FIGURE 12.22 A p–V diagram and work**
See Exercises 26 and 27.

27. ■■ Suppose that after the final process shown in Fig. 12.22 (see Exercise 26), the pressure of the gas is decreased isometrically from 1.0×10^5 Pa to 0.70×10^5 Pa, and then the gas is compressed isobarically from 1.0 m³ to 0.80 m³. What is the total work done in all of these processes, including 1 through 5?

28. IE ■■■ A gas is compressed as shown on the p–V diagram in ▼Fig. 12.23. (a) Is the work done by the gas (1) positive, (2) zero, or (3) negative? Why? (b) What is the magnitude of the work done by the gas?

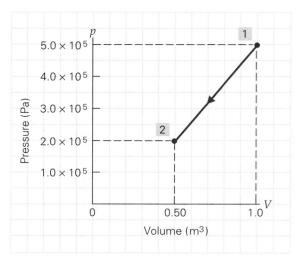

▲ **FIGURE 12.23 A variable p–V process and work**
See Exercise 28.

29. ■■■ One mole of ideal gas is taken through the cyclic process shown in ▼Fig. 12.24. (a) Compute the work involved (W) for each of the four processes. (b) Find ΔU, W, and Q for the complete cycle. (c) What is T_3?

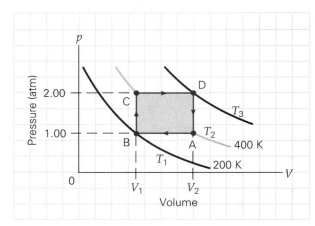

▲ **FIGURE 12.24 A cyclic process** See Exercise 29.

12.4 The Second Law of Thermodynamics and Entropy

30. The second law of thermodynamics (a) describes the state of a system, (b) applies only when the first law is satisfied, (c) precludes perpetual motion machines, or (d) does not apply to an isolated system.

31. In any natural process, the overall change in the entropy of the universe is (a) negative, (b) positive, (c) zero, or (d) the same as the change in the internal energy.

32. CQ Do the entropies of the *objects* in the following processes increase or decrease? (a) *Ice* melts; (b) *water vapor* condenses; (c) *water* is heated on a stove; (d) *food* is cooled in a refrigerator.

33. CQ In Example 12.6, what would be the implication if the total change in entropy had been negative?

34. CQ When a quantity of hot water is mixed with a quantity of cold water, the combined system comes to thermal equilibrium at some intermediate temperature. How does the entropy of the system change?

35. CQ A student challenges the second law of thermodynamics by saying that entropy does not have to increase in all situations, such as when water freezes to ice. Is this challenge valid? Why or why not?

36. CQ An ideal gas expands isothermally. What happens to its entropy, and why?

37. ■ What change in entropy is associated with the reversible phase change of 1.0 kg of ice to water at 0°C?

38. **IE** ■ A process involves 0.50 kg of steam condensing to water at 100°C. (a) The change in entropy of the steam (water) is (1) positive, (2) zero, or (3) negative. Why? (b) What is the change in entropy of the steam (water) associated with the process?

39. ■ What is the change in entropy of mercury vapor ($L_v = 2.7 \times 10^5$ J/kg) when 0.50 kg of it condenses to a liquid at its boiling point of 357°C?

40. ■ Which process has the greater change in entropy, 0.75 kg of ice changing to water at 0°C or 0.25 kg of water changing to steam at 100°C? Comment on the entropy changes in terms of order and disorder

41. ■ One mole of ideal gas goes through an isothermal compression at room temperature. If 7.5×10^3 J of work is done in compressing the gas, what is the change in entropy of the gas?

42. **IE** ■■ A quantity of an ideal gas initially at STP undergoes a reversible isothermal expansion and does 3.0×10^3 J of work on its surroundings in the process. (a) Will the entropy of the gas (1) increase, (2) remain the same, or (3) decrease? Why? (b) What is the change in the entropy of the gas?

43. ■■ During a liquid-to-solid phase change of a substance, its change in entropy is -4.19×10^3 J/K. If 1.67×10^6 J of heat is removed in the process, what is the freezing point of the substance in degrees Celsius?

44. ■■ An isolated system consists of two very large thermal reservoirs, one hot and one cold, with constant temperatures of 373 K and 273 K, respectively. If 1000 J of heat were to flow from the cold reservoir to the hot reservoir spontaneously, (a) what would be the change in entropy of the hot reservoir? (b) of the cold reservoir? (c) What would be the total change in entropy of the isolated system? (d) Could the process take place naturally?

45. ■■ Two heat reservoirs at temperatures 200°C and 60°C, respectively, are brought into thermal contact, and 1.50×10^3 J of heat spontaneously flows from one to the other with no significant temperature change. What is the change in the entropy of the two-reservoir system?

46. ■■ In the winter, the heat in a house with an inside temperature of 18°C leaks out at a rate of 2.0×10^4 J per second. If the outside temperature is 0°C, by how much does the entropy of the house decrease per second?

47. ■■ (a) How much heat is transferred in the processes shown on the *T–S* diagram in ▶Fig. 12.25? (b) What type of process takes place as the system goes from state 2 to state 3?

48. ■■ Suppose that the system described by the *T–S* diagram in Fig. 12.25 is returned to its original state, state 1, by a reversible process depicted by a straight line from

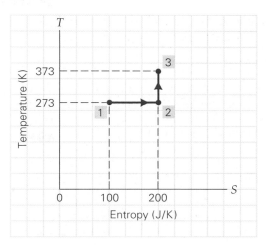

▲ **FIGURE 12.25 Entropy and heat** See Exercises 47 and 48.

states 3 to 1. What is the change in entropy of system after the cycle is over? How much heat is transferred in the cyclic process?

49. ■■■ A 50.0-g ice cube at 0°C is placed in 500 mL of water at 20°C. Estimate the change in entropy when all the ice has melted (a) for the ice, (b) for the water, and (c) for the ice–water system.

12.5 Heat Engines and Thermal Pumps*

50. For a cyclic heat engine, (a) $\varepsilon > 1$, (b) $Q_h = W_c$, (c) $\Delta U = W_{net}$, or (d) $Q_h > Q_c$.

51. A thermal pump (a) is rated by thermal efficiency, (b) requires work input, (c) is not consistent with the second law of thermodynamics, or (d) violates the first law of thermodynamics.

52. Which one of the following determines the thermal efficiency of a heat engine? (a) Q_h/Q_c; (b) Q_c/Q_h; (c) $Q_h - Q_c$; (d) $Q_h + Q_c$.

53. **CQ** What happens to the internal energy of a cyclic heat engine after a complete cycle?

54. **CQ** Is leaving a refrigerator door open a practical way to air-condition a room? Explain.

55. **CQ** Lord Kelvin's statement of the second law of thermodynamics as applied to heat engines ("No heat engine operating in a cycle can convert its heat input completely to work") refers to their operation *in a cycle*. Why is the phrase "in a cycle" included?

56. **CQ** The energy output of a thermal pump is greater than the energy used to operate the pump. Is this device a violation of the first law of thermodynamics?

*Consider efficiencies to be exact.

57. CQ In normal atmospheric convection cycles, colder air from a higher altitude is transferred to a lower, warmer level. Does this process violate the second law of thermodynamics? Explain.

58. ■ A gasoline engine has a thermal efficiency of 28%. If the engine absorbs 2000 J of heat in each cycle, (a) what is the net work output in each cycle? (b) How much heat is exhausted in each cycle?

59. ■ If an engine does 200 J of net work and exhausts 600 J of heat per cycle, what is its thermal efficiency?

60. ■ A heat engine with a thermal efficiency of 20% does 800 J of net work each cycle. How much heat per cycle is lost to the surroundings (the low-temperature reservoir)?

61. ■ An internal combustion engine with a thermal efficiency of 15.0% does 2.60×10^4 J of net work each cycle. How much heat is lost by the engine in each cycle?

62. ■ The heat output of a particular engine is 7.5×10^3 J per cycle, and the net work out is 4.0×10^3 J per cycle. What is the thermal efficiency of the engine?

63. ■■ A steam engine has a thermal efficiency of 10.0% and does 4500 J of useful work each cycle. What is the heat lost to the environment in each cycle?

64. IE ■■ An engineer redesigns a heat engine and improves its thermal efficiency from 20% to 25%. (a) Does the ratio of the heat output to heat input (1) increase, (2) remain the same, or (3) decrease? Why? (b) What is the change in Q_c/Q_h?

65. ■■ A gasoline engine burns gas that releases 3.3×10^8 J of heat per hour. (a) What is the energy input during a period of 2.0 h? (b) If the engine delivers 25 kW of power during this time, what is its thermal efficiency?

66. ■■ A refrigerator takes heat from its cold interior at a rate of 7.5 kW when the required input work is done at a rate of 2.5 kW. At what rate is heat exhausted to the kitchen?

67. ■■ A refrigerator with a COP of 2.2 removes 4.2×10^5 J of heat from its storage area each cycle. (a) How much heat is exhausted each cycle? (b) What is the total work input in joules for 10 cycles?

68. ■■ A heat pump removes 2.0×10^3 J of heat from the outdoors and delivers 3.5×10^3 J of heat to the inside of a house each cycle. (a) How much work is required per cycle? (b) What is the COP of this pump?

69. ■■ An air conditioner has a COP of 2.75. What is the power rating of the unit if it is to remove 1.00×10^7 J of heat in 20 min?

70. ■■ A heat engine has a thermal efficiency of 30.0%. If its heat input for each cycle is supplied by the condensation

of 8.00 kg of steam at 100°C, (a) what is the net work output per cycle, and (b) how much heat is lost to the surroundings each cycle?

71. ■■ A heat pump that uses an underground water reservoir as a heat source extracts 2.1×10^5 J each cycle, while requiring 3.0×10^4 J of work to do so. (a) How much heat is delivered to the inside of the house? (b) What is the temperature drop of the water when it is returned to the reservoir if 5.0 kg is used each cycle?

72. ■■■ A coal-fired power plant produces 900 MW of electricity power and operates at an overall thermal efficiency of 25%. (a) What is the rate of heat input to the plant? (b) What is the rate of heat discharge from the plant? (c) Why is the water heated by the discharge cooled in a cooling tower before being ejected into a nearby river?

73. ■■■ A four-stroke engine runs on the Otto cycle. (See the related Insight on p. 436.) It delivers 150 hp at 3600 rpm. (a) How many cycles are there in 1 min? (b) If the thermal efficiency of the engine is 20%, what is the heat input per minute? (c) How much heat is wasted (per minute) to the environment?

12.6 The Carnot Cycle and Ideal Heat Engines

74. The Carnot cycle consists of (a) two isobaric and two isothermal processes, (b) two isometric and two adiabatic processes, (c) two adiabatic and two isothermal processes, or (d) four arbitrary processes that return the system to its initial state.

75. CQ Which of the following temperature-reservoir relationships would yield the highest efficiency for a Carnot engine? (a) $T_c = 0.15\ T_h$, (b) $T_c = 0.25\ T_h$, (c) $T_c = 0.50\ T_h$, or (d) $T_c = 0.90\ T_h$.

76. CQ For a heat engine that operates between two reservoirs of temperatures T_c and T_h, the Carnot efficiency is the (a) highest possible value, (b) lowest possible value, (c) average value, or (d) none of the preceding.

77. CQ An idealized cycle for a two-stroke internal combustion engine is shown in ▶ Fig. 12.26. It consists of an adiabatic expansion and compression (legs 1 and 3, respectively) and an isometric decompression and compression (legs 2 and 4, respectively). This chapter has stated that in a Carnot cycle, heat transfers occur at constant temperatures. For the cycle in Fig. 12.26, (a) is heat taken in at a constant temperature? (b) Is heat rejected at a constant temperature? (c) Draw a sketch that indicates the areas representing work (W, with signs) during expansion and compression, and draw another sketch for the net work.

78. CQ In Fig. 12.26, legs 1 and 3 are adiabatic processes. Suppose these processes were isothermal instead. (a) Does this change make the cycle a Carnot cycle? Explain. (b) In

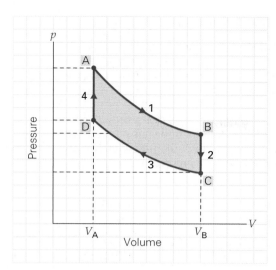

▲ **FIGURE 12.26 Non-Carnot cycle** Two adiabats and two isomets. See Exercises 77 and 78. (Not drawn to scale.)

which legs are there heat transfers? [Indicate directions by signs.] Do these transfers occur at constant temperatures?

79. **CQ** Automobile engines can be either air cooled or water cooled. Which type of engine would you expect to be more efficient, and why?

80. **CQ** If you have the choice of running your heat engine between the following two sets of temperatures for the cold and hot reservoirs, which would you choose, and why? 100°C and 300°C; 50°C and 250°C.

81. **CQ** Diesel engines are much more efficient than gasoline engines. Which type of engine runs hotter? Why?

82. ■ A steam engine operates between 100°C and 30°C. What is the Carnot efficiency of the ideal engine that operates between these temperatures?

83. ■ A Carnot engine has an efficiency of 35% and takes in heat from a high-temperature reservoir at 147°C. What is the Celsius temperature of the engine's low-temperature reservoir?

84. ■ What is the temperature of the hot reservoir of a Carnot engine that is 30% efficient and has a 20°C cold reservoir?

85. ■ It has been proposed that temperature differences in the ocean could be used to run a heat engine to generate electricity. In tropical regions, the water temperature is about 25°C at the surface and about 5°C at very large depths. (a) What would be the maximum theoretical efficiency of such an engine? (b) Would a heat engine with such a low efficiency be practical? Explain.

86. ■ An engineer wants. to run a heat engine with an efficiency of 40% between a high-temperature reservoir at 350°C and a low-temperature reservoir. Below what tem-

perature must the low-temperature reservoir be for practical operation of the engine?

87. ■■ A Carnot engine takes 2.7×10^4 J of heat per cycle from a high-temperature reservoir at 320°C and exhausts some of it to a low-temperature reservoir at 120°C. How much net work is done by the engine per cycle?

88. ■■ A Carnot engine with an efficiency of 40% operates with a low-temperature reservoir at 50°C and exhausts 1200 J of heat each cycle. What are (a) the heat input per cycle and (b) the Celsius temperature of the high-temperature reservoir?

89. ■■ A Carnot engine takes in heat from a reservoir at 327°C and has an efficiency of 30%. If the exhaust temperature is not changed and the efficiency is increased to 40%, what is the increase in the temperature of the hot reservoir?

90. ■■ An inventor claims to have developed a heat engine that, on each cycle, takes in 5.0×10^5 J of heat from a high-temperature reservoir at 400°C and exhausts 2.0×10^5 J to the surroundings at 125°C. Would you invest your money in the production of this engine? Explain.

91. **IE** ■■ Equation 12.15 shows that the greater the temperature difference between the reservoirs of a heat engine, the greater is the engine's Carnot efficiency. Suppose you had the choice of raising the high temperature reservoir by a certain number of kelvins or lowering the low temperature reservoir by the same number of kelvins. (a) You should choose (1) to raise the high-temperature reservoir, (2) to lower the low-temperature reservoir, or (3) both (1) or (2) produce the same change in efficiency, so it does not matter which you choose. Why? (b) Prove your answer to (a) numerically.

92. ■■ A heat engine operates at a thermal efficiency that is 45% of the Carnot efficiency. If the temperatures of the high temperature and low temperature reservoirs are 400°C and 100°C, respectively, what are the Carnot efficiency and the thermal efficiency of the engine?

93. ■■ A heat engine has half the thermal efficiency of a Carnot engine operating between temperatures of 100°C and 375°C. (a) What is the Carnot efficiency of the heat engine? (b) If the heat engine absorbs heat at a rate of 50 kW, at what rate is heat exhausted?

94. ■■ The working substance of a cyclic heat engine is 0.75 kg of an ideal gas. The cycle consists of two isobaric processes and two isometric processes, as shown in ▶Fig. 12.27. What would be the efficiency of a Carnot engine operating with the same high temperature and low temperature reservoirs?

95. ■■ In each cycle, a Carnot engine takes 800 J of heat from a high temperature reservoir and discharges 600 J to a low

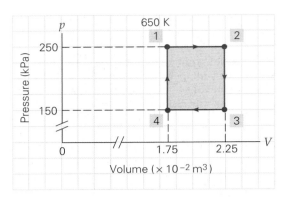

▲ **FIGURE 12.27 Thermal efficiency** See Exercise 94.

temperature reservoir. (a) What is the Carnot efficiency of the engine? (b) What is the ratio of the temperature of the high temperature reservoir to that of the low temperature reservoir?

96. ■■ A Carnot engine operating between reservoirs at 27°C and 227°C does 1500 J of work each cycle. (a) What is the efficiency of the engine? (b) What is the change in entropy for the engine for each cycle?

97. ■■ Because of limitations on materials, the maximum temperature of the superheated steam used in a turbine for the generation of electricity is about 540°C. (a) If the steam condenser operates at 20°C, what is the ideal efficiency? (b) The actual efficiency is about 35–40%. What does this range tell you?

98. ■■■ There is a Carnot coefficient of performance (COP$_c$) for an ideal, or Carnot, heat pump. (a) Show that this quantity is given by

$$COP_C = \frac{T_h}{T_h - T_c}$$

(b) What does this quantity tell you about adjusting the temperatures for maximum efficiency of a heat pump? (Can you guess the equation for the COP$_C$ for a refrigerator?)

99. ■■■ A salesperson tells you that a new refrigerator with a high COP removes 2.6×10^3 J each cycle from the inside of the refrigerator at a temperature of 5.0°C and rejects 2.8×10^3 J into the 30°C kitchen. (a) What is the refrigerator's COP? (b) Is this scenario possible? Justify your answer.

100. ■■■ A Carnot engine has an efficiency of 40%. Suppose it could be run in reverse as a heat pump. What would be the COP$_C$ of the pump? (See Exercise 98.)

101. ■■■ An ideal heat pump is equivalent to a Carnot engine running in reverse. Show that the input work required for an ideal heat pump is given by

$$W = Q_c \left[\left(\frac{T_h}{T_c} \right) - 1 \right]$$

Additional Exercises

102. A thermally isolated quantity of gas with an initial volume of 10 L is compressed isobarically at a pressure of 300 kPa. If 900 J of work is done on the gas, what is the final volume of the gas?

103. For the two-step reversible process illustrated in ▼ Fig. 12.28, the same amount of heat is added to and removed from the system. Prove that the total entropy of the universe increases in the process.

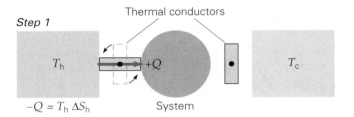

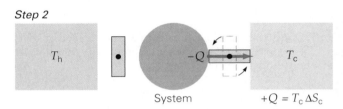

▲ **FIGURE 12.28 Entropy increase?** See Exercise 103.

104. Use the general definition of the thermal efficiency of a heat engine, $\varepsilon = W_{net}/Q_{in}$, to derive the Carnot efficiency. [*Hint*: Consider the cycle in terms of entropy, and see Fig. 12.18b.]

105. A gas is enclosed in a cylindrical piston with a 12.0-cm radius. Heat is slowly added to the gas while the pressure is maintained at 1.00 atm. During the process, the piston moves 6.00 cm. (a) What type of process is this? (b) If the heat transferred to the gas during the expansion is 420 J, what is the change in the internal energy of the gas?

106. A heat engine operates between a reservoir at 20°C and one at 250°C. What is the Carnot efficiency of the engine? Could this efficiency ever be achieved? Why?

107. A monatomic ideal gas is compressed adiabatically from a pressure of 1.00×10^5 Pa and volume of 240 L to a volume of 40.0 L. (a) What is the new pressure of the gas? (b) How much work is done on the gas?

108. The energy or useful work lost as a result of a change in the entropy of the universe is given by $W = T_c \Delta S_u$, where T_c is the temperature of the cold reservoir for the process and S_u is the entropy of the universe. (a) Show that for heat (Q) transferred from a high-temperature reservoir to a low-temperature reservoir, $W = Q[1 - (T_c/T_h)]$. (b) What is the quantity within the brackets?

109. In an isothermal expansion at 27°C, an ideal gas does 30 J of work. What is the change in entropy of the gas?

110. An ideal gas in a cylinder is compressed adiabatically while 1500 J of work is done in moving the piston. What is the change in the internal energy of the gas? Does the temperature of the gas increase or decrease?

111. In a thermodynamic process, 2500 J of heat is transferred to a system, and 1000 J of work is done on the system. What is the change in the internal energy of the system?

112. A Carnot engine operates between 70°C and 400°C. If the engine does 5.0×10^4 J of work per cycle, what is the heat exhausted in each cycle?

113. A gram of water (1.00 cm^3) at 100°C is converted to a volume of 1671 cm^3 of steam at atmospheric pressure. (a) How much work is done by the water against the atmosphere in its expansion? (b) What is the change in the internal energy of the system?

114. In 1834, Joule performed an experiment that produced some important information about the internal energy of a gas. In this experiment, two containers connected by a valve are submerged in a water bath (see ▼Fig. 12.29). Container A is filled with a gas at high pressure, and container B is empty (evacuated). If the valve is opened quickly, the gas in container A undergoes "free expansion" into the evacuated container. (In other words, in pushing against "a vacuum," the gas does no work.) The temperature of the water bath is the same after the gas has again come to equilibrium. Does the internal energy of the gas change in the process?

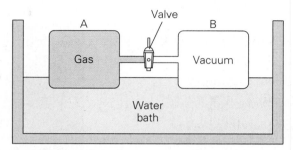

▲ **FIGURE 12.29 Free expansion** See Exercise 114.

Vibrations and Waves

INSIGHTS

- Earthquakes, Seismic Waves, and Seismology
- Desirable and Undesirable Resonances

Learn by Drawing

- Oscillating in a Parabolic Potential Well

The picture you see is what a lot of people probably first think of when they hear the word *wave*. We're all familiar with ocean waves or their smaller relatives, the ripples that form on the surface of a lake or pond when something disturbs the surface. Yet in many ways, the waves that are most important to us, as well as most interesting to physicists, either are invisible or don't look like waves. Sound, for example, is a wave. Because they travel through a medium, sound waves share many properties with water waves. Perhaps most surprisingly, light is a wave. In fact, all electromagnetic radiations are waves— radio waves, microwaves, X rays, and so on. Whenever you peer through a microscope, put on a pair of glasses, or look up at a rainbow, you are experiencing wave energy in the form of light. In Chapter 28, you'll learn how even moving particles have wavelike properties. But first, we need to look at the basic description of waves.

What is a wave, and what properties do waves of *all* kinds have in common? You will find that, to understand wave motion, it is necessary to study some other kinds of motion which at first might not seem closely related to waves. It turns out, however, that the motion of a weight bouncing up and down on a spring or the swinging of a pendulum in a grandfather clock provides important keys to an understanding of wave motion, whether it be that of an electron, a Slinky®, or the perfect wave that surfers dream about.

A vibration, or an oscillation, involves back-and-forth motion, such as that of a swinging pendulum. Another example is a ball in a round-bottomed bowl. If the ball is displaced from its equilibrium position, it will roll back and forth, or oscillate, and finally come to rest at the equilibrium position, where its potential energy is lowest.

Recall from Chapter 8 that the ball must be oscillating about a point of *stable* equilibrium, for which there is a *restoring* force or torque. This is true in general for objects that undergo vibrating or oscillating motions. In a material medium, the restoring force is provided by intermolecular forces. If a molecule is disturbed, restoring forces exerted by its neighbors tend to return the molecule to its original position, and it begins to oscillate. In so doing, it affects adjacent molecules, which are in turn set into oscillation. This is referred to as *propagation*. But what is propagated by the molecules in a material? The answer is, energy. A single disturbance, which happens when you give the end of a stretched rope a quick shake, gives rise to what is referred to as a *wave pulse*. A continuous, repetitive disturbance gives rise to a continuous propagation of energy that we call *wave motion*. But before looking at waves in media, it is helpful to analyze the oscillations of a single mass.

13.1 Simple Harmonic Motion

OBJECTIVES: To (a) describe simple harmonic motion and (b) describe how energy and speed vary in such motion.

The motion of an oscillating object depends on the restoring force that makes the object go back and forth. It is convenient to begin to study such motion by considering the simplest type of force acting along the x-axis: a force that is directly proportional to the object's displacement from equilibrium. A common example is the (ideal) spring force, described by **Hooke's law** (Section 5.2),

Note: Hooke's law (for an ideal spring) is discussed in Section 5.2.

$$F_s = -kx \qquad (13.1)$$

where k is the spring constant. Recall from Chapter 5 that the minus sign indicates that the force is always in the direction opposite that of the displacement. That is, the force always tends to restore the spring to its equilibrium position.

Suppose that an object on a horizontal frictionless surface is connected to a spring as shown in ▶ Fig. 13.1. When the object is displaced to one side of its equilibrium position and released, it will move back and forth—that is, it will vibrate, or oscillate. Here, an oscillation or a vibration is clearly a *periodic motion*—a motion that repeats itself again and again along the same path. For linear oscillations, like those of an object attached to a spring, the path may be back and forth or up and down. For the angular oscillation of a pendulum, the path is back and forth along a circular arc.

Motion under the influence of the type of force described by Hooke's law is called **simple harmonic motion (SHM)**, because the force is the simplest restoring force and because the motion can be described by harmonic functions (sines and cosines), as you will see later in the chapter. The directed distance of an object in SHM from its equilibrium position is the object's **displacement**. Note in Fig. 13.1 that the displacement can be either positive or negative ($+x$ or $-x$), which indicates direction. The maximum displacements are $+A$ and $-A$ (Fig. 13.1b,d).

The magnitude of the maximum displacement, or the maximum distance of an object from its equilibrium position, is called the object's **amplitude** (A), a scalar quantity that expresses the distance of both extreme displacements from the equilibrium position.

Besides the amplitude, two other important quantities used in describing an oscillation are its period and frequency. The **period** (T) is the time it takes the object to complete one cycle of motion. A cycle is a *complete* round-trip, or motion through a complete oscillation. For example, if an object starts at $x = A$ (Fig. 13.1b), then when it returns to $x = A$ (as in Fig. 13.1f), it will have completed one cycle in a time we call one period. If an object were initially at $x = 0$ when disturbed, then its second return to this point would mark a cycle. In either case, the object would travel a distance of $4A$ during one cycle. Can you show this? (Why a *second* return?)

PHYSLET®
ILLUSTRATION

Spring Constant

Note: Here, $+x$ signifies spring extension and $-x$ means compression.

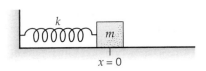

(a) Equilibrium

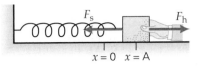

$x = 0$ $x = A$

(b) $t = 0$ Just before release

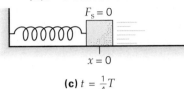

$x = 0$

(c) $t = \frac{1}{4}T$

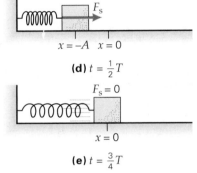

$x = -A$ $x = 0$

(d) $t = \frac{1}{2}T$

$F_s = 0$

$x = 0$

(e) $t = \frac{3}{4}T$

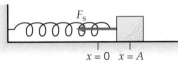

$x = 0$ $x = A$

(f) $t = T$

▲ **FIGURE 13.1 Simple harmonic motion (SHM)** When an object on a spring **(a)** is displaced from its equilibrium position $x = 0$ and **(b)** is released, the object undergoes SHM (assuming no frictional losses). The time it takes to complete one cycle is the period of oscillation (T). (Here, F_s is the spring force and F_h is the force the hand applies.) **(c)** At $t = T/4$, the object is back at its equilibrium position; **(d)** at $t = T/2$, it is at $x = -A$. **(e)** During the next half cycle, the motion is to the right; **(f)** at $t = T$, the object is back at its initial ($t = 0$) starting position.

The **frequency** (f) is the number of cycles per second. The frequency and the period are related by

$$f = \frac{1}{T} \quad \begin{array}{l} \textit{frequency} \\ \textit{and period} \end{array} \qquad (13.2)$$

SI unit of frequency: hertz (Hz), or cycle per second (cycle/s)

The inverse relationship is reflected in the units. The period is the number of seconds per cycle, and the frequency is the number of cycles per second. For example, if $T = \frac{1}{2}$ s/cycle, then $f = 2$ cycles/s.

The standard unit of frequency is the **hertz** (Hz), which is one cycle per second.* From Eq. 13.2, frequency has the unit inverse seconds ($1/s$, or s^{-1}), since the period is a measure of time. Although cycle is not really a unit, you might find it convenient at times to express frequency in cycles per second to help with unit analysis. This is similar to the way the radian (rad) is used in the description of circular motion in Sections 7.1 and 7.2.

The preceding terms used to describe SHM are summarized in Table 13.1.

Energy and Speed of a Mass–Spring System in SHM

Recall from Chapter 5 that the potential energy stored in a spring that is stretched or compressed a distance $\pm x$ from equilibrium (chosen to be $x = 0$) is

$$U = \frac{1}{2}kx^2 \qquad (13.3)$$

The *change* in potential energy of an object oscillating on a spring is related to the work done by the spring force. An object with mass m oscillating on a spring also has kinetic energy. The kinetic and potential energies together give the total mechanical energy of the system:

$$E = K + U = \frac{1}{2}mv^2 + \frac{1}{2}kx^2 \qquad (13.4)$$

When the object is at one of its maximum displacements, $+A$ or $-A$, it is instantaneously at rest, $v = 0$ (▶Fig. 13.2). Thus, all the energy is in the form of potential energy (U_{max}) at this location; that is,

$$E = \frac{1}{2}m(0)^2 + \frac{1}{2}k(\pm A)^2 = \frac{1}{2}kA^2$$

or

$$E = \frac{1}{2}kA^2 \quad \begin{array}{l} \textit{total energy of an} \\ \textit{object in SHM on a spring} \end{array} \qquad (13.5)$$

This outcome is a general result for SHM:

The total energy of an object in simple harmonic motion is directly proportional to the square of the amplitude of the object's motion.

TABLE 13.1 Terms Used to Describe Simple Harmonic Motion

displacement—the directed distance of an object ($\pm x$) from its equilibrium position.
amplitude (A)—the magnitude of the maximum displacement, or the maximum distance, of an object from its equilibrium position.
period (T)—the time for one complete cycle of motion.
frequency (f)—the number of cycles per second (in hertz or inverse seconds, where $f = 1/T$).

---

*The unit is named for Heinrich Hertz (1857–1894), a German physicist and early investigator of electromagnetic waves.

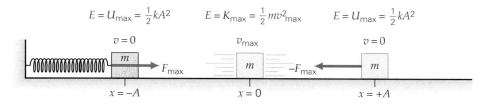

$$E = U_{max} = \tfrac{1}{2}kA^2 \qquad E = K_{max} = \tfrac{1}{2}mv^2_{max} \qquad E = U_{max} = \tfrac{1}{2}kA^2$$

$v = 0$ v_{max} $v = 0$

$x = -A$ $x = 0$ $x = +A$

◀ **FIGURE 13.2 Oscillations and energy** For a mass oscillating in SHM on a spring (on a frictionless surface), the total energy at the amplitude positions ($\pm A$) is all potential energy (U_{max}), and $E = \tfrac{1}{2}kx^2 = \tfrac{1}{2}kA^2$, which is the total energy of the system. At the center position ($x = 0$), the total energy is all kinetic energy ($E = \tfrac{1}{2}mv^2_{max}$, where m is the mass of the block). How is the total energy divided between $x = 0$ and $x = \pm A$?

Note: This discussion will be limited to light springs, the mass of which can be considered negligible.

Equation 13.5 allows us to express the velocity of an object oscillating on a spring as a function of position:

$$E = K + U \qquad \text{or} \qquad \tfrac{1}{2}kA^2 = \tfrac{1}{2}mv^2 + \tfrac{1}{2}kx^2$$

Rearranging terms, we get

$$v^2 = \frac{k}{m}(A^2 - x^2)$$

so

$$v = \pm\sqrt{\frac{k}{m}(A^2 - x^2)} \quad \text{velocity of an object in SHM} \quad (13.6)$$

where the positive and negative signs indicate the direction of the velocity. Note that at $x = \pm A$ the velocity is zero, since the object is instantaneously at rest at its maximum displacement from equilibrium.

Note also that when the oscillating object passes through its equilibrium position ($x = 0$), its potential energy is zero. At that instant, all the energy is kinetic, and the object is traveling at its maximum speed v_{max}. The expression for the energy in this case is

$$E = \tfrac{1}{2}kA^2 = \tfrac{1}{2}mv^2_{max}$$

and thus,

$$v_{max} = \sqrt{\frac{k}{m}}\,A \quad \begin{array}{l}\textit{maximum speed of}\\ \textit{mass on a spring}\end{array} \quad (13.7)$$

In the next Example, as well as in the Learn by Drawing feature on p. 448, you can visualize the continuous trade-off between kinetic and potential energy.

Example 13.1 ■ A Block and a Spring: Simple Harmonic Motion

A block with a mass of 0.25 kg sitting on a frictionless surface is connected to a light spring that has a spring constant of 180 N/m (see Fig. 13.1). If the block is displaced 15 cm from its equilibrium position and released, what are (a) the total energy of the system and (b) the speed of the block when it is 10 cm from its equilibrium position?

Thinking It Through. The total energy depends on the spring constant (k) and the amplitude (A), which are given. At $x = 10$ cm, the speed should be less than the maximum speed. (Why?)

Solution. First we list the given data, as usual, and what is to be found. The initial displacement is the amplitude. (Why?)

Given: $m = 0.25$ kg   **Find:** (a) E (total energy)
$k = 180$ N/m   (b) v (speed)
$A = 15$ cm $= 0.15$ m
$x = 10$ cm $= 0.10$ m

(a) The total energy is given by Eq. 13.5:

$$E = \tfrac{1}{2}kA^2 = \tfrac{1}{2}(180\ \text{N/m})(0.15\ \text{m})^2 = 2.0\ \text{J}$$

Learn by Drawing

Oscillating in a Parabolic Potential Well

A way to visualize the conservation of energy in simple harmonic motion is shown in Fig. 1. The potential energy of a mass–spring system can be sketched on a plot of energy (E) versus position (x). Since $U = kx^2 \propto x^2$, the graph is a *parabola*.

In the absence of nonconservative forces, the total energy of the system, E, is constant. But E is the sum of the kinetic and potential energies. During the oscillations, there is a continuous trade-off between the two types of energies, but their sum remains constant. Mathematically, this relationship is written as $E = K + U$. In Fig. 2, U (indicated by a blue arrow) is represented by the vertical distance from the x-axis. Since E is constant and independent of x, it will plot as a horizontal line (shown in green). The kinetic energy is the part of the total energy that is *not* potential energy; that is, $K = E - U$; it can be graphically interpreted (purple arrow) as the vertical distance between the potential-energy parabola and the horizontal green total-energy line. As the object oscillates

on the x-axis, the energy trade-offs can be visualized as the lengths of the two arrows change.

A general location, x_1, is shown in Fig. 2. Neither the kinetic energy nor the potential energy is at its maximum value of E there. These maximum values occur instead at $x = 0$ and $x = \pm A$, respectively. The motion cannot exceed $x = \pm A$, because that would imply a negative kinetic energy, which is physically impossible. (Why?) The amplitude positions are sometimes called the *endpoints* of the motion, because they are the locations where the speed is instantaneously zero and the object reverses direction.

Try using the graphical approach to answer the following questions (and make up some of your own): What do you have to do to E to increase the amplitude of oscillation, and how might you do this? What happens to the amplitude of a real-life system in SHM in the presence of a force such as friction when E decays with time?

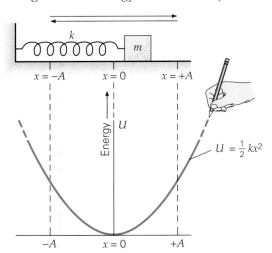

FIGURE 1 The potential energy "well" of a spring–mass system The potential energy of a spring that is stretched or compressed ($\pm x$) from its equilibrium position ($x = 0$) is a parabola, since $U \propto x^2$. At $x = \pm A$, all of the system's energy is potential.

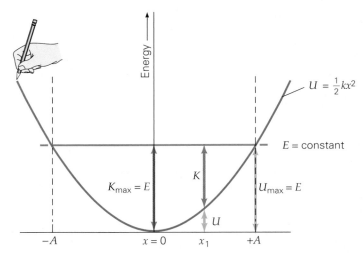

FIGURE 2 Energy transfers as the spring–mass system oscillates The vertical distance from the x-axis to the parabola is the system's potential energy. The remainder—the vertical distance between the parabola and the horizontal line representing the system's constant total energy E—is the system's kinetic energy.

Energy and Mass of a Spring

(b) The instantaneous speed of the block at a distance 10 cm from the equilibrium position is given by Eq. 13.6 without directional signs:

$$v = \sqrt{\frac{k}{m}(A^2 - x^2)} = \sqrt{\frac{180 \text{ N/m}}{0.25 \text{ kg}} [(0.15 \text{ m})^2 - (0.10 \text{ m})^2]} = \sqrt{9.0 \text{ m}^2/\text{s}^2} = 3.0 \text{ m/s}$$

What would the speed be at $x = -10$ cm ?

Follow-up Exercise. In part (b) of this Example, the block at $x = 10$ cm is at two-thirds, or 67%, of its maximum displacement. Is its speed at that position therefore 67%

of its maximum speed? Prove your answer mathematically. (*Answers to all Follow-up Exercises are at the back of the text.*)

The spring constant is commonly determined by placing an object of known mass on the end of a spring and letting it settle vertically to a new equilibrium position. The next Example shows some typical results.

Example 13.2 ■ The Spring Constant: Experimental Determination

When a 0.50-kg mass is suspended from a spring, the spring stretches a distance of 10 cm to a new equilibrium position (▼Fig. 13.3a). (a) What is the spring constant of the spring? (b) The mass is then pulled down another 5.0 cm and released. What is the highest position of the oscillating mass?

Thinking It Through. At the equilibrium position, the net force on the mass is zero because $a = 0$. In (b), negative y will be used to designate "downward," as is customary in problems involving vertical motion.

Solution.

Given: $m = 0.50$ kg
$y_o = 10$ cm $= 0.10$ m
$y = -5.0$ cm $= -0.050$ m (new reference point)

Find: (a) k (spring constant)
(b) A (amplitude)

(a) When the suspended mass is in equilibrium (Fig. 13.3a), the net force on the mass is zero. Thus, the weight of the mass and the spring force are equal and opposite. Then, equating their magnitudes,

$$F_s = w$$

or

$$ky_o = mg$$

Hence,

$$k = \frac{mg}{y_o} = \frac{(0.50 \text{ kg})(9.8 \text{ m/s}^2)}{0.10 \text{ m}} = 49 \text{ N/m}$$

(b) Once set into motion, the mass oscillates up and down through the equilibrium position. Since the motion is symmetric about this point, it is designated as the zero reference point of the oscillation (Fig. 13.3b). The initial displacement is $-A$, so the highest position of the mass will be 5.0 cm above the equilibrium position $(+A)$.

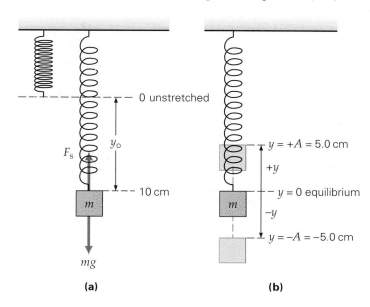

(a)

(b)

◀ **FIGURE 13.3 Determination of the spring constant** (a) When an object suspended on a spring is in equilibrium, the two forces on the object cancel, so that $F_s = w$, or $ky_o = mg$. Thus, the spring constant k can be computed: $k = mg/y_o$. (b) The zero reference point of an object in SHM and suspended on a spring is conveniently taken to be the new equilibrium position, as the motion is symmetric about that point. (See Example 13.2.)

Follow-up Exercise. How much more potential energy does the spring in this Example have at the bottom of its oscillation than at the top?

13.2 Equations of Motion

OBJECTIVES: To (a) understand the equation of SHM and (b) explain what is meant by phase and phase differences.

We refer to the **equation of motion** of an object as the equation that gives the object's position as a function of time. For example, the equation of motion with a constant linear acceleration is $x = v_0 t + \frac{1}{2} at^2$, where v_0 is the initial velocity (Chapter 2). However, the acceleration is not constant in simple harmonic motion, so the kinematic equations of Chapter 2 do not apply to this case.

The equation of motion for an object in SHM can be found from a relationship between simple harmonic and uniform circular motions. SHM can be simulated by a component of uniform circular motion, as illustrated in ▼ Fig. 13.4. As the illuminated object moves in uniform circular motion (with constant angular speed ω) in a vertical plane, its shadow moves back and forth vertically, following the same path as the object on the spring, which is in simple harmonic motion. Since the shadow and the object have the same position at any time, it follows that the equation of motion for the shadow of the object in circular motion is the same as the equation of motion for the oscillating object on the spring.

From the reference circle in Fig. 13.4b, the y-coordinate (position) of the object is given by

$$y = A \sin \theta$$

But the object moves with a constant angular velocity of magnitude ω. In terms of the angular distance θ, assuming that $\theta = 0°$ at $t = 0$, we have $\theta = \omega t$, so

$$y = A \sin \omega t \quad (SHM \; for \; y_o = 0, \; initial \; upward \; motion) \quad (13.8)$$

▼ **FIGURE 13.4 Reference circle for vertical motion** (a) The shadow of an object in uniform circular motion has the same vertical motion as an object oscillating on a spring in simple harmonic motion. (b) The motion can thus be described by $y = A \sin \theta = A \sin \omega t$ (assuming that $y = 0$ at $t = 0$).

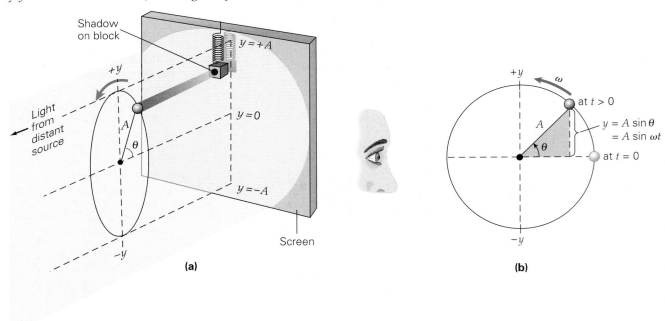

Note that as t increases from zero, y increases, in the positive direction, so the equation describes initial upward motion.

With Eq. 13.8 as the equation of motion, the mass *must* always be initially at $y_o = 0$. But what if the mass on the spring were initially at the amplitude position $+A$? In that case, the sine equation would not describe the motion, because it does *not* describe the *initial condition*—that is, $y_o = +A$ at $t_o = 0$. So another equation of motion is needed, and $y = A \cos \omega t$ applies. By this equation, at $t_o = 0$, the mass is at $y_o = A \cos \omega t = A \cos \omega (0) = +A$, and the cosine equation correctly describes the initial conditions (▼Fig. 13.5):

$$y = A \cos \omega t \quad \text{(initial downward motion with } y_o = +A) \quad (13.9)$$

Here, the initial motion is downward, because, for times shortly after $t_o = 0$, the value of y decreases. If the amplitude were $-A$, the mass would initially be at that position and the initial motion would be upward.

Thus, the equation of motion for an oscillating object may be either a sine or a cosine function. Both of these functions are referred to as being *sinusoidal*. That is, simple harmonic motion is described by a sinusoidal function of time.

The angular speed ω (in rad/s) of the *reference circle object* is called the *angular frequency* of the oscillating object, since $\omega = 2\pi f$, where f is the frequency of revolution or rotation of the object (Section 7.2). Figure 13.4 shows that the frequency of the "orbiting" object is the same as the frequency of oscillation of the object on the spring. Thus, using $f = 1/T$, we can write Eq. 13.8 as

$$y = A \sin (2\pi ft) = A \sin \left(\frac{2\pi t}{T}\right) \quad \begin{array}{l}\text{(SHM for } y_o = 0, \\ \text{initial upward motion)}\end{array} \quad (13.10)$$

Note that this equation is for initial upward motion, because after $t_o = 0$, the value of y increases positively. For initial downward motion, the amplitude term would be $-A$.

Equations 13.8 and 13.10 give three equivalent forms of the equation of motion for an object in SHM. Any one of them can be used for convenience, depending on the known parameters. For example, suppose you are given the time t in terms of the period T—say, $t_o = 0$, $t_1 = T/4$, and $t_2 = 3T/4$—and are asked to find the position of an object in SHM at these times. In this case, it is convenient to use Eq. 13.10, and

$$t_o = 0 \qquad y_o = A \sin [2\pi(0)/T] = A \sin 0 = 0$$

$$t_1 = \frac{T}{4} \qquad y_1 = A \sin [2\pi(T/4)/T] = A \sin \pi/2 = A$$

$$t_2 = \frac{3T}{4} \qquad y_2 = A \sin [2\pi(3T/4)/T] = A \sin 3\pi/2 = -A$$

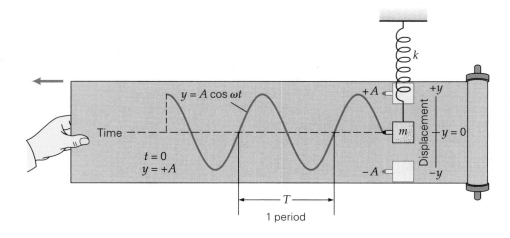

◀ FIGURE 13.5 Sinusoidal equation of motion As time passes, the oscillating object traces out a sinusoidal curve on the moving paper. In this case, $y = A \cos \omega t$, because the object's initial displacement is $y_o = +A$.

The results tell us that the object was initially at $y = 0$, as we knew. One-quarter period later, it was at $y = A$, or the amplitude of its oscillation; and at a time of three quarters of a period $(3T/4)$, it was at the $-A$ position, which is to be expected, since the motion is periodic. (Where would the object be at $T/2$ and T?)

Hence, we may write in general,

$$y = \pm A \sin \omega t = \pm A \sin (2\pi f t) = \pm A \sin \left(\frac{2\pi t}{T} \right)$$

(+A for initial motion upward with $y_o = 0$
−A for initial motion downward with $y_o = 0$)

By a similar development, Eq. 13.9 has the general form

$$y = \pm A \cos \omega t = \pm A \cos (2\pi f t) = \pm A \cos \left(\frac{2\pi t}{T} \right)$$

(+A for initial motion downward with $y_o = +A$
−A for initial motion upward with $y_o = -A$)

To show the power of the reference circle, let us use it to compute the period of the spring–object system. Note that the time for the object in the reference circle to make one complete "orbit" is exactly the time it takes for the oscillating object to make one complete cycle. (See Fig. 13.4.) Thus, all we need is the time for one orbit around the reference circle, and we have the period of oscillation. Because the object "orbiting" the reference circle is in uniform circular motion at a constant speed equal to the maximum speed of oscillation v_{max}, the object travels a distance of one circumference in one period. Then $t = d/v$, where $t = T$, d is the circumference, and v is v_{max} given by Equation 13.7; that is,

$$T = \frac{d}{v_{max}} = \frac{2\pi A}{\sqrt{k/m}\, A}$$

or

$$T = 2\pi \cdot \sqrt{\frac{m}{k}} \qquad \textit{period of object oscillating on a spring} \qquad (13.11)$$

Note: Period and frequency are independent of amplitude in SHM.

Because the amplitudes canceled out in Eq. 13.11, *the period (and frequency) are independent of the amplitude of the motion.* This statement is a general feature of simple harmonic oscillators—that is, oscillators driven by a linear restoring force, such as a spring obeying Hooke's law.

We see from Eq. 13.11 that the greater the mass, the longer is the period, and the greater the spring constant (or the stiffer the spring), the shorter is the period. It is the *ratio* of mass to stiffness that determines the period. Thus, you can offset an increase in mass by using a stiffer spring.

Since $f = 1/T$,

$$f = \frac{1}{2\pi} \sqrt{\frac{k}{m}} \qquad \textit{frequency of object oscillating on a spring} \qquad (13.12)$$

Thus, the greater the spring constant (the stiffer the spring), the more frequently the system vibrates, as you might expect.

Also, note that since $\omega = 2\pi f$, we may write

$$\omega = \sqrt{\frac{k}{m}} \qquad \textit{angular frequency of object oscillating on a spring} \qquad (13.13)$$

As another example, a simple pendulum (a small, heavy object on a string) will undergo simple harmonic motion for small angles of oscillation. The period of a simple pendulum oscillating through a small angle $\theta \lesssim 10°$ is given, to a good approximation, by

$$T = 2\pi \sqrt{\frac{L}{g}} \quad \begin{array}{l}\textit{period of a} \\ \textit{simple pendulum}\end{array} \quad (13.14)$$

where L is the length of the pendulum and g is the acceleration due to gravity. A pendulum-driven clock that is not properly rewound and is running down would still keep correct time, because the period would remain unchanged as the amplitude decreased. As shown by Eq. 13.11, the period is independent of amplitude.

An important difference between the period of the mass–spring system and that of the pendulum is that the latter is independent of the mass of the bob. (See Eq. 13.11 and Eq. 13.14.) Can you explain why? Think about what supplies the restoring force for the pendulum's oscillations. It is gravity. Hence, the acceleration (along with the velocity and period) is expected to be independent of mass. That is, the gravitational force automatically provides the same acceleration to different masses of bobs on pendulums with the same length. We have seen that similar effects occur in free fall (Chapter 2) and with blocks sliding and cylinders rolling down inclines (Chapters 4 and 8, respectively). The next Example demonstrates the usage of the equation of motion for SHM.

Example 13.3 ■ An Oscillating Mass: Applying the Equation of Motion

A mass on a spring oscillates vertically with an amplitude of 15 cm, a frequency of 0.20 Hz, and an equation of motion given by Eq. 13.8, with $y_o = 0$ at $t_o = 0$ and initial upward motion. (a) What are the position and direction of motion of the mass at $t = 3.1$ s ? (b) How many oscillations (cycles) does the mass make in a time of 12 s?

Thinking It Through. Part (a) is a straightforward application of Eq. 13.8. In part (b), the number of oscillations means the number of cycles, and recall that frequency is sometimes expressed in cycles per second. Hence, multiplying the frequency by the time would give the number of cycles or oscillations.

Solution.

Given: $A = 15$ cm $= 0.15$ m   *Find:* (a) y (position) and direction of motion
         $f = 0.20$ Hz   (b) n (number of oscillations or cycles)
         $y = A \sin \omega t$ (Eq. 13.8)
    (a) $t = 3.1$ s (b) $t = 12$ s

(a) First, since the frequency f is given, it is convenient to use the equation of motion in the form $y = A \sin 2\pi ft$ (Eq. 13.10). As can be seen from the equation, at $t_o = 0$, $y_o = 0$, so initially the mass is at the zero position. Then, at $t = 3.1$ s,

$$y = A \sin 2\pi ft = (0.15 \text{ m}) \sin [2\pi(0.20 \text{ s}^{-1})(3.1 \text{ s})] = (0.15 \text{ m}) \sin (3.9 \text{ rad}) = -0.10 \text{ m}$$

So the mass is at $y = -0.10$ m at $t = 3.1$ s. But what is its direction of motion? Let's take a look at the period (T) and see what part of its cycle the mass is in. By Eq. 13.2,

$$T = \frac{1}{f} = \frac{1}{0.20 \text{ Hz}} = 5.0 \text{ s}$$

In $t = 3.1$ s, the mass has gone through 3.1 s/5.0 s $= 0.62$, or 62%, of a period or cycle, so it is moving downward [up ($\frac{1}{4}$ cycle) and back ($\frac{1}{4}$ cycle) to $y_o = 0$ in $\frac{1}{2}$, or 50%, of the cycle, and downward during the next $\frac{1}{4}$ cycle].
(b) The number of oscillations (cycles) is equal to the product of the frequency (cycles/s) and the elapsed time (s), both of which are given:

$$n = ft = (0.20 \text{ cycles/s})(12 \text{ s}) = 2.4 \text{ cycle}$$

Thus, the mass has gone through two complete cycles and 0.4 of another, which means that it is on its way back to $y_0 = 0$ from its amplitude position of $+A$. (Why?)

Follow-up Exercise. Find what is asked for in Exercise 13.3 at times (1) $t = 4.5$ s and (2) $t = 7.5$ s.

Problem-Solving Hint

Note that in the calculation in part (a) of Example 13.3, where we have sin (3.9), the angle is in radians, *not* degrees. Don't forget to set your calculator to radians (rather than degrees) when finding the value of a trigonometric function in equations for simple harmonic or circular motion.

Example 13.4 ■ Fun with a Pendulum: Frequency and Period

A helpful older brother takes his sister to play on the swings in the park. He pushes her from behind on each return. Assuming that the swing behaves as a simple pendulum with a length of 2.50 m, (a) what would be the frequency of the oscillations, and (b) what would be the interval between the brother's pushes?

Thinking It Through. (a) The period is given by Eq. 13.14, and the frequency and period are inversely related: $f = 1/T$. (b) Since the brother pushes from one side on each return, he must push once every cycle that is completed, so the time between his pushes is equal to the swing's period.

Solution.

Given: $L = 2.50$ m *Find:* (a) f (frequency)
(b) T (period)

(a) We can take the reciprocal of Eq. 13.14 to solve directly for the frequency:

$$f = \frac{1}{T} = \frac{1}{2\pi}\sqrt{\frac{g}{L}} = \frac{1}{2\pi}\sqrt{\frac{9.80\ \text{m/s}^2}{2.50\ \text{m}}} = 0.315\ \text{Hz}$$

(b) The period is then found from the frequency:

$$T = 1/f = 1/(0.315\ \text{Hz}) = 3.17\ \text{s}$$

The brother must push every 3.17 s to maintain a steady swing (and to keep his sister from complaining).

Follow-up Exercise. In this Example, the older brother, a physics buff, carefully measures the period of the swing to be 3.18 s, not 3.17 s. If the length of 2.50 m is accurate, what is the acceleration due to gravity at the location of the park? Considering this accurate value of g, do you think the park is at sea level?

Initial Conditions and Phase

You may be wondering how to decide whether to use a sine or cosine function to describe a particular case of simple harmonic motion. In general, the form of the function is determined by the initial displacement and velocity of the object: the *initial conditions* of the system. These initial conditions are the values of the displacement and velocity at $t = 0$; taken together, they tell how the system is initially set into motion.

Let's look at four special cases. If an object in vertical SHM has an initial displacement of $y = 0$ at $t = 0$ and moves initially upward, the equation of motion is $y = A \sin \omega t$ (► Fig. 13.6a). Note that $y = A \cos \omega t$ does not satisfy the initial condition, because $y_0 = A \cos \omega t = A \cos \omega (0) = A$, since $\cos 0 = 1$.

Suppose that the object is initially released ($t = 0$) from its positive amplitude position ($+A$), as in the case of the object on a spring shown in Fig. 13.5. Here, the equation of motion is $y = A \cos \omega t$ (Fig. 13.6b). This expression satisfies the initial condition: $y_0 = A \cos \omega(0) = A$.

Note: Initial conditions include both the displacement y_0 and velocity v_0 at $t = 0$.

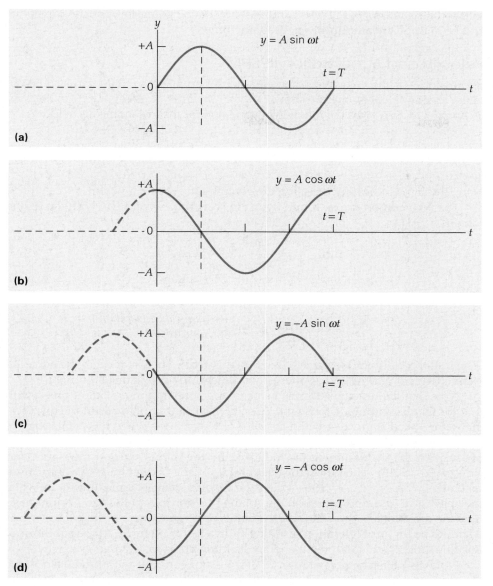

(a)

$y = A \sin \omega t$

$t = T$

(b)

$y = A \cos \omega t$

$t = T$

(c)

$y = -A \sin \omega t$

$t = T$

(d)

$y = -A \cos \omega t$

$t = T$

◀ **FIGURE 13.6 Initial conditions and equations of motion** The initial conditions (y_o and t_o) determine the form of the equation of motion—for the cases shown here, either a sine or a cosine. For $t_o = 0$, the initial displacements are **(a)** $y_o = 0$, **(b)** $y_o = +A$, **(c)** $y_o = 0$, and **(d)** $y_o = -A$. The equations of motion must match the initial conditions. (See text for description.)

The other two cases are (1) $y = 0$ at $t = 0$, with motion initially downward (for an object on a spring) or in the negative direction (for horizontal SHM), and (2) $y = -A$ at $t = 0$, meaning that the object is initially at its negative-amplitude position. These motions are described by $y = -A \sin \omega t$ and $y = -A \cos \omega t$, respectively, as illustrated in Figs. 13.6c and 13.6d.

Only these four initial conditions will be considered in our study. Should y_o have a value other than 0 or $\pm A$, the equation of motion is complicated. Note in Fig. 13.6 that if the curves are extended in the negative direction of the horizontal axis (dashed purple lines), they all have the same shape, but have been "shifted," so to speak. In (a) and (b), one curve is ahead of the other by 90°, or $\frac{1}{4}$ cycle; that is, the two curves shifted by a quarter cycle with respect to one another. The oscillations are then said to have a *phase difference* of 90°. In (a) and (c), the curves are shifted 180° and are 180° out of phase. (Note in this case that the oscillations are opposite: When one mass is going up, the other is going down.) What about the oscillations in (a) and (d)?

A figure with a 360° (or 0°) phase shift is not shown, because this would be the same as that in (a). When two objects in SHM have the same equation of motion, they are said to be oscillating *in phase*, which means that they are oscillating together with identical motions. Objects with a 180° phase shift or difference are said to be *completely out of phase* and will always be going in opposite directions and be at opposite amplitudes at the same time.

Velocity and Acceleration in SHM

Expressions for the velocity and acceleration of an object in SHM can also be obtained. Using calculus, one can show that $v = \Delta y/\Delta t = \Delta(A \sin \omega t)/\Delta t$ in the limit as Δt goes to zero gives the following expression for the instantaneous velocity:

Note: Maximum speed $v = \omega A$.

$$v = \omega A \cos \omega t \quad \begin{array}{l}\textit{(vertical velocity if } v_o \\ \textit{is upward at } t_o = 0, \ y_o = 0)\end{array} \quad (13.15)$$

Here, the signs indicating direction are given by the cosine function.

The acceleration can be found by using Newton's second law with the spring force $F_s = -ky$:

$$a = \frac{F_s}{m} = \frac{-ky}{m} = -\frac{k}{m} A \sin \omega t$$

Since $\omega = \sqrt{k/m}$,

$$a = -\omega^2 A \sin \omega t = -\omega^2 y \quad \begin{array}{l}\textit{(vertical acceleration if } v_o \text{ is} \\ \textit{upward at } t_o = 0, \ y_o = 0)\end{array} \quad (13.16)$$

(The initial conditions are for the sine function term. The last term in the equation is independent of time and applies at any time. Convince yourself of this.)

Note that the functions for the velocity and acceleration are out of phase with that for the displacement. Since the velocity is 90° out of phase with the displacement, the speed is greatest when $\cos \omega t = \pm 1$ at $y = 0$—when the oscillating object is passing through its equilibrium position. The acceleration is 180° out of phase with the displacement (as indicated by the minus sign on the right-hand side of Eq. 13.16). Therefore, the magnitude of the acceleration is a maximum when $\sin \omega t = \pm 1$ at $y = \pm A$—when the displacement is a maximum, or when the object is at an amplitude position. At any position except the equilibrium position, the directional sign of the acceleration is opposite that of the displacement, as it should be for an acceleration resulting from a restoring force. At the equilibrium position, both the displacement and acceleration are zero. (Can you see why?)

Note: Maximum acceleration magnitude $a = \omega^2 A$.

Note also that the acceleration in SHM is not constant with time. Hence, the kinematic equations for acceleration (Chapter 2) *cannot* be used, since they describe constant acceleration.

Damped Harmonic Motion

PHYSLET® ILLUSTRATION

Simple Harmonic Motion

Simple harmonic motion with constant amplitude implies that there are no losses of energy, but in practical applications there are always some frictional losses. Therefore, to maintain a constant amplitude motion, energy must be added to a system by some external driving force, such as someone pushing a swing. Without a driving force, the amplitude and the energy of an oscillator decrease with time, giving rise to **damped harmonic motion** (▶Fig. 13.7a). The time required for the oscillations to cease, or damp out, depends on the magnitude and type of the damping force (e.g., air resistance).

In many applications involving continuous periodic motion, damping is unwanted and necessitates an energy input. However, in some instances, damping is desirable. For example, the dial in a spring-operated bathroom scale oscillates briefly before stopping at a weight reading. If not properly damped, these oscillations would continue for some time, and you would have to wait before you could read your weight. Shock absorbers provide damping in the suspension systems of automobiles (Fig. 13.7b; also see Fig. 9.9b). Without "shocks" to dissipate energy

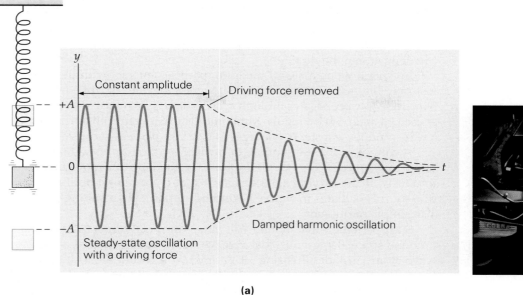

(a) **(b)**

▲ **FIGURE 13.7 Damped harmonic motion** **(a)** When a driving force adds energy to a system in an amount equal to the energy losses of the system, the oscillation is steady with a constant amplitude. When the driving force is removed, the oscillations decay (i.e., they are damped), and the amplitude decreases nonlinearly with time. **(b)** In some applications, damping is desirable and even promoted, as with shock absorbers in automobile suspension systems. Otherwise, the passengers would be in for a bouncy ride.

after hitting a bump, the ride would be bouncy. In California, many new buildings incorporate damping mechanisms (giant shock absorbers) to dampen their oscillatory motion after they are set in motion by earthquake waves.

13.3 Wave Motion

OBJECTIVES: **To (a) describe wave motion in terms of various parameters and (b) identify different types of waves.**

The world is full of waves of various types; some examples are water waves, sound waves, waves generated by earthquakes, and light waves. All waves result from a disturbance, the source of the wave. In this chapter, we will be concerned with mechanical waves—those which are propagated in some medium. (Light waves, which do not require a propagating medium, will be considered in more detail in later chapters.)

When a medium is disturbed, energy is imparted to it. Suppose that energy is added to a material mechanically, such as by impact or (in the case of a gas) by compression. The addition of the energy sets some of the particles in the medium vibrating. Because the particles are linked by intermolecular forces, the oscillation of each particle affects that of its neighbors. The added energy propagates, or spreads, by means of interactions among the particles of the medium. An analogy to this process is shown in ►Fig. 13.8, where the "particles" are dominoes. As each domino falls, it topples the one next to it. Thus, energy is transferred from domino to domino, and the disturbance propagates through the medium.

In this case, there is no restoring force between the dominoes, so they do not oscillate, as do particles in a continuous material medium. Therefore, the disturbance moves in space, but it does not repeat itself in time at any one location.

Similarly, if the end of a stretched rope is given a quick shake, the disturbance transfers energy from the hand to the rope, as illustrated in ►Fig. 13.9. The forces acting between the "particles" in the rope cause them to move in response to the motion of the hand, and a *wave pulse* travels down the rope. Each "particle" goes up and then back down as the pulse passes by. This motion of individual particles

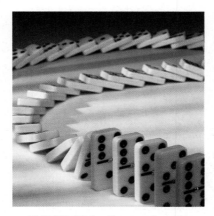

▲ **FIGURE 13.8 Energy transfer** The propagation of a disturbance, or a transfer of energy through space, is seen in a row of falling dominoes.

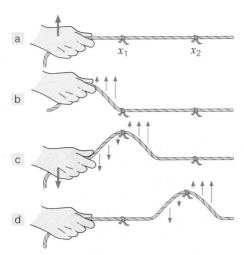

▶ **FIGURE 13.9 Wave pulse** The hand disturbs the stretched rope in a quick up-and-down motion, and a wave pulse propagates along the rope. (The red arrows represent the velocities of the hand and of pieces of the rope at different times and locations.) The rope "particles" move up and down as the pulse passes. The energy in the pulse is thus *both* kinetic (elastic) and potential (gravitational).

Note: A wave is a combination of oscillations in space and time.

▶ **FIGURE 13.10 Periodic wave** A continuous harmonic disturbance can set up a sinusoidal wave in a stretched rope, and the wave travels down the rope with wave speed v. Note that the "particles" in the rope oscillate vertically in simple harmonic motion. The distance between two successive points that are in phase (e.g., at two crests) on the waveform is the wavelength λ of the wave. Can you tell how much time has elapsed, as a fraction of the period T, between the first (red) and last (blue) waves?

and the propagation of the wave pulse as a whole can be observed by tying pieces of ribbon onto the rope (at x_1 and x_2 in the figure). As the disturbance passes point x_1, the ribbon rises and falls, as do the rope's "particles." Later, the same happens to the ribbon at x_2, which indicates that the energy disturbance is propagating, or traveling, along the rope.

In a continuous material medium, particles interact with their neighbors, and restoring forces cause them to oscillate when they are disturbed. Thus, any disturbance not only propagates through space, but may be repeated over and over in time at each position. Such a regular, rhythmic disturbance in both time and space is called a **wave**, and the transfer of energy is said to take place by means of **wave motion**.

A continuous wave motion, or *periodic wave*, requires a disturbance from an oscillating source (▼Fig. 13.10). In this case, the particles move up and down continuously. If the driving source is such that a constant amplitude is maintained (the source oscillates in simple harmonic motion), the resulting particle motion is also simple harmonic.

Such periodic wave motion will have sinusoidal forms (sine or cosine) in both time and space. Being sinusoidal in space means that if you took a photograph of the wave at any instant ("freezing" it in time), you would see a sinusoidal waveform (such as one of the curves in Fig. 13.10). However, if you looked at a single point in space as a wave passed by, you would see a particle of the medium oscillating up and down sinusoidally with time, like the mass on a spring discussed in Section 13.2. (For example, imagine looking through a thin slit at a fixed location on the moving paper in Fig. 13.5. The wave trace would be seen rising and falling like a particle.)

Wave Characteristics

Specific quantities are used to describe sinusoidal waves. As with a particle in simple harmonic motion, the *amplitude* (A) of a wave is the magnitude of the maximum displacement, or the maximum distance, from the particle's equilibrium position (Fig. 13.10). This quantity corresponds to the height of a wave crest or the depth of a trough. Recall from Section 13.2 that, in SHM, the total energy of the oscillator is proportional to the square of the amplitude. Similarly, the energy *transported* by a wave is proportional to the square of its amplitude ($E \propto A^2$). Note the difference, though: A wave is one way of *transmitting* energy through space, whereas an oscillator's energy is localized in space.

For a periodic wave, the distance between two successive crests (or troughs) is called the **wavelength** (λ) (Fig. 13.10). Actually, it is the distance between any two successive parts of the wave that are in phase (i.e., that are at identical points on the waveform). The crest and trough positions are usually used for convenience. Note that a wavelength corresponds spatially to one cycle. Keep in mind that the wave, not the medium or material, is traveling.

The *frequency* (f) of a periodic wave is the number of cycles per second—that is, the number of complete waveforms, or wavelengths, that pass by a given point during each second. The frequency of the wave is the same as the frequency of the SHM source that created it.

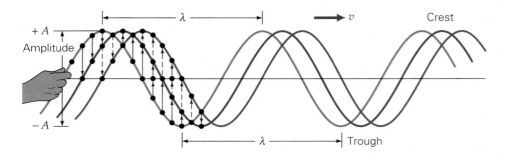

A periodic wave is said to possess a *period* (*T*). The period $T = 1/f$ is the time for one complete waveform (a wavelength) to pass by a given point. Since a wave moves, it also has a **wave speed** (or velocity if the wave's direction is specified). Any particular point on the wave (e.g., a crest) travels a distance of one wavelength λ in a time of one period *T*. Then, since $v = d/t$ and $f = 1/T$, we have

$$v = \frac{\lambda}{T} = \lambda f \qquad wave\ speed \qquad (13.17)$$

Note that the dimensions of *v* are correct (length/time). In general, the wave speed depends on the nature of the medium, in addition to the source frequency *f*.

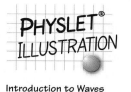

PHYSLET® ILLUSTRATION

Introduction to Waves

Example 13.5 ■ Dock of the Bay: Finding Wave Speed

A person on a pier observes a set of incoming waves that have a sinusoidal form with a distance of 1.6 m between the crests. If a wave laps against the pier every 4.0 s, what are (a) the frequency and (b) the speed of the waves?

Thinking It Through. We know the period and wavelength, so we can use the definition of frequency and Eq. 13.17 for wave speed.

Solution. The distance between crests is the wavelength, so we have the following information:

Given: $\lambda = 1.6$ m *Find:* (a) *f* (frequency)
$\qquad\quad\ T = 4.0$ s $\qquad\quad$ (b) *v* (wave speed)

(a) The lapping indicates the arrival of a wave crest; hence, 4.0 s is the wave period—the time it takes to travel one wavelength (the crest-to-crest distance). Then

$$f = \frac{1}{T} = \frac{1}{4.0\ \text{s}} = 0.25\ \text{s}^{-1} = 0.25\ \text{Hz}$$

(b) The frequency or the period can be used in Eq. 13.17 to find the wave speed:

$$v = \lambda f = (1.6\ \text{m})(0.25\ \text{s}^{-1}) = 0.40\ \text{m/s}$$

Alternatively,

$$v = \frac{\lambda}{T} = \frac{1.6\ \text{m}}{4.0\ \text{s}} = 0.40\ \text{m/s}$$

Follow-up Exercise. On another day, the person measures the speed of sinusoidal water waves at 0.25 m/s. (a) How far does a wave crest travel in 2.0 s? (b) If the distance between successive crests is 2.5 m, what is the frequency of these waves?

Types of Waves

In general, waves may be divided into two types, based on the direction of the particles' oscillations relative to that of the wave velocity. In a **transverse wave**, the particle motion is perpendicular to the direction of the wave velocity. The wave produced in a stretched string (Fig. 13.10) is an example of a transverse wave, as is the wave shown in ▶Fig. 13.11a. A transverse wave is sometimes called a *shear wave*, because the disturbance supplies a force that tends to shear the medium—to separate layers of that medium at a right angle to the direction of the wave velocity. Shear waves can propagate only in solids, since a liquid or a gas cannot support a shear. That is, a liquid or a gas does not have sufficient restoring forces between its particles to propagate a transverse wave.

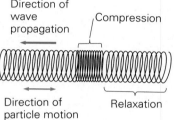

Direction of wave propagation

Direction of particle motion

(a)

Direction of wave propagation

Compression

Direction of particle motion

Relaxation

(b)

▲ **FIGURE 13.11 Transverse and longitudinal waves** (Wave pulses shown here for simplicity) **(a)** In a transverse wave, the motion of the particles is perpendicular to the direction of the wave velocity, as shown here in a spring for a wave moving to the left. A transverse wave is sometimes called a *shear wave*, because it supplies a force that tends to shear the medium. Transverse shear waves can propagate only in solids. (Why?) **(b)** In a longitudinal wave, the particle motion is parallel to (or *along*) the direction of the wave velocity. Here, a wave pulse also moves to the left. A longitudinal wave is sometimes called a *compressional wave*, because the force tends to compress the medium. Longitudinal compressional waves can propagate in all media—solid, liquid, and gas. Can you explain the motion of the wave *source* for both types of waves?

In a **longitudinal wave**, the particle oscillation is parallel to the direction of the wave velocity. A longitudinal wave can be produced in a stretched spring by moving the coils back and forth along the spring axis (Fig. 13.11b). Alternating pulses of compression and relaxation travel along the spring. A longitudinal wave is sometimes called a *compressional wave*.

Sound waves in air are another example of longitudinal waves. A periodic disturbance produces compressions in the air. Between the compressions are *rarefactions*—regions where the density of the air is reduced, or rarefied. A loudspeaker oscillating back and forth, for example, can create these compressions and rarefactions, which travel out into the air as sound waves. (Sound is discussed in detail in Chapter 14.)

Longitudinal waves can propagate in solids, liquids, and gases, since all phases of matter can be compressed to some extent. The propagations of transverse and longitudinal waves in different media give information about the Earth's interior structure, as discussed in the Insight on p. 462.

The sinusoidal profile of water waves might make you think that they are transverse waves. Actually, they reflect a combination of longitudinal and transverse motions (▶ Fig. 13.12). The particle motion may be nearly circular at the surface, but becomes more elliptical with depth, eventually becoming longitudinal. A hundred meters or so below the surface of a large body of water, the wave disturbances have little effect. For example, a submarine at these depths is undisturbed by large waves on the ocean's surface. As a wave approaches shallower water near shore, the water particles have difficulty completing their elliptical paths. When the water becomes too shallow, the particles can no longer move through the bottom parts of their paths, and the wave breaks. Its crest falls forward to form breaking surf as the waves' kinetic energy is transformed into potential energy—a water "hill" that eventually topples over.

13.4 Wave Properties

OBJECTIVE: **To explain various wave properties and resulting phenomena.**

Among the properties exhibited by all waves are interference, superposition, reflection, refraction, dispersion, and diffraction. Particles do not exhibit these properties.

Superposition and Interference

When two or more waves meet or pass through the same region of a medium, they pass through each other and proceed without being altered. While they are in the same region, the waves are said to be interfering.

What happens during interference? That is, what does the combined waveform look like? The relatively simple answer is given by the **principle of superposition**:

> At any time, the combined waveform of two or more interfering waves is given by the sum of the displacements of the individual waves at each point in the medium.

The principle of **interference** is illustrated in ▶ Fig. 13.13. The displacement of the combined waveform at any point is $y = y_1 + y_2$, where y_1 and y_2 are the displacements of the individual pulses at that point. (Directions are indicated by plus and minus signs.) Interference, then, is the physical addition of waves. In adding waves, we must take into account the possibility that they are producing disturbances in opposite directions. In other words, we must treat the disturbances in terms of vector addition.

In the figure, the vertical displacements of the two pulses are in the same direction, and the amplitude of the combined waveform is greater than that of either pulse. This situation is called **constructive interference**. Conversely, if one pulse

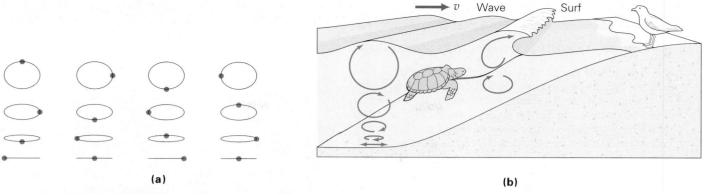

(a)

(b)

▲ **FIGURE 13.12 Water waves** Water waves are a combination of longitudinal and transverse motions. **(a)** At the surface, the water particles move in circles, but their motions become more longitudinal with depth. **(b)** When a wave approaches the shore, the lower particles are forced into steeper paths until, finally, the wave breaks or falls over to form surf.

has a negative displacement, the two pulses tend to cancel each other when they overlap, and the amplitude of the combined waveform is smaller than that of either pulse. This situation is called **destructive interference**.

The special cases of total constructive and total destructive interference for traveling wave pulses of the same width and amplitude are shown in ▸ Fig. 13.14. At the instant these interfering waves exactly overlap (crest coinciding with crest), the amplitude of the combined waveform is twice that of either individual wave. This case is referred to as **total constructive interference**. When the interfering pulses have opposite displacements and are exactly superimposed (crest coinciding with trough), the waveforms momentarily disappear; that is, the amplitude of the combined wave is zero. This case is called **total destructive interference**.

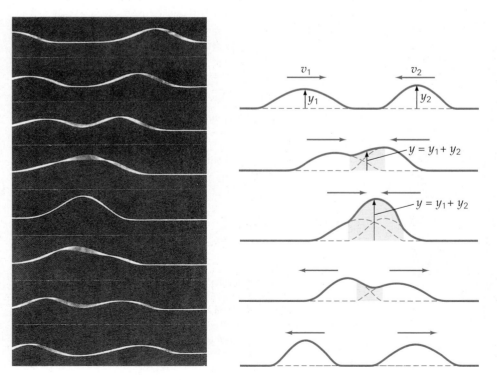

◀ **FIGURE 13.13 Principle of superposition** When two waves meet, they interfere. The beige tint marks the area where the two waves, moving in opposite directions, overlap and combine. The displacement at any point on the combined wave is equal to the sum of the displacements on the individual waves: $y = y_1 + y_2$.

Earthquakes, Seismic Waves, and Seismology

The structure of the Earth's interior is still something of a mystery. The deepest mine shafts and drillings extend only a few kilometers into the Earth, compared with a depth of about 6400 km to the Earth's center. Using waves to probe the Earth's structure is one way to investigate it further. Waves generated by earthquakes have proved to be especially useful for this purpose. Seismology is the study of these waves, called *seismic waves*.

Earthquakes are caused by the sudden release of built-up stress along cracks and faults, such as the famous San Andreas Fault in California (Fig. 1). According to the geological theory of plate tectonics, the outer layer of the Earth consists of rigid plates—huge slabs of rock that move very slowly relative to one another. Stresses are continuously built up, particularly along boundaries between plates.

When slippage finally occurs, the energy from this stress-relieving event propagates outward as (seismic) waves from a site below the surface called the *focus*. Seismic waves are of two general types: surface waves and body waves. *Surface waves*, which move along the Earth's surface, account for most earthquake damage (Fig. 2). *Body waves* travel through the Earth and are both longitudinal and transverse. The compressional (longitudinal) waves are called P *waves*, and the shear (transverse) waves are called S *waves* (Fig. 3). The P and S stand for *primary* and *secondary* and indicate the waves' relative speeds (actually, their arrival times at monitoring stations). In general, primary waves travel through materials faster than do secondary waves and are detected first. An earthquake's rating on the Richter scale is related to the energy released in the form of seismic waves.

Seismic stations around the world monitor P and S waves with sensitive detecting instruments called *seismographs* (Fig. 4). From the data gathered, we can map the paths of the waves through the Earth and thereby learn about the structure of the interior of our planet. The Earth's interior seems to be divided into three general regions: the crust, the man-

FIGURE 2 Bad vibrations Earthquake damage caused by the major shock that struck Kobe, Japan, in January 1995.

tle, and the core, which itself has a solid inner region and a liquid outer region.*

The locations of these regions' boundaries are determined in part by *shadow zones*, regions where no waves of a particular type are detected. These zones appear because, al-

*In most places, the crust is about 24–30 km (15–20 mi) thick; the mantle is 2900 km (1800 mi) thick; and the core has a radius of 3450 km (2150 mi). The solid inner core has a radius of about 1200 km (750 mi).

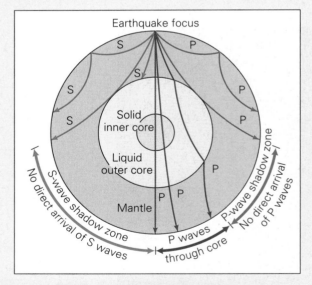

FIGURE 3 Compressional and shear waves Earthquakes produce waves that travel through the Earth. Because transverse S waves are not detected on the opposite side of the Earth, scientists believe that at least part of the Earth's core is a viscous liquid under high pressures and temperatures. The waves bend continuously, or refract, because their speed varies with depth.

FIGURE 1 The San Andreas Fault A small section of the fault, which runs through the San Francisco Bay area as well as across the more rural regions of California, is shown here.

though longitudinal waves can travel through solids *or* liquids, transverse waves can travel only through solids. When an earthquake occurs at a particular location, P waves are detected on the other side of the Earth, but S waves are not. (See Fig. 3.) The absence of S waves in a shadow zone leads to the conclusion that the Earth must have a region near its center that is in the liquid phase. This region is a highly viscous metallic liquid—but definitely a liquid, since it does not support a shear. (Transverse waves are not propagated.)

When the transmitted P waves enter and leave the liquid region, they are refracted (bent). This refraction gives rise to a P-wave shadow zone, which indicates that only the outer part of the core is liquid. As you will learn in Chapter 19, the combination of a liquid outer core and the Earth's rotation may be responsible for the Earth's magnetic field.

Related Exercises: 73 and 110

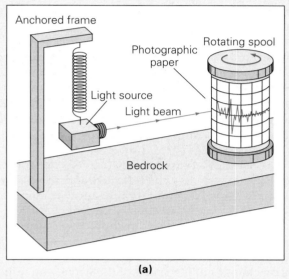

(a)

(b)

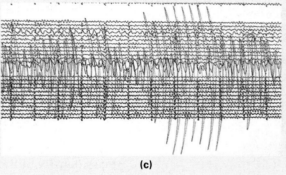

(c)

FIGURE 4 The seismograph **(a)** A simple seismograph. The device records the amplitudes of ground vibrations. The energy of seismic waves is proportional to the square of their amplitude. **(b)** A U.S. Geological Survey scientist monitors seismograph readings. **(c)** Seismograph trace from the destructive Kobe, Japan, earthquake of 1995.

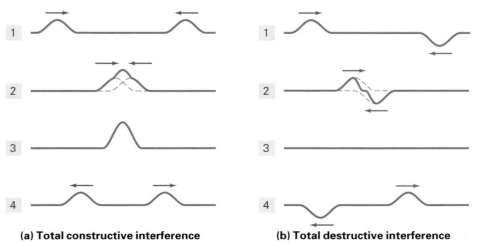

(a) Total constructive interference   **(b) Total destructive interference**

◀ **FIGURE 13.14 Interference** **(a)** When two wave pulses of the same amplitude meet and are in phase, they interfere constructively. When the pulses are exactly superimposed (3), total constructive interference occurs. **(b)** When the interfering pulses have opposite amplitudes and are exactly superimposed (3), total destructive interference occurs.

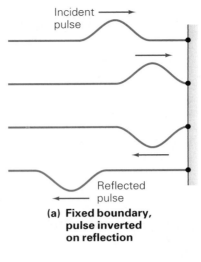

(a) **Fixed boundary, pulse inverted on reflection**

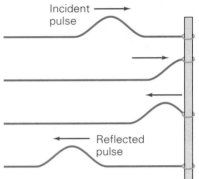

(b) **Free (movable) boundary, pulse not inverted on reflection**

▲ **FIGURE 13.15 Reflection**
(a) When a wave (pulse) on a string is reflected from a fixed boundary, the reflected wave is inverted. **(b)** If the string is free to move at the boundary, the phase of the reflected wave is not shifted from that of the incident wave.

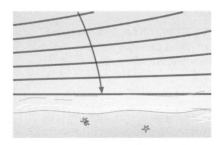

▲ **FIGURE 13.16 Refraction**
The refraction of water waves is shown from overhead. As the crests approach the beach, their left edge slows because it enters shallow water first. Thus, the whole crest rotates, approaching the beach more or less head-on.

The word *destructive* unfortunately tends to imply that the energy, as well as the form of the waves, are destroyed. This is not the case. At the point of total destructive interference, when the net waveshape and, hence potential energy, are zero, the wave energy is stored in the medium completely in the form of kinetic energy. That is, the straight string has instantaneous velocity.

Reflection, Refraction, Dispersion, and Diffraction

Besides meeting other waves, waves can (and do) meet objects or a boundary with another medium. In such cases, several things may occur. One of these is **reflection**, which occurs when a wave strikes an object or comes to a boundary with another medium and is at least partly diverted back into the original medium. An echo is the reflection of sound waves, and mirrors reflect light waves.

Two cases of reflection are illustrated in ◄Fig. 13.15. If the end of the string is fixed, the reflected pulse is inverted (Fig. 13.15a). This is because the pulse causes the string to exert an upward force on the wall, and the wall exerts an equal and opposite downward force on the string (by Newton's third law). The downward force creates the downward, or inverted, reflected pulse. If the end of the string is free to move, then the reflected pulse is not inverted. (There is no phase shift.) This is illustrated in Fig. 13.15b, which shows the string attached to a light ring that can move freely on a smooth pole. The ring is accelerated upward by the front portion of the incoming pulse and then comes back down, thus creating a noninverted reflected pulse.

More generally, when a wave strikes a boundary, the wave is not completely reflected. Instead, some of the wave's energy is reflected and some is transmitted or absorbed. When a wave crosses a boundary into another medium, its speed generally changes because the new material has different characteristics. Entering the medium obliquely (at an angle), the transmitted wave moves in a direction different from that of the incident wave. This phenomenon is called **refraction** (◄Fig. 13.16).

Since refraction depends on changes in the speed of the wave, you might be wondering which physical parameters determine the wave speed. Generally, there are two types of situations. The simplest kind of wave is one whose speed does *not* depend on its wavelength (or frequency). All such waves travel at the same speed, determined solely by the properties of the medium. These waves are called *nondispersive waves*, because they do not disperse, or spread apart from one another. An example of a nondispersive transverse wave is a wave on a string, whose speed, as we shall see, is determined only by the tension and mass density of the string (Section 13.5). Sound is a nondispersive longitudinal wave; the speed of sound (in air) is determined only by the compressibility and density of the air. Indeed, if the speed of sound did depend on the frequency, at the back of the symphony hall you might hear the violins well before the clarinets, even though the two sound waves were in perfect synchronization when they left the orchestra pit.

When the wave speed *does* depend on wavelength (or frequency), the waves are said to exhibit **dispersion**: Waves of different frequencies spread apart from one another. Although waves of light are nondispersive in a vacuum, when they enter some media, they are spread out or dispersed. This is the basis for prisms separating sunlight into a color spectrum and for the formation of a rainbow, as we shall see in Chapter 22. Dispersion will be most important for us in our study of light, but you should remember that waves other than light can also be dispersive under the right conditions.

Diffraction refers to the bending of waves around an edge of an object and is not related to refraction. For example, if you stand along an outside wall of a building near the corner of the street, you can hear people talking around the corner. Assuming that there are no reflections or air motion (wind), this would not be

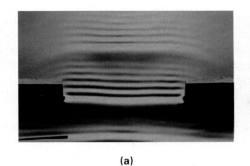
(a)

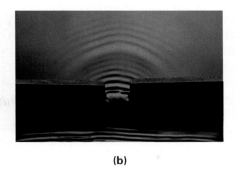
(b)

◀ **FIGURE 13.17 Diffraction**
Diffraction effects are greatest when the opening (or object) is about the same size as or smaller than the wavelength of the waves. **(a)** With an opening much larger than the wavelength of these plane water waves, diffraction is noticeable only near the edges. **(b)** With an opening about the same size as the wavelength of the waves, diffraction produces nearly semicircular waves.

possible if the sound waves traveled in a straight line. As the sound waves pass the corner, instead of being sharply cut off, they "wrap around" the edge; thus, you can hear the sound.

In general, the effects of diffraction are evident only when the size of the diffracting object or opening is about the same as or smaller than the wavelength of the waves. The dependence of diffraction on the wavelength and size of the object or opening is illustrated in ▲Fig. 13.17. For many waves, diffraction is negligible under normal circumstances. For instance, visible light has wavelengths on the order of 10^{-6} m. Such wavelengths are much too small to exhibit diffraction when they pass through common-sized openings, such as an eyeglass lens.

Reflection, refraction, dispersion, and diffraction will be considered in more detail when we study light waves in Chapters 22 and 24.

13.5 Standing Waves and Resonance

OBJECTIVES: **To (a) describe the formation and characteristics of standing waves and (b) explain the phenomenon of resonance.**

If you shake one end of a stretched rope, waves travel down it to the fixed end and are reflected back. The waves going down and back interfere. In most cases, the combined waveforms have a changing, jumbled appearance. But if the rope is shaken at just the right frequency, a steady waveform, or series of uniform loops, appears to stand in place along the rope. Appropriately, this phenomenon is called a **standing wave** (▼Fig. 13.18a). It arises because of interference with the reflected waves, which have the same wavelength, amplitude, and speed as the incident waves. Since the two identical waves travel in opposite directions, the net energy flow down the rope is zero. In effect, the energy is standing in the loops.

Some points on the rope remain stationary at all times and are called **nodes**. At these points, the displacements of the interfering waves are *always* equal and

▶ **FIGURE 13.18 Standing waves** **(a)** Standing waves are formed by interfering waves traveling in opposite directions. **(b)** Conditions of destructive and constructive interference recur as each wave travels a distance of $\lambda/4$ in a time $t = T/4$. The velocities of the rope's particles are indicated by the arrows. This motion gives rise to standing waves with stationary nodes and maximum-amplitude antinodes.

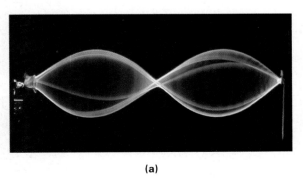
(a)

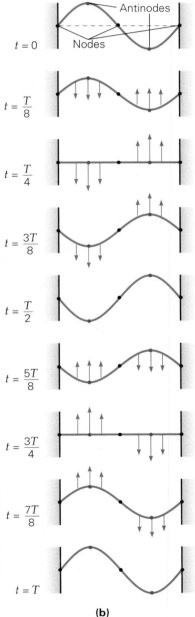
(b)

opposite. Thus, by the principle of superposition, the interfering waves cancel each other completely at these points, and the rope does not undergo displacement there. At all other points, the rope oscillates back and forth at the same frequency. The points of maximum amplitude, where constructive interference is greatest, are called **antinodes**. As you can see in Fig. 13.18b, adjacent antinodes are separated by a half-wavelength ($\lambda/2$), or one loop; adjacent nodes are also separated by a half-wavelength.

Standing waves can be generated in a rope by more than one driving frequency; the higher the frequency, the more oscillating half-wavelength loops are in the rope. The only requirement is that the half-wavelengths "fit" the length of the rope. The frequencies at which large amplitude standing waves are produced are called **natural frequencies**, or **resonant frequencies**. The resulting standing-wave patterns are called *normal*, or *resonant, modes of vibration*. In general, all systems that oscillate have one or more natural frequencies, which depend on such factors as mass, elasticity or restoring force, and geometry (boundary conditions). The natural frequencies of a system are sometimes called its *characteristic* frequencies.

A stretched string or rope can be analyzed to determine its natural frequencies. The boundary condition is that the ends are fixed; thus, there must be a node at each end. The number of closed segments or loops of a standing wave that will fit between the nodes at the ends (along the length of the string) is equal to an integral number of *half* wavelengths (▼Fig. 13.19). Note that $L = \lambda_1/2$, $L = 2(\lambda_2/2)$, $L = 3(\lambda_3/2)$, $L = 4(\lambda_4/2)$, and so on. In general,

$$L = n\left(\frac{\lambda_n}{2}\right) \quad \text{or} \quad \lambda_n = \frac{2L}{n} \quad \text{(for } n = 1, 2, 3, \dots)$$

The natural frequencies of oscillation are

$$f_n = \frac{v}{\lambda_n} = n\left(\frac{v}{2L}\right) = nf_1 \quad \text{for } n = 1, 2, 3, \dots \qquad \begin{array}{l}\textit{natural frequencies}\\ \textit{for a stretched string}\end{array} \quad (13.18)$$

▼ **FIGURE 13.19 Natural frequencies** A stretched string can have standing waves only at certain frequencies. These correspond to the numbers of half-wavelength loops that will fit along the length of string between the nodes at the fixed ends.

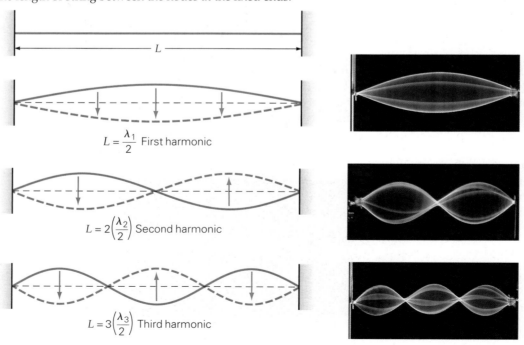

$L = \dfrac{\lambda_1}{2}$ First harmonic

$L = 2\left(\dfrac{\lambda_2}{2}\right)$ Second harmonic

$L = 3\left(\dfrac{\lambda_3}{2}\right)$ Third harmonic

where v is the speed of waves on a string. The lowest natural frequency ($f_1 = v/2L$ for $n = 1$) is called the **fundamental frequency**. All of the other natural frequencies are integral multiples of the fundamental frequency: $f_n = nf_1$ (for $n = 1, 2, 3, \ldots$). The set of frequencies f_1, $f_2 = 2f_1$, $f_3 = 3f_1, \ldots$ is called a **harmonic series**: f_1 (the fundamental frequency) is the *first harmonic*, f_2 the *second harmonic*, and so on.

Strings that are fixed at each end are found in stringed musical instruments such as violins, pianos, and guitars. When such a string is excited, the resulting vibration generally includes several harmonics in addition to the fundamental frequency. The number of harmonics depends on how and where the string is excited—that is, whether it is plucked, struck, or bowed. It is the combination of harmonic frequencies that gives a particular instrument its characteristic sound quality (more on this in Chapter 14). As Eq. 13.18 shows, the fundamental frequency of a stretched string, as well as the other harmonics, depends on the length of the string. Think of how different notes are obtained on a particular string of a violin or a guitar (▶ Fig. 13.20).

Natural frequencies also depend on other parameters, such as mass and force, which affect the wave speed in the string. For a stretched string, the wave speed (v) can be shown to be

$$v = \sqrt{\frac{F_T}{\mu}} \quad \text{(wave speed in a stretched string)} \quad (13.19)$$

where F_T is the tension in the string and μ is the linear mass density (mass per unit length, $\mu = m/L$). (We use F_T, rather than the T of previous chapters, so as not to confuse the tension with the period T.) Thus, Eq. 13.18 can be written as

$$f_n = n\left(\frac{v}{2L}\right) = \frac{n}{2L}\sqrt{\frac{F_T}{\mu}} = nf_1 \quad \text{(for } n = 1, 2, 3, \ldots) \quad (13.20)$$

Note that the greater the linear mass density of a string, the lower are its natural frequencies. As you may know, the low-note strings on a violin or guitar are thicker, or more massive, than the high-note strings. By tightening a string, we increase all frequencies of that string. Changing the tension in the string is how violinists, for example, tune their instruments before a performance.

▲ **FIGURE 13.20 Fundamental frequencies** Performers of a stringed instrument such as the violin or guitar use their fingers to stop or fret the strings. By pressing a string against the fingerboard, the player reduces the amount of its length that is free to vibrate. This reduction changes the resonant frequency of the string and thus the pitch of the tone it produces.

Example 13.6 ■ A Piano String: Fundamental Frequency and Harmonics

A piano string with a length of 1.15 m and a mass of 20.0 g is under a tension of 6.30×10^3 N. (a) What is the fundamental frequency of the string when it is struck? (b) What are the frequencies of the next two harmonics?

Thinking It Through. We have the tension and can calculate the linear mass density from the data. This will allow us to find the fundamental frequency, and from that, the harmonics can be calculated.

Solution.

Given: $L = 1.15$ m
$m = 20.0$ g $= 0.0200$ kg
$F_T = 6.30 \times 10^3$ N

Find: (a) f_1 (fundamental frequency)
(b) f_2 and f_3 (frequencies of next two harmonics)

(a) The linear mass density of the string is

$$\mu = \frac{m}{L} = \frac{0.0200 \text{ kg}}{1.15 \text{ m}} = 0.0174 \text{ kg/m}$$

Then, using Eq. 13.20, we have

$$f_1 = \frac{1}{2L}\sqrt{\frac{F_T}{\mu}} = \frac{1}{2(1.15\text{ m})}\sqrt{\frac{6.30 \times 10^3\text{ N}}{0.0174\text{ kg/m}}} = 262\text{ Hz}$$

This is approximately the frequency of middle C (C_4) on a piano.

(b) Since $f_2 = 2f_1$ and $f_3 = 3f_1$, it follows that

$$f_2 = 2f_1 = 2(262\text{ Hz}) = 524\text{ Hz}$$

and

$$f_3 = 3f_1 = 3(262\text{ Hz}) = 786\text{ Hz}$$

The second harmonic corresponds approximately to C_5 on a piano, since, by definition, the frequency doubles with each octave (every eighth white key).

Note: Don't be confused by the language: *First* overtone means the first frequency above the fundamental frequency—that is, the *second* harmonic.

Follow-up Exercise. A musical note is referenced to the fundamental vibrational frequency, or first harmonic. In musical terms, the second harmonic is the first overtone, the third harmonic is the second overtone, and so on. If an instrument has a third overtone with a frequency of 880 Hz, what is the frequency of the first overtone?

Integrated Example 13.7 ■ Tuning Up: Raising the Frequency of a Guitar String

Suppose you wish to raise the fundamental frequency of a guitar string from the A note (220 Hz) below middle C to the A note (440 Hz) above middle C. (a) Would you (1) loosen the string to halve its tension, (2) tighten the string to double its tension, (3) use another string of the same material with half the diameter at the same tension, or (4) use another string of the same material with twice the diameter at the same tension? (b) If the guitar strings are made of steel ($\rho = 7.8 \times 10^3\text{ kg/m}^3$, Table 9.2), and an initial thicker string has a diameter of 0.30 cm, show that a string with half the diameter will have double the fundamental frequency.

(a) Conceptual Reasoning. The fundamental frequency of a stretched string is given by Eq. 13.20:

$$f = \frac{1}{2L}\sqrt{\frac{F_T}{\mu}} \quad (for\ n = 1)$$

The frequency of the string is thus proportional to the *square root* of the tension force F_T, so loosening the string—that is, decreasing F_T—would not increase the frequency. Nor would doubling the tension double the frequency (because $\sqrt{2F_T} \neq 2\sqrt{F_T}$). Thus, neither (1) nor (2) is the correct answer.

In using another string, the question is, then, how does the frequency vary with the linear mass density μ of the string? If the strings are of the same material (same density ρ), then the greater the diameter of a string, the greater is its mass per unit length (greater μ). Hence, a thinner string, with a smaller μ, will vibrate at a higher frequency, and the answer is (3).

(b) Thinking It Through. At first, one might think of computing the frequencies directly, using Eq. 13 20. But this can't be done, because not enough data are given, which usually implies the use of a ratio. To show the difference in frequencies with different diameter strings, the frequency equation needs to be expressed in terms of the diameter of a string. Inspecting Eq. 13.20, we see that the diameter of the string does not depend on the length L (assumed constant between the bridge and neck of the guitar) or on the constant tension force F_T (assuming negligible stretching). This leads to a consideration of the linear mass density $\mu = m/L$.

Recall that the mass of a string depends on its density and volume; that is, $\rho = m/V$, or $m = \rho V$. Then, the volume V of a length (L) of wire, which may be thought of as that of a long cylinder with circular cross section A, can be determined by $V = AL$. The circular area is proportional to the square of the diameter of the wire, so this is the key to our proof.

Given: $f = \dfrac{1}{2L}\sqrt{\dfrac{F_T}{\mu}}$ (for $n = 1$, Eq. 13.20)   **Find:** Prove that a string with a diameter d_2 that is half that of a thicker string (diameter d_1) will have twice the fundamental frequency of the thicker string.

$\rho = 7.8 \times 10^3\,\text{kg/m}^3$ (steel)
$d_1 = 0.30\,\text{cm}$
$d_2 = d_1/2 = 0.15\,\text{cm}$

As noted in the "Thinking It Through" section, the linear mass density of the wire string can be expressed in terms of its density and volume, the latter of which is proportional to the string's diameter:

$$\mu = \frac{m}{L} = \frac{\rho V}{L} = \frac{\rho AL}{L} = \rho\left(\frac{\pi d^2}{4}\right)$$

Putting this into Eq. 13.20 yields

$$f = \frac{1}{2L}\sqrt{\frac{F_T}{\mu}} = \frac{1}{2L}\sqrt{\frac{4F_T}{\rho\pi d^2}} = \left(\frac{1}{L}\sqrt{\frac{F_T}{\rho\pi}}\right)\frac{1}{d}$$

The quantities in the brackets are constant, and $f \propto 1/d$, so, in ratio form,

$$\frac{f_2}{f_1} = \frac{d_1}{d_2} = \frac{0.30\,\text{cm}}{0.15\,\text{cm}} = 2 \quad \text{and} \quad f_2 = 2f_1$$

Follow-up Exercise. The fundamental frequency of a violin string is A below middle C (220 Hz). How could you tune this string to middle C (264 Hz) without changing strings as was done in this Example?

When an oscillating system is driven at one of its natural, or resonant, frequencies, the maximum amount of energy is transferred to the system. The natural frequencies of a system are the frequencies at which the system "wants" to vibrate, so to speak. The condition of driving a system at a natural frequency is referred to as **resonance**.

A common example of a system in mechanical resonance is someone being pushed on a swing. Basically, a swing is a simple pendulum and has only one resonant frequency for a given length $[f = 1/(2\pi)\sqrt{g/L}]$. If you push the swing with this frequency and in phase with its motion, its amplitude and energy increase ($\blacktriangleright$ Fig. 13.21). If you push at a slightly different frequency, the energy transfer is no longer a maximum. (What do you think happens if you push with the resonant frequency, but 180° out of phase with the swing's motion?)

Unlike a simple pendulum, a stretched string has many natural frequencies. Almost any driving frequency will cause a disturbance in the string. However, if the frequency of the driving force is not equal to one of the natural frequencies, the resulting wave will be relatively small and jumbled. By contrast, when the frequency of the driving force matches one of the natural frequencies, the maximum amount of energy is transferred to the string. A steady standing-wave pattern results, with the amplitude at the antinodes becoming relatively large.

Mechanical resonance is not the only type of resonance. When you tune a radio, you are changing the resonant frequency of an electrical circuit (Chapter 21) so that it will be driven by, or will pick up, a signal at the frequency of the station you want. Other examples of resonance are described in the Insight on the next page.

$\blacktriangle$ **FIGURE 13.21 Resonance in the playground** The swing behaves like a pendulum in SHM. To transfer energy efficiently, the man must time his pushes to the natural frequency of the swing.

Desirable and Undesirable Resonances

When we hold a large seashell to our ear, we say we hear the ocean or a sound like that of the ocean. The cause of this sound is a resonance effect—a desirable one because it is pleasant. Ambient sounds enter the shell, and the frequencies of some of these are at the resonant frequencies of the shell, which acts as a resonant cavity. Standing sound waves are set up that can be heard when the shell is held close to the ear. (We shall learn more about sound in Chapter 14.)

The air in the shell resonates at the shell's natural frequencies. The variations in sound arise from the different ambient sounds, or the coming and going, of different resonant frequencies. "The coming and going of these resonant frequencies gives the listener the illusion of hearing the ocean waves come and go."* Basically, the brain processes the sound, looking for a pattern that has been previously experienced. Most of us have heard the sound of ocean waves before, so this is what we associate with when we listen to a seashell. You can also hear a similar "ocean" sound by placing an empty glass or cupping your hand over your ear.

When a large number of soldiers march over a small bridge, they are generally ordered to break step. The reason is that the marching frequency may correspond to one of the natural frequencies of the bridge and set it into resonant vibration, which could cause it to collapse. This actually occurred on a suspension bridge in England in 1831. The bridge was weak and in need of repair, but the resonance vibrations induced by the marching soldiers crossing the bridge caused it to collapse—with some injuries.

Another incident of a bridge vibrating was due, not to marching soldiers, but to the driving force of the wind. On the morning of November 7, 1940, winds with speeds of 40 to 45 mi/h started the main span of the Tacoma Narrows Bridge (in Washington State) vibrating. The bridge, 2800 ft (855 m) long and 39 ft (12 m) wide, had been opened to traffic only four months earlier.

During the first month that the bridge was used, small transverse modes of vibration had been observed. But on November 7, special wind effects drove the bridge in near

*Jearl Walker, *The Flying Circus of Physics*, John Wiley & Sons, 1977.

resonance, and the main span vibrated at a frequency of 36 vib/min and an amplitude of 1.5 ft. At 10 a.m., the main span began to vibrate in a torsional (twisting) mode in two segments at a frequency of 14 vib/min. The wind continued to drive the bridge in resonance, and the vibrational amplitude increased. Shortly after 11 a.m., the main span collapsed (Fig. 1).*

"Galloping Gertie" (the nickname given to the bridge) was rebuilt on the same tower foundations. However, the new design made the structure stiffer, to increase its resonant frequency so that high winds could not produce unwanted resonance.

*It is doubtful that the gusting of the wind set the bridge into vibration. The wind velocity was moderately steady, and gust fluctuations are normally random. One explanation for the driving source of the oscillations involves the formation of vortices as the wind blew past the bridge. Vortices are like the eddies that form in water at the end of an oar when a boat is rowed. The wind blowing over and under the bridge formed vortices that rotated in opposite directions. The formation and "shedding" of the vortices (like eddies coming off oars) would have imparted energy to the bridge, and if the frequency of this action approximated a natural frequency, a standing wave would have been set up.

FIGURE 1   Galloping Gertie   The collapse of the Tacoma Narrows Bridge on November 7, 1940, is captured in this frame from a movie camera.

Chapter Review

Important Concepts and Equations

- **Simple harmonic motion (SHM)** requires a restoring force directly proportional to the displacement, such as an ideal spring force, which is given by Hooke's law.

  Hooke's law:
  $$F_s = -kx \qquad (13.1)$$

- The **frequency** (f) and **period** (T) for SMH are the inverse of each other.

  Frequency and period for SHM:
  $$f = \frac{1}{T} \qquad (13.2)$$

- In general, the total energy of an object in SHM is directly proportional to the square of the amplitude.

  Total energy of a spring and mass in SHM:
  $$E = \tfrac{1}{2}kA^2 = \tfrac{1}{2}mv^2 + \tfrac{1}{2}kx^2 \qquad (13.4\text{–}5)$$

- The form of an equation of motion for an object in SHM depends on the object's initial (y_o) displacement.

Equations of motion for SHM:

$$y = \pm A \sin \omega t = \pm A \sin (2\pi f t) = \pm A \sin \left(\frac{2\pi t}{T} \right) \quad \text{(13.8,}$$
$$\text{13.10)}$$

$+A$ for initial motion upward with $y_o = 0$
$-A$ for initial motion downward with $y_o = 0$

$$y = \pm A \cos \omega t = \pm A \cos (2\pi f t) = \pm A \cos \left(\frac{2\pi t}{T} \right) \quad \text{(13.9)}$$

$+A$ for initial motion downward with $y_o = +A$
$-A$ for initial motion upward with $y_o = -A$

Velocity of a mass oscillating on a spring:

$$v = \pm \sqrt{\frac{k}{m}(A^2 - x^2)} \quad \text{(13.6)}$$

Period of a mass oscillating on a spring:

$$T = 2\pi \sqrt{\frac{m}{k}} \quad \text{(13.11)}$$

Angular frequency of a mass oscillating on a spring:

$$\omega = 2\pi f = \sqrt{\frac{k}{m}} \quad \text{(13.13)}$$

Period of a simple pendulum (small-angle approximation):

$$T = 2\pi \sqrt{\frac{L}{g}} \quad \text{(13.14)}$$

Velocity of a mass in SHM:

$$v = \omega A \cos \omega t \quad \text{(if } v_o \text{ is upward at } y_o = 0) \quad \text{(13.15)}$$

Acceleration of a mass in SHM:

$$a = -\omega^2 A \sin \omega t = -\omega^2 y \quad \begin{array}{l} \text{(if } v_o \text{ is upward} \\ \text{at } y_o = 0) \end{array} \quad \text{(13.16)}$$

- A wave is a disturbance in time and space; energy is transferred or propagated by wave motion.

Wave speed:

$$v = \frac{\lambda}{T} = \lambda f \quad \text{(13.17)}$$

- At any time, the combined waveform of two or more interfering waves is given by the sum of the displacements of the individual waves at each point in the medium.

- At natural frequencies, standing waves can form on a string as a result of the interference of two waves of identical wavelength, amplitude, and speed traveling in opposite directions on a string.

Natural frequencies for a stretched string:

$$f_n = n\frac{v}{2L} = \frac{n}{2L}\sqrt{\frac{F_T}{\mu}} = nf_1 \quad \text{(13.20)}$$
$$\text{(for } n = 1, 2, 3, \dots)$$

Exercises

13.1 Simple Harmonic Motion

1. A particle in SHM has (a) variable amplitude, (b) a restoring force in the form of Hooke's law, (c) a frequency directly proportional to its period, or (d) a position that is represented graphically by $x(t) = at + b$.

2. The maximum kinetic energy of a mass–spring system in SHM is equal to (a) A, (b) A^2, (c) kA, (d) $kA^2/2$.

3. If the period of a system in SHM is doubled, the frequency of the system is (a) doubled, (b) halved, (c) four times as large, or (d) one-quarter as large.

4. When a particle in SHM is at the equilibrium position, the potential energy of the system is (a) zero, (b) maximum, (c) negative, or (d) none of the above.

5. CQ If the amplitude of a mass in SHM is doubled, how are (a) the energy and (b) the maximum speed affected?

6. CQ How does the speed of a mass in SHM change as the mass approaches its equilibrium position? Explain.

7. CQ A mass–spring system in SHM has an amplitude A and period T. How long does it take for the mass to travel a distance A? How about $2A$?

8. CQ A tennis player uses a racket to bounce a ball up and down with a constant period. Is this a simple harmonic motion? Explain.

9. ■ A particle oscillates in SHM with an amplitude A. What is the total *distance* the particle travels in one period?

10. ■ A 1.0-kg toy oscillating on a spring completes a cycle every 0.50 s. What is the frequency of this oscillation?

11. ■ A particle in simple harmonic motion has a frequency of 40 Hz. What is the period of this oscillation?

12. ■ The frequency of a simple harmonic oscillator is doubled from 0.25 Hz to 0.50 Hz. What is the change in its period?

13. ■ What is the spring constant of a spring scale that stretches 6.0 cm when a basket of vegetables of mass 0.25 kg is suspended from it?

14. ■ An object of mass 0.50 kg is attached to a spring with spring constant 10 N/m. If the object is pulled down 0.050 m from the equilibrium position and released, what is its maximum speed?

15. ■■ Atoms in a solid are in continuous vibrational motion due to thermal energy. At room temperature, the amplitude of these atomic vibrations is typically about 10^{-9} cm,

and their frequency is on the order of 10^{12} Hz. (a) What is the approximate period of oscillation of a typical atom? (b) What is its maximum speed of such an atom?

16. **IE ■■** (a) At what position is the magnitude of the force on a mass in a mass–spring system maximum? (1) $x = 0$, (2) $x = -A$, or (3) $x = +A$. Why? (b) If $m = 0.500$ kg, $k = 150$ N/m, and $A = 0.150$ m, what are the magnitude of the force on the mass and the acceleration of the mass at $x = 0$, 0.050 m, and 0.150 m?

17. **IE ■■** (a) At what position is the speed of a mass in a mass–spring system maximum? (1) $x = 0$, (2) $x = -A$, or (3) $x = +A$. Why? (b) If $m = 0.250$ kg, $k = 100$ N/m, and $A = 0.10$ m, what is the maximum speed?

18. **■■** A mass–spring system is in SHM in the horizontal direction. If the mass is 0.25 kg, the spring constant is 12 N/m, and the amplitude is 15 cm, (a) what would be the maximum speed of the mass, and (b) where would this occur? (c) What would be the speed at a half-amplitude position?

19. **■■** In Exercise 18, (a) what is the speed of the mass if it is at $x = 10$ cm ? (b) what is the magnitude of the force exerted by the spring on the mass?

20. **■■** A 0.25-kg object suspended on a light spring is released from a position 15 cm above the stretched equilibrium position. The spring has a spring constant of 80 N/m. (a) What is the total energy of the system? (Neglect gravitational potential energy.) (b) Does this energy depend on the mass of the object? Explain.

21. **■■** What is the speed of the object in Exercise 20 when the object is (a) 5.0 cm above its equilibrium position and (b) 5.0 cm below its equilibrium position? (c) What is the object's maximum speed, and where does this occur?

22. **■■** The mass of a mass–spring system in SHM is at $x = A/2$. What fraction of the mechanical energy of the mass is potential? What fraction is kinetic?

23. **■■■** A 75-kg circus performer jumps from a 5.0-m height onto a trampoline and stretches it downward 0.30 m. Assuming that the trampoline obeys Hooke's law, (a) how far will it stretch if the performer jumps from a height of 8.0 m? (b) How far will the trampoline stretch if the performer stands still on it while taking a bow?

24. **■■■** A 0.250-kg ball is dropped from a height of 10.0 cm onto a spring, as illustrated in ▶ Fig. 13.22. If the spring has a spring constant of 60.0 N/m, (a) what distance will the spring be compressed? (Neglect energy loss during collision.) (b) On recoiling upward, how high will the ball go?

13.2 Equations of Motion

25. The equation of motion for a particle in SHM (a) is always a cosine function, (b) reflects damping action, (c) is inde-

▲ **FIGURE 13.22 How far down?** See Exercise 24.

pendent of the initial conditions, or (d) gives the position of the particle as a function of time.

26. **CQ** If the length of a pendulum is doubled, what is the ratio of the new period to the old one?

27. **CQ** The apparatus in Fig. 13.5 demonstrates that the motion of a mass on a spring can be described by a sinusoidal function of time. How could this same relationship be demonstrated for a pendulum?

28. **CQ** Could simple harmonic motion be described by a tangent function? Explain.

29. **CQ** Would the period of a pendulum in an upward-accelerating elevator be increased or decreased compared with its period in a nonaccelerating elevator? Explain.

30. **CQ** Automobiles use shock absorbers and springs to damp oscillations upon hitting a bump. Compare the frequency of oscillation between a car with just a driver and one with a driver and three passengers.

31. **CQ** If a mass–spring system is taken to the Moon, will the period of the system change? How about the period of a pendulum taken to the Moon? Explain.

32. **■** A 0.50-kg mass oscillates in simple harmonic motion on a spring with a spring constant of 200 N/m. What are (a) the period and (b) the frequency of the oscillation?

33. **■** The simple pendulum in a grandfather clock is 1.0 m long. What are (a) the period and (b) the frequency of this pendulum?

34. **■** What mass on a spring with a spring constant of 100 N/m will oscillate with a period of 2.0 s?

35. ■ A breeze sets a suspended lamp into oscillation. If the period is 1.0 s, what is the distance from the ceiling to the lamp at the lowest point? Assume that the lamp acts as a simple pendulum.

36. ■ Write the general equation of motion for a mass that is on a horizontal frictionless surface and is connected to a spring at equilibrium (a) if the mass is initially given a quick push away from the spring and (b) if the mass is pulled away from the spring and released.

37. ■ The equation of motion for an oscillator in vertical SHM is given by $y = (0.10 \text{ m}) \sin 100\, t$. What are the (a) amplitude, (b) frequency, and (c) period of this motion?

38. ■ The displacement of an object is given by $y = (5.0 \text{ cm}) \sin 20\pi t$. What are the object's (a) amplitude, (b) frequency, and (c) period of oscillation?

39. ■ If the displacement of an oscillator in SHM is described by the equation $y = (0.25 \text{ m}) \cos 314t$, where y is in meters and t is in seconds, what is the position of the oscillator at (a) $t = 0$, (b) $t = 5.0$ s, and (c) $t = 15$ s ?

40. IE ■■ The oscillations of two oscillating mass–spring systems are graphed in ▼Fig. 13.23. The mass in System A is four times that in System B. (a) Compared with System B, System A has (1) more, (2) the same, or (3) less energy. Why? (b) Calculate the ratio of energy between System B and System A.

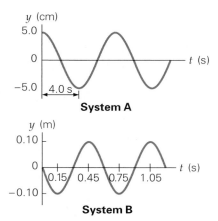

▲ **FIGURE 13.23   Wave energy and equation of motion**
See Exercises 40, 53, and 54.

41. ■■ Show that the total energy of a mass–spring system in simple harmonic motion is given by $\frac{1}{2} m\omega^2 A^2$.

42. IE ■■ (a) If the mass in a mass–spring system is doubled, the new period is (1) 2, (2) $\sqrt{2}$, or (3) $1/\sqrt{2}$ times the old period. Why? (b) If the initial period is 3.0 s and the mass is reduced to 1/3 of its initial value, what is the new period?

43. IE ■■ (a) If the spring constant in a mass–spring system is tripled, the new period is (1) 3, (2) $\sqrt{3}$, or (3) $1/\sqrt{3}$

times the old period. Why? (b) If the initial period is 2.0 s and the spring constant is halved, what is the new period?

44. ■■ Show that, for a pendulum to oscillate at the same frequency as a mass on a spring, the pendulum's length must be $L = mg/k$.

45. ■■ In a lab experiment, you are given a spring with a spring constant of 12 N/m. What mass would you suspend on the spring to have an oscillation period of 0.91 s when the mass is in SHM?

46. IE ■■ Since there is little gravity in outer space, astronauts cannot measure their mass on a spring scale as we do on the Earth. (a) Can you design a method whereby they could measure their mass on a spring scale in space? (b) If the period of oscillation of an astronaut on a 3000-N/m spring is 1.0 s, what is the astronaut's mass?

47. ■■ Students use a simple pendulum with a length of 36.90 cm to measure the acceleration of gravity at the location of their school. If the period of the pendulum is 1.220 s, what is the experimental value of g at the school?

48. ■■ The equation of motion of a particle in vertical SHM is given by $y = (10 \text{ cm}) \sin 0.50t$. What are the particle's (a) displacement, (b) velocity, and (c) acceleration at $t = 1.0$ s?

49. ■■ A 0.15-kg mass oscillates vertically in SHM on a spring with spring constant 6.0 N/m. The mass was initially raised 8.0 cm from equilibrium. (a) Write the equation of motion of the system. (b) What is the displacement of the mass at $t = 0.50$ s?

50. ■■ Two equal masses oscillate on light springs, the second with a spring constant twice that of the first. Which system will have the greater frequency and how many times greater?

51. IE ■■ (a) If a grandfather clock (see Exercise 33) were taken to the Moon, where the acceleration due to gravity is only one-sixth (assume the figure to be exact) that on the Earth, will the period of vibration (1) increase, (2) remain the same, or (3) decrease? Why? (b) If the period on the Earth is 2.0 s, what is the period on the Moon?

52. ■■ A 0.075-kg mass oscillates on a light spring, and another 0.075-kg mass oscillates as the bob of a simple pendulum in SHM. If the periods of the oscillations are equal and the length of the pendulum is 0.30 m, what is the spring constant of the spring?

53. ■■ The motion of a particle is described by the curve for System A in Fig. 13.23. Write the equation of motion in terms of a cosine function.

54. ■■ The motion of a 0.35-kg mass oscillating on a light spring is described by the curve for System B in Fig. 13.23.

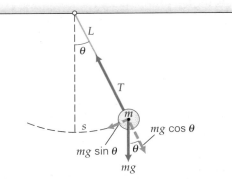

▲ **FIGURE 13.24 SHM of a pendulum**   See Exercise 55.

(a) Write the equation for the displacement of the mass as a function of time. (b) What is the spring constant of the spring?

55. ■■■ The forces acting on a simple pendulum are shown in ▲ Fig. 13.24. (a) Show that, for the small-angle approximation ($\sin \theta \approx \theta$), the force producing the motion has the same form as Hooke's law. (b) Show by analogy with a mass on a spring that the period of a simple pendulum is given by $T = 2\pi \sqrt{L/g}$. [*Hint*: Think of the effective spring constant.]

56. ■■■ A clock uses a pendulum that is 75 cm long. The clock is accidentally broken, and when it is repaired, the length of the pendulum is shortened by 2.0 mm. Consider the pendulum to be a simple pendulum. (a) Will the repaired clock gain or lose time? (b) By how much will the time indicated by the repaired clock differ from the correct time (taken to be the time determined by the original pendulum in 24 h)? (c) If the pendulum string were metal, would the surrounding temperature make a difference in the timekeeping of the clock? Explain.

13.3 Wave Motion

57. Wave motion in a material medium involves (a) the propagation of a disturbance, (b) interparticle interactions, (c) the transfer of energy, or (d) all of the preceding.

58. For a periodic wave propagating at a speed v, the following relationship holds: (a) $\lambda = v/f$, (b) $v = \lambda/f$, (c) $v = \lambda f^2$, or (d) $f = \lambda/v$.

59. A water wave is (a) transverse, (b) longitudinal, (c) a combination of transverse and longitudinal, or (d) none of the above.

60. **CQ** What type(s) of wave(s), transverse or longitudinal, will propagate through (a) solids, (b) liquids, and (c) gases?

61. **CQ** ▶ Figure 13.25 shows snapshots of two waves. Identify each as transverse or longitudinal.

▲ **FIGURE 13.25 Transverse or longitudinal?**
See Exercise 61.

62. **CQ** Standing on a hill and looking at a tall wheat field, you see a beautiful wave traveling across the field when there is a breeze. What type of wave is this? Explain.

63. ■ A longitudinal sound wave has a speed of 340 m/s in air. If this wave produces a tone with a frequency of 1000 Hz, what is its wavelength?

64. ■ A transverse wave has a wavelength of 0.50 m and a frequency of 20 Hz. What is the wave speed?

65. ■ A student reading his physics book on a lake dock notices that the distance between two incoming wave crests is about 2.4 m, and he then measures the time of arrival between the crests to be 1.6 s. What is the approximate speed of the waves?

66. ■ Light waves travel in a vacuum at a speed of 300 000 km/s. The frequency of visible light is about 5×10^{14} Hz. What is the approximate wavelength of the light?

67. ■ A certain laser emits light of wavelength 633×10^{-9} m. What is the frequency of this light in a vacuum?

68. ■■ The range of sound frequencies audible to the human ear extends from about 20 Hz to 20 kHz. If the speed of sound in air is 345 m/s, what are the limits of this audible range, expressed in wavelengths?

69. **IE** ■■ The AM frequencies on a radio dial range from 550 kHz to 1600 kHz, and the FM frequencies range from 88.0 MHz to 108 MHz. All of these radio waves travel at a speed of 3.00×10^8 m/s (speed of light). (a) Compared with the FM frequencies, the AM frequencies

have (1) longer, (2) the same, or (3) shorter wavelengths. Why? (b) What are the wavelength ranges of the AM band and the FM band?

70. ■■ A sonar generator on a submarine produces periodic ultrasonic waves at a frequency of 2.50 MHz. The wavelength of the waves in seawater is 4.80×10^{-4} m. When the generator is directed downward, an echo reflected from the ocean floor is received 10.0 s later. How deep is the ocean at that point? (Assume that wavelength is constant at all depths.)

71. ■■ In watching a transverse wave go by, a woman in a boat notes that 13 crests pass by in a time of 3.0 s. If she measures the distance between two successive crests to be 0.75 m, and if the first and the last points that pass her are crests, what is the speed of the wave?

72. ■■ A wave traveling in the $+x$-direction is shown in ▼ Fig. 13.26a. The particle displacement at a particular location in the medium through which the wave travels is shown in Fig. 13.26b. (a) What is the amplitude of the traveling wave? (b) What is the wave speed?

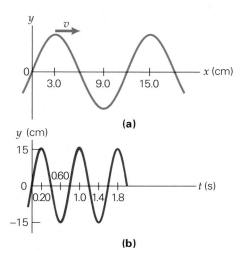

(a)

(b)

▲ **FIGURE 13.26 How high and how fast?**   See Exercise 72.

73. ■■ Assume that P and S (primary and secondary) waves from an earthquake with a focus near the Earth's surface travel through the Earth at nearly constant average speeds of 8.0 km/s and 6.0 km/s, respectively. Assume that there is no deflection or refraction of the waves. (a) How long is the delay between the arrivals of successive waves at a seismic monitoring station located 90° in latitude from the epicenter (the spot on the surface directly above the focus) of the quake? (b) Do the waves cross the boundary of the mantle? (c) How long does it take for the waves to arrive at a monitoring station on the opposite side of the Earth?

74. ■■■ The speed of longitudinal waves traveling in a long, solid rod is given by $v = \sqrt{Y/\rho}$, where Y is Young's modulus and ρ is the density of the solid. If a disturbance has

a frequency of 40 Hz, what is the wavelength of the waves it produces in (a) an aluminum rod and (b) a copper rod? [*Hint*: See Tables 9.1 and 9.2.]

75. ■■■ Fred strikes a steel train rail with a hammer at a frequency of 0.50 Hz, and Wilma puts her ear to the rail 1.0 km away. (a) How long after the first strike does Wilma hear the sound? (b) What is the time interval between the successive sound pulses she hears? [*Hint*: See Tables 9.1 and 9.2 and Exercise 74.]

13.4 Wave Properties

76. When two waves meet each other and interfere, the resultant waveform is determined by (a) reflection, (b) refraction, (c) diffraction, or (d) superposition.

77. Refraction (a) involves constructive interference, (b) refers to a change in direction at media interfaces, (c) is identical to diffraction, (d) occurs only for media or mechanical waves.

78. You can often hear people talking from around a corner of a building. This is due primarily to (a) reflection, (b) refraction, (c) interference, or (d) diffraction.

79. CQ What is destroyed when destructive interference occurs? What happens to energy? Explain.

80. CQ Dolphins and bats determine the location of their prey by emitting ultrasonic sound waves. Which wave phenomenon is involved?

81. CQ If sound waves were dispersive (i.e., if the speed of sound depended on its frequency), what consequences would you experience if you were listening to an orchestra in a concert hall?

13.5 Standing Waves and Resonance

82. For two traveling waves to form standing waves, the waves must have the same (a) frequency, (b) amplitude, (c) speed, or (d) all of the preceding.

83. When a stretched violin string oscillates in its third harmonic mode, the standing wave in the string will exhibit (a) 3 wavelengths, (b) 1/3 wavelength, (c) 3/2 wavelengths, or (d) 2 wavelengths.

84. CQ A child's swing (a pendulum) has only one natural frequency f_1, yet it can be driven or pushed smoothly at frequencies of $f_1/2$, $f_1/3$, $2f_1$, and $3f_1$. How is this possible?

85. CQ A guitar string fixed at both ends is vibrating in its fourth harmonic frequency. How many nodes are there on the string?

86. CQ A wave travels in a string with a fixed tension. How does the wavelength change if the frequency is increased? How does the wave speed change?

87. **CQ** You feel that one string on your old violin sounds a bit low in frequency (pitch). How would you adjust the string Explain.

88. **CQ** Given the same tension and length, will a thicker or thinner guitar string sound higher in frequency? Why?

89. ■ A standing wave is formed in a stretched string that is 3.0 m long. What are the wavelengths of (a) the first harmonic and (b) the third harmonic?

90. ■ The fundamental frequency of a stretched string is 100 Hz. What are the frequencies of (a) the second harmonic and (b) the third harmonic?

91. ■ If the frequency of the third harmonic of a vibrating string is 450 Hz, what is the frequency of the first harmonic?

92. ■■ Will a standing wave be formed in a 4.0-m length of stretched string that transmits waves at a speed of 12 m/s if it is driven at a frequency of (a) 15 Hz or (b) 20 Hz?

93. ■■ Two waves of equal amplitude and wavelength of 0.80 m travel in opposite directions at a speed of 250 m/s in a string. If the string is 2.0 m long, for which harmonic mode is the standing wave set up in the string?

94. **IE** ■■ A piece of rubber tubing is stretched by a force. (a) If the force doubles, the transverse wave speed (1) doubles, (2) halves, (3) increases by $\sqrt{2}$, or (4) decreases by $\sqrt{2}$. Why? (b) If the linear mass density of a 10.0-m tubing is 0.125 kg/m and it is stretched by a force of 9.00 N, what is the transverse wave speed in the tubing? (c) What are its waves' natural frequencies?

95. ■■ Find the first four harmonics for a string that is 2.0 m long, has a linear mass density of 2.5×10^{-2} kg/m, and is under a tension of 40 N.

96. ■■ Two stretched strings A and B have the same tension and linear mass density. Are any of the first six harmonics of the strings equal if the string lengths are (a) 1.0 m and 3.0 m or (b) 1.5 m and 2.0 m, respectively?

97. **IE** ■■ A violin string is tuned to a certain frequency (the fundamental frequency, or first harmonic). (a) If a violinist wants a higher frequency, should the string be (1) lengthened, (2) kept the same length, or (3) shortened? Why? (b) If the string is tuned to 440 Hz and the violinist puts a finger down on the string one-eighth of the string length from the neck end, what is the frequency of the string when the instrument is played this way?

98. ■■■ A thin, flexible metal rod is 1.0 m long. It is clamped at one end to a table, and the other end can vibrate freely. What are the natural frequencies of the rod if the wave speed in the material is 3.5×10^3 m/s?

99. ■■■ In a common laboratory experiment on standing waves, the waves are produced in a stretched string by an electrical vibrator that oscillates at 60 Hz (▼Fig. 13.27). The string runs over a pulley, and a hanger is suspended from the end. The tension in the string is varied by adding weights to the hanger. If the active length of the string (the part that vibrates) is 1.5 m and this length of the string has a mass of 0.10 g, what masses must be suspended to produce the first four harmonics in that length?

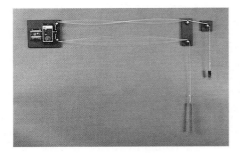

▲ **FIGURE 13.27 Standing waves in strings** Twin vibrating strings with standing waves. This demonstration model allows you to vary string's tension, length, and type (linear mass density). Also, the vibration frequency can be adjusted. See Exercise 99.

Additional Exercises

100. A 0.10-kg mass suspended on a spring is pulled to 8.0 cm below its equilibrium position and released. When the mass passes through the equilibrium position, it has a speed of 0.40 m/s. What is the speed of the mass when it is 4.0 cm from the equilibrium position?

101. A standing wave has nodes at $x = 0$ cm, $x = 6.0$ cm, $x = 12$ cm, and $x = 18$ cm. (a) What is the wavelength of the waves that are interfering to produce this standing wave? (b) At what positions are the antinodes?

102. A simple pendulum makes 5.0 complete cycles in 10 s. (a) What are the frequency and period of this pendulum? (b) What is the length of the pendulum?

103. The motion of an object is described by $x = (0.10$ m$) \cos 50t$. What are the maximum magnitudes of the object's velocity and acceleration?

104. A steel piano wire is 60 cm long and has a mass of 3.0 g. If the tension in the wire is 550 N, what are the wire's fundamental frequency and wavelength?

105. When driven at a frequency of 420 Hz, a stretched string has four equal loops in a standing wave. What driving frequency will set up a wave with two equal segments?

106. On a violin, a correctly tuned A string has a frequency of 440 Hz. If an A string produces sound at 450 Hz under a

tension of 500 N, what should the tension be to produce the correct frequency?

107. If the tension in a fixed string is doubled, what will happen to the (a) wavelength and (b) frequency of the first harmonic?

108. Show that Equation 13.19, $v = \sqrt{\dfrac{F_T}{\mu}}$, is dimensionally correct.

109. An object in vertical SHM is described by $y = (0.20 \text{ cm}) \sin 1.8\pi t$. What is the speed of the object at $t = 10$ s?

110. During an earthquake, the corner of a tall building oscillates with an amplitude of 20 cm at 0.50 Hz. What are the magnitudes of (a) the maximum displacement, (b) the maximum velocity, and (c) the maximum acceleration of the corner of the building? (Assume SHM.)

111. A mass resting on a horizontal frictionless surface is connected to a fixed spring. The mass is displaced 16 cm from its equilibrium position and released. At $t = 0.50$ s, the mass is 8.0 cm from its equilibrium position (and has not passed through it yet). What is the period of oscillation of the mass?

CHAPTER 14

Sound

Good vibrations, clearly! We owe a lot to sound waves. Not only do they provide us with one of our main sources of enjoyment in the form of music, but they also bring us a wealth of vital information about our environment, from the chime of a doorbell to the warning shrill of a police siren to the song of a mockingbird. Indeed, sound waves are the basis for our major form of communication: speech. They can also constitute a highly irritating distraction (noise). But sound waves become music, speech, or noise only when our ears perceive them. Physically, sound is simply waves that propagate in solids, liquids, and gases. Without a medium, there can be no sound; in a vacuum such as outer space, there is utter silence.

This distinction between the sensory and physical meanings of sound gives you a way to answer the old philosophical question, If a tree falls in the forest where there is no one to hear it, does it make a sound? The answer depends on how sound is defined—it is no if we are thinking in terms of sensory hearing, but yes if we are considering only the physical waves. Since sound waves are all around us most of the time, we are exposed to many interesting sound phenomena. You'll explore some of the most important of these in this chapter.

14.1 Sound Waves

OBJECTIVES: **To (a) define sound and (b) explain the sound frequency spectrum.**

For sound waves to exist, there must be a disturbance or vibrations in some medium. This disturbance may be the clapping of hands or the skidding of tires as a car comes to a sudden stop. Under water, you can hear the click of rocks against one another. If you put your ear to a thin wall, you can hear sounds from the other side of the wall. **Sound waves** in gases and liquids (both are fluids) are primarily longitudinal waves. However, sound disturbances moving through solids can have both longitudinal and transverse components. The intermolecular interactions in solids are much stronger than in fluids and allow transverse components to propagate.

The characteristics of sound waves can be visualized by considering those produced by a tuning fork, essentially a metal bar bent into a U shape (▼Fig. 14.1). The prongs, or tines, vibrate when struck. The fork vibrates at its fundamental frequency (with an antinode at the end of each tine), so a single tone is heard. (A *tone* is sound with a definite frequency.) The vibrations disturb the air, producing alternating high-pressure regions called *condensations* and low-pressure regions called *rarefactions*. As the fork vibrates, these disturbances propagate outward, and a series of them can be described by a sinusoidal wave.

As the disturbances traveling through the air reach the ear, the eardrum (a thin membrane) is set into vibration by the pressure variations. On the other side of the eardrum, tiny bones (the hammer, anvil, and stirrup) carry the vibrations to the inner ear, where they are picked up by the auditory nerve. (See the Insight on p. 486.)

Characteristics of the ear limit the perception of sound. Only sound waves with frequencies between about 20 Hz and 20 kHz (kilohertz) initiate nerve impulses that are interpreted by the human brain as sound. This frequency range is called the **audible region** of the **sound frequency spectrum** (▷Fig. 14.2).

Frequencies lower than 20 Hz are in the **infrasonic region**. Waves in this region, which humans are unable to hear, are found in nature. Longitudinal waves generated by earthquakes have infrasonic frequencies, and we use these waves to study the Earth's interior (Chapter 13 Insight, p. 462). Infrasonic waves, or *infrasound*, are also generated by wind and weather patterns. Elephants and cattle have hearing response in the infrasonic region and may even give early warnings of earthquakes and weather disturbances. Aircraft, automobiles, and other rapidly moving objects also can produce infrasound.

Above 20 kHz is the **ultrasonic region**. Ultrasonic waves can be generated by high-frequency vibrations in crystals. Ultrasonic waves, or *ultrasound*, cannot be detected by humans, but can be by other animals. The audible region for dogs extends to about 45 kHz, so ultrasonic whistles can be used to call dogs without disturbing people. Cats and bats have even higher audible ranges, up to about 70 kHz and 100 kHz, respectively.

▼ **FIGURE 14.1   Vibrations make waves**   **(a)** A vibrating tuning fork disturbs the air, producing alternating high-pressure regions (condensations) and low-pressure regions (rarefactions), which form sound waves. **(b)** After being picked up by a microphone, the pressure variations are converted to electrical signals. When these signals are displayed on an oscilloscope, the sinusoidal waveform is evident.

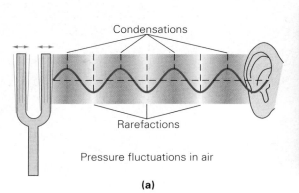

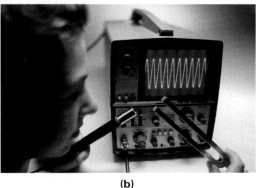

(a)   (b)

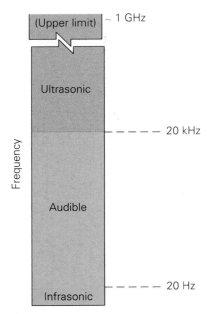

▲ **FIGURE 14.2 Sound frequency spectrum** The audible region of sound for humans lies between about 20 Hz and 20 kHz. Below this is the infrasonic region, and above it is the ultrasonic region. The upper limit is about 1 GHz, because of the elastic limitations of materials.

There are many practical applications of ultrasound. Since ultrasound can travel for kilometers in water, it is used in sonar, the ultrasound counterpart of radar, which employs radio waves for the detection and ranging of objects. Sound pulses generated by the sonar apparatus are reflected by underwater objects, and the resulting echoes are picked up by a detector. The time required for a sound pulse to make one round-trip, together with the speed of sound in water, gives the distance or range of the object. Sonar also is widely used by fishermen to detect schools of fish, and in a similar manner, ultrasound is used in autofocus cameras. Distance measurement allows focal adjustments to be made. An application of ultrasonic sonar from the natural world is given in the Insight on page 481.

In medicine, ultrasound is used to clean teeth (e.g., there are ultrasonic toothbrushes) and to examine internal tissues and organs that are nearly invisible to X rays. Perhaps the best-known medical application of ultrasound is its use to view a fetus without exposing it to the dangerous effects of X rays. Ultrasonic generators (transducers) made of quartz crystals produce high-frequency waves that are used to scan a designated region of the body from several angles. Reflections from the scanned areas are monitored, and a computer constructs an image from the reflected signals. Images are recorded several times each second. The series of images provides a "moving picture" of an internal structure, such as the heart of a fetus. A still shot, or echogram, is shown in ▼Fig. 14.3.

In industrial and home applications, ultrasonic baths are used to clean metal machine parts, dentures, and jewelry. The high-frequency (short wavelength) ultrasound vibrations loosen particles in otherwise inaccessible places.

Ultrasonic frequencies extend into the megahertz (MHz) range, but the sound frequency spectrum does not continue indefinitely. There is an upper limit of about 10^9 Hz, or 1 GHz (gigahertz), which is determined by the upper limit of the elasticity of the materials through which the sound propagates.

14.2 The Speed of Sound

OBJECTIVES: To (a) tell how the speed of sound differs in different media and (b) describe the temperature dependence of the speed of sound in air.

In general, the speed at which a disturbance moves through a medium depends on the elasticity and density of the medium. For example, as you learned in Chapter 13, the wave speed in a stretched string is given by $v = \sqrt{F_T/\mu}$, where F_T is the tension in the string and μ is the linear mass density of the string.

Similar expressions describe wave speeds in solids and liquids, for which the elasticity is expressed in terms of moduli (Chapter 9). In general, the speed

▶ **FIGURE 14.3 Ultrasound in use** Ultrasound generated by transducers, which convert electrical oscillations into mechanical vibrations and vice versa, is transmitted through tissue and is reflected from internal structures. The reflected waves are detected by the transducers, and the signals are used to construct an image, or echogram, shown here for a well-developed fetus.

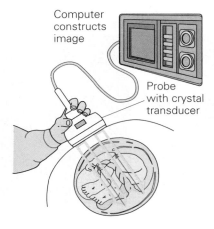

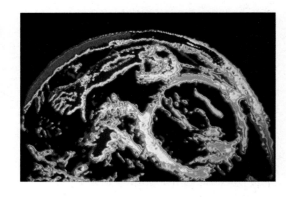

Audible and Ultrasonic: White Noise and Echolocation

White Noise

In an application of audible sound, one kind of noise is used to eliminate another. Noise is generally defined as unwanted sound, and you can probably think of occasions when you would have liked some distracting noise reduced or eliminated. If a radio or TV is too loud, the volume can be turned down. But what about people talking and ambient noises? It has been found that these noises can be reduced by countering them with *white noise*—sound that contains the complete range of audible frequencies. White noise is analogous to *white light*, which is light that contains all of the different colors or frequencies (Section 22.5). An electric fan produces a good approximation of white noise, as does the sound of cymbals and snare drums.

White noise can be used to mask other sounds. For example, if someone in another room is talking and distracting you from doing your homework, you might turn on a fan to mask the noise. What you are doing is "fooling" your brain so that it can get back to the homework. This effect is explained as follows: If two people nearby are talking simultaneously, the brain "selects" one of the voices, usually the loudest, and listens to it so as to understand what is being said. The same holds for three people talking at the same time. But suppose 1000 people were talking. The brain would not be able to pick out one voice. The talking from 1000 people sounds like white noise. So when you turn on a fan to create white noise, similar to that of the 1000 talking people, the voice distracting you from your homework makes it 1001, and the brain doesn't concentrate on it anymore.

Sound masking is available commercially for use in large offices where normal operations can produce distracting noise, such as several employees talking. As just described, the brain focuses on one voice and processes what is being said, which is distracting. Sound masking works by producing sound electronically, similar to that of soft blowing air, an approximation to the white noise produced by a fan. Speakers are located so as to put the masking sound in the appropriate areas.

There is also a musical application of white noise: Electronically synthesized white noise can be filtered (blocking some frequencies) so as to produce sounds with combinations of frequencies that are not heard from typical musical instruments. Maybe even music can be made with pure white noise? How would that sound?

Echolocaton

Sonar appeared in the animal kingdom long before it was developed by human engineers. On their nocturnal hunting flights, bats use a kind of natural sonar to navigate in and out of their caves and to locate and catch flying insects (Fig. 1). The bats emit pulses of ultrasound and track their prey by means of the reflected echoes. The technique is known as *echolocation*. Certain species of cave-dwelling birds have evolved the same ability.

The frequencies of bat cries vary with the species, in the range from 30 000 Hz to 80 000 Hz (30 to 80 kHz). The use of such high frequencies is essential to the bat's sonar system. To determine the nature of objects by means of reflected sounds, it is necessary that the wavelengths of the sounds be on the order of the dimensions of the object, and even shorter for finer details. When the bat's sound pulses strike a small object such as a flying insect, the reflected sound has only a small fraction of the original sound energy. Traveling to the bat through the air, the sound is further weakened. The auditory system and data-processing capabilities of bats are truly amazing. (Note the size of the bat's ears in Fig. 1.)

On the basis of the intensity of the echo, a bat can tell how big an insect is: The smaller the insect, the less intense is the echo. The direction of motion of an insect is sensed by the frequency of the echo. If an insect is moving away from the bat, the returning echo will have a lower frequency. If the insect is moving toward the bat, the echo will have a higher frequency. The change in frequency is known as the *Doppler effect*, which is presented in more detail in Section 14.5.

The bat, the only mammal to have evolved true flight, is a much maligned and feared creature. However, because they feed on tons of insects yearly, bats save the environment from a lot of insecticides. "Blind as a bat" is a common expression, yet bats have fairly good vision, which complements their use of echolocation. Finally, do you know why bats roost and hang upside down (Fig. 1)? That is their takeoff position. Unlike birds, bats can't launch themselves from the ground. Their wings don't produce enough lift to allow takeoff directly from the ground, and their legs are so small and underdeveloped that they can't run to build up takeoff speed. So, they use their claws to hang, and then fall into flight when they are ready to fly.

Related Exercises: 13, 23, 74

FIGURE 1 Echolocation results With the aid of their own natural sonar systems, bats hunt flying insects. The bats emit pulses of ultrasonic waves, which lie within their audible region, and use the echoes reflected from their prey to guide their attack. (Note the size of the ears.)

TABLE 14.1 Speed of Sound in Various Media (typical values)	
Medium	Speed (m/s)
Solids	
Aluminum	5100
Copper	3500
Iron	4500
Glass	5200
Polystyrene	1850
Zinc	3200
Liquids	
Alcohol, ethyl	1125
Mercury	1400
Water	1500
Gases	
Air (0°C)	331
Air (100°C)	387
Helium (0°C)	965
Hydrogen (0°C)	1284
Oxygen (0°C)	316

of sound in a solid and in a liquid is given by $v = \sqrt{Y/\rho}$ and $v = \sqrt{B/\rho}$, respectively, where Y is Young's modulus, B is the bulk modulus, and ρ is the density. The speed of sound in a gas is inversely proportional to the square root of the molecular mass, but the equation is more complicated and will not be presented here.

Solids are generally more elastic than liquids, which in turn are more elastic than gases. In a highly elastic material, the restoring forces between the atoms or molecules cause a disturbance to propagate faster. Thus, the speed of sound is generally about 2 to 4 times faster in solids than in liquids and about 10 to 15 times faster in solids than in gases such as air (Table 14.1).

Although not expressed explicitly in the preceding equations, the speed of sound generally depends on the temperature of the medium. In dry air, for example, the speed of sound is 331 m/s (about 740 mi/h) at 0°C. As the temperature increases, so does the speed of sound. For *normal environmental temperatures*, the speed of sound in air increases by about 0.6 m/s for each degree Celsius above 0°C. Thus, a good approximation of the speed of sound in air for a particular (environmental) temperature is given by

$$v = (331 + 0.6T_C) \text{ m/s} \quad \text{\textit{speed of sound in dry air}} \quad (14.1)$$

where T_C is the air temperature in degrees Celsius.* Although not written explicitly, the units associated with the factor 0.6 are meters per second per Celsius degree $[\text{m/(s·C°)}]$.

Let's take a comparative look at the speed of sound in different media.

Example 14.1 ■ Solid, Liquid, Gas: Speed of Sound in Different Media

From their material properties, find the speed of sound in (a) a solid copper rod, (b) liquid water, and (c) air at room temperature (20°C).

Thinking It Through. We know that the speed of sound in a solid or a liquid depends on the elastic modulus and the density of the solid or liquid. These values are available in Tables 9.1 and 9.2. The speed of sound in air is given by Eq. 14.1.

Solution.

Given: $Y_{Cu} = 11 \times 10^{10} \text{ N/m}^2$
$B_{H_2O} = 2.2 \times 10^9 \text{ N/m}^2$
$\rho_{Cu} = 8.9 \times 10^3 \text{ kg/m}^3$
$\rho_{H_2O} = 1.0 \times 10^3 \text{ kg/m}^3$
(all values from Tables 9.1 and 9.2)
$T_C = 20°C$ (for air)

Find: (a) v_{Cu} (speed in copper)
(b) v_{H_2O} (speed in water)
(c) v_{air} (speed in air)

(a) To find the speed of sound in a copper rod, we use the expression $v = \sqrt{Y/\rho}$:

$$v_{Cu} = \sqrt{\frac{Y}{\rho}} = \sqrt{\frac{11 \times 10^{10} \text{ N/m}^2}{8.9 \times 10^3 \text{ kg/m}^3}} = 3.5 \times 10^3 \text{ m/s}$$

(b) For water, $v = \sqrt{B/\rho}$:

$$v_{H_2O} = \sqrt{\frac{B}{\rho}} = \sqrt{\frac{2.2 \times 10^9 \text{ N/m}^2}{1.0 \times 10^3 \text{ kg/m}^3}} = 1.5 \times 10^3 \text{ m/s}$$

*A better approximation of these and higher temperatures is given by the expression

$$v = \left(331\sqrt{1 + \frac{T_C}{273}}\right) \text{ m/s}$$

In Table 14.1, see v for air at 100°C, which is outside the normal environmental temperature range.

(c)   For air at 20°C, by Eq. 14.1, we have

$$v_{\text{air}} = (331 + 0.6T_{\text{C}})\,\text{m/s} = [331 + 0.6(20)]\,\text{m/s} = 343\,\text{m/s} = 3.43 \times 10^2\,\text{m/s}$$

Follow-up Exercise.   In this Example, how many times faster is the speed of sound in copper (a) than in water and (b) than in air (at room temperature)? Compare your results with the values given at the beginning of the section. *(Answers to all Follow-up Exercises are at the back of the text.)*

A generally useful value for the speed of sound in air is $\frac{1}{3}$ km/s (or $\frac{1}{5}$ mi/s). Using this value, you can, for example, estimate how far away lightning is by counting the number of seconds between the time you observe the flash and the time you hear the associated thunder. Because the speed of light is so fast, you see the lightning flash almost instantaneously. The sound waves of the thunder travel relatively slowly, at about $\frac{1}{3}$ km/s. For example, if the interval between the two events is measured to be 6 s (often by counting "one thousand one, one thousand two, ... "), the lightning stroke was about 2 km away ($\frac{1}{3}$ km/s $\times$ 6 s).

You may also have noticed the delay in the arrival of sound relative to that of light at a baseball game. If you're sitting in the outfield stands, you see the batter hit the ball before you hear the crack of the bat.

Integrated Example 14.2 ■ Safe or Out? You Make the Call

On a cool October afternoon (air temperature = 15°C), from your seat in the centerfield stands 113 m from first base, you witness the play that will decide the World Series. You see the runner's foot touch the bag; half a second later, straining your ears, you hear the faint thud of the ball in the first baseman's glove. The umpire signals safe; half the fans boo loudly. (a) How could you determine whether the umpire was correct? (b) Did the ump blow it? Apply the physics from part (a) to find out.

(a) Conceptual Reasoning.   This situation has to do with the delay in hearing the sound caused by an action; hence, it involves the speed of sound. You can see the action almost instantaneously, yet the sound takes an appreciable time to reach you, because the speed of light is much faster than that of sound. If the travel time of sound is less than $\frac{1}{2}$ s, then the runner had already touched the base when the first baseman caught the ball, and the runner is safe. If the travel time is greater than $\frac{1}{2}$ s, he's out.

(b) Thinking It Through.   To see if the ump was wrong, we need to calculate the time it took for the sound to travel to you.

Given: $d = 113\,\text{m}$ *Find:* t (travel time of sound)
 $T = 15°\text{C}$
 $t = \frac{1}{2}\,\text{s}$

From Eq. 14.1, with a temperature of 15°C, the speed of sound is

$$v = (331 + 0.6T_{\text{C}})\,\text{m/s} = [331 + 0.6\,(15°\text{C})]\,\text{m/s} = 340\,\text{m/s}$$

For a constant speed, the general distance–time relation is $d = vt$, so the time for the sound to travel the distance to your seat is

$$t = \frac{d}{v} = \frac{113\,\text{m}}{340\,\text{m/s}} = 0.332\,\text{s}$$

or just under one-third of a second. The runner reached the base about $\frac{1}{2}\,\text{s} - \frac{1}{3}\,\text{s} = \frac{1}{6}\,\text{s}$ before the ball arrived. The ump was right!

To show why it is justified to say that the visual observation takes place almost instantaneously, let's see how long it takes for light to travel from first base to your seat. For practical purposes, the speed of light in air is the same as that in a vacuum,

$c = 3.00 \times 10^8$ m/s, which is on the order of 10^6 (a million) times greater than the speed of sound. The time for light to travel 113 m is then

$$t = \frac{d}{c} = \frac{113 \text{ m}}{3.00 \times 10^8 \text{ m/s}} = 3.77 \times 10^{-7} \text{ s}$$

or 0.377μs—not a very long time.

Follow-up Exercise. On a day when the air temperature is 20°C, a hiker shouts and hears the echo from the face of a vertical stone cliff 5.00 s later. (a) How far away is the cliff? (b) If the air temperature were 15°C, what would be the difference in the time it takes to hear the echoes?

The speed of sound in air depends on various factors. Temperature is the most important, but there are other considerations, such as the homogeneity and composition of the air. For example, the air composition may not be "normal" in a polluted area. These effects are relatively small and will not be considered, except conceptually in the next Example.

Conceptual Example 14.3 ■ Speed of Sound: Sound Traveling Far and Wide

Note that the speed of sound in *dry* air for a given temperature is given to a good approximation by Eq. 14.1. However, the moisture content of the air (humidity) varies, and this variation affects the speed of sound. At the same temperature, would sound travel faster in (a) dry air or (b) moist air?

Reasoning and Answer. According to an old folklore saying, "Sound traveling far and wide, a stormy day will betide." This saying implies that sound travels faster on a highly humid day, when a storm or precipitation is likely. But is the saying true?

Near the beginning of this section, we learned that the speed of sound in a gas is inversely proportional to the square root of the molecular mass of the gas. So, at constant pressure, is moist air more or less dense than dry air?

In a volume of moist air, a large number of water (H_2O) molecules occupy the space normally occupied by either nitrogen (N_2) or oxygen (O_2) molecules, which make up 98% of the air. Water molecules are less massive than both nitrogen and oxygen molecules. [From Section 10.3, the molecular (formula) masses are H_2O, 18 g; N_2, 28 g; and O_2, 32 g.] Thus, the average molecular mass of a volume of moist air is less than that of dry air, and the speed of sound is greater in moist air.

We can look at this situation another way: Since water molecules are less massive, they have less inertia and respond to the sound wave faster than nitrogen or oxygen molecules do. The water molecules therefore propagate the disturbance faster.

Follow-up Exercise. Considering only molecular masses, would you expect the speed of sound to be greatest in nitrogen, oxygen, or helium (at the same temperature and pressure)? Explain.

Always keep in mind that our discussion generally assumes ideal conditions for the propagation of sound. Actually, the speed of sound depends on many things, one of which is humidity, as the previous Conceptual Example shows. A variety of other properties affect the propagation of sound. As an example, let's ask the question, Why do ships' foghorns have such a low pitch or frequency? The answer is, Because low-frequency sound waves travel farther than high-frequency ones under identical conditions. This effect is explained by a couple of characteristics of sound waves. First, sound waves are attenuated (i.e., they lose energy) because of the vis-

Note: Humidity was included here as an interesting consideration for the speed of sound in air. However, henceforth, in computing the speed of sound in air at a certain temperature, we will consider only dry air (Eq. 14.1) unless otherwise stated.

cosity of the air (Ch. 9.5). Second, sound waves tend to interact with oxygen and water molecules in the air. The combined result of these two properties is that the total attenuation of sound in air depends on the frequency of the sound: the higher the frequency, the more is the attenuation and the shorter is the distance traveled. For example, a 200-Hz sound will travel much farther than an 800-Hz sound. Thus, low-frequency foghorns are used. In accordance with this dependence on frequency, you might notice that when a storm's lightning is farther away, the thunder you generally hear is a low-frequency rumble. Sound is one of our major means of communicating. (For more on our speech and hearing, see the Insight on page 486.)

14.3 Sound Intensity and Sound Intensity Level

OBJECTIVES: To (a) define sound intensity and explain how it varies with distance from a point source and (b) calculate sound intensity levels on the decibel scale.

Wave motion involves the propagation of energy. The rate of energy transfer is expressed in terms of **intensity**, which is the energy transported per unit time across a unit area. Since energy divided by time is power, intensity is power divided by area:

$$\text{intensity} = \frac{\text{energy/time}}{\text{area}} = \frac{\text{power}}{\text{area}} \quad \left[I = \frac{E/t}{A} = \frac{P}{A} \right]$$

The standard units of intensity (power/area) are watts per square meter (W/m^2).

Consider a point source that sends out spherical sound waves, as shown in ▼Fig. 14.4. If there are no losses, the sound intensity at a distance R from the source is

$$I = \frac{P}{A} = \frac{P}{4\pi R^2} \tag{14.2}$$

where P is the power of the source and $4\pi R^2$ is the area of a sphere of radius R, through which the sound energy passes perpendicularly.

The intensity of a point source of sound is therefore *inversely proportional to the square of the distance from the source* (an inverse-square relationship). Two intensities at different distances from a source of constant power can be compared as a ratio:

$$\frac{I_2}{I_1} = \frac{P/(4\pi R_2^2)}{P/(4\pi R_1^2)} = \frac{R_1^2}{R_2^2}$$

or

$$\frac{I_2}{I_1} = \left(\frac{R_1}{R_2} \right)^2 \tag{14.3}$$

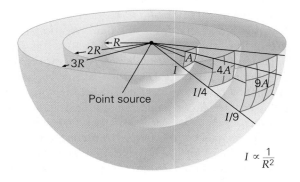

◀ **FIGURE 14.4 Intensity of a point source** The energy emitted from a point source spreads out equally in all directions. Since intensity is power divided by area, $I = P/A = P/(4\pi R^2)$, where the area is that of a spherical surface. The intensity then decreases with the distance from the source as $1/R^2$ (figure not to scale).

Speech and Hearing

Sound is one of our most important means of interpersonal communication, and our most important source of sound is the human voice. Let's look at the anatomy and physics of the voice and the ear, the latter of which serves as our receiver of sound.

The Human Voice

The energy for sounds associated with the human voice originates in the muscle action of the diaphragm, which forces air up from the lungs. To produce variations in sound, this steady stream of air must be periodically disturbed, or "modulated." The fundamental modulating organ is the larynx (the "voice box"), across which are stretched membranelike folds called the vocal cords. The opening of the vocal cords modulates the airstream to produce sounds (Fig. 1).

As illustrated in the figure, there are two sets of vocal cords: (1) the *false vocal cords*, which do not produce sound, but help close the larynx during swallowing, and (2) the *true vocal cords*, which are elastic and are responsible for vocal sounds.

Changing the tension of the vocal cords controls the pitch (perceived frequency) of the vocal sounds (as we shall see in Section 14.6). The loudness of the sound is related to the pressure of the air passing over the vocal cords. A strong blast of air results in a greater amplitude of the vibrations and hence a louder sound.

The waveforms of voice sounds are quite specific to individuals and provide a "voiceprint" that can be used for iden-

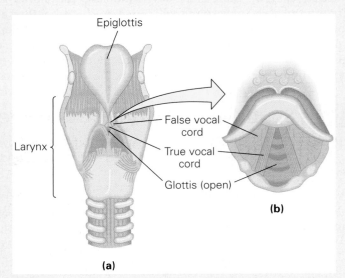

FIGURE 1 Anatomy of the voice box (a) A cross section of the larynx. (b) A down-the-throat view of the vocal cords.

tification, just as fingerprints are. The validity of voiceprints for legal identification, however, is highly controversial.

Hearing

The anatomy of the human ear is illustrated in Fig. 2. Sound enters the outer ear and travels through the ear (or auditory)

Suppose that the distance from a point source is doubled; that is, $R_2 = 2R_1$, or $R_1/R_2 = \frac{1}{2}$. Then

$$\frac{I_2}{I_1} = \left(\frac{R_1}{R_2}\right)^2 = \left(\frac{1}{2}\right)^2 = \frac{1}{4}$$

and

$$I_2 = \frac{I_1}{4}$$

Since the intensity decreases by a factor of $1/R^2$, doubling the distance decreases the intensity to a quarter of its original value.

A good way to understand this inverse-square relationship intuitively is to look at the geometry of the situation. As Fig. 14.4 shows, the greater the distance from the source, the larger the area over which a given amount of sound energy is spread, and thus the lower its intensity. (Imagine having to paint two walls of different areas. If you had the same amount of paint to use on each, you'd have to spread it more thinly over the larger wall.) Since this area increases as the square of the radius R, the intensity decreases accordingly—that is, as $1/R^2$.

Sound intensity is perceived by the ear as **loudness**. On the average, the human ear can detect sound waves (at 1 kHz) with an intensity as low as 10^{-12} W/m^2. This intensity (I_o) is referred to as the *threshold of hearing*. Thus, for us to hear a sound, it must not only have a frequency in the audible range, but also be of sufficient inten-

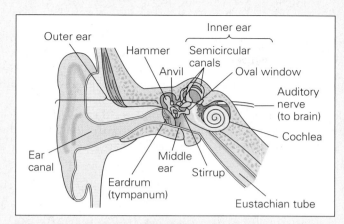

FIGURE 2   Anatomy of the human ear   The ear converts pressure in the air into electrical nerve impulses that are interpreted as sounds by the brain.

canal to the *eardrum* (the tympanum), which separates the outer ear from the middle ear. The eardrum is a membrane that vibrates in response to the impinging sound waves. The vibrations are transmitted through the middle ear, which contains an intricate set of connected bones, commonly called the *hammer* (malleus), *anvil* (incus), and *stirrup* (stapes), because of their shapes.

The inner ear includes the semicircular canals, which are important in controlling balance, and the *cochlea*. It is in

the cochlea that sound waves are translated into nerve impulses and that pitch or frequency discrimination is made. The cochlea consists of a series of liquid-filled tubes or ducts, coiled into a spiral shape that resembles a snail shell. The *basilar membrane* supports the cochlea and forms the floor of the cochlear ducts. Within the ducts are thousands of special receptor cells called *hair cells*.

The hair cells are specialized nerve receptors. When a region of the basilar membrane is set vibrating by sound of a particular frequency, hair cells in that region are stimulated and nerve impulses are sent to the brain, where they are interpreted as sound. The brain translates this positional information (impulses from particular fibers originating in specific regions of the basilar membrane) back into frequency information (the subjective physiological sensation of pitch).

Incidentally, the middle ear is connected to the throat by the Eustachian tube, the end of which is normally closed. It opens during swallowing and yawning to permit air to enter and leave, so that internal and external pressures are equalized. You have probably experienced a "stopping up" of your ears with a sudden change in atmospheric pressure (e.g., during rapid ascents or descents in elevators or airplanes). Swallowing opens the Eustachian tubes and relieves the excess pressure difference on the middle ear.

Related Exercises: 24, 31, 84

sity. As the intensity is increased, the perceived sound becomes louder. At an intensity of 1.0 W/m², the sound is uncomfortably loud and may be painful to the ear. This intensity (I_p) is called the *threshold of pain*.

Note that the thresholds of pain and hearing differ by a factor of 10^{12}:

$$\frac{I_P}{I_o} = \frac{1.0\ \text{W/m}^2}{10^{-12}\ \text{W/m}^2} = 10^{12}$$

That is, the intensity at the threshold of pain is a *trillion* times greater than that at the threshold of hearing. Within this enormous range, the perceived loudness is not directly proportional to the intensity. Thus, if the intensity is doubled, the perceived loudness does not double. In fact, a doubling of perceived loudness corresponds approximately to an increase in intensity by a factor of 10. For example, a sound with an intensity of 10^{-5} W/m² would be perceived to be twice as loud as one with an intensity of 10^{-6} W/m². (The smaller the negative exponent, the larger is the number.)

Sound Intensity Level: The Bel and the Decibel

It is convenient to compress the large range of sound intensities by using a logarithmic scale (base 10) to express intensity levels. The intensity level of a sound must be referenced to a standard intensity, which is taken to be that of the threshold of hearing, $I_o = 10^{-12}$ W/m². Then, for any intensity I, the intensity level is the logarithm (or log) of the ratio of I to I_o, that is, $\log I/I_o$. For example, if a sound has an intensity of $I = 10^{-6}\ W/m^2$,

$$\log \frac{I}{I_o} = \log \frac{10^{-6}\ W/m^2}{10^{-12}\ W/m^2} = \log 10^6 = 6\ B$$

Note: The bel was named in honor of Alexander Graham Bell, the inventor of the telephone.

(Recall that $\log_{10} 10^x = x$.) The exponent of the power of 10 in the final log term is taken to have a unit called the **bel** (B). Thus, a sound with an intensity of $10^{-6}\ W/m^2$ has an intensity level of 6 B on this scale. That way, the intensity range from $10^{-12}\ W/m^2$ to $1.0\ W/m^2$ is compressed into a scale of intensity levels ranging from 0 B to 12 B.

A finer intensity scale is obtained by using a smaller unit, the **decibel** (dB), which is a tenth of a bel. The range from 0 to 12 B corresponds to 0 to 120 dB. In this case, the equation for the relative **sound intensity level**, or **decibel level** (β), is

Sound intensity level in decibels

$$\beta = 10 \log \frac{I}{I_o} \quad \text{(where } I_o = 10^{-12}\ W/m^2) \qquad (14.4)$$

Note that the sound intensity level (in decibels, which are dimensionless) is *not* the same as the sound intensity (in watts per square meter).

Threshold of hearing: $I_o = 10^{-12}\ W/m^2$

Threshold of pain: $I_p = 1.0\ W/m^2$

The decibel intensity scale and familiar sounds at some intensity levels are shown in ▼ Fig. 14.5. Sound intensities can have detrimental effects on hearing, and because of this, the U.S. government has set occupational noise-exposure limits.

Example 14.4 ■ Sound Intensity Levels: Using Logarithms

What are the intensity levels of sounds with intensities of (a) $10^{-12}\ W/m^2$ and (b) $5.0 \times 10^{-6}\ W/m^2$?

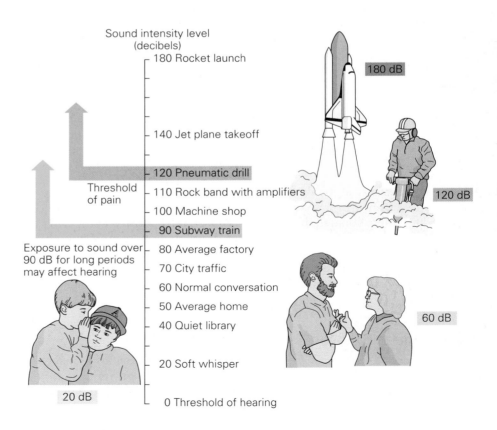

▶ **FIGURE 14.5 Sound intensity levels and the decibel scale**
The intensity levels of some common sounds on the decibel (dB) scale.

Thinking It Through.   The sound intensity levels can be found by using Eq. 14.4.

Solution.

Given: (a) $I = 10^{-12} \text{ W/m}^2$ *Find:* (a) β (sound intensity level)
   (b) $I = 5.0 \times 10^{-6} \text{ W/m}^2$ (b) β

(a)   Using Eq. 14.4, we have

$$\beta = 10 \log \frac{I}{I_o} = 10 \log \left(\frac{10^{-12} \text{ W/m}^2}{10^{-12} \text{ W/m}^2} \right) = 10 \log 1 = 0 \text{ dB}$$

The intensity is the same as that at the threshold of hearing. (Recall that $\log 1 = 0$, since $1 = 10^0$ and $\log 10^0 = 0$.) Note that an intensity level of 0 dB does not mean that there is no sound.

(b) $\beta = 10 \log \dfrac{I}{I_o} = 10 \log \left(\dfrac{5.0 \times 10^{-6} \text{ W/m}^2}{10^{-12} \text{ W/m}^2} \right)$

$$= 10 \log (5.0 \times 10^6) = 10(\log 5.0 + \log 10^6) = 10(0.70 + 6.0) = 67 \text{ dB}$$

Follow-up Exercise.   Note in this Example that the intensity of $5.0 \times 10^{-6} \text{ W/m}^2$ is halfway between 10^{-6} and 10^{-5} (or 60 and 70 dB), yet this intensity does not correspond to a midway value of 65 dB. (a) Why? (b) What intensity *does* correspond to 65 dB? (Compute it to three significant figures.)

Example 14.5 ■ Intensity Level Differences: Using Ratios

(a) What is the difference in the intensity levels if the intensity of a sound is doubled?
(b) By what factors does the intensity increase for intensity level *differences* of 10 dB and 20 dB?

Thinking It Through.   (a) If the intensity is doubled, then $I_2 = 2I_1$, or $I_2/I_1 = 2$. We can then use Eq. 14.4 to find the intensity difference. Recall that $\log a - \log b = \log a/b$. (b) Here, it is important to note that these values are intensity level *differences*, $\Delta\beta = \beta_2 - \beta_1$, *not* intensity *levels*. The equation developed in (a) will work. (Why?)

Solution.   Listing the data, we have.

Given: (a) $I_2 = 2I_1$ *Find:* (a) $\Delta\beta$ (intensity level difference)
   (b) $\Delta\beta = 10$ dB   (b) I_2/I_1 (factors of increase)
    $\Delta\beta = 20$ dB

(a)   Using Eq. 14.4 and the relationship $\log a - \log b = \log a/b$, we have, for difference $\Delta\beta = \beta_2 - \beta_1 = 10[\log (I_2/I_o) - \log (I_1/I_o)] = 10 \log [(I_2/I_o)/(I_1/I_o)] = 10 \log I_2/I_1$. Then,

$$\Delta\beta = 10 \log \frac{I_2}{I_1} = 10 \log 2 = 3 \text{ dB}$$

Thus, doubling the intensity increases the intensity level by 3 dB (e.g., an increase from 55 dB to 58 dB).

(b)   For a 10-dB difference,

$$\Delta\beta = 10 \text{ dB} = 10 \log \frac{I_2}{I_1} \quad \text{and} \quad \log \frac{I_2}{I_1} = 1.0$$

Since $\log 10^1 = 1$, the intensity ratio is 10:1 because

$$\frac{I_2}{I_1} = 10^1 \quad \text{and} \quad I_2 = 10 \, I_1$$

Similarly, for a 20-dB difference,

$$\Delta\beta = 20 \text{ dB} = 10 \log \frac{I_2}{I_1} \quad \text{and} \quad \log \frac{I_2}{I_1} = 2.0$$

Since $\log 10^2 = 2$,

$$\frac{I_2}{I_1} = 10^2 \quad \text{and} \quad I_2 = 100\, I_1$$

Thus, an intensity level difference of 10 dB corresponds to changing (increasing or decreasing) the intensity by a factor of 10. An intensity level difference of 20 dB corresponds to changing the intensity by a factor of 100.

You should be able to guess the factor that corresponds to an intensity level difference of 30 dB. In general, the factor of the intensity change is $10^{\Delta B}$, where ΔB is the difference in levels of bels. Since 30 dB = 3 B and $10^3 = 1000$, the intensity changes by a factor of 1000 for an intensity level difference of 30 dB.

Follow-up Exercise. A $\Delta\beta$ of 20 dB and a $\Delta\beta$ of 30 dB correspond to factors of 100 and 1000, respectively, in intensity changes. Does a $\Delta\beta$ of 25 dB correspond to an intensity change factor of 500? Explain.

Example 14.6 ■ Combined Sound Levels: Adding Intensities

Sitting at a sidewalk restaurant table, a friend talks to you in normal conversation (60 dB). At the same time, the intensity level of the street traffic reaching you is also 60 dB. What is the total intensity level of the combined sounds?

Thinking It Through. It is tempting simply to add the two sound intensity levels together and say that the total is 120 dB. But intensity levels in decibels are logarithmic, so you can't add them in the normal way. However, intensities (I) can be added arithmetically, since energy and power are scalar quantities. Then the combined intensity level can be found from the sum of the intensities.

Solution. We have the following information:

Given: $\beta_1 = 60 \text{ dB}$ *Find:* Total β
$\beta_2 = 60 \text{ dB}$

Let's find the intensities associated with the intensity levels:

$$\beta_1 = 60 \text{ dB} = 10 \log \frac{I_1}{I_o} = 10 \log \left(\frac{I_1}{10^{-12} \text{ W/m}^2} \right)$$

By inspection, we have

$$I_1 = 10^{-6} \text{ W/m}^2$$

Similarly, $I_2 = 10^{-6} \text{ W/m}^2$, since both intensity levels are 60 dB. So the total intensity is

$$I_{\text{total}} = I_1 + I_2 = 1.0 \times 10^{-6} \text{ W/m}^2 + 1.0 \times 10^{-6} \text{ W/m}^2 = 2.0 \times 10^{-6} \text{ W/m}^2$$

Then, converting back to intensity level, we get

$$\beta = 10 \log \frac{I_{\text{total}}}{I_o} = 10 \log \left(\frac{2.0 \times 10^{-6} \text{ W/m}^2}{10^{-12} \text{ W/m}^2} \right) = 10 \log (2.0 \times 10^6)$$
$$= 10(\log 2.0 + \log 10^6) = 10(0.30 + 6.0) = 63 \text{ dB}$$

This value is a long way from 120 dB! Notice that the combined intensities doubled the intensity value, and the intensity level increased by 3 dB, in agreement with our finding in part (a) of Example 14.5.

Follow-up Exercise. In this Example, suppose the added noise gave a total that *tripled* the intensity level of the conversation. What would be the total combined intensity level in this case?

Protect Your Hearing

Hearing may be damaged by excessive noise, so our ears sometimes need protection from continuous loud sounds (▼Fig. 14.6). Hearing damage depends on the sound intensity level (decibel level) and the exposure time. The exact combinations vary for different people, but a general guide to noise levels is given in Table 14.2. Studies have shown that sound levels of 90 dB and above will damage

◀ **FIGURE 14.6 Protect your hearing** Continuous loud sounds can damage hearing, so our ears may need protection as shown here. Note in Table 14.2 that the intensity level of lawn mowers is on the order of 90 dB.

TABLE 14.2 Sound Levels and Ear Damage Exposure Times

	Decibels (dB)	Examples	Damage Can Occur with Nonstop Exposure
Faint	30	Quiet library, whispering	
Moderate	60	Normal conversation, sewing machine	
Very loud	80	Heavy traffic, noisy restaurant, screaming child	10 hours
	90	Lawn mower, motorcycle, loud party	Less than 8 hours
	100	Chain saw, subway train, snowmobile	Less than 2 hours
Extremely loud	110	Stereo headset at full blast, rock concert	30 minutes
	120	Dance clubs, car stereos, action movies, some musical toys	15 minutes
	130	Jackhammer, loud computer games, loud sporting events	Less than 15 minutes
Painful	140	Boom stereos, gunshot blast, firecrackers	Any length (for example, hearing loss can occur from a few shots of a high-powered gun if protection is not worn)

Courtesy of The EAR Foundation.

receptor nerves in the ear, resulting in a loss of hearing. At 90 dB, it takes 8 hours or less for damage to occur. In general, if the sound level is increased by 5 decibels, the safe exposure time is cut in half. For example, if a sound level of 95 dB (that of a loud lawn mower or motorcycle) takes 4 hours to damage your hearing, then a sound level of 105 dB takes only 1 hour to do damage.

14.4 Sound Phenomena

OBJECTIVES: To (a) explain the reflection, refraction, and diffraction of sound and (b) distinguish between constructive and destructive interference.

Reflection, Refraction, and Diffraction

An echo is a familiar example of the *reflection* of sound—sound "bouncing" off a surface. Sound *refraction* is less common than reflection, but you may have experienced it on a calm summer evening, when it is possible to hear distant voices or other sounds that ordinarily would not be audible. This effect is due to the refraction, or bending (change in direction), of the sound waves as they pass from one region into a region where the air density is different. The effect is similar to what would happen if the sound passed into another medium.

The required conditions for sound to be refracted are a layer of cooler air near the ground or water and a layer of warmer air above it. These conditions occur frequently over bodies of water, which cool after sunset (▼Fig. 14.7). As a result of the cooling, the waves are refracted in an arc that may allow a distant person to receive an increased intensity of sound.

Another bending phenomenon is *diffraction*, described in Section 13.4. Sound may be diffracted, or bent, around corners or around an object. We usually think of waves as traveling in straight lines. However, you can hear someone you cannot see standing around a corner. This bending is different from that of refraction, in which no obstacle causes the bending.

Reflection, refraction, and diffraction are described in a general sense here for sound. These phenomena are important considerations for light waves as well and will be discussed more fully in Chapters 22 and 24.

▼ **FIGURE 14.7 Sound refraction** Sound travels more slowly in the cool air near the water surface than in the upper, warmer air. As a result, the waves are refracted, or bent. This bending increases the intensity of the sound at a distance where it otherwise might not be heard.

Interference

Like waves of any kind, sound waves *interfere* when they meet. Suppose that two loudspeakers separated by some distance emit sound waves in phase at the same frequency. If we consider the speakers to be point sources, then the waves will spread out spherically and interfere (▼Fig. 14.8a). The lines from a particular speaker represent wave crests (or condensations), and the troughs (or rarefactions) lie in the intervening white areas.

In particular regions of space, there will be constructive or destructive interference. For example, if two waves meet in a region where they are exactly in phase (two crests or two troughs coincide), there will be total **constructive interference** (Fig. 14.8b). Notice that the waves have the same motion at point C in the figure. If, instead, the waves meet such that the crest of one coincides with the trough of the other (at point D), the two waves will cancel each other out (Fig. 14.8c). The result will be total **destructive interference**.

It is convenient to describe the path lengths traveled by the waves in terms of wavelength (λ) to determine whether they arrive in phase. Consider the waves arriving at point C in Fig. 14.8b. The path lengths in this case are $L_{AC} = 4\lambda$ and $L_{BC} = 3\lambda$. The **phase difference** ($\Delta\theta$) is related to the **path-length difference** (ΔL) by the simple relationship

$$\Delta\theta = \frac{2\pi}{\lambda}(\Delta L) \qquad \begin{array}{l}\textit{phase difference and}\\ \textit{path-length difference}\end{array} \qquad (14.5)$$

Since 2π rad is equivalent, in angular terms, to a full wave cycle or wavelength, multiplying the path-length difference by $2\pi/\lambda$ gives the phase difference in radians. For the example illustrated in Fig. 14.8b, we have

$$\Delta\theta = \frac{2\pi}{\lambda}(L_{AC} - L_{BC}) = \frac{2\pi}{\lambda}(4\lambda - 3\lambda) = 2\pi \text{ rad}$$

When $\Delta\theta = 2\pi$ rad, the waves are shifted by one wavelength. This is the same as $\Delta\theta = 0°$, so the waves are actually back in phase. Thus, the waves interfere constructively in the region of point C, increasing the intensity, or loudness, of the sound detected there.

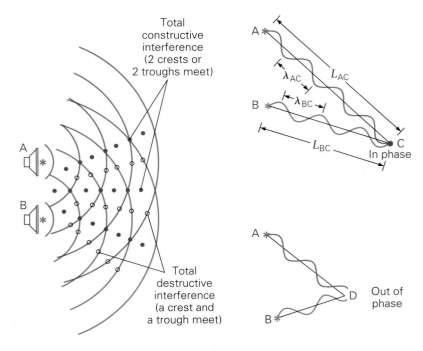

◀ **FIGURE 14.8 Interference**
(a) Sound waves from two point sources spread out and interfere. **(b)** At points where the waves arrive in phase (with a zero phase difference), such as point C, constructive interference occurs. **(c)** At points where the waves arrive completely out of phase (with a phase difference of 180°), such as point D, destructive interference occurs. The phase difference at a particular point depends on the path lengths the waves travel to reach that point.

From Eq. 14.5, we see that the sound waves are in phase at any point where the path-length difference is zero or an integral multiple of the wavelength. That is,

$$\Delta L = n\lambda \quad (n = 0, 1, 2, 3, \ldots) \qquad \text{condition for constructive interference} \qquad (14.6)$$

A similar analysis of the situation in Fig. 14.8c, where $L_{AD} = 2\frac{3}{4}\lambda$ and $L_{BD} = 2\frac{1}{4}\lambda$, gives

$$\Delta\theta = \frac{2\pi}{\lambda}\left(2\tfrac{3}{4}\lambda - 2\tfrac{1}{4}\lambda\right) = \pi \text{ rad}$$

or $\Delta\theta = 180°$. At point D, the waves are completely out of phase, and destructive interference occurs in this region.

Sound waves will be out of phase at any point where the path-length difference is an odd number of half-wavelengths ($\lambda/2$), or

$$\Delta L = m\left(\frac{\lambda}{2}\right) \quad (m = 1, 3, 5, \ldots) \qquad \text{condition for destructive interference} \qquad (14.7)$$

At these points, a softer, or less intense, sound will be heard or detected. If the amplitudes of the waves are exactly equal, the destructive interference is total and no sound is heard.

An application of destructive interference in sound waves is to reduce loud noises. For example, engine noise in a helicopter can be distracting and cause hearing discomfort. Special headsets, with a microphone near the ear, pick up the loud noise and supply sound with a phase difference that cancels the original sound as much as possible. Ideally, the sound that is supplied is $180°$ out of phase with the undesirable noise.

Example 14.7 ■ Pump Up the Volume: Sound Interference

At an open-air concert on a hot day (with an air temperature of 25°C), you sit 7.00 m and 9.10 m, respectively, from a pair of speakers, one at each side of the stage. A musician, warming up, plays a single 494-Hz tone. What do you hear? (Consider the speakers to be point sources.)

Thinking It Through. The sound waves from the speakers interfere. Is the interference constructive, destructive, or something in between? It depends on the path-length difference, which we can compute from the given distances.

Solution.

Given: $d_1 = 7.00$ m and $d_2 = 9.10$ m   *Find:* ΔL (path-length difference
$\qquad\quad f = 494$ Hz $\qquad\qquad\qquad\qquad\qquad$ in wavelength units)
$\qquad\quad T = 25°C$

The path-length difference (2.10 m) between the waves arriving at your location must be expressed in terms of the wavelength of the sound. To do this, we first need to know the wavelength. Given the frequency, we can find the wavelength from the relationship $\lambda = v/f$, provided that the speed of sound, v, at the given temperature is known. The speed v can be found by using Eq. 14.1:

$$v = 331 + 0.6T_C = 331 + 0.6(25) = 346 \text{ m/s}$$

The wavelength of the sound waves is then

$$\lambda = \frac{v}{f} = \frac{346 \text{ m/s}}{494 \text{ Hz}} = 0.700 \text{ m}$$

Thus, the distances in terms of wavelength are

$$d_1 = (7.00 \text{ m})\left(\frac{\lambda}{0.700 \text{ m}}\right) = 10.0\lambda \quad \text{and} \quad d_2 = (9.10 \text{ m})\left(\frac{\lambda}{0.700 \text{ m}}\right) = 13.0\lambda$$

The path-length difference in terms of wavelengths is

$$\Delta L = d_2 - d_1 = 13.0\lambda - 10.0\lambda = 3.0\lambda$$

This is an integral number of wavelengths ($n = 3$), so constructive interference occurs. The sounds of the two speakers reinforce each other, and you hear an intense tone at 494 Hz.

Follow-up Exercise. Suppose that in this Example the tone traveled to a person sitting 7.00 m and 8.75 m, respectively, from the two speakers. What would be the situation in that case?

Another interesting interference effect occurs when two tones of nearly the same frequency ($f_1 \approx f_2$) are sounded simultaneously. The ear senses pulsations in loudness known as **beats**. The human ear can detect as many as 7 beats per second before they sound "smooth" (continuous, without any pulsations).

Suppose that two sinusoidal waves with the same amplitude, but slightly different frequencies, interfere (▼Fig. 14.9a). Figure 14.9b represents the resulting sound wave. The amplitude of the combined wave varies sinusoidally, as shown by the black curves (known as *envelopes*) that outline the wave.

What does this variation in amplitude mean in terms of what the listener perceives? A listener will hear a pulsating sound (beats), as determined by the envelopes. The maximum amplitude is $2A$ (at the point where the maxima of the two original waves interfere constructively). Detailed mathematics shows that a listener will hear the beats at a frequency called the **beat frequency** (f_b), given by

$$f_b = |f_1 - f_2| \tag{14.8}$$

The absolute value is taken because the frequency f_b cannot be negative, even if $f_2 > f_1$. A negative beat frequency would be meaningless.

Beats can be produced when tuning forks of nearly the same frequency are vibrating at the same time. For example, using forks with frequencies of 516 Hz and

▼ **FIGURE 14.9 Beats** Two traveling waves of equal amplitude and slightly different frequencies interfere and give rise to pulsating tones called beats. The beat frequency is given by $f_b = |f_1 - f_2|$.

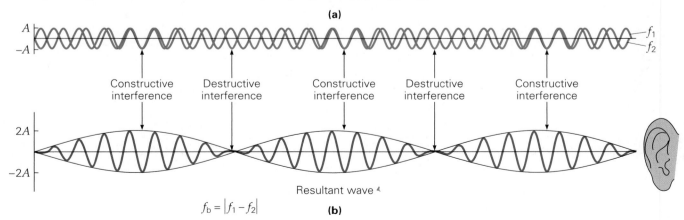

513 Hz, one can generate a beat frequency of $f_b = 516\text{ Hz} - 513\text{ Hz} = 3\text{ Hz}$, and three beats are heard each second. Musicians tune two stringed instruments to the same note by adjusting the tensions in the strings until the beats disappear ($f_1 = f_2$).

14.5 The Doppler Effect

OBJECTIVES: **To (a) describe and explain the Doppler effect and (b) give some examples of its occurrences and applications.**

The Austrian physicist Christian Doppler (1803–1853) first described what we now call the Doppler effect.

If you stand along a highway and a car or truck approaches you with its horn blowing, the **pitch** (the perceived frequency) of the sound is higher as the vehicle approaches and lower as it recedes. You can also hear variations in the frequency of the motor noise when you watch a race car going around a track. A variation in the perceived sound frequency due to the motion of the source is an example of the **Doppler effect**.

As ▼Fig. 14.10 shows, the sound waves emitted by a moving source tend to bunch up in front of the source and spread out in back. The Doppler shift in frequency can be found by assuming that the air is at rest in a reference frame such as that depicted in ▶Fig. 14.11. The speed of sound in air is v, and the speed of the moving source is v_s. The frequency of the sound produced by the source is f_s. In one period, $T = 1/f_s$, a wave crest moves a distance $d = vT = \lambda$. (The sound wave would travel this distance in still air in any case, regardless of whether the source is moving.) But in one period, the source travels a distance $d_s = v_s T$ before emitting another wave crest. The distance between the successive wave crests is thus shortened to a wavelength λ':

$$\lambda' = d - d_s = vT - v_s T = (v - v_s)T = \frac{v - v_s}{f_s}$$

▼ **FIGURE 14.10 The Doppler effect** The sound waves bunch up in front of a moving source—the whistle—giving a higher frequency there. They trail out behind the source, giving a lower frequency there. (The figure is not drawn to scale. Why?)

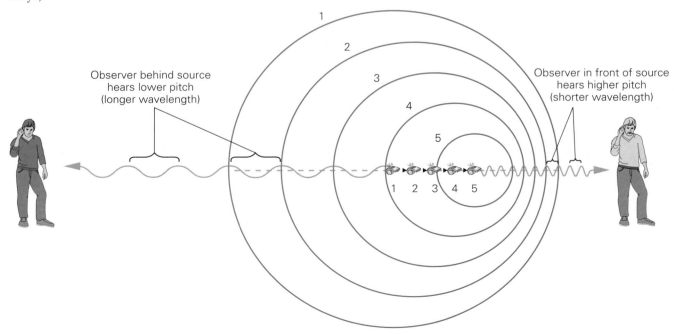

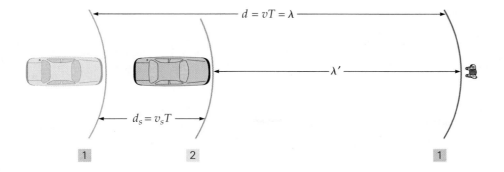

◀ **FIGURE 14.11 The Doppler effect and wavelength** Sound from a moving car's horn travels a distance d in a time T. During this time, the car (the source) travels a distance d_s before putting out a second pulse, thereby shortening the observed wavelength of the sound in the approaching direction.

The frequency heard by the observer (f_o) is related to the shortened wavelength by $f_o = v/\lambda'$, and substituting λ' gives

$$f_o = \frac{v}{\lambda'} = \left(\frac{v}{v - v_s}\right)f_s$$

or

$$f_o = \left(\frac{1}{1 - \dfrac{v_s}{v}}\right)f_s \qquad \begin{array}{l}\textit{(source moving toward}\\ \textit{a stationary observer)}\end{array} \qquad (14.9)$$

$$\begin{array}{ll}\textit{where} & v_s = \textit{speed of source}\\ \textit{and} & v = \textit{speed of sound}\end{array}$$

Since $1 - (v_s/v)$ is less than 1, f_o is greater than f_s in this situation. For example, suppose that the speed of the source is a tenth of the speed of sound; that is, $v_s = v/10$, or $v_s/v = \frac{1}{10}$. Then, by Eq. 14.9, $f_o = \frac{10}{9}f_s$.

Similarly, when the source is moving away from the observer ($\lambda' = d + d_s$), the observed frequency is given by

$$f_o = \left(\frac{v}{v + v_s}\right)f_s = \left(\frac{1}{1 + \dfrac{v_s}{v}}\right)f_s \qquad \begin{array}{l}\textit{(source moving}\\ \textit{away from a stationary}\\ \textit{observer)}\end{array} \qquad (14.10)$$

Here, f_o is less than f_s. (Why?)

Combining Eqs. 14.9 and 14.10 yields a general equation for the observed frequency with a moving source and a stationary observer:

$$f_o = \left(\frac{v}{v \pm v_s}\right)f_s = \left(\frac{1}{1 \pm \dfrac{v_s}{v}}\right)f_s \qquad \left\{\begin{array}{l}- \textit{ for source moving}\\ \textit{toward stationary observer}\\ + \textit{ for source moving}\\ \textit{away from stationary observer}\end{array}\right. \qquad (14.11)$$

As you might expect, the Doppler effect also occurs with a moving observer and a stationary source, although this situation is a bit different. As the observer moves toward the source, the distance between successive wave crests is the normal wavelength (or $\lambda = v/f_s$), but the measured wave speed is different. Relative to the approaching observer, the sound from the stationary source has a wave speed of $v' = v + v_o$, where v_o is the speed of the observer and v is the speed of sound in still air. (The observer moving toward the source is moving in a direction opposite that of the propagating waves and thus meets more wave crests in a given time.)

With $\lambda = v/f_s$, the observed frequency is then

$$f_o = \frac{v'}{\lambda} = \left(\frac{v + v_o}{v}\right)f_s$$

PHYSLET® ILLUSTRATION

Doppler Effect

or

$$f_o = \left(1 + \frac{v_o}{v}\right)f_s \quad \begin{array}{l} \text{(observed moving} \\ \text{toward a stationary source)} \end{array} \quad (14.12)$$

$$\textit{where} \quad v_o = \textit{speed of observer}$$
$$\textit{and} \quad v = \textit{speed of sound}$$

Similarly, for an observer moving away from a stationary source, the perceived wave speed is $v' = v - v_o$, and

$$f_o = \frac{v'}{\lambda} = \left(\frac{v - v_o}{v}\right)f_s$$

or

$$f_o = \left(1 - \frac{v_o}{v}\right)f_s \quad \begin{array}{l} \text{(observer moving} \\ \text{away from a stationary source)} \end{array} \quad (14.13)$$

Equations 14.12 and 14.13 can be combined into a general equation for a moving observer and a stationary source:

$$f_o = \left(\frac{v \pm v_o}{v}\right)f_s = \left(1 \pm \frac{v_o}{v}\right)f_s \quad \left\{ \begin{array}{l} + \textit{for observer moving} \\ \textit{toward stationary source} \\ - \textit{for observer moving} \\ \textit{away from stationary source} \end{array} \right. \quad (14.14)$$

Example 14.8 ■ On the Road Again: The Doppler Effect

As a truck traveling at 96 km/h approaches and passes a person standing along the highway, the driver sounds the horn. If the horn has a frequency of 400 Hz, what are the frequencies of the sound waves heard by the person (a) as the truck approaches and (b) after it has passed? (Assume that the speed of sound is 346 m/s.)

Thinking It Through. This situation is an application of the Doppler effect, Eq. 14.11, with a moving source and a stationary observer. In such problems, it is important to identify the data correctly.

Solution.

Given: $v_s = 96$ km/h $= 27$ m/s   *Find:* (a) f_o (observed frequency while truck
$\quad\quad\quad f_s = 400$ Hz $\quad\quad\quad\quad\quad\quad\quad$ is approaching)
$\quad\quad\quad v = 346$ m/s $\quad\quad\quad\quad\quad\quad$ (b) f_o (observed frequency while truck
$\quad\quad\quad\quad\quad\quad\quad\quad\quad\quad\quad\quad\quad\quad\quad\quad\quad$ is moving away)

(a) From Eq. 14.11 with a minus sign (source approaching stationary observer),

$$f_o = \left(\frac{v}{v - v_s}\right)f_s = \left(\frac{346 \text{ m/s}}{346 \text{ m/s} - 27 \text{ m/s}}\right)(400 \text{ Hz}) = 434 \text{ Hz}$$

(b) A plus sign is used in Eq. 14.11 when the source is moving away:

$$f_o = \left(\frac{v}{v + v_s}\right)f_s = \left(\frac{346 \text{ m/s}}{346 \text{ m/s} + 27 \text{ m/s}}\right)(400 \text{ Hz}) = 371 \text{ Hz}$$

Follow-up Exercise. Suppose that the observer in this Example were initially moving toward and then past a stationary 400-Hz source at a speed of 96 km/h. What would be the observed frequencies? (Would they differ from those for the moving source?)

There are also cases in which both the source and the observer are moving, either toward or away from one another (see Exercise 102). We will not consider them mathematically here (but will do so conceptually in the next Example).

Conceptual Example 14.9 ■ It's All Relative: Moving Source and Moving Observer

Suppose a sound source and an observer are moving away from one another in opposite directions, each at half the speed of sound in air. Then the observer would (a) receive sound with a frequency higher than the source frequency, (b) receive sound with a frequency lower than the source frequency, (c) receive sound with the same frequency as the source frequency, or (d) receive no sound from the source.

Reasoning and Answer. As we know, when a source moves away from a stationary observer, the observed frequency is lower (Eq. 14.10). Similarly, when an observer moves away from a stationary source, the observed frequency is also lower (Eq. 14.13). With both source and observer moving away from each other in opposite directions, the combined effect would make the observed frequency even less, so neither (a) nor (c) is the answer.

It would appear that (b) is the correct answer, but we must logically eliminate (d) for completeness. Remember that the speed of sound relative to the air is constant. Therefore, (d) would be correct *only if the observer is moving faster than the speed of sound* relative to the air. Since the observer is moving at only half the speed of sound, (b) is the correct answer.

Think about it this way: Regardless of how fast the source is moving, the sound from the source is moving at the speed of sound through the air toward the observer. The observer is moving at only half the speed of sound through the air, so the sound from the source can easily reach and pass the observer. (For a mathematical expression for a moving source and a moving observer, see Exercise 102.)

Follow-up Exercise. What would be the situation if both the source and the observer were traveling in the same direction with the same subsonic speed? (*Subsonic*, as opposed to *supersonic*, refers to a speed that is less than the speed of sound in air.)

Problem-Solving Hint

You may find it difficult to remember whether a plus or minus sign is used in the general equations for the Doppler effect. Let your experience help you. For the common case of a stationary observer, the frequency of the sound increases when the source approaches, so the denominator in Eq. 14.11 must be smaller than the numerator. Accordingly, in this case you use the minus sign. When the source is receding, the frequency is lower. The denominator in Eq. 14.11 must then be larger than the numerator, and you use the plus sign. Similar reasoning will help you choose a plus or minus sign for the numerator in Eq. 14.14.

The Doppler effect also applies to light waves, although the equations describing the effect are different from those just given. When a distant light source such as a star moves away from us, the frequency of the light we receive from it is lowered. That is, the light is shifted toward the red (long wavelength) end of the spectrum, an effect known as a *Doppler red shift*. Similarly, the frequency of light from an object approaching us is increased—the light is shifted toward the blue (short wavelength) end of the spectrum, producing a *Doppler blue shift*. The magnitude of the shift is related to the speed of the source.

The Doppler shift of light from astronomical objects is very useful to astronomers. The rotation of a planet, a star, or some other body can be established by looking at the Doppler shifts of light from opposite sides of the object; because of the rotation, one side is receding (and hence is red shifted) and the other is approaching (and thus is blue shifted). Similarly, the Doppler shifts of light from stars in different regions of our galaxy, the Milky Way, indicate that the galaxy is rotating.

You have been subjected to a practical application of the Doppler effect if you have ever been caught speeding in your car by police radar, which uses reflected radio waves. (*Radar* stands for *ra*dio *d*etecting *a*nd *r*anging and is similar to underwater sonar, which uses ultrasound.) If radio waves are reflected from a parked car, the reflected waves return to the source with the same frequency. But for a car that is moving toward a patrol car, the reflected waves have a higher frequency, or are Doppler shifted. Actually, there is a double Doppler shift: In receiving the wave, the moving car acts like a moving observer (the first Doppler shift), and in reflecting the wave, the car acts like a moving source emitting a wave (the second Doppler shift). The magnitudes of the shifts depend on the speed of the car. A computer quickly calculates this speed and displays it for the police officer.

For other important applications of the Doppler effect, see the Insight on p. 502.

Sonic Booms

Consider a jet plane that can travel at supersonic speeds. As the speed of a moving source of sound approaches the speed of sound, the waves ahead of the source come close together (▼Fig. 14.12a). When a plane is traveling at the speed of sound,

▶ **FIGURE 14.12 Bow waves and sonic booms** **(a)** When an aircraft exceeds the speed of sound in air, v_s, the sound waves form a pressure ridge, or shock wave. As the trailing shock wave passes over the ground, observers hear a sonic boom (actually, two booms, because shock waves are formed at the front and tail of the plane). **(b)** A bullet traveling at a speed of 500 m/s. Note the shock waves produced (and the turbulence behind the bullet). The image was made by using interferometry with polarized light and a pulsed laser, with an exposure time of 20 ns.

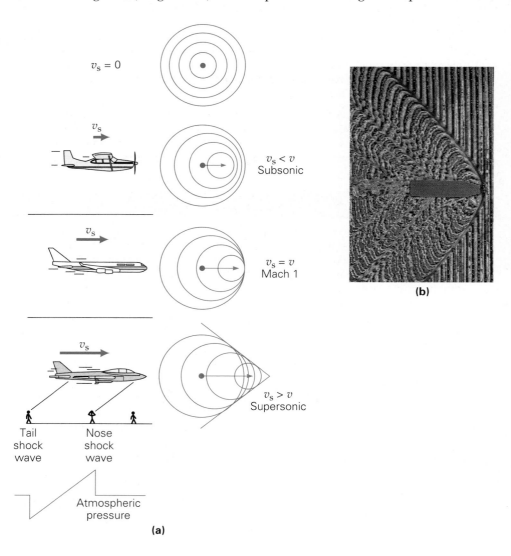

$v_s = 0$

v_s
$v_s < v$
Subsonic

v_s
$v_s = v$
Mach 1

v_s
$v_s > v$
Supersonic

Tail shock wave

Nose shock wave

Atmospheric pressure

(a)

(b)

the waves can't outrun it, and they pile up in front. At supersonic speeds, the waves overlap. This overlapping of a large number of waves produces many points of constructive interference, forming a large pressure ridge, or *shock wave*. This kind of wave is sometimes called a *bow wave* because it is analogous to the wave produced by the bow of a boat moving through water at a speed greater than the speed of the water waves. Figure 14.12b shows the shock wave of a bullet traveling at 500 m/s.

From aircraft traveling at supersonic speed, the shock wave trails out to the sides and downward. When this pressure ridge passes over an observer on the ground, the large concentration of energy produces what is known as a **sonic boom**. There is really a double boom, because shock waves are formed at both ends of the aircraft. Under certain conditions, the shock waves can break windows and cause other damage to structures on the ground. (Sonic booms are no longer heard as frequently as in the past. Pilots are now instructed to fly supersonically only at high altitudes and away from populated areas.)

On a smaller scale, you have probably heard a "mini" sonic boom, the "crack" of a whip. This type of sonic boom is created by the transonic speed of the whip's tip; that is, the speed of the tip changes from subsonic to supersonic (and back).

A common misconception is that a sonic boom is heard only when a plane breaks the sound barrier. As an aircraft approaches the speed of sound, the pressure ridge in front of it is essentially a barrier that must be overcome with extra power. However, once supersonic speed is reached, the barrier is no longer there, and the shock waves, continuously created, trail behind the plane, producing booms along its ground path.

Ideally, the sound waves produced by a supersonic aircraft form a cone-shaped shock wave (▶ Fig. 14.13). The waves travel outward with a speed v, and the speed of the source (plane) is v_s. Note from the figure that the angle between a line tangent to the spherical waves and the line along which the plane is moving is given by

$$\sin \theta = \frac{vt}{v_s t} = \frac{v}{v_s} = \frac{1}{M} \qquad (14.15)$$

The inverse ratio of the speeds is called the **Mach number** (M), named after Ernst Mach (1838–1916), an Austrian physicist, who used it in studying supersonics, and is given by

$$M = \frac{v_s}{v} = \frac{1}{\sin \theta} \qquad (14.16)$$

If v equals v_s, the plane is flying at the speed of sound, and the Mach number is 1 (i.e., $v_s/v = 1$). Therefore, a Mach number less than 1 indicates a subsonic speed, and a Mach number greater than 1 indicates a supersonic speed. In the latter case, the Mach number tells the speed of the aircraft in terms of a multiple of the speed of sound. A Mach number of 2, for instance, indicates a speed twice the speed of sound. Note that since $\sin \theta \le 1$, no shock wave can exist unless $M \ge 1$.

14.6 Musical Instruments and Sound Characteristics

OBJECTIVE: To explain some of the sound characteristics of musical instruments in physical terms.

Musical instruments provide good examples of standing waves and boundary conditions. On some stringed instruments, different notes are produced by using finger pressure to vary the lengths of the strings (▶ Fig. 14.14). As learned in Chapter 13, the natural frequencies of a stretched string (fixed at each end, as is the case

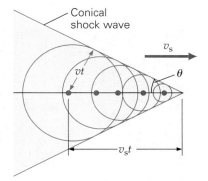

▲ **FIGURE 14.13 Shock wave cone and Mach number**   When the speed of the source (v_s) is greater than the speed of sound in air (v), the interfering spherical sound waves form a conical shock wave that appears as a V-shaped pressure ridge when viewed in two dimensions. The angle θ is given by $\sin \theta = v/v_s$, and the inverse ratio v_s/v is called the Mach number.

▲ **FIGURE 14.14 A shorter vibrating string, a higher frequency**   Different notes are produced on stringed instruments such as guitars, violins, or cellos by placing a finger on a string to change its effective, or vibrating, length.

INSIGHT

Doppler Applications: Blood Cells and Raindrops

Blood cells Besides its well-known role of producing an image of a fetus (Fig. 14.3), ultrasound provides a variety of other uses in the medical field. Since the Doppler effect can detect and provide information on moving objects, it can be used to examine blood flow in the major arteries and veins in the arms and legs (Fig. 1). The reflectors here are red blood cells. The tests provide physicians information that helps them diag-

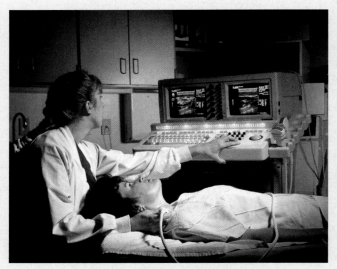

FIGURE 1 Medical application of the Doppler effect The Doppler effect is used to examine blood flow in major arteries and veins, here the carotid artery in the neck. Turbulence in the blood flow can be detected, which might reveal a narrowing of blood vessels, clots, or aneurysms (abnormal blood-filled dilatation of blook vessels).

nose such things as blood clots, arterial occlusion (closing), and venous insufficiency. Ultrasound procedures offer a less invasive alternative to other diagnostic procedures, such as arteriography (X-ray pictures of an artery after the injection of a dye).

Another medical use of ultrasound is the electrocardiogram, which is an examination of the heart. On a monitor, this ultrasonic procedure can display the beating movements of the heart, and the physician can see the heart's chambers, valves, and blood flow as it makes its way in and out of the organ.

While we are focusing on the body and sound, here is something for you to try: In a quiet room, put both thumbs in your ears firmly and listen. Do you hear a low pulsating sound? Why is this? Would you believe that you are hearing the sound, at about 25 Hz, made by the contracting and relaxing of the muscle fibers in your hands and arms? Although in the audible range, these sounds are not normally heard, because the human ear is relatively insensitive to low-frequency sounds.

Raindrops Radar has been used since the early 1940s to provide information about rainstorms and other forms of precipitation. This information is obtained from the intensity of the reflected signal. Such conventional radars can also detect the hooked (rotational) "signature" of a tornado, but only after the storm is well developed.

A major improvement in weather forecasting came about with the development of a radar system that could measure the Doppler frequency shift in addition to the magnitude of the echo signal reflected from precipitation (usually raindrops). The Doppler shift is related to the velocity of the precipitation blown by the wind.

A Doppler-based radar system (Fig. 2a) can penetrate a storm and monitor its wind speeds. The direction of a storm's

for the strings on an instrument) are $f_n = n(v/2L)$, from Eq. 13.20, where the speed of the wave in the string is given by $v = \sqrt{F_T/\mu}$. Initially adjusting the tension in a string tunes it to a particular (fundamental) frequency. Then the effective length of the string is varied by finger location and pressure.

Standing waves are also set up in wind instruments. For example, consider a pipe organ with fixed lengths of pipe, which may be open or closed (▶Fig. 14.15). An open pipe is open at both ends, and a closed pipe is closed at one end and open at the other (the end with the antinode). Analysis similar to that done in Chapter 13 for a stretched string with the proper boundary conditions shows that the natural frequencies of the pipes are

$$f_n = \frac{v}{\lambda_n} = n\left(\frac{v}{2L}\right) = nf_1 \qquad n = 1, 2, 3, \ldots \qquad \text{(natural frequencies for a pipe open on both ends)} \quad (14.17)$$

and

$$f_m = \frac{v}{\lambda_m} = m\left(\frac{v}{4L}\right) = mf_1 \qquad m = 1, 3, 5, \ldots \qquad \text{(natural frequencies for a pipe closed on both ends)} \quad (14.18)$$

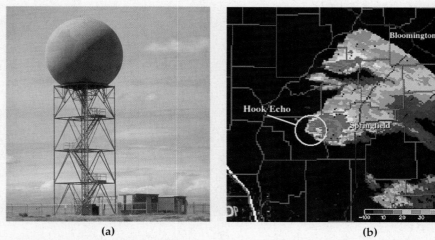

FIGURE 2   Doppler radar   (a) A Doppler radar installation. **(b)** Doppler radar depicts the precipitation inside a thunderstorm. A hook echo is a signature of a possible tornado.

wind-driven rain gives a wind "field" map of the affected region. Such maps provide strong clues of developing tornadoes, so meteorologists can detect them much earlier than was ever before possible (Fig. 2b). With Doppler radar, forecasters have been able to predict tornadoes as much as 20 min before they touch down, as compared with just over 2 min for conventional radar. Doppler radar has saved many lives with this increased warning time. The National Weather Service has a network of Doppler radars around the United States, and Doppler radar scans are now common on both TV weather forecasts and the Internet.

Doppler radars installed at major airports have another use. Several airplane crashes and near crashes have been attributed to downward wind bursts (also known as microbursts and downbursts). Such strong down drafts cause wind shears capable of forcing landing aircraft to crash.

Wind bursts generally result from high-speed downdrafts in the turbulence of thunderstorms, but they can also occur in clear air when rain evaporates high above ground. Since Doppler radar can detect the wind speed and the direction of raindrops in clouds, as well as dust and other objects floating in the air, it can provide an early warning against dangerous wind shear conditions. Two or three radar sites are needed to detect motions in two or three directions (dimensions), respectively.

Related Exercises: 58, 59

where v is the speed of sound in air. Note that the natural frequencies depend on the length of the pipe. This is an important consideration in a pipe organ (Fig. 14.15c), particularly in selecting the dominant or fundamental frequency. (The diameter of the pipe is also a factor, but is not considered in this simple analysis.)

The same physical principles apply to wind and brass instruments. In all of these, the human breath is used to create standing waves in an open tube. Most such instruments allow the player to vary the effective length of the tube and thus the pitch produced—either with the help of slides or valves that vary the actual length of tubing in which the air can resonate, as in most brasses, or by opening and closing holes in the tube, as in woodwinds (▶Fig. 14.16).

Recall from Chapter 13 that a musical note or tone is referenced to the fundamental vibrational frequency of an instrument. In musical terms, the first overtone is the second harmonic, the second overtone is the third harmonic, and so on. Note that for a closed organ pipe (Eq. 14.18), the even harmonics are missing.

Example 14.10 ■ Pipe Dreams: Fundamental Frequency

A particular open organ pipe has a length of 0.653 m. Taking the speed of sound in air to be 345 m/s, what is the fundamental frequency of this pipe?

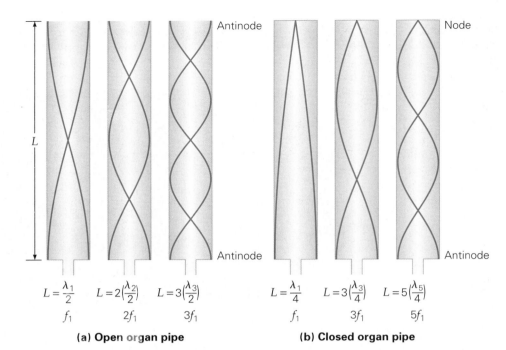

$L = \dfrac{\lambda_1}{2}$ $L = 2\left(\dfrac{\lambda_2}{2}\right)$ $L = 3\left(\dfrac{\lambda_3}{2}\right)$ $L = \dfrac{\lambda_1}{4}$ $L = 3\left(\dfrac{\lambda_3}{4}\right)$ $L = 5\left(\dfrac{\lambda_5}{4}\right)$

f_1 $2f_1$ $3f_1$ f_1 $3f_1$ $5f_1$

(a) Open organ pipe   **(b) Closed organ pipe**

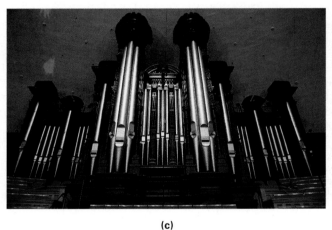

(c)

▶ **FIGURE 14.15   Organ pipes**
Longitudinal standing waves
(illustrated here as sinusoidal
curves) are formed in vibrating air
columns in pipes. **(a)** An open pipe
has antinodes at both ends. **(b)** A
closed pipe has a closed (node) end
and an open (antinode) end. **(c)** A
modern pipe organ. The pipes can
be open or closed.

Thinking It Through.   The fundamental frequency ($n = 1$) of an open pipe is given
directly by Eq. 14.17. Physically, there is a half-wavelength ($\lambda/2$) in the length of the
pipe, so $\lambda = 2L$.

Solution.

Given: $L = 0.653$ m   *Find:* f_1 (fundamental frequency)
            $v = 345$ m/s (speed of sound)

With $n = 1$,

$$f_1 = \frac{v}{2L} = \frac{345 \text{ m/s}}{2(0.653 \text{ m})} = 264 \text{ Hz}$$

This frequency is middle C (C_4).

Follow-up Exercise.   A closed organ pipe has a fundamental frequency of 256 Hz.
What would be the frequency of its first overtone? Is this frequency audible?

___

Perceived sounds are described by terms whose meanings are similar to those
used to describe the physical properties of sound waves. Physically, a wave is gen-

erally characterized by intensity, frequency, and waveform (harmonics). The corresponding terms used to describe the sensations of the ear are loudness, pitch, and quality (or timbre). These general correlations are shown in Table 14.3. However, the correspondence is not perfect. The physical properties are objective and can be measured directly. The sensory effects are subjective and vary from person to person. (Think of temperature as measured by a thermometer and by the sense of touch.)

Sound intensity and its measurement on the decibel scale were covered in Section 14.3. Loudness is related to intensity, but the human ear responds differently to sounds of different frequencies. For example, two tones with the same intensity (in watts per square meter), but different frequencies, might be judged by the ear to be different in loudness.

Frequency and *pitch* are often used synonymously, but again there is an objective–subjective difference: If the same low-frequency tone is sounded at two intensity levels, most people will say that the more intense sound has a lower pitch, or perceived frequency.

The curves in the graph of intensity level versus frequency shown in ▾Fig. 14.17 are called *equal-loudness contours* (or Fletcher–Munson curves, after the researchers who generated them). These contours join points representing intensity–frequency combinations that a person with average hearing judges to be equally loud. The top curve shows that the decibel level of the threshold of pain (120 dB) does not vary a great deal over the normal hearing range, regardless of the frequency of the sound. In contrast, the threshold of hearing, represented by the lowest contour, varies widely with frequency. For a tone with a frequency of 2000 Hz, the threshold of hearing is 0 dB, but a 20-Hz tone would have to have an intensity level of over 70 dB (the extrapolated *y*-intercept of the lowest curve) just to be heard.

It is interesting to note the dips (or minima) in the curves. These indicate that the ear is most sensitive to sounds with frequencies around 4000 Hz and 12 000 Hz. A tone with a frequency of 4000 Hz can be heard at intensity levels *below* 0 dB. The minima occur as a result of resonance in a closed cavity in the auditory canal (similar to a closed pipe). The length of the cavity is such that it has a fundamental resonance frequency of about 4000 Hz, resulting in extra sensitivity. As in a closed cavity, the next natural frequency is the third harmonic (see Eq. 14.18), which is three times the fundamental frequency, or about 12 000 Hz.

▾ **FIGURE 14.17 Equal-loudness contours** The curves indicate tones that are judged to be equally loud, although they have different frequencies and intensity levels. For example, on the lowest contour, a 1000-Hz tone at 0 dB sounds as loud as a 50-Hz tone at 40 dB. Note that the frequency scale is logarithmic to compress the large frequency range.

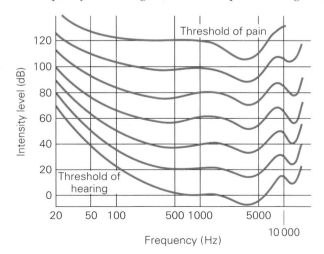

(a)

All holes covered

First five holes covered   Higher *f*

First three holes covered   Still higher *f*

$$f \mu \frac{1}{L}$$

(b)

▴ **FIGURE 14.16 Wind instruments** (a) Wind instruments are essentially open tubes. (Here, Kenny G performs on a soprano saxophone at a Macy's Thanksgiving Day Parade.) (b) The effective length of the air column, and hence the pitch of the sound, is varied by opening and closing holes along the tube. The frequency *f* is inversely proportional to the effective length *L* of the air column.

TABLE 14.3 General Correlation between Perceptual and Physical Characteristics of Sound	
Sensory Effect	Physical Wave Property
Loudness	Intensity
Pitch	Frequency
Quality (timbre)	Waveform (harmonics)

The **quality** of a tone is the characteristic that enables it to be distinguished from another tone of basically the same intensity and frequency. Tone quality depends on the waveform—specifically, the number of harmonics (overtones) present and their relative intensities (▼Fig. 14.18). The tone of a voice depends in large part on the vocal resonance cavities. One person can sing a tone with the same basic frequency and intensity as another, but different combinations of overtones give the two voices different qualities.

The notes of a musical scale correspond to certain frequencies; as we saw in Example 14.10, middle C (C_4) has a frequency of 264 Hz. When a note is played on an instrument, its assigned frequency is that of the first harmonic, which is the fundamental frequency. (The second harmonic is the first overtone, the third harmonic is the second overtone, and so on.) The fundamental frequency is dominant over the accompanying overtones that determine the sound quality of the instrument. Recall from Chapter 13 that the overtones which are produced depend on how an instrument is played. Whether a violin string is plucked or bowed, for example, can be discerned from the quality of identical notes.

▶ **FIGURE 14.18 Waveform and quality** (a) The superposition of sounds of different frequencies and amplitudes gives a complex waveform. The harmonics, or overtones, determine the quality of the sound. (b) The waveform of a violin tone is displayed on an oscilloscope.

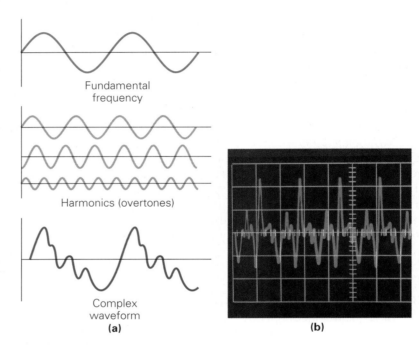

Fundamental frequency

Harmonics (overtones)

Complex waveform

(a)

(b)

Chapter Review

Important Concepts and Equations

- The sound frequency spectrum is divided into infrasonic ($f < 20$ Hz), audible (20 Hz $< f < 20$ kHz), and ultrasonic ($f > 20$ kHz) frequency regions.

- The speed of sound in a medium depends on the elasticity of the medium and its density. In general, $v_{solids} > v_{liquids} > v_{gases}$.

Speed of sound in air (meters per second):

$$v = (331 + 0.6T_C) \text{ m/s} \qquad (14.1)$$

- The intensity of a point source is inversely proportional to the square of the distance from the source.

Intensity of a point source:

$$I = \frac{P}{4\pi R^2} \quad \text{and} \quad \frac{I_2}{I_1} = \left(\frac{R_1}{R_2}\right)^2 \qquad (14.2, 14.3)$$

- The sound intensity level is a logarithmic function of the sound intensity and is expressed in decibels (dB).

Intensity level (in decibels, dB):

$$\beta = 10 \log \frac{I}{I_o} \quad \text{where} \quad I_o = 10^{-12} \text{ W/m}^2 \qquad (14.4)$$

- Sound wave interference of two point sources depends on phase difference as related to path-length difference. Sound waves that arrive at a point in phase reinforce each other (constructive interference); sound waves that arrive at a point out of phase cancel each other (destructive interference).

Phase difference (where ΔL is the path-length difference):

$$\Delta\theta = \frac{2\pi}{\lambda}(\Delta L) \qquad (14.5)$$

Condition for constructive interference:

$$\Delta L = n\lambda \qquad (n = 0, 1, 2, 3, \ldots) \qquad (14.6)$$

Condition for destructive interference:

$$\Delta L = m\left(\frac{\lambda}{2}\right) \qquad (m = 1, 3, 5, \ldots) \qquad (14.7)$$

Beat frequency:

$$f_b = |f_1 - f_2| \qquad (14.8)$$

- The Doppler effect depends on the velocities of the sound source and observer relative to still air. When the relative motion of the source and observer is toward each other, the observed pitch increases; when the relative motion of source and observer is away from each other, the observed pitch decreases.

Doppler effect:

$$f_o = \left(\frac{v}{v \pm v_s}\right)f_s = \left(\frac{1}{1 \pm \dfrac{v_s}{v}}\right)f_s \qquad (14.11)$$

$$\text{where} \quad v_s = speed\ of\ source$$
$$\text{and} \quad v = speed\ of\ sound$$

$\begin{cases} -for\ source\ moving\ toward\ stationary\ observer \\ +for\ source\ moving\ away\ from\ stationary\ observer \end{cases}$

$$f_o = \left(\frac{v \pm v_o}{v}\right)f_s = \left(1 \pm \frac{v_o}{v}\right)f_s \qquad (14.14)$$

$$\text{where} \quad v_o = speed\ of\ observer$$
$$\text{and} \quad v = speed\ of\ sound$$

$\begin{cases} +for\ observer\ moving\ toward\ stationary\ source \\ -for\ observer\ moving\ away\ from\ stationary\ source \end{cases}$

Angle for conical shock wave:

$$\sin\theta = \frac{vt}{v_s t} = \frac{v}{v_s} = \frac{1}{M} \qquad (14.15)$$

Mach number:

$$M = \frac{v_s}{v} = \frac{1}{\sin\theta} \qquad (14.16)$$

Natural frequencies of organ pipe open on both ends:

$$f_n = n\left(\frac{v}{2L}\right) = nf_1 \qquad (n = 1, 2, 3, \ldots) \qquad (14.17)$$

Natural frequencies of organ pipe closed on one end:

$$f_m = m\left(\frac{v}{4L}\right) = mf_1 \qquad (m = 1, 3, 5, \ldots) \qquad (14.18)$$

Exercises

14.1 Sound Waves
and
14.2 The Speed of Sound

1. A sound wave with a frequency of 15 Hz is in what region of the sound spectrum? (a) audible, (b) infrasonic, (c) ultrasonic, or (d) supersonic.

2. A sound wave in air (a) is longitudinal, (b) is transverse, (c) has longitudinal and transverse components, or (d) travels faster than a sound wave through a liquid.

3. The speed of sound is generally greatest in (a) solids, (b) liquids, (c) gases, or (d) a vacuum.

4. The speed of sound in air (a) is about 1/3 km/s, (b) is about 1/5 mi/s, (c) depends on temperature, or (d) all of these.

5. CQ Suggest a possible explanation of why some flying insects produce buzzing sounds and some do not.

6. CQ Explain why sound travels faster in warmer than in colder air.

7. CQ Two sounds that differ in frequency are emitted from a single loudspeaker. Which sound will reach your ear first, the one with the lower or the higher frequency?

8. CQ The speed of sound in air depends on temperature. What effect, if any, should humidity have?

9. CQ When dogs sleep, they usually put their ear on the floor. Why do you suppose this is so? Is it related to people putting their ears on railroad tracks in Western movies?

10. ■ What is the speed of sound in air at (a) 10°C and (b) 20°C?

11. ■ The speed of sound in air on a summer day is 350 m/s. What is the air temperature?

12. ■ The thunder from a lightning flash is heard by an observer 3.0 s after she sees the flash. What is the approximate distance to the lightning strike in (a) kilometers and (b) miles?

13. ■ Sonar is used to map the ocean floor. If an ultrasonic signal is received 2.0 s after it is emitted, how deep is the ocean floor at that location?

14. ■ The wave speed in a liquid is given by $v = \sqrt{B/\rho}$, where B is the bulk modulus of the liquid and ρ is its density. Show that this equation is dimensionally correct. What about $v = \sqrt{Y/\rho}$ for a solid? (Y is Young's modulus.)

15. IE ■■ A tuning fork vibrates at a frequency of 256 Hz. (a) When the air temperature increases, the wavelength of the sound from the tuning fork (1) increases, (2) remains the same, or (3) decreases. Why? (b) If the temperature rises from 0°C to 20°C, what is the change in the wavelength?

16. ■■ Particles approximately 3.0×10^{-2} cm in diameter are to be scrubbed loose from machine parts in an aqueous ultrasonic cleaning bath. Above what frequency should the bath be operated to produce wavelengths of this size and smaller?

17. ■■ Medical ultrasound uses a frequency of around 20 MHz to diagnose human conditions and ailments. (a) If the speed of sound in tissue is 1500 m/s, what is the smallest detectable object? (b) If the penetration depth is about 200 wavelengths, how deep can this instrument penetrate?

18. ■■ Brass is an alloy of copper and zinc. Does the addition of zinc to copper cause an increase or a decrease in the speed of sound in brass rods compared to copper rods? Explain.

19. ■■ The speed of sound in steel is about 4.5 km/s. A steel rail is struck with a hammer, and an observer 0.30 km away has one ear to the rail. (a) How much time will elapse from the time the sound is heard through the rail until the time it is heard through the air? Assume that the air temperature is 20°C and that no wind is blowing. (b) How much time would elapse if the wind were blowing toward the observer at 36 km/h from where the rail was struck?

20. ■■ At a baseball game on a cool day (with an air temperature of 16°C), a fan hears the crack of the bat 0.25 s after observing the batter hit the ball. How far is the fan from home plate?

21. ■■ A 2000-Hz tone is sounded when the air temperature is 20°C and then again at 10°C. What is the percentage change in the wavelength of the sound between the higher and the lower temperature?

22. ■■ A person holds a rifle horizontally and fires at a target. The bullet has a muzzle speed of 200 m/s, and the person hears the bullet strike the target 1.00 s after firing it. The air temperature is 72°F. What is the distance to the target?

23. ■■ A freshwater dolphin sends ultrasonic sound to locate a prey. If the echo off the prey is received by the dolphin 0.12 second after being sent, how far is the prey from the dolphin?

24. ■■ The size of your eardrum (the tympanum; see Fig. 2 in the Insight on page 486) partially determines the upper frequency limit of your audible region, usually between 16 000 Hz and 20 000 Hz. If the wavelength is on the order of twice the diameter of the eardrum and the air temperature is 20°C, how wide is your eardrum? Is your answer reasonable?

25. IE ■■■ On hiking up a mountain that has several overhanging cliffs, a climber drops a stone at the first cliff to determine its height by measuring the time it takes to hear the stone hit the ground. (a) At a second cliff that is twice the height of the first, the measured time of the sound from the dropped stone is (1) less than double, (2) double, or (3) more than double that of the first. Why? (b) If the measured time is 4.8 s for the stone dropping from the first cliff, and the air temperature is 20°C, how high is the cliff? (c) If the height of a third cliff is three times that of the first one, what would be the measured time for a stone dropped there to reach the ground?

26. ■■■ Sound propagating through air at 30°C passes through a vertical cold front into air that is 4.0°C. If the sound has a frequency of 2400 Hz, by what percentage does its wavelength change in crossing the boundary?

14.3 Sound Intensity and Sound Intensity Level

27. If the air temperature increases, would the sound intensity from a constant-output point source (a) increase, (b) decrease, or (c) remain unchanged?

28. The decibel scale is referenced to a standard intensity of (a) 1.0 W/m², (b) 10^{-12} W/m², (c) normal conversation, or (d) the threshold of pain.

29. CQ The Richter scale, used to measure the intensity level of earthquakes, is a logarithmic scale, as is the decibel scale. Why are such scales used?

30. **CQ** Can there be negative decibel levels, such as −10 dB? If so, what would these mean?

31. ■ Assuming that the diameter of your eardrum is 1 cm (see Exercise 24), what is the sound power received by the eardrum at the threshold of (a) hearing and (b) pain?

32. ■ Calculate the intensity generated by a 1.0-W point source of sound at a location (a) 3.0 m and (b) 6.0 m from it.

33. **IE** ■ (a) If the distance from a point sound source triples, the sound intensity will be (1) 3, (2) 1/3, (3) 9, or (4) 1/9 times the original value. Why? (b) By how many times must the distance from a point source be increased to reduce the sound intensity by half?

34. ■ Calculate the intensity level for (a) the threshold of hearing and (b) the threshold of pain.

35. ■ Find the intensity levels in decibels for sounds with intensities of (a) $10^{-2}\,\text{W/m}^2$, (b) $10^{-6}\,\text{W/m}^2$, and (c) $10^{-15}\,\text{W/m}^2$.

36. ■ If the intensity of one sound is $10^{-4}\,\text{W/m}^2$ and the intensity of another is $10^{-2}\,\text{W/m}^2$, what is the difference in their intensity levels?

37. **IE** ■■ (a) If the power of a sound source doubles, the intensity level at a certain distance from the source (1) increases, (2) exactly doubles, or (3) decreases. Why? (b) What are the intensity levels at a distance of 10 m from a 5.0-W and a 10-W source, respectively?

38. ■■ What is the intensity of a sound that has an intensity level of (a) 50 dB or (b) 90 dB? (a)

39. ■■ Noise levels for some common aircraft are given in Table 14.4. What are the lowest and highest intensities for (a) takeoff and (b) landing?

TABLE 14.4 Takeoff and Landing Noise Levels for Some Common Commercial Jet Aircraft*

Aircraft	Takeoff noise (dB)	Landing noise (dB)
737	85.7–97.7	99.8–105.3
747	89.5–110.0	103.8–107.8
DC-10	98.4–103.0	103.8–106.6
L-1011	95.9–99.3	101.4–102.8

*Noise level readings are taken from 198 m (650 ft). The range depends on the aircraft model and the type of engine used.

40. **IE** ■■ If the distance to a sound source is halved, (a) will the sound intensity level change by a factor of (1) 2, (2) 1/2, (3) 4, (4) 1/4, or (5) none of the above? Why? (b) What is the change in the sound intensity level?

41. ■■ What is the intensity level of a 23-dB sound after being amplified (a) 10 thousand times, (b) a million times, (c) a billion times?

42. ■■ A tape player has a signal-to-noise ratio of 53 dB. How many times larger is the intensity of the signal than that of the background noise?

43. ■■ The sound intensity levels for a machine shop and a quiet library are 90 dB and 40 dB, respectively. (a) What is each intensity? (b) How many times greater is the intensity of the sound in the machine shop than that in the library?

44. **IE** ■■ A dog's bark has a sound intensity level of 40 dB. (a) If two of the same dogs are barking, the intensity level is (1) less than 40 dB, (2) between 40 dB and 80 dB, or (3) 80 dB. (b) What would be the intensity level?

45. ■■ At a rock concert, the average sound intensity level for a person in a front-row seat is 110 dB for a single band. If all the bands scheduled to play produce sound of that same intensity, how many of them would have to play simultaneously for the sound level to be at or above the threshold of pain?

46. ■■ At a distance of 10.0 m from a point source, the intensity level is measured to be 70 dB. At what distance from the source will the intensity level be 40 dB?

47. ■■ At a 4th of July celebration, a firecracker explodes (▼Fig. 14.19). Considering the firecracker to be a point source, what are the intensities heard by observers at points B, C, and D, relative to that heard by the observer at A?

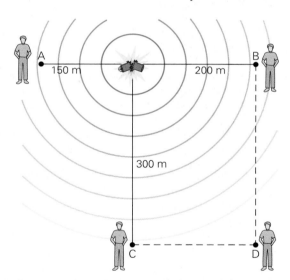

▲ **FIGURE 14.19 A big bang** See Exercise 47.

48. ■■ A person standing 4.0 m from a wall shouts such that the sound strikes the wall with an intensity of $2.5 \times 10^{-4}\,\text{W/m}^2$. Assuming that the wall absorbs 20% of the incident energy and reflects the rest, what is the sound intensity level just before and after the sound is reflected?

49. ■■ A gas-powered lawn mower is rated at 95 dB. (a) What is the sound intensity for this mower? (b) How

many times more intense is the sound of this mower than that of an electric-powered mower rated at 83 dB?

50. ■■■ A 1000-Hz tone from a loudspeaker has an intensity level of 100 dB at a distance of 2.5 m. If the speaker is assumed to be a point source, how far from the speaker will the sound have intensity levels (a) of 60 dB and (b) barely high enough to be heard?

51. ■■■ A bee produces a buzzing sound that is barely audible to a person 3.0 m away. How many bees would have to be buzzing at that distance to produce a sound with an intensity level of 50 dB?

14.4 Sound Phenomena
and
14.5 The Doppler Effect

52. Beats are the direct result of (a) interference, (b) refraction, (c) diffraction, or (d) the Doppler effect.

53. Police radar makes use of (a) beats, (b) Doppler effect, (c) interference, or (d) sonic boom.

54. CQ Do interference beats have anything to do with the "beat" of music? Explain.

55. CQ (a) Is there a Doppler effect if a sound source and an observer are moving with the same velocity? (b) What would be the effect if a moving source accelerated toward a stationary observer?

56. CQ How fast would a "jet fish" have to swim to create an aquatic sonic boom?

57. CQ As a person walks *in between* a pair of loudspeakers that produce tones of the same amplitude and frequency, he hears a varying sound intensity. Explain.

58. CQ How can a Doppler radar used in weather forecasting measure both the location and motion of the clouds?

59. CQ In Chapter 20, you will learn that red light has a lower frequency than blue light. If an orange star appears red to us (this is the famous redshift in astronomy), what can you conclude about the motion of the star? Also, propose a method of measuring the velocity of the star.

60. ■ Two sound waves with the same wavelength, 0.50 m, arrive at a point after having traveled (a) 2.50 m and 3.75 m and (b) 3.25 m and 8.25 m, respectively. What type of interference occurs in each case?

61. ■ Two adjacent point sources, A and B, are directly in front of an observer and emit identical 1000-Hz tones. To what closest distance behind source B would source A have to be moved for the observer to hear no sound? (Assume that the air temperature is 20°C and ignore the falling off of intensity with distance.)

62. ■ A violinist and a pianist simultaneously sound notes with frequencies of 436 Hz and 440 Hz, respectively. What beat frequency will the musicians hear?

63. IE ■ A violinist tuning her instrument to a piano note of 264 Hz detects three beats per second. (a) The frequency of the violin could be (1) less than 264 Hz, (2) equal to 264 Hz, (3) greater than 264 Hz, or (4) both (1) and (3). Why? (b) What are the possible frequencies of the violin tone?

64. ■ What is the frequency heard by a person driving 50 km/h directly toward a factory whistle ($f = 800$ Hz) if the air temperature is 0°C?

65. IE ■ On a day with a temperature of 20°C and no wind blowing, the frequency heard by a moving person from a 500-Hz stationary siren is 520 Hz. (a) The person is (1) moving toward, (2) moving away from, or (3) stationary relative to the siren. Why? (b) What is the person's speed?

66. ■■ While standing near a railroad crossing, you hear a train horn. The frequency emitted by the horn is 400 Hz. If the train is traveling at 90.0 km/h and the air temperature is 25°C, what is the frequency you hear (a) when the train is approaching and (b) after it has passed?

67. ■■ Two identical strings on different cellos are tuned to the 440-Hz A note. The peg holding one of the strings slips, so its tension is decreased by 1.5%. What is the beat frequency heard when the strings are then played together?

68. ■■ How fast, in kilometers per hour, must a sound source be moving toward you to make the observed frequency 5.0% greater than the true frequency? (Assume that the speed of sound is 340 m/s.)

69. ■■ What is the half-angle of the shock wave of a jet aircraft just as it breaks the sound barrier?

70. IE ■■ On transatlantic flights, the supersonic transport (SST) Concorde flies at a speed of Mach 1.5. (a) If the Concorde were to fly faster than Mach 1.5, the half-angle of the conical shock wave would (1) increase, (2) remain the same, or (3) decrease. Why? (b) What is the half-angle of the conical shock wave formed by the Concorde at Mach 1.5?

71. ■■ The half-angle of the conical shock wave formed by a supersonic jet is 35°. What are (a) the Mach number of the aircraft and (b) the actual speed of the aircraft if the air temperature is −20°C?

72. ■■■ Two point-source loudspeakers are a certain distance apart, and a person stands 12.0 m in front of one of them on a line perpendicular to the baseline of the speakers. If the speakers emit identical 1000-Hz tones, what is their minimum nonzero separation so that the observer hears little or no sound? (Take the speed of sound to be exactly 340 m/s.)

73. ■■■ A bystander hears a siren vary in frequency from 476 Hz to 404 Hz as a fire truck approaches, passes by, and moves away on a straight street (▼Fig. 14.20). What is the speed of the truck? (Take the speed of sound in air to be 343 m/s.)

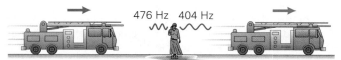

476 Hz   404 Hz

▲ **FIGURE 14.20   The siren's wail**   See Exercise 73.

74. ■■■ Bats emit sounds of frequencies around 35.0 kHz and use echolocation to find their prey. If a bat is moving with a speed of 12.0 m/s toward an insect at an air temperature of 20.0°C, (a) what frequency is heard by the insect? (b) What frequency is heard by the bat from the reflected sound? (c) Would the speed of the bat affect the answers?

14.6 Musical Instruments and Sound Characteristics

75. The human ear can hear tones best at (a) 1000 Hz, (b) 4000 Hz, (c) 6000 Hz, or (d) all frequencies.

76. The quality of sound depends on its (a) waveform, (b) frequency, (c) speed, or (d) intensity.

77. CQ (a) After a snowfall, why does it seem particularly quiet? (b) Why do empty rooms sound hollow? (c) Why do people's voices sound fuller or richer when they sing in the shower?

78. CQ Why aren't the frets on a guitar evenly spaced?

79. CQ Is it possible for an open organ pipe and an organ pipe closed at one end, each of the same length, to produce notes of the same frequency? Justify your answer.

80. CQ When you blow across the top of a bottle with water in it, why does the frequency of the sound increase with increasing levels of water?

81. CQ Why are there no even harmonics in a pipe that is closed on one end?

82. ■ The first three natural frequencies of an organ pipe are 126 Hz, 378 Hz, and 630 Hz. (a) Is the pipe an open or a closed one? (b) (b) Taking the speed of sound in air to be 340 m/s, find the length of the pipe.

83. ■ A closed organ pipe has a fundamental frequency of 528 Hz (a C note) at 20°C. What is the fundamental frequency of the pipe when the temperature is 0°C?

84. ■ The human ear canal is about 2.5 cm long. It is open at one end and closed at the other. (See Fig. 2 in the Insight on p. 495.) (a) What is the fundamental frequency of the

ear canal at 20°C? (b) To what frequency is the ear most sensitive? (c) If a person's ear canal is longer than 2.5 cm, is the fundamental frequency higher or lower than that in (a)? Explain.

85. ■■ An organ pipe that is closed at one end has a length of 0.90 m. At 20°C, what is the distance between a node and an adjacent antinode for (a) the second harmonic and (b) the third harmonic?

86. ■■ An open organ pipe and an organ pipe that is closed at one end both have lengths of 0.52 m at 20°C. What is the fundamental frequency of each pipe?

87. IE ■■ When all of its holes are closed, a flute is essentially a tube that is open at both ends, with the length being from the mouthpiece to the far end (as in Fig. 14.16b). If a hole is open, then the length of the tube is effectively from the mouthpiece to the hole. (a) Is the position at the mouthpiece (1) a node, (2) an antinode, or (c) neither a node nor an antinode? Why? (b) If the lowest fundamental frequency on a flute is 262 Hz, what is the minimum length of the flute at 20°C? (c) If a note of frequency 440 Hz is to be played, which hole should be open? Express your answer as a distance from the hole to the mouthpiece.

88. ■■ A tuning fork with a frequency of 440 Hz is held above a resonance tube that is partially filled with water. Assuming that the speed of sound in air is 342 m/s, for what three smallest heights of the air column will resonance occur? Draw the longitudinal standing waves in the air columns for each height. [*Hint*: For resonance to occur, the frequency of the tuning fork must match that of the tube.]

89. ■■■ An organ pipe that is closed at one end is filled with helium. The pipe has a fundamental frequency of 660 Hz in air at 0°C. What is the pipe's fundamental frequency with the helium in it?

Additional Exercises

90. Two identical sources producing 440-Hz tones are located 6.97 m and 8.90 m, respectively, from an observation point. If the air temperature that day is 15°C, how do the waves interfere at the point?

91. If a person standing 30.0 m from a 550-Hz point source moves 5.0 m closer to the source, by what factor does the sound intensity change?

92. An office in an e-commerce company has 50 computers, which generate a sound intensity level of 40 dB (from the keyboards). The office manager tries to cut the noise to half as loud by removing 25 computers. Does he achieve his goal? What is the intensity level generated by 25 computers?

93. How fast must an observer be moving toward a 500-Hz siren so that she hears double the frequency?

94. The intensity levels of two people holding a conversation are 60.0 dB and 65.0 dB, respectively. What is the intensity level of the combined sounds?

95. The intensity level of a sound is 90 dB at a certain distance from its source. How much energy falls on an eardrum with a diameter of 1.0 cm in 5.0 s?

96. A jet flies at a speed of Mach 2.0. What is the half-angle of the conical shock wave formed by the aircraft? Can you tell the speed of the shock wave?

97. The note A (440 Hz) is produced by a stringed musical instrument. The air temperature is 20°C. Approximately how many vibrations does the string make before the sound reaches a person 30 m away?

98. A person drops a stone into a deep well and hears it splash 3.16 s later. How deep is the well? (Assume the air temperature in the well to be 10°C.)

99. The frequency of an ambulance siren is 700 Hz. What are the frequencies heard by a stationary pedestrian as the ambulance approaches and passes her at a speed of 90.0 km/h? (Assume that the air temperature is 20°C).

100. One hunter sees another who is 300 km away fire his rifle. (Smoke comes out of the barrel.) If the air temperature is 5°C, how long will it be until the first hunter hears the shot? (Assume no wind.)

101. An open organ pipe has a length of 0.75 m. What would be the length of an organ pipe closed at one end whose third harmonic ($m = 3$) is the same as the fundamental frequency of the open pipe?

102. ■■ Show that the general equation of the Doppler effect for a moving source and a moving observer is given by

$$f_o = f_s \left(\frac{v \pm v_o}{v \mp v_s} \right)$$

using the sign convention of Eqs. 14.11 and 14.14.

103. ■■ A fire truck travels at a speed of 90 km/h with its siren emitting sound at a frequency of 500 Hz. What is the frequency heard by a passenger in a car traveling at 65 km/h in the oncoming lane to the fire truck (a) approaching it and (b) moving away from it? [*Hint*: See Exercise 102, and take the speed of sound to be 354 m/s. (It is a hot day.)]

Electric Charge, Forces, and Fields

INSIGHTS

- Xerography, Electrostatic Copiers, and Laser Printers
- Lightning and Lightning Rods

Learn by Drawing

- Using the Superposition Principle to Determine the Electric Field Direction
- Sketching Electric Lines of Force

Few natural processes deliver such an enormous amount of energy in a fraction of a second as a lightning bolt. Yet many people have never experienced its tremendous power at close range; only a few hundred people are actually struck by lightning each year in the United States.

It might surprise you, therefore, to realize that you have almost certainly had a similar experience, at least from a physicist's point of view. Have you ever walked across a carpeted room and gotten a tiny shock when you reached for a metallic doorknob? Although the scale is dramatically different, the physical processes involved are much the same as being struck by lightning.

Electricity seems to have a flair for the dramatic. For most of us, the word conjures up striking images: a blaze of floodlights, crackling live wires, the skyline of a great city at night, or a stroke of lightning. There is often a hint of anxiety in these associations, for we know that electric-ity can be dangerous. We also know, however, that electricity can be tamed—even "domesticated." In the home or office, we now take it for granted. Indeed, the extent to which we depend on electricity becomes evident only when the power goes off unexpectedly, giving us a dramatic reminder of the role that it plays in our daily lives. Yet less than a century ago there were no power lines crossing the country, no electric lights or appliances—none of the seemingly endless electrical applications around us today.

We now know that electricity and magnetism are, in fact, related. Today, this "unified" electromagnetic force is recognized as one of the four fundamental forces (along with gravity, discussed in Chapter 7, and the strong and weak nuclear forces, discussed in Chapters 29 and 30). It is customary in introductory physics courses to consider the electrical part of the electromagnetic force before the magnetic part and only then to combine

513

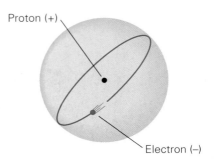

Proton (+)

Electron (−)

(a) Hydrogen atom

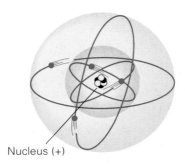

Nucleus (+)

(b) Beryllium atom

▲ **FIGURE 15.1 Simplistic model of atoms** The so-called solar system model of **(a)** a hydrogen atom and **(b)** a beryllium atom views the electrons (negatively charged) as orbiting the nucleus (positively charged), analogously to the planets orbiting the Sun. The electronic structure of atoms is actually much more complicated than this.

Note: Recall the discussion of Newton's third law in Section 4.4.

electricity and magnetism into electromagnetism. This will be our approach. Among other things, in this chapter you'll learn what the lightning bolt and the spark from the doorknob have in common.

15.1 Electric Charge

OBJECTIVES: To **(a)** distinguish between the two types of electric charge, **(b)** state the charge–force law that operates between charged objects, and **(c)** understand and use the law of charge conservation.

What is electricity? Perhaps the broadest answer is that electricity is a collective term describing phenomena associated with the interaction between *electrically charged* objects. Our study will start with the simplest situation, electro*statics*. As the name implies, the charged objects are *at rest.*

Like mass, **electric charge** is a fundamental property of matter. Electric charge is associated with particles that make up the atom: the electron and the proton. The simplistic solar system model of the atom, shown in ◄Fig. 15.1, likens its structure to planets orbiting the Sun. The *electrons* are viewed as orbiting a nucleus, a core containing most of the atom's mass in the form of *protons* and electrically neutral particles called *neutrons*. As we saw in Section 7.5, the centripetal force that keeps the planets in orbit about the Sun is supplied by gravity. Similarly, the force that keeps the electrons in orbit around the nucleus is the electrical force. However, there are important distinctions between gravitational and electrical forces.

A basic one is that there is only one type of mass in nature, and gravitational forces are known to be only attractive. Electric charge, however, comes in two types, distinguished by the labels positive (+) and negative (−). Protons carry a positive charge, and electrons carry a negative charge. Different combinations of the two types of charge can produce *either* attractive *or* repulsive electrical forces.

The directions of the electric forces when charges interact with one another are given by the following principle, called the **law of charges** or the **charge–force law**:

| Like charges repel each other, and unlike charges attract each other. |

That is, two negatively charged particles or two positively charged particles repel each other, whereas particles with opposite charges attract each other (▼ Fig. 15.2). The repulsive and attractive forces are equal and opposite, and they act on different objects, in keeping with Newton's third law (action–reaction).

The charge on an electron and that on a proton are equal in magnitude, but opposite in sign. The magnitude of the charge on an electron is abbreviated as *e* and is the fundamental unit of charge, since it is the smallest charge observed in nature.*

▶ **FIGURE 15.2 The charge–force law, or law of charges** **(a)** Like charges repel. **(b)** Unlike charges attract.

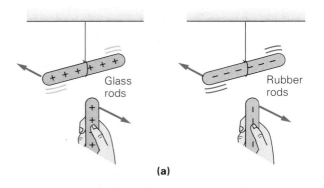

Glass rods

Rubber rods

(a)

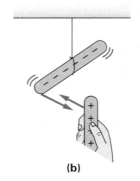

(b)

*According to relatively recent experiments, protons (as well as neutrons and other particles) are made up of particles called *quarks*, which carry charges of $\pm\frac{1}{3}$ and $\pm\frac{2}{3}$ of the electronic charge. There is experimental evidence of the existence of quarks within the nucleus, but free quarks have not been detected. Current theory implies that that may be impossible (Chapter 30).

The SI unit of charge is the **coulomb (C)**, named for the French physicist Charles A. Coulomb (1736–1806), who discovered a relationship between electric force and charge (considered in Section 15.3). The charges and masses of the electron, proton, and neutron are given in Table 15.1, where we see that $e = 1.6 \times 10^{-19}$ C. Thus, the charge on one electron is written as $-e = -1.60 \times 10^{-19}$ C and on one proton is $+e = +1.60 \times 10^{-19}$ C.

We frequently use several terms when we discuss charged objects. Saying that an object has a **net charge** means that the object has an excess of either positive or negative charges. (It is common, however, to ask about the "charge" of an object when we really mean the net charge.) As you will see in Section 15.2, excess charge is most commonly produced by a transfer of electrons, *not* protons. (Protons are bound into the nucleus and, under most common situations, are therefore not the charges that move.) For example, if an object has a (net) charge of $+1.6 \times 10^{-18}$ C, then it must have a deficiency of 10 electrons, since $10(1.6 \times 10^{-19}$ C$) = 1.6 \times 10^{-18}$ C. That is, 10 electrons have been removed, and the total number of electrons in the object no longer completely cancels out the positive charge of all the protons—resulting in a net positive charge. On an atomic level, some of the atoms that make up the object would be deficient in electrons. Such positively charged atoms are called *positive ions*. Atoms with an excess of electrons are called *negative ions*.

Since the charge on the electron is such a tiny fraction of a coulomb, a coulomb's worth of net charge on an object is rarely seen in everyday situations. Therefore, it is common to express charges on objects using *microcoulombs* (μC, or 10^{-6} C), *nanocoulombs* (nC, or 10^{-9} C), and *picocoulombs* (pC, or 10^{-12} C).

Because the (net) electric charge on an object is caused by a deficiency or an excess of electrons, it must always be an integer multiple of the charge on an electron. Our symbol for charge will be q (or Q), and a plus sign or a minus sign will indicate whether the object has a deficiency ($+q$) or an excess ($-q$) of electrons. Thus, for the (net) charge of an object, we can write

$$q = \pm ne \qquad (15.1)$$

SI unit of charge: coulomb (C)

where $n = 1, 2, 3, \ldots$. We sometimes say that charge is "quantized," which means that it occurs only in integral multiples of the fundamental electronic charge.

In dealing with any electrical phenomena, another important principle is **conservation of charge**:

The net charge of an isolated system remains constant.

The net charge may be other than zero, but it remains constant. Suppose, for example, that a system consists initially of two electrically neutral objects, and one million electrons are transferred from one to the other. The object with the added electrons will then have a net negative charge, and the object with the reduced number of electrons will have a net positive charge of equal magnitude. (See Example 15.1.) Thus, the net charge of the *system* remains zero. If the universe is considered as a whole, conservation of charge means that the net charge *of the universe* is constant.

Note that this principle doesn't prohibit the creation or destruction of charged particles. In fact, physicists have known for a long time that charged particles can be created and destroyed on the atomic and nuclear levels. However, because of

TABLE 15.1	Particles and Electric Charge	
Particle	*Electric Charge**	*Mass**
Electron	-1.602×10^{-19} C	$m_e = 9.109 \times 10^{-31}$ kg
Proton	$+1.602 \times 10^{-19}$ C	$m_p = 1.673 \times 10^{-27}$ kg
Neutron	0	$m_n = 1.675 \times 10^{-27}$ kg

*Even though the values are displayed to four significant figures, we will usually use only two or three in our calculations.

charge conservation, charged particles are created or destroyed only in pairs with equal and opposite charges.

Integrated Example 15.1 ■ On the Carpet: Conservation of Quantized Charge

You shuffle across a carpeted floor on a dry day and the carpet acquires a net positive charge (for details on how objects can acquire charge, see Section 15.2). (a) Will you have a (1) deficiency or (2) an excess of electrons? (b) If the charge the carpet acquired has a magnitude of 2.0 nC, how many missing or extra electrons will you have?

(a) Conceptual Reasoning. (a) Since the carpet has a net positive charge, it must have lost electrons. By charge conservation, then, you must have gained all of them. Thus, your charge is negative, indicating an excess of electrons, and the correct answer is (2).

(b) Thinking It Through. By knowing the charge on one electron, you can quantify the excess of electrons.

Solution. Express the charge in coulombs, and state what is to be found.

Given: $q_c = +2.0\,nC\left(\dfrac{10^{-9}\,C}{1\,nC}\right)$ 　　　　　*Find:* n, number of missing or
　　　　　　　　　　　　　　　　　　　　　　　　　　　　excess electrons on you
　　　$= +2.0 \times 10^{-9}\,C$
　$q_e = -1.6 \times 10^{-19}\,C$ (from Table 15.1)

The net charge on you is

$$q = -q_c = -2.0 \times 10^{-9}\,C$$

Using this result in Eq. 15.1, we have

$$n = \frac{q}{q_e} = \frac{-2.0 \times 10^{-9}\,C}{-1.60 \times 10^{-19}\,C/electron} = 1.3 \times 10^{10}\ electrons$$

As can be seen from this familiar situation, net charges usually involve huge numbers of electrons (here, more than 10 billion).

Follow-up Exercise. In this Example, if your mass is 80 kg, by what percentage has your mass increased due to the excess electrons? *(Answers to all Follow-up Exercises are at the back of the text.)*

15.2 Electrostatic Charging

OBJECTIVES: To (a) distinguish between conductors and insulators, (b) explain the operation of the electroscope, and (c) distinguish among charging by friction, conduction, induction, and polarization.

The existence of two types of electric charge (and thus both attractive and repulsive electrical forces) can be demonstrated easily. Before learning how this is done, you need to be able to distinguish between electrical conductors and insulators. What distinguishes these broad groups of substances is their ability to conduct, or transmit, electric charge. Some materials, particularly metals, are good **conductors** of electric charge. Others, such as glass, rubber, and most plastics, are **insulators**, or poor electrical conductors. A comparison of the relative magnitudes of the conductivities of some materials is given in ▸ Fig. 15.3.

In conductors, the *valence* electrons of the atoms—the electrons in the outermost orbits—are loosely bound. As a result, they can be easily removed from the atom and moved about in the conductor, or they can leave the conductor altogether. That is, the valence electrons are not permanently bound to a particular atom. In insulators, however, most of the electrons are tightly bound. Thus, charge does not move easily through, nor is it easily removed from, an insulator.

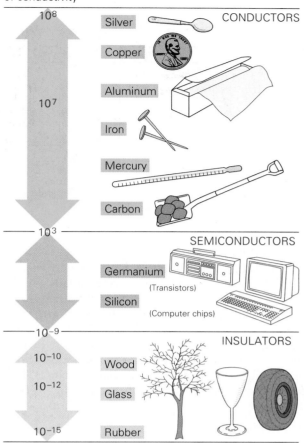

Relative magnitude   Material
of conductivity

10^8 Silver   CONDUCTORS

Copper

10^7 Aluminum

Iron

Mercury

Carbon

10^3 SEMICONDUCTORS

Germanium
(Transistors)

Silicon
(Computer chips)

10^{-9} INSULATORS

10^{-10} Wood

10^{-12} Glass

10^{-15} Rubber

◄ **FIGURE 15.3 Conductors, semiconductors, and insulators** A comparison of the relative magnitudes of the electrical conductivities of various materials (not drawn to scale).

As Fig. 15.3 shows, there is also a class of materials called **semiconductors**. Their ability to conduct charge is intermediate between that of insulators and conductors. The movement of electrons in semiconductors is much more difficult to describe than the simple valence electron approach used for insulators and conductors. In fact, the details of semiconductor properties can be understood only with the aid of quantum mechanics, which is beyond the scope of this book.

However, it is interesting to note that the conductivity of semiconductors can be adjusted by adding certain types of atomic impurities in varying concentrations. Beginning in the 1940s, scientists undertook research into the properties of semiconductors to create applications for such materials. Scientists used semiconductors to create transistors, then solid-state circuits, and, eventually, our modern-day computer microchips. The microchip is one of the major developments responsible for the high-speed computer technology we enjoy today.

Now that we know a bit about conductors and insulators, let's learn about one way of determining the signs of charged objects. The *electroscope* is one of the simplest devices used to demonstrate the characteristics of electric charge (▶Fig. 15.4). In its simplest form, it consists of a metal rod with a metallic bulb at one end. The rod is attached to a solid, rectangular piece of metal that has an attached foil "leaf," usually made of gold or aluminum. This arrangement is insulated from its protective glass container by an insulating frame. When charged objects are brought close to the bulb, electrons in the bulb are either attracted to or repelled by the charged objects, in accordance with the charge–force law. For example, if a negatively charged rod is brought near the bulb, electrons in the bulb are repelled, and the bulb is left with a positive charge. The electrons are conducted down to

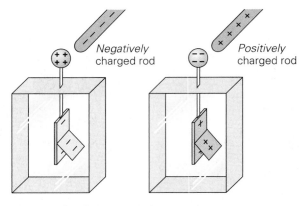

Bulb

Negatively charged rod

Positively charged rod

(a) Neutral electroscope has charges evenly distributed; leaf is vertical.

(b) Electrostatic forces cause leaf to separate away.

▶ **FIGURE 15.4 The electroscope** An electroscope can be used to determine whether an object is electrically charged. When a charged object is brought near the bulb, the leaf separates away from the metal piece.

Note: An uncharged electroscope can detect only whether an object is electrically charged. If the electroscope is charged with a known sign, it can then also feel the sign of the charge on the object.

Note: From an external viewpoint, you cannot tell whether the rubber rod gained negative charges or the fur gained positive charges. In other words, moving electrons to the rubber rod results in the same physical situation as moving positive charges to the fur. However, because the rubber is an insulator and its electrons are therefore tightly bound, we might suspect that the fur lost electrons and the rubber gained them. In solids, the protons, being in the nuclei of the atoms, do not move; only electrons move. It is just a question of which material most easily loses electrons.

the metal rectangle and its attached foil leaf, which then will swing away, since they have like charges (Fig. 15.4b). Similarly, if a positively charged rod is brought near the bulb, the leaf also swings away. (Can you explain why?)

Notice that the net charge on the electroscope remains zero in these instances, because the device is isolated (insulated); only the *distribution* of charge is altered. However, it is possible to give an electroscope (and other objects) a net charge by **electrostatic charging**. The term refers to any process by which an insulator or an insulated conductor receives a net charge. In this case, the leaf remains swung out from the metal even without the presence of a charged object. To identify the various types of electrostatic charging, consider the following processes.

Charging by Friction

In one charging process, when certain insulator materials are rubbed with cloth or fur, they become electrically charged. For example, if a hard rubber rod is rubbed with fur, the rod will acquire a net negative charge; rubbing a glass rod with silk will give the rod a net positive charge. The rubbing process is called **charging by friction**. The transfer of charge is due to the contact between the materials, and the amount of charge transferred depends, as you might expect, on the nature of the materials involved.

Example 15.1 was actually an example of frictional charging, in which you picked up a net charge from the carpet. If you reached for a metal object, such as a doorknob, you might get "zapped" by a spark. Your negative charge produces an electric force great enough to *ionize* (i.e., free electrons from) the air molecules between your hand and the knob. The resulting flow of freed electrons gives rise to the spark discharge you sometimes see (and always feel) between hand and metal. This phenomenon usually doesn't occur on humid days, because, with adequate humidity, a thin film of moisture on objects prevents the buildup of charge by conducting it away.

Charging by Conduction (Contact)

Bringing a charged rod close to an electroscope will reveal that the rod is charged, but it does not tell you what type of charge the rod has (positive or negative). The sign of the charge can be determined, however, if the electroscope is first given a known type of (net) charge. For example, electrons can be transferred to the electroscope from a negatively charged object, as illustrated in ▶ Fig. 15.5a. The electrons in the rod repel one another, and some will transfer onto the electroscope. Notice that the leaf is now permanently diverged from the

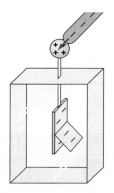 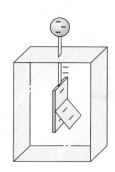 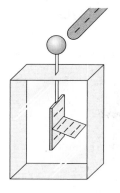 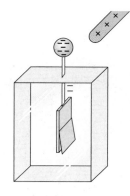

(a) Neutral electroscope is touched with negatively charged rod.

(b) Charges are transferred to bulb; electroscope has net negative charge.

(c) Negatively charged rod repels electrons; leaf separates further.

(d) Positively charged rod attracts electrons; leaves collapse.

▲ **FIGURE 15.5 Charging by conduction (a)** The electroscope is initially neutral (but the charges are separated), and a nearby charged rod is then touched to the bulb. **(b)** Charge is transferred to the electroscope. **(c)** When a rod of the same charge is brought near the bulb, the leaf separates further. **(d)** When an oppositely charged rod is brought nearby, the leaf collapses.

metal piece. In this case, we say that the electroscope has been **charged by contact** or by **conduction** (Fig. 15.5b). "Conduction" refers to the flow of charge during the short period of time the electrons are transferred.

If a negatively charged rod is brought close to the now negatively charged electroscope, the leaf will diverge even further as more electrons are repelled from the bulb (Fig. 15.5c). An oppositely (positively) charged rod will cause the leaf to collapse by attracting some electrons up to the bulb and away from the leaf area (Fig. 15.5d).

Charging by Induction

Using a negatively charged rubber rod (charged by friction), you might ask whether it is possible to create an electroscope that is positively charged. The answer is yes, and doing so involves a process called **charging by induction**. Starting with an uncharged electroscope, you touch the bulb with a finger, which *grounds* the electroscope—that is, provides a path by which electrons can escape from the bulb (▶Fig. 15.6). Then, when a negatively charged rod is brought close to (but is not touching) the bulb, the rod repels electrons from the bulb onto your finger and down into the Earth. Removing your finger *while the charged rod is kept nearby* leaves the electroscope with a net positive charge. This is because when the negative rod is removed, the electrons that traveled to the Earth (ground) have no way of coming back. (Their return path has been removed.)

Charge Separation by Polarization

Charging by contact and charging by induction both involve the removal of charge from an object. However, an object can have charge moved *within it* and yet keep a net charge of zero. In this case, induction brings about **polarization**, or separation of charge (▶Fig. 15.7). If an object is electrically neutral overall, but has equal, yet opposite, amounts of charge at both ends, we say it acts as an *electric dipole*. (See Section 15.4.) Examples of molecular electric dipoles are shown in Fig. 15.7. Now you can understand why, when you rub a balloon on your hair or sweater, the balloon can stick to the wall. The balloon is charged by friction, inducing an opposite charge on the wall's surface, thereby creating an attractive electric force. (See Section 15.3 for details on the electric force.)

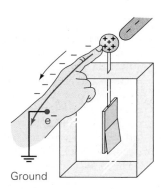

(a) Repelled by the negatively charged rod, electrons are transferred to ground through hand.

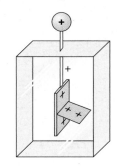

(b) Electroscope is left positively charged.

▲ **FIGURE 15.6 Charging by induction (a)** Touching the bulb with a finger provides a path to ground for charge transfer. The symbol e⁻ stands for "electron." **(b)** When the finger is removed, the electroscope has a net positive charge, opposite that of the rod.

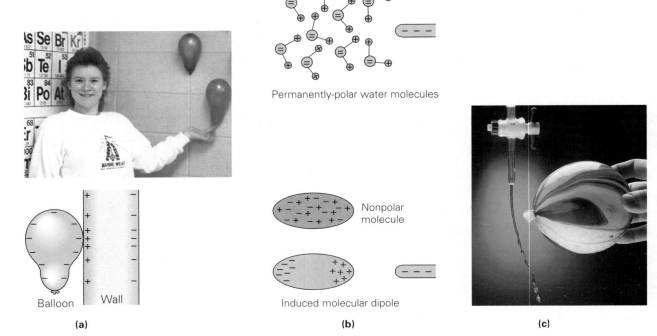

▲ FIGURE 15.7 Polarization   (a) When the balloons are charged by friction and placed in contact with the wall, an opposite charge is induced on the wall's surface, to which the balloons then stick by the force of electrostatic attraction. **(b)** Some molecules, such as those of water, are polar in nature; that is, they have separated regions of positive and negative charge. But even some molecules that are not normally dipolar in nature can be polarized temporarily by the presence of a nearby charged object. The electric force induces a separation of charge and, consequently, temporary molecular dipoles. **(c)** A stream of water bends toward a charged balloon. No polarization needs to be induced in the water molecules, because they are already natural dipoles. The charged balloon simply attracts the ends of the molecules that carry the opposite charge.

Electrostatic charging can be annoying, as when static cling causes clothes and papers to stick together, or even dangerous, such as when electrostatic spark discharges start a fire or cause an explosion in the presence of a flammable gas. But electrostatic charges can also be beneficial in a variety of practical applications. For example, the air we breathe is cleaner because of electrostatic precipitators used in smokestacks. In these devices, electrical discharges cause the particles (by-products of fuel combustion) to acquire a net charge. The charged particles can then be removed from the flue gases by attracting them to electrically charged surfaces. On a smaller scale, electrostatic air cleaners are available for the home. Another, now almost indispensable, application is the electrostatic copier. (see the Insight on p. 522.)

15.3 Electric Force

OBJECTIVES: **To (a) understand Coulomb's law and (b) use it to calculate the electric force between charged particles.**

The relative directions of the electric forces on mutually interacting charges are given by the charge–force law. However, what about the *magnitude*, or strength, of the electric force? This was investigated by Coulomb, who found that the magnitude of the electric force between two "point" (very small) charges q_1 and q_2 depended directly on the product of the charges (magnitude only) and inversely on the square of the distance between them. That is, $F_e \propto q_1 q_2/r^2$. (This inverse-square relationship is mathematically similar to that for the force of gravity between two point masses: $F_g \propto m_1 m_2/r^2$; see Chapter 7.)

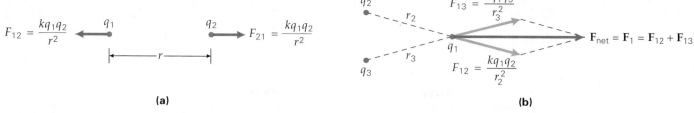

$$F_{12} = \frac{kq_1q_2}{r^2} \qquad q_1 \qquad\qquad q_2 \qquad F_{21} = \frac{kq_1q_2}{r^2}$$

$$\longleftarrow r \longrightarrow$$

(a)

$$F_{13} = \frac{kq_1q_3}{r_3^2}$$

$$\mathbf{F}_{net} = \mathbf{F}_1 = \mathbf{F}_{12} + \mathbf{F}_{13}$$

$$F_{12} = \frac{kq_1q_2}{r_2^2}$$

(b)

▲ **FIGURE 15.8 Coulomb's law   (a)** The mutual electrostatic forces on two point charges are equal and opposite. **(b)** For a configuration of two or more point charges, the force on a particular charge is the vector sum of the forces on it due to all the other charges. (Note: In each of these situations, all of the charges are of the same sign. How can we tell that this is true? Can you tell their *sign*? What is the direction of the force on q_2 due to q_3?)

Like Cavendish's measurements to determine the universal gravitational constant G (Section 7.5), Coulomb's measurements provided a constant of proportionality, k, so that the electric force could be written in equation form. Thus, the magnitude of the electric force between two point charges is described by an equation called **Coulomb's law** (remember, the q's here refer to the magnitude of the charge):

Note: Coulomb's law gives the electric force, but only between point charges.

$$F_e = \frac{kq_1q_2}{r^2} \quad (point\ charges\ only) \qquad (15.2)$$

magnitude of the electric force
between two point charges

Here, r is the distance between the charges (▲Fig. 15.8a) and k a constant with a value of

$$k = 8.988 \times 10^9\ \mathrm{N \cdot m^2/C^2} \approx 9.00 \times 10^9\ \mathrm{N \cdot m^2/C^2}$$

There are always equal and opposite forces between any charges, in accordance with Newton's third law (Fig. 15.8a). The charge–force law determines whether they are mutually repulsive or attractive. In some instances, we are concerned with the force on a particular charge in a system of more than two charges. The net electric force on any particular charge in the system is simply the vector sum of the forces on that charge due to all the other charges (Fig. 15.8b), as we shall see in the next two Examples.

Note: In calculations, we will take k to be exact at $9.00 \times 10^9\ \mathrm{N \cdot m^2/C^2}$ for significant-figure purposes.

Conceptual Example 15.2 ■ Free of Charge: Electric Forces

A rubber comb pulled through dry hair can acquire a net negative charge. The charged comb can then be used to attract and pick up small pieces of uncharged paper. This would seem to violate Coulomb's force law. Since the paper has no net charge, you might expect there to be no electric force on it. Which charging mechanism explains this phenomenon, and how does it explain it? (a) conduction, (b) friction, or (c) polarization.

Reasoning and Answer. Since the comb doesn't touch the paper when it starts to attract it, the paper cannot be charged by either conduction or friction, because both of these mechanisms require contact. Thus, the answer must be (c). When the charged comb is near the paper, the paper becomes polarized (▶Fig. 15.9). The key to understanding the attraction is to note that the charged ends of the paper are not the same distance from the comb. The positive end of the paper is closer to the comb than the negative end. Since the electric force decreases with distance, the attraction ($\mathbf{F}_1$) between the comb and the positive end of the paper is greater than the repulsion ($\mathbf{F}_2$) between the comb and the paper's negative end. Therefore the net force on the paper is toward the comb, and the paper accelerates in that direction.

Follow-up Exercise. Does the phenomenon just described tell you the sign of the charge on the comb? Explain why or why not? (*Answers to all Follow-up Exercises are at the back of the text.*)

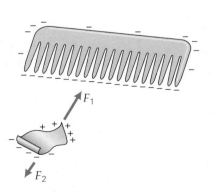

▲ **FIGURE 15.9 Comb and paper** See Conceptual Example 15.2.

INSIGHT

Xerography, Electrostatic Copiers, and Laser Printers

Xerography (from the Greek *xeros*, meaning "dry," and *graphein*, meaning "to write") refers to a dry process by which almost any printed material can be copied. This process uses a photoconductor, which is a light-sensitive semiconductor, such as amorphous silica. When kept in darkness, a photoconductor is an insulator and can be electrostatically charged. However, when light strikes the material, it becomes conductive and the electric charge can be removed.

In transfer xerography, a photoconductor-coated plate, drum, or belt is electrostatically charged and then receives a projected image of the page to be copied (Fig. 1). The illuminated portions of the photoconductor coating become con-

FIGURE 1 Printing by xerography (a) A typical xerographic copier in cutaway front view. ① The drum is charged positively at charging electrode C_1. ② The light reflected from the original page is focused onto the drum, creating a "positive-charge" image. ③ Then, negatively charged toner is sprayed onto the drum, making a true-ink positive of the original. The paper is carried by conveyor belt from the blank paper feed to a positive charging electrode C_2 and then onto the selenium-coated drum. ④ The positive toner image on the drum is transferred to the paper. ⑤ The paper is then carried through hot rollers, which press and "set" the toner, and the copy is delivered to the output tray. **(b)** Details of the processes occurring at the numbered locations in (a).

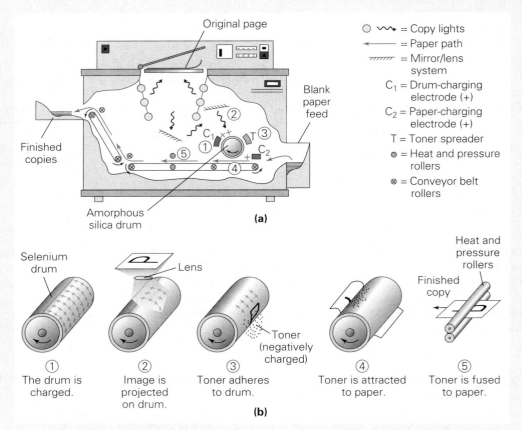

(a)

(b)
① The drum is charged.
② Image is projected on drum.
③ Toner adheres to drum.
④ Toner is attracted to paper.
⑤ Toner is fused to paper.

Example 15.3 ■ Coulomb's Law: Reviewing Vector Addition

(a) Two point charges of -1.0 nC and $+2.0$ nC are separated by a distance of 0.30 m (▶ Fig. 15.10a). What is the electric force on each particle? (b) A configuration of three charges is shown in Fig. 15.10b. What is the electrostatic force on q_3?

Thinking It Through. Adding electric forces is no different from adding any other type of force. The only difference here is that we use Coulomb's law to calculate their magnitudes. Then it is just a matter of computing components. (a) For the two point charges, we use Coulomb's law (Eq. 15.2), noting that the forces are attractive. (Why?) (b) Here we must use components to vectorially add the two forces acting on q_3 due to q_1 and q_2. We can find θ from the distances between charges. This angle is necessary to calculate the x and y force components. (See the Problem-Solving Hint on pg. 524.)

ducting and are discharged, but the areas corresponding to the dark print remain charged. Essentially, this creates an electric-charge copy of the original sheet. Then the photoconductor copy comes into contact with a negatively charged powder called toner, or dry ink. The toner is attracted to, and adheres to, the charged regions. Paper is placed over the inked photoconductor and is given a positive charge. The toner is then attracted to the paper, and heating causes it to be permanently fused to the paper. All of this takes place quickly, and out comes your copy.

The laser printer now commonly used with computers is basically a xerographic machine in which there is no original copy, since the information to be printed is actually stored in the computer. A laser (Chapter 27) scans back and forth across a rotating, charged drum in the printer (Fig. 2). The laser beam passes through a device called a modulator, which turns the beam on or off, according to the signals received from the computer. Wherever the light beam strikes the drum, that point is discharged, while unilluminated areas, corresponding to the letters on the document being produced, remain charged. In this way, a charged image of what is to be printed is produced. The rest of the printing process then takes place as in xerography.

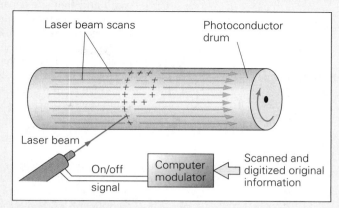

FIGURE 2 Laser printer A computer-controlled laser scans the charged photoconductor drum, causing the charge to bleed off where the beam strikes. When the laser beam is turned off, charged regions remain, which can be reproduced as in the xerography process.

Solution. Listing the data and converting nanocoulombs to coulombs, we have

Given: (a) $q_1 = -1.0 \text{ nC} \left(\dfrac{10^{-9} \text{ C}}{1 \text{ nC}} \right) = -1.0 \times 10^{-9} \text{ C}$ **Find:** (a) $\mathbf{F}_{12}$ and $\mathbf{F}_{21}$
(b) $\mathbf{F}_3$

$q_2 = +2.0 \text{ nC} \left(\dfrac{10^{-9} \text{ C}}{1 \text{ nC}} \right) = +2.0 \times 10^{-9} \text{ C}$

$r = 0.30 \text{ m}$

(b) Data given in Figure 15.10b. Convert charges to coulombs as in (a).

(a) Equation 15.2 gives the magnitude of the force acting on each point charge:

$$F_{12} = F_{21} = \frac{kq_1q_2}{r^2} = \frac{(9.00 \times 10^9 \text{ N} \cdot \text{m}^2/\text{C}^2)(1.0 \times 10^{-9} \text{ C})(2.0 \times 10^{-9} \text{ C})}{(0.30 \text{ m})^2}$$

$$= 0.20 \times 10^{-6} \text{ N} = 0.20 \text{ } \mu\text{N}$$

▼ **FIGURE 15.10 Coulomb's law and electrostatic forces** See Example 15.3.

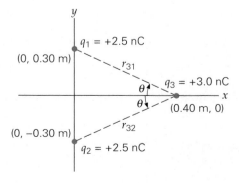

(a)

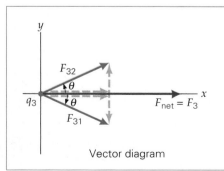

(b)

Note again that *only the charge magnitudes are used*, since Coulomb's law gives only the force magnitude. However, because the charges are of opposite sign, we know by the charge–force law that the force is attractive. Note that the forces are extremely small compared with everyday forces. This is because the charges involved are extremely small and they are separated by large distances. (See Example 15.4 for what happens when the distances become on the order of atomic and nuclear dimensions.)

(b) The forces $\mathbf{F}_{31}$ and $\mathbf{F}_{32}$ must be added vectorially to find the net force. Since all the charges are positive, the forces are repulsive, as shown in the vector diagram in Fig. 15.10b. Since $q_1 = q_2$ and the charges are equidistant from q_3, it follows that $\mathbf{F}_{31}$ and $\mathbf{F}_{32}$ have the same magnitude.

Note from the figure that $r_{31} = r_{32} = 0.50$ m. (Why?) With data from the figure, we again use Eq. 15.2:

$$F_{32} = \frac{kq_2q_3}{r_{32}^2} = \frac{(9.00 \times 10^9 \,\text{N} \cdot \text{m}^2/\text{C}^2)(2.5 \times 10^{-9}\,\text{C})(3.0 \times 10^{-9}\,\text{C})}{(0.50\,\text{m})^2}$$

$$= 0.27 \times 10^{-6}\,\text{N} = 0.27\,\mu\text{N}$$

Taking into account the directions of $\mathbf{F}_{31}$ and $\mathbf{F}_{32}$, we see by symmetry that the y-components of the vectors cancel. Thus, $\mathbf{F}_3$ (the net force on charge 3) acts along the positive x-axis and has a magnitude of

$$F_3 = F_{31_x} + F_{32_x} = 2\,F_{31_x}$$

since $F_{31} = F_{32}$.

The angle θ can be determined from the distance triangles; that is,

$$\theta = \tan^{-1}\left(\frac{0.30\,\text{m}}{0.40\,\text{m}}\right) = 37°.$$ So one x-component has a value of

$$F_{32_x} = F_{32}\cos\theta = (0.27\,\mu\text{N})\cos 37° = 0.22\,\mu\text{N}$$

and therefore,

$$F_3 = 2\,F_{32_x} = 2(0.22\,\mu\text{N}) = 0.44\,\mu\text{N}$$

in the positive x-direction.

Follow-up Exercise. In part (b) of this Example, calculate the force $\mathbf{F}_1$ on q_1. *(Answers to all Follow-up Exercises are at the back of the text.)*

Problem-Solving Hint

The signs of the charges can be used explicitly in Eq. 15.2. If they are so used, a positive value for F indicates a repulsive force and a negative value an attractive force. However, such an approach is *not* recommended, because this sign convention is useful only in the case of one-dimensional forces (forces that have only one nonzero component, as in Example 15.3a). More generally, when forces are two dimensional, Eq. 15.2 is used to calculate the magnitude of the force, which means only the magnitude of the charges (see Example 15.3b). Then the charge–force law is used to determine the direction of the force between each pair of charges. (Draw a sketch and put in the angles.) Last comes calculating each force's components by using trigonometry and combining them appropriately. This latter approach will be the one we use.

The magnitudes of the charges in Example 15.3 are typical of static charges that can be produced by frictional rubbing; that is, they are tiny. Thus, the forces involved are also tiny by everyday standards, much smaller than any force we have studied so far. However, on the atomic scale, even tiny forces can produce huge accelerations, because the particles (such as electrons and protons) have extremely small mass. Consider the following Example compared with the answers in Example 15.3.

Example 15.4 ■ Inside the Nucleus: Repulsive Electrostatic Forces

(a) What is the magnitude of the repulsive electrostatic force between two protons in a nucleus? Take the distance from center to center of the nuclear protons to be 3.0×10^{-15} m. (b) If the protons were released from rest, how would the magnitude of their initial acceleration compare with that of the acceleration due to gravity on the Earth's surface, g?

Thinking It Through. (a) We must apply Coulomb's law to find the repulsive force. (b) To find the initial acceleration, we use Newton's second law ($\mathbf{F}_{net} = m\mathbf{a}$).

Solution. Listing the known quantities, we have the following:

Given: $r = 3.0 \times 10^{-15}$ m   *Find:* (a) F_e (magnitude of force)
$\qquad q_1 = q_2 = +1.60 \times 10^{-19}$ C (from Table 15.1)
$\qquad m_p = 1.67 \times 10^{-27}$ kg (from Table 15.1)   (b) $\dfrac{a}{g}$ (magnitude of accel eration compared with g)

(a) Using Coulomb's law (Eq. 15.2), we have

$$F_e = \frac{kq_1q_2}{r^2} = \frac{(9.00 \times 10^9\ \text{N} \cdot \text{m}^2/\text{C}^2)(1.60 \times 10^{-19}\ \text{C})(1.60 \times 10^{-19}\ \text{C})}{(3.0 \times 10^{-15}\ \text{m})^2} = 26\ \text{N}$$

(Notice that this force is much larger than that in the previous Example and is equivalent to the weight of an object with a mass of about 3 kg. Thus, with its small mass, we expect the proton to experience a huge acceleration.)

(b) If it acted alone on a proton, this force would produce an acceleration of

$$a = \frac{F_e}{m_p} = \frac{26\ \text{N}}{1.67 \times 10^{-27}\ \text{kg}} = 1.6 \times 10^{28}\ \text{m/s}^2$$

Then

$$\frac{a}{g} = \frac{1.6 \times 10^{28}\ \text{m/s}^2}{9.8\ \text{m/s}^2} = 1.6 \times 10^{27}$$

That is, $a = (1.6 \times 10^{27})g \approx 10^{27}g$ (a whole lot bigger than the gravitational acceleration).

Atoms above helium in the periodic chart contain more than two protons in their nuclei. With these enormous mutually repulsive forces, you would expect nuclei to fly apart. Since this doesn't generally happen, there must be a stronger attractive force holding the nucleus together. This is the nuclear (or strong) force, which will be discussed in Chapters 29 and 30.

Follow-up Exercise. Suppose you could anchor a proton to the ground and you wished to place a second one directly above the first so that the weight of the second proton would be exactly balanced by the electric repulsion between them. How far apart must the protons be? *(Answers to all Follow-up Exercises are at the back of the text.)*

Although there is a striking similarity between the mathematical form of the expressions for the electric and gravitational forces, there is a huge difference in the relative strengths of the two forces, as is shown in the next Example.

Example 15.5 ■ Inside the Atom: Electric versus Gravitational Force

Compare the magnitudes of the electric and gravitational forces between a proton and an electron. Express your answer as a ratio of electric force to gravitational force. In other words, how many times larger is the electric force than the gravitational force?

Thinking It Through. The distance between the proton and electron is not given. However, both the electrical force (for point charges) and the gravitational force (for point masses) vary as the inverse square of the distance, so the distance will cancel out in a ratio. By using Coulomb's law and Newton's law of gravitation (Chapter 7), the ratio can be determined if one knows the charges, masses, and appropriate electric and gravitational constants.

Solution. The charges and masses of the particles are known (Table 15.1), as are the electrical constant k and the universal gravitational constant G.

Given: $q_e = -1.60 \times 10^{-19}\,\text{C}$ *Find:* $\dfrac{F_e}{F_g}$ (ratio of forces)

$q_p = +1.60 \times 10^{-19}\,\text{C}$

$m_e = 9.11 \times 10^{-31}\,\text{kg}$

$m_p = 1.67 \times 10^{-27}\,\text{kg}$

The expressions for the forces are

$$F_e = \frac{k q_e q_p}{r^2} \quad \text{and} \quad F_g = \frac{G m_e m_p}{r^2}$$

Forming a ratio of magnitudes for comparison purposes (and to cancel r) gives

$$\frac{F_e}{F_g} = \frac{k q_e q_p}{G m_e m_p}$$

$$= \frac{(9.00 \times 10^9\,\text{N} \cdot \text{m}^2/\text{C}^2)(1.60 \times 10^{-19}\,\text{C})^2}{(6.67 \times 10^{-11}\,\text{N} \cdot \text{m}^2/\text{kg}^2)(9.11 \times 10^{-31}\,\text{kg})(1.67 \times 10^{-27}\,\text{kg})} = 2.27 \times 10^{39}$$

or

$$F_e = (2.27 \times 10^{39}) F_g$$

The magnitude of the electrostatic force between a proton and an electron is more than 10^{39} times greater than the magnitude of the gravitational force! For this reason, the gravitational force between charged particles is usually neglected in solving problems in electrostatics. (Note that 10^{39} is 1000 trillion trillion trillion. In comparison, our national debt is on the order of trillions of dollars.)

Follow-up Exercise. Show that gravity is even more negligible compared with the electric repulsive force between two electrons by repeating the calculation in this Example, but using the mass of, and charge on, an electron. Explain why this is so. *(Answers to all Follow-up Exercises are at the back of the text.)*

15.4 Electric Field

OBJECTIVES: **To (a) understand the definition of the electric field and (b) plot electric field lines and calculate electric fields for simple charge distributions.**

The electric force, like the gravitational force, is an "action-at-a-distance" force. In fact, the range of the electric force between two point charges is infinite, since $F_e \propto 1/r^2$ and so approaches zero only if r approaches infinity. Thus, a particular arrangement, or configuration, of charges can have an effect on an additional charge placed anywhere nearby.

Note: Charges set up an electric field, which then acts on other charges placed in that field.

The idea of a force acting across space was difficult for early investigators to accept, and the more modern concept of a *force field*, or simply a field, was introduced. An *electric field* is envisioned as surrounding every arrangement of charges. Thus, the electric field represents the *physical effect* of a particular configuration of charges on the nearby space. The field is the way to represent what is different about the nearby space because those charges are there. The concept allows us to think of charges as interacting with the electric field created by other charges, rather than with the other charges "at a distance." In sum, the logic of the electric field is as follows: Think of a configuration of charges creating an electric field in the nearby space. If another charge is now placed in this electric field, a force will be exerted on it by the field. Charges create fields, and the fields exert forces on other charges.

Thus, an electric field is a *vector field* (since it must include direction) that enables us to determine the force exerted on a charge at a particular position in space. However, the electric field is *not* a force. Instead, the magnitude (or strength, as we sometimes say) of the electric field is expressed as force per unit charge. Measuring an electric field's strength may be visualized as placing a small *test charge* at various locations, measuring the force acting on the charge, dividing by the charge, and finding the number of newtons of force that would be exerted on a charge of one coulomb. Then the test charge is removed and the force disappears (why?), but the field remains, because it is created by the nearby charges. With the strength of the electric field measured in various locations near the charges that create it, we now have a partial "map" of the electric field strength. The map is partial because we still lack the direction of the field.

The direction of the electric field is specified by the direction of the force on the test charge. However, the direction of this force depends on whether the test charge is positive or negative. The sign convention is that a *positive test charge* $(+q_0)$ is used for measuring electric field direction (see ▼Fig. 15.11). That is,

Note: A *test charge* (q_0) is small and positive.

| the electric field direction is in the direction of the force experienced by a positive test charge.

Once you know the electric field mapping (magnitude and direction) due to a charge configuration, you can ignore the "source" charges and talk solely in terms of the field they have produced. This procedure often greatly facilitates calculations.

The **electric field (E)** at any point in space is then formally defined as

$$\mathbf{E} = \frac{\mathbf{F}_{on\,q_0}}{q_0} \qquad (15.3)$$

electric field definition

SI unit of electric field: newton/coulomb (N/C)

Remember that **E** is a vector field. At any location, it has the direction of the net force acting on a *positive* (test) charge q_0 placed at that location.

For the special case of the electric field (magnitude) at a distance r from a single point charge of q coulombs, we can use Coulomb's force law to calculate the force between the point charge and the imaginary test charge and then divide by the test charge:

$$E = \frac{F_e}{q_0} = \frac{(kq_0 q/r^2)}{q_0} = \frac{kq}{r^2}$$

That is,

$$E = \frac{kq}{r^2} \quad \begin{array}{l}\textit{(magnitude of electric field} \\ \textit{due to point charge q)}\end{array} \qquad (15.4)$$

It is important to note that in the derivation of Eq. 15.4, q_0 canceled out. *This must always happen*, because the field is produced by the nearby arrangement of charges, *not* by the test charge.

Some electric field vectors in the vicinity of a positive charge are illustrated in ▶Fig. 15.12a. Note that the vectors point away from the positive charge, in the direction of the force exerted on a positive test charge, and that the magnitude of the vectors decreases with distance from the charge (an inverse-square relationship).

If there is more than one charge creating an electric field, then the total, or net, electric field at any point is found by using the **superposition principle for electric fields**, which can be stated as follows.

| For a configuration of charges, the total, or net, electric field at any point is the vector sum of the electric fields due to the individual charges.

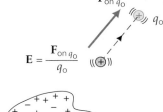

▲ **FIGURE 15.11 Checking for electric field direction** By convention, the direction of the electric field (**E**) at any location in space is in the same direction as the force experienced by an imaginary small positive test charge placed at that location. **E** is easily determined by asking yourself which way the test charge would accelerate if you imagine releasing it. In the situation pictured, the "system of charges" produces a net electric field upwards and to the right at the location of the test charge. To understand why, look at the signs of the charges in the system. Which type predominates?

▶ **FIGURE 15.12 Electric field**
(a) The electric field points away from a positive point charge, in the direction a force would be exerted on a small positive test charge. Note that the field's magnitude (as indicated by the lengths of vectors) becomes smaller as the distance from the source charge increases, reflecting the inverse-square distance relationship characteristic of the field produced by a point charge. **(b)** In this simple case, the vectors are easily connected to give an electric field line pattern from a positive point charge.

Note: Think of the electric field definition as being useful in the same way that the price per pound is for food items. Knowing how much you want of an item, you can compute how much it will cost if you know the price per pound. Similarly, given the magnitude of a charge placed in an electric field, you can compute the force on it if you know the field strength in newtons per coulomb.

Note: Total electric field: $E = \Sigma E_i$.

Electric Fields

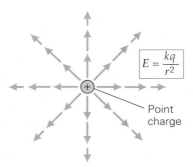

$$E = \frac{kq}{r^2}$$

Point charge

(a) Electric field vectors

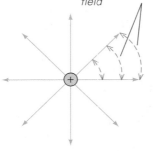

The closer together the lines of force, the stronger the field

(b) Electric field lines (lines of force)

The use of this principle is demonstrated in the next two Examples, and a way to qualitatively determine the direction of the electric field from a group of charges is shown in the Learn By Drawing Feature entitled "Using the Superposition Principle to Determine the Electric Field Direction" on p. 529.

Example 15.6 ■ Electric Fields in One Dimension: Zero Field

Two point charges are placed on the x-axis as shown in ▼Fig. 15.13. Find the locations on the axis where the electric field is zero.

Thinking It Through. Each point charge produces its own electric field. By the superposition principle, the total field is the vector sum of the two fields. We want places on the x-axis where these fields are equal and opposite, to cancel and give no total (or *net*) electric field.

Solution. Let us specify the location as a distance x from q_1 (assumed to be located at $x = 0$). (Convert charges from microcoulombs to coulombs.)

Given: $d = 0.60$ m (distance between charges) *Find:* x [the location(s) of zero **E**]
$q_1 = +1.5 \ \mu C = +1.5 \times 10^{-6}$ C
$q_2 = +6.0 \ \mu C = +6.0 \times 10^{-6}$ C

Since both charges are positive, their fields point to the right at all points to the right of q_2. (Why?) Therefore, the fields cannot cancel in that region. Similarly, to the left of q_1, both fields point to the left and cannot cancel. The only possibility of cancellation is *between* the charges. In that region, the two fields will cancel if their magnitudes are equal, since they are oppositely directed. Thus:

$$E_1 = E_2 \quad \text{or} \quad \frac{kq_1}{x^2} = \frac{kq_2}{(d-x)^2}$$

Rearranging this expression and canceling the constant k, we can write

$$\frac{1}{x^2} = \frac{(q_2/q_1)}{(d-x)^2}$$

Since $q_2/q_1 = 4$, we can take the square root of both sides:

$$\sqrt{\frac{1}{x^2}} = \sqrt{\frac{q_2/q_1}{(d-x)^2}} = \sqrt{\frac{4}{(d-x)^2}} \quad \text{or} \quad \frac{1}{x} = \frac{2}{d-x}$$

▶ **FIGURE 15.13 Electric field in one dimension** See Example 15.6.

Where is **E** = 0?

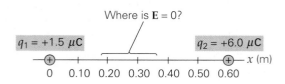

$q_1 = +1.5 \ \mu C$ $q_2 = +6.0 \ \mu C$ x (m)

0   0.10 0.20 0.30 0.40 0.50 0.60

Solving for x, we get $x = d/3 = 0.60\,\text{m}/3 = 0.20$ m. (Why don't we use the negative square root? Try it.) This location makes sense physically. Because q_2 is the larger of the two positive charges, the only way to have the two fields be equal in magnitude at that location is for it to be closer to q_1 than to q_2.

Follow-up Example. Repeat this Example, changing the sign of the right-hand charge. How is the situation physically different? *(Answers to all Follow-up Exercises are at the back of the text.)*

Integrated Example 15.7 ■ Electric Fields in Two Dimensions: Using Vector Components

▼Fig. 15.14a shows a configuration of three point charges. **(a)** In what quadrant is the net electric field at the origin? (1) the first quadrant, (2) the second quadrant, or (3) the third quadrant. Explain your reasoning, using a sketch of the individual electric fields and the superposition principle. **(b)** Calculate the magnitude and direction of the electric field at the origin due to this arrangement of charges.

(a) Conceptual Reasoning. In general, the electric field points towards a negative point charge and away from a positive point charge. Therefore, $\mathbf{E}_1$ and $\mathbf{E}_2$ point in the positive x-direction. Similarly, $\mathbf{E}_3$ points along the positive y-axis. Since the electric field is the sum of these vectors, both of its components are positive. Because of this, $\mathbf{E}$ must be in the first quadrant, making an angle of less than 90° above the positive x-axis (Fig. 15.14b). Thus, the correct answer is (1).

(b) Thinking It Through. The directions of the electric fields due to the individual charges are shown in the sketch in part (a). All that needs to be done is to add the fields vectorially. That is, the electric field is the vector sum of the individual fields, or $\mathbf{E} = \mathbf{E}_1 + \mathbf{E}_2 + \mathbf{E}_3$.

Solution. Listing the data given and converting the charges into coulombs, we have the following information:

Given: $q_1 = -1.00\,\mu\text{C} = -1.00 \times 10^{-6}\,\text{C}$ *Find:* $\mathbf{E}$ (total electric field at origin)
$q_2 = +2.00\,\mu\text{C} = +2.00 \times 10^{-6}\,\text{C}$
$q_3 = -1.50\,\mu\text{C} = -1.50 \times 10^{-6}\,\text{C}$
$r_1 = 3.50$ m
$r_2 = 5.00$ m
$r_3 = 4.00$ m

The superposition principle is used to determine the electric field $\mathbf{E}$ by adding the individual fields. Note from the sketch that $\mathbf{E}_y$ is entirely due to $\mathbf{E}_3$ and $\mathbf{E}_x$ is the sum of $\mathbf{E}_1$

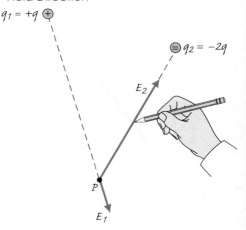

Learn by Drawing

Using the Superposition Principle to Determine the Electric Field Direction

To qualitatively determine the direction of the electric field at any point P, simply sketch the individual electric fields and add them vectorially, taking into account their relative magnitudes if you can. In the specific situation shown here, $\mathbf{E}_1$ has been drawn much smaller than $\mathbf{E}_2$ because of both distance and charge factors. Can you explain why $\mathbf{E}_2$, if drawn very accurately, would be about eight times longer than $\mathbf{E}_1$? The final step would be to complete the vector addition, thus finding $\mathbf{E}$ at P. You should be able to do that.

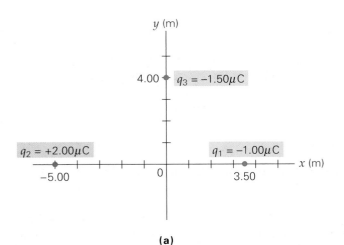

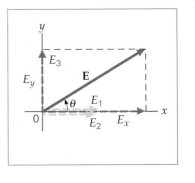

(a)　　　　　　　**(b)**

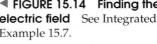

◄ **FIGURE 15.14 Finding the electric field** See Integrated Example 15.7.

and E_2. To calculate the magnitudes of the three fields that make up the total field, use Eq. 15.4:

$$E_1 = \frac{kq_1}{r_1^2} = \frac{(9.00 \times 10^9 \, \text{N} \cdot \text{m}^2/\text{C}^2)(1.00 \times 10^{-6} \, \text{C})}{(3.50 \, \text{m})^2} = 7.35 \times 10^2 \, \text{N/C}$$

$$E_2 = \frac{kq_2}{r_2^2} = \frac{(9.00 \times 10^9 \, \text{N} \cdot \text{m}^2/\text{C}^2)(2.00 \times 10^{-6} \, \text{C})}{(5.00 \, \text{m})^2} = 7.20 \times 10^2 \, \text{N/C}$$

$$E_3 = \frac{kq_3}{r_3^2} = \frac{(9.00 \times 10^9 \, \text{N} \cdot \text{m}^2/\text{C}^2)(1.50 \times 10^{-6} \, \text{C})}{(4.00 \, \text{m})^2} = 8.44 \times 10^2 \, \text{N/C}$$

The magnitudes of the x and y components of the total field are then

$$E_x = E_1 + E_2 = 7.35 \times 10^2 \, \text{N/C} + 7.20 \times 10^2 \, \text{N/C} = 1.46 \times 10^3 \, \text{N/C}$$

and

$$E_y = E_3 = 8.44 \times 10^2 \, \text{N/C}$$

In component form,

$$\mathbf{E} = E_x\hat{\mathbf{x}} + E_x\hat{\mathbf{y}} = (1.46 \times 10^3 \, \text{N/C})\hat{\mathbf{x}} + (8.44 \times 10^2 \, \text{N/C})\hat{\mathbf{y}}$$

You should be able to show that, in magnitude–angle form, this is equivalent to

$$E = 1.69 \times 10^3 \, \text{N/C at } \theta = 30.0° \quad \text{relative to the positive } x\text{-axis}$$
$$(\theta \text{ is in the first quadrant})$$

Follow-up Exercise. In this Example, suppose q_1 was removed. Find the electric field at q_1's former location due to the two remaining charges. *(Answers to all Follow-up Exercises are at the back of the text.)*

Learn by Drawing

Sketching Electric Lines of Force

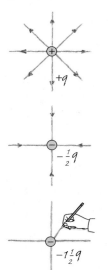

How many lines should be drawn for $-1\frac{1}{2}\,q$, and what should their direction be?

Electric Lines of Force

A convenient way of *graphically* representing the electric field pattern produced by a configuration of charges is by using *electric lines of force*, or **electric field lines**. For simplicity, consider the electric field vectors near a positive point charge, as in Fig. 15.12a. The vectors are connected in Fig. 15.12b. This is called "constructing the electric field lines" due to a point charge. Notice that the electric field is stronger nearer the charge, where the field lines are closer together. Also, at any point on a field line, the electric field direction is tangent to the line. (The lines usually have arrows attached to them that indicate the general field direction.) Lastly, note that electric field lines can't cross. If they did, it would mean that at the crossing spot there would be two directions for the force on a charge if one were placed there—a physically unreasonable result.

The general rules for sketching and interpreting electric field lines are as follows:

1. The closer together the field lines, the stronger is the electric field.
2. At any point, the direction of the electric field is tangent to the field lines.
3. The electric field lines start at positive charges and end at negative charges.
4. The number of lines leaving or entering a charge is proportional to the magnitude of that charge.
5. Electric field lines can never cross.

These rules enable us to "map" the pattern of electric lines of force due to various charge configurations. (See the Learn by Drawing feature, "Sketching Electric Lines of Force pg. 531.")

Let's now apply these rules and the superposition principle to map out the electric field line pattern due to an *electric dipole*. An **electric dipole** consists of two equal, but opposite, electric charges (or "poles," as they were known historically). Even though the net charge on the dipole is zero, it creates an electric field because the charges are separated. If they were not, the charges would cancel, and the net field would be zero everywhere.

In addition to learning how to determine electric field lines, it is important to study dipoles, because they occur in nature. For example, electric dipoles can serve as a model for permanently polarized molecules, such as the water molecule. (See Fig. 15.7.) Similarly, they can help us describe the electric properties of living organisms, such as the electric eel (Fig. 15.15a). In the next Example, we qualitatively construct the electric field pattern from a dipole.

Example 15.8 ■ Constructing the Electric Field Pattern Due to a Dipole

Using the superposition principle and the electric field line rules, construct a typical electric field line for an electric dipole.

Thinking It Through. The construction involves vector addition of the individual electric fields from the two opposite ends of the dipole.

Solution.

Given: an electric dipole of two equal and opposite charges separated by a distance, d

Find: a typical electric field line

An electric dipole is shown in ▼Fig. 15.15b. To keep track of the two fields, let's label the positive charge q_+ and the negative charge q_-. Their individual fields, $\mathbf{E}_+$ and $\mathbf{E}_-$, will be designated with the same subscripts.

Since electric fields (and thus field lines) start at positive charges, let's begin at location A, near charge q_+. Because this location is much closer to q_+ than to q_-, it follows that $E_+ > E_-$. We know that $\mathbf{E}_+$ will *always* point away from q_+ and $\mathbf{E}_-$ will *always* point toward q_-. Putting these two facts together constructs the two fields shown at A. The parallelogram method determines their vector sum: the electric field at A.

Since we are trying to map out the electric field line, and since the electric field $\mathbf{E}$ is tangent to the lines, the general direction of the electric field at A points us approximately to our next location, B. At B, there is a reduced magnitude (why?) and slight directional change in both $\mathbf{E}_+$ and $\mathbf{E}_-$.

▼ **FIGURE 15.15 Mapping the electric field due to a dipole** **(a)** Electric field pattern generated by an electric eel. **(b)** The construction of one electric field line from a dipole is shown. The electric field is the vector sum of the two fields produced by the two ends of the dipole. (See Example 15.8 for details.) **(c)** The full electric dipole field is determined by following the procedure in part (b) at other locations near the dipole.

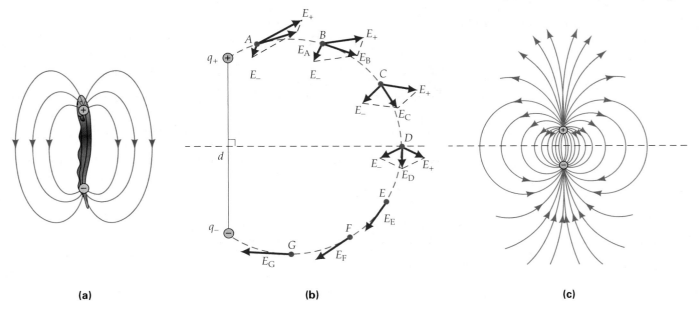

(a)   (b)   (c)

By now you should be able to see how the fields at C and D are determined. Location D is special because it is on the perpendicular bisector of the dipole axis—the imaginary line d that connects the two charges. The electric field points downward anywhere on this line. You should be able to show the construction at points E, F and G.

Lastly, we want to construct the electric field line; let's start at the positive end of the dipole, because the field lines leave that end. Since the electric field vectors are tangent to the field lines, we continue to fulfill this requirement. You should be able to fill in the other lines and understand the complete dipole field pattern shown in Fig. 15.15c.

Follow-up Exercise. Using the techniques in this Example, construct the field lines that start (a) just above the positive charge, (b) just below the negative charge, and (c) just below the positive charge. *(Answers to all Follow-up Exercises are at the back of the text.)*

▸Figures 15.16a and b show the use of the superposition principle to construct the electric field lines created by a single large charged plate. This is then extended to create the electric field pattern due to two oppositely charged parallel plates (Fig. 15.16c). Once we understand what the field looks like near one large plate, putting two together is easy. Due to the symmetric cancellation of the horizontal field components (as long as we stay away from the edges of the plate), the field is constant in direction and magnitude. Thus, between the two oppositely charged plates, the electric field is, to a good approximation, uniform and points from the positive plate to the negative plate. (Think of the direction of the force acting on a positive test charge).

The mathematical expression for the electric field magnitude between two closely spaced plates is too complicated to derive here, but the result is

$$E = \frac{4\pi k Q}{A} \quad (between\ parallel\ plates) \quad (15.5)$$

where Q is the magnitude of the total charge on *one* of the plates and A is the area of *one* plate. Parallel plates are common in electronic applications. For example, in Chapter 16 we shall see that an important electric circuit element is a device called a capacitor, which, in its simplest form, is just a set of parallel plates. Capacitors play a crucial role in lifesaving devices such as *heart defibrillators.* These instruments can restart a heart by passing an electric charge through it.

Cloud-to-ground lightning can be approximated by closely spaced parallel plates as in the next Example. (See the Insight "Lightning and Lightning Rods" on p. 534.)

PHYSLET®
ILLUSTRATION

Electric Field in a Parallel-Plate Capacitor

Example 15.9 ■ Parallel Plates: Estimating the Charge on Storm Clouds

The electric field E required to ionize moist air is about 1.0×10^6 N/C. When the field reaches this value, the least bound atoms begin leaving their molecules, eventually creating lightning. Assume that the existing value for E between the negatively charged lower cloud surface and the positively charged ground is 1.00% of the ionization value, or 1.0×10^4 N/C. (See Fig. 1a. of Insight on p. 534) Take the clouds to be squares 10 mi on each side. Estimate the magnitude of the total negative charge on the lower surface.

Thinking It Through. The electric field is given, so Eq. 15.5 can be used to estimate Q. First we must convert the cloud area A (one of the "plates") into square meters.

Solution.

Given: $E = 1.0 \times 10^4$ N/C   *Find:* Q (the magnitude of the charge
        $d = 10$ mi $\approx 1.6 \times 10^4$ m   on the lower cloud surface)

Using $A = d^2$ for the area of a square, we solve Eq. 15.5 for the magnitude of the charge (the cloud surface is negative):

$$Q = \frac{EA}{4\pi k} = \frac{(1.0 \times 10^4 \text{ N/C})(1.6 \times 10^4 \text{ m})^2}{4\pi(9.0 \times 10^9 \text{ N} \cdot \text{m}^2/\text{C}^2)} = 23 \text{ C}$$

This expression is justified only if the distance between the clouds and the ground is much less than their size. (Why?) Such an assumption is equivalent to assuming that the clouds are less than several miles from the Earth's surface.

Note, that the charge here is huge compared with the frictional static charges we develop when walking on a carpet. However, the cloud charge is spread out over a very large area, and any one small area does not contain a lot of charge.

Follow-up Exercise. In this Example, (a) what is the direction of the electric field between the cloud and the Earth? and (b) how many excess electrons are on the lower cloud surface per square meter? (*Answers to all Follow-up Exercises are at the back of the text.*)

The complete electric field patterns for some other common charge configurations are shown in ▼Fig. 15.17. Note that the electric field lines begin on positive charges and end on negative charges (or at infinity when there is no nearby negative charge). Be sure to choose the number of lines emanating from or ending at a charge in proportion to the magnitude of that charge. (See the Learn by Drawing feature "Sketching Electric Lines of Force" on p. 530.)

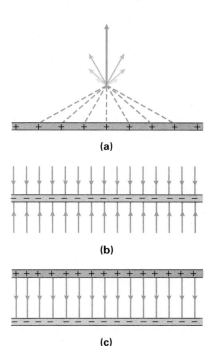

▲ **FIGURE 15.16 Electric field due to very large parallel plates** (a) Above a positively charged plate, the net electric field points upward. Here, the horizontal components of the electric fields from various locations on the plate cancel out. Below the plate, *E* points downward (not shown). (b) For a negatively charged plate, the electric field direction (shown on both sides of the plate) is reversed. (c) Superimposing the fields from both plates results in cancellation outside the plates and an approximately uniform field between them.

15.5 Conductors and Electric Fields

OBJECTIVES: To (a) describe the electric field near the surface and in the interior of a conductor, (b) determine where the highest concentration of excess charge accumulates on a charged conductor, and (c) sketch the electric field line pattern outside a charged conductor.

The electric fields associated with charged conductors (that are isolated or insulated) have several interesting properties. By definition, in a static situation, for the charges to remain at rest, the net electric force on each of them must be zero, so

The electric field is zero everywhere *inside* a charged conductor.

Also, as you might expect, excess charges on a conductor tend to get as far away from each other as possible, since they are highly mobile. Thus,

Any *excess* charge on an isolated conductor resides entirely on the surface of the conductor.

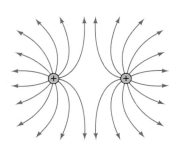

(a) Like point charges

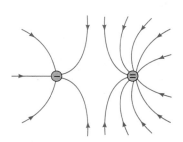

(b) Unequal like point charges

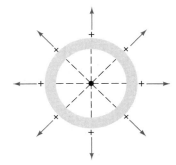

(c) Positively charged metal sphere

▲ **FIGURE 15.17 Electric fields** Electric fields for: **(a)** Like point charges. **(b)** Unequal like point charges. **(c)** A positively charged metal sphere. The electric field outside the sphere is as though all the charge on the sphere were concentrated at its center. The electric field inside the sphere is zero.

Lightning and Lightning Rods

We are all familiar with the violent release of electrical energy in the form of lightning. Although it is a common occurrence, we still have a lot to learn about the formation of lightning. It is known that during the development of a cumulonimbus (storm) cloud, a separation of charge occurs. The cloud acquires regions of different charge, with the bottom generally negatively charged. As a result, an opposite charge is induced on the surface of the Earth (Fig. 1a). Eventually, lightning may reduce this charge difference by ionizing the air, allowing a flow of charge between cloud and ground. However, air is a good insulator, so the electric field must be quite strong for ionization to occur. (See Example 15.9 for a quantitative estimate of the charge on a cloud.)

How the separation of charge takes place in a cloud is not fully understood, but it must be associated somehow with the rapid vertical movement of air and moisture within storm

FIGURE 1   Lightning and lightning rods   (a) Cloud polarization induces a charge on the Earth's surface. **(b)** When the field becomes large enough, an electrical discharge results, which we call lightning. **(c)** A lightning rod provides a path to ground so as to prevent damage.

Another property of static electric fields and conductors is that there cannot be any tangential component of the electric field at the surface of the conductor. If this were not true, charges would move *along* the surface, contrary to our assumption of a static situation. Thus,

| The electric field at the surface of a charged conductor is perpendicular to the surface.

Lastly, the excess charge on a conductor of irregular shape is most closely packed where the surface is highly curved (i.e., at the sharpest points). Since the charge is densest there, the electric field will be the greatest just above these locations. That is,

| Excess charge tends to accumulate at sharp points, or locations of highest curvature, on charged conductors. As a result, the electric field due to this excess charge is greatest at such locations.

These results are illustrated in ▶ Fig. 15.18. *Note that they are true only for conductors* and, even then, only under static conditions. Electric fields *can* exist inside nonconducting materials and also inside conductors when conditions vary with time.

To understand *why* most of the charge accumulates in the highly curved surface regions, consider the forces acting *between* charges on the surface of the conductor.

clouds. Water is a polar molecule: It has separated regions of opposite charge, and under certain circumstances water molecules can break apart to produce positively charged and negatively charged ions. Some ionization may occur as a result of frictional forces between water droplets. Another possibility is that charges are separated during the formation of ice pellets. It has been shown experimentally that as water droplets freeze, positively charged ions become concentrated in the colder, outer regions of the droplets, whereas negatively charged ions concentrate in the warmer, interior regions. Thus, a freezing droplet has a positively charged outer ice shell and a negatively charged liquid interior.

As the interior of a droplet begins to freeze, it expands and shatters the outer shell, giving rise to positively charged ice fragments, which are carried upward by turbulence in the interior of the cloud. This occurs on a large scale, and the remaining, relatively heavy negatively charged droplets eventually settle to the base of the cloud.

Most lightning occurs entirely within a cloud (intracloud discharges), where it cannot be seen directly. However, familiar visible discharges take place between clouds (cloud-to-cloud discharges) and between a cloud and the Earth (cloud-to-ground discharges). Pictures of cloud-to-ground discharges taken with special high-speed cameras reveal a nearly invisible downward ionization path. The lightning discharges in a series of jumps or steps and so is called a *stepped leader*. As the leader nears the ground, positively charged ions in the form of a *streamer* rise from trees, tall buildings, or the ground to meet it.

When a streamer and a leader make contact, the electrons along the leader channel begin to flow downward. The initial flow is near the ground, and as it continues, electrons positioned successively higher begin to migrate downward. Hence, the path of electron flow is extended upward in a *return stroke*. The surge of charge in the return stroke causes the conductive path to be illuminated, producing the bright flash seen by the eye and recorded in time-exposure photographs of lightning (Fig. 1b). Most lightning flashes have a duration of less than 0.50 s. Usually, after the initial discharge ionization again takes place along the original channel, and another return stroke occurs. Typical lightning events have three or four return strokes.

Ben Franklin is often said to have been the first to demonstrate the electrical nature of lightning. In 1750, he suggested an experiment using a metal rod on a tall building. However, a Frenchman named d'Alibard set up the experiment and drew sparks from a rod during a thunderstorm. Franklin later performed a similar experiment with a kite he flew, also during a thunderstorm. Both were extremely lucky not to have been electrocuted. *Under no circumstances should you try to duplicate this experiment.* On average, lightning kills 200 people a year in the United States and injures another 550.

A practical outcome of Franklin's work was the lightning rod. It consists of a pointed metal rod connected by a wire to a metal rod that is driven into the Earth, or "grounded" (Fig. 1c). Franklin wrote that the rod "either prevents the stroke from the cloud or, if the stroke is made, conducts the stroke to the Earth with safety of the building."

The latter comment accurately describes the principle of operation of a lightning rod. (See Section 15.5.) The elevated rod intercepts the downward ionized stepped leader, discharging it harmlessly to the ground before it reaches a structure or makes contact with an upward streamer. This prevents the formation of a damaging electrical surge associated with a return stroke. (See related exercises 70 and 72 at the end of the chapter.)

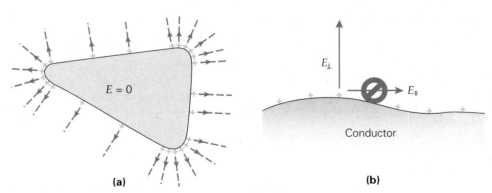

(a)   **(b)**

▲ **FIGURE 15.18 Electric fields and conductors** **(a)** Under static conditions, the electric field is zero inside a conductor. Any excess charge resides on the conductor's surface. For an irregularly shaped conductor, the excess charge accumulates in the regions of highest curvature (the sharpest points), as shown. The electric field near the surface is perpendicular to that surface and strongest where the charge is densest. **(b)** Under static conditions, the electric field must *not* have a component tangential to the conductor's surface.

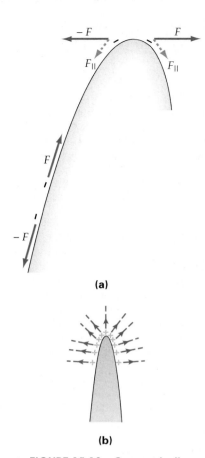

(a)

(b)

▲ **FIGURE 15.19 Concentration of charge on a curved surface**
(a) On a flat surface, the repulsive forces between excess charges are parallel to the surface and tend to push the charges apart. On a sharply curved surface, in contrast, these forces are directed at an angle to the surface. Their components parallel to the surface are smaller, allowing more charge to concentrate in such areas. **(b)** Taken to the extreme, a sharply pointed metallic needle has a dense concentration of charge at the tip. This produces a large electric field in the region above the tip, which is the principle of the lightning rod.

(See ◂Fig. 15.19a.) Where the surface is fairly flat, these forces will be directed nearly parallel to the surface. The charges will spread out until the parallel forces from neighboring charges in opposite directions cancel out. At a sharp end, the forces between charges will be directed more nearly perpendicular to the surface. Here there will be little tendency for the charges to move parallel to the surface, since the force component parallel to the surface will be very small. Thus, the highly curved regions of the surface accumulate a higher concentration of charge.

An interesting situation occurs if there is a large concentration of charge on a conductor with a sharp point (Fig. 15.19b). The electric field strength in the region above the point may be high enough to start the ionization of air molecules (i.e., to pull or push electrons off the molecules). The resulting free electrons are further accelerated by the electric field and can cause secondary ionizations by striking other air molecules, resulting in an avalanche of electrons, visible as a spark discharge. More charge can be placed on a gently curved conductor, such as a sphere, before a spark discharge will occur. The concentration of charge at the sharp point of a conductor is one reason for the effectiveness of lightning rods. (See the Insight on p. 534.)

To further examine the properties of conductors with regard to excess charge and external electric fields, consider the next two conceptual examples.

Conceptual Example 15.10 ■ The Classic Ice Pail Experiment

A positively charged rod is held inside an isolated metal container that has uncharged electroscopes conductively attached to its inside and outside surfaces (▸Fig. 15.20a). What will happen to the leaves of the electroscopes? (Justify your answer.) (a) Neither electroscope's leaf will show a deflection. (b) Only the outside-connected electroscope's leaf will show a deflection. (c) Only the inside-connected electroscope's leaf will show a deflection. (d) The leaves of both electroscopes will show deflections.

Reasoning and Answer. The positively charged rod will attract negative charges, causing the inside of the metal container to become negatively charged. The outside electroscope will thus acquire a positive charge. Hence, both electroscopes will be charged (though with opposite signs) and show deflections, so the answer is (d). A similar experiment was performed by the 19th-century English physicist Michael Faraday using ice pails, so this setup is often called Faraday's ice pail experiment.

Follow-up Exercise. Suppose that the positively charged rod *touched* the metal container as shown in Fig. 15.20b. What would be the effect on the electroscopes? *(Answers to all Follow-up Exercises are at the back of the text.)*

Conceptual Example 15.11 ■ A Practical Use for Conductors: Shielding From External Electric Fields

An uncharged solid metal sphere is placed in a uniform electric field (▸Fig. 15.21a) created by two oppositely charged plates not shown in the figure. (a) Describe the electric field line pattern after the sphere is inserted. (b) The sphere is now "hollowed out" as shown in Fig. 15.21b. What is the electric field inside the hollow region? Explain your reasoning.

Reasoning and Answer. (a) Remember that in a metal, there are many electrons ready to move about. When static conditions are reached, the electric field lines must be perpendicular to the surface of the sphere, and since the field is zero inside, the lines must end or start at the metal surface. For the field lines on the left to end at the spherical surface, a negative charge must be induced on that side. For the field lines to continue to the negatively charged plate, they must start at the right-hand side of the sphere, which therefore must have a positive induced charge on it. Since the total charge on the sphere is zero, by charge conservation, the induced charges are equal in magnitude, but opposite in sign. Putting it all together, we can meet all the electric field requirements with the field lines configured as in Fig. 15.21b.

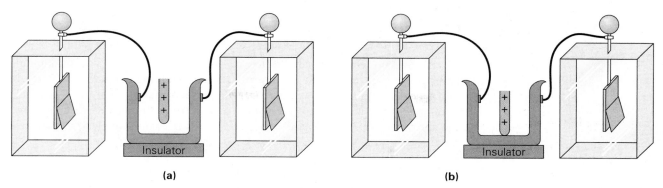

(a)   (b)

(b) Since there is no electric field inside the solid metal in part (a), removing most of the interior leaves the same result: a hollow region with no electric field (Fig. 15.21c). This effect, *electric shielding*, is commonly used to protect delicate instruments from stray electric fields from, say, nearby electric power outlets. Typically, the apparatus is put inside a metallic screened cage, named a *Faraday cage*, after the scientist Michael Faraday.

Follow-up Exercise. A single negative point charge is placed at the center of an initially uncharged metallic spherical shell. Describe the electric field line patterns (a) in the interior hollow section of the shell, (b) in the shell itself, and (c) outside the shell. (*Hint*: charges are induced on both surfaces of the spherical shell. Why?) (*Answers to all Follow-up Exercises are at the back of the text.*)

▲ **FIGURE 15.20 An ice pail experiment** **(a)** See Conceptual Example 15.10. **(b)** See Follow-up Exercise 15.10.

---

*15.6 Gauss's Law for Electric Fields: A Qualitative Approach

OBJECTIVES: To (a) state the physical basis of Gauss's law and (b) use the law to make qualitative predictions.

One of the fundamental laws that govern the behavior of electric fields was discovered by Karl Friedrich Gauss (1777–1855), a German mathematician. In its mathematical form, this relationship can be used to calculate electric field strengths, but doing that involves techniques beyond the scope of the book. However, even a conceptual look at Gauss's law can teach us some interesting physics.

Consider the single positive electric charge in ▶Fig. 15.22a. Now picture an *imaginary closed surface* surrounding the charge. Such a surface is called a **Gaussian surface**. Now, let us designate electric field lines that pass through the surface and point outward as positive and inward pointing ones as negative. If we count the lines of both types and total them up (i.e., subtract the number of negative lines from the number of positive ones), we find that the total is positive. (Of course, in this case there are *only* positive lines!) That is, there is a net number of outward-pointing electric field lines passing through the surface. Similarly, for a negative charge (Fig. 15.22b), the count would yield a negative total, indicating a net number of inward-pointing lines passing through the surface. Note that these results would be true for *any* closed surface surrounding the charge, regardless of its shape. If we double the magnitude of the negative charge (Fig. 15.22c), our negative field line count would also double. (Why?)

Figure 15.22d shows a dipole with four different imaginary closed surfaces. As we have seen for single charges, Surface 1 encloses a net positive charge and therefore has a positive field line count. Similarly, Surface 2, which encloses a net negative charge, has a negative field line count. The more interesting cases are Surfaces 3 and 4. Note that both include zero net charge—Surface 3 because it includes no charges at all and Surface 4 because it includes equal and opposite charges that cancel out. Note that both Surfaces 3 and 4 have a net field line count of zero, correlating with no net charge enclosed.

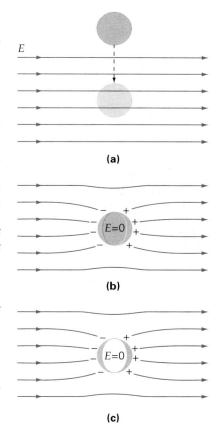

(a)

(b)

(c)

▲ **FIGURE 15.21 Electric shielding** See Conceptual Example 15.11.

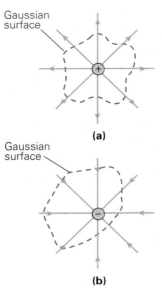

(a)

(b)

(c)

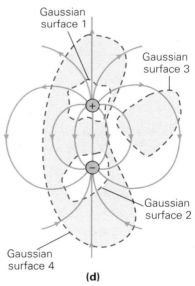

(d)

▲ **FIGURE 15.22 Various Gaussian surfaces and lines of force** **(a)** surrounding a single positive point charge, **(b)** surrounding a single negative point charge, and **(c)** surrounding a larger negative point charge. **(d)** Four different surfaces surrounding various parts of an electric dipole.

These situations illustrate the fundamental (conceptual) physical idea of **Gauss's law:***

> The net number of electric field lines passing through an imaginary closed surface is proportional to the amount of net charge enclosed within that surface.

An everyday analogy illustrated in ▼ Fig. 15.23 may help you understand this principle. If you surround a lawn sprinkler with an imaginary surface, you will find that there is a net flow of water out through that surface—clear evidence that what you have inside is a "source" of water (disregarding the pipe's bringing water into the sprinkler). In an analogous way, a net outward-pointing electric field indicates the presence of a net positive charge inside the surface, since positive charges, as we know, are "sources" of the electric field. Similarly, a drain inside our imaginary surface would give itself away because there would be a net inward flow of water through the surface (again neglecting the pipe's carrying the water away). The following Example illustrates the power of Gauss's law, even in its qualitative form.

Conceptual Example 15.12 ■ Charged Conductors Revisited: Gauss's Law

A net charge Q is placed on a conductor of arbitrary shape (▼ Fig. 15.24). Use the qualitative version of Gauss's law to prove that all the charge must lie on the conductor's surface under electrostatic conditions.

Reasoning and Answer. Since the situation is static equilibrium, there can be no electric field inside the volume of the conductor; otherwise, the almost-free electrons would be moving around. Let us take a Gaussian surface that follows the shape of the actual conductor, but is *just barely* inside the actual surface. Since there are no electric field lines inside the conductor, there are no electric field lines passing through our imaginary surface in either direction. So the net result is zero electric field lines penetrating the Gaussian surface. But, by Gauss's law, the net number of field lines is proportional to the amount of charge enclosed inside the surface. Thus, there must be no net charge within the surface!

Since our imaginary surface can be made as close as we wish to the true surface, it follows that the excess charge, if it cannot be inside the volume of the conductor, must be on the conductor's surface. This is a very elegant proof of what we determined before by considering repulsion and forces between charges.

Follow-up Exercise. In this Example, if the net charge on the conductor is negative, what is the sign of the net number of lines through a Gaussian surface that completely encloses the conductor, excess charge and all? Explain your reasoning. *(Answers to all Follow-up Exercises are at the back of the text.)*

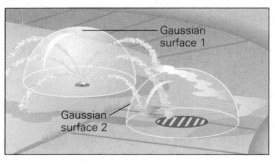

▲ **FIGURE 15.23 Water analogy to Gauss's law** A net outward flow of water indicates a source of water inside closed surface 1. A net inward flow of water indicates a water drain inside closed surface 2.

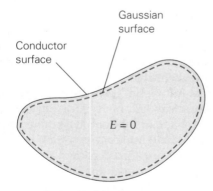

▲ **FIGURE 15.24 Gauss's law: Excess charge on a conductor** See Conceptual Example 15.12.

*Strictly speaking, this is Gauss's law for electric fields. There is also a version of Gauss's law for magnetic fields, which we will not discuss.

Chapter Review

Important Concepts and Equations

- The **law of charges, or charge–force law**, states that like charges repel, and opposite charges attract.
- The principle of **charge conservation** means that the net charge of an isolated system remains constant.
- **Conductors** are materials (usually metallic) that conduct electric charge readily because their atoms have one or more loosely bound electrons.
- **Insulators** are materials that do not easily gain, lose, or conduct electric charge.
- **Electrostatic charging** involves processes that enable an object to gain a net charge. Among these processes are charging by friction, contact (conduction), and induction.
- The **electric polarization** of an object involves creating separate and equal amounts of positive and negative charge in different locations on that object.
- **Coulomb's law** expresses the magnitude of the force between two point charges:

$$F_e = \frac{kq_1q_2}{r^2} \quad \text{(two point charges)} \quad (15.2)$$

where $k \approx 9.00 \times 10^9 \, \text{N} \cdot \text{m}^2/\text{C}^2$

- The **electric field** is a vector field that describes how charges modify the space around them. It is defined as the electric force per unit positive charge, or

$$\mathbf{E} = \frac{\mathbf{F}_{\text{on } q_0}}{q_0} \quad (15.3)$$

- According to the **superposition principle for electric fields**, the (net) electric field at any location due to a configuration of charges is the vector sum of the individual electric fields from individual charges that make up that configuration.
- **Electric field lines** are a graphical visualization of the electric field and are created by connecting electric field vectors. The spacing between lines is inversely related to the field strength, and tangents to the lines give the direction of the electric field.
- Under static conditions, the **electric fields associated with conductors** have the following properties:

  The electric field is zero everywhere *inside* a charged conductor.

  Any *excess* charge on an isolated conductor resides entirely on its surface.

  The electric field at the surface of a charged conductor is perpendicular to its surface.

  Excess charge on a conductor tends to accumulate at locations of highest surface curvature.

  The electric field due to the excess charge on a conductor is greatest at locations of highest surface curvature.

Exercises*

15.1 Electric Charge

1. A combination of two electrons and three protons would have a net charge of (a) +1, (b) −1, (c) $+1.6 \times 10^{-19}$ C, or (d) -1.6×10^{-19} C.

2. The directions of the electric forces on two interacting charges are given by (a) the conservation of charge, (b) the charge–force law, (c) the magnitude of the charges, or (d) the distance between the charges.

3. CQ (a) How do we know that there are two types of electric charge? (b) What would be the effect of designating the charge on the electron as positive and the charge on the proton as negative?

4. CQ An electrically neutral object can be given a net charge by several means. Does this violate the conservation of charge? Explain.

*Take k to be *exact* at a $9.00 \times 10^9 \, \text{N} \cdot \text{m}^2/\text{C}^2$ and e to be exact at 1.60×10^{-19} C for significant-figure purposes.

5. CQ If a solid neutral object becomes positively charged, does its mass increase or decrease? What if it becomes negatively charged?

6. CQ How can you determine the type of charge on an object? Explain?

7. CQ If two objects electrically repel each other, are both necessarily charged? What if they attract each other?

8. ■ What would be the net electric charge of an object with 1.0 million excess electrons?

9. ■ In walking across a carpet, you acquire a net negative charge of 50 μC. How many excess electrons do you have?

10. IE ■ A glass rod rubbed with silk acquires a charge of $+8.0 \times 10^{-10}$ C. (a) Is the charge on the silk (1) positive, (2) zero, or (3) negative? Why? (b) What is the charge on the silk, and how many electrons have been transferred to the silk?

11. **IE ■** A rubber rod rubbed with fur acquires a charge of -4.8×10^{-9} C. (a) Is the charge on the fur (1) positive, (2) zero, or (3) negative? Why? (b) What is the charge on the fur, and how much mass is transferred to the rod?

12. **■■** An alpha particle is the nucleus of a helium atom with no electrons. What would be the charge on two alpha particles?

15.2 Electrostatic Charging

13. **CQ** A rubber rod is rubbed with fur. The fur is then quickly brought near the bulb of an uncharged electroscope. What is the sign of the charge on the leaves of the electroscope and why? [Refer to Exercise 11.]

14. **CQ** Explain why a stream of water is deflected, or bent, when an electrically charged object is brought close to it. (Try this yourself with a charged plastic pen or hard rubber comb.)

15. **CQ** A balloon is charged by rubbing and then clings to a wall. Does this mean that the wall is also charged? Explain.

16. **CQ** Fuel trucks often have metal chains reaching from their frames to the ground. Why?

17. **CQ** Is there a gain or loss of electrons when an object is electrically polarized? Explain.

18. **CQ** How could an electroscope be positively charged by induction? How could you prove that it was positively charged?

19. **CQ** Two metal spheres mounted on insulated supports are in contact. How could both spheres be given a net electric charge without being touched by a charged object? Would the spheres be charged positively or negatively?

15.3 Electric Force

20. The magnitude of the electric force between two point charges is given by (a) the charge–force law, (b) conservation of charge, (c) Coulomb's law, or (d) both (a) and (b).

21. Compared with the electric force, the gravitational force between two protons is (a) about the same, (b) somewhat larger, (c) very much larger, or (d) very much smaller.

22. **CQ** The Earth attracts us by its gravitational force, but the electric force is much greater than the gravitational force. Why don't we experience an electric force?

23. **CQ** In calculating planetary orbits around the Sun, why can astronomers safely ignore the electric force?

24. **CQ** Would the net forces on two fixed charged objects be affected by the presence of a third charged object?

25. **CQ** Coulomb's law is also called an inverse square law. What does that tell you about the relationship between force and distance?

26. **IE ■** An electron that is a certain distance from a proton is acted upon by an electrical force. (a) If the electron were moved twice that distance away from the proton, would the electrical force be (1) 2, (2) $\frac{1}{2}$, (3) 4, or (4) $\frac{1}{4}$ times the original force? Why? (b) If the initial electric force is F, and the electron were moved to one-third the original distance toward the proton, what would be the new electrical force?

27. **■** Two identical point charges are a fixed distance apart. By what factor would the electric force be affected if (a) one of their charges were doubled and the other were halved, (b) both their charges were halved, and (c) one charge were halved and the other were left unchanged?

28. **■** In a certain organic molecule, the nuclei of two carbon atoms are separated by a distance of 0.25 nm. What is the magnitude of the electric repulsion between them?

29. **■** An electron and a proton are separated by 2.0 nm. (a) What is the magnitude of the force on the electron? (b) What is the net force on the system?

30. **IE ■** Two charges originally separated by a certain distance are moved farther apart, until the force between them has decreased by a factor of 10. (a) Is the new distance (1) less than 10, (2) equal to 10, or (3) greater than 10 times the original distance. Why? (b) If the original distance was 30 cm, how far apart are the charges?

31. **■** Two charges are separated until they are a distance of 100 cm apart, causing the electric force between them to increase by a factor of exactly 5. What was their initial separation distance?

32. **■■** The closest distance between the singly charged sodium and chlorine ions in crystals of table salt is 2.82×10^{-10} m. What is the attractive electric force between the ions?

33. **■■** Two point charges of $-2.0\ \mu C$ are fixed at opposite ends of a meterstick. Where on the meterstick could (a) a free electron and (b) a free proton be in electrostatic equilibrium?

34. **■■** Two point charges of $-1.0\ \mu C$ and $+1.0\ \mu C$ are fixed at opposite ends of a meterstick. Where could (a) a free electron and (b) a free proton be in electrostatic equilibrium?

35. **■■** Two charges, q_1 and q_2, are located at the origin and at $(0.50$ m, $0)$, respectively. Where on the x-axis must a third charge, q_3, of arbitrary sign be placed to be in electrostatic equilibrium if (a) q_1 and q_2 are like charges of equal magnitude, (b) q_1 and q_2 are unlike charges of equal magnitude, and (c) $q_1 = +3.0\ \mu C$ and $q_2 = -7.0\ \mu C$?

36. **■■** On average, the electron and proton in a hydrogen atom are separated by a distance of 5.3×10^{-11} m

(▼Fig. 15.25). Assuming the orbit of the electron to be circular, (a) what is the electric force on the electron? (b) What is the electron's orbital speed? (c) What is the magnitude of the electron's centripetal acceleration in units of g?

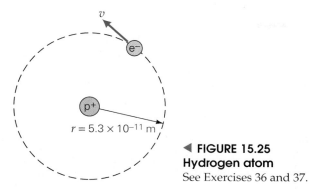

◀ **FIGURE 15.25**
Hydrogen atom
See Exercises 36 and 37.

37. ■■ Compute the gravitational force between the electron and proton in the hydrogen atom (Fig. 15.25), and then calculate the ratio of the magnitudes of the electric force to the gravitational force, using the result of Exercise 36a.

38. ■■■ Three charges are located at the corners of an equilateral triangle, as depicted in ▼Fig. 15.26. What are the magnitude and the direction of the force on q_1?

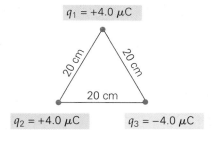

◀ **FIGURE 15.26**
Charge triangle
See Exercises 38, 59, and 60.

39. ■■■ Four charges are located at the corners of a square, as illustrated in ▼Fig. 15.27. What are the magnitude and the direction of the force (a) on charge q_2 and (b) on charge q_4?

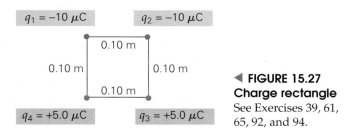

◀ **FIGURE 15.27**
Charge rectangle
See Exercises 39, 61, 65, 92, and 94.

40. ■■■ Two 0.10-g pith balls are suspended from the same point by threads 30 cm long. (Pith is a light insulating material once used to make helmets worn in tropical climates.) When the balls are given equal charges, they

come to rest 18 cm apart, as shown in ▼Fig. 15.28. What is the magnitude of the charge on each ball? (Neglect the mass of the thread.)

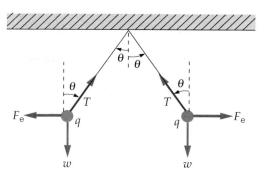

▲ **FIGURE 15.28 Repelling pith balls** See Exercise 40.

15.4 Electric Field

41. The electric field due to a positive point charge (a) varies as $1/r$, (b) points toward the charge, (c) points away from the charge, or (d) has a finite range.

42. The units of electric field are (a) C, (b) N/C, (c) N, or (d) J.

43. At a point in space, an electric force acts vertically downward on an electron. The direction of the electric field at that point is (a) down, (b) up, (c) zero, or (d) undetermined by the data.

44. CQ How is the relative magnitude of the electric field in different regions determined from a field vector diagram?

45. CQ How can the relative magnitudes of the field in different regions be determined from an electric field line diagram?

46. CQ Can electric field lines ever cross? Explain.

47. CQ A positive charge is inside an isolated metal sphere, as shown in ▼Fig. 15.29. Describe the situation in terms of the electric field and the charge on the sphere. How would the situation change if the charge were negative?

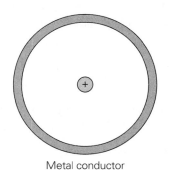

Metal conductor

◀ **FIGURE 15.29 A point charge inside a thick metal spherical shell**
See Exercise 47.

48. CQ If the electric field due to the excess charge on the Earth points downward, what type of charge does the Earth carry? Why?

49. CQ Could the electric field due to two charges ever be zero at some location(s) nearby? If yes, describe and sketch the situation.

50. IE ■ (a) If the distance from a charge is doubled, is the magnitude of the electric field (1) 2, (2) $\frac{1}{2}$, (3) 4, or (4) $\frac{1}{4}$ times as large? Why? (b) If the original electric field is 1.0×10^{-4} N/C and the distance is reduced to one-third the original distance, what is the magnitude of the new electric field?

51. ■ An isolated electron is acted on by an electric force of 3.2×10^{-14} N. What is the magnitude of the electric field at the electron's location?

52. ■ What is the magnitude of the electric field at a point 0.25 cm away from a point charge of $+2.0\ \mu$C?

53. ■ At what distance from a proton is the magnitude of the electric field 1.0×10^5 N/C?

54. IE ■■ Two fixed charges, $-4.0\ \mu$C and $-5.0\ \mu$C, are separated by a certain distance. (a) Is the net electric field at a location halfway between the two charges (1) directed toward the $-4.0\ \mu$C charge, (2) zero, or (3) directed toward the $-5.0\ \mu$ C charge? Why? (b) If the charges are separated by 20 cm, calculate the magnitude of the net electric field halfway between the charges?

55. ■■ What would be the magnitude and the direction of a vertical electric field that would just support the weight of a proton on the surface of the Earth? of an electron?

56. IE ■■ Two charges, $-3.0\ \mu$C and $-4.0\ \mu$C, are located at $(-0.50$ m, 0) and (0.75 m, 0), respectively. (a) Is the location of a point on the x-axis between the two charges where the electric field is zero (1) nearer the $-3.0\ \mu$C charge, (2) at the origin, or (3) nearer the $-4.0\ \mu$C charge? Why? (b) Find the location of the point where the electric field is zero.

57. ■■ Three charges, $+2.5\ \mu$C, $-4.8\ \mu$C, and $-6.3\ \mu$C, are located at $(-0.20$ m, 0.15 m), (0.50 m, -0.35 m), and $(-0.42$ m, -0.32 m), respectively. What is the electric field at the origin?

58. ■■ Two charges of $+4.0\ \mu$C and $+9.0\ \mu$C are 30 cm apart. Where on the line joining the charges is the electric field zero?

59. ■■ What is the electric field at the center of the triangle in Fig. 15.26?

60. ■■ Compute the electric field at a point midway between charges q_1 and q_2 in Fig. 15.26.

61. ■■ What is the electric field at the center of the square in Fig. 15.27?

62. ■■ A particle with a mass of 2.0×10^{-5} kg and a charge of $+2.0\ \mu$C is released in a (parallel-plate) uniform horizontal electric field of 12 N/C. (a) How far horizontally does the particle travel in 0.50 s? (b) What is the horizontal component of its velocity at that point? (c) If the plates are 5.0 cm on each side, how much charge is on each?

63. IE ■■ Two very large parallel plates are oppositely and uniformly charged. (a) The electric field between the plates does *not* depend on (1) the charge on the plates, (2) the area of the plates, or (3) the separation of the plates. Why? (b) If the field between the plates is 1.7×10^6 N/C, how dense is the charge on each plate (in μC/m^2)?

64. ■■ Two square, oppositely charged conducting plates measure 20 cm on each side. They are placed close together and parallel to each other. They have charges of $+4.0$ nC and -4.0 nC, respectively. (a) What is the electric field between the plates? (b) What force is exerted on an electron in this region?

65. ■■■ Compute the electric field at a point 4.0 cm from q_2 along a line running toward q_3 in Fig. 15.27.

66. ■■■ Two equal and opposite charges form a dipole, as shown in ▼Fig. 15.30. (a) Draw the electric field vectors to determine the direction of the net electric field at point P. (b) What is the magnitude of the electric field at point P, in terms of k, q, d, and x? (c) If point P is very far away, show that the result is approximately $E = kqd/x^3$.

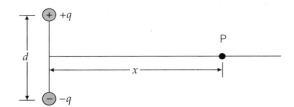

▲ **FIGURE 15.30 Electric dipole field** See Exercise 66.

15.5 Conductors and Electric Fields

67. In electrostatic equilibrium, is the electric field just below the surface of a charged conductor (a) the same value as the field just above the surface (b) zero, (c) infinite, or (d) kq/R^2?

68. An uncharged thin metal slab is placed in an external electric field that points horizontally to the left. What is the electric field inside the slab? (a) Zero, (b) the same value as the original external field, or (c) somewhere between zero and the original external field value.

69. The direction of the electric field at the surface of a charged conductor under electrostatic conditions (a) is parallel to the surface, (b) is perpendicular to the surface, (c) is at a 45° angle to the surface, or (d) depends on the charge on the conductor.

70. **CQ** Is it safe to stay in a car in a lightning storm (▼Fig. 15.31)? Explain.

▲ **FIGURE 15.31 Safe inside a car?** See Exercise 70.

71. **CQ** Under electrostatic conditions, the excess charge on a conductor is on its surface. Does this mean that all of the conduction electrons in a conductor are on the surface? Explain.

72. **CQ** Tall buildings have lightning rods to protect them from lightning strikes. Explain why the rods are pointed in shape and taller than the buildings.

73. **IE ■** A solid conducting sphere is surrounded by a thick, spherical conducting shell. Assume that a total charge $+Q$ is initially placed at the center of the inner sphere and then released. After equilibrium is reached, the inner surface of the shell will have (1) negative, (2) zero, or (3) positive charge. How much charge is on (b) the interior of the solid sphere, (c) the surface of the solid sphere, (d) the inner surface of the shell, and (e) the outer surface of the shell?

74. **■** In Exercise 73, what is the electric field direction (a) in the interior of the solid sphere, (b) between the sphere and the shell, (c) inside the shell, and (d) outside the shell?

75. **■■** In Exercise 73, write expressions for the electric field magnitude (a) in the interior of the solid sphere, (b) between the sphere and the shell, (c) inside the shell, and (d) outside the shell. Your answer should be in terms of Q, r (the distance from the center of the sphere), and k.

76. **■■** A flat, triangular piece of metal with rounded corners has a net positive charge on it. Sketch the charge distribution on the surface and the electric field lines near the surface of the metal (including their direction).

77. **■■■** Approximate a metal needle as a long cylinder with a very pointed, but slightly rounded, end. Sketch the charge distribution and outside electric field lines if the needle has an excess of electrons on it.

*15.6 Gauss's Law for Electric Fields: A Qualitative Approach

78. A Gaussian surface surrounds an object with a net charge of $+5.0\,\mu C$. Which of the following is true? (a) More electric field lines will point outward than inward. (b) More

electric field lines will point inward than outward. (c) The net number of field lines through the surface is zero.

79. **CQ** How many net electric field lines would pass through a Gaussian surface located completely within the region between a set of oppositely charged parallel plates?

80. **CQ** Two concentric spherical surfaces enclose a point charge. The radius of the outer sphere is twice that of the inner one. Which sphere will have more electric field lines penetrating its surface?

81. **■** The same Gaussian surface is used to surround two charged objects separately. The net number of field lines penetrating the surface is the same in both cases, but the lines are oppositely directed. What can you say about the net charges on the two objects?

82. **IE ■■** A Gaussian surface has 16 field lines leaving it when it surrounds a point charge of $+10.0\,\mu C$ and 75 field lines entering it when it surrounds an unknown point charge. (a) The magnitude of the unknown charge is (1) greater than $10.0\,\mu C$, (b) equal to $10.0\,\mu C$, or (3) less than $10.0\,\mu C$. Why? (b) What is the unknown charge?

83. **■■** If 10 field lines leave a Gaussian surface when it completely surrounds the positive end of an electric dipole, what would the count be if the surface surrounded just the other end?

84. **■■** Suppose a Gaussian surface encloses both a positive point charge that has six field lines leaving it and a negative point charge with twice the magnitude of charge as the positive one. What is the net number of field lines passing through the Gaussian surface?

85. **■■** If a net number of electric field lines point outward from a Gaussian surface, does that necessarily mean there are no negative charges in the interior? Explain with an example.

Additional Exercises

86. An electron is in a constant electric field of $3.5 \times 10^3\,N/C$ directed along the negative y-axis. (a) What is the force on the electron? (b) If the electron is released, how much kinetic energy does it have after it has moved 10 cm?

87. A 2.5-g copper penny is given a charge of $-0.50\,\mu C$. (a) How many excess electrons make up this charge? (b) What is the percentage of the mass of these excess electrons to the mass of the penny?

88. Two fixed charges, $-3.0\,\mu C$ and $-5.0\,\mu C$, are 0.40 m apart. (a) Where should a third charge of $-1.0\,\mu C$ be placed to put the system in electrostatic equilibrium? (b) What about a third charge of $+1.0\,\mu C$?

89. An electron is released from rest at the origin in a uniform electric field that has a magnitude of 450 N/C in the positive x-direction. (a) How long will it take the electron to travel 1.0 m? (b) What will its position be (in x–y coordinates) at half that time?

90. Find the electric field at point O for the charge configuration shown in ▼Fig. 15.32.

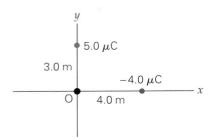

▲ **FIGURE 15.32 Electric field**   See Exercise 90.

91. A charge of $+6.0\ \mu C$ at the origin of a set of coordinate axes is acted upon by a force of 1.8 N in the positive y-direction due to another charge located 0.30 m away. What are the magnitude and the position of the other charge?

92. Compute the electric field at a point midway between charges q_1 and q_4 in Fig. 15.27.

93. A solid spherical (uncharged) conductor is placed in a uniform electric field pointing to the right. Sketch the charges on the conductor and the electric field after electrostatic conditions are attained.

94. If a Gaussian surface completely surrounds the charge arrangement in Fig. 15.27, what is the sign of the net number of lines passing through the surface?

95. For an electric dipole, the product qd is the *dipole moment*. It is taken to be the vector **p** by assuming that it points from the negative end of the dipole toward the positive end. With a sketch, show that a dipole in a uniform field will rotate until its dipole moment lines up with the field.

96. What happens to the electric dipole in Exercise 95 if the field is not uniform (▼Fig. 15.33), assuming that the dipole is free to move?

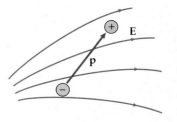

◀ **FIGURE 15.33 Dipole moment, dipoles in electric field**   An electric dipole placed in a nonuniform electric field. See Exercise 96.

97. CQ In result (c) of Exercise 66, the dipole moment qd (see definition in Exercise 95) is in the numerator. Explain why the electric field on the axis perpendicular to the dipole goes to zero if d is zero.

98. A uniform metal slab is inserted between and parallel to a pair of oppositely charged parallel plates. Sketch the resulting electric field.

99. At a speed of $6.0 \times 10^7\,\text{m/s}$, an electron in a computer monitor enters midway between two parallel oppositely charged plates, as shown in ▼Fig. 15.34. If the electric field between the plates is $2.0 \times 10^4\,\text{N/C}$, what is the vertical deflection d of the electron when it leaves the plates?

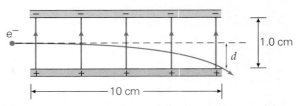

▲ **FIGURE 15.34 Electron in a computer monitor**   See Exercise 99.

Electric Potential, Energy, and Capacitance

INSIGHT

■ Electric Potential and Nerve Signal Transmission

Learn by Drawing

✎ ΔV Is Independent of Reference Point

✎ Graphical Relationship between Electric Field Lines and Equipotentials

The girl in the photo is experiencing some electrical effects. In fact, she is charged to an electric potential of several thousand volts! But household circuits operate at only 120 volts and can still give you a nasty and potentially dangerous shock. Yet the girl doesn't seem to be having a problem. What's going on? You'll find the explanation of this and many other electrical phenomena in this and the next two chapters. This chapter introduces the basic concept of electric potential and examines its properties and its usefulness in describing some electrical phenomena and instruments.

16.1 Electric Potential Energy and Electric Potential Difference

OBJECTIVES: To (a) understand the concept of electric potential difference ("voltage") and its relationship to electric potential energy and (b) calculate electric potential differences.

In Chapter 15, electrical effects were analyzed in terms of electric field vectors and electric lines of force. Recall that studying mechanics initiated the use of Newton's laws (Chapter 4) with free-body diagrams and vectors. Then,

545

a search for a simpler approach led to the introduction of *scalar* quantities such as work, kinetic energy, and potential energy (Chapter 5). With these concepts, energy (conservation) methods could be employed to solve many problems that would have been much more difficult using the vector (force) approach. It turns out to be extremely useful, both conceptually and from a problem-solving standpoint, to extend these energy methods to the study of electric fields.

Electric Potential Energy

Let's start with one of the simplest electric field patterns: the field between two large, oppositely charged parallel plates. As we learned in Chapter 15, near the center of the plates the field is uniform in magnitude and direction (◀Fig. 16.1a). Suppose a small positive charge q_0 is moved at constant velocity against the electric field, $\mathbf{E}$, directly from the negative plate (A) to the positive plate (B). An external force with the same magnitude as the electric force, $F_e = q_0 E$, is required to accomplish this. The work done by the external force is positive, since the force and the displacement are in the same direction ($\theta = 0°$ in Eq. 5.2). Thus, the amount of work done by the external force is $W = F_e d \cos 0° = +q_0 Ed$.

If the test charge is released when it reaches the positive plate, it will accelerate back towards the negative plate, gaining kinetic energy. By moving the charge from plate A to plate B, the external force has increased the charge's **electric potential energy**, U_e, by an amount equal to the work done on the charge $(U_B > U_A)$:

$$\Delta U_e = U_B - U_A = q_0 Ed$$

Note: From Eq. 15.3, $E = F/q_0$, we get $F = q_0 E$.

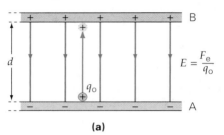

$$E = \frac{F_e}{q_0}$$

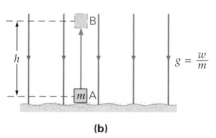

$$g = \frac{w}{m}$$

(b)

▲ **FIGURE 16.1 Changes in potential energy in uniform electric and gravitational fields** **(a)** Moving a positive charge q_0 against the electric field requires positive work and increases the electric potential energy. **(b)** Moving a mass m against the gravitational field requires positive work and increases the gravitational potential energy.

The gravitational analogy to the parallel-plate electric field is the gravitational field near the Earth's surface, where that field is uniform in direction and magnitude. When an object is raised a vertical distance h at constant velocity, the change in its potential energy is positive $(U_B > U_A)$ and equal to the work done by the external (lifting) force. This force is equal to the weight of the object: $F = w = mg$ (Fig. 16.1b). Thus, the increase in gravitational potential energy is

$$\Delta U_g = U_B - U_A = Fh = mgh$$

To distinguish clearly between the two situations, different symbols are used (h and d) to represent distances in gravitational and electric fields, respectively.

Electric Potential Difference

In dealing with electric forces, defining the electric field as the electric force per unit positive test charge eliminated the dependence on the test charge. Then, with knowledge of the electric field, the force on *any* point charge placed at that location could be determined (from $F_e = q_0 E$). Similarly, the **electric potential difference**, ΔV, between any two points in space is defined as *the change in potential energy, or work done, per unit positive test charge*. Thus, the definition of electric potential difference is

Definition of: electric potential difference or voltage

$$\Delta V = \frac{\Delta U_e}{q_0} = \frac{W}{q_0} \tag{16.1}$$

SI unit of electric potential difference: joule/coulomb (J/C), or volt (V)

The SI unit of electric potential difference is the joule per coulomb. This unit has been renamed the **volt** (V) in honor of Alessandro Volta (1745–1827), an Italian scientist who constructed the first battery. (See Chapter 17.) Thus, 1 V = 1 J/C. Potential difference is commonly called **voltage**, and the symbol for potential difference is usually changed from ΔV to just V, as is done later in this chapter.

Potential difference, although based on potential-energy difference, is *not* the same thing. Since potential difference is defined *per unit charge*, it does *not* depend on the amount of charge moved. Like the electric field, potential difference is a useful property, since, from ΔV, we can calculate ΔU_e for any amount of charge moved. To illustrate this point, let us calculate the potential difference associated with the uniform field between two parallel plates:

$$\Delta V = \frac{\Delta U_e}{q_o} = \frac{q_o E d}{q_o} = Ed \qquad \begin{array}{l} \textit{potential difference} \\ \textit{(parallel plates only)} \end{array} \quad (16.2)$$

Notice that the amount of charge moved, q_o, cancels out. Thus, the potential difference ΔV depends only on the characteristics of the charged plates—that is, the field they produce (E) and their separation (d). We say that the positively charged plate is at a *higher electric potential* than the negatively charged one, by an amount ΔV.

Notice that electric potential *difference* was introduced without first defining electric potential. Although this approach may seem backward, there is a good reason for it. Electric potential *differences* are the physically meaningful quantities that are actually measured. (Voltages are measured using voltmeters, which will be discussed in detail in Chapter 18.). The electric potential value at a specific location, in contrast, isn't definable in an absolute way—it depends entirely on the choice of a reference point. A constant can be added to, or subtracted from, all potentials and not alter potential *differences*.

The same idea was encountered during our study of mechanical forms of potential energy associated with springs and gravitation in Chapter 5. There, only *changes* in potential energy were important. Definite values of potential energy could be determined, but only after the zero reference point was defined. Thus, in the case of gravity, zero gravitational potential energy is sometimes chosen at the Earth's surface, for convenience. However, it is just as correct (and sometimes more convenient) to define the zero value at an infinite distance from Earth, as was done in Chapter 7.

This property also holds for electric potential energy and electric potential. The electric potential may be defined or chosen as zero at the negative plate of a pair of charged parallel plates. However, it is sometimes convenient to locate the zero value at infinity, as we shall see shortly in the case of a point charge. Either way, the *difference* between two given points will be unaffected. This idea is brought home in the Learn by Drawing figure on this page. Suppose that, with a certain choice for the location of zero electric potential, point A turns out to be at a potential of +100 V and point B at +300 V. With a different zero point, the potential at A might be +1100 V. In that case, the value of the electric potential at B would be +1300 V. Regardless of the zero point, B will always be 200 V higher in potential than A.

The following Example illustrates the relationship between electric potential energy and electric potential difference. It also shows how energy methods can be extended to electric fields.

Note: We commonly assign a value of zero for the electric potential of the negative plate, but this is arbitrary. Only potential *differences* are meaningful.

Learn by Drawing

ΔV Is Independent of Reference Point

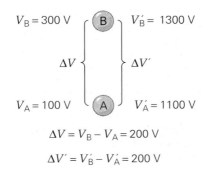

$V_B = 300\ V \qquad V'_B = 1300\ V$

$\Delta V \qquad \Delta V'$

$V_A = 100\ V \qquad V'_A = 1100\ V$

$\Delta V = V_B - V_A = 200\ V$

$\Delta V' = V'_B - V'_A = 200\ V$

$\Delta V = \Delta V'$

Example 16.1 ■ Moving a Proton: Parallel Plates and Potential Difference

Imagine moving a proton from the negative plate to the positive plate of the parallel-plate arrangement in ▶ Fig. 16.2a. The plates are 1.50 cm apart, and the electric field is uniform with a magnitude of 1500 N/C. (a) What is the change in the proton's electric potential energy? (b) What is the electric potential difference (voltage) between the plates? (c) If the proton is released from rest at the positive plate (Fig. 16.2b), what speed will it have just before it hits the negative plate?

Thinking It Through. (a) The change in potential energy can be computed from the work that is done to move the charge. (b) The electric potential difference between the two plates can then be determined by dividing the work done by the charge moved. (c) When the proton is released, its electric potential energy is converted into kinetic energy. Knowing this, one can calculate the proton's speed.

PHYSLET®
ILLUSTRATION

Work and Potential Difference

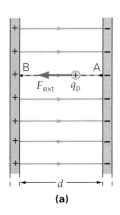

(a)

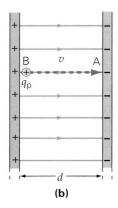

(b)

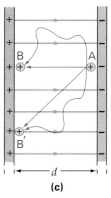

(c)

▲ **FIGURE 16.2   Accelerating a charge** (a) Moving a proton from the negative to the positive plate increases the proton's potential energy. (See Example 16.1.) **(b)** When it is released from the positive plate, the proton accelerates back toward the negative plate, gaining kinetic energy at the expense of electric potential energy. **(c)** The work done to move a proton between any two points in an electric field, such as A and B or A and B′, is independent of the path.

Solution.   The magnitude of the electric field, E, is given. Since a proton is involved, its mass and charge are known (Table 15.1).

Given: $E = 1500 \text{ N/C}$
$q_p = +1.60 \times 10^{-19} \text{ C}$
$m_p = 1.67 \times 10^{-27} \text{ kg}$
$d = 1.50 \text{ cm} = 1.50 \times 10^{-2} \text{ m}$

Find: (a) ΔU_e (potential-energy change)
(b) ΔV (potential difference between plates)
(c) v (speed of released proton just before it reaches negative plate)

(a)   Electric potential energy is increased, because positive work must be done on the proton to move it against the electric field, toward the positive plate. Thus,

$$\Delta U_e = (q_p E)d = (+1.60 \times 10^{-19} \text{ C})(1500 \text{ N/C})(1.50 \times 10^{-2} \text{ m})$$
$$= +3.60 \times 10^{-18} \text{ J}$$

(b)   The potential difference, or voltage, is the potential-energy change per unit charge, as expressed in Eq. 16.1:

$$\Delta V = \frac{\Delta U_e}{q_p} = \frac{+3.60 \times 10^{-18} \text{ J}}{+1.60 \times 10^{-19} \text{ C}} = +22.5 \text{ V}$$

The positive plate is 22.5 V higher in electric potential than the negative plate.
(c)   The total energy of the proton is constant; therefore, $\Delta K + \Delta U_e = 0$. (The loss in potential energy is equal in magnitude to the gain in kinetic energy.) However, the proton has no initial kinetic energy ($K_o = 0$). Hence, $\Delta K = K - K_o = K$. From this relationship, the speed of the proton can be calculated as follows:

$$\Delta K = K = -\Delta U_e$$

or

$$\tfrac{1}{2} m_p v^2 = -\Delta U_e$$

Thus, $\Delta U_e = -3.60 \times 10^{-18}$ J, and the proton's speed just before it hits the negative plate is

$$v = \sqrt{\frac{2(-\Delta U_e)}{m_p}} = \sqrt{\frac{2[-(-3.60 \times 10^{-18} \text{ J})]}{1.67 \times 10^{-27} \text{ kg}}} = 6.57 \times 10^4 \text{ m/s}$$

Even though the kinetic energy gained is very small, the proton acquires such a high speed because its mass is extremely small.

Follow-up Exercise.   How would your answers change in this Example if an alpha particle were moved instead of a proton? (An alpha particle is the nucleus of a helium atom and has a charge of $+2e$ with a mass approximately four times that of a proton.) (*Answers to all Follow-up Exercises are at the back of the text.*)

Example 16.1 can also be used to show that electric potential-energy changes (and therefore, electric potential differences) are independent of the path taken. This is true for all conservative forces, of which the electrostatic force is one. In Fig. 16.2c, the work done in moving the proton from A to B is the same, *regardless of the route* taken. Thus, the alternative wiggly paths from A to B and A to B′ require the same work as do their respective straight-line paths. This is because movement at right angles to the field requires no work. (Why?)

The gravitational analogy to moving the proton in Example 16.1 is, of course, raising an object in a uniform gravitational field. When the object is raised, its gravitational potential energy increases, because the force of gravity is downward. However, with electricity, there are two types of charges, and the force between them can also be repulsive. At this point, the simple analogy to gravity breaks down.

To understand why the analogy fails, consider how the energy discussion in Example 16.1 would differ if an electron had been used instead of a proton. The electron, being negatively charged, would be attracted to plate B, so the external force would have to be in a direction *opposite* the electron's displacement (i.e., directed backwards, to prevent the electron from gaining speed). Thus, for an electron, the external force would do *negative* work, thereby *decreasing* the electric potential energy. Unlike the proton, the electron is attracted to the positive plate (the plate with the higher electric potential). If allowed to move freely, the electron would "fall" toward the region of higher potential (just as the proton "fell" toward the region of lower potential), losing potential energy and gaining kinetic energy.

In sum, we can say this about the behavior of charged particles in electric fields:

| Positive charges, when released, tend to move toward regions of lower electric potential.

| Negative charges, when released, tend to move toward regions of higher electric potential.

Consider the following medical application involving the creation of X rays from fast-moving electrons, accelerated by large electric potential differences (voltages).

Note: Potential differences, like electric fields, are defined in terms of positive charges. Electrons are subject to the same potential difference, but the opposite potential energy change. To determine whether the potential increases or decreases, decide whether an external force does positive or negative work on a *positive* test charge.

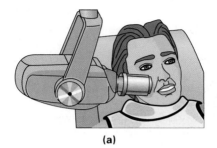
(a)

Example 16.2 ■ Creating X rays: Accelerating Electrons

Most modern dental offices have X-ray machines for diagnosing hidden dental problems (▶ Fig. 16.3a). Typically, electrons are accelerated through electric potential differences (voltages) of 25 000 V. (Large voltages like this are created by transformers, which are examined in Chapter 20.) When the electrons hit the positively charged plate, their kinetic energy is converted into high-energy light particles sometimes called X-ray photons (Fig. 16.3b). (Photons are particles of light and are discussed in detail in Ch 27.) Suppose a single electron's kinetic energy is distributed equally among five X rays. How much energy would one X ray have?

Thinking It Through. From energy conservation, the kinetic energy gained by any one electron is equal in magnitude to the electric potential energy it loses. From this kinetic energy, the energy of one X ray can be calculated.

Solution. The charge of an electron is known (from Table 15.1), and the accelerating voltage is given.

Given: $q = -1.60 \times 10^{-19}\,C$ *Find:* Energy (E) of one X ray
     $\Delta V = 2.50 \times 10^4\,V$

The electron leaves the negatively charged plate and moves toward the region of highest electric potential (i.e., "uphill"). Thus, the change in its electric potential energy is

$$\Delta U_e = q\,\Delta V = (-1.60 \times 10^{-19}\,C)(+2.50 \times 10^4\,V) = -4.00 \times 10^{-15}\,J$$

The gain in kinetic energy comes from this loss in electric potential energy. Since the electrons have no appreciable kinetic energy when they start,

$$K = -\Delta U_e = +4.00 \times 10^{-15}\,J$$

Therefore, one of the five X rays will have an energy of

$$E = \frac{K}{5} = 8.00 \times 10^{-16}\,J$$

Follow-up Exercise. In this Example, use energy methods to determine the speed of one of the electrons when it is one-tenth of the way to the positive plate.

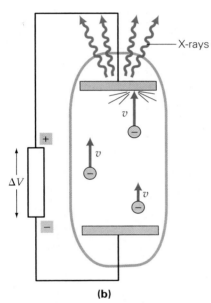
▲ **FIGURE 16.3 X-ray production (a)** A dental X-ray machine. **(b)** A schematic diagram of the X-ray production.

Electric Potential Difference Due to a Point Charge

In nonuniform electric fields, the potential difference between two points is determined by applying the definition given in Eq. 16.1. However, when the field strength varies, the mathematics is complicated because the work is done by a varying force. The only nonuniform field we will consider in any detail is that due

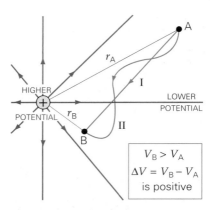

▲ FIGURE 16.4 Electric field and potential due to a point charge Electric potential increases as you move closer to a positive charge. Thus, B is at a higher potential than A.

$$V_B > V_A$$
$$\Delta V = V_B - V_A$$
$$\text{is positive}$$

to a point charge (◄Fig. 16.4). The expression for the potential difference (voltage) between two points at distances r_A and r_B from a point charge q requires calculus to derive, so here we simply state it:

$$\Delta V = \frac{kq}{r_B} - \frac{kq}{r_A} \quad \begin{array}{l} \textit{electric potential difference} \\ \textit{(point charge only)} \end{array} \quad (16.3)$$

In Fig. 16.4, the point charge is positive. Since location B is closer to the point charge than A, the potential difference is positive: B is at a higher potential than A. This is because it takes *positive* external work to move a positive test charge from A to B.

From this analysis, we see that the electric potential increases as we consider locations nearer to a positive charge. Since the electric force is a conservative force, the potential difference is the same, regardless of the path. (The work done along path II is the same as that done along path I in Figure 16.4.)

Now consider what would happen if the point charge were negative. In this case, B would be at a *lower* potential than A. (Isn't the work required to move a positive test charge closer now *negative*?)

To summarize, the potential changes according to the following rules:

| Electric potential increases as we consider locations nearer to positive charges or farther from negative charges.

| Electric potential decreases as we consider locations farther from positive charges or nearer to negative charges.

The potential at a very large distance from a point charge is usually chosen to be zero (much as we chose for the gravitational potential energy for a point mass in Chapter 7). For this choice, the *electric potential V* at a distance r from a point charge is given by

$$V = \frac{kq}{r} \quad \begin{array}{l} \textit{electric potential} \\ \textit{(point charge only)} \end{array} \quad (16.4)$$

Keep in mind, however, that only electric potential *differences* are important, as the next Example illustrates.

Integrated Example 16.3 ■ Describing the Hydrogen Atom: Potential Differences near a Proton

In the Bohr model of the hydrogen atom (Chapter 27), the electron in orbit around the proton can exist only in certain circular orbits. The smallest has a radius of 0.0529 nm, and the next largest has a radius of 0.212 nm. (a) What can you say about the electric potential associated with each orbit? (1) The smaller one has a higher potential, (2) the larger one has a higher potential, or (3) they both have the same value of electric potential. Explain your reasoning. (b) Verify your answer to part (a) by calculating the values of the electric potential at the locations of the two orbits.

(a) Conceptual Reasoning. The electron orbits in the field of a proton, whose charge is positive. Since electric potential increases with decreasing distance from a positive charge, the answer must be (1), namely, that the smaller orbit is associated with the higher value of electric potential.

(b) Thinking It Through. We know the charge on the proton, so Eq. 16.4 can be used to check the actual potential values. We list the values as follows:

Given: $q_p = +1.60 \times 10^{-19}\,\text{C}$
$r_1 = 0.0529\,\text{nm} = 5.29 \times 10^{-11}\,\text{m}$
$r_2 = 0.212\,\text{nm} = 2.12 \times 10^{-10}\,\text{m}$ (using the fact that 1 nm = 10^{-9} m)

Find: The value of the electric potential (V) for each orbit

Equation 16.4 allows us to determine the electric potential at both distances. Thus,

$$V_1 = \frac{kq_\mathrm{p}}{r_1} = \frac{(9.00 \times 10^9\,\mathrm{N \cdot m^2/C^2})(+1.60 \times 10^{-19}\,\mathrm{C})}{5.29 \times 10^{-11}\,\mathrm{m}} = +27.2\,\mathrm{V}$$

For the larger orbit,

$$V_2 = \frac{kq_\mathrm{p}}{r_2} = \frac{(9.00 \times 10^9\,\mathrm{N \cdot m^2/C^2})(+1.60 \times 10^{-19}\,\mathrm{C})}{2.12 \times 10^{-10}\,\mathrm{m}} = +6.79\,\mathrm{V}$$

Follow-up Exercise. In this Example, suppose the electron were moved from the smallest to the next orbit. (a) Has it moved to a region of higher or lower electrical potential? (b) What would be the change in the electric potential *energy* of the electron?

Electric Potential Energy of Various Charge Configurations

The discussion of gravitational energy (Chapter 7) considered the gravitational potential energy of systems of masses in some detail. The expressions for electric force and gravitational force are mathematically similar, and so are those for the corresponding potential energies, except for the use of charge in place of mass and the two signs associated with charge. In the gravitational case of two masses, the mutual gravitational potential energy is negative, because the force is always *attractive*. For electric potential energy, the result can be either positive or negative, because the electric force can be either *repulsive* or *attractive*.

If a positive point charge q_1 is fixed in space and a second positive charge q_2 is brought toward it from a very large distance $(r \rightarrow \infty)$ to a distance r_{12} (▼Fig. 16.5a), the work done is positive. Therefore, the system of two charges gains electric potential energy. The potential at a large distance (V_∞) is chosen to be zero. (Recall that the zero reference point is arbitrary.) Thus, from Eq. 16.3, we see that the change in potential energy is

$$\Delta U_\mathrm{e} = q_2 \Delta V = q_2(V_1 - V_\infty) = q_2\left(\frac{kq_1}{r_{12}} - 0\right) = \frac{kq_1 q_2}{r_{12}}$$

Since the large-distance value of the electric potential energy is chosen as zero, it follows that $\Delta U_\mathrm{e} = U_{12} - U_\infty = U_{12}$. With this choice of reference value, the potential energy of a two-charge system is

$$U_{12} = \frac{kq_1 q_2}{r_{12}} \quad \begin{array}{l}\textit{electric potential energy}\\ \textit{(two charges)}\end{array} \quad (16.5)$$

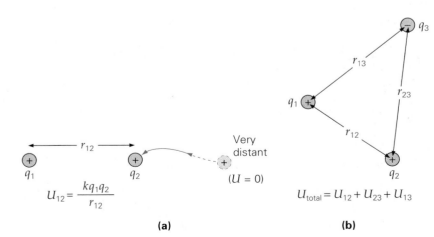

(a)

(b)

◀ **FIGURE 16.5   Mutual electric potential energy of point charges** (a) If a positive charge is moved from a large distance to a distance r_{12} from another positive charge, there is an increase in potential energy because positive work must be done to bring the mutually repelling charges closer. (b) For more than two charges, the system's electric potential energy is the sum of the mutual potential energies of each pair.

For *unlike* charges, the electrical force is attractive and the electric potential energy is negative. For *like* charges, the potential energy is positive. Thus, if the two charges are of the same sign, when released, they will move apart, gaining kinetic energy as they lose potential energy. Conversely, it takes positive work to increase the separation of two opposite charges, such as the proton and the electron. (See Integrated Example 16.3.)

Because energy is a scalar quantity, for a configuration of charges, the *total* potential energy U_{total} is the algebraic sum of the mutual potential energies of all pairs of charges:

$$U_{total} = U_{12} + U_{23} + U_{13} + \cdots \qquad (16.6)$$

Only the first three terms of Eq. 16.6 would be needed for the configuration shown in Fig. 16.5b. Note that the signs of the charges keep things straight mathematically, as Example 16.4 shows.

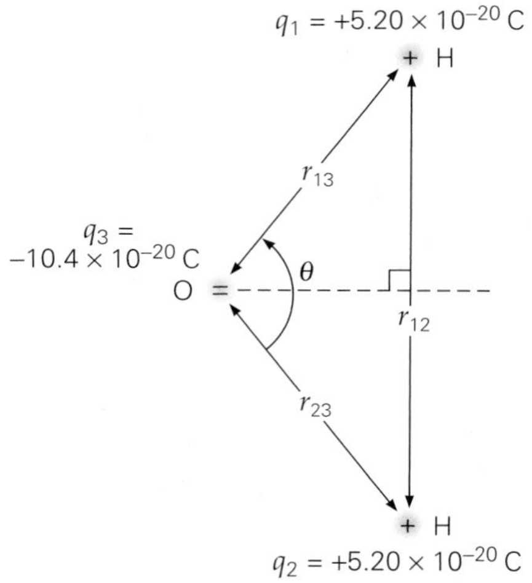

$q_1 = +5.20 \times 10^{-20}$ C

$q_3 = -10.4 \times 10^{-20}$ C

$q_2 = +5.20 \times 10^{-20}$ C

▲ **FIGURE 16.6 Electrostatic potential energy of a water molecule** A charge configuration has electric potential energy, since work is required to bring the charges together from large distances. The charges shown on the water molecule are net average charges because the atoms within the molecule share electrons. Thus, the charges on the ends of the water molecule can be smaller than the charge on the electron or proton. (See Example 16.4 for details.)

Example 16.4 ■ The Molecule of Life: The Electric Potential Energy of a Water Molecule

The water molecule is the foundation of life as we know it. Many of its important properties (e.g., the reason it is a liquid on the Earth's surface) are related to the fact that it is a permanent polar molecule (an electric dipole; see Section 15.4). A simple picture of the water molecule, including the charges, is given in ◄Fig. 16.6. The distance from each hydrogen atom to the oxygen atom is 9.60×10^{-11} m, and the angle (θ) between the two hydrogen–oxygen bond directions is 104°. What is the total electrostatic energy of the water molecule?

Thinking It Through. The model of the water molecule involves three charges. Thus, the sum in Eq. 16.6 will have three terms. The charges are given, but the distance between the hydrogen atoms must be calculated by trigonometry. The total electrostatic potential energy is the algebraic sum of the potential energies of the three pairs.

Solution. The following data are taken from Fig. 16.6.

Given: $q_1 = q_2 = +5.20 \times 10^{-20}$ C   *Find:* U_{total} (total electrostatic potential
$q_3 = -10.4 \times 10^{-20}$ C   of water molecule)
$r_{13} = r_{23} = 9.60 \times 10^{-11}$ m
$\theta = 104°$

Notice that $(r_{12}/2)/r_{13} = \sin(\theta/2)$. Hence, we can solve for r_{12}:

$$r_{12} = 2r_{13}\sin\frac{\theta}{2} = 2(9.60 \times 10^{-11}\,\text{m})(\sin 52°) = 1.51 \times 10^{-10}\,\text{m}$$

To determine the algebraic sum of the potential energies of the three pairs, let's calculate each pair's contribution to the total separately. Note that $U_{13} = U_{23}$. (Why?) Applying Eq. 16.5, we have

$$U_{12} = \frac{kq_1q_2}{r_{12}} = \frac{(9.00 \times 10^9\,\text{N}\cdot\text{m}^2/\text{C}^2)(+5.20 \times 10^{-20}\,\text{C})(+5.20 \times 10^{-20}\,\text{C})}{1.51 \times 10^{-10}\,\text{m}}$$
$$= +1.61 \times 10^{-19}\,\text{J}$$

and

$$U_{13} = U_{23} = \frac{kq_2q_3}{r_{23}} = \frac{(9.00 \times 10^9\,\text{N}\cdot\text{m}^2/\text{C}^2)(+5.20 \times 10^{-20}\,\text{C})(-10.4 \times 10^{-20}\,\text{C})}{9.60 \times 10^{-11}\,\text{m}}$$
$$= -5.07 \times 10^{-19}\,\text{J}$$

The total electrostatic potential energy is

$$U_{total} = U_{12} + U_{13} + U_{23} = (+1.61 \times 10^{-19}\,J) + (-5.07 \times 10^{-19}\,J) + (-5.07 \times 10^{-19}\,J)$$
$$= -8.53 \times 10^{-19}\,J$$

The negative result indicates that the molecule requires positive external work to break it apart. (That is, it must be pulled apart.)

Follow-up Exercise. Another common polar molecule is carbon monoxide (CO), a toxic gas commonly produced in automobiles when the combustion of fuel is incomplete. The carbon atom is positively charged and the oxygen atom negatively charged. The distance between the carbon atom and the oxygen atom is $1.20 \times 10^{-10}\,m$, and the (average) charge on each is $6.60 \times 10^{-20}\,C$. Determine the electrostatic energy of this molecule. Is it more or less electrically stable than a water molecule in this Example?

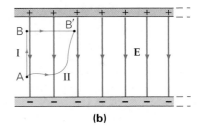

(a)

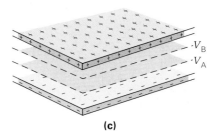

(b)

16.2 Equipotential Surfaces and the Electric Field

OBJECTIVES: To (a) explain what is meant by an equipotential surface, (b) sketch equipotential surfaces for simple charge configurations, and (c) explain the relationship between equipotential surfaces and electric fields.

Equipotential Surfaces

Suppose that a positive charge is moved perpendicularly to an electric field (such as path I of ▶Fig. 16.7a). As the charge moves from A to A′ between parallel charged plates, *no work* is done on it by the electric field, because the angle between the electric force and the charge's displacement is 90°. If no work is done, then the potential energy of the charge does not change between A and A′, so $\Delta U_{AA'} = 0$. From this result, we conclude that these two points—and *all* other points on path I—are at the same potential V; that is,

$$\Delta V_{AA'} = V_{A'} - V_A = \frac{\Delta U_{AA'}}{q} = 0 \qquad \text{or} \qquad V_A = V_{A'}$$

This property also holds for all points on the *plane* parallel to the plates and containing path I. Such a plane, on which the electric potential is constant, is an **equipotential surface** (sometimes referred to as an *equipotential*). The word *equipotential* means "same potential." An equipotential surface need not be a flat plane, nor is it necessarily parallel to the plates or charges that created the field, as we see in the special parallel-plate case we are considering here.

Since no work is required to move a charge along an equipotential surface, it must be generally true that

| Equipotential surfaces are always at right angles to the electric field.

Moreover, because the electric field is a conservative field, the work will be the same whether we take path I, path II, or any other path from A to A′ (Fig. 16.7a). As long as the charge returns to the same equipotential surface from which it started, the work done is zero and its electric potential energy stays the same.

If the positive charge is moved in the opposite direction of **E** (path I in Fig. 16.7b)—at right angles to the equipotentials—the electric potential energy, and hence the electric potential, will increase. (Why?) When B is reached, the charge is on a different equipotential surface—one of a higher potential than A. If, instead, the charge had been moved from A to B′, the work required would be the same as in moving from A to B. Hence, B and B′ are also on an equipotential surface. Thus, for charged parallel plates, the equipotential surfaces are planes parallel to the plates, as shown in Fig. 16.7c.

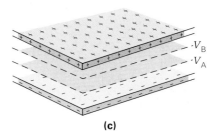

(c)

▲ **FIGURE 16.7 Construction of equipotential surfaces between parallel plates** **(a)** The work done in moving a charge is zero as long as you start and stop on the same equipotential surface. (Compare paths I and II.) **(b)** Once you move to a higher potential (for example, from point A to point B), you can stay on that new equipotential surface by moving perpendicularly to the electric field (B to B′). The change in potential is independent of the path, since the same change occurs whether path I or path II is used. (Why?) **(c)** The actual equipotential surfaces within the parallel plates are planes parallel to those plates. Two such plates are shown, with $V_B > V_A$.

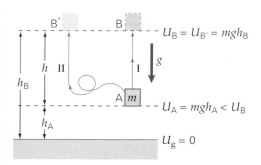

▲ FIGURE 16.8 Gravitational potential energy analogy Raising an object in a uniform gravitational field results in an increase in gravitational potential energy. Thus, $U_B > U_A$. At a given height, the object's potential energy is constant as long as it remains on that (gravitational) equipotential surface. Here, **g** points downward, like **E** in Fig. 16.7.

To help understand the concept of an electric equipotential surface, consider a gravitational analogy. If the gravitational potential energy is designated as zero at ground level and an object is raised a height $h = h_B - h_A$ (from A to B in ◄Fig. 16.8), then the work done by an external force is mgh and is positive. For horizontal movement, the potential energy does not change. This means that the dashed plane at height h_B is a gravitational equipotential surface—and so is the plane at h_A, but at a lower potential. Thus, surfaces of constant gravitational potential energy are planes parallel to the Earth's surface. Topographic maps, which display land contours by plotting lines of constant elevation (usually relative to sea level), are actually maps of constant gravitational potential (▼Fig. 16.9a,b). Note how the equipotentials near a point charge are qualitatively similar to the contours of a hill (Fig. 16.9c,d).

It is extremely useful to know how to sketch equipotential surfaces, because they are intimately related to the electric field and to practical aspects such as voltage. The Learn by Drawing feature "Graphical Relationship between Electric Field Lines and Equipotentials" on p. 556 summarizes a qualitative method that is useful for sketching equipotential surfaces, given an electric field-line pattern. As this feature shows, the method is also useful for the converse problem: sketching the electric field lines if the equipotential surfaces are given. Can you see how these ideas were used to construct the equipotentials of an electric dipole in ►Fig. 16.10?

▼ FIGURE 16.9 Topographic maps—a gravitational analogy to equipotential surfaces **(a)** A symmetrical hill with slices at different elevations. Each slice is a plane of constant gravitational potential. **(b)** A topographic map of the slices in (a). The contours, where the planes intersect the surface, represent increasingly larger values of gravitational potential as one goes up the hill. **(c)** The electric potential V near a point charge q forms a similar symmetrical hill. V is constant at fixed distances from q. **(d)** Electrical equipotentials around a point charge are spherical (in two dimensions they are circles) centered on the charge. The closer the equipotential is to the positive charge, the larger is its electric potential.

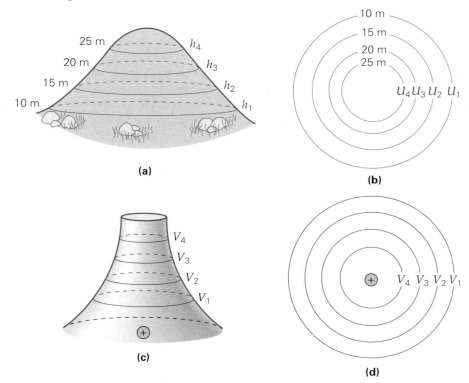

To determine the relationship between the electric field (E) and the electric potential (V), let's first consider the special case of the uniform electric field (▼Fig. 16.11). The potential difference (ΔV) between any two equipotential planes (e.g., those labeled 1 and 3 in the figure) can be calculated with the same technique used to derive Eq. 16.2. The result is

$$\Delta V = V_3 - V_1 = E \Delta x \qquad (16.7)$$

Thus, if you start on equipotential surface 1 and move *perpendicularly away* from it and *against* the electric field to equipotential surface 3, there is a potential *increase* (ΔV) that depends on the electric field strength (E) and the distance traveled (Δx).

For a given travel distance Δx, this perpendicular movement yields the maximum possible gain in potential. Think of taking one step of length Δx in *any* direction, starting from surface 1. The way to maximize the increase would be to step onto surface 3. A step in any direction not perpendicular to surface 1 (for example, ending on surface 2) yields a smaller increase in potential.

By finding the direction of the *maximum* potential increase, we are finding the direction opposite that of **E**. This means that

the direction of the electric field **E** is the direction in which the electric potential decreases the most rapidly.

Thus, at any location, the magnitude of the electric field is the maximum rate of change of the potential with distance, or

$$E = \left| \frac{\Delta V}{\Delta x} \right|_{max} \qquad (16.8)$$

The units of electric field are volts per meter (V/m). In Chapter 15, E was expressed in newtons per coulomb (N/C; see Section 15.4). You should show, through dimensional analysis, that 1 V/m = 1 N/C. A graphical interpretation of the relationship between **E** and V is shown in the Learn by Drawing feature on pg. 556.

In most practical situations, the potential difference (*voltage*), rather than the electric field, is specified. For example, a D-cell flashlight battery has a terminal voltage of 1.5 V, meaning that it can maintain a potential difference of 1.5 V between its terminals. Most automotive batteries have a terminal voltage of about 12 V. Some of the common potential differences, or voltages, are listed in Table 16.1.

Whether you know it or not, you live in an electric field near the Earth's surface. This field varies with weather conditions and, consequently, can be an indicator of approaching storms. Example 16.5 applies the equipotential surface concept to help in understanding the Earth's electric field.

TABLE 16.1	Common Electric Potential Differences (Voltages)
Source	Approximate Voltage (ΔV)
Across nerve membranes	100 mV
Small-appliance batteries	about 1 V
Automotive batteries	12 V
Household outlet (United States)	110 to 120 V
Outlets (Europe)	220 to 240 V
Automotive ignitions (spark plug firing)	10 000 V
Laboratory Van de Graaff generators	25 000 V
High-voltage electric power delivery lines	300 kV or more
Cloud-to-Earth surface during thunderstorm	100 MV or more

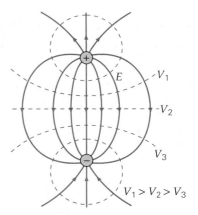

▲ **FIGURE 16.10 Equipotentials of an electric dipole**
Equipotentials are perpendicular to electric field lines. $V_1 > V_2$, since equipotential surface 1 is closer to the positive charge than is surface 2. To understand how equipotentials are constructed, see the Learn by Drawing feature on p. 556.

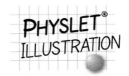

Equipotential Surfaces

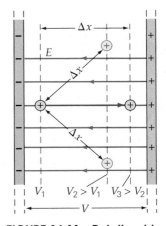

▲ **FIGURE 16.11 Relationship between the potential change (ΔV) and the electric field (E)**
The electric field direction is that of maximum decrease in potential, or opposite the direction of maximum increase in potential (here, in the direction of the solid blue arrow, not the dashed ones; why?). The electric field magnitude is given by the rate at which the potential changes over distance (usually in volts per meter).

Learn by Drawing

Graphical Relationship between Electric Field Lines and Equipotentials

Since it takes no work to move a charge along an equipotential surface, such surfaces must always be perpendicular to the electric field lines. Also, the electric field has a magnitude equal to the change in potential per unit distance (V/m) and points in the direction in which the potential decreases most rapidly. We can use these facts to construct equipotentials if we know the electric field line pattern. The reverse is also true: Given the equipotentials, we can construct the electric field lines. Furthermore, if we know the potential (in volts) associated with each equipotential, the strength and direction of the field can be estimated from the rate at which the potential changes with distance (Eq. 16.8).

A couple of examples should provide a graphical insight into the connection between equipotential surfaces and their associated electric fields. Consider Fig. 1, in which you are given the electric field lines and want to determine the shape of the equipotentials. Pick any point, such as A, and begin moving at right angles to the electric field lines. Keep moving so as to always maintain this same perpendicular orientation to the field lines. Between lines you may have to approximate, but plan ahead to the next field line so that you can cross it at a right angle. To find another equipotential, start at another point, such as B, and proceed the same way. Sketch as many equipotentials as you need to map the area of interest. The figure shows the result of sketching four such equipotentials, from A (at the highest potential—can you tell why?) to D (at the lowest potential).

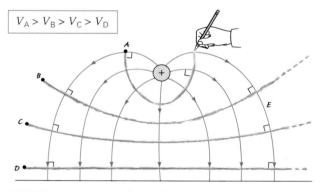

$$V_A > V_B > V_C > V_D$$

FIGURE 1 **Sketching equipotentials from electric field lines** If you know the electric field pattern, pick a point in the region of interest and move so that your path is always perpendicular to the next field line. Keep your path as smooth as possible, planning ahead so that each succeeding field line is also crossed at right angles. To map surface with a higher (or lower) potential, move in the opposite (or the same) direction as the electric field and repeat the process. Here, $V_A > V_B$, and so on.

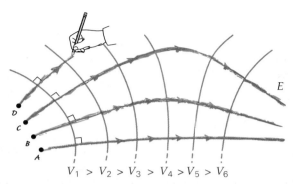

$$V_1 > V_2 > V_3 > V_4 > V_5 > V_6$$

FIGURE 2 **Mapping the electric field from equipotentials** Start at a convenient point, and trace a line that crosses each equipotential at a right angle. Repeat the process as often as needed to reveal the field pattern, adding arrows to indicate the direction of the field lines from high to low potential. In going from one potential to the next, plan ahead so that each succeeding equipotential is also crossed at right angles.

Now suppose you are given the equipotentials instead of the field lines (Fig. 2). The electric field lines point in the direction of decreasing V and are perpendicular to the equipotential surfaces. Thus, to map the field, start at any point, and move in such a way that your path intersects each equipotential surface at a right angle. The resulting field line is shown in Fig. 2, beginning at point A. Starting at points B, C, and D provides additional field lines that suggest the complete electric field pattern; you need only add the arrows in the direction of decreasing potential.

Lastly, suppose you want to estimate the magnitude of **E** at some point P, knowing the values of the equipotentials 1.0 cm on either side of it (Fig. 3). From this information, you can tell that the field points roughly from A to B. (How do you know that?) Its approximate magnitude is

$$E = \left| \frac{\Delta V}{\Delta x} \right|_{max} = \frac{(1000 \text{ V} - 950 \text{ V})}{2.0 \times 10^{-2} \text{ m}}$$
$$= 2.5 \times 10^3 \text{ V/m}$$

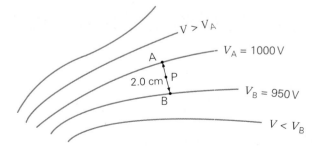

FIGURE 3 **Estimating the magnitude of the electric field** The magnitude of the potential change per meter at any point gives the strength of the electric field at that point.

Example 16.5 ■ The Earth's Electric Field and Equipotential Surfaces: Electric Barometers?

Under normal atmospheric conditions, the Earth's surface is electrically charged. This creates an approximately constant electric field of about 150 V/m pointing *down* near the surface. (a) Under these conditions, what is the shape of the equipotential surfaces, and in what direction does the electric potential decrease the most rapidly? (b) How far apart are two equipotential surfaces that have a 1000-V difference between them? Which has a higher potential, the one farther from the Earth or the one closer?

Thinking It Through. (a) Near the Earth's surface, the electric field is approximately uniform, so the equipotentials are similar to those of charged parallel plates. The discussion surrounding Eqs. 16.7 and 16.8 enables us to determine which way the potential increases. (b) Equation 16.8 can then be used to determine how far apart the equipotential surfaces are.

Solution. Listing the data, we have:

Given: $E = 150$ V/m, downwards *Find:* (a) Shape of equipotential surfaces and
    $\Delta V = 1000$ V   direction of decrease in potential
                      (b) Δx [distance between equipotentials]

(a) Uniform electric fields are associated with plane equipotential surfaces. In this case, the planes are parallel to the Earth's surface. The electric field points downward. This is the direction in which the potential decreases most rapidly.

(b) To determine the distance between the two equipotentials, think of moving vertically so that $\Delta V/\Delta x$ has its maximum value. (Why?) Solving Eq. 16.8 for Δx yields

$$\Delta x = \frac{\Delta V}{E} = \frac{1000 \text{ V}}{150 \text{ V/m}} = 6.67 \text{ m}$$

Since the potential decreases as we move downward (in the direction of **E**), the higher potential is associated with the surface that is 6.67 m *farther* from the ground.

Follow-up Exercise. Re-examine the preceding Example, but under storm conditions. During a lightning storm, the electric field can rise to many times the normal value. (a) Under these conditions, if the field is 900 V/m and points *upward*, how far apart are two equipotential surfaces that differ by 2000 V? (b) Which surface is at a higher potential, the one closer to the Earth or the one farther away? (c) Can you tell where the two surfaces are located relative to the ground? Explain.

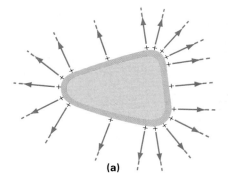

(a)

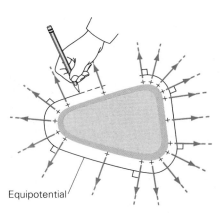

Equipotential

(b)

Equipotential surfaces can also be useful for describing the field near a charged conductor, as the next Conceptual Example shows.

Conceptual Example 16.6 ■ The Equipotential Surfaces Outside a Charged Conductor

A solid conductor with an excess positive charge on it is shown in ▶Fig. 16.12a. Which of the following best describes the shape of the equipotential surfaces just outside the conductor's surface? Explain your reasoning. The surfaces are (a) flat planes, (b) spheres, or (c) approximately in the shape of the conductor's surface.

Reasoning and Answer. Choice (a) can be eliminated immediately, because flat equipotential surfaces are associated with flat plates. While it might be tempting to pick answer (b) because of symmetry, a quick look at the electric field pattern, in conjunction with the Learn by Drawing feature on pg. 556, shows that the correct answer is (c). To verify that (c) is the correct answer, recall that the excess charge accumulates preferentially at the sharp corners and thus the electric field is strongest there (Fig. 16.12a). More importantly, just above the surface, the electric field must be perpendicular to that surface. Since

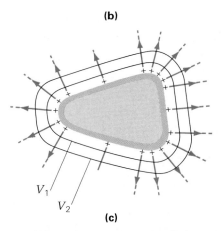

V_1

V_2

(c)

▲ **FIGURE 16.12 Equipotential surfaces near a charged conductor** See Conceptual Example 16.6.

the equipotential surfaces are themselves perpendicular to the electric field lines, they must, therefore, follow the contour of the conductor's surface (Fig. 16.12b).

Follow-up Exercise. In this Example, (a) which of the two equipotentials (1 or 2) shown in Fig. 16.12c is at a higher potential? (b) What is the shape of the equipotential surfaces *very far away from* the conductor? Explain your reasoning. (*Hint*: What does the charged conductor look like when you are very far away from it?)

The Electron Volt

The concept of electric potential provides a convenient unit of energy that is particularly useful in molecular, atomic, nuclear, and elementary particle physics. The **electron volt** (eV) is defined as the kinetic energy acquired by an electron or proton accelerated through a potential difference, or voltage, of exactly 1 V. The electron's gain in kinetic energy is equal to its loss in electric potential energy. Using this relationship, we can express the gain in kinetic energy in joules:

$$\Delta K = -\Delta U_e = -(e\Delta V) = -(-1.60 \times 10^{-19}\,C)(1.00\,V) = +1.60 \times 10^{-19}\,J$$

Since this is what is meant by 1 electron volt, the conversion factor between the electron volt and the joule (to three significant figures) is

$$1\,eV = 1.60 \times 10^{-19}\,J$$

Note: The electron volt is a unit of energy. The volt is not. Do not confuse the two!

The electron volt is typical of energies on the atomic scale. Thus, it is convenient to express atomic energies in terms of electron volts instead of the joule. The energy of *any* charged particle accelerated through *any* potential difference can be expressed in electron volts. For example, if an electron is accelerated through a potential difference of 1000 V, its gain in kinetic energy (ΔK) is one thousand times larger than a 1-eV electron, or

$$\Delta K = e\Delta V = (1\,e)(1000\,V) = 1000\,eV = 1\,keV$$

The "keV" is the abbreviation for the *kilo*electron volt.

The electron volt is defined in terms of a particle with the minimum unit of charge (the electron or proton). However, the energy of a particle with *any* amount of charge can also be expressed in electron volts. Thus, if a particle with a charge of $+2e$, such as an alpha particle (a helium nucleus, consisting of two protons and two neutrons), were accelerated through a potential difference (voltage) of 1000 volts, it would gain a kinetic energy of $\Delta K = e\Delta V = (2\,e)(1000\,V) = 2000\,eV = 2\,keV$. Note how easy it is to compute the kinetic energy of an accelerated charged particle if you work in electron volts.

Still, larger units of energy are sometimes needed. For example, in nuclear and elementary particle physics, it is not uncommon to find particles with energies of *mega*electron volts (MeV) and *giga*electron volts (GeV). ($1\,MeV = 10^6\,eV$, and $1\,GeV = 10^9\,eV$, respectively.)*

In working out numerical problems, it is especially important to be aware that the electron volt (eV) is *not* an SI unit. Hence, in using energies to calculate speeds, you must first convert from electron volts to joules. For example, to calculate the speed of an electron accelerated from rest through 10.0 V, first convert the electron's kinetic energy (10.0 eV) to joules: $K = (10.0\,eV)(1.60 \times 10^{-19}\,J/eV) = 1.60 \times 10^{-18}\,J$. To make sure that everything is in the SI system, the mass of the electron is expressed in kg. Then the speed is calculated as follows:

$$v = \sqrt{2K/m} = \sqrt{2(1.60 \times 10^{-18}\,J)/(9.11 \times 10^{-31}\,kg)} = 1.87 \times 10^6\,m/s$$

*At one time, a billion electron volts was referred to as BeV, but this usage was abandoned because confusion arose. In some countries, such as Great Britain and Germany, a billion means 10^{12} (which is called a trillion in the United States).

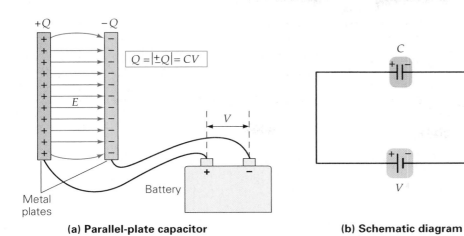

$$Q = |\pm Q| = CV$$

(a) Parallel-plate capacitor

(b) Schematic diagram

◀ **FIGURE 16.13 Capacitor and circuit diagram** **(a)** Two parallel metal plates are charged by a battery that moves electrons from the positive plate to the negative one through the wire. Work is done while charging the capacitor, and energy is stored in the electric field. **(b)** This diagram represents the charging situation shown in part (a). It also shows the symbols commonly used for a battery (*V*) and a capacitor (*C*). The longer line of the battery symbol is the positive terminal, and the shorter line represents the negative terminal. The symbol for a capacitor is similar, but the lines are of equal length.

Note: Capacitors store energy in their electric fields.

16.3 Capacitance

OBJECTIVES: To (a) define capacitance and explain what it means physically and (b) calculate the charge, voltage, electric field, and energy storage for parallel-plate capacitors.

A pair of parallel plates, if charged, stores electrical energy (▲Fig. 16.13). Such an arrangement of conductors is an example of a **capacitor**. (Actually, any pair of conductors qualifies as a capacitor.) The energy storage occurs because it takes work to transfer the charge from one plate to the other. Imagine that one electron is moved between a pair of initially uncharged plates. Once that is done, transferring a *second* electron would be more difficult, because it is not only repelled by the first electron on the negative plate, but also attracted by a double positive charge on the positive plate. Thus, to separate the charges requires more and more work as more and more charge accumulates on the plates. (This is analogous to stretching a spring. The more you stretch it, the harder it is to stretch it further.)

The work needed to charge parallel plates can be done quickly (usually in a few microseconds) by a battery. Although we won't discuss battery action in detail until the next chapter, all you need to know now is that a battery removes electrons from the positive plate and transfers, or "pumps," them through a wire to the negative plate. In the process of doing work on them, it loses some of its internal chemical potential energy. Of primary interest here is the result: a separation of charge and the creation of an electric field in the capacitor. The battery will continue to charge the capacitor until the potential difference between the plates is equal to the terminal voltage of the battery—for example, 12 volts if you use a standard automotive battery. When the capacitor is disconnected from the battery, it becomes a storage "reservoir" of electrical energy.

For a capacitor, the potential difference across the plates is proportional to the charge Q on the plates, or $Q \propto V$. (Here, Q denotes the magnitude of the charge on *either* plate, *not* the net charge on the whole capacitor, which is zero.) This proportionality can be made into an equation by using a constant, C, called *capacitance*:

Note: The charges on the plates are $+Q$ and $-Q$, but it is customary to refer in general to the magnitude of these charges, Q (meaning $|\pm Q|$), on a capacitor.

Note: Recall that our notation for potential difference, or voltage (ΔV), will be replaced by V for convenience.

$$Q = CV \qquad \text{or} \qquad C = \frac{Q}{V} \qquad (16.9)$$

SI unit of capacitance: coulomb per volt (C/V), or farad (F)

Capacitance means the charge stored *per volt*. When a capacitor has a "large capacitance," it holds a large amount of charge *per volt*, compared with one of "smaller capacitance." If you connected the same battery to two different capacitors, the one with the larger capacitance would store more charge and more energy.

Note: The farad was named for the English scientist Michael Faraday (1791–1867), an early investigator of electrical phenomena who first introduced the concept of the electric field.

The unit of capacitance, the coulomb per volt (C/V), is called a **farad** (F). The farad is a large unit (see Example 16.7), so the *microfarad* ($1 \, \mu\text{F} = 10^{-6} \, \text{F}$), the *nanofarad* ($1 \, \text{nF} = 10^{-9} \, \text{F}$), and the *picofarad* ($1 \, \text{pF} = 10^{-12} \, \text{F}$) are commonly used.

Capacitance depends *only* on the geometry (size, shape, and spacing) of the plates (and the material between the plates, Section 16.5). Consider the parallel-plate capacitor, which has a uniform electric field given by Eq. 15.5:

$$E = \frac{4\pi k Q}{A}$$

The voltage across the plates can be computed from Eq. 16.2:

$$V = Ed = \frac{4\pi k Q d}{A}$$

The capacitance of a parallel-plate arrangement is then

$$C = \frac{Q}{V} = \left(\frac{1}{4\pi k} \right) \frac{A}{d} \quad (\textit{parallel plates only}) \quad (16.10)$$

It is common to replace the expression in the parentheses in Eq. 16.10 with a single symbol, ε_o, called the **permittivity of free space**. The value of this constant (to three significant figures) is

$$\varepsilon_\text{o} = \frac{1}{4\pi k} = 8.85 \times 10^{-12} \, \text{C}^2/(\text{N} \cdot \text{m}^2) \quad \begin{array}{c} \textit{permittivity} \\ \textit{of free space} \end{array} \quad (16.11)$$

The constant ε_o describes the electrical properties of free space (a vacuum), but its value in air is only 0.05% larger. In our calculations, we will take them to be the same.

It is common to rewrite Eq. 16.10 in terms of ε_o:

$$C = \frac{\varepsilon_\text{o} A}{d} \quad (\textit{parallel plates only}) \quad (16.12)$$

Let us use Eq. 16.12 in the next Example to show just how unrealistically large an air-filled capacitor with a capacitance of 1.0 F would be.

Example 16.7 ■ Parallel-Plate Capacitors: How Large Is a Farad?

What would be the plate area of an air-filled 1.0-F parallel-plate capacitor if the plate separation were 1.0 mm? Would it be realistic to consider building such a capacitor?

Thinking It Through. The area can be calculated directly from Eq. 16.12. Remember to keep all quantities in SI units, so that the answer will be in square meters. We can use the vacuum value of ε_o for air without creating a significant error.

Solution.

Given: $C = 1.0 \, \text{F}$ *Find:* A (area of one of the plates)
   $d = 1.0 \, \text{mm} = 1.0 \times 10^{-3} \, \text{m}$

Solving Eq. 16.12 for the area gives

$$A = \frac{Cd}{\varepsilon_\text{o}} = \frac{(1.0 \, \text{F})(1.0 \times 10^{-3} \, \text{m})}{8.85 \times 10^{-12} \, \text{C}^2/(\text{N} \cdot \text{m}^2)} = 1.1 \times 10^8 \, \text{m}^2$$

This is about 100 km² (about 40 mi²), or a square 10 km (over 6 mi) on a side. It is unrealistic to build a capacitor that big; 1.0 F is therefore a very large value of capacitance! There are ways, however, to make high-capacity capacitors (Section 16.4).

Follow-up Exercise. In this Example, what would the plate spacing have to be if you wanted the capacitor to have a plate area of 1 cm²? Compare your answer with a typical atomic diameter of 10^{-9} to 10^{-10} m. Is it feasible to build this capacitor?

The expression for the energy stored in a charged capacitor can be obtained by graphical analysis, since both Q and V vary during *charging*—the process of applying a charge, such as by a battery. A plot of voltage versus charge for charging a capacitor is a straight line with a slope of $1/C$, since $V = (1/C)Q$ (▶Fig. 16.14). The graph represents the charging of an initially uncharged capacitor ($V_o = 0$) to a final voltage (V). The work done is equivalent to transferring the total charge, using an average voltage $\overline{V}$. Since the voltage varies linearly with charge, the average voltage is half the final voltage V:

$$\overline{V} = \frac{V_{\text{final}} + V_{\text{initial}}}{2} = \frac{V + 0}{2} = \frac{V}{2}$$

Thus, the energy stored in the capacitor (equal to the work done by the battery) is

$$U_C = W = Q\overline{V} = \tfrac{1}{2}QV$$

Since $Q = CV$, this equation can be written in several equivalent forms:

$$U_C = \tfrac{1}{2}QV = \frac{Q^2}{2C} = \tfrac{1}{2}CV^2 \qquad \begin{array}{l}\textit{energy storage}\\ \textit{in a capacitor}\end{array} \qquad (16.13)$$

Typically, the form $U_C = \tfrac{1}{2}CV^2$ is the most practical, since the capacitance and the voltage are usually the known quantities. A very important medical application of the capacitor is in the *cardiac defibrillator* discussed in the next Example.

Example 16.8 ■ Capacitors to the Rescue: Energy Storage in a Cardiac Defibrillator

Following a heart attack, the heart sometimes beats in an erratic (and useless) fashion, called *fibrillation*. One way to get the heart back into its normal rhythm is to shock it with electrical energy supplied by an instrument called a *cardiac defibrillator* (▶Fig. 16.15). About 300 J of energy is required to produce the desired effect. Typically, a defibrillator stores this energy in a capacitor charged by a 5000-V power supply. (a) What capacitance is required? (b) What is the charge on the capacitor's plates?

Thinking It Through. (a) To find the capacitance, solve for C in Eq. 16.13. (b) The charge then follows from the definition of capacitance (Eq. 16.9).

Solution. We list the given data:

Given: $U_C = 300\text{ J}$ *Find:* (a) C (the capacitance)
$V = 5000\text{ V}$ (b) Q (charge on capacitor)

(a) The most appropriate form of Eq. 16.13 is $U_C = \tfrac{1}{2}CV^2$. (Why?) Solving for C, we get

$$C = \frac{2U_C}{V^2} = \frac{2(300\text{ J})}{(5000\text{ V})^2} = 2.40 \times 10^{-5}\text{ F} = 24.0\ \mu\text{F}$$

(b) The charge (magnitude on either plate) is then

$$Q = CV = (2.40 \times 10^{-5}\text{ F})(5000\text{ V}) = 0.120\text{ C}$$

Follow-up Exercise. In this Example, if the maximum allowable energy for any one defibrillation attempt is 750 J, what is the maximum voltage that should be used?

Sometimes capacitors can successfully model real-life phenomena. For example, a lightning storm can be considered to be the discharge of a negatively charged cloud to the positively charged ground—in effect, a "cloud-ground" capacitor. (See the related Insight in Chapter 15.) Another interesting application of electric potential treats nerve membranes as cylindrical capacitors to help explain nerve signal transmission. (See the Insight on p. 562.)

Note: Do not confuse U_C, the energy stored in a capacitor, with ΔU_e, the change in electric potential energy of a charged particle. (See Section 16.1.)

Note: Generally, we will use the lowercase letter q to represent charges on single particles and the uppercase letter Q for the larger amounts of charges on capacitor plates.

Note: Practice using the various forms of capacitor energy. In a pinch, you need recall only one, together with the definition of capacitance, $C = Q/V$.

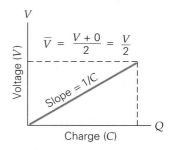

▲ **FIGURE 16.14 Capacitor voltage versus charge** A plot of voltage (V) versus charge (Q) for a capacitor is a straight line with slope $1/C$ (since $V = (1/C)Q$). The average voltage is $\overline{V} = \tfrac{1}{2}V$, and the total work done is equivalent to transferring the charge through $\overline{V}$. Thus, $U_C = W = Q\overline{V} = \tfrac{1}{2}QV$, the area under the curve (a triangle).

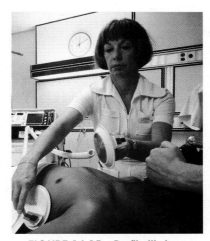

▲ **FIGURE 16.15 Defibrillator** A burst of electric current (flow of charge) from a defibrillator may restore a normal heartbeat in persons who have suffered cardiac arrest. Capacitors store the electrical energy that the device depends on.

Electric Potential and Nerve Signal Transmission

One of the most important systems in the human body is the nervous system, which is responsible for our reception of external stimuli through our senses (e.g., touch). Nerves also provide communication between the brain and our organs and muscles. If you touch something hot, your brain sends the signal "Move!" through the nervous system to your hand. But what is this signal, and how does it work?

A typical nerve consists of a bundle of nerve cells called *neurons*, much like individual telephone wires bundled into a single cable. The structure of a typical neuron is shown in Fig. 1a. The cell body, or *soma*, has long treelike structures called *dendrites* that receive the input signal. The soma is responsible for processing the signal and transmitting it down

the *axon*. At the other end of the axon are projections with knobs called *synaptic terminals*. At these knobs, the electrical signal is transmitted to another neuron across a gap called the *synapse*. The human body contains on the order of 100 billion neurons, and each neuron can have several hundred connections! Running the nervous system costs the body about 25% of its energy intake each day.

To understand the electrical nature of nerve signal transmission, let us focus on the axon. A vital component of the axon is its cell membrane, which is typically about 10 nm thick and consists of lipids (polarized hydrocarbon molecules) and embedded protein molecules (Fig. 1b). The membrane has pores called *ion channels*, where large protein molecules regulate the flow of ions (primarily sodium) across

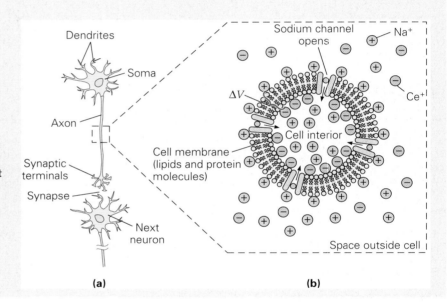

FIGURE 1   **(a)** The structure of a typical neuron. **(b)** An enlargement of the axon membrane, showing the membrane (about 10 nm thick) and the concentration of ions inside and outside the cell. The charge polarization across the membrane leads to a voltage, or membrane potential. When triggered by an external stimulus, the sodium ion channels open, allowing sodium ions into the cell. This influx changes the membrane potential.

(a)   **(b)**

TABLE 16.2 Dielectric Constants for Some Materials	
Material	*Dielectric Constant (κ)*
Vacuum	1.000 00
Air	1.000 59
Paper	3.7
Polyethylene	2.3
Polystyrene	2.6
Teflon	2.1
Glass (range)	3–7
Pyrex glass	5.6
Bakelite	4.9
Silicon oil	2.6
Water	80
Strontium titanate	233

16.4 Dielectrics

OBJECTIVES: To (a) understand what a dielectric is and (b) understand how it affects the physical properties of a capacitor.

In most capacitors, a sheet of insulating material, such as paper or plastic, is placed between the plates. Such an insulating material, called a **dielectric**, serves several purposes. For one, it keeps the plates from coming into contact. Contact would allow the electrons to flow back onto the positive plate, thereby neutralizing the charge on the capacitor and the energy stored. A dielectric also allows flexible plates of metallic foil to be rolled into a cylinder, giving the capacitor a more compact (and therefore more practical) size. Finally, a dielectric increases the charge storage capacity of the capacitor and therefore, under the right conditions, the energy stored in the capacitor. This capability depends on the type of material and is characterized by the **dielectric constant** (κ). Values of the dielectric constant for some common materials are given in Table 16.2.

the membrane. The key to nerve signal transmission is the fact that these ion channels are selective: They allow only certain types of ions to cross the membrane; others cannot.

The fluid outside the axon, although electrically neutral, contains sodium ions (Na^+) and chlorine ions (Cl^-) in solution. By contrast, the axon's internal fluid is rich in potassium ions (K^+) and negatively charged protein molecules. If it were not for the selective nature of the cell membrane, the Na^+ concentration would be equal on both sides. Under normal (or *resting*) conditions, it is difficult for Na^+ to penetrate the interior of the nerve cell. This process gives rise to a polarization of charge across the membrane. The exterior is positive (with Na^+ trying to enter the region of lower concentration), which attracts the negative proteins to the inner surface of the membrane (Fig. 1b). Thus a cylindrical capacitorlike charge-storage system exists across an axon membrane when it is resting. The *resting membrane potential* (the voltage across the membrane) is defined as $\Delta V = V_{in} - V_{out}$. Since the outside is positively charged, as defined, the resting potential is a negative quantity; it ranges from about -40 to -90 mV (millivolts), with a typical value of -70 mV.

Signal conduction occurs when the cell membrane receives a stimulus from the dendrites. Only then does the membrane potential change, and this change is propagated down the axon. The stimulus triggers Na^+ channels in the membrane (closed while resting, like a gate) to open and temporarily allows sodium ions to enter the cell (Fig. 1b). These positive ions are attracted to the negative charge layer on the interior and are driven by the difference in concentration. In about 0.001 s, enough sodium ions have passed through the gated channel to cause a reversal of polarity, and the membrane potential rises, typically to $+30$ mV. This time sequence for the change in membrane potential is shown in Fig. 2. When the difference in Na^+ concentration causes the membrane voltage to become positive, the sodium gates close. A chemical process known as the *Na/K ATPase molecular pump*

then reestablishes the resting potential at -70 mV by selectively transporting the excess Na^+ back out to the cell exterior.

This variation in membrane potential (a total of 100 mV, from -70 mV to $+30$ mV) is called the cell's *action potential*. The action potential *is* the signal that is actually transmitted down the axon. The voltage "wave" travels at speeds of from 1 to 100 m/s on its way to triggering another such pulse in the adjacent neuron. This speed, along with other factors such as time delays in the synapse region, is responsible for typical human reaction times totaling a few tenths of a second.

Related Exercises: 39 and 104.

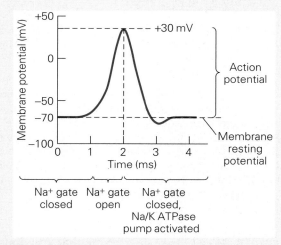

FIGURE 2 As the sodium channels open like gates and sodium ions rush to the cell's interior, the membrane potential changes quickly from its resting value of -70 mV to about $+30$ mV. The resting potential is restored (about 4 ms later) by a pump process that chemically removes the excess Na^+ after the sodium gates have closed (at 2 ms).

How a dielectric affects the electrical properties of a capacitor is illustrated in ▶Fig. 16.16. The capacitor is fully charged (creating a field E_o) and carefully disconnected from the battery, after which a dielectric is inserted (Fig. 16.16a). In the dielectric material, work is done on molecular dipoles by the existing electric field, aligning them with that field (Fig. 16.16b). (The molecular polarization may be permanent or temporarily induced by the electric field. In either case, the effect is the same.) Work is also done on the dielectric slab as a whole, since the charged plates pull it into the space between them.

The result is that the dielectric creates a "reverse" electric field (shown as E_d in Fig. 16.16c) that partially cancels the field between the plates. This means that the *net* field (E) between the plates is reduced, and therefore, so is the voltage across the plates (since $V = Ed$). The dielectric constant κ of the material is defined as the ratio of the voltage with the material in place (V) to the vacuum voltage (V_o). Since V is proportional to E, the voltage ratio is the same as the electric field ratio:

$$\kappa = \frac{V_o}{V} = \frac{E_o}{E} \quad \text{(only when the capacitor charge is constant)} \quad (16.14)$$

Note: Equation 16.14 holds only if the battery is disconnected.

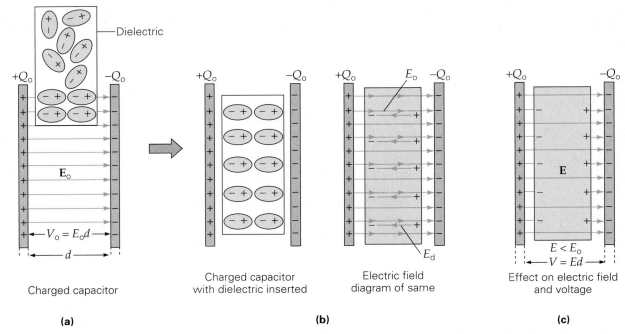

Dielectric

+Q_o −Q_o

E_o

$V_o = E_o d$

d

Charged capacitor

(a)

+Q_o −Q_o

Charged capacitor
with dielectric inserted

+Q_o E_o −Q_o

E_d

Electric field
diagram of same

(b)

+Q_o −Q_o

E

$E < E_o$

$V = Ed$

Effect on electric field
and voltage

(c)

▲ **FIGURE 16.16 The effects of a dielectric on an isolated capacitor** **(a)** A dielectric material with randomly oriented permanent molecular dipoles (or dipoles induced by the electric field) is inserted between the plates of an isolated charged capacitor. As the dielectric is inserted, the capacitor tends to pull it in, thus doing work on it. (Note the attractive forces between the plate charges and those induced on the dielectric surfaces.) **(b)** When the material is in the capacitor's electric field, the dipoles orient themselves with the field, giving rise to an opposing electric field E_d. **(c)** The dipole field partially cancels the field due to the plate charges. The net effect is a decrease in both the electric field and the voltage. Since the stored charge remains the same, the capacitance increases.

Note that κ is dimensionless and is greater than 1, since $V < V_o$. From Eq. 16.14, we can see that one way of determining the dielectric constant is by measuring the two voltages. (Voltmeters are discussed in detail in Chapter 18.) Since the battery was disconnected and the capacitor isolated, the charge on the plates, Q_o, is unaffected. Because $V = V_o/\kappa$, the value of the capacitance with the dielectric inserted is larger than the vacuum value by a factor of κ. In effect, the same amount of charge is now being stored at a lower voltage, and the result is an increase in capacitance. To understand this effect, apply the definition of capacitance:

$$C = \frac{Q}{V} = \frac{Q_o}{(V_o/\kappa)} = \kappa\left(\frac{Q_o}{V_0}\right) \quad \text{or} \quad C = \kappa C_o \quad (16.15)$$

Thus, inserting a dielectric into an isolated capacitor results in a larger capacitance. But what about energy storage? Since there is no energy input (the battery is disconnected) and the capacitor does work on the dielectric by pulling it into the region between the plates, the stored energy *drops* by a factor of κ (▸Fig. 16.17a), as the following equation shows:

$$U_C = \frac{Q^2}{2C} = \frac{Q_o^2}{2\kappa C_o} = \frac{Q_o^2/2C_o}{\kappa} = \frac{U_o}{\kappa} < U_o \quad (\textit{battery disconnected})$$

A different situation occurs, however, if the dielectric is inserted *and the battery remains connected*. In this case, the original voltage is maintained and the battery is able to supply (pump) more charge—and therefore do work (Fig. 16.17b). Since the battery does additional work, we expect the energy stored in the capacitor

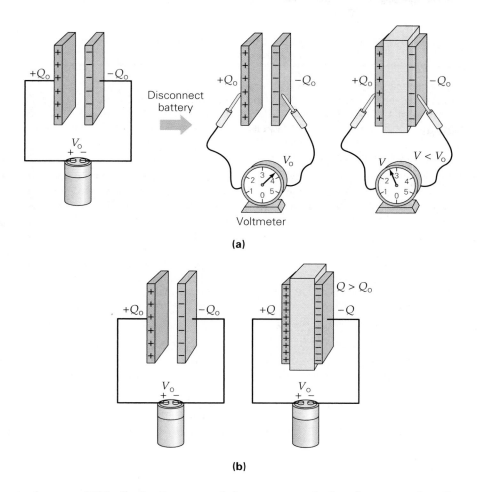

► FIGURE 16.17 **Dielectrics and**
capacitance (a) A parallel-plate capacitor in air (no dielectric) is charged by a battery to a charge Q_o and a voltage V_o (left). If the battery is disconnected and the potential across the capacitor is measured by a voltmeter, a reading of V_o is obtained (center). But if a dielectric is now inserted between the capacitor plates, the voltage drops to $V = V_o/\kappa$ (right), so the stored energy decreases. (Can you estimate the dielectric constant from the voltage readings?) (b) A capacitor is charged as in part (a), but the battery is left connected. When a dielectric is inserted into the capacitor, the voltage is maintained at V_o. (Why?) However, the charge on the plates increases to $Q = \kappa Q_o$. Therefore, more energy is now stored in the capacitor. In both cases, the capacitance increases by a factor of κ.

to *increase*. With the battery remaining connected, the charge on the plates increases by a factor κ, or $Q = \kappa Q_o$. Once again the capacitance increases, but now it is because more charge is stored at the same voltage. From the definition of capacitance, the result is the same as Eq. 16.15, since $C = Q/V = \kappa Q_o/V_o = \kappa(Q_o/V_o) = \kappa C_o$. Thus,

| the effect of a dielectric is to increase the capacitance by a factor of κ, regardless of the conditions under which the dielectric is inserted.

In the case of a capacitor kept at constant voltage, the energy storage of the capacitor increases at the expense of the battery. To see this, let's recalculate the energy with the dielectric in place:

$$U_C = \tfrac{1}{2}CV^2 = \tfrac{1}{2}\kappa C_o V_o^2 = \kappa(\tfrac{1}{2}C_o V_o^2) = \kappa U_o > U_o \quad (battery\ connected)$$

For a parallel-plate capacitor with a dielectric, the capacitance is increased over its (air) value in Eq. 16.12 by a factor of κ:

$$C = \kappa C_o = \frac{\kappa \varepsilon_o A}{d} \quad (parallel\ plates\ only) \quad (16.16)$$

This relationship is sometimes written as $C = \varepsilon A/d$, where $\varepsilon = \kappa \varepsilon_o$ is called the **dielectric permittivity** of the material, which is always greater than ε_o.

A sketch of the inside of a typical cylindrical capacitor and an assortment of real capacitors is shown in ►Fig. 16.18. Changes in capacitance can be used to monitor motion in our technological world, as the next Example shows.

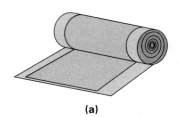

(a)

(b)

▲ **FIGURE 16.18 Capacitors in use** **(a)** The dielectric material between the capacitor plates enables the plates to be constructed so that they are close together, thus increasing the capacitance. In addition, the plates can then be rolled up into a compact, more practical capacitor. **(b)** Capacitors (flat, brown circles and purple cylinders) among other circuit elements in a microcomputer.

▶ **FIGURE 16.19 Capacitors in use** Capacitors can be used to convert movement into electrical signals that can be measured and analyzed by computer. As the distance between the plates changes, so does the capacitance, which causes a change in the charge on the capacitor. Some computer keyboards themselves operate in this way, as do other instruments, such as seismographs. (See Chapter 13.) See Example 16.9.

Example 16.9 ■ The Capacitor as a Motion Detector: Computer Keyboards

Consider a capacitor (with dielectric) underneath a computer key (▼Fig. 16.19). The capacitor is connected to a 12.0-volt battery and has a normal (uncompressed) plate separation of 3.00 mm and a plate area of 0.750 cm^2. (a) What is the required dielectric constant if the capacitance is 1.10 pF? (b) How much charge is stored on the plates under normal conditions? (c) How much charge flows onto the plates (i.e., what is the change in their charge) if they are compressed to a separation of 2.00 mm?

Thinking It Through. (a) The capacitance of air-filled plates can be found from Eq. 16.12, and then the dielectric constant can be determined from Eq. 16.15. (b) The charge follows from Eq. 16.9. (c) The compressed-plate separation distance must be used to recompute the capacitance. Then the new charge can be found as in (b).

Solution. The given data are as follows:

Given: $V = 12.0 \text{ V}$
$d = 3.00 \text{ mm} = 3.00 \times 10^{-3} \text{ m}$
$A = 0.750 \text{ cm}^2 = 7.50 \times 10^{-5} \text{ m}^2$
$C = 1.10 \text{ pF} = 1.10 \times 10^{-12} \text{ F}$
$d' = 2.00 \text{ mm} = 2.00 \times 10^{-3} \text{ m}$

Find: (a) κ (dielectric constant)
(b) Q (initial capacitor charge)
(c) ΔQ (change in capacitor charge)

(a) From Eq. 16.12, the capacitance if the plates were separated by air would be

$$C_o = \frac{\varepsilon_o A}{d} = \frac{(8.85 \times 10^{-12} \text{ C}^2/\text{N} \cdot \text{m}^2)(7.50 \times 10^{-5} \text{ m}^2)}{3.00 \times 10^{-3} \text{ m}} = 2.21 \times 10^{-13} \text{ F}$$

Since the dielectric increases the capacitance, its value is

$$\kappa = \frac{C}{C_o} = \frac{1.10 \times 10^{-12} \text{ F}}{2.21 \times 10^{-13} \text{ F}} = 4.98$$

(b) The initial charge is then

$$Q = CV = (1.10 \times 10^{-12} \text{ F})(12.0 \text{ V}) = 1.32 \times 10^{-11} \text{C}$$

(c) Under compressed conditions, the capacitance is

$$C' = \frac{\kappa \varepsilon_o A}{d'} = \frac{(4.98)(8.85 \times 10^{-12} \text{ C}^2/\text{N} \cdot \text{m}^2)(7.50 \times 10^{-5} \text{ m}^2)}{2.00 \times 10^{-3} \text{ m}} = 1.65 \times 10^{-12} \text{ F}$$

The voltage remains the same, $Q' = C'V = (1.65 \times 10^{-12} \text{ F})(12.0 \text{ V}) = 1.98 \times 10^{-11} \text{ C}$ Since the capacitance increased, the charge increased by

$$\Delta Q = Q' - Q = (1.98 \times 10^{-11} \text{ C}) - (1.32 \times 10^{-11} \text{ C}) = +6.60 \times 10^{-12} \text{ C}$$

As the key is depressed, a charge, whose magnitude is related to the displacement, flows onto the capacitor, providing a way of measuring the movement electrically.

Follow-up Exercise. In this Example, suppose instead that the spacing between the plates were *increased* by 1.00 mm from the normal value of 3.00 mm. Would charge flow onto or away from the capacitor? How much charge would this be?

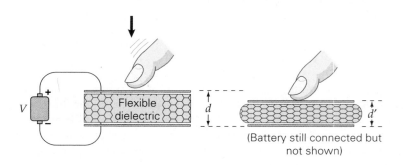

(Battery still connected but not shown)

16.5 Capacitors in Series and in Parallel

OBJECTIVES: To (a) find the equivalent capacitance of capacitors connected in series and in parallel, (b) calculate the charges, voltages, and energy storage of individual capacitors in series and parallel configurations and (c) analyze capacitor networks that include both series and parallel arrangements.

Capacitors can be connected in two basic ways: *in series* or *in parallel*. In series, the capacitors are connected head to tail (▼Fig. 16.20a). When connected in parallel, all the leads on one side of the capacitors have a common connection. (Think of all the "tails" connected together and all the "heads" connected together; Fig. 16.20b.)

Note: For so-called sandwiched dielectric parallel-plate capacitors, there is no head or tail distinction between the leads. Some types of capacitors do have particular positive and negative sides, and then the distinction must be made.

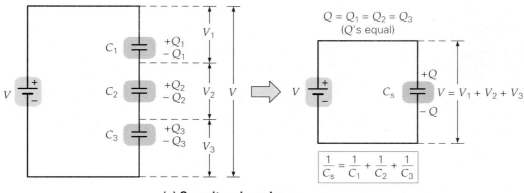

(a) Capacitors in series

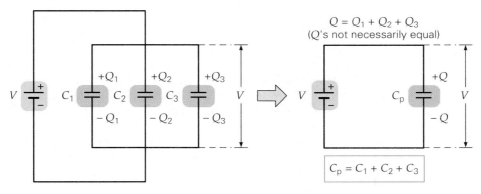

(b) Capacitors in parallel

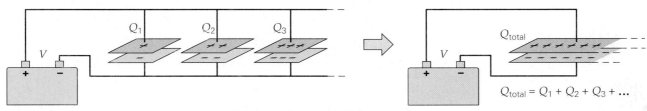

(c) Capacitors in parallel

▲ **FIGURE 16.20 Capacitors in series and in parallel** (a) All capacitors connected in series have the same charge, and the sum of the voltage drops is equal to the voltage of the battery. The total series capacitance is equivalent to the value of C_s. **(b)** When capacitors are connected in parallel, the voltage drops across the capacitors are the same, and the total charge is equal to the sum of the charges on the individual capacitors. The total parallel capacitance is equivalent to the value of C_p. **(c)** In a parallel connection, thinking of the plates makes it easier to see why the total charge is the sum of the individual charges. In effect, this arrangement represents a capacitor with two large plates.

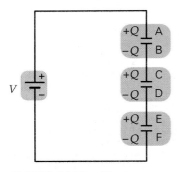

▲ FIGURE 16.21 Charges on capacitors in series Plates B and C together had zero net charge to start. When the battery placed $+Q$ on plate A, charge $-Q$ was induced on B; thus, C must have acquired $+Q$ for the BC combination to remain neutral. Continuing this way through the string, we see that all the charges must be the same in magnitude.

Capacitors in Series

When capacitors are wired in series, the charge Q must be the same on all the plates:

$$Q = Q_1 = Q_2 = Q_3 = \cdots$$

To see why this must be true, examine ◀Fig. 16.21. Note that only plates A and F are actually connected to the battery. Since the plates labelled B and C are isolated, the total charge on them must always be zero. Thus, if the battery puts a charge of $+Q$ on plate A, then $-Q$ is induced on B at the expense of plate C, which acquires a charge of $+Q$. This charge in turn induces $-Q$ on D, and so on down the line.

As we have seen, "voltage drop" is just another name for "change in electrical potential energy per unit charge." Thus, when we add up all the series capacitor voltage drops (see Fig 16.20a), we must get the same value as the voltage across the battery terminals. Thus, the sum of the individual voltage drops across all the capacitors is equal to the voltage of the source:

$$V = V_1 + V_2 + V_3 + \cdots$$

The **equivalent series capacitance**, C_s, is defined as the value of a single capacitor that could replace the series combination and store the same charge at the same voltage. Since the combination of capacitors stores a charge of Q at a voltage of V, it follows that $C_s = Q/V$, or $V = Q/C_s$. However, the individual voltages are related to the individual charges by $V_1 = Q/C_1, V_2 = Q/C_2, V_3 = Q/C_3, \ldots$.

Substituting these expressions into the voltage equation, we have

$$\frac{Q}{C_s} = \frac{Q}{C_1} + \frac{Q}{C_2} + \frac{Q}{C_3} + \cdots$$

Canceling the common Q's, we get

$$\frac{1}{C_s} = \frac{1}{C_1} + \frac{1}{C_2} + \frac{1}{C_3} + \cdots \qquad \textit{equivalent series} \atop \textit{capacitance} \qquad (16.17)$$

Note that the value of C_s is always smaller than the smallest capacitance in the combination. Since, in series, all the capacitors have the same charge, the charge stored by this arrangement is $Q = C_i V_i$ (where the subscript i refers to *any* of the individual capacitors in the string). Since $V_i < V$, the series arrangement stores *less* charge than any individual capacitor connected by itself to the same battery.

It makes sense that in series the smallest capacitance receives the largest voltage. A small value of C means less charge stored per volt. In order for the charge on all the capacitors to be the same, the smaller the value of capacitance, the larger is the fraction of the total voltage required ($Q = CV$).

Capacitors in Parallel

With a parallel arrangement (Fig. 16.20b), the voltages across the capacitors are the same (why?), and each individual voltage is equal to that of the battery:

$$V = V_1 = V_2 = V_3 = \cdots$$

The total charge is the sum of the charges on each capacitor (Fig 16.20c):

$$Q_{\text{total}} = Q_1 + Q_2 + Q_3 + \cdots$$

We expect the equivalent capacitance in parallel to be larger than the largest capacitance, because more charge per volt can be stored in this way than if any one capacitor were connected to the battery by itself. The individual charges are given by $Q_1 = C_1 V, Q_2 = C_2 V, \ldots$ A capacitor with the **equivalent parallel capacitance**, C_p, would hold this same total charge when connected to the battery, so $C_p = Q_{\text{total}}/V$, or $Q_{\text{total}} = C_p V$. Substituting these expressions into the previous equation, we have

$$C_p V = C_1 V + C_2 V + C_3 V + \cdots$$

and, canceling the common V, we obtain

$$C_p = C_1 + C_2 + C_3 + \cdots \quad \begin{array}{l} \textit{equivalent parallel} \\ \textit{capacitance} \end{array} \quad (16.18)$$

Thus, in the parallel case, the equivalent capacitance C_p is the sum of the individual capacitances. In this case, the equivalent capacitance is larger than the largest individual capacitance. Since capacitors in parallel have the same voltage, the largest capacitance will store the most charge. As a comparison of capacitors in series and in parallel, consider the next Example.

Example 16.10 ■ A Direct Comparison of Charge: Capacitors in Series and in Parallel

Given two capacitors, one with a capacitance of $2.50\ \mu$F and the other of $5.00\ \mu$F, what are the charge on each and the total charge stored if they are connected across a 12.0-volt battery (a) in series and (b) in parallel?

Thinking It Through. (a) Capacitors in series have the same charge; therefore, the smaller capacitor will have most of the voltage. ($Q = CV =$ constant; if C is smaller, it must be that V is larger.) Eq. 16.17 enables us to find the equivalent capacitance. From that, the charge on each capacitor can then be calculated. (b) Capacitors in parallel have the same voltage; hence, the larger capacitor will have the most charge. ($Q = CV$, thus if C is larger, Q must increase in proportion.)

Capacitors in Parallel and in Series

Solution. Listing the data, we have the following:

Given: $C_1 = 2.50\ \mu$F $= 2.50 \times 10^{-6}$ F   *Find:* (a) Q on each capacitor in series and
       $C_2 = 5.00\ \mu$F $= 5.00 \times 10^{-6}$ F Q_{total} (total charge)
       $V = 12.0$ V   (b) Q on each capacitor in parallel
                                     and Q_{total} (total charge)

(a) In series, the total (equivalent) capacitance is $1/C_s = 1/C_1 + 1/C_2$. Thus, we have

$$\frac{1}{C_s} = \frac{1}{2.50 \times 10^{-6}\ \text{F}} + \frac{1}{5.00 \times 10^{-6}\ \text{F}} = \frac{3}{5.00 \times 10^{-6}\ \text{F}}$$

so

$$C_s = 1.67 \times 10^{-6}\ \text{F}$$

(Note C_s is less than the smallest capacitance in the series chain, as expected.)
    Since the charge on each capacitor is the same in series and the same as the total, we have

$$Q_{total} = Q_1 = Q_2 = C_s V = (1.67 \times 10^{-6}\ \text{F})(12.0\ \text{V}) = 2.00 \times 10^{-5}\ \text{C}$$

(b) Here, the parallel equivalent capacitance relationship is used:

$$C_p = C_1 + C_2 = 2.50 \times 10^{-6}\ \text{F} + 5.00 \times 10^{-6}\ \text{F} = 7.50 \times 10^{-6}\ \text{F}.$$

(Note again a double check: This result is reasonable because it is greater than the largest individual value of C in the parallel arrangement.)
    Therefore,

$$Q_{total} = C_p V = (7.50 \times 10^{-6}\ \text{F})(12.0\ \text{V}) = 9.00 \times 10^{-5}\ \text{C}$$

In parallel, each capacitor has the full 12.0 V across it; hence,

$$Q_1 = C_1 V = (2.50 \times 10^{-6}\ \text{F})(12.0\ \text{V}) = 3.00 \times 10^{-5}\ \text{C}$$
$$Q_2 = C_2 V = (5.00 \times 10^{-6}\ \text{F})(12.0\ \text{V}) = 6.00 \times 10^{-5}\ \text{C}$$

As a final double check, notice that the total stored charge is equal to the sum of the charges on both capacitors.

Follow-up Exercise. In part (a) of this Example, what is the voltage across each capacitor, and how much energy is stored in each?

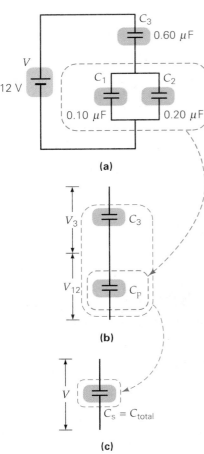

(a)

(b)

(c)

▲ **FIGURE 16.22 Circuit reduction** By combining capacitances, the combination of capacitors is reduced to a single equivalent capacitance. See Example 16.11.

Capacitor arrangements typically involve *both* series and parallel connections, as shown in the next Example. In this situation, you simplify the circuit, using the equivalent parallel and series capacitance expressions, until you end up with one single, overall equivalent capacitance. To find the results for each individual capacitor, you work backward until you get to the original arrangement.

Example 16.11 ■ One Step at a Time: Capacitors in Series–Parallel Combination

Three capacitors are connected in a circuit as shown in ◄Fig. 16.22a. What is the voltage across each capacitor?

Thinking It Through. The voltage across each capacitor could be found from $V = Q/C$ if the charge on each capacitor were known. The total charge on the capacitors is found by reducing the series–parallel combination to a single equivalent capacitance. Two of the capacitors are in parallel. Their single equivalent capacitance (C_p) is in series with the last capacitor—a fact that enables the total capacitance to be found. Working backward will allow the voltage across each capacitor to be found.

Solution.

Given: Values of capacitance and voltage from figure *Find:* V_1, V_2, and V_3 (voltages across capacitors)

Starting with the parallel combination, we have

$$C_p = C_1 + C_2 = 0.10\,\mu F + 0.20\,\mu F = 0.30\,\mu F$$

Now the arrangement is partially reduced, as shown in Fig. 16.22b. Next, considering C_p in series with C_3, we can find the total, or overall, equivalent capacitance of the original arrangement:

$$\frac{1}{C_s} = \frac{1}{C_3} + \frac{1}{C_p} = \frac{1}{0.60\,\mu F} + \frac{1}{0.30\,\mu F} = \frac{1}{0.60\,\mu F} + \frac{2}{0.60\,\mu F} = \frac{1}{0.20\,\mu F}$$

Therefore,

$$C_s = 0.20\,\mu F = 2.0 \times 10^{-7}\,F$$

This is the total equivalent capacitance of the arrangement (Fig. 16.22c). Treating the problem as one for a single capacitor, we can find the charge on that equivalent capacitance:

$$Q = C_s V = (2.0 \times 10^{-7}\,F)(12\,V) = 2.4 \times 10^{-6}\,C$$

This is the charge on C_3 and C_p, since they are in series. We can use this fact to calculate the voltage across C_3:

$$V_3 = \frac{Q}{C_3} = \frac{2.4 \times 10^{-6}\,C}{6.0 \times 10^{-7}\,F} = 4.0\,V$$

The sum of the voltages across the capacitors equals the voltage across the battery terminals. The voltages across C_1 and C_2 are the same because they are in parallel. Since the voltage across C_1 (or C_2) plus the voltage across C_3 equals the total voltage (the battery voltage), we can write $V = V_{12} + V_3 = 12\,V$. (See Fig. 16.22a.) Here, V_{12} represents the voltage across either C_1 or C_2. Solving for V_{12}, we have

$$V_{12} = V - V_3 = 12\,V - 4.0\,V = 8.0\,V$$

It is reasonable that the parallel combination of C_1 and C_2 sustains most of the voltage drop, because they are equivalent to a capacitance that is less than C_3. Since C_p and C_3 are in series, it follows that C_p (and therefore both C_1 and C_2) have most of the voltage.

Follow-up Exercise. In this Example, find (a) the charge stored on each capacitor and (b) the energy stored in each.

Chapter Review

Important Concepts

- The **electric potential difference** (or **voltage**) between two points is the work done per unit positive charge between those two points, or the change in electric potential energy per unit positive charge. Expressed in equation form, this relationship is

$$\Delta V = \frac{\Delta U_e}{q_o} = \frac{W}{q_o} \qquad (16.1)$$

- **Equipotential surfaces** (surfaces of constant electric potential, also called **equipotentials**) are surfaces on which a charge has a constant electric potential energy. Alternatively, we say that it takes no work to move a charge from one point to another on an equipotential surface. These surfaces are everywhere perpendicular to the electric field.

- The expression for the **electric potential due to a point charge** (choosing $V = 0$ at $r = \infty$) is

$$V = \frac{kq}{r} \qquad (16.4)$$

- The **electric potential energy for a pair of point charges** is given by (choosing $U = 0$ at $r = \infty$)

$$U_{12} = \frac{kq_1 q_2}{r_{12}} \qquad (16.5)$$

- The **electric potential energy of a configuration of more than two point charges** is given by a sum of point-charge pair terms from Eq. 16.5:

$$U_{\text{total}} = U_{12} + U_{23} + U_{13} + \cdots \qquad (16.6)$$

- The electric field is related to how rapidly the electric potential changes with distance. The electric field (**E**) points in the direction of the most rapid decrease in electric potential (V). The electric field magnitude (E) is the rate of change of the potential with distance, or

$$E = \left| \frac{\Delta V}{\Delta x} \right|_{\text{max}} \qquad (16.8)$$

- The **electron volt (eV)** is a unit of energy commonly used in atomic and nuclear physics. It is defined as the kinetic energy gained by an electron or a proton accelerated through a potential difference of 1 volt.

- A **capacitor** is any arrangement of two metallic plates. Capacitors store charge on their plates, and therefore electric energy in their electric fields.

- **Capacitance** is a quantitative measure of how effective a capacitor is in storing charge. It is defined as

the magnitude of the charge stored on either plate per volt, or

$$C = \frac{Q}{V} \qquad (16.9)$$

- The **capacitance of a parallel-plate capacitor** (in air) is

$$C = \frac{\varepsilon_o A}{d} \qquad (16.12)$$

where $\varepsilon_o = 8.85 \times 10^{-12}\, C^2/(N \cdot m^2)$ is called the **permittivity of free space**.

- The **energy stored in a capacitor** depends on the capacitor's capacitance and the amount of charge the capacitor stores (or, equivalently, the voltage across its plates). There are three equivalent expressions for this energy:

$$U_C = \tfrac{1}{2}QV = \frac{Q^2}{2C} = \tfrac{1}{2}CV^2 \qquad (16.13)$$

- A **dielectric** is a nonconducting material that increases the capacitance value upon insertion between the plates of a capacitor.

- The **dielectric constant** κ describes the effect of a dielectric on capacitance. A dielectric increases the capacitor's capacitance over its value with air between the plates, or

$$C = \kappa C_o \qquad (16.15)$$

- When capacitors are connected in series, they can be thought of as being equivalent to one capacitor, with a capacitance called the **equivalent series capacitance** C_s. In series, all the capacitors have the same charge. The equivalent series capacitance is always less than that of the smallest capacitor in the combination and is given by

$$\frac{1}{C_s} = \frac{1}{C_1} + \frac{1}{C_2} + \frac{1}{C_3} + \cdots \qquad (16.17)$$

- When capacitors are connected in parallel, they can be thought of as being equivalent to one capacitor, with a capacitance called the **equivalent parallel capacitance** C_p. In parallel, all the capacitors have the same voltage. The equivalent parallel capacitance is always greater than that of the largest capacitor in the combination and is given by

$$C_p = C_1 + C_2 + C_3 + \cdots \qquad (16.18)$$

Exercises

16.1 Electric Potential Energy and Electric Potential Difference

1. The SI unit of electric potential difference is the (a) joule, (b) newton, (c) newton-meter, or (d) joule per coulomb.

2. The electrostatic potential energy of two point charges (a) is inversely proportional to their separation distance, (b) is a vector quantity, (c) is always positive, (d) has units of N/C.

3. CQ What is the difference (a) between electrostatic potential energy and electric potential and (b) between electric potential difference and voltage?

4. CQ When a proton approaches another proton, what happens to the electric potential energy of the first proton?

5. CQ In terms of the way they react to differences in electrical potential, describe why positive charges speed up as they approach negative charges.

6. CQ An electron is released in a region where there is a varying electric potential. Will the electron move toward the lower potential region or the higher potential region? Explain.

7. CQ If the electric potential is constant inside and throughout an object, what is the electric field in the object? Explain.

8. For charged parallel plates, where does a proton have the lowest electric potential energy? (a) Near the positive plate, (b) near the negative plate, or (c) midway between the plates.

9. CQ If two locations are at the same potential, does it take net work to move a charge from one location to the other? Explain.

10. ■ A pair of parallel plates is charged by a 24-V battery. How much work is required to move a $-4.0\text{-}\mu\text{C}$ charge from the positive to the negative plate?

11. ■ If it takes $+1.6 \times 10^{-5}$ J to move a positive charge between two charged parallel plates, (a) what is the magnitude of the charge if the plates are connected to a 6.0-V battery? (b) Taking the charge to be negative, was it moved from the negative to the positive plate or from the positive to the negative plate?

12. ■ What are the magnitude and direction of the electric field between the two charged parallel plates in Exercise 11 if the plates are separated by 4.0 mm?

13. ■ A proton is accelerated by a potential difference of 10 kV. How fast is the proton moving if it started from rest?

14. ■ An electron is accelerated by a uniform electric field (1000 V/m) pointing vertically upward. Use Newton's laws to determine the electron's velocity after it moves 0.50 cm from rest.

15. ■ (a) Repeat Exercise 14, but find the speed by using energy methods. Get the direction in which the electron is moving by considering electric potential-energy changes. (b) Does the electron gain or lose potential energy?

16. IE ■ Consider two points at different distances from a positive point charge. (a) The point closer to the charge has a (1) higher, (2) equal, or (3) lower potential than the point farther away from the charge. Why? (b) What is the difference in potential between two points 20 cm and 40 cm from a charge of $5.5\,\mu\text{C}$?

17. IE ■■ (a) At one-third the original distance from a positive point charge, by what factor is the electric potential changed? (1) 1/3, (2) 3, (3) 1/9, or (4) 9. Why? (b) How far from a $+1.0\text{-}\mu\text{C}$ charge is a point with an electric potential value of 10 kV? (c) How much of a change in potential would occur if the point were moved to three times that distance?

18. IE ■■ In the Bohr model of the hydrogen atom, the electron can exist only in circular orbits of certain radii. (a) Will a larger orbit have a (1) higher, (2) equal, or (3) lower electric potential than a smaller orbit? Why? (b) Determine the potential difference between two orbits of radii 0.21 nm and 0.48 nm.

19. ■■ In Exercise 18, by how much does the potential energy of the atom change if the electron goes (a) from the lower to the higher orbit, (b) from the higher to the lower orbit, and (c) from the larger orbit to a very large distance?

20. ■■ How much work is required to completely separate two charges (each $-1.4\,\mu\text{C}$) at constant velocity if they were initially 8.0 mm apart?

21. ■■ In Exercise 20, if the two charges are released at their initial separation distance, how much kinetic energy would each have when they are very distant from one another?

22. ■■ It takes $+6.0$ J of work to move two charges from a large distance apart to 1.0 cm from one another. If the charges have the same magnitude, (a) how large is each charge, and (b) what can you tell about their signs?

23. ■■ A $+2.0\text{-}\mu\text{C}$ charge is initially 0.20 m from a fixed $-5.0\text{-}\mu\text{C}$ charge and is then moved to a position 0.50 m from the fixed charge. (a) What work was required to move the charge? (b) Does the work depend on the path through which the charge is moved?

24. ■■ An electron is moved from point A to point B and then to point C along two legs of an equilateral triangle with sides of length 0.25 m (▼Fig. 16.23). If the horizontal electric field is 15 V/m, (a) what is the magnitude of the work required? (b) What is the potential difference between points A and C? (c) Which point is at a higher potential?

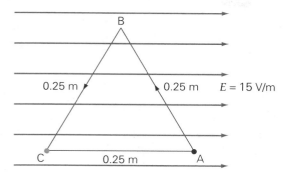

▲ **FIGURE 16.23 Work and energy** See Exercise 24.

25. ■■ Compute the energy necessary to bring together the charges in the configuration shown in ▼Fig. 16.24.

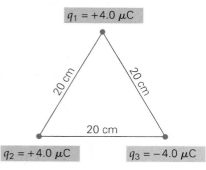

▲ **FIGURE 16.24 A charge triangle** See Exercises 25, 27, and 28.

26. ■■ Compute the energy necessary to bring together the charges in the configuration shown in ▼Fig. 16.25.

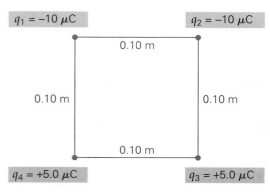

▲ **FIGURE 16.25 A charge rectangle** See Exercises 26, 29, 30, and 32.

27. ■■ What is the electric potential at the center of the triangle in Fig. 16.24?

28. ■■ Compute the electric potential at a point midway between q_2 and q_3 in Fig. 16.24.

29. ■■■ What is the electric potential at the center of the square in Fig. 16.25?

30. ■■■ Compute the electric potential at a point midway between q_1 and q_4 in Fig. 16.25.

31. IE ■■■ In a computer monitor, electrons are accelerated from rest through a potential difference in an electron gun arrangement (▼Fig. 16.26). (a) Should the left side of the gun be at (1) a higher, (2) an equal, or (3) a lower potential than the right side? Why? (b) If the potential difference in the gun is 10 kV, what is the "muzzle speed" of the electrons emerging from the gun? (c) If the gun is directed at a screen 35 cm away, how long does it take the electrons to reach the screen?

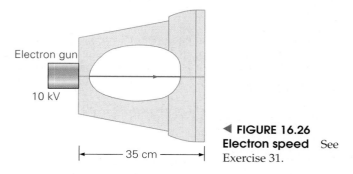

◀ **FIGURE 16.26 Electron speed** See Exercise 31.

32. ■■■ Compute the electric potential at a point 4.0 cm from q_4 along the line joining it to q_1 in Fig. 16.25.

16.2 Equipotential Surfaces and the Electric Field

33. Equipotential surfaces are those surfaces on which (a) the potential is constant, (b) the electric field is zero, or (c) the potential is zero.

34. Equipotential surfaces are (a) parallel to the electric field, (b) perpendicular to the electric field, or (c) at any angle with respect to the electric field.

35. Sketch the topographic map you would expect as you walk away from the ocean up a gently sloping uniform beach. Label the gravitational equipotentials as to relative height and potential value. Show how to predict, from the map, which way a ball would accelerate if it is initially rolled up the beach away from the water.

36. CQ Explain why two equipotential surfaces cannot intersect.

37. CQ Suppose that you start with a charge at rest on an equipotential surface, move it off the surface, then return it to the surface, and, finally, bring it to rest. Is it still true that the net work done on the charge is zero? Explain.

38. **CQ** What geometrical shape are the equipotential surfaces between two charged parallel plates?

39. **CQ** (a) What is the approximate shape of the equipotential surfaces inside the axon cell membrane? (See Fig. 1, p. 562.) (b) Under resting-potential conditions, where is the region of highest electric potential inside the membrane? (c) What about during reversed polarity conditions?

40. **CQ** The equipotential surfaces outside a point charge are what geometrical shape?

41. **CQ** Near a fixed positive point charge, if you go from one equipotential surface to another one with a smaller radius, what happens to the value of the potential?

42. Show that the unit V/m is the same as the unit N/C.

43. **CQ** Explain why an electron volt is a unit of energy. Which is larger, a GeV or a MeV?

44. At a given point on an equipotential surface, the electric field points directly (a) to the next highest equipotential, (b) to the next lowest equipotential, (c) parallel to the equipotential surface?

45. **CQ** Can electric field lines ever cross? Explain, using the result of Exercise 36.

46. **CQ** Can the electric field at a point be zero, yet there exist a nonzero electric potential at that point? Explain.

47. ■ For a $+3.50$-μC point charge, what is the radius of the equipotential surface that is at a potential of 2.50 kV?

48. ■ A uniform electric field of 10 kV/m points vertically upward. How far apart are the equipotential planes that differ by 100 V?

49. ■ In Exercise 48, if the ground is called zero potential, how far above the ground is the equipotential surface corresponding to 7.0 kV?

50. ■ Determine the potential 2.5 mm from the negative plate of a pair of parallel plates separated by 10 mm and connected to a 24-V battery.

51. ■ Relative to the positive plate in Exercise 50, where is the point with a potential of 20 V?

52. ■ If the radius of the equipotential surface of point charge is 14.3 m at a potential of 2.20 kV, what is the magnitude of the point charge creating the potential?

53. **IE** ■ (a) The shape of an equipotential surface at a certain distance from a point charge consists of (1) concentric spheres, (2) concentric cylinders, or (3) planes. Why? (b) Calculate the amount of work (in eV and in J) it would take to move an electron from 12.6 m to 14.3 m away from a $+3.50$-μC point charge.

54. ■ The potential difference involved in a typical lightning discharge may be up to 100 MV (million volts). What would be the gain in kinetic energy of an electron after moving through this potential difference? Give your answer in (a) eV and (b) joules. (Assume that there are no collisions.)

55. ■ In a typical Van de Graaff linear accelerator, protons can be accelerated through a potential difference of 20 MV. What is their kinetic energy if they started from rest? Give your answer in (a) eV, (b) keV, (c) MeV, (d) GeV, and (e) joules.

56. **CQ** ■ In Exercise 55, how do your answers change if a doubly charged ($+2e$) alpha particle is accelerated instead? (Recall that an alpha particle consists of two neutrons and two protons.)

57. ■■ In Exercises 55 and 56, compute the speed of the proton and alpha particle on being accelerated through the potential.

58. ■■ Calculate the voltage required to accelerate a beam of protons initially at rest, and calculate their speed if they have a kinetic energy of (a) 3.5 eV, (b) 4.1 keV, and (c) 8.0×10^{-16} J.

59. ■■ Repeat the calculation in Exercise 58 for electrons instead of protons.

60. ■■■ Two large parallel plates are separated by 3.0 cm and connected to a 12-V battery. Starting at the negative plate and moving 1.0 cm toward the positive plate at a 45° angle (▼Fig. 16.27), what value of potential would be reached, assuming the negative plate were defined as zero potential?

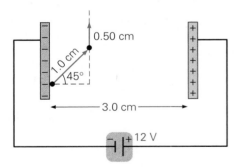

▲ **FIGURE 16.27 Reaching our potential** See Exercises 60 and 61.

61. ■■■ In Exercise 60, what would be the value of the potential if you then moved 0.50 cm parallel to the plates?

16.3 Capacitance

62. Capacitance has units of (a) farads, (b) joules, (c) coulomb per volt, (d) both (a) and (c).

63. **CQ** How can you increase the capacitance and the energy-storage capability of an air-filled parallel-plate capacitor?

64. **CQ** If the plates of an isolated parallel-plate capacitor are moved closer to each other, does the energy storage increase, decrease, or remain the same? Explain.

65. **CQ** If the potential difference across a capacitor is doubled, what happens to (a) the charge on the capacitor and (b) the energy stored in the capacitor?

66. **CQ** A capacitor is connected to a 12-V battery. If the plate separation increases, what happens to the charge on the capacitor?

67. ■ How much charge flows through a 12-V battery when a 2.0-μF capacitor is connected across its terminals?

68. ■ A parallel-plate capacitor has a plate area of 0.50 m^2 and a plate separation of 2.0 mm. What is its capacitance?

69. ■ What plate separation is required for a parallel-plate capacitor to have a capacitance of 5.0×10^{-9} F if the plate area is 0.40 m^2?

70. **IE** ■ (a) For a parallel-plate capacitor, a larger plate area results in (1) a larger, (2) an equal, or (3) a smaller capacitance. Why? (b) How about a larger plate separation? Why? (c) A 2.5×10^{-9} F parallel-plate capacitor has a plate area of 0.425 m^2. If the capacitance is to double, what is the required plate area?

71. ■■ A 12-V battery is connected to a parallel-plate capacitor with a plate area of 0.20 m^2 and a plate separation of 5.0 mm. (a) What is the resulting charge on the capacitor? (b) How much energy is stored in the capacitor?

72. ■■ If the plate separation of the capacitor in Exercise 71 changed to 10 mm after the capacitor was disconnected from the battery, how do your answers to that Exercise change?

73. ■■■ Current state-of-the-art capacitors are capable of storing many times the energy of older ones. Such a capacitor, with a capacitance of 1.0 F, is able to light a small 0.50-W bulb at full power for 5.0 s before it quits. What was the terminal voltage of the battery that charged the capacitor?

16.4 Dielectrics

74. Putting a dielectric in a charged (but not connected to a battery) parallel-plate capacitor (a) decreases the capacitance, (b) decreases the voltage, (c) increases the charge, or (d) causes a discharge because the dielectric is a conductor.

75. A parallel-plate capacitor is connected to a battery. If a dielectric is inserted between the plates, (a) the capacitance decreases, (b) the voltage increases, (c) the voltage decreases, or (d) the charge increases.

76. **CQ** Give several reasons why a conductor would not be a good choice as a dielectric for a capacitor.

77. **CQ** A parallel-plate capacitor is connected to a battery. If a dielectric is inserted between the plates, what happens to (a) the capacitance and (b) the voltage?

78. ■ A capacitor has a capacitance of 50 pF, which increases to 150 pF when a dielectric material is between its plates. What is the dielectric constant of the material?

79. ■ A 50-pF capacitor is immersed in silicone oil ($\kappa = 2.6$). When the capacitor is connected to a 24-V battery, what will be the charge on the capacitor and the amount of stored energy?

80. ■ The dielectric of a parallel-plate capacitor is to be constructed from glass that completely fills the volume between the plates. The area of each plate is 0.50 m^2. (a) What thickness should the glass have if the capacitance is to be 0.10 μF? (b) What is the charge on the capacitor if it is connected to a 12-V battery?

81. ■■ A parallel-plate capacitor has a capacitance of 1.5 μF with air between the plates. The capacitor is connected to a 12-V battery and charged. The battery is then removed. When a dielectric is placed between the plates, a potential difference of 5.0 V is measured across the plates. What is the dielectric constant of the material?

82. ■■ A parallel-plate capacitor has rectangular plates with dimensions of 6.0 cm $\times$ 8.0 cm. If the plates are separated by a sheet of 1.5-mm-thick Teflon™ ($\kappa = 2.1$), how much energy is stored in the capacitor when it is connected to a 12-V battery?

16.5 Capacitors in Series and in Parallel

83. Capacitors in series have the same (a) voltage, (b) charge, or (c) energy storage.

84. Capacitors in parallel have the same (a) voltage, (b) charge, or (c) energy storage.

85. **CQ** Under what conditions would two capacitors in series have the same voltage?

86. **CQ** Under what conditions would two capacitors in parallel have the same charge?

87. **CQ** If you are given two capacitors, how should you connect them to get (a) maximum equivalent capacitance and (b) minimum equivalent capacitance?

88. ■ What is the equivalent capacitance of two capacitors with capacitances of 0.40 μF and 0.60 μF when they are connected (a) in series and (b) in parallel?

89. **IE** ■ (a) Two capacitors can be connected to a battery in either a series or parallel combination. The parallel combination will draw (1) more, (2) equal, or (3) less energy from a battery than the series combination. Why? (b) When a series combination of two uncharged capacitors is connected to a 12-V battery, 173 μJ of energy is drawn from the bat-

tery. If one of the capacitors has a capacitance of $4.0\,\mu F$, what is the capacitance of the other?

90. ■■ For the arrangement of three capacitors in ▼Fig. 16.28, what value of C_1 will give a total equivalent capacitance of $1.7\,\mu F$?

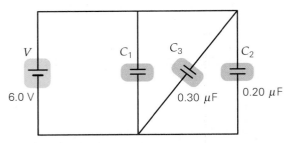

▲ **FIGURE 16.28 A capacitor triad** See Exercises 90 and 94.

91. IE ■■ (a) Three capacitors of equal capacitance are connected in parallel to a battery, and together they draw a certain amount of charge Q from that battery. Will the charge on each capacitor be (1) Q, (2) $3Q$, or (3) $Q/3$? (b) Three capacitors of $0.25\,\mu F$ each are connected in parallel to a 12-V battery. What is the charge on each capacitor? (c) How much charge is drawn from the battery?

92. IE ■■ (a) If you are given three identical capacitors, you can obtain (1) three, (2) five, or (3) seven different capacitance values. (b) If the three capacitors each have a capacitance of $1.0\,\mu F$, what are the different values of equivalent capacitance?

93. ■■ What are the maximum and minimum equivalent capacitances that can be obtained by combinations of three capacitors of $1.5\,\mu F$, $2.0\,\mu F$, and $3.0\,\mu F$?

94. ■■■ If the capacitance $C_1 = 0.10\,\mu F$, what is the charge on each of the capacitors in the circuit in Fig. 16.28?

95. ■■■ Four capacitors are connected in a circuit as illustrated in ▼Fig. 16.29. Find the charge on, and the voltage difference across, each of the capacitors.

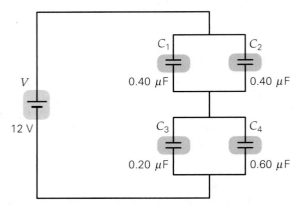

▲ **FIGURE 16.29 Double parallel in series** See Exercise 95.

Additional Exercises

96. When you go vertically upward 10.0 mm, you find that you are moving in a direction in which the electric potential decreases the most rapidly. (a) If the change in potential is 200 V, estimate the direction and magnitude of the electric field. (b) If the ground is 10 cm below, by how much does the potential there differ from that at your location?

97. An electric dipole consists of two point charges $\pm q$, separated by a distance d. How much work (in terms of k, q, and d) did it take to bring these charges together from a large distance?

98. Sketch the equipotential surfaces and the electric field line pattern outside a uniformly (negatively) charged long wire. Label the surfaces with relative potential value, and indicate the electric field direction.

99. A helium atom with one electron already removed (a positive helium ion) consists of a single orbiting electron and a nucleus of two protons. If the electron is in its minimum orbital radius of 0.027 nm, what is the potential energy of the system? Give your answer in eV.

100. Suppose that the three capacitors in Figure 16.22 have the following values: $C_1 = 0.15\,\mu F$, $C_2 = 0.25\,\mu F$, and $C_3 = 0.30\,\mu F$. (a) What is the equivalent capacitance of this arrangement? (b) How much charge will be drawn from the battery? (c) What is the voltage across each capacitor?

101. IE Two horizontal, conductive parallel plates are separated by a vertical distance. An electron is to be suspended in midair between them. (a) The top plate should be at (1) a higher, (2) an equal, or (3) a lower potential compared with the bottom plate. Why? (b) If the vertical distance between the plates is 1.5 cm, what voltage across the plates is required?

102. A 1.0 F parallel-plate capacitor with a distance of 0.50 mm between the plates is to be built. (a) What is the required area of the plates without a dielectric between them? (b) What is the required plate area if, instead, the plates are filled with polystyrene ($\kappa = 2.6$) 0.50 mm thick between them?

103. The electric field between two parallel plates separated by 3.0 cm has a magnitude of 5.5×10^4 V/m. If the electron is released from the negative plate, how much kinetic energy has it gained when it reaches the positive plate?

104. Assuming that the axon cell membrane is 10 nm thick (Fig. 1, Nerve Signal Transmission Insight, pg. 562), estimate the (a) direction and (b) magnitude of the electric field inside the membrane under resting-potential conditions. (*Hint*: See Learn by Drawing on sketching electric field lines, Fig. 3, p. 556.)

Electric Current and Resistance

If you were asked to think of electricity and its uses, many favorable images would probably come to mind, including such diverse applications as lamps, television remote controls, and electric leaf blowers. You might also think of some unfavorable images, such as dangerous lightning, or sparks you may have experienced from an overloaded electric outlet.

Common to all of these images is the concept of electric energy. For an electric appliance, energy is supplied by electric current in wires; for lightning or a spark, it is conducted through the air. In either case, the light, heat, or mechanical energy given off is simply electric energy converted to a different form. In this photograph, for example, the light given off by the spark is emitted by air molecules.

In this chapter, we are concerned with the fundamental principles governing electric circuits. These principles will enable us to answer questions such as: What is electric current and how does it travel? What

causes an electric current to move through an appliance when we flick a switch? Why does the electric current cause the filament in a bulb to glow brightly, but not affect the connecting wires in the same way? We can apply the principles to gain an understanding of a wide range of phenomena, from the operation of house-hold appliances to the workings of Nature's spectacular fireworks—lightning.

17.1 Batteries and Direct Current

OBJECTIVES: To (a) introduce the properties of a battery, (b) explain how a battery produces a direct current in a circuit, and (c) learn various circuit symbols for sketching schematic circuit diagrams.

After studying electric force and energy in Chapters 15 and 16, you can probably guess what is required to produce an *electric current*, or a flow of charge. Here are

Note: Recall that *voltage* is used to mean "difference in electric potential."

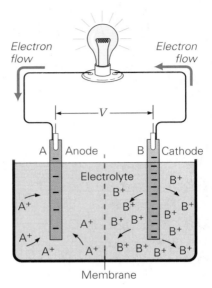

▲ **FIGURE 17.1   Battery action in a chemical battery or cell** Chemical processes involving an electrolyte and two unlike metal electrodes cause ions of both metals to dissolve into the solution at different rates. Thus, one electrode (the cathode) becomes more negatively charged than the other (the anode). The anode is at a higher potential than the cathode. By convention, the anode is designated the positive terminal and the cathode the negative. This potential difference (V) can cause a current, or a flow of charge (electrons), in the wire. The positive ions migrate as shown. (A membrane is necessary to prevent mixing of the two types of ion; why?)

some analogies to help you. Water flows downhill, from areas of higher to lower gravitational potential energy. Heat flows spontaneously only when there is a temperature *difference*. For electricity, a flow of electric charge is caused by an electric potential *difference*—which is what we call "voltage."

In solid conductors—particularly metals—the outer electrons of atoms are relatively free to move. (In liquid conductors and charged gases called *plasmas*, both positive and negative ions, as well as electrons, can move.) Energy is required to move electric charge. Electric energy is generated through the conversion of other forms of energy, giving rise to a potential difference, or voltage. Any device that can produce potential differences fits under the general name of a *power supply*.

Battery Action

One common type of power supply is the battery. A **battery** converts stored *chemical* potential energy into electrical energy. The Italian scientist Allesandro Volta is credited with constructing one of the first practical batteries. A simple battery consists of two unlike metal *electrodes* in an *electrolyte*, a solution that conducts electricity. With the appropriate electrodes and electrolyte, a potential difference develops across the electrodes as a result of chemical action (◄Fig. 17.1).

When a complete circuit is formed—for example, by connecting a lightbulb and wires (Fig. 17.1)—electrons from the more negative electrode (B) will move through the wire and bulb to the less negative electrode (A).* The result is a flow of electrons in the wire. As electrons move through the bulb's filament, colliding with and transferring energy to its atoms (typically tungsten), the filament reaches a sufficient temperature to give off visible light (glow). Since electrons tend to move to regions of higher potential, electrode A must be at a higher electric potential than electrode B. In other words, there is now a potential *difference* across the terminals of the battery. Electrode A is called the **anode** and labeled with a plus (+) sign. Electrode B is called the **cathode** and labeled with a negative (−) sign. It is easy to keep track of their relative signs by recalling that electrons themselves are negatively charged and thus will move through the wire from B (−) to A (+).

For the purpose of studying electric circuits, a battery may be represented as a "black box" that maintains a constant potential difference between its terminals. Inserted into a circuit, a battery is capable of giving energy to the electrons in the wire (at the expense of its own internal chemical energy), which in turn deliver that energy to circuit elements external to the battery. At those elements, the energy is converted into other forms, such as mechanical motion (electric lawn mowers), heat (immersion heaters) and light (lightbulbs). Other sources of voltage, such as generators and photocells, will be considered later.

To help understand the role of a battery in a circuit, consider the gravitational analogy shown in ▶Fig. 17.2. A gasoline-fueled pump lifts water, using the chemical energy stored in the gasoline to do work on the water. Therefore, the water gains gravitational potential energy. The water then returns to the pump by flowing down the trough (analogous to the wire) into the pond. On the way down, the water does work on the wheel, resulting in its rotational kinetic energy, analogous to the electrons transferring energy to a lightbulb.

Battery EMF and Terminal Voltage

The potential difference across the terminals of a battery *when it is not connected* to a circuit is called the battery's **electromotive force** (**emf**) ($\mathscr{E}$). The name is somewhat misleading, because electromotive force is *not* a force, but a potential difference, or

---

*As we shall see soon, a *complete circuit* is any complete loop consisting of wires and electrical devices (such as batteries and lightbulbs).

voltage. To avoid confusion with the force concept, we will call electromotive force just "emf." A battery's emf is the work done by the battery (taken from stored chemical energy) per coulomb of charge that passes through it. If a battery does 1 joule of work on 1 coulomb of charge passing through it, then its emf is 1 joule per coulomb (1 J/C), or 1 volt (1 V).

The emf actually represents the theoretical maximum potential difference across the battery's terminals (▼Fig. 17.3a). When a battery is connected to a circuit and charge flows, the voltage across the terminals is always slightly less than the emf. This "operating voltage" (V) of a battery (the symbol for a battery is the pair of unequal-length parallel lines in Fig. 17.3b) is called its **terminal voltage**. Because batteries in actual operation are of most interest to us, it is the terminal voltage that is most important.

Under many conditions, the emf and terminal voltage are almost the same. Any difference is due to the fact that the battery itself offers some resistance to the flow of charge in the circuit. Conceptually, we represent this so-called *internal resistance* (symbolized by r) explicitly in a circuit diagram (Fig. 17.3b). (Resistance, discussed in detail in Section 17.3, represents the opposition to a flow of charge.) Internal resistances are typically small, so the terminal voltage of a battery is essentially the same as the emf ($V \approx \mathcal{E}$). However, when a battery supplies a large current or when its internal resistance is high (older batteries), the terminal voltage may drop appreciably below the emf. The reason is that it takes some voltage just to produce a current in the battery because of its internal resistance. Mathematically, the difference between the emf and the terminal voltage is given by $\mathcal{E} - V = Ir$, where I is called the *electric current* (Section 17.2) in the battery. Thus, the terminal voltage, and not the emf, is a true indication of the state of the battery. Unless otherwise specified, we will assume batteries to have negligible internal resistance, so that $V = \mathcal{E}$ when the battery is in a circuit.

There is a wide variety of batteries in use. One of the most common is the 12-V automobile battery, consisting of six 2-V cells connected in *series*.* That is, the

---

*Chemical energy is converted to electrical energy in a chemical *cell*. The term *battery* generally refers to a collection, or "battery," of cells.

▲ **FIGURE 17.2 Gravitational analog to a battery and lightbulb** A gasoline-powered pump lifts water from the pond, increasing the potential energy of the water. As the water flows downhill, it transfers energy to (or does work on) a waterwheel, causing the wheel to spin. This action is analogous to the delivery of energy to a lightbulb by an electrical current (for example, as in Fig. 17.1).

Note: A battery's emf is sometimes called its *open-circuit terminal* voltage, signifying that the battery is connected to nothing.

Note: We reserve the symbol r to represent internal battery resistance. R is used to represent *any* resistance external to the battery, such as a lightbulb.

▼ **FIGURE 17.3 Electromotive force (emf) and terminal voltage** (a) The emf ($\mathcal{E}$) of a battery is the maximum potential difference across its terminals. This maximum occurs when the battery is not connected to an external circuit. **(b)** Because of internal resistance (r), the terminal voltage V when the battery is in operation is less than the emf $\mathcal{E}$. Here, R is the resistance of the lightbulb.

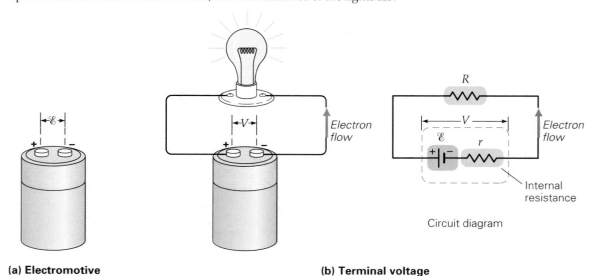

(a) Electromotive force (emf)

(b) Terminal voltage

Circuit diagram

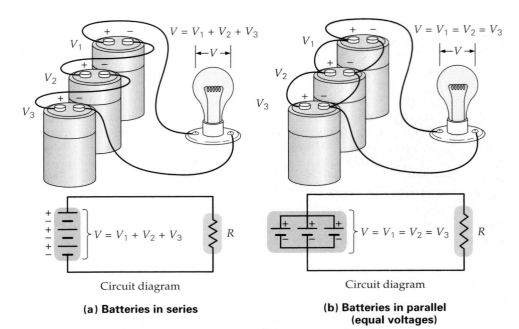

$$V = V_1 + V_2 + V_3$$

$$V = V_1 = V_2 = V_3$$

Circuit diagram

(a) Batteries in series

Circuit diagram

(b) Batteries in parallel (equal voltages)

▲ **Figure 17.4   Batteries in series and in parallel**   **(a)** When batteries are connected in series, their voltages add, and the voltage across the resistance R is the sum of the voltages. **(b)** When batteries of the same voltage are connected in parallel, the voltage across the resistance is the same, as if only a single battery were present. In this case, each battery supplies part of the total current.

Learn by Drawing

Sketching Circuits

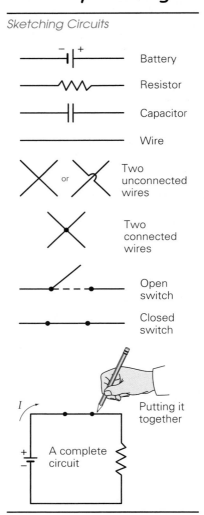

positive terminal of each cell is connected to the negative terminal of the next cell (shown for three cells in ▲Fig. 17.4a). When batteries or cells are connected in this fashion, their voltages add. If cells are connected in *parallel*, all their positive terminals have a common connection, as do their negative terminals (Fig. 17.4b). When identical batteries are connected in this way, the potential difference is the same across all of them, and each one supplies a fraction of the current to the circuit. For three batteries with equal voltages, each one supplies one-third of the current. Parallel connections of batteries are familiar to people who have had their car jump-started. For a jump start, the weak (high r) battery is connected in parallel to a normal (low r) battery, which delivers most of the current to get the car started.

Circuit Diagrams and Symbols

To analyze circuits, it is common to draw circuit diagrams that are schematic representations of the wires, batteries, and appliances, as they are connected. Each element that makes up a circuit is represented by its own symbol in the circuit diagram. As we have seen in Fig. 17.3b and Fig. 17.4, the symbol for a battery is two parallel lines, the longer of which represents the positive ($+$) terminal and the shorter the negative ($-$) terminal. Any circuit element that opposes the flow of charge is generally represented by the symbol ⌇⌇⌇, which stands for resistance R. (Electrical resistance is considered in detail in section 17.3; at this point, we merely introduce the circuit symbol.) Connecting wires are drawn as unbroken lines. Where lines cross, it is assumed that they do not contact one another, unless they have a heavy dot at their intersection. Lastly, switches are shown as "drawbridges" in wires, capable of going up (to open the circuit and thus stop the current) and down (closed to complete the circuit and allow current). These symbols, along with that of the capacitor (from Chapter 16), are summarized in the Learn by Drawing feature on this page.

Conceptual Example 17.1 ■ Asleep at the Switch?

▶Fig. 17.5 shows a circuit diagram that represents two identical batteries (each having terminal voltage V) connected in parallel to a lightbulb (represented by a resistor). The bulb has a voltage of V across it (neglecting the resistance of the wire). We know that before the switch S_1 is opened, the voltage across the lightbulb equals V (i.e., $V_{AB} = V$). What happens to the voltage across the lightbulb when S_1 is opened? (a) The voltage remains the same as before the switch is opened, a value of V. (b) The voltage drops to $V/2$, since only one battery is now connected to the bulb. (c) The voltage drops to zero.

Reasoning and Answer. It might be tempting to choose answer (b), because there is only one battery, not two. But look again. The remaining battery is still connected to the lightbulb. This means that there must be *some* voltage across the lightbulb, so the answer certainly cannot be (c). But it also means that the answer cannot be (b), because the remaining (identical) battery will maintain a voltage of V across the lightbulb. Hence, the answer is (a).

Follow-up Exercise. In this Example, what would the correct answer be if, instead of opening S_1, switch S_2 were opened. Explain your answer and reasoning. (*Answers to all Follow-up Exercises are at the back of the text.*)

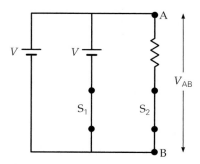

▲ **FIGURE 17.5 What happens to the resistor?** See Example 17.1.

17.2 Current and Drift Velocity

OBJECTIVES: To (a) define electric current, (b) distinguish between electron flow and conventional current, and (c) explain the concept of drift velocity and electric energy transmission.

As we saw in the previous section, a battery or some other voltage source connected to a continuous conducting path forms a **complete circuit**. In ordinary applications, *a sustained electric current requires a voltage source and a complete circuit.* Recall that a practical circuit will likely have a switch that is used to open or close the circuit.

Electric Current

Since it is electrons that move in the circuit's wires, the flow of charge is away from the negative terminal of the battery. Historically, however, circuit analysis has been done in terms of **conventional current**, which is in the direction in which positive charges would flow, or the direction *opposite* to the electron flow (▶Fig. 17.6). (Some situations exist in which a positive charge flow *is* responsible for the current—for example, in semiconductors.)

The battery is said to *deliver* current to a circuit or a component. Alternatively, we say that the circuit (or its components) *draws* current from the battery. The current eventually returns to the battery. A battery can only cause a current in one direction, from the negative ($-$) terminal (the cathode) to the positive ($+$) terminal (the anode). This type of one-directional charge flow is called **direct current (dc)**.

Quantitatively, the **electric current** (I) is defined as the time rate of flow of net charge. In this chapter, we will be concerned primarily with steady charge flow. In such a flow, if a net charge q passes through a cross-sectional area in a time interval t (▶Fig. 17.7), the electric current is defined as

$$I = \frac{q}{t} \quad \text{electric current} \qquad (17.1)$$

SI unit of current: coulomb per second (C/s) or ampere (A)

The coulomb per second has been designated the **ampere** (**A**) in honor of the French physicist André Ampére (1775–1836), an early investigator of electrical

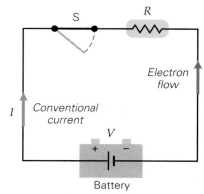

▲ **FIGURE 17.6 Conventional current** For historical reasons, circuit analysis is usually done with conventional current. Conventional current is in the direction in which positive charges would flow, or opposite to the electron flow.

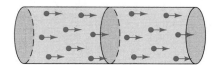

▲ **FIGURE 17.7 Electric current** Electric current (I) in a wire is defined as the rate at which the net charge (q) passes through the wire's cross-sectional area: $I = q/t$. The units of I are amperes (A), or "amps" for short.

and magnetic phenomena. In everyday usage, the ampere is commonly shortened to "amp." Thus, a current of 10 A is read as "ten amps." Small currents may be expressed in *milliamperes* (mA, or 10^{-3} A) or *microamperes* (μA, or 10^{-6} A), which are shortened to milliamps and microamps. In a typical household circuit, it is not unusual for the wires to carry several amps of current. To understand the relationship between charge and current, consider the next example.

Example 17.2 ■ Counting Electrons: Current and Charge

There is a current of 0.50 A in a flashlight bulb for 2.0 min. How much charge passes through the bulb during this time? How many electrons does this represent?

Thinking It Through. The current and time elapsed are given. The definition of current (Eq. 17.1) allows us to find the charge q. Since each electron has a charge with a magnitude of 1.6×10^{-19} C, charge can be converted into a specific number of electrons.

Solution. Listing the data given and converting the time interval into seconds:

Given: $I = 0.50$ A   *Find:* q (amount of charge)
       $t = 2.0$ min $= 1.2 \times 10^2$ s n (number of electrons)

By Eq. 17.1, $I = q/t$, so the magnitude of the charge is given by

$$q = It = (0.50 \text{ A})(1.2 \times 10^2 \text{ s}) = (0.50 \text{ C/s})(1.2 \times 10^2 \text{ s}) = 60 \text{ C}$$

The charge q is made up of n electron charges. Solving for n and using the magnitude of the charge on the electron, we have

$$n = \frac{q}{e} = \frac{60 \text{ C}}{1.6 \times 10^{-19} \text{ C/electron}} = 3.8 \times 10^{20} \text{ electrons}$$

Follow-up Exercise. Many sensitive laboratory instruments can easily measure currents in the nanoampere range, or even smaller. How long, in years, would it take for 1.0 C of charge to flow past a given point in a wire that carries a current of 1.0 nA?

Drift Velocity, Electron Flow, and Electric Energy Transmission

Although we frequently mention charge flow in analogy to water flow, electric charge does not flow in a conductor in exactly the same way that water flows through a pipe. In the absence of a potential difference, the free electrons in a metal wire move randomly at high speeds, colliding many times per second with the metal atoms. As a result, there is no average net flow of charge, since equal amounts of charge pass through a given point in opposite directions during any interval.

However, when a potential difference (voltage) is applied across the ends of the wire (e.g., by a battery), an electric field appears in one direction. Then a flow of electrons occurs opposite that direction. This does *not* mean that electrons are moving directly from one end of the wire to the other. They still move about in all directions as they collide with the atoms of the conductor (◀Fig. 17.8), but there is an *added* component (in one direction) to their velocities. The overall result is that their motions are less random, and more of them, on average, move toward the positive terminal of the battery than away from it.

This net electron flow is characterized by an average velocity called the **drift velocity**, which is much smaller than the random velocities of the electrons themselves. The magnitude of the drift velocity is on the order of 1 mm/s. A quick calculation shows that it would take an electron about 17 min to travel 1 m along a wire. Yet a lamp comes on almost instantaneously when you flip the switch (complete the circuit), and the electronic signals carrying telephone conversations travel almost instantaneously over miles of wire. How can that be?

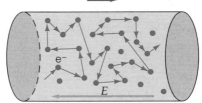

Drift velocity v_d

▲ **FIGURE 17.8 Drift velocity** Because of collisions with the atoms of the conductor, electron motion is random. However, when the conductor is connected, for example, to a battery to form a complete circuit, there is a small net motion in the direction opposite the electric field (toward the high-potential (positive) terminal, or anode). The speed and direction of this net motion form the drift velocity of the electrons.

Evidently, *something* must be moving faster than the "drifting" electrons. Indeed, something is: the electric field. When a potential difference is applied, the associated electric field in the conductor travels at a speed close to that of light (roughly 10^8 m/s). The electric field therefore influences the motion of electrons *throughout the conductor* almost instantaneously. Thus, there is current everywhere in the circuit almost simultaneously. You don't have to wait for electrons to "get there" from elsewhere. The electrons already in the filament of the bulb begin to move almost immediately due to the electric force, delivering energy and creating light essentially with no delay.

Sometimes this effect is explained in analogy to a row of dominos. When you topple a domino at one end, that *signal* or energy is transmitted rapidly down the row. At the other end, the domino that finally topples (and delivers the energy) is clearly not the one that you pushed. Thus, it was the energy, not the dominos, that traveled rapidly down the row.

17.3 Resistance and Ohm's Law

OBJECTIVES: To (a) define electrical resistance and explain what is meant by an ohmic resistor, (b) summarize the factors that determine resistance, and (c) calculate the effect of these factors in simple situations.

If you are given the applied voltage (potential difference) across the two ends of any conducting material, how do you determine the current? As you might expect, usually the greater the voltage, the greater is the current.

However, another factor besides voltage influences the current. Just as internal friction (viscosity) affects fluid flow in pipes (Chapter 9), the resistance of the material from which the wire is made will affect the flow of charge. Any object that offers significant resistance to the flow of charge is called a *resistor* and is represented in circuit diagrams by the zigzag symbol (Section 17.1). This symbol is used to represent a wide range of "resistors," from the cylindrical color-coded ones on printed circuit boards (▶Fig. 17.9) to electrical devices and appliances such as hair dryers and lightbulbs.

The **resistance** (R) of any object is defined as the ratio of the voltage across the object to the resulting current through that object. This definition makes physical sense. For example, if a huge voltage is applied across a resistor and the result is a very small current, then that resistor has a high resistance. Resistance is therefore defined as

$$R = \frac{V}{I} \quad \text{electrical resistance} \quad (17.2)$$

SI unit of resistance: volt per ampere (V/A), or ohm (Ω)

The units of resistance are volts per ampere (V/A), called the **ohm** (Ω) in honor of the German physicist Georg Ohm (1789–1854), who investigated the relationship between current and voltage in many different materials. Large values of resistance are expressed as kilohms (kΩ) and megohms (MΩ). A schematic circuit diagram showing how, in principle, resistance is determined is illustrated in ▶Fig. 17.10a. (In Chapter 18, we will study the instruments that are used to measure electrical currents and voltages, called ammeters and voltmeters, respectively.)

For some materials, the resistance may be constant over a range of voltages. A resistor that exhibits constant resistance is said to obey **Ohm's law**, or to be *ohmic*. The law was named after Ohm, who found materials possessing that property. A plot of voltage versus current for a material with an ohmic resistance gives a straight line with a slope equal to its resistance R (Fig. 17.10b). A common and practical form of Ohm's law is $V = IR$, where R is constant.

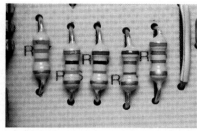

▲ **FIGURE 17.9 Resistors in use** A printed circuit board, typically used in computers, includes resistors of different values. The large, striped cylinders are resistors; their four-band color code indicates their resistance in ohms.

Note: *Resistor* is a generic term for any object that possesses significant electrical resistance.

PHYSLET®
ILLUSTRATION

Ohm's Law

Note: *Ohmic* means "having a constant resistance."

Note: Remember, *V* stands for ΔV.

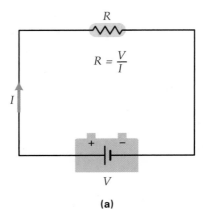

$$R = \frac{V}{I}$$

(a)

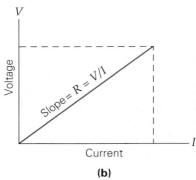

(b)

▲ **FIGURE 17.10 Resistance and Ohm's law** **(a)** In principle, any object's electrical resistance can be determined by dividing the voltage across it by the current through it. **(b)** If the element obeys Ohm's law (applicable only to a constant resistance), then a plot of voltage versus current gives a straight line with a slope equal to R, the element's resistance. (Its resistance does not change with voltage.)

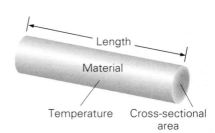

▲ **FIGURE 17.11 Resistance factors** Factors directly affecting the electrical resistance of a cylindrical conductor are the type of material it is made of, its length (L), its cross-sectional area (A), and its temperature (T).

It is important to realize that Ohm's law is not a fundamental law in the same sense as, for example, Newton's laws of motion are. There is no "law" which states that materials *must* have constant resistance. Indeed, many of our technological advances in computers and communication are based on semiconductors, which can have *nonlinear* voltage–current relationships.

Unless specified otherwise, we will assume that resistors are ohmic. Keep in mind, however, that many materials are nonohmic. For instance, the tungsten filaments of lightbulbs have more resistance at their high operating temperatures than at room temperature. The following Example shows how the resistance of the human body can make the difference between life and death.

Example 17.3 ■ Danger in the House: Human Resistance

Any room in the house that is exposed to water and electrical voltage can present hazards. (See the discussion of electrical safety in Section 18.5.) For example, suppose a person steps out of a shower and inadvertently touches an exposed 120-V wire with his finger. The human body, when wet, can have an electrical resistance as low as 300 Ω across its whole length. Estimate the current in that person's body.

Thinking It Through. The wire has an electric potential of 120 V. By definition, the floor is at 0 V. Therefore the voltage across the person's body is 120 V. To determine the current, we can use Eq. 17.2.

Solution. Listing the data and converting the current to amps we have

Given: $V = 120$ V   *Find:* I (current in the body)
         $R = 300$ Ω

Using Eq. 17.2, we have

$$I = \frac{V}{R} = \frac{120 \text{ V}}{300 \text{ Ω}} = 0.400 \text{ A} = 400 \text{ mA}$$

While this is a small current by everyday standards, it is a large current for the human body. A current over 10 mA can cause severe muscle contractions, and currents on the order of 100 mA can stop the heart. So this current is potentially deadly. (See the Insight on Electricity and Personal Safety, Table 1, p. 626, Chapter 18.)

Follow-up Exercise. When the human body is dry, its resistance (over its length) can be as high as 100 kΩ. What voltage would be required to produce a current of 1.0 mA (the value that people can barely feel)?

Factors That Influence Resistance

On the atomic level, resistance arises when electrons collide with the atoms that make up a material. Thus, resistance partially depends on the type of material an object is composed of. However, geometrical factors also influence an object's resistance. Overall, the resistance of an object of uniform cross section, such as a length of wire, depends on four properties of the object (◄Fig. 17.11): (1) the type of material (including any impurities), (2) its length, (3) its cross-sectional area, and (4) its temperature.

As you might expect, the resistance of an object (such as an everyday strand of wire) is *directly* proportional to the length (L) of the object and *inversely* proportional to its cross-sectional area (A): $R \propto L/A$. For example, a uniform metal wire 4 m long offers twice as much resistance as a similar wire 2 m long, but a wire with a cross-sectional area of 0.50 mm² has only half the resistance of one with an area of 0.25 mm². These geometrical conditions are analogous to those for liquid flow in a pipe. The longer the pipe, the more is its resistance (drag), but the larger the cross-sectional area of the pipe, the more liquid it can carry.

Resistivity

The resistance of a material is partly determined by intrinsic atomic properties, described by the material's **resistivity** (ρ). The resistance of uniform cross-section object thus includes the resistivity and is given by

Note: Do not confuse resistivity with mass density, which has the same symbol (ρ).

$$R = \rho \frac{L}{A} \qquad (17.3)$$

SI unit of resistivity: ohm-meter ($\Omega \cdot m$)

Solving Eq. 17.3 for the resistivity, we obtain $\rho = RA/L$, from which it can be seen that the units of resistivity are ohm-square meters per meter, or ohm-meters ($\Omega \cdot m$). Thus, knowing its resistivity allows us to calculate the resistance of any material with a constant cross section.

The resistivities of some common conductors, semiconductors, and insulators are given in Table 17.1. The values apply at 20°C, because resistivity can depend on temperature.

An interesting and potentially important medical application involves the measurement of human body resistance and its relationship to body fat. (See the Insight on Bioelectrical Impedance Analysis on p. 586.)

Resistivity and Resistance

Example 17.4 ■ A Longer Cord: Resistance of a Wire

A 1.5-m insulated extension cord consists of two parallel copper wires. If each copper wire has a diameter of 2.3 mm (equivalent to American Wire Gauge, or AWG, No. 12), what is the resistance of *one* of the wires at room temperature?

Thinking It Through.   The wire is a cylinder, and we know its dimensions and the type of material it is made of. Thus, we can apply Eq. 17.3 after we look up the resistivity of copper from Table 17.1.

Solution.   We list the geometrical data and the resistivity we found:

Given: $L = 1.5$ m   *Find:* R (resistance)
$\quad\quad\quad d = 2.3$ mm $= 2.3 \times 10^{-3}$ m
$\quad\quad\quad \rho = 1.70 \times 10^{-8}\ \Omega \cdot m$ (Table 17.1)

The cross-sectional area of the wire is

$$A = \pi r^2 = \pi \left(\frac{d}{2}\right)^2 = \frac{\pi (2.3 \times 10^{-3}\,m)^2}{4} = 4.2 \times 10^{-6}\ m^2$$

TABLE 17.1   Resistivities (at 20°C) and Temperature Coefficients of Resistivity for Various Materials*

	$\rho\ (\Omega \cdot m)$	$\alpha\ (C^{\circ-1})$		$\rho\ (\Omega \cdot m)$	$\alpha\ (C^{\circ-1})$
Conductors			*Semiconductors*		
Aluminum	2.82×10^{-8}	4.29×10^{-3}	Carbon	3.6×10^{-5}	-5.0×10^{-4}
Copper	1.70×10^{-8}	6.80×10^{-3}	Germanium	4.6×10^{-1}	-5.0×10^{-2}
Iron	10×10^{-8}	6.51×10^{-3}	Silicon	2.5×10^{2}	-7.0×10^{-2}
Mercury	98.4×10^{-8}	0.89×10^{-3}			
Nichrome (alloy of nickel	100×10^{-8}	0.40×10^{-3}	*Insulators*		
and chromium)			Glass	10^{12}	
Nickel	7.8×10^{-8}	6.0×10^{-3}	Rubber	10^{15}	
Platinum	10×10^{-8}	3.93×10^{-3}	Wood	10^{10}	
Silver	1.59×10^{-8}	6.1×10^{-3}			
Tungsten	5.6×10^{-8}	4.5×10^{-3}			

*Values for semiconductors are general ones, and resistivities for insulators are typical orders of magnitude.

Then, using Eq. 17.3, we get

$$R = \frac{\rho L}{A} = \frac{(1.70 \times 10^{-8}\,\Omega \cdot m)(1.5\,m)}{4.2 \times 10^{-6}\,m^2} = 6.1 \times 10^{-3}\,\Omega = 6.1\,m\Omega$$

This result demonstrates why the resistances of the connecting wires in an electrical circuit can usually be neglected: The resistances of the circuit *components* are generally much larger than those of the connecting wires.

Follow-up Exercise.   Old fashioned lightbulbs were constructed from carbon filaments. Estimate the length of the cylindrical carbon filament with a diameter of 0.50 mm that would have a total resistance of 150 Ω at room temperature.

For many materials, the temperature dependence of resistivity is nearly linear if the temperature change is not too great. That is, the resistivity at a temperature T after a temperature change $\Delta T = T - T_o$ is given by

$$\rho = \rho_o(1 + \alpha \Delta T) \qquad \begin{array}{l}\textit{temperature variation}\\ \textit{of resistivity}\end{array} \qquad (17.4)$$

where α is a constant (usually only over a certain temperature range) called the **temperature coefficient of resistivity** and ρ_o is a reference resistivity at T_o (usually 0°C). Eq. 17.4 can be rewritten as

$$\Delta \rho = \rho_o \alpha \Delta T \qquad (17.5)$$

INSIGHT

Bioelectrical Impedance Analysis (BIA)

Traditional (and very approximate) methods for determining body fat percentages involve the use of buoyancy tanks (for density measurements), or calipers to pinch the flesh. However, in recent years, electrical resistance experiments have been designed to measure the body fat of the human body.* In theory, these measurements—termed *bioelectrical impedance analysis* (BIA)—have the potential to determine, with more accuracy, a patient's total body water, fat-free mass, and body fat (adipose tissue).

The principle of BIA is based upon the water content of the human body. Water in the human body is a relatively good conductor of electric current, due to the presence of ions such as potassium (K^+) and sodium (Na^+). Since muscle tissue holds more water per kilogram than does fat, it is a better conductor. Because muscle tissue is less resistive than fat, for a given voltage the difference in currents should be a good indicator of the fat-to-muscle percentage.

In practice, to carry out BIA, one electrode of a low-voltage power supply is connected to the wrist and the other to the opposite ankle. The current is kept below 1 mA for safety purposes, with typical currents being about 800μA. (See the Insight, Chapter 18, p. 626, on the effects of current on humans.) The subject cannot feel this small current. Typical resistance values are about 250 Ω. From Ohm's

*Technically, this technique measures the body's *impedance*, which includes effects of capacitance and magnetic effects as well. (See Chapter 21.) However, these contributions are about 10% of the total. Hence, we will use the word *resistance* here.

law, we can estimate the required voltage: $V = IR = (8 \times 10^{-4}\,A)(250\,\Omega) = 0.200\,V$, or about 200 mV. In actuality, the voltage alternates in polarity (ac) at a frequency of 50 kHz, because this frequency range is known not to trigger electrically excitable tissues, such as nerves and cardiac muscle.

From what has been presented in this chapter (e.g., Eq. 17.3), you should be able to understand some of the factors involved in interpreting the results of human resistance measurements. The resistance that is measured is actually the total resistance. However, the current travels, not through a uniform conductor, but rather through the arm, trunk, and leg. Not only does each of these parts of the body have a different fat-to-muscle ratio, which affects resistivity (ρ), but they are all of a different length (L) and cross-sectional area (A). Thus, the arm and the leg, usually dominated by muscle and with a small cross-sectional area, offer the most resistance. Conversely, the trunk, which usually contains a relatively high percentage of fat and has a large cross-sectional area, has a low resistance.

By subjecting BIAs to statistical analysis, researchers hope to understand how the wide range of physical and genetic parameters present in humans affects resistance measurements. Among these parameters are height, weight, body type, and ethnicity. Once the correlations are understood, BIA may well become a valuable medical tool not only in routine physicals, but also in the diagnosis of various diseases.

Related Exercise: 36

where $\Delta\rho = \rho - \rho_\mathrm{o}$ is the change in resistivity that occurs when the temperature changes by ΔT. The ratio $\Delta\rho/\rho_\mathrm{o}$ is dimensionless, so α must have units of inverse Celsius degrees ($\mathrm{C^{\circ -1}}$, or $1/\mathrm{C^\circ}$). Physically, α represents the fractional change in resistivity ($\Delta\rho/\rho_\mathrm{o}$) per Celsius degree. The temperature coefficients of resistivity for some materials are listed in Table 17.1. These coefficients are assumed to be constant over the temperature ranges used in the Examples and Exercises.

Note: Compare the form of Eq. 17.5 with Eq. 10.10 for the linear expansion of a solid.

We can use the fact that resistance is directly proportional to resistivity (Eq. 17.3), together with Eqs. 17.4 and 17.5, to derive an expression for the resistance of a conductor of uniform cross section as a function of temperature:

$$R = R_\mathrm{o}(1 + \alpha\Delta T) \quad \text{or} \quad \Delta R = R_\mathrm{o}\alpha\Delta T \quad \begin{array}{l}\textit{temperature variation} \\ \textit{of resistance}\end{array} \quad (17.6)$$

Here, $\Delta R = R - R_\mathrm{o}$, the change in resistance relative to its value R_o, usually taken to be at 0°C. The variation of resistance with temperature provides a means of measuring temperature in the form of an *electrical resistance thermometer*, as illustrated in the next Example.

Example 17.5 ■ An Electrical Thermometer: Variation of Resistance with Temperature

A platinum wire has a resistance of 0.50 Ω at 0°C. It is placed in a water bath, where its resistance rises to a final value of 0.60 Ω. What is the temperature of the bath?

Thinking It Through. From the temperature coefficient of resistivity for platinum from Table 17.1, we can solve for ΔT from Eq. 17.6 and add it to 0°C, the initial temperature, to find the temperature of the bath.

Solution.

Given: $T_\mathrm{o} = 0°\mathrm{C}$
$\qquad\quad R_\mathrm{o} = 0.50\ \Omega$
$\qquad\quad R = 0.60\ \Omega$
$\qquad\quad \alpha = 3.93 \times 10^{-3}\ \mathrm{C^{\circ -1}}$ (Table 17.1)

Find: T (temperature of the bath)

The ratio $\Delta R/R_\mathrm{o}$ is the fractional change in the initial resistance R_o (at 0°C). We solve Eq. 17.6 for ΔT, using the given values:

$$\Delta T = \frac{\Delta R}{\alpha R_\mathrm{o}} = \frac{R - R_\mathrm{o}}{\alpha R_\mathrm{o}} = \frac{0.60\ \Omega - 0.50\ \Omega}{(3.93 \times 10^{-3}\ \mathrm{C^{\circ -1}})(0.50\ \Omega)} = 51\ \mathrm{C^\circ}$$

Thus, the bath is at $T = T_\mathrm{o} + \Delta T = 0°\mathrm{C} + 51\ \mathrm{C^\circ} = 51°\mathrm{C}$.

Follow-up Exercise. In this Example, if the material had been copper, for which $R_\mathrm{o} = 0.50\ \Omega$, rather than platinum, what would its resistance be at the same bath temperature? What conclusions can you draw about whether you would choose a material with a high or low temperature coefficient of resistivity to make a sensitive "resistance" thermometer?

Carbon and other semiconductors have negative temperature coefficients of resistivity. (See Table 17.1.) This means that the resistance of a semiconductor actually increases with decreasing temperature. However, many materials have positive temperature coefficients of resistivity, so their resistances decrease as temperature decreases. You might wonder how far electrical resistance can be reduced by lowering the temperature. In certain cases, the resistance can reach zero—not just close to zero, but, as accurately as we can measure, *exactly* to zero! This condition, called **superconductivity**, which has led to technological applications of great importance, is discussed in the Insight on p. 588.

Superconductivity

The electrical resistance of metals and alloys generally decreases with decreasing temperature. At relatively low temperatures, some materials exhibit what is called *superconductivity*; that is, their electrical resistance drops to zero.

Superconductivity was discovered in 1911 by Heike Kamerlingh Onnes, a Dutch physicist. The phenomenon was first observed in solid mercury at a temperature of about 4 K ($-269°C$). The mercury was cooled to this temperature by using liquid helium. The boiling point of helium (the temperature at which it condenses to a liquid) is about $-267°C$ at 1 atm. Lead also exhibits superconductivity when cooled to such a temperature.

Theoretically, an electric current established in a superconducting loop should persist *indefinitely*, with no resistive losses. Currents introduced into superconducting loops have been observed to remain constant for several years. In 1957, a theory was presented by the American physicists John Bardeen (a coinventor of the transistor), Leon Cooper, and Robert Schrieffer in an attempt to explain various aspects of metallic superconductors. Known as the BCS theory, it presents a quantum mechanical model in which electrons are viewed as waves traveling through a material. According to this theory, under normal conditions lattice vibrations and imperfections scatter electrons and give rise to resistance and energy loss. The theory predicts that these mechanisms have no effect on the electrons in superconductors. In the absence of this scattering, the resistance is zero, and a current can persist as long as the metal is in a superconducting state.

Keeping a superconductor at the low temperatures of liquid helium is somewhat difficult and costly. Such low temperatures have continued to preclude the use of superconductors in everyday life. However, an ongoing search has revealed materials that become superconducting at higher temperatures. Other superconducting metals and alloys were found for which the critical temperature (the temperature at which the material becomes superconductive) was about 18 K ($-255°C$). In 1973, a material with a critical temperature of 23 K ($-250°C$) was discovered.

In 1986, a major breakthrough was made with the discovery of a new class of superconductors, the so-called ceramic alloys of rare earth elements such as lanthanum and yttrium. These superconductors were prepared by grinding ordinary metallic and rare earth elements together and heating the mixture to a high temperature to produce a ceramic material. One such mixture consisted of barium, yttrium, and copper oxide. The critical temperature for these ceramic mixtures was about 57 K ($-216°C$).

In 1987, a ceramic material with a critical temperature of 98 K ($-175°C$) was discovered. This was a major advance, because it meant that superconductivity could be obtained by using liquid nitrogen, which has a boiling point of 77 K ($-196°C$), as a coolant. Liquid nitrogen is relatively plentiful (nitrogen is the chief constituent of air) and inexpensive: It costs cents per liter (much less than milk), compared to dollars per liter for liquid helium.

More recent reports indicate higher critical temperatures and even suggest that certain copper oxide compounds may lose their resistance to electrical current at room temperature and above. As scientists study the new superconductors, they will gain a better understanding of them, and no doubt the critical temperature will rise. One problem scientists face is the difficulty of confirming a superconducting state in small samples or regions of samples. Another is related to the BCS theory: Although many of its fundamental elements may help explain high-temperature superconductivity, some reinterpretation and modification will likely be necessary.

There are many possible applications for superconductors. One is superconducting magnets. The strength of an electromagnet depends on the magnitude of the current in its wire coil (Chapter 19). If there were no resistance, there would be greater current and no losses. Used in motors or engines, superconducting electromagnets would provide more power for the same energy input. (Superconducting magnets cooled by liquid helium have been utilized in ships' engines and particle accelerators.) Such magnets have been used to levitate (Fig. 1) and propel trains. Another application of superconductors might be underground transmission cables with no resistive losses. However, among the many technological problems still to be overcome is how to form wires out of ceramic materials, which are generally brittle. The application most likely to be realized soon will be in making faster computer circuits.

Imagine what it would mean if someone discovered a material that was superconducting at refrigerator or room temperature. The absence of electrical resistance opens many possibilities. You're likely to hear more about superconductor applications in the near future.

FIGURE 1 Magnetic levitation A magnetic cube levitates above a piece of "high temperature" superconducting material that is cooled with liquid nitrogen. When the material becomes cold enough to superconduct, it acquires certain magnetic properties that repel the magnetic cube and cause it to levitate.

17.4 Electric Power

OBJECTIVES: To (a) define electric power, (b) calculate the power delivery of simple electric circuits, and (c) explain joule heating and its significance.

When a sustained current exists in a circuit, the electrons are given energy by the voltage source, such as a battery. As these charge carriers then pass through circuit components, they collide with the atoms of the material (i.e., they encounter resistance) and lose energy. The energy transferred in the collisions can result in an increase in the temperature of the components. In this way, electrical energy can be transformed, at least partially, into thermal energy.

However, electrical energy can also be converted into other forms of energy. Not only is it useful for generating heat (as in electric stoves), but also light (as in lightbulbs) and mechanical motion or work (as in electric drills). According to the law of conservation of energy, whatever forms energy may take, the total energy delivered to the charge carriers by the battery must be transferred to the circuit elements. That is, in traveling around a complete circuit, any charge carrier loses all the electric potential energy it gained from the battery by the time it returns to the battery.

The energy gained by an amount of charge q from a voltage source (voltage V) is qV. Over a time interval t, the *rate* at which energy is delivered by a power supply may not be constant. The average rate of energy delivery is called the average **electric power**, $\overline{P}$, and is given by

$$\overline{P} = \frac{W}{t} = \frac{qV}{t}$$

In the special case when the current and voltage are steady with time (as with a battery), then the average power is the same as the power at all times. For steady (dc) currents, $I = q/t$ (Eq. 17.1). Thus, we can rewrite the preceding power equation as:

$$P = IV \quad \textit{electric power} \quad (17.7a)$$

As you learned in Chapter 5, the SI unit of power is the watt (W). The ampere (the unit of current I) times the volt (the unit of voltage V) gives the joule per second (J/s), or watt. (You should check this.)

A visual mechanical analogy to help explain Eq. 17.7a is given in ▼Fig. 17.12. The figure depicts a simple electric circuit as a system for transferring energy, in analogy to a conveyor belt delivery system.

▼ **FIGURE 17.12 Electric power analogy**   Electric circuits can be thought of as energy delivery systems much like a conveyor belt. **(a)** Imagine the current made up of consecutive segments of charge $q = 1.0$ C, each carrying $qV = 12$ J of energy supplied by the battery. The current is $I = V/R = 6.0$ A, or 6.0 C/s. Thus, the power (or energy delivery rate) to the resistor is $(6.0 \text{ C/s})(12 \text{ J/C}) = 72 \text{ J/s} = 72$ W. **(b)** The conveyor belt comprises a series of buckets, each carrying 12 kg of sand (analogous to the energy carried by each charge q). Arriving at one bucket every 6.0 s (analogous to current I).

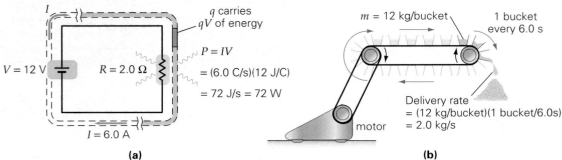

(a)   (b)

Since $R = V/I$, power can be written in three equivalent forms:

$$P = IV = \frac{V^2}{R} = I^2R \quad \textit{electric power} \quad (17.7b)$$

Joule Heat

The thermal energy expended in a current-carrying resistor is referred to as **joule heat**, or I^2R (pronounced "*I* squared *R*") **losses**. In many instances (e.g., in electrical transmission lines), joule heating is an undesirable side effect. However, in other situations, the conversion of electrical energy to thermal energy is the main purpose. Heating applications include the heating elements (burners) of electric stoves, hair dryers, immersion heaters, and toasters.

Electric lightbulbs are rated in watts (power)—for example, 100 W or 60 W (▶Fig. 17.13a). Incandescent lamps are relatively inefficient as light sources. Typically, less than 5% of the electrical energy is converted to visible light; most of the energy produced is invisible infrared radiation and heat.

Electrical appliances are tagged or stamped with their power ratings. Either the voltage and power requirements or the voltage and current requirements are given (Fig. 17.13b). In either case, the current, power, and effective resistance can be calculated. The power requirements of some household appliances are given in Table 17.2.

TABLE 17.2 Typical Power and Current Requirements for Various Household Appliances (120 V)

Appliance	Power	Current	Appliance	Power	Current
Air conditioner, room	1500 W	12.5 A	Heater, portable	1500 W	12.5 A
Air-conditioning, central	5000 W	41.7 A*	Microwave oven	900 W	5.2 A
Blender	800 W	6.7 A	Radio–cassette player	14 W	0.12 A
Clothes dryer	6000 W	50 A*	Refrigerator, frost-free	500 W	4.2 A
Clothes washer	840 W	7.0 A	Stove, top burners	6000 W	50.0 A*
Coffeemaker	1625 W	13.5 A	Stove, oven	4500 W	37.5 A*
Dishwasher	1200 W	10.0 A	Television, color	100 W	0.83 A
Electric blanket	180 W	1.5 A	Toaster	950 W	7.9 A
Hair dryer	1200 W	10.0 A	Water heater	4500 W	37.5 A*

*A high-power appliance such as this is typically wired to a 240-V house supply to reduce the current to half these values (Section 18.5).

Integrated Example 17.6 ■ A Modern "Appliance": The Power to Compute

(a) Consider two appliances that operate at the same voltage. Appliance A has a higher power rating than appliance B. (a) How does the resistance of A compare with that of B? (1) It is larger, (2) it is smaller, (3) it is the same. (b) A desktop computer system includes a color monitor and a central processing unit (CPU) with keyboard, each operating at 120 V. The color monitor has a power requirement of 130 W and the CPU's requirement is 240 W. What is the resistance of each component under these operating conditions?

(a) Conceptual Reasoning. Power depends on current and voltage. Since the two appliances operate at the same voltage, they can't carry the same current. Therefore, answer (3) cannot be correct. Because both appliances operate at the same voltage, the one with higher power (A) must carry the most current. For A to carry more current at the same voltage as B, it must have less resistance than B. Therefore, the correct answer is (2), A has less resistance than B.

(b) Thinking It Through. The definition of resistance is $R = V/I$ (Eq. 17.2). To use this definition, we need the current, which can be determined from Eq. 17.7 ($P = IV$).

We list the data, using the subscript m for monitor and C for CPU:

Given: $P_m = 130$ W   **Find:** R (resistance of each computer component)
$P_C = 240$ W
$V = 120$ V

Solving for the current from Eq. 17.7, we get

$$I_m = \frac{P_m}{V} = \frac{130 \text{ W}}{120 \text{ V}} = 1.08 \text{ A}$$

and

$$I_C = \frac{P_C}{V} = \frac{240 \text{ W}}{120 \text{ V}} = 2.00 \text{ A}$$

The resistances then follow:

$$R_m = \frac{V}{I_m} = \frac{120 \text{ V}}{1.08 \text{ A}} = 111 \ \Omega$$

and

$$R_C = \frac{V}{I_C} = \frac{120 \text{ V}}{2.00 \text{ A}} = 60.0 \ \Omega$$

Hence, an appliance's resistance is inversely related to its power requirement.

Follow-up Exercise. Many desktop computer systems employ a laser printer. While printing, such printers typically have a power requirement of 175 W. When they are not printing, they are in "standby mode," which takes only 8.00 W. Determine the resistance of the printer in each of its two modes.

(a)

(b)

▲ **FIGURE 17.13   Power ratings**
(a) Lightbulbs are rated in watts. Operated at 120 V, this 60-W bulb uses 60 J of energy each second. **(b)** Appliance ratings list either voltage and power or voltage and current. From either, the current, power, and effective resistance can be found. Here, one appliance is rated at 120 V and 18 W and the other at 120 V and 300 mA. Can you compute the current and resistance for the former and the power required and resistance for the latter?

Example 17.7 ■ A Potentially Dangerous Repair: Joule Heat Revisited

A hair dryer is rated at 1200 W for a 115-V operating voltage. The uniform wire filament breaks near one end, and the owner repairs it by removing one section near the break and simply reconnecting it. The filament is then 10.0% shorter than its original length. What will be the change in the heater's power output?

Thinking It Through. Both before and after the breakage, the wire will have the same voltage. However, shortening the wire will decrease its resistance, resulting in a larger current. With this increase in current, we expect the power output to go up.

Solution. We list the numerical values, using the subscript 1 to indicate "before breakage" and the subscript 2 for "after the repair":

Given: $P_1 = 1200$ W   **Find:** P_2 (power output after the repair)
$V_1 = V_2 = 115$ V
$L_2 = 0.900 L_1$

Because 10.0% of the filament's length is gone, it will have 90.0% of its original resistance (R). From Eq. 17.3, the original resistance is

$$R_1 = \rho \frac{L_1}{A}$$

Thus,

$$R_2 = \rho \frac{L_2}{A} = \rho \frac{0.900 L_1}{A} = 0.900 \left(\rho \frac{L_1}{A} \right) = 0.900 R_1$$

This means that the current will increase, since the applied voltage is the same ($V_2 = V_1$). We can write this requirement as $V_2 = I_2 R_2 = V_1 = I_1 R_1$. Thus, the current increases by about 11%, since

$$I_2 = \left(\frac{R_1}{R_2}\right) I_1 = \left(\frac{R_1}{0.900 R_1}\right) I_1 = 1.11 I_1$$

Under the initial conditions, the current was

$$I_1 = \frac{P_1}{V_1} = \frac{1200 \text{ W}}{115 \text{ V}} = 10.4 \text{ A}$$

Then the new current will be

$$I_2 = 1.11 I_1 = (1.11)(10.4 \text{ A}) = 11.5 \text{ A}$$

Using $P = I^2 R$ in Eq. 17.7, we get the original resistance

$$R_1 = \frac{V_1}{I_1} = \frac{115 \text{ V}}{10.4 \text{ A}} = 11.1 \text{ }\Omega$$

The new resistance is therefore

$$R_2 = 0.900 R_1 = (0.900)(11.1 \text{ }\Omega) = 9.99 \text{ }\Omega$$

For the repaired dryer, the power output is then

$$P_2 = I_2^2 R_2 = (11.5 \text{ A})^2 (9.99 \text{ }\Omega) = 1.32 \times 10^3 \text{ W}$$

The power output of the dryer has been increased by 120 W. *Do not attempt such a repair job!* With reduced resistance and increased current, the filament could overheat and start a fire.

Follow-up Exercise. Suppose that the owner of the hair dryer in this Example instead replaced the coil with one of the same length and material, but made with wire that was 10.0% thicker. What would be the new power of the dryer?

We often complain about our electric bills, but what do we actually buy from the electric or power company? What we pay for is electric energy measured in the unit of **kilowatt-hour** (**kWh**). Let's see for which physical quantity this is a unit. Power is the time rate at which work is done ($P = W/t$ or $W = Pt$), so work has units of watt-seconds (power × time). Converting this unit to the larger unit of kilowatt-hours (kWh), we see that the kilowatt-hour is a unit of work (or energy), equivalent to 3.6 million joules:

$$1 \text{ kWh} = (1000 \text{ W})(3600 \text{ s}) = (1000 \text{ J/s})(3600 \text{ s}) = 3.6 \times 10^6 \text{ J}$$

Thus, we pay the "power" company for electrical energy that we use to do work. Let's take a look at the cost of electricity.

Example 17.8 ■ The Price of Coolness: Electric Power Cost

If the motor of a frost-free refrigerator runs 15% of the time, how much does it cost to operate per month (to the nearest penny) if the power company charges 11¢ per kilowatt-hour? (Assume 30 days in a month.)

Thinking It Through. From the power (*rate* of energy usage) and the amount of time the motor is on per day, we can compute the electrical energy the refrigerator consumes *daily* and then project to a 30-day month. The cost is based on the typical cost of electrical energy.

Solution. In talking about practical electrical energy delivery, kilowatt-hours are used because the joule is a relatively tiny unit. Listing the data,

Given: $P = 500$ W (Table 17.2)   *Find:* Operating cost per month
          Cost = \$0.11/kWh

Since the refrigerator motor operates 15% of the time, in one day it runs

$$t = (0.15)(24 \text{ h}) = 3.60 \text{ h}$$

Since $P = W/t$, the electrical work done, or the average energy expended *per day*, is

$$W = Pt = (500 \text{ W})(3.60 \text{ h}) = 1.80 \times 10^3 \text{ Wh} = 1.80 \text{ kWh}$$

Then, as a per-day cost, we have

$$\left(\frac{1.80 \text{ kWh}}{\text{day}}\right)\left(\frac{\$0.11}{\text{kWh}}\right) = \frac{\$0.198}{\text{day}}$$

or 19.8¢ per day. Thus, for a 30-day month, the cost is

$$\left(\frac{\$0.198}{\text{day}}\right)\left(\frac{30 \text{ day}}{\text{month}}\right) = \$5.94 \text{ per month}$$

Follow-up Exercise. How long would you have to leave a 60-W lightbulb on to use the same amount of electrical energy that the refrigerator motor in this Example uses each hour it is on?

The cost of electricity varies with location. On the average in the United States, it ranges from a low of about 8¢ (per kilowatt-hour) to many times that in certain areas. In the last few years, electric energy rates have been deregulated. Coupled with increasing demand (without a corresponding increase in supply), deregulation has given rise to skyrocketing rates in places like California. Do you know the price of electricity in your locality? Check an electric bill to find out, especially if you are in one of the areas affected by dramatically rising rates. The Insight on electrical costs (p. 594) gives another type of cost comparison that you might find interesting.

Electrical Efficiency and Natural Resources

The demand for electricity in the United States is continuously increasing. About 25% of the electricity generated in the United States goes into lighting (▶Fig. 17.14). This percentage is roughly equivalent to the output of 100 electricity-generating (power) plants. Refrigerators consume about 7% of the electricity produced in the United States (the output of about 28 power plants).

This huge (and increasing) consumption of electricity has prompted the federal government and many state governments to set minimum efficiency limits for refrigerators, freezers, air conditioners, water heaters, dishwashers, heat pumps, and so forth (▶Fig. 17.15). Also, more efficient fluorescent lighting has been developed. The most efficient fluorescent lamp now in use consumes about 25–30% less energy than the average fluorescent lamp and roughly 75% less energy than incandescent lamps with an equivalent light output.

The result of all these measures has been significant energy savings as new, more efficient appliances gradually replace inefficient older models. Energy saved translates directly into savings of fuels and other natural resources, as well as a reduction in environmental hazards, such as chemical pollution and global warming. To see what kind of results can be achieved by applying just one energy efficiency standard, consider the next Example.

▲ **FIGURE 17.14   All lit up**   A satellite image of the Americas at night. Can you identify the major population centers in the United States and elsewhere? The red spots across part of South America indicate large-scale burning of vegetation. The small yellow spot in Central America shows burning gas flares at oil production sites. At the top right edge, you can just glimpse a few of the white city lights of Europe. This image was recorded by a visible–infrared system.

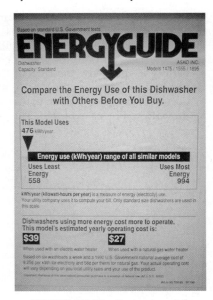

▲ **FIGURE 17.15   Energy guide** Consumers are made aware of the efficiencies of appliances in terms of the average yearly cost of their operation. Sometimes the yearly cost is given for different kilowatt-hour (kWh) rates, which vary around the United States.

The Cost of Portable Electricity

In the United States, the cost of household electricity ranges from about 8¢ per kilowatt-hour to many times that value, depending on the region. Even though electric energy rates have soared in recent years (especially in certain areas of the country, such as California), they still are a bargain, considering the work that electricity does and the convenience it provides.

Batteries are also widely used to supply our electrical energy needs. For example, they power toys, radios, TV remote controls, and a variety of other portable devices. Have you ever thought about the cost of this "portable" electricity, compared with that of household electricity? Let's take a look and make a general cost comparison. To do this, we need some data on batteries.

For one thing, we need to know the cost of a battery. This varies a great deal with location and with quantity purchased. Focusing on just the AA-cell and the D-cell (purchased in August 2001 in California) revealed the average costs listed in Table 1.

TABLE 1 Battery Data (California, August 2001)

	Cost	Rating	Voltage
AA-cell	$1.00	2450 mA·h = 2.45 A·h	1.5 V
D-cell	$1.50	14 250 mA·h = 14.24 A·h	1.5 V

We also need to know the amount of electric energy expected from a battery over its lifetime. Manufacturers rate their batteries in a mixed unit called the ampere-hour $(A \cdot h)$. This unit allows you to estimate the lifetime of a battery if you know the current (rate) it will have to supply; that is, battery life (h) = amp-hour rating/current output. Thus, the greater the current output, the shorter the battery life, as expected. The amp-hour ratings are not usually listed on batteries or their packages, but are available from manufacturers. Some typical values are given in Table 1.

Notice that the amp-hour is current × time (It) and is therefore a unit of charge. Hence, the battery rating is a measure of the charge supplied over the lifetime of a battery. The amp-hour is equivalent to 3600 coulombs of charge, which can be seen by converting the time in hours to seconds: $1 A \cdot h = (1 C/s)(3600s) = 3600 C$. Thus, the total lifetime charge q each battery can supply to a circuit is as follows (using the conversion factor between coulombs and $A \cdot h$):

AA-cell: $q = (2.45 A \cdot h)[3.60 \times 10^3 C/(A \cdot h)] = 8.82 \times 10^3 C$
D-cell: $q = (14.25 A \cdot h)[3.60 \times 10^3 C/(A \cdot h)]$
$= 5.13 \times 10^4 C$

The energy supplied by a battery is given by $W = qV$, so the lifetime energy outputs of each of these batteries are (in kWh)

AA - cell: $qV = (8.82 \times 10^3 C)(1.5 V)$
$= 1.3 \times 10^4 J = 0.0036 kWh$

and

D - cell: $qV = (5.13 \times 10^4 C)(1.5 V)$
$= 7.7 \times 10^4 J = 0.021 kWh$

The cost per kWh of this portable (battery) electricity is then

AA-cell: price/kWh = $1.00/0.0036 kWh = $278/kWh

and

D-cell: price kWh = $1.50/0.021 kWh = $71/kWh

Clearly, compared with household electrical energy rates, we pay a great deal for the convenience of batteries.

Related Exercise: 68 and 99

Example 17.9 ■ What We Can Save: Increasing Electrical Efficiency

Most modern power plants produce electricity at a rate of about 1.0 GW (electric power output). Estimate, to two significant figures, how many fewer power plants the state of California would need if all its households switched from the 500-W refrigerators of Example 17.9 to more efficient 400-W refrigerators. (Assume that there are about 6 million homes with an average of 1.2 refrigerators operating per home.)

Thinking It Through. We want an estimate of the effect of switching to the more efficient refrigerators. The results from Example 17.9 can be used to calculate the overall effect.

Solution.

Given: Plant rate = 1.0 GW = 1.0×10^6 kW
Energy requirement, 500-watt model
= 1.80 kWh/day (Example 17.8)
Number of homes = 6.0×10^6
Number of refrigerators per home = 1.2

Find: How many fewer power plants are required by switching to more efficient refrigerators

For the entire state, the energy usage per day with older refrigerators is

$$\left(\frac{1.80 \text{ kWh/day}}{\text{refrig}}\right)(6.0 \times 10^6 \text{ homes})\left(\frac{1.2 \text{ refrigerators}}{\text{home}}\right) = 1.3 \times 10^7 \frac{\text{kWh}}{\text{day}}$$

The more energy-efficient refrigerators would use only 80% (400 W/500 W = 0.80) of this figure, or approximately 1.0×10^7 kWh/day. The difference, 3.0×10^6 kWh/day, is the rate at which electric energy is saved. One 1.0-gigawatt power plant produces

$$(1.0 \times 10^6 \text{ kW/plant})\left(24\frac{\text{h}}{\text{day}}\right) = 2.4 \times 10^7 \frac{(\text{kWh/plant})}{\text{day}}$$

So the replacement refrigerators would save about

$$\frac{3.0 \times 10^6 \text{ kWh/day}}{2.4 \times 10^7 \text{ kWh/(plant-day)}} = 0.13 \text{ plant}$$

or about one-eighth of the output of a typical plant. This saving would result from a change in a single domestic appliance. Developing (and using!) more efficient electrical appliances is one way to avoid having to build new electric energy generating plants.

Follow-up Exercise. Electric and gas water heaters are often said to be equally efficient—typically, about 95%. In reality, while gas water heaters are capable of 95% efficiency, it might be more accurate to describe electric water heaters as only about 33% efficient, even though approximately 95% of the electrical energy they consume is transferred to the water in the form of heat. Explain. [*Hint:* What is the source of energy for an electric water heater? Compare this to the energy delivery of natural gas. Recall the discussion of electrical generation in Section 12.4 and Carnot efficiency in Section 12.5.]

Chapter Review

Important Concepts and Equations

- A **battery** is capable of generating an **electromotive force (emf)**, or a voltage, across its two terminals. The high-voltage terminal of a battery is the **anode**, and the low-voltage terminal is the **cathode**.

- **Emf** ($\mathscr{E}$) is measured in volts and is equal to the number of joules of energy that a battery (or any power supply) delivers to 1 coulomb of charge that passes through it (1 J/C = 1 V). For a battery, this energy comes from a depletion of chemical potential energy.

- **Electric current** (I) is the rate at which charge flows. Its direction is expressed in terms of the **conventional current**, which is in the direction in which positive charge actually flows or appears to flow. In metals, the charge flow is due to the electrons, and hence, the conventional current direction is opposite the electron flow direction. Current is defined in amperes (1 A = 1 C/s) as

$$I = \frac{q}{t} \tag{17.1}$$

- An electric current that is steady in direction is called a **direct current**.

- For an electric current to exist in a circuit, it must be a **complete circuit**—that is, a circuit that continuously connects both terminals of a battery or power supply.

- **Terminal voltage** is the voltage across the terminals of a battery or power supply when there is a current flowing through it. Because of internal resistance, a battery's terminal voltage is less than its emf.

- The **electrical resistance** (R) of an object is the voltage across the object divided by the current through it, or

$$R = \frac{V}{I} \quad \text{or} \quad V = IR \tag{17.2}$$

- A circuit element obeys **Ohm's law** if that element exhibits constant electrical resistance. Ohm's law is commonly written $V = IR$, where R is constant.

- The resistance of an object depends on the **resistivity** (ρ) of the material from which it is made (atomic properties), the cross-sectional area A, and the length L. For objects with a uniform cross section, the resistance is

$$R = \rho\frac{L}{A} \tag{17.3}$$

- **Electric power** (P) is the rate at which work is done by a battery or power supply, or the rate at which energy is transferred to a circuit element. The power delivered to a circuit element depends on the element's resistance and the voltage applied to it and can be expressed in three equivalent ways:

$$P = IV = \frac{V^2}{R} = I^2R \qquad (17.7b)$$

- **Joule heat** is the term used for the rate at which electrical energy is delivered as heat.
- The **kilowatt-hour** (kWh) is the unit used commercially to measure electric energy. It can be thought of as the energy used by a 1-kilowatt appliance running for 1 hour. The kilowatt-hour is equivalent to 3.6 million joules.

Exercises

In this chapter, assume that all batteries have negligible internal resistance unless otherwise indicated.

17.1 Batteries and Direct Current

1. When a battery is placed into a complete circuit, the voltage across its terminals is its (a) emf, (b) terminal voltage, (c) power, (d) all of these.

2. As a battery ages, its (a) emf increases, (b) emf decreases, (c) terminal voltage increases, (d) terminal voltage decreases.

3. When four 1.5-V batteries are connected in parallel, the output voltage of the combination is (a) 1.5 V, (b) 3.0 V, (c) 6.0 V, (d) none of these.

4. CQ Why does the battery design described on p. 593 require a chemical membrane?

5. CQ Is it possible for a battery with an emf of 12 V to measure exactly 12 V when it is in use? Explain.

6. CQ Discuss the difference between emf and terminal voltage. Which one is greater?

7. ■ (a) Two 1.5-V dry cells are connected in series. What is the total voltage of the combination? (b) What would be the total voltage if the cells were connected in parallel?

8. ■ What is the voltage across six 1.5-V batteries when they are connected (a) in series and (b) in parallel?

9. ■■ Two 6.0-V batteries and one 12-V battery are connected in series. (a) What is the voltage across the whole arrangement? (b) What arrangement of these three batteries would give a total voltage of 12 V?

10. ■■ Given three batteries with voltages of 1.0 V, 3.0 V, and 12 V, respectively, how many different voltages could be obtained by connecting one or more of the batteries in series or parallel, and what are these voltages?

11. IE ■■ You are given four AA batteries that are rated at 1.5 V each. The batteries are grouped in pairs. In arrangement A, the two batteries in each pair are in series, and then the pairs are connected in parallel. In arrangement B, the two batteries in each pair are in parallel, and then the pairs are connected in series. (a) Compared with arrangement B, will arrangement A have (1) a higher, (2) the same, or (3) a lower total voltage? (b) What are the total voltages for each arrangement?

17.2 Current and Drift Velocity

12. The unit of electrical current is (a) C, (b) C/s, (c) A, (d) both (b) and (c).

13. CQ In a circuit such as that in Fig. 17.4, in which direction, from positive to negative or from negative to positive, is the current inside the battery? Explain.

14. CQ Drift velocity, the average velocity at which electrons travel in a circuit, is on the order of 1 mm/s. Yet a lamp comes on instantaneously when you flip a switch. Explain.

15. ■ A net charge of 30 C passes through the cross-sectional area of a wire in 2.0 min. What is the current in the wire?

16. ■ How long would it take for a net 2.5-C charge to pass through the cross-sectional area of a wire to produce a steady current of 5.0 mA?

17. ■ A small toy car draws a 0.50-mA current from a 3.0-V nicad (nickel–cadmium) battery. In 10 min of operation, (a) how much charge flows through the toy car, and (b) how much energy is lost by the battery?

18. ■■ A car's starter motor draws 50 A from the car's battery when starting up. If the start-up time is 1.5 s, how many electrons pass a location in the circuit during that time?

19. IE ■■ Suppose a greater net charge passes a given location in wire A than in wire B. (a) Compared with the current in wire B, the current in wire A (1) is higher, (2) is the same, (3) is lower, or (4) none of the preceding. Why? (b) A net 20-C charge passes a location in a wire in 1.25 min; a net 30-C charge passes a location in another wire in 1.52 min. Which wire carries more current? How much more?

20. ■■ A fully charged, heavy-duty battery is rated at 100 A · h at 5.0 A. Thus, the life of the battery when delivering 5.0 A is 20 h (100 A · h/5.0 A). How much charge will the battery deliver in this time?

21. **IE** ■■ Imagine that some protons are moving to the left at the same time that some electrons are moving to the right past the same location. (a) Will the net current be (1) to the right, (2) to the left, (3) zero, or (4) none of the preceding? Why? (b) In 4.5 s, 6.7 C of electrons flow to the right at the same time that 8.3 C of protons flow to the left. What is the magnitude of the total current?

22. ■■ In a proton linear accelerator, a 9.5-mA proton current hits a target. (a) How many protons hit the target each second? (b) What is the energy delivered to the target each second if the protons each have a kinetic energy of 20 MeV and come to a complete stop in the target?

17.3 Resistance and Ohm's Law*

23. The unit of resistance is the (a) V/A, (b) A/V, (c) W, (d) V.

24. For an ohmic resistor, current and resistance (a) vary with temperature, (b) are directly proportional, (c) are independent of voltage, (d) none of these.

25. **CQ** If the voltage (V) were plotted on the same graph versus current (I) for two ohmic conductors with different resistances, how could you tell which is less resistive?

26. **CQ** The filaments in bulbs usually fail just after the bulbs are turned on rather than when they have already been on for some time. Why?

27. **CQ** Nichrome, an alloy of nickel and chromium, has high resistivity and a low temperature coefficient of resistivity, compared with copper and aluminum (Table 17.1). How does this affect nichrome's performance as a heating element?

28. **CQ** A wire is connected across a steady voltage source. (a) If that wire is replaced with one of the same material that is twice as long and has twice the cross-sectional area, how will the current in the wire be affected? (b) How will the current be affected if, instead, the new wire has the same length, but half the diameter, as the old one?

29. **CQ** In a semiconductor, the number of conducting charges can vary drastically with temperature. What does a negative temperature coefficient of resistivity for a semiconductor imply about how the number of charge carriers varies with temperature in such a material?

30. **CQ** A real battery always has some internal resistance r (▶Fig. 17.16) that increases with the battery's age. Explain why the terminal voltage drops as the internal

*Assume that the temperature coefficients of resistivity given in Table 17.1 apply over large temperature ranges.

resistance increases even though the emf remains constant.

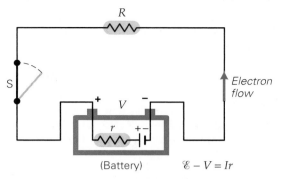

▲ **FIGURE 17.16 Emf and terminal voltage** See Exercises 30 and 31.

31. ■ A battery labeled 12.0 V supplies 1.90 A to a 6.00-Ω resistor (Fig. 17.16). (a) What is the terminal voltage of the battery? (b) What is its internal resistance?

32. ■ What is the emf of a battery with a 0.15-Ω internal resistance if the battery delivers 1.5 A to an externally connected 5.0-Ω resistor?

33. **IE** ■ Some states allow the use of aluminum wire in houses in place of copper. (a) If you wanted the resistance of your aluminum wires to be the same as that of copper, would the aluminum wire have to have (1) a greater diameter than, (2) a smaller diameter than, or (3) the same diameter as the copper wire? Why? (b) Calculate the thickness ratio of aluminum to that of copper?

34. ■ How much current is drawn from a 12-V battery when a 15-Ω resistor is connected across its terminals?

35. ■ What voltage must a battery have to produce a 0.50-A current through a 2.0-Ω resistor?

36. ■ During a research experiment on the conduction of current in the human body, a medical technician attaches one electrode to the wrist and a second to the shoulder. If 100 mV are applied across the two electrodes and the resulting current is 12.5 mA, what is the overall resistance of the patient's arm?

37. ■■ A 0.60-m-long copper wire has a diameter of 0.10 cm. What is the resistance of the wire?

38. ■■ A 100-Ω resistor is placed in a circuit with two identical batteries connected in series. If the resistor draws 1.5 A, what is the terminal voltage of each battery?

39. ■■ An automobile starter is connected to a 12-V battery. If the starter has an effective resistance of 2.5 Ω, what is the rate at which electrons drift past a point in the circuit?

40. ■■ A material is formed into a long rod with a square cross section 0.50 cm on each side. When a 100-V voltage is applied across a 20-m length of the rod, a 5.0-A current is carried. (a) What is the resistivity of the material? (b) Is the material a conductor, an insulator, or a semiconductor?

41. ■■ Two copper wires have equal cross-sectional areas and lengths of 2.0 m and 0.50 m, respectively. What is the ratio of the current in the shorter wire to that in the longer one if they are connected to the same power supply?

42. IE ■■ Two copper wires have equal lengths, but the diameter of one is three times that of the other. (a) The resistance of the thinner wire is (1) 3, (2) 1/3, (3) 9, or (4) 1/9 times that of the resistance of the thicker wire. Why? (b) If the thicker wire has a resistance of 1.0 Ω, what is the resistance of the thinner wire?

43. ■■ The wire in a heating element of an electric stove burner has a 0.75-m effective length and a 2.0×10^{-6}-m^2 cross-sectional area. If the wire is made of iron and operates at a 380°C temperature, what is its operating resistance?

44. ■■ What is the percentage variation of the resistivity of copper over the temperature range from room temperature (20°C) to 100°C?

45. ■■ A copper wire has a 25-mΩ resistance at 20°C. When the wire is carrying a current, heat produced by the current causes the temperature of the wire to increase by 27 C°. What is the change in the wire's resistance?

46. ■■ What is the change in resistance of a 2000-m-long aluminum wire with a 10.00-Ω resistance (at 0°C) as the temperature of the wire changes from the freezing point to the boiling point of water?

47. ■■ When a resistor is connected to a 12-V source, it draws a 185-mA current. The same resistor connected to a 90-V source draws a 1.25-A current. Is the resistor ohmic? Justify your answer mathematically.

48. ■■ What length of wire is required to make a 20.0-Ω resistor by winding AWG (American Wire Gauge) No. 10 nichrome wire in a coil? (AWG No. 10 wire has a diameter of 2.588 mm.)

49. IE ■■ As a wire is stretched out so that its length increases, its cross-sectional area decreases, while the total volume of the wire remains constant. (a) Will the resistance after the stretch be (1) greater than, (2) the same as, or (3) less than that before the stretch? Why? (b) A 1.0-m length of copper wire with a 2.0-mm diameter is stretched out; its length increases by 25% while its cross-sectional area decreases, but remains uniform. Compute the resistance ratio.

50. ■■ A particular application requires a 20-m length of aluminum wire to have a 0.25-mΩ resistance at 20°C. What must be the wire's diameter?

51. ■■ If the resistance of the wire in Exercise 50 cannot vary by more than ±5.0%, what is the wire's operating temperature range?

52. ■■ ▼Figure 17.17 shows data on the dependence of the current through a resistor on the voltage across that resistor. (a) Is the resistor ohmic? (b) What is its resistance?

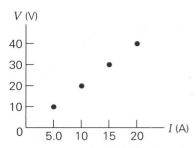

▲ **FIGURE 17.17 An ohmic resistor?** See Exercise 52.

53. ■■■ At 20°C, a silicon rod is connected to a battery with a terminal voltage of 6.0 V. A 0.50-A current results. If the temperature of the rod is increased to 25°C, how much current does the rod then carry?

54. IE ■■■ A platinum wire is connected to a battery. (a) If the temperature increases, will the current in the wire (1) increase, (2) remain the same, or (3) decrease? Why? (b) An electrical resistance thermometer is made of platinum wire that has a 5.0-Ω resistance at 20°C. The wire is connected to a 1.5-V battery. When the thermometer is heated to 2020°C, by how much does the current change?

17.4 Electric Power

55. Electric power has units of (a) $A^2 \cdot \Omega$, (b) J/s, (c) V^2/Ω, (d) all of these.

56. If the voltage across an ohmic resistor is doubled, the power expended in the resistor (a) increases by a factor of 2, (b) increases by a factor of 4, (c) decreases by half, (d) none of these.

57. If the current through an ohmic resistor is halved, the power expended in the resistor (a) increases by a factor of 2, (b) increases by a factor of 4, (c) decreases by half, (d) decreases by a factor of 4.

58. CQ Assuming that the resistance of your hair dryer obeys Ohm's law, what would happen to its power output if you plugged it directly into a 240-V outlet in Europe if it is designed to be used in the 120-V outlets of the United States?

59. CQ Most lightbulb filaments are made of tungsten and are about the same length. What would be different about the filament in a 60-W bulb compared with that in a 40-W bulb?

60. CQ Which one consumes more power from a 12-V battery, a 5.0-Ω resistor or a 10-Ω resistor? Why?

61. ■ A digital video disc (DVD) player is rated at 100 W at 120 V. What is its resistance?

62. ■ A freezer of resistance 10 Ω is connected to a 110-V source. What is the power delivered when this freezer is on?

63. ■ The current through a refrigerator with a resistance of 12 Ω is 13 A (when the refrigerator is on). What is the power delivered to the refrigerator?

64. ■ Show that the quantity volts squared per ohm (V^2/Ω) has SI units of power.

65. ■ An electric water heater is designed to produce 50 kW of heat when it is connected to a 240-V source. What must be the resistance of the heater?

66. ■■ Assuming that the heater in Exercise 65 is 90% efficient, how long would it take to heat 50 gal of water from 20°C to 80°C?

67. IE ■■ An ohmic resistor in a circuit is designed to operate at 120 V. If you connect the resistor to a 60-V power source, will the resistor dissipate heat at (1) 2, (2) 4, (3) 1/2, or (4) 1/4 times the designed power? Why? (b) If the designed power is 90 W at 120 V, but the resistor is connected to 40 V, what is the power delivered to the resistor at the lower voltage?

68. ■■ An electric toy with a resistance of 2.50 Ω is operated by a 1.50-V battery. (a) What current does the toy draw? (b) Assuming that the battery delivers a steady current for its lifetime of 6.00 h, how much charge passed through the toy? (c) How much energy was delivered to the toy?

69. ■■ A welding machine draws 18 A of current at 240 V. (a) How much energy does the machine use each second? (b) What is its resistance?

70. ■■ On average, an electric water heater operates for 2.0 h each day. (a) If the cost of electricity is $0.15/kWh, what is the cost of operating the heater during a 30-day month? (b) What is the resistance of a typical water heater? [*Hint*: See Table 17.2.]

71. ■■ What is the resistance of a heating coil if it is to generate 15 kJ of heat per minute when it is connected to a 120-V source?

72. ■■ A 200-W computer power supply is on 10 h per day. If the cost of electricity is $0.15/kWh, what is the cost (to the nearest dollar) of using the computer for a year (365 days)?

73. ■■ A 120-V air conditioner unit draws 15 A of current. If it operates for 20 min, (a) how much energy in kilowatt-hours does it use? (b) If the cost of electricity is $0.15/kWh, what is the cost (to the nearest penny) of operating the unit for 20 min?

74. ■■ Two resistors, 100 Ω and 25 kΩ, are rated for a maximum power output of 1.5 W and 0.25 W, respectively.

What is the maximum voltage that can be safely applied to each resistor?

75. ■■ A wire 5.0 m long and 3.0 mm in diameter has a resistance of 100 Ω. A 15-V potential difference is applied across the wire. Find (a) the current in the wire, (b) the resistivity of its material, and (c) the rate at which heat is being produced in the wire.

76. IE ■■ When connected to a voltage source, a coil of tungsten wire initially dissipates 500 W of power. In a short time, the temperature of the coil increases by 150 C° because of joule heating. (a) Will the dissipated power (1) increase, (2) remain the same, or (3) decrease? Why? (b) What is the corresponding change in the power?

77. ■■ A 20-Ω resistor is connected to four 1.5-V batteries. What is the joule heat loss per minute in the resistor if the batteries are connected (a) in series and (b) in parallel?

78. ■■ A 5.5-kW water heater operates at 240 V. (a) Should the heater circuit have a 20-A or a 30-A circuit breaker? (A circuit breaker is a safety device that opens the circuit at its rated current.) (b) Assuming 85% efficiency, how long will the heater take to heat the water in a 55-gal tank from 20° to 80°C?

79. ■■ A student uses an immersion heater to heat 0.30 kg of water from 20°C to 80°C for tea. If the heater is 75% efficient and takes 2.5 min, what is its resistance? (Assume 120-V household voltage.)

80. ■■ An ohmic appliance is rated at 100 W when it is connected to a 120-V source. If the power company cuts the voltage by 5.0% to conserve energy, what is the power consumed by the appliance after the voltage drop?

81. ■■ A lightbulb's output is 60 W when it operates at 120 V. If the voltage were cut in half and the power dropped to 20 W during a brownout, what is the ratio of the bulb's resistance at full power to its resistance during the brownout?

82. ■■ To empty a flooded basement, a water pump must do work (lift the water) at a rate of 2.00 kW. If the pump is wired to a 240-V source and is 84% efficient, (a) how much current does it draw and (b) what is its resistance?

83. ■■■ Find the total monthly (30-day) electric bill (to the nearest dollar) for the following household appliance usage if the utility rate is $0.12/kWh: Central air conditioning runs 30% of the time; a blender is used 0.50 h/month; a dishwasher is used 8.0h/month; a microwave oven is used 15 min/day; the motor of a frost-free refrigerator runs 15% of the time; a stove (burners plus oven) is used a total of 10 h/month; and a color television is operated 120 h/month. (Use the information given in Table 17.2.)

Additional Exercises

84. **IE** A piece of carbon and a piece of copper have the same resistance at room temperature. (a) If the temperature of each piece is increased by 10.0 C°, will the copper piece have (1) a higher resistance than, (2) the same resistance as, or (3) a lower resistance than the carbon piece? Why? (b) Calculate the ratio of the resistance of copper to that of carbon.

85. A resistance thermometer made of platinum shows a 25% change in resistance over a certain temperature range. What is the change in temperature for this range?

86. An iron conductor with a resistance of 1.75 Ω at 20°C is connected to a 12-V source. If the conductor is heated in an oven to 300°C, what is the change in the current through it?

87. A computer CD-ROM drive that operates on 120 V is rated at 40 W when it is operating. (a) How much current does the drive draw? (b) What is the drive's resistance?

88. If the price of electricity is $0.15/kWh, how much would it cost (to the nearest penny) to operate ten 75-W light-bulbs for 8.0 h?

89. A toaster oven is rated for 1.60 kW at 120 V. When the filament ribbon of the oven burns out, the owner makes a repair that reduces the length of the filament by 10%. How does this reduction affect the power output of the oven, and what is the new value?

90. The tungsten filament of an incandescent lamp has a resistance of 200 Ω at room temperature. What would the resistance be at an operating temperature of 1600°C?

91. Two pieces of aluminum and copper wire are identical in length and diameter. At some temperature, one of the wires will have the same resistance that the other has at 20°C. What is that temperature? (Is there more than one temperature?)

92. (a) What is the current in an aluminum wire of length 100 m and radius 1.0 mm if the wire is connected to a 1.5-V battery? (b) How many joules of energy does the battery dissipate in 10 min?

93. An electric iron with a 15-Ω heating element operates at 120 V. How many joules of energy does the iron convert to heat in 1.0 h?

94. When a wire 1.0 mm in diameter and 2.0 m in length is connected to a 3.00-V source, an 11.8-A current results. (a) What is the resistivity of the wire? (b) What material is the wire likely to be made of?

95. A 100-Ω resistor is rated at 0.25 W. (a) What is the maximum current the resistor can carry? (b) What is the maximum operating voltage that should be applied across the resistor?

96. A wire carries a 75-mA current. How many electrons pass a given point in the wire in 2.0 s?

97. A 0.10-A current exists in a single resistor in a circuit with a 40-V voltage source. (a) What is the resistance of the resistor? (b) How much power is dissipated by the resistor? (c) How much energy is dissipated in 2.0 min?

98. Find the ratio of the resistance of a copper wire to the resistance of an aluminum wire (a) if the wires have the same length and diameter and (b) if the copper wire is twice as long as the aluminum wire and has a diameter half that of the aluminum wire.

99. A 3.00-V battery is connected to a 5.00-W lightbulb in a small flashlight. (a) How much current is in the bulb when the flashlight is on? (b) Assuming that the flashlight is left on and the current is constant for 3.00 h until the battery quits, how much charge moved through the bulb? (c) How much energy was initially stored in the battery?

100. A copper bus bar, used to conduct large amounts of current at a power station, is 1.5 m long and 8.0 cm by 4.0 cm in cross section. What voltage applied to the ends of the bar will produce a 3000-A current through it?

101. **CQ** Why is electric power delivered over long distances at high voltages when we know that high voltages can be dangerous?

102. ▼ Figure 17.18 shows free charge carriers that each have a charge q and move with a speed v_d (drift speed) in a conductor of cross-sectional area A. Let n be the number of free charge carriers per unit volume. (a) Prove that the total charge (ΔQ) free to move in the volume element shown is given by $\Delta Q = (nAx)q$. (b) Prove that the current in the conductor is given by $I = nqv_dA$.

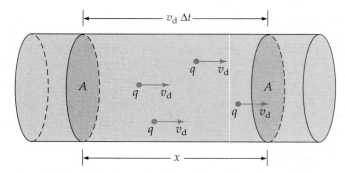

▲ **FIGURE 17.18 Total charge and current** See Exercises 102 and 103.

103. A copper wire with a cross-sectional area of 13.3 mm² (AWG No. 6) carries a 1.2-A current. If the wire contains 8.5×10^{22} free electrons per cubic centimeter, what is the drift velocity of the electrons? [*Hint*: See Exercise 102 and Fig. 17.18.]

Basic Electric Circuits

INSIGHTS

- RC Circuits in Action: Studsensors™ Help around the House
- Electricity and Personal Safety

Learn by Drawing

- Kirchhoff Plots: A Graphical Interpretation of Kirchhoff's Loop Theorem

W e usually think of metallic wires as the "connectors" between resistors in a circuit. However, wires are not the only conductors of electricity, as the photo shows. Since the bulb is lit, the circuit must be complete. We can conclude, therefore, that the "lead" in a pencil (actually a form of carbon called *graphite*) is a good conductor of electricity. The same must be true for the liquid in the beaker—in this case, a solution of water and ordinary table salt.

Electric circuits are of many kinds and can be designed for many purposes, from boiling water to lighting a Christmas tree. Circuits containing "liquid" conductors, such as shown in this photo, have practical applications in the laboratory and in industry; for example, they can be used to synthesize or purify chemical substances and to electroplate metals (such as in making silver plate). Armed with the principles learned in Chapters 15, 16, and 17, you are now ready to analyze some electric circuits. This analysis will give you an appreciation of how electricity actually works.

Circuit analysis most often deals with voltage, current, and power requirements. A circuit may be analyzed theoretically before being assembled. The analysis might show that the circuit would not function properly as designed or that there could be a safety problem (such as overheating due to joule heat). To help in this analysis, we will rely heavily on diagrams of circuits to visualize and understand their function. A few of these diagrams were included in Chapter 17.

Let's begin our analysis of circuits by looking at arrangements of resistive elements, such as lightbulbs, toasters, and immersion heaters.

18.1 Resistances in Series, Parallel, and Series–Parallel Combinations

OBJECTIVES: To **(a)** determine the equivalent resistance of resistors in series, parallel, and series–parallel combinations and **(b)** use equivalent resistances to analyze simple circuits.

The resistance symbol ($-\!\!\bigwedge\!\!\bigwedge\!\!-$) can represent *any* type of circuit element such as a lightbulb. Here, it is assumed that all elements are ohmic (they have constant resistance), unless otherwise stated. Also, the resistance of the wires will be neglected.

Resistors in Series

In analyzing a circuit, keep in mind that the *sum of the voltages around a circuit loop* (that is, the gains and losses, represented by $+$ and $-$ signs, respectively) *is zero*. For example, for the circuit in ▼ Fig. 18.1a, the individual voltages (V_i, where $i = 1$, 2, or 3) across the resistors combine to equal the voltage increase (V) across the battery terminals. Each resistor in series carries the same current (I). Therefore, summing the voltage gains and losses, we have $V - \Sigma V_i = 0$. Finally, Ohm's law tells us that for each resistor, $V_i = IR_i$. Substituting this expression into the previous equation, we obtain

Note: For resistors, $V = IR$

$$V - \Sigma IR_i = 0 \qquad \text{or} \qquad V = \Sigma IR_i \qquad (18.1)$$

The fact that the sum of the voltages around the circuit loop is zero is really a statement of conservation of energy. We saw in Chapter 17 that for the charge carriers traversing the circuit, the energy given to them by a voltage source must be completely transferred, or delivered, to the other circuit elements.

The circuit elements in Fig. 18.1a are said to be connected in series, or connected end to end. *When resistors are in series, the current is the same through all the resistors*, as required by the conservation of charge. If this condition were not true, then charge would build up or disappear, which cannot happen. ▶ Figure 18.2 shows the analogous flow of water along a smooth stream bed punctuated by a series of rapids.

▼ **FIGURE 18.1   Resistors in series**   **(a)** When resistors (representing the resistances of lightbulbs) are in series, the current through each is the same. ΣV_i, the sum of the voltage drops across the resistors, is equal to V, the voltage of the battery. **(b)** The equivalent resistance R_s of the resistors in series.

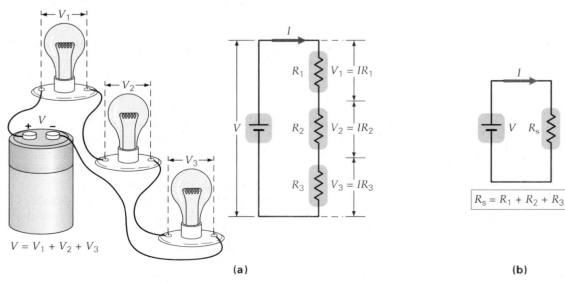

(a)   (b)

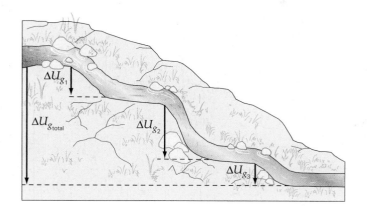

Even though a different amount of gravitational potential energy is lost as the water flows through each set of rapids, the current of the water is the same in each set. The total loss of gravitational potential energy is the sum of the three losses. (To make this circuit complete, some external agent, such as a pump, would have to do work to return the water to the top of the hill, restoring its original gravitational potential energy.)

If we label the common current in the resistors as I, then Eq. 18.1 can be written explicitly for three resistors (such as the three lightbulbs in Fig. 18.1a):

$$V = V_1 + V_2 + V_3$$
$$= IR_1 + IR_2 + IR_3 = I(R_1 + R_2 + R_3)$$

To replace the three resistors by one resistor R_s (called the **equivalent series resistance**) and maintain the same current, we need $V = IR_s$, or $R_s = V/I$. Hence, from the previous equation, the three resistors in series have an equivalent resistance

Note: To better understand resistor connections, review the discussion of capacitors connected in series and parallel arrangements in Chapter 16.

$$R_s = \frac{V}{I} = R_1 + R_2 + R_3$$

That is, the equivalent resistance of resistors in series is the sum of the individual resistances. The three resistors (bulbs) in Fig. 18.1a could be replaced by a single resistor of resistance R_s (Fig. 18.1b) without affecting the current. For example, if each resistor in Fig. 18.1a had a value of 10 Ω, then R_s would be 30 Ω. *Note that the equivalent series resistance is larger than the resistance of the largest resistor in the series.*

Note: R_s is larger than the largest resistance in a series arrangement.

This result can be extended to any number of resistors in series:

$$R_s = R_1 + R_2 + R_3 + \cdots = \Sigma R_i \qquad \begin{array}{l}\textit{equivalent series}\\ \textit{resistance}\end{array} \qquad (18.2)$$

Series connections are not normally used in everyday circuits such as house wiring, because they present two major disadvantages. The first disadvantage is clear if you consider what happens if one of the lightbulbs in Fig. 18.1a burns out. What would happen? *All* of the bulbs would go out, because the circuit would no longer be complete, or continuous. If the filament of one of the bulbs is broken, the circuit is said to be *open*. Think of an open circuit as equivalent to an infinite resistance; thus, the current in an open circuit is zero.

A second disadvantage of series connections is that each resistor operates at less than the battery voltage (V). If a fourth resistor were added, then the voltage across each of the first three bulbs (and their currents) would decrease, resulting in reduced power delivered to all the bulbs. Clearly, this condition is not acceptable in a household setting.

Resistors in Parallel

We can also connect resistors to a battery in parallel (▶ Fig. 18.3a). In this case, all the resistors have common connections—that is, all the leads on one side of the resistors are attached together to one terminal of the battery. All the leads on the other side are attached to the other terminal. *When resistors are connected in parallel*

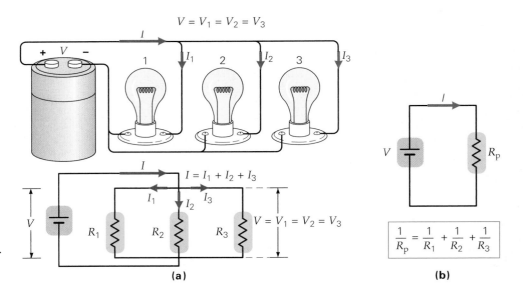

$$V = V_1 = V_2 = V_3$$

$$I = I_1 + I_2 + I_3$$

$$V = V_1 = V_2 = V_3$$

$$\frac{1}{R_p} = \frac{1}{R_1} + \frac{1}{R_2} + \frac{1}{R_3}$$

(a)

(b)

▶ **FIGURE 18.3 Resistors in parallel** **(a)** When resistors are connected in parallel, the voltage drop across each of the resistors is the same. The current from the battery divides among the resistors. **(b)** The equivalent resistance R_p of the resistors in parallel.

Note: In actuality, the wires of the circuit are not ordinarily arranged in the neat rectangular pattern of a circuit diagram. The rectangular form is simply a convention that provides a neater presentation and easier visualization of the circuit arrangement.

to a source of emf (Section 17.1), *the voltage drop across each resistor is the same*. It may not surprise you to learn that household circuits are wired in parallel. (See Section 18.5.) This is because parallel circuits do not have the disadvantages of series circuits. When wired in parallel, each appliance operates at the full voltage, and turning one appliance off or on does not affect the others.

However, unlike the case of resistors in series, the current in a parallel circuit divides into different paths (Fig. 18.3a). The current divides when it comes to a *junction* (a location where several wires come together), much like traffic on a highway divides when it reaches a fork in the road (▼Fig. 18.4a). The total current out of the battery is equal to the sum of these currents. Specifically, for three resistors in parallel, we can express this relationship as $I = I_1 + I_2 + I_3$. Notice that if the resistances are equal, the current will divide equally. However, in general, the resistances will not be equal. Then the current will divide among the resistors in inverse proportion to their resistances—that is, the greatest current will take the path of least resistance.

The **equivalent parallel resistance** (R_p) is the value of a single resistor that could replace all the resistors and maintain the same current. Thus, $R_p = V/I$, or $I = V/R_p$. The voltage drop (V) is the same across each resistor. To visualize this sit-

▼ **FIGURE 18.4 Analogies to resistors in parallel** **(a)** When a road forks, the total number of cars entering the two branches each minute is equal to the number of cars arriving at the fork each minute. Movement of charge at a junction can be considered in the same way. **(b)** When water flows from behind a dam at a given height, the amount of gravitational potential energy that it loses in falling to the stream bed below is the same regardless of the path of descent. This situation is analogous to voltages across parallel resistors.

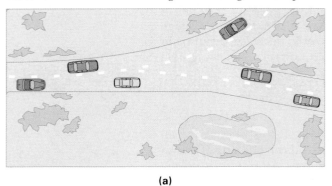

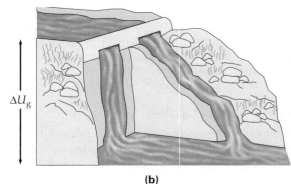

(a)

(b)

uation, imagine a water analogy. Consider two separate paths for the water, each leading from the top of a dam to the bottom. The water loses the same amount of gravitational potential energy (analogous to V) regardless of the path (Fig. 18.4b). For electricity, a given amount of charge loses the same amount of electrical potential energy, regardless of which parallel resistor it passes through.

Thus, the current through each resistor is $I_i = V/R_i$. (Here, the subscript i represents any of the resistors: $1, 2, 3, \ldots$.) Then, substituting for each current, we obtain

$$I = I_1 + I_2 + I_3 = \frac{V}{R_1} + \frac{V}{R_2} + \frac{V}{R_3}$$

Therefore,

$$\frac{V}{R_p} = V\left(\frac{1}{R_p}\right) = V\left(\frac{1}{R_1} + \frac{1}{R_2} + \frac{1}{R_3}\right)$$

By equating the two resistance expressions in the parentheses, we see that the equivalent resistance R_p (of three resistors) in parallel is given (in reciprocal form) by

$$\frac{1}{R_p} = \frac{1}{R_1} + \frac{1}{R_2} + \frac{1}{R_3}$$

That is, the reciprocal of the parallel equivalent resistance is equal to the sum of the reciprocals of the resistances of the individual resistors.

This result can be generalized to include any number of resistors in parallel:

$$\frac{1}{R_p} = \frac{1}{R_1} + \frac{1}{R_2} + \frac{1}{R_3} + \cdots = \Sigma \frac{1}{R_i} \quad \begin{array}{l} \textit{equivalent parallel} \\ \textit{resistance} \end{array} \quad (18.3)$$

If there are only two resistors in parallel, the equivalent resistance can be written in nonreciprocal form, which can be more convenient to use than Eq. 18.3. In this case, Eq 18.3 reads

$$\frac{1}{R_p} = \frac{1}{R_1} + \frac{1}{R_2} = \frac{R_1 + R_2}{R_1 R_2}$$

or

$$R_p = \frac{R_1 R_2}{R_1 + R_2} \quad \begin{array}{l} \textit{(only for two} \\ \textit{resistors in parallel)} \end{array} \quad (18.3a)$$

Problem-Solving Hint

Note that Eq. 18.3 gives $1/R_p$, *not* R_p. You must take the reciprocal to solve for R_p. The units are not ohms until you invert for the final answer. (If you carry units along in your calculations, errors of this type will be less likely to occur.)

The equivalent resistance of resistors in parallel is always less than the smallest resistance in the arrangement. For example, two parallel resistors—say, of resistances 6.0 Ω and 12.0 Ω—are equivalent to a single with a resistance of 4.0 Ω.

Physically, the reason that the equivalent resistance is smaller when you combine resistors in parallel can be seen by considering a 12-V battery in a complete circuit with a single 6.0-Ω resistor. The current in the circuit is 2.0 A ($I = V/R$). Now imagine connecting a 12.0-Ω resistor in parallel to the 6.0-Ω resistor. The current through the 6.0-Ω resistor will be unaffected—it will remain

Note: R_p is smaller than the smallest resistance in a parallel arrangement.

PHYSLET® ILLUSTRATION

Resistors in Series and Parallel

at 2.0 A. (Why?) However, the new resistor will have a current of 1.0 A, since $I = V/R = 12\,V/12\,\Omega = 1.0\,A$. Since the current divides, the *total* current in the circuit is 1.0 A + 2.0 A, or 3.0 A. Now look at the overall picture of what happened. The current increased while the voltage remained the same, so the equivalent resistance of the circuit *must have decreased below* 6.0 Ω *when the* 12-Ω *resistor was attached.* Due to the extra path available there is more total current. Therefore, the equivalent circuit resistance has been reduced.

Notice that this argument does not depend on the value of the added resistor. All that matters is that another path with some resistance is added.

The general rules of thumb are as follows:

| Series connections provide a way to increase total resistance.
| Parallel connections provide a way to decrease total resistance.

To see how these ideas work consider the Example 18.1.

Example 18.1 ■ Connections Count: Resistors in Series and in Parallel

---

What is the equivalent resistance of three resistors (1.0 Ω, 2.0 Ω, and 3.0 Ω) when they are connected (a) in series (Fig. 18.1a) and (b) in parallel (Fig. 18.3a)? (c) How much current will be delivered by a 12-V battery in each of these arrangements?

Thinking It Through. To find the equivalent resistances for (a) and (b), apply Eqs. 18.2 and 18.3, respectively. To find the series current for (c), calculate the current through the battery by treating the battery as if it were connected to a single resistor—the series equivalent resistance. For the parallel arrangement, the total current can be determined by using the parallel equivalent resistance. From the knowledge that each resistor in parallel has the same voltage across it, the individual currents can be calculated.

Solution. Listing the data, we have

Given: $R_1 = 1.0\,\Omega$ *Find:* (a) R_s (series resistance)
$R_2 = 2.0\,\Omega$ (b) R_p (parallel resistance)
$R_3 = 3.0\,\Omega$ (c) I (total current for each case)
$V = 12\,V$

(a) The equivalent series resistance (Eq. 18.2) is

$$R_s = R_1 + R_2 + R_3 = 1.0\,\Omega + 2.0\,\Omega + 3.0\,\Omega = 6.0\,\Omega$$

Our result is larger than the largest resistance, as expected.

(b) The equivalent parallel resistance is determined from Eq. 18.3

$$\frac{1}{R_p} = \frac{1}{R_1} + \frac{1}{R_2} + \frac{1}{R_3} = \frac{1}{1.0\,\Omega} + \frac{1}{2.0\,\Omega} + \frac{1}{3.0\,\Omega}$$

$$= \frac{6.0}{6.0\,\Omega} + \frac{3.0}{6.0\,\Omega} + \frac{2.0}{6.0\,\Omega} = \frac{11}{6.0\,\Omega}$$

or

$$R_p = \frac{6.0\,\Omega}{11} = 0.55\,\Omega$$

(c) From the equivalent series resistance and the battery voltage we get the current:

$$I = \frac{V}{R_s} = \frac{12\,V}{6.0\,\Omega} = 2.0\,A$$

Let's calculate the voltage drop across each resistor:

$$V_1 = IR_1 = (2.0\,A)(1.0\,\Omega) = 2.0\,V$$
$$V_2 = IR_2 = (2.0\,A)(2.0\,\Omega) = 4.0\,V$$
$$V_3 = IR_3 = (2.0\,A)(3.0\,\Omega) = 6.0\,V$$

Notice that to ensure that the current through each resistor is the same, it must be that *in series, the larger resistors require more voltage.* As a check, note that the sum of the resistor voltage drops $(V_1 + V_2 + V_3)$ does equal the battery voltage.

For the parallel arrangement, the total current is determined by using the parallel equivalent resistance:

$$I = \frac{V}{R_p} = \frac{12\ V}{0.55\ \Omega} = 22\ A$$

Note that the current for the parallel combination is much larger than that for the series combination. (Why?) Here, the current through each of the resistors can be determined, since each resistor has a voltage of 12 V across it. Therefore,

$$I_1 = \frac{V}{R_1} = \frac{12\ V}{1.0\ \Omega} = 12\ A$$

$$I_2 = \frac{V}{R_2} = \frac{12\ V}{2.0\ \Omega} = 6.0\ A$$

$$I_3 = \frac{V}{R_3} = \frac{12\ V}{3.0\ \Omega} = 4.0\ A$$

Notice that the total of the three currents is indeed equal to the current through the battery.

As you can see, for resistors in parallel, the resistor with the smallest resistance gets most of the current, giving rise to the adage "*Current takes the path of least resistance.*" This adage is true because all resistors in parallel experience the same voltage.

Follow-up Exercise.   (a) Calculate the power delivered to each resistor for each arrangement in this Example. (b) What generalizations can you make? For instance, which resistor gets the most power in series? in parallel? (c) Does the series arrangement require more or less total power than the parallel case? (*Answers to all Follow-up Exercises are at the back of the text.*)

As a wiring application, consider strings of Christmas tree lights. In the past, such strings of lights had large bulbs connected in series. When one bulb blew out, all the others on that string also went out, leaving you to hunt for the faulty bulb. You had a real problem if more than one bulb blew out at the same time! Now, with newer strings that have smaller bulbs, one or more bulbs may burn out, but the others remain lit. Does this mean that the bulbs are now wired in parallel? No, parallel wiring would yield a large current, which could be dangerous.

Instead, an insulated jumper, or "shunt," is wired in parallel with each bulb filament (▶ Fig. 18.5). In normal operation, the shunt is insulated from the filament wires and thus does not conduct current. When the filament breaks and the bulb "burns out," there is momentarily no current anywhere in the string. Thus, the voltage across the open circuit at the broken filament will be the full 120-V household voltage. This voltage causes sparking that burns off the shunt's insulating material. Now the shunt makes electrical contact with the filament wires, thus completing the circuit. The rest of the lights in the string continue to glow. (The shunt, a wire with very little resistance compared with that of a bulb filament, is indicated by the small resistance symbol in the circuit diagram of Fig. 18.5. Note that under normal operation, there is a gap—the insulation—between the shunt and the filament wire.) To understand the effect (if any) of a burnt-out bulb on the string, consider the following Example.

Conceptual Example 18.2 ■ Oh, Tannenbaum! Christmas Tree Lights Revisited

In a string of Christmas tree lights composed of bulbs with jumper shunts, if the filament of one bulb burns out, will the other bulbs (a) glow a little more brightly, (b) glow a little more dimly, or (c) be unaffected?

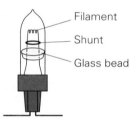
Filament
Shunt
Glass bead

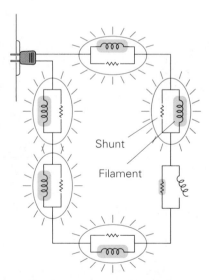
Shunt
Filament

▲ **FIGURE 18.5   Shunt-wired Christmas tree lights**   A shunt, or jumper, in parallel with the bulb filament reestablishes a complete circuit when one of the bulb filaments burns out. Without the shunt, if one bulb were to burn out, all of the bulbs would go out.

Reasoning and Answer.   If one bulb filament burns out and its shunt completes the circuit, there will be less total resistance in the circuit, because the shunt's resistance is much less than the filament's resistance. (Note that the filaments of the good bulbs and the shunt of the burnt-out bulb are in series, so the resistances just add.)

 With less total resistance, there will be more current in the circuit, and the remaining good bulbs will glow a little brighter. This effect occurs because the light output of a bulb is directly related to the power delivered to that bulb (and electrical power is related to the current by $P = I^2R$). So the answer is (a). For example, suppose the string of lights has 18 identical bulbs. Then the voltage drop across each bulb would be $(120\ \text{V})/18 = 6.7\ \text{V}$. (The voltage drops across the bulbs must add up to the voltage from the electrical socket into which the string of lights is plugged: 120 V.) If one bulb is burnt out, the voltage across each lighted bulb would be $(120\ \text{V})/17 = 7.1\ \text{V}$. This increased voltage causes the current through each bulb to increase. Both increases contribute to brighter lights.

Follow-up Exercise.   In this Example, if you were to remove one bulb, what would be the voltage across (a) the empty socket and (b) any of the remaining bulbs? Explain.

Series–Parallel Resistor Combinations

Resistors may be connected in a circuit in a variety of series–parallel combinations. As shown in ▾Fig. 18.6, circuits with only one voltage source can sometimes be reduced to a single equivalent loop, containing just the voltage source and one equivalent resistance, by applying the series and parallel results.

 A procedure for analyzing circuits that have only series and parallel resistor combinations is given next. This procedure allows us to find the voltages across, and the currents in, the various resistors. The steps are as follows:

1. Determine which blocks of resistors are in series and which are in parallel, and reduce all blocks to equivalent resistances, using Eqs. 18.2 and 18.3.

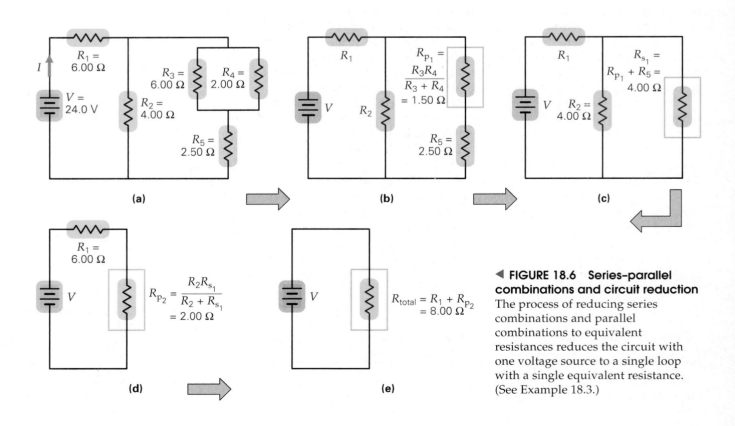

◀ **FIGURE 18.6   Series–parallel combinations and circuit reduction** The process of reducing series combinations and parallel combinations to equivalent resistances reduces the circuit with one voltage source to a single loop with a single equivalent resistance. (See Example 18.3.)

2. Reduce the circuit further by treating the separate equivalent resistances (from Step 1) as individual resistors. Proceed until you get to a single loop with one total (overall or equivalent) resistance value.

3. Find the current delivered to the reduced circuit by using $I = V/R_{\text{total}}$.

4. Expand the reduced circuit back to the actual circuit by reversing the reduction steps, one at a time. Use the current of the reduced circuit to find the currents and voltages for the resistors in each step.

To see this procedure in use, consider the next Example.

Example 18.3 ■ Series–Parallel Combination of Resistors: Same Voltage or Same Current?

What are the voltage across and the current in each of the resistors R_1 through R_5 in Fig. 18.6a?

Thinking It Through. We apply the steps described previously. It is important to identify parallel and series combinations before starting. It is clear that R_3 is in parallel with R_4. This parallel combination is itself in series with R_5. Furthermore, the R_3–R_4–R_5 leg is in parallel with R_2. Lastly, this parallel combination is in series with R_1. Combining the resistors step by step should enable us to determine the total equivalent circuit resistance (Step 2). From that value, the total current can be calculated. Then, working backwards, we can find the current in, and voltage across, each resistor.

Solution. To avoid rounding errors, the results will be carried to two decimal places.

Given: Values in Fig. 18.6a   *Find:* Current and voltage for each resistor (Fig. 18.6a)

The parallel combination at the right-hand side of the circuit diagram can be reduced to the equivalent resistance R_{P_1} (see Fig. 18.6b), using Eq. 18.3:

$$\frac{1}{R_{\text{P}_1}} = \frac{1}{R_3} + \frac{1}{R_4} = \frac{1}{6.00\ \Omega} + \frac{1}{2.00\ \Omega} = \frac{4}{6.00\ \Omega}$$

This expression is equivalent to

$$R_{\text{P}_1} = 1.50\ \Omega$$

This operation leaves a series combination of R_{P_1} and R_5 along that side, which is reduced to R_{S_1}, using Eq. 18.2 (Fig. 18.6c):

$$R_{\text{S}_1} = R_{\text{P}_1} + R_5 = 1.50\ \Omega + 2.50\ \Omega = 4.00\ \Omega$$

Then, R_2 and R_{S_1} are in parallel and can be reduced (again using Eq. 18.3) to R_{P_2} (Fig. 18.6d):

$$\frac{1}{R_{\text{P}_2}} = \frac{1}{R_2} + \frac{1}{R_{\text{S}_1}} = \frac{1}{4.00\ \Omega} + \frac{1}{4.00\ \Omega} = \frac{2}{4.00\ \Omega}$$

This expression is equivalent to

$$R_{\text{P}_2} = 2.00\ \Omega$$

This operation leaves two resistances (R_1 and R_{P_2}) in series. These resistances combine to give the total equivalent resistance (R_{total}) of the circuit (Fig. 18.6e):

$$R_{\text{total}} = R_1 + R_{\text{P}_2} = 6.00\ \Omega + 2.00\ \Omega = 8.00\ \Omega$$

Thus, the battery delivers a current of

$$I = \frac{V}{R_{\text{total}}} = \frac{24.0\ \text{V}}{8.00\ \Omega} = 3.00\ \text{A}$$

Now let's work backward and "rebuild" the actual circuit. Note that the battery current is the same as the current through R_1 and R_{p_2}, since they are in series. (In Fig. 18.6d, $I = I_1 = 3.00$ A and $I = I_{p_2} = 3.00$ A.) Therefore, the voltages across these resistors are

$$V_1 = I_1R_1 = (3.00 \text{ A})(6.00 \text{ }\Omega) = 18.0 \text{ V}$$

and

$$V_{p_2} = I_{p_2}R_{p_2} = (3.00 \text{ A})(2.00 \text{ }\Omega) = 6.00 \text{ V}$$

Since R_{p_2} is made up of R_2 and R_{s_1} (Fig. 18.6c and d), there must be a 6.00-V drop across both of these resistors. We can use this fact to calculate the current through each. Since the resistors have the same voltage and resistance, their currents are therefore

$$I_2 = \frac{V_2}{R_2} = \frac{6.00 \text{ V}}{4.00 \text{ }\Omega} = 1.50 \text{ A} \quad \text{and} \quad I_{s_1} = \frac{V_{s_1}}{R_{s_1}} = \frac{6.00 \text{ V}}{4.00 \text{ }\Omega} = 1.50 \text{ A}$$

respectively. Now notice that I_{s_1} is also the current in R_{p_1} and R_5, because they are in series. (In Fig. 18.6b, $I_{s_1} = I_{p_1} = I_5 = 1.50$ A.)

The resistors' individual voltages are therefore

$$V_{p_1} = I_{s_1}R_{p_1} = (1.50 \text{ A})(1.50 \text{ }\Omega) = 2.25 \text{ V}$$

and

$$V_5 = I_{s_1}R_5 = (1.50 \text{ A})(2.50 \text{ }\Omega) = 3.75 \text{ V}$$

respectively. (As a check, note that the voltages do, in fact, add to 6.00 V.)

Finally, the voltage across R_3 and R_4 is the same as V_{p_1} (why?), and thus

$$V_{p_1} = V_3 = V_4 = 2.25 \text{ V}$$

With the voltages and resistances we have obtained, the last two currents, I_3 and I_4, are

$$I_3 = \frac{V_3}{R_3} = \frac{2.25 \text{ V}}{6.00 \text{ }\Omega} = 0.38 \text{ A}$$

and

$$I_4 = \frac{V_4}{R_4} = \frac{2.25 \text{ V}}{2.00 \text{ }\Omega} = 1.13 \text{ A}$$

The current (I_{s_1}) is expected to divide at the R_3–R_4 junction. Thus, a check is available: $I_3 + I_4$ does equal I_{s_1}, within rounding errors.

Follow-up Exercise. In this Example, verify that the total power delivered to all of the resistors is the same as the power output of the battery.

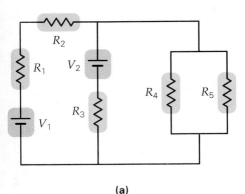

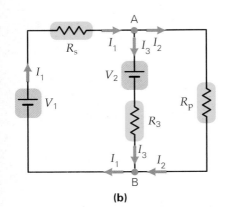

▲ **FIGURE 18.7 Multiloop circuit** In general, a circuit that contains voltage sources in more than one loop cannot be further reduced by series and parallel reductions. However, some reductions within each loop may be possible, such as from part **(a)** to part **(b)**. At a circuit junction, where three or more wires come together, the current divides or currents come together, as at junctions A and B in part **(b)**, respectively. The path between two junctions is called a *branch*. In part **(b)**, there are three branches—that is, three different paths between junctions A and B.

18.2 Multiloop Circuits and Kirchhoff's Rules

OBJECTIVES: To (a) understand the physical principles that underlie Kirchhoff's circuit rules and (b) apply these rules in the analysis of actual circuits.

Series–parallel circuits with a single voltage source can always be reduced to a single loop, as we have seen in Example 18.3. However, circuits may contain several loops, each one having several voltage sources, resistances, or both. In many cases, resistors may not be connected either in series or in parallel. A multiloop circuit, which does not lend itself to the analysis method described in Section 18.1, is shown in ◀Fig. 18.7a. Even though some combinations of resistors can be replaced by their equivalent resistances (Fig. 18.7b), this circuit can be reduced only so far by using parallel and series procedures.

Analyzing these types of circuits requires a more general approach—that is, the application of **Kirchhoff's rules**. These rules embody the conservation of charge and the conservation of energy. (Although they were not stated specifically, Kirchhoff's rules were applied to the simple circuits analyzed in Section 18.1.) First, it is useful to introduce some terminology that will help describe complex circuits:

Note: Kirchhoff's rules were developed by the German physicist Gustav Kirchhoff (1824–1887).

- A point at which three or more wires are joined is called a **junction**, or **node**—for example, point A in Fig. 18.7b.
- A path connecting two junctions is called a **branch**. A branch may contain one or more circuit elements.

Kirchhoff's Junction Theorem

Kirchhoff's first rule, or **junction theorem**, states that the algebraic sum of the currents at any junction is zero:

$$\Sigma I_i = 0 \qquad \begin{array}{l}\textit{sum of currents}\\ \textit{at a junction}\end{array} \qquad (18.4)$$

Thus, the sum of the currents entering at a junction (taken as positive) and the currents leaving the junction (taken as negative) is zero. This rule is a statement of charge conservation. For the junction at point A in Fig. 18.7b, the algebraic sum of the currents is $I_1 - I_2 - I_3 = 0$; equivalently,

$$I_1 = I_2 + I_3$$
$$\textit{current in} = \textit{current out}$$

(This rule was applied in analyzing parallel resistances in Section 18.1.)

Problem-Solving Hint

Sometimes you cannot tell whether a particular current is directed into or out of a junction simply by looking at a circuit diagram. If this is the case, you simply *assume* directions of currents at that junction. This assumption helps in problem solving, because it means that you do not have to worry about directions of currents. For example, if one of your assumptions turns out to be opposite to the actual direction of the current, then a negative answer for that current will result. This outcome tells you that the direction of the current is opposite to the initially chosen direction.

Kirchhoff's Loop Theorem

Kirchhoff's second rule, or **loop theorem**, states that the algebraic sum of the potential differences (voltages) across all of the elements of any *closed loop* is zero:

$$\Sigma V_i = 0 \qquad \begin{array}{l}\textit{sum of voltages}\\ \textit{around a closed loop}\end{array} \qquad (18.5)$$

This expression means that the sum of the voltage rises (an increase in potential) equals the sum of the voltage drops (a decrease in potential) around a closed loop, which must be true if energy is conserved. (This rule was used in analyzing series resistances in Section 18.1.)

Notice that traversing a circuit loop in different directions will yield either a voltage rise or a voltage drop across each circuit element. Thus, it is important to establish a sign convention for voltages. We will use the convention illustrated in ▶Fig. 18.8. The voltage across a battery is taken to be positive (a voltage rise) if the

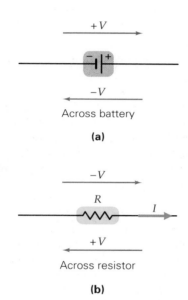

▲ **FIGURE 18.8 Sign convention for Kirchhoff's rules** **(a)** When Kirchhoff's rules are applied in going around a circuit loop, the voltage across a battery is taken to be positive $(+)$ if a battery is traversed from the negative terminal to the positive terminal and negative $(-)$ if the battery is traversed from the positive terminal to the negative terminal. **(b)** The voltage across a resistor is taken to be negative $(-)$ if the resistance is traversed in the direction of the assigned branch current and positive $(+)$ if the resistance is traversed in the direction opposite that of the assigned branch current.

loop is traversed from the negative terminal toward the positive terminal of the battery (Fig. 18.8a). The voltage across a battery is taken to be negative if the loop is traversed in the opposite direction, from the positive to the negative terminal. (Notice that the direction of the currents in the circuit has nothing to do with the sign of the voltage across a given battery. The sign of the voltage depends only on the direction in which we choose to cross the battery—positive to negative terminal or vice versa.)

The voltage across a resistor is taken to be negative (a decrease) if the resistor is traversed in the same direction as the assigned current in that branch (Fig. 18.8b) and positive if the resistor is traversed in the opposite direction. Together, these sign conventions allow you to sum the voltages around a closed loop, regardless of the direction chosen to do the sum. This is because Eq. 18.5 is mathematically the same in either case. To see this condition, note that reversing the direction in which you choose to sum the voltages simply amounts to multiplying Eq. 18.5 for the original direction by -1. This operation, of course, does not change the equation.

Learn by Drawing

Kirchhoff Plots: A Graphical Interpretation of Kirchhoff's Loop Theorem

Kirchhoff's loop theorem is usually stated in mathematical terms. However, there is a geometrical interpretation of it that may help you develop better insight into its meaning. This graphical approach allows you to visualize the potential changes in a circuit, either to anticipate the results of mathematical analysis or to confirm the order of magnitude of your results. (Of course, don't forget that to analyze most circuits numerically, Kirchhoff's junction theorem must also be used—see Example 18.5.)

The idea is to make a three-dimensional plot out of the circuit diagram. The wires and elements of the circuit form the basis for the x-y plane, or the diagram's "floor." Plotted perpendicularly to this plane, along the z-axis, is the value of the electric potential, with an appropriate choice for zero. Such a diagram is called a *Kirchhoff plot* (Fig. 1).

The rules for constructing a Kirchhoff plot are simple: Start at a known potential value, and go around a complete loop, finishing where you started. Since you come back to the same location, the sum of all the rises in potential (positive voltages) must be balanced out by the sum of the drops (negative voltages). This requirement is the geometrical expression of the conservation of energy, as embodied mathematically by Kirchhoff's loop theorem.

Thus, if the potential increases (say, in traversing a battery from cathode to anode—that is, from the negative terminal to the positive terminal), draw a rise in the z-direction. In this instance, the rise represents the terminal voltage of the battery. Similarly, if the potential decreases (for example, in traversing a resistor in the direction of the current), make sure that the potential drops. If possible, try to draw the rises and drops (the voltages) to scale. That is, if there is a large rise in potential (such as you would have across a high-voltage battery), then draw that rise to be large in proportion to the others on the diagram.

For elaborate circuits, this graphical method may prove to be too complicated for practical use. Neverthe-

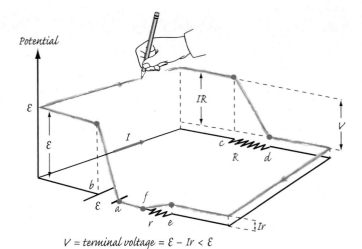

$$V = terminal\ voltage = \mathcal{E} - Ir < \mathcal{E}$$

FIGURE 1   Kirchhoff plots: A graphical problem-solving strategy   The schematic of the circuit is laid out in the x–y plane, and the electric potential is plotted perpendicularly along the z-axis. Usually, the zero of the potential is taken to be the negative terminal of the battery. A direction for current is assigned, and the value of the potential is plotted around the circuit, following the rules for gains and losses in potential. This Kirchhoff plot shows a rise in potential when the battery is traversed from cathode to anode, a drop in potential across the external resistor, and smaller drop in potential across the internal resistance of the battery.

In applying Kirchhoff's loop theorem, notice that the sign of a voltage across a resistor (which tells you whether electric potential increases or decreases) is determined by the direction of the current in that resistor. Recall, however, that there can be situations in which the direction of the current is not obvious. How do you handle the voltage signs in such cases? The answer is simple: After assuming a direction for the current you follow the voltage sign convention *based on this assumed direction*. This technique makes the two sign conventions consistent with one another.

A graphical interpretation of Kirchhoff's loop theorem that you may find helpful is presented in the Learn by Drawing feature on pp. 612–613. Integrated Example 18.4, in which a simple parallel circuit is reexamined from the point of view of Kirchhoff's rules, shows that these rules are equivalent to our previous series–parallel considerations. At the same time, notice how important it can be to draw a correct circuit diagram—it can give you hints as to how to proceed.

PHYSLET® ILLUSTRATION

Kirchhoff's Laws

Learn by Drawing

less, it is always good to keep this concept in mind, as it illustrates the fundamental idea behind the loop theorem.

As an example of this method, consider the circuit in Fig. 1: a battery with internal resistance r wired to a single external resistor R. The direction of the current is from the anode to the cathode through the external resistor as shown. We choose the potential of the battery's cathode as zero and start there. Then, traversing the circuit in the direction of the current, we show a rise in potential going from the cathode to the anode. Next, we show that the potential is constant as we follow the current through the wires until it meets the external resistor. That is, we indicate no voltage *drop* along connecting wires.

At the resistor, there must be a significant drop in potential. However, we do not make the drop all the way to zero, because there must be some voltage left to produce a current through the internal resistance. Thus, the terminal voltage of the battery, V, must be less than its emf (the rise between points a and b).

Figure 2 shows two resistors in series, and that combination in parallel with a third resistor. For simplicity, all three resistors are assumed to have the same resistance R, and the internal resistance of the battery is assumed to be zero. Starting at point a, there is a rise in potential corresponding to the voltage of the battery. Then, as we trace the loop through the single resistor, there must be a single drop in potential equal in magnitude to $\mathscr{E}$.

If we follow the loop that includes the two resistors, we see that each has only half the total voltage drop. Thus, each will carry only half the current of the single resistor. We know from Section 18.1 that in parallel circuits, the largest resistance carries the least current. But notice

how this geometrical approach can help develop your intuition and allow you to anticipate numerical results. Here, since all the resistances are equal, we would expect one third of the total current through the two resistors and two thirds through the single resistor.

As an exercise, try redrawing Fig. 2 as it would appear if the two resistors in series had resistances of $0.5R$ and $1.5R$. Which resistor now has the largest voltage across it, and how do the currents in the branches compare? Analyze the circuit mathematically to see if your expectations are confirmed.

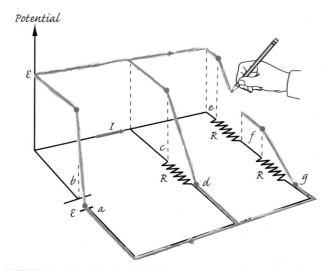

FIGURE 2 Kirchhoff plot of a more complex circuit
Imagine how the plot would change if you were to vary the values of the three resistors. Then analyze the circuit mathematically to see whether your plots allowed you to anticipate the voltages and currents.

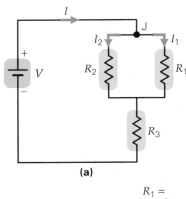

(a)

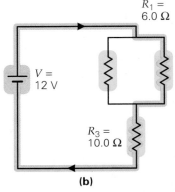

(b)

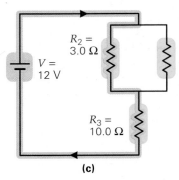

(c)

▲ **FIGURE 18.9 Sketching circuit diagrams by using Kirchhoff's rules** **(a)** Sketching the circuit diagram from the written description in Integrated Example 18.4. **(b)** and **(c)** The two different loops used in the analysis of Integrated Example 18.4, by applying Kirchhoff's loop theorem to this circuit.

Integrated Example 18.4 ■ A Simple Circuit: Using Kirchhoff's Rules

Two resistors R_1 and R_2 are connected in parallel. This combination is then connected in series to a third resistor R_3 that has the largest resistance of the three. A battery is also attached to complete the circuit, with one electrode at the beginning and the other at the end of this network. (a) Which resistor will carry the most current, (1) R_1, (2) R_2, or (3) R_3? Explain. (b) In the actual circuit, assume that $R_1 = 6.0\ \Omega$, $R_2 = 3.0\ \Omega$, $R_3 = 10.0\ \Omega$, and the battery's terminal voltage is 12.0 V. Apply Kirchhoff's rules to determine the current in each resistor and the voltage across each resistor.

(a) Conceptual Reasoning. It is easiest to see what is happening if a schematic circuit diagram is made (◄Fig. 18.9). You might initially think that the resistor with the least resistance would carry the most current. But be careful; this holds only if all the resistors are in parallel. This is *not* the case here. The two parallel resistors each carry only a fraction of the total current. However, R_3 is not in parallel with them and, *does* carry the total current. Hence, the correct answer is (3): R_3 carries the most current.

(b) Thinking It Through.

Given: $R_1 = 6.0\ \Omega$ *Find:* the current in each resistor and the voltage
     $R_2 = 3.0\ \Omega$ across each resistor
     $R_3 = 10.0\ \Omega$
     $V = 12.0\ V$

There are three unknown currents: the total current (I) and the currents in each of the parallel resistors (labeled as I_1 and I_2). Since there is only one battery, the current must be clockwise (as shown in the figure). Applying Kirchhoff's junction theorem to the first junction (J in Fig.18.9a), we have

$$\Sigma I_i = 0 \quad \text{or} \quad I - I_1 - I_2 = 0 \tag{1}$$

Using the loop theorem in the clockwise direction in Fig. 18.9b, we cross the battery from the negative to the positive terminal and then traverse R_1 and R_3 to complete the loop. The resulting equation (showing the voltage signs explicitly) is

$$\Sigma V_i = 0 \quad \text{or} \quad +V + (-I_1R_1) + (-IR_3) = 0 \tag{2}$$

A third equation can be obtained by applying the loop theorem as done previously, except this time going through R_2 instead of R_1 (Fig. 18.9c). This yields

$$\Sigma V_i = 0 \quad \text{or} \quad +V + (-I_2R_2) + (-IR_3) = 0 \tag{3}$$

Putting in the battery voltage (in volts) and resistances (in ohms) and rearranging these equations, we then have

$$I = I_1 + I_2 \tag{1a}$$
$$12 - 6I_1 - 10I = 0 \quad \text{or} \quad 6 - 3I_1 - 5I = 0 \tag{2a}$$
$$12 - 3I_2 - 10I = 0 \tag{3a}$$

Adding Eqs. (2a) and (3a) yields $18 - 3(I_1 + I_2) - 15I = 0$. However, from Eq. (1a), $I = I_1 + I_2$. Therefore,

$$18 - 3I - 15I = 0 \quad \text{or} \quad I = 1.00\ A$$

How do we know that the answer is in amperes? We know this because we consistently used volts and ohms. If you stay within this set of units, the results will be evident and consistent, and you don't need to carry units. (Of course, it is always a good idea to double-check your units at each step.)

Eqs. (3a) and (1a) can then be solved for the remaining currents:

$$I_2 = \tfrac{2}{3}\ A \quad \text{and} \quad I_1 = \tfrac{1}{3}\ A$$

These answers are consistent with our circuit-diagram reasoning in part (a).

Since the currents and resistances are known, the voltages are obtained from Ohm's law, $V = IR$. Thus,

$$V_1 = I_1R_1 = (\tfrac{1}{3}\ A)(6.0\ \Omega) = 2.0\ V$$

$$V_2 = I_2 R_2 = (\tfrac{2}{3} \text{ A})(3.0 \text{ }\Omega) = 2.0 \text{ V}$$
$$V_3 = I_3 R_3 = (1.0 \text{ A})(10.0 \text{ }\Omega) = 10.0 \text{ V}$$

Note that, as expected, the two voltage drops across the parallel resistors are the same. Because of that, two thirds of the total current is in the resistor with the least resistance (the 3.00-Ω resistor). Notice also that the total voltage across the network is 12.0 V.

Follow-up Exercise. (a) In part (b) of this Example, using what you know about series and parallel combinations, predict what will happen to the various currents if R_2 is increased. Explain your reasoning. (b) Rework part (b) of this Example, changing R_2 to 8.0 Ω, and compare your numerical results with your expectations from part (a).

Application of Kirchhoff's Rules

Integrated Example 18.4 is a relatively simple application of Kirchhoff's rules and could have been carried out using the expressions for equivalent resistances. More complicated, multiloop circuits require a more structured approach. In this book, we will use the following general steps in applying Kirchhoff's rules:

1. Assign a current and direction of current for each branch in the circuit. This assignment is done most conveniently at junctions.
2. Indicate the loops and the arbitrarily chosen directions in which they are to be traversed (▶ Fig. 18.10). Every branch *must* be in at least one loop.
3. Apply Kirchhoff's first rule to write the equations for the currents, one for each junction that gives a unique equation. (This step gives a set of equations that includes all branch currents, and you may have redundant junctions.)
4. Traverse the number of loops necessary to include all branches. In traversing a loop, apply Kirchhoff's second rule (remember that $V = IR$ for each resistor), and write the equations, using our sign conventions.

If this procedure is applied properly, Steps 3 and 4 give a set of N equations with N unknown currents. These equations may then be solved for the currents. If more loops are traversed than necessary, you will have redundant equations. Only the number of loops that includes each branch once is needed.

This procedure may seem complicated, but it's generally straightforward, as the following Example shows.

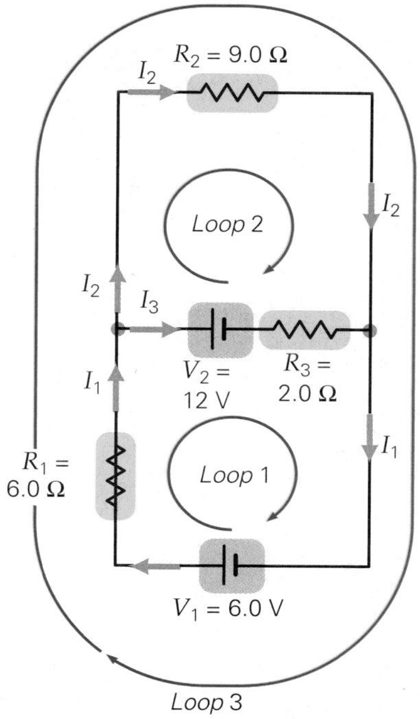

▲ FIGURE 18.10 Application of Kirchhoff's rules To analyze a circuit such as the one shown, assign a current and direction of current for each branch in the circuit (most conveniently done at junctions). Identify each loop and the arbitrary direction of traversal. Then write equations for each independent junction (using Kirchhoff's first rule) and for as many loops as needed to include every branch (using Kirchhoff's second rule). Be careful to observe sign conventions. See Example 18.5.

Example 18.5 ■ Branch Currents: Using Kirchhoff's Rules

For the circuit diagrammed in Fig. 18.10, find the current in each branch.

Thinking It Through. Series or parallel calculations cannot be used here. (Why?) Instead, the solution is begun by assigning directions of current ("best guesses") in each loop, and then the junction theorem and the loop theorem are used (twice— once for each inner loop) to generate three equations, since there are three currents.

Solution.

Given: Values in Fig. 18.10 *Find:* The current in each of the three branches

The currents in the branches and their directions, as well as the loops, have been assigned as shown in the figure. There is a current in every branch, and every branch is in at least one loop. (Some branches are in more than one loop, which is acceptable.)

Applying Kirchhoff's first rule at the left-hand junction gives

$$I_1 = I_2 + I_3 \qquad (1)$$

For the other junction, we could write $I_2 + I_3 = I_1$ (current in = current out), but this equation is equivalent to Eq. (1).

Going around loop 1 as in Fig. 18.10 and applying Kirchhoff's second rule with the sign convention gives

$$+V_1 + (-I_1 R_1) + (-V_2) + (-I_3 R_3) = 0$$

Then, putting in the numerical values from the figure yields

$$+6 - 6I_1 - 12 - 2I_3 = 0$$

or, after rearrangement of terms,

$$6I_1 + 2I_3 = -6 \quad \text{or} \quad 3I_1 + I_3 = -3 \qquad (2)$$

For convenience, units are omitted, and resistances are written to one significant figure.

Similarly, for loop 2,

$$+V_2 + (-I_2 R_2) + (+I_3 R_3) = 0$$

and

$$+12 - 9I_2 + 2I_3 = 0$$

Thus,

$$9I_2 - 2I_3 = 12 \qquad (3)$$

Equations (1), (2), and (3) form a set of three equations with three unknowns. You can solve for the I's in several ways. For example, first substitute Eq. (1) into Eq. (2) to eliminate I_1:

$$3(I_2 + I_3) + I_3 = -3$$

This expression simplifies to

$$3I_2 + 4I_3 = -3, \quad \text{or} \quad I_2 = -1 - \tfrac{4}{3} I_3 \qquad (4)$$

Then, substituting Eq. (4) into Eq. (3) eliminates I_2:

$$9(-1 - \tfrac{4}{3} I_3) - 2I_3 = 12$$

That is,

$$-14I_3 = 21 \quad \text{or} \quad I_3 = -1.5 \text{ A}$$

The minus sign in the result tells you that the wrong direction was assumed for I_3.

Putting the value of I_3 into Eq. (4) gives the value of I_2:

$$I_2 = -1 - \tfrac{4}{3}(-1.5 \text{ A}) = 1.0 \text{ A}$$

Then, from Eq. (1),

$$I_1 = I_2 + I_3 = 1.0 \text{ A} - 1.5 \text{ A} = -0.5 \text{ A}$$

The minus sign indicates that I_1 was also assigned the opposite of its true direction. (See the Problem-Solving Hint on p. 613.) You might have suspected this, because of the larger of the two batteries, the 12-V one is in the I_3 branch.

Note that this analysis did not use loop 3. The equation for this loop would be redundant, giving four equations and three unknowns.

Follow-up Exercise. Rework this Example, using the junction theorem and loops 3 and 1 instead of loops 1 and 2.

18.3 RC Circuits

OBJECTIVES: **To (a) understand the charging and discharging of a capacitor through a resistor and (b) calculate the current and voltage at specific times during these processes.**

The previous sections have dealt with circuits that have constant currents. In some direct-current (dc) circuits, the current can vary with time while maintaining a constant direction (thus it still is "dc"). This is the case with **RC circuits**, which have a resistor (R), or other resistive components, and a capacitor (C).

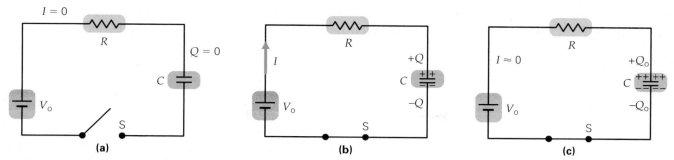

▲ **FIGURE 18.11 A series RC circuit** (a) Even though there is a space between the plates of the capacitor, (b) there is a current when the switch is closed until the capacitor is charged to its maximum value. The rate of charging (and discharging) depends on the time constant (τ) of the circuit: $\tau = RC$. (c) For times much larger than τ, the current is essentially zero, and the capacitor is essentially fully charged.

Charging a Capacitor through a Resistor

The charging of an initially uncharged capacitor by a battery is depicted in time sequence in ▲Fig. 18.11. After the switch is closed, even though there is a separation between the plates of the capacitor, charge does flow while the capacitor is charging.

The maximum charge (Q_o) that the capacitor can accumulate depends upon the capacitance (C) and the voltage of the battery (V_o): $Q_o = CV_o$. At $t = 0$, there is no charge on the capacitor and thus no voltage across it. Therefore, the full voltage of the battery appears across the resistor, resulting in an initial current $I_o = V_o/R$. As charge increases on the capacitor's plates, so does the voltage across the plates. Eventually, the capacitor is charged to the maximum, and the current diminishes to zero as the resistor's voltage drops to zero and the capacitor's voltage climbs to V_o.

The resistance helps determine how fast the capacitor is charged, since the larger its value, the greater is the resistance to charge flow. The capacitance also influences the speed of charging—it takes longer to charge a larger capacitor. Analysis of this type of circuit requires mathematics beyond the level of this book. However, it can be shown that the voltage across the capacitor increases exponentially with time according to the equation

$$V_C = V_o[1 - e^{-t/(RC)}] \quad \begin{array}{l}\textit{(charging capacitor's}\\ \textit{voltage in an RC circuit)}\end{array} \quad (18.6)$$

where e has an approximate value of 2.718. (The irrational number e is the base of the system of *natural logarithms*.) A graph of V_C versus t is presented in ►Fig. 18.12a. As expected, V_C approaches V_o, the capacitor's maximum voltage, after a long time.

A graph of I versus t is given in Fig. 18.12b. The current varies with time according to the equation

$$I = I_o e^{-t/(RC)} \quad (18.7)$$

The current decreases exponentially with time and is greatest initially.

According to Eq. 18.6, it would theoretically take an infinite amount of time for the capacitor to become fully charged. However, in practice, capacitors become almost completely charged in relatively short times. It is customary to use a special value to express the charging time. This value, called the **time constant** (τ), is expressed as

$$\tau = RC \quad \textit{time constant for RC circuits} \quad (18.8)$$

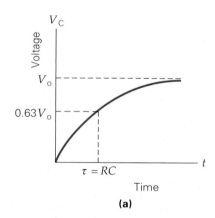

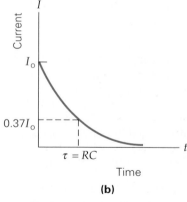

▲ **FIGURE 18.12 Capacitor charging** (a) In a series RC circuit, the voltage across the capacitor increases nonlinearly with time, reaching 63% of its maximum voltage (V_o) in one time constant: $t = \tau = RC$. (b) The current is initially a maximum ($I_o = V_o/R$) and decays exponentially, falling to 37% of its initial value in one time constant.

After an elapsed time equal to one time constant, $t = \tau = RC$, the voltage across the charging capacitor has risen to 63% of the maximum possible. This can be seen by evaluating V_C (Eq. 18.6) after a length of time equal to $\tau(=RC)$ has elapsed:

$$V_C = V_o(1 - e^{-\tau/\tau}) = V_o(1 - e^{-1})$$

$$\approx V_o\left(1 - \frac{1}{2.718}\right) = 0.63V_o$$

This implies that the capacitor is 63% charged, since $Q \propto V_C$. Note that at that time, the current is only 37% of its initial maximum value (I_o).

At the end of two time constants, $t = 2\tau = 2RC$, the capacitor is charged to more than 86% of its maximum value; at $t = 3\tau = 3RC$, the capacitor is charged to 95% of its maximum value; and so on. The capacitor is usually considered to be "fully charged" after several time constants have elapsed.

Discharging a Capacitor through a Resistor

◀Figure 18.13a shows a capacitor being *discharged* through a resistor. In this case, the voltage across the capacitor decreases exponentially with time, as does the current. The expression for the decay of the capacitor's voltage (from its maximum voltage of V_o) is

$$V_C = V_o e^{-t/(RC)} = V_o e^{-t/\tau} \qquad \begin{array}{l}\textit{(discharging capacitor's}\\ \textit{voltage in an RC circuit)}\end{array} \qquad (18.9)$$

For example, in one time constant, the voltage across the capacitor falls to 37% of its original value (Fig. 18.13b). The current in the circuit decays exponentially, following Eq. 18.7. As a practical example consider Example 18.6.

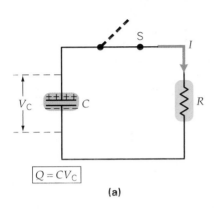

$$Q = CV_C$$

(a)

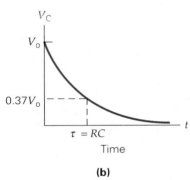

(b)

▲ **FIGURE 18.13 Capacitor discharging** (a) As the switch is closed, current appears in the circuit as the capacitor discharges. (b) When a charged capacitor is discharged through a resistance, the voltage across the capacitor (and the current in the circuit) decays exponentially with time, falling to 37% of its initial value in one time constant: $t = \tau = RC$.

Example 18.6 ■ RC Circuits in Cameras: "Flash Photography Is as Easy as Falling Off a Log(arithm)"

In many small cameras, the built-in flashbulb is fired from energy stored in a capacitor. The capacitor is kept charged using long-life batteries with voltages of typically 9.00 V. Once the bulb is fired, the capacitor must charge up quickly, through an internal RC circuit. If the capacitor has a value of 0.100 F, what must be the resistance so that the capacitor is charged to 80% of its maximum charge (the minimum amount of charge to fire the bulb again) in 5.00 s?

Thinking It Through. After one time constant, the capacitor is charged to 63% of its maximum charge. Since the capacitor needs 80% of its maximum voltage to have 80% of its full charge, it follows that the time constant must be less than 5.00 s. We can use Eq. 18.6 (along with our handy calculator) to determine the time constant.

Solution. The data given include the final voltage across the capacitor, V_C, which is 80% of the battery's voltage which means Q is 80% of the maximum charge.

Given: $C = 0.100$ F   *Find:* R (the resistance required so that the
$\qquad\quad V_B = V_o = 9.00$ V $\qquad\qquad\qquad$ capacitor is 80% charged in 5.00 s)
$\qquad\quad V_C = 0.80V_o = 7.20$ V
$\qquad\quad t = 5.00$ s

Putting the data into Eq. 18.6, $V_C = V_o(1 - e^{-t/\tau})$, we have

$$7.20 = 9.00\,(1 - e^{-5.00/\tau})$$

Rearranging this equation yields $e^{-5.00/\tau} = 0.20$, and the reciprocal of this expression (to make the exponent positive) is

$$e^{5.00/\tau} = 5.00$$

To solve for the time constant, recall that if $e^a = b$, then a is the *natural logarithm* (ln) of b. Thus, in our case, $5.00/\tau$ is the natural logarithm of 5.00. Using a calculator, we find that $\ln 5.00 = 1.61$. Therefore, we have

$$\frac{5.00}{\tau} = \ln 5.00 = 1.61$$

or

$$\tau = RC = \frac{5.00}{1.61} = 3.11 \text{ s}$$

Solving for R yields

$$R = \frac{3.11 \text{ s}}{C} = \frac{3.11 \text{ s}}{0.10 \text{ F}} = 31 \; \Omega$$

As expected, the time constant is less than 5.0 s, because achieving 80% of the maximum voltage requires longer than one time constant.

Follow-up Exercise. (a) In this Example, how does the energy stored in the capacitor (after 5.00 s) compare with the maximum energy storage? Explain why it isn't 80%. (b) If you waited 10.00 s to charge the capacitor, what would its voltage be? Why isn't it twice the voltage that exists across the capacitor after 5.00 s?

An application of an RC circuit is diagrammed in ▶Fig. 18.14a. This circuit is called a *blinker circuit* (or a *neon-tube relaxation oscillator*). The resistor and capacitor are initially wired in series, and then a neon tube is connected in parallel with the capacitor. (These neon tubes are the size of miniature Christmas tree lights.)

When the circuit is closed, the voltage across the capacitor (and the neon tube) rises from 0 to V_b, which is the *breakdown voltage* of the neon gas in the tube (about 80 V). At that voltage, the gas is ionized and begins to conduct electricity, and the tube lights up. When the tube is in a conducting state, the capacitor discharges through it, and the voltage falls rapidly (Figure 18.14b). When the voltage drops below V_m, called the *maintaining voltage*, the discharge of the tube can no longer be sustained, and the tube stops conducting. The capacitor begins charging again, the voltage rises from V_m to V_b, and the cycle repeats. The continual repetition of this cycle causes the tube to blink on and off.

Other applications of RC circuits are discussed in the Insight on p. 620.

PHYSLET®
ILLUSTRATION

RC Circuit

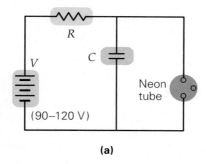

(a)

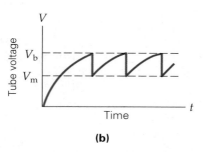

(b)

▲ **FIGURE 18.14 Blinker circuit** (a) When a neon tube is connected across the capacitor in a series RC circuit that has the proper voltage source, the voltage across the tube (and capacitor) will relax, or oscillate, with time. As a result, the tube periodically flashes or blinks. (b) A graph of voltage versus time shows the oscillating effect between V_b, the breakdown voltage, and V_m, the maintaining voltage.

18.4 Ammeters and Voltmeters

OBJECTIVES: To understand (a) how galvanometers are used as ammeters and voltmeters, (b) how multirange versions of these devices are constructed, and (c) how they are connected to measure current and voltage in real circuits.

As the names imply, an **ammeter** measures current *through* circuit elements and a **voltmeter** measures voltages *across* circuit elements. A basic component of both of these types of meters is a **galvanometer** (▶Fig. 18.15a). The galvanometer operates on magnetic principles that will be covered in Chapter 19. In this chapter, it will simply be treated as a circuit element that has an internal resistance r (typically about 50 Ω) and whose needle deflection is directly proportional to the current in it (Fig. 18.15b).

Ammeters

A galvanometer measures current, but because of its small coil resistance, only currents in the microampere range can be measured without burning out the coil wires. However there is a way to construct an ammeter to measure larger currents and still use a galvanometer. To do so, we connect a small *shunt resistor*

RC Circuits in Action: Studsensors™ Help around the House

To install wall shelving or to hang a picture securely, it is usually necessary to locate the studs in the walls. Studs are the vertical wooden supports to which plasterboard or paneling is nailed, and they can be very elusive. The traditional way of finding them, by tapping on the wall, generally leads to more frustration than success. Commercially available stud finders employ pivoted magnets, which move when they pass over a nail head in the stud behind the wall, but these devices, too, are far from infallible.

An electronic device called Studsensor™ is much easier to use and relies on capacitive sensing. The capacitor forms part of a simple RC charging circuit. However, for this application, the metallic plates of the capacitor are arranged end to end rather than facing one another as in a parallel-plate capacitor (Fig. 1). In this configuration, the fringe electric field extends a considerable distance beyond the plane of the plates, and any nearby material acts as a dielectric in this field. The unit is initially calibrated by placing it against the wall and balancing the charging time of the RC circuit against a reference standard. When the sensor containing the capacitor plates is moved over a stud, the character of the dielectric changes, and the capacitance of the plates increases. The greater the capacitance, the longer is the charging time of the circuit relative to the calibration standard. The changes in charging time are sensed electronically and conveyed to the user by lighted indicators (Fig. 2).

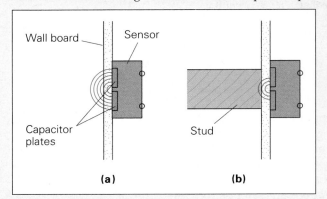

FIGURE 1 RC sensing **(a)** The fringe electric field of the capacitor plates in a Studsensor™ permeates the wall board, which acts as a dielectric. **(b)** When the sensor is over a stud, the capacitance changes. The sensor detects this change as a change in an RC circuit.

FIGURE 2 Studsensor™ The lower green light indicates that the unit is calibrated. Vertical red lights signal the approach to a stud, and the top red light indicates that the sensor is directly over a stud. Some models have both light and sound indicators.

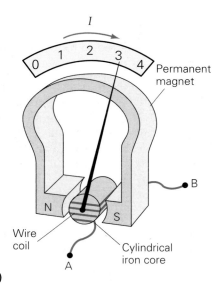

(a)

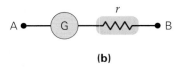

(b)

◀ **FIGURE 18.15 The galvanometer** **(a)** A galvanometer is a current-sensitive device whose needle deflection is proportional to the current through its coil. **(b)** The circuit symbol for a galvanometer is a circle containing a "G." The internal resistance (r) of the meter is indicated explicitly.

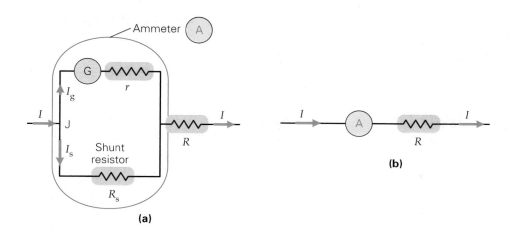

(a)

(with a resistance of R_s) in parallel with a galvanometer to take most of the current (▲Fig. 18.16). This requires that the shunt have much less resistance than the galvanometer ($R_s \ll r$). The following Example illustrates how the resistance of the shunt is determined in the design of an ammeter.

Example 18.7 ■ Ammeter Design Using Kirchhoff's Rules: Choosing a Shunt Resistor

Suppose you have a galvanometer that can safely carry a maximum coil current of 200 μA (called its *full-scale sensitivity*) and that has a coil resistance of 50 Ω. It is to be used in an ammeter designed to read currents up to 3.0 A (at full scale). What is the required shunt resistance? (See Fig. 18.16a).

Thinking It Through.   The galvanometer itself can carry only a very small current. Therefore, most of the current will have to be diverted, or "shunted," through the shunt resistor. Thus, we expect the shunt resistance to be much less than the galvanometer's internal resistance ($R_s \ll r$). The two resistors are in parallel, which means that they have the same voltage. This information should enable us to determine the required value of R_s.

Solution.   Listing the data, we have

Given: $I_g = 200\ \mu A = 2.00 \times 10^{-4}$ A   *Find:* R_s (shunt resistance)
  $r = 50\ \Omega$
  $I_{max} = 3.0$ A

Because the voltages across the galvanometer and the shunt resistor are equal, we can write (using subscripts "g" for galvanometer and "s" for shunt—see Fig. 18.16a)

$$V_g = V_s \quad \text{or} \quad I_g r = I_s R_s$$

Using Kirchhoff's junction rule at point J, the current I in the external circuit is, $I = I_g + I_s$, or $I_s = I - I_g$. Then substituting into the previous equation for I_s, we have

$$I_g r = (I - I_g)R_s$$

Solving this equation symbolically, we find that the shunt's resistance R_s is

$$R_s = \frac{I_g r}{I_{max} - I_g} = \frac{(2.00 \times 10^{-4}\,\text{A})(50\ \Omega)}{3.0\,\text{A} - 2.00 \times 10^{-4}\,\text{A}}$$
$$= 3.3 \times 10^{-3}\ \Omega = 3.3\ \text{m}\Omega$$

Note that the shunt's resistance is very small compared with the coil's resistance which allows most of the current (2.9998 A at full scale) to pass through the shunt-resistor branch. The ammeter will read currents linearly up to 3.0 A. For example, if a current of 1.5 A were to flow into the ammeter, there would be a current of 100 μA

Note: Ammeters are connected in series with the element whose current they are measuring (Fig. 18.16b).

Note: Ammeters are galvanometers with *small resistors in parallel.* Voltmeters are galvanometers with *large resistors in series.*

Note: Voltmeters are connected in parallel, or across, the element whose voltage they are measuring (Fig. 18.17b).

(half the allowed maximum) in the coil of the galvanometer, which would give a half-scale reading.

Follow-up Exercise. In this Example, if we had used a shunt resistance of $1.0 \text{ m}\Omega$, what would be the full-scale reading (maximum current reading) of the ammeter?

Voltmeters

A voltmeter that is capable of reading voltages higher than the microvolt range (anything higher would burn out the galvanometer if it were alone) is constructed by connecting a large *multiplier resistor* in series with a galvanometer (◄Fig. 18.17). Because the voltmeter has a large resistance, due to the multiplier resistor, it draws little current from the circuit element whose voltage it measures. However, the current that exists in the voltmeter is proportional to the voltage across the circuit element. Thus, the voltmeter can be calibrated in volts. To better understand this configuration, consider Example 18.8.

Example 18.8 ■ Voltmeter Design: Using Kirchhoff's Rules to Choose a Multiplier Resistor

Suppose that the galvanometer in Example 18.7 is to be used instead in a voltmeter with a full-scale reading of 3.0 V. What is the required multiplier resistance?

Thinking It Through. To turn a galvanometer into a voltmeter, we need a reduction in current—accomplished by adding a large "multiplier resistor" in series. All the data necessary to calculate the multiplier resistance are given here and in Example 18.7.

Solution. First, we list the data given:

Given: $I_g = 200 \ \mu\text{A} = 2.00 \times 10^{-4} \text{ A}$ *Find:* R_m (multiplier resistance)
        (from Example 18.7)
        $r = 50 \ \Omega$ (from Example 18.7)
     $V_{max} = 3.0 \text{ V}$

The resistances of the galvanometer and multiplier are in series. This combination is itself in parallel with the external circuit element (R). Therefore, the voltage across the external circuit element is the sum of the voltages across the galvanometer and multiplier (Fig. 18.17):

$$V = V_g + V_m$$

The voltages across the galvanometer and multiplier resistors can be written as

$$V_g = I_g r \quad \text{and} \quad V_m = I_g R_m$$

respectively. Combining these three equations, we have

$$V = V_g + V_m = I_g r + I_g R_m = I_g(r + R_m)$$

Solving for the resistance of the multiplier, we find that the result is

$$R_m = \frac{V - I_g r}{I_g} = \frac{3.0 \text{ V} - (2.00 \times 10^{-4} \text{ A})(50 \ \Omega)}{2.00 \times 10^{-4} \text{ A}} = 1.5 \times 10^4 \ \Omega = 15 \text{ k}\Omega$$

Notice that the second term in the numerator ($I_g r$) is negligible compared with the full-scale reading of 3.0 V. Thus, to a good approximation, $R_m \approx V/I_g$, or $V \propto I_g$. Therefore, the measured voltage is indeed proportional to the current in the galvanometer.

Follow-up Exercise. The voltmeter in this Example is used to measure the voltage of a resistor in a circuit. A current of 3.00 A flows through the resistor ($1.00 \ \Omega$) *before* the voltmeter is connected. Assuming that the total *incoming* current (I in Fig. 18.17b) remains the same after the voltmeter is connected, calculate the current in the galvanometer.

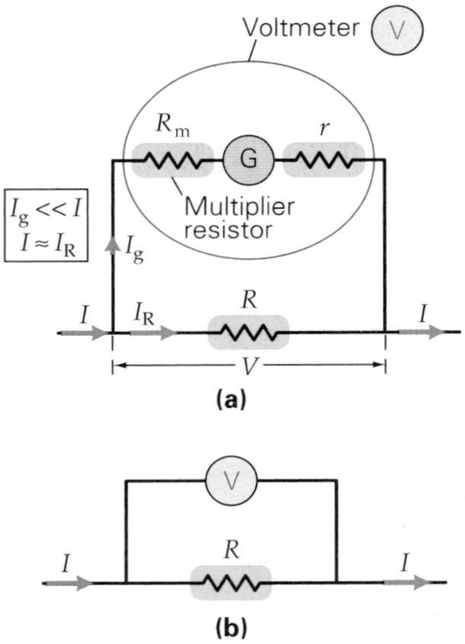

▲ **FIGURE 18.17 A dc voltmeter**
Here, R is the resistance of the resistor whose voltage is being measured. **(a)** A galvanometer in series with a multiplier resistor (R_m) is a voltmeter capable of measuring various ranges of voltage, depending on the value of R_m. **(b)** The circuit symbol for a voltmeter is a circle with a "V" inside it. (See Example 18.8 for a discussion of voltmeter design.)

For versatility, ammeters and voltmeters may be multiranged. This task is accomplished by providing the user with a choice of several shunt or multiplier resistors (▶Fig. 18.18a and b). Such meters can also be combined into *multimeters*, which measure voltage, current, and often resistance. Electronic digital multimeters are common (Fig. 18.18c). In place of mechanical galvanometers, these devices use electronic circuits that analyze digital signals to calculate voltages, currents, and resistances.

18.5 Household Circuits and Electrical Safety

OBJECTIVES: To understand (a) how household circuits are wired and (b) the underlying principles that govern electrical safety devices.

Although household circuits generally use alternating current, which has not yet been discussed, they include practical applications of some of the principles we have already studied.

For example, would you expect the elements (lamps, appliances, and so on) in a household circuit to be connected in series or in parallel? From the discussion of Christmas tree lights in Section 18.1, it should be apparent that household elements must be connected in parallel. For example, when the bulb in a lamp blows out, other elements in the circuit continue to work. Moreover, household appliances and lamps are generally rated for 120 V. If these elements were connected in series, none of the individual circuit elements would have its required 120 V.

Electrical power is supplied to a house by a three-wire system (▼Fig. 18.19). There is a difference in potential of 240 V between the two hot, or high-potential, wires, and each of these wires has a 120-V difference in potential with respect to the ground. The third wire is grounded at the point where the wires enter the house, usually by a metal rod driven into the ground. This wire is defined as the zero potential and is called the *ground*, or *neutral, wire*.

▼ **FIGURE 18.19 Household wiring** A 120-V circuit is obtained by connecting between the +120-V (or −120-V) line and the ground line. A difference in potential of 240 V for large appliances such as electric stoves, central air conditioners, and hot-water heaters is obtained by connecting the +120-V line and the −120-V line.

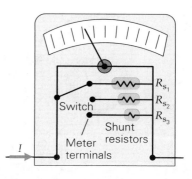

(a) Multirange ammeter

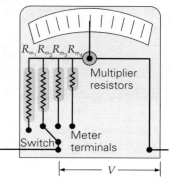

(b) Multirange voltmeter

(c)

▲ **FIGURE 18.18   Multirange meters** **(a)** An ammeter or **(b)** a voltmeter measures several ranges of current and voltage by switching among different shunt or multiplier resistors, respectively. (Instead of a switch, there may be an exterior terminal for each range.) **(c)** Both functions can be combined in a multimeter.

Note: Household voltage can fluctuate, under normal conditions, between 110 and 120 V. Similarly, 240-V connections can be as low as 220 V and still be considered normal.

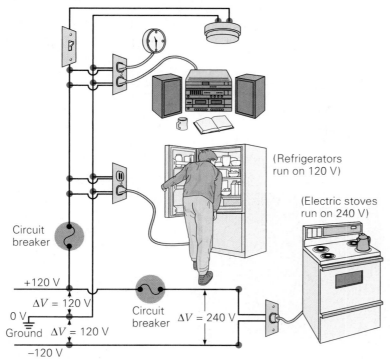

(Refrigerators run on 120 V)

(Electric stoves run on 240 V)

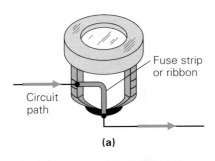

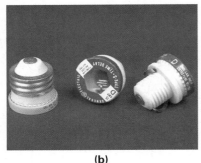

(b)

▲ **FIGURE 18.20   Fuses**
(a) A fuse contains a metallic strip, or ribbon, that melts when the current exceeds a rated value. This action opens the circuit and prevents the occurrence of overheating. (b) Edison-base fuses (left) have threads similar to those on lightbulbs; fuses with different ampere ratings can be interchanged. Type-S fuses (right) have different threads for different ratings. They fit specific adapters in the fuse sockets and thus cannot be interchanged.

▶ **FIGURE 18.21   Circuit breaker**
(a) A diagram of a thermal trip element. With increased current and joule heating, the element bends until it opens the circuit at some preset value of current. Magnetic trip elements are also used. (b) A bank of typical household circuit breakers.

The difference in potential of 120 V needed for most household appliances is obtained by connecting them between the ground wire and either high-potential wire: $\Delta V = 120\,V - 0\,V = 120\,V$ or $\Delta V = 0\,V - (-120\,V) = 120\,V$. (See Fig. 18.19.) Even though the ground wire has zero potential, it is a current-carrying wire, because it is part of the complete circuit. Large appliances such as central air conditioners, ovens, and hot-water heaters need 240 V, and this voltage is obtained by connecting them between the two hot wires: $\Delta V = 120\,V - (-120\,V) = 240\,V$. Although the current through an appliance may be given on a rating tag, it can also be determined from the power rating (using $P = IV$). For example, a stereo rated at 180 W at 120 V would draw an average current of 1.5 A (from $I = P/V$).

There are limitations on the number of elements that can be put in a circuit and the total current in that circuit. The joule heat (or I^2R loss) in the wires must be considered. Generally, the more elements (appliances and, therefore, resistances) connected in parallel, the smaller is the equivalent resistance of the circuit. Adding elements increases the total current. Remember that real wires have some resistance and could be subject to significant joule heating if the current is large enough. Therefore, by adding too many current-carrying elements, it is possible to overload a household circuit so that it carries too much current and produces too much heat *in the wires*. This heat could melt the insulation and start a fire.

Overloading is prevented by limiting the current in a circuit by means of two types of devices: fuses and circuit breakers. **Fuses** (◀Fig. 18.20) are common in older homes. An Edison-base fuse has threads like those on the base of a lightbulb. (See Fig. 18.20b.) Inside the fuse is a metal strip that melts because of joule heat when the current is larger than the rated value (which is typically 15 A for a 120-V circuit). The melting of the strip breaks (or opens) the circuit, and the current drops to zero. This *open circuit*, in effect, has an infinite resistance.

Circuit breakers are now used exclusively in wiring new homes. One type (▼ Fig. 18.21) uses a bimetallic strip (see Chapter 10). As the current in the strip increases, the strip becomes warmer and bends. At the rated value of the current, the strip will bend sufficiently to open the circuit. The strip then quickly cools, so the breaker can be reset. However, a blown fuse or a tripped circuit breaker indicates that the circuit is drawing or attempting to draw too much current. *Find and correct the problem before replacing the fuse or resetting the circuit breaker.*

Switches, fuses, and circuit breakers are placed in the hot (high-potential) side of the line. They would work in the grounded side as a heating-overload sensor.

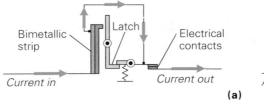

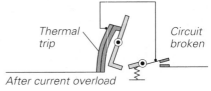

(a)

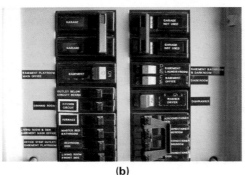

(b)

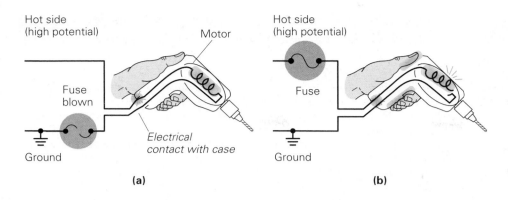

(a)   **(b)**

◀ **FIGURE 18.22 Electrical safety**
(a) Switches and fuses or circuit breakers should always be wired in the hot side of the line, *not* in the grounded side as shown. If these elements are wired in the grounded side, the line is at a high potential even when the fuse is blown or a switch is open. **(b)** Even if the fuse or circuit breaker is wired in the hot side, a potentially dangerous situation exists. If an internal wire comes in contact with the metal casing of an appliance or power tool, a person touching the casing, which is at high voltage, can get a shock.

But even if the switch were open, the fuse blown, or the breaker tripped, the circuit and any elements in it would still be connected to a high potential, which could be dangerous if a person made electrical contact (▲Fig. 18.22a).

Even with fuses or circuit breakers in the hot side of the line, there is still a possibility of getting an electrical shock from a defective appliance that has a metal casing, such as a hand drill. If a wire comes loose inside, it could make contact with the casing, which would then be "hot," or at a high potential (Fig. 18.22b). A person could provide a path to the ground and become part of the circuit, thus receiving a shock. For a discussion of the effects of electric shock, see the Insight on p. 626.

To prevent a shock, a third, dedicated grounding wire is added to the circuit that grounds the metal casing of appliances or power tools (▶Fig. 18.23). This wire provides a path of very low resistance, bypassing the tool. Such a bypass route is a type of *short circuit*, analogous to taking a shortcut on a trip. This wire does not normally carry current. If a hot wire comes in contact with the casing, the circuit is completed to this grounded, or shorting, wire. The fuse is blown or the circuit breaker tripped, as most of the current is in the third wire to the ground (a low-resistance path) rather than in you (a high-resistance path).

On three-prong **grounded plugs**, the large, round prong connects with the dedicated grounding wire. Adapters can be used between a three-prong plug and a two-prong socket. Such an adapter has a grounding lug or grounding wire (▶Fig. 18.24a) that should be fastened to the receptacle box by the plate-fastening screw or some other means. The receptacle box is grounded by means of the grounding wire. If the adapter lug or wire is not connected, the system is left unprotected, which defeats the purpose of the dedicated grounding safety feature.

You may have noticed another type of plug, a two-prong plug that fits in the socket only in one orientation, because one prong is wider than the other and one of the slits of the receptacle is also larger (Fig. 18.24b). This type of plug is called a **polarized plug**. *Polarizing* in the electrical sense is a method of identifying the hot and grounded sides of the line so that particular connections can be made.

Such polarized plugs and sockets are now a common safety feature. Wall receptacles are wired so that the small slit connects to the hot side and the large slit connects to the neutral, or ground, side. Having the hot side identified in this way makes two safeguards possible. First, the manufacturer of an electrical appliance can design it so that the switch is always in the hot side of the line. Thus, all of the wiring of the appliance beyond the switch is safely neutral when the switch is open and the appliance off. Moreover, the casing of an appliance is connected by the manufacturer to the ground side by means of a polarized plug. Should a hot wire inside the appliance come loose and contact the metal casing, the effect would be similar to that with a dedicated grounding system: The hot side of the line would be shorted to the ground, which would blow a fuse or trip a circuit breaker.

Another type of electrical safety device, the ground fault circuit interrupter, or GFCI, is discussed in Chapter 20.

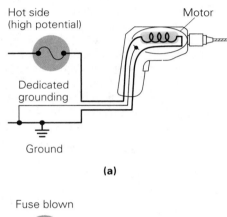

(a)

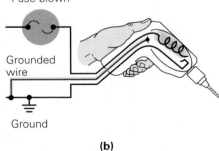

(b)

▲ **FIGURE 18.23 Dedicated grounding** **(a)** For safety, a third wire is connected from an appliance or power tool to ground. This dedicated grounding wire normally carries no current (as opposed to the grounded wire of the circuit). **(b)** If a loose internal wire comes in contact with the grounded casing, the shorting to ground through the dedicated grounding wire blows the protective fuse or circuit breaker. This action prevents the creation of a potentially dangerous high voltage between the case and ground.

Electricity and Personal Safety

Safety precautions are necessary to prevent injuries when people work with and use electricity. Electrical conductors (such as wires) are coated with insulating materials so they can be handled safely. However, when a person comes in contact with a charged conductor, a difference in potential may exist across part of the person's body. A bird can sit on a high-voltage line without any problem because the bird's two feet are at essentially the same potential, and thus there is no difference in potential to generate a current in the bird. But if a person carrying an aluminum (conducting) ladder touches it to an electrical line, a difference in potential exists from the line to the ground, and the ladder and the person are then part of a current-carrying circuit.

The extent of personal injury in such a case depends on the amount of electric current through the body and on the exact path through the body. From the relationship between voltage, current, and resistance, we know that the current in the body is given by $I = V/R_{body}$, where R_{body} is the resistance of the body. Thus, for a given voltage, the current depends on the body's resistance.

However, the body's resistance varies, depending on the environment. If the skin is dry, the resistance can be $0.50 \, M\Omega$ ($0.50 \times 10^6 \, \Omega$) or more. For a difference in potential of 120 V in such a case, there would be a current of

$$I = \frac{V}{R_{body}} = \frac{120 \, V}{0.50 \times 10^6 \, \Omega} = 0.24 \times 10^{-3} \, A = 0.24 \, mA$$

This current is almost too weak to be felt (Table 1).

Suppose, however, that the skin is wet with perspiration. Then R_{body} is only about $5.0 \, k\Omega$ ($5.0 \times 10^3 \, \Omega$), and the current is

$$I = \frac{V}{R_{body}} = \frac{120 \, V}{5.0 \times 10^3 \, \Omega} = 24 \times 10^{-3} \, A = 24 \, mA$$

which could be very dangerous. (See Table 1 again.)

A basic precaution is to avoid contact with an electrical conductor that might cause a difference in potential across your body or part of it. The physical damage resulting from such contact depends on the path of the current. If that path is from the little finger to the thumb on one hand, probably only a burn would result from a large current. However, if the path is from hand to hand through the chest (and therefore likely through the heart), the effect can be much worse. Some of the possible effects of this latter circuit path are given in Table 1.

Injury results because the current interferes with muscle functions and/or causes burns. Muscle functions are regulated by electrical impulses through the nerves, and these impulses can be influenced by external currents. Muscle reaction and pain can occur from a current of just a few milliamperes. At about 10 mA, muscle paralysis can prevent a person from releasing the conductor. At about 20 mA, contraction of the chest muscles occurs, which can cause impairment or stoppage of breathing. Death can occur in a few minutes. At 100 mA, there are rapid uncoordinated movements of the heart muscles (called ventricular fibrillation), which prevent the proper pumping action from taking place and can be fatal in a matter of seconds.

Working safely with electricity requires a knowledge of fundamental electrical principles *and* common sense. Electricity must be treated with respect.

Related Exercises: 88 and 89

TABLE 1 Effects of Electric Current on the Human Body*

Current (approximate)	Effect
2.0 mA (0.002 A)	Mild shock or heating
10 mA (0.01 A)	Paralysis of motor muscles
20 mA (0.02 A)	Paralysis of chest muscles, causing respiratory arrest; fatal in a few minutes
100 mA (0.1 A)	Ventricular fibrillation, preventing coordination of the heart's beating; fatal in a few seconds
1000 mA (1 A)	Serious burns; fatal almost instantly

*The effect on the human body of a given amount of current depends on a variety of conditions. This table gives only general and relative descriptions. The descriptions assume a circuit path that includes the upper chest.

▶ FIGURE 18.24 Plugging into ground (a) A three-prong plug and adapter for a two-prong socket. The grounding lug (loop) on the adapter should be connected to the plate-fastening screw on the grounded receptacle box—otherwise, the safety feature is lost. (b) A polarized plug. The differently sized prongs permit prewired identification of the high and ground sides of the line.

(a)

(b)

Chapter Review

Important Concepts and Equations

- When resistors are wired in **series**, the current through each of them is the same. The **equivalent resistance** of resistors in series is

$$R_s = R_1 + R_2 + R_3 + \cdots = \Sigma R_i \qquad (18.2)$$

- When resistors are wired in **parallel**, the voltage across each of them is the same. The **equivalent resistance** is

$$\frac{1}{R_p} = \frac{1}{R_1} + \frac{1}{R_2} + \frac{1}{R_3} + \cdots = \Sigma \frac{1}{R_i} \qquad (18.3)$$

- **Kirchhoff's junction theorem** states that the total current into any **junction** equals the total current out of that junction (conservation of electric charge).

$$\Sigma I_i = 0 \quad \text{sum of currents at a junction} \qquad (18.4)$$

- **Kirchhoff's loop theorem** states that in traversing a complete circuit loop, the algebraic sum of the voltage gains and losses is zero, or the sum of the voltage gains equals the sum of the voltage losses (conserva-

tion of energy in an electric circuit). In terms of voltages, this can be written as

$$\Sigma V_i = 0 \quad \begin{array}{l}\text{sum of voltages} \\ \text{around a closed loop}\end{array} \qquad (18.5)$$

- The **time constant** (τ) for an RC circuit is a characteristic time by which we measure the capacitor's charging and discharging rate. τ is given by

$$\tau = RC \qquad (18.8)$$

- An **ammeter** is a device for measuring current; it consists of a galvanometer and a shunt resistor in parallel. Ammeters are connected in series, with the circuit element carrying the current to be measured, and have very little resistance.

- A **voltmeter** is a device for measuring voltage; it consists of a galvanometer and a multiplier resistor wired in series. Voltmeters are connected in parallel, with the circuit element experiencing the voltage to be measured, and have large resistance.

Exercises

Assume that all resistors are ohmic unless otherwise stated.

18.1 Resistances in Series, Parallel, and Series–Parallel Combinations

1. Which of the following is always the same for resistors in series? (a) voltage; (b) current; (c) power; (d) energy.

2. Which of the following is always the same for resistors in parallel? (a) voltage; (b) current; (c) power; (d) energy.

3. CQ Are the voltage drops across resistors in series generally the same? If not, under what circumstance(s) could they be the same?

4. CQ Are the currents in resistors in parallel generally the same? If not, under what circumstance(s) could they be the same?

5. CQ If a large resistor and a small resistor are connected in series, will the effective resistance be closer in value to that of the large resistance or the small one? What if they are connected in parallel?

6. CQ In some strings of Christmas lights, if one bulb burns out, all the remaining bulbs will not remain lit. How are these bulbs connected?

7. CQ Three resistors that have values of $5\,\Omega, 2\,\Omega$, and $1\,\Omega$, respectively, are connected in series to a battery. Which resistor gets the most power, and why?

8. CQ Three resistors that have values of $5\,\Omega, 2\,\Omega$, and $1\,\Omega$, respectively, are connected in parallel to a battery. Which resistor gets the most power, and why?

9. ■ Three resistors that have values of $10\,\Omega, 20\,\Omega$, and $30\,\Omega$, respectively, are to be connected. (a) How should you connect them to get the maximum equivalent resistance, and what is this maximum value? (b) How should you connect them to get the minimum equivalent resistance, and what is this minimum value?

10. ■ Two identical resistors (R) are connected in series and then wired in parallel to a 20-Ω resistor. If the total equivalent resistance is 10 Ω, what is the value of R?

11. ■ Two identical resistors (R) are connected in parallel and then wired in series to a 40-Ω resistor. If the total equivalent resistance is 55 Ω, what is the value of R?

12. IE ■ For three 4.0-Ω resistors, (a) how many values of equivalent resistance can you get, (1) three, (2) five, or (3) seven? (b) List the equivalent resistance for each.

13. ■ Three resistors with values of 5.0 Ω, 10 Ω, and 15 Ω, respectively, are connected in series in a circuit with a 9.0-V battery. (a) What is the total equivalent resistance? (b) What is the current in each resistor? (c) At what rate is energy delivered to the 15-Ω resistor?

14. ■ Find the equivalent resistances for all possible combinations of two or more of the three resistors in Exercise 13.

15. ■ Three resistors with values 1.0 Ω, 2.0 Ω, and 4.0 Ω, respectively, are connected in parallel in a circuit with a 6.0-V battery. What are (a) the total equivalent resistance, (b) the voltage across each resistor, and (c) the power delivered to the 4.0-Ω resistor?

16. IE ■■ (a) If you had only an infinite supply of 1.0-Ω resistors, what is the minimum number of resistors required to make an equivalent resistance of 1.5 Ω, (1) two, (2) three, or (3) four? (b) Describe or show by a sketch how the resistors should be connected.

17. IE ■■ A length of wire with a resistance *R* is cut into two equal segments. The segments are then twisted together to form a conductor half as long as the original wire. (a) The resistance of the shortened conductor is (1) $R/4$, (2) $R/2$, or (3) R. (b) If the resistance of the original wire is 27 $\mu\Omega$ and the wire is cut into three equal segments, what is the resistance of the shortened conductor?

18. IE ■■ You have four 5.00-Ω resistors. (a) Can you connect all the resistors to produce an effective total resistance of 3.75 Ω? (b) Describe how you would connect them.

19. ■■ Three resistors with values of 2.0 Ω, 4.0 Ω, and 6.0 Ω, respectively, are connected in series in a circuit with a 12-V battery. (a) How much current is delivered to the circuit by the battery? (b) What is the current in each resistor? (c) How much power is delivered to each resistor? (d) How does this power compare with the power delivered to the total equivalent resistance?

20. ■■ Suppose that the resistors in Exercise 19 are connected in parallel. (a) How much current is delivered to the circuit by the battery? (b) What is the current in each resistor? (c) How much power is delivered to each resistor? (d) How does this power compare with the power delivered to the total equivalent resistance?

21. ■■ Two 8.0-Ω resistors are connected in parallel, as are two 4.0-Ω resistors. These two combinations are then connected in series in a circuit with a 12-V battery. What is the current in each resistor and the voltage across each resistor?

22. What is the equivalent resistance of the resistors in ▼Fig. 18.25?

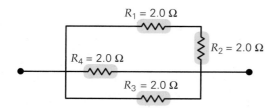

▲ **FIGURE 18.25 Series–parallel combination**
See Exercises 22 and 32.

23. ■■ What is the equivalent resistance between points A and B in ▼Fig. 18.26?

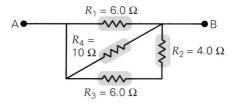

▲ **FIGURE 18.26 Series–parallel combination**
See Exercises 23 and 34.

24. ■ What is the equivalent resistance of the arrangement of resistors shown in ▼Fig. 18.27?

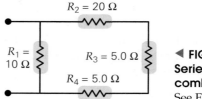

◄ **FIGURE 18.27 Series–parallel combination**
See Exercise 24.

25. ■■ Several 60-W lightbulbs are connected in parallel with a 120-V source. The very last bulb blows a 15-A fuse in the circuit. (a) Sketch a schematic circuit diagram to show the fuse in relation to the bulbs. (b) How many lightbulbs are in the circuit (including the very last bulb)?

26. ■■ Three 50-Ω resistors and a 120-V source may be used in a circuit. (a) What arrangement yields the maximum power? Sketch the circuit diagram. (b) What arrangement yields the minimum power? Sketch its circuit diagram.

27. ■■ Find the current in and voltage of the 10-Ω resistor shown in ▼Fig. 18.28.

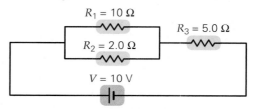

▲ **FIGURE 18.28 Current and voltage drop of a resistor**
See Exercises 27 and 48.

28. ■■ A three-way lightbulb can produce 50 W, 100 W, or 150 W of power at 120 V. For each power setting, find the bulb's (a) current and (b) resistance.

29. ■■ For the circuit shown in ▼Fig. 18.29, find (a) the current in each resistor, (b) the voltage across each resistor, and (c) the total power delivered.

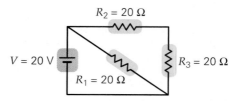

▲ **FIGURE 18.29 Circuit reduction** See Exercises 29 and 49.

30. ■■ A 120-V circuit has a circuit breaker rated to "trip" (to create an open circuit) at 15 A. How many 300-Ω resistors could be connected in parallel without tripping the breaker?

31. ■■ In your dorm room, you have two 100-W lights, a 150-W color TV, a 300-W refrigerator, a 900-W hairdryer, and a 200-W computer (including the monitor). If there is a 15-A circuit breaker in the 120-V power line, will the breaker "trip" (i.e., create an open circuit)?

32. ■■ Suppose that the resistor arrangement in Fig. 18.25 is connected to a 12-V battery. What will be (a) the current in each resistor, (b) the voltage drop across each resistor, and (c) the total power delivered?

33. ■■ To make hot tea, you use a 500-W heater connected to a 120-V line to heat 0.20 kg of water from 20°C to 80°C. Assuming that there is no heat loss other than that delivered to the water, how long does this process take?

34. ■■ The terminals of a 6.0-V battery are connected to points A and B in Fig. 18.26. (a) How much current is in each resistor? (b) How much power is delivered to each? (c) Compare the sum of the individual powers with the power delivered to the equivalent resistance for the circuit.

35. ■■ Lightbulbs with the power ratings (expressed in watts) given in ▼Fig. 18.30 are connected in a circuit as shown. (a) What current does the voltage source deliver to the circuit? (b) Find the power delivered to each bulb. (Take the bulbs' resistances to be the same as at their normal operating voltage.)

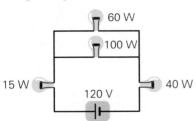

▲ **FIGURE 18.30 Watt's up?** See Exercise 35.

36. ■■■ Two resistors R_1 and R_2 are in series with a 7.0-V battery. If R_1 has a resistance of 2.0 Ω and R_2 receives energy at the rate of 6.0 W, what is (are) the circuit's current(s)?

37. ■■■ For the circuit in ▼Fig. 18.31, find (a) the current in each resistor and (b) the voltage across each resistor.

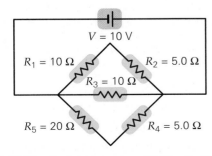

▲ **FIGURE 18.31   Resistors and currents**   See Exercise 37.

38. ■■■ What is the total power delivered to the circuit shown in ▼Fig. 18.32?

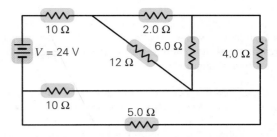

▲ **FIGURE 18.32   Power dissipation**   See Exercise 38.

39. ■■■ What is the equivalent resistance of the arrangement shown in ▼Fig. 18.33?

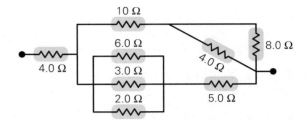

▲ **FIGURE 18.33   Equivalent resistance**   See Exercise 39.

40. ■■■ The circuit shown in ▼Fig. 18.34, named *Wheatstone bridge*, after Sir Charles Wheatstone (1802–1875), is used to measure resistance without the corrections sometimes necessary when using ammeter–voltmeter measurements. (See, for example, Exercises 82 and 83.) The resistances R_1, R_2, and R_s are known, and R_x is the unknown resistance to be measured; R_s is variable and is adjusted until the bridge circuit is balanced, when the galvanometer (G) shows a zero reading (i.e., no current through the galvanometer branch). Show that when the bridge is balanced, R_x is given by

$$R_x = \left(\frac{R_2}{R_1}\right)R_s$$

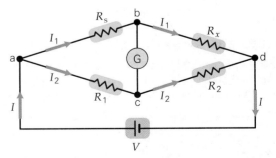

▲ **FIGURE 18.34   Wheatstone bridge**   See Exercise 40.

18.2 Multiloop Circuits and Kirchhoff's Rules

41. A multiloop circuit has more than one (a) junction, (b) branch, (c) current, or (d) all of the preceding.

42. By our sign convention, if a resistor is traversed in the direction of the current, (a) the current is negative, (b) the current is positive, (c) the voltage is negative, or (d) the voltage is positive.

43. **CQ** If you traverse a battery from its negative terminal (cathode) to its positive terminal (anode), is the voltage positive or negative?

44. **CQ** Kirchhoff's junction theorem is based on the conservation of charge. Explain this theorem with a simple example.

45. **CQ** Kirchhoff's loop theorem is based on the conservation of energy. Explain this theorem with a simple example.

46. ■ Traverse loop 3 *opposite* to the direction shown in Fig. 18.10, and demonstrate that the resulting equation is the same as if you had followed the direction of the arrows.

47. ■ For the circuit shown in Fig. 18.10, reverse the directions of loops 1 and 2, and demonstrate that equations equivalent to those in Example 18.5 are obtained.

48. ■ Use Kirchhoff's loop theorem to find the current in each resistor in Fig. 18.28.

49. ■ Apply Kirchhoff's rules to the circuit in Fig. 18.29 to find the current in each resistor.

50. **IE** ■ Two batteries, with terminal voltages of 10 V and 4 V, respectively, are connected with their positive terminals together. A 12-Ω resistor is wired between their negative terminals. (a) The current in the resistor is (1) 0 A, (2) between 0 A and 1.0 A, or (3) greater than 1.0 A. Why? (b) Use Kirchhoff's loop theorem to find the current in the circuit and the power delivered to the resistor. (c) Compare this result with the power output of each battery.

51. ■■ Using Kirchhoff's rules, find the current in each resistor in ▼ Fig. 18.35.

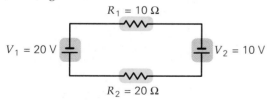

▲ **FIGURE 18.35 Single-loop circuit** See Exercise 51.

52. ■■ Apply Kirchhoff's rules to the circuit in ▼ Fig. 18.36, and find (a) the current in each resistor and (b) the rate at which energy is being delivered to the 8.0-Ω resistor.

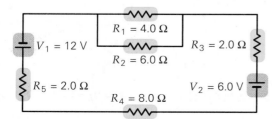

▲ **FIGURE 18.36 A loop in a loop** See Exercise 52.

53. ■■ Find the current in each resistor in the circuit shown in ▼ Fig. 18.37.

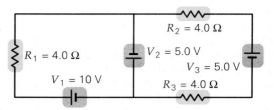

▲ **FIGURE 18.37 Double-loop circuit** See Exercise 53.

54. ■■ Find the currents in the circuit branches in ▼ Fig. 18.38.

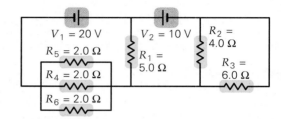

▲ **FIGURE 18.38 How many loops?** See Exercise 54.

55. ■ For the multiloop circuit shown in ▼ Fig. 18.39, what is the current in each branch?

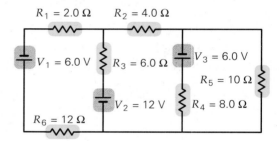

▲ **FIGURE 18.39 Triple-loop circuit** See Exercise 55.

18.3 RC Circuits

56. When a capacitor discharges through a resistor, the voltage across the capacitor is a maximum (a) at the beginning of the process, (b) near the middle of the process, (c) at the end of the process, or (d) after one time constant.

57. When a capacitor discharges through a resistor, the current in the circuit is a minimum (a) at the beginning of the process, (b) near the middle of the process, (c) at the end of the process, or (d) after one time constant.

58. **CQ** In what ways can you increase the time constant of an RC series circuit?

59. **CQ** Explain why the voltage across a charging capacitor is at its minimum at the beginning of the charging process.

60. **CQ** Explain why the current in a charging RC circuit decreases as the capacitor is being charged.

61. ■ In Fig. 18.11b, the switch is closed at $t = 0$, and the capacitor begins to charge. What is the voltage across the resistor and across the capacitor, expressed as fractions of V_0 (to two significant figures), (a) just after the switch is closed, (b) after one time constant has elapsed, and (c) after many time constants have lapsed?

62. ■ A capacitor in a single-loop RC circuit is charged to 63% of its final voltage in 1.5 s. Find (a) the time constant for the circuit and (b) the percentage of the circuit's final voltage after 15 s. (Use five significant figures to show how close this percentage is to 100%.)

63. IE ■ In a flashing neon light, a certain time constant is desired. (a) To increase this time constant, you should (1) increase the capacitance, (2) decrease the capacitance, or (3) not use a capacitor. Why? (b) If a 2.0-s time constant is desired and you have a 1.0-μF capacitor, what resistance should you use in the circuit?

64. ■■ How many time constants will it take for an initially charged capacitor to be discharged to half of its initial voltage?

65. ■■ A 1.00-μF capacitor, initially charged to 12 V, is connected in series with a resistor. (a) What resistance is necessary to cause the capacitor to have only 37% of its initial charge 1.50 s after starting to discharge? (b) What is the voltage across the capacitor at $t = 3\tau$ if the capacitor is charged by the same battery through the same resistor?

66. ■■ An RC circuit with $C = 40\ \mu$F and $R = 6.0\ \Omega$ has a 24-V source in it. With the capacitor initially uncharged, an open switch in the circuit is closed. (a) What is the voltage across the resistor immediately afterward? (b) What is the voltage across the capacitor at that time? (c) What is the current in the resistor at that time?

67. ■■ (a) For the circuit in Exercise 66, after the switch has been closed for $t = 4\tau$, what is the charge on the capacitor? (b) After a long time has passed, what are the voltages across the capacitor and the resistor?

68. ■■■ An RC circuit with a 5.0-MΩ resistor and a 0.40-μF capacitor is connected to a 12-V source. If the capacitor is initially uncharged, what is the change in voltage across it between $t = 2\tau$ and $t = 4\tau$?

69. ■■■ A 3.0-MΩ resistor is connected in series with a 0.28-μF capacitor. This arrangement is then connected across four 1.5-V batteries (also in series). (a) What are the initial voltage across the capacitor and the initial current in the circuit? (b) How much charge is on the capacitor after 4.0 s?

18.4 Ammeters and Voltmeters

70. A voltmeter has a (a) large shunt resistance, (b) large multiplier resistance, (c) small shunt resistance, or (d) small multiplier resistance.

71. An ammeter has a (a) large shunt resistance, (b) large multiplier resistance, (c) small shunt resistance, or (d) small multiplier resistance.

72. To correctly measure the voltage across a circuit element, a voltmeter should be connected (a) in series with the element, (b) in parallel with the element, (c) between the high potential side of the element and ground, or (d) none of the preceding.

73. CQ (a) What would happen if an ammeter were connected in parallel with a current-carrying circuit element? (b) What would happen if a voltmeter were connected in series with a current-carrying circuit element?

74. CQ What is the resistance of an ideal voltmeter, and why?

75. CQ If designed properly, should an ammeter have a large or small resistance? Why?

76. IE ■ A galvanometer with a full-scale sensitivity of 2000 μA has a coil resistance of 100 Ω. It is to be used in an ammeter with a full-scale reading of 30 A. (a) Should you use (1) a shunt resistor, (2) a zero-resistor, or (3) a multiplier resistor? Why? (b) What is the necessary resistance?

77. IE ■ The galvanometer in Exercise 76 is to be used in a voltmeter with a full-scale reading of 15 V. (a) Should you use (1) a shunt resistor, (2) a zero-resistor, or (3) a multiplier resistor? Why? (b) What is the required resistance?

78. ■ A galvanometer with a full-scale sensitivity of 600 μA and a coil resistance of 50 Ω is to be used to build an ammeter designed to read 5.0 A at full scale. What is the required shunt resistance?

79. ■■ A galvanometer has a coil resistance of 20 Ω. A current of 200 μA deflects the needle through 10 divisions at full scale. What resistance is needed to convert the galvanometer to a full-scale 10-V voltmeter?

80. ■■ An ammeter has a resistance of 1.0 mΩ. Find the current in the ammeter when it is properly connected to a 10-Ω resistor and a 6.0-V source. (Express your answer to five significant figures to show how it differs from 0.60 A.)

81. ■■ A voltmeter has a resistance of 30 kΩ. What is the current in the meter when it is properly connected across a 10-Ω resistor that is hooked to a 6.0-V source?

82. IE ■■■ An ammeter and a voltmeter can measure the value of a resistor. Suppose that the ammeter is connected in series with the resistor and that the voltmeter is placed across the resistor only. (a) For accurate measurement, the internal resistance of the voltmeter should be (1) zero, (2) equal to the resistance to be measured, or (3) infinite. Why? (b) Show that the resistance is then

$$R = \frac{V}{I - (V/R_V)}$$

where V is the voltage measured by the voltmeter, I is the current measured by the ammeter, and R_V is the resistance of the voltmeter.

83. **IE ■■■** An ammeter and a voltmeter can measure the value of a resistor. Suppose that the ammeter is connected in series with the resistor and that the voltmeter is placed across both the ammeter and the resistor. (a) For accurate measurement, the internal resistance of the ammeter should be (1) zero, (2) equal to the resistance to be measured, or (3) infinite. Why? (b) Show that the resistance is then

$$R = (V/I) - R_A$$

where V is the voltage measured by the voltmeter, I is the current measured by the ammeter, and R_A is the resistance of the ammeter.

18.5 Household Circuits and Electrical Safety

84. The ground wire in household wiring (a) is a current-carrying wire, (b) is at a voltage of 240 V from one of the "hot" wires, (c) carries no current, or (d) none of the preceding.

85. A dedicated grounding wire (a) is the basis for the polarized plug, (b) is necessary for a circuit breaker, (c) normally carries no current, or (d) none of the preceding.

86. **CQ** In terms of electrical safety, what is wrong with the circuit in ▼ Fig. 18.40, and why?

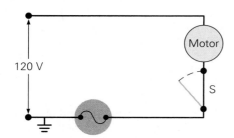

▲ **FIGURE 18.40 A safety problem?** See Exercise 86.

87. **CQ** The severity of bodily injury from electrocution depends on the magnitude of the current and its path, yet you commonly see signs that warn "Danger: High Voltage" (▼ Fig. 18.41). Shouldn't such signs be changed to refer to high current? Explain.

▲ **FIGURE 18.41 Danger—high voltage** Shouldn't it be "high current" instead of "high voltage?" See Exercise 87.

88. **CQ** Explain why it is perfectly safe for birds to perch with both feet on the same high-voltage wire, even if the insulation is worn through.

89. **CQ** After a collision with a power pole, you are trapped in your car, with a high-voltage line (frayed insulation) in contact with the hood of the car. Is it safer to step out of the car one foot at a time or to jump with both feet leaving the car at the same time? Explain your reasoning.

90. **CQ** Most electrical codes require the metal case of an electric clothes dryer to have a wire running from the case to a nearby faucet (or any metal plumbing). Explain why.

Additional Exercises

91. Two 50-Ω resistors are connected in parallel. That arrangement is then connected to a third 50-Ω resistor in series. What is the effective resistance?

92. Two 100-Ω resistors are connected in series. That arrangement is then connected to a third 100-Ω resistor in parallel. What is the effective resistance?

93. A 4.0-Ω resistor and a 6.0-Ω resistor are connected in series. A third resistor is connected in parallel with the 6.0-Ω resistor. The whole arrangement gives a total equivalent resistance of 7.0 Ω. What is the value of the third resistor?

94. Find the current in each resistor in the circuit in ▼ Fig. 18.42.

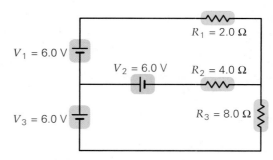

▲ **FIGURE 18.42 Kirchhoff's rules** See Exercise 94.

95. Resistors of 30 Ω and 15 Ω are connected in parallel in a circuit with a 9.0-V battery. How much energy is delivered each second to both resistors (together)?

96. Four resistors are connected to a 90-V source as shown in ▼ Fig. 18.43. (a) Which resistor(s) receive(s) the most power, and how much? (b) What is the total power delivered to the circuit from the power supply?

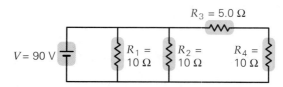

▲ **FIGURE 18.43 How much power is delivered?** See Exercise 96.

97. Four resistors are connected in a circuit with a 110-V source as shown in ▼Fig. 18.44. (a) What is the current in each resistor? (b) How much power is delivered to each resistor?

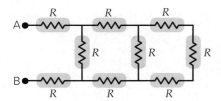

▲ **FIGURE 18.44 Joule-heat losses** See Exercise 97.

98. Nine resistors, each of value R, are connected in a "ladder" fashion as shown in ▼Fig. 18.45. What is the effective resistance of this network between points A and B?

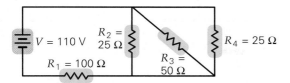

▲ **FIGURE 18.45 A resistance ladder** See Exercise 98.

99. In Exercise 98, if $R = 10\ \Omega$ and a 12.0-V battery is connected from point A to point B, how much current is in each resistor?

100. The time between flashes of the neon tube, or the period of the oscillator circuit, in Fig. 18.14 is the time it takes for the voltage to rise from V_m to V_b. Show that the period of such an oscillator is given by

$$T = t_b - t_m = RC \ln\left(\frac{V_o - V_m}{V_o - V_b}\right)$$

where V_o is the maximum voltage, or the voltage of the battery in the circuit.

101. ▼Fig. 18.46 shows the workings of a *potentiometer*, a very accurate device for determining emf. It consists of three batteries, an ammeter, and several resistors, including a uniform long wire that can be "tapped" for a specific fraction of its total resistance. $\mathscr{E}_0$ is the emf of a working battery, $\mathscr{E}_1$ designates a battery with a precisely known emf, and $\mathscr{E}_2$ a battery whose emf is unknown. The switch S is thrown toward battery 1, and the point T (for "tapped") is moved along the resistor until the ammeter reads zero. The resistance of this arrangement is R_1. This procedure is repeated with the switch thrown toward battery 2, and the point T is moved to T' until the ammeter again reads zero. The resistance of this arrangement is R_2. Show that

$$\mathscr{E}_2 = \frac{R_2}{R_1}\mathscr{E}_1$$

102. Two 100-W lightbulbs are connected to a 110-V power source. What are the current in each bulb and the power delivered if the bulbs are connected (a) in series and (b) in parallel?

103. If a 120-V source were connected across the leads (open ends) of the circuit shown in ▼Fig. 18.47, how much current would be drawn from it?

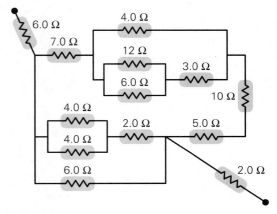

▲ **FIGURE 18.47 Connect to voltage source**
See Exercise 103.

104. If a combination of three 30-Ω resistors receives energy at the rate of 3.2 W when connected to a 12-V battery, how are the resistors connected in the circuit?

105. A battery has three cells, each with an internal resistance of 0.020 Ω and an emf of 1.50 V. The battery is connected in parallel with a 10.0-Ω resistor. (a) Determine the voltage across the resistor. (b) How much current is in each cell? (The cells in a battery are in series.)

106. A galvanometer with an internal resistance of 50 Ω and a full-scale sensitivity of 200 μA is used to construct a multirange voltmeter. What values of multiplier resistors allow for three full-scale voltage readings of 20 V, 100 V, and 200 V? (See Fig. 18.18b.)

107. A galvanometer with an internal resistance of 100 Ω and a full-scale sensitivity of 100 μA is used to construct a multirange ammeter. What values of shunt resistors allow for three full-scale current readings of 1.0 A, 5.0 A and 10 A? (See Fig. 18.18a.)

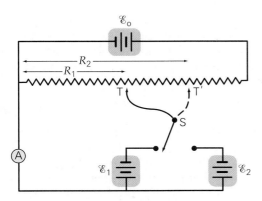

▲ **FIGURE 18.46 The potentiometer** See Exercise 101.

CHAPTER 19

Magnetism

INSIGHT
- Magnetism in Nature

When one thinks of magnetism, one tends to think of an attraction. For example, it is well known that certain things can be picked up with a magnet. Most of us have probably encountered magnetic latches that hold cabinet doors shut or used magnets to stick notes to the refrigerator. It is less likely that one thinks of repulsion. Yet repulsive magnetic forces exist—and they can be just as useful as attractive ones.

In that regard, the opening photo for this chapter shows an interesting example. At first glance, the vehicle looks like an ordinary locomotive. But where are the wheels? In fact, it isn't a conventional train at all, but a high-speed magnetically levitated train. The train doesn't physically touch the "rails." Rather it "floats" above them, supported by repulsive forces produced by powerful magnets. The advantages are obvious: with no

wheels, there is no rolling friction and no bearings to lubricate—in fact, there are very few moving parts of any kind. Magnetic forces are also used to accelerate and decelerate the train.

But where do the magnetic forces come from? For centuries, the attractive properties of magnets were attributed to supernatural forces. Magnetism is now associated with electricity, because physicists discovered that both are really different aspects of a single force: the electromagnetic force. Electromagnetism is put to use in motors, generators, radios, telephones, magnetically driven trains, and many other familiar applications. In the future, the further development of high-temperature superconducting materials (Chapter 17) may open the way for the practical application of many more devices that are now found only in the laboratory or as experimental prototypes.

Although electricity and magnetism are manifestations of the same fundamental force, it is instructive first to consider them individually and then put them together, so to speak (as electromagnetism). This chapter and the next will investigate magnetism and its intimate relationship to electricity.

19.1 Magnets, Magnetic Poles, and Magnetic Field Direction

OBJECTIVES: To (a) learn the force rule between magnetic poles and (b) explain how the direction of a magnetic field is determined with a compass.

One of the features of a common bar magnet is that it has two "centers" of force, called *poles*, at or near each end of the magnet (▶Fig. 19.1). To avoid confusion with the electric charge notation of positive and negative, these poles are called north (N) and south (S). This terminology comes from the early use of the magnetic compass; that is, to determine direction. The north pole of a compass magnet was historically defined as the *north-seeking* end—the end that tends to point *north* on the Earth.

By using two bar magnets, the attractive and repulsive forces acting between the ends of the magnets can be determined experimentally. Each pole of a bar magnet is attracted to the opposite pole of the other and repelled by the same pole of the other. Analogous to the charge–force law is the **pole–force law**, or **law of poles**:

Like magnetic poles repel each other, and unlike magnetic poles attract each other (▼Fig. 19.2).

One immediate (and sometimes confusing) result of the historical definition of a north pole has to do with the Earth's magnetic field. Since the north pole of a bar magnet is attracted to the Earth's north polar region, that region must be acting, magnetically speaking, as a south pole. (See Section 19.8.)

Even though, at first glance, a bar magnet may look like the exact magnetic analogue of the electric dipole, there are significant and fundamental differences between the two. For example, magnetic poles always occur in pairs, never singly. Two opposite magnetic poles, such as those of a bar magnet, form a *magnetic dipole*. You might think that breaking a bar magnet in half would yield two isolated poles. However, the resulting pieces of the magnet always turn out to be just two shorter magnets, *each with its own set of north and south poles*. While a single magnetic pole (a *magnetic monopole*) could exist in theory, it has yet to be found experimentally.

The fact that there is no magnetic analogy to electric charge provides a strong hint about the differences between electric and magnetic fields, even though, superficially, magnetism appears to be similar to electricity. For example, the source of magnetism is actually the electric charge. However, as we will see in Sections 19.6 and 19.7, one way magnetism is produced is when electric charges are *in motion*, such as electric currents in circuits and orbiting (or spinning) atomic electrons. The latter is actually the source of the bar magnet's field (Section 19.3).

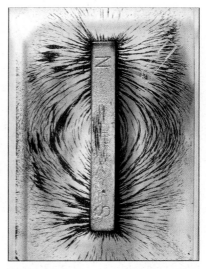

▲ **FIGURE 19.1 Bar magnet**
The iron filings indicate the poles, or centers of force, of a common bar magnet. The compass direction designates these poles as north (N) and south (S). (See Fig. 19.3.)

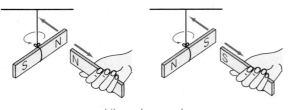 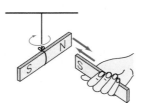

Like poles repel   Unlike poles attract

◀ **FIGURE 19.2 The pole–force law, or law of poles** Like poles (N–N or S–S) repel, and unlike poles (N–S) attract.

▼ FIGURE 19.3 Magnetic fields
(a) Magnetic field lines can be traced and outlined by using iron filings or a compass, as shown in the case of the magnetic field due to a bar magnet. The filings behave like tiny compasses and line up with the field. The closer together the field lines, the stronger is the magnetic field. **(b)** (Below) Iron filing pattern for the magnetic field between unlike poles; the field lines converge. **(c)** Iron filing pattern for the magnetic field between like poles; the field lines diverge.

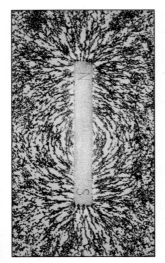

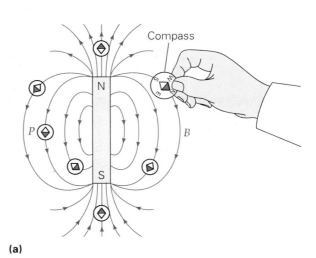

(a)

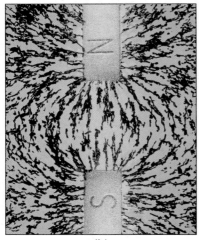

(b)

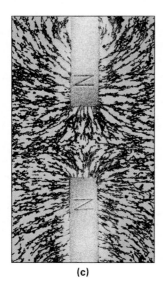

(c)

PHYSLET® ILLUSTRATION

The Field of a Bar Magnet

Magnetic Field Direction

The historical approach to analyzing a bar magnet's magnetic field was to view either pole as a magnetic version of "charge" and to try to express the magnetic force between poles in a mathematical form similar to Coulomb's law for the electric force (Chapter 15). In fact, Coulomb did develop such a law, using magnetic pole strengths in place of electric charge. However, this law is now rarely used, because it does not fit our modern understanding that is single magnetic poles have never been found. Instead, the modern description uses the concept of the *magnetic field*.

Recall that electric charges produce an electric field, which can be represented by electric field lines. The electric field (vector) is defined as the force per unit charge, $E = F_e/q_o$. Similarly, magnetic interactions can be described in terms of the **magnetic field**, a vector quantity represented by the symbol **B**. Just as electric fields exist in the neighborhood of electric charges, magnetic fields surround permanent magnets. The pattern of magnetic field lines surrounding a magnet can be made visible by sprinkling iron filings over a magnet that is covered with a piece of paper or glass (Fig. 19.1). Because of the magnetic field, the iron filings become induced magnets and line up in the direction of the field, behaving like small compass needles.

To describe the magnetic field, both magnitude (sometimes called "strength") and direction must be specified. The direction of a magnetic field (usually referred to as a "*B* field") is defined in terms of a compass that has been calibrated for direction in the Earth's magnetic field:

> The direction of a magnetic field (**B**) at any location is the direction that the north pole of a compass would point if placed at that location.

This definition provides a method for mapping a magnetic field by moving a small compass to various locations in the field. At any location, the compass needle will line up in the direction of the *B* field that exists there. If the compass is then moved in the direction in which its needle (the north end) points, the path of the needle traces out a *magnetic field line*, as illustrated in ▲Fig. 19.3a.

Because the north end of a compass points away from the north pole of a bar magnet, the field lines of a bar magnet point away from its north pole and towards its south pole. The rules that govern the interpretation of the magnetic field lines are the same that apply to electric field lines:

> The closer together the field lines, the stronger is the field. At any location, the direction tangent to the field line in which the north end of a compass points is the direction of the field.

Note that the concentration of iron filings in the pole regions (Fig. 19.3b,c) indicates closely spaced field lines and therefore a relatively strong magnetic field compared with the field in other places around the magnet. As for the field direction, note that just outside the middle of the magnet, for example, the field points vertically down, tangent to the field line at that point, as expected (Fig. 19.3a, point P).

You might think that the magnitude of **B** would be defined as the magnetic force per unit pole strength, in analogy to **E**. However, since magnetic monopoles don't exist, the magnitude of the magnetic field is defined in terms of the magnetic force exerted on a moving electric charge, as discussed in the next section.

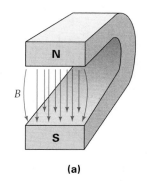

(a)

19.2 Magnetic Field Strength and Magnetic Force

OBJECTIVES: To (a) define magnetic field strength and (b) determine the magnetic force exerted by a magnetic field on a moving charged particle.

Experiments indicate that one of the important quantities in determining the *magnetic* force on a particle is the *electric* charge on that particle. That is, there seems to be some connection between *electrical* properties of objects and how they respond to *magnetic* fields. The study of interactions between electrically charged particles and magnetic fields is called **electromagnetism**.

Let us consider the following electromagnetic interaction. Suppose a positively charged particle ($+q$) is moving at a constant velocity as it enters a uniform magnetic field so that its velocity is perpendicular to the field. (A fairly uniform magnetic field exists between the poles of a "horseshoe" magnet, as shown in ▶ Fig. 19.4a.) When it enters the field, the particle is *deflected* into a curved path, which would be a circular arc if the field were perfectly uniform (Fig. 19.4b).

From our study of circular motion (Section 7.3), for a particle to move in a circular orbit, there must exist a centripetal force that is always perpendicular to the particle's velocity. But what gives rise to this force here? No electric field is present. The gravitational force is too weak to cause such a deflection; furthermore, it would deflect the particle into a parabolic arc, not a circular one. Evidently, the force is due to the interaction of the moving charge and the magnetic field. This indicates that *a magnetic field can exert a force on a moving electrically charged particle.*

From experimental observation, the magnitude of the magnetic force is directly proportional to the particle's charge and speed. For the special case in which the particle's velocity (**v**) is perpendicular to the magnetic field (**B**), the magnitude of the field (the magnetic field strength) is defined as

$$B = \frac{F}{qv} \quad \begin{array}{l} \textit{(valid only for } \mathbf{v} \\ \textit{perpendicular to } \mathbf{B}) \end{array} \quad (19.1)$$

SI unit of magnetic field: newton per ampere-meter [N/(A·m), or tesla (T)]

Thus *B* means the magnetic force exerted on a charged particle per unit charge (coulomb) and per unit speed (meter per second). From Eq. 19.1, the magnetic field strength has units of N/(C·m/s) or N/(A·m), since the ampere is the coulomb per second. This combination of units is named the **tesla** (T) after Nikola Tesla (1856–1943), a famous early researcher in magnetic fields. Thus, 1 T = 1 N/(A·m). Most everyday magnetic field strengths, such as those from permanent magnets, are much smaller than 1 T. In these situations, it is common to express magnetic field strengths in terms of a much smaller (non-SI) unit called the *gauss* (G), which is one ten-thousandth of a tesla. Hence, 1 G = 0.0001 T = 10^{-4} T. For example, the Earth's magnetic field is on the order of several tenths of a gauss. On the other hand, conventional laboratory magnets have produced fields as high as 3 T, and superconducting magnets up to 25 T.

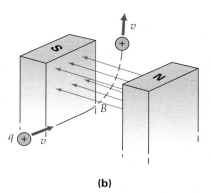

(b)

▲ **FIGURE 19.4 Force on a moving charged particle** (a) A horseshoe magnet, created by bending a permanent bar magnet, produces a fairly uniform field between its poles. (b) When a charged particle enters a magnetic field, the particle is acted on by a force whose direction is obvious by the deflection of the particle from its original path.

Note: The magnetic field plays a vital role in magnetic resonance imaging (MRI), a technique widely used in medical diagnostics. (See Chapter 28.)

Once the magnetic field strength in a location has been determined (from Eq. 19.1), the force on any charged particle moving at any speed can be found (if **v** is perpendicular to **B**).* Eq. 19.1 can be solved to find the force:

$$F = qvB \qquad \text{(\textit{valid only for }\textbf{v}\\ \textit{perpendicular to }\textbf{B})} \qquad (19.2)$$

In the more general case, the particle's velocity may *not* be perpendicular to the magnetic field. Then, the magnitude of the force depends also on the sine of the angle (called θ) between the velocity vector and the magnetic field vector. In general, the magnitude of the magnetic force is

$$F = qvB \sin\theta \qquad \text{\textit{magnetic force on}\\ \textit{a charged particle}} \qquad (19.3)$$

Thus, the magnetic force is zero when **v** and **B** are parallel ($\theta = 0°$) or oppositely directed ($\theta = 180°$), since $\sin 0° = \sin 180° = 0$. The force attains its maximum value when these two vectors are perpendicular. With $\theta = 90°$ ($\sin 90° = 1$), this maximum value is $F = qvB \sin 90° = qvB$.

The Right-Hand Force Rule for Moving Charges

The *direction* of the magnetic force on any moving charged particle is determined by the orientation of the particle's velocity relative to the magnetic field. The magnetic force direction is given by the **right-hand force rule** (◄Fig. 19.5a):

> When the fingers of the right hand are pointed in the direction of a charged particle's velocity **v** and then curled (through the smallest angle) toward the vector **B**, the extended thumb points in the direction of the force **F** that acts on a *positive* charge. If the particle has a negative charge, the magnetic force is in the direction opposite to that of the thumb.

You might imagine the fingers of the right hand to be physically turning or rotating the vector **v** into **B** so that **v** and **B** are aligned.

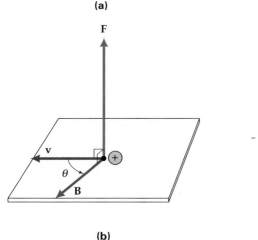

(a)

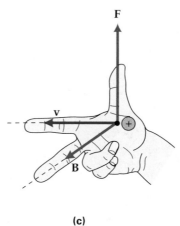

(b)　　　　　　　　　　(c)

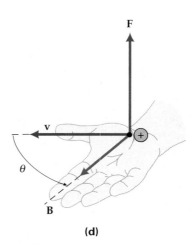

(d)

▲ **FIGURE 19.5 Right-hand rules for magnetic force** (a) When the fingers of the right hand are pointed in the direction of **v** and then curled toward the direction of **B**, the extended thumb points in the direction of the force **F** on a *positive* charge. (b) The magnetic force is always perpendicular to the plane of **B** and **v** and thus is always perpendicular to the direction of the particle's motion. (c) When the extended forefinger of the right hand points in the direction of **v** and the middle finger points in the direction of **B**, the extended right thumb points in the direction of **F** on a *positive* charge. (d) When the fingers of the right hand are pointed in the direction of **B** and the thumb in the direction of **v**, the palm points in the direction of the force **F** on a *positive* charge. (Regardless of the rule you employ, remember to reverse the direction for a negative charge.)

*Strictly speaking, the speeds must be considerably less than the speed of light to avoid relativistic complications (Chapter 26).

Notice that the magnetic force is always *perpendicular to the plane formed by the vectors* **v** *and* **B** (Fig. 19.5b). Because the force is perpendicular to the particle's direction of motion (**v**), it cannot do any work on the particle. (From the definition of work, $W = Fd \cos 90° = 0$.) Therefore, a magnetic field alone cannot change the speed (kinetic energy) of a particle—only its direction.

Several common alternative equivalent right-hand force rules are shown in Fig. 19.5c. For negative charges, it is suggested that you start by assuming that the charge is positive. Then determine the direction of the force for that case, using the right-hand force rule. Lastly, *reverse* this direction to determine the direction of the actual force. To see how this rule is applied to both signs of charges, consider the next conceptual example.

Conceptual Example 19.1 ■ Even "Lefties" Use the Right-Hand Rule

In a linear particle accelerator, a beam of protons travels horizontally northward. To deflect the protons eastward in a uniform magnetic field, the field should point in which direction? (a) vertically downward, (b) west, (c) vertically upward, or (d) south.

Reasoning and Answer. Since the force on a moving charged particle is perpendicular to the plane of **v** and **B**, the magnetic field cannot be in the horizontal plane; if it were, it would deflect the protons down or up. (Verify this with the right-hand force rule.) Thus, we can eliminate choices (b) and (d). Next, use the right-hand force rule (assuming a positive charge) to see if **B** could be downward (answer a). You should verify (make a sketch) that for a downward magnetic field, the force would be to the west. Hence, the answer must be (c). The magnetic field must point upward to deflect the protons to the east. You should verify that this answer is correct, using the right-hand force rule.

Follow-up Exercise. What direction would the beam deflect if it were actually composed of electrons moving in the opposite direction as the protons in this Example? (*Answers to all Follow-up Exercises are at the back of the text.*)

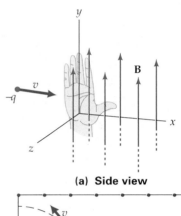

(a) Side view

As discussed, charged particles entering uniform magnetic fields travel in circular arcs. Consider the situation in Example 19.2 carefully.

Example 19.2 ■ Going Around in Circles: Force on a Moving Charge

A particle with a charge of -5.0×10^{-4} C and a mass of 2.0×10^{-9} kg moves at a speed of 1.0×10^3 m/s in the $+x$ direction toward a uniform magnetic field of 0.20 T in the $+y$ direction (▶Fig. 19.6a). (a) Which way will the particle start to deflect as it enters the field? (b) What is the force on the particle just as it enters the magnetic field? (c) What is the radius of the circular orbit of the particle while it is in the field?

Thinking It Through. The particle's initial deflection is in the direction of the initial magnetic force. Note that the particle is negatively charged, so care must be taken in applying the right-hand force rule. A circular arc trajectory is expected, since the magnetic force is always perpendicular to the particle's velocity. The magnitude of the magnetic force on a single charge is given by Eq. 19.3. Newton's second law, which relates net force to acceleration, should provide a way to determine the radius of the circular orbit from the magnetic force.

Solution. We list the given data.

Given: $q = -5.0 \times 10^{-4}$ C
$v = 1.0 \times 10^3$ m/s ($+x$ direction)
$m = 2.0 \times 10^{-9}$ kg
$B = 0.20$ T ($+y$ direction)

Find: (a) Initial deflection direction
(b) Initial magnetic force **F**
(c) Radius r of the orbit

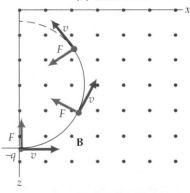

(b) Top view

▲ **FIGURE 19.6 Path of a charged particle in a magnetic field** **(a)** A charged particle entering a uniform magnetic field will be deflected—here toward the *xy*-plane by the right-hand rule, because the charge is negative. **(b)** In the field, the force is always perpendicular to the particle's velocity. The particle moves in a circular path if the field is constant and the particle enters the field perpendicularly to its direction. (See Example 19.2.)

(a) By the right-hand force rule, the force on a positive charge would be in the $+z$ direction (see direction of palm, Fig 19.6a). Since the charge is negative, the force is in the opposite direction; thus, the particle will begin to deflect in the $-z$ direction.

(b) The magnitude of the force can be determined from Eq. 19.3. Because only the magnitude is of interest, the sign of q can be neglected. We have

$$F = qvB \sin \theta$$
$$= (5.0 \times 10^{-4}\,\text{C})(1.0 \times 10^3\,\text{m/s})(0.20\,\text{T})(\sin 90°) = 0.10\,\text{N}$$

(c) Since the magnetic force is the only force acting on the particle, it is also the net (centripetal) force that causes the circular path (Fig. 19.6b). For circular motion, Newton's second law reads

$$F_{\text{net}} = ma_c$$

Now, we substitute the expression for the magnetic force (from Eq. 19.2, since $\theta = 90°$) for the net force and the expression for centripetal acceleration ($a_c = v^2/r$; see Section 7.3):

$$qvB = \frac{mv^2}{r} \quad \text{or} \quad r = \frac{mv}{qB}$$

Finally, inserting the numerical values, we have

$$r = \frac{mv}{qB} = \frac{(2.0 \times 10^{-9}\,\text{kg})(1.0 \times 10^3\,\text{m/s})}{(5.0 \times 10^{-4}\,\text{C})(0.20\,\text{T})} = 2.0 \times 10^{-2}\,\text{m} = 2.0\,\text{cm}$$

Follow-up Exercise. In this Example, if the particle had been a proton traveling initially in the $+z$ direction, (a) in what direction would it initially be deflected? (b) If the radius of its circular path were 10 cm and its speed were 1.0×10^6 m/s, what would be the magnetic field strength?

19.3 Applications: Charged Particles in Magnetic Fields

OBJECTIVES: To understand how the magnetic force is used in modern appliances and instruments.

In general, whenever a moving charged particle enters a magnetic field, it is acted upon by a magnetic force. This force deflects the particle in a manner that depends on the particle's mass, charge, and velocity (speed and direction), as well as the strength of the field. This deflective force is of primary importance in many of our common appliances, machines, and instruments.

The Cathode-Ray Tube (CRT): Oscilloscope Screens, Television Sets, and Computer Monitors

The **cathode-ray tube** (**CRT**) is a vacuum tube that is used as a display screen for a laboratory instrument called an *oscilloscope* (◄Fig. 19.7). The basic operation of both the oscilloscope and the television picture tube is shown in ►Fig. 19.8. Electrons are emitted from a hot metallic filament and are accelerated by a voltage applied between the cathode ($-$) and the anode ($+$) in an "electron gun" arrangement. In one kind of design, these instruments use current-carrying coils to produce a magnetic field (Section 19.6), which controls the deflection of the electron beam. As the field strength is varied, the electron beam scans the fluorescent screen in a fraction of a second. When the electrons hit the fluorescent material, they cause the atoms to emit light (Section 27.4). For a black-and-white TV, the signals transmitted from a video camera reproduce an image on the

▲ **FIGURE 19.7 Cathode-ray tube (CRT)** The motion of the deflected beam traces a pattern on a fluorescent screen.

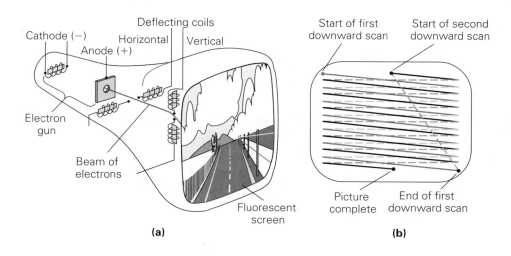

Cathode (−)
Deflecting coils
Horizontal | Vertical
Anode (+)

Electron
gun

Beam of
electrons

Fluorescent
screen

(a)

Start of first
downward scan
Start of second
downward scan

Picture
complete
End of first
downward scan

(b)

◀ **FIGURE 19.8 Television tube**
(a) A television picture tube is a
cathode-ray (electron) tube, or CRT.
The electrons are accelerated
between the cathode and anode and
are then deflected to the proper
location on a fluorescent screen by
magnetic fields produced by
current-carrying coils. **(b)** The beam
scans every other line on the screen
in one downward pass that takes
1/60 s and then scans the lines in
between in a second pass of 1/60 s.
This yields a complete picture of
525 lines in 1/30 s.

screen as a mosaic of light and dark dots, depending on whether the intermittent
beam is on or off at a particular instant.

Producing the images on a color TV or a color computer monitor is somewhat
more involved. A common color picture tube has three beams, one for each of the
primary colors (red, green, and blue; Chapter 25). Phosphor dots on the screen are
arranged in groups of three (triads), with one dot for each primary color. The excitation of the appropriate dots and the resulting emission (fluorescence) of a combination of colors produce a color picture.

The Velocity Selector and the Mass Spectrometer

Have you ever thought about how the mass of an atom or molecule is measured? Electric and magnetic fields provide a way in a **mass spectrometer**
("mass spec" for short). Mass spectrometers perform many functions in modern laboratories. For example, they can be used to track short-lived molecules
in studies of the biochemistry of living organisms. They can also determine the
structure of large organic molecules, to analyze the composition of complex
mixtures, such as a sample of smog-laden air. In criminal cases, forensic
chemists can use the mass spectrometer to identify traces of materials, such as a
streak of paint left over in a car accident. In physics, these instruments can be
used to separate out atoms to establish the age of ancient rocks and human arti-
facts. In modern hospitals, mass spectrometers are essential for measuring and
maintaining the proper balance of gaseous medications, such as anesthetic
gases administered during an operation.

Actually, it is the masses of *ions*, or charged molecules, that are measured.*
Ions with a known charge $(+q)$ are produced by removing electrons from atoms
and molecules. At this point, the resulting beam of ions would have a distribution
of speeds, rather than a single speed. Therefore, before they enter the mass spectrometer, ions with a specific velocity are selected by a *velocity selector*. A velocity
selector consists of an electric field and a magnetic field at right angles. This
arrangement allows particles traveling only at a unique velocity to pass through
undeflected.

The selected velocity is determined by the magnitudes of the crossed electric
and magnetic fields. To see this, consider a positive ion approaching the crossed-
field arrangement at right angles to both fields. The electric field produces a

Note: The difference in mass between
a neutral atom or molecule and its
charged counterparts (ions) amounts
to only the mass of an electron or two
and is therefore negligible under most
circumstances.

*Recall that removing or adding electrons to an atom or molecule produces an ion. However, an
ion's mass is negligibly different from that of its neutral atom, because the electron's mass is very small
compared to the masses of the protons and neutrons in the atomic nuclei.

downward force ($F_e = qE$), and the magnetic field produces an upward force ($F_m = qvB_1$). (You should show each force's direction in ▼ Fig. 19.9.)

If the beam is not to be deflected, the resultant force must be zero; thus, the two forces must be equal and opposite to cancel. Hence,

$$F_e = F_m \quad \text{or} \quad qE = qvB_1$$

which can then be solved for a unique speed that depends on the two field strengths:

$$v = \frac{E}{B_1}$$

If the plates are parallel, the electric field between them is given by $E = V/d$. Thus a more practical version of the previous equation is

$$v = \frac{V}{B_1 d} \quad \begin{array}{l} \textit{selected speed of a} \\ \textit{velocity selector} \end{array} \tag{19.4}$$

Thus, a speed can be selected by choosing the values of V, B_1, and d.

Beyond the velocity selector, the beam passes through a slit into another magnetic field ($\mathbf{B}_2$), which is perpendicular to the direction of the beam. Here the particle beam is bent into a circular arc. The analysis is identical to that in Example 19.2. Therefore,

$$F_{net} = ma_c \quad \text{or} \quad qvB_2 = m\frac{v^2}{r}$$

Using Eq. 19.4, the mass of the particle is given by

$$m = \left(\frac{qdB_1 B_2}{V}\right)r \quad \textit{mass spectrometer} \tag{19.5}$$

The quantity inside the parentheses is a constant (assuming that all ions have the same charge q). Hence, the greater the mass of an ion, the greater is the radius of the ion's circular path. Two circular paths of different radii are shown in Fig. 19.9, which indicates that the beam actually contains ions of two different masses. If the radius of curvature r is measured (by recording the position of the beam with a detector), the mass of the ion can be calculated from Eq. 19.5.

In a mass spectrometer of slightly different design, the detector is at a fixed position (r). In this case, the instrument is operated by varying the magnetic field magnitude (B_2) with time and recording the detector reading as time goes on. Since, with this design, m is proportional to B_2 (note that $m = (qdB_1 r/V)B_2$, and

▶ **FIGURE 19.9 Principle of the mass spectrometer** Ions pass through the velocity selector; only those with a particular velocity ($v = E/B_1$) then enter a magnetic field ($\mathbf{B}_2$). These ions are deflected, with the radius of the circular path depending on the mass and charge of the ion. Paths of two different radii indicate that the beam contains ions of two different masses (assuming that they have the same charge).

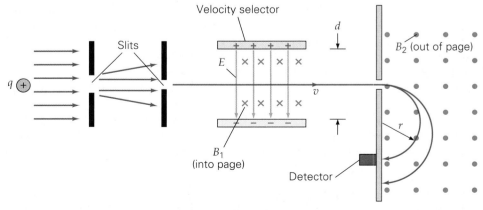

Top view

therefore, $m \propto B_2$), as B_2 varies the detector records the masses and number of ions that enter it as a function of time. Regardless of the design, the result—called a *mass spectrum* (the number of ions plotted against their molecular mass)—is typically displayed on an oscilloscope or computer screen and digitized for storage and analysis purposes (►Fig. 19.10). Consider the following Example of a mass spectrometer arrangement.

Example 19.3 ■ The Mass of a Molecule: a Mass Spectrometer

One electron is removed from a methane molecule before the molecule enters the mass spectrometer arrangement shown in Fig. 19.9. After passing through the velocity selector, the charged molecule has a speed of 1.00×10^3 m/s. It then enters the final magnetic field region, in which the field is 6.70×10^{-3} T, and follows a circular path. The molecule lands on the detector 5.00 cm from the entrance to the field. Determine the mass of this molecule. (Neglect the mass of the electron that is removed.)

Thinking It Through. The centripetal force for the circular arc is provided by the magnetic force acting on the charged molecule. Since the velocity and magnetic field are at right angles, the magnetic force is given by Eq. 19.2. Applying Newton's second law to circular motion, we see that the only unknown is the mass of the molecule.

Solution. First, convert the given data into the SI system.

Given: $q = 1.60 \times 10^{-19}$ C (why?) *Find:* m (mass of a methane molecule)
$r = d/2 = (5.00 \text{ cm})/2 = 0.0250$ m
$B_2 = 6.70 \times 10^{-3}$ T
$v = 1.00 \times 10^3$ m/s

The centripetal force on the molecule ($F_c = mv^2/r$) is provided by the magnetic force ($F_m = qvB_2$); thus,

$$\frac{mv^2}{r} = qvB_2$$

The mass is determined by solving this equation for m and putting in the numerical values:

$$m = \frac{qB_2r}{v} = \frac{(1.60 \times 10^{-19} \text{ C})(6.70 \times 10^{-3} \text{ T})(0.0250 \text{ m})}{1.00 \times 1.0^3 \text{ m/s}} = 2.73 \times 10^{-26} \text{ kg}$$

Follow-up Exercise. In this Example, if the magnetic field between the velocity selector's parallel plates, which are 10.0 mm apart, is 5.00×10^{-2} T, what voltage must be applied to the plates?

Silent Propulsion: Magnetohydrodynamics

In a search for quiet and efficient methods of propulsion at sea, engineers have invented a system based on *magnetohydrodynamics*—the study of the interactions of moving fluids and magnetic fields. This method of propulsion relies on the magnetic force and does not require moving parts—such as motors, bearings, and shafts—that are common to most ships and submarines. To avoid detection, the "silent-running" feature is of particular importance to modern submarine design.

Essentially, seawater enters the front of the unit and is expelled at high speeds from the rear (►Fig. 19.11). A superconducting electromagnet (see Section 19.7) is used to produce a large magnetic field. At the same time, an electric generator produces a large dc voltage, sending a current through the seawater. (Recall that seawater is a good conductor since it has a large concentration of ions such as sodium (Na^+) and chlorine (Cl^-)). The magnetic force on the current pushes the water backward, expelling a jet of water. By Newton's third law, a reaction force pushes forward on the submarine, which then accelerates silently in that direction.

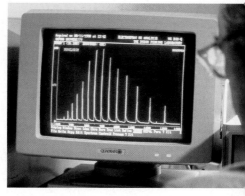

▲ **FIGURE 19.10 Mass spectrometer** Display of a mass spectrometer, with the number of molecules plotted vertically and the molecular mass horizontally. The molecule being analyzed is myoglobin, a protein that stores oxygen in muscle tissue. For myoglobin, each peak on the display represents the mass of an ionized fragment. Such patterns, *mass spectra*, help determine the composition and structure of large molecules. The mass spectrometer can also be used to identify tiny amounts of a molecule in a complex mixture.

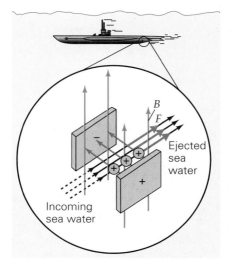

▲ **FIGURE 19.11 Propulsion via magnetohydrodynamics** In magneto-hydrodynamic propulsion, seawater has an electric current passed through it by a dc voltage. A magnetic field exerts a force on the current, pushing the water out of the submarine or boat. The reaction force pushes the vessel in the opposite direction.

19.4 Magnetic Forces on Current-Carrying Wires

OBJECTIVES: **To (a) calculate the magnetic force on a current-carrying wire and the torque on a current-carrying loop and (b) explain the concept of the magnetic moment of a loop or coil.**

An electric charge moving in a magnetic field is acted on by a force, unless it is moving parallel to or directly opposite to the direction of the field. Thus, a current-carrying wire in a magnetic field would be expected to be subject to such a force, because a current is composed of moving charges. The sum of the individual magnetic forces on the moving charges should give the total magnetic force on the wire.

Note: Remember that thinking of a current in terms of a positive charge is simply a useful convention. In reality, negative electrons are the charge carriers in ordinary electric current.

Recall that "conventional current" is thought of as positive charges moving in a wire, as depicted in ▼Fig. 19.12.* Here, the magnetic force is at its maximum, since $\theta = 90°$. In a time t, a charge q would move, on the average, through a length $L = vt$, where v is the average drift velocity. Since all the moving charges (Σq_i) in this length of wire are acted on by a magnetic force in the same direction, the magnitude of the total force on this length of wire, from Eq. 19.2, is

$$F = (\Sigma q_i)vB$$

Substituting L/t for v and rearranging gives

$$F = (\Sigma q_i)\left(\frac{L}{t}\right)B = \left(\frac{\Sigma q_i}{t}\right)LB$$

But $\Sigma q_i/t$ is just the current (I). Therefore,

$$F = ILB \qquad \begin{array}{l}\textit{(current and magnetic field}\\ \textit{are perpendicular)}\end{array} \qquad (19.6)$$

Equation 19.6 gives the maximum force on the wire, because the current is at an angle of 90° to the magnetic field. If the current in the wire makes some angle θ with respect to the field, then the force on the wire will be less. In general, the force on a length of current-carrying wire in a uniform magnetic field is

$$F = ILB \sin \theta \qquad \begin{array}{l}\textit{magnetic force on a}\\ \textit{current-carrying wire}\end{array} \qquad (19.7)$$

Note that if the current in the wire is parallel or oppositely directed to the field, the force on the wire is zero.

The direction of the magnetic force on a current-carrying wire is also given by a right-hand rule. As was the case for individual charged particles, there are sev-

▲ **FIGURE 19.12 Force on a wire segment** Magnetic fields exert forces on wires that carry currents, because electric current is composed of many moving charged particles. The maximum force on a length L of current-carrying wire in a magnetic field is shown, since the angle between the velocity of the charge and the magnetic field is 90°.

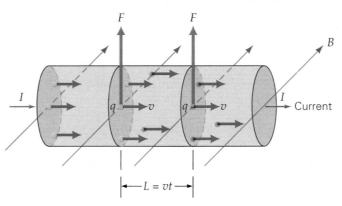

*Use the right-hand force rule to convince yourself that electrons traveling to the left would give the same magnetic force direction.

eral equivalent versions of the **right-hand force rule for a current-carrying wire**, the most common being:

> When the fingers of the right hand are pointed in the direction of the conventional current *I* and then curled toward the vector **B**, the extended thumb points in the direction of the magnetic force on the wire (see ▼Figs. 19.13a and b).

An equivalent alternative is shown in ▶Fig. 19.14.

> When the fingers of the right hand are extended in the direction of the magnetic field and the thumb pointed in the direction of the conventional current *I* carried by the wire, the palm of the right hand points in the direction of the magnetic force on the wire.

Regardless of which rule you choose, they must give the same direction, because they are extensions of the right-hand force rules for individual charges. To see how two current-carrying wires interact magnetically, consider the next Example.

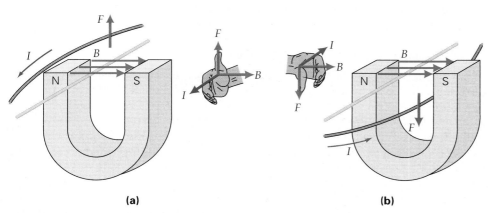

(a)   **(b)**

▲ **FIGURE 19.13   A right-hand force rule for current-carrying wires**   The direction of the force is given by pointing the fingers of the right hand in the direction of the conventional current *I* and then curling them toward **B**. The extended thumb points in the direction of **F**. The force is **(a)** upward and **(b)** downward.

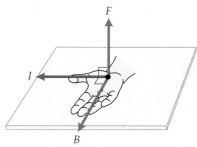

▲ **FIGURE 19.14   An alternative right-hand force rule**   When the fingers of the right hand are extended in the direction of the magnetic field **B** and the thumb is pointed in the direction of the conventional current *I*, then the palm points in the direction of **F**. You should check that this gives the same direction as the equivalent rule shown in Fig. 19.13.

Integrated Example 19.4 ■ Magnetic Forces on Wires Suspended at the Equator

Since a current-carrying wire is acted on by a magnetic force, it would seem possible to suspend such a wire at rest above the ground using the Earth's own magnetic field. (a) In what direction would the current have to be in a long, straight wire (located at the equator) to perform such a feat? (1) up, (2) down, (3) east, or (4) west. Draw a sketch to help you decide. (b) Calculate the current required to suspend the wire, assuming that the Earth's magnetic field is 0.40 G at the equator and the wire is 1.0 m long with a mass of 30 g.

(a) Conceptual Reasoning.   The required magnetic force direction is up, since gravity acts downward (▶Fig. 19.15). The Earth's magnetic field at the equator is parallel to the ground and points north. Because the magnetic force is perpendicular to both the current and the magnetic field, the current cannot be up or down, which eliminates the first two choices. To decide between east and west, simply choose one and see if it works (or doesn't work). Suppose the current is to the west. Using the right-hand force rule shows that the magnetic force acts downward. Since this is incorrect, the only correct answer is (3), easterly. You should verify that this is correct by using the right-hand rule directly.

(b) Thinking It Through.   The mass of the wire is known, and therefore we can calculate its weight. This must be equal and opposite to the magnetic force. The current and the field are at right angles to each other; hence, the magnetic force is given by Eq. 19.6, and from that, the current can be determined.

First we list the data (and convert to SI units at the same time):

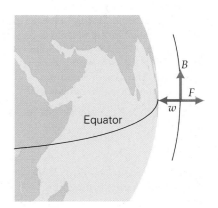

▲ **FIGURE 19.15   Defying gravity by using a magnetic field?**   Near the Earth's equator it is theoretically possible to cancel the pull of gravity with an upward magnetic force on a wire. What must be the direction and magnitude of the current? (See Integrated Example 19.4.)

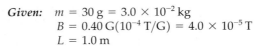
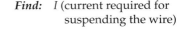

Given: $m = 30\,\text{g} = 3.0 \times 10^{-2}\,\text{kg}$
$\quad\quad\quad B = 0.40\,\text{G}(10^{-4}\,\text{T/G}) = 4.0 \times 10^{-5}\,\text{T}$
$\quad\quad\quad L = 1.0\,\text{m}$

Find: I (current required for suspending the wire)

The wire's weight is $w = mg = 0.29\,\text{N}$. For balance, this equals the magnetic force, or

$$w = ILB$$

Therefore,

$$I = \frac{w}{LB} = \frac{0.29\,\text{N}}{(1.0\,\text{m})(4.0 \times 10^{-5}\,\text{T})} = 7.4 \times 10^{3}\,\text{A}$$

This is a huge current, so suspending the wire in this manner is not a practical idea.

Follow-up Exercise. (a) Using the right-hand force rule, show that the idea of suspending a wire, as shown in this Example, could not work at either the south or north pole of the Earth. (b) In this Example, what would the mass of the wire have to be in order for it to be suspended when carrying a more reasonable current of 10 A? Does this seem like a reasonable mass for a 1-m length of wire?

Torque on a Current-Carrying Loop

Another important use of magnetic forces are forces exerted on a current-carrying loop, such as the rectangular loop in ◄Fig. 19.16a. (The wires connecting the loop to a voltage source are not shown.) Suppose that the loop is free to rotate about an axis passing through opposite sides, as shown. There is no net force or torque due to the forces acting on the pivoted sides of the loop (the sides through which the axis of rotation passes). The forces on them are equal and opposite and are in the plane of the loop and therefore produce no net torque or force. However, the equal and opposite forces on the two sides of the loop *parallel* to the axis of rotation, while not creating a net force, *do* create a net torque.

To see how this works, consider Fig. 19.16b, which is a side view of Fig.19.16a. The magnitude of the magnetic force on each of the nonpivoted sides (length L) is $F = ILB$. Recall that the torque produced by a force (Section 8.2) is given by $\tau = r_\perp F$, where $r_\perp$ is the perpendicular distance (the lever arm) from the axis of rotation to the line of action of the force. From Fig. 19.16b, $r_\perp = \frac{1}{2}w \sin\theta$, where w is the width of the loop and θ is the angle between the normal to the plane of the loop and the direction of the magnetic field. The net torque on the loop from both forces is the sum of the two torques, or

$$\tau = r_\perp F + r_\perp F = (\tfrac{1}{2}w \sin\theta)F + (\tfrac{1}{2}w \sin\theta)F = wF \sin\theta = w(ILB)\sin\theta$$

Then, since wL is the area (A) of the loop, we can express the magnitude of the torque on a single pivoted, current-carrying loop as

$$\tau = IAB \sin\theta \qquad \text{\textit{torque on current-carrying loop}} \qquad (19.8)$$

Although derived for a rectangular loop, Eq. 19.8 is actually valid for a flat loop of *any* shape and area. A *coil* is composed of N loops (where $N = 2, 3, \ldots$). Thus, for a coil, the torque is N times that on one loop (since the current is the same in all the loops), or

$$\tau = NIAB \sin\theta \qquad \text{\textit{torque on current-carrying coil}} \qquad (19.9)$$

The magnitude of the **magnetic moment** vector of a coil, m, is defined as

$$m = NIA \qquad \text{\textit{magnetic moment of a coil}} \qquad (19.10)$$

(SI units of magnetic moment: ampere · meter² or A · m²)

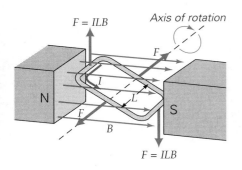

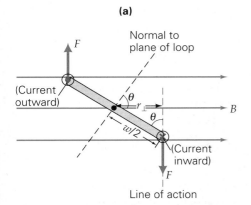

▲ **FIGURE 19.16 Force and torque on a current-carrying pivoted loop** (a) A current-carrying rectangular loop oriented in a magnetic field as shown is acted on by a force on each of its sides. Only the forces on the sides parallel to the axis of rotation produce a torque that causes the loop to rotate. (b) A side view shows the geometry for determining the torque. (See text for details.)

Note: A *coil* consists of N loops of the same size, all carrying the same current *I* in series.

The direction of the magnetic moment is determined by circling the fingers of the right hand in the direction of the (conventional) current. The thumb gives the direction of the magnetic moment vector **m**, always perpendicular to the plane of the coil (▶Fig. 19.17a). Eq. 19.10 can be rewritten in terms of the magnetic moment:

$$\tau = mB \sin \theta \qquad (19.11)$$

The magnetic torque tends to align the magnetic moment vector (**m**) with the magnetic field direction. To see this, notice that a "current loop" in a magnetic field is subject to a torque until $\sin \theta = 0$ (i.e., $\theta = 0°$), at which point the forces producing the torque are parallel to the plane of the loop (see Fig. 19.17b). This situation exists when the plane of the loop is perpendicular to the field. If the loop is started at rest such that its magnetic moment makes some angle with the magnetic field, the loop will undergo an angular acceleration that will rotate it to the zero angle position. Rotational inertia will carry it through the equilibrium position (zero angle, Fig. 19.17c) to the other side. On that side, the torque will slow the loop down, stop it, and then reaccelerate it back toward equilibrium. In other words, the torque on the current loop is a *restoring* torque and tends to cause the loop's magnetic moment to oscillate about the field direction, much like a compass needle as it settles down to point north.

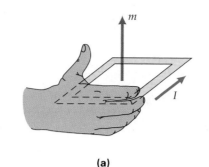

(a)

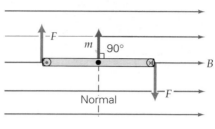

(b)

Example 19.5 ■ Magnetic Torque: Doing the Twist?

A laboratory technician makes a circular coil out of 100 loops of thin copper wire with a resistance of 0.50 Ω. The coil diameter is 10 cm and the coil is connected to a 6.0-V battery. (a) Determine the magnetic moment of the coil. (b) Determine the maximum torque on the coil if it were placed between the pole faces of a magnet where the magnetic field strength was 0.40 T.

Thinking It Through. The magnetic moment includes not only the number of loops and the area of the coil, but also the current in the wires. Ohm's law can be used to find the current. The maximum torque occurs when the angle between the magnetic moment vector and the B-field is 90°, as given by Eq. 19.11.

Solution. The given data are listed, with the radius of the circle expressed in SI units.

Given: $N = 100$ loops   *Find:* (a) m (coil magnetic moment)
$r = d/2 = 5.0 \text{ cm} = 5.0 \times 10^{-2} \text{ m}$ (b) τ (maximum torque on the coil)
$R = 0.50 \text{ }\Omega$
$V = 6.0 \text{ V}$

(a) The magnetic moment is given by Eq. 19.10, so the area and current are needed:

$$A = \pi r^2 = (3.14)(5.0 \times 10^{-2} \text{ m})^2 = 7.9 \times 10^{-3} \text{ m}^2$$

and

$$I = \frac{V}{R} = \frac{6.0 \text{ V}}{0.50 \text{ }\Omega} = 12 \text{ A}$$

Therefore,

$$m = NIA = (100)(12 \text{ A})(7.9 \times 10^{-3} \text{ m}^2) = 9.5 \text{ A} \cdot \text{m}^2$$

(b) The maximum torque is at $\theta = 90°$ in Eq. 19.11:

$$\tau = mB \sin \theta = (9.5 \text{ A} \cdot \text{m}^2)(0.40 \text{ T})(\sin 90°) = 3.8 \text{ m} \cdot \text{N}$$

Follow-up Exercise. In this Example, (a) show that if the coil were rotated so that its magnetic moment vector was at 45°, the torque would *not* be half the maximum torque. (b) At what angle would the torque be half the maximum torque?

▲ **FIGURE 19.17 Magnetic moment of a current-carrying loop** **(a)** A right-hand rule determines the direction of the loop's magnetic moment vector **m**. The fingers wrap around the loop in the direction of the current, and the thumb gives the direction of **m**. **(b)** Condition of maximum torque. **(c)** Condition of zero torque. If the loop is free to rotate, the magnetic moment vector of the loop will tend to align with the direction of the external magnetic field.

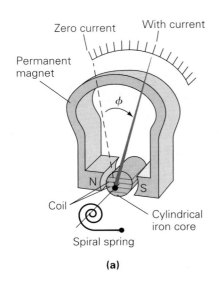

Zero current   With current

Permanent magnet

ϕ

Coil

Cylindrical iron core

Spiral spring

N   S

(a)

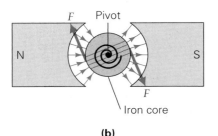

F   Pivot

N   S

F

Iron core

(b)

▲ **FIGURE 19.18 The galvanometer** **(a)** The deflection (ϕ) of the needle from its zero-current position is proportional to the current in the coil. A galvanometer can therefore detect and measure currents. **(b)** A magnet with curved pole faces is used so that the field lines are always perpendicular to the core surface and the torque does not vary with ϕ.

19.5 Applications: Current-Carrying Wires in Magnetic Fields

OBJECTIVE: To explain the operation of various instruments whose functions depend on electromagnetic interactions.

With the principles of electromagnetic interactions learned so far, you can understand the operation of some familiar laboratory measuring devices and dc motors.

The Galvanometer: The Foundation of the Ammeter and Voltmeter

Recall from Chapter 18 that both ammeters and voltmeters used the *galvanometer* as an essential part in their design.* The galvanometer was previously treated as a "black box" that measured small currents, but now you are ready to see how it actually works. As ◄Fig. 19.18a shows, a galvanometer consists of a coil of wire loops on an iron core that pivots between the pole faces of a permanent magnet. When a current exists in the coil, a torque is exerted on the coil. A small spring supplies a counter-, or restoring, torque, and in equilibrium a pointer indicates a deflection ϕ that is proportional to the current in the coil.

A practical problem arises if the galvanometer's magnetic field is not shaped correctly. If the coil rotated from its position of maximum torque ($\theta = 90°$), the torque would become less, and the pointer deflection ϕ, would *not* then be proportional to the current. This problem is avoided by making the pole faces curved and by wrapping the coil on a cylindrical iron core. The core tends to concentrate the field lines such that **B** is always perpendicular to the nonpivoted side of the coil (Fig. 19.18b). With this design, the deflection angle is proportional to the current through the galvanometer ($\phi \propto I$), as required.

The dc Motor

In general, *an electrical motor is a device that converts electrical energy into mechanical energy.* Such a conversion actually occurs during the movement of a galvanometer needle. However, a galvanometer is not considered a motor, because a practical **dc motor** must have continuous rotation for continuous energy output.

A pivoted, current-carrying coil in a magnetic field will rotate, but for only a half-cycle—that is, through a maximum angle of 180°. When the magnetic field and the coil's magnetic moment are lined up ($\sin \theta = 0$), the torque on the coil is zero and the coil is in equilibrium.

To provide for continuous rotation, the current is reversed every half-turn so that the torque-producing forces are reversed. This is done by means of a *split-ring commutator*, an arrangement of two metal half-rings insulated from each other (►Fig. 19.19a). The ends of the wire that forms the coil are fixed to the half-rings so that they rotate together. The current is supplied to the coil through the commutator by means of contact brushes. Then, with one half-ring electrically positive ($+$) and the other negative ($-$), the coil and ring rotate. When they have gone through half a rotation, the half-rings come in contact with the opposite brushes. Since their polarity is suddenly reversed, the current in the coil is now in the opposite direction. This reverses the directions of the magnetic forces, keeping the torque in the same direction (clockwise in Fig. 19.19b). Even though the torque is zero at the equilibrium position, the coil is in unstable equilibrium and

---

*Even though most voltmeters and ammeters are now digital, it is useful to understand how the mechanical versions use magnetic forces to make electrical measurements.

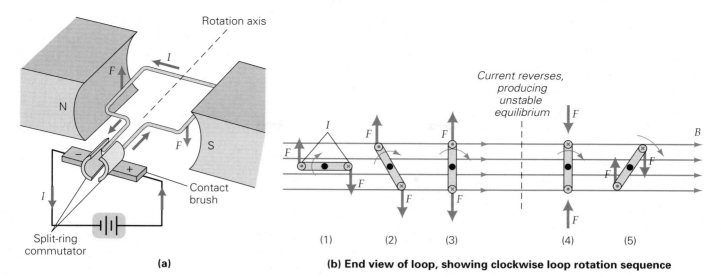

(a)

(b) End view of loop, showing clockwise loop rotation sequence

▲ **FIGURE 19.19 A dc motor**
(a) A split-ring commutator reverses the polarity and current each half-cycle, so the coil rotates continuously. **(b)** An end view shows the forces on the coil and its orientation during a half-cycle. (For simplicity, we depict a single loop, but the coil actually has N loops.) Notice the current reversal (shown by the dot and cross notation) between situations (3) and (4).

has enough rotational momentum to continue through the equilibrium point, whereupon the torque takes over and the coil rotates another half-cycle. The process repeats in continuous operation. In a real motor, the rotating shaft is called the *armature*.

The Electronic Balance

Traditional laboratory balances measure mass by balancing the weight force of an unknown mass against that of a known mass. The newer, digital electronic balances (▼Fig. 19.20a) work on a different principle. There is still a suspended beam with a pan on one end that holds the object to be weighed (or, more correctly, "massed"), but there is no known mass. The balancing downward force is supplied by current-carrying coils of wire in the field of a permanent magnet (Fig. 19.20b). The coils move up and down in the cylindrical gap of the magnet, and the downward force is proportional to the current in the coils. The mass of the object in the pan is determined from the coil current, which produces a force just sufficient to balance the beam.

The current required to produce balance is controlled automatically by photosensors and an electronic feedback loop. When the beam is balanced and horizontal,

▼ **FIGURE 19.20 Electronic balance** **(a)** A digital electronic balance. **(b)** Diagram of the principle of an electronic balance. The balance force is supplied by electromagnetism.

(a)

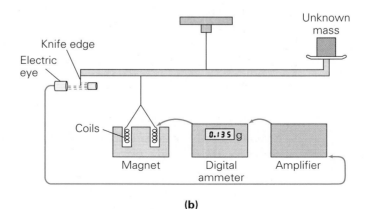

(b)

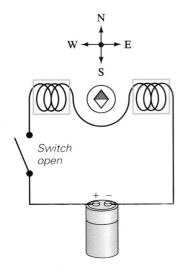

N
W ← → E
S

Switch open

+ −

(a) No current

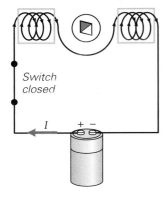

Switch closed

I   + −

(b) Current

▲ **FIGURE 19.21 Electric current and magnetic field** **(a)** With no current in the wire, the compass needle points north. **(b)** With a current in the wire, the compass needle is deflected, indicating the presence of an additional magnetic field superimposed on that of the Earth. In this case, the strength of the additional field is roughly equal in magnitude to that of the Earth. How can you tell?

a knife-edge obstruction cuts off part of the light from a source that falls on a photosensitive "electric eye," the resistance of which depends on the amount of light falling on it. This resistance controls the current that an amplifier sends through the coil. For example, if the beam tilts so that the knife edge rises and more light strikes the eye, the current in the coil is increased to counterbalance the tilting. In this manner, the beam is electronically maintained in nearly horizontal equilibrium. The current that keeps the beam in the horizontal position is read out on a digital ammeter calibrated in grams or milligrams instead of amperes.

19.6 Electromagnetism: The Source of Magnetic Fields

OBJECTIVES: To (a) understand the production of a magnetic field by electric currents, (b) calculate the strength of the magnetic field in simple cases, and (c) use the right-hand source rule to determine the direction of the magnetic field from the direction of the current that produces it.

Electric and magnetic phenomena, although clearly different in their details, are closely and fundamentally related. As we have seen, the magnetic force on a particle depends on the particle's electrical properties. But where does the magnetic field really come from? Danish physicist Hans Christian Oersted discovered the answer to this question in 1820, when he found that *electric currents produce magnetic fields*. His studies marked the beginnings of the discipline now called **electromagnetism**, part of which involves the relationship between electric currents and magnetic fields.

In particular, Oersted first noted that an electric current could produce a deflection of a compass needle. This property can be demonstrated with an arrangement like that in ◄Fig. 19.21. When the circuit is open and there is no current, the compass needle points, as usual, in the northerly direction. However, when the switch is closed and there is current in the circuit, the compass needle points in a different direction, indicating that an additional magnetic field (due to the current) must be affecting the needle. When the switch is reopened and the current drops to zero, the compass needle goes back to pointing north.

Developing expressions for the magnitude of the magnetic field near a current-carrying wire requires mathematics beyond the scope of this book. Thus, this section will simply present the results for the magnitude of the **B** field in several different current configurations.

Magnetic Field near a Long, Straight, Current-Carrying Wire At a perpendicular distance *d* from a long, straight wire carrying a current *I* (▶Fig. 19.22), the magnitude of **B** is

$$B = \frac{\mu_\text{o} I}{2\pi d} \quad \begin{array}{l}\textit{magnetic field due}\\ \textit{to long, straight wire}\end{array} \quad (19.12)$$

where $\mu_\text{o} = 4\pi \times 10^{-7} \, \text{T} \cdot \text{m/A}$ is a proportionality constant called the *magnetic permeability of free space*. For long, straight wires only, the field lines are closed circles centered on the wire (Fig. 19.22a). Note in Fig. 19.22b that the direction of **B** is given by a right-hand rule that is convenient for use with straight wires, namely, one of the **right-hand source rules**:

> If a straight current-carrying wire is grasped with the right hand with the extended thumb pointing in the direction of the current (*I*), the curled fingers indicate the circular sense of the magnetic field direction.

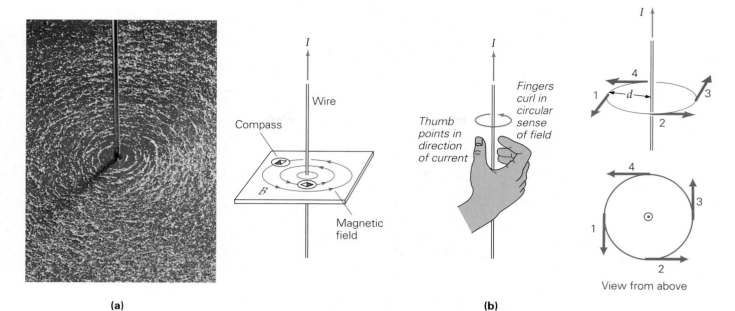

(a)

(b)

▲ FIGURE 19.22 Magnetic field around a long, straight current-carrying wire (a) The field lines form concentric circles around the wire, as revealed by the pattern of iron filings. (b) The circular sense of the field lines is given by the right-hand source rule, and the magnetic field vector is tangent to the circular field line at any point.

Example 19.6 ■ Common Fields: Magnetic Field from a Current-Carrying Wire

The maximum household current in a wire is about 15 A. Assume that this current exists in a long straight wire in a west-to-east direction (▼Fig. 19.23). What are the magnitude and direction of the magnetic field the current produces 1.0 cm directly below the wire?

Thinking It Through. To find the magnitude of the field, we should use Eq. 19.12, for a long, straight wire. The direction of the field is given by the right-hand source rule.

Solution. We list the data.

Given: $I = 15$ A
$d = 1.0$ cm $= 0.010$ m

Find: **B** (magnitude and direction)

From Eq. 19.12, the magnitude of the field at the point 1.0 cm directly below the wire is

$$B = \frac{\mu_{\circ} I}{2\pi d} = \frac{(4\pi \times 10^{-7}\ \text{T.m/A})(15\ \text{A})}{2\pi(0.010\text{m})} = 3.0 \times 10^{-4}\ \text{T}$$

Note that this field is about 10 times larger than the Earth's field. By the right-hand source rule, shown in Fig. 19.23, the direction of the field at that point is to the north.

Follow-up Exercise. (a) In this Example, what is the direction of the magnetic field due to the wire 5.0 cm above it? (b) Calculate the current needed to produce a magnetic field of the same strength as that in the Example, but 5.0 cm from the wire.

PHYSLET®
ILLUSTRATION

The Magnetic Field Around a Current-Carying Wire

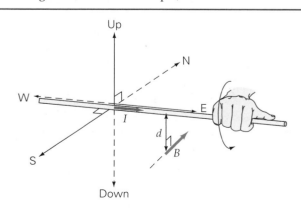

◀ FIGURE 19.23 Magnetic field Finding the magnitude and direction of the magnetic field produced by a straight current-carrying wire. (See Example 19.6.)

Magnetic Field at the Center of a Circular Current-Carrying Wire Loop At the *center* of a flat circular coil of wire consisting of N loops, each of radius r and each carrying current I (▼Fig. 19.24a shows one such loop), the magnitude of **B** is

$$B = \frac{\mu_o N I}{2r} \qquad \begin{array}{l} \textit{magnetic field at center of} \\ \textit{circular coil of N loops} \end{array} \qquad (19.13)$$

In this case (and all circular current arrangements, such as solenoids, discussed next), it is more convenient to determine the direction of **B** using a right-hand source rule that is slightly different from (but equivalent to) the one for straight wires:

> If a circular loop of current-carrying wires is grasped with the right hand so that the fingers are curled in the direction of the current, the magnetic field direction in the area enclosed by the loop is given by the direction in which the extended thumb is pointing (Fig. 19.24b).

Of course, you should know that in all cases the magnetic field lines form closed loops. Thus, the field direction depends crucially on the particular location (Fig. 19.24c). Finally, notice that the overall field of the loop is geometrically similar to that of a bar magnet.

Magnetic Field in a Current-Carrying Solenoid A *solenoid* is constructed by winding a long wire into a tight coil, or helix, with many circular loops, as shown in ▶Fig. 19.25. If the radius of the loops is small compared with the length (L) of the coil, the interior magnetic field is parallel to the solenoid's longitudinal axis and constant in magnitude. The direction of the interior solenoid field is given by the right-hand source rule for circular geometry. If the solenoid has N turns (loops) and carries a current I, the magnitude of the magnetic field at the center is

$$B = \frac{\mu_o N I}{L} \qquad \begin{array}{l} \textit{magnetic field near} \\ \textit{the center of a solenoid} \end{array} \qquad (19.14)$$

Notice that the solenoid field depends crucially on how densely, or closely packed, the turns of wire are wound. The expression $n = N/L$, or the number of turns per meter, is sometimes called the *linear turn density*, and Eq. 19.14 is sometimes rewritten as $B = \mu_o n I$.

▼ **FIGURE 19.24 Magnetic field due to a circular current-carrying loop** **(a)** Pattern of iron filings for a current-carrying loop. Note that the magnetic field at the loop's center is perpendicular to the loop's plane. **(b)** The direction of the field in the area enclosed by the loop is given by a right-hand source rule. With the fingers wrapped around the loop in the direction of the conventional current, the thumb indicates the direction of **B** in the plane of the loop. **(c)** The overall magnetic field of a current-carrying circular loop is similar to that of a bar magnet.

(a)

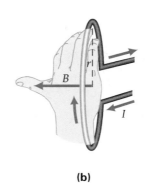

(b)

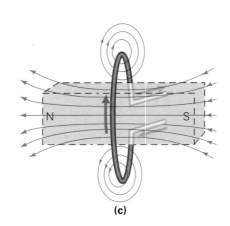

(c)

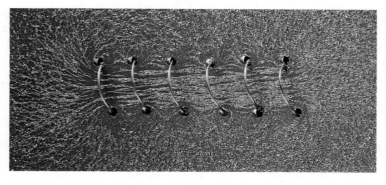

(a)

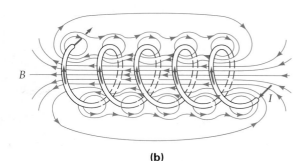

(b)

▲ FIGURE 19.25 Magnetic field of a solenoid (a) The magnetic field of a current-carrying solenoid is fairly uniform near the central axis of the solenoid, as seen in this pattern of iron filings. (b) The direction of the field in the interior can be determined by applying the right-hand source rule to any of the loops. Notice its resemblance to the field near a bar magnet.

The longer the solenoid, the more uniform is the magnetic field across the cross-sectional area within its coil. An ideal, or infinitely long, solenoid would have a perfectly uniform internal field. Although this ideal condition cannot be realized, a long solenoid does produce a relatively uniform magnetic field. Notice how the field created by a solenoid (Fig. 19.25) resembles that of a permanent bar magnet.

To see why the solenoid might be best suited for magnetic applications requiring a large magnetic field, consider the next Example.

Example 19.7 ■ Wire versus Solenoid: Concentrating the Magnetic Field

A solenoid is 0.30 m long with 300 turns and carries a current of 5.0 A. (a) What is the magnitude of the magnetic field at the center of this solenoid? (b) Compare your result with the field near the single wire in Example 19.6 (notice that the wire in that example carried *three times* the current of the solenoid in this Example), and comment.

Thinking It Through. The B field depends on the number of turns (N), the solenoid length (L), and the current (I). This is a direct application of Eq. 19.14.

Solution.

Given: $I = 5.0$ A *Find:* (a) B (magnitude of magnetic field near
$N = 300$ turns   the solenoid center)
$L = 0.30$ m   (b) Compare the answer in part (a) with that
                from a long straight wire in Example 19.6

From Eq. 19.14,

$$B = \frac{\mu_0 NI}{L} = \frac{(4\pi \times 10^{-7}\,\text{T}\cdot\text{m/A})(300)(5.0\,\text{A})}{0.30\,\text{m}} = 2\pi \times 10^{-3}\,\text{T} \approx 6.3\,\text{mT}$$

Notice that this is about *20 times larger* than the field near the wire in Example 19.6 and *several hundred* times larger than the Earth's magnetic field. The reason is that the solenoid's field is the vector sum of the fields from 300 loops.

Follow-up Exercise. In this Example, if the current were reduced to 1.0 A and the solenoid shortened to 0.10 m, how many turns would be needed to create the same magnetic field near the center of the solenoid?

In the next Integrated Example, all aspects of electromagnetism are involved: forces on electric currents and the production of magnetic fields by electric currents. Study the example carefully, especially the use of the appropriate right-hand rule.

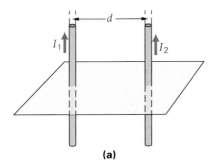

(a)

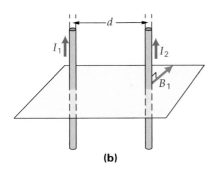

(b)

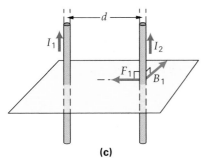

(c)

▲ **FIGURE 19.26   Mutual interaction of parallel current-carrying wires   (a)** Two parallel wires carry current in the same direction. **(b)** Wire 1 creates a magnetic field at the site of wire 2. **(c)** Wire 2 is pulled toward wire 1 by a force. (See Integrated Example 19.8 for details.)

Integrated Example 19.8 ■ Attraction or Repulsion: Magnetic Force between Two Parallel Wires

Two long, parallel wires carry currents in the same direction, as illustrated in ◄Fig. 19.26a. (a) Is the magnetic force between these wires (1) attractive or (2) repulsive? Make a sketch to show how you obtained your result. (b) If both wires carry a current of 5.0 A, have a length of 50 cm, and are separated by 3.0 mm, determine the magnitude of the force on each wire.

(a) Conceptual Reasoning.   Choose one wire, and determine the direction of the magnetic field it produces (using the right-hand source rule for straight wires) at the other wire. (In Fig. 19.26b, we have chosen wire 1.) This is the magnetic field in which the other wire (wire 2) is placed. Then use the right-hand force rule to determine the direction of the force on wire 2. The result, shown in Fig. 19.26c, is an attractive force, so (1) is the correct answer. You should be able to show that wire 2 exerts an attractive force on wire 1, in keeping with Newton's third law.

(b) Thinking It Through.   To find the magnetic field strength produced by wire 1, use Eq. 19.12. Since the magnetic field is at right angles to the current in wire 2, the force on it is just ILB. Be careful to determine which field and which current are involved.

The symbols are given in Fig. 19.26. We list the given data and convert to SI units.

Given: $I_1 = I_2 = 5.0$ A   *Find:* F (the magnitude of the force
          $d = 3.0$ mm $= 3.0 \times 10^{-3}$ m   between the wires)
          $L = 50$ cm $= 5.0 \times 10^{-1}$ m

The magnetic field due to I_1 is

$$B_1 = \frac{\mu_o I_1}{2\pi d} = \frac{(4\pi \times 10^{-7}\,\text{T·m/A})(5.0\,\text{A})}{2\pi(3.0 \times 10^{-3}\,\text{m})} = 3.3 \times 10^{-4}\,\text{T}$$

The magnitude of the magnetic force on wire 2 is

$$F_2 = I_2 L B_1 = (5.0\,\text{A})(0.50\,\text{m})(3.3 \times 10^{-4}\,\text{T}) = 8.3 \times 10^{-4}\,\text{N}$$

Follow-up Exercise.   (a) In this Example, determine the direction of the force between the wires if the current in both of them is reversed in direction. (b) If the magnitude of the force between the wires is one-fourth of the value in the Example, how far apart are the wires if the currents remain the same?

The magnetic force between parallel wires in an arrangement like that analyzed in Integrated Example 19.8 provides the basis for the definition of the ampere. The National Institute of Standards and Technology (NIST) defines the ampere as

> that current which, if maintained in each of two long parallel wires separated by a distance of 1 m in free space, would produce a magnetic force between the wires of exactly 2×10^{-7} N for each meter of wire.

19.7 Magnetic Materials

OBJECTIVES: To explain **(a)** how ferromagnetic materials enhance external magnetic fields, **(b)** how "permanent" magnets are produced, and **(c)** how "permanent" magnetism can be destroyed.

Why are some materials magnetic or easily magnetized, whereas others are not? How can a bar magnet create a magnetic field when it carries no obvious current? First, you know that a current is needed to produce a magnetic field. Now, if you compare the magnetic fields of a bar magnet and a long solenoid (e.g., compare Figs. 19.1 and 19.25), you might begin to think that the magnetic field of the bar magnet is due to *internal* currents—caused perhaps by electrons orbiting the atoms or by electron spin. However, detailed analysis of atomic structure shows that the net magnetic field produced by orbital motion is zero or very small.

What, then, *is* the source of the magnetism produced by magnetic materials? Modern atomic quantum theory tells us that the permanent type of magnetism, like that exhibited by an iron bar magnet, is produced by electron *spin*. Classical physics likens a spinning electron to the Earth rotating on its axis. However, this is *not* the case. Electron spin is strictly a quantum mechanical effect with no direct classical analogue. Nonetheless, the mechanical picture of spinning electrons creating magnetic fields is useful for qualitative thinking and reasoning. In effect, each "spinning" electron produces a field similar to that of a current loop (Fig. 19.24c). This pattern, resembling that from a small bar magnet, leads us to picture the spinning electrons as tiny compass needles.

In multielectron atoms, the electrons *usually* are arranged in pairs with their spins oppositely aligned. The magnetic fields then cancel each other, and the material is not magnetic. Aluminum is such a material. However, in certain strongly magnetic materials, known as **ferromagnetic materials**, the fields due to electron spins do *not* cancel. In fact, there is a strong interaction between neighboring spins and atoms that leads to the formation of large groups of atoms called **magnetic domains**. In a given domain, many of the electron spins are actually aligned in approximately the same direction, producing a relatively strong (net) magnetic field. Not many ferromagnetic materials occur naturally. The most common are iron, nickel, and cobalt. Gadolinium and certain manufactured alloys are also ferromagnetic.

In an unmagnetized ferromagnetic material, the domains are randomly oriented and there is no net magnetization (▼Fig. 19.27a). But when a ferromagnetic material (like an iron bar magnet) is placed in an external magnetic field, the domains begin to change their orientation and size (Fig. 19.27b). These effects can be understood physically by representing the electron spin by a small compass needle, ready to "line up" in an external magnetic field. As the iron bar is introduced into the field, the material exhibits the following two effects:

1. Domain boundaries change, and the domains with magnetic orientations in the direction of the external field grow at the expense of the other domains.

2. The magnetic orientation of some domains may change slightly so as to be more aligned with the field.

You now can understand why an unmagnetized piece of iron is attracted to a magnet and why iron filings line up with a magnetic field. Essentially, the pieces of iron become induced magnets (Fig. 19.27c).

▼ **FIGURE 19.27 Magnetic domains** (a) With no external magnetic field, the magnetic domains of a ferromagnetic material are randomly oriented and the material is unmagnetized. (b) In an external magnetic field, domains with orientations parallel to the field may grow at the expense of other domains, and the orientations of some domains may become more aligned with the field. (c) As a result, the material becomes magnetized, or exhibits magnetic properties.

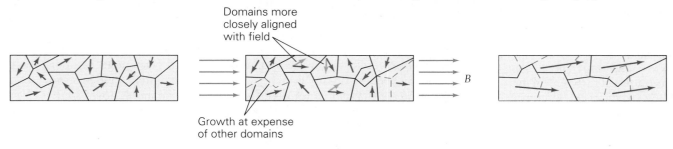

Domains more closely aligned with field

Growth at expense of other domains

B

(a) No external magnetic field   **(b) With external magnetic field**   **(c) Resulting bar magnet**

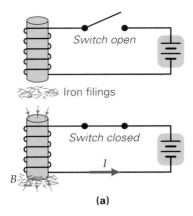

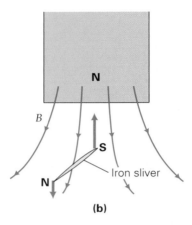

(c)

▲ **FIGURE 19.28 Electromagnet**
(a) (*top*) With no current in the circuit, there is no magnetic force. (*bottom*) However, with a current in the coil, there is a magnetic field and the iron core becomes magnetized. **(b)** Detail of the lower end of the electromagnet in (a). The sliver of iron is attracted to the end of the electromagnet. **(c)** An electromagnet picking up scrap metal.

Electromagnets and Magnetic Permeability

Ferromagnetic materials are used to make electromagnets, usually by wrapping a wire around an iron core (◄Fig. 19.28a). The current in the coil creates a magnetic field in the iron, which in turn creates its own field many times larger than the coil's field. By turning the current on and off, we can turn the magnetic field on and off at will. When the electromagnet is on, it can induce magnetism in ferromagnetic materials (e.g., the iron sliver in Fig. 19.28b) and pick up large amounts of scrap iron (Fig. 19.28c).

The iron used in an electromagnet is called *soft iron*. It is manufactured in such a way that when the external field is removed, the magnetic domains quickly become unaligned and the iron is again demagnetized. The adjective "soft" refers not to the metal's mechanical hardness, but to its magnetic properties.

When an electromagnet is on (lower drawing in Fig. 19.28a), the iron core is magnetized and adds to the field of the solenoid. The total field is expressed as

$$B = \frac{\mu N I}{L} \qquad \begin{array}{l} \textit{magnetic field at the center} \\ \textit{of the iron-core solenoid} \end{array} \qquad (19.15)$$

Notice that this equation is identical to that for the magnetic field of an air-core solenoid (Eq. 19.14), except that it contains μ instead of μ_o, the permeability of free space. Here, μ represents the **magnetic permeability** of the *core material*, not free space. The role permeability plays in magnetism is similar to that of the permittivity ε in electricity (Chapter 16). For magnetic materials, the magnetic permeability is defined in terms of its value in free space; thus,

$$\mu = \kappa_m \mu_o \qquad (19.16)$$

where κ_m is called the *relative* permeability (dimensionless) and is the magnetic analogue of the dielectric constant κ.

The value of κ_m for a vacuum is equal to unity. (Why?) Since, for ferromagnetic materials, the total magnetic field far exceeds that from the wire wrapping, it follows that $\mu \gg \mu_o$ and $\kappa_m \gg 1$. A core of a ferromagnetic material with a large permeability in an electromagnet can enhance its field thousands of times compared with an air core. In other words, ferromagnetic materials have values of κ_m on the order of thousands.

From Eq. 19.15, it can be seen that the magnetic field strength of an electromagnet also depends on the current in the wires. Large currents produce large fields, but this generation of fields is accompanied by much greater joule heating (I^2R losses) in the wires, which therefore require water-cooling lines. The problem can be alleviated by using a superconductor for the wires, because the superconductor has zero resistance and thus no joule heating. In fact, this technology exists on a small scale in laboratories. However, for commercial use, superconducting magnets are currently not yet practical because of the tremendous cooling required to keep the conductors in their low-temperature superconducting state.

The type of iron that retains some of its magnetism after being in an external magnetic field is called *hard iron* and is used to make so-called permanent magnets. You may have noticed that a paper clip or a screwdriver blade becomes slightly magnetized after being near a magnet. Permanent magnets are produced by heating pieces of some ferromagnetic material in an oven and then cooling them in a strong magnetic field to get the maximum effect. In permanent magnets, the domains do *not* become unaligned when the external magnetic field is removed.

A *permanent* magnet is not necessarily truly permanent, however, because its magnetism can be destroyed. Hitting such a magnet with a hard object or dropping it on the floor can cause a loss of some or all of the domain alignment, thus reducing or eliminating the magnet's overall magnetic field. Heating, too, can cause a loss of magnetism, since the resulting increase in random (thermal) motions of atoms also tends to disrupt the alignment of domains. One of the worst things you can do to an audio or video magnetic tape cassette is to leave it on the dashboard of a car on a hot day. The thermal motion caused by the heat partially destroys the magnetic voice or video signal imprinted on the tape. Above a certain critical temperature, called the **Curie temperature** (or Curie point), domain coupling is destroyed by the increased thermal oscillations, and a ferromagnetic material loses its ferromagnetism. This effect was discovered by the French physicist Pierre Curie (1859–1906), husband of Madame Marie Curie. The Curie temperature for iron is 770°C.

Ferromagnetic domain alignment plays an important role in certain areas of geology and geophysics. For example, it is well known that, when cooled to everyday temperatures, lava flows from volcanoes that initially contain iron above its Curie temperature can retain some magnetism due to the Earth's magnetic field. Measuring the strength and orientation of these older lava flows at various locations around the world has enabled geophysicists to map the changes in the Earth's magnetic field and polarity over time.

Some of the first evidence in support of plate tectonic motion came from measuring the direction of the magnetic polarity of seafloor samples containing iron.* The seafloor near the mid-Atlantic ridge, for example, is composed of lava flows from underwater volcanoes. These solidified flows were found to exhibit permanent magnetism, but the polarity varied with time (older samples are farther out from the ridge) as the Earth's magnetic polarity changed.

*19.8 Geomagnetism: The Earth's Magnetic Field

OBJECTIVES: To (a) state some of the general characteristics of the Earth's magnetic field, (b) explain some theories about its possible source, and (c) discuss some of the ways in which the Earth's magnetic field affects our planet's local environment.

The magnetic field of the Earth was used for centuries before people had any clues about its origin. In ancient times, navigators used lodestones or magnetized needles to locate north. Some other forms of life, including certain bacteria and homing pigeons, also use the Earth's magnetic field for navigation. (See Insight on page 658.)

An early study of magnetism was carried out by the English scientist Sir William Gilbert about 1600. In investigating the magnetic field of a specially cut spherical lodestone that simulated the Earth, he came to believe that the Earth as a whole acts as a magnet. Gilbert thought that a large body of permanently magnetized material within the Earth might produce its magnetic field.

In fact, the Earth's external magnetic field, or the *geomagnetic field*, as it is called, does have a configuration similar to that which would be produced by a large interior bar magnet with the south pole of the magnet pointing north (▶Fig. 19.29). The magnitude of the horizontal component of the Earth's magnetic field at the magnetic equator is about 10^{-5} T (approximately 0.1 G), and the vertical component at the geomagnetic poles is about 10^{-4} T (roughly, 1 G). It has been calculated that, for a ferromagnetic material of maximum magnetization to produce this field, it would have to occupy only about 0.01% of the Earth's volume.

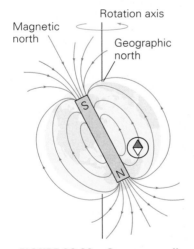

▲ **FIGURE 19.29 Geomagnetic field** The Earth's magnetic field is similar to that of a bar magnet. However, a permanent solid magnet could not exist within the Earth because of the high temperatures there. The Earth's magnetic field is believed to be associated with motions in the liquid outer core deep within the planet.

---

*The outermost solid layer of the Earth's crust is known to be composed of sections or "plates." These plates are in constant, but very slow, motion—a centimeter per year is a typical speed. At certain intersections, such as that where the Pacific Plate meets the southern coastline of Alaska, the plates collide, giving rise to volcanism and earthquakes. At other intersections, such as the mid-Atlantic ridge, the plates are receding from one another, with new material coming up from the Earth's interior in the form of hot lava.

INSIGHT

Magnetism in Nature

For centuries, humans have relied on compasses to provide directional information enabling them to navigate according to the Earth's magnetic field (Fig. 19.29). In contrast, certain living creatures seem to have their own built-in directional sensors. For example, some species of bacteria have been determined to be *magnetotactic*—that is, able to sense or feel the presence and direction of the Earth's magnetic field.

Initially, experiments were done in the 1980s on bacteria commonly found in the mud of bogs, swamps, and ponds.* In a laboratory magnetic field, when a droplet of muddy water was viewed under a microscope, one species of bacteria always migrated in the direction of the field (Fig. 1)—just as these bacteria do in their natural environment with the Earth's field. Furthermore, when these bacteria died and therefore could no longer migrate, they maintained their alignment with the magnetic field even when its direction changed. It became apparent that members of this species act like magnetic dipoles, or compass needles. Once aligned with the field, they migrate along magnetic field lines simply by moving their *flagella*, which are whiplike appendages (Fig. 2).

What makes these bacteria living compasses? Even among known magnetotactic species, "new" bacteria (formed

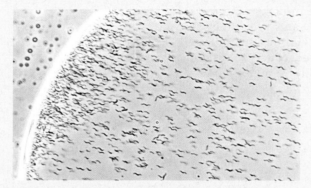

FIGURE 1 Magnetotactic bacteria in migration
Bacteria in a drop of muddy water, as viewed under a microscope, align along the direction of the applied magnetic field (north to the left) and accumulate at the edge. When the field is reversed, so is the direction of migration.

*See, for example, R. P. Blakemore and R. B. Frankel, "Magnetic Navigation in Bacteria," *Scientific American*, December 1981. We are indebted to Professor Frankel for several interesting discussions of this topic.

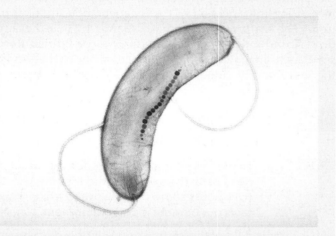

FIGURE 2 An elliptical magnetotactic bacterium A freshwater magnetotactic bacterium, shown in an electron micrograph. Two whiplike appendages, or flagella, are clearly visible, along with a chain of magnetite particles.

The idea of a ferromagnet of this size within the Earth may not seem unreasonable at first, but this cannot be a correct model. The interior temperature of the Earth is well above the Curie temperatures of iron and nickel, the ferromagnetic materials believed to be the most abundant in the Earth's interior. For iron, the Curie temperature is 770°C, which is attained at a depth of only 100 km. The temperature of the Earth is higher at greater depths. Thus, the existence of a permanent internal magnet is not possible.

The fact that an electric current produces a magnetic field has led scientists to speculate that the Earth's magnetic field is associated with motions in the liquid outer core, which, in turn, may be connected some way with the Earth's rotation. We know that Jupiter, a planet that is largely gaseous and rotates very rapidly, has a magnetic field much larger than that of Earth. Mercury and Venus have very weak magnetic fields; these planets are more like Earth in density and rotate relatively slowly.

Several theoretical models have been proposed to explain the origin of the Earth's magnetic field. For example, it has been suggested that it arises from currents associated with thermal convection cycles in the liquid outer core caused by heat from the inner core. But the details of the mechanism are still not clear.

The axis of the Earth's magnetic field does not lie along the planet's rotational axis, which defines the geographic poles. Hence, the north magnetic pole and the

by cell division) do not have this magnetotactic sense. However, if they live in a solution containing a minimum concentration of iron, they are able to synthesize a chain of small magnetic particles (Fig. 2). Oddly enough, these internal compasses have the same chemical composition as the original slivers of naturally occurring ore used as compasses by ancient sailors: magnetite (chemical symbol Fe_3O_4). The individual particles in the chain are approximately 50 nm across, and the chain of a mature bacterium typically contains about 20 such particles, each of which is a single magnetic domain.

Thus, the bacteria are passively steered by their internal compasses. But why is it biologically important for these bacteria to follow the Earth's magnetic field direction? An important piece of the puzzle was found while investigators were studying the same species from Southern Hemisphere waters. These bacteria migrate *opposite* the direction of the

Earth's field. Recall that in the Northern Hemisphere the Earth's field inclines downward and that the reverse is true in the Southern Hemisphere. These facts lead scientists to believe that the bacteria are using the field direction for survival (Fig. 3). Since oxygen is toxic to them, they are most likely to survive in the muddy, nutrient-rich depths, and the Earth's magnetic field direction points them that way. Their directional sense also aids them near the equator; there it does not direct them downward, but instead keeps them at a constant depth, thus avoiding an upward migration to the deadly oxygen-rich surface waters. Evidence of magnetic field navigation has been found not only in bacteria, but also in such diverse organisms as bees, butterflies, homing pigeons, and dolphins. It is hoped that continued research in this area will provide insight into magnetic sensing in other organisms.

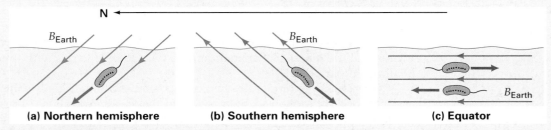

(a) Northern hemisphere **(b) Southern hemisphere** **(c) Equator**

FIGURE 3 Survival of the fittest? **(a)** In the Northern Hemisphere, where the Earth's magnetic field inclines downward, magnetotactic bacteria follow the field to the nutrient-rich depths. **(b)** In the Southern Hemisphere, the Earth's field is inclined upward, but the bacteria migrate opposite the field and so are able to stay in deep waters. **(c)** Around the equator, the bacteria move parallel to the water surface and thus are kept away from the shallow, oxygen-rich (and hence poisonous) waters.

geographic North Pole do not coincide (see Fig. 19.29): The magnetic pole is several thousand kilometers south of the geographic North Pole (true north). The south magnetic pole is displaced even more from its geographic pole, so the magnetic axis is not even a straight line through the center of the Earth.

A compass indicates the direction of *magnetic north*, not "true," or geographic, north. The angular difference in these two directions is called the *magnetic declination* (▶Fig. 19.30). As shown on the map, the magnetic declination varies for different locations. Knowing the variations in the magnetic declination is particularly important in the accurate navigation of airplanes and ships, as you can imagine.

The Earth's magnetic field also exhibits a variety of fluctuations with time. As discussed in the previous section, the permanent magnetism created in iron-rich rocks as they cooled in the Earth's magnetic field has provided us with much evidence of these fluctuations over long time scales. For example, the Earth's magnetic poles have switched polarity at various times in the past, most recently about 700 000 years ago. During a period of reversed polarity, the south magnetic pole is near the south geographic pole—the opposite of today's polarity. We are not certain why such pole reversals have occurred.

On a shorter time scale, the magnetic poles do not always remain in the same locations, but tend to "wander." The north magnetic pole has recently been moving about 1° of latitude (roughly 110 km or 70 mi) per decade. For some unknown

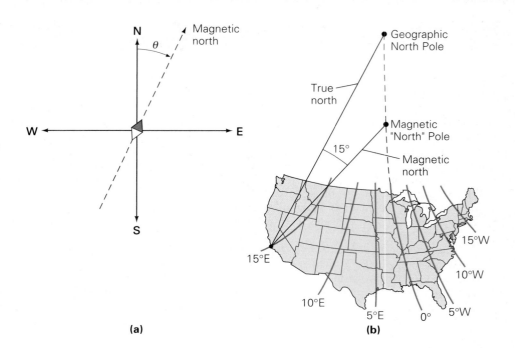

▶ **FIGURE 19.30 Magnetic declination** **(a)** The angular difference between magnetic north and "true," or geographic, north is called the magnetic declination. **(b)** The magnetic declination varies with location and time. The map shows *isogonic* lines (lines with the same magnetic declination) for the continental United States. For locations on the 0° line, magnetic north is in the same direction as true (geographic) north. On either side of this line, a compass has an easterly or westerly variation. For example, on a 15°E line, a compass has an easterly declination of 15°. (Magnetic north is 15° east of true north.)

reason, it has moved consistently northward from its 1904 latitude of 69°N, and westward, crossing the 100°W longitudinal meridian. This long-term polar drift means that the magnetic declination map (Fig. 19.30b) varies with time and must be updated periodically.

On a still shorter time scale, there are sometimes dramatic daily shifts of as much as 80 km (50 mi), followed by a return to the starting positions. These shifts are thought to be caused by charged particles from the Sun that reach the Earth's upper atmosphere and set up currents that change the planet's overall magnetic field.

Charged particles from the Sun entering the Earth's magnetic field give rise to other phenomena. A charged particle that enters a uniform magnetic field at an angle that is *not* perpendicular to the field spirals in a helix (▼Fig. 19.31a). This is because the component of the particle's velocity parallel to the field does not change. (Recall that a magnetic field acts only on the perpendicular component of the veloc-

▼ **FIGURE 19.31 Magnetic confinement** **(a)** A charged particle entering a uniform magnetic field at an angle other than 90° moves in a spiraling path. **(b)** In a nonuniform, bulging magnetic field, particles spiral back and forth as though confined in a magnetic bottle. **(c)** Charged particles are trapped in the Earth's magnetic field, and the regions where they are concentrated are called Van Allen belts.

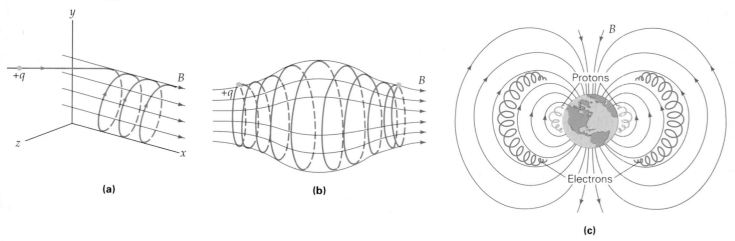

ity.) The motions of charged particles in a nonuniform field are quite complex. However, for a bulging field such as that depicted in Fig. 19.31b, the particles spiral back and forth as though in a "magnetic bottle."

An analogous phenomenon occurs in the Earth's magnetic field, giving rise to regions with concentrations of charged particles. Two large donut-shaped regions at altitudes of several thousand kilometers are called the *Van Allen radiation belts* (Fig. 19.31c). In the lower Van Allen belt, light emissions called *aurorae* occur—the aurora borealis, or northern lights, in the Northern Hemisphere and the aurora australis, or southern lights, in the Southern Hemisphere. These eerie, flickering lights are most commonly observed in the Earth's polar regions, but have been seen at all latitudes (▶Fig. 19.32).

It is believed that an aurora is created when charged solar particles become trapped in the Earth's magnetic field. Maximum aurora activity occurs after a solar disturbance, such as a solar flare—a violent magnetic storm on the Sun that spews out enormous quantities of charged particles and radiation. Trapped in the Earth's magnetic field, the charged particles are guided toward the polar regions, where they excite or ionize oxygen and nitrogen atoms in the atmosphere. When the excited atoms return to their normal state and the ions regain their normal number of electrons, light is emitted (Chapter 27), producing the beautiful glow of the aurora.

▲ **FIGURE 19.32 Aurora borealis: the northern lights** This spectacular display is caused by energetic solar particles trapped in the Earth's magnetic field. The particles excite or ionize air atoms; on de-excitation (or recombination) of the atoms, light is emitted.

Chapter Review

Important Concepts and Equations

- The **pole–force law**, or **law of poles**, states that opposite magnetic poles attract and like poles repel.

- The **magnetic field (B)** is expressed in units of the tesla (T) and $1\,T = 1\,N/(A \cdot m)$. Magnetic fields can exert forces on moving charged particles and electric currents. The magnetic force on a charged particle is given by

$$F = qvB \sin \theta \qquad (19.3)$$

The magnetic force on a current-carrying wire is given by

$$F = ILB \sin \theta \qquad (19.7)$$

- **Right-hand force rules** are used to determine the direction of a *magnetic force* on moving charged particles and current-carrying wires.

- A series of N current-carrying circular loops, each with a plane area A and carrying a current I can be subject to a **magnetic torque** when placed in a B field. The expression for the magnitude of the torque on such an arrangement is

$$\tau = NIAB \sin \theta \qquad (19.9)$$

- The magnitude of the **magnetic moment** vector of a coil has units of $A \cdot m^2$ and is defined as

$$m = NIA \qquad (19.10)$$

- The magnitude of the **magnetic field** produced by a long, straight wire is given by

$$B = \frac{\mu_o I}{2\pi d} \qquad (19.12)$$

$$\mu_o = 4\pi \times 10^{-7}\,T \cdot m/A$$

where is the **magnetic permeability of free space**. For long, straight wires only, the field lines are closed circles centered on the wire.

- The magnitude of the **magnetic field** produced at the center of an arrangement of N circular current-carrying loops of radius r is

$$B = \frac{\mu_o NI}{2r} \qquad (19.13)$$

- The magnitude of the **magnetic field** produced near the center of the interior of a solenoid with N windings and a length L is given by

$$B = \frac{\mu_o NI}{L} \qquad (19.14)$$

- **Right-hand source rules** are used to determine the direction of the magnetic field from various current configurations.

- In **ferromagnetic materials**, the electron spins align, creating **domains**. When an external field is applied, the effect is to increase the size of those domains which already point in the direction of the field at the expense of the other domains. When the external magnetic field is removed, a **permanent magnet** remains.

Exercises

19.1 Magnets, Magnetic Poles, and Magnetic Field Direction

1. When the ends of two bar magnets are near each other, they attract one another. The ends must be (a) one north, the other south, (b) one south, the other north, (c) both north, (d) both south, or (e) either (a) or (b).

2. A compass placed in a magnetic field directed from east to west will point (a) perpendicularly to the field lines, (b) with the south pole to the east, (c) with the south pole to the west, or (d) at a nonzero angle ($\theta < 90°$) to the field direction.

3. If you look directly down on the south pole of a bar magnet, the magnetic field points (a) to the right, (b) to the left, (c) away from you, or (d) toward you.

4. CQ Given two identical iron bars, one of which is a permanent magnet and the other unmagnetized, how could you tell which is which by using only the two bars?

5. List some of the similarities and differences between electric and magnetic forces.

19.2 Magnetic Field Strength and Magnetic Force

6. A proton moves vertically upward perpendicularly to a uniform magnetic field and deflects to the right as you watch it. What is the magnetic field direction? (a) Directly away from you, (b) directly toward you, (c) to the right, or (d) to the left.

7. Can a constant magnetic field set a stationary electron in motion? If so, how?

8. If a negatively charged particle were moving downward along the right edge of this page, which way should a magnetic field (known to be perpendicular to the plane of the paper) be oriented so that the particle would initially be deflected to the left?

9. An electron passes through a magnetic field without being deflected. What do you conclude about the orientation between the magnetic field and the velocity of the electron, assuming that no other forces act on the electron?

10. CQ A proton and an electron are moving at the same velocity perpendicularly to a constant magnetic field. (a) Compare the magnitude of the magnetic force on them. (b) Compare the magnitude of their accelerations.

11. CQ If a charged particle moves in a straight line and there are no other forces on it except possibly from a magnetic field, can you say with certainty that no magnetic field is present? Explain.

12. CQ Three particles enter a uniform magnetic field as shown in ▶Fig. 19.33a. Particles 1 and 3 have equal speeds and charges of the same magnitude. What can you say about (a) the charges of the particles and (b) their masses?

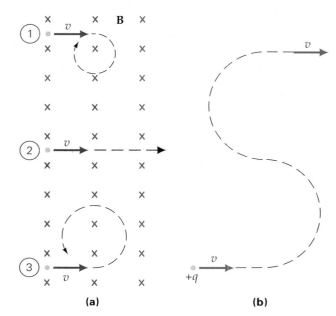

(a)   (b)

▲ **FIGURE 19.33 Charges in motion** See Exercises 12 and 13.

13. CQ You want to deflect a positively charged particle in an "S" path, as shown in Fig. 19.33b, using only magnetic fields. (a) Explain how this could be done by using magnetic fields perpendicular to the plane of the page. (b) How does the magnitude of an emerging particle's velocity compare with the particle's initial velocity?

14. IE ■ A positive charge moves horizontally to the right across this page and enters a magnetic field directed vertically downward. (a) Is the magnetic force on the charge directed into or out of the page? Explain. (b) If the charge is 0.25 C, its speed is 2.0×10^2 m/s, and it is acted on by a force of 20 N, what is the magnetic field strength?

15. ■ A charge of 0.050 C moves vertically in a field of 0.080 T that is oriented horizontally. What speed must the charge have such that the force acting on it is 10 N?

16. ■ A magnetic field can be used to determine the sign of charge carriers in a current-carrying wire. Consider a wide conducting strip in a magnetic field oriented as shown in ▼Fig. 19.34. The charge carriers are deflected

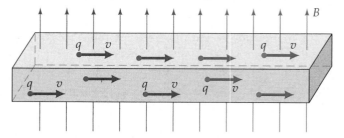

▲ **FIGURE 19.34 The Hall effect** See Exercise 16.

by the magnetic force and accumulate on one side of the strip, giving rise to a measurable voltage across it. (This phenomenon is known as the *Hall effect*.) If the sign of the charge carriers is unknown (they are either positive charges moving as indicated by the arrows in the figure or negative charges moving in the opposite direction), how does polarity or sign of the measured voltage allow the sign of the charge to be determined? Assume that only one type of charge carrier is responsible for the current.

17. ■■ A beam of protons is accelerated to a speed of 5.0×10^6 m/s in a particle accelerator and emerges horizontally from the accelerator into a uniform magnetic field. What **B** field perpendicular to the velocity of the proton would cancel the force of gravity and keep the beam moving exactly horizontally?

18. IE ■■ An electron travels in the $+x$ direction in a magnetic field and is acted upon by a magnetic force in the $-y$ direction. (a) Is the magnetic field directed in the (1) $-x$, (2) $+y$, (3) $+z$, or (4) $-z$ direction? Explain. (b) If the electron speed is 3.0×10^6 m/s and the magnitude of the force is 5.0×10^{-19} N, what is the magnetic field strength?

19. ■■ An electron travels at a speed of 2.0×10^4 m/s through a uniform magnetic field whose magnitude is 1.2×10^{-3} T. What is the magnitude of the magnetic force on the electron if its velocity and the magnetic field (a) are perpendicular, (b) make an angle of 45°, or (c) are parallel?

20. ■■ What angle(s) does a particle's velocity have to make with the direction of the magnetic field for the particle to be subjected to half the maximum possible magnetic force?

21. ■■■ A proton beam is first accelerated to a speed of 3.0×10^5 m/s in a particle accelerator. The beam then enters a uniform magnetic field of 0.50 T, which is oriented at an upward angle of 37° relative to the direction of the beam. (a) What is the initial acceleration of a proton in the accelerated beam? (b) If the beam were made of electrons rather than protons, what would be the difference in the force on the particles as the beam entered the magnetic field?

19.3 Applications: Charged Particles in Magnetic Fields

22. A mass spectrometer (a) can be used to determine the masses of atoms and molecules, (b) requires charged particles, (c) can be used to determine relative abundances of isotopes, or (d) all of these.

23. Why can a nearby magnet distort the display of a computer monitor or television picture tube (▶ Fig. 19.35)?

24. ■ An ionized deuteron (a particle with a $+e$ charge) passes through a velocity selector whose perpendicular magnetic and electric fields have magnitudes of 40 mT and 8.0 kV/m, respectively. Find the speed of the ion.

25. ■ In a velocity selector, the uniform magnetic field of 1.5 T is produced by a large magnet. Two parallel plates with a separation of 1.5 cm produce the perpendicular electric field. What voltage should be applied across the plates so that (a) a singly charged ion traveling at 8.0×10^4 m/s will pass through undeflected or (b) a doubly charged ion traveling at the same speed will pass through undeflected?

26. ■ A charged particle travels undeflected through perpendicular electric and magnetic fields whose magnitudes are 3000 N/C and 30 mT, respectively. Find the speed of the particle if it is (a) a proton or (b) an alpha particle. (An alpha particle is a helium nucleus—a positive ion with a double positive charge.)

27. ■■ In an experimental technique for treating deep tumors, unstable positively charged pions (π, elementary particles with a mass of 2.25×10^{-28} kg) are aimed to penetrate the flesh and to disintegrate at the tumor site, releasing energy to kill cancer cells. If pions with a kinetic energy of 10 keV are required and if a velocity selector with an electric field strength of 2.0×10^3 V/m is used, what must be the magnetic field strength?

28. ■■ In a mass spectrometer, a singly charged ion having a particular velocity is selected by using a magnetic field of 0.10 T perpendicular to an electric field of 1.0×10^3 V/m. This same magnetic field is then used to deflect the ion, which moves in a circular path with a radius of 1.2 cm. What is the mass of the ion?

29. ■■ In a mass spectrometer, a doubly charged ion having a particular velocity is selected by using a magnetic field of 100 mT perpendicular to an electric field of 1.0 kV/m. This same magnetic field is then used to deflect the ion in a circular path with a radius of 15 mm. Find (a) the mass of the ion and (b) the kinetic energy of the ion. (c) Does the kinetic energy of the ion increase in the circular path? Explain.

30. ■■■ In a mass spectrometer, a beam of protons enters a magnetic field. Some protons make exactly a one-quarter circular arc of radius 0.50 m. If the field is always perpendicular to the proton's velocity, what is the field's magnitude if exiting protons have a kinetic energy of 10 keV?

◀ **FIGURE 19.35 Magnetic disturbance** See Exercise 23.

19.4 Magnetic Forces on Current-Carrying Wires

and

19.5 Applications: Current-Carrying Wires in Magnetic Fields

31. A long, straight, horizontal wire located on the equator carries a current directed toward the east. What is the direction of the force on the wire due to the Earth's magnetic field? (a) East, (b) west, (c) south, or (d) upward?

32. **CQ** Two straight wires are parallel to each other. If the currents in the wires are in the same direction, will the wires attract or repel each other? What if the currents are opposite?

33. **CQ** What happens to the length of a spring when a large current passes through it? (*Hint:* Consider the current direction in the neighboring spring coils.)

34. **CQ** Is it possible to orient a current loop in a uniform magnetic field such that there is no torque on the loop? If yes, describe the orientation.

35. A magnetic field exists in the east–west direction, but it is unknown whether it points east or west. Can you determine the direction of the magnetic field, using a current-carrying wire with a known current direction? Explain.

36. In a galvanometer, the needle's deflection is directly proportional to (a) the current in the wire coil, (b) the number of turns in the coil, (c) the magnetic field, or (d) all of these.

37. Explain the operation of the doorbell and door chimes illustrated in ▼Fig. 19.36.

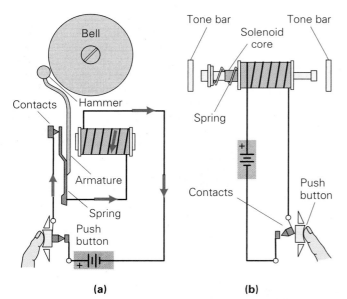

(a)   **(b)**

▲ **FIGURE 19.36 Electromagnetic applications** Both **(a)** a doorbell and **(b)** door chimes have electromagnets. See Exercise 37.

38. **IE** ■ A straight, horizontal segment of wire carries a current in the $+x$ direction in a magnetic field that is directed in the $-z$ direction. (a) Is the magnetic force on the wire directed in the (1) $-x$, (2) $+z$, (3) $+y$, or (4) $-y$ direction? Explain. (b) If the wire is 1.0 m long and carries a current of 5.0 A and the magnitude of the magnetic field is 0.30 T, what is the magnitude of the force on the wire?

39. ■ A 2.0-m length of straight wire carries a current of 20 A in a uniform magnetic field of 50 mT whose direction is at an angle of 37° from the direction of the current. Find the force on the wire.

40. ■ Show how you can use a right-hand force rule to find the direction of the current in a wire in a magnetic field if you know the force on the wire. The forces on some specific wires are shown in ▼Fig. 19.37.

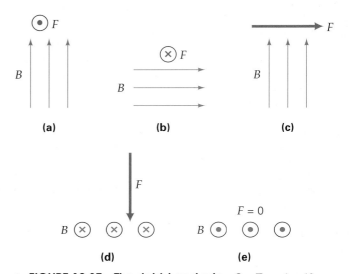

(a)   **(b)**   **(c)**

(d)   **(e)**

▲ **FIGURE 19.37 The right-hand rule** See Exercise 40.

41. ■■ A straight wire 50 cm long conducts a current of 4.0 A directed vertically upward. If the wire is acted on by a force of 1.0×10^{-2} N in the eastward direction due to a magnetic field at right angles to the length of the wire, what are the magnitude and direction of the magnetic field?

42. ■■ A horizontal magnetic field of 1.0×10^{-4} T is at an angle of 30° to the direction of the current in a straight, horizontal wire 75 cm long. If the wire carries a current of 15 A, what is the magnitude of the force on the wire?

43. ■■ A wire carries a current of 10 A in the $+x$ direction in a uniform magnetic field of 0.40 T. Find the magnitude of the force per unit length and the direction of the force on the wire if the magnetic field points in (a) the $+x$ direction, (b) the $+y$ direction, (c) the $+z$ direction, (d) the $-y$ direction, and (e) the $-z$ direction.

44. ■■ A straight wire 25 cm long is oriented vertically in a uniform horizontal magnetic field of 0.30 T pointing in the $-x$ direction. What current (including direction) would cause the wire to be subject to a force of 0.050 N in the $+y$ direction?

45. ■■ A wire carries a current of 10 A in the $+x$ direction. Find the force per unit length on the wire if it is in a magnetic field that has components of $B_x = 0.020$ T, $B_y = 0.040$ T, and $B_z = 0$ T.

46. ■■ A set of jumper cables used to start a car from another car's battery is connected to the terminals of both batteries. If 15 A of current exists in the cables during the starting procedure and the cables are parallel and 15 cm apart, what is the force per unit length on the cables?

47. IE ■■ Two long, straight, parallel wires carry current in the same direction. (a) Use the right-hand source and force rules to determine whether the forces on the wires are (1) attractive or (2) repulsive. (b) If the wires are 24 cm apart and carry currents of 2.0 A and 4.0 A, respectively, find the force per unit length on each wire.

48. IE ■■ Two long, straight, parallel wires 10 cm apart carry currents in opposite directions. (a) Use the right-hand source and force rules to determine whether the forces on the wires are (1) attractive or (2) repulsive. (b) If the wires carry equal currents of 3.0 A, what is the force per unit length on the wires?

49. ■■ A nearly horizontal dc power line on the equator carries a current of 1000 A directly eastward. If the Earth's magnetic field at the location of the power line is 5.0×10^{-5} T due north, what are the magnitude and direction of the magnetic force on a 15-m section of the line?

50. ■■ What is the force per unit length on wire 1 in ▼Fig. 19.38?

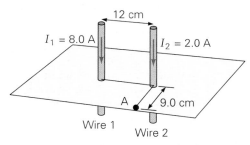

▲ **FIGURE 19.38 Parallel current-carrying wires** See Exercises 50, 51, 67, 70, and 71.

51. ■■ What is the force per unit length on wire 2 in Fig. 19.38?

52. IE ■■ A long wire is placed 2.0 cm directly below a rigidly mounted second wire (▶Fig. 19.39). (a) Use the right-hand source and force rules to determine whether the currents in the wires should be in (1) the same or (2) the opposite direction so that the lower wire is in equilibrium. (It "floats.") (b) If the lower wire has a linear mass density of 1.5×10^{-3} kg/m and the wires carry the same current, what should be the current?

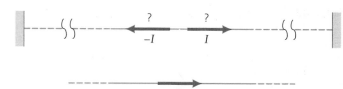

▲ **FIGURE 19.39 Magnetic suspension** The bottom wire is magnetically attracted to the top (rigidly fixed) wire. See Exercise 52.

53. ■■ A wire is bent as shown in ▼Fig. 19.40 and placed in a magnetic field with a magnitude of 1.0 T in the indicated direction. Find the magnitude of the force on each segment of the wire if $x = 50$ cm and the wire carries a current of 5.0 A to the left.

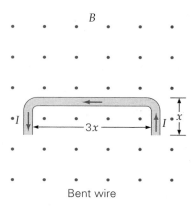

▲ **FIGURE 19.40 Current-carrying wire in a magnetic field** See Exercise 53.

54. IE ■■ A loop of current-carrying wire is in a 1.6-T magnetic field. (a) For the magnetic torque on the loop to be at maximum, should the plane of the coil be (1) parallel, (2) perpendicular, or (3) at a 45° angle to the magnetic field? Explain. (b) If the loop is rectangular with dimensions 20 cm by 30 cm and carries a current of 1.5 A, what is the magnitude of the magnetic moment of the loop, and what is the maximum torque?

55. ■■■ Two straight wires are positioned at right angles to each other as in ▼Fig. 19.41. What is the net force on each wire? Is there a net torque on each wire?

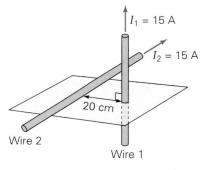

▲ **FIGURE 19.41 Perpendicular current-carrying wires** See Exercises 55 and 78.

56. ■■■ A rectangular wire loop with a cross-sectional area of 0.20 m² carries a current of 0.25 A. The loop is free to rotate about an axis that is perpendicular to a uniform magnetic field with strength 0.30 T. The plane of the loop is at an angle of 30° to the direction of the magnetic field. What is the magnitude of the torque on the loop?

19.6 Electromagnetism: The Source of Magnetic Fields

57. A long, straight wire is parallel to the ground and carries a steady current to the east. At a point directly below the wire, what is the direction of the magnetic field the wire produces? (a) North, (b) east, (c) south, or (d) west.

58. You are looking directly into one end of a long solenoid. The magnetic field at its center points at you. What is the direction of the current in the solenoid, as viewed by you? (a) Clockwise, (b) counterclockwise, (c) directly toward you, or (d) directly away from you.

59. A long, straight current-carrying wire is oriented vertically. On its east side, the field it creates points south. What is the direction of the current? (a) Up, or (b) down.

60. CQ A circular current-carrying loop is lying flat on a table and creates a field at the loop's center that points vertically upward. If you look straight down on the loop, what is the direction of the current?

61. CQ If the distance from a long current-carrying wire is doubled, what happens to the magnetic field strength?

62. CQ There are two solenoids, one with 100 turns and the other with 200 turns. If both carry the same current, will the one with more turns necessarily produce a stronger magnetic field at its center? Explain.

63. ■ The magnetic field at the center of a 50-turn coil of radius 15 cm is 0.80 mT. Find the current in the coil.

64. ■ A long, straight wire carries a current of 2.5 A. Find the magnitude of the magnetic field 25 cm from the wire.

65. ■ In a physics lab, a student discovers that the magnitude of the magnetic field at a certain distance from a long wire is 4.0 μT. If the wire carries a current of 5.0 A, what is the distance of the magnetic field from the wire?

66. ■ A solenoid is 0.20 m long and consists of 100 turns of wire. At its center, the solenoid produces a magnetic field with a strength of 1.5 mT. Find the current in the coil.

67. ■■ Two long, parallel wires carry currents of 8.0 A and 2.0 A (Fig. 19.38). (a) What is the magnitude of the magnetic field midway between the wires? (b) Where on a line perpendicular to and joining the wires is the magnetic field zero?

68. ■■ Two long, parallel wires separated by 50 cm each carry currents of 4.0 A in a horizontal direction. Find the magnetic field midway between the wires if the currents are (a) in the same direction and (b) in opposite directions.

69. ■■ Two long, parallel wires separated by 0.20 m carry equal currents of 1.5 A in the same direction. Find the magnitude of the magnetic field 0.15 m away from each wire on the side opposite the other wire (▼Fig. 19.42).

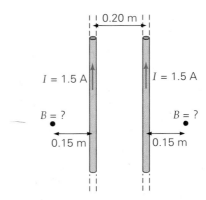

▲ FIGURE 19.42 Magnetic field summation
See Exercise 69.

70. ■■ In Fig. 19.38, find the magnetic field (magnitude and direction) at point A, which is located 9.0 cm away from wire 2 on a line perpendicular to the line joining the wires.

71. ■■ Suppose that the current in wire 1 in Fig. 19.38 were in the opposite direction. What would be the magnetic field midway between the wires?

72. ■■ How much current must flow in a circular loop of radius 10 cm to produce a magnetic field at the center of the loop that is the same magnitude as the horizontal component of the Earth's magnetic field at the equator?

73. ■■ A coil of four circular loops of radius 5.0 cm carries a current of 2.0 A clockwise, as viewed from above the coil's plane. What is the magnetic field at the center of the coil?

74. IE ■■ A circular loop of wire in the horizontal plane carries a counterclockwise current, as viewed from above. (a) Use the right-hand source rule to determine whether the direction of the magnetic field at the center of the loop is (1) toward or (2) away from the observer. (b) If the diameter of the loop is 12 cm and the current is 1.8 A, what is the magnitude of the magnetic field at the center of the loop?

75. ■■ A circular loop of wire with a radius of 5.0 cm carries a current of 1.0 A. Another circular loop of wire is concentric with (that is, has a common center with) the first and carries a current of 2.0 A. The magnetic field at the center of the loops is zero. What is the radius of the second loop?

76. ■■ A solenoid 10 cm long is wound with 1000 turns of wire. How much current must exist in the windings to produce a magnetic field of 4.0×10^{-4} T at the solenoid's center?

77. ■■ A solenoid is wound with 200 turns per centimeter. An outer layer of insulated wire with 180 turns per centimeter is wound over the solenoid's first layer of wire. When the solenoid is operating, the inner coil carries a current of 10 A and the outer coil carries a current of 15 A in the direction opposite to that of the current in the inner coil (▼Fig. 19.43). What is the magnitude of the magnetic field at the center of the doubly wound solenoid?

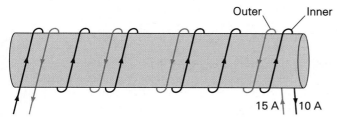

15 A   10 A

▲ FIGURE 19.43 Double it up? See Exercise 77.

78. ■■■ Two long, perpendicular wires carry currents of 15 A, as illustrated in Fig. 19.41. What is the magnitude of the magnetic field at the midpoint of the line joining the wires?

79. ■■■ Four wires running through the corners of a square with sides of length a, as shown in ▼Fig. 19.44, carry equal currents I. Calculate the magnetic field at the center of the square in terms of these parameters.

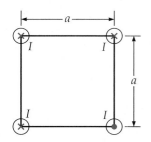

◀ FIGURE 19.44 Current-carrying wires in a square array See Exercise 79.

80. ■■■ A particle with charge q and mass m moves in a horizontal plane at right angles to a uniform vertical magnetic field B. (a) What is the frequency f of the particle's circular motion in terms of q, B, and m? (This frequency is called the *cyclotron frequency*.) (b) Show that the time required for any charged particle to make one complete revolution is independent of its speed and radius. (c) Compute the path radius and the cyclotron frequency if the particle is an electron with speed $v = 10^5$ m/s and the field strength is $B = 10^{-4}$ T.

19.7 Magnetic Materials

81. The main source of magnetism in magnetic materials is from (a) electron orbits, (b) electron spin, (c) magnetic poles, or (d) nuclear properties.

82. When a ferromagnetic material is placed in an external magnetic field, (a) the domain orientation may change, (b) the domain boundaries may change, (c) new domains are created, or (d) both (a) and (b).

83. CQ If you are looking down on the orbital plane of the electron in a hydrogen atom and the electron orbits counterclockwise, what is the direction of the magnetic field the electron produces at the proton?

84. CQ What is the purpose of the iron core often used at the center of a solenoid?

85. CQ How can you destroy or reduce the magnetic field of a permanent magnet?

86. ■■ A solenoid with 100 turns per centimeter has an iron core with a relative permeability of 2000. The solenoid carries a current of 1.2 A. (a) What is the magnetic field at the center of the solenoid? (b) How much greater is the magnetic field with the iron core than it would be without it?

87. ■■■ What is the magnetic field (due to the electron only) at the center of the circular orbit of the electron in a hydrogen atom? The orbital radius is 0.0529 nm. [*Hint*: Find the electron's period by considering the centripetal force.]

*19.8 Geomagnetism: The Earth's Magnetic Field

88. The Earth's magnetic field (a) has poles that coincide with the geographic poles, (b) only exists at the poles, (c) reverses polarity every few hundred years, or (d) none of these.

89. Aurorae (see Fig. 19.32) (a) occur only in the Northern Hemisphere, (b) are related to the lower Van Allen belt, (c) occur because of Earth's magnetic pole reversals, or (d) happen predominantly when there are no solar disturbances.

90. CQ What is the direction of the force due to the Earth's magnetic field on an electron near the equator if the electron's velocity is directed (a) due south, (b) northwest, or (c) upward?

91. CQ What is the polarity of the magnetic pole near the Earth's geographic South Pole?

Additional Exercises

92. A proton is accelerated from rest through a potential difference of 3.0 kV. It then enters a region where its velocity is initially perpendicular to an electric field. The field is created by two parallel plates separated by 10 cm with a potential difference of 250 V across them. Find the magnitude of the magnetic field (perpendicular to E) needed to allow the proton to pass undeflected between the plates.

93. A solenoid 10 cm long has 3000 turns of wire and carries a current of 5.0 A. A 2000-turn coil of wire of the same length as the solenoid surrounds it and is concentric (shares a common central axis) with it. The outer coil carries a current of 10 A in the same direction as that of the current in the solenoid. Find the magnetic field at the common center.

94. **IE** A horizontal beam of electrons travels from north to south in a discharge tube, where the Earth has a downward component of its magnetic field. (a) Is the direction of the magnetic force on the electron directed to (1) west, (2) east, (3) south, or (4) north? Explain. (b) If the electron's speed is 1.0×10^3 m/s and the downward component of the Earth's magnetic field is 5.0×10^{-5} T, what is the magnitude of the force on the electrons?

95. A proton enters a uniform magnetic field that is at a right angle to its velocity. The field strength is 0.80 T and the proton follows a circular path with a radius of 4.6 cm. What is the kinetic energy of the proton?

96. A square loop carries a current of 10 A. When placed in a magnetic field of 500 mT at an angle of 50° relative to the loop's plane, the loop experiences a torque of 0.15 m·N about an axis through the loop's center and parallel to one of the sides. Find the length of a side of the square.

97. A proton is accelerated from rest through a potential difference of 1.0 kV. It enters a uniform magnetic field of 4.5 mT that is initially perpendicular to its velocity.

(a) Find the radius of the proton's circular path. (b) Calculate the period of revolution of the proton.

98. **IE** A beam of protons traveling north is required to pass undeflected through two horizontal parallel plates, where a constant electric field and a magnetic field exist at right angles to each other. The electric field is directed vertically upward from the bottom to the top plate. (a) Should the magnetic field be directed (1) west, (2) east, (3) south, or (4) north? Explain. (b) If the speed of the protons is 2.0×10^2 m/s and the electric field has a magnitude of 100 V/m, what is the magnitude of the magnetic field?

99. How fast should an electron travel at right angles to a magnetic field of 0.010 T so that the magnetic force just balances the gravitational force on the electron? Ignore the Earth's field.

100. A particle with a charge of 4.0×10^{-8} C moves at a speed of 3.0×10^2 m/s through a magnetic field in the direction at which the magnetic force on the particle is maximum. If the force on the particle is 1.8×10^{-6} N, what is the magnitude of the magnetic field?

101. A current-carrying loop is rectangular with dimensions of 20 cm by 30 cm. The loop carries a current of 10 A and is in a uniform magnetic field of 50 mT directed parallel to the plane of the loop. Find the torque on the loop.

Electromagnetic Induction and Waves

20.1 Induced emf: Faraday's Law and Lenz's Law

20.2 Electric Generators and Back emf

20.3 Transformers and Power Transmission

20.4 Electromagnetic Waves

INSIGHT

■ Electromagnetic Induction in Action

As we saw in Chapter 19, an electric current produces a magnetic field. But the mutual relationship between electricity and magnetism does not stop there. In this chapter, you will learn that under the right conditions, a magnetic field can produce an electric current. How is this done? Chapter 19 considered only *constant* magnetic fields. No current is induced in a loop of wire that is stationary in a constant magnetic field. However, if the magnetic field changes with time, or if the wire loop moves across, or is rotated in, the field, a current *is* induced in the wire.

The practical uses of this interrelationship of electricity and magnetism are many. One example happens during the playing of music recorded on an audiocassette, which is actually a magnetic tape that has information encoded on it as variations in its magnetism. These variations can be used to produce electrical currents, which, in turn, are amplified and drive speakers, from which sound waves are emitted. Similar processes are involved when information is stored on or retrieved from a disk in your computer.

On a larger scale, consider the generation of the electric energy that provides the basis for our modern civilization. A great deal of the energy that we extract from the environment is converted to electricity before being put to use. At hydroelectric plants such as that in the photo, one of the oldest and simplest energy sources on Earth—falling water—is used to generate the electric energy. The gravitational potential energy of the water is converted into kinetic energy, and some of this kinetic energy is transformed, eventually, into electric energy. But how does this last step take place?

Regardless of the ultimate source of the energy—the burning of oil, coal, or gas; a nuclear reactor; or falling water—the actual conversion to electric energy is accomplished by means of magnetic fields and electromagnetic induction. This chapter not only examines the underlying electromagnetic principles that make such conversion possible, but also discusses several practical applications. Moreover, we will also see that the creation and propagation of electromagnetic radiation (or "light") is intimately related to electromagnetic induction.

20.1 Induced emf: Faraday's Law and Lenz's Law

OBJECTIVES: **To (a) define magnetic flux and explain how an induced emf is created and (b) determine induced emfs and currents.**

It is observed experimentally that a magnet held stationary near a conducting wire loop does *not* induce a current in that loop (▼ Fig. 20.1a). If the magnet is moved toward the loop, however, as shown in Fig. 20.1b, the deflection of the galvanometer needle indicates that current exists in the loop, but only during the motion. Furthermore, if the magnet is moved away from the loop, as shown in Fig. 20.1c, the galvanometer needle is deflected in the opposite direction, which indicates a reversal of the current's direction, but, again, only during the motion.

Deflections of the galvanometer needle, indicating the presence of *induced currents*, also occur if it is the loop that is moved toward or away from the stationary magnet. The effect thus depends on *relative* motion of the loop and magnet. The induced current depends on the speed of that relative motion. However, experimen-

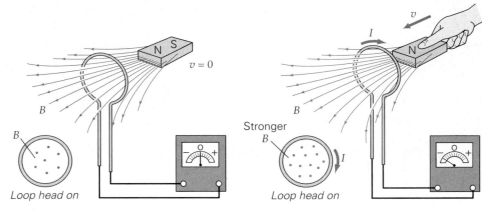

(a) No motion between magnet and loop   **(b) Magnet is moved toward loop**

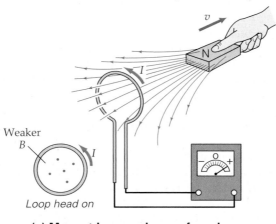

(c) Magnet is moved away from loop

▲ **FIGURE 20.1   Electromagnetic induction   (a)** When there is no relative motion between the magnet and the wire loop, the number of field lines through the loop (in this case, 7) is constant, and the galvanometer shows no deflection. **(b)** Moving the magnet toward the loop increases the number of field lines passing through the loop (now 12), and an induced current is detected. **(c)** Moving the magnet away from the loop decreases the number of field lines passing through the loop (to 5). The induced current is now in the opposite direction. (See the needle.)

tally, there is a noteworthy exception. If a loop is moved (but not rotated) in a *uniform* magnetic field, as shown in ▶Fig. 20.2, no current is induced.

Yet another way to induce a current in a stationary wire loop is to vary the current in another, nearby loop. When the switch in the battery-powered circuit in ▼Fig. 20.3a is closed, the current in the right-hand loop goes from zero to some constant maximum value in a short time. Only during the time of buildup of current does the magnetic field caused by the current in this loop (and therefore also in the region of the left-hand loop) increase. The galvanometer needle deflects, indicating current in the left-hand loop. When the current in the right-hand loop attains its maximum value, the magnetic field it produces becomes constant, and the current in the left-hand loop drops to zero. Similarly, when the switch in the right-hand loop is opened (Fig. 20.3b), its current and field decrease quickly to zero, and the galvanometer deflects in the opposite direction, indicating a reversal in direction of the current induced in the left-hand loop. Note that in both cases, *the induced current in the left-hand loop occurs only while the magnetic field through that loop changes.*

In Fig. 20.1, moving the magnet changes the magnetic environment in a loop, causing an induced emf that, in turn, causes an induced current. For the case of *two* stationary loops (Fig. 20.3), a changing current in the right-hand loop produced a changing magnetic environment in the left-hand loop, thereby inducing an emf and a current in the left-hand loop.* There is a convenient way of summarizing what is happening in both Fig. 20.1 and Fig. 20.3: To induce currents in a loop or complete circuit, a process called **electromagnetic induction**, all that matters is whether the magnetic field through the loop or circuit is changing.

Detailed experiments on electromagnetic induction were done independently by Michael Faraday in England and Joseph Henry in the United States around 1830. Instead of using the subjective term "magnetic environment," Faraday found that the important factor in electromagnetic induction was the time rate of change of the number of magnetic field lines passing through the loop or circuit area. That is, he discovered that

> an induced emf is produced in a loop or complete circuit whenever the number of magnetic field lines passing through the plane of the loop or circuit changes.

Magnetic Flux

Since the induced emf (and induced current) in a loop depends on the rate of change of the number of magnetic field lines passing through it, determining induced emf demands that the number of field lines through the loop be quantified. Consider a loop of wire in a uniform magnetic field (▶Fig. 20.4a). The number of field lines through the loop depends on the loop's area, its orientation relative to the B field, and the strength of that field. To describe the loop's orientation, the concept of an *area vector* (**A**) is employed. Its direction is normal to the loop's plane, and its magnitude is equal to the area of the loop. An angle θ, which is the angle between the magnetic field (**B**) and the area vector (**A**), is used as a measure of their relative orientation. For example, in Fig. 20.4a, $\theta = 0°$, meaning that the two vectors are in the same direction, or that the area plane is perpendicular to the magnetic field.

For the case of a magnetic field that does not vary over the area, the number of magnetic field lines passing through a particular area (the area within a loop, in our case) is proportional to the **magnetic flux** (Φ), which is defined as

$$\Phi = BA \cos \theta \quad \begin{array}{c} \textit{magnetic flux} \\ \textit{(constant magnetic field)} \end{array} \quad (20.1)$$

SI unit of magnetic flux: tesla-meter squared ($\text{T} \cdot \text{m}^2$), or weber (Wb)

---

*The term *mutual induction* is used to describe the situation when emfs and currents are induced between two (or more) loops.

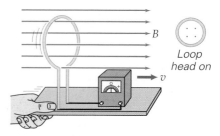

▲ **FIGURE 20.2 Relative motion and no induction** When a loop is moved parallel to a uniform magnetic field, there is no change in the number of field lines passing through the loop, and there is no induced current.

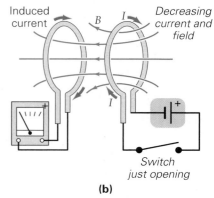

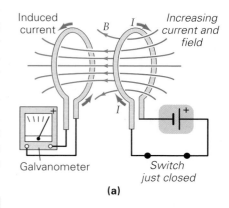

▲ **FIGURE 20.3 Mutual induction** **(a)** When the switch is closing in the right-loop circuit, the buildup of current produces a changing magnetic field in the other loop, inducing a current in it. **(b)** When the switch is opened, the magnetic field collapses, and the magnetic field in the left-hand loop decreases. The induced current in this loop is then in the opposite direction. The induced currents occur only when the magnetic field passing through a loop changes.

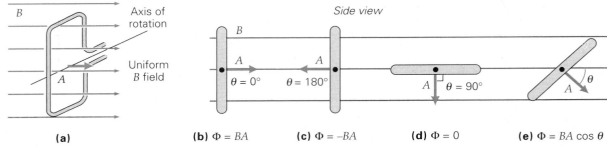

(a)
            (b) $\Phi = BA$ **(c)** $\Phi = -BA$ **(d)** $\Phi = 0$ **(e)** $\Phi = BA \cos \theta$

▲ **FIGURE 20.4 Magnetic flux** (a) Magnetic flux (Φ) is a measure of the number of field lines passing through an area (A). The area can be represented by a vector **A** perpendicular to the plane of the area. (b) When the plane of a loop is perpendicular to the field and $\theta = 0°$, then $\Phi = \Phi_{max} = +BA$. (c) When $\theta = 180°$, the magnetic flux has the same magnitude, but is opposite in direction: $\Phi = -\Phi_{max} = -BA$. (d) When $\theta = 90°$, then $\Phi = 0$. (e) As the loop plane is changed from an orientation perpendicular to the field to one more parallel to the field, less area is open to the field lines, and the flux decreases. In general, $\Phi = BA \cos \theta$.

The SI unit of the magnetic field B is the tesla $[1\text{ T} = 1\text{ N}/(\text{A} \cdot \text{m})]$; thus, magnetic flux has SI units of $\text{T} \cdot \text{m}^2$. However, when the emphasis is on flux, the tesla is sometimes replaced by the weber (Wb) per square meter ($1\text{ T} = 1\text{ Wb/m}^2$). Thus, the weber is a unit of flux (Φ). To see this relationship, note that the units of BA are $(\text{Wb/m}^2)(\text{m}^2)$, or Wb. (The cosine is dimensionless.)

The orientation of the loop with respect to the magnetic field affects the number of field lines passing through it, and this factor is accounted for by the cosine term in Eq. 20.1. Let us consider several possible orientations:

- If **B** and **A** are parallel ($\theta = 0°$), then the magnetic flux is positive and has a maximum value of $\Phi_{max} = BA \cos 0° = +BA$. The maximum possible number of magnetic field lines pass through the loop (Fig. 20.4b).
- If **B** and **A** are oppositely directed ($\theta = 180°$), then the magnitude of the magnetic flux is a maximum again, but of opposite sign: $\Phi_{180°} = BA \cos 180° = -BA = -\Phi_{max}$ (Fig. 20.4c).
- If **B** and **A** are perpendicular, then there are no field lines passing through the plane of the loop, and the flux is zero: $\Phi_{90°} = BA \cos 90° = 0$ (Fig. 20.4d).
- For situations at intermediate angles, the flux is less than the maximum value, but nonzero (Fig. 20.4e). We can interpret $A \cos \theta$ as the effective area of the loop perpendicular to the field lines (◄Fig. 20.5a). Alternatively, $B \cos \theta$ can be viewed as the perpendicular component of the field through the full area of the loop, A, as shown in Fig. 20.5b. Thus, Eq. 20.1 can be thought of as either $\Phi = (B \cos \theta)A$ or $\Phi = B(A \cos \theta)$, depending on the interpretation. In either case, the answer is the same.

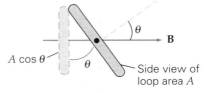

$\Phi = B (A \cos \theta)$

(a)

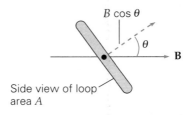

Side view of loop area A

$\Phi = (B \cos \theta) A$

(b)

▲ **FIGURE 20.5 Magnetic flux through a loop: An alternative interpretation** Instead of defining the flux (Φ) **(a)** in terms of the magnetic field (B) passing through a reduced area ($A \cos \theta$), we can define it **(b)** in terms of the perpendicular component of the magnetic field ($B \cos \theta$) passing through A. Either way, Φ is a measure of the number of field lines passing through A and is given by $\Phi = BA \cos \theta$ (Eq. 20.1).

Faraday's Law of Induction and Lenz's Law

From quantitative experiments, Faraday determined that the emf induced in a coil (consisting of N loops) depends on the time rate of change of the number of magnetic field lines through all the loops, or the *time rate of change of the magnetic flux through all the loops*. This dependence, known as **Faraday's law of induction**, is expressed mathematically as

$$\mathcal{E} = -N\frac{\Delta \Phi}{\Delta t} \quad \textit{induced emf} \qquad (20.2)$$

where $\Delta \Phi$ is the change in flux through *one loop*. In a coil of N loops of wire, the total change of flux is $N\Delta \Phi$. Note that the induced emf $\mathcal{E}$ in Eq. 20.2 is an *average* value over the time interval Δt. (Why?)

The minus sign is included in Eq. 20.2 to give an indication of the *polarity*, or *direction*, of the induced emf, which is found by considering the resulting induced current and its effect. A Russian physicist, Heinrich Lenz (1804–1865), discovered the law that governs the direction of the induced emf. **Lenz's law** is stated as follows:

> An induced emf in a wire loop or coil has a direction, or polarity, such that the current it creates produces its own magnetic field that opposes the *change* in magnetic flux through that loop or coil.

This law means that the magnetic field *due to the induced current* is in a direction that tries to keep the flux through the loop from changing. For example, if the flux increases in the $+x$ direction, the magnetic field due to the induced current will be in the $-x$ direction (▼Fig. 20.6a). This effect tends to cancel the increase in the flux, or *oppose the change*. Essentially, the magnetic field due to the induced current tries to maintain the existing magnetic flux. This effect is sometimes called "electromagnetic inertia," by analogy to the tendency of objects to resist changes in their velocity. In the long run, the induced current cannot prevent the magnetic flux from changing. However, during the time that the flux through the coil is changing, the induced magnetic field will oppose the change.

The direction of the induced current is given by the **induced-current right-hand rule**:

> With the thumb of the right hand pointing in the direction of the induced field, the fingers point in the direction of the induced current. (See Fig. 20.6b and Integrated Example 20.1.)

You might recognize this rule as a version of the right-hand rules used to find the direction of a magnetic field produced by a current (Chapter 19). Here it is used in reverse. Typically, you know the induced *B*-field direction (for example, $-x$ in Fig. 20.6b) and want the direction of the current that produces it. An application of Lenz's law is illustrated in Integrated Example 20.1.

▼ **FIGURE 20.6   Finding the direction of the induced current**   **(a)** An external magnetic field is shown increasing to the right. The induced current creates its own magnetic field to try to counteract the flux change that is occurring. **(b)** The induced current right-hand rule determines the direction of the induced current. Given that the direction of the induced magnetic field is to the left, with the thumb of the right hand pointing that way, the fingers give the current direction as shown.

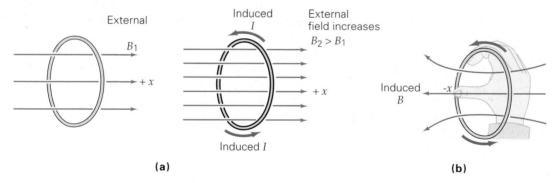

(a)   (b)

Integrated Example 20.1 ■ Lenz's Law and Induced Currents

(a) The south end of a bar magnet is pulled far away from a tiny wire coil. (See ▶Fig. 20.7a.) Looking from behind the coil (Fig. 20.7b), which way is the induced current: (1) counterclockwise, (2) clockwise, or (3) there is no induced current? (b) Suppose that the magnetic field across the area of the coil is initially 0.040 T, the coil's radius is 2.0 mm, and there are 100 loops in the coil. Determine the magnitude of the average induced emf in the coil if the bar magnet is removed in 0.75 s. Assume that the magnetic field is constant over the coil area.

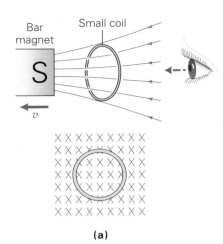

(a)

(b)

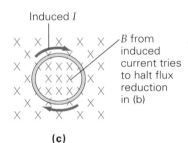

Induced I

B from induced current tries to halt flux reduction in (b)

(c)

▲ **FIGURE 20.7 Using a bar magnet to induce currents**
(a) The south end of a bar magnet is pulled rapidly away from a small wire loop. **(b)** The view from the right of the loop shows the magnetic field pointing away from the observer, or into the page, and decreasing. **(c)** To try to counteract the loss of magnetic flux into the page, current is induced in the clockwise direction, so as to provide its own magnetic field into the page. See Integrated Example 20.1.

(a) Conceptual Reasoning. There is initially magnetic flux *into* the plane of the coil (Fig. 20.7b), and at the end of the experiment, when the magnet is very far away from the coil, there is no flux through the coil; thus, the flux has changed. Therefore, there must be some induced emf, so answer (3) cannot be correct. As the bar magnet pulls away, the field weakens, but it maintains this direction. The induced emf will thus produce an (induced) current that, in turn, will produce a magnetic field into the page so as to try to prevent this decrease in flux. Therefore, the induced current (as well as the polarity of the induced emf) is in the clockwise direction, as found using the induced-current right-hand rule (Fig. 20.7c).

(b) Thinking It Through. This example is a straightforward application of Eq. 20.2. The initial flux is the maximum possible. The following data are given:

Given: $B_i = 0.040$ T   *Find:* Induced emf $\mathscr{E}$ (magnitude)
$\quad\quad\quad r = 2.00$ mm $= 2.00 \times 10^{-3}$ m
$\quad\quad\quad N = 100$ loops
$\quad\quad\quad \Delta t = 0.75$ s

To find the initial magnetic flux through one loop of the coil, use Eq. 20.1 with an angle of $\theta = 0°$. (Why?) First, the area must be computed:

$$A = \pi r^2 = \pi(2.00 \times 10^{-3}\,\text{m})^2 = 1.26 \times 10^{-5}\,\text{m}^2$$

Therefore, the initial flux, Φ_i, through one loop is

$$\Phi_i = B_i A \cos\theta = (0.040\,\text{T})(1.26 \times 10^{-5}\,\text{m}^2)\cos 0° = 5.03 \times 10^{-7}\,\text{T}\cdot\text{m}^2$$

Since the final flux is zero, $\Delta\Phi = \Phi_f - \Phi_i = 0 - \Phi_i = -\Phi_i$. Therefore, the average induced emf (magnitude) is

$$|\mathscr{E}| = N\frac{|\Delta\Phi|}{\Delta t} = (100\,\text{loops})\frac{(5.03 \times 10^{-7}\,\text{T}\cdot\text{m}^2\,\text{loop})}{(0.75\,\text{s})} = 6.70 \times 10^{-5}\,\text{V} = 67.0\,\mu\text{V}$$

Follow-up Exercise. In this Example, (a) in which direction is the induced current if instead a north magnetic pole approaches the coil quickly? Explain. (b) In this example, what would be the average induced current if the coil had a total resistance of 0.2 Ω? *(Answers to all Follow-up Exercises are at the back of the text.)*

Lenz's law incorporates the principle of energy conservation. Consider a situation where a wire loop has an increasing magnetic flux through its area. Contrary to Lenz's law, suppose instead that the magnetic field from the induced current *added* to the flux instead of keeping it at its original value. This increased flux would then lead to an even greater induced current. In turn, this greater induced current would produce a still greater magnetic flux, which in turn would give a greater induced current, and so on. Such a something-for-nothing energy situation would violate the law of the conservation of energy.

To understand the polarity of the induced emf in a loop in terms of forces, consider the case of the moving magnet (for example, Fig. 20.1b). Recall that a current-carrying loop creates its own magnetic field similar to that of a bar magnet. (See Figs. 19.3 and 19.25.) Thus, the induced current sets up a magnetic field in the loop, and that loop acts like a bar magnet with a polarity that opposes the motion of the real bar magnet (▶Fig. 20.8). You should be able to show that if the bar magnet is pulled away from the loop, the loop exerts a magnetic attraction so as to keep the magnet from leaving—electromagnetic inertia in action.

Substituting the expression for the magnetic flux (Φ) given by Eq. 20.1 into Eq. 20.2, we have

$$\mathscr{E} = -\frac{N\Delta\Phi}{\Delta t} = -\frac{N\Delta(BA\cos\theta)}{\Delta t} \tag{20.3}$$

Thus, an induced emf results if the

1. the strength of the magnetic field, B, changes and/or,
2. the loop area, A, changes and/or,
3. the angle between the normal to the loop area and the field, θ, changes.

In situation (1), a flux change is created because of a time-varying magnetic field, such as can be obtained from a time-varying current in a nearby circuit or by moving a magnet near a coil, as in Fig. 20.1 (or by moving the coil near the magnet).

In situation (2), a flux change results because of a time-varying loop area. This situation might occur if a loop were stretched or flattened or had an adjustable circumference (such as an adjustable loop around a balloon as it is being blown up; see Exercise 21 at the end of the chapter).

Finally, in situation (3), a change in flux can result from a *change in orientation of the loop with time*. This situation can occur when a coil is rotated in a magnetic field. The change in the number of field lines through a single loop is evident in the sequential views of the loop in Fig. 20.4. Rotating a coil in a magnetic field is a common way of inducing an emf and will be considered separately in Section 20.2. The emfs that result from changing the field strength and loop area are analyzed in the next three Examples. (Also, see the Insight on p. 677 for some applications of electromagnetic induction.)

Conceptual Example 20.2 ■ Fields in the Fields: Electromagnetic Induction

In rural areas where electric power lines carry electricity to big cities, it is possible to generate small electric currents by means of induction in a conducting loop. The overhead power lines carry relatively large alternating currents that periodically reverse in direction 60 times per second. How would you orient the plane of the loop to best produce an induced current if the power lines run north to south, (a) parallel to the Earth's surface, (b) perpendicular to the Earth's surface in the north–south direction, or (c) perpendicular to the Earth's surface in the east–west direction? (See ▸ Fig. 20.9a.)

Reasoning and Answer. Magnetic field lines from long wires are circular in shape and centered on the wire. (See Fig. 19.23.) By the source right-hand rule, the direction of the magnetic field at ground level is parallel to the Earth's surface, but alternates in direction. The orientation choices are shown in Fig. 20.9b. Neither answer (a) nor (c) cannot be correct, because in this orientation, there would *never* be any magnetic flux passing through the loop. With a constant flux (in this case, zero), there is no induced emf. Hence, the answer must be (b). If the loop is oriented perpendicular to the Earth's surface with its plane in the north–south direction, the flux through it would vary from zero to its maximum value and back 60 times per second, and there would be an induced emf in the loop.

Follow-up Exercise. Suggest possible ways of increasing the amount of current produced by the arrangement discussed in this Example by changing only properties of the loop and not of the overhead wire.

Example 20.3 ■ Induced Currents: A Potential Hazard to Equipment?

Electrical instruments can be damaged or destroyed if they are in a rapidly changing magnetic field [situation (1) below Eq. 20.3]. This might occur if an instrument were located near an electromagnet operating under alternating-current conditions. It is possible that the electromagnet's external field could produce a changing magnetic flux within the instrument. If the resulting induced currents are large enough, they could damage the instrument. Consider a laptop computer's speaker that is near such an electromagnet (▸ Fig. 20.10). Suppose an electromagnet exposes the speaker to a maximum magnetic field of 1.00×10^{-3} T which reverses direction every $\frac{1}{120}$ s.

Assume that the speaker's coil consists of 100 circular wire loops (each with a radius of 3.00 cm) with a total resistance of 1.00 Ω. According to the manufacturer of the

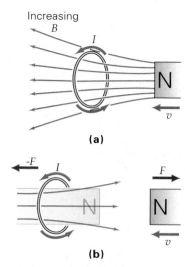

▲ **FIGURE 20.8 Lenz's law described in terms of forces** (a) If the north end of a bar magnet is moved rapidly toward a wire loop, current is induced in the direction shown. (b) While the induced current exists, the loop acts like a small bar magnet with its "north end" close to the north end of the real bar magnet. Thus, there is a magnetic repulsion. This is an alternative way of viewing Lenz's law: Induce a current so as to try to keep the flux from changing.

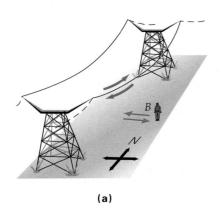

(a)

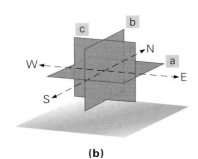

(b)

▲ **FIGURE 20.9 Induced emfs below electric power lines** **(a)** If current-carrying wires run in the north–south direction, then the alternating current in them produces a magnetic field on the ground below that periodically points east and then west. **(b)** There are three choices for loop orientation in Conceptual Example 20.2.

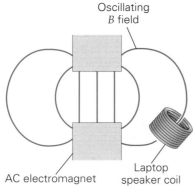

▲ **FIGURE 20.10 Instrument hazard?** The small coil of a computer speaker system is placed close to an alternating-current electromagnet. The changing flux in the coil produces an induced emf and, thus, an induced current that depends on the resistance of the coil. See Example 20.3.

speaker, the current in the coil should not exceed 25.0 mA. (a) Calculate the magnitude of the average induced emf in the coil during the $\frac{1}{120}$ s interval. (b) Does the average induced current exceed the manufacturer's limit?

Thinking It Through. (a) The flux goes from a (maximum) positive to a (maximum) negative value in $\frac{1}{120}$ s. The magnetic fluxes and flux change can be determined from Eq. 20.1 with $\theta = 0°$ and $\theta = 180°$. The average induced emf can then be calculated from Eq. 20.2. (b) Once we know the emf, the induced current can be calculated from $I = \mathcal{E}/R$.

Solution. Listing the data, we have the following:

Given: $B_i = +1.00 \times 10^{-3}$ T (pointing one way)   *Find:* (a) $\mathcal{E}$ (average induced
      $B_f = -1.00 \times 10^{-3}$ T (pointing the   emf in coil)
              opposite way)   (b) I (average induced
      $\Delta t = \frac{1}{120}$ s $= 8.33 \times 10^{-3}$ s   current in coil)
      $N = 100$ loops
      $R = 1.00 \ \Omega$
      $r = 3.00$ cm $= 3.00 \times 10^{-2}$ m
      $I_{max} = 25.0$ mA

(a) First, we compute the circular loop area:

$$A = \pi r^2 = \pi(3.00 \times 10^{-2} \text{ m})^2 = 2.83 \times 10^{-3} \text{ m}^2$$

Then the initial flux through *one* loop is given by Eq. 20.1:

$$\Phi_i = B_i A \cos \theta = (1.00 \times 10^{-3} \text{ T})(2.83 \times 10^{-3} \text{ m}^2)(\cos 0°) = 2.83 \times 10^{-6} \text{ T} \cdot \text{m}^2/\text{loop}$$

Since the final flux is the negative of this, the change in flux through one loop is

$$\Delta \Phi = \Phi_f - \Phi_i = -\Phi_i - \Phi_i = -2\Phi_i = -5.66 \times 10^{-6} \text{ T} \cdot \text{m}^2/\text{loop}$$

From Eq. 20.2, we obtain

$$\mathcal{E} = N\frac{|\Delta \Phi|}{\Delta t} = (100 \text{ loops})\left(\frac{5.66 \times 10^{-6} \text{ T} \cdot \text{m}^2/\text{loop}}{8.33 \times 10^{-3} \text{ s}}\right) = 6.79 \times 10^{-2} \text{ V}$$

(b) This voltage is small by everyday standards, but keep in mind that the speaker coil's resistance is also small. The induced emf will create an induced current in the speaker coil. To determine the current, we have

$$I = \frac{\mathcal{E}}{R} = \frac{6.79 \times 10^{-2} \text{ V}}{1.00 \ \Omega} = 6.79 \times 10^{-2} \text{ A} = 67.9 \text{ mA}$$

This value exceeds the allowed speaker current of 25.0 mA.

Follow-up Exercise. In this example, if the speaker coil were moved farther from the electromagnet, it could reach a point where the induced average current was equal to 25.0 mA. Determine the magnetic field strength, B_{max} at this point.

Integrated Example 20.4 ■ **The Essence of Electric-Energy Generation: Mechanical Work into Electrical Current**

Changes in loop area, as depicted in ▶ Fig. 20.11a, can be used to study the basic principles behind the generation of electrical energy. An external force does work as the moveable bar is pulled upward, and this work is converted to electrical energy. Since the "circuit" (wires, resistor, and bar in the figure) is in a magnetic field, the flux through it changes with time, thereby inducing a current. (a) In Fig. 20.11a, what is the direction of the induced current in the resistor, (1) from 1 to 2 or (2) from 2 to 1? (b) If the bar is 20 cm long and is pulled at a steady speed of 10 cm/s, what would be the induced current if the resistor has a value of 5.0 Ω and the circuit is in a uniform magnetic field of 0.25 T?

INSIGHT

Electromagnetic Induction in Action

You have probably used applications of electromagnetic induction without knowing it. One such application is magnetic tape—either audio or video. In an audiotape recorder, a plastic tape coated with a film of iron oxide or chromium oxide runs past a recording head, which consists of a coil wound around an iron core with a gap (Fig. 1). Current in the coil produces a magnetic field in the gap, magnetizing the tape's film. The strength and direction of the magnetization in the film are determined by the gap field, which is determined by the pulses of current that are generated by sound from a microphone.

To reproduce the sound, the tape is run by the same head or one similar to it. As the tape passes through the head area, the changing flux due to the magnetization of the film on the tape induces an emf in the coil, matching the original fluctuations in voltage or current. These fluctuations are amplified and converted to sound in a speaker (see Exercise 9 at the end of this chapter).

Electromagnetic induction is also used in an electrical safety device called a *ground-fault circuit interrupter* (GFCI). A GFCI can be plugged into a wall outlet or installed as part of a home circuit. It senses any abrupt change in current in an electrical appliance plugged into the outlet or circuit and turns off the appliance, protecting you from electrical shock.

The principle of the GFCI is illustrated in Fig. 2. It consists of a sensing coil and a circuit breaker. The coil is wrapped on an iron ring, through which pass the wires carrying current to and from the protected outlet or circuit. This arrangement is called a *differential transformer*, because it can sense a difference in the currents carried by the two wires.

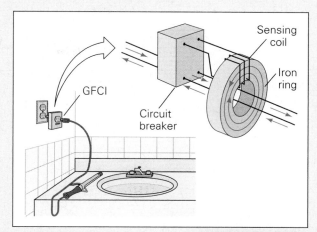

FIGURE 2 Ground-fault circuit interrupter (GFCI) A safety device that quickly detects currents to ground and opens the circuit.

The opposite currents in the two wires produce oppositely directed magnetic fields concentrated in the iron ring. Since the fields are always equal and opposite, the total field is normally zero. Hence, the net flux through the coil is zero, and no emf is induced in the sensing coil.

However, suppose that a wire breaks inside an appliance and touches its metal casing (Section 18.5) or a plugged-in hair dryer falls into water. If you touch the appliance or put your hand in the water, some of the current will pass through your body to ground (a fault to ground). The currents in the wires passing through the ring are then not equal, giving rise to a nonzero magnetic flux. A changing flux causes an induced emf in the sensing coil; the resulting current trips the circuit breaker, opening the circuit. All of this happens in about 30 ms (30 thousandths of a second) in response to an induced current of 4 to 6 mA. Notice that this induced current is much less than is required to trip a protective circuit breaker, usually 20 A.

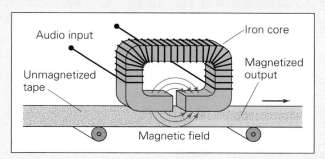

FIGURE 1 Tape-recorder head Pulses of current in the coil wrapped around an iron core produce magnetic fields that magnetize the metallic coating on the tape as it passes by.

(a) Conceptual Reasoning.   As illustrated in Fig. 20.11a, the magnetic flux points to the left and increases with time. According to Lenz's law, the field from the induced current must be to the right. Using the induced-current right-hand rule, we find that the induced current is from 1 to 2 (Fig. 20.11b).

(b) Thinking It Through.   The flux change is due to an area change. The area changes because the bar is pulled upward. If we can express the area change in terms of the bar's speed, we could determine the flux change. From the flux change, the induced emf is determined from Eq. 20.2. Finally, the induced current can be found by using Ohm's law.

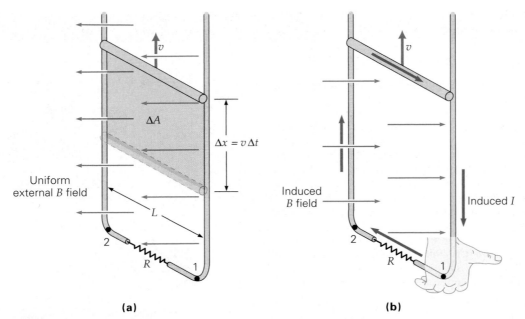

(a)   (b)

▲ **FIGURE 20.11 Motional emf**   (a) As the metal rod is pulled on the metal frame, the area
of the rectangular loop varies with time. A current is induced in the loop as a result of the
changing flux. **(b)** To counteract the upward increase of the flux to the left, an induced current
must create a magnetic field to the right. See Integrated Example 20.4.

**PHYSLET®
ILLUSTRATION**

Faraday's Law

Listing the data, we have the following:

Given: $B = 0.25$ T   **Find:** Induced current in the resistor
$L = 20$ cm $= 0.20$ m
$v = 10$ cm/s $= 0.10$ m/s
$R = 5.0 \, \Omega$

As the bar moves upward, the area of the circuit increases by an amount $\Delta A = L\Delta x$.
(See Fig. 20.11a.) At constant speed, the distance traveled by the bar in a time interval Δt
is $\Delta x = v\Delta t$. Therefore,

$$\Delta A = Lv\Delta t$$

Since the angle between the magnetic field and the normal to the area (θ) is $0°$, the flux
is given by $\Phi = BA \cos 0° = BA$. Since the magnetic field is constant, the change in flux
can be written as $\Delta\Phi = B\Delta A$, or

$$\Delta\Phi = BLv\Delta t$$

The magnitude of the induced emf $\mathscr{E}$, from Eq. 20.2, is

$$|\mathscr{E}| = \frac{|\Delta\Phi|}{\Delta t} = \frac{BLv\Delta t}{\Delta t} = BLv$$

Putting in given values, we find that the magnitude of the emf is

$$|\mathscr{E}| = BLv = (0.25 \text{ T})(0.20 \text{ m})(0.10 \text{ m/s}) = 5.0 \times 10^{-3} \text{ V}$$

Lastly, the induced current is

$$I = \frac{\mathscr{E}}{R} = \frac{5.0 \times 10^{-3} \text{ V}}{5.0 \, \Omega} = 1.0 \times 10^{-3} \text{A}$$

Follow-up Exercise.   In this Example, if the magnetic field were increased by 10
times, what would the speed of the bar have to be to generate an induced current of 0.1 A?

The emf produced by the movement of the bar in Integrated Example 20.4 is known as *motional emf*. You can tell by the numerical result of Integrated Example 20.4 that this particular arrangement isn't a very practical way to generate huge amounts of electrical energy from mechanical energy. In that example, the power dissipated in the resistor is only 5.0×10^{-6} W, or $5.0\,\mu$W. (You should verify this result.) However, the basic principle is of extreme importance to our modern society—through induced currents, mechanical energy is converted into electrical energy in large amounts and on demand.

20.2 Electric Generators and Back emf

OBJECTIVES: To **(a)** understand the operation of electrical generators and calculate the emf produced by an ac generator and **(b)** explain the origin of back emf and its effect on the behavior of motors.

One method to induce an emf or current in a loop is through a change in the loop's orientation (Fig. 20.4). This method of producing a flux change is the operational principle behind electric generators.

Electric Generators

An *electric generator* is a device that converts mechanical energy into electrical energy. Basically, the function of a generator is the reverse of that of a motor.

Recall that a simple chemical battery supplies direct current (dc). That is, the polarity of the voltage and the direction of the current do not change. However, most generators produce alternating current (ac), so named because the polarity of the voltage and the direction of the current change periodically. Thus, the electric energy used in homes and industry is delivered in the form of alternating voltage and current. (See Chapter 21 for analysis of ac circuits.)

An **ac generator** is sometimes called an *alternator*. The elements of a simple ac generator are shown in ▶Fig. 20.12. A wire loop called an *armature*, is mechanically rotated in a magnetic field by some external means, such as water flow or steam rushing over turbine blades. The rotation of the loop causes the magnetic flux through the loop to change, and a current is induced in the loop. The ends of the loop are connected to an external circuit by means of slip rings and brushes. In practice, generators have many loops, or windings, on their armatures.

When the loop is rotated at a constant angular speed (ω), the angle (θ) between the magnetic-field vector and the area vector of the loop (which is perpendicular to the plane of the loop) changes with time: $\theta = \omega t$ (assuming that $\theta = 0°$ at $t = 0$). As a result, the number of field lines through the loop changes with time, causing an induced emf. From Eq. 20.1, the flux (for one loop) varies as

$$\Phi = BA \cos \theta = BA \cos \omega t$$

By Faraday's law, the induced emf will also vary with time. For a rotating coil of N loops, Faraday's law becomes

$$\mathscr{E} = -N\frac{\Delta \Phi}{\Delta t} = -NBA\left(\frac{\Delta(\cos \omega t)}{\Delta t}\right)$$

Here, we have removed B and A from the time rate of change, because they are constant. By using methods beyond the scope of this book, it can be shown that the term in the parentheses can be rewritten as $\Delta(\cos \omega t)/\Delta t = -\omega \sin \omega t$. Thus, the induced emf can be expressed as

$$\mathscr{E} = (NBA\omega) \sin \omega t$$

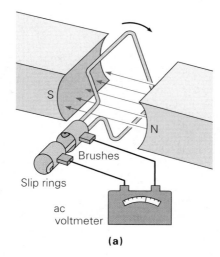

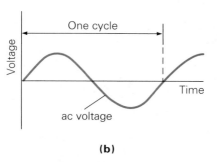

(a)

(b)

▲ **FIGURE 20.12 A simple ac generator** **(a)** The rotation of a wire loop in a magnetic field produces **(b)** a voltage output whose polarity reverses with each half-cycle. This alternating voltage gives rise to an alternating current if the wire loop is connected to a complete circuit.

Notice that the product of terms, $NBA\omega$, represents the magnitude of the maximum emf, which occurs whenever $\sin \omega t = \pm 1$. If $NBA\omega$ is called $\mathscr{E}_0$, the maximum value of the emf, then the foregoing equation can be rewritten compactly as

$$\mathscr{E} = \mathscr{E}_0 \sin \omega t \qquad (20.4)$$

Since the value of the sine function varies between $+1$ and -1, the sign, or polarity, of the emf changes with time (▼Fig. 20.13). Note from the figure that the emf has its maximum value $\mathscr{E}_0$ when $\theta = 90°$ or $\theta = 270°$, that is, when the plane of the rotating loop is parallel to field lines, and the magnetic flux is zero at some instant in time. The change in flux is greatest at these angles, because, although the flux is momentarily zero, it is changing rapidly, due to a *sign* change. Near the flux's largest value ($\theta = 0°$ or $\theta = 180°$), the flux is approximately constant.

The direction of the current produced by this alternating induced emf also changes periodically. In everyday applications, it is common to refer to the frequency (f) of the armature (in hertz or rotations per second), rather than the angular frequency (ω). Since they are related by $\omega = 2\pi f$, Eq. 20.4 can be rewritten as

$$\mathscr{E} = \mathscr{E}_0 \sin (2\pi f t) \quad \textit{alternator emf} \qquad (20.5)$$

The ac frequency in the United States and most of the western hemisphere is 60 Hz. A frequency of 50 Hz is common in Europe and other areas.

Keep in mind that Eqs. 20.4 and 20.5 give the instantaneous value of the emf and that $\mathscr{E}$ varies between $+\mathscr{E}_0$ and $-\mathscr{E}_0$ over one-half of an armature rotational period ($\frac{1}{120}$ of a second in the United States). For practical ac electrical circuits, time-averaged values for ac voltage and current are more important. This concept will be developed in Chapter 21. To see how various factors influence the generator's output, examine the next Example closely.

Example 20.5 ■ An ac Generator: Renewable Electric Energy

A farmer decides to use a waterfall to create a small hydroelectric power plant. He builds a coil consisting of 1500 circular loops of wire with a radius of 20 cm, which rotates on the generator's armature at 60 Hz in a magnetic field. To average 120 V, he wants to generate a maximum emf of 170 V. What must be the magnitude of the magnetic field in the generator?

Thinking It Through. We can determine the magnetic field from the expression for $\mathscr{E}_0$.

Solution.

Given: $\mathscr{E}_0 = 170$ V   *Find:* Magnitude of the magnetic field (B)
$N = 1500$ loops
$r = 20$ cm $= 0.20$ m
$f = 60$ Hz

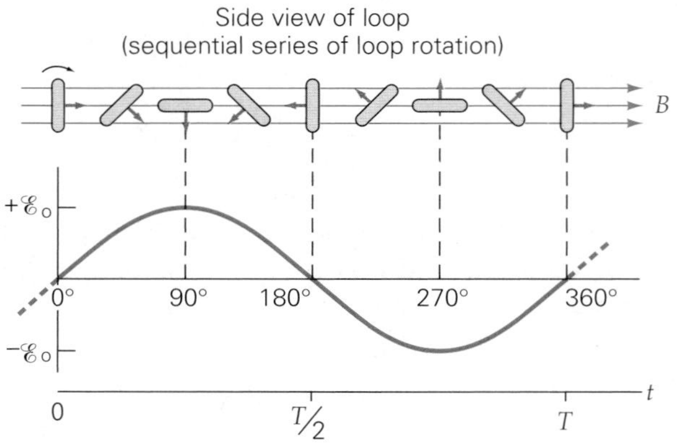

▶ FIGURE 20.13   An ac generator output   A graph of the sinusoidal output of the generator, together with a side view of the corresponding orientations of the loop during a cycle, showing the flux variation with time. The emf is a maximum when the flux changes the most rapidly, as it passes through zero and changes in sign.

The generator's maximum (or peak) emf is given by $\mathscr{E}_o = NBA\omega$. Because $\omega = 2\pi f$ and, for a circle, $A = \pi r^2$, this expression can be rewritten as

$$\mathscr{E}_o = NB(\pi r^2)(2\pi f) = 2\pi^2 NBr^2 f$$

Solving for B, we obtain

$$B = \frac{\mathscr{E}_o}{2\pi^2 Nr^2 f} = \frac{170 \text{ V}}{2\pi^2(1500)(0.20 \text{ m})^2(60 \text{ Hz})} = 2.4 \times 10^{-3} \text{ T}$$

Follow-up Exercise. In this Example, suppose that the farmer wanted to generate an emf with an average of 240 V, which requires a maximum emf of 340 V. If he chose to do so by changing the size of the coils, what would their new radius have to be?

In most large-scale ac generators (power plants), the armature is actually stationary, and magnets revolve about it. The revolving magnetic field produces a time-varying flux through the coils of the armature and thus an ac output. A turbine supplies the mechanical energy required to spin the magnets in the generator (▼Fig. 20.14a). Turbines are typically powered by steam generated from the heat of combustion of fossil fuels or by nuclear reactions, but they can also be rotated by falling water (hydroelectricity), as shown in Fig. 20.14b. Thus, the basic difference between the various types of power plants is the source of the energy that turns the turbines.

Back emf

Although their main job is to convert electric energy into mechanical energy, motors also generate emfs at the same time. Like a generator, a motor has a rotating armature in a magnetic field. The induced emf in this case is called a **back emf** (or *counter emf*), $\mathscr{E}_b$, because its polarity is opposite to that of the line voltage and tends to reduce the current in the armature coils.

Note: Motors are discussed in Section 19.5.

If V is the line voltage, then the net voltage driving the motor is less than V (since the line voltage and the back emf are of opposite polarity). The net voltage is given by $V_{net} = V - \mathscr{E}_b$. For a motor with an armature of internal resistance R, the current the motor draws while in operation is

$$I = \frac{V_{net}}{R} = \frac{V - \mathscr{E}_b}{R}$$

or

$$\mathscr{E}_b = V - IR \quad \text{(back emf of a motor)} \quad (20.6)$$

where V is the line voltage.

◀ **FIGURE 20.14 Electrical generation** **(a)** Turbines such as those depicted here, generate electric energy in much larger quantities than can the hydroelectric plant in the opening photo of this chapter. **(b)** Gravitational potential energy of water, here trapped behind the Glen Canyon dam on the Colorado River in Arizona, is converted into electric energy.

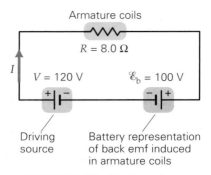

Armature coils

$R = 8.0 \ \Omega$

I

$V = 120 \ V$ $\mathscr{E}_b = 100 \ V$

Driving source   Battery representation of back emf induced in armature coils

▲ **FIGURE 20.15 Back emf** The back emf in the armature of a dc motor can be represented as a battery with polarity opposite that of the driving source. See Example 20.6

The back emf of a motor depends on the rotational speed of the armature and builds up from zero to some maximum value as the armature goes from rest to its normal operating speed. On start-up, the back emf is zero (why?), so the starting current is a maximum (Eq. 20.6 with $\mathscr{E}_b = 0$). Ordinarily, a motor turns something; that is, it has a mechanical load. Without a load, the armature speed will increase until the back emf almost equals the line voltage. The result is a small current in the coils, just enough to overcome friction and joule heat losses. Under normal load conditions, the back emf is less than the line voltage. The larger the load, the slower the motor rotates and the smaller the back emf is. If a motor is overloaded and turns very slowly, the back emf may be reduced so much that the current becomes very large (since V_{net} increases as $\mathscr{E}_b$ decreases) and may burn out the coils. Thus, the back emf plays a vital role in the regulation of a motor's operation by limiting the current in it.

Schematically, a back emf in a dc motor circuit can be represented as an "induced battery" with polarity opposite that of the driving voltage (◄Fig. 20.15). To see how the back emf determines the current in a motor, consider the next example.

Example 20.6 ■ Getting up to Speed: Back emf in a dc Motor

A dc motor is built with windings that have a resistance of 8.00 Ω and operates at a line voltage of 120 V. With a normal load, there is a back emf of 100 V when the motor reaches full speed. (See Fig. 20.15.) Determine (a) the starting current drawn by the motor and (b) the armature current at operating speed under a normal load.

Thinking It Through. (a) The only difference between start-up and full speed is that there is no back emf at start-up. The net voltage and resistance determine the current, so Eq. 20.6 can be applied. (b) At operating speed, the back emf increases and is opposite in polarity to the line voltage. Equation 20.6 can again be used to determine the current.

Solution. Let's list the data as usual:

Given: $R = 8.00 \ \Omega$ *Find:* (a) I_s (starting current)
$V = 120 \ V$ (b) I (operating current)
$\mathscr{E}_b = 100 \ V$

(a) From Eq. 20.6, the current in the windings is

$$I_s = \frac{V}{R} = \frac{120 \ V}{8.00 \ \Omega} = 15.0 \ A$$

(b) When the motor is at full speed, the back emf is 100 V; thus, the current is

$$I = \frac{V - \mathscr{E}_b}{R} = \frac{120 \ V - 100 \ V}{8.00 \ \Omega} = 2.50 \ A$$

Note that, with no back emf, the starting current of the motor is relatively large. Thus, when a big motor, such as that of a central air-conditioning unit for a building, starts up, the lights in the building might momentarily dim, because of the large starting current that the motor draws. In some motor designs, resistors are temporarily connected in series with a motor's coil to protect the windings from burning out as a result of large starting currents.

Follow-up Exercise. In this Example, (a) how much energy is required to bring the motor to operating speed if it takes 10 s and the back emf averages 50 V during that time? (b) Compare this amount with the amount of energy required to keep the motor running for 10 s once it reaches its operating conditions.

Since motors and generators are opposites, so to speak, and a back emf develops in a motor, you may be wondering whether a back force develops in a generator. The answer is yes. When an operating generator is not connected to an external circuit, no current exists, and therefore there is no force on the coils of the armature, due to the magnetic field. However, when the generator delivers power to a circuit and current *is* in the coils, there is a force that produces a *countertorque*, which opposes the rotation of the armature. As more current is drawn, the countertorque increases, and a greater driving force is needed to turn the armature. Therefore, the higher the output of current of a generator, the greater is the energy expended (fuel consumed) in overcoming the countertorque.

20.3 Transformers and Power Transmission

OBJECTIVES: To (a) explain transformer action in terms of Faraday's law, (b) calculate the output of step-up and step-down transformers, and (c) understand the importance of transformers in electric energy delivery systems.

Electric energy is transmitted by power lines over long distances. It is desirable to minimize I^2R losses (joule heat) that can occur in these transmission lines. Since the resistance of a line is fixed, reducing I^2R losses means reducing current. However, the power output of a generator is determined by its outputs of current and voltage ($P = IV$), and for a fixed voltage, such as 120 V, a reduction in current would mean a reduced power output. It might appear that there is no way to reduce the current while maintaining the power level. Fortunately, electromagnetic induction can reduce power-transmission losses by increasing the voltage and simultaneously reducing the current in such a way that the delivered *power* is essentially unchanged. This is done with a device called a **transformer**.

A simple transformer consists of two coils of insulated wire wound on the same (closed) iron core ($\blacktriangleright$ Fig. 20.16a). When ac voltage is applied to the input coil, or *primary coil*, the alternating current produces an alternating magnetic flux that is concentrated in the iron core, without any leakage of flux outside the core. Thus, under these conditions, the same changing flux also passes through the output coil, or *secondary coil*, inducing an alternating voltage and current in it. (Note that it is common in transformer design to refer to emfs as "voltages," as we did in Chapter 18. We will use this language here also.)

The difference between the induced voltage in the secondary coil and the voltage in the primary coil depends on the ratio of the numbers of turns in the two coils. By Faraday's law, the induced voltage in the secondary coil is

$$V_s = -N_s \frac{\Delta \Phi}{\Delta t}$$

where N_s is the number of turns in the secondary coil. The changing flux in the primary coil produces a back emf of

$$V_p = -N_p \frac{\Delta \Phi}{\Delta t}$$

where N_p is the number of turns in the primary coil. If the resistance of the primary coil is neglected, this back emf is equal in magnitude to the external voltage applied to the primary coil. Then, forming a ratio of the output voltage (secondary) to the input voltage (primary) yields

$$\frac{V_s}{V_p} = \frac{-N_s(\Delta \Phi / \Delta t)}{-N_p(\Delta \Phi / \Delta t)}$$

or

$$\frac{V_s}{V_p} = \frac{N_s}{N_p} \qquad (20.7)$$

Note: When discussing transformers, it is customary to use the term *voltage* rather than *emf*.

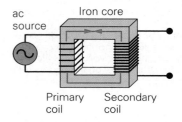

(a) Step-up transformer: high-voltage (low-current) output

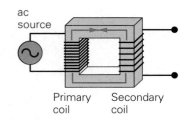

(b) Step-down transformer: low-voltage (high-current) output

$\blacktriangle$ **FIGURE 20.16 Transformers** **(a)** A step-up transformer has more turns in the secondary coil than in the primary coil. **(b)** A step-down transformer has more turns in the primary coil than in the secondary coil.

If the transformer is assumed to be 100% efficient (i.e., there are no energy losses), then the power input is equal to the power output. Since $P = IV$, we have

$$I_p V_p = I_s V_s \qquad (20.8)$$

Although some energy is always lost, this equation is a good approximation, since a well-designed transformer can have an efficiency of more than 95%. (The sources of energy losses will be discussed shortly.) Then, from Eq. 20.8, the transformer currents and voltages are related to the turn ratio by

$$\frac{I_p}{I_s} = \frac{V_s}{V_p} = \frac{N_s}{N_p} \qquad (20.9)$$

With these equations, it is easy to see how a transformer affects the voltage and current. In terms of the output,

$$V_s = \left(\frac{N_s}{N_p}\right)V_p \qquad (20.10a)$$

and

(ideal relationship between transformer voltage and current)

$$I_s = \left(\frac{N_p}{N_s}\right)I_p \qquad (20.10b)$$

Note: Step-up transformer: $N_s > N_p$, or $N_s/N_p > 1$ (more turns in secondary coil than in primary coil).

Note: Step-down transformer: $N_s < N_p$, or $N_s/N_p < 1$ (more turns in primary coil than in secondary coil).

Note: The terms "step-up" and "step-down" refer to what happens to the primary, or input, *voltage*, not to the current. The effect on the current is the reverse of the effect on the voltage.

That is, if the secondary coil has more windings than the primary coil does (i.e., $N_s > N_p$, or $N_s/N_p > 1$), as in Fig. 20.16a, the voltage is "stepped up," since $V_s > V_p$. This design is called a *step-up transformer*. Notice that there is less current in the secondary coil than in the primary coil ($N_p/N_s < 1$ and $I_s < I_p$).

The opposite situation, in which the secondary coil has fewer turns than the primary coil does, characterizes a *step-down transformer* (Fig. 20.16b). In this case, the voltage is "stepped down," and the current is increased. Depending upon the design details, a step-up transformer may be used as a step-down transformer by reversing the output and input connections.

Example 20.7 ■ Transformer Orientation: Step-Up or Step-Down Configuration?

An ideal 600-W transformer has 50 turns on its primary coil and 100 turns on its secondary coil. (a) Is this configuration a step-up or step-down arrangement? Explain. (b) If the primary coil is connected to a 120-V source, what is the output voltage of the secondary coil? (c) If the transformer is operated in reverse and the 120-V input is applied to the 100-turn coil, what is the output current?

Thinking It Through. (a) The type of transformer is determined by the turn ratio. (b) The output voltage can be determined from Eq. 20.10a, once the turn ratio is established. (c) Here, the situation is reversed. The input current can be determined from the power delivered. Once the turn ratio is established, the output current can then be determined from Eq. 20.10b.

Solution.

Given: (b) $N_p = 50$
$N_s = 100$
$V_p = 120$ V
$P_p = P_s = 600$ W
(c) $N_p = 100$
$N_s = 50$
$V_p = 120$ V
$P_p = P_s = 600$ W

Find: (a) Determine if the transformer is a step-up or step-down transformer
(b) V_s (output, or secondary, voltage)
(c) I_s (output, or secondary current) under reversed conditions

(a) Since $N_s > N_p$, $V_s > V_p$. Therefore, it is a step-up transformer.

(b) To find the output, or secondary, voltage, apply Eq. 20.10a with the turn ratio of 2, since $N_s = 2N_p$. Thus,

$$V_s = \left(\frac{N_s}{N_p}\right)V_p = (2)(120\text{ V}) = 240\text{ V}$$

(c) With the reverse connections, it is a step-down transformer. Since the roles of primary and secondary are reversed, we have $N_p > N_s$ and $N_s/N_p = \frac{50}{100} = \frac{1}{2}$. Therefore,

$$V_s = \left(\frac{N_s}{N_p}\right)V_p = \left(\frac{1}{2}\right)(120\text{ V}) = 60\text{ V}$$

To find the output current, remember that the transformer is "ideal," and the input power is the same as the output power. Equating the two powers, we have

$$P_p = I_p V_p = I_s V_s = P_s = 600\text{ W}$$

From this result, the input current is

$$I_p = \frac{600\text{ W}}{V_p} = \frac{600\text{ W}}{120\text{ V}} = 5.00\text{ A}$$

Since the voltage is stepped down by a factor of two, the output current should be stepped up by a factor of two. From Eq. 20.10b,

$$I_s = \left(\frac{N_p}{N_s}\right)I_p = 2 \times 5.00\text{ A} = 10.0\text{ A}$$

as expected.

Follow-up Exercise. (a) When traveling in Europe (where average ac voltages are 240 V), what type of transformer would enable you to use 120-V appliances? Explain. (b) For a 1500-W hair dryer (assumed ohmic) designed for use at 120 V, what would be the transformer's output current in Europe, assuming it to be ideal?

The preceding relationships strictly apply only to ideal (or "lossless") transformers; actual transformers have energy losses. Although well-designed transformers generally have an internal-energy loss of less than 5%, there is no such thing as an ideal transformer (0% loss) in reality. There are many factors that determine how close a real transformer comes to performing like an ideal one.

First, there is flux leakage; that is, not all of the flux passes through the secondary coil. In some transformer designs, one of the insulated coils is wound directly on top (interlocking) of the other rather than having the two coils on separate "legs." This configuration helps minimize flux leakage while reducing transformer size. Second, when ac current flows in the primary coil, the changing magnetic flux through the loops gives rise to an induced emf in that coil. This phenomenon is called *self-induction*. By Lenz's law, the self-induced emf will oppose the change in current and will thus limit the current (similar to a back emf in a motor). A third reason that transformers are less than ideal is joule heating (I^2R losses) due to the resistance of the coil wires. Usually, this loss is small, because the wires have little resistance. A fourth reason for energy loss is *eddy currents*, or swirling movements of charge, in the transformer core itself. To increase the density of the magnetic flux, the core is made of a highly permeable material (such as iron), but such materials are usually good conductors. The changing magnetic flux in the core induces emfs, which, in turn, create eddy currents in the core material. These eddy currents then dissipate energy by heating the core (I^2R losses again).

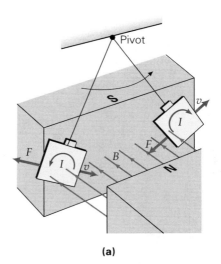

(a)

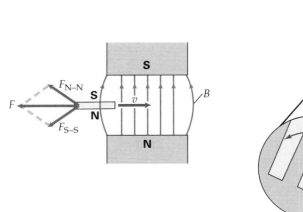

(b)

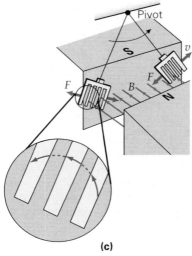

(c)

▲ **FIGURE 20.17   Eddy currents**   (a) Eddy currents are induced in a nonmagnetic conductive plate moving in a magnetic field. The induced currents oppose the change in flux, and a retarding force opposes the motion. Note that the currents reverse direction as the plate swings through the field. (b) An overhead view as the plate swings toward the field from the left. The retarding force can be seen to be a result (F_{net}) of the two repulsive forces (F_{N-N} and F_{S-S}) acting on like magnetic poles. The side of the plate closest to the north pole of the permanent magnet acts as a north pole, and the other side acts as a south pole. (c) If the plate has slits, the eddy currents and magnetic forces are drastically reduced.

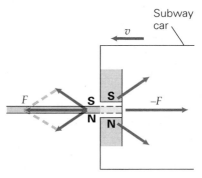

▲ **FIGURE 20.18**
Electromagnetic braking and mass transit   Magnetic braking is based on Faraday's and Lenz's laws. When braking, a train energizes an electromagnet onboard. This electromagnet straddles a long metal rail. The induced currents in the rail produce a mutually repulsive force between the rail and the train, thereby slowing the train.

To reduce this effect, transformer cores are made of thin sheets of material (usually iron) laminated with an insulating glue between them. The insulating layers between the sheets break up the eddy currents or confine them to the sheets, greatly reducing energy loss due to them.

The effects of eddy currents can be demonstrated by allowing a plate made of a conductive, but nonmagnetic, metal, such as aluminum, to swing through a magnetic field (▲ Fig. 20.17a). Eddy currents are set up in the plate as a result of its motion in the field, which changes the magnetic flux through the plate's area. By Lenz's law, eddy currents are induced, opposing the flux change.

When the plate enters the field (the position of the left-hand plate in Fig. 20.17a), a counterclockwise current is induced. (Apply Lenz's law to show this effect.) This situation is equivalent to the plate's having a north magnetic pole near the permanent magnet's north pole and a south magnetic pole near the permanent magnet's south pole (Fig. 20.17b). Thus, there are two repulsive magnetic forces on the plate, and the net force tends to slow it down as it enters the field. The eddy current in the plate is reversed in direction as it leaves the field, producing attractive magnetic forces. In both cases, the induced emfs act to slow the plate's motion.

The reduction of eddy currents (similar to how the laminated layers in a transformer work) can be demonstrated by using a plate with slits (Fig. 20.17c). When this plate swings between the magnet's poles, it swings relatively freely, because the eddy currents are greatly reduced by the gaps (slits). Thus, the magnetic force on the plate is also reduced.

The damping effect of eddy currents has been applied in the braking systems of rapid-transit railcars. When an electromagnet (housed in the car) is turned on, it applies a magnetic field to a rail. The repulsive force due to the induced eddy currents in the rail acts as a braking force (◀Fig. 20.18). As the car slows, the eddy currents in the rail decrease, allowing a smooth braking action.

Power Transmission and Transformers

For power transmission over long distances, transformers provide a means to increase (step up) the voltage and reduce the current of a generator's output in order to cut down the resistive I^2R losses in the wires carrying the current. A schematic diagram of an ac power distribution system is shown in ▶ Fig. 20.19. The voltage output of the generator is stepped up, and the energy is transmitted over long distances to an area substation near the consumers. There, the voltage is stepped down. There are further voltage step-downs at distributing substations and utility poles before 120-V and 240-V electricity is supplied to homes and businesses.

The following Example illustrates the benefits of being able to step up the voltage (and step down the current) for electrical power transmission.

Example 20.8 ■ Cutting Your Losses: Power Transmission at High Voltage

A small hydroelectric power plant produces energy in the form of electric current at 10 A and a voltage of 440 V. The voltage is stepped up to 4400 V (by an ideal transformer) for transmission over 40 km of power line, which has a total resistance of 20 Ω. (a) What percentage of the original energy would have been lost in transmission if the voltage had not been stepped up? (b) What percentage of the original energy is actually lost when the voltage is stepped up?

Thinking It Through. (a) The power output can be computed from $P = IV$ and compared with the power lost in the wire, $P = I^2R$. (b) Equations 20.10a and 20.10b should be used to determine the stepped-up voltage and stepped-down currents, respectively. Then the calculation is repeated, and the results are compared with those of part (a).

Solution.

Given: $I_p = 10$ A *Find:* (a) Percentage energy loss without voltage step-up
$V_p = 440$ V   (b) Percentage energy loss with voltage step-up
$V_s = 4400$ V
$R = 20\ \Omega$

(a) The power output by the generator is

$$P = I_pV_p = (10\ \text{A})(440\ \text{V}) = 4400\ \text{W}$$

The rate of energy loss of the wire (joules per second, or watts) in transmitting a current of 10 A is

$$P_{loss} = I^2R = (10\ \text{A})^2(20\ \Omega) = 2000\ \text{W}$$

Thus, the percentage of the produced energy lost to joule heat in the wires is

$$\% \text{ loss} = \frac{P_{loss}}{P} \times 100\% = \frac{2000\ \text{W}}{4400\ \text{W}} \times 100\% = 45\%$$

(b) When the voltage is stepped up to 4400 V, the transmitted current is reduced by a factor of 10:

$$I_s = \left(\frac{V_p}{V_s}\right)I_p = \left(\frac{440\ \text{V}}{4400\ \text{V}}\right)(10\ \text{A}) = 1.0\ \text{A}$$

The power is thus reduced by a factor of 100, since it varies as the square of the current:

$$P_{loss} = I^2R = (1.0\ \text{A})^2(20\ \Omega) = 20\ \text{W}$$

Therefore, the percentage of power lost is also reduced by a factor of 100:

$$\% \text{ loss} = \frac{P_{loss}}{P} \times 100\% = \frac{20\ \text{W}}{4400\ \text{W}} \times 100\% = 0.45\%$$

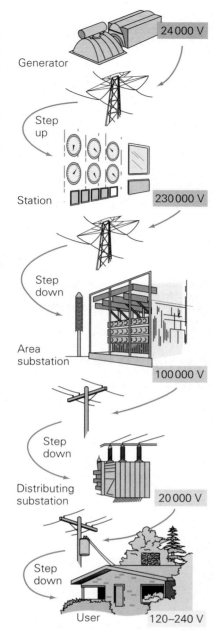

▲ **FIGURE 20.19 Power transmission** A diagram of a typical electrical-power distribution system.

Follow-up Exercise. Some heavy-duty electrical appliances, such as water pumps, can be wired to 240 V or 120 V. Their power rating is the same regardless of the voltage at which they run. (a) Explain the efficiency advantage of operating such appliances at the higher voltage. (b) For a 1.00-hp pump (746 W), estimate the ratio of the power lost in the wires at 240 V to the power lost at 120 V (assuming that all resistances are ohmic and the connecting wires are the same).

20.4 Electromagnetic Waves

OBJECTIVES: To (a) explain the physical nature, origin, and means of propagation of electromagnetic waves and (b) describe the properties and uses of various types of electromagnetic waves.

Electromagnetic waves (or *electromagnetic radiation*) were considered as a means of heat transfer in Section 11.4. Now you are ready to understand the production and characteristics of electromagnetic radiation. These waves can be described in terms of the electric and magnetic fields.

James Clerk Maxwell showed that four fundamental relationships could describe all observed electromagnetic phenomena. He used these equations to predict the existence of electromagnetic waves. Because of his contributions, the set of equations is known as *Maxwell's equations*, although the equations were, for the most part, developed by other scientists (for example, Faraday's law of induction).

Essentially, Maxwell's equations combine the electric field and the magnetic field into a single electromagnetic field. The apparently separate fields are symmetrically related in the sense that either one can create the other under the proper conditions. This symmetry is evident in the equations as presented in their advanced mathematical form (not shown). For us, a qualitative description is sufficient:

> A time-varying magnetic field produces a time-varying electric field.
> A time-varying electric field produces a time-varying magnetic field.

The first statement is simply the observation that, as we saw in Section 20.1, a changing magnetic flux gives rise to an induced emf and current in a wire. The second statement is crucial to understanding the self-propagating characteristic of electromagnetic waves. Together, these two phenomena enable these waves to travel through a vacuum, whereas all other waves require a supporting medium.

According to Maxwell's theory, *accelerating* electric charges, such as an oscillating electron, produce electromagnetic waves. The electron in question could be one of the many nearly free electrons in the metal antenna of a radio transmitter, driven by an electrical (voltage) oscillator at a frequency of 10^6 Hz (1 MHz). As each electron oscillates, it continually accelerates and decelerates and thus radiates an electromagnetic wave (▶Fig. 20.20a). The continual oscillations of many such charges due to the alternating voltage in the transmitter produce time-varying electric and magnetic fields in the immediate vicinity of the antenna. The electric field, shown in red in Fig. 20.20a, is in the plane of the page and continually changes direction, as does the magnetic field (shown in blue and going into and out of the paper).

Both the electric and the magnetic fields carry energy and propagate outward with the speed of light. This speed is symbolized by the letter c and is, to three significant figures, $c = 3.00 \times 10^8$ m/s. Maxwell's equations show that at large distances from the source, these electromagnetic waves become plane waves. (Figure 20.20b shows a wave at an instant in time.) Here, the electric field (**E**) is perpendicular to the magnetic field (**B**), and each varies sinusoidally with time. Both **E** and **B** are perpendicular to the direction of wave propagation. Thus, electromagnetic waves are *transverse* waves, with the *fields* oscillating perpendicularly to the direction of propagation. According to Maxwell's theory, as one field changes, it creates the other field. This process, repeated again and again, gives rise to the traveling electromagnetic wave we call light.

Note: The Scottish physicist James Clerk Maxwell (1831–1879) unified the concepts of electric and magnetic fields.

Note: A changing B field produces a changing E field. A changing E field produces a changing B field.

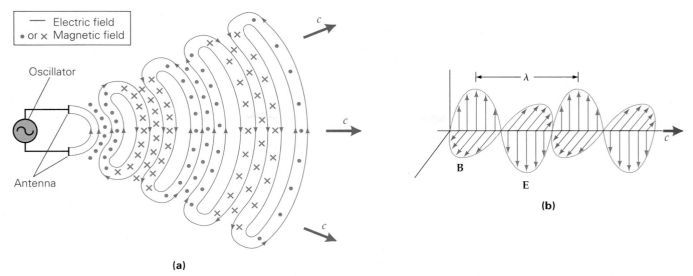

(a)

▲ **FIGURE 20.20 Source of electromagnetic waves** Electromagnetic waves are produced, fundamentally, by accelerating electric charges. **(a)** The charges are moved in simple harmonic motion by an oscillating voltage source connected to a metal antenna. As the polarity of the antenna and the direction of the current in the wire periodically change, alternating electric and magnetic fields propagate outward. The electric field **E** and the magnetic field **B** are in phase and are perpendicular to one another and to the direction of propagation of the wave. Thus, electromagnetic waves are transverse waves. **(b)** At large distances from the source, these curved wavefronts become almost plane.

It is important to realize the following fact about the transmission of electromagnetic waves in a vacuum:

> In a vacuum, all electromagnetic waves, regardless of their frequency or wavelength, travel at the same speed, c. To three significant figures, $c = 3.00 \times 10^8$ m/s.*

This was predicted by Maxwell's equations and has been verified experimentally.

Since everyday distances are quite short, the time delay due to the travel of light can usually be neglected. However, for interplanetary trips, this delay can be a problem. Consider the following Example.

PHYSLET®
ILLUSTRATION

Electromagnetic Waves

Example 20.9 ■ Long-Distance Guidance: The Speed of Electromagnetic Waves in a Vacuum

Viking space probes landed on Mars in 1976 and sent radio and TV signals (both are electromagnetic waves) to the Earth. How much longer did it take for a signal to reach us when Mars was farthest from the Earth than when it was closest to us? The average distances of Mars and the Earth from the Sun are 229 million km (d_M) and 150 million km (d_E), respectively. Assume that both planets have circular orbits, and use the average distances as the radii of the circles.

Thinking It Through. This situation calls for a time–distance calculation. The planets are farthest apart when they are on opposite sides of the Sun and are thus separated by a distance of $d_M + d_E$. (This arrangement requires signals to be sent through the Sun, which, of course, is not possible. However, it does serve to determine the upper limit on transmission times.) The planets are closest when they are aligned on the same side of the Sun; their separation distance is then $d_M - d_E$. (Draw a diagram to help you visualize this

*See the discussion on pp. 691 through 694 of this section for descriptions of the different kinds of electromagnetic waves. They all have the same fundamental electric and magnetic field structure (Fig. 20.20), but differ in frequency and wavelength. In common usage, "light" refers to "visible light," that is, the frequencies to which our eyes are sensitive. "Electromagnetic radiation" ("light" in the broader sense) refers to all frequencies, visible and invisible.

configuration.) Since the speed of electromagnetic waves in a vacuum is known, the times can be found from $t = d/c$.

Solution. Listing the data, and converting the distances to meters, we have:

Given: $d_M = 229 \times 10^6 \, km = 2.29 \times 10^{11} \, m$ **Find:** Δt (difference in time for light
$d_E = 150 \times 10^6 \, km = 1.50 \times 10^{11} \, m$ to travel the longest and
shortest distances)

Radio and TV waves travel at the speed, c. Thus, the longest travel time t_L is

$$t_L = \frac{d_M + d_E}{c} = \frac{3.79 \times 10^{11} \, m}{3.00 \times 10^8 \, m/s} = 1.26 \times 10^3 \, s \quad (or \ 21.1 \, min)$$

For the shortest distance the shortest travel time t_s is

$$t_s = \frac{d_M - d_E}{c} = \frac{7.90 \times 10^{10} \, m}{3.00 \times 10^8 \, m/s} = 2.63 \times 10^2 \, s \quad (or \ 4.39 \, min)$$

The difference in the times is thus

$$\Delta t = t_L - t_s = 1.00 \times 10^3 \, s \quad (or \ 16.7 \, min)$$

Follow-up Exercise. Assume that the *Sojourner* roving Martian vehicle (see Example 2.1, on p. 34) is heading for a collision with a rock 2.0 m ahead of it. When it is at that distance, the vehicle sends a picture of the rock to controllers on the Earth. If Mars is at the closest point in its orbit to the Earth, what is the maximum speed that the *Sojourner* could have and still avoid a collision? Assume that the video signal from the *Sojourner* reaches the Earth and that the signal for the *Sojourner* to stop is sent back immediately.

Radiation Pressure

An electromagnetic wave carries energy. Consequently, it can do work and can exert a force on a material it strikes. Consider light striking an electron at rest on a surface (◄Fig. 20.21). The electric field of the electromagnetic wave exerts a force on the electron, giving it a downward velocity (**v**) as shown in the figure. Since a charged particle moving in a magnetic field experiences a force, there is a magnetic force on the electron, due to the magnetic-field component of the light wave. By the right-hand rule, this force is initially in the direction that the wave is propagating (Fig. 20.21a). Since the electromagnetic wave produces the same force on many electrons, it exerts a force on the surface in the direction in which it is traveling.

The radiation force per area is called the **radiation pressure**. Radiation pressure is negligibly small for most everyday situations, but it can be important in atmospheric and astronomical phenomena, as well as in atomic and nuclear physics, where masses are small and there is no friction. For example, radiation pressure plays a key role in determining the direction in which the tail of a comet points. Sunlight delivers energy to the comet's "head," which consists of frozen ices and dust. Some of this material evaporates as the comet nears the Sun, and the evaporated gases are pushed away from the Sun by radiation pressure. Thus, the tail generally points away from the Sun, no matter whether the comet is approaching or leaving the Sun's vicinity.

Another potential use of radiation pressure from sunlight is to propel interplanetary "sailing" satellites outward from the Sun in a slowly enlarging and spiraling orbit, eventually reaching the outer planets (▶ Fig. 20.22a). To create enough force, using the extremely low pressure of the sunlight, the sails would have to be very large in area, and the satellite would have to have as little mass as possible. The payoff is that no fuel (except for small amounts for course corrections) would be needed once the satellite was launched. Consider the following Conceptual Example, which concerns radiation pressure and space travel.

Note: Even the light from your desk lamp exerts a very, very small force on your desk. However, all such effects are negligible in everyday settings.

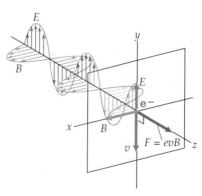

▲ **FIGURE 20.21 Radiation pressure** The electric field of an electromagnetic wave that strikes a surface acts on an electron, giving it a velocity (**v**). The magnetic field then exerts a force on the moving charge in the direction of propagation of the incident light. (Verify this direction, using the magnetic right-hand force rule.)

Conceptual Example 20.10 ■ Sailing the Sea of Space: Radiation Pressure in Action

Some scientists have proposed that a relatively light spacecraft with a huge sail could be built and launched from a space station in the Earth's orbit. With little or no power of its own, it would use the pressure of sunlight to propel it to the outer planets. To get the maximum propulsive force, what kind of surface should the sail have, (a) shiny and reflective, (b) dark and absorptive, or (c) surface characteristics would not matter?

Reasoning and Answer. At first glance, you might think that the answer is (c). However, as we have seen, radiation possesses energy and is capable of exerting force, and it transfers momentum to whatever it strikes. Thus, the interaction between the radiation and the sail must be thought of in terms of conservation of momentum, as shown in Fig. 20.22b. (See Section 6.3.) If the radiation is absorbed, the situation is analogous to a completely inelastic collision (like a putty wad sticking to a door), and the sail would acquire all of the momentum (**p**) originally possessed by the radiation.

However, if the radiation is *reflected*, the situation is analogous to a completely elastic collision (like a Superball™ bouncing off a wall). Since the momentum of the radiation after the collision would be equal to its original momentum in magnitude, but opposite in direction, its momentum would thus be reversed (from **p** to −**p**). To conserve momentum, the momentum transferred to the shiny sail would be twice as great (2**p**) as that for the dark sail. Because force is the rate of change of momentum, reflective sails would experience, on average, twice as much force as absorptive ones. Thus, the answer is (a).

Follow-up Exercise. (a) Would the sail in this Example provide less or more acceleration as the interplanetary sailing ship moves farther from the Sun? (b) How could this change in acceleration be counteracted, assuming that you wanted to keep it constant?

Types of Electromagnetic Waves

Electromagnetic waves are classified by ranges of frequencies or wavelengths in a spectrum. Recall from Chapter 13 that frequency and wavelength are inversely related by the traveling-wave relationship $\lambda = c/f$, where the speed of light, c, has been substituted for the general wave speed v. The higher the frequency, the shorter is the wavelength, and vice versa. The electromagnetic spectrum is continuous, so the limits of the various ranges are approximate (▼Fig. 20.23). Table 20.1 lists these frequency and wavelength ranges for the general types of electromagnetic waves.

Power Waves Electromagnetic waves with a frequency of 60 Hz result from alternating currents in electrical circuits. These power waves have a wavelength of 5.0×10^6 m, or 5000 km (more than 3000 mi). Waves of such low frequency are of little practical use. They may occasionally produce a so-called 60-Hz hum on your stereo or introduce unwanted electrical noise in delicate

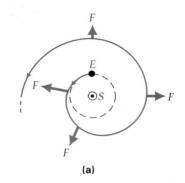
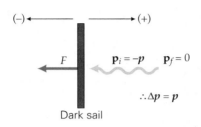
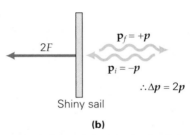

(a)

(b)

▲ **FIGURE 20.22** "Sailing" the solar system (a) A satellite equipped with a large sail would be acted on by radiation pressure from sunlight. This cost-free force would cause the satellite to spiral outward. With planning, the craft would get to outer planets with little or no extra fuel. **(b)** Is it better for the sail to be black or shiny? See Conceptual Example 20.10.

◀ **FIGURE 20.23** The electromagnetic spectrum The spectrum of frequencies or wavelengths is divided into regions, or ranges. Note that the region of visible light is a very small part of the total electromagnetic spectrum. The wavelengths are given in nanometers: 1 nm = 10^{-9} m. (Wavelengths illustrated at the top of the figure are not drawn to scale.)

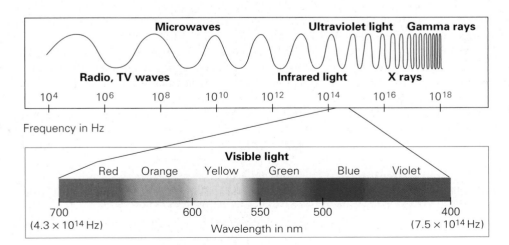

TABLE 20.1 Classification of Electromagnetic Waves

Type of Wave	Approximate Frequency Range (Hz)	Approximate Wavelength Range (m)	Source
Power waves	60	5.0×10^6	Electric currents
Radio waves			Electric circuits
AM	$(0.53 \times 10^6)-(1.7 \times 10^6)$	570–186	
FM	$(88 \times 10^6)-(108 \times 10^6)$	3.4–2.8	
TV	$(54 \times 10^6)-(890 \times 10^6)$	5.6–0.34	
Microwaves	10^9-10^{11}	$10^{-1}-10^{-3}$	Special vacuum tubes
Infrared radiation	$10^{11}-10^{14}$	$10^{-3}-10^{-7}$	Warm and hot bodies
Visible light	$(4.0 \times 10^{14})-(7.0 \times 10^{14})$	10^{-7}	The Sun and lamps
Ultraviolet radiation	$10^{14}-10^{17}$	$10^{-7}-10^{-10}$	Very hot bodies and special lamps
X rays	$10^{17}-10^{19}$	$10^{-10}-10^{-12}$	High-speed electron collisions and atomic processes
Gamma rays	above 10^{19}	below 10^{-12}	Nuclear reactions, processes in particle accelerators, and natural processes

instruments. More serious concerns have been expressed about the possible effects of these waves on health. Some early research tended to suggest that very low-frequency fields may have potentially harmful biological effects on cells and tissues. However, recent surveys indicate that this is not the case in daily life.

Radio and TV Waves Radio and TV waves are generally in the frequency range from 500 kHz to about 1000 MHz. The AM (amplitude-modulated) band runs from 530 to 1710 kHz (1.71 MHz). Higher frequencies, up to 54 MHz, are used for "shortwave" bands. TV bands range from 54 MHz to 890 MHz. The FM (frequency-modulated) radio band runs from 88 to 108 MHz, which lies in a gap between channels 6 and 7 of the range of TV bands. Cellular phones use radio waves to transmit voice communication in the ultrahigh frequency (UHF) band, with frequencies similar to those used for TV channels 13 and higher.

Early global communications used the "shortwave" bands, as do amateur (ham) radio operators today. But how are the normally straight-line radio waves transmitted around the curvature of the Earth? This feat is accomplished by reflection off ionic layers in the upper atmosphere. Energetic particles from the Sun ionize gas molecules, giving rise to several ion layers. Certain of these layers reflect radio waves. Thus, by "bouncing" radio waves off these layers, radio transmissions can be sent beyond the horizon, to any region of the Earth.

Such reflection of radio waves requires the ionic layers to have uniform density. When, from time to time, a solar disturbance produces a shower of energetic particles that upsets this uniformity, a communications "blackout" can occur as the radio waves are scattered in many directions rather than reflected in straight lines. To avoid such disruptions, global communications have, in the past, relied largely on transoceanic cables. Now we also have communications satellites, which can provide line-of-sight transmission to any point on the globe.

Microwaves Microwaves, with frequencies in the gigahertz (GHz) range, are produced by special vacuum tubes (called *klystrons* and *magnetrons*). Microwaves are commonly used in communications and radar applications. In addition to its roles in navigation and guidance, radar provides the basis for the speed guns used to time such things as baseball pitches, tennis serves, and speeding motorists. When radar waves are reflected from a moving object, their wavelength is shifted by the Doppler effect (Section 14.5). The magnitude and sign of the shift indicate the velocity of the object toward or away from the observer.

Infrared Radiation (IR) The infrared region of the electromagnetic spectrum lies adjacent to the low-frequency, or long-wavelength, end of the visible spectrum. The frequency at which a warm body emits IR radiation depends on that body's temperature. (See Chapter 27.) Such a body emits electromagnetic waves of many frequencies, but the frequency of the maximum intensity characterizes the radiation. A body at about room temperature emits radiation in the far infrared region. ("Far" here is taken as relative to the visible region.)

Recall from Section 11.4 that infrared radiation is sometimes referred to as "heat rays." This is because water molecules, which are present in most materials, readily absorb electromagnetic radiation in the infrared-wavelength region. When they do, their random thermal motion is increased—they "heat up," as well as heat their surroundings. Infrared lamps are used in therapeutic applications, such as for easing pain in strained muscles, and to keep food warm in cafeterias. IR is also associated with maintaining the Earth's temperature through the *greenhouse effect*. In this effect, incoming visible light (which passes relatively easily through the atmosphere) is absorbed by the Earth's surface and reradiated as infrared (longer wavelength) radiation, which is trapped by greenhouse gases such as carbon dioxide and water vapor. Its name comes from greenhouses that use glass to trap the energy (rather than atmospheric gases).

Visible Light The region of visible light occupies only a small portion of the electromagnetic spectrum. It runs from a frequency of about 4×10^{14} Hz to about 7×10^{14} Hz, or a wavelength range of about 700 to 400 nm, (Fig. 20.23). Only the radiation in this wavelength/frequency region activates the receptors in human eyes. Visible light emitted or reflected from the objects around us provides us with our visual information about our world. Visible light and optics will be discussed in Chapters 22 to 25.

Note: Recall that 1 nanometer (nm) $= 10^{-9}$ m.

It is interesting to note that not all animals are sensitive to the same range of wavelengths. For example, snakes can detect infrared radiation, and the visible range of many insects extends well into the ultraviolet range. The sensitivity range of human eyes conforms closely to the spectrum of wavelengths emitted by the Sun, which has its maximum emissions in the yellow–green region.

Ultraviolet Radiation (UV) The Sun's spectrum, although consisting mostly of visible light, has a small component of ultraviolet (UV) light, whose frequency range lies beyond the violet end of the visible region. Ultraviolet radiation is also produced by special lamps and very hot bodies. In addition to causing tanning of the skin, UV radiation can cause sunburn and/or skin cancer if exposure to it is too high.

Upon its arrival at the Earth, most of the Sun's ultraviolet emission is absorbed in the ozone (O_3) layer in the atmosphere, at an altitude of about 30 to 50 km. Because the ozone layer plays a protective role, there is concern about its depletion by chlorofluorocarbon gases (such as Freon, once commonly used in refrigerators) that drift upward and react with the ozone.

Most ultraviolet radiation is absorbed by certain molecules in ordinary glass. Therefore, you cannot get much of a tan through glass windows. Sunglasses are now labeled to indicate the UV protection standards they meet in shielding the eyes from this potentially harmful radiation. There are certain types of high-tech glass (called "photo-gray" glass) that darken when exposed to UV. These materials form the basis for sunglasses that darken when exposed to sunlight. Of course, these sunglasses aren't very useful when worn by someone while driving a car. (Why?) Welders wear special glass goggles or facemasks to protect their eyes from the large amounts of UV radiation produced by the arcs of welding torches. Similarly, it is important to shield your eyes from sunlamps or from snow-covered surfaces. The ultraviolet component of sunlight reflected from snow-covered surfaces can produce snowblindness in unprotected eyes.

X rays Beyond the ultraviolet region of the electromagnetic spectrum is the important X-ray region. We are familiar with X rays primarily through medical applications. X rays were discovered accidentally in 1895 by the German physicist Wilhelm Roentgen (1845–1923) when he noted the glow of a piece of fluorescent paper caused by some mysterious radiation coming from a cathode-ray tube. Because of the apparent mystery involved, this radiation was named *x-radiation* or *X rays*.

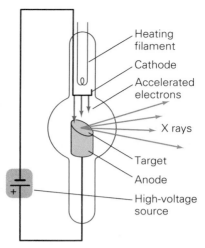

▲ FIGURE 20.24   The X-ray tube
Electrons accelerated through a
large voltage strike a target
electrode. There, they slow down
and interact with (excite) the atomic
electrons of the target material.
Energy is emitted in the form of
X rays during both the "braking"
process and the deexcitation of the
atoms in the material.

The basic elements of an X-ray tube are shown in ◀Fig. 20.24. An accelerating voltage, typically several thousand volts, is applied across the electrodes in a sealed, evacuated tube. Electrons emitted from the heated negative electrode (cathode) are accelerated toward the positive electrode (anode). When they strike the anode, some of their kinetic-energy loss is converted to electromagnetic energy in the form of X rays.

A similar process takes place in color-television picture tubes, which use high voltages and electron beams. When the high-speed electrons hit the screen, they can emit X rays into the environment. Fortunately, all modern televisions have the shielding necessary to protect viewers from exposure to this radiation. In the early days of color television, this was not always the case—hence the warning that came with the set: "Do not to sit too close to the screen."

As you will learn in Chapter 27, the energy carried by electromagnetic radiation depends on its frequency. High-frequency X rays have very high energies and can cause cancer, skin burns, and other harmful effects. However, at low intensities, X rays can be used with relative safety to view the internal structure of the human body and other opaque objects.* X rays can pass through materials that are opaque to other types of radiation. The denser the material, the greater is its absorption of X rays and the less intense the transmitted radiation will be. For example, as X rays pass through the human body, many more X rays are absorbed or scattered by bone than by tissue. If the transmitted radiation is directed onto a photographic plate or film, the exposed areas show variations in intensity—a picture of internal structures.

The combination of the computer with modern X-ray machines permits the formation of three-dimensional images by means of a technique called *computerized tomography*, or *CT* (▼Fig. 20.25).

Gamma Rays   The electromagnetic waves of the uppermost frequency range of the known electromagnetic spectrum are called *gamma rays* (γ rays). This high-frequency radiation is produced in nuclear reactions, in particle accelerators, and also as a result of certain types of nuclear decay (radioactivity). Gamma rays will be discussed in more detail in Chapter 29.

▼ FIGURE 20.25   CT scan   In an ordinary X-ray image, the entire thickness of the body is projected onto the film. However, internal structures often overlap, making details hard to distinguish. In CT—computerized tomography (from the Greek words *tomo*, meaning "slice," and *graph*, meaning "picture")—X-ray beams scan across a slice of the body. **(a)** The transmitted radiation is recorded by a series of detectors and processed by a computer. Using information from multiple slices, the computer can construct a three-dimensional image. Any single slice can also be displayed for study, as on the monitor. CT scans typically provide physicians with much more information than can be obtained from a conventional X-ray image. **(b)** CT image of a brain with a benign tumor.

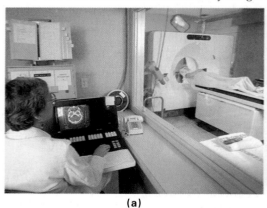

(a)

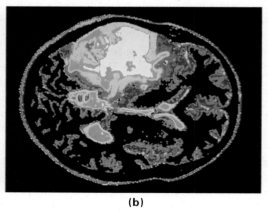

(b)

*Most health scientists believe that there is no safe "threshold" level for X rays or other energetic radiation—that is, no level of exposure that is completely risk free—and that some of the dangerous effects are cumulative over a lifetime. People should therefore avoid unnecessary medical X rays or any other unwarranted exposure to "hard" radiation (Chapter 29). However, when properly used, X rays can be an extremely useful diagnostic tool capable of saving lives.

Chapter Review

Important Concepts and Equations

- **Magnetic flux** (Φ) is a measure of the number of magnetic field lines that pass through an area.

$$\Phi = BA \cos \theta \qquad (20.1)$$

where B is the strength of the magnetic field (assumed constant across the area), A is the area, and θ is the angle between the direction of the magnetic field and the direction of the normal to the plane of the area.

- **Electromagnetic induction** refers to the creation of induced emfs whenever the magnetic flux through a coil, loop, or circuit is changed.

- **Faraday's law of induction** states that the magnitude of the induced emf is equal to the time rate of change of the magnetic flux. In equation terms, this law is

$$\mathcal{E} = -N \frac{\Delta \Phi}{\Delta t} \quad induced\ emf \quad (20.2)$$

where $\Delta \Phi$ is the change in flux through *one loop* and there are N total loops.

- **Lenz's law** states that when a change in magnetic flux induces an emf in a coil, loop, or circuit, the resulting current is in such a direction as to create its own magnetic field to oppose the change in flux.

- An **ac generator** converts mechanical energy into electrical energy. The generator's emf as a function of time is

$$\mathcal{E} = \mathcal{E}_o \sin \omega t \qquad (20.4)$$

where $\mathcal{E}_o$, the maximum emf.

- **Back emf** is an induced reverse emf created by induction in motors when their armature rotates in their magnetic field.

- A **transformer** is a device that changes the voltage supplied to it by means of induction. The voltage applied to the input, or primary (p), side of the transformer is changed into the output, or secondary (s), voltage according to

$$V_s = \left(\frac{N_s}{N_p} \right) V_p \qquad (20.10a)$$

In a step-up transformer, $N_s > N_p$, hence the voltage is increased. In this case, the output current is decreased. In a step-down transformer, the reverse is true since $N_p > N_s$.

- An **electromagnetic wave** consists of time-varying electric and magnetic fields that propagate at a constant speed in a vacuum ($c = 3.00 \times 10^8$ m/s). The various types of electromagnetic radiation (such as UV, radio waves, and visible light) differ in frequency and wavelength.

- Electromagnetic radiation carries energy and momentum and can exert a force called **radiation pressure**.

Exercises

20.1 Induced emf: Faraday's Law and Lenz's Law

1. A unit of magnetic flux is (a) Wb, (b) $T \cdot m^2$, (c) $T \cdot m/A$, or (d) both (a) and (b).

2. The magnetic flux through a loop can change due to a change in (a) the area of the coil, (b) the strength of the magnetic field, (c) the orientation of the loop, or (d) all of the preceding.

3. For an induced current to appear in a wire loop, (a) there must be a large magnetic flux through the loop, (b) the loop's plane must be parallel to the magnetic field, (c) the loop's plane must be perpendicular to the magnetic field, or (d) the magnetic flux through the loop must vary with time.

4. CQ A bar magnet is dropped through a coil of wire as shown in ▶ Fig. 20.26. (a) Describe what is observed on the galvanometer by sketching a graph of $\mathcal{E}$ versus t. (b) Does the magnet fall freely? Explain.

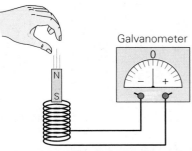

▲ **FIGURE 20.26   A time-varying magnetic field**   What will the galvanometer measure? See Exercise 4.

5. CQ In Fig. 20.1b, what would be the direction of the induced current in the loop if the south pole of the magnet were approaching instead of the north pole?

6. CQ What are the factors that affect the magnitude of the induced emf in a solenoid?

7. CQ Does the induced emf in a closed loop depend on the value of the magnetic flux in the loop? Explain.

8. CQ Two identical strong magnets are dropped simultaneously by two students into two vertical tubes of the same dimensions (▼Fig. 20.27). One tube is made of copper, and the other is made of plastic. From which tube will the magnet emerge first? Why?

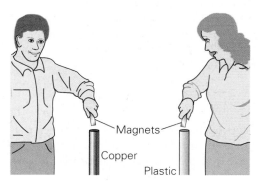

▲ **FIGURE 20.27 Free fall?** See Exercise 8.

9. CQ A basic telephone has both a speaker–transmitter and a receiver (▼Fig. 20.28). Until the advent of digital phones in the 1990s, the transmitter had a diaphragm coupled to a carbon chamber (called the *button*), which contained loosely packed granules of carbon. As the diaphragm vibrated because of incident sound waves, the pressure on the granules varied, causing them to be more or less closely packed. As a result, the resistance of the button changed. The receiver converted these electrical impulses to sound. Applying the principles of electricity and magnetism that you have learned, explain the basic operation of this type of telephone.

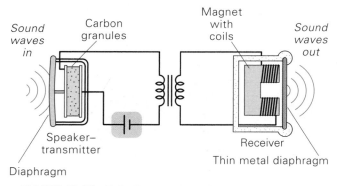

▲ **FIGURE 20.28 Telephone operation** See Exercise 9.

10. ■ A circular loop with an area of 0.015 m² is in a uniform magnetic field of 0.30 T. What is the flux through the loop's plane if it is (a) parallel to the field, (b) at an angle of 37° to the field, and (c) perpendicular to the field?

11. ■ A circular loop (radius of 20 cm) is positioned at various orientations in a uniform magnetic field of 0.15 T. Find the magnetic flux if the normal to the plane of the loop is (a) perpendicular to the field, (b) parallel to the field, and (c) at an angle of 40° to the field.

12. ■ The plane of a conductive loop with an area of 0.020 m² is perpendicular to a uniform magnetic field of 0.30 T. If the field drops to zero in 0.0045 s, what is the magnitude of the average emf induced in the loop?

13. ■ A loop in the form of a right triangle with one side of 40.0 cm and a hypotenuse of 50.0 cm lies in a plane perpendicular to a uniform magnetic field of 550 mT. What is the flux through the loop?

14. ■ A square coil of wire with 10 turns is in a magnetic field of 0.25 T. The total flux through the coil is 0.50 T·m². Find the area of one turn if the field (a) is perpendicular to the plane of the coil and (b) makes an angle of 60° with the plane of the coil.

15. ■■ An ideal solenoid with a current of 1.5 A has a radius of 3.0 cm and a turn density of 250 turns/m. What is the magnetic flux (due to its own field) through its cross-sectional area at its center?

16. ■■ A magnetic field exists at right angles to the plane of a wire loop with an area of 0.40 m². If the field decreases by 0.20 T in 10^{-3} s, what is the magnitude of the average emf induced in the loop?

17. ■■ A square loop of wire with sides of length 40 cm experiences a uniform magnetic field perpendicular to its area. If the field's strength is initially 100 mT and it decays to zero in 0.010 s, what is the magnitude of the average emf induced in the loop?

18. ■■ The magnetic flux through one turn of a 60-turn coil of wire is reduced from 35 Wb to 5.0 Wb in 0.10 s. The average induced current in the coil is 3.6 mA. Find the total resistance of the wire.

19. ■■ When the magnetic flux through a single loop of wire increases by 30 T·m², an average current of 40 A is induced in the wire. Assuming that the wire has a resistance of 2.5 Ω, over what period of time did the flux increase?

20. ■■ In 0.20 s, a coil of wire with 50 loops experiences an average induced emf of 9.0 V, due to a changing magnetic field perpendicular to the plane of the coil. The radius of the coil is 10 cm, and the initial strength of the magnetic field is 1.5 T. Assuming that the strength of the field decreased with time, what is the final strength of the field?

21. IE ■■ A single strand of wire of adjustable length is wound around the circumference of a round balloon. A uniform magnetic field is perpendicular to the plane of the loop (▶Fig. 20.29). (a) If the balloon is inflated, in which direction is the induced current, looking down from above,

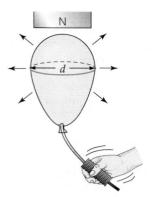

▲ FIGURE 20.29 **Pumping energy** See Exercise 21.

(1) counterclockwise, (2) clockwise, or (3) there is no induced current? (b) If the magnitude of the magnetic field is 0.15 T and the wire's diameter increases from 20 cm to 40 cm in 0.040 s, what is the magnitude of the average value of the emf induced in the loop?

22. ■■ The magnetic field perpendicular to the plane of a wire loop with an area of 0.10 m² changes with time as shown in ▼Fig. 20.30. What is the magnitude of the average emf induced in the loop for each segment of the graph (for example, from 0 to 2.0 ms)?

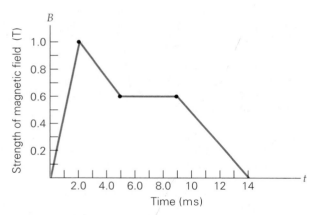

▲ FIGURE 20.30 **Magnetic field versus time** See Exercise 22.

23. IE ■■ A boy is traveling due north at a constant speed while carrying a metal rod. The rod's length is oriented in the east–west direction and is parallel to the ground. (a) There will be no induced emf when the rod is (1) at the equator of the Earth, (2) near the Earth's magnetic poles, or (3) somewhere between the equator and the poles. Why? (b) If the vertical component of the Earth's magnetic field is 1.0×10^{-5} T at a particular location, the boy runs with a speed of 5.0 m/s at that location, and the rod is 1.0 m long, what is the induced emf in the rod?

24. ■■ A metal airplane with a wingspan of 30 m flies horizontally at a constant speed of 320 km/h in a region where the vertical component of the Earth's magnetic field is 5.0×10^{-5} T. What is the induced motional emf between the tips of the plane's wings?

25. ■■ Suppose that the metal rod in Fig. 20.11 is 20 cm long and is moving at a speed of 10 m/s in a magnetic field of 0.30 T and that the metal frame is covered with an insulating material. Find (a) the magnitude of the induced emf across the rod and (b) the current in the rod.

26. ■■ The flux through a fixed loop of wire changes uniformly from +40 Wb to −20 Wb in 1.5 ms. (a) What is the significance of the negative flux? (b) What is the average induced emf in the loop?

27. ■■■ A coil of wire with 10 turns and an area of 0.055 m² is placed in a magnetic field of 1.8 T and oriented such that the area is perpendicular to the field. The coil is then flipped by 90° in 0.25 s and ends up with the area parallel to the field (▼Fig. 20.31). What is the magnitude of the average emf induced in the coil?

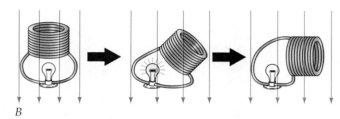

▲ FIGURE 20.31 **Flipping the coil** See Exercises 27 and 28.

28. IE ■■■ In Fig 20.31, the coil is flipped by 180° in the same time interval as in Exercise 27. (a) How does the magnitude of the average emf compare with that in Exercise 27, where the coil was flipped only 90°? (1) It is higher, (2) it is the same, or (3) it is lower. Why? (b) What is the magnitude of the average emf in this case?

29. ■■■ A uniform magnetic field of 0.50 T penetrates a double-incline block as shown in ▼Fig. 20.32. Determine the magnetic flux through each surface of the block.

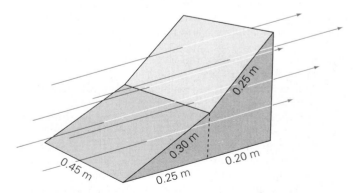

▲ FIGURE 20.32 **Magnetic flux** See Exercise 29. (Not drawn to scale.)

20.2 Electric Generators and Back emf

30. Increasing the coil area in an ac generator (a) increases the frequency of rotation, (b) decreases the maximum induced emf, or (c) increases the maximum induced emf.

31. The back emf of a motor depends on (a) the input voltage, (b) the input current, (c) the armature's rotational speed, or (d) none of the preceding.

32. **CQ** What is the orientation of the armature loop in a simple ac generator when the value of (a) the emf is a minimum and (b) the magnetic flux is a minimum? Explain your reasoning for each situation, and discuss why they do not occur at the same orientation.

33. **CQ** A student has a bright idea for a generator: For the arrangement shown in ▼ Fig. 20.33, the magnet is pulled down and released. With a highly elastic spring, the student thinks that there should be a relatively continuous electrical output. What is wrong with this idea?

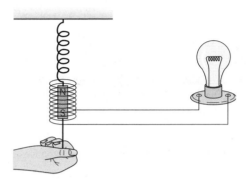

▲ **FIGURE 20.33 Inventive genius?** See Exercise 33.

34. **CQ** In a dc motor, if the armature is jammed or turns very slowly under a heavy load, the coils in the motor may burn out. Why?

35. **CQ** Discuss the factors that affect the maximum induced emf in an ac generator.

36. ■ An emergency ac generator operates at a rotation frequency of 60 Hz. If the output voltage is a maximum (in magnitude) at $t = 0$, when is it next (a) a maximum (in magnitude), (b) zero, and (c) at its initial value?

37. ■ A student makes a simple generator by using a single square loop 10 cm on each side. The loop is then rotated at a frequency of 60 Hz in a magnetic field of 0.015 T. (a) What is the maximum emf output? (b) What would be the maximum emf output if 10 such loops were used instead?

38. ■ A simple ac generator consists of a coil with 10 turns (each turn has an area of 50 cm²). The coil rotates in a uniform magnetic field of 350 mT with a frequency of 60 Hz. (a) Write an expression in the form of Eq. 20.5 for the generator's emf variation with time. (b) Compute the maximum emf.

39. **IE** ■■ A 60-Hz ac voltage source has a maximum voltage of 120 V. A student wants to determine the voltage $\frac{1}{180}$ s after it is zero. (a) How many possible voltages are there, (1) one, (2) two, or (3) three? Why? (b) Determine all possible voltages.

40. ■■ An ac generator with a maximum emf of 400 V is to be constructed from wire loops (radius 0.15 m). It will operate at a frequency of 60 Hz and use a magnetic field of 200 mT. How many loops will be needed?

41. ■■ An ac generator operates at a rotational frequency of 60 Hz and has a maximum emf of 100 V. Assume that it has zero emf at start-up. What is the instantaneous emf (a) $\frac{1}{240}$ s after start-up and (b) $\frac{1}{120}$ s after the emf passes through zero as it begins to reverse polarity?

42. ■■ The armature of a simple ac generator has 20 circular loops of wire, each with a radius of 10 cm. It is rotated with a frequency of 60 Hz in a uniform magnetic field of 800 mT. What is the maximum emf induced in the loops, and how often is this value attained?

43. ■■ The armature of an ac generator has 100 turns. Each turn is a rectangular loop measuring 8.0 cm by 12 cm. The generator has a sinusoidal voltage output with an amplitude of 24 V. If the magnetic field of the generator is 250 mT, with what frequency does the armature turn?

44. **IE** ■■ (a) To increase the output of an ac generator, a student has the choice of *either* doubling the generator's magnetic field *or* its frequency. To maximize the increase in emf output, he should (1) double the magnetic field, (2) double the frequency, or (3) it does not matter which one he doubles. Explain. (b) Two students display their ac generators at a science fair. The generator made by student A has a loop area of 100 cm² rotating in a magnetic field of 20 mT at 60 Hz. The one made by student B has a loop area of 75 cm² rotating in a magnetic field of 200 mT at 120 Hz. Which one generates the largest maximum emf? Justify your answer mathematically.

45. **IE** ■■ A motor has a resistance of 2.50 Ω and is connected to a 110-V line. (a) Is the operating current of the motor (1) higher than 44 A, (2) 44 A, or (3) lower than 44 A? Why? (b) If the back emf of the motor at operating speed is 100 V, what is its operating current?

46. ■■ The starter motor in an automobile has a resistance of 0.40 Ω in its armature windings. The motor operates on 12 V and has a back emf of 10 V when running at normal operating speed. How much current does the motor draw (a) when running at its operating speed and (b) when starting up?

47. ■■ A 240-V dc motor has an armature whose resistance is 1.50 Ω. When running at its operating speed, it draws a current of 16.0 A. (a) What is the back emf of the motor when it is operating normally? (b) What is the starting current? (Assume that there is no additional resistance in the circuit.) (c) What series resistance would be required to limit the starting current to 25 A?

20.3 Transformers and Power Transmission

48. A step-up transformer has (a) more windings in the primary coil, (b) more windings in the secondary coil, or

(c) the same number of windings in the primary and secondary coils.

49. The power delivered by an ideal step-down transformer is (a) greater in the primary coil, (b) greater in the secondary coil, or (c) the same in the primary and secondary coils.

50. CQ To reduce resistance losses, should electric power be transmitted at low voltage or at high voltage? Explain.

51. CQ Can a step-up transformer be used as a step-down transformer? If so, how?

52. CQ Can a transformer work using dc voltages? Explain.

53. CQ The voltage used to fire a spark plug in an automobile is supplied by a *coil*, which is actually a pair of induction coils (▼Fig. 20.34). Explain how the voltage of the car's battery (12 V) is raised as high as 25 kV by this device. Also explain the function of the distributor.

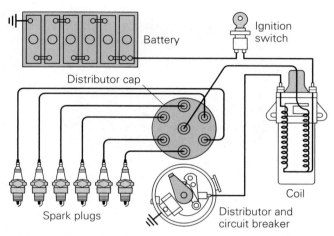

▲ **FIGURE 20.34 Auto ignition with coil** See Exercise 53.

54. IE ■ The secondary coil of an ideal transformer has 450 turns, and the primary coil has 75 turns. (a) Is this transformer a (1) step-up or (2) step-down transformer? Why? (b) What is the ratio of the current in the primary coil to the current in the secondary coil? (c) What is the ratio of the voltage across the primary coil to the voltage in the secondary coil?

55. ■ An ideal transformer steps 8.0 V up to 2000 V, and the 4000-turn secondary coil carries 2.0 A. (a) Find the number of turns in the primary coil. (b) Find the current in the primary coil.

56. ■ The primary coil of an ideal transformer has 720 turns, and the secondary coil has 180 turns. If the primary coil carries 15 A at a voltage of 120 V, what are (a) the voltage and (b) the output current of the secondary coil?

57. ■ The transformer in the power supply for a computer's 250-MB ZIP drive changes a 120-V input to a 5.0-V output. Find the ratio of the number of turns in the primary coil to the number of turns in the secondary coil.

58. ■ The primary coil of an ideal transformer is connected to a 120-V source and draws 10 A. The secondary coil has 800 turns and a current of 4.0 A. (a) What is the voltage across the secondary coil? (b) How many turns are in the primary coil?

59. ■ An ideal transformer has 840 turns in its primary coil and 120 turns in its secondary coil. If the primary coil draws 2.50 A at 110 V, what are (a) the current and (b) the output voltage of the secondary coil?

60. ■■ The efficiency $\mathscr{E}$ of a transformer is defined as the ratio of the power output to the power input: $\mathscr{E} = I_s V_s/(I_p V_p)$. Show that in terms of the ratios of currents and voltages given in Eq. 20.10 (for an ideal transformer), an efficiency of 100% is obtained. What does this finding imply?

61. IE ■■ The specifications of a transformer used with a small appliance read as follows: Input, 120 V, 6.0 W; Output, 9.0 V, 300 mA. (a) Is this transformer (1) an ideal or (2) a nonideal transformer? Why? (b) What is its efficiency? [See Exercise 60.]

62. ■■ A circuit component operates at 20 V and 0.50 A. A transformer with 300 turns in its primary coil is used to convert 120-V household electricity to the proper voltage. (a) How many turns must the secondary coil have? (b) How much current is in the primary coil?

63. ■■ A transformer in a door chime steps down the voltage from 120 V to 6.0 V and supplies a current of 0.50 A to the chime mechanism. (a) What is the turn ratio of the transformer? (b) What is the current input to the transformer?

64. ■■ An ac generator supplies 20 A at 440 V to a 10 000-V power line. If the step-up transformer has 150 turns in its primary coil, how many turns are in its primary coil, how many turns are in the secondary coil?

65. ■■ The electricity supplied in Exercise 64 is transmitted over a line 80.0 km long with a resistance of 0.80 Ω/ km. (a) How many kilowatt-hours are saved in 5.00 h by stepping up the voltage? (b) At $0.10/kWh, how much of a savings (to the nearest $10) is this to the consumer in a 30-day month, assuming that the energy is supplied continuously?

66. ■■ At an area substation, the power-line voltage is stepped down from 100 000 V to 20 000 V. If 10 MW of power is delivered to the 20 000-V circuit, what are the primary and secondary currents in the transformer?

67. ■■ A voltage of 200 000 V in a transmission line is reduced to 100 000 V at an area substation, to 7200 V at a distributing substation, and finally to 240 V at a utility pole outside of a house. (a) What turn ratio N_s/N_p is required for each reduction step? (b) By what factor is the transmission-line current stepped up in each voltage step-down? (c) What is the overall factor by which the current is stepped up from transmission line to utility pole?

68. ■■ A plant produces electric energy at 50 A and 20 kV. The energy is transmitted 25 km over transmission lines whose resistance is 1.2 Ω/km. (a) What is the power loss in the lines if the energy were transmitted at 20 kV? (b) What should be the output voltage of the generator to decrease the power loss by a factor of 15?

69. ■■ Electrical power is transmitted through a power line 175 km long with a resistance of 1.2 Ω/km. The generator's output is 50 A at its operating voltage of 440 V. This voltage has a single step-up for transmission at 44 kV. (a) How much power is lost as joule heat during the transmission? (b) What must be the turn ratio of a transformer at the delivery point in order to provide an output voltage of 220 V? (Neglect the voltage drop in the line.)

20.4 Electromagnetic Waves

70. Relative to the blue end of the visible spectrum, the yellow and green regions have (a) higher frequencies, (b) longer wavelengths, (c) shorter wavelengths, or (d) both (a) and (c).

71. Which of the following electromagnetic waves has the lowest frequency? (a) UV; (b) IR; (c) X ray; (d) microwave.

72. CQ An antenna is connected to a battery. Will the antenna emit an electromagnetic wave? Why?

73. CQ On a cloudy summer day, you work outside and feel cool, yet that evening, you find that you have a sunburn. Why? Explain in terms of infrared and ultraviolet radiation.

74. CQ Radiation can exert pressure on surfaces (radiation pressure). Will the pressure be greater on a shiny surface or a dark surface? Why?

75. CQ Radar operates at wavelengths of a few centimeters, whereas FM radio operates at wavelengths on the order of meters. How do radar frequencies compare with the frequencies of the FM band on your radio?

76. ■ Find the frequencies of electromagnetic waves with wavelengths of (a) 2.0 m, (b) 25 m, and (c) 75 m.

77. ■ The frequency ranges of the radio AM band, radio FM band, and TV band are 530 kHz to 1710 kHz, 88 MHz to 108 MHz, and 54 MHz to 890 MHz, respectively. What are the corresponding wavelength ranges for these bands?

78. ■ A meteorologist in a TV station is using radar to determine the distance to a cloud. He notes that a time of 0.24 ms elapses between the sending and the return of a radar pulse. How far away is the cloud?

79. ■ How long does it take for a laser beam to travel from the Earth to a reflector on the Moon and back? Take the distance from the Earth to the Moon to be 2.4×10^5 mi. (This experiment was done when the *Apollo* flights of the early 1970s left laser reflectors on the lunar surface.)

80. ■■ Orange light has a wavelength of 600 nm, and green light has a wavelength of 510 nm. What is the difference in frequency between the two types of light?

81. ■■ A certain type of radio antenna is called a *quarter-wavelength antenna*, because its length is equal to one quarter of the wavelength to be received. If you were going to make such antennae for the AM and FM radio bands by using the middle frequencies of each band, what lengths of wire would you use?

82. IE ■■ Microwave ovens can have cold spots and hot spots, due to standing electromagnetic waves, analogous to standing wave nodes and antinodes in strings (▼ Fig. 20.35). (a) The longer the distance between the cold spots, (1) the higher is the frequency, (2) the lower is the frequency, or (3) the frequency is independent of this distance. Why? (b) If your microwave has cold spots (nodes) approximately every 5.0 cm, what is the frequency of the waves it uses to cook?

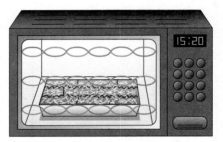

▲ **FIGURE 20.35 Cold spots?** See Exercise 82.

Additional Exercises

83. A 120-V ac motor has a resistance of 4.0 Ω. When operating at normal speed, the motor develops a back emf of 110 V. What are (a) the start-up current of the motor and (b) the current when the motor is operating at normal speed?

84. An electric doorbell operates at 4.5 V. If this voltage is obtained from standard 120-V household voltage, which of the windings in the bell's transformer has more turns, and how many times more?

85. A circular wire loop of 10 turns is in a uniform magnetic field of 150 mT. The flux through the loop is 0.12 T·m². What is the radius of the loop if the normal to the plane of the loop makes an angle of 45° with the field?

86. A power line transmits electric energy at a voltage of 1.50 kV to a residential area. There, a transformer with 500 turns in its primary coil steps the voltage down to 240 V. How many turns are in the transformer's secondary coil?

87. A pivoted coil of wire is rotated 50 times per second in a uniform magnetic field. How often is the induced emf in the coil zero?

88. A 120-V dc motor draws a current of 6.0 A and has a back emf of 96 V at its operating speed. (a) What starting current does the motor draw? (Assume that there are no additional resistances in the circuit.) (b) What series resistance would be required to limit the starting current to 15 A?

89. IE In ▼Figure 20.36, a metal bar of length L moves in a region of constant magnetic field. That field is directed into the page. (a) The direction of the induced current through the resistor is (1) up, (2) down, or (3) there is no current. Why? (b) If the magnitude of the magnetic field is 250 mT, what is the current?

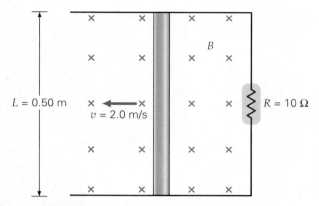

▲ **FIGURE 20.36 Motional emf** See Exercise 89.

90. An ideal transformer has 120 turns in its primary coil and 840 turns in its secondary coil. If the current in the primary coil is 14 A, with a voltage of 120 V, what are (a) the voltage and (b) the current in the secondary coil?

91. A square loop of wire has a length of 1.0 m on each side. Its plane is perpendicular to a uniform magnetic field of 0.75 T. What is the flux through the loop?

92. An ideal transformer has 80 turns in its primary coil and 360 turns in its secondary coil. When 12 V is applied to the primary coil, a 3.0-A current results in that coil. What are the (a) current, (b) voltage output, and (c) power output of the secondary coil?

93. A square conductive loop 50 cm on each side has a resistance of 0.200 Ω. Find the rate at which a magnetic field perpendicular to the plane of the loop must change with time in order to cause an average joule heating of 5.0 J/s in the loop.

94. A coil with 150 concentric loops of wire and a radius of 12 cm is used as the armature of a generator with a magnetic field of 0.80 T. With what frequency should the armature be rotated to make the polarity change every 0.010 s?

95. The transformer on a utility pole steps the voltage down from 20 000 V to 220 V for use in a college science building. If the building uses 6.6 kW of power, what are the primary and secondary currents in the transformer?

96. Gamma rays have typical wavelengths on the order of 3×10^{-15} m (nuclear diameter size). What is the frequency of typical gamma-ray radiation?

97. In some countries, the voltage is delivered at 50 Hz instead of the 60 Hz delivered in the United States. (a) What is the rotation rate of the turbines that drive the electric generators for such voltage delivery? (b) Would different transformers be needed if the step-up and step-down requirements of the delivery systems were to be the same? Explain carefully, considering both ideal and nonideal transformers.

CHAPTER

2 1 (stylized chapter number)

AC Circuits

Direct-current circuits have many uses, but the nuclear reactor control panel in the photo operates many devices that use alternating current (ac). The electric power delivered to our homes and offices is also ac, and most everyday devices and appliances require alternating current.

There are several reasons for this heavy reliance on alternating current. For one thing, almost all electric energy is produced by generators using electromagnetic induction, and such generators (Chapter 20) produce ac outputs. Furthermore, electrical energy produced in ac fashion can be transmitted economically over long distances through the use of transformers. But perhaps the most important reason that ac is used so universally is that the alternation of the current produces electromagnetic effects that can be exploited in a wide variety of

versatile devices. For example, every time you tune a radio to a favorite station, you take advantage of a special property of the ac circuits in the radio.

In analyzing common dc circuits, only electrical resistance is of concern. There also is resistance in ac circuits, of course, but other factors affect the flow of charge as well. For instance, a capacitor in a dc circuit offers infinite resistance (an open circuit). However, in an ac circuit, this is not the case. Alternating voltage continually charges and discharges a capacitor. Under these conditions, current can exist in a branch of a circuit even if it contains a capacitor. Moreover, wrapped coils of wire will oppose an ac current through electromagnetic induction (Lenz's law; Section 20.1).

In this chapter, the basic principles of ac circuits are discussed. More generalized forms of Ohm's law and expressions for power are developed for ac circuits. Finally, we explore the phenomenon of circuit resonance—the condition for maximum energy transfer in an ac circuit—and some of its applications.

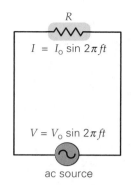

21.1 Resistance in an AC Circuit

OBJECTIVES: To (a) specify how voltage, current, and power vary with time in an ac circuit, (b) understand the concepts of rms and peak values, and (c) learn how resistors respond under ac conditions.

An ac circuit contains an ac voltage source (such as a small generator or simply a household outlet) and one or more other elements. The diagram for an ac circuit with a single resistive element is shown in ▸Fig. 21.1. If the source's output voltage varies sinusoidally, as is the case for a simple generator (see Section 20.2), the voltage across the resistor varies with time in accordance with the equation

$$V = V_o \sin \omega t = V_o \sin 2\pi ft \tag{21.1}$$

where ω is the angular frequency (in rad/s) and is related to the frequency f (in Hz) by $\omega = 2\pi f$. The voltage oscillates between $+V_o$ and $-V_o$ as $\sin 2\pi ft$ oscillates between ± 1. The voltage V_o, called the **peak** (or *maximum*) **voltage**, represents the amplitude of the voltage oscillations.

▲ **FIGURE 21.1   A purely resistive circuit**   The ac source delivers a sinusoidal voltage to a circuit consisting of a single resistor. The voltage across, and current in, the resistor vary sinusoidally at the frequency of the applied ac voltage.

AC Current and Power

Under ac conditions, the current through the resistor oscillates in direction and magnitude. From Ohm's law the current in the resistor, as a function of time, is

$$I = \frac{V}{R} = \left(\frac{V_o}{R}\right)\sin 2\pi ft$$

Since V_o represents the peak or maximum voltage across the resistor, the fraction in parentheses must represent the maximum current in the resistor. Thus, this equation can be rewritten as

$$I = I_o \sin 2\pi ft \tag{21.2}$$

where the amplitude of the current is $I_o = V_o/R$ and is called the **peak** (or *maximum*) **current**.

▸Fig. 21.2 shows both current and voltage as functions of time for a resistor. Note that they are in step, or *in phase*. That is, both reach their zero values, minima, and maxima at the same time. The current oscillates and has positive and negative values (indicating reversals of direction) during each cycle. *Thus, the average current is zero*. Mathematically, this reflects the fact that the time-averaged value of the sine function over one or more *complete* (360°) cycles is zero. Using overbars to denote a time-averaged value, we can write $\overline{\sin \theta} = \overline{\sin 2\pi ft} = 0$. Similarly, $\overline{\cos \theta} = 0$.

The fact that the *average* current is zero, however, does not mean that there is no joule heating (I^2R losses). The dissipation of electrical energy in a resistor does not depend on the direction of the current. The instantaneous power as a function of time is obtained from the instantaneous current (Eq. 21.2). Thus,

$$P = I^2R = (I_o^2 R)\sin^2 2\pi ft \tag{21.3}$$

Even though the current changes sign, the *square* of the current, I^2, is always positive. Thus, the average value of I^2R is *not* zero. The average, or mean, value of I^2 is

$$\overline{I^2} = \overline{I_o^2 \sin^2 2\pi ft} = I_o^2 \overline{\sin^2 2\pi ft}$$

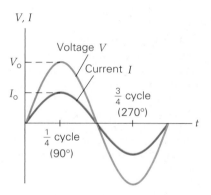

▲ **FIGURE 21.2   Voltage and current in phase**   In a purely resistive ac circuit, the voltage and current are in step, or in phase.

Using the trigonometric identity $\sin^2 \theta = \frac{1}{2}(1 - \cos 2\theta)$, we obtain $\overline{\sin^2 \theta} = \frac{1}{2}(1 - \overline{\cos 2\theta})$. Since $\overline{\cos 2\theta} = 0$ (just as $\overline{\cos \theta} = 0$), it follows that $\overline{\sin^2 \theta} = \frac{1}{2}$. Thus, the previous expression for $\overline{I^2}$ can be rewritten as

$$\overline{I^2} = I_o^2 \overline{\sin^2 2\pi ft} = \frac{1}{2}I_o^2 \qquad (21.4)$$

The average power is therefore

$$\overline{P} = \overline{I^2}R = \frac{1}{2}I_o^2 R \qquad (21.5)$$

It is customary to write ac power in the same form as dc power ($P = I^2 R$). To do this, we define the special current

$$I_{\text{rms}} = \sqrt{\overline{I^2}} = \sqrt{\frac{1}{2}I_o^2} = \frac{I_o}{\sqrt{2}} = \frac{\sqrt{2}}{2}I_o = 0.707 I_o \qquad (21.6)$$

I_{rms} is the **rms current**, or **effective current**. (Here, *rms* stands for *root-mean-square*, indicating the square *root* of the *mean* value of the *square* of the current.) Using $I_{\text{rms}}^2 = (I_o/\sqrt{2})^2 = \frac{1}{2}I_o^2$, we can rewrite the average power as

$$\overline{P} = \frac{1}{2}I_o^2 R = I_{\text{rms}}^2 R \qquad (21.7)$$

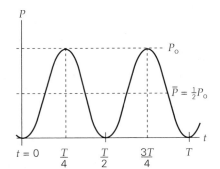

▲ **FIGURE 21.3 Power variation with time in a resistor** Although both current and voltage oscillate in direction (sign), their product (power) is always a positive oscillating quantity. Because of the way in which the rms values of current and voltage are defined, the average power is one-half the peak power.

The average power is equivalent to the time-varying (oscillating) power averaged over time (◄Fig. 21.3).

AC Voltage

The peak values of voltage and current for a resistor are related by $V_o = I_o R$. Using a development similar to that for rms current, we define the **rms voltage**, or **effective voltage**, by

$$V_{\text{rms}} = \frac{V_o}{\sqrt{2}} = \frac{\sqrt{2}}{2}V_o = 0.707 V_o \qquad (21.8)$$

For resistors under ac conditions, then, dc ideas can be used—as long as we realize that the quantities represent rms values. Thus, for ac situations involving only a resistor, the relationship between rms values of current and voltage is

$$V_{\text{rms}} = I_{\text{rms}}R \qquad \begin{array}{l}\textit{voltage across} \\ \textit{a resistor}\end{array} \qquad (21.9)$$

Combining Eqs. 21.9 and 21.7, we have equivalent expressions for power under ac conditions:

$$\overline{P} = I_{\text{rms}}^2 R \qquad \textit{ac power} \qquad (21.10)$$

$$= I_{\text{rms}}V_{\text{rms}}$$

$$= \frac{V_{\text{rms}}^2}{R}$$

It is customary to measure and specify rms values for ac quantities. For example, the household line voltage of 120 V, as you may have guessed, is really the rms value of the voltage. It has a peak or maximum value of

$$V_o = \sqrt{2}V_{\text{rms}} = 1.414(120 \text{ V}) = 170 \text{ V}$$

The graphical interpretations of peak and rms values of current and voltage are shown in ▶Fig. 21.4.

Example 21.1 ■ A Lightbulb: RMS Values versus Peak Values

A lamp with a 60-W bulb is plugged into a 120-V outlet. (a) What are the rms and peak currents through the lamp? (b) What is the resistance of the bulb under these conditions?

Thinking It Through. (a) Since the average power and rms voltage are known, we can find the rms current from Eq. 21.10. From the rms current, Eq. 21.6 can be used to calculate the peak current. (b) The resistance is found from Eq. 21.9.

Solution. The average power and the rms voltage of the source are given.

Given: $\overline{P} = 60\text{ W}$ *Find:*   (a) I_{rms} and I_o (rms and peak currents)
        $V_{rms} = 120\text{ V}$ (b) R (bulb resistance)

(a) The rms current is

$$I_{rms} = \frac{\overline{P}}{V_{rms}} = \frac{60\text{ W}}{120\text{ V}} = 0.50\text{ A}$$

and the peak current is determined by rearranging Eq. 21.6:

$$I_o = \sqrt{2}I_{rms} = \sqrt{2}(0.50\text{ A}) = 0.71\text{ A}$$

(b) The resistance of the bulb is

$$R = \frac{V_{rms}}{I_{rms}} = \frac{120\text{ V}}{0.50\text{ A}} = 240\ \Omega$$

Follow-up Exercise. What would be the (a) rms current and (b) peak current in a 60-W lightbulb in Great Britain, where the house rms voltage is 240 V at 50 Hz? (c) How would be the resistance of a 60-W bulb in Great Britain, compared with one designed for operation at 120 V? Why are the two resistances different? *(Answers to all Follow-up Exercises are at the back of the text.)*

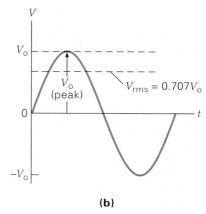

▲ **FIGURE 21.4 Root-mean-square (rms) current and voltage** The rms values of **(a)** current and **(b)** voltage are 0.707, or $1/\sqrt{2}$, times the peak (maximum) values.

Conceptual Example 21.2 ■ Across the Pond: British versus American Electrical Systems

In many countries, the line voltage is 240 V. If a British tourist visiting in the United States plugged in a hair dryer/blower brought from home (where the voltage is 240 V), you would expect it (a) not to operate, (b) to operate normally, (c) to operate poorly, or (d) to burn out.

Reasoning and Answer. British appliances operate at 240 V. At a decreased voltage, there would be decreased current ($I = V/R$) and reduced joule heating (since $P = IV$). If the resistance of the appliance were constant, then, at half the voltage, there would be only one-fourth the power output. Thus, the heating element of the hair dryer/blower might get warm, but it would not work as expected, so the answer is (c). In addition, the decreased current could cause the blower motor to run slower than normal.

     Fortunately, most people do not make this mistake, because plugs and sockets vary from country to country. On foreign travel with appliances from home, a converter/adapter kit can be useful (▶Fig. 21.5). This kit contains a selection of plugs for adapting to the new sockets, as well as a voltage converter to fit into a plug. The converter is a solid-state device that converts 240 V to 120 V for U.S. travelers and vice versa for tourists visiting the United States who are used to 240 V at home.

Follow-up Exercise. What happens if an American tourist inadvertently plugs a 120-V appliance into a British 240-V outlet without a converter? Explain.

21.2 Capacitive Reactance

OBJECTIVES: To (a) explain the behavior of capacitors in ac circuits and (b) calculate the effect of a capacitor on ac current (capacitive reactance).

In Chapter 16, we learned that when a capacitor is connected to a dc voltage source, current exists only for the short time required to charge the capacitor. As charge accumulates on the capacitor's plates, the voltage across them increases, opposing the current. That is, a capacitor in a dc circuit will limit or oppose the current as the capacitor charges. When the capacitor is fully charged, the current drops to zero.

Things are different when a capacitor is driven by an ac voltage source (▼Fig. 21.6a). Under these conditions, the capacitor limits the current, but doesn't completely prevent the flow of charge. This is because the capacitor is alternately charged and discharged as the current and voltage reverse each half-cycle.

Plots of ac current and voltage versus time for a circuit with a capacitor are shown in Fig. 21.6b. Let's look at the changing conditions of the capacitor with time (▶Fig. 21.7).

- $t = 0$ is arbitrarily chosen as the time of maximum voltage* (Fig. 21.7a). At the start, the capacitor is fully charged ($Q_o = CV_o$) with the polarity shown. Since the plates cannot accommodate more charge, there is no current in the circuit.
- As the voltage decreases, the capacitor begins to discharge, giving rise to a counterclockwise current (negative, compare Fig. 21.6b to Fig. 21.7b).
- The current reaches its maximum value when the voltage drops to zero and the capacitor plates are completely discharged (Fig. 21.7c). This occurs exactly one-quarter of the way through the cycle ($t = T/4$).
- The ac voltage source now reverses polarity and starts to increase in magnitude. The capacitor begins to charge, this time with the opposite polarity (Fig. 21.7d). With the plates uncharged, there is no opposition to the current, so the current is at its maximum value. However, as the plates accumulate charge, they begin to inhibit the current; thus, the current decreases in magnitude.
- Halfway through the cycle ($t = T/2$), the capacitor is fully charged, but opposite in polarity to its starting condition (Fig. 21.7e). The current is zero, and the voltage is at its maximum magnitude, but opposite the initial polarity.

During the next half-cycle (not shown in Fig. 21.7), the process is reversed and the circuit returns to its initial condition.

Note that the current and voltage are *not* in step (i.e., not in phase) in this situation. The current reaches its maximum a quarter-cycle *ahead* of the voltage. The relationship between the current and the voltage is commonly stated in this way:

> In a purely capacitive ac circuit, the *current* leads the *voltage* by 90°, or one-quarter ($\frac{1}{4}$) cycle.

As in a dc circuit, a capacitor provides opposition to the charging process in an ac circuit, but it is not totally limiting (as it is in a dc circuit, where it behaves as an open circuit). The quantitative measure of the opposition of a capacitor to current is referred to as the capacitor's **capacitive reactance** (X_C). In an ac circuit, capacitive reactance is given by

$$X_C = \frac{1}{\omega C} = \frac{1}{2\pi f C} \quad \textit{capacitive reactance} \quad (21.11)$$

SI unit of capacitive reactance: ohm (Ω), or second per farad (s/F)

*We have arbitrarily chosen the polarity of initial capacitor voltage as positive (Fig 21.6b and Fig. 21.7a).

▲ **FIGURE 21.5 Converter and adapters** In countries that have 240-V line voltages, U.S. tourists need conversion to 120 V to operate normal U.S. appliances properly. Note the different types of plugs for different countries. The small plugs go into the foreign sockets, and the converter prongs fit into the back of a socket. A U.S. standard two-prong plug fits into the converter, which has a 120-V output. (See Conceptual Example 21.2.)

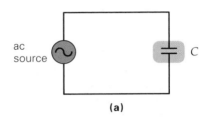

(a)

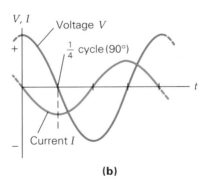

(b)

▲ **FIGURE 21.6 A purely capacitive circuit** (a) In a circuit with only capacitance, (b) the current leads the voltage by 90°, or one-quarter cycle. Half of a cycle of voltage and current, shown, corresponds to Fig. 21.7.

where $\omega = 2\pi f$, C is the capacitance (in farads), and f is the frequency (in Hz). Like resistance, reactance is measured in ohms (Ω). Can you show that the ohm is equivalent to the second per farad?

Equation 21.11 shows that the reactance is inversely proportional to both the capacitance (C) and the voltage frequency (f). Both of these dependencies can be understood as follows. Recall that *capacitance* means "charge stored per volt" ($C = Q/V$). For a particular voltage, therefore, the greater the capacitance, the more charge the capacitor can accommodate for a chosen frequency. Putting more charge on the plates requires a larger charge flow (i.e., a greater current), which means that the opposition to the current, or the reactance, is smaller. Similarly, the greater the frequency of the ac voltage source, the shorter is the charging time during each cycle. A shorter charging time means less charge accumulated on the plates and therefore less opposition to the current. Thus, the capacitive reactance is inversely proportional to *both* frequency and capacitance. As a special case, note that in Eq. 21.11, if $f = 0$, (nonoscillating dc conditions) the capacitive reactance is infinite and there is no current, as expected.

The capacitive reactance is related to the voltage across the capacitor and the current by an equation that has the same form as $V = IR$ for pure resistances:

$$V_{rms} = I_{rms}X_C \qquad \begin{array}{l}\textit{voltage across}\\ \textit{a capacitor}\end{array} \qquad (21.12)$$

Consider the next Example: a capacitor connected to an ac voltage source.

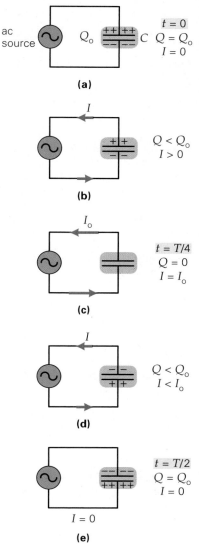

(a)

(b)

(c)

(d)

(e)

▲ **FIGURE 21.7 A capacitor under ac conditions** This sequence shows the voltage, charge, and current in a circuit containing only a capacitor and an ac voltage source. All five circuit diagrams taken together represent physically what is plotted in the first half of the cycle (from $t = 0$ to $t = T/2$) in the graph shown in Fig. 21.6b.

Example 21.3 ■ Current Under AC Conditions: Capacitive Reactance

A 15.0-μF capacitor is connected to a 120-V, 60-Hz source. What are (a) the capacitive reactance and (b) the current (rms and peak) in the circuit?

Thinking It Through. The capacitive reactance can be obtained from the capacitance and the frequency by using Eq. 21.11. (b) We can then calculate the rms current from the reactance and rms voltage via Eq. 21.12. Finally, Eq. 21.6 gives the peak current.

Solution. Assuming the 60-Hz frequency to be exact, our answers are to three significant figures. We list the data:

Given: $C = 15.0\,\mu F = 15.0 \times 10^{-6}\,F$ *Find:* (a) X_C (capacitive reactance)
$\qquad V_{rms} = 120\,V$ (b) I_o (peak current)
$\qquad\quad f = 60\,Hz$ (exact) I_{rms} (rms current)

(a) The capacitive reactance is

$$X_C = \frac{1}{2\pi fC} = \frac{1}{2\pi(60\,Hz)(15.0 \times 10^{-6}\,F)} = 177\,\Omega$$

(b) Then the rms current is

$$I_{rms} = \frac{V_{rms}}{X_C} = \frac{120\,V}{177\,\Omega} = 0.678\,A$$

and therefore, the peak current is

$$I_o = \sqrt{2}I_{rms} = \sqrt{2}(0.678\,A) = 0.959\,A$$

The current oscillates at 60 cycles per second with a magnitude of 0.959 A.

Follow-up Exercise. In this Example, (a) what is the peak voltage, and (b) what frequency would give the same current if the capacitance were twice as large?

PHYSLET®
ILLUSTRATION

Capacitors in ac Circuits

21.3 Inductive Reactance

OBJECTIVES: To (a) explain what an inductor is, (b) explain the behavior of inductors in ac circuits, and (c) calculate the effect of inductors on ac current (inductive reactance).

Note: Lenz's law is given in Section 20.1.

Inductance is a measure of the opposition a circuit element presents to a time-varying current (by Lenz's law). In principle, all circuit elements (even resistors) have some inductance. However, a coil of wire with negligible resistance has only inductance. When placed in a circuit with a time-varying current, such a coil, called an **inductor**, exhibits a reverse voltage, or back emf, in opposition to the changing current. The changing current through the coil produces a changing magnetic field and flux. The back emf is the induced emf in opposition to this changing flux. Since the back emf is induced in the inductor as a result of its own changing magnetic field, this phenomenon is called *self-induction*.

The self-induced emf is given by Faraday's law (Eq. 20.2): $\mathscr{E} = -N\Delta\Phi/\Delta t$. The time rate of change of the flux, $\Delta\Phi/\Delta t$, is proportional to the rate of change of the current in the coil, $\Delta I/\Delta t$, since the current produces the magnetic field responsible for the changing flux. Thus, the back emf is proportional, and directed opposite, to the rate of change of the current. This relationship is expressed mathematically with a constant of proportionality, L:

$$\mathscr{E} = -L\left(\frac{\Delta I}{\Delta t}\right) \tag{21.13}$$

L is called the *inductance* of the coil (more properly, its *self*-inductance). From the equation, it can be seen that inductance has units of volt-seconds per ampere $(\text{V}\cdot\text{s/A})$. This combination is called a **henry** $(1\text{ H} = 1\text{ V}\cdot\text{s/A})$, in honor of Joseph Henry (1797–1878), an American physicist and early investigator of electromagnetic induction. The smaller millihenry (mH) is commonly used $(1\text{ mH} = 10^{-3}\text{ H})$.

The opposition that an inductor presents to the current in an ac circuit depends on the value of the inductance and on the frequency of the voltage. This relationship is expressed quantitatively by the circuit's **inductive reactance**, given by

$$X_{\text{L}} = \omega L = 2\pi fL \qquad \textit{inductive reactance} \tag{21.14}$$

SI unit of inductive reactance: ohm (Ω), or henry per second (H/s)

where f is the frequency of the driving voltage, $\omega = 2\pi f$, and L is the inductance. Like capacitive reactance, inductive reactance is measured in ohms (Ω), which you should be able to show are equivalent to henrys per second.

Note that the inductive reactance is directly proportional to both the inductance (L) of the coil and the frequency (f) of the voltage source. The inductance of a coil is a property of the coil that depends on the number of turns, the coil's diameter, its length, and the material of the core (if any). The frequency of the source plays a role because the more rapidly the current in the coil changes, the greater is the rate of change in its magnetic flux. This implies a larger self-induced (reverse) emf to oppose the changes in current.

In terms of X_{L}, the equation for the voltage across an inductor has a mathematical form similar to that for resistors:

$$V_{\text{rms}} = I_{\text{rms}}X_{\text{L}} \qquad \begin{array}{l}\textit{voltage across}\\ \textit{an inductor}\end{array} \tag{21.15}$$

PHYSLET® ILLUSTRATION

Inductors in ac Circuits

The circuit symbol for an inductor and the graphs of the voltage across the inductor and the current in the circuit are shown in ▶ Fig. 21.8. When an inductor is connected to an ac voltage source, maximum voltage corresponds to zero current. When the voltage drops to zero, the current is maximum. This happens because, as the voltage changes polarity (causing the magnetic flux through the inductor to drop to zero), the inductor acts to prevent the change in accordance with Lenz's law, so the induced emf creates a current. Thus, the current *lags* one-quarter cycle behind the voltage, a relationship commonly expressed as follows:

> In a purely inductive ac circuit, the *voltage* leads the *current* by 90°, or one-quarter ($\frac{1}{4}$) cycle.

The phase relationships of current and voltage for purely inductive and purely capacitive circuits are opposite. A phrase that may help you remember this is

> *ELI* the *ICE* man

With E representing voltage (for *emf*) and I representing current, *ELI* indicates that, with an inductance (L), the voltage leads the current (I). Similarly, *ICE* tells you that, with a capacitance (C), the current leads the voltage.

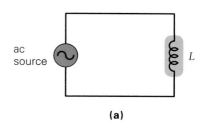

(a)

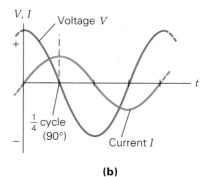

(b)

▲ **FIGURE 21.8 A purely inductive circuit (a)** In a circuit with only inductance, **(b)** the voltage leads the current by 90°, or one-quarter cycle.

Example 21.4 ■ Current Opposition without Resistance: Inductive Reactance

A 125-mH inductor is connected to a 120-V, 60-Hz source. What are (a) the inductive reactance and (b) the rms current in the circuit?

Thinking It Through.   Since the inductance and frequency are known, we can compute the inductive reactance from Eq. 21.14 and the current from Eq. 21.15.

Solution.   Listing the given data, we have the following:

Given: $L = 125\ \text{mH} = 0.125\ \text{H}$ **Find:**   (a) X_L (inductive reactance)
$V_{\text{rms}} = 120\ \text{V}$ (b) I_{rms}
$f = 60\ \text{Hz}$ (exact)

(a)   The inductive reactance is

$$X_L = 2\pi f L = 2\pi (60\ \text{Hz})(0.125\ \text{H}) = 47.1\ \Omega$$

(b)   The rms current is then

$$I_{\text{rms}} = \frac{V_{\text{rms}}}{X_L} = \frac{120\ \text{V}}{47.1\ \Omega} = 2.55\ \text{A}$$

Follow-up Exercise.   In this Example, (a) what is the peak current, and (b) what voltage frequency would give the same current for an inductor with triple the inductance?

21.4 Impedance: RLC Circuits

OBJECTIVES: To **(a)** calculate currents and voltages when various reactive circuit elements are present in ac circuits, **(b)** use phase diagrams to calculate overall impedance and rms currents, and **(c)** understand and use the concept of the power factor in ac circuits.

The previous sections considered purely capacitive or purely inductive circuits separately and without resistance present. However, in the real world, it is impossible to have purely reactive circuits, since there is always some resistance—at a minimum, that from the connecting wires or the inductor's wire. Thus, resistances, capacitive reactances, and inductive reactances *combine* to impede the current in ac circuits. An analysis of some simple combination circuits illustrates these effects.

Series RC Circuit

Suppose that an ac circuit consists of a voltage source, a resistor, and a capacitive element connected in series (▶ Fig. 21.9a). The phase relationship between the current and the voltage is different for each circuit element. As a result, a special

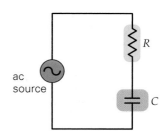

(a) RC Circuit diagram

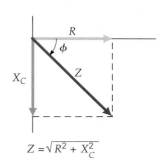

$$Z = \sqrt{R^2 + X_C^2}$$

(b) Phase diagram

▲ **FIGURE 21.9 A series RC circuit (a)** In a series RC circuit, **(b)** the impedance Z is the phasor sum of the resistance R and the capacitive reactance X_C.

graphical method is used to find the overall opposition to the current in the circuit. This method employs a *phase diagram*.

In a *phase diagram* such as the one in Fig. 21.9b for an RC circuit, the resistance and reactance in the circuit are endowed with vectorlike properties and their magnitudes represented by arrows called *phasors*. On a set of *x–y* coordinate axes, the resistance is plotted on the positive *x*-axis (that is, at 0°), since the voltage–current phase difference for a resistor is zero. The capacitive reactance is plotted along the negative *y*-axis, to reflect a phase difference (ϕ) of $-90°$. The negative phase angle symbolizes the fact that, for a capacitor, the voltage lags behind the current.

The phasor sum is the effective, or net, opposition to the current, which we call the **impedance** (Z). Phasors are added in the same way as vectors (because the effects of the resistor and capacitor are not in phase). For the series RC circuit,

$$Z = \sqrt{R^2 + X_C^2} \quad \begin{array}{l} series\ RC\ circuit \\ impedance \end{array} \quad (21.16)$$

The unit of impedance is the ohm.

The generalization of Ohm's law to circuits containing capacitors and inductors along with resistors is

$$V_{rms} = I_{rms}Z \quad \begin{array}{l} Ohm's\ law \\ for\ ac\ circuits \end{array} \quad (21.17)$$

To see how phasors are used to analyze an RC circuit, consider the next Example. Take particular note of part b, in which there is an *apparent* violation of Kirchhoff's loop theorem—explained by phase differences between the voltages across the two elements of the RC circuit.

Note: The word *impedance* (denoted by Z) refers to the overall circuit opposition to current. We reserve the words *reactance* and *resistance* for opposition in individual elements.

Note: $V_{rms} = I_{rms}Z$ can be applied to any circuit, as long as the impedance Z is properly calculated, using phasors.

Example 21.5 ■ RC Impedance and Kirchhoff's Loop Theorem

A series RC circuit has a resistance of 100 Ω and a capacitance of 15.0 μF. (a) What is the (rms) current in the circuit when it is driven by a 120-V, 60-Hz source? (b) Compute the (rms) voltage across each circuit element and the two elements combined. Compare it with that of the voltage source. Is Kirchhoff's loop theorem satisfied?

Thinking It Through. (a) Note that a resistor has been added to the circuit in Example 21.3. By using phasors, the capacitive reactance and the resistance are combined to determine the overall impedance (Eq. 21.16). From Eq. 21.17, the impedance and voltage can be used to find the current. (b) Since the current is the same everywhere at any given time in a series circuit, the result of part (a) can be used to calculate the voltages. The rms voltage across both elements together is found by recalling that the individual voltages are out of phase by 90°. What this means physically is that they reach their peak values *not* at the same time, but rather one-fourth of a period apart. Thus, we *cannot* simply add the voltages.

Solution.

Given: $R = 100\ \Omega$
$C = 15.0\ \mu F = 15.0 \times 10^{-6}\ F$
$V_{rms} = 120\ V$
$f = 60\ Hz$ (exact)

Find: (a) I (rms current)
(b) V_C (rms voltage across capacitor)
V_R (rms voltage across resistor)
$V_{(R+C)}$ (combined rms voltage)

(a) In Example 21.3, we calculated the reactance for the capacitor at the frequency f to be $V_C = 177\ \Omega$. Then Eq. 21.16 can be used to get the circuit impedance:

$$Z = \sqrt{R^2 + X_C^2} = \sqrt{(100\ \Omega)^2 + (177\ \Omega)^2} = 203\ \Omega$$

Since $V_{rms} = I_{rms}Z$, the rms current is

$$I_{rms} = \frac{V_{rms}}{Z} = \frac{120\ V}{203\ \Omega} = 0.591\ A$$

(b) Using Eq. 21.17 first for the rms voltage across resistor alone $(Z = R)$,

$$V_R = I_{rms}R = (0.591\ A)(100\ \Omega) = 59.1\ V$$

For the capacitor alone $(Z = X_C)$, the rms voltage across the capacitor is

$$V_C = I_{rms}X_C = (0.591\ A)(177\ \Omega) = 105\ V$$

The algebraic sum of these two rms voltages is 164 V, which is *not* the same as the rms value of the voltage source. This does *not* mean that Kirchhoff's loop theorem no longer holds. In fact, the source voltage does equal the combined voltages across the capacitor and resistor *if you account for phase differences*. The combined voltage must be calculated properly; that is, it must take into account the 90° phase difference between the two voltages. Thus, using the Pythagorean theorem to get the total voltage, we have

$$V_{(R+C)} = \sqrt{V_R^2 + V_C^2} = \sqrt{(59.1\ V)^2 + (105\ V)^2} = 120\ V$$

When the individual voltage differences are combined properly, it is seen that Kirchhoff's law remains valid: The total rms voltage across both elements is equal to the rms voltage supplied by the source.

Follow-up Exercise. (a) How would the result in part (a) of this Example change if the circuit were driven by a voltage source with the same rms voltage, but oscillating at 120 Hz? (b) Is the resistor or the capacitor responsible for the change?

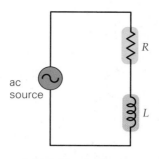

(a) RL Circuit diagram

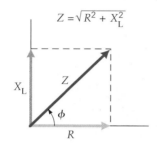

(b) Phase diagram

▲ **FIGURE 21.10   A series RL circuit** (a) In a series RL circuit, (b) the impedance Z is the phasor sum of the resistance R and the inductive reactance X_L.

Series RL Circuit

The analysis of a series RL circuit (▶Fig. 21.10) is similar to that of a series RC circuit. However, the inductive reactance is plotted along the positive y-axis in the phase diagram, to reflect a phase difference of $+90°$ with respect to the resistance. Remember that a positive phase angle means that the voltage *leads* the current, as is the case for an inductor.

The impedance is

$$Z = \sqrt{R^2 + X_L^2} \qquad \begin{array}{l} series\ RL\ circuit \\ impedance \end{array} \qquad (21.18)$$

Series RLC Circuit

More generally, an ac circuit may contain all three circuit elements—a resistor, an inductor, and a capacitor—shown in series in ▶Fig. 21.11. Again, phasor addition determines the overall circuit impedance. Combining the vertical components (the inductive and capacitive reactances) gives the total reactance, $X_L - X_C$. Subtraction is used because the phase difference between X_L and X_C is 180°. The overall circuit impedance is the phasor sum of the resistance and the total reactance. The phasor diagram shows that the impedance is

$$Z = \sqrt{R^2 + (X_L - X_C)^2} \qquad \begin{array}{l} series\ RLC\ circuit \\ impedance \end{array} \qquad (21.19)$$

Impedance in an RLC Circuit

The **phase angle** (ϕ) between the source voltage and the current in the circuit is the angle between the overall impedance phasor (Z) and the +x-axis (Fig. 21.11b), or

$$\tan \phi = \frac{X_L - X_C}{R} \qquad \begin{array}{l} phase\ angle\ in\ series \\ RLC\ circuit \end{array} \qquad (21.20)$$

Notice that if X_L is greater than X_C (as in Fig. 21.11b), the phase angle is positive $(+\phi)$, and the circuit is said to be *inductive*, since the nonresistive part of the impedance (the reactance) is dominated by the inductor. If X_C is greater than X_L, the phase angle is negative $(-\phi)$, and the circuit is said to be *capacitive*, because the capacitor dominates the reactance.

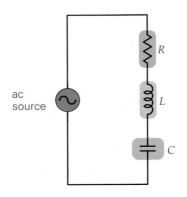

(a) RLC Circuit diagram

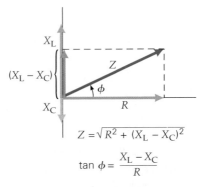

$$Z = \sqrt{R^2 + (X_L - X_C)^2}$$

$$\tan \phi = \frac{X_L - X_C}{R}$$

(b) Phase diagram

▲ **FIGURE 21.11 A series RLC circuit** (a) In a series RLC circuit, (b) the impedance Z is the phasor sum of the resistance R and the total (or net) reactance $(X_L - X_C)$.

TABLE 21.1 Impedances and Phase Angles for Series Circuits

Circuit element(s)	Impedance Z (in Ω)	Phase angle ϕ
R	R	$0°$
C	X_C	$-90°$
L	X_L	$+90°$
RC	$\sqrt{R^2 + X_C^2}$	negative (meaning that ϕ is between $0°$ and $-90°$)
RL	$\sqrt{R^2 + X_L^2}$	positive (meaning that ϕ is between $0°$ and $+90°$)
RLC	$\sqrt{R^2 + (X_L - X_C)^2}$	positive if $X_L > X_C$ negative if $X_C > X_L$

A summary of impedances and phase angles for the three circuit elements and various combinations is given in Table 21.1. Example 21.6 discusses the RLC circuit.

Example 21.6 ■ All Together Now: Impedance in an RLC Circuit

A series RLC circuit has a resistance of 25.0 Ω, a capacitance of 50.0 μF, and an inductance of 0.300 H. If the circuit is driven by a 120-V, 60-Hz source, what are (a) the total impedance of the circuit, (b) the rms current in the circuit, and (c) the phase angle between the current and the voltage?

Thinking It Through. (a) To calculate the overall impedance from Eq. 21.19, the individual reactances must be determined. (b) The current is computed from the generalization of Ohm's law, $V_{rms} = I_{rms}Z$ (Eq. 21.17). (c) The phase angle is calculated from Eq. 21.20.

Solution. We are given all the necessary data, which we list:

Given: $R = 25.0 \ \Omega$
$C = 50.0 \ \mu\text{F} = 5.00 \times 10^{-5} \text{ F}$
$L = 0.300 \text{ H}$
$V_{rms} = 120 \text{ V}$
$f = 60 \text{ Hz (exact)}$

Find: (a) Z (overall circuit impedance)
(b) I_{rms}
(c) ϕ (phase angle)

(a) The individual reactances are

$$X_C = \frac{1}{2\pi f C} = \frac{1}{2\pi(60 \text{ Hz})(5.00 \times 10^{-5} \text{ F})} = 53.1 \ \Omega$$

and

$$X_L = 2\pi f L = 2\pi(60 \text{ Hz})(0.300 \text{ H}) = 113 \ \Omega$$

Then,

$$Z = \sqrt{R^2 + (X_L - X_C)^2} = \sqrt{(25.0 \ \Omega)^2 + (113\Omega - 53.1 \ \Omega)^2} = 64.9 \ \Omega$$

(b) Since $V_{rms} = I_{rms}Z$, we have

$$I_{rms} = \frac{V_{rms}}{Z} = \frac{120 \text{ V}}{64.9 \ \Omega} = 1.85 \text{ A}$$

(c) Solving $\tan \phi = (X_L - X_C)/R$ for the phase angle gives

$$\phi = \tan^{-1}\left(\frac{X_L - X_C}{R}\right) = \tan^{-1}\left(\frac{113 \ \Omega - 53.1 \ \Omega}{25.0 \ \Omega}\right) = +67.3°$$

We should have expected a positive phase angle, since the inductive reactance is greater than the capacitive reactance. (See the results of part (a).)

Follow-up Exercise. (a) Consider the RLC circuit in this Example, but with the driving frequency doubled. Reasoning conceptually, should the phase angle ϕ be greater or less than the +67.3° with the increase in frequency? (b) Compute the new phase angle to show that your reasoning is correct.

You should now appreciate the importance of phasor diagrams in finding impedances, voltages, and currents in ac circuits. However, you might still be wondering about the usefulness of the phase angle ϕ. To illustrate its use and physical importance, let's examine the power loss for an RLC circuit. Notice that power loss analysis depends heavily on the proper use of phasor diagrams.

Power Factor for a Series RLC Circuit

It is interesting to consider the power (energy dissipated per second) in an RLC circuit. The first thing to realize is that the only power loss (joule heating) takes place in the resistor. *There are no power losses associated with capacitors and inductors.* Capacitors and inductors simply store energy and give it back, without loss. Ideally, neither has any resistance, and thus their joule heating is zero.

As we have seen, the average (rms) power dissipated by a resistor is $P_{rms} = I_{rms}^2 R$. The rms power can also be expressed in terms of the rms current and voltage, but the voltage in this case *must be that across the resistor* (V_R), since this is the only dissipative element. That is, the average power dissipated in a series RLC circuit is

$$\overline{P} = P_R = I_{rms}V_R$$

The voltage across the resistor is conveniently found from a voltage triangle that corresponds to the phasor triangle (▶ Fig. 21.12). The rms voltages across the individual components in an RLC circuit are $V_R = I_{rms}R$, $V_L = I_{rms}X_L$, and $V_C = I_{rms}X_C$. Combining the last two voltages, we can write $(V_L - V_C) = I_{rms}(X_L - X_C)$. Therefore, if we multiply each leg of the phasor triangle (Fig. 21.12a) by the rms current, an equivalent voltage triangle results (Fig. 21.12b). As this figure shows, the voltage across the resistor is

$$V_R = V_{rms} \cos \phi \qquad (21.21)$$

The term $\cos \phi$ is called the **power factor**. From Fig. 21.11,

$$\cos \phi = \frac{R}{Z} \qquad series\ RLC\ power\ factor \qquad (21.22)$$

The average power, rewritten in terms of the power factor, is

$$\overline{P} = I_{rms}V_{rms} \cos \phi \qquad series\ RLC\ power \qquad (21.23)$$

Since power is dissipated only in the resistance ($\overline{P} = I_{rms}^2 R$), Eq. 21.22 enables us to express the average power as

$$\overline{P} = I_{rms}^2 Z \cos \phi \qquad series\ RLC\ power \qquad (21.24)$$

Note that $\cos \phi$ varies from a maximum of +1 (when $\phi = 0°$) to a minimum of 0 (when $\phi = \pm\pi/2$). When $\phi = 0°$, the circuit is said to be *completely resistive*. That is, there is maximum power dissipation (as though the circuit contained only a resistor). The power factor decreases as the phase angle increases in either direction [$+\phi$ or $-\phi$, since $\cos(-\phi) = \cos \phi$]—in other words, as the circuit becomes inductive or capacitive. At $\phi = +90°$, the circuit is *completely inductive*; at $\phi = -90°$, it is *completely capacitive*. In these cases, the circuit contains only an inductor or a capacitor, respectively, so no power is dissipated. In practice, since there is always

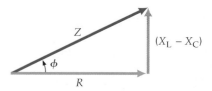

(a) Phasor triangle

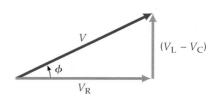

(b) Equivalent voltage triangle

▲ **FIGURE 21.12 Phasor and voltage triangles** The rms voltages across the components of a series RLC circuit are given by $V_R = I_{rms}R$, $V_L = I_{rms}X_L$, and $V_C = I_{rms}X_C$. Since the current is the same through each, **(a)** the phasor triangle can be converted to **(b)** a voltage triangle. Note that $V_R = V \cos \phi$.

some resistance, a circuit can never be completely inductive or capacitive. It is possible, however, for an RLC circuit to appear to be completely resistive even if it contains both a capacitor and an inductor, as we shall see when we study resonance in Section 21.5. Let's look at our previous RLC Example with an emphasis on power.

Example 21.7 ■ Power Factor Revisited

How much power is dissipated in the circuit described in Example 21.6?

Thinking It Through. The power factor can be determined because the resistance (R) and impedance (Z) are known. Once the power factor is known, the actual power can be calculated.

Solution.

Given: See Example 21.6   *Find:* $\overline{P}$

In Example 21.6, the circuit was found to have an overall impedance of $Z = 64.9\ \Omega$, and its resistance was $R = 25.0\ \Omega$. The power factor of the circuit is therefore

$$\cos\phi = \frac{R}{Z} = \frac{25.0\ \Omega}{64.9\ \Omega} = 0.385$$

Using other data from Example 21.6 and Eq. 21.23 gives

$$\overline{P} = I_{\text{rms}}V_{\text{rms}}\cos\phi = (1.85\ \text{A})(120\ \text{V})(0.385) = 85.5\ \text{W}$$

This is less than the power that would be dissipated without a capacitor and an inductor.

Follow-up Exercise. In this Example, (a) at what frequency would the dissipated power be at its maximum value, and (b) what would that maximum power be?

21.5 Circuit Resonance

OBJECTIVES: **To (a) understand the concept of resonance in ac circuits and (b) calculate the resonance frequency of an RLC circuit.**

PHYSLET®
ILLUSTRATION

Resonance in an RLC Circuit

When the power factor ($\cos\phi$) of an RLC series circuit is equal to unity, maximum power is transferred to the circuit. For a particular voltage, the current in the circuit must be at its maximum, since the impedance is at its minimum. This occurs when the inductive and capacitive reactances effectively cancel each other—when they are equal in magnitude and 180° out of phase. This can be made to happen in any RLC circuit by choosing the appropriate source frequency.

The key to finding the correct frequency is to realize that both the inductive and capacitive reactances are frequency dependent, so the overall impedance is also a function of the frequency. From the expression for the impedance of an RLC circuit, $Z = \sqrt{R^2 + (X_L - X_C)^2}$, we can see that the impedance is at a minimum when the inductive and capacitive reactances are equal in magnitude. This occurs at a particular frequency, designated as f_o, which can be found by setting $X_L = X_C$. Using the expressions for the two reactances, we have

$$2\pi f_o L = \frac{1}{2\pi f_o C}$$

Solving for this frequency yields

$$f_o = \frac{1}{2\pi\sqrt{LC}} \qquad \textit{RLC resonance frequency} \quad (21.25)$$

Equation 21.25 gives the frequency for the condition of minimum impedance—that is, the **resonance frequency** of the circuit. A graph of reactance versus frequency is shown in ▼Fig. 21.13a. The curves X_C and X_L intersect at f_o, where their values are equal.

A Physical Explanation of Resonance

The physical explanation of resonance in a series RLC circuit is worth exploring. We know that the voltage across the resistor is in phase with the current. We also know that the voltage across the capacitor *lags* the current (and thus the voltage across the resistor) by 90°. The voltage across the inductor *leads* the current (and thus the voltage across the resistor) by 90°. Hence, *the capacitor and inductor voltages are always 180° out of phase, or have opposite polarity.* In other words, they tend to cancel out. This is embodied graphically in the phasor diagram of Fig. 21.11b.

Normally, the capacitive and inductive voltages are *not* of the same magnitude and so do *not* cancel completely. Thus, the voltage across the resistor is generally less than the voltage of the source. Therefore, the power dissipated in the resistor (joule heat) is less than its maximum possible value. However, if the capacitive and inductive voltages do cancel exactly, then the full source voltage appears across the resistor, the power factor becomes equal to unity, and the resistor then dissipates the maximum power output. In other words, the circuit is being driven (at frequency f_o) "at resonance."

Applications of Resonance

When a series RLC circuit is driven at its resonance frequency, the power transfer from the voltage source to the circuit and the current in the circuit are a maximum. A graph of rms current versus driving frequency is shown in Fig. 21.13b for several different values of resistance. Note that the maximum current values occur at the resonance frequency f_o and also the curve becomes sharper and narrower as the resistance decreases.

Resonant circuits have a wide variety of applications. One application is in the tuning mechanism of a radio. Each radio station has an assigned broadcast frequency at which radio waves are transmitted (see Insight on p. 717). The oscillating electric and magnetic fields of the radio waves set electrons in the receiving antenna into regular back-and-forth motion (i.e., they produce an alternating current), as a voltage source in the circuit would do. (This reception process is the reverse of the *production* of electromagnetic waves by oscillating electrons in a transmitting antenna.)

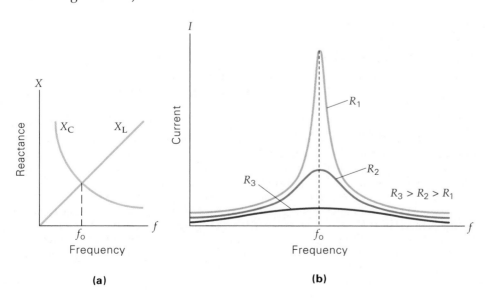

◀ **FIGURE 21.13 Resonance frequency for a series RLC circuit** **(a)** At the resonance frequency (f_o), the capacitive and inductive reactances are equal ($X_L = X_C$). On a graph of X versus f, this is the frequency at which the curves of X_C and X_L intersect. **(b)** On a graph of I versus f, the current is a maximum at f_o. The curve becomes sharper and narrower as the resistance in the circuit decreases.

(a)

(b)

▲ **FIGURE 21.14   Variable air capacitor**   Rotating the movable plates between the fixed plates changes the overlap area and thus the capacitance. Such capacitors were common in tuning circuits in older radios.

Signals broadcast by many different stations (at different frequencies) generally reach a radio together, but the receiver circuit selectively picks up only the one with a frequency at or near its resonance frequency. Most radios allow you to alter this resonance frequency to "tune in" different stations. In the early days of radio, variable air capacitors were used for this purpose (◄Fig. 21.14). Today, more compact variable capacitors in smaller radios have a polymer dielectric between thin plates. The polymer sheets help maintain the plate separation and increase the capacitance, allowing makers to use plates of a much smaller area. (Recall from Chapter 16 that $C = K\varepsilon_0 A/d$.) In most modern radios, solid-state devices replace variable capacitors.

Integrated Example 21.8 ■ AM versus FM: Resonance in Radio Reception

(a) When you switch from AM frequencies (the "AM band") to the FM band in some radios, you are effectively changing the capacitance of the receiving circuit, assuming constant inductance. Is the capacitance (1) increased or (2) decreased when you make this change? (b) Suppose you were listening to music on an FM station at 99.7 MHz and then switched over to a program on the AM band at 920 kHz. By what factor would you have changed the capacitance of the receiving circuit in the radio?

(a) Conceptual Reasoning.   Since FM stations broadcast at significantly higher frequencies than do AM stations (see Table 20.1), the resonance frequency of the receiver must be increased to receive power in the FM band. An increase of the resonance frequency requires reducing the capacitance since the inductance is fixed. Thus, the correct answer is (2).

(b) Thinking It Through.   The resonance frequency (Eq. 21.25) depends on the inductance and the capacitance. A ratio can be formed. Since the question asks for a "factor," it should be clear that it is asking for a ratio of the new capacitance to the original capacitance. The frequencies have to be expressed in the same units, so let's convert MHz into kHz.

Given: $f_o = 99.7\,\text{MHz} = 99.7 \times 10^3\,\text{kHz}$ **Find:** C'/C (ratio of new capacitance to
$f_o' = 920\,\text{kHz}$ original capacitance)

From Eq. 21.25, the two resonant frequencies are given by

$$f_o = \frac{1}{2\pi\sqrt{LC}} \quad \text{and} \quad f_o' = \frac{1}{2\pi\sqrt{LC'}}$$

Dividing the first of these equations by the second gives

$$\frac{f_o}{f_o'} = \frac{2\pi\sqrt{LC'}}{2\pi\sqrt{LC}} = \sqrt{\frac{C'}{C}}$$

Solving for the capacitance ratio by squaring, and substituting the numbers, we have

$$\frac{C'}{C} = \left(\frac{f_o}{f_o'}\right)^2 = \left(\frac{99.7 \times 10^3\,\text{kHz}}{920\,\text{kHz}}\right)^2 = 1.17 \times 10^4$$

Thus, $C' = 1.17 \times 10^4\,C$, and the capacitance was increased by a factor of more than 10 000 when you switched stations.

Follow-up Exercise.   (a) Based on the resonance curves shown in Fig. 21.13b, can you explain how it is possible to pick up two radio stations *simultaneously* on your radio? (You may have encountered this phenomenon, particularly between two cities located far a part from one another. Two stations are sometimes granted licenses for broadcasting at closely spaced frequencies under the assumption that they won't both be received by the same radio. However, under certain atmospheric conditions, this may not be true.) (b) In part (b) of this Example, if you next decreased the capacitance by a factor of two (starting with the news at 920 kHz) to listen to a hockey game, to what new frequency on the AM band would you now be tuned?

Oscillator Circuits: Broadcasters of Electromagnetic Radiation

To generate the high-frequency electromagnetic waves used in radio communications and television (Fig. 1), electric current must be made to oscillate at high frequencies in electronic circuits. This can be accomplished with RLC circuits. Such circuits are called *oscillator circuits*, because the current in them oscillates at a frequency determined by their inductive and capacitive elements.

When the resistance in an RLC circuit is very small, the circuit is essentially an LC circuit. Energy in such a circuit oscillates back and forth between the inductor and capacitor at a frequency f. This is the natural frequency of the circuit and is equal to its resonance frequency (Eq. 21.25). The small resistance in the circuit dissipates some energy. However, in an ideal LC circuit with no ohmic resistance, the oscillation would continue indefinitely.

To understand this energy oscillation better, consider an ideal LC parallel circuit as shown in Fig. 2a. Let's assume that the capacitor is initially charged and the switch is then closed. The following sequence of events occurs:

1. The capacitor would discharge instantaneously (since $RC = 0$) were it not for the current having to pass through the coil. When the switch is closed (at $t = 0$), the current through the coil is zero (Fig. 2a). As the current builds up, so does the magnetic field in the coil. By Lenz's law, the increasing magnetic field and change of flux through the coil induce an emf in such a direction as to oppose the increase in current. Because of this opposition, the capacitor takes some time to discharge.

2. When the capacitor is in a fully discharged state (Fig. 2b), all of its initial energy has been transferred to the inductor in the form of a magnetic field. (Since $R = 0$, no energy is lost to joule heating.) At this time (one-quarter of a cycle or period), the magnetic field and the current in the coil are at their maximum, and all the energy is now stored in the inductor.

3. As the magnetic field collapses from its maximum value, an emf that opposes the collapse is induced in the coil. This emf is in a direction that would tend to continue the current in the coil even as the current is decreasing (Lenz's law again). The polarity of the emf is now opposite what it was in step 1. Thus, current continues to transport charge to the

capacitor plates and charges the capacitor with a polarity that is the reverse of its initial value.

4. When the capacitor is fully charged (at reverse polarity), it has the same energy that it had initially (no joule losses; see Fig.2c). This occurs halfway through the cycle, or half a period from the start. The magnetic field in the coil is now zero again, as is the current in the circuit.

5. The capacitor then begins to discharge and the previous four steps are repeated, continuing the "oscillation" in the circuit.

For an ideal case of a circuit without resistance, the oscillations would continue indefinitely. Note that, a small ac voltage of the proper frequency applied to the circuit is sufficient to maintain strong oscillations between the electric field of the capacitor and the magnetic field of the inductor. An LC circuit is sometimes called a "tank circuit," which suggests its energy-storage capability. Without an input of voltage, however, the oscillator would run down as a result of I^2R losses and radiation losses (just as its mechanical counterpart would be slowed down by frictional losses).

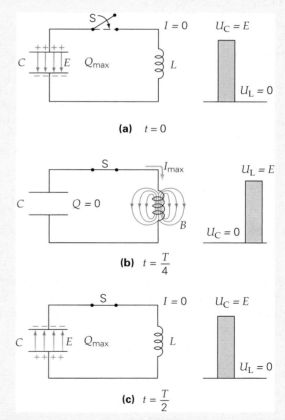

(a) $t = 0$

(b) $t = \dfrac{T}{4}$

(c) $t = \dfrac{T}{2}$

FIGURE 2   An oscillating LC circuit   If the resistance is negligible, this circuit will oscillate indefinitely. Half a cycle is shown. Energy is transferred back and forth between magnetic and electric types of energy. The oscillating electrons in the wire will give off electromagnetic radiation at the circuit's oscillating frequency.

FIGURE 1   Broadcast antennas

Chapter Review

Important Concepts

- An **ac voltage** is described by

$$V = V_o \sin \omega t = V_o \sin 2\pi f t \qquad (21.1)$$

- For a sinusoidally varying current, called **ac current,** the **peak current** I_o and the **rms** (root-mean-square or effective) **current** I_{rms} are related by

$$I_{rms} = \frac{I_o}{\sqrt{2}} = 0.707 I_o \qquad (21.6)$$

- For an ac voltage, the **peak voltage** V_o is related to its the **rms** (root-mean-square) **voltage** V_{rms} by

$$V_{rms} = \frac{V_o}{\sqrt{2}} = 0.707 V_o \qquad (21.8)$$

- The current in a resistor is in phase with the voltage across it. For a capacitor, the current is 90° (one-quarter of a cycle) ahead of the voltage. For an inductor, the current lags the voltage by 90°.

- In ac circuits, joule heating is due entirely to the resistive elements, and the time-averaged **power dissipation** is

$$\overline{P} = I_{rms}^2 R \qquad (21.7)$$

- In an ac circuit, capacitors and inductors allow current and create opposition to current. This opposition is characterized by **capacitive reactance** and **inductive reactance**, respectively. The capacitive reactance is given by

$$X_C = \frac{1}{\omega C} = \frac{1}{2\pi f C} \qquad (21.11)$$

The inductive reactance is given by

$$X_L = \omega L = 2\pi f L \qquad (21.14)$$

- Ohm's law, as applied to each type of circuit element, is a generalization of the version from dc circuits. The relationship between rms current and the rms voltage for a resistor is:

$$V_{rms} = I_{rms} R \qquad (21.9)$$

The relationship between rms current and the rms voltage for a capacitor is

$$V_{rms} = I_{rms} X_C \qquad (21.12)$$

The relationship between rms current and the rms voltage for an inductor is

$$V_{rms} = I_{rms} X_L \qquad (21.15)$$

- **Phasors** are vectorlike quantities that allow resistances and reactances to be represented graphically.

- **Impedance (Z)** is the total, or effective, opposition to current that takes into account both resistances and reactances. Impedance is related to current and circuit voltage by a generalization of Ohm's law:

$$V_{rms} = I_{rms} Z \qquad (21.17)$$

- The **impedance for a series RLC circuit** is

$$Z = \sqrt{R^2 + (X_L - X_C)^2} \qquad (21.19)$$

- The **phase angle (ϕ)** between the rms voltage and the rms current in a series RLC circuit is

$$\tan \phi = \frac{X_L - X_C}{R} \qquad (21.20)$$

- The **power factor (cos ϕ)** for a series RLC circuit is a measure of how close to the maximum power dissipation the circuit is. The power factor is

$$\cos \phi = \frac{R}{Z} \qquad (21.22)$$

The average power dissipated (joule heating in the resistor) is given by

$$\overline{P} = I_{rms} V_{rms} \cos \phi \qquad (21.23)$$

or

$$\overline{P} = I_{rms}^2 Z \cos \phi \qquad (21.24)$$

- The **resonance frequency (f_o)** of an RLC circuit is the frequency at which the circuit dissipates maximum power. This frequency is

$$f_o = \frac{1}{2\pi\sqrt{LC}} \qquad (21.25)$$

Exercises

21.1 Resistance in an AC Circuit

1. Which of the following voltages is larger in a sinusoidally varying ac voltage? (a) V_o, (b) V_{rms}, or (c) they have the same value.

2. CQ If the average current in an ac circuit is zero, why isn't the average power zero?

3. CQ The voltage and current of a resistor in an ac circuit are *in phase*. What does that mean?

4. **CQ** A 60-W lightbulb designed to work at 240 V is instead connected to a 120-V source. Discuss the changes in the bulb's rms current and power when it is at 120 V compared with 240 V.

5. **CQ** If the ac voltage and current for a circuit element are given respectively by $V = 120 \sin(120\pi t)$ and $I = 30 \sin(120\pi t + \pi/2)$, could the circuit element be a resistor? Explain.

6. ■ What are the peak voltages of a 120-V ac line and a 240-V ac line?

7. ■ An ac circuit has an rms current of 5.0 A. What is the peak current?

8. ■ The maximum voltage across a resistor in an ac circuit is 156 V. Find the resistor's rms voltage.

9. ■ How much ac rms current must be in a 10-Ω resistor to produce an average power of 15 W?

10. ■ An ac circuit contains a resistor with a resistance of 5.0 Ω. The resistor has an rms current of 0.75 A. (a) Find its rms voltage and peak voltage. (b) Find the average power delivered to the resistor.

11. ■ A hair dryer is rated at 1200 W when plugged into a 120-V outlet. Find (a) its rms current; (b) its peak current; and (c) its resistance.

12. **IE** ■■ The voltage across a 10-Ω resistor varies as $V = (170 \text{ V}) \sin(120t)$. (a) Will the current in the resistor (1) be in phase with the voltage, (2) lead the voltage by 90°, or (3) lag the voltage by 90°? (b) Write the expression for the current in the resistor as a function of time.

13. ■■ An ac voltage is applied to a 25-Ω resistor that it dissipates 500 W of power. Find the resistor's (a) rms and peak currents and (b) rms and peak voltages.

14. **IE** ■■ An ac voltage source has a peak voltage of 85 V and a frequency of 60 Hz. The voltage at $t = 0$ is zero. (a) A student wants to calculate the voltage at $t = 1/240$ s. How many possible answers are there, (1) one, (2) two, or (3) three? Why? (b) List all possible answers.

15. ■■ An ac voltage source has an rms value of 120 V. The voltage goes from zero to its maximum value in 4.20 ms. Write an expression for the voltage as a function of time.

16. ■■ What are the resistance and rms current of a 100-W, 120-V computer monitor?

17. ■■ Find the rms and peak currents through a 40-W, 120-V lightbulb.

18. ■■ A 50-kW heater is designed to run using a 240-V ac source. Find its (a) peak current and (b) peak voltage.

19. ■■ A circuit has a current given by $I = (8.0 \text{ A}) \sin(40\pi t)$ with an applied voltage $V = (60 \text{ V}) \sin(40\pi t)$. What is the average power delivered to the circuit?

20. ■■ The current and voltage outputs of an operating ac generator have peak values of 2.5 A and 16 V, respectively. What is the average power output of the generator?

21. ■■■ The current in a 60-Ω resistor is given by $I = (2.0 \text{ A}) \sin(380t)$. (a) What is the frequency of the current? (b) What is the rms current? (c) How much average power is delivered to the resistor? (d) Write an equation for the voltage across the resistor as a function of time. (e) Write an equation for the power delivered to the resistor as a function of time. (f) Show that the rms power obtained from (e) is the same as your answer to (c).

21.2 Capacitive Reactance
and
21.3 Inductive Reactance

22. In a purely capacitive ac circuit, (a) the current and voltage are in phase, (b) the current leads the voltage, (c) the current lags the voltage, or (d) none of the above.

23. The unit of capacitive reactance is (a) F, (b) $(\text{F} \cdot \text{s})^{-1}$, (c) Ω, or (d) $\Omega \cdot \text{m}$.

24. The unit of inductance is (a) H/s, (b) H, (c) Ω, or (d) $\Omega \cdot \text{s}$.

25. **CQ** Explain why a capacitor acts as an ac open circuit while an inductor acts as a short circuit at low frequencies.

26. **CQ** In an ac circuit, what circuit elements can oppose current other than ohmic resistances? Explain how they do this.

27. **CQ** Explain these statements: "The voltage across an inductor leads the current by 90°," and "The voltage across a capacitor lags the current by 90°."

28. **CQ** Can an inductor oppose dc current? Explain.

29. **CQ** If the current on a 10-μF capacitor is described by $I = (120 \text{ A}) \sin(120\pi t + \pi/2)$, what is the instantaneous voltage across the capacitor at $t = 0$? Explain.

30. ■ Find the frequency at which a 25-μF capacitor has a reactance of 25 Ω.

31. ■ A single 2.0-μF capacitor is connected across the terminals of a 60-Hz voltage source, and a current of 2.0 mA is measured on an ac ammeter. What is the capacitive reactance of the capacitor?

32. ■ What capacitance would have a reactance of 100 Ω in a 60-Hz ac circuit?

33. ■ A single 50-mH inductor forms a complete circuit by being connected to an ac voltage source at 120 V and 60 Hz. (a) What is the inductive reactance of the circuit? (b) How much current is in the circuit? (c) What is the phase angle between the current and the applied voltage? (Assume negligible resistance.)

34. ■ How much current is in a circuit containing only a 50-μF capacitor connected to an ac generator with an output of 120 V and 60 Hz?

35. ■■ A variable capacitor in a circuit with a 120-V, 60-Hz source initially has a capacitance of 0.25 μF. The capacitance is then increased to 0.40 μF. What is the percentage change in the current in the circuit?

36. ■■ An inductor has a reactance of 90 Ω in a 60-Hz ac circuit. What is its inductance?

37. ■■ Find the frequency at which a 250-mH inductor has a reactance of 400 Ω.

38. IE ■■ A capacitor is connected to a variable-frequency ac voltage source. (a) If the frequency increases by a factor of 3, the capacitive reactance will be (1) 3, (2) 1/3, (3) 9, or (4) 1/9 times the original reactance. Why? (b) If the capacitive reactance of a capacitor at 120 Hz is 100 Ω, what is its reactance if the frequency is changed to 60 Hz?

39. ■■ With a single 150-mH inductor in a circuit with a 60-Hz voltage source, a current of 1.6 A is measured on an ac ammeter. (a) What is the rms voltage of the source? (b) What is the phase angle between the current and that voltage?

40. ■■ What inductance has the same reactance in a 120-V, 60-Hz circuit as a capacitance of 10 μF?

41. ■■ A circuit with a single capacitor is connected to a 120-V, 60-Hz source. What is its capacitance if there is a current of 0.20 A in the circuit?

42. IE ■■ An inductor is connected to a variable-frequency ac voltage source. (a) If the frequency decreases by a factor of 2, the rms current will be (1) 2, (2) 1/2, (3) 4, or (4) 1/4 times the original rms current. Why? (b) If the rms current in an inductor at 40 Hz is 9.0 A, what is its rms current if the frequency is changed to 120 Hz?

21.4 Impedance: RLC Circuits
and
21.5 Circuit Resonance

43. The impedance of an RLC circuit depends on (a) frequency, (b) inductance, (c) capacitance, or (d) all of these.

44. If the capacitance of a series RLC circuit is decreased, (a) the capacitive reactance increases, (b) the inductive reactance increases, (c) the current remains constant, or (d) the power factor remains constant.

45. When a series RLC circuit is driven at its resonance frequency, (a) energy is dissipated only by the resistive element, (b) the power factor has a value of one, (c) there is maximum power delivered to the circuit, or (d) all of these.

46. CQ What is the impedance of an RLC circuit at resonance and why?

47. CQ Is there any power delivered to capacitors or inductors in an ac circuit? Why or why not?

48. CQ What are the factors that determine the resonant frequency of an RLC circuit? Is resistance a factor? Explain?

49. ■ A coil in a 60-Hz circuit has a resistance of 100 Ω and an inductance of 0.45 H. Calculate (a) the coil's reactance and (b) the circuit's impedance.

50. ■ A series RC circuit has a resistance of 200 Ω and a capacitance of 25 μF and is driven by a 120-V, 60-Hz source. (a) Find the capacitive reactance and impedance of the circuit. (b) How much current is drawn from the source?

51. ■ A series RL circuit has a resistance of 100 Ω and a inductance of 100 mH and is driven by a 120-V, 60-Hz source. (a) Find the inductive reactance and the impedance of the circuit. (b) How much current is drawn from the source?

52. ■ An RC circuit has a resistance of 250 Ω and a capacitance of 6.0 μF. If the circuit is driven by a 60-Hz source, find (a) the inductance and (b) the impedance of the circuit.

53. IE ■ An RC circuit has a resistance of 100 Ω and a capacitive capacitance of 50 Ω. (a) Will the phase angle be (1) positive, (2) zero, or (3) negative? Why? (b) What is the phase angle of this circuit?

54. ■■ A series RLC circuit has a resistance of 25 Ω, an inductance of 0.30 H, and a capacitance of 8.0 μF. (a) At what frequency should the circuit be driven for the maximum power to be transferred from the driving source? (b) What is the impedance at that frequency?

55. IE ■■ In a series RLC circuit, $R = X_C = X_L = 40$ Ω for a particular driving frequency. (a) This circuit is (1) inductive, (2) capacitive, or (3) in resonance. Why? (b) If the driving frequency is doubled, what will be the impedance of the circuit?

56. IE ■■ (a) An RLC series circuit is in resonance. Which one of the following can you change without upsetting the resonance? (1) resistance, (2) capacitance, (3) inductance, or (4) frequency. Why? (b) A resistor, an inductor, and a capacitor have values of 500 Ω, 500 mH, and 3.5 μF, respectively. They are connected in series to a power supply of 240 V with a frequency of 60 Hz. What values of resistance and inductance would be required for this circuit to be in resonance?

57. ■■ How much power is dissipated in the circuit described in Exercise 56b?

58. ■■ What is the resonant frequency of an RLC circuit with a resistance of 100 Ω, an inductance of 100 mH, and a capacitance of 5.00 μF?

59. ■■ A tuning circuit in an antique radio receiver has a fixed inductance of 0.50 mH and a variable capacitor (▼Fig. 21.15). If the circuit is tuned to a radio station broadcasting at 980 kHz on the AM dial, what is the capacitance of the capacitor?

▲ **FIGURE 21.15   Stay tuned**   See Exercises 59, 60, and 73.

60. ■■ What would be the range of the variable capacitor in Exercise 59 required for tuning over the complete AM band? [*Hint*: See Table 20.1.]

61. ■■ Find the currents supplied by the ac source for all possible connections in ▼Fig. 21.16.

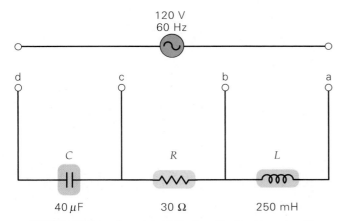

▲ **FIGURE 21.16   A series RLC circuit**   See Exercise 61.

62. IE ■■ A coil with a resistance of 30 Ω and an inductance of 0.15 H is connected to a 120-V, 60-Hz source. (a) Is the phase angle of this circuit (1) positive, (2) zero, or (3) negative? Why? (b) What is the phase angle of the circuit? (c) How much rms current is in the circuit? (d) What is the average power delivered to the circuit?

63. ■■ A small welder uses a voltage source of 120 V at 60 Hz. When the source is operating, it requires 1200 W of power, and the power factor is 0.75. Find the rms current in the welder.

64. ■■ A series circuit is connected to a 220-V, 60-Hz power supply. The circuit has the following components: a 10-Ω resistor, a coil with an inductive reactance of 120 Ω, and a capacitor with a reactance of 120 Ω. Compute the rms voltage across (a) the resistor, (b) the inductor, and (c) the capacitor.

65. ■■ A series RLC circuit has a resistance of 25 Ω, a capacitance of 0.80 μF, and an inductance of 250 mH. The circuit is connected to a variable-frequency source with a fixed rms voltage output of 12 V. If the frequency that is supplied is set at the circuit's resonance frequency, what is the rms voltage across each of the circuit elements?

66. CQ ■■ In Exercises 64 and 65, $V_R + V_C + V_L$ is much greater than the source voltage. How do you explain the discrepancy?

67. IE ■■ (a) If the circuit in ▼Fig. 21.17 is in resonance, the impedance of the circuit is (1) greater than 25 Ω, (2) equal to 25 Ω, or (3) less than 25 Ω. Why? (b) If the driving frequency is 60 Hz, what is the circuit's impedance?

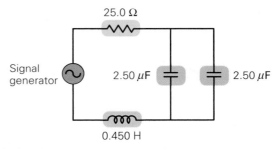

▲ **FIGURE 21.17   Tune to resonance**   See Exercise 67.

68. ■■■ A series RLC circuit with a resistance of 400 Ω has capacitive and inductive reactances of 300 Ω and 500 Ω, respectively. (a) What is the power factor of the circuit? (b) If the circuit operates at 60 Hz, what additional capacitance should be connected to the original capacitance to give a power factor of unity, and how should the capacitors be connected?

69. ■■■ A series RLC circuit has components with $R = 50$ Ω, $L = 0.15$ H, and $C = 20$ μF. The circuit is driven by a 120-V, 60-Hz source. What is the power delivered to the circuit, expressed as a percentage of the power delivered when the circuit is in resonance?

Additional Exercises

70. A 60.0-W lightbulb operates on 120-V household voltage. What are (a) the peak voltage across the bulb and (b) the peak current in the bulb?

71. The rms current in an ac circuit is 0.25 A. What is the peak current?

72. A 120-V, 60-Hz source is connected across a 0.40-H inductor. Find (a) the current in the inductor and (b) the phase angle between the current and the applied voltage.

73. A radio receiver circuit with an inductance of 1.50 μH is tuned to an FM station at 98.9 MHz by adjusting a variable capacitor. What is the capacitance of the capacitor when the circuit is tuned to this station?

74. A circuit connected to a 110-V, 60-Hz source contains a 50-Ω resistor and a coil with an inductance of 100 mH. Find (a) the reactance of the coil, (b) the impedance of the circuit, (c) the current in the circuit, and (d) the power dissipated by the coil.

75. Calculate the phase angle between the current and the applied voltage in Exercise 74.

76. A 1.0-μF capacitor is connected to a 120-V, 60-Hz source. (a) What is the capacitive reactance of the circuit? (b) How much current is in the circuit? (c) What is the phase angle between the current and the applied voltage?

77. The circuit in ▼Fig. 21.18a is called a *low-pass filter* because a large current is delivered to the load (RL circuit) only by a low-frequency source. The circuit in Fig. 21.18b is called a *high-pass filter* because a large current is delivered to the load only by a high-frequency source. Show mathematically why the circuits have these characteristics.

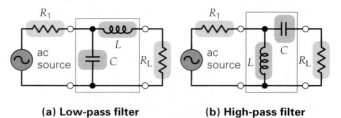

(a) Low-pass filter   **(b) High-pass filter**

▲ **FIGURE 21.18 Low-pass and high-pass filters**
See Exercise 77.

78. IE (a) If an RLC circuit is in resonance, the phase angle of the circuit is (1) positive, (2) zero, or (3) negative. Why? (b) A circuit has an inductive reactance of 280 Ω at 60 Hz. What value of capacitance would set this circuit into resonance?

Reflection and Refraction of Light

INSIGHTS

- Dark, Rainy Night
- Understanding Refraction: A Marching Analogy
- Fiber Optics: Medical Applications
- The Rainbow

Learn by Drawing

✎ Tracing the Reflecting Rays

We live in a visual world, surrounded by eye-catching images such as that shown in the photo. How these images are formed is something we take largely for granted—until we see something that we can't easily explain. Optics is the study of light and vision. Human vision requires light, specifically what we call visible light. The broad definition of *visible light* is "any radiation in or near the visible region of the electromagnetic spectrum." (See Fig. 20.23.) For our study, this definition includes infrared and ultraviolet radiation. Similar optical properties, such as reflection and refraction, are shared by all electromagnetic waves. Light acts like a wave in its propagation (Chapter 24) and as a particle when it interacts with matter (Chapter 27).

In this chapter, we will investigate the basic optical phenomena of reflection, refraction, total internal reflection, and dispersion. The principles that govern reflection explain the behavior of mirrors, while those that govern refraction explain the properties of lenses. With the aid of these and other optical principles, we can understand many optical phenomena that we experience every day: why a glass prism spreads out light into a

spectrum of colors, what causes mirages, how rainbows are formed, and why the legs of a person standing in a lake or swimming pool seem to shorten. We will also explore some less familiar territory, including the fascinating field of fiber optics.

A simple geometrical approach, involving straight lines and angles, can be used to investigate many aspects of the properties of light, especially how light propagates. For these purposes, we need not be concerned with the physical (wave) nature of electromagnetic waves described in Chapter 20. The principles of geometrical optics will be introduced here and applied in greater detail in the study of mirrors and lenses in Chapter 23.

22.1 Wave Fronts and Rays

OBJECTIVE: To define and explain the concepts of wave fronts and rays.

Waves, electromagnetic or otherwise, are conveniently described in terms of wave fronts. A **wave front** is the line or surface defined by adjacent portions of a wave that are in phase. If an arc is drawn along one of the crests of a circular water wave moving out from a point

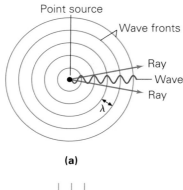

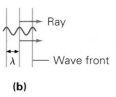

(a)

(b)

▲ **FIGURE 22.1   Wave fronts and rays**   A wave front is defined by adjacent points on a wave that are in phase, such as those along wave crests or troughs. **(a)** Near a point source, the wave fronts are circular in two dimensions and spherical in three dimensions. **(b)** Very far from a point source, the wave fronts are approximately linear or planar. A line perpendicular to a wave front in the direction of the wave's propagation is called a ray.

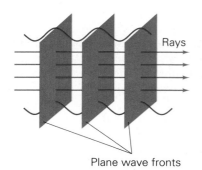

▲ **FIGURE 22.2   Light rays**   A plane wave travels in a direction perpendicular to its wave fronts. A beam of light can be represented by a group of parallel rays (or by a single ray).

PHYSLET®
ILLUSTRATION

Law of Reflection

source, all the particles on the arc will be in phase (◄Fig. 22.1a). An arc along a wave trough would work equally well. For a three-dimensional spherical wave, such as a sound or light wave emitted from a point source, the wave front is a spherical surface rather than a circle.

Very far from the source, the curvature of a short segment of a circular or spherical wave front is extremely small. Such a segment may be approximated as a linear wave front (in two dimensions) or a **plane wave front** (in three dimensions), just as we take the surface of the Earth to be locally flat (Fig. 22.1b). A plane wave front can also be produced directly by a luminous flat surface. In a uniform medium, wave fronts propagate outward from the source at a speed characteristic of the medium. We saw this for sound waves in Chapter 14, and the same occurs for light, although at a much faster speed. The speed of light is greatest in a vacuum: $c = 3.00 \times 10^8$ m/s.

The geometrical description of a wave in terms of wave fronts tends to neglect the fact that the wave is actually oscillating, like those studied in Chapter 13. This simplification is carried a step further with the concept of a ray. As illustrated in Fig. 22.1, a line drawn perpendicular to a series of wave fronts and pointing in the direction of propagation is called a **ray**. Note that a ray points in the direction of the energy flow of a wave. A plane wave is assumed to travel in a straight line in a medium in the direction of its rays, perpendicular to its plane wave fronts. A beam of light can be represented by a group of rays or simply as a single ray (◄Fig. 22.2). The representation of light as rays is adequate and convenient for describing many optical phenomena.

The use of the geometrical representations of wave fronts and rays to explain phenomena such as the reflection and refraction of light is called **geometrical optics**. However, certain other phenomena, such as the interference of light, cannot be treated in this manner and must be explained in terms of actual wave characteristics. These phenomena will be considered in Chapter 24.

22.2   Reflection

OBJECTIVES: To **(a)** explain the law of reflection and **(b)** distinguish between regular (specular) and irregular (diffuse) reflection.

The reflection of light is an optical phenomenon of enormous importance: If light were not reflected to our eyes by objects around us, we wouldn't see the objects at all. **Reflection** involves the absorption and reemission of light by means of complex electromagnetic vibrations in the atoms of the reflecting medium. However, the phenomenon is easily described by using rays.

A light ray incident on a surface is described by an **angle of incidence** (θ_i). This angle is measured relative to a *normal*—a line perpendicular to the reflecting surface (►Fig. 22.3). Similarly, the reflected ray is described by an **angle of reflection** (θ_r), also measured from the normal. The relationship between these angles is given by the **law of reflection**:

The angle of incidence is equal to the angle of reflection, or

$$\theta_i = \theta_r \quad \text{law of reflection} \quad (22.1)$$

Two other attributes of reflection are that the incident ray, the reflected ray, and the normal all lie in the same plane, which is sometimes called the plane of incidence, and that the incident and the reflected rays are on opposite sides of the normal.

When the reflecting surface is smooth, the reflected rays from parallel incident rays are also parallel (►Fig. 22.4a). This type of reflection is called **regular**, or **specular**, **reflection** (Fig. 22.4b). The reflection from a flat mirror is specular or regular reflection. If the reflecting surface is rough, however, the reflected rays are not parallel, because of the irregular nature of the surface (►Fig. 22.5). This type of reflection is termed **irregular**, or **diffuse**, **reflection**. The reflection of light from

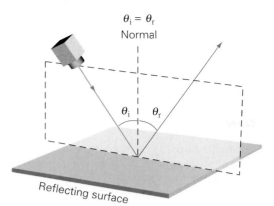

$\theta_i = \theta_r$

Normal

Reflecting surface

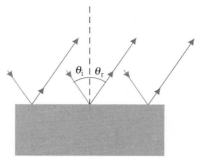

$\theta_i \mid \theta_r$

◀ FIGURE 22.3 The law of reflection According to the law of reflection, the angle of incidence (θ_i) is equal to the angle of reflection (θ_r). Note that the angles are measured relative to a normal (a line perpendicular to the reflecting surface). The normal and the incident and reflected rays always lie in the same plane.

(a) Regular, or specular, reflection

(b)

▲ FIGURE 22.4 Regular (specular) reflection (a) When a light beam is reflected from a smooth surface and the reflected rays are parallel, the reflection is said to be regular or specular. **(b)** Regular, or specular, reflection from a smooth water surface produces an almost perfect mirror image of salt mounds at this Australian salt mine.

this page is an example of diffuse reflection. The Insight "Dark, Rainy Night," on p. 727, discusses more about the difference between specular and diffuse reflection in a real-life situation.

Note in Fig. 22.4a and Fig. 22.5 that the law of reflection still applies locally to both specular and diffuse reflection. However, the type of reflection involved determines whether we see images from a reflecting surface. In specular reflection, the reflected, parallel rays produce an image when they are viewed by an optical system such as an eye or a camera (Fig. 22.4b). Diffuse reflection does not produce an image, because the light is reflected in various directions.

Both experience with friction and direct investigations show that all surfaces are rough on a microscopic scale. What, then, determines whether reflection is specular or diffuse? In general, if the dimensions of the surface irregularities are greater than the wavelength of the light, the reflection is diffuse. Therefore, to make a good mirror, glass (with a metal coating) or metal must be polished at least until the surface irregularities are about the same size as the wavelength of light. Recall from Chapter 20 that the wavelength of visible light is on the order of 10^{-7} m. (You will learn more about reflection from a mirror in the Learn by Drawing exercise presented in Example 22.1.)

It is because of diffuse reflection that we are able to see illuminated objects, such as the Moon. If the Moon's spherical surface were smooth, only the reflected sunlight from a small region would come to an observer on the Earth, and only that small illuminated area would be seen. Also, you can see the beam of light from a flashlight or spotlight because of diffuse reflection from dust and particles in the air.

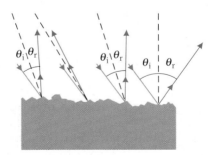

$\theta_i \mid \theta_r$ $\theta_i \mid \theta_r$ $\theta_i \mid \theta_r$

◀ FIGURE 22.5 Irregular (diffuse) reflection Reflected rays from a relatively rough surface, such as this page, are not parallel; the reflection is said to be irregular or diffuse. (Note that the law of reflection still applies locally to each individual ray.)

Example 22.1 ■ Learn by Drawing: Tracing the Reflecting Rays

Two mirrors, M_1 and M_2, are perpendicular to each other, and a light ray is incident on one of the mirrors as shown in ▶ Fig. 22.6. (a) Sketch a diagram to trace the path of the light ray. (b) Find the direction of the ray after it is reflected by M_2.

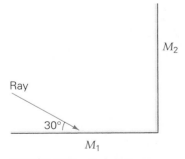

M_2

Ray

$30°$

M_1

▲ FIGURE 22.6 Trace the ray See Example 22.1.

Learn by Drawing

Tracing the Reflecting Rays
(see Example 22.1)

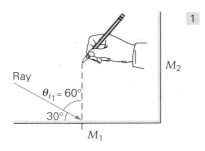

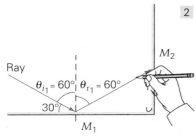

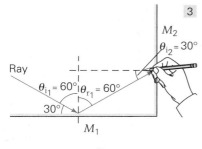

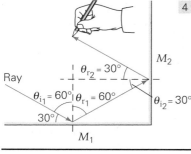

Thinking It Through. The law of reflection can be used to determine the direction of the ray after it reaches both mirrors.

Solution.

Given: $\theta = 30°$ (angle relative to M_1)   *Find:* (a) Sketch a diagram tracing the light ray
                                                   (b) θ_{r_2} (angle of reflection from M_2)

(a) and (b) follow steps 1–4 on the Learn by Drawing illustration:

(a) 1. Since the incident and reflected rays are all measured from the normal (a line perpendicular to the reflecting surface), we draw the normal to mirror M_1 at the point the incident ray hits M_1. From geometry, it can be seen that the angle of incidence on M_1 is $\theta_{i_1} = 60°$.
    2. According to the law of reflection, the angle of reflection from M_1 is also $\theta_{r_1} = 60°$. Next, draw this reflected ray with an angle of reflection of $60°$, and extend it until it hits M_2.
    3. Draw another normal to M_2 at the point where the ray hits M_2. Also from geometry (focus on the triangle in the diagram), the angle of incidence on M_2 is $\theta_{i_2} = 30°$. (Why?)
(b) 4. The angle of reflection off of M_2 is $\theta_{r_2} = \theta_{i_2} = 30°$. This is the final ray reflected after reaching both mirrors.

What if the directions of the rays are reversed? In other words, if a ray is first incident on M_2, in the direction opposite that of the one we drew for (b), will all the rays reverse their directions? Draw another diagram to prove that this is indeed the case. Thus, light rays are reversible.

Follow-up Exercise. When following an 18-wheel truck, you may see a sign on the back stating, "If you can't see my mirror, I can't see you." What does this mean? *(Answers to all Follow-up Exercises are at the back of the text.)*

22.3 Refraction

OBJECTIVES: To (a) explain refraction in terms of Snell's law and the index of refraction and (b) give examples of refractive phenomena.

Refraction refers to the change in direction of a wave at a boundary where the wave passes from one transparent medium into another. In general, when a wave is incident on a boundary between media, some of the wave's energy is reflected and some is transmitted. For example, when light traveling in air is incident on a transparent material such as glass, it is partially reflected and partially transmitted (▶Fig. 22.7). But the direction of the transmitted light is different from the direction of the incident light, so the light is said to have been refracted; in other words, it has changed direction. This change in direction is caused by the fact that light travels with different speeds in different media. Intuitively, you might expect the passage of light to take longer through a medium with more atoms per volume, and the speed of light is, in fact, generally less in denser media. For example, the speed of light in water is about 75% of that in air or a vacuum. ▶Fig. 22.8a shows the refraction of light at an air–water boundary. You can read the Insight "Understanding Refraction: A Marching Analogy," on p. 729, to help you understand why a change in the speed of light from one medium to another results in a change in the direction of light propagation.

The change in the direction of wave propagation is described by the **angle of refraction**. In Fig. 22.8b, θ_1 is the angle of incidence and θ_2 is the angle of refraction. Willebord Snell (1591–1626), a Dutch physicist, discovered a relationship between the angles (θ) and the speeds (v) of light in two media, as shown in the figure:

$$\frac{\sin \theta_1}{\sin \theta_2} = \frac{v_1}{v_2} \qquad \textit{Snell's law} \qquad (22.2)$$

This expression is known as **Snell's law**. Note that θ_1 and θ_2 are always taken with respect to the normal.

INSIGHT

Dark, Rainy Night

When you drive on a dry night, you can see the road and the street signs ahead of you clearly. However, here is a familiar scene on a dark, rainy night: Even though you have your headlights on, you can hardly see the road ahead. When a car approaches, the situation becomes even worse. You see the reflections of the approaching car's headlights on the surface of the road, and they appear brighter than usual. You may be blinded and cannot see anything except the reflective glare of the oncoming headlights.

What causes these conditions and how can they be explained? When the road surface is dry, the reflection off of the road is irregular or diffuse because the surface is rough. Light from your headlights hitting the road in front of you reflects in all directions; some of it reflects back to you, and the road may be clearly seen (just as the page of this book can be read because the paper is microscopically rough). However, when the road surface is wet, water fills the crevices, turning the road into a relatively smooth reflecting surface (Fig. 1a). Light from your headlights then reflects ahead. The normally diffuse reflection is gone and is replaced by specular reflection. Images of lighted buildings and road lights form, confusing your view of the surface, and the specular reflection of oncoming cars' headlights may cause difficulty in seeing the road (Fig. 1b).

Besides wet, slippery surfaces, specular reflection is a major cause of accidents on rainy nights; thus, extra caution is advised under such conditions.

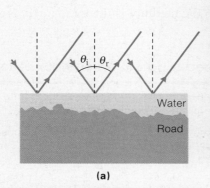

(a)   (b)

FIGURE 1 Diffuse to specular **(a)** Water on the road's surface turns the diffuse reflection from a dry road into specular reflection. **(b)** Thus, instead of seeing the road, a driver sees the reflected images of lights, buildings, etc.

Thus, light is refracted when passing from one medium into another because the speed of light is different in the two media. The speed of light is greatest in a vacuum, and it is therefore convenient to compare the speed of light in other media with this constant value (c). We do so by defining a ratio called the **index of refraction** (n):

$$n = \frac{c}{v} = \frac{\text{speed of light in a vacuum}}{\text{speed of light in a medium}} \qquad (22.3)$$

As a ratio of speeds, the index of refraction is a unitless quantity. The indices of refraction of several substances are given in Table 22.1. Note that these values are for a specific wavelength of light. The wavelength is specified because v, and consequently n, are slightly different for different wavelengths. (This is the cause of dispersion, to be discussed later in the chapter.). The values of n given in the table will be used in examples and exercises in this chapter for all wavelengths of light in the visible region, unless otherwise noted. Observe that n is always greater than 1, because the speed of light in a vacuum is greater than the speed of light in any material ($c > v$).

The frequency (f) of light does not change when the light enters another medium, but the wavelength of light in a material differs from the wavelength of that light in a vacuum, as can be easily shown:

$$n = \frac{c}{v} = \frac{\lambda f}{\lambda_m f}$$

▲ **FIGURE 22.7 Reflection and refraction** A beam of light is incident on a trapezoidal prism from the left. Part of the beam is reflected, and part is refracted. The refracted beam is partially reflected and partially refracted at the bottom glass–air surface. (What happens to the reflected portion?)

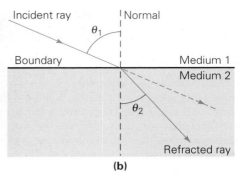

(a)

(b)

► FIGURE 22.8 Refraction
(a) Light changes direction upon entering a different medium. **(b)** The refracted ray is described by the angle of refraction, θ_2, measured from the normal.

Note: When light is refracted,
• its speed and wavelength are changed;
• its frequency remains unchanged.

or

$$n = \frac{\lambda}{\lambda_m} \qquad (22.4)$$

The wavelength of light in the medium is then $\lambda_m = \lambda/n$. Since $n > 1$, it follows that $\lambda_m < \lambda$.

TABLE 22.1 Indices of Refraction (at $\lambda = 590\,nm$)*

Substance	n
Air	1.000 29
Water	1.33
Human eye	1.336–1.406
Ethyl alcohol	1.36
Fused quartz	1.46
Glycerine	1.47
Polystyrene	1.49
Oil (typical value)	1.50
Glass (by type)†	1.45–1.70
crown	1.52
flint	1.66
Zircon	1.92
Diamond	2.42

*One nanometer (nm) is 10^{-9} m.
†Crown glass is a soda–lime silicate glass; flint glass is a lead–alkali silicate glass. Flint glass is more dispersive than crown glass (Section 22.5).

Example 22.2 ■ The Speed of Light in Water: Index of Refraction

Light from a laser with a wavelength of 632.8 nm travels from air into water. What are the speed and wavelength of the laser light in water?

Thinking It Through. If we know the index of refraction (n) of a medium, the speed and wavelength of light in the medium can be obtained from Eq. 22.3 and Eq. 22.4.

Solution.

Given: $n_{water} = 1.33$ (from Table 22.1)
      $\lambda = 632.8$ nm
      $c = 3.00 \times 10^8$ m/s (speed of light in air)

Find: v and λ_m (speed and wavelength of light in water)

Since $n = c/v$,

$$v = \frac{c}{n} = \frac{3.00 \times 10^8 \text{ m/s}}{1.33} = 2.26 \times 10^8 \text{ m/s}$$

Note that $1/n = v/c = 1/1.33 = 0.75$; therefore, v is 75% of the speed of light in a vacuum. Also, $n = \lambda/\lambda_m$, so

$$\lambda_m = \frac{\lambda}{n} = \frac{632.8 \text{ nm}}{1.33} = 475.8 \text{ nm}$$

Follow-up Exercise. The speed of light in a particular liquid is 2.40×10^8 m/s. What is the index of refraction of the liquid?

The index of refraction, n, is a measure of the speed of light in a transparent material, or technically, a measure of the *optical density* of the material. For example, the speed of light in water is less than that in air, so water is said to be optically denser than air. (Optical density in general correlates with mass density. However, in some instances, a material with a greater optical density than another can have a lower mass density.) Thus, the greater the index of refraction of a material, the greater is the material's optical density and the smaller is the speed of light in the material.

For practical purposes, the index of refraction is measured in air rather than in a vacuum, since the speed of light in air is very close to c, and

$$n_{air} = \frac{c}{v_{air}} \approx \frac{c}{c} = 1$$

INSIGHT

Understanding Refraction: A Marching Analogy

The refraction of light is not as easy to understand or visualize as its reflection. We say that the direction of the light changes when it enters a new medium because light has different speeds in different media. The transmission of light through a transparent medium involves complex atomic absorption and emission processes, but it makes sense to suppose that these processes would take longer in a denser medium, and that is indeed the case. What is difficult to visualize is why light changes direction because of a change in speed.

To afford some insight into this phenomenon, let's consider an analogy of a band marching across a field (Fig. 1). Part of the field is wet and muddy, and the marching column enters this region obliquely (at a nonzero angle of incidence). As the marchers enter the wet, slippery region, they keep marching with the same cadence (frequency). However, slipping in the mud, the stride (wavelength) of the marchers is shorter, so they are slowed down.

The band members in the far end of the same row are still on dry ground and continue on with their original stride. The effect of the change in speed is a change in direction when the band enters the second medium. We might think of the marching rows as wave fronts. As in refraction, the frequency (cadence) remains the same, but the wavelength, speed, and direction change (to keep a row aligned) as the band enters another medium.

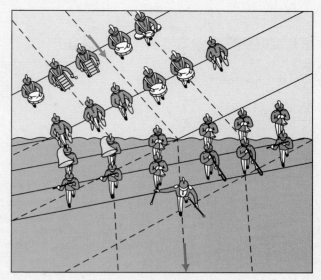

FIGURE 1 Marching analogy for refraction On obliquely entering a muddy field, a marching row changes direction slightly, analogous to the refraction of a wave front.

(From Table 22.1, $n_{air} = 1.00029$.)

A more practical form of Snell's law can be rewritten as

$$\frac{\sin \theta_1}{\sin \theta_2} = \frac{v_1}{v_2} = \frac{c/n_1}{c/n_2} = \frac{n_2}{n_1}$$

or

$$n_1 \sin \theta_1 = n_2 \sin \theta_2 \qquad \begin{array}{l}\textit{Snell's law} \\ \textit{(another form)}\end{array} \qquad (22.5)$$

where n_1 and n_2 are the indices of refraction for the first and second media, respectively.

Note that Eq. 22.5 can be used to measure the index of refraction. If the first medium is air, then $n_1 \approx 1$ and $n_2 \approx \sin \theta_1 / \sin \theta_2$. Thus, only the angles of incidence and refraction need to be measured to determine the index of refraction of a material experimentally. On the other hand, if the index of refraction of a material is known, it can be used in Snell's law to find the angle of refraction for any angle of incidence.

Note also that the sine of the refraction angle is inversely proportional to the index of refraction: $\sin \theta_2 \approx \sin \theta_1 / n_2$. Hence, for a given angle of incidence, the greater the index of refraction, the smaller is $\sin \theta_2$ and the smaller is the angle of refraction, θ_2.

More generally, the following relationships hold:

- If the second medium is more optically dense than the first medium ($n_2 > n_1$), the ray is refracted *toward* the normal ($\theta_2 < \theta_1$), as illustrated in ▸Fig. 22.9a.
- If the second medium is less optically dense than the first medium ($n_2 < n_1$), the ray is refracted *away from* the normal ($\theta_2 > \theta_1$), as illustrated in Fig. 22.9b.

PHYSLET® ILLUSTRATION

Law of Refraction

Note: The product of $n \sin \theta$ is a constant in any medium.

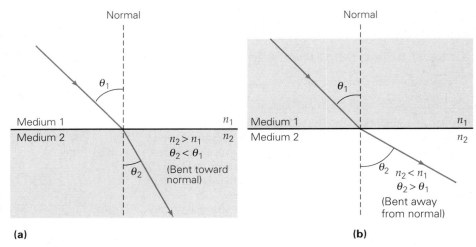

▲ FIGURE 22.9 Index of refraction and ray deviation **(a)** When the second medium is more optically dense than the first ($n_2 > n_1$), the ray is refracted toward the normal, as in the case of light entering water from air. **(b)** When the second medium is less optically dense than the first ($n_2 < n_1$), the ray is refracted away from the normal. [This is the case if the ray in part (a) is traced in reverse, going from medium 2 to medium 1.]

Integrated Example 22.3 ■ Angle of Refraction: Snell's Law

Light in air is incident on a piece of crown glass at an angle of 37.0° (relative to the normal). (a) Will the refracted ray bend (1) toward or (2) away from the normal? Use a diagram to illustrate. (b) What is the angle of refraction?

(a) Conceptual Reasoning. According to the alternative form of Snell's law (Eq. 22.5), $n_1 \sin \theta_1 = n_2 \sin \theta_2$, (1) is the correct answer. Since $n_2 > n_1$, the angle of refraction must be smaller than the angle of incidence ($\theta_2 < \theta_1$). Because both θ_1 and θ_2 are measured from the normal, the refracted ray will bend toward the normal. The ray diagram in this case is identical to Fig. 22.9a. (Remember the marching band analogy?)

(b) Thinking It Through. Again, the alternative form of Snell's law (Eq. 22.5) is most practical in this case. (Why?) Listing the given quantities, we have

Given: $\theta_1 = 37.0°$ *Find:* (b) θ_2 (angle of refraction)
        $n_1 = 1.00$ (air)
        $n_2 = 1.52$ (crown glass, from Table 22.1)

To find the angle of refraction, we use

$$\sin \theta_2 = \frac{n_1 \sin \theta_1}{n_2} = \frac{(1.00)(\sin 37.0°)}{1.52} = 0.396$$

and we obtain

$$\theta_2 = \sin^{-1}(0.396) = 23.3°$$

Follow-up Exercise. It is found experimentally that a beam of light entering a liquid from air at an angle of incidence of 37.0° exhibits an angle of refraction of 28.8° in the liquid. What is the speed of light in the liquid?

Example 22.4 ■ A Glass Tabletop: More about Refraction

A beam of light traveling in air strikes the glass top of a coffee table at an angle of incidence of 45° (▶ Fig. 22.10). The glass has an index of refraction of 1.5. (a) What is the angle of refraction for the light transmitted into the glass? (b) Prove that the emergent beam is parallel to the incident beam—that is, that $\theta_4 = \theta_1$. (c) If the glass is 2.0 cm thick,

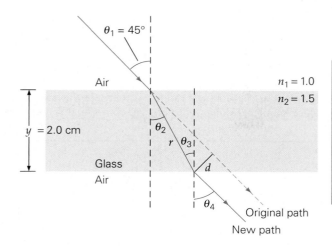

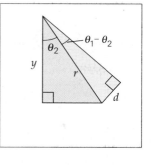

◀ **FIGURE 22.10   Two refractions**
In the glass, the refracted ray is
displaced laterally (sideways) a
distance d from the incident ray,
and the emergent ray is parallel to
the original ray. (See Example 22.4.)

what is the lateral displacement between the ray entering and the ray emerging from the
glass (the perpendicular distance between the two rays—d in the figure)?

Thinking It Through.   Since there are two refractions involved in this example, we
use Snell's law in (a), again in (b), and then some geometry and trigonometry in (c). [See
Appendix I.]

Solution.   Listing the data, we have the following:

Given: $\theta_1 = 45°$ *Find:*   (a) θ_2 (angle of refraction)
            $n_1 = 1.0$ (air)   (b) Show that $\theta_4 = \theta_1$
            $n_2 = 1.5$ (c) d (lateral displacement)
            $y = 2.0$ cm

(a)   Using the practical form of Snell's law, Eq. 22.5, with $n_1 = 1.0$ for air gives

$$\sin \theta_2 = \frac{\sin \theta_1}{n_2} = \frac{\sin 45°}{1.5} = \frac{0.707}{1.5} = 0.47$$

Thus,

$$\theta_2 = \sin^{-1}(0.47) = 28°$$

Note that the beam is refracted toward the normal.
(b)   If $\theta_1 = \theta_4$, then the emergent ray is parallel to the incident ray. Applying Snell's law
to the beam at both surfaces gives

$$n_1 \sin \theta_1 = n_2 \sin \theta_2$$

and

$$n_2 \sin \theta_3 = n_1 \sin \theta_4$$

From the figure, we see that $\theta_2 = \theta_3$. Therefore,

$$n_1 \sin \theta_1 = n_1 \sin \theta_4$$

or

$$\theta_1 = \theta_4$$

Thus, the emergent beam is parallel to the incident beam, but displaced laterally or perpendicularly at a distance d.
(c)   It can be seen from the inset in Fig. 22.10 that, to find d, we need to first find r from
the known information in the pink right triangle. We have

$$\frac{y}{r} = \cos \theta_2 \quad \text{or} \quad r = \frac{y}{\cos \theta_2}$$

In the yellow right triangle, $d = r \sin(\theta_1 - \theta_2)$. Substituting r from the previous step yields

$$d = \frac{y \sin(\theta_1 - \theta_2)}{\cos \theta_2} = \frac{(2.0 \text{ cm}) \sin(45° - 28°)}{\cos 28°} = 0.66 \text{ cm}$$

Follow-up Exercise. If the glass in this Example had $n = 1.6$, would the lateral displacement be the same, larger, or smaller? Explain your answer conceptually, and then calculate the actual value to verify your reasoning.

Example 22.5 ■ The Human Eye: Refraction and Wavelength

A simplified representation of the crystalline lens in a human eye shows it to have a cortex (an outer layer) of $n_{cortex} = 1.386$ and a nucleus (a core) of $n_{nucleus} = 1.406$. See Fig. 25.1b. (Note that both refraction indices are within the range listed for the human eye in Table 22.1.) If a beam of monochromatic (single-frequency or wavelength) light of wavelength 590 nm is directed from air through the front of the eye and into the crystalline lens, qualitatively compare and list the frequency, speed, and wavelength of light in air, the nucleus, and the cortex. First do the comparison part without numbers, and then calculate the actual values to verify your reasoning.

Reasoning and Answer. First, we need the relative magnitudes of the indices of refraction, where $n_{air} < n_{cortex} < n_{nucleus}$.

As you learned earlier in this section, the frequency (f) of light is the same in all three media: air, the cortex, and the nucleus. Thus, the frequency can be calculated by using the speed and the wavelength of light in any of these materials, but it is easiest in air. (Why?) From the wave relationship $c = \lambda f$ (Eq. 13.17), we have

$$f = f_{air} = f_{cortex} = f_{nucleus} = \frac{c}{\lambda} = \frac{3.00 \times 10^8 \text{ m/s}}{590 \times 10^{-9} \text{ m}} = 5.08 \times 10^{14} \text{ Hz}$$

The speed of light in a medium depends on its index of refraction, since $v = c/n$. The smaller the index of refraction, the higher the speed. Therefore, the speed of light is the highest in air ($n = 1.00$ and $v = c = 3.00 \times 10^8$ m/s) and lowest in the nucleus ($n = 1.406$).

The speed of light in the cortex is

$$v_{cortex} = \frac{c}{n_{cortex}} = \frac{3.00 \times 10^8 \text{ m/s}}{1.386} = 2.16 \times 10^8 \text{ m/s}$$

and the speed of light in the nucleus is

$$v_{nucleus} = \frac{3.00 \times 10^8 \text{ m/s}}{1.406} = 2.13 \times 10^8 \text{ m/s}$$

We also know that the wavelength of light in a medium depends on the index of refraction of the medium ($\lambda_m = \lambda/n$.) The smaller the index of refraction, the longer is the wavelength. Therefore, the wavelength of light is the longest in air ($n = 1$ and $\lambda = 590$ nm) and shortest in the nucleus ($n = 1.406$).

The wavelength in the cortex can be calculated from Eq. 22.4:

$$\lambda_{cortex} = \frac{\lambda}{n_{cortex}} = \frac{590 \text{ nm}}{1.386} = 426 \text{ nm}$$

and the wavelength in the nucleus is

$$\lambda_{nucleus} = \frac{590 \text{ nm}}{1.406} = 420 \text{ nm}$$

Finally, we can construct a table to more easily compare the frequency, speed, and wavelength of light in the three media:

	Frequency (Hz)	Speed (m/s)	Wavelength (nm)
Air	5.08×10^{14}	3.00×10^8	590
Cortex	5.08×10^{14}	2.16×10^8	426
Nucleus	5.08×10^{14}	2.13×10^8	420

Follow-up Exercise. A light source of a single frequency is submerged in water in a special fish tank. The beam travels in the water, through double glass panes at the side of the tank (each glass pane has a different n), and into air. In general, what happens to (a) the frequency and (b) the wavelength of the light when it emerges into the outside air?

Refraction is common in everyday life and explains many things we observe. Let's look at refraction in action.

Mirage: A common example of this phenomenon sometimes occurs on a highway on a hot summer day. The refraction of light is caused by layers of air that are at different temperatures (the layer closer to the road is at a higher temperature, lower density, and lower index of refraction). This variation in indices of refraction gives rise to the observed "wet" spot and an inverted image of an object such as a car (▼ Fig. 22.11a). The term *mirage* generally brings to mind a thirsty person in the desert "seeing" a pool of water that really isn't there. This optical illusion plays tricks on the mind, with the image usually seen as in a pool of water and our eye's past experience unconsciously leading us to conclude that there is water on the road.

In Fig. 22.11b, there are two ways to see the car. First, the horizontal rays come directly from the car to our eyes, so we see the car above the ground. Also, the rays from the car that travel toward the road surface will be refracted by the layered air. After hitting the surface, these rays will be refracted again and travel toward our eyes. (See the inset in the figure.) Cooler air has a higher density and so a higher index of refraction. A ray traveling toward the road surface will be gradually refracted with a larger angle of refraction until it hits the surface. It will then be refracted again with a smaller angle of refraction, going toward our eyes. As a consequence, we also see an inverted image of the car, below the road surface. In other words, the surface of the road acts almost like a mirror.

Not where it should be: You may have experienced a refractive effect while trying to reach for something underwater, such as a fish (▶ Fig. 22.12a). We are used to

▼ **FIGURE 22.11 Refraction in action** (a) An inverted car on a "wet" road, a mirage. (b) The mirage is formed when light from the object is refracted by layers of air at different temperatures near the surface of the road.

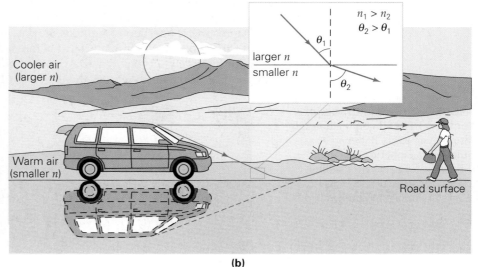

(a)

(b)

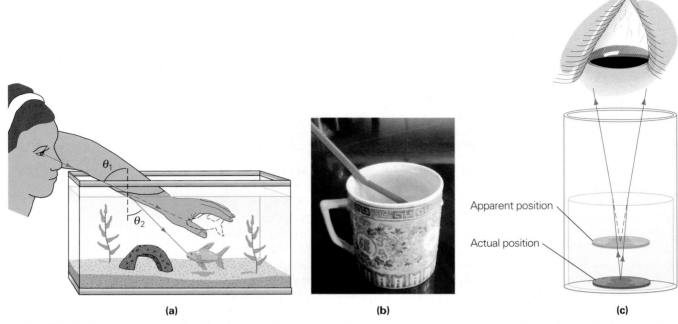

| (a) | (b) | (c) |

▲ **FIGURE 22.12 Refractive effects** **(a)** The light is refracted, and because we tend to think of light as traveling in straight lines, the fish is not where we think it is. **(b)** The chopstick appears bent at the air–water boundary. If the cup is transparent, we see another type of refraction. (See Exercise 27.) **(c)** Because of refraction, the coin appears to be closer than it actually is.

PHYSLET® ILLUSTRATION

Apparent Depth

light traveling in straight lines from objects to our eyes, but the light reaching our eyes from a submerged object has a directional change at the air–water interface. (Note in the figure that the ray is refracted away from the normal.) As a result, the object appears to be closer to the surface than it actually is, and therefore we tend to miss the object when reaching for it. For the same reason, a chopstick in a cup appears bent (Fig. 22.12b), a coin in a glass of water will appear closer than it really is (Fig. 22.12c), and the legs of a person standing in water seem shorter than their actual length. The relationship between the true depth and the apparent depth can be calculated. (See Exercise 45.)

Atmospheric effects: The Sun on the horizon sometimes appears to be flattened, with its horizontal dimension greater than its vertical dimension (▼Fig. 22.13a). This effect is the result of temperature and density variations in the denser air along the horizon. These variations occur predominantly vertically, so light from the top and light from the bottom portions of the Sun are refracted differently as the two sets of beams pass through different atmospheric densities with different indices of refraction.

Atmospheric refraction lengthens the day, so to speak, by allowing us to see the Sun (or the Moon, for that matter) just before it actually rises above the horizon and just after it actually sets below the horizon (as much as 20 minutes on both ends). The denser air near the Earth refracts the light over the horizon toward us (Fig. 22.13b).

▶ **FIGURE 22.13 Atmospheric effects** **(a)** The Sun on the horizon commonly appears flattened as a result of atmospheric refraction. **(b)** Before rising and after setting, the Sun can be seen briefly also because of atmospheric refraction.

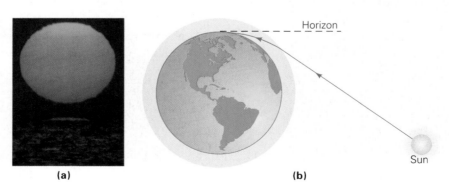

| (a) | (b) |

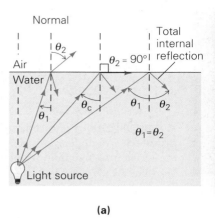

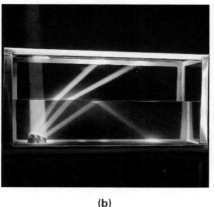

(a)   (b)

◀ **FIGURE 22.14 Internal reflection** (a) When light enters a less optically dense medium, it is refracted away from the normal. At a critical angle (θ_c), the light is refracted along the interface (common boundary) of the media. At an angle greater than the critical angle ($\theta_1 > \theta_c$), there is total internal reflection. (b) Can you estimate the critical angle in the photograph?

22.4 Total Internal Reflection and Fiber Optics

OBJECTIVES: To (a) describe total internal reflection and (b) understand fiber-optic applications.

An interesting phenomenon occurs when light travels from a more optically dense medium into a less optically dense one, such as when light goes *from* water *into* air. As you know, in such a case a ray will be refracted away from the normal. (The angle of refraction is larger than the angle of incidence.) Furthermore, Snell's law states that the greater the angle of incidence, the greater is the angle of refraction. That is, as the angle of incidence increases, the farther the refracted ray diverges from the normal.

However, there is a limit. For a certain angle of incidence called the **critical angle** (θ_c), the angle of refraction is 90°, and the refracted ray is directed along the boundary between the media. But what happens if the angle of incidence is even larger? If the angle of incidence is greater than the critical angle ($\theta_1 > \theta_c$), the light isn't refracted at all, but is internally reflected (▲Fig. 22.14). This condition is called **total internal reflection**. The reflection process is almost 100% efficient. (There is still some absorption of light in the materials.) Because of total internal reflection, glass prisms can be used as mirrors (▶Fig. 22.15). In summary, reflection and refraction occur at all angles for $\theta_1 \leq \theta_c$, but the refracted or transmitted ray disappears at $\theta_1 > \theta_c$.

An expression for the critical angle can be obtained from Snell's law. If $\theta_1 = \theta_c$ in the optically denser medium, $\theta_2 = 90°$, and it follows that

$$\frac{\sin \theta_1}{\sin \theta_2} = \frac{\sin \theta_c}{\sin 90°} = \frac{n_2}{n_1}$$

Since $\sin 90° = 1$,

$$\sin \theta_c = \frac{n_2}{n_1} \quad \text{where } n_1 > n_2 \tag{22.6}$$

If the second medium is air, $n_2 \approx 1$, and the critical angle at the boundary from a medium into air is given by $\sin \theta_c = 1/n$, where n is the index of refraction of the medium.

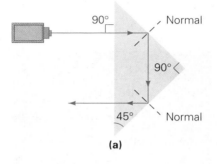

(a)

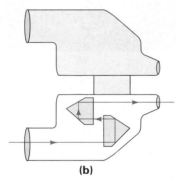

(b)

▲ **FIGURE 22.15 Internal reflection in a prism** (a) Because the critical angle of glass is less than 45°, prisms with 45° and 90° angles can be used to reflect light through 180°. (b) Internal reflection of light by prisms in binoculars makes this instrument much shorter than a telescope.

Example 22.6 ■ A View from the Pool: Critical Angle

(a) What is the critical angle for light traveling in water and incident on a water–air boundary? (b) If a diver submerged in a pool looked up at the surface of the water at an angle of $\theta < \theta_c$, what would she see? (Neglect any thermal or motional effects.)

◀ **FIGURE 22.16 Panoramic and distorted** An underwater view of the surface of a swimming pool in Hawaii.

Thinking It Through. (a) The critical angle is given by Eq. 22.6. (b) As shown in Fig. 22.14a, θ_c forms a cone of vision for viewing from below the water.

Solution.

Given: $n_1 = 1.33$ (for water, from Table 22.1)   *Find:* (a) θ_c (critical angle)
$n_2 \approx 1$ (why?)   (b) View for $\theta < \theta_c$

(a) The critical angle is

$$\theta_c = \sin^{-1}\left(\frac{n_2}{n_1}\right) = \sin^{-1}\left(\frac{1}{1.33}\right) = 48.8°$$

(b) Using Fig. 22.14a, trace the rays in reverse for light coming from all angles outside the pool. Light coming from the above-water 180° panorama could be viewed only in a cone with a half-angle of 48.8°. As a result, objects above the surface would also appear distorted. An underwater panoramic view is seen in ▲ Fig. 22.16. Now can you explain why wading birds like herons usually keep their locations low before trying to catch a fish?

Follow-up Exercise. What would the diver see when looking up at the water surface at an angle of $\theta > \theta_c$?

Internal reflections enhance the brilliance of cut diamonds. (Brilliance is a measure of the amount of light returning straight back to the viewer. Brilliance is reduced if light leaks out the back of a diamond—the reflection is not total.) The critical angle for a diamond–air surface is

$$\theta_c = \sin^{-1}\left(\frac{1}{n}\right) = \sin^{-1}\left(\frac{1}{2.42}\right) = 24.4°$$

A so-called brilliant-cut diamond has many facets, or faces (58 in all—33 on the upper face and 25 on the lower). Light entering the lower facets from the upper facets above the critical angle is internally reflected in the diamond. The light then emerges from the upper facets, giving rise to the diamond's brilliance (◀Fig. 22.17).

(a)

(b)

▲ **FIGURE 22.17 Diamond brilliance** **(a)** Internal reflection gives rise to a diamond's brilliance. **(b)** The "cut," or the depth proportions, of the facets is critical. If a stone is too shallow or too deep, light will be lost (refracted out) through the lower facets.

Fiber Optics

When a fountain is illuminated from below, the light is transmitted along the curved streams of water. This phenomenon was first demonstrated in 1870 by the British scientist John Tyndall (1820–1893), who showed that light was "conducted" along the curved path of a stream of water flowing from a hole in the side of a container. The phenomenon is observed because light undergoes total internal reflection along the stream.

Total internal reflection forms the basis of **fiber optics**, a fascinating field centered on the use of transparent fibers to transmit light. Multiple total internal reflections make it possible to "pipe" light along a transparent rod (just like streams of water), even if the rod is curved (▶ Fig. 22.18). Note from the figure that the smaller the diameter of the light pipe, the more total internal reflections it has. In a small fiber, there can be as many as several hundred total internal reflections per centimeter.

Total internal reflection is an exceptionally efficient process. Optical fibers can be used to transmit light over very long distances with losses of only about 25% per kilometer. These losses are due primarily to impurities in the fiber, which scatter the light. Transparent materials have different degrees of transmission. Fibers are made of special plastics and glasses for maximum transmission efficiency. The greatest efficiency is achieved with infrared radiation, because there is less scattering, as you will learn in Section 24.5.

The greater efficiency of multiple total internal reflections compared with multiple mirror reflections can be illustrated by a good reflecting plane mirror, which has at best a reflectivity of about 95%. After each reflection, the beam intensity is 95% of that of the incident beam from the preceding reflection ($I_1 = 0.95\,I_o$; $I_2 = 0.95\,I_1 = 0.95^2\,I_o; \dots$). Therefore, the intensity I of the reflected beam after n reflections is given by

$$I = 0.95^n\,I_o$$

where I_o is the initial intensity of the beam before the first reflection. Thus, after 14 reflections,

$$I = 0.95^{14}\,I_o = 0.49\,I_o$$

In other words, after 14 reflections, the intensity is reduced to less than half (49%). For 100 reflections, $I = 0.006\,I_o$, and the intensity is only 0.6% of the initial intensity! Compare this to about 75% of the initial intensity in optical fibers over a kilometer in length with *thousands and thousands* of reflections, and you can see the advantage of total internal reflection.

Fibers whose diameters are about 10 μm (10^{-5} m) are grouped together in flexible bundles that are 4 to 10 mm in diameter and up to several meters in length, depending on the application (▼ Fig. 22.19). A fiber bundle with a cross-sectional area of 1 cm^2 can contain as many as 50 000 individual fibers. (A coating is needed to keep the fibers from touching.)

There are many important and interesting applications of fiber optics, ranging from communications and computer networking to medical applications. (See the

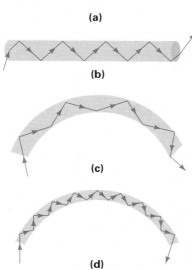

(a)

(b)

(c)

(d)

▲ **FIGURE 22.18 Light pipes**
(a) Total internal reflection in an optical fiber. **(b)** When light is incident on the end of a cylindrical form of transparent material such that the internal angle of incidence is greater than the critical angle of the material, the light undergoes total internal reflection down the length of the light pipe. **(c)** Light is also transmitted along curved light pipes by total internal reflection. **(d)** As the diameter of the rod or fiber becomes smaller, the number of reflections per unit length increases.

▼ **FIGURE 22.19 Fiber-optic bundle** **(a)** Hundreds or even thousands of extremely thin fibers are grouped together **(b)** to make an optical fiber, colored blue by a laser.

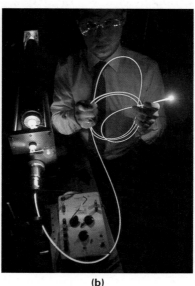

(a)

(b)

Fiber Optics: Medical Applications

Before fiber optics, *endoscopes*—instruments used to view internal portions of the human body—consisted of lens systems in long, narrow tubes. Some contained a dozen or more lenses and produced relatively poor images. Also, because the lenses had to be aligned in certain ways, the tubes had to have rigid sections, which limited the endoscope's maneuverability. Such an endoscope could be inserted down the patient's throat into the stomach to observe the stomach lining. However, there would be blind spots due to the curvature of the stomach and the inflexibility of the instrument.

Fiber-optic bundles have eliminated these problems. Lenses placed at the end of the fiber bundles focus the light, and a prism is used to change the direction for its return. The incident light is usually transmitted by an outer layer of fiber bundles, and the image is returned through a central core of fibers. Mechanical linkages allow maneuverability. The end of a fiber endoscope can be equipped with devices to obtain specimens of the viewed tissues for biopsy (diagnostic examination) or even to perform surgical procedures (Fig. 1). For example, arthroscopic surgery is performed on the knees of injured athletes. The arthroscope that is now routinely used for inspecting *and* repairing damaged joints is simply a fiber endoscope fitted with appropriate surgical implements.

A fiber-optic *cardioscope* (for direct observation of heart valves) typically is a fiber bundle about 4 mm in diameter and 30 cm long. Such a cardioscope passes easily to the heart through the jugular vein in the neck, which is about 15 mm in diameter. To displace the blood and provide a clear field of view for observing and photographing, a transparent balloon at the tip of the cardioscope is inflated with saline (saltwater) solution.

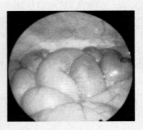

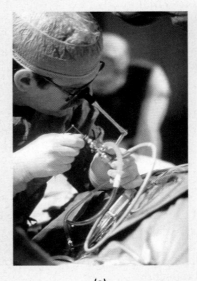

(b)

(a)

FIGURE 1 Endoscopy (a) Surgeons use a fiber-optic endoscope to perform surgery. **(b)** An endoscopic view of the intestines.

Insight on this page.) Light signals, converted from electrical signals, are transmitted through optical telephone lines and computer networks. At the other end, they are converted back to electrical signals. Optical fibers have lower energy losses than current-carrying wires, particularly at higher frequencies, and can carry far more data. Also, optical fibers are lighter than metal wires, have greater flexibility, and are not affected by electromagnetic disturbances (electric and magnetic fields), since they are made of materials that are electrical insulators.

22.5 Dispersion

OBJECTIVE: To explain dispersion and some of its effects.

Light of a single frequency, and consequently a single wavelength, is called *monochromatic light* (from the Greek *mono*, meaning "one," and *chroma*, meaning "color"). Visible light that contains all the component frequencies, or colors, at about the same intensities (e.g., sunlight) is termed *white light*. When a beam of white light passes through a glass prism, as shown in ▸Fig. 22.20a, it is spread out, or dispersed, into a spectrum of colors. This phenomenon led Newton to believe that sunlight is a mixture of colors. When the beam enters the prism, the component colors, corresponding to different wavelengths of light, are refracted at slightly different angles, so they spread out into a spectrum (Fig. 22.20b).

The emergence of a spectrum indicates that the index of refraction of glass is slightly different for different wavelengths, which is true of many transparent media (Fig. 22.20c). The reason has to do with the fact that in a dispersive medi-

Note: You can remember the sequence of the colors of the visible spectrum (from the long-wavelength end to the short-wavelength end) by using the name ROY G. BIV, which is an acronym for *red, orange, yellow, green, blue, indigo,* and *violet.*

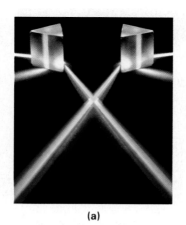
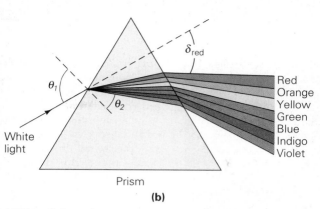

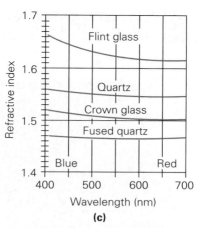

▲ FIGURE 22.20 Dispersion (a) White light is dispersed into a spectrum of colors by glass prisms. **(b)** In a dispersive medium, the index of refraction varies slightly with wavelength. Red light, longest in wavelength, has the smallest index of refraction and is refracted least. The angle between the incident beam and an emergent ray is the angle of deviation (δ) for that ray. (The angles are exaggerated for clarity.) **(c)** Variation in the index of refraction with wavelength for some common transparent media.

um the speed of light is slightly different for different wavelengths. Since the index of refraction n of a medium is a function of the speed of light in that medium ($n = c/v$), the index of refraction is different for different wavelengths. It follows from Snell's law that light of different wavelengths will be refracted at different angles.

We can summarize the preceding discussion by saying that in a transparent material with different indices of refraction for different wavelengths of light, refraction causes a separation of light according to wavelengths, and the material is said to exhibit **dispersion**. Dispersion varies with different media. Also, because the differences in the indices of refraction for different wavelengths are small, a representative value at some specified wavelength can be used for general purposes. (See Table 22.1.)

PHYSLET®
ILLUSTRATION

Dispersion in a Prism

Example 22.7 ■ Forming a Spectrum: Dispersion

The index of refraction of a particular transparent material is 1.4503 for the red end ($\lambda_r = 700$ nm) of the visible spectrum and 1.4698 for the blue end ($\lambda_b = 400$ nm). If white light is incident on a prism of this material as in Fig. 22.20b at an angle of incidence of 45°, what is the angular separation of the visible spectrum inside the prism?

Thinking It Through.   The angle of refraction is given by Snell's law, and we can compute the angle of refraction for the red and blue ends of the visible spectrum. The angular separation of the light inside the prism is the difference in these angles of refraction.

Solution.

Given: (red) $n_r = 1.4503$ for $\lambda_r = 700$ nm   *Find:* $\Delta\theta_2$ (angular separation)
    (blue) $n_b = 1.4698$ for $\lambda_b = 400$ nm
        $\theta_1 = 45°$

Using Eq. 22.5 with $n_1 = 1.00$ (air), we get

$$\sin \theta_{2_r} = \frac{\sin \theta_1}{n_{2_r}} = \frac{\sin 45°}{1.4503} = 0.48756 \quad \text{and} \quad \theta_{2_r} = 29.180°$$

Similarly,

$$\sin \theta_{2_b} = \frac{\sin \theta_1}{n_{2_b}} = \frac{\sin 45°}{1.4698} = 0.48109 \quad \text{and} \quad \theta_{2_b} = 28.757°$$

INSIGHT

The Rainbow

FIGURE 1 Rainbow Notice that the colors of the primary rainbow run vertically from red (top) to blue (bottom).

We have all been fascinated by the beautiful array of colors of a rainbow (Fig. 1). With the optical principles learned in this chapter, we are now in a position to understand the formation of this spectacular display.

A rainbow is formed by refraction, dispersion, and internal reflection of light within water droplets. When sunlight shines on millions of water droplets in the air during and after a rainstorm, we see a multicolored arc whose colors run from violet along the lower part of the spectrum (in order of wavelength) to red along the upper. Occasionally, more than one rainbow is seen: The main, or primary, rain-

bow is sometimes accompanied by a fainter and higher secondary rainbow (Fig. 1) or even a third rainbow. These higher order rainbows are caused by more than one total internal reflection within the water droplets.

The light that forms the primary rainbow is first refracted and dispersed in the water droplet, then reflected once inside each water droplet, and finally refracted and dispersed again upon exiting the water droplet, resulting in the light being spread out into a spectrum of colors (Fig. 2a). However, because of the conditions for refraction and total internal reflection in water, the angles between incoming and outgoing rays for violet to red light lie within a narrow range of 40°–42°. This means that you can see a rainbow only when the Sun is behind you, so that the dispersed light is reflected to you through these angles.

Red appears on the top of the rainbow because light of shorter wavelengths from those water droplets will pass over our eyes (Fig.2b). Similarly, blue is at the bottom of the rainbow because light of longer wavelengths passes under our eyes.

We generally see rainbows only as arcs, because their formation by water droplets is cut off at the ground. If you were on a cliff or on an airplane, you might see a complete circular rainbow (Fig.2b). Also, the higher the Sun is in the sky, the less of a rainbow you will be able to see from the ground. In fact, you won't see a primary rainbow if the Sun's angle above the horizon is greater than 42°. The primary rainbow can still be seen from a height, however. As an observer's elevation increases, more of the arc becomes visible. You may also have seen a circular rainbow in the spray from a garden hose.

So

$$\Delta\theta_2 = \theta_{2_r} - \theta_{2_b} = 29.180° - 28.757° = 0.423°$$

This is not much of a deviation, but as the light travels, it is refracted and dispersed again by the second boundary to spread out the colors further. Finally, the light emerges from the prism, and the dispersion becomes evident (Fig. 22.20a).

Follow-up Exercise. If the green light exhibits an angular separation of 0.156° from the red light, what is the index of refraction for green light in the hypothetical material of the example? Will the green light refract more or less than the red light? Explain.

A good example of a dispersive material is diamond, which is about five times more dispersive than glass. In addition to revealing the brilliance resulting from internal reflections off many facets, a cut diamond shows a display of colors, or "fire," resulting from the dispersion of the refracted light (Fig. 22.17).

Dispersion is a cause of chromatic aberration in lenses, which is described fully in Chapter 23. Optical systems in cameras often consist of several lenses to minimize this problem (see Section 23.4).

Another dramatic example of dispersion is the production of a rainbow, as discussed in the Insight on this page.

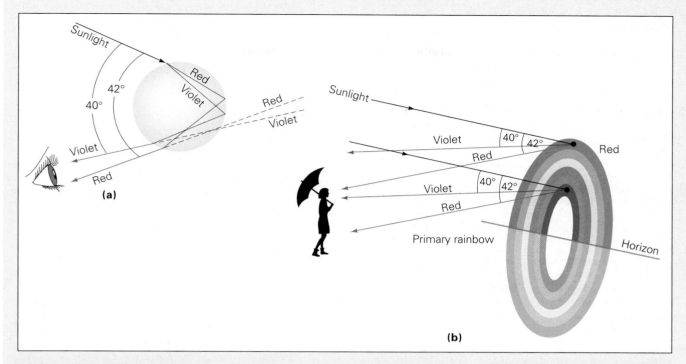

FIGURE 2 The rainbow   Rainbows are created by the refraction, dispersion, and internal reflection of sunlight. **(a)** Light of different color emerges from the water droplet in different directions. **(b)** An observer sees red light at the top of the bow and violet or blue at the bottom.

Chapter Review

Important Concepts and Equations

- **Law of reflection:** The angle of incidence equals the angle of reflection (as measured from the normal to the reflecting surface):

$$\theta_i = \theta_r \qquad (22.1)$$

- The **index of refraction** of any medium is the ratio of the speed of light in a vacuum to its speed in that medium:

$$n = \frac{c}{v} = \frac{\lambda}{\lambda_m} \qquad (22.3, 22.4)$$

- The angle of refraction as a ray of light moves from one medium to another is given by **Snell's law**. If the second medium is more optically dense, the ray is refract-

ed toward the normal; if the medium is less dense, the ray is refracted away from the normal. Snell's law is

$$\frac{\sin \theta_1}{\sin \theta_2} = \frac{v_1}{v_2} \quad \text{or} \quad n_1 \sin \theta_1 = n_2 \sin \theta_2 \quad (22.2, 22.5)$$

- **Total internal reflection** occurs if the second medium is less dense that the first and the angle of incidence exceeds the critical angle:

$$\sin \theta_c = \frac{n_2}{n_1} \quad (n_1 > n_2) \qquad (22.6)$$

- The refractive **dispersion** of light occurs in some media because different wavelengths have slightly different indices of refraction and hence different speeds.

Exercises*

22.1 Wave Fronts and Rays
and
22.2 Reflection

1. A wave front is (a) always circular, (b) parallel to a ray, (c) described by a surface of equal phase, or (d) none of these.

2. A ray (a) is perpendicular to the direction of energy flow, (b) is always parallel to other rays, (c) is perpendicular to a series of wave fronts, or (d) illustrates the wave nature of light.

3. For regular, or specular, reflection, (a) the angle of incidence equals the angle of reflection, (b) the rays of a reflected beam are parallel, (c) the incident ray, the reflected ray, and the normal lie in the same plane, or (d) all of the above.

4. For irregular, or diffuse, reflection, (a) the angle of incidence equals the angle of reflection, (b) the rays of a reflected beam are not parallel, (c) the incident ray, the reflected ray, and the local normal lie in the same plane, or (d) all of these.

5. CQ When you see the Sun over a lake or the ocean, you often observe a long swath of light (▼ Fig. 22.21). What causes this effect, sometimes called a "glitter path?"

▲ **FIGURE 22.21   A glitter path**   See Exercise 5.

6. CQ Under what circumstances will the angle of reflection be smaller than the angle of incidence?

7. CQ If a mirror is perfect (reflecting 100%), can you see its surface?

8. CQ The book you are reading does not have a light source, so it must be reflecting light from other sources. What type of reflection is this?

9. CQ It is difficult to drive during the night after a rain because of the glare. Instead of seeing the road, you see the images of buildings, trees, and so on. What causes this glare? Why is there no such glare when the road is dry? (See Insight on p. 727.)

10. ■ The angle of incidence of a light ray on a mirrored surface is 35°. What is the angle between the incident and reflected rays?

11. ■ A beam of light is incident on a plane mirror at an angle of 32° relative to the normal. What is the angle between the reflected rays and the surface of the mirror?

12. IE ■ A beam of light is incident on a plane mirror at an angle α relative to the surface of the mirror. (a) Will the angle between the reflected ray and the normal be (1) α, (2) $90° - \alpha$, or (3) 2α? (b) If $\alpha = 43°$, what is the angle between the reflected ray and the normal?

13. ■ A light ray incident on a plane mirror is at an angle of 55° relative to the surface of the mirror. At what angle of reflection is the reflected ray?

14. IE ■■ Two upright plane mirrors touch along one edge, where their planes make an angle of α. A beam of light is directed onto one of the mirrors at an angle of incidence $\beta < \alpha$ and is reflected onto the other mirror. (a) Will the angle of reflection of the beam from the second mirror be (1) α, (2) β, (3) $\alpha + \beta$, or (4) $\alpha - \beta$? (b) If $\alpha = 60°$ and $\beta = 40°$, what will be the angle of reflection of the beam from the second mirror?

15. IE ■■ Two identical plane mirrors of width w are placed a distance d apart with their mirrored surfaces parallel and facing each other. (a) A beam of light is incident at one end of one mirror so that the light just strikes the far end of the other mirror after reflection. Will the angle of incidence be (1) $\sin^{-1} w/d$, (2) $\cos^{-1} w/d$, or (3) $\tan^{-1} w/d$? (b) If $d = 50$ cm and $w = 25$ cm, what is the angle of incidence?

16. ■■ Two people stand 3.0 m away from a large plane mirror and spaced 5.0 m apart in a dark room. At what angle of incidence should one of them shine a flashlight on the mirror so that the reflected beam directly strikes the other person?

17. ■■ If you hold a 900-cm² square plane mirror 45 cm from your eyes and can just see the full length of an 8.5-m flagpole behind you, how far are you from the pole? [*Hint*: A diagram is helpful here.].

18. ■■■ Two plane mirrors, M_1 and M_2, are placed together as illustrated in ▶ Fig. 22.22. (a) If the angle α between the mirrors is 70° and the angle of incidence, θ_{i_1}, of a light ray

*Assume angles to be exact.

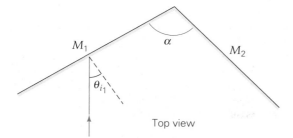

▲ **FIGURE 22.22 Plane mirrors together** See Exercises 18 and 19.

◀ **FIGURE 22.23 Refraction effect** See Exercise 27.

incident on M_1 is 35°, what is the angle of reflection, θ_{r_2}, from M_2? (b) If $\alpha = 115°$ and $\theta_{i_1} = 60°$, what is θ_{r_2}?

19. ■■■ For the plane mirrors in Fig. 22.22, what angles α and θ_{i_1} would allow a ray to be reflected back in the direction from which it came (parallel to the incident ray)?

22.3 Refraction
and
22.4 Total Internal Reflection and Fiber Optics

20. Light refracted at the boundary of two different media (a) is bent toward the normal when $n_1 > n_2$, (b) is bent away from the normal when $n_1 > n_2$, (c) has the same angle of refraction as the angle of incidence, or (d) always decreases in speed.

21. The index of refraction (a) is always greater than or equal to 1, (b) is inversely proportional to the speed of light in a medium, (c) is inversely proportional to the wavelength of light in the medium, or (d) all of these.

22. **CQ** What is the fundamental physical reason for refraction?

23. **CQ** As light travels from one medium to another, does its wavelength change? its frequency? its speed?

24. **CQ** Ice has an index of reflection of 1.31, smaller than the value for water at 1.33. Why?

25. The critical angle for total internal reflection at a medium–air boundary (a) is independent of the wavelength of the light in the medium, (b) is greater for a medium with a smaller index of refraction, (c) may be greater than 90°, or (d) none of these.

26. **CQ** Under what conditions will total internal reflection occur?

27. **CQ** Explain why the pencil in ▶Fig. 22.23 appears almost severed. Also, compare this figure with Fig. 22.12b.

28. **CQ** The photos in ▶Fig. 22.24 were taken with a camera on a tripod at a fixed angle. There is a penny in the container, but only its tip is seen initially. However, when water is added, more of the coin is seen. Why? Use a diagram to explain.

▲ **FIGURE 22.24 You barely see it, but then you do** See Exercises 28 and 74.

29. **CQ** Two hunters, one with bow and arrow and the other with a laser gun, see a fish under water. They both aim directly where they see it. Which one, the arrow or the laser beam, has a better chance of hitting the fish? Explain.

30. **CQ** Will total internal reflection occur if light is traveling from air to glass?

31. **CQ** Why is fiber optics so useful in medicine?

32. ■ The speed of light in the core of the crystalline lens in a human eye is 2.13×10^8 m/s. What is the index of refraction of the core?

33. ■ Is the speed of light greater in diamond or in zircon? Express the difference as a percentage.

34. **IE** ■ A beam of light enters water. (a) Will the angle of refraction be (1) greater than, (2) equal to, or (3) less than the angle of incidence? Why? (b) If the beam enters the water at an angle of 60° relative to the normal of the surface, find the angle of refraction

35. ■ Light passes from air into water. If the angle of refraction is 20°, what is the angle of incidence?

36. ■ A beam of light traveling in air is incident on a transparent plastic material at an angle of incidence of 50°. The angle of refraction is 35°. What is the index of refraction of the plastic?

37. IE ■ (a) For total internal reflection to occur, should the light be directed from air to a diamond or from a diamond to air? Why? (b) What is the critical angle of the diamond in air?

38. ■ The critical angle for a certain type of glass in air is 41.8°. What is the index of refraction of the glass?

39. ■■ A beam of light in air is incident on the surface of a slab of fused quartz. Part of the beam is transmitted into the quartz at an angle of refraction of 30° relative to a normal to the surface, and part is reflected. What is the angle of reflection?

40. ■■ A beam of light is incident on a flat piece of polystyrene at an angle of 55° relative to a normal to the surface. What angle does the refracted ray make with the plane of the surface?

41. ■■ Monochromatic blue light that has a frequency of 6.5×10^{14} Hz enters a piece of flint glass. What are the frequency and wavelength of the light in the glass?

42. ■■ Light passes from material A, which has an index of refraction of $\frac{4}{3}$, into material B, which has an index of refraction of $\frac{5}{4}$. Find the ratio of the speed of light in material B to the speed of light in material A.

43. ■■ In Exercise 42, what is the ratio of the light's wavelength in material B to that in material A?

44. ■■ The laser used in cornea surgery to treat corneal disease is the excimer laser, which emits ultraviolet light with a wavelength of 193 nm in air. The index of refraction of the cornea is 1.376. What are the wavelength and frequency of the light in the cornea?

45. IE ■■ (a) An object immersed in water appears closer to the surface than it actually is. What is the cause of this illusion? (b) Using ▼Fig. 22.25, show that the apparent

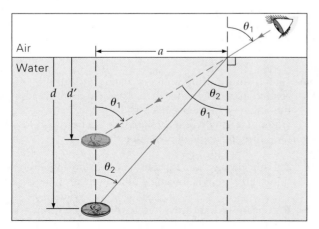

▲ **FIGURE 22.25** Apparent depth? See Exercise 45. (For small angles only; angles enlarged for clarity.)

depth for small angles of refraction is d/n, where n is the index of refraction of the water. [*Hint*: Recall that for small angles, $\tan\theta \approx \sin\theta$.]

46. ■■ A fish tank is made of glass with an index of refraction of 1.50. A person shines a light beam on the glass at an incident angle of 40° to see a fish inside. Is the fish illuminated? Justify your answer.

47. IE ■■ (a) A beam of light is to undergo total internal reflection through a 45°–90°–45° prism (▼Fig. 22.26). Will this arrangement depend on (1) the index of refraction of the prism, (2) the index of refraction of the surrounding medium, or (3) the indices of refraction of both? Why? (b) Calculate the index of refraction of the prism if the surrounding medium is air or water.

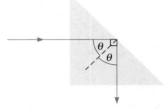

▲ **FIGURE 22.26** Total internal reflection in a prism See Exercises 47 and 48.

48. ■■ A 45°–90°–45° prism (Fig. 22.26) is made of a material with an index of refraction of 1.85. Can the prism be used to deflect a beam of light by 90° (a) in air or (b) in water?

49. ■■ A light ray in air is incident on a glass plate 10.0 cm thick at an angle of incidence of 40°. The glass has an index of refraction of 1.65. The emerging ray on the other side of the plate is parallel to the incident ray, but is laterally displaced. What is the perpendicular distance between the original direction of the ray and the direction of the emerging ray? [*Hint*: See Example 22.4.]

50. ■■ A person lying at poolside looks over the edge of the pool and sees a bottle cap on the bottom directly below, where the depth is 3.2 m. How far below the water surface does the bottle cap appear to be? (See Exercise 45b.)

51. ■■ What percentage of the actual depth is the apparent depth of an object submerged in water if the observer is looking almost straight downward. (See Exercise 45b.)

52. ■■ At what angle to the surface must a diver submerged in a lake look toward the surface to see the setting Sun?

53. ■■ A submerged diver shines a light toward the surface of a body of water at angles of incidence of 40° and 50°. Can a person on the shore see a beam of light emerging from the surface in either case? Justify your answer mathematically.

54. IE ■■ To a submerged diver looking upward through the water, the altitude of the Sun (the angle between the Sun and the horizon) appears to be 45°. (a) Is 45° the actual altitude of the Sun? (b) If not, what is the Sun's actual altitude?

55. ■■ A coin lies on the bottom of a pool under 1.5 m of water and 0.90 m from the sidewall (▼ Fig. 22.27). If a light beam is incident on the water surface at the wall, at what angle θ relative to the wall must the beam be directed so that it will illuminate the coin?

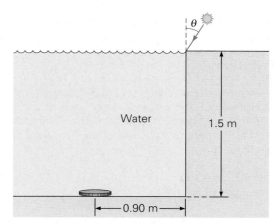

▲ FIGURE 22.27 Find the coin See Exercise 55 (not drawn to scale).

56. ■■ Describe a method for measuring the index of refraction of the fluid in Fig. 22.8a. Determine the index of refraction of the liquid.

57. IE ■■ A light beam traveling upward in a plastic material with an index of refraction of 1.60 is incident on an upper horizontal air interface at an angle of 45°. (a) Is the beam transmitted? (b) Suppose the upper surface of the plastic material is covered with a layer of liquid with an index of refraction of 1.20. What happens in this case?

58. ■■ A crown-glass plate 2.5 cm thick is placed over a newspaper. How far beneath the top surface of the plate would the print appear to be if you were looking almost vertically downward through the plate? (See Exercise 45b.)

59. ■■■ An outdoor circular fish pond has a diameter of 4.0 m and a uniform full depth of 1.50 m. A fish halfway down in the pond and 0.50 m from the near side can just see the full height of a 1.8-m-tall person. How far away from the edge of the pond is the person?

60. ■■■ A cube of flint glass sits on a newspaper on a table. By looking into one of the vertical sides of the cube, is it possible to see the portion of the newspaper covered by the glass?

61. ■■■ Two glass prisms are placed together (▶ Fig. 22.28). (a) If a beam of light strikes the face of one of the prisms at normal incidence as shown, at what angle θ does the

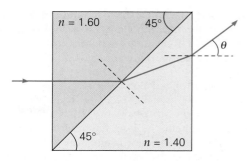

▲ FIGURE 22.28 Joined prisms See Exercise 61.

beam emerge from the other prism? (b) At what angle of incidence would the beam be refracted along the interface of the prisms?

22.5 Dispersion

62. Dispersion can occur only if the light is (a) monochromatic, (b) polychromatic, (c) white light, or (d) both (b) and (c).

63. Dispersion can occur only in (a) reflection, (b) refraction, (c) total internal reflection, or (d) all of the above.

64. CQ What causes light of different frequencies to separate upon being refracted?

65. CQ Why is dispersion more prominent in a triangular-shaped prism than a square block?

66. CQ A glass prism disperses white light into a spectrum. Can a second glass prism be used to recombine the spectral components? Explain.

67. CQ You can never walk under a rainbow. Explain why.

68. CQ A light beam consisting of two colors, A and B, is sent through a prism. Color A is refracted more than color B. Which color has a longer wavelength? Explain.

69. CQ (a) If glass is dispersive, why don't we see a spectrum of colors when sunlight passes through a windowpane? (b) Does dispersion occur for polychromatic light incident on a dispersive medium at an angle of 0°? Explain. (Are the speeds of each color of light the same in the medium?)

70. IE ■■ The index of refraction of crown glass is 1.515 for red light and 1.523 for blue light. (a) If light is incident on crown class from air, which color, red or blue, will be refracted more? Why? (b) Find the angle separating rays of the two colors in a piece of crown glass if their angle of incidence is 37°.

71. ■■ White light passes through a prism made of crown glass and strikes an interface with air at an angle of 41.15°. Using the indices of refraction given in Exercise 70, describe what happens.

72. ■■ A beam of light with red and blue components of wavelengths 670 nm and 425 nm, respectively, strikes a

slab of fused quartz at an incident angle of 30°. On refraction, the different components are separated by an angle of 0.001 31 rad. If the index of refraction of the red light is 1.4925, what is the index of refraction of the blue light?

73. ■■■ A beam of red light is incident on an equilateral prism as shown in ▼Fig. 22.29. (a) If the index of refraction of red light on the prism is 1.400, at what angle θ does the beam emerge from the other face of the prism? (b) Suppose the incident beam were white light. What would be the angular separation of the red and blue components in the emergent beam if the index of refraction of blue light were 1.403? (c) If the index of refraction of blue light were 1.405?

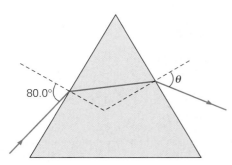

▲ FIGURE 22.29 **Prism revisited** See Exercise 73.

Additional Exercises

74. An opaque container that is empty except for a single coin is 15 cm deep. When looking into the container at a viewing angle of 50° relative to the vertical side of the container, you see nothing on the bottom. When the container is filled with water, you see the coin (from the same viewing angle) on the bottom of, and just beyond, the side of the container. (See Fig. 22.24.) How far is the coin from the side of the container?

75. Light travels from a material whose index of refraction is $n_1 = 2.0$ into another material whose index of refraction is $n_2 = 1.7$. The opposite parallel surface of the second material is exposed to the air. (a) Find the critical angle at which the light will be reflected at the interface of the materials. (b) Find the angle at which the light will pass through the interface and then be reflected at the opposite surface of the second material.

76. Light strikes a surface at an angle of 50° relative to the surface. What is the angle of reflection?

77. A beam of light traveling in water strikes a surface of a transparent material at an angle of incidence of 45°. If the angle of refraction in the material is 35°, what is the index of refraction of the material?

78. Yellow-green light of wavelength 550 nm is incident on the surface of a flat piece of crown glass at an angle of 40°. What is (a) the angle of refraction of the light? (b) the speed of the light in the glass? (c) the wavelength of the light in the glass?

79. Light strikes water perpendicularly to the surface. What is the angle of refraction?

80. In Fig. 22.20b, if the glass prism has an index of refraction of 1.5 and the experiment is done in water rather than in air, what happens to the spectrum emerging from the prism? How about in a liquid that also has an index of refraction of 1.5? Explain.

81. Light passes from medium A into medium B at an angle of incidence of 30°. The index of refraction of A is 1.5 times that of B. (a) What is the angle of refraction? (b) What is the ratio of the speed of light in B to the speed of light in A? (c) What is the ratio of the frequency of the light in B to the frequency of light in A? (d) At what angle of incidence would the light be internally reflected?

Mirrors and Lenses

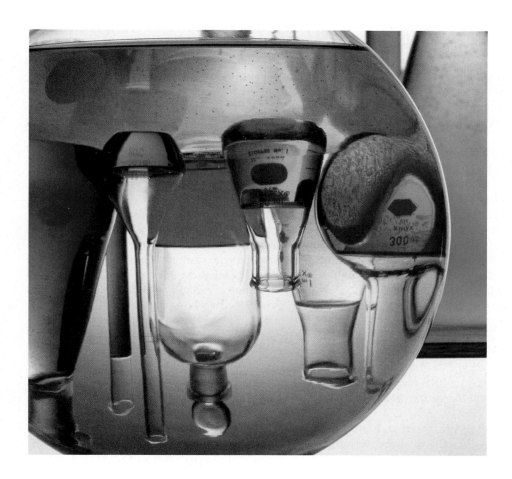

What would life be like if there were no mirrors in bathrooms or on cars and if no one could get glasses? Imagine a world without optical images of any kind—no photographs, no movies, no TV. Think about how little we'd know about the universe if there were no telescopes with which to observe distant planets and stars, how little we'd know about biology and medicine if there were no microscopes with which to see bacteria and our own cells. We sometimes forget how dependent we are on mirrors and lenses.

The first mirror was probably the reflecting surface of a pool of water. Later, people discovered that polished metals and glass have reflective properties. They must also have noticed that when they looked at things through glass, the objects looked different than when viewed directly, depending on the shape of the glass. In some cases, the objects appeared to be enlarged (magnified) or inverted, as in the photo shown on this page. In time, people learned to shape glass purposefully into lenses, paving the way for the eventual development of the many optical devices we now take so much for granted.

The optical properties of mirrors and lenses are based on the principles of reflection and refraction of light, introduced in Chapter 22. In this chapter, you'll learn how mirrors and lenses work. Among other things, you'll discover why the image in the photo is upside down, whereas your image in an ordinary flat mirror is right-side up—but it doesn't seem to comb its hair with the same hand you use!

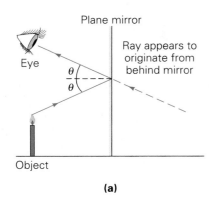

Plane mirror

Ray appears to
originate from
behind mirror

Eye

θ
θ

Object

(a)

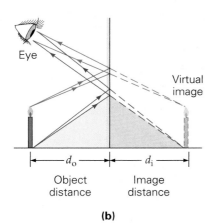

Eye

Virtual
image

d_o | d_i

Object
distance

Image
distance

(b)

▲ **FIGURE 23.1 Image formed by a plane mirror** (a) A ray from a point on the object is reflected in the mirror according to the law of reflection. **(b)** Rays from various points on the object produce an extended image. Because the two shaded triangles are identical, the image distance d_i (the distance of the image from the mirror) is equal to the object distance d_o. That is, the image appears to be the same distance behind the mirror as the object is in front of the mirror. The rays appear to diverge from the image position. In this case, the image is said to be virtual.

23.1 Plane Mirrors

OBJECTIVES: To (a) understand how images are formed and (b) describe the characteristics of images formed by plane mirrors.

Mirrors are smooth reflecting surfaces, usually made of polished metal or glass that has been coated with some metallic substance. As you know, even an uncoated piece of glass, such as a window pane, can act as a mirror. However, when one side of a piece of glass is coated with a compound of tin, mercury, or silver, the reflectivity of the glass is increased, as light is not transmitted through the coating. A mirror may be front coated or back coated, depending on the application.

When you look directly into a mirror, you see the reflected images of yourself and objects around you (apparently on the other side of the surface of the mirror). The geometry of a mirror's surface affects the size, orientation, and type of image. In general, an *image* is the visual counterpart of an object, produced by reflection (mirrors), refraction (lenses), or the passage of rays through a small hole.

A mirror with a flat surface is called a **plane mirror**. How images are formed by a plane mirror is illustrated by the ray diagram in ◄Fig. 23.1. An image appears to be behind or "inside" the mirror. This is because, when the mirror reflects a ray of light from the object to the eye (Fig. 23.1a), the ray appears to originate from behind the mirror. Reflected rays from the top and bottom of an object are shown in Fig. 23.1b. In actuality, light rays coming from all points on the side of the object facing the mirror are reflected, and an image of the complete object is observed.

The image formed in this way *appears* to be behind the mirror. Such an image is called a **virtual image**. Light rays appear to diverge from virtual images, but do not actually do so. However, spherical mirrors (discussed in Section 23.2) can project images in front of the mirror. This type of image is called a **real image**. An example of a real image is the image produced by an overhead projector in a classroom.

Notice in Fig. 23.1b the positions or distances of the object and image from the mirror. Quite logically, the distance of an object from a mirror is called the *object distance* (d_o), and the distance its image appears to be behind the mirror is called the *image distance* (d_i). By geometry of identical triangles, it can be shown that $d_o = d_i$, which means that *the image formed by a plane mirror appears to be at a distance behind the mirror that is equal to the distance between the object and the front of the mirror.* (See Exercise 17.)

We are interested in various characteristics of images. Two of these features are the size and orientation of an image compared with those of its object. Both are expressed in terms of the **lateral magnification factor** (*M*), which is defined as a ratio:

$$M = \frac{\text{image height}}{\text{object height}} = \frac{h_i}{h_o} \tag{23.1}$$

Note that *M* is a dimensionless quantity, as it is a ratio of heights or lengths. In ►Fig. 23.2, you should be able to see that the image and object are pointing in the same direction, so $h_i = h_o$. Therefore, $M = +1$ for a plane mirror, the image is upright, and there is no magnification. That is, you and your image in a plane mirror are the same size.

We use a lighted candle as an object to allow us to address another image characteristic: orientation—that is, whether the image is upright or inverted with respect to the orientation of the object. (In sketching ray diagrams, an arrow is a convenient object for this purpose.) For a plane mirror, the image is always upright (or erect). This means that the image is oriented in the same vertical direction as the object. We then say that h_i and h_o have the same sign, so *M* is positive. Therefore, for a plane mirror, $M = +1$, as we have already calculated.

With other types of mirrors, such as spherical mirrors (which we will consider shortly), it is possible to have inverted images. In summary, the sign of M tells us the orientation of the image relative to the object, and the absolute value of M gives the magnification.

Another characteristic of reflected images is the so-called right–left reversal. When you look at yourself in a mirror and raise your right hand, it appears that your image is raising its left hand. However, this right–left reversal is an apparent one that is actually caused by the front–back reversal. That is, if your front faces south, your back "faces" north. Your image, on the other hand, has its front to the north and back to the south—a front–back reversal. You can demonstrate this reversal by asking one of your friends to stand facing you (without a mirror). If your friend raises his right hand, you can see that that hand is actually on your left side.

The main characteristics of an image formed by a plane mirror are summarized in Table 23.1. See also the Insight on page 751, A View to a Curve.

▲ **FIGURE 23.2 Magnification** The lateral, or height-distance, magnification factor is given by $M = h_i/h_o$. For a plane mirror, $M = +1$, which means that $h_i = h_o$—the image is the same height as the object, and the image is upright.

TABLE 23.1 Characteristics of Images Formed by Plane Mirrors

The image distance is equal to the object distance ($d_i = d_o$). That is, the image appears to be as far behind the mirror as the object is in front.

The image is virtual, upright, and unmagnified ($M = +1$).

Example 23.1 ■ All of Me: Minimum Mirror Length

What is the minimum vertical length of a plane mirror needed for a person to be able to see a complete (head-to-toe) image of him- or herself (▼Fig. 23.3)?

Thinking It Through. Applying the law of reflection, we see in the figure that two triangles are formed by the rays needed for the image to be complete. These triangles relate the person's height to the minimum mirror length.

Solution. To determine this length, consider the situation shown in Fig. 23.3. With a mirror of minimum length, a ray from the top of the person's head would be reflected at the top of the mirror, and a ray from the person's feet would be reflected at the bottom of the mirror. The length L of the mirror is then the distance between the dashed horizontal lines perpendicular to the mirror at its top and bottom.

However, these lines are also the normals for the ray reflections. By the law of reflection, they bisect the angles between incident and reflected rays; that is, $\theta_i = \theta_r$. Then, because their respective triangles on each side of the dashed normal are similar, the length

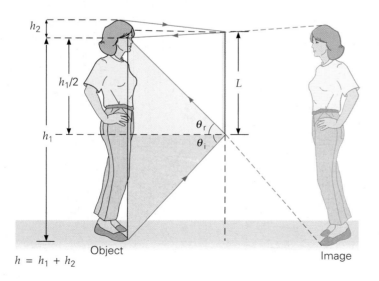

◀ **FIGURE 23.3 Seeing it all** The minimum height, or vertical length, of a plane mirror needed for a person to see his or her complete (head-to-toe) image turns out to be half the person's height. See Example 23.1.

of the mirror from its bottom to a point even with the person's eyes is $h_1/2$, where h_1 is the person's height from the feet to the eyes. Similarly, the small upper length of the mirror is $h_2/2$ (the vertical distance between the person's eyes and the top of mirror). Thus,

$$L = \frac{h_1}{2} + \frac{h_2}{2} = \frac{h_1 + h_2}{2} = \frac{h}{2}$$

where h is the person's total height.

Hence, for a person to see his or her complete image in a plane mirror, the minimum height, or vertical length, of the mirror must be half the height of the person.

You can do a simple experiment to prove this conclusion. Get some newspaper and tape, and find a full-length mirror. Gradually cover parts of the mirror with the newspaper until you cannot see your complete image. You will find you need only a mirror length that is half your height.

Follow-up Exercise. What effect does a person's distance from the mirror have on the minimum mirror length required to produce his or her complete image? *(Answers to all Follow-up Exercises are at the back of the text.)*

23.2 Spherical Mirrors

OBJECTIVES: To **(a)** distinguish between converging and diverging spherical mirrors, **(b)** describe images and their characteristics, and **(c)** determine these image characteristics from ray diagrams and the spherical-mirror equation.

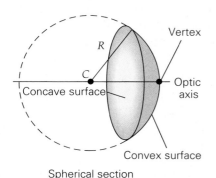

▲ **FIGURE 23.4 Spherical mirrors** A spherical mirror is a section of a sphere. Either the outside (convex) surface or the inside (concave) surface of the spherical section may be the reflecting surface.

As the name implies, a **spherical mirror** is a reflecting surface with spherical geometry. ◄Figure 23.4 shows that if a portion of a sphere of radius R is sliced off along a plane, the severed section has the shape of a spherical mirror. Either the inside or outside of such a section can be the reflecting surface. For reflections on the inside surface, the section behaves as a **concave mirror**. (Think of looking into a *cave* in order to help yourself remember that a con*cave* mirror has an indented surface.) For reflections from the outside surface, the section behaves as a **convex mirror**.

The radial line through the center of the spherical mirror is called the *optic axis*, and it intersects the surface of the mirror at the *vertex* of the spherical section (Fig. 23.4). The point on the optic axis that corresponds to the center of the sphere of which the mirror forms a section is called the **center of curvature** (C). The distance between the vertex and the center of curvature is equal to the radius of the sphere and is called the **radius of curvature** (R).

When rays parallel and close to the optic axis are incident on a concave mirror, the reflected rays intersect, or converge, at a common point called the **focal point** (F). As a result, a concave mirror is called a **converging mirror** (▼ Fig. 23.5a). Note that the law of reflection, $\theta_i = \theta_r$, is satisfied for each ray.

▶ **FIGURE 23.5 Focal point**
(a) Rays parallel and close to the optic axis of a concave spherical mirror converge at the focal point F. **(b)** Rays parallel and close to the optic axis of a convex spherical mirror are reflected along paths as though they came from a focal point behind the mirror. Note that the law of reflection, $\theta_i = \theta_r$, is satisfied for each ray in each diagram.

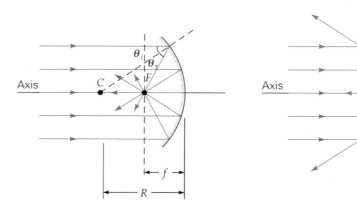

(a) Concave, or converging, mirror

(b) Convex, or diverging, mirror

INSIGHT

A View to a Curve

When you walk into a room that has large mirrors on opposite walls (as in a dance studio), you see many images of yourself that appear to stretch far into the distance. In such a situation, do you see an infinite number of images that stretch into infinity?

In theory, the answer is yes. Consider the case of an object between two parallel mirrors as illustrated in Fig. 1. Mirror 1 (M_1) and mirror 2 (M_2) form the image of the object (O) at I_1 and I_2, respectively. Then M_1 forms an image of I_2 at I_3 and M_2 forms an image of I_1 at I_4, and so on. In each image, the "object" distance is always equal to the image distance. Note that the image formed by one mirror serves as an "object" for another mirror, and secondary images can also form.

However, in reality, you won't see an infinite number of images, and often the images will appear to curve in one direction. Why?

No mirror reflects 100% of the light incident on it. Good mirrors reflect about 95%. After about 14 reflections, only about 50% of the light remains ($0.95 \times 0.95 \times \cdots \times 0.95 = 0.95^{14} =$ 0.48 = 48%). After 27 reflections, only about 25% of the light is left, and so on, until none of the light remains to travel between the mirrors.

Even if perfect mirrors (100% reflection) are used, you still won't see an infinite number of images. Since your eyes are in the middle of your head, you won't see those images that are smaller (farther away from you) and hidden behind the larger images (closer to you).

It is, however, possible to see a large number of images, but the images lie along a curve and eventually are lost "around the curve" (Fig. 2a). This is because two mirrors can never be perfectly parallel. If there is a slight deviation from a perfectly parallel setup, this imperfection will actually multiply after numerous reflections (Fig. 2b), which leads you to a view to a curve.

Related Exercises: 14 and 15

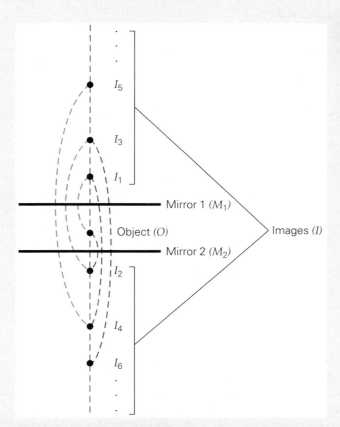

FIGURE 1   Infinite to infinity?   Theoretically, an infinite number of images is formed for an object in between two parallel perfect mirrors.

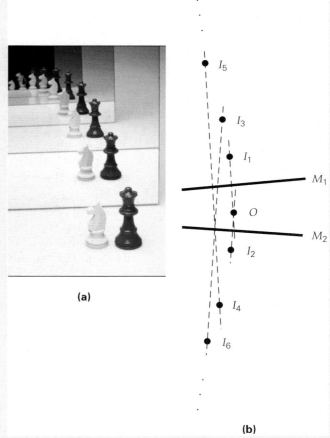

(a)

(b)

FIGURE 2   A view to a curve   (a) Multiple images from two almost-parallel mirrors. **(b)** Notice how successive images are moved to the side, because the mirrors are not perfectly parallel, giving the curve effect. (Top view.)

Note:
Concave mirror = converging mirror
Convex mirror = diverging mirror

▲ **FIGURE 23.6 Diverging mirror**
Note by reverse-ray tracing in
Fig. 23.5b that a diverging (convex)
spherical mirror gives an expanded
field of view, as can be seen in this
store-monitoring mirror.

Similarly, a beam parallel and close to the optic axis of a convex mirror diverges on reflection, as though the reflected rays came from a focal point behind the mirror's surface (Fig. 23.5b). Thus, a convex mirror is called a **diverging mirror** (◄Fig. 23.6). When you see diverging rays, your brain interprets or assumes that there is an object from which the rays *appear* to diverge, even though no such object is there.

The distance from the vertex to the focal point of parallel rays near the axis of a spherical mirror is called the **focal length** *f*. (See Fig. 23.5.) The focal length is related to the radius of curvature by the following simple equation:

$$f = \frac{R}{2} \qquad \begin{array}{l} \textit{focal length,} \\ \textit{spherical mirror} \end{array} \qquad (23.2)$$

Ray Diagrams

The characteristics of images formed by spherical mirrors can be determined from geometrical optics (introduced in Chapter 22). The method involves drawing rays emanating from one or more points on an object. The law of reflection ($\theta_i = \theta_r$) applies, and three key rays are defined with respect to the mirror's geometry as follows:

1. A **parallel ray** is a ray that is incident along a path parallel to the optic axis and is reflected through (or appears to go through) the focal point (as do all rays near and parallel to the axis).

2. A **chief ray**, or **radial ray**, is a ray that is incident through the center of curvature (*C*) of the spherical mirror. Since the chief ray is incident normal to the mirror's surface, this ray is reflected back along its incident path, through point *C*.

3. A **focal ray** is a ray that passes through (or appears to go through) the focal point and is reflected parallel to the optic axis. (It is a reversed parallel ray, so to speak.)

Using any two of these three key rays, we can locate the image (image distance) and determine its size (magnified or reduced), orientation (upright or inverted), and type (real or virtual). It is customary to use the tip of the object (for example, the head of an arrow or the flame of a candle) as the origin point of the rays. This method makes it easy to see whether the image is upright or inverted. The corresponding point of the image is at the point of intersection of the rays.

Keep in mind, however, that *properly traced rays from any point on the object can be used to find the image*. Every point on a visible object acts as an emitter of light. For example, for a candle, the flame emits its own light, but every point on the candle reflects light.

Example 23.2 ■ Learn by Drawing: A Mirror Ray Diagram

An object is placed 39.0 cm in front of a concave spherical mirror of radius 24.0 cm. (a) Use a ray diagram to locate the image formed by this mirror. (b) Discuss the characteristics of the image.

Thinking It Through. A ray diagram, drawn accurately, can by itself provide information about image location and image characteristics that might otherwise be determined mathematically.

Solution.

Given: $R = 24.0$ cm   *Find:* (a) image location
$d_o = 39.0$ cm   (b) image characteristics

(a)   Since we have been asked to use a ray diagram (drawing) to locate the image, the first thing we need to decide on is a scale for the drawing. If we use a scale of 1 cm (on the drawing) to represent 10 cm, the object would be drawn 3.90 cm in front of the mirror.

First we draw the optic axis, the mirror, the object (a lighted candle), and the center of curvature (C). From Eq. 23.2, $f = 24.0$ cm/2 $= 12.0$ cm. So we can draw the focal point (F) halfway from the vertex to the center of curvature.

To locate the image, follow steps 1–4 on the Learn by Drawing illustration on this page:

1.   The first ray we draw here is the parallel ray [(1) in the drawing]. From the tip of the flame, we draw a horizontal ray (parallel to the optic axis). After reflecting, this ray goes through the focal point, F.
2.   We can then draw the chief ray [(2) in the drawing]. From the tip of the flame, we draw a ray going through the center of curvature, C. This ray will be reflected back in the original direction. (Why?)
3.   We can clearly see that the two rays that are reflected from the parallel and chief rays intersect. The point of intersection is the tip of the image of the candle. From this point, we can draw the image by extending the tip of the flame to the optic axis.
4.   Only two rays are needed to locate the image. However, if we do draw the third ray, the focal ray in this case [(3) in the drawing], it must go through the same point on the image at which the other two rays intersect (if we are drawing the diagram carefully). The ray from the tip of the flame going through the focal point, F, after reflection, will travel parallel to the optic axis.

(b)   From the ray diagram we drew in part (a), it can clearly be seen that the image is real (because the rays intersect in front of the mirror). The reflected rays converge and pass through the image (at a point). As a result, the real image could be seen on a screen (for example, a piece of white paper) that is positioned at a distance d_i from the concave mirror. The image is also inverted (the image of the candle points downward) and is smaller than the object.

Follow-up Exercise.   In this example, what would the characteristics of the image be if the object were 15.0 cm in front of the mirror? Locate the image, and discuss its characteristics.

An example of a ray diagram using the same three rays for a convex (diverging) mirror will be shown in Integrated Example 23.4.

A converging mirror does *not* always form a real image. For a converging spherical mirror, the characteristics of the image change with the distance of the object from the mirror. Dramatic changes take place at two points: C (the center of curvature) and F (the focal point). These points divide the optic axis into three regions (▸Fig. 23.7a): $d_o > R$, $R > d_o > f$, and $d_o < f$.*

Let's start with an object in the region farthest from the mirror ($d_o > R$) and move toward the mirror:

- The case of $d_o > R$ has already been dealt with in Example 23.2.
- When $d_o = R = 2f$, the image is real, inverted, and the same size as the object.
- When $R > d_o > f$, an enlarged, inverted, real image is formed (Fig. 23.7b). The image is magnified when the object is inside the center of curvature, C.

*This is the case when the rays are close to the optic axis—that is, for small angles of reflection.

Learn by Drawing

A Mirror Ray Diagram
(See Example 23.2)

Converging (concave) mirror

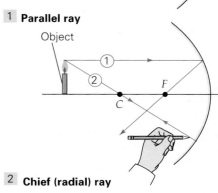

1 **Parallel ray**

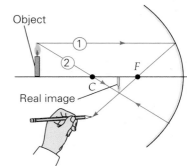

2 **Chief (radial) ray**

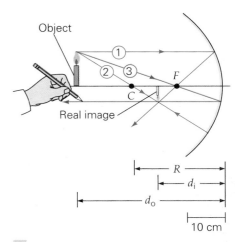

3 **Locating image**

4 **Can also use focal ray**

- When $d_o = f$, the object is at the focal point (Fig. 23.7c). The reflected rays are parallel, and the image is said to be formed at infinity. The focal point F is a special "crossover" point, as it divides the space in front of the mirror into two regions.
- When $d_o < f$, the object is inside the focal point (between the focal point and the mirror's surface). A virtual, enlarged, and upright image is formed (Fig. 23.7d).

When $d_o > f$, the image is real; when $d_o < f$, the image is virtual. For $d_o = f$, we can't see an image formed at infinity, but we say that the image is formed there because of symmetry with the case in Fig. 23.7c. When an object is at "infinity"—that is, it is so far away that rays emanating from it and falling on the mirror are essentially parallel—its image is formed at F. By reverse ray tracing, rays from an object at a great ("infinite") distance from the mirror are shown to be essentially parallel when near the mirror and to form an image on a screen aligned in the focal plane. This fact provides an easy method for determining the focal length of a concave mirror.

▼ **FIGURE 23.7 Concave mirrors** **(a)** For a concave, or converging, mirror, the object is located within one of three regions defined by the center of curvature (C) and the focal point (F), or at one of these two points. For $d_o > R$, the image is real, inverted, and smaller than the object, as shown by the ray diagrams in Example 23.2. **(b)** For $R > d_o > f$, the image will also be real and inverted but enlarged, or magnified. **(c)** For an object at the focal point F, or $d_o = f$, the image is said to be formed at infinity. **(d)** For $d_o < f$, the image will be virtual, upright, and enlarged.

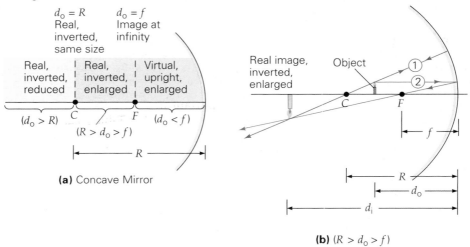

(a) Concave Mirror

(b) ($R > d_o > f$)

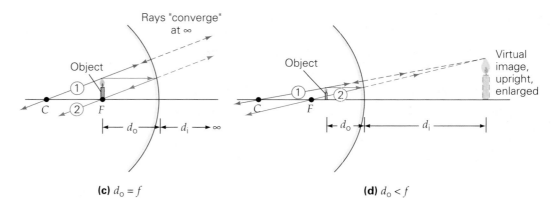

(c) $d_o = f$

(d) $d_o < f$

The position, orientation, and size of the image can be determined graphically from ray diagrams drawn to scale. However, these characteristics can be determined more quickly and accurately by analytical methods. It can be shown by means of geometry that the object distance (d_o), the image distance (d_i) and the focal length (f) are related. That relationship is known as the **spherical-mirror equation**:

$$\frac{1}{d_o} + \frac{1}{d_i} = \frac{1}{f} = \frac{2}{R} \quad \text{or} \quad d_i = \frac{d_o f}{d_o - f} \quad \textit{spherical-mirror equation} \quad (23.3)$$

Note that this equation can be written in terms of either the radius of curvature, R, or the focal distance f, since by Eq. 23.2, $f = R/2$.

The **lateral magnification factor** M defined in Eq. 23.1 can also be found analytically for a spherical mirror. Again, by using geometry, it can be expressed in terms of the image and object distances:

$$M = -\frac{d_i}{d_o} \quad \textit{magnification equation} \quad (23.4)$$

The minus sign is added by convention to indicate the orientation of the image: A positive value for M indicates an upright image, whereas a negative M implies an inverted image. Also, if $|M| > 1$, the image is magnified, or larger than the object. If $|M| < 1$, the image is reduced, or smaller than the object. Note that, for mirrors, the lateral magnification M, also called the *magnification factor*, or simply *magnification*, is conveniently expressed in terms of the image distance d_i and the object distance d_o rather than in terms of the image and object heights used in Eq. 23.1. (A description of the origin of Eqs. 23.3 and 23.4 follows as optional coverage.)

The signs of the various quantities are very important in the application of Eqs. 23.3 and 23.4. We will use the sign conventions summarized in Table 23.2.

Example 23.3 and Integrated Example 23.4 show how these equations and sign conventions are used for spherical mirrors. In general, this approach usually

---

TABLE 23.2 Sign Conventions for Spherical Mirrors

Focal length
Concave (converging) mirror: f(or R) is positive
Convex (diverging) mirror: f(or R) is negative

Object distance d_o (same for both concave and convex mirrors)
d_o is positive when the object is in front of the mirror (real object)
d_o is negative when the object is behind the mirror (virtual object)*

Image distance d_i and **image type** (same for both mirrors)
d_i is positive when the image is formed in front of the mirror (real image)
d_i is negative when the image is formed behind the mirror (virtual image)

Image orientation M (same for both mirrors)
M is positive when the image is upright with respect to the object
M is negative when the image is inverted with respect to the object

*In a combination of two (or more) mirrors, the image formed by the first mirror is the object of the second mirror (and so on). If this image–object falls behind the second mirror, it is referred to as a *virtual* object, and the object distance is taken to be negative. This concept is more important for lens combinations, as we will see in Section 23.3, and is mentioned here only for completeness.

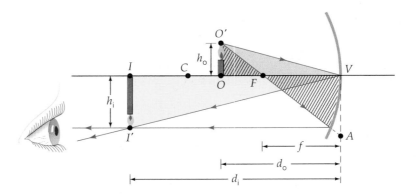

▶ **FIGURE 23.8 Spherical-mirror equation** The rays provide the geometry, through similar triangles, for the derivation of the spherical-mirror equation.

Note: A helpful hint for remembering that the magnification is d_i over d_o is that the ratio is in alphabetical order ("i" over "o").

Note: $|M|$ is the absolute value of M: its magnitude without regard to sign. For example, $|+2| = |-2| = 2$.

involves finding the image of an object; you will be asked where the image is formed (d_i) and what the image characteristics are (M). These characteristics tell whether the image is real or virtual, upright or inverted, and larger or smaller than the object (magnified or reduced).

*Derivation of the Spherical-Mirror Equation You might often wonder where various equations originate. The spherical-mirror equation can be derived with the aid of a little geometry. Consider the ray diagram in ▲Fig. 23.8. The object and image distances (d_o and d_i) and the heights of the object and image (h_o and h_i) are shown. Note that these lengths make up the bases and heights of triangles formed by the ray reflected at the vertex (V). These triangles ($O'VO$ and $I'VI$) are similar, since, by the law of reflection, their angles at V are equal. Hence, we can write

$$\frac{h_i}{h_o} = -\frac{d_i}{d_o} \tag{1}$$

This equation is Eq. 23.4, from the definition of Eq. 23.1. The negative sign inserted here signifies the fact that the image is inverted, so h_i is negative.

The (focal) ray through F also forms similar triangles, $O'FO$ and VFA (in the approximation that the mirror is small compared with its radius). (Why are the triangles similar?) The bases of these triangles are $VF = f$ and $OF = d_o - f$. Then, if VA is taken to be h_i,

$$\frac{h_i}{h_o} = -\frac{VF}{OF} = -\frac{f}{d_o - f} \tag{2}$$

Again, the negative sign inserted here signifies the fact that the image is inverted, so h_i is negative.

Equating Eqs. 1 and 2, we have

$$\frac{d_i}{d_o} = \frac{f}{d_o - f} \tag{3}$$

Cross multiplying gives

$$d_i d_o - d_i f = d_o f \quad\text{or}\quad d_i f + d_o f = d_i d_o \tag{4}$$

Dividing by $d_o d_i f$ yields

$$\frac{1}{d_o} + \frac{1}{d_i} = \frac{1}{f}$$

which is the spherical-mirror equation (Eq. 23.3).

Example 23.3 ■ What Kind of Image? Behavior of a Concave Mirror

A concave mirror has a radius of curvature of 30 cm. If an object is placed (a) 45 cm, (b) 20 cm, and (c) 10 cm from the mirror, where is the image formed, and what are its characteristics? (Specify whether each image is real or virtual, upright or inverted, and magnified or reduced.)

Thinking It Through. Here, we are given R, from which we can compute $f = R/2$. We are also given three different object distances, which we can apply in Eqs. 23.3 and 23.4 to determine the image location and characteristics.

Solution.

Given: $R = 30$ cm, so $f = R/2 = 15$ cm **Find:** d_i, M, and image characteristics
  (a) $d_o = 45$ cm   for given object distances
  (b) $d_o = 20$ cm
  (c) $d_o = 10$ cm

PHYSLET®
ILLUSTRATION

Concave Mirror

Note that the given object distances correspond to the regions shown in Fig. 23.7a. There is no need to convert the distances to meters as long as all distances are in the same unit (centimeters in this case). You could draw representative ray diagrams for each of these cases in order to find the characteristics of each image.

(a) In this case, the object distance is greater than the radius of curvature ($d_o > R$), and

$$\frac{1}{d_o} + \frac{1}{d_i} = \frac{1}{f} \quad \text{or} \quad \frac{1}{45 \text{ cm}} + \frac{1}{d_i} = \frac{1}{15 \text{ cm}}$$

Then

$$\frac{1}{d_i} = \frac{1}{15 \text{ cm}} - \frac{1}{45 \text{ cm}} = \frac{2}{45 \text{ cm}} \quad \text{or} \quad d_i = \frac{45 \text{ cm}}{2} = +22.5 \text{ cm}$$

and

$$M = -\frac{d_i}{d_o} = -\frac{22.5 \text{ cm}}{45 \text{ cm}} = -\frac{1}{2}$$

Thus, the image is real (positive d_i), inverted (negative M), and half as large as the object ($|M| = \frac{1}{2}$).

(b) Here, $R > d_o > f$, and the object is between the focal point and the center of curvature:

$$\frac{1}{20 \text{ cm}} + \frac{1}{d_i} = \frac{1}{15 \text{ cm}} \quad \text{or} \quad \frac{1}{d_i} = \frac{1}{15 \text{ cm}} - \frac{1}{20 \text{ cm}} = \frac{1}{60 \text{ cm}}$$

Thus,

$$d_i = +60 \text{ cm}$$

and

$$M = -\frac{60 \text{ cm}}{20 \text{ cm}} = -3.0$$

In this case, the image is real (positive d_i), inverted (negative M), and three times the size of the object ($|M| = 3$).

(c) For this case, $d_o < f$, and the object is inside the focal point. We use the alternate form of Eq. 23.3:

$$d_i = \frac{d_o f}{d_o - f} = \frac{(10 \text{ cm})(15 \text{ cm})}{10 \text{ cm} - 15 \text{ cm}} = -30 \text{ cm}$$

Then

$$M = -\frac{d_i}{d_o} = -\frac{(-30\text{ cm})}{10\text{ cm}} = +3.0$$

In this case, the image is virtual (negative d_i), upright (positive M), and three times the size of the object ($|M| = 3$).

From the denominator of the expression for d_i, you can see that d_i will always be negative when d_o is less than f. Therefore, a virtual image is always formed for an object inside the focal point of a converging mirror.

Follow-up Exercise. For the converging mirror in this Example, where is the image formed and what are its characteristics if the object is at 30 cm, or $d_o = R$?

Problem-Solving Hint

When using the spherical-mirror equations to find image characteristics, it is helpful to first make a quick sketch (approximate, not necessarily to scale) of the ray diagram for the situation. This sketch shows you the image characteristics and helps you avoid making mistakes when applying the sign conventions. *The ray diagram and the mathematical solution must agree.*

Convex Mirror

Integrated Example 23.4 ■ Similarities and Differences: Behavior of a Convex Mirror

An object (in this case, a candle) is 20 cm in front of a diverging mirror that has a focal length of −15 cm (see the sign conventions in Table 23.2). (a) Use a ray diagram to determine if the image formed is (1) real or virtual, (2) upright or inverted, and (3) magnified or reduced. (b) Find the location and characteristics of the image by using the mirror equations.

(a) Conceptual Reasoning. Since we know the object distance and the focal length of the convex mirror, a ray diagram can be drawn and the image characteristics can, therefore, be determined. The first thing we need to decide on is a scale for the ray diagram. In this example, we could use a scale of 1 cm (on the drawing) to represent 10 cm. That way, the object would be 2.0 cm in front of the mirror in our drawing. We draw the optic axis, the mirror, the object (a lighted candle), and the focal point (F). Since this mirror is convex, the focal point (F) and the center of curvature (C) are behind the mirror. From Eq. 23.2, $R = 2f = 2(-15\text{ cm}) = -30$ cm. So we can draw C at twice the distance from the vertex than is F.

Only two out of the three key rays are necessary to locate the image (▶Fig. 23.9). The parallel ray (1) starts from the tip of the flame, travels parallel to the optic axis, and then diverges from the mirror after reflection, appearing to come from F. The chief ray (2) originates from the tip of the flame, appears to go through C, and then reflects straight back, but appears to come from C. It is clearly seen that these two rays, after reflection, diverge from each other, and there is no chance for them to intersect. However, they both appear to start from a common point behind the mirror: the image point of the tip of the flame. We can also draw the focal ray (3) to verify that all three rays appear to emanate from the same image point.

Therefore, the image is virtual (the reflecting rays don't actually come from a point behind the mirror), upright, and smaller than the object. Measuring from the diagram (keep in mind the drawing scale we are using), we find that $d_i \approx 9.0$ cm, and the magnification $\approx 0.5/1.2 = +0.4$.

(b) Thinking It Through. The object distance and focal length are given. The image position and characteristics can be calculated by using the mirror equations.

Given: $d_o = 20$ cm *Find:* d_i, M, and image characteristics
 $f = -15$ cm

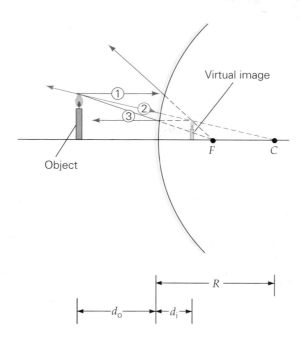

◀ **FIGURE 23.9 Diverging mirror**
Ray diagram of a diverging mirror.
See Integrated Example 23.4.

Note that the focal length is negative for a convex mirror. (See Table 23.2.) Using Eq. 23.3, we have

$$\frac{1}{20 \text{ cm}} + \frac{1}{d_i} = \frac{1}{-15 \text{ cm}}$$

or

$$\frac{1}{d_i} = \frac{1}{-15 \text{ cm}} - \frac{1}{20 \text{ cm}} = -\frac{7}{60 \text{ cm}}$$

so

$$d_i = -\frac{60 \text{ cm}}{7} = -8.6 \text{ cm}$$

Then

$$M = -\frac{d_i}{d_o} = -\frac{(-8.6 \text{ cm})}{20 \text{ cm}} = +0.43$$

Thus, the image is virtual (negative d_i), upright (positive M), and 0.43 times the size (height) of the object. Since f is negative, the image of a real object is always virtual for a diverging (convex) mirror. (Why?)

Follow-up Exercise. As has been pointed out, a diverging mirror always forms a virtual image of a real object. What about the other characteristics of the image—its orientation and magnification? Can any general statements be made about them?

Spherical Mirror Aberrations

Technically, our descriptions of image characteristics for spherical mirrors are true only for objects near the optic axis—that is, only for small angles of incidence and reflection. If these conditions do not hold, the images will be blurred (out of focus) or distorted, because not all of the rays will converge in the same plane. As illustrated in ▶ Fig. 23.10, incident parallel rays far from the optic axis do not converge at the focal point. The farther the incident ray is from the axis, the more distant is its reflected ray from the focal point. This effect is called **spherical aberration**.

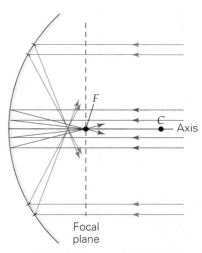

▲ **FIGURE 23.10 Spherical aberration for a mirror**
According to the small-angle approximation, rays parallel to and near the mirror's axis converge at the focal point. However, when parallel rays not near the axis are reflected, they converge in front of the focal point. This effect, called *spherical aberration*, gives rise to blurred images.

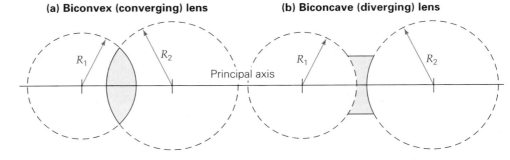

(a) Biconvex (converging) lens   (b) Biconcave (diverging) lens

► **FIGURE 23.11 Spherical lenses** Spherical lenses have surfaces defined by two spheres, and the surfaces are either convex or concave. **(a)** Biconvex and **(b)** biconcave lenses are shown here. If $R_1 = R_2$, a lens is spherically symmetric.

Spherical aberration does not occur with a parabolic mirror. (As the name *parabolic mirror* implies, a section through the center of a parabolic mirror has the form of a parabola.) All of the incident rays parallel to the optic axis of such a mirror have a common focal point. For this reason, parabolic mirrors are used in most astronomical telescopes, as we will see in Chapter 24. However, these mirrors are more difficult to make than spherical mirrors (and are therefore more expensive).

23.3 Lenses

OBJECTIVES: To **(a)** distinguish between converging and diverging lenses, **(b)** describe images and their characteristics, and **(c)** find image locations and characteristics by using ray diagrams and the thin-lens equation.

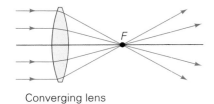

Converging lens

(a) Biconvex (converging) lens

(b)

▲ **FIGURE 23.12 (a) Converging lens** For a thin biconvex lens, rays parallel to the axis converge at the focal point F. **(b) Burning glass** A magnifying glass (converging lens) can be used to focus the Sun's rays to a spot—with incendiary results. Do not try this at home!

The word *lens* comes from the Latin word for lentil, a seed whose shape is similar to that of a common lens. An optical **lens** is made from some transparent material (most commonly glass, but sometimes plastic or crystal). One or both surfaces usually have a spherical contour. *Biconvex* spherical lenses (with both surfaces convex) and *biconcave* spherical lenses (with both surfaces concave) are illustrated in ▲Fig. 23.11.

The properties of lenses are due to the refraction of light passing through them. When light rays pass through a lens, they are refracted, or deviated from their original paths, according to the law of refraction (see Chapter 22).

A biconvex lens is a **converging lens**: Incident light rays parallel to the axis of the lens converge at a focal point on the opposite side of the lens (◄Fig. 23.12a). You may have focused the Sun's rays with a magnifying glass (a biconvex, or converging, lens) and thereby witnessed the concentration of radiant energy that results (Fig. 23.12b). The parallel rays coming from the Sun or some other distant object (at infinity) converge at the focal point of the lens. This fact provides a way to experimentally determine the focal length of a converging lens.

Conversely, a biconcave lens is a **diverging lens**: Incident parallel rays emerge from the lens as though they emanated from a focal point on the incident side of the lens (►Fig. 23.13). (See Exercise 98 for more information on how a lens can converge or diverge rays.)

There are several types of converging and diverging lenses (►Fig. 23.14). Meniscus lenses are the type most commonly used for corrective eyeglasses. In general, a converging lens is thicker at its center than at its periphery, and a diverging lens is thinner at its center than at its periphery. This discussion will be limited to spherically symmetric biconvex and biconcave lenses, for which both surfaces have the same radius of curvature.

When light travels inside a lens, it is refracted and displaced laterally, as shown in Example 22.4 for a piece of glass (Fig. 22.10). If a lens is thick, this displacement may be fairly large and can complicate analysis of the lens's characteristics. This problem does not arise with thin lenses, for which the refractive displacement of transmitted light is negligible. Our discussion will be limited to thin lenses.

Like a spherical mirror, a lens with spherical geometry has, *for each lens surface*, a center of curvature, a radius of curvature, a focal point, and a focal length. The focal points are at equal distances on either side of a thin lens. However, for a spherical lens, $f \neq R/2$, as it does for a spherical mirror (Eq. 23.2). Usually, only the focal length of a lens is specified, rather than the radius of curvature.

The general rules for drawing ray diagrams for lenses are similar to those for spherical mirrors, but some modifications are necessary, since light passes through, rather than reflects from, a lens. Opposite sides of a lens are generally distinguished as the *object side* and the *image side*. The object side is the side on which an object is positioned, and the image side is the *opposite* side of the lens (where a real image would be formed). The three rays from a point on an object are drawn as follows:

1. A **parallel ray** is a ray that is parallel to the lens's optic axis on incidence and that, after refraction, either (a) passes through the focal point on the image side of a converging lens *or* (b) appears to diverge from the focal point on the object side of a diverging lens.
2. A **chief ray**, or **central ray**, is a ray that passes through the center of the lens and is undeviated.*
3. A **focal ray** is a ray that (a) passes through the focal point on the object side of a converging lens *or* (b) appears to pass through the focal point on the image side of a diverging lens and, after refraction, is parallel to the lens's optic axis.

As with spherical mirrors, only two rays are needed to determine the image; we will use the parallel and chief rays. (As in the case of mirrors, however, it is generally a good idea to include a third ray, the focal ray, in your diagrams as a check.)

Example 23.5 ■ Learn by Drawing: A Lens Diagram

An object is placed 30 cm in front of a thin biconvex lens of focal length 20 cm. (a) Use a ray diagram to locate the image. (b) Discuss the characteristics of the image.

Thinking It Through. Remember the steps we followed for the previous ray diagram, for a concave spherical mirror (Example 23.2).

Solution.

Given: $d_o = 30$ cm   *Find:*   (a) location of the image (using a ray diagram)
        $f = 20$ cm   (b) the image's characteristics

(a) Since we have been asked to use a ray diagram (see Learn by Drawing on page 762 to locate the image, the first thing we need to decide on is a scale for the drawing. In this example, we could use a scale of 1 cm (as shown in the accompanying drawing) to represent 10 cm. That way, the object would be 3.00 cm in front of the mirror in our drawing.

First we draw the optic axis, the lens, the object (a lighted candle), and the focal points (F). A vertical dashed line through the center of the lens is drawn as a reference for the rays. For simplicity, the refraction of the rays is depicted as if it occurs at the center of each lens. In reality, it would occur at the air–glass and glass–air surfaces of each lens.

Follow steps 1–4 on the Learn by Drawing illustration:

1. The first ray we draw is the parallel ray [(1) in the drawing]. From the tip of the flame, we draw a horizontal ray (parallel to the optic axis). After passing through the lens, this ray goes through the focal point F on the image side.
2. We then draw the chief ray [(2) in the drawing]. From the tip of the flame, we draw a ray passing through the center of the lens. This ray will go undeviated through the thin lens to the image side.
3. It can be clearly seen that these two rays intersect on the image side. The point of intersection is the image point of the tip of the candle. From this point, we can draw the image by extending the tip of the flame to the optic axis.

---

*Only two rays are needed to determine the image, and we will use only two—the parallel and chief rays—for clarity in our illustration. However, you may draw the third type, a focal ray, as a check.

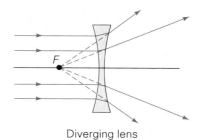

Diverging lens

Biconcave (diverging) lens

▲ **FIGURE 23.13 Diverging lens** Rays parallel to the axis of a biconcave, or diverging, lens appear to diverge from a focal point on the incident side of the lens.

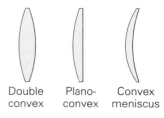

Double convex   Plano-convex   Convex meniscus

Converging lenses

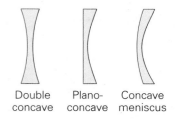

Double concave   Plano-concave   Concave meniscus

Diverging lenses

▲ **FIGURE 23.14 Lens shapes** Lens shapes vary widely and are normally categorized as converging or diverging. In general, a converging lens is thicker at its center than at the periphery, and a diverging lens is thinner at its center than at the periphery.

Learn by Drawing

A Lens Ray Diagram (See Example 23.5.)

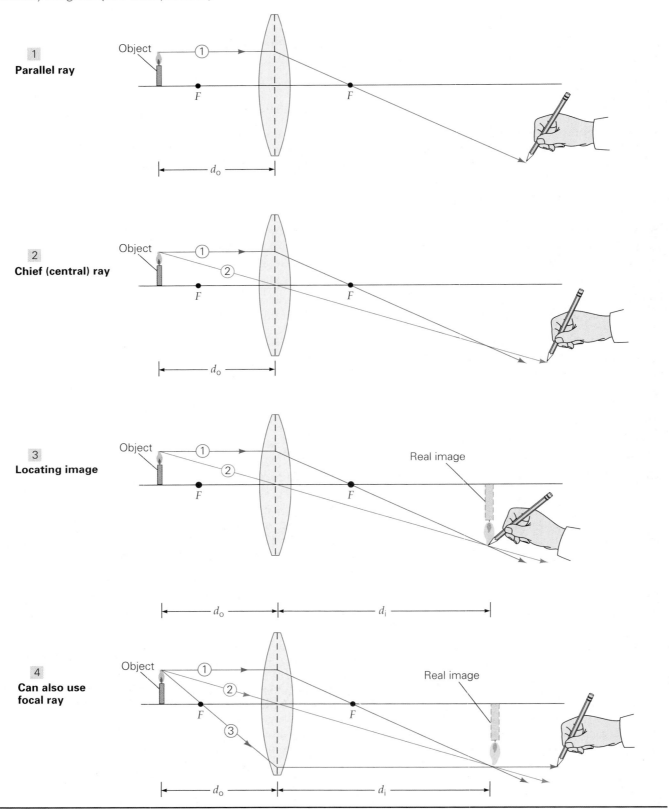

1 Parallel ray

2 Chief (central) ray

3 Locating image

4 Can also use focal ray

4. Only two rays are needed to locate the image. However, if we do draw the third ray, in this case the focal ray [(3) in the drawing], it must go through the same point on the image at which the other two rays intersect (if you are drawing the diagram carefully). The ray from the tip of the flame passing through the focal point F on the object side will travel parallel to the optic axis on the image side.

(b) From the ray diagram we drew in part (a), it can be clearly seen that the image is real (because the rays intersect on the image side). The rays converge and pass through the image (at a point). As a result, the real image could be seen on a screen (for example, a piece of white paper) that is positioned at a distance d_i from the converging lens. The image is also inverted (the image of the candle points downward) and is larger than the object.

In this case, $d_o = 30$ cm and $f = 20$ cm, so $2f > d_o > f$. Using similar ray diagrams, you can prove that for any d_o in this range, the image is always real, enlarged, and inverted. Actually, the overhead projector in your classroom uses this particular arrangement.

Follow-up Exercise. In this example, what does the image look like if the object is 10 cm in front of the lens? Locate the image graphically, and discuss the characteristics of the image.

▼ Fig. 23.15 shows other ray diagrams with different object distances for a converging lens, along with real-life applications.

For a lens, the image of an object is real when formed or projected on the side of the lens opposite the object's location (on the image side; see Fig. 23.15a) and virtual when formed on the same side of the lens as the object is located (on the object side; see Fig. 23.15b). Another way of looking at these properties is that rays converge at a real image and appear to diverge from a virtual image. Also, as with spherical mirrors, a real image from a lens can be formed on a screen, but a virtual image cannot.

Regions could be similarly defined for the object distance for a converging lens as was done for a converging mirror in Fig. 23.7a. Here, an object distance of

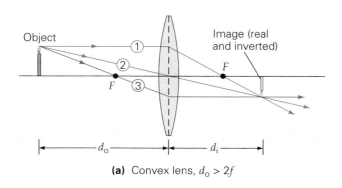

(a) Convex lens, $d_o > 2f$

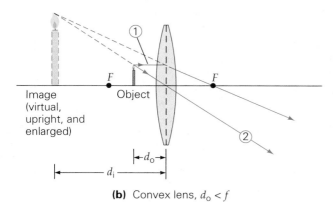

(b) Convex lens, $d_o < f$

◀ **FIGURE 23.15 Ray diagrams for lenses** (a) A converging biconvex lens forms a real object when $d_o > 2f$. The image is real, inverted, and reduced. **(b)** Ray diagrams for a converging lens with $d_o < f$. The image is virtual, upright, and magnified. Practical examples are shown for both cases.

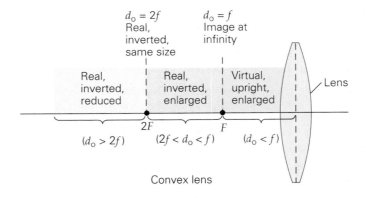

▶ **FIGURE 23.16 Convex lens**
For a convex, or converging, lens,
the object is located within one of
three regions defined by the focal
point (F) and twice the focal point
($2F$) or at one of these two points.
For $d_o > 2f$, the image is real,
inverted, and reduced (Fig. 23.15a).
For $2f > d_o > f$, the image will
also be real and inverted, but
enlarged, or magnified, as shown
by the ray diagrams in Example 23.5.
For $d_o < f$, the image will be
virtual, upright, and enlarged
(Fig. 23.15b).

$d_o = 2f$ for a converging lens has a significance similar to that of $d_o = R = 2f$ for
a converging mirror (▲ Fig. 23.16).

The ray diagram for a diverging lens will be discussed in Example 23.8. Like
diverging mirrors, diverging lenses can form only virtual images.

The image distances and characteristics for a lens can also be found analytically. The equations for thin lenses are identical to those for spherical mirrors. (See
Exercise 68.) The **thin-lens equation** is

$$\frac{1}{d_o} + \frac{1}{d_i} = \frac{1}{f} \qquad \text{or} \qquad d_i = \frac{d_o f}{d_o - f} \qquad \text{thin-lens equation} \quad (23.5)$$

The **magnification factor**, like that for spherical mirrors, is given by

$$M = -\frac{d_i}{d_o} \qquad (23.6)$$

(A thin lens is a lens for which the thickness of the lens is assumed to be negligible
compared with the lens's focal length.) The sign conventions for these thin-lens
equations are given in Table 23.3.

Just as when working with mirrors, it is helpful to sketch a ray diagram before
working a lens problem analytically.

PHYSLET® ILLUSTRATION

Convex Lens

TABLE 23.3 Sign Conventions for Thin Lenses

Focal length
Converging lens (sometimes called a *positive* lens): f is positive
Diverging lens (sometimes called a *negative* lens): f is negative
Object distance (d_o)
d_o is positive when the object is in front of the lens (real object)
d_o is negative when the object is behind the lens (virtual object)*
Image distance d_i and **Image type**
d_i is positive when the image is formed on the image side of the lens—opposite to the
object (real image)
d_i is negative when the image is formed on the object side of the lens—same side as the
object (virtual image)
Image orientation M
M is positive when the image is upright with respect to the object
M is negative when the image is inverted with respect to the object

---

*In a combination of two (or more) lenses, the image formed by the first lens is taken as the ob-
ject of the second lens (and so on). If this image–object falls behind the second lens, it is referred to as
a virtual object, and the object distance is taken to be negative ($-$).

Example 23.6 ■ Three Images: Behavior of a Converging Lens

A biconvex lens has a focal length of 12 cm. For an object (a) 60 cm, (b) 15 cm, and (c) 8 cm from the lens, where is the image formed, and what are its characteristics?

Thinking It Through. With the focal length (f) and the object distances (d_o), we can apply Eq. 23.5 to find the image distances (d_i) and Eq. 23.6 to determine the image characteristics. Sketch ray diagrams for all these cases to get an idea of the image characteristics.

Solution.

Given: $f = 12$ cm   *Find:* d_i and the image characteristics for all three cases
 (a) $d_o = 60$ cm
 (b) $d_o = 15$ cm
 (c) $d_o = 8$ cm
 (All three distances are taken to be exact for simplicity.)

(a) The object distance is greater than twice the focal length ($d_o > 2f$). Using Eq. 23.5,

$$\frac{1}{60 \text{ cm}} + \frac{1}{d_i} = \frac{1}{12 \text{ cm}} \quad \text{or} \quad \frac{1}{d_i} = \frac{1}{12 \text{ cm}} - \frac{1}{60 \text{ cm}} = \frac{5}{60 \text{ cm}} - \frac{1}{60 \text{ cm}} = \frac{4}{60 \text{ cm}} = \frac{1}{15 \text{ cm}}$$

Then

$$d_i = +15 \text{ cm} \quad \text{and}$$

$$M = -\frac{d_i}{d_o} = -\frac{15 \text{ cm}}{60 \text{ cm}} = -\frac{1}{4}$$

The image is real (positive d_i), inverted (negative M), and one-fourth the object's size ($|M| = \frac{1}{4}$). A camera uses this arrangement when the object distance is greater than $2f (d_o > 2f)$.

(b) Here, $2f > d_o > f$. Using Eq. 23.5, we have

$$\frac{1}{d_i} = \frac{1}{12 \text{ cm}} - \frac{1}{15 \text{ cm}} = \frac{5}{60 \text{ cm}} - \frac{4}{60 \text{ cm}} = \frac{1}{60 \text{ cm}}$$

Then

$$d_i = +60 \text{ cm} \quad \text{and}$$

$$M = -\frac{d_i}{d_o} = -\frac{60 \text{ cm}}{15 \text{ cm}} = -4$$

The image is real (positive d_i), inverted (negative M), and four times the object 's size ($|M| = 4$). This situation applies to the overhead projector and slide projector ($2f > d_o > f$).

(c) For this case, $d_o < f$. Using Eq. 23.5, we have

$$\frac{1}{d_i} = \frac{1}{12 \text{ cm}} - \frac{1}{8 \text{ cm}} = \frac{4}{48 \text{ cm}} - \frac{6}{48 \text{ cm}} = -\frac{2}{48 \text{ cm}} = -\frac{1}{24 \text{ cm}}$$

Then

$$d_i = -24 \text{ cm} \quad \text{and}$$

$$M = -\frac{d_i}{d_o} = -\frac{(-24 \text{ cm})}{8 \text{ cm}} = +3$$

The image is virtual (negative d_i), upright (positive M), and three times the object's size ($|M| = 3$). This situation is an example of a simple microscope or magnifying glass ($d_o < f$).

As you can see, a converging lens is versatile. Depending on the object distance (relative to the focal length), the lens can be used as a camera, projector, or magnifying glass.

Follow-up Exercise. If the object distance of a convex lens is allowed to vary, at what object distance does the real image change from being reduced to being magnified?

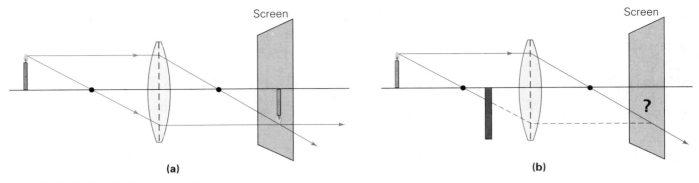

(a)

(b)

▲ **FIGURE 23.17 Half a lens, half an image?** (a) A converging lens forms an image on a screen. (b) The lower half of the lens is blocked. What happens to the image? See Conceptual Example 23.7.

Conceptual Example 23.7 ■ Half an Image?

A converging lens forms an image on a screen, as shown in ▲Fig.23.17a. Then the lower half of the lens is blocked, as shown in Fig. 23.17b. As a result, (a) only the top half of the original image will be visible on the screen; (b) only the bottom half of the original image will be visible on the screen; or (c) the entire image will be visible.

Reasoning and Answer. At first thought, you might imagine that blocking off half of the lens would eliminate half of the image. However, rays from *every* point on the object pass through *all parts* of the lens. Thus, the upper half of the lens can form a total image (as could the lower half), so the answer is (c).

You might confirm this conclusion by drawing a chief ray in Fig. 23.17b. Or, you might use the scientific method and experiment—particularly if you wear eyeglasses. Block off the bottom part of your glasses, and you will find that you can still read through the top part (unless you wear bifocals).

Follow-up Exercise. Can you think of any property of the image that *would* be affected by blocking off half of the lens? Explain.

Example 23.8 ■ Time for a Change: Behavior of a Diverging Lens

An object is 24 cm in front of a diverging lens that has a focal length of -15 cm. Find the location and characteristics of the image through (a) a ray diagram (b) calculations with the thin-lens equations.

Thinking It Through. See the sign conventions in Table 23.3.

Solution.

Given: $d_o = 24$ cm
$f = -15$ cm (diverging lens)

Find: (a) Image location and characteristics through a ray diagram
(b) d_i, M, and image characteristics (mathematically)

(a) We once again use a scale of 1 cm (in our drawing) to represent 10 cm. That way, the object will be 2.4 cm in front of the lens in our drawing. We draw the optic axis, the lens, the object (in this case, a lighted candle), the focal point (F), and a vertical dashed line through the center of the lens.

The parallel ray (1) starts from the tip of the flame, travels parallel to the optic axis, diverges from the lens after refraction, and appears to diverge from the F on the object side. The chief ray (2) originates from the tip of the flame and goes through the center of the lens, with no direction change. It is clearly seen that these two rays, after refraction, diverge and do not intersect. However, they appear to come from in front of the lens (object side), and that apparent intersection is the image point of the tip of the flame (◄Fig. 23.18). We can also draw the focal ray (3) to verify that these rays appear to come from the same image point. The focal ray appears to go through the focal point on the image side and travels parallel to the optic axis after refraction from the lens.

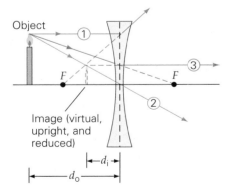

▲ **FIGURE 23.18 Diverging lens** Ray diagram of a diverging lens. Here, the image is virtual and in front of the lens, upright, and smaller than the object.

This image is virtual (why?), upright, and smaller than the object. Measuring from the diagram (keeping in mind the drawing scale we are using), we find that $d_i \approx 9$ cm and the magnification $\approx \frac{5}{14} = 0.4$.

(b) Note that the focal length is negative for a diverging lens. (See Table 23.3.) From Eq. 23.5,

$$\frac{1}{24 \text{ cm}} + \frac{1}{d_i} = \frac{1}{-15 \text{ cm}} \quad \text{or} \quad \frac{1}{d_i} = \frac{1}{-15 \text{ cm}} - \frac{1}{24 \text{ cm}} = -\frac{13}{120 \text{ cm}}$$

so

$$d_i = -9.2 \text{ cm}$$

Then

$$M = -\frac{d_i}{d_o} = -\frac{(-9.2 \text{ cm})}{24 \text{ cm}} = +0.38$$

Thus, the image is virtual (negative d_i) and upright (positive M), and it is 0.38 times the size (height) of the object. Since f is negative, the image of a real object is always virtual for a diverging lens. (Why?)

Follow-up Exercise.   A diverging lens always forms a virtual image of a real object. What general statements can be made about the image's orientation and magnification?

A special type of lens that you may have encountered is discussed in the Insight on p. 768.

Combinations of Lenses

Many optical instruments, such as microscopes and telescopes (Chapter 25), use a combination of lenses, or a compound-lens system. When two or more lenses are used in combination, we can determine the overall image produced by considering the lenses individually in sequence. That is, the image formed by the first lens becomes the object for the second lens, and so on. For this reason, we present the principles of lens combinations before considering the specifics of their real-life applications.

If the first lens produces an image in front of the second lens, that image is treated as a real object for the second lens (▼Fig. 23.19a). If, however, the lenses are

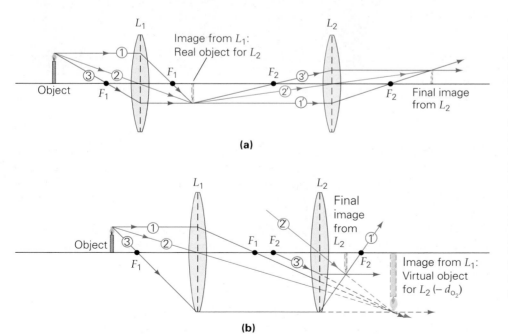

(a)

(b)

◀ **FIGURE 23.19 Lens combinations** The final image produced by a compound-lens system can be found by treating the image of one lens as the object for the adjacent lens. **(a)** If the image of the first lens (L_1) is formed in front of the second lens (L_2), the object for the second lens is said to be real. (Note that rays 1′, 2′, and 3′ are the parallel, chief, and focal rays, respectively, for L_2. They are *not* continuations of rays 1, 2, and 3, the parallel, chief, and focal rays, respectively, for L_1.) **(b)** If the rays pass through the second lens before the image is formed, the object for the second lens is said to be virtual, and the object distance for the second lens is taken to be negative.

Fresnel Lenses

To focus parallel light or to produce a large beam of parallel light rays, a sizable converging lens is necessary. The large mass of glass necessary to form such a lens is bulky and heavy; moreover, the thick lens absorbs some of the light and is likely to show distortions. A French physicist named Augustin Fresnel (Fre-nel'; 1788–1827) developed a solution to this problem for lenses used in lighthouses.

Fresnel recognized that the refraction of light takes place at the surfaces of a lens. Hence, a lens could be made thinner—even flat—by removing glass from the interior, as long as the refracting properties of the surfaces were not changed.

This can be accomplished by cutting a series of concentric grooves in the surface of the lens (Fig. 1a). Note that the surface of each remaining curved segment is nearly parallel to the corresponding surface of the original lens. Together,

the concentric segments refract light as does the original biconvex lens (Fig. 1b). In effect, the lens has simply been slimmed down by the removal of unnecessary glass between the refracting surfaces.

A lens with such a series of concentric curved surfaces is called a *Fresnel lens*. Such lenses are widely used in overhead projectors and in beacons (Fig. 1c). A Fresnel lens is very thin and therefore much lighter than a conventional biconvex lens with the same optical properties. Also, Fresnel lenses are easily molded from plastic—often with one flat side (plano-convex) so that the lens can be attached to a glass surface.

One disadvantage of Fresnel lenses is that concentric circles are visible when an observer is looking through such a lens and when an image produced by one is projected on a screen, as when we use an overhead projector.

Related Exercises: 56 and 64

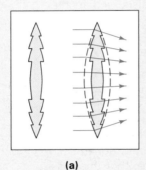

(a)

(b)

(c)

FIGURE 1 Fresnel lens   (a) The focusing action of a lens comes from refraction at its surfaces. It is therefore possible to reduce the thickness of a lens by cutting away glass in concentric grooves, leaving a set of curved surfaces with the same refractive properties as the lens from which they were derived. **(b)** A flat Fresnel lens with concentric curved surfaces magnifies like a biconvex converging lens. **(c)** An array of Fresnel lenses produces focused beams in this Boston Harbor light. (Fresnel lenses were, in fact, developed for use in lighthouses.)

close enough so that the image from the first lens is not formed before the rays pass through the second lens (Fig. 23.19b), then a modification must be made in the sign conventions. In this case, the image from the first lens is treated as a *virtual* object for the second lens, and the object distance for it is taken to be *negative* in the lens equation (Table 23.3).

The total magnification (M_{total}) of a compound-lens system is the product of the individual magnification factors of the component lenses. For example, for a two-lens system, as in Fig. 23.19,

$$M_{\text{total}} = M_1 M_2 \qquad (23.7)$$

The conventional signs for M_1 and M_2 carry through the product to indicate, from the sign of M_{total}, whether the final image is upright or inverted. (See Exercise 76.)

Example 23.9 ■ A Special Offer: A Lens Combo and a Virtual Object

Consider two lenses similar to those illustrated in Fig. 23.19b. Suppose the object is 20 cm in front of lens L_1, which has a focal length of 15 cm. Lens L_2, with a focal length of 12 cm, is 26 cm from L_1. What is the location of the final image, and what are its characteristics?

Thinking It Through. This is a double application of the thin-lens equation. The lenses are treated successively. The image of lens L_1 becomes the object of lens L_2. We must keep the quantities distinctly labeled and the distances appropriately referenced.

Solution. We have

Given: $d_{o_1} = 20$ cm   *Find:* d_{i_2} and image characteristics
$f_1 = 15$ cm
$f_2 = 12$ cm
$D = 26$ cm (distance between lenses)

The first step is to apply the thin-lens equation (Eq. 23.5) and the magnification factor for thin lenses (Eq. 23.6) to L_1:

$$\frac{1}{d_{i_1}} = \frac{1}{f_1} - \frac{1}{d_{o_1}} = \frac{1}{15 \text{ cm}} - \frac{1}{20 \text{ cm}} = \frac{4}{60 \text{ cm}} - \frac{3}{60 \text{ cm}} = \frac{1}{60 \text{ cm}}$$

or

$$d_{i_1} = 60 \text{ cm (real image)}$$

and

$$M_1 = -\frac{d_{i_1}}{d_{o_1}} = -\frac{60 \text{ cm}}{20 \text{ cm}} = -3.0 \text{ (inverted and magnified)}$$

The image from lens L_1 becomes the object for lens L_2. This image is then $d_{i_1} - D = 60$ cm $- 26$ cm $= 34$ cm on the right, or image, side of L_2. Therefore, it is a virtual object (see Table 23.3), and $d_{o_2} = -34$ cm. (The d_o for virtual objects is taken to be negative.)
Then we apply the equations to the second lens, L_2:

$$\frac{1}{d_{i_1}} = \frac{1}{f_2} - \frac{1}{d_{o_1}} = \frac{1}{12 \text{ cm}} - \frac{1}{(-34 \text{ cm})} = \frac{23}{204 \text{ cm}}$$

or

$$d_{i_2} = 8.9 \text{ cm (real image)}$$

and

$$M_2 = -\frac{d_{i_1}}{d_{o_2}} = -\frac{8.9 \text{ cm}}{(-34 \text{ cm})} = 0.26 \text{ (upright and reduced)}$$

(*Note*: The virtual object for L_2 was inverted, and thus the term "upright" means that the final image is also inverted.) The total magnification M_{total} is then

$$M_{\text{total}} = M_1 M_2 = (-3.0)(0.26) = -0.78$$

The sign is carried through with the magnifications. We determine that the final real image is located 8.9 cm on the right (image) side of L_2 and that it is inverted (negative sign) relative to the initial object and reduced.

Follow-up Exercise. Suppose the object in Fig. 23.19b were located 30 cm in front of L_1. Where would the final image be formed in this case, and what would be its characteristics?

23.4 Lens Aberrations

OBJECTIVES: To (a) describe some common lens aberrations and (b) explain how they can be reduced or corrected.

Lenses, like mirrors, can also have aberrations. We now discuss some common aberrations.

Spherical Aberration

The discussion of lenses thus far has concentrated on rays that are near the optic axis. Like spherical mirrors, however, converging lenses may show **spherical aberration**, an effect that occurs when parallel rays passing through different regions of a lens do not come together on a common focal plane. In general, rays close to the axis of a converging lens are refracted less and come together at a point farther from the lens than do rays passing through the periphery of the lens (▼Fig. 23.20a).

Spherical aberration can be minimized by using an aperture to reduce the effective area of the lens, so that only light rays near the axis are transmitted. Also, combinations of converging and diverging lenses can be used, as the aberration of one lens can be compensated for by the optical properties of another lens.

Chromatic Aberration

Chromatic aberration is an effect that occurs because the index of refraction of the material making up a lens is not the same for all wavelengths of light (that is, the material is dispersive). When white light is incident on a lens, the transmitted rays of different wavelengths (colors) do not have a common focal point, and images of different colors are produced at different locations (Fig. 23.20b).

This dispersive aberration can be minimized, but not eliminated, by using a compound-lens system consisting of lenses of different materials, such as crown glass and flint glass. The lenses are chosen so that the dispersion produced by one is compensated for by opposite dispersion produced by the other. With a properly constructed two-component lens system, called an *achromatic doublet* (*achromatic* means "without color"), the images of any two selected colors can be made to coincide.

▼ **FIGURE 23.20 Lens aberrations** (a) Spherical aberration. In general, rays closer to the axis of a lens are refracted less and come together at a point farther from the lens than do rays passing through the periphery of the lens. **(b)** Chromatic aberration. Because of dispersion, different wavelengths (colors) of light are focused in different planes, which results in distortion of the overall image.

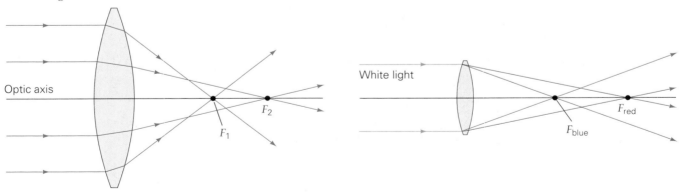

(a) Spherical aberration

(b) Chromatic aberration

Astigmatism

A circular beam of light along the lens axis forms a circular illuminated area on the lens. When incident on a converging lens, the parallel beam converges at the focal point. However, when a circular cone of light from an off-axis source falls on the convex spherical surface of a lens some distance away, the light forms an elliptical illuminated area on the lens. The rays entering along the major and minor axes of the ellipse then focus at different points after passing through the lens. This condition is called **astigmatism**.

With different focal points in different planes, the images in both planes are blurred. For example, the image of a point is no longer a point, but rather two separated short-line images (blurred points). Astigmatism can be reduced by decreasing the effective area of the lens with an aperture or by adding a cylindrical lens to compensate.

*23.5 The Lens Maker's Equation

OBJECTIVES: To (a) describe the lens maker's equation, (b) explain how its application differs from that of the thin-lens equation, and (c) understand lens power in diopters.

The biconvex and biconcave thin lenses considered so far in this chapter have been relatively easy to analyze. However, there is a variety of other shapes of lenses, as illustrated in Fig. 23.14. For them, the analysis becomes more involved, but it is important to know the focal lengths of such lenses for optical considerations, because lenses are ground for specific purposes or applications.

Lens refraction depends on the shapes of the lens's surfaces and on the index of refraction of the lens. These properties are related to the focal length of a thin lens by the **lens maker's equation**, which gives the focal length of a thin lens in air ($n_{air} = 1$) as

$$\frac{1}{f} = (n - 1)\left(\frac{1}{R_1} - \frac{1}{R_2}\right) \quad \begin{array}{l}(\textit{for thin lens}\\ \textit{in air})\end{array} \quad (23.8)$$

where n is the index of refraction of the lens material and R_1 and R_2 are the radii of curvature of the first (front side) and second (back side) lens surfaces, respectively. (The first surface is the one on which light from an object is first incident.)

A sign convention is required for the lens maker's equation, and a common one is summarized in Table 23.4. The signs depend on the location of a surface's center of curvature (C) relative to the side of the lens on which light is incident or to the side from which light emerges (▶Fig. 23.21).

If the lens is surrounded by a medium other than air, then the first term in parentheses in Eq. 23.8 becomes $(n/n_m) - 1$, where n and n_m are the indices of refraction of the lens material and the surrounding medium, respectively. Now we can see why some converging lenses in air become diverging when submerged in water: If $n_m > n$, then f is negative, and the lens is diverging.

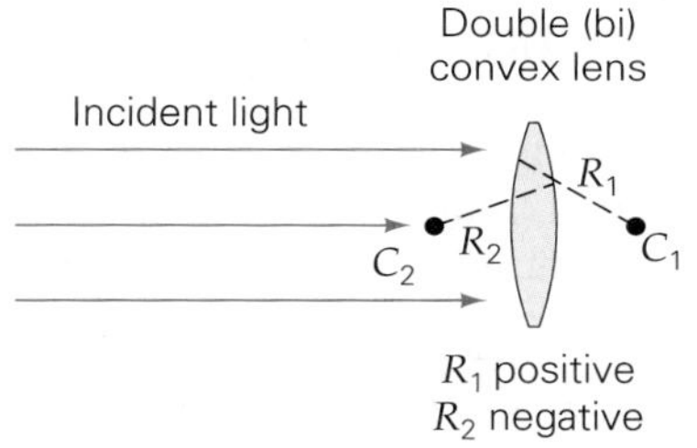

▲ **FIGURE 23.21 Centers of curvature** Lenses, such as this biconvex lens, have two centers of curvature, which define the signs of the radii of curvature. See Table 23.4.

TABLE 23.4 Sign Conventions for Lens Maker's Equation

R is positive when C is on the side of the lens from which light emerges, or on the back side of the lens

R is negative when C is on the side of the lens on which light is incident, or on the front side of the lens

$R = \infty$ for a plane (flat) surface

f is positive for a converging (positive) lens

f is negative for a diverging (negative) lens

Lens Power: Diopters

Notice that the lens maker's equation (Eq. 23.8) gives the inverse focal length $1/f$. Optometrists use this inverse relationship to express the *lens power* (P) of a lens in units called **diopters** (abbreviated as "D"). The lens power is the reciprocal of the focal length of the lens expressed in *meters*:

$$P(\text{expressed in diopters}) = \frac{1}{f(\text{expressed in meters})} \qquad (23.9)$$

So, $1\,D = 1\,m^{-1}$. The lens maker's equation gives a lens's power ($1/f$) in diopters if the radii of curvature are expressed in meters.

If you wear glasses, you may have noticed that the prescription the optometrist gave you for your eyeglass lenses was written in terms of diopters. Converging and diverging lenses are referred to as positive ($+$) and negative ($-$) lenses, respectively. Thus, if an optometrist prescribes a corrective lens with a power of $+2$ diopters, it is a converging lens with a focal length of

$$f = \frac{1}{P} = \frac{1}{+2\,D} = \frac{1}{2\,m^{-1}} = 0.50\,m = +50\,cm$$

The greater the power of the lens in diopters, the shorter its focal length is and the more strongly converging or diverging it is. Thus, a "stronger" prescription lens (greater lens power) has a smaller f than does a "weaker" prescription lens (lesser lens power).

Chapter Review

Important Concepts and Equations

- **Plane mirrors** form virtual, upright, and unmagnified images. The object distance is equal to the image distance ($d_o = d_i$).

- The **lateral magnification factor** for all mirrors and lenses is

$$M = -\frac{d_i}{d_o} \qquad (23.4, 23.6)$$

- **Spherical mirrors** are either concave (converging) or convex (diverging). Diverging spherical mirrors always form upright, reduced, virtual images.

The **focal length of a spherical mirror**:

$$f = \frac{R}{2} \qquad (23.2)$$

Spherical-mirror equation:

$$\frac{1}{d_o} + \frac{1}{d_i} = \frac{1}{f} = \frac{2}{R} \qquad (23.3)$$

- Bispherical lenses are either convex (converging) or concave (diverging). Diverging spherical lenses always form upright, reduced, virtual images.

- The **thin-lens equation** relates focal length, object distance, and image distance:

$$\frac{1}{d_o} + \frac{1}{d_i} = \frac{1}{f} \qquad (23.5)$$

- *The **lens maker's equation** is used to compute the grinding radii for the desired focal length of a lens:

$$\frac{1}{f} = (n - 1)\left(\frac{1}{R_1} - \frac{1}{R_2}\right) \quad \begin{array}{l}(\textit{thin lens in}\\ \textit{air only})\end{array} \qquad (23.8)$$

- ***Lens power in diopters** (where f is in meters) is given by

$$P = \frac{1}{f} \qquad (23.9)$$

*This definition incorporates the thin-lens approximation (negligible lateral deviation).

Exercises

23.1 Plane Mirrors

1. A plane mirror (a) has a greater image distance than object distance; (b) produces a virtual, upright, unmagnified image; (c) changes the vertical orientation of an object; or (d) reverses an object's top and bottom.

2. A plane mirror (a) can be used to magnify, (b) produces both real and virtual images, (c) always produces a virtual image, or (d) forms images by diffuse reflection.

3. **CQ** (a) A transparent windowpane can serve as a mirror, but usually only when it is dusk or dark outside. Why? (b) When looking at a reflecting windowpane at night, you may see two similar images. Why? (c) One-way, half-silvered mirrors reflect on one side and can be seen through on the other. (This effect is sometimes used on sunglasses as well.) What is the principle used in making a one-way mirror? [*Hint*: At night, a windowpane may be a one-way mirror.]

4. **CQ** Day–night rearview mirrors are common in cars. At night, you tilt the mirror backward, and the intensity and glare of headlights behind you are reduced (▼Fig. 23.22). The mirror is wedge shaped and is silvered on the back. The effect has to do with front-surface and back-surface reflections. The unsilvered front surface reflects about 5% of incident light; the silvered back surface reflects about 90% of the incident light. Explain how the day–night mirror works.

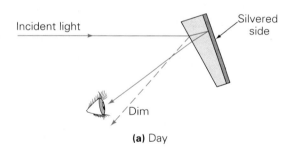

(a) Day

(b) Night

▲ **FIGURE 23.22 Automobile day–night mirror** See Exercise 4.

5. **CQ** When you stand in front of a plane mirror, there is a right–left reversal. (a) Why is there not a top–bottom reversal of your body? (b) Could you effect an apparent top–bottom reversal by positioning your body differently?

6. **CQ** What is the magnification of a plane mirror? What does it signify?

7. **CQ** Why do some emergency vehicles have AMBULANCE (▼Fig. 23.23) printed on the front?

▲ **FIGURE 23.23 Backward and reversed** See Exercise 7.

8. **CQ** Can a virtual image be projected onto a screen? Why or why not?

9. **CQ** What is the focal length of a plane mirror? Why?

10. ■ A person stands 2.0 m away from the reflecting surface of a plane mirror. (a) What is the apparent distance between the person and his or her image? (b) What are the image characteristics?

11. ■ An object 5.0 cm tall is placed 40 cm from a plane mirror. Find (a) the distance from the object to the image, (b) the height of the image, and (c) the image's magnification.

12. ■ Standing 2.5 m in front of a plane mirror with your camera, you decide to take a picture of yourself. To what distance should the camera be manually focused so that you will get a complete image?

13. ■■ A small dog sits at a distance of 1.5 m in front of a plane mirror. (a) Where is the dog's image? (b) If the dog jumps at the mirror with a speed of 0.50 m/s, how fast does the dog approach its image?

14. ■■ A woman fixing the hair on the back of her head holds a plane mirror 30 cm in front of her face so as to look into a plane mirror on the bathroom wall behind her. If she is 90 cm from the wall mirror, approximately how far does the image of the back of her head appear in front of her?

15. **IE** ■■ (a) When you stand between two plane mirrors on opposite walls in a dance studio, you observe (1) one, (2) two, or (3) multiple images. Why? (b) If you stand 3.0 m from the mirror on the north wall and 5.0 m from the mirror on the south wall, what are the image distances for the first two images in both mirrors?

16. ■■ A woman 1.70 m tall stands 3.0 m in front of a plane mirror. (a) What is the minimum height the mirror must be to allow the woman to view her complete image from head to foot? Assume that her eyes are 10 cm below the top of her head. (b) What would be the required minimum height of the mirror if she were to stand 5.0 m away?

17. ■■ Prove that $d_o = d_i$ for a plane mirror. [*Hint*: Refer to Fig. 23.2, and use similar and identical triangles.]

18. CQ ■■■ Draw ray diagrams that show how three images of an object are formed in two plane mirrors at right angles as shown in ▼Fig. 23.24a. [*Hint*: Consider rays from both ends of the object in the drawing for each image.] Figure 23.24b shows a similar situation from a different point of view that gives four images. Explain the extra image in this case.

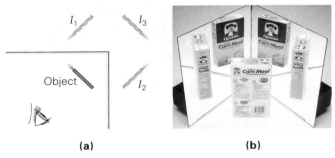

(a)   (b)

▲ FIGURE 23.24 Two mirrors—multiple images
See Exercise 18.

23.2 Spherical Mirrors

19. Which one of the following statements concerning spherical mirrors is correct? (a) A converging mirror alone can produce an inverted virtual image. (b) A diverging mirror alone can produce an inverted virtual image. (c) A diverging mirror can produce an inverted real image. (d) A converging mirror can produce an inverted real image.

20. The image produced by a convex mirror is always (a) virtual and upright, (b) real and upright, (c) virtual and inverted, or (d) real and inverted.

21. CQ (a) What is the purpose of using a dual mirror on a car or truck, such as the one shown in ▼Fig. 23.25? (b) Some

▲ FIGURE 23.25 Mirror applications See Exercise 21.

rearview mirrors on the passenger side of automobiles have the warning "OBJECTS IN MIRROR ARE CLOSER THAN THEY APPEAR." Explain why this is so. (c) Could a TV satellite dish be considered a converging mirror? Explain.

22. CQ (a) If you look into a shiny spoon, you see an inverted image on one side and an upright image on the other (▼Fig. 23.26). (Try it.) Why? (b) Could you see upright images on both sides ? Explain.

Outside   Inside
Bowl   Bowl

▲ FIGURE 23.26 Reflections from concave and convex surfaces See Exercise 22.

23. CQ (a) A 10-cm-tall mirror bears the following advertisement: "Full-view mini mirror. See your full body in 10 cm." How can this be? (b) A popular novelty item consists of a concave mirror with a ball suspended at or slightly inside the center of curvature (▼Fig. 23.27). When the ball swings toward the mirror, its image grows larger and suddenly fills the whole mirror. The image appears to be jumping out of the mirror. Explain what is happening.

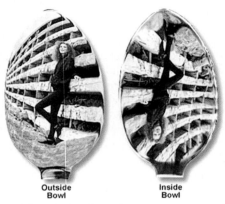

◄ FIGURE 23.27
Spherical-mirror toy
See Exercise 23.

24. CQ How can the focal length be determined experimentally for a concave mirror? Can you do the same thing for a convex mirror?

25. CQ Can a convex mirror produce an image that is taller than the object? Why or why not?

26. CQ Should a shaving/makeup mirror be concave or convex? Why?

27. ■ A spherical mirror has a radius of curvature of 10 cm. What is the focal length of this mirror?

28. ■ An object 3.0 cm tall is placed 20 cm from the front of a concave mirror with a radius of curvature of 30 cm. Where is the image formed, and how tall is it?

29. ■ If the object in Exercise 28 is moved to a position 10 cm from the front of the mirror, what will be the characteristics of the image?

30. IE ■ An object is 30 cm in front of a convex mirror that has a focal length of 60 cm. (a) Use a ray diagram to determine if the image is (1) real or virtual, (2) upright or inverted, and (3) magnified or smaller than the object. (b) Calculate the image distance and image height.

31. ■ A candle with a flame 1.5 cm tall is placed 5.0 cm from the front of a concave mirror. A virtual image is produced that is 10 cm from the vertex of the mirror. (a) Find the focal length and radius of curvature of the mirror. (b) How tall is the image of the flame?

32. ■■ An object 3.0 cm tall is placed at different locations in front of a concave mirror whose radius of curvature is 30 cm. Determine the location of the image and its characteristics when the object distance is 40 cm, 30 cm, 15 cm, and 5.0 cm, using (a) a ray diagram and (b) the mirror equation.

33. ■■ Use the mirror equation and the magnification factor to show that when $d_o = R = 2f$ for a concave mirror, the image is real, inverted, and the same size as the object.

34. ■■ Using the spherical-mirror equation and the magnification factor, show that for a concave mirror with $d_o < f$, the image of an object is always virtual, upright, and magnified.

35. ■■ Using the spherical-mirror equation and the magnification factor, show that for a convex mirror, the image of an object is always virtual, upright, and reduced.

36. IE ■■ The erect image of an object 18 cm in front of a mirror is half the size of the object. (a) The mirror is (1) convex, (2) concave, or (3) plane. Why? (b) What is the focal length of the mirror?

37. ■■ A concave shaving mirror is constructed so that a man at a distance of 20 cm from the mirror sees his image magnified 1.5 times. What is the radius of curvature of the mirror?

38. ■■ A concave mirror has a magnification of +3.0 for an object placed 50 cm in front of it. (a) What type of image is produced? (b) Find the radius of curvature of the mirror.

39. ■■ A child looks at a reflective Christmas treeball ornament that has a diameter of 9.0 cm and sees an image of her face that is half the real size. How far is the child's face from the ball?

40. IE ■■ A dentist uses a spherical mirror that produces an upright image of a tooth which is magnified four times. (a) The mirror is (1) converging, (2) diverging, or (3) flat. Why? (b) What is the mirror's focal length in terms of the object distance?

41. ■■ A 15-cm-long pencil is placed with its eraser on the optic axis of a concave mirror and its point directed upward at a distance of 20 cm in front of the mirror. The radius of curvature of the mirror is 30 cm. Use (a) a ray diagram and (b) the mirror equation to locate the image and determine the image characteristics.

42. IE ■■ A pill bottle 3.0 cm tall is placed 12 cm in front of a mirror. A 9.0-cm-tall upright image is formed. (a) The mirror is (1) convex, (2) concave, or (3) flat. Why? (b) What is its radius of curvature?

43. ■■ A spherical mirror at an amusement park shows anyone who stands 2.5 m in front of it an upright image two times the person's height. What is the mirror's radius of curvature?

44. ■■ For values of d_o from 0 to ∞, (a) sketch graphs of (1) d_i versus d_o and (2) M versus d_o for a converging mirror, and (b) sketch similar graphs for a diverging mirror.

45. ■■■ The front surface of a wooden cube 5.0 cm on each side is placed a distance of 30 cm in front of a converging mirror that has a focal length of 20 cm. Where is the image of the front surface of the cube located, and what are the image characteristics?

46. ■■■ A section of a sphere is mirrored on both sides. If the magnification of an object is +1.8 when the section is used as a concave mirror, what is the magnification of an object at the same distance in front of the convex side?

47. ■■■ A concave mirror has a radius of curvature of 20 cm. For what two object distances will the image have twice the height of the object?

48. ■■■ A convex mirror is on the exterior of the passenger side of many trucks. If the focal length of such a mirror is −40.0 cm, what will be the location and height of the image of a car that is 2.0 m tall and (a) 100 m behind the mirror and (b) 10.0 m in front of the mirror? (See Exercise 21a.)

49. ■■■ Two students in a physics laboratory each have a concave mirror with the same radius of curvature, 40 cm. Each student places an object in front of her mirror. The image in both mirrors is three times the size of the object. However, when the students compare notes, they find that the object distances are not the same. Is this possible? If so, what are the object distances?

23.3 Lenses
and
23.4 Lens Aberrations

50. A diverging lens (a) must have at least one concave surface, (b) always produces a virtual image, (c) is thinner at its center than at the periphery, or (d) all of the preceding.

51. The image produced by a diverging lens is always (a) virtual and magnified, (b) real and magnified, (c) virtual and reduced, or (d) real and reduced.

52. A lens aberration that is caused by dispersion is called (a) spherical aberration, (b) chromatic aberration, (c) refractive aberration, or (d) none of the preceding.

53. CQ Explain why a fish in a spherical fish bowl, viewed from the side, appears larger than it really is.

54. CQ Can a converging lens ever form a virtual image of a real object? If yes, under what conditions?

55. CQ How can you quickly determine the focal length of a converging lens? Will the same method work for a diverging lens?

56. CQ If you want to use a converging lens to design a simple overhead projector so as to project the magnified image of some small writing onto a screen on a wall, how far should you place the object in front of the lens?

57. ■ An object is placed 50.0 cm in front of a converging lens of focal length 10.0 cm. What are the image distance and the lateral magnification?

58. ■ An object placed 30 cm in front of a converging lens forms an image 15 cm behind the lens. What is the focal length of the lens?

59. ■ A converging lens with a focal length of 20 cm is used to produce an image on a screen that is 2.0 m from the lens. What is the object distance?

60. IE ■■ An object 4.0 cm tall is in front of a converging lens of focal length 22 cm. The object is 15 cm away from the lens. (a) Use a ray diagram to determine if the image is (1) real or virtual, (2) upright or inverted, and (3) magnified or smaller than the object. (b) Calculate the image distance and lateral magnification.

61. ■■ An object is placed in front of a biconcave lens whose focal length is 18 cm. Where is the image located, and what are its characteristics, if the object distance is (a) 10 cm and (b) 25 cm? Sketch ray diagrams for each case.

62. ■■ Using the thin-lens equation and the magnification factor, show that for a spherical diverging lens, the image of a real object is always virtual, upright, and reduced.

63. ■■ A biconvex lens has a focal length of 0.12 m. Where on the lens axis should an object be placed in order to get

(a) a real, enlarged image with a magnification of 2.0 and (b) a virtual, enlarged image with a magnification of 2.0?

64. ■■ (a) Design the lens in a single-lens slide projector that will form a sharp image on a screen 4.0 m away with the transparent slides 6.0 cm from the lens. (b) If the object on a slide is 1.0 cm tall, how tall will the image on the screen be, and how should the slide be placed in the projector?

65. ■■ A simple single-lens camera (biconvex lens) is used to photograph a man 1.7 m tall who is standing 4.0 m from the camera. If the man's image fills the length of a frame of film (35 mm), what is the focal length of the lens?

66. ■■ An object 5.0 cm tall is 10 cm from a concave lens. The resulting image is one-fifth as large as the object. What is the focal length of the lens?

67. ■■ An object is placed 40 cm from a screen. (a) At what point between the object and the screen should a converging lens with a focal length of 10 cm be placed so that it will produce a sharp image on the screen? (b) What is the lens's magnification?

68. ■■ Using ▼Fig. 23.28, derive (a) the thin-lens equation and (b) the magnification equation for a thin lens. [*Hint*: Use similar triangles.]

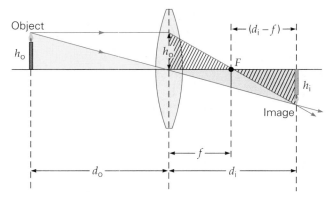

▲ **FIGURE 23.28 The thin-lens equation** The geometry for deriving the thin-lens equation (and magnification factor). Note the two sets of similar triangles. See Exercise 68.

69. ■■ (a) For a biconvex lens, what is the *minimum* distance between an object and its image if the image is real? (b) What is the distance if the image is virtual?

70. ■■ With a magnifying glass, a biology student on a field trip views a small insect. If she sees the insect magnified by a factor of 3.5 when the glass is held 3.0 cm from it, what is the focal length of the lens?

71. ■■ (a) If a book is held 30 cm from an eyeglass lens with a focal length of −45 cm, where is the image of the print formed? (b) If an eyeglass lens with a focal length of +57 cm is used, where is the image formed?

72. ■■ For the arrangement shown in ▼Fig. 23.29, an object is placed 0.40 m in front of the converging lens, which has a focal length of 0.15 m. If the concave mirror has a focal length of 0.13 m, where is the final image formed, and what are its characteristics?

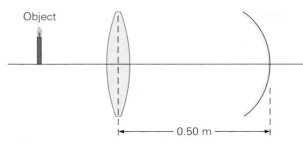

▲ FIGURE 23.29 Lens—mirror combination
See Exercise 72.

73. ■■ A simple camera uses a single converging lens with a focal length of 4.5 cm. A picture is taken of a 26-cm-tall physics book standing on a table at a distance of 1.5 m from the camera. If a sharp image is formed on the film, how far is the film from the lens? (b) What is the height and orientation of the image of the book on the film?

74. ■■■ The geometry of a compound microscope, which consists of two converging lenses, is shown in ▼Fig. 23.30. (More detail on microscopes is given in Chapter 25.) The objective lens and the eyepiece lens have focal lengths of 2.8 mm and 3.3 cm, respectively. If an object is located 3.0 mm from the objective lens, where is the final image located, and what type of image is it?

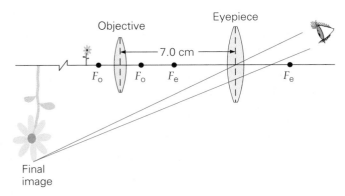

▲ FIGURE 23.30 Compound microscope
See Exercise 74.

75. ■■■ Two converging lenses L_1 and L_2 have focal lengths of 30 cm and 20 cm, respectively. The lenses are placed 60 cm apart along the same axis, and an object is placed 50 cm from L_1 on the side opposite L_2. Where is the image formed relative to L_2, and what are its characteristics?

76. CQ ■■■ For a lens combination, why is the total magnification $M_{\text{total}} = M_1 M_2$? [Hint: Think about the definition of magnification.]

77. ■■■ Show that for thin lenses that have focal lengths f_1 and f_2 and are in contact, the effective focal length (f) is given by

$$\frac{1}{f} = \frac{1}{f_1} + \frac{1}{f_2}$$

*23.5 The Lens Maker's Equation

78. The power of a lens is expressed in units of (a) watts, (b) diopters, (c) meters, or (d) both (b) and (c).

79. CQ What is the focal length of a rectangular glass block?

80. CQ (a) When a lens with $n = 1.60$ is immersed in water, is there a change in the focal length of the lens? If so, which way? (b) What would be the case for a submerged lens whose index of refraction is less than that of the fluid?

81. CQ A lens that is converging in air is submerged in a fluid whose index of refraction is greater than that of the lens. Is the lens still converging?

82. CQ When you open your eyes underwater, everything is blurry. However, when you wear goggles, you can see clearly. Explain.

83. ■ A nearsighted student was prescribed by an optometrist to wear glasses with a power of −2.0 D. What is the focal length of the glass lenses?

84. ■■ An optometrist prescribes a corrective lens with a power of +1.5 D. The lens maker will start with a glass blank that has an index of refraction of 1.6 and a convex front surface whose radius of curvature is 20 cm. To what radius of curvature should the other surface be ground?

85. ■■ A plastic plano-concave lens has a radius of curvature of 50 cm for its concave surface. If the index of refraction of the plastic is 1.35, what is the power of the lens?

86. IE ■■■ A biconvex lens is made of glass whose index of refraction is 1.6. The lens has a radius of curvature of 30 cm for one surface and 40 cm for the other. (a) Will the focal length of this lens (1) increase, (2) remain the same, or (3) decrease if the lens is moved from air to under water? Why? (b) Calculate the focal length of this lens as used in air and under water.

Additional Exercises

87. To photograph a full Moon, a photographer uses a single-lens camera having a focal length of 60 mm. What will be the diameter of the Moon's image on the film? [Note: Data

about the Moon are inside the back cover of this text.]

88. **IE** A virtual image of magnification +0.50 is produced when an object is placed in front of a spherical mirror. (a) The mirror is (1) convex, (2) concave, or (3) flat. Why? (b) Find the radius of curvature of the mirror if the object is 7.0 cm in front of it.

89. A bottle 6.0 cm tall is located 75 cm from the concave surface of a mirror with a radius of curvature of 50 cm. Where is the image located, and what are its characteristics?

90. **IE** A shaving mirror has a magnification of +4.00. (a) The mirror is (1) convex, (2) concave, or (3) flat. Why? (b) What is the focal length of the mirror if your face is 10.0 cm in front of the mirror?

91. **CQ** A method of determining the focal length of a diverging lens is called *autocollimation*. As ▼ Fig. 23.31 shows, first a sharp image of a light source is projected on a screen by a converging lens. Second, the screen is replaced with a plane mirror. Third, a diverging lens is placed between the converging lens and the mirror. Light will then be reflected by the mirror back through the compound-lens system, and an image will be formed on a screen near the light source. This image is made sharp by adjusting the distance

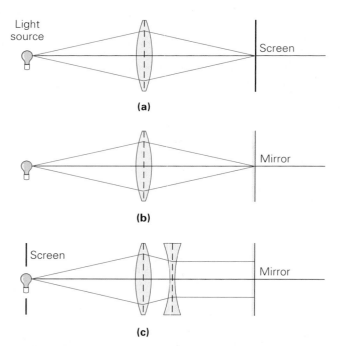

(a)

(b)

(c)

▲ **FIGURE 23.31 Autocollimation** See Exercise 91.

between the diverging lens and the mirror. The distance at which the image is clearest is equal to the focal length of the lens. Explain why this method works.

92. Show that the magnification for objects near the optic axis of a convex mirror is given by $|M| = d_i/d_o$. [*Hint*: Use a ray diagram with rays reflected at the mirror's vertex.]

93. A biconvex lens produces a real, inverted image of an object that is magnified 2.5 times when the object is 20 cm from the lens. What is the focal length of the lens?

94. (a) For values of d_o from 0 to ∞, sketch graphs of (1) d_i versus d_o and (2) M versus d_o for a converging lens. (b) Sketch similar graphs for a diverging lens.

95. Two lenses, each having a power of +10 D, are placed 20 cm apart along the same axis. If an object is 60 cm from the first lens on the side opposite the second lens, where is the final image relative to the first lens, and what are its characteristics?

96. An object is 15 cm from a converging lens whose focal length is 10 cm. On the opposite side of that lens, at a distance of 60 cm, is a converging lens with a focal length of 20 cm. Where is the final image formed, and what are its characteristics?

97. **IE** A biology student wants to examine a bug at a magnification of +5.00. (a) The lens should be (1) convex, (2) concave, or (3) flat. Why? (b) If the bug is 5.00 cm from the lens, what is the focal length of the lens?

98. (a) Use ray diagrams to show that a ray parallel to the optic axis of a biconvex lens is refracted toward the axis at the incident surface and again at the exit surface. (b) Show that the refraction effect also holds for a biconcave lens, but with deflections away from the axis.

99. A converging glass lens with an index of refraction of 1.62 has a focal length of 30 cm in air. What is the focal length when the lens is submerged in water?

100. The image of an object located 30 cm from a concave mirror is formed on a screen located 20 cm from the mirror. What is the mirror's radius of curvature?

101. A concave cosmetic mirror produces a virtual image 1.5 times the size of a person whose face is 20 cm from the mirror. (a) Draw a ray diagram of this situation. (b) What is the focal length of the mirror?

Physical Optics: The Wave Nature of Light

It's always intriguing to see brilliant colors produced by objects that we know don't ordinarily have any colors of their own. The glass of a prism, for example, clear and transparent by itself, nevertheless gives rise to a whole array of colors when white light passes through it. Prisms, like the water droplets that produce the rainbow, don't create anything that isn't already there. They merely separate the different wavelengths that make up white light and spread them out.

The phenomena of reflection and refraction are conveniently analyzed by using geometrical optics (Chapter 22). Ray diagrams show what happens when light is reflected from a mirror or passed through a lens. However, some other phenomena involving light, such as the interference patterns in the opening photo, cannot be adequately explained or described using the ray concept, since this technique ignores the wave nature of light. Other wave phenomena include diffraction and polarization.

Physical optics, or **wave optics**, takes into account those wave properties that geometrical optics ignores. The wave theory of light leads to satisfactory explanations of those phenomena that cannot be analyzed with rays. Thus, in this chapter, the wave nature of light must be used to analyze phenomena such as interference and diffraction.

Wave optics must be used to explain how light propagates around small objects or through small openings. An object or opening is considered small if it is in the order of magnitude of the wavelength of light.

779

24.1 Young's Double-Slit Experiment

OBJECTIVES: To (a) explain how Young's experiment demonstrated the wave nature of light and (b) compute the wavelength of light from experimental results.

It has been stated that light behaves like a wave, but no proof of this assertion has been discussed. How would you go about demonstrating the wave nature of light? One method that involves the use of interference was first devised in 1801 by the English scientist Thomas Young (1773–1829). **Young's double-slit experiment** not only demonstrates the wave nature of light, but also allows the measurement of its wavelengths. Essentially, light can be shown to be a wave if it exhibits wave properties such as interference and diffraction.

Recall from the discussion of wave interference in Sections 13.4 and 14.4 that superimposed waves may interfere constructively or destructively. Total constructive interference occurs when two crests are superimposed, and total destructive interference occurs when a crest and a trough of two identical waves are superimposed. Interference can be observed with water waves (◄Fig. 24.1), for which constructive and destructive interference produce obvious interference patterns.

The interference of (visible) light waves is not as easily observed, because of their relatively short wavelengths ($\approx 10^{-7}$ m) and their not being monochromatic (single frequency). Also, stationary interference patterns are produced only with *coherent sources*—sources that produce light waves having a constant phase relationship to one another. For example, for constructive interference to occur at some point, the waves meeting at that point must be in phase. As the waves meet, a crest must *always* overlap a crest, and a trough must *always* overlap a trough. If a phase difference develops between the waves over time, the interference pattern changes, and a stable or stationary pattern will not be established.

In an ordinary light source, the atoms are excited randomly, and the emitted light waves fluctuate in amplitude and frequency. Thus, light from two such sources is *incoherent* and cannot produce a stationary interference pattern. Interference does occur, but the phase difference between the interfering waves changes so fast that the interference effects are not discernible. To obtain two coherent sources, a barrier with one narrow slit is placed in front of a light source, and a barrier with two very narrow slits is positioned symmetrically in front of the first barrier (►Fig. 24.2a). Waves propagating out from the single slit are in phase, and the double slits then act as two coherent sources by separating each wave into two parts. Any random changes in the light from the original source will thus occur for the light passing through both slits, and the phase difference will be constant. The modern laser beam, a coherent light source, makes the observation of a stable interference pattern much easier. A series of bright lines is then observed on a screen placed relatively far from the slits (Fig. 24.2b). This pattern results from *wave interference.*

To help analyze Young's experiment, let's imagine that light with a single wavelength (monochromatic light) is used. Because of diffraction (see Sections 13.4 and 14.4 and, in this chapter, Section 24.3), or the spreading of light as it passes through a slit, the waves spread out and interfere as illustrated in Fig. 24.2a. Coming from two coherent "sources," the interfering waves produce a stable interference pattern on the screen. The pattern consists of a bright central maximum (►Fig. 24.3a) and a series of symmetrical dark (Fig. 24.3b) and bright (Fig. 24.3c) side fringes, which mark the positions at which destructive and constructive interference occur. The existence of this interference pattern clearly demonstrates the wave nature of light. The bright fringes decrease in intensity on either side of this central maximum.

Measuring the wavelength of light requires us to look at the geometry of Young's experiment, as shown in ►Fig. 24.4. Let the screen be a distance L from the

Note: Compare Fig. 24.1 with Fig. 14.8a.

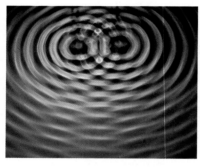

▲ **FIGURE 24.1 Water-wave interference** The constructive and destructive interference of water waves from two coherent sources in a ripple tank produces interference patterns.

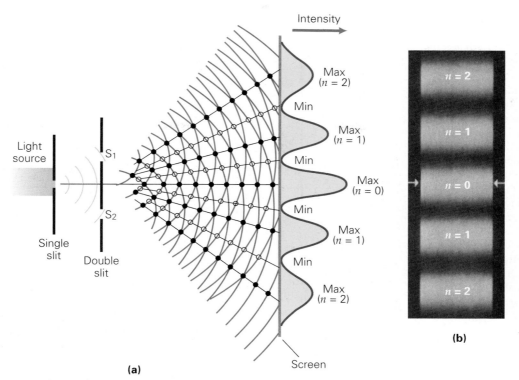

(b)

(a)

▲ **FIGURE 24.2 Double-slit interference** **(a)** The coherent waves from two slits are shown in blue (top slit) and red (bottom slit). The waves spread out as a result of diffraction from narrow slits. The waves interfere, producing alternating maxima and minima, or bright and dark fringes, on the screen. **(b)** An interference pattern. Note the symmetry of the pattern about the central maximum ($n = 0$).

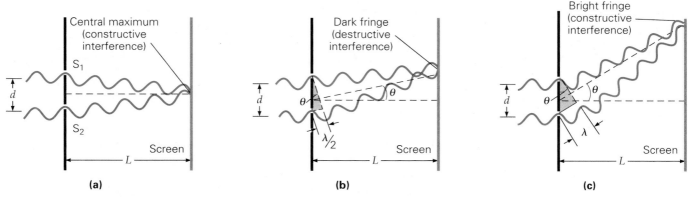

(a) **(b)** **(c)**

▲ **FIGURE 24.3 Interference** The interference that produces bright or dark fringes depends on the difference in the path lengths of the light from the two slits. **(a)** The path-length difference at the position of the central maximum is zero, so the waves arrive in phase and interfere constructively. **(b)** At the position of the first dark fringe, the path-length difference is $\lambda/2$, and the waves interfere destructively. **(c)** At the position of the first bright fringe, the path-length difference is λ, and the interference is constructive.

slits and P be an arbitrary point on the screen. P is located a distance y from the center of the central maximum and at an angle θ relative to a normal line between the slits. The slits S_1 and S_2 are separated by a distance d. Note that the light path from slit S_2 to P is longer than the path from slit S_1 to P. As the figure shows, the path-length difference (ΔL) is approximately

$$\Delta L = d \sin \theta$$

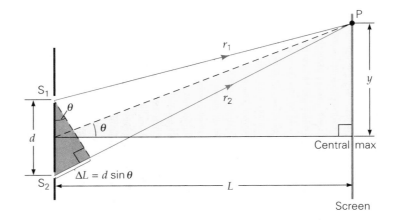

► FIGURE 24.4 Geometry of Young's double-slit experiment The difference in the path lengths for light traveling from the two slits to a point P is $r_2 - r_1 = \Delta L$, which forms a side of the small shaded triangle. Because the barrier with the slits is parallel to the screen, the angle between r_2 and the barrier (at S_2, in the small shaded triangle) is equal to the angle between r_2 and the screen. When L is much greater than y, that angle is almost identical to the angle between the screen and the dashed line, which is an angle in the large shaded triangle. The two shaded triangles are then almost exactly similar, and the angle at S_1 in the small triangle is almost exactly equal to θ. Thus, $\Delta L = d \sin \theta$. By the condition for constructive interference, $d \sin \theta = n\lambda$, where $n = 0, 1, 2, 3, \ldots$ is the order of interference. (Not drawn to scale. Assume that $d \ll L$.)

PHYSLET®
ILLUSTRATION

Two-Slit Interference

The fact that the angle in the small shaded triangle is almost equal to θ can be shown by a simple geometrical argument involving similar triangles when $d \ll L$, as described in the caption of Fig. 24.4.

The relationship of the phase difference of two waves to their path-length difference was discussed in Chapter 14 for the interference of sound waves. The conditions for interference hold for any type of wave, including light waves. Constructive interference occurs at any point where the path-length difference between the two waves from the slits is an integral number of wavelengths:

$$\Delta L = n\lambda \quad \text{for } n = 0, 1, 2, 3, \ldots \quad \begin{array}{l}\textit{condition for}\\ \textit{constructive interference}\end{array} \quad (24.1)$$

Similarly, for destructive interference, the path-length difference is an odd number of half-wavelengths:

$$\Delta L = \frac{m\lambda}{2} \quad \text{for } m = 1, 3, 5, \ldots \quad \begin{array}{l}\textit{condition for}\\ \textit{destructive interference}\end{array} \quad (24.2)$$

Thus, in Fig. 24.4, the position of a bright fringe (constructive interference) satisfies

$$d \sin \theta = n\lambda \quad \text{for } n = 0, 1, 2, 3, \ldots \quad \begin{array}{l}\textit{condition for}\\ \textit{bright fringes}\end{array} \quad (24.3)$$

where n is called the *order number*. The zeroth-order fringe ($n = 0$) corresponds to the central maximum; the first-order fringe ($n = 1$) is the first bright fringe on either side of the central maximum; and so on. As the path-length difference varies from point to point, so does the phase difference and the resulting type of interference (constructive or destructive).

The wavelength can therefore be determined by measuring d and θ for a bright fringe of a particular order (other than the central maximum) from $\lambda = (d \sin \theta)/n$.

The angle θ is related to the location of a fringe relative to the central maximum (y), which can be conveniently measured from a photograph of the interference pattern, such as Fig. 24.2b. If θ is small ($y \ll L$), $\sin \theta \approx \tan \theta = y/L$. Substituting this expression for $\sin \theta$ into Eq. 24.3 and then solving for y gives a good approximation of the distance of the nth bright fringe (y_n) from the central maximum on either side:

$$y_n \approx \frac{nL\lambda}{d} \quad \begin{array}{l}\textit{(lateral distance to bright fringe,}\\ \textit{for } n = 0, 1, 2, 3, \ldots, \textit{for small } \theta \textit{ only)}\end{array} \quad (24.4)$$

A similar analysis gives the locations of the dark fringes. (See Exercise 12a.)

From Eq. 24.3, we see that, except for the zeroth-order fringe, $n = 0$ (the central maximum), the positions of the fringes depend on wavelength—different wavelengths (λ) give different values of sin θ and therefore of θ. Hence, when the experimenter uses white light, as Young did, the central fringe is white, but the other orders contain a spectrum of colors. Since y is proportional to λ ($y \propto \lambda$), we expect to see that red is farther out than blue for a particular order of bright fringe.

By measuring the positions of the color fringes within a particular order, Young was able to determine the wavelengths of the colors of visible light. It could also be seen that the size of the interference pattern, y_n, depends on the inverse of the slit separation d. The smaller the d, the more spread out is the interference pattern. For a large d, the interference pattern is so compressed that it appears to us as a single fringe.

Example 24.1 ■ Measuring the Wavelength of Light: Young's Double-Slit Experiment

In a lab experiment similar to the one shown in Fig. 24.4, monochromatic light (light with only one wavelength or frequency) passes through two narrow slits that are 0.050 mm apart. The interference pattern is observed on a white wall 1.0 m from the slits, and the second-order bright fringe is at an angle of $\theta = 1.5°$. (a) What is the wavelength of the light? (b) What is the distance between the second-order and third-order bright fringes?

Thinking It Through. (a) We can use Eq. 24.3 to find the wavelength. (b) Since $L \gg d$ (1.0 m $\gg$ 0.050 mm), the angle θ is small. We could compute y_2 and y_3 from Eq. 24.4 and then get the distance between the second-order and third-order fringes ($y_3 - y_2$). However, the bright fringes for a given wavelength of light are evenly spaced; in general, the distance between adjacent bright fringes is constant (for a small θ).

Solution. From the information given in this Example, we have

Given: $d = 0.050$ mm $= 5.0 \times 10^{-5}$ m   *Find:* (a) λ (wavelength)
$L = 1.0$ m   (b) $y_3 - y_2$ (distance between
$\theta_2 = 1.5°$ $n = 2$ and $n = 3$)
$n = 2$

(a) Using Eq. 24.3 gives

$$\lambda = \frac{d \sin \theta}{n} = \frac{(5.0 \times 10^{-5} \text{ m}) \sin 1.5°}{2} = 6.5 \times 10^{-7} \text{ m} = 650 \text{ nm}$$

This value is 650 nm, which is the wavelength of orange-red light (see Fig. 20.23).
(b) Using a general approach for n and $n + 1$, we get

$$y_{n+1} - y_n = \frac{(n + 1)L\lambda}{d} - \frac{nL\lambda}{d} = \frac{L\lambda}{d}$$

In this case, the distance between successive fringes is

$$y_3 - y_2 = \frac{L\lambda}{d} = \frac{(1.0 \text{ m})(6.5 \times 10^{-7} \text{ m})}{5.0 \times 10^{-5} \text{ m}} = 1.3 \times 10^{-2} \text{ m} = 1.3 \text{ cm}$$

Follow-up Exercise. Suppose white light were used instead of monochromatic light in this Example. What would be the separation distance of the red ($\lambda = 700$ nm) and blue ($\lambda = 400$ nm) components in the second-order fringe? (*Answers to all Follow-up Exercises are at the back of the text.*)

24.2 Thin-Film Interference

OBJECTIVES: To (a) describe how thin films can produce colorful displays and (b) give some examples of practical applications of thin-film interference.

Have you ever wondered what causes the rainbowlike colors when white light is reflected from a thin film of oil or a soap bubble? This effect—known as thin-film interference—is a result of the interference of light reflected from opposite surfaces of the film and may be readily understood in terms of wave interference.

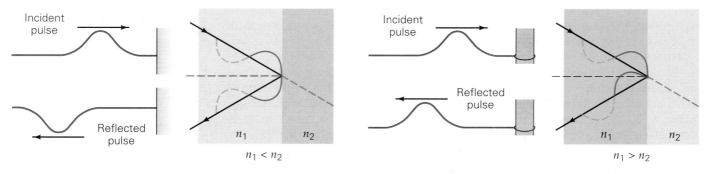

(a) Fixed end: 180° phase shift   **(b)** Free end: zero phase shift

▲ **FIGURE 24.5 Reflection and phase shifts**   The phase changes that light waves undergo on reflection are analogous to those for pulses in strings. **(a)** The phase of a pulse in a string is shifted by 180° on reflection from a fixed end, and so is the phase of a light wave when it is reflected from a more optically dense medium. **(b)** A pulse in a string has a phase shift of zero (it is not shifted) when reflected from a free end. Analogously, a light wave is not phase shifted when reflected from a less optically dense medium.

First, however, you need to see how the phase of a light wave is affected by reflection. Recall from Chapter 13 that a wave pulse on a rope undergoes a 180° phase change, or a half wave ($\lambda/2$) shift, when reflected from a rigid support and no phase shift when reflected from a free support (▲Fig. 24.5). Similarly, as the figure shows, the phase change for the reflection of light waves at a boundary depends on the optical densities, or the indices of refraction (n), of the two materials:

- A light wave traveling in one medium and reflected from the boundary of a second medium whose index of refraction is greater than that of the first medium ($n_2 > n_1$) undergoes a 180° phase change.
- If the reflecting medium has the smaller index of refraction ($n_2 < n_1$), there is no phase change.

To understand why you see colors from an oil film (for example, floating on water or on a wet road), consider the reflection of monochromatic light from a thin film, as illustrated in ▼Fig. 24.6. The path length of the wave in the film depends

▼ **FIGURE 24.6 Thin-film interference**   For an oil film on water, there is a 180° phase shift for light reflected from the air–oil interface and a zero phase shift at the oil–water interface. λ' is the wavelenth in the oil. **(a)** Destructive interference occurs if the oil film has a minimum thickness of $\lambda'/2$ for normal incidence. (Waves are displaced and angled for clarity.) **(b)** Constructive interference occurs with a minimum film thickness of $\lambda'/4$. **(c)** Thin-film interference in an oil slick. Different film thicknesses give rise to the reflections of different colors.

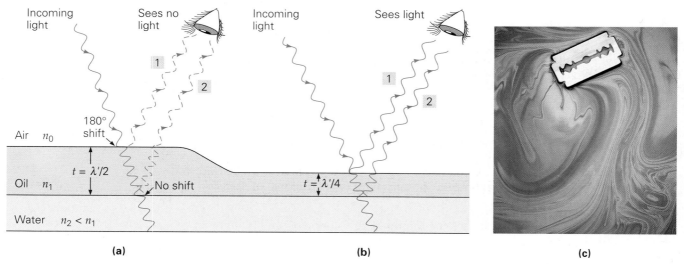

(a)   **(b)**   **(c)**

on the angle of incidence (why?), but, for simplicity, we will assume normal (perpendicular) incidence for the light, even though the rays are drawn at an angle in the figure, for clarity.

PHYSLET®
ILLUSTRATION

Thin-Film Interference

The oil film has a greater index of refraction than that of air, and the light reflected from the air–oil interface (wave 1 in the figure) undergoes a 180° phase shift. The transmitted waves pass through the oil film and are reflected at the oil–water interface. In general, the index of refraction of oil is greater than that of water (see Table 22.1)—that is, $n_2 < n_1$—so a reflected wave in this instance (wave 2) is *not* phase shifted.

You might think that if the path length of the wave in the oil film ($2t$, twice the thickness—down and back) were an integral number of wavelengths [$2t = 2(\lambda'/2) = \lambda'$, or $t = \lambda'/2$, in Fig. 24.6a], then the waves reflected from the two surfaces would interfere constructively. But keep in mind that the wave reflected from the top surface (wave 1) undergoes a 180° phase shift. The reflected waves from the two surfaces are therefore actually *out of phase* for this film thickness and interfere destructively. This means that reflected light of this wavelength is not seen.

Similarly, if the path length of the wave in the film were an odd number of half-wavelengths [$2t = 2(\lambda'/4) = \lambda'/2$, or $t = \lambda'/4$, in Fig. 24.6b], then the reflected waves would actually be *in phase* (as a result of a 180° phase shift of wave 1) and would interfere constructively. Light of this wavelength would be observed when reflected from the oil film.

Keep in mind that the path length is expressed in terms of the wavelength of light *in the film*, λ'. Recall from Section 22.3 that the wavelength is different in different media. Here, $\lambda' = \lambda/n$, where λ' is the wavelength in the oil, λ the wavelength in air, and n the index of refraction of the oil. This relationship can be seen from $n = c/v = f\lambda/(f\lambda') = \lambda/\lambda'$, or $\lambda' = \lambda/n$, since the frequency of light, f, is the same in both materials (why?).

Because oil and soap films generally have different thicknesses in different regions, particular wavelengths (colors) of white light interfere constructively in different regions after reflection. As a result, a vivid display of various colors appears (Fig. 24.6c), which may change if the film thickness changes with time. Thin-film interference may be seen when two glass slides are stuck together with an air film between them (►Fig. 24.7a). The bright colors of a peacock, an example of colorful interference in nature, are a result of layers of fibers in its feathers. Light reflected from successive layers interferes constructively, giving bright colors, even though the feather has no pigment of its own. Since the condition for constructive interference depends on the angle of incidence, the color pattern changes somewhat with the viewing angle and motion of the bird (Fig. 24.7b).

From this analysis of thin-film interference, you can see that the word "destructive" in this context does not imply that energy is destroyed. Destructive interference is simply a description of a physical fact—that light energy is not present at a particular location, but is somewhere else. When destructive interference occurs for light incident on the surfaces of a thin film, you know that the light energy is not reflected, but that the incident beam (energy) is, instead, transmitted. Similarly, the mathematical description of Young's double-slit experiment in Section 24.1 tells you that you *should not expect* to find light energy at the locations of the dark fringes. The light energy is at the locations of the bright fringes, and the interference analysis shows that the light energy is merely redistributed. In other words, energy is conserved, and it is simply "moved around."

A practical application of thin-film interference is nonreflective coatings for lenses. (See the Insight on p. 788 for more detailed information.) In this situation, a film coating is used to create destructive interference between the reflected waves so as to increase the light transmission into the glass (►Fig. 24.8). The index

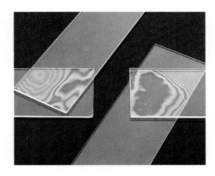

(a)

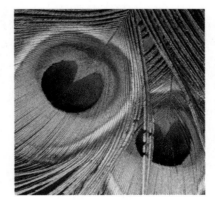

(b)

▲ **FIGURE 24.7 Thin-film interference** (a) A thin air film between microscope slides gives colorful patterns. (b) Multilayer interference in a peacock's feathers gives rise to bright colors. The brilliant throat colors of hummingbirds are produced in the same way.

For a thin film on a glass lens, there is a 180° phase shift at each interface when the index of refraction of the film is less than that of the glass. The waves reflected off the top and bottom surfaces of the film interfere. For clarity, the angle of incidence is drawn to be large, but, in reality, it is almost zero.

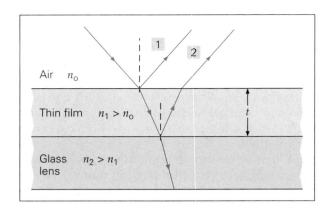

of refraction of the film is between that of air and glass ($n_o < n_1 < n_2$). Consequently, phase shifts of incident light take place at the surfaces of both the film and the glass.

In such a case, the condition for constructive interference of the reflected light is

$$\Delta L = 2t = m\lambda' \quad \text{or} \quad t = \frac{m\lambda'}{2} = \frac{m\lambda}{2n_1} \quad m = 0, 1, 2, \ldots \qquad \begin{array}{l}\textit{condition for}\\ \textit{constructive}\\ \textit{interference when}\\ n_o < n_1 < n_2 \end{array} \quad (24.5)$$

and the condition for destructive interference is

$$\Delta L = 2t = \frac{m\lambda'}{2} \quad \text{or} \quad t = \frac{m\lambda'}{4} = \frac{m\lambda}{4n_1} \quad m = 1, 3, 5, \ldots \qquad \begin{array}{l}\textit{condition for}\\ \textit{destructive}\\ \textit{interference when}\\ n_o < n_1 < n_2 \end{array} \quad (24.6)$$

Here, the wavelength of the light in the film (λ') is related to that in air (λ) by $\lambda' = \lambda/n_1$. Therefore, the minimum film thickness for destructive interference occurs when $m = 1$, so

$$t_{\min} = \frac{\lambda}{4n_1} \quad \begin{array}{l}\textit{minimum film thickness}\\ \textit{(for } n_o < n_1 < n_2)\end{array} \qquad (24.7)$$

If the index of refraction of the film is greater than that of air and glass ($n_o < n_1 > n_2$), then only the reflection at the n_o–n_1 interface has the phase shift. Therefore, $2t = m\lambda'$ will actually be for destructive interference and $2t = m\lambda'/2$ for constructive interference. (Why?)

Example 24.2 ■ Nonreflective Coatings: Thin-Film Interference

A glass lens ($n = 1.60$) is coated with a thin, transparent film of magnesium fluoride ($n = 1.38$) to make the lens nonreflecting. (a) What is the minimum film thickness for the lens to be nonreflecting for incident light of wavelength 550 nm? (b) Will a film thickness of 996 nm make the lens nonreflecting?

Thinking It Through. (a) We can use Eq. 24.7 directly to get an idea of the minimum film thickness for a nonreflective coating. (b) We need to determine whether 996 nm satisfies the condition in Eq. 24.6.

Solution.

Given: $n_o = 1$ (air)   **Find:** (a) t_{min} (minimum film thickness)
$n_1 = 1.38$ (for film)   (b) determine if $t = 998$ nm gives nonreflecting
$n_2 = 1.60$ (for lens)
$\lambda = 550$ nm

(a) Since $n_2 > n_1 > n_o$,

$$t_{min} = \frac{\lambda}{4n_1} = \frac{550 \text{ nm}}{4(1.38)} = 99.6 \text{ nm}$$

which is quite thin ($\approx 10^{-5}$ cm). In terms of atoms, which have diameters on the order of 10^{-10} m, or 10^{-1} nm, the film is 10^3 atoms thick.

(b)

$$t = 996 \text{ nm} = 10(99.6 \text{ nm}) = 10t_{min} = 10\left(\frac{\lambda}{4n_1}\right) = 5\left(\frac{\lambda}{2n_1}\right)$$

So this film thickness does not satisfy the condition required for the lens to be nonreflective (destructive interference). Actually, it satisfies the requirement for constructive interference (Eq. 24.5) with $m = 5$. Such a coating specific for infrared radiation on car and house windows could be useful in hot climates, because it maximizes reflection and minimizes transmission.

Follow-up Exercise.   For the glass lens in this Example to reflect, rather than transmit, the incident light through the lens, what would be the minimum film thickness?

Optical Flats and Newton's Rings

The phenomenon of thin-film interference can be used to check the smoothness and uniformity of optical components such as mirrors and lenses. *Optical flats* are made by grinding and polishing glass plates until they are as flat and smooth as possible. (The surface roughness is usually in the order of $\lambda/20$.) The degree of flatness can be checked by putting two such plates together at a slight angle so that a very thin air wedge is between them (▶ Fig. 24.9a). The reflected waves off the top (wave 1) and bottom (wave 2) plates interfere. Note that wave 2 in Fig. 24.9a has a 180° phase shift as it is reflected from an air–plate interface. Therefore, at a certain point from where the two plates touch (point O), the condition for constructive interference is $2t = m\lambda/2$ ($m = 1, 3, 5, \ldots$), and the condition for destructive interference is $2t = m\lambda$ ($m = 0, 1, 2, \ldots$). The thickness t determines the type of interference (constructive or destructive). If the plates are smooth and flat, a regular interference pattern of bright and dark fringes, or bands, appears (Fig. 24.9b). This pattern is a result of the uniformly varying differences in path lengths between the plates. Any irregularity in the pattern indicates an irregularity in at least one plate. Once a good optical flat is verified, it can be used to check the flatness of a reflecting surface, such as that of a precision mirror.

Direct evidence of the 180° phase shift discussed earlier can be clearly seen in Fig. 24.9. At the point where the two plates touch ($t = 0$), we see a dark band. If there were no 180° phase shift, $t = 0$ would correspond to $\Delta L = 0$, and a bright band would appear. The fact that a dark band is seen at this point proves that there is a 180° phase shift in reflection from a more optically dense material.

A similar technique is used to check the smoothness and symmetry of lenses. When a curved lens is placed on an optical flat, a radially symmetric air wedge is formed between the lens and the optical flat (▶ Fig. 24.10a). Since the thickness of the air wedge determines the condition for constructive and destructive interference, the regular interference pattern in this case is a set of concentric bright and dark circular fringes (Fig. 24.10b). They are called *Newton's rings*, after Isaac Newton,

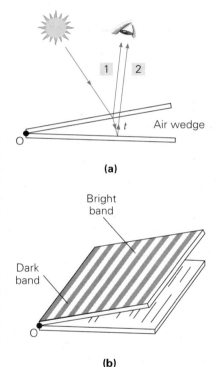

(a)

▲ FIGURE 24.9   Optical flatness
(a) An optical flat is used to check the smoothness of a reflecting surface. The flat is placed so that there is an air wedge between it and the surface. The waves reflected from the two plates interfere, and the thickness of the air wedge at certain points determines if bright or dark bands are seen. **(b)** If the surface is smooth, a regular or symmetrical interference pattern is seen. Note that a dark band is at point O where $t = 0$.

Nonreflecting Lenses

You may have noticed the blue–purple tint of the coated optical lenses used in cameras and binoculars. The coating makes the lenses practically "nonreflecting." If a lens is a nonreflecting type, the incident light is mostly transmitted. Maximum transmission of light is desirable for the exposure of photographic film and for viewing objects with binoculars.

For a typical reflection off an air–glass interface, about 4% of the light is reflected and 96% is transmitted. A modern camera lens, for example, is actually made up of a group of lenses (elements) in order to minimize aberrations and improve image quality. For instance, a certain 35 mm–70 mm zoom lens consists of 13 elements, or there are 26 reflective surfaces.

After one reflection, 0.96 = 96% of the light is transmitted. After two reflections, or one element, the transmitted light is $0.96 \times 0.96 = 0.96^2 = 0.92 = 92\%$. After 26 such reflections, the transmitted light is only $0.96^{26} = 0.35 = 35\%$. Therefore, almost all modern lenses are coated with nonreflecting film to reduce loss of reflection.

A lens is made to be nonreflecting by coating it with a thin film of a material that has an index of refraction between the indices of refraction of air and glass (Fig. 24.8). If the coating is a quarter-wavelength ($\lambda'/4$) thick, the difference in path length between the reflected rays is $\lambda'/2$, where λ' is the wavelength of light in the coating. In this case, both reflected waves undergo a 180° phase shift, and therefore they are out of phase for a path-length difference of $\lambda'/2$ and interfere destructively. That is, the incident light is transmitted, and the coated lens is nonreflecting.

Note that the actual thickness of a quarter-wavelength thickness of film is specific to the particular wavelength of light. The thickness is usually chosen to be a quarter-wavelength

of yellow–green light ($\lambda \approx 550$ nm), to which the human eye is most sensitive. The wavelengths at the red and blue ends of the visible region are still partially reflected, giving the coated lens its bluish–purple tint (Fig. 1). Sometimes other quarter-wavelength thicknesses are chosen, giving rise to other hues, such as amber or reddish purple, depending on the application of the lenses.

Nonreflective coatings are also applied to the surfaces of solar cells, which convert light into electrical energy (Chapter 27). Since the thickness of such a coating is wavelength dependent, like that on a nonreflecting lens, not all of the light is transmitted. However, the losses of reflection can be decreased from around 30% to 10%, making the cell more efficient.

Related Exercises: 25, 31, and 34

FIGURE 1 Coated lenses The nonreflective coating on binocular and camera lenses generally produces a characteristic bluish–purple hue. (Why?)

who first described this interference effect. Note that at the point where the lens and the optical flat touch ($t = 0$), there is, once again, a dark spot. (Why?) Lens irregularities give rise to a distorted fringe pattern, and the radii of these rings can be used to calculate the radius of curvature of the lens.

▶ **FIGURE 24.10 Newton's rings** **(a)** A lens placed on an optical flat forms a ring-shaped air wedge, which gives rise to interference of the waves reflected from the top (wave 1) and the bottom (wave 2) of the air wedge. **(b)** The resulting interference pattern is a set of concentric rings called *Newton's rings*. Note that at the center of the pattern is a dark spot. Lens irregularities produce a distorted pattern.

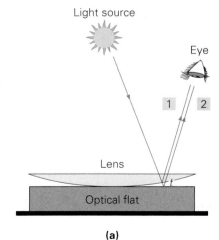

(a)

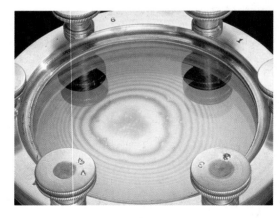

(b)

24.3 Diffraction

OBJECTIVES: **To (a) define diffraction and (b) give examples of diffractive effects.**

In geometrical optics, light is represented by rays and pictured as traveling in straight lines. If this model were to represent the real nature of light, however, there would be no interference effects in Young's double-slit experiment. Instead, there would be only two bright images of slits on the screen, with a well-defined shadow area where no light enters. But we do see interference patterns, which means that the light must deviate from a straight-line path and enter the regions that would otherwise be in shadow. The waves spread out as they pass through the slits; this spreading of the light wave is called **diffraction**. Diffraction generally occurs when waves pass through small openings or around sharp edges or corners. The diffraction of water waves is shown in ▶Fig. 24.11. (See also Fig. 13.17.)

As Fig. 13.17 shows, there are different degrees of spreading out or diffraction. The amount of diffraction depends on the wavelength of the wave and corresponds to the size of the opening or object. In general, *the longer the wavelength compared to the width of the opening or object, the greater the diffraction.* This principle is also in ▼Fig. 24.12.

In Fig. 24.12a, the width of the opening is much greater than the wavelength ($w \gg \lambda$), and there is little diffraction—the wave keeps traveling in its original direction without much spreading out. (There is some degree of diffraction around the edges of the opening.) In Fig. 24.12b, with the wavelength and width of the opening being about equal or in the same order of magnitude ($w \sim \lambda$), there is noticeable diffraction—the wave spreads out and deviates from its original direction of travel. Part of the wave keeps traveling in its original direction but the rest of the wave bends around the opening and spreads out in many directions that are not in the same original direction.

The diffraction of sound (Chapter 14) is quite evident. Someone can talk to you from another room or around the corner of a building, and even in the absence of reflections, you can easily hear the person. Audible sound waves have wavelengths on the order of centimeters to meters. Thus, the widths of ordinary objects and openings are about the same as or narrower than the wavelengths of sound, and diffraction readily occurs for sound.

▲ **FIGURE 24.11 Water-wave diffraction** This photograph of an Israeli beach dramatically shows single-slit diffraction of ocean waves through the barrier openings. Note the beach has been shaped by the circular wave fronts.

▼ **FIGURE 24.12 Wavelength and opening dimensions** In general, the narrower the opening compared to the wavelength, the greater the diffraction. (a) Without much diffraction ($w \gg \lambda$), the wave would keep traveling in its original direction. (b) With noticeable diffraction ($w \approx \lambda$), the wave bends around the opening and spreads out.

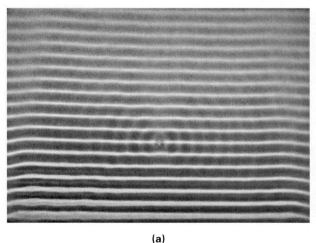

(a)

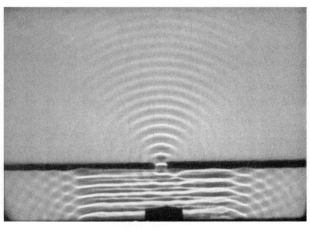

(b)

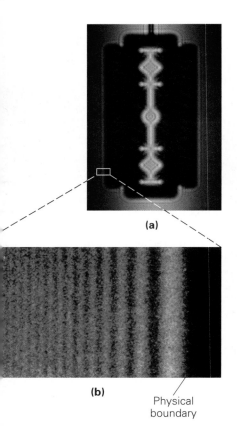

(a)

(b)

Physical
boundary

▲ **FIGURE 24.13 Diffraction in action** (a) Diffraction patterns produced by a razor blade. (b) A close-up view of the fringes formed at the edge of the blade.

Visible light waves, however, have wavelengths on the order of 10^{-7} m (Chapter 20); therefore diffraction phenomena for these waves often go unnoticed. Careful observation will reveal that a shadow boundary is blurred or fuzzy. On close inspection of the area around a sharp razor blade, for example, you can see a pattern of bright and dark fringes (◀Fig. 24.13). These interference patterns are evidence of the diffraction of the light around the edge of the object.

As an illustration of "single-slit" diffraction, consider a slit in a barrier (▼Fig. 24.14). Suppose that the slit (of width w) is illuminated with monochromatic light. A diffraction pattern consisting of a bright central maximum and a symmetrical array of maxima (regions of constructive interference) on both sides is observed on a screen at a distance L from the slit (where $L \gg w$).

A diffraction pattern resulting from an infinite number of wave interferences can be considered an interference pattern. Various points on the wave front passing through the slit can be considered to be small point sources. Then the interference of those waves can be analyzed much as we analyzed double-slit interference earlier.

The analysis is not done here, however. From the geometry, the dark fringes (regions of destructive interference), located at θ, satisfy the relationship

$$w \sin \theta = m\lambda \quad \text{for } m = 1, 2, 3, \ldots \qquad \begin{array}{c} \textit{condition for} \\ \textit{dark fringes} \end{array} \quad (24.8)$$

where θ is the angle of a particular minimum, designated by $m = 1, 2, 3, \ldots$, on either side of the central maximum and m is called the order number. (There is no $m = 0$. Why?)

Although the foregoing result is very similar in form to that for Young's double-slit experiment (Eq. 24.3), it is extremely important to point out that for the single-slit experiment, dark fringes, rather than bright fringes, are analyzed. Also, note that the width of the slit (w) is used in diffraction.

The small-angle approximation, $\sin \theta \approx y/L$, can be made in some instances ($y \ll L$) in obtaining a formula for the location, relative to the central maximum, of a particular diffraction minimum on the screen. This relationship gives a good approximation of the distances of the dark fringes on either side of the center of the central maximum:

$$y_m = m\left(\frac{L\lambda}{w}\right) \quad \text{for } m = 1, 2, 3, \ldots \qquad \begin{array}{c} \textit{location for} \\ \textit{dark fringes} \end{array} \quad (24.9)$$

▶ **FIGURE 24.14 Single-slit diffraction** The diffraction of light by a single slit gives rise to a diffraction pattern consisting of a large and bright central maximum and a symmetric array of side fringes. The order number m corresponds to the minima or dark fringes. (See text for description.)

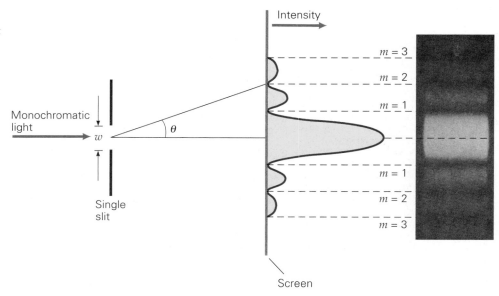

The qualitative predictions from Eq. 24.9 are quite interesting and instructive:

- For a given slit width (w), the longer the wavelength (λ), the wider the diffraction pattern is.
- For a given wavelength (λ), the narrower the slit width (w), the wider the diffraction pattern is.
- The width of the central maximum (between the two dark fringes with $m = 1$) is twice the width of the side maxima (each maximum other than the central one).

Let's consider each of these predictions.

As the slit is made narrower, the central maximum and the side fringes spread out and become larger. Equation 24.9 is not applicable to very small slit widths (because of the small-angle approximation). If the slit width is decreased until it is of the same order of magnitude as the wavelength of the light, then the central maximum spreads out over the whole screen. That is, diffraction becomes dramatically evident when the width of the slit is about the same as the wavelength of the light used. Diffraction effects are most easily observed when $\lambda/w \approx 1$, or $w \approx \lambda$.

Conversely, if the slit is made wider for a given wavelength of light, then the diffraction pattern becomes narrower. The fringes move closer together and eventually become difficult to distinguish when w is much wider than λ ($w \gg \lambda$). The pattern then appears as a fuzzy shadow around the central maximum, which is the illuminated image of the slit. This type of pattern is observed for the image produced by sunlight entering a dark room through a hole in a curtain. Such an observation led early experimenters to investigate the wave nature of light. The acceptance of this concept was, in large part, due to the explanation of diffraction offered by physical optics.

Conceptual Example 24.3 ■ Diffraction and Radio Reception

Driving with the car radio on in the city or in mountainous areas of the country, you have probably noticed that on certain broadcast bands, the quality of your radio reception can vary sharply from place to place, with stations seeming to fade out and reappear. What do you think might cause this variation, and which of the following bands would you expect to be least affected by it? (a) Weather (162 MHz); (b) FM (88–108 MHz); (c) AM (525–1610 kHz).

Reasoning and Answer. Radio waves, like light, are electromagnetic waves and so tend to travel in straight lines for long distances from their sources. They can be blocked by objects in their path—especially if the objects are massive (such as hills) or are made largely of metal (such as buildings).

However, because of diffraction, radio waves can also "wrap around" obstacles or "fan out" as they pass through obstacles and openings, just as sound waves do, provided that their wavelength is at least roughly the size of the obstacle or opening. The longer the wavelength, the greater will be the diffraction, and so the less likely the radio waves are to be obstructed.

To determine which band benefits most by such diffraction, we need to find the wavelengths corresponding to the given frequencies, using the relationship $c = \lambda f$. When we do this, we find that AM radio waves, with $\lambda = 186$ to 571 m, are the longest of the three bands (by a factor of about 100). We thereby conclude that AM broadcasts are more likely to be diffracted around such objects as mountains or through the openings between them. Thus, the answer is (c).

Follow-up Exercise. Woodwind instruments, such as the clarinet and the flute, are usually smaller than brass instruments, like the trumpet and trombone. During halftime at a football game, when a marching band faces you, you can easily hear both the woodwind instruments and the brass instruments. Yet, when the band marches away from you, the brass instruments sound muted, but you can hear the woodwinds quite well. Why?

Notice from Fig. 24.14 that the central maximum differs in width from the side maxima. The central maximum is twice as wide, which can be shown as follows: Taking the width of the central maximum to be the distance between the bounding minima on each side ($m = 1$), or a value of $2y_1$, we obtain, from Eq. 24.9 with $y_1 = L\lambda/w$,

$$2y_1 = \frac{2L\lambda}{w} \quad \text{width of central maximum} \quad (24.10)$$

Similarly, the width of the bright fringes on the sides is given by

$$y_2 - y_1 = y_3 - y_2 = \frac{L\lambda}{w}$$

Or, in general,

$$y_{m+1} - y_m = \frac{L\lambda}{w} \quad \text{width of side fringes} \quad (24.11)$$

Thus, the width of the central maximum is twice that of the side fringes.

Example 24.4 ■ Width of the Central Maximum: Single-Slit Diffraction

---

Monochromatic light (yellow–green) of wavelength $\lambda = 550$ nm passes through a slit whose width is 0.050 mm. (a) At what angle will the third minimum be seen? (b) What is the width of the central maximum on a screen located 1.0 m from the slit?

Thinking It Through. This situation is a direct application of (a) Eq. 24.8 and (b) Eq. 24.10.

Solution.

Given: $\lambda = 550$ nm $= 5.50 \times 10^{-7}$ m   *Find:* (a) θ_3
       $w = 0.050$ mm $= 5.0 \times 10^{-5}$ m   (b) $2y_1$ (width of the central
   (a) $m = 3$ maximum)
   (b) $L = 1.0$ m

(a) From Eq. 24.8, we have

$$\sin \theta_3 = \frac{m\lambda}{w} = \frac{3(5.50 \times 10^{-7}\,\text{m})}{5.0 \times 10^{-5}\,\text{m}} = 0.033$$

So,

$$\theta_3 = \sin^{-1} 0.033 = 1.9°$$

(b) Equation 24.10 gives

$$2y_1 = \frac{2L\lambda}{w} = \frac{2(1.0\,\text{m})(5.50 \times 10^{-7}\,\text{m})}{5.0 \times 10^{-5}\,\text{m}} = 2.2 \times 10^{-2}\,\text{m} = 2.2\,\text{cm}$$

Follow-up Exercise. By what factor would the width of the central maximum change if red light ($\lambda = 700$ nm) were used in this Example?

Diffraction Gratings

Bright and dark fringes result from diffraction followed by interference when monochromatic light passes through a set of double slits. As the number of slits is increased, the bright fringes become sharper (narrower) and the dark fringes wider. The sharp fringes are very useful in optical analysis of light sources and other applications. ▶Fig. 24.15 shows a typical experiment with monochromatic light incident on a **diffraction grating**, which consists of large numbers of par-

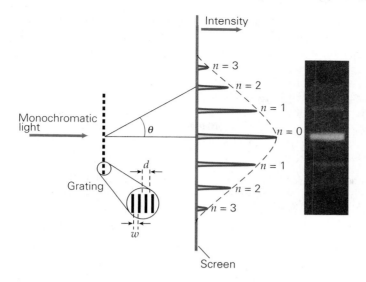

◀ FIGURE 24.15 Diffraction grating A diffraction grating produces a sharply defined interference/diffraction pattern. Two parameters define a grating: the slit separation d and the slit width w. The combination of multiple-slit interference and single-slit diffraction determine the intensity distribution of the various orders of maxima.

allel, closely spaced slits. Two parameters define a diffraction grating: the slit separation between successive slits, d, and the individual slit width, w. The resulting pattern is the result of interference and diffraction. ▾Fig. 24.16 shows the intensity distribution of these fringes as a result of both interference and diffraction.

Diffraction gratings were first made of fine strands of wire. Their effects were similar to what can be seen by viewing a candle flame through a feather held close to the eye. Better gratings have a large number of fine lines or grooves on glass or metal surfaces. If light is transmitted through a grating, it is called a *transmission grating*. However, *reflection gratings* are also common. The closely spaced grooves of a compact disc act as a reflection grating, giving rise to their familiar iridescent sheen (▶Fig. 24.17). Commercial master gratings are made by depositing a thin film of aluminum on an optically flat surface and then removing some of the reflecting metal by cutting regularly spaced, parallel lines. Precision diffraction gratings are made using two coherent laser beams that intersect at an angle. The beams expose a layer of photosensitive material, which is then etched. The spacing of the grating lines is determined by the intersection angle of the

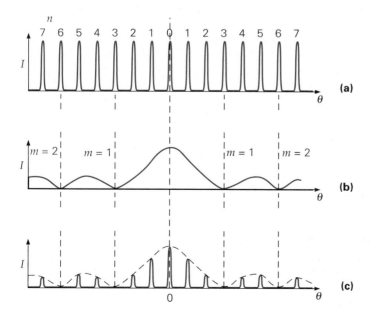

◀ FIGURE 24.16 Intensity distribution of interference and diffraction (a) Interference determines the positions of the interference maxima:
$d \sin \theta = n\lambda, n = 0, 1, 2, 3, \ldots$.
(b) Diffraction locates the positions of the diffraction minima:
$w \sin \theta = m\lambda, m = 1, 2, 3, \ldots$.
(c) The combination (product) of interference and diffraction determine the intensity distribution of the fringes.

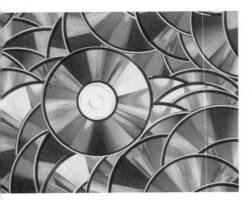

▲ FIGURE 24.17 Diffraction effects The narrow grooves of compact discs (CDs) act as reflection diffraction gratings, producing colorful displays.

Note: *d* is the distance between adjacent slits.

beams. Precision gratings may have 30 000 lines per centimeter or more and are therefore expensive and difficult to fabricate. Most gratings used in laboratory instruments are replica gratings, which are plastic castings of high-precision master gratings.

It can be shown that the condition for interference maxima for a grating illuminated with monochromatic light is identical to that for a double slit. The expression is

$$d \sin \theta = n\lambda \quad \text{for } n = 0, 1, 2, 3, \ldots \qquad \begin{array}{l}\textit{interference maxima} \\ \textit{for a grating}\end{array} \quad (24.12)$$

where *n* is called the *order-of-interference maximum* and θ is the angle at which that maximum occurs for a particular wavelength. The zeroth-order maximum is coincident with the central maximum of the diffraction pattern. The spacing between adjacent slits (*d*) is obtained from the number of lines or slits per unit length of the grating: $d = 1/N$. For example, if $N = 5000$ lines/cm, then

$$d = \frac{1}{N} = \frac{1}{5000/\text{cm}} = 2.0 \times 10^{-4} \text{ cm}$$

If the light incident on a grating is white light or polychromatic, then the fringes are multicolored (▼Fig. 24.18a). There is no deviation of the components of the light for the zeroth order ($\sin \theta = 0$ for all wavelengths), so the central maximum is white. However, the colors separate for higher orders, since the position of the maximum depends on wavelength (Eq. 24.12), with the longer wavelength having a larger θ, which produces a spectrum. Note that it is possible for higher orders produced by a diffraction grating to overlap. That is, the angles for different orders may be the same for two different wavelengths.

The sharp spectra produced by diffraction gratings are used in instruments called *spectrometers* (Fig. 24.18b). In a spectrometer, materials are illuminated with light of various wavelengths to find which wavelengths are strongly transmitted or reflected. The diffraction pattern helps identify the material, because materials usually absorb or emit unique and characteristic wavelengths. The grating is rotated so that the sample is illuminated with a succession of different wavelengths. The wavelengths may also be in the infrared and ultraviolet regions.

▼ FIGURE 24.18 Spectroscopy **(a)** In each side fringe, components of different wavelengths are separated, since the deviation depends on wavelength: $\theta = \sin^{-1}(n\lambda/d)$. **(b)** As a result, gratings are used in spectrometers to determine the wavelengths present in a beam of light by measuring their angles of diffraction and to separate the various wavelengths for further analysis.

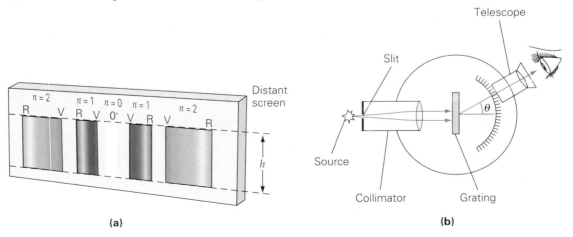

(a)   **(b)**

Only a limited number of spectral orders can be seen when using a diffraction grating. The number depends on the wavelength of the light and on the grating's spacing (d). From Eq. 24.12, since $\sin \theta$ cannot exceed 1 ($\sin \theta \leq 1$),

$$\sin \theta = \frac{n\lambda}{d} \leq 1 \quad \text{or} \quad n \leq \frac{d}{\lambda}$$

Diffraction gratings have almost completely replaced prisms in spectroscopy. The creation of a spectrum and the measurement of wavelengths by a grating depend only on geometrical measurements such as lengths and/or angles. Wavelength determination with a prism, in contrast, depends on the dispersive characteristics of the glass or other material of which the prism is made. Thus, it is crucial to know how the index of refraction depends on the wavelength of light. In contrast to a prism, which deviates red light least and violet light most, a diffraction grating produces the smallest angle for violet light (short λ) and the greatest for red light (long λ). Also, a prism disperses white light into a single spectrum. A diffraction grating, however, produces a number of spectra, one for each order other than $n = 0$.

Example 24.5 ■ A Diffraction Grating: Line Spacing and Spectral Orders

A particular diffraction grating produces an $n = 2$ spectral order at an angle of 30° for light with a wavelength of 500 nm. (a) How many lines per centimeter does the grating have? (b) At what angle can the $n = 3$ spectral order be seen?

Thinking It Through. (a) To find the number of lines per centimeter (N) the grating has, we need to know the grating spacing (d), since $N = 1/d$. With the given data, we can find d from Eq. 24.12. (b) Using Eq. 24.12 again, we can find θ for $n = 3$.

Solution.

Given: $\lambda = 500 \text{ nm} = 5.00 \times 10^{-7} \text{ m}$ *Find:* (a) N (lines/cm)
$\quad\quad\quad n = 2$ $\quad\quad\quad\quad\quad\quad\quad\quad\quad\quad\quad\quad\quad$ (b) θ for $n = 3$
$\quad\quad\quad \theta = 30°$ for $n = 2$

(a) Using Eq. 24.12, we get the grating spacing:

$$d = \frac{n\lambda}{\sin \theta} = \frac{2(5.00 \times 10^{-7} \text{ m})}{\sin 30°} = 2.00 \times 10^{-6} \text{ m} = 2.00 \times 10^{-4} \text{ cm}$$

Then

$$N = \frac{1}{d} = \frac{1}{2.00 \times 10^{-4} \text{ cm}} = 5000 \text{ lines/cm}$$

(b)

$$\sin \theta = \frac{n\lambda}{d} = \frac{3(5.00 \times 10^{-7} \text{ m})}{2.00 \times 10^{-6} \text{ m}} = 0.75$$

so

$$\theta = \sin^{-1} 0.75 = 48.6°$$

Follow-up Exercise. Can you determine the answer to part (b) without first calculating the grating spacing d?

X-ray Diffraction

In principle, the wavelength of any electromagnetic wave can be determined by using a diffraction grating with the appropriate spacing. Diffraction was used to determine the wavelengths of X rays early in the 20th century. Experimental evidence indicated that the wavelengths of X rays were probably around 10^{-8} cm, but it is impossible to construct a diffraction grating with this spacing. Around 1913, Max von Laue (1879–1960), a German physicist, suggested that the regular spacing of the atoms in a crystalline solid might make the crystal act as a diffraction grating for X rays, since the atomic spacing in crystals is on the order of 10^{-8} cm (▼Fig. 24.19). X rays were directed at crystals, and diffraction patterns were indeed observed. (See Fig. 24.19b).

Figure 24.19a illustrates diffraction by the planes of atoms in a crystal such as sodium chloride. The path-length difference is $2d \sin \theta$, where d is the distance between the crystal's internal planes. Thus, the condition for constructive interference is

$$2d \sin \theta = n\lambda \quad \text{for} \quad n = 1, 2, 3, \ldots \quad \textit{X-ray diffraction} \quad (24.13)$$

This relationship is known as **Bragg's law**, after W. L. Bragg (1890–1971), the British physicist who first derived it. Note that θ is *not* measured from the normal, as is the convention in optics.

The first X-ray wavelengths were experimentally determined by this means, and X-ray diffraction is now used to investigate the internal structure not only of simple crystals, but also of large, complex biological molecules such as proteins and DNA (Fig. 24.19c). Because of their short wavelengths, which are comparable with atomic spacings in molecules, X rays provide a method for investigating the atomic arrangement of molecules.

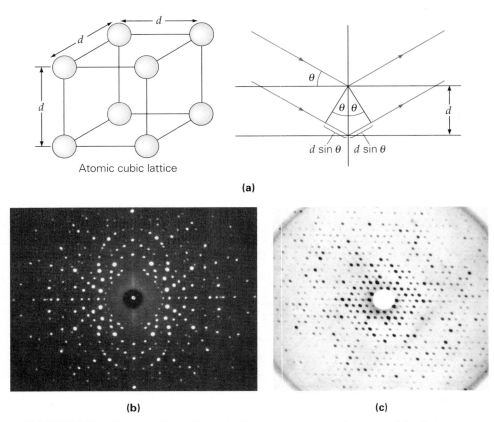

Atomic cubic lattice

(a)

(b)   (c)

▲ **FIGURE 24.19 Crystal diffraction** **(a)** The array of atoms in a crystal-lattice structure acts as a diffraction grating, and X rays are diffracted from the planes of atoms. With a lattice spacing of d, the path-length difference for the X rays diffracted from adjacent planes is $2d \sin \theta$. **(b)** X-ray diffraction pattern of a crystal of potassium sulfate. By analyzing the geometry of such patterns, investigators can deduce the structure of the crystal and the position of its various atoms. **(c)** X-ray diffraction pattern of the protein hemoglobin, which carries oxygen in blood.

24.4 Polarization

OBJECTIVES: To (a) explain light polarization and (b) give examples of polarization, both in the environment and in commercial applications.

When you think of polarized light, you may visualize polarizing (or Polaroid™) sunglasses, since this is one of the more common applications of polarization. When something is polarized, it has a preferential direction, or orientation. (Think of a polarized electrical plug, as described in Chapter 18.) In terms of light waves, **polarization** refers to the orientation of the transverse wave oscillations.

Note: See Fig. 18.24.

Recall from Chapter 20 that light is an electromagnetic wave with oscillating electric and magnetic field vectors (**E** and **B**, respectively) perpendicular (transverse) to the direction of propagation. Light from most sources consists of a large number of electromagnetic waves emitted by the atoms of the source. Each atom produces a wave with a particular orientation, corresponding to the direction of the atomic vibration. However, since electromagnetic waves are produced by numerous atoms, all orientations of the **E** and **B** fields are possible in the composite light emitted. When the field vectors are randomly oriented, the light is said to be *unpolarized*. This situation is commonly represented schematically in terms of the electric field vector as shown in ▼Fig. 24.20a. As viewed along the direction of propagation, the electric field is randomly or equally distributed in all directions. However, as viewed parallel to the direction of propagation, this random or equal distribution can be represented by two directions (such as the *x*- and *y*-directions in a two-dimensional coordinate system). Here, the vertical arrows denote the electric field components in that direction, and the dots represent the component going in and out of the paper. This notation will be used throughout this section.

Note: Refer to Figure 20.20.

If there is some preferential orientation of the field vectors, the light is said to be *partially polarized*. Both representations in Fig. 24.20b show that there are more electric field vectors in the vertical direction than in the horizontal direction. If the field vectors oscillate in only *one* plane, the light is *plane polarized*, or *linearly polarized* (Fig. 24.20c). Note that polarization is evidence that light is a transverse wave. True longitudinal waves, such as sound waves in air, cannot be polarized, because there are no vibrations perpendicular to the direction of propagation.

PHYSLET® ILLUSTRATION

Polarization

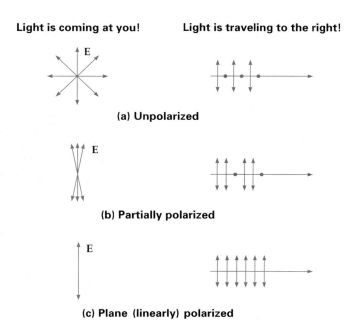

Light is coming at you! **Light is traveling to the right!**

(a) Unpolarized

(b) Partially polarized

(c) Plane (linearly) polarized

◀ **FIGURE 24.20   Polarization**
Polarization is represented by the orientation of the plane of vibration of the electric field vectors.
(a) When the vectors are randomly oriented, the light is unpolarized. The dots represent a direction perpendicular to the paper of the electric field, and the vertical arrows denote an up-and-down direction of the electric field. Equal dots and arrows are used to represent unpolarized light.
(b) With preferential orientation of the field vectors, the light is partially polarized. Here, there are fewer dots than arrows. **(c)** When the vectors are in one plane, the light is plane polarized, or linearly polarized. No dots are seen here.

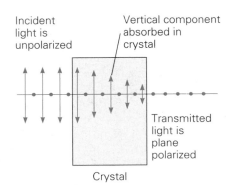

▲ FIGURE 24.21 Selective absorption (dichroism) Dichroic crystals selectively absorb one polarized component (the vertical component) more than the other. If the crystal is thick enough, the emerging beam is linearly polarized.

Light can be polarized in several ways. Polarization by selective absorption, reflection, and double refraction will be discussed here. Polarization by scattering will be considered in Section 24.5.

Polarization by Selective Absorption (Dichroism)

Some crystals, such as those of the mineral tourmaline, exhibit the interesting property of absorbing one of the electric field components more than the other. This property is called **dichroism**. If a dichroic crystal is sufficiently thick, the more strongly absorbed component may be completely absorbed. In that case, the emerging beam is plane polarized (◄Fig. 24.21).

Another dichroic crystal is quinine sulfide periodide (commonly called *herapathite*, after W. Herapath, an English physician who discovered its polarizing properties in 1852). This crystal was of great practical importance in the development of modern polarizers. Around 1930, Edwin H. Land (1909–1991), an American scientist, found a way to align tiny, needle-shaped dichroic crystals in sheets of transparent celluloid. The result was a thin sheet of polarizing material that was given the commercial name *Polaroid™*.

Better polarizing films have been developed that use synthetic polymer materials instead of celluloid. During the manufacturing process, this kind of film is stretched to align the long molecular chains of the polymer. With proper treatment, the outer (valence) electrons of the molecules can move along the oriented chains. As a result, light with **E** vectors parallel to the oriented chains is readily absorbed, but light with **E** vectors perpendicular to the chains is transmitted. The direction perpendicular to the orientation of the molecular chains is called the **transmission axis**, or the **polarization direction**. Thus, when unpolarized light falls on a polarizing sheet, the sheet acts as a polarizer and transmits polarized light (▼Fig. 24.22). Since one of the two electric field components is absorbed, the light intensity after the polarizer is half of the intensity incident on it $(I_o/2)$. The human eye cannot distinguish between polarized and unpolarized light. To tell whether light is polarized, we must use an *analyzer*, which may be another sheet of polarizing film. As shown in Fig. 24.22, if the transmission axis of an analyzer is parallel to the plane of polarization of polarized light, there is maximum trans-

▼ FIGURE 24.22 Polarizing sheets (a) When polarizing sheets are oriented so that their transmission axes are the same, the emerging light is polarized. The first sheet acts as a polarizer, and the second acts as an analyzer. (b) When one of the sheets is rotated 90° and the transmission axes are perpendicular (crossed polarizers), little light (ideally, none) is transmitted. (c) Crossed polarizers made using polarizing sunglasses.

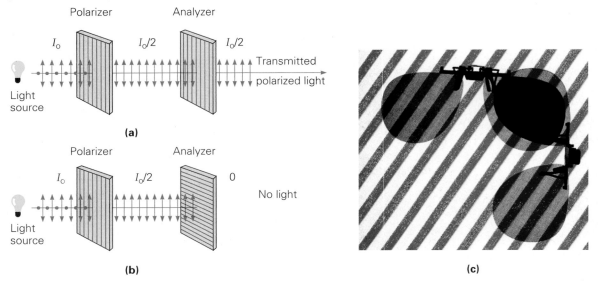

mission. If the transmission axis of the analyzer is perpendicular to the plane of polarization, little light (ideally, none) will be transmitted.

In general, the intensity of the transmitted light is given by

$$I = I_o \cos^2 \theta \qquad \text{Malus's Law} \qquad (24.14)$$

where θ is the angle between the transmission axes of the polarizer and analyzer. This expression is known as *Malus's law*, after its discoverer, French physicist E. L. Malus (1775–1812).

Polarizing glasses whose lenses have different transmission axes are used to view some 3D movies. The pictures are projected on the screen by two projectors that transmit slightly different images, photographed by cameras a short distance apart. The projected light from each projector is linearly polarized, but in mutually perpendicular directions. The lenses of the 3D glasses also have transmission axes that are perpendicular. Thus, one eye sees the image from one projector, and the other eye sees the image from the other projector. The brain interprets the slight difference in perspective of the two images as depth, or a third dimension, just as it does in normal vision.

Integrated Example 24.6 ■ Make Something Out of Nothing: Three Polarizers

In Figs. 24.22b and c, there is no light transmitted after the analyzer, because the transmission axes of the polarizer and analyzer are perpendicular. Assume that the unpolarized light incident on the first polarizer has an intensity of I_o. A second polarizer is inserted in between the first polarizer and analyzer, and the transmission axis of the second polarizer makes an angle of θ with the first polarizer. (a) Is it possible for some light to go through this arrangement? If yes, does it occur at (1) $\theta = 0°$, (2) $\theta = 45°$, or (3) $\theta = 90°$? Why? What happens if the second polarizer is rotated? (b) When $\theta = 30°$, what is the light intensity transmitted?

(a) Conceptual Reasoning. Yes, it is possible for some light to go through this arrangement at (2) $\theta = 45°$, or any other angle that is not $0°$ or $90°$, for that matter. The Learn by Drawing feature on p. 800 can help us understand this situation.

With just the first polarizer and the analyzer, no light is transmitted, according to Malus's law (Eq. 24.14), because the angle between the transmission axes is $90°$. However, when a second polarizer is inserted in between the first polarizer and the analyzer, some light can actually pass through the system. For example, if the transmission axis of the second polarizer makes an angle of θ with that of the first polarizer, then the angle between the transmission axes of the second polarizer and the analyzer will be $90° - \theta$. (Why?)

When unpolarized light of intensity I_o is incident on the first polarizer, the transmitted light after the first polarizer is $I_o/2$, because only one of the two electric field components is transmitted. After the second polarizer, the intensity is decreased by a factor of $\cos^2 \theta$. After the analyzer, the intensity is decreased further by a factor of $\cos^2 (90° - \theta) = \sin^2 \theta$. So the transmitted intensity is $I = (I_o/2)(\cos^2 \theta)(\sin^2 \theta)$. Therefore, as long as θ is not $0°$ or $90°$, there will be some transmitted light through the system.

Since the transmitted light depends on the angle θ, rotating the second polarizer will change the intensity transmitted.

(b) Thinking It Through. Once we understand this situation, part (b) is a straightforward calculation.

Given: $\theta = 30°$ *Find:* (b) I after three polarizers

When $\theta = 30°$,

$$I = \frac{I_o}{2}(\cos^2 30°)(\sin^2 30°) = \frac{3I_o}{32}$$

Follow-up Exercise. For what value of θ will the transmitted intensity be a maximum in this example?

Learn by Drawing

Three Polarizers (See Integrated Example 24.6.)

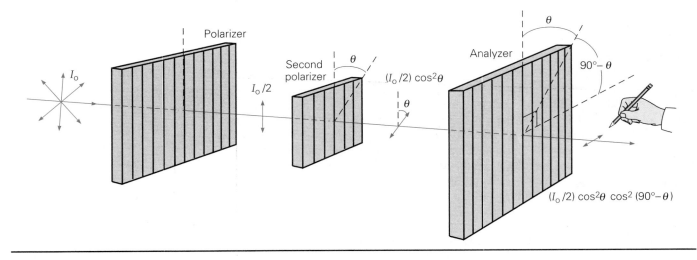

Polarization by Reflection

When a beam of unpolarized light strikes a smooth, transparent medium such as glass, the beam is partially reflected and partially transmitted. The reflected light may be completely polarized, partially polarized, or unpolarized, depending on the angle of incidence. The unpolarized case occurs for 0°, or normal incidence. As the angle of incidence is varied, both the reflected and refracted light are partially polarized. The electric field components normal to the surface are reflected more strongly, producing partial polarization (▼ Fig. 24.23a). However, at one particular angle of incidence, the reflected beam is completely polarized (Fig. 24.23b). At this angle, though, the refracted beam is still only partially polarized.

David Brewster (1781–1868), a Scottish physicist, found that the complete polarization of the reflected beam occurs when the reflected and refracted beams are perpendicular. The incident angle at which this polarization occurs is called the **polarizing angle** (θ_p), or the **Brewster angle**, and it depends on the indices of re-

▶ **FIGURE 24.23 Polarization by reflection** **(a)** When light is incident on an air-medium boundary, the reflected and refracted light are normally partially polarized. **(b)** When the reflected and refracted components of a beam of light are 90° apart, the reflected component is linearly polarized, and the refracted component is partially polarized. This situation occurs when

$$\theta_1 = \theta_p = \tan^{-1}\left(\frac{n_2}{n_1}\right).$$

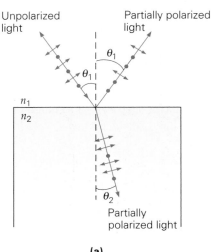

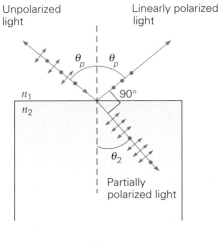

(a)   (b)

fraction of the media. In Fig. 24.23b, the reflected and refracted beams are at 90° and the angle θ_1 is the polarizing angle θ_p; thus, $\theta_1 = \theta_p$, so

$$\theta_1 + 90° + \theta_2 = 180°$$

Then

$$\theta_1 + \theta_2 = 90° \quad \text{or} \quad \theta_2 = 90° - \theta_1$$

By Snell's law (Chapter 22),

$$n_1 \sin \theta_1 = n_2 \sin \theta_2$$

In this case, $\sin \theta_2 = \sin(90° - \theta_1) = \cos \theta_1$. Therefore,

$$\frac{\sin \theta_1}{\sin \theta_2} = \frac{\sin \theta_1}{\cos \theta_1} = \tan \theta_1 = \frac{n_2}{n_1}$$

With $\theta_1 = \theta_p$, we get

$$\tan \theta_p = \frac{n_2}{n_1} \quad \text{or} \quad \theta_p = \tan^{-1} \frac{n_2}{n_1} \qquad (24.15)$$

θ_p (polarizing or Brewster angle): incident angle for complete polarization of reflected light

If the first medium is air ($n_1 = 1$), then $\tan \theta_p = \frac{n_2}{1} = n_2 = n$, where n is the index of refraction of the second medium.

Now you can understand the principle behind polarizing sunglasses. As you have seen, light reflected from a smooth surface is partially polarized. The direction of polarization is mostly parallel to the surface, or horizontal. (See Fig. 24.23b.) Light reflected from the surface of a road or water can be so intense that it gives rise to visual glare (▼Fig. 24.24a). To reduce this effect, the polarizing lenses of glasses are oriented with their transmission axes vertical so that some of the partially polarized light from reflective surfaces is absorbed. Polarizing filters also enable cameras to take "clean" pictures without glare (Fig. 24.24b).

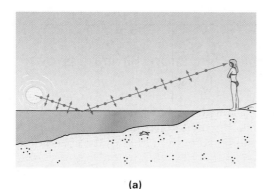

(a)

(b)

◄ **FIGURE 24.24 Glare reduction**
(a) Light reflected from a horizontal surface is partially polarized in the horizontal plane. When sunglasses are oriented so that their transmission axis is vertical, the horizontally polarized component of such light is not transmitted, so glare is reduced. **(b)** Polarizing filters for cameras use the same principle. The photo at right was taken with such a filter. Note the reduction in reflections from the store window.

It is interesting to note that light from a rainbow is also partially polarized. This polarization occurs because the reflection angle inside a water droplet is somewhat close to the Brewster angle. The plane of polarization is tangential to the rainbow at each point.

Example 24.7 ■ Sunlight on a Pond: Polarization by Reflection

Sunlight is reflected from the smooth surface of a pond. What is the Sun's altitude (the angle between the Sun and the horizon) when the polarization of the reflected light is greatest?

Thinking It Through. Incident light at the Brewster angle has the greatest polarization upon reflection, so the altitude of the Sun needs to be formed at this angle.

Solution. The index of refraction of water is listed in Table 22.1.

Given: $n_1 = 1$ *Find:* θ (altitude angle for greatest polarization)
        $n_2 = 1.33$ (Table 22.1)

Since the angle of incidence is measured from the normal and the altitude angle is measured from the horizon, the angle of incidence is the angle complementary to the altitude angle (draw a sketch to help yourself visualize this situation), so

$$\theta = 90° - \theta_p$$

where θ_p is the Brewster angle.
Using Eq. 24.15, we find that

$$\theta_p = \tan^{-1}\frac{n_2}{n_1} = \tan^{-1}\left(\frac{1.33}{1}\right) = 53.1°$$

So,

$$\theta = 90° - \theta_p = 90° - 53.1° = 36.9°$$

Follow-up Exercise. Light is incident on a flat, transparent material with an index of refraction of 1.38. At what angle of refraction would the transmitted light have the greatest polarization?

Polarization by Double Refraction (Birefringence)

When monochromatic light travels in glass, its speed is the same in all directions and is characterized by a single index of refraction. Any material that has such a property is said to be *isotropic*, meaning that it has the same optical characteristics in all directions. Some crystalline materials, such as quartz, calcite, and ice, are *anisotropic* with respect to the speed of light; that is, the speed of light, and therefore the index of refraction, is different in different directions within the material. Anisotropy gives rise to some unique optical properties, one of which is that the index of refraction may vary with the direction of propagation. Such materials are said to be doubly refracting, or to exhibit **birefringence**, and polarization is involved.

For example, a beam of unpolarized light incident on a birefringent crystal of calcite ($CaCO_3$, calcium carbonate) is illustrated in ▶Fig. 24.25. When the beam propagates at an angle to a particular crystal axis, the beam is doubly refracted and separated into two components, or rays, upon refraction. These two rays are linearly polarized in mutually perpendicular directions. One ray, called the *ordinary* (o) *ray*, passes through the crystal in an undeflected path and is characterized by an index of refraction n_o. The second ray, called the *extraordinary* (e) *ray*, is refracted and is characterized by an index of refraction n_e. The particular axis direction indicated by dashed lines in Fig. 24.25a is called the *optic axis*. Along this direction, $n_o = n_e$, and nothing extraordinary is noted about the transmitted light.

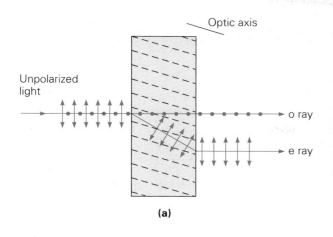

Optic axis

Unpolarized
light

o ray

e ray

(a)

(b)

◀ **FIGURE 24.25 Double refraction, or birefringence**
(a) Unpolarized light incident normal to the surface of a birefringent crystal and at an angle to a particular direction in the crystal (dashed lines) is separated into two components. The ordinary (o) ray and the extraordinary (e) ray are plane polarized in mutually perpendicular directions.
(b) Birefringence seen through a calcite crystal.

Some transparent materials have the ability to rotate the plane of polarization of linearly polarized light. This property, called **optical activity**, is due to the molecular structure of the material (▼Fig. 24.26a). The rotation may be clockwise or counterclockwise, depending on the molecular orientation. Optically active molecules include those of certain proteins, amino acids, and sugars.

Glasses and plastics become optically active when under stress. The greatest rotation of the direction of polarization of the transmitted light occurs in the regions where the stress is the greatest. Viewing the stressed piece of material through crossed polarizers allows the points of greatest stress to be identified. This determination is called *optical stress analysis* (Fig. 24.26b). Another use of polarizing films, the liquid crystal display (LCD), is described in the Insight on p. 804.

▼ **FIGURE 24.26 Optical activity and stress detection**   **(a)** Some substances have the property of rotating the polarization direction of linearly polarized light. This ability, which depends on the molecular structure of the substance, is called *optical activity*. **(b)** Glasses and plastics become optically active under stress, and the points of greatest stress are apparent when the material is viewed through crossed polarizers. Engineers can thus test plastic models of structural elements to see where the greatest stresses will occur when the models are "loaded." Here, a model of a suspension-bridge strut is being analyzed.

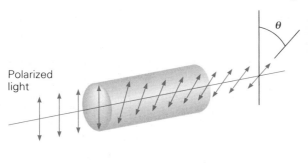

Polarized
light

θ

(a)

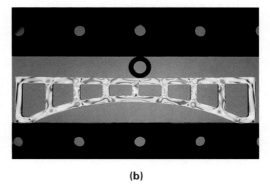

(b)

*24.5 Atmospheric Scattering of Light

OBJECTIVES: To **(a)** discuss scattering and **(b)** explain why the sky is blue and sunsets are red.

When light is incident on a suspension of particles, such as the molecules of air, some of the light may be absorbed and reradiated. This process is called *scattering*. The scattering of sunlight in the atmosphere produces some interesting effects,

LCDs and Polarized Light

LCDs (liquid crystal displays) are commonplace on watches, calculators, gas pumps, and even some television and computer monitors. The name "liquid crystal" may seem self-contradictory. Normally, when a crystalline solid melts, the resulting liquid no longer has an orderly atomic or molecular arrangement. Some organic compounds, however, pass through an intermediate state in which the molecules may rearrange somewhat, but still maintain the overall order that is characteristic of a crystal.

A liquid crystal flows like a liquid, but its optical properties may depend on the order of its molecules. For example, certain liquid crystals with orderly arrangements are transparent, but the crystalline order can easily be disturbed by applied electrical forces. The disordered liquid then scatters light and is opaque.

A common type of LCD, called a *twisted-nematic display*, makes use of the effect of a liquid-crystal on polarized light (Fig. 1). A display such as those commonly used on wristwatches and calculators is made by sandwiching a layer of liquid-crystal material between two glass plates that have fine parallel grooves, or channels, in their surfaces. One of the plates is then rotated or twisted 90°. The molecules in contact with the plates remain parallel to the grooves, and the result is a molecular orientation that rotates through 90° between the plates. In this configuration, the liquid crystal is optically active and will rotate in the direction of polarization of linearly polarized light.

The plates are then placed between crossed polarizing sheets and backed with a mirrored surface. Light entering and passing through the LCD is polarized, rotated 90°, re-

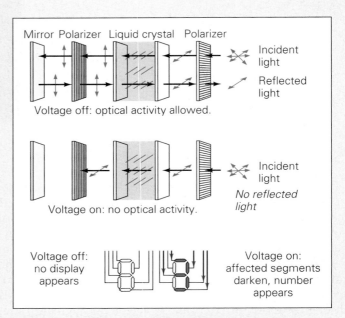

FIGURE 1 **Liquid-crystal display (LCD)** A twisted-nematic display is an application involving the optical activity of a liquid crystal and crossed polarizers. When the crystalline order is disoriented by an electric field from an applied voltage, the liquid crystal loses its optical activity in that region, and light is not transmitted and reflected. Numerals and letters are formed by applying voltages to segments of a block display.

including the polarization of skylight (that is, sunlight that has been scattered by the atmosphere), the blueness of the sky, and the redness of sunsets and sunrises.

Atmospheric scattering causes the skylight to be polarized. When unpolarized sunlight is incident on air molecules, the electric field of the light wave sets electrons of the molecules into vibration. The vibrations are complex, but these accelerated charges emit radiation, like the vibrating electrons in the antenna of a radio broadcast station (see Section 20.4). The intensity of this emitted radiation is strongest along a line perpendicular to the oscillation, and, as illustrated in ▶ Fig. 24.27, an observer viewing from an angle of 90° with respect to the direction of the sunlight will receive linearly polarized light, because of the charge oscillations normal to the surface. At other viewing angles, both components are present, and skylight seen through a polarizing filter appears partially polarized, because of the stronger component.

Since the scattering of light with the greatest degree of polarization occurs at a right angle to the direction of the Sun, at sunrise and sunset the scattered light from directly overhead has the greatest degree of polarization. The polarization of skylight can be observed by viewing the sky through a polarizing filter (or a polarizing sunglass lens) and rotating the filter. Light from different regions of the sky will be transmitted in different degrees, depending on its degree of polariza-

flected, and again rotated 90° by its components. After the return trip through the liquid crystal, the direction of polarization of the light is the same as that of the initial polarizer. Thus, the light is transmitted and leaves the display unit. Because of the reflection and transmission, the display appears to be of a light color (usually light gray) when illuminated with white, unpolarized light.

By using an analyzer, we can readily show the light from the bright regions of an LCD to be polarized. The whole display appears dark when the analyzer is properly oriented (Fig. 1). You may have noticed this effect if you have ever tried to see the time on the LCD of a wristwatch while you were wearing polarizing sunglasses (Fig. 2).

The dark numbers or letters on an LCD are formed by applying an electric field to parts of the liquid-crystal layer.

The molecules of the liquid crystal are polar, so an electric field can disorient them. Transparent, electrically conductive film coatings arranged in a seven-block pattern are put onto the glass plates. Each block, or display segment, has a separate electrical connection. When a voltage is applied across one or more of the segments, the electric field disorients the molecules of the liquid crystal in that area, and their optical activity is lost. (Note that all the numerals 0 through 9 can be formed by the segmented display.) The incident polarized light passing through the disoriented regions of the liquid crystal is absorbed by the second polarizer. Thus, these regions are opaque and appear dark. To produce images on small TV monitors, the display segments of the LCD are coupled so that many of them can be energized with a single lead.

One of the major advantages of LCDs is their low power consumption. Other similar displays, such as those using red light-emitting diodes (LEDs), produce light themselves, using relatively large amounts of power. LCDs produce no light, but instead use reflected light.

Color flat-panel computer monitors, which rely on LCD technology, are becoming more and more popular among mainstream computer users. As compared with traditional cathode ray tube (CRT) monitors of the same screen size, they are about one-quarter the physical size, consume about half the electric energy, and are much easier on the eyes. Computer displays usually are measured in pixels, much like a small square on a graph paper. To produce color, the three LCD segments (red, green, and blue) of a flat-panel monitor are grouped together on each pixel. By controlling the intensities of the three colors, each pixel can generate every color in the visible spectrum.

 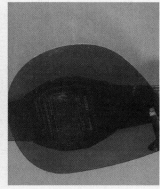

FIGURE 2 Polarized light   The light from an LCD is polarized, as can be shown by using polarizing sunglasses as an analyzer.

Related Exercise: 67

tion. It is believed that some insects, such as bees, use polarized skylight to determine navigational directions relative to the Sun.

Why the Sky Is Blue

The scattering of sunlight by air molecules also causes the sky to look blue. This effect is not due to polarization, but is caused by the selective absorption of light. As oscillators, air molecules have resonant frequencies (at which they scatter most efficiently) in the ultraviolet region. Consequently, when sunlight is scattered, the light in the blue end of the visible region is scattered more than in the red end.

For particles such as air molecules, which are much smaller than the wavelength of light, the scattering intensity S (of unit W/m^2) is inversely proportional to the wavelength to the fourth power (that is, $S \propto 1/\lambda^4$). This relationship between wavelength and scattering is called **Rayleigh scattering**, after Lord Rayleigh (1842–1919), a British physicist who derived it. This inverse relationship predicts that light of the shorter wavelength, or blue, end of the spectrum will be scattered more than light of the longer wavelength, or red, end. The scattered blue light is rescattered in the atmosphere and eventually is directed toward the ground. This is why the sky appears blue.

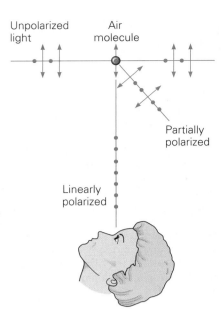

▲ FIGURE 24.27 Polarization by scattering When incident unpolarized sunlight is scattered by a gas molecule in the air, the light perpendicular to the direction of the incident ray is linearly polarized. Light scattered at some arbitrary angle is partially polarized. An observer at a right angle (90°) to the direction of the incident sunlight receives linearly polarized light.

▲ FIGURE 24.28 Red sky at night A spectacular red sunset over a mountaintop observatory in Chile. The red sky results from the scattering of sunlight by atmospheric gases and small solid particles. A directly observed reddening Sun is due to the scattering of the wavelengths toward the blue end of the spectrum in the direct line of sight.

Keep in mind that all colors are actually present in skylight, but that the dominant color is blue. (The sky doesn't appear violet because our eyes are more sensitive to blue light, and there is more blue than violet light in sunlight.) You may have noticed that the sky is more blue directly overhead than toward the horizon and appears white just above the horizon. (Look for this effect the next time you are outside on a clear day.) This is because there are relatively fewer air molecules—scatterers—directly overhead than toward the horizon. Multiple scatterings occur in the denser air near the horizon, and the recombination of scattered light gives rise to the white appearance. (Analogously, if you add a drop or two of milk to a glass of water and illuminate the suspension with intense white light, the scattered light has a bluish hue, while a glass of undiluted milk is white, because of multiple scatterings.) In the same way, atmospheric pollution may impart a milky white appearance to most or all of the sky.

Example 24.8 ■ The Red and the Blue: Rayleigh Scattering

How much more is light at the blue end of the visible spectrum scattered by air molecules than is light at the red end?

Thinking It Through. We know that Rayleigh scattering is proportional to $1/\lambda^4$ and that light from the blue end of the spectrum (shorter wavelength) is scattered more than light from the red end. The wording "how much more" implies a factor or ratio.

Solution. The Rayleigh scattering relationship is $S \propto 1/\lambda^4$, where S is the amount, or intensity, of scattering for a particular wavelength. Thus, you can form the ratio

$$\frac{S_{\text{blue}}}{S_{\text{red}}} = \left(\frac{\lambda_{\text{red}}}{\lambda_{\text{blue}}}\right)^4$$

The blue end of the spectrum (violet light) has a wavelength of about $\lambda_{\text{blue}} = 400$ nm, and red light has a wavelength of about $\lambda_{\text{red}} = 700$ nm. Inserting these values gives

$$\frac{S_{\text{blue}}}{S_{\text{red}}} = \left(\frac{\lambda_{\text{red}}}{\lambda_{\text{blue}}}\right)^4 = \left(\frac{700 \text{ nm}}{400 \text{ nm}}\right)^4 = 9.4 \qquad \text{or} \qquad S_{\text{blue}} = 9.4 S_{\text{red}}$$

Thus, blue light is scattered almost 10 times as much as is red light.

Follow-up Exercise. What wavelength of light is scattered twice as much as red light? What color light is this?

Why Sunsets and Sunrises Are Red

When the Sun is near the horizon, sunlight travels a greater distance through the denser air near the Earth's surface. Since the light therefore undergoes a great deal of scattering, you might think that only the least scattered light, the red light, would reach observers on the Earth's surface. This would explain red sunsets. However, it has been shown that the dominant color of white light after only molecular scattering is orange. Thus, there must be other types of scattering that shifts the light from the setting (or rising) Sun toward the red end of the spectrum (◄Fig. 24.28).

Red sunsets have been found to result from the scattering of sunlight by atmospheric gases *and* by small foreign particles. These particles are not necessary for the blueness of the sky, but are compulsory for deep-red sunsets and sunrises. (This is why spectacular red sunsets are observed in the months after a large volcanic eruption that puts tons of particulate matter into the atmosphere.) Red sunsets occur most often when there is a high-pressure air mass to the west, since the concentration of particles is generally greater in high-pressure air masses than in low-pressure air masses. Similarly, red sunrises occur most often when there is a high-pressure air mass to the east.

Now you can understand the old saying "Red sky at night, sailors' delight. Red sky in the morning, sailors take warning." Fair weather generally accompanies high-pressure air masses, because they are associated with reduced cloud formation. Most of the United States lies in the Westerlies wind zone, in which air masses generally move from west to east. A red sky at night is thus likely to indicate a fair-weather, high-pressure air mass to the west that will be coming your way. A red sky in the morning means that the high-pressure air mass has passed and poor weather may set in.

As a final note, how would you like a sky that is normally *red*? Then try Mars, the "red planet." The thin Martian atmosphere is about 95% carbon dioxide (CO_2). The CO_2 molecule is more massive than an oxygen (O_2) or a nitrogen (N_2) molecule. As a result, CO_2 molecules have a lower resonant frequency (longer wavelength) and preferentially scatter the red end of the visible spectrum. Hence, the Martian sky is red during the day. And what of the color of sunrises and sunsets on Mars? Think about it . . . .

Chapter Review

Important Concepts and Equations

- **Young's double-slit experiment** provides evidence of the wave nature of light and a way to measure the wavelength of light ($\approx 10^{-7}$ m).

  The angular position (θ) of the **bright fringes** satisfies the condition

  $$d \sin \theta = n\lambda \quad \text{for } n = 0, 1, 2, 3, \ldots \quad (24.3)$$

  where d is the slit separation.

  For small θ, the distance between the nth bright fringe and the central maximum is

  $$y_n \approx \frac{nL\lambda}{d} \quad \text{for } n = 0, 1, 2, 3, \ldots \quad (24.4)$$

- Light reflected at a media boundary for which $n_2 > n_1$ undergoes a 180° **phase change**. If $n_2 < n_1$, there is no phase change on reflection. The phase changes affect thin-film interference, which also depends on film thickness and index of refraction.

  The **minimum thickness for a nonreflecting film** is

  $$t_{min} = \frac{\lambda}{4n_1} \quad (\text{for } n_2 > n_1 > n_o) \quad (24.7)$$

- In a **single-slit diffraction** experiment, the **dark fringes** at location θ satisfy

  $$w \sin \theta = m\lambda \quad \text{for } m = 1, 2, 3, \ldots \quad (24.8)$$

  where w is the slit width. In general, the longer the wavelength as compared with the width of an opening or object, the greater is the diffraction.

- For a **diffraction grating**, the maxima (bright fringes) satisfy

  $$d \sin \theta = n\lambda \quad \text{for } n = 0, 1, 2, \ldots \quad (24.12)$$

  where $d = 1/N$ and N is the number of lines per unit length.

- **Polarization** is the preferential orientation of the electric field vectors that make up a light wave and is evidence that light is a transverse wave. Light can be polarized by selective absorption, reflection, double refraction (birefringence), and scattering.

  When the transmission axes of a polarizer and an analyzer make an angle of θ, the intensity of the transmitted light is given by **Malus' law**:

  $$I = I_o \cos^2 \theta \quad (24.14)$$

  In reflection, if the angle of incidence is equal to the **Brewster (polarizing) angle** θ_p, then the reflected light is linearly polarized:

  $$\tan \theta_p = \frac{n_2}{n_1} \quad \text{or} \quad \theta_p = \tan^{-1} \frac{n_2}{n_1} \quad (24.15)$$

- The intensity of Rayleigh scattering is inversely proportional to the fourth power of the wavelength of the light. The blueness of the Earth's sky results from the preferential scattering of sunlight by air molecules.

Exercises

24.1 Young's Double-Slit Experiment

1. In a Young's experiment using monochromatic light, if the slit spacing d decreases, the interference fringe spacing will (a) decrease, (b) increase, (c) remain unchanged, or (d) disappear.

2. If the path-length difference between two identical and coherent beams is 2λ when they arrive at a point on a screen, the point will be (a) bright, (b) dark, (c) multicolored, or (d) gray.

3. CQ When white light is used in Young's experiment, many bright fringes with a spectrum of colors are seen. In a given fringe, is the red end or the blue end closer to the central maximum? Explain.

4. CQ What would be observed in Young's experiment if the light passing through the slit were not coherent?

5. CQ Television pictures often flutter when an airplane passes by (▼Fig. 24.29). Explain a possible cause of this fluttering, based on interference effects.

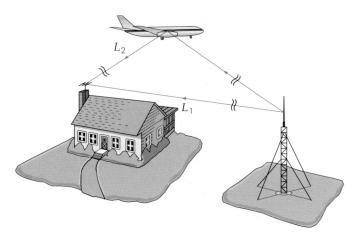

▲ **FIGURE 24.29 Interference** See Exercise 5.

6. CQ Will two flashlights held close to each other produce a stationary interference fringe pattern in a Young's double-slit experiment? Explain.

7. CQ Describe what would happen to the interference pattern in Young's double-slit experiment if the wavelength of the monochromatic light were to increase.

8. CQ The intensity of the central bright fringe in the interference pattern of a Young's double-slit experiment is about four times that of either light wave. Is this a violation of the conservation of energy? Explain.

9. ■ In the development of Young's experiment, a small-angle approximation ($\tan \theta \approx \sin \theta$) was used to find the lateral displacements of the bright and dark fringes. How good is this approximation? For example, what is the percentage error for $\theta = 15°$?

10. ■ To study wave interference, a student uses two speakers driven by the same sound wave of wavelength 0.50 m. If the distances from a point to the speakers differ by 0.75 m, will the waves interfere constructively or destructively at that point? What if the distances differ by 1.0 m?

11. ■ Two point sources of coherent sound are located at (2.0 m, 5.0 m) and (4.0 m, 3.0 m) in the x–y plane. What is the longest possible wavelength of the sound waves at the point (15 m, 20 m) if (a) they interfere destructively and (b) they have a path-length difference of $\lambda/8$?

12. ■■ (a) Derive a relationship that gives the locations of the dark fringes in a Young's double-slit experiment. What is the distance between the dark fringes? (b) For a third-order destructive interference (the third dark-side fringe from the central maximum), what is the path-length difference between that location and the two slits?

13. ■■ An interference pattern is formed on a screen when light of wavelength 550 nm is incident on two parallel slits 50 μm apart. At what angle can the second-order bright fringe be seen?

14. IE ■■ Monochromatic light passes through two narrow slits and forms an interference pattern on a screen. (a) If the wavelength of light used increases, will the distance between the bright fringes (1) increase, (2) remain the same, or (3) decrease? Why? (b) If the slit separation is 0.25 mm, the screen is 1.5 m away from the slits, and light of wavelength 680 nm is used, what is the distance from the center of the central maximum to the center of the third-order bright fringe?

15. ■■ In a double-slit experiment using monochromatic light, a screen is placed 1.25 m away from the slits, which have a separation distance of 0.0250 mm. The third-order bright fringe is 6.60 cm from the center of the central maximum. Find (a) the wavelength of the light and (b) the lateral displacement of the second-order bright fringe.

16. ■■ Monochromatic light illuminates two parallel slits that are 0.20 mm apart. The adjacent bright lines of the interference pattern on a screen 1.5 m away from the slits are 0.45 cm apart. What is the wavelength and color of the light?

17. IE ■■ Two parallel slits are illuminated with monochromatic light, and an interference pattern is observed on a screen. (a) If the distance between the slits were increased, would the distance between the bright fringes (1) increase, (2) remain the same, or (3) decrease? Why? (b) If the slit separation is 1.0 mm, the wavelength is 640 nm, and the distance from the slits to the screen is 3.00 m, what is the separation between adjacent interference maxima?

18. ■■ Yellow–green light ($\lambda = 550$ nm) is used in a double-slit experiment in which the slit separation distance is 1.75×10^{-4} m. If the screen is located 2.00 m from the slits, determine (a) the angular separation between the central maximum and the second-order bright fringe and (b) the lateral displacement of this fringe.

19. ■■ In a double-slit experiment with monochromatic light and a screen at a distance of 1.50 m from the slits, the angle between the second-order bright fringe and the central maximum is 0.0230 rad. If the separation distance of the slits is 0.0350 mm, what are (a) the wavelength and color of the light and (b) the lateral displacement of the fringe?

20. IE ■■■ (a) If the apparatus for a Young's double-slit experiment is completely immersed in water will the spacing of the interference fringes (1) increase, (2) remain the same, or (3) decrease? Why? (b) What would the lateral displacement in Exercise 14 be if the entire system were immersed in still water?

21. ■■■ Light of two different wavelengths is used in a double-slit experiment. The location of the third-order bright fringe for the first light, yellow–orange light ($\lambda = 600$ nm), coincides with the location of the fourth-order bright fringe for the other color's light. What is the wavelength of the other light?

24.2 Thin-Film Interference

22. For a thin film with $n_1 > n_o$ and $n_1 > n_2$, where n_1 is the index of refraction of the film, a film thickness for constructive interference of the reflected light is (a) $\lambda'/4$, (b) $\lambda'/2$, (c) λ', or (d) both (a) and (b).

23. For a thin film with $n_o < n_1 < n_2$, where n_1 is the index of refraction of the film, the minimum film thickness for destructive interference of the reflected light is (a) $\lambda'/4$, (b) $\lambda'/2$, or (c) λ'.

24. CQ What would be the effect on thin-film interference if the incident light were not perpendicular to the film?

25. CQ Most lenses used in cameras are coated with thin films and appear bluish purple when viewed with reflected light. What wavelengths are not visible in the reflected light?

26. CQ When destructive interference of two waves occurs at a certain location, there is no energy at that location. Is this situation a violation of the conservation of energy? Explain.

27. CQ At the center of a Newton's rings arrangement (Fig. 24.10a), the air wedge has a thickness of zero. Why is this area always dark?

28. CQ When a thin film of kerosene spreads out on water, the thinnest part looks bright. Is the index of refraction of kerosene greater than or less than that of water? Explain.

29. ■ Light of wavelength 550 nm in air is normally incident on a glass plate ($n = 1.5$) whose thickness is 1.1×10^{-5} m. (a) What is the thickness of the glass in terms of the wavelength of light in glass? (b) Will the light interfere constructively or destructively?

30. ■ A lens with an index of refraction of 1.60 is to be coated with a material ($n = 1.40$) that will make the lens nonreflecting for red light ($\lambda = 700$ nm) normally incident on the lens. What is the minimum required thickness of the coating?

31. ■ Magnesium fluoride ($n = 1.38$) is frequently used as a lens coating to make nonreflecting lenses. What is the difference in the minimum film thickness required for maximum transmission of blue light ($\lambda = 400$ nm) and of red light ($\lambda = 700$ nm)?

32. IE ■■ A solar cell is designed to have a nonreflective coating of a transparent material. (a) Will the thickness of the coating depend on the index of refraction of the underlying material in the solar cell? Discuss the possible scenarios. (b) If $n_{solar} > n_{film}$ and $n_{film} = 1.22$, what is the minimum thickness of the film for light with a wavelength of 550 nm?

33. ■■ A thin layer of oil ($n = 1.50$) floats on water. Destructive interference is observed for light of wavelengths 480 nm and 600 nm, each at a different location. Find the two minimum thicknesses of the oil film, assuming normal incidence.

34. ■■ A scientist wants to make a nonreflecting lens for an infrared-radiation detector. If the coating material has an index of refraction of 1.20, what should be the film thickness for radiation with a frequency of 3.75×10^{14} Hz?

35. ■■ Two parallel plates are separated by a small distance as illustrated in ▼Fig. 24.30. If the top plate is illuminated with light from a He–Ne laser ($\lambda = 632.8$ nm), for what minimum separation distances will the light be (a) reflected back to the observer from air-layer surfaces and (b) transmitted through the plates? [*Note*: $t = 0$ is *not* an answer for (b).]

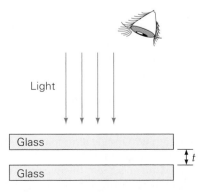

▲ **FIGURE 24.30 Reflection or transmission?**
See Exercise 35.

36. IE ■■■ An air wedge such as that shown in ▼Fig. 24.31 can be used to measure small dimensions, such as the diameter of a thin wire. (a) If the top glass plate is illuminated with monochromatic light, what kind of interference pattern will be observed? Why? (b) Express the locations of the bright interference fringes in terms of wedge thickness measured from the apex of the wedge.

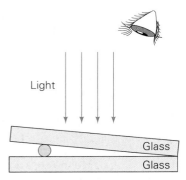

Light

Glass
Glass

▲ **FIGURE 24.31 Air wedge** See Exercises 36 and 37.

37. ■■■ The glass plates in Fig. 24.31 are separated by a thin, round filament. When the top plate is illuminated normally with light of wavelength 550 nm, the filament lies directly below the sixth bright fringe. What is the diameter of the filament?

24.3 Diffraction

38. In a single-slit diffraction pattern, (a) all maxima have the same width, (b) the central maximum is twice as wide as the side maxima, (c) the side maxima are twice as wide as the central maximum, or (d) none of the preceding.

39. As the number of lines per unit length of a diffraction grating increases, the spacing between bright fringes (a) increases, (b) decreases, or (c) remains unchanged.

40. CQ From Eq. 24.8, can the $m = 2$ dark fringes be seen if $w = \lambda$? How about the $m = 1$ dark fringe?

41. CQ In our discussion of single-slit diffraction, the length of the slit was assumed to be much greater than the width. What changes would be observed in the diffraction pattern if the length were comparable with the width of the slit?

42. CQ How does the width of the central maximum in a single-slit diffraction pattern change if the wavelength of light increases? Explain.

43. CQ In a diffraction grating, the slits are very closely spaced. What is the advantage of this design?

44. ■ A slit of width 0.20 mm is illuminated with monochromatic light of wavelength 480 nm, and a diffraction pattern is formed on a screen 1.0 m away from the slit. (a) What is the width of the central maximum? (b) What are the widths of the second-order and third-order bright fringes?

45. ■ A slit 0.025 mm wide is illuminated with red light ($\lambda = 680$ nm). How wide are (a) the central maximum and (b) the side maxima of the diffraction pattern formed on a screen 1.0 m from the slit?

46. ■ At what angle will the second-order diffraction maximum be seen from a diffraction grating of spacing 1.25 μm when illuminated by light of wavelength 550 nm?

47. ■ A venetian blind is essentially a diffraction grating—not for visible light, but for waves with longer wavelengths. If the spacing between the slats of a blind is 2.5 cm, (a) for what wavelength is there a first-order maximum at an angle of 10°, and (b) what type of radiation is this?

48. IE ■■ A single slit is illuminated with monochromatic light, and a screen is placed behind the slit to observe the diffraction pattern. (a) If the width of the slit is increased, will the width of the central maximum (1) increase, (2) remain the same, or (3) decrease? Why? (b) If the width of the slit is 0.50 mm, the wavelength is 680 nm, and the screen is 1.80 m from the slit, what is the angle between the second dark fringe ($m = 2$) and the central maximum, and what is the lateral displacement of this dark fringe?

49. ■■ What is the width of the central maximum for the single-slit arrangement in Exercise 48b?

50. ■■ A certain crystal gives a deflection angle of 25° for the first-order diffraction of monochromatic X rays with a frequency of 5.0×10^{17} Hz. What is the lattice spacing of the crystal?

51. ■■ Find the angles of the blue ($\lambda = 420$ nm) and red ($\lambda = 680$ nm) components of the first- and second-order spectra in a pattern produced by a diffraction grating with 7500 lines/cm.

52. IE ■■ (a) Discuss what limits the number of bright fringes that can be observed with a diffraction grating. (b) How many bright fringes appear when monochromatic light of wavelength 560 nm illuminates a diffraction grating that has 10 000 lines/cm, and what are their order numbers?

53. ■■ In a particular diffraction pattern, the red component (700 nm) in the second-order spectrum is deviated at an angle of 20°. (a) How many lines per centimeter does the grating have? (b) If the grating is illuminated with white light, how many bright fringes of the complete visible spectrum are produced?

54. ■■ White light whose components have wavelengths from 400 nm to 700 nm illuminates a diffraction grating with 4000 lines/cm. Do the first- and second-order spectra overlap? Justify your answer.

55. IE ■■ White light ranging from blue (400 nm) to red (700 nm) illuminates a diffraction grating with 8000 lines/cm. (a) For the first spectral order, which color, blue or red, will be closer to the central maximum? Why?

(b) What is the angular width of the first-order spectrum produced?

56. ■■ A diffraction grating with 8000 lines/cm is illuminated with a beam of monochromatic red light from a He–Ne laser ($\lambda = 632.8$ nm). How many side maxima are formed in the diffraction pattern, and at what angles are they observed?

57. ■■■ Show that for a diffraction grating, the violet ($\lambda = 400$ nm) portion of the third-order spectrum overlaps the yellow–orange ($\lambda = 600$ nm) portion of the second-order spectrum, regardless of the grating's spacing.

58. IE ■■■ A teacher standing in a doorway 1.0 m wide blows a whistle with a frequency of 1000 Hz to summon children from the playground (▼Fig. 24.32). Two boys are playing on the swings 100 m away from the school building. The boy on the left tells the other that they had better go, but the boy on the right, at an angle of 19.6° from a line normal to the doorway, asks "Why?"—he has not heard the whistle. (a) Is this situation possible? Why? (b) Taking the speed of sound in air to be 335 m/s, justify your answer mathematically.

1.0 m

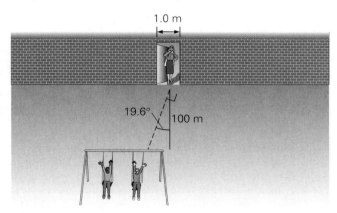

▲ FIGURE 24.32 **Moment of truth** See Exercise 58. (Not drawn to scale.)

24.4 Polarization

59. Light can be polarized by (a) reflection, (b) refraction, (c) scattering, or (d) all of the preceding.

60. The Brewster angle depends on (a) the index of refraction of a material, (b) Bragg's law, (c) internal reflection, or (d) interference.

61. CQ Given two pairs of sunglasses, could you tell if one or both were polarizing?

62. CQ How would you hold an analyzer to detect the polarization of a rainbow? Could you block out the polarized light completely? Explain.

63. CQ Suppose that you held two polarizing sheets in front of you and looked through both of them. How many times

would you see the sheets lighten and darken (a) if one were rotated through one complete rotation, (b) if both were rotated through one complete rotation at the same rate in opposite directions, (c) if both were rotated through one complete rotation at the same rate in the same direction, and (d) if both were rotated through one complete rotation in opposite directions, but one twice as fast as the other?

64. CQ How can you use the Brewster angle to determine the index of refraction of a material?

65. CQ Can a sound wave be polarized? Why or why not?

66. CQ How does selective absorption produce polarized light?

67. CQ If you place a pair of polarizing sunglasses in front of your calculator's LCD display and rotate them, what do you observe?

68. ■ Some types of glass have a range of indices of refraction of about 1.4 to 1.7. What is the range of the polarizing (Brewster) angle for these glasses?

69. ■ If the polarizing (Brewster) angle for a certain material in air is 58°, what is the index of refraction of the material?

70. ■ A polarizer–analyzer pair can have their transmission axes at either 30° or 45° angles. Which angle will allow more light to be transmitted?

71. ■■ Unpolarized light of intensity I_o is incident on a polarizer–analyzer pair with their transmission axes at a 30° angle. (a) What light intensity is transmitted through the polarizer? (b) How about after the analyzer?

72. ■■ A beam of light is incident on a glass plate ($n = 1.62$), and the reflected ray is completely polarized. What is the angle of refraction for the beam?

73. ■■ The critical angle for internal reflection in a certain medium is 45°. What is the polarizing (Brewster) angle for light externally incident on the medium?

74. IE ■■ The angle of incidence is adjusted so there is maximum linear polarization for the reflection of light from a transparent piece of plastic. (a) Will there be transmitted light? Why? (b) If the answer to (a) is yes, what is the angle of refraction in the plastic if it has an index of refraction of 1.22?

75. ■■ Sunlight is reflected off a vertical plate-glass window ($n = 1.55$). What would be the Sun's altitude (angle above the horizon) have to be for the reflected light to be completely polarized?

76. ■■ Find the polarizing (Brewster) angle for a piece of glass ($n = 1.60$) that is submerged in water.

77. ■■■ A plate of crown glass is covered with a layer of water. A beam of light traveling in air is incident on the

water and partially transmitted. Is there any angle of incidence for which the light reflected from the water–glass
interface will have maximum linear polarization? Justify
your answer mathematically.

*24.5 Atmospheric Scattering of Light

78. Which of the following colors is scattered the most in the
atmosphere? (a) blue; (b) yellow; (c) red; (d) color makes
no difference.

79. CQ Explain why the sky is red in the morning and
evening and blue during the day.

80. CQ (a) Why does the sky not have a uniform blueness?
(b) What color would an astronaut on the Moon see when
looking at the sky or into space?

Additional Exercises

81. If the Brewster angle for a glass plate is 1.05 rad, what is
the index of refraction of the glass?

82. If the slit width in a single-slit experiment were doubled,
the distance to the screen reduced by one-third, and the
wavelength of the light changed from 600 nm to 450 nm,
how would the width of the bright fringes be affected?

83. A glass plate has an index of refraction of 1.5. At what
angle of incidence will light be reflected from the plate
with the maximum linear polarization?

84. When illuminated with monochromatic light, a diffraction grating with 1000 lines/cm produces a first-order
maximum at 2.6° from the central maximum. What is the
wavelength of the light?

85. A thin air wedge between two flat glass plates forms bright
and dark interference bands when illuminated with
normally incident monochromatic light. (See Fig. 24.9.)
(a) Show that the thickness of the air wedge changes by
$\lambda/2$ from one bright fringe to the next, where λ is the
wavelength of the light. (b) What would be the change
in the thickness of the wedge between bright fringes if
the space were filled with a liquid with an index of refraction n?

86. A diffraction grating is designed to have the second-order
maxima at 10° from the central maximum for the red end
($\lambda = 700\text{ nm}$) of the visible spectrum. How many lines
per centimeter does the grating have?

87. Two parallel slits 0.075 mm apart are illuminated with
monochromatic light of wavelength 480 nm. Find the
angle between the center of the central maximum and the
center of an adjacent bright fringe.

88. Monochromatic light of wavelength 500 nm falls on two
slits separated by a distance of 40 μm. What is the distance
between the first-order and third-order bright fringes
formed on a screen 1.0 m away from the slits?

89. A film on a lens is 1.0×10^{-7} m thick and is illuminated
with white light. The index of refraction of the film is 1.4.
For what wavelength of light will the lens be nonreflecting?

90. What is the highest spectral order that can be seen in a diffraction grating with 9000 lines/cm when the grating is illuminated by white light?

91. In a double-slit experiment that uses monochromatic
light, the angular separation between the central maximum and the second-order bright fringe is 0.160°. What is
the wavelength of the light if the distance between the
slits is 0.350 mm?

92. A camera lens is coated with a thin layer of a material
that has an index of refraction of 1.35. This coating
makes the lens nonreflecting for light of wavelength
450 nm (in air) that is normally incident on the lens.
What is the thickness of the thinnest film that will make
the lens nonreflecting?

Vision and Optical Instruments

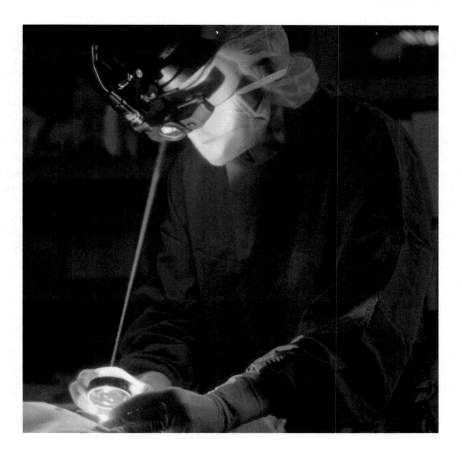

INSIGHTS

- Cornea Surgery and Bioengineering
- Telescopes Using Nonvisible Radiation

Vision is one of our chief means of acquiring information about the world around us. However, the images seen by many eyes are not clear or in focus, and glasses or some other remedy is needed. Great progress has been made in the last decade in surgical improvement or correction of vision defects. A popular procedure is laser surgery, as shown in the opening photo. Laser surgery can be used for such procedures as repairing torn retinas, destroying eye tumors, and stopping abnormal growth of blood vessels that can endanger vision.

Optical instruments, the basic function of which is to improve and extend the power of observation beyond that of the human eye, augment our vision. Mirrors and lenses are used in a variety of optical instruments, including microscopes and telescopes.

The earliest magnifying lenses were drops of water captured in a small hole. By the 17th century, craftsmen were able to grind fair-quality lenses for simple microscopes or magnifying glasses, which were used primarily for botanical studies. (These early lenses also found a use in spectacles.) Soon, the basic compound microscope, which uses two lenses, was developed. Modern compound microscopes, which can magnify an object up to 200 times, extended our vision into the microbe world.

Around 1609, Galileo used lenses to construct an astronomical telescope that allowed him to observe valleys and mountains on the Moon, sunspots, and the four largest satellites (moons) of Jupiter. Now, huge telescopes that use lenses and mirrors extend our vision light-years into the history of the universe.

813

What would our knowledge of the universe and our world be if these instruments and their modern versions had never been invented? Bacteria would still be unknown, and planets, stars, and galaxies would have remained nothing to us but mysterious points of light.

Mirrors and lenses were discussed in Chapter 23 and other optical phenomena in Chapter 24. The principles developed in those chapters can be applied to the study of vision and optical instruments. In this chapter, you will learn about the fundamental optical instrument, without which all others would be of little use—the human eye. You will also learn more about the design of microscopes and telescopes and about the factors that limit viewing with these devices.

25.1 The Human Eye

OBJECTIVES: **To (a) describe the optical workings of the eye and (b) explain some common vision defects and how they are corrected.**

Note: Image formation by a converging lens is discussed in Section 23.3; see Figure 23.15a.

The human eye is the most fundamental optical instrument, since without it, the field of optics would not exist. The human eye is analogous to a simple camera in several respects (▼Fig. 25.1). A simple camera consists of a converging lens, which is used to focus images on light-sensitive film at the back of the camera's interior chamber. (Recall from Chapter 23 that for relatively distant objects, a converging lens produces a smaller, inverted, real image.) There is an adjustable diaphragm opening, or aperture, and a shutter to control the amount of light entering the camera.

The eye, too, has a converging lens that focuses images on the light-sensitive lining (called the *retina*) on the rear surface of the eyeball. (Other transparent media in the eye also help create the image by refracting incoming light.) The eyelid might be thought of as a shutter; however, the shutter of a camera, which controls the expo-

▶ **FIGURE 25.1 Camera and eye analogy** In some respects, a simple camera is similar to the human eye. An image is formed on the film in a camera and on the retina of the eye. (The complex refractive properties of the eye are not shown here, because multiple refractive media are involved.)

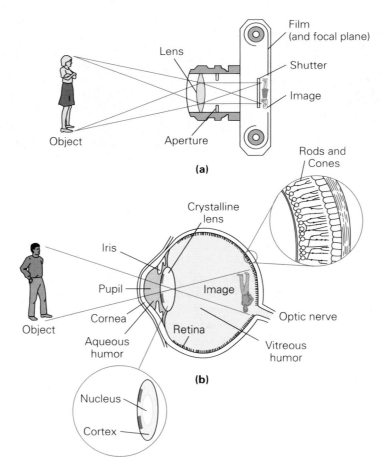

sure time, is generally opened only for a fraction of a second, while the eyelid is normally open for continuous exposure. The human nervous system performs a function analogous to that of a shutter in analyzing image signals from the eye at a rate of 20 to 30 times per second. The eye might therefore be better likened to a movie or video camera, which exposes a similar number of frames (images) per second.

Although the optical functions of the eye are relatively simple, its physiological functions are quite complex. As Fig. 25.1b shows, the eyeball is a nearly spherical chamber. It has an internal diameter of about 1.5 cm and is filled with a transparent jellylike substance called the *vitreous humor*. The eyeball has a white outer covering called the *sclera*, part of which is visible as the "white" of the eye. Light enters the eye through a curved, transparent tissue called the *cornea* and passes into a clear fluid known as the *aqueous humor*. Behind the cornea is a circular diaphragm, the *iris*, whose central hole is called the *pupil*. The iris contains the pigment that determines eye color. Through muscle action, the iris can change the area of the pupil (from 2 to 8 mm in diameter), thereby controlling the amount of light entering the eye.

Behind the iris is a *crystalline lens*, a converging lens composed of microscopic glassy fibers. A simplified model of the crystalline lens has a core (nucleus) with a uniform index of refraction of $n = 1.406$ and an outer cortex with an uniform index of refraction of $n = 1.386$. When tension is exerted on the lens by attached muscles, the glassy fibers slide over each other, causing the shape and focal length of the lens to change, thus focusing the image properly. Notice that an inverted image is formed. We do not see an inverted image, however, because somehow the brain interprets this image as being right side up.

On the back interior wall of the eyeball is a light-sensitive surface called the **retina**. From the retina, the optic nerve relays signals to the brain. The retina is composed of nerves and two types of light receptors, or photosensitive cells, called **rods** and **cones**, because of their shapes. The rods are more sensitive to light than are the cones and distinguish light from dark in low light intensities (twilight vision). The cones can distinguish frequency ranges of sufficiently intense light, which the brain interprets as colors (color vision). Most of the cones are clustered together around a central region of the retina. The rods, which are more numerous than the cones, are outside this region and are distributed nonuniformly over the retina.

The focusing adjustment of the eye differs from that of a simple camera. A camera lens has a constant focal length, and the image distance is varied by moving the lens relative to the film in order to produce sharp images on the film for different object distances. In the eye, the image distance is constant, and the focal length of the lens is varied (as the attached muscles act to change the eye's shape) in order to produce sharp images on the retina for different object distances. When the eye is focused on distant objects, the muscles are relaxed, and the crystalline lens is thinnest and has a power of about 20 D (diopters). Recall that the power (P) of a lens in diopters (D) is the reciprocal of its focal length *in meters*. So 20 D corresponds to a focal length of $f = 1/(20\,\text{D}) = 0.050$ m. When the eye is focused on closer objects, the lens becomes thicker, and the radius of curvature and the focal length are decreased. The lens power may increase to 30 D ($f = 0.033$ m), or even more in young children. The adjustment of the focal length of the crystalline lens is called *accommodation*. (Look at a nearby object and then at an object in the distance, and notice how fast accommodation takes place. It's practically instantaneous.)

Note: This relationship is presented in Eq. 23.9, in Section 23.5.

The extremes of the range over which distinct vision (sharp focus) is possible are known as the *far point* and the *near point*. The *far point* is the greatest distance at which the eye can see objects clearly and is taken to be infinity. The *near point* is the position closest to the eye at which objects can be seen clearly. This position depends on the extent to which the lens can be deformed (thickened) by accommodation. The range of accommodation gradually diminishes with age as the crystalline lens loses its elasticity. That is, the near point gradually recedes with age. The approximate positions of the near point at various ages are listed in Table 25.1.

Note: The eye sees clearly between its far point and near point.

TABLE 25.1 Approximate Near Points of the Normal Eye at Different Ages	
Age (years)	Near Point (centimeters)
10	10
20	12
30	15
40	25
50	40
60	100

PHYSLET® ILLUSTRATION

Nearsightedness

PHYSLET® ILLUSTRATION

Farsightedness

Children can see sharp images of objects that are within 10 cm of their eyes, and the crystalline lens of a normal young-adult eye can be deformed to produce sharp images of objects as close as 12 to 15 cm. However, adults at about the age of 40 normally experience a shift in the near point to beyond 25 cm. You may have noticed people over the age of 40 holding reading material away from their eyes to move it out to within the range of accommodation. When the print gets too small or the arms too short, corrective reading glasses are the solution. The recession of the near point with age is not considered an abnormal defect of vision, since it proceeds at about the same rate in most normal eyes.

Vision Defects

Speaking of the "normal" eye (▾Fig. 25.2a) implies that some eyes produce defective vision. This is indeed the case, as is quite apparent from the number of people who wear glasses or contact lenses. Many people's eyes cannot accommodate within the normal range of 25 cm to infinity (near point to far point). These people have one of the two most common visual defects: nearsightedness (myopia) or farsightedness (hyperopia). Both of these conditions can usually be corrected with glasses.

Nearsightedness (or *myopia*) is the ability to see nearby objects clearly, but not distant objects. That is, the far point is not infinity, but some nearer point. When an object beyond the far point is viewed, the rays come to focus in front of the retina (Fig. 25.2b). As a result, the image on the retina is blurred, or out of focus. As the object is moved closer to the eye, its image moves back toward the retina. If the object is moved to within the far point, a sharp image is seen, formed on the retina.

Nearsightedness arises because the eyeball is too long or perhaps because the curvature of the cornea is too great. Whatever the reason, the images of distant objects are focused in front of the retina. Appropriate diverging lenses correct this condition. Such a lens causes the rays to diverge, and the eye focuses the image farther back so that it falls on the retina.

Farsightedness (or *hyperopia*) is the ability to see distant objects clearly, but not nearby objects. That is, the near point is not at the normal position, but at some point farther from the eye. The image of an object that is closer to the eye than the near point is formed behind the retina (Fig. 25.2c). Farsightedness arises because the eyeball is too short or perhaps because of insufficient curvature of the cornea. A similar farsighted condition (*presbyopia*) occurs when the near point recedes with age, as discussed previously.

Farsightedness is usually corrected with appropriate converging lenses. Such a lens causes the rays to converge on the retina, and the eye is then able to focus the image on the retina. Converging lenses are also used to correct the farsightedness associated with the natural recession of the near point with age. Thus, people middle aged or older often wear reading glasses, which employ converging lenses.

▾ **FIGURE 25.2 Nearsightedness and farsightedness** (a) The normal eye produces sharp images on the retina for objects located between its near point and its far point. The image is real, inverted, and always smaller than the object. (Why?) Here, the object is a distant, upward-pointing arrow (not shown) and the light rays come from its tip. (b) In a nearsighted eye, the image of a *distant* object is focused *in front of* the retina. This defect is corrected with a diverging lens. (c) In a farsighted eye, the image of a *nearby* object is focused *behind* the retina. This defect is corrected with a converging lens. (Not drawn to scale.)

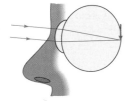

(a) Normal

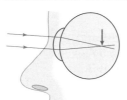

Uncorrected

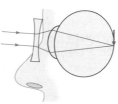

Corrected

(b) Nearsightedness (myopia)

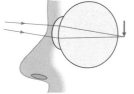

Uncorrected

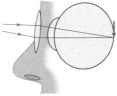

Corrected

(c) Farsightedness (hyperopia)

Integrated Example 25.1 ■ Correcting Nearsightedness: Use of Diverging Lenses

(a) An optometrist has a choice to give a patient either regular glasses or contact lenses to correct his nearsightedness. Usually, regular glasses sit a few centimeters in front of the eye and contact lenses right on the eye. Should the power of the contact lenses prescribed be (1) the same as, (2) greater than, or (3) less than that of the regular glasses? Why? (b) A certain nearsighted person cannot see objects clearly when they are more than 78 cm from either eye. What power must corrective lenses have, for both regular glasses and contact lenses, if this person is to see distant objects clearly? Assume that the glasses are 3.0 cm in front of the eye.

(a) Conceptual Reasoning. For nearsightedness, the corrective lens is diverging (▼Fig. 25.3). The lens must effectively put the image of a distant object ($d_o = \infty$) at the far point d_f. The image, which acts as an object for the eye, is then within the range of accommodation. Since the image distance is *measured from the lens*, a contact lens will have a *longer* image distance. For a contact lens, $d_i = -|d_f|$, and for regular glasses, $d_i = -|d_f - d|$, where d is the distance between the regular glasses and the eye. A minus sign and absolute values are used for image distance because the image is virtual, being on the object side of the lens. (You may recall from Chapter 23 that diverging lenses can form only virtual images.)

Recall that the power of a lens is $P = 1/f$ (Eq. 23.9). We can use the thin-lens equation (Eq. 23.5) to find P if we can determine the object and image distances, d_o and d_i:

$$P = \frac{1}{f} = \frac{1}{d_o} + \frac{1}{d_i} = \frac{1}{\infty} + \frac{1}{d_i} = \frac{1}{d_i} = -\frac{1}{|d_i|}$$

That is, a longer $|d_i|$ will yield a smaller P, so the contact lenses should have a lower power than the regular glasses. Thus, the answer is (3).

(b) Thinking It Through. Once we understand how corrective lens work, the calculation for part (b) is straightforward.

Given: $d_f = 78\text{ cm} = 0.78\text{ m}$ (far point)   *Find:* P (in diopters) for regular glasses
$d = 3.0\text{ cm} = 0.030\text{ m}$ P (in diopters) for contact lenses

For regular glasses,

$$|d_i| = |d_f - d| = 0.78\text{ m} - 0.030\text{ m} = 0.75\text{ m}$$

(See Fig 25.3, which is not drawn to scale.) So, $d_i = -0.75$ m.
Then, using the thin-lens equation, we get

$$P = \frac{1}{f} = \frac{1}{d_o} + \frac{1}{d_i} = \frac{1}{\infty} + \frac{1}{-0.75\text{ m}} = -\frac{1}{0.75\text{ m}} = -1.33\text{ D}$$

or

$$f = \frac{1}{P} = \frac{1}{-1.33\text{ D}} = -0.75\text{ m}$$

Note: Review Example 23.6.

Note: Image formation by a diverging lens is discussed in Section 23.3; see Figure 23.18.

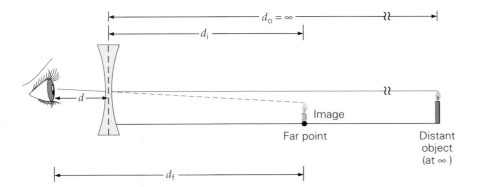

◀ FIGURE 25.3 Correcting nearsightedness A diverging lens is used. See Integrated Example 25.1. Only regular glasses are shown. For contact lenses, the lens is immediately in front of the eye ($d = 0$).

A negative, or diverging, lens with a power of 1.33 D is needed.
For contact lenses,

$$|d_i| = |d_f| = 0.78 \text{ m}$$

($d = 0$.) So, $d_i = -0.78$ m.
Then, using the thin-lens equation, we get

$$P = \frac{1}{\infty} + \frac{1}{-0.78 \text{ m}} = -\frac{1}{0.78 \text{ m}} = -1.28 \text{ D}$$

or

$$f = \frac{1}{-1.28 \text{ D}} = -0.78 \text{ m}$$

Follow-up Exercise. Suppose a mistake were made for regular glasses in this Example such that a "corrective" lens of +1.33 D were used. What effect would this error have? (*Answers to all Follow-up Exercises are at the back of the text.*)

If the far point is changed by corrective diverging lenses, as in Integrated Example 25.1, the near point will be affected as well. This condition causes another viewing problem by making close-up vision worse, but bifocal lenses can be used in this situation to address the problem. Bifocals were invented by Benjamin Franklin, who glued two lenses together. They are now made by grinding or molding lenses with different curvatures in two different regions. Both nearsightedness and farsightedness can be treated at the same time with bifocals. (Trifocals are also available, with lenses having three different curvatures.)

A more modern technique that involves the use of a laser to correct nearsightedness is shown in the Insight entitled "Cornea Surgery and Bioengineering," on page 819. In this procedure, a flap of material on the surface of the cornea is cut and pulled up. The laser is then used to shape the exposed surface of the cornea, which changes its refractive characteristics so as to have the image of a distant object fall on the retina. The corneal flap is then replaced. This procedure eliminates the need for corrective lenses. It is relatively painless, and the time to recover clear vision is on the order of 24 hours.

Example 25.2 ■ Correcting Farsightedness: Use of a Converging Lens

A farsighted person has a near point of 75 cm for one eye and a near point of 100 cm for the other. What powers should contact lenses have to allow the person to see an object clearly at a distance of 25 cm?

Thinking It Through. For farsightedness, the corrective lens must be converging and must form the image at its eye's near point of an object at the normal eye's near point.

Solution. Let us distinguish the eyes as 1 and 2. The image distances are negative. (Why?)

Given: $d_{i_1} = -75 \text{ cm} = -0.75 \text{ m}$ *Find:* P_1 and P_2 (lens power for each eye)
$d_{i_2} = -100 \text{ cm} = -1.0 \text{ m}$
$d_o = 25 \text{ cm} = 0.25 \text{ m}$

The optics of a person's eyes are usually different, as in this problem, and a different lens prescription is usually required for each eye. In this case, each lens is to form an image at its eye's near point of an object that is at a distance (d_o) of 0.25 m. The image will then act as an object within the eye's range of accommodation. This situation corresponds to a person wearing reading glasses (▶Fig. 25.4). (For the sake of clarity, the lens in Fig. 25.4a is not in contact with the eye.)

Cornea Surgery and Bioengineering

The imperfect shapes or surfaces of the cornea often cause refractive errors that cause vision defects. For example, a cornea that is curved too much causes nearsightness, a flatter-than-normal cornea causes farsightness, and irregular cornea surfaces cause astigmatism.

Recently, laser surgery to reshape the cornea has become very popular (Fig. 1). This procedure corrects the defective shape or irregular surface of the cornea so that it can better focus light on the retina, thereby reducing or even eliminating vision defects.

First, a very precise instrument called a *microkeratome* is used to create a thin corneal flap with a hinge on one side

of the cornea (Fig. 2a.). Once the flap is made and folded back, a tightly focused ultraviolet pulsed laser is used to perform the procedure in the deeper layers of the cornea. Each laser pulse accurately removes a microscopic layer of the inner cornea in the targeted area, thus reshaping the cornea to correct vision defects (Fig. 2b). Once the procedure is performed, the flap is placed back in its original position without the need for stitches (Fig. 2c.). The procedure is usually painless, and patients typically have only minimal discomfort.

Further exciting advances in the treatment of vision defects are on the horizon. For example, researchers have developed techniques for replacing a damaged cornea with freshly bioengineered tissues. If the patient has one healthy eye, stem cells from it are harvested and then planted. The cells will grow into a sturdy layer of tissue that can be used to replace the bad corneal tissues by stitching the new tissue onto the damaged eye. If both of the patient's eyes are damaged, donor tissues can be collected from close relatives.

Related Exercise: 2

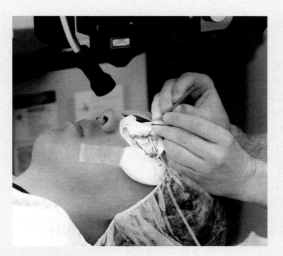

FIGURE 1   Eye surgery   Laser surgery is performed to reshape the cornea. Some patients achieve corrected vision within a day after this procedure. Notice that the surgeon is wearing no latex gloves. The fine chalk dust used on the gloves as a lubricant could contaminate the eye.

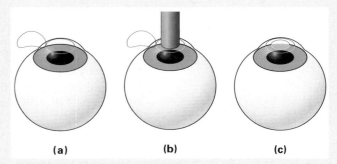

(a)   **(b)**   **(c)**

FIGURE 2   Cornea reshaping   **(a)** A flap is made on the corneal surface. **(b)** A laser beam is used to reshape the cornea. **(c)** The flap is placed back.

▼ **FIGURE 25.4   Reading glasses and correcting farsightedness**   **(a)** When an object at the normal near point (25 cm) is viewed through reading glasses with converging lenses, the image is formed farther away, but within the eye's range of accommodation (beyond the receded near point). See Example 25.2. **(b)** Small print as viewed through the lens of reading glasses. The camera used to take this picture is focused past this page onto where the virtual image is.

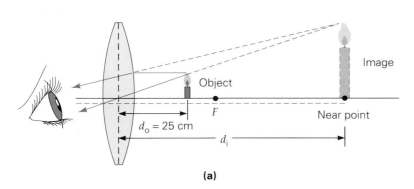

(a)

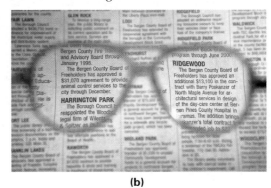

(b)

The image distances are negative, since the images are virtual (i.e., the image is on the same side as the object). With contact lenses, the distance from the eye to the object and the distance from the lens to the object are taken to be the same. Then

$$P_1 = \frac{1}{f_1} = \frac{1}{d_o} + \frac{1}{d_{i_1}} = \frac{1}{0.25 \text{ m}} - \frac{1}{0.75 \text{ m}} = \frac{2}{0.75 \text{ m}} = +2.7 \text{ D}$$

and

$$P_2 = \frac{1}{f_2} = \frac{1}{d_o} + \frac{1}{d_{i_2}} = \frac{1}{0.25 \text{ m}} - \frac{1}{1.0 \text{ m}} = \frac{3}{1.0 \text{ m}} = +3.0 \text{ D}$$

Note that these lenses are positive, or converging, as expected.

Follow-up Exercise. A mistake is made in grinding or molding the corrective lenses in this Example such that the left lens is made to the prescription intended for the right eye, and vice versa. What effect would this error have?

Another common defect of vision is **astigmatism**, which is usually due to a refractive surface, normally the cornea or crystalline lens, being out of round (nonspherical). As a result, the eye has different focal lengths in different planes (▼Fig. 25.5a). Points may appear as lines, and the image of a line may be distinct in one direction and blurred in another or blurred in both directions. A test for astigmatism is given in Fig. 25.5b.

Astigmatism can be corrected with lenses that have greater curvature in the plane in which the cornea or crystalline lens has deficient curvature (Fig. 25.5c). Astigmatism is lessened in bright light, because the pupil of the eye becomes smaller, so only rays near the axis are entering the eye, thus avoiding the outer edges of the cornea.

You have probably heard of 20/20 vision. But what is it? *Visual acuity* is a measure of how vision is affected by object distance. This quantity is commonly determined by using a chart of letters placed at a given distance from the eyes. The result is usually expressed as a fraction: The *numerator* is the distance at which the test eye sees a standard symbol, such as the letter "E," clearly; the *denominator* is the distance at which the letter is seen clearly by a *normal* eye. A 20/20 (test/normal) rating, which is sometimes called "perfect" vision, means that at a distance of 20 ft, the eye being tested can see standard-sized letters as clearly as can a normal eye.

A person's eyes can have differing visual acuity; for example, one eye may have 20/20 vision and the other 20/30 vision. The latter means that the eye can read standard-sized letters from a chart at a maximum distance of 20 ft, while someone with normal vision could do so at 30 ft. That is, the eye in question has to

▼ **FIGURE 25.5 Astigmatism** When one of the eye's refracting components is not spherical, the eye has different focal lengths in different planes. **(a)** The effect occurs because rays in the vertical plane (red) and horizontal plane (blue) are focused at different points: F_v and F_h, respectively. **(b)** To someone with eyes that are astigmatic, some or all of the lines in this diagram will appear blurred. **(c)** Nonspherical lenses, such as plano-convex cylindrical lenses, are used to correct astigmatism.

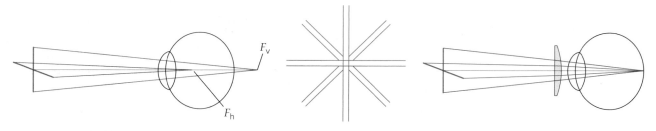

(a) Uncorrected astigmatism   **(b) Test for astigmatism**   **(c) Corrected by lens**

be closer to the chart than a normal eye would to see the letters clearly; hence, the eye is *less* sharp than a normal eye. Similarly, a person with 20/15 vision would have a particularly sharp eye relative to a normal eye. The test eye could see an object clearly from a distance of 20 ft, whereas the normal eye would have to be at a distance of 15 ft in order to do so. Thus, the greater the ratio or fraction, the better is the visual acuity of the eye.

25.2 Microscopes

OBJECTIVES: To (a) distinguish between lateral and angular magnification and (b) describe simple and compound microscopes and their magnifications.

Microscopes are used to magnify objects so that we can see more detail or see features that are normally indiscernible. Two basic types of microscope will be considered here.

The Magnifying Glass (A Simple Microscope)

When we look at an object in the distance, it appears to be very small. As it is brought or comes closer to our eyes, it appears larger. How large an object appears depends on the size of the image on the retina. This size is related to the angle subtended by the object (▼Fig. 25.6): The greater the angle, the larger is the image.

When we want to examine detail or look at something closely, we bring it close to our eyes so that it subtends a greater angle. For example, you may examine the detail of a figure in this book by bringing it closer to your eyes. You'll see the greatest amount of detail when the book is at your near point (assuming that you are not wearing glasses). If your eyes were able to accommodate to shorter distances, an object brought very close to them would appear even larger. However, as you can easily prove by bringing this book very close to your eyes, images are blurred when objects are at distances inside the near point.

A **magnifying glass**, which is simply a single convex lens (sometimes called a *simple microscope*), forms a clear image of an object when it is closer than the near point. In such a position, the image of an object subtends a greater angle and therefore appears larger, or magnified (▶Fig. 25.7). The lens produces a virtual image beyond the near point on which the eye focuses. If a handheld magnifying glass is used, its position is usually adjusted until this image is seen clearly.

As illustrated in Fig. 25.7, the angle subtended by the virtual image of an object is much greater when a magnifying glass is used. The magnification of an object *viewed through a magnifying glass* is expressed in terms of this angle. This **angular magnification**, or magnifying power, is designated by the symbol m. The angular magnification is defined as the ratio of the angular size of the object as viewed through the magnifying glass (θ) to the angular size of the object as viewed without the magnifying glass (θ_o):

$$m = \frac{\theta}{\theta_o} \quad \textit{angular magnification} \quad (25.1)$$

Note: Angular magnification is not the same as lateral magnification, which is discussed in Section 23.1. (See Eq. 23.1.)

(This m is not the same as M, the lateral magnification, which is a ratio of heights: $M = h_i/h_o$.)

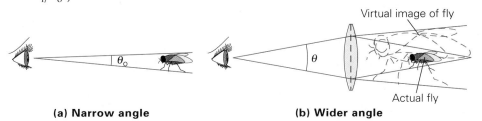

(a) **Narrow angle**

(b) **Wider angle**

◀ **FIGURE 25.6 Magnification and angle** (a) How large an object appears is related to the angle subtended by the object. (b) The angle and the size of the virtual image of an object are increased with a converging lens.

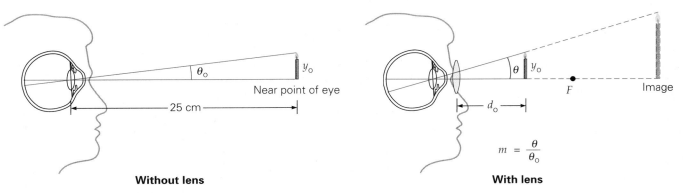

Without lens   With lens

$$m = \frac{\theta}{\theta_o}$$

▲ FIGURE 25.7 **Angular magnification** The angular magnification (m) of a lens is defined as the ratio of the angular size of an object viewed through the lens to the angular size of the object viewed without the lens: $m = \theta/\theta_o$.

The maximum angular magnification occurs when the image seen through the glass is at the eye's near point, $d_i = -25$ cm, since this position is as close as it can be seen clearly. (A value of 25 cm will be assumed to be typical for the near point in this discussion. The minus sign is used because the image is virtual.) The corresponding object distance can be calculated from the thin-lens equation, Eq. 23.5, as

$$d_o = \frac{d_i f}{d_i - f} = \frac{(-25 \text{ cm}) f}{-25 \text{ cm} - f}$$

or

$$d_o = \frac{25f}{25 + f} \qquad (25.2)$$

where f must be in centimeters.

The angular sizes of the object are related to its height by

$$\tan \theta_o = \frac{y_o}{25} \qquad \text{and} \qquad \tan \theta = \frac{y_o}{d_o}$$

(See Fig. 25.7.) Assuming that a small-angle approximation $(\tan \theta \approx \theta)$ is valid gives

$$\theta_o \approx \frac{y_o}{25} \qquad \text{and} \qquad \theta \approx \frac{y_o}{d_o}$$

Then the maximum angular magnification can be expressed as

$$m = \frac{\theta}{\theta_o} = \frac{y_o/d_o}{y_o/25} = \frac{25}{d_o}$$

Substituting for d_o from Eq. 25.2 gives

$$m = \frac{25}{25f/(25 + f)}$$

which simplifies to

$$m = 1 + \frac{25 \text{ cm}}{f} \qquad \begin{array}{l} \textit{angular magnification for} \\ \textit{image at near point (25 cm)} \end{array} \qquad (25.3)$$

Thus, lenses with shorter focal lengths give greater angular magnifications.

In the derivation of Eq. 25.3, the object being viewed by the unaided eye was taken to be at the near point, as was the image viewed through the lens. Actually, the normal eye can focus on an image located anywhere between the near point and infinity. At the extreme where the image is at infinity, the eye is more relaxed—the muscles attached to the crystalline lens are relaxed, and the lens is thin. For the image to be at infinity, the object must be at the focal point of the lens. In this case,

$$\theta \approx \frac{y_o}{f}$$

and the angular magnification is

$$m = \frac{25 \text{ cm}}{f} \qquad \begin{array}{l} \textit{angular magnification} \\ \textit{for image at infinity} \end{array} \qquad (25.4)$$

Mathematically, it seems that the magnifying power can be increased to any desired value by using lenses that have sufficiently short focal lengths. Physically, however, lens aberrations limit the practical range of a single magnifying glass to about $3\times$ or $4\times$ (read as "three ex" and "four ex"), or a sharp image magnification of three or four times the size of the object when used normally.

Example 25.3 ■ Elementary: Angular Magnification of a Magnifying Glass

Sherlock Holmes uses a converging lens with a focal length of 12 cm to examine the fine detail of some cloth fibers found at the scene of a crime. (a) What is the maximum magnification given by the lens? (b) What is the magnification for relaxed-eye viewing?

Thinking It Through. Equations 25.3 and 25.4 apply here. Part (a) asks for the maximum magnification, which is discussed in the derivation of Eq. 25.3 and occurs when the image formed by the lens is at the near point of the eye. For part (b), note that the eye is most relaxed when viewing distant objects.

Solution.

Given: $f = 12$ cm   Find: (a) m (d_i = near point)
                              (b) m ($d_i = \infty$)

(a)   For Equation 25.3, the near point was taken to be 25 cm:

$$m = 1 + \frac{25 \text{ cm}}{f} = 1 + \frac{25 \text{ cm}}{12 \text{ cm}} = 3.1\times$$

(b)   Equation 25.4 gives the magnification for the image formed by the lens at infinity:

$$m = \frac{25 \text{ cm}}{f} = \frac{25 \text{ cm}}{12 \text{ cm}} = 2.1\times$$

Follow-up Exercise.   Taking the maximum practical magnification of a magnifying glass to be $4\times$, which would have the longer focal length, a glass for near-point viewing or one for distant viewing, and how much longer?

The Compound Microscope

A compound microscope provides greater magnification than is attained with a single lens, or a simple microscope. A basic **compound microscope** consists of a pair of converging lenses, each of which contributes to the magnification (▶Fig. 25.8a). The converging lens with a relatively short focal length ($f_o < 1$ cm) is known as the **objective**. It produces a real, inverted, and enlarged image of an object positioned

Note: It might be useful to review Section 23.3 and Fig. 23.15.

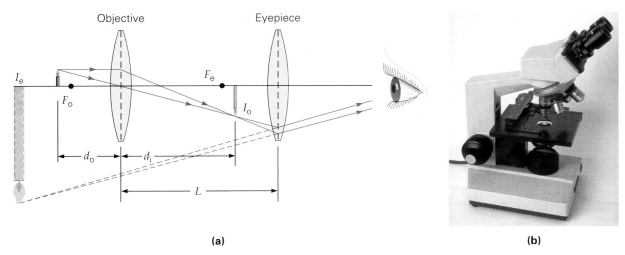

Objective | **Eyepiece**

(a)

(b)

▲ **FIGURE 25.8 The compound microscope** (a) In the optical system of a compound microscope, the real image formed by the objective falls just within the focal point of the eyepiece (F_e) and acts as an object for this lens. An observer looking through the eyepiece sees an enlarged image. **(b)** A compound microscope.

slightly beyond its focal point. The other lens, called the **eyepiece**, or **ocular**, has a longer focal length (f_e is a few centimeters) and is positioned so that the image formed by the objective falls just *inside* its focal point. This lens forms a magnified, inverted, and virtual image that is viewed by the observer. In essence, the objective gives a magnified real image, and the eyepiece is a simple magnifying glass.

The total magnification (m_{total}) of a lens combination is the product of the magnifications produced by the two lenses. The image formed by the objective is larger than its object by a factor M_o equal to the lateral magnification ($M_o = -d_i/d_o$). In Fig. 25.8a, note that the image distance for the objective lens is approximately equal to L, the distance between the lenses—that is, $d_i \approx L$. (The image I_o is formed by the objective just inside the focal point of the eyepiece, which has a short focal length.) Also, since the object is very close to the focal point of the objective, $d_o \approx f_o$. With these approximations,

$$M_o \approx -\frac{L}{f_o}$$

Eq. 25.4 gives the angular magnification of the eyepiece for an image at infinity.

$$m_e = \frac{25 \text{ cm}}{f_e}$$

Since the object for the eyepiece (the image formed by the objective) is very near the focal point of the eyepiece, a good approximation is

$$m_{total} = M_o m_e = -\left(\frac{L}{f_o}\right)\left(\frac{25 \text{ cm}}{f_e}\right)$$

or

$$m_{total} = -\frac{(25 \text{ cm})L}{f_o f_e} \qquad \begin{array}{l} \textit{angular magnification} \\ \textit{of compound microscope} \end{array} \qquad (25.5)$$

where f_o, f_e, and L are in centimeters.

The angular magnification of a compound microscope is negative, indicating that the final image is inverted compared to the initial orientation of the object. However, we often state only the magnification ($100\times$ microscope, not $-100\times$).

Example 25.4 ■ A Compound Microscope: Finding the Magnification

A microscope has an objective with a focal length of 10 mm and an eyepiece with a focal length of 4.0 cm. The lenses are positioned 20 cm apart in the barrel. Determine the approximate total magnification of the microscope.

Thinking It Through. This is a direct application of Eq. 25.5.

Solution.

Given: $f_o = 10 \text{ mm} = 1.0 \text{ cm}$ *Find:* m_{total} (total magnification)
       $f_e = 4.0 \text{ cm}$
       $L = 20 \text{ cm}$

Using Eq. 25.5, we get

$$m_{total} = -\frac{(25 \text{ cm})L}{f_o f_e} = -\frac{(25 \text{ cm})(20 \text{ cm})}{(1.0 \text{ cm})(4.0 \text{ cm})} = -125\times$$

Note the relatively short focal length of the objective. The negative sign indicates that the final image is inverted.

Follow-up Exercise. If the focal length of the eyepiece in this Example were doubled, how would the length of the microscope be affected for the same magnification? (Express the change as a percentage.)

A modern compound microscope is shown in Fig. 25.8b. Interchangeable eyepieces with magnifications from about $5\times$ to over $100\times$ are available. For standard microscopic work in biology or medical laboratories, $5\times$ and $10\times$ eyepieces are normally used. Microscopes are often equipped with rotating turrets, which usually contain three objectives for different magnifications, such as $10\times, 43\times$, and $97\times$. These objectives and the $5\times$ and $10\times$ eyepieces can be used in various combinations to provide magnifying powers from $50\times$ to $970\times$. The maximum magnification obtained from a compound microscope is about $2000\times$.

Opaque objects are usually illuminated with a light source placed above them. Specimens that are transparent, such as cells or thin sections of tissues on glass slides, are illuminated with a light source beneath the microscope stage so that light passes through the specimen. A modern microscope is usually equipped with a light condenser (converging lens) and diaphragm below the stage, which are used to concentrate the light and control its intensity. A microscope may have an internal light source. The light is reflected into the condenser from a mirror. Older microscopes have two mirrors with reflecting surfaces: One is a plane mirror for reflecting light from a high-intensity external source, and the other is a concave mirror for converging low-intensity light such as skylight.

25.3 Telescopes

OBJECTIVES: To (a) distinguish between refractive and reflective telescopes and (b) describe the advantages of each.

Telescopes apply the optical principles of mirrors and lenses to improve our ability to see distant objects. Used for both terrestrial and astronomical observations, telescopes allow some objects to be viewed in greater detail and other, fainter or more distant objects simply to be seen. Basically, there are two types of telescopes—refracting and reflecting—characterized by the gathering and converging of light by lenses or mirrors, respectively.

Refracting Telescope

The principle underlying one type of **refracting telescope** is similar to that of a compound microscope. The major components of a refracting telescope are objective and eyepiece lenses, as illustrated in ▾Fig. 25.9. The objective is a large converging lens with a long focal length, and the movable eyepiece has a relatively short focal length. Rays from a distant object are essentially parallel and form an image (I_o) at the focal point (F_o) of the objective. This image acts as an object for the eyepiece, which is moved until the image lies just inside its focal point (F_e). A large, inverted, virtual image (I_e) is seen by an observer.

For relaxed viewing, the eyepiece is adjusted so that its image (I_e) is at infinity, which means that the objective image (I_o) is at the focal point of the eyepiece (f_e). As Fig. 25.9 shows, the distance between the lenses is then the sum of the focal lengths ($f_o + f_e$), which is the length of the telescope tube. The magnifying power of a telescope focused for the final image at infinity can be shown to be

$$m = -\frac{f_o}{f_e} \qquad \begin{array}{l} \textit{angular magnification} \\ \textit{of refracting telescope} \end{array} \qquad (25.6)$$

(see Exercise 57), where the minus sign is inserted to indicate that the image is inverted, as in our lens sign convention described in Section 23.3. Thus, to achieve the greatest magnification, the focal length of the objective should be made as long as possible and the focal length of the eyepiece as short as possible.

The telescope illustrated in Fig. 25.9 is called an **astronomical telescope**. The final image produced by an astronomical telescope is inverted, but this condition poses little problem to astronomers. (Why?) However, someone viewing an object on Earth through a telescope finds it more convenient to have an upright image. A telescope in which the final image is upright is called a **terrestrial telescope**. An upright final image can be obtained in several ways; two are illustrated in ▸Fig. 25.10.

In the telescope diagrammed in Fig. 25.10a, a diverging lens is used as an eyepiece. This type of terrestrial telescope is referred to as a *Galilean telescope*, because Galileo built one in 1609. A real image is formed by the objective to the left of the eyepiece, and this image acts as a "virtual" object for the eyepiece. (See Section 23.3.) An observer sees a magnified, upright, virtual image. (Note that with a diverging lens and negative focal length, Eq. 25.6 gives a $+m$, indicating an upright image.)

Galilean telescopes have several disadvantages, most notably very narrow fields of view and limited magnification. A better type of terrestrial telescope, il-

Note: Astronomical telescopes give an inverted image.

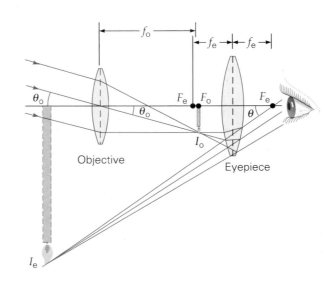

▸ **FIGURE 25.9 The refracting astronomical telescope** In an astronomical telescope, rays from a distant object form an intermediate image (I_o) at the focal point of the objective (F_o). The eyepiece is moved so that the image is at or slightly inside its focal point (F_e). An observer sees an enlarged image at infinity (I_e, shown at a finite distance here for illustration).

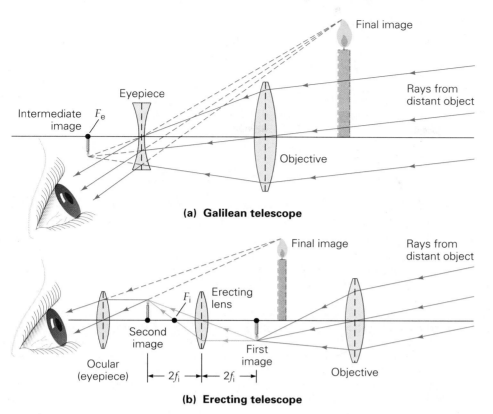

(a) Galilean telescope

(b) Erecting telescope

◀ **FIGURE 25.10 Terrestrial telescopes** **(a)** A Galilean telescope uses a diverging lens as an eyepiece, producing upright, virtual images. **(b)** Another way to produce upright images is to use a converging "erecting" lens (focal length f_i) between the objective and eyepiece in an astronomical telescope. This addition elongates the telescope, but the length can be shortened by using internally reflecting prisms.

lustrated in Fig. 25.10b, uses a third lens, called the *erecting lens*, or *inverting lens*, between converging objective and eyepiece lenses. If the image is formed by the objective at a distance that is twice the focal length of the intermediate erecting lens ($2f_i$), then the lens merely inverts the image without magnification, and the telescope magnification is still given by Eq. 25.6.

However, to achieve the upright image in this way requires a greater telescope length. Using the intermediate erecting lens to invert the image increases the length of the telescope by four times the focal length of the erecting lens ($2f_i$ on each side). The inconvenient length can be avoided by using internally reflecting prisms. This is the principle behind prism binoculars, which are double telescopes—one for each eye (▶Fig. 25.11).

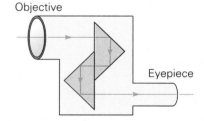

▲ **FIGURE 25.11 Prism binoculars** A schematic cutaway view of one ocular (one-half of a pair of prism binoculars), showing the internal reflections in the prisms, which reduce the overall physical length.

Example 25.5 ■ Astronomical Telescope—and Longer Terrestrial Telescope

An astronomical telescope has an objective lens with a focal length of 30 cm and an eyepiece with a focal length of 9.0 cm. (a) What is the magnification of the telescope? (b) If an erecting lens with a focal length of 7.5 cm is used to convert the telescope to a terrestrial type, what is the overall length of the telescope tube?

Thinking It Through. Equation 25.6 applies directly in part (a). In part (b), the erecting lens elongates the telescope by four times the focal length of the lens ($4f_i$) to the length of the scope (Fig. 25.10b).

Solution. Listing the data, we have

Given: $f_o = 30$ cm $\qquad$ *Find:* (a) m (magnification)
$\qquad f_e = 9.0$ cm $\qquad\qquad\qquad$ (b) L (length of
$\qquad f_i = 7.5$ cm (intermediate erecting lens) $\qquad\qquad$ telescope tube)

(a) The magnification is given by Eq. 25.6 as

$$m = -\frac{f_o}{f_e} = -\frac{30 \text{ cm}}{9.0 \text{ cm}} = -3.3\times$$

where the minus sign indicates that the final image is inverted.

(b) Taking the length of the astronomical tube to be the distance between the lenses, we find that this length is just the sum of the lenses' focal lengths:

$$L_1 = f_o + f_e = 30 \text{ cm} + 9.0 \text{ cm} = 39 \text{ cm}$$

The overall length is then

$$L = L_1 + L_2 = 39 \text{ cm} + 4f_i = 39 \text{ cm} + 4(7.5 \text{ cm}) = 69 \text{ cm}$$

Hence, the telescope length is over two-thirds of a meter, with an upright image, but the same magnification, 3.3× (why?).

Follow-up Exercise. A terrestrial telescope 66 cm in length has an intermediate erecting lens with a focal length of 12 cm. What is the focal length of an erecting lens that would reduce the telescope length to a more manageable 50 cm?

Conceptual Example 25.6 ■ Constructing a Telescope

A student is given two converging spherical lenses, one with a focal length of 5.0 cm and the other with a focal length of 20 cm. To construct a telescope to best view distant objects with these lenses, the student should hold the lenses (a) more than 25 cm apart; (b) less than 25 cm, but more than 20 cm, apart; (c) less than 20 cm, but more than 5.0 cm, apart; or (d) less than 5.0 cm apart. Specify which lens should be used as the eyepiece.

Reasoning and Answer. First let's see which lens should be used as the eyepiece. The only type of telescope that can be constructed with two converging lenses is an astronomical telescope. In this type of telescope, the lens with the longer focal length is used as an objective lens to produce a real image of a distant object. That image is then viewed with the lens with the shorter focal length, the eyepiece, used as a simple magnifier.

If the object is at a great distance, a real image is formed by the objective lens in the focal plane of the lens (Fig. 25.9). This image acts as the object for the eyepiece, which is positioned so that the image/object lies just inside its focal point so as to produce a large, inverted second image.

However, the two lenses must be *slightly* less than 25 cm apart, so answer (a) is not correct. Answers (c) and (d) are also not correct, because the eyepiece would be too close to the objective to produce the large secondary image needed for optimal viewing of a distant object. In these cases, the rays would pass through the second lens before the image was formed, and a *reduced* image might be produced. (See Section 23.3.) Thus, answer (b), with the objective image just inside the eyepiece's focal point, is the correct answer.

Follow-up Exercise. A third converging lens with a focal length of 4.0 cm is used with the aforementioned two lenses to produce a terrestrial telescope in which the third lens does nothing more than invert the image. How should the lenses be positioned and how far apart should they be for the final image to be of maximum size and upright?

Reflecting Telescope

For viewing the Sun, Moon, and nearby planets, large magnifications are important to see details. However, even with the highest feasible magnification, stars appear only as faint points of light. For distant stars and galaxies, it is more important to gather enough light than to increase the magnification, so that the object can be seen at all. The intensity of light from a distant source is very low. In many instances, such a source can be detected only when the light is gathered and focused on a photographic plate over a long period of time.

Recall from Section 14.3 that intensity is energy per unit time per unit *area*. Thus, more light can be gathered if the size of the objective is increased. This increases the distance at which the telescope can detect faint objects, such as distant galaxies. (Recall that the light intensity of a point source is inversely proportional to the *square* of the distance between the source and the observer.) However, producing a large lens involves difficulties associated with glass quality, grinding, and polishing. Compound-lens systems are required in order to reduce aberrations, and a very large lens may sag under its own weight, producing further aberrations. The largest objective lens in use has a diameter of 40 in. (102 cm) and is part of the refracting telescope of the Yerkes Observatory at Williams Bay, Wisconsin.

The previously mentioned problems are reduced with a **reflecting telescope**, which uses a large, concave, front-surface parabolic mirror (▼Fig. 25.12). A parabolic mirror does not exhibit spherical aberration, and a mirror has no inherent chromatic aberration. (Why?) High-quality glass is not needed, since the light is reflected by a mirrored front surface. And only one surface has to be ground, polished, and silvered.

The largest single-mirror telescope, with a mirror 8.2 m (323 in.) in diameter, is at the European Southern Observatory in Chile (▶Fig 25.13a). The largest reflecting telescope in the United States, with a mirror 5.1 m (200 in.) in diameter, is the Hale Observatories reflecting telescope, on Palomar Mountain in California.

Even though reflecting telescopes have advantages over refracting telescopes, they also have their own problems. Like a large lens, a large mirror may sag under its own weight, and the weight necessarily increases with the size of the mirror. The weight factor also increases the costs of constructing such a telescope, since the supporting elements for a heavier mirror must be more massive.

These problems are being addressed by new technologies. One approach is to use an array of small mirrors, coordinated so as to function as a single large mirror. Examples include the twin Keck telescopes at Mauna Kea in Hawaii. Each has a mirror consisting of 36 hexagonal segments that are computer positioned to give the equivalence of a 10-m mirror. The European Southern Observatory plans to

▼ **FIGURE 25.12 Reflecting telescopes** A concave mirror can be used in a telescope to converge light to form an image of a distant object. **(a)** The image may be at the prime focus, or **(b)** a small mirror and lens can be used to focus the image outside the telescope, a configuration called a *Newtonian focus*. **(c)** Another arrangement, called a *Cassegrain focus*, uses a mirror to form the image below the main mirror.

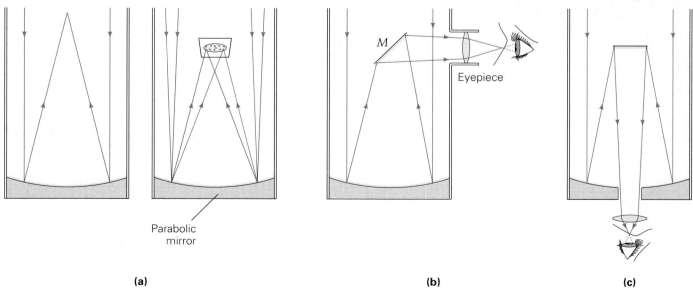

Parabolic mirror

Eyepiece

(a)   (b)   (c)

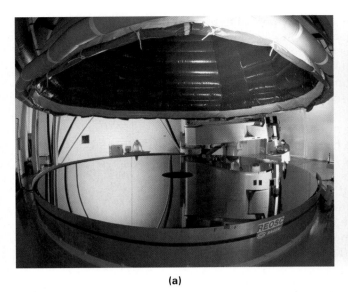

(a)

(b)

▲ **FIGURE 25.13 European Southern Observatory, near Paranal, Chile** **(a)** An 8.2-m-diameter mirror is undergoing the final phase of polishing. **(b)** Four 8.2-m telescopes will form a VLT (Very Large Telescope) with an equivalent diameter of 16 m.

▲ **FIGURE 25.14 Hubble Space Telescope (HST)** Late in 1993, astronauts from the Space Shuttle *Endeavor* visited the HST in orbit. They installed corrective equipment that compensated for many of the telescope's optical flaws and repaired or replaced other malfunctioning systems. Since then, the HST has produced a wealth of spectacular images and helped astronomers make many important discoveries. The latest servicing mission was performed between March 1 and 12, 2002.

have four 8.2-m-diameter mirrors to form a VLT (Very Large Telescope,) with an equivalent diameter of 16 m (Fig. 25.13b).

Another way of extending our view into space is to put telescopes into orbit around the Earth. Above the atmosphere, the view is unaffected by the twinkling effect of atmospheric turbulence and refraction, and there is no background problem from city lights. In 1990, the optical Hubble Space Telescope (HST) was launched into orbit (◄Fig. 25.14). Even with a mirror diameter of only 2.4 m, its privileged position has allowed the HST to produce images seven times clearer than those formed by Earthbound telescopes.

However, not all telescopes are optical, as illustrated by the Insight on p. 833.

25.4 Diffraction and Resolution

OBJECTIVES: To **(a)** describe the relationship of diffraction and resolution and **(b)** state and explain Rayleigh's criterion.

The diffraction of light places a limitation on our ability to distinguish objects that are close together when we use microscopes or telescopes. This effect can be understood by considering two point sources located far from a narrow slit of width w (▶Fig. 25.15). The sources could be distant stars, for example. In the absence of diffraction, two bright spots, or images, would be observed on a screen. As you know from Section 24.3, however, the slit diffracts the light, and each image consists of a central maximum with a pattern of weaker bright and dark fringes on either side. If the sources are close together, the two central maxima may overlap. In this case, the images cannot be distinguished, or are said to be *unresolved*. For the images to be *resolved*, the central maxima must not overlap appreciably.

In general, images of two sources can be resolved if the central maximum of one falls at or beyond the first minimum (dark fringes) of the other. This generally accepted limiting condition for the **resolution** of two diffracted images—that is, the ability to distinguish both images as separate—was first proposed by Lord Rayleigh (1842–1919), a British physicist. The condition is known as the **Rayleigh criterion**:

Two images are said to be just resolved when the central maximum of one image falls on the first minimum of the diffraction pattern of the other image.

The Rayleigh criterion can be expressed in terms of the angular separation (θ) of the sources. (See Fig. 25.15.) The first minimum ($m = 1$) for a single-slit diffraction pattern satisfies this relationship:

$$w \sin \theta = m\lambda = \lambda$$

or

$$\sin \theta = \frac{\lambda}{w}$$

This equation gives the minimum angular separation for two images to be just resolved according to the Rayleigh criterion. In general, for visible light, the wavelength is much smaller than the slit width ($\lambda \ll w$), so a good approximation is $\sin \theta \approx \theta$. The limiting, or minimum, angle of resolution (θ_{min}) for a slit of width w is then

$$\theta_{min} = \frac{\lambda}{w} \qquad \begin{array}{l} \textit{minimum angle of} \\ \textit{resolution (for a slit)} \end{array} \qquad (25.7)$$

(Note that θ_{min} is a pure number and therefore must be expressed in radians.) Thus, the images of two sources will be *distinctly* resolved if the angular separation of the sources is greater than λ/w.

The apertures (openings) of cameras, microscopes, and telescopes are generally circular. Thus, there is a circular diffraction pattern around the central maximum, in the form of a bright circular disk (▶Fig. 25.16). Detailed analysis for a circular aperture shows that the minimum angular separation for the images of two objects to be just resolved is

$$\theta_{min} = \frac{1.22\lambda}{D} \qquad \begin{array}{l} \textit{minimum angle of resolution} \\ \textit{(for a circular aperture)} \end{array} \qquad (25.8)$$

where D is the diameter of the aperture and θ_{min} is in radians.

Equation 25.8 applies to the objective lens of a microscope or telescope or the iris of the eye, which may be considered to be a circular aperture for light. According to Eqs. 25.7 and 25.8, the smaller the θ_{min}, the better is the resolution. The minimum angle of resolution, θ_{min}, should be small so that objects close together can be resolved; therefore, the aperture should be as large as possible. This is another reason for using large lenses (and mirrors) in telescopes, in addition to their greater light-gathering power.

Example 25.7 ■ Eye and Telescope: Evaluating Resolution with the Rayleigh Criterion

Determine the minimum angle of resolution by the Rayleigh criterion for (a) the pupil of the eye (daytime diameter of about 4.0 mm) for visible light with a wavelength of 660 nm; (b) the European Southern Observatory refracting telescope (diameter of 8.2 m), for visible light of the same wavelength as in part (a); and (c) a radio telescope 25 m in diameter for radiation with a wavelength of 21 cm.

Thinking It Through. This is a comparison of θ_{min} for apertures with different diameters—a direct application of Eq. 25.8.

Solution.

Given: (a) $D = 4.0$ mm $= 4.0 \times 10^{-3}$ m
$\qquad \lambda = 660$ nm $= 6.60 \times 10^{-7}$ m
$\qquad$ (b) $D = 8.2$ m
$\qquad \lambda = 660$ nm $= 6.60 \times 10^{-7}$ m
$\qquad$ (c) $D = 25$ m
$\qquad \lambda = 21$ cm $= 0.21$ m

Find: (a) θ_{min} (minimum angles of resolution)
$\qquad$ (b) θ_{min}
$\qquad$ (c) θ_{min}

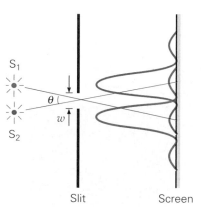

(a) Resolved

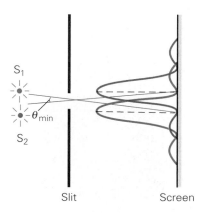

(b) Just resolved

▲ **FIGURE 25.15   Resolution** Two light sources in front of a slit produce diffraction patterns. **(a)** When the angle subtended by the sources at the slit is large enough for the diffraction patterns to be distinguishable, the images are said to be resolved. **(b)** At smaller angles, the central maxima are closer together. At θ_{min}, the central maximum of one image's diffraction pattern falls on the first dark fringe of the other image's pattern, and the images are said to be just resolved. For smaller angles, the patterns are unresolved.

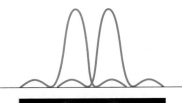

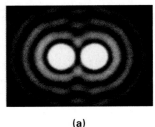

(a)

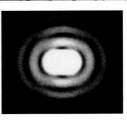

(b)

▲ **FIGURE 25.16 Circular-aperture resolution** **(a)** When the angular separation of two objects is large enough, the images are well resolved. (Compare with Fig. 25.15a.) **(b)** Rayleigh criterion: The central maximum of the diffraction pattern of one image falls on the first minimum of the diffraction pattern of the other image. (Compare with Fig. 25.15b.) The images of objects with smaller angular separations cannot be clearly distinguished as individual images.

(a) For the eye,

$$\theta_{min} = \frac{1.22\lambda}{D} = \frac{1.22(6.60 \times 10^{-7}\,\text{m})}{4.0 \times 10^{-3}\,\text{m}} = 2.0 \times 10^{-4}\,\text{rad}$$

(b) For the light telescope,

$$\theta_{min} = \frac{1.22(6.60 \times 10^{-7}\,\text{m})}{8.2\,\text{m}} = 9.8 \times 10^{-8}\,\text{rad}$$

(*Note*: The resolution of Earthbound telescopes with large-diameter objectives is usually not limited by diffraction, but rather by other effects, such as atmospheric turbulence. These telescopes have a θ_{min} on the order of 10^{-6} rad.)
(c) For the radio telescope,

$$\theta_{min} = \frac{1.22(0.21\,\text{m})}{25\,\text{m}} = 0.010\,\text{rad}$$

The smaller the angular separation, the better is the resolution. What do the results tell you?

Follow-up Exercise. As noted in Section 25.3, the Hubble Space Telescope has a mirror diameter of 2.4 m. How does its resolution compare with that of the largest Earthbound telescopes? [See the note in part (b) of this Example.]

For a microscope, it is more convenient to specify the actual separation (*s*) between two point sources. Since the objects are usually near the focal point of the objective, to a good approximation,

$$\theta_{min} = \frac{s}{f} \quad \text{or} \quad s = f\theta_{min}$$

where *f* is the focal length of the lens. (Here, *s* is taken as the arc length subtended by θ_{min}, and $s = r\theta_{min} = f\theta_{min}$.) Then, using Eq. 25.8, we get

$$s = f\theta_{min} = \frac{1.22\lambda f}{D} \qquad \begin{array}{l}\textit{resolving power}\\ \textit{of a microscope}\end{array} \qquad (25.9)$$

This minimum distance between two points whose images can be just resolved is called the **resolving power** of the microscope. In practice, the resolving power of a microscope indicates the ability of the objective to distinguish fine detail in specimens' structures. For another real-life example of resolution, see ▼ Fig. 25.17.

▼ **FIGURE 25.17 Real-life resolution** **(a)**, **(b)**, **(c)** A sequence of an approaching automobile's headlights. In (a), the headlights are almost unresolved through the circular aperture of the camera (or your eye).

(a)

(b)

(c)

Telescopes Using Nonvisible Radiation

The word *telescope* usually brings to mind visual observations. However, the visible region is a very small part of the electromagnetic spectrum, and celestial objects emit radiation of many other types, including radio waves. This fact was discovered accidentally in 1931 by an electrical engineer named Carl Jansky while he was working on the problem of static interference with intercontinental radio communications. Jansky found an annoying static hiss that came from a fixed direction in space, apparently from a celestial source. It was soon apparent that radio waves are another source of astronomical information, and radio telescopes were built to investigate this source.

A radio telescope operates similarly to a reflecting light telescope. A reflector with a large area collects and focuses the radio waves at a point where a detector picks up the signal (Fig. 1). The parabolic collector, called a *dish*, is covered with metal wire mesh or metal plates. Since the wavelengths of radio waves range from a few millimeters to several meters, wire mesh is "smooth" enough and a good reflecting surface for such waves.

Radio telescopes supplement optical telescopes and provide some definite advantages. Radio waves pass freely through the huge clouds of dust that hide a large part of our galaxy from visual observation. Also, radio waves easily penetrate the Earth's atmosphere, which reflects and scatters a large percentage of the incoming visible light.

The Earth's atmosphere absorbs many wavelengths of electromagnetic radiation and completely blocks some from reaching the ground. Water vapor, a strong absorber of infrared radiation, is concentrated in the lower part of the atmosphere, near the surface of the Earth. Thus, observations with infrared telescopes are made from high-flying aircraft or from orbiting spacecraft.

The first orbiting infrared observatory was launched in 1983. Not only are atmospheric interferences eliminated in space, but also a telescope may be cooled to a very low temperature without becoming coated with condensed water vapor from the atmosphere. Cooling the telescope helps eliminate interfering infrared radiation generated by the telescope itself. The orbiting telescope launched in 1983 was cooled with liquid helium to about 10 K; it carried out an infrared survey of the entire sky.

The atmosphere is virtually opaque to ultraviolet radiation, X rays, and gamma rays from distant sources. Orbiting satellites with telescopes sensitive to these types of radiation have mapped out portions of the sky, and other surveys are planned.

At the altitudes reached by orbiting satellites, even observations in the visible region are not affected by air motion and temperature refraction. Perhaps in the not-too-distant future, a permanently staffed orbiting observatory carrying a variety of telescopes will help expand our knowledge of the universe.

Related Exercise: 81

FIGURE 1   Radio telescopes   Several of the dish antennae that make up the Very Large Array (VLA) radio telescope near Socorro, New Mexico. There are 27 movable dishes, each 25 m in diameter, forming the array along a Y-shaped railway network. The data from all the antennae are combined to produce a single radio image. In this way, it is possible to attain a resolution equivalent to that of one giant radio dish.

Conceptual Example 25.8   ■   Viewing from Space: The Great Wall of China

The Great Wall of China was originally about 2400 km (1500 mi) long, with a base width of about 6.0 m and a top width of about 3.7 m. Several hundred kilometers of the Wall remain intact (▶Fig. 25.18). The Wall is said to be the only human construction that can be seen with the unaided eye by an astronaut orbiting the Earth. Using the result from part (a) of Example 25.7, see if this is true. (Neglect any atmospheric effects.)

Reasoning and Answer.   Despite the length of the Wall, it would not be visible from space unless its *width* subtends the minimum angle of resolution for the eye of an observing astronaut ($\theta_{min} = 2.0 \times 10^{-4}$ rad from Example 25.7). Guard towers with roofs wider than 6.0 m were located every 180 m along the Wall, so let's take the maximum observable width dimension to be 7.0 m. [Actually, it is the circular arc length that subtends the angle, but at such a radius, the chord length (width) is very nearly equal to the circular arc length. Refer to Example 7.2, and make yourself a sketch.]

Let's assume that the astronaut is just able to distinguish the Wall. Recall that $s = r\theta$ (Eq. 7.3), where s is an approximation of the maximum observable width of the Wall and r is the radial (height) distance. Then, the astronaut would have to be at a radial distance of

$$r = \frac{s}{\theta} = \frac{7.0 \text{ m}}{2.0 \times 10^{-4} \text{ (rad)}} = 3.5 \times 10^4 \text{ m} = 35 \text{ km } (= 22 \text{ mi})$$

So, above 35 km, the Wall would not be able to be seen with the unaided eye. Orbiting satellites are 300 km (190 mi) or more above the Earth so as to be above the denser part of the atmosphere. (Why?) Hence, for an astronaut to see the Great Wall, he or she would have to come back very near to Earth. The statement about the ability to see the Wall from space is false.

Follow-up Exercise. What would be the minimum diameter of the objective of a telescope that would allow an astronaut orbiting the Earth at an altitude of 300 km to see the Great Wall? (Take all conditions to be the same as stated in this Example, and assume that the wavelength of light is 660 nm.)

▲ **FIGURE 25.18 The Great Wall** The walkway of the Great Wall of China, which was built as a fortification along China's northern border.

Note: The relationship between wavelength and index of refraction is given in Section 22.3; see Eq. 22.4.

Note: The relationship between pitch and frequency for sound is discussed in Section 14.6.

Note from Eq. 25.8 that higher resolution can be gained by using radiation of a shorter wavelength. Thus, a telescope with an objective of a given size will have greater resolution with violet light than with red light. For microscopes, it is possible to increase resolving power by shortening the wavelengths of the light used to create the image. This can be done with a specialized objective called an *oil immersion lens*. When such a lens is used, a drop of transparent oil fills the space between the objective and the specimen. Recall that the wavelength of light in oil is $\lambda_m = \lambda/n$, where n is the index of refraction of the oil and λ is the wavelength of light in air. With values of n of about 1.50 or higher, the wavelength is significantly reduced, and the resolving power is increased proportionally.

*25.5 Color

OBJECTIVE: To relate color vision and light.

In general, physical properties are fixed or absolute. For example, a particular type of electromagnetic radiation has a certain frequency or wavelength. However, visual perception of this radiation may vary from person to person. How we "see" radiation gives rise to what we call *color vision*.

Color Vision

Color is perceived because of a physiological response to excitation by light of the cone receptors in the retina of the human eye. (Many animals have no cone cells and thus live in a black-and-white world.) The cones are sensitive to light with frequencies approximately between 7.5×10^{14} Hz and 4.3×10^{14} Hz (wavelengths between 400 nm and 700 nm). Different frequencies of light are perceived by the brain as different colors. The association of a color with a particular frequency is subjective and may vary from person to person. As pitch is to sound and hearing, color is to light and vision.

Color vision is not well understood. One of the most widely accepted theories is that there are three types of cones in the retina, responding to different parts of the visible spectrum, particularly in the red, green, and blue regions (▶ Fig. 25.19). Presumably, the types of cones absorb light of specific ranges of frequencies and functionally overlap one another to form combinations that are interpreted by the brain as various colors of the spectrum. For example, when red and green cones

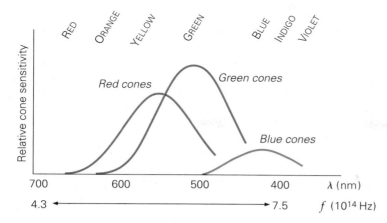

are stimulated equally by light of a particular frequency, the brain interprets the color as yellow. But when the red cones are stimulated more strongly than the green cones, the brain senses orange. *Color blindness* is believed to result when one or more type of cone is missing.

As Fig. 25.19 shows, the human eye is not equally sensitive to all colors. Some colors evoke a greater brightness response than others do and therefore appear brighter for the same intensity. The wavelength of maximum visual sensitivity is about 550 nm, in the yellow–green region.

The foregoing theory of color vision is based on the experimental fact that beams of varying intensities of red, green, and blue light will produce most colors. The red, blue, and green from which we interpret a full spectrum of colors are called the **additive primary colors**. When light beams of the additive primaries are projected on a white screen so that they overlap, other colors will be produced, as illustrated in ▶ Fig. 25.20. This technique is called the **additive method of color production**. Triad dots consisting of three phosphors that emit the additive primary colors when excited are used in television picture tubes to produce colored images.

Note in Fig. 25.20 that a certain combination of the primary colors appears white to the eye. Also, many *pairs* of colors appear white to the eye when combined. The colors of such pairs are said to be **complementary colors**. The complement of blue is yellow, that of red is cyan, and that of green is magenta. As the figure shows, the complementary color of a particular primary is the combination, or sum, of the other two primaries. Hence, the primary and its complement together appear white.

Edwin H. Land (the developer of Polaroid™ film) showed that when the proper mixtures of only two wavelengths (colors) of light are passed through black and white transparencies (no color), the wavelengths produce images of various colors. Land wrote, "*In this experiment we are forced to the astonishing conclusion that the rays are not in themselves color-making. Rather they are bearers of information that the eye uses to assign appropriate colors to various objects in an image.*"* From Land's experiments, it seems that information about colors other than for the two wavelengths used is developed in the brain via the cone cells. However, our knowledge about color vision is far from complete.

Objects have a color when they are illuminated with white light because they reflect (scatter) or transmit light predominately of the frequency of that color. The other frequencies of the white light are mostly absorbed. For example, when white light strikes a ripe red apple, mostly the waves in the red portion of the spectrum are reflected—all others (and thus all other colors) are mostly absorbed. Similarly, when white light passes through a piece of transparent red glass, or a filter, mostly the red rays are transmitted. This occurs because the color pigments in the glass are selective absorbers.

*From Edwin H. Land, "Experiments in Color Vision," in *Scientific American*, May 1959, pp. 84–99.

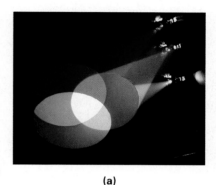

(a)

(b)

▲ FIGURE 25.20 Additive method of color production When light beams of the primary colors (red, blue, and green) are projected onto a white screen, mixtures of them produce other colors. Varying the intensities of the beams allows most colors to be produced.

Note: Recall from Chapter 22 that the colors of the visible spectrum may be remembered by the acronym ROY G. BIV.

Pigments are mixed to form various colors, such as in the production of paints and dyes. You are probably aware that mixing yellow and blue paints produces green. This is because the yellow pigment absorbs most of the wavelengths except those in the yellow and nearby regions of the visible spectrum, and the blue pigment absorbs most of the wavelengths except those in the blue and nearby regions. The wavelengths in the intermediate green region, adjacent to both the yellow and blue wavelengths, are not strongly absorbed by either pigment, and therefore the mixture appears green. The same effect can be accomplished by passing white light through stacked yellow and blue filters. The light coming through both filters appears green.

Mixing pigments results in the subtraction of colors. The color that is perceived is created by whatever is *not* absorbed by the pigment—that is, not subtracted from the original beam. This is the principle of what is called the **subtractive method of color production**. Three particular pigments—cyan, magenta, and yellow—are called the **subtractive primary pigments**. Various combinations of two of the three subtractive primaries produce the three additive primary colors (red, blue, and green), as illustrated in ▼Fig. 25.21. When the subtractive primaries are mixed in the proper proportions, the mixture appears black (because all wavelengths are absorbed). Painters often refer to the subtractive primaries as red, yellow, and blue. They are loosely referring to magenta (purplish red), yellow, and cyan ("true" blue). Mixing these paints in the proper proportions produces a broad spectrum of colors.

Note in Fig. 25.21 that the magenta pigment essentially subtracts the color green where it overlaps with cyan and yellow. As a result, magenta is sometimes referred to as "minus green." If a magenta filter were placed in front of a green light, no light would be transmitted. Similarly, cyan is called "minus red," and yellow is called "minus blue." An example of subtractive color mixing is a photographer's use of a yellow filter to bring out white clouds on black and white film. This filter absorbs blue from the sky, darkening it relative to the clouds, which reflect white light. Hence, the contrast between the two is enhanced. What type of filter would you use to darken green vegetation on black and white film? to lighten it?

▼ **FIGURE 25.21 Subtractive method of color production** **(a)** When the primary pigments (cyan, magenta, and yellow) are mixed, different colors are produced by subtractive absorption; for example, the mixing of yellow and magenta produces red. When all three pigments are mixed and all the wavelengths of visible light are absorbed, the mixture appears black. **(b)** Subtractive color mixing, using filters. The principle is the same as in part (a). Each pigment selectively absorbs certain colors, removing them from the white light. The colors that remain are what we see.

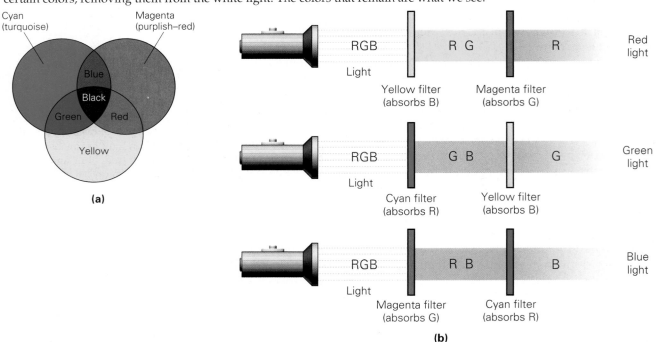

Chapter Review

Important Concepts and Equations

- Nearsighted people cannot see distant objects clearly. Farsighted people cannot see nearby objects clearly. These conditions may be corrected by diverging and converging lenses, respectively.

- The magnification of a magnifying glass (or simple microscope) is expressed in terms of **angular magnification** (m), as distinguished from the lateral magnification (M; see Chapter 23):

$$m = \frac{\theta}{\theta_o} \qquad (25.1)$$

Magnification of a magnifying glass with the image at the near point (25 cm) is expressed as

$$m = 1 + \frac{25 \text{ cm}}{f} \qquad (25.3)$$

Magnification of a magnifying glass with the image at infinity is expressed as

$$m = \frac{25 \text{ cm}}{f} \qquad (25.4)$$

- The objective of a compound microscope has a relatively short focal length, and the eyepiece, or ocular, has a longer focal length. Both contribute to the **total magnification**, m_{total}, given by

$$m_{total} = M_o m_e = -\frac{(25)L}{f_o f_e} \qquad (25.5)$$

where L, f_o, and f_e are in centimeters.

- A refracting telescope uses a converging lens to gather light, and a reflecting telescope uses a converging mirror. The image created by either one is magnified by the eyepiece. The **magnification of a refracting telescope** is

$$m = -\frac{f_o}{f_e} \qquad (25.6)$$

- Diffraction places a limit on the ability to resolve, or distinguish, objects that are close together. Two images are said to be just resolved when the central maximum of one image falls on the first minimum of the diffraction pattern of the other image (Rayleigh criterion).

  For a rectangular slit, the **minimum angle of resolution is**

$$\theta_{min} = \frac{\lambda}{w} \qquad (25.7)$$

The **minimum angle of resolution for a circular aperture** of diameter D is

$$\theta_{min} = \frac{1.22\lambda}{D} \qquad (25.8)$$

The **resolving power** of a microscope is

$$s = f\theta_{min} = \frac{1.22\lambda f}{D} \qquad (25.9)$$

- Color is our brain's interpretation of the frequency of light.

Exercises

25.1 The Human Eye*

1. The rods of the retina (a) are responsible for 20/20 vision, (b) are responsible for black-and-white twilight vision, (c) are responsible for color vision, or (d) focus light.

2. An imperfect cornea can cause (a) astigmatism, (b) nearsightedness, (c) farsightedness, or (d) all of the preceding.

3. CQ The focal length of the crystalline lens of the human eye varies with muscle action. Discuss the shape (curvature) of the lens when it is looking at distant and close objects, respectively.

4. CQ Is the image formed on the retina inverted or upright? Why?

*Assume that corrective lenses are in contact with the eye (contact lenses) unless otherwise stated.

5. CQ People and other animals often exhibit "red eye" when photographed with a flash camera. Light reflected from the retina is red because of blood vessels near the surface. Some cameras have an anti-red-eye option, which, when activated, gives a quick flash before the longer picture-taking flash. Explain how this option reduces red eye.

6. CQ Which parts of the camera correspond to the iris, crystalline lens, and retina of the eye?

7. CQ (a) If an eye has a far point of 15 m and a near point of 25 cm, is that eye nearsighted or farsighted? (b) How about an eye with a far point at infinity and a near point at 50 cm? (c) What type of corrective lenses (converging or diverging) would you use to correct the vision defects in parts (a) and (b)?

8. CQ Will wearing glasses to correct nearsightedness and farsightedness, respectively, affect the size of the image on the retina? Explain.

9. ■ What are the powers of (a) a converging lens of focal length 20 cm and (b) a diverging lens of focal length −50 cm?

10. IE ■ The far point of a certain nearsighted person is 90 cm. (a) Which type of contact lens, (1) converging, (2) diverging, or (3) bifocal, should be prescribed to enable the person to see more distant objects clearly? Why? (b) What would the power of the lens be, in diopters?

11. IE ■ A certain farsighted person has a near point of 50 cm. (a) Which type of contact lens, (1) converging, (2) diverging, or (3) bifocal, should an optometrist prescribe to enable the person to see objects clearly as close as 25 cm? Why? (b) no need the (in diopter) What is the power of the lens, in diopters?

12. ■■ A woman cannot see objects clearly when they are farther than 12.5 m away. (a) Is she nearsighted or farsighted? (b) Which type of lens will allow her to see distant objects clearly, and of what power should the lens be?

13. ■■ A nearsighted woman has an uncorrected far point of 200 cm. Which type of contact lens would correct this condition, and of what power should it be?

14. ■■ A farsighted professor can *just* see the print in a book clearly when she holds the book at arm's length (0.80 m from the eyes). (a) Which type of lens will allow her to read the text at the normal near point? (b) What is the lens's focal length?

15. IE ■■ To correct a case of hyperopia, an optometrist prescribes positive contact lenses that effectively move the patient's near point from 100 cm to 25 cm. (a) Will the patient be able to see distant objects clearly with the corrective glasses on, or will she have to take them out? Why? (a) What is the power of the lenses?

16. ■■ A farsighted person with a near point of 0.95 m gets contact lenses and can then read a newspaper held at a distance of 25 cm. What is the power of the lenses? (Assume that the lenses are the same for both eyes.)

17. ■■ A farsighted man is unable to focus on objects nearer than 1.5 m. (a) Which type of contact lens will allow him to focus on the print of a book held 25 cm from his eyes? (b) Of what power should it be?

18. ■■ A certain myopic man has a far point of 150 cm. (a) What power must a contact lens have to allow him to see distant objects clearly? (b) If he is able to read print at 25 cm while wearing his contacts, is his near point less than 25 cm? If so, what is it? (c) Give an approximation of the man's age, based on the normal rate of recession of the near point.

19. ■■ A middle-aged man starts to wear eyeglasses with lenses of +2.0 D that allow him to read a book held as closely as 25 cm. Several years later, he finds that he must hold a book no closer than 33 cm to read it clearly with the same glasses, so he gets new glasses. What is the power of the new lenses? (Assume that both lenses are the same.)

20. ■■ A college professor can see objects clearly only if they are between 70 and 500 cm from her eyes. Her optometrist prescribes bifocals (▼Fig. 25.22) that enable her to see distant objects through the top half of the lenses and read students' papers at a distance of 25 cm through the lower half. What are the respective powers of the top and bottom lenses? [Assume that both lenses (right and left) are the same.]

Nearsightedness correction

Farsightedness correction

▲ **FIGURE 25.22** **Bifocals** See Exercises 20 and 23.

21. ■■ A nearsighted student wears contact lenses to correct for a far point that is 4.00 m from her eyes. When she is not wearing her contact lenses, her near point is 20 cm. What is her near point when she is wearing her contacts?

22. ■■■ A nearsighted woman has a far point located 7.5 m from one eye. (a) If a corrective lens is worn 2.0 cm from the eye, what would be the necessary power of the lens for her to see distant objects? (b) What would be the necessary power if a contact lens were used?

23. ■■■ Bifocal glasses are used to correct both nearsightedness and farsightedness at the same time (Fig. 25.22). If the near points in the right and left eyes are 35.0 cm and 45.0 cm, respectively, and the far point is 220 cm for both eyes, what are the powers of the lenses prescribed for the glasses? (Assume that the glasses are worn 3.00 cm from the eyes.)

25.2 Microscopes*

24. A magnifying glass (a) is a concave lens, (b) forms a clear image of an object that is closer than the near point, (c) magnifies by effectively increasing the angle the object subtends, or (d) both (b) and (c).

25. A compound microscope has (a) unlimited magnification, (b) two lenses of the same focal length, (c) a diverging objective lens, or (d) an eyepiece of relatively long focal length.

26. CQ With an object at the focal point of a magnifying glass, the magnification is given by $m = (25 \text{ cm})/f$ (Eq. 25.4).

*The normal near point should be taken as 25 cm unless otherwise specified.

According to this equation, the magnification could be increased indefinitely by using lenses with shorter focal lengths. Why, then, do we need compound microscopes?

27. **CQ** When you use a simple convex lens as a magnifying glass, where should you put the object, farther away than the focal length or inside the focal length? Explain.

28. ■ Using the small-angle approximation, compare the angular sizes of a car 1.0 m in height when at distances of (a) 500 m and (b) 1025 m.

29. ■ An object is placed 10 cm in front of a converging lens with a focal length of 18 cm. What are (a) the lateral magnification and (b) the angular magnification?

30. ■ A biology student uses a converging lens to examine the details of a small insect. If the focal length of the lens is 12 cm, what is the maximum angular magnification?

31. ■ A physics student uses a converging lens with a focal length of 15 cm to read a small measurement scale. (a) What is the maximum magnification that can be obtained? (b) What is the magnification for viewing with the relaxed eye?

32. ■ A student uses a magnifying glass to examine the details of a microcircuit in the lab. If the lens has a focal length of 8.0 cm and a virtual image is formed at the student's near point (25 cm), (a) how far from the circuit is the lens held, and (b) what is the magnification?

33. ■■ A detective looks at a fingerprint with a magnifying glass whose power is +3.5 D. What is the maximum magnification of the print?

34. **IE** ■■ A lens with a power of +10 D is used as a simple microscope. (a) For the image of an object to be seen clearly, can the object be placed infinitely close to the lens, or is there a limit on how close it can be? Why? (b) Calculate how close an object can be brought to the lens. (c) What is the angular magnification at this point?

35. ■■ What is the maximum magnification of a magnifying glass with a power of +3.0 D for (a) a person with a near point of 25 cm and (b) a person with a near point of 10 cm?

36. ■■ A compound microscope has a distance of 15 cm between lenses and an ocular with a focal length of 8.0 mm. What power should the objective have to give a total magnification of −360×?

37. **IE** ■■ Two lenses of focal length 3.0 cm and 0.35 cm are available for a compound microscope, and the distance between the lenses has to be 15 cm. (a) Which lens, the one with the longer focal length or the one with the shorter focal length, should be used as the objective? (b) What is the total magnification of the microscope.

38. ■■ The focal length of the objective lens of a compound microscope is 4.5 mm. The eyepiece has a focal length of 3.0 cm. If the distance between the lenses is 18 cm, what is the magnification of a viewed image?

39. ■■ A compound microscope has an objective lens with a focal length of 0.50 cm and an ocular with a focal length of 3.25 cm. The separation distance between the lenses is 22 cm. A student with a normal near point uses the microscope. (a) What is the total magnification? (b) Compare the total magnification (as a percentage) with the magnification of the eyepiece alone as a simple magnifying glass.

40. ■■ A −150× microscope has an objective whose focal length is 0.75 cm. If the distance between the lenses is 20 cm, find the focal length of the eyepiece.

41. ■■ A specimen is 5.0 mm from the objective of a compound microscope that has a power of +250 D. What must be the magnifying power of the eyepiece if the total magnification of the specimen is −100×?

42. ■■■ Referring to ▼Fig. 25.23, show that the magnifying power of a magnifying glass held at a distance d from the eye is given by

$$m = \left(\frac{25}{f}\right)\left(1 - \frac{d}{D}\right) + \frac{25}{D}$$

when the actual object is located at the near point (25 cm). [*Hint*: Use a small-angle approximation, and note that $y_i/y_o = d_i/d_o$, by similar triangles.]

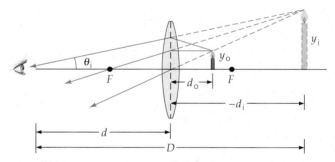

▲ **FIGURE 25.23 Power of a magnifying glass** See Exercise 42.

43. ■■■ A magnifying glass with a focal length of 10 cm is held 4.0 cm from the eyes to view the small print of a book. What is the magnification if the magnifying glass is held 5.0 cm from the book? [*Hint*: See Exercise 42.]

44. **IE** ■■■ A modern microscope is equipped with a turret that has three objectives with focal lengths of 16 mm, 4.0 mm, and 1.6 mm and interchangeable eyepieces of 5.0× and 10×. A specimen is positioned such that each objective produces an image 150 mm from the objective. (a) Which objective-and-eyepiece combination would you use if you want to have the greatest magnification? How about the least magnification? Why? (b) What are the greatest and least magnifications possible?

25.3 Telescopes

45. An inverted image is produced by (a) a terrestrial telescope, (b) an astronomical telescope, (c) a Galilean telescope, or (d) all of the proceding.

46. Compared with large refracting telescopes, large reflecting telescopes have the advantage of (a) greater light-gathering capability, (b) freedom from chromatic aberration, (c) lower cost, or (d) all of the preceding.

47. CQ In Fig. 25.12b, part of the light entering the concave mirror is obstructed by a small plane mirror that is used to redirect the rays to a viewer. Does this mean that only a portion of a star can be seen? How does the size of the obstruction affect the image?

48. CQ Why is chromatic aberration an important factor in refracting telescopes, but not in reflecting telescopes?

49. CQ If you are given two lenses with different focal lengths, which one should you use as the objective for a telescope? Why?

50. ■ Find the magnification of a telescope whose objective has a focal length of 50 cm and whose eyepiece has a focal length of 2.5 cm.

51. ■ An astronomical telescope has an objective and an eyepiece whose focal lengths are 60 cm and 15 cm, respectively. What are the telescope's (a) magnifying power and (b) length?

52. ■■ A telescope has an eyepiece with a focal length of 10 mm. If the length of the tube is 1.5 m, what is the angular magnification of the telescope when it is focused for an object at infinity?

53. ■■ An astronomical telescope has objective and eyepiece lenses with focal lengths of 87.5 cm and 8.00 mm, respectively. (a) What is the magnification of the telescope, and (b) what is its approximate length?

54. IE ■■ A terrestrial telescope has three lenses: an objective, an erecting lens, and an eyepiece. (a) Does the erecting lens (1) increase the magnification, (2) increase the physical length of the telescope, (3) decrease the magnification, or (4) decrease the physical length of the telescope? Why? (b) The three lenses of this terrestrial telescope have focal lengths of 40 cm, 20 cm, and 15 cm for the objective, erecting lens, and eyepiece, respectively. What is the magnification of the telescope for an object at infinity? (c) What is the length of the telescope barrel?

55. IE ■■ You are given two objectives and two eyepieces and are instructed to make a telescope with them. The focal lengths of the objectives are 60.0 cm and 40.0 cm, and the focal lengths of the eyepieces are 0.90 cm and 0.80 cm, respectively. (a) Which lens combination would you pick if you want to have maximum magnification?

How about minimum magnification? Why? (b) Calculate the maximum and minimum magnifications.

56. ■■ A telescope has an angular magnification of $-50\times$ and a barrel 1.02 m long. What are the focal lengths of the objective and the eyepiece?

57. ■■■ Referring to ▼Fig. 25.24, show that the angular magnification of a refracting telescope focused for the final image at infinity is $m = f_o/f_e$. (Since telescopes are designed for viewing distant objects, the angular size of an object viewed with the unaided eye is the angular size of the object at its actual location rather than at the near point, as is true for a microscope.)

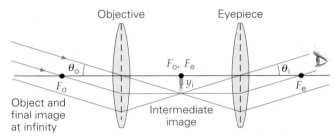

▲ **FIGURE 25.24 Angular modification of a refracting telescope** See Exercise 57.

25.4 Diffraction and Resolution*

58. The images of two sources are said to be resolved when (a) the central maxima of the diffraction patterns fall on each other, (b) the first bright fringes of the diffraction patterns fall on each other, (c) the central maximum of one diffraction pattern falls on the first dark fringe of the other, or (d) none of the preceeding.

59. For a telescope with a circular aperture, the minimum angle of resolution is (a) greater for red light than for blue light, (b) independent of the frequency of the light, (c) directly proportional to the radius of the aperture, or (d) independent of the area of the aperture.

60. The purpose of using oil immersion lenses on microscopes is to (a) reduce the size of the microscope, (b) increase the magnification, (c) increase the wavelength of light so as to increase the resolving power, or (d) reduce the wavelength of light so as to increase the resolving power.

61. CQ When an optical instrument is designed, high resolution is often desired so that the instrument may be used to observe fine details. Does higher resolution mean a smaller or larger minimum angle of resolution? Explain.

62. CQ A reflecting telescope with a large objective mirror can collect more light from stars than a reflecting telescope with a smaller objective mirror. What other advantage is gained with a large mirror? Why?

*Ignore atmospheric blurring unless otherwise stated.

63. ■ According to the Rayleigh criterion, what is the minimum angle of resolution for two point sources of red light ($\lambda = 680$ nm) in the diffraction pattern produced by a single slit with a width of 0.55 mm?

64. ■ The minimum angular separation of the images of two identical monochromatic point sources in a single-slit diffraction pattern is 0.0055 rad. If a slit width of 0.10 mm is used, what is the wavelength of the sources?

65. ■ What is the resolution limit due to diffraction for the European Southern Observatory reflecting telescope (8.20-m, or 323-in., diameter) for light with a wavelength of 550 nm?

66. ■ What is the resolution due to diffraction for the Hale telescope at Mount Palomar, with its 200-in.-diameter mirror, for light with a wavelength of 550 nm? Compare this value with the resolution limit for the European Southern Observatory telescope found in Exercise 65.

67. IE ■■ A human eye views small objects of different colors, and the eye's resolution is measured. (a) The eye obtains the maximum resolution and sees the finest details for objects of which color, (1) red, (2) yellow, (3) blue, or (4) it does not matter? Why? (b) The maximum diameter of the eye's pupil at night is about 7.0 mm. What is the minimum angle of separation for two 550-nm sources from which light passes through the pupil of the eye?

68. ■■ Some African tribesmen claim to be able to see the moons of Jupiter with the unaided eye. If two moons of Jupiter are at a minimum distance of 3.1×10^8 km away from Earth and at a maximum separation distance of 3.0×10^6 km, is this possible? Explain. Assume that the moons reflect sufficient light and that their observation is not restricted by Jupiter. [*Hint*: See Exercise 67b.]

69. ■■ Assuming that the headlights of a car are point sources 1.7 m apart, what is the maximum distance from an observer to the car at which the headlights are distinguishable from each other? [*Hint*: See Exercise 67b.]

70. ■■ A refracting telescope with a lens whose diameter is 30.0 cm is used to view a binary star system that emits light in the visible region. (a) What is the minimum angular separation of the two stars for them to be barely resolved? (b) If the binary star is a distance of 6.00×10^{20} km from the Earth, what is the lateral separation between the two stars? (Assume that a line joining the stars is perpendicular to our line of sight.)

71. IE ■■ Two astronomical telescopes have the characteristics shown in the following table:

Telescope	Objective focal length (cm)	Eyepiece focal length (cm)	Objective diameter (cm)
A	90.0	0.84	75
B	85.0	0.77	60

(a) Which telescope would you choose (1) for best magnification? Why? (2) for best resolution? Why? (b) Calculate the maximum magnification and the minimum resolving angle for a wavelength of 550 nm.

72. ■■ The objective of a microscope is 2.50 cm in diameter and has a focal length of 30.0 mm. (a) If yellow light with a wavelength of 570 nm is used to illuminate a specimen, what is the minimum angular separation of two fine details of the specimen for them to be just resolved? (b) What is the resolving power of the lens?

73. ■■■ A microscope with an objective 1.20 cm in diameter is used to view a specimen via light from a mercury source with a wavelength of 546.1 nm. (a) What is the limiting angle of resolution? (b) If details finer than those observable in part (a) are to be observed, what color of light in the visible spectrum would have to be used? (c) If an oil immersion lens were used ($n_{oil} = 1.50$), what would be the change (expressed as a percentage) in the resolving power?

*25.5 Color

74. An additive primary color is (a) blue, (b) green, (c) red, or (d) all of the preceding.

75. A subtractive primary color is (a) cyan, (b) yellow, (c) magenta, or (d) all of the preceding.

76. CQ Describe how the American flag would appear if it were illuminated with light of each of the primary colors.

77. CQ Can white be obtained by the subtractive method of color production? Explain. It is sometimes said that black is the absence of all color or that a black object absorbs all incident light. If so, why do we see black objects?

78. CQ Several beverages, such as root beer, develop a "head" of foam when poured into a glass. Why is the foam generally white or light colored, whereas the liquid is dark?

79. CQ White light is incident on two filters as shown in ▼Fig. 25.25. Complete the color rays to show which color of light emerges from the yellow filter.

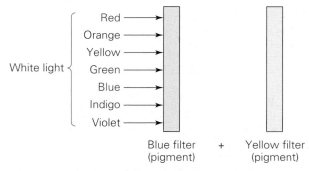

▲ FIGURE 25.25 **Color absorption** See Exercise 79.

Additional Exercises

80. A nearsighted woman has eyeglasses with lenses that correct for a far point of 130 cm. What is the power of the lenses? (Assume that the lenses are the same for both eyes.)

81. A radio telescope has a diameter of 300 m and can be tuned to wavelengths as short as 4.0 m. What is the minimum angular separation of two stars that can be distinguished by the telescope?

82. A terrestrial telescope uses an objective and eyepiece with focal lengths of 45 cm and 15 cm, respectively. What should the focal length of the erecting lens be if the overall length of the telescope is to be 0.80 m?

83. A nearsighted man wears eyeglasses whose lenses have a power of -0.15 D. How far away is his far point?

84. When viewing an object with a magnifying glass whose focal length is 15 cm, a student positions the lens so that there is minimum eyestrain. What is the observed magnification?

85. Light from two monochromatic point sources, both of wavelength 550 nm, is incident on a slit that is 0.050 mm wide. What is the minimum angle of resolution in the diffraction pattern formed by the slit?

86. A compound microscope has an objective with a focal length of 4.00 mm and an eyepiece with a magnification of $10.0\times$. If the objective and eyepiece are 15.0 cm apart, what is the total magnification of the microscope?

87. A person using a magnifying glass with a focal length of 12 cm views an object at the focal point of the lens. What is the magnification?

88. A refracting telescope has an objective with a focal length of 50 cm and an eyepiece with a focal length of 15 mm.

The telescope is used to view an object that is 10 cm high and located 50 m away. What is the apparent angular height of the object as viewed through the telescope?

89. An eyeglass lens with a power of $+2.8$ D allows a farsighted child to read a book held at a distance of 25 cm from her eyes. At what distance must she hold the book to read it without glasses?

90. A student views the details of a dollar bill with a magnifying glass, achieving its maximum magnification of $3\times$. What is the focal length of the magnifying glass?

91. A microscope has objective and eyepiece lenses that are 15 cm apart and have focal lengths of 7.5 mm and 10 mm, respectively. What is the total magnification of the microscope?

92. The amount of light that reaches the film in a camera depends on the lens aperture (the effective area) as controlled by the diaphragm. The f-number is the ratio of the focal length of the lens to its effective diameter. For example, an f/8 setting means that the diameter of the aperture is one-eighth of the focal length of the lens. The lens setting is commonly referred to as the *f-stop*. (a) Determine how much light each of the following lens settings admits to the camera as compared with f/8: (1) f/3.2 and (2) f/16. (b) The exposure time of a camera is controlled by the shutter speed. If a photographer correctly uses a lens setting of f/8 with a film exposure time of 1/60 s, what exposure time should he use to get the same amount of light exposure if he sets the f-stop at f/5.6?

93. From a spacecraft in orbit 150 km above the Earth's surface, an astronaut wishes to observe her hometown as she passes over it. What size features will she be able to identify with the unaided eye, neglecting atmospheric effects? [*Hint:* Estimate the diameter of the human iris.]

Relativity

Y ou might not think so at first glance, but this photograph tells us something very remarkable about the universe. The bright shapes are galaxies, each consisting of many billions of stars—trillions, probably, for the larger ones. They are very far from us—billions of light years away. The faint arcs that make the photograph resemble a spider's web represent light from galaxies far more distant still!

What is remarkable about these fine wisps of light, however, is not the billions of years that they have been traveling to reach us, but the path they have taken. Light usually travels in straight lines, yet the light from these distant galaxies has had its direction changed by the gravitational field of the galaxies in the foreground, creating the arcs in the photo.

The fact that light can be affected by gravity was predicted by Albert Einstein as a consequence of his theory of general relativity. The subject of relativity originated from the analysis of physical phenomena when speeds approach that of light. Indeed, relativity caused us to rethink our basic understanding of space, time, and gravitation. It successfully challenged Newtonian concepts that had dominated scientific thinking for 250 years.

The impact of relativity has been especially great in those branches of science concerned with two extremes of physical reality: the subatomic realm of nuclear and particle physics (Chapters 29 and 30), where time intervals and distances are almost inconceivably small, and the cosmic realm, where time intervals and distances are almost unimaginably large. All modern theories about the birth, evolution, and ultimate fate of our universe are inextricably linked to our understanding of relativity.

In this chapter, you will learn how Einstein's theory of relativity explains the changes in length and time that are observed in rapidly moving objects, the equivalence of energy and mass, and the bending of light rays by gravitational fields—phenomena that appear very strange from the classical Newtonian viewpoint.

26.1 Classical Relativity and the Michelson–Morley Experiment

OBJECTIVES: To (a) summarize the concepts of classical relativity, (b) define inertial and noninertial reference frames, and (c) explain the logic behind the ether hypothesis and the reasons for its demise.

Physics is concerned with the description of the world around us and depends on observations and measurements. We expect some aspects of nature to be consistent and unvarying; that is, the ground rules by which nature plays should be consistent, and our descriptions of nature or physical principles should not change from observation to observation. We emphasize this constancy by referring to such principles as "laws"—for example, the laws of motion. Not only have physical laws proved valid over time, but they are the same for all observers.

To put it another way, a physical principle or law should not depend upon the observer's frame of reference. When a measurement is made or an experiment performed, reference is usually made to a particular frame or coordinate system, most often the laboratory, which is considered to be "at rest." Now envision the experiment as observed by a passerby (moving relative to the laboratory). On comparing notes, the experimenter and the observer should find the results of the experiment and the physical principles involved to be the same. Physicists believe that the laws of nature are the same, regardless of the observer. Measured quantities may vary and descriptions may be different, but the laws that these quantities obey must be the same for all observers.

Suppose that you are at rest and observe two cars traveling in the same direction on a straight road at speeds of 60 km/h and 90 km/h. Even though we rarely say it, it is assumed that these speeds are measured relative to your reference frame—the ground. However, a person in the car traveling at 60 km/h observes the other car traveling at 30 km/h *relative to her reference frame*—the car in which she is riding. (What does someone in the car traveling at 90 km/h observe?) That is, each person observes a *relative* velocity—the velocity relative to his or her reference frame.

In measuring relative velocity, there seems to be no "true" rest frame. Any reference frame can be considered to be at rest if the observer moves along with it. We can, however, make a distinction between what are called *inertial* and *noninertial* reference frames. An **inertial reference frame** is defined as a reference frame in which Newton's first law of motion holds. That is, in an inertial system, an object on which there is no net force remains stationary or moves with a constant velocity. Since Newton's first law holds in this reference frame, the second law of motion ($\mathbf{F}_{net} = m\mathbf{a}$) also holds.

Conversely, in a noninertial (or accelerating) reference frame, an object with no net force on it would appear to accelerate. Note, however, that it is the frame that is accelerating, not the object. If such an object is observed from an accelerated reference frame, then Newton's second law will not correctly describe its motion. One example of a noninertial reference frame is an automobile accelerating forward from a stop sign. A cup on the dashboard may appear, when viewed from the car's (noninertial) reference frame, to accelerate backward without being acted on by any force. In fact, the noninertial observer would have to invoke a *fictitious* back-

Note: Review the discussion of relative velocities in Section 3.3.

Note: The relationship between Newton's first and second laws is discussed in Section 4.3.

ward force to explain the cup's apparent acceleration! From the inertial frame of a sidewalk observer, however, the cup simply tends to stay put (in accordance with the first law, since, ignoring friction, no net force acts on the cup) as the car accelerates out from under it.

Any reference frame moving with a constant velocity relative to an inertial reference frame is itself an inertial frame. Given a constant relative velocity, no acceleration effects are introduced in comparing one frame with another. In such cases, $\mathbf{F}_{net} = m\mathbf{a}$ can be used by observers in *either* frame to analyze a situation, and both observers will come to the same conclusions. That is, Newton's second law holds in both frames. Thus, with respect to the laws of mechanics, no inertial frame is preferred over another. This precept is sometimes called the *principle of classical, or Newtonian, relativity*:

| The laws of mechanics are the same in all inertial reference frames. |

The Absolute Reference Frame: The Ether

In the 1800s, with the development of the theories of electricity and magnetism, some serious questions arose. Maxwell's equations predicted light to be an electromagnetic wave that travels with a speed of $c = 3.00 \times 10^8$ m/s in a vacuum. But relative to what reference frame does light have this speed? Classically, the speed of light, measured from different frames of reference, would be expected to differ. For example, by vector addition of velocities (relative velocity, see Chapter 3), it would be greater than 3.00×10^8 m/s if you were approaching the light wave.

Consider ▶ Fig. 26.1: A person is in a reference frame (the truck) moving relative to another frame (the ground) with a constant velocity $\mathbf{v}'$ and throws a ball with a velocity $\mathbf{v}_b$. Then the so-called stationary observer would say that the ball had a velocity of $\mathbf{v} = \mathbf{v}' + \mathbf{v}_b$ relative to the ground (his frame). For example, suppose a truck were moving at 20 m/s east relative to the ground and a ball were thrown at 10 m/s (relative to the truck), also easterly. The ball would then have a speed of 20 m/s + 10 m/s = 30 m/s to the east as observed from the ground reference frame.

Now suppose that the person on the truck turned on a flashlight, projecting a beam of light to the east. In this case, according to Newtonian relativity, $\mathbf{v} = \mathbf{v}' + \mathbf{c}$, and the speed of light measured by the observer would be greater than 3.00×10^8 m/s. According to Newtonian or classical relativity then, the speed of light can take on any value, depending on the observer's reference frame.

Assuming Newtonian relativity to be true, it followed that the particular light speed of 3.00×10^8 m/s must be referenced to some unique frame, in analogy to other waves. Thus, the assumption of a unique reference frame for light seemed quite natural. Since the Earth receives light from the Sun and from distant stars it was thought that a light-transporting medium must permeate all space. This medium was called the *luminiferous ether*, or simply, **ether**.

The idea of an undetected ether became popular in the latter part of the 19th century. Maxwell, whose work laid the foundations for our understanding of electromagnetic waves (Chapter 20), believed in the existence of an etherlike substance, as is evidenced by a quote from his writings:

> Whatever difficulties we may have in forming a consistent idea of the constitution of the ether, there can be no doubt that the interplanetary and interstellar spaces are not empty, but are occupied by a material substance or body which is certainly the largest, and probably the most uniform body of which we have any knowledge.

It would seem, then, that Maxwell's equations (the laws of electromagnetism, which describe the propagation of light) did *not* satisfy the Newtonian relativity principle, as did the laws of mechanics. On the basis of the preceding discussion, a preferential or special inertial reference frame would appear to exist—one that could be considered absolutely at rest—the ether frame.

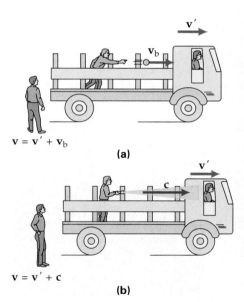

▲ **FIGURE 26.1 Relative velocity**
(a) According to a stationary observer on the ground, $\mathbf{v} = \mathbf{v}' + \mathbf{v}_b$. **(b)** Similarly, the velocity of light would classically be measured to be $\mathbf{v} = \mathbf{v}' + \mathbf{c}$, with a magnitude greater than c. (Velocity vectors are not drawn to scale—can you tell why?)

This was the state of affairs toward the end of the 19th century, when scientists set out to investigate whether they had come upon a new dimension of physics or perhaps a flaw in what were considered to be established principles. One of the first attempts to resolve the situation was to test the ether theory. If it could be proven that the ether existed, then, presumably, a truly absolute rest frame could finally be identified. This was the purpose of the famous *Michelson–Morley experiment*.

During the 1880s, two American scientists, A. A. Michelson and E. W. Morley, carried out a series of experiments designed to see if they could measure the Earth's velocity relative to the ether. They sought to do this, in effect, by measuring differences in the speed of light due to the Earth's orbital velocity. According to the ether theory of those times, if you were moving relative to the ether (the absolute reference frame), then you would measure the speed of light to be different from *c*. The scientists' experimental apparatus, while crude by today's standards, was capable of making such a measurement.

Michelson and Morley performed their experiment over several years at various times during the year, taking advantage of the variation in the direction of the Earth's velocity to try all orientations. The experiment showed that the speed of light was *always c*. Over the next century, Michelson and Morley's experiment has been refined and become much more accurate. Yet the results are always the same: *The speed of light in a vacuum is the same (c), regardless of the observer's motion.*

In essence, then, the ether did not exist. There was no absolute frame of rest. These null results forced physicists to give up the idea of the absolute ether frame. Simply put, light could propagate through a vacuum, and its speed (in a vacuum) was the same for all observers. As we shall see in the remainder of the chapter, this seemingly simple statement radically changed our notions of space and time.

26.2 The Postulates of Special Relativity and the Relativity of Simultaneity

OBJECTIVES: To explain (a) the two postulates of relativity and (b) how they lead to the relativity of simultaneity.

The failure of the Michelson–Morley experiment to detect the ether left the scientific community in a quandary. The inconsistencies between Newtonian mechanics and electromagnetic theory remained unexplained. Many physicists were convinced that the experiment needed to be more accurate; they could not believe that light did not need a medium in which to propagate. These problems were finally resolved by a theory introduced in 1905 by Albert Einstein (◄Fig. 26.2). Interestingly, Einstein was apparently *not* motivated by the Michelson–Morley experiment in the development of his theory of relativity. Einstein later could not recall whether he had even known about the experiment at the time he formulated his theory (while working as a clerk in the Swiss Patent Office!).

Einstein's key insight was based on his idea that the laws of mechanics should not be the only ones to obey the relativity principle. He reasoned that nature should be more symmetrical than that. *All* physical laws should be encompassed by the relativity principle. In Einstein's view, the inconsistencies in electromagnetic theory were due to the assumption that an absolute rest frame (the ether) existed. His theory did away with the need for an ether by placing all laws of physics on an equal footing, thus eliminating any way of measuring the absolute speed of an inertial reference frame.

The first of the two postulates on which relativity is based is thus an extension (generalization) of Newtonian relativity. Einstein's **principle of relativity** applies to *all* the fundamental laws of physics, including those of electricity and magnetism:

> Postulate I (principle of relativity): All the laws of physics are the same in all inertial reference frames.

This postulate implies that all inertial reference frames are equivalent, with all physical laws, not just those involving mechanics, being the same in all of them.

▲ **FIGURE 26.2 Einstein and Michelson** A 1931 photo shows Einstein (right) with Michelson during a meeting in Pasadena, California.

As a consequence, no type of experiment performed entirely within a reference frame would enable an observer in that frame to detect its motion. *Thus, there is no such entity as an absolute reference frame.* In hindsight, the first postulate seems reasonable: We have no reason to think that nature would play favorites by picking the laws of mechanics over the laws from other areas of physics.

Einstein's second postulate involves the speed of light. According to Newtonian relativity, the observed speed of light could have any value. In fact, if a reference system were traveling in the same direction as the light beam and with the same speed, the speed of light in that frame would be zero! This possibility was the source of Einstein's question, "What would I see if I rode a beam of light?" In the same frame as the electromagnetic wave, the electric and magnetic field vectors would be static; that is, they would not vary with time. According to electromagnetic induction, the time variation of electric and magnetic fields is crucial to the propagation of light. Hence, such static fields were inconsistent with the electromagnetic theory of the propagation of light.

To avoid this inconsistency, Einstein formulated his second postulate, called the **constancy of the speed of light**, which can be stated as follows:

Postulate II (constancy of the speed of light): The speed of light in a vacuum has the same value in all inertial systems.

These two postulates form the basis of Einstein's **special theory of relativity**. The "special" designation indicates that the theory deals only with the special case of inertial reference frames. The *general* theory of relativity, discussed later in the chapter, deals with the more general case of noninertial, or accelerating, frames.

The second postulate is perhaps more difficult to accept than the first. It states that two observers in different inertial reference frames measure the speed of light to be *c*, *independently of the speed of the source or the observer*. For example, if a person moving toward you with a constant velocity turned on a light, you and that person would measure the speed of light to be *c*, regardless of your relative velocity (▶ Fig. 26.3).

The second postulate is essential to the validity of the first and consistent with the null result of the Michelson–Morley experiment. By doing away with the ideas of the ether and an absolute reference frame, Einstein could reconcile the apparently fundamental differences between mechanics and electromagnetism. Even so, the second postulate seemed to go against common sense. However, keep in mind that we have very little everyday experience dealing with speeds near that of light! The ultimate test of any theory is provided by the scientific method. What does Einstein's theory predict, and has it been experimentally verified? As we shall see, the answer to the last part of the question is a resounding "yes" for every experiment ever performed. We will now explore some of the implications of Einstein's theory.

The postulates of special relativity can be better understood by imagining simple situations to see what these postulates predict. Einstein did this himself using what he termed *gedanken*, or "thought," experiments. Let us begin with a series of famous Einstein gedanken experiments related to simultaneity and the fundamental idea of how we measure length and time. We will then look at some experimental evidence that supports the implications of the theory.

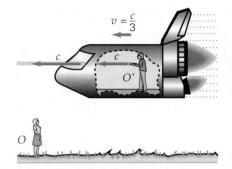

▲ **FIGURE 26.3 Constancy of the speed of light** Two observers in different inertial frames measure the speed of the same beam of light. The observer in frame O' measures a speed of c inside the ship. According to Newtonian mechanics, the observer in frame O would measure a speed of $c + c/3 = 4c/3$ as the beam passes her, but instead, according to Einsteinian mechanics, she measures c.

The Relativity of Simultaneity

In everyday life, we think of two events that are simultaneous to one person as being simultaneous to everyone. What could be more obvious? Simultaneous events are those that occur "at the same time," and doesn't that mean the same thing to all observers? The answer is "no"—at least, not at high speeds.

Think of an inertial reference frame (called O) in which two events are designed to be simultaneous. For example, two firecrackers (located at points A and B on the *x*-axis) could be arranged to explode when a switch controlling a voltage source, placed midway between them, is flipped to the "on" position (▶ Fig. 26.4a).

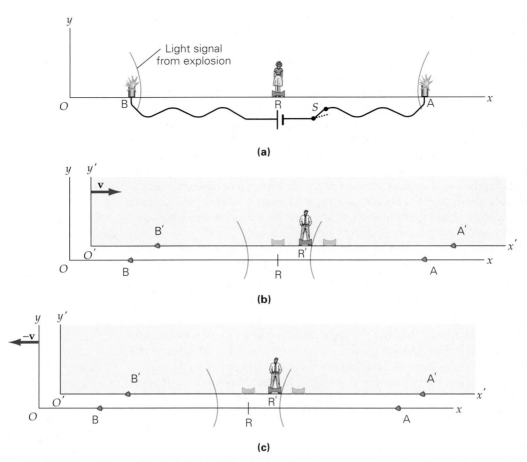

▲ **FIGURE 26.4 The relativity of simultaneity** (a) An observer in reference frame O triggers two explosions (at A and B) simultaneously. A light receptor R, located midway between them, records the two light signals as arriving at the same time. **(b)** An observer midway between the two explosions, but in frame O', moving with respect to O, sees the burn marks made by the two explosions on the x'-axis, but sees the explosion at A happen before that at B. **(c)** The situation as viewed from O'. The observer in O' sees the explosion at A before that at B. To him, O is moving to the left.

Let us equip the observer in this frame with a light receptor at point R, exactly midway between the firecrackers. This detector is capable of detecting whether two light flashes do, in fact, arrive at the same time. (Actually, the light receptor could be placed at any location, but then the reception times would have to be corrected for differences due to the unequal distances the light would travel. To avoid these complications, all simultaneity detectors will be placed at the midpoint.)

After detonation, the light receptor records that the two explosions went off simultaneously *in the O frame*. But consider the same two explosions as seen by another observer in a different inertial frame, O'. Viewed from O, O' is moving to the right at a speed v. The observer in O' has equipped himself with a series of light receptors on his x'-axis, because he is not sure which one will end up midway between the explosion points.

After the explosions, there are burn marks on both the x-axis (at A and B) and the x'-axis (at A' and B'), as shown in Fig. 26.4b. These marks can be used to identify the particular O' light receptor (call it R') that was, in fact, located midway between A' and B'. But when the observer in O' reviews the data from this receptor, he finds that it did *not* record the explosions simultaneously. (The speed of light, c, is the same in both reference frames; why?) This fact does not cause the observer in O to doubt her conclusion, however; she has a simple explanation of what happened. As she sees the situation, during the time it took for the light to get to R', that receptor had moved some distance toward A and away from B. Consequently, the receptor in O' received the flash from A before that from B.

Which observer is correct? It's hard to find any objection to the conclusion reached by the observer in O—but isn't the observer in O' making a mistake? It should be obvious to him that he is moving with respect to the firecrackers. Why doesn't he realize this and take his motion into account? After all, wasn't his light receptor moving toward B and away from B? If so, then it should not surprise him that it recorded the flash from A before the flash from B. All he has to do is to allow for this motion in his calculations, and he will conclude that the flashes "really" were simultaneous.

But if we reason this way, we are ignoring the postulates of relativity. This reasoning amounts to assuming that when the situation is viewed from the O frame (as in Figs. 26.4a and b), we are looking at what "really" happened from the vantage point of the frame that is "really" at rest. But according to the first postulate, no inertial reference frame is more valid than any other, and none can be considered absolutely at rest. The observer in O' doesn't think of himself as moving. To him, the O frame is moving, and O' is the rest frame. He observes the firecrackers moving at a speed v to the left (Fig. 26.4c), but this motion would not affect his conclusions. To him, the explosions, equally distant from R', were recorded at different times; therefore they were not simultaneous.

You might wonder whether it could be arranged so that the observer in O' would agree that the explosions were simultaneous. To accomplish this, the observer in O could delay the setting off of firecracker A relative to B so that R' would receive the two signals at the same time. However, the two observers would still disagree as to whether the events were simultaneous—because now they would no longer be simultaneous in O.

What are we to make of this curious situation? In a nonrelativistic world, one of the observers would have to be wrong. But as we have seen, both observers performed the measurements correctly. Neither one used faulty instruments, or made any errors in logic. So the conclusion must be that *both* are correct.

Notice that there is nothing special about a firecracker explosion. Any "happening" at a particular point in space at a particular time—a karate kick, a clock striking the hour, a soap bubble bursting—would have done just as well. Such a happening is called an **event**, in the language of relativity. In sum, then

> Events that are simultaneous in one inertial reference frame may not be simultaneous in a different inertial frame.

This kind of gedanken experiment convinced Einstein to give up on simultaneity as an absolute concept.

Notice that if the relative speed of the two reference frames is slow, then the lack of simultaneity is completely undetectable. Thus, since everyday speeds are very low compared with that of light, we conclude, erroneously, that simultaneity is absolute. Most relativistic effects have this property. That is, their departure from familiar experience is not apparent when the speeds involved are much less than the speed of light. Since everyday speeds never approach that of light, it is hardly surprising that these conclusions and results seem strange to us.

Conceptual Example 26.1 ■ Agreeing to Disagree: The Relativity of Simultaneity

From Fig. 26.4, estimate the relative speed of the two observers. If the relative speed were only 10 m/s, would there be better agreement on simultaneity? Why?

Reasoning and Answer.   To estimate their relative speed, compare the distance between B and B′ in Fig. 26.4b with the distance the light has traveled from B. The figure indicates that the O' frame has moved about 25% as far as the light wave has. Therefore, the relative speed between the two reference frames is approximately 25% of the speed of light, or $v \approx 0.25c$. By contrast, at 10 m/s, the two frames would *not* have moved a noticeable distance; thus, both observers would agree on simultaneity.

Follow-up Exercise. Show that two events that occur simultaneously on the y-axis of O are perceived as simultaneous by an observer in O', regardless of the relative speed, as long as the relative motion is along their common x–x' axes. *(Answers to all Follow-up Exercises are at the back of the text.)*

Now think about how important simultaneity is to measuring the lengths of moving objects. To measure the length of an object that is moving relative to us, the positions of both ends of it must be marked *simultaneously*. However, two inertial observers in relative motion *disagree* on simultaneity! Thus, they will also disagree on the length of the object. It also turns out that they will disagree on time intervals. These two effects, time dilation and relativistic length contraction, are the subjects of the next section.

26.3 The Relativity of Length and Time: Time Dilation and Length Contraction

OBJECTIVES: To (a) understand the concepts of time dilation and length contraction and (b) calculate the relationship between time intervals and lengths observed in different inertial frames.

Time Dilation

Another of Einstein's thought experiments pertained to the measurement of time intervals in different inertial frames. To compare time intervals in different inertial reference frames, he envisioned a *light pulse clock*, illustrated in ▼Fig. 26.5a. A tick (time interval) on the clock corresponds to the time a light pulse would take to make a round-trip between the source and the mirror. The observers in the two reference frames, O and O', have identical light clocks, and the clocks run at the same rate when they are at rest relative to one another. For an observer at rest with

▼ **FIGURE 26.5 Time dilation** **(a)** A light clock that measures time in units of round-trip reflections of light pulses. The time for light to travel up and back is $\Delta t_o = 2L/c$. **(b)** An observer in O measures a time interval of $\Delta t = (2L/c)[1/\sqrt{1 - (v/c)^2}]$ on the clock in the O' frame. Thus, the moving clock appears to run slowly to the observer in O.

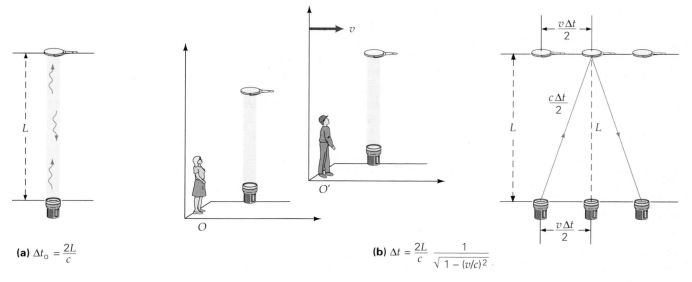

(a) $\Delta t_o = \dfrac{2L}{c}$

(b) $\Delta t = \dfrac{2L}{c} \dfrac{1}{\sqrt{1 - (v/c)^2}}$

respect to one of these clocks, the time interval (Δt_o) for a round-trip of a light pulse is the total distance traveled, divided by the speed of light, or

$$\Delta t_o = \frac{2L}{c} \qquad (26.1)$$

Now, suppose O' is in motion relative to O with a constant velocity $\mathbf{v}$ to the right. With his identical clock (at rest in O') the observer in O' measures the same time interval for his clock, Δt_o. But, according to the observer in O, the clock in the O' system is moving, and the path of its light pulse forms the sides of two right triangles (Fig. 26.5b). Thus, the observer in O sees the light pulse from the clock in O' take a longer path (and therefore a longer time interval Δt) than the light from her own clock. From the O frame, we can apply the Pythagorean theorem:

$$\left(\frac{c\Delta t}{2}\right)^2 = \left(\frac{v\Delta t}{2}\right)^2 + L^2$$

(Here, Δt is the time interval of the O' clock as measured by the observer in O.) Since the speed of light is the same for all observers, the light in the "moving" clock's reference frame takes a longer time to cover the path, *according to the observer in O*. From the O reference frame, the moving clock runs slowly, that is, the ticks occur at a slower rate. To determine the relationship between the time intervals, solve the preceding equation for Δt:

$$\Delta t = \frac{2L}{c}\left[\frac{1}{\sqrt{1 - (v/c)^2}}\right] \qquad (26.2)$$

But the time interval measured by an observer at rest with respect to a clock is given by $\Delta t_o = 2L/c$ (Eq. 26.1). So the measured time interval on the clock moving with respect to an inertial frame is longer than the interval on a clock which is at rest in that frame. Combining the equations, we obtain

$$\Delta t = \frac{\Delta t_o}{\sqrt{1 - (v/c)^2}} \qquad \textit{relativistic time dilation} \qquad (26.3)$$

Since $\sqrt{1 - (v/c)^2}$ is less than one (why?), $\Delta t > \Delta t_o$. Thus, an observer in O measures a longer time interval on the O' clock than does the observer in O' on the same clock. This effect is called **time dilation**. With longer time between ticks, the O' clock appears to O to run more slowly than the O clock. The situation is symmetric and relative: The observer in O' would say that the clock in the O frame ran slowly relative to the O' clock.

PHYSLET®
ILLUSTRATION

Time Dilation

| Moving clocks are observed to run more slowly than clocks that are at rest in the observer's own frame of reference.

This effect, like all relativistic effects, is significant only if the relative speeds are close to that of light.

To distinguish between the two time intervals, the term **proper time interval** is used. As with most measurements, it is usually "proper" or normal to be at rest with respect to a clock when a time interval is measured. In the preceding development, the proper time interval is Δt_o. Stated another way, the proper time interval between two events is that interval measured by an observer who is at rest relative to the events and who sees the events occur *at the same location in space*. (What are the two events for the light-pulse clock? Are they at the same location in O' for the O' clock?) In Fig. 26.5, the observer in O sees the events by which the time interval of the O' clock is measured at *different* locations. Because the clock is moving, the starting event (the light pulse leaving) occurs at a different location in

Note: The proper time interval Δt_o is always less than the dilated time interval Δt.

TABLE 26.1	Some Values

of $\gamma = \dfrac{1}{\sqrt{1-(v/c)^2}}$

v	γ*
0	1.00
0.100c	1.01
0.200c	1.02
0.300c	1.05
0.400c	1.09
0.500c	1.15
0.600c	1.25
0.700c	1.40
0.800c	1.67
0.900c	2.29
0.950c	3.20
0.990c	7.09
0.995c	10.0
0.999c	22.4
c	∞

*To three significant figures.

O from that of the ending event (the light pulse returning). Thus, Δt, the time interval measured by the observer in O, is *not* the proper time interval.

Many of the equations of special relativity can be written more compactly if the expression $1/\sqrt{1-(v/c)^2}$ is replaced by the symbol γ (Greek letter gamma); that is,

$$\gamma = \frac{1}{\sqrt{1-(v/c)^2}} \tag{26.4}$$

Note that γ is always greater than or equal to one. (When is it equal to one?) Note also that as v approaches c, γ approaches infinity. Since an infinite time interval is not physically possible, *relative speeds equal to or greater than that of light are not possible*. The values of γ for several values of v (expressed as fractions of c) are listed in Table 26.1. Notice how the speed of an object must be an appreciable fraction of c before relativistic effects can be observed. For example, at a speed $0.10c$, γ differs from 1.00 by 1%. Using Eq. 26.4, we can express the time dilation relationship (Eq. 26.3) more compactly as

$$\Delta t = \gamma \Delta t_o \tag{26.5}$$

Suppose you observed a clock at rest in a system moving past you at a constant velocity ($v = 0.600c$) relative to you. For that relative speed, γ is 1.25. Because of time dilation, when 20 min have elapsed on that clock, you observe an interval of $\Delta t = \gamma \Delta t_o = (1.25)(20\text{ min}) = 25\text{ min}$ on *your* clock. The 20 min is the proper time, since the events defining the 20-min interval took place at the same location (that of the "moving" clock—for example, for readings at 8:00 and 8:20). Thus, the "moving" clock runs more slowly (20 min elapsed, as opposed to 25 min on your clock) when viewed by an observer (you) moving relative to it.

Finally, note that the time-dilation effect cannot apply solely to the artificial light-pulse clock. It must be true for all clocks and hence all time intervals (i.e., anything that keeps to a rhythm or frequency, including the heart). If this were not the case—if mechanical watches, for example, did not exhibit time dilation—then a watch and a light-pulse clock would run at different rates in a given inertial frame. Then an observer in that frame would have a way of telling whether he or she was moving by making a comparison *solely within their frame*. Since this violates Postulate I, all moving clocks, regardless of their nature, must exhibit the same time-dilation effect. Let's take a look at an actual situation in nature.

Example 26.2 ■ Muon Decay Viewed from the Ground: Time Dilation Verified by Experiment

There are many subatomic particles in nature. One of these, a muon, is created in the Earth's atmosphere when cosmic rays (mostly protons) collide with the nuclei of the atoms that compose air molecules. Once created, muons approach the Earth's surface with a speed near to c (typically, about $0.998c$). However, muons are unstable and decay into other particles. The average lifetime of a muon *at rest* is 2.20×10^{-6} s. During this time, the muon would travel a distance of $d = v_o \Delta t = 0.998c(2.20 \times 10^{-6}\text{ s}) = [0.998(3.00 \times 10^8\text{ m/s})](2.20 \times 10^{-6}\text{ s}) = 659$ m. This is 0.659 km, or less than half a mile. Since muons are created at altitudes of 5 to 15 km, we should expect that very few of them would reach the Earth's surface. However, an appreciable number actually do reach the Earth. Using time dilation, explain this apparent paradox.

Thinking It Through. The apparent paradox arises because the preceding calculation does not take time dilation into account. That is, a muon decays by its own "internal clock," as measured in its own reference frame. (What is measured is a proper time interval, since, in the rest frame of the muon, "birth" and "death" events take place at the same location.) Thus, to an observer on Earth, any clock in the muon's reference frame would appear to run more slowly than a clock on Earth (◄Fig. 26.6).

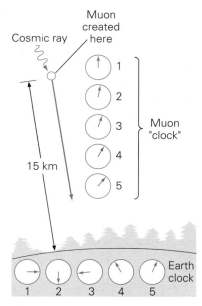

▲ **FIGURE 26.6 Experimental evidence of time dilation** Muons are observed at the surface of the Earth, as predicted by special relativity. (See Example 26.2.)

Solution. We list the given quantities:

Given: $v = 0.998c$
   $\Delta t_o = 2.20 \times 10^{-6}\,$s (muon proper lifetime)

Find: An explanation, based on time dilation, of why muons make it to the Earth's surface

Instead of the proper time interval $\Delta t_o = 2.20 \times 10^{-6}\,$s, an observer on Earth would measure a time interval longer by a factor of γ. From Eq. 26.4,

$$\gamma = \frac{1}{\sqrt{1 - (v/c)^2}} = \frac{1}{\sqrt{1 - (0.998c/c)^2}} = 15.8$$

From Eq. 26.5, the lifetime of the muon, according to an observer on Earth, is

$$\Delta t = \gamma \Delta t_o = (15.8)(2.20 \times 10^{-6}\,\text{s}) = 3.48 \times 10^{-5}\,\text{s}$$

The distance the muon travels, according to an observer on Earth (using the dilated time interval), is therefore

$$d = v\Delta t = 0.998(3.0 \times 10^8\,\text{m/s})(3.48 \times 10^{-5}\,\text{s})$$
$$= 1.04 \times 10^4\,\text{m} = 10.4\,\text{km}$$

This distance is approximately the same altitude at which muons are created. Hence, the detection of more muons than expected is an experimental confirmation of time dilation.

Follow-up Exercise. In this Example, what speed would enable the muons to travel twice as far relative to the Earth? Would they have to travel twice as fast? Explain.

Problem-Solving Hint

In working time-dilation problems, the proper time interval Δt_o must be identified. To do this, first identify (1) the events that define the beginning and end of the interval and (2) a clock that is present at both events. This clock, and the observer at rest with respect to it, measures the proper time interval Δt_o. Since $\Delta t = \gamma \Delta t_o$ and $\gamma > 1$, the proper time interval Δt_o is always less than a time interval measured by any other inertial observer moving relative to this "proper" clock.

Length Contraction

To measure the length of an object that is not at rest in our reference frame, we must take care to mark both ends simultaneously. Consider again the two inertial reference frames used in the discussion of simultaneity, and imagine a measuring stick lying on the x-axis at rest in O (▶Fig. 26.7a). If the observer in O marks the ends simultaneously, the observer in O′ will observe end A marked before B. Thus, for the observer in O′ to make a correct length measurement, the observer in O must delay the marking of A relative to that of B (Fig. 26.7b). Picture the observer in O setting off explosions that make burn marks in both reference frames (i.e., on both the x- and x′-axes), with the information about these two events traveling at the speed c in all directions as visible light. When the ends are marked so that the observer in O′ agrees that they were done simultaneously, all that needs to be done is to subtract the two positions to get the length of the stick. Notice that it is *less* than that measured by O.

Again, it might be asked, "Which observer makes the correct measurement?" Both do, since *both have measured the length correctly* in their own frames. Neither thinks that the other observer has done things correctly, but each is satisfied with his or her own measurement. The observer in O′ would have measured the same length as that in O *if he were willing to overlook the fact that the burn marks were not made simultaneously*, as in Fig. 26.7a. However, this is *not* the correct way to measure the length of the moving stick. The lack of agreement on simultaneity leads to the following qualitative statement about length contraction:

Length Contraction

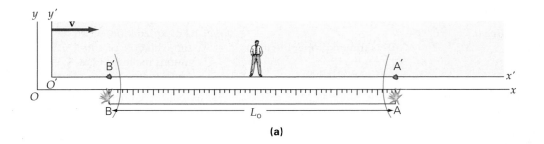

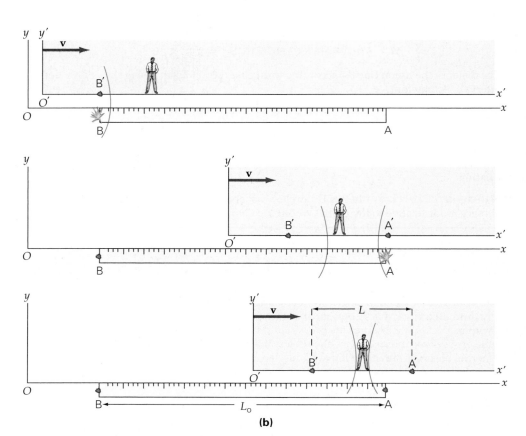

▲ FIGURE 26.7 Measuring lengths correctly To measure the length of a moving object correctly, the ends must be marked simultaneously. **(a)** When the observer in frame O marks the ends simultaneously, the observer in frame O' does not agree and observes A marked before B, resulting in too long a length from the viewpoint of O'. **(b)** When O delays the marking of A relative to that of B by just the correct amount (how do we know from the sketch?), the observer in O' measures the correct length from his point of view. The length measured by the observer in O' is less than the rest (or proper) length measured by the observer in O.

> An object's length is largest as measured by an observer at rest with respect to it (the proper observer). If the object is moving relative to an inertial observer, that observer measures a smaller length than the proper observer.*

Once again, this effect is entirely negligible at speeds that are low compared with c.

Note: The proper length L_o is always greater than the contracted length L.

The distance between two points as measured by the observer at rest with respect to them is designated by L_o and is called the **proper length**. The proper length (also known as the *rest length*) is the largest possible length. The term "prop-

---

*Here, "length" refers to that dimension of the object which is in the direction of its relative velocity. Thus, if a cylindrical stick is moving parallel to its long axis, then only its long-axis "length"—and not its diameter—would exhibit length contraction.

er" has nothing to do with the correctness of the measurement, because each observer is measuring correctly from his or her point of view.

A gedanken experiment can help us develop an expression for length contraction. Consider a rod at rest in frame O. This means that the observer in frame O is the proper length measurer for this rod. Thus, the length of the rod that he measures is L_o. An observer in O', traveling at a constant speed v in the direction parallel to the stick, also measures the length of the rod ($\blacktriangleright$ Fig. 26.8). She does this by using the clock she is holding to measure the time interval required for the two ends of the rod to pass her. Since she measures the proper time interval (how do we know this?), the time interval she measures is Δt_o. In her reference frame, the rod is moving to the left with speed v. With that information, she can determine the length L, since $L = v\Delta t_o$ (speed × time).

The observer in O could also measure the length of the rod by the same means. To him, the observer in O' is moving to the right past the rod. If he notes the times on his clock when O' passes the ends of the rod, he measures a time interval Δt. For him, the length of the stick is the proper length, therefore, $L_o = v\Delta t$. Then, dividing one length by the other, we get

$$\frac{L}{L_o} = \frac{v\Delta t_o}{v\Delta t} = \frac{\Delta t_o}{\Delta t} \qquad (26.6)$$

But we know that $\Delta t = \gamma\Delta t_o$ (Eq. 26.5), or $\Delta t_o/\Delta t = 1/\gamma$. Equation 26.6 then becomes

$$L = \frac{L_o}{\gamma} = L_o\sqrt{1 - (v/c)^2} \quad \textit{relativistic length contraction} \quad (26.7)$$

Since γ is always greater than one, $L < L_o$. This difference in length is called **relativistic length contraction**.

Thus, measurements of time and space intervals (i.e., lengths and distances) are affected by relative motion. To see the effects of length contraction and time dilation, consider the following two high-speed Examples.

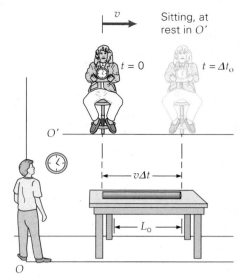

▲ **FIGURE 26.8 Derivation of length contraction** The observer in O measures the time it takes for the observer in O' to move past the ends of the rod. Similarly, the observer in O' measures the time it takes for the ends of the rod to pass her. The observer in O' is the *proper time measurer*; she measures the shortest possible time between these two events. The measured lengths of the rod are not the same. The observer in O is the *proper length measurer*; he measures the longest possible length.

Example 26.3 ■ Warp Speed? Length Contraction and Time Dilation

An observer sees a spaceship, measured to be 100 m long when at rest, pass completely by (that is, nose to tail) in uniform motion with a speed of $0.500c$ ($\blacktriangleright$ Fig. 26.9). While this observer is watching the ship, a time of 2.00 s elapses on a clock onboard the ship. (a) What is the length of the ship as measured by the observer? (b) What time interval elapses on the observer's clock during the 2.00-s interval on the ship's clock?

Thinking It Through. One hundred meters is the proper length. (Why?) The time interval of 2.00 s is the proper time interval, because the same clock, in the same location, measures the beginning and end of the interval. In part (a), Eq. 26.7 can be used to calculate the contracted length, and in part b, Eq. 26.5 can be employed to determine the dilated time interval.

Solution. We list the given quantities:

Given: $L_o = 100$ m (proper length)   *Find:* (a) L (contracted length)
$\quad\quad\quad v = 0.500c$ $\quad\quad\quad\quad\quad\quad$ (b) Δt (dilated time interval)
$\quad\quad\quad \Delta t_o = 2.00$ s (proper time interval)

(a) By calculation or from Table 26.1, $\gamma = 1.15$ for $v = 0.500c$, and the length contraction is given by Eq. 26.7:

$$L = \frac{L_o}{\gamma} = \frac{100\text{ m}}{1.15} = 87.0\text{ m}$$

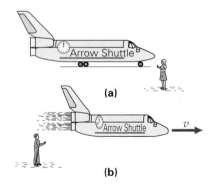

(a)

(b)

▲ **FIGURE 26.9 Length contraction and time dilation** As a result of length contraction, moving objects are observed to be shorter, or contracted, in the direction of motion, and moving clocks are observed to run more slowly because of time dilation. (See Example 26.3.)

(b) The time interval Δt measured by the observer is longer than the proper time interval Δt_o and is given by Eq. 26.5:

$$\Delta t = \gamma \Delta t_o = 1.15(2.00 \text{ s}) = 2.30 \text{ s}$$

Follow-up Exercise. In this Example, calculate the time it takes the spaceship to pass a given point in the observer's reference frame, as seen by (a) a person in the spaceship and (b) the observer watching the ship move by. (c) Explain why these time intervals are not the same.

Example 26.4 ■ Muon Decay Revisited: Alternative Explanations

Example 26.2 showed that many more muons reach the Earth's surface than can be accounted for without relativistic considerations. Experimenters on the Earth explain this phenomenon in terms of relativistic time dilation. Since a hypothetical observer *on the muon* could not use this argument (why not?), how would she explain the fact that the average muon makes it to the surface? Which explanation is correct?

Thinking It Through. A muon traveling at $v = 0.998c$ decays by its own clock (proper time) in $\Delta t_o = 2.20 \times 10^{-6}$ s. In that time, it would travel a distance of only 659 m. Yet muons *are* observed on Earth and so must travel more than 10 km (on average) before decaying. In Example 26.2, this apparent paradox was explained (for the Earth observer) by a time-dilation effect. According to the Earth observer, the muon's "clock" runs slowly, thus enabling it to travel farther (before decay) than expected. But how is the "paradox" explained by a hypothetical observer on the muon? For such an observer, the muon clock is correct—and it registers a time interval that is not sufficient for the muon to reach the Earth's surface! How can this be reconciled with the observation of the Earth-bound observer? After all, two observers cannot disagree on the end result that a significant fraction of the muons do, indeed, make it to sea level. To the observer on the muon, it is the distance to the surface that moves by quickly, so the reasoning will involve length contraction.

Solution.

Given: See Example 26.2   *Find:* the explanation for muons making it to the Earth's surface from the muon reference frame

The apparent paradox disappears when length contraction is taken into account. For the imaginary observer on the muon, the muon "clock" reads correctly ($\Delta t_o = 2.20 \times 10^{-6}$ s), but the observed travel distance is shorter because of length contraction. With $\gamma = 15.8$ for $v = 0.998c$, a travel length of 10.0 km in the Earth frame, which is the proper length L_o (why?), is measured by the observer on the muon to be

$$L = \frac{L_o}{\gamma} = \frac{10.0 \text{ km}}{15.8} = 0.633 \text{ km} = 633 \text{ m}$$

To travel this distance would take a time (according to the muon observer) of

$$\Delta t = \frac{L}{v} = \frac{L}{0.998c} = \frac{633 \text{ m}}{0.998(3.00 \times 10^8 \text{ m/s})} = 2.11 \times 10^{-6} \text{ s}$$

This is roughly the average muon lifetime *in the muon's reference frame*. Thus, through relativistic considerations, both observers agree that many muons reach the Earth (the experimental result). The Earth observer explains this result by saying that the muon clock is running slow (time dilation). The observer on the muon says, "No, the clock is fine, but the distance we have to travel is considerably less than you claim" (length contraction). Who is correct? Both are. The reasoning is different for different observers, but the result is the same.

Follow-up Exercise. Muons are actually created with a range of speeds. What is the speed of a muon if it decays 5.00 km from its creation point as measured by an Earth observer?

The Twin Paradox

Time dilation gives rise to a popular relativistic topic: the **twin paradox**, or **clock paradox**. According to special relativity, a clock moving relative to an observer runs more slowly than one in that observer's frame. Since heartbeat intervals and ages are proper time intervals, the question arises, Do you age more quickly than does a person moving relative to you?

One way to explore this question is through another gedankenexperiment. Consider identical twins, one of whom goes on a high-speed journey into space. Will the space traveler come back younger than his Earth-bound twin? Or will the space twin see the Earth twin age more slowly? *Both can't be right*, and therein lies the apparent paradox.

The resolution of this apparent paradox lies in the fact that in leaving and returning to Earth, the space twin *must* experience accelerations and so is not always in an inertial reference frame. The stay-at-home twin does not feel the forces associated with speed and directional changes that the traveling twin experiences. Thus, the two twins are individually "marked," and their experiences are not symmetrical. However, if the acceleration periods (at the start, at turnaround, and at the return) occupy only a small part of the total time taken for the trip, special relativity can be applied. When it is, the result is that the space-traveling twin does indeed return younger than the Earth-bound twin, and both agree on that fact. (Einstein's general theory, which considers accelerating systems, confirms this result.)

During the time the spaceship travels at a constant velocity on the outward and return trips, the proper length of the trip is measured in the Earth twin's reference frame with fixed beginning and end points. The traveling twin thus measures a length contraction, or a shorter distance for the trip. Traveling at the same relative speed, the space twin travels a shorter time and returns home younger than his Earth-bound twin, according to the traveling twin. According to the Earth twin, the traveling twin's heart beats more slowly, so the traveling twin returns home younger than the Earth-bound twin. Thus, there is no disagreement. The traveler returns having aged less than the nontraveler. At extremely high speeds, it would be possible for the traveler to arrive back and find many generations of Earthlings long since gone.

The twin paradox has been experimentally verified with extremely accurate atomic clocks flown around the world. Extremely accurate clocks were necessary because relativistic effects are extremely small at low speeds. The time recorded by the traveling clock was compared with that measured by an identical clock that stayed at home. The clocks were accurate enough to measure the predicted effect (with corrections made for accelerations on landings and takeoffs), and time dilation in our everyday world, although extremely tiny, was experimentally verified.* To see the calculations for a hypothetical twin-paradox situation, consider the next Example.

Example 26.5 ■ To the Stars: Relativity and Space Travel

Consider a high-speed round-trip into space taken by one of a pair of twins. Ignore accelerations at the start, end, and turnaround points, and assume that a negligible amount of time is spent at the turnaround. (a) Find the speed at which a space explorer would need to travel to make the round-trip to a star 100 light-years away in only 20.0 years of traveler time. (The light-year is defined as the distance light travels in a vacuum in 1 year.) (b) How much time elapses on Earth during this trip?

*Hafele and Keating, "Around the World Atomic Clocks: Relativistic Time Gains Observed," *Science*, vol. 117, no. 4044 (July 14, 1972), pp. 166–170.

Thinking It Through. Let us calculate a one-way trip. The explorer measures the one-way proper time (10.0 years), because his clock is present at the start and end of the trip. People on Earth measure the one-way proper length, 100 light-years, since the beginning and end markers (the Earth and star) of the trip are at rest with respect to the Earth. (a) To find the speed, either the Earth-bound explanation (time dilation) or that of the traveler (length contraction) can be used. (b) To determine the time from the Earth's viewpoint, we use the proper distance and the explorer's speed (from part a).

Solution. We list the given quantities:

Given: $L_o = 100$ light-years (proper length)   *Find:* (a) v (speed of the traveler)
$\quad\quad\quad \Delta t_o = 10.0$ years (one-way proper time)   (b) Δt (time elapsed on Earth)

(a) For the traveler, the length to use is the contracted version of 100 light-years. According to the traveler, a one-way trip takes

$$\Delta t_o = \frac{\text{distance traveled}}{\text{speed}} = \frac{L}{v} = \frac{L_o\sqrt{1 - \left(\dfrac{v}{c}\right)^2}}{v}$$

This equation can be solved for v:

$$v = \frac{c}{\sqrt{1 + \left(\dfrac{c\Delta t_o}{L_o}\right)^2}} = \frac{c}{\sqrt{1 + \left(\dfrac{10.0 \text{ light-years}}{100 \text{ light-years}}\right)^2}} = 0.995c$$

Thus, v is 99.5% the speed of light. Notice that we have used the fact that the distance light travels in 10.0 years ($c\Delta t_o$) is, by definition, 10.0 light-years. We did not have to convert to meters because both distances ($c\Delta t_o$ and L_o) were expressed in light-years.
(b) The people on Earth observe the traveler covering a total of 200 light-years at $0.995c$, so the round-trip time would be 200 light-years/$0.995c$, or about 201 years, compared with 20.0 years for the traveler. The traveler would come back to find that everyone who was alive on Earth when he left had long since died.

Follow-up Exercise. Verify the same result as that obtained in part (b) of this Example, but from time-dilation considerations. Show, using that approach, that the Earth observers find 201 of their years elapsed during the trip. (*Hint*: Carry intermediate results to five decimal places, and round to three significant figures at the end.)

26.4 Relativistic Kinetic Energy, Momentum, Total Energy, and Mass–Energy Equivalence

OBJECTIVES: To (a) understand the relativistically correct expressions for kinetic energy, momentum, and total energy, (b) understand the equivalence of mass and energy, and (c) use the relativistically correct expressions to calculate energy and momentum in particle interactions.

The ramifications of special relativity are particularly important in particle physics, in which speeds can be appreciable fractions of the speed of light. Such speeds can be created, for example, in particle accelerators. Many of the results from classical mechanics are incorrect when they are applied to high-speed particles. Kinetic energy, for example, is different from the expression used in our studies of classical mechanics ($K = \frac{1}{2}mv^2$) as we will now see.

Relativistic Kinetic Energy

Einstein showed that the expression for kinetic energy (K) at high speeds is indeed quite different from the one used in classical mechanics. Kinetic energy still increases with speed, but in a different way. The **relativistic kinetic energy** of a particle of mass m moving with speed v is

$$K = \left[\frac{1}{\sqrt{1 - (v/c)^2}} - 1 \right] mc^2 = (\gamma - 1)mc^2 \quad \begin{array}{l} \textit{relativistic kinetic} \\ \textit{energy} \end{array} \quad (26.8)$$

It can be shown that this expression becomes the more familiar $K = \frac{1}{2}mv^2$ when $v \ll c$.

According to Eq. 26.8, as v approaches c, the kinetic energy of an object becomes infinite. In other words, to accelerate an object to $v = c$ would require an infinite amount of energy or work, which is not possible. Thus, no object can travel as fast as, or faster than, the speed of light.

Particle accelerators can accelerate charged particles to very high speeds. There is complete agreement between the experimentally measured kinetic energies of these charged particles and Eq. 26.8. A graphical comparison of the relativistically correct expression for kinetic energy and the classical (low-speed) expression is shown in ▶Fig. 26.10. Note that the relativistic and classical expressions are in agreement at low speeds.

Relativistic Momentum

In special relativity, the definition of momentum must also be changed from the low-speed expression $\mathbf{p} = m\mathbf{v}$. **Relativistic momentum** is given by

$$\mathbf{p} = \frac{m\mathbf{v}}{\sqrt{1 - \left(\dfrac{v}{c}\right)^2}} = \gamma m\mathbf{v} \quad \begin{array}{l} \textit{relativistic} \\ \textit{momentum} \end{array} \quad (26.9)$$

Note that momentum retains its vector character—that is, the momentum vector ($\mathbf{p}$) is still in the direction of the particle's velocity vector. Linear momentum is still conserved under the proper conditions. (See Chapter 6.)

Relativistic Total Energy and Rest Energy: the Equivalence of Mass and Energy

In classical mechanics, the total mechanical energy of an object is the sum of its kinetic and potential energies ($E = K + U$). When there is no potential energy, this expression reduces to $E = K$; that is, the total energy is all in the form of kinetic energy. However, Einstein was able to show that the **relativistic total energy** of an object is given by

$$E = \frac{mc^2}{\sqrt{1 - \left(\dfrac{v}{c}\right)^2}} = \gamma mc^2 \quad \textit{relativistic total energy} \quad (26.10)$$

This means that even when a body is at rest and thus has no kinetic energy ($K = 0$), it *still has an energy* of mc^2. This minimum energy that an object always possesses is called its **rest energy** and is given by

$$E_o = mc^2 \quad \textit{rest energy} \quad (26.11)$$

According to Eq. 26.10, the energy of an object decreases as v decreases—which also happens according to the classical equation, since, with no potential energy, $E = K = \frac{1}{2}mv^2$. But unlike the classical mechanical situation, when $v = 0$, the relativistic total energy of the particle is *not* zero, but E_o. Since $K = (\gamma - 1)mc^2$, the total energy of a particle can be expressed as the sum of its kinetic and rest

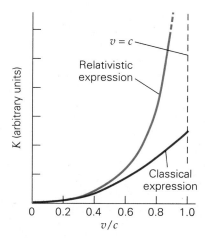

▲ **FIGURE 26.10 Relativistic versus classical kinetic energy** The variation in kinetic energy with particle speed, expressed as a fraction of c, is shown for the relativistically correct expression and for the classical expression. The classical expression becomes negligibly different from the relativistic one for speeds less than about $0.2c$. As objects approach the speed c, their kinetic energy becomes very large. Objects cannot have a speed of exactly c, because their kinetic energy would be infinite.

energies. To see this, note that the relativistic kinetic energy expression (Eq. 26.8) can be rewritten as

$$K = (\gamma - 1)mc^2 = \gamma mc^2 - mc^2 = E - E_o$$

Thus, an alternative to Eq. 26.10 is

$$E = K + E_o = K + mc^2 \qquad \begin{array}{l} \textit{relativistic total energy} \\ \textit{with no potential energy} \end{array} \qquad (26.12)$$

(When a particle has potential energy U, its total energy is $E = K + U + mc^2$.)

A direct relationship between a particle's total energy and rest energy can be obtained by replacing mc^2 with E_o in Eq. 26.10:

$$E = \gamma E_o \qquad (26.13)$$

For $v = 0$ (or $K = 0$) and thus $\gamma = 1$, $E = E_o$, as noted earlier.

Equation 26.11 expresses Einstein's famous **mass–energy equivalence**. An object has energy even at rest—its rest energy. Consequently, *mass is a form of energy*. In nuclear and particle physics, it is impossible to keep the law of conservation of energy if this idea is not included.

Mass–energy equivalence does not mean that mass can be converted into useful energy at will. If so, our energy problems would be solved, since there is a lot of mass on the Earth! However, significant practical conversion of mass into other forms of energy, such as heat, does take place in nuclear reactors that generate electric energy (Chapter 30).

The (rest) energy equivalent of an object depends on its mass. For example, the mass of an electron is 9.109×10^{-31} kg, and its energy is

$$E_o = mc^2 = (9.109 \times 10^{-31}\,\text{kg})(2.998 \times 10^8\,\text{m/s})^2 = 8.187 \times 10^{-14}\,\text{J}$$

or, converted into electron-volts (eV) and mega-electron-volts (MeV)

$$E_o = (8.187 \times 10^{-14}\,\text{J})\left(\frac{1\,\text{eV}}{1.602 \times 10^{-19}\,\text{J}}\right) = 5.110 \times 10^5\,\text{eV} = 0.5110\,\text{MeV}$$

In high-energy, particle, and nuclear physics, it is common to express rest energies in terms of the electron-volt (eV) or multiples of it, such as the MeV. For example, the rest energy of an electron is said to be 511 keV or 0.511 MeV. This is very convenient because, often, the particles are accelerated by potential differences measured in volts.

How do you know whether you need to use the relativistically correct formulas or whether you can "get away" with the classical ones? The general rule of thumb is that if an object's speed is 10% of the speed of light or less, then the error in using the nonrelativistic kinetic energy expression is less than 1%. (The classical expression is lower than the relativistic one). At $v < 0.1c$, the object's kinetic energy is less than 0.5% of its rest energy. Thus, a commonly accepted practice is to use this speed and kinetic-energy region as a dividing line:

For speeds below 10% of the speed of light or kinetic energies less than 0.5% of an object's rest energy, the error in using the nonrelativistic formulas is less than 1%, and it is then usually acceptable to use the nonrelativistic expressions.

For example, an electron with a kinetic energy of 50 eV would qualify as a "nonrelativistic electron" because 50 eV is 10000 times (0.01%) smaller than the electron's rest energy. An electron with a kinetic energy of 0.511 MeV would, however, be *highly* relativistic, because its kinetic energy is the same as its rest energy. Thus, what counts is the kinetic energy relative to the rest energy of the particle, or the speed of the particle relative to that of light—even the dividing line is relative! The following Example shows how particle energies are calculated using relativistically correct expressions.

Example 26.6 ■ A Speedy Electron: Energy Required for Acceleration

(a) How much work is required to accelerate an electron from rest to a speed of $0.900c$?
(b) How much error is made by using the nonrelativistic expression?

Thinking It Through. (a) By the work–energy theorem, the work needed is equal to the electron's gain in kinetic energy. The gain in kinetic energy is the final kinetic energy, since the initial kinetic energy is zero. The rest energy of the electron is known: $E_o = mc^2 = 0.5110$ MeV. From its speed, its kinetic energy can be determined with Eq. 26.8. (b) The nonrelativistic kinetic energy is given by $K = \frac{1}{2}mv^2$.

Solution. The given data are as follows:

Given: $v = 0.900c$ *Find:* (a) W (work required)
$E_o = 0.5110$ MeV (from text) (b) K (nonrelativistic kinetic energy)

(a) The relativistic kinetic energy is given by Eq. 26.8:

$$K = (\gamma - 1)mc^2 = (\gamma - 1)E_o$$

With $v = 0.900c$, by calculation or from Table 26.1, $\gamma = 2.29$; therefore,

$$K = (\gamma - 1)E_o = (2.29 - 1)(0.511 \text{ MeV}) = 0.659 \text{ MeV}$$

Thus, by the work–energy theorem, 0.659 MeV of work is required to accelerate an electron to a speed of $0.900c$.

(b) Many times, even with nonrelativistic expressions, a shortcut can be applied as follows: Essentially, you can work with mass expressed in energy units. The technique involves first multiplying and then dividing by c^2, which enables you to use E_o:

$$K_{nonrel} = \tfrac{1}{2}mv^2 = \tfrac{1}{2}(mc^2)\left(\frac{v}{c}\right)^2 = \tfrac{1}{2}(0.5110 \text{ MeV})(0.900)^2 = 0.207 \text{ MeV}$$

In this case, the kinetic energy is low by a factor of more than three.

Follow-up Exercise. In this Example, (a) what is the relativistic total energy of the electron? (b) What would be the electron's total energy if it were treated nonrelativistically?

To see how relativistic momentum is employed at relativistic speeds, consider the next Example.

Integrated Example 26.7 ■ When 1 + 1 Doesn't Equal 2: Conservation of Relativistic Momentum and Energy

A particle of mass m, initially moving, collides with an identical particle initially at rest. The two stick together, forming a single particle of mass m'. (a) Do you expect that (1) $m' > 2m$, (2) $m' < 2m$, or (3) $m' = 2m$? Explain. (b) If the incoming particle is initially moving at a speed $v = 0.800c$ to the right, what is m' in terms of m?

(a) Conceptual Reasoning. This is an example of an *inelastic* collision (Chapter 6). In such a collision, kinetic energy is not conserved. Some, but not all, of the initial kinetic energy is lost. All of it cannot be lost, because that would mean the combined particle would be at rest after the collision, a situation that would violate conservation of momentum. Since total relativistic energy is conserved, any loss of kinetic energy is converted into mass. If this weren't true, then (3) would be the correct answer. However, the combined mass must include the mass equivalent of the "lost" kinetic energy, or $m' = 2m +$ "some kinetic energy in the form of mass." Therefore, the correct answer is (1): $m' > 2m$.

(b) Thinking It Through. Collisions are usually analyzed using conservation of momentum and total energy. After the collision, the combined particle must be moving to the right to conserve the direction (vector) of the total momentum. The magnitude of the moving particle's momentum before the collision must equal the magnitude of the single combined particle's momentum afterward. The total relativistic energy is also conserved. From these considerations, we can determine the combined particle's mass.

Solution.

Given: $v = 0.800c$ *Find:* m' (mass of combined particle)
  $m = $ mass of one particle

For the incoming particle,

$$\gamma = 1/\sqrt{1 - (v/c)^2} = 1/\sqrt{1 - (0.800)^2} = 1.67$$

The total system momentum **P** is conserved; thus,

$$\mathbf{P}_i = \mathbf{P}_f$$

Using the expression for relativistic momentum (Eq. 26.9) and equating the magnitude of the momentum of the incoming particle to that of the combined particle, we have

$$\gamma m v = \gamma' m' v'$$

where γ' refers to the combined particle after the collision. Putting in the numbers,

$$1.67m(0.800c) = \gamma' m' v'$$

Next, we equate the total relativistic energy before the collision to that after the collision. Remember that the total energy of a single particle is related to its rest energy by $E = \gamma E_o$. Therefore, the initial total energy is the sum of the energy due to the moving particle and the rest energy of the "target," or $E_i = 1.67mc^2 + mc^2$. The final total energy is that of the combined particle, or $E_f = \gamma' m' c^2$. Thus, energy conservation requires that

$$1.67mc^2 + mc^2 = \gamma' m' c^2 \quad \text{or} \quad 2.67m = \gamma' m'$$

(In the last step, the speed of light cancels out of both sides of the equation.) Dividing the last result into the momentum result, we obtain

$$\frac{1.67m(0.800\,c)}{2.67m} = \frac{\gamma' m' v'}{\gamma' m'} = v'$$

or

$$v' = 0.500\,c$$

Using this result, we find that $\gamma' = 1/\sqrt{1 - (0.500)^2} = 1.15$. Now the energy equation can be used to solve for the mass of the combined particle m':

$$2.67mc^2 = 1.15m'c^2 \quad \text{or} \quad m' = 2.32m$$

As expected, the mass of the combined particle is greater than $2m$ because some kinetic energy is converted into mass (energy).

Follow-up Exercise.   (a) In this Example, how much kinetic energy is lost? (b) What would be the mass of the combined particle if the two particles initially approached head-on each with a speed of 0.800c and stuck? (Your answers should be in terms of m and c.)

26.5 The General Theory of Relativity

OBJECTIVES: To (a) explain the principle of equivalence and (b) specify some of the predictions of general relativity.

Special relativity applies to inertial systems, but not to accelerating systems. Accelerating systems require an extremely complex theory, which was described by Einstein in several papers published about 1915. Called the **general theory of relativity**, it is essentially a theory about gravity.

The Principle of Equivalence

An important aspect of general relativity was first stated by Einstein (after yet another gedankenexperiment). He stated it as the **principle of equivalence**:

> An inertial reference frame in a uniform gravitational field is physically equivalent to a reference frame that is not in a gravitational field, but that is in uniform linear acceleration.

What this means physically is that

> No experiment performed in a closed system can distinguish between the effects of a gravitational field and the effects of an acceleration.

Thus, an observer in an accelerating system would find the effects of a gravitational field and those of their acceleration to be equivalent and indistinguishable. For simplicity, we will restrict ourselves to systems that are accelerating linearly (i.e., no rotational acceleration).

To understand the principle of equivalence, consider the situations illustrated in ▶Fig. 26.11. Imagine yourself as an astronaut in a closed spaceship. You drop a pencil, and it accelerates to the floor. What does this mean? According to the principle of equivalence, it could mean (a) that you are in a gravitational field or (b) that you are in an accelerating system (Fig. 26.11a). Any experiment performed entirely inside the ship cannot determine one from the other. Whether the spaceship is in free space and accelerating with an acceleration $a = g$ or whether it is in a gravitational field with $g = -a$, the pencil would accelerate to the floor with the same observed acceleration. In your closed system (you are not allowed to look outside!), there is no experiment you could perform to distinguish whether the pencil's fall is a gravitational or an acceleration effect. According to the principle of equivalence, the two are physically indistinguishable.

Suppose that the pencil does not begin to fall, but instead remains suspended in midair next to you. This could mean that (a) you are in an inertial frame with no gravity or (b) you are in free fall in a gravitational field. Once again, in the closed spaceship (Fig. 26.11b), you have no way of distinguishing between these two possibilities. Thus, they are *equivalent*.

Light and Gravitation

The principle of equivalence leads to an important prediction: *that a gravitational field bends light*. To see how this prediction arises, let's use another gedankenexperiment. Suppose a beam of light traverses a spaceship that is accelerating rapidly upward. If the spaceship were stationary, light entering the ship at point A would arrive at point B on the far wall (▶Fig. 26.12a). However, because the spaceship is accelerating during the time it takes the light to traverse the ship, the light bends into an arc and arrives at point C. (Notice that if the spaceship were simply moving up without accelerating, the light path would be a straight line. This would not be consistent with a gravitational trajectory. Why?)

Now consider the situation as observed by a person onboard the spaceship (Fig. 26.12b). From her point of view, the light enters at A and arrives at C after traversing a downward-curving path—as though the light path were bent by the gravitational field, much as a baseball's path would be. Although it is the acceleration of the rocket that produces this effect, we conclude by the principle of equivalence that light paths should be bent by a gravitational field.

Under most conditions, this effect must be very small, since we don't observe any everyday evidence of light bending in the Earth's gravitational field. However, this prediction of Einstein's theory was experimentally verified in 1919 during a solar eclipse. At night, when they can be clearly seen, distant stars are measured

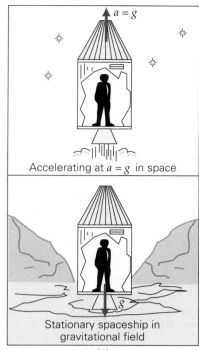

Accelerating at $a = g$ in space

Stationary spaceship in gravitational field

(a)

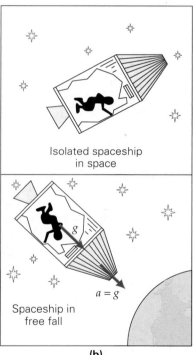

Isolated spaceship in space

Spaceship in free fall

(b)

▲ **FIGURE 26.11 The principle of equivalence** (a) In a closed spaceship, the astronaut can perform no experiment that would determine whether he was in a gravitational field or an accelerating system. (b) Similarly, an inertial frame without gravity cannot be distinguished from free fall in a gravitational field.

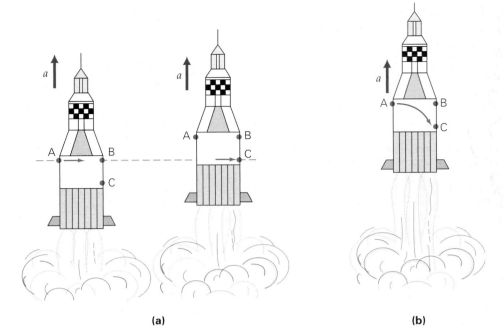

◄ FIGURE 26.12 Light bending
(a) Light traversing an accelerating, closed rocket from point A arrives at point C. **(b)** In the accelerating system, the light path would appear to be bent. Since an acceleration produces this effect, by the principle of equivalence, light should also be bent by a gravitational field.

(a)   **(b)**

to have a constant angular separation (▼Fig. 26.13a). Light from one of the stars may pass near the Sun, which has a relatively strong gravitational field. However, any bending would not normally be observed on Earth because the starlight is usually masked by the strong glare of the Sun and by the sunlight scattered in the atmosphere (which is why stars aren't seen during the day).

However, conditions in which stars may be seen clearly can occur during a total solar eclipse. When the Moon comes between the Earth and the Sun, an observer in the Moon's shadow (the umbra) can see stars that are not normally observed (Fig. 26.13b). If the light from a star passing near the Sun is bent, then the star will have an *apparent* location that is different from its normal location. As a result, the angular distance between stars will be measured as slightly

▼ FIGURE 26.13 Gravitational attraction of light **(a)** Normally, two distant stars are observed to have a certain angular separation. **(b)** During a solar eclipse, the star behind the Sun can still be seen because of the effect of solar gravity on starlight. The star has a larger measured angular separation than it has in the absence of such an eclipse. **(c)** General relativity views a gravitational field as a warping of space and time. A simplified analogy is the surface of a warped rubber sheet.

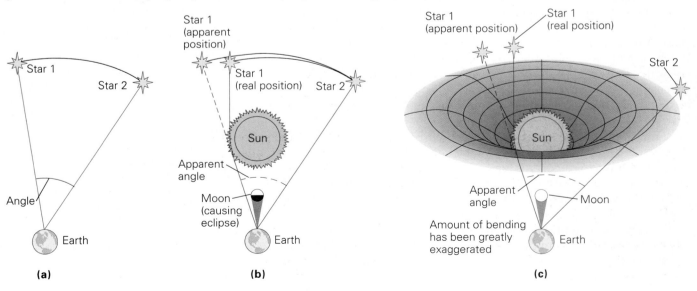

larger than normal. Einstein's theory predicted that the angular difference between the apparent positions of the stars during the eclipse of 1919 and their positions in the absence of such an eclipse would be 1.75 seconds of arc, or a tiny angle of about 0.00005°. The experimental angular distance was found to be 1.61 ± 0.30 seconds of arc, in agreement within the level of uncertainty of the measurement.

General relativity describes a gravitational field as a "warping" of space and time, as illustrated in Fig. 26.13c. A light beam follows the curvature of space–time like a ball rolling on a curved surface. The bending of light was verified during subsequent solar eclipses and also by signals sent back to Earth by various space probes that pass near the Sun.

Gravitational Lensing Another effect of the gravitational bending of light is known as *gravitational lensing*. In the late 1970s, a double quasar was discovered. (A quasar is a powerful astronomical radio source.) The fact that it was a double quasar was not unusual, but everything about the two quasars seemed to be *exactly* the same, except that one was fainter than the other. It was suggested that perhaps there was only one quasar and that, somewhere between it and the Earth, a massive, but optically faint, object had bent its electromagnetic radiation, producing multiple images. The subsequent detection of a faint galaxy between the two quasars confirmed this hypothesis, and other examples have since been discovered (▼Fig. 26.14).

The discovery of gravitational lenses gave general relativity a new role in modern astronomy. By examining multiple images of a distant galaxy or quasar and their relative brightness, astronomers can gain information about an intervening galaxy or cluster of galaxies whose gravitational field causes the bending of light.

Black Holes The idea that gravity can affect light finds its most extreme application in the concept of a black hole. A **black hole** is thought to form from the gravitationally collapsed remnant of a massive star, typically many times the mass of our Sun.* Such an object has a density so great and a gravitational field so

▼ **FIGURE 26.14 Gravitational lensing** **(a)** The bending of light by a massive object such as a galaxy or a cluster of galaxies can give rise to multiple images of a more distant object. **(b)** The discovery of what appeared to be four images of the same quasar (the "Einstein Cross") suggested the possibility of gravitational lensing. On investigation, a faint intervening galaxy was found.

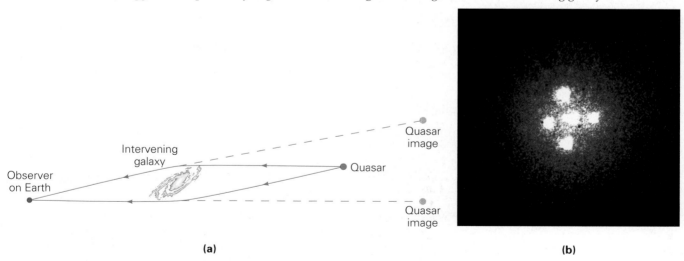

(a)

(b)

*It is speculated that black holes also may originate in other ways, such as from the collapse of entire star clusters in the center of a galaxy or at the beginning of the universe during the big bang. Some current theories (combining quantum mechanics and general relativity) hint that black holes might emit material particles and radiation and thus "evaporate," but none have been discovered to be evaporating. In this book, we will limit our discussion to stellar collapse.

intense that nothing can escape it. In terms of our space–time warp analogy, a black hole is graphically represented as a bottomless pit in the fabric of space–time. Even light can't escape the intense gravitational field of a black hole—hence the blackness.

An estimate of the size of a black hole can be obtained by using the concept of escape speed. Recall from Section 7.6 that the escape speed from the surface of a spherical body of mass M and radius R is given by

$$v_{esc} = \sqrt{\frac{2GM}{R}} \qquad (26.14)$$

If light does not escape from a black hole, its escape speed must exceed the speed of light. The critical radius of a sphere around a black hole from which light cannot escape is obtained by substituting $v_{esc} = c$ into Eq. 26.14. Solving for R, we obtain

$$R = \frac{2GM}{c^2} \qquad \textit{Schwarzschild radius} \qquad (26.15)$$

Note: A collapsing star becomes a black hole if it collapses beyond its Schwarzschild radius, or event horizon.

The quantity R is called the **Schwarzschild radius**, after Karl Schwarzschild (1873–1916), a German astronomer who developed the concept. The boundary of a sphere of radius R defines the black hole's **event horizon**. Any event occurring within this horizon is invisible to an outside observer, since light cannot escape. The event horizon thus gives the limiting distance within which light cannot escape from a black hole. Information about what goes on inside the event horizon can never reach us, so questions such as "What does it look like inside the event horizon?" are unanswerable (at least, given the current state of knowledge in physics).

Example 26.8 ■ If the Sun Were a Black Hole: Schwarzschild Radius

What would be the Schwarzschild radius if our Sun collapsed to a black hole? (The mass of the Sun is $M_S = 2.0 \times 10^{30}$ kg.)

Thinking It Through. This is a straightforward calculation using Eq. 26.15. We will need the gravitational constant and speed of light.

Solution.

Given: $M_S = 2.0 \times 10^{30}$ kg   *Find:* R (Schwarzschild radius)
$G = 6.67 \times 10^{-11}$ N·m²/kg²
$c = 3.00 \times 10^8$ m/s

Keeping all the quantities in SI units, we have

$$R = \frac{2GM_S}{c^2} = \frac{2(6.67 \times 10^{-11}\, \text{N·m}^2/\text{kg}^2)(2.0 \times 10^{30}\, \text{kg})}{(3.00 \times 10^8\, \text{m/s})^2} = 3.0 \times 10^3\, \text{m} = 3.0\, \text{km}$$

This is less than 2 miles! The Sun's radius is about 7×10^5 km. However, our Sun will *not* become a black hole. This fate befalls only stars much more massive than the Sun.

Follow-up Exercise. Once black holes form, they continuously draw in matter, increasing their Schwarzschild radius. How many times more massive would our Sun have to be for its Schwarzschild radius to extend to, and swallow up, Mercury, the innermost planet? (The average distance from the Sun to Mercury is 5.79×10^{10} m.)

If nothing, including radiation, escapes a black hole from inside its event horizon, how, then, might we observe or even merely locate a black hole? This question is addressed in the Insight on p. 867.

Black Holes

The idea of black holes has caught the public's fancy. Try to imagine something so dense that nothing—not even light—can escape from it. It is so dense that a tablespoon full of its matter would contain more mass than Mt. Everest. According to the theory of stellar evolution, black holes could result from the collapse of stars having much greater mass than the Sun. But gathering experimental proof of the existence of a black hole is another matter.

The boundary of the "blackness" of the hole is the surface of the event horizon at the Schwarzschild radius. The boundary of a black hole is thus not a sphere of matter, but rather the distance at which the gravitational force is sufficiently strong to keep even light from escaping. What form or size matter takes inside a black hole is not known and in principle can never be known. Even if a probe could be sent "inside" a black hole (i.e., closer than its Schwarzschild radius), the probe could not send data back to us.

How, then, might a black hole be detected? Light passing nearby it would be bent, but it is unlikely that we would ever observe that, considering the vastness of space. The most likely possibility comes from observing binary star systems, which are common. (A binary star system consists of two stars that orbit each other about a common center of mass.) Cygnus X-1, the first X-ray source discovered in the constellation Cygnus, provides the best evidence so far for a black hole in our galaxy. (X-ray sources are observed by means of rockets and satellites, since X rays do not penetrate our atmosphere. Most astronomical X-ray sources, however, are not massive enough to qualify as black holes.)

Looking at the location in the sky where Cygnus X-1 is located, in visible light (Fig. 1) we observe a giant star (visible light cannot detect two separate stars) whose spectrum shows periodic Doppler redshifts and blueshifts, indicating a periodic orbital motion away from us and toward us, respectively. Measuring the precise orbital data of both stars could, in principle, allow us to compute their masses. In Cygnus X-1, there is an unseen companion to the giant star that *seems* to have more than enough mass to qualify as a black hole.

It is speculated that, in general, in binary star systems such as Cygnus X-1, one member has become a black hole. Matter drawn from the other, more normal, star would create an *accretion disk* of spiraling matter around the black hole (Fig. 2). The matter falling into the disk would be accelerated and heated. Collisions and deceleration would produce a characteristic spectrum of X rays, and the black hole would appear as an X-ray source. Note that the X rays must be emitted by the hot matter before it gets closer than the Schwarzschild radius. (Why?)

A similar mechanism may explain the enormous energy output of *active galaxies* and *quasars*. It has been proposed that the centers of these brilliant objects, which produce vast amounts of radiation, might contain black holes with masses millions or even billions of times that of our Sun. In fact, recent observations suggest that even many normal galaxies, including our own Milky Way galaxy, may harbor enormously massive black holes in their cores.

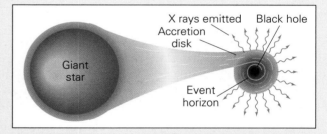

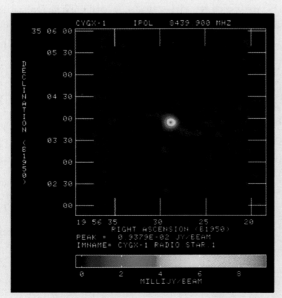

FIGURE 1 Cygnus X-1 The overexposed large dark spot at the center is a giant star believed to be a companion of the X-ray source and possible black hole Cygnus X-1.

FIGURE 2 X rays and black holes Matter drawn from the "normal" member of a binary star system forms a spiraling accretion disk around the hole. Matter falling into the disk is accelerated, and collisions give rise to the emission of X rays.

*26.6 Relativistic Velocity Addition

OBJECTIVES: To (a) understand the necessity for a relativistic velocity addition equation and (b) apply it to simple relative velocity calculations.

As we have seen, the postulates of special relativity affect our classical concepts of distance and time. Since velocity involves these two quantities, relative velocities should also be affected, and that is indeed the case.

Recall from Section 26.1 that, according to Newtonian relativity, there is a problem with vector addition when light is involved. In Fig. 26.1, the vector addition of velocities predicts a speed greater than c for the beam of light. Such a speed violates the second postulate of relativity.

The same problem occurs with objects moving at appreciable fractions of the speed of light. Consider the rocket separation shown in ▼ Fig. 26.15. After separation, the jettisoned stage has a velocity **v** with respect to the Earth, and the rocket has a velocity **u′** with respect to the jettisoned stage. Then, from Newtonian relativity and vector addition, the velocity of the rocket payload with respect to the Earth is **u** = **v** + **u′**.

However, at relativistic speeds, this addition law gives a result contradictory to special relativity. Suppose, for example, that the velocities are all in the same direction with magnitudes of $v = 0.50c$ and $u' = 0.60c$. Then $u = v + u' = 0.50c + 0.60c = 1.10c$. Thus, classical relativity predicts that an observer on the Earth would measure the rocket's speed to be greater than c; however, according to the argument in Section 26.3, no object can travel faster than the speed of light.

Einstein recognized the problem with the Newtonian approach. Since length and time differ depending on the observer's reference frame, classical velocity vector addition is not correct at relativistic speeds. He showed the correct equation (for motion in a straight line) to be

$$u = \frac{v + u'}{1 + \dfrac{vu'}{c^2}} \qquad \begin{array}{l} \textit{relativistic} \\ \textit{velocity addition} \\ \textit{(one dimension)} \end{array} \qquad (26.16)$$

where the velocities have the same meanings as in the preceding paragraph and sign notation is used to indicate velocity directions. Notice that the observed velocity u is reduced by a factor of $1/[1 + (vu'/c^2)]$ from that of the classical result. At very low speeds, (i.e., $v \ll 1$ and $u' \ll 1$), $\dfrac{vu'}{c^2} = \dfrac{v}{c} \cdot \dfrac{u'}{c} \ll 1$. Then the denominator in Eq. 26.16 is, for all practical purposes, equal to one, and the Newtonian result, $u = v + u'$, is obtained.

In working out problems, it is important to identify the velocities clearly:

v = velocity of object 1 with respect to an inertial observer

u′ = velocity of object 2 with respect to object 1

u = velocity of object 2 with respect to an inertial observer

Since we will consider only one-dimensional-motion problems, signs are used to indicate velocity directions. Let's see in the next Example how Eq. 26.16 gives results that do not violate the second postulate.

Note: It is usually convenient to express velocities as fractions of c.

▶ **FIGURE 26.15 Relativistic velocity addition** After jettisoning, the rocket payload has a velocity **u′** with respect to the jettisoned stage, and the jettisoned stage has a velocity **v** with respect to the Earth. For relativistic velocities, the velocity **u** of the rocket with respect to the Earth is obtained from a relativistically correct version.

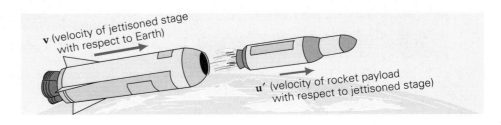

v (velocity of jettisoned stage with respect to Earth)

u′ (velocity of rocket payload with respect to jettisoned stage)

Example 26.9 ■ Faster than the Speed of Light?
Relativistic Velocity Addition

For the rocket separation in Fig. 26.15, let the speeds be $v = 0.50c$ and $u' = 0.60c$ (with directions shown in the figure). What is the velocity of the payload, as measured by an observer on Earth?

Thinking It Through. Clearly, Eq. 26.16 needs to be applied here because the speeds are relativistic, and u must be less than c.

Solution. The velocities are taken to be in the positive direction.

Given: $v = +0.50c$ (speed of jettisoned stage with respect to Earth)
$u' = +0.60c$ (speed of rocket with respect to jettisoned stage)

Find: u (speed of rocket with respect to Earth)

Since both velocities are to the right, they are designated with plus signs. From Eq. 26.16, we have

$$u = \frac{v + u'}{1 + \dfrac{vu'}{c^2}} = \frac{+0.50c + 0.60c}{1 + \dfrac{(+0.50c)(+0.60c)}{c^2}} = \frac{+1.1c}{1.3} = +0.85c$$

As expected, u is less than c and is directed to the right, as indicated by the plus sign.

Follow-up Exercise. Repeat this Example with both objects moving to the right at $0.40c$. According to Newtonian relativity, the velocity relative to the Earth is $0.80c$, which does *not* violate the second postulate. What is the correct result? (*Hint*: It should be lower than the Newtonian result of $0.80c$. Why?)

Chapter Review

Important Concepts and Equations

- **Newtonian or classical relativity** supported the belief in the existence of an absolute inertial reference frame somewhere in the universe that was "at rest." In this reference frame was the material called the **ether**, the medium through which light could propagate.

- **The Michelson–Morley experiment** attempted to measure the speed of the Earth relative to the ether frame. The results were always zero. Thus, physicists had to give up the idea of a reference frame that was absolutely at rest.

- **Special theory of relativity** involves inertial reference frames moving relative to one another and is based on two postulates:

  Principle of relativity: All the laws of physics are the same in all inertial reference frames.

  Principle of the constancy of the speed of light: The speed of light in a vacuum has the same value in all inertial systems.

- A time interval measured by a clock that is present at both the starting and stopping events of the interval is a **proper time interval** Δt_o. The time inter-

val Δt of an observer in any other inertial frame is larger than the proper time interval. The two intervals are related by

$$\Delta t = \frac{\Delta t_o}{\sqrt{1 - (v/c)^2}} \qquad (26.3)$$

For convenience, the factor γ (pronounced "gamma") is defined as

$$\gamma = \frac{1}{\sqrt{1 - (v/c)^2}} \qquad (26.4)$$

Equation 26.3 is written more compactly as

$$\Delta t = \gamma \Delta t_o \qquad (26.5)$$

This effect on time intervals is called **time dilation**.

- The length of an object, as measured by an observer at rest with respect to it, is called the object's **proper length** L_o. To an observer in any other inertial frame, the length L is smaller than the proper length by a factor of $1/\gamma$, or

$$L = \frac{L_o}{\gamma} = L_o\sqrt{1 - (v/c)^2} \qquad (26.7)$$

This phenomenon is known as **length contraction**.

- At speeds near the speed of light, an object's **relativistic kinetic energy** is

$$K = \left[\frac{1}{\sqrt{1 - (v/c)^2}} - 1 \right] mc^2 = (\gamma - 1)mc^2 \quad (26.8)$$

- At speeds near the speed of light, an object's **relativistic momentum** is

$$\mathbf{p} = \frac{m\mathbf{v}}{\sqrt{1 - \left(\dfrac{v}{c}\right)^2}} = \gamma m\mathbf{v} \quad (26.9)$$

- At speeds near the speed of light, the **relativistic total energy** of an object, is

$$E = \frac{mc^2}{\sqrt{1 - \left(\dfrac{v}{c}\right)^2}} = \gamma mc^2 \quad (26.10)$$

- Even when an object is at rest, it still has an energy of mc^2. This minimum energy called its **rest energy** and is

$$E_o = mc^2 \quad (26.11)$$

- An object's relativistic total energy can be written in terms of its rest energy as

$$E = K + E_o = K + mc^2 \quad (26.12)$$

or

$$E = \gamma E_o \quad (26.13)$$

- The **general theory of relativity** expresses how physical quantities are measured by observers in reference frames that are accelerating. The **principle of equivalence**, which states that

  an inertial reference frame in a uniform gravitational field is physically equivalent to a reference frame that is not in a gravitational field, but that is in uniform linear acceleration.

  This means that

  no experiment performed in a closed system can distinguish between the effects of a gravitational field and the effects of an acceleration.

- General relativity predicts many interesting phenomena that have been experimentally verified such as stars called **black holes**, and the bending of light by a gravitational field.

Exercises

Note: Assume c to be exact at 3.00×10^8 m/s. Consider speeds given in terms of c as having the same number of significant figures as the accompanying numerical coefficient. (For example, $0.85c$ has two significant figures.)

26.1 Classical Relativity and the Michelson–Morley Experiment

1. An object free of all forces exhibits a changing velocity in a certain reference frame. It follows that (a) the frame is inertial, (b) $\mathbf{F} = m\mathbf{a}$ applies in this frame, (c) the laws of mechanics are the same in this reference frame as in all inertial frames, or (d) none of these.

2. The principle of classical relativity (a) did not seem to apply to electricity and magnetism, (b) required a constant speed of light different from that predicted by Maxwell's equations, or (c) implied that nothing can travel as fast as light.

3. The existence of the ether (a) would provide a special or absolute reference frame, (b) is necessary for the propagation of light, (c) made Maxwell's equations relativistically correct, or (d) both (a) and (b).

4. The Michelson–Morley experiment showed that the speed of light in a vacuum (a) depends on the observer's motion, (b) proved the existence of the ether, (c) is a constant, or (d) none of the above.

5. CQ We live on a rotating Earth and therefore are in an accelerating, noninertial system. How, then, can we apply Newton's laws of motion on the Earth?

6. ■ Car A is traveling eastward at 85 km/h. Car B is traveling with a speed of 65 km/h. Find the relative velocity of car B with respect to car A if car B is traveling (a) westward and (b) eastward.

7. ■ A person 1.20 km away from you fires a gun, and you observe the flash from the muzzle. A wind of 10.0 m/s is blowing. How long will it take the sound of gunfire to reach you if the wind is (a) toward you and (b) toward the person who fired the gun? (Take the speed of sound to be 345 m/s.)

8. ■ A small airplane has an airspeed (speed with respect to air) of 200 km/h. Find the airplane's ground speed if there is (a) a head wind of 35 km/h and (b) a tailwind of 25 km/h.

9. ■ A speedboat can travel with a speed of 50 m/s in still water. If the boat is in a river that has a flow speed of 5.0 m/s, find the maximum and minimum values of the boat's speed relative to an observer on the riverbank.

10. IE ■■ A boat can make a round-trip between two locations, A and B, on the same side of a river in a time t if there is no current in the river. (a) If there is a constant current in the river, the time it takes the boat to make the same round-trip will be (1) longer, (2) the same, or (3) shorter. Why? (b) If the boat can travel with a speed of 20 m/s in still water, the speed of the river current is 5.0 m/s, and the distance between points A and B is 1.0 km, calculate the times when there is no current and when there is current.

11. ■■■ The apparatus used by the French scientist Armand Fizeau in 1849 for measuring the speed of light is illus-

trated in ▼Fig. 26.16. Teeth on a rotating wheel periodically interrupt a beam of light. The flashes of light travel to a plane mirror and are reflected back to an observer. Show that if the wheel is rotated so that light passing through one gap reaches the mirror and is reflected to the observer through the adjacent gap, the speed of light is given by $c = 2fNL$, where N is the number of gaps in the wheel; f is the frequency of, or number of revolutions per second made by, the rotating wheel; and L is the distance between the wheel and the mirror.

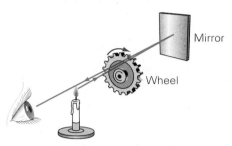

▲ **FIGURE 26.16 Fizeau's apparatus** See Exercise 11.

26.2 The Postulates of Special Relativity and the Relativity of Simultaneity

12. The special theory of relativity (a) applies to inertial systems, (b) applies to noninertial systems, (c) applies to both inertial and noninertial systems, or (d) disproves the constancy of the speed of light.

13. Events that are simultaneous in one inertial reference frame are (a) always simultaneous in other inertial reference frames, (b) never simultaneous in other inertial reference frames, (c) sometimes simultaneous in other inertial reference frames, or (d) none of these.

14. CQ In a space war, a warrior traveling in a spaceship at a speed of $0.85c$ heads directly toward a laser. At what speed does the warrior see the light from the laser approach him?

15. CQ Is it possible for either observer on two approaching rocket ships to observe the other ship with a velocity greater than c? Why?

16. CQ In the gedanken experiment shown in ▼Fig. 26.17, two events in the same inertial reference frame O are related by cause and effect: (1) A gun at the origin fires a bullet along the x-axis with a speed of 300 m/s. (Assume that there are

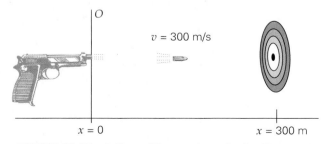

▲ **FIGURE 26.17 A thought experiment** See Exercise 16.

no gravitational or frictional forces.) (2) The bullet hits a target at $x = +300$ m. (a) What is the velocity of the inertial reference frame that would determine the proper time between these two events? (b) Use qualitative arguments to show that the two events cannot be viewed simultaneously by any inertial observer. (*Note:* This shows that special relativity preserves the time *sequence* of two events if they are related as cause and effect. Thus, to all observers, the gun fires before the bullet hits the target.)

17. CQ In a gedanken experiment (▼Fig. 26.18), two events that cannot be related by cause and effect occur in the same inertial reference frame O: (1) Strobe light A, at the origin of the x-axis, flashes; (2) strobe light B, located at $x = +600$ m, flashes 1.0 μs later. B's flash cannot be caused by A's flash. (Light travels only 300 m in the time between the two events.) (a) Show that there exists another inertial reference frame, traveling at less than c, in which these two events would be observed to occur simultaneously. (b) What is the direction of the velocity of the reference frame in (a)? Relativity does *not* preserve the time sequence of two events if they are not related as cause and effect. (*Note:* Since one event doesn't cause the other, no physical principles are violated if observers see them in reversed order.)

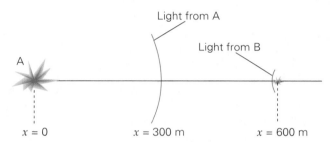

▲ **FIGURE 26.18 Another thought experiment** See Exercise 17.

26.3 The Relativity of Length and Time: Time Dilation and Length Contraction

18. An observer sees a friend passing by her in a rocket ship that has a uniform velocity with a magnitude of an appreciable fraction of the speed of light. The observer knows her friend to be 5 ft, 8 in., tall, and he is standing such that his length is perpendicular to their relative velocity. To the observer, her friend will appear (a) taller than 5 ft, 8 in., (b) shorter than 5 ft, 8 in., or (c) exactly 5 ft, 8 in., tall.

19. CQ A farm boy who knows some physics wants to store a 5-m-long pole in a storage shed that is only 4 m long. He claims that if he runs sufficiently fast, the pole will fit in the shed as a result of length contraction. Is this possible from his viewpoint? Explain.

20. CQ On returning from a round-trip through space at high speeds (close to c), would you find yourself younger than your twin brother who stayed home? Explain.

21. CQ You are standing on a road and observe a rocket-powered car as it passes by you. Sketch the shape of the

car when it is stationary and when it is moving at the very high speed of 0.50*c*.

22. **CQ** You are standing on the Earth and observe a fast-moving spacecraft go by with your professor on board. (a) If both you and your professor are observing your wristwatch, who is measuring the proper time? (b) Who measures the proper length of the spacecraft?

23. ■ A spacecraft moves past a student with a relative velocity of 0.90*c*. If the pilot of the spacecraft observes 10 min to elapse on his watch, how much time has elapsed according to the student's watch?

24. **IE** ■ You have a pulse rate of 80 beats/min, and your physics professor in a spacecraft is moving with a speed of 0.85*c* relative to you. (a) According to your professor, your pulse rate is (1) greater than 80 beats/min, (2) equal to 80 beats/min, or (3) less than 80 beats/min. Why? (b) What is your pulse rate, according to your professor?

25. ■ You fly your 15.0-m-long spaceship with a speed of *c*/3 relative to your friend. Your velocity is parallel to the ship's length. How long is your spaceship, as observed by your friend?

26. **IE** ■ An astronaut in a spacecraft moves past a field 100 m long (according to a person standing on the field) and parallel to the field's length, with a speed of 0.75*c*. (a) Will the length of the field, according to the astronaut, be (1) longer than 100 m, (2) equal to 100 m, or (3) shorter than 100 m? Why? (b) What is the length as measured by the astronaut? (c) Which length is the proper length?

27. ■■ The proper lifetime of a muon is 2.20 μs. If the muon has a lifetime of 34.8 μs according to an observer on Earth, what is the muon's speed, as a fraction of *c*, relative to the observer?

28. ■■ One of a pair of 25-year-old twins takes a round-trip through space while the other twin remains on Earth. The traveling twin moves at a speed of 0.95*c* for 39 years, according to Earth time. Assuming that special relativity applies, what are the twins' ages when the traveling twin returns to Earth?

29. **IE** ■■ Alpha Centauri, a binary star close to our solar system, is about 4.3 light-years away. Suppose a spaceship traveled this distance with a constant speed of 0.60*c* relative to Earth. (a) Compared with a clock on the spaceship, an Earth-based clock will measure (1) a longer time, (2) an equal time, or (3) a shorter time. Why? (b) How much time would elapse on an Earth-based clock and on a clock on the spaceship?

30. ■■ A cylindrical spaceship of length 35.0 m and diameter 8.35 m is traveling in the direction of its cylindrical axis (length). It passes by the Earth with a relative speed of 2.44 $\times$ 10^8 m/s. What are the dimensions of the ship, as measured by an Earth observer?

31. ■■ A pole-vaulter at the Relativistic Olympics sprints past you to do a vault with a speed of 0.65*c*. When he is at rest, his pole is 7.0 m long. What length do you perceive the pole to be as he passes you? (What assumption do you need to make to arrive at your answer?)

32. ■■ A "flying wedge" spaceship is a right triangle in side view. When the ship is at rest, its base and altitude (which form the 90° angle) measure 40.0 m and 15.0 m, respectively. As the ship moves with a speed of 0.900*c* past an observer on Earth, what does she (a) measure the area of the side of the ship to be? (b) calculate the angle of the triangular "nose" to be?

33. ■■ How fast must a meterstick be moving, parallel to its length, relative to an observer so that he measures its length to be 50 cm?

34. ■■ The distance to Planet *X* is 1.00 light-year. How long does it take a spaceship to reach *X*, according to the pilot of the spaceship, if the speed of the ship is 0.700*c* relative to *X*?

35. ■■ What is the length contraction (ΔL) of an automobile 5.00 m long when it is traveling at 100 km/h? [*Hint*: For $x = 1$, $\sqrt{1 - x^2} \approx 1 - (x^2/2)$.]

36. **IE** ■■ A student in a certain reference frame cannot understand why there is a problem in converting to the metric system, because she notes that the length of her professor's meterstick in another reference frame is the same length as her yardstick (parallel to the meterstick). (a) Which of the following is true? (1) the student is moving relative to the professor, (2) the professor is moving relative to the student, or (3) the only thing that matters is that they are moving relative to one another. Why? (b) What is their relative speed (assuming them to be moving in a direction parallel to their respective sticks)?

37. ■■■ Sirius is about 9.0 light-years from Earth. To reach the star by spaceship in 12 years (ship time), how fast must you travel?

26.4 Relativistic Kinetic Energy, Momentum, Total Energy, and Mass–Energy Equivalence

38. The special theory of relativity predicts relativistic effect(s) for the fundamental properties of (a) length, (b) kinetic energy, (c) time, or (d) all of these.

39. The total energy of a relativistic moving particle is (a) $E = mv^2$, (b) $E = \gamma E_o$, (c) $E = K + \gamma mc^2$, or (d) $E = mc^2$.

40. **CQ** The special theory of relativity places an upper limit on the speed an object can have. Are there similar limits on energy and momentum? Explain.

41. **CQ** Is it physically possible to accelerate a mass to the speed of light if a constant net force continuously acts on the mass? Why?

42. **CQ** Energy can be expressed in the unit of MeV, the mega electron volt (Chapter 16). Physicists often use the MeV as a unit of mass in atomic and nuclear physics. Is this wrong? Why?

43. ■ If an electron has a kinetic energy of 2 keV, could the classical expression for kinetic energy be used to compute its speed accurately? What if its kinetic energy was 2 MeV? Explain.

44. ■ An electron travels at a speed of $0.600c$. What is its total energy?

45. ■ An electron is accelerated from rest through a potential difference of 2.50 MV. Find the electron's (a) speed, (b) kinetic energy, and (c) momentum.

46. ■ How fast must an object travel for its total energy to be (a) 1% more than its rest energy and (b) 99% more than its rest energy?

47. ■ An average home uses about 1.5×10^4 kWh of electricity per year. How much matter would have to be converted to energy (assuming 100% efficiency) to supply energy for one year to a city with 250 000 such homes? (Are you surprised by the answer?)

48. ■ The United States uses approximately 3.0 trillion kWh of electricity annually. If this electrical energy were supplied by nuclear generating plants, how much nuclear mass would have to be converted to energy, assuming a production efficiency of 25%?

49. ■ To travel to a nearby star, a spaceship is accelerated to $0.99c$ in order to take advantage of time dilation. If the ship has a mass of 3.0×10^6 kg, how much work must be done to get it up to speed from rest? Compare this value with the annual electricity usage of the United States. (See Exercise 48.)

50. ■■ An electron has a total energy of 2.8 MeV. What is its momentum?

51. ■■ How much energy, in keV, is required to accelerate an electron from rest to $0.50c$?

52. ■■ The kinetic energy of an electron is 60% of its total energy. Find the (a) speed and (b) momentum of the electron.

53. ■■ Sketch graphs of the (a) momentum and (b) energy of an object as a function of speed between zero and c. (Use the relativistically correct expressions.)

54. ■■ A proton moves with a speed of $0.35c$. What are its (a) total energy (b) kinetic energy, and (c) momentum?

55. ■■ A proton moving with a constant speed has a total energy 2.5 times its rest energy. What are the proton's (a) speed and (b) kinetic energy?

56. **IE** ■■ The Sun has a mass of 1.989×10^{30} kg and radiates at a rate of 3.827×10^{23} kW. (a) Over time, will the mass of the Sun (1) increase, (2) remain the same, or (3) decrease? Why? (b) Estimate the lifetime of the Sun from this data.

57. ■■ A nickel has a mass of 5.00 g. If this mass could be completely converted to electric energy, how long would it keep a 100-W lightbulb lit?

58. **IE** ■■ Phase changes require energy in the form of latent heat (Chapter 11). (a) If 1 kg of ice at 0°C is converted to water at 0°C, will the water have (1) more, (2) the same, or (3) less mass than the ice? Why? (b) What is the difference in mass between the ice and the water? Do you think this difference would be detectable?

59. ■■ A beam of electrons is accelerated from rest to a speed of $0.950c$ in a particle accelerator. In MeV, what are the (a) kinetic energy and (b) total energy of the electrons?

60. **IE** ■■■ A particle of mass m, initially moving with a speed v, collides head-on elastically with an identical particle initially at rest. (a) Do you expect the total mass of the two particles after the collision to be (1) greater than $2m$, (2) equal to $2m$, or (3) less than $2m$? Why? (b) What are the total energy and momentum of the two particles after the collision, in terms of m, v, and c?

61. ■■■ Using the relationship $E = \gamma mc^2$, show that the total energy E and the magnitude of the relativistic momentum p are related by $E^2 = p^2c^2 + (mc^2)^2$.

62. ■■■ In a linear accelerator, protons are accelerated to an energy of 600 MeV. (a) How do we know that all of this energy must be kinetic energy and not total energy? (b) What is the speed of the protons? (c) What is their momentum?

26.5 The General Theory of Relativity

63. The general theory of relativity (a) provides a theoretical basis for explaining the gravitational force, (b) applies only to rotating systems, (c) applies only to inertial systems, or (d) refutes the principle of equivalence.

64. One of the predictions of the general theory of relativity is (a) mass–energy equivalence, (b) the constancy of the speed of light, (c) the twin paradox, or (d) the bending of light in a gravitational field.

65. The Schwarzschild radius is the radius of (a) a black hole, (b) a star, (c) an event horizon, or (d) none of the above.

66. **CQ** Suppose a meterstick were dropped toward the event horizon of a black hole. Describe what the effect(s) might be on the meterstick.

67. **IE** ■ The mass of the planet Jupiter is about 318 times that of the Earth ($M_J = 318 M_E$). (a) If these two planets could develop into black holes, would the event horizon for Jupiter be (1) larger, (2) the same, or (3) smaller than the Earth's? Why? (b) The mass of the Earth is $M_E = 6.0 \times 10^{24}$ kg; what are the radii of the event horizons for the Earth and Jupiter?

68. ■■ If the Sun became a black hole, what would be its average density? (See Example 26.8.)

69. ■■ A black hole has an event horizon of 5.00×10^3 m. (a) What is its mass? (b) Determine the lower limit on the density of the black hole.

70. **CQ** ■■ An apparatus like that in ▼Fig. 26.19 was given to Albert Einstein on his 76th birthday by Eric M. Rogers, a physics professor at Princeton University. The goal is to get the ball into the cup without touching the ball. (Jiggling the pole up and down will not do it.) Einstein solved the puzzle immediately and then confirmed his answer with an experiment. How did Einstein get the ball into the cup? [*Hint*: He used a fundamental concept of general relativity.][†]

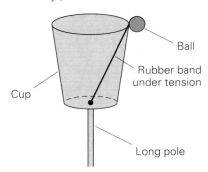

Ball

Rubber band under tension

Cup

Long pole

▲ **FIGURE 26.19   How to get the ball into the cup**   See Exercise 70.

*26.6 Relativistic Velocity Addition

71. ■ After jettisoning a stage, a rocket has a velocity of $+0.20c$ relative to the jettisoned stage. An observer on Earth sees the jettisoned stage moving with a velocity of $+7.5 \times 10^7$ m/s, relative to her, in the same direction as the rocket. What is the velocity of the rocket relative to the Earth observer?

72. ■ In moving away from Planet Z, a spacecraft fires a probe with a speed of $0.15c$, relative to the spacecraft, back toward Z. If the speed of the spacecraft is $0.40c$ relative to Z, what is the velocity of the probe as seen by an observer on Z?

73. ■■ A rocket launched outward from Earth has a speed of $0.100c$ relative to Earth. The rocket is directed toward an incoming meteor that may hit the planet. If the meteor moves with a speed of $0.250c$ relative to the rocket and di-

[†]This problem is adapted from R. T. Weidner, *Physics*, Allyn and Bacon, 1985, p. 333.

rectly toward it, what is the velocity of the meteor as observed from Earth?

74. **IE** ■■ Two spaceships, each with a speed of $0.60c$ relative to Earth, approach each other head-on. (a) The speed of one ship relative to the other is (1) greater than c, (2) equal to c, or (3) less than c. Why? (b) What is the speed of one ship relative to the other?

75. ■■■ In a colliding-beam apparatus, two beams of protons are aimed directly at each other. The first beam contains protons moving with a speed of $0.800c$ to the right, and the second beam's protons have a speed of $0.900c$ to the left. Both speeds are measured relative to the laboratory frame. What are (a) the velocity of the second beam's protons relative to that of the first, and (b) the velocity of the first beam's protons relative to that of the second?

Additional Exercises

76. Compute the rest energy of a proton (in MeV).

77. A spaceship travels with a speed of $0.60c$ relative to an observer in an inertial frame. How much time does a clock onboard the spaceship appear to lose in a day, according to that observer?

78. An electron travels at a speed of 9.5×10^7 m/s. What is its total energy?

79. A relativistic rocket is measured to be 50 m long, 2.5 m high, and 2.0 m wide by the pilot. To an inertial observer, what are its dimensions when the rocket travels at $0.65c$ in the direction of its length?

80. If the mass of 1.0 kg of coal (or any substance) could be completely converted into energy, how many kilowatt-hours of energy would be produced?

81. Imagine that you are moving with a speed of $0.80c$ past a person who is reading this chapter in his textbook. If he takes 30 min to read the chapter by his clock, how much time would you observe to elapse on your clock?

82. In electron–positron annihilation, an electron and a positron (which has the same mass as the electron, but carries a positive charge) collide, and both masses are completely converted to electromagnetic radiation. Assuming that both particles are at rest when they are annihilated, what is the total energy of the radiation?

83. A spaceship has a length of 150 m in its rest frame. An observer in another, inertial frame measures the length of the spaceship as 110 m, with a relative velocity parallel to that length. What is the speed of the rocket relative to the observer?

84. At a typical nuclear power plant, refueling occurs about every 18 months. Assuming that a plant has operated continuously since the last refueling and produces 1.2 GW of electric power at an efficiency of 33%, how much less massive are the fuel rods at the end of the 18 months than at the start? (Assume 30-day months.)

Quantum Physics

Lasers are used in a variety of everyday applications—in bar-code scanners at store checkout counters, in computer printers, in various types of surgery, and in laboratory situations, such as the one shown in the opening photo. You own a laser yourself if you have a CD or DVD player.

The laser is a practical application of principles that revolutionized physics. These principles were first developed in the early 20th century, one of the most productive eras in the history of physics. For example, special relativity (Chapter 26) helped resolve problems faced by classical (Newtonian) relativity in describing objects moving at speeds comparable with that of light. However, there were other troublesome areas in which classical theories did not agree with experimental results. To address these issues, scientists devised new hypotheses based on non-

traditional approaches, ushering in a profound revolution in our understanding of the physical world. Chief among these new theories was the idea that light is *quantized* into discrete amounts of energy. This concept and others like it led to the formulation of a new set of principles and a new branch of physics, *quantum mechanics*.

Quantum theory demonstrated that particles often exhibit wave properties and that waves frequently behave like particles. Thus was born the *wave–particle duality* of matter. As a result of quantum theory, calculations in the realm of the very small—dimensions the sizes of atoms and smaller—must deal with *probabilities* rather than in the precisely determined values associated with classical theory.

A detailed treatment of quantum mechanics requires extremely complex mathematics. However, a general overview of the important results of such a treatment is essential to an understanding of physics as it is known today. Thus, the important developments of "quantum" physics are presented in this chapter, and an introduction to quantum mechanics is provided in Chapter 28.

27.1 Quantization: Planck's Hypothesis

OBJECTIVES: To (a) define blackbody radiation and use Wien's law and (b) understand how Planck's hypothesis paved the way for quantum ideas.

One of the problems scientists faced at the end of the 19th century was how to explain the spectra of electromagnetic radiation emitted by hot objects—solids, liquids, and dense gases. This radiation is sometimes called **thermal radiation**. You learned in Chapter 11 that the total intensity of the emitted radiation from such objects is proportional to the fourth power of the absolute Kelvin temperature (T^4) of the object. Thus, all objects emit thermal radiation to some degree.

However, at everyday temperatures, this radiation is almost all in the infrared region (IR) and thus not visible to our eyes. At a temperature of about 1000 K, an object begins to emit an appreciable amount of radiation in the long-wavelength end of the visible spectrum, observed as a reddish glow. A hot electric stove burner is a good example of this condition. Still higher temperatures cause the radiation to shift to even shorter wavelengths and the color to change to yellow-orange. Above a temperature of about 2000 K, an object glows yellowish-white, like the filament of a light bulb, and gives off appreciable amounts of all the visible colors (wavelengths), but with different percentages.

Although we observe a dominant color (wavelength) with our eyes, in actuality there is a *continuous spectrum*. A spectrum shows how the intensity of emitted energy depends upon wavelength, as illustrated in ▶ Fig. 27.1a for a hot object. Notice that practically all wavelengths are present, but there is a dominant color (wavelength region), which depends upon the object's temperature.

Note: The concept of a blackbody was introduced in Section 11.4.

The curves shown in Fig. 27.1a are for an idealized **blackbody**. A blackbody is an ideal system or object that absorbs and emits all radiation that is incident on it. Although an ideal blackbody is not attainable, it can be experimentally approximated quite well by a small hole that leads to an internal cavity inside a block of material (Fig. 27.1b). Radiation falling on the hole enters the cavity and is reflected back and forth by the cavity walls. If the hole is very small in comparison with the surface area of the cavity, only a small amount of radiation will make its way out of the hole. Since nearly all radiation incident on the hole is absorbed, as viewed from the outside, the hole is a good approximation of the surface of a blackbody.

Two things happen to the spectrum as the temperature increases (Fig. 27.1a). Clearly, more radiation is emitted at every wavelength, but also the wavelength of the maximum-intensity component (λ_{max}) becomes shorter. This wavelength shift is described by **Wien's displacement law**

$$\lambda_{max}T = 2.90 \times 10^{-3} \, \text{m} \cdot \text{K} \qquad (27.1)$$

where λ_{max} is the wavelength of the radiation (in meters) at which maximum intensity occurs and T is the temperature of the body (in kelvins).

Wien's law can be used to determine the wavelength of the maximum spectral component if the temperature of the emitter is known, or the temperature of the emitter if the wavelength of the strongest emission is known. Thus, it can be used to estimate the temperatures of stars (dense gases) from their radiation spectrum, as the following Example shows.

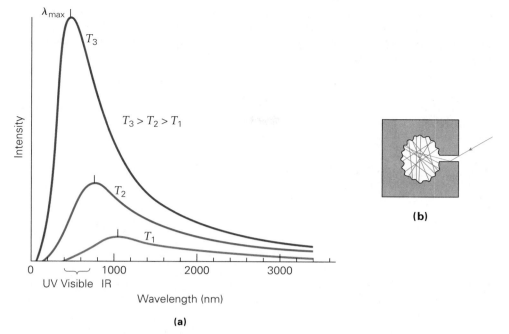

(b)

▲ **FIGURE 27.1   Thermal radiation**   **(a)** Intensity-versus-wavelength curves for the thermal radiation from a blackbody at different temperatures. The wavelength associated with the maximum intensity (λ_{max}) becomes shorter with increasing temperature. **(b)** A blackbody can be approximated by a small hole leading to an interior cavity in a block of material.

Example 27.1 ■ Solar Colors: Using Wien's Law

The visible surface of the Sun is the gaseous photosphere from which radiation escapes. At the top of the photosphere, the temperature is 4500 K; at a depth of about 260 km, the temperature is 6800 K. To a good approximation, the Sun radiates energy as if it were a blackbody. Assuming this condition, (a) what are the wavelengths of the radiation of maximum intensity for these temperatures, and (b) to what colors do these wavelengths correspond?

Thinking It Through.   Wien's displacement law (Eq. 27.1) enables us to determine the wavelengths.

Solution.

Given: T_1 = 4500 K   *Find:*   (a) λ_{max} (for the two different temperatures)
        T_2 = 6800 K   (b) Colors corresponding to these λ_{max} values

(a)   At the top of the photosphere,

$$\lambda_{max} = \frac{2.90 \times 10^{-3} \text{ m} \cdot \text{K}}{4500 \text{ K}} = (6.44 \times 10^{-7} \text{ m})(10^9 \text{ nm/m}) = 644 \text{ nm}$$

and at the 260-km depth,

$$\lambda_{max} = \frac{2.90 \times 10^{-3} \text{ m} \cdot \text{K}}{6800 \text{ K}} = (4.26 \times 10^{-7} \text{ m})(10^9 \text{ nm/m}) = 426 \text{ nm}$$

(b)   As the temperature increases with photosphere depth, the wavelength of the radiation of maximum intensity shifts toward the blue end of the spectrum. Hence, at the Sun's surface, the emitted radiation of maximum intensity is in the orange–red region of the visible spectrum. (For a discussion of the visible spectrum and color in relation to wavelength, see Section 20.4 and Fig. 20.23.) At a depth of 260 km, λ_{max} is near the violet

PHYSLET®
ILLUSTRATION

Blackbody Spectrum

end of the spectrum. Combining all the depths and temperatures in between, we have a spectrum that shows all wavelengths (colors), but is dominated by yellow. Notice that some of the emitted radiation will be in the ultraviolet region. (How do we know this?) This UV component is ordinarily absorbed by the ozone layer in the Earth's atmosphere.

Follow-up Exercise. What would you expect to be the dominant wavelengths (and corresponding colors) from the following stars? (a) Betelgeuse, with an average surface temperature of 3.00×10^3 K; (b) Rigel, with an average surface temperature of 1.00×10^4 K. (*Answers to all Follow-up Exercises are at the back of the text.*)

The Ultraviolet Catastrophe and the Planck Hypothesis

Classically, thermal radiation results from the thermal oscillations of electric charges near the surface of an object. Since these electric charges oscillate at many different frequencies, a continuous spectrum of emitted radiation is expected.

Classical calculations describing the spectrum of radiation emitted by a black-body predict an intensity inversely related to wavelength (actually $I \propto 1/\lambda^4$). At long wavelengths, the classical theory agrees fairly well with experimental data. However, at short wavelengths, the agreement disappears. Contrary to experimental observations, the classical theory predicts that the radiation intensity should continue to increase without bound as the wavelength gets smaller. These results are illustrated in ◄Fig. 27.2. The classical prediction is sometimes called the *ultraviolet catastrophe*—"ultraviolet" because the difficulty occurs for short wavelengths beyond the violet end of the visible spectrum, and "catastrophe" because it predicts that the emitted energy grows without limits at these wavelengths.

The failure of classical electromagnetic theory to explain the characteristics of thermal radiation led Max Planck (1858–1947), a German physicist, to reexamine the phenomenon. In 1900 Planck formulated a theory that correctly predicted the observed distribution of the blackbody radiation spectrum. (Compare the solid blue curve with the data points in Fig. 27.2.) However, his theory depended upon a radical new idea. He had to assume that the thermal oscillators (the atoms emitting the radiation) have only *discrete*, or particular, amounts of energy rather than a continuous distribution of possible energies. Only with this assumption did his theory agree with experimental results.

Planck found that these discrete amounts of energy were related to the frequency f of the atomic oscillations by

$$E_n = n(hf) \quad \text{for } n = 1, 2, 3, \dots \qquad \begin{array}{l}\textit{Planck's quantization}\\ \textit{hypothesis}\end{array} \qquad (27.2)$$

That is, the energy occurs only in integral multiples of hf. The symbol h represents a constant known as **Planck's constant** and has a value (to three significant figures) of

$$h = 6.63 \times 10^{-34}\,\text{J} \cdot \text{s}$$

The idea expressed in Eq. 27.2 is called **Planck's hypothesis**. Rather than the oscillator energy being able to have any value, as is true classically, Planck's hypothesis states that the energy is *quantized*, that is, it occurs only in discrete amounts. The smallest possible amount of oscillator energy, according to Eq. 27.2 with $n = 1$, is

$$E_1 = hf \qquad (27.3)$$

All other permitted values of the energy are integral multiples of hf. The quantity hf is called a **quantum** of energy (from the Latin *quantus*, meaning "how much"). As a result, the energy of each atom can change only by the absorption or emission of energy in discrete, or *quantum*, amounts.

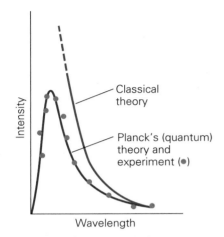

▲ **FIGURE 27.2   The ultraviolet catastrophe** Classical theory predicts that the intensity of thermal radiation emitted by a blackbody should be inversely related to the wavelength of the emitted radiation. If this were true, the intensity would become infinite as the wavelength approaches zero. In contrast, Planck's quantum theory agrees with the observed radiation distribution (solid dots).

Note: As with other fundamental constants, we will take h to be *exact* at 6.63×10^{-34} J · s, for calculational purposes.

Although the theoretical predictions agreed with experiment, Planck himself was not convinced of the validity of his quantum hypothesis. However, the concept of quantization was then extended to explain other phenomena that could not be explained classically. Thus, despite Planck's hesitation, the quantum hypothesis earned him a Nobel Prize in 1918.

27.2 Quanta of Light: Photons and the Photoelectric Effect

OBJECTIVES: To (a) describe the photoelectric effect, (b) explain how it can be understood by assuming that light energy is carried by particles, and (c) summarize the properties of photons.

The concept of the quantization of light was introduced in 1905 by Albert Einstein in a paper about light absorption and emission, at about the same time he published his famous paper on special relativity. Einstein reasoned that energy quantization should be a fundamental property of electromagnetic waves (light). He suggested that if the energy of the thermal oscillators in a hot substance is quantized, then it necessarily followed that, to conserve energy, *the emitted radiation should also be quantized*. For example, suppose an atom initially had an energy of $3hf$ ($n = 3$ in Eq. 27.2) and ended up with a final energy of $2hf$ ($n = 2$ in Eq. 27.2). Einstein proposed that the atom *must* emit a specific amount (or *quantum*) of light energy—in this case, $3hf - 2hf$, or hf. His quantum, or packet, of light is today referred to as a **photon**. Each photon has a definite amount of energy E that depends on the frequency f of the light according to

$$E = hf \qquad \begin{array}{l}\textit{photon energy} \\ \textit{and light frequency}\end{array} \qquad (27.4)$$

This idea suggests that light can behave as discrete quanta (plural of "quantum"), or "particles," of energy rather than as waves. Scientists interpret Eq. 27.4 as a mathematical "connection" between the *wave nature* of light (a wave of frequency f) and the *particle nature* of light (photons each with an energy E). Given light of a certain frequency or wavelength, we can use this equation to calculate the amount of energy in each photon, or vice versa.

Einstein used the photon concept to explain the **photoelectric effect**, another phenomenon for which the classical description was inadequate. Certain metallic materials are *photosensitive*, that is, when light strikes their surface, electrons may be emitted. The radiant energy supplies the work necessary to free the electrons from the material's surface. A schematic representation of a typical photoelectric-effect experiment is shown in ▶Fig. 27.3a. A voltage is maintained between the anode and the cathode. When light strikes the cathode ($-$), which is photosensitive, electrons are emitted. Because they are released by absorption of light energy, these emitted electrons are given the special name of *photoelectrons*. As they travel to the anode ($+$) and around the complete circuit, a current is registered on the ammeter.

When a photocell is illuminated with monochromatic (single-wavelength) light of different intensities, characteristic curves are obtained as a function of the applied voltage (Fig. 27.3b). For positive voltages, the anode is positive and attracts the electrons. Under these conditions, the *photocurrent* I_p, which signifies the flow of electrons released by the light, does not vary with voltage. This is because, under a positive voltage, the anode is positively charged, and thus *all* the electrons reach it. As expected classically, I_p is proportional to the incident-light intensity—the greater the intensity (see $I_2 > I_1$ in Fig. 27.3b), the more energy is available to free additional electrons.

The kinetic energy of the photoelectrons can be measured by reversing the voltage across the electrodes ($V < 0$), creating *retarding-voltage* conditions. The

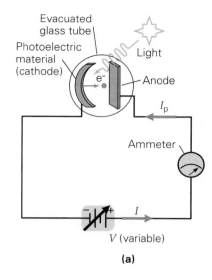

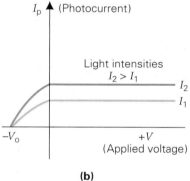

(b)

▲ **FIGURE 27.3 The photoelectric effect and characteristic curves**
(a) Incident monochromatic light on the photoelectric material in a photocell (or phototube) causes the emission of electrons, which results in a current in the circuit. The applied voltage is variable. **(b)** As the plots of photocurrent versus voltage for two intensities of light show, the current stays constant as the voltage is increased. However, for negative voltages (using the battery with reversed polarity), the current goes to zero at the stopping potential with a magnitude of V_o, which depends on the type of material, but is independent of intensity.

name "retarding voltage" comes from the fact that the electrons are now repelled instead of being attracted. As the retarding voltage is made more and more negative, the photocurrent decreases. Only the electrons with an initial kinetic energy greater than $e|V|$ can make it to the negative anode and be recorded as the part of the photocurrent by the ammeter. The electrons' initial kinetic energy is converted into electric potential energy as they approach the now negatively charged plate. At some value of retarding voltage, with a magnitude of V_o, called the **stopping potential**, the photocurrent is zero. There are no electrons collected—even the fastest electrons turn around. Hence, the maximum kinetic energy (K_{max}) of the photoelectrons is related to the magnitude of the stopping potential (V_o) by

$$K_{max} = eV_o \qquad (27.5)$$

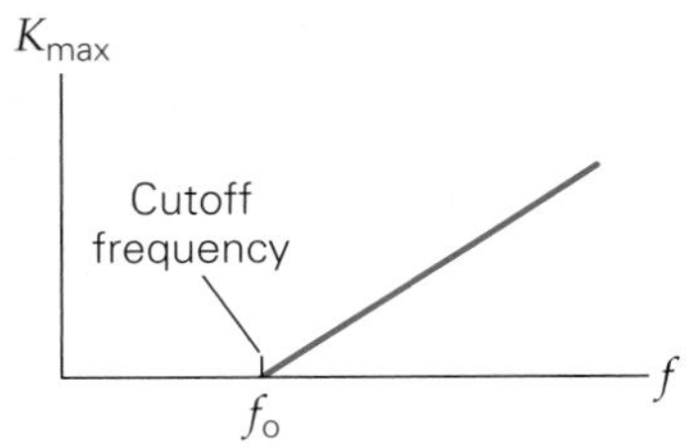

When the frequency of the incident light varies, the maximum kinetic energy of the electrons increases linearly with the frequency (◀Fig. 27.4). No emission of electrons is observed for light with a frequency below a certain *cutoff frequency* f_o. Even if the light intensity is very low, the current begins essentially instantaneously, with no observable time delay, as long as a photomaterial is being illuminated by light with a frequency $f > f_o$.

The important characteristics of the photoelectric effect are summarized in Table 27.1. Notice that only one of the characteristics is predicted correctly by classical wave theory, whereas Einstein's photon concept explains all of the results. He proposed that light energy occurs in quantum "packages," or photons, each with energy $E = hf$. When an electron absorbs a quantum of light energy, some of the photon's energy goes to free the electron, with the remainder showing up as kinetic energy. The work required to free the electron is designated by ϕ. So, when a photon of energy E is absorbed, conservation of energy requires that $E = K + \phi$, or, using Eq. 27.4 to replace E, we have

$$hf = K + \phi \qquad (27.6)$$

Since the energies are very small, the commonly used energy unit is the electron volt (eV; see Chapter 16). To three significant figures, $1.00 \text{ eV} = 1.60 \times 10^{-19} \text{ J}$.

The least tightly bound electron will have the maximum kinetic energy K_{max}. (Why?) The energy needed to free the least-bound electron is called the **work function** (ϕ_o) of the material. For this situation, Eq. 27.6 becomes

$$\underset{\substack{\text{incident} \\ \text{photon} \\ \text{energy}}}{hf} = \underset{\substack{\text{maximum} \\ \text{kinetic energy} \\ \text{of freed} \\ \text{electron}}}{K_{max}} + \underset{\substack{\text{minimum work} \\ \text{needed to} \\ \text{free the electron}}}{\phi_o} \qquad (27.7)$$

▲ FIGURE 27.4 Maximum kinetic energy versus light frequency in the photoelectric effect The maximum kinetic energy (K_{max}) of the photoelectrons is a linear function of the incident light frequency. Below a certain cutoff frequency f_o, no photoemission occurs, regardless of the intensity of the light.

TABLE 27.1 Characteristics of the Photoelectric Effect

Characteristic	Predicted by wave theory?
1. The photocurrent is proportional to the intensity of the light.	Yes
2. The maximum kinetic energy of the emitted electrons depends upon the frequency of the light, but not on its intensity.	No
3. No photoemission occurs for light with a frequency below a certain cutoff frequency f_o, regardless of the light intensity.	No
4. A photocurrent is observed immediately when the light frequency is greater than f_o, even if the light intensity is extremely low.	No

Learn by Drawing

The Photoelectric Effect and Energy Conservation

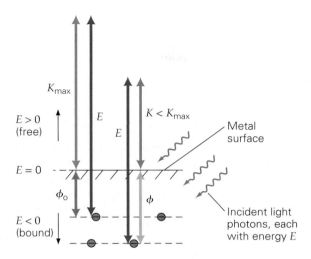

Other electrons require more energy than the minimum in order to be freed, so their kinetic energy will be less than K_{max}. This concept is explored visually in the Learn by Drawing feature on the photoelectric effect and energy conservation on this page. Some typical numerical values are shown in the next Example.

Example 27.2 ■ The Photoelectric Effect: Electron Speed and Stopping Potential

The work function of a particular metal is known to be 2.00 eV. If the metal is illuminated with light of wavelength 550 nm, what will be (a) the maximum kinetic energy of the emitted electrons and (b) their maximum speed? (c) What is the stopping potential?

Thinking It Through. (a) By energy conservation (Eq. 27.7), the maximum kinetic energy is the difference between the incoming photon energy and the work function. (b) Speed can be determined from kinetic energy, since the mass of an electron is known ($m = 9.11 \times 10^{-31}$ kg). (c) The stopping potential is found by requiring that all of the kinetic energy be converted to electric potential energy (Eq. 27.5).

Solution. First, we convert the data into SI units.

Given: $\phi_o = (2.00 \text{ eV})(1.60 \times 10^{-19} \text{ J/eV})$ *Find:* (a) K_{max} (maximum kinetic energy)
$ = 3.20 \times 10^{-19} \text{ J}$ (b) v_{max} (maximum speed)
$ \lambda = 550 \text{ nm} = 5.50 \times 10^{-7} \text{ m}$ (c) V_o (stopping potential)

(a) Using $\lambda f = c$, we find that the photon energy of light with the given wavelength is

$$E = hf = \frac{hc}{\lambda} = \frac{(6.63 \times 10^{-34} \text{ J} \cdot \text{s})(3.00 \times 10^8 \text{ m/s})}{5.50 \times 10^{-7} \text{ m}} = 3.62 \times 10^{-19} \text{ J}$$

Then

$$K_{max} = E - \phi_o = 3.62 \times 10^{-19} \text{ J} - 3.20 \times 10^{-19} \text{ J}$$

$$= (4.20 \times 10^{-20} \text{ J})\left(\frac{1 \text{ eV}}{1.60 \times 10^{-19} \text{ J}}\right) = 0.263 \text{ eV}$$

(b) v_{max} can be found from $K_{max} = \frac{1}{2}mv_{max}^2$:

$$v_{max} = \sqrt{\frac{2K_{max}}{m}} = \sqrt{\frac{2(4.20 \times 10^{-20}\,\text{J})}{9.11 \times 10^{-31}\,\text{kg}}} = 3.04 \times 10^5\,\text{m/s}$$

(c) The stopping potential is related to K_{max} by $K_{max} = eV_o$, therefore,

$$V_o = \frac{K_{max}}{e} = \frac{0.42 \times 10^{-19}\,\text{J}}{1.60 \times 10^{-19}\,\text{C}} = 0.26\,\text{V}$$

Follow-up Exercise. In this Example, suppose that a different wavelength of light is used and that the result is a new stopping voltage of 0.50 V. What is the wavelength of this new light? Explain why this wavelength requires a larger stopping voltage.

Einstein's quantum theory is consistent with *all* the experimental results of the photoelectric effect. In the photon model, an increase in light intensity means an increase in the number of photons and therefore in the number of electrons freed (the photocurrent). However, an increase in intensity would *not* mean a change in the energy of an individual photon, since that energy depends only on the light frequency ($E = hf$). Therefore, K_{max} should be independent of intensity, but linearly dependent on the frequency of the incident light—as is observed experimentally.

Einstein's theory also explains the existence of a cut off frequency. In the Einstein interpretation, since photon energy depends on frequency, below a certain (cutoff) frequency (f_o) the photons simply don't have enough energy to dislodge even the most loosely bound electrons. Therefore, no current is observed for those frequencies. Since, at the cutoff frequency, no electrons are emitted, the cutoff frequency can be found by setting $K_{max} = 0$ in Eq. 27.7:

$$hf_o = K_{max} + \phi_o = 0 + \phi_o$$

or

$$f_o = \frac{\phi_o}{h} \qquad \textit{threshold frequency} \qquad (27.8)$$

The cutoff frequency f_o is sometimes called the **threshold frequency**. If the incident light has a frequency that is less than f_o, then no matter how many photons are available (that is, no matter how intense the light), the photons will not have enough energy to free the least-bound electrons from the material. In this situation, the binding energy of the least-bound electron exceeds the photon energy. (How would you explain this scenario, using a sketch such as the one in the Learn by Drawing feature on p. 881?)

(How would you explain this scenario, using a sketch such as the one in the Learn by Drawing feature on p. 881?)

> **Note**: A photon of energy hf_o will barely free an electron from a solid, but the electron will have essentially no kinetic energy.

Example 27.3 ■ The Photoelectric Effect: Threshold Frequency and Wavelength

What are the threshold frequency and corresponding wavelength for the metal described in Example 27.2?

Thinking It Through. Example 27.2 gives the work function, so Eq. 27.8 allows us to determine the threshold frequency. The threshold wavelength can then be computed from $\lambda = c/f$.

Solution. Listing the data, we have

Given: $\phi_o = 2.00\,\text{eV}$ *Find:* f_o (threshold frequency)
       $= 3.20 \times 10^{-19}\,\text{J}$ λ_o (wavelength at threshold frequency)
       (from Example 27.2)

Solving for the threshold frequency f_o from $\phi_o = hf_o$ (Eq. 27.8), we get

$$f_o = \frac{\phi_o}{h} = \frac{3.20 \times 10^{-19}\,\text{J}}{6.63 \times 10^{-34}\,\text{J}\cdot\text{s}} = 4.83 \times 10^{14}\,\text{Hz}$$

The wavelength (the threshold wavelength) corresponding to this frequency is

$$\lambda_o = \frac{c}{f_o} = \frac{3.00 \times 10^8 \text{ m/s}}{4.83 \times 10^{14} \text{ Hz}} = 6.21 \times 10^{-7} \text{ m} = 621 \text{ nm}$$

Any frequency lower than 4.83×10^{14} Hz, or, alternatively, any wavelength longer than 621 nm, would not yield photoelectrons. Notice that this wavelength lies in the red end of the electromagnetic spectrum, so yellow light, for example, would dislodge electrons, but infrared would not.

Follow-up Exercise. In this Example, what would be the stopping voltage if the frequency of the light were twice the cutoff frequency?

Problem-Solving Hint

In photon calculations, we are often given the wavelength of the light rather than the frequency. Typically, what is needed is the photon energy. Instead of first calculating the frequency ($f = c/\lambda$), then the energy in joules ($E = hf$), and finally converting to electron volts, we can do this all in one step. To do so, we combine these two equations to form $E = hf = hc/\lambda$ and express the product hc in electron-volt nanometers (eV·nm). The value of this useful constant is

$$hc = (6.63 \times 10^{-34} \text{ J·s})(3.00 \times 10^8 \text{ m/s}) = 1.99 \times 10^{-25} \text{ J·m}$$

$$= \frac{(1.99 \times 10^{-25} \text{ J·m})(10^9 \text{ nm/m})}{1.60 \times 10^{-19} \text{ J/eV}}$$

$$= 1.24 \times 10^3 \text{ eV·nm}$$

This shortcut can save time and effort in working problems and allows estimation of the photon energy associated with light of a given wavelength (or vice versa). Thus, if orange light ($\lambda = 600$ nm) is used, you need only divide to find that each photon carries approximately 2 eV of energy:

$$E = \frac{hc}{\lambda} = \frac{1.24 \times 10^3 \text{ eV·nm}}{600 \text{ nm}} = 2.07 \text{ eV}$$

There are many applications of the photoelectric effect. The fact that the current produced by photocells is proportional to the intensity light makes them ideal for use in photographers' light meters. Photocells are also used in solar-energy applications to convert sunlight to electricity.

Another common application of the photocell is the electric eye (▶Fig. 27.5a). As long as light strikes the photocell, there is current in the circuit. Blocking the light opens the circuit in the relay (magnetic switch), which, in turn, controls some device. A common application of the electric eye is to turn on streetlights automatically at night. A safety application of the electric eye is shown in Fig. 27.5b. The threshold frequencies of some materials lie outside the visible region. Hence, nonvisible infrared light can be used in such applications as burglar alarms and home protection systems.

PHYSLET®
ILLUSTRATION

Photoelectric Effect

27.3 Quantum "Particles": The Compton Effect

OBJECTIVES: To (a) understand how the photon model of light explains scattering of light from electrons (the Compton effect) and (b) calculate the wavelength of the scattered light in the Compton effect.

In 1923, the American physicist Arthur H. Compton (1892–1962) explained the scattering of X rays from a graphite (carbon) block by assuming the radiation to be composed of quanta. His explanation of the observed effect provided additional convincing evidence that, at least in certain types of experiments, light (electromagnetic) energy is carried by *photons*.

Compton had observed that when a beam of monochromatic (single wavelength) X rays was scattered by various materials, the wavelength of the scattered

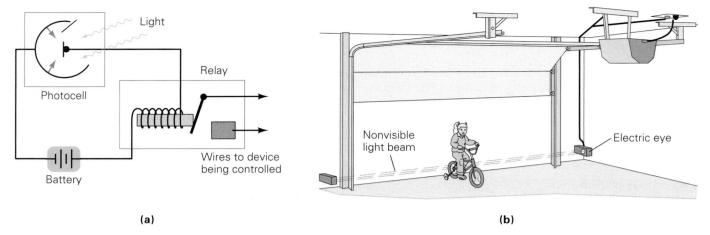

(a)　　　　　　　　　　　　　　　　　　　　　　　　　　**(b)**

▲ **FIGURE 27.5　Photoelectric applications: The electric eye**　**(a)** A diagram of an electric-eye circuit. When light strikes a photocell, there is a current in the circuit. Interruption of the light beam opens the circuit in the relay (a magnetic switch) that controls the particular device. **(b)** Electric-eye circuits are used in automatic garage-door openers. When the door starts to move downward, any interruption of the electric-eye beam (usually IR light) causes the door to stop, protecting anything that may be under the descending door.

Note: Elastic collisions are discussed in Section 6.4.

X ray was longer than the wavelength of the incident X ray. In addition, he noted that the change in the wavelength depended on the angle θ through which the X rays were scattered, but *not* on the nature of the scattering material (▶Fig. 27.6). This phenomenon came to be known as the **Compton effect**.

According to the wave model, any scattered radiation should have the same frequency (and wavelength) as the incident radiation. In this model, the electrons in the atoms of the scattering material are accelerated by the oscillating electric field of the radiation and therefore oscillate at the same frequency as the incident wave. The scattered radiation should then have the same frequency, regardless of direction.

According to Einstein's photon picture, the energy E of each photon is proportional to the frequency f of the light. Then a change in frequency or wavelength would indicate a change in photon energy. Since the wavelength increased (and the frequency decreased, since $f = c/\lambda$), the scattered photons had *less* energy than the incident ones. Moreover, the change in photon energy increased with the scattering angle—which reminded Compton of an elastic collision of two particles. Could the same principles apply in the scattering of quantum "particles"?

Pursuing this idea, Compton assumed that a photon behaves like a particle when it collides with electrons. He reasoned that if an incident photon collides with an electron initially at rest, the photon should transfer some energy and momentum to that electron. In agreement with the experiment, the energy and frequency of the scattered photon should both decrease (since $E = hf$). Applying conservation of energy and linear momentum, Compton showed that the shift in the wavelength of the photon scattered at an angle θ is given by

$$\Delta\lambda = \lambda - \lambda_\mathrm{o} = \lambda_\mathrm{C}(1 - \cos\theta) \qquad \textit{Compton scattering} \quad (27.9)$$

where λ_o is the wavelength of the incident photon and λ is that of the scattered photon. The constant λ_C is called the **Compton wavelength of the electron**. It is inversely related to the mass of the electron by $\lambda_\mathrm{C} = h/m_e c$. λ_C has a numerical value of 2.43×10^{-12} m $= 2.43 \times 10^{-3}$ nm.* Equation 27.9 correctly predicts the observed wavelength shift. For his work, Compton was awarded a Nobel Prize in 1927.

*This numerical value is the Compton wavelength of the *electron*. Compton scattering can occur from any particle; hence, there is a Compton wavelength of the proton, the neutron, etc. These values are much smaller than the electron's Compton wavelength, because other particles are much more massive than electrons.

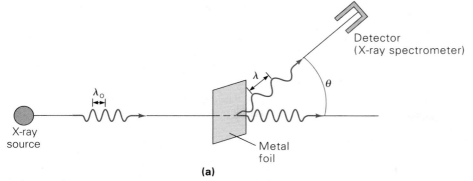

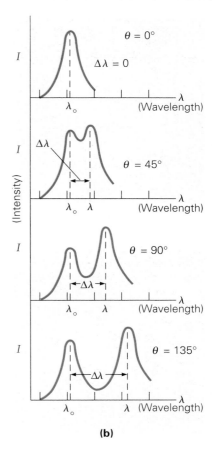

▲ FIGURE 27.6 X-ray scattering **(a)** When X rays of a single wavelength are scattered by the electrons in metal foil, the scattered wavelength (λ) is longer than the incident wavelength (λ_o). Most of the incident X rays pass through without interacting (and therefore undergo no change in wavelength). The scattered electrons are not shown, because they remain in the sample. **(b)** The change in wavelength increases with the scattering angle (θ).

Note that the *maximum* wavelength increase occurs when $\theta = 180°$ and has a value of $\Delta\lambda_{max} = 2\lambda_C = 4.86 \times 10^{-3}$ nm. (To see this, use Eq. 27.9 and note that for $\theta = 180°$, $\cos\theta = -1$ and $1 - \cos\theta = 2$.) The scattered wavelength is longest when the photon reverses direction, and the electron goes forward with the maximum amount of kinetic energy. Since this value is the maximum wavelength *change*, it will be hard to measure for incident wavelengths several thousand times this value or greater, such as for wavelengths larger than several nanometers (strong UV). In other words, the Compton effect is negligible for UV light and other types of light with longer wavelengths, such as visible, IR, etc. It is significant, percentagewise, only for X-ray and gamma-ray scattering.

Example 27.4 ■ X-Ray Scattering: The Compton Effect

A monochromatic beam of X rays of wavelength 1.35×10^{-10} m is scattered by the electrons in a metal foil. By what percentage is the wavelength shifted if the scattered X rays are observed at an angle of 90°?

Thinking It Through. The change, or shift, in the wavelength, $\Delta\lambda$, is given by Eq. 27.9 with $\theta = 90°$, and the fractional change is $\Delta\lambda/\lambda_o$. The change is positive, because the scattered light has a longer wavelength than that of the incident light.

Solution. Listing the data, we have

Given: $\lambda_o = 1.35 \times 10^{-10}$ m   *Find:* Percentage change in wavelength
    $\theta = 90°$

Starting from Eq. 27.9, we can compute the fractional change directly, since $\cos 90° = 0$:

$$\frac{\Delta\lambda}{\lambda_o} = \frac{\lambda_C}{\lambda_o}\left(1 - \cos\theta\right) = \frac{2.43 \times 10^{-12}\,\text{m}}{1.35 \times 10^{-10}\,\text{m}}(1 - \cos 90°) = 1.80 \times 10^{-2}$$

So

$$\frac{\Delta\lambda}{\lambda_o} \times 100\% = 1.80\%$$

Follow-up Exercise. In this Example, (a) what would be the maximum percentage change if gamma rays with a wavelength of 1.50×10^{-14} m were used instead? (b) Is it larger or smaller than the maximum possible percentage change for the X rays in the Example? Why?

(a)

(b)

▲ **FIGURE 27.7 Gas-discharge tubes** **(a)** These luminous glass tubes are gas-discharge tubes, in which atoms of various gases emit light when electrically excited. Each gas radiates its own characteristic wavelengths. **(b)** Only some "neon lights" actually contain neon, which glows with a red hue; other gases produce other colors.

Einstein's and Compton's successes with photons left scientists with two apparently competing theories. Classically, light is pictured as a traveling wave, and this theory satisfactorily explains such phenomena as interference and diffraction. Conversely, quantum theory is necessary to explain the photoelectric and Compton effects. The two theories combined to give rise to a description called the **dual nature** (or **wave–particle duality**) **of light**. That is, to explain all electromagnetic phenomena, light has to be considered sometimes as a wave and other times as a beam of photons.

27.4 The Bohr Theory of the Hydrogen Atom

OBJECTIVES: To (a) understand how the Bohr model of the hydrogen atom explains the atom's emission and absorption spectra, (b) calculate the energies and wavelengths of emitted and absorbed photons for transitions in atomic hydrogen and (c) understand how the generalized concept of atomic energy levels can explain other atomic phenomena.

In the 1800s, much experimental work was done with gas discharge tubes—for example, those containing hydrogen, neon, and mercury vapor. Common neon "lights" are actually gas discharge tubes (◀Fig. 27.7). Light from an incandescent source, such as a light bulb's hot filament, exhibits a *continuous spectrum* in which all wavelengths are present. However, when light emissions from gas discharge tubes were analyzed, discrete spectra with only certain wavelengths present were observed (▶Fig. 27.8). The spectrum of light coming from such a tube is called a *bright-line spectrum*, or **emission spectrum**, because only certain wavelengths are emitted. In general, the wavelengths present in an emission spectrum are characteristic of the individual atoms or molecules of the particular gas.

Atoms can absorb light as well as emit it. If white light is passed through a relatively cool gas, certain frequencies or wavelengths are missing, or absorbed. The result is a *dark-line spectrum*, or **absorption spectrum**—a series of dark lines superimposed on a continuous spectrum (Fig. 27.8c). Just as in emission spectra, the missing wavelengths are uniquely related to the type of atom or molecule doing the absorbing. By determining the unique pattern of emitted and/or absorbed wavelengths, the type of atoms or molecules present in a sample can be identified. This method is called *spectroscopic analysis* and is widely used in physics, astrophysics, biology, and chemistry. For example, the element helium was first discovered to exist on the Sun when scientists found that its absorption-line pattern did not match any known pattern on the Earth.

Although the reason for line spectra was not understood in the 1800s, they provided an important clue to the electron structure of atoms. Hydrogen, with its relatively simple visible spectrum, received much of the attention. It is also the simplest atom, consisting of only one electron and one proton. In the late 19th century, the Swiss physicist J. J. Balmer found an empirical formula that gives the wavelengths of the four spectral lines of hydrogen in the visible region:

$$\frac{1}{\lambda} = R\left(\frac{1}{2^2} - \frac{1}{n^2}\right) \quad \text{for } n = 3, 4, 5, \text{ and } 6 \quad \textit{visible spectrum of hydrogen} \quad (27.10)$$

In this equation, R is called the *Rydberg constant* and has a value of $1.097 \times 10^{-2}\,\text{nm}^{-1}$. The four spectral lines of hydrogen in the visible region, are part of the **Balmer series**. They were found to fit the formula, but it was not understood why. Similar formulas were found to fit other spectral-line series that were completely in the ultraviolet and infrared regions.

An explanation of the spectral lines was given in a theory of the hydrogen atom put forth in 1913 by the Danish physicist Niels Bohr (1885–1962). Bohr assumed that the electron of the hydrogen atom orbits the proton in a circular orbit analogous to a planet orbiting the Sun. The attractive electrical force between the electron and pro-

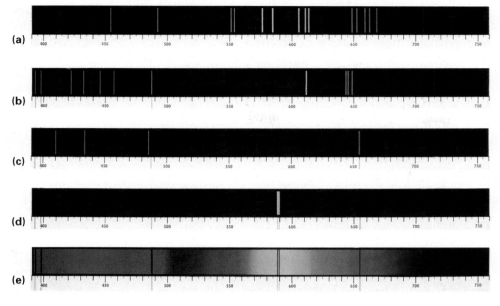

(a)

(b)

(c)

(d)

(e)

◀ **FIGURE 27.8 Emission and absorption spectra of gases**
When a gas is excited by heat or electricity, the light it emits can be separated into its various wavelengths by a prism or diffraction grating; the result is a bright-line, or emission, spectrum, such as the ones shown here from **(a)** barium, **(b)** calcium, **(c)** hydrogen, and **(d)** sodium, each with its own characteristic pattern. **(e)** When a continuous spectrum from a hot solid or dense gas is viewed after passing through a cool gas, a dark-line, or absorption, spectrum is observed. Each line represents a particular wavelength the gas has absorbed. The absorption spectrum of the Sun provided here shows several prominent absorption lines produced by the gases of the solar atmosphere before the sunlight makes it to the Earth. In fact, the inert gas helium was first discovered to exist on the Sun by this very method.

ton supplies the necessary centripetal force for the circular motion. Recall that the centripetal force is given by $F_c = mv^2/r$, where v is the electron's orbital speed, m is its mass, and r is the radius of its orbit. The force between the proton and electron is given by Coulomb's law as $F_e = kq_1q_2/r^2 = ke^2/r^2$, where e is the magnitude of the charge of the proton and the electron (▶Fig. 27.9). Equating these two forces, we have

$$\frac{mv^2}{r} = \frac{ke^2}{r^2} \tag{27.11}$$

The total energy of the atom is the sum of its kinetic and potential energies. Recall from Chapter 16 that the electric potential energy of two point charges is given by $U_e = kq_1q_2/r$. Since the electron and proton are oppositely charged, $U_e = -ke^2/r$. Thus, the expression for the total energy becomes

$$E = K + U_e = \tfrac{1}{2}mv^2 - \frac{ke^2}{r}$$

From Eq. 27.11, the kinetic energy can be written as $\tfrac{1}{2}mv^2 = ke^2/2r$. With this relationship, the total energy becomes

$$E = \frac{ke^2}{2r} - \frac{ke^2}{r} = -\frac{ke^2}{2r} \tag{27.12}$$

Note that E is negative, indicating that the system is bound. As the radius gets very large, E approaches zero. With $E = 0$, the electron would no longer be bound to the proton, and the atom, having lost its electron, would be ionized.

Up to this point, only classical principles have been applied. At this step in the theory, Bohr made a radical assumption—radical in the sense that he introduced a quantum concept to attempt to explain atomic line spectra:

> Bohr assumed that the angular momentum of the electron was quantized and could have only discrete values that were integral multiples of $h/2\pi$, where h is Planck's constant.

Recall that in a circular orbit of radius r, the angular momentum L is given by mvr (Eq. 8.14). Therefore, Bohr's assumption translates into

$$mvr = n\left(\frac{h}{2\pi}\right) \quad \text{for } n = 1, 2, 3, 4, \ldots \tag{27.13}$$

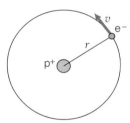

▲ **FIGURE 27.9 The Bohr model of the hydrogen atom** The electron is pictured as revolving around the much more massive proton in a circular orbit. The electric force of attraction provides the centripetal force.

Note: Review the concept of potential-energy wells, presented in Sections 5.4 and 7.5. (See Figs. 5.14 and 7.18.)

The integer n is called a *quantum number*, specifically, the atom's **principal quantum number**.*With this assumption, the orbital speed of the electron (v) can be found. Its quantized values are given by

$$v_n = \frac{nh}{2\pi mr} \quad \text{for } n = 1, 2, 3, 4, \ldots$$

Putting this expression for v into Eq. 27.11 and solving for r, we obtain

$$r_n = \left(\frac{h^2}{4\pi^2 ke^2 m}\right)n^2 \quad \text{for } n = 1, 2, 3, 4, \ldots \tag{27.14}$$

Here, the subscript n on r is used to indicate that only certain radii are possible—that is, the size of the orbit is quantized. The energy for an orbit can be found by substituting this expression for r into Eq. 27.12, which gives

$$E_n = -\left(\frac{2\pi^2 k^2 e^4 m}{h^2}\right)\frac{1}{n^2} \quad \text{for } n = 1, 2, 3, 4, \ldots \tag{27.15}$$

where the energy is also written with a subscript of n to show its dependence on n. The quantities in the parentheses on the right-hand sides of Eqs. 27.14 and 27.15 are constants and can be evaluated numerically. Because the radii are so small, they are typically expressed in nanometers (nm); similarly, energies are expressed in electron-volts (eV).

$$r_n = 0.0529 n^2 \text{ nm} \quad \text{for } n = 1, 2, 3, 4, \ldots \qquad \text{\textit{orbital radii and}} \tag{27.16}$$
$$E_n = \frac{-13.6}{n^2}\text{ eV} \quad \text{for } n = 1, 2, 3, 4, \ldots \qquad \begin{array}{l}\textit{energies for the}\\ \textit{hydrogen atom}\end{array} \tag{27.17}$$

The use of these expressions is shown in the next Example.

Example 27.5 ■ A Bohr Orbit: Radius and Energy

Find the orbital radius and energy of an electron in a hydrogen atom characterized by the principal quantum number $n = 2$.

Thinking It Through. Eqs. 27.16 and 27.17 are used with $n = 2$.

Solution. For $n = 2$,

$$r_2 = 0.0529 n^2 \text{ nm} = 0.0529(2)^2 \text{ nm} = 0.212 \text{ nm}$$

and

$$E_2 = \frac{-13.6}{n^2}\text{ eV} = \frac{-13.6}{2^2}\text{ eV} = -3.40 \text{ eV}$$

Follow-up Exercise. In this Example, what is the speed of the orbiting electron?

However, there was still a problem with Bohr's theory. Classically, any accelerating charge should radiate electromagnetic energy. For the Bohr circular orbits, the electron is accelerating centripetally. Thus, the orbiting electron should lose energy and spiral into the nucleus. Clearly, this doesn't happen in the hydrogen atom, so Bohr had to make another nonclassical assumption:

> Bohr postulated that the hydrogen electron does *not* radiate energy when it is in a bound, discrete orbit. It radiates energy only when it makes a *downward transition* to an orbit of lower energy. It makes an *upward transition* to an orbit of higher energy by absorbing energy.

*The principal quantum number is only one of four quantum numbers necessary to completely describe each electron in an atom. See Chapter 28.

Energy Levels

The "allowed" orbits of the electron in a hydrogen atom are commonly expressed in terms of their energy (▼Fig. 27.10). In this context, the electron is referred to as being in a particular "energy level" or state. The principal quantum number designates the energy level. The lowest energy level ($n = 1$) is called the **ground state**. The energy levels above the ground state are called **excited states**. For example, $n = 2$ is the *first excited state* (see Example 27.5), and so on.

The electron is normally in the ground state and must be given energy in order to raise it to an excited state. The electron can be excited only by absorbing discrete amounts of energy. The energy levels are somewhat analogous to the rungs of a ladder. When a person goes up and down a ladder, he or she changes his or her gravitational potential energy by discrete amounts. Similarly, an electron goes up and down its own energy ladder in discrete steps. Notice, however, that the energy levels of the hydrogen atom are not evenly spaced, as they are on a ladder.

If enough energy is absorbed, it is possible for the electron to no longer be bound to the atom. In this case, we say that the atom is *ionized*. For example, to ionize a hydrogen atom initially in its ground state requires a minimum of 13.6 eV of energy. This process makes the final energy of the electron zero (since it is free), and it has a principle quantum number of $n = \infty$. However, if the electron is already in an excited state, then less energy is needed to ionize the atom. Since the energy of the electron in any state is E_n, the energy needed to free it from the atom is $-E_n$. This energy is called the **binding energy** of the electron. Note that $-E_n$ is positive (why?) and represents the energy required to ionize the atom if the electron initially is in a state with a principal quantum number of n.

An electron generally does not remain in an excited state for long; it decays, or makes a downward transition to a lower energy level, in a very short time. The time an electron spends in an excited state is called the **lifetime** of the excited state. For many states, the lifetime is about 10^{-8} s. In making a transition to a lower state, the electron emits energy in the form of a photon of light (▶Fig. 27.11). The energy ΔE of the photon is equal in magnitude to the energy *difference* of the levels:

$$\Delta E = E_{n_i} - E_{n_f} = \left[\frac{-13.6}{n_i^2} \text{ eV} \right] - \left[\frac{-13.6}{n_f^2} \text{ eV} \right]$$

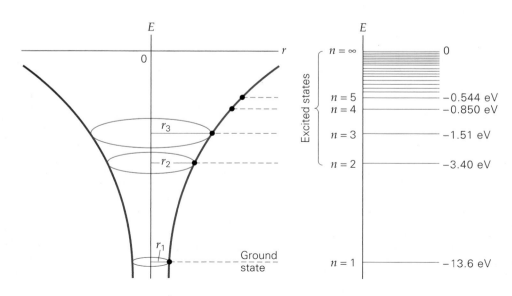

◀ **FIGURE 27.10 Orbits and energy levels of the hydrogen electron** The Bohr theory predicts that the hydrogen electron can occupy only certain orbits having discrete radii. Each allowed orbit has a corresponding total energy, conveniently displayed as an energy-level diagram. The lowest energy level ($n = 1$) is the ground state; those levels above it ($n > 1$) are excited states. The orbits are shown on the left, with orbital plotted against the $1/r$ electrical potential of the proton. The electron in the ground state is deepest in the potential-energy well, analogous to the gravitational potential-energy well of Fig. 7.18. (Neither r nor the energy levels are drawn to scale—can you tell why?)

Excited atom

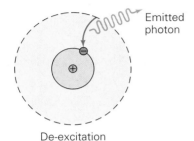

De-excitation

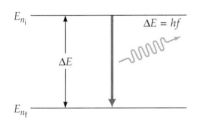

▲ **FIGURE 27.11 Electron transitions and photon emission** When a hydrogen atom emits light, its electron makes a downward transition to a lower orbit (with less energy), and a photon is emitted. The photon's energy is equal to the energy difference between the two levels.

or

$$\Delta E = 13.6\left(\frac{1}{n_f^2} - \frac{1}{n_i^2}\right)\text{eV} \qquad \begin{array}{l}\textit{photon energy emitted} \\ \textit{by an H atom}\end{array} \qquad (27.18)$$

Here, the subscripts "i" and "f" refer to the initial and final states, respectively. According to the Bohr theory, this energy difference is emitted as a photon with an energy E. Therefore, $E = \Delta E = hf = hc/\lambda$, and only photons of particular frequencies and wavelengths are emitted. These particular frequencies of wavelengths correspond to the various transitions between energy levels and explain the existence of the emission type of spectrum.

The final principal quantum number n_f refers to the energy level in which the electron ends up during the emission process. The original *Balmer* emission series in the visible region corresponds to $n_f = 2$ and $n_i = 3, 4, 5$ and 6. There is only one emission series entirely in the ultraviolet range, called the *Lyman* series, in which all the transitions end in the $n_f = 1$ state (the ground state). There are many series entirely in the infrared region, most notably the *Paschen* series, which ends with the electron in the second excited state, $n_f = 3$.

Usually, the wavelength of the light is what is measured during the emission process. Since photon energy and light wavelength are related by Einstein's equation (Eq. 27.4), the wavelength of the emitted light, λ, can be obtained from

$$\lambda = \frac{hc}{\Delta E} = \frac{1.24 \times 10^3}{\Delta E(\text{in eV})}\,\text{nm} \qquad (27.19)$$

(See the Problem-Solving Hint on p. 883.) Consider Example 27.6.

Example 27.6 ■ Investigating the Balmer Series: Visible Light from Hydrogen

What is the wavelength of the emitted light when an electron in a hydrogen atom undergoes a transition from the $n = 3$ energy level to the $n = 2$ energy level?

Thinking It Through. The emitted photon has an energy equal to the energy difference between the two energy levels (Eq. 27.18). The wavelength of the light can then be obtained by using Eq. 27.19.

Solution.

Given: $n_i = 3$ *Find:* λ (wavelength of emitted light)
       $n_f = 2$

The energy of the emitted photon is equal to the magnitude of the atom's change in energy. Thus,

$$\Delta E = 13.6\left(\frac{1}{n_f^2} - \frac{1}{n_i^2}\right)\text{eV} = 13.6\left(\frac{1}{4} - \frac{1}{9}\right)\text{eV} = 1.89\text{ eV}$$

Using Eq. 27.19 (and making sure that ΔE is expressed in electron volts), we obtain

$$\lambda = \frac{1.24 \times 10^3 \text{ eV}\cdot\text{nm}}{\Delta E} = \frac{1.24 \times 10^3 \text{ eV}\cdot\text{nm}}{1.89\text{ eV}} = 656\text{ nm}$$

which is in the red portion of the visible spectrum.

Follow-up Exercise. Light of what wavelength would be just sufficient to ionize a hydrogen atom if it started in its first excited state? What type of light is it, visible, UV, or IR?

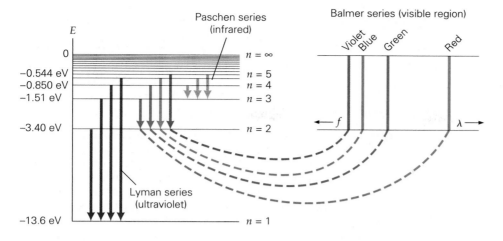

Paschen series
(infrared)

Balmer series (visible region)

FIGURE 27.12 Hydrogen spectrum Transitions may occur between two or more energy levels as the electron returns to the ground state. Transitions to the $n = 2$ state give spectral lines with wavelengths in the visible region (the Balmer series). Transitions to other levels give rise to other series (not in the visible region), as shown.

In summary, since the Bohr model of hydrogen requires that the electron make transitions only between discrete energy levels, the atom emits photons of discrete energies (or light of discrete wavelengths), which results in emission spectra. This process is summarized in ▲Fig. 27.12.

Integrated Example 27.7 ■ The Balmer Series: Entirely Visible?

We know that four wavelengths of the Balmer series are in the visible range (Fig. 27.12). (a) Are there more wavelengths than these four in the series? If so, what type of light are they likely to be, (1) infrared, (2) visible, or (3) ultraviolet? (b) What is the longest wavelength of nonvisible light in the Balmer series?

(a) Conceptual Reasoning. The Balmer series of emission lines is given off when the electron ends in the first excited state, that is, $n_f = 2$ (Fig. 27.12). There are four distinct visible wavelengths, corresponding to four different starting states: $n_i = 3, 4, 5,$ and 6. Any other wavelengths in this series must start with $n_i = 7$ or higher and therefore represent a larger energy difference than those for the visible lines. This means that these photons carry more energy than visible-light photons (or, alternatively, light of a shorter wavelength than that of visible light). Thus the answer is (3), since the other Balmer lines must be in the ultraviolet (UV) region of the spectrum.

(b) Thinking It Through. The longest nonvisible Balmer-series wavelength corresponds to the smallest photon energy above the $n = 6 \rightarrow 2$ transition. (Why?). Hence, we are talking about $n_i = 7$ and $n_f = 2$. The energy difference can be computed from Eq. 27.18. Then λ can be calculated from Eq. 27.19.

Given: The Balmer emission series for hydrogen   *Find:* The longest nonvisible wavelength in the Balmer series

From Eq. 27.18,

$$\Delta E = 13.6\left(\frac{1}{n_f^2} - \frac{1}{n_i^2}\right)eV = 13.6\left(\frac{1}{4} - \frac{1}{49}\right)eV = 3.12 \text{ eV}$$

This energy difference corresponds to light of wavelength

$$\lambda = \frac{1.24 \times 10^3 \text{ eV} \cdot \text{nm}}{\Delta E} = \frac{1.24 \times 10^3 \text{ eV} \cdot \text{nm}}{3.12 \text{ eV}} = 397 \text{ nm}$$

This is just below the lower limit of the visible spectrum which ends at 400 nm.

Follow-up Exercise. In hydrogen, what is the longest wavelength of light emitted in the Lyman series? In what region of the spectrum is this light?

Assume that a hydrogen atom, initially in its ground state, absorbs a photon. In general, how many emitted photons would you expect to be associated with the deexcitation process back to the ground state? Can there be (a) more than one photon, or must there be (b) only one photon?

Reasoning and Answer. Since the difference between light emission and absorption is the direction of the transition (down for emission, up for absorption), you might be tempted to answer (b), since there was only one photon required for the excitation process. However, if you look at the details of the two processes, you will realize that they are not necessarily symmetrical. In absorbing a photon's energy, a hydrogen atom will jump from its ground state to an excited state—let's say from $n = 1$ to $n = 4$. This process requires a single photon of a unique energy (that is, light of a unique wavelength).

However, in dropping back to the ground state, the atom may take any one of *several* possible routes. For example, the atom could go from the $n = 4$ state to the $n = 3$ state, followed by a transition to the ground state ($n = 1$). This process would involve the emission of two photons. Therefore, the answer is (a). In general, following the absorption of a single photon, several photons can be emitted.

Follow-up Exercise. (a) How many different energy photons may a hydrogen atom emit in deexciting from the third excited state to the ground state? (b) Starting from the ground state, which excitation transition *must* result in only one emitted photon when the hydrogen atom de-excites? Explain your reasoning.

PHYSLET® ILLUSTRATION

The Bohr Atom

Bohr's theory gave excellent agreement with experiment for hydrogen gas as well as for other ions with just one electron, such as singly ionized helium. However, it could not successfully describe multielectron atoms. Bohr's theory was incomplete in the sense that it patched new quantum ideas into a basically classical framework. The theory contains some correct concepts, but a complete description of the atom did not come until the development of quantum mechanics (Chapter 28). Nevertheless, the idea of discrete energy levels in atoms enables us to qualitatively understand phenomena such as fluorescence.

In **fluorescence**, an electron in an excited state returns to the ground state in two or more steps, like a ball bouncing down a flight of stairs. At each step, a photon is emitted. Each such step represents a smaller energy transition than the original energy required for the upward transition. Therefore, each emitted photon must have a lower energy and a longer wavelength than the original exciting photon. For example, the atoms of many minerals can be excited by absorbing ultraviolet (UV) light and fluoresce, or glow, in the visible region when they deexcite (◄Fig. 27.13). A variety of living organisms, from corals to butterflies, manufacture fluorescent pigments that emit visible light.

(a) Visible illumination

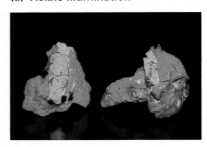

(b) UV illumination

▲ **FIGURE 27.13 Fluorescence** Many minerals emit light of visible wavelengths when illuminated by invisible ultraviolet light (so-called black light). The visible light is produced when atoms excited by the UV light deexcite to lower energy levels in several smaller steps, yielding photons of less energy and longer (visible) wavelengths.

27.5 A Quantum Success: The Laser

OBJECTIVE: To understand some of the practical applications of the quantum hypothesis—in particular, the laser.

The development of the laser was a major technological success. Unlike the numerous inventions that have come about by trial and error or accident, including Roentgen's discovery of X rays and Edison's electric lamp, the laser was developed on theoretical grounds. Through the use of quantum-physics ideas, the principle of the **laser** was first predicted and then designed, built, and, finally, applied. (The word *laser* is an acronym that stands for *light amplification by stimulated emission of radiation*. Stimulated emission will be discussed shortly.)

The laser has found widespread applications, some of which are discussed at the end of this section.

The existence of atomic energy levels is of prime importance in understanding the laser's operation. Usually, an electron makes a transition to a lower energy level almost immediately, remaining in an excited state for only about 10^{-8} s. However, the lifetimes of some excited states are appreciably longer than this. An atomic state with a relatively long lifetime is called a **metastable state**. For example, **phosphorescent** materials are composed of atoms that have such metastable states. These materials are used on luminous watch dials, toys, and other items that "glow in the dark." When a phosphorescent material is exposed to light, the atoms are excited to higher energy levels. Many of the atoms return to their normal state very quickly. However, there are also metastable states in which the atoms may remain for seconds, minutes, or even more than an hour. Consequently, the material can emit light and glow for some time (▶Fig. 27.14).

A major consideration in laser operation is the emission process. As ▼Fig. 27.15 shows, absorption and spontaneous emission of radiation can occur between two energy levels. That is, a photon is absorbed and a photon is emitted almost immediately. However, when the higher energy state is metastable, there is another possible emission process, called **stimulated emission**. Einstein first proposed this process in 1919. If a photon with an energy equal to an allowed transition strikes an atom already in a metastable state, it may stimulate that atom to make a transition to a lower energy level. This transition yields a second photon that is identical to the first one. Thus, *two* photons with the same frequency and phase will go off in the same direction. Notice that stimulated emission is an amplification process—one photon in, two out. But, this process is not a case of getting something for nothing, since the atom must be initially excited, and energy is required to do this.

Ordinarily, when light passes through a material, photons are more likely to be absorbed than to give rise to stimulated emission. This is because there normally are many more atoms in their ground state than in excited states. However, it is possible to prepare a material so that more of its atoms are in an excited metastable state than in the ground state. This condition is known as a **population inversion**. In this case, there may be more stimulated emission than absorption, and the net result is amplification. With the proper instrumentation (see the upcoming discussion), the result is a laser.

Today, there are many types of lasers capable of producing light of different wavelengths. The helium–neon (He–Ne) gas laser is probably the most familiar, since it is used for classroom demonstrations and laboratory experiments. The characteristic reddish–pink light produced by the He–Ne laser ($\lambda = 632.8$ nm) is also used in the optical scanning systems at supermarket checkouts. The gas mixture is about 85% helium and 15% neon. Essentially, the helium is used for

▲ **FIGURE 27.14**
Phosphorescence and metastable states When atoms in a phosphorescent material are excited, some of them do not immediately return to the ground state, but remain in metastable states for longer than normal periods of time. In this exhibit at the San Francisco Exploratorium, the phosphorescent walls and floor continue to glow for about 30 seconds after being illuminated, retaining the shadows of children who were present when the phosphors were initially exposed to light.

▼ **FIGURE 27.15 Photon absorption and emission** (a) In the absorption of light, once a photon is absorbed, the atom is excited to a higher energy level. (b) After a short time, the atom spontaneously decays to a lower energy level with the emission of a photon. (c) If another photon with an energy equal to that of the *downward* transition strikes an excited atom, stimulated emission can occur. The result is *two* photons (the original plus the new one) with the same frequency. They travel in the same direction as that of the incident photon and are in phase with one another.

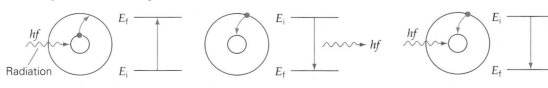

(a) **Absorption**   (b) **Spontaneous emission**   (c) **Stimulated emission**

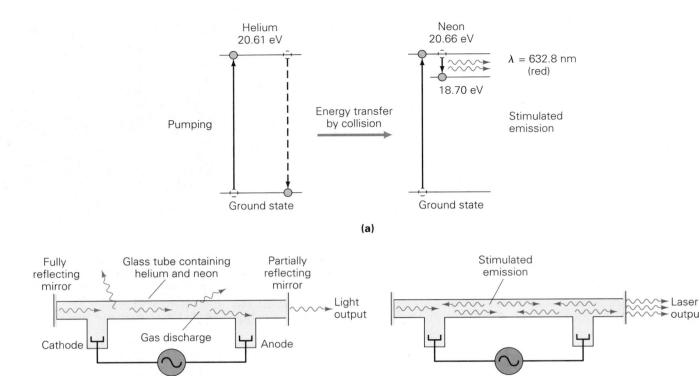

Helium
20.61 eV

Neon
20.66 eV

$\lambda = 632.8$ nm
(red)

18.70 eV

Pumping

Energy transfer
by collision

Stimulated
emission

Ground state

Ground state

(a)

Fully
reflecting
mirror

Glass tube containing
helium and neon

Partially
reflecting
mirror

Light
output

Cathode

Gas discharge

Anode

Radio-frequency
generator

Stimulated
emission

Laser
output

Radio-frequency
generator

(b)

▲ **FIGURE 27.16 The helium–neon laser** (a) Helium atoms are first excited (or "pumped") by collision with electrons. This energy is then transferred from the helium atoms to the neon atoms in a metastable state. All that is needed is one photon from a downward transition to stimulate emission from the other excited neon atoms. When this process occurs, the result is the emission of an intense beam of red laser light. **(b)** End mirrors on the laser tube are used to enhance the light beam (from stimulated emission) in a direction along the tube's axis. One of the mirrors is only partially silvered, allowing for some of the light to come out of the tube and resulting in the beam of red light we observe.

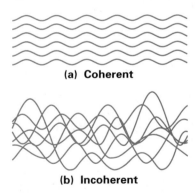

(a) Coherent

(b) Incoherent

▲ **FIGURE 27.17 Coherent light**
(a) Laser light is monochromatic (single frequency and single wavelength or color) and coherent, meaning that all the light waves are locked in phase. **(b)** Light waves from sources such as a lightbulb filament are emitted randomly. They consist of many different wavelengths (colors) and are incoherent or, on average, out of phase.

energizing and the neon for amplification. The gas mixture is subjected to a high-voltage discharge of electrons produced by a radio-frequency power supply converted to dc. The helium atoms are excited by collision with these electrons (▲Fig. 27.16a). This process is referred to as *pumping*. Energy is pumped into the system, and the helium atoms are pumped into an excited state that is 20.61 eV above its ground state.

This excited state in helium has a relatively long lifetime of about 10^{-4} s and has almost the same energy as an excited state in the Ne atom at 20.66 eV. Because this lifetime is so long, there is a good chance that before an excited He atom can spontaneously emit a photon, it will collide with a Ne atom in its ground state. When such a collision occurs, energy can be transferred to the Ne atom. The lifetime of the 20.66-eV neon state is also relatively long. The delay in this metastable state of neon causes a population inversion in the neon atoms—a higher percentage of atoms in the 20.66-eV state than in ones below it.

The stimulated emission of the light emitted by these neon atoms is enhanced by reflections from mirrors placed at each end of the laser tube (Fig. 27.16b). Some excited Ne atoms spontaneously emit photons in all directions, and these photons, in turn, induce stimulated emissions. In stimulated emission, the two photons leave the atom in the same direction as that of the incident photon. Photons traveling in the direction of the tube axis are reflected back through the tube by the end mirrors. The photons, in reflecting back and forth, cause even more stimulated emissions. The result is an intense, highly directional, coherent (in phase), monochromatic (single-wavelength) beam of light traveling back and forth along the tube axis. Part of the beam emerges through one of the end mirrors, because it is only partially silvered.

The monochromatic, coherent, and directional properties of laser light are responsible for its unique properties (◄Fig. 27.17). Light from sources such as incandescent lamps is emitted from the atoms randomly and at different frequencies (a result of many different transitions). As a result, the light is out of phase, or incoherent. Such beams spread out and become less intense. The properties of laser light allow the formation of a very narrow beam, which with amplification can be very intense.

Some laser applications are shown in ►Fig. 27.18, and another is discussed in the Insight on Compact Disc Players (p. 896). Although laser beams are used to weld detached retinas in the eye, viewing a laser beam directly can be quite hazardous, because the beam is focused on the retina in a very small area. If the beam intensity and the viewing time are sufficient, the photosensitive cells of the retina can be burned and destroyed by this concentrated energy.

Another interesting application of laser light is the production of three-dimensional images in a process called **holography**. The process does not use lenses, as ordinary image-forming processes do, yet it recreates the original scene in three dimensions. The key to holography is the coherent property of laser light, which gives the light waves a definite spatial relationship to each other.

In the photographic process of making a *hologram*, an arrangement such as that illustrated in ▼Fig. 27.19 is used. Part of the light from the laser (the object beam) passes through a partially reflecting mirror to the object. The other part, or reference beam, is reflected to the film. The light incident on the object is also reflected to the film, and it interferes with the reference beam. The film records the interference pattern of the two light beams, which essentially imprints on the film the information carried from the object by the light's wave fronts.

When the film is developed, the interference pattern bears no resemblance to the object and appears as a meaningless pattern of light and dark areas. However, when the wave-front information is reconstructed by passing light through the film, a three-dimensional image is seen (►Fig. 27.20). If part of the three-dimensional image is hidden from view, you can see it by moving your head to one side, just as you would to see a hidden part of a real object.

(a)

(b)

▲ **FIGURE 27.18 Some laser applications (a)** Laser eye surgeries, which minimize bleeding and scarring, can be used to repair detached retinas, treat glaucoma, correct retinal abnormalities, and alleviate various other problems. **(b)** An industrial laser cuts through a steel plate.

▼ **FIGURE 27.19 Holography (a)** The coherent light from a laser is split into reference and object beams. The interference pattern between these beams is recorded on a photographic plate. **(b)** When the developed plate, or hologram, is illuminated by normal light, the viewer sees a reconstructed three-dimensional image of the object.

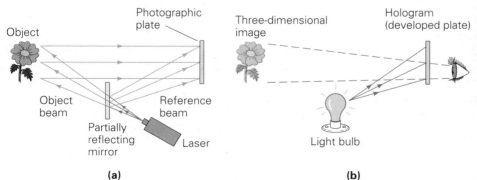

▲ **FIGURE 27.20 Holographic images** A three-dimensional hologram can be viewed from any angle, just as if it were a real object.

The Compact Disc

The compact disc (CD) represents a method of storing data, such as musical recordings. The CD system was introduced around 1980. Since then, musical CDs have almost completely supplanted tapes and long-playing vinyl records. The disc itself is 12 cm in diameter and can store more than 6 billion bits of information. This capacity is the equivalent of more than 1000 floppy disks or over 275000 pages of text! For audio use, a CD can store 74 min of music, and the sound reproduction is virtually unaffected by dust, scratches, or fingerprints on the disc.

Information on the disc is in the form of raised areas called "pits," which are separated by flat areas called "land." The surface of the disc is coated with a thin layer of aluminum to reflect the laser beam that "reads" the information (Fig. 1). The pits are arranged in a spiral track like the grooves in a phonograph record; however, CD tracks are about 60 times closer together than are the grooves on a record.

In the readout system, a laser beam, which comes from a small semiconductor (solid-state) laser, is applied from below the disc and focused on the aluminum coating of the track. The disc rotates at about 3.5 to 8 revolutions per second as the laser beam follows the spiral track. When the beam strikes a land area between two pits, it is reflected. When the beam spot overlaps a land area and a pit, the light reflected from the different areas interferes, causing fluctuations in the reflected beam. To make the fluctuations more distinct, the raised-pit thickness (t) is made to be one-fourth of the wavelength of the laser light. Then, reflected light from land areas travels an additional path length of one half of a wavelength. Destructive interference occurs when the two parts of the reflected beam combine. (See Section 24.1.) As a result, there is less reflected intensity when the beam passes over the edge of a pit than when it passes over a land area alone.

Then the reflected beam of varying intensity strikes a photodiode (a solid-state photocell), which reads the information. The fluctuations of the reflected light convey the coded information as a series of binary numbers (zeros and ones). A pit represents a binary 1, and a land area is read as a binary 0. The signals are then electronically converted into sound.

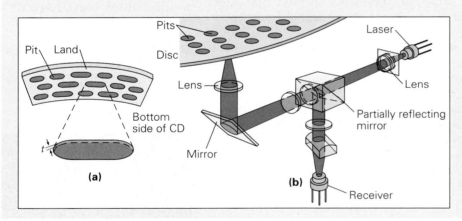

◀ FIGURE 1 The compact disc (a) The information on the disc is recorded in the form of raised areas called *pits*, which are separated by flat areas called *land*. The pits are on the bottom of the disc. (b) The surface is coated with a thin layer of aluminum to reflect the laser beam, which "reads" the information.

Chapter Review

Important Concepts

- All hot solids, liquids, and dense gases emit a **continuous spectrum** of electromagnetic (thermal) radiation. The maximum amount of energy is emitted at a wavelength (λ_{max}) determined by the material's absolute temperature T. **Wein's displacement law** gives the (inverse) relationship of wavelength and absolute temperature:

$$\lambda_{max}T = 2.90 \times 10^{-3}\,\text{m}\cdot\text{K} \qquad (27.1)$$

- The classical theory of atomic vibrations could not explain the thermal radiation spectra. In particular, it predicted that an infinite amount of energy is emitted at the short wavelengths, a phenomenon called the **ultraviolet catastrophe**, which did not agree with experiment. To explain the observed spectrum from such objects, **Planck's hypothesis** stated that the energy of the atoms in the material was quantized in multiples of their vibrational frequency, or

$$E_n = n(hf) \quad \text{for } n = 1, 2, 3, \ldots \qquad (27.2)$$

where h, **Planck's constant**, has a numerical value of $6.63 \times 10^{-34}\,\text{J}\cdot\text{s}$.

- In the **photoelectric effect**, light falling on a surface causes electrons to come off that surface. To explain the experimental details, Einstein had to assume that light consisted of discrete quanta of energy, called **photons**, or particles of light. The energy of a photon associated with light of frequency f is

$$E = hf \qquad (27.4)$$

- Light can also interact with electrons via a scattering process called **Compton scattering**. To explain the details of such scattering, light (frequency f) must be treated as a photon carrying a specific amount of energy, hf. When photons scatter off electrons, they impart kinetic energy to the electrons. Therefore, the scattered photon has less energy than the incident photons, or, in terms of waves, the scattered light has a longer wavelength λ than the wavelength of the incident light, λ_o. The relationship between the two wavelengths is given by the Compton scattering equation

$$\Delta \lambda = \lambda - \lambda_o = \lambda_C(1 - \cos \theta) \qquad (27.9)$$

where $\lambda_C = h/m_e c = 2.43 \times 10^{-12}\,\text{m} = 2.43 \times 10^{-3}\,\text{nm}$ is the **Compton wavelength** of the electron.

- The **wave–particle duality of light** means that light must be thought of as having both particle and wave natures.

- The **Bohr theory** of hydrogen treats the electron as a classical particle held in circular orbit around the proton by the electrical force of attraction. In his theory, Bohr made two nonclassical, quantum postulates:

    The angular momentum of the electron is quantized and can have only discrete values that are integral multiples of $h/2\pi$, where h is Planck's constant;

and

    the electron does *not* radiate energy when it is in a bound, discrete orbit. It radiates energy only when it makes a downward transition to an orbit of lower energy. It makes an upward transition to an orbit of higher energy by absorbing energy.

- The Bohr model of hydrogen led directly to the quantization of the radius and energy of the atom. Their values are

$$r_n = 0.0529\,n^2\,\text{nm} \quad n = 1, 2, 3, 4, \ldots \qquad (27.16)$$

and

$$E_n = \frac{-13.6}{n^2}\,\text{eV} \quad n = 1, 2, 3, 4, \ldots \qquad (27.17)$$

where n is the **principal quantum number**. When the hydrogen atom is in the $n = 1$ state, it has its smallest size and lowest energy and is in its **ground state**. States of larger size and higher energy are **excited states**.

- Atoms emit light in an **emission spectrum**, which shows light emitted only at certain wavelengths characteristic of the atom. Since atomic energies are quantized, the emitted light must come off in quanta (photons) with fixed amounts of energy. These amounts of energy correspond to light of only those certain wavelengths.

- Atoms absorb light in an **absorption spectrum**, which shows light absorbed only at certain wavelengths characteristic of the atom. Since atomic energies are quantized, the absorbed light must be in the form of quanta (photons) with fixed amounts of energy. These amounts of energy correspond to absorbed light of only those certain wavelengths.

- A **metastable state** is a relatively long-lived excited atomic state.

- **Stimulated emission** can occur when a photon prematurely causes a downward atomic transition, yielding a photon identical to itself.

- A **laser** uses **population inversion**. This phenomenon is caused by metastable excited atomic states and results in light amplification through stimulated emission.

Exercises

Note: Take h to be exact at $6.63 \times 10^{-34}\,\text{J} \cdot \text{s}$ for significant figure purposes, and use $hc = 1.24 \times 10^3\,\text{eV} \cdot \text{nm}$ (three significant figures).

27.1 Quantization: Planck's Hypothesis

1. A blackbody (a) absorbs all radiation incident on it, (b) re-emits all radiation incident on it, (c) has a wavelength at which maximum radiation intensity occurs that is inversely proportional to its temperature, or (d) all of the preceding.

2. Planck's hypothesis (a) was the basis of Wien's law, (b) called for the intensity of blackbody radiation to be proportional to $1/\lambda^4$, (c) quantized the energy of thermal atomic oscillators, or (d) predicted the ultraviolet catastrophe.

3. **CQ** Does a blackbody at 200 K emit twice as much total radiation as when its temperature is 100 K? Explain.

4. **CQ** Some stars appear reddish, and some others are blue. Which of these stars have the lower surface temperature?

5. ■ Sketch a graph of temperature versus (a) frequency and (b) wavelength of the most intense radiation component of blackbody radiation.

6. ■ Find the approximate temperature of a red star that emits light with a wavelength of maximum emission of 700 nm (deep red).

7. ■ What are the wavelength and frequency of the most intense radiation component from a blackbody with a temperature of 0°C?

8. IE ■ (a) If you have a fever, will the wavelength of the radiation component of maximum intensity emitted by your body (1) increase, (2) remain the same, or (3) decrease as compared with its value when your temperature is normal? Why? (b) Assume that human skin has a temperature of 32°C. What is the wavelength of the radiation component of maximum intensity emitted by our bodies? In what region of the EM spectrum is this wavelength?

9. ■ The walls of a blackbody cavity are at a temperature of 27°C. What is the frequency of the radiation of maximum intensity?

10. IE ■■ The temperature of a blackbody increases from 200°C to 400°C. (a) Will the frequency of the most intense spectral component emitted by this blackbody (1) increase, but not double; (2) double; (3) be reduced in half; or (4) decrease, but not in half? Why? (b) What is the change in the frequency of the most intense spectral component of this blackbody?

11. ■■ What is the minimum energy of a thermal oscillator in a blackbody producing radiation at λ_{max} at a temperature of 212°F?

12. ■■ The temperature of a blackbody is 1000 K. If the intensity of the emitted radiation, 2.0 W/m^2, were due entirely to the most intense frequency component, how many quanta of radiation would be emitted per second per square meter?*

27.2 Quanta of Light: Photons and the Photoelectric Effect

13. In the photoelectric effect, classical theory predicted that (a) no photoemission can occur below a threshold frequency; (b) the photocurrent is proportional to the light intensity; (c) the kinetic energy of the emitted electron is dependent on frequency, but not on intensity; or (d) a photocurrent is observed immediately.

14. A photocurrent is observed when (a) the light frequency is above the threshold frequency, (b) the energy of the photons is greater than the work function, (c) the light frequency is below the threshold frequency, or (d) both (a) and (b).

*Assume that each quantum has the minimum allowed energy.

15. CQ If electromagnetic radiation is made up of quanta, why don't we detect the discrete packets of energy when listening to a radio?

16. CQ With respect to ionization (which can cause biological damage), is it more dangerous to stand in front of a very low energy beam of X-ray radiation or a much higher energy beam of red light? How does the photon model of light explain this apparent paradox?

17. CQ In the photoelectric effect, if the frequency of the radiation is below a certain cutoff frequency, no photoelectrons will be observed *no matter how intense* the radiation is. Why does this fact favor a particle theory of light over a wave theory?

18. CQ In the photoelectric effect, if the intensity of the incident light is increased, will the photoelectrons be ejected in a shorter time? Explain.

19. CQ In the photoelectric effect, if the frequency of the incident light is increased, but the intensity is kept constant, will more photoelectrons be ejected? Explain.

20. ■ What is the energy of a quantum of light that has a frequency of 5.0×10^{14} Hz?

21. ■ A photon has an energy of 3.3×10^{-15} J. What type of radiation is this photon?

22. IE ■ (a) Compared with a quantum of red light ($\lambda = 700$ nm), a quantum of violet light ($\lambda = 400$ nm) has (1) more, (2) the same amount of, or (3) less energy. Why? (b) Calculate the ratio of energy of a photon associated with violet light to the energy of a photon related to red light.

23. ■ A certain source of ultraviolet light has a wavelength of 150 nm. How much energy does one of its photons have in (a) joules and (b) electron-volts?

24. IE ■ The work function of metal A is greater than that of metal B. (a) The threshold wavelength for metal A is (1) shorter than, (2) the same as, or (3) longer than that of metal B. Why? (b) If the threshold wavelength for metal B is 620 nm and the work function of metal A is twice that of metal B, what is the threshold wavelength for metal A?

25. ■ The photoelectrons ejected from a surface have a maximum kinetic energy of 3.0 eV. If the intensity of the light producing the photoelectrons is tripled, what does the maximum kinetic energy of photoelectrons become?

26. ■■ Assume that a 100-W lightbulb gives off 2.50% of its energy as visible light. How many photons of visible light are given off in 1.00 minute? (Use an average visible wavelength of 550 nm.)

27. ■■ ▶ Figure 27.21 shows a graph of stopping potential versus frequency for a photoelectric material. Deter-

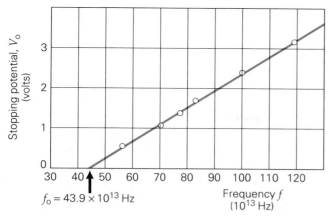

▲ FIGURE 27.21 Stopping potential versus frequency
See Exercise 27.

mine (a) Planck's constant and (b) the work function of the material from the data contained in the graph.

28. ■■ A metal with a work function of 2.40 eV is illuminated with a beam of monochromatic light. It is found that the stopping potential for the emitted electrons is 2.50 V. What is the wavelength of the light in nanometers?

29. ■■ What is the lowest frequency of light that can cause the release of electrons from a metal that has a work function of 2.8 eV?

30. ■■ The threshold wavelength for emission of electrons from a metallic surface is 500 nm. Calculate the maximum speed of photoelectrons if we use light having a wavelength of (a) 400 nm, (b) 500 nm, and (c) 600 nm.

31. ■■ In Exercise 30, what is the work function of the metal in eV?

32. ■■ The work function of a photoelectric material is 3.5 eV. If the material is illuminated with monochromatic light ($\lambda = 300$ nm), what are (a) the stopping potential of emitted photoelectrons and (b) the cutoff frequency?

33. IE ■■ Blue light with a wavelength of 420 nm is incident on a certain material and observed to cause the emission of photoelectrons with a maximum kinetic energy of 1.00×10^{-19} J. (a) When red light ($\lambda = 700$ nm) is used, there will be (1) more, (2) the same number of, (3) fewer, or (4) no photoelectrons. (b) Justify your answer to part (a) mathematically.

34. ■■ Gold has a work function of 4.82 eV. If a gold surface is illuminated with ultraviolet light ($\lambda = 160$ nm), what are (a) the maximum kinetic energy (in eV) of the emitted photoelectrons and (b) the threshold frequency for gold?

35. ■■ Sodium and silver have work functions of 2.46 eV and 4.73 eV, respectively. (a) If the surfaces of the metals are both illuminated with the same monochromatic light, the photoelectrons from which metal will have the greater maximum speed upon emission? (b) What wavelengths (in nm) are required for photoelectrons to be emitted from each metal surface?

36. ■■■ When a certain photoelectric material is illuminated with red light ($\lambda = 700$ nm) and then blue light ($\lambda = 400$ nm), it is found that the maximum kinetic energy of the photoelectrons resulting from the blue light is twice that of those from red light. What is the work function of the material?

37. ■■■ When the surface of a particular material is illuminated with monochromatic light of various frequencies, the stopping potentials for the photoelectrons are determined to be the following:

Frequency (in Hz)

| 9.9×10^{14} | 7.6×10^{14} | 6.2×10^{14} | 5.0×10^{14} |

Stopping potential (in V)

| 2.6 | 1.6 | 1.0 | 0.60 |

Plot these data, and determine the values of Planck's constant and the work function of the metal.

27.3 Quantum "Particles": The Compton Effect

38. The Compton effect was first observed using (a) visible light, (b) infrared radiation, (c) ultraviolet light, or (d) X rays.

39. The wavelength shift for Compton scattering is maximum when (a) the photon scattering angle is 90°, (b) the electron scattering angle is 90°, (c) the shift is equal to the Compton wavelength, or (d) the incident photon is backscattered.

40. CQ A photon can undergo Compton scattering from a molecule. How does the maximum wavelength shift for Compton scattering from a molecule compare with that from a free electron? Explain.

41. CQ In the center of the Sun, most of the electromagnetic energy is in the form of X rays. By the time it reaches the surface, however, it is mostly in the visible range. Estimate how many successive Compton scatterings are needed to effect this wavelength change.

42. CQ In Compton scattering, why does the scattered light always have a longer wavelength than the incident light?

43. ■ Calculate the maximum wavelength shift for Compton scattering from a free electron.

44. ■ What is the change in wavelength when monochromatic X rays are scattered by electrons through an angle of 30°?

45. ■ A monochromatic beam of X rays with a wavelength of 0.280 nm is scattered by a metal foil. What is the wavelength of the scattered X rays observed at an angle of 45° from the direction of the incident beam?

46. ■ X rays with a wavelength of 0.0045 nm are used in a Compton-scattering experiment. If the X rays are scattered

through an angle of 53°, what is the wavelength of the scattered radiation?

47. **IE ▪▪** A photon with an energy of 5.0 keV is scattered by a free electron. (a) The scattered photon could have an energy of (1) zero; (2) less than 5.0 keV, but not zero; or (3) equal to 5.0 keV. Why? (b) If the wavelength of the scattered photon is 0.25 nm, what is the recoiling electron's kinetic energy?

48. **▪▪** X rays scattered from a carbon atom show a wavelength shift of 0.000326 nm. What is their scattering angle?

49. **▪▪** If the Compton shift for a photon scattered by an electron is 1.25×10^{-4} nm, what is the scattering angle?

50. **IE ▪▪▪** The Compton effect can occur for scattering from any particle—for example, from protons. (a) Compared with the Compton wavelength for an electron, the Compton wavelength for a proton is (1) longer, (2) the same size, or (3) shorter. Why? (b) What is the value of the Compton wavelength for a proton? (c) Determine the ratio of the maximum Compton wavelength shift for scattering by an electron to that for scattering by a proton.

27.4 The Bohr Theory of the Hydrogen Atom

51. In his theory of the hydrogen atom, Bohr postulated the quantization of (a) energy, (b) centripetal acceleration, (c) light, or (d) angular momentum.

52. An excited hydrogen atom emits light when its electron (a) makes a transition to a lower energy level, (b) is excited to a higher energy level, or (c) is in the ground state.

53. **CQ** Explain why the Bohr theory is applicable only to the hydrogen atom and hydrogenlike atoms, such as singly ionized helium, doubly ionized lithium, and other one-electron systems.

54. **CQ** Will it take more energy to ionize (free) the electron in a hydrogen atom if the electron is in an excited state than if it is in the ground state? Explain.

55. **CQ** Very accurate measurements of the wavelengths emitted by a hydrogen atom indicate that they are all slightly longer than expected from the Bohr theory. Explain how conservation-of-momentum ideas might help explain this observation. [*Hint*: Photons carry momentum and energy, both of which need to be conserved.]

56. **CQ** If the principal quantum number of a hydrogen atom doubles, what happens to the radius of the electron orbit? Why?

57. If the principal quantum number of a hydrogen atom doubles, what happens to the energy of the electron?

58. **▪** Find the energy required to excite a hydrogen electron from (a) the ground state to the first excited state

and (b) from the first excited state to the second excited state.

59. **▪** In Exercise 58, classify the type of light needed to create each of the transitions.

60. **▪** Find the energy needed to ionize a hydrogen atom whose electron is in the (a) $n = 2$ state and (b) $n = 3$ state.

61. **▪** What is the frequency of light that would excite the electron of a hydrogen atom from (a) a state with a principal quantum number of $n = 2$ to that with a principle quantum number of $n = 5$? (b) What about from $n = 2$ to $n = \infty$?

62. **▪** Find the radius of the electron orbit in a hydrogen atom for states with the following principal quantum numbers: (a) $n = 2$; (b) $n = 4$; (c) $n = 5$.

63. **▪** Scientists are now beginning to study "large" atoms, that is, atoms with orbits that are almost large enough to be measured by our everyday units of measurement. For what excited state (give an approximate principal quantum number) of a hydrogen atom would the diameter of the orbit be a micron (10^{-6} m)—that is, close to the size of a dust particle?

64. **▪** Use the results of the Bohr theory to find the value of the Rydberg constant in Balmer's empirical formula (Eq. 27.10).

65. **▪** Find the energy of the hydrogen electron corresponding to states with the following principal quantum numbers: (a) $n = 3$; (b) $n = 6$; (c) $n = 10$.

66. **▪▪** Find the binding energy of the hydrogen electron for states with the following principal quantum numbers: (a) $n = 3$; (b) $n = 5$; (c) $n = 10$.

67. **IE ▪▪** A hydrogen atom has an ionization energy of 13.6 eV. When it absorbs a photon with an energy greater than this energy, the electron will be emitted with some kinetic energy. (a) If the energy of such a photon is doubled, the kinetic energy of the emitted electron will (1) increase, but not double; (2) remain the same; (3) exactly double; or (4) decrease. Why? (b) A photon associated with light of a frequency of 7.00×10^{15} Hz is absorbed by a hydrogen atom. What is the kinetic energy of the emitted electron?

68. **▪▪** A hydrogen atom in its ground state is excited to the $n = 5$ energy level. The electron then makes a transition directly to the $n = 2$ level before returning to the ground state. (a) What are the wavelengths of the emitted photons? (b) Would any of the emitted light be in the visible region?

69. **IE ▪▪** (a) For which one of the following transitions in a hydrogen atom is the photon of greatest energy emitted? (1) $n = 5$ to $n = 3$; (2) $n = 6$ to $n = 2$; (3) $n = 2$ to $n = 1$. (b) Justify your answer mathematically.

70. ■■ The hydrogen spectrum has a series of lines called the *Lyman series*, which results from transitions to the ground state. What is the longest wavelength in this series, and in what region of the EM spectrum does it lie?

71. ■■ What is the binding energy for an electron in the ground state in the following hydrogenlike ions? (a) He^+; (b) Li^{2+}.

72. ■■ A hydrogen atom absorbs light of wavelength 486 nm. (a) How much energy did the atom absorb? (b) What are the values of the principal quantum numbers of the initial and final states of this transition?

73. ■■■ Show that the speeds of an electron in the Bohr orbits are given (to two significant figures) by $v_n = (2.2 \times 10^6 \text{ m/s})/n$.

74. ■■■ Show, by adding the kinetic and electric potential energies, that the total energy of an electron in a hydrogen atom in an orbit with a radius of 0.0529 nm is -13.6 eV.

75. ■■■ In Exercise 74, (a) how much of the energy is potential, and how much is kinetic? (b) How do the magnitudes of the two forms of energy compare?

27.5 A Quantum Success: The Laser

76. Which of the following is not essential for laser action? (a) population inversion; (b) phosphorescence; (c) pumping; (d) stimulated emission.

77. CQ Why does a laser emit only one or a few colors, whereas a lightbulb emits a continuous spectrum?

78. CQ In what sense is a laser an "amplifier" of energy? Explain why this concept does not violate conservation of energy.

79. CQ Explain the main difference between spontaneous emission and stimulated emission.

Additional Exercises

80. An FM radio station broadcasts at a frequency of 98.9 MHz and radiates 750 kW of power. How many photons are radiated by the station's antenna each second?

81. A photon is a "massless" particle, meaning that it has no mass. Show that the momentum of a photon is given by $p = h/\lambda$. [*Hint*: Use the relativistic relationship $E^2 = p^2c^2 + (mc^2)^2$, which relates the total energy, mass, and momentum of a particle. (See Exercise 61 in Chapter 26.).]

82. Light of wavelength 340 nm is incident on a metal surface and ejects electrons that have a maximum speed of 3.5×10^5 m/s. What is the work function of the metal?

83. What are the energies of the photons required to excite a hydrogen atom in its ground state to states with principal quantum numbers of (a) $n = 3$ and (b) $n = 5$?

84. How many quanta of monochromatic red light ($\lambda = 700$ nm) would it take to have 1.0 J of energy?

85. For what scattering angle would the wavelength shift for Compton scattering from free electrons be 25% of the electron Compton wavelength?

86. IE (a) How many transitions in a hydrogen atom result in the emission of light in the visible region of the spectrum (400–700 nm), (1) one, (2) two, (3) three, or (4) four? (b) Calculate these wavelengths. [*Hint*: Consider the Balmer series.]

Quantum Mechanics and Atomic Physics

CHAPTER 28

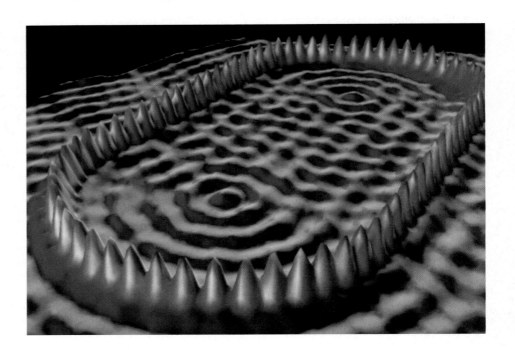

Just a few decades ago, if someone had claimed to have a photograph of an atom, people would have laughed. Today, a device called the scanning tunneling microscope (STM) routinely produces images such as the opening photograph. The blue shapes represent iron atoms, neatly arranged on a copper surface.

The STM operates on a quantum-mechanical phenomenon called *tunneling*. Tunneling reflects some of the fundamental features of the subatomic realm: the probabilistic character of quantum processes and the wave nature of particles. These features explain how particles may turn up in places where, according to classical notions, they should not be.

In the 1920s, a new kind of physics, based on the synthesis of wave and quantum ideas, was introduced. This new theory, called **quantum mechanics**, combines the wave–particle duality of matter into a single consistent description. It revolutionized scientific thought and provides the basis of our understanding of phenomena that occur on the scale of molecular sizes and smaller.

In this chapter, some of the basic ideas of quantum mechanics are presented in order to show how they describe matter. Practical applications made possible by the quantum-mechanical view of nature, such as the electron microscope and magnetic resonance imaging (MRI), are also discussed.

28.1 Matter Waves: The de Broglie Hypothesis

OBJECTIVES: To (a) explain de Broglie's hypothesis, (b) calculate the wavelength of a matter wave and (c) specify under what circumstances the wave nature of matter will be observable.

Since a photon travels at the speed of light, it must be treated relativistically as a particle with no mass. If this were not the case (i.e., if it were to have mass), it would have infinite energy. This is because, at $v = c$, the dimensionless relativistic factor γ becomes infinite, and its total energy $E = \gamma mc^2$ would become infinite unless $m = 0$. A photon's energy E and momentum p are related by $p = E/c$. Recall from the Einstein relationship (Eq. 27.3) that the energy of a photon can be written in terms of wave frequency and wavelength as $E = hf = hc/\lambda$. Combining these two results, we find that the momentum of a photon is inversely related to its wavelength by

Note: These relationships are discussed in Section 26.4.

$$p = \frac{E}{c} = \frac{hf}{c} = \frac{h}{\lambda} \quad (photon\ momentum) \qquad (28.1)$$

In the early part of the 20th century, French physicist Louis de Broglie (1892–1987) considered that there might be a symmetry between waves and particles. He reasoned that if light sometimes behaves like a particle, then perhaps material particles, such as electrons, also have wave properties. In 1924, de Broglie hypothesized that a moving particle has a wave associated with it. He proposed that the particle's wavelength is related to the magnitude of its momentum ($p = mv$) by an equation similar to that for a photon (Eq. 28.1). The **de Broglie hypothesis** states the following:

A particle with momentum of magnitude p has a wave associated with it. This wave's wavelength is given by

$$\lambda = \frac{h}{p} = \frac{h}{mv} \quad (material\ particles) \qquad (28.2)$$

The waves associated with moving particles were called **matter waves**, or, more commonly, **de Broglie waves**, and were thought to somehow influence or guide the particle's motion. The electromagnetic wave is the associated wave for a photon. However, de Broglie waves associated with particles such as electrons and protons are *not* electromagnetic waves. The de Broglie equation for the wavelength gives no clue as to the nature of the wave associated with a particle that has mass.

Needless to say, de Broglie's hypothesis met with great skepticism. The idea that the motions of photons were somehow governed by the wave properties of light seemed reasonable. But the extension of this idea to the motion of a particle *with mass* was difficult to accept. Moreover, there was no evidence at the time that particles exhibited any wave properties, such as interference and diffraction.

In support of his hypothesis, de Broglie showed how it could give an interpretation of the quantization of the angular momentum postulated by Bohr in his theory of the hydrogen atom. Recall that Bohr had to hypothesize that the angular momentum of the orbiting electron was quantized in integer multiples of $h/2\pi$ (Chapter 27). De Broglie argued that for a *free* particle, the associated wave would be a *traveling* wave. However, the *bound* electron of a hydrogen atom travels repeatedly in discrete circular orbits. The associated matter wave might therefore be expected to be a *standing* wave. For a standing wave to be produced, an integral number of wavelengths has to fit into the orbital circumference. The wave must reinforce itself constructively, much as a standing wave in a circular string (▶ Fig. 28.1).

Note: The Bohr model is outlined in Section 27.4.

Note: Compare this standing wave with that in the discussion of standing waves in Section 13.5.

The circumference of a Bohr orbit of radius r_n is $2\pi r_n$, where n is the quantum number. De Broglie equated this circumference to an integer number of electron "wavelengths" in the following manner:

$$2\pi r_n = n\lambda \quad \text{for } n = 1, 2, 3, \ldots$$

Using the de Broglie relationship, $\lambda = h/mv$, and recalling that, for a circular orbit, the angular momentum $L = mvr$, we have

Note: See Eq. 8.14 for a reminder about angular momentum.

$$2\pi r_n = \frac{nh}{mv} \quad \text{or} \quad L_n = mvr_n = n\left(\frac{h}{2\pi}\right) \quad \text{for } n = 1, 2, 3, \ldots$$

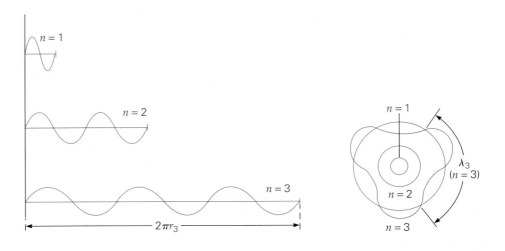

FIGURE 28.1 de Broglie waves and Bohr orbits Similar to standing waves in a stretched string, de Broglie waves form circular standing waves on the circumferences of the Bohr orbits. The number of wavelengths in a particular orbit of radius r, shown here for $n = 3$, is equal to the principal quantum number of that orbit. (Not drawn to scale.)

Thus, Bohr's angular-momentum quantization is equivalent to the de Broglie assumption that, in some way, the electron behaves like a wave.

For orbits other than those allowed by the Bohr theory, the de Broglie wave for the orbiting electron would not close on itself. This observation is consistent with the Bohr postulate of the electron being only in certain "allowed" orbits. This implied that *the amplitude of the de Broglie wave might be related to the location of the electron*. This idea is actually a fundamental cornerstone of modern quantum mechanics, as we shall see.

If particles really have wavelike properties, why don't we observe their wave nature in everyday living? It is because effects such as diffraction are significant only when the wavelength λ is on the order of the size of the object or the opening it meets. (See Section 24.3.) If λ is much smaller than these dimensions, then diffraction is negligible. The numbers in Examples 28.1 and 28.2 should convince you about the difference between the atomic world and our everyday world.

Example 28.1 ■ Should Ballplayers Worry About Diffraction: de Broglie Wavelength

A pitcher throws a fastball to his catcher at 40 m/s through a square strike-zone practice target. The square opening is cut in a sheet of canvas and is 50 cm on each side. If the ball's mass is 0.15 kg, (a) what is the wavelength of the de Broglie wave associated with the ball? (b) Should the catcher expect diffraction to occur as the ball passes through the opening to his mitt?

Thinking It Through.   (a) The de Broglie relationship (Eq. 28.2) can be used to calculate the wavelength of the matter wave associated with the ball. (b) All that matters is whether the wavelength is much larger than, much smaller than, or approximately the same size as the opening.

Solution.

Given: $v = 40$ m/s   *Find:* (a) λ (de Broglie wavelength)
           $d = 50$ cm $= 0.50$ m   (b) Whether diffraction is likely
           $m = 0.15$ kg

(a)   Eq. 28.2 gives the wavelength of the baseball:

$$\lambda = \frac{h}{mv} = \frac{6.63 \times 10^{-34}\,\text{J}\cdot\text{s}}{(0.15\,\text{kg})(40\,\text{m/s})} = 1.1 \times 10^{-34}\,\text{m}\quad(\textit{extremely small!})$$

(b)   For significant diffraction to occur when a wave passes through an opening, the wave must have a wavelength similar in size to that of the opening. Since

1.1×10^{-34} m $\ll 0.50$ m, the baseball travels straight into the catcher's mitt, with no noticeable wave diffraction.

Follow-up Exercise. In this Example, (a) how fast would the ball have to be thrown for diffractive effects to become important? (b) At that speed, how long would the ball take to travel the 20 m to the plate? (Compare your answer with the age of the universe—about 15 billion years. Would someone watching this ball think that it is moving?) (*Answers to all Follow-up Exercises are at the back of the text.*)

From Example 28.1, it is little wonder that the wave nature of matter isn't observed in our everyday lives. Notice, however, that a particle's de Broglie wavelength varies inversely with its mass and speed. So particles with very small masses traveling at low speeds might be another story, as the following Example shows.

Note: Single-slit diffraction is discussed in Section 24.3.

Example 28.2 ■ A Whole Different Ball Game: de Broglie Wavelength of an Electron

(a) What is the de Broglie wavelength of the wave associated with an electron that has been accelerated from rest through a potential of 50.0 V? (b) Compare your answer with the typical distance between atoms in a solid crystal, about 10^{-10} m. Would you expect diffraction to occur as these electrons pass between such atoms?

Thinking It Through. (a) We can use the de Broglie relationship (Eq. 28.2), but the electron's speed must first be calculated, using the given accelerating voltage. This computation involves consideration of energy conservation—the conversion of electric potential energy into kinetic energy. (b) For diffractive effects to be important, λ must be on the order of 10^{-10} m.

Solution. We list the data, as well as the mass of an electron, which can be found in Table 15.1.

Given: $V = 50.0$ V $\qquad\qquad$ *Find:* (a) λ (de Broglie wavelength)
$\qquad\;\;\, m = 9.11 \times 10^{-31}$ kg (from Table 15.1) $\qquad\;\;\;$ (b) Whether diffraction is likely

(a) The magnitude of the potential energy lost by the electron is $|\Delta U_e| = eV$ and is equal to its gain in kinetic energy ($\Delta K = \frac{1}{2}mv^2$, since $K_o = 0$). Equating these two quantities enables us to calculate the speed:

$$\tfrac{1}{2}mv^2 = eV \qquad \text{or} \qquad v = \sqrt{\frac{2eV}{m}}$$

So,

$$v = \sqrt{\frac{2(1.60 \times 10^{-19}\,\text{C})(50.0\,\text{V})}{9.11 \times 10^{-31}\,\text{kg}}} = 4.19 \times 10^6 \text{ m/s}$$

Thus, the electron's de Broglie wavelength is

$$\lambda = \frac{h}{mv} = \frac{6.63 \times 10^{-34}\,\text{J}\cdot\text{s}}{(9.11 \times 10^{-31}\,\text{kg})(4.19 \times 10^6\,\text{m/s})} = 1.74 \times 10^{-10} \text{ m}$$

(b) Since this result is the same order of magnitude as the opening between atoms, diffraction should be observed. Thus, passing electrons through a crystal lattice could prove de Broglie's hypothesis.

Follow-up Exercise. In this Example, what would change if the particle were a proton? In other words, would the de Broglie wave of a proton be more or less likely to exhibit diffraction effects than would an electron under the same conditions? Explain, and give a numerical answer.

In Example 28.2, if the accelerating voltage were changed, the calculation of v and λ would have to be repeated. This process would again involve the use of constants such as the electron's mass, its charge, and Planck's constant. In problems involving accelerating voltages, therefore, it is convenient to use the numerical values of m, e, and h to derive a nonrelativistic expression for the de Broglie wavelength of an electron when it is accelerated through a difference in potential, V. To begin, the kinetic energy is expressed in terms of momentum. Since $p = mv$, the kinetic energy, $\frac{1}{2}mv^2$, can also be written as $p^2/2m$. Then, by energy conservation,

$$\frac{p^2}{2m} = eV \quad \text{or} \quad p = \sqrt{2meV}$$

Thus, the de Broglie wavelength is

$$\lambda = \frac{h}{p} = \frac{h}{\sqrt{2meV}} = \sqrt{\frac{h^2}{2meV}}$$

Note: This alternative de Broglie wavelength expression for electrons accelerated through a potential difference, in which V must be in volts, applies only to nonrelativistic electrons initially at rest.

Inserting the values of h, e, and m and rounding the result to three significant figures, we have

$$\lambda = \sqrt{\frac{1.50}{V}} \times 10^{-9}\ \text{m} = \sqrt{\frac{1.50}{V}}\ \text{nm} \quad \begin{array}{l}(\textit{nonrelativistic electrons} \\ \textit{initially at rest; V in volts})\end{array} \quad (28.3)$$

Thus, if $V = 50.0$ V,

$$\lambda = \sqrt{\frac{1.50}{50.0}} \times 10^{-9}\ \text{m} = 0.173\ \text{nm}$$

Note that this answer differs in the last digit from that found in Example 28.2, because there, the values for h, e, and m were rounded to three significant figures *before* the calculations were performed.

You may wish to derive a comparable formula for a proton, to use in solving similar problems. What quantities would have to be changed?

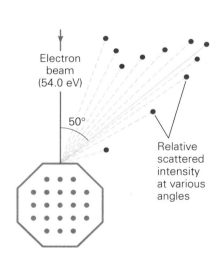

Electron beam (54.0 eV)

50°

Relative scattered intensity at various angles

▲ **FIGURE 28.2 The Davisson–Germer experiment**
When a beam of electrons of kinetic energy 54.0 eV is incident on the face of a nickel crystal, a maximum in the scattering intensity is observed at an angle of 50°.

In 1927, two physicists in the United States, C. J. Davisson and L. H. Germer, used a crystal to diffract a beam of electrons, thereby demonstrating a wavelike property of particles. A single crystal of nickel was cut to expose a spacing of $d = 0.215$ nm between the planes of atoms. When a beam of electrons was directed normally onto the crystal face, a maximum in the intensity of scattered electrons was observed at an angle of 50° relative to the surface normal (◀Fig. 28.2). The scattering was most intense for an accelerating potential of 54.0 V.

According to wave theory, constructive interference due to waves reflected from two lattice planes (a distance d apart) should occur at certain scattering angles θ. The theory predicts that the first-order maximum should be observed at an angle given by

$$\sin\theta = \frac{\lambda}{d}$$

This condition for the experimental setup in Fig. 28.2 requires a wavelength of

$$\lambda = d\sin\theta = (0.215\ \text{nm})\sin 50° = 0.165\ \text{nm}$$

Using Eq. 28.3 to determine the de Broglie wavelength of the electrons after being accelerated through 54.0 V, we obtain

$$\lambda = \sqrt{\frac{1.50}{V}}\ \text{nm} = \sqrt{\frac{1.50}{54.0}}\ \text{nm} = 0.167\ \text{nm}$$

The agreement was well within the experimental uncertainty. The Davisson–Germer experiment gave convincing proof of de Broglie's matter-wave hypothesis. A practical application of the wavelike properties of electrons is discussed in the Insight on electron microscopes on p. 908.

Note: Diffraction from crystal lattices is discussed in Section 24.3.

28.2 The Schrödinger Wave Equation

OBJECTIVES: To understand qualitatively (a) the reasoning that underlies the Schrödinger wave equation and (b) the equation's use in finding particle wave functions.

De Broglie's hypothesis predicts that moving particles have associated waves that somehow govern their behavior. However, it does not tell us the *form* of these waves, only their wavelengths. To have a useful theory, we need an equation that will give the mathematical form of these matter waves. We also need to know how these waves govern particle motion. In 1926, Erwin Schrödinger, an Austrian physicist, presented a general equation that describes the de Broglie matter waves and their interpretation.

A traveling de Broglie wave varies with location and time in a similar way to everyday wave motion (Section 13.2). The de Broglie wave is denoted by ψ (the Greek letter psi, pronounced "sigh") and is called the **wave function**. ψ is associated with the particle's kinetic, potential, and total energy. Recall that, for a conservative mechanical system (Section 5.5), the total mechanical energy E, which is the sum of the kinetic and potential energies, is a constant; that is $K + U = E$. Schrödinger proposed a similar equation for the de Broglie matter waves, involving the wave function ψ. **Schrödinger's wave equation*** has the general form

$$(K + U)\psi = E\psi \tag{28.4}$$

Equation 28.4 can be solved for ψ, but doing so involves complex mathematical operations well beyond the scope of this text. For us, the more important question is related to ψ's physical significance. During the early development of quantum mechanics, it was not at all clear how ψ should be interpreted. After much thought and investigation, Schrödinger and his colleagues hypothesized the following:

> The square of a particle's wave function is proportional to the probability of finding that particle at a given location.

Note: More precisely, the square of the absolute value of the wave-function solution to Schrödinger's equation, $|\psi|^2$, gives the probability of finding a particle at a location.

The interpretation of ψ^2 as a *probability* altered the idea that the electron in a hydrogen atom could be found only in orbits at discrete distances from the nucleus, as described in the Bohr theory. When the Schrödinger equation was solved for the hydrogen atom, there was a nonzero probability of finding the electron at almost any distance from the nucleus! The relative probability of finding an electron [in the ground state $(n = 1)$] at a given distance from the proton is shown in ▶Fig. 28.3a. The *maximum* probability coincides with the Bohr radius of 0.0529 nm, but it is possible that, for instance, the electron could even be inside the nucleus. Notice that, although the ground-state wave function exists for distances well beyond 0.20 nm from the proton, there is little chance of finding an electron beyond this distance. The probability density distribution gives rise to the idea of an *electron cloud* around the nucleus (Fig. 28.3b). The cloud is actually a probability density cloud. It does *not* mean that the electron is somehow smeared out over space.

---

*Although Eq. 28.4 looks like a multiplication of $E = K + U$ by ψ, it is much more complex. For example, K is no longer $\frac{1}{2}mv^2$, but is instead replaced by a quantity that enables us to *extract* the kinetic energy from ψ.

The Electron Microscope

The de Broglie hypothesis led to the development of an important practical application—the *electron microscope*. As the Davisson–Germer experiment demonstrated, under the right conditions, electrons experience diffraction, as do light waves. Does this mean that electrons might also be focused like light waves? The answer is yes. In fact, electron "waves" can be focused to form images, though the focusing mechanism is different than those used with light in a regular microscope. As Example 28.2 shows, moving electrons have very short de Broglie wavelengths. With such short wavelengths, it is possible to obtain greater magnification and finer resolution than can be provided by any light microscope. Recall the inverse relationship between resolving power and wavelength—a smaller wavelength means greater resolving power, or the ability to see details. (See Section 25.4.) In fact, the resolving power of standard electron microscopes is on the order of a few nanometers, which is only about 10 times larger than atomic sizes.

Knowledge of how to focus electron beams by using magnetic coils permitted the construction of the first electron microscope in Germany in 1931. In a *transmission electron microscope* (TEM), an electron beam is directed onto a very thin specimen. Different numbers of electrons pass through different parts of the specimen, depending on its structure. The transmitted beam is then brought into focus by a magnetic objective coil. The general components of electron and light microscopes are analogous (Fig. 1), but an electron microscope must be housed in a high-vacuum chamber in order to

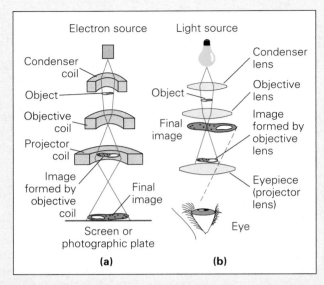

Figure 1   Electron and light microscopes   A comparison of the elements of **(a)** an electron microscope and **(b)** a compound light microscope. The light microscope is drawn upside down for a better comparison.

prevent the electrons from colliding with air molecules. As a result, an electron microscope looks nothing like a light microscope (Fig. 2). A normal light-microscope image (Fig. 3a)

▶ **FIGURE 28.3   Electron probability for hydrogen-atom orbits** **(a)** The square of the wave function is the probability of finding the hydrogen electron at a particular location. Here, it is assumed to be in the ground state ($n = 1$), and its probability is plotted as a function of radial distance from the proton. The electron has the greatest probability of being at a distance of 0.0529 nm, which matches the radius of the first Bohr orbit. **(b)** The probability distribution gives rise to the idea of an electron probability cloud around the nucleus. The cloud's density reflects the probability density.

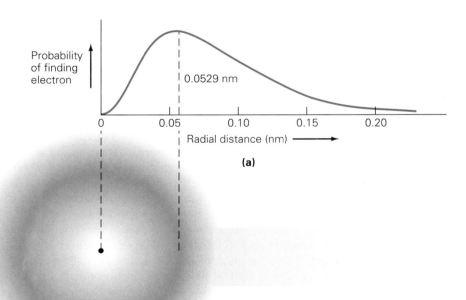

(b) Electron "cloud"

Figure 2　An electron microscope　The microscope is housed in a cylindrical vacuum chamber, to the left.

is limited to a magnification of about 2000×, while magnifications up to 100 000× can be achieved with an electron microscope (Fig. 3b).

Another difference between light and electron microscopes is that the final lens in an electron microscope, called the *projector coil*, has to project a real image onto a fluorescent screen or photographic film, since the eye cannot perceive an electron image directly. Because specimens for transmission electron microscopy must be very thin, special techniques for specimen preparation are required. These methods allow for specimen sections to be sliced as thin as 10 nm (only about 100 atoms thick).

The surfaces of thicker objects can be examined by the *reflection* of the electron beam from the surface. This process is accomplished with the more recently developed *scanning electron microscope* (*SEM*). A beam spot is scanned across the specimen by means of deflecting coils, much as is done in a television tube. Surface irregularities cause directional variations in the intensity of the reflected electrons, which gives contrast to the image. The specimens must be coated with a thin layer of metal (such as gold or aluminum) in order to make them conducting. Otherwise, they would be charged nonuniformly by the electron beam, resulting in a distortion of the final image. Through such techniques, a scanning electron microscope gives pictures with a remarkable three-dimensional quality, such as those in Fig. 3c.

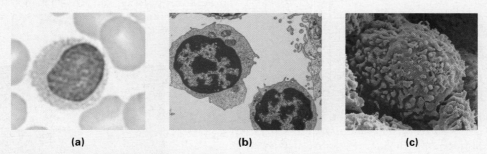

(a)　　　　　　　　　(b)　　　　　　　　　(c)

Figure 3　Lymphocytes (white blood cells)　Images produced by **(a)** a light microscope, **(b)** a transmission electron microscope (TEM), and **(c)** a scanning electron microscope (SEM). Notice the enormous increase in magnification produced by the electron microscope.

An interesting quantum-mechanical result that runs counter to our everyday experiences is *tunneling*. In classical physics, there are regions forbidden to particles by energy considerations. These regions are areas where a particle's potential energy would be greater than its total energy. Classically, the particle is not allowed in such regions because it would have a negative kinetic energy there $(E - U = K < 0)$, which is impossible. In such situations, we say that the particle's location is limited by a *potential-energy barrier*.

In certain instances, however, quantum mechanics predicts a small, but finite, probability of the particle's wave function penetrating the barrier and thus of the particle being found on the other side of the barrier. Thus, there is a certain probability (which is practically zero for everyday objects) of the particle "tunneling" through the barrier, especially on the atomic level, where the wave nature of particles is exhibited. Such tunneling forms the basis of the scanning tunneling microscope (known by the acronym "STM"; see the Insight on the scanning tunneling microscope on p. 910), which creates images with a resolution on the order of the size of a single atom. Barrier penetration also explains certain nuclear decay processes (Chapter 29).

The Scanning Tunneling Microscope (STM)

The *scanning tunneling microscope* (*STM*) was invented in the late 1970s and promptly revolutionized the field of surface physics. STMs use quantum-mechanical tunneling to produce stunning images of atoms, such as this chapter's opening photograph.

The STM produces atomic-sized images by positioning its sharp tip very close (about 1 nm) to a conducting surface, such as aluminum. A voltage is applied between the tip and the surface, causing electron tunneling through the vacuum gap (Fig. 1). This tunneling current is extremely sensitive to the separation distance. A feedback circuit monitors this current and moves the probe vertically to keep the current constant. The separation distance is digitized, recorded, and processed by computers for display. By passing the probe over the surface of the specimen in successive nearby parallel movements, a three-dimensional image of the surface can be displayed. Typical vertical resolution for modern STMs is 0.001 nm, and lateral resolutions are about 0.1 nm. Because the diameters of individual atoms are on the order of several tenths of a nanometer, STMs allow detailed images of atoms to be created. Figure 2 shows the result of a surface scan on gallium arsenide, a semiconductor material.

Note that for an STM to work properly, the surface of the material must be electrically conductive. More recently, an *atomic force microscope* (*AFM*) has been developed to analyze nonconducting surfaces. It uses a probe to scan the surface, as does the STM, but instead of creating and measuring a tunneling current, it measures the force between the tip and surface. This force, which depends sensitively on the tip–surface separation, can be used to monitor the separation distance and produce surface images with resolution similar to that of the STM.

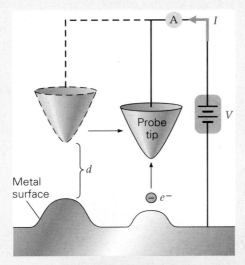

Figure 1 **Schematic representation of the STM** The tip of a metal probe is moved across the contours of a metal-coated specimen surface. Applying a small voltage between the probe and the surface causes electrons to tunnel across the vacuum gap. The tunneling current is extremely sensitive to the separation distance d between the probe tip and the surface. A feedback circuit (not shown) moves the probe up and down so as to keep constant the tunneling current (measured by the ammeter A). Thus, when the probe is scanned across the sample, surface features smaller than atoms (up to 0.001 nm vertically) can be detected. By passing the probe over the surface in many successive and nearby parallel paths, the resulting data can be processed to produce three-dimensional images.

Figure 2 **Scanning tunneling microscopy results** Atoms of the semiconductor gallium arsenide.

28.3 Atomic Quantum Numbers and the Periodic Table

OBJECTIVE: **To understand the structure of the periodic table in terms of quantum-mechanical electron orbits and the Pauli exclusion principle.**

The Hydrogen Atom

When the Schrödinger equation was solved for the hydrogen atom, the results predicted the energy levels to be the same as those from the Bohr theory (Section 27.4). Recall that the Bohr-model energy values depended only on the principal quantum number n. However, in addition, the solution to the Schrödinger equation gave two other quantum numbers, designated as ℓ and m_ℓ. Three quantum numbers are needed because the electron can move in three dimensions.

The quantum number ℓ is called the **orbital quantum number**. It is associated with the orbital angular momentum of the electron. For each value of n, the ℓ quantum number has integer values from zero up to a maximum value of $n - 1$. For example, if $n = 3$, the three possible values of ℓ are 0, 1, and 2. ▶Figure 28.4 shows three orbits with different angular momenta, but the same energy. The number of different ℓ values for a given n value is equal to n. In the hydrogen atom, the energy of the electron depends only on n. Thus, orbits with the same n value, but different ℓ values, have the same energy and are said to be *degenerate*.

The quantum number m_ℓ is called the **magnetic quantum number**. The name originated from experiments in which an external *magnetic* field was applied to a sample. They showed that a particular energy level (one with given values of n and ℓ) of a hydrogen atom actually consists of several orbits that differ slightly in energy *only when in a magnetic field*. Thus, in the absence of a field, there was additional energy degeneracy. Clearly, there must be more to the description of the orbit than just n and ℓ. The quantum number m_ℓ was introduced to enumerate the number of levels that existed for a given orbital quantum number ℓ. Under zero-magnetic-field conditions, the energy of the atom does not depend on either of these quantum numbers.

The magnetic quantum number m_ℓ is associated with the orientation of the orbital angular-momentum vector **L** in space (▼Fig. 28.5). If there is no external magnetic field, then all orientations of **L** have the same energy. For each value of ℓ, m_ℓ is an integer that can range from zero to $\pm\ell$. That is, $m_\ell = 0, \pm1, \pm2, \ldots, \pm\ell$. For example, an orbit described by $n = 3$ and $\ell = 2$ can have m_ℓ values of $-2, -1, 0, +1$, and $+2$. In this case, the orbital angular-momentum vector **L** has five possible orientations, all with the same energy if no magnetic field is present. In general, for a given value of ℓ, there are $2\ell + 1$ possible values of m_ℓ. For example, with $\ell = 2$, there are five values of m_ℓ, since $2\ell + 1 = (2 \times 2) + 1 = 5$.

However, this finding was not the end of the story. The use of high-resolution optical spectrometers showed that each emission line of hydrogen is, in fact, two very closely spaced lines. Thus, each emitted wavelength is actually two. This splitting is called *spectral fine structure*. Hence, a fourth quantum number was necessary in order to describe each atomic state completely. This number is called the **spin quantum number** m_s of the electron. It is associated with the *intrinsic* angular momentum of the electron. This property, called *electron spin*, is sometimes described by analogy to the angular momentum associated with a spinning object (▶Fig. 28.6). Because each energy level is split into only two levels, the electron's intrinsic angular momentum (or, more simply, its "spin") can possess only two orientations, called "spin up" and "spin down."

Thus, the fine structure of an atom's energy levels results from the electron spin's having two orientations with respect to the atom's *internal* magnetic field, produced by the electron's orbital motion. Analogous to a magnetic moment (such as a compass) in a magnetic field, the atom possesses slightly less

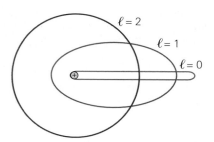

▲ **FIGURE 28.4 The orbital quantum number (ℓ)** The orbits of an electron are shown for the second excited state in hydrogen. For the principal quantum number $n = 3$, there are three different values of angular momentum (corresponding to the three differently shaped orbits and three different values of ℓ), but they all have the same total energy. The circular orbit has the maximum angular momentum; the narrowest orbit would actually pass through the proton classically and thus has zero angular momentum.

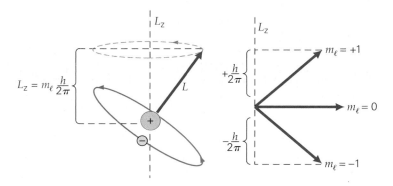

◀ **FIGURE 28.5 The magnetic quantum number (m_ℓ)** Here, **L** is the vector angular momentum associated with the orbit of the electron. Because the energy of an orbit is independent of the orientation of its plane, orbits with the same ℓ that differ only in m_ℓ have the same energy. There are $2\ell + 1$ possible orientations for a given ℓ. The value of m_ℓ tells the *component* of the angular-momentum vector in a given direction, as shown (right) for $\ell = 1$.

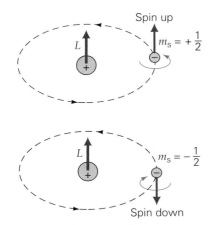

Spin up

L

$m_s = +\frac{1}{2}$

L

$m_s = -\frac{1}{2}$

Spin down

▲ **FIGURE 28.6 The electron-spin quantum number (m_s)** The electron spin can be either up or down. Electron spin is strictly a quantum mechanical property and should not be identified with the physical spin of a macroscopic body, except with respect to conceptual reasoning.

Note: The n quantum number designates orbital shells; the ℓ quantum number designates orbital subshells.

PHYSLET® ILLUSTRATION

Hydrogenic Atoms

energy when its electron's spin is "lined up" with the field than when the spin is aligned opposite to the field. However, keep in mind that spin is fundamentally a purely quantum-mechanical concept; it is not really analogous to a spinning top. This is because, as far as we know now, the electron possesses no size—that is, it is truly a point particle.

Thus, for the excited state of hydrogen with $n = 3$ and $\ell = 2$, each value of m_ℓ also has two possible spin orientations. For example, when $m_\ell = +1$, there are two possible sets of the four quantum numbers: $n = 3$, $\ell = 2$, and $m_\ell = +1$, with $m_s = +\frac{1}{2}$; and $n = 3$, $\ell = 2$, and $m_\ell = +1$, with $m_s = -\frac{1}{2}$. Both sets would have nearly the same energy. The orbit's energy is therefore almost independent of the electron's spin direction. This condition results in yet another (approximate) energy degeneracy. In summary, the energy of the various states of the hydrogen atom are, to a very high degree, determined solely by the primary quantum number n.

The four quantum numbers for the hydrogen atom are summarized in Table 28.1. Other particles, such a protons, neuturons, and composites of them called *atomic nuclei* also possess spin. A particularly useful property of the spin of a proton, and of atomic nuclei in general, is discussed in the Insight on magnetic resonance imaging on p. 914.

Multielectron Atoms

The Schrödinger equation cannot be solved exactly for atoms with more than one electron (multielectron atoms). However, a solution can be found, to a workable approximation, in which each electron occupies a state characterized by a set of quantum numbers similar to those for hydrogen. Because of the repulsive forces between the electrons, the description of a multielectron atom is much more complicated. For one, the energy depends not only on the principal quantum number n, but also on the orbital quantum number ℓ. This condition gives rise to a subdivision (or "splitting") of the degeneracy seen in hydrogen atoms. In multielectron atoms, the energies of the atomic levels generally depend on all four quantum numbers.

It is common to refer to all the electron levels that share the same n value as making up a **shell** and to all the electron levels that share the same ℓ values within that shell as **subshells**. That is, electron levels with the same n value are said to be "in the same shell." Similarly, electrons with the same n and ℓ values are said to be "in the same subshell."

The ℓ subshells can be designated by integers; however, it is common to use letters instead. The letters $s, p, d, f, g, \ldots$ correspond to the values of $\ell = 0, 1, 2, 3, 4, \ldots$ respectively. After f, the letters go alphabetically (Table 28.2).

Since, in multielectron atoms an electron's energy depends on both n and ℓ both quantum numbers are used to label atomic energy levels (a shell and subshell, respectively). The labeling convention is as follows: n is written as a number, followed by the letter that stands for the value of ℓ. For example, 1s denotes an energy level with $n = 1$ and $\ell = 0$; 2p is for $n = 2$ and $\ell = 1$; 3d for $n = 3$ and $\ell = 2$; and so on. Also, it is common to refer to the m_ℓ values as representing *orbitals*. For example, a 2p energy level has three orbitals, corresponding to the m_ℓ values of -1, 0, and $+1$ (since $\ell = 1$).

The hydrogen-atom energy levels are not evenly spaced, but do increase sequentially. In multielectron atoms, not only are the energy levels unevenly

TABLE 28.1 Quantum Numbers for the Hydrogen Atom

Quantum Number	Symbol	Allowed Values	Number of Allowed Values
Principal	n	$1, 2, 3, \ldots$	no limit
Orbital angular momentum	ℓ	$0, 1, 2, 3, \ldots, (n-1)$	n (for each n)
Orbital magnetic	m_ℓ	$0, \pm 1, \pm 2, \pm 3, \ldots, \pm \ell$	$2\ell + 1$ (for each ℓ)
Spin	m_s	$\pm \frac{1}{2}$	2 (for each m_ℓ)

spaced, but also their numerical sequence is out of order. The shell–subshell ($n-\ell$ notation) energy-level sequence for a multielectron atom is shown in ▼Fig. 28.7a. Notice, for example, that the 4s level is, energywise, below the 3d level. Such variations result in part from electrical forces between the electrons. Furthermore, note that the electrons in the outer orbits are "shielded" from the attractive force of the nucleus by the electrons that are closer to the nucleus. For example, consider the highly elliptical orbit (Fig. 28.4) of an electron in the 4s ($\ell = 0$) orbit. The electron clearly spends more time near the nucleus, and hence is more tightly bound, than if it were in the more circular 3d ($\ell = 2$) orbit. A convenient way to remember the order of the levels is given in Fig. 28.7b.

The ground state of a multielectron atom has some similarities to that of the hydrogen atom, with its one electron in the 1s, or lowest energy, level. In a multielectron atom, the ground state still has the lowest total energy, but is a combination of energy levels in this case. That is, the electrons are in the lowest possible energy levels. But to identify how the electrons fill these levels, we must know how many electrons can occupy a particular energy level. For example, the lithium (Li) atom has three electrons. Can they all be in the 1s level? As we shall see, the answer, given by the Pauli exclusion principle, is no.

TABLE 28.2 Subshell Designations	
Value of ℓ	Designation
$\ell = 0$	s
$\ell = 1$	p
$\ell = 2$	d
$\ell = 3$	f
$\ell = 4$	g
$\ell = 5$	h
$\cdots$	$\cdots$

The Pauli Exclusion Principle

Exactly how the electrons of a multielectron atom distribute themselves in the ground-state energy levels is governed by a principle set forth in 1928 by the Austrian physicist Wolfgang Pauli. The **Pauli exclusion principle** states the following:

No two electrons in an atom can have the same set of quantum numbers (n, ℓ, m_ℓ, m_s). That is, no two electrons in an atom can be in the same quantum state.

This principle limits the number of electrons that can occupy a given energy level. For example, the 1s ($n = 1$ and $\ell = 0$) level can have only one m_ℓ value, $m_\ell = 0$, along with only two m_s values, $m_s = \pm\frac{1}{2}$. Thus, there are only two unique sets of quantum numbers (n, ℓ, m_ℓ, m_s) for the 1s level—$(1, 0, 0, +\frac{1}{2})$ and $(1, 0, 0, -\frac{1}{2})$—so only two electrons can occupy the 1s level. If this situation is indeed the case, then

▼ **FIGURE 28.7 Energy levels of a multielectron atom**   **(a)** The shell–subshell ($n-\ell$) sequence shows that the energy levels are not evenly spaced and that the sequence of energy levels has numbers out of order. For example, the 4s level lies below the 3d level. The maximum number of electrons for a subshell, $2(2\ell + 1)$, is shown in parentheses on representative levels. (The vertical energy differences may not be drawn to scale.) **(b)** A convenient way to remember the energy-level order of a multielectron atom is to list the n-versus-ℓ values as shown here. The diagonal lines then give the energy levels in ascending order.

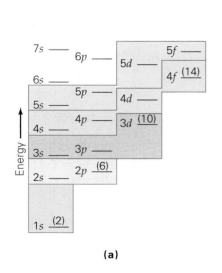

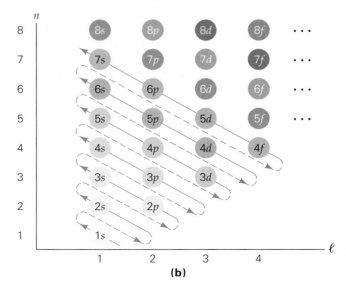

(a)

(b)

INSIGHT

Magnetic Resonance Imaging (MRI)

Magnetic resonance imaging, or MRI, originally known as NMR (for "nuclear magnetic resonance"), has become an increasingly common and important medical technique for the noninvasive examination of the human body (Fig. 1). It is based on the quantum-mechanical concept of spin.

A current-carrying loop in a magnetic field experiences a torque that tends to orient the loop's magnetic moment parallel to the field, much like a compass (Chapter 20). Electrons possess an intrinsic angular momentum called *spin*. This condition gives rise to a *spin magnetic moment*. When atoms are placed in a magnetic field, each of their energy levels is split into two levels, each with a slightly different energy, which results in the fine structure in the atoms' emission spectra.

Nuclei exhibit similar spin properties when placed in magnetic fields. For example, nuclei possess magnetic moments. To understand the basic principles of MRI operation, let us concentrate on the simplest atom, hydrogen, with a single proton for a nucleus. The magnetic resonance of hydrogen is commonly measured in an MRI apparatus, since hydrogen is the most abundant element in the human body. The spin angular momentum of the proton can take on only two values, similar to that of electron spin, called "spin up" and "spin down." They describe the orientation of the proton's magnetic moment relative to the direction of an external magnetic field (Fig. 2a). With spin up, the magnetic moment is parallel to the field, and with spin down, it points opposite to the field. The spin-down orientation has a slightly higher energy, and the energy difference between the two levels is proportional to the magnitude of the magnetic field. The transition to the higher level can be made by allowing the proton to absorb a photon with just this amount of energy, ΔE.

In a typical MRI apparatus, the sample (usually a region of the human body) is placed in a magnetic field **B**. The magnitude of the field determines the energy E (or frequency f) of the photon needed to cause a transition. This is

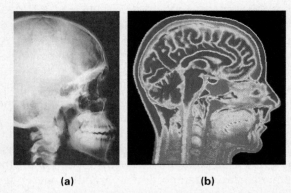

(a)   **(b)**

Figure 1 Diagnostic images **(a)** An X ray of a human head. **(b)** A magnetic resonance image (MRI) of a human head. The amount of detail captured, especially in the soft tissues of the brain, makes such images very useful for medical diagnosis.

because the absorption requires that $E = hf = \Delta E$, which is proportional to the strength of the magnetic field, B. The photons that trigger this transition are supplied by a pulsed beam of radio-frequency (RF) radiation applied to the sample. If the frequency of the radiation is adjusted so that the photon energy equals the energy difference between the levels, then many nuclei will absorb the photon energy and be excited into the higher, spin-down energy level. The frequency that creates such excitation is known as the *resonance frequency* ($f = \Delta E/h$), hence the name magnetic *resonance* imaging.

A diagram and photo of a typical MRI device are shown in Fig. 3. Large coils produce the magnetic field. (Notice the solenoid arrangement in Fig. 3a.) Other coils produce the RF signals ("RF photons") that cause the nuclei to "flip their spin," or be excited from the lower to the

---

we say that such a shell is *full*; all other electrons are excluded from it by Pauli's principle. Thus, for a Li atom, with three electrons, the third electron must occupy the next higher level (2s) when the atom is in the ground state. This case is illustrated in ▸Fig. 28.8, along with the ground-state energy levels for some other atoms.

Integrated Example 28.3 ■ The Quantum Shell Game: How Many States?

(a) How does the number of possible electron states compare between the 2p and the 4p subshells? (1) The number of states in the 2p subshell is greater than that in the 4p subshell; (2) the number of states in the 2p subshell is less than that in the 4p subshell; or (3) the number of states in the 2p subshell is the same as that in the 4p subshell. (b) Compare the number of possible electron states in the 3p subshell to the number in the 4d subshell.

(a) Conceptual Reasoning. The term "subshell" refers to the states that share the same ℓ quantum number within a shell. That is, they all have the same principal

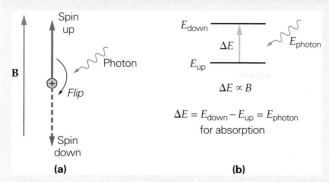

Figure 2 Nuclear spin **(a)** In a uniform magnetic field, the spin angular momentum, or spin magnetic moment, of a hydrogen nucleus (a proton) can have only two values—called "spin up" and "spin down," in reference to the direction of the external magnetic field. **(b)** This condition gives rise to two energy levels for the nucleus. Energy must be absorbed in order to "flip" the proton spin.

upper energy level. The resulting absorption of energy is detected, as is emitted radiation coming from a return transition to the lower state. Regions that produce the greatest absorption (or reemission) are those areas with the greatest concentration of the particular nucleus to which the apparatus is "tuned" by the choice of B and f. (Other nuclei besides hydrogen have their own characteristic intrinsic spins and can be imaged by changing the tuning to match their resonance, or "spin-flip," frequency.) Images are produced by means of computerized tomography, similar to that used in X-ray CT scans (see Fig. 20.25). The result is a two-

or three-dimensional image (Fig. 1b), which can provide a great deal of diagnostic medical information.

Although a variety of atoms or nuclei exhibits nuclear magnetic resonance, most MRI work is done with hydrogen, because of the varying water content of tissue. For example, muscle tissue has more water than does fatty tissue, so there is a distinct contrast in the radiation intensity of the two materials. Similarly, fatty deposits in blood vessels are perceived distinctly from the tissue of the vessel walls. A tumor with a water content different than that of the surrounding tissue would also show up in an MRI image.

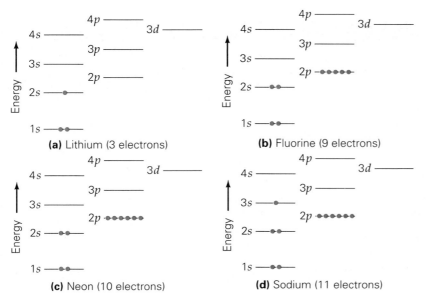

Figure 3 MRI **(a)** A diagram and **(b)** a photograph of the apparatus used for magnetic resonance imaging.

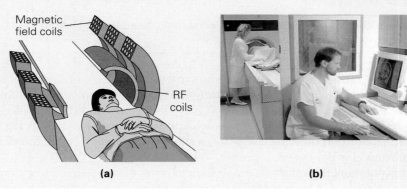

(a) Lithium (3 electrons)

(b) Fluorine (9 electrons)

(c) Neon (10 electrons)

(d) Sodium (11 electrons)

◄ **FIGURE 28.8 Filling subshells** The electron subshell distributions for several unexcited atoms (in their ground state) according to the Pauli exclusion principle. Because of spin, any s subshell can hold a maximum of two electrons, and any p subshell can hold a maximum of six electrons. What can you say about the d subshells?

quantum number n. All that matters is that their ℓ values are the same. Therefore, the number of states in the two subshells is the same, so the correct answer is (3).

(b) Thinking It Through. This part is a matter of following the quantum-mechanical counting rules. Also remember that each letter stands for a specific value of ℓ. The p level means that $\ell = 1$, and the d level means that $\ell = 2$. In each subshell, we replace the letter designation for ℓ by its number. Thus, the data given are as follows:

Given: $3p$ level means $n = 3$ and $\ell = 1$ *Find:*   the number of quantum states in $3p$
            $4d$ level means $n = 4$ and $\ell = 2$ subshell as compared with the
                                                    number in the $4d$ subshell

For a particular subshell, it is the ℓ value that determines the number of states. Recall that there are $(2\ell + 1)$ possible m_ℓ values for a given ℓ. Thus, for $\ell = 1$, there are $[(2 \times 1) + 1] = 3$ values for m_ℓ. (They are $+1$, 0, and -1.) Each of these values can have two m_s values $(\pm\frac{1}{2})$, making six different combinations of (n, ℓ, m_ℓ, m_s), or six states.

In general, the number of possible states for a given ℓ value is $2(2\ell + 1)$, taking into account the two possible "spin states" for each orbital state:

$$\text{Number of electron states for } \ell = 1 \text{ is } 2(2\ell + 1) = 2[(2 \times 1) + 1] = 6$$
$$\text{Number of electron states for } \ell = 2 \text{ is } 2(2\ell + 1) = 2[(2 \times 2) + 1] = 10$$

Comparison shows that the d subshell has more possible electron states than does the p subshell, regardless of which shell they are in (i.e., their n value).

These results are summarized in Table 28.3. Notice that the total number of states in a given *shell* (designated by n) is $2n^2$. For example, for the $n = 2$ shell, the total number of states for its combined s and p subshells ($\ell = 0, 1$) is $2n^2 = 2(2)^2 = 8$. This means that up to eight electrons can be accommodated in the $n = 2$ shell: two in the $2s$ subshell and six in the $2p$ subshell.

Follow-up Exercise. How many electrons could be accommodated in the $3d$ subshell if there were no spin quantum number?

Electron Configurations The electron structure of the ground state of atoms can be determined, so to speak, by putting an increasing number of electrons in the lower energy subshells [hydrogen (H), 1 electron; helium (He), 2 electrons; lithium (Li), 3 electrons; and so on], as was done for four elements in Fig. 28.8. However, rather than drawing diagrams, a shorthand notation called the **electron configuration** is widely used.

In this notation, the subshells are written in order of increasing energy, and the number of electrons in each subshell designated with a superscript. For example,

TABLE 28.3   Possible Sets of Quantum Numbers and States

Electron Shell n	Subshell ℓ	Subshell Notation	Orbitals (m_ℓ)	Number of Orbitals (m_ℓ) in Subshell $(2\ell + 1)$	Number of States m_s in Subshell $2(2\ell + 1)$	Total Electron States for Shell $2n^2$
1	0	$1s$	0	1	2	2
2	0	$2s$	0	1	2	8
	1	$2p$	$1, 0, -1$	3	6	
3	0	$3s$	0	1	2	
	1	$3p$	$1, 0, -1$	3	6	18
	2	$3d$	$2, 1, 0, -1, -2$	5	10	
4	0	$4s$	0	1	2	
	1	$4p$	$1, 0, -1$	3	6	32
	2	$4d$	$2, 1, 0, -1, -2$	5	10	
	3	$4f$	$3, 2, 1, 0, -1, -2, -3$	7	14	

$3p^5$ means that a $3p$ subshell is occupied by five electrons. The electron configurations for the atoms shown in Fig. 28.8 can thus be written as follows:

Li	(3 electrons)	$1s^2 2s^1$
F	(9 electrons)	$1s^2 2s^2 2p^5$
Ne	(10 electrons)	$1s^2 2s^2 2p^6$
Na	(11 electrons)	$1s^2 2s^2 2p^6 3s^1$

In writing an electron configuration, when one subshell is filled, you go on to the next higher one. The total of all the superscripts in any configuration must add up to the number of electrons in the atom.

The energy spacing between adjacent subshells is not uniform, as Figs. 28.7a and 28.8 show. In general, there are relatively large energy gaps between the s subshells and the subshells immediately below them. (Compare the $4s$ subshell with the $3p$ one in Fig. 28.7a.) The subshells just below the s subshells are usually p subshells, with the exception of the lowest subshell—the $1s$ subshell is below the $2s$ subshell. The gaps between other subshells—for example, between the $3s$ subshell and the $3p$ subshell above it, or between the $4d$ and $5p$ subshells—are considerably smaller.

This unevenness in energy differences gives rise to periodic large energy gaps, represented by vertical lines between certain subshells in the electron configuration:

$$1s^2 \mid 2s^2 2p^6 \mid 3s^2 3p^6 \mid 4s^2 3d^{10} 4p^6 \mid 5s^2 4d^{10} 5p^6 \mid 6s^2 4f^{14} 5d^{10} 6p^6 \mid \ldots$$

(*number of states*) (2) (8) (8) (18) (18) (32)

The subshells *between* the lines have only slightly different energies. The grouping of subshells (for example, $2s^2 2p^6$) that have about the same energy is referred to as an **electron period**.

Electron periods are the basis of the periodic table of elements. With your present knowledge of electron configurations, you are now in a position to understand the periodic table better than the person who originally developed it.

The Periodic Table of Elements

By 1860, over 60 chemical elements had been discovered. Several attempts had been made to classify the elements into some orderly arrangement, but none were satisfactory. It had been noted in the early 1800s that the elements could be listed in such a way that similar chemical properties recurred periodically throughout the list. With this idea, in 1869, a Russian chemist, Dmitri Mendeleev (pronounced men-duh-*lay*-eff), created an arrangement of the elements, based on this periodic property. The modern version of his **periodic table of elements** is used today and can be seen on the walls of just about every science building (▶ Fig. 28.9).

Mendeleev arranged the known elements in rows, called **periods**, in order of increasing atomic mass. When he came to an element that had chemical properties similar to those of one of the previous elements, he put this element below the previous similar one. In this manner, he formed both horizontal rows of elements and vertical columns called **groups**, or families of elements with similar chemical properties. The table was later rearranged in order of increasing atomic, or proton, number (the number of protons in the nucleus of an atom is the number at the top left of each of the element boxes in Fig. 28.9) in order to resolve some inconsistencies. Notice that if atomic masses were used, cobalt and nickel, atomic numbers 27 and 28, respectively, would fall in reversed columns.

With only 65 elements, there were vacant spaces in Mendeleev's table. The elements for these spaces were yet to be discovered. Because the missing elements were part of a sequence and had properties similar to those of other elements in a group, Mendeleev could predict their masses and chemical properties. Less than 20 years after Mendeleev devised his table, which showed chemists what to look for in order to find the undiscovered elements, three of the missing elements were, in fact, discovered.

▲ **FIGURE 28.9 The periodic table of elements** The elements are arranged in order of increasing atomic, or proton, number. Horizontal rows are called *periods*, and vertical columns are called *groups*. The elements in a group have similar chemical properties. Each atomic mass represents an average of that element's isotopes, weighted to reflect their relative abundance in our immediate environment. The masses have been rounded to two decimal places; more precise values are given in Appendices IV and V. (A value in parentheses represents the mass number of the best-known or longest-lived isotope of an unstable element. See Appendix IV for an alphabetical listing of elements.

Key:
- Atomic number — 26
- Symbol — Fe
- Atomic mass — 58.85
- Outer electron configuration — $3d^6 4s^2$

Transition elements (d block)

GROUP I	GROUP II											GROUP III	GROUP IV	GROUP V	GROUP VI	GROUP VII	GROUP VIII
1 H 1.01 $1s^1$																	2 He 4.00 $1s^2$
3 Li 6.94 $2s^1$	4 Be 9.01 $2s^2$											5 B 10.81 $2p^1$	6 C 12.01 $2p^2$	7 N 14.01 $2p^3$	8 O 16.00 $2p^4$	9 F 19.00 $2p^5$	10 Ne 20.18 $2p^6$
11 Na 22.99 $3s^1$	12 Mg 24.31 $3s^2$											13 Al 26.98 $3p^1$	14 Si 28.09 $3p^2$	15 P 30.97 $3p^3$	16 S 32.07 $3p^4$	17 Cl 35.45 $3p^5$	18 Ar 39.95 $3p^6$
19 K 39.10 $4s^1$	20 Ca 40.08 $4s^2$	21 Sc 44.96 $3d^1 4s^2$	22 Ti 47.88 $3d^2 4s^2$	23 V 50.94 $3d^3 4s^2$	24 Cr 52.00 $3d^5 4s^1$	25 Mn 54.94 $3d^5 4s^2$	26 Fe 55.85 $3d^6 4s^2$	27 Co 58.93 $3d^7 4s^2$	28 Ni 58.69 $3d^8 4s^2$	29 Cu 63.55 $3d^{10} 4s^1$	30 Zn 65.39 $3d^{10} 4s^2$	31 Ga 69.72 $4p^1$	32 Ge 72.61 $4p^2$	33 As 74.92 $4p^3$	34 Se 78.96 $4p^4$	35 Br 79.90 $4p^5$	36 Kr 83.80 $4p^6$
37 Rb 85.47 $5s^1$	38 Sr 87.62 $5s^2$	39 Y 88.96 $4d^1 5s^2$	40 Zr 91.22 $4d^2 5s^2$	41 Nb 92.91 $4d^4 5s^1$	42 Mo 95.94 $4d^5 5s^1$	43 Tc (98) $4d^5 5s^2$	44 Ru 101.07 $4d^7 5s^1$	45 Rh 102.91 $4d^8 5s^1$	46 Pd 106.42 $4d^{10} 5s^6$	47 Ag 107.87 $4d^{10} 5s^1$	48 Cd 112.41 $4d^{10} 5s^2$	49 In 114.82 $5p^1$	50 Sn 118.71 $5p^2$	51 Sb 121.76 $5p^3$	52 Te 127.60 $5p^4$	53 I 126.90 $5p^5$	54 Xe 131.29 $5p^6$
55 Cs 132.91 $6s^1$	56 Ba 137.33 $6s^2$	57 La 138.91 $5d^1 6s^2$ *	72 Hf 178.49 $5d^2 6s^2$	73 Ta 180.95 $5d^3 6s^2$	74 W 183.85 $5d^4 6s^2$	75 Re 186.21 $5d^5 6s^2$	76 Os 190.2 $5d^6 6s^2$	77 Ir 192.22 $5d^7 6s^2$	78 Pt 195.08 $5d^9 6s^1$	79 Au 196.97 $5d^{10} 6s^1$	80 Hg 200.59 $5d^{10} 6s^2$	81 Tl 204.36 $6p^1$	82 Pb 207.2 $6p^2$	83 Bi 208.98 $6p^3$	84 Po (209) $6p^4$	85 At (210) $6p^5$	86 Rn (222) $6p^6$
87 Fr (223) $7s^1$	88 Ra 226.03 $7s^2$	89 Ac 227.03 $6d^1 7s^2$ †	104 Rf (261) $6d^2 7s^2$	105 Db (262) $6d^3 7s^2$	106 Sg (263) $6d^4 7s^2$	107 Bh (264) $6d^5 7s^2$	108 Hs (265) $6d^6 7s^2$	109 Mt (268) $6d^7 7s^2$	110 (269)	111 (272)	112 (277)						

* (Lanthanides) — f block

58 Ce 140.12 $5d^1 4f^1 6s^2$	59 Pr 140.91 $4f^3 6s^2$	60 Nd 144.24 $4f^4 6s^2$	61 Pm (145) $4f^5 6s^2$	62 Sm 150.36 $4f^6 6s^2$	63 Eu 151.96 $4f^7 6s^2$	64 Gd 157.25 $5d^1 4f^7 6s^2$	65 Tb 158.93 $5d^1 4f^8 6s^2$	66 Dy 162.50 $4f^{10} 6s^2$	67 Ho 164.93 $4f^{11} 6s^2$	68 Er 167.26 $4f^{12} 6s^2$	69 Tm 168.93 $4f^{13} 6s^2$	70 Yb 173.04 $4f^{14} 6s^2$	71 Lu 174.97 $5d^1 4f^{14} 6s^2$

† (Actinides)

90 Th 232.04 $6d^2 7s^2$	91 Pa 231.04 $5f^2 6d^1 7s^2$	92 U 238.03 $5f^3 6d^1 7s^2$	93 Np 237.05 $5f^4 6d^1 7s^2$	94 Pu (244) $5f^6 6d^0 7s^2$	95 Am (243) $5f^7 6d^0 7s^2$	96 Cm (247) $5f^7 6d^1 7s^2$	97 Bk (247) $5f^9 6d^0 7s^2$	98 Cf (251) $5f^{10} 6d^0 7s^2$	99 Es (252) $5s^{11} 6d^0 7s^2$	100 Fm (257) $5f^{12} 6d^0 7s^2$	101 Md (258) $5f^{13} 6d^0 7s^2$	102 No (259) $5f^{14} 6d^0 7s^2$	103 Lr (260) $5f^{14} 6d^1 7s^2$

PERIODS 1–7

The periodic table puts the elements into seven horizontal rows, or periods. The first period has only two elements. Periods 2 and 3 each have 8 elements, and periods 4 and 5 each have 18 elements. Recall that the s, p, d, and f subshells can contain a maximum of 2, 6, 10, and 14 electrons $[2(2\ell + 1)]$, respectively. You should begin to see a correlation between these numbers and the arrangements of elements in the periodic table.

The periodicity of the periodic table can be understood in terms of the electron configurations of the atoms. For $n = 1$, the electrons are in one of two s states ($1s$); for $n = 2$, electrons can fill the $2s$ and $2p$ states, which gives a total of 10 electrons; and so on. Thus, the period number for a given element is equal to the highest n shell containing electrons in the atom. Notice the electron configurations for the elements in Fig. 28.9. Also, compare the electron periods given earlier, as defined by energy gaps, and the periods in the periodic table ($\blacktriangledown$Fig. 28.10). There is a one-to-one correlation, so the periodicity comes from energy-level considerations in atoms.

Chemists refer to *representative elements* as elements in which the last (least bound) electron enters an s or p subshell. In *transition elements*, the last electron enters a d subshell; and in *inner transition elements*, the last electron enters an f subshell. So that the periodic table is not unmanageably wide, the f subshell elements are usually placed in two rows at the bottom of the table. Each of the two rows is given a name— the *lanthanide series* and the *actinide series*—based on its position within the period.

Finally, we can also understand why elements in vertical columns, or groups, have similar chemical properties. The chemical properties of an atom, such as its ability to react and form compounds, depend almost entirely on the atom's outermost electrons—that is, those electrons in the outermost *unfilled* shell. It is these electrons, called *valence electrons*, that form chemical bonds with other atoms. Because of the way in which the elements are arranged in the table, the outermost electron configurations of all the atoms in any one group are similar. The atoms in such a group would thus be expected to have similar chemical properties, and they do. For example, notice the first two groups at the left of the table. They have one and two outermost electrons in an s subshell, respectively. These elements are all highly reactive metals that form compounds which have many similarities.

Shell (last to be filled)	Subshells	Number of electrons in subshell, $2(2\ell + 1)$	Corresponding period in periodic table
$n = 7$	$7p$	6	
	$6d$	10	Period 7
	$5f$	14	(32 elements)
	$7s$	2	
$n = 6$	$6p$	6	
	$5d$	10	Period 6
	$4f$	14	(32 elements)
	$6s$	2	
$n = 5$	$5p$	6	
	$4d$	10	Period 5
	$5s$	2	(18 elements)
$n = 4$	$4p$	6	
	$3d$	10	Period 4
	$4s$	2	(18 elements)
$n = 3$	$3p$	6	Period 3
	$3s$	2	(8 elements)
$n = 2$	$2p$	6	Period 2
	$2s$	2	(8 elements)
$n = 1$	$1s$	2	Period 1 (2 elements)

Energy ↑

◀ **FIGURE 28.10 Electron periods** The periods of the periodic table are related to electron configurations. The last n shell to be filled is equal to the period number. The electron periods and the corresponding periods of the table are defined by relatively large energy gaps between successive subshells (such as between $4s$ and $3p$) of the atoms.

The group at the far right, the noble gases, includes elements with completely filled subshells (and thus a full shell). These elements are at the ends of electron periods, or just before a large energy gap. These gases are nonreactive and can form compounds (by chemical bonding) only under very special conditions.

Conceptual Example 28.4 ■ Combining Atoms: Performing Chemistry on the Periodic Table

Combinations of atoms, called *molecules*, can form if atoms come together and share outer electrons. This sharing process is called *covalent bonding*. In this "shared-custody" scheme, both atoms find it energetically beneficial (that is, they lower their combined total energies) to have the equivalent of a filled outer shell of electrons, if only on a part-time basis. Using your knowledge of electron shells and the periodic chart, determine which of the following atoms would most likely form a covalent arrangement with oxygen: (a) neon (Ne); (b) calcium (Ca); or (c) hydrogen (H).

Reasoning and Answer. Choice (a), neon, with a total of 10 electrons, can be eliminated immediately, because it has a full outer shell of 8 electrons and, as such, has nothing to be gained by losing or adding electrons. Looking at the periodic table, we see that oxygen, with 6 outer-shell electrons, is 2 electrons shy of having a full complement of 8 electrons. It *could* occasionally have those 2 electrons by sharing electrons with another atom or atoms. Choice (b), calcium, with its 20 electrons, is 2 electrons beyond the previous full shell of 10. You might think, therefore, that calcium is a possible covalent partner. However, you must remember that the covalent arrangement is a two-way street. In other words, the arrangement would also require calcium sometimes to have two *more* electrons than normal.

This situation would put the calcium atom in the awkward position of being 4 electrons beyond the full shell of 10 and 14 electrons away from the next complete shell. Hence, even though this attempt at covalent bonding might seem to work for oxygen, it certainly won't work for calcium. The two species do, however, combine to form calcium oxide (CaO). The bonding that keeps calcium oxide together is based on the electrical attraction between the positive calcium ion, Ca^{+2}, and the negative oxygen ion, O^{-2}. In this type of bonding, the two atoms permanently exchange two electrons, making each a doubly charged ion of opposite signs. This bond is called an *ionic bond* and is not covalent. Thus, answer (b), calcium, is not correct.

The remaining candidate, hydrogen, has one electron fewer than a full shell of two. Thus, if a hydrogen atom could add one electron, it would attain an electron configuration like that of the lightest inert gas, helium. Since each hydrogen atom needs to share only one electron, two of them can accomplish this by sharing with a single oxygen atom. Part of the time, the hydrogen atoms must share their electron with the oxygen atom in order to create the latter's full outer shell of eight electrons. So, the correct answer is (c)—two hydrogen atoms covalently bound to a single oxygen atom. This combination has the molecular formula H_2O—water.

Follow-up Exercise. In this Example, (a) what would be the electron configuration of the oxygen in the water molecule at some instant when it has "custody" of the electrons from both hydrogen atoms? (b) What would be the net charge on the oxygen in this case?

28.4 The Heisenberg Uncertainty Principle

OBJECTIVE: To understand the inherent quantum-mechanical limits on the accuracy of physical observations.

An important aspect of quantum mechanics has to do with measurement and accuracy. In classical mechanics, there is no limit to the accuracy of a measurement. Theoretically, by continual refinement of a measurement instrument and procedure, the accuracy could be improved to any degree so as to give *exact* values. This theoretical approach results in a *deterministic* view of nature. For example, if both the position and the velocity of an object are known *exactly* at a particular time, you can determine exactly where the object will be in the future and where it was in the past (assuming that you know all the forces that act on it).

However, quantum theory predicts otherwise and sets limits on the accuracy of measurements. This idea was introduced in 1927 by the German physicist Werner Heisenberg, who had developed another approach to quantum mechanics that complemented Schrödinger's wave theory. The **Heisenberg uncertainty principle** as applied to position and momentum (or velocity) can be stated as follows:

It is impossible to know simultaneously an object's exact position and momentum.

This concept is often illustrated with a simple thought experiment. Suppose that you want to measure the position and momentum (actually, the velocity) of an electron. In order for you to "see," or locate, the electron, at least one photon must bounce off the electron and come to your eye (or detector), as illustrated in ▶Fig. 28.11. However, in the collision process, some of the photon's energy and momentum are transferred to the electron (as in the Compton effect, described in Section 27.3).

After the collision, the electron recoils. Thus, in the very process of locating the electron's position accurately, uncertainty is introduced into the electron's velocity (or momentum, since $\Delta\mathbf{p} = m\Delta\mathbf{v}$). This effect isn't noticed in our everyday macroscopic world, because the recoil produced by viewing an object with light is negligible. This is because the pressure exerted by the light cannot appreciably alter the motion or position of an object of everyday mass.

According to wave optics, the position of an electron can be measured *at best* to an uncertainty Δx of about the wavelength λ of the light used—that is, $\Delta x \approx \lambda$. The photon "particle" used for this location has a momentum of $p = h/\lambda$. Since the amount of momentum transferred during collision isn't determined, the final momentum of the electron would have an uncertainty on the order of the momentum of the photon, or $\Delta p \approx h/\lambda$.

Notice that the product of these two uncertainties is *at least* as large as h, since

$$(\Delta p)(\Delta x) \approx \left(\frac{h}{\lambda}\right)(\lambda) = h$$

This equation relates the *minimum* uncertainties, or maximum accuracies, of *simultaneous* measurements of the momentum and position. In actuality, the uncertainties could be worse, depending on the amount of light (number of photons) used, the apparatus, and the technique. Using more detailed considerations, Heisenberg found that the product of the two uncertainties would equal, *at a minimum*, $h/2\pi$. However, it could be higher. Hence, we can write the following:

$$(\Delta p)(\Delta x) \geq \frac{h}{2\pi} \qquad (28.5)$$

That is, the product of the *minimum* uncertainties of simultaneous momentum and position measurements is on the order of Planck's constant divided by 2π (that is, about 10^{-34} J·s).

In order to locate the position of the particle accurately (that is, to make Δx as small as possible), a photon with a very short wavelength must be used. However, this type of photon carries a lot of momentum, which results in an increased uncertainty in the momentum. To take the extreme case, if the location of a particle could be measured exactly (that is, $\Delta x \to 0$), we would have no idea about its momentum ($\Delta p \to \infty$). Thus, it is the measurement process itself that limits the accuracy to which position and momentum can be measured simultaneously. In Heisenberg's words, "Since the measuring device has been constructed by the observer ... we have to remember that what we observe is not nature in itself but nature exposed to our method of questioning."

To see how the Heisenberg uncertainty principle affects the microscopic and macroscopic worlds, consider the following Integrated Example.

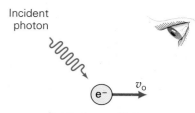

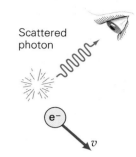

(a) Before collision

(b) After collision

▲ **FIGURE 28.11 Measurement-induced uncertainty** **(a)** To measure the position and momentum (or velocity) of an electron, at least one photon must collide with the electron and be scattered toward the eye or detector. **(b)** In the collision process, energy and momentum are transferred to the electron, which induces uncertainty in its velocity.

Heisenberg Uncertainty Principle for momentum and position

Integrated Example 28.5 ■ Electron versus Bullet: The Uncertainty Principle

An electron and a bullet both have the same speed, measured to the same accuracy. (a) How would their minimum uncertainties in position compare? (1) The electron's location would be more uncertain than that of the bullet; (2) the bullet's location would be more uncertain than that of the electron; or (3) their location uncertainties would be the same. (b) If the bullet's mass is 20.0 grams and both the bullet and the electron have a speed of 300 m/s, with an uncertainty of $\pm0.010\%$, determine the minimum uncertainty in the position of each.

(a) Conceptual Reasoning. The minimum uncertainty in location is related to the minimum uncertainty in momentum. Since both the bullet and the electron have the same uncertainty in velocity, the bullet's momentum is much more uncertain (momentum is proportional to mass, as is $\Delta \mathbf{p}$, since $\Delta \mathbf{p} = m\Delta \mathbf{v}$). Therefore, the bullet's location will be much less uncertain than that of the electron, so the answer is (1).

(b) Thinking It Through. The uncertainty principle (Eq. 28.5) can be solved for Δx in each case, because $\Delta \mathbf{p}$ can be determined from the uncertainty in velocity, $\Delta \mathbf{v}$. The electron is affected much more, because of its very small mass, as seen in part (a).

Listing the quantities, including the known electron mass, we have

Given: $m_e = 9.11 \times 10^{-31}$ kg
$m_b = 20$ g $= 0.020$ kg
$v_b = v_e = 300$ m/s $\pm 0.010\%$

Find: Δx_e and Δx_b (minimum uncertainties in position)

The uncertainty in speed is 0.01% (or 0.00010) for both the electron and the bullet. This uncertainty is

$$(300 \text{ m/s})(0.00010) = 0.030 \text{ m/s}$$

so

$$v = 300 \text{ m/s} \pm 0.010\% = 300 \text{ m/s} \pm 0.030 \text{ m/s}$$

The *total* uncertainty in speed is *twice* this amount, because the measurements can be off by 0.010% above or below the measured values; hence, $\Delta v = 0.060$ m/s.

For the electron, the minimum uncertainty in position is

$$\Delta x_e = \frac{h}{2\pi\Delta p} = \frac{h}{2\pi m_e \Delta v} = \frac{6.63 \times 10^{-34} \text{ J} \cdot \text{s}}{2\pi(9.11 \times 10^{-31} \text{ kg})(0.060 \text{ m/s})} = 0.0019 \text{ m} = 1.9 \text{ mm}$$

Similarly, the bullet's minimum uncertainty in position is

$$\Delta x_b = \frac{h}{2\pi m_b \Delta v} = \frac{6.63 \times 10^{-34} \text{ J} \cdot \text{s}}{2\pi(0.020 \text{ kg})(0.060 \text{ m/s})} = 8.8 \times 10^{-32} \text{ m}$$

Notice that the uncertainty in the bullet's position is many times smaller than the diameter of a nucleus. Its position uncertainly is also many orders of magnitude less than that of the electron. The lesson is that uncertainty in location for everyday objects traveling at ordinary speeds is negligible. However, for electrons, the 1.9 mm is significant and measurable.

Follow-up Exercise. In this Example, what would the minimum uncertainty in the electron's speed have to be for the minimum uncertainty in its position to be on the order of atomic dimensions, or 0.10 nm?

An equivalent form of the uncertainty principle relates uncertainties in energy and time. As with the position and momentum, a detailed analysis shows that, *at best*, the product of the uncertainties in energy and time is $h/2\pi$. Of course, it could be larger; hence, this form of the uncertainty principle is written as

Heisenberg Uncertainty Principle
for energy and time

$$(\Delta E)(\Delta t) \geq \frac{h}{2\pi} \tag{28.6}$$

This form of the Heisenberg uncertainty principle shows that the energy of an object may be uncertain by an amount ΔE, depending on the time taken to measure it, Δt. For longer times, the energy measurement becomes increasingly more accurate. Once again, these uncertainties are important only for very light objects. Such uncertainties are of particular importance in nuclear physics and elementary particle interactions (Chapter 30).

Notice that the energy of a particle cannot be measured exactly unless an infinite amount of time is taken to do so. If a measurement of energy is carried out in a time Δt, then the energy is uncertain by an amount ΔE. For example, the measurement of the frequency of light emitted by an atom is really the measurement of the energy of the photon associated with the transition from an excited state to the ground state. The measurement must be carried out in an amount of time comparable with the lifetime of the excited state. As a result, the observed emission line (Sect. 27.4) has a nonzero energy width, since $\Delta E = h\Delta f$ is *not* zero (▶Fig. 28.12). This *natural broadening* is generally small for atomic emissions and was therefore ignored in Chapter 27, where spectral lines were considered to have exact frequencies.

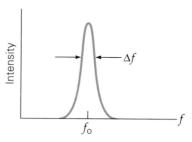

▲ **FIGURE 28.12 Natural line broadening** Because a measurement must be carried out in an amount of time comparable with the lifetime (Δt) of an excited atomic state, the energy of that state is uncertain by an amount $\Delta E = h\Delta f$. The observed emission line has a width of Δf, rather than being a line of single frequency f_o with zero width.

28.5 Particles and Antiparticles

OBJECTIVES: To understand (a) the relationship between particles and antiparticles and (b) the energy requirements for pair production.

When the British physicist Paul A. M. Dirac, in 1928, extended quantum mechanics to include relativistic considerations, something new and very different was predicted—a particle called the **positron**. The positron was predicted to have the same mass as the electron, but to carry a *positive* charge. The oppositely charged positron is the **antiparticle** of the electron. (Antiparticles will be discussed in more depth in Chapters 29 and 30.)

The positron was first observed experimentally in 1932 by the American physicist C. D. Anderson in cloud-chamber experiments with cosmic rays. The curvature of the particle tracks in a magnetic field showed two types of particles, with opposite charge and the same mass (▶Fig. 28.13). Anderson had, in fact, discovered the positron.

Because electric charge is conserved, a positron can be created only with the simultaneous creation of an electron (so that the net charge created is zero). This process is called **pair production**. In Anderson's experiment, positrons were observed to be emitted from a thin lead plate exposed to cosmic rays from outer space, which contain highly energetic X rays. Pair production occurs when an X-ray photon comes near a nucleus—the nuclei of the lead atoms of the plate in the Anderson experiment. In this process, the photon goes out of existence, and an "electron–positron pair" (an electron and a positron) is created, as illustrated in ▶Fig. 28.14. This result represents a direct conversion of electromagnetic (photon) energy into mass. By the conservation of energy (neglecting the small recoil kinetic energy of the massive nearly nucleus),

$$hf = 2m_e c^2 + K_{e^-} + K_{e^+}$$

where hf is energy of the photon, $2m_e c^2$ is the total mass–energy equivalent of the electron–positron pair, and the K's are the kinetic energies of the particles and the recoil nucleus.

The minimum energy to produce such a pair occurs when they are produced at rest—when K_{e^-} and K_{e^+} are zero. Thus, the *minimum* photon energy to produce an electron—positron pair is

$$E_{min} = hf = 2m_e c^2 = 1.022 \text{ MeV} \tag{28.7}$$

(Here, we have used the result from Section 26.5 which says that the mass of an electron is equivalent to an energy of $m_e c^2 = 0.511$ MeV.) This minimum photon energy is called the **threshold energy** for pair production.

But if they are created by cosmic rays, why aren't positrons commonly found in nature? This is because, almost immediately after their creation, positrons go out of existence by a process called **pair annihilation**. When an energetic positron appears, it loses kinetic energy by collision as it passes through matter. Finally, almost at rest, it combines with an electron and forms a hydrogenlike

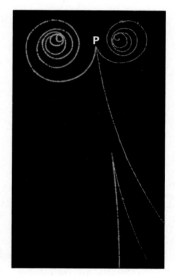

▲ **FIGURE 28.13 Cloud-chamber photograph of pair production** In this false-color cloud-chamber photograph, a gamma-ray photon (not visible, but it enters the region in a downward direction) interacts with a nearby atomic nucleus (point P) to produce an electron and a positron (green and red spiral tracks, respectively, at the top). In the process, the photon also dislodges an orbital electron (the nearly straight and vertical green track). An external magnetic field causes the electron and positron to be deflected in paths of opposite curvature. A similar event is recorded in the bottom half of the photo. (Why might the paths of the particles created in this case show less deflection?)

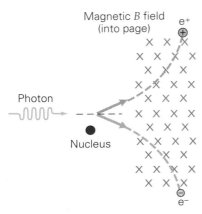

▲ FIGURE 28.14 Pair production
An electron (e⁻) and a positron (e⁺) can be created when an energetic photon passes near a heavy nucleus.

atom, called a *positronium atom*, in which a positron substitutes for a proton. The positronium atom is unstable and quickly decays (≈10^{-10} s) into two photons, each with an energy of 0.511 MeV (▼Fig. 28.15). Pair annihilation is then a direct conversion of mass into electromagnetic energy—the inverse of pair production, so to speak. Pair annihilation is the basis for a medical diagnostic tool called a *positron emission tomography* (or PET) *scan*. This application will be discussed in more detail in Chapter 29 after you learn more about nuclear decay processes.

More generally, all particles have antiparticles. For example, there is an antiproton with the same mass as a proton, but with a negative charge. Even a neutral particle such as the neutron has an antiparticle—the antineutron. It is even conceivable that antiparticles predominate in some parts of the universe. If so, atoms made of the **antimatter** in these regions would consist of negatively charged nuclei composed of antiprotons and antineutrons, surrounded by orbiting positively charged positrons (antielectrons). It would be difficult to distinguish a region of antimatter visibly, since the physical behavior of antimatter atoms would presumably be the same as that of ordinary matter.

▼ FIGURE 28.15 Pair annihilation A slow positron and electron can form a system called a *positronium atom*. The disappearance of a positronium atom is signaled by the appearance of two photons, each with an energy of 0.511 MeV. Why would we not expect one photon with an energy of 1.022 MeV? ("CM" stands for "center of mass.")

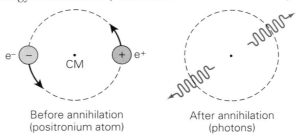

Before annihilation
(positronium atom)

After annihilation
(photons)

Chapter Review

Important Concepts

- The **de Broglie hypothesis** assigns a wavelength to material particles, by analogy with the assignment of momentum to light particles (photons). The **de Broglie wavelength** of a particle is inversely related to its momentum by

$$\lambda = \frac{h}{p} = \frac{h}{mv} \qquad (28.2)$$

- A quantum-mechanical **wave function** ψ is a "probability wave" associated with a particle. The Schrödinger wave equation enables us to calculate the wave function in different situations. The probability density, the square of the wave function, gives the relative probability of finding a particle at a particular location.

- Electron-orbital energies are determined primarily by the **principal quantum number** n, where $n = 1, 2, 3, \ldots$.

- The quantum number ℓ is called the **orbital quantum number** and is associated with the orbital an-

gular momentum of the electron's orbit. For each value of n, the ℓ quantum number has one of n possible integer values from zero up to a maximum value of $n - 1$.

- The quantum number m_ℓ is called the **magnetic quantum number** and is associated with the z-component of the electron's orbital angular momentum. For a given ℓ, there are $2\ell + 1$ possible m_ℓ values (integers), given by $m_\ell = 0, \pm 1, \pm 2, \ldots, \pm \ell$.

- The quantum number that describes an electron's intrinsic angular momentum is the **spin quantum number** m_s, which, for electrons, protons, and neutrons, can have only two values: $m_s = \pm\frac{1}{2}$. These values correspond to the two angular momentum directions of spin: up and down.

- Orbits with different quantum numbers that have the same energy are said to be **degenerate**.

- Orbits that share a common principal quantum number n are said to be in the same **shell**.

- Orbits that share a common orbital quantum number ℓ are said to be in the same **subshell**.

- The **Pauli exclusion principle** states that in a given atom, no two electrons can have exactly the same set of quantum numbers.

- The **Heisenberg uncertainty principle** states that you cannot simultaneously measure both the position and the momentum (or velocity) of a particle exactly. The same condition holds true for the particle's energy and the period of time during which that energy is measured. The uncertainties satisfy

$$(\Delta p)(\Delta x) \geq \frac{h}{2\pi} \qquad (28.5)$$

and

$$(\Delta E)(\Delta t) \geq \frac{h}{2\pi} \qquad (28.6)$$

- **Pair production** refers to the creation of a particle and its **antiparticle**. The reverse process is **pair annihilation**, in which a particle and antiparticle annihilate and their energy is converted into two photons.

Exercises*

28.1 Matter Waves: The de Broglie Hypothesis

1. The momentum of a photon is (a) zero, (b) equal to c, (c) proportional to its frequency, (d) proportional to its wavelength, or (e) given by the de Broglie hypothesis.

2. The Davisson–Germer experiment (a) was concerned with X-ray spectra, (b) verified the Heisenberg principle, (c) supported the Pauli exclusion principle, or (d) demonstrated the wavelike properties of electrons.

3. **CQ** The de Broglie hypothesis predicts that there is a wave associated with any object that has momentum. Why don't we observe the wave associated with a moving car?

4. **CQ** If a baseball and a bowling ball were traveling at the same speed, which one would have a longer de Broglie wavelength? Why?

5. **CQ** An electron is accelerated from rest through an electric-potential difference. Will increasing the difference in potential result in a longer or shorter de Broglie wavelength? Why?

6. ■ What is the de Broglie wavelength associated with a 1000-kg car moving at 25 m/s?

7. **IE** ■ An electron and a proton are moving with the same speed. (a) Compared with the proton, will the electron have (1) a shorter, (2) an equal, or (3) a longer de Broglie wavelength? Why? (b) If the speed of the electron and proton is 100 m/s, what are their de Broglie wavelengths?

8. ■ Calculate the de Broglie wavelength of a 70-kg person running at a speed of 2.0 m/s.

9. ■■ A proton and an electron are accelerated from rest through the same difference in potential, V. What is

*Note: Use Eq. 28.3 for λ where appropriate.

ratio of the de Broglie wavelength of an electron to that of a proton (to two significant figures)?

10. **IE** ■■ Electrons are accelerated from rest through an electric-potential difference. If this potential difference increases to nine times the original value, the new de Broglie wavelength will be (1) nine, (2) three, (3) one-ninth, or (4) one-third times that of the original. Why? (b) If the original potential is 250 kV and the new potential is 600 kV, what is the ratio of the new de Broglie wavelength to the original?

11. ■■ An electron is accelerated from rest through a potential difference so that its de Broglie wavelength is 0.010 nm. What is the potential difference?

12. ■■ A charged particle is accelerated through a potential difference V. By what factor would its de Broglie wavelength change if the voltage were doubled?

13. **IE** ■■ A proton traveling at a speed of 4.5 ... Why? accelerated through a potential difference ...ngth of its de Broglie wavelength (1) increase, (2 ... or (3) decrease, due to the potenti... (b) By what percentage does the ...ctrons that exthe proton change? ...ngle of 25° when ...a lattice plane spac-

14. ■■ What is the energ... hibits a first-order... diffracted by a ...avelength of the Earth in its ing of 0.15 n... ...me a circular orbit.)

...e Bohr theory of the hydrogen atom,

15. ■■ W... electron in the first Bohr orbit is or... (a) What is the wavelength of the matter ...ated with the electron? (b) How does this ...th compare with the circumference of the first ...bit?

17. ■■ A scientist wants to use an electron microscope to observe details on the order of 0.25 nm. Through what potential difference must the electrons be accelerated from rest so that they have a de Broglie wavelength of this magnitude?

28.2 The Schrödinger Wave Equation

18. The wave-function solution to the Schrödinger equation (a) can never be found, (b) is the probability of finding a particle, (c) functionally describes the de Broglie wave of a particle, or (d) none of the preceding.

19. The square of a particle's wave function is interpreted as being (a) the energy of the particle, (b) the probability of locating the particle, (c) the quantum number of a state, or (d) the basis of the Pauli exclusion principle.

20. ■■■ A particle in a box is constrained to move in one dimension, like a bead on a wire, as illustrated in ▼Fig. 28.16. Assuming that no forces act on the particle in the interval $0 < x < L$ and that it hits a perfectly rigid wall, it may be thought of as a particle in an infinitely deep well with $U = 0$. (a) Show that the spatial wave function for the particle is $\psi_n = A \sin(n\pi x/L)$ for $n = 1, 2, 3, \ldots$, where A is the wave function amplitude. (b) Show that the kinetic energy of the particle is given by $K_n = n^2[h^2/(8mL^2)]$, where m is the particles mass. [Hint: (a) Consider boundary conditions like those of a standing wave in a string tied at both ends. (b) Recall that $K_n = p^2/(2m)$ and that the wave function involves the de Broglie wavelength.]

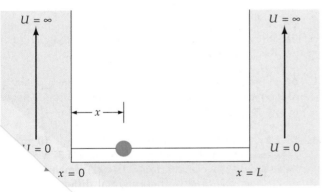

21. ■■ Particle in a box See Exercises 20 and 21.
dens.
states.
the $n = $ ᵇ wave functions and the probability
ᵉ in Exercise 20 for the first three
rticle most likely to be found for

28.3 Atomic Q. Periodic Table

22. The ℓ quantum num. mines the total energy **ers and the** angular momentum of the the orientation of the angula. associated with the electron spᵗom (a) deter-
ᵈd with the
'ed with
(d) is

23. The quantum number m_ℓ (a) determines the energy of the electron, (b) tells whether the electron is spinning up or down, (c) tells the orientation of the angular-momentum vector of the electron in orbit, or (d) all of the preceding.

24. The quantum number m_s (a) is purely a quantum-mechanical concept, (b) arises from the orbital motion of an electron, (c) is due to actual electron spin, or (d) all of the preceding.

25. CQ What information does the quantum number n give for a hydrogen atom?

26. CQ Niels Bohr set forth a *correspondence principle*, which states that the results of quantum mechanics and classical physics become more and more in agreement when quantum numbers become very large. Discuss this principle in terms of the hydrogen atom.

27. CQ What is the basis of the periodic table of elements in terms of quantum theory, and what do the elements in a particular group have in common?

28. ■ (a) How many possible sets of quantum numbers are there for $n = 2$ and $n = 3$ shells? (b) Write the explicit values of all the quantum numbers (n, ℓ, m_ℓ, m_s) for these levels.

29. ■ How many possible sets of quantum numbers are there for the subshells with (a) $\ell = 0$ and (b) $\ell = 3$?

30. IE ■ (a) Which has more possible sets of quantum numbers associated with it, $n = 2$ or $\ell = 2$? (b) Prove your answer to part (a).

31. ■ An electron in an atom is in an orbit that has a magnetic quantum number of $m_\ell = 2$. What are the minimum values that (a) ℓ and (b) n could be for that orbit?

32. ■■ Draw the ground-state energy-level diagrams like those in Fig. 28.8 for (a) nitrogen (N) and (b) potassium (K).

33. ■■ Draw schematic diagrams for the electrons in the subshells of (a) sodium (Na) and (b) argon (Ar) atoms in the ground state.

34. ■■ Identify the atoms of each of the following ground-state electron configurations: (a) $1s^2 2s^2$; (b) $1s^2 2s^2 2p^3$; (c) $1s^2 2s^2 2p^6$; (d) $1s^2 2s^2 2p^6 3s^2 3p^4$.

35. ■■ Write the ground-state electron configurations for each of the following atoms: (a) boron (B), (b) calcium (Ca), (c) zinc (Zn), and (d) tin (Sn).

36. IE ■■■ (a) If there were no electron spin, the $1s$ state would contain a maximum of (1) zero, (2) one, or (3) two electrons. Why? (b) What would be the first two inert or noble gases if there were no electron spin?

37. ■■■ How would the electronic structure of lithium differ if electron spin were to have three possible orientations instead of just two?

28.4 The Heisenberg Uncertainty Principle

38. If the uncertainty in the position of a moving particle increases, (a) the particle may be located more exactly, (b) the uncertainty in its momentum decreases, (c) the uncertainty in its velocity increases, or (d) none of the preceding.

39. According to the uncertainty principle, measurement of the exact energy of a particle requires (a) special equipment, (b) an infinite amount of time, (c) uncertainty in the momentum, or (d) none of the preceding.

40. CQ Why is it impossible to simultaneously and accurately measure the position and velocity of a particle?

41. CQ A bowling ball has well-defined position and speed, whereas an electron does not. Why?

42. ■ A 0.50-kg ball has a position of 5.0 m ± 0.01 m. To what minimum uncertainty can its momentum be measured?

43. IE ■ An electron and a proton each have a momentum of 3.28470×10^{-30} kg·m/s ± 0.00025×10^{-30} kg·m/s. (a) The minimum uncertainty in the position of the electron compared with that of the proton will be (1) larger, (2) the same, or (3) smaller. Why? (b) Calculate the minimum uncertainty in the position for each.

44. ■ What is the minimum uncertainty in the velocity of an electron that is known to be somewhere between 0.050 nm and 0.10 nm from a proton?

45. ■ What is the minimum uncertainty in the velocity of a 0.50-kg ball that is known to be at 1.0000 cm ± 0.0005 cm from the edge of a table?

46. ■■ The energy of a 2.00-keV electron is known to within ±3.00%. How accurately can its position be measured?

47. ■■ If an excited state of an atom has a lifetime of 1.0×10^{-7} s, what is the minimum error associated with the measurement of the energy of this state?

48. ■■ The energy of the first excited state of a hydrogen atom is −0.34 eV ± 0.0003 eV. What is the average lifetime for this state?

49. IE ■■ (a) If the lifetime of excited state A is longer than that of state B, then the width of a spectral line due to natural broadening for a transition from state A to state B will be (1) smaller than, (2) the same as, or (3) greater than that for a transition from state B. Why? (b) Calculate the ratio of the width of a spectral line due to natural broadening for a transition from an excited state with a lifetime of 10^{-12} s to that for a state with a lifetime of 10^{-8} s.

28.5 Particles and Antiparticles

50. Pair production involves (a) the production of two electrons, (b) the production of two positrons, (c) a positronium atom, or (d) a certain threshold energy.

51. Due to momentum considerations, pair annihilation cannot result in the emission of how many photons, (a) one, (b) two, or (c) several?

52. CQ It has been suggested in science fiction that matter and antimatter could be combined as a source of energy. (The starship Enterprise on Star Trek had antimatter engines, with the antimatter being stored in "pods.") Speculate if this scenario is possible.

53. IE ■ A photon with an energy of 1.04 MeV comes near a heavy nucleus. (a) Will pair production result? (b) What is the frequency of the photon?

54. ■ What is the energy of the photons produced in electron–positron pair annihilation, assuming that both particles are essentially at rest initially?

55. ■ What is the threshold energy for the production of a proton–antiproton pair?

56. IE ■■ A muon, or μ meson, has the same charge as an electron, but is 207 times more massive. (a) Compared with electron–position pair production, the pair production of a muon and an antimuon requires a photon of (1) more, (2) the same amount of, or (3) less energy. Why? (b) What would be the minimum energy for such a photon?

Additional Exercises

57. If the spacing between the lattice planes of a crystal is 0.19 nm, what voltage is needed to accelerate electrons (from rest) so that they exhibit a first-order diffraction maximum at an angle of 45°?

58. With what accuracy would you have to measure the position of a moving electron so that its velocity is uncertain by 0.500 cm/s?

59. A gamma-ray photon has an energy of 7.5 MeV. What are (a) its momentum and (b) the wavelength of the light wave with which it is associated?

60. An electron is accelerated from rest through a potential of 5.00 kV ± 3.0%. What is the minimum uncertainty in the position of the electron after the acceleration?

61. Show that, for a particle moving in one dimension between fixed boundaries $-L \le x \le L$, the wave functions are given by $\psi_n = A \cos(n\pi x/2L)$, where $n = 1, 3, 5, \ldots$, and $\psi_n = A \sin(n\pi x/2L)$, where $n = 2, 4, 6, \ldots$, and A is the wave function's amplitude.

62. Show that the allowed kinetic energies in Exercise 61 are four times smaller than those in Exercise 20. Why is this the case?

63. In Exercise 61, where is the particle most likely to be found in its ground state?

64. What is the de Broglie wavelength for the matter wave associated with (a) a 150-g ball thrown at 20 m/s and (b) a 1200-kg automobile traveling at 90 km/h?

The Nucleus

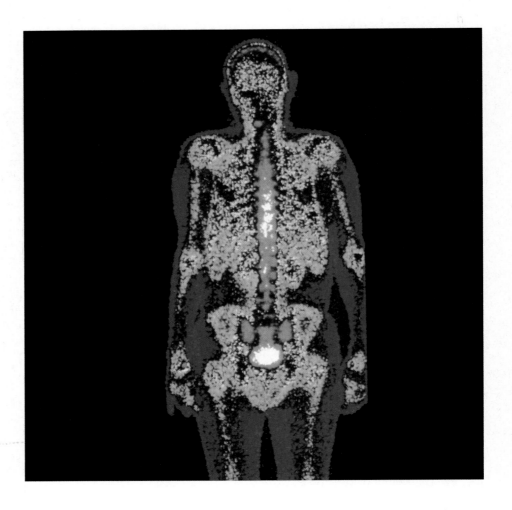

This skeletal image (a bone scan) was created by radiation from a radioactive source. *Radiation* and *radioactivity* are words that produce anxiety, but we often overlook the many beneficial uses of radiation. For instance, exposure to high-energy radiation can cause cancer—yet, precisely the same sort of radiation can be useful in the diagnosis and treatment of cancer.

The bone scan shown in the opening photo was created by radiation released in the body when unstable nuclei spontaneously broke apart—a process we call *radioactive decay*. But what makes some nuclei stable, while others are able to decay? For radioactive nu-

clei, what determines the rate at which they break down and the particles that they emit when they do decay? These are some of the questions we'll explore in this chapter. We'll also learn how radiation is detected and measured, as well as more about its dangers and uses.

In addition, the study of radioactivity and nuclear stability gives us some general ideas about the nature of the nucleus, the energy it possesses, and how this energy can be released. The release of nuclear energy has become one of our major energy sources, which will be considered in Chapter 30. First, let's take a look at the nucleus itself.

29.1 Nuclear Structure and the Nuclear Force

OBJECTIVES: To **(a)** distinguish between the Thomson and Rutherford–Bohr models of the atom, **(b)** specify some of the basic properties of the strong nuclear force, and **(c)** understand nuclear notation.

It is evident from the emission of electrons from heated filaments (*thermionic emission*) and the photoelectric effect that atoms contain electrons. Since an atom is normally electrically neutral, it must contain a positive charge equal in magnitude to the total charge on all the electrons in the atom. Also, since the electron's mass is small compared with the mass of even the lightest of atoms, most of the atomic mass appears to be associated with that positive charge.

Based on these observations, Sir J. J. Thomson (1856–1940), a British physicist who had experimentally proven the existence of the electron in 1897, proposed a model of the atom. In the Thomson model, the negatively charged electrons are uniformly distributed within a continuous sphere of positive charge. It was called the "plum pudding" model, because the electrons in the positive charge are analogous to the raisins in a plum pudding. The region of positive charge was assumed to have a radius on the order of 10^{-10} m, or 0.1 nm, based on calculations of the bulk properties of matter.

Our modern model of atomic structure is quite different. This model pictures all of an atom's positive charge, and practically all of its mass, as concentrated in a central "nucleus," which is surrounded by the orbiting negatively charged electrons. The concept of an atomic nucleus was proposed by the British physicist Ernest Rutherford (1871–1937). Combined with Bohr's theory of orbiting electrons (Section 27.4), this idea led to the simplistic "solar system" model, or **Rutherford–Bohr model**, of the atom.

Rutherford's insight came from the results of alpha-particle scattering experiments performed in his laboratory about 1911. An alpha (α) particle is a doubly positively charged particle ($q_\alpha = +2e$) that is naturally emitted from some radioactive materials. (See Section 29.2.) A beam of these alpha particles was directed at a thin gold-foil "target," and the deflection angles of the scattered particles were observed (▼Fig. 29.1).

An alpha particle is over 7000 times more massive than an electron. Thus, the Thomson model predicts only small deflections—the result of collisions with the very light electrons as an alpha particle passes through a gold atom (▶Fig. 29.2). Surprisingly, however, Rutherford and his colleagues observed alpha particles scattered at appreciable angles. In about 1 in every 8000 scatterings, the alpha particles were actually *backscattered*; that is, they were scattered through angles greater than 90° (▶Fig. 29.3).

PHYSLET®
ILLUSTRATION

Rutherford Scattering

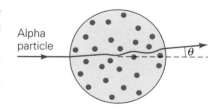

▲ **FIGURE 29.2   The plumpudding model**   In Thomson's plum-pudding model of the atom, massive alpha particles were expected to be only slightly deflected by collisions with the electrons (blue dots) in the atom. The experimental results were quite different.

▼ **FIGURE 29.1   Rutherford's scattering experiment**   A beam of alpha particles from a radioactive source was scattered by gold nuclei in a thin foil, and the scattering was observed as a function of the scattering angle θ. The observer detects the light (viewed through a lens) given off by a scintillating phosphorescent screen.

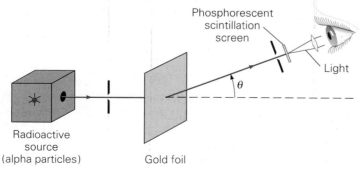

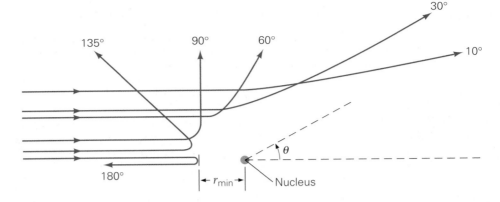

◀ **FIGURE 29.3 Rutherford scattering** A compact, dense atomic nucleus with a positive charge accounts for the observed scattering. An alpha particle in a head-on collision with the nucleus would be scattered directly backward ($\theta = 180°$) after coming within a distance r_{min} of the nucleus. At this scale, the electron orbits (about the nucleus) are too far away to be seen.

Calculations showed that the probability of backscattering taking place with a Thomson model was miniscule—much, much less than 1 in 8000. As Rutherford described the backscattering, "It was almost as incredible as if you had fired a 15-inch shell at a piece of tissue paper and it came back and hit you."

The experimental results led Rutherford to the concept of an atomic nucleus: "On consideration, I realized that this scattering backward must be the result of a single collision, and when I made calculations I saw that it was impossible to get anything of that order of magnitude unless you took a system in which the greater part of the mass of the atom was concentrated in a minute nucleus. It was then that I had the idea of an atom with a minute massive center carrying a charge."

If all of the positive charge of a target atom were concentrated in a very small region, then an alpha particle coming close to this region would experience a large deflecting (electrical repulsion) force. The mass of this "nucleus" of positive charge would be larger than that of the alpha particle, since most of the atomic mass is associated with the positive charge. Thus, in this model, backscattering is more likely to occur, because the positive charge is concentrated in a small nucleus rather than being spread throughout the atom.

A simple estimate can give an idea of the size of the nucleus. It is during a head-on collision that an alpha particle comes closest to the nucleus (a distance labeled as r_{min} in Fig. 29.3). That is, the alpha particle approaching the nucleus stops at r_{min} and is accelerated back along its original path. Assuming a spherical charge distribution, the electric potential energy of the alpha particle (α) and nucleus (n) when separated by a center-to-center distance r is $U_e = kq_\alpha q_n/r = k(2e)(Ze)/r$ (Eq. 16.5). Here, Z is the **atomic number**, or the number of protons in the nucleus, and therefore the total charge of the nucleus is $q_n = +Ze$. By conservation of energy, the kinetic energy of the incoming alpha particle is completely converted into electric potential energy at the turnaround point, r_{min}. Using $q_\alpha = +2e$, we equate the two energies, which results in

$$\tfrac{1}{2}mv^2 = \frac{k(2e)Ze}{r_{min}}$$

or, solving for r_{min}, we obtain

$$r_{min} = \frac{4kZe^2}{mv^2} \tag{29.1}$$

In Rutherford's experiment, the kinetic energy of the alpha particles from the particular source had been measured, and Z was known to be 79 for gold. Using these values, along with the constants in Eq. 29.1, Rutherford found r_{min} to be on the order of 10^{-14} m.

Although the nuclear model of the atom is useful, the nucleus is much more than a simple volume of positive charge. The nucleus is actually composed of two types of particles—protons and neutrons—collectively referred to as **nucleons**. The nucleus of the hydrogen atom is a single proton. Rutherford suggested that the hydrogen nucleus be named *proton* (from the Greek word meaning "first") after he became convinced that no nucleus could be less massive than the hydrogen nucleus. A neutron is an electrically neutral particle with a mass slightly greater than that of a proton. The existence of the neutron was not experimentally verified until 1932.

The Nuclear Force

Of the forces in the nucleus, there is certainly the attractive gravitational force between nucleons. But in Chapter 15, this gravitational force was shown to be negligible compared with the repulsive electrical force between the positively charged protons. Taking only these repulsive forces into account, we would predict that the nucleus will fly apart. Yet, the nuclei of most atoms are stable. Therefore, there must be an additional attractive force between nucleons that holds the nucleus together. This strong attractive force is called the **strong nuclear force**, or simply the *nuclear force*.

The exact expression for the nuclear force is extremely complex. However, some general features of this force are as follows:

- The nuclear force is strongly attractive and much larger in magnitude than both the electrostatic force and the gravitational force between nucleons.
- The nuclear force is very short ranged; that is, a nucleon interacts only with its nearest neighbors, over distances on the order of 10^{-15} m.
- The nuclear force is independent of electric charge; that is, it acts between *any* two nucleons—two protons, a proton and a neutron, or two neutrons.

Thus, nearby protons repel each other electrically, but attract each other (and nearby neutrons) by the strong nuclear force, with the latter winning the battle. Having no electric charge, neutrons only attract nearby protons and neutrons.

Nuclear Notation

To describe the nuclei of different atoms, it is convenient to use the notation illustrated in ▶ Fig. 29.4a. The chemical symbol of the element is used with subscripts and a superscript. The subscript on the left is called the *atomic number* (Z), which indicates the number of protons in the nucleus. A more descriptive name for the symbol Z is **proton number**, which will be used in this book. For electrically neutral atoms, Z is equal to the number of orbital electrons. (Why?)

The number of protons in the nucleus of an atom determines the species of the atom—the element to which the atom belongs. In Fig. 29.4b, the proton number $Z = 6$ indicates that the nucleus belongs to a carbon atom. The proton number thus defines which chemical symbol is used. Electrons can be removed from (or added to) an atom to form an ion, *but this does not change the atom's species*. For example, a nitrogen atom with an electron removed, N^+, is still nitrogen—a nitrogen *ion*. It is the proton number, rather than the electron number, that determines the species of atom.

The superscript to the left of the chemical symbol is called the **mass number** (A)—the total number of protons and neutrons in the nucleus. Since protons and neutrons have roughly equal masses, the mass numbers of nuclei give a relative comparison of nuclear masses. For the carbon nucleus in Fig. 29.4b, the mass number is $A = 12$, because there are six protons and six neutrons. The number of neutrons, called the **neutron number** (N), is sometimes indicated by a subscript on the right side of the chemical symbol. However, this subscript is usually omitted, because it can be calculated from A and Z; that is, $N = A - Z$. Similarly, the proton number is routinely omitted, because the chemical symbol uniquely specifies the value of Z.

Even though the atoms of an element all have the same number of protons in their nuclei, they may have different numbers of neutrons. For example, nuclei of different carbon atoms ($Z = 6$) may contain six, seven, or eight neutrons. In nuclear

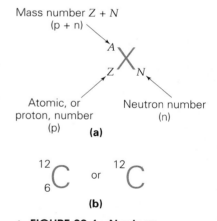

▲ **FIGURE 29.4 Nuclear notation** (a) The composition of a nucleus is shown by the chemical symbol of the element with the mass number A (sum of protons and neutrons) as a left superscript and the proton (atomic) number Z as a left subscript. The neutron number N may be shown as a right subscript, but both Z and N are routinely omitted, because the letter symbol tells you Z, and $N = A - Z$. (b) The two most common nuclear notations for a nucleus of one of the stable isotopes of carbon.

notation, these atoms would be written as $^{12}_{6}C_6$, $^{13}_{6}C_7$, and $^{14}_{6}C_8$, respectively. Atoms whose nuclei have the same number of protons, but different numbers of neutrons, are called **isotopes**. The three atoms we just listed are three isotopes of carbon.

Note: All the isotopes in one family have almost the same orbital-electron structure and thus very similar chemical properties.

Isotopes are like members of a family. They all have the same Z number and the same surname (element name), but they are distinguishable by the number of neutrons in their nuclei and, therefore, by their mass. Isotopes are referred to by their mass numbers; for example, these isotopes of carbon are called *carbon-12, carbon-13*, and *carbon-14*, respectively. There are other isotopes of carbon that are unstable: ^{11}C, ^{15}C, and ^{16}C. A particular nuclear species or isotope of any element is also called a **nuclide**. So far, we have mentioned six nuclides of carbon. Generally, only a few isotopes of a given species are stable. But this number can vary widely from none to several or more. In fact, in our carbon example, ^{14}C is unstable, although it is long lived. Only ^{12}C and ^{13}C, in fact, are truly stable isotopes of carbon.

Another important family of isotopes is that of hydrogen, which has three isotopes: ^{1}H, ^{2}H, and ^{3}H. These isotopes are given special names. ^{1}H is called *ordinary hydrogen*, or simply *hydrogen*; ^{2}H is called *deuterium*. Deuterium, which is stable, is sometimes known as *heavy hydrogen*. It can combine with oxygen to form heavy water (written D_2O). The third isotope of hydrogen, ^{3}H, called *tritium*, is unstable.

29.2 Radioactivity

OBJECTIVES: To (a) define the term *radioactivity*, (b) distinguish among alpha, beta, and gamma decay and (c) write nuclear-decay equations.

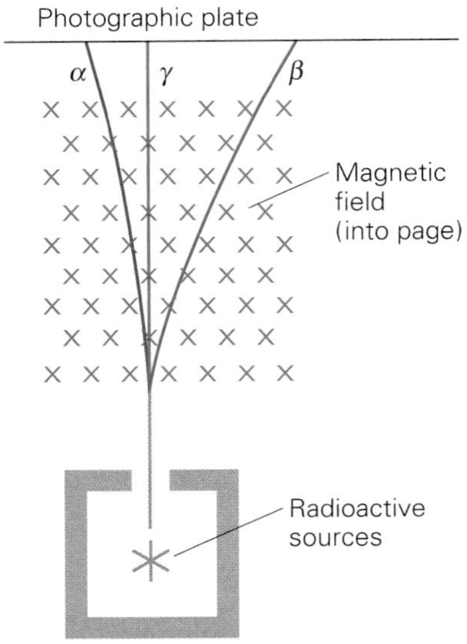

▲ **FIGURE 29.5 Nuclear radiation** Different types of radiation from radioactive sources can be distinguished by passing them through a magnetic field. Alpha and beta particles are deflected. From the right-hand magnetic-force rule, alpha particles are positively charged and beta particles are negatively charged. The radii of curvature (not drawn to scale) allow the particles to be distinguished by mass. Gamma rays are not deflected and thus are uncharged; they are quanta of electromagnetic energy.

Most elements have at least one stable isotope. It is atoms with stable nuclides with which we are most familiar in the environment. However, some nuclei are unstable and disintegrate spontaneously (or decay), emitting energetic particles and photons. Unstable isotopes are said to be *radioactive* or to exhibit **radioactivity**. For example, tritium has a radioactive nucleus. Of all the unstable nuclides, only a small number occur naturally. Others can be produced artificially (Chapter 30).

Radioactivity is unaffected by normal physical or chemical processes, such as heat, pressure, and chemical reactions. This is because processes such as these do *not* affect the source of the radioactivity—the nucleus. Nor can nuclear instability be explained by a simple imbalance of attractive and repulsive forces within the nucleus. This is because, experimentally, nuclear disintegrations (of a given isotope) occur at a fixed rate. That is, the nuclei in a given sample do not all decay at the same time. According to classical theories, identical nuclei *should* decay at the same time. Therefore, radioactive decay suggests probability effects of quantum mechanics.

The discovery of radioactivity is credited to the French scientist Henri Becquerel. In 1896, while studying the fluorescence of a uranium compound, he discovered that a photographic plate near a sample had been darkened, even though the compound had not been activated by exposure to light and was not fluorescing. Apparently, this darkening was caused by some new type of radiation emitted from the compound itself. In 1898, Pierre and Marie Curie announced the discovery of two radioactive elements, radium and polonium, which they had isolated from uranium pitchblende ore.

Experiment shows that the radiation emitted by radioactive isotopes is of three different kinds. When a radioactive isotope is placed in a chamber so that the emitted radiation passes through a magnetic field to a photographic plate (◂Fig. 29.5), the various types of radiation expose the plate, producing characteristic spots by which the types of radiation may be identified. The positions of the spots show that some isotopes emit radiation that is deflected to the left; some emit radiation that is deflected to the right; and some emit radiation that is undeflected. These spots are characteristic of what came to be known as *alpha, beta*, and *gamma* radiations.

From the opposite deflections of two of the types of radiation in the magnetic field, it is evident that positively charged particles are associated with alpha decay and that negatively charged particles are emitted during beta decay. Because of their much smaller deflection, alpha particles have to be considerably more massive than beta particles. The undeflected gamma radiation must be electrically neutral. (Why?)

Detailed investigations of the three different radiation types revealed the following:

- **Alpha particles** are actually doubly charged ($+2e$) particles that contain two protons and two neutrons. They are identical to the nucleus of the helium atom ($^4_2$He).
- **Beta particles** are electrons (positive electrons or *positrons* were discovered later).
- **Gamma rays** are high-energy quanta of electromagnetic energy (photons).

Radiation in a Magnetic Field

For a few radioactive elements, two spots are found on the film, indicating that the elements decay by two different modes. Let's now look at some details of each of these three modes of decay.

Alpha Decay

When an alpha particle is ejected from a radioactive nucleus, the nucleus loses two protons and two neutrons, so the mass number (A) is decreased by four ($\Delta A = -4$). The proton number (Z) is also decreased by two ($\Delta Z = -2$). Because the *parent nucleus* (the original nucleus) loses two protons, the *daughter nucleus* (the resulting nucleus) must be the nucleus of another element, defined by the new proton number. Thus, the **alpha-decay** process is a process of nuclear *transmutation*, in which the nuclei of one element change into the nuclei of a lighter element.

An example of an isotope, or nuclide, that undergoes alpha decay is polonium-214. The decay process is represented as a nuclear equation (usually written without neutron numbers):

$$^{214}_{84}\text{Po} \longrightarrow ^{210}_{82}\text{Pb} + ^{4}_{2}\text{He}$$

polonium   lead   alpha particle
(helium nucleus)

Notice that both the mass-number and proton-number totals must be equal on each side of the equation: ($214 = 210 + 4$) and ($84 = 82 + 2$), respectively. This condition reflects the fact that *two conservation laws apply to all nuclear processes*. The first is the **conservation of nucleons**:

| The total number of nucleons (A) remains constant in any nuclear process. |

The second is the familiar **conservation of charge**:

| The total charge remains constant in any nuclear process. |

These conservation laws allow us to predict the composition of the daughter nucleus, as the following Example illustrates.

Example 29.1 ■ Uranium's Daughter: Alpha Decay

A $^{238}_{92}$U nucleus undergoes alpha decay. What is the resulting daughter nucleus?

Thinking It Through. Nucleon conservation allows the prediction of the daughter's proton number. From that information, the element's name can be determined from the periodic table.

Solution. Since $\Delta Z = -2$ for alpha decay, the uranium-238 (^{238}U) nucleus loses two protons, and the daughter nucleus has a proton number $Z = 92 - 2 = 90$, which is thorium's proton number (see the periodic table, Fig. 28.9). The equation for this decay can therefore be written as

$$^{238}_{92}\text{U} \rightarrow ^{234}_{90}\text{Th} + ^{4}_{2}\text{He} \quad \text{or} \quad ^{238}_{92}\text{U} \rightarrow ^{234}_{90}\text{Th} + \alpha$$

where the helium nucleus is written as α, or sometimes as $^{4}_{2}\alpha$.

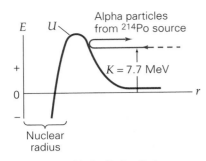

▲ FIGURE 29.6   Potential-energy barrier for alpha particles Alpha particles from radioactive polonium with energies of 7.7 MeV do not have enough energy to overcome the electrostatic potential-energy barrier of the ^{238}U nucleus and are scattered.

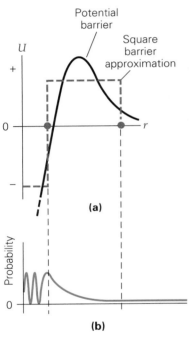

▲ FIGURE 29.7   Tunneling effect, or barrier penetration   (a) The potential-energy barrier presented by a nucleus to an alpha particle can be approximated by a rectangular barrier. **(b)** The probability of finding the alpha particle at a given location, according to quantum-mechanical calculations, is shown. If the particle is initially inside the nucleus, it has a likelihood of "tunneling" through the barrier and appearing outside the nucleus. Typically, this event has a very small, but nonzero, probability of occurring for elements above lead on the periodic table.

Follow-up Exercise.   Using high-energy accelerators, it is possible to *add* an alpha particle to a nucleus—essentially the reverse of the reaction in this Example. Write the equation for this nuclear reaction, and predict the identity of the resulting nucleus if an alpha particle is added to a ^{12}C nucleus. (*Answers to all Follow-up Exercises are at the back of the text.*)

The energies of the alpha particles from radioactive sources are typically a few MeV. (See Section 16.2.) For example, the energy of the alpha particle from the decay of ^{214}Po is about 7.7 MeV, and that from ^{238}U decay is about 4.14 MeV. Alpha particles from such sources were used in the scattering experiments that led to the Rutherford nuclear model described in the previous section.

Outside the nucleus, the repulsive electric force increases as an alpha particle approaches the nucleus. Inside the nucleus, however, the strongly attractive nuclear force dominates. These conditions are depicted in ◄Fig. 29.6, which shows a graph of the potential energy U as a function of r, the distance from the center of the nucleus. Consider alpha particles (with kinetic energy of 7.7 MeV) from a ^{214}Po source incident on ^{238}U (Fig. 29.6). The alpha particles don't have enough kinetic energy to overcome the electric potential-energy "barrier," whose height exceeds 7.7 MeV. Thus, Rutherford scattering occurs. On the other side of the coin, so to speak, we do know that the ^{238}U nucleus does undergo alpha decay, emitting an alpha particle with an energy of 4.4 MeV, which is below the height of the barrier. How can these lower energy alpha particles cross a potential-energy barrier from the inside when higher energy alpha particles cannot cross from the outside? According to classical theory, this situation is impossible, since it violates the conservation of energy. However, quantum mechanics offers an explanation.

Quantum mechanics predicts a finite probability of finding the alpha particle *outside* the nucleus (◄Fig 29.7). This phenomenon is called **tunneling**, or **barrier penetration**, since the alpha particle has a nonzero probability of tunneling through the barrier. (Recall that electrons do this in the process employed by the scanning tunneling microscope [STM], described in Chapter 28.)

Beta Decay

The emission of an electron (a beta particle) in a nuclear-decay process might seem contradictory to the proton–neutron model of the nucleus. Note, however, that the electron emitted in **beta decay** is *not* part of the original nucleus. *During the decay, the electron is created in the nucleus.* There are several types of beta decay. When a negative electron is emitted, the process is called β^- **decay**. An example of this type of beta decay is that of ^{14}C:

$$\underset{\text{carbon}}{^{14}_{6}\text{C}} \longrightarrow \underset{\text{nitrogen}}{^{14}_{7}\text{N}} + \underset{\substack{\text{beta particle} \\ \text{(electron)}}}{^{0}_{-1}\text{e}}$$

The parent carbon's nucleus has six protons and eight neutrons, whereas the daughter nucleus (nitrogen) has seven protons and seven neutrons. Notice that the electron symbol has a nucleon number of zero (because the electron is not a nucleon) and a charge number of -1. Thus, both nucleon number (14) and electric charge ($+6$) are conserved.

In this type of beta decay, the neutron number of the parent nucleus decreases by one, and the proton number of the daughter nucleus increases by one. Thus, the nucleon number remains unchanged. In essence, *a neutron within the nucleus decays into a proton and an electron (which is emitted):*

$$\underset{\text{neutron}}{^{1}_{0}\text{n}} \longrightarrow \underset{\text{proton}}{^{1}_{1}\text{p}} + \underset{\text{electron}}{^{0}_{-1}\text{e}} \qquad (\textit{basic } \beta^- \textit{ decay})$$

Beta decay generally happens when a nucleus is unstable because it has too many neutrons compared with the number of protons. (See Section 29.5, which discusses nuclear stability.) In the carbon-14 example, the most massive stable isotope of carbon is ^{13}C, with only seven neutrons. Therefore, ^{14}C has too many neutrons. Since beta decay both decreases the neutron number and increases the proton number, the product is more stable. In this case, the product is the nucleus ^{14}N, which is stable. For completeness, we note that there is another elementary particle, called a *neutrino*, emitted in beta decay. For simplicity, it will not be shown in the nuclear-decay equations here. Its important role in beta decay will be discussed in Chapter 30.

There are actually two modes of beta decay, β^- and β^+, as well as a competing process called *electron capture*. As we have seen, β^- decay involves the emission of an electron. **β^+ decay**, or *positron decay*, involves the emission of a positron (see Section 28.5). The positron symbol is $_{+1}^{0}e$. Nuclei that undergo β^+ decay have too many protons relative to the number of neutrons. The net effect of β^+ decay is to convert a proton into a neutron. As in β^- decay, this process serves to create a more stable daughter nucleus. An example of β^+ decay is the following:

$$^{15}_{8}O_7 \quad \rightarrow \quad ^{15}_{7}N_8 \quad + \quad ^{0}_{+1}e$$

oxygen   *nitrogen*   *positron*

Positron emission is also accompanied by a neutrino (but a different type than that associated with β^- decay), which we again omit until Chapter 30.

The competing process of **electron capture** (abbreviated as **EC**) involves the absorption of *orbital* electrons by a nucleus. The net result is a daughter nucleus that would have been produced by positron decay. Therefore, these two types of decay compete; that is, there is a probability for both to happen. A specific example of electron capture is as follows:

$$^{0}_{-1}e \quad + \quad ^{7}_{4}Be \quad \rightarrow \quad ^{7}_{3}Li$$

orbital electron   *beryllium*   *lithium*

As in β^+ decay, a proton changes into a neutron, but no beta particle is emitted in electron capture.

Gamma Decay

In **gamma decay**, the nucleus emits a gamma (γ) ray, a high-energy photon of electromagnetic energy. The emission of a gamma ray by a nucleus in an excited state is analogous to the emission of a photon by an excited atom. Most commonly, the nucleus emitting the gamma ray is a daughter nucleus left in an excited state after alpha decay, beta decay, or electron capture.

Nuclei possess energy levels analogous to those of atoms. However, nuclear energy levels are much farther apart and more complicated than those of an atom. The nuclear energy levels are typically separated by *kilo*electron volts (keV) and *mega*electron volts (MeV), rather than the few electron volts that usually separate atomic levels. As a result, gamma rays are very energetic, having frequencies greater than those of X rays and, thus, much shorter wavelengths. It is common to indicate a nucleus in an excited state with a superscript asterisk. Consider the decay of ^{61}Ni from an excited nuclear state (indicated by the asterisk) to one of lesser energy:

$$^{61}_{28}Ni^* \quad \rightarrow \quad ^{61}_{28}Ni \quad + \quad \gamma$$

nickel (excited)   *nickel*   *gamma ray*

Note that *in gamma decay, the mass and proton numbers do not change*. The daughter nucleus in this case is simply the parent nucleus with less energy (in a lower energy state). As an example of gamma emission following beta decay, consider the following Integrated Example.

Note: Remember that the positron or electron emitted during $\beta^\pm$ decay is *not* initially present in the neutron or proton that decays. Among other things, its presence before decay would violate conservation of energy, the uncertainty principle, and conservation of angular momentum. The electron or positron is *created at the time of the decay* and does not exist before that. See Chapter 30 for further details.

Naturally occurring cesium has only one stable isotope, $^{133}_{55}$Cs. However, the unstable isotope $^{137}_{55}$Cs is a common nucleus found in used nuclear fuel rods at power plants after their original uranium fuel has decayed. (See Chapter 30.) When $^{137}_{55}$Cs decays, its daughter nucleus is sometimes left in an excited state. After the initial decay, the daughter emits a gamma ray to produce a final stable nucleus. (a) Does $^{137}_{55}$Cs first decay by (1) β^+ decay, (2) β^- decay, or (3) electron capture? Explain. (b) Find the final daughter product by writing the chain of decay equations. Show all the steps leading to the final stable nucleus.

(a) Conceptual Reasoning. The $^{137}_{55}$Cs isotope has too many neutrons to be stable, as $^{133}_{55}$Cs, with four fewer neutrons, is stable. Choices (1) and (3) both increase the number of neutrons relative to the number of protons. Lowering the number of neutrons calls for β^- decay, so the correct choice is (2), β^- decay.

(b) Thinking It Through. Since we know that the cesium-137 nucleus must decay by emitting a β^- particle, its daughter (in an excited state) can be determined from charge and nucleon conservation. The final state of the daughter will result after a gamma-ray photon is emitted.
 The data is as follows:

Given: Initial nucleus of $^{137}_{55}$Cs   *Find:*   The decay schemes that lead to
the stable nucleus

 During β^- decay, the proton number increases by one; thus, the daughter will be barium ($Z = 56$). (See the periodic table, Fig. 28.9.) The decay equation should indicate that barium is left in an excited state that is ready to decay via gamma emission. (As usual in this chapter, we omit the neutrino.) Thus, the decay equation is

$$^{137}_{55}\text{Cs} \quad \longrightarrow \quad ^{137}_{56}\text{Ba*} \quad + \quad ^{0}_{-1}\text{e}$$

$$\text{\textit{cesium}} \qquad\qquad \text{\textit{barium}} \qquad \text{\textit{electron}}$$

$$\text{\textit{(excited)}}$$

This process is then quickly followed by the emission of a gamma ray from the excited barium nucleus:

$$^{137}_{55}\text{Ba*} \quad \longrightarrow \quad ^{137}_{55}\text{Ba} \quad + \quad \gamma$$

$$\text{\textit{barium}} \qquad\qquad \text{\textit{barium}} \qquad \text{\textit{gamma ray}}$$

$$\text{\textit{(excited)}}$$

Sometimes, this whole process is written as a combined equation in order to show the sequential behavior:

$$^{137}_{55}\text{Cs} \quad \longrightarrow \quad ^{137}_{56}\text{Ba*} \quad + \quad ^{0}_{-1}\text{e}$$

$$\text{\textit{cesium}} \qquad\qquad \text{\textit{barium}} \qquad \text{\textit{electron}}$$

$$\text{\textit{(excited)}}$$

$$\longmapsto \quad ^{137}_{56}\text{Ba} \quad + \quad \gamma$$

$$\qquad\quad \text{\textit{barium}} \qquad \text{\textit{gamma ray}}$$

Follow-up Exercise. An unstable isotope of sodium, ^{22}Na, can be produced in nuclear reactors. The only stable isotope of sodium is ^{23}Na. ^{22}Na is known to decay by one type of beta decay. (a) Which type of beta decay is it? Explain. (b) Write down the beta-decay scheme, and predict the daughter nucleus.

Radiation Penetration

The absorption, or degree of penetration, of nuclear radiation is an important consideration in applications. A familiar one is the radioisotope treatment of cancer. However, radiation penetration is also important in determining the necessary amount of nuclear shielding around a nuclear reactor. In addition, gamma radiation is being explored as a quick and convenient way to kill bacteria and sterilize food, and absorption of nuclear radiation is regularly used to monitor and control the thickness of metal and plastic sheets in fabrication processes.

The three types of radiation (alpha, beta, and gamma) are absorbed quite differently. As they move along their paths, the electrically charged alpha and beta particles interact with the electrons of the atoms of a material and may ionize some of them. The charge and speed of the particle determine the rate at which it loses energy along its path (remember that ionizing an atom takes energy) and, thus, the degree of penetration. Degree of penetration also depends on properties of the material, such as its density. In general, the following is what happens when radiation enters a material:

- Alpha particles are doubly charged, have a relatively large mass, and generally move relatively slowly. A few centimeters of air or a sheet of paper will usually completely stop them.

- Beta particles are much less massive and are singly charged. They can travel a few meters in air or a few millimeters in aluminum before being stopped.

- Gamma rays are uncharged and are therefore more penetrating than alpha and beta particles. A significant portion of a beam of high-energy gamma rays can penetrate a centimeter or more of a dense material, such as lead. Lead is commonly used as shielding against harmful X and gamma rays. Photons lose energy or are removed from a beam of gamma rays by a combination of Compton scattering, the photoelectric effect, and pair production (if their energy is above about 1 MeV).

Radiation passing through matter can do considerable damage. Structural materials can become brittle and lose their strength when exposed to strong radiation, such as that in nuclear reactors (Chapter 30) and the intense cosmic radiation to which space vehicles are exposed. In biological tissue, the radiation damage is chiefly due to ionizations in living cells (Section 29.5). We are exposed to normal background radiation from radioisotopes in the environment and cosmic radiation from outer space. The energy we absorb and the damage inflicted to cells from exposure to everyday levels of such radiation is usually too low to be harmful. However, concern has been expressed about the radiation exposure of people employed in several different types of jobs where radiation levels may be considerably higher. For example, workers at nuclear power plants are constantly monitored for absorbed radiation and subject to strict rules that govern the amount of time for which they can work in a given period. Also, flight crews who spend many hours aboard high-flying jet aircraft may receive above-average exposure to radiation from cosmic rays.

Of the many unstable nuclides, only a small number occur naturally. Most of the radioactive nuclides found in nature are products of the decay series of heavy nuclei. There is continual radioactive decay progressing in a series into successively lighter elements. The ^{238}U decay series is shown in ▶ Fig. 29.8. It stops when the stable isotope of lead, ^{206}Pb, is reached. Note that some nuclides in the series decay by two modes and that radon (^{222}Rn) is part of this decay series. This radioactive gas has received a great deal of attention, because it can accumulate in significant amounts in poorly ventilated buildings.

Note: See also Figs. 29.20 (^{237}Np) and 29.21 (^{239}Pu).

29.3 Decay Rate and Half-Life

OBJECTIVES: To (a) explain the concepts of activity, decay constant, and half-life of a radioactive sample and (b) use radioactive decay to find the age of objects.

The nuclei in a sample of radioactive material do *not* decay all at once, but rather do so randomly at a characteristic rate unaffected by external forces. It is impossible to tell exactly when a *particular* unstable nucleus will decay. What can be

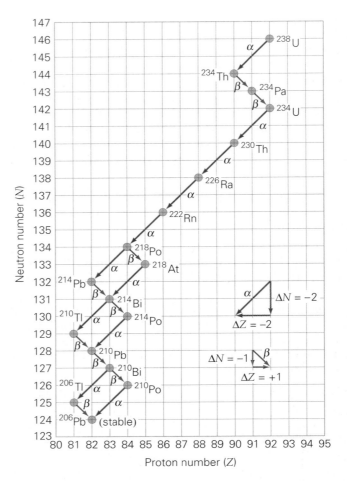

FIGURE 29.8 Decay series of uranium-238 On this plot of N versus Z, a diagonal transition from right to left is an alpha-decay process, and a diagonal transition from left to right is a β^--decay process. (How can you tell?) The decay series continues until the stable nucleus ^{206}Pb is reached.

determined, however, is how many nuclei in a sample will decay during a given period of time.

The **activity** of a sample of radioactive nuclide is defined as the number of nuclear disintegrations, or decays, per second. For a given amount of material, activity decreases with time, as fewer and fewer radioactive nuclei remain. Each nuclide has its own characteristic rate of decrease. The rate at which the number of parent nuclei (N) decreases is proportional to the number present, or $\Delta N/\Delta t \propto N$. This rate can be written in equation form with a constant of proportionality as

$$\frac{\Delta N}{\Delta t} = -\lambda N$$

Note: Activity = number of decays per second = $|\Delta N/\Delta t|$ and is a positive number.

where the constant λ is called the **decay constant**. λ has SI units of s^{-1} and varies from nuclide to nuclide. The larger the decay constant, the greater is the rate of decay. The minus sign in the previous equation indicates that N is decreasing. The **activity (R)** of a radioactive sample is the magnitude of $\Delta N/\Delta t$, in decays per second, without a minus sign (see the usage in Example 29.3):

$$R = \text{activity} = \left|\frac{\Delta N}{\Delta t}\right| = \lambda N \qquad (29.2)$$

The way in which the number of parent nuclei N decreases with time (assuming an initial number N_o) is illustrated in ▸Fig. 29.9. The number N decays *exponentially*, and the graph follows an exponential function ($e^{-\lambda t}$) with time. Here, λ is the decay constant, and $e \approx 2.718$ is the base of natural logarithms. By using the

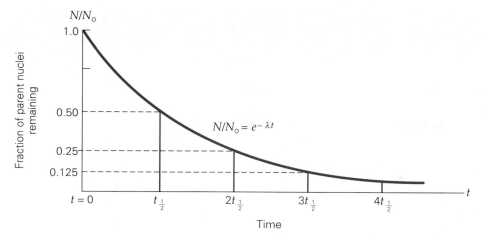

◀ FIGURE 29.9 Radioactive
decay The fraction of the remaining parent nuclei (N/N_o) in a radioactive sample plotted as a function of time follows an exponential-decay curve. The curve's shape and steepness depend upon the decay constant λ or the half-life $t_{1/2}$.

techniques of calculus, Eq. 29.2 can be solved for the number of the remaining (or undecayed) parent nuclei (N) at a time t compared with the number at $t = 0$:

$$N = N_o e^{-\lambda t} \tag{29.3}$$

(Here, N_o represents the initial number of nuclei present at $t = 0$.)

The decay rate of a nuclide is commonly expressed in terms of its half-life rather than the decay constant. The **half-life** ($t_{1/2}$) is defined as the time it takes for half of the radioactive nuclei in a sample to decay. This is the time corresponding to $N_o/2$ parent nuclei in Fig. 29.9—that is, when $N/N_o = \frac{1}{2}$. In the same amount of time, activity (decays per second) also is cut in half, since the activity is proportional to the number of parent nuclei present. Because of this proportionality, it is the decay rate that is usually measured to determine half-life. What is actually measured is the rate at which the decay particles are emitted.

For example, ▼Fig. 29.10 shows that the half-life of strontium-90 (^{90}Sr) is 28 years. An alternative way to view the concept of half-life is to consider the amount, or mass, of the parent material. Thus, if there were initially 100 micrograms (μg) of ^{90}Sr, only 50 μg would remain after 28 years. The other 50 μg would have decayed by the following beta-decay process:

$$\underset{\text{strontium}}{^{90}_{38}\text{Sr}} \longrightarrow \underset{\text{yttrium}}{^{90}_{39}\text{Y}} + \underset{\text{electron}}{^{0}_{-1}\text{e}}$$

Thus, the sample would contain a mixture of both strontium and yttrium. After another 28 years, another half of the strontium nuclei would decay, leaving 25 μg of strontium-90, and so on.

The half-lives of radioactive nuclides vary greatly, as Table 29.1 shows. Nuclides with very short half-lives are generally created in nuclear reactions (Chapter 30).

PHYSLET®
ILLUSTRATION

Half-Life

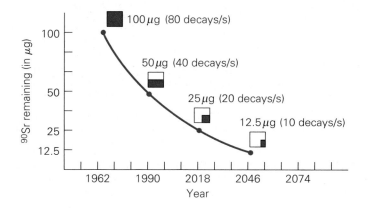

◀ FIGURE 29.10 Radioactive
decay and half-life As shown here for strontium-90, after each half-life ($t_{1/2} = 28$ y), only half of the amount of ^{90}Sr present at the start of that period of time remains, with the other half having decayed to ^{90}Y via beta decay. Similarly, the activity (decays per second) has also decreased by half after 28 years.

TABLE 29.1 The Half-Lives of Some Radioactive Nuclides (In Order of Increasing Half-Life)

Nuclide	Primary Decay Mode	Half-Life of Decay Mode
Beryllium-8 ($^{8}_{4}$Be)	α	1×10^{-16} s
Polonium-213 ($^{213}_{84}$Po)	α	4×10^{-16} s
Oxygen-19 ($^{19}_{8}$O)	β^{-}	27 s
Fluorine-17 ($^{17}_{9}$F)	β^{+}, EC	66 s
Polonium-218 ($^{218}_{84}$Po)	α, β^{-}	3.05 min
Technetium-104 ($^{104}_{43}$Tc)	β^{-}	18 min
Krypton-76 ($^{76}_{36}$Kr)	EC	14.8 h
Magnesium-28 ($^{28}_{12}$Mg)	β^{-}	21 h
Iodine-123 ($^{123}_{53}$I)	EC	13.3 h
Radon-222 ($^{222}_{86}$Rn)	α	3.82 d
Iodine-131 ($^{131}_{53}$I)	β^{-}	8.0 d
Cobalt-60 ($^{60}_{27}$Co)	β^{-}	5.3 y
Strontium-90 ($^{90}_{38}$Sr)	β^{-}	28 y
Radium-226 ($^{226}_{88}$Ra)	α	1600 y
Carbon-14 ($^{14}_{6}$C)	β^{-}	5730 y
Plutonium-239 ($^{239}_{94}$Pu)	α	2.4×10^{4} y
Uranium-238 ($^{238}_{92}$U)	α	4.5×10^{9} y
Rubidium-87 ($^{87}_{37}$Rb)	β^{-}	4.7×10^{10} y

If these nuclides had existed when the Earth was formed (about 4.7 billion years ago), they would have long since decayed. In fact, this is the case for technetium (Tc) and promethium (Pm). These elements do not exist naturally, but can be produced in laboratories. Conversely, the half-life of the naturally occurring ^{238}U isotope is about 4.5 billion years. That is, about half of the original ^{238}U present when the Earth was formed exists today. The longer the half-life of a nuclide, the more slowly it decays and the smaller is the decay constant λ. The half-life and the decay constant have an inverse relationship, or $t_{1/2} \propto 1/\lambda$. To show the numerical relationship, consider Eq. 29.3. When $t = t_{1/2}$, then $N/N_{o} = \frac{1}{2}$. Therefore,

$$\frac{N}{N_{o}} = \tfrac{1}{2} = e^{-\lambda t_{1/2}}$$

But because $e^{-0.693} \approx \frac{1}{2}$ (you should check this, using your calculator) by comparison of the exponents, we have, to three significant figures,

$$t_{1/2} = \frac{0.693}{\lambda} \tag{29.4}$$

The concept of half-life is important in medical applications, as is shown in Example 29.3.

Example 29.3 ■ Check Before You Send the Patient Home: Half-Life and Activity

The half-life of iodine-131, used in thyroid treatments, is 8.0 days. At a certain time, 4.0×10^{14} iodine-131 nuclei are in a hospital patient's thyroid gland. (a) What is the ^{131}I activity in the thyroid at that time? (b) How many ^{131}I nuclei remain after 1.0 day?

Thinking It Through. (a) Equation 29.4 enables us to determine the decay constant λ from the half-life. We then use Eq. 29.2 to find the initial activity. (b) To get N, Eq. 29.3 can be used in connection with the e^{x} button on a calculator.

Solution. Listing the data and converting the half-life into seconds, we have

Given: $t_{1/2} = 8.0$ days $= 6.9 \times 10^5$s   **Find:** (a) R_o (activity at $t = 0$)
$N_o = 4.0 \times 10^{14}$ nuclei (initially)   (b) N (number of undecayed
$t = 1.0$ day   nuclei after 1.0 day)

(a) The decay constant is determined from its relationship to the half-life (Eq. 29.4) as follows:

$$\lambda = \frac{0.693}{t_{1/2}} = \frac{0.693}{6.9 \times 10^5 \text{s}} = 1.0 \times 10^{-6} \text{ s}^{-1}$$

Using the initial number of undecayed nuclei, N_o, we find that the initial activity R_o is

$$R_o = \left| \frac{\Delta N}{\Delta t} \right| = \lambda N_o = (1.0 \times 10^{-6} \text{ s}^{-1})(4.0 \times 10^{14}) = 4.0 \times 10^8 \text{ decays/s}$$

(b) With $t = 1.0$ day and $\lambda = 0.693/t_{1/2} = 0.693/8.0$ days $= 0.087$ day^{-1}, we have

$N = N_o e^{-\lambda t} = (4.0 \times 10^{14} \text{ nuclei}) e^{-(0.087 \text{ day}^{-1}/8.0 \text{ day}^{-1})(1.0 \text{ day})}$

$= (4.0 \times 10^{14} \text{ nuclei}) e^{-0.087} = (4.0 \times 10^{14} \text{ nuclei})(0.917) = 3.7 \times 10^{14} \text{ nuclei}$

The e^x-function calculator button is sometimes labeled as the inverse of the ln x function. Become familiar with it. Here, it told us that $e^{-0.087} \approx 0.917$ to three significant figures.

Follow-up Exercise. In this Example, suppose that the attending physician will not allow the patient to go home until the activity is $\frac{1}{64}$ of its original level. (a) How long would the patient have to remain in observation? (b) In practice, the amount of time is much shorter. Can you think of a possible biological reason for this?

The "strength" of a radioactive sample is specified by its activity R. A common unit of radioactivity is named in honor of Pierre and Marie Curie.* One **curie (Ci)** is defined as

$$1 \text{ Ci} \equiv 3.70 \times 10^{10} \text{decays/s}$$

This definition is a historical and traditional one and is based on the activity of 1.00 g of pure radium. However, the SI unit is the **becquerel (Bq)**, which is defined as

$$1 \text{ Bq} \equiv 1 \text{ decay/s}$$

Therefore,

$$1 \text{ Ci} = 3.70 \times 10^{10} \text{ Bq}$$

Even with the present-day emphasis on SI units, radioactive sources are still commonly specified in curies. The curie is a relatively large unit, however, so the *millicurie* (mCi), the *microcurie* (μCi), and sometimes even smaller multiples such as the *nanocurie* (nCi) and *picocurie* (pCi) are used. Teaching laboratories, for example, typically use samples with activities of one microcurie or less. The strength of a source is calculated in the following Example.

Example 29.4 ■ Declining Source Strength: Get a Half-Life!

A ^{90}Sr beta source has an initial activity of 10.0 mCi. How many decays per second will be taking place after 84.0 years?

---

*Marie Sklodowska Curie (1867–1934) was born in Poland and studied in France, where she met and married physicist Pierre Curie (1859–1906). In 1903, Madame Curie (as she is commonly known) and Pierre shared the Nobel Prize in physics with Henri Becquerel (1852–1908) for their work on radioactivity. She was also awarded the Nobel Prize in chemistry in 1911 for the discovery of radium.

Thinking It Through. Table 29.1 lists the half-life for the source. In this Example, we can make use of the fact that in each successive half-life, the activity decreases by half from what it was at the start of that interval. Thus, Eq. 29.3. need not be used, because the elapsed time is exactly three half-lives. (This approach is advisable only when the elapsed time is an integral multiple of the half-life, as it is here.)

Solution.

Given: Initial activity = 10.0 mCi   **Find:** R (activity after 84.0 years)
t = 84.0 years
$t_{1/2}$ = 28.0 years
(from Table 29.1)

Since 84 years is exactly three half-lives, the activity after that amount of time has elapsed will be one-eighth as great ($\frac{1}{2} \times \frac{1}{2} \times \frac{1}{2} = \frac{1}{8}$), and the strength of the source will then be

$$R = \left| \frac{\Delta N}{\Delta t} \right| = 10.0 \text{ mCi} \times \frac{1}{8} = 1.25 \text{ mCi} = 1.25 \times 10^{-3} \text{ Ci}$$

In terms of decays per second, or becquerels, we have

$$R = \left| \frac{\Delta N}{\Delta t} \right| = (1.25 \times 10^{-3} \text{ Ci})\left(3.70 \times 10^{10} \frac{\text{decays/s}}{\text{Ci}} \right)$$
$$= 4.63 \times 10^7 \text{ decays/s} = 4.63 \times 10^7 \text{ Bq}$$

Follow-up Exercise. For the material in this Example, suppose a radiation safety officer tells you that this sample can go into a low-level waste disposal canister only when its activity drops to one-millionth of its current activity (in other words, when it drops from 10.0 mCi to 10.0 pCi, a factor of a million). Using your calculator, estimate, to two significant figures, how long you have to keep the sample before you can throw it out. [*Hint*: On your calculator, 2 raised to what power produces about a million?]

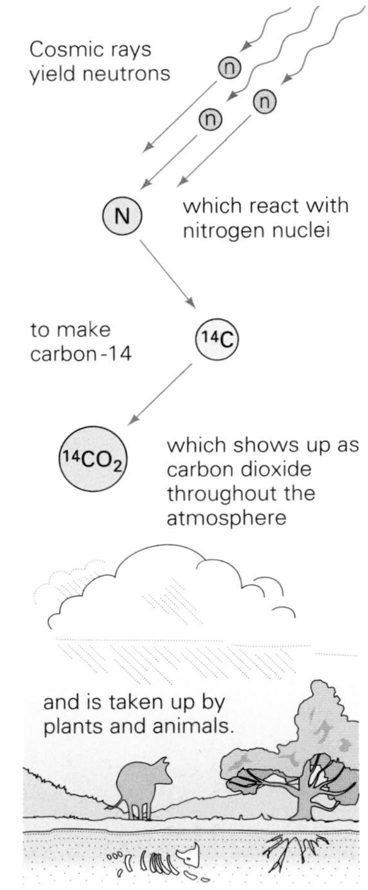

Radioactive Dating

Because their decay rates are constant, radioactive nuclides can be used as nuclear clocks. In the previous Example, the half-life of a radioactive nuclide was used to determine how much of the sample will exist in the future. Similarly, by using the half-life to calculate backward in time, scientists can determine the age of objects that contain such radioactive nuclides. As you might surmise, some idea of the initial composition or initial amount of a nuclide must first be known.

To illustrate the principle of radioactive dating, let's look at how it is done with ^{14}C. **Carbon-14 dating** is used on materials that were once part of living things and on the remnants of objects made from or containing such materials (such as wood, bone, leather, or parchment). The process depends on the fact that living things (including yourself) contain a known amount of radioactive ^{14}C. The concentration is very small—about one ^{14}C atom for every 7.2×10^{11} atoms of ordinary ^{12}C. Even so, the ^{14}C in our bodies *cannot* be due to any ^{14}C that was present when the Earth was formed. This is because the half-life of ^{14}C is $t_{1/2}$ = 5730 years, which is short in comparison with the age of the Earth (about 4.7×10^9 years).

Concentrations of ^{14}C are found in living things because the isotope is continuously produced in the upper atmosphere. Cosmic rays from outer space cause reactions that produce neutrons (◀Fig. 29.11). These neutrons are then absorbed by the nuclei of the nitrogen atoms of the air, which, in turn, decay by emitting a proton to produce ^{14}C by the reaction

$$^{14}_{7}\text{N} + ^{1}_{0}\text{n} \rightarrow ^{14}_{6}\text{C} + ^{1}_{1}\text{H}$$

^{14}C eventually decays by β^- decay ($^{14}_{6}\text{C} \rightarrow ^{14}_{7}\text{N} + ^{0}_{-1}\text{e}$), because it is neutron rich. Although the intensity of incident cosmic rays may not be constant over time, the

Cosmic rays yield neutrons

which react with nitrogen nuclei

to make carbon-14

which shows up as carbon dioxide throughout the atmosphere

and is taken up by plants and animals.

But when an organism dies, no fresh carbon-14 replaces the carbon-14 decaying in its tissues, and the carbon-14 radioactivity decreases by half every 5730 years.

▲ **FIGURE 29.11 Carbon-14 radioactive dating** The formation of carbon-14 in the atmosphere and its entry into the biosphere.

Note: The cosmic-ray production of ^{14}C is an example of a nuclear reaction that induces a nuclear transmutation. Such reactions will be studied in more detail in Chapter 30.

concentration of ^{14}C in the atmosphere is relatively constant, because of atmospheric mixing and the fixed decay rate.

The ^{14}C atoms are oxidized into carbon dioxide (CO_2), so a small fraction of the CO_2 molecules in the air is radioactive. Plants take in this radioactive CO_2 by photosynthesis, and animals ingest the plant material. As a result, the concentration of ^{14}C in living organic matter is the same as the concentration in the atmosphere, one part in 7.2×10^{11}. However, once an organism dies, the ^{14}C is *not* replenished, and the concentration decreases. *Thus, the concentration of ^{14}C in dead matter relative to that in living things can be used to establish the time that the organism died.* Since radioactivity is generally measured in terms of activity, the ^{14}C activity in organisms now alive must somehow be found. The following Example shows how this is done.

Example 29.5 ■ Living Organisms: Natural Carbon-14 Activity

Determine the average ^{14}C activity R in decays per minute per gram of natural carbon, found in living organisms, if the concentration of ^{14}C in the organisms is the same as that in the atmosphere.

Thinking It Through. From the previous discussion, we know the concentration of ^{14}C relative to that of ^{12}C. To calculate the ^{14}C activity, we need the decay constant (λ), which can be computed from the half-life of ^{14}C ($t_{1/2} = 5730$ years; see Table 29.1) and the number of ^{14}C atoms (N) per gram. Carbon has an atomic mass of 12.0, so N can be found from Avogadro's number (recall that $N_A = 6.02 \times 10^{23}$ atoms/mole) and the number of moles, $n = N/N_A$ (see Section 10.3).

Solution. Listing the known ratio and the half-life from Table 29.1 (and converting the half-life into minutes), we have

Given: $\dfrac{^{14}C}{^{12}C} = \dfrac{1}{7.2 \times 10^{11}} = 1.4 \times 10^{-12}$ *Find:* Average activity R per gram

$t_{1/2} = (5730 \text{ years})(5.26 \times 10^5 \text{ min/year})$
$\qquad = 3.01 \times 10^9 \text{ min}$
Carbon has 12.0 g per mole

From the half-life, the decay constant is

$$\lambda = \frac{0.693}{t_{1/2}} = \frac{0.693}{3.01 \times 10^9 \text{ min}} = 2.30 \times 10^{-10} \text{ min}^{-1}$$

For 1.0 g of carbon, the number of moles is $n = 1.0 \text{ g}/(12\text{g/mol}) = \frac{1}{12}$ mol, so the number of atoms (N) is

$$N = nN_A = \left(\frac{1}{12}\text{mol}\right)(6.02 \times 10^{23} \text{ C nuclei/mol}) = 5.0 \times 10^{22} \text{ C nuclei (per gram)}$$

The number of ^{14}C nuclei per gram is given by the concentration factor

$$N\left(\frac{^{14}C}{^{12}C}\right) = (5.0 \times 10^{22} \text{ C nuclei/g})\left(1.4 \times 10^{-12} \frac{^{14}C \text{ nuclei}}{\text{C nuclei}}\right) = 7.0 \times 10^{10} \text{ }^{14}C \text{ nuclei/g}$$

The activity in decays per gram of carbon per minute (to two significant figures), is

$$\left|\frac{\Delta N}{\Delta t}\right| = \lambda N = (2.30 \times 10^{-10} \text{ min}^{-1})(7.0 \times 10^{10} \text{ }^{14}C/\text{g}) = 16 \frac{^{14}C \text{ decays}}{\text{g} \cdot \text{min}}$$

Thus, if an artifact such as a bone or a piece of cloth has a current activity of 8.0 counts per gram of carbon per minute, then the original living organism would have died about one half-life, or about 5700 years, ago. This would put the date of the artifact near 3700 B.C.

Follow-up Exercise. Suppose that your instruments could measure ^{14}C beta emissions only down to 1.0 decay/g · min. How far back (to two significant figures) could you estimate the ages of dead organisms?

Now consider how the activity calculated in Example 29.5 can be used to date ancient organic finds.

Example 29.6 ■ Old Bones: Carbon-14 Dating

A bone is unearthed in an archeological dig. Laboratory analysis determines that there are 20 beta emissions per minute from 10 g of carbon in the bone. What is the approximate age of the bone?

Thinking It Through. Since the initial activity of a living sample is known (Example 29.5), we can work backward to determine the amount of time elapsed.

Solution. For comparison purposes, the activity *per gram* is the relevant number.

Given: Activity = 20 decays/min in 10 g of carbon   *Find:* Age of the bone
$$= 2.0 \text{ decays/g} \cdot \text{min}$$

Assuming that the bone had the normal concentration of ^{14}C when the organism died, the ^{14}C activity at the time of death would have been 16 decays/g·min (Example 29.5). Afterward, the decay rate would decrease by half for each half-life:

$$16 \xrightarrow{t_{1/2}} 8 \xrightarrow{t_{1/2}} 4 \xrightarrow{t_{1/2}} 2 \text{ decay/g} \cdot \text{min}$$

So, with an observed activity of 2.0 decays/g·min, the ^{14}C in the bone has gone through approximately three half-lives. Thus, the bone is three half-lives old, or, to two significant figures,

$$\text{Age} \approx 3.0 t_{1/2} = (3.0)(5730 \text{ years}) = 1.7 \times 10^4 \text{ years}$$

Follow-up Exercise. Isotopic data indicate that the stable nuclide ^{39}K and the long-lived nuclide ^{40}K are the only isotopes of potassium that currently exist on the Earth. The ^{40}K nuclide has a half-life of 1.28×10^9 years. (See Appendix V.) The present abundance ratio of the two potassium isotopes is ^{39}K/^{40}K = 14. At what time in the past would the two isotopes have had equal abundances? (According to present theories of the formation of the elements and of our solar system, this date should approximate the age of the Earth. Compare your answer with the age of the Earth, which is estimated at 4.7×10^9 years. Is the calculated time close to this number?)

The limit of radioactive carbon dating depends on the ability to measure the very low activity in old samples. Current techniques give an age-dating limit of about 40 000–50 000 years, depending on the size of the sample. After about nine half-lives, the radioactivity of a carbon sample has decreased so much that it is barely measurable (less than two decays per gram per *hour*).

Another radioactive dating process uses lead-206 (^{206}Pb) and uranium-238 (^{238}U). This dating method is used extensively in geology, because of the long half-life of ^{238}U. Lead-206 is the stable end isotope of the ^{238}U decay series. (See Fig. 29.8.) If a rock sample contains both of these isotopes, the lead is assumed to be a decay product of the uranium that was there when the rock first formed. Thus, the ratio of ^{206}Pb/^{238}U is a measure of the geologic age of the rock.

29.4 Nuclear Stability and Binding Energy

OBJECTIVES: To (a) state which proton- and neutron-number combinations result in stable nuclei, (b) explain the pairing effect and magic numbers in relation to nuclear stability and (c) calculate nuclear binding energies.

Now that we have considered some of the properties of unstable isotopes, let's turn to the stable isotopes. Stable isotopes exist naturally for all elements having proton numbers from 1 to 83, except for those with $Z = 43$ (technetium) and $Z = 61$ (promethium). The nuclear interactions that give rise to nuclear stability are extremely complicated. However, by looking at some of the fundamental properties of stable nuclei, it is possible to obtain general criteria for nuclear stability.

Nucleon Populations

One of the first considerations is the relative number of protons and neutrons in stable nuclei. Nuclear stability depends on the dominance of the attractive nuclear force between nucleons over the repulsive Coulomb force between protons. This force dominance depends on the ratio of protons to neutrons.

For stable nuclei of low mass numbers (about $A < 40$), the ratio of neutrons to protons (N/Z) is approximately 1. That is, the number of protons and the number of neutrons are equal or nearly equal. For example, ^4_2He, $^{12}_6\text{C}$, $^{23}_{11}\text{Na}$, and $^{27}_{13}\text{Al}$ have the same or about the same number of protons as neutrons. For stable nuclei of higher mass numbers ($A > 40$), the number of neutrons exceeds the number of protons ($N/Z > 1$). The heavier the nuclei, the higher is this ratio, that is, the more the neutrons outnumber the protons.

This trend is illustrated in ▼Fig. 29.12, a plot of neutron number (N) versus proton number (Z) for stable nuclei. The heavier stable nuclei lie above the $N = Z$ line ($N > Z$). Examples of heavy stable nuclei include $^{62}_{28}\text{Ni}$, $^{114}_{50}\text{Sn}$, $^{208}_{82}\text{Pb}$, and $^{209}_{83}\text{Bi}$. In fact, bismuth-209 is the heaviest element that has a stable isotope.*

Radioactive decay "adjusts" the proton and neutron numbers of an unstable nuclide until a stable nuclide is produced—that is, the product lands on the stability curve in Fig. 29.12. Since alpha decay decreases the numbers of protons and neutrons by equal amounts, alpha decay alone would give nuclei with neutron populations that are *larger* than those of the stable nuclides on the curve. However, β^- decay *following* alpha decay can lead to a stable combination, since the effect of β^- decay is to convert a neutron into a proton. Thus, very heavy nuclei undergo a chain, or sequence, of alpha and beta decays until a stable nucleus is reached, as was illustrated in Fig. 29.8 for ^{238}U.

Pairing

Many stable nuclei have even numbers of both protons and neutrons, and very few have odd numbers of *both* protons and neutrons. A survey of the stable isotopes (Table 29.2) shows that 168 stable nuclei have this even–even combination,

TABLE 29.2 Pairing Effect of Stable Nuclei

Proton Number	Neutron Number	Number of Stable Nuclei
Even	Even	168
Even	Odd }	
Odd	Even }	107
Odd	Odd	4

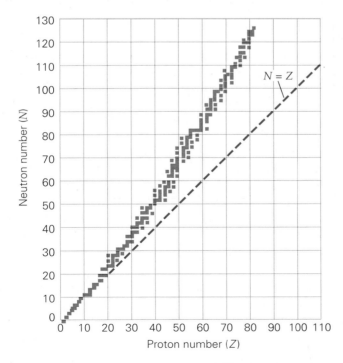

◀ **FIGURE 29.12 A plot of *N* versus *Z* for stable nuclei** For nuclei with mass numbers $A < 40$ ($Z < 20$ and $N < 20$), the number of protons and the number of neutrons are equal or nearly equal. For nuclei with $A > 40$, the number of neutrons exceeds the number of protons, so these nuclei lie above the $N = Z$ line.

*Bismuth-209 alpha decays, but with a half-life of 2×10^{18} years; for practical purposes, it is considered to be stable.

107 have even–odd or odd–even arrangements, and only 4 contain odd numbers of both protons and neutrons. These four are isotopes of the elements with the four lowest odd proton numbers: $_1^2H$, $_3^6Li$, $_5^{10}B$, and $_7^{14}N$.

The dominance of even–even combinations indicates that the protons and neutrons in nuclei tend to "pair up." That is, two protons pair up and, separately, two neutrons pair up. Aside from the four exceptions, all odd–odd nuclei are unstable. Also, odd–even and even–odd nuclei tend to be less stable than the even–even variety.

This **pairing effect** provides a qualitative criterion for stability. For example, you might expect the aluminum isotope $_{13}^{27}Al$ to be stable (even–odd), but not $_{13}^{26}Al$ (odd–odd). This is actually the case.

The *general criteria for nuclear stability* can be summarized as follows:

1. All isotopes with a proton number greater than $83 (Z > 83)$ are unstable.
2. (a) Most even–even nuclei are stable.
   (b) Many odd–even and even–odd nuclei are stable.
   (c) Only four odd–odd nuclei are stable ($_1^2H$, $_3^6Li$, $_5^{10}B$, and $_7^{14}N$).
3. (a) Stable nuclei with mass numbers less than $40 (A < 40)$ have approximately the same number of protons and neutrons.
   (b) Stable nuclei with mass numbers greater than $40 (A > 40)$ have more neutrons than protons.

Conceptual Example 29.7 ■ Running Down the Checklist: Nuclear Stability

Is the sulfur isotope $_{16}^{38}S$ likely to be stable?

Reasoning and Answer. We use the general criteria for nuclear stability to analyze this case:

1. *Satisfied.* Isotopes with $Z > 83$ are unstable. With $Z = 16$, this criterion is satisfied.
2. *Satisfied.* The isotope $_{16}^{38}S_{22}$ has an even–even nucleus, so it could be stable.
3. *Not satisfied.* Here, $A < 40$, but $Z = 16$ and $N = 22$ are not approximately equal.

Therefore, the ^{38}S isotope is likely to be unstable. (The nucleus is unstable and decays by β^- emission, since it is neutron rich.)

Follow-up Exercise. (a) List likely isotopes of copper ($Z = 29$). (b) Apply the criteria for nuclear stability to see which of those isotopes should be stable. Use Appendix V to check your conclusions.

Binding Energy

Note: The concept of binding energy was introduced in Section 27.4.

An important quantitative aspect of nuclear stability is the binding energy of the nucleons. Binding energy can be calculated by considering the mass–energy equivalence of the nuclear masses. Since nuclear masses are so small in relation to the SI standard kilogram, another standard, the **atomic mass unit (u)**, is used to measure them. The conversion factor (to six significant figures) between the atomic mass unit and the kilogram is

$$1\,u = 1.66054 \times 10^{-27}\,kg$$

The masses of the various particles are typically expressed in atomic mass units. (See Table 29.3.) The listed energy equivalents reflect Einstein's $E = mc^2$ mass–energy equivalence relationship from Eq. 26.11. Thus, a mass of 1 u has an energy equivalent of

$$mc^2 = (1.66054 \times 10^{-27}\,kg)(2.9977 \times 10^8\,m/s)^2 = 1.4922 \times 10^{-10}\,J$$
$$= \frac{1.4922 \times 10^{-10}\,J}{1.602 \times 10^{-13}\,J/MeV} = 931.5\,MeV$$

TABLE 29.3 The Atomic Mass Unit (u), Particle Masses, and Their Energy Equivalents

Particle	u	Mass (kg)	Equivalent Energy (MeV)
	1 (exact)	1.66054×10^{-27}	931.5
Electron	5.48578×10^{-4}	9.10935×10^{-31}	0.511
Proton	1.007276	1.67262×10^{-27}	938.27
Hydrogen atom	1.007825	1.67356×10^{-27}	938.79
Neutron	1.008665	1.67500×10^{-27}	939.57

as given in the first entry of Table 29.3. We will use 931.5 MeV/u (to four significant figures) as a handy conversion factor in order to avoid having to multiply by c^2.

Note that in Table 29.3, the proton and the hydrogen atom ($^1_1$H) are listed separately, as they have slightly different masses. This difference is due to the mass of the atomic electron. We measure experimentally the masses of neutral atoms (nucleons plus Z electrons) rather than of nuclei. Keep this factor in mind. Since nuclear-energy calculations usually involve very small differences in mass, the mass of the electron can be significant.

Nuclear stability can be looked at in terms of energy. For example, if the mass of a helium-4 nucleus is compared with the total mass of nucleons that compose it, a significant inequity emerges: A neutral helium atom (including its *two* electrons) has a mass of 4.002 603 u. (Atomic masses of various atoms are given in Appendix V.) The total mass of two hydrogen atoms (^{1}H) (including *two* electrons) and two neutrons is

$$2m(^1\text{H}) = 2.015650 \text{ u}$$
$$2m_\text{n} = \underline{2.017330 \text{ u}}$$
$$\text{Total} = 4.032980 \text{ u}$$

This total is greater than the mass of the helium atom (4.002 603 u). The helium nucleus is less massive than the sum of its parts by an amount

$$\Delta m = [2m(^1\text{H}) + 2m_\text{n}] - m(^4\text{He})$$
$$= 4.032980 \text{ u} - 4.002603 \text{ u} = 0.030377 \text{ u}$$

Notice that the two electronic masses of helium subtract out, since the mass of two hydrogen atoms included two electrons. This difference in mass called the **mass defect**, has an energy equivalent of

$$(0.030377 \text{ u})(931.5 \text{ MeV/u}) = 28.30 \text{ MeV}$$

This energy is the total binding energy (E_b) of the ^{4}He nucleus.

In general, for any nucleus, the **total binding energy** is related to the mass defect by

$$E_\text{b} = (\Delta m)c^2 \quad \text{total binding energy} \quad (29.5)$$

where Δm is the mass defect. An alternative interpretation of the binding energy is that it represents the energy required to separate the constituent nucleons completely into free particles. This concept is illustrated in ▸Fig. 29.13 for the helium nucleus, where 28.30 MeV of energy is necessary to separate it into four nucleons.

We can gain some insight into the nature of the nuclear force by considering the *average binding energy per nucleon* for stable nuclei. This quantity is the total binding energy of a nucleus, divided by the total number of nucleons, or E_b/A, where A is the mass number. For the helium nucleus (^{4}He) in Fig. 29.13, the average binding energy per nucleon is

$$\frac{E_\text{b}}{A} = \frac{28.30 \text{ MeV}}{4} = 7.075 \text{ MeV/nucleon}$$

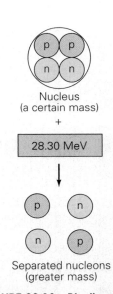

▲ **FIGURE 29.13 Binding energy** 28.30 MeV is required to separate a helium nucleus into its constituent protons and neutrons. Conversely, if two protons and two neutrons combine to form a helium nucleus, 28.30 MeV of energy would be released.

Compared with the binding energy of atomic electrons (13.6 eV for a hydrogen electron in the ground state), nuclear binding energies are millions of times larger, indicative of a very strong binding force.

Example 29.8 ■ The Stablest of the Stable: Binding Energy per Nucleon

Compute the average binding energy per nucleon of the iron-56 nucleus ($^{56}_{26}$Fe).

Thinking It Through. The atomic mass of iron 56 is found in Appendix V, and the other needed masses are in Table 29.3. We can then compute the mass defect Δm, the total binding energy E_b, and, lastly, the average binding energy per nucleon, E_b/A.

Solution.

Given: $^{56}_{26}$Fe mass = 55.934 939 u   *Find:* E_b/A (average binding energy per
$^{1}_{1}$H mass = $m(^1\text{H})$ = 1.007 825 u   nucleon)
$^{1}_{0}$n mass = m_n = 1.008 665 u

(Notice the use of the masses of the iron *atom* and the hydrogen *atom* rather than the nuclear masses.)

The mass defect is the difference between the mass of the iron atom and the mass of its separated constituents. The total mass of the constituents (here, 26 hydrogen atoms and 30 neutrons) is found as follows:

$$26m(^1\text{H}) = 26(1.007\,825\text{ u}) = 26.203\,450\text{ u}$$
$$30m_n = 30(1.008\,665\text{ u}) = 30.259\,950\text{ u}$$
$$\text{Total} = 56.463\,400\text{ u}$$

Thus, the mass defect is

$$\Delta m = [26m(^1\text{H}) + 30m_n] - m(^{56}\text{Fe}) = 56.463\,400\text{ u} - 55.934\,939\text{ u} = 0.528\,461\text{ u}$$

The total binding energy is easily calculated using the energy equivalence of 1 u:

$$E_b = (\Delta m)c^2 = (0.528\,461\text{ u})(931.5\text{ MeV/u}) = 492.3\text{ MeV}$$

This iron nuclide has 56 nucleons, so the average binding energy per nucleon is

$$\frac{E_b}{A} = \frac{492.3\text{ MeV}}{56} = 8.791\text{ MeV/nucleon}$$

Follow-up Exercise. (a) To illustrate the pairing effect, compare the average binding energy per nucleon for ^{4}He (calculated previously to be 7.075 MeV) with that of ^{3}He. (Find the atomic masses in Appendix V.) (b) How does your answer reflect the pairing effect?

If E_b/A is calculated for various nuclei and plotted versus mass number, the values generally lie along the curve shown in ▶Fig. 29.14. The value of E_b/A rises rapidly with increasing A for light nuclei and starts to level off (around $A = 15$) at about 8.0 MeV/nucleon, with a maximum value of about 8.8 MeV/nucleon in the vicinity of iron, which has the most stable nucleus. (See Example 29.8.) For $A > 60$, the E_b/A values decrease slowly, indicating that the nucleons are, on average, less tightly bound.

The importance of the maximum in the curve cannot be understated. Consider what would happen on either side of it. If a massive nucleus split, or *fissioned*, into two lighter nuclei, the nucleons would be more tightly bound, and energy would be released. On the low-mass side of the maximum, if two nuclei could be fused, in a process called *fusion*, a more tightly bound nucleus would be created, and energy would be released. The details and application of these processes will be discussed in detail in Chapter 30.

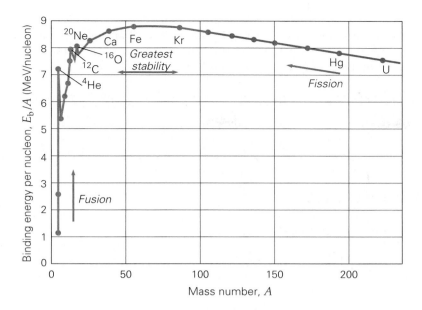

◀ **FIGURE 29.14   A plot of binding energy per nucleon versus mass number**   If the binding energy per nucleon (E_b/A) is plotted versus mass number (A), the curve has a maximum near iron (Fe). This indicates that the nuclei in this region are, on average, the most tightly bound and have the greatest stability. Extremely heavy nuclei can release energy by splitting (fissioning). Extremely light nuclei can combine by fusion to release energy.

The E_b/A curve shows that, except for very light nuclei, the binding energy per nucleon does not change a great deal and has a value of $E_b/A \approx 8$ MeV/nucleon. This means that, to a good approximation, we can write $E_b \propto A$. In other words, the *total* binding energy is (approximately) proportional to the total number of nucleons.

This proportionality indicates a characteristic of the nuclear force that is quite different than the electrical force. Suppose that the attractive nuclear force did act between *all* the pairs of nucleons in a nucleus. Each pair of nucleons would then contribute to the total binding energy. In a nucleus containing A nucleons, there are $A(A - 1)/2$ pairs, so there would be $A(A - 1)/2$ contributions to the total binding energy. For nuclei for which $A \gg 1$ (heavier nuclei), $A(A - 1) \approx A^2$, and it would be expected that the binding energy would be proportional to the square of A, or $E_b \propto A^2$ *if the nucleon–nucleon force were to act over a long range.* But as we have seen, in actuality, $E_b \propto A$. This relationship indicates that a given nucleon is *not* bound to all the other nucleons. This phenomenon, called *saturation*, implies that the nuclear forces act over a very short range and that a particular nucleon interacts only with its nearest neighbors.

Magic Numbers

We are familiar with the concept of filled shells in atoms. There is an analogous effect in the nucleus. Although the concept of individual nucleon "orbits" inside the nucleus is hard to visualize, experimental evidence does indicate the existence of "closed nuclear shells" when the number of protons *or* neutrons is 2, 8, 20, 28, 50, 82, or 126. Important work on the nuclear-shell model was done by Maria Goeppert-Mayer (1906–1972), a German-born physicist.

The number of stable isotopes of various elements provides solid evidence of the existence of such **magic numbers**. If an element has a magic number of protons, it has an unusually high number of stable isotopes. Elements whose proton number is far away numerically from a magic number may have only 1 or 2 (or even no) stable isotopes. Aluminum, for example, with 13 protons, has just 1 stable isotope, ^{27}Al. But tin, with $Z = 50$ (a magic number), has 10 stable isotopes, ranging from $N = 62$ to $N = 74$. Neighboring indium, with $Z = 49$, has only 2 stable isotopes, and antimony, with $Z = 51$, also has only 2.

Another piece of experimental evidence for magic numbers is related to binding energies. High-energy gamma-ray photons can be used to knock out single nucleons from a nucleus (a phenomenon called the *photonuclear effect*) in a manner analogous to the photoelectric effect in metals. Experimentally, a nuclide with a proton magic number, such as tin, requires about 2 MeV *more* photon energy to eject a proton from its nucleus than does a nuclide that does not have a magic number of protons. Thus, magic numbers are associated with extra large binding energies, another sure sign of higher stability.

29.5 Radiation Detection and Applications

OBJECTIVES: To (a) gain insight into the operating principles of various nuclear-radiation detectors, (b) investigate the medical and biological effects of radiation exposure and (c) study some of the practical uses and applications of radiation.

Detecting Radiation

Since our senses cannot detect radioactive decay directly, detection must be accomplished through indirect means. For example, people who work with radioactive materials or in nuclear reactors usually wear film badges that indicate cumulative exposure to radiation by the degree of darkening of the film when developed. If more immediate ("real time") and quantitative methods are needed to detect radiation, a variety of instruments is available.

These instruments, as a group, are known as **radiation detectors**. Fundamentally, they are all based on the ionization or excitation of atoms, a phenomenon caused by the passage of energetic particles through matter. The electrically charged alpha and beta particles transfer energy to atoms by electrical interactions, removing electrons and creating ions. Gamma-ray photons can produce ionization by the photoelectric effect and Compton scattering (Section 27.3). They may also produce electrons and positrons by pair production (Section 28.5), if their energy is large enough. Regardless of the source, the particles produced by these interactions, not the actual radiated particles, are the objects "detected" by a radiation detector.

One of the most common radiation detectors is the *Geiger counter*, developed by Hans Geiger, a student and then colleague of Ernest Rutherford. The principle of the Geiger counter is illustrated in ◄Fig. 29.15. A voltage of about 1000 V is applied across the wire electrode and outer electrode (a metallic tube) of the Geiger tube. The tube contains a gas (such as argon) at low pressure. When an ionizing particle enters the tube through a thin window at one end, it ionizes some of the gas atoms. The freed electrons are accelerated toward the positive anode. On their way, they strike and ionize other gas atoms. This process snowballs, and the resulting "avalanche" discharge produces a current pulse. The pulse is amplified and sent to an electronic counter that counts the pulses, or the number of particles detected. The pulses are sometimes used to drive a loudspeaker so that particle detection is heard as a click.

Another method of detection, one of the oldest, is the *scintillation counter* (▼Fig. 29.16). Here, the atoms of a phosphor material (such as sodium iodide

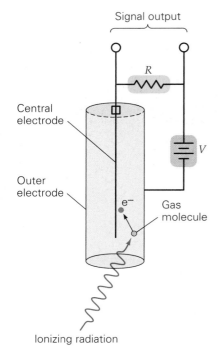

▲ FIGURE 29.15 The Geiger counter Incident radiation ionizes a gas atom, freeing an electron that is, in turn, accelerated toward the central (positive) electrode. On the way, this electron produces additional electrons through ionization, resulting in a current pulse that is detected as a voltage across the external resistor R.

► FIGURE 29.16 The scintillation counter A photon emitted by a phosphor atom excited by an incoming particle causes the emission of a photoelectron from the photocathode. Accelerated through a difference in potential in a photomultiplier tube, the photoelectrons free secondary electrons when they collide with successive electrodes at higher potentials. After several steps, a relatively weak scintillation is converted into a measurable electric pulse.

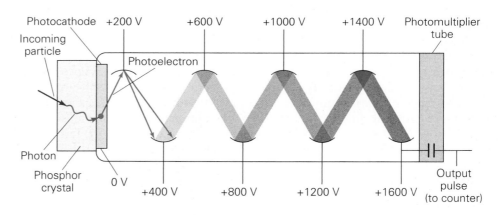

[NaI]) are excited by an incident particle. A visible light pulse is emitted when the atoms return to their ground state. The light pulse is converted to an electrical pulse by a photoelectric material. The pulse is then amplified in a *photomultiplier tube*, which consists of a series of electrodes of successively higher potential. The photoelectrons are accelerated toward the first electrode and acquire sufficient energy to cause several secondary electrons from ionization to be emitted when they strike the electrode. This process continues, and relatively weak scintillations are converted into sizable electrical pulses, which are then counted electronically.

In a *solid-state*, or *semiconductor, detector*, charged particles passing through a semiconductor material produce electrons, because of ionization. When a voltage is applied across the material, the electrons are collected as an electric current, which can be amplified and counted.

The three aforementioned detectors determine the number of particles produced by a sample radioactive isotope. Other methods of detection allow the actual trajectory, or tracks, of charged particles to be seen or recorded. Among this type of detector are the cloud chamber, the bubble chamber, and the spark chamber. In the first two, vapors and liquids are supercooled and superheated, respectively, by suddenly varying the volume and pressure. The *cloud chamber* was developed early in the 1900s by C. T. R. Wilson, a British atmospheric physicist. In the chamber, supercooled vapor condenses into droplets on ionized molecules created along the path of an energetic particle. When the chamber is illuminated, the droplets scatter the light, making the path visible (▶Fig. 29.17).

The *bubble chamber*, which was invented by the American physicist D. A. Glazer in 1952, uses a similar principle. A reduction in pressure causes a liquid to be superheated and able to boil. Ions produced along the path of an energetic particle become sites for bubble formation, and a trail of bubbles is created. Since the bubble chamber uses a liquid, commonly liquid hydrogen, the density of atoms in it is much greater than in the vapor of a cloud chamber. Thus, tracks are more readily observable in bubble chambers, and hence bubble chambers have largely replaced cloud chambers.

In a *spark chamber*, the path of a charged particle is registered by a series of sparks. The charged particle passes between a pair of electrodes that have a high difference in potential and are immersed in an inert (noble) gas. The charged particle causes the ionization of gas molecules, giving rise to a visible spark (flash of light) between the electrodes as the released electrons travel to the positive electrode. A spark chamber is merely an array of such electrodes in the form of parallel plates or wires. A series of sparks, which can be photographed, then marks the particle's path.

Once the particle's trajectory is displayed, the particle's energy can be determined. Typically, a magnetic field is applied across any of the three types of chambers, charged particles are deflected, and the energy of a particle is calculated from the radius of curvature of the particle's path. Gamma rays do not leave visible tracks in any of these types of chamber detectors. However, their presence can be detected indirectly, because they are able to produce electrons by such processes as the photoelectric effect and pair production. The gamma-ray energy can be determined from the measured energy of these electrons.

▲ **FIGURE 29.17 Cloud-chamber tracks** The circular track in this photograph of a cloud chamber was made by a positron in a strong magnetic field. (Can you explain the approximately circular path of the particle in terms of the orientation of the magnetic field relative to the positron's velocity?)

Biological Effects and Medical Applications of Radiation

In medicine, nuclear radiation can be used beneficially in the diagnosis and treatment of some diseases, but it also is potentially harmful if not handled and administered properly. Nuclear radiation and X rays can penetrate human tissue without pain or any other sensation. However, early investigators quickly learned that large doses or repeated small doses lead to red skin, lesions, and other conditions. It is now known that certain types of cancers can be caused by excessive exposure to radiation.

The chief hazard of radiation is damage to living cells, due primarily to ionization. Ions, particularly complex ions or radicals produced by radiation, may be highly reactive (for example, a hydroxyl ion [OH⁻] produced from water). Such reactive ions interfere with the normal chemical operations of the cell. If enough cells are damaged or killed, cell reproduction might not be fast enough, and the irradiated tissue could eventually die. In other instances, genetic damage, or mutation, may occur in a chromosome in the cell nucleus. If the affected cells are sperm or egg cells (or their precursors), any children that they produce may suffer from various birth defects. If the damaged cells are ordinary body cells, they may become cancerous, reproducing in a rapid and uncontrolled manner and eventually becoming a malignant tumor. The human cells most susceptible to radiation damage are those of the reproductive organs, bone marrow, and lymph nodes. To begin our discussion of radiation damage and applications, let us investigate how the radiation "dose" is quantified.

Radiation Dosage An important consideration in radiation therapy and radiation safety is the amount, or *dose*, of radiation. Several quantities are used to describe this amount in terms of *exposure*, *absorbed dose*, or *equivalent dose*. The earliest unit of dosage, the **roentgen (R)**, based on exposure and defined in terms of ionization produced in air. One roentgen is the quantity of X rays or gamma rays required to produce an ionization charge of 2.58×10^{-4} C/kg in air.

The **rad** (radiation absorbed dose) is an *absorbed* dose unit. One rad is an absorbed dose of radiation energy of 10^{-2} J per kilogram of absorbing *material*. Note that the rad is based on energy absorbed from the radiation rather than simply ionization caused by the radiation in air (as the roentgen is). As such, it is more directly related to the biological damage caused by the radiation. Because of this characteristic, the rad has largely replaced the roentgen.

The rad is not an SI unit. The SI unit for absorbed dose is the **gray (Gy)**, defined as

$$1 \text{ Gy} \equiv 1 \text{ J/kg} \equiv 100 \text{ rad}$$

However, the most meaningful assessment of the effects of radiation must involve measuring the *biological damage* produced, because it is well known that equal doses (in rads) of different types of radiation produce *different* effects. For example, a relatively massive alpha particle with a charge of $+2e$ moves through the tissue rather slowly, with a great deal of electrical interaction. The ionizing collisions thus occur close together along a short penetration path and are more localized. Therefore, potentially more dangerous damage is done by alpha particles than by electrons or gamma rays.

This *effective dose* is measured in terms of the **rem** (rad equivalent man). The various degrees of effectiveness of different particles are characterized by a factor called **relative biological effectivenss** (RBE), or *quality factor* (QF), which has been tabulated for various particles in Table 29.4. (Note in Table 29.4 that X rays and gamma rays have, by definition, an RBE of 1.)

The effective dose is given by the product of the dose in rads and the appropriate RBE:

$$\text{effective dose (in rems)} = \text{dose (in rads)} \times \text{RBE} \qquad (29.6)$$

Thus, 1 rem of *any* type of radiation does approximately the same amount of biological damage. For example, a 20-rem effective dose of alpha particles does the same amount of damage as a 20-rem effective dose of X rays. However, note that to administer these doses, 20 rad of X rays is needed, compared with only 1 rad of alpha particles.

Remember that the SI unit of *absorbed dose* is the gray. The SI unit of *effective dose* is the **sievert (Sv)**:

$$\text{effective dose (in sieverts)} = \text{dose (in grays)} \times \text{RBE} \qquad (29.7)$$

Since 1 Gy $\equiv$ 100 rad, it follows that 1 Sv $\equiv$ 100 rem.

TABLE 29.4 Typical Relative Biological Effectiveness (RBE) Values of Various Types of Radiations

Type	RBE (or QF)
X rays and gamma rays	1
Beta particles	1.2
Slow neutrons	4
Fast neutrons and protons	10
Alpha particles	20

Note: The gray was named in honor of Louis Harold Gray, a British radiobiologist whose studies laid the foundation for measuring absorbed dose.

It is difficult to set a maximum permissible radiation dosage, but the general standard for humans is an average dose of 5 rem/y after the age of 18, with no more than 3 rem in any three-month period. In the United States, the normal average annual dose per capita is about 200 mrem (millirem). About 125 mrem comes from the natural background of cosmic rays and naturally occurring radioactive isotopes in soil, building materials, and so on. The remainder is chiefly from diagnostic medical applications, mostly X rays.

Medical Treatment Using Radiation Some radioactive isotopes can be used for medical treatment, typically of cancerous conditions. Since a radioactive isotope, or a *radioisotope* as it is sometimes called, behaves chemically like a stable isotope of the element, it can participate in chemical reactions associated with normal bodily functions. One such radioisotope, used to treat thyroid cancer, is ^{131}I. Under usual conditions, the thyroid gland absorbs normal iodine. However, if ^{131}I is absorbed in a large enough dose, it can kill cancer cells. To see how a dose of radiation to the thyroid from ^{131}I can be estimated, consider Example 29.9. For a discussion of further uses of radioisotopes, see the Insight on applications of radiation on p. 954.

Example 29.9 ■ Radiation Dosage: Iodine-131 and Thyroid Cancer

One method of treating a cancerous thyroid is to administer a hefty amount of the radioactive isotope ^{131}I. The thyroid absorbs this iodine, and the iodine's gamma rays kill cells in the thyroid. (For data on ^{131}I, see Example 29.3.) (a) Write down the decay scheme of ^{131}I, and predict the identity of the daughter nucleus, which, in this case, is stable after emitting a gamma ray. (b) The charged particle (part (a) tells the type) has an average kinetic energy of 200 keV. Assume that the patient was given 0.0500 mCi of ^{131}I and that the thyroid absorbs only 25% of this amount. Further assume that only 40% of that 25% actually decays in the thyroid. If all of the energy carried by the charged particles is deposited in the thyroid, estimate the dose received by the thyroid (mass of 50.0 g). (Do not include the gamma rays.)

Thinking It Through. (a) The decay will be β^- decay, because the initial nucleus contains too many neutrons (78 neutrons compared with the 74 for stable iodine). The daughter nucleus is determined by its proton number. The daughter is left in an excited state and emits a gamma ray in order to become stable. (b) The dose depends on the energy deposited per kilogram of thyroid. Hence, we need to know how many β^- particles are emitted (and therefore absorbed). This number is determined by the initial number of ^{131}I nuclei that are actually present in the thyroid. The effective dose will depend on the RBE for β^- particles, found in Table 29.4.

Solution.

Given: 0.0500 mCi of ^{131}I ingested
25% of the ^{131}I makes it to the thyroid
40% of the ^{131}I that makes it to the thyroid decays there
$\overline{K}_\beta = 200$ keV
$m = 50.0$ g $= 0.0500$ kg
RBE (see Table 29.4)

Find: (a) the decay scheme for ^{131}I
(b) the dose (in rem) from the emitted particles

(a) Looking up the element with $Z = 54$, we find that it is xenon (Xe). The decay scheme is then

$$\begin{array}{ccccc}
^{131}_{53}\text{I} & \longrightarrow & ^{131}_{54}\text{Xe}^* & + & \beta^- \\
\text{iodine} & & \text{xenon} & & \text{beta} \\
& & \text{(excited)} & & \\
\end{array}$$

$$\quad\quad\quad\quad \longmapsto \quad ^{131}_{54}\text{Xe} \quad + \quad \gamma$$
$$\quad\quad\quad\quad\quad\quad\quad\quad \text{xenon} \quad\quad \text{gamma ray}$$

INSIGHT

Biological and Medical Applications of Radiation

Radiation has always been a double-edged sword. We know of its potentially harmful side; yet, sources of radiation can also provide solutions to problems. Exposure to high levels of gamma radiation is now an approved method of food sterilization in the United States. Food such as chicken and beef are commonly sterilized this way, thus reducing the threat of salmonella and *E. coli* contamination. In the aftermath of the terrorist attacks on the United States in the fall of 2001, some of this gamma technology is being retooled so that it can kill anthrax spores and other bioweapons.

External radiation sources such as ^{60}Co are also commonly used to treat cancer. ^{60}Co is a source of energetic gamma rays. After it beta decays (can you figure out the type of beta decay?) to ^{60}Ni in an excited state, two consecutive gamma rays of energy 1.17 and 1.33 MeV are emitted. These energies are much higher than provided by conventional X-ray machines; hence, a sample of ^{60}Co, with its relatively long half-life of about five years, can provide an inexpensive and convenient source of penetrating radiation. Essentially, all you need is the ^{60}Co sample in a lead box with a small hole to allow for gamma rays to exit in one direction. One problem, however, with this single-beam method is that the gamma rays deposit energy in normal flesh both in front of and behind the targeted tumor.

An improved version of ^{60}Co treatment, called the *gamma knife*, is in use at several research hospitals (Fig. 1). Once the tumor is located, a ring of gamma rays from ^{60}Co sources is accurately aimed at it. Any one source is relatively weak, thus not doing too much damage outside the tumor. However, where the beams meet at the tumor site, a lot of energy is deposited. Thus, a large dose can be deposited in the tumor, with minimal damage to the surrounding tissue. This instrument requires careful computer calculations and is still under development.

Even more exotic techniques are being tried in an effort to kill inoperable tumors. One method is called *pion therapy*. A pion is an unstable elementary particle (Chapter 30) that can be produced in accelerators by bombarding a target, like carbon, with high-energy protons. Pions can be charged or uncharged. Of interest in this area are the negatively charged pions. Pions of a specific kinetic energy can be selected and focused by magnetic fields (Chapter 19) onto a region of the body where a tumor exists. Unlike photons (gamma rays and X rays), charged particles create most of their ionization "damage" at the end of their path, when they are moving slowly (Fig. 2). By adjusting their kinetic energy, the pions can be made to end up at the tumor site, thus causing maximum damage to the cancer cells. As a bonus, since the pions are unstable, they will give off gamma rays that will do even more damage inside the tumor (Fig. 3). Research on this technology-intensive technique is ongoing, with some limited success.

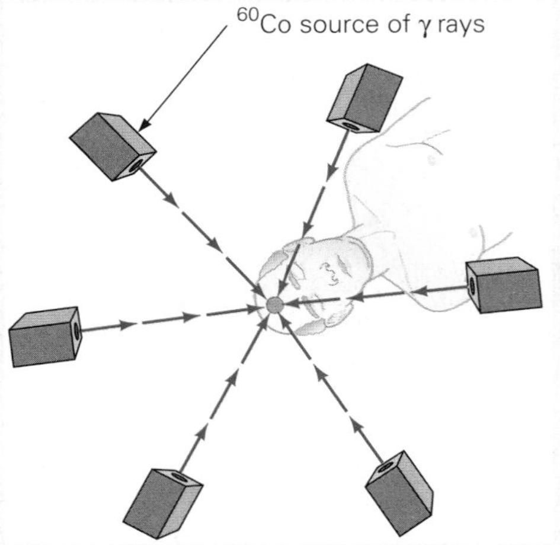

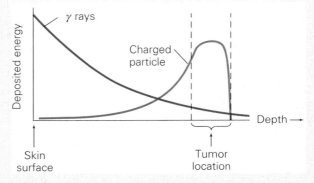

Figure 2　A comparison of the energy deposited for a gamma-ray beam with that of a charged particle passing through tissue.

Figure 1　The gamma knife consists of many relatively weak beams of gamma rays (usually from ^{60}Co) that are calculated to converge on the location of an inoperable tumor. Thus, unlike in the traditional single-beam treatment, much more of the energy ends up at the tumor site, and less damage is done in front of, or behind, the tumor.

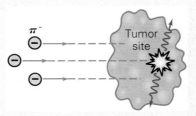

Figure 3　Pionic cancer therapy consists of focusing a beam of negative pions onto a tumor. Unlike with beams of gamma rays, most of the pion energy is deposited at the tumor site. In addition, when pions decay, gamma rays are released at the tumor causing further destruction of cancer cells.

(b) Of the 0.0500 mCi, only 0.0125 mCi makes it to the thyroid. Of that, only 40%, or .00500 mCi (5.00×10^{-6} Ci), actually decays in the thyroid. From this information and Eq. 29.2 the number of ^{131}I nuclei that decay in the thyroid can be found. From Example 29.3, the decay constant for ^{131}I is $\lambda = 1.0 \times 10^{-6}$ s^{-1}. Thus, the number of ^{131}I nuclei, N, that decays in the thyroid is

$$N = \frac{R}{\lambda} = \frac{(5.00 \times 10^{-6} \text{ Ci})\left(3.7 \times 10^{10} \dfrac{\text{nuclei/sec}}{\text{Ci}}\right)}{1.0 \times 10^{-6} \text{ s}^{-1}} = 1.85 \times 10^{11} \,^{131}\text{I nuclei}$$

Each ^{131}I nucleus releases one β^- particle, with an average kinetic energy of 200 keV. Remembering that there is 1.60×10^{-19} J/eV, or 1.60×10^{-16} J/keV, we find that the energy, E, deposited in the thyroid is

$$E = (1.85 \times 10^{11} \,^{131}\text{I nuclei})\left(200 \frac{\text{keV}}{^{131}\text{I nuclei}}\right)\left(1.60 \times 10^{-16} \frac{\text{J}}{\text{keV}}\right)$$
$$= 5.92 \times 10^{-3} \text{ J}$$

The absorbed dose is

$$\text{absorbed dose} = \frac{5.92 \times 10^{-3} \text{ J}}{0.0500 \text{ kg}} = 0.118 \frac{\text{J}}{\text{kg}}$$
$$= 0.118 \text{ Gy or } 11.8 \text{ rad}$$

The effective dose (in sieverts and rems) is this multiplied by the RBE for beta particles (1.2):

$$\text{effective dose} = (0.118 \text{ Gy})(1.2) = 0.142 \text{ Sv} = 14.2 \text{ rem}$$

Follow-up Exercise. In this Example, determine the absorbed and effective dose from the gamma rays, assuming that 10% of the gamma rays are absorbed in the thyroid tissue and that their energy is 364 keV. Compare these results with the dose from the β^- radiation.

Medical Diagnosis Applications That Use Radiation Besides theraputic use, such as that previously described for iodine-131, radioactive isotopes can be used for diagnostic procedures. Since the radioisotope behaves chemically like a stable isotope, attaching radioisotopes to molecules enables the molecules to be used as tracers as they travel to different organs and regions of the body.

Many bodily functions can be studied by monitoring the location and activity of tracer molecules as they are absorbed during body processes. For example, the activity of the thyroid gland can be determined by monitoring its iodine uptake with small amounts of radioactive iodine-123. This isotope emits gamma rays and has a half-life of 13.3 h. The uptake of radioactive iodine by a person's thyroid can be monitored by a gamma detector and compared with the function of a normal thyroid to see if there is any gland abnormality. Similarly, radioactive solutions of iodine and gold are quickly absorbed by the liver.

One of the most commonly used diagnostic tracers is technetium-99 (^{99}Tc). It has a convenient half-life of 6 h, emits gamma rays, and combines with a large variety of compounds. When injected into the blood stream, ^{99}Tc will not be absorbed by the brain, because of the blood–brain barrier. However, tumors do not have this barrier. Thus, brain tumors readily absorb the ^{99}Tc. These tumors then show as gamma-ray emitters on detectors external to the body. Similarly, other areas of the body can be scanned and unusual activities noted and measured.

It is possible to image gamma-ray activity in a single plane, or "slice," through the body. A gamma detector is moved around the patient to measure the emission intensity from many angles. A complete image can then be constructed by using computer-assisted tomography, as in X-ray CT. This process is referred to as *single-photon emission tomography* (SPET). Another technique, *positron emission tomography* (PET; introduced in Chapter 28), uses tracers that are positron emitters, such as ^{11}C and ^{15}O. When a positron is emitted, it is quickly annihilated, and two gamma rays are produced, and travel in opposite directions. The gamma rays are recorded simultaneously by a ring of detectors surrounding the patient (▶Fig. 29.18).

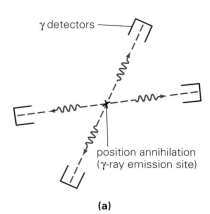

γ detectors

position annihilation (γ-ray emission site)

(a)

(b)

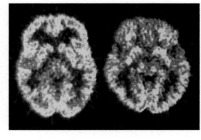

(c)

▲ **FIGURE 29.18   PET scan**
(a) and (b) A PET scanner can, for example, monitor brain activity after the administration of glucose-containing radioactive isotopes. Note the use of an array of detectors for the gamma radiation produced when a positron is annihilated. Any one detector pair pinpoints a line on which the source of the gamma emission was located. With many such pairs working together, the site can be determined to an accuracy of about a centimeter. **(c)** PET scans of a normal brain (left) and the brain of a schizophrenic patient (right).

In a common application, PET technology is used to detect fast-growing cancer cells. The positron emitter ^{18}F is chemically attached to glucose molecules and administered to the patient. Actively growing cells absorb glucose under normal conditions, but very active cancer cells absorb considerably more than average. By comparing the emissions coming from a given region on a potentially sick patient with that from a normal, healthy person, these "overactive" cancer cells can be detected. Such PET scans are now routinely done, for example, as a follow-up to chemotherapy treatment of lymphoma (cancer of the lymph system). A PET scan can detect even tiny leftover active tumors that can then be targeted for further treatment.

Domestic and Industrial Applications of Radiation

A common application of radioactivity in the home is the smoke detector. In this detector, a weak radioactive source ionizes air molecules, setting up a small current in the detector circuit. If smoke enters the detector, the ions there become attached to the smoke particles, causing a reduction in the current. The drop in current is sensed electronically, which triggers an alarm (▼ Fig. 29.19).

Industry also makes good use of radioactive isotopes. Radioactive tracers are used to determine flow rates in pipes, to detect leaks, and to study corrosion and wear. Also, it is possible to radioactivate certain compounds at a particular stage in a process by irradiating them with particles, generally neutrons. This technique is called **neutron activation analysis** and is an important method of identifying elements in a sample. Before the development of this procedure, the chief methods of identification were chemical and spectral analyses. In both of these methods, a fairly large amount of a sample has to be destroyed in the analysis procedure. As a result, a sample may not be large enough for analysis, or small traces of elements in a sample may go undetected. Neutron activation analysis has the advantage over these methods on both scores. Only minute samples are needed, and the method can detect just trace amounts of an element.

A typical neutron activation process might start with californium-252, an unstable neutron emitter:

$$\text{(source)} \qquad ^{252}_{98}\text{Cf} \rightarrow\ ^{251}_{98}\text{Cf} +\ ^{1}_{0}\text{n}$$

The neutrons emitted are used to bombard a sample and create characteristic gamma rays. A common target is nitrogen, consisting mostly of ^{14}N. When ^{14}N absorbs a neutron, the ^{15}N nucleus is usually in an excited state. The excited nitrogen-15 nucleus decays with the emission of a gamma ray with a distinctive energy. This reaction is shown as follows:

$$^{1}_{0}\text{n} +\ ^{14}_{7}\text{N} \rightarrow\ ^{15}_{7}\text{N}^{*} \rightarrow\ ^{15}_{7}\text{N} + \gamma$$

$$\quad\quad\quad\quad\ \ \text{(excited} \qquad\qquad \text{(gamma}$$
$$\quad\quad\quad\quad\ \ \text{nucleus)} \qquad\qquad\ \text{ray)}$$

By placing an energy-sensitive gamma-ray detector to the side of the sample, the presence of nitrogen can be ascertained.

Nitrogen activation is commonly used as an important antiterrorist tool at airports. Virtually all explosives contain nitrogen. Thus, by using neutron activation

▶ **FIGURE 29.19 Smoke detector** A weak radioactive source ionizes the air and sets up a small current. Smoke particles in the detector reduce the current, causing an alarm to sound.

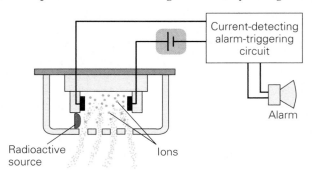

and analyzing the energy of any gamma-ray emission coming from a suitcase, we can check for an explosive device in the suitcase. Other materials in the suitcase may contain nitrogen too, so manual checks are made to confirm any suspicious findings.

Recently, the U.S. government has given permission for the use of gamma radiation in the processing of poultry. The radiation kills bacteria, helps preserve the food, and in no way makes the food radioactive. There are, however, continuing concerns with this process from food health professionals. Even though the gamma-ray emission cannot make the meat radioactive, it *can* change some of the *chemical* bonding through ionization effects. This possibility has prompted enough concerns about whether this process can affect the chemical structure of the meat—making it unsafe to eat—to warrant further study.

Chapter Review

Important Concepts

- **The Rutherford–Bohr model** of the atom is a planetary model, with negative electrons orbiting the positively charged nucleus, where almost all the mass of the atom is located.

- The **strong nuclear force** is the short-range attractive force between nucleons that is responsible for holding the nucleus together against the repulsive electric force of its protons.

- The **nucleus** of an atom contains protons and neutrons, collectively named **nucleons**. The nucleus is characterized by its **proton number** Z and its **neutron number** N. Its mass number, or **atomic number**, A, is the total number of nucleons, so $A = N + Z$.

- **Isotopes** of an element differ only in the number of neutrons they contain in their nucleus.

- **Radioactivity** refers to any type of nuclear instability that results in the emission of particles and photons.

- Nuclei may undergo radioactive decay by the emission of an alpha particle, or helium nucleus ($\boldsymbol{\alpha}$ **decay**); a beta particle, either an electron ($\boldsymbol{\beta^-}$ **decay**) or a positron ($\boldsymbol{\beta^+}$ **decay**); or a gamma ray, a high-energy photon of electromagnetic radiation ($\boldsymbol{\gamma}$ **decay**). Nuclei can also change by capturing one of the atom's orbital electrons (electron capture, or **EC**).

- In any nuclear reaction or decay, two conservation rules pertain: **Total charge** and **total nucleon number** are conserved.

- The **half-life** of a nuclide is the time required for the number of undecayed nuclei in a sample to fall to half of its initial value. The number of undecayed nuclei remaining after a time t is given by the exponential-decay relationship

$$N = N_0 e^{-\lambda t} \qquad (29.3)$$

where the **decay constant** (λ) is inversely related to the half-life by

$$t_{1/2} = \frac{0.693}{\lambda} \qquad (29.4)$$

The **activity** (R) of a radioactive sample is the rate at which the nuclei in the sample decay. It is directly proportional to the number of nuclei remaining undecayed and is given by

$$R = \text{activity} = \left| \frac{\Delta N}{\Delta t} \right| = \lambda N \qquad (29.2)$$

Activity is measured in units of the **curie (Ci)** or the **becquerel (Bq)**. $1\ \text{Ci} \equiv 3.70 \times 10^{10}$ decays/s, and $1\ \text{Bq} \equiv 1$ decay/s; thus, $1\ \text{Ci} = 3.70 \times 10^{10}$ Bq.

- **Nuclear stability** depends primarily on the proton-to-neutron ratio and the pairing effect. The latter makes nuclei with even numbers of protons and neutrons (even–even nuclei) more stable and abundant than even–odd or odd–odd nuclei.

- Nuclear masses are most conveniently measured in units of the **atomic mass unit (u)**. In terms of Einstein's mass–energy equivalence, the annihilation of 1 u of mass releases 931.5 MeV of energy. The atomic mass unit is related to the kilogram by $1\ \text{u} = 1.66054 \times 10^{-27}$ kg.

- **Total binding energy** (E_b) is the minimum amount of energy needed to separate a nucleus into its constituent nucleons:

$$E_b = (\Delta m)c^2 \qquad (29.5)$$

Δm is the **mass defect** and is the difference between the sum of the masses of the constituent nucleons and the mass of the nucleus as a whole.

- The effective dose of radiation (**rem** or **sievert**) is determined by the energy deposited per kilogram of material (**rad** or **gray**) and the type of particle depositing the energy (**RBE**):

Effective Dose:

Dose (in rem) = dose (in rad) × RBE $\qquad (29.6)$

Dose (in Sv) = dose (in Gy) × RBE $\qquad (29.7)$

The RBE is a factor that depends on the type of particle (radiation) that deposits the energy.

Exercises

29.1 Nuclear Structure and the Nuclear Force

1. The concept of the atomic nucleus resulted from the work of (a) Bohr, (b) Thomson, (c) Rutherford, or (d) Geiger.

2. Carbon-12 (a) has the same number of protons and neutrons, (b) is a nuclide, (c) is an isotope, or (d) all of the preceding.

3. The strong nuclear force (a) binds the orbital electrons to the atomic nucleus, (b) has a longer range than that of the electrostatic force, (c) acts only between identical particles, or (d) overcomes the force of repulsion between protons in the nucleus.

4. CQ Is the nuclear force attractive or repulsive? What particles can it act between? Is it a long- or short-ranged force?

5. CQ What do the isotopes of an element have in common? How do they differ?

6. ■ Determine the number of protons, neutrons, and electrons in a neutral atom with the following nuclei: (a) $^{43}_{20}$Ca and (b) $^{58}_{29}$Cu.

7. ■ (a) Write the nuclear notations for the isotopes of hydrogen, using the symbols H, D, and T. (b) Write the chemical symbols for six different forms of water, using these symbols. (c) Identify the radioactive forms.

8. ■ Oxygen has three stable isotopes, with 8, 9, and 10 neutrons, respectively. Write these isotopes in nuclear notation.

9. ■ An isotope of potassium has the same number of neutrons as the nuclide argon-40. Write the nuclear notation for this potassium isotope.

10. ■ ^{24}Mg and ^{25}Mg are two isotopes of magnesium. What are the numbers of protons, neutrons, and electrons in each if (a) the atom is electrically neutral, (b) the ion has a −2 charge, and (c) the ion has a +1 charge?

11. ■ One isotope of uranium has a mass number of 238. What are the numbers of protons, neutrons, and electrons in a neutral atom of this isotope?

12. IE ■■ (a) Isotopes of an element have the same (1) atomic number, (2) neutron number, or (3) mass number. (b) Write two possible isotopes for the element ^A_ZX.

13. ■■ An approximate expression for the nuclear radius (R) is $R = R_o A^{1/3}$, where $R_o = 1.2 \times 10^{-15}$ m and A is the mass number of the nucleus. (a) Find the nuclear radii of atoms of the noble gases: He, Ne, Ar, Kr, Xe, and Rn. (b) How does the order of magnitude of your answers compare with the approximate size of an atom, roughly 0.1 nm?

29.2 Radioactivity

14. The conservation of nucleons and the conservation of charge apply to (a) only alpha decay, (b) only beta decay, (c) only gamma decay, or (d) all nuclear processes.

15. CQ The neutron number is not conserved in a beta decay. Is this a violation of the conservation of nucleons? Explain

16. CQ What are the main differences between alpha, beta, and gamma rays?

17. IE ■ Tritium is radioactive. (a) Would you expect it to (1) β^+, (2) β^-, or (3) alpha decay? Why? (b) What is the identity of the daughter nucleus? Is it stable? [Hint: Write the nuclear equation for each.]

18. ■ Write the nuclear equations expressing (a) the beta decay of $^{60}_{27}$Co and (b) the alpha decay of $^{226}_{88}$Ra.

19. ■ Write the nuclear equations for (a) the alpha decay of neptunium-237, (b) the β^- decay of phosphorus-32, (c) the β^+ decay of cobalt-56, (d) electron capture in cobalt-56, and (e) the γ decay of potassium-42.

20. IE ■ Polonium-214 can decay by alpha decay. (a) The product of its decay has how many fewer protons than polonium-214, (1) zero, (2) one, (3) two, or (4) four? Why? (b) Write the nuclear equation for this decay.

21. ■ A lead-209 nucleus results from both alpha–beta sequential decays and beta–alpha sequential decays. What was the grandparent nucleus? (Show this result for both decay routes by writing the nuclear equations for both decay processes.)

22. ■■ Complete the following nuclear-decay equations:
    (a) $^8_4$Be $\rightarrow$ $^4_2$He +
    (b) $^{240}_{94}$Po $\rightarrow$ $^{97}_{38}$Sr + $^{139}_{56}$Ba +
    (c) $^{47}_{21}$Sc* $\rightarrow$ $^{47}_{21}$Sc +
    (d) $^{29}_{11}$Na $\rightarrow$ $^0_{-1}$e +

23. ■■ Complete the following nuclear-decay equations:
    (a) $^{238}_{92}$U $\rightarrow$ $^{234}_{90}$Th +
    (b) $^{40}_{19}$K $\rightarrow$ $^{40}_{20}$Ca +
    (c) $^{236}_{92}$U $\rightarrow$ $^{131}_{53}$I + 3($^1_0$n) +
    (d) $^{23}_{11}$Na + γ $\rightarrow$
    (e) $^{22}_{11}$Na + $\rightarrow$ $^{22}_{10}$Ne

24. ■■ Actinium-227 ($^{227}_{89}$Ac) decays by alpha decay or by beta decay and is part of a decay sequence like the one shown in Fig. 29.8. Each daughter nucleus then decays to radium-223 ($^{223}_{88}$Ra), which subsequently decays to polonium-215 ($^{215}_{84}$Po). Write the nuclear equations for the decay process in the decay series from ^{227}Ac to ^{215}Po.

25. ■■ The decay series for neptunium-237 is shown in ▼ Fig. 29.20. Identify the decay modes and each of the nuclei in the sequence.

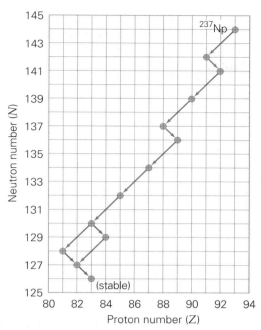

▲ **FIGURE 29.20** **Neptunium-237 decay series** See Exercise 25.

29.3 Decay Rate and Half-life

26. After one half-life, a sample of a particular radioactive material (a) is half as massive, (b) has its half-life reduced by half, (c) is no longer radioactive, or (d) has its activity reduced by half.

27. In two half-lives, the activity of a radioactive sample will have decreased by what percent, (a) 25%, (b) 50%, (c) 75%, or (d) 87.5%?

28. CQ What physical or chemical properties affect the decay rate, or half-life, of a radioactive isotope?

29. CQ Nuclide A has a decay constant that is half that of nuclide B. If the two nuclides start with the same number of undecayed nuclei, will twice as many of nuclide A's nuclei as the number of nuclide B's nuclei decay in a given time? Explain.

30. CQ What are the (a) half-life and (b) decay constant for a stable isotope?

31. ■ A particular radioactive sample undergoes 2.50×10^6 decays/s. What is the activity of the sample in (a) curies and (b) becquerels?

32. ■ At present, a laboratory radioactive beta source has an activity of 20 mCi. (a) What is the present decay rate in decays per second? (b) Assuming that one beta particle is emitted per decay, how many are currently emitted per minute?

33. IE ■■ The half-life of a radioactive isotope is 1 h. (a) What fraction of a sample would be left after 3 h, (1) one-third, (2) one-eighth, or (3) one-ninth? Why? (b) What fraction of a sample would be left after 1 day?

34. ■■ A sample of technetium-104 has an activity of 10 mCi. Estimate the activity of the sample after one hour has elapsed ($t_{1/2} = 18$ min).

35. ■■ What period of time is required for a sample of radioactive tritium (^{3}H) to lose 80.0% of its activity? Tritium has a half-life of 12.3 years.

36. IE ■■ The half-lives for $^{28}_{12}$Mg and $^{104}_{32}$Tc are 21 h and 18 min, respectively. (a) The decay constant of $^{104}_{32}$Tc will be (1) larger than, (2) equal to, or (3) smaller than that of $^{28}_{12}$Mg. Why? (b) Calculate the ratio of the decay constant of the technicium isotope to that of the magnesium isotope.

37. ■■ A certain amount of ^{131}I is introduced into a patient in a medical diagnostic procedure for her thyroid. What percentage of the sample remains after 1.00 day, assuming that all of the ^{131}I is retained in the patient's thyroid gland? (Answer to three significant figures.)

38. IE ■■ Carbon-14 dating is used to determine the age of some buried bones. (a) If the activity of bone A is higher than that of bone B, then bone A is (1) older than, (2) younger than, or (3) the same age as bone B. Why? (b) A sample of old bone is found to have 4.0 beta decays/min for each gram of carbon. Approximately how old is the bone?

39. ■■ Show that the number N of radioactive nuclei remaining in a sample after n half-lives is given by

$$N = \frac{N_o}{2^n} = \left(\tfrac{1}{2}\right)^n N_o$$

where N_o is the initial number of nuclei.

40. ■■ Some ancient writings on parchment are found sealed in a jar in a cave. If carbon-14 dating shows the parchment to be 28 650 years old, what percentage of the carbon-14 atoms still remains in the sample compared with the number that was present when the parchment was made?

41. ■■ The activity of an unknown beta source is observed to decrease by 87.5% in 54 min. What is the most likely identity of this isotope?

42. ■■ How long would it take (to the nearest whole year) for the activity of a sample of cobalt-60 to reduce to 20% of the original activity?

43. ■■ A soil sample contains 40 μg of ^{90}Sr. Approximately how much ^{90}Sr will be in the sample 150 years from now?

44. ■■ (a) What is the decay constant of fluorine-17 ($t_{1/2} = 66.0$ s)? (b) How long will it take for the activity of a sample of ^{17}F to decrease to 10% of its initial value?

45. ■■ Francium-223 ($^{223}_{87}$Fr) has a half-life of 21.8 min. (a) How many nuclei are initially present in a 25.0-mg sample of this isotope? (b) How many nuclei will be present 1 h and 49 min later?

46. ■■ A basement room containing radon gas ($t_{1/2} =$ 3.82 days) is sealed to be airtight. (a) If there are 7.50×10^{10} radon atoms trapped in the room, estimate how many radon atoms remain in the room after one week. (b) Radon undergoes alpha decay. After 30 days, is the number of its daughter nuclei equal to the number of radon parents that have decayed? Explain.

47. ■■ In 1898, Pierre and Marie Curie isolated about 10 mg of radium-226 from eight tons of uranium ore. If this sample had been placed in a museum, how much of the radium would remain in the year 2100?

48. ■■■ An ancient artifact is found to contain 250 g of carbon and has an activity of 475 decays per minute. What is the approximate age of the artifact, to the nearest thousand years?

49. ■■■ The recoverable U.S. reserves of high-grade uranium-238 ore (high-grade ore contains about 10 kg of $^{238}U_3O_8$ per ton) are estimated to be about 500 000 tons. Neglecting any geological changes, what mass of ^{238}U existed in this high-grade ore when the Earth was formed, about 4.6 billion years ago? [*Hint*: See Appendix V.]

50. **IE** ■■■ Nitrogen-13, with a half-life of 10 min, decays by positron emission. (a) The end product is (1) ^{13}N, (2) ^{13}C, or (3) ^{13}O. Why? (b) If a pure sample of ^{13}N has a mass of 1.5 g at a certain time, what is the activity 35 min later? (c) What percentage of the sample is ^{13}N at this time?

29.4 Nuclear Stability and Binding Energy

51. For nuclei with a mass number greater than 40, which of the following statements is correct? (a) The number of protons is approximately equal to the number of neutrons; (b) the number of protons exceeds the number of neutrons; (c) all such nuclei are stable up to $Z = 92$; (d) none of the preceding.

52. The average binding energy per nucleon of the daughter nucleus in a decay process is (a) greater than, (b) less than, or (c) equal to that of the parent nucleus.

53. ■ Determine whether each of the three isotopes of hydrogen is likely to be stable or unstable.

54. ■ Using Appendix V, list those stable nuclei that are "doubly magic." That is, make a list of nuclei that are likely to have magic numbers for both neutrons and protons.

55. ■ From which one of the following pairs of nuclei would you expect it to be easier to remove a neutron? (a) $^{16}_8$O or $^{17}_8$O; (b) $^{40}_{20}$Ca or $^{42}_{20}$Ca; (c) $^{10}_5$B or $^{11}_5$B; (d) $^{208}_{82}$Pb or $^{209}_{83}$Bi. State your reasoning for your choice in each case.

56. ■ Write in nuclear notation the most abundant stable isotope of the nuclide with a proton magic number of 28. (See Appendix V.)

57. ■ Only two isotopes of Sb (antimony, $Z = 51$) are stable. Pick the two stable isotopes from the following list: (a) ^{120}Sb; (b) ^{121}Sb; (c) ^{122}Sb; (d) ^{123}Sb; (e) ^{124}Sb.

58. ■ Determine which of the following isotopes are likely to be unstable: (a) ^{23}Na, (b) ^{50}V; (c) ^{209}Bi; (d) ^{209}Po; (e) ^{44}Ca.

59. ■ The total binding energy of $^2_1$H is 2.224 MeV. Use this information to compute the mass of a ^{2}H nucleus from the known mass of a proton and a neutron.

60. ■■ Use Avogadro's number (see Section 10.3), $N_A = 6.02 \times 10^{23}$ atoms/mole, to show that $1\,u = 1.66 \times 10^{-27}$ kg. (Recall that a ^{12}C atom has a mass of exactly 12 u.)

61. ■■ The mass of $^{12}_6$C is exactly 12 u. (a) What is the total binding energy of this nucleus? (b) What is the average binding energy per nucleon?

62. ■■ The mass of $^{16}_8$O is 15.994915 u. What is the average binding energy per nucleon (E_b/A) for this nucleus?

63. ■■ Which isotope of hydrogen has the lower average binding energy per nucleon, deuterium or tritium? Justify your answer mathematically.

64. ■■ Near high-neutron areas, such as a nuclear reactor, neutrons will be absorbed by protons (the hydrogen nucleus in water molecules) and will give off a gamma ray of a characteristic energy in the process. What is the energy of the gamma ray (to three significant figures)?

65. ■■ How much energy (to four significant figures) would be required to completely separate all the nucleons of a nitrogen-14 nucleus, the atom of which has a mass of 14.003074 u?

66. ■■ Calculate the binding energy of the last neutron in the $^{40}_{19}$K nucleus. [*Hint*: Compare the mass of $^{40}_{19}$K with the mass of $^{39}_{19}$K plus the mass of a neutron.]

67. ■■ If an alpha particle could be removed intact from an aluminum-27 nucleus ($m = 26.981541$ u), a sodium-23 nucleus ($m = 22.989770$ u) would remain. How much energy would be required to do this operation?

68. ■■ On average, are nucleons more tightly bound in an ^{27}Al nucleus or in a ^{23}Na nucleus?

69. ■■ The atomic mass of $^{235}_{92}$U is 235.043925 u. Find the average binding energy per nucleon for this isotope.

70. **IE** ■■ The total binding energy of $^6_3$Li is 32.0 MeV. (a) Compared with the masses of three protons and three neutrons taken together, the mass of $^6_3$Li is (1) greater, (2) the

same, or (3) less. Why? (b) What is the mass (in u) of the ^{6}Li nucleus?

71. ■■■ The mass of $^8_4$Be is 8.005305 u. (a) Which is less, the total mass of two alpha particles or the mass of the ^{8}Be nucleus? (b) Which is greater, the total binding energy of the ^{8}Be nucleus or the total binding energy of two alpha particles? (c) On the basis of your answers to parts (a) and (b) alone, do you expect the ^{8}Be nucleus to decay spontaneously into two alpha particles?

29.5 Radiation Detection and Applications

72. Which type of detector can record the trajectory of charged particles? (a) Geiger counter, (b) scintillation counter, (c) solid-state detector, or (d) spark chamber?

73. A unit of radiation dosage is the (a) rad, (b) gray, (c) sievert, or (d) all of the preceding.

74. CQ A basic assumption of radiocarbon dating is that the cosmic-ray intensity has been generally constant for the last 40 000 years or so. Suppose it were found that the intensity was much less 100 000 years ago than it is today. How would this finding affect the process of carbon-14 dating?

75. CQ Will X ray and alpha particles of the same dose have the same effective dose? Why?

76. ■ A patient receives a 1.0-rad dose of radiation from X rays and a 1.0-rad dose from an alpha source. (a) Which dose has the greater biological effectiveness? (b) What are the doses in rems?

77. ■ In a diagnostic procedure, a patient in a hospital ingests 80 mCi of gold-198 ($t_{1/2}$ = 2.7 days). What is the activity at the end of one month if none of the gold is eliminated from the body by biological functions?

78. ■ A technician working at a nuclear-reactor facility is exposed to a slow neutron radiation and receives a dose of 1.25 rad. (a) How much energy is absorbed by 200 g of the worker's tissue? (b) Was the maximum permissible radiation dosage exceeded?

79. ■ A person working with nuclear isotopes for a two-month period receives a 0.5-rad dose from a gamma source, a 0.3-rad dose from a slow-neutron source, and a 0.1-rad dose from an alpha source. Was the maximum permissible radiation dosage exceeded?

80. ■■ A cancer treatment called the *gamma knife* (see Insight p. 954) uses a large number of ^{60}Co sources to treat tumors. ^{60}Co emits two gamma rays of energy 1.33 MeV and 1.17 MeV in quick succession. Assume that 50.0% of the total gamma-ray energy is absorbed by a tumor. The total activity of the ^{60}Co sources is 1.00 mCi, the tumor's mass is 0.100 kg and the patient is exposed for an hour. Calculate the effective radiation dose received by the tumor. (Since the ^{60}Co half-life is 5.3 y, changes in its activity during treatment can be neglected.)

81. ■■ Neutron activation analysis was performed on small pieces of hair that had been taken from the exiled Napoleon after he died on the island of St. Helena in 1821. The samples were found to contain abnormally high levels of arsenic, which supported a theory that his death was not due to natural causes. If this evidence was derived by studying beta emissions coming from arsenic-76 nuclei, what was the arsenic isotope present in the hair, and what was the final nucleus after the beta decay?

Additional Exercises

82. A fossil specimen is believed to be about 18 000 years old. Explain how this hypothesis could be confirmed.

83. The binding energy per nucleon of $^4_2$He is 7.075 MeV. (a) What is the nucleus's total binding energy? (b) What is the mass of a ^{4}He nucleus?

84. Starting with uranium-234 ($^{234}_{92}$U), there is a decay sequence of four alpha decays and two β^- decays. What is the nucleus that remains at the end of this sequence?

85. A sample of ^{215}Bi, which beta decays ($t_{1/2}$ = 2.4 min), contains Avogadro's number of nuclei. How many bismuth nuclei are present after (a) 10 min and (b) 1.0 h? (c) What are the activities, in curies and becquerels, at these times?

86. An old bone yields, on average, 1.00 beta emissions per minute per gram of carbon. Approximately how old is the bone?

87. Complete the following nuclear-decay equations:
    (a) $^{47}_{21}$Sc → $^{47}_{22}$Ti +
    (b) $^{226}_{88}$Ra → $^4_2$He +
    (c) $^{237}_{93}$Np → $^0_{-1}$e +
    (d) $^{210}_{84}$Po* → $^{210}_{84}$Po +
    (e) $^{11}_6$C → $^{11}_5$B +

88. ▶Figure 29.21 shows the decay series for plutonium-239. Identify the first four nuclei in this decay process.

89. The value of E_b/A for $^{238}_{92}$U is 7.58 MeV. (a) What is the total binding energy of this nucleus? (b) What is its mass?

90. What are the rest energies of (a) a carbon-12 atom and (b) an alpha particle?

91. Determine which of the following isotopes are likely to be stable: (a) ^{20}C; (b) ^{100}Sn; (c) ^{6}Li; (d) ^{9}Be; (e) ^{22}Na.

92. Thorium-229 ($^{229}_{90}$Th) undergoes alpha decay. (a) What is the daughter nucleus of this process? Write the decay equation. (b) This daughter nucleus then undergoes beta decay. What is the resulting granddaughter nucleus? Write the decay equation.

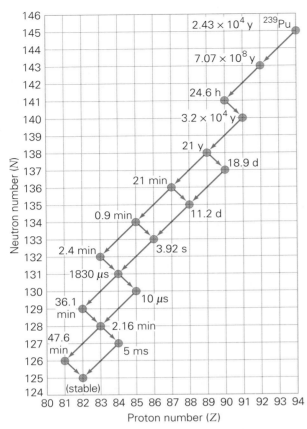

▲ **FIGURE 29.21 Plutonium-239 decay series** See Exercise 88.

93. When a $^6_3$Li nucleus is struck by a proton, an alpha particle and another nucleus are released. What is this other nucleus?

94. Gold-198 beta decays with a half-life of 2.7 days. (a) What is the activity of a sample of 1.0 g of pure $^{198}_{79}$Au? (b) What is the activity of the sample at the end of one month?

95. From Exercise 13, we know that the approximate radius of a nucleus is given by $R = R_0 A^{1/3}$, where $R_0 = 1.2 \times 10^{-15}$ m and A is the mass number of the nucleus. Assuming that nuclei are spherical (they are approximately so in many cases), (a) show that the average nucleon density in a nucleus is 1.4×10^{44} nucleons/m^3 and (b) estimate the nuclear density in kilograms per cubic meter.

96. $^3_1$H has a half-life of 12.33 years. What percentage of a sample containing $^3_1$H will remain after 6.00 years?

97. **IE** The nuclear mass and the atomic mass of hydrogen are essentially the same. (a) The nuclear density must be (1) greater than, (2) the same as, or (3) less than the average density of the atom. Why? (b) Prove your answer to part (a) by comparing the average nuclear density results of Exercise 95(b), to the average density of a hydrogen atom with a radius of 0.0529 nm, assuming that atom is spherical. (c) Does your number for hydrogen agree with the density of hydrogen gas at normal temperatures and pressures? Why or why not?

98. Calculate the kinetic energy of an alpha particle emitted by $^{232}_{92}$U. Neglect the recoil velocity of the daughter nucleus, $^{228}_{90}$Th.

Nuclear Reactions and Elementary Particles

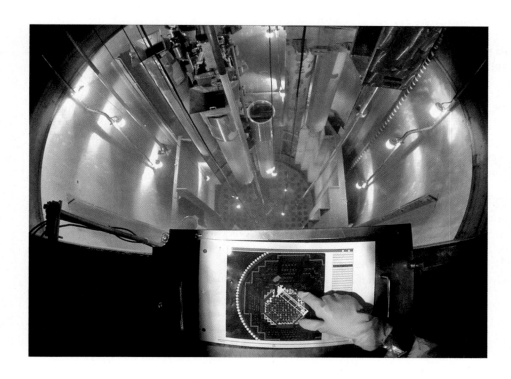

30.1 Nuclear Reactions

30.2 Nuclear Fission

30.3 Nuclear Fusion

30.4 Beta Decay and the Neutrino

30.5 Fundamental Forces and Exchange Particles

30.6 Elementary Particles

30.7 The Quark Model

30.8 Force Unification Theories, the Standard Model, and the Early Universe

Today, more than half a century after the first nuclear reactor was built, these facilities remain a center of controversy. Some people see them as essential to solving the world's energy problems. To others, they embody all that is wrong and dangerous about modern technology. The reactor shown in the opening photo is a symbol of this ongoing debate. It is part of the Three Mile Island (TMI) plant in Pennsylvania, the site of one of the largest and most publicized nuclear accidents in the world, second perhaps only to Chernobyl (more on that later). In this photo, you are looking into the heart of one of the two remaining TMI reactors, both of which are still running safely.

In this chapter, we'll study the nuclear fission reactions that power such reactors. We will also learn how scientists are trying to harness fusion reactions, which occur within the Sun and other stars and might one day provide a safe and virtually inexhaustible supply of clean energy for our planet.

We will also look at elementary particles and their interactions. Investigations have shown that a variety of particles other than the proton and neutron is associated with the nucleus. The discovery of these particles has given us insight into the world of the nucleus and provided a clearer understanding of the evolution of the universe.

30.1 Nuclear Reactions

OBJECTIVES: To (a) use charge and nucleon conservation to write nuclear reaction equations and (b) understand and use the concepts of Q value and threshold energy to analyze nuclear reactions.

Ordinary chemical reactions between atoms and molecules involve only orbital electrons. The nuclei of these atoms do not participate in the process, so the atoms retain their identity. Conversely, in **nuclear reactions**, the original nuclei are converted into the nuclei of other elements. Scientists first became aware of this type of reaction during experimental study of the nucleus conducted by bombarding nuclei with energetic particles.

The first artificially induced nuclear reaction was produced by Ernest Rutherford in 1919. Nitrogen was bombarded with alpha particles from a natural source (^{214}Bi). The particles that were produced by the reactions were identified as protons. Rutherford reasoned that an alpha particle colliding with a nitrogen nucleus must sometimes induce a reaction that produces a proton. As a result, the nitrogen nucleus is *artificially transmuted* into an oxygen nucleus:

$$\underset{\substack{nitrogen \\ (14.003\,074\ u)}}{^{14}_{7}\text{N}} \quad + \quad \underset{\substack{alpha\ particle \\ (4.002\,603\ u)}}{^{4}_{2}\text{He}} \quad \rightarrow \quad \underset{\substack{oxygen \\ (16.991\,33\ u)}}{^{17}_{8}\text{O}} \quad + \quad \underset{\substack{proton \\ (1.007\,825\ u)}}{^{1}_{1}\text{H}}$$

(The atomic masses are given for later use.)

This reaction and many others like it actually form a short-lived (intermediate or temporary) *compound nucleus* in an excited state. For example, the preceding reaction can be written more precisely as

$$^{14}_{7}\text{N} + {}^{4}_{2}\text{He} \rightarrow ({}^{18}_{9}\text{F}^{*}) \rightarrow {}^{17}_{8}\text{O} + {}^{1}_{1}\text{H}$$

The intermediate nucleus is the fluorine nucleus, $^{18}_{9}\text{F}^{*}$, formed in an excited state, which is indicated, as usual, by the asterisk. A compound nucleus typically loses excess energy by ejecting a particle or particles—in this case, a proton. Since a compound nucleus lasts only a very short time, it is commonly omitted from the nuclear-reaction equation.

What Rutherford discovered was a way to change one element into another. This was the age-old dream of alchemists, although their main goal was to change common metals, such as mercury and lead, into gold. This seemingly profitable metamorphosis and many other transmutations can be initiated today with *particle accelerators*, machines that accelerate charged particles to very high speeds and energies. When these particles strike target nuclei, they can initiate nuclear reactions. One possible reaction that is initiated when protons strike the nuclei of mercury is

$$\underset{\substack{mercury \\ (199.968\,321\ u)}}{^{200}_{80}\text{Hg}} \quad + \quad \underset{\substack{proton \\ (1.007\,825\ u)}}{^{1}_{1}\text{H}} \quad \rightarrow \quad \underset{\substack{gold \\ (196.966\,56\ u)}}{^{197}_{79}\text{Au}} \quad + \quad \underset{\substack{alpha\ particle \\ (4.002\,603\ u)}}{^{4}_{2}\text{He}}$$

In this reaction, mercury is converted into gold, so it would seem that modern physics has fulfilled the alchemists' dream. However, making such small amounts of gold in an accelerator costs far more than the gold is worth.

Reactions such as the foregoing ones have the general form

$$A + a \rightarrow B + b$$

where the uppercase letters represent the nuclei and the lowercase letters represent the particles. Such reactions are often written in a shorthand notation:

$$A(a, b)B$$

For example, in this form, the two previous reactions can be rewritten more compactly as

$$^{14}\text{N}(\alpha, p)^{17}\text{O} \quad \text{and} \quad {}^{200}\text{Hg}(p, \alpha)^{197}\text{Au}$$

The periodic table (see Fig. 28.9 and rear cover of this text) lists over 100 elements, but only 90 elements occur naturally on Earth. Elements with proton numbers greater than uranium ($Z = 92$), as well as technetium ($Z = 43$) and promethium ($Z = 61$), are created artificially by nuclear reactions. The name *technetium* comes from the Greek word *technetos*, meaning "artificial"; technetium was the first unknown element to be created by artificial means. Elements of all Z values up to $Z = 112$ have been created artificially.*

*Beyond $Z = 112$, there are gaps. Extremely short-lived nuclei with $Z = 114$, 116, and 118 have been discovered, but are yet unnamed. Can you explain why the even values have been discovered and not the odd ones in between?

Conservation of Mass–Energy and the Q Value

In every nuclear reaction, total (relativistic) energy ($E = K + mc^2$) must be conserved. (See Chapter 26.) Consider the reaction in which nitrogen is converted into oxygen: $^{14}N(\alpha, p)^{17}O$. By the conservation of total relativistic energy,

$$(K_N + m_N c^2) + (K_\alpha + m_\alpha c^2) = (K_O + m_O c^2) + (K_p + m_p c^2)$$

where the subscripts refer to the particular particle or nucleus. Rearranging the equation, we have

$$K_O + K_p - (K_N + K_\alpha) = (m_N + m_\alpha - m_O - m_p)c^2$$

The **Q value** of the reaction is defined as the change in kinetic energy. Thus,

$$Q = \Delta K = (K_O + K_p) - (K_N + K_\alpha) \tag{30.1}$$

Q can be positive or negative, depending on whether the total kinetic energy of the system increases or decreases. Thus, the Q value is a measure of the kinetic energy released or absorbed in a reaction. Equation 30.1 can alternatively be expressed in terms of the masses:

$$Q = (m_N + m_\alpha - m_O - m_p)c^2 \tag{30.2}$$

In terms of a general reaction of the form $A + a \rightarrow B + b$,

$$Q = \Delta K = (m_A + m_a - m_B - m_b)c^2 = (\Delta m)c^2 \tag{30.3}$$

Thus, as alternative interpretation of Q is that it is the difference in the mass-equivalent energies of the reactants (initial) and the products (final) of a reaction. This is a reflection of the fact that mass can be converted into kinetic energy and vice versa. Note that the mass difference Δm can be positive or negative. Notice also that $\Delta m = m_i - m_f$, the opposite of our usual convention. This is to guarantee that the value of Q is correctly related to ΔK. That is, if the total mass of the system increases during the reaction and the kinetic energy decreases, Q must be negative. Similarly, if the total mass decreases and the kinetic energy increases, then Q is positive.

Note: Here $\Delta m = m_i - m_f$.

If Q is negative, the reaction requires a minimum amount of kinetic energy before the mass of the products can be attained and the reaction can happen. To see this, let us look at Rutherford's original reaction in some detail. Using the masses given under the reactants and products in the equation on p. 964 for the $^{14}N(\alpha, p)^{17}O$ reaction, we have

$$
\begin{aligned}
Q &= (m_N + m_\alpha - m_O - m_p)c^2 \\
&= [(14.003\,074\ u + 4.002\,603\ u) - (16.999\,133\ u + 1.007\,825\ u)]c^2 \\
&= (-0.001\,281\ u)c^2
\end{aligned}
$$

or, using the mass–energy equivalence factor from Section 29.4, we obtain

$$Q = (-0.001\,281\ u)(931.5\ \text{MeV/u}) = -1.193\ \text{MeV}$$

The negative Q value indicates that some kinetic energy was lost in the reaction. In this case, the reaction is said to be **endoergic** (or *endothermic*). In endoergic reactions, the kinetic energy of the reacting particles is partially converted into mass.

When the Q value of a reaction is positive, energy is released, and the reaction is said to be **exoergic** (or *exothermic*). That is, energy is produced by (*exo*) the reaction. In this case, some mass is converted into energy in the form of increased kinetic energy of the reaction products.

Example 30.1 ■ A Possible Energy Source: Q Value of a Reaction

Determine if the following reaction is endoergic or exoergic, and calculate its Q value.

$$\underset{\substack{deuteron \\ (2.014\,102\ u)}}{^{2}_{1}\text{H}} \quad + \quad \underset{\substack{deuteron \\ (2.014\,102\ u)}}{^{2}_{1}\text{H}} \quad \rightarrow \quad \underset{\substack{helium \\ (3.016\,029\ u)}}{^{3}_{2}\text{He}} \quad + \quad \underset{\substack{neutron \\ (1.008\,665\ u)}}{^{1}_{0}\text{n}}$$

Thinking It Through. The reaction is endoergic if $Q > 0$, and exoergic if $Q < 0$. We need the mass difference (Δm) to determine Q from Eq. 30.3.

Solution. Δm is calculated by subtracting the final masses from the initial masses. Therefore,

$$\Delta m = 2m_{\text{D}} - m_{\text{He}} - m_{\text{n}}$$
$$= 2(2.014\,102\ u) - 3.016\,029\ u - 1.008\,665\ u = +0.003\,51\ u$$

Thus, mass has been lost, the total kinetic energy has increased, and the reaction is exoergic. The Q value is

$$Q = (0.003\,51\ u)(931.5\ \text{MeV/u}) = +3.27\ \text{MeV}$$

Follow-up Exercise. Determine if the following reaction is endoergic or exoergic, and calculate its Q value:

$$\underset{\substack{carbon \\ (12.000\,000\ u)}}{^{12}_{6}\text{C}} \quad + \quad \underset{\substack{helium \\ (4.002\,603\ u)}}{^{4}_{2}\text{He}} \quad \rightarrow \quad \underset{\substack{carbon \\ (13.003\,355\ u)}}{^{13}_{6}\text{C}} \quad + \quad \underset{\substack{helium \\ (3.016\,029\ u)}}{^{3}_{2}\text{He}}$$

(Answers to all Follow-up Exercises are at the back of the text.)

Problem-Solving Hint

Note in Example 30.1 that Q values are computed from the mass difference, expressed in atomic mass units, by using the mass–energy conversion factor derived in Chapter 29. (See Table 29.3.) This method eliminates the need to use c^2 and gives Q directly in MeV.

TABLE 30.1 Interpretation of Q Values

Q value	Effect
Positive $(Q > 0)$	Exoergic, some mass converted into energy (mass of reactants greater than mass of products)
Negative $(Q < 0)$	Endoergic, some kinetic energy converted into mass (mass of products greater than mass of reactants)

Radioactive decay (Chapter 29) is a special type of nuclear reaction with one reactant nucleus and two (or more) products. The Q value of radioactive decay is always positive, because there is a gain in kinetic energy. For decay reactions, Q is called the *disintegration energy*. The meaning of the sign of the Q value is summarized in Table 30.1.

When a reaction's Q value is negative, you might think that the reaction could occur if the incident particle had a kinetic energy at least equal to Q—that is, if it were to have $K_{\text{min}} = |Q|$.* However, if all the kinetic energy were converted to mass, the particles would be at rest after the reaction, which violates conservation of linear momentum.

Hence, in an endoergic reaction, to conserve linear momentum, the kinetic energy of the incident particle must be *greater* than $|Q|$. The minimum kinetic energy (K_{min}) that a particle needs in order to initiate an endoergic reaction is called the **threshold energy**. For nonrelativistic energies, the threshold energy is

$$K_{\text{min}} = \left(1 + \frac{m_{\text{a}}}{M_{\text{A}}}\right)|Q| \qquad \begin{array}{l}(stationary \\ target\ only)\end{array} \qquad (30.4)$$

where m_{a} and M_{A} are the masses of the incident particle and the stationary target nucleus, respectively. In Eq. 30.4, the factor by which $|Q|$ is multiplied is greater

---

*Kinetic energy is written in terms of $|Q|$, the absolute value of Q, because kinetic energy cannot be negative. The sign of Q arises from the mass difference and indicates the gain or loss of mass during the reaction. If Q is negative, we want K_{min} to be positive—hence, the use of absolute value.

than 1 (why?), so, as expected, $K_{min} > |Q|$. The calculation of a threshold energy is shown in the next Example.

Example 30.2 ■ Nitrogen into Oxygen: Threshold Energy

What is the threshold energy for the reaction $^{14}N(\alpha, p)^{17}O$?

Thinking It Through. The Q value for this reaction was calculated previously in the text. To get the threshold energy, we use Eq. 30.4.

Solution. The following data are taken from the text:

Given: $m_a = m_\alpha = 4.002\,603$ u $\quad$ *Find:* K_{min} (threshold energy)
$\quad\quad M_A = m_N = 14.003\,074$ u
$\quad\quad Q = -1.193$ MeV

From Eq. 30.4,

$$K_{min} = \left(1 + \frac{m_\alpha}{M_N}\right)|Q| = \left(1 + \frac{4.002\,603\text{ u}}{14.003\,074\text{ u}}\right)|-1.193\text{ MeV}| = 1.534\text{ MeV}$$

Follow-up Exercise. In this Example, how much of the threshold kinetic energy goes into increasing the mass of the system, and how much shows up as kinetic energy in the final state? Explain your reasoning.

Reaction Cross-sections

In an endoergic reaction, when the incident particle has more than the threshold energies of several reactions, any of the reactions may occur, usually with differing probabilities. A measure of the probability that a particular reaction will occur is called the **cross-section** for that reaction. The probability for a particular reaction depends on many factors. Usually, it depends on the kinetic energy of the initiating particle, sometimes very dramatically. For positively charged incident particles, the presence of the Coulomb barrier means that the probability of a given reaction occurring generally increases with the kinetic energy of the incident particle.

Being electrically neutral, neutrons are unaffected by the Coulomb barrier. As a result, the cross-section for a given reaction involving neutrons can be quite large, even for low-energy neutrons. Reactions involving neutrons such as $^{27}Al(n, \gamma)^{28}Al$ are called *neutron-capture reactions*. As the energy of the neutron increases, the cross-section can vary a great deal, as ▶ Fig. 30.1 shows. The peaks in the curve, called *resonances*, are associated with nuclear energy levels in the nucleus being formed. If the neutron's energy is "just right" to create the final nucleus in one of its energy levels, there is a relatively high probability that neutron absorption will occur.

30.2 Nuclear Fission

OBJECTIVES: To understand (a) the process of nuclear fission, (b) the nature and cause of a nuclear chain reaction and (c) the basic principles involved in the operation of nuclear reactors.

In early attempts to make heavier elements artificially, uranium, the heaviest element known at the time, was bombarded with neutrons. An unexpected result was that the uranium nuclei sometimes split into fragments. These fragments were identified as the nuclei of lighter elements. The process was dubbed *nuclear fission*, after the biological fission process of cell division.

In a **fission reaction**, a heavy nucleus divides into two lighter nuclei with the emission of neutrons. Mass is converted into kinetic energy of the neutron and fragments during this process. Some heavy nuclei undergo *spontaneous fission*, but at very slow rates. However, fission can be *induced*, and this is the important

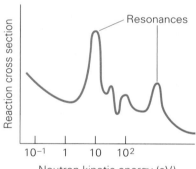

▲ **FIGURE 30.1 Reaction cross-section** A typical graph of a neutron reaction's cross-section versus energy. The peaks where the probabilities of reactions are greatest are called *resonances*. They correspond to energy levels in the compound nucleus formed when the neutron is temporarily captured.

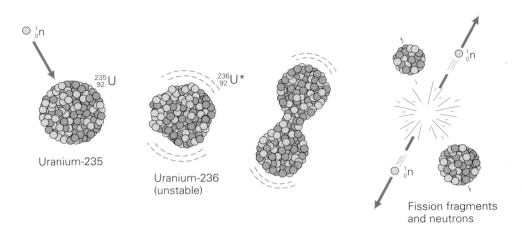

► FIGURE 30.2 Liquid-drop model of fission When an incident neutron is absorbed by a fissionable nucleus, such as ^{235}U, the unstable compound nucleus (^{236}U) undergoes violent oscillations and breaks apart like a liquid drop, typically emitting two or more neutrons and yielding two radioactive fragments.

Uranium-235

Uranium-236 (unstable)

Fission fragments and neutrons

process in energy production. For example, when a ^{235}U nucleus absorbs a neutron, it may fission into xenon and strontium by the reaction

$$^{235}_{92}\text{U} + {}^{1}_{0}\text{n} \rightarrow ({}^{236}_{92}\text{U}^*) \rightarrow {}^{140}_{54}\text{Xe} + {}^{94}_{38}\text{Sr} + 2({}^{1}_{0}\text{n})$$

According to the *liquid-drop model*, this intermediate nucleus (^{236}U) undergoes oscillations and becomes distorted like a liquid drop (▲ Fig. 30.2). The separation of the nucleons into different parts of the "drop" weakens the nuclear force, and the repulsive electrical force between the two parts of the "nuclear drop" causes it to split, or fission.

Note that the preceding reaction involving ^{235}U is *not* unique. Other outcomes include the following (note that compound nuclei are omitted):

$$^{1}_{0}\text{n} + {}^{235}_{92}\text{U} \rightarrow {}^{141}_{56}\text{Ba} + {}^{92}_{36}\text{Kr} + 3({}^{1}_{0}\text{n})$$

and

$$^{1}_{0}\text{n} + {}^{235}_{92}\text{U} \rightarrow {}^{150}_{60}\text{Nd} + {}^{81}_{32}\text{Ge} + 5({}^{1}_{0}\text{n})$$

Only certain nuclei undergo fission. For them, the probability of fissioning depends on the energy of the incident neutrons. For example, the largest probabilities for fission of ^{235}U and ^{239}Pu occur for "slow" neutrons, that is, neutrons with kinetic energies less than about 1 eV. However, for ^{232}Th, "fast" neutrons with energies of 1 MeV or greater are more likely to trigger a fission reaction.

An estimate of the energy released in a fission reaction can be obtained by considering the E_b/A curve for stable nuclei (Fig. 29.14). When a nucleus with a high mass number (A), such as uranium, splits into two nuclei, it is, in effect, moving inward and upward along the sloping tail of this curve toward more stable nuclei. As a result, the average binding energy per nucleon increases from about 7.8 MeV to approximately 8.8 MeV. The energy liberated is about 1 MeV per nucleon in the fission products. In the reaction at the top of this page, $140 + 94 = 234$ nucleons are bound in the products. Thus, the energy release is approximately $(1\text{ MeV/nucleon}) \times 234\text{ nucleons} \approx 234\text{ MeV}$.

At first glance, this amount might not seem like much energy. 234 MeV is only about 3.7×10^{-11} J, which pales in comparison with everyday energies. In fact, 200 MeV is less than 0.1 percent of the energy equivalent of the mass of the ^{235}U nucleus, which is approximately $(235\text{ nucleons})(939\text{ MeV/nucleon}) = 2.2 \times 10^5$ MeV. Nevertheless, on a percentage basis, it is many times larger than the amount of energy released in ordinary chemical reactions, such as the burning of oil or coal.

Practical amounts of energy from fission can, however, be obtained when huge numbers of these fissions occur per second. One way of accomplishing this is by a **chain reaction**. For example, suppose a ^{235}U nucleus fissions (on its own or triggered by an external neutron) with the release of two neutrons (► Fig. 30.3). Ideally, the released neutrons can then initiate two more fission reactions, a process

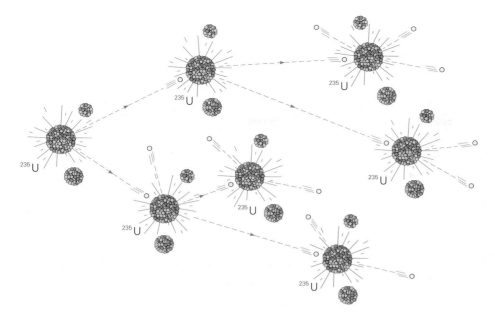

◀ **FIGURE 30.3 Fission chain reaction** The neutrons that result from one fission event can initiate other fission reactions, which, in turn, initiate further fission reactions, and so on. When enough fissionable material is present, the sequence of reactions can be adjusted to be self-sustaining (a chain reaction).

Note: A chain reaction requires a critical mass.

Nuclear Chain Reaction

that, in turn, releases four neutrons. These neutrons may initiate more reactions, and so on. Thus, the process can multiply, with the number of neutrons doubling with each generation. When this occurs, the neutron production rate (and, hence, the energy released from the sample) grows exponentially.

To maintain a sustained chain reaction, there must be an adequate quantity of fissionable material. The minimum mass required to produce a sustained chain reaction is called the **critical mass**. When the critical mass is attained, there is enough fissionable material such that at least one neutron from each fission event, on average, goes on to fission another nucleus.

Several factors determine critical mass. Most evident is the amount of fissionable material. If the quantity of material is small, many neutrons will escape from the sample (through the surface) before inducing a fission, and the chain reaction will die out. Also, nuclides other than ^{235}U in the sample may absorb neutrons, thereby limiting the chain reaction. As a result, the purity of the fissionable isotope affects the critical mass.

Natural uranium is made up of the isotopes ^{238}U and ^{235}U. The natural concentration of the fissionable ^{235}U isotope is only about 0.7%. The remaining 99.3% is ^{238}U, which can absorb neutrons without fissioning, thereby preventing those neutrons from continuing the chain reaction of ^{235}U fission. To have more fissionable ^{235}U nuclei in a sample and thus reduce the critical mass, the ^{235}U can be concentrated. This enrichment varies from 3–5% ^{235}U for reactor-grade material used in electrical generation to over 99% for weapons-grade material. This difference is important, since it is highly desirable that a nuclear reactor *not* explode like an atomic bomb nor use fuel capable of being easily made into a bomb.

Chain reactions take place almost instantly. If such a reaction proceeds uncontrolled, the quick and enormous release of energy can cause an explosion. This is the principle of the so-called atomic bomb. (A better name is the *fission bomb*.) In such a bomb, several subcritical pieces of fuel are suddenly imploded to form a critical mass. The resulting chain reaction is then out of control, releasing an enormous amount of energy in a short period of time. For the steady production of energy from the fission process, the chain-reaction process must be controlled. We shall now see how this task is accomplished in **nuclear reactors**.

Nuclear Reactors

The Power Reactor Currently, the only practical nuclear reactor design for generating electrical power is based on the fission chain reaction. A typical design for a

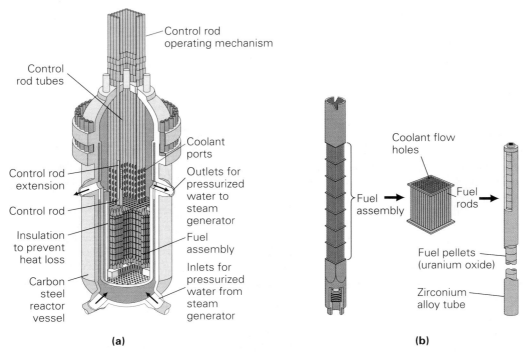

▲ **FIGURE 30.4 Nuclear reactor**   (a) A schematic diagram of a reactor vessel. (b) A fuel rod and its assembly.

nuclear reactor is shown in ▲ Fig. 30.4. There are five key elements to a reactor: fuel rods, core, coolant, control rods, and moderator.

Tubes packed with pellets of uranium oxide form the **fuel rods**, which are located in the reactor **core**, the central portion of the reactor. A typical commercial reactor contains fuel rods bundled in fuel assemblies of approximately 200 rods each. Coolant flows around the rods to remove the heat energy from the chain reaction. The reactors used in the United States are light-water reactors, which means that ordinary water is used as a **coolant** to remove heat. However, the hydrogen nuclei of ordinary water can capture neutrons to form deuterium, thus removing neutrons from the chain reaction. Hence, enriched uranium with 3–5% ^{235}U must be used.*

The chain-reaction rate and, therefore, the energy output of a reactor are controlled by boron or cadmium **control rods**, which can be inserted into, or withdrawn from, the reactor core. Cadmium and boron have a very high probability for absorbing neutrons. By inserting the control rods between the fuel-rod assemblies, neutrons are removed from the chain reaction. The control rods are adjusted so that the chain reaction proceeds at a steady rate. The idea is to create a sustainable fission chain reaction in which the average fission produces only one more fission. For refueling, or in an emergency, the control rods can be fully inserted, and enough neutrons are removed to curtail the chain reaction and shut down the reactor. However, even with the chain reaction shut down, water must continue to circulate in order to prevent heat buildup due to the continuing decay of radioactive fission products in the fuel rods. If not, damage to the fuel rods can result. Melting and cracking of fuel rods due to inadequate cooling was the cause of the accident at Three Mile Island in 1979. The core of that reactor is still highly radioactive, because fission fragments were released from the rods.

*Many Canadian reactors use heavy water, D_2O, as a coolant in place of light water. The advantage is that the deuterium does not readily absorb neutrons, so the fuel can be of lower uranium enrichment. However, D_2O must first be separated from normal water, H_2O, an operation that takes energy.

The water flowing through the fuel-rod assemblies acts not only as a coolant, but also as a **moderator**. The fission cross-section for ^{235}U is largest for slow neutrons (sometimes called *thermal* neutrons with kinetic energies less than 1 eV). However, neutrons emitted from a fission are fast neutrons (with kinetic energies of about 2 MeV). Their speed is reduced, or moderated, by collisions with the water molecules. It takes about 20 collisions to moderate fast neutrons down to energies of about 1 eV.

The Breeder Reactor In a commercial power reactor, the ^{238}U goes along for the ride, so to speak. However, while it is unlikely to fission by absorbing a slow neutron, ^{238}U can be involved in reactions caused by *fast* neutrons. If a neutron's energy has not been completely moderated, reactions with ^{238}U can occur. For example, a conversion of ^{238}U to ^{239}Pu via successive beta decays can happen as follows:

$$\underset{(fast)}{^{1}_{0}\text{n} + {}^{238}_{92}\text{U}} \rightarrow ({}^{239}_{92}\text{U}^{*}) \rightarrow {}^{239}_{93}\text{Np} + \beta^{-}$$
$$\qquad\qquad\qquad\qquad\qquad \hookrightarrow {}^{239}_{94}\text{Pu} + \beta^{-}$$

^{239}Pu, with a half-life of 24 000 years, *is* fissionable. Since it is possible to actively promote the conversion of ^{238}U to ^{239}Pu in a reactor by reducing the degree of moderation, the same amount (or more!) of fissionable fuel can be produced (^{239}Pu) as is consumed (^{235}U). This is the principle behind the **breeder reactor**. Notice this isn't a case of something for nothing. Rather, the reactor is converting the unfissionable ^{238}U part of the fuel to the fissionable ^{239}Pu, while continuing to produce power by ^{235}U fission chain reactions.

Developmental work on the breeder reactor in the United States was essentially stopped in the 1970s. However, France went on to develop operational breeder reactors that provide nuclear fuel (^{239}Pu) for power reactors. France and other nations are highly dependent on nuclear energy and breeder reactors for their electrical energy needs.

Electricity Generation The components of a typical pressurized water reactor used in the United States are shown in ▼Fig. 30.5. The heat generated by the controlled chain reaction is carried away by the water passing through the rods in the fuel assembly. The water is pressurized to several hundred atmospheres so that it can reach temperatures over 300°C for more efficient heat removal. The hot water is then pumped to a heat exchanger, where the heat energy is transferred to the water of a steam generator. Notice that the reactor coolant and the exchanger water are two separate and distinct closed systems. (Why?)

Next, high-pressure steam turns a turbine that operates an electrical-energy generator, as is the case for any nonnuclear power plant. The steam is then cooled

Note: A breeder reactor doesn't produce something from nothing. It uses the kinetic energy of the neutrons, wasted in a nonbreeder reactor, to create fissionable ^{239}Pu from the nonfissionable ^{238}U component of uranium.

Note: Recall from Section 11.3 that the boiling point of water increases with increasing pressure.

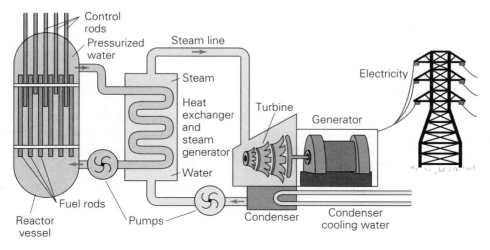

◀ **FIGURE 30.5 Pressurized water reactor** The components of a pressurized water reactor. The heat energy from the reactor core is carried away by the circulating water. The water in the reactor is pressurized so that it can be heated to high temperatures for more efficient heat removal. The energy is used to generate steam, which drives the turbine that turns the generator to produce electrical energy.

and condensed after turning the turbine. This final loop of coolant water typically carries the heat to a nearby ocean, lake, or river.

Nuclear Reactor Safety

Nuclear energy is used to generate a substantial amount of the electricity in the world. More than 25 countries now produce electricity by this method. Several hundred nuclear reactor units are in operation throughout the world, with more than 100 units in the United States. With the increasing number of nuclear facilities comes the fear of nuclear accidents and the subsequent release of radioactive materials into the environment.

If the coolant of a light-water reactor is lost, the chain reaction stops, because the coolant is also the moderator. However, the decay of the fission fragments, some with half-lives of hundreds of years, continue. In such a **LOCA** (loss-of-coolant accident), the fuel rods might become hot enough (several thousand degrees Celsius) for the cladding (the outside covering) to melt and fracture. Once this occurs, the hot, fissioning mass could fall into the water on the floor of the containment vessel and cause a steam explosion, a hydrogen explosion, or both. This explosion could rupture the walls of the containment vessel and allow radioactive fragments into the environment. Even if the walls were not breached, the hot "melt" of the fuel rods could burn through the floor of the building, eventually reaching ground level and the atmosphere (a situation called the *China syndrome*). A partial **meltdown** did occur at the TMI generating plant in 1979. This meltdown was a LOCA, and a small amount of radioactive steam was vented to the atmosphere. Inside one reactor vessel, now sealed, electronic robots have discovered heavy damage to the fuel rods.

The April 1986 nuclear accident at Chernobyl was a meltdown caused by human error, magnified by the instability resulting from the use of carbon as a moderator instead of water. When the flow of cooling water was inadvertently removed, the chain reaction went out of control—something that could not happen in a light-water reactor—producing a huge rise in temperature. The resulting explosions blew the top off the building (◄Fig. 30.6). When the fuel rods melted, the graphite blocks burned like a massive charcoal barbecue, spewing radioactive smoke into the air. Winds carried this radioactive smoke over much of Europe and over the North Pole into Canada and the United States, where significant amounts of core fission fragments (such as ^{131}I, ^{90}Sr, and ^{137}Cs) were detected.

Even if nuclear reactors operate safely (and their safety record, particularly in the United States, is a very good one), there remains the problem of radioactive waste. Fission fragments are radioactive and have long half-lives. As a result, the safe handling of nuclear waste will be a problem for centuries to come. In the 1990s, the United States opened its first nuclear-waste site, in New Mexico. Where and how to seal, safely transport, bury, and guard nuclear waste will be important decisions for many generations to come.

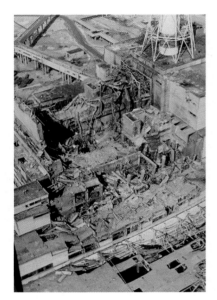

▲ **FIGURE 30.6   Chernobyl, April 1986** An aerial photo showing damage to the reactor at Chernobyl. This accident released large amounts of radioactive materials into the environment, with dire consequences.

30.3 Nuclear Fusion

OBJECTIVES: To (a) explain the fundamental difference between fusion and fission, (b) calculate energy releases in fusion reactions and (c) understand how fusion might eventually provide a source of electric energy.

Another type of nuclear reaction that can produce energy is *fusion*. In a **fusion reaction**, light nuclei fuse to form a more massive nucleus, releasing energy in the process ($Q > 0$). A simple fusion reaction—the fusion of two deuterium nuclei

($^2_1$H), sometimes called a D–D reaction–was examined in Example 30.1. There, it was shown that this reaction releases 3.27 MeV of energy per fusion.

Another example is the fusion of deuterium and tritium (a D–T reaction):

$$\underset{(2.014\,102\ u)}{^2_1\text{H}} \quad + \quad \underset{(3.016\,049\ u)}{^3_1\text{H}} \quad \longrightarrow \quad \underset{(4.002\,603\ u)}{^4_2\text{He}} \quad + \quad \underset{(1.008\,665\ u)}{^1_0\text{n}}$$

Using the given masses, you should be able to show that this reaction involves a release of 17.6 MeV per fusion.

A fusion reaction releases much less energy in comparison with the more than 200 MeV released from a typical single fission. However, equal-mass samples of hydrogen and uranium have many, many more hydrogen nuclei than uranium nuclei. As a result, *per kilogram*, the fusion of hydrogen gives almost three times the energy released from uranium fission.

In a sense, our lives depend crucially on nuclear fusion, because it is the source of energy for stars, including our Sun. One sequence of fusion reactions that is believed to be responsible for the Sun's energy output is as follows. First, we have proton–proton fusion,

$$^1_1\text{H} + {}^1_1\text{H} \rightarrow {}^2_1\text{H} + \beta^+ + \nu$$

(where ν represents a neutrino, a particle to be discussed in the next section). Then another proton fuses with the deuteron:

$$^1_1\text{H} + {}^2_1\text{H} \rightarrow {}^3_2\text{He} + \gamma$$

Finally, two of the ^{3}He nuclei fuse:

$$^3_2\text{He} + {}^3_2\text{He} \rightarrow {}^4_2\text{He} + {}^1_1\text{H} + {}^1_1\text{H}$$

The net effect of this sequence, called the *proton–proton cycle*, is that four protons ($^1_1$H) combine to form one helium nucleus ($^4_2$He) plus two positrons (β^+), two gamma rays (γ), and two neutrinos (ν) with a release of energy:

$$4(^1_1\text{H}) \rightarrow {}^4_2\text{He} + 2\beta^+ + 2\gamma + 2\nu + Q \qquad (Q = 24.7\ \text{MeV})$$

In a star such as our Sun, gamma-ray photons scatter off nuclei on their way to the surface. Each scattering results in a reduction in energy, until each photon has only a few electron volts of energy. Thus, on reaching the surface, the photons are mostly visible-light photons. In our Sun, fusion involves only the central 10% of the Sun's mass. It has been going on for about 5 billion years and should continue, approximately as is, for another 5 billion years. As the next Example shows, an enormous number of fusion reactions per second is required to power the Sun.

Example 30.3 ■ Still Going: The Fusion Power of the Sun

Incoming sunlight falls on the Earth at the rate of $1.40 \times 10^3\ \text{W/m}^2$. Assume that the Sun's energy is produced by the proton–proton cycle and that 10% of its total hydrogen is available for fusion. Calculate the mass lost by the Sun per second.

Thinking It Through. To find the mass-loss rate, we need to know the Sun's total power output. Imagine the power flow through a sphere centered on the Sun, with a radius the size of the Earth's orbit, and then calculate that sphere's area. The total power can then be determined from the power per m^2 and the total area in m^2. The Sun's mass-loss rate can be calculated from the total power, based on the mass–energy equivalence (Table 29.3).

Solution. The data are as follows (using some solar system data given in Appendix III):

Given: $R_{E-S} = 1.50 \times 10^8$ km $= 1.50 \times 10^{11}$ m *Find:* $\Delta m/\Delta t$ (overall mass-loss rate)
$M_S = 2.00 \times 10^{30}$ kg
$P_S/A = 1.40 \times 10^3$ W/m^2

The surface area of the imaginary sphere that intercepts all the Sun's energy is

$$A = 4\pi R_{E-S}^2 = 4\pi(1.5 \times 10^{11}\text{ m})^2 = 2.83 \times 10^{23}\text{ m}^2$$

Thus, the total power output of the Sun is

$$P_S = (1.40 \times 10^3\text{ W/m}^2)(2.83 \times 10^{23}\text{ m}^2) = 3.96 \times 10^{26}\text{ W} = 3.96 \times 10^{26}\text{ J/s}$$

To find the equivalent mass-loss rate, we convert this power to MeV per second:

$$\frac{3.96 \times 10^{26}\text{ J/s}}{1.60 \times 10^{-13}\text{ J/MeV}} = 2.48 \times 10^{39}\text{ MeV/s}$$

Then, using the mass–energy equivalence (931.5 MeV/u), we find that the mass-loss rate is

$$\frac{\Delta m}{\Delta t} = \frac{(2.48 \times 10^{39}\text{ MeV/s})(1.66 \times 10^{-27}\text{ kg/u})}{931.5\text{ MeV/u}} = 4.42 \times 10^9\text{ kg/s}$$

Follow-up Exercise. In this Example, how many proton–proton cycles happen per second? [*Hint*: Each cycle releases 24.7 MeV.]

Fusion as a Source of Energy Produced on the Earth

In several ways, fusion appears to be an ideal energy source for the future. Enough deuterium exists in the oceans, in the form of heavy water, to supply our needs for centuries. In addition, fusion does not depend on a chain reaction, so there is less danger of the release of radioactive material. Also, fusion products tend to have relatively short half-lives. For example, tritium has a half-life of only 12.3 years, compared with hundreds or thousands of years for fission products.

However, there are many unresolved technical problems to be solved before controlled fusion can be used commercially to produce electric energy. A primary problem is that very high temperatures are needed to *initiate* fusion reactions, because of the electrical repulsion between the nuclei. Temperatures on the order of millions of degrees are needed to initiate these *thermonuclear fusion reactions*. The problem is in confining sufficient energy in a reaction region to maintain these high temperatures. Because of this difficulty, practical fusion reactors have not yet been achieved. However, uncontrolled fusion has been demonstrated in the form of the hydrogen (H) bomb. In this case, the fusion reaction is initiated by an implosion created by a small atomic (fission) bomb. This implosion provides the necessary density and temperatures to begin the fusion process.

At the high temperatures required for fusion, electrons are stripped from their nuclei, and a gas is created that consists of positively charged ions and free, negatively charged electrons. Such a "gas" of charged particles is called a **plasma**.* Plasmas have a number of special physical properties and are sometimes referred to as the *fourth phase of matter*.

The technological problems involved in confining a plasma are being approached in at least two different ways: magnetic confinement and inertial confinement. Since a plasma is a gas of charged particles, it can be controlled and manipulated by using electric and magnetic fields. In **magnetic confinement**, magnetic fields are used to hold the plasma in a confined space, a so-called magnetic bottle. (See Fig.19.33b.)

*Plasmas exist in our everyday world, such as in fluorescent lamps and lightning strokes.

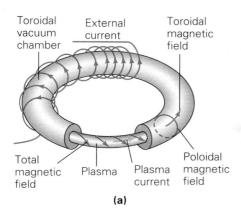

◀ **FIGURE 30.7 Magnetic confinement** Magnetic confinement is one method by which controlled nuclear fusion might be achieved. **(a)** Tokamak configuration, showing the *B* field generated by external currents. The magnetic field confines the plasma in the ring. **(b)** The Princeton Tokamak Fusion Test Reactor (TFTR).

Once a plasma is confined, electric fields can be used to produce electric currents in it. The currents, in turn, raise the plasma's temperature. Temperatures of 100 million kelvins have been achieved in a design called a *tokomak* (▲Fig. 30.7). This design uses a magnetic field arranged in a donut shape to trap the charged particles.

In addition to high temperatures, the initiation of fusion has minimum requirements on plasma density and confinement time. The trick is to meet all these requirements at the same time. The generation of several megawatts of power for less than a second in a magnetically confined plasma is typical of the best results so far. Clearly, for commercial applications, much higher power levels must be attained at a continuous level.

Inertial confinement depends on implosion techniques. Hydrogen fuel pellets are either dropped or positioned in a reactor chamber. Pulses of laser, electron, or ion beams are then used to implode the pellet, producing compression and high densities and temperatures. Fusion can occur if the pellet stays together for a sufficient time, which depends on its inertia (hence the name "inertial confinement"). At present, laser and particle beams are not powerful enough to produce sustainable fusion by inertial confinement. In summary, practical energy production from fusion is not expected to be accomplished until well into this century, if at all.

30.4 Beta Decay and the Neutrino

OBJECTIVES: To (a) explain why the neutrino is necessary in order to account for observed beta decay data, (b) specify some of the physical properties of neutrinos, and (c) write complete beta decay equations.

At first glance, beta decay *appears* to be a two-body decay process in which unstable nuclei emit an electron ($_{-1}^{0}e$) or a positron ($_{+1}^{0}e$). Examples of both types of decay are

$$\underset{(14.003\,242\ u)}{_{6}^{14}\text{C}} \quad \longrightarrow \quad \underset{(14.003\,074\ u)}{_{7}^{14}\text{N}} \quad + \quad _{-1}^{0}e \qquad\qquad \beta^{-}\ decay$$

$$\underset{(13.005\,739\ u)}{_{7}^{13}\text{N}} \quad \longrightarrow \quad \underset{(13.003\,355\ u)}{_{6}^{13}\text{C}} \quad + \quad _{+1}^{0}e \qquad\qquad \beta^{+}\ decay$$

However, when analyzed in detail, these equations appear to violate the conservation of energy and linear momentum, amongst other conservation laws.

The energy released (or disintegration energy) in the foregoing β^{-} process, as calculated from the mass defect (using the given masses*), is $Q = 0.156$ MeV.

---

*Use of the atomic mass of the daughter ^{14}N (with seven electrons) is necessary in order to take into account the emitted electron, since the ^{14}N resulting from beta decay would have only the six electrons that orbit the parent ^{14}C nucleus.

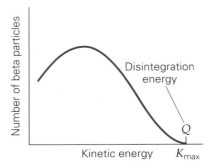

▲ FIGURE 30.8 Beta ray spectrum For a typical beta-decay process, all beta particles are emitted with $K < Q$, leaving unaccounted-for energy.

(You should show this.) Therefore, if the decay involves only two particles (electron and daughter), the electron, being much lighter than the daughter, should always have a kinetic energy of just slightly less than 0.156 MeV. However, this is *not* what happens. When the electron's kinetic energies are measured, a *continuous spectrum* of energies is observed up to $K_{max} \approx Q$ (◄Fig. 30.8). (See Conceptual Example 30.4.) That is, not all of the released energy is accounted for by the electron's kinetic energy. Nor is this the only difficulty. The emitted β^- and the daughter nucleus hardly ever leave the disintegration site in opposite directions. Thus, linear-momentum conservation appears to be violated as well.*

What, then, is the problem? We would hope that our conservation laws are not invalid. An alternative explanation is that these apparent violations are telling us something about nature that we do not yet recognize. All of the apparent difficulties can be resolved if it is assumed that an unobserved particle is also emitted during the decay. This explanation and the existence of such a particle were first proposed in 1930 by Wolfgang Pauli. Fermi christened this particle the **neutrino** (meaning "little neutral one" in Italian). For charge to be conserved, the neutrino has to be electrically neutral. Because the neutrino had been virtually impossible to observe, it must interact very weakly with matter. In fact, scientists eventually discovered that the neutrino interacts with matter through a second nuclear force, much weaker than the strong force, called the *weak interaction*, or the *weak nuclear force*. (See Section 30.5.)

Details of the initial experimental observations of beta decay suggested that the neutrino had zero mass and therefore traveled with the speed of light. It also had linear momentum p related to its total energy E by $E = pc$. Furthermore, it had a spin quantum number of $\frac{1}{2}$. In 1956, a particle with these properties was finally detected, and the neutrino's existence was firmly established.

The previous beta-decay equations can now be written correctly and completely as

$$^{14}_{6}\text{C} \rightarrow {}^{14}_{7}\text{N} + \beta^- + \bar{\nu}_e$$

and

$$^{13}_{7}\text{N} \rightarrow {}^{13}_{6}\text{C} + \beta^+ + \nu_e$$

where the Greek letter ν (nu) symbolizes the neutrino. The symbol with a bar over it represents an *antineutrino*. The "overbar" notation is a common way of indicating an antiparticle. Two *different* neutrinos are associated with beta decay. A neutrino is emitted in β^+ decay, and an antineutrino is emitted in β^- decay. The subscript "e" identifies the neutrinos as associated with electron–positron beta decay. As we will see in Section 30.5, there are additional types of neutrinos associated with other decays triggered by the weak interaction.

Conceptual Example 30.4 ■ Having It All? Maximum Kinetic Energy in Beta Decay

Consider the decay of ^{14}C initially at rest:

$$^{14}_{6}\text{C} \rightarrow {}^{14}_{7}\text{N} + \beta^- + \bar{\nu}_e$$

What can you say about the maximum possible kinetic energy (K_{max}) of the beta particle, (a) $K_{max} > Q$, (b) $K_{max} = Q$, or (c) $K_{max} < Q$? Explain your reasoning clearly.

---

*This process also violates the conservation of angular momentum. Careful analysis of the nuclear spin before the decay shows that it does *not* match the total spin (angular momentum) of the daughter plus the emitted electron.

Reasoning and Answer. Answer (a) certainly cannot be correct, as it violates energy conservation. No one particle can have more than the total amount of energy available. So, can answer (b) be correct? Suppose that the beta particle did get all the released energy. That would mean that neither the neutrino nor the daughter nucleus had any energy, and therefore, neither would have any momentum. But to conserve momentum, at least one of these two particles must move off in the direction opposite to that of the beta particle. Hence, at least the neutrino or the daughter must have some energy. Therefore, the beta particle can't have it all, and answer (c) must be the correct one: $K_{max} < Q$.

Follow-up Exercise. A typical energy spectrum of emitted beta particles is shown in Fig. 30.8. This spectrum indicates that there is a small probability of the beta particle having almost no kinetic energy. What would the decay products' trajectories look like after the decay if this were indeed the case?

30.5 Fundamental Forces and Exchange Particles

OBJECTIVES: **To (a) understand the quantum-mechanical description of forces and (b) classify the various forces according to their strengths, properties, ranges, and virtual particles.**

The forces involved in everyday activities are complicated, because of the large numbers of atoms that make up ordinary objects. Contact forces between two hard objects, for example, are due to the repulsive electromagnetic forces between the atomic electrons. Looking at the *fundamental* interactions between particles makes things simpler. On this level, there are only four known **fundamental forces**: the *gravitational force*, the *electromagnetic force*, the *strong nuclear force*, and the *weak nuclear force*.

The most familiar of these forces are the gravitational and electromagnetic forces. Gravity acts between all particles, while the electromagnetic force is restricted to charged particles.* Both forces decrease with increasing particle separation distance and have a very large range.

To describe these forces classically, we employed the concept of a field. Modern physics, however, provides an alternative, more fundamental, description of how forces are transmitted. The force transmittal process is viewed as an exchange of particles. For example, a repulsive force would be analogous to you and another person interacting by tossing a ball back and forth. As you throw the ball and the other person catches it, for example, each of you feels a backward force. An observer who couldn't see the ball might conclude that there was a repulsive force between you and the other person.

The creation of such force-carrying particles would seem to violate energy conservation. However, because of the uncertainty principle (Section 28.4), a particle can be created for a *short time* with no outside energy input without violating energy conservation. Thus, over long time intervals, energy is conserved. For *extremely* short time intervals, the uncertainty principle *permits* a large *uncertainty* in energy ($\Delta E \propto 1/\Delta t$); creation of a particle is allowed; and energy conservation can be briefly violated. However, the created particle is absorbed before it is ever detected, so we never *observe* energy nonconservation. A particle created and absorbed in such a manner is called a **virtual particle**. In this sense, *virtual* means "undetected."

Thus, in the modern view, the fundamental forces are carried, or transmitted, by virtual **exchange particles**. The exchange particles for the four forces

*Relativity shows that both the electric and magnetic forces are components of a single force—the electromagnetic force.

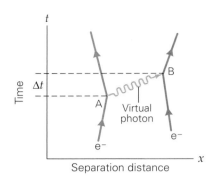

▲ FIGURE 30.9 Feynman diagram of an electron–electron interaction The interacting electrons undergo a change in energy and momentum, due to the exchange of a virtual photon, which is created at A and absorbed at B in an amount of time (Δt) that is consistent with the uncertainty principle.

must differ in mass. Recall that the greater the mass of the particle, the greater is the energy ΔE required to create it, and thus the shorter is the time Δt for which it exists. Since a massive particle can exist for only a short time, the distance it can travel, and hence the range of the associated force, must be small. That is, *the range of a force associated with an exchange particle is inversely proportional to the mass of that exchange particle.*

The Electromagnetic Force and the Photon

The exchange particle of the electromagnetic force is a (virtual) **photon**. As a "massless" particle, it has an infinite range, as is required for the electromagnetic force. A particle exchange can be visualized graphically by using a *Feynman diagram*, as in ◄Fig. 30.9. This graph shows the specific example of how the exchange idea can explain the electrical repulsion between two electrons. Such *space–time diagrams* are named after American scientist and Nobel Prize winner Richard Feynman (1918–1988), who used them to analyze electromagnetic interactions in his *quantum electrodynamics* theory. The important points are the vertices of the diagram. One electron creates a virtual photon at point A, and the other electron absorbs it at point B. Each of the electrons undergoes a change in energy and in momentum (including direction) by virtue of the photon exchange and resulting force.

The Strong Nuclear Force and Mesons

Japanese physicist Hideki Yukawa (1907–1981) proposed in 1935 that the short-range strong nuclear force between two nucleons is associated with an exchange particle called the **meson**. An estimate of the meson's mass can be made from the uncertainty principle. If a nucleon were to create a meson, conservation of energy would be (undetectably) violated by an amount of energy at least equal to the energy equivalent of the meson's mass, or

$$\Delta E = (\Delta m)c^2 = m_m c^2$$

where m_m is the mass of the meson.

By the uncertainty principle, the meson would have to be absorbed in the exchange process in an amount of time on the order of

$$\Delta t \approx \frac{h}{2\pi \Delta E} = \frac{h}{2\pi m_m c^2}$$

In this amount of time, the meson could travel a distance R (which stands for "range") that must be less than that traveled by light in the same time; thus,

$$R < c\Delta t = \frac{h}{2\pi m_m c} \tag{30.5}$$

Taking this distance to be the experimentally known range of the nuclear force ($R \approx 1.4 \times 10^{-15}$ m) and solving for m_m gives $m_m \approx 270 m_e$, where m_e is the electron's mass. Thus, if Yukawa's meson exists, it should have a mass on the order of 270 times that of an electron.

Virtual mesons in an exchange process cannot be directly observed, because they are emitted and reabsorbed during the nucleon–nucleon interaction. Physicists reasoned, however, that if sufficient energy were involved in the *collision* of nucleons, real mesons might be created from the energy available in the collision. These mesons could then be detectable. At the time of Yukawa's prediction, there were no known particles with masses between that of the electron (m_e) and the proton ($m_p = 1836 m_e$).

In 1936, Yukawa's prediction seemed to come true when a new particle with a mass of about $200m_e$ was discovered in cosmic rays. Originally called the μ (mu) meson, and now just the **muon**, it was shown to have two charge varieties, $\pm e$, with a mass of $m_{\mu^{\pm}} = 207m_e$. However, further investigations showed that the muon did *not* behave like the strongly interacting particle of Yukawa's theory. In fact, muon interaction with matter was found to be very weak. The muons from cosmic radiation were found to penetrate a large amount of material and thus were detected in very deep mines.

Note: Muon decay was treated in Examples 26.2 and 26.4 as evidence of time dilation.

This situation was a source of controversy and confusion for years. But, in 1947, more particles in this mass range were discovered in cosmic radiation. These particles (one positive, one negative, and one with no charge) were called π (pi) mesons (for *primary mesons*) and now are more commonly called **pions**. Measurement showed the masses of the pions to be $m_{\pi^{\pm}} = 273m_e$ and $m_{\pi^{\circ}} = 264m_e$. Moreover, pions were found to interact strongly with matter. The pion fulfilled the requirements of Yukawa's theory. This meson was generally accepted as the particle primarily responsible for the transmission of the strong nuclear force. Feynman diagrams for some nucleon–nucleon interactions are shown in ▼ Fig. 30.10.

Free pions and muons are unstable. For example, the π^+ particle decays in about 10^{-8} s into a muon and another type of neutrino:

$$\pi^+ \rightarrow \mu^+ + \nu_{\mu}$$

The ν_{μ} is called a *muon neutrino*, and it differs from the electron neutrino (ν_e) produced in beta decay. The muons can also decay into positrons and electrons with the emission of both types of neutrinos. For example, the positive-muon decay scheme is

$$\mu^+ \rightarrow \beta^+ + \nu_e + \bar{\nu}_{\mu}$$

The Weak Nuclear Force and the W particle

The discrepancies in beta decay discussed in the preceding section led to another discovery. Electrons and neutrinos are emitted from unstable nuclei, but they do *not* exist inside the nucleus before the decay takes place. Enrico Fermi proposed that these particles are actually created when the radioactive nucleus decays. For β^- decay, this means that a neutron is in some way changed, or *transmuted*, into three particles:

$$n \rightarrow p^+ + \beta^- + \bar{\nu}_e$$

Experiments confirmed that free neutrons disintegrate by this decay scheme, with a half-life of about 10.4 min. But which force could cause a neutron

▼ **FIGURE 30.10 Feynman diagrams of nucleon–nucleon (strong nuclear force) interactions** Some possible nucleon–nucleon interactions that occur through the exchange of virtual pions. **(a)** An n–p reaction. **(b)** A p–n reaction. **(c)** A p–p scattering. (Diagrams (a) and (b) are called "charge exchange" interactions. Why?)

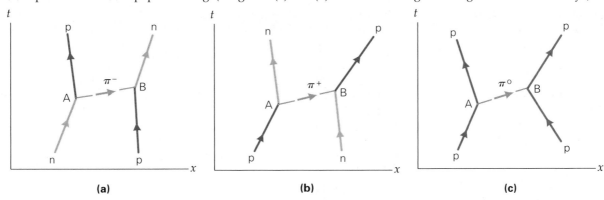

(a)   (b)   (c)

to disintegrate in this manner? Since the neutron is outside the nucleus (free) when it decays, none of the known forces, including the strong nuclear force, seemed applicable. Thus, some other fundamental force must be acting in beta decay. Decay-rate measurements indicated that the force was extraordinarily weak—weaker than the electromagnetic force, but still much stronger than the gravitational force. This force was dubbed the **weak nuclear force**.

Originally, it was thought that the weak interaction is extremely localized, without any measurable range. We now know that the weak force has a range of about 10^{-17} m. While this range is much smaller than that of the strong force, it isn't zero. This means that the exchange particles associated with the weak force (the virtual force carriers) must be much more massive than the pions of the strong force. The virtual exchange particles associated with the weak force were named **W particles**.* W (weak) particles have masses about 100 times that of a proton, a fact that correlates with the extremely short range of the weak force. The existence of the W particle was confirmed in the 1980s when accelerators were built with enough energy to create the first real (nonvirtual) W particles.

The weak force is the only force that acts on neutrinos, which explains why these particles are so difficult to detect. Research has shown that the weak force is involved in the transmutation of other subatomic particles as well. In general, the weak force is limited to transmuting the identities of particles within the nucleus. The only way it manifests its existence in the outside world is through the emitted neutrinos. For example, the Sun's fusion reactions create neutrinos that continuously pass through the Earth.

One highly noticeable, but infrequent, announcement of the weak force at work occurs during the explosion of a stellar supernova. In a *supernova*, the collapse of the core of an aging star gives rise to a huge energy release, accompanied by a great number of neutrinos. In a relatively "nearby" supernova (a mere 1.5×10^{18} km away) observed in 1987, a burst of neutrinos was detected at essentially the same time as the light flash. Since both signals arrived after traveling extremely long distances, this result can be used to set an upper limit on the mass of the neutrino. If the neutrino had mass, its speed would have been less than that of light, and it would thus have arrived later. At present, some very delicate experiments have indicated that neutrinos might have a very small amount of mass. It is hoped that ongoing research will eventually determine if this is true and, if it is, the actual mass of neutrinos.

The Gravitational Force and Gravitons

The exchange particles associated with the gravitational force are called **gravitons**. There is still no firm evidence of the existence of this massless particle. (Why must gravitons be massless?) Ongoing experiments to detect the graviton have as yet proven unsuccessful, due to the relative weakness of its interaction.

A comparison of the relative strengths of the four fundamental forces is given in Table 30.2.

TABLE 30.2 Fundamental Forces

Force	Relative Strength	Action Distance	Exchange Particle	Particles That Experience the Interaction
Strong nuclear	1	Short range ($\approx 10^{-15}$ m)	Pion (π meson)	Hadrons*
Electromagnetic	10^{-3}	Inverse square (infinite)	Photon	Electrically charged
Weak nuclear	10^{-8}	Extremely short range ($\approx 10^{-17}$ m)	W particle†	All
Gravitational	10^{-45}	Inverse square (infinite)	Graviton	All

*Hadrons are discussed in Section 30.6.

†Three particles are involved, as described in Section 30.8.

*The weak force is actually carried by *three* exchange particles: W^+, W^-, and Z^0 (neutral).

30.6 Elementary Particles

OBJECTIVES: To (a) classify the elementary particles into families and (b) understand the different properties of the various families of elementary particles.

The fundamental building blocks of matter are referred to as **elementary particles**. Simplicity reigned when it was thought that an atom was an indivisible particle and therefore was *the* elementary particle. Early in the 20th century, the proton, neutron, and electron were discovered to be constituents of atoms. It was then hoped that these three particles are Nature's elementary particles. However, scientists now know of a huge variety of subatomic particles and are working to simplify and reduce this list to a smaller set of truly elementary particles—building blocks for all the other "composite" particles—if this task is indeed possible.

Several systems classify elementary particles on the basis of their various properties. One classification uses the distinction of nuclear-force interactions. Particles that interact via the weak nuclear force, but not the strong force, are called **leptons** ("light ones"). The lepton family includes the electron, the muon, and their neutrinos. Other particles, called **hadrons**, are the only particles to interact by the strong nuclear force. These particles include the proton, neutron, and pion. Let's look briefly at the lepton and hadron families.

Leptons

The most familiar lepton is the electron. It is the only lepton that exists naturally in atoms. There is no evidence that it has any internal structure, at least down to 10^{-17} m. Thus, at present, the electron is considered to be a point particle.

Muons were first observed in cosmic rays. They appear not to have any internal structure and are therefore sometimes referred to as *heavy electrons*. Muons are unstable and decay in about 2×10^{-6} s, according to the following scheme:

$$\mu^- \rightarrow \beta^- + \bar{\nu}_e + \nu_\mu$$

A third charged lepton is known as a tau (τ^-) particle, or **tauon**. It has a mass about twice that of a proton. The electron, muon, and tauon are all negatively charged, have no apparent internal structure, and have positively charged antiparticles.

The remaining leptons are neutrinos, which are present in cosmic rays and are emitted by the Sun and, as we have seen, in some radioactive decays. Neutrinos have no mass (or close to none) and thus travel at the speed of light (or very near to it). They feel neither the electromagnetic force nor the strong force and pass through matter with very little interaction.

There are three types of neutrinos, each associated with a different charged lepton $(e^\pm, \mu^\pm, \tau^\pm)$. They are named, not surprisingly, the *electron neutrino* (ν_e), the *muon neutrino* (ν_μ), and the *tau neutrino* (ν_τ). There is also an antineutrino for each of them, for a total of six different neutrinos. This completes the list of leptons. With a total of 6 leptons plus their antiparticles, there are 12 different leptons in all. Current theories predict that there should be no others.

Hadrons

Another family of elementary particles is the hadrons. All hadrons interact by the strong force, the weak force, and gravity. The electrically charged members can also interact by the electromagnetic force.

The hadrons are subdivided into *baryons* and *mesons*. **Baryons** include the familiar nucleons—the proton and neutron. They are distinguished from mesons in that they possess half-integer intrinsic spin values $(\frac{1}{2}, \frac{3}{2}, \dots)$. Except for the stable proton, baryons decay into products that eventually include a proton. (Recall from Section 29.2 the beta decay of a neutron into a proton, for example.)

Mesons, which include pions, have integer spin values (0, 1, 2, ...) and eventually decay into leptons and photons. For example, the neutral pion decays into two gamma rays. (Why must there be two?)

The large number of hadrons suggests that they may be composites of other truly elementary particles. Some help in sorting out the hadron "zoo" came in 1963 when Murray Gell-Mann and George Zweig of the California Institute of Technology put forth the *quark theory*, which is discussed in the next section.

Leptons and hadrons and their properties are summarized in Table 30.3.

30.7 The Quark Model

OBJECTIVES: To (a) become familiar with the quark model and properties of quarks and (b) understand how the quark model accounts for the properties of baryons and mesons.

Gell-Mann and Zweig proposed that, in fact, hadrons are *not* elementary particles and therefore are not fundamental building blocks. Hadrons, they theorized, are composite particles composed of truly elementary (fundamental) particles. They named these particles **quarks** (taken from James Joyce's novel *Finnegan's Wake**). However, Gell-Mann and Zweig noted that because leptons and photons are es-

TABLE 30.3 Some Elementary Particles and Their Properties

Family Name	Particle Type	Particle Symbol	Antiparticle Symbol	Rest Energy (MeV)	Lifetime* (s)
Lepton	Electron	e^-	e^+	0.511	stable
	Muon	μ^-	μ^+	105.7	2.2×10^{-6}
	Tauon	τ^-	τ^+	1784	$\approx 3 \times 10^{-13}$
	Electron neutrino	ν_e	$\bar{\nu}_e$	$0^\dagger$	stable
	Muon neutrino	ν_μ	$\bar{\nu}_\mu$	$0^\dagger$	stable
	Tauon neutrino	ν_τ	$\bar{\nu}_\tau$	$0^\dagger$	stable
Hadron					
Mesons	Pion	π^+	π^-	139.6	2.6×10^{-8}
		π^0	same	135.0	8.4×10^{-17}
	Kaon	K^+	K^-	493.7	1.2×10^{-8}
		K^0	$\overline{K^0}$	497.7	8.9×10^{-11}
Baryons	Proton	p	$\bar{\text{p}}$	938.3	stable (?)§
	Neutron	n	$\bar{\text{n}}$	939.6	0.9×10^2
	Lambda	Λ^0	$\overline{\Lambda^0}$	1116	2.6×10^{-10}
	Sigma	Σ^+	$\overline{\Sigma^-}$	1189	8.0×10^{-10}
		Σ^0	$\overline{\Sigma^0}$	1192	0.6×10^{-20}
		Σ^-	$\overline{\Sigma^+}$	1197	1.5×10^{-10}
	Xi	Ξ^0	$\overline{\Xi^0}$	1315	2.9×10^{-10}
		Ξ^-	$\overline{\Xi^+}$	1321	1.6×10^{-10}
	Omega	Ω^-	Ω^+	1672	8.2×10^{-11}

*Lifetimes are expressed to two significant figures or fewer.

†Neutrinos have either zero mass or very small masses. Experiments yield only upper limits; for example, it is known that the mass of the electron neutrino is less than 7×10^{-6} MeV.

§Electroweak theory predicts that the proton is unstable, with a half-life of 1000 trillion times the age of the universe.

*A line in the novel exclaims, "Three quarks for Muster Mark!" The "three quarks" denote the three children of a character in the novel, Mister (Muster) Mark, who is also known as Mr. Finn.

sentially point particles, they *are* likely to be truly elementary particles with no internal structure.

Noting that some hadrons are electrically charged, Gell-Mann and Zweig reasoned that quarks must also possess charge. Their initial **quark model** consisted of three different quarks (with fractional charges) and their antiparticles (called *antiquarks*). Table 30.4 shows that, to account for hadrons that were undiscovered at the time, the list of quarks eventually had to be expanded to include six types. The original quark idea required only three quarks (plus three antiquarks) to construct the hadrons known at that time. These quarks were named the *up* quark (u), the *down* quark (d), and the *strange* quark (s). By using various combinations of these three types of quarks, the relatively heavy hadrons, the baryons (whose name means "heavy ones"), could be built. In addition, quark–antiquark pairs could account for the lighter hadrons, the mesons. Thus, Gell-Mann and Zweig had proposed a radical new idea:

| Quarks are the fundamental particles of the hadron family.

Since several quarks have to combine to give the charge on the hadron, the quarks must have fractions of an electron charge e. The theory proposed that u, d, and s quarks have charges of $+\frac{2}{3}e$, $-\frac{1}{3}e$, and $-\frac{1}{3}e$, respectively. The antiquarks, designated by overbars, such as $\bar{u}$, have opposite charges. Thus, three quark combinations could produce any baryon. For instance, the quark composition of the proton and neutron would be uud and udd, respectively. Mesons could be constructed from various pairs of a quark and an antiquark, such as $u\bar{d}$ for the positive pion, π^+ (▼Fig. 30.11).

Note: Quark names (such as up, down, and strange) are arbitrary and therefore should not be taken literally. They serve simply to identify the quarks.

TABLE 30.4 Types of Quarks*

Name	Symbol	Charge
Up	u	$+\frac{2}{3}e$
Down	d	$-\frac{1}{3}e$
Strange	s	$-\frac{1}{3}e$
Charm	c	$+\frac{2}{3}e$
Top (Truth)	t	$+\frac{2}{3}e$
Bottom (Beauty)	b	$-\frac{1}{3}e$

*Antiquarks are designated by an overbar and have opposite charges compared with those of the corresponding quarks.

Note: Three quarks in proper combination account for all known baryons. Quark–antiquark pairs in proper combination account for all known mesons.

▼ **FIGURE 30.11 Hadronic quark structure** **(a)** Three combinations of quarks can be used to construct all the baryons, such as the proton and neutron. **(b)** Quark–antiquark combinations can be used to construct all the mesons, such as the positive pion.

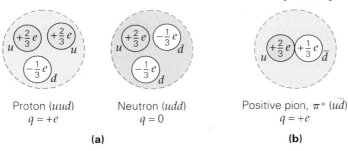

Proton (uud)
$q = +e$

Neutron (udd)
$q = 0$

Positive pion, π^+ ($u\bar{d}$)
$q = +e$

(a)

(b)

Conceptual Example 30.5 ■ Building Mesons: Quark Engineering, Inc.

Using Tables 30.3 and 30.4, explain why it is not possible to build a meson from two quarks (in other words, without using any antiquarks).

Reasoning and Answer. It can be seen from the list of mesons in Table 30.3 that mesons are either neutral or have a charge that is an integral multiple of e. In Table 30.4, note that all the positively charged quarks have a charge of $+\frac{2}{3}e$. Thus, any two of them would add up to a meson charge of $+\frac{4}{3}e$, which is contrary to observation. Similar reasoning holds if we use two negatively charged quarks: Since they all have the same charge ($-\frac{1}{3}e$), any two of them add up to a total charge of $-\frac{2}{3}e$, again in disagreement with experiment. Finally, consider combining one positively charged quark and one negatively charged quark. The net charge of this combination is $+\frac{1}{3}e$, again not in agreement with observation. Thus, no combination of two quarks (or two antiquarks) can be used to produce a meson. The combinations must include at least one antiquark.

Follow-up Exercise. The antiparticle of the positive pion (π^+) is the negative pion (π^-). What is the quark structure of the negative pion? Show that it is composed of the antiquarks of the quarks that make up the positive pion.

The discovery of new subatomic particles in the 1970s led to the addition of the last three quark types: *charm* (*c*); *top*, or *truth* (*t*); and *bottom*, or *beauty* (*b*). Today, there is firm experimental evidence of the existence of all six quarks and their six antiquarks. How many more such particles will be needed to keep up with our growing "zoo" of elementary particles? The hope is, of course, that this list will not need to be expanded. In summary, the present picture includes the following truly elementary particles: leptons (and antileptons), quarks (and antiquarks), and exchange particles such as the photon.

Quark Confinement, Color Charge, and Gluons

Thus far in our discussion on quarks, one thing has been missing—direct experimental observation of a quark. Unfortunately, even in the most energetic particle collisions, *a free quark has never been observed*. Physicists now believe that quarks are permanently confined within their particles by a springlike force. That is, a force exists between quarks that grows as they separate from one another. This force grows very rapidly with distance and prevents the ejection of a quark from its particle. This phenomenon is called **quark confinement**.

In order explain the force between quarks and to clear up some problems with apparent violation of the Pauli exclusion principle (see the discussion of atomic structure, in Section 28.3), quarks were endowed with another characteristic called **color charge**, or simply *color*. There are three types of color charge: red, green, and blue. (These names have nothing to do with visual color.) In analogy to electric charge, the quark confinement force exists because different color charges attract each other. Recall from Section 30.5 that the electromagnetic force is due to virtual photon exchanges between charged particles. Similarly, the force between quarks of different color is due to exchanges of virtual particles called **gluons** (▼Fig. 30.12). This force is sometimes called the **color force**. This theory is named *quantum chromodynamics* (*QCD*) (*chromo* for "color"), in analogy to Feynman's quantum *electro*dynamics (QED), which successfully explains the electromagnetic force.

The concept of the force between quarks can be extended to explain the force between hadrons—the strong nuclear force. Consider our previous explanation of the strong force between a neutron and proton as shown in Fig. 30.10a:

▼ **FIGURE 30.12 Quarks, color charge, confinement, and gluons** (a) Quarks of different color charge attract each other via the color force, which keeps them confined (here, inside a baryon—how do you know this is a baryon?). (b) Gluons (wiggly arrows) are exchanged between quarks of different color charge, creating the color force—analogous to virtual photons and the electric force.

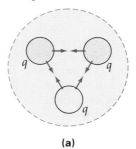
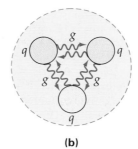

(a)   (b)

the exchange of a virtual negative pion. More fundamentally, we can qualitatively depict this force in terms of an exchange of quarks between the hadrons (▶Fig. 30.13).

30.8 Force Unification Theories, the Standard Model, and the Early Universe

OBJECTIVES: To (a) become familiar with current attempts to unify the four fundamental forces and (b) understand why elementary-particle interactions might hold the key to understanding the very early evolution of the universe.

Unification Theories

Early in the 20th century, Einstein was one of the first theorists to conjecture that it might be possible to unify the fundamental forces of nature—that these four apparently very different forces might really be just different manifestations of one force. Each manifestation would appear under a different set of physical conditions. For example, an electrically neutral particle would not exhibit the electromagnetic portion. Since then, it has been the dream of physicists to understand the "fundamental interactions" in the universe, using just one force. Attempts to unify the various forces are called **unification theories**.

Actually, the first step toward unification was taken in the 19th century by Maxwell when he combined the electric and magnetic forces into a single electromagnetic force. Einstein later showed that the two are connected by relativity. The next major step occurred in the 1960s when Sheldon Glashow, Abdus Salam, and Steven Weinberg successfully combined the electromagnetic force with the weak force into a single **electroweak force**. For their efforts, they were awarded a Nobel Prize in 1979.

How can such apparently very different forces be unified? After all, their exchange particles are so different—recall the massless photon (γ) for the electromagnetic force and the massive W particle for the weak force. However, it was reasoned that, at extremely high energies, the mass–energy of the W particle is negligible compared with its total energy, in effect making it massless, like the photon. To understand this reasoning, recall from Section 26.5 that the total energy of a particle is the sum of its kinetic and rest energies ($E = K + mc^2$). When $K \gg mc^2$ (that is, at high energies), the particle's rest energy is negligible compared with its kinetic energy—perhaps, Glashow, Salam, and Weinberg reasoned, making the W particle more "photonlike" than was previously thought.

In the electroweak theory, weakly interacting particles such as electrons and neutrinos carry a *weak charge*. This weak charge is responsible for the exchange of particles, thus creating the combined electroweak force. The electroweak unification theory predicted the existence of three electroweak exchange particles: W^+ and W^- when there is charge exchange, and the neutral Z^0 when there is no charge transfer (▶Fig. 30.14). The eventual discovery of these particles led to a Nobel Prize for Carlo Rubbia and Simon van der Meer in 1984.

Scientists are now attempting to unify the electroweak force with the strong nuclear force. If successful, this **grand unified theory (GUT)** would reduce the number of fundamental forces to two. Most GUT attempts require several dozen exchange particles. In addition, leptons and quarks are combined into one family and can change into each other through the exchange of these particles. Scientists are mildly optimistic that experimental verification of one of the GUT candidates might occur early in the 21st century. For example, most of the current GUT theories predict that the proton should be unstable and decay with a half-life of about 10^{32} years, which is 10^{22} times the age of the universe! Experiments looking for

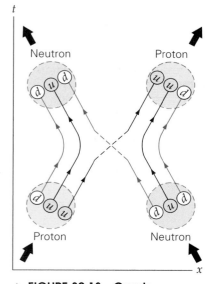

▲ **FIGURE 30.13 Quark depiction of the strong nuclear force** Instead of envisioning the n–p interaction as an exchange of a virtual π^- particle, we can use a model of quark exchange. In this exchange, a pair of quarks (equivalent to a π^- particle) is transferred. A proton becomes a neutron, and vice versa.

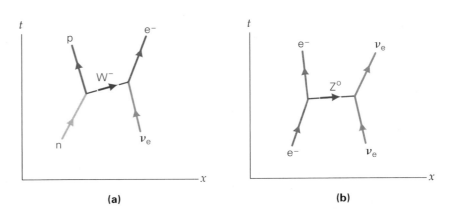

FIGURE 30.14 Weak-force interactions The Feynman diagrams for **(a)** the neutrino-induced conversion of a neutron into a proton and an electron through the exchange of a W⁻ particle and **(b)** the scattering of a neutrino by an electron through the exchange of a neutral Z° particle.

this decay are currently underway, so far with no success. The experiments require huge amounts of water (protons) in order to make up for the expected very slow decay rate.

The ultimate unification would be to fold the gravitational force into the GUT, creating a single **superforce**. How this would be done, or even *if* it can ever be done, is not clear at this point. One problem is that, although the three components of the GUT can be represented as force fields in space and time, in our current view, gravity *is* space and time!

The Standard Model

Taken together, the electroweak theory and the QCD model for the strong interaction are referred to as the **Standard Model**. In this model, the gluons carry the strong force. This force keeps quarks together to form composite particles such as protons and pions. Leptons do not experience the strong force and participate only in the gravitational and electroweak interactions. Presumably, the gravitational interaction is carried by the graviton. The electromagnetic part of the electroweak interaction is carried by the photon, and the W and Z bosons carry the weak portion of the electroweak force.

Evolution of the Universe and the Superforce

An interesting connection now exists between elementary particle physicists and astrophysicists interested in the evolution of the universe—in particular, its very early evolution. This connection occurs because, according to present ideas, the universe began 10 to 15 billion years ago with the *Big Bang*. It is theorized that temperatures during the first 10^{-45} second of the universe were on the order of 10^{32} K. This corresponds to particle kinetic energies of about 10^{22} MeV, high enough that the rest energy of even the most massive elementary particles would be negligible. In effect, the particles would be massless, perhaps placing the four forces on an equal footing.

As the universe expanded and cooled, these early elementary particles condensed into what we see today. First, protons, neutrons, and electrons formed; in turn, they combined into atoms and, eventually, molecules. Over the billions of years, the average temperature of the universe has cooled off to its present value of about 3 K. In the process, scientists think that the superforce symmetry (the equal footing of all four forces) has been lost, leaving us with four very different-looking forces that are really components of a superforce.

It is hoped that future experiments might tell us more about the early moments of the universe and thus about the superforce. The ultimate goal of physics might even be within reach—to understand the basic interactions that govern the universe. While great strides are being made, it is likely to take well into this century, if not further, to achieve this goal.

Chapter Review

Important Concepts

- A **nuclear reaction** (including decay) involves the interactions of nuclei and particles, usually resulting in different product nuclei or particles.

- The **Q value** of a reaction (or decay) is the energy released or absorbed in the process; it appears as a change in the total mass of the system. If the total mass increases, energy is required. This situation is called an **endoergic reaction**. If the total mass decreases, energy is released in an **exoergic reaction**. For a two-body reaction of the form A + a → B + b,

$$Q = (m_A + m_a - m_B - m_b)c^2 = (\Delta m)c^2 \qquad (30.3)$$

- In **fission**, an unstable heavy nucleus decays by splitting into two fragments and several neutrons, which together have less total mass than that of the original nucleus; kinetic energy is thus released. A **chain reaction** occurs when the neutrons released from one fission trigger other fissions, which trigger further fissions, and so on.

- In **a nuclear power reactor**, the fission chain reaction is kept from going out of control by a set of control rods. The heat energy released is usually used to create steam, which eventually turns generator turbine blades to create electricity.

- In **fusion**, two light nuclei fuse, producing a nucleus with less total mass than that of the original nuclei; energy is thus released. Controlled fusion reactions are, at present, not commercially feasible, because no one has achieved confinement of the gas of ionized atoms and free electrons, called a **plasma**, at the proper density and temperature.

- The **beta decay** of an unstable nucleus produces a daughter nucleus, an electron or positron, and an antineutrino or neutrino.

- **Exchange particles** are virtual particles believed to be associated with various forces. The **pion** (a meson) is the exchange particle primarily responsible for the strong nuclear force.

- The **weak nuclear force**, transmitted by the **W particle**, is primarily responsible for beta decay and the instability of the neutron.

- **Hadrons** are the family of elementary particles that interact by the strong force, the weak force, and gravity. If electrically charged, hadrons also interact by the electromagnetic force. The hadrons are subdivided into baryons and mesons. **Baryons** include the familiar nucleons—the proton and neutron. Except for the stable proton, baryons decay into products that eventually include a proton. **Mesons**, which include pions, eventually decay into leptons and photons.

- **Leptons** are the family of elementary particles that interact through the weak force, electromagnetism, and gravity, but not the strong force. Electrons, muons, and neutrinos make up the lepton family.

- **Quarks** are elementary, fractionally charged particles that make up hadrons.

- **Charge color** describes the color force between quarks. Color force explains why a free quark is never likely to be seen, a phenomenon called **quark confinement**. Quark exchange is the fundamental explanation for the strong nuclear force.

- The **electroweak force** is the name given to the unified electromagnetic and weak forces.

- The **grand unified theory** (**GUT**) is an attempt to unify the electroweak force with the strong nuclear force.

- The **superforce** is the single force that will result if the long-pursued unification of the "fundamental" forces ever materializes.

- In the **Standard Model**, gluons carry the strong force, which keeps quarks together in composite particles such as protons. Leptons participate only in the gravitational and electroweak interactions. The former is carried by the graviton. The electroweak interaction is carried by the photon and the W and Z bosons.

Exercises

30.1 Nuclear Reactions

1. The Q value of a reaction (a) is the difference in the kinetic energies of the products and reactants, (b) is equal to the mass–energy difference between the initial and final systems, (c) may be positive or negative, or (d) all of the preceding.

2. To initiate an endoergic reaction, a particle incident on a stationary nucleus must have (a) a minimum disintegration energy; (b) a kinetic energy equal to the Q value; (c) a kinetic energy less than a certain threshold energy; or (d) a kinetic energy larger than the Q value, which depends on the ratio of incident-particle mass to target mass.

3. **CQ** What is the main difference between the results of a chemical reaction and those of a nuclear reaction?

4. **CQ** How does the threshold energy of an incident particle in a nuclear reaction vary with the mass of the target nuclei? (Sketch a graph of threshold energy versus target mass if you can.) What is the interpretation of $Q = 0$?

5. **CQ** In an exoergic reaction, which has more mass, the reactants or the products? Why?

6. ■ Complete the following nuclear reactions:
   (a) $^1_0n + ^{40}_{18}Ar \rightarrow \quad + ^{\ 0}_{-1}e$
   (b) $^1_0n + ^{235}_{92}U \rightarrow ^{98}_{40}Zr + \quad + 3(^1_0n)$
   (c) $^1_0n + ^{235}_{92}U \rightarrow ^{133}_{51}Sb + ^{99}_{41}Nb +$
   (d) $\quad (\alpha, p)^{17}_8O$
   (e) $^{137}_{56}Ba(n, \quad)^{137}_{55}Cs$

7. ■ Complete the following nuclear reactions:
   (a) $^{13}_6C + ^1_1H \rightarrow \gamma +$
   (b) $^{10}_5B + ^4_2He \rightarrow ^{12}_6C +$
   (c) $^{27}_{13}Al(\alpha, n)$
   (d) $^{14}_7N(\alpha, p)$
   (e) $^{13}_6C(p, \alpha)$

8. ■ Determine the compound nuclei for the reactions in Exercise 6.

9. ■ Determine the compound nuclei for the reactions in Exercise 7.

10. **IE** ■ (a) In the reaction $^{13}_6C + ^1_1H \rightarrow ^4_2He + ^{10}_5B$, add up the initial and final masses to determine if it is endoergic or exoergic. (b) If it is exoergic, find the amount of energy released; if it is endoergic, find the threshold energy.

11. ■ What would the daughter nuclei in the following decay equations be? Will the reactions occur spontaneously?
   (a) $^{22}_{10}Ne \rightarrow \quad + ^{\ 0}_{-1}e$
   (b) $^{226}_{88}Ra \rightarrow \quad + ^4_2He$
   (c) $^{16}_8O \rightarrow \quad + ^4_2He$

12. ■ Show that the Q value for the reaction $^1_1H + ^2_1H \rightarrow ^3_2He + \gamma$ is 5.49 MeV.

13. ■ **IE** Uranium-238 undergoes alpha decay as follows:
$$^{238}_{92}U \rightarrow ^{234}_{90}Th + ^4_2He$$
$$(238.050\,786\,u) \quad (234.043\,583\,u) \quad (4.002\,603\,u)$$
   (a) Would you expect the Q value to be (1) positive, (2) negative, or (3) zero? Why? (b) Find the Q value.

14. ■■ Find the threshold energy for the following reaction:
$$^{16}_8O + ^1_0n \rightarrow ^{13}_6C + ^4_2He$$
$$(15.994\,915\,u\ ) \quad (1.008\,665\,u\ ) \quad (13.003\,355\,u) \quad (4.002\,603\,u)$$

15. ■■ Find the threshold energy for the following reaction:
$$^3_2He + ^1_0n \rightarrow ^2_1H + ^2_1H$$
$$(3.016\,029\,u) \quad (1.008\,665\,u) \quad (2.014\,102\,u) \quad (2.014\,102\,u)$$

16. ■■ Find the threshold energy for the following reaction:
$$^{13}_6C + ^1_1H \rightarrow ^1_0n + ^{13}_7N$$
$$(13.003\,355\,u) \quad (1.007\,825\,u) \quad (1.008\,665\,u) \quad (13.005\,739\,u)$$

17. **IE** ■■ (a) If a nuclear reaction requires the incident particle to have a minimum kinetic energy to occur, the reaction is (1) endoergic, (2) exoergic, or (3) atomic. (b) Determine the minimum kinetic energy the incident alpha particle must have so that it can initiate the following reaction:
$$^{14}_7N + ^4_2He \rightarrow ^{17}_8O + ^1_1H$$
$$(14.003\,074\,u\ ) \quad (4.002\,603\,u\ ) \quad (16.999\,131\,u\ ) \quad (1.007\,825\,u\ )$$

18. ■■ Is the given reaction endoergic or exoergic? Prove your answer.
$$^7_3Li + ^1_1H \rightarrow ^4_2He + ^4_2He$$
$$(7.016\,005\,u) \quad (1.007\,825\,u) \quad (4.002\,603\,u) \quad (4.002\,603\,u)$$

19. ■■ Is the reaction $^{200}Hg(p, \alpha)^{197}$ Au endoergic or exoergic? Prove your answer. (The reaction on p. xxx gives the mass values.)

20. ■■ Determine the Q value of the following reaction:
$$^9_4Be + ^4_2He \rightarrow ^{12}_6C + ^1_0n$$
$$(9.012\,183\,u) \quad (4.002\,603\,u) \quad (12.000\,000\,u) \quad (1.008\,665\,u)$$

21. ■■ What is the minimum kinetic energy a proton must have in order to initiate the reaction $^3_1H(p, d)^2_1H$? ("d" stands for a deuterium nucleus and is called the *deuteron*.)

22. ■■ ^{226}Ra decays and emits a 4.706-MeV alpha particle. Find the kinetic energy of the recoiling daughter nucleus from the decay of a stationary radium-226 nucleus.

23. **IE** ■■■ The same type of incident particle is used for two endoergic reactions. In one reaction, the mass of the target nucleus is 15 times greater than that of the incident particle, and in the other reaction, it is 20 times greater. The Q value of the first reaction is known to be three times that of the second. (a) Compared with the second reaction, the first reaction has (1) greater, (2) the same amount of, or (3) less minimum threshold energy. (b) Prove your answer to part (a) by calculating the ratio of the minimum threshold energy for the first reaction to that for the second reaction.

24. ■■■ Consider n (where n is an integer ≥ 1) ceramic pie plates of radius R randomly fixed (but none overlapping) on a rectangular wall with dimensions L and W. If you were to throw a very small (point) object at the wall, what would be the percent probability of hitting a pie plate, in terms of the given parameters? [*Hint*: Think in terms of area and your "reaction" cross-section.]

25. ■■■ As you might have guessed, the SI units of cross-section are m^2. When neutron resonances were first discovered, however, they were much, much larger than any cross-section previously measured, and someone said they were "as big as a barn." Hence, a more common unit for cross-section is the barn (b), defined as $1 \text{ b} \equiv 10^{-28} \text{ m}^2$. Using the empirical equation for the radius of a nucleus, $R = R_0 A^{1/3}$, where $R_0 \approx 1.2 \times 10^{-15}$ m = 1.2 fm, estimate the geometrical cross-section (in barns) for (a) carbon-12, (b) iron-56, (c) lead-208, and (d) uranium-238.

26. ■■■ Assume that the average kinetic energy of ions in a plasma is given by the equation for the kinetic energy of the atoms in an ideal gas $(\frac{1}{2}mv^2 = \frac{3}{2}k_B T)$. If fusion can occur when the ions approach each other to within a distance of the upper limit for the nuclear diameter $(R = 10^{-12}$ cm), calculate the temperature required for fusion of two deuterium ions. (Recall that Boltzmann's constant is $k_B = 1.38 \times 10^{-23}$ J/K.)

30.2 Nuclear Fission
and
30.3 Nuclear Fusion

27. Nuclear fission (a) is endoergic, (b) occurs only for uranium-235, (c) releases about 500 MeV per fission, or (d) requires a critical mass for a sustained reaction.

28. A nuclear reactor (a) can operate on natural (unenriched) uranium, (b) has its chain reaction controlled by neutron-absorbing materials, (c) can be partially controlled by the amount of moderator, or (d) all of the preceding.

29. A nuclear fusion reaction (a) has a negative Q value, (b) may occur spontaneously, (c) is an example of "splitting" the atom, or (d) releases less than 50 MeV of energy per fusion process.

30. Controlled fusion requires (a) no critical mass, (b) confinement, (c) formation of a plasma, or (d) all of the preceding.

31. CQ What is the fundamental difference between a fission and a fusion?

32. CQ In a nuclear fission reaction, which has more mass, the original nuclei or the products? How about fusion? Why?

33. CQ The energy produced in fission reactions is carried off as kinetic energies of the products. How is this energy converted to heat in a nuclear reactor?

34. ■ Find the approximate energy released in the following fission reactions: (a) $^{235}_{92}\text{U} + ^{1}_{0}\text{n} \rightarrow$ fission products plus the release of five neutrons and (b) $^{235}_{94}\text{Pu} + ^{1}_{0}\text{n} \rightarrow$ fission products plus the release of three neutrons.

35. ■ Calculate the amounts of energy released in the following fusion reactions: (a) $^{1}_{1}\text{H} + ^{1}_{0}\text{n} \rightarrow ^{2}_{1}\text{H} + \gamma$ and (b) $^{3}_{2}\text{He} + ^{3}_{2}\text{He} \rightarrow ^{4}_{2}\text{He} + 2^{1}_{1}\text{H}$.

36. ■ Calculate the amounts of energy released in the following fusion reactions: (a) $^{2}_{1}\text{H} + ^{2}_{1}\text{H} \rightarrow ^{3}_{2}\text{He} + ^{1}_{0}\text{n}$ and (b) $^{2}_{1}\text{H} + ^{3}_{1}\text{H} \rightarrow ^{4}_{2}\text{He} + ^{1}_{0}\text{n}$.

37. ■■ In power reactors, using water as a moderator works well, because the proton and neutron have nearly the same mass. For a head-on elastic collision, we might expect a neutron to lose all of its kinetic energy in one collision, whereas for an "almost miss," we might expect it to lose essentially none. Assume that, on average, the neutron loses 60% of its kinetic energy during each collision. Estimate how many collisions are needed to reduce a 2.0-MeV neutron to a neutron with a kinetic energy of only 0.02 eV (approximately "thermal").

30.4 Beta Decay and the Neutrino

38. In the absence of a neutrino, what is not conserved in beta decay, (a) energy, (b) linear momentum, (c) angular momentum, or (d) all of the preceding?

39. A neutrino interacts with matter by (a) the electromagnetic interaction, (b) the strong interaction, (c) the weak interaction, or (d) both (b) and (c).

40. CQ Why is it so difficult to detect neutrinos experimentally?

41. CQ What are the fundamental properties of a neutrino?

42. ■ A neutrino created in a beta-decay process has an energy of 2.65 MeV. What is the de Broglie wavelength of the neutrino?

43. ■■ In Exercise 42, what is the sum of the linear momentums of the beta particle and of the daughter nucleus? Include the direction of this sum and your reasoning.

44. IE ■■ In Exercise 42, if the disintegration energy were 3.51 MeV, (a) what would be the maximum possible kinetic energy of the beta particle, (1) zero, (2) 0.86 MeV, (3) 2.65 MeV, or (4) 3.51 MeV? Why? (b) What would the beta particle's momentum (including direction) be in this case? (c) What would be the kinetic energy and momentum of the daughter nucleus in this case?

45. ■■ Show that the disintegration energy for β^- decay is $Q = (m_P - m_D - m_e)c^2 = (M_P - M_D)c^2$, where the m's represent the masses of the parent and daughter *nuclei* and the M's represent the masses of the neutral *atoms*.

46. ■■ What is the maximum kinetic energy of the electron emitted when a ^{12}B nucleus beta decays into a ^{12}C nucleus? (See Exercise 45.)

47. ■■ The kinetic energy of an electron emitted from a ^{32}P nucleus that beta decays into a ^{32}S nucleus is observed to be 1.00 MeV. What is the energy of the accompanying neutrino of the decay process? Neglect the recoil energy of the daughter nucleus. (See Exercise 45.)

48. ■■ Show that the disintegration energy for β^+ decay is $Q = (m_P - m_D - m_e)c^2 = (M_P - M_D - 2m_e)c^2$, where the m's represent the masses of the parent and daughter *nuclei* and the M's represent the masses of the neutral *atoms*.

49. ■■■ The kinetic energy of a positron emitted from the β^+ decay of a ^{13}N nucleus into a ^{13}C nucleus is measured to be 1.190 MeV. What is the energy of the accompanying neutrino of the decay process? Neglect the recoil energy of the daughter nucleus. (See Exercise 48.)

50. IE ■■■ The expressions for the Q values associated with both β^- and β^+ decay are given in Exercises 45 and 48. Assume that the daughter's atomic mass M_D is the same in both processes. (a) If both expressions are to be energetically possible, the mass of the parent atom, M_P, for β^- decay must be (1) greater than, (2) equal to, or (3) less than that for β^+ decay. Why? (b) Write the mass requirements of the parent atoms for β^- and β^+ processes, respectively, in terms of the atomic masses of the daughter, parent, and electron.

30.5 Fundamental Forces and Exchange Particles

51. Virtual particles (a) form virtual images, (b) exist only in an amount of time as permitted by the uncertainty principle, (c) make up positrons, or (d) can be observed in exchange processes.

52. The exchange particle for the strong nuclear force is the (a) pion, (b) W particle, (c) muon, or (d) positron.

53. CQ If virtual exchange particles are unobservable by themselves, how is their existence verified?

54. CQ When a proton interacts with another proton, which of the four fundamental forces would be involved? How about when an electron interacts with another electron?

55. ■ Assuming the range of the nuclear force to be on the order of 10^{-15} m, predict the mass of the exchange particle.

56. ■■ In a certain type of reaction, a high-speed proton collides with a nucleus and travels, on average, a distance of 5.0×10^{-16} m within the nucleus before the reaction takes place. What type of interaction is this, and how much time elapses before the interaction takes place?

57. IE ■■ (a) In an interaction using the virtual-particle model, the range of the interaction (1) increases, (2) remains the same, or (3) decreases as energy of the exchange particle increases. Why? (b) A W particle in a weak interaction is found to have rest energy of 1.00 GeV. What is the approximate range for the weak interaction with this exchange particle?

58. ■■ By what minimum amount of energy is the conservation of energy "violated" during a neutral pi-meson exchange process?

59. ■■ How long is the conservation of energy "violated" in a neutral pi-meson exchange process?

30.6 Elementary Particles *and* 30.7 The Quark Model

60. Particles that interact by the strong nuclear force are called (a) *muons*, (b) *hadrons*, (c) *W particles*, or (d) *leptons*.

61. Quarks make up which of the following particles? (a) hadrons; (b) muons; (c) Z particles; (d) all of the preceding.

62. CQ What is meant by quark flavor and color? Can these attributes be changed? Explain.

63. CQ With so many types of hadrons, why aren't fractional electronic charges observed?

64. CQ Describe the differences between baryons and mesons.

65. ■ Out of the following pairs, which particle has more mass? (a) π^+ and π^0; (b) K^+ and K^0; (c) Σ^+ and Σ^0; (d) Ξ^0 and Ξ^-? [*Hint*: See Table 30.3.]

66. IE ■ (a) The quark combination for a proton is (1) *uuu*, (2) *uud*, (3) *udd*, or (4) *ddd*. (b) Prove your answer to part (a).

67. IE ■ (a) The quark combination for a neutron is (1) *uuu*, (2) *uud*, (3) *udd*, or (4) *ddd*. (b) Prove your answer to part (a).

30.8 Force Unification Theories, the Standard Model, and the Early Universe

68. The *grand unified theory* would reduce the number of fundamental forces to (a) one, (b) two, (c) three, or (d) four.

69. The magnetic force is part of the (a) electroweak force, (b) weak force, (c) strong force, or (d) superforce.

Additional Exercises

70. Complete the following nuclear reactions:
    (a) $^6_3$Li $+ {}^1_1$H $\rightarrow {}^3_2$He $+$
    (b) $^{58}_{28}$Ni $+ {}^2_1$H $\rightarrow {}^{59}_{28}$Ni $+$
    (c) $^{235}_{92}$U $+ {}^1_0$n $\rightarrow {}^{138}_{54}$Xe $+ 5{}^1_0$n $+$
    (d) $^9_4$Be(α, n)
    (e) $^{16}_8$O(n, p)

71. Give the compound nuclei for the reactions in Exercise 70.

72. Compute the Q value (to three significant figures) of the following fusion reaction: $^2_1$H $+ {}^2_1$H $\rightarrow {}^3_1$H $+ {}^1_1$H.

73. Find the energy released in the beta decay of ^{14}C.

74. Show that the Q value for electron capture is given by $Q = (m_P + m_e - m_D)c^2 = (M_P - M_D)c^2$, where the m's represent the masses of the parent and daughter *nuclei* and the M's represent the masses of the neutral *atoms*.

75. **IE** A ^{7}Be nucleus can be converted into a ^{7}Li nucleus by capturing an electron (see Exercise 74) or through β^+ decay (see Exercise 48). (a) From an energy point of view, which one is more likely to occur, (1) electron capture, (2) β^+ decay, or (3) both are equally likely to occur. Why? (b) What is the energy of the emitted neutrino (neglect daughter-nucleus recoil) in electron capture?

76. Complete the following nuclear reactions:
    (a) $^4_2$He + $^{14}_7$N → $^{17}_8$O +
    (b) $^{13}_7$N + $^1_0$n → $^4_2$He +
    (c) $^{92}_{40}$Zr(p, α)
    (d) $^3_2$He(γ, p)

77. Determine the threshold energy of the following reaction:

$$\begin{array}{ccccccc} ^{16}_{8}\text{O} & + & ^{1}_{0}\text{n} & \longrightarrow & ^{13}_{6}\text{C} & + & ^{4}_{2}\text{He} \\ (15.994\,915\ u) & & (1.008\,665\ u) & & (13.003\,355\ u) & & (4.002\,603\ u) \end{array}$$

Appendices

Appendix I Mathematical Relationships

Algebraic Relationships

$(a + b)^2 = a^2 + 2ab + b^2$
$(a - b)^2 = a^2 - 2ab + b^2$
$(a^2 - b^2) = (a + b)(a - b)$

Quadratic Formula

If $ax^2 + bx + c = 0$, then $x = \dfrac{-b \pm \sqrt{b^2 - 4ac}}{2a}$

Powers and Exponents

$x^0 = 1$

$x^1 = x$ $\qquad x^{-1} = \dfrac{1}{x}$

$x^2 = x \cdot x$ $\qquad x^{-2} = \dfrac{1}{x^2}$ $\qquad x^{\frac{1}{2}} = \sqrt{x}$

$x^3 = x \cdot x \cdot x$ $\qquad x^{-3} = \dfrac{1}{x^3}$ $\qquad x^{\frac{1}{3}} = \sqrt[3]{x}$

etc. $\qquad$ etc. $\qquad$ etc.

$x^a \cdot x^b = x^{(a+b)}$
$x^a / x^b = x^{(a-b)}$
$(x^a)^b = x^{ab}$

Logarithms

If $x = a^n$, then $n = \log_a x$.

common logarithms: base 10
(assumed when abbreviation "log" is used, unless another base is specified)

$\log 10^x = x$
$\log xy = \log x + \log y$
$\log x/y = \log x - \log y$
$\log x^y = y \log x$

natural logarithms: base $e = 2.71828\ldots$ (abbreviated "ln")

$\ln e^x = x$
$\log x = 0.43429 \ln x$
$\ln x = 2.3026 \log x$

Geometric and Trigonometric Relationships

Areas and Volumes of Some Common Shapes

Circle: $\qquad A = \pi r^2 = \dfrac{\pi d^2}{4}$ (area)

$\qquad\qquad c = 2\pi r = \pi d$ (circumference)

Triangle: $\qquad A = \frac{1}{2} ab$

Sphere: $\qquad A = 4\pi r^2$
$\qquad\qquad V = \frac{4}{3}\pi r^3$

Cylinder: $\qquad A = \pi r^2$ (end)
$\qquad\qquad A = 2\pi rh$ (body)
$\qquad\qquad A = 2(\pi r^2) + 2\pi rh$ (total)
$\qquad\qquad V = \pi r^2 h$

Definitions of Trigonometric Functions

$\sin \theta = \dfrac{y}{r} \qquad \cos \theta = \dfrac{x}{r} \qquad \tan \theta = \dfrac{\sin \theta}{\cos \theta} = \dfrac{y}{x}$

$\theta°$ (rad)	$\sin \theta$	$\cos \theta$	$\tan \theta$
0° (0)	0	1	0
30° ($\pi/6$)	0.500	0.866	0.577
45° ($\pi/4$)	0.707	0.707	1.00
60° ($\pi/3$)	0.866	0.500	1.73
90° ($\pi/2$)	1	0	$\rightarrow \infty$

For very small angles,

$\cos \theta \approx 1 \qquad \sin \theta \approx \theta$ (radians)

$\tan \theta = \dfrac{\sin \theta}{\cos \theta} \approx \theta$ (radians)

The sign of a trigonometric function depends on the quadrant, or the signs of x and y; for example, in the second quadrant

$(-x, y)$, $-x/r = \cos\theta$ and $y/r = \sin\theta$. The sign can also be assigned by using the reduction formulas.

Reduction Formulas

(θ in second quadrant)	(θ in third quadrant)	(θ in fourth quadrant)

$\sin\theta = \cos(\theta - 90°) = -\sin(\theta - 180°) = -\cos(\theta - 270°)$
$\cos\theta = -\sin(\theta - 90°) = -\cos(\theta - 180°) = \sin(\theta - 270°)$

Fundamental Identities

$\sin^2\theta + \cos^2\theta = 1$
$\sin 2\theta = 2\sin\theta\cos\theta$
$\cos 2\theta = \cos^2\theta - \sin^2\theta = 2\cos^2\theta - 1 = 1 - 2\sin^2\theta$
$\sin^2\theta = \frac{1}{2}(1 - \cos 2\theta)$
$\cos^2\theta = \frac{1}{2}(1 + \cos 2\theta)$

For half-angle $(\theta/2)$ identities, replace θ with $\theta/2$; for example,
$\sin^2\theta/2 = \frac{1}{2}(1 - \cos\theta)$
$\cos^2\theta/2 = \frac{1}{2}(1 + \cos\theta)$

$\sin(\alpha \pm \beta) = \sin\alpha\cos\beta \pm \cos\alpha\sin\beta$
$\cos(\alpha \pm \beta) = \cos\alpha\cos\beta \mp \sin\alpha\sin\beta$

$$\tan(\alpha \mp \beta) = \frac{\tan\alpha \pm \tan\beta}{1 \mp \tan\alpha\tan\beta}$$

Law of Cosines

For a triangle with angles A, B, and C with opposite sides a, b, and c, respectively:
$a^2 = b^2 + c^2 - 2bc\cos A$

(with similar results for $b^2 = \cdots$ and for $c^2 = \cdots$).

If $A = 90°$, this equation reduces to the Pythagorean theorem:
$a^2 = b^2 + c^2$ (of the form $r^2 = x^2 + y^2$)

Law of Sines

For a triangle with angles A, B, and C with opposite sides a, b, and c, respectively:
$$\frac{a}{\sin A} = \frac{b}{\sin B} = \frac{c}{\sin C}$$

Powers-of-10 (Scientific) Notation

In physics, many numbers are very big or very small. To express them, **powers-of-10 (scientific) notation** is frequently used. When the number 10 is squared or cubed, we get

$10^2 = 10 \times 10 = 100$
$10^3 = 10 \times 10 \times 10 = 1000$

You can see that the number of zeros is just equal to the power of 10. As an example, 10^{23} is a 1 followed by 23 zeros.

Negative powers of 10 also can be used. For example,

$$10^{-2} = \frac{1}{10^2} = \frac{1}{100} = 0.01$$

Thus, if a power of ten has a negative exponent, we shift the decimal place to the left once for each power of 10. For example, 1 centimeter (cm) is 1/100 m, or 10^{-2} m, which is 0.01 m.

A number can be represented in powers-of-10 notation in many different ways—all correct. For example, the distance from Earth to the Sun is 93 million miles. This value can be represented as 93 000 000 miles, 93×10^6 miles, 9.3×10^7 miles, or 0.93×10^8 miles. Any of the given representations of 93 million miles is correct, although 9.3×10^7 is preferred. (In expressing powers-of-10 notation, it is customary to have one digit to the left of the decimal point, unless significant figures are involved.)

Thus it can be seen that the exponent, or power of 10, changes when the decimal point of the prefix number is shifted. General rules for this notation are as follows:

Rules for Using Powers-of-10 Notation

1. The exponent, or power of 10, is *increased* by 1 for every place the decimal point is shifted to the *left*.

2. The exponent, or power of 10, is *decreased* by 1 for every place the decimal point is shifted to the *right*.

This is simply a way of saying that if the coefficient (prefix number) gets smaller, the exponent gets correspondingly larger, and vice versa. Overall, the number is the same.

Example 1. ■ Expressing Numbers in Powers-of-10 Notation

Express the following numbers in powers-of-10 notation.

(a) 360 000 (b) 246.7 (c) 0.0694 (d) 0.000 011

Solution. Applying the preceding rules, we obtain the following:

(a) 360 000 = 3.6×10^5 (shift to left, rule 1)

(b) 246.7 = 2.467×10^2 (shift to left, rule 1)

(c) 0.0694 = 6.94×10^{-2} (shift to right, rule 2)

(d) 0.000 011 = 1.1×10^{-5} (shift to right, rule 2)

Powers-of-10 notation is useful in expressing the results of mathematical operations with the proper number of significant figures. For example, consider the operation

$$325 \times 45 = 14625$$

Expressing the result with two significant figures, we obtain

$$325 \times 45 = 1.5 \times 10^4$$

Example 2. ■ Using Powers-of-10 Notation to Express Calculation Results

Perform the following mathematical operation on a calculator, and express the result properly, using significant figures and scientific notation:

$$\frac{0.0024}{8.05} = ?$$

Solution. Doing this operation on a calculator gives

$$\frac{0.0024}{8.05} = 0.000\,298\,136$$

(*Note*: The number of digits in the result may vary with different calculators.)

The number 0.0024 has two significant figures, so

$$\frac{0.0024}{8.05} = 3.0 \times 10^{-4}$$

Arithmetic procedures of multiplication and division, as well as addition and subtraction, can be done in powers-of-10 notation.

Multiplication of Powers of 10

In multiplication, the exponents are added.

Example 3. ■

$$(2 \times 10^4)(4 \times 10^3) = 8 \times 10^7$$

and

$$(1.2 \times 10^{-2})(3 \times 10^6) = 3.6 \times 10^4$$

Division of Powers of 10

In division, the denominator exponent is subtracted from the numerator for exponent.

Example 4. ■

$$\frac{4.8 \times 10^8}{2.4 \times 10^2} = 2.0 \times 10^6$$

and

$$\frac{3.4 \times 10^{-8}}{1.7 \times 10^{-2}} = 2.0 \times 10^{-6}$$

An alternative method for division is to transfer all powers of 10 from the denominator to the numerator by changing the sign of the exponent. Then, the exponents of the powers of 10 may be added, because they are now being multiplied. The decimal parts are not transferred; they are divided in the usual manner. This method requires an additional step, but many students find that it leads to the correct answer more consistently. Thus,

$$\frac{4.8 \times 10^8}{2.4 \times 10^2} = \frac{4.8 \times 10^8 \times 10^{-2}}{2.4} = 2.0 \times 10^6$$

Squaring Powers of 10

When squaring exponential numbers, multiply the exponent by 2. The decimal part is multiplied by itself.

Example 5. ■

$$(3 \times 10^4)^2 = 9 \times 10^8$$
$$(4 \times 10^{-7})^2 = 16 \times 10^{-14}$$

Finding the Square Root of Powers of 10

To find the square root of an exponential number, follow the rule $\sqrt{10^a} = 10^{\left(\frac{a}{2}\right)}$. Then, $10^{\left(\frac{a}{2}\right)}$ is easily found with a calculator (using a 10^x function). At times you may want to get an idea of the order of magnitude by doing the square root in your head. In this case, express the $\sqrt{10^a}$ such that the power of ten is an even number so it can be evenly divided by 2, and $10^{\left(\frac{a}{2}\right)}$ is the square root. The square root of a prefix of the power of ten can be estimated. Here are a couple of examples where the square roots of the prefixes are evident.

Example 6. ■

$$\sqrt{9 \times 10^8} = 3 \times 10^4$$
$$\sqrt{2.5 \times 10^{-17}} = \sqrt{25 \times 10^{-18}} = 5 \times 10^{-9}$$

Addition and Subtraction of Powers of 10

In addition or subtraction, the exponents of 10 must be the same value.

Example 7. ■

$$\begin{array}{r} 4.6 \times 10^{-8} \\ + 1.2 \times 10^{-8} \\ \hline 5.8 \times 10^{-8} \end{array}$$

and

$$\begin{array}{r} 4.8 \times 10^7 \\ - 2.5 \times 10^7 \\ \hline 2.3 \times 10^7 \end{array}$$

Appendix II   Kinetic Theory of Gases

The basic assumptions are as follows:

1. All the molecules of a pure gas have the same mass (m) and are in continuous and completely random motion. (The mass of each molecule is so small that the effect of gravity on it is negligible.)

2. The gas molecules are separated by large distances and occupy a volume that is negligible compared with these distances.

3. The molecules exert no forces on each other except when they collide.

4. Collisions of the molecules with one another and with the walls of the container are perfectly elastic.

The magnitude of the force exerted on the wall of the container by a gas molecule colliding with it is $F = \Delta p/\Delta t$. Assuming that the direction of the velocity (v_x) is normal to the wall, the magnitude of the average force is

$$F = \frac{\Delta(mv)}{\Delta t} = \frac{mv_x - (-mv_x)}{\Delta t} = \frac{2mv_x}{\Delta t} \qquad (1)$$

After striking one wall of the container, which, for convenience, is assumed to be a cube with sides of dimensions L, the molecule recoils in a straight line. Suppose that the molecule reaches the opposite wall without colliding with any other molecules along the way. The molecule then travels the distance L in a time equal to L/v_x. After the collision with that wall, again assuming no collisions on the return trip, the round trip will take $\Delta t = 2L/v_x$. Thus, the number of collisions per unit time a molecule makes with a particular wall is $v_x/2L$, and the average force of the wall from successive collisions is

$$F = \frac{2mv_x}{\Delta t} = \frac{2mv_x}{2L/v_x} = \frac{mv_x^2}{L} \qquad (2)$$

The random motions of the many molecules produce a relatively constant force on the walls, and the pressure (p) is the total force on a wall divided by the wall's area:

$$p = \frac{\Sigma F_i}{L^2} = \frac{m(v_{x_1}^2 + v_{x_2}^2 + v_{x_3}^2 + \cdots)}{L^3} \qquad (3)$$

The subscripts refer to individual molecules.

The average of the squares of the speeds is given by

$$\overline{v_x^2} = \frac{v_{x_1}^2 + v_{x_2}^2 + v_{x_3}^2 + \cdots}{N}$$

where N is the number of molecules in the container. In terms of this average, Eq. 3 can be written as

$$p = \frac{Nm\overline{v_x^2}}{L^3} \qquad (4)$$

However, the molecules' motions occur with equal frequency along any one of the three axes, so $\overline{v_x^2} = \overline{v_y^2} = \overline{v_z^2}$ and $\overline{v^2} = \overline{v_x^2} + \overline{v_y^2} + \overline{v_z^2} = 3\overline{v_x^2}$. Then

$$\sqrt{\overline{v^2}} = v_{rms}$$

where v_{rms} is called the root-mean-square (rms) speed. Substituting this result into Eq. 4 and replacing L^3 with V (since L^3 is the volume of the cubical container) gives

$$pV = \tfrac{1}{3}Nmv_{rms}^2 \qquad (5)$$

This result is correct even though collisions between molecules were ignored. Statistically, these collisions average out, so the number of collisions with each wall is as described. This result is also independent of the shape of the container. A cube merely simplifies the derivation.

We now combine this result with the empirical perfect gas law:

$$pV = Nk_BT = \tfrac{1}{3}Nmv_{rms}^2$$

The average kinetic energy per gas molecule is thus proportional to the absolute temperature of the gas:

$$\overline{K} = \tfrac{1}{2}mv_{rms}^2 = \tfrac{3}{2}k_BT \qquad (6)$$

The collision time is negligible compared with the time between collisions. Some kinetic energy will be momentarily converted to potential energy during a collision; however, this potential energy can be ignored, because each molecule spends a negligible amount of time in collisions. Therefore, by this approximation, the total kinetic energy is the internal energy of the gas, and the internal energy of a perfect gas is directly proportional to its absolute temperature.

Appendix III   Planetary Data

Name	Equatorial Radius (km)	Mass (Compared with Earth's)*	Mean Density (× 10³ kg/m³)	Surface Gravity (Compared with Earth's)	Semimajor Axis × 10⁶ km	AU†	Orbital Period Years	Days	Eccentricity	Inclination to Ecliptic
Mercury	2439	0.0553	5.43	0.378	57.9	0.3871	0.24084	87.96	0.2056	7°00′26″
Venus	6052	0.8150	5.24	0.894	108.2	0.7233	0.61515	224.68	0.0068	3°23′40″
Earth	6378.140	1	5.515	1	149.6	1	1.00004	365.25	0.0167	0°00′14″
Mars	3397.2	0.1074	3.93	0.379	227.9	1.5237	1.8808	686.95	0.0934	1°51′09″
Jupiter	71398	317.89	1.36	2.54	778.3	5.2028	11.862	4337	0.0483	1°18′29″
Saturn	60000	95.17	0.71	1.07	1427.0	9.5388	29.456	10760	0.0560	2°29′17″
Uranus	26145	14.56	1.30	0.8	2871.0	19.1914	84.07	30700	0.0461	0°48′26″
Neptune	24300	17.24	1.8	1.2	4497.1	30.0611	164.81	60200	0.0100	1°46′27″
Pluto	1500–1800	0.02	0.5–0.8	~0.03	5913.5	39.5294	248.53	90780	0.2484	17°09′03″

*Planet's mass/Earth's mass, where $M_E = 6.0 \times 10^{24}$ kg.
†Astronomical unit: 1 AU = 1.5×10^8 km, the average distance between the Earth and the Sun.

Appendix IV Alphabetical Listing of the Chemical Elements (The perodic table is provided inside the back cover.)

Element	Symbol	Atomic Number (Proton Number)	Atomic Mass	Element	Symbol	Atomic Number (Proton Number)	Atomic Mass	Element	Symbol	Atomic Number (Proton Number)	Atomic Mass
Actinium	Ac	89	227.0278	Hafnium	Hf	72	178.49	Praseodymium	Pr	159	140.9077
Aluminum	Al	13	26.98154	Hahnium	Ha	105	(262)	Promethium	Pm	61	(145)
Americium	Am	95	(243)	Hassium	Hs	108	(265)	Protactinium	Pa	91	231.0359
Antimony	Sb	51	121.757	Helium	He	2	4.00260	Radium	Ra	88	226.0254
Argon	Ar	18	39.948	Holmium	Ho	67	164.9304	Radon	Rn	86	(222)
Arsenic	As	33	74.9216	Hydrogen	H	1	1.00794	Rhenium	Re	75	186.207
Astatine	At	85	(210)	Indium	In	49	114.82	Rhodium	Rh	45	102.9055
Barium	Ba	56	137.33	Iodine	I	53	126.9045	Rubidium	Rb	37	85.4678
Berkelium	Bk	97	(247)	Iridium	Ir	77	192.22	Ruthenium	Ru	44	101.07
Beryllium	Be	4	9.01218	Iron	Fe	26	55.847	Rutherfordium	Rf	104	(261)
Bismuth	Bi	83	208.9804	Krypton	Kr	36	83.80	Samarium	Sm	62	150.36
Bohrium	Bh	107	(264)	Lanthanum	La	57	138.9055	Scandium	Sc	21	44.9559
Boron	B	5	10.81	Lawrencium	Lr	103	(260)	Seaborgium	Sg	106	(263)
Bromine	Br	35	79.904	Lead	Pb	82	207.2	Selenium	Se	34	78.96
Cadmium	Cd	48	112.41	Lithium	Li	3	6.941	Silicon	Si	14	28.0855
Calcium	Ca	20	40.078	Lutetium	Lu	71	174.967	Silver	Ag	47	107.8682
Californium	Cf	98	(251)	Magnesium	Mg	12	24.305	Sodium	Na	11	22.98977
Carbon	C	6	12.011	Manganese	Mn	25	54.9380	Strontium	Sr	38	87.62
Cerium	Ce	58	140.12	Meitnerium	Mt	109	(268)	Sulfur	S	16	32.066
Cesium	Cs	55	132.9054	Mendelevium	Md	101	(258)	Tantalum	Ta	73	180.9479
Chlorine	Cl	17	35.453	Mercury	Hg	80	200.59	Technetium	Tc	43	(98)
Chromium	Cr	24	51.996	Molybdenum	Mo	42	95.94	Tellurium	Te	52	127.60
Cobalt	Co	27	58.9332	Neodymium	Nd	60	144.24	Terbium	Tb	65	158.9254
Copper	Cu	29	63.546	Neon	Ne	10	20.1797	Thallium	Tl	81	204.383
Curium	Cm	96	(247)	Neptunium	Np	93	237.048	Thorium	Th	90	232.0381
Dubnium	Db	105	(262)	Nickel	Ni	28	58.69	Thulium	Tm	69	168.9342
Dysprosium	Dy	66	162.50	Niobium	Nb	41	92.9064	Tin	Sn	50	118.710
Einsteinium	Es	99	(252)	Nitrogen	N	7	14.0067	Titanium	Ti	22	47.88
Erbium	Er	68	167.26	Nobelium	No	102	(259)	Tungsten	W	74	183.85
Europium	Eu	63	151.96	Osmium	Os	76	190.2	Uranium	U	92	238.0289
Fermium	Fm	100	(257)	Oxygen	O	8	15.9994	Vanadium	V	23	50.9415
Fluorine	F	9	18.998403	Palladium	Pd	46	106.42	Xenon	Xe	54	131.29
Francium	Fr	87	(223)	Phosphorus	P	15	30.97376	Ytterbium	Yb	70	173.04
Gadolinium	Gd	64	157.25	Platinum	Pt	78	195.08	Yttrium	Y	39	88.9059
Gallium	Ga	31	69.72	Plutonium	Pu	94	(244)	Zinc	Zn	30	65.39
Germanium	Ge	32	72.561	Polonium	Po	84	(209)	Zirconium	Zr	40	91.22
Gold	Au	79	196.9665	Potassium	K	19	39.0983				

Appendix V Properties of Selected Isotopes

Atomic Number (Z)	Element	Symbol	Mass Number (A)	Atomic Mass*	Abundance (%) or Decay Mode† (if radioactive)	Half-life (if radioactive)
0	(Neutron)	n	1	1.008665	β^-	10.6 min
1	Hydrogen	H	1	1.007825	99.985	
	Deuterium	D	2	2.014102	0.015	
	Tritium	T	3	3.016049	β^-	12.33 y
2	Helium	He	3	3.016029	0.00014	
			4	4.002603	≈100	
3	Lithium	Li	6	6.015123	7.5	
			7	7.016005	92.5	
4	Beryllium	Be	7	7.016930	EC, γ	53.3 d

Atomic Number (Z)	Element	Symbol	Mass Number (A)	Atomic Mass*	Abundance (%) or Decay Mode† (if radioactive)	Half-life (if radioactive)
			8	8.005305	2α	6.7×10^{-17} s
			9	9.012183	100	
5	Boron	B	10	10.012938	19.8	
			11	11.009305	80.2	
			12	12.014353	β^-	20.4 ms
6	Carbon	C	11	11.011433	β^+, EC	20.4 ms
			12	12.000000	98.89	
			13	13.003355	1.11	
			14	14.003242	β^-	5730 y
7	Nitrogen	N	13	13.005739	β^-	9.96 min
			14	14.003074	99.63	
			15	15.000109	0.37	
8	Oxygen	O	15	15.003065	β^+, EC	122 s
			16	15.994915	99.76	
			18	17.999159	0.204	
9	Fluorine	F	19	18.998403	100	
10	Neon	Ne	20	19.992439	90.51	
			22	21.991384	9.22	
11	Sodium	Na	22	21.994435	β^+, EC, γ	2.602 y
			23	22.989770	100	
			24	23.990964	β^-, γ	15.0 h
12	Magnesium	Mg	24	23.985045	78.99	
13	Aluminum	Al	27	26.981541	100	
14	Silicon	Si	28	27.976928	92.23	
			31	30.975364	β^-, γ	2.62 h
15	Phosphorus	P	31	30.973763	100	
			32	31.973908	β^-	14.28 d
16	Sulfur	S	32	31.972072	95.0	
			35	34.969033	β^-	87.4 d
17	Chlorine	Cl	35	34.968853	75.77	
			37	36.965903	24.23	
18	Argon	Ar	40	39.962383	99.60	
19	Potassium	K	39	38.963708	93.26	
			40	39.964000	β^-, EC, γ, β^+	1.28×10^9 y
20	Calcium	Ca	40	39.962591	96.94	
24	Chromium	Cr	52	51.940510	83.79	
25	Manganese	Mn	55	54.938046	100	
26	Iron	Fe	56	55.934939	91.8	
27	Cobalt	Co	59	58.933198	100	
			60	59.933820	β^-, γ	5.271 y
28	Nickel	Ni	58	57.935347	68.3	
			60	59.930789	26.1	
			64	63.927968	0.91	
29	Copper	Cu	63	62.929599	69.2	
			64	63.929766	β^-, β^+	12.7 h
			65	64.927792	30.8	
30	Zinc	Zn	64	63.929145	48.6	
			66	65.926035	27.9	
33	Arsenic	As	75	74.921596	100	
35	Bromine	Br	79	78.918336	50.69	
36	Krypton	Kr	84	83.911506	57.0	
			89	88.917563	β^-	3.2 min
38	Strontium	Sr	86	85.909273	9.8	
			88	87.905625	82.6	
			90	89.907746	β^-	28.8 y
39	Yttrium	Y	89	89.905856	100	
43	Technetium	Tc	98	97.907210	β^-, γ	4.2×10^6 y
47	Silver	Ag	107	106.905095	51.83	

Atomic Number (Z)	Element	Symbol	Mass Number (A)	Atomic Mass*	Abundance (%) or Decay Mode† (if radioactive)	Half-life (if radioactive)
			109	108.904754	48.17	
48	Cadmium	Cd	114	113.903361	28.7	
49	Indium	In	115	114.90388	95.7; β^-	5.1×10^{14} y
50	Tin	Sn	120	119.902199	32.4	
53	Iodine	I	127	126.904477	100	
			131	130.906118	β^-, γ	8.04 d
54	Xenon	Xe	132	131.90415	26.9	
			136	135.90722	8.9	
55	Cesium	Cs	133	132.90543	100	
56	Barium	Ba	137	136.90582	11.2	
			138	137.90524	71.7	
			144	143.92273	β^-	11.9 s
61	Promethium	Pm	145	144.91275	EC, α, γ	17.7 y
74	Tungsten (Wolfram)	W	184	183.95095	30.7	
76	Osmium	Os	191	190.96094	β^-, γ	15.4 d
			192	191.96149	41.0	
78	Platinum	Pt	195	194.96479	33.8	
79	Gold	Au	197	196.96656	100	
80	Mercury	Hg	202	201.97063	29.8	
81	Thallium	Tl	205	204.97441	70.5	
			210	209.990069	β^-	1.3 min
82	Lead	Pb	204	203.973044	β^-, 1.48	1.4×10^{17} y
			206	205.97446	24.1	
			207	206.97589	22.1	
			208	207.97664	52.3	
			210	209.98418	α, β^-, γ	22.3 y
			211	210.98874	β^-, γ	36.1 min
			212	211.99188	β^-, γ	10.64 h
			214	213.99980	β^-, γ	26.8 min
83	Bismuth	Bi	209	208.98039	100	
			211	210.98726	α, β^-, γ	2.15 min
84	Polonium	Po	210	209.98286	α, γ	138.38 d
			214	213.99519	α, γ	164 μs
86	Radon	Rn	222	222.017574	α, β	3.8235 d
87	Francium	Fr	223	223.019734	α, β^-, γ	21.8 min
88	Radium	Ra	226	226.025406	α, γ	1.60×10^3 y
			228	228.031069	β^-	5.76 y
89	Actinium	Ac	227	227.027751	α, β^-, γ	21.773 y
90	Thorium	Th	228	228.02873	α, γ	1.9131 y
			232	232.038054	100; α, γ	1.41×10^{10} y
92	Uranium	U	232	232.03714	α, γ	72 y
			233	233.039629	α, γ	1.592×10^5 y
			235	235.043925	0.72; α, γ	7.038×10^8 y
			236	236.045563	α, γ	2.342×10^7 y
			238	238.050786	99.275; α, γ	4.468×10^9 y
			239	239.054291	β^-, γ	23.5 min
93	Neptunium	Np	239	239.052932	β^-, γ	2.35 d
94	Plutonium	Pu	239	239.052158	α, γ	2.41×10^4 y
95	Americium	Am	243	243.061374	α, γ	7.37×10^3 y
96	Curium	Cm	245	245.065487	α, γ	8.5×10^3 y
97	Berkelium	Bk	247	247.07003	α, γ	1.4×10^3 y
98	Californium	Cf	249	249.074849	α, γ	351 y
99	Einsteinium	Es	254	254.08802	α, γ, β^-	276 d
100	Fermium	Fm	253	253.08518	EC, α, γ	3.0 d

*The masses given throughout this table are those for the neutral atom, including the Z electrons.
†"EC" stands for electron capture.

Answers to Follow-up Exercises

Chapter 1

1.1 A mass of 1000 kg has a weight of 2200 lb. The weight of a metric ton is equivalent to the British *long* ton (2200 lb), or 200 lb greater than the British *short* ton (2000 lb).

1.2 Yes, $[L] = [L]$, or m = m.

1.3 (a) 50 mi/h $[(0.447 \text{ m/s})/(\text{mi/h})] = 22$ m/s.
(b) $(1 \text{ mi/h})(1609 \text{ km/mi})(1 \text{ h}/3600 \text{ s}) = 0.477$ m/s.

1.4 13.3 times.

1.5 1 m^3 = 10^6 cm^3.

1.6 European. 10 mi/gal $\approx$ 16 km/4L = 4 km/L, as compared with 10 km/L.

1.7 (a) 7.0×10^5 kg^2. (b) 3.02×10^2 (no units).

1.8 (a) 23.70. (b) 22.09.

1.9 2.3×10^{-3} m^3, or 2.3×10^3 cm^3.

1.10 16.9 m.

1.11 23.5 N or 23.5 S (Tropic of Cancer and Tropic of Capricorn, respectively).

1.12 750 cm^3 = 7.50×10^{-4} m^3 $\approx 10^{-3}$ m^3, $m = \rho V \approx$ $(10^3 \text{ kg/m}^3)(10^{-3} \text{ m}^3) = 1$ kg. (By direct calculation, $m = 0.79$ kg.)

1.13 $V \approx 10^{-2}$ m^3, cells/vol $\approx 10^4$ cells/mm^3 $(10^9$ mm^3/m$^3) =$ 10^{13} cells/m^3, and (cells/vol)(vol) $\approx 10^{11}$ white cells.

Chapter 2

2.1 (a) No. If meters per minute is used, the answer will be in minutes. (b) 0.200 m/min (or 0.00333 m/s).

2.2 (a) $s_1 = 2.00$ m/s; $s_2 = 1.52$ m/s; $s_3 = 1.72$ m/s $\neq 0$, although the velocity is zero.

2.3 No. If the velocity is also in the negative direction, the object will speed up.

2.4 9.0 m/s in the direction of the original motion.

2.5 Yes, 96 m. (A lot quicker, isn't it?)

2.6 No, changes x_o positions, but the separation distance is the same.

2.7 $v_B = 2v_A$.

2.8 $x = v^2/2a$, $x_B = 48.6$ m, and $x_C = 39.6$ m; the Blazer should not tailgate within at least 9.0 m.

2.9 1.16 s longer.

2.10 Time for bill to fall its length = 0.179 s. This time is less than the average reaction time (0.192 s) computed in the Example, so most people cannot catch the bill.

2.11 $y_u = y_d = 5.12$ m, as measured from reference $y = 0$ at the release point.

2.12 Eq. 2.8′, $t = 4.6$ s; Eq. 2.10′, $t = 4.6$ s.

Chapter 3

3.1 $v_x = -0.40$ m/s, $v_y = +0.30$ m/s; the distance is unchanged.

3.2 $x = 9.00$ m, $y = 12.6$ m (same).

3.3 $\mathbf{v} = (0)\hat{\mathbf{x}} + (3.7 \text{ m/s})\hat{\mathbf{y}}$.

3.4 No, the boat would curve back and forth, with both components acting.

3.5 $\mathbf{C} = (-7.7 \text{ m})\hat{\mathbf{x}} + (-4.3 \text{ m})\hat{\mathbf{y}}$.

3.6 $d = 524.976$ m = 525 m (rounding differences).

3.7 14.5° W of N.

3.8 (a) $y_o = +25$ m and $y = 0$; the equation is the same.
(b) $\mathbf{v} = (8.25 \text{ m/s})\hat{\mathbf{x}} + (-22.1 \text{ m/s})\hat{\mathbf{y}}$.

3.9 Both increase sixfold.

3.10 (a) If not, the stone would hit to the side of the block. (b) Eq. 3.11 does not apply; the initial and final heights are not the same. $R = 15$ m, which is way off the 27-m answer.

3.11 The ball thrown at 45°. It would have a greater initial velocity.

3.12 At the top of the parabolic arc, the player's vertical motion is zero and is very small on either side of this maximum height. Here, the player's horizontal velocity component dominates, and he moves horizontally, with little motion in the vertical direction. This gives the illusion of "hanging" in the air.

3.13 4.15 m from the net.

Chapter 4

4.1 6.0 m/s in the direction of the net force.

4.2 (a) 11 lb. (b) Weight in pounds $\approx$ 2.2 lb/kg.

4.3 8.3 N

4.4 (a) 50° above the +x-axis. (b) x- and y-components reversed: $\mathbf{v} = (9.8 \text{ m/s})\hat{\mathbf{x}} + (4.5 \text{ m/s})\hat{\mathbf{y}}$.

4.5 2000 N.

4.6 (a) $m_2 > 1.7$ kg. (b) $\theta < 17.5°$.

4.7 (a) 7.35 N. (b) Neglecting air resistance, 7.35 N, downward.

4.8 $T_1 = 25$ N.

4.9 (a) $F_1 = 3.5w$. Even greater than F_2. (b) $\Sigma F_y = ma$, and F_1 and F_2 would both increase.

4.10 $\mu_s = 1.41\mu_k$ (for three cases in Table 4.1).

4.11 No. F varies with angle, with the angle for minimum applied force being around 33° in this case. (Greater forces are required for 20° and 50°.) In general, the optimum angle depends on the coefficient of friction.

4.12 The deceleration of the crate would be less than that of the truck. That is, the truck slows down more than the crate, and the crate slides forward.

4.13 Air resistance depends not only on speed, but also on size and shape. If the heavier ball were larger, it would have more exposed area to collide with air molecules, and the retarding force would increase faster. Depending on the size difference, the heavier ball might reach terminal velocity first, and the lighter ball would strike the ground first. Alternatively, the balls might reach terminal velocity together.

Chapter 5

5.1 -2.0 J

5.2 Work is done initially in lifting the handles and load. But when you move the wheelbarrow forward at a constant height, no work is done by the vertical component because that component is perpendicular to the motion.

5.3 No, speed would decrease and it would stop moving.

5.4 $W_{x_1} = 0.034$ J, $W_x = 0.64$ J (measured from x_o)

5.5 No, $W_2/W_1 = 4$, or 4 times as much

5.6 Here we have $m_s = m_g/2$ as before. However, $v_s/v_g =$ $(6.0 \text{ m/s})/(4.0 \text{ m/s}) = \frac{3}{2}$. Using a ratio, $K_s/K_g = \frac{9}{8}$, and the safety still has more kinetic energy than the guard. (Answer could also be obtained from direct calculations of kinetic energies, but for a relative comparison, a ratio is usually quicker.)

5.7 $W_3/W_2 = 1.4$, or 40% larger. More work, but a smaller percentage increase.

5.8 $\Delta K_{total} = 0$, $\Delta U_{total} = 0$

5.9 9.9 m/s

5.10 No. $E_o = E$ or $\frac{1}{2}mv_o^2 + mgh = \frac{1}{2}mv^2$. The mass cancels and the speed is independent of mass. (Recall that in free fall, all objects or projectiles fall with the same vertical acceleration g—see Section 2.5.)

5.11 0.025 m

5.12 (a) 59% (b) $E_{loss}/t = mg(y/t) = mgv = (60\,mg)\,$J/s

5.13 Block would stop in rough area.

5.14 52%

5.15 (a) Same work in twice the time. (b) Same work in half the time.

5.16 (a) No. (b) Creation of energy.

Chapter 6

6.1 5.0 m/s. Yes, this is 18 km/h or 11 mi/h, a speed at which humans can run.

6.2 (1) Ship the greatest KE. (2) Bullet the least KE.

6.3 $(-3.0\ \text{kg} \cdot \text{m/s})\hat{\mathbf{x}} + (4.0\ \text{kg} \cdot \text{m/s})\hat{\mathbf{y}}$

6.4 It would increase to 60 m/s: greater speed, longer drive, ideally. (There is also a directional consideration.)

6.5 (a) No, for the m_1/m_2 system, external force on block. Yes, for the m_1/m_2 Earth system. But with m_2 attached to the Earth, the mass of this part of the system would be vastly greater than that of m_2, so its change in velocity would be negligible. (b) Assuming the ball is tossed in the + direction: for the tosser, $v_t = -0.50$ m/s; for the catcher, $v_c = 0.48$ m/s. For the ball: $p = 0$, $+25$ kg·m/s, $+1.2$ kg·m/s.

6.6 No. Energy went into work of breaking the brick, and some lost as heat and sound.

6.7 No.

6.8 No; all of the kinetic energy cannot be lost to make the dent. The momentum after the collision cannot be zero, since it was not zero initially. Thus, the balls must be moving and have kinetic energy. This can also be seen from Eq. 6.11: $K_f/K_i = m_1/(m_1 + m_2)$, and K_f cannot be zero (unless m_1 is zero, which is not possible).

6.9 5.0 m

6.10 No. If $v_1 = v_2$, then $m_1 - m_2 = 2m_1$, which requires that $m_2 = -m_1$. Negative mass is not possible, so $m_1 m_2$ is always less than $2m_1$ and the velocities cannot be equal.

6.11 Same as reverse collision, 0.0022.

6.12 All of the balls swing out, but to different degrees. With $m_1 > m_2$, the stationary ball (m_2) moves off with a greater speed after collision than the incoming, heavier ball (m_1), and the heavier ball's speed is reduced after collision, in accordance with Eq. 6.16 (see Fig. 6.14b). Hence, a "shot" of momentum is passed along the row of balls with equal mass (see Fig. 6.14a), and the end ball swings out with the same speed as was imparted to m_2. Then, the process is repeated: m_1, *now moving more slowly*, collides again with the initial ball in the row (m_2), and another, but smaller, shot of momentum is passed down the row. The new end ball in the row receives less kinetic energy than the one that swung out just a moment previously, and so doesn't swing as high. This process repeats itself instantaneously for each ball, with the observed result that all of the balls swing out to different degrees.

6.13 $M = 11$ kg, initially at origin

6.14 $(X_{CM}, Y_{CM}) = (0.47\ \text{m}, 0.10\ \text{m})$; same location as in Example, two-thirds of the length of the bar from m_1. Note: The location of the CM does not depend on the frame of reference.

6.15 Yes, the CM does not move.

Chapter 7

7.1 1.61×10^3 m = 1.61 km (about a mile)

7.2 (a) 0.35% for 10° (b) 1.2% for 20°

7.3 (a) 4.7 rad/s, 0.38 m/s; 4.7 rad/s, 0.24 m/s (b) To equalize the running distances, because the curved sections of the track have different radii and thus different lengths.

7.4 120 rpm

7.5 (a) 86% (b) No; the astronaut still has mass and thus, with gravity acting, has weight by definition. (See the end of Section 7.6.)

7.6 (a) 106 rpm (b) $a = \sqrt{2}\,g = 13.9$ m/s^2, at 45° below plane of centrifuge.

7.7 The string cannot be exactly horizontal; it must make some small angle to the horizontal so that there will be an upward component of the tension force to balance the ball's weight.

7.8 No; it depends on mass: $F_c = \mu_s mg$.

7.9 No. Both masses have the same angular frequency or speed ω, and $a_c = r\omega^2$, so actually $a_c \propto r$. Remember, $v = 2\pi r/T$, and note that $v_2 > v_1$, with $a_c = v^2/r$.

7.10 $T = 5.2$ N

7.11 (a) The directions of ω and α would be downward, perpendicular to the plane of the CD. (b) Negative α, which means it is the opposite direction of ω.

7.12 -0.031 rad/s^2

7.13 2.8×10^{-3} m/s^2 (a large force, but a small acceleration)

7.14 (a) No, they do not vary linearly; $\Delta U = 2.4 \times 10^9$ J, only a 9.1% increase. (b) A smaller negative value means a greater energy in the Earth's negative potential well (higher up in the well).

7.15 This is the amount of *negative* work done by an external force or agent when the masses are brought together. To separate the masses by infinite distances, an equal amount of positive work (against gravity) would have to be done.

7.16 4.7 times

7.17 Because of air resistance, the spacecraft would lose more energy and continue its downward spiral. With this frictional effect, the spacecraft would heat up. (Recall that meteors burn up in the atmosphere.) To keep a spacecraft from overheating, a material is used that evaporates from the outer surface tiles. This requires work, which is supplied by heat energy. Enough heat is carried off for the spacecraft to remain relatively cool. (See latent heat, Chapter 11.)

7.18 You wouldn't want to apply a forward thrust as in this Example, since the radius of the orbit of your spacecraft would increase and the spacecraft travel more slowly, falling farther behind. By using a reverse thrust (Example 7.16), you would move to a lower orbit and speed up. Once you had overtaken the station, you could apply an appropriate forward thrust to put the spacecraft back into the proper orbit for docking.

Chapter 8

8.1 $s = r\omega = 5(0.12\ \text{m})(1.7) = 0.20$ m; $s = v_{CM}t = (0.10\ \text{m/s})(2.00\ \text{s}) = 0.20$ m

8.2 No.

8.3 The weights of the balls and the forearm produce torques that tend to cause rotation in the direction opposite that of the applied torque.

8.4 $T \propto 1/\sin \theta$, and as θ gets smaller, so does $\sin \theta$ and T increases. In the limit, $\sin \theta \to 0$ and $T \to$ infinity (unrealistic).

8.5 $\Sigma \tau$: $Nx - m_1 g x_1 - m_2 g x_2 - m_3 g x_3 = (200 \text{ g})g(50 \text{ cm}) - (25 \text{ g})g(0 \text{ cm}) - (75 \text{ g})g(20 \text{ cm}) - (100 \text{ g})g(85 \text{ cm}) = 0$, where $N = Mg$.

8.6 No. With f_{s_1}, the reaction force N would not generally be the same (f_{s_2} and N are perpendicular components of the force exerted on the ladder by the wall). In this case, we still have $N = f_{s_1}$, but $Ny - (m_1 g)x_1 - (m_m g)x_m - f_{s_2}$, and $x_3 = 0$.

8.7 5 bricks

8.8 (d) no (equal masses) (e) Yes; with larger mass farther from axis of rotation, $I = 360 \text{ kg} \cdot \text{m}^2$.

8.9 The long pole (or your extended arms) increases the moment of inertia by placing more mass farther from the axis of rotation (the tightrope or rail). When the walker leans to the side, a gravitational torque tends to produce a rotation about the axis of rotation, causing a fall. However, with a greater rotational inertia (greater I), the walker has time to shift his or her body so that the center of gravity is again over the rope or rail and thus again in (unstable) equilibrium. With very flexible poles, the CG may be below the wire, thus ensuring stability.

8.10 $t = 0.63 \text{ s}$

8.11

$$\alpha = \frac{2 \, mg - (2\tau_f R)}{(2 \, m + M)R} \cdot \frac{N}{\text{kg} \cdot \text{m}}; \quad \text{and} \quad \frac{N}{\text{kg} \cdot \text{m}} = \frac{\text{kg} \cdot \text{m/s}^2}{\text{kg} \cdot \text{m}} = \frac{1}{\text{s}^2}$$

8.12 With an initial velocity, the yo-yo would have more kinetic energy, some of which goes into rotational kinetic energy so the yo-yo spins faster and longer in the "sleeper" mode.

8.13 The yo-yo would roll back and forth, oscillating about the critical angle.

8.14 (a) 0.24 m (b) The force of *static* friction, f_s, acts at the point of contact, which is always instantaneously at rest and so does no work. Some frictional work may be done due to rolling friction, but this is considered negligible for hard objects and surfaces.

8.15 $v_{CM} = 2.2 \text{ m/s}$; using a ratio, 1.4 times greater; no rotational energy

8.16 You already know the answer: 5.6 m/s. (It doesn't depend on the mass of the ball.)

Chapter 9

9.1 (a) +0.10% (b) 39 kg

9.2 2.3×10^{-4} L, or 2.3×10^{-7} m^3

9.3 (1) Having enough nails, and (2) having them all of equal height and not so sharp a point. This could be achieved by filing off the tips of the nails so as to have a "uniform" surface. Also, this would increase the effective area.

9.4 3.03×10^4 N (or 6.82×10^3 lb—about 3.4 tons!) This is roughly the force on your back right now. Our bodies don't collapse under atmospheric pressure because cells are filled with incompressible fluids (mostly water), bone, and muscle, which react with an equal outward pressure (equal and opposite forces). As with forces, it is a pressure *difference* that gives rise to dynamic effects.

9.5 $d_o = \sqrt{\dfrac{F_o}{F_i}} \, d_i = \sqrt{\dfrac{1}{10}} \, (8.0 \text{ cm}) = 2.5 \text{ cm}$

9.6 10.3 m (about 34 ft). You can see why we don't use water barometers.

9.7 Pressure in veins is lower than that in arteries (120/80).

9.8 greater, by 7.75×10^2

9.9 (a) The object would sink, so the buoyant force is less than the object's weight. Hence, the scale would have a reading greater than 40 N. Note that with a greater density, the object would not be as large and less water would be displaced. (b) 41.8 N.

9.10 11%

9.11 −18%

9.12 69%

9.13 As the water falls, speed (v) increases and area (A) must decrease to have $Av =$ a constant.

9.14 0.38 m

Chapter 10

10.1 (a) 40°C (b) You should immediately know the answer—this is the temperature at which the Fahrenheit and Celsius temperatures are numerically equal.

10.2 (a) $T_R = T_F + 460$ (b) $T_R = \frac{9}{5} T_C + 492$ (c) $T_R = \frac{9}{5} T_K$

10.3 96°C

10.4 273°C; no, not on Earth

10.5 50 C°

10.6 It depends on the metal of the bar. If the thermal expansion coefficient (α) of the bar is less than that of iron, it will not expand as much and not be as long as the diameter of the circular ring after heating. However, if the bar's α is greater than that of iron, the bar will expand more than the ring and the ring will be distorted.

10.7 Basically, the situations would be reversed. Faster cooling would be achieved by submerging the ice in Example 10.7—the cooler water would be less dense and would rise, promoting mixing. For a lake with cooling at the surface, cooler, less-dense water would remain at the surface until minimum density was achieved. With further cooling, the denser water would sink and freezing would occur from the bottom up.

10.8 v_{rms}, 1.69%; K, 3.41%

10.9 The rotational kinetic energy for oxygen is the difference between the total energies, 2.44×10^3 J. The oxygen is less massive and so has the higher v_{rms}.

Chapter 11

11.1 2.44×10^3 m

11.2 (a) 44°C (b) 6.7×10^5 J

11.3 (a) The ratio will be smaller because the specific heat of aluminum is greater than that of copper. (b) $Q_w/Q_p = 15.2$

11.4 The final temperature (T_f) is expected to be higher because the water was at a higher initial temperature. $T_f = 34.4°C$

11.5 The heat from the copier will raise the room temperature an additional 1.0 C°.

11.6 -1.09×10^5 J (negative because heat is lost)

11.7 (a) 2.64×10^{-2} kg or 26.4 g of ice melts. (b) The final temperature is still 0°C because the liver cannot lose enough heat to melt all the ice, even if the ice started at 0°C. The final result is an ice/water/liver system at 0°C, but with more water than in the Example.

11.8 1.1×10^5 J/s (difference due to rounding)

11.9 No, since the air spaces provide good insulation because air is a poor conductor. The many small pockets of air between the body and the outer garment form an insulating layer that minimizes conduction and so retards the loss of body heat. (There is little convection in the small spaces.)

11.10 -30 J/s or -30 W

11.11 Drapes reduce heat loss by limiting radiation through the window and by keeping convection currents away from the glass.

Chapter 12

12.1 It will take him about 19 days less to reach his goal.

12.2 In both cases, heat flow is into the gas. During the isothermal expansion $Q = W = +3.14 \times 10^3$ J. During the isobaric expansion $W = +4.53 \times 10^3$ J and $\Delta U = +6.80 \times 10^3$ J, therefore $Q = \Delta U + W = +1.13 \times 10^4$ J.

12.3 (a) $\Delta U = +464$ J (b) $753°$C (c) $Q = +316$J

12.4 As the air comes to lower elevations and higher pressures, it quickly compresses. This is approximately an adiabatic process, resulting in a temperature rise of the air.

12.5 (a) 142 K or $-131°$C (b) For a monatomic gas, $\Delta U = \frac{3}{2} nR\Delta T = -3.76 \times 10^3$ J. This should be the same as $-W$ since, for an adiabatic process $Q = 0 = \Delta U + W$; therefore, $\Delta U = -W$. The slight difference is due to rounding.

12.6 -1.22×10^3 J/K

12.7 Overall zero entropy change requires $|\Delta S_w| = |\Delta S_m|$ or $|Q_w/T_w| = |Q_m/T_m|$. Because the system is isolated, the magnitudes of the two heat flows *must* be the same $|Q_w| = |Q_m|$. Thus no overall entropy change requires the water and the metal to have the same average temperature $\bar{T}_w = \bar{T}_m$. This is not possible, unless they are initially at the *same* temperature. Thus, this can only happen if there is no heat flow.

12.8 As the blades of grass break down, they become more disordered. Thus they experience an entropy increase. The overall universe entropy must also increase, since this degradation is a naturally occurring process.

12.9 If the basics of the cycle are kept, that is, the triangular shape and the volume doubling, then a way to increase the net work (the area inside the cycle) is to drop the pressure even further at the end of the isometric leg. If you allow the volume to more than double during the isobaric expansion, that would achieve the same end. Anything that increases the net area (work) will do.

12.10 (a) 150 J/cycle (b) 850 J/cycle

12.11 $Q_{34} = 610$ J and $Q_{23} = 730$ J, therefore $Q_c = Q_{23} + Q_{34} = 1.34 \times 10^3$ J. This agrees with $Q_c = Q_h - W_{net} = 59 \times 10^3$ J $- 245$ J $= 1.35 \times 10^3$ J (to within rounding errors).

12.12 (a) The new values are: $COP_{ref} = 3.3$ and $COP_{hp} = 4.3$. (b) The COP of the air conditioner has the largest percentage increase.

12.13 It would show an increase of 7.5%.

Chapter 13

13.1 No, its maximum speed is $\left(\sqrt{k/m}\right)A = 4.0$ m/s. Thus it is traveling at 75% of the maximum speed.

13.2 0.49 J

13.3 (1) $y = -0.0881$ m, up. $n = 0.90$. (2) $y = 0$, going up. $n = 1.5$.

13.4 9.76 m/s^2; no. Since this is less than the accepted value at sea level, the park is probably at an altitude above sea level.

13.5 (a) 0.50 m (b) 0.10 Hz

13.6 440 Hz

13.7 increase the tension (by 44% as can be calculated).

Chapter 14

14.1 (a) 2.3 (b) 10.2

14.2 (a) 858 m (b) $\Delta t = 0.05$ s

14.3 It would be greatest in He, because it has the smallest molecular mass. (It would be lowest in oxygen, which has the largest molecular mass.)

14.4 (a) The dB scale is logarithmic, not linear.
(b) 3.16×10^{-6} W/m^2

14.5 No, $I_2 = (316)I_1$

14.6 65 dB

14.7 destructive interference: $\Delta L = 2.5\lambda = 5(\lambda/2)$, and $m = 5$. No sound would be heard if the waves from the speakers had equal amplitudes. Of course, during a concert the sound would not be single-frequency tones but would have a variety of frequencies and amplitudes. Listeners at certain locations might not hear certain parts of the audible spectrum, but this probably wouldn't be noticed.

14.8 toward, 431 Hz; past, 369 Hz

14.9 With the source and the observer traveling in the same direction at the same speed, their relative velocity would be zero. That is, the observer would consider the source to be stationary. Since the speed of the source and observer is subsonic, the sound from the source would overtake the observer without a shift in frequency. Generally, for motions involved in a Doppler shift, the word *toward* is associated with an *increase* in frequency and *away* with a *decrease* in frequency. Here, the source and observer remain a constant distance apart. (What would be the case if the speeds were supersonic?)

14.10 768 Hz; yes

Chapter 15

15.1 1.5×10^{-20}%.

15.2 No; if the comb were positive, it would polarize the paper in the reverse direction and still attract it.

15.3 $F_1 = 3.8 \times 10^{-7}$ N at an angle of $56°$ above the negative x axis, or $(-0.22 \mu N)\hat{x} + (0.32 \mu N)\hat{y}$

15.4 0.12 m

15.5 $F_e/F_g = 4.16 \times 10^{42}$; F_g is even *less* than F_e because the electron's mass is considerably smaller than a proton's, and F_e (magnitude) doesn't change.

15.6 The field is zero to the left of q_1 at $x = -0.60$ m. The location cannot be between the charges, because both fields point to the left there. Since q_1 is the smaller of the two charges, the location must be closer to it than to q_2.

15.7 $\mathbf{E} = (-65.8$ N/C$)\hat{x} + (360$ N/C$)\hat{y}$, or 366 N/C at $79.6°$ above the negative x axis.

15.8 (a) There are two fields, a large one pointing upward due to the nearby positive charge and a smaller one pointing downward from the more distant negative charge, therefore the net field points upward. (b) There are two fields, a large one pointing upward from the nearby negative charge and a smaller one pointing downward from the more distant positive charge, therefore the net field points upward. (c) There are two fields, a large one pointing downward from the nearby positive charge and a smaller one also pointing downward from the more distant negative charge, thus the net field points downward.

15.9 (a) $\mathbf{E}$ points upward, from ground to cloud. (b) 5.6×10^{11} electrons/m^2.

15.10 Positive charge is completely on the outside surface. Thus, only the electroscope attached to the outside would show deflection.

15.11 A positive charge equal in magnitude to the negative point charge is induced on the shell's interior surface. A negative charge equal to the negative point charge is induced on the exterior surface. The fields in the three regions are as follows. (a) The field lines point radially inward towards the negative point charge. (b) The field is zero. (c) The lines are radially inward toward the negative point charge, ending at the negative charge on the outside surface.

15.12 Their sign is negative since electric field lines points toward negative charges.

Chapter 16

16.1 (a) $\Delta U_e = +7.20 \times 10^{-19}$ J (b) ΔV is unchanged (why?) (c) $v = 4.65 \times 10^4$ m/s.

16.2 $v = 2.96 \times 10^7$ m/s

16.3 (a) It moved to a region of lower potential. (b) $\Delta U_e = q\Delta V = +3.27 \times 10^{-18}$ J.

16.4 $U_{12} = -3.27 \times 10^{-19}$ J. It is less stable because it would take less work ($+3.27 \times 10^{-19}$ J) to break it apart than for the water molecule ($+8.53 \times 10^{-19}$ J).

16.5 (a) 2.22 m. (b) The one closest to the Earth's surface is at a higher potential. (c) No, you cannot tell how far from the surface they are, only the separation between the two surfaces.

16.6 (a) Surface 1 is at a higher potential than surface 2. (b) Very far away, any charged conductor "looks like" a point charge, hence the equipotential surfaces gradually become spherical.

16.7 $d = 8.9 \times 10^{-16}$ m which is much smaller than an atomic diameter. Therefore such a capacitor is not practical.

16.8 $V = 7.91 \times 10^3$ V.

16.9 If the spacing were increased, C would decrease since $C \propto 1/d$. Because $Q = CV$, then $Q \propto C$ (the voltage across the capacitor is constant). Hence in this situation because C decreases, Q would also decrease. Thus, charge would flow away from the capacitor: $\Delta Q = -3.30 \times 10^{-12}$ C.

16.10 $V_1 = 8.00$ V; $V_2 = 4.00$ V; $U_{C_1} = 8.00 \times 10^{-5}$; $U_{C_2} = 4.00 \times 10^{-5}$ J.

16.11 (a) $Q_1 = 8.0 \times 10^{-7}$ C; $Q_2 = 1.6 \times 10^{-6}$ C; $Q_3 = 2.4 \times 10^{-6}$ C; (b) $U_{C_1} = 3.2 \times 10^{-6}$ J; $U_{C_2} = 6.4 \times 10^{-6}$ J; $U_{C_3} = 4.8 \times 10^{-6}$ J.

Chapter 17

17.1 The result is the same, that is, $V_{AB} = V$.

17.2 32 years

17.3 100 V

17.4 0.82 m

17.5 $R = 0.67$ Ω. For a sensitive electrical thermometer, use material with a high temperature coefficient of resistivity to get a large resistance change over a given temperature change.

17.6 print mode: $R = 82.3$ Ω; standby mode: $R = 1.80$ kΩ.

17.7 1.45×10^3 W.

17.8 8.3 h

17.9 At best, power plants produce electric energy with efficiencies of 35% (ignoring joule heat losses during transmission). Thus, in terms of primary fuels, 35% is the maximum efficiency of *any electrical appliance*. The electric hot water heater would then convert 95% of the delivered electric energy to heat. The net result is that 95% of 35%, or only 33%, of the original fuel energy content is actually used to heat water. Natural gas is delivered to the water heater's burner at es-

sentially no energy cost. Thus approximately 95% of the gas's heat content goes into heating the water. Gas heating systems are almost three times more efficient than electric systems in terms of primary fuel energy content.

Chapter 18

18.1 (a) series: $P_1 = 4.0$ W, $P_2 = 8.0$ W, $P_3 = 12.0$ W; parallel: $P_1 = 144$ W, $P_2 = 72$ W, $P_3 = 48$ W. (b) In series, the most power is dissipated in the largest resistance. In parallel, the most power is dissipated in the smallest resistance. (c) The series arrangement requires less power, 24 W as compared to 264 W for the parallel arrangement. It dissipates less total power, because it has a larger equivalent resistance and carries a much smaller current at the same voltage.

18.2 (a) The voltage across the open socket will be 120 V. (b) The voltage across the remaining bulbs will be 0 V.

18.3 $P_1 = I_1^2 R_1 = 54.0$ W, $P_2 = I_2^2 R_2 = 9.0$ W, $P_3 = I_3^2 R_3 = 0.87$ W, $P_4 = I_4^2 R_4 = 2.55$ W, and $P_5 = I_5^2 R_5 = 5.63$ W. The sum of these is 72.1 W, rounded to three significant figures. For the battery, we have agreement to within rounding error since $P_b = I_b V_b = 72.0$ W.

18.4 (a) If R_2 is increased, the equivalent parallel resistance of R_2 and R_1 increases. Thus, the total circuit resistance increases, resulting in a reduction in the total circuit current. Since the current in R_3 is the same as the total current, I_3 should decrease. From this, V_3 should decrease. Thus V_1 and V_2 should both increase (since they are equal and $V = V_2 + V_3 = $ constant). Since R_1 has not changed, due to the voltage increase, I_1 should increase. Since I_3 decreases and I_1 increases, it must be, from $I_3 = I_1 + I_2$, that I_2 must decrease. (b) Recalculating confirms these predictions: $I_1 = 0.51$ A (increase), $I_2 = 0.38$ A (decrease), and $I_3 = 0.89$ A (decrease).

18.5 At the junction, we still have $I_1 = I_2 + I_3$ (Eq. 1). Using the loop theorem around loop 3 in the clockwise direction (all terms in volts): $6 - 6I_1 - 9I_2 = 0$ (Eq. 2). For loop 1, the result is $6 - 6I_1 - 12 - 2I_3 = 0$ (Eq. 3). Solve Eq. 1 for I_2 and substitute into Eq. 2. Then solve Eq. 2 and Eq. 3 simultaneously for I_1 and I_3. All answers are the same as those in the Example, as they should be.

18.6 (a) The maximum energy storage (at 9.00 V) is 4.05 J. At 7.20 V, the capacitor stores only $U_C = \frac{1}{2}CV^2 = 2.59$ J, or 64% of the maximum. This is because the energy storage varies as the square of the voltage across the capacitor, and $0.8^2 = 0.64$. (b) 8.64 V, because the voltage rises to a maximum of 9.00 V exponentially, not linearly.

18.7 10 A

18.8 0.20 mA

Chapter 19

19.1 East, since reversing both the velocity direction and the sign of the charge leaves the direction the same.

19.2 (a) Using the force right-hand rule, the proton would initially deflect in the negative x direction. (b) 0.10 T

19.3 0.500 V

19.4 (a) At the poles the magnetic field is perpendicular to the ground. Since the current is parallel to the ground, according to the force right-hand rule the force on the wire would be in a plane parallel to the ground. Thus it would not be able to cancel the downward force of gravity. (b) The wire's mass is 0.041 g, which is unrealistically low.

19.5 (a) At 45°, the torque is 0.269 m · N or 70.7% of the maximum torque. (b) 30°.
19.6 (a) South (b) 75 A.
19.7 500 turns
19.8 (a) The force is still attractive, since the currents are in the same direction. You should be able to show this by using the right-hand source and force rules. (b) 0.012 m or 12 mm.

Chapter 20

20.1 (a) Clockwise. (b) 0.335 mA.
20.2 Any way that will increase the flux, such as increasing the loop area or the number of loops. Changing to a lower resistance would also help.
20.3 7.36×10^{-4} T
20.4 1.0 m/s
20.5 0.28 m
20.6 (a) 6.1×10^3 J (b) 5.0×10^3 J so about 12 times more energy is used during startup.
20.7 (a) You would use it as a step-down transformer because U.S. appliances are designed to work at half of the European voltage. (b) 12.5 A.
20.8 (a) Higher voltages allow for lower current usage. This in turn reduces joule heat losses in the delivery wires and in the motor windings, making more energy available for doing mechanical work and therefore a higher efficiency. (b) Since the voltage is doubled, the current is halved. The heat loss in the wire is proportional to the *square* of the current. Thus, losses will be cut by a factor of 4 or be reduced to 25% of their value at 120 V.
20.9 0.38 cm/s.
20.10 (a) With increasing distance, the Sun's light intensity (energy per second per unit area) drops. Therefore so would the force due to the light pressure on the sail. In turn, the ship's acceleration would be reduced. (b) You would need to somehow enlarge the sail area to catch more light.

Chapter 21

21.1 (a) 0.25A (b) 0.35 A (c) $9.6 \times 10^2 \, \Omega$, larger than the 240 Ω required for a bulb of the same power in the United States. The voltage in Great Britain is larger than that in the United States. Thus to keep the power constant, the current must be reduced by using a larger resistance.
21.2 If the resistance of the appliance is constant, the power will quadruple since $P \propto V^2$. Even if the resistance increased, the power would probably be much more than the appliance was designed for and it would likely burn out, or at least blow a fuse.
21.3 (a) $\sqrt{2}(120 \text{ V}) = 170$ V (b) 30 Hz
21.4 (a) $\sqrt{2}(2.55 \text{ A}) = 3.61$ A (b) 20 Hz
21.5 (a) The current would increase to 0.896 A. (b) The capacitor is responsible; with a frequency increase, X_C decreases. Since resistance is independent of frequency, it remains constant and overall Z decreases.
21.6 (a) In an RLC circuit, the phase angle ϕ depends on the difference $X_L - X_C$. If you increase the frequency, X_L would increase and X_C would decrease, thus their difference would increase and so would ϕ. (b) $\phi = 84.0°$, an increase as expected.
21.7 (a) 41.1 Hz (b) 576 W

21.8 (a) If you have a receiver tuned to a frequency *between* the two station frequencies, you would not receive the maximum strength signal from either station but there might be enough power from each to hear them simultaneously. (b) 1300 kHz

Chapter 22

22.1 Light travels in straight lines and is reversible. If you can see someone in a mirror, that person can see you. Conversely, if you can't see the trucker's mirror, then he or she can't see your image in that mirror and won't know that your car is behind the truck.
22.2 $n = 1.25$
22.3 By Snell's law, $n_2 = 1.25$, so $v = c/n_2 = 2.40 \times 10^8$ m/s.
22.4 With a greater n, θ_2 is smaller so the refracted light inside the glass is toward the lower-left. Therefore the lateral displacement is larger. 0.72 cm.
22.5 (a) The frequency of the light is unchanged in the different media, so the emerging light has the same frequency as that of the source. (b) The wavelength in air is independent of the water and glass media, as can be shown by adding another step (medium) to the Example solution. By reverse analysis, $\lambda_{air} = n_{water}\lambda_{water} = (c/v_{water})\lambda_{water} = c/f$. Thus, the wavelength in air is c/f.
22.6 Because of total internal reflections, the diver could not see anything above water. Instead, he would see the reflection of something on the sides and/or bottom of the pool. (Use reverse ray tracing.)
22.7 $n = 1.4574$. Green light will be refracted more than red light as green has a shorter wavelength, thus greater n than red light. By Snell's law, green will have a smaller angle of refraction so it is refracted more.

Chapter 23

23.1 No effect. Note that the solution to the Example does not include the distance. The geometry of the situation is the same regardless of the distance from the mirror.
23.2 $d_i \approx 60$ cm; real, inverted, and magnified.
23.3 $d_i = d_o$ and $M = -1$; real, inverted, and same size
23.4 The image is also always upright and smaller than the object.
23.5 $d_i = -20$ cm (in front of the lens); virtual, upright, and magnified.
23.6 $d_o = 2f = 24$ cm
23.7 Blocking off half of the lens would result in half the *amount* of light focused at the image plane, so the resulting image would be less bright but still full size.
23.8 The image is also always upright and smaller than the object
23.9 3 cm behind L_2; real, inverted, and smaller than the object ($M_{total} = -0.75$)

Chapter 24

24.1 $\Delta y = y_r - y_b = 1.2 \times 10^{-2}$ m = 1.2 cm
24.2 twice as thick, $t = 199$ nm
24.3 In brass instruments, the sound comes from a relatively large, flared opening. Thus there is little diffraction, so most

of the energy is radiated in the forward direction. In wood-wind instruments, much of the sound comes from tone holes along the column of the instrument. These holes are small compared to the wavelength of the sound, so there is appreciable diffraction. As a result, the sound is radiated in nearly all directions, even backward.

24.4 The width would increase by a factor of $700/550 = 1.27$.
24.5 Yes. $d \sin \theta_2 = 2\lambda$ and $d \sin \theta_3 = 3\lambda$, $\sin \theta_3 = 3 \sin \theta_2/2 = 3 \sin 30°/2 = 0.75$. So $\theta_3 = \sin^2 0.75 = 48.6°$.
24.6 $45°$
24.7 $\theta_2 = 35.9°$
24.8 $\lambda = 589$ nm; yellow

Chapter 25

25.1 It wouldn't work, a real image would form on person's side of lens ($d_i = +0.75$ m).
25.2 For an object at $d_o = 25$ cm, the image for eye 1 would be formed at 1.0 m; this is beyond the near point for that eye, so the object could be seen clearly. The image for eye 2 would be formed at 0.77 m; this is inside the near point for that eye, so the object would not be seen clearly.
25.3 glass for near-point viewing, 2.0 cm longer
25.4 length doubles
25.5 $f_i = 8.0$ cm
25.6 The erecting lens (of focal length f_e) should go between the objective and the eyepiece, positioned a distance of $2f_e$ from the image formed by the objective, which acts as an object. The erecting lens then produces an inverted image of the same size at $2f_e$ on the opposite side of the lens, which acts as an object for the eyepiece. The use of the erecting lens lengthens the telescope by $4f_e$.
25.7 3.4×10^{-7} rad, an order of magnitude better than the typical 10^{-6} rad
25.8 3.5 cm

Chapter 26

26.1 Light waves from two simultaneous events on the y-axis meet at some midpoint receptor on the y-axis. Since there is no relative motion along that axis, a simultaneous recording of the two events will also be recorded along the y'-axis. Hence the two observers agree on simultaneity for this situation.
26.2 $v = 0.9995c$; no, not twice as fast. Travel is limited to less than c, so this is only about a 0.15% increase.
26.3 (a) 0.667 μs (b) 0.580 μs (c) The observer watching the ship measures the proper time interval. To the person on the ship, that time interval is dilated.
26.4 $v = 0.991c$
26.5 The traveler measures the proper time interval of 20.0 y, but Earth inhabitants measure the dilated version of this (why?). The gamma factor is based on a recalculated value for traveler speed $v = 0.905\,04c$. Keeping five places after the decimal and then rounding to three significant figures we get $\gamma = 1/\sqrt{1 - (0.995\,04c/c)^2} = 10.049\,88$ and $\Delta t = \gamma \Delta t_o = (10.049\,88)(20.0 \text{ y}) = 200.99 \text{ y} \approx 201 \text{ y}$.
26.6 (a) 1.17 MeV (b) 0.207 MeV
26.7 (a) $0.319mc^2$ (b) $3.33mc^2$
26.8 1.95×10^7 more massive
26.9 $u = +0.69c$ which, as expected, is lower in magnitude than the nonrelativistic (and wrong) result of $0.80c$.

Chapter 27

27.1 (a) The wavelength is 967 nm, which is infrared. With some emissions in the red region, it would appear reddish. (b) The wavelength is 290 nm, which is ultraviolet. With significant visible emissions, it would appear bluish-white.
27.2 496 nm, which is shorter than the Example, indicating a higher photon energy. Thus the maximum kinetic energy of the photoelectrons is higher, requiring an increased stopping voltage.
27.3 2.00 V
27.4 (a) The ratio is $\Delta\lambda/\lambda_o = 324$ meaning a wavelength increase of $3.24 \times 10^4 \%$. (b) Percentagewise this is a much larger increase than in the Example because the wavelength of the incoming light (λ_o) is much smaller for gamma rays.
27.5 1.09×10^6 m/s
27.6 365 nm (UV).
27.7 The minimum energy required for ionization from the $n = 2$ (first excited state) in hydrogen is 10.2 eV. If an incoming photon carried exactly 10.2 eV, it could be absorbed. Then the electron would escape the atom, and the atom would "barely" be ionized. To carry 10.2 eV, the light needs a wavelength of 122 nm, which is ultraviolet.
27.8 (a) There are six possible transitions from the $n = 4$ state to the ground state, and thus the emitted light has six different possible wavelengths. (b) If the atom is excited from the ground state to the first excited state ($n = 1$ to $n = 2$), then it has no choice but to emit a single photon during de-excitation state ($n = 2$ to $n = 1$) because there are no intermediate states.

Chapter 28

28.1 (a) 8.8×10^{-33} m/s (b) 2.3×10^{33} s, or 7.2×10^{25} y. This is about 4.8×10^{15} times longer than the age of the universe. This movement would definitely not be noticeable.
28.2 The proton's de Broglie wavelength is 4.1×10^{-12} m, or about twenty times smaller than atomic spacing distances. With a wavelength much smaller than the atomic spacing, these protons would *not* be expected to exhibit significant diffraction effects.
28.3 Only five electrons could be accommodated in the 3d sub-shell if there were no spin (with spin there can be ten).
28.4 (a) $1s^2\ 2s^2\ 2p^6$ (b) $-2e$ or -3.2×10^{-19} C
28.5 1.2×10^6 m/s

Chapter 29

29.1 $^{12}_{6}C + ^{4}_{2}He \rightarrow ^{16}_{8}O$, thus the resulting nucleus is oxygen-16.
29.2 (a) Since $^{23}_{11}Na$ is the stable isotope with 11 protons and 12 neutrons, $^{22}_{11}Na$ is one neutron shy of being stable. In other words, it is proton-rich or neutron-poor. Thus, the expected decay mode is β^+ or positron decay. (b) Neglecting the emitted neutrino, the decay is $^{22}_{11}Na \rightarrow ^{22}_{10}Ne + ^{0}_{+1}e$. The daughter nucleus is neon-22.
29.3 (a) 48 d because reducing the activity by a factor of 64 requires six half lives; $1/2^6 = 1/64$. (b) The process of excretion from the body can also remove ^{131}I.
29.4 The closest integer is 20, since $2^{20} \approx 1.05 \times 10^6$. Thus it takes 20 half-lives, or about 560 y.
29.5 The measurement can be made to four ^{14}C half-lives, or 2.3×10^4 y.

29.6 Using a calculator you can easily focus in on about 3.8 half-lives, since $1/2^{3.81} \approx 1/14$. To two significant figures, this gives 4.9×10^9 y, which is in decent agreement with the age of the Earth.

29.7 (a) Starting with 29 protons and 29 neutrons (why?), we have the following candidates: $^{58}_{29}Cu$, $^{59}_{29}Cu$, $^{60}_{29}Cu$, $^{61}_{29}Cu$, $^{62}_{29}Cu$, $^{63}_{29}Cu$, $^{64}_{29}Cu$, etc. Now delete the odd–odd isotopes (why?) to get the most likely (stable) isotopes: $^{59}_{29}Cu$, $^{61}_{29}Cu$, $^{63}_{29}Cu$, $^{65}_{29}Cu$, etc. (b) Further trimming of the list can be done by deleting those with $N \approx Z$ (why?) and those with N significantly larger than Z (why?). Since Z should be just a bit smaller than N in this mass region, we expect neutron numbers in the mid-30s. Hence a good guess would be just $^{63}_{29}Cu$ and $^{65}_{29}Cu$. According to Appendix V, these are, in fact, the only two stable isotopes of copper.

29.8 (a) The result for ^3He is 2.573 MeV/nucleon, which is considerably smaller than the 7.075 MeV/nucleon for ^4He. Thus ^4He is the more tightly bound of the two. (b) Unlike ^3He, all protons and neutrons in ^4He *are* paired, resulting in a more tightly bound nucleus.

29.9 The absorbed dose is 0.0215 Gy or 2.15 rad. Since the RBE for gamma rays is 1, the effective dose is 0.0215 Sv or 2.15 rem. Both of these are about one-seventh of the dose from the beta radiation.

Chapter 30

30.1 $Q = -15.63$ MeV, so it is endoergic (takes energy to happen).

30.2 The increase in mass has an energy equivalent of 1.193 MeV. The rest of the incident kinetic energy (1.534 MeV − 1.193 MeV, or 0.341 MeV) must be distributed between the kinetic energies of the proton and of the oxygen-17.

30.3 There are 1.00×10^{38} proton cycles per second.

30.4 Because the beta particle has little energy and therefore little momentum, the neutrino and the daughter nucleus would have to recoil in opposite directions in order to conserve linear momentum. This assumes the original nucleus had zero linear momentum.

30.5 Since the negative pion (π^-) has a charge of $-e$, its quark structure must be $\bar{u}d$. Using similar reasoning, the quark structure of the positive pion (π^+) is $u\bar{d}$. Thus the antiparticle of the π^+ would have the composition of $\bar{u}\bar{\bar{d}}$. However, the antiquark of an antiquark is just the original quark—for example $(\bar{\bar{u}}) = u$. Hence the antiparticle of the π^+ has the quark structure $\bar{u}\bar{\bar{d}} = \bar{u}d$, which is the π^- quark structure, as expected.

Answers to Odd-Numbered Exercises

Chapter 1

1. (b)
3. (c)
5. (b)
7. Decimal (base 10) has a dime worth 10¢ and a dollar worth 10 dimes or 100¢. By analogy a duodecimal system would have a dime worth 12¢ and a dollar worth 12 "dimes" or $1.44 in current dollars.
9. 1 nautical mile = 6076 ft
11. (c)
13. (c)
15. Dimensional analysis uses the fundamental dimensions of physical quantities such as length ([L]), mass ([M]), time ([T]), etc. Unit analysis uses a specific system of units. For example, if the mks system is used, then meter (m), kilogram (kg), and second (s) are used in unit analysis.
17. (d)
19. $m^2 = m^2$
21. no; $V = \pi d^3/6$
23. no; $m/s \neq m/s - m$
25. yes; $[L]^2 = [L]^2 + [L]^2$
27. $1/s$, or s^{-1}
29. (a) $kg \cdot m^2/s^2$ (b) yes
31. (a)
33. (a) cm as it is the smallest unit among the ones listed (b) 183 cm
35. 37 000 000 times
37. 73.0 m, 79.8 m, 24.1 m
39. 68 L
41. metric; $9.4 \times 10^2 \, m^2$
43. (a) 341 m/s (b) 0.268 s
45. (a) 16 km/h for each 10 mi/h (b) 113 km/h
47. (a) 59.1 mL (b) 3.53 oz
49. 6.1 cm
51. (a) $26 \, m^3$ (b) $9.2 \times 10^2 \, ft^3$
53. (a) $62.3 \, lb/ft^3$ (b) 8.34 lb
55. (a)
57. 5.05 cm; 5.05×10^{-1} dm; 5.05×10^{-2} m
59. no; only one doubtful digit; best 25.48 cm
61. (a) 1.0 m (b) 8.0 cm (c) 16 kg (d) $1.5 \times 10^{-2} \, \mu s$
63. (a) 10.1 m (b) 775 km (c) 2.55×10^{-3} kg (d) 9.30×10^7 mi
65. $1.1 \times 10^{-2} \, m^2$
67. (a) three, since the height has only 3 significant figures (b) $470 \, cm^2$
69. (a) $2.0 \, kg \cdot m/s$ (b) $2.1 \, kg \cdot m/s$ (c) no, rounding difference
71. (c)
73. 56 m
75. 100 kg
77. (a) 52% (b) 64 g, 20 g
79. 9.47×10^{15} m
81. 0.87 m
83. same area for both, $1.3 \, cm^2$
85. 25 min
87. 17 m
89. $0.32, \$0.32/L = \$1.21/gal$
91. $42 \, cm^2$
93. about $10^2 \, cm^3$
95. 0.60 L
97. not reasonable (56 mi/h)
99. (a) Since $d = (13 \, mi) \tan 25°$ and $\tan 25° < 1(\tan 45° = 1)$, d is less than 13 mi. (b) 6.1 mi
101. 7.2 kg

Chapter 2

1. (a) scalar, (b) vector, (c) scalar, (d) vector
3. (a)
5. Yes, for a round trip. No; distance is always greater than or equal to the magnitude of displacement.
7. Speed is the magnitude of the velocity.
9. The distance traveled is greater than or equal to 300 m.
11. 1.65 m down
13. 30 km, no
15. (a) 0.50 m/s (b) 8.3 min
17. (a) between 40 m and 60 m, as any side of a triangle cannot be greater than the sum of the other two sides. The magnitude of the displacement is the hypotenuse of the right triangle so it cannot be smaller than the longer of the sides perpendicular to each other. (b) 45 m at 27° west of north
19. (a) 2.7 cm/s (b) 1.9 cm/s
21. (a) 7.35 s (b) no
23. (a) $\bar{s}_{0-2.0s} = 1.0 \, m/s; \bar{s}_{2.0s-3.0s} = 0;$ $\bar{s}_{3.0s-4.5s} = 1.3 \, m/s; \bar{s}_{4.5s-6.5s} = 2.8 \, m/s;$ $\bar{s}_{6.5s-7.5s} = 0; \bar{s}_{7.5s-9.0s} = 1.0 \, m/s$ (b) $\bar{v}_{0-2.0s} = 1.0 \, m/s; \bar{v}_{2.0s-3.0s} = 0;$ $\bar{v}_{3.0s-4.5s} = 1.3 \, m/s; \bar{v}_{4.5s-6.5s} = -2.8 \, m/s;$ $\bar{v}_{6.5s-7.5s} = 0; \bar{v}_{7.5s-9.0s} = 1.0 \, m/s$ (c) $v_{1.0s} = s_{0-2.0s} = 1.0 \, m/s;$ $v_{2.5s} = s_{2.0s-3.0s} = 0; v_{4.5s} = 0;$ $v_{6.0s} = \bar{s}_{4.5s-6.5s} = -2.8 \, m/s$ (d) $v_{4.5s-9.0s} = -0.89 \, m/s$
25. 8.7×10^2 km
27. 59.9 mi/h; no
29. impossible, because the average speed to home would have to be infinity
31. Yes, although the speed of the car is constant, its velocity is not because of the change in direction.
33. v_o
35. yes. For example, an object moving in the +x-axis (positive velocity) slows down (negative acceleration or acceleration in the −x-axis).
37. $1.85 \, m/s^2$
39. 3.7 s
41. 36 m
43. $a_{0-4} = 2.0 \, m/s^2; a_{4-10} = 0;$ $a_{10-18} = -1.0 \, m/s^2$
45. 150 s
47. (d)
49. (a)
51. no, $9.9 \, m/s^2$
53. (a) $3.5 \, m/s^2$ (b) 4.5 s
55. (a) 81.4 km/h (b) 0.794 s
57. 3.09 s and 13.7 s. After 3.09 s, it is 175 m from where the reverse thrust was applied. However, if the reverse thrust is continuously applied (possible, but not likely) it will reverse its direction and be back to 175 m from the point where the initial reverse thrust was applied, it had taken 13.7 s.
59. (a) travels in the +x direction and then reverses. The object has initial velocity in the +x direction and it takes time for the object to decelerate and stops, and then reverse direction. (b) 23 s (c) 40 m/s in −x
61. 1.43×10^{-4} s
63. (a) $30 \, m/s^2$ (b) 75 m/s
65. no, $x = 13.3$ m
67. (a) −12 m/s; −4.0 m/s (b) −18 m (c) 50 m
69. (a) x_o—initial position, x—final position, v_o—initial velocity, v—final velocity, a—acceleration, t—time interval (b) The condition under which these equations hold is that the motion needs to be motion with constant acceleration.
71. (d)
73. zero and $9.8 \, m/s^2$ downward
75. The ball moves with a constant velocity.
77. 78.4 m
79. (a) 4 times, as displacement is proportional to the time squared (b) 15.9 m and 4.0 m
81. no, not a good deal.
83. 67 m
85. slightly less than 8.0 m/s
87. (a) 48 m (b) 38 m/s downward
89. (a) same, 1.14 m/s (b) 11.2 m/s, −11.2 m/s (equal but opposite)
91. (a) less than 95%, as the height depends on the velocity squared (b) 3.61 m
93. hits 14 cm in front of the prof
95. (a) $\sqrt{6}$, as time interval depends on the square root of acceleration (b) Earth: 16.5 m, 3.67 s; Moon: 99.2 m, 22.0 s
97. (a) −20.9 m/s (b) 2.87 s
99. (a) 21 m/s (b) 19 m
101. 45 mi/h
103. (a) greater than R but less than $2R$. For any right triangle, the hypotenuse is always greater than any one of the two sides that are perpendicular to each other (R) and less than the sum of the two sides perpendicular to each other ($R + R = 2R$). (b) 71 m
105. 16.6 m/s
107. 1.2×10^2 m
109. (a) 27 m/s (b) 4.3 s
111. (a) It takes sound time to travel from the bottom of the well to the person. (b) 51.7 m
113. (a) $5.0 \, m/s^2$ (b) 13 m/s (c)

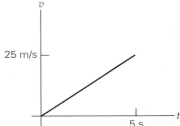

(d)

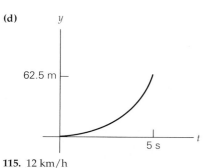

115. 12 km/h

Chapter 3

1. (a)

3. yes, e.g. circular motion

5. (a) Linear velocity increases or decreases in magnitude only. **(b)** parabolic path **(c)** circle

7. (a) between 4.0 m/s² and 7.0 m/s² as the hypotenuse of a right triangle can never be smaller than either of the two sides perpendicular to each other (so it must be greater than 4.0 m/s²) and greater than the sum of the two sides perpendicular to each other (so it must be less than 7.0 m/s²) **(b)** 5.0 m/s², 53° above +x-axis

9. (a) 6.0 m/s **(b)** 3.6 m/s

11. (a) 70 m **(b)** 0.57 min, 0.43 min

13. $x = 1.75$ m; $y = -1.75$ m

15. (a) $x = 6.56$ m; $y = 3.25$ m **(b)** 5.41 m/s at 13.9° above +x-axis

17. 1.0 m/s

19. (a) 1.5 m/s **(b)** 70 m

21. opposite direction, same direction, at right angle

23. yes when the vector is in the y-direction

25. $v_x = 8.66$ m/s; $v_y = 5.00$ m/s

27. (a) Yes, vector addition is associative.

(b)

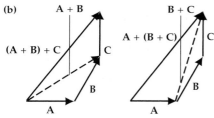

29. (a) +15 km/h **(b)** −75 km/h

31. 109 mi/h

33. (a)

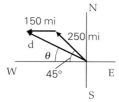

(b) 372 mi at 28.4° north of west

35. (a) $(-3.4$ cm$)\,\hat{x} + (-2.9$ cm$)\,\hat{y}$
(b) 4.5 cm, 63° above −x-axis **(c)** $(4.0$ cm$)\,\hat{x} + (-6.9$ cm$)\,\hat{y}$

37. (a) $(14.4$ N$)\,\hat{y}$ **(b)** 12.7 N at 85.0° above +x-axis

39. 16 m/s at 79° above the −x-axis

41. opposite

43. 8.5 N at 21° below −x-axis

45. parallel 30 N, perpendicular 40 N

47. 27 m at 72° above −x-axis

49. (a) west of north **(b)** 102 mi/h at 61.1° north of west

51. 242 N at 48° below −x-axis

53. Since the rain is coming down at an angle relative to you, you should hold the umbrella so it is tilted forward (perpendicular to the velocity of the rain relative to you).

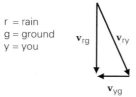

r = rain
g = ground
y = you

55. straight up. Both you and the ball has the same horizontal velocity relative to the ground or they have zero horizontal velocity relative to each other so the object returns to your hand.

57. 13 s

59. (a) +85 km/h **(b)** −5 km/h

61. 4.0 min

63. 146 s = 2.43 min

65. (a) 68 m **(b)** 0.76 m/s, 11° relative to shore

67. (a) also increases **(b)** 4.7 m/s

69. (a) 24° east of south **(b)** 1.5 h

71. 45°

73. The vertical motion does not affect the horizontal motion.

75. 2.7×10^{-13} m; no

77. 40 m

79. horizontal: 26 m/s, vertical: 15 m/s

81. (a) Ball B collides with ball A because they have the same horizontal velocity. **(b)** 0.11 m, 0.11 m

83. (a) 0.77 m **(b)** The ball would not fall back in.

85. (a) 1.15 km **(b)** 30.7 s **(c)** 6.13 km

87. 35° or 55°

89. (a) 26 m **(b)** 23 m/s at 68° below horizontal

91. 40.9 m/s, 11.9° above horizontal

93. The pass is short.

95. (a)

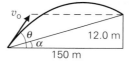

(b) 66.0 m/s **(c)** too long

97. $(4.0$ m/s$)\,\hat{x} + (2.0$ m/s$)\,\hat{y}$; 4.5 m/s at 27° above +x-axis

99. 90 miles at 45° north of east

101. 6.02 m/s

103. 63°

105. (a) 30.3° or 59.7° **(b)** no

107. (a) 21.7 m/s **(b)** 33.3 m/s, 49.3° below horizontal

109. (a) 53° above or below +x-axis **(b)** ±8.0 m/s

111. 6.1×10^2 km

Chapter 4

1. no. If an object remains at rest, the *net force* is zero. There could still be forces acting on it as long as the net force is zero.

3. Your tendency is to remain at rest or move with constant velocity. However, the plane is accelerating to a velocity faster than yours so you are "behind" and feel being "pushed" into the seat. The seat actually supplies a forward force to accelerate you to the same velocity as the plane.

5. (c)

7. no, same mass, same inertia

9. balloon moves **(a)** forward **(b)** backward

11. The dishes at rest tend to remain at rest.

13. $m_{Al} = 1.35 m_{water}$

15. $\mathbf{F}_3 = (-7.6$ N$)\,\hat{x}$

17. (b)

19. No, both mass and gravity decrease.

21. "Soft hands" here result in longer contact time between the ball and the hands. The increase in contact time decreases the magnitude of acceleration. From Newton's second law, this in turn decreases the force required to stop the ball and its reaction force, the force on the hands.

23. 1.7 kg

25. 7.0×10^5 N

27. 75.5 kg

29. 78 N, 18 lb

31. (a) on the Earth **(b)** 5.4 kg (2 lb)

33. (a) 1.2 m/s² **(b)** same

35. 2.40 m/s²

37. (a) 0.133 m/s² **(b)** Only 300 N needed to maintain constant v.

39. 8.9×10^4 N

41. (c)

43. The forces on different objects (one on horse, one on cart) cannot cancel.

45. Yes, the forces on different objects (one on ball, one on bat) cannot cancel.

47. brick hits the fist hard with a force of 800 N, no

49. 1.5 m/s², opposite to hers

51.

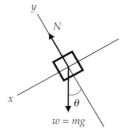

mg-gravitational force
N-normal force by ramp on car

53. (a) all of the preceding, depends on the acceleration **(b)** $a = 1.8$ m/s² downward

55. (a) 735 N **(b)** 735 N **(c)** 585 N

57. (a) 2.0 m/s², 19° north of east **(b)** 1.3 m/s² 30° south of east

59. (a) 0.96 m/s² **(b)** 2.6×10^2 N

61. 64 m

63. (a) both the tree separation and sag **(b)** 6.1×10^2 N

65. (a) 0.711 m/s² **(b)** 1067 N

67. (a) 2.5 m/s² right **(b)** 2.0 m/s² left
69. 1.1 m/s², up
71. 3.9 × 10² N
73. 1.2 m/s², m_1 up and m_2 down **(b)** 21 N
75. (a) no friction **(b)** opposite its direction of velocity **(c)** sideways **(d)** forward, in direction of velocity
77. Kinetic friction (sliding) is less than static friction (rolling).
79. increasing normal force, therefore friction; yes; more difficult to accelerate
81. The treads are designed to displace water so cars with regular tires can drive in the rain. However, the wide and smooth drag race tires increase friction because it uses a softer compound therefore large coefficient of friction.
83. 2.7 × 10² N
85. (a) zero **(b)** 3.1 m/s²
87. 0.064
89. 0.44 m/s²
91. (a) 30° **(b)** 22°
93. 0.73
95. (a) 26 N **(b)** 21 N
97. 0.33
99. (a) between 0.72 kg and 1.7 kg **(b)** between 0.88 kg and 1.5 kg
101. 1.50 × 10³ N
103. no
105. (a) 75 N **(b)** 0.097
107. 0.32 N
109. 5.2 m
111. no; 24.6 m < 26 m

Chapter 5

1. (d)
3. negative, as force and displacement are opposite
5. (a) No, the weight is not moving, so there is no displacement and therefore, no work. **(b)** Yes, positive work is done by the force exerted by the weightlifter. **(c)** No, as in (a), no work is done. **(d)** Yes, but the positive work is done by gravity, not the weightlifter.
7. Positive on the way down and negative on the way up. No, it is not constant. Maximum at points B and D ($\theta = 0°$ or $180°$) and minimum at points A and C ($\theta = 90°$).

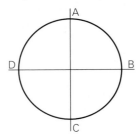

9. −98 J
11. 6.65 m
13. 3.7 J
15. (a) negative because the direction of force is opposite (180°) to the displacement (up). **(b)** 1.47 × 10⁵ J
17. 2.3 × 10³ J

19. (a) 1.48 × 10⁵ J **(b)** −1.23 × 10⁵ J **(c)** 2.50 × 10⁴ J
21. (c)
23. three times as much
25. 2.5 cm
27. 1.25 × 10⁵ N/m
29. 0.21 J
31. (a) 15 kg **(b)** 5.1 kg more
33. −3.0 J
35. Because the heel can lift from the blade, the blade will be in contact with the ice longer. This will make the displacement a bit greater (longer stride) so that the work done is greater. Greater work translates to faster speed according to the work-energy theorem.
37. reducing speed. Reducing the speed by half will reduce K by $\frac{3}{4}$ while reducing the mass by half will only reduce K by half.
39. $\sqrt{2}\,v$
41. (a) 75% **(b)** 7.5 J
43. (a) 45 J **(b)** 21 m/s
45.
$$W_{net} = -fd = \tfrac{1}{2}mv^2 - \tfrac{1}{2}mv_o^2 = 0 - \tfrac{1}{2}mv_o^2.$$
$$d = \frac{v_o^2}{2f} \propto v_o^2; \ 200 \text{ m}$$
47. 8.3 × 10³ J
49. $x^2 - x_o^2$
51. 4.9 × 10² J
53. 5.9 J
55. (a) 8.0 J **(b)** 5.6 J
57. (a) all the same as the change in potential energy is independent of the reference level **(b)** $U_b = -44$ J; $U_a = 66$ J **(c)** −1.1 × 10² J
59. (d)
61. final height = initial height
63. Each time you land on the trampoline, you coil your legs on the way down, then push down on the trampoline as you land, thereby compressing it more; storing more energy, you rebound and go higher. The limit is determined by the "spring constant" of the trampoline and how much you can compress it.
65. no
67. (a) 15.0 J, zero, 15.0 J **(b)** 7.65 J, 7.35 J, 15.0 J **(c)** zero, 15.0 J, 15.0 J
69. (a) 20 J; 60 J **(b)** 8.9 m/s **(c)** 80 J
71. (a) 1.03 m **(b)** 0.841 m **(c)** 2.32 m/s
73. (a) 11 m/s **(b)** no **(c)** 7.7 m/s
75. (a) 2.7 m/s **(b)** 0.38 m **(c)** 29°
77. −7.4 × 10² J
79. 12 m/s
81. no, energy; 9.0 × 10⁶ J
83. They are doing the same amount of work (same mass, same height). So the one that arrives first will have expended more power due to shorter time interval.
85. 97 W
87. 5.7 × 10⁻⁵ W
89. 6.7 × 10² J
91. 2.3 × 10⁶ J
93. (a) 2.3 hp **(b)** 10 hp
95. $0.36
97. (a) more **(b)** 0 − 10 cm: 0.25 J; 10 − 20 cm: 0.75 J

99. (a) 4.9 m **(b)** independent of mass
101. (a) 8.17 × 10⁵ J **(b)** 3.90 × 10³ W
103. (a) 0.166 m **(b)** 4.70 m/s
105. (a) 5.5 × 10² W **(b)** 0.74 hp

Chapter 6

1. (b)
3. no, mass is also a factor
5. not necessarily; different mass can still have different kinetic energies
7. 75 kg
9. (a) no, as velocity is also a factor in calculating momentum **(b)** the running back, 38 kg · m/s more
11. 24 J
13. 4.05 kg · m/s in direction opposite to initial velocity
15. (a) 3.7 kg · m/s down **(b)** 7.0 kg · m/s down
17. 3.5 m/s in the same direction or 8.2 m/s in opposite direction
19.
$$\Delta \mathbf{p} = (-1.1 \text{ kg} \cdot \text{m/s})\,\hat{\mathbf{x}} + (-2.8 \text{ kg} \cdot \text{m/s})\,\hat{\mathbf{y}}$$
21. 6.5 × 10³ N
23. (a) 491 kg · m/s downward **(b)** yes; 519 kg · m/s downward
25. increase contact time so to increase impulse
27. (a) Drive: large impulse (large $\bar{F}$ and Δt); chip shot: small impulse (small $\bar{F}$ and Δt). **(b)** Jab: small impulse (small $\bar{F}$ and Δt); knock-out punch: large impulse (large $\bar{F}$ and Δt). **(c)** Bunting: small impulse (small $\bar{F}$ and Δt); home-run swing: large impulse (large $\bar{F}$ and Δt).
29. Not necessarily, contact time is also important.
31. 6.0 × 10³ N
33. (a) 1.2 × 10³ N **(b)** 1.2 × 10⁴ N
35. (a) hitting it back. When a ball changes its direction, the change in momentum is greater. **(b)** 1.2 × 10² N in direction opposite to v_o
37. 1.1 × 10³ N and 4.7 × 10² N
39. 8.2 N upward
41. 0.057 s
43. (a)
45. Throw something or even blow breath (there is no friction so you cannot walk).
47. Golf ball has a much smaller mass.
49. 1.0 m/s westward
51. 1.67 m/s in original direction of bullet
53. 7.6 m/s, 12° above the +x-axis
55. (a) 11 m/s to the right **(b)** 22 m/s to the right **(c)** at rest
57. 0.78 m/s
59. 0.61 m/s, 0.73 m/s
61. (c)
63. Momentum is a vector and kinetic energy is a scalar.
65. elastic, as the kinetic energy is conserved
67. $v_1 = -0.48$ m/s; $v_2 = +0.020$ m/s
69. $v_p = -1.8 × 10^6$ m/s; $v_a = 1.2 × 10^6$ m/s

71. (a) completely inelastic, because the fish and the bird are combined after the "collision" (catch) **(b)** 5.6 m/s

73. (a) south of east **(b)** 13.9 m/s, 53.1° south of east

75. no, 1.1 kg·m/s at 69° below x-axis or 249°

77. (a) $m_1/m_2 = 1/2$ **(b)** $v_2 = (2/3)v_{1o}$

79. (a) $v = v_o/90$ **(b)** 2.5×10^2 m/s **(c)** no, 99%

81. (b) $e_{\text{elastic}} = 1.0$; $e_{\text{inelastic}} = 0$

83. (d)

85. directly above the foot on the ground

87. (a) $(0, -0.45$ m$)$ **(b)** no, only that they are equidistant from CM

89. (a) 4.6×10^6 m from the center of Earth **(b)** 1.8×10^6 m below the surface of Earth

91. 82.8 m

93. still at center of sheet

97. (a) the 65 kg travels 3.3 m and the 45 kg travels 4.7 m **(b)** same distances as in part (a).

99. (a) no **(b)** in the same direction

101. 2.1×10^5 J

103. $v_1 = v_2 = 2.0$ m/s

105. (a) 36 km/h, 56° north of east **(b)** 49% lost

107. (a) inelastic **(b)** 2.1 N·s **(c)** 21 N

109. (a) no **(b)** 4.5 m/s

111. 7.7×10^{-23} m/s

Chapter 7

1. (c)

3. no, yes

5. (4.5 m, 2.8 m)

7. (a) 0.26 rad **(b)** 0.79 rad **(c)** 1.6 rad **(d)** 2.1 rad

9. 1.4×10^9 m

11. (a) 4.00 rad **(b)** 229°

15. 10.7 rad

17. (a) 9.2×10^{-3} rad $= 0.53°$ **(b)** 3.4×10^{-2} rad $= 1.9°$

19. (a) no **(b)** 6 such pieces and one 0.28 rad piece

21. (b)

23. Yes, they all sweep through the same angle. No, they do not have the same tangential speed as the distances to the center of the wheel are different.

25. farthest and closest to the axis; all the same

27. 2.4 rev/day

29. 1.8 s

31. B has greater angular speed.

33. (a) 0.84 rad/s **(b)** 3.4 m/s, 4.2 m/s

35. (a) rotating angular speed, as the time is less and the angular displacement is the same. **(b)** 7.27×10^{-5} rad/s, 1.99×10^{-7} rad/s

37. (a) 60 rad **(b)** 3.6×10^3 m

39. (d)

41. No. When an object is in uniform circular motion, it always has a non-zero centripetal acceleration and a zero tangential acceleration.

43. in direction of acceleration, inward; no

45. Centripetal force is required for a car to maintain its circular path. When a car is on a banked turn, the horizontal component of the normal force on the car is pointing toward the center of the circular path. This component will enable the car to negotiate the turn even when there is no friction.

47. 0.049 rad/s, or about 680 rev/day

49. 1.3 m/s

51. no; 1.33 m/s^2 > 1.25 m/s^2

53. 11.3°

55. 9.9 m/s

57. (a) $v = \sqrt{rg}$ **(b)** $h = (5/2)r$

59. (b) independent of mass

(c) $\tan\theta = \dfrac{v^2}{gr} - \mu_s$

61. (d)

63. no. Any car in traveling on a circular track always has centripetal acceleration. It could have angular acceleration if its speed changes.

65. 1.1×10^{-3} rad/s^2

67. 7.5 rev

69. (a) 1.82 rad/s^2 **(b)** 28.7 rev

71. (a) both. There is always centripetal acceleration for any car in circular motion. When the car increases its speed on a circular track, there is also angular acceleration. **(b)** 53 s **(c)** $\mathbf{a} = (8.5$ m/s$^2)$ $\hat{\mathbf{r}}$ + $(1.4$ m/s$^2)$ $\hat{\mathbf{t}}$

73. (c)

75. No, these terms are not correct. Gravity acts on the astronauts and the spacecraft, providing the necessary centripetal force for the orbit, so g is not zero and there is weight by definition $(w = mg)$. The "floating" occurs because the spacecraft and astronauts are "falling" ("accelerating" toward Earth at the same rate).

77. Yes, if we also know the radius of the Earth. The acceleration due to gravity near the surface of the Earth can be written as $a_g = GM_E/R_E^2$. By simply measuring a_g, we can determine $M_E = a_g R_E^2/G$.

79. 1.6 m/s^2

81. (a) lunar eclipse, because the forces by the Sun and the Earth on the Moon are in the same direction so they add up. **(b)** solar: 2.4×10^{20} N; lunar: 6.4×10^{20} N, both toward Sun

83. 8.0×10^{-10} N, toward opposite corner

85. 3.4×10^5 m

87. $F_{E-S} = (1.7 \times 10^2)F_{E-M}$

89. (a) -2.5×10^{-10} J **(b)** 0

91. (c)

93. (a) Rockets are launched eastward to get more velocity relative to space because the Earth rotates toward the east. **(b)** The tangential speed of the Earth is higher in Florida because Florida is closer to the equator than California so hence a greater distance from the axis of rotation. Also, the launch is over the ocean for safety.

95. (a) 3.7×10^3 m/s **(b)** 34%

97. 1.53×10^9 m

99. 2.0 min

101. (a) 0 **(b)** 1.2×10^3 N **(c)** 1.8×10^3 N

103. (a) 1.10×10^{10} m/s^2 **(b)** 1.12×10^9 g

105. 1.1×10^4 m/s

107. 1 y^2/AU^3

Chapter 8

1. (a)

3. yes; rolling motion

5. (b)

7. zero

9. 0.10 m

11. (a) 0.50 m/s **(b)** 1.0 m/s

13. (a) equal to **(b)** yes

15. 0.58 rotations

17. (a)

19. back muscles have to exert greater torque

21. Yes, the toy clown is in stable equilibrium. Its center of gravity is directly below the tightrope. If the clown leans to one side, its own weight will restore its equilibrium position. If the weights are removed, the clown will be in an unstable equilibrium and it will fall.

23. 3.3×10^2 N

25. 6 stable (faces) and 20 unstable (12 edges and 8 corners)

27. (a) yes if the lever arms are appropriate for the weights of the children because torque is equal to force times lever arm **(b)** 2.3 m

31. 1.6×10^2 N

33. 3.6 N

35. $m_2 = 0.20$ kg, $m_3 = 0.50$ kg, $m_4 = 0.40$ kg

37. Yes. The center of gravity of every stick is at or to the left of the edge of the table

39. 1.2 m from left end of board

41. (a) equal to **(b)** 0.19 **(c)** 3.5 m/s

43. (a)

45. depends on how mass is distributed about an axis

47. When the paper is pulled quickly (a large force is required to accelerate the roll), the force the paper can provide is not great enough to accelerate the paper roll. However, if the paper is pulled slowly, the paper is strong enough to accelerate the roll because the force required is smaller. The more paper the roll has, the greater the moment of inertia, the greater the force required to accelerate the roll, and therefore the easier to tear.

49. yes, increases

51. (a) 22.5 kg·m^2 **(b)** 62.5 kg·m^2 **(c)** 85.0 kg·m^2

53. 0.64 m·N

55. (a) minimum according to the parallel axis theorem **(b)** 8.0 kg·m^2; 7.5 kg·m^2

57. 1.1 rad

59. 3.5 times

61. 1.2 m/s^2

63. 2.4 N

65. (a) 1.5g **(b)** 67-cm position

67. 6.5 m/s^2

69. (a) $7\mu_s/2$ **(b)** 63.8°

71. (c)

73. small total mass with more mass near center

75. (a) 28 J (b) 14 W
77. 0.47 m · N
79. 0.16 m
81. cylinder goes higher by 7.1%
83. (a) 1.31×10^8 J (b) 1.46×10^6 W
85. (a) 29% (b) 40% (c) 50%
87. (a) $v = \sqrt{gh}$ (b) $h = 2.7R$
(c) weightlessness
89. Walking toward the center decreases the moment of inertia and so increases the rotational speed.
91. The arms and legs are put onto these positions to decrease the moment of inertia. This decrease in moment of inertia increases the rotational speed.
93. The cat manipulates its body to change the moment of inertia to rotate or flip over. It is done by twisting one way with part of the body and then the desired part (the feet) may be rotated the other way.
95. The two rotors, rotating in opposite directions, compensate each other so the total angular momentum of the whole system is zero.
97. 1.4 rad/s
99. $L_{rot} = 2.4 \times 10^{29}$ kg · m^2/s;
$L_{rev} = 2.8 \times 10^{34}$ kg · m^2/s
101. $T = 3.3T_o$
103. (a) 4.3 rad/s (b) $K = 1.1K_o$
(c) work done by skater
105. (a) in opposite direction, from angular momentum conservation (b) 0.56 rad/s
(c) no, 2.1 rad
107. 52 J
109. (a) 4.9 m/s^2 (b) 9.8 rad/s^2
111. zero
113. (a) 2.5 cm (b) 36 cm
115. 1.3 rad/s
117. $T_1 = 21$ N; $T_2 = 15$ N
119. smaller side, 1.4 times

Chapter 9

1. (c)
3. steel wire
5. N/m
7. 0.020
9. 1.1×10^{11} N/m^2
11. 47 N
13. 3.92×10^{-5} m
15. 2.0×10^{-7} m
17. 6.7×10^3 N/m^2
19. (a) ethyl alcohol as it has the smallest bulk modulus (b) $\Delta p_w / \Delta p_{ea} = 2.2$
21. 5.4×10^{-5}
23. (c)
25. When the bowl is full, the atmospheric pressure on the water does not allow any water out of the bottle. When the water level in the bowl decreases below the neck of the bottle, air bubbles in and water flows out until the pressures are equalized again. No, the height does not depend on the surface area.
27. Bicycle tires have a much smaller contact area with the ground so they need a higher pressure to balance the weight of the bicycle and the rider.

29. Pressure inside balances pressure outside.
31. (a) higher due to its low density (b) 10 m
33. (a) higher as it has a lower density
(b) 22 cm
35. 6.39×10^{-4} m^2
37. (a) a lower pressure inside the can as stream condenses (b) 1.6×10^4 N = 3600 lb
39. Air density decreases rapidly with altitude.
41. 2.2×10^5 N (about 50 000 lb)
43. (a) 1.1×10^8 Pa (b) 1.9×10^6 N
45. 470 N; 1.2×10^6 Pa
47. 2.6 mm
49. (a)
51. The level does not change. As the ice melts, the volume of the newly converted water decreases; however, the ice, which was initially above the water surface, is now under the water. This compensated for the decrease in volume. It does not matter whether the ice is hollow or not. Both can be proved mathematically.
53. They will receive the same buoyant force as it depends only on the volume of the fluids displaced and independent of the mass of the object.
55. Yes, reading increases due to the reaction force of the buoyant force.
57. (a) water displacement (b) no;
18.8×10^3 kg/m$^3 < \rho_g = 19.3 \times 10^3$ kg/m^3
59. 33 N, same
61. no,
14.5×10^3 kg/m$^3 < \rho_g = 19.3 \times 10^3$ kg/m^3
63. 0.33 m
65. 17.7 m
67. 2.3 m/s^2
69. There are more capillaries than arteries.
71. (c)
73. The concave bottom makes the air travel faster under the car. This increase in speed will reduce the pressure under the car. The pressure difference forces the car more on the ground to provide a greater normal force and friction for traction.
75. the side facing the batter spinning downward so to generate the extra downward force caused by the pressure difference between the top and bottom of the ball
77. (a) increase by a factor of 4 from the equation of contunity (b) decrease by a factor of 9.
79. (a) 3.5 cm^3/s (b) 0.031% (c) It is a physiological need. The slow speed is needed to give time for the exchange of substances such as oxygen between the blood and the tissues.
81. (a) 0.99 m/s; 1.7 m/s; 2.2 m/s; 2.6 m/s
(b) 0.44 m, from $y = 20$ cm
83. 2.2 Pa
85. (b)
87. (a)
89. 3.5×10^2 Pa
91. 99.99 cm
93. (a) 0.09 m (b) 8.1 kg
95. 1.32×10^5 N or 2.96×10^4 lb
97. 8.5×10^2 kg/m^3
99. 0.45 m
101. 2.2 m/s
103. (a) $v_s = 11$ cm/s; $v_n = 66$ cm/s

(b) 1.2×10^3 Pa
105. (a) 3.3×10^4 N (b) 3.8×10^4 N
107. (a) 9.8×10^{-4} m^3 (b) 1.5×10^3 kg/m^3
(c) Yes, but less precisely—the sphere would float, and you would have to estimate the portion of its volume that was submerged.

Chapter 10

1. (d)
3. not necessarily. Internal energy does not solely depend on temperature. It also depends on mass.
5. Air has water and it may freeze, potentially at high altitudes.
7. (a) 538°C (b) −18°C (c) −29°C
(d) −40°C
9. −70°C
11. 56.7°C and −62°C
13. 136°F; −128°F
15. (a) −101 F° (b) 558 F°
17. (c)
19. (a) increases (b) constant
21. The pressure of the gas is held constant. So if the temperature increases, so does the volume and vice versa, according to the ideal gas law. Therefore temperature is determined from volume.
23. The balloons collapsed. Due to the decrease in temperature, the volume also decreases.
25. a higher atmospheric pressure at lower altitude, which causes the bag not to inflate as its pressure is lower.
27. (a) −273°C (b) −23°C (c) 0°C (d) 52°C
29. 300K
31. (a) 2.2 mol (b) 2.5 mol (c) 3.0 mol
(d) 2.5 mol
33. 1.2×10^5 Pa
35. 0.0247 m^3
37. (a) increase from the ideal gas law
(b) $p_2 = 4p_1$
39. 33.4 lb/in^2
41. (a) increase from the ideal gas law
(b) 10.6%
43. (d)
45. (a) Ice moves upward. (b) Ice moves downward. (c) copper
47. Volume increases.
49. No, it will not be distorted because both the ring and the bar are made of iron so they will expand at the same rate as one single piece. Yes, the circular ring will be distorted, if the bar is made of aluminum.
51. Lid expands more than glass.
53. (a) high because the tape shrinks
(b) 0.06%
55. 0.0027 cm
57. (a) 60.07 cm (b) 3.91×10^{-3} cm^2; yes
59. (a) The ring, so it expands and then the ball can go through. (b) 353°C
61. (a) spill because the coefficient of volume expansion is greater for gasoline than steel. (b) 0.48 gal
63. 13.4×10^3 kg/m^3
65. (a) 1.54×10^4 °C (b) no
67. (c)

69. The gases diffuse through the porous membrane, but the helium gas diffuses faster because its atoms have a smaller mass. Eventually there will be equal concentrations of gases on both sides of the container.

71. (a) 6.1×10^{-21} J (b) 7.7×10^{-21} J

73. (a) 6.21×10^{-21} J (b) 1.37×10^3 m/s

75. increases by a factor of $\sqrt{2}$

77. 1.12 times

79. (a)

81. it has more degrees of freedom

83. 6.1×10^3 J

85. 0.46 gal

87. 6.8×10^{-4} m

89. 7.8 cm

91. 33 C°

93. (a) 3.3×10^{-2} mol (b) 2.0×10^{22} molecules (c) 0.94 g

95. (a) 18 F° (b) 5.6 C°

97. 9.9×10^{20}

101. 3.3 cm

103. 2.026×10^5 Pa

Chapter 11

1. (d)

3. 2.54×10^5 J

5. 5.86×10^3 W

7. (a) 60 000 times (b) 83 h

9. Water has a high specific heat and takes longer to cool off.

11. yes; losing heat

13. just the temperature difference

15. specific heat and mass

17. Yes, c_p is always greater than c_v for gases. At constant volume, no work is done when adding heat to the gases. Under constant pressure, more heat is needed because of the work done as the gas expands.

19. 1.7×10^6 J

21. 1.27 kg

23. 0.13 kg

25. 55 L

27. (a) higher. If some water splashed out, there will be less water to absorb the heat. So the final temperature will be higher and the measured specific heat value will be in error and appear to be higher than accepted value. (b) 3.1×10^2 J/(kg·C°)

29. 21.5°C

31. (d)

33. Different substances have different internal energies, different molecular structures, and different intrinsic thermal properties. These quantities affect the temperature at which phase changes take place due to the different effects the addition or removal of heat has on different substances. Latent heats will also be different for different substances because of different molecular structures, or bonds. The latent heat energy goes into breaking these bonds.

35. This is due to the high value of the latent heat of vaporization. When steam condenses, it releases 2.26×10^6 J/kg of heat. When 100°C water drops its temperature by 1 C°, it releases only 4186 J/kg.

37. Fan blows relatively dry air, promoting evaporation, and thereby keeping you cool.

39. (a) More because the heat of vaporization L_v is greater than the heat of fusion, L_f. (b) vaporization by 1.8×10^6 J more

41. 8.0×10^4 J

43. 4.1×10^3 J

45. 0.81 kg

47. 11°C

49. 0.94 L or 0.94 kg

51. (a) Water has high heat of evaporation. (b) 1.1×10^6 J

53. 0.17 L

55. Because of convection, the warm (less dense) water rises to the top and the cooler (more dense) water sinks to the bottom.

57. (a) This convects the heat from the hot soup to the cooler air. (b) No. The ice blocks air flow and cooling of the air. (Also, the air conditioner is less efficient and runs more, increasing electric cost.) Also, ice is a poor conductor.

59. It is to increase the surface area for better conduction and radiation.

61. The double-walled and partially evacuated container is to counteract conduction and convection because both processes depend on a medium to transfer the heat (the double-walls are more for holding the partially evacuated region than for reducing conduction and convection). The mirrored interior minimizes the loss by radiation.

63. Air is a poor heat conductor.

65. (a) faster as tile has a higher thermal conductivity (b) 4.5

67. 4.54×10^6 J

69. (a) 5.5×10^5 W (b) 73 kg, no; this answer is not reasonable because a lot of heat does not go into the water.

71. 7.8 h

73. (a) 4.9 in. (b) 6.9 in.

75. (a) less since power radiated depends on the emissivity and a blackbody has an emissivity of 1 (b) 1.7

77. 2.3 cm

79. 7.8×10^5 J

81. 1.0×10^8 J

83. 84°C

85. 88.7°C

87. 1.9 g; into heating skates and loss to the environment.

89. (a) more, because aluminum has a higher specific heat. (b) Al by 1.8×10^4 J more

91. 2.3×10^6 J

93. It will stay hot longer if cream is added right away because the heat loss is proportional to the temperature difference. By adding the cream right away, the temperature difference is less and so is the heat loss.

Chapter 12

1. (c)

3. (a)

5.

7. (b)

9. (d)

11. This is an adiabatic compression. When the plunger is pushed in, the work done goes into increasing of the internal energy of the air. The increase in internal energy increases the temperature of the air and causes the paper to catch fire.

13. through work

15. 3, 2, 1

17. (a) the same, because $\Delta U = 0$ (b) added, 400 J

19. (a) decreases, as $Q = 0$ and W is positive, therefore ΔU and ΔT are negative. (b) −500 J

21. (a) 3.3×10^3 J (b) yes, 5.1×10^3 J

23. (a) zero because the initial and final temperatures are the same so $\Delta T = 0$. (b) $-p_1 V_1$ (on the gas) (c) $-p_1 V_1$ (out of the gas)

25. (a) by the gas (b) 3.5×10^3 J

27. 3.6×10^4 J

29. (a) path AB: -1.66×10^3 J; path BC: 0; path CD: 3.31×10^3 J; path DA: 0 (b) 0, $Q = W = 1.65 \times 10^3$ J (c) 800K

31. (b)

33. There would be energy created.

35. No, ice or water itself is not an isolated system.

37. 1.2×10^3 J/K

39. -2.1×10^2 J/K

41. −26 J/K

43. 126°C

45. 1.33 J/K

47. (a) 2.73×10^4 J (b) isotropic

49. (a) 61.0 J/K (b) −57.8 J/K (c) 3.2 J/K

51. (b)

53. It is unchanged since it returns to its original value.

55. This is important because heat can be completely converted to work for a single process (not a cycle) such as an isothermal expansion process of an ideal gas.

57. No, as the warm air rises to the higher altitude, both gravity and buoyancy forces do work. Since it is a natural process with work input, the entropy increases and the second law is not violated.

59. 25%

61. 1.47×10^5 J

63. 4.05×10^4 J

65. (a) 6.6×10^8 J (b) 27%

67. (a) 6.1×10^5 J (b) 1.9×10^6 J

69. 3.0 kW

71. (a) 2.4×10^5 J (b) 10 C°

73. (a) 1800 (b) 3.4×10^7 J (c) 2.7×10^7 J

75. (a)

77. (a) no, $Q_h = +Q_4$ (b) no, $Q_c = -Q_2$ (c) Expansion, positive work.

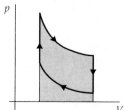

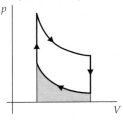

Compression, negative work.

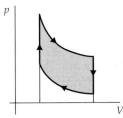

Net work

79. Efficiency depends on ΔT; water can maintain a large ΔT.

81. Diesel engines run hotter as diesel fuel has a higher spontaneous combustion temperature. According to Carnot efficiency, the higher the hot reservoir temperature, and the higher the efficiency, for a fixed low temperature reservoir.

83. 0°C

85. (a) 6.7% (b) probably not at the moment, due to low efficiency and relatively cheap fossil fuels.

87. 9.1×10^3 J

89. 100 C°

91. lowering the low temperature reservoir

93. (a) 42% (b) 39 kW

95. (a) 25% (b) $T_h = 1.3T_c$

97. (a) 64% (b) A lot more energy is lost than in the ideal situation.

99. (a) 13 (b) no, $COP_C = 11$

105. (a) isobaric expansion (b) 146 J

107. (a) 1.99×10^6 Pa (b) -8.30×10^4 J

109. 0.10 J/K

111. 3500 J

113. (a) 169 J (b) 2.09×10^3 J

Chapter 13

1. (b)
3. (b)
5. (a) four times greater (b) twice as large
7. $T/4$, $T/2$
9. $4A$
11. 0.025 s
13. 41 N/m
15. (a) 10^{-12} s (b) 63 m/s
17. (a) $x = 0$ (b) 2.0 m/s
19. (a) 0.77 m/s (b) 1.2 N
21. (a) 2.5 m/s (b) 2.5 m/s (c) 2.7 m/s, equilibrium position
23. (a) 0.38 m (b) 8.5×10^{-3} m
25. (d)
27. trace out path on scrolling horizontal paper
29. In an upward accelerating elevator, the effective gravitational acceleration increases so the period would be decreased.

31. no, as the period is independent on the gravitational acceleration for a mass-spring system; yes, as the period depends on the gravitational acceleration

33. (a) 2.0 s (b) 0.50 Hz
35. 0.25 m
37. (a) 0.10 m (b) 16 Hz (c) 0.063 s
39. (a) 0.25 m (b) 0.17 m (c) −0.18 m
43. (a) $1/\sqrt{3}$, as the period is inversely proportional to the square root of the spring constant (b) 2.8 s
45. 0.25 kg
47. 9.787 m/s²
49. (a) $y = (8.0 \text{ cm}) \cos (6.3 \text{ rad/s})t$ (b) −8.0 cm, or at the other amplitude.
51. (a) increase as the period is inversely proportional to the gravitational acceleration (b) 4.8 s
53. $y = (5.0 \text{ cm}) \cos (\pi t/4)$
57. (d)
59. (c)
61. The one on the left is transverse and the one on the right is longitudinal.
63. 0.34 m
65. 1.5 m/s
67. 4.7×10^{-14} Hz
69. (a) longer as low frequency corresponds to longer wavelength, according to $v = \lambda f$ (b) AM: 1.88×10^2 m to 5.45×10^2 m; FM: 2.78 m to 3.41 m
71. 3.0 m/s
73. (a) 3.8×10^2 s (b) yes, 1.9×10^3 km > 30 km (c) 1.6×10^3 s, S waves do not go through the liquid core
75. (a) 0.20 s (b) 2.0 s
77. (b)
79. Only waveform; energy is not destroyed but redistributed.
81. Sounds from different instruments would arrive at different times.
83. (c)
85. 5
87. increase the string tension to increase the speed therefore frequency
89. (a) 6.0 m (b) 2.0 m
91. 150 Hz
93. $n = 5$
95. 10 Hz; 20 Hz; 30 Hz; 40 Hz
97. (a) shortened because a shorter string has a shorter wavelength, therefore a higher frequency given the speed is a constant. (b) 503 Hz
99. 0.22 kg; 0.055 kg; 0.024 kg; 0.014 kg
101. (a) 12 cm (b) 3.0 cm, 9.0 cm, 15 cm
103. 5.0 m/s, 2.5×10^2 m/s²
105. 210 Hz
107. (a) unchanged (b) increases by $\sqrt{2}$ times
109. 1.1 cm/s
111. 3.0 s

Chapter 14

1. (b)
3. (a)
5. sounds not all in our audible range
7. same time
9. Sound travels considerably faster in solids than in air. Dogs can hear better by

putting their ear on the floor. Yes, it is related to people putting their ears on railroad tracks.

11. 32°C
13. 1.5×10^3 m
15. (a) increases as the speed of sound increases with temperature and $v = \lambda f$. So if v increases and f remains the same, λ increases (b) +0.047 m
17. (a) 7.5×10^{-5} m (b) 1.5×10^{-2} m
19. (a) 0.81 s (b) 0.78 s
21. −1.75%
23. 90 m
25. (a) less than double because the total time is the sum of the time it takes for the stone to hit the ground (free fall motion) and the time it takes sound to travel back that distance (b) 1.0×10^2 m (c) 8.7 s
27. (c)
29. to compress a large physical range into a smaller scale
31. (a) 8×10^{-17} W (b) 8×10^{-5} W
33. (a) 1/9 as I is inversely proportional to the square of R. Tripling R will reduce I to $1/3^2 = 1/9$ (b) 1.4 times
35. (a) 100 dB (b) 60 dB (c) −30 dB
37. (a) increases but will not double. Doubling the power will double the intensity but not the intensity level. (b) 96 dB; 99 dB
39. (a) 3.72×10^{-4} W/m²; 1.00×10^{-1} W/m² (b) 9.55×10^{-3} W/m²; 6.03×10^{-2} W/m²
41. (a) 63 dB (b) 83 dB (c) 113 dB
43. (a) 10^{-3} W/m², 10^{-8} W/m² (b) $I_M/I_L = 10^5$
45. 10 bands
47. $I_B = 0.56I_A$, $I_C = 0.25I_A$, $I_D = 0.17I_A$
49. (a) 3.2×10^{-3} W/m² (b) 16
51. 10^5 bees
53. (b)
55. (a) no (b) increasing frequency
57. The varying sound intensity is caused by the interference effect. At certain locations there is constructive interference and at other locations, there are destructive interference.
59. moving away from us
61. 0.172 m
63. (a) both (1) and (3) because the beat frequency only measure the frequency difference between the two and it does not specify which frequency is higher. So the frequency of the violin can be either higher or lower than that of the instrument. (b) 267 Hz and 261 Hz
65. (a) Since the heard frequency is higher than the siren frequency, the person is moving toward the siren. (b) 13.7 m/s
67. 3.3 Hz
69. 90°
71. (a) 1.74 (b) 555 m/s
73. 28 m/s
75. (b)
77. (a) The snow absorbs sound so there is little reflection. (b) In an empty room, there is less absorption. So the reflections die out more slowly and therefore the sound sounds hollow and echoing. (c) Sound is reflected by the shower walls and standing waves are set

up, giving rise to more harmonics and therefore richer sound quality.

79. no

81. the closed end must be a node

83. 510 Hz

85. **(a)** does not exist, only odd harmonics **(b)** 0.30 m

87. **(a)** The position at the mouthpiece is an antinode because it has the maximum vibration. **(b)** 0.655 m **(c)** 0.390 m

89. 1.92×10^3 Hz

91. $l_2 = 1.44 l_1$

93. speed of sound

95. 3.9×10^{-7} J

97. 38 vibrations

99. 755 Hz approaching and 652 moving away

101. 1.1 m

103. **(a)** 566 Hz **(b)** 443 Hz

Chapter 15

1. **(c)**

3. **(a)** because attractive and repulsive forces can be produced by different combinations of just two types of charges. **(b)** no effect as it is simply a sign convention.

5. decrease; increase

7. yes, not necessarily (polarization)

9. 3.1×10^{14} electrons

11. **(a)** positive because of the conservation of charge **(b)** $+4.8 \times 10^{-9}$ C, 2.7×10^{-20} kg

13. positive as the leaves are charged by polarization when the positively charged fur is brought near an electroscope

15. No, the wall is neutral but polarized.

17. No, charges simply reorient.

19. induction; opposite charges

21. **(d)**

23. Objects are electrically neutral.

25. Force depends on the inverse square of the distance between the charges.

27. **(a)** 1 **(b)** 1/4 **(c)** 1/2

29. **(a)** 5.8×10^{-11} N **(b)** zero

31. 2.24 m

33. **(a)** 50 cm **(b)** 50 cm

35. **(a)** $x = 0.25$ m **(b)** nowhere **(c)** $x = -0.94$ m for $\pm q_3$

37. 3.6×10^{-47} N, 2.3×10^{39}

39. **(a)** 96 N, 39° below $+x$ axis **(b)** 61 N, 84° above $-x$ axis

41. **(c)**

43. **(b)**

45. by the relative density or spacing of the field lines

47. If a positive charge is at center of the spherical shell, the electric field is *not* zero inside. The field lines run radially outward to the inside surface of the shell where they stop at the induced negative charges on this surface. The field lines reappear on the outside shell surface (positively charged), and continue radially outward as if emanating from the point charge at center. If the charge were negative, the field lines would reverse their directions.

49. Yes, this is possible, for example, when the electric fields created by the two charges

are equal in magnitude and opposite in direction at a location.

51. 2.0×10^5 N/C

53. 1.2×10^{-7} m

55. 1.0×10^{-7} N/C upward; 5.6×10^{-11} N/C downward

57. $\mathbf{E} = (2.2 \times 10^5 \text{ N/C}) \, \hat{\mathbf{x}} + (-4.1 \times 10^5 \text{ N/C}) \, \hat{\mathbf{y}}$

59. 5.4×10^6 N/C toward the charge of $-4.0 \, \mu$C

61. 3.8×10^7 N/C in the $+y$ direction

63. **(a)** the separation of the plates because $E = 4\pi kQ/A$ **(b)** 15 μC/m^2

65. $\mathbf{E} = (-4.4 \times 10^6 \text{ N/C}) \, \hat{\mathbf{x}} + (7.3 \times 10^7 \text{ N/C}) \, \hat{\mathbf{y}}$

67. **(b)**

69. **(b)**

71. no. There are conduction electrons inside a conductor. However, the protons inside neutralize these conduction electrons.

73. **(a)** negative due to induction **(b)** zero (c) $+Q$ **(d)** $-Q$ **(e)** $+Q$

75. **(a)** zero **(b)** kQ/r^2 **(c)** zero **(d)** kQ/r^2

77.

79. zero

81. The net charges are equal and opposite in sign.

83. 10 field lines entering (negative)

85. no, just more positives than negatives: net positive charge

87. **(a)** 3.1×10^{12} electrons **(b)** 1.1×10^{-13} %

89. **(a)** 1.6×10^{-7} s **(b)** $x = -0.25$ m; $y = 0$

91. $+3.0 \, \mu$C on the $-y$ axis or $-3.0 \, \mu$C on the $+y$ axis

93.

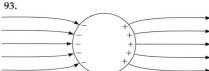

97. Two charges ($+$ and $-$) would produce equal and opposite electric fields that cancel out.

99. 4.9 mm

Chapter 16

1. **(d)**

3. **(a)** Electrical potential is the electrostatic potential energy *per unit charge*, i.e., $V = U/q_o$. **(b)** no difference

5. Approaching a negative charge means moving towards a region of lower electric potential that is losing potential. Positive charges tend to move towards regions of lower potential, thus losing potential energy and gaining kinetic energy (speeding up).

7. Electric field is related to the change in potential in space. If there is no change in potential, the electric field is zero.

9. No, it does not. Since $W = q_o \Delta V$, if $\Delta V = 0$, $W = 0$.

11. **(a)** 2.7 μC **(b)** positive to negative

13. 1.4×10^6 m/s

15. **(a)** 1.3×10^6 m/s, downward **(b)** It loses potential energy.

17. **(a)** 3 because electric potential is inversely proportional to the distance, $V = kq/r$ **(b)** 0.90 m **(c)** -6.7 kV

19. **(a)** gains 6.2×10^{-19} J **(b)** loses 6.2×10^{-19} J **(c)** gains 4.8×10^{-19} J

21. 1.1 J

23. **(a)** $+0.27$ J **(b)** no

25. -0.72 J

27. 3.1×10^5 V

29. -1.3×10^6 V

31. **(a)** It should be at a lower potential because electrons have a negative charge. They move toward higher potential regions where they have lower potential energy. **(b)** 5.9×10^7 m/s **(c)** 5.9×10^{-9} s

33. **(a)**

35.

37. Yes, because there is no change in kinetic or potential energy. So the net work done is zero.

39. **(a)** cylindrical **(b)** near the inner surface **(c)** near the outer surface

41. It increases to a larger positive value.

43. The electron volt unit is gotten from $e \Delta V = q \Delta V = \Delta U_e$. It is also the kinetic energy gained by an electron when it goes through a potential difference of 1 V. So it is a unit of energy. A GeV is larger than a MeV by 1000 times.

45. no. From Exercise 16.36, equipotential surfaces cannot cross. Since electric field lines are perpendicular to the equipotential surfaces, electric field lines cannot cross either.

47. 12.6 m

49. 70 cm

51. 1.7 mm away from the positive plate, toward the negative plate

53. **(a)** concentric spheres because the electric potential depends only on the distance from the charge, $V = kq/r$. So it is radially symmetric. **(b)** $+297$ eV $= +4.76 \times 10^{-17}$ J

55. **(a)** 2.0×10^7 eV **(b)** 2.0×10^4 keV **(c)** 20 MeV **(d)** 2.0×10^{-2} GeV **(e)** 3.2×10^{-12} J

57. 6.2×10^7 m/s (proton); 4.4×10^7 m/s (alpha)

59. **(a)** 3.5 V; 1.1×10^6 m/s **(b)** 4.1 kV; 3.8×10^7 m/s **(c)** 5.0 kV; 4.2×10^7 m/s

61. $+2.8$ V

63. increase plate area and/or decrease plate separation

65. **(a)** It doubles. **(b)** It quadruples.

67. 2.4×10^{-5} C
69. 0.71 mm
71. (a) 4.2×10^{-9} C (b) 2.5×10^{-8} J
73. 2.2 V
75. (d)
77. (a) increases (b) remains the same
79. 3.1×10^{-9} C; 3.7×10^{-8} J
81. $\kappa = 2.4$
83. (b)
85. equal capacitance
87. (a) parallel (b) series
89. (a) more because the equivalent capacitance is higher and the energy drawn is proportional to capacitance. (b) 6.0μF
91. (a) $Q/3$ because $Q_t = Q_1 + Q_2 + Q_3$. Also $Q_1 = Q_2 = Q_3$ because the capacitors have the same capacitance. (b) 3.0μC (c) 9.0μC
93. max. 6.5μF; min. 0.67μF
95. C_1: 2.4μC, 6.0 V; C_2: 2.4μC, 6.0 V; C_3: 1.2μC, 6.0 V; C_4: 3.6μC, 6.0 V
97. $-kq^2/d$
99. -1.1×10^2 eV
101. (a) higher because electron has negative charge. It will experience an upward force if the potential is higher at the top.
(b) 8.4×10^{-13} V
103. 2.6×10^{-16} J

Chapter 17

1. (b)
3. (a)
5. No; terminal voltage is lower than emf.
7. (a) 3.0 V (b) 1.5 V
9. (a) 24 V (b) two 6.0-V in series, together in parallel with the 12-V
11. (a) The total voltage of identical batteries in parallel is the same as the voltage of each individual battery and the total voltage of the batteries in series is the sum of the voltages of each individual battery. Each arrangement has one parallel and one series so they have the same total voltage. (b) 3.0 V, 3.0 V
13. negative to positive
15. 0.25 A
17. (a) 0.30 C (b) 0.90 J
19. (a) None of the preceding, since current depends on both the charge and the time interval the charge flows past a certain location. (b) 2nd wire by 0.062 A
21. (a) to the left. The current due to the protons will be to the left and the current due to the electrons will also be to the left because electrons have negative charge. (b) 3.3 A
23. (a)
25. the one with the smaller slope
27. Performance is more temperature independent and compact in size.
29. The number rises with temperature.
31. (a) 11.4 V (b) 0.32 Ω
33. (a) a greater diameter. Since aluminum has a higher resistivity, its area (diameter) must be greater, if the length of the wire is the same, to have the same resistance as copper according to $R = \rho L/A$. (b) 1.29
35. 1.0 V

37. 1.3×10^{-2} Ω
39. 3.0×10^{19} electrons/s
41. The shorter carries 4 times the current.
43. 0.13 Ω
45. 4.6 mΩ
47. not ohmic because $R_1 = 65$ Ω and $R_2 = 72$ Ω
49. (a) greater. After the stretch, the length L increases and the cross-sectional area A decreases, so R increases according to $R = \rho L/A$. (b) 1.6
51. 8.2°C to 32°C
53. 0.77 A
55. (d)
57. (d)
59. The wire in the 60-W bulb would be thicker.
61. 144 Ω
63. 2.0×10^3 W
65. 1.2 Ω
67. (a) If the voltage is halved, the current is also halved. Since power is equal to voltage times current, power becomes 1/4 as large. (b) 10 W
69. (a) 4.3×10^3 J (b) 13 Ω
71. 58Ω
73. (a) 0.60 kWh (b) $0.09
75. (a) 0.15 A (b) 1.4×10^{-4} $\Omega \cdot$ m (c) 2.3 W
77. (a) 1.1×10^2 J (b) 6.8 J
79. 21 Ω
81. $R_{120}/R_{60} = 4/3$
83. $151
85. 64 C°
87. (a) 0.33 A (b) 3.6×10^2 Ω
89. It increases to 1.78 kW.
91. copper 117°C or aluminum −73°C
93. 3.5×10^6 J
95. (a) 5.0×10^{-2} A (b) 5.0 V
97. (a) 4.0×10^2 Ω (b) 4.0 W (c) 4.8×10^2 J
99. (a) 1.67 A (b) 1.80×10^4 C (c) 5.40×10^4 J
101. to reduce I^2R losses. Since $P = IV$, a higher voltage means lower current if P is kept constant. The resistance of the transmission lines is approximately a constant. With a lower current, the power loss in the lines is drastically reduced since joule heating varies as the square of the current ($P = I^2R$).
103. 6.6×10^{-6} m/s

Chapter 18

1. (b)
3. No; only if all resistances are equal.
5. large; small
7. 5 Ω; it gets the most voltage for the same current
9. (a) in series, 60 Ω (b) in parallel, 5.5 Ω
11. 30 Ω
13. (a) 30 Ω (b) 0.30 A (c) 1.4 W
15. (a) 0.57 Ω (b) 6.0 V (c) 9.0 W
17. (a) Each shortened segment has a resistance of $R/2$ because resistance in proportional to length (Chapter 17). Then two $R/2$ resistors in parallel gives $R/4$. (b) $3.0 \mu\Omega$
19. (a) 1.0 A (b) 1.0 A in all (c) $P_{2\Omega} = 2.0$ W; $P_{4\Omega} = 4.0$ W; $P_{6\Omega} = 6.0$ W

(d) $P_{sum} = P_{total} = 12$ W
21. I(for all) = 1.0 A; $V_{8.0} = 8.0$ V; $V_{4.0} = 4.0$ V
23. 2.7 Ω
25. (a)

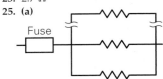

(b) 31
27. 0.25 A; 2.5 V
29. (a) 1.0 A; 0.50 A; 0.50 A (b) 20 V; 10 V; 10 V (c) 30 W
31. no, 14.6 A < 15 A
33. 100 s = 1.7 min
35. (a) 0.085 A (b) $P_{15} = 7.0$ W, $P_{40} = 2.6$ W, $P_{60} = 0.24$ W, $P_{100} = 0.41$ W
37. (a) $I_1 = I_2 = 0.67$ A, $I_3 = 1.0$ A, $I_4 = I_5 = 0.40$ A (b) $V_1 = 6.7$ V, $V_2 = 3.3$ V, $V_3 = 10$ V, $V_4 = 2.0$ V, $V_5 = 8.0$ V
39. 8.1 Ω
41. (d)
43. positive
45. The loop theorem basically says that the total voltage around a complete loop is equal to zero. Since voltage is defined as potential energy difference per unit charge, it is equivalent to that the potential energy difference around a complete loop is equal to zero. This is essentially the conservation of energy as the same point in a complete loop can only have the same potential energy value so the potential energy difference is equal to zero.
47. Around loop 1 (reverse), $-V_1 + I_3R_3 + V_2 + I_1R_1 = 0$. If we multiply -1 on both sides, it is the same as the equation for loop 1 (forward). Around loop 2 (reverse), $I_2R_2 - V_2 - I_3R_3 = 0$. Again if we multiply by -1 on both sides, it is the same as the equation for loop 2 (forward).
49. $I_1 = 1.0$ A; $I_2 = I_3 = 0.50$ A
51. $I_1 = 0.33$ A left; $I_2 = 0.33$ A right
53. $I_1 = 3.75$ A (up); $I_2 = 1.25$ A (left); $I_3 = 1.25$ A (right)
55. $I_1 = 0.664$ A left, $I_2 = 0.786$ A right, $I_3 = 1.450$ A up, $I_4 = 0.770$ A down, $I_5 = 0.016$ A down, $I_6 = 0.664$ A down
57. (c)
59. there is little charge on the capacitor
61. (a) $V_R = V_o$, $V_C = 0$ (b) $V_R = 0.37 V_o$, $V_C = 0.63 V_o$ (c) $V_R = 0$, $V_C = V_o$
63. (a) increase capacitance to increase the time constant as $\tau = RC$ (b) 2.0 MΩ
65. (a) 1.50 MΩ (b) 11.4 V
67. (a) 9.4×10^{-4} C (b) $V_C = 24$ V; $V_R = 0$
69. (a) 0 V, 2.0 μA (b) 1.7×10^{-6} C
71. (c)
73. (a) An ammeter has very low resistance, so if it were connected in parallel in a circuit, the circuit current would be very high and the galvanometer could burn out.
(b) A voltmeter has very high resistance, so if it were connected in series in a circuit, it would read the voltage of the source because it has the highest resistance (most probably)

and therefore the most voltage drop among the circuit elements.

75. small resistance. An ammeter is used to measure current when it is connected in series to a circuit element. If it has small resistance, there will be little voltage across it so it will not affect the voltage across the circuit element, therefore its current.

77. (a) multiplier resistor. Since a galvanometer cannot have a large voltage across it, the large voltage has to be across a series resistor (multiplier). **(b)** 7.4 kΩ

79. 50 kΩ

81. 0.20 mA

83. (a) zero. An ammeter is connected in series with a circuit element. If its resistance is zero, it will not affect the current through the circuit element.

85. (c)

87. No, high voltage could produce harmful current even if resistance is high.

89. safer to jump. If you step off the car one foot at a time, there will be a high voltage between your feet. If you jump, the voltage between your feet is zero because your feet will be at the same potential all the time.

91. 75 Ω

93. 6.0 Ω

95. 8.1 J/s

97. (a) $I = I_1 = 1.0$ A, $I_2 = I_4 = 0.40$ A, $I_3 = 0.20$ A **(b)** $P_1 = 100$ W, $P_2 = P_4 = 4.0$ W, $P_3 = 2.0$ W

99. $I_1 = I_2 = 0.440$ A, $I_3 = 0.323$ A, $I_4 = I_5 = 0.117$ A, $I_6 = 0.0878$ A, $I_7 = I_8 = I_9 = 0.0293$ A

103. 11.8 A

105. (a) 4.47 V **(b)** 0.447 A

107. 10 mΩ, 2.0 mΩ, and 1.0 mΩ

Chapter 19

1. (e)

3. (c)

5. Similarities: Two kinds of poles, north and south (charges, positive and negative); like poles (charges) repel and unlike poles (charges) attract.
Differences: Poles come in pairs (single charge can exist).

7. No, charge must be in motion to experience magnetic force.

9. They are either parallel or opposite.

11. not necessarily, because there still could be a magnetic force ($\theta = 0°$)

13. (a) The bottom half would have a magnetic field directed into the page and the top half would have a magnetic field directed out of the page.
(b) They are the same since centripetal force does not change the speed of the particle.

15. 2.5×10^3 m/s

17. 2.0×10^{-14} T, left, looking in the direction of the velocity

19. (a) 3.8×10^{-18} N **(b)** 2.7×10^{-18} N **(c)** zero

21. (a) 8.6×10^{12} m/s², horizontal toward right **(b)** same magnitude, opposite direction

23. magnetic force on electron beam

25. (a) 1.8×10^3 V **(b)** same voltage, independent of the charge

27. 5.3×10^{-4} T

29. (a) 4.8×10^{-26} kg **(b)** 2.4×10^{-18} J **(c)** no, work equals zero

31. (d)

33. it shortens

35. yes. Orient the current-carrying wire in a direction that is not east-west and then observe the magnetic force on the current-carrying wire. Then use the right-hand force rule to determine the direction of the magnetic field.

37. Pushing the button in both cases completes the circuit. The current in the wires activates the electromagnet, causing the clapper to be attracted and ring the bell. However, this breaks the armature contact and opens the circuit. Holding the button causes this to repeat and the bell rings continuously. For the chimes, when the circuit is completed, the electromagnet attracts the core and compresses the spring. Inertia causes it to hit one tone bar and the spring force then sends the core in the opposite direction to strike the other bar.

39. 1.2 N perpendicular to the plane of **B** and I

41. 5.0×10^{-3} T north to south

43. (a) 0 **(b)** 4.0 N/m in $+z$ **(c)** 4.0 N/m in $-y$ **(d)** 4.0 N/m in $-z$ **(e)** 4.0 N/m in $+y$

45. 0.40 N/m; $+z$

47. (a) attractive **(b)** 6.7×10^{-6} N/m

49. 0.75 N upward

51. 2.7×10^{-5} N/m toward wire 1

53. Left: 2.5 N to the left; right: 2.5 N to the right; top: 7.5 N to the top

55. zero, yes

57. (a)

59. (b)

61. halved

63. 3.8 A

65. 0.25 m

67. (a) 2.0×10^{-5} T **(b)** 9.6 cm from wire 1

69. both 2.9×10^{-6} T

71. 3.3×10^{-5} T

73. 1.0×10^{-4} T, away from the observer

75. 10 cm

77. 8.8×10^{-2} T

79. $B = \sqrt{2}\mu_o I/(\pi a)$, at 45° toward lower left wire

81. (b)

83. away from you

85. by hitting or heating it

87. 12 T

89. (b)

91. north magnetic pole

93. 0.44 T

95. 1.0×10^{-14} J

97. (a) 1.0 m **(b)** 1.5×10^{-5} s

99. 5.6×10^{-9} m/s

101. 3.0×10^{-2} m·N

Chapter 20

1. (d)

3. (d)

5. counterclockwise (in head-on view)

7. No, it depends on the rate of the flux change with time.

9. Sound waves cause the resistance of the button to change as described. This results in change in the current, so the sound waves produce electrical pulses. These pulses travel through the phone lines, and to a receiver. The receiver has a coil wrapped around a magnet, and the pulses create a varying magnetic field as they pass through the coil causing the diaphragm to vibrate and thus produce sound waves as the diaphragm vibrates in the air.

11. (a) 0 **(b)** 1.9×10^{-2} T·m² **(c)** 1.4×10^{-2} T·m²

13. 3.3×10^{-2} T·m²

15. 1.3×10^{-6} T·m²

17. 1.6 V

19. 0.30 s

21. (a) counterclockwise **(b)** 0.35 V

23. (a) at the equator, because the metal rod is parallel to the magnetic field there. **(b)** 50 μV

25. (a) 0.60 V **(b)** 0 A

27. 4.0 V

29. 0.037 T·m² for the lower incline, 0.034 T·m² for the upper incline, 0.071 T·m² for the back side, zero through all other surfaces.

31. (c)

33. The magnet moving through the coils produces a current. As the magnet moves up and down in the coil it will induce a current in the coil which will light the bulb. However, the magnet produces the current (Faraday's law of induction) at the expense of its kinetic energy and potential energy. The magnet's motion will therefore damp out (Lenz's law).

35. N, B, A, and ω, as $\mathscr{E} = \mathscr{E}_o \sin 2\pi ft$, where $\mathscr{E}_o = NBA\omega$.

37. (a) 0.057 V **(b)** 0.57 V

39. (a) Two, because the initial voltage direction was not specified, so there are two possible directions. **(b)** ±104 V

41. (a) 100 V **(b)** 0

43. 16 Hz

45. (a) lower than 44 A, because of the back emf induced when the motor turns. The back emf lowers the effective voltage on the motor and so the current is lower than 44 A. **(b)** 4.00 A

47. (a) 216 V **(b)** 160 A **(c)** 8.1 Ω

49. (c)

51. Yes, reverse the primary and secondary coils.

53. First of all, the dc voltage has to be converted to a time varying voltage. The points in older cars and electronic devices in newer cars achieved this. The coil functions as a step-up transformer to give high voltage pulses. The distributor distributes the high voltage pulses to different spark plugs to ignite the gas–air mixture.

55. (a) 16 **(b)** 5.0×10^2 A

57. 24:1

59. (a) 17.5 A **(b)** 15.7 V

61. (a) non-ideal, because the power in the secondary is lower than that in the primary, $P_s < P_p$ **(b)** 45%

63. (a) N_s/N_p is 1:20 **(b)** 2.5×10^{-2} A

65. (a) 128 kWh **(b)** $1840

67. (a) 1:2, 1:14, 1:30 **(b)** 2.0, 14, 30 **(c)** 833

69. (a) 53 W **(b)** $N_p/N_s = 200$

71. (d)

73. Infrared (heat) radiation is absorbed by clouds (water molecules), but the sunburning ultraviolet is not.

75. Radar frequencies are much higher because wavelengths are much shorter.

77. 1.8×10^2 m to 5.7×10^2 m; 2.8 m to 3.4 m; 0.34 m to 5.6 m

79. 2.6 s

81. AM: 67 m; FM: 0.77 m

83. (a) 30 A **(b)** 2.5 A

85. 0.60 m

87. once every 0.01 s

89. (a) up, according to Lenz's law **(b)** 25 mA

91. 0.75 T·m^2

93. 4.0 T/s

95. 0.33 A; 30 A

97. (a) 50 rotations per second **(b)** No, if they are ideal (lossless) the same turn ratios would work. However in reality the 50-Hz delivery would have fewer losses because the eddy currents are less due to the fact that the magnetic flux in the transformer is changing less rapidly.

Chapter 21

1. (a)

3. Both reach maximum or minimum at the same time.

5. No, voltage and current should be in phase for a resistor.

7. 7.1 A

9. 1.2 A

11. (a) 10 A **(b)** 14 A **(c)** 12 Ω

13. (a) 4.5 A, 6.3 A **(b)** 112 V, 158 V

15. $V = (170\text{ V}) \sin(119\pi t)$

17. 0.33 A; 0.47 A

19. 2.4×10^2 W

21. (a) 60 Hz **(b)** 1.4 A **(c)** 1.2×10^2 W **(d)** $V = (120\text{ V}) \sin(380t)$ **(e)** $P = (240\text{ W}) \sin^2(380t)$

23. (c)

25. For a capacitor, the *lower the frequency*, the longer the charging time in each cycle. If the frequency is very low (dc), then the charging time is very long so it acts as an ac open circuit. For an inductor, the *lower the frequency*, the more slowly the current changes in the inductor. If the frequency is very low (dc), then the current in the inductor won't change for a long time so it acts as a short circuit.

27. One cycle is equal to 2π radians or 360°. "The voltage across an inductor leads the current by 90°" means the voltage leads the current by a quarter-cycle or reaches maximum a quarter-cycle ahead of the current. "The voltage across a capacitor lags the current by 90°" means the voltage lags the current by a quarter-cycle or reaches maximum a quarter-cycle after the current.

29. Zero, current leads voltage by 90°.

31. $1.3 \times 10^3\,\Omega$

33. (a) 19 Ω **(b)** 6.4 A **(c)** Voltage leads current by 90°.

35. an increase of 60%

37. 255 Hz

39. (a) 90 V **(b)** Voltage leads current by 90°.

41. 4.4 μF

43. (d)

45. (d)

47. no, $\phi = 90°$ so $\cos\phi = 0$

49. (a) 1.7×10^2 Ω **(b)** 2.0×10^2 Ω

51. (a) 38 Ω; 1.1×10^2 Ω **(b)** 1.1 A

53. (a) negative, because this is a capacitive circuit **(b)** −27°

55. (a) in resonance, because $X_L = X_C$ so $Z = R$ **(b)** 72 Ω

57. 50 W

59. 5.3×10^{-11} F

61. ab: 1.3 A, ac: 1.2 A, bc: 4.0 A, cd: 1.8 A, bd: 1.6 A, ad: 2.9 A

63. 13 A

65. $V_R = 12$ V; $V_L = 2.7 \times 10^2$ V; $V_C = 2.7 \times 10^2$ V

67. (a) equal to 25 Ω. At resonance, $X_L = X_C$, so $Z = R$. **(b)** 362 Ω

69. 30%

71. 0.35 A

73. 1.7×10^{-12} F

75. 37°

77. From $X_C = 1/(2\pi f C)$ and $X_L = 2\pi f L$, we can see that the capacitor opposes current more strongly at lower frequencies and the inductor opposes current more strongly at high frequencies. In Fig. 21.18a, the inductor in series with R_L filters out the high frequency current and so only the low frequency current reaches R_L. In Fig. 21.18b, the capacitor in series with R_L filters out the low frequency current and so only the high-frequency current reaches R_L.

Chapter 22

1. (c)

3. (d)

5. Water surface has waves, only certain segments of the surface are oriented so as to reflect the Sun's image toward us at any instant. It is like reflections from a series of little mirrors oriented in different directions. The reflecting facets change almost randomly from moment to moment as the water surface undulates.

7. No; if it is reflecting 100%, then you cannot see its surface but the reflections of other objects.

9. After rain, water fills the unevenness of the road and the reflection off the road is specular so images of buildings, trees, and so on are formed. When the road is dry, the unevenness of the road cause diffuse reflection so there are no images of buildings, tree, etc.

11. 58°

13. 35°

15. (a) $\tan^{-1} w/d$ **(b)** 27°

17. 12 m

19. 90°, any θ_{i1}

21. (d)

23. yes, no, yes

25. (b)

27. This severed look is because the angle of refraction is different for air-glass interface than the angle of refraction for water-glass interface. The top portion refracts from air to glass and the bottom portion refracts from water to glass. This is different from what's on Fig. 22.12b. In that figure, we see the top portion directly in air and the bottom portion in water through refraction from water to air. The angle of refraction made the pencil appears to be bent.

29. The laser beam has a better chance to hit the fish. The fish appears to the hunter at a location different from its true location due to refraction. The laser beam obeys the same law of refraction and retraces the light the hunter sees to the fish. The arrow goes into the water in a near-straight line path.

31. It is so useful due to some of its unique properties. An fiber optic is very flexible, small in size, and can be inserted in almost everywhere, compared to the old mechanical based devices with mirrors and lenses.

33. 26% greater in zircon

35. 27°

37. (a) diamond to air, because diamond has a higher index of refraction **(b)** 24°

39. 47°

41. 6.5×10^{14} Hz; 2.8×10^{-7} m

43. $\frac{16}{15}$

45. (a) This is caused by refraction of light in the water-air interface. The angle of refraction in air is greater than the angle of incidence in water so the object immersed in water appears closer to the surface.

47. (a) both, because $\theta_c \geq \sin^{-1}(n_2/n_1)$ **(b)** 1.41; 1.88

49. 3.2 cm

51. 75%

53. seen for 40° but not for 50°; $\theta_c = 49°$

55. 43°

57. (a) no, $\theta_c = 38.7°$ **(b)** transmitted, $\theta_c = 48.6°$

59. 2.0 m

61. (a) 12.5° **(b)** 26.2°

63. (b)

65. two refractions and two dispersions in a prism

67. To see a rainbow, the light has to be behind you. Actually, you won't see a primary rainbow if the Sun's angle above the horizon is greater than 42°. Therefore you cannot look up to find a rainbow thus you cannot walk under a rainbow.

69. (a) $\theta \approx 0°$ **(b)** No (no); the speeds are different.

71. Red is transmitted and blue is internally reflected.

73. (a) 21.7° **(b)** 0.22° **(c)** 0.37°

75. (b) 58° **(c)** 30°

77. 1.64

79. 0°: no direction change

81. **(a)** 49° **(b)** 1.5 **(c)** 1 **(d)** 42°

Chapter 23

1. **(b)**

3. **(a)** Reflections from the window are seen clearly against a dark background, During the day, light passes both ways through the pane, and although some is reflected, it is difficult to see the reflection due to the light coming through. At night, there is little light coming through the pane, so the reflections are seen much more clearly. **(b)** The two images are due to reflections on both sides of the pane glass, producing two similar images. **(c)** This works on a combination of half-silvering and bright light on one side and dark on the other. For example, at night people inside the house cannot see things outside because there is little light coming through the window from the outside and they see their own reflections.

5. **(a)** The left-right reversal is an apparent one. It is caused by the front-back reversal. Right and left are directional senses like clockwise and counterclockwise rather than fixed directions referenced to a coordinate system. **(b)** No. If you rotate your body 90°, it is still a left-right reversal.

7. When viewed by a driver through a rear view mirror, the right-left reversal property of the image formed by a plane mirror will make it read "AMBULANCE."

9. infinite, because it cannot focus light to a point

11. **(a)** 0.80 m **(b)** 5.0 cm **(c)** +1.0

13. **(a)** 1.5 m behind the mirror **(b)** 1.0 m/s

15. **(a)** multiple images **(b)** 3.0 m and 13 m behind north mirror; 5.0 m and 11 m behind south mirror

19. **(d)**

21. **(a)** The plane mirror gives a large view of the area immediately around that side of the truck. The small convex mirror gives a wide-angle perspective of the road in back of both sides of the truck (but the image is smaller). **(b)** These are convex mirrors that give a better field of view, but also images are smaller than objects (so the image distances are smaller than object distances) and hence appear closer than they actually are. **(c)** Yes, it can be considered as a converging mirror, because it collects a large amount of radio waves and focuses them onto a small area.

23. **(a)** The image is smaller than the object and it is possible to "see your full body in 10 cm" in a diverging mirror. **(b)** As the ball swings toward the mirror and approaches the focal point, the image enlarges. An enlarged image appears to be closer to our eyes and so it appears to move toward the observer and therefore produces the effect of appearing to "jump" out of the mirror as the ball swings through the focal point.

25. No, convex mirror produces only reduced images.

27. 5.0 cm

29. $d_i = -30$ cm; $h_i = 9.0$ cm; image is virtual, upright, and magnified.

31. **(a)** $f = 10$ cm, $R = 20$ cm **(b)** 3.0 cm

37. 120 cm

39. 2.3 cm

41. **(a)**

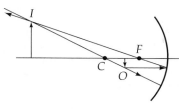

(b) $d_i = 60$ cm, $M = -3.0$, real and inverted

43. 10 m

45. $d_i = 60$ cm; image is real, inverted, and magnified.

47. 5.0 cm, 15 cm

49. yes: 13 cm; 27 cm

51. **(c)**

53. When the fish is inside the focal point, the image is upright, virtual, and magnified.

55. locate the image of a distant object; no

57. $d_i = 12.5$ cm; $M = -0.25$

59. 22 cm

61. **(a)** $d_i = -6.4$ cm; $M = +0.64$ **(b)** $d_i = -10.5$ cm; $M = +0.42$

63. **(a)** 18 cm **(b)** 6.0 cm

65. 8.1 cm

67. **(a)** 20 cm **(b)** $M = -1$

69. **(a)** $d = 4f$ **(b)** approaches 0

71. **(a)** -18 cm **(b)** -63 cm

73. **(a)** 4.6 cm **(b)** 0.80 cm, inverted

75. 8.6 cm; real, inverted, $M_{total} = -0.86$

79. ∞

81. no

83. -0.50 m

85. -0.70 D

87. 0.55 mm

89. $d_i = 37.5$ cm, $h_i = 3.0$ cm; real and inverted

91. The image formed by the converging lens is at the mirror. This image is the object for the diverging lens. If the mirror is at the focal point of the diverging lens, the rays refracted after the diverging lens will be parallel to the axis. These rays will be reflected back parallel to the axis by the mirror and will form another image at the mirror. This second image is now the object for the converging lens. By reversing the rays, a sharp image is formed on the screen located where the original object is. Therefore the distance from the diverging lens to the mirror is the focal length of the diverging lens.

93. 14 cm

95. 20 cm on object side of first lens; virtual, inverted, $M_{total} = -1.0$

97. **(a)** convex **(b)** 6.25 cm

99. 85 cm

101. **(a)**

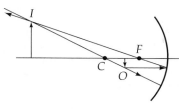

(b) 60 cm

Chapter 24

1. **(b)**

3. blue, since angle increases with wavelength

5. Path-length difference changes; so does interference.

7. The spacing between the maxima would increase.

9. 3.4%

11. **(a)** 0.80 m **(b)** 3.2 m

13. 1.3°

15. **(a)** 440 nm **(b)** 4.40 cm

17. **(a)** decrease, since it is inversely proportional to the distance between slits **(b)** 1.9 mm

19. **(a)** $\lambda = 402$ nm, violet **(b)** 3.45 cm

21. 450 nm

23. **(a)**

25. all wavelengths except bluish purple

27. destructive interference due to 180° phase shift

29. **(a)** 30λ **(b)** destructively

31. 54 nm

33. 160 nm; 200 nm

35. **(a)** 158.2 nm **(b)** 316.4 nm

37. 1.51×10^{-6} m

39. **(a)**

41. a second diffraction pattern perpendicular to the first

43. more spread-out diffraction pattern, because d is small

45. **(a)** 5.4 cm **(b)** 2.7 cm

47. **(a)** 4.3 mm **(b)** microwave

49. 4.9 mm

51. blue: 18.4°, 39.2°; red: 30.7°, not possible

53. **(a)** 2.44×10^3 lines/cm **(b)** 11

55. **(a)** blue, because it has a shorter wavelength **(b)** 15.4°

57. $\theta_{3v} = \theta_{1y}$

59. **(d)**

61. Looking through the lens of one pair while rotate the lens of the other pair in front of the first pair. If the intensity changes as the lenses are rotated, both pairs are polarized. If the intensity does not change, either both are not polarized or one of the two pairs are not polarized.

63. **(a)** twice **(b)** four times **(c)** none **(d)** six times

65. No; sound is a longitudinal wave.

67. The numbers appear and disappear as the sunglasses are rotated.

69. 1.6

71. **(a)** $0.50I_o$ **(b)** $0.375I_o$

73. 55°

75. 57.2°

77. no; $\sin \theta > 1$ which is impossible

79. Blue scatters more than red so red is seen.

81. 1.74
83. 56°
85. (b) $\lambda/(2n)$
87. 0.37°
89. 560 nm
91. 489 nm

Chapter 25

1. (b)
3. distant: large radius; close: small radius
5. The pre-flash occurs before the aperture is open and the film exposed. The bright light causes the iris to reduce down (giving a small pupil) so that when the second flash comes momentarily, you don't have a wide opening through which you get the red-eye reflection from the retina.
7. (a) nearsighted (b) farsighted (c) for (a), diverging; for (b), converging
9. (a) +5.0 D (b) −2.0 D
11. (a) converging, as the person is farsighted (b) +2.0 D
13. diverging, −0.50 D
15. (a) Take them out, because the positive contact lens will alter the person's far point. (b) +3.0 D
17. (a) converging (b) +3.3 D
19. +3.0 D
21. 21 cm
23. right: +1.42 D, −0.46 D; left: + 2.16 D, −0.46 D
25. (d)
27. Inside; when the object in inside the focal length, the image is virtual, upright, and magnified.
29. (a) 2.3× (b) 2.5 ×
31. (a) 2.7 × (b) 1.7×
33. 1.9×
35. (a) 1.8× (b) 1.3×
37. (a) the one with the shorter focal length (b) −360×
39. (a) −340× (b) 3900%
41. 25×
43. 3.6×
45. (b)
47. No, the whole star can still be seen. The obstruction will reduce the intensity or brightness of the image.
49. The one with the longer focal length; the magnification of the telescope is proportional to the focal length of the objective. $(m = -f_o/f_e)$.
51. (a) −4.0× (b) 75 cm
53. (a) −110× (b) 88.3 cm
55. (a) 60.0 cm and 0.80 cm; 40.0 cm and 0.90 cm (b) −75×;−44×
57. $\theta_i \approx \tan\theta_i = \dfrac{y_i}{f_e}$ and $\theta_o = \dfrac{y_i}{f_o}$.
So $m = \dfrac{\theta_i}{\theta_o} = \dfrac{y_i/f_e}{y_i/f_o} = \dfrac{f_o}{f_e}$.
59. (a)
61. Smaller; minimum angle of resolution corresponds to higher resolution because smaller details can be resolved at smaller angle of resolution.
63. 1.2×10^{-3} rad
65. 8.18×10^{-8} rad

67. (a) Blue, because the minimum angle of resolution is proportional to wavelength, $\theta_{min} = 1.22\lambda/D$, and blue has the shorter wavelength. (b) 9.6×10^{-5} rad
69. 18 km
71. (a) (1) B, (2) A; for best magnification, the high f_o/f_e ratio is desired. For best resolution, the large objective diameter should be used. (b) −110×, 8.9×10^{-7} rad
73. (a) 5.55×10^{-5} rad (b) blue (c) 33.3%
75. (d)
77. Since white is obtained by adding colors, it cannot be obtained by the subtractive method. That method subtracts colors, and the one we see is the one that is not absorbed. Black objects do not absorb all wavelengths of light. We see the objects because we perceive the extremely faint light as black. (Think of twilight vision.)
79. green
81. 1.6×10^{-2} rad
83. 6.7 m
85. 1.1×10^{-2} rad
87. 2.1×
89. 83 cm
91. −500×
93. objects as large as typical houses

Chapter 26

1. (d)
3. (a)
5. The centripetal acceleration due to the rotation of the Earth is very small for most purposes, so we can ignore it and therefore we can treat the Earth as an inertial frame of reference.
7. (a) 3.38 s (b) 3.58 s
9. 55 m/s and 45 m/s
13. (c)
15. No, any inertial observer cannot measure an object with mass to travel at a speed equal to or greater than c.
17. (a) An inertial frame O' moving from A to B could see the two flashes simultaneously. (b) From A to B.
19. no. From the boy's view, the barn is moving at the same speed so it would appear to contract and be even shorter than 4.0 m
21.

Stationary

At very high speed (0.5c)

23. 23 min
25. 14.1 m
27. 0.998c
29. (a) a longer time due to time dilation (b) 7.2 years and 5.7 years
31. 5.3 m; assume relative velocity is parallel to length of pole
33. 0.87c
35. 2.14×10^{-14} m

37. 0.60c
39. (b)
41. no. To get to the speed of light means the object must have an infinite amount of energy, since when $v = c$, $\gamma = \infty$, and $E = \infty$. Since it takes an infinite amount of work to do this, no force can accomplish it.
43. Yes, because the kinetic energy is much less than the rest-energy of the electron (2 keV ≪ 511 keV). No, since the kinetic energy is considerably greater than the rest-energy of the electron (2 MeV ≫ 0.511 MeV).
45. (a) 0.985c (b) 2.50 MeV (c) 1.56×10^{-21} kg · m/s
47. 0.15 kg
49. 1.6×10^{24} J, or 150 000 times more!
51. 79 keV
53. (a)

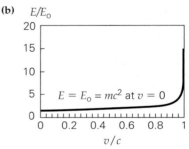

(b) E/E_o

55. (a) 0.92c (b) 1.4×10^3 MeV
57. 1.43×10^5 years
59. (a) 1.13 MeV (b) 1.64 MeV
63. (a)
65. (c)
67. (a) larger, as it has more mass and the radius of the event horizon is directly proportion to the mass, $R = 2GM/c^2$. (b) 8.9 mm and 2.8 m
69. (a) 3.37×10^{30} kg (b) 6.44×10^{18} kg/m³
71. 0.43c
73. −0.154c (toward Earth)
75. (a) 0.988c to the left (b) 0.988c to the right
77. 4.8 h
79. length, 38 m; height, 2.5 m; width, 2.0 m
81. 50 min
83. 0.68c

Chapter 27

1. (d)
3. No, it actually emits 2^4 or 16 times the total radiation since the radiation rate depends on T^4.

5. (a)

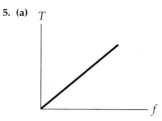

(b)

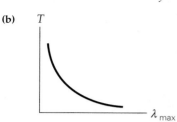

7. 1.06×10^{-5} m, 2.83×10^{13} Hz

9. 3.1×10^{13} Hz

11. 2.56×10^{-20} J

13. (b)

15. The energy packets of radio photons arrive in such quantity and so quickly that when they are converted to sound our ear cannot distinguish between discrete arrivals. It is somewhat analogous to a movie, where the picture is actually single frames passing in front of us at a rapid speed, but we see it as a smooth and continuous moving picture. Likewise, although the radio signals are single frames, we hear the radio signals continuously.

17. The energy of radiation is proportional to its intensity, according to the wave theory, and its frequency, according to the particle theory. It takes a certain amount of energy to eject a photoelectron. Since only the frequency, not the intensity, matters in this case, it favors the particle theory.

19. No, the number of photoelectrons ejected depends on the intensity of light.

21. $\lambda = 6.0 \times 10^{-11}$ m; X ray

23. (a) 1.32×10^{-18} J **(b)** 8.27 eV

25. 3.0 eV

27. (a) 6.7×10^{-34} J·s **(b)** 2.9×10^{-19} J

29. 6.8×10^{14} Hz

31. 2.48 eV

33. (a) no photoelectrons
(b) $hf = 2.84 \times 10^{-19}$ J $< \phi_o = 3.74 \times 10^{-19}$ J

35. (a) sodium
(b) $\lambda_{silver} = 262$ nm; $\lambda_{sodium} = 504$ nm

37.
K_{max} (eV)

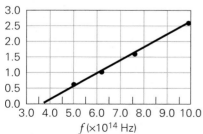

$h \approx 6.6 \times 10^{-34}$ J·s, $\phi_o = 1.5$ eV

39. (d)

41. 200 000

43. 4.86×10^{-3} nm

45. 0.281 nm

47. (a) less than 5.0 keV but not zero. According to the conservation of energy and momentum, the free electron will carry some kinetic energy and the scattered photon will also be moving away after the scattering.
(b) 40 eV

49. 18.5°

51. (d)

53. The theory does not include electron–electron interactions.

55. The recoiling atom carries some kinetic energy away.

57. It increases (becoming less negative) by a factor of 4.

59. (a) ultraviolet **(b)** visible (red)

61. (a) 6.89×10^{14} Hz **(b)** 8.21×10^{14} Hz

63. $n \approx 100$

65. (a) -1.51 eV **(b)** -0.378 eV **(c)** -0.136 eV

67. (a) increase but not double **(b)** 15.4 eV

69. (a) $n = 2$ to $n = 1$ **(b)** $\Delta E_{53} = 0.967$ eV; $\Delta E_{62} = 2.97$ eV; $\Delta E_{21} = 10.2$ eV.

71. (a) 54.4 eV **(b)** 122 eV

75. (a) potential is -27.2 eV and kinetic is $+13.6$ eV **(b)** $|U| = 2K$

77. Through stimulated emission, the photon from a laser is caused by electron transitions between two discrete energy levels and so there are only a few colors (frequencies). A light bulb emits thermal radiation at many different frequencies.

79. Spontaneous emission is really a random emission. Electrons jump from a higher energy orbit to a lower energy orbit and a photon is released in the process. Stimulated emission is kind of an induced or controlled emission. The electron in the higher energy orbit can jump to a lower energy orbit when a photon of energy equal to the difference of the energy between the two orbits is introduced. In this respect, the process is not random but is totally controllable. Once there are enough electrons in the higher energy orbit, the photons are introduced and all the electrons will eventually jump to the lower energy orbit. For each introduced photon, two photons of the same frequency are produced (the other one is from the emission of the electron transition).

83. (a) 12.1 eV **(b)** 13.1 eV

85. 41°

Chapter 28

1. (c)

3. Its wavelength is too short compared to everyday dimensions.

5. It is shorter, since the higher potential difference causes a higher momentum.

7. (a) It will have a longer wavelength, since the de Broglie wavelength is inversely proportional to the mass, $\lambda = h/(mv)$.
(b) electron: 7.28×10^{-6} m, proton: 3.97×10^{-9} m

9. $\lambda_e/\lambda_p = 43$

11. 1.5×10^4 V

13. (a) It will decrease, since the proton will gain speed from the potential difference and the de Broglie wavelength is inversely proportional to speed, $\lambda = h/(mv)$.
(b) -53% (a decrease)

15. 3.7×10^{-63} m

17. 24 V

19. (b)

21. (a) Wave function

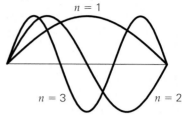

$n = 1$

$n = 3$ $n = 2$

Probability Density

$n = 3$ $n = 1$ $n = 2$

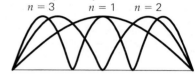

(b) $x = L/2$

23. (c)

25. total energy and orbital radius

27. The periodical table groups elements according to the values of quantum numbers n and l. Within a group, the elements have the same or very similar electronic configurations for the outmost electrons.

29. (a) 2 **(b)** 14

31. (a) $\ell = 2$ **(b)** $n = 3$

33. (a)

Na has 11 electrons

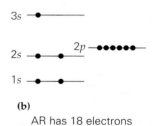

(b)

AR has 18 electrons

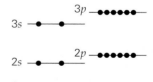

35. (a) $1s^2\, 2s^2\, 2p^1$
(b) $1s^2\, 2s^2\, 2p^6\, 3s^2\, 3p^6\, 4s^2$
(c) $1s^2\, 2s^2\, 2p^6\, 3s^2\, 3p^6\, 3d^{10}\, 4s^2$
(d) $1s^2\, 2s^2\, 2p^6\, 3s^2\, 3p^6\, 3d^{10}\, 4s^2\, 4p^6\, 4d^{10}\, 5s^2\, 5p^2$

37. It would have a $1s^3$ and would be the first closed shell inert gas. With three spins orientations, s states can contain three electrons without violating the Pauli exclusion principle.

39. (b)

41. According to the uncertainty principle, the product of uncertainty in position and the uncertainty in momentum is in the order of Planck's constant. A bowling ball's diameter and mass are so large that the uncertainty in them (determined by the extremely small value of Planck's constant) is undetectable. However for an electron, with its very small mass, this is not true.

43. (a) They are the same, since both have the same momentum.
(b) both 0.21 m
45. 2.1×10^{-29} m/s
47. 1.1×10^{-27} J
49. (a) smaller. According to the uncertainty principle, $\Delta E \Delta t \geq h/(2\pi)$, the greater the lifetime, Δt, the smaller the energy difference, therefore the smaller the width of spectral line.
(b) 10^4
51. (a)
53. (a) yes **(b)** 2.5×10^{20} Hz
55. 1.9 GeV
57. 84 V
59. (a) 4.0×10^{-21} kg·m/s **(b)** 1.7×10^{-13} m
63. $x = 0$

Chapter 29

1. (c)
3. (d)
5. They have the same number of protons, different number of neutrons.
7. (a) ^{1}H, ^{2}D, ^{3}T
(b) H_2O, D_2O, T_2O, HDO, HTO, DTO
(c) T_2O, HTO, DTO
9. $^{41}_{19}$K
11. 92 p, 146 n, and 92 e
13. (a) $R_{\text{He}} = 1.9 \times 10^{-15}$ m; $R_{\text{Ne}} = 3.3 \times 10^{-15}$ m; $R_{\text{Ar}} = 4.1 \times 10^{-15}$ m; $R_{\text{Kr}} = 5.3 \times 10^{-15}$ m; $R_{\text{Xe}} = 6.1 \times 10^{-15}$ m; $R_{\text{Rn}} = 7.3 \times 10^{-15}$ m
(b) They are roughly from ten thousand to fifty thousand times smaller than an atom.
15. No, this is not a violation. In a negative beta decay for example, a proton, which is a nucleon, is created to make up for the loss of a neutron.
17. (a) Since tritium, $^3_1$H, has an extra neutron compared to the stable deuterium ($^2_1$H), we expect β^- decay. **(b)** Helium-3; yes, it's stable.
19. (a) $^{237}_{93}$Np $\rightarrow$ $^{233}_{91}$Pa + $^4_2$He
(b) $^{32}_{15}$P $\rightarrow$ $^{32}_{16}$S + $^0_{-1}$e
(c) $^{56}_{27}$Co $\rightarrow$ $^{56}_{26}$Fe + $^0_{+1}$e
(d) $^{56}_{27}$Co + $^0_{-1}$e $\rightarrow$ $^{56}_{26}$Fe
(e) $^{42}_{19}$K* $\rightarrow$ $^{42}_{19}$K + γ
21. $^{213}_{83}$Bi
23. (a) $^4_2$He **(b)** $^0_{-1}$e **(c)** $^{102}_{39}$Y **(d)** $^{23}_{11}$Na* **(e)** $^0_{-1}$e
25. α to ^{233}Pa; β^- to ^{233}U; α to ^{229}Th; α to ^{225}Ra; β^- to ^{225}Ac; α to ^{221}Fr; α to ^{217}At; α to ^{213}Bi; α to ^{209}Tl; β^- to ^{209}Pb; β to ^{209}Bi.

Or
α to ^{233}Pa; β^- to ^{233}U; α to ^{229}Th; α to ^{225}Ra; β^- to ^{225}Ac; α to ^{221}Fr; α to ^{217}At; α to ^{213}Bi; β^- to ^{213}Po; α tò ^{209}Pb; β^- to ^{209}Bi.
27. (c)
29. No, the decay is exponential.
31. (a) 67.6μCi **(b)** 2.50×10^6Bq
33. (a) Since 3 h is 3 half-lives, then $(\frac{1}{2})^3$ or 1/8 would be left. **(b)** $1/2^{24}$ or approximately 6×10^{-6} % of the original
35. 28.6 years
37. 91.7%
41. ^{104}Tc
43. 0.98μg
45. (a) 6.75×10^{19} nuclei
(b) 2.11×10^{18} nuclei
47. 9.2 mg
49. 7.6×10^6 kg
51. (d)
53. ^{1}H and ^{2}H: stable; ^{3}H: unstable
55. (a) $^{17}_8$O **(b)** $^{42}_{20}$Ca **(c)** $^{10}_5$B **(d)** approximately the same
57. (b) and **(d)**; others are odd–odd
59. 2.013553 u
61. (a) 92.2 MeV **(b)** 7.68 MeV/nucleon
63. deuterium (1.11 MeV/nucleon); tritium (2.83 MeV/nucleon)
65. 104.7 MeV
67. 10.1 MeV
69. 7.59 MeV/nucleon
71. (a) two alphas **(b)** two alphas **(c)** yes
73. (d)
75. No, the relative biological effectiveness (RBE) values are different.
77. 36μCi
79. yes (3.7 rem in two months)
81. $^{75}_{33}$As, $^{76}_{34}$Se
83. (a) 28.3 MeV **(b)** 4.001501 u
85. (a) 3.4×10^{22} nuclei **(b)** 1.8×10^{16} nuclei **(c)** after 10 min: 9.7×10^{19} decays/min; after 1 h: 5.2×10^{15}decays/min
87. (a) $^0_{-1}$e **(b)** $^{222}_{86}$Rn **(c)** $^{237}_{94}$Pu **(d)** γ **(e)** $^0_{+1}$e
89. (a) 1.80×10^3MeV **(b)** 237.997821 u
91. (c) and **(d)**
93. $^3_2$He
95. (b) 2.3×10^{17}kg/m^3
97. (a) It is greater, since the nucleus occupies a much smaller volume than the atom and both have about the same mass. **(b)** The nuclear density is greater by 10^{14} times. **(c)** no

Chapter 30

1. (d)
3. Nuclei are changed in a nuclear reaction.
5. The reactants, because energy is released in an exoergic reaction.
7. (a) $^{14}_7$N **(b)** $^2_1$H **(c)** $^{30}_{15}$P **(d)** $^{17}_8$O **(e)** $^{10}_5$B
9. (a) $^{14}_7$N* **(b)** $^{14}_7$N* **(c)** $^{31}_{15}$P* **(d)** $^{18}_9$F* **(e)** $^{14}_7$N*
11. (a) $^{22}_{11}$Na, no($Q < 0$)
(b) $^{222}_{86}$Rn, yes ($Q > 0$) **(c)** $^{12}_6$C, no ($Q < 0$)

13. (a) Positive, because this is a decay (spontaneous) process. **(b)** +4.28 MeV
15. 4.36 MeV
17. (a) endoergic **(b)** 1.53 MeV
19. exoergic; $Q = +6.50$ MeV
21. 5.38 MeV
23. (a) greater **(b)** 3.05
25. (a) 0.24 b **(b)** 0.66 b **(c)** 1.6 b **(d)** 1.7 b
27. (d)
29. (d)
31. fission: heavy nuclei split into lighter nuclei; fusion: light nuclei fuse into heavier nuclei
33. The reaction products collide with the core materials and coolant and in the process heat the water which converts it to steam that powers a turbine.
35. (a) 2.22 MeV **(b)** 12.9 MeV
37. 36 collisions
39. (c)
41. $m = 0, v = c$, spin $= \frac{1}{2}$
43. 1.41×10^{-21} kg·m/s, opposite the neutrino direction
47. 0.71 MeV
49. 9 keV
51. (b)
53. The effects of the virtual exchange particles can be predicted and these predictions confirmed experimentally (scientific method).
55. 3.5×10^{-28} kg
57. (a) It decreases, because the mass increases and the range is inversely proportional to the mass, $R = h/(2\pi mc)$.
(b) 1.98×10^{-16} m
59. 4.71×10^{-24} s
61. (a)
63. All hadrons contain quarks and/or antiquarks. Quarks are not believed to exist freely outside the nucleus.
65. (a) π^+ **(b)** K° **(c)** Σ° **(d)** Ξ^-
67. (a) udd **(b)** The neutron has no charge: $\frac{2}{3}e - \frac{1}{3}e - \frac{1}{3}e = 0$.
69. (a)
71. (a) $^7_4$Be* **(b)** $^{60}_{29}$Cu* **(c)** $^{236}_{92}$U* **(d)** $^{13}_6$C* **(e)** $^{17}_8$O*
73. 0.156 MeV
75. (a) Electron capture since β^- decay is not energetically possible. **(b)** 0.86 MeV
77. 2.35 MeV

Photo Credits

Chapter 15—CO.15 Zefa/London/Corbis/Stock Market Fig. 15.7a Jerry Wilson/Jerry Wilson Fig. 15.7c Charles D. Winters/Photo Researchers, Inc. Fig. 15.19.1b Keith Kent/Peter Arnold, Inc. Fig. 15.19.1c Philippe Wojazer/Reuters/CORBIS

Chapter 16—CO.16 Richard Megna/Fundamental Photographs Fig. 16.15 Spencer Grant/Photo Researchers, Inc. Fig. 16.18b © 1989 Paul Silverman, Fundamental Photographs

Chapter 17—CO.17 Arthur S. Aubry/Getty Images, Inc. Fig. 17.9 Tom Pantages Fig. 17.13a Ronald Brown/Arnold & Brown Fig. 17.13b1 Frank Labua/Pearson Education/PH College Fig. 17.13b2 Frank Labua/Pearson Education/PH College Fig. 17.14 NASA/Mark Marten/Photo Researchers, Inc. Fig. 17.15 Frank Labua/Pearson Education/PH College Fig. 17.13.2a Frank LaBua/Pearson Education/PH College Fig. 17.13.2b Frank LaBua/Pearson Education/PH College

Chapter 18—CO.18 Richard Megna/Fundamental Photographs Fig. 18.15a Courtesy of Cenco/Sargent-Welch/VWR International Fig. 18.18c Richard Megna/Fundamental Photographs Fig. 18.20b Frank Labua/Pearson Education/PH College Fig. 18.21b Frank LaBua/Pearson Education/PH College Fig. 18.24a Frank Labua/Pearson Education/PH College Fig. 18.24b Frank Labua/Pearson Education/PH College Fig. 18.41 M. Antman/The Image Works

Chapter 19—CO.19 Peter Menzel/Stock Boston Fig. 19.1 Courtesy of Cenco Fig. 19.3a L/Richard Megna/Fundamental Photographs Fig. 19.3b Richard Megna/Fundamental Photographs Fig. 19.3c Richard Megna/Fundamental Photographs Fig. 19.22a Richard Megna/Fundamental Photographs Fig. 19.24a Richard Megna/Fundamental Photographs Fig. 19.25a Richard Megna/Fundamental Photographs Fig. 19.7 Jerry Wilson Fig. 19.10 James Holmes/Oxford Centre for Molecular Sciences/SPL Fig. 19.20 Courtesy of Cenco/Sargent-Welch/VWR International Fig. 19.29.2 Richard Blakemore, University of New Hampshire Fig. 19.32 Pekka Parviainen/Science Photo Library/Photo Researchers, Inc. Fig. 19.29.1 Richard Blakemore, University of New Hampshire Fig. 19.29.2 Richard Blakemore, University of New Hampshire

Chapter 20—CO.20 Tim Barnwell/Stock Boston Fig. 20.14b U.S. Department of the Interior Fig. 20.14a Photo by Tim Matsui/Liaison Fig. 20.25a Larry Mulvehill/Science Source/Photo Researchers, Inc. Fig. 20.25b GCA/CNRI/Phototake NYC Fig. 20.25.1 Figure prepared by Dr. Jay Herman, Laboratory for Atmospheres, NASA - Goddard Space Flight Center. Fig. 20.9.3 Richard Megna/Fundamental Photographs

Chapter 21—CO.21 Getty Images, Inc.

Chapter 22—CO.22 Martin Harvey/Peter Arnold, Inc. Fig. 22.4b Peter M. Fisher/Corbis/Stock Market Fig. 22.5.1 E. R. Degginger/Color-Pic, Inc. Fig. 22.21 Photri/Corbis/Stock Market Fig. 22.7 Richard Megna/Fundamental Photographs Fig. 22.9 Richard Megna/Fundamental Photographs Fig. 22.11a Kent Wood/Photo Researchers, Inc. Fig. 22.12b Jerry Wilson Fig. 22.14b Ken Kay/Fundamental Photographs Fig. 22.16 Stuart Westmorland/CORBIS Fig. 22.17a © Gemological Institute of America Fig. 22.18a Courtesy of Cenco Fig. 22.19a Nick Koudis/Getty Images, Inc. Fig. 22.19b Hank Morgan/Photo Researchers, Inc. Fig. 22.20a David Parker/Photo Researchers, Inc. Fig. 22.23a SIU/Photo Researchers, Inc. Fig. 22.24a Frank LaBua/Pearson Education/PH College Fig. 22.24b Frank LaBua/Pearson Education/PH College Fig. 22.13.1a © Edward Pascuzzi Fig. 22.19.1a Southern Illinois University/Photo Researchers, Inc. Fig. 22.19.1b Charles Lightdale/ Photo Researchers, Inc. Fig. 22.20.1 Doug Johnson/Photo Researchers, Inc.

Chapter 23—CO.23 Peticolas/Megna/Fundamental Photographs Fig. 23.6 Paul Silverman/Fundamental Photographs Fig. 23.13 Richard Hutchings/Photo Researchers, Inc. Fig. 23.15a John Smith/Jerry Wilson Fig. 23.15b Richard Megna/Fundamental Photographs Fig. 23.22 Frank Labua/Pearson Education/PH College Fig. 23.24b Michael Freeman/Michael Freeman Fig. 23.25 Tom Tracy/Corbis/Stock Market Fig. 23.27 John Smith/Jerry Wilson Fig. 23.26 Michael W. Davidson, National High Magnetic Field Laboratory, The Florida State University. Fig. 23.18.1b John Smith/Jerry Wilson Fig. 23.18.1c Bohdan Hrynewych/Stock Boston

Chapter 24—CO.24 Adrienne Hart Davis/ Science Photo Library/Photo Researchers, Inc. Fig. 24.1 Richard Megna/Fundamental Photographs Fig. 24.2b From the "Atlas of Optical Phenomena," Michel Cagnet, Maurice Francon, Jean Claude Thrierr. © by Springer-Verlag OHG, Berlin, 1962. Fig. 24.6c Peter Aprahamian/Photo Researchers, Inc. Fig. 24.7a David Parker/Science Photo Library/Photo Researchers, Inc. Fig. 24.7b Gregory G. Dimijian, MD/Photo Researchers, Inc. Fig. 24.9b Ken Kay/Fundamental Photographs Fig. 24.11 Sabina Zigman, sophomore at Cornell University Fig. 24.12a Education Development Center, Inc. Fig. 24.13a Ken Kay/Fundamental Photographs Fig. 24.12b Education Development Center, Inc. Fig. 24.13b Ken Kay/Fundamental Photographs Fig. 24.17 Dan McCoy/Rainbow Fig. 24.19b A.J. Stosick Fig. 24.19c Courtesy of Dr. M.F. Perutz/Medical Research Council Fig. 24.25b Ed Degginger/Color-Pic, Inc. Fig. 24.22c Diane Schiumo/Fundamental Photographs Fig. 24.24b1 Nina Barnett Photography Fig. 24.24b2 Nina Barnett Photography Fig. 24.26b Peter Aprahamian/Sharples Stress Engineers Ltd./Science Photo Library/Photo Researchers, Inc. Fig. 24.28 Roger Ressmeyer © 1994 CORBIS

Chapter 25—CO.25 Joe McNally Photography Fig. 25.5.1 Robert Dyer/Robert Dyer Fig. 25.4b Frank Labua/Pearson Education/PH College Fig. 25.8b Rocher/Jerrican/Photo Researchers, Inc. Fig. 25.13a SAGEM/European Southern Observatory Fig. 25.13b SAGEM/European Southern Observatory Fig. 25.14 NASA Headquarters Fig. 25.16a Reproduced by permission from Michel Cagnet, Maurice Franzon, and Jean Claude Thierr, Atlas of Optical Phenomena. New York: Springer-Verlag, 1962. © 1962 by Springer-Verlag GmbH & Co. Fig. 25.16b Reproduced by permission from Michel Cagnet, Maurice Franzon, and Jean Claude Thierr, Atlas of Optical Phenomena. New York: Springer-Verlag, 1962. © 1962 by Springer-Verlag GmbH & Co. Fig. 25.17a The Image Finders Fig. 25.17b The Image Finders Fig. 25.17c The Image Finders Fig. 25.18 Bill Bachmann/Photo Researchers, Inc. Fig. 25.20a Fritz Goro, Life Magazine © TimePix Fig. 25.18.1 Tony Craddock/Science Photo Library/Photo Researchers, Inc.

Chapter 26—CO.26 W. Couch, University of New South Wales, Australia/Space Telescope Science Institute OPO/NASA Fig. 26.2 Science Photo Library/Photo Researchers, Inc. Fig. 26.14b Space Telescope Science Institute/Photo Researchers, Inc. Fig. 26.15a Space Telescope Science Institute Fig. 26.16.1 National Radio Astronomy Observatory Fig. 26.16.2 David Hardy/Science Photo Library/Photo Researchers, Inc.

Chapter 27—CO.27 Lawrence Livermore National Laboratory Fig. 27.7b Ann Purcell/Photo Researchers, Inc. Fig. 27.8a Wabash Instrument Corp./Fundamental Photographs Fig. 27.8b Wabash Instrument Corp./Fundamental Photographs Fig. 27.8c Wabash Instrument Corp./Fundamental Photographs Fig. 27.13a © Mark A. Schneider/Photo Researchers, Inc. Fig. 27.13b © Mark A. Schneider/Photo Researchers, Inc. Fig. 27.14 Dan McCoy/Rainbow Fig. 27.18a © Alexander Tsiaras/Science Source/Photo Researchers, Inc. Fig. 27.18b Rosenfeld Images Ltd/Science Photo Library/Photo Researchers, Inc. Fig. 27.20 Philippe Plailly/Photo Researchers, Inc.

Chapter 28—CO.28 Courtesy of International Business Machines Corporation; Almaden Research Center. Unauthorized use not permitted. Fig. 28.3.2 Bob Thomason/Stone/Getty Images Inc Fig. 28.3.5 R.J. Driscoll, M.G. Youngquist, & J.D. Baldeschwieler, California Institute of Technology/Science Photo Library/Photo Researchers, Inc. Fig. 28.3.3a Manfred Kage/Peter Arnold, Inc. Fig. 28.3.3b Dr. R. Kessel/Peter Arnold, Inc. Fig. 28.6.1L Omikron/Science Source/Photo Researchers, Inc. Fig. 28.6.1R Mehau Kulyk/Science Photo Library/Photo Researchers, Inc. Fig. 28.3.6 Will & Deni McIntyre/Photo Researchers, Inc.

Chapter 29—CO.29 Roger Tully/Getty Images Inc Fig. 29.17 Lawrence Berkeley National Laboratory/Photo Researchers, Inc. Fig. 29.18a Dan McCoy/Rainbow Fig. 29.18c Monte S. Buchsbaum, M.D., Mount Sinai School of Medicine, New York, NY

Chapter 30—CO.30 Alexander Tsiaras/Stock Boston Fig. 30.6 Shone/Getty Images, Inc, Fig. 30.7b Plasma Physics Laboratory, Princeton University

Index

Induction. *See* Electromagnetic induction
Inductive circuits, 711
Inductive reactance, 708–9
Inductor(s), 708, 713
Inelastic collisions, 194–97, 861
Inertia
  electromagnetic, 673
  force and, 106–7
  moment of, 274–78, 290, 291
  in Newton's first law of motion, 105–6, 107n
  pulley, 279–81
  rotational, 276, 293
Inertial confinement, 975
Inertial reference frame, 844–45, 846–47
  relativity of simultaneity and, 847–50
Infrared radiation, 393–94, 395, 692–93
Infrasonic region, 479
Infrasonic waves (infrasound), 479
Inner transition elements, 919
Input (primary) coil, 683
Instantaneous
  acceleration, 40, 116–17
  angular speed, 223
  angular velocity, 223
  axis of rotation, 262
  force, 116–17
  speed, 34
  velocity, 37–38, 39
Insulation, 389–91
  R-values (thermal resistance values), 392
Insulator(s), 516–17
  foam, 393
  thermal, 388, 389–91
Intensity, of sound, 485–92, 505
Interference, 460–64
  constructive, 460–61, 493, 781, 782, 784
  destructive, 461, 493, 494, 781, 782, 784, 785, 790
  sound, 493–96
  thin-film, 783–88
  total constructive, 461, 463, 493
  total destructive, 461, 463, 464
  Young's double-slit experiment, 780–81
Interference maxima, 794
Interference minima, 790
Internal combustion engines, 425, 428
Internal energy, 346
  of diatomic gases, 365–66
  of monatomic gases, 362
  in thermodynamics, 406–18
Internal motion, 205–6
Internal reflection, and fiber optics, 736–37
Internal resistance, 579
International System of Units (SI), 2–5
Intravenous (IV) blood transfusions, 335–36
Inverse-square law, 236, 244
Inversion, population, 893
Iodine-123, 955

Iodine-131, 953–55
Ion(s), 641
Ion channels, 562–63
Ionization, 518
Ionized atoms, 889
Iris, 814, 815
Irregular (diffuse) reflection, 724–25
Irreversible process, 406
Irrotational flow, 326
Isobaric expansion, 412
Isobaric process, 411–14
Isobars, 411, 425
Isochoric process, 414–15
Isogonic lines, 660
Isolated (closed) systems, 158, 188, 194, 420–21
Isomet, 414, 425
Isometric process, 414–15
Isotherm(s), 410–11, 412, 417–18, 433
Isothermal process, 410–11, 420
Isotope(s), 932. *See also* Half-life
  properties of, A5–9
Isotropic expansion, 357
Isotropic material, 802
IV (intravenous injection) under gravity flow, 320–21

Jansky, Carl, 833
Jet propulsion, 115, 116, 208–10
Joule, James, 143, 374–75, 443
Joule heat, 590–93, 624
  electrical resistance and, 591–92
  loss, 590
joule (J), 143, 374, 375, 558
joule per coulomb, 579
Jumper, 607–8
Junctions (nodes), electrical, 611
Junction theorem, 611, 621–22

Karate chops, 311
Keck telescope, 829
Kelvin, Lord, 353, 427, 433
kelvin (K), 5, 353
Kelvin temperature scale, 351, 352–56
Kepler, Johannes, 237, 244
Kepler's laws of planetary motion, 237, 243–45
  first law (law of orbits), 244
  second law (law of areas), 244–45, 290
  third law (law of periods), 245
Keyboards, computer, 566
Kidneys, 363
kilocalorie (kcal), 374
kiloelectron volt (keV), 558, 582
kilogram (kg), 4, 7
kilowatt-hour (kWh), 592
kilowatt (kW), 167
Kinematic equations, 45–49, 67–70. *See also* Constant acceleration; Free fall

Kinematics, 32–63. *See also* Acceleration; Displacement; Distance; Free fall; Speed; Velocity
  rotational, 235
Kinetic energy
  in beta decay, 976–77
  collisions and, 193–202
  defined, 150
  gravitational potential energy and, 155–56
  momentum vs., 181–82, 185, 186
  of orbiting satellite, 248
  Q value and, 965–67
  relativistic, 858–59
  rotational work and, 283–87
  translational, 285–86, 346, 360, 365, 366
  work-energy theorem and, 150–54, 284–87
Kinetic (sliding) friction, 124
  coefficient of, 125–28
Kinetic theory of gases, 360–66, 387, A3–4
  absolute temperature in, 361
  diffusion in, 362–64
  equipartition theorem in, 364–65
Kirchhoff, Gustav, 611
Kirchhoff plots, 612–13
Kirchhoff's rules, 610–16
  application of, 614–16
  first rule (junction theorem), 611, 621–22
  second rule (loop theorem), 611–13, 622, 710–11
Klystrons, 692

Laminar flow, 333–34
Land, Edwin H., 798, 835
Land, of compact disc, 896
Lanthanide series, 919
Lapse rate, atmospheric, 372
Large numbers, writing, 3n
Larynx (voice box), 486
Laser(s), 813, 875, 892–96
  for correcting nearsightedness, 818
  in eye surgery, 818, 819, 895
  helium-neon, 893–94
  in holography, 895
Laser printers, 522–23
Latent heat, 381–86
  of fusion, 382–83
  of sublimation, 382
  of vaporization, 382–83, 387
Lateral magnification, 748
Latitudes, distance around Earth at different, 23–24
Law
  of areas (Kepler's second law), 244–45, 290
  of charges (charge-force law), 514
  of conservation of linear momentum, 187
  of conservation of mechanical energy, 160
  of conservation of total energy, 158

  of cosines, A2
  of inertia. *See* Newton's laws of motion, first
  of orbits (Kepler's first law), 244
  of periods (Kepler's third law), 245
  of poles (pole-force law), 635
  of reflection, 724, 725
  of sines, A2
LC circuit, oscillating, 717
LCDs (liquid crystal displays), 381, 803, 804–5
Lead-206 dating, 944
Leaning Tower of Pisa, 50, 51, 273
LEDs (light-emitting diodes), 805
Length
  arc, 220–22
  British unit, 9
  equivalent, 13
  focal, 752, 755, 761, 764
  proper, 854–55
  SI unit, 3, 9
  Young's modulus and, 307–9
Length contraction, 853–56
Lens(es), 747, 760–72
  aberrations in, 770–71
  blocking off half, 766
  combinations of, 767–69
  converging (biconvex), 760, 761, 763–65, 814, 816, 818–20
  crystalline, 814, 815
  diopters, 772
  diverging (biconcave), 760, 761, 766–67, 817–18
  erecting, 827
  Fresnel, 768
  image side, 761
  meniscus, 760
  nonreflecting, 788
  object side, 761
  oil immersion, 834
  optical flats, 787–88
  ray diagrams for, 762, 763
Lens maker's equation, 771–72
Lens power, 772
Lenz, Heinrich, 673
Lenz's law, 672–79, 685, 686, 708
Leptons, 981, 982, 986
Lever (moment) arm, 264
Life, and second law of thermodynamics, 423
Lifetime, of excited state, 889
Lift, airplane, 330
Lift, hydraulic, 315–17
Lift force, 116
Light
  atmospheric scattering of, 803–7
  determination of color, 835–36
  dual nature of, 886
  in general theory of relativity, 863–66
  monochromatic, 738
  quantization of, 879–83
  relativistic velocity addition for, 869
  speed of. *See* Speed, of light